P9-CDL-962

Rational (Reciprocal) Function

$f(x) = \dfrac{1}{x}$

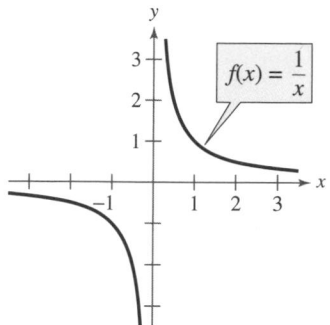

Domain: $(-\infty, 0) \cup (0, \infty)$
Range: $(-\infty, 0) \cup (0, \infty)$
No intercepts
Decreasing on $(-\infty, 0)$ and $(0, \infty)$
Odd function
Origin symmetry
Vertical asymptote: y-axis
Horizontal asymptote: x-axis

Exponential Function

$f(x) = a^x, \ a > 0, \ a \neq 1$

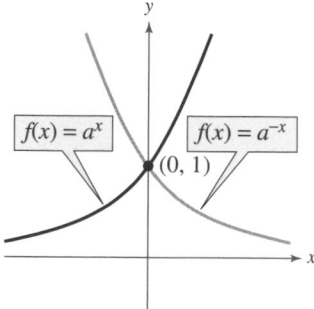

Domain: $(-\infty, \infty)$
Range: $(0, \infty)$
Intercept: $(0, 1)$
Increasing on $(-\infty, \infty)$
    for $f(x) = a^x$
Decreasing on $(-\infty, \infty)$
    for $f(x) = a^{-x}$
x-axis is a horizontal asymptote
Continuous

Logarithmic Function

$f(x) = \log_a x, \ a > 0, \ a \neq 1$

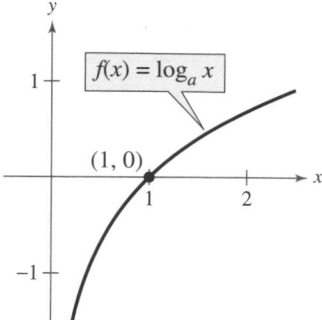

Domain: $(0, \infty)$
Range: $(-\infty, \infty)$
Intercept: $(1, 0)$
Increasing on $(0, \infty)$
y-axis is a vertical asymptote
Continuous
Reflection of graph of $f(x) = a^x$
    in the line $y = x$

Sine Function

$f(x) = \sin x$

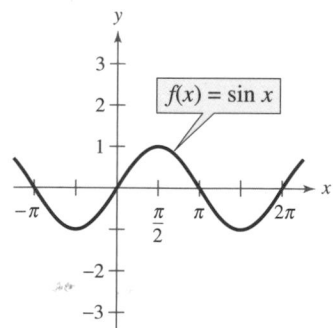

Domain: $(-\infty, \infty)$
Range: $[-1, 1]$
Period: 2π
x-intercepts: $(n\pi, 0)$
y-intercept: $(0, 0)$
Odd function
Origin symmetry

Cosine Function

$f(x) = \cos x$

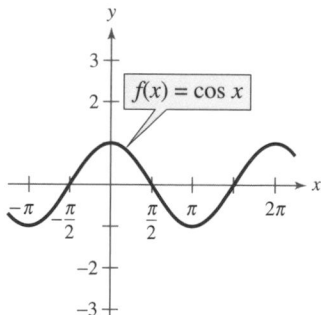

Domain: $(-\infty, \infty)$
Range: $[-1, 1]$
Period: 2π
x-intercepts: $\left(\dfrac{\pi}{2} + n\pi, 0\right)$
y-intercept: $(0, 1)$
Even function
y-axis symmetry

Tangent Function

$f(x) = \tan x$

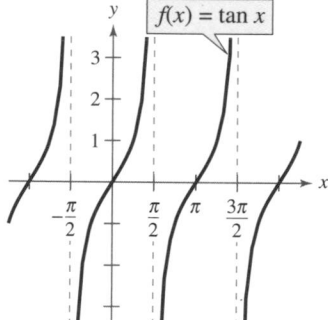

Domain: $x \neq \dfrac{\pi}{2} + n\pi$
Range: $(-\infty, \infty)$
Period: π
x-intercepts: $(n\pi, 0)$
y-intercept: $(0, 0)$
Vertical asymptotes:
    $x = \dfrac{\pi}{2} + n\pi$
Odd function
Origin symmetry

Cosecant Function

$f(x) = \csc x$

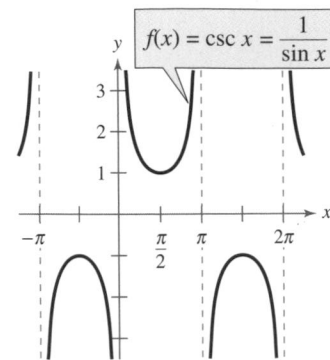

$f(x) = \csc x = \dfrac{1}{\sin x}$

Domain: $x \neq n\pi$
Range: $(-\infty, -1] \cup [1, \infty)$
Period: 2π
No intercepts
Vertical asymptotes: $x = n\pi$
Odd function
Origin symmetry

Secant Function

$f(x) = \sec x$

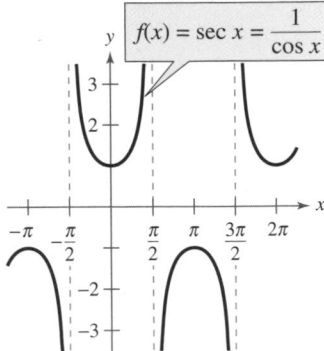

$f(x) = \sec x = \dfrac{1}{\cos x}$

Domain: $x \neq \dfrac{\pi}{2} + n\pi$
Range: $(-\infty, -1] \cup [1, \infty)$
Period: 2π
y-intercept: $(0, 1)$
Vertical asymptotes:

$$x = \dfrac{\pi}{2} + n\pi$$

Even function
y-axis symmetry

Cotangent Function

$f(x) = \cot x$

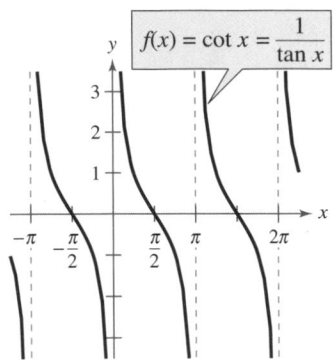

$f(x) = \cot x = \dfrac{1}{\tan x}$

Domain: $x \neq n\pi$
Range: $(-\infty, \infty)$
Period: π
x-intercepts: $\left(\dfrac{\pi}{2} + n\pi, 0 \right)$

Vertical asymptotes: $x = n\pi$
Odd function
Origin symmetry

Inverse Sine Function

$f(x) = \arcsin x$

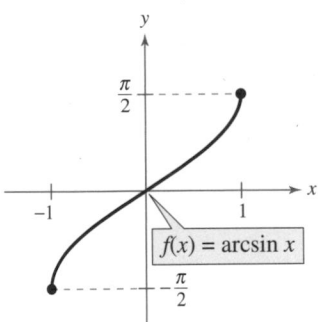

$f(x) = \arcsin x$

Domain: $[-1, 1]$

Range: $\left[-\dfrac{\pi}{2}, \dfrac{\pi}{2} \right]$

Intercept: $(0, 0)$
Odd function
Origin symmetry

Inverse Cosine Function

$f(x) = \arccos x$

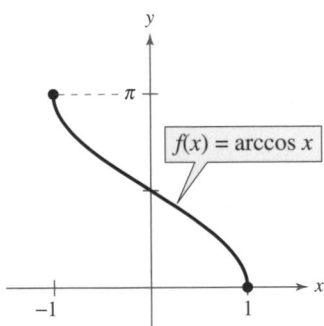

$f(x) = \arccos x$

Domain: $[-1, 1]$
Range: $[0, \pi]$

y-intercept: $\left(0, \dfrac{\pi}{2} \right)$

Inverse Tangent Function

$f(x) = \arctan x$

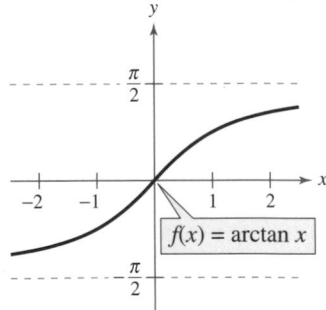

$f(x) = \arctan x$

Domain: $(-\infty, \infty)$

Range: $\left(-\dfrac{\pi}{2}, \dfrac{\pi}{2} \right)$

Intercept: $(0, 0)$
Horizontal asymptotes:

$$y = \pm \dfrac{\pi}{2}$$

Odd function
Origin symmetry

FIFTH EDITION

Precalculus Functions and Graphs
A Graphing Approach

Ron Larson
The Pennsylvania State University
The Behrend College

Robert Hostetler
The Pennsylvania State University
The Behrend College

Bruce H. Edwards
University of Florida

With the assistance of David C. Falvo
The Pennsylvania State University
The Behrend College

Houghton Mifflin Company   Boston   New York

Publisher: Richard Stratton
Sponsoring Editor: Cathy Cantin
Senior Marketing Manager: Jennifer Jones
Development Editor: Lisa Collette
Supervising Editor: Karen Carter
Senior Project Editor: Patty Bergin
Art and Design Manager: Gary Crespo
Cover Design Manager: Anne S. Katzeff
Photo Editor: Jennifer Meyer Dare
Composition Buyer: Chuck Dutton
New Title Project Manager: James Lonergan
Editorial Associate: Jeannine Lawless
Marketing Associate: Mary Legere
Editorial Assistant: Jill Clark
Composition and Art: Larson Texts, Inc.

Cover photograph © Jeff Foott/Getty Images

We have included examples and exercises that use real-life data as well as technology output from a variety of software. This would not have been possible without the help of many people and organizations. Our wholehearted thanks go to all their time and effort.

Printed in the U.S.A.

Library of Congress Catalog Card Number: 2006931641

Instructor's exam copy:
ISBN13: 978-0-618-85155-3
ISBN10: 0-618-85155-0

For orders, use student text ISBNs:
ISBN13: 978-0-618-85150-8
ISBN10: 0-618-85150-X

123456789–DOW–11 10 09 08 07

Contents

A Word from the Authors   vii
Features Highlights   xii

Chapter 1   **Functions and Their Graphs**   1
Introduction to Library of Parent Functions   2
1.1   Lines in the Plane   3
1.2   Functions   16
1.3   Graphs of Functions   30
1.4   Shifting, Reflecting, and Stretching Graphs   42
1.5   Combinations of Functions   51
1.6   Inverse Functions   62
1.7   Linear Models and Scatter Plots   73
Chapter Summary   82   **Review Exercises**   83
Chapter Test   88   **Proofs in Mathematics**   89

Chapter 2   **Polynomial and Rational Functions**   91
2.1   Quadratic Functions   92
2.2   Polynomial Functions of Higher Degree   103
2.3   Real Zeros of Polynomial Functions   116
2.4   Complex Numbers   131
2.5   The Fundamental Theorem of Algebra   139
2.6   Rational Functions and Asymptotes   146
2.7   Graphs of Rational Functions   156
2.8   Quadratic Models   165
Chapter Summary   172   **Review Exercises**   173
Chapter Test   179   **Proofs in Mathematics**   180
Progressive Summary 1 and 2   182

Chapter 3   **Exponential and Logarithmic Functions**   183
3.1   Exponential Functions and Their Graphs   184
3.2   Logarithmic Functions and Their Graphs   193
3.3   Properties of Logarithms   207
3.4   Solving Exponential and Logarithmic Equations   214
3.5   Exponential and Logarithmic Models   225
3.6   Nonlinear Models   237
Chapter Summary   246   **Review Exercises**   247
Chapter Test   252   **Cumulative Test 1–3**   253
Proofs in Mathematics   255
Progressive Summary 1–3   256

Chapter 4   **Trigonometric Functions**   257
4.1   Radian and Degree Measure   258
4.2   Trigonometric Functions: The Unit Circle   269
4.3   Right Triangle Trigonometry   277
4.4   Trigonometric Functions of Any Angle   288
4.5   Graphs of Sine and Cosine Functions   297
4.6   Graphs of Other Trigonometric Functions   309
4.7   Inverse Trigonometric Functions   320
4.8   Applications and Models   331
      Chapter Summary   343   **Review Exercises**   344
      Chapter Test   349   **Proofs in Mathematics**   350

Chapter 5   **Analytic Trigonometry**   351
5.1   Using Fundamental Identities   352
5.2   Verifying Trigonometric Identities   360
5.3   Solving Trigonometric Equations   368
5.4   Sum and Difference Formulas   380
5.5   Multiple-Angle and Product-to-Sum Formulas   387
      Chapter Summary   399   **Review Exercises**   400
      Chapter Test   403   **Proofs in Mathematics**   404

Chapter 6   **Additional Topics in Trigonometry**   407
6.1   Law of Sines   408
6.2   Law of Cosines   417
6.3   Vectors in the Plane   424
6.4   Vectors and Dot Products   438
6.5   Trigonometric Form of a Complex Number   448
      Chapter Summary   460   **Review Exercises**   461
      Chapter Test   465   **Cumulative Test 4–6**   466
      Proofs in Mathematics   468   **Progressive Summary 1–6**   472

Chapter 7   **Linear Systems and Matrices**   473
7.1   Solving Systems of Equations   474
7.2   Systems of Linear Equations in Two Variables   485
7.3   Multivariable Linear Systems   495
7.4   Matrices and Systems of Equations   511
7.5   Operations with Matrices   526
7.6   The Inverse of a Square Matrix   541
7.7   The Determinant of a Square Matrix   551
7.8   Applications of Matrices and Determinants   559
      Chapter Summary   569   **Review Exercises**   570
      Chapter Test   576   **Proofs in Mathematics**   577

Chapter 8   Sequences, Series, and Probability   579

8.1   Sequences and Series   580

8.2   Arithmetic Sequences and Partial Sums   592

8.3   Geometric Sequences and Series   601

8.4   Mathematical Induction   611

8.5   The Binomial Theorem   619

8.6   Counting Principles   627

8.7   Probability   637

Chapter Summary   650   **Review Exercises**   651

Chapter Test   655   **Proofs in Mathematics**   656

Chapter 9   Topics in Analytic Geometry   659

9.1   Circles and Parabolas   660

9.2   Ellipses   671

9.3   Hyperbolas   680

9.4   Rotation and Systems of Quadratic Equations   690

9.5   Parametric Equations   699

9.6   Polar Coordinates   707

9.7   Graphs of Polar Equations   713

9.8   Polar Equations of Conics   722

Chapter Summary   729   **Review Exercises**   730

Chapter Test   734   **Cumulative Test 7–9**   735

Proofs in Mathematics   737

Progressive Summary 3–9   740

Chapter 10   Analytic Geometry in Three Dimensions   741

10.1   The Three-Dimensional Coordinate System   742

10.2   Vectors in Space   750

10.3   The Cross Product of Two Vectors   757

10.4   Lines and Planes in Space   764

Chapter Summary   773   **Review Exercises**   774

Chapter Test   776   **Proofs in Mathematics**   777

Chapter 11   Limits and an Introduction to Calculus   779

11.1   Introduction to Limits   780

11.2   Techniques for Evaluating Limits   791

11.3   The Tangent Line Problem   801

11.4   Limits at Infinity and Limits of Sequences   811

11.5   The Area Problem   820

Chapter Summary   828   **Review Exercises**   829

Chapter Test   832   **Cumulative Test 10–11**   833

Proofs in Mathematics   835

Progressive Summary 3–11   836

Appendices   **Appendix A   Technology Support Guide   A1**

Appendix B   Review of Graphs, Equations, and Inequalities   A25

B.1   The Cartesian Plane   A25

B.2   Graphs of Equations   A36

B.3   Solving Equations Algebraically and Graphically   A47

B.4   Solving Inequalities Algebraically and Graphically   A63

B.5   Representing Data Graphically   A76

Appendix C   Concepts in Statistics   A85

C.1   Measures of Central Tendency and Dispersion   A85

C.2   Least Squares Regression   A94

Appendix D   Variation   A96

Appendix E   Solving Linear Equations and Inequalities   A103

Appendix F   Systems of Inequalities   A106

F.1   Solving Systems of Inequalities   A106

F.2   Linear Programming   A116

Appendix G   Study Capsules   A125

Answers to Odd-Numbered Exercises and Tests   A137

Index of Selected Applications   A289

Index   A291

A Word from the Authors

Welcome to *Precalculus Functions and Graphs: A Graphing Approach*, Fifth Edition. We are pleased to present this new edition of our textbook in which we focus on making the mathematics accessible, supporting student success, and offering instructors flexible teaching options.

Accessible to Students

We have taken care to write this text with the student in mind. Paying careful attention to the presentation, we use precise mathematical language and a clear writing style to develop an effective learning tool. We believe that every student can learn mathematics, and we are committed to providing a text that makes the mathematics of the precalculus course accessible to all students.

Throughout the text, solutions to many examples are presented from multiple perspectives—algebraically, graphically, and numerically. The side-by-side format of this pedagogical feature helps students to see that a problem can be solved in more than one way and to see that different methods yield the same result. The side-by-side format also addresses many different learning styles.

We have found that many precalculus students grasp mathematical concepts more easily when they work with them in the context of real-life situations. Students have numerous opportunities to do this throughout this text. The *Make a Decision* feature further connects real-life data and applications and motivates students. It also offers students the opportunity to generate and analyze mathematical models from large data sets. To reinforce the concept of functions, we have compiled all the elementary functions as a *Library of Parent Functions*, presented in a summary on the endpapers of the text for convenient reference. Each function is introduced at the first point of use in the text with a definition and description of basic characteristics.

We have carefully written and designed each page to make the book more readable and accessible to students. For example, to avoid unnecessary page turning and disruptions to students' thought processes, each example and corresponding solution begins and ends on the same page.

Supports Student Success

During more than 30 years of teaching and writing, we have learned many things about the teaching and learning of mathematics. We have found that students are most successful when they know what they are expected to learn and why it is important to learn the concepts. With that in mind, we have incorporated a thematic study thread throughout this textbook.

Each chapter begins with a list of applications that are covered in the chapter and serve as a motivational tool by connecting section content to real-life situations. Using the same pedagogical theme, each section begins with a set of section learning objectives—*What You Should Learn*. These are followed by an engaging real-life application—*Why You Should Learn It*—that motivates students and illustrates an area where the mathematical concepts will be applied in an example or exercise in the section. The *Chapter Summary—What Did You Learn?*—at the end of each chapter includes *Key Term*s with page references and *Key Concepts*, organized by section, that were covered throughout the chapter. The *Chapter Summary* serves as a useful study aid for students.

Throughout the text, other features further improve accessibility. *Study Tips* are provided throughout the text at point-of-use to reinforce concepts and to help students learn how to study mathematics. *Explorations* reinforce mathematical concepts. Each example with worked-out solution is followed by a *Checkpoint*, which directs the student to work a similar exercise from the exercise set. The *Section Exercises* begin with a *Vocabulary Check*, which gives the students an opportunity to test their understanding of the important terms in the section. A *Prerequisites Skills* is offered in margin notes throughout the textbook exposition. Reviewing the prerequisite skills will enable students to master new concepts more quickly. *Synthesis Exercises* check students' conceptual understanding of the topics in each section. *Skills Review Exercises* provide additional practice with the concepts in the chapter or previous chapters. *Review Exercises*, *Chapter Tests*, and periodic *Cumulative Tests* offer students frequent opportunities for self-assessment and to develop strong study and test-taking skills. The *Progressive Summaries* and the *Study Capsules* serve as a quick reference when working on homework or as a cumulative study aid.

The use of technology also supports students with different learning styles, and graphing calculators are fully integrated into the text presentation. The *Technology Support Appendix* makes it easier for students to use technology. *Technology Support* notes are provided throughout the text at point-of-use. These notes guide students to the *Technology Support Appendix*, where they can learn how to use specific graphing calculator features to enhance their understanding of the concepts presented in the text. These notes also direct students to the *Graphing Technology Guide*, in the *Online Study Center*, for keystroke support that is available for numerous calculator models. *Technology Tips* are provided in the text at point-of-use to call attention to the strengths and weaknesses of graphing technology, as well as to offer alternative methods for solving or checking a problem using technology. Because students are often misled by the limitations of graphing calculators, we have, where appropriate, used color to enhance the graphing calculator displays in the textbook. This enables students to visualize the mathematical concepts clearly and accurately and avoid common misunderstandings.

Numerous additional text-specific resources are available to help students succeed in the precalculus course. These include "live" online tutoring, instructional DVDs, and a variety of other resources, such as tutorial support and self-assessment, which are available on the Web and in Eduspace®. In addition, the *Online Notetaking Guide* is a notetaking guide that helps students organize their class notes and create an effective study and review tool.

Flexible Options for Instructors

From the time we first began writing textbooks in the early 1970s, we have always considered it a critical part of our role as authors to provide instructors with flexible programs. In addition to addressing a variety of learning styles, the optional features within the text allow instructors to design their courses to meet their instructional needs and the needs of their students. For example, the *Explorations* throughout the text can be used as a quick introduction to concepts or as a way to reinforce student understanding.

Our goal when developing the exercise sets was to address a wide variety of learning styles and teaching preferences. The *Vocabulary Check* questions are provided at the beginning of every exercise set to help students learn proper mathematical terminology. In each exercise set we have included a variety of exercise types, including questions requiring writing and critical thinking, as well as real-data applications. The problems are carefully graded in difficulty from mastery of basic skills to more challenging exercises. Some of the more challenging exercises include the *Synthesis Exercises* that combine skills and are used to check for conceptual understanding, and the *Make a Decision* exercises that further connect real-life data and applications and motivate students. *Skills Review Exercises*, placed at the end of each exercise set, reinforce previously learned skills. The *Proofs in Mathematics*, at the end of each chapter, are proofs of important mathematical properties and theorems and illustrate various proof techniques. This feature gives the instructors the opportunity to incorporate more rigor into their course. In addition, Houghton Mifflin's Eduspace® website offers instructors the option to assign homework and tests online—and also includes the ability to grade these assignments automatically.

Several other print and media resources are available to support instructors. The *Online Instructor Success Organizer* includes suggested lesson plans and is an especially useful tool for larger departments that want all sections of a course to follow the same outline. The *Instructor's Edition* of the *Online Student Notetaking Guide* can be used as a lecture outline for every section of the text and includes additional examples for classroom discussion and important definitions. This is another valuable resource for schools trying to have consistent instruction and it can be used as a resource to support less experienced instructors. When used in conjunction with the *Online Student Notetaking Guide* these resources can save instructors preparation time and help students concentrate on important concepts.

Instructors who stress applications and problem solving and integrate technology into their course will be able to use this text successfully.

We hope you enjoy the Fifth Edition.

Ron Larson

Robert Hostetler

Bruce H. Edwards

PREFACE

Acknowledgments

We would like to thank the many people who have helped us prepare the text and supplements package, including all those reviewers who have contributed to this and previous editions of the text. Their encouragement, criticisms, and suggestions have been invaluable to us.

Reviewers

Tony Homayoon Akhlaghi
Bellevue Community College

Daniel D. Anderson
University of Iowa

Bruce Armbrust
Lake Tahoe Community College

Jamie Whitehead Ashby
Texarkana College

Teresa Barton
Western New England College

Kimberly Bennekin
Georgia Perimeter College

Charles M. Biles
Humboldt State University

Phyllis Barsch Bolin
Oklahoma Christian University

Khristo Boyadzhiev
Ohio Northern University

Dave Bregenzer
Utah State University

Anne E. Brown
Indiana University-South Bend

Diane Burleson
Central Piedmont Community College

Alexander Burstein
University of Rhode Island

Marilyn Carlson
University of Kansas

Victor M. Cornell
Mesa Community College

John Dersh
Grand Rapids Community College

Jennifer Dollar
Grand Rapids Community College

Marcia Drost
Texas A & M University

Cameron English
Rio Hondo College

Susan E. Enyart
Otterbein College

Patricia J. Ernst
St. Cloud State University

Eunice Everett
Seminole Community College

Kenny Fister
Murray State University

Susan C. Fleming
Virginia Highlands Community College

Jeff Frost
Johnson County Community College

James R. Fryxell
College of Lake County

Khadiga H. Gamgoum
Northern Virginia Community College

Nicholas E. Geller
Collin County Community College

Betty Givan
Eastern Kentucky University

Patricia K. Gramling
Trident Technical College

Michele Greenfield
Middlesex County College

Bernard Greenspan
University of Akron

Zenas Hartvigson
University of Colorado at Denver

Rodger Hergert
Rock Valley College

Allen Hesse
Rochester Community College

Rodney Holke-Farnam
Hawkeye Community College

Lynda Hollingsworth
Northwest Missouri State University

Jean M. Horn
Northern Virginia Community College

Spencer Hurd
The Citadel

Bill Huston
Missouri Western State College

Deborah Johnson
Cambridge South Dorchester High School

Francine Winston Johnson
Howard Community College

Luella Johnson
State University of New York, College at Buffalo

Susan Kellicut
Seminole Community College

John Kendall
Shelby State Community College

Donna M. Krawczyk
University of Arizona

Peter A. Lappan
Michigan State University

Charles G. Laws
Cleveland State Community College

JoAnn Lewin
Edison Community College

Richard J. Maher
Loyola University

Carl Main
Florida College

Marilyn McCollum
North Carolina State University

Judy McInerney
Sandhills Community College

David E. Meel
Bowling Green University

Beverly Michael
University of Pittsburgh

Roger B. Nelsen
Lewis and Clark College

Jon Odell
Richland Community College

Paul Oswood
Ridgewater College

Wing M. Park
College of Lake County

Rupa M. Patel
University of Portland

Robert Pearce
South Plains College

David R. Peterson
University of Central Arkansas

James Pommersheim
Reed College

Antonio Quesada
University of Akron

Laura Reger
Milwaukee Area Technical College

Jennifer Rhinehart
Mars Hill College

Lila F. Roberts
Georgia Southern University

Keith Schwingendorf
Purdue University North Central

George W. Shultz
St. Petersburg Junior College

Stephen Slack
Kenyon College

Judith Smalling
St. Petersburg Junior College

Pamela K. M. Smith
Fort Lewis College

Cathryn U. Stark
Collin County Community College

Craig M. Steenberg
Lewis-Clark State College

Mary Jane Sterling
Bradley University

G. Bryan Stewart
Tarrant County Junior College

Mahbobeh Vezvaei
Kent State University

Ellen Vilas
York Technical College

Hayat Weiss
Middlesex Community College

Howard L. Wilson
Oregon State University

Joel E. Wilson
Eastern Kentucky University

Michelle Wilson
Franklin University

Fred Worth
Henderson State University

Karl M. Zilm
Lewis and Clark Community College

We would like to thank the staff of Larson Texts, Inc. who assisted in preparing the manuscript, rendering the art package, and typesetting and proofreading the pages and supplements.

On a personal level, we are grateful to our wives, Deanna Gilbert Larson, Eloise Hostetler, and Consuelo Edwards for their love, patience, and support. Also, a special thanks goes to R. Scott O'Neil.

If you have suggestions for improving this text, please feel free to write us. Over the past two decades we have received many useful comments from both instructors and students, and we value these very much.

Ron Larson
Robert Hostetler
Bruce H. Edwards

ACKNOWLEDGMENTS

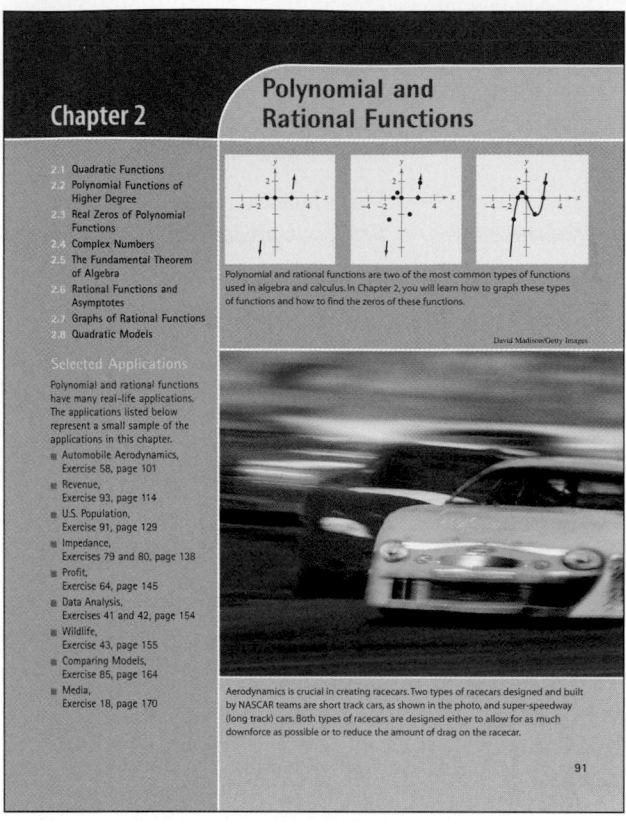

Chapter 2

2.1 Quadratic Functions
2.2 Polynomial Functions of Higher Degree
2.3 Real Zeros of Polynomial Functions
2.4 Complex Numbers
2.5 The Fundamental Theorem of Algebra
2.6 Rational Functions and Asymptotes
2.7 Graphs of Rational Functions
2.8 Quadratic Models

Selected Applications

Polynomial and rational functions have many real-life applications. The applications listed below represent a small sample of the applications in this chapter.

- Automobile Aerodynamics, Exercise 58, page 101
- Revenue, Exercise 93, page 114
- U.S. Population, Exercise 91, page 129
- Impedance, Exercises 79 and 80, page 138
- Profit, Exercise 64, page 145
- Data Analysis, Exercises 41 and 42, page 154
- Wildlife, Exercise 43, page 155
- Comparing Models, Exercise 85, page 164
- Media, Exercise 18, page 170

Polynomial and Rational Functions

Polynomial and rational functions are two of the most common types of functions used in algebra and calculus. In Chapter 2, you will learn how to graph these types of functions and how to find the zeros of these functions.

David Madison/Getty Images

Aerodynamics is crucial in creating racecars. Two types of racecars designed and built by NASCAR teams are short track cars, as shown in the photo, and super-speedway (long track) cars. Both types of racecars are designed either to allow for as much downforce as possible or to reduce the amount of drag on the racecar.

91

Chapter Opener

Each chapter begins with a comprehensive overview of the chapter concepts. The photograph and caption illustrate a real-life application of a key concept. Section references help students prepare for the chapter.

Applications List

An abridged list of applications, covered in the chapter, serve as a motivational tool by connecting section content to real-life situations.

"What You Should Learn" and "Why You Should Learn It"

Sections begin with *What You Should Learn*, an outline of the main concepts covered in the section, and *Why You Should Learn It*, a real-life application or mathematical reference that illustrates the relevance of the section content.

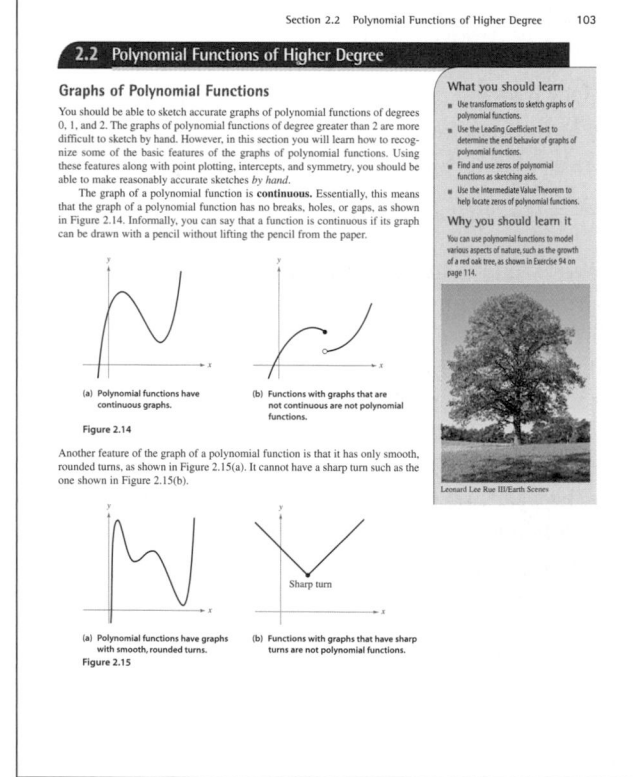

Section 2.2 Polynomial Functions of Higher Degree 103

2.2 Polynomial Functions of Higher Degree

Graphs of Polynomial Functions

You should be able to sketch accurate graphs of polynomial functions of degrees 0, 1, and 2. The graphs of polynomial functions of degree greater than 2 are more difficult to sketch by hand. However, in this section you will learn how to recognize some of the basic features of the graphs of polynomial functions. Using these features along with point plotting, intercepts, and symmetry, you should be able to make reasonably accurate sketches *by hand*.

The graph of a polynomial function is **continuous**. Essentially, this means that the graph of a polynomial function has no breaks, holes, or gaps, as shown in Figure 2.14. Informally, you can say that a function is continuous if its graph can be drawn with a pencil without lifting the pencil from the paper.

What you should learn
- Use transformations to sketch graphs of polynomial functions.
- Use the Leading Coefficient Test to determine the end behavior of graphs of polynomial functions.
- Find and use zeros of polynomial functions as sketching aids.
- Use the Intermediate Value Theorem to help locate zeros of polynomial functions.

Why you should learn it
You can use polynomial functions to model various aspects of nature, such as the growth of a red oak tree, as shown in Exercise 94 on page 114.

(a) Polynomial functions have continuous graphs.

(b) Functions with graphs that are not continuous are not polynomial functions.

Figure 2.14

Another feature of the graph of a polynomial function is that it has only smooth, rounded turns, as shown in Figure 2.15(a). It cannot have a sharp turn such as the one shown in Figure 2.15(b).

Sharp turn

(a) Polynomial functions have graphs with smooth, rounded turns.

(b) Functions with graphs that have sharp turns are not polynomial functions.

Figure 2.15

Leonard Lee Rue III/Earth Scenes

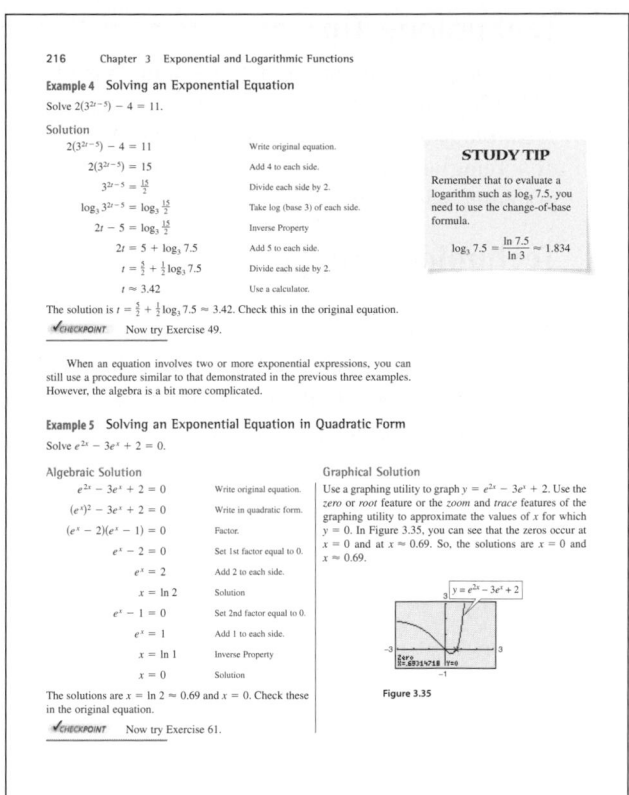

Examples

Many examples present side-by-side solutions with multiple approaches—algebraic, graphical, and numerical. This format addresses a variety of learning styles and shows students that different solution methods yield the same result.

Checkpoint

The *Checkpoint* directs students to work a similar problem in the exercise set for extra practice.

Study Tips

Study Tips reinforce concepts and help students learn how to study mathematics.

Library of Parent Functions

The *Library of Parent Functions* feature defines each elementary function and its characteristics at first point of use. The *Study Capsules* are also referenced for further review of each elementary function.

Explorations

The *Explorations* engage students in active discovery of mathematical concepts, strengthen critical thinking skills, and help them to develop an intuitive understanding of theoretical concepts.

New! Prerequisite Skills

A review of algebra skills needed to complete the examples is offered to the students at point of use throughout the text.

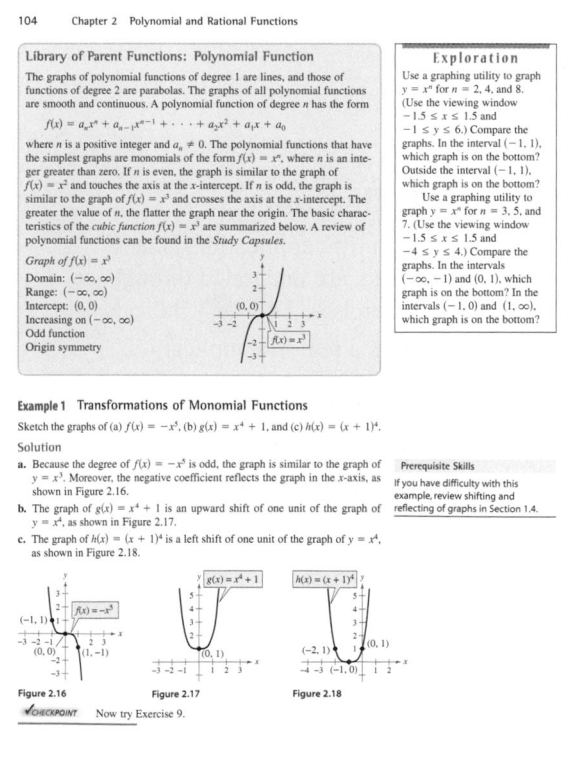

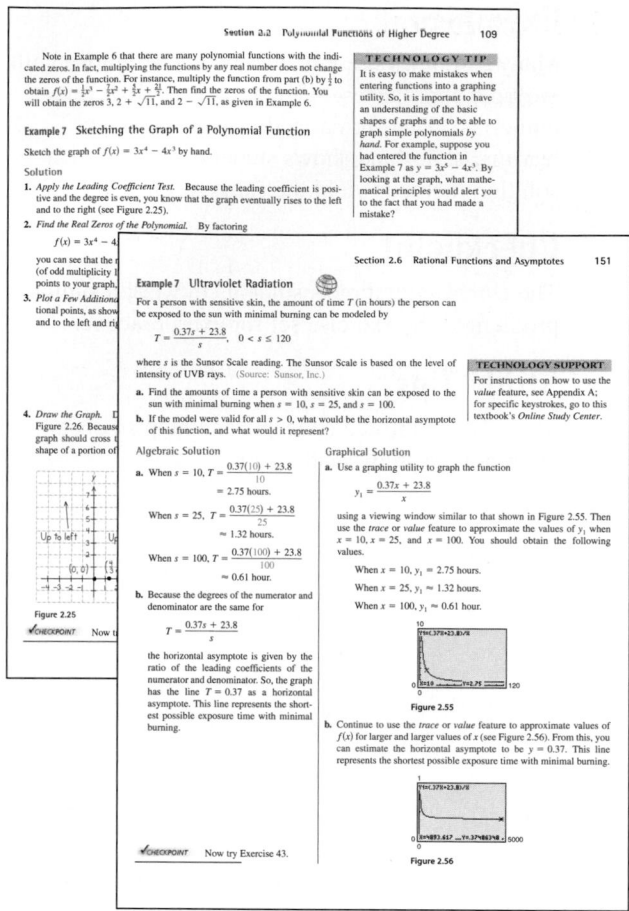

Technology Tip

Technology Tips point out the pros and cons of technology use in certain mathematical situations. *Technology Tips* also provide alternative methods of solving or checking a problem by the use of a graphing calculator.

Technology Support

The *Technology Support* feature guides students to the *Technology Support Appendix* if they need to reference a specific calculator feature. These notes also direct students to the *Graphing Technology Guide*, in the *Online Study Center*, for keystroke support that is available for numerous calculator models.

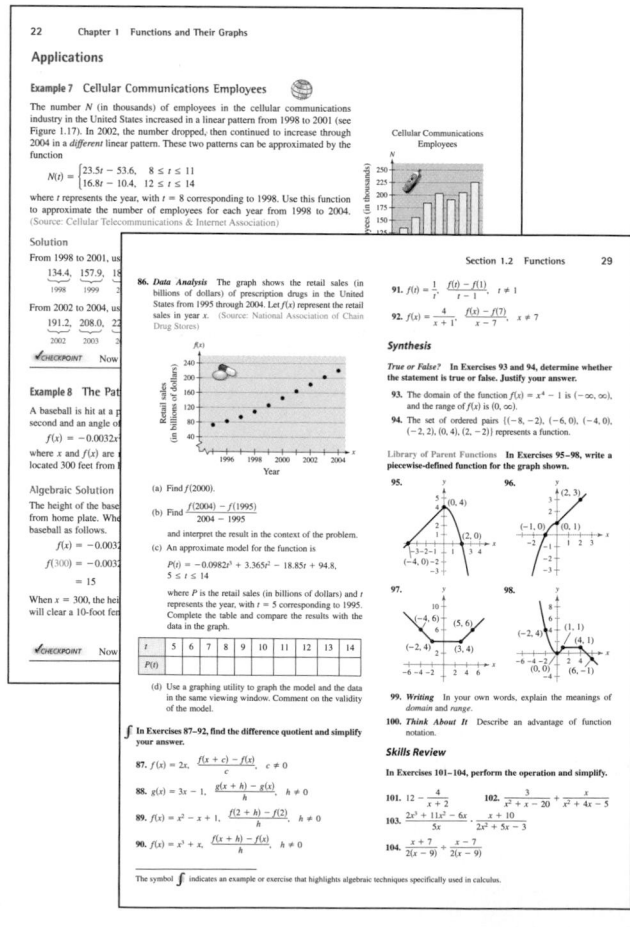

Real-Life Applications

A wide variety of real-life applications, many using current real data, are integrated throughout the examples and exercises. The 🌐 indicates an example that involves a real-life application.

Algebra of Calculus

Throughout the text, special emphasis is given to the algebraic techniques used in calculus. *Algebra of Calculus* examples and exercises are integrated throughout the text and are identified by the symbol $\int$.

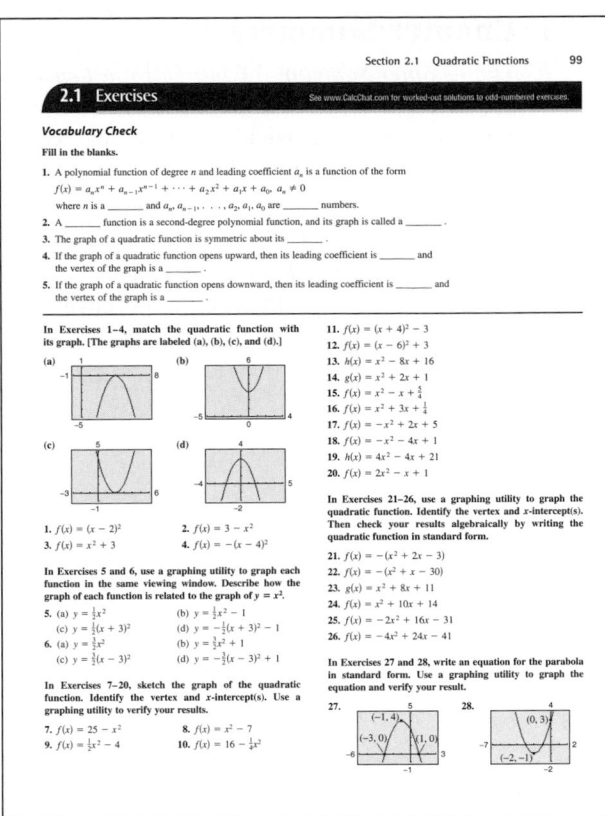

Section 2.1   Quadratic Functions   99

2.1 Exercises   See www.CalcChat.com for worked-out solutions to odd-numbered exercises.

Vocabulary Check

Fill in the blanks.

1. A polynomial function of degree n and leading coefficient a_n is a function of the form
$$f(x) = a_n x^n + a_{n-1} x^{n-1} + \cdots + a_2 x^2 + a_1 x + a_0, \quad a_n \neq 0$$
where n is a _____ and $a_n, a_{n-1}, \ldots, a_2, a_1, a_0$ are _____ numbers.
2. A _____ function is a second-degree polynomial function, and its graph is called a _____.
3. The graph of a quadratic function is symmetric about its _____.
4. If the graph of a quadratic function opens upward, then its leading coefficient is _____ and the vertex of the graph is a _____.
5. If the graph of a quadratic function opens downward, then its leading coefficient is _____ and the vertex of the graph is a _____.

In Exercises 1–4, match the quadratic function with its graph. [The graphs are labeled (a), (b), (c), and (d).]

(a)   (b)   (c)   (d)

1. $f(x) = (x - 2)^2$
2. $f(x) = 3 - x^2$
3. $f(x) = x^2 + 3$
4. $f(x) = -(x - 4)^2$

In Exercises 5 and 6, use a graphing utility to graph each function in the same viewing window. Describe how the graph of each function is related to the graph of $y = x^2$.

5. (a) $y = \frac{1}{2}x^2$ (b) $y = \frac{1}{2}x^2 - 1$
   (c) $y = \frac{1}{2}(x + 3)^2$ (d) $y = -\frac{1}{2}(x + 3)^2 - 1$
6. (a) $y = \frac{3}{2}x^2$ (b) $y = \frac{3}{2}x^2 + 1$
   (c) $y = \frac{3}{2}(x - 3)^2$ (d) $y = -\frac{3}{2}(x - 3)^2 + 1$

In Exercises 7–20, sketch the graph of the quadratic function. Identify the vertex and x-intercept(s). Use a graphing utility to verify your results.

7. $f(x) = 25 - x^2$
8. $f(x) = x^2 - 7$
9. $f(x) = \frac{1}{2}x^2 - 4$
10. $f(x) = 16 - \frac{1}{4}x^2$

11. $f(x) = (x + 4)^2 - 3$
12. $f(x) = (x - 6)^2 + 3$
13. $h(x) = x^2 - 8x + 16$
14. $g(x) = x^2 + 2x + 1$
15. $f(x) = x^2 - x + \frac{5}{4}$
16. $f(x) = x^2 + 3x + \frac{1}{4}$
17. $f(x) = -x^2 + 2x + 5$
18. $f(x) = -x^2 - 4x + 1$
19. $h(x) = 4x^2 - 4x + 21$
20. $f(x) = 2x^2 - x + 1$

In Exercises 21–26, use a graphing utility to graph the quadratic function. Identify the vertex and x-intercept(s). Then check your results algebraically by writing the quadratic function in standard form.

21. $f(x) = -(x^2 + 2x - 3)$
22. $f(x) = -(x^2 + x - 30)$
23. $g(x) = x^2 + 8x + 11$
24. $f(x) = x^2 + 10x + 14$
25. $f(x) = -2x^2 + 16x - 31$
26. $f(x) = -4x^2 + 24x - 41$

In Exercises 27 and 28, write an equation for the parabola in standard form. Use a graphing utility to graph the equation and verify your result.

27.   28.

Section Exercises

The section exercise sets consist of a variety of computational, conceptual, and applied problems.

Vocabulary Check

Section exercises begin with a *Vocabulary Check* that serves as a review of the important mathematical terms in each section.

New! Calc Chat

The worked-out solutions to the odd-numbered text exercises are now available at *www.CalcChat.com*.

Synthesis and Skills Review Exercises

Each exercise set concludes with three types of exercises.

Synthesis exercises promote further exploration of mathematical concepts, critical thinking skills, and writing about mathematics. The exercises require students to show their understanding of the relationships between many concepts in the section.

Skills Review Exercises reinforce previously learned skills and concepts.

New! *Make a Decision* exercises, found in selected sections, further connect real-life data and applications and motivate students. They also offer students the opportunity to generate and analyze mathematical models from large data sets.

---

102   Chapter 2   Polynomial and Rational Functions

62. **Data Analysis** The factory sales S of VCRs (in millions of dollars) in the United States from 1990 to 2004 can be modeled by $S = -28.40t^2 + 218.1t + 2435$, for $0 \le t \le 14$, where t is the year, with $t = 0$ corresponding to 1990. (Source: Consumer Electronics Association)
   (a) According to the model, when did the maximum value of factory sales of VCRs occur?
   (b) According to the model, what was the value of the factory sales in 2004? Explain your result.
   (c) Would you use the model to predict the value of the factory sales for years beyond 2004? Explain.

Synthesis

True or False? In Exercises 63 and 64, determine whether the statement is true or false. Justify your answer.

63. The function $f(x) = -12x^2 - 1$ has no x-intercepts.
64. The graphs of $f(x) = -4x^2 - 10x + 7$ and $g(x) = 12x^2 + 30x + 1$ have the same axis of symmetry.

Library of Parent Functions In Exercises 65 and 66, determine which equation(s) may be represented by the graph shown. (There may be more than one correct answer.)

65. (a) $f(x) = -(x - 4)^2 + 2$
   (b) $f(x) = -(x + 2)^2 + 4$
   (c) $f(x) = -(x + 2)^2 - 4$
   (d) $f(x) = -x^2 - 4x - 8$
   (e) $f(x) = -(x - 2)^2 - 4$
   (f) $f(x) = -x^2 + 4x - 8$

66. (a) $f(x) = (x - 1)^2 + 3$
   (b) $f(x) = (x + 1)^2 + 3$
   (c) $f(x) = (x - 3)^2 + 1$
   (d) $f(x) = x^2 + 2x + 4$
   (e) $f(x) = (x + 3)^2 + 1$
   (f) $f(x) = x^2 + 6x + 10$

Think About It In Exercises 67–70, find the value of b such that the function has the given maximum or minimum value.

67. $f(x) = -x^2 + bx - 75$; Maximum value: 25
68. $f(x) = -x^2 + bx - 16$; Maximum value: 48
69. $f(x) = x^2 + bx + 26$; Minimum value: 10
70. $f(x) = x^2 + bx - 25$; Minimum value: -50

71. **Profit** The profit P (in millions of dollars) for a recreational vehicle retailer is modeled by a quadratic function of the form $P = at^2 + bt + c$, where t represents the year. If you were president of the company, which of the following models would you prefer? Explain your reasoning.
   (a) a is positive and $t \ge -b/(2a)$.
   (b) a is positive and $t \le -b/(2a)$.
   (c) a is negative and $t \ge -b/(2a)$.
   (d) a is negative and $t \le -b/(2a)$.

72. **Writing** The parabola in the figure below has an equation of the form $y = ax^2 + bx - 4$. Find the equation of this parabola in two different ways, by hand and with technology (graphing utility or computer software). Write a paragraph describing the methods you used and comparing the results of the two methods.

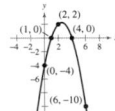

Skills Review

In Exercises 73–76, determine algebraically any point(s) of intersection of the graphs of the equations. Verify your results using the *intersect* feature of a graphing utility.

73. $x + y = 8$
   $-\frac{2}{3}x + y = 6$
74. $y = 3x - 10$
   $y = \frac{1}{4}x + 1$
75. $y = 9 - x^2$
   $y = x + 3$
76. $y = x^3 + 2x - 1$
   $y = -2x + 15$

77. **Make a Decision** To work an extended application analyzing the height of a basketball after it has been dropped, visit this textbook's *Online Study Center*.

FEATURES

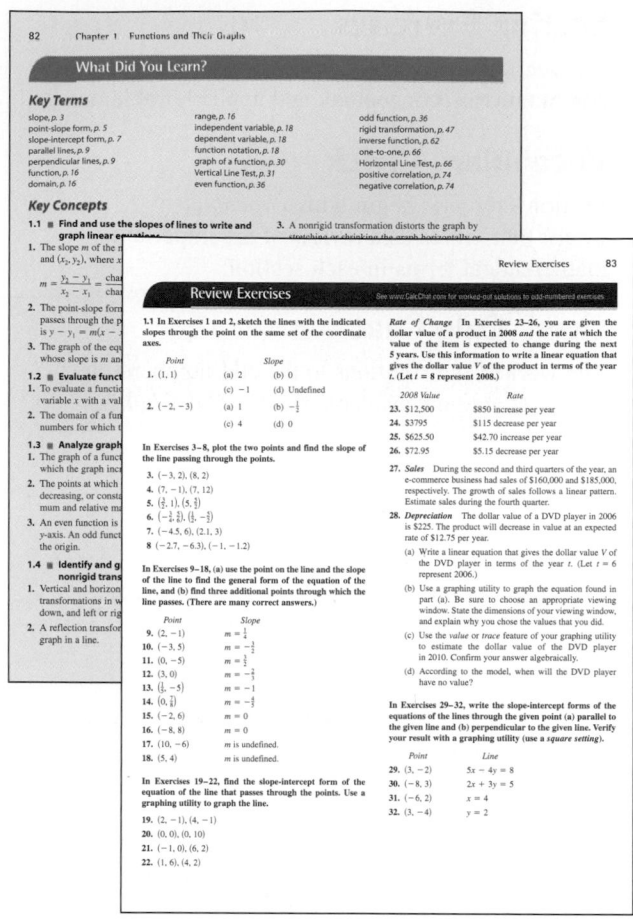

Chapter Summary

The *Chapter Summary* "*What Did You Learn?*" includes *Key Terms* with page references and *Key Concepts*, organized by section, that were covered throughout the chapter.

Review Exercises

The chapter *Review Exercises* provide additional practice with the concepts covered in the chapter.

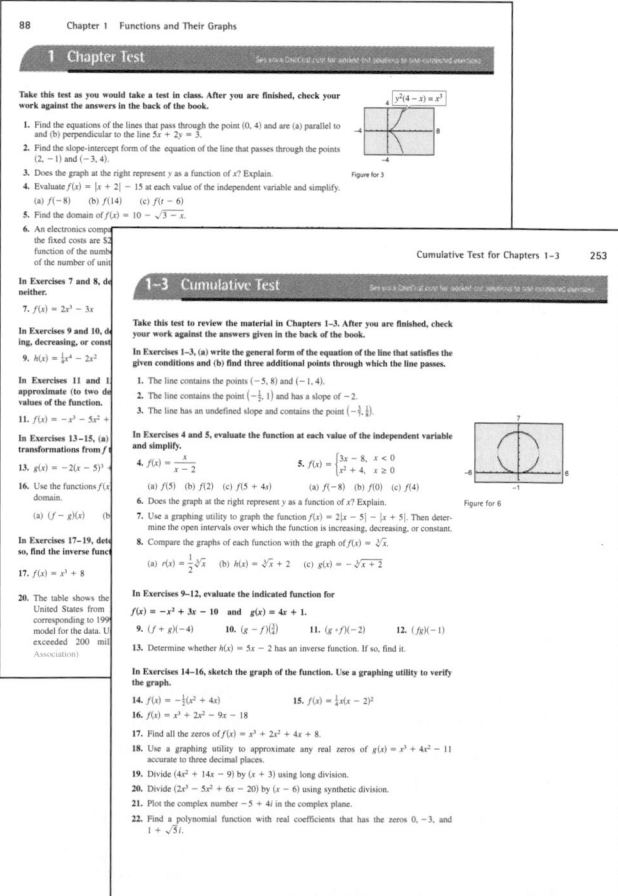

Chapter Tests and Cumulative Tests

Chapter Tests, at the end of each chapter, and periodic *Cumulative Tests* offer students frequent opportunities for self-assessment and to develop strong study and test-taking skills.

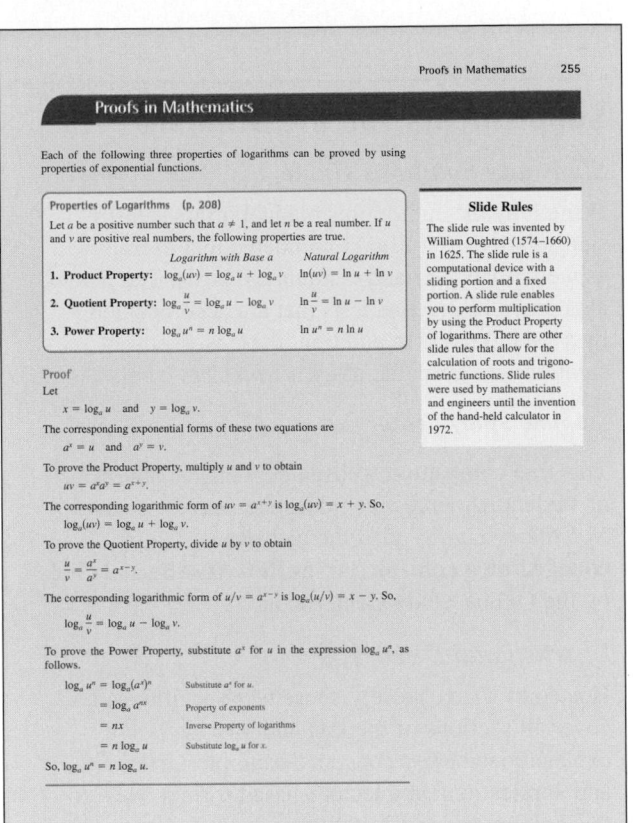

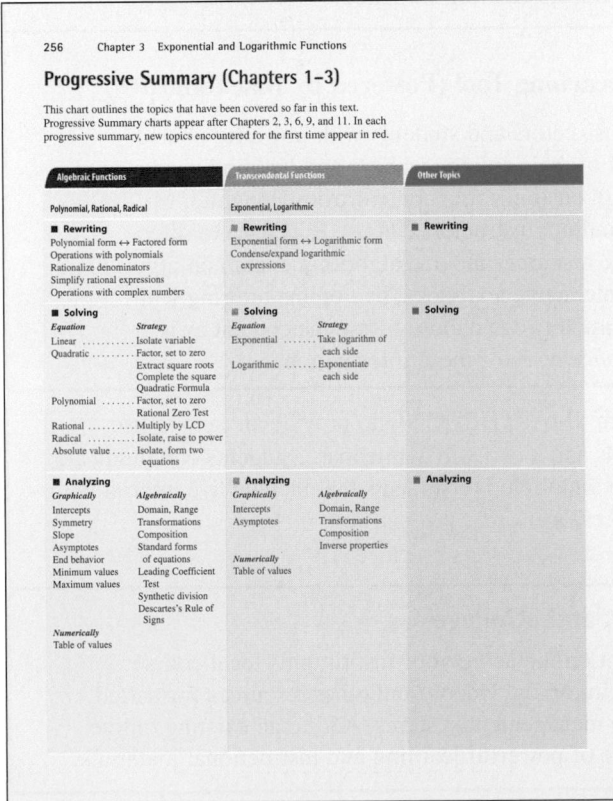

Proofs in Mathematics

At the end of every chapter, proofs of important mathematical properties and theorems are presented as well as discussions of various proof techniques.

New! Progressive Summaries

The *Progressive Summaries* are a series of charts that are usually placed at the end of every third chapter. Each *Progressive Summary* is completed in a gradual manner as new concepts are covered. Students can use the *Progressive Summaries* as a cumulative study aid and to see the connection between concepts and skills.

New! Study Capsules

Each *Study Capsule* in Appendix G summarizes many of the key concepts covered in previous chapters. A *Study Capsule* provides definitions, examples, and procedures for solving, simplifying, and graphing functions. Students can use this appendix as a quick reference when working on homework or studying for a test.

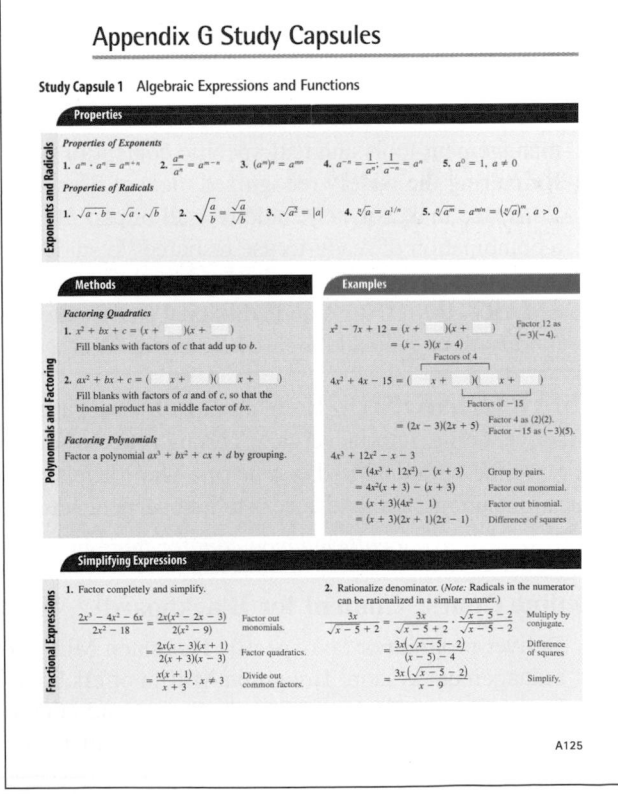

FEATURES

Supplements for the Instructor

Instructor's Annotated Edition (IAE)

Online Complete Solutions Guide

Online Instructor Success Organizer

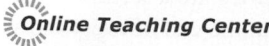 **Online Teaching Center**

This free companion website contains an abundance of instructors resources. Visit **college.hmco.com/pic/larsonPFGAGA5e** and click on the Online Teaching Center icon.

HM Testing™ (Powered by Diploma™) "Testing the way you want it"

HM Testing provides instructors all the tools they need to **create, author/edit, customize,** and **deliver** multiple types of tests. Instructors can use existing test bank content, edit the content, and author new static or algorithmic questions—all within Diploma's powerful electronic platform.

Supplements for the Student

Study and Solutions Guide

Written by the author, this manual offers step-by-step solutions for all odd-numbered text exercises as well as Chapter and Cumulative Tests. The manual also provides practice tests that are accompanied by a solution key. In addition, these worked-out solutions are available at *www.CalcChat.com*.

 **Online Study Center**

This free companion website contains an abundance of student resources including the *Online Student Notetaking Guide*. Visit the website **college.hmco.com/pic/larsonPFGAGA5e** and click on the Online Study Center icon.

Instructional DVDs

Hosted by Dana Mosely, these text-specific DVDs cover all sections of the text and provide key explanations of key concepts, examples, exercises, and applications in a lecture-based format. New to this edition, the DVDs will now include Captioning for the Hearing Impaired.

 **Eduspace®: Houghton Mifflin's Online Learning Tool (Powered by Blackboard™)**

Eduspace is a web-based learning system that provides instructors and students with powerful course management tools and text-specific content to support all of their online teaching and learning needs. By pairing the widely recognized tools of Blackboard with customizable content from Houghton Mifflin, Eduspace makes it easy to deliver all or part of a course online. Instructors can use Eduspace to offer a combination of ready-to-use resources to students. These resources include algorithmic and non-algorithmic homework, quizzes, tests, tutorials, instructional videos, interactive textbooks, live online tutoring with **SMARTHINKING®**, and additional study materials. Instructors can choose to use the content as is, modify it, or even add their own content. Visit *www.eduspace.com* for more information.

SMARTHINKING®: Houghton Mifflin has partnered with SMARTHINKING to provide an easy-to-use, effective, online tutorial service. Through state-of-the-art tools and a two-way whiteboard, students communicate in real-time with qualified e-structors. Three levels of service are offered to students that include live tutorial help, question submission, and access to independent study resources.

Visit *smarthinking.college.hmco.com* for more information.

Online Course Content for Blackboard™, WebCT®, and eCollege®

Deliver program or text-specific Houghton Mifflin content online using your institution's local course management system. Houghton Mifflin offers homework, tutorials, videos, and other resources formatted for Blackboard™, WebCT®, eCollege®, and other course management systems. Add to an existing online course or create a new one by selecting from a wide range of powerful learning and instructional materials.

SUPPLEMENTS

Chapter 1

Functions and Their Graphs

1.1 Lines in the Plane
1.2 Functions
1.3 Graphs of Functions
1.4 Shifting, Reflecting, and Stretching Graphs
1.5 Combinations of Functions
1.6 Inverse Functions
1.7 Linear Models and Scatter Plots

Selected Applications

Functions have many real-life applications. The applications listed below represent a small sample of the applications in this chapter.

- Rental Demand,
  Exercise 86, page 14
- Postal Regulations,
  Exercise 77, page 27
- Motor Vehicles,
  Exercise 83, page 28
- Fluid Flow,
  Exercise 92, page 40
- Finance,
  Exercise 58, page 50
- Bacteria,
  Exercise 81, page 61
- Consumer Awareness,
  Exercise 84, page 61
- Shoe Sizes,
  Exercises 103 and 104, page 71
- Cell Phones,
  Exercise 12, page 79

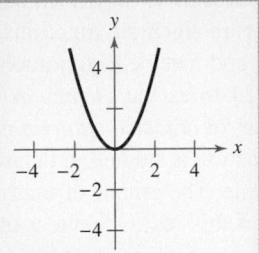 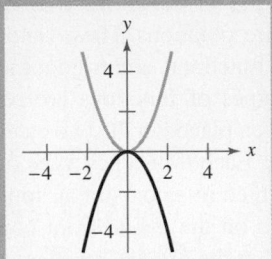 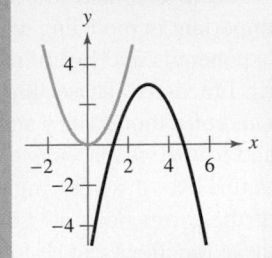

An equation in *x* and *y* defines a relationship between the two variables. The equation may be represented as a graph, providing another perspective on the relationship between *x* and *y*. In Chapter 1, you will learn how to write and graph linear equations, how to evaluate and find the domains and ranges of functions, and how to graph functions and their transformations.

© Index Stock Imagery

Refrigeration slows down the activity of bacteria in food so that it takes longer for the bacteria to spoil the food. The number of bacteria in a refrigerated food is a function of the amount of time the food has been out of refrigeration.

Introduction to Library of Parent Functions

In Chapter 1, you will be introduced to the concept of a *function*. As you proceed through the text, you will see that functions play a primary role in modeling real-life situations.

There are three basic types of functions that have proven to be the most important in modeling real-life situations. These functions are algebraic functions, exponential and logarithmic functions, and trigonometric and inverse trigonometric functions. These three types of functions are referred to as the *elementary functions*, though they are often placed in the two categories of *algebraic functions* and *transcendental functions*. Each time a new type of function is studied in detail in this text, it will be highlighted in a box similar to this one. The graphs of many of these functions are shown on the inside front cover of this text. A review of these functions can be found in the *Study Capsules*.

Algebraic Functions

These functions are formed by applying algebraic operations to the identity function $f(x) = x$.

Name	Function	Location
Linear	$f(x) = ax + b$	Section 1.1
Quadratic	$f(x) = ax^2 + bx + c$	Section 2.1
Cubic	$f(x) = ax^3 + bx^2 + cx + d$	Section 2.2
Polynomial	$P(x) = a_n x^n + a_{n-1}x^{n-1} + \cdots + a_2 x^2 + a_1 x + a_0$	Section 2.2
Rational	$f(x) = \dfrac{N(x)}{D(x)}$, $N(x)$ and $D(x)$ are polynomial functions	Section 2.6
Radical	$f(x) = \sqrt[n]{P(x)}$	Section 1.2

Transcendental Functions

These functions cannot be formed from the identity function by using algebraic operations.

Name	Function	Location
Exponential	$f(x) = a^x, a > 0, a \neq 1$	Section 3.1
Logarithmic	$f(x) = \log_a x, x > 0, a > 0, a \neq 1$	Section 3.2
Trigonometric	$f(x) = \sin x, f(x) = \cos x, f(x) = \tan x,$ $f(x) = \csc x, f(x) = \sec x, f(x) = \cot x$	Section 4.4
Inverse Trigonometric	$f(x) = \arcsin x, f(x) = \arccos x, f(x) = \arctan x$	Section 4.7

Nonelementary Functions

Some useful nonelementary functions include the following.

Name	Function	Location
Absolute value	$f(x) = |g(x)|$, $g(x)$ is an elementary function	Section 1.2
Piecewise-defined	$f(x) = \begin{cases} 3x + 2, & x \geq 1 \\ -2x + 4, & x < 1 \end{cases}$	Section 1.2
Greatest integer	$f(x) = [\![g(x)]\!]$, $g(x)$ is an elementary function	Section 1.3
Data defined	Formula for temperature: $F = \dfrac{9}{5}C + 32$	Section 1.2

1.1 Lines in the Plane

The Slope of a Line

In this section, you will study lines and their equations. The **slope** of a nonvertical line represents the number of units the line rises or falls vertically for each unit of horizontal change from left to right. For instance, consider the two points (x_1, y_1) and (x_2, y_2) on the line shown in Figure 1.1. As you move from left to right along this line, a change of $(y_2 - y_1)$ units in the vertical direction corresponds to a change of $(x_2 - x_1)$ units in the horizontal direction. That is,

$$y_2 - y_1 = \text{the change in } y$$

and

$$x_2 - x_1 = \text{the change in } x.$$

The slope of the line is given by the ratio of these two changes.

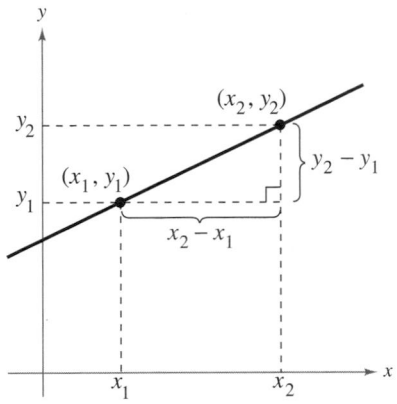

Figure 1.1

Definition of the Slope of a Line

The **slope** m of the nonvertical line through (x_1, y_1) and (x_2, y_2) is

$$m = \frac{y_2 - y_1}{x_2 - x_1} = \frac{\text{change in } y}{\text{change in } x}$$

where $x_1 \neq x_2$.

When this formula for slope is used, the *order of subtraction* is important. Given two points on a line, you are free to label either one of them as (x_1, y_1) and the other as (x_2, y_2). However, once you have done this, you must form the numerator and denominator using the same order of subtraction.

$$m = \frac{y_2 - y_1}{x_2 - x_1} \qquad\qquad m = \frac{y_1 - y_2}{x_1 - x_2} \qquad\qquad m = \frac{y_2 - y_1}{x_1 - x_2}$$

  Correct   Correct   Incorrect

Throughout this text, the term *line* always means a *straight* line.

Example 1   Finding the Slope of a Line

Find the slope of the line passing through each pair of points.

a. $(-2, 0)$ and $(3, 1)$ **b.** $(-1, 2)$ and $(2, 2)$ **c.** $(0, 4)$ and $(1, -1)$

Solution

Difference in y-values

a. $m = \dfrac{\overbrace{y_2 - y_1}}{\underbrace{x_2 - x_1}} = \dfrac{1 - 0}{3 - (-2)} = \dfrac{1}{3 + 2} = \dfrac{1}{5}$

Difference in x-values

b. $m = \dfrac{2 - 2}{2 - (-1)} = \dfrac{0}{3} = 0$

c. $m = \dfrac{-1 - 4}{1 - 0} = \dfrac{-5}{1} = -5$

The graphs of the three lines are shown in Figure 1.2. Note that the *square setting* gives the correct "steepness" of the lines.

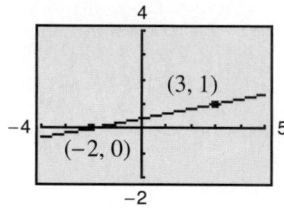

(a)

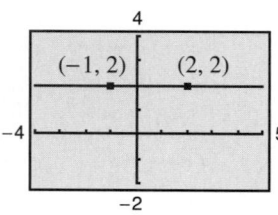

(b)

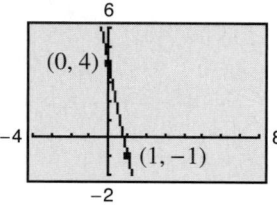

(c)

Figure 1.2

✓CHECKPOINT   Now try Exercise 9.

> The definition of slope does not apply to vertical lines. For instance, consider the points $(3, 4)$ and $(3, 1)$ on the vertical line shown in Figure 1.3. Applying the formula for slope, you obtain

$$m = \frac{4 - 1}{3 - 3} = \frac{3}{0}. \qquad \text{Undefined}$$

Because division by zero is undefined, the slope of a vertical line is undefined.

> From the slopes of the lines shown in Figures 1.2 and 1.3, you can make the following generalizations about the slope of a line.

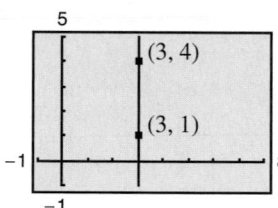

Figure 1.3

The Slope of a Line

1. A line with positive slope $(m > 0)$ *rises* from left to right.

2. A line with negative slope $(m < 0)$ *falls* from left to right.

3. A line with zero slope $(m = 0)$ is *horizontal*.

4. A line with undefined slope is *vertical*.

Exploration

Use a graphing utility to compare the slopes of the lines $y = 0.5x$, $y = x$, $y = 2x$, and $y = 4x$. What do you observe about these lines? Compare the slopes of the lines $y = -0.5x$, $y = -x$, $y = -2x$, and $y = -4x$. What do you observe about these lines? (*Hint:* Use a *square setting* to guarantee a true geometric perspective.)

The Point–Slope Form of the Equation of a Line

If you know the slope of a line *and* you also know the coordinates of one point on the line, you can find an equation for the line. For instance, in Figure 1.4, let (x_1, y_1) be a point on the line whose slope is m. If (x, y) is any *other* point on the line, it follows that

$$\frac{y - y_1}{x - x_1} = m.$$

This equation in the variables x and y can be rewritten in the **point-slope form** of the equation of a line.

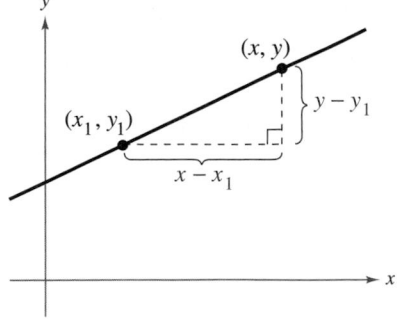

Figure 1.4

> Point-Slope Form of the Equation of a Line
>
> The **point-slope form** of the equation of the line that passes through the point (x_1, y_1) and has a slope of m is
>
> $$y - y_1 = m(x - x_1).$$

The point-slope form is most useful for finding the equation of a line if you know at least one point that the line passes through and the slope of the line. You should remember this form of the equation of a line.

Example 2   The Point–Slope Form of the Equation of a Line

Find an equation of the line that passes through the point $(1, -2)$ and has a slope of 3.

Solution

$$\begin{aligned} y - y_1 &= m(x - x_1) && \text{Point-slope form} \\ y - (-2) &= 3(x - 1) && \text{Substitute for } y_1, m, \text{ and } x_1. \\ y + 2 &= 3x - 3 && \text{Simplify.} \\ y &= 3x - 5 && \text{Solve for } y. \end{aligned}$$

The line is shown in Figure 1.5.

✓CHECKPOINT   Now try Exercise 25.

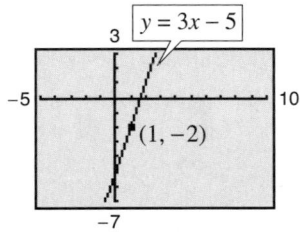

Figure 1.5

The point-slope form can be used to find an equation of a nonvertical line passing through two points (x_1, y_1) and (x_2, y_2). First, find the slope of the line.

$$m = \frac{y_2 - y_1}{x_2 - x_1}, \; x_1 \neq x_2$$

Then use the point-slope form to obtain the equation

$$y - y_1 = \frac{y_2 - y_1}{x_2 - x_1}(x - x_1).$$

This is sometimes called the **two-point form** of the equation of a line.

STUDY TIP

When you find an equation of the line that passes through two given points, you need to substitute the coordinates of only one of the points into the point-slope form. It does not matter which point you choose because both points will yield the same result.

Example 3 A Linear Model for Sales Prediction

During 2004, Nike's net sales were \$12.25 billion, and in 2005 net sales were \$13.74 billion. Write a linear equation giving the net sales y in terms of the year x. Then use the equation to predict the net sales for 2006.   (Source: Nike, Inc.)

Solution

Let $x = 0$ represent 2000. In Figure 1.6, let $(4, 12.25)$ and $(5, 13.74)$ be two points on the line representing the net sales. The slope of this line is

$$m = \frac{13.74 - 12.25}{5 - 4} = 1.49. \qquad m = \frac{y_2 - y_1}{x_2 - x_1}$$

By the point-slope form, the equation of the line is as follows.

$$y - 12.25 = 1.49(x - 4) \qquad \text{Write in point-slope form.}$$

$$y = 1.49x + 6.29 \qquad \text{Simplify.}$$

Now, using this equation, you can predict the 2006 net sales $(x = 6)$ to be

$$y = 1.49(6) + 6.29 = 8.94 + 6.29 = \$15.23 \text{ billion.}$$

✓**CHECKPOINT**   Now try Exercise 45.

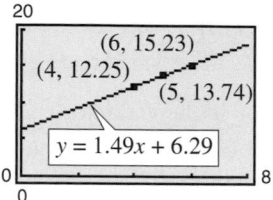

Figure 1.6

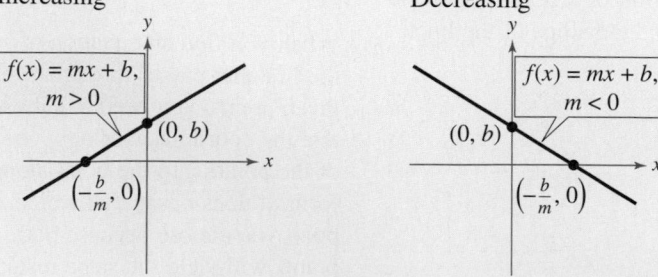

Library of Parent Functions: Linear Function

In the next section, you will be introduced to the precise meaning of the term *function*. The simplest type of function is a *linear function* of the form

$$f(x) = mx + b.$$

As its name implies, the graph of a linear function is a line that has a slope of m and a y-intercept at $(0, b)$. The basic characteristics of a linear function are summarized below. (Note that some of the terms below will be defined later in the text.) A review of linear functions can be found in the *Study Capsules.*

Graph of $f(x) = mx + b, m > 0$
Domain: $(-\infty, \infty)$
Range: $(-\infty, \infty)$
x-intercept: $(-b/m, 0)$
y-intercept: $(0, b)$
Increasing

Graph of $f(x) = mx + b, m < 0$
Domain: $(-\infty, \infty)$
Range: $(-\infty, \infty)$
x-intercept: $(-b/m, 0)$
y-intercept: $(0, b)$
Decreasing

When $m = 0$, the function $f(x) = b$ is called a *constant function* and its graph is a horizontal line.

STUDY TIP

The prediction method illustrated in Example 3 is called **linear extrapolation.** Note in the top figure below that an extrapolated point does not lie between the given points. When the estimated point lies between two given points, as shown in the bottom figure, the procedure used to predict the point is called **linear interpolation.**

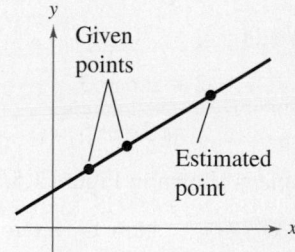

Linear Extrapolation

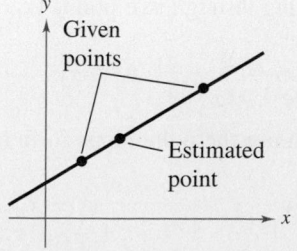

Linear Interpolation

Sketching Graphs of Lines

Many problems in coordinate geometry can be classified as follows.

1. Given a graph (or parts of it), find its equation.

2. Given an equation, sketch its graph.

For lines, the first problem is solved easily by using the point-slope form. This formula, however, is not particularly useful for solving the second type of problem. The form that is better suited to graphing linear equations is the **slope-intercept form** of the equation of a line, $y = mx + b$.

Slope-Intercept Form of the Equation of a Line

The graph of the equation

$$y = mx + b$$

is a line whose slope is m and whose y-intercept is $(0, b)$.

Example 4   Using the Slope-Intercept Form

Determine the slope and y-intercept of each linear equation. Then describe its graph.

a. $x + y = 2$ **b.** $y = 2$

Algebraic Solution

a. Begin by writing the equation in slope-intercept form.

$x + y = 2$	Write original equation.
$y = 2 - x$	Subtract x from each side.
$y = -x + 2$	Write in slope-intercept form.

From the slope-intercept form of the equation, the slope is -1 and the y-intercept is $(0, 2)$. Because the slope is negative, you know that the graph of the equation is a line that falls one unit for every unit it moves to the right.

b. By writing the equation $y = 2$ in slope-intercept form

$$y = (0)x + 2$$

you can see that the slope is 0 and the y-intercept is $(0, 2)$. A zero slope implies that the line is horizontal.

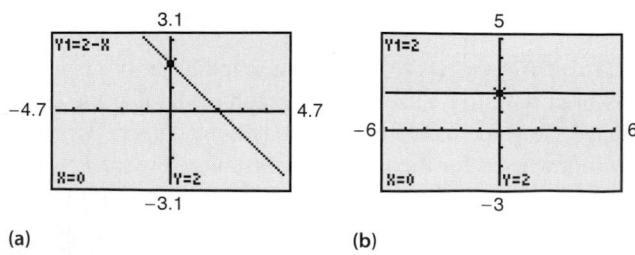

✓CHECKPOINT   Now try Exercise 47.

Graphical Solution

a. Solve the equation for y to obtain $y = 2 - x$. Enter this equation in your graphing utility. Use a decimal viewing window to graph the equation. To find the y-intercept, use the *value* or *trace* feature. When $x = 0$, $y = 2$, as shown in Figure 1.7(a). So, the y-intercept is $(0, 2)$. To find the slope, continue to use the *trace* feature. Move the cursor along the line until $x = 1$. At this point, $y = 1$. So the graph falls 1 unit for every unit it moves to the right, and the slope is -1.

b. Enter the equation $y = 2$ in your graphing utility and graph the equation. Use the *trace* feature to verify the y-intercept $(0, 2)$, as shown in Figure 1.7(b), and to see that the value of y is the same for all values of x. So, the slope of the horizontal line is 0.

(a)   **(b)**

Figure 1.7

From the slope-intercept form of the equation of a line, you can see that a horizontal line ($m = 0$) has an equation of the form $y = b$. This is consistent with the fact that each point on a horizontal line through $(0, b)$ has a y-coordinate of b. Similarly, each point on a vertical line through $(a, 0)$ has an x-coordinate of a. So, a vertical line has an equation of the form $x = a$. This equation cannot be written in slope-intercept form because the slope of a vertical line is undefined. However, *every* line has an equation that can be written in the **general form**

$$Ax + By + C = 0 \qquad \text{General form of the equation of a line}$$

where A and B are not *both* zero.

> **Summary of Equations of Lines**
>
> 1. General form: $Ax + By + C = 0$
> 2. Vertical line: $x = a$
> 3. Horizontal line: $y = b$
> 4. Slope-intercept form: $y = mx + b$
> 5. Point-slope form: $y - y_1 = m(x - x_1)$

> ### Exploration
>
> Graph the lines $y_1 = 2x + 1$, $y_2 = \frac{1}{2}x + 1$, and $y_3 = -2x + 1$ in the same viewing window. What do you observe?
>
> Graph the lines $y_1 = 2x + 1$, $y_2 = 2x$, and $y_3 = 2x - 1$ in the same viewing window. What do you observe?

Example 5 Different Viewing Windows

The graphs of the two lines

$$y = -x - 1 \qquad \text{and} \qquad y = -10x - 1$$

are shown in Figure 1.8. Even though the slopes of these lines are quite different (-1 and -10, respectively), the graphs seem misleadingly similar because the viewing windows are different.

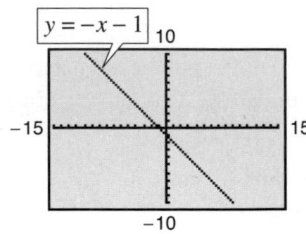   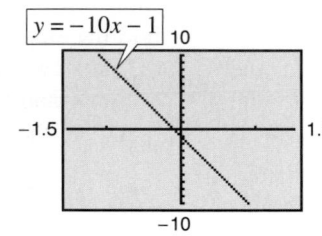

Figure 1.8

 ✓CHECKPOINT   Now try Exercise 51.

TECHNOLOGY TIP   When a graphing utility is used to graph a line, it is important to realize that the graph of the line may not visually appear to have the slope indicated by its equation. This occurs because of the viewing window used for the graph. For instance, Figure 1.9 shows graphs of $y = 2x + 1$ produced on a graphing utility using three different viewing windows. Notice that the slopes in Figures 1.9(a) and (b) do not visually appear to be equal to 2. However, if you use a *square setting*, as in Figure 1.9(c), the slope visually appears to be 2.

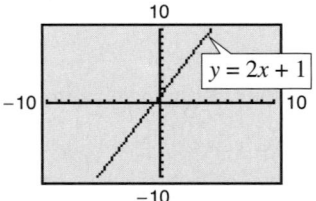

(a)

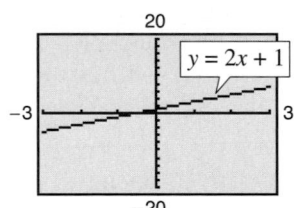

(b)

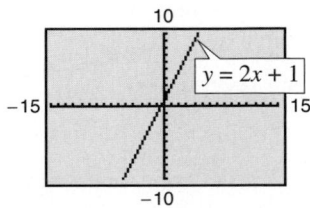

(c)
Figure 1.9

Parallel and Perpendicular Lines

The slope of a line is a convenient tool for determining whether two lines are parallel or perpendicular.

Parallel Lines

Two distinct nonvertical lines are **parallel** if and only if their slopes are equal. That is,

$$m_1 = m_2.$$

Example 6   Equations of Parallel Lines

Find the slope-intercept form of the equation of the line that passes through the point $(2, -1)$ and is parallel to the line $2x - 3y = 5$.

Solution

Begin by writing the equation of the given line in slope-intercept form.

$2x - 3y = 5$ Write original equation.

$-2x + 3y = -5$ Multiply by -1.

$3y = 2x - 5$ Add $2x$ to each side.

$y = \dfrac{2}{3}x - \dfrac{5}{3}$ Write in slope-intercept form.

Therefore, the given line has a slope of $m = \frac{2}{3}$. Any line parallel to the given line must also have a slope of $\frac{2}{3}$. So, the line through $(2, -1)$ has the following equation.

$y - (-1) = \dfrac{2}{3}(x - 2)$ Write in point-slope form.

$y + 1 = \dfrac{2}{3}x - \dfrac{4}{3}$ Simplify.

$y = \dfrac{2}{3}x - \dfrac{7}{3}$ Write in slope-intercept form.

Notice the similarity between the slope-intercept form of the original equation and the slope-intercept form of the parallel equation. The graphs of both equations are shown in Figure 1.10.

✔CHECKPOINT   Now try Exercise 57(a).

Perpendicular Lines

Two nonvertical lines are **perpendicular** if and only if their slopes are negative reciprocals of each other. That is,

$$m_1 = -\dfrac{1}{m_2}.$$

TECHNOLOGY TIP

Be careful when you graph equations such as $y = \frac{2}{3}x - \frac{7}{3}$ with your graphing utility. A common mistake is to type in the equation as

$$Y1 = 2/3X - 7/3$$

which may not be interpreted by your graphing utility as the original equation. You should use one of the following formulas.

$$Y1 = 2X/3 - 7/3$$
$$Y1 = (2/3)X - 7/3$$

Do you see why?

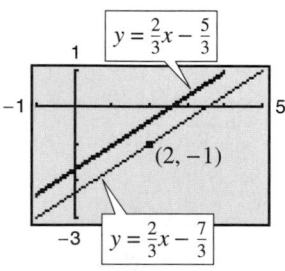

Figure 1.10

Example 7   Equations of Perpendicular Lines

Find the slope-intercept form of the equation of the line that passes through the point $(2, -1)$ and is perpendicular to the line

$$2x - 3y = 5.$$

Solution

From Example 6, you know that the equation can be written in the slope-intercept form $y = \frac{2}{3}x - \frac{5}{3}$. You can see that the line has a slope of $\frac{2}{3}$. So, any line perpendicular to this line must have a slope of $-\frac{3}{2}$ (because $-\frac{3}{2}$ is the negative reciprocal of $\frac{2}{3}$). So, the line through the point $(2, -1)$ has the following equation.

$$y - (-1) = -\frac{3}{2}(x - 2) \qquad \text{Write in point-slope form.}$$

$$y + 1 = -\frac{3}{2}x + 3 \qquad \text{Simplify.}$$

$$y = -\frac{3}{2}x + 2 \qquad \text{Write in slope-intercept form.}$$

The graphs of both equations are shown in Figure 1.11.

✓**CHECKPOINT**   Now try Exercise 57(b).

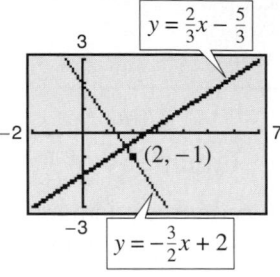

Figure 1.11

Example 8   Graphs of Perpendicular Lines

Use a graphing utility to graph the lines

$$y = x + 1$$

and

$$y = -x + 3$$

in the same viewing window. The lines are supposed to be perpendicular (they have slopes of $m_1 = 1$ and $m_2 = -1$). Do they appear to be perpendicular on the display?

Solution

If the viewing window is nonsquare, as in Figure 1.12, the two lines will not appear perpendicular. If, however, the viewing window is square, as in Figure 1.13, the lines will appear perpendicular.

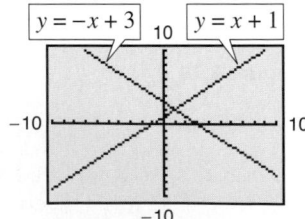

Figure 1.12

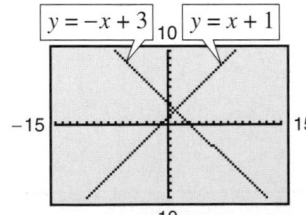

Figure 1.13

✓**CHECKPOINT**   Now try Exercise 67.

1.1 Exercises

See www.CalcChat.com for worked-out solutions to odd-numbered exercises.

Vocabulary Check

1. Match each equation with its form.

(a) $Ax + By + C = 0$ (i) vertical line

(b) $x = a$ (ii) slope-intercept form

(c) $y = b$ (iii) general form

(d) $y = mx + b$ (iv) point-slope form

(e) $y - y_1 = m(x - x_1)$ (v) horizontal line

In Exercises 2–5, fill in the blanks.

2. For a line, the ratio of the change in y to the change in x is called the _____ of the line.

3. Two lines are _____ if and only if their slopes are equal.

4. Two lines are _____ if and only if their slopes are negative reciprocals of each other.

5. The prediction method _____ is the method used to estimate a point on a line that does not lie between the given points.

In Exercises 1 and 2, identify the line that has the indicated slope.

1. (a) $m = \frac{2}{3}$ (b) m is undefined.   (c) $m = -2$

2. (a) $m = 0$ (b) $m = -\frac{3}{4}$ (c) $m = 1$

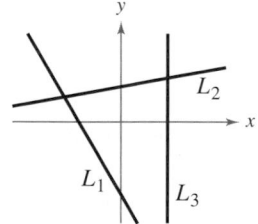

Figure for 1

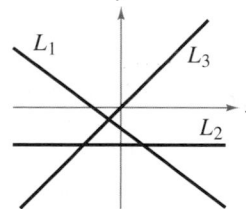

Figure for 2

In Exercises 3 and 4, sketch the lines through the point with the indicated slopes on the same set of coordinate axes.

	Point		*Slopes*		
3.	$(2, 3)$	(a) 0	(b) 1	(c) 2	(d) -3
4.	$(-4, 1)$	(a) 3	(b) -3	(c) $\frac{1}{2}$	(d) Undefined

In Exercises 5 and 6, estimate the slope of the line.

5.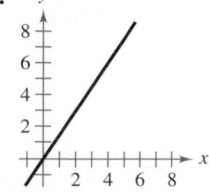

6.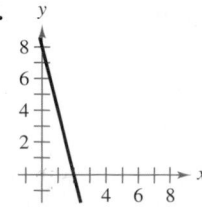

In Exercises 7–10, find the slope of the line passing through the pair of points. Then use a graphing utility to plot the points and use the *draw* feature to graph the line segment connecting the two points. (Use a *square setting*.)

7. $(0, -10), (-4, 0)$ **8.** $(2, 4), (4, -4)$

9. $(-6, -1), (-6, 4)$ **10.** $(-3, -2), (1, 6)$

In Exercises 11–18, use the point on the line and the slope of the line to find three additional points through which the line passes. (There are many correct answers.)

	Point	*Slope*
11.	$(2, 1)$	$m = 0$
12.	$(3, -2)$	$m = 0$
13.	$(1, 5)$	m is undefined.
14.	$(-4, 1)$	m is undefined.
15.	$(0, -9)$	$m = -2$
16.	$(-5, 4)$	$m = 2$
17.	$(7, -2)$	$m = \frac{1}{2}$
18.	$(-1, -6)$	$m = -\frac{1}{2}$

In Exercises 19–24, (a) find the slope and y-intercept (if possible) of the equation of the line algebraically, and (b) sketch the line by hand. Use a graphing utility to verify your answers to parts (a) and (b).

19. $5x - y + 3 = 0$ **20.** $2x + 3y - 9 = 0$

21. $5x - 2 = 0$ **22.** $3x + 7 = 0$

23. $3y + 5 = 0$ **24.** $-11 - 8y = 0$

In Exercises 25–32, find the general form of the equation of the line that passes through the given point and has the indicated slope. Sketch the line by hand. Use a graphing utility to verify your sketch, if possible.

Point	Slope
25. $(0, -2)$	$m = 3$
26. $(-3, 6)$	$m = -2$
27. $(2, -3)$	$m = -\frac{1}{2}$
28. $(-2, -5)$	$m = \frac{3}{4}$
29. $(6, -1)$	m is undefined.
30. $(-10, 4)$	m is undefined.
31. $\left(-\frac{1}{2}, \frac{3}{2}\right)$	$m = 0$
32. $(2.3, -8.5)$	$m = 0$

In Exercises 33–42, find the slope-intercept form of the equation of the line that passes through the points. Use a graphing utility to graph the line.

33. $(5, -1), (-5, 5)$

34. $(4, 3), (-4, -4)$

35. $(-8, 1), (-8, 7)$

36. $(-1, 4), (6, 4)$

37. $\left(2, \frac{1}{2}\right), \left(\frac{1}{2}, \frac{5}{4}\right)$

38. $(1, 1), \left(6, -\frac{2}{3}\right)$

39. $\left(-\frac{1}{10}, -\frac{3}{5}\right), \left(\frac{9}{10}, -\frac{9}{5}\right)$

40. $\left(\frac{3}{4}, \frac{3}{2}\right), \left(-\frac{4}{3}, \frac{7}{4}\right)$

41. $(1, 0.6), (-2, -0.6)$

42. $(-8, 0.6), (2, -2.4)$

In Exercises 43 and 44, find the slope-intercept form of the equation of the line shown.

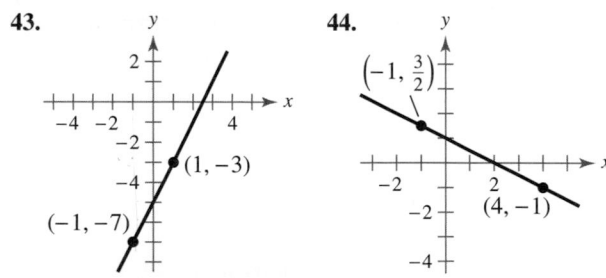

43.

44.

45. *Annual Salary* A jeweler's salary was $28,500 in 2004 and $32,900 in 2006. The jeweler's salary follows a linear growth pattern. What will the jeweler's salary be in 2008?

46. *Annual Salary* A librarian's salary was $25,000 in 2004 and $27,500 in 2006. The librarian's salary follows a linear growth pattern. What will the librarian's salary be in 2008?

In Exercises 47–50, determine the slope and y-intercept of the linear equation. Then describe its graph.

47. $x - 2y = 4$

48. $3x + 4y = 1$

49. $x = -6$

50. $y = 12$

In Exercises 51 and 52, use a graphing utility to graph the equation using each of the suggested viewing windows. Describe the difference between the two graphs.

51. $y = 0.5x - 3$

Xmin = -5	Xmin = -2
Xmax = 10	Xmax = 10
Xscl = 1	Xscl = 1
Ymin = -1	Ymin = -4
Ymax = 10	Ymax = 1
Yscl = 1	Yscl = 1

52. $y = -8x + 5$

Xmin = -5	Xmin = -5
Xmax = 5	Xmax = 10
Xscl = 1	Xscl = 1
Ymin = -10	Ymin = -80
Ymax = 10	Ymax = 80
Yscl = 1	Yscl = 20

In Exercises 53–56, determine whether the lines L_1 and L_2 passing through the pairs of points are parallel, perpendicular, or neither.

53. L_1: $(0, -1), (5, 9)$
L_2: $(0, 3), (4, 1)$

54. L_1: $(-2, -1), (1, 5)$
L_2: $(1, 3), (5, -5)$

55. L_1: $(3, 6), (-6, 0)$
L_2: $(0, -1), \left(5, \frac{7}{3}\right)$

56. L_1: $(4, 8), (-4, 2)$
L_2: $(3, -5), \left(-1, \frac{1}{3}\right)$

In Exercises 57–62, write the slope-intercept forms of the equations of the lines through the given point (a) parallel to the given line and (b) perpendicular to the given line.

Point	Line
57. $(2, 1)$	$4x - 2y = 3$
58. $(-3, 2)$	$x + y = 7$
59. $\left(-\frac{2}{3}, \frac{7}{8}\right)$	$3x + 4y = 7$
60. $(-3.9, -1.4)$	$6x + 2y = 9$
61. $(3, -2)$	$x - 4 = 0$
62. $(-4, 1)$	$y + 2 = 0$

In Exercises 63 and 64, the lines are parallel. Find the slope-intercept form of the equation of line y_2.

63.

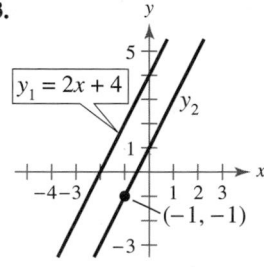

64.

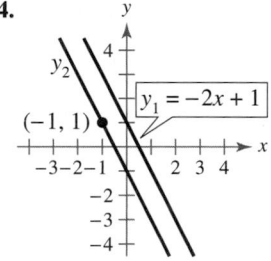

In Exercises 65 and 66, the lines are perpendicular. Find the slope-intercept form of the equation of line y_2.

65.   **66.**

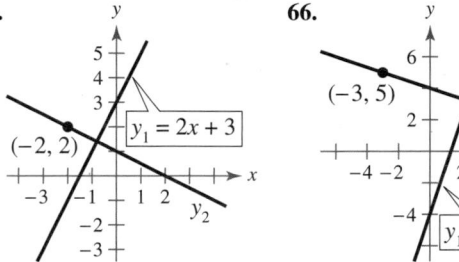

Graphical Analysis   **In Exercises 67–70, identify any relationships that exist among the lines, and then use a graphing utility to graph the three equations in the same viewing window. Adjust the viewing window so that each slope appears visually correct. Use the slopes of the lines to verify your results.**

67. (a) $y = 2x$ (b) $y = -2x$ (c) $y = \frac{1}{2}x$

68. (a) $y = \frac{2}{3}x$ (b) $y = -\frac{3}{2}x$ (c) $y = \frac{2}{3}x + 2$

69. (a) $y = -\frac{1}{2}x$ (b) $y = -\frac{1}{2}x + 3$ (c) $y = 2x - 4$

70. (a) $y = x - 8$ (b) $y = x + 1$ (c) $y = -x + 3$

71. *Earnings per Share*   The graph shows the earnings per share of stock for Circuit City for the years 1995 through 2004.   (Source: Circuit City Stores, Inc.)

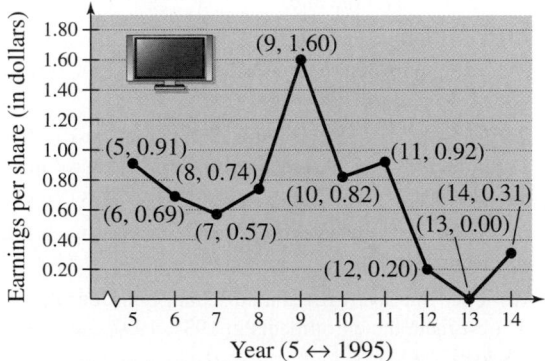

(a) Use the slopes to determine the years in which the earnings per share of stock showed the greatest increase and greatest decrease.

(b) Find the equation of the line between the years 1995 and 2004.

(c) Interpret the meaning of the slope of the equation from part (b) in the context of the problem.

(d) Use the equation from part (b) to estimate the earnings per share of stock in the year 2010. Do you think this is an accurate estimation? Explain.

72. *Sales*   The graph shows the sales (in billions of dollars) for Goodyear Tire for the years 1995 through 2004, where $t = 5$ represents 1995.   (Source: Goodyear Tire)

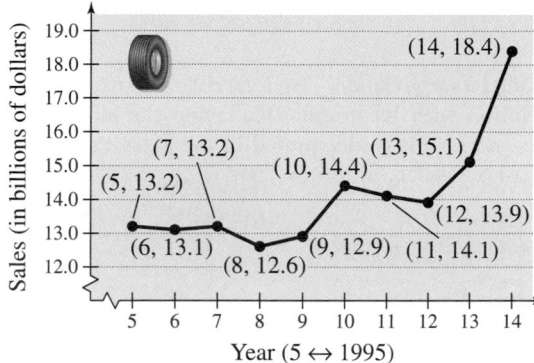

(a) Use the slopes to determine the years in which the sales for Goodyear Tire showed the greatest increase and the smallest increase.

(b) Find the equation of the line between the years 1995 and 2004.

(c) Interpret the meaning of the slope of the equation from part (b) in the context of the problem.

(d) Use the equation from part (b) to estimate the sales for Goodyear Tire in the year 2010. Do you think this is an accurate estimation? Explain.

73. *Height*   The "rise to run" ratio of the roof of a house determines the steepness of the roof. The rise to run ratio of the roof in the figure is 3 to 4. Determine the maximum height in the attic of the house if the house is 32 feet wide.

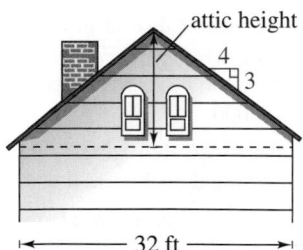

74. *Road Grade*   When driving down a mountain road, you notice warning signs indicating that it is a "12% grade." This means that the slope of the road is $-\frac{12}{100}$. Approximate the amount of horizontal change in your position if you note from elevation markers that you have descended 2000 feet vertically.

Rate of Change In Exercises 75–78, you are given the dollar value of a product in 2006 and the rate at which the value of the product is expected to change during the next 5 years. Write a linear equation that gives the dollar value *V* of the product in terms of the year *t*. (Let $t = 6$ represent 2006.)

	2006 Value	Rate
75.	$2540	$125 increase per year
76.	$156	$4.50 increase per year
77.	$20,400	$2000 decrease per year
78.	$245,000	$5600 decrease per year

Graphical Interpretation In Exercises 79–82, match the description with its graph. Determine the slope of each graph and how it is interpreted in the given context. [The graphs are labeled (a), (b), (c), and (d).]

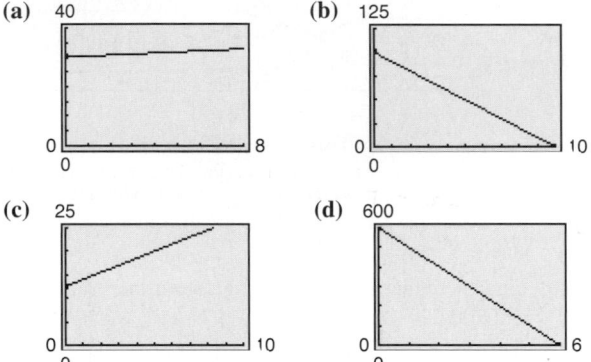

(a)   40   **(b)**   125

(c)   25   **(d)**   600

79. You are paying $10 per week to repay a $100 loan.

80. An employee is paid $12.50 per hour plus $1.50 for each unit produced per hour.

81. A sales representative receives $30 per day for food plus $.35 for each mile traveled.

82. A computer that was purchased for $600 depreciates $100 per year.

83. ***Depreciation*** A school district purchases a high-volume printer, copier, and scanner for $25,000. After 10 years, the equipment will have to be replaced. Its value at that time is expected to be $2000.

(a) Write a linear equation giving the value *V* of the equipment during the 10 years it will be used.

(b) Use a graphing utility to graph the linear equation representing the depreciation of the equipment, and use the *value* or *trace* feature to complete the table.

t	0	1	2	3	4	5	6	7	8	9	10
V											

(c) Verify your answers in part (b) algebraically by using the equation you found in part (a).

84. ***Meteorology*** Recall that water freezes at 0°C (32°F) and boils at 100°C (212°F).

(a) Find an equation of the line that shows the relationship between the temperature in degrees Celsius *C* and degrees Fahrenheit *F*.

(b) Use the result of part (a) to complete the table.

C		−10°	10°			177°
F	0°			68°	90°	

85. ***Cost, Revenue, and Profit*** A contractor purchases a bulldozer for $36,500. The bulldozer requires an average expenditure of $5.25 per hour for fuel and maintenance, and the operator is paid $11.50 per hour.

(a) Write a linear equation giving the total cost *C* of operating the bulldozer for *t* hours. (Include the purchase cost of the bulldozer.)

(b) Assuming that customers are charged $27 per hour of bulldozer use, write an equation for the revenue *R* derived from *t* hours of use.

(c) Use the profit formula $(P = R - C)$ to write an equation for the profit derived from *t* hours of use.

(d) Use the result of part (c) to find the break-even point (the number of hours the bulldozer must be used to yield a profit of 0 dollars).

86. ***Rental Demand*** A real estate office handles an apartment complex with 50 units. When the rent per unit is $580 per month, all 50 units are occupied. However, when the rent is $625 per month, the average number of occupied units drops to 47. Assume that the relationship between the monthly rent *p* and the demand *x* is linear.

(a) Write the equation of the line giving the demand *x* in terms of the rent *p*.

(b) Use a graphing utility to graph the demand equation and use the *trace* feature to estimate the number of units occupied when the rent is $655. Verify your answer algebraically.

(c) Use the demand equation to predict the number of units occupied when the rent is lowered to $595. Verify your answer graphically.

87. ***Education*** In 1991, Penn State University had an enrollment of 75,349 students. By 2005, the enrollment had increased to 80,124.   (Source: Penn State Fact Book)

(a) What was the average annual change in enrollment from 1991 to 2005?

(b) Use the average annual change in enrollment to estimate the enrollments in 1984, 1997, and 2000.

(c) Write the equation of a line that represents the given data. What is its slope? Interpret the slope in the context of the problem.

88. *Writing*   Using the results of Exercise 87, write a short paragraph discussing the concepts of *slope* and *average rate of change*.

Synthesis

True or False?   **In Exercises 89 and 90, determine whether the statement is true or false. Justify your answer.**

89. The line through $(-8, 2)$ and $(-1, 4)$ and the line through $(0, -4)$ and $(-7, 7)$ are parallel.

90. If the points $(10, -3)$ and $(2, -9)$ lie on the same line, then the point $\left(-12, -\frac{37}{2}\right)$ also lies on that line.

Exploration   **In Exercises 91–94, use a graphing utility to graph the equation of the line in the form**

$$\frac{x}{a} + \frac{y}{b} = 1, \quad a \neq 0, b \neq 0.$$

Use the graphs to make a conjecture about what a and b represent. Verify your conjecture.

91. $\dfrac{x}{5} + \dfrac{y}{-3} = 1$ **92.** $\dfrac{x}{-6} + \dfrac{y}{2} = 1$

93. $\dfrac{x}{4} + \dfrac{y}{-\frac{2}{3}} = 1$ **94.** $\dfrac{x}{\frac{1}{2}} + \dfrac{y}{5} = 1$

In Exercises 95–98, use the results of Exercises 91–94 to write an equation of the line that passes through the points.

95. x-intercept: $(2, 0)$ **96.** x-intercept: $(-5, 0)$
    y-intercept: $(0, 3)$ y-intercept: $(0, -4)$
97. x-intercept: $\left(-\frac{1}{6}, 0\right)$ **98.** x-intercept: $\left(\frac{3}{4}, 0\right)$
    y-intercept: $\left(0, -\frac{2}{3}\right)$ y-intercept: $\left(0, \frac{4}{5}\right)$

Library of Parent Functions   **In Exercises 99 and 100, determine which equation(s) may be represented by the graph shown. (There may be more than one correct answer.)**

99. 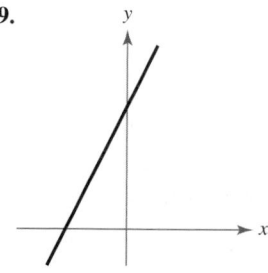   **100.**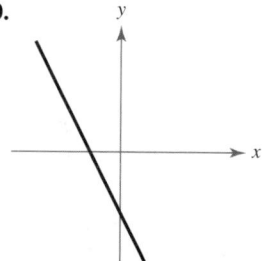

(a) $2x - y = -10$ (a) $2x + y = 5$
(b) $2x + y = 10$ (b) $2x + y = -5$
(c) $x - 2y = 10$ (c) $x - 2y = 5$
(d) $x + 2y = 10$ (d) $x - 2y = -5$

Library of Parent Functions   **In Exercises 101 and 102, determine which pair of equations may be represented by the graphs shown.**

101. 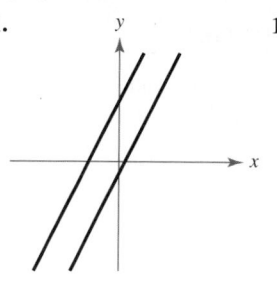   **102.**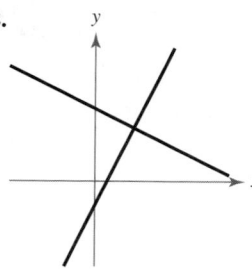

(a) $2x - y = 5$ (a) $2x - y = 2$
    $2x - y = 1$ $x + 2y = 12$
(b) $2x + y = -5$ (b) $x - y = 1$
    $2x + y = 1$ $x + y = 6$
(c) $2x - y = -5$ (c) $2x + y = 2$
    $2x - y = 1$ $x - 2y = 12$
(d) $x - 2y = -5$ (d) $x - 2y = 2$
    $x - 2y = -1$ $x + 2y = 12$

103. *Think About It*   Does every line have both an x-intercept and a y-intercept? Explain.

104. *Think About It*   Can every line be written in slope-intercept form? Explain.

105. *Think About It*   Does every line have an infinite number of lines that are parallel to the given line? Explain.

106. *Think About It*   Does every line have an infinite number of lines that are perpendicular to the given line? Explain.

Skills Review

In Exercises 107–112, determine whether the expression is a polynomial. If it is, write the polynomial in standard form.

107. $x + 20$ **108.** $3x - 10x^2 + 1$
109. $4x^2 + x^{-1} - 3$ **110.** $2x^2 - 2x^4 - x^3 + 2$
111. $\dfrac{x^2 + 3x + 4}{x^2 - 9}$ **112.** $\sqrt{x^2 + 7x + 6}$

In Exercises 113–116, factor the trinomial.

113. $x^2 - 6x - 27$ **114.** $x^2 - 11x + 28$
115. $2x^2 + 11x - 40$ **116.** $3x^2 - 16x + 5$

117. **Make a Decision** To work an extended application analyzing the numbers of bachelor's degrees earned by women in the United States from 1985 to 2005, visit this textbook's *Online Study Center*. (Data Source: U.S. Census Bureau)

The *Make a Decision* exercise indicates a multipart exercise using large data sets. Go to this textbook's *Online Study Center* to view these exercises.

1.2 Functions

Introduction to Functions

Many everyday phenomena involve pairs of quantities that are related to each other by some rule of correspondence. The mathematical term for such a rule of correspondence is a **relation.** Here are two examples.

1. The simple interest I earned on an investment of $1000 for 1 year is related to the annual interest rate r by the formula $I = 1000r$.

2. The area A of a circle is related to its radius r by the formula $A = \pi r^2$.

Not all relations have simple mathematical formulas. For instance, people commonly match up NFL starting quarterbacks with touchdown passes, and hours of the day with temperature. In each of these cases, there is some relation that matches each item from one set with exactly one item from a different set. Such a relation is called a **function.**

> **Definition of a Function**
>
> A **function** f from a set A to a set B is a relation that assigns to each element x in the set A exactly one element y in the set B. The set A is the **domain** (or set of inputs) of the function f, and the set B contains the **range** (or set of outputs).

To help understand this definition, look at the function that relates the time of day to the temperature in Figure 1.14.

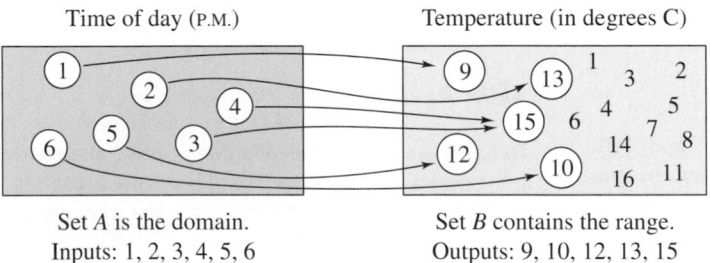

Set A is the domain.
Inputs: 1, 2, 3, 4, 5, 6

Set B contains the range.
Outputs: 9, 10, 12, 13, 15

Figure 1.14

This function can be represented by the ordered pairs $\{(1, 9°), (2, 13°), (3, 15°), (4, 15°), (5, 12°), (6, 10°)\}$. In each ordered pair, the first coordinate (x-value) is the **input** and the second coordinate (y-value) is the **output.**

> **Characteristics of a Function from Set A to Set B**
>
> 1. Each element of A must be matched with an element of B.
> 2. Some elements of B may not be matched with any element of A.
> 3. Two or more elements of A may be matched with the same element of B.
> 4. An element of A (the domain) cannot be matched with two different elements of B.

Library of Functions: Data Defined Function

Many functions do not have simple mathematical formulas, but are defined by real-life data. Such functions arise when you are using collections of data to model real-life applications. Functions can be represented in four ways.

1. *Verbally* by a sentence that describes how the input variables are related to the output variables

   Example: The input value x is the election year from 1952 to 2004 and the output value y is the elected president of the United States.

2. *Numerically* by a table or a list of ordered pairs that matches input values with output values

   Example: In the set of ordered pairs {(2, 34), (4, 40), (6, 45), (8, 50), (10, 54)}, the input value is the age of a male child in years and the output value is the height of the child in inches.

3. *Graphically* by points on a graph in a coordinate plane in which the input values are represented by the horizontal axis and the output values are represented by the vertical axis

   Example: See Figure 1.15.

4. *Algebraically* by an equation in two variables

   Example: The formula for temperature, $F = \frac{9}{5}C + 32$, where F is the temperature in degrees Fahrenheit and C is the temperature in degrees Celsius, is an equation that represents a function. You will see that it is often convenient to approximate data using a mathematical model or formula.

STUDY TIP

To determine whether or not a relation is a function, you must decide whether each input value is matched with exactly one output value. If any input value is matched with two or more output values, the relation is not a function.

Example 1 Testing for Functions

Decide whether the relation represents y as a function of x.

a.

Input, x	2	2	3	4	5
Output, y	11	10	8	5	1

b.

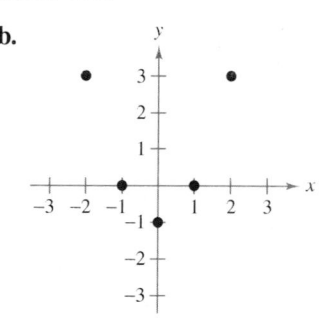

Figure 1.15

Prerequisite Skills

When plotting points in a coordinate plane, the x-coordinate is the directed distance from the y-axis to the point, and the y-coordinate is the directed distance from the x-axis to the point. To review point plotting, see Appendix B.1.

Solution

a. This table *does not* describe y as a function of x. The input value 2 is matched with two different y-values.

b. The graph in Figure 1.15 *does* describe y as a function of x. Each input value is matched with exactly one output value.

✔*CHECKPOINT*   Now try Exercise 5.

STUDY TIP

Be sure you see that the *range* of a function is not the same as the use of *range* relating to the viewing window of a graphing utility.

In algebra, it is common to represent functions by equations or formulas involving two variables. For instance, the equation $y = x^2$ represents the variable y as a function of the variable x. In this equation, x is the **independent variable** and y is the **dependent variable.** The domain of the function is the set of all values taken on by the independent variable x, and the range of the function is the set of all values taken on by the dependent variable y.

Example 2   Testing for Functions Represented Algebraically

Which of the equations represent(s) y as a function of x?

a. $x^2 + y = 1$ **b.** $-x + y^2 = 1$

Solution

To determine whether y is a function of x, try to solve for y in terms of x.

a. Solving for y yields

$$x^2 + y = 1$$ Write original equation.

$$y = 1 - x^2.$$ Solve for y.

Each value of x corresponds to exactly one value of y. So, y is a function of x.

b. Solving for y yields

$$-x + y^2 = 1$$ Write original equation.

$$y^2 = 1 + x$$ Add x to each side.

$$y = \pm\sqrt{1 + x}.$$ Solve for y.

The $\pm$ indicates that for a given value of x there correspond two values of y. For instance, when $x = 3$, $y = 2$ or $y = -2$. So, y is not a function of x.

✓CHECKPOINT   Now try Exercise 19.

Function Notation

When an equation is used to represent a function, it is convenient to name the function so that it can be referenced easily. For example, you know that the equation $y = 1 - x^2$ describes y as a function of x. Suppose you give this function the name "f." Then you can use the following **function notation.**

Input	Output	Equation
x	$f(x)$	$f(x) = 1 - x^2$

The symbol $f(x)$ is read as the *value of f at x* or simply *f of x.* The symbol $f(x)$ corresponds to the y-value for a given x. So, you can write $y = f(x)$. Keep in mind that f is the *name* of the function, whereas $f(x)$ is the *output value* of the function at the *input value x.* In function notation, the *input* is the independent variable and the *output* is the dependent variable. For instance, the function $f(x) = 3 - 2x$ has *function values* denoted by $f(-1), f(0)$, and so on. To find these values, substitute the specified input values into the given equation.

input

For $x = -1$, $f(-1) = 3 - 2(-1) = 3 + 2 = 5.$

For $x = 0$, $f(0) = 3 - 2(0) = 3 - 0 = 3.$

Exploration

Use a graphing utility to graph $x^2 + y = 1$. Then use the graph to write a convincing argument that each x-value has at most one y-value.

Use a graphing utility to graph $-x + y^2 = 1$. (*Hint:* You will need to use two equations.) Does the graph represent y as a function of x? Explain.

TECHNOLOGY TIP

You can use a graphing utility to evaluate a function. Go to this textbook's *Online Study Center* and use the Evaluating an Algebraic Expression program. The program will prompt you for a value of x, and then evaluate the expression in the equation editor for that value of x. Try using the program to evaluate several different functions of x.

f is (this input) = 3 - 2x

Although f is often used as a convenient function name and x is often used as the independent variable, you can use other letters. For instance,

$$f(x) = x^2 - 4x + 7, \quad f(t) = t^2 - 4t + 7, \quad \text{and} \quad g(s) = s^2 - 4s + 7$$

all define the same function. In fact, the role of the independent variable is that of a "placeholder." Consequently, the function could be written as

$$f(\quad) = (\quad)^2 - 4(\quad) + 7.$$

Example 3   Evaluating a Function

Let $g(x) = -x^2 + 4x + 1$. Find (a) $g(2)$, (b) $g(t)$, and (c) $g(x + 2)$.

Solution

a. Replacing x with 2 in $g(x) = -x^2 + 4x + 1$ yields the following.

$$g(2) = -(2)^2 + 4(2) + 1 = -4 + 8 + 1 = 5$$

b. Replacing x with t yields the following.

$$g(t) = -(t)^2 + 4(t) + 1 = -t^2 + 4t + 1$$

c. Replacing x with $x + 2$ yields the following.

$$g(x + 2) = -(x + 2)^2 + 4(x + 2) + 1 \qquad \text{Substitute } x + 2 \text{ for } x.$$

$$= -(x^2 + 4x + 4) + 4x + 8 + 1 \qquad \text{Multiply.}$$

$$= -x^2 - 4x - 4 + 4x + 8 + 1 \qquad \text{Distributive Property}$$

$$= -x^2 + 5 \qquad \text{Simplify.}$$

✓*CHECKPOINT*   Now try Exercise 29.

In Example 3, note that $g(x + 2)$ is not equal to $g(x) + g(2)$. In general, $g(u + v) \neq g(u) + g(v)$.

Library of Parent Functions: Piecewise-Defined Function

A *piecewise-defined function* is a function that is defined by two or more equations over a specified domain. The *absolute value function* given by $f(x) = |x|$ can be written as a piecewise-defined function. The basic characteristics of the absolute value function are summarized below. A review of piecewise-defined functions can be found in the *Study Capsules*.

Graph of $f(x) = |x| = \begin{cases} x, & x \geq 0 \\ -x, & x < 0 \end{cases}$

Domain: $(-\infty, \infty)$
Range: $[0, \infty)$
Intercept: $(0, 0)$
Decreasing on $(-\infty, 0)$
Increasing on $(0, \infty)$

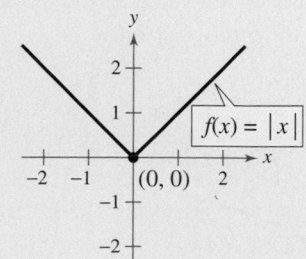

Example 4   A Piecewise–Defined Function

Evaluate the function when $x = -1$ and $x = 0$.

$$f(x) = \begin{cases} x^2 + 1, & x < 0 \\ x - 1, & x \geq 0 \end{cases}$$

Solution

Because $x = -1$ is less than 0, use $f(x) = x^2 + 1$ to obtain

$$f(-1) = (-1)^2 + 1 = 2.$$

For $x = 0$, use $f(x) = x - 1$ to obtain

$$f(0) = 0 - 1 = -1.$$

✓**CHECKPOINT**   Now try Exercise 37.

> **TECHNOLOGY TIP**
>
> Most graphing utilities can graph piecewise-defined functions. For instructions on how to enter a piecewise-defined function into your graphing utility, consult your user's manual. You may find it helpful to set your graphing utility to *dot mode* before graphing such functions.

The Domain of a Function

The domain of a function can be described explicitly or it can be *implied* by the expression used to define the function. The **implied domain** is the set of all real numbers for which the expression is defined. For instance, the function

$$f(x) = \frac{1}{x^2 - 4}$$
    Domain excludes x-values that result in division by zero.

has an implied domain that consists of all real x other than $x = \pm 2$. These two values are excluded from the domain because division by zero is undefined. Another common type of implied domain is that used to avoid even roots of negative numbers. For example, the function

$$f(x) = \sqrt{x}$$
    Domain excludes x-values that result in even roots of negative numbers.

is defined only for $x \geq 0$. So, its implied domain is the interval $[0, \infty)$. In general, the domain of a function *excludes* values that would cause division by zero *or* result in the even root of a negative number.

> **Exploration**
>
> Use a graphing utility to graph $y = \sqrt{4 - x^2}$. What is the domain of this function? Then graph $y = \sqrt{x^2 - 4}$. What is the domain of this function? Do the domains of these two functions overlap? If so, for what values?

Library of Parent Functions: Radical Function

Radical functions arise from the use of rational exponents. The most common radical function is the *square root function* given by $f(x) = \sqrt{x}$. The basic characteristics of the square root function are summarized below. A review of radical functions can be found in the *Study Capsules*.

Graph of $f(x) = \sqrt{x}$

Domain: $[0, \infty)$
Range: $[0, \infty)$
Intercept: $(0, 0)$
Increasing on $(0, \infty)$

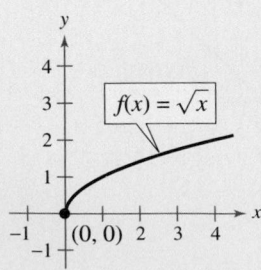

> **STUDY TIP**
>
> Because the square root function is not defined for $x < 0$, you must be careful when analyzing the domains of complicated functions involving the square root symbol.

Example 5   Finding the Domain of a Function

Find the domain of each function.

a. $f: \{(-3, 0), (-1, 4), (0, 2), (2, 2), (4, -1)\}$

b. $g(x) = -3x^2 + 4x + 5$

c. $h(x) = \dfrac{1}{x + 5}$

d. Volume of a sphere: $V = \frac{4}{3}\pi r^3$

e. $k(x) = \sqrt{4 - 3x}$

Prerequisite Skills

In Example 5(e), $4 - 3x \geq 0$ is a *linear inequality*. To review solving of linear inequalities, see Appendix E.

Solution

a. The domain of f consists of all first coordinates in the set of ordered pairs.

$$\text{Domain} = \{-3, -1, 0, 2, 4\}$$

b. The domain of g is the set of all *real* numbers.

c. Excluding x-values that yield zero in the denominator, the domain of h is the set of all real numbers x except $x = -5$.

d. Because this function represents the volume of a sphere, the values of the radius r must be positive. So, the domain is the set of all real numbers r such that $r > 0$.

e. This function is defined only for x-values for which $4 - 3x \geq 0$. By solving this inequality, you will find that the domain of k is all real numbers that are less than or equal to $\frac{4}{3}$.

✔**CHECKPOINT**   Now try Exercise 59.

In Example 5(d), note that the *domain of a function may be implied by the physical context*. For instance, from the equation $V = \frac{4}{3}\pi r^3$, you would have no reason to restrict r to positive values, but the physical context implies that a sphere cannot have a negative or zero radius.

For some functions, it may be easier to find the domain and range of the function by examining its graph.

Example 6   Finding the Domain and Range of a Function

Use a graphing utility to find the domain and range of the function

$$f(x) = \sqrt{9 - x^2}.$$

Solution

Graph the function as $y = \sqrt{9 - x^2}$, as shown in Figure 1.16. Using the *trace* feature of a graphing utility, you can determine that the x-values extend from -3 to 3 and the y-values extend from 0 to 3. So, the domain of the function f is all real numbers such that $-3 \leq x \leq 3$ and the range of f is all real numbers such that $0 \leq y \leq 3$.

✔**CHECKPOINT**   Now try Exercise 67.

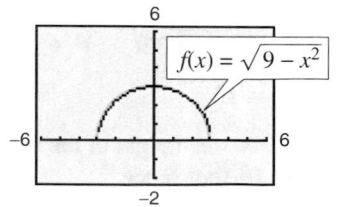

Figure 1.16

Applications

Example 7 Cellular Communications Employees

The number N (in thousands) of employees in the cellular communications industry in the United States increased in a linear pattern from 1998 to 2001 (see Figure 1.17). In 2002, the number dropped, then continued to increase through 2004 in a *different* linear pattern. These two patterns can be approximated by the function

$$N(t) = \begin{cases} 23.5t - 53.6, & 8 \le t \le 11 \\ 16.8t - 10.4, & 12 \le t \le 14 \end{cases}$$

where t represents the year, with $t = 8$ corresponding to 1998. Use this function to approximate the number of employees for each year from 1998 to 2004. (Source: Cellular Telecommunications & Internet Association)

Cellular Communications Employees

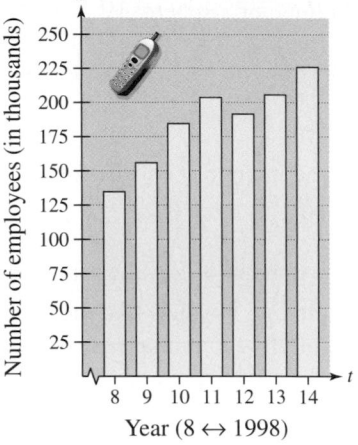

Figure 1.17

Solution

From 1998 to 2001, use $N(t) = 23.5t - 53.6$.

$$\underbrace{134.4,}_{1998} \quad \underbrace{157.9,}_{1999} \quad \underbrace{181.4,}_{2000} \quad \underbrace{204.9}_{2001}$$

From 2002 to 2004, use $N(t) = 16.8t - 10.4$.

$$\underbrace{191.2,}_{2002} \quad \underbrace{208.0,}_{2003} \quad \underbrace{224.8}_{2004}$$

✓CHECKPOINT   Now try Exercise 87.

Example 8 The Path of a Baseball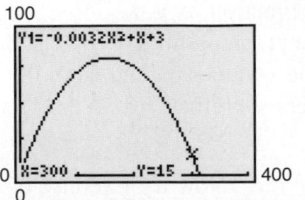

A baseball is hit at a point 3 feet above the ground at a velocity of 100 feet per second and an angle of 45°. The path of the baseball is given by the function

$$f(x) = -0.0032x^2 + x + 3$$

where x and $f(x)$ are measured in feet. Will the baseball clear a 10-foot fence located 300 feet from home plate?

Algebraic Solution

The height of the baseball is a function of the horizontal distance from home plate. When $x = 300$, you can find the height of the baseball as follows.

$$f(x) = -0.0032x^2 + x + 3 \qquad \text{Write original function.}$$

$$f(300) = -0.0032(300)^2 + 300 + 3 \qquad \text{Substitute 300 for } x.$$

$$= 15 \qquad \text{Simplify.}$$

When $x = 300$, the height of the baseball is 15 feet, so the baseball will clear a 10-foot fence.

✓CHECKPOINT   Now try Exercise 89.

Graphical Solution

Use a graphing utility to graph the function $y = -0.0032x^2 + x + 3$. Use the *value* feature or the *zoom* and *trace* features of the graphing utility to estimate that $y = 15$ when $x = 300$, as shown in Figure 1.18. So, the ball will clear a 10-foot fence.

Figure 1.18

Difference Quotients

One of the basic definitions in calculus employs the ratio

$$\frac{f(x + h) - f(x)}{h}, \qquad h \neq 0.$$

This ratio is called a **difference quotient,** as illustrated in Example 9.

Example 9   Evaluating a Difference Quotient

For $f(x) = x^2 - 4x + 7$, find $\dfrac{f(x + h) - f(x)}{h}$.

Solution

$$
\begin{aligned}
\frac{f(x + h) - f(x)}{h} &= \frac{[(x + h)^2 - 4(x + h) + 7] - (x^2 - 4x + 7)}{h} \\[2mm]
&= \frac{x^2 + 2xh + h^2 - 4x - 4h + 7 - x^2 + 4x - 7}{h} \\[2mm]
&= \frac{2xh + h^2 - 4h}{h} \\[2mm]
&= \frac{h(2x + h - 4)}{h} = 2x + h - 4, \ h \neq 0
\end{aligned}
$$

✔**CHECKPOINT**   Now try Exercise 93.

Summary of Function Terminology

Function: A **function** is a relationship between two variables such that to each value of the independent variable there corresponds exactly one value of the dependent variable.

Function Notation: $y = f(x)$

  f is the *name* of the function.
  y is the **dependent variable,** or output value.
  x is the **independent variable,** or input value.
  $f(x)$ is the *value of the function at x.*

Domain: The **domain** of a function is the set of all values (inputs) of the independent variable for which the function is defined. If x is in the domain of f, f is said to be *defined* at x. If x is not in the domain of f, f is said to be *undefined* at x.

Range: The **range** of a function is the set of all values (outputs) assumed by the dependent variable (that is, the set of all function values).

Implied Domain: If f is defined by an algebraic expression and the domain is not specified, the **implied domain** consists of all real numbers for which the expression is defined.

STUDY TIP

Notice in Example 9 that h cannot be zero in the original expression. Therefore, you must restrict the domain of the simplified expression by adding $h \neq 0$ so that the simplified expression is equivalent to the original expression.

The symbol ∫ indicates an example or exercise that highlights algebraic techniques specifically used in calculus.

1.2 Exercises

See www.CalcChat.com for worked-out solutions to odd-numbered exercises.

Vocabulary Check

Fill in the blanks.

1. A relation that assigns to each element x from a set of inputs, or _____ , exactly one element y in a set of outputs, or _____ , is called a _____ .

2. For an equation that represents y as a function of x, the _____ variable is the set of all x in the domain, and the _____ variable is the set of all y in the range.

3. The function $f(x) = \begin{cases} x^2 - 4, & x \le 0 \\ 2x + 1, & x > 0 \end{cases}$ is an example of a _____ function.

4. If the domain of the function f is not given, then the set of values of the independent variable for which the expression is defined is called the _____ .

5. In calculus, one of the basic definitions is that of a _____ , given by $\dfrac{f(x + h) - f(x)}{h}$, $h \neq 0$.

In Exercises 1–4, does the relation describe a function? Explain your reasoning.

1. *Domain   Range*

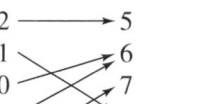

2. *Domain   Range*

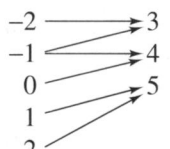

3. *Domain   Range*

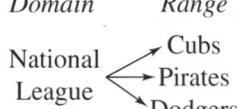

4. *Domain   Range*
(Year) (Number of North Atlantic tropical storms and hurricanes)

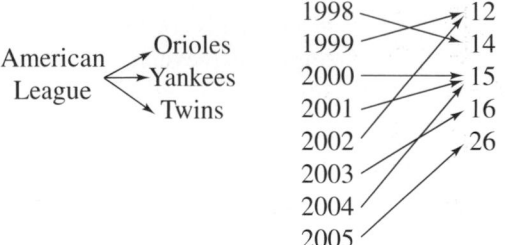

In Exercises 5–8, decide whether the relation represents y as a function of x. Explain your reasoning.

5.

Input, x	−3	−1	0	1	3
Output, y	−9	−1	0	1	9

6.

Input, x	0	1	2	1	0
Output, y	−4	−2	0	2	4

7.

Input, x	10	7	4	7	10
Output, y	3	6	9	12	15

8.

Input, x	0	3	9	12	15
Output, y	3	3	3	3	3

In Exercises 9 and 10, which sets of ordered pairs represent functions from A to B? Explain.

9. $A = \{0, 1, 2, 3\}$ and $B = \{-2, -1, 0, 1, 2\}$
  (a) $\{(0, 1), (1, -2), (2, 0), (3, 2)\}$
  (b) $\{(0, -1), (2, 2), (1, -2), (3, 0), (1, 1)\}$
  (c) $\{(0, 0), (1, 0), (2, 0), (3, 0)\}$
  (d) $\{(0, 2), (3, 0), (1, 1)\}$

10. $A = \{a, b, c\}$ and $B = \{0, 1, 2, 3\}$
  (a) $\{(a, 1), (c, 2), (c, 3), (b, 3)\}$
  (b) $\{(a, 1), (b, 2), (c, 3)\}$
  (c) $\{(1, a), (0, a), (2, c), (3, b)\}$
  (d) $\{(c, 0), (b, 0), (a, 3)\}$

Circulation of Newspapers **In Exercises 11 and 12, use the graph, which shows the circulation (in millions) of daily newspapers in the United States.** (Source: Editor & Publisher Company)

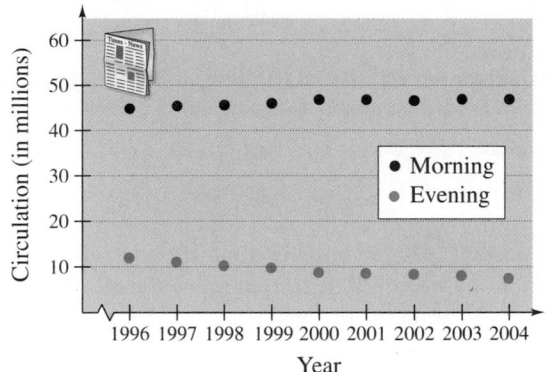

11. Is the circulation of morning newspapers a function of the year? Is the circulation of evening newspapers a function of the year? Explain.

12. Let $f(x)$ represent the circulation of evening newspapers in year x. Find $f(2004)$.

In Exercises 13–24, determine whether the equation represents y as a function of x.

13. $x^2 + y^2 = 4$ 14. $x = y^2 + 1$
15. $y = \sqrt{x^2 - 1}$ 16. $y = \sqrt{x + 5}$
17. $2x + 3y = 4$ 18. $x = -y + 5$
19. $y^2 = x^2 - 1$ 20. $x + y^2 = 3$
21. $y = |4 - x|$ 22. $|y| = 4 - x$
23. $x = -7$ 24. $y = 8$

In Exercises 25 and 26, fill in the blanks using the specified function and the given values of the independent variable. Simplify the result.

25. $f(x) = \dfrac{1}{x + 1}$

(a) $f(4) = \dfrac{1}{(\quad) + 1}$ (b) $f(0) = \dfrac{1}{(\quad) + 1}$

(c) $f(4t) = \dfrac{1}{(\quad) + 1}$ (d) $f(x + c) = \dfrac{1}{(\quad) + 1}$

26. $g(x) = x^2 - 2x$

(a) $g(2) = (\quad)^2 - 2(\quad)$
(b) $g(-3) = (\quad)^2 - 2(\quad)$
(c) $g(t + 1) = (\quad)^2 - 2(\quad)$
(d) $g(x + c) = (\quad)^2 - 2(\quad)$

In Exercises 27–42, evaluate the function at each specified value of the independent variable and simplify.

27. $f(t) = 3t + 1$
   (a) $f(2)$ (b) $f(-4)$ (c) $f(t + 2)$
28. $g(y) = 7 - 3y$
   (a) $g(0)$ (b) $g(\frac{7}{3})$ (c) $g(s + 2)$
29. $h(t) = t^2 - 2t$
   (a) $h(2)$ (b) $h(1.5)$ (c) $h(x + 2)$
30. $V(r) = \frac{4}{3}\pi r^3$
   (a) $V(3)$ (b) $V(\frac{3}{2})$ (c) $V(2r)$
31. $f(y) = 3 - \sqrt{y}$
   (a) $f(4)$ (b) $f(0.25)$ (c) $f(4x^2)$
32. $f(x) = \sqrt{x + 8} + 2$
   (a) $f(-8)$ (b) $f(1)$ (c) $f(x - 8)$
33. $q(x) = \dfrac{1}{x^2 - 9}$
   (a) $q(0)$ (b) $q(3)$ (c) $q(y + 3)$
34. $q(t) = \dfrac{2t^2 + 3}{t^2}$
   (a) $q(2)$ (b) $q(0)$ (c) $q(-x)$
35. $f(x) = \dfrac{|x|}{x}$
   (a) $f(3)$ (b) $f(-3)$ (c) $f(t)$
36. $f(x) = |x| + 4$
   (a) $f(4)$ (b) $f(-4)$ (c) $f(t)$
37. $f(x) = \begin{cases} 2x + 1, & x < 0 \\ 2x + 2, & x \geq 0 \end{cases}$
   (a) $f(-1)$ (b) $f(0)$ (c) $f(2)$
38. $f(x) = \begin{cases} 2x + 5, & x \leq 0 \\ 2 - x^2, & x > 0 \end{cases}$
   (a) $f(-2)$ (b) $f(0)$ (c) $f(1)$
39. $f(x) = \begin{cases} x^2 + 2, & x \leq 1 \\ 2x^2 + 2, & x > 1 \end{cases}$
   (a) $f(-2)$ (b) $f(1)$ (c) $f(2)$
40. $f(x) = \begin{cases} x^2 - 4, & x \leq 0 \\ 1 - 2x^2, & x > 0 \end{cases}$
   (a) $f(-2)$ (b) $f(0)$ (c) $f(1)$
41. $f(x) = \begin{cases} x + 2, & x < 0 \\ 4, & 0 \leq x < 2 \\ x^2 + 1, & x \geq 2 \end{cases}$
   (a) $f(-2)$ (b) $f(1)$ (c) $f(4)$

42. $f(x) = \begin{cases} 5 - 2x, & x < 0 \\ 5, & 0 \le x < 1 \\ 4x + 1, & x \ge 1 \end{cases}$

(a) $f(-2)$ (b) $f\left(\frac{1}{2}\right)$ (c) $f(1)$

In Exercises 43–46, complete the table.

43. $h(t) = \frac{1}{2}|t + 3|$

t	-5	-4	-3	-2	-1
$h(t)$					

44. $f(s) = \dfrac{|s - 2|}{s - 2}$

s	0	1	$\frac{3}{2}$	$\frac{5}{2}$	4
$f(s)$					

45. $f(x) = \begin{cases} -\frac{1}{2}x + 4, & x \le 0 \\ (x - 2)^2, & x > 0 \end{cases}$

x	-2	-1	0	1	2
$f(x)$					

46. $h(x) = \begin{cases} 9 - x^2, & x < 3 \\ x - 3, & x \ge 3 \end{cases}$

x	1	2	3	4	5
$h(x)$					

In Exercises 47–50, find all real values of x such that $f(x) = 0$.

47. $f(x) = 15 - 3x$ **48.** $f(x) = 5x + 1$

49. $f(x) = \dfrac{3x - 4}{5}$ **50.** $f(x) = \dfrac{2x - 3}{7}$

In Exercises 51 and 52, find the value(s) of x for which $f(x) = g(x)$.

51. $f(x) = x^2, \quad g(x) = x + 2$

52. $f(x) = x^2 + 2x + 1, \quad g(x) = 7x - 5$

In Exercises 53–62, find the domain of the function.

53. $f(x) = 5x^2 + 2x - 1$ **54.** $g(x) = 1 - 2x^2$

55. $h(t) = \dfrac{4}{t}$ **56.** $s(y) = \dfrac{3y}{y + 5}$

57. $f(x) = \sqrt[3]{x - 4}$ **58.** $f(x) = \sqrt[4]{x^2 + 3x}$

59. $g(x) = \dfrac{1}{x} - \dfrac{3}{x + 2}$ **60.** $h(x) = \dfrac{10}{x^2 - 2x}$

61. $g(y) = \dfrac{y + 2}{\sqrt{y - 10}}$ **62.** $f(x) = \dfrac{\sqrt{x + 6}}{6 + x}$

In Exercises 63–66, use a graphing utility to graph the function. Find the domain and range of the function.

63. $f(x) = \sqrt{4 - x^2}$ **64.** $f(x) = \sqrt{x^2 + 1}$

65. $g(x) = |2x + 3|$ **66.** $g(x) = |x - 5|$

In Exercises 67–70, assume that the domain of f is the set $A = \{-2, -1, 0, 1, 2\}$. Determine the set of ordered pairs representing the function f.

67. $f(x) = x^2$ **68.** $f(x) = x^2 - 3$

69. $f(x) = |x| + 2$ **70.** $f(x) = |x + 1|$

71. *Geometry*   Write the area A of a circle as a function of its circumference C.

72. *Geometry*   Write the area A of an equilateral triangle as a function of the length s of its sides.

73. *Exploration*   The cost per unit to produce a radio model is $60. The manufacturer charges $90 per unit for orders of 100 or less. To encourage large orders, the manufacturer reduces the charge by $0.15 per radio for each unit ordered in excess of 100 (for example, there would be a charge of $87 per radio for an order size of 120).

(a) The table shows the profit P (in dollars) for various numbers of units ordered, x. Use the table to estimate the maximum profit.

Units, x	Profit, P
110	3135
120	3240
130	3315
140	3360
150	3375
160	3360
170	3315

(b) Plot the points (x, P) from the table in part (a). Does the relation defined by the ordered pairs represent P as a function of x?

(c) If P is a function of x, write the function and determine its domain.

74. *Exploration* An open box of maximum volume is to be made from a square piece of material, 24 centimeters on a side, by cutting equal squares from the corners and turning up the sides (see figure).

(a) The table shows the volume V (in cubic centimeters) of the box for various heights x (in centimeters). Use the table to estimate the maximum volume.

Height, x	Volume, V
1	484
2	800
3	972
4	1024
5	980
6	864

(b) Plot the points (x, V) from the table in part (a). Does the relation defined by the ordered pairs represent V as a function of x?

(c) If V is a function of x, write the function and determine its domain.

(d) Use a graphing utility to plot the point from the table in part (a) with the function from part (c). How closely does the function represent the data? Explain.

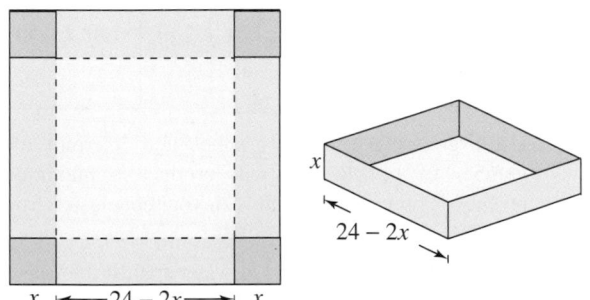

75. *Geometry* A right triangle is formed in the first quadrant by the x- and y-axes and a line through the point $(2, 1)$ (see figure). Write the area A of the triangle as a function of x, and determine the domain of the function.

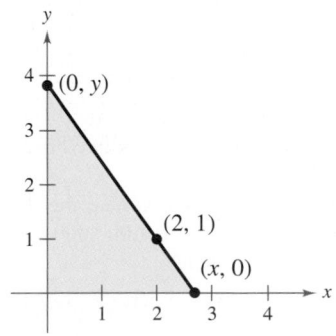

76. *Geometry* A rectangle is bounded by the x-axis and the semicircle $y = \sqrt{36 - x^2}$ (see figure). Write the area A of the rectangle as a function of x, and determine the domain of the function.

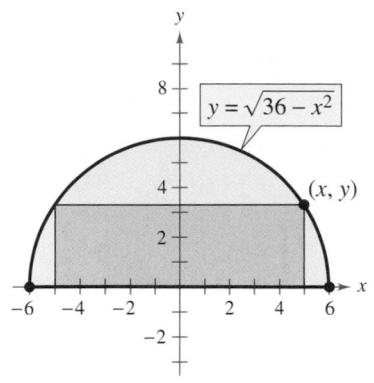

77. *Postal Regulations* A rectangular package to be sent by the U.S. Postal Service can have a maximum combined length and girth (perimeter of a cross section) of 108 inches (see figure).

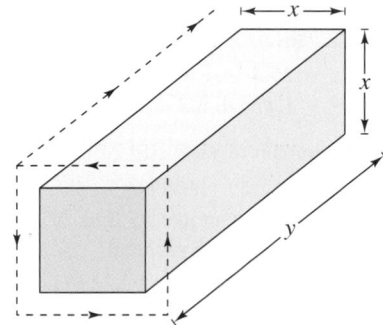

(a) Write the volume V of the package as a function of x. What is the domain of the function?

(b) Use a graphing utility to graph the function. Be sure to use an appropriate viewing window.

(c) What dimensions will maximize the volume of the package? Explain.

78. *Cost, Revenue, and Profit* A company produces a toy for which the variable cost is \$12.30 per unit and the fixed costs are \$98,000. The toy sells for \$17.98. Let x be the number of units produced and sold.

(a) The total cost for a business is the sum of the variable cost and the fixed costs. Write the total cost C as a function of the number of units produced.

(b) Write the revenue R as a function of the number of units sold.

(c) Write the profit P as a function of the number of units sold. (*Note:* $P = R - C$.)

Revenue In Exercises 79–82, use the table, which shows the monthly revenue y (in thousands of dollars) of a landscaping business for each month of 2006, with $x = 1$ representing January.

Month, x	Revenue, y
1	5.2
2	5.6
3	6.6
4	8.3
5	11.5
6	15.8
7	12.8
8	10.1
9	8.6
10	6.9
11	4.5
12	2.7

A mathematical model that represents the data is

$$f(x) = \begin{cases} -1.97x + 26.3 \\ 0.505x^2 - 1.47x + 6.3 \end{cases}.$$

79. What is the domain of each part of the piecewise-defined function? Explain your reasoning.

80. Use the mathematical model to find $f(5)$. Interpret your result in the context of the problem.

81. Use the mathematical model to find $f(11)$. Interpret your result in the context of the problem.

82. How do the values obtained from the model in Exercises 80 and 81 compare with the actual data values?

83. *Motor Vehicles* The numbers n (in billions) of miles traveled by vans, pickup trucks, and sport utility vehicles in the United States from 1990 to 2003 can be approximated by the model

$$n(t) = \begin{cases} -6.13t^2 + 75.8t + 577, & 0 \le t \le 6 \\ 24.9t + 672, & 6 < t \le 13 \end{cases}$$

where t represents the year, with $t = 0$ corresponding to 1990. Use the *table* feature of a graphing utility to approximate the number of miles traveled by vans, pickup trucks, and sport utility vehicles for each year from 1990 to 2003. (Source: U.S. Federal Highway Administration)

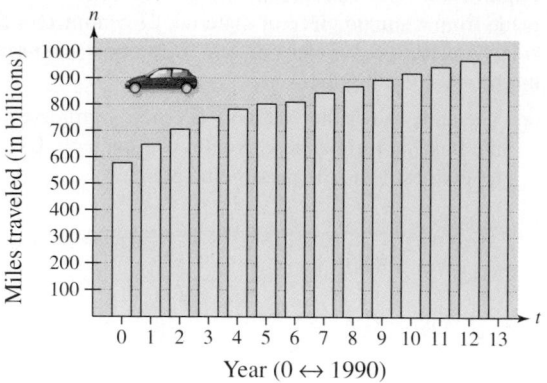

Figure for 83

84. *Transportation* For groups of 80 or more people, a charter bus company determines the rate per person according to the formula

Rate $= 8 - 0.05(n - 80)$, $n \ge 80$

where the rate is given in dollars and n is the number of people.

(a) Write the revenue R of the bus company as a function of n.

(b) Use the function from part (a) to complete the table. What can you conclude?

n	90	100	110	120	130	140	150
$R(n)$							

(c) Use a graphing utility to graph R and determine the number of people that will produce a maximum revenue. Compare the result with your conclusion from part (b).

85. *Physics* The force F (in tons) of water against the face of a dam is estimated by the function

$$F(y) = 149.76\sqrt{10}\,y^{5/2}$$

where y is the depth of the water (in feet).

(a) Complete the table. What can you conclude from it?

y	5	10	20	30	40
$F(y)$					

(b) Use a graphing utility to graph the function. Describe your viewing window.

(c) Use the table to approximate the depth at which the force against the dam is 1,000,000 tons. How could you find a better estimate?

(d) Verify your answer in part (c) graphically.

86. *Data Analysis* The graph shows the retail sales (in billions of dollars) of prescription drugs in the United States from 1995 through 2004. Let $f(x)$ represent the retail sales in year x.   (Source: National Association of Chain Drug Stores)

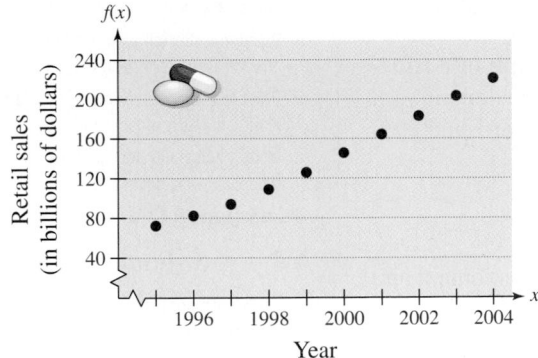

(a) Find $f(2000)$.

(b) Find $\dfrac{f(2004) - f(1995)}{2004 - 1995}$

and interpret the result in the context of the problem.

(c) An approximate model for the function is

$$P(t) = -0.0982t^3 + 3.365t^2 - 18.85t + 94.8,$$
$$5 \le t \le 14$$

where P is the retail sales (in billions of dollars) and t represents the year, with $t = 5$ corresponding to 1995. Complete the table and compare the results with the data in the graph.

t	5	6	7	8	9	10	11	12	13	14
$P(t)$										

(d) Use a graphing utility to graph the model and the data in the same viewing window. Comment on the validity of the model.

∫ **In Exercises 87–92, find the difference quotient and simplify your answer.**

87. $f(x) = 2x$, $\dfrac{f(x + c) - f(x)}{c}$, $c \ne 0$

88. $g(x) = 3x - 1$, $\dfrac{g(x + h) - g(x)}{h}$, $h \ne 0$

89. $f(x) = x^2 - x + 1$, $\dfrac{f(2 + h) - f(2)}{h}$, $h \ne 0$

90. $f(x) = x^3 + x$, $\dfrac{f(x + h) - f(x)}{h}$, $h \ne 0$

91. $f(t) = \dfrac{1}{t}$, $\dfrac{f(t) - f(1)}{t - 1}$, $t \ne 1$

92. $f(x) = \dfrac{4}{x + 1}$, $\dfrac{f(x) - f(7)}{x - 7}$, $x \ne 7$

Synthesis

True or False?   **In Exercises 93 and 94, determine whether the statement is true or false. Justify your answer.**

93. The domain of the function $f(x) = x^4 - 1$ is $(-\infty, \infty)$, and the range of $f(x)$ is $(0, \infty)$.

94. The set of ordered pairs $\{(-8, -2), (-6, 0), (-4, 0), (-2, 2), (0, 4), (2, -2)\}$ represents a function.

Library of Parent Functions   **In Exercises 95–98, write a piecewise-defined function for the graph shown.**

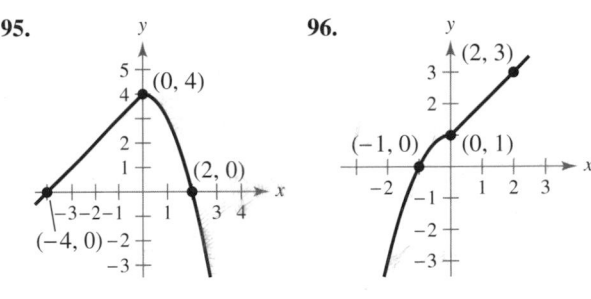

95.   **96.**

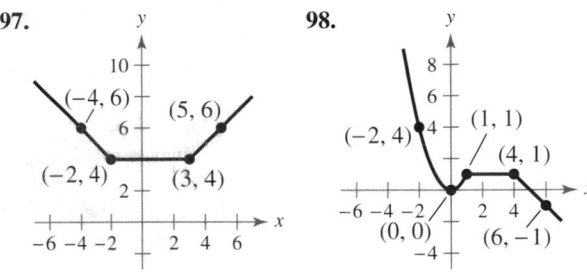

97.   **98.**

99. *Writing*   In your own words, explain the meanings of *domain* and *range*.

100. *Think About It*   Describe an advantage of function notation.

Skills Review

In Exercises 101–104, perform the operation and simplify.

101. $12 - \dfrac{4}{x + 2}$

102. $\dfrac{3}{x^2 + x - 20} + \dfrac{x}{x^2 + 4x - 5}$

103. $\dfrac{2x^3 + 11x^2 - 6x}{5x} \cdot \dfrac{x + 10}{2x^2 + 5x - 3}$

104. $\dfrac{x + 7}{2(x - 9)} \div \dfrac{x - 7}{2(x - 9)}$

The symbol ∫ indicates an example or exercise that highlights algebraic techniques specifically used in calculus.

1.3   Graphs of Functions

The Graph of a Function

In Section 1.2, functions were represented graphically by points on a graph in a coordinate plane in which the input values are represented by the horizontal axis and the output values are represented by the vertical axis. The **graph of a function** f is the collection of ordered pairs $(x, f(x))$ such that x is in the domain of f. As you study this section, remember the geometric interpretations of x and $f(x)$.

$x =$ the directed distance from the y-axis

$f(x) =$ the directed distance from the x-axis

Example 1 shows how to use the graph of a function to find the domain and range of the function.

Example 1   Finding the Domain and Range of a Function

Use the graph of the function f shown in Figure 1.19 to find (a) the domain of f, (b) the function values $f(-1)$ and $f(2)$, and (c) the range of f.

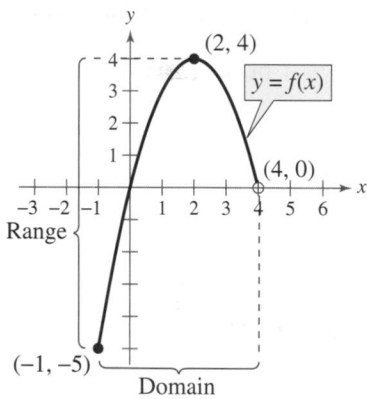

Figure 1.19

What you should learn

- Find the domains and ranges of functions and use the Vertical Line Test for functions.
- Determine intervals on which functions are increasing, decreasing, or constant.
- Determine relative maximum and relative minimum values of functions.
- Identify and graph step functions and other piecewise-defined functions.
- Identify even and odd functions.

Why you should learn it

Graphs of functions provide a visual relationship between two variables. For example, in Exercise 88 on page 40, you will use the graph of a step function to model the cost of sending a package.

Stephen Chernin/Getty Images

Solution

a. The closed dot at $(-1, -5)$ indicates that $x = -1$ is in the domain of f, whereas the open dot at $(4, 0)$ indicates that $x = 4$ is not in the domain. So, the domain of f is all x in the interval $[-1, 4)$.

b. Because $(-1, -5)$ is a point on the graph of f, it follows that

$f(-1) = -5.$

Similarly, because $(2, 4)$ is a point on the graph of f, it follows that

$f(2) = 4.$

c. Because the graph does not extend below $f(-1) = -5$ or above $f(2) = 4$, the range of f is the interval $[-5, 4]$.

✓**CHECKPOINT**   Now try Exercise 3.

STUDY TIP

The use of dots (open or closed) at the extreme left and right points of a graph indicates that the graph does not extend beyond these points. If no such dots are shown, assume that the graph extends beyond these points. ∞

Example 2   Finding the Domain and Range of a Function

Find the domain and range of

$$f(x) = \sqrt{x - 4}.$$

Algebraic Solution

Because the expression under a radical cannot be negative, the domain of $f(x) = \sqrt{x - 4}$ is the set of all real numbers such that $x - 4 \geq 0$. Solve this linear inequality for x as follows. (For help with solving linear inequalities, see Appendix E.)

$x - 4 \geq 0$	Write original inequality.
$x \geq 4$	Add 4 to each side.

So, the domain is the set of all real numbers greater than or equal to 4. Because the value of a radical expression is never negative, the range of $f(x) = \sqrt{x - 4}$ is the set of all nonnegative real numbers.

Graphical Solution

Use a graphing utility to graph the equation $y = \sqrt{x - 4}$, as shown in Figure 1.20. Use the *trace* feature to determine that the x-coordinates of points on the graph extend from 4 to the right. When x is greater than or equal to 4, the expression under the radical is nonnegative. So, you can conclude that the domain is the set of all real numbers greater than or equal to 4. From the graph, you can see that the y-coordinates of points on the graph extend from 0 upwards. So you can estimate the range to be the set of all nonnegative real numbers.

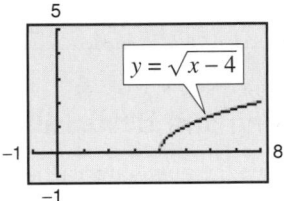

Figure 1.20

CHECKPOINT   Now try Exercise 7.

By the definition of a function, at most one y-value corresponds to a given x-value. It follows, then, that a vertical line can intersect the graph of a function at most once. This leads to the **Vertical Line Test** for functions.

> **Vertical Line Test for Functions**
>
> A set of points in a coordinate plane is the graph of y as a function of x if and only if no vertical line intersects the graph at more than one point.

Example 3   Vertical Line Test for Functions

Use the Vertical Line Test to decide whether the graphs in Figure 1.21 represent y as a function of x.

Solution

a. This is *not* a graph of y as a function of x because you can find a vertical line that intersects the graph twice.

b. This *is* a graph of y as a function of x because every vertical line intersects the graph at most once.

CHECKPOINT   Now try Exercise 17.

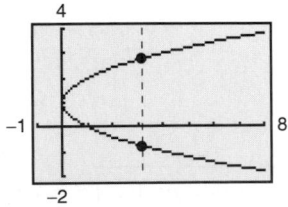

(a)

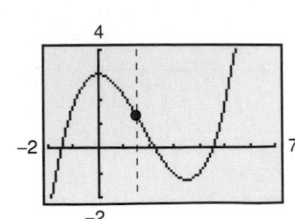

(b)

Figure 1.21

Increasing and Decreasing Functions

The more you know about the graph of a function, the more you know about the function itself. Consider the graph shown in Figure 1.22. Moving from *left to right*, this graph falls from $x = -2$ to $x = 0$, is constant from $x = 0$ to $x = 2$, and rises from $x = 2$ to $x = 4$.

Increasing, Decreasing, and Constant Functions

A function f is **increasing** on an interval if, for any x_1 and x_2 in the interval,

$$x_1 < x_2 \text{ implies } f(x_1) < f(x_2).$$

A function f is **decreasing** on an interval if, for any x_1 and x_2 in the interval,

$$x_1 < x_2 \text{ implies } f(x_1) > f(x_2).$$

A function f is **constant** on an interval if, for any x_1 and x_2 in the interval,

$$f(x_1) = f(x_2).$$

Example 4   Increasing and Decreasing Functions

In Figure 1.23, determine the open intervals on which each function is increasing, decreasing, or constant.

Solution

a. Although it might appear that there is an interval in which this function is constant, you can see that if $x_1 < x_2$, then $(x_1)^3 < (x_2)^3$, which implies that $f(x_1) < f(x_2)$. So, the function is increasing over the entire real line.

b. This function is increasing on the interval $(-\infty, -1)$, decreasing on the interval $(-1, 1)$, and increasing on the interval $(1, \infty)$.

c. This function is increasing on the interval $(-\infty, 0)$, constant on the interval $(0, 2)$, and decreasing on the interval $(2, \infty)$.

TECHNOLOGY TIP

Most graphing utilities are designed to graph functions of x more easily than other types of equations. For instance, the graph shown in Figure 1.23(a) represents the equation $x - (y - 1)^2 = 0$. To use a graphing utility to duplicate this graph you must first solve the equation for y to obtain $y = 1 \pm \sqrt{x}$, and then graph the two equations $y_1 = 1 + \sqrt{x}$ and $y_2 = 1 - \sqrt{x}$ in the same viewing window.

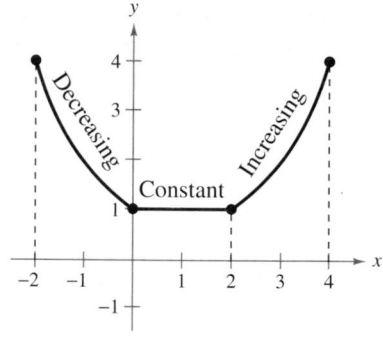

Figure 1.22

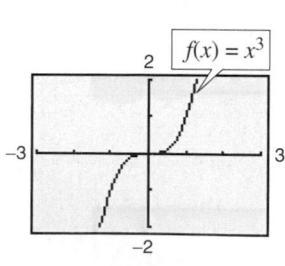

(a)

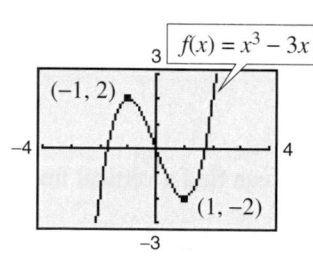

(b)

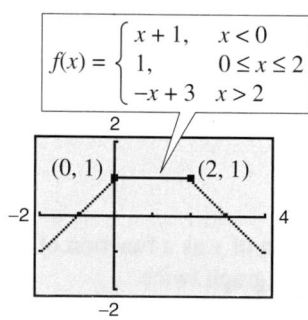

(c)

Figure 1.23

✓CHECKPOINT   Now try Exercise 21.

Relative Minimum and Maximum Values

The points at which a function changes its increasing, decreasing, or constant behavior are helpful in determining the relative maximum or relative minimum values of the function.

Definitions of Relative Minimum and Relative Maximum

A function value $f(a)$ is called a **relative minimum** of f if there exists an interval (x_1, x_2) that contains a such that

$$x_1 < x < x_2 \quad \text{implies} \quad f(a) \leq f(x).$$

A function value $f(a)$ is called a **relative maximum** of f if there exists an interval (x_1, x_2) that contains a such that

$$x_1 < x < x_2 \quad \text{implies} \quad f(a) \geq f(x).$$

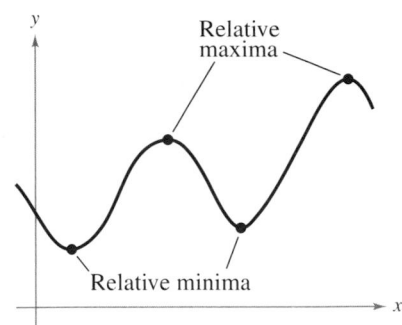

Figure 1.24

Figure 1.24 shows several different examples of relative minima and relative maxima. In Section 2.1, you will study a technique for finding the *exact points* at which a second-degree polynomial function has a relative minimum or relative maximum. For the time being, however, you can use a graphing utility to find reasonable approximations of these points.

Example 5 Approximating a Relative Minimum

Use a graphing utility to approximate the relative minimum of the function given by $f(x) = 3x^2 - 4x - 2$.

Solution

The graph of f is shown in Figure 1.25. By using the *zoom* and *trace* features of a graphing utility, you can estimate that the function has a relative minimum at the point

$$(0.67, -3.33). \qquad \text{See Figure 1.26.}$$

Later, in Section 2.1, you will be able to determine that the exact point at which the relative minimum occurs is $\left(\frac{2}{3}, -\frac{10}{3}\right)$.

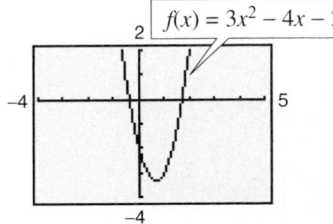

Figure 1.25

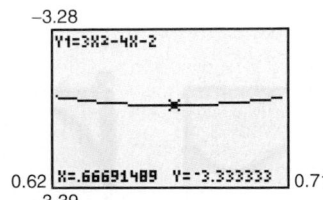

Figure 1.26

✔CHECKPOINT   Now try Exercise 31.

TECHNOLOGY TIP   Some graphing utilities have built-in programs that will find minimum or maximum values. These features are demonstrated in Example 6.

TECHNOLOGY TIP

When you use a graphing utility to estimate the x- and y-values of a relative minimum or relative maximum, the *zoom* feature will often produce graphs that are nearly flat, as shown in Figure 1.26. To overcome this problem, you can manually change the vertical setting of the viewing window. The graph will stretch vertically if the values of Ymin and Ymax are closer together.

Example 6 Approximating Relative Minima and Maxima

Use a graphing utility to approximate the relative minimum and relative maximum of the function given by $f(x) = -x^3 + x$.

Solution

The graph of f is shown in Figure 1.27. By using the *zoom* and *trace* features or the *minimum* and *maximum* features of the graphing utility, you can estimate that the function has a relative minimum at the point

$(-0.58, -0.38)$ See Figure 1.28.

and a relative maximum at the point

$(0.58, 0.38)$.   See Figure 1.29.

If you take a course in calculus, you will learn a technique for finding the exact points at which this function has a relative minimum and a relative maximum.

 **CHECKPOINT** Now try Exercise 33.

Example 7 Temperature

During a 24-hour period, the temperature y (in degrees Fahrenheit) of a certain city can be approximated by the model

$$y = 0.026x^3 - 1.03x^2 + 10.2x + 34, \qquad 0 \le x \le 24$$

where x represents the time of day, with $x = 0$ corresponding to 6 A.M. Approximate the maximum and minimum temperatures during this 24-hour period.

Solution

To solve this problem, graph the function as shown in Figure 1.30. Using the *zoom* and *trace* features or the *maximum* feature of a graphing utility, you can determine that the maximum temperature during the 24-hour period was approximately 64°F. This temperature occurred at about 12:36 P.M. ($x \approx 6.6$), as shown in Figure 1.31. Using the *zoom* and *trace* features or the *minimum* feature, you can determine that the minimum temperature during the 24-hour period was approximately 34°F, which occurred at about 1:48 A.M. ($x \approx 19.8$), as shown in Figure 1.32.

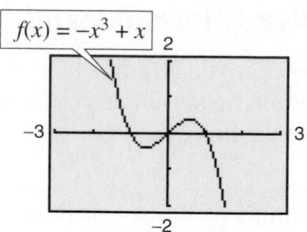

Figure 1.27

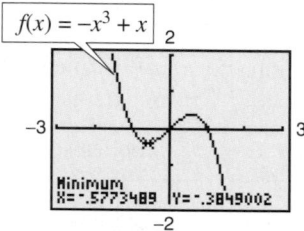

Figure 1.28

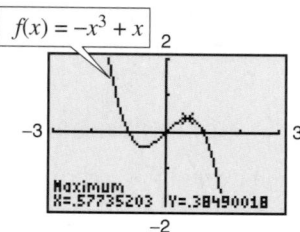

Figure 1.29

TECHNOLOGY SUPPORT

For instructions on how to use the *minimum* and *maximum* features, see Appendix A; for specific keystrokes, go to this textbook's *Online Study Center*.

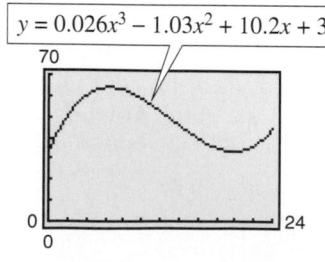

Figure 1.30

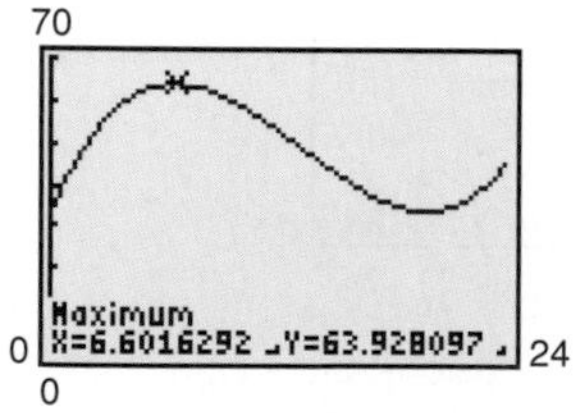

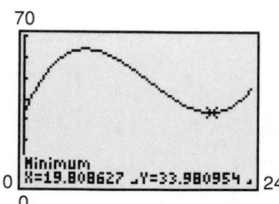

Figure 1.31

Figure 1.32

 **CHECKPOINT** Now try Exercise 91.

Graphing Step Functions and Piecewise-Defined Functions

Library of Parent Functions: Greatest Integer Function

The *greatest integer function*, denoted by $[\![x]\!]$ and defined as the greatest integer less than or equal to x, has an infinite number of breaks or steps—one at each integer value in its domain. The basic characteristics of the greatest integer function are summarized below. A review of the greatest integer function can be found in the *Study Capsules*.

Graph of $f(x) = [\![x]\!]$

Domain: $(-\infty, \infty)$

Range: the set of integers

x-intercepts: in the interval $[0, 1)$

y-intercept: $(0, 0)$

Constant between each pair of consecutive integers

Jumps vertically one unit at each integer value

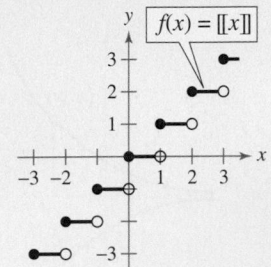

$f(x) = [\![x]\!]$

Could you describe the greatest integer function using a piecewise-defined function? How does the graph of the greatest integer function differ from the graph of a line with a slope of zero?

TECHNOLOGY TIP

Most graphing utilities display graphs in *connected mode*, which means that the graph has no breaks. When you are sketching graphs that do have breaks, it is better to use *dot mode*. Graph the greatest integer function [often called Int (x)] in *connected* and *dot modes*, and compare the two results.

Because of the vertical jumps described above, the greatest integer function is an example of a **step function** whose graph resembles a set of stairsteps. Some values of the greatest integer function are as follows.

$$[\![-1]\!] = (\text{greatest integer} \leq -1) = -1$$

$$\left[\!\left[\tfrac{1}{10}\right]\!\right] = \left(\text{greatest integer} \leq \tfrac{1}{10}\right) = 0$$

$$[\![1.5]\!] = (\text{greatest integer} \leq 1.5) = 1$$

In Section 1.2, you learned that a piecewise-defined function is a function that is defined by two or more equations over a specified domain. To sketch the graph of a piecewise-defined function, you need to sketch the graph of each equation on the appropriate portion of the domain.

Example 8   Graphing a Piecewise-Defined Function

Sketch the graph of $f(x) = \begin{cases} 2x + 3, & x \leq 1 \\ -x + 4, & x > 1 \end{cases}$ by hand.

Solution

This piecewise-defined function is composed of two linear functions. At and to the left of $x = 1$, the graph is the line given by $y = 2x + 3$. To the right of $x = 1$, the graph is the line given by $y = -x + 4$ (see Figure 1.33). Notice that the point $(1, 5)$ is a solid dot and the point $(1, 3)$ is an open dot. This is because $f(1) = 5$.

✓CHECKPOINT   Now try Exercise 43.

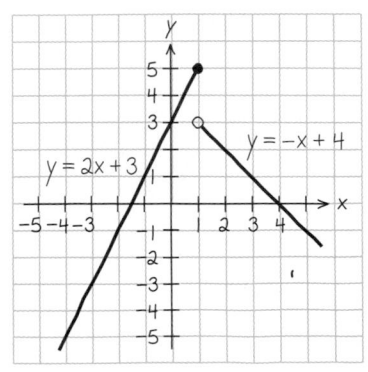

$y = 2x + 3$

$y = -x + 4$

Figure 1.33

Even and Odd Functions

A graph has *symmetry with respect to the y-axis* if whenever (x, y) is on the graph, so is the point $(-x, y)$. A graph has *symmetry with respect to the origin* if whenever (x, y) is on the graph, so is the point $(-x, -y)$. A graph has *symmetry with respect to the x-axis* if whenever (x, y) is on the graph, so is the point $(x, -y)$. A function whose graph is symmetric with respect to the y-axis is an **even function.** A function whose graph is symmetric with respect to the origin is an **odd function.** A graph that is symmetric with respect to the x-axis is not the graph of a function (except for the graph of $y = 0$). These three types of symmetry are illustrated in Figure 1.34.

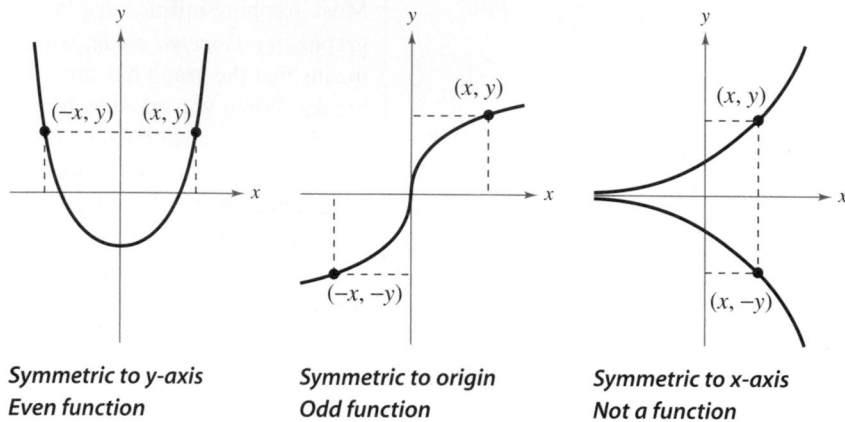

Symmetric to y-axis
Even function
Figure 1.34

Symmetric to origin
Odd function

Symmetric to x-axis
Not a function

Test for Even and Odd Functions

A function f is **even** if, for each x in the domain of f, $f(-x) = f(x)$.
A function f is **odd** if, for each x in the domain of f, $f(-x) = -f(x)$.

Example 9   Testing for Evenness and Oddness

Is the function given by $f(x) = |x|$ even, odd, or neither?

Algebraic Solution

This function is even because

$$f(-x) = |-x|$$
$$= |x|$$
$$= f(x).$$

Graphical Solution

Use a graphing utility to enter $y = |x|$ in the *equation editor*, as shown in Figure 1.35. Then graph the function using a standard viewing window, as shown in Figure 1.36. You can see that the graph appears to be symmetric about the y-axis. So, the function is even.

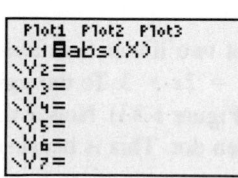

Figure 1.35

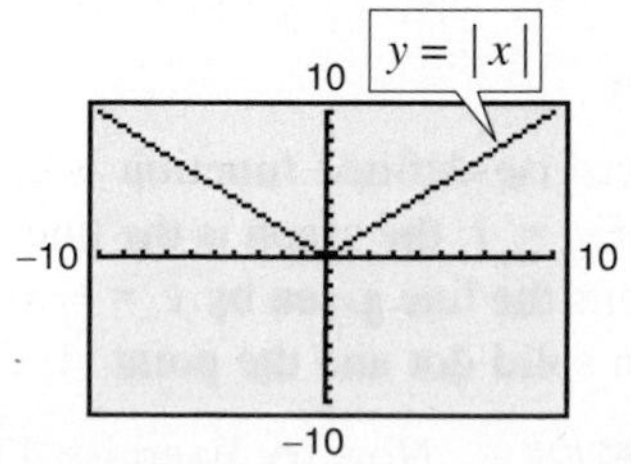

Figure 1.36

✔CHECKPOINT   Now try Exercise 59.

Example 10   Even and Odd Functions

Determine whether each function is even, odd, or neither.

a. $g(x) = x^3 - x$

b. $h(x) = x^2 + 1$

c. $f(x) = x^3 - 1$

Algebraic Solution

a. This function is odd because

$$g(-x) = (-x)^3 - (-x)$$

$$= -x^3 + x$$

$$= -(x^3 - x)$$

$$= -g(x).$$

b. This function is even because

$$h(-x) = (-x)^2 + 1$$

$$= x^2 + 1$$

$$= h(x).$$

c. Substituting $-x$ for x produces

$$f(-x) = (-x)^3 - 1$$

$$= -x^3 - 1.$$

Because $f(x) = x^3 - 1$ and $-f(x) = -x^3 + 1$, you can conclude that $f(-x) \neq f(x)$ and $f(-x) \neq -f(x)$. So, the function is neither even nor odd.

✓CHECKPOINT   Now try Exercise 61.

Graphical Solution

a. In Figure 1.37, the graph is symmetric with respect to the origin. So, this function is odd.

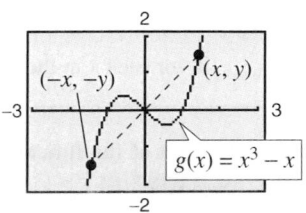

Figure 1.37

b. In Figure 1.38, the graph is symmetric with respect to the y-axis. So, this function is even.

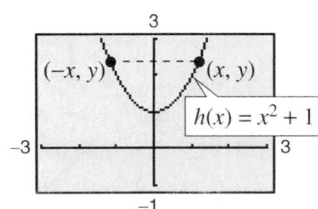

Figure 1.38

c. In Figure 1.39, the graph is neither symmetric with respect to the origin nor with respect to the y-axis. So, this function is neither even nor odd.

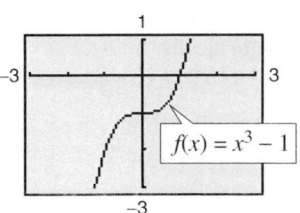

Figure 1.39

To help visualize symmetry with respect to the origin, place a pin at the origin of a graph and rotate the graph 180°. If the result after rotation coincides with the original graph, the graph is symmetric with respect to the origin.

1.3 Exercises

See www.CalcChat.com for worked-out solutions to odd-numbered exercises.

Vocabulary Check

Fill in the blanks.

1. The graph of a function f is a collection of _____ (x, y) such that x is in the domain of f.

2. The _____ is used to determine whether the graph of an equation is a function of y in terms of x.

3. A function f is _____ on an interval if, for any x_1 and x_2 in the interval, $x_1 < x_2$ implies $f(x_1) > f(x_2)$.

4. A function value $f(a)$ is a relative _____ of f if there exists an interval (x_1, x_2) containing a such that $x_1 < x < x_2$ implies $f(a) \le f(x)$.

5. The function $f(x) = [\![x]\!]$ is called the _____ function, and is an example of a step function.

6. A function f is _____ if, for each x in the domain of f, $f(-x) = f(x)$.

In Exercises 1–4, use the graph of the function to find the domain and range of f. Then find $f(0)$.

1.
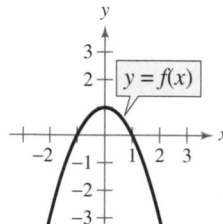

2.
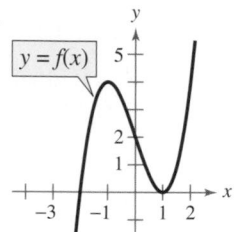

3.
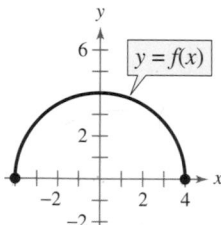

4.
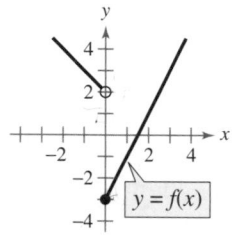

In Exercises 5–10, use a graphing utility to graph the function and estimate its domain and range. Then find the domain and range algebraically.

5. $f(x) = 2x^2 + 3$

6. $f(x) = -x^2 - 1$

7. $f(x) = \sqrt{x - 1}$

8. $h(t) = \sqrt{4 - t^2}$

9. $f(x) = |x + 3|$

10. $f(x) = -\frac{1}{4}|x - 5|$

In Exercises 11–14, use the given function to answer the questions.

(a) Determine the domain of the function.

(b) Find the value(s) of x such that $f(x) = 0$.

(c) The values of x from part (b) are referred to as what graphically?

(d) Find $f(0)$, if possible.

(e) The value from part (d) is referred to as what graphically?

(f) What is the value of f at $x = 1$? What are the coordinates of the point?

(g) What is the value of f at $x = -1$? What are the coordinates of the point?

(h) The coordinates of the point on the graph of f at which $x = -3$. can be labeled $(-3, f(-3))$ or $(-3, \boxed{})$.

11. $f(x) = x^2 - x - 6$

12. $f(x) = x^3 - 4x$

13.
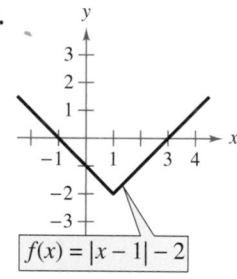
$f(x) = |x - 1| - 2$

14.
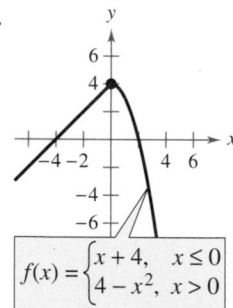
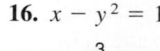
$f(x) = \begin{cases} x + 4, & x \le 0 \\ 4 - x^2, & x > 0 \end{cases}$

In Exercises 15–18, use the Vertical Line Test to determine whether y is a function of x. Describe how you can use a graphing utility to produce the given graph.

15. $y = \frac{1}{2}x^2$

16. $x - y^2 = 1$

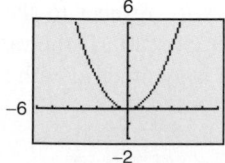

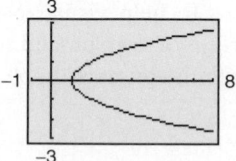

17. $x^2 + y^2 = 25$ **18.** $x^2 = 2xy - 1$

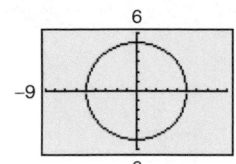   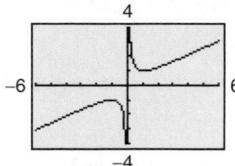

In Exercises 19–22, determine the open intervals over which the function is increasing, decreasing, or constant.

19. $f(x) = \frac{3}{2}x$ **20.** $f(x) = x^2 - 4x$

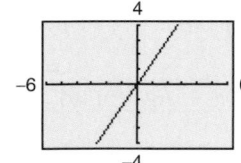   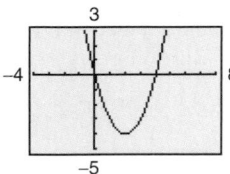

21. $f(x) = x^3 - 3x^2 + 2$ **22.** $f(x) = \sqrt{x^2 - 1}$

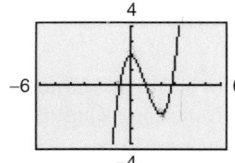   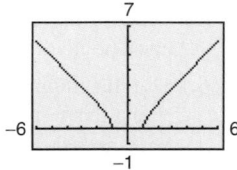

In Exercises 23–30, (a) use a graphing utility to graph the function and (b) determine the open intervals on which the function is increasing, decreasing, or constant.

23. $f(x) = 3$

24. $f(x) = x$

25. $f(x) = x^{2/3}$

26. $f(x) = -x^{3/4}$

27. $f(x) = x\sqrt{x + 3}$

28. $f(x) = \sqrt{1 - x}$

29. $f(x) = |x + 1| + |x - 1|$

30. $f(x) = -|x + 4| - |x + 1|$

In Exercises 31–36, use a graphing utility to approximate any relative minimum or relative maximum values of the function.

31. $f(x) = x^2 - 6x$

32. $f(x) = 3x^2 - 2x - 5$

33. $y = 2x^3 + 3x^2 - 12x$

34. $y = x^3 - 6x^2 + 15$

35. $h(x) = (x - 1)\sqrt{x}$

36. $g(x) = x\sqrt{4 - x}$

In Exercises 37–42, (a) approximate the relative minimum or relative maximum values of the function by sketching its graph using the point-plotting method, (b) use a graphing utility to approximate any relative minimum or relative maximum values, and (c) compare your answers from parts (a) and (b).

37. $f(x) = x^2 - 4x - 5$ **38.** $f(x) = 3x^2 - 12x$

39. $f(x) = x^3 - 3x$ **40.** $f(x) = -x^3 + 3x^2$

41. $f(x) = 3x^2 - 6x + 1$ **42.** $f(x) = 8x - 4x^2$

In Exercises 43–50, sketch the graph of the piecewise-defined function by hand.

43. $f(x) = \begin{cases} 2x + 3, & x < 0 \\ 3 - x, & x \geq 0 \end{cases}$

44. $f(x) = \begin{cases} x + 6, & x \leq -4 \\ 2x - 4, & x > -4 \end{cases}$

45. $f(x) = \begin{cases} \sqrt{4 + x}, & x < 0 \\ \sqrt{4 - x}, & x \geq 0 \end{cases}$

46. $f(x) = \begin{cases} 1 - (x - 1)^2, & x \leq 2 \\ \sqrt{x - 2}, & x > 2 \end{cases}$

47. $f(x) = \begin{cases} x + 3, & x \leq 0 \\ 3, & 0 < x \leq 2 \\ 2x - 1, & x > 2 \end{cases}$

48. $g(x) = \begin{cases} x + 5, & x \leq -3 \\ -2, & -3 < x < 1 \\ 5x - 4, & x \geq 1 \end{cases}$

49. $f(x) = \begin{cases} 2x + 1, & x \leq -1 \\ x^2 - 2, & x > -1 \end{cases}$

50. $h(x) = \begin{cases} 3 + x, & x < 0 \\ x^2 + 1, & x \geq 0 \end{cases}$

Library of Parent Functions In Exercises 51–56, sketch the graph of the function by hand. Then use a graphing utility to verify the graph.

51. $f(x) = [\![x]\!] + 2$

52. $f(x) = [\![x]\!] - 3$

53. $f(x) = [\![x - 1]\!] + 2$

54. $f(x) = [\![x - 2]\!] + 1$

55. $f(x) = [\![2x]\!]$

56. $f(x) = [\![4x]\!]$

In Exercises 57 and 58, use a graphing utility to graph the function. State the domain and range of the function. Describe the pattern of the graph.

57. $s(x) = 2\left(\frac{1}{4}x - [\![\frac{1}{4}x]\!]\right)$

58. $g(x) = 2\left(\frac{1}{4}x - [\![\frac{1}{4}x]\!]\right)^2$

In Exercises 59–66, **algebraically** determine whether the function is even, odd, or neither. Verify your answer using a graphing utility.

59. $f(t) = t^2 + 2t - 3$ **60.** $f(x) = x^6 - 2x^2 + 3$

61. $g(x) = x^3 - 5x$ **62.** $h(x) = x^3 - 5$

63. $f(x) = x\sqrt{1 - x^2}$ **64.** $f(x) = x\sqrt{x + 5}$

65. $g(s) = 4s^{2/3}$ **66.** $f(s) = 4s^{3/2}$

Think About It   In Exercises 67–72, find the coordinates of a second point on the graph of a function f if the given point is on the graph and the function is (a) even and (b) odd.

67. $\left(-\frac{3}{2}, 4\right)$ **68.** $\left(-\frac{5}{3}, -7\right)$

69. $(4, 9)$ **70.** $(5, -1)$

71. $(x, -y)$ **72.** $(2a, 2c)$

In Exercises 73–82, use a graphing utility to graph the function and determine whether it is even, odd, or neither. Verify your answer algebraically.

73. $f(x) = 5$ **74.** $f(x) = -9$

75. $f(x) = 3x - 2$ **76.** $f(x) = 5 - 3x$

77. $h(x) = x^2 - 4$ **78.** $f(x) = -x^2 - 8$

79. $f(x) = \sqrt{1 - x}$ **80.** $g(t) = \sqrt[3]{t - 1}$

81. $f(x) = |x + 2|$ **82.** $f(x) = -|x - 5|$

In Exercises 83–86, graph the function and determine the interval(s) (if any) on the real axis for which $f(x) \geq 0$. Use a graphing utility to verify your results.

83. $f(x) = 4 - x$ **84.** $f(x) = 4x + 2$

85. $f(x) = x^2 - 9$ **86.** $f(x) = x^2 - 4x$

87. *Communications*   The cost of using a telephone calling card is \$1.05 for the first minute and \$0.38 for each additional minute or portion of a minute.

(a) A customer needs a model for the cost C of using the calling card for a call lasting t minutes. Which of the following is the appropriate model?

$$C_1(t) = 1.05 + 0.38[\![t - 1]\!]$$

$$C_2(t) = 1.05 - 0.38[\![-(t - 1)]\!]$$

(b) Use a graphing utility to graph the appropriate model. Use the *value* feature or the *zoom* and *trace* features to estimate the cost of a call lasting 18 minutes and 45 seconds.

88. *Delivery Charges*   The cost of sending an overnight package from New York to Atlanta is \$9.80 for a package weighing up to but not including 1 pound and \$2.50 for each additional pound or portion of a pound. Use the greatest integer function to create a model for the cost C of overnight delivery of a package weighing x pounds, where $x > 0$. Sketch the graph of the function.

$\int$ In Exercises 89 and 90, write the height h of the rectangle as a function of x.

89.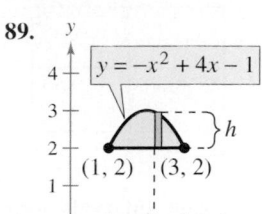

90.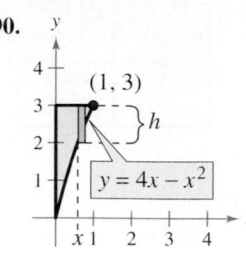

91. *Population*   During a 14 year period from 1990 to 2004, the population P (in thousands) of West Virginia fluctuated according to the model

$$P = 0.0108t^4 - 0.211t^3 + 0.40t^2 + 7.9t + 1791,$$

$$0 \leq t \leq 14$$

where t represents the year, with $t = 0$ corresponding to 1990.   (Source: U.S. Census Bureau)

(a) Use a graphing utility to graph the model over the appropriate domain.

(b) Use the graph from part (a) to determine during which years the population was increasing. During which years was the population decreasing?

(c) Approximate the maximum population between 1990 and 2004.

92. *Fluid Flow*   The intake pipe of a 100-gallon tank has a flow rate of 10 gallons per minute, and two drain pipes have a flow rate of 5 gallons per minute each. The graph shows the volume V of fluid in the tank as a function of time t. Determine in which pipes the fluid is flowing in specific subintervals of the one-hour interval of time shown on the graph. (There are many correct answers.)

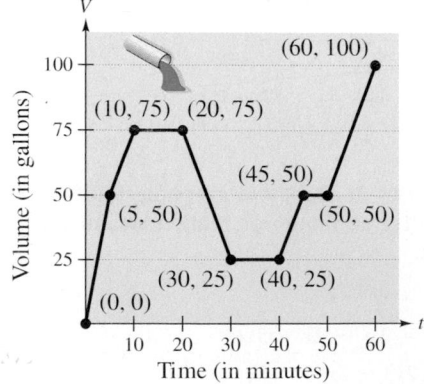

Synthesis

True or False?   **In Exercises 93 and 94, determine whether the statement is true or false. Justify your answer.**

93. A function with a square root cannot have a domain that is the set of all real numbers.

94. It is possible for an odd function to have the interval $[0, \infty)$ as its domain.

Think About It   **In Exercises 95–100, match the graph of the function with the best choice that describes the situation.**

(a) **The air temperature at a beach on a sunny day**

(b) **The height of a football kicked in a field goal attempt**

(c) **The number of children in a family over time**

(d) **The population of California as a function of time**

(e) **The depth of the tide at a beach over a 24-hour period**

(f) **The number of cupcakes on a tray at a party**

95.
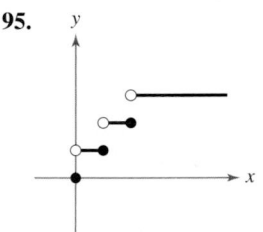

96.
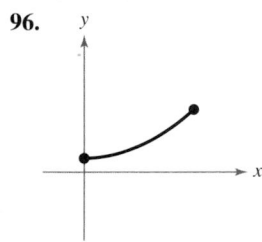

97.
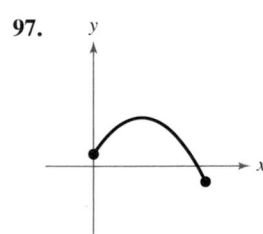

98.
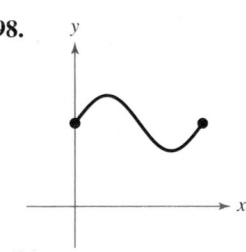

99.
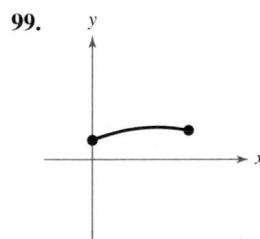

100.
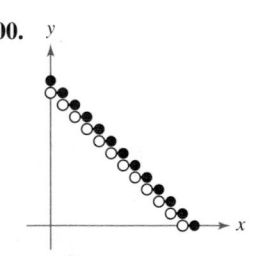

101. *Proof*   Prove that a function of the following form is odd.
$$y = a_{2n+1}x^{2n+1} + a_{2n-1}x^{2n-1} + \cdots + a_3x^3 + a_1x$$

102. *Proof*   Prove that a function of the following form is even.
$$y = a_{2n}x^{2n} + a_{2n-2}x^{2n-2} + \cdots + a_2x^2 + a_0$$

103. If f is an even function, determine if g is even, odd, or neither. Explain.

(a) $g(x) = -f(x)$ (b) $g(x) = f(-x)$

(c) $g(x) = f(x) - 2$ (d) $g(x) = -f(x - 2)$

104. *Think About It*   Does the graph in Exercise 16 represent x as a function of y? Explain.

105. *Think About It*   Does the graph in Exercise 17 represent x as a function of y? Explain.

106. *Writing*   Write a short paragraph describing three different functions that represent the behaviors of quantities between 1995 and 2006. Describe one quantity that decreased during this time, one that increased, and one that was constant. Present your results graphically.

Skills Review

In Exercises 107–110, identify the terms. Then identify the coefficients of the variable terms of the expression.

107. $-2x^2 + 8x$

108. $10 + 3x$

109. $\dfrac{x}{3} - 5x^2 + x^3$

110. $7x^4 + \sqrt{2}x^2$

In Exercises 111–114, find (a) the distance between the two points and (b) the midpoint of the line segment joining the points.

111. $(-2, 7), (6, 3)$

112. $(-5, 0), (3, 6)$

113. $\left(\frac{5}{2}, -1\right), \left(-\frac{3}{2}, 4\right)$

114. $\left(-6, \frac{2}{3}\right), \left(\frac{3}{4}, \frac{1}{6}\right)$

In Exercises 115–118, evaluate the function at each specified value of the independent variable and simplify.

115. $f(x) = 5x - 1$

(a) $f(6)$ (b) $f(-1)$ (c) $f(x - 3)$

116. $f(x) = -x^2 - x + 3$

(a) $f(4)$ (b) $f(-2)$ (c) $f(x - 2)$

117. $f(x) = x\sqrt{x - 3}$

(a) $f(3)$ (b) $f(12)$ (c) $f(6)$

118. $f(x) = -\frac{1}{2}x|x + 1|$

(a) $f(-4)$ (b) $f(10)$ (c) $f\left(-\frac{2}{3}\right)$

$\int$ **In Exercises 119 and 120, find the difference quotient and simplify your answer.**

119. $f(x) = x^2 - 2x + 9,\ \dfrac{f(3 + h) - f(3)}{h},\ h \neq 0$

120. $f(x) = 5 + 6x - x^2,\ \dfrac{f(6 + h) - f(6)}{h},\ h \neq 0$

1.4 Shifting, Reflecting, and Stretching Graphs

Summary of Graphs of Parent Functions

One of the goals of this text is to enable you to build your intuition for the basic shapes of the graphs of different types of functions. For instance, from your study of lines in Section 1.1, you can determine the basic shape of the graph of the linear function $f(x) = mx + b$. Specifically, you know that the graph of this function is a line whose slope is m and whose y-intercept is $(0, b)$.

The six graphs shown in Figure 1.40 represent the most commonly used functions in algebra. Familiarity with the basic characteristics of these simple graphs will help you analyze the shapes of more complicated graphs.

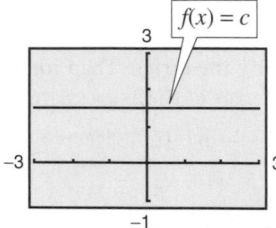

(a) Constant Function

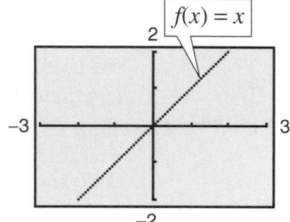

(b) Identity Function

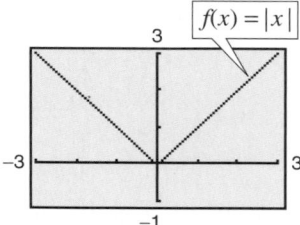

(c) Absolute Value Function

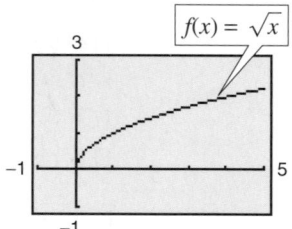

(d) Square Root Function

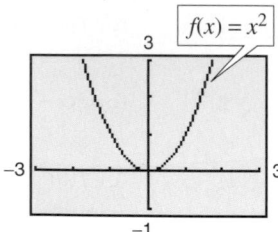

(e) Quadratic Function

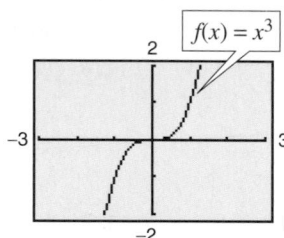

(f) Cubic Function

Figure 1.40

Throughout this section, you will discover how many complicated graphs are derived by shifting, stretching, shrinking, or reflecting the parent graphs shown above. Shifts, stretches, shrinks, and reflections are called *transformations*. Many graphs of functions can be created from combinations of these transformations.

What you should learn

- Recognize graphs of parent functions.
- Use vertical and horizontal shifts and reflections to graph functions.
- Use nonrigid transformations to graph functions.

Why you should learn it

Recognizing the graphs of parent functions and knowing how to shift, reflect, and stretch graphs of functions can help you sketch a wide variety of simple functions by hand. This skill is useful in sketching graphs of functions that model real-life data. For example, in Exercise 57 on page 49, you are asked to sketch a function that models the amount of fuel used by vans, pickups, and sport utility vehicles from 1990 through 2003.

Tim Boyle/Getty Images

Vertical and Horizontal Shifts

Many functions have graphs that are simple transformations of the graphs of parent functions summarized in Figure 1.40. For example, you can obtain the graph of

$$h(x) = x^2 + 2$$

by shifting the graph of $f(x) = x^2$ two units *upward,* as shown in Figure 1.41. In function notation, h and f are related as follows.

$$h(x) = x^2 + 2$$
$$= f(x) + 2 \qquad \text{Upward shift of two units}$$

Similarly, you can obtain the graph of

$$g(x) = (x - 2)^2$$

by shifting the graph of $f(x) = x^2$ two units to the *right,* as shown in Figure 1.42. In this case, the functions g and f have the following relationship.

$$g(x) = (x - 2)^2$$
$$= f(x - 2) \qquad \text{Right shift of two units}$$

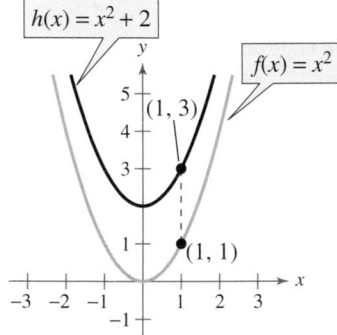

Figure 1.41   *Vertical shift upward: two units*

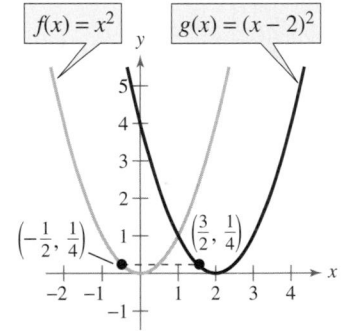

Figure 1.42   *Horizontal shift to the right: two units*

> **Exploration**
>
> Use a graphing utility to display (in the same viewing window) the graphs of $y = x^2 + c$, where $c = -2, 0, 2,$ and 4. Use the results to describe the effect that c has on the graph.
>
> Use a graphing utility to display (in the same viewing window) the graphs of $y = (x + c)^2$, where $c = -2, 0, 2,$ and 4. Use the results to describe the effect that c has on the graph.

The following list summarizes vertical and horizontal shifts.

Vertical and Horizontal Shifts

Let c be a positive real number. **Vertical and horizontal shifts** in the graph of $y = f(x)$ are represented as follows.

  1. Vertical shift c units *upward*: $h(x) = f(x) + c$

  2. Vertical shift c units *downward*: $h(x) = f(x) - c$

  3. Horizontal shift c units to the *right*: $h(x) = f(x - c)$

  4. Horizontal shift c units to the *left*: $h(x) = f(x + c)$

In items 3 and 4, be sure you see that $h(x) = f(x - c)$ corresponds to a *right* shift and $h(x) = f(x + c)$ corresponds to a *left* shift for $c > 0$.

Example 1   Shifts in the Graph of a Function

Compare the graph of each function with the graph of $f(x) = x^3$.

a. $g(x) = x^3 - 1$ **b.** $h(x) = (x - 1)^3$ **c.** $k(x) = (x + 2)^3 + 1$

Solution

a. Graph $f(x) = x^3$ and $g(x) = x^3 - 1$ [see Figure 1.43(a)]. You can obtain the graph of g by shifting the graph of f one unit downward.

b. Graph $f(x) = x^3$ and $h(x) = (x - 1)^3$ [see Figure 1.43(b)]. You can obtain the graph of h by shifting the graph of f one unit to the right.

c. Graph $f(x) = x^3$ and $k(x) = (x + 2)^3 + 1$ [see Figure 1.43(c)]. You can obtain the graph of k by shifting the graph of f two units to the left and then one unit upward.

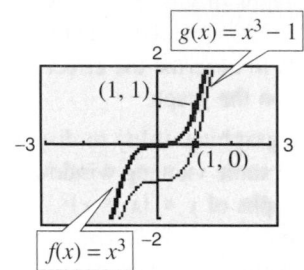
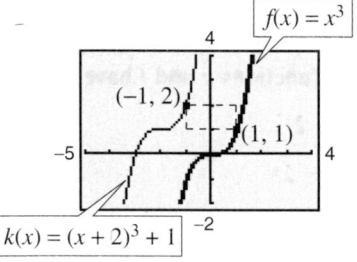

(a) Vertical shift: one unit downward

(b) Horizontal shift: one unit right

(c) Two units left and one unit upward

Figure 1.43

✓CHECKPOINT   Now try Exercise 23.

Example 2   Finding Equations from Graphs

The graph of $f(x) = x^2$ is shown in Figure 1.44. Each of the graphs in Figure 1.45 is a transformation of the graph of f. Find an equation for each function.

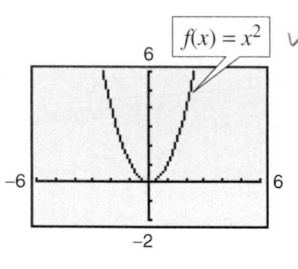
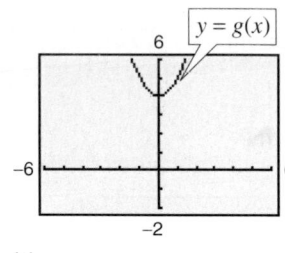
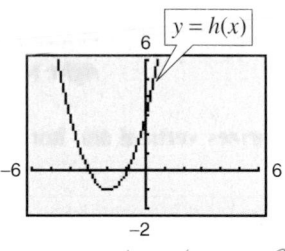

Figure 1.44

Figure 1.45

(a) $g(x) = x^2 + 4$

(b) $h(x) = (x + 2)^2 - 1$

Solution

a. The graph of g is a vertical shift of four units upward of the graph of $f(x) = x^2$. So, the equation for g is $g(x) = x^2 + 4$.

b. The graph of h is a horizontal shift of two units to the left, and a vertical shift of one unit downward, of the graph of $f(x) = x^2$. So, the equation for h is $h(x) = (x + 2)^2 - 1$.

✓CHECKPOINT   Now try Exercise 17.

Reflecting Graphs

Another common type of transformation is called a **reflection.** For instance, if you consider the x-axis to be a mirror, the graph of $h(x) = -x^2$ is the mirror image (or reflection) of the graph of $f(x) = x^2$ (see Figure 1.46).

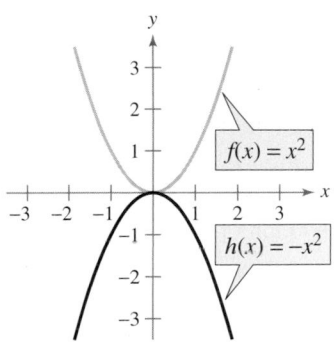

Figure 1.46

> **Exploration**
>
> Compare the graph of each function with the graph of $f(x) = x^2$ by using a graphing utility to graph the function and f in the same viewing window. Describe the transformation.
>
> **a.** $g(x) = -x^2$
>
> **b.** $h(x) = (-x)^2$

Reflections in the Coordinate Axes

Reflections in the coordinate axes of the graph of $y = f(x)$ are represented as follows.

1. Reflection in the x-axis: $h(x) = -f(x)$

2. Reflection in the y-axis: $h(x) = f(-x)$

Example 3   Finding Equations from Graphs

The graph of $f(x) = x^4$ is shown in Figure 1.47. Each of the graphs in Figure 1.48 is a transformation of the graph of f. Find an equation for each function.

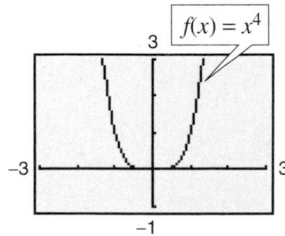

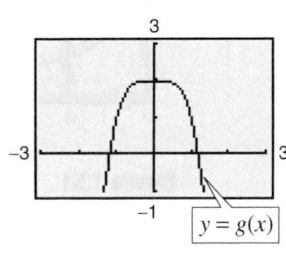

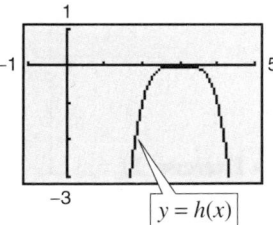

(a)   **(b)**

Figure 1.47   **Figure 1.48** $g(x) = -(x)^4 + 2$

Solution

a. The graph of g is a reflection in the x-axis *followed by* an upward shift of two units of the graph of $f(x) = x^4$. So, the equation for g is $g(x) = -x^4 + 2$.

b. The graph of h is a horizontal shift of three units to the right *followed by* a reflection in the x-axis of the graph of $f(x) = x^4$. So, the equation for h is $h(x) = -(x - 3)^4$.

✓*CHECKPOINT*   Now try Exercise 19.

Example 4   Reflections and Shifts

Compare the graph of each function with the graph of $f(x) = \sqrt{x}$.

a. $g(x) = -\sqrt{x}$ **b.** $h(x) = \sqrt{-x}$ **c.** $k(x) = -\sqrt{x+2}$

Algebraic Solution

a. Relative to the graph of $f(x) = \sqrt{x}$, the graph of g is a reflection in the x-axis because

$$g(x) = -\sqrt{x}$$
$$= -f(x).$$

b. The graph of h is a reflection of the graph of $f(x) = \sqrt{x}$ in the y-axis because

$$h(x) = \sqrt{-x}$$
$$= f(-x).$$

c. From the equation

$$k(x) = -\sqrt{x+2}$$
$$= -f(x+2)$$

you can conclude that the graph of k is a left shift of two units, followed by a reflection in the x-axis, of the graph of $f(x) = \sqrt{x}$.

Graphical Solution

a. Use a graphing utility to graph f and g in the same viewing window. From the graph in Figure 1.49, you can see that the graph of g is a reflection of the graph of f in the x-axis.

b. Use a graphing utility to graph f and h in the same viewing window. From the graph in Figure 1.50, you can see that the graph of h is a reflection of the graph of f in the y-axis.

c. Use a graphing utility to graph f and k in the same viewing window. From the graph in Figure 1.51, you can see that the graph of k is a left shift of two units of the graph of f, followed by a reflection in the x-axis.

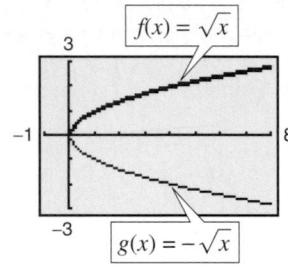

Figure 1.49

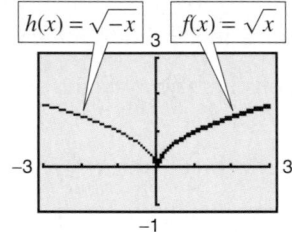

Figure 1.50

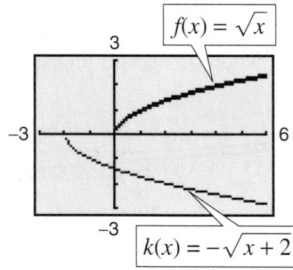

Figure 1.51

✓CHECKPOINT   Now try Exercise 21.

When graphing functions involving square roots, remember that the domain must be restricted to exclude negative numbers inside the radical. For instance, here are the domains of the functions in Example 4.

Domain of $g(x) = -\sqrt{x}$: $x \geq 0$

Domain of $h(x) = \sqrt{-x}$: $x \leq 0$

Domain of $k(x) = -\sqrt{x+2}$: $x \geq -2$

Nonrigid Transformations

Horizontal shifts, vertical shifts, and reflections are called **rigid transformations** because the basic shape of the graph is unchanged. These transformations change only the *position* of the graph in the coordinate plane. **Nonrigid transformations** are those that cause a *distortion*—a change in the shape of the original graph. For instance, a nonrigid transformation of the graph of $y = f(x)$ is represented by $y = cf(x)$, where the transformation is a **vertical stretch** if $c > 1$ and a **vertical shrink** if $0 < c < 1$. Another nonrigid transformation of the graph of $y = f(x)$ is represented by $h(x) = f(cx)$, where the transformation is a **horizontal shrink** if $c > 1$ and a **horizontal stretch** if $0 < c < 1$.

Example 5   Nonrigid Transformations

Compare the graph of each function with the graph of $f(x) = |x|$.

a. $h(x) = 3|x|$

b. $g(x) = \frac{1}{3}|x|$

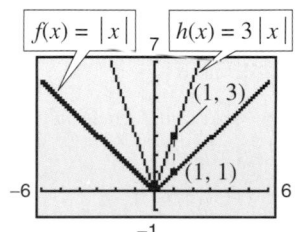

Figure 1.52

Solution

a. Relative to the graph of $f(x) = |x|$, the graph of

$$h(x) = 3|x|$$
$$= 3f(x)$$

is a vertical stretch (each y-value is multiplied by 3) of the graph of f. (See Figure 1.52.)

b. Similarly, the graph of

$$g(x) = \frac{1}{3}|x|$$
$$= \frac{1}{3}f(x)$$

is a vertical shrink $\left(\text{each } y\text{-value is multiplied by } \frac{1}{3}\right)$ of the graph of f. (See Figure 1.53.)

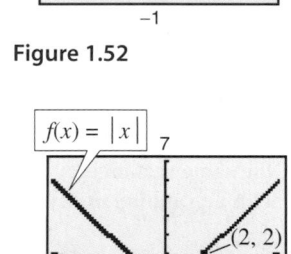

Figure 1.53

 Now try Exercise 31.

Example 6   Nonrigid Transformations

Compare the graph of $h(x) = f\left(\frac{1}{2}x\right)$ with the graph of $f(x) = 2 - x^3$.

Solution

Relative to the graph of $f(x) = 2 - x^3$, the graph of

$$h(x) = f\left(\tfrac{1}{2}x\right) = 2 - \left(\tfrac{1}{2}x\right)^3 = 2 - \tfrac{1}{8}x^3$$

is a horizontal stretch (each x-value is multiplied by 2) of the graph of f. (See Figure 1.54.)

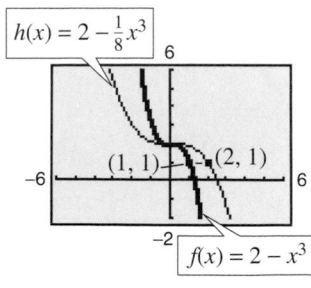

Figure 1.54

 Now try Exercise 39.

1.4 Exercises

See www.CalcChat.com for worked-out solutions to odd-numbered exercises.

Vocabulary Check

In Exercises 1–5, fill in the blanks.

1. The graph of a _____ is U-shaped.

2. The graph of an _____ is V-shaped.

3. Horizontal shifts, vertical shifts, and reflections are called _____ .

4. A reflection in the x-axis of $y = f(x)$ is represented by $h(x) =$ _____ , while a reflection in the y-axis of $y = f(x)$ is represented by $h(x) =$ _____ .

5. A nonrigid transformation of $y = f(x)$ represented by $cf(x)$ is a vertical stretch if _____ and a vertical shrink if _____ .

6. Match the rigid transformation of $y = f(x)$ with the correct representation, where $c > 0$.

   (a) $h(x) = f(x) + c$ (i) horizontal shift c units to the left
   (b) $h(x) = f(x) - c$ (ii) vertical shift c units upward
   (c) $h(x) = f(x - c)$ (iii) horizontal shift c units to the right
   (d) $h(x) = f(x + c)$ (iv) vertical shift c units downward

---

In Exercises 1–12, sketch the graphs of the three functions by hand on the same rectangular coordinate system. Verify your result with a graphing utility.

1. $f(x) = x$
   $g(x) = x - 4$
   $h(x) = 3x$

2. $f(x) = \frac{1}{2}x$
   $g(x) = \frac{1}{2}x + 2$
   $h(x) = \frac{1}{2}(x - 2)$

3. $f(x) = x^2$
   $g(x) = x^2 + 2$
   $h(x) = (x - 2)^2$

4. $f(x) = x^2$
   $g(x) = x^2 - 4$
   $h(x) = (x + 2)^2 + 1$

5. $f(x) = -x^2$
   $g(x) = -x^2 + 1$
   $h(x) = -(x - 2)^2$

6. $f(x) = (x - 2)^2$
   $g(x) = (x + 2)^2 + 2$
   $h(x) = -(x - 2)^2 - 1$

7. $f(x) = x^2$
   $g(x) = \frac{1}{2}x^2$
   $h(x) = (2x)^2$

8. $f(x) = x^2$
   $g(x) = \frac{1}{4}x^2 + 2$
   $h(x) = -\frac{1}{4}x^2$

9. $f(x) = |x|$
   $g(x) = |x| - 1$
   $h(x) = |x - 3|$

10. $f(x) = |x|$
    $g(x) = |x + 3|$
    $h(x) = -2|x + 2| - 1$

11. $f(x) = \sqrt{x}$
    $g(x) = \sqrt{x + 1}$
    $h(x) = \sqrt{x - 2} + 1$

12. $f(x) = \sqrt{x}$
    $g(x) = \frac{1}{2}\sqrt{x}$
    $h(x) = -\sqrt{x + 4}$

13. Use the graph of f to sketch each graph. To print an enlarged copy of the graph, go to the website *www.mathgraphs.com.*

    (a) $y = f(x) + 2$
    (b) $y = -f(x)$
    (c) $y = f(x - 2)$
    (d) $y = f(x + 3)$
    (e) $y = 2f(x)$
    (f) $y = f(-x)$
    (g) $y = f\left(\frac{1}{2}x\right)$

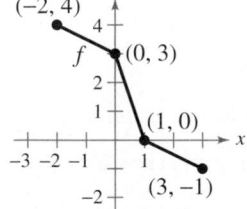

14. Use the graph of f to sketch each graph. To print an enlarged copy of the graph, go to the website *www.mathgraphs.com.*

    (a) $y = f(x) - 1$
    (b) $y = f(x + 1)$
    (c) $y = f(x - 1)$
    (d) $y = -f(x - 2)$
    (e) $y = f(-x)$
    (f) $y = \frac{1}{2}f(x)$
    (g) $y = f(2x)$

In Exercises 15–20, identify the parent function and describe the transformation shown in the graph. Write an equation for the graphed function.

15.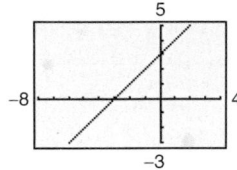

16.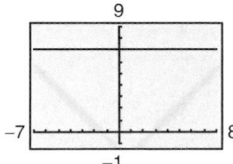

17.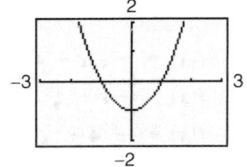

18.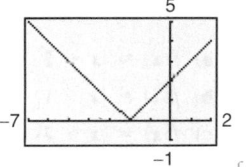

19.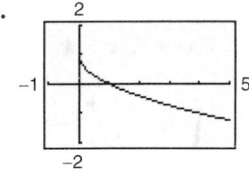

20.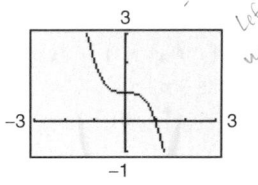

In Exercises 21–26, compare the graph of the function with the graph of $f(x) = \sqrt{x}$.

21. $y = -\sqrt{x} - 1$ **22.** $y = \sqrt{x} + 2$

23. $y = \sqrt{x - 2}$ **24.** $y = \sqrt{x + 4}$

25. $y = 2\sqrt{x}$ **26.** $y = \sqrt{-x + 3}$

In Exercises 27–32, compare the graph of the function with the graph of $f(x) = |x|$.

27. $y = |x + 5|$ **28.** $y = |x| - 3$

29. $y = -|x|$ **30.** $y = |-x|$

31. $y = 4|x|$ **32.** $y = \left|\frac{1}{2}x\right|$

In Exercises 33–38, compare the graph of the function with the graph of $f(x) = x^3$.

33. $g(x) = 4 - x^3$ **34.** $g(x) = -(x - 1)^3$

35. $h(x) = \frac{1}{4}(x + 2)^3$ **36.** $h(x) = -2(x - 1)^3 + 3$

37. $p(x) = \left(\frac{1}{3}x\right)^3 + 2$ **38.** $p(x) = [3(x - 2)]^3$

In Exercises 39–42, use a graphing utility to graph the three functions in the same viewing window. Describe the graphs of g and h relative to the graph of f.

39. $f(x) = x^3 - 3x^2$
$g(x) = f(x + 2)$
$h(x) = \frac{1}{2}f(x)$

40. $f(x) = x^3 - 3x^2 + 2$
$g(x) = f(x - 1)$
$h(x) = f(3x)$

41. $f(x) = x^3 - 3x^2$
$g(x) = -\frac{1}{3}f(x)$
$h(x) = f(-x)$

42. $f(x) = x^3 - 3x^2 + 2$
$g(x) = -f(x)$
$h(x) = f(2x)$

In Exercises 43–56, g is related to one of the six parent functions on page 42. (a) Identify the parent function f. (b) Describe the sequence of transformations from f to g. (c) Sketch the graph of g by hand. (d) Use function notation to write g in terms of the parent function f.

43. $g(x) = 2 - (x + 5)^2$ **44.** $g(x) = -(x + 10)^2 + 5$

45. $g(x) = 3 + 2(x - 4)^2$ **46.** $g(x) = -\frac{1}{4}(x + 2)^2 - 2$

47. $g(x) = 3(x - 2)^3$ **48.** $g(x) = -\frac{1}{2}(x + 1)^3$

49. $g(x) = (x - 1)^3 + 2$ **50.** $g(x) = -(x + 3)^3 - 10$

51. $g(x) = |x + 4| + 8$ **52.** $g(x) = |x + 3| + 9$

53. $g(x) = -2|x - 1| - 4$ **54.** $g(x) = \frac{1}{2}|x - 2| - 3$

55. $g(x) = -\frac{1}{2}\sqrt{x + 3} - 1$ **56.** $g(x) = -\sqrt{x + 1} - 6$

57. *Fuel Use*   The amounts of fuel F (in billions of gallons) used by vans, pickups, and SUVs (sport utility vehicles) from 1990 through 2003 are shown in the table. A model for the data can be approximated by the function $F(t) = 33.0 + 6.2\sqrt{t}$, where $t = 0$ represents 1990. (Source: U.S. Federal Highway Administration)

Year	Annual fuel use, F (in billions of gallons)
1990	35.6
1991	38.2
1992	40.9
1993	42.9
1994	44.1
1995	45.6
1996	47.4
1997	49.4
1998	50.5
1999	52.8
2000	52.9
2001	53.5
2002	55.2
2003	56.3

(a) Describe the transformation of the parent function $f(t) = \sqrt{t}$.

(b) Use a graphing utility to graph the model and the data in the same viewing window.

(c) Rewrite the function so that $t = 0$ represents 2003. Explain how you got your answer.

58. *Finance* The amounts M (in billions of dollars) of home mortgage debt outstanding in the United States from 1990 through 2004 can be approximated by the function

$$M(t) = 32.3t^2 + 3769$$

where $t = 0$ represents 1990. (Source: Board of Governors of the Federal Reserve System)

(a) Describe the transformation of the parent function $f(t) = t^2$.

(b) Use a graphing utility to graph the model over the interval $0 \le t \le 14$.

(c) According to the model, when will the amount of debt exceed 10 trillion dollars?

(d) Rewrite the function so that $t = 0$ represents 2000. Explain how you got your answer.

Synthesis

True or False? **In Exercises 59 and 60, determine whether the statement is true or false. Justify your answer.**

59. The graph of $y = f(-x)$ is a reflection of the graph of $y = f(x)$ in the x-axis.

60. The graph of $y = -f(x)$ is a reflection of the graph of $y = f(x)$ in the y-axis.

61. *Exploration* Use the fact that the graph of $y = f(x)$ has x-intercepts at $x = 2$ and $x = -3$ to find the x-intercepts of the given graph. If not possible, state the reason.

(a) $y = f(-x)$ (b) $y = -f(x)$ (c) $y = 2f(x)$

(d) $y = f(x) + 2$ (e) $y = f(x - 3)$

62. *Exploration* Use the fact that the graph of $y = f(x)$ has x-intercepts at $x = -1$ and $x = 4$ to find the x-intercepts of the given graph. If not possible, state the reason.

(a) $y = f(-x)$ (b) $y = -f(x)$ (c) $y = 2f(x)$

(d) $y = f(x) - 1$ (e) $y = f(x - 2)$

63. *Exploration* Use the fact that the graph of $y = f(x)$ is increasing on the interval $(-\infty, 2)$ and decreasing on the interval $(2, \infty)$ to find the intervals on which the graph is increasing and decreasing. If not possible, state the reason.

(a) $y = f(-x)$ (b) $y = -f(x)$ (c) $y = 2f(x)$

(d) $y = f(x) - 3$ (e) $y = f(x + 1)$

64. *Exploration* Use the fact that the graph of $y = f(x)$ is increasing on the intervals $(-\infty, -1)$ and $(2, \infty)$ and decreasing on the interval $(-1, 2)$ to find the intervals on which the graph is increasing and decreasing. If not possible, state the reason.

(a) $y = f(-x)$ (b) $y = -f(x)$ (c) $y = \frac{1}{2}f(x)$

(d) $y = -f(x - 1)$ (e) $y = f(x - 2) + 1$

Library of Parent Functions In Exercises 65–68, determine which equation(s) may be represented by the graph shown. There may be more than one correct answer.

65.
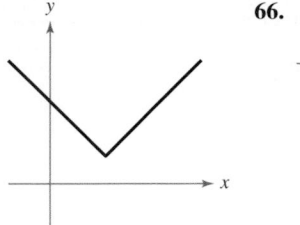

66.
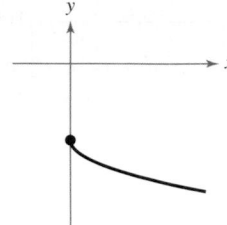

(a) $f(x) = |x + 2| + 1$ (a) $f(x) = -\sqrt{x} - 4$

(b) $f(x) = |x - 1| + 2$ (b) $f(x) = -4 - \sqrt{x}$

(c) $f(x) = |x - 2| + 1$ (c) $f(x) = -4 - \sqrt{-x}$

(d) $f(x) = 2 + |x - 2|$ (d) $f(x) = \sqrt{-x} - 4$

(e) $f(x) = |(x - 2) + 1|$ (e) $f(x) = \sqrt{-x} + 4$

(f) $f(x) = 1 - |x - 2|$ (f) $f(x) = \sqrt{x} - 4$

67.
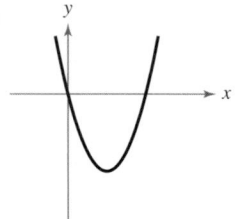

68.
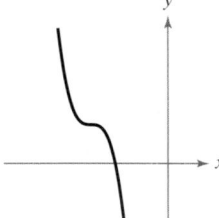

(a) $f(x) = (x - 2)^2 - 2$ (a) $f(x) = -(x - 4)^3 + 2$

(b) $f(x) = (x + 4)^2 - 4$ (b) $f(x) = -(x + 4)^3 + 2$

(c) $f(x) = (x - 2)^2 - 4$ (c) $f(x) = -(x - 2)^3 + 4$

(d) $f(x) = (x + 2)^2 - 4$ (d) $f(x) = (-x - 4)^3 + 2$

(e) $f(x) = 4 - (x - 2)^2$ (e) $f(x) = (x + 4)^3 + 2$

(f) $f(x) = 4 - (x + 2)^2$ (f) $f(x) = (-x + 4)^3 + 2$

Skills Review

In Exercises 69 and 70, determine whether the lines L_1 and L_2 passing through the pairs of points are parallel, perpendicular, or neither.

69. L_1: $(-2, -2)$, $(2, 10)$

    L_2: $(-1, 3)$, $(3, 9)$

70. L_1: $(-1, -7)$, $(4, 3)$

    L_2: $(1, 5)$, $(-2, -7)$

In Exercises 71–74, find the domain of the function.

71. $f(x) = \dfrac{4}{9 - x}$

72. $f(x) = \dfrac{\sqrt{x - 5}}{x - 7}$

73. $f(x) = \sqrt{100 - x^2}$

74. $f(x) = \sqrt[3]{16 - x^2}$

1.5 Combinations of Functions

Arithmetic Combinations of Functions

Just as two real numbers can be combined by the operations of addition, subtraction, multiplication, and division to form other real numbers, two *functions* can be combined to create new functions. If $f(x) = 2x - 3$ and $g(x) = x^2 - 1$, you can form the sum, difference, product, and quotient of f and g as follows.

$$f(x) + g(x) = (2x - 3) + (x^2 - 1)$$
$$= x^2 + 2x - 4 \qquad \text{Sum}$$
$$f(x) - g(x) = (2x - 3) - (x^2 - 1)$$
$$= -x^2 + 2x - 2 \qquad \text{Difference}$$
$$f(x) \cdot g(x) = (2x - 3)(x^2 - 1)$$
$$= 2x^3 - 3x^2 - 2x + 3 \qquad \text{Product}$$
$$\frac{f(x)}{g(x)} = \frac{2x - 3}{x^2 - 1}, \qquad x \neq \pm 1 \qquad \text{Quotient}$$

The domain of an **arithmetic combination** of functions f and g consists of all real numbers that are common to the domains of f and g. In the case of the quotient $f(x)/g(x)$, there is the further restriction that $g(x) \neq 0$.

What you should learn

- Add, subtract, multiply, and divide functions.
- Find compositions of one function with another function.
- Use combinations of functions to model and solve real-life problems.

Why you should learn it

Combining functions can sometimes help you better understand the big picture. For instance, Exercises 75 and 76 on page 60 illustrate how to use combinations of functions to analyze U.S. health care expenditures.

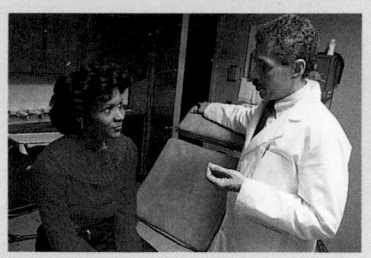

SuperStock

Sum, Difference, Product, and Quotient of Functions

Let f and g be two functions with overlapping domains. Then, for all x common to both domains, the sum, difference, product, and quotient of f and g are defined as follows.

1. **Sum:** $\qquad (f + g)(x) = f(x) + g(x)$

2. **Difference:** $\qquad (f - g)(x) = f(x) - g(x)$

3. **Product:** $\qquad (fg)(x) = f(x) \cdot g(x)$

4. **Quotient:** $\qquad \left(\dfrac{f}{g}\right)(x) = \dfrac{f(x)}{g(x)}, \quad g(x) \neq 0$

Example 1 Finding the Sum of Two Functions

Given $f(x) = 2x + 1$ and $g(x) = x^2 + 2x - 1$, find $(f + g)(x)$. Then evaluate the sum when $x = 2$.

Solution

$$(f + g)(x) = f(x) + g(x) = (2x + 1) + (x^2 + 2x - 1) = x^2 + 4x$$

When $x = 2$, the value of this sum is $(f + g)(2) = 2^2 + 4(2) = 12$.

✓CHECKPOINT   Now try Exercise 7(a).

Example 2   Finding the Difference of Two Functions

Given $f(x) = 2x + 1$ and $g(x) = x^2 + 2x - 1$, find $(f - g)(x)$. Then evaluate the difference when $x = 2$.

Algebraic Solution

The difference of the functions f and g is

$$(f - g)(x) = f(x) - g(x)$$
$$= (2x + 1) - (x^2 + 2x - 1)$$
$$= -x^2 + 2.$$

When $x = 2$, the value of this difference is

$$(f - g)(2) = -(2)^2 + 2$$
$$= -2.$$

Note that $(f - g)(2)$ can also be evaluated as follows.

$$(f - g)(2) = f(2) - g(2)$$
$$= [2(2) + 1] - [2^2 + 2(2) - 1]$$
$$= 5 - 7$$
$$= -2$$

✓*CHECKPOINT*   Now try Exercise 7(b).

Graphical Solution

You can use a graphing utility to graph the difference of two functions. Enter the functions as follows (see Figure 1.55).

$$y_1 = 2x + 1$$
$$y_2 = x^2 + 2x - 1$$
$$y_3 = y_1 - y_2$$

Graph y_3 as shown in Figure 1.56. Then use the *value* feature or the *zoom* and *trace* features to estimate that the value of the difference when $x = 2$ is -2.

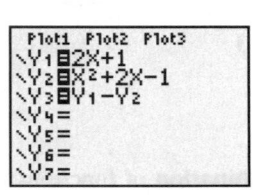

Figure 1.55

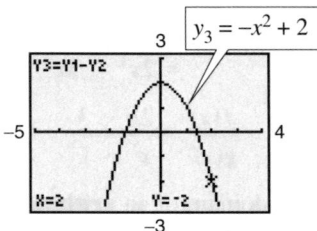

Figure 1.56

In Examples 1 and 2, both f and g have domains that consist of all real numbers. So, the domain of both $(f + g)$ and $(f - g)$ is also the set of all real numbers. Remember that any restrictions on the domains of f or g must be considered when forming the sum, difference, product, or quotient of f and g. For instance, the domain of $f(x) = 1/x$ is all $x \neq 0$, and the domain of $g(x) = \sqrt{x}$ is $[0, \infty)$. This implies that the domain of $(f + g)$ is $(0, \infty)$.

Example 3   Finding the Product of Two Functions

Given $f(x) = x^2$ and $g(x) = x - 3$, find $(fg)(x)$. Then evaluate the product when $x = 4$.

Solution

$$(fg)(x) = f(x)g(x)$$
$$= (x^2)(x - 3)$$
$$= x^3 - 3x^2$$

When $x = 4$, the value of this product is

$$(fg)(4) = 4^3 - 3(4)^2 = 16.$$

✓*CHECKPOINT*   Now try Exercise 7(c).

Example 4   Finding the Quotient of Two Functions

Find $(f/g)(x)$ and $(g/f)(x)$ for the functions given by $f(x) = \sqrt{x}$ and $g(x) = \sqrt{4 - x^2}$. Then find the domains of f/g and g/f.

Solution

The quotient of f and g is

$$\left(\frac{f}{g}\right)(x) = \frac{f(x)}{g(x)} = \frac{\sqrt{x}}{\sqrt{4 - x^2}},$$

and the quotient of g and f is

$$\left(\frac{g}{f}\right)(x) = \frac{g(x)}{f(x)} = \frac{\sqrt{4 - x^2}}{\sqrt{x}}.$$

The domain of f is $[0, \infty)$ and the domain of g is $[-2, 2]$. The intersection of these domains is $[0, 2]$. So, the domains for f/g and g/f are as follows.

Domain of (f/g): $[0, 2)$ Domain of (g/f): $(0, 2]$

✔CHECKPOINT   Now try Exercise 7(d).

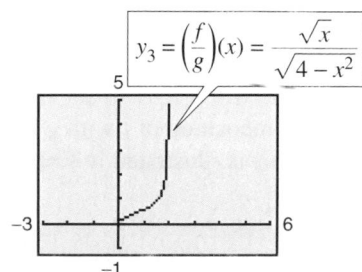

Figure 1.57

TECHNOLOGY TIP   You can confirm the domain of f/g in Example 4 with your graphing utility by entering the three functions $y_1 = \sqrt{x}$, $y_2 = \sqrt{4 - x^2}$, and $y_3 = y_1/y_2$, and graphing y_3, as shown in Figure 1.57. Use the *trace* feature to determine that the x-coordinates of points on the graph extend from 0 to 2 but do not include 2. So, you can estimate the domain of f/g to be $[0, 2)$. You can confirm the domain of g/f in Example 4 by entering $y_4 = y_2/y_1$ and graphing y_4, as shown in Figure 1.58. Use the *trace* feature to determine that the x-coordinates of points on the graph extend from 0 to 2 but do not include 0. So, you can estimate the domain of g/f to be $(0, 2]$.

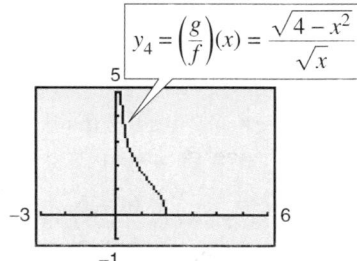

Figure 1.58

Compositions of Functions

Another way of combining two functions is to form the **composition** of one with the other. For instance, if $f(x) = x^2$ and $g(x) = x + 1$, the composition of f with g is

$$f(g(x)) = f(x + 1) = (x + 1)^2.$$

This composition is denoted as $f \circ g$ and is read as "f composed with g."

Definition of Composition of Two Functions

The **composition** of the function f with the function g is

$$(f \circ g)(x) = f(g(x)).$$

The domain of $f \circ g$ is the set of all x in the domain of g such that $g(x)$ is in the domain of f. (See Figure 1.59.)

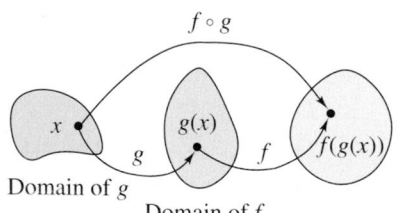

Figure 1.59

Example 5   Forming the Composition of f with g

Find $(f \circ g)(x)$ for $f(x) = \sqrt{x}$, $x \geq 0$, and $g(x) = x - 1$, $x \geq 1$. If possible, find $(f \circ g)(2)$ and $(f \circ g)(0)$.

Solution

$$(f \circ g)(x) = f(g(x)) \qquad \text{Definition of } f \circ g$$

$$= f(x - 1) \qquad \text{Definition of } g(x)$$

$$= \sqrt{x - 1}, \quad x \geq 1 \qquad \text{Definition of } f(x)$$

The domain of $f \circ g$ is $[1, \infty)$. So, $(f \circ g)(2) = \sqrt{2 - 1} = 1$ is defined, but $(f \circ g)(0)$ is not defined because 0 is not in the domain of $f \circ g$.

✔CHECKPOINT   Now try Exercise 35.

The composition of f with g is generally not the same as the composition of g with f. This is illustrated in Example 6.

> **Exploration**
>
> Let $f(x) = x + 2$ and $g(x) = 4 - x^2$. Are the compositions $f \circ g$ and $g \circ f$ equal? You can use your graphing utility to answer this question by entering and graphing the following functions.
>
> $$y_1 = (4 - x^2) + 2$$
>
> $$y_2 = 4 - (x + 2)^2$$
>
> What do you observe? Which function represents $f \circ g$ and which represents $g \circ f$?

Example 6   Compositions of Functions

Given $f(x) = x + 2$ and $g(x) = 4 - x^2$, evaluate (a) $(f \circ g)(x)$ and (b) $(g \circ f)(x)$ when $x = 0, 1, 2,$ and 3.

Algebraic Solution

a. $(f \circ g)(x) = f(g(x))$ Definition of $f \circ g$

$\qquad = f(4 - x^2)$ Definition of $g(x)$

$\qquad = (4 - x^2) + 2$ Definition of $f(x)$

$\qquad = -x^2 + 6$

$(f \circ g)(0) = -0^2 + 6 = 6$

$(f \circ g)(1) = -1^2 + 6 = 5$

$(f \circ g)(2) = -2^2 + 6 = 2$

$(f \circ g)(3) = -3^2 + 6 = -3$

b. $(g \circ f)(x) = g(f(x))$ Definition of $g \circ f$

$\qquad = g(x + 2)$ Definition of $f(x)$

$\qquad = 4 - (x + 2)^2$ Definition of $g(x)$

$\qquad = 4 - (x^2 + 4x + 4)$

$\qquad = -x^2 - 4x$

$(g \circ f)(0) = -0^2 - 4(0) = 0$

$(g \circ f)(1) = -1^2 - 4(1) = -5$

$(g \circ f)(2) = -2^2 - 4(2) = -12$

$(g \circ f)(3) = -3^2 - 4(3) = -21$

Note that $f \circ g \neq g \circ f$.

✔CHECKPOINT   Now try Exercise 37.

Numerical Solution

a. You can use the *table* feature of a graphing utility to evaluate $f \circ g$ when $x = 0, 1, 2,$ and 3. Enter $y_1 = g(x)$ and $y_2 = f(g(x))$ in the *equation editor* (see Figure 1.60). Then set the table to *ask* mode to find the desired function values (see Figure 1.61). Finally, display the table, as shown in Figure 1.62.

b. You can evaluate $g \circ f$ when $x = 0, 1, 2,$ and 3 by using a procedure similar to that of part (a). You should obtain the table shown in Figure 1.63.

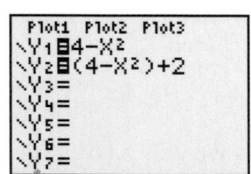

Figure 1.60

Figure 1.61

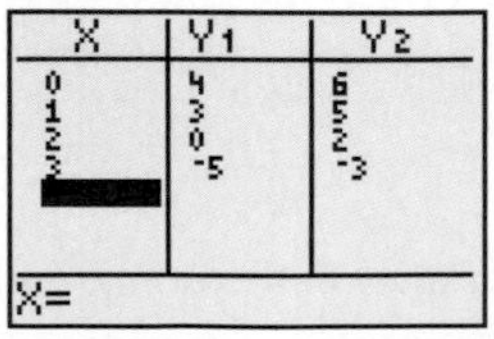

Figure 1.62

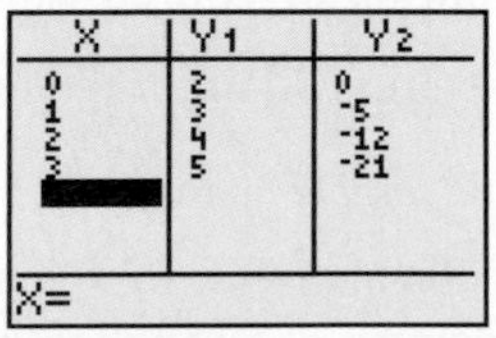

Figure 1.63

From the tables you can see that $f \circ g \neq g \circ f$.

To determine the domain of a composite function $f \circ g$, you need to restrict the outputs of g so that they are in the domain of f. For instance, to find the domain of $f \circ g$ given that $f(x) = 1/x$ and $g(x) = x + 1$, consider the outputs of g. These can be any real number. However, the domain of f is restricted to all real numbers except 0. So, the outputs of g must be restricted to all real numbers except 0. This means that $g(x) \neq 0$, or $x \neq -1$. So, the domain of $f \circ g$ is all real numbers except $x = -1$.

Example 7   Finding the Domain of a Composite Function

Find the domain of the composition $(f \circ g)(x)$ for the functions given by

$$f(x) = x^2 - 9 \qquad \text{and} \qquad g(x) = \sqrt{9 - x^2}.$$

Algebraic Solution

The composition of the functions is as follows.

$$(f \circ g)(x) = f(g(x))$$

$$= f\left(\sqrt{9 - x^2}\right)$$

$$= \left(\sqrt{9 - x^2}\right)^2 - 9$$

$$= 9 - x^2 - 9$$

$$= -x^2$$

From this, it might appear that the domain of the composition is the set of all real numbers. This, however, is not true. Because the domain of f is the set of all real numbers and the domain of g is $[-3, 3]$, the domain of $(f \circ g)$ is $[-3, 3]$.

✓CHECKPOINT   Now try Exercise 39.

Graphical Solution

You can use a graphing utility to graph the composition of the functions $(f \circ g)(x)$ as $y = \left(\sqrt{9 - x^2}\right)^2 - 9$. Enter the functions as follows.

$$y_1 = \sqrt{9 - x^2} \qquad y_2 = y_1^2 - 9$$

Graph y_2, as shown in Figure 1.64. Use the *trace* feature to determine that the x-coordinates of points on the graph extend from -3 to 3. So, you can graphically estimate the domain of $(f \circ g)(x)$ to be $[-3, 3]$.

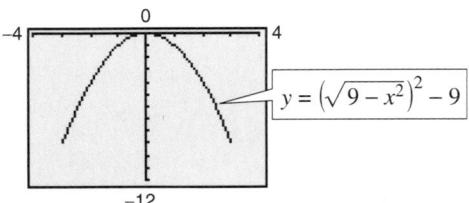

$$y = \left(\sqrt{9 - x^2}\right)^2 - 9$$

Figure 1.64

Example 8   A Case in Which $f \circ g = g \circ f$

Given $f(x) = 2x + 3$ and $g(x) = \frac{1}{2}(x - 3)$, find each composition.

a. $(f \circ g)(x)$ **b.** $(g \circ f)(x)$

Solution

a. $(f \circ g)(x) = f(g(x))$

$$= f\left(\frac{1}{2}(x - 3)\right)$$

$$= 2\left[\frac{1}{2}(x - 3)\right] + 3$$

$$= x - 3 + 3$$

$$= x$$

b. $(g \circ f)(x) = g(f(x))$

$$= g(2x + 3)$$

$$= \frac{1}{2}\left[(2x + 3) - 3\right]$$

$$= \frac{1}{2}(2x)$$

$$= x$$

✓CHECKPOINT   Now try Exercise 51.

STUDY TIP

In Example 8, note that the two composite functions $f \circ g$ and $g \circ f$ are equal, and both represent the identity function. That is, $(f \circ g)(x) = x$ and $(g \circ f)(x) = x$. You will study this special case in the next section.

In Examples 5, 6, 7, and 8, you formed the composition of two given functions. In calculus, it is also important to be able to identify two functions that make up a given composite function. Basically, to "decompose" a composite function, look for an "inner" and an "outer" function.

Example 9   Identifying a Composite Function

Write the function $h(x) = (3x - 5)^3$ as a composition of two functions.

Solution

One way to write h as a composition of two functions is to take the inner function to be $g(x) = 3x - 5$ and the outer function to be $f(x) = x^3$. Then you can write

$$h(x) = (3x - 5)^3$$

$$= f(3x - 5)$$

$$= f(g(x)).$$

✓CHECKPOINT   Now try Exercise 65.

Example 10   Identifying a Composite Function

Write the function

$$h(x) = \frac{1}{(x - 2)^2}$$

as a composition of two functions.

Solution

One way to write h as a composition of two functions is to take the inner function to be $g(x) = x - 2$ and the outer function to be

$$f(x) = \frac{1}{x^2}$$

$$= x^{-2}.$$

Then you can write

$$h(x) = \frac{1}{(x - 2)^2}$$

$$= (x - 2)^{-2}$$

$$= f(x - 2)$$

$$= f(g(x)).$$

✓CHECKPOINT   Now try Exercise 69.

Exploration

Write each function as a composition of two functions.

a. $h(x) = |x^3 - 2|$

b. $r(x) = |x^3| - 2$

What do you notice about the inner and outer functions?

Exploration

The function in Example 10 can be decomposed in other ways. For which of the following pairs of functions is $h(x)$ equal to $f(g(x))$?

a. $g(x) = \dfrac{1}{x - 2}$ and

   $f(x) = x^2$

b. $g(x) = x^2$ and

   $f(x) = \dfrac{1}{x - 2}$

c. $g(x) = \dfrac{1}{x}$ and

   $f(x) = (x - 2)^2$

Example 11   Bacteria Count

The number N of bacteria in a refrigerated food is given by

$$N(T) = 20T^2 - 80T + 500, \qquad 2 \le T \le 14$$

where T is the temperature of the food (in degrees Celsius). When the food is removed from refrigeration, the temperature of the food is given by

$$T(t) = 4t + 2, \qquad 0 \le t \le 3$$

where t is the time (in hours).

a. Find the composition $N(T(t))$ and interpret its meaning in context.

b. Find the number of bacteria in the food when $t = 2$ hours.

c. Find the time when the bacterial count reaches 2000.

Solution

a. $N(T(t)) = 20(4t + 2)^2 - 80(4t + 2) + 500$

$$= 20(16t^2 + 16t + 4) - 320t - 160 + 500$$

$$= 320t^2 + 320t + 80 - 320t - 160 + 500$$

$$= 320t^2 + 420$$

The composite function $N(T(t))$ represents the number of bacteria as a function of the amount of time the food has been out of refrigeration.

b. When $t = 2$, the number of bacteria is

$$N = 320(2)^2 + 420$$

$$= 1280 + 420$$

$$= 1700.$$

c. The bacterial count will reach $N = 2000$ when $320t^2 + 420 = 2000$. You can solve this equation for t algebraically as follows.

$$320t^2 + 420 = 2000$$

$$320t^2 = 1580$$

$$t^2 = \frac{79}{16}$$

$$t = \frac{\sqrt{79}}{4}$$

$$t \approx 2.22 \text{ hours}$$

So, the count will reach 2000 when $t \approx 2.22$ hours. When you solve this equation, note that the negative value is rejected because it is not in the domain of the composite function. You can use a graphing utility to confirm your solution. First graph the equation $N = 320t^2 + 420$, as shown in Figure 1.65. Then use the *zoom* and *trace* features to approximate $N = 2000$ when $t \approx 2.22$, as shown in Figure 1.66.

✓CHECKPOINT   Now try Exercise 81.

Exploration

Use a graphing utility to graph $y_1 = 320x^2 + 420$ and $y_2 = 2000$ in the same viewing window. (Use a viewing window in which $0 \le x \le 3$ and $400 \le y \le 4000$.) Explain how the graphs can be used to answer the question asked in Example 11(c). Compare your answer with that given in part (c). When will the bacteria count reach 3200?

Notice that the model for this bacteria count situation is valid only for a span of 3 hours. Now suppose that the minimum number of bacteria in the food is reduced from 420 to 100. Will the number of bacteria still reach a level of 2000 within the three-hour time span? Will the number of bacteria reach a level of 3200 within 3 hours?

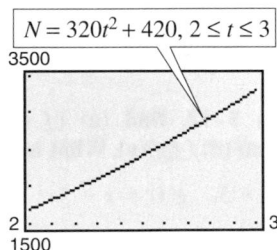

Figure 1.65

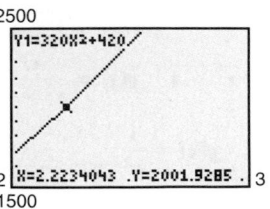

Figure 1.66

1.5 Exercises

See www.CalcChat.com for worked-out solutions to odd-numbered exercises.

Vocabulary Check

Fill in the blanks.

1. Two functions f and g can be combined by the arithmetic operations of _____ , _____ , _____ , and _____ to create new functions.

2. The _____ of the function f with the function g is $(f \circ g)(x) = f(g(x))$.

3. The domain of $f \circ g$ is the set of all x in the domain of g such that _____ is in the domain of f.

4. To decompose a composite function, look for an _____ and an _____ function.

In Exercises 1–4, use the graphs of f and g to graph $h(x) = (f + g)(x)$. To print an enlarged copy of the graph, go to the website www.mathgraphs.com.

1.
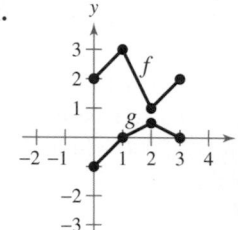

2.
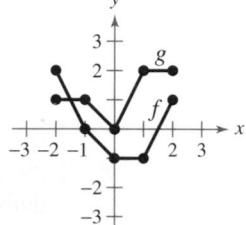

3.
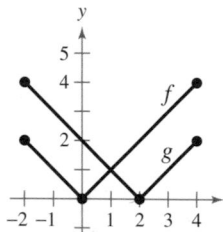

4.
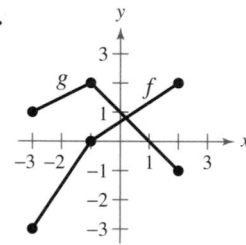

In Exercises 5–12, find (a) $(f + g)(x)$, (b) $(f - g)(x)$, (c) $(fg)(x)$, and (d) $(f/g)(x)$. What is the domain of f/g?

5. $f(x) = x + 3$, $g(x) = x - 3$

6. $f(x) = 2x - 5$, $g(x) = 1 - x$

7. $f(x) = x^2$, $g(x) = 1 - x$

8. $f(x) = 2x - 5$, $g(x) = 4$

9. $f(x) = x^2 + 5$, $g(x) = \sqrt{1 - x}$

10. $f(x) = \sqrt{x^2 - 4}$, $g(x) = \dfrac{x^2}{x^2 + 1}$

11. $f(x) = \dfrac{1}{x}$, $g(x) = \dfrac{1}{x^2}$

12. $f(x) = \dfrac{x}{x + 1}$, $g(x) = x^3$

In Exercises 13–26, evaluate the indicated function for $f(x) = x^2 - 1$ and $g(x) = x - 2$ algebraically. If possible, use a graphing utility to verify your answer.

13. $(f + g)(3)$

14. $(f - g)(-2)$

15. $(f - g)(0)$

16. $(f + g)(1)$

17. $(fg)(4)$

18. $(fg)(-6)$

19. $\left(\dfrac{f}{g}\right)(-5)$

20. $\left(\dfrac{f}{g}\right)(0)$

21. $(f - g)(2t)$

22. $(f + g)(t - 4)$

23. $(fg)(-5t)$

24. $(fg)(3t^2)$

25. $\left(\dfrac{f}{g}\right)(-t)$

26. $\left(\dfrac{f}{g}\right)(t + 2)$

In Exercises 27–30, use a graphing utility to graph the functions f, g, and h in the same viewing window.

27. $f(x) = \frac{1}{2}x$, $g(x) = x - 1$, $h(x) = f(x) + g(x)$

28. $f(x) = \frac{1}{3}x$, $g(x) = -x + 4$, $h(x) = f(x) - g(x)$

29. $f(x) = x^2$, $g(x) = -2x$, $h(x) = f(x) \cdot g(x)$

30. $f(x) = 4 - x^2$, $g(x) = x$, $h(x) = f(x)/g(x)$

In Exercises 31–34, use a graphing utility to graph f, g, and $f + g$ in the same viewing window. Which function contributes most to the magnitude of the sum when $0 \le x \le 2$? Which function contributes most to the magnitude of the sum when $x > 6$?

31. $f(x) = 3x$, $g(x) = -\dfrac{x^3}{10}$

32. $f(x) = \dfrac{x}{2}$, $g(x) = \sqrt{x}$

33. $f(x) = 3x + 2$, $g(x) = -\sqrt{x + 5}$

34. $f(x) = x^2 - \frac{1}{2}$, $g(x) = -3x^2 - 1$

In Exercises 35–38, find (a) $f \circ g$, **(b)** $g \circ f$, **and, if possible,**
(c) $(f \circ g)(0)$.

35. $f(x) = x^2$, $g(x) = x - 1$

36. $f(x) = \sqrt[3]{x - 1}$, $g(x) = x^3 + 1$

37. $f(x) = 3x + 5$, $g(x) = 5 - x$

38. $f(x) = x^3$, $g(x) = \dfrac{1}{x}$

In Exercises 39–48, determine the domains of (a) f,
(b) g, **and (c)** $f \circ g$. **Use a graphing utility to verify your**
results.

39. $f(x) = \sqrt{x + 4}$, $g(x) = x^2$

40. $f(x) = \sqrt{x + 3}$, $g(x) = \dfrac{x}{2}$

41. $f(x) = x^2 + 1$, $g(x) = \sqrt{x}$

42. $f(x) = x^{1/4}$, $g(x) = x^4$

43. $f(x) = \dfrac{1}{x}$, $g(x) = x + 3$

44. $f(x) = \dfrac{1}{x}$, $g(x) = \dfrac{1}{2x}$

45. $f(x) = |x - 4|$, $g(x) = 3 - x$

46. $f(x) = \dfrac{2}{|x|}$, $g(x) = x - 1$

47. $f(x) = x + 2$, $g(x) = \dfrac{1}{x^2 - 4}$

48. $f(x) = \dfrac{3}{x^2 - 1}$, $g(x) = x + 1$

In Exercises 49–54, (a) find $f \circ g$, $g \circ f$, **and the domain of**
$f \circ g$. **(b) Use a graphing utility to graph** $f \circ g$ **and** $g \circ f$.
Determine whether $f \circ g = g \circ f$.

49. $f(x) = \sqrt{x + 4}$, $g(x) = x^2$

50. $f(x) = \sqrt[3]{x + 1}$, $g(x) = x^3 - 1$

51. $f(x) = \tfrac{1}{3}x - 3$, $g(x) = 3x + 9$

52. $f(x) = \sqrt{x}$, $g(x) = \sqrt{x}$

53. $f(x) = x^{2/3}$, $g(x) = x^6$

54. $f(x) = |x|$, $g(x) = -x^2 + 1$

In Exercises 55–60, (a) find $(f \circ g)(x)$ **and** $(g \circ f)(x)$, **(b)**
determine algebraically whether $(f \circ g)(x) = (g \circ f)(x)$, **and**
(c) use a graphing utility to complete a table of values for
the two compositions to confirm your answers to part (b).

55. $f(x) = 5x + 4$, $g(x) = 4 - x$

56. $f(x) = \tfrac{1}{4}(x - 1)$, $g(x) = 4x + 1$

57. $f(x) = \sqrt{x + 6}$, $g(x) = x^2 - 5$

58. $f(x) = x^3 - 4$, $g(x) = \sqrt[3]{x + 10}$

59. $f(x) = |x|$, $g(x) = 2x^3$

60. $f(x) = \dfrac{6}{3x - 5}$, $g(x) = -x$

In Exercises 61–64, use the graphs of f **and** g **to evaluate the**
functions.

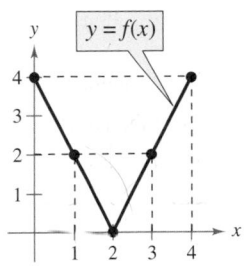
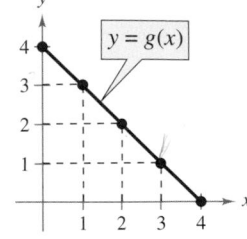

61. (a) $(f + g)(3)$ (b) $(f/g)(2)$

62. (a) $(f - g)(1)$ (b) $(fg)(4)$

63. (a) $(f \circ g)(2)$ (b) $(g \circ f)(2)$

64. (a) $(f \circ g)(1)$ (b) $(g \circ f)(3)$

∫ **In Exercises 65–72, find two functions** f **and** g **such that**
$(f \circ g)(x) = h(x)$. **(There are many correct answers.)**

65. $h(x) = (2x + 1)^2$ **66.** $h(x) = (1 - x)^3$

67. $h(x) = \sqrt[3]{x^2 - 4}$ **68.** $h(x) = \sqrt{9 - x}$

69. $h(x) = \dfrac{1}{x + 2}$

70. $h(x) = \dfrac{4}{(5x + 2)^2}$

71. $h(x) = (x + 4)^2 + 2(x + 4)$

72. $h(x) = (x + 3)^{3/2} + 4(x + 3)^{1/2}$

73. *Stopping Distance*   The research and development
department of an automobile manufacturer has determined
that when required to stop quickly to avoid an accident, the
distance (in feet) a car travels during the driver's reaction
time is given by

$$R(x) = \tfrac{3}{4}x$$

where x is the speed of the car in miles per hour. The
distance (in feet) traveled while the driver is braking is
given by

$$B(x) = \tfrac{1}{15}x^2.$$

(a) Find the function that represents the total stopping
distance T.

(b) Use a graphing utility to graph the functions R, B, and
T in the same viewing window for $0 \le x \le 60$.

(c) Which function contributes most to the magnitude of
the sum at higher speeds? Explain.

74. Sales From 2000 to 2006, the sales R_1 (in thousands of dollars) for one of two restaurants owned by the same parent company can be modeled by $R_1 = 480 - 8t - 0.8t^2$, for $t = 0, 1, 2, 3, 4, 5, 6$, where $t = 0$ represents 2000. During the same seven-year period, the sales R_2 (in thousands of dollars) for the second restaurant can be modeled by $R_2 = 254 + 0.78t$, for $t = 0, 1, 2, 3, 4, 5, 6$.

(a) Write a function R_3 that represents the total sales for the two restaurants.

(b) Use a graphing utility to graph R_1, R_2, and R_3 (the total sales function) in the same viewing window.

Data Analysis **In Exercises 75 and 76, use the table, which shows the total amounts spent (in billions of dollars) on health services and supplies in the United States and Puerto Rico for the years 1995 through 2005. The variables y_1, y_2, and y_3 represent out-of-pocket payments, insurance premiums, and other types of payments, respectively.** (Source: U.S. Centers for Medicare and Medicaid Services)

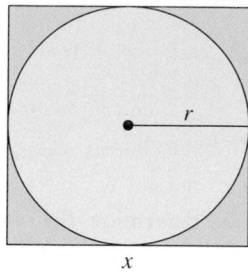

Year	y_1	y_2	y_3
1995	146	330	457
1996	152	344	483
1997	162	361	503
1998	176	385	520
1999	185	414	550
2000	193	451	592
2001	202	497	655
2002	214	550	718
2003	231	601	766
2004	246	647	824
2005	262	691	891

The models for this data are $y_1 = 11.4t + 83$, $y_2 = 2.31t^2 - 8.4t + 310$, and $y_3 = 3.03t^2 - 16.8t + 467$, where t represents the year, with $t = 5$ corresponding to 1995.

75. Use the models and the *table* feature of a graphing utility to create a table showing the values of y_1, y_2, and y_3 for each year from 1995 to 2005. Compare these values with the original data. Are the models a good fit? Explain.

76. Use a graphing utility to graph y_1, y_2, y_3, and $y_T = y_1 + y_2 + y_3$ in the same viewing window. What does the function y_T represent? Explain.

77. Ripples A pebble is dropped into a calm pond, causing ripples in the form of concentric circles. The radius (in feet) of the outermost ripple is given by $r(t) = 0.6t$, where t is the time (in seconds) after the pebble strikes the water. The area of the circle is given by $A(r) = \pi r^2$. Find and interpret $(A \circ r)(t)$.

78. Geometry A square concrete foundation was prepared as a base for a large cylindrical gasoline tank (see figure).

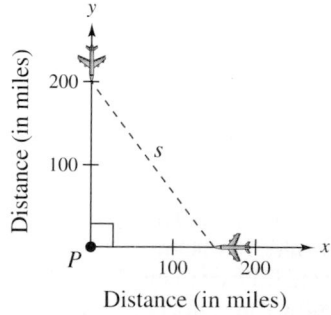

(a) Write the radius r of the tank as a function of the length x of the sides of the square.

(b) Write the area A of the circular base of the tank as a function of the radius r.

(c) Find and interpret $(A \circ r)(x)$.

79. Cost The weekly cost C of producing x units in a manufacturing process is given by

$$C(x) = 60x + 750.$$

The number of units x produced in t hours is $x(t) = 50t$.

(a) Find and interpret $C(x(t))$.

(b) Find the number of units produced in 4 hours.

(c) Use a graphing utility to graph the cost as a function of time. Use the *trace* feature to estimate (to two-decimal-place accuracy) the time that must elapse until the cost increases to $15,000.

80. Air Traffic Control An air traffic controller spots two planes at the same altitude flying toward each other. Their flight paths form a right angle at point P. One plane is 150 miles from point P and is moving at 450 miles per hour. The other plane is 200 miles from point P and is moving at 450 miles per hour. Write the distance s between the planes as a function of time t.

81. Bacteria The number of bacteria in a refrigerated food product is given by $N(T) = 10T^2 - 20T + 600$, for $1 \le T \le 20$, where T is the temperature of the food in degrees Celsius. When the food is removed from the refrigerator, the temperature of the food is given by $T(t) = 2t + 1$, where t is the time in hours.

(a) Find the composite function $N(T(t))$ or $(N \circ T)(t)$ and interpret its meaning in the context of the situation.

(b) Find $(N \circ T)(6)$ and interpret its meaning.

(c) Find the time when the bacteria count reaches 800.

82. Pollution The spread of a contaminant is increasing in a circular pattern on the surface of a lake. The radius of the contaminant can be modeled by $r(t) = 5.25\sqrt{t}$, where r is the radius in meters and t is time in hours since contamination.

(a) Find a function that gives the area A of the circular leak in terms of the time t since the spread began.

(b) Find the size of the contaminated area after 36 hours.

(c) Find when the size of the contaminated area is 6250 square meters.

83. Salary You are a sales representative for an automobile manufacturer. You are paid an annual salary plus a bonus of 3% of your sales over $500,000. Consider the two functions

$$f(x) = x - 500,000 \quad \text{and} \quad g(x) = 0.03x.$$

If x is greater than $500,000, which of the following represents your bonus? Explain.

(a) $f(g(x))$ (b) $g(f(x))$

84. Consumer Awareness The suggested retail price of a new car is p dollars. The dealership advertised a factory rebate of $1200 and an 8% discount.

(a) Write a function R in terms of p giving the cost of the car after receiving the rebate from the factory.

(b) Write a function S in terms of p giving the cost of the car after receiving the dealership discount.

(c) Form the composite functions $(R \circ S)(p)$ and $(S \circ R)(p)$ and interpret each.

(d) Find $(R \circ S)(18,400)$ and $(S \circ R)(18,400)$. Which yields the lower cost for the car? Explain.

Synthesis

True or False? In Exercises 85 and 86, determine whether the statement is true or false. Justify your answer.

85. If $f(x) = x + 1$ and $g(x) = 6x$, then

$$(f \circ g)(x) = (g \circ f)(x).$$

86. If you are given two functions $f(x)$ and $g(x)$, you can calculate $(f \circ g)(x)$ if and only if the range of g is a subset of the domain of f.

Exploration In Exercises 87 and 88, three siblings are of three different ages. The oldest is twice the age of the middle sibling, and the middle sibling is six years older than one-half the age of the youngest.

87. (a) Write a composite function that gives the oldest sibling's age in terms of the youngest. Explain how you arrived at your answer.

(b) If the oldest sibling is 16 years old, find the ages of the other two siblings.

88. (a) Write a composite function that gives the youngest sibling's age in terms of the oldest. Explain how you arrived at your answer.

(b) If the youngest sibling is two years old, find the ages of the other two siblings.

89. Proof Prove that the product of two odd functions is an even function, and that the product of two even functions is an even function.

90. Conjecture Use examples to hypothesize whether the product of an odd function and an even function is even or odd. Then prove your hypothesis.

91. Proof Given a function f, prove that $g(x)$ is even and $h(x)$ is odd, where $g(x) = \frac{1}{2}[f(x) + f(-x)]$ and $h(x) = \frac{1}{2}[f(x) - f(-x)]$.

92. (a) Use the result of Exercise 91 to prove that any function can be written as a sum of even and odd functions. (*Hint:* Add the two equations in Exercise 91.)

(b) Use the result of part (a) to write each function as a sum of even and odd functions.

$$f(x) = x^2 - 2x + 1, \quad g(x) = \frac{1}{x + 1}$$

Skills Review

In Exercises 93–96, find three points that lie on the graph of the equation. (There are many correct answers.)

93. $y = -x^2 + x - 5$ **94.** $y = \frac{1}{5}x^3 - 4x^2 + 1$

95. $x^2 + y^2 = 24$ **96.** $y = \frac{x}{x^2 - 5}$

In Exercises 97–100, find an equation of the line that passes through the two points.

97. $(-4, -2), (-3, 8)$ **98.** $(1, 5), (-8, 2)$

99. $\left(\frac{3}{2}, -1\right), \left(-\frac{1}{3}, 4\right)$ **100.** $(0, 1.1), (-4, 3.1)$

1.6 Inverse Functions

Inverse Functions

Recall from Section 1.2 that a function can be represented by a set of ordered pairs. For instance, the function $f(x) = x + 4$ from the set $A = \{1, 2, 3, 4\}$ to the set $B = \{5, 6, 7, 8\}$ can be written as follows.

$$f(x) = x + 4: \ \{(1, 5), (2, 6), (3, 7), (4, 8)\}$$

In this case, by interchanging the first and second coordinates of each of these ordered pairs, you can form the **inverse function** of f, which is denoted by f^{-1}. It is a function from the set B to the set A, and can be written as follows.

$$f^{-1}(x) = x - 4: \ \{(5, 1), (6, 2), (7, 3), (8, 4)\}$$

Note that the domain of f is equal to the range of f^{-1}, and vice versa, as shown in Figure 1.67. Also note that the functions f and f^{-1} have the effect of "undoing" each other. In other words, when you form the composition of f with f^{-1} or the composition of f^{-1} with f, you obtain the identity function.

$$f(f^{-1}(x)) = f(x - 4) = (x - 4) + 4 = x$$

$$f^{-1}(f(x)) = f^{-1}(x + 4) = (x + 4) - 4 = x$$

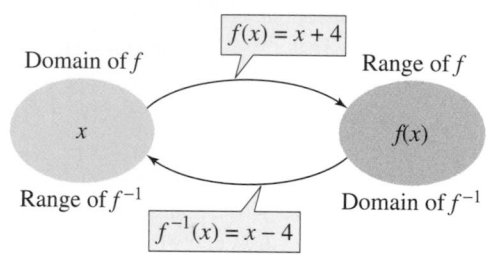

Domain of f $\boxed{f(x) = x + 4}$ Range of f

x $f(x)$

Range of f^{-1} $\boxed{f^{-1}(x) = x - 4}$ Domain of f^{-1}

Figure 1.67

What you should learn

- Find inverse functions informally and verify that two functions are inverse functions of each other.
- Use graphs of functions to decide whether functions have inverse functions.
- Determine if functions are one-to-one.
- Find inverse functions algebraically.

Why you should learn it

Inverse functions can be helpful in further exploring how two variables relate to each other. For example, in Exercises 103 and 104 on page 71, you will use inverse functions to find the European shoe sizes from the corresponding U.S. shoe sizes.

Example 1   Finding Inverse Functions Informally

Find the inverse function of $f(x) = 4x$. Then verify that both $f(f^{-1}(x))$ and $f^{-1}(f(x))$ are equal to the identity function.

Solution

The function f *multiplies* each input by 4. To "undo" this function, you need to *divide* each input by 4. So, the inverse function of $f(x) = 4x$ is given by

$$f^{-1}(x) = \frac{x}{4}.$$

You can verify that both $f(f^{-1}(x))$ and $f^{-1}(f(x))$ are equal to the identity function as follows.

$$f(f^{-1}(x)) = f\left(\frac{x}{4}\right) = 4\left(\frac{x}{4}\right) = x \qquad f^{-1}(f(x)) = f^{-1}(4x) = \frac{4x}{4} = x$$

✓CHECKPOINT   Now try Exercise 1.

STUDY TIP

Don't be confused by the use of the exponent -1 to denote the inverse function f^{-1}. In this text, whenever f^{-1} is written, it always refers to the inverse function of the function f and not to the reciprocal of $f(x)$, which is given by

$$\frac{1}{f(x)}.$$

Example 2 Finding Inverse Functions Informally

Find the inverse function of $f(x) = x - 6$. Then verify that both $f(f^{-1}(x))$ and $f^{-1}(f(x))$ are equal to the identity function.

Solution

The function f *subtracts* 6 from each input. To "undo" this function, you need to *add* 6 to each input. So, the inverse function of $f(x) = x - 6$ is given by

$$f^{-1}(x) = x + 6.$$

You can verify that both $f(f^{-1}(x))$ and $f^{-1}(f(x))$ are equal to the identity function as follows.

$$f(f^{-1}(x)) = f(x + 6) = (x + 6) - 6 = x$$

$$f^{-1}(f(x)) = f^{-1}(x - 6) = (x - 6) + 6 = x$$

✓CHECKPOINT   Now try Exercise 3.

A table of values can help you understand inverse functions. For instance, the following table shows several values of the function in Example 2. Interchange the rows of this table to obtain values of the inverse function.

x	-2	-1	0	1	2
$f(x)$	-8	-7	-6	-5	-4

x	-8	-7	-6	-5	-4
$f^{-1}(x)$	-2	-1	0	1	2

In the table at the left, each output is 6 less than the input, and in the table at the right, each output is 6 more than the input.

The formal definition of an inverse function is as follows.

Definition of Inverse Function

Let f and g be two functions such that

$$f(g(x)) = x \qquad \text{for every } x \text{ in the domain of } g$$

and

$$g(f(x)) = x \qquad \text{for every } x \text{ in the domain of } f.$$

Under these conditions, the function g is the **inverse function** of the function f. The function g is denoted by f^{-1} (read "f-inverse"). So,

$$f(f^{-1}(x)) = x \qquad \text{and} \qquad f^{-1}(f(x)) = x.$$

The domain of f must be equal to the range of f^{-1}, and the range of f must be equal to the domain of f^{-1}.

If the function g is the inverse function of the function f, it must also be true that the function f is the inverse function of the function g. For this reason, you can say that the functions f and g are *inverse functions of each other*.

Example 3   Verifying Inverse Functions Algebraically

Show that the functions are inverse functions of each other.

$$f(x) = 2x^3 - 1 \qquad \text{and} \qquad g(x) = \sqrt[3]{\dfrac{x+1}{2}}$$

Solution

$$f(g(x)) = f\left(\sqrt[3]{\dfrac{x+1}{2}}\right) = 2\left(\sqrt[3]{\dfrac{x+1}{2}}\right)^3 - 1$$

$$= 2\left(\dfrac{x+1}{2}\right) - 1$$

$$= x + 1 - 1$$

$$= x$$

$$g(f(x)) = g(2x^3 - 1) = \sqrt[3]{\dfrac{(2x^3 - 1) + 1}{2}}$$

$$= \sqrt[3]{\dfrac{2x^3}{2}}$$

$$= \sqrt[3]{x^3}$$

$$= x$$

✓**CHECKPOINT**   Now try Exercise 15.

Example 4   Verifying Inverse Functions Algebraically

Which of the functions is the inverse function of $f(x) = \dfrac{5}{x-2}$?

$$g(x) = \dfrac{x-2}{5} \qquad \text{or} \qquad h(x) = \dfrac{5}{x} + 2$$

Solution

By forming the composition of f with g, you have

$$f(g(x)) = f\left(\dfrac{x-2}{5}\right) = \dfrac{5}{\left(\dfrac{x-2}{5}\right) - 2} = \dfrac{25}{x - 12} \neq x.$$

Because this composition is not equal to the identity function x, it follows that g *is not* the inverse function of f. By forming the composition of f with h, you have

$$f(h(x)) = f\left(\dfrac{5}{x} + 2\right) = \dfrac{5}{\left(\dfrac{5}{x} + 2\right) - 2} = \dfrac{5}{\left(\dfrac{5}{x}\right)} = x.$$

So, it appears that h is the inverse function of f. You can confirm this by showing that the composition of h with f is also equal to the identity function.

✓**CHECKPOINT**   Now try Exercise 19.

TECHNOLOGY TIP

Most graphing utilities can graph $y = x^{1/3}$ in two ways:

$$y_1 = x \wedge (1/3) \quad \text{or}$$

$$y_1 = \sqrt[3]{x}.$$

However, you may not be able to obtain the complete graph of $y = x^{2/3}$ by entering $y_1 = x \wedge (2/3)$. If not, you should use

$$y_1 = (x \wedge (1/3))^2 \quad \text{or}$$

$$y_1 = \sqrt[3]{x^2}.$$

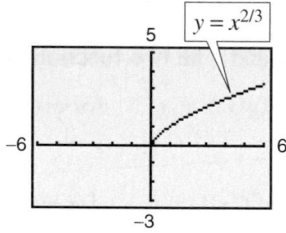

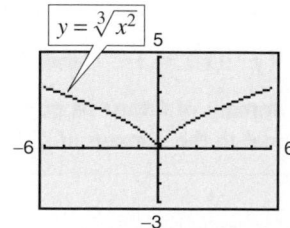

The Graph of an Inverse Function

The graphs of a function f and its inverse function f^{-1} are related to each other in the following way. If the point (a, b) lies on the graph of f, then the point (b, a) must lie on the graph of f^{-1}, and vice versa. This means that the graph of f^{-1} is a *reflection* of the graph of f in the line $y = x$, as shown in Figure 1.68.

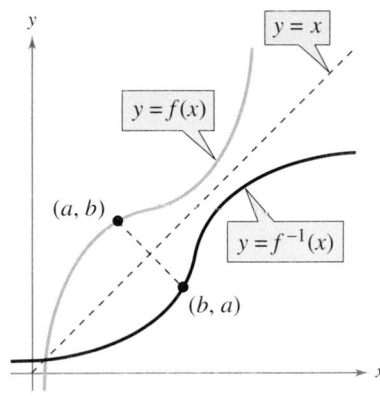

Figure 1.68

> **TECHNOLOGY TIP**
>
> In Examples 3 and 4, inverse functions were verified algebraically. A graphing utility can also be helpful in checking whether one function is the inverse function of another function. Use the Graph Reflection Program found at this textbook's *Online Study Center* to verify Example 4 graphically.

Example 5   Verifying Inverse Functions Graphically and Numerically

Verify that the functions f and g from Example 3 are inverse functions of each other graphically and numerically.

Graphical Solution

You can verify that f and g are inverse functions of each other *graphically* by using a graphing utility to graph f and g in the same viewing window. (Be sure to use a *square setting*.) From the graph in Figure 1.69, you can verify that the graph of g is the reflection of the graph of f in the line $y = x$.

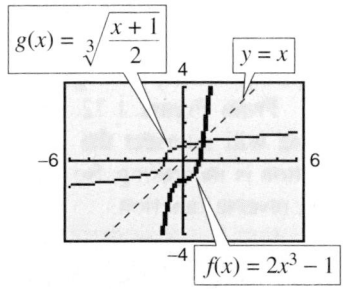

Figure 1.69

Numerical Solution

You can verify that f and g are inverse functions of each other *numerically*. Begin by entering the compositions $f(g(x))$ and $g(f(x))$ into a graphing utility as follows.

$$y_1 = f(g(x)) = 2\left(\sqrt[3]{\frac{x+1}{2}}\right)^3 - 1$$

$$y_2 = g(f(x)) = \sqrt[3]{\frac{(2x^3 - 1) + 1}{2}}$$

Then use the *table* feature of the graphing utility to create a table, as shown in Figure 1.70. Note that the entries for x, y_1, and y_2 are the same. So, $f(g(x)) = x$ and $g(f(x)) = x$. You can now conclude that f and g are inverse functions of each other.

X	Y₁	Y₂
-3	-3	-3
-2	-2	-2
-1	-1	-1
0	0	0
1	1	1
2	2	2
3	3	3

X = -3

Figure 1.70

✓*CHECKPOINT*   Now try Exercise 25.

The Existence of an Inverse Function

Consider the function $f(x) = x^2$. The first table at the right is a table of values for $f(x) = x^2$. The second table was created by interchanging the rows of the first table. The second table does not represent a function because the input $x = 4$ is matched with two different outputs: $y = -2$ and $y = 2$. So, $f(x) = x^2$ does not have an inverse function.

To have an inverse function, a function must be **one-to-one**, which means that no two elements in the domain of f correspond to the same element in the range of f.

x	-2	-1	0	1	2
$f(x)$	4	1	0	1	4

x	4	1	0	1	4
$g(x)$	-2	-1	0	1	2

Definition of a One-to-One Function

A function f is **one-to-one** if, for a and b in its domain, $f(a) = f(b)$ implies that $a = b$.

Existence of an Inverse Function

A function f has an inverse function f^{-1} if and only if f is one-to-one.

From its graph, it is easy to tell whether a function of x is one-to-one. Simply check to see that every horizontal line intersects the graph of the function at most once. This is called the **Horizontal Line Test.** For instance, Figure 1.71 shows the graph of $y = x^2$. On the graph, you can find a horizontal line that intersects the graph twice.

Two special types of functions that pass the Horizontal Line Test are those that are increasing or decreasing on their entire domains.

1. If f is *increasing* on its entire domain, then f is one-to-one.

2. If f is *decreasing* on its entire domain, then f is one-to-one.

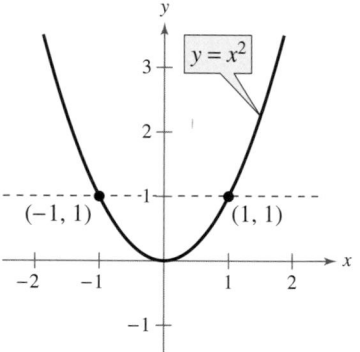

Figure 1.71 $f(x) = x^2$ *is not one-to-one.*

Example 6   Testing for One-to-One Functions

Is the function $f(x) = \sqrt{x} + 1$ one-to-one?

Algebraic Solution

Let a and b be nonnegative real numbers with $f(a) = f(b)$.

$$\sqrt{a} + 1 = \sqrt{b} + 1 \qquad \text{Set } f(a) = f(b).$$
$$\sqrt{a} = \sqrt{b}$$
$$a = b$$

So, $f(a) = f(b)$ implies that $a = b$. You can conclude that f is one-to-one and *does* have an inverse function.

Graphical Solution

Use a graphing utility to graph the function $y = \sqrt{x} + 1$. From Figure 1.72, you can see that a horizontal line will intersect the graph at most once and the function is increasing. So, f is one-to-one and *does* have an inverse function.

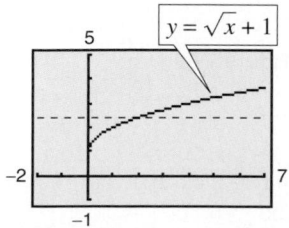

Figure 1.72

✓CHECKPOINT   Now try Exercise 55.

Finding Inverse Functions Algebraically

For simple functions, you can find inverse functions by inspection. For more complicated functions, however, it is best to use the following guidelines.

Finding an Inverse Function

1. Use the Horizontal Line Test to decide whether f has an inverse function.

2. In the equation for $f(x)$, replace $f(x)$ by y.

3. Interchange the roles of x and y, and solve for y.

4. Replace y by $f^{-1}(x)$ in the new equation.

5. Verify that f and f^{-1} are inverse functions of each other by showing that the domain of f is equal to the range of f^{-1}, the range of f is equal to the domain of f^{-1}, and $f(f^{-1}(x)) = x$ and $f^{-1}(f(x)) = x$.

The function f with an implied domain of all real numbers may not pass the Horizontal Line Test. In this case, the domain of f may be restricted so that f does have an inverse function. For instance, if the domain of $f(x) = x^2$ is restricted to the nonnegative real numbers, then f does have an inverse function.

Example 7   Finding an Inverse Function Algebraically

Find the inverse function of $f(x) = \dfrac{5 - 3x}{2}$.

Solution

The graph of f in Figure 1.73 passes the Horizontal Line Test. So you know that f is one-to-one and has an inverse function.

$$f(x) = \frac{5 - 3x}{2} \qquad \text{Write original function.}$$

$$y = \frac{5 - 3x}{2} \qquad \text{Replace } f(x) \text{ by } y.$$

$$x = \frac{5 - 3y}{2} \qquad \text{Interchange } x \text{ and } y.$$

$$2x = 5 - 3y \qquad \text{Multiply each side by 2.}$$

$$3y = 5 - 2x \qquad \text{Isolate the } y\text{-term.}$$

$$y = \frac{5 - 2x}{3} \qquad \text{Solve for } y.$$

$$f^{-1}(x) = \frac{5 - 2x}{3} \qquad \text{Replace } y \text{ by } f^{-1}(x).$$

The domains and ranges of f and f^{-1} consist of all real numbers. Verify that $f(f^{-1}(x)) = x$ and $f^{-1}(f(x)) = x$.

✓CHECKPOINT   Now try Exercise 59.

TECHNOLOGY TIP

Many graphing utilities have a built-in feature for drawing an inverse function. To see how this works, consider the function $f(x) = \sqrt{x}$. The inverse function of f is given by $f^{-1}(x) = x^2$, $x \geq 0$. Enter the function $y_1 = \sqrt{x}$. Then graph it in the standard viewing window and use the *draw inverse* feature. You should obtain the figure below, which shows both f and its inverse function f^{-1}. For instructions on how to use the *draw inverse* feature, see Appendix A; for specific keystrokes, go to this textbook's *Online Study Center*.

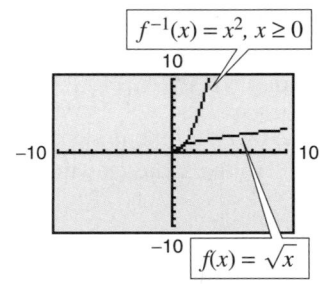

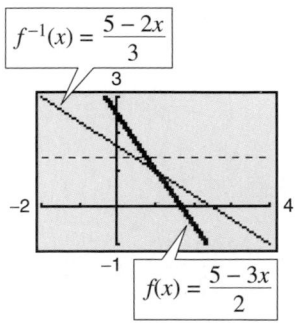

Figure 1.73

Example 8 Finding an Inverse Function Algebraically

Find the inverse function of $f(x) = x^3 - 4$ and use a graphing utility to graph f and f^{-1} in the same viewing window.

Solution

$f(x) = x^3 - 4$	Write original function.
$y = x^3 - 4$	Replace $f(x)$ by y.
$x = y^3 - 4$	Interchange x and y.
$y^3 = x + 4$	Isolate y.
$y = \sqrt[3]{x + 4}$	Solve for y.
$f^{-1}(x) = \sqrt[3]{x + 4}$	Replace y by $f^{-1}(x)$.

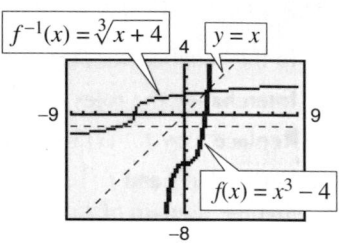

The graph of f in Figure 1.74 passes the Horizontal Line Test. So, you know that f is one-to-one and has an inverse function. The graph of f^{-1} in Figure 1.74 is the reflection of the graph of f in the line $y = x$. Verify that $f(f^{-1}(x)) = x$ and $f^{-1}(f(x)) = x$.

Figure 1.74

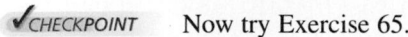CHECKPOINT   Now try Exercise 61.

Example 9 Finding an Inverse Function Algebraically

Find the inverse function of $f(x) = \sqrt{2x - 3}$ and use a graphing utility to graph f and f^{-1} in the same viewing window.

Solution

$f(x) = \sqrt{2x - 3}$	Write original function.
$y = \sqrt{2x - 3}$	Replace $f(x)$ by y.
$x = \sqrt{2y - 3}$	Interchange x and y.
$x^2 = 2y - 3$	Square each side.
$2y = x^2 + 3$	Isolate y.
$y = \dfrac{x^2 + 3}{2}$	Solve for y.
$f^{-1}(x) = \dfrac{x^2 + 3}{2}, \quad x \geq 0$	Replace y by $f^{-1}(x)$.

The graph of f in Figure 1.75 passes the Horizontal Line Test. So you know that f is one-to-one and has an inverse function. The graph of f^{-1} in Figure 1.75 is the reflection of the graph of f in the line $y = x$. Note that the range of f is the interval $[0, \infty)$, which implies that the domain of f^{-1} is the interval $[0, \infty)$. Moreover, the domain of f is the interval $\left[\frac{3}{2}, \infty\right)$, which implies that the range of f^{-1} is the interval $\left[\frac{3}{2}, \infty\right)$. Verify that $f(f^{-1}(x)) = x$ and $f^{-1}(f(x)) = x$.

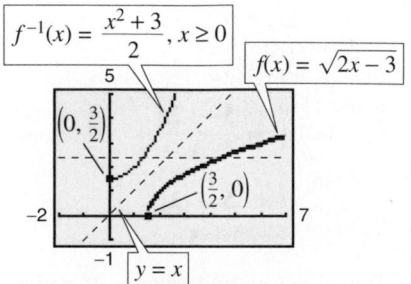

CHECKPOINT   Now try Exercise 65.

Figure 1.75

1.6 Exercises

See www.CalcChat.com for worked-out solutions to odd-numbered exercises.

Vocabulary Check

Fill in the blanks.

1. If the composite functions $f(g(x)) = x$ and $g(f(x)) = x$, then the function g is the _____ function of f, and is denoted by _____ .

2. The domain of f is the _____ of f^{-1}, and the _____ of f^{-1} is the range of f.

3. The graphs of f and f^{-1} are reflections of each other in the line _____ .

4. To have an inverse function, a function f must be _____ ; that is, $f(a) = f(b)$ implies $a = b$.

5. A graphical test for the existence of an inverse function is called the _____ Line Test.

In Exercises 1–8, find the inverse function of f informally. Verify that $f(f^{-1}(x)) = x$ and $f^{-1}(f(x)) = x$.

1. $f(x) = 6x$
2. $f(x) = \frac{1}{3}x$
3. $f(x) = x + 7$
4. $f(x) = x - 3$
5. $f(x) = 2x + 1$
6. $f(x) = \dfrac{x - 1}{4}$
7. $f(x) = \sqrt[3]{x}$
8. $f(x) = x^5$

In Exercises 9–14, (a) show that f and g are inverse functions algebraically and (b) use a graphing utility to create a table of values for each function to numerically show that f and g are inverse functions.

9. $f(x) = -\dfrac{7}{2}x - 3, \quad g(x) = -\dfrac{2x + 6}{7}$

10. $f(x) = \dfrac{x - 9}{4}, \quad g(x) = 4x + 9$

11. $f(x) = x^3 + 5, \quad g(x) = \sqrt[3]{x - 5}$

12. $f(x) = \dfrac{x^3}{2}, \quad g(x) = \sqrt[3]{2x}$

13. $f(x) = -\sqrt{x - 8}, \quad g(x) = 8 + x^2, \quad x \leq 0$

14. $f(x) = \sqrt[3]{3x - 10}, \quad g(x) = \dfrac{x^3 + 10}{3}$

In Exercises 15–20, show that f and g are inverse functions algebraically. Use a graphing utility to graph f and g in the same viewing window. Describe the relationship between the graphs.

15. $f(x) = x^3, \quad g(x) = \sqrt[3]{x}$

16. $f(x) = \dfrac{1}{x}, \quad g(x) = \dfrac{1}{x}$

17. $f(x) = \sqrt{x - 4}; \quad g(x) = x^2 + 4, \quad x \geq 0$

18. $f(x) = 9 - x^2, \quad x \geq 0; \quad g(x) = \sqrt{9 - x}$

19. $f(x) = 1 - x^3, \quad g(x) = \sqrt[3]{1 - x}$

20. $f(x) = \dfrac{1}{1 + x}, \quad x \geq 0; \quad g(x) = \dfrac{1 - x}{x}, \quad 0 < x \leq 1$

In Exercises 21–24, match the graph of the function with the graph of its inverse function. [The graphs of the inverse functions are labeled (a), (b), (c), and (d).]

(a)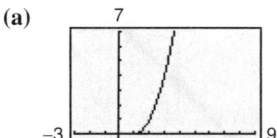

(b)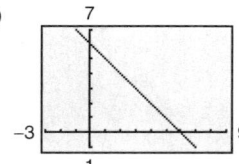

(c)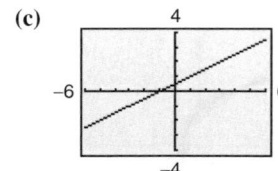

(d)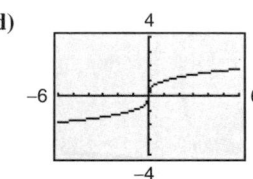

21.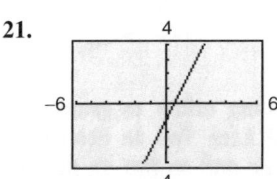

22.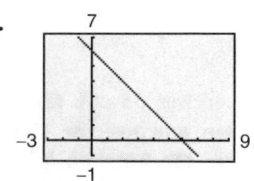

23.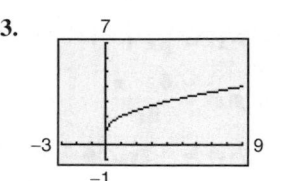

24.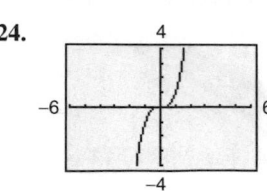

In Exercises 25–28, show that f and g are inverse functions (a) graphically and (b) numerically.

25. $f(x) = 2x, \quad g(x) = \dfrac{x}{2}$

26. $f(x) = x - 5, \quad g(x) = x + 5$

27. $f(x) = \dfrac{x-1}{x+5}, \quad g(x) = -\dfrac{5x+1}{x-1}$

28. $f(x) = \dfrac{x+3}{x-2}, \quad g(x) = \dfrac{2x+3}{x-1}$

In Exercises 29–34, determine if the graph is that of a function. If so, determine if the function is one-to-one.

29.
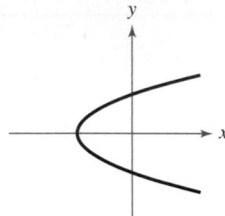

30.
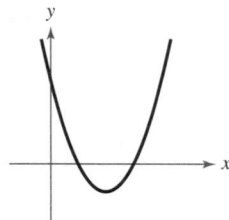

31.
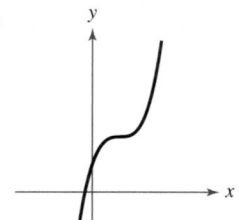

32.
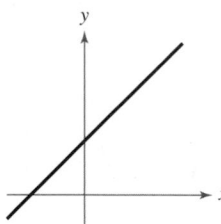

33.
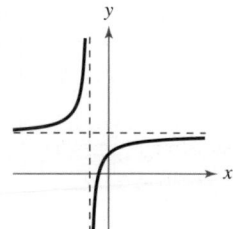

34.
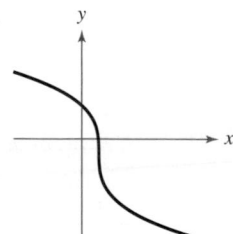

In Exercises 35–46, use a graphing utility to graph the function and use the Horizontal Line Test to determine whether the function is one-to-one and so has an inverse funtion exists.

35. $f(x) = 3 - \tfrac{1}{2}x$

36. $f(x) = \tfrac{1}{4}(x + 2)^2 - 1$

37. $h(x) = \dfrac{x^2}{x^2 + 1}$

38. $g(x) = \dfrac{4 - x}{6x^2}$

39. $h(x) = \sqrt{16 - x^2}$

40. $f(x) = -2x\sqrt{16 - x^2}$

41. $f(x) = 10$

42. $f(x) = -0.65$

43. $g(x) = (x + 5)^3$

44. $f(x) = x^5 - 7$

45. $h(x) = |x + 4| - |x - 4|$

46. $f(x) = -\dfrac{|x - 6|}{|x + 6|}$

In Exercises 47–58, determine algebraically whether the function is one-to-one. Verify your answer graphically.

47. $f(x) = x^4$

48. $g(x) = x^2 - x^4$

49. $f(x) = \dfrac{3x + 4}{5}$

50. $f(x) = 3x + 5$

51. $f(x) = \dfrac{1}{x^2}$

52. $h(x) = \dfrac{4}{x^2}$

53. $f(x) = (x + 3)^2, \quad x \geq -3$

54. $q(x) = (x - 5)^2, \quad x \leq 5$

55. $f(x) = \sqrt{2x + 3}$

56. $f(x) = \sqrt{x - 2}$

57. $f(x) = |x - 2|, \quad x \leq 2$

58. $f(x) = \dfrac{x^2}{x^2 + 1}$

In Exercises 59–68, find the inverse function of f algebraically. Use a graphing utility to graph both f and f^{-1} in the same viewing window. Describe the relationship between the graphs.

59. $f(x) = 2x - 3$

60. $f(x) = 3x$

61. $f(x) = x^5$

62. $f(x) = x^3 + 1$

63. $f(x) = x^{3/5}$

64. $f(x) = x^2, \quad x \geq 0$

65. $f(x) = \sqrt{4 - x^2}, \quad 0 \leq x \leq 2$

66. $f(x) = \sqrt{16 - x^2}, \quad -4 \leq x \leq 0$

67. $f(x) = \dfrac{4}{x}$

68. $f(x) = \dfrac{6}{\sqrt{x}}$

Think About It   In Exercises 69–78, restrict the domain of the function f so that the function is one-to-one and has an inverse function. Then find the inverse function f^{-1}. State the domains and ranges of f and f^{-1}. Explain your results. (There are many correct answers.)

69. $f(x) = (x - 2)^2$

70. $f(x) = 1 - x^4$

71. $f(x) = |x + 2|$

72. $f(x) = |x - 2|$

73. $f(x) = (x + 3)^2$

74. $f(x) = (x - 4)^2$

75. $f(x) = -2x^2 + 5$

76. $f(x) = \tfrac{1}{2}x^2 - 1$

77. $f(x) = |x - 4| + 1$

78. $f(x) = -|x - 1| - 2$

In Exercises 79 and 80, use the graph of the function f to complete the table and sketch the graph of f^{-1}.

79.

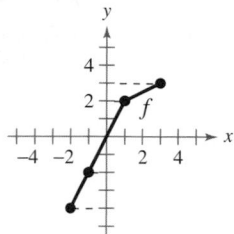

x	$f^{-1}(x)$
-4	
-2	
2	
3	

80.

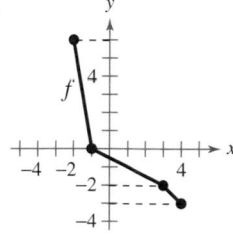

x	$f^{-1}(x)$
-3	
-2	
0	
6	

In Exercises 81–88, use the graphs of $y = f(x)$ and $y = g(x)$ to evaluate the function.

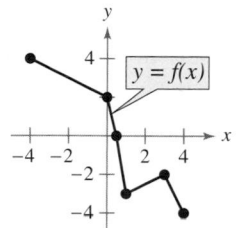 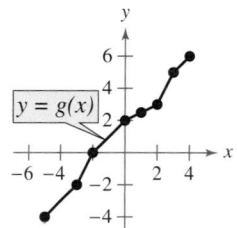

81. $f^{-1}(0)$ **82.** $g^{-1}(0)$

83. $(f \circ g)(2)$ **84.** $g(f(-4))$

85. $f^{-1}(g(0))$ **86.** $(g^{-1} \circ f)(3)$

87. $(g \circ f^{-1})(2)$ **88.** $(f^{-1} \circ g^{-1})(-2)$

Graphical Reasoning **In Exercises 89–92, (a) use a graphing utility to graph the function, (b) use the *draw inverse* feature of the graphing utility to draw the inverse of the function, and (c) determine whether the graph of the inverse relation is an inverse function, explaining your reasoning.**

89. $f(x) = x^3 + x + 1$ **90.** $h(x) = x\sqrt{4 - x^2}$

91. $g(x) = \dfrac{3x^2}{x^2 + 1}$ **92.** $f(x) = \dfrac{4x}{\sqrt{x^2 + 15}}$

In Exercises 93–98, use the functions $f(x) = \frac{1}{8}x - 3$ and $g(x) = x^3$ to find the indicated value or function.

93. $(f^{-1} \circ g^{-1})(1)$ **94.** $(g^{-1} \circ f^{-1})(-3)$

95. $(f^{-1} \circ f^{-1})(6)$ **96.** $(g^{-1} \circ g^{-1})(-4)$

97. $(f \circ g)^{-1}$ **98.** $g^{-1} \circ f^{-1}$

In Exercises 99–102, use the functions $f(x) = x + 4$ and $g(x) = 2x - 5$ to find the specified function.

99. $g^{-1} \circ f^{-1}$ **100.** $f^{-1} \circ g^{-1}$

101. $(f \circ g)^{-1}$ **102.** $(g \circ f)^{-1}$

103. *Shoe Sizes*   The table shows men's shoe sizes in the United States and the corresponding European shoe sizes. Let $y = f(x)$ represent the function that gives the men's European shoe size in terms of x, the men's U.S. size.

Men's U.S. shoe size	Men's European shoe size
8	41
9	42
10	43
11	45
12	46
13	47

(a) Is f one-to-one? Explain.

(b) Find $f(11)$.

(c) Find $f^{-1}(43)$, if possible.

(d) Find $f(f^{-1}(41))$.

(e) Find $f^{-1}(f(13))$.

104. *Shoe Sizes*   The table shows women's shoe sizes in the United States and the corresponding European shoe sizes. Let $y = g(x)$ epresent the function that gives the women's European shoe size in terms of x, the women's U.S. size.

Women's U.S. shoe size	Women's European shoe size
4	35
5	37
6	38
7	39
8	40
9	42

(a) Is g one-to-one? Explain.

(b) Find $g(6)$.

(c) Find $g^{-1}(42)$.

(d) Find $g(g^{-1}(39))$.

(e) Find $g^{-1}(g(5))$.

105. *Transportation*   The total values of new car sales f (in billions of dollars) in the United States from 1995 through 2004 are shown in the table. The time (in years) is given by t, with $t = 5$ corresponding to 1995.   (Source: National Automobile Dealers Association)

Year, t	Sales, $f(t)$
5	456.2
6	490.0
7	507.5
8	546.3
9	606.5
10	650.3
11	690.4
12	679.5
13	699.2
14	714.3

(a) Does f^{-1} exist?

(b) If f^{-1} exists, what does it mean in the context of the problem?

(c) If f^{-1} exists, find $f^{-1}(650.3)$.

(d) If the table above were extended to 2005 and if the total value of new car sales for that year were \$690.4 billion, would f^{-1} exist? Explain.

106. *Hourly Wage*   Your wage is \$8.00 per hour plus \$0.75 for each unit produced per hour. So, your hourly wage y in terms of the number of units produced x is $y = 8 + 0.75x$.

(a) Find the inverse function. What does each variable in the inverse function represent?

(b) Use a graphing utility to graph the function and its inverse function.

(c) Use the *trace* feature of a graphing utility to find the hourly wage when 10 units are produced per hour.

(d) Use the *trace* feature of a graphing utility to find the number of units produced when your hourly wage is \$22.25.

Synthesis

True or False?   **In Exercises 107 and 108, determine whether the statement is true or false. Justify your answer.**

107. If f is an even function, f^{-1} exists.

108. If the inverse function of f exists, and the graph of f has a y-intercept, the y-intercept of f is an x-intercept of f^{-1}.

109. *Proof*   Prove that if f and g are one-to-one functions, $(f \circ g)^{-1}(x) = (g^{-1} \circ f^{-1})(x)$.

110. *Proof*   Prove that if f is a one-to-one odd function, f^{-1} is an odd function.

In Exercises 111–114, decide whether the two functions shown in the graph appear to be inverse functions of each other. Explain your reasoning.

111.
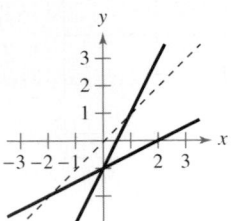

112.
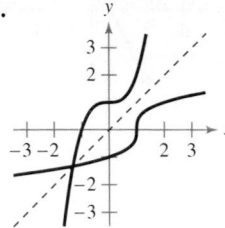

113.
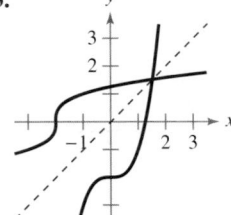

114.
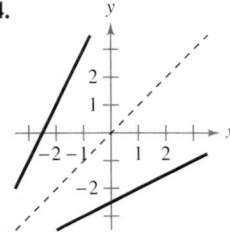

In Exercises 115–118, determine if the situation could be represented by a one-to-one function. If so, write a statement that describes the inverse function.

115. The number of miles n a marathon runner has completed in terms of the time t in hours

116. The population p of South Carolina in terms of the year t from 1960 to 2005

117. The depth of the tide d at a beach in terms of the time t over a 24-hour period

118. The height h in inches of a human born in the year 2000 in terms of his or her age n in years

Skills Review

In Exercises 119–122, write the rational expression in simplest form.

119. $\dfrac{27x^3}{3x^2}$

120. $\dfrac{5x^2 y}{xy + 5x}$

121. $\dfrac{x^2 - 36}{6 - x}$

122. $\dfrac{x^2 + 3x - 40}{x^2 - 3x - 10}$

In Exercises 123–128, determine whether the equation represents y as a function of x.

123. $4x - y = 3$

124. $x = 5$

125. $x^2 + y^2 = 9$

126. $x^2 + y = 8$

127. $y = \sqrt{x + 2}$

128. $x - y^2 = 0$

1.7   Linear Models and Scatter Plots

Scatter Plots and Correlation

Many real-life situations involve finding relationships between two variables, such as the year and the outstanding household credit market debt. In a typical situation, data is collected and written as a set of ordered pairs. The graph of such a set is called a *scatter plot*. (For a brief discussion of scatter plots, see Appendix B.1.)

Example 1   Constructing a Scatter Plot

The data in the table shows the outstanding household credit market debt D (in trillions of dollars) from 1998 through 2004. Construct a scatter plot of the data. (Source: Board of Governors of the Federal Reserve System)

Year	Household credit market debt, D (in trillions of dollars)
1998	6.0
1999	6.4
2000	7.0
2001	7.6
2002	8.4
2003	9.2
2004	10.3

Solution

Begin by representing the data with a set of ordered pairs. Let t represent the year, with $t = 8$ corresponding to 1998.

$(8, 6.0), (9, 6.4), (10, 7.0), (11, 7.6), (12, 8.4), (13, 9.2), (14, 10.3)$

Then plot each point in a coordinate plane, as shown in Figure 1.76.

✔CHECKPOINT   Now try Exercise 1.

From the scatter plot in Figure 1.76, it appears that the points describe a relationship that is nearly linear. The relationship is not *exactly* linear because the household credit market debt did not increase by precisely the same amount each year.

A mathematical equation that approximates the relationship between t and D is a *mathematical model*. When developing a mathematical model to describe a set of data, you strive for two (often conflicting) goals—accuracy and simplicity. For the data above, a linear model of the form

$$D = at + b$$

appears to be best. It is simple and relatively accurate.

What you should learn

- Construct scatter plots and interpret correlation.
- Use scatter plots and a graphing utility to find linear models for data.

Why you should learn it

Real-life data often follows a linear pattern. For instance, in Exercise 20 on page 81, you will find a linear model for the winning times in the women's 400-meter freestyle Olympic swimming event.

Nick Wilson/Getty Images

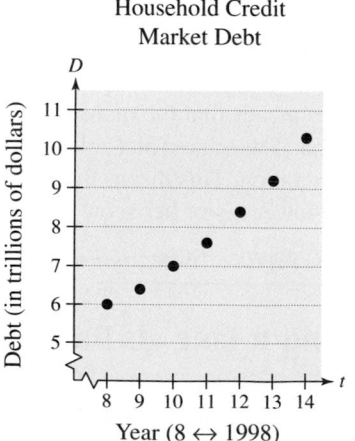

Figure 1.76

Consider a collection of ordered pairs of the form (x, y). If y tends to increase as x increases, the collection is said to have a **positive correlation.** If y tends to decrease as x increases, the collection is said to have a **negative correlation.** Figure 1.77 shows three examples: one with a positive correlation, one with a negative correlation, and one with no (discernible) correlation.

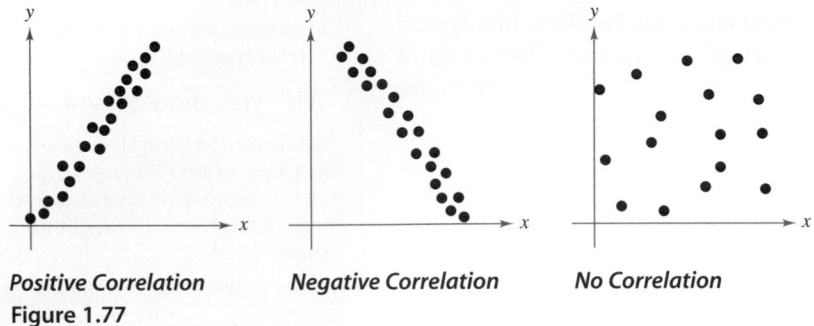

Positive Correlation   *Negative Correlation*   *No Correlation*
Figure 1.77

Example 2   Interpreting Correlation

On a Friday, 22 students in a class were asked to record the numbers of hours they spent studying for a test on Monday and the numbers of hours they spent watching television. The results are shown below. (The first coordinate is the number of hours and the second coordinate is the score obtained on the test.)

Study Hours: $(0, 40)$, $(1, 41)$, $(2, 51)$, $(3, 58)$, $(3, 49)$, $(4, 48)$, $(4, 64)$, $(5, 55)$, $(5, 69)$, $(5, 58)$, $(5, 75)$, $(6, 68)$, $(6, 63)$, $(6, 93)$, $(7, 84)$, $(7, 67)$, $(8, 90)$, $(8, 76)$, $(9, 95)$, $(9, 72)$, $(9, 85)$, $(10, 98)$

TV Hours: $(0, 98)$, $(1, 85)$, $(2, 72)$, $(2, 90)$, $(3, 67)$, $(3, 93)$, $(3, 95)$, $(4, 68)$, $(4, 84)$, $(5, 76)$, $(7, 75)$, $(7, 58)$, $(9, 63)$, $(9, 69)$, $(11, 55)$, $(12, 58)$, $(14, 64)$, $(16, 48)$, $(17, 51)$, $(18, 41)$, $(19, 49)$, $(20, 40)$

a. Construct a scatter plot for each set of data.

b. Determine whether the points are positively correlated, are negatively correlated, or have no discernible correlation. What can you conclude?

Solution

a. Scatter plots for the two sets of data are shown in Figure 1.78.

b. The scatter plot relating study hours and test scores has a positive correlation. This means that the more a student studied, the higher his or her score tended to be. The scatter plot relating television hours and test scores has a negative correlation. This means that the more time a student spent watching television, the lower his or her score tended to be.

✓**CHECKPOINT**   Now try Exercise 3.

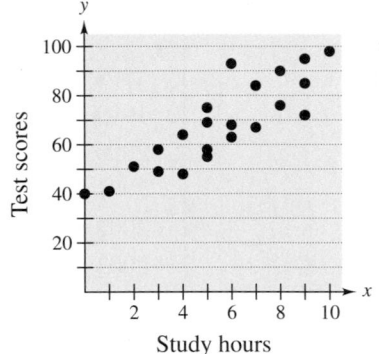

Study hours

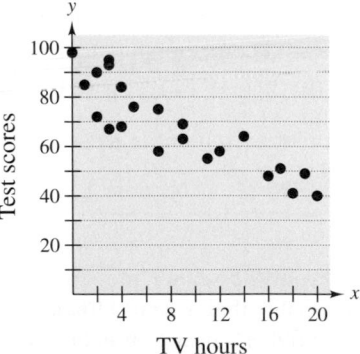

TV hours

Figure 1.78

Fitting a Line to Data

Finding a linear model to represent the relationship described by a scatter plot is called **fitting a line to data.** You can do this graphically by simply sketching the line that appears to fit the points, finding two points on the line, and then finding the equation of the line that passes through the two points.

Example 3 Fitting a Line to Data

Find a linear model that relates the year to the outstanding household credit market debt. (See Example 1.)

Year	Household credit market debt, D (in trillions of dollars)
1998	6.0
1999	6.4
2000	7.0
2001	7.6
2002	8.4
2003	9.2
2004	10.3

Solution

Let t represent the year, with $t = 8$ corresponding to 1998. After plotting the data in the table, draw the line that you think best represents the data, as shown in Figure 1.79. Two points that lie on this line are $(9, 6.4)$ and $(13, 9.2)$. Using the point-slope form, you can find the equation of the line to be

$$D = 0.7(t - 9) + 6.4$$
$$= 0.7t + 0.1. \qquad \text{Linear model}$$

✓CHECKPOINT   Now try Exercise 11(a) and (b).

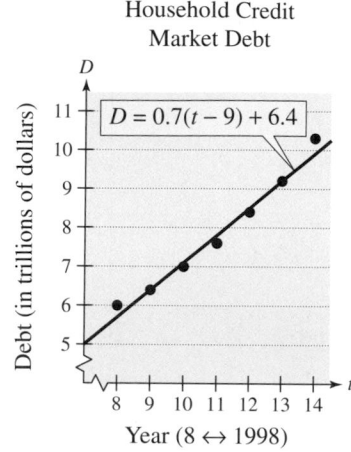

Household Credit Market Debt

$D = 0.7(t - 9) + 6.4$

Debt (in trillions of dollars)

Year (8 ↔ 1998)

Figure 1.79

Once you have found a model, you can measure how well the model fits the data by comparing the actual values with the values given by the model, as shown in the following table.

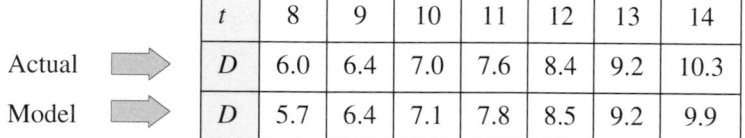

	t	8	9	10	11	12	13	14
Actual	D	6.0	6.4	7.0	7.6	8.4	9.2	10.3
Model	D	5.7	6.4	7.1	7.8	8.5	9.2	9.9

The sum of the squares of the differences between the actual values and the model values is the **sum of the squared differences.** The model that has the least sum is the **least squares regression line** for the data. For the model in Example 3, the sum of the squared differences is 0.31. The least squares regression line for the data is

$$D = 0.71t. \qquad \text{Best-fitting linear model}$$

Its sum of squared differences is 0.3015. See Appendix C for more on the least squares regression line.

STUDY TIP

The model in Example 3 is based on the two data points chosen. If different points are chosen, the model may change somewhat. For instance, if you choose $(8, 6)$ and $(14, 10.3)$, the new model is

$$D = 0.72(t - 8) + 6$$
$$= 0.72t + 0.24.$$

Example 4   A Mathematical Model

The numbers S (in billions) of shares listed on the New York Stock Exchange for the years 1995 through 2004 are shown in the table.   (Source: New York Stock Exchange, Inc.)

Year	Shares, S
1995	154.7
1996	176.9
1997	207.1
1998	239.3
1999	280.9
2000	313.9
2001	341.5
2002	349.9
2003	359.7
2004	380.8

TECHNOLOGY SUPPORT

For instructions on how to use the *regression* feature, see Appendix A; for specific keystrokes, go to this textbook's *Online Study Center.*

a. Use the *regression* feature of a graphing utility to find a linear model for the data. Let t represent the year, with $t = 5$ corresponding to 1995.

b. How closely does the model represent the data?

Graphical Solution

a. Enter the data into the graphing utility's list editor. Then use the *linear regression* feature to obtain the model shown in Figure 1.80. You can approximate the model to be $S = 26.47t + 29.0$.

b. You can use a graphing utility to graph the actual data and the model in the same viewing window. In Figure 1.81, it appears that the model is a fairly good fit for the actual data.

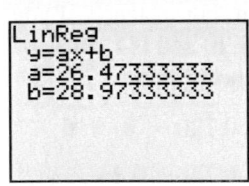

Figure 1.80

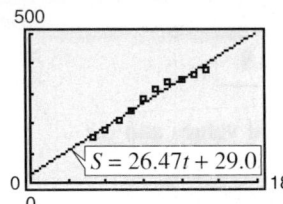

Figure 1.81

✔CHECKPOINT   Now try Exercise 9.

Numerical Solution

a. Using the *linear regression* feature of a graphing utility, you can find that a linear model for the data is $S = 26.47t + 29.0$.

b. You can see how well the model fits the data by comparing the actual values of S with the values of S given by the model, which are labeled S^* in the table below. From the table, you can see that the model appears to be a good fit for the actual data.

Year	S	S^*
1995	154.7	161.4
1996	176.9	187.8
1997	207.1	214.3
1998	239.3	240.8
1999	280.9	267.2
2000	313.9	293.7
2001	341.5	320.2
2002	349.9	346.6
2003	359.7	373.1
2004	380.8	399.6

When you use the *regression* feature of a graphing calculator or computer program to find a linear model for data, you will notice that the program may also output an "*r*-value." For instance, the *r*-value from Example 4 was $r \approx 0.985$. This *r*-value is the **correlation coefficient** of the data and gives a measure of how well the model fits the data. Correlation coefficients vary between -1 and 1. Basically, the closer $|r|$ is to 1, the better the points can be described by a line. Three examples are shown in Figure 1.82.

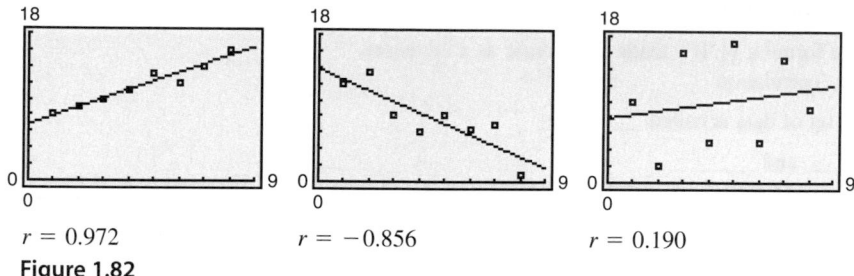

$r = 0.972$ $r = -0.856$ $r = 0.190$

Figure 1.82

Example 5 Finding a Least Squares Regression Line

The following ordered pairs (w, h) represent the shoe sizes w and the heights h (in inches) of 25 men. Use the *regression* feature of a graphing utility to find the least squares regression line for the data.

(10.0, 70.5)	(10.5, 71.0)	(9.5, 69.0)	(11.0, 72.0)	(12.0, 74.0)
(8.5, 67.0)	(9.0, 68.5)	(13.0, 76.0)	(10.5, 71.5)	(10.5, 70.5)
(10.0, 71.0)	(9.5, 70.0)	(10.0, 71.0)	(10.5, 71.0)	(11.0, 71.5)
(12.0, 73.5)	(12.5, 75.0)	(11.0, 72.0)	(9.0, 68.0)	(10.0, 70.0)
(13.0, 75.5)	(10.5, 72.0)	(10.5, 71.0)	(11.0, 73.0)	(8.5, 67.5)

Solution

After entering the data into a graphing utility (see Figure 1.83), you obtain the model shown in Figure 1.84. So, the least squares regression line for the data is

$$h = 1.84w + 51.9.$$

In Figure 1.85, this line is plotted with the data. Note that the plot does not have 25 points because some of the ordered pairs graph as the same point. The correlation coefficient for this model is $r \approx 0.981$, which implies that the model is a good fit for the data.

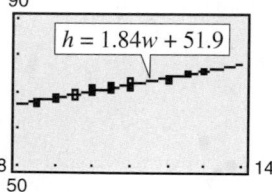

Figure 1.83   **Figure 1.84**   **Figure 1.85**

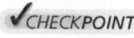 *CHECKPOINT*   Now try Exercise 20.

1.7 Exercises

See www.CalcChat.com for worked-out solutions to odd-numbered exercises.

Vocabulary Check

Fill in the blanks.

1. Consider a collection of ordered pairs of the form (x, y). If y tends to increase as x increases, then the collection is said to have a _____ correlation.

2. Consider a collection of ordered pairs of the form (x, y). If y tends to decrease as x increases, then the collection is said to have a _____ correlation.

3. The process of finding a linear model for a set of data is called _____ .

4. Correlation coefficients vary between _____ and _____ .

---

1. *Sales* The following ordered pairs give the years of experience x for 15 sales representatives and the monthly sales y (in thousands of dollars).

$(1.5, 41.7)$, $(1.0, 32.4)$, $(0.3, 19.2)$, $(3.0, 48.4)$, $(4.0, 51.2)$,
$(0.5, 28.5)$, $(2.5, 50.4)$, $(1.8, 35.5)$, $(2.0, 36.0)$,
$(1.5, 40.0)$, $(3.5, 50.3)$, $(4.0, 55.2)$, $(0.5, 29.1)$, $(2.2, 43.2)$,
$(2.0, 41.6)$

(a) Create a scatter plot of the data.

(b) Does the relationship between x and y appear to be approximately linear? Explain.

2. *Quiz Scores* The following ordered pairs give the scores on two consecutive 15-point quizzes for a class of 18 students.

$(7, 13)$, $(9, 7)$, $(14, 14)$, $(15, 15)$, $(10, 15)$, $(9, 7)$,
$(14, 11)$, $(14, 15)$, $(8, 10)$, $(9, 10)$, $(15, 9)$, $(10, 11)$,
$(11, 14)$, $(7, 14)$, $(11, 10)$, $(14, 11)$, $(10, 15)$, $(9, 6)$

(a) Create a scatter plot of the data.

(b) Does the relationship between consecutive quiz scores appear to be approximately linear? If not, give some possible explanations.

In Exercises 3–6, the scatter plots of sets of data are shown. Determine whether there is positive correlation, negative correlation, or no discernible correlation between the variables.

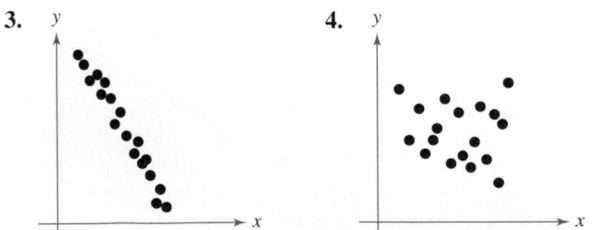

3.   **4.**

5.   **6.**

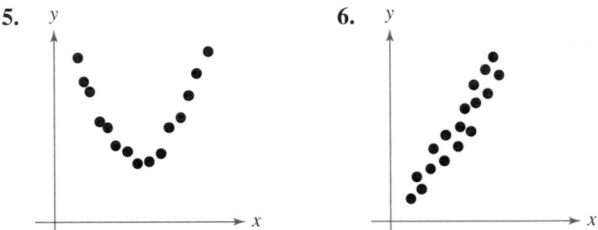

In Exercises 7–10, (a) for the data points given, draw a line of best fit through two of the points and find the equation of the line through the points, (b) use the *regression* feature of a graphing utility to find a linear model for the data, and to identify the correlation coefficient, (c) graph the data points and the lines obtained in parts (a) and (b) in the same viewing window, and (d) comment on the validity of both models. To print an enlarged copy of the graph, go to the website *www.mathgraphs.com*.

7.   **8.**

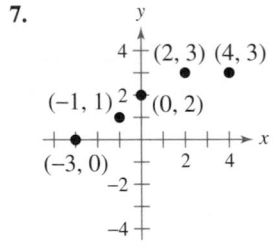   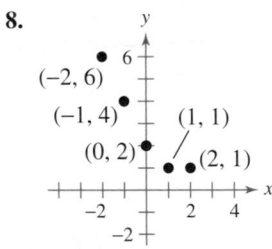

9.   **10.**

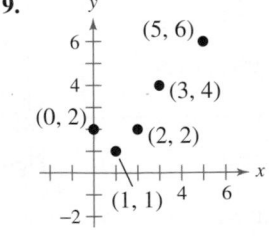   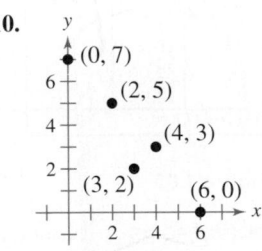

11. Hooke's Law   Hooke's Law states that the force F required to compress or stretch a spring (within its elastic limits) is proportional to the distance d that the spring is compressed or stretched from its original length. That is, $F = kd$, where k is the measure of the stiffness of the spring and is called the *spring constant*. The table shows the elongation d in centimeters of a spring when a force of F kilograms is applied.

Force, F	Elongation, d
20	1.4
40	2.5
60	4.0
80	5.3
100	6.6

(a) Sketch a scatter plot of the data.

(b) Find the equation of the line that seems to best fit the data.

(c) Use the *regression* feature of a graphing utility to find a linear model for the data. Compare this model with the model from part (b).

(d) Use the model from part (c) to estimate the elongation of the spring when a force of 55 kilograms is applied.

12. Cell Phones   The average lengths L of cellular phone calls in minutes from 1999 to 2004 are shown in the table. (Source: Cellular Telecommunications & Internet Association)

Year	Average length, L (in minutes)
1999	2.38
2000	2.56
2001	2.74
2002	2.73
2003	2.87
2004	3.05

(a) Use a graphing utility to create a scatter plot of the data, with $t = 9$ corresponding to 1999.

(b) Use the *regression* feature of a graphing utility to find a linear model for the data. Let t represent the year, with $t = 9$ corresponding to 1999.

(c) Use a graphing utility to plot the data and graph the model in the same viewing window. Is the model a good fit? Explain.

(d) Use the model to predict the average lengths of cellular phone calls for the years 2010 and 2015. Do your answers seem reasonable? Explain.

13. Sports   The mean salaries S (in thousands of dollars) for professional football players in the United States from 2000 to 2004 are shown in the table.   (Source: National Collegiate Athletic Assn.)

Year	Mean salary, S (in thousands of dollars)
2000	787
2001	986
2002	1180
2003	1259
2004	1331

(a) Use a graphing utility to create a scatter plot of the data, with $t = 0$ corresponding to 2000.

(b) Use the *regression* feature of a graphing utility to find a linear model for the data. Let t represent the year, with $t = 0$ corresponding to 2000.

(c) Use a graphing utility to plot the data and graph the model in the same viewing window. Is the model a good fit? Explain.

(d) Use the model to predict the mean salaries for professional football players in 2005 and 2010. Do the results seem reasonable? Explain.

(e) What is the slope of your model? What does it tell you about the mean salaries of professional football players?

14. Teacher's Salaries   The mean salaries S (in thousands of dollars) of public school teachers in the United States from 1999 to 2004 are shown in the table. (Source: Educational Research Service)

Year	Mean salary, S (in thousands of dollars)
1999	41.4
2000	42.2
2001	43.7
2002	43.8
2003	45.0
2004	45.6

(a) Use a graphing utility to create a scatter plot of the data, with $t = 9$ corresponding to 1999.

(b) Use the *regression* feature of a graphing utility to find a linear model for the data. Let t represent the year, with $t = 9$ corresponding to 1999.

(c) Use a graphing utility to plot the data and graph the model in the same viewing window. Is the model a good fit? Explain.

(d) Use the model to predict the mean salaries for teachers in 2005 and 2010. Do the results seem reasonable? Explain.

15. Cable Television The average monthly cable television bills C (in dollars) for a basic plan from 1990 to 2004 are shown in the table.   (Source: Kagan Research, LLC)

Year	Monthly bill, C (in dollars)
1990	16.78
1991	18.10
1992	19.08
1993	19.39
1994	21.62
1995	23.07
1996	24.41
1997	26.48
1998	27.81
1999	28.92
2000	30.37
2001	32.87
2002	34.71
2003	36.59
2004	38.23

(a) Use a graphing utility to create a scatter plot of the data, with $t = 0$ corresponding to 1990.

(b) Use the *regression* feature of a graphing utility to find a linear model for the data and to identify the correlation coefficient. Let t represent the year, with $t = 0$ corresponding to 1990.

(c) Graph the model with the data in the same viewing window.

(d) Is the model a good fit for the data? Explain.

(e) Use the model to predict the average monthly cable bills for the years 2005 and 2010.

(f) Do you believe the model would be accurate to predict the average monthly cable bills for future years? Explain.

16. State Population The projected populations P (in thousands) for selected years for New Jersey based on the 2000 census are shown in the table.   (Source: U.S. Census Bureau)

Year	Population, P (in thousands)
2005	8745
2010	9018
2015	9256
2020	9462
2025	9637
2030	9802

(a) Use a graphing utility to create a scatter plot of the data, with $t = 5$ corresponding to 2005.

(b) Use the *regression* feature of a graphing utility to find a linear model for the data. Let t represent the year, with $t = 5$ corresponding to 2005.

(c) Use a graphing utility to plot the data and graph the model in the same viewing window. Is the model a good fit? Explain.

(d) Use the model to predict the population of New Jersey in 2050. Does the result seem reasonable? Explain.

17. State Population The projected populations P (in thousands) for selected years for Wyoming based on the 2000 census are shown in the table.   (Source: U.S. Census Bureau)

Year	Population, P (in thousands)
2005	507
2010	520
2015	528
2020	531
2025	529
2030	523

(a) Use a graphing utility to create a scatter plot of the data, with $t = 5$ corresponding to 2005.

(b) Use the *regression* feature of a graphing utility to find a linear model for the data. Let t represent the year, with $t = 5$ corresponding to 2005.

(c) Use a graphing utility to plot the data and graph the model in the same viewing window. Is the model a good fit? Explain.

(d) Use the model to predict the population of Wyoming in 2050. Does the result seem reasonable? Explain.

18. Advertising and Sales The table shows the advertising expenditures x and sales volumes y for a company for seven randomly selected months. Both are measured in thousands of dollars.

Month	Advertising expenditures, x	Sales volume, y
1	2.4	202
2	1.6	184
3	2.0	220
4	2.6	240
5	1.4	180
6	1.6	164
7	2.0	186

(a) Use the *regression* feature of a graphing utility to find a linear model for the data and to identify the correlation coefficient.

(b) Use a graphing utility to plot the data and graph the model in the same viewing window.

(c) Interpret the slope of the model in the context of the problem.

(d) Use the model to estimate sales for advertising expenditures of $1500.

19. *Number of Stores*   The table shows the numbers T of Target stores from 1997 to 2006.   (Source: Target Corp.)

Year	Number of stores, T
1997	1130
1998	1182
1999	1243
2000	1307
2001	1381
2002	1475
2003	1553
2004	1308
2005	1400
2006	1505

(a) Use the *regression* feature of a graphing utility to find a linear model for the data and to identify the correlation coefficient. Let t represent the year, with $t = 7$ corresponding to 1997.

(b) Use a graphing utility to plot the data and graph the model in the same viewing window.

(c) Interpret the slope of the model in the context of the problem.

(d) Use the model to find the year in which the number of Target stores will exceed 1800.

(e) Create a table showing the actual values of T and the values of T given by the model. How closely does the model fit the data?

20. *Sports*   The following ordered pairs (t, T) represent the Olympic year t and the winning time T (in minutes) in the women's 400-meter freestyle swimming event.   (Source: *The World Almanac* 2005)

(1948, 5.30)	(1968, 4.53)	(1988, 4.06)
(1952, 5.20)	(1972, 4.32)	(1992, 4.12)
(1956, 4.91)	(1976, 4.16)	(1996, 4.12)
(1960, 4.84)	(1980, 4.15)	(2000, 4.10)
(1964, 4.72)	(1984, 4.12)	(2004, 4.09)

(a) Use the *regression* feature of a graphing utility to find a linear model for the data. Let t represent the year, with $t = 0$ corresponding to 1950.

(b) What information is given by the sign of the slope of the model?

(c) Use a graphing utility to plot the data and graph the model in the same viewing window.

(d) Create a table showing the actual values of y and the values of y given by the model. How closely does the model fit the data?

(e) Can the model be used to predict the winning times in the future? Explain.

Synthesis

True or False?   In Exercises 21 and 22, determine whether the statement is true or false. Justify your answer.

21. A linear regression model with a positive correlation will have a slope that is greater than 0.

22. If the correlation coefficient for a linear regression model is close to -1, the regression line cannot be used to describe the data.

23. *Writing*   A linear mathematical model for predicting prize winnings at a race is based on data for 3 years. Write a paragraph discussing the potential accuracy or inaccuracy of such a model.

24. *Research Project*   Use your school's library, the Internet, or some other reference source to locate data that you think describes a linear relationship. Create a scatter plot of the data and find the least squares regression line that represents the points. Interpret the slope and y-intercept in the context of the data. Write a summary of your findings.

Skills Review

In Exercises 25–28, evaluate the function at each value of the independent variable and simplify.

25. $f(x) = 2x^2 - 3x + 5$

(a) $f(-1)$ (b) $f(w + 2)$

26. $g(x) = 5x^2 - 6x + 1$

(a) $g(-2)$ (b) $g(z - 2)$

27. $h(x) = \begin{cases} 1 - x^2, & x \le 0 \\ 2x + 3, & x > 0 \end{cases}$

(a) $h(1)$ (b) $h(0)$

28. $k(x) = \begin{cases} 5 - 2x, & x < -1 \\ x^2 + 4, & x \ge -1 \end{cases}$

(a) $k(-3)$ (b) $k(-1)$

In Exercises 29–34, solve the equation algebraically. Check your solution graphically.

29. $6x + 1 = -9x - 8$ **30.** $3(x - 3) = 7x + 2$

31. $8x^2 - 10x - 3 = 0$ **32.** $10x^2 - 23x - 5 = 0$

33. $2x^2 - 7x + 4 = 0$ **34.** $2x^2 - 8x + 5 = 0$

What Did You Learn?

Key Terms

slope, *p. 3*
point-slope form, *p. 5*
slope-intercept form, *p. 7*
parallel lines, *p. 9*
perpendicular lines, *p. 9*
function, *p. 16*
domain, *p. 16*

range, *p. 16*
independent variable, *p. 18*
dependent variable, *p. 18*
function notation, *p. 18*
graph of a function, *p. 30*
Vertical Line Test, *p. 31*
even function, *p. 36*

odd function, *p. 36*
rigid transformation, *p. 47*
inverse function, *p. 62*
one-to-one, *p. 66*
Horizontal Line Test, *p. 66*
positive correlation, *p. 74*
negative correlation, *p. 74*

Key Concepts

1.1 ■ Find and use the slopes of lines to write and graph linear equations

1. The slope m of the nonvertical line through (x_1, y_1) and (x_2, y_2), where $x_1 \neq x_2$, is

$$m = \frac{y_2 - y_1}{x_2 - x_1} = \frac{\text{change in } y}{\text{change in } x}.$$

2. The point-slope form of the equation of the line that passes through the point (x_1, y_1) and has a slope of m is $y - y_1 = m(x - x_1)$.

3. The graph of the equation $y = mx + b$ is a line whose slope is m and whose y-intercept is $(0, b)$.

1.2 ■ Evaluate functions and find their domains

1. To evaluate a function $f(x)$, replace the independent variable x with a value and simplify the expression.

2. The domain of a function is the set of all real numbers for which the function is defined.

1.3 ■ Analyze graphs of functions

1. The graph of a function may have intervals over which the graph increases, decreases, or is constant.

2. The points at which a function changes its increasing, decreasing, or constant behavior are the relative minimum and relative maximum values of the function.

3. An even function is symmetric with respect to the y-axis. An odd function is symmetric with respect to the origin.

1.4 ■ Identify and graph shifts, reflections, and nonrigid transformations of functions

1. Vertical and horizontal shifts of a graph are transformations in which the graph is shifted up or down, and left or right.

2. A reflection transformation is a mirror image of a graph in a line.

3. A nonrigid transformation distorts the graph by stretching or shrinking the graph horizontally or vertically.

1.5 ■ Find arithmetic combinations and compositions of functions

1. An arithmetic combination of functions is the sum, difference, product, or quotient of two functions. The domain of the arithmetic combination is the set of all real numbers that are common to the two functions.

2. The composition of the function f with the function g is

$$(f \circ g)(x) = f(g(x)).$$

The domain of $f \circ g$ is the set of all x in the domain of g such that $g(x)$ is in the domain of f.

1.6 ■ Find inverse functions

1. If the point (a, b) lies on the graph of f, then the point (b, a) must lie on the graph of its inverse function f^{-1}, and vice versa. This means that the graph of f^{-1} is a reflection of the graph of f in the line $y = x$.

2. Use the Horizontal Line Test to decide if f has an inverse function. To find an inverse function algebraically, replace $f(x)$ by y, interchange the roles of x and y and solve for y, and replace y by $f^{-1}(x)$ in the new equation.

1.7 ■ Use scatter plots and find linear models

1. A scatter plot is a graphical representation of data written as a set of ordered pairs.

2. The best-fitting linear model can be found using the *linear regression* feature of a graphing utility or a computer program.

Review Exercises

See www.CalcChat.com for worked-out solutions to odd-numbered exercises.

1.1 In Exercises 1 and 2, sketch the lines with the indicated slopes through the point on the same set of the coordinate axes.

Point		Slope
1. $(1, 1)$	(a) 2	(b) 0
	(c) -1	(d) Undefined
2. $(-2, -3)$	(a) 1	(b) $-\frac{1}{2}$
	(c) 4	(d) 0

In Exercises 3–8, plot the two points and find the slope of the line passing through the points.

3. $(-3, 2), (8, 2)$

4. $(7, -1), (7, 12)$

5. $\left(\frac{3}{2}, 1\right), \left(5, \frac{5}{2}\right)$

6. $\left(-\frac{3}{4}, \frac{5}{6}\right), \left(\frac{1}{2}, -\frac{5}{2}\right)$

7. $(-4.5, 6), (2.1, 3)$

8 $(-2.7, -6.3), (-1, -1.2)$

In Exercises 9–18, (a) use the point on the line and the slope of the line to find the general form of the equation of the line, and (b) find three additional points through which the line passes. (There are many correct answers.)

Point	Slope
9. $(2, -1)$	$m = \frac{1}{4}$
10. $(-3, 5)$	$m = -\frac{3}{2}$
11. $(0, -5)$	$m = \frac{3}{2}$
12. $(3, 0)$	$m = -\frac{2}{3}$
13. $\left(\frac{1}{5}, -5\right)$	$m = -1$
14. $\left(0, \frac{7}{8}\right)$	$m = -\frac{4}{5}$
15. $(-2, 6)$	$m = 0$
16. $(-8, 8)$	$m = 0$
17. $(10, -6)$	m is undefined.
18. $(5, 4)$	m is undefined.

In Exercises 19–22, find the slope-intercept form of the equation of the line that passes through the points. Use a graphing utility to graph the line.

19. $(2, -1), (4, -1)$

20. $(0, 0), (0, 10)$

21. $(-1, 0), (6, 2)$

22. $(1, 6), (4, 2)$

Rate of Change In Exercises 23–26, you are given the dollar value of a product in 2008 *and* the rate at which the value of the item is expected to change during the next 5 years. Use this information to write a linear equation that gives the dollar value V of the product in terms of the year t. (Let $t = 8$ represent 2008.)

2008 Value	Rate
23. $12,500	$850 increase per year
24. $3795	$115 decrease per year
25. $625.50	$42.70 increase per year
26. $72.95	$5.15 decrease per year

27. *Sales* During the second and third quarters of the year, an e-commerce business had sales of $160,000 and $185,000, respectively. The growth of sales follows a linear pattern. Estimate sales during the fourth quarter.

28. *Depreciation* The dollar value of a DVD player in 2006 is $225. The product will decrease in value at an expected rate of $12.75 per year.

(a) Write a linear equation that gives the dollar value V of the DVD player in terms of the year t. (Let $t = 6$ represent 2006.)

(b) Use a graphing utility to graph the equation found in part (a). Be sure to choose an appropriate viewing window. State the dimensions of your viewing window, and explain why you chose the values that you did.

(c) Use the *value* or *trace* feature of your graphing utility to estimate the dollar value of the DVD player in 2010. Confirm your answer algebraically.

(d) According to the model, when will the DVD player have no value?

In Exercises 29–32, write the slope-intercept forms of the equations of the lines through the given point (a) parallel to the given line and (b) perpendicular to the given line. Verify your result with a graphing utility (use a *square setting*).

Point	Line
29. $(3, -2)$	$5x - 4y = 8$
30. $(-8, 3)$	$2x + 3y = 5$
31. $(-6, 2)$	$x = 4$
32. $(3, -4)$	$y = 2$

1.2 In Exercises 33 and 34, which sets of ordered pairs represent functions from A to B? Explain.

33. $A = \{10, 20, 30, 40\}$ and $B = \{0, 2, 4, 6\}$

(a) $\{(20, 4), (40, 0), (20, 6), (30, 2)\}$

(b) $\{(10, 4), (20, 4), (30, 4), (40, 4)\}$

(c) $\{(40, 0), (30, 2), (20, 4), (10, 6)\}$

(d) $\{(20, 2), (10, 0), (40, 4)\}$

34. $A = \{u, v, w\}$ and $B = \{-2, -1, 0, 1, 2\}$

(a) $\{(v, -1), (u, 2), (w, 0), (u, -2)\}$

(b) $\{(u, -2), (v, 2), (w, 1)\}$

(c) $\{(u, 2), (v, 2), (w, 1), (w, 1)\}$

(d) $\{(w, -2), (v, 0), (w, 2)\}$

In Exercises 35–38, determine whether the equation represents y as a function of x.

35. $16x^2 - y^2 = 0$

36. $2x - y - 3 = 0$

37. $y = \sqrt{1 - x}$

38. $|y| = x + 2$

In Exercises 39–42, evaluate the function at each specified value of the independent variable, and simplify.

39. $f(x) = x^2 + 1$

(a) $f(1)$ (b) $f(-3)$

(c) $f(b^3)$ (d) $f(x - 1)$

40. $g(x) = x^{4/3}$

(a) $g(8)$ (b) $g(t + 1)$

(c) $g(-27)$ (d) $g(-x)$

41. $h(x) = \begin{cases} 2x + 1, & x \le -1 \\ x^2 + 2, & x > -1 \end{cases}$

(a) $h(-2)$ (b) $h(-1)$

(c) $h(0)$ (d) $h(2)$

42. $f(x) = \dfrac{3}{2x - 5}$

(a) $f(1)$ (b) $f(-2)$

(c) $f(t)$ (d) $f(10)$

In Exercises 43–48, find the domain of the function.

43. $f(x) = \dfrac{x - 1}{x + 2}$

44. $f(x) = \dfrac{x^2}{x^2 + 1}$

45. $f(x) = \sqrt{25 - x^2}$

46. $f(x) = \sqrt{x^2 - 16}$

47. $g(s) = \dfrac{5s + 5}{3s - 9}$ **48.** $f(x) = \dfrac{2x + 1}{3x + 4}$

49. *Cost* A hand tool manufacturer produces a product for which the variable cost is \$5.35 per unit and the fixed costs are \$16,000. The company sells the product for \$8.20 and can sell all that it produces.

(a) Write the total cost C as a function of x, the number of units produced.

(b) Write the profit P as a function of x.

50. *Consumerism* The retail sales R (in billions of dollars) of lawn care products and services in the United States from 1997 to 2004 can be approximated by the model

$$R(t) = \begin{cases} 0.126t^2 - 0.89t + 6.8, & 7 \le t < 11 \\ 0.1442t^3 - 5.611t^2 + 71.10t - 282.4, & 11 \le t \le 14 \end{cases}$$

where t represents the year, with $t = 7$ corresponding to 1997. Use the *table* feature of a graphing utility to approximate the retail sales of lawn care products and services for each year from 1997 to 2004. (Source: The National Gardening Association)

∫ In Exercises 51 and 52, find the difference quotient and simplify your answer.

51. $f(x) = 2x^2 + 3x - 1$, $\dfrac{f(x + h) - f(x)}{h}$, $h \ne 0$

52. $f(x) = x^3 - 5x^2 + x$, $\dfrac{f(x + h) - f(x)}{h}$, $h \ne 0$

1.3 In Exercises 53–56, use a graphing utility to graph the function and estimate its domain and range. Then find the domain and range algebraically.

53. $f(x) = 3 - 2x^2$

54. $f(x) = \sqrt{2x^2 - 1}$

55. $h(x) = \sqrt{36 - x^2}$

56. $g(x) = |x + 5|$

In Exercises 57–60, (a) use a graphing utility to graph the equation and (b) use the Vertical Line Test to determine whether y is a function of x.

57. $y = \dfrac{x^2 + 3x}{6}$

58. $y = -\dfrac{2}{3}|x + 5|$

59. $3x + y^2 = 2$

60. $x^2 + y^2 = 49$

In Exercises 61–64, (a) use a graphing utility to graph the function and (b) determine the open intervals on which the function is increasing, decreasing, or constant.

61. $f(x) = x^3 - 3x$

62. $f(x) = \sqrt{x^2 - 9}$

63. $f(x) = x\sqrt{x - 6}$

64. $f(x) = \dfrac{|x + 8|}{2}$

In Exercises 65–68, use a graphing utility to approximate (to two decimal places) any relative minimum or relative maximum values of the function.

65. $f(x) = (x^2 - 4)^2$

66. $f(x) = x^2 - x - 1$

67. $h(x) = 4x^3 - x^4$

68. $f(x) = x^3 - 4x^2 - 1$

In Exercises 69–72, sketch the graph of the function by hand.

69. $f(x) = \begin{cases} 3x + 5, & x < 0 \\ x - 4, & x \geq 0 \end{cases}$

70. $f(x) = \begin{cases} x^2 + 7, & x < 1 \\ x^2 - 5x + 6, & x \geq 1 \end{cases}$

71. $f(x) = [\![x]\!] + 3$

72. $f(x) = [\![x + 2]\!]$

In Exercises 73–78, determine algebraically whether the function is even, odd, or neither. Verify your answer using a graphing utility.

73. $f(x) = x^2 + 6$

74. $f(x) = x^2 - x - 1$

75. $f(x) = (x^2 - 8)^2$

76. $f(x) = 2x^3 - x^2$

77. $f(x) = 3x^{5/2}$

78. $f(x) = 3x^{2/5}$

1.4 In Exercises 79–84, identify the parent function and describe the transformation shown in the graph. Write an equation for the graphed function.

79.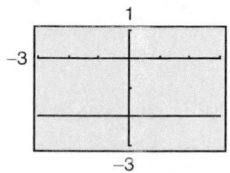

80.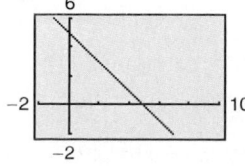

81.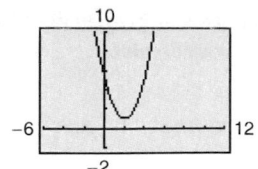

82.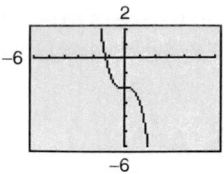

83.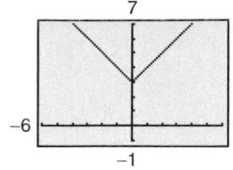

84.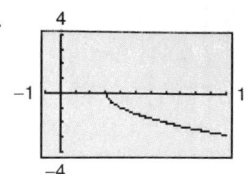

In Exercises 85–88, use the graph of $y = f(x)$ to graph the function.

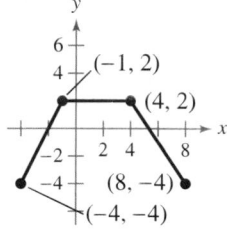

85. $y = f(-x)$

86. $y = -f(x)$

87. $y = f(x) - 2$

88. $y = f(x - 1)$

Library of Parent Functions In Exercises 89–100, h is related to one of the six parent functions on page 42. (a) Identify the parent function f. (b) Describe the sequence of transformations from f to h. (c) Sketch the graph of h by hand. (d) Use function notation to write h in terms of the parent function f.

89. $h(x) = x^2 - 6$

90. $h(x) = -x^2 - 3$

91. $h(x) = (x - 2)^3 + 5$

92. $h(x) = -(x + 2)^2 - 8$

93. $h(x) = -(x - 2)^2 - 8$

94. $h(x) = \frac{1}{2}(x - 3)^2 - 6$

95. $h(x) = -\sqrt{x} + 5$

96. $h(x) = 2\sqrt{x} + 5$

97. $h(x) = \sqrt{x - 1} + 3$

98. $h(x) = |x| + 9$

99. $h(x) = -\frac{1}{2}|x| + 9$

100. $h(x) = |x + 8| - 1$

1.5 In Exercises 101–110, let $f(x) = 3 - 2x$, $g(x) = \sqrt{x}$, and $h(x) = 3x^2 + 2$, and find the indicated values.

101. $(f - g)(4)$

102. $(f + h)(5)$

103. $(f + g)(25)$

104. $(g - h)(1)$

105. $(fh)(1)$

106. $\left(\dfrac{g}{h}\right)(1)$

107. $(h \circ g)(7)$

108. $(g \circ f)(-2)$

109. $(f \circ g)(-4)$

110. $(g \circ h)(6)$

∫ **In Exercises 111–116, find two functions f and g such that $(f \circ g)(x) = h(x)$. (There are many correct answers.)**

111. $h(x) = (x + 3)^2$

112. $h(x) = (1 - 2x)^3$

113. $h(x) = \sqrt{4x + 2}$

114. $h(x) = \sqrt[3]{(x + 2)^2}$

115. $h(x) = \dfrac{4}{x + 2}$

116. $h(x) = \dfrac{6}{(3x + 1)^3}$

Data Analysis **In Exercises 117 and 118, the numbers (in millions) of students taking the SAT (y_1) and ACT (y_2) for the years 1990 through 2004 can be modeled by**

$$y_1 = 0.00204t^2 + 0.0015t + 1.021$$

and

$$y_2 = 0.0274t + 0.785$$

where t represents the year, with $t = 0$ corresponding to 1990. (Source: College Entrance Examination Board and ACT, Inc.)

117. Use a graphing utility to graph y_1, y_2, and $y_1 + y_2$ in the same viewing window.

118. Use the model $y_1 + y_2$ to estimate the total number of students taking the SAT and ACT in 2008.

1.6 In Exercises 119–122, find the inverse function of f informally. Verify that $f(f^{-1}(x)) = x$ and $f^{-1}f(x)) = x$.

119. $f(x) = 6x$

120. $f(x) = x + 5$

121. $f(x) = \dfrac{1}{2}x + 3$

122. $f(x) = \dfrac{x - 4}{5}$

In Exercises 123 and 124, show that f and g are inverse functions (a) graphically and (b) numerically.

123. $f(x) = 3 - 4x$, $g(x) = \dfrac{3 - x}{4}$

124. $f(x) = \sqrt{x + 1}$, $g(x) = x^2 - 1$, $x \geq 0$

In Exercises 125–128, use a graphing utility to graph the function and use the Horizontal Line Test to determine whether the function is one-to-one and an inverse function exists.

125. $f(x) = \frac{1}{2}x - 3$

126. $f(x) = (x - 1)^2$

127. $h(t) = \dfrac{2}{t - 3}$

128. $g(x) = \sqrt{x + 6}$

In Exercises 129–134, find the inverse function of f algebraically.

129. $f(x) = \frac{1}{2}x - 5$

130. $f(x) = \dfrac{7x + 3}{8}$

131. $f(x) = 4x^3 - 3$

132. $f(x) = 5x^3 + 2$

133. $f(x) = \sqrt{x + 10}$

134. $f(x) = 4\sqrt{6 - x}$

1.7 In Exercises 135 and 136, the scatter plots of sets of data are shown. Determine whether there is positive correlation, negative correlation, or no discernible correlation between the variables.

135.   **136.**

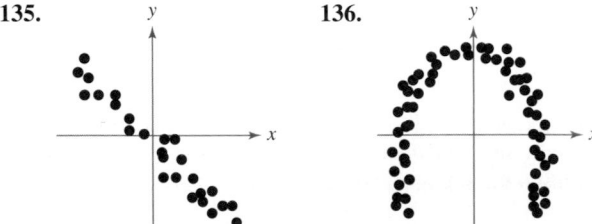

137. *Education*   The following ordered pairs give the entrance exam scores x and the grade-point averages y after 1 year of college for 10 students.

(75, 2.3), (82, 3.0), (90, 3.6), (65, 2.0), (70, 2.1), (88, 3.5), (93, 3.9), (69, 2.0), (80, 2.8), (85, 3.3)

(a) Create a scatter plot of the data.

(b) Does the relationship between x and y appear to be approximately linear? Explain.

138. *Stress Test*   A machine part was tested by bending it x centimeters 10 times per minute until it failed (y equals the time to failure in hours). The results are given as the following ordered pairs.

(3, 61), (6, 56), (9, 53), (12, 55), (15, 48), (18, 35), (21, 36), (24, 33), (27, 44), (30, 23)

(a) Create a scatter plot of the data.

(b) Does the relationship between x and y appear to be approximately linear? If not, give some possible explanations.

139. *Falling Object*   In an experiment, students measured the speed s (in meters per second) of a ball t seconds after it was released. The results are shown in the table.

Time, t	Speed, s
0	0
1	11.0
2	19.4
3	29.2
4	39.4

(a) Sketch a scatter plot of the data.

(b) Find the equation of the line that seems to fit the data best.

(c) Use the *regression* feature of a graphing utility to find a linear model for the data and identify the correlation coefficient. Compare this model with the model from part (b).

(d) Use the model from part (c) to estimate the speed of the ball after 2.5 seconds.

140. *Sports*   The following ordered pairs (x, y) represent the Olympic year x and the winning time y (in minutes) in the men's 400-meter freestyle swimming event. (Source: *The World Almanac* 2005)

(1964, 4.203)	(1980, 3.855)	(1996, 3.800)
(1968, 4.150)	(1984, 3.854)	(2000, 3.677)
(1972, 4.005)	(1988, 3.783)	(2004, 3.718)
(1976, 3.866)	(1992, 3.750)	

(a) Use the *regression* feature of a graphing utility to find a linear model for the data. Let x represent the year, with $x = 4$ corresponding to 1964. Identify the correlation coefficient for the model.

(b) Use a graphing utility to create a scatter plot of the data.

(c) Graph the model with the data in the same viewing window.

(d) Does the model appear to be a good fit for the data? Explain.

(e) Would this model be appropriate for predicting the winning times in future Olympics? Explain.

Height   In Exercises 141–144, the following ordered pairs (x, y) represent the percent y of women between the ages of 20 and 29 who are under a certain height x (in feet). (Source: U.S. National Center for Health Statistics)

(4.67, 0.6)	(5.17, 21.8)	(5.67, 92.4)
(4.75, 0.7)	(5.25, 34.3)	(5.75, 96.2)
(4.83, 1.2)	(5.33, 48.9)	(5.83, 98.6)
(4.92, 3.1)	(5.42, 62.7)	(5.92, 99.5)
(5.00, 6.0)	(5.50, 74.0)	(6.00, 100.0)
(5.08, 11.5)	(5.58, 84.7)	

141. Use the *regression* feature of a graphing utility to find a linear model for the data.

142. Use a graphing utility to plot the data and graph the model in the same viewing window.

143. How closely does the model fit the data?

144. Can the model be used to estimate the percent of women who are under a height of 6.3 feet?

Synthesis

True or False?   In Exercises 145–148, determine whether the statement is true or false. Justify your answer.

145. If the graph of the parent function $f(x) = x^2$ is moved six units to the right, moved three units upward, and reflected in the x-axis, then the point $(-1, 28)$ will lie on the graph of the transformation.

146. If $f(x) = x^n$ where n is odd, f^{-1} exists.

147. There exists no function f such that $f = f^{-1}$.

148. The sign of the slope of a regression line is always positive.

88   Chapter 1   Functions and Their Graphs

1   Chapter Test   See www.CalcChat.com for worked-out solutions to odd-numbered exercises.

Take this test as you would take a test in class. After you are finished, check your work against the answers in the back of the book.

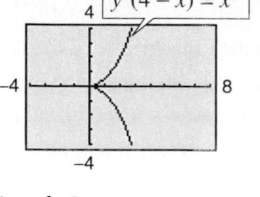

Figure for 3

1. Find the equations of the lines that pass through the point $(0, 4)$ and are (a) parallel to and (b) perpendicular to the line $5x + 2y = 3$.

2. Find the slope-intercept form of the equation of the line that passes through the points $(2, -1)$ and $(-3, 4)$.

3. Does the graph at the right represent y as a function of x? Explain.

4. Evaluate $f(x) = |x + 2| - 15$ at each value of the independent variable and simplify.

   (a) $f(-8)$ (b) $f(14)$ (c) $f(t - 6)$

5. Find the domain of $f(x) = 10 - \sqrt{3 - x}$.

6. An electronics company produces a car stereo for which the variable cost is $5.60 and the fixed costs are $24,000. The product sells for $99.50. Write the total cost C as a function of the number of units produced and sold, x. Write the profit P as a function of the number of units produced and sold, x.

In Exercises 7 and 8, determine algebraically whether the function is even, odd, or neither.

7. $f(x) = 2x^3 - 3x$ 8. $f(x) = 3x^4 + 5x^2$

In Exercises 9 and 10, determine the open intervals on which the function is increasing, decreasing, or constant.

9. $h(x) = \frac{1}{4}x^4 - 2x^2$ 10. $g(t) = |t + 2| - |t - 2|$

In Exercises 11 and 12, use a graphing utility to graph the functions and to approximate (to two decimal places) any relative minimum or relative maximum values of the function.

11. $f(x) = -x^3 - 5x^2 + 12$ 12. $f(x) = x^5 - x^3 + 2$

In Exercises 13–15, (a) identify the parent function f, (b) describe the sequence of transformations from f to g, and (c) sketch the graph of g.

13. $g(x) = -2(x - 5)^3 + 3$ 14. $g(x) = \sqrt{-x - 7}$ 15. $g(x) = 4|-x| - 7$

16. Use the functions $f(x) = x^2$ and $g(x) = \sqrt{2 - x}$ to find the specified function and its domain.

   (a) $(f - g)(x)$ (b) $\left(\dfrac{f}{g}\right)(x)$ (c) $(f \circ g)(x)$ (d) $(g \circ f)(x)$

In Exercises 17–19, determine whether the function has an inverse function, and if so, find the inverse function.

17. $f(x) = x^3 + 8$ 18. $f(x) = x^2 + 6$ 19. $f(x) = \dfrac{3x\sqrt{x}}{8}$

20. The table shows the numbers of cellular phone subscribers S (in millions) in the United States from 1999 through 2004, where t represents the year, with $t = 9$ corresponding to 1999. Use the *regression* feature of a graphing utility to find a linear model for the data. Use the model to find the year in which the number of subscribers exceeded 200 million. (Source: Cellular Telecommunications & Internet Association)

Year, t	Subscribers, S
9	86.0
10	109.5
11	128.4
12	140.8
13	158.7
14	182.1

Table for 20

Proofs in Mathematics

Conditional Statements

Many theorems are written in the **if-then form** "if p, then q," which is denoted by

$p \rightarrow q$ Conditional statement

where p is the **hypothesis** and q is the **conclusion.** Here are some other ways to express the conditional statement $p \rightarrow q$.

p implies q. p, only if q. p is sufficient for q.

Conditional statements can be either true or false. The conditional statement $p \rightarrow q$ is false only when p is true and q is false. To show that a conditional statement is true, you must prove that the conclusion follows for all cases that fulfill the hypothesis. To show that a conditional statement is false, you need only to describe a single **counterexample** that shows that the statement is not always true.

For instance, $x = -4$ is a counterexample that shows that the following statement is false.

If $x^2 = 16$, then $x = 4$.

The hypothesis "$x^2 = 16$" is true because $(-4)^2 = 16$. However, the conclusion "$x = 4$" is false. This implies that the given conditional statement is false.

For the conditional statement $p \rightarrow q$, there are three important associated conditional statements.

1. The **converse** of $p \rightarrow q$: $q \rightarrow p$
2. The **inverse** of $p \rightarrow q$: $\sim p \rightarrow \sim q$
3. The **contrapositive** of $p \rightarrow q$: $\sim q \rightarrow \sim p$

The symbol $\sim$ means the **negation** of a statement. For instance, the negation of "The engine is running" is "The engine is not running."

Example 1   Writing the Converse, Inverse, and Contrapositive

Write the converse, inverse, and contrapositive of the conditional statement "If I get a B on my test, then I will pass the course."

Solution

a. *Converse:* If I pass the course, then I got a B on my test.

b. *Inverse:* If I do not get a B on my test, then I will not pass the course.

c. *Contrapositive:* If I do not pass the course, then I did not get a B on my test.

In the example above, notice that neither the converse nor the inverse is logically equivalent to the original conditional statement. On the other hand, the contrapositive *is* logically equivalent to the original conditional statement.

Biconditional Statements

Recall that a conditional statement is a statement of the form "if p, then q." A statement of the form "p if and only if q" is called a **biconditional statement.** A biconditional statement, denoted by

$p \leftrightarrow q$ Biconditional statement

is the conjunction of the conditional statement $p \rightarrow q$ and its converse $q \rightarrow p$.

A biconditional statement can be either true or false. To be true, *both* the conditional statement and its converse must be true.

Example 2   Analyzing a Biconditional Statement

Consider the statement $x = 3$ if and only if $x^2 = 9$.

a. Is the statement a biconditional statement?   **b.** Is the statement true?

Solution

a. The statement is a biconditional statement because it is of the form "p if and only if q."

b. The statement can be rewritten as the following conditional statement and its converse.

> *Conditional statement:* If $x = 3$, then $x^2 = 9$.
> *Converse:* If $x^2 = 9$, then $x = 3$.

The first of these statements is true, but the second is false because x could also equal -3. So, the biconditional statement is false.

Knowing how to use biconditional statements is an important tool for reasoning in mathematics.

Example 3   Analyzing a Biconditional Statement

Determine whether the biconditional statement is true or false. If it is false, provide a counterexample.

A number is divisible by 5 if and only if it ends in 0.

Solution

The biconditional statement can be rewritten as the following conditional statement and its converse.

> *Conditional statement:* If a number is divisible by 5, then it ends in 0.
> *Converse:* If a number ends in 0, then it is divisible by 5.

The conditional statement is false. A counterexample is the number 15, which is divisible by 5 but does not end in 0. So, the biconditional statement is false.

Chapter 2

Polynomial and Rational Functions

2.1 Quadratic Functions

2.2 Polynomial Functions of Higher Degree

2.3 Real Zeros of Polynomial Functions

2.4 Complex Numbers

2.5 The Fundamental Theorem of Algebra

2.6 Rational Functions and Asymptotes

2.7 Graphs of Rational Functions

2.8 Quadratic Models

Selected Applications

Polynomial and rational functions have many real-life applications. The applications listed below represent a small sample of the applications in this chapter.

- Automobile Aerodynamics, Exercise 58, page 101
- Revenue, Exercise 93, page 114
- U.S. Population, Exercise 91, page 129
- Impedance, Exercises 79 and 80, page 138
- Profit, Exercise 64, page 145
- Data Analysis, Exercises 41 and 42, page 154
- Wildlife, Exercise 43, page 155
- Comparing Models, Exercise 85, page 164
- Media, Exercise 18, page 170

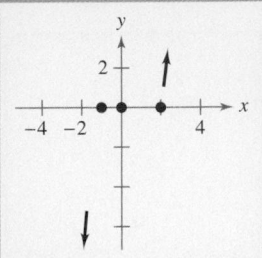 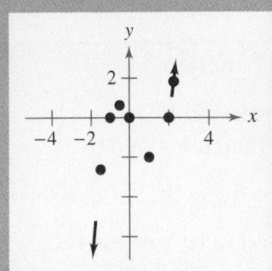 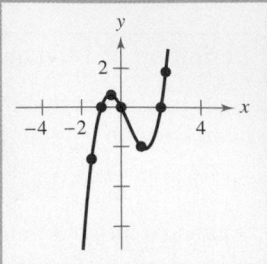

Polynomial and rational functions are two of the most common types of functions used in algebra and calculus. In Chapter 2, you will learn how to graph these types of functions and how to find the zeros of these functions.

David Madison/Getty Images

Aerodynamics is crucial in creating racecars. Two types of racecars designed and built by NASCAR teams are short track cars, as shown in the photo, and super-speedway (long track) cars. Both types of racecars are designed either to allow for as much downforce as possible or to reduce the amount of drag on the racecar.

2.1 Quadratic Functions

The Graph of a Quadratic Function

In this and the next section, you will study the graphs of polynomial functions.

> ### Definition of Polynomial Function
>
> Let n be a nonnegative integer and let $a_n, a_{n-1}, \ldots, a_2, a_1, a_0$ be real numbers with $a_n \neq 0$. The function given by
>
> $$f(x) = a_n x^n + a_{n-1} x^{n-1} + \cdots + a_2 x^2 + a_1 x + a_0$$
>
> is called a **polynomial function in x of degree n.**

Polynomial functions are classified by degree. For instance, the polynomial function

$$f(x) = a \qquad \text{Constant function}$$

has degree 0 and is called a **constant function.** In Chapter 1, you learned that the graph of this type of function is a horizontal line. The polynomial function

$$f(x) = mx + b, \quad m \neq 0 \qquad \text{Linear function}$$

has degree 1 and is called a **linear function.** You also learned in Chapter 1 that the graph of the linear function $f(x) = mx + b$ is a line whose slope is m and whose y-intercept is $(0, b)$. In this section, you will study second-degree polynomial functions, which are called **quadratic functions.**

> ### Definition of Quadratic Function
>
> Let a, b, and c be real numbers with $a \neq 0$. The function given by
>
> $$f(x) = ax^2 + bx + c \qquad \text{Quadratic function}$$
>
> is called a **quadratic function.**

Often real-life data can be modeled by quadratic functions. For instance, the table at the right shows the height h (in feet) of a projectile fired from a height of 6 feet with an initial velocity of 256 feet per second at any time t (in seconds). A quadratic model for the data in the table is $h(t) = -16t^2 + 256t + 6$ for $0 \leq t \leq 16$.

The graph of a quadratic function is a special type of U-shaped curve called a **parabola.** Parabolas occur in many real-life applications, especially those involving reflective properties, such as satellite dishes or flashlight reflectors. You will study these properties in a later chapter.

All parabolas are symmetric with respect to a line called the **axis of symmetry,** or simply the **axis** of the parabola. The point where the axis intersects the parabola is called the **vertex** of the parabola.

What you should learn

- Analyze graphs of quadratic functions.
- Write quadratic functions in standard form and use the results to sketch graphs of functions.
- Find minimum and maximum values of quadratic functions in real-life applications.

Why you should learn it

Quadratic functions can be used to model the design of a room. For instance, Exercise 53 on page 100 shows how the size of an indoor fitness room with a running track can be modeled.

Dwight Cendrowski

t	h
0	6
2	454
4	774
6	966
8	1030
10	966
12	774
14	454
16	6

Library of Parent Functions: Quadratic Function

The simplest type of *quadratic function* is $f(x) = ax^2$, also known as the *squaring function* when $a = 1$. The basic characteristics of a quadratic function are summarized below. A review of quadratic functions can be found in the *Study Capsules*.

Graph of $f(x) = ax^2$, $a > 0$
Domain: $(-\infty, \infty)$ ⟷
Range: $[0, \infty)$ ↕
Intercept: $(0, 0)$
Decreasing on $(-\infty, 0)$
Increasing on $(0, \infty)$
Even function
Axis of symmetry: $x = 0$
Relative minimum or vertex: $(0, 0)$

Graph of $f(x) = ax^2$, $a < 0$
Domain: $(-\infty, \infty)$
Range: $(-\infty, 0]$
Intercept: $(0, 0)$
Increasing on $(-\infty, 0)$
Decreasing on $(0, \infty)$
Even function
Axis of symmetry: $x = 0$
Relative maximum or vertex: $(0, 0)$

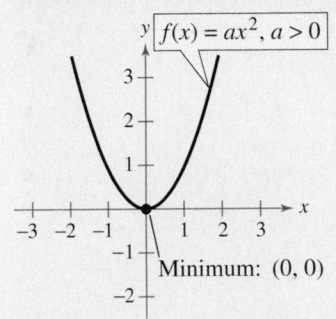

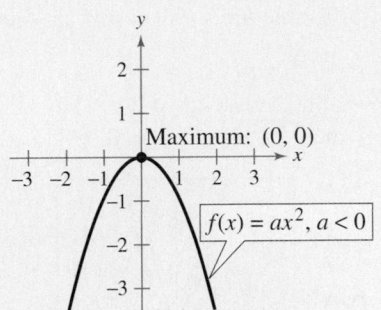

For the general quadratic form $f(x) = ax^2 + bx + c$, if the leading coefficient a is positive, the parabola opens upward; and if the leading coefficient a is negative, the parabola opens downward. Later in this section you will learn ways to find the coordinates of the vertex of a parabola.

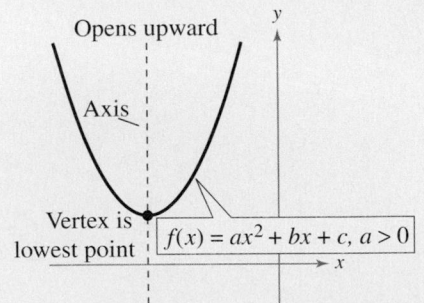

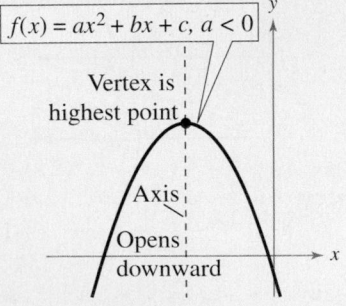

When sketching the graph of $f(x) = ax^2$, it is helpful to use the graph of $y = x^2$ as a reference, as discussed in Section 1.4. There you saw that when $a > 1$, the graph of $y = af(x)$ is a vertical stretch of the graph of $y = f(x)$. When $0 < a < 1$, the graph of $y = af(x)$ is a vertical shrink of the graph of $y = f(x)$. This is demonstrated again in Example 1.

Example 1   Graphing Simple Quadratic Functions

Describe how the graph of each function is related to the graph of $y = x^2$.

a. $f(x) = \dfrac{1}{3}x^2$ **b.** $g(x) = 2x^2$

c. $h(x) = -x^2 + 1$ **d.** $k(x) = (x + 2)^2 - 3$

Solution

a. Compared with $y = x^2$, each output of f "shrinks" by a factor of $\frac{1}{3}$. The result is a parabola that opens upward and is broader than the parabola represented by $y = x^2$, as shown in Figure 2.1.

b. Compared with $y = x^2$, each output of g "stretches" by a factor of 2, creating a narrower parabola, as shown in Figure 2.2.

c. With respect to the graph of $y = x^2$, the graph of h is obtained by a *reflection* in the x-axis and a vertical shift one unit *upward*, as shown in Figure 2.3.

d. With respect to the graph of $y = x^2$, the graph of k is obtained by a horizontal shift two units *to the left* and a vertical shift three units *downward*, as shown in Figure 2.4.

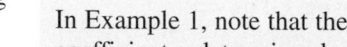

STUDY TIP

In Example 1, note that the coefficient a determines how widely the parabola given by $f(x) = ax^2$ opens. If $|a|$ is small, the parabola opens more widely than if $|a|$ is large.

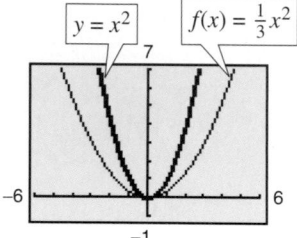

Figure 2.1

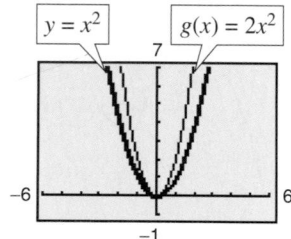

Figure 2.2

Prerequisite Skills

If you have difficulty with this example, review shifting, reflecting, and stretching of graphs in Section 1.4.

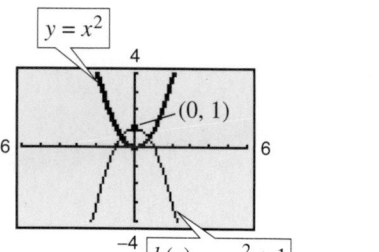

Figure 2.3

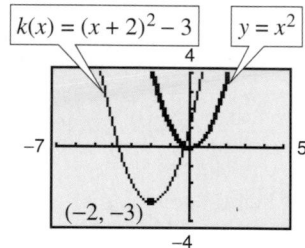

Figure 2.4

✓CHECKPOINT   Now try Exercise 5.

Recall from Section 1.4 that the graphs of $y = f(x \pm c)$, $y = f(x) \pm c$, $y = -f(x)$, and $y = f(-x)$ are rigid transformations of the graph of $y = f(x)$.

$y = f(x \pm c)$ Horizontal shift $y = -f(x)$ Reflection in x-axis

$y = f(x) \pm c$ Vertical shift $y = f(-x)$ Reflection in y-axis

The Standard Form of a Quadratic Function

The equation in Example 1(d) is written in the **standard form**

$$f(x) = a(x - h)^2 + k.$$

This form is especially convenient for sketching a parabola because it identifies the vertex of the parabola as (h, k).

> **Standard Form of a Quadratic Function**
>
> The quadratic function given by
>
> $$f(x) = a(x - h)^2 + k, \qquad a \neq 0$$
>
> is in **standard form.** The graph of f is a parabola whose axis is the vertical line $x = h$ and whose vertex is the point (h, k). If $a > 0$, the parabola opens upward, and if $a < 0$, the parabola opens downward.

Exploration

Use a graphing utility to graph $y = ax^2$ with $a = -2, -1,$ $-0.5, 0.5, 1,$ and 2. How does changing the value of a affect the graph?

Use a graphing utility to graph $y = (x - h)^2$ with $h = -4, -2, 2,$ and 4. How does changing the value of h affect the graph?

Use a graphing utility to graph $y = x^2 + k$ with $k = -4, -2, 2,$ and 4. How does changing the value of k affect the graph?

Example 2   Identifying the Vertex of a Quadratic Function

Describe the graph of $f(x) = 2x^2 + 8x + 7$ and identify the vertex.

Solution

Write the quadratic function in standard form by completing the square. Recall that the first step is to factor out any coefficient of x^2 that is not 1.

$$f(x) = 2x^2 + 8x + 7 \qquad \text{Write original function.}$$

$$= (2x^2 + 8x) + 7 \qquad \text{Group } x\text{-terms.}$$

$$= 2(x^2 + 4x) + 7 \qquad \text{Factor 2 out of } x\text{-terms.}$$

$$= 2(x^2 + 4x + 4 - 4) + 7 \qquad \underline{\text{Add and subtract } (4/2)^2 = 4 \text{ within}}$$
$$\text{parentheses to complete the square.}$$

$$\left(\frac{4}{2}\right)^2$$

$$= 2(x^2 + 4x + 4) - 2(4) + 7 \qquad \text{Regroup terms.}$$

$$= 2(x + 2)^2 - 1 \qquad \text{Write in standard form.}$$

why?

From the standard form, you can see that the graph of f is a parabola that opens upward with vertex $(-2, -1)$, as shown in Figure 2.5. This corresponds to a left shift of two units and a downward shift of one unit relative to the graph of $y = 2x^2$.

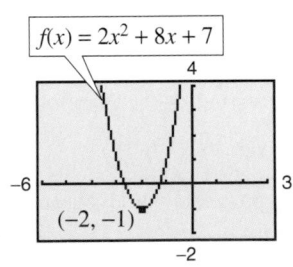

$f(x) = 2x^2 + 8x + 7$

$(-2, -1)$

Figure 2.5

✓*CHECKPOINT*   Now try Exercise 13.

To find the x-intercepts of the graph of $f(x) = ax^2 + bx + c$, solve the equation $ax^2 + bx + c = 0$. If $ax^2 + bx + c$ does not factor, you can use the Quadratic Formula to find the x-intercepts, or a graphing utility to approximate the x-intercepts. Remember, however, that a parabola may not have x-intercepts.

Example 3   Identifying *x*-Intercepts of a Quadratic Function

Describe the graph of $f(x) = -x^2 + 6x - 8$ and identify any *x*-intercepts.

Solution

$$f(x) = -x^2 + 6x - 8$$ Write original function.

$$= -(x^2 - 6x) - 8$$ Factor -1 out of *x*-terms.

$$= -(x^2 - 6x + 9 - 9) - 8$$ Because $b = 6$, add and subtract $(6/2)^2 = 9$ within parentheses.

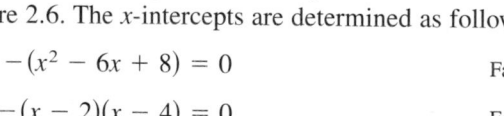

$$= -(x^2 - 6x + 9) - (-9) - 8$$ Regroup terms.

$$= -(x - 3)^2 + 1$$ Write in standard form.

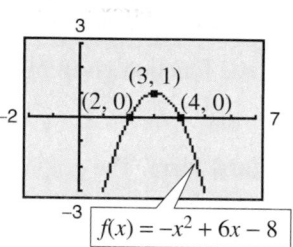

Figure 2.6

The graph of *f* is a parabola that opens downward with vertex $(3, 1)$, as shown in Figure 2.6. The *x*-intercepts are determined as follows.

$$-(x^2 - 6x + 8) = 0$$ Factor out -1.

$$-(x - 2)(x - 4) = 0$$ Factor.

$$x - 2 = 0 \quad \Longrightarrow \quad x = 2$$ Set 1st factor equal to 0.

$$x - 4 = 0 \quad \Longrightarrow \quad x = 4$$ Set 2nd factor equal to 0.

So, the *x*-intercepts are $(2, 0)$ and $(4, 0)$, as shown in Figure 2.6.

✓CHECKPOINT   Now try Exercise 17.

Example 4   Writing the Equation of a Parabola in Standard Form

Write the standard form of the equation of the parabola whose vertex is $(1, 2)$ and that passes through the point $(3, -6)$.

Solution

Because the vertex of the parabola is $(h, k) = (1, 2)$, the equation has the form

$$f(x) = a(x - 1)^2 + 2.$$ Substitute for *h* and *k* in standard form.

Because the parabola passes through the point $(3, -6)$, it follows that $f(3) = -6$. So, you obtain

$$-6 = a(3 - 1)^2 + 2$$

$$-6 = 4a + 2$$

$$-2 = a.$$

The equation in standard form is $f(x) = -2(x - 1)^2 + 2$. You can confirm this answer by graphing $f(x) = -2(x - 1)^2 + 2$ with a graphing utility, as shown in Figure 2.7 Use the *zoom* and *trace* features or the *maximum* and *value* features to confirm that its vertex is $(1, 2)$ and that it passes through the point $(3, -6)$.

✓CHECKPOINT   Now try Exercise 29.

STUDY TIP

In Example 4, there are infinitely many different parabolas that have a vertex at $(1, 2)$. Of these, however, the only one that passes through the point $(3, -6)$ is the one given by

$$f(x) = -2(x - 1)^2 + 2.$$

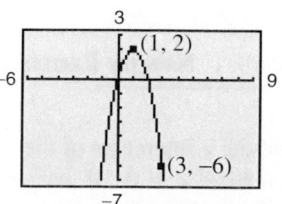

Figure 2.7

Finding Minimum and Maximum Values

Many applications involve finding the maximum or minimum value of a quadratic function. By completing the square of the quadratic function $f(x) = ax^2 + bx + c$, you can rewrite the function in standard form.

$$f(x) = a\left(x + \frac{b}{2a}\right)^2 + \cdot\left(c - \frac{b^2}{4a}\right) \qquad \text{Standard form}$$

You can see that the vertex occurs at $x = -b/(2a)$, which implies the following.

Minimum and Maximum Values of Quadratic Functions

1. If $a > 0$, f has a *minimum* value at $x = -\dfrac{b}{2a}$.

2. If $a < 0$, f has a *maximum* value at $x = -\dfrac{b}{2a}$.

Example 5   The Maximum Height of a Baseball

A baseball is hit at a point 3 feet above the ground at a velocity of 100 feet per second and at an angle of 45° with respect to the ground. The path of the baseball is given by the function $f(x) = -0.0032x^2 + x + 3$, where $f(x)$ is the height of the baseball (in feet) and x is the horizontal distance from home plate (in feet). What is the maximum height reached by the baseball?

Algebraic Solution

For this quadratic function, you have

$$f(x) = ax^2 + bx + c = -0.0032x^2 + x + 3$$

which implies that $a = -0.0032$ and $b = 1$. Because the function has a maximum when $x = -b/(2a)$, you can conclude that the baseball reaches its maximum height when it is x feet from home plate, where x is

$$x = -\frac{b}{2a} = -\frac{1}{2(-0.0032)}$$

$$= 156.25 \text{ feet.}$$

At this distance, the maximum height is

$$f(156.25) = -0.0032(156.25)^2 + 156.25 + 3$$

$$= 81.125 \text{ feet.}$$

✔CHECKPOINT   Now try Exercise 55.

Graphical Solution

Use a graphing utility to graph $y = -0.0032x^2 + x + 3$ so that you can see the important features of the parabola. Use the *maximum* feature (see Figure 2.8) or the *zoom* and *trace* features (see Figure 2.9) of the graphing utility to approximate the maximum height on the graph to be $y \approx 81.125$ feet at $x \approx 156.25$.

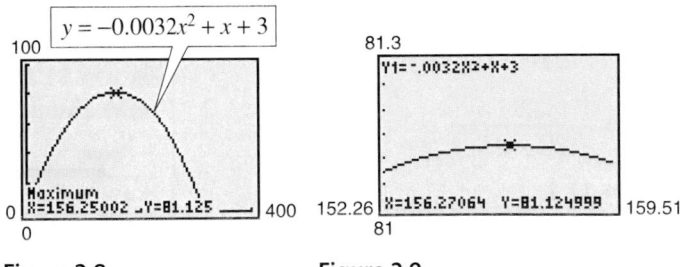

Figure 2.8   Figure 2.9

> **TECHNOLOGY TIP**
>
> Note in the graphical solution for Example 5, that when using the *zoom* and *trace* features, you might have to change the y-scale in order to avoid a graph that is "too flat."

TECHNOLOGY SUPPORT   For instructions on how to use the *maximum*, the *minimum*, the *table*, and the *zoom* and *trace* features, see Appendix A; for specific keystrokes, go to this textbooks's *Online Study Center*.

Example 6   Cost

A soft drink manufacturer has daily production costs of

$$C(x) = 70{,}000 - 120x + 0.055x^2$$

where C is the total cost (in dollars) and x is the number of units produced. Estimate numerically the number of units that should be produced each day to yield a minimum cost.

X	Y₁
800	9200
900	6550
1000	5000
1100	4550
1200	5200
1300	6950
1400	9800

X=1100

Figure 2.10

Solution

Enter the function $y = 70{,}000 - 120x + 0.055x^2$ into your graphing utility. Then use the *table* feature of the graphing utility to create a table. Set the table to start at $x = 0$ and set the table step to 100. By scrolling through the table you can see that the minimum cost is between 1000 units and 1200 units, as shown in Figure 2.10. You can improve this estimate by starting the table at $x = 1000$ and setting the table step to 10. From the table in Figure 2.11, you can see that approximately 1090 units should be produced to yield a minimum cost of \$4545.50.

✓CHECKPOINT   Now try Exercise 57.

X	Y₁
1060	4598
1070	4569.5
1080	4552
1090	4545.5
1100	4550
1110	4565.5
1120	4592

X=1090

Figure 2.11

Example 7   Grants

The numbers g of grants awarded from the National Endowment for the Humanities fund from 1999 to 2003 can be approximated by the model

$$g(t) = -99.14t^2 + 2{,}201.1t - 10{,}896, \qquad 9 \le t \le 13$$

where t represents the year, with $t = 9$ corresponding to 1999. Using this model, determine the year in which the number of grants awarded was greatest. (Source: U.S. National Endowment for the Arts)

Algebraic Solution

Use the fact that the maximum point of the parabola occurs when $t = -b/(2a)$. For this function, you have $a = -99.14$ and $b = 2201.1$. So,

$$t = -\frac{b}{2a}$$

$$= -\frac{2201.1}{2(-99.14)}$$

$$\approx 11.1.$$

From this t-value and the fact that $t = 9$ represents 1999, you can conclude that the greatest number of grants were awarded during 2001.

✓CHECKPOINT   Now try Exercise 61.

Graphical Solution

Use a graphing utility to graph

$$y = -99.14x^2 + 2{,}201.1x - 10{,}896$$

for $9 \le x \le 13$, as shown in Figure 2.12. Use the *maximum* feature (see Figure 2.12) or the *zoom* and *trace* features (see Figure 2.13) of the graphing utility to approximate the maximum point of the parabola to be $x \approx 11.1$. So, you can conclude that the greatest number of grants were awarded during 2001.

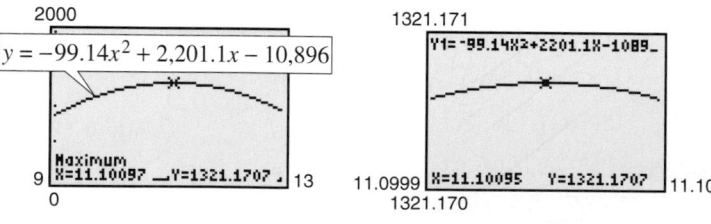

Figure 2.12   **Figure 2.13**

2.1 Exercises

Vocabulary Check

Fill in the blanks.

1. A polynomial function of degree n and leading coefficient a_n is a function of the form
$$f(x) = a_n x^n + a_{n-1}x^{n-1} + \cdots + a_2 x^2 + a_1 x + a_0, \ a_n \neq 0$$
where n is a _____ and $a_n, a_{n-1}, \ldots, a_2, a_1, a_0$ are _____ numbers.

2. A _____ function is a second-degree polynomial function, and its graph is called a _____ .

3. The graph of a quadratic function is symmetric about its _____ .

4. If the graph of a quadratic function opens upward, then its leading coefficient is _____ and the vertex of the graph is a _____ .

5. If the graph of a quadratic function opens downward, then its leading coefficient is _____ and the vertex of the graph is a _____ .

In Exercises 1–4, match the quadratic function with its graph. [The graphs are labeled (a), (b), (c), and (d).]

(a)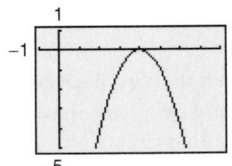

(b)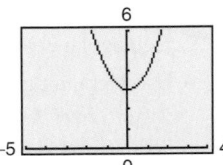

(c)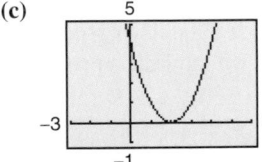

(d)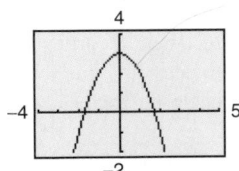

1. $f(x) = (x - 2)^2$
2. $f(x) = 3 - x^2$
3. $f(x) = x^2 + 3$
4. $f(x) = -(x - 4)^2$

In Exercises 5 and 6, use a graphing utility to graph each function in the same viewing window. Describe how the graph of each function is related to the graph of $y = x^2$.

5. (a) $y = \frac{1}{2}x^2$ (b) $y = \frac{1}{2}x^2 - 1$
   (c) $y = \frac{1}{2}(x + 3)^2$ (d) $y = -\frac{1}{2}(x + 3)^2 - 1$
6. (a) $y = \frac{3}{2}x^2$ (b) $y = \frac{3}{2}x^2 + 1$
   (c) $y = \frac{3}{2}(x - 3)^2$ (d) $y = -\frac{3}{2}(x - 3)^2 + 1$

In Exercises 7–20, sketch the graph of the quadratic function. Identify the vertex and x-intercept(s). Use a graphing utility to verify your results.

7. $f(x) = 25 - x^2$
8. $f(x) = x^2 - 7$
9. $f(x) = \frac{1}{2}x^2 - 4$
10. $f(x) = 16 - \frac{1}{4}x^2$

11. $f(x) = (x + 4)^2 - 3$
12. $f(x) = (x - 6)^2 + 3$
13. $h(x) = x^2 - 8x + 16$
14. $g(x) = x^2 + 2x + 1$
15. $f(x) = x^2 - x + \frac{5}{4}$
16. $f(x) = x^2 + 3x + \frac{1}{4}$
17. $f(x) = -x^2 + 2x + 5$
18. $f(x) = -x^2 - 4x + 1$
19. $h(x) = 4x^2 - 4x + 21$
20. $f(x) = 2x^2 - x + 1$

In Exercises 21–26, use a graphing utility to graph the quadratic function. Identify the vertex and x-intercept(s). Then check your results algebraically by writing the quadratic function in standard form.

21. $f(x) = -(x^2 + 2x - 3)$
22. $f(x) = -(x^2 + x - 30)$
23. $g(x) = x^2 + 8x + 11$
24. $f(x) = x^2 + 10x + 14$
25. $f(x) = -2x^2 + 16x - 31$
26. $f(x) = -4x^2 + 24x - 41$

In Exercises 27 and 28, write an equation for the parabola in standard form. Use a graphing utility to graph the equation and verify your result.

27.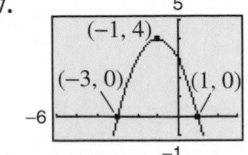

28.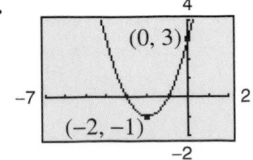

In Exercises 29–34, write the standard form of the quadratic function that has the indicated vertex and whose graph passes through the given point. Verify your result with a graphing utility.

29. Vertex: $(-2, 5)$;   Point: $(0, 9)$

30. Vertex: $(4, 1)$;   Point: $(6, -7)$

31. Vertex: $(1, -2)$;   Point: $(-1, 14)$

32. Vertex: $(-4, -1)$;   Point: $(-2, 4)$

33. Vertex: $\left(\frac{1}{2}, 1\right)$;   Point: $\left(-2, -\frac{21}{5}\right)$

34. Vertex: $\left(-\frac{1}{4}, -1\right)$;   Point: $\left(0, -\frac{17}{16}\right)$

Graphical Reasoning   In Exercises 35–38, determine the *x*-intercept(s) of the graph visually. How do the *x*-intercepts correspond to the solutions of the quadratic equation when $y = 0$?

35.
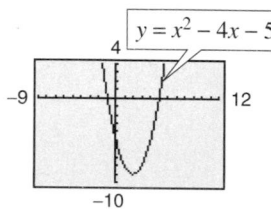
$y = x^2 - 4x - 5$

36.
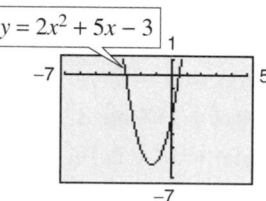
$y = 2x^2 + 5x - 3$

37.
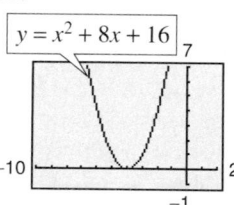
$y = x^2 + 8x + 16$

38.
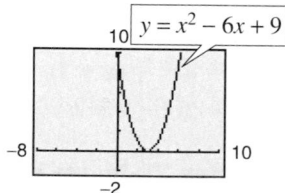
$y = x^2 - 6x + 9$

In Exercises 39–44, use a graphing utility to graph the quadratic function. Find the *x*-intercepts of the graph and compare them with the solutions of the corresponding quadratic equation when $y = 0$.

39. $y = x^2 - 4x$

40. $y = -2x^2 + 10x$

41. $y = 2x^2 - 7x - 30$

42. $y = 4x^2 + 25x - 21$

43. $y = -\frac{1}{2}(x^2 - 6x - 7)$

44. $y = \frac{7}{10}(x^2 + 12x - 45)$

In Exercises 45–48, find two quadratic functions, one that opens upward and one that opens downward, whose graphs have the given *x*-intercepts. (There are many correct answers.)

45. $(-1, 0), (3, 0)$

46. $(0, 0), (10, 0)$

47. $(-3, 0), \left(-\frac{1}{2}, 0\right)$

48. $\left(-\frac{5}{2}, 0\right), (2, 0)$

In Exercises 49–52, find two positive real numbers whose product is a maximum.

49. The sum is 110.

50. The sum is *S*.

51. The sum of the first and twice the second is 24.

52. The sum of the first and three times the second is 42.

53. *Geometry*   An indoor physical fitness room consists of a rectangular region with a semicircle on each end. The perimeter of the room is to be a 200-meter single-lane running track.

(a) Draw a diagram that illustrates the problem. Let *x* and *y* represent the length and width of the rectangular region, respectively.

(b) Determine the radius of the semicircular ends of the track. Determine the distance, in terms of *y*, around the inside edge of the two semicircular parts of the track.

(c) Use the result of part (b) to write an equation, in terms of *x* and *y*, for the distance traveled in one lap around the track. Solve for *y*.

(d) Use the result of part (c) to write the area *A* of the rectangular region as a function of *x*.

(e) Use a graphing utility to graph the area function from part (d). Use the graph to approximate the dimensions that will produce a rectangle of maximum area.

54. *Numerical, Graphical, and Analytical Analysis*   A rancher has 200 feet of fencing to enclose two adjacent rectangular corrals (see figure). Use the following methods to determine the dimensions that will produce a maximum enclosed area.

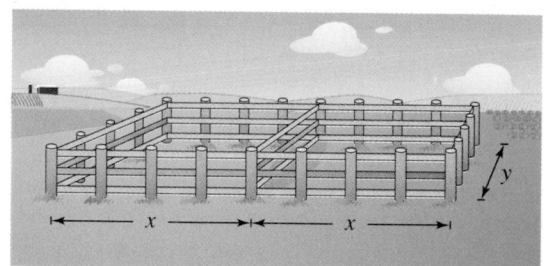

(a) Write the area *A* of the corral as a function of *x*.

(b) Use the *table* feature of a graphing utility to create a table showing possible values of *x* and the corresponding areas *A* of the corral. Use the table to estimate the dimensions that will produce the maximum enclosed area.

(c) Use a graphing utility to graph the area function. Use the graph to approximate the dimensions that will produce the maximum enclosed area.

(d) Write the area function in standard form to find algebraically the dimensions that will produce the maximum area.

(e) Compare your results from parts (b), (c), and (d).

55. *Height of a Ball*   The height y (in feet) of a punted football is approximated by

$$y = -\frac{16}{2025}x^2 + \frac{9}{5}x + \frac{3}{2}$$

where x is the horizontal distance (in feet) from where the football is punted.

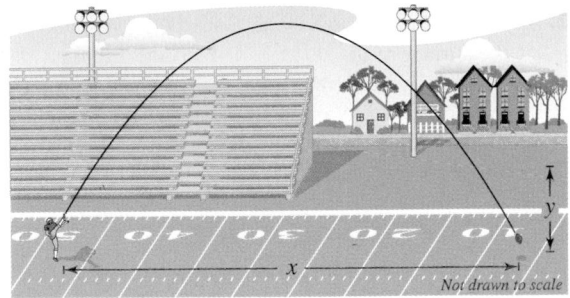

(a) Use a graphing utility to graph the path of the football.

(b) How high is the football when it is punted? (*Hint:* Find y when $x = 0$.)

(c) What is the maximum height of the football?

(d) How far from the punter does the football strike the ground?

56. *Path of a Diver*   The path of a diver is approximated by

$$y = -\frac{4}{9}x^2 + \frac{24}{9}x + 12$$

where y is the height (in feet) and x is the horizontal distance (in feet) from the end of the diving board (see figure). What is the maximum height of the diver? Verify your answer using a graphing utility.

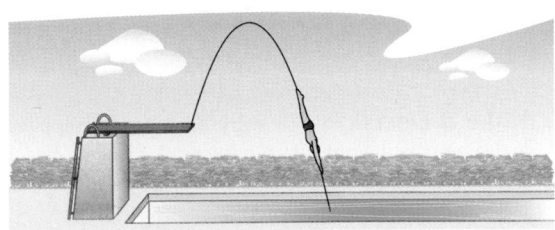

57. *Cost*   A manufacturer of lighting fixtures has daily production costs of

$$C(x) = 800 - 10x + 0.25x^2$$

where C is the total cost (in dollars) and x is the number of units produced. Use the *table* feature of a graphing utility to determine how many fixtures should be produced each day to yield a minimum cost.

58. *Automobile Aerodynamics*   The number of horsepower H required to overcome wind drag on a certain automobile is approximated by

$$H(s) = 0.002s^2 + 0.05s - 0.029, \quad 0 \le s \le 100$$

where s is the speed of the car (in miles per hour).

(a) Use a graphing utility to graph the function.

(b) Graphically estimate the maximum speed of the car if the power required to overcome wind drag is not to exceed 10 horsepower. Verify your result algebraically.

59. *Revenue*   The total revenue R (in thousands of dollars) earned from manufacturing and selling hand-held video games is given by

$$R(p) = -25p^2 + 1200p$$

where p is the price per unit (in dollars).

(a) Find the revenue when the price per unit is $20, $25, and $30.

(b) Find the unit price that will yield a maximum revenue.

(c) What is the maximum revenue?

(d) Explain your results.

60. *Revenue*   The total revenue R (in dollars) earned by a dog walking service is given by

$$R(p) = -12p^2 + 150p$$

where p is the price charged per dog (in dollars).

(a) Find the revenue when the price per dog is $4, $6, and $8.

(b) Find the price that will yield a maximum revenue.

(c) What is the maximum revenue?

(d) Explain your results.

61. *Graphical Analysis*   From 1960 to 2004, the annual per capita consumption C of cigarettes by Americans (age 18 and older) can be modeled by

$$C(t) = 4306 - 3.4t - 1.32t^2, \quad 0 \le t \le 44$$

where t is the year, with $t = 0$ corresponding to 1960. (Source: U.S. Department of Agriculture)

(a) Use a graphing utility to graph the model.

(b) Use the graph of the model to approximate the year when the maximum annual consumption of cigarettes occurred. Approximate the maximum average annual consumption. Beginning in 1966, all cigarette packages were required by law to carry a health warning. Do you think the warning had any effect? Explain.

(c) In 2000, the U.S. population (age 18 and older) was 209,117,000. Of those, about 48,306,000 were smokers. What was the average annual cigarette consumption *per smoker* in 2000? What was the average daily cigarette consumption *per smoker*?

62. *Data Analysis*   The factory sales S of VCRs (in millions of dollars) in the United States from 1990 to 2004 can be modeled by $S = -28.40t^2 + 218.1t + 2435$, for $0 \le t \le 14$, where t is the year, with $t = 0$ corresponding to 1990.   (Source: Consumer Electronics Association)

(a) According to the model, when did the maximum value of factory sales of VCRs occur?

(b) According to the model, what was the value of the factory sales in 2004? Explain your result.

(c) Would you use the model to predict the value of the factory sales for years beyond 2004? Explain.

Synthesis

True or False?   **In Exercises 63 and 64, determine whether the statement is true or false. Justify your answer.**

63. The function $f(x) = -12x^2 - 1$ has no x-intercepts.

64. The graphs of $f(x) = -4x^2 - 10x + 7$ and $g(x) = 12x^2 + 30x + 1$ have the same axis of symmetry.

Library of Parent Functions   **In Exercises 65 and 66, determine which equation(s) may be represented by the graph shown. (There may be more than one correct answer.)**

65. (a) $f(x) = -(x - 4)^2 + 2$

(b) $f(x) = -(x + 2)^2 + 4$

(c) $f(x) = -(x + 2)^2 - 4$

(d) $f(x) = -x^2 - 4x - 8$

(e) $f(x) = -(x - 2)^2 - 4$

(f) $f(x) = -x^2 + 4x - 8$

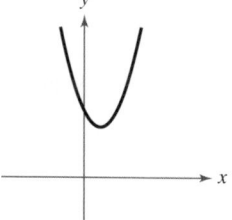

66. (a) $f(x) = (x - 1)^2 + 3$

(b) $f(x) = (x + 1)^2 + 3$

(c) $f(x) = (x - 3)^2 + 1$

(d) $f(x) = x^2 + 2x + 4$

(e) $f(x) = (x + 3)^2 + 1$

(f) $f(x) = x^2 + 6x + 10$

Think About It   **In Exercises 67–70, find the value of b such that the function has the given maximum or minimum value.**

67. $f(x) = -x^2 + bx - 75$; Maximum value: 25

68. $f(x) = -x^2 + bx - 16$; Maximum value: 48

69. $f(x) = x^2 + bx + 26$; Minimum value: 10

70. $f(x) = x^2 + bx - 25$; Minimum value: -50

71. *Profit*   The profit P (in millions of dollars) for a recreational vehicle retailer is modeled by a quadratic function of the form $P = at^2 + bt + c$, where t represents the year. If you were president of the company, which of the following models would you prefer? Explain your reasoning.

(a) a is positive and $t \ge -b/(2a)$.

(b) a is positive and $t \le -b/(2a)$.

(c) a is negative and $t \ge -b/(2a)$.

(d) a is negative and $t \le -b/(2a)$.

72. *Writing*   The parabola in the figure below has an equation of the form $y = ax^2 + bx - 4$. Find the equation of this parabola in two different ways, by hand and with technology (graphing utility or computer software). Write a paragraph describing the methods you used and comparing the results of the two methods.

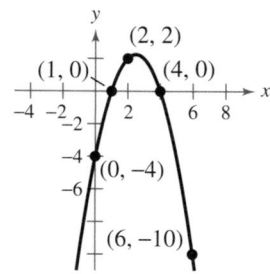

Skills Review

In Exercises 73–76, determine algebraically any point(s) of intersection of the graphs of the equations. Verify your results using the *intersect* feature of a graphing utility.

73. $x + y = 8$
   $-\frac{2}{3}x + y = 6$

74. $y = 3x - 10$
   $y = \frac{1}{4}x + 1$

75. $y = 9 - x^2$
   $y = x + 3$

76. $y = x^3 + 2x - 1$
   $y = -2x + 15$

77. *Make a Decision*   To work an extended application analyzing the height of a basketball after it has been dropped, visit this textbook's *Online Study Center.*

2.2 Polynomial Functions of Higher Degree

Graphs of Polynomial Functions

You should be able to sketch accurate graphs of polynomial functions of degrees 0, 1, and 2. The graphs of polynomial functions of degree greater than 2 are more difficult to sketch by hand. However, in this section you will learn how to recognize some of the basic features of the graphs of polynomial functions. Using these features along with point plotting, intercepts, and symmetry, you should be able to make reasonably accurate sketches *by hand*.

The graph of a polynomial function is **continuous.** Essentially, this means that the graph of a polynomial function has no breaks, holes, or gaps, as shown in Figure 2.14. Informally, you can say that a function is continuous if its graph can be drawn with a pencil without lifting the pencil from the paper.

What you should learn

■ Use transformations to sketch graphs of polynomial functions.

■ Use the Leading Coefficient Test to determine the end behavior of graphs of polynomial functions.

■ Find and use zeros of polynomial functions as sketching aids.

■ Use the Intermediate Value Theorem to help locate zeros of polynomial functions.

Why you should learn it

You can use polynomial functions to model various aspects of nature, such as the growth of a red oak tree, as shown in Exercise 94 on page 114.

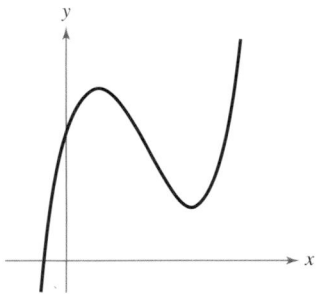

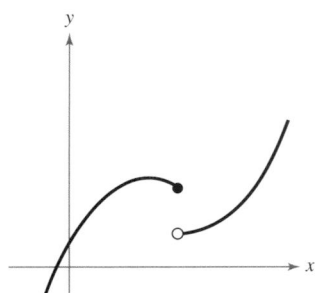

(a) Polynomial functions have continuous graphs.

(b) Functions with graphs that are not continuous are not polynomial functions.

Figure 2.14

Leonard Lee Rue III/Earth Scenes

Another feature of the graph of a polynomial function is that it has only smooth, rounded turns, as shown in Figure 2.15(a). It cannot have a sharp turn such as the one shown in Figure 2.15(b).

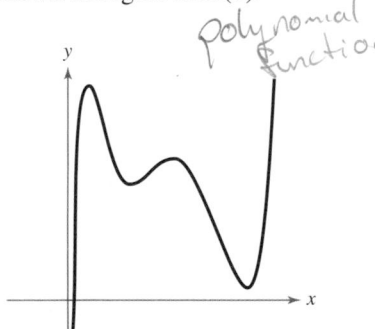

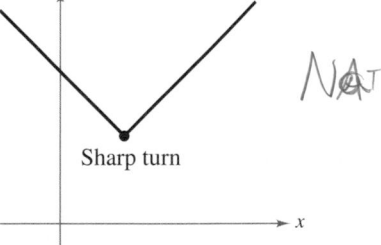

Sharp turn

(a) Polynomial functions have graphs with smooth, rounded turns.

(b) Functions with graphs that have sharp turns are not polynomial functions.

Figure 2.15

Library of Parent Functions: Polynomial Function

The graphs of polynomial functions of degree 1 are lines, and those of functions of degree 2 are parabolas. The graphs of all polynomial functions are smooth and continuous. A polynomial function of degree n has the form

$$f(x) = a_n x^n + a_{n-1} x^{n-1} + \cdots + a_2 x^2 + a_1 x + a_0$$

where n is a positive integer and $a_n \neq 0$. The polynomial functions that have the simplest graphs are monomials of the form $f(x) = x^n$, where n is an integer greater than zero. If n is even, the graph is similar to the graph of $f(x) = x^2$ and touches the axis at the x-intercept. If n is odd, the graph is similar to the graph of $f(x) = x^3$ and crosses the axis at the x-intercept. The greater the value of n, the flatter the graph near the origin. The basic characteristics of the *cubic function* $f(x) = x^3$ are summarized below. A review of polynomial functions can be found in the *Study Capsules*.

Graph of $f(x) = x^3$

Domain: $(-\infty, \infty)$
Range: $(-\infty, \infty)$
Intercept: $(0, 0)$
Increasing on $(-\infty, \infty)$
Odd function
Origin symmetry

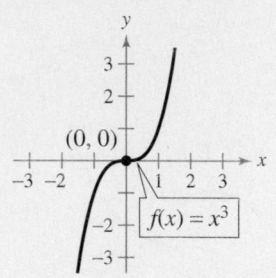

Exploration

Use a graphing utility to graph $y = x^n$ for $n = 2$, 4, and 8. (Use the viewing window $-1.5 \le x \le 1.5$ and $-1 \le y \le 6$.) Compare the graphs. In the interval $(-1, 1)$, which graph is on the bottom? Outside the interval $(-1, 1)$, which graph is on the bottom?

Use a graphing utility to graph $y = x^n$ for $n = 3$, 5, and 7. (Use the viewing window $-1.5 \le x \le 1.5$ and $-4 \le y \le 4$.) Compare the graphs. In the intervals $(-\infty, -1)$ and $(0, 1)$, which graph is on the bottom? In the intervals $(-1, 0)$ and $(1, \infty)$, which graph is on the bottom?

Example 1   Transformations of Monomial Functions

Sketch the graphs of (a) $f(x) = -x^5$, (b) $g(x) = x^4 + 1$, and (c) $h(x) = (x + 1)^4$.

Solution

a. Because the degree of $f(x) = -x^5$ is odd, the graph is similar to the graph of $y = x^3$. Moreover, the negative coefficient reflects the graph in the x-axis, as shown in Figure 2.16.

b. The graph of $g(x) = x^4 + 1$ is an upward shift of one unit of the graph of $y = x^4$, as shown in Figure 2.17.

c. The graph of $h(x) = (x + 1)^4$ is a left shift of one unit of the graph of $y = x^4$, as shown in Figure 2.18.

Prerequisite Skills

If you have difficulty with this example, review shifting and reflecting of graphs in Section 1.4.

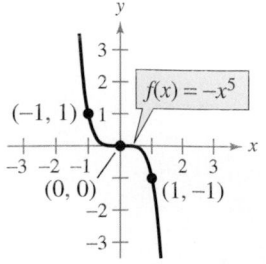

Figure 2.16

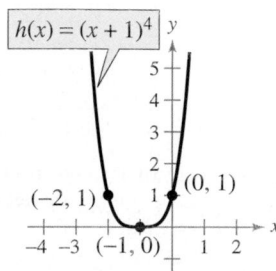

Figure 2.17   **Figure 2.18**

✓CHECKPOINT   Now try Exercise 9.

The Leading Coefficient Test

In Example 1, note that all three graphs eventually rise or fall without bound as x moves to the right. Whether the graph of a polynomial eventually rises or falls can be determined by the polynomial function's degree (even or odd) and by its leading coefficient, as indicated in the **Leading Coefficient Test.**

Leading Coefficient Test

As x moves without bound to the left or to the right, the graph of the polynomial function $f(x) = a_n x^n + \cdots + a_1 x + a_0$, $a_n \neq 0$, eventually rises or falls in the following manner.

1. When n is odd:

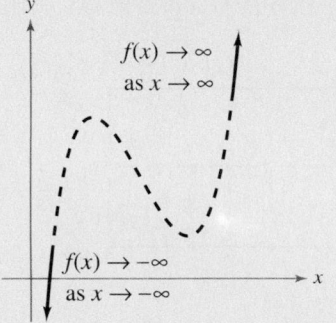

If the leading coefficient is positive $(a_n > 0)$, the graph falls to the left and rises to the right.

If the leading coefficient is negative $(a_n < 0)$, the graph rises to the left and falls to the right.

2. When n is even:

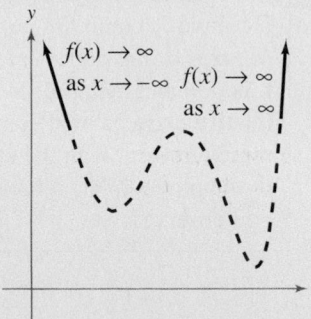
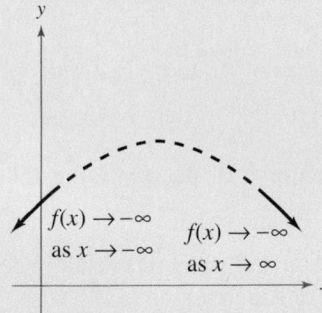

If the leading coefficient is positive $(a_n > 0)$, the graph rises to the left and right.

If the leading coefficient is negative $(a_n < 0)$, the graph falls to the left and right.

Note that the dashed portions of the graphs indicate that the test determines only the right-hand and left-hand behavior of the graph.

As you continue to study polynomial functions and their graphs, you will notice that the degree of a polynomial plays an important role in determining other characteristics of the polynomial function and its graph.

Exploration

For each function, identify the degree of the function and whether the degree of the function is even or odd. Identify the leading coefficient and whether the leading coefficient is positive or negative. Use a graphing utility to graph each function. Describe the relationship between the degree and sign of the leading coefficient of the function and the right- and left-hand behavior of the graph of the function.

a. $y = x^3 - 2x^2 - x + 1$
b. $y = 2x^5 + 2x^2 - 5x + 1$
c. $y = -2x^5 - x^2 + 5x + 3$
d. $y = -x^3 + 5x - 2$
e. $y = 2x^2 + 3x - 4$
f. $y = x^4 - 3x^2 + 2x - 1$
g. $y = -x^2 + 3x + 2$
h. $y = -x^6 - x^2 - 5x + 4$

STUDY TIP

The notation "$f(x) \to -\infty$ as $x \to -\infty$" indicates that the graph falls to the left. The notation "$f(x) \to \infty$ as $x \to \infty$" indicates that the graph rises to the right.

Example 2 Applying the Leading Coefficient Test

Use the Leading Coefficient Test to describe the right-hand and left-hand behavior of the graph of each polynomial function.

a. $f(x) = -x^3 + 4x$ **b.** $f(x) = x^4 - 5x^2 + 4$ **c.** $f(x) = x^5 - x$

Solution

a. Because the degree is odd and the leading coefficient is negative, the graph rises to the left and falls to the right, as shown in Figure 2.19.

b. Because the degree is even and the leading coefficient is positive, the graph rises to the left and right, as shown in Figure 2.20.

c. Because the degree is odd and the leading coefficient is positive, the graph falls to the left and rises to the right, as shown in Figure 2.21.

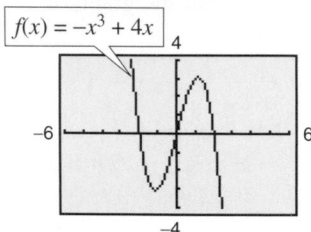

Figure 2.19

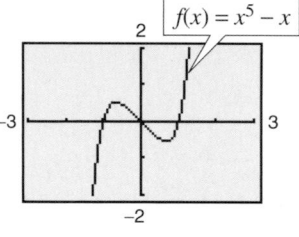

Figure 2.20

Figure 2.21

   Now try Exercise 15.

In Example 2, note that the Leading Coefficient Test only tells you whether the graph *eventually* rises or falls to the right or left. Other characteristics of the graph, such as intercepts and minimum and maximum points, must be determined by other tests.

Zeros of Polynomial Functions

It can be shown that for a polynomial function f of degree n, the following statements are true.

1. The function f has at most n real zeros. (You will study this result in detail in Section 2.5 on the Fundamental Theorem of Algebra.)

2. The graph of f has at most $n - 1$ relative **extrema** (relative **minima** or **maxima**).

Recall that a **zero** of a function f is a number x for which $f(x) = 0$. Finding the zeros of polynomial functions is one of the most important problems in algebra. You have already seen that there is a strong interplay between graphical and algebraic approaches to this problem. Sometimes you can use information about the graph of a function to help find its zeros. In other cases, you can use information about the zeros of a function to find a good viewing window.

> **Exploration**
>
> For each of the graphs in Example 2, count the number of zeros of the polynomial function and the number of relative extrema, and compare these numbers with the degree of the polynomial. What do you observe?

Real Zeros of Polynomial Functions

If f is a polynomial function and a is a real number, the following statements are equivalent.

   1. $x = a$ is a *zero* of the function f.

   2. $x = a$ is a *solution* of the polynomial equation $f(x) = 0$.

   3. $(x - a)$ is a *factor* of the polynomial $f(x)$.

   4. $(a, 0)$ is an x-intercept of the graph of f.

TECHNOLOGY SUPPORT

For instructions on how to use the *zero* or *root* feature, see Appendix A; for specific keystrokes, go to this textbook's *Online Study Center*.

Finding zeros of polynomial functions is closely related to factoring and finding x-intercepts, as demonstrated in Examples 3, 4, and 5.

Example 3   Finding Zeros of a Polynomial Function

Find all real zeros of $f(x) = x^3 - x^2 - 2x$.

Algebraic Solution

$$f(x) = x^3 - x^2 - 2x \qquad \text{Write original function.}$$
$$0 = x^3 - x^2 - 2x \qquad \text{Substitute 0 for } f(x).$$
$$0 = x(x^2 - x - 2) \qquad \text{Remove common monomial factor.}$$
$$0 = x(x - 2)(x + 1) \qquad \text{Factor completely.}$$

So, the real zeros are $x = 0$, $x = 2$, and $x = -1$, and the corresponding x-intercepts are $(0, 0)$, $(2, 0)$, and $(-1, 0)$.

Check

$$(0)^3 - (0)^2 - 2(0) = 0 \qquad x = 0 \text{ is a zero. } \checkmark$$
$$(2)^3 - (2)^2 - 2(2) = 0 \qquad x = 2 \text{ is a zero. } \checkmark$$
$$(-1)^3 - (-1)^2 - 2(-1) = 0 \qquad x = -1 \text{ is a zero. } \checkmark$$

✓*CHECKPOINT*   Now try Exercise 33.

Graphical Solution

Use a graphing utility to graph $y = x^3 - x^2 - 2x$. In Figure 2.22, the graph appears to have the x-intercepts $(0, 0)$, $(2, 0)$, and $(-1, 0)$. Use the *zero* or *root* feature, or the *zoom* and *trace* features, of the graphing utility to verify these intercepts. Note that this third-degree polynomial has two relative extrema, at $(-0.55, 0.63)$ and $(1.22, -2.11)$.

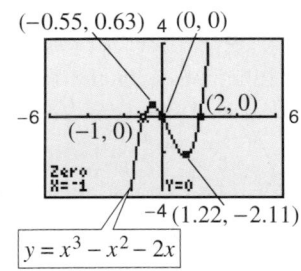

Figure 2.22

Example 4   Analyzing a Polynomial Function

Find all real zeros and relative extrema of $f(x) = -2x^4 + 2x^2$.

Solution

$$0 = -2x^4 + 2x^2 \qquad \text{Substitute 0 for } f(x).$$
$$0 = -2x^2(x^2 - 1) \qquad \text{Remove common monomial factor.}$$
$$0 = -2x^2(x - 1)(x + 1) \qquad \text{Factor completely.}$$

So, the real zeros are $x = 0$, $x = 1$, and $x = -1$, and the corresponding x-intercepts are $(0, 0)$, $(1, 0)$, and $(-1, 0)$, as shown in Figure 2.23. Using the *minimum* and *maximum* features of a graphing utility, you can approximate the three relative extrema to be $(-0.71, 0.5)$, $(0, 0)$, and $(0.71, 0.5)$.

✓*CHECKPOINT*   Now try Exercise 45.

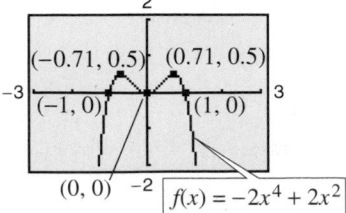

Figure 2.23

Repeated Zeros

For a polynomial function, a factor of $(x - a)^k$, $k > 1$, yields a **repeated zero** $x = a$ of **multiplicity** k.

1. If k is odd, the graph *crosses* the x-axis at $x = a$.
2. If k is even, the graph *touches* the x-axis (but does not cross the x-axis) at $x = a$.

STUDY TIP

In Example 4, note that because k is even, the factor $-2x^2$ yields the repeated zero $x = 0$. The graph touches (but does not cross) the x-axis at $x = 0$, as shown in Figure 2.23.

Example 5 Finding Zeros of a Polynomial Function

Find all real zeros of $f(x) = x^5 - 3x^3 - x^2 - 4x - 1$.

Solution

Use a graphing utility to obtain the graph shown in Figure 2.24. From the graph, you can see that there are three zeros. Using the *zero* or *root* feature, you can determine that the zeros are approximately $x \approx -1.86$, $x \approx -0.25$, and $x \approx 2.11$. It should be noted that this fifth-degree polynomial factors as

$$f(x) = x^5 - 3x^3 - x^2 - 4x - 1 = (x^2 + 1)(x^3 - 4x - 1).$$

The three zeros obtained above are the zeros of the cubic factor $x^3 - 4x - 1$ (the quadratic factor $x^2 + 1$ has two complex zeros and so no *real* zeros).

✔CHECKPOINT   Now try Exercise 47.

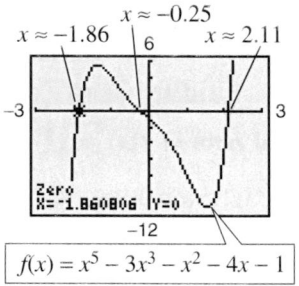

$f(x) = x^5 - 3x^3 - x^2 - 4x - 1$

Figure 2.24

Example 6 Finding a Polynomial Function with Given Zeros

Find polynomial functions with the following zeros. (There are many correct solutions.)

a. $-\dfrac{1}{2}, 3, 3$ **b.** $3, 2 + \sqrt{11}, 2 - \sqrt{11}$

Solution

a. Note that the zero $x = -\frac{1}{2}$ corresponds to either $\left(x + \frac{1}{2}\right)$ or $(2x + 1)$. To avoid fractions, choose the second factor and write

$$f(x) = (2x + 1)(x - 3)^2$$
$$= (2x + 1)(x^2 - 6x + 9) = 2x^3 - 11x^2 + 12x + 9.$$

b. For each of the given zeros, form a corresponding factor and write

$$f(x) = (x - 3)\left[x - \left(2 + \sqrt{11}\right)\right]\left[x - \left(2 - \sqrt{11}\right)\right]$$
$$= (x - 3)\left[(x - 2) - \sqrt{11}\right]\left[(x - 2) + \sqrt{11}\right]$$
$$= (x - 3)\left[(x - 2)^2 - \left(\sqrt{11}\right)^2\right]$$
$$= (x - 3)(x^2 - 4x + 4 - 11)$$
$$= (x - 3)(x^2 - 4x - 7) = x^3 - 7x^2 + 5x + 21.$$

✔CHECKPOINT   Now try Exercise 55.

Exploration

Use a graphing utility to graph

$$y_1 = x + 2$$
$$y_2 = (x + 2)(x - 1).$$

Predict the shape of the curve $y = (x + 2)(x - 1)(x - 3)$, and verify your answer with a graphing utility.

Note in Example 6 that there are many polynomial functions with the indicated zeros. In fact, multiplying the functions by any real number does not change the zeros of the function. For instance, multiply the function from part (b) by $\frac{1}{2}$ to obtain $f(x) = \frac{1}{2}x^3 - \frac{7}{2}x^2 + \frac{5}{2}x + \frac{21}{2}$. Then find the zeros of the function. You will obtain the zeros 3, $2 + \sqrt{11}$, and $2 - \sqrt{11}$, as given in Example 6.

TECHNOLOGY TIP

It is easy to make mistakes when entering functions into a graphing utility. So, it is important to have an understanding of the basic shapes of graphs and to be able to graph simple polynomials *by hand*. For example, suppose you had entered the function in Example 7 as $y = 3x^5 - 4x^3$. By looking at the graph, what mathematical principles would alert you to the fact that you had made a mistake?

Example 7 Sketching the Graph of a Polynomial Function

Sketch the graph of $f(x) = 3x^4 - 4x^3$ by hand.

Solution

1. *Apply the Leading Coefficient Test.* Because the leading coefficient is positive and the degree is even, you know that the graph eventually rises to the left and to the right (see Figure 2.25).
2. *Find the Real Zeros of the Polynomial.* By factoring

$$f(x) = 3x^4 - 4x^3 = x^3(3x - 4)$$

you can see that the real zeros of f are $x = 0$ (of odd multiplicity 3) and $x = \frac{4}{3}$ (of odd multiplicity 1). So, the x-intercepts occur at $(0, 0)$ and $\left(\frac{4}{3}, 0\right)$. Add these points to your graph, as shown in Figure 2.25.

3. *Plot a Few Additional Points.* To sketch the graph by hand, find a few additional points, as shown in the table. Be sure to choose points between the zeros and to the left and right of the zeros. Then plot the points (see Figure 2.26).

x	-1	0.5	1	1.5
$f(x)$	7	-0.31	-1	1.69

4. *Draw the Graph.* Draw a continuous curve through the points, as shown in Figure 2.26. Because both zeros are of odd multiplicity, you know that the graph should cross the x-axis at $x = 0$ and $x = \frac{4}{3}$. If you are unsure of the shape of a portion of the graph, plot some additional points.

Exploration

Partner Activity Multiply three, four, or five distinct linear factors to obtain the equation of a polynomial function of degree 3, 4, or 5. Exchange equations with your partner and sketch, *by hand*, the graph of the equation that your partner wrote. When you are finished, use a graphing utility to check each other's work.

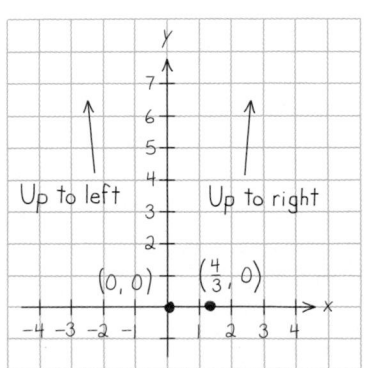

Figure 2.25

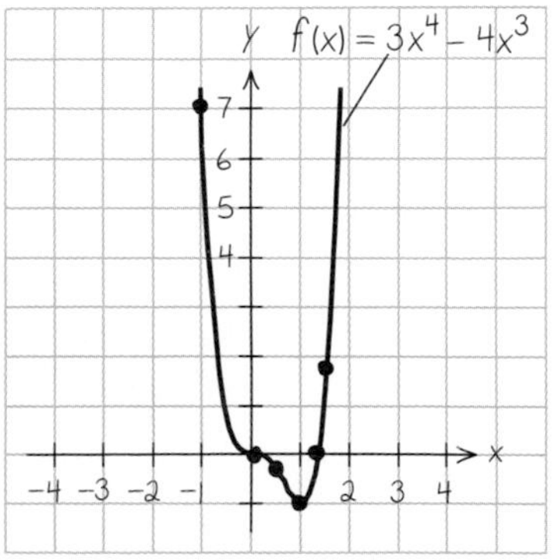

Figure 2.26

✓CHECKPOINT Now try Exercise 71.

Example 8   Sketching the Graph of a Polynomial Function

Sketch the graph of $f(x) = -2x^3 + 6x^2 - \frac{9}{2}x$.

Solution

1. *Apply the Leading Coefficient Test.*   Because the leading coefficient is negative and the degree is odd, you know that the graph eventually rises to the left and falls to the right (see Figure 2.27).

2. *Find the Real Zeros of the Polynomial.*   By factoring

$$f(x) = -2x^3 + 6x^2 - \frac{9}{2}x$$
$$= -\frac{1}{2}x(4x^2 - 12x + 9)$$
$$= -\frac{1}{2}x(2x - 3)^2$$

   you can see that the real zeros of f are $x = 0$ (of odd multiplicity 1) and $x = \frac{3}{2}$ (of even multiplicity 2). So, the x-intercepts occur at $(0, 0)$ and $\left(\frac{3}{2}, 0\right)$. Add these points to your graph, as shown in Figure 2.27.

3. *Plot a Few Additional Points.*   To sketch the graph by hand, find a few additional points, as shown in the table. Then plot the points (see Figure 2.28.)

x	-0.5	0.5	1	2
$f(x)$	4	-1	-0.5	-1

4. *Draw the Graph.*   Draw a continuous curve through the points, as shown in Figure 2.28. As indicated by the multiplicities of the zeros, the graph crosses the x-axis at $(0, 0)$ and touches (but does not cross) the x-axis at $\left(\frac{3}{2}, 0\right)$.

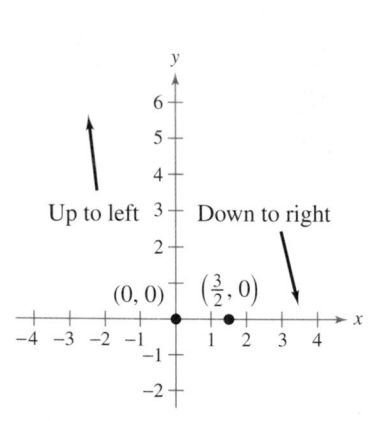

Figure 2.27

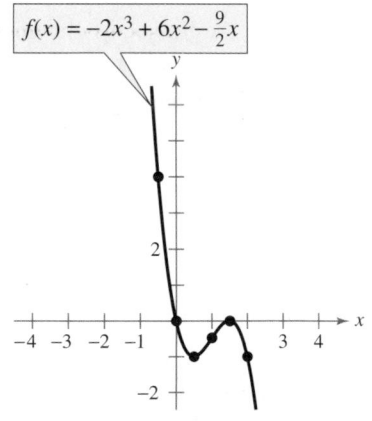

$f(x) = -2x^3 + 6x^2 - \frac{9}{2}x$

Figure 2.28

✔CHECKPOINT   Now try Exercise 73.

TECHNOLOGY TIP   Remember that when using a graphing utility to verify your graphs, you may need to adjust your viewing window in order to see all the features of the graph.

The Intermediate Value Theorem

The **Intermediate Value Theorem** concerns the existence of real zeros of polynomial functions. The theorem states that if $(a, f(a))$ and $(b, f(b))$ are two points on the graph of a polynomial function such that $f(a) \neq f(b)$, then for any number d between $f(a)$ and $f(b)$ there must be a number c between a and b such that $f(c) = d$. (See Figure 2.29.)

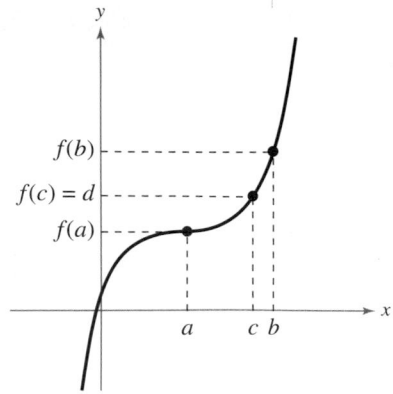

> **Intermediate Value Theorem**
>
> Let a and b be real numbers such that $a < b$. If f is a polynomial function such that $f(a) \neq f(b)$, then in the interval $[a, b]$, f takes on every value between $f(a)$ and $f(b)$.

Figure 2.29

This theorem helps you locate the real zeros of a polynomial function in the following way. If you can find a value $x = a$ at which a polynomial function is positive, and another value $x = b$ at which it is negative, you can conclude that the function has at least one real zero between these two values. For example, the function $f(x) = x^3 + x^2 + 1$ is negative when $x = -2$ and positive when $x = -1$. Therefore, it follows from the Intermediate Value Theorem that f must have a real zero somewhere between -2 and -1.

Example 9   Approximating the Zeros of a Function

Find three intervals of length 1 in which the polynomial $f(x) = 12x^3 - 32x^2 + 3x + 5$ is guaranteed to have a zero.

Graphical Solution

Use a graphing utility to graph

$$y = 12x^3 - 32x^2 + 3x + 5$$

as shown in Figure 2.30.

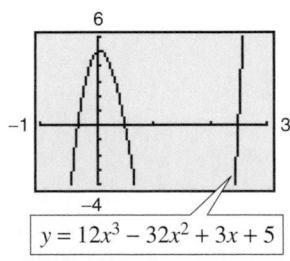

$$y = 12x^3 - 32x^2 + 3x + 5$$

Figure 2.30

From the figure, you can see that the graph crosses the x-axis three times—between -1 and 0, between 0 and 1, and between 2 and 3. So, you can conclude that the function has zeros in the intervals $(-1, 0)$, $(0, 1)$, and $(2, 3)$.

✔CHECKPOINT   Now try Exercise 79.

Numerical Solution

Use the *table* feature of a graphing utility to create a table of function values. Scroll through the table looking for consecutive function values that differ in sign. For instance, from the table in Figure 2.31 you can see that $f(-1)$ and $f(0)$ differ in sign. So, you can conclude from the Intermediate Value Theorem that the function has a zero between -1 and 0. Similarly, $f(0)$ and $f(1)$ differ in sign, so the function has a zero between 0 and 1. Likewise, $f(2)$ and $f(3)$ differ in sign, so the function has a zero between 2 and 3. So, you can conclude that the function has zeros in the intervals $(-1, 0)$, $(0, 1)$, and $(2, 3)$.

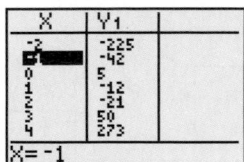

Figure 2.31

2.2 Exercises

See www.CalcChat.com for worked-out solutions to odd-numbered exercises.

Vocabulary Check

Fill in the blanks.

1. The graphs of all polynomial functions are _____ , which means that the graphs have no breaks, holes, or gaps.

2. The _____ is used to determine the left-hand and right-hand behavior of the graph of a polynomial function.

3. A polynomial function of degree n has at most _____ real zeros and at most _____ turning points, called _____ .

4. If $x = a$ is a zero of a polynomial function f, then the following statements are true.
   (a) $x = a$ is a _____ of the polynomial equation $f(x) = 0$.
   (b) _____ is a factor of the polynomial $f(x)$.
   (c) $(a, 0)$ is an _____ of the graph of f.

5. If a zero of a polynomial function is of even multiplicity, then the graph of f _____ the x-axis, and if the zero is of odd multiplicity, then the graph of f _____ the x-axis.

6. The _____ Theorem states that if f is a polynomial function such that $f(a) \neq f(b)$, then in the interval $[a, b]$, f takes on every value between $f(a)$ and $f(b)$.

In Exercises 1–8, match the polynomial function with its graph. [The graphs are labeled (a) through (h).]

(a)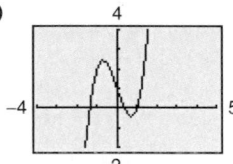

(b)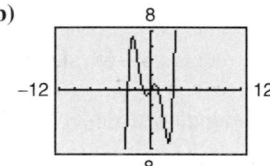

(c)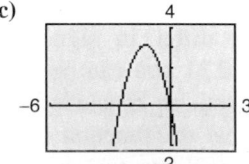

(d)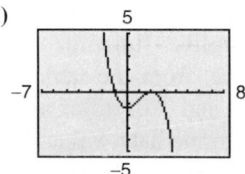

(e)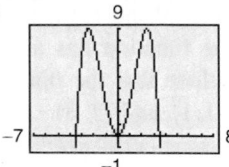

(f)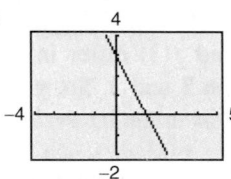

(g)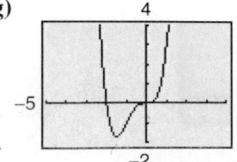

(h)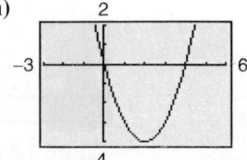

1. $f(x) = -2x + 3$
2. $f(x) = x^2 - 4x$
3. $f(x) = -2x^2 - 5x$
4. $f(x) = 2x^3 - 3x + 1$
5. $f(x) = -\frac{1}{4}x^4 + 3x^2$
6. $f(x) = -\frac{1}{3}x^3 + x^2 - \frac{4}{3}$
7. $f(x) = x^4 + 2x^3$
8. $f(x) = \frac{1}{5}x^5 - 2x^3 + \frac{9}{5}x$

In Exercises 9 and 10, sketch the graph of $y = x^n$ and each specified transformation.

9. $y = x^3$
   (a) $f(x) = (x - 2)^3$
   (b) $f(x) = x^3 - 2$
   (c) $f(x) = -\frac{1}{2}x^3$
   (d) $f(x) = (x - 2)^3 - 2$

10. $y = x^4$
   (a) $f(x) = (x + 5)^4$
   (b) $f(x) = x^4 - 5$
   (c) $f(x) = 4 - x^4$
   (d) $f(x) = \frac{1}{2}(x - 1)^4$

Graphical Analysis **In Exercises 11–14, use a graphing utility to graph the functions f and g in the same viewing window. Zoom out far enough so that the right-hand and left-hand behaviors of f and g appear identical. Show both graphs.**

11. $f(x) = 3x^3 - 9x + 1$, $g(x) = 3x^3$
12. $f(x) = -\frac{1}{3}(x^3 - 3x + 2)$, $g(x) = -\frac{1}{3}x^3$
13. $f(x) = -(x^4 - 4x^3 + 16x)$, $g(x) = -x^4$
14. $f(x) = 3x^4 - 6x^2$, $g(x) = 3x^4$

In Exercises 15–22, use the Leading Coefficient Test to describe the right-hand and left-hand behavior of the graph of the polynomial function. Use a graphing utility to verify your result.

15. $f(x) = 2x^4 - 3x + 1$ 16. $h(x) = 1 - x^6$

17. $g(x) = 5 - \frac{7}{2}x - 3x^2$ 18. $f(x) = \frac{1}{3}x^3 + 5x$

19. $f(x) = \dfrac{6x^5 - 2x^4 + 4x^2 - 5x}{3}$

20. $f(x) = \dfrac{3x^7 - 2x^5 + 5x^3 + 6x^2}{4}$

21. $h(t) = -\frac{2}{3}(t^2 - 5t + 3)$

22. $f(s) = -\frac{7}{8}(s^3 + 5s^2 - 7s + 1)$

In Exercises 23–32, find all the real zeros of the polynomial function. Determine the multiplicity of each zero. Use a graphing utility to verify your result.

23. $f(x) = x^2 - 25$ 24. $f(x) = 49 - x^2$

25. $h(t) = t^2 - 6t + 9$ 26. $f(x) = x^2 + 10x + 25$

27. $f(x) = x^2 + x - 2$ 28. $f(x) = 2x^2 - 14x + 24$

29. $f(t) = t^3 - 4t^2 + 4t$ 30. $f(x) = x^4 - x^3 - 20x^2$

31. $f(x) = \frac{1}{2}x^2 + \frac{5}{2}x - \frac{3}{2}$ 32. $f(x) = \frac{5}{3}x^2 + \frac{8}{3}x - \frac{4}{3}$

Graphical Analysis   In Exercises 33–44, (a) find the zeros algebraically, (b) use a graphing utility to graph the function, and (c) use the graph to approximate any zeros and compare them with those from part (a).

33. $f(x) = 3x^2 - 12x + 3$

34. $g(x) = 5x^2 - 10x - 5$

35. $g(t) = \frac{1}{2}t^4 - \frac{1}{2}$

36. $y = \frac{1}{4}x^3(x^2 - 9)$

37. $f(x) = x^5 + x^3 - 6x$

38. $g(t) = t^5 - 6t^3 + 9t$

39. $f(x) = 2x^4 - 2x^2 - 40$

40. $f(x) = 5x^4 + 15x^2 + 10$

41. $f(x) = x^3 - 4x^2 - 25x + 100$

42. $y = 4x^3 + 4x^2 - 7x + 2$

43. $y = 4x^3 - 20x^2 + 25x$

44. $y = x^5 - 5x^3 + 4x$

In Exercises 45–48, use a graphing utility to graph the function and approximate (accurate to three decimal places) any real zeros and relative extrema.

45. $f(x) = 2x^4 - 6x^2 + 1$

46. $f(x) = -\frac{3}{8}x^4 - x^3 + 2x^2 + 5$

47. $f(x) = x^5 + 3x^3 - x + 6$

48. $f(x) = -3x^3 - 4x^2 + x - 3$

In Exercises 49–58, find a polynomial function that has the given zeros. (There are many correct answers.)

49. $0, 4$ 50. $-7, 2$

51. $0, -2, -3$ 52. $0, 2, 5$

53. $4, -3, 3, 0$ 54. $-2, -1, 0, 1, 2$

55. $1 + \sqrt{3}, 1 - \sqrt{3}$ 56. $6 + \sqrt{3}, 6 - \sqrt{3}$

57. $2, 4 + \sqrt{5}, 4 - \sqrt{5}$ 58. $4, 2 + \sqrt{7}, 2 - \sqrt{7}$

In Exercises 59–64, find a polynomial function with the given zeros, multiplicities, and degree. (There are many correct answers.)

59. Zero: -2, multiplicity: 2   60. Zero: 3, multiplicity: 1
    Zero: -1, multiplicity: 1   Zero: 2, multiplicity: 3
    Degree: 3   Degree: 4

61. Zero: -4, multiplicity: 2   62. Zero: -5, multiplicity: 3
    Zero: 3, multiplicity: 2   Zero: 0, multiplicity: 2
    Degree: 4   Degree: 5

63. Zero: -1, multiplicity: 2   64. Zero: -1, multiplicity: 2
    Zero: -2, multiplicity: 1   Zero: 4, multiplicity: 2
    Degree: 3   Degree: 4
    Rises to the left,   Falls to the left,
    Falls to the right   Falls to the right

In Exercises 65–68, sketch the graph of a polynomial function that satisfies the given conditions. If not possible, explain your reasoning. (There are many correct answers.)

65. Third-degree polynomial with two real zeros and a negative leading coefficient

66. Fourth-degree polynomial with three real zeros and a positive leading coefficient

67. Fifth-degree polynomial with three real zeros and a positive leading coefficient

68. Fourth-degree polynomial with two real zeros and a negative leading coefficient

In Exercises 69–78, sketch the graph of the function by (a) applying the Leading Coefficient Test, (b) finding the zeros of the polynomial, (c) plotting sufficient solution points, and (d) drawing a continuous curve through the points.

69. $f(x) = x^3 - 9x$ 70. $g(x) = x^4 - 4x^2$

71. $f(x) = x^3 - 3x^2$ 72. $f(x) = 3x^3 - 24x^2$

73. $f(x) = -x^4 + 9x^2 - 20$ 74. $f(x) = -x^6 + 7x^3 + 8$

75. $f(x) = x^3 + 3x^2 - 9x - 27$

76. $h(x) = x^5 - 4x^3 + 8x^2 - 32$

77. $g(t) = -\frac{1}{4}t^4 + 2t^2 - 4$

78. $g(x) = \frac{1}{10}(x^4 - 4x^3 - 2x^2 + 12x + 9)$

In Exercises 79–82, (a) use the Intermediate Value Theorem and a graphing utility to find graphically any intervals of length 1 in which the polynomial function is guaranteed to have a zero, and (b) use the *zero* or *root* feature of the graphing utility to approximate the real zeros of the function. Verify your answers in part (a) by using the *table* feature of the graphing utility.

79. $f(x) = x^3 - 3x^2 + 3$ **80.** $f(x) = -2x^3 - 6x^2 + 3$

81. $g(x) = 3x^4 + 4x^3 - 3$ **82.** $h(x) = x^4 - 10x^2 + 2$

In Exercises 83–90, use a graphing utility to graph the function. Identify any symmetry with respect to the *x*-axis, *y*-axis, or origin. Determine the number of *x*-intercepts of the graph.

83. $f(x) = x^2(x + 6)$ **84.** $h(x) = x^3(x - 4)^2$

85. $g(t) = -\frac{1}{2}(t - 4)^2(t + 4)^2$

86. $g(x) = \frac{1}{8}(x + 1)^2(x - 3)^3$

87. $f(x) = x^3 - 4x$ **88.** $f(x) = x^4 - 2x^2$

89. $g(x) = \frac{1}{5}(x + 1)^2(x - 3)(2x - 9)$

90. $h(x) = \frac{1}{5}(x + 2)^2(3x - 5)^2$

91. *Numerical and Graphical Analysis* An open box is to be made from a square piece of material 36 centimeters on a side by cutting equal squares with sides of length *x* from the corners and turning up the sides (see figure).

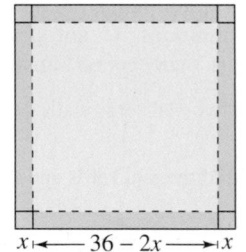
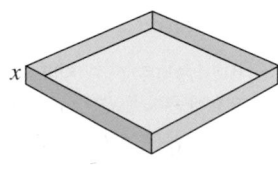

(a) Verify that the volume of the box is given by the function $V(x) = x(36 - 2x)^2$.

(b) Determine the domain of the function *V*.

(c) Use the *table* feature of a graphing utility to create a table that shows various box heights *x* and the corresponding volumes *V*. Use the table to estimate a range of dimensions within which the maximum volume is produced.

(d) Use a graphing utility to graph *V* and use the range of dimensions from part (c) to find the *x*-value for which $V(x)$ is maximum.

92. *Geometry* An open box with locking tabs is to be made from a square piece of material 24 inches on a side. This is done by cutting equal squares from the corners and folding along the dashed lines, as shown in the figure.

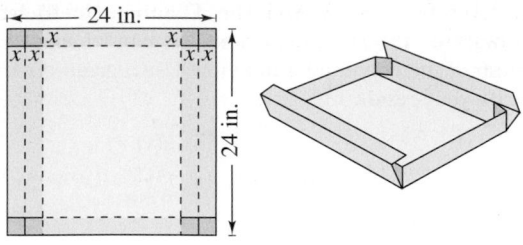

Figure for 92

(a) Verify that the volume of the box is given by the function $V(x) = 8x(6 - x)(12 - x)$.

(b) Determine the domain of the function *V*.

(c) Sketch the graph of the function and estimate the value of *x* for which $V(x)$ is maximum.

93. *Revenue* The total revenue *R* (in millions of dollars) for a company is related to its advertising expense by the function

$$R = 0.00001(-x^3 + 600x^2), \quad 0 \le x \le 400$$

where *x* is the amount spent on advertising (in tens of thousands of dollars). Use the graph of the function shown in the figure to estimate the point on the graph at which the function is increasing most rapidly. This point is called the **point of diminishing returns** because any expense above this amount will yield less return per dollar invested in advertising.

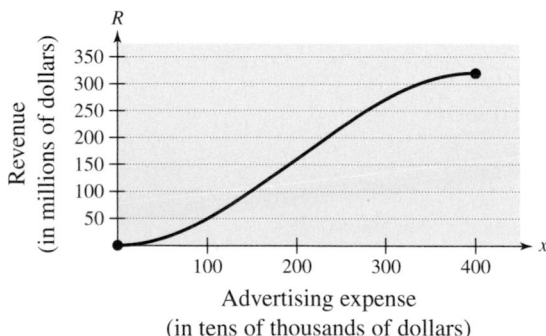

94. *Environment* The growth of a red oak tree is approximated by the function

$$G = -0.003t^3 + 0.137t^2 + 0.458t - 0.839$$

where *G* is the height of the tree (in feet) and *t* ($2 \le t \le 34$) is its age (in years). Use a graphing utility to graph the function and estimate the age of the tree when it is growing most rapidly. This point is called the **point of diminishing returns** because the increase in growth will be less with each additional year. (*Hint:* Use a viewing window in which $0 \le x \le 35$ and $0 \le y \le 60$.)

Data Analysis   In Exercises 95–98, use the table, which shows the median prices (in thousands of dollars) of new privately owned U.S. homes in the Northeast y_1 and in the South y_2 for the years 1995 through 2004. The data can be approximated by the following models.

$$y_1 = 0.3050t^3 - 6.949t^2 + 53.93t - 8.8$$

$$y_2 = 0.0330t^3 - 0.528t^2 + 8.35t + 65.2$$

In the models, t represents the year, with $t = 5$ corresponding to 1995.   (Sources: National Association of Realtors)

Year, t	y_1	y_2
5	126.7	97.7
6	127.8	103.4
7	131.8	109.6
8	135.9	116.2
9	139.0	120.3
10	139.4	128.3
11	146.5	137.4
12	164.3	147.3
13	190.5	157.1
14	220.0	169.0

95. Use a graphing utility to plot the data and graph the model for y_1 in the same viewing window. How closely does the model represent the data?

96. Use a graphing utility to plot the data and graph the model for y_2 in the same viewing window. How closely does the model represent the data?

97. Use the models to predict the median prices of new privately owned homes in both regions in 2010. Do your answers seem reasonable? Explain.

98. Use the graphs of the models in Exercises 95 and 96 to write a short paragraph about the relationship between the median prices of homes in the two regions.

Synthesis

True or False?   In Exercises 99–104, determine whether the statement is true or false. Justify your answer.

99. It is possible for a sixth-degree polynomial to have only one zero.

100. The graph of the function

$$f(x) = 2 + x - x^2 + x^3 - x^4 + x^5 + x^6 - x^7$$

rises to the left and falls to the right.

101. The graph of the function $f(x) = 2x(x - 1)^2(x + 3)^3$ crosses the x-axis at $x = 1$.

102. The graph of the function $f(x) = 2x(x - 1)^2(x + 3)^3$ touches, but does not cross, the x-axis.

103. The graph of the function $f(x) = -x^2(x + 2)^3(x - 4)^2$ crosses the x-axis at $x = -2$.

104. The graph of the function $f(x) = 2x(x - 1)^2(x + 3)^3$ rises to the left and falls to the right.

Library of Parent Functions   In Exercises 105–107, determine which polynomial function(s) may be represented by the graph shown. There may be more than one correct answer.

105. (a) $f(x) = x(x + 1)^2$
    (b) $f(x) = x(x - 1)^2$
    (c) $f(x) = x^2(x - 1)$
    (d) $f(x) = -x(x + 1)^2$
    (e) $f(x) = -x(x - 1)^2$

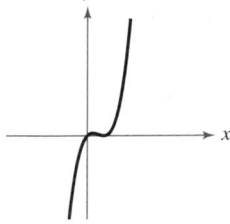

106. (a) $f(x) = x^2(x - 2)^2$
    (b) $f(x) = x^2(x + 2)^2$
    (c) $f(x) = x(x - 2)$
    (d) $f(x) = -x^2(x + 2)^2$
    (e) $f(x) = -x^2(x - 2)^2$

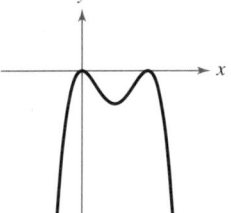

107. (a) $f(x) = (x - 1)^2(x + 2)^2$
    (b) $f(x) = (x - 1)(x + 2)$
    (c) $f(x) = (x + 1)^2(x - 2)^2$
    (d) $f(x) = -(x - 1)^2(x + 2)^2$
    (e) $f(x) = -(x + 1)^2(x - 2)^2$

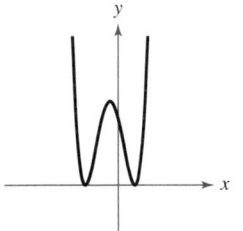

Skills Review

In Exercises 108–113, let $f(x) = 14x - 3$ and $g(x) = 8x^2$. Find the indicated value.

108. $(f + g)(-4)$

109. $(g - f)(3)$

110. $(fg)\left(-\dfrac{4}{7}\right)$

111. $\left(\dfrac{f}{g}\right)(-1.5)$

112. $(f \circ g)(-1)$

113. $(g \circ f)(0)$

In Exercises 114–117, solve the inequality and sketch the solution on the real number line. Use a graphing utility to verify your solution graphically.

114. $3(x - 5) < 4x - 7$

115. $2x^2 - x \geq 1$

116. $\dfrac{5x - 2}{x - 7} \leq 4$

117. $|x + 8| - 1 \geq 15$

2.3   Real Zeros of Polynomial Functions

Long Division of Polynomials

Consider the graph of

$$f(x) = 6x^3 - 19x^2 + 16x - 4.$$

Notice in Figure 2.32 that $x = 2$ appears to be a zero of f. Because $f(2) = 0$, you know that $x = 2$ is a zero of the polynomial function f, and that $(x - 2)$ is a factor of $f(x)$. This means that there exists a second-degree polynomial $q(x)$ such that $f(x) = (x - 2) \cdot q(x)$. To find $q(x)$, you can use **long division of polynomials.**

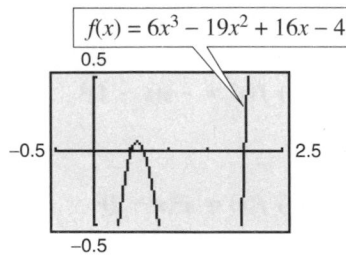

Figure 2.32

Example 1   Long Division of Polynomials

Divide $6x^3 - 19x^2 + 16x - 4$ by $x - 2$, and use the result to factor the polynomial completely.

Solution

$$
\begin{array}{r}
6x^2 - 7x + 2 \\
x - 2 \overline{)\, 6x^3 - 19x^2 + 16x - 4} \\
\end{array}
$$

$-(6x^3 + 12x^2)$	Multiply: $6x^2(x - 2)$.
$-7x^2 + 16x$	Subtract.
$-(+7x^2 + {}^-14x)$	Multiply: $-7x(x - 2)$.
$2x - 4$	Subtract.
$-2x + 4$	Multiply: $2(x - 2)$.
0	Subtract.

You can see that

$$6x^3 - 19x^2 + 16x - 4 = (x - 2)(6x^2 - 7x + 2)$$
$$= (x - 2)(2x - 1)(3x - 2).$$

Note that this factorization agrees with the graph of f (see Figure 2.32) in that the three x-intercepts occur at $x = 2$, $x = \frac{1}{2}$, and $x = \frac{2}{3}$.

✔CHECKPOINT   Now try Exercise 1.

What you should learn

- Use long division to divide polynomials by other polynomials.
- Use synthetic division to divide polynomials by binomials of the form $(x - k)$.
- Use the Remainder and Factor Theorems.
- Use the Rational Zero Test to determine possible rational zeros of polynomial functions.
- Use Descartes's Rule of Signs and the Upper and Lower Bound Rules to find zeros of polynomials.

Why you should learn it

The Remainder Theorem can be used to determine the number of coal mines in the United States in a given year based on a polynomial model, as shown in Exercise 92 on page 130.

Ian Murphy/Getty Images

STUDY TIP

Note that in Example 1, the division process requires $-7x^2 + 14x$ to be subtracted from $-7x^2 + 16x$. Therefore it is implied that

$$
\begin{array}{r}
-7x^2 + 16x \\
-(-7x^2 + 14x) \\
\end{array}
=
\begin{array}{r}
-7x^2 + 16x \\
7x^2 - 14x \\
\hline
\end{array}
$$

and instead is written simply as

$$
\begin{array}{r}
-7x^2 + 16x \\
-7x^2 + 14x. \\
\hline
2x \\
\end{array}
$$

In Example 1, $x - 2$ is a factor of the polynomial $6x^3 - 19x^2 + 16x - 4$, and the long division process produces a remainder of zero. Often, long division will produce a nonzero remainder. For instance, if you divide $x^2 + 3x + 5$ by $x + 1$, you obtain the following.

Divisor $\Rightarrow$ $x + 1\overline{)\,x^2 + 3x + 5}$ ← Quotient $x + 2$ / Dividend

$\underline{x^2 + \;x}$

$2x + 5$

$\underline{2x + 2}$

3 ← Remainder

In fractional form, you can write this result as follows.

$$\underbrace{\frac{x^2 + 3x + 5}{x + 1}}_{\text{Dividend / Divisor}} = \underbrace{x + 2}_{\text{Quotient}} + \underbrace{\frac{3}{x + 1}}_{\text{Remainder / Divisor}}$$

This implies that

$$x^2 + 3x + 5 = (x + 1)(x + 2) + 3 \qquad \text{Multiply each side by } (x + 1).$$

which illustrates the following theorem, called the **Division Algorithm.**

The Division Algorithm

If $f(x)$ and $d(x)$ are polynomials such that $d(x) \neq 0$, and the degree of $d(x)$ is less than or equal to the degree of $f(x)$, there exist unique polynomials $q(x)$ and $r(x)$ such that

$$f(x) = d(x)q(x) + r(x)$$

Dividend | Quotient | Divisor | Remainder

where $r(x) = 0$ *or* the degree of $r(x)$ is less than the degree of $d(x)$. If the remainder $r(x)$ is zero, $d(x)$ *divides evenly* into $f(x)$.

The Division Algorithm can also be written as

$$\frac{f(x)}{d(x)} = q(x) + \frac{r(x)}{d(x)}.$$

In the Division Algorithm, the rational expression $f(x)/d(x)$ is **improper** because the degree of $f(x)$ is greater than or equal to the degree of $d(x)$. On the other hand, the rational expression $r(x)/d(x)$ is **proper** because the degree of $r(x)$ is less than the degree of $d(x)$.

Before you apply the Division Algorithm, follow these steps.

1. Write the dividend and divisor in descending powers of the variable.

2. Insert placeholders with zero coefficients for missing powers of the variable.

Example 2   Long Division of Polynomials

Divide $8x^3 - 1$ by $2x - 1$.

Solution

Because there is no x^2-term or x-term in the dividend, you need to line up the subtraction by using zero coefficients (or leaving spaces) for the missing terms.

$$
\begin{array}{r}
4x^2 + 2x + 1 \\
2x - 1 \overline{)\, 8x^3 + 0x^2 + 0x - 1} \\
\underline{8x^3 - 4x^2} \\
4x^2 + 0x \\
\underline{4x^2 - 2x} \\
2x - 1 \\
\underline{2x - 1} \\
0
\end{array}
$$

So, $2x - 1$ divides evenly into $8x^3 - 1$, and you can write

$$
\frac{8x^3 - 1}{2x - 1} = 4x^2 + 2x + 1, \quad x \neq \frac{1}{2}.
$$

✓CHECKPOINT   Now try Exercise 7.

You can check the result of Example 2 by multiplying.

$$(2x - 1)(4x^2 + 2x + 1) = 8x^3 + 4x^2 + 2x - 4x^2 - 2x - 1$$
$$= 8x^3 - 1$$

Example 3   Long Division of Polynomials

Divide $-2 + 3x - 5x^2 + 4x^3 + 2x^4$ by $x^2 + 2x - 3$.

Solution

Begin by writing the dividend in descending powers of x.

$$
\begin{array}{r}
2x^2 \qquad + 1 \\
x^2 + 2x - 3 \overline{)\, 2x^4 + 4x^3 - 5x^2 + 3x - 2} \\
\underline{2x^4 + 4x^3 - 6x^2} \\
x^2 + 3x - 2 \\
\underline{x^2 + 2x - 3} \\
x + 1
\end{array}
$$

Note that the first subtraction eliminated two terms from the dividend. When this happens, the quotient skips a term. You can write the result as

$$
\frac{2x^4 + 4x^3 - 5x^2 + 3x - 2}{x^2 + 2x - 3} = 2x^2 + 1 + \frac{x + 1}{x^2 + 2x - 3}.
$$

✓CHECKPOINT   Now try Exercise 9.

Synthetic Division

There is a nice shortcut for long division of polynomials when dividing by divisors of the form $x - k$. The shortcut is called **synthetic division.** The pattern for synthetic division of a cubic polynomial is summarized as follows. (The pattern for higher-degree polynomials is similar.)

Synthetic Division (of a Cubic Polynomial)

To divide $ax^3 + bx^2 + cx + d$ by $x - k$, use the following pattern.

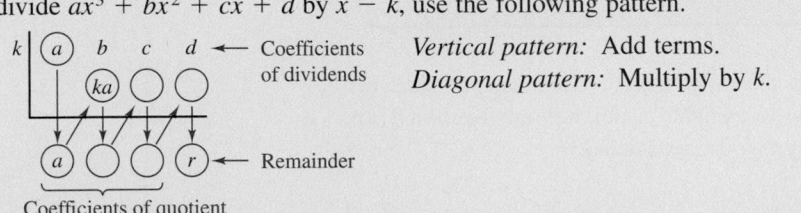

Vertical pattern: Add terms.
Diagonal pattern: Multiply by k.

This algorithm for synthetic division works *only* for divisors of the form $x - k$. Remember that $x + k = x - (-k)$.

Example 4   Using Synthetic Division

Use synthetic division to divide $x^4 - 10x^2 - 2x + 4$ by $x + 3$.

Solution

You should set up the array as follows. Note that a zero is included for each missing term in the dividend.

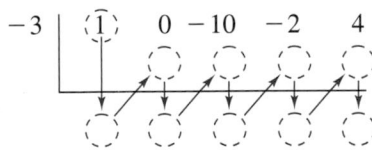

Then, use the synthetic division pattern by adding terms in columns and multiplying the results by -3.

Divisor: $x + 3$ Dividend: $x^4 - 10x^2 - 2x + 4$

$$
\begin{array}{r|rrrrr}
-3 & 1 & 0 & -10 & -2 & 4 \\
   &   & -3 & 9 & 3 & -3 \\
\hline
   & 1 & -3 & -1 & 1 & 1 \\
\end{array}
$$
← Remainder: 1

Quotient: $x^3 - 3x^2 - x + 1$

So, you have

$$\frac{x^4 - 10x^2 - 2x + 4}{x + 3} = x^3 - 3x^2 - x + 1 + \frac{1}{x + 3}.$$

✓CHECKPOINT   Now try Exercise 15.

Exploration

Evaluate the polynomial $x^4 - 10x^2 - 2x + 4$ at $x = -3$. What do you observe?

The Remainder and Factor Theorems

The remainder obtained in the synthetic division process has an important interpretation, as described in the **Remainder Theorem.**

The Remainder Theorem   (See the proof on page 180.)

If a polynomial $f(x)$ is divided by $x - k$, the remainder is

$\quad r = f(k).$

The Remainder Theorem tells you that synthetic division can be used to evaluate a polynomial function. That is, to evaluate a polynomial function $f(x)$ when $x = k$, divide $f(x)$ by $x - k$. The remainder will be $f(k)$.

Example 5   Using the Remainder Theorem

Use the Remainder Theorem to evaluate the following function at $x = -2$.

$\quad f(x) = 3x^3 + 8x^2 + 5x - 7$

Solution

Using synthetic division, you obtain the following.

$$
\begin{array}{r|rrrr}
-2 & 3 & 8 & 5 & -7 \\
   &   & -6 & -4 & -2 \\
\hline
   & 3 & 2 & 1 & -9
\end{array}
$$

Because the remainder is $r = -9$, you can conclude that

$\quad f(-2) = -9. \qquad r = f(k)$

This means that $(-2, -9)$ is a point on the graph of f. You can check this by substituting $x = -2$ in the original function.

Check

$$
\begin{aligned}
f(-2) &= 3(-2)^3 + 8(-2)^2 + 5(-2) - 7 \\
      &= 3(-8) + 8(4) - 10 - 7 \\
      &= -24 + 32 - 10 - 7 = -9
\end{aligned}
$$

✓**CHECKPOINT**   Now try Exercise 35.

Another important theorem is the **Factor Theorem.** This theorem states that you can test whether a polynomial has $(x - k)$ as a factor by evaluating the polynomial at $x = k$. If the result is 0, $(x - k)$ is a factor.

The Factor Theorem   (See the proof on page 180.)

A polynomial $f(x)$ has a factor $(x - k)$ if and only if $f(k) = 0$.

Example 6   Factoring a Polynomial: Repeated Division

Show that $(x - 2)$ and $(x + 3)$ are factors of

$$f(x) = 2x^4 + 7x^3 - 4x^2 - 27x - 18.$$

Then find the remaining factors of $f(x)$.

Algebraic Solution

Using synthetic division with the factor $(x - 2)$, you obtain the following.

$$
\begin{array}{r|rrrrr}
2 & 2 & 7 & -4 & -27 & -18 \\
  &   & 4 & 22 & 36 & 18 \\
\hline
  & 2 & 11 & 18 & 9 & 0
\end{array}
$$

0 remainder; $(x - 2)$ is a factor.

Take the result of this division and perform synthetic division again using the factor $(x + 3)$.

$$
\begin{array}{r|rrrr}
-3 & 2 & 11 & 18 & 9 \\
   &   & -6 & -15 & -9 \\
\hline
   & 2 & 5 & 3 & 0
\end{array}
$$

$\underbrace{}$
$2x^2 + 5x + 3$

0 remainder; $(x + 3)$ is a factor.

Because the resulting quadratic factors as

$$2x^2 + 5x + 3 = (2x + 3)(x + 1)$$

the complete factorization of $f(x)$ is

$$f(x) = (x - 2)(x + 3)(2x + 3)(x + 1).$$

✓CHECKPOINT   Now try Exercise 45.

Graphical Solution

The graph of a polynomial with factors of $(x - 2)$ and $(x + 3)$ has x-intercepts at $x = 2$ and $x = -3$. Use a graphing utility to graph

$$y = 2x^4 + 7x^3 - 4x^2 - 27x - 18.$$

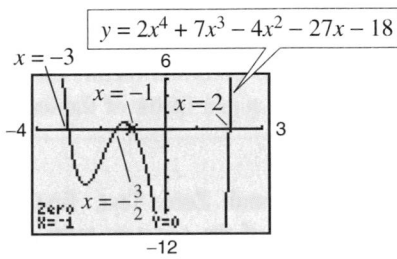

Figure 2.33

From Figure 2.33, you can see that the graph appears to cross the x-axis in two other places, near $x = -1$ and $x = -\frac{3}{2}$. Use the *zero* or *root* feature or the *zoom* and *trace* features to approximate the other two intercepts to be $x = -1$ and $x = -\frac{3}{2}$. So, the factors of f are $(x - 2)$, $(x + 3)$, $\left(x + \frac{3}{2}\right)$, and $(x + 1)$. You can rewrite the factor $\left(x + \frac{3}{2}\right)$ as $(2x + 3)$, so the complete factorization of f is $f(x) = (x - 2)(x + 3)(2x + 3)(x + 1)$.

Using the Remainder in Synthetic Division

In summary, the remainder r, obtained in the synthetic division of $f(x)$ by $x - k$, provides the following information.

1. The remainder r gives the value of f at $x = k$. That is, $r = f(k)$.

2. If $r = 0$, $(x - k)$ is a factor of $f(x)$.

3. If $r = 0$, $(k, 0)$ is an x-intercept of the graph of f.

   Throughout this text, the importance of developing several problem-solving strategies is emphasized. In the exercises for this section, try using more than one strategy to solve several of the exercises. For instance, if you find that $x - k$ divides evenly into $f(x)$, try sketching the graph of f. You should find that $(k, 0)$ is an x-intercept of the graph.

The Rational Zero Test

The **Rational Zero Test** relates the possible rational zeros of a polynomial (having integer coefficients) to the leading coefficient and to the constant term of the polynomial.

> **The Rational Zero Test**
>
> If the polynomial
>
> $$f(x) = a_n x^n + a_{n-1} x^{n-1} + \cdots + a_2 x^2 + a_1 x + a_0$$
>
> has integer coefficients, every rational zero of f has the form
>
> $$\text{Rational zero} = \frac{p}{q}$$
>
> where p and q have no common factors other than 1, p is a factor of the constant term a_0, and q is a factor of the leading coefficient a_n.

To use the Rational Zero Test, first list all rational numbers whose numerators are factors of the constant term and whose denominators are factors of the leading coefficient.

$$\text{Possible rational zeros} = \frac{\text{factors of constant term}}{\text{factors of leading coefficient}}$$

Now that you have formed this list of *possible rational zeros*, use a trial-and-error method to determine which, if any, are actual zeros of the polynomial. Note that when the leading coefficient is 1, the possible rational zeros are simply the factors of the constant term. This case is illustrated in Example 7.

Example 7   Rational Zero Test with Leading Coefficient of 1

Find the rational zeros of $f(x) = x^3 + x + 1$.

Solution

Because the leading coefficient is 1, the possible rational zeros are simply the factors of the constant term.

> *Possible rational zeros:* ± 1

By testing these possible zeros, you can see that neither works.

$$f(1) = (1)^3 + 1 + 1 = 3$$

$$f(-1) = (-1)^3 + (-1) + 1 = -1$$

So, you can conclude that the polynomial has *no* rational zeros. Note from the graph of f in Figure 2.34 that f does have one real zero between -1 and 0. However, by the Rational Zero Test, you know that this real zero is *not* a rational number.

✔**CHECKPOINT**   Now try Exercise 49.

Graph the polynomial $y = x^3 - 53x^2 + 103x - 51$ in the standard viewing window. From the graph alone, it appears that there is only one zero. From the Leading Coefficient Test, you know that because the degree of the polynomial is odd and the leading coefficient is positive, the graph falls to the left and rises to the right. So, the function must have another zero. From the Rational Zero Test, you know that ± 51 might be zeros of the function. If you zoom out several times, you will see a more complete picture of the graph. Your graph should confirm that $x = 51$ is a zero of f.

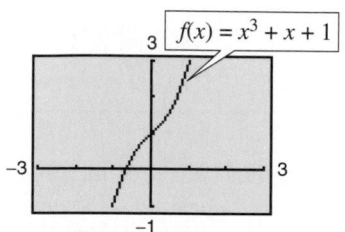

Figure 2.34

If the leading coefficient of a polynomial is not 1, the list of possible rational zeros can increase dramatically. In such cases, the search can be shortened in several ways.

1. A programmable calculator can be used to speed up the calculations.

2. A graphing utility can give a good estimate of the locations of the zeros.

3. The Intermediate Value Theorem, along with a table generated by a graphing utility, can give approximations of zeros.

4. The Factor Theorem and synthetic division can be used to test the possible rational zeros.

Finding the first zero is often the most difficult part. After that, the search is simplified by working with the lower-degree polynomial obtained in synthetic division.

Example 8 Using the Rational Zero Test

Find the rational zeros of $f(x) = 2x^3 + 3x^2 - 8x + 3$.

Solution

The leading coefficient is 2 and the constant term is 3.

Possible rational zeros:

$$\frac{\text{Factors of 3}}{\text{Factors of 2}} = \frac{\pm 1, \pm 3}{\pm 1, \pm 2} = \pm 1, \pm 3, \pm \frac{1}{2}, \pm \frac{3}{2}$$

By synthetic division, you can determine that $x = 1$ is a rational zero.

$$
\begin{array}{r|rrrr}
1 & 2 & 3 & -8 & 3 \\
  &   & 2 & 5 & -3 \\
\hline
  & 2 & 5 & -3 & 0
\end{array}
$$

So, $f(x)$ factors as

$$f(x) = (x - 1)(2x^2 + 5x - 3) = (x - 1)(2x - 1)(x + 3)$$

and you can conclude that the rational zeros of f are $x = 1$, $x = \frac{1}{2}$, and $x = -3$, as shown in Figure 2.35.

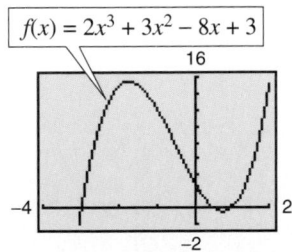

$$f(x) = 2x^3 + 3x^2 - 8x + 3$$

Figure 2.35

✓CHECKPOINT   Now try Exercise 51.

A graphing utility can help you determine which possible rational zeros to test, as demonstrated in Example 9.

Example 9 Finding Real Zeros of a Polynomial Function

Find all the real zeros of $f(x) = 10x^3 - 15x^2 - 16x + 12$.

Solution

Because the leading coefficient is 10 and the constant term is 12, there is a long list of possible rational zeros.

Possible rational zeros:

$$\frac{\text{Factors of } 12}{\text{Factors of } 10} = \frac{\pm 1, \pm 2, \pm 3, \pm 4, \pm 6, \pm 12}{\pm 1, \pm 2, \pm 5, \pm 10}$$

With so many possibilities (32, in fact), it is worth your time to use a graphing utility to focus on just a few. By using the *trace* feature of a graphing utility, it looks like three reasonable choices are $x = -\frac{6}{5}$, $x = \frac{1}{2}$, and $x = 2$ (see Figure 2.36). Synthetic division shows that only $x = 2$ works. (You could also use the Factor Theorem to test these choices.)

$$
\begin{array}{r|rrrr}
2 & 10 & -15 & -16 & 12 \\
  &   & 20 & 10 & -12 \\
\hline
  & 10 & 5 & -6 & 0
\end{array}
$$

So, $x = 2$ is one zero and you have

$$f(x) = (x - 2)(10x^2 + 5x - 6).$$

Using the Quadratic Formula, you find that the two additional zeros are irrational numbers.

$$x = \frac{-5 + \sqrt{265}}{20} \approx 0.56 \quad \text{and} \quad x = \frac{-5 - \sqrt{265}}{20} \approx -1.06$$

✓**CHECKPOINT** Now try Exercise 55.

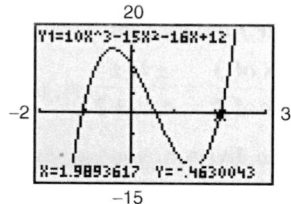

Figure 2.36

Other Tests for Zeros of Polynomials

You know that an nth-degree polynomial function can have *at most* n real zeros. Of course, many nth-degree polynomials do not have that many real zeros. For instance, $f(x) = x^2 + 1$ has no real zeros, and $f(x) = x^3 + 1$ has only one real zero. The following theorem, called **Descartes's Rule of Signs,** sheds more light on the number of real zeros of a polynomial.

Descartes's Rule of Signs

Let $f(x) = a_n x^n + a_{n-1}x^{n-1} + \cdots + a_2 x^2 + a_1 x + a_0$ be a polynomial with real coefficients and $a_0 \neq 0$.

1. The number of *positive real zeros* of f is either equal to the number of variations in sign of $f(x)$ or less than that number by an even integer.
2. The number of *negative real zeros* of f is either equal to the number of variations in sign of $f(-x)$ or less than that number by an even integer.

A **variation in sign** means that two consecutive (nonzero) coefficients have opposite signs.

When using Descartes's Rule of Signs, a zero of multiplicity k should be counted as k zeros. For instance, the polynomial $x^3 - 3x + 2$ has two variations in sign, and so has either two positive or no positive real zeros. Because

$$x^3 - 3x + 2 = (x - 1)(x - 1)(x + 2)$$

you can see that the two positive real zeros are $x = 1$ of multiplicity 2.

Example 10   Using Descartes's Rule of Signs

Describe the possible real zeros of $f(x) = 3x^3 - 5x^2 + 6x - 4$.

Solution

The original polynomial has *three* variations in sign.

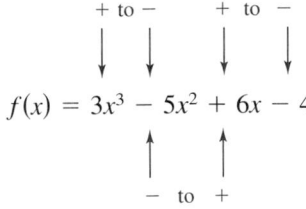

The polynomial

$$f(-x) = 3(-x)^3 - 5(-x)^2 + 6(-x) - 4 = -3x^3 - 5x^2 - 6x - 4$$

has no variations in sign. So, from Descartes's Rule of Signs, the polynomial $f(x) = 3x^3 - 5x^2 + 6x - 4$ has either three positive real zeros or one positive real zero, and has no negative real zeros. By using the *trace* feature of a graphing utility, you can see that the function has only one real zero (it is a positive number near $x = 1$), as shown in Figure 2.37.

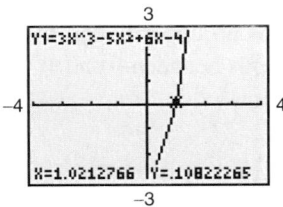

Figure 2.37

✓**CHECKPOINT**   Now try Exercise 65.

Another test for zeros of a polynomial function is related to the sign pattern in the last row of the synthetic division array. This test can give you an upper or lower bound of the real zeros of f, which can help you eliminate possible real zeros. A real number b is an **upper bound** for the real zeros of f if no zeros are greater than b. Similarly, b is a **lower bound** if no real zeros of f are less than b.

Upper and Lower Bound Rules

Let $f(x)$ be a polynomial with real coefficients and a positive leading coefficient. Suppose $f(x)$ is divided by $x - c$, using synthetic division.

1. If $c > 0$ and each number in the last row is either positive or zero, c is an **upper bound** for the real zeros of f.

2. If $c < 0$ and the numbers in the last row are alternately positive and negative (zero entries count as positive or negative), c is a **lower bound** for the real zeros of f.

Example 11 Finding the Zeros of a Polynomial Function

Find the real zeros of $f(x) = 6x^3 - 4x^2 + 3x - 2$.

Solution

The possible real zeros are as follows.

$$\frac{\text{Factors of 2}}{\text{Factors of 6}} = \frac{\pm 1, \pm 2}{\pm 1, \pm 2, \pm 3, \pm 6} = \pm 1, \pm \frac{1}{2}, \pm \frac{1}{3}, \pm \frac{1}{6}, \pm \frac{2}{3}, \pm 2$$

The original polynomial $f(x)$ has three variations in sign. The polynomial

$$f(-x) = 6(-x)^3 - 4(-x)^2 + 3(-x) - 2$$

$$= -6x^3 - 4x^2 - 3x - 2$$

has no variations in sign. As a result of these two findings, you can apply Descartes's Rule of Signs to conclude that there are three positive real zeros or one positive real zero, and no negative real zeros. Trying $x = 1$ produces the following.

$$
\begin{array}{r|rrrr}
1 & 6 & -4 & 3 & -2 \\
  &   & 6 & 2 & 5 \\
\hline
  & 6 & 2 & 5 & 3
\end{array}
$$

So, $x = 1$ is not a zero, but because the last row has all positive entries, you know that $x = 1$ is an upper bound for the real zeros. Therefore, you can restrict the search to zeros between 0 and 1. By trial and error, you can determine that $x = \frac{2}{3}$ is a zero. So,

$$f(x) = \left(x - \frac{2}{3}\right)(6x^2 + 3).$$

Because $6x^2 + 3$ has no real zeros, it follows that $x = \frac{2}{3}$ is the only real zero.

✓CHECKPOINT   Now try Exercise 75.

Before concluding this section, here are two additional hints that can help you find the real zeros of a polynomial.

1. If the terms of $f(x)$ have a common monomial factor, it should be factored out before applying the tests in this section. For instance, by writing

$$f(x) = x^4 - 5x^3 + 3x^2 + x = x(x^3 - 5x^2 + 3x + 1)$$

   you can see that $x = 0$ is a zero of f and that the remaining zeros can be obtained by analyzing the cubic factor.

2. If you are able to find all but two zeros of $f(x)$, you can always use the Quadratic Formula on the remaining quadratic factor. For instance, if you succeeded in writing

$$f(x) = x^4 - 5x^3 + 3x^2 + x = x(x - 1)(x^2 - 4x - 1)$$

   you can apply the Quadratic Formula to $x^2 - 4x - 1$ to conclude that the two remaining zeros are $x = 2 + \sqrt{5}$ and $x = 2 - \sqrt{5}$.

Exploration

Use a graphing utility to graph

$$y_1 = 6x^3 - 4x^2 + 3x - 2.$$

Notice that the graph intersects the x-axis at the point $\left(\frac{2}{3}, 0\right)$. How does this information relate to the real zero found in Example 11? Use a graphing utility to graph

$$y_2 = x^4 - 5x^3 + 3x^2 + x.$$

How many times does the graph intersect the x-axis? How many real zeros does y_2 have?

Exploration

Use a graphing utility to graph

$$y = x^3 + 4.9x^2 - 126x$$
$$+ 382.5$$

in the standard viewing window. From the graph, what do the real zeros appear to be? Discuss how the mathematical tools of this section might help you realize that the graph does not show all the important features of the polynomial function. Now use the *zoom* feature to find all the zeros of this function.

2.3 Exercises

See www.CalcChat.com for worked-out solutions to odd-numbered exercises.

Vocabulary Check

1. Two forms of the Division Algorithm are shown below. Identify and label each part.

$$f(x) = d(x)q(x) + r(x) \qquad \frac{f(x)}{d(x)} = q(x) + \frac{r(x)}{d(x)}$$

In Exercises 2–7, fill in the blanks.

2. The rational expression $p(x)/q(x)$ is called _____ if the degree of the numerator is greater than or equal to that of the denominator, and is called _____ if the degree of the numerator is less than that of the denominator.

3. An alternative method to long division of polynomials is called _____ , in which the divisor must be of the form $x - k$.

4. The test that gives a list of the possible rational zeros of a polynomial function is known as the _____ Test.

5. The theorem that can be used to determine the possible numbers of positive real zeros and negative real zeros of a function is called _____ of _____ .

6. The _____ states that if a polynomial $f(x)$ is divided by $x - k$, then the remainder is $r = f(k)$.

7. A real number b is an _____ for the real zeros of f if no zeros are greater than b, and is a _____ if no real zeros of f are less than b.

In Exercises 1–14, use long division to divide.

1. Divide $2x^2 + 10x + 12$ by $x + 3$.
2. Divide $5x^2 - 17x - 12$ by $x - 4$.
3. Divide $x^4 + 5x^3 + 6x^2 - x - 2$ by $x + 2$.
4. Divide $x^3 - 4x^2 - 17x + 6$ by $x - 3$.
5. Divide $4x^3 - 7x^2 - 11x + 5$ by $4x + 5$.
6. Divide $2x^3 - 3x^2 - 50x + 75$ by $2x - 3$.
7. Divide $7x^3 + 3$ by $x + 2$.
8. Divide $8x^4 - 5$ by $2x + 1$.
9. $(x + 8 + 6x^3 + 10x^2) \div (2x^2 + 1)$
10. $(1 + 3x^2 + x^4) \div (3 - 2x + x^2)$
11. $(x^3 - 9) \div (x^2 + 1)$
12. $(x^5 + 7) \div (x^3 - 1)$
13. $\dfrac{2x^3 - 4x^2 - 15x + 5}{(x - 1)^2}$
14. $\dfrac{x^4}{(x - 1)^3}$

In Exercises 15–24, use synthetic division to divide.

15. $(3x^3 - 17x^2 + 15x - 25) \div (x - 5)$
16. $(5x^3 + 18x^2 + 7x - 6) \div (x + 3)$
17. $(6x^3 + 7x^2 - x + 26) \div (x - 3)$
18. $(2x^3 + 14x^2 - 20x + 7) \div (x + 6)$
19. $(9x^3 - 18x^2 - 16x + 32) \div (x - 2)$
20. $(5x^3 + 6x + 8) \div (x + 2)$
21. $(x^3 + 512) \div (x + 8)$
22. $(x^3 - 729) \div (x - 9)$
23. $\dfrac{4x^3 + 16x^2 - 23x - 15}{x + \frac{1}{2}}$
24. $\dfrac{3x^3 - 4x^2 + 5}{x - \frac{3}{2}}$

Graphical Analysis **In Exercises 25–28, use a graphing utility to graph the two equations in the same viewing window. Use the graphs to verify that the expressions are equivalent. Verify the results algebraically.**

25. $y_1 = \dfrac{x^2}{x + 2}$, $y_2 = x - 2 + \dfrac{4}{x + 2}$
26. $y_1 = \dfrac{x^2 + 2x - 1}{x + 3}$, $y_2 = x - 1 + \dfrac{2}{x + 3}$
27. $y_1 = \dfrac{x^4 - 3x^2 - 1}{x^2 + 5}$, $y_2 = x^2 - 8 + \dfrac{39}{x^2 + 5}$
28. $y_1 = \dfrac{x^4 + x^2 - 1}{x^2 + 1}$, $y_2 = x^2 - \dfrac{1}{x^2 + 1}$

In Exercises 29–34, write the function in the form $f(x) = (x - k)q(x) + r(x)$ for the given value of k. Use a graphing utility to demonstrate that $f(k) = r$.

Function	Value of k
29. $f(x) = x^3 - x^2 - 14x + 11$	$k = 4$
30. $f(x) = 15x^4 + 10x^3 - 6x^2 + 14$	$k = -\frac{2}{3}$
31. $f(x) = x^3 + 3x^2 - 2x - 14$	$k = \sqrt{2}$
32. $f(x) = x^3 + 2x^2 - 5x - 4$	$k = -\sqrt{5}$
33. $f(x) = 4x^3 - 6x^2 - 12x - 4$	$k = 1 - \sqrt{3}$
34. $f(x) = -3x^3 + 8x^2 + 10x - 8$	$k = 2 + \sqrt{2}$

In Exercises 35–38, use the Remainder Theorem and synthetic division to evaluate the function at each given value. Use a graphing utility to verify your results.

35. $f(x) = 2x^3 - 7x + 3$

    (a) $f(1)$ (b) $f(-2)$ (c) $f\left(\frac{1}{2}\right)$ (d) $f(2)$

36. $g(x) = 2x^6 + 3x^4 - x^2 + 3$

    (a) $g(2)$ (b) $g(1)$ (c) $g(3)$ (d) $g(-1)$

37. $h(x) = x^3 - 5x^2 - 7x + 4$

    (a) $h(3)$ (b) $h(2)$ (c) $h(-2)$ (d) $h(-5)$

38. $f(x) = 4x^4 - 16x^3 + 7x^2 + 20$

    (a) $f(1)$ (b) $f(-2)$ (c) $f(5)$ (d) $f(-10)$

In Exercises 39–42, use synthetic division to show that x is a solution of the third-degree polynomial equation, and use the result to factor the polynomial completely. List all the real zeros of the function.

Polynomial Equation	Value of x
39. $x^3 - 7x + 6 = 0$	$x = 2$
40. $x^3 - 28x - 48 = 0$	$x = -4$
41. $2x^3 - 15x^2 + 27x - 10 = 0$	$x = \frac{1}{2}$
42. $48x^3 - 80x^2 + 41x - 6 = 0$	$x = \frac{2}{3}$

In Exercises 43–48, (a) verify the given factors of the function f, (b) find the remaining factors of f, (c) use your results to write the complete factorization of f, and (d) list all real zeros of f. Confirm your results by using a graphing utility to graph the function.

Function	Factor(s)
43. $f(x) = 2x^3 + x^2 - 5x + 2$	$(x + 2)$
44. $f(x) = 3x^3 + 2x^2 - 19x + 6$	$(x + 3)$
45. $f(x) = x^4 - 4x^3 - 15x^2$ $+ 58x - 40$	$(x - 5), (x + 4)$
46. $f(x) = 8x^4 - 14x^3 - 71x^2$ $- 10x + 24$	$(x + 2), (x - 4)$
47. $f(x) = 6x^3 + 41x^2 - 9x - 14$	$(2x + 1)$
48. $f(x) = 2x^3 - x^2 - 10x + 5$	$(2x - 1)$

In Exercises 49–52, use the Rational Zero Test to list all possible rational zeros of f. Then find the rational zeros.

49. $f(x) = x^3 + 3x^2 - x - 3$

50. $f(x) = x^3 - 4x^2 - 4x + 16$

51. $f(x) = 2x^4 - 17x^3 + 35x^2 + 9x - 45$

52. $f(x) = 4x^5 - 8x^4 - 5x^3 + 10x^2 + x - 2$

In Exercises 53–60, find all real zeros of the polynomial function.

53. $f(z) = z^4 - z^3 - 2z - 4$

54. $f(x) = x^4 - x^3 - 29x^2 - x - 30$

55. $g(y) = 2y^4 + 7y^3 - 26y^2 + 23y - 6$

56. $h(x) = x^5 - x^4 - 3x^3 + 5x^2 - 2x$

57. $f(x) = 4x^4 - 55x^2 - 45x + 36$

58. $z(x) = 4x^4 - 43x^2 - 9x + 90$

59. $f(x) = 4x^5 + 12x^4 - 11x^3 - 42x^2 + 7x + 30$

60. $g(x) = 4x^5 + 8x^4 - 15x^3 - 23x^2 + 11x + 15$

Graphical Analysis   **In Exercises 61–64, (a) use the *zero* or *root* feature of a graphing utility to approximate (accurate to the nearest thousandth) the zeros of the function, (b) determine one of the exact zeros and use synthetic division to verify your result, and (c) factor the polynomial completely.**

61. $h(t) = t^3 - 2t^2 - 7t + 2$

62. $f(s) = s^3 - 12s^2 + 40s - 24$

63. $h(x) = x^5 - 7x^4 + 10x^3 + 14x^2 - 24x$

64. $g(x) = 6x^4 - 11x^3 - 51x^2 + 99x - 27$

In Exercises 65–68, use Descartes's Rule of Signs to determine the possible numbers of positive and negative real zeros of the function.

65. $f(x) = 2x^4 - x^3 + 6x^2 - x + 5$

66. $f(x) = 3x^4 + 5x^3 - 6x^2 + 8x - 3$

67. $g(x) = 4x^3 - 5x + 8$

68. $g(x) = 2x^3 - 4x^2 - 5$

In Exercises 69–74, (a) use Descartes's Rule of Signs to determine the possible numbers of positive and negative real zeros of f, (b) list the possible rational zeros of f, (c) use a graphing utility to graph f so that some of the possible zeros in parts (a) and (b) can be disregarded, and (d) determine all the real zeros of f.

69. $f(x) = x^3 + x^2 - 4x - 4$

70. $f(x) = -3x^3 + 20x^2 - 36x + 16$

71. $f(x) = -2x^4 + 13x^3 - 21x^2 + 2x + 8$

72. $f(x) = 4x^4 - 17x^2 + 4$

73. $f(x) = 32x^3 - 52x^2 + 17x + 3$

74. $f(x) = 4x^3 + 7x^2 - 11x - 18$

Occasionally, throughout this text, you will be asked to round to a place value rather than to a number of decimal places.

In Exercises 75–78, use synthetic division to verify the upper and lower bounds of the real zeros of f. Then find the real zeros of the function.

75. $f(x) = x^4 - 4x^3 + 15$

   Upper bound: $x = 4$;   Lower bound: $x = -1$

76. $f(x) = 2x^3 - 3x^2 - 12x + 8$

   Upper bound: $x = 4$;   Lower bound: $x = -3$

77. $f(x) = x^4 - 4x^3 + 16x - 16$

   Upper bound: $x = 5$;   Lower bound: $x = -3$

78. $f(x) = 2x^4 - 8x + 3$

   Upper bound: $x = 3$;   Lower bound: $x = -4$

In Exercises 79–82, find the rational zeros of the polynomial function.

79. $P(x) = x^4 - \frac{25}{4}x^2 + 9 = \frac{1}{4}(4x^4 - 25x^2 + 36)$

80. $f(x) = x^3 - \frac{3}{2}x^2 - \frac{23}{2}x + 6 = \frac{1}{2}(2x^3 - 3x^2 - 23x + 12)$

81. $f(x) = x^3 - \frac{1}{4}x^2 - x + \frac{1}{4} = \frac{1}{4}(4x^3 - x^2 - 4x + 1)$

82. $f(z) = z^3 + \frac{11}{6}z^2 - \frac{1}{2}z - \frac{1}{3} = \frac{1}{6}(6z^3 + 11z^2 - 3z - 2)$

In Exercises 83–86, match the cubic function with the correct number of rational and irrational zeros.

(a) **Rational zeros: 0;**   **Irrational zeros: 1**

(b) **Rational zeros: 3;**   **Irrational zeros: 0**

(c) **Rational zeros: 1;**   **Irrational zeros: 2**

(d) **Rational zeros: 1;**   **Irrational zeros: 0**

83. $f(x) = x^3 - 1$ **84.** $f(x) = x^3 - 2$

85. $f(x) = x^3 - x$ **86.** $f(x) = x^3 - 2x$

In Exercises 87–90, the graph of $y = f(x)$ is shown. Use the graph as an aid to find all the real zeros of the function.

87. $y = 2x^4 - 9x^3 + 5x^2$
       $+ 3x - 1$

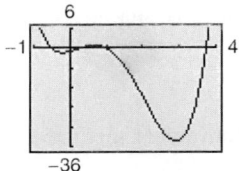

88. $y = x^4 - 5x^3 - 7x^2$
       $+ 13x - 2$

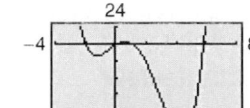

89. $y = -2x^4 + 17x^3$
       $- 3x^2 - 25x - 3$

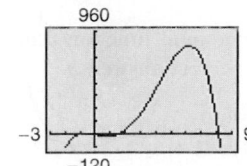

90. $y = -x^4 + 5x^3$
       $- 10x - 4$

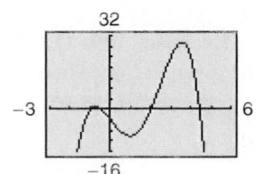

91. *U.S. Population*   The table shows the populations P of the United States (in millions) from 1790 to 2000.   (Source: U.S. Census Bureau)

Year	Population (in millions)
1790	3.9
1800	5.3
1810	7.2
1820	9.6
1830	12.9
1840	17.1
1850	23.2
1860	31.4
1870	39.8
1880	50.2
1890	63.0
1900	76.2
1910	92.2
1920	106.0
1930	123.2
1940	132.2
1950	151.3
1960	179.3
1970	203.3
1980	226.5
1990	248.7
2000	281.4

The population can be approximated by the equation

$$P = 0.0058t^3 + 0.500t^2 + 1.38t + 4.6, \quad -1 \le t \le 20$$

where t represents the year, with $t = -1$ corresponding to 1790, $t = 0$ corresponding to 1800, and so on.

(a) Use a graphing utility to graph the data and the equation in the same viewing window.

(b) How well does the model fit the data?

(c) Use the Remainder Theorem to evaluate the model for the year 2010. Do you believe this value is reasonable? Explain.

92. *Energy* The number of coal mines C in the United States from 1980 to 2004 can be approximated by the equation $C = 0.232t^3 - 2.11t^2 - 261.8t + 5699$, for $0 \le t \le 24$, where t is the year, with $t = 0$ corresponding to 1980. (Source: U.S. Energy Information Administration)

(a) Use a graphing utility to graph the model over the domain.

(b) Find the number of mines in 1980. Use the Remainder Theorem to find the number of mines in 1990.

(c) Could you use this model to predict the number of coal mines in the United States in the future? Explain.

93. *Geometry* A rectangular package sent by a delivery service can have a maximum combined length and girth (perimeter of a cross section) of 120 inches (see figure).

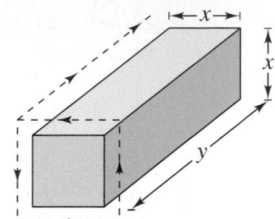

(a) Show that the volume of the package is given by the function $V(x) = 4x^2(30 - x)$.

(b) Use a graphing utility to graph the function and approximate the dimensions of the package that yield a maximum volume.

(c) Find values of x such that $V = 13,500$. Which of these values is a physical impossibility in the construction of the package? Explain.

94. *Automobile Emissions* The number of parts per million of nitric oxide emissions y from a car engine is approximated by the model $y = -5.05x^3 + 3,857x - 38,411.25$, for $13 \le x \le 18$, where x is the air-fuel ratio.

(a) Use a graphing utility to graph the model.

(b) It is observed from the graph that two air-fuel ratios produce 2400 parts per million of nitric oxide, with one being 15. Use the graph to approximate the second air-fuel ratio.

(c) Algebraically approximate the second air-fuel ratio that produces 2400 parts per million of nitric oxide. (*Hint:* Because you know that an air-fuel ratio of 15 produces the specified nitric oxide emission, you can use synthetic division.)

Synthesis

True or False? **In Exercises 95 and 96, determine whether the statement is true or false. Justify your answer.**

95. If $(7x + 4)$ is a factor of some polynomial function f, then $\frac{4}{7}$ is a zero of f.

96. $(2x - 1)$ is a factor of the polynomial

$$6x^6 + x^5 - 92x^4 + 45x^3 + 184x^2 + 4x - 48.$$

Think About It **In Exercises 97 and 98, the graph of a cubic polynomial function $y = f(x)$ with integer zeros is shown. Find the factored form of f.**

97.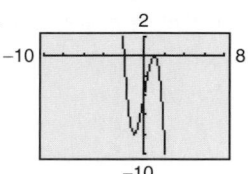

98.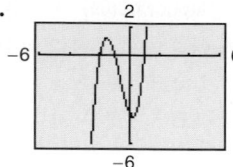

99. *Think About It* Let $y = f(x)$ be a quartic polynomial with leading coefficient $a = -1$ and $f(\pm 1) = f(\pm 2) = 0$. Find the factored form of f.

100. *Think About It* Let $y = f(x)$ be a cubic polynomial with leading coefficient $a = 2$ and $f(-2) = f(1) = f(2) = 0$. Find the factored form of f.

101. *Think About It* Find the value of k such that $x - 4$ is a factor of $x^3 - kx^2 + 2kx - 8$.

102. *Think About It* Find the value of k such that $x - 3$ is a factor of $x^3 - kx^2 + 2kx - 12$.

103. *Writing* Complete each polynomial division. Write a brief description of the pattern that you obtain, and use your result to find a formula for the polynomial division $(x^n - 1)/(x - 1)$. Create a numerical example to test your formula.

(a) $\dfrac{x^2 - 1}{x - 1} = $ ▨

(b) $\dfrac{x^3 - 1}{x - 1} = $ ▨

(c) $\dfrac{x^4 - 1}{x - 1} = $ ▨

104. *Writing* Write a short paragraph explaining how you can check polynomial division. Give an example.

Skills Review

In Exercises 105–108, use any convenient method to solve the quadratic equation.

105. $9x^2 - 25 = 0$

106. $16x^2 - 21 = 0$

107. $2x^2 + 6x + 3 = 0$

108. $8x^2 - 22x + 15 = 0$

In Exercises 109–112, find a polynomial function that has the given zeros. (There are many correct answers.)

109. $0, -12$

110. $1, -3, 8$

111. $0, -1, 2, 5$

112. $2 + \sqrt{3}, 2 - \sqrt{3}$

2.4   Complex Numbers

The Imaginary Unit *i*

Some quadratic equations have no real solutions. For instance, the quadratic equation $x^2 + 1 = 0$ has no real solution because there is no real number x that can be squared to produce -1. To overcome this deficiency, mathematicians created an expanded system of numbers using the **imaginary unit *i*,** defined as

$$i = \sqrt{-1} \qquad \text{Imaginary unit}$$

where $i^2 = -1$. By adding real numbers to real multiples of this imaginary unit, you obtain the set of **complex numbers.** Each complex number can be written in the **standard form** $a + bi$. For instance, the standard form of the complex number $\sqrt{-9} - 5$ is $-5 + 3i$ because

$$\sqrt{-9} - 5 = \sqrt{3^2(-1)} - 5 = 3\sqrt{-1} - 5 = 3i - 5 = -5 + 3i.$$

In the standard form $a + bi$, the real number a is called the **real part** of the **complex number** $a + bi$, and the number bi (where b is a real number) is called the **imaginary part** of the complex number.

Definition of a Complex Number

If a and b are real numbers, the number $a + bi$ is a **complex number,** and it is said to be written in **standard form.** If $b = 0$, the number $a + bi = a$ is a real number. If $b \neq 0$, the number $a + bi$ is called an **imaginary number.** A number of the form bi, where $b \neq 0$, is called a **pure imaginary number.**

The set of real numbers is a subset of the set of complex numbers, as shown in Figure 2.38. This is true because every real number a can be written as a complex number using $b = 0$. That is, for every real number a, you can write $a = a + 0i$.

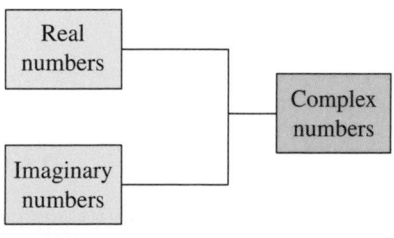

Figure 2.38

Equality of Complex Numbers

Two complex numbers $a + bi$ and $c + di$, written in standard form, are equal to each other

$$a + bi = c + di \qquad \text{Equality of two complex numbers}$$

if and only if $a = c$ and $b = d$.

What you should learn

- Use the imaginary unit *i* to write complex numbers.
- Add, subtract, and multiply complex numbers.
- Use complex conjugates to write the quotient of two complex numbers in standard form.
- Plot complex numbers in the complex plane.

Why you should learn it

Complex numbers are used to model numerous aspects of the natural world, such as the impedance of an electrical circuit, as shown in Exercises 79 and 80 on page 138.

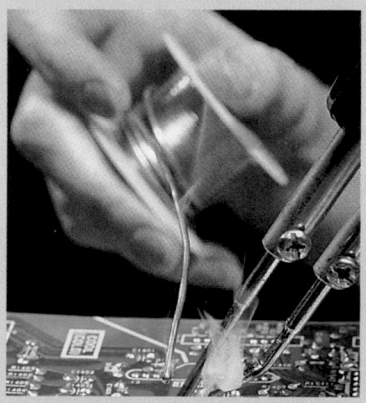

Phil Degginger/Getty Images

Operations with Complex Numbers

To add (or subtract) two complex numbers, you add (or subtract) the real and imaginary parts of the numbers separately.

Addition and Subtraction of Complex Numbers

If $a + bi$ and $c + di$ are two complex numbers written in standard form, their sum and difference are defined as follows.

Sum: $(a + bi) + (c + di) = (a + c) + (b + d)i$

Difference: $(a + bi) - (c + di) = (a - c) + (b - d)i$

The **additive identity** in the complex number system is zero (the same as in the real number system). Furthermore, the **additive inverse** of the complex number $a + bi$ is

$-(a + bi) = -a - bi.$ Additive inverse

So, you have $(a + bi) + (-a - bi) = 0 + 0i = 0$.

Example 1 Adding and Subtracting Complex Numbers

a.
$$(3 - i) + (2 + 3i) = 3 - i + 2 + 3i \qquad \text{Remove parentheses.}$$
$$= 3 + 2 - i + 3i \qquad \text{Group like terms.}$$
$$= (3 + 2) + (-1 + 3)i$$
$$= 5 + 2i \qquad \text{Write in standard form.}$$

b.
$$\sqrt{-4} + \left(-4 - \sqrt{-4}\right) = 2i + (-4 - 2i) \qquad \text{Write in } i\text{-form.}$$
$$= 2i - 4 - 2i \qquad \text{Remove parentheses.}$$
$$= -4 + 2i - 2i \qquad \text{Group like terms.}$$
$$= -4 \qquad \text{Write in standard form.}$$

c.
$$3 - (-2 + 3i) + (-5 + i) = 3 + 2 - 3i - 5 + i$$
$$= 3 + 2 - 5 - 3i + i$$
$$= 0 - 2i$$
$$= -2i$$

d.
$$(3 + 2i) + (4 - i) - (7 + i) = 3 + 2i + 4 - i - 7 - i$$
$$= 3 + 4 - 7 + 2i - i - i$$
$$= 0 + 0i$$
$$= 0$$

✓CHECKPOINT   Now try Exercise 19.

STUDY TIP

In Examples 1(b) and 1(d), note that the sum of complex numbers can be a real number.

Many of the properties of real numbers are valid for complex numbers as well. Here are some examples.

Associative Properties of Addition and Multiplication
Commutative Properties of Addition and Multiplication
Distributive Property of Multiplication over Addition

Notice how these properties are used when two complex numbers are multiplied.

$$(a + bi)(c + di) = a(c + di) + bi(c + di) \qquad \text{Distributive Property}$$

$$= ac + (ad)i + (bc)i + (bd)i^2 \qquad \text{Distributive Property}$$

$$= ac + (ad)i + (bc)i + (bd)(-1) \qquad i^2 = -1$$

$$= ac - bd + (ad)i + (bc)i \qquad \text{Commutative Property}$$

$$= (ac - bd) + (ad + bc)i \qquad \text{Associative Property}$$

The procedure above is similar to multiplying two polynomials and combining like terms, as in the FOIL Method.

> ### Exploration
>
> Complete the following:
>
> | | |
> |---|---|
> | $i^1 = i$ | $i^7 = $ ▢ |
> | $i^2 = -1$ | $i^8 = $ ▢ |
> | $i^3 = -i$ | $i^9 = $ ▢ |
> | $i^4 = 1$ | $i^{10} = $ ▢ |
> | $i^5 = $ ▢ | $i^{11} = $ ▢ |
> | $i^6 = $ ▢ | $i^{12} = $ ▢ |
>
> What pattern do you see? Write a brief description of how you would find i raised to any positive integer power.

Example 2 Multiplying Complex Numbers

a. $\sqrt{-4} \cdot \sqrt{-16} = (2i)(4i)$ Write each factor in i-form.

$\qquad\qquad\qquad = 8i^2$ Multiply.

$\qquad\qquad\qquad = 8(-1)$ $i^2 = -1$

$\qquad\qquad\qquad = -8$ Simplify.

b. $(2 - i)(4 + 3i) = 8 + 6i - 4i - 3i^2$ Product of binomials

$\qquad\qquad\qquad = 8 + 6i - 4i - 3(-1)$ $i^2 = -1$

$\qquad\qquad\qquad = 8 + 3 + 6i - 4i$ Group like terms.

$\qquad\qquad\qquad = 11 + 2i$ Write in standard form.

c. $(3 + 2i)(3 - 2i) = 9 - 6i + 6i - 4i^2$ Product of binomials

$\qquad\qquad\qquad = 9 - 4(-1)$ $i^2 = -1$

$\qquad\qquad\qquad = 9 + 4$ Simplify.

$\qquad\qquad\qquad = 13$ Write in standard form.

d. $4i(-1 + 5i) = 4i(-1) + 4i(5i)$ Distributive Property

$\qquad\qquad\qquad = -4i + 20i^2$ Simplify.

$\qquad\qquad\qquad = -4i + 20(-1)$ $i^2 = -1$

$\qquad\qquad\qquad = -20 - 4i$ Write in standard form.

e. $(3 + 2i)^2 = 9 + 6i + 6i + 4i^2$ Product of binomials

$\qquad\qquad\qquad = 9 + 12i + 4(-1)$ $i^2 = -1$

$\qquad\qquad\qquad = 9 - 4 + 12i$ Group like terms.

$\qquad\qquad\qquad = 5 + 12i$ Write in standard form.

✓CHECKPOINT   Now try Exercise 29.

> ### STUDY TIP
>
> Before you perform operations with complex numbers, be sure to rewrite the terms or factors in i-form first and then proceed with the operations, as shown in Example 2(a).

Complex Conjugates

Notice in Example 2(c) that the product of two complex numbers can be a real number. This occurs with pairs of complex numbers of the forms $a + bi$ and $a - bi$, called **complex conjugates.**

$$(a + bi)(a - bi) = a^2 - abi + abi - b^2i^2 = a^2 - b^2(-1) = a^2 + b^2$$

Example 3   Multiplying Conjugates

Multiply $3 - 5i$ by its complex conjugate.

Solution

The complex conjugate of $3 - 5i$ is $3 + 5i$.

$$(3 - 5i)(3 + 5i) = 3^2 - (5i)^2 = 9 - 25i^2 = 9 - 25(-1) = 34$$

✓CHECKPOINT   Now try Exercise 37.

To write the quotient of $a + bi$ and $c + di$ in standard form, where c and d are not both zero, multiply the numerator and denominator by the complex conjugate of the *denominator* to obtain

$$\frac{a + bi}{c + di} = \frac{a + bi}{c + di}\left(\frac{c - di}{c - di}\right)$$ Multiply numerator and denominator by complex conjugate of denominator.

$$= \frac{ac + bd}{c^2 + d^2} + \left(\frac{bc - ad}{c^2 + d^2}\right)i.$$ Standard form

Example 4   Writing a Quotient of Complex Numbers in Standard Form

Write the quotient $\dfrac{2 + 3i}{4 - 2i}$ in standard form.

Solution

$$\frac{2 + 3i}{4 - 2i} = \frac{2 + 3i}{4 - 2i}\left(\frac{4 + 2i}{4 + 2i}\right)$$ Multiply numerator and denominator by complex conjugate of denominator.

$$= \frac{8 + 4i + 12i + 6i^2}{16 - 4i^2}$$ Expand.

$$= \frac{8 - 6 + 16i}{16 + 4}$$ $i^2 = -1$

$$= \frac{2 + 16i}{20}$$ Simplify.

$$= \frac{1}{10} + \frac{4}{5}i$$ Write in standard form.

✓CHECKPOINT   Now try Exercise 49.

TECHNOLOGY TIP

Some graphing utilities can perform operations with complex numbers. For instance, on some graphing utilities, to divide $2 + 3i$ by $4 - 2i$, use the following keystrokes.

(2 + 3 i) ÷

(4 − 2 i) ENTER

The display will be as follows.

$$.1 + .8i \quad \text{or} \quad \frac{1}{10} + \frac{4}{5}i$$

Fractals and the Mandelbrot Set

Most applications involving complex numbers are either theoretical or very technical, and are therefore not appropriate for inclusion in this text. However, to give you some idea of how complex numbers can be used in applications, a general description of their use in **fractal geometry** is presented.

To begin, consider a coordinate system called the **complex plane.** Just as every real number corresponds to a point on the real number line, every complex number corresponds to a point in the complex plane, as shown in Figure 2.39. In this figure, note that the vertical axis is called the **imaginary axis** and the horizontal axis is called the **real axis.** The point that corresponds to the complex number $a + bi$ is (a, b).

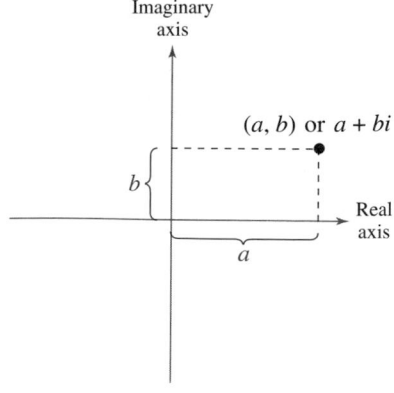

Figure 2.39

Example 5   Plotting Complex Numbers

Plot each complex number in the complex plane.

a. $2 + 3i$ **b.** $-1 + 2i$ **c.** 4 **d.** $-3i$

Solution

a. To plot the complex number $2 + 3i$, move (from the origin) two units to the right on the real axis and then three units upward, as shown in Figure 2.40. In other words, plotting the complex number $2 + 3i$ in the complex plane is comparable to plotting the point $(2, 3)$ in the Cartesian plane. (Note that in Figure 2.40, i is called the imaginary unit because it is located one unit from the origin on the imaginary axis of the complex plane.)

b. The complex number $-1 + 2i$ corresponds to the point $(-1, 2)$, as shown in Figure 2.40.

c. The complex number 4 corresponds to the point $(4, 0)$, as shown in Figure 2.40.

d. The complex number $-3i$ corresponds to the point $(0, -3)$, as shown in Figure 2.40.

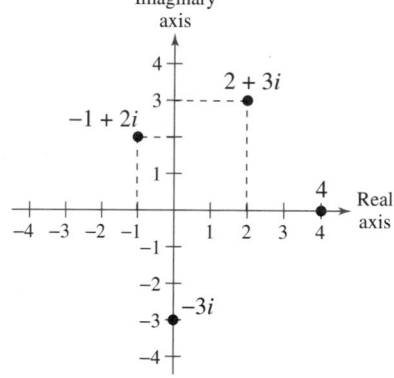

Figure 2.40

✓**CHECKPOINT**   Now try Exercise 71.

In the hands of a person who understands fractal geometry, the complex plane can become an easel on which stunning pictures, called **fractals,** can be drawn. The most famous such picture is called the **Mandelbrot Set,** named after the Polish-born mathematician Benoit Mandelbrot. To draw the Mandelbrot Set, consider the following sequence of numbers.

$$c, c^2 + c, (c^2 + c)^2 + c, [(c^2 + c)^2 + c]^2 + c, \ldots$$

The behavior of this sequence depends on the value of the complex number c. For some values of c this sequence is **bounded,** which means that the absolute value of each number $\left(|a + bi| = \sqrt{a^2 + b^2}\right)$ in the sequence is less than some fixed number N. For other values of c, the sequence is **unbounded,** which means that the absolute values of the terms of the sequence become infinitely large. If the sequence is bounded, the complex number c is in the Mandelbrot Set; if the sequence is unbounded, the complex number c is not in the Mandelbrot Set.

Example 6   Members of the Mandelbrot Set

a. The complex number -2 is in the Mandelbrot Set, because for $c = -2$, the corresponding Mandelbrot sequence is $-2, 2, 2, 2, 2, 2, \ldots$, which is bounded.

b. The complex number i is also in the Mandelbrot Set, because for $c = i$, the corresponding Mandelbrot sequence is

$$i, \quad -1 + i, \quad -i, \quad -1 + i, \quad -i, \quad -1 + i, \quad \ldots$$

which is bounded.

c. The complex number $1 + i$ is not in the Mandelbrot Set, because for $c = 1 + i$, the corresponding Mandelbrot sequence is

$$1 + i, \quad 1 + 3i, \quad -7 + 7i, \quad 1 - 97i, \quad -9407 - 193i,$$

$$88,454,401 + 3,631,103i, \quad \ldots$$

which is unbounded.

✓**CHECKPOINT**   Now try Exercise 77.

With this definition, a picture of the Mandelbrot Set would have only two colors: one color for points that are in the set (the sequence is bounded), and one color for points that are outside the set (the sequence is unbounded). Figure 2.41 shows a black and yellow picture of the Mandelbrot Set. The points that are black are in the Mandelbrot Set and the points that are yellow are not.

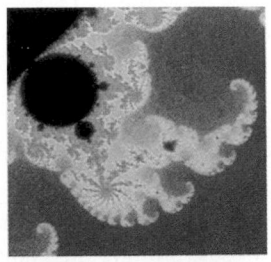

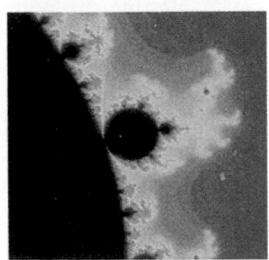

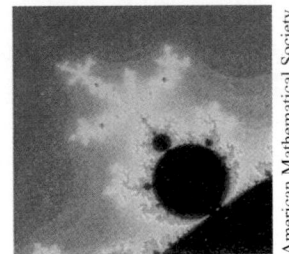

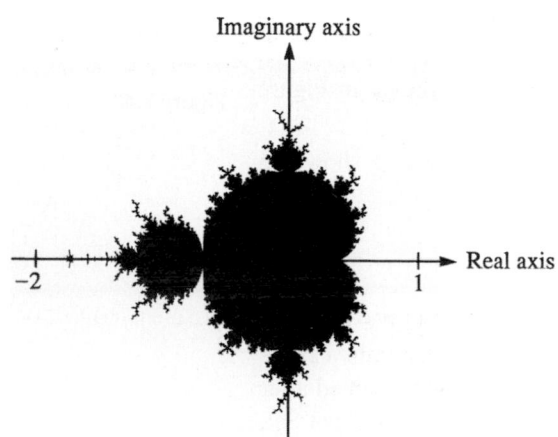

Figure 2.41

American Mathematical Society

Figure 2.42

To add more interest to the picture, computer scientists discovered that the points that are not in the Mandelbrot Set can be assigned a variety of colors, depending on "how quickly" their sequences diverge (become infinitely large). Figure 2.42 shows three different appendages of the Mandelbrot Set. (The black portions of the picture represent points that are in the Mandelbrot Set.)

Figure 2.43 shows another type of fractal. From this picture, you can see why fractals have fascinated people since their discovery (around 1980). The program for creating the fractal fern on a graphing utility is available at this textbook's *Online Study Center.*

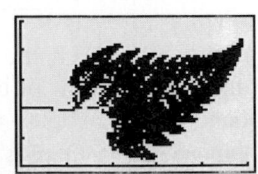

Figure 2.43   *A Fractal Fern*

2.4 Exercises

See www.CalcChat.com for worked-out solutions to odd-numbered exercises.

Vocabulary Check

1. Match the type of complex number with its definition.

(a) real number   (i) $a + bi, a = 0, b \neq 0$

(b) imaginary number   (ii) $a + bi, b = 0$

(c) pure imaginary number   (iii) $a + bi, a \neq 0, b \neq 0$

In Exercises 2–5, fill in the blanks.

2. The imaginary unit i is defined as $i =$ _____ , where $i^2 =$ _____ .

3. The set of real multiples of the imaginary unit i combined with the set of real numbers is called the set of _____ numbers, which are written in the standard form _____ .

4. Complex numbers can be plotted in the complex plane, where the horizontal axis is the _____ axis and the vertical axis is the _____ axis.

5. The most famous fractal is called the _____ .

In Exercises 1–4, find real numbers a and b such that the equation is true.

1. $a + bi = -9 + 4i$ **2.** $a + bi = 12 + 5i$

3. $(a - 1) + (b + 3)i = 5 + 8i$

4. $(a + 6) + 2bi = 6 - 5i$

In Exercises 5–14, write the complex number in standard form.

5. $5 + \sqrt{-16}$ **6.** $2 - \sqrt{-9}$

7. -6 **8.** 8

9. $-5i + i^2$ **10.** $-3i^2 + i$

11. $\left(\sqrt{-75}\right)^2$ **12.** $\left(\sqrt{-4}\right)^2 - 7$

13. $\sqrt{-0.09}$ **14.** $\sqrt{-0.0004}$

In Exercises 15–24, perform the addition or subtraction and write the result in standard form.

15. $(4 + i) - (7 - 2i)$ **16.** $(11 - 2i) - (-3 + 6i)$

17. $\left(-1 + \sqrt{-8}\right) + \left(8 - \sqrt{-50}\right)$

18. $\left(7 + \sqrt{-18}\right) + \left(3 + \sqrt{-32}\right)$

19. $13i - (14 - 7i)$ **20.** $22 + (-5 + 8i) - 10i$

21. $\left(\frac{3}{2} + \frac{5}{2}i\right) + \left(\frac{5}{3} + \frac{11}{3}i\right)$ **22.** $\left(\frac{3}{4} + \frac{7}{5}i\right) - \left(\frac{5}{6} - \frac{1}{6}i\right)$

23. $(1.6 + 3.2i) + (-5.8 + 4.3i)$

24. $-(-3.7 - 12.8i) - \left(6.1 - \sqrt{-24.5}\right)$

In Exercises 25–36, perform the operation and write the result in standard form.

25. $\sqrt{-6} \cdot \sqrt{-2}$ **26.** $\sqrt{-5} \cdot \sqrt{-10}$

27. $\left(\sqrt{-10}\right)^2$ **28.** $\left(\sqrt{-75}\right)^2$

29. $(1 + i)(3 - 2i)$ **30.** $(6 - 2i)(2 - 3i)$

31. $4i(8 + 5i)$ **32.** $-3i(6 - i)$

33. $\left(\sqrt{14} + \sqrt{10}i\right)\left(\sqrt{14} - \sqrt{10}i\right)$

34. $\left(3 + \sqrt{-5}\right)\left(7 - \sqrt{-10}\right)$

35. $(4 + 5i)^2 - (4 - 5i)^2$ **36.** $(1 - 2i)^2 - (1 + 2i)^2$

In Exercises 37–44, write the complex conjugate of the complex number. Then multiply the number by its complex conjugate.

37. $4 + 3i$ **38.** $7 - 5i$

39. $-6 - \sqrt{5}i$ **40.** $-3 + \sqrt{2}i$

41. $\sqrt{-20}$ **42.** $\sqrt{-13}$

43. $3 - \sqrt{-2}$ **44.** $1 + \sqrt{-8}$

In Exercises 45–52, write the quotient in standard form.

45. $\dfrac{6}{i}$ **46.** $-\dfrac{5}{2i}$

47. $\dfrac{2}{4 - 5i}$ **48.** $\dfrac{3}{1 - i}$

49. $\dfrac{2 + i}{2 - i}$ **50.** $\dfrac{8 - 7i}{1 - 2i}$

51. $\dfrac{i}{(4 - 5i)^2}$ **52.** $\dfrac{5i}{(2 + 3i)^2}$

In Exercises 53–56, perform the operation and write the result in standard form.

53. $\dfrac{2}{1 + i} - \dfrac{3}{1 - i}$ **54.** $\dfrac{2i}{2 + i} + \dfrac{5}{2 - i}$

55. $\dfrac{i}{3-2i} + \dfrac{2i}{3+8i}$ **56.** $\dfrac{1+i}{i} - \dfrac{3}{4-i}$

In Exercises 57–62, simplify the complex number and write it in standard form.

57. $-6i^3 + i^2$ **58.** $4i^2 - 2i^3$

59. $\left(\sqrt{-75}\right)^3$ **60.** $\left(\sqrt{-2}\right)^6$

61. $\dfrac{1}{i^3}$ **62.** $\dfrac{1}{(2i)^3}$

63. Cube each complex number. What do you notice?

  (a) 2 (b) $-1+\sqrt{3}i$ (c) $-1-\sqrt{3}i$

64. Raise each complex number to the fourth power and simplify.

  (a) 2 (b) -2 (c) $2i$ (d) $-2i$

In Exercises 65–70, determine the complex number shown in the complex plane.

65.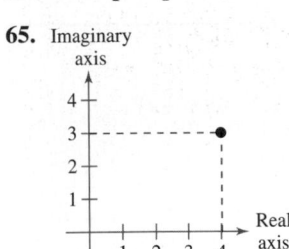

66.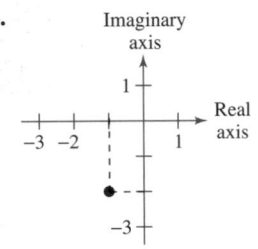

67.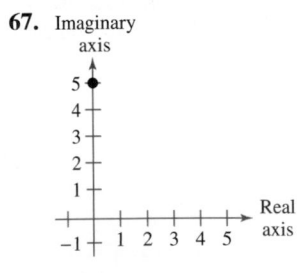

68.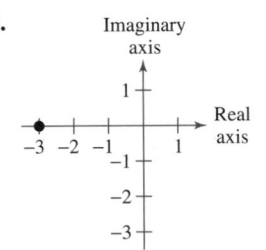

69.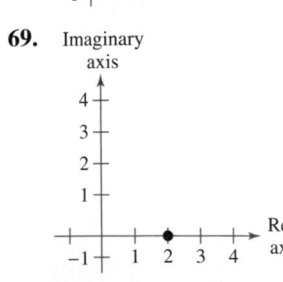

70.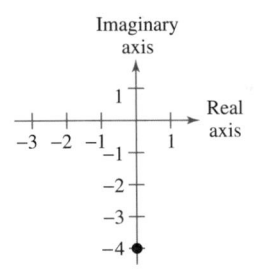

In Exercises 71–76, plot the complex number in the complex plane.

71. $4-5i$ **72.** $-7+2i$

73. $3i$ **74.** $-5i$

75. 1 **76.** -6

Fractals   In Exercises 77 and 78, find the first six terms of the sequence given on page 135. From the terms, do you think the given complex number is in the Mandelbrot Set? Explain your reasoning.

77. $c = \frac{1}{2}i$ **78.** $c = 2$

Impedance   In Exercises 79 and 80, use the following information. The opposition to current in an electrical circuit is called its impedance. The impedance z in a parallel circuit with two pathways satisfies the equation $1/z = 1/z_1 + 1/z_2$ where z_1 is the impedance (in ohms) of pathway 1 and z_2 is the impedance (in ohms) of pathway 2. Use the table to determine the impedance of the parallel circuit. (*Hint:* You can find the impedance of each pathway in a parallel circuit by adding the impedances of all components in the pathway.)

	Resistor	Inductor	Capacitor
Symbol	$a\ \Omega$	$b\ \Omega$	$c\ \Omega$
Impedance	a	bi	$-ci$

79. 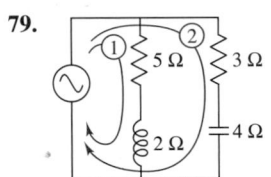   **80.**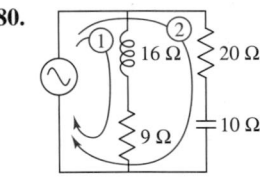

Synthesis

True or False?   In Exercises 81–86, determine whether the statement is true or false. Justify your answer.

81. There is no complex number that is equal to its conjugate.

82. $i^{44} + i^{150} - i^{74} - i^{109} + i^{61} = -1$

83. The sum of two imaginary numbers is always an imaginary number.

84. The product of two imaginary numbers is always an imaginary number.

85. The conjugate of the product of two complex numbers is equal to the product of the conjugates of the two complex numbers.

86. The conjugate of the sum of two complex numbers is equal to the sum of the conjugates of the two complex numbers.

Skills Review

In Exercises 87–90, perform the operation and write the result in standard form.

87. $(4x-5)(4x+5)$ **88.** $(x+2)^3$

89. $\left(3x-\frac{1}{2}\right)(x+4)$ **90.** $(2x-5)^2$

2.5   The Fundamental Theorem of Algebra

The Fundamental Theorem of Algebra

You know that an nth-degree polynomial can have at most n real zeros. In the complex number system, this statement can be improved. That is, in the complex number system, every nth-degree polynomial function has *precisely n zeros*. This important result is derived from the **Fundamental Theorem of Algebra,** first proved by the German mathematician Carl Friedrich Gauss (1777–1855).

> **The Fundamental Theorem of Algebra**
>
> If $f(x)$ is a polynomial of degree n, where $n > 0$, then f has at least one zero in the complex number system.

Using the Fundamental Theorem of Algebra and the equivalence of zeros and factors, you obtain the **Linear Factorization Theorem.**

> **Linear Factorization Theorem**   (See the proof on page 181.)
>
> If $f(x)$ is a polynomial of degree n, where $n > 0$, f has precisely n linear factors
>
> $$f(x) = a_n(x - c_1)(x - c_2) \cdot \cdot \cdot (x - c_n)$$
>
> where $c_1, c_2, \dots, c_n$ are complex numbers.

Note that neither the Fundamental Theorem of Algebra nor the Linear Factorization Theorem tells you *how* to find the zeros or factors of a polynomial. Such theorems are called *existence theorems*. To find the zeros of a polynomial function, you still must rely on other techniques.

Remember that the n zeros of a polynomial function can be real or complex, and they may be repeated. Examples 1 and 2 illustrate several cases.

Example 1   Real Zeros of a Polynomial Function

Counting multiplicity, confirm that the second-degree polynomial function

$$f(x) = x^2 - 6x + 9$$

has exactly *two* zeros: $x = 3$ and $x = 3$.

Solution

$$x^2 - 6x + 9 = (x - 3)^2 = 0$$

$$x - 3 = 0 \quad \Longrightarrow \quad x = 3 \qquad \text{Repeated solution}$$

The graph in Figure 2.44 touches the x-axis at $x = 3$.

✓CHECKPOINT   Now try Exercise 1.

What you should learn

- Use the Fundamental Theorem of Algebra to determine the number of zeros of a polynomial function.
- Find all zeros of polynomial functions, including complex zeros.
- Find conjugate pairs of complex zeros.
- Find zeros of polynomials by factoring.

Why you should learn it

Being able to find zeros of polynomial functions is an important part of modeling real-life problems. For instance, Exercise 63 on page 145 shows how to determine whether a ball thrown with a given velocity can reach a certain height.

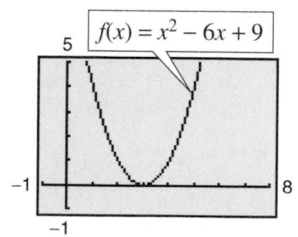

Jed Jacobsohn/Getty Images

$f(x) = x^2 - 6x + 9$

Figure 2.44

Example 2   Real and Complex Zeros of a Polynomial Function

Confirm that the third-degree polynomial function

$$f(x) = x^3 + 4x$$

has exactly three zeros: $x = 0$, $x = 2i$, and $x = -2i$.

Solution

Factor the polynomial completely as $x(x - 2i)(x + 2i)$. So, the zeros are

$$x(x - 2i)(x + 2i) = 0$$
$$x = 0$$
$$x - 2i = 0 \quad \Longrightarrow \quad x = 2i$$
$$x + 2i = 0 \quad \Longrightarrow \quad x = -2i.$$

In the graph in Figure 2.45, only the real zero $x = 0$ appears as an x-intercept.

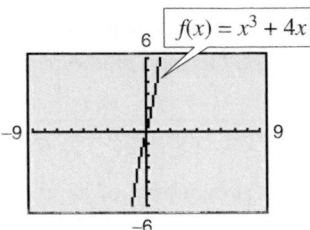

Figure 2.45

✔CHECKPOINT   Now try Exercise 3.

Example 3 shows how to use the methods described in Sections 2.2 and 2.3 (the Rational Zero Test, synthetic division, and factoring) to find all the zeros of a polynomial function, including complex zeros.

Example 3   Finding the Zeros of a Polynomial Function

Write $f(x) = x^5 + x^3 + 2x^2 - 12x + 8$ as the product of linear factors, and list all the zeros of f.

Solution

The possible rational zeros are ± 1, ± 2, ± 4, and ± 8. The graph shown in Figure 2.46 indicates that 1 and -2 are likely zeros, and that 1 is possibly a repeated zero because it appears that the graph touches (but does not cross) the x-axis at this point. Using synthetic division, you can determine that -2 is a zero and 1 is a repeated zero of f. So, you have

$$f(x) = x^5 + x^3 + 2x^2 - 12x + 8 = (x - 1)(x - 1)(x + 2)(x^2 + 4).$$

By factoring $x^2 + 4$ as

$$x^2 - (-4) = \left(x - \sqrt{-4}\right)\left(x + \sqrt{-4}\right) = (x - 2i)(x + 2i)$$

you obtain

$$f(x) = (x - 1)(x - 1)(x + 2)(x - 2i)(x + 2i)$$

which gives the following five zeros of f.

$$x = 1, x = 1, x = -2, x = 2i, \text{ and } x = -2i$$

Note from the graph of f shown in Figure 2.46 that the *real* zeros are the only ones that appear as x-intercepts.

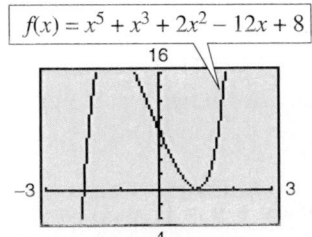

Figure 2.46

✔CHECKPOINT   Now try Exercise 27.

Conjugate Pairs

In Example 3, note that the two complex zeros are **conjugates.** That is, they are of the forms $a + bi$ and $a - bi$.

> **Complex Zeros Occur in Conjugate Pairs**
>
> Let $f(x)$ be a polynomial function that has *real coefficients*. If $a + bi$, where $b \neq 0$, is a zero of the function, the conjugate $a - bi$ is also a zero of the function.

Be sure you see that this result is true only if the polynomial function has *real coefficients*. For instance, the result applies to the function $f(x) = x^2 + 1$, but not to the function $g(x) = x - i$.

Example 4 Finding a Polynomial with Given Zeros

Find a *fourth-degree* polynomial function with real coefficients that has $-1, -1,$ and $3i$ as zeros.

Solution

Because $3i$ is a zero *and* the polynomial is stated to have real coefficients, you know that the conjugate $-3i$ must also be a zero. So, from the Linear Factorization Theorem, $f(x)$ can be written as

$$f(x) = a(x + 1)(x + 1)(x - 3i)(x + 3i).$$

For simplicity, let $a = 1$ to obtain

$$f(x) = (x^2 + 2x + 1)(x^2 + 9) = x^4 + 2x^3 + 10x^2 + 18x + 9.$$

✓**CHECKPOINT**   Now try Exercise 39.

Factoring a Polynomial

The Linear Factorization Theorem states that you can write any nth-degree polynomial as the product of n linear factors.

$$f(x) = a_n(x - c_1)(x - c_2)(x - c_3) \cdots (x - c_n)$$

However, this result includes the possibility that some of the values of c_i are complex. The following theorem states that even if you do not want to get involved with "complex factors," you can still write $f(x)$ as the product of linear and/or quadratic factors.

> **Factors of a Polynomial**   (See the proof on page 181.)
>
> Every polynomial of degree $n > 0$ with real coefficients can be written as the product of linear and quadratic factors with real coefficients, where the quadratic factors have no real zeros.

A quadratic factor with no real zeros is said to be **prime** or **irreducible over the reals.** Be sure you see that this is not the same as being *irreducible over the rationals*. For example, the quadratic

$$x^2 + 1 = (x - i)(x + i)$$

is irreducible over the reals (and therefore over the rationals). On the other hand, the quadratic

$$x^2 - 2 = \left(x - \sqrt{2}\right)\left(x + \sqrt{2}\right)$$

is irreducible over the rationals, but *reducible* over the reals.

> **STUDY TIP**
>
> Recall that irrational and rational numbers are subsets of the set of real numbers, and the real numbers are a subset of the set of complex numbers.

Example 5 Factoring a Polynomial

Write the polynomial

$$f(x) = x^4 - x^2 - 20$$

a. as the product of factors that are irreducible over the *rationals*,

b. as the product of linear factors and quadratic factors that are irreducible over the *reals*, and

c. in completely factored form.

Solution

a. Begin by factoring the polynomial into the product of two quadratic polynomials.

$$x^4 - x^2 - 20 = (x^2 - 5)(x^2 + 4)$$

Both of these factors are irreducible over the rationals.

b. By factoring over the reals, you have

$$x^4 - x^2 - 20 = \left(x + \sqrt{5}\right)\left(x - \sqrt{5}\right)(x^2 + 4)$$

where the quadratic factor is irreducible over the reals.

c. In completely factored form, you have

$$x^4 - x^2 - 20 = \left(x + \sqrt{5}\right)\left(x - \sqrt{5}\right)(x - 2i)(x + 2i).$$

✓**CHECKPOINT**   Now try Exercise 47.

In Example 5, notice from the completely factored form that the *fourth-degree* polynomial has *four* zeros.

Throughout this chapter, the results and theorems have been stated in terms of zeros of polynomial functions. Be sure you see that the same results could have been stated in terms of solutions of polynomial equations. This is true because the zeros of the polynomial function

$$f(x) = a_n x^n + a_{n-1} x^{n-1} + \cdots + a_2 x^2 + a_1 x + a_0$$

are precisely the solutions of the polynomial equation

$$a_n x^n + a_{n-1} x^{n-1} + \cdots + a_2 x^2 + a_1 x + a_0 = 0.$$

Example 6   Finding the Zeros of a Polynomial Function

Find all the zeros of

$$f(x) = x^4 - 3x^3 + 6x^2 + 2x - 60$$

given that $1 + 3i$ is a zero of f.

Algebraic Solution

Because complex zeros occur in conjugate pairs, you know that $1 - 3i$ is also a zero of f. This means that both

$$x - (1 + 3i) \qquad \text{and} \qquad x - (1 - 3i)$$

are factors of f. Multiplying these two factors produces

$$[x - (1 + 3i)][x - (1 - 3i)] = [(x - 1) - 3i][(x - 1) + 3i]$$

$$= (x - 1)^2 - 9i^2$$

$$= x^2 - 2x + 10.$$

Using long division, you can divide $x^2 - 2x + 10$ into f to obtain the following.

$$
\begin{array}{r}
x^2 - x - 6 \\
x^2 - 2x + 10 \overline{\smash{\big)}\ x^4 - 3x^3 + 6x^2 + 2x - 60} \\
\underline{x^4 - 2x^3 + 10x^2} \\
-x^3 - 4x^2 + 2x \\
\underline{-x^3 + 2x^2 - 10x} \\
-6x^2 + 12x - 60 \\
\underline{-6x^2 + 12x - 60} \\
0
\end{array}
$$

So, you have

$$f(x) = (x^2 - 2x + 10)(x^2 - x - 6)$$

$$= (x^2 - 2x + 10)(x - 3)(x + 2)$$

and you can conclude that the zeros of f are $x = 1 + 3i$, $x = 1 - 3i$, $x = 3$, and $x = -2$.

✓**CHECKPOINT**   Now try Exercise 53.

Graphical Solution

Because complex zeros always occur in conjugate pairs, you know that $1 - 3i$ is also a zero of f. Because the polynomial is a fourth-degree polynomial, you know that there are at most two other zeros of the function. Use a graphing utility to graph

$$y = x^4 - 3x^3 + 6x^2 + 2x - 60$$

as shown in Figure 2.47.

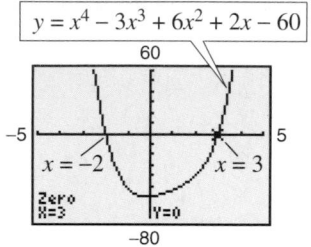

Figure 2.47

You can see that -2 and 3 appear to be x-intercepts of the graph of the function. Use the *zero* or *root* feature or the *zoom* and *trace* features of the graphing utility to confirm that $x = -2$ and $x = 3$ are x-intercepts of the graph. So, you can conclude that the zeros of f are

$$x = 1 + 3i, \ x = 1 - 3i, \ x = 3, \text{ and}$$
$$x = -2.$$

In Example 6, if you were not told that $1 + 3i$ is a zero of f, you could still find all zeros of the function by using synthetic division to find the real zeros -2 and 3. Then, you could factor the polynomial as $(x + 2)(x - 3)(x^2 - 2x + 10)$. Finally, by using the Quadratic Formula, you could determine that the zeros are $x = 1 + 3i$, $x = 1 - 3i$, $x = 3$, and $x = -2$.

2.5 Exercises

See www.CalcChat.com for worked-out solutions to odd-numbered exercises.

Vocabulary Check

Fill in the blanks.

1. The _____ of _____ states that if $f(x)$ is a polynomial function of degree n $(n > 0)$, then f has at least one zero in the complex number system.

2. The _____ states that if $f(x)$ is a polynomial of degree n, then f has precisely n linear factors

$$f(x) = a_n(x - c_1)(x - c_2) \cdots (x - c_n)$$

where $c_1, c_2, \cdots, c_n$ are complex numbers.

3. A quadratic factor that cannot be factored further as a product of linear factors containing real numbers is said to be _____ over the _____ .

4. If $a + bi$ is a complex zero of a polynomial with real coefficients, then so is its _____ .

In Exercises 1–4, find all the zeros of the function.

1. $f(x) = x^2(x + 3)$

2. $g(x) = (x - 2)(x + 4)^3$

3. $f(x) = (x + 9)(x + 4i)(x - 4i)$

4. $h(t) = (t - 3)(t - 2)(t - 3i)(t + 3i)$

Graphical and Analytical Analysis **In Exercises 5–8, find all the zeros of the function. Is there a relationship between the number of real zeros and the number of x-intercepts of the graph? Explain.**

5. $f(x) = x^3 - 4x^2$
   $\quad + x - 4$

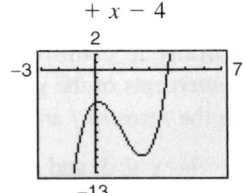

6. $f(x) = x^3 - 4x^2$
   $\quad - 4x + 16$

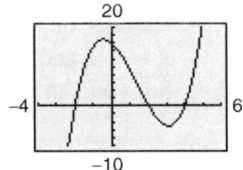

7. $f(x) = x^4 + 4x^2 + 4$

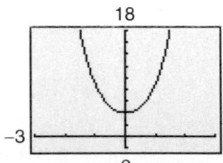

8. $f(x) = x^4 - 3x^2 - 4$

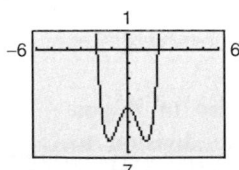

In Exercises 9–28, find all the zeros of the function and write the polynomial as a product of linear factors. Use a graphing utility to graph the function to verify your results graphically. (If possible, use your graphing utility to verify the complex zeros.)

9. $h(x) = x^2 - 4x + 1$

10. $g(x) = x^2 + 10x + 23$

11. $f(x) = x^2 - 12x + 26$

12. $f(x) = x^2 + 6x - 2$

13. $f(x) = x^2 + 25$

14. $f(x) = x^2 + 36$

15. $f(x) = 16x^4 - 81$

16. $f(y) = 81y^4 - 625$

17. $f(z) = z^2 - z + 56$

18. $h(x) = x^2 - 4x - 3$

19. $f(x) = x^4 + 10x^2 + 9$

20. $f(x) = x^4 + 29x^2 + 100$

21. $f(x) = 3x^3 - 5x^2 + 48x - 80$

22. $f(x) = 3x^3 - 2x^2 + 75x - 50$

23. $f(t) = t^3 - 3t^2 - 15t + 125$

24. $f(x) = x^3 + 11x^2 + 39x + 29$

25. $f(x) = 5x^3 - 9x^2 + 28x + 6$

26. $f(s) = 3s^3 - 4s^2 + 8s + 8$

27. $g(x) = x^4 - 4x^3 + 8x^2 - 16x + 16$

28. $h(x) = x^4 + 6x^3 + 10x^2 + 6x + 9$

In Exercises 29–36, (a) find all zeros of the function, (b) write the polynomial as a product of linear factors, and (c) use your factorization to determine the x-intercepts of the graph of the function. Use a graphing utility to verify that the real zeros are the only x-intercepts.

29. $f(x) = x^2 - 14x + 46$

30. $f(x) = x^2 - 12x + 34$

31. $f(x) = 2x^3 - 3x^2 + 8x - 12$

32. $f(x) = 2x^3 - 5x^2 + 18x - 45$

33. $f(x) = x^3 - 11x + 150$

34. $f(x) = x^3 + 10x^2 + 33x + 34$

35. $f(x) = x^4 + 25x^2 + 144$

36. $f(x) = x^4 - 8x^3 + 17x^2 - 8x + 16$

In Exercises 37–42, find a polynomial function with real coefficients that has the given zeros. (There are many correct answers.)

37. $2, i, -i$

38. $3, 4i, -4i$

39. $2, 2, 4 - i$

40. $-1, -1, 2 + 5i$

41. $0, -5, 1 + \sqrt{2}i$

42. $0, 4, 1 + \sqrt{2}i$

In Exercises 43–46, the degree, the zeros, and a solution point of a polynomial function f are given. Write f (a) in completely factored form and (b) in expanded form.

	Degree	Zeros	Solution Point
43.	4	$1, -2, 2i$	$f(-1) = 10$
44.	4	$-1, 2, i$	$f(1) = 8$
45.	3	$-1, 2 + \sqrt{5}i$	$f(-2) = 42$
46.	3	$-2, 2 + 2\sqrt{2}i$	$f(-1) = -34$

In Exercises 47–50, write the polynomial (a) as the product of factors that are irreducible over the *rationals*, (b) as the product of linear and quadratic factors that are irreducible over the *reals*, and (c) in completely factored form.

47. $f(x) = x^4 - 6x^2 - 7$ **48.** $f(x) = x^4 + 6x^2 - 27$

49. $f(x) = x^4 - 2x^3 - 3x^2 + 12x - 18$

(*Hint:* One factor is $x^2 - 6$.)

50. $f(x) = x^4 - 3x^3 - x^2 - 12x - 20$

(*Hint:* One factor is $x^2 + 4$.)

In Exercises 51–58, use the given zero to find all the zeros of the function.

	Function	Zero
51.	$f(x) = 2x^3 + 3x^2 + 50x + 75$	$5i$
52.	$f(x) = x^3 + x^2 + 9x + 9$	$3i$
53.	$g(x) = x^3 - 7x^2 - x + 87$	$5 + 2i$
54.	$g(x) = 4x^3 + 23x^2 + 34x - 10$	$-3 + i$
55.	$h(x) = 3x^3 - 4x^2 + 8x + 8$	$1 - \sqrt{3}i$
56.	$f(x) = x^3 + 4x^2 + 14x + 20$	$-1 - 3i$
57.	$h(x) = 8x^3 - 14x^2 + 18x - 9$	$\frac{1}{2}(1 - \sqrt{5}i)$
58.	$f(x) = 25x^3 - 55x^2 - 54x - 18$	$\frac{1}{5}(-2 + \sqrt{2}i)$

Graphical Analysis   **In Exercises 59–62, (a) use the *zero* or *root* feature of a graphing utility to approximate the zeros of the function accurate to three decimal places and (b) find the exact values of the remaining zeros.**

59. $f(x) = x^4 + 3x^3 - 5x^2 - 21x + 22$

60. $f(x) = x^3 + 4x^2 + 14x + 20$

61. $h(x) = 8x^3 - 14x^2 + 18x - 9$

62. $f(x) = 25x^3 - 55x^2 - 54x - 18$

63. *Height*   A baseball is thrown upward from ground level with an initial velocity of 48 feet per second, and its height h (in feet) is given by

$$h(t) = -16t^2 + 48t, \quad 0 \le t \le 3$$

where t is the time (in seconds). You are told that the ball reaches a height of 64 feet. Is this possible? Explain.

64. *Profit*   The demand equation for a microwave is $p = 140 - 0.0001x$, where p is the unit price (in dollars) of the microwave and x is the number of units produced and sold. The cost equation for the microwave is $C = 80x + 150{,}000$, where C is the total cost (in dollars) and x is the number of units produced. The total profit obtained by producing and selling x units is given by $P = R - C = xp - C$. You are working in the marketing department of the company that produces this microwave, and you are asked to determine a price p that would yield a profit of $9 million. Is this possible? Explain.

Synthesis

True or False?   **In Exercises 65 and 66, decide whether the statement is true or false. Justify your answer.**

65. It is possible for a third-degree polynomial function with integer coefficients to have no real zeros.

66. If $x = 4 + 3i$ is a zero of the function

$$f(x) = x^4 - 7x^3 - 13x^2 + 265x - 750$$

then $x = 4 - 3i$ must also be a zero of f.

67. *Exploration*   Use a graphing utility to graph the function $f(x) = x^4 - 4x^2 + k$ for different values of k. Find values of k such that the zeros of f satisfy the specified characteristics. (Some parts have many correct answers.)

(a) Two real zeros, each of multiplicity 2

(b) Two real zeros and two complex zeros

68. *Writing*   Compile a list of all the various techniques for factoring a polynomial that have been covered so far in the text. Give an example illustrating each technique, and write a paragraph discussing when the use of each technique is appropriate.

Skills Review

In Exercises 69–72, sketch the graph of the quadratic function. Identify the vertex and any intercepts. Use a graphing utility to verify your results.

69. $f(x) = x^2 - 7x - 8$

70. $f(x) = -x^2 + x + 6$

71. $f(x) = 6x^2 + 5x - 6$

72. $f(x) = 4x^2 + 2x - 12$

2.6 Rational Functions and Asymptotes

Introduction to Rational Functions

A **rational function** can be written in the form

$$f(x) = \frac{N(x)}{D(x)}$$

where $N(x)$ and $D(x)$ are polynomials and $D(x)$ is not the zero polynomial.

In general, the *domain* of a rational function of x includes all real numbers except x-values that make the denominator zero. Much of the discussion of rational functions will focus on their graphical behavior near these x-values.

Example 1 Finding the Domain of a Rational Function

Find the domain of $f(x) = 1/x$ and discuss the behavior of f near any excluded x-values.

Solution

Because the denominator is zero when $x = 0$, the domain of f is all real numbers except $x = 0$. To determine the behavior of f near this excluded value, evaluate $f(x)$ to the left and right of $x = 0$, as indicated in the following tables.

x	-1	-0.5	-0.1	-0.01	-0.001	$\to 0$
$f(x)$	-1	-2	-10	-100	-1000	$\to -\infty$

x	$0 \leftarrow$	0.001	0.01	0.1	0.5	1
$f(x)$	$\infty \leftarrow$	1000	100	10	2	1

From the table, note that as x approaches 0 *from the left*, $f(x)$ decreases without bound. In contrast, as x approaches 0 *from the right*, $f(x)$ increases without bound. Because $f(x)$ decreases without bound from the left and increases without bound from the right, you can conclude that f is not continuous. The graph of f is shown in Figure 2.48.

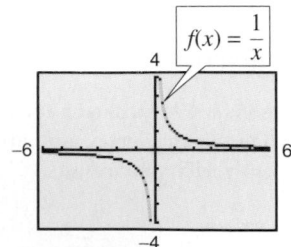

$$f(x) = \frac{1}{x}$$

Figure 2.48

✔CHECKPOINT   Now try Exercise 1.

Library of Parent Functions: Rational Function

A *rational function* $f(x)$ is the quotient of two polynomials,

$$f(x) = \frac{N(x)}{D(x)}.$$

A rational function is not defined at values of x for which $D(x) = 0$. Near these values the graph of the rational function may increase or decrease without bound. The simplest type of rational function is the *reciprocal function* $f(x) = 1/x$. The basic characteristics of the reciprocal function are summarized below. A review of rational functions can be found in the *Study Capsules*.

Graph of $f(x) = \dfrac{1}{x}$

Domain: $(-\infty, 0) \cup (0, \infty)$
Range: $(-\infty, 0) \cup (0, \infty)$
No intercepts
Decreasing on $(-\infty, 0)$ and $(0, \infty)$
Odd function
Origin symmetry
Vertical asymptote: y-axis
Horizontal asymptote: x-axis

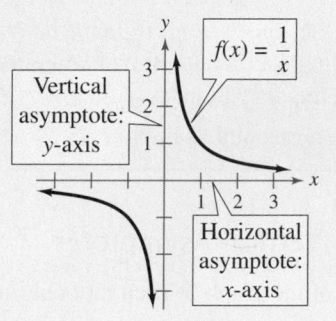

Exploration

Use a table of values to determine whether the functions in Figure 2.49 are continuous. If the graph of a function has an asymptote, can you conclude that the function is not continuous? Explain.

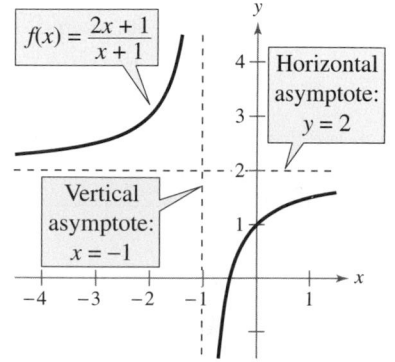

Horizontal and Vertical Asymptotes

In Example 1, the behavior of f near $x = 0$ is denoted as follows.

$\underbrace{f(x) \to -\infty \text{ as } x \to 0^-}$

$f(x)$ decreases without bound as x approaches 0 from the left.

$\underbrace{f(x) \to \infty \text{ as } x \to 0^+}$

$f(x)$ increases without bound as x approaches 0 from the right.

The line $x = 0$ is a **vertical asymptote** of the graph of f, as shown in the figure above. The graph of f has a **horizontal asymptote**—the line $y = 0$. This means the values of $f(x) = 1/x$ approach zero as x increases or decreases without bound.

$\underbrace{f(x) \to 0 \text{ as } x \to -\infty}$

$f(x)$ approaches 0 as x decreases without bound.

$\underbrace{f(x) \to 0 \text{ as } x \to \infty}$

$f(x)$ approaches 0 as x increases without bound.

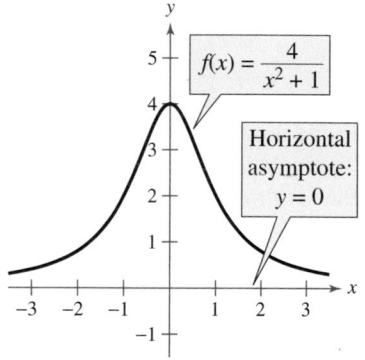

Definition of Vertical and Horizontal Asymptotes

1. The line $x = a$ is a **vertical asymptote** of the graph of f if $f(x) \to \infty$ or $f(x) \to -\infty$ as $x \to a$, either from the right or from the left.

2. The line $y = b$ is a **horizontal asymptote** of the graph of f if $f(x) \to b$ as $x \to \infty$ or $x \to -\infty$.

Figure 2.49 shows the horizontal and vertical asymptotes of the graphs of three rational functions.

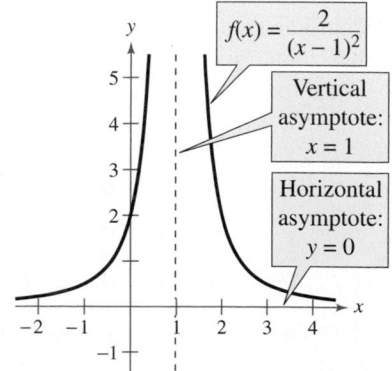

Figure 2.49

Vertical and Horizontal Asymptotes of a Rational Function

Let f be the rational function

$$f(x) = \frac{N(x)}{D(x)} = \frac{a_n x^n + a_{n-1} x^{n-1} + \cdots + a_1 x + a_0}{b_m x^m + b_{m-1} x^{m-1} + \cdots + b_1 x + b_0}$$

where $N(x)$ and $D(x)$ have no common factors.

1. The graph of f has *vertical* asymptotes at the zeros of $D(x)$.

2. The graph of f has at most one *horizontal* asymptote determined by comparing the degrees of $N(x)$ and $D(x)$.

   a. If $n < m$, the graph of f has the line $y = 0$ (the x-axis) as a horizontal asymptote.

   b. If $n = m$, the graph of f has the line $y = a_n/b_m$ as a horizontal asymptote, where a_n is the leading coefficient of the numerator and b_m is the leading coefficient of the denominator.

   c. If $n > m$, the graph of f has no horizontal asymptote.

> **Exploration**
>
> Use a graphing utility to compare the graphs of y_1 and y_2.
>
> $$y_1 = \frac{3x^3 - 5x^2 + 4x - 5}{2x^2 - 6x + 7}$$
>
> $$y_2 = \frac{3x^3}{2x^2}$$
>
> Start with a viewing window in which $-5 \le x \le 5$ and $-10 \le y \le 10$, then zoom out. Write a convincing argument that the shape of the graph of a rational function eventually behaves like the graph of $y = a_n x^n / b_m x^m$, where $a_n x^n$ is the leading term of the numerator and $b_m x^m$ is the leading term of the denominator.

Example 2   Finding Horizontal and Vertical Asymptotes

Find all horizontal and vertical asymptotes of the graph of each rational function.

a. $f(x) = \dfrac{2x}{3x^2 + 1}$ **b.** $f(x) = \dfrac{2x^2}{x^2 - 1}$

Solution

a. For this rational function, the degree of the numerator is *less than* the degree of the denominator, so the graph has the line $y = 0$ as a horizontal asymptote. To find any vertical asymptotes, set the denominator equal to zero and solve the resulting equation for x.

$$3x^2 + 1 = 0 \qquad \text{Set denominator equal to zero.}$$

Because this equation has no real solutions, you can conclude that the graph has no vertical asymptote. The graph of the function is shown in Figure 2.50.

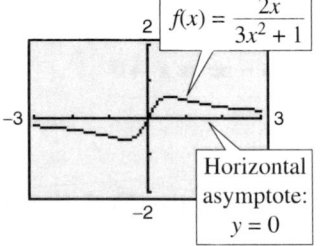

Figure 2.50

b. For this rational function, the degree of the numerator is *equal to* the degree of the denominator. The leading coefficient of the numerator is 2 and the leading coefficient of the denominator is 1, so the graph has the line $y = 2$ as a horizontal asymptote. To find any vertical asymptotes, set the denominator equal to zero and solve the resulting equation for x.

$$x^2 - 1 = 0 \qquad \text{Set denominator equal to zero.}$$

$$(x + 1)(x - 1) = 0 \qquad \text{Factor.}$$

$$x + 1 = 0 \implies x = -1 \qquad \text{Set 1st factor equal to 0.}$$

$$x - 1 = 0 \implies x = 1 \qquad \text{Set 2nd factor equal to 0.}$$

This equation has two real solutions, $x = -1$ and $x = 1$, so the graph has the lines $x = -1$ and $x = 1$ as vertical asymptotes, as shown in Figure 2.51.

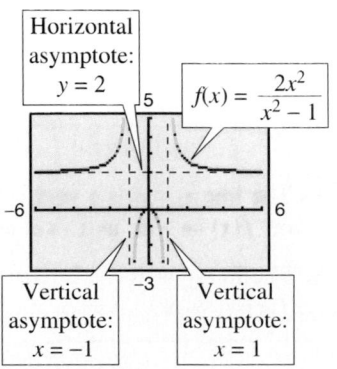

Figure 2.51

✓CHECKPOINT   Now try Exercise 13.

Values for which a rational function is undefined (the denominator is zero) result in a vertical asymptote or a hole in the graph, as shown in Example 3.

Example 3   Finding Horizontal and Vertical Asymptotes and Holes

Find all horizontal and vertical asymptotes and holes in the graph of

$$f(x) = \frac{x^2 + x - 2}{x^2 - x - 6}.$$

 *used in class*

Solution

For this rational function the degree of the numerator is *equal to* the degree of the denominator. The leading coefficients of the numerator and denominator are both 1, so the graph has the line $y = 1$ as a horizontal asymptote. To find any vertical asymptotes, first factor the numerator and denominator as follows.

$$f(x) = \frac{x^2 + x - 2}{x^2 - x - 6} = \frac{(x - 1)(x + 2)}{(x + 2)(x - 3)} = \frac{x - 1}{x - 3}, \quad x \neq -2$$

By setting the denominator $x - 3$ (of the simplified function) equal to zero, you can determine that the graph has the line $x = 3$ as a vertical asymptote, as shown in Figure 2.52. To find any holes in the graph, note that the function is undefined at $x = -2$ and $x = 3$. Because $x = -2$ is not a vertical asymptote of the function, there is a hole in the graph at $x = -2$. To find the y-coordinate of the hole, substitute $x = -2$ into the simplified form of the function.

$$y = \frac{x - 1}{x - 3} = \frac{-2 - 1}{-2 - 3} = \frac{3}{5}$$

So, the graph of the rational function has a hole at $\left(-2, \frac{3}{5}\right)$.

✓CHECKPOINT   Now try Exercise 17.

Example 4   Finding a Function's Domain and Asymptotes

For the function f, find (a) the domain of f, (b) the vertical asymptote of f, and (c) the horizontal asymptote of f.

$$f(x) = \frac{3x^3 + 7x^2 + 2}{-4x^3 + 5}$$

Solution

a. Because the denominator is zero when $-4x^3 + 5 = 0$, solve this equation to determine that the domain of f is all real numbers except $x = \sqrt[3]{\frac{5}{4}}$.

b. Because the denominator of f has a zero at $x = \sqrt[3]{\frac{5}{4}}$, and $\sqrt[3]{\frac{5}{4}}$ is not a zero of the numerator, the graph of f has the vertical asymptote $x = \sqrt[3]{\frac{5}{4}} \approx 1.08$.

c. Because the degrees of the numerator and denominator are the same, and the leading coefficient of the numerator is 3 and the leading coefficient of the denominator is -4, the horizontal asymptote of f is $y = -\frac{3}{4}$.

✓CHECKPOINT   Now try Exercise 19.

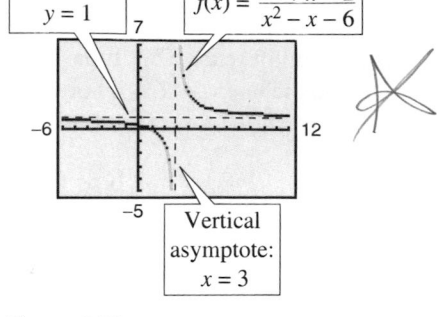

Figure 2.52

TECHNOLOGY TIP

Notice in Figure 2.52 that the function appears to be defined at $x = -2$. Because the domain of the function is all real numbers except $x = -2$ and $x = 3$, you know this is not true. Graphing utilities are limited in their resolution and therefore may not show a break or hole in the graph. Using the *table* feature of a graphing utility, you can verify that the function is not defined at $x = -2$.

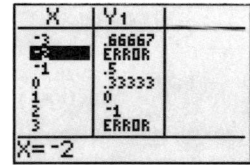

Example 5 A Graph with Two Horizontal Asymptotes

A function that is not rational can have two horizontal asymptotes—one to the left and one to the right. For instance, the graph of

$$f(x) = \frac{x + 10}{|x| + 2}$$

is shown in Figure 2.53. It has the line $y = -1$ as a horizontal asymptote to the left and the line $y = 1$ as a horizontal asymptote to the right. You can confirm this by rewriting the function as follows.

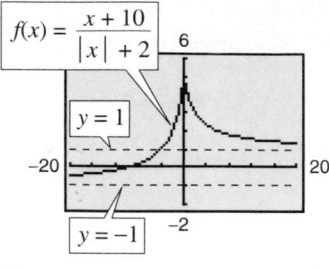

$$f(x) = \begin{cases} \dfrac{x + 10}{-x + 2}, & x < 0 \\[2mm] \dfrac{x + 10}{x + 2}, & x \geq 0 \end{cases}$$

$|x| = -x$ for $x < 0$

$|x| = x$ for $x \geq 0$

Figure 2.53

✓CHECKPOINT   Now try Exercise 21.

Applications

There are many examples of asymptotic behavior in real life. For instance, Example 6 shows how a vertical asymptote can be used to analyze the cost of removing pollutants from smokestack emissions.

Example 6 Cost-Benefit Model

A utility company burns coal to generate electricity. The cost C (in dollars) of removing $p\%$ of the smokestack pollutants is given by $C = 80{,}000p/(100 - p)$ for $0 \leq p < 100$. Use a graphing utility to graph this function. You are a member of a state legislature that is considering a law that would require utility companies to remove 90% of the pollutants from their smokestack emissions. The current law requires 85% removal. How much additional cost would the utility company incur as a result of the new law?

Solution

The graph of this function is shown in Figure 2.54. Note that the graph has a vertical asymptote at $p = 100$. Because the current law requires 85% removal, the current cost to the utility company is

$$C = \frac{80{,}000(85)}{100 - 85} \approx \$453{,}333.$$ Evaluate C at $p = 85$.

If the new law increases the percent removal to 90%, the cost will be

$$C = \frac{80{,}000(90)}{100 - 90} = \$720{,}000.$$ Evaluate C at $p = 90$.

So, the new law would require the utility company to spend an additional

$$720{,}000 - 453{,}333 = \$266{,}667.$$ Subtract 85% removal cost from 90% removal cost.

✓CHECKPOINT   Now try Exercise 39.

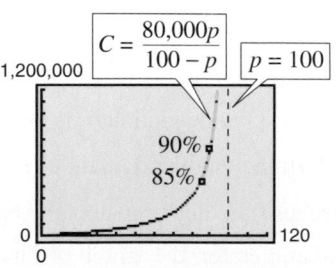

Figure 2.54

Example 7   Ultraviolet Radiation

For a person with sensitive skin, the amount of time T (in hours) the person can be exposed to the sun with minimal burning can be modeled by

$$T = \frac{0.37s + 23.8}{s}, \quad 0 < s \leq 120$$

where s is the Sunsor Scale reading. The Sunsor Scale is based on the level of intensity of UVB rays.   (Source: Sunsor, Inc.)

a. Find the amounts of time a person with sensitive skin can be exposed to the sun with minimal burning when $s = 10$, $s = 25$, and $s = 100$.

b. If the model were valid for all $s > 0$, what would be the horizontal asymptote of this function, and what would it represent?

> **TECHNOLOGY SUPPORT**
>
> For instructions on how to use the *value* feature, see Appendix A; for specific keystrokes, go to this textbook's *Online Study Center*.

Algebraic Solution

a. When $s = 10$, $T = \dfrac{0.37(10) + 23.8}{10}$

$= 2.75$ hours.

When $s = 25$, $T = \dfrac{0.37(25) + 23.8}{25}$

≈ 1.32 hours.

When $s = 100$, $T = \dfrac{0.37(100) + 23.8}{100}$

≈ 0.61 hour.

b. Because the degrees of the numerator and denominator are the same for

$$T = \frac{0.37s + 23.8}{s}$$

the horizontal asymptote is given by the ratio of the leading coefficients of the numerator and denominator. So, the graph has the line $T = 0.37$ as a horizontal asymptote. This line represents the shortest possible exposure time with minimal burning.

Graphical Solution

a. Use a graphing utility to graph the function

$$y_1 = \frac{0.37x + 23.8}{x}$$

using a viewing window similar to that shown in Figure 2.55. Then use the *trace* or *value* feature to approximate the values of y_1 when $x = 10$, $x = 25$, and $x = 100$. You should obtain the following values.

When $x = 10$, $y_1 = 2.75$ hours.

When $x = 25$, $y_1 \approx 1.32$ hours.

When $x = 100$, $y_1 \approx 0.61$ hour.

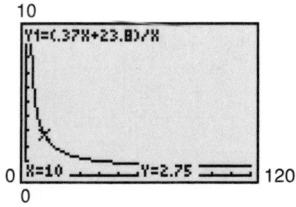

Figure 2.55

b. Continue to use the *trace* or *value* feature to approximate values of $f(x)$ for larger and larger values of x (see Figure 2.56). From this, you can estimate the horizontal asymptote to be $y = 0.37$. This line represents the shortest possible exposure time with minimal burning.

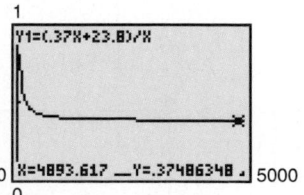

Figure 2.56

✓CHECKPOINT   Now try Exercise 43.

2.6 Exercises

See www.CalcChat.com for worked-out solutions to odd-numbered exercises.

Vocabulary Check

Fill in the blanks.

1. Functions of the form $f(x) = N(x)/D(x)$, where $N(x)$ and $D(x)$ are polynomials and $D(x)$ is not the zero polynomial, are called _____ .

2. If $f(x) \to \pm\infty$ as $x \to a$ from the left (or right), then $x = a$ is a _____ of the graph of f.

3. If $f(x) \to b$ as $x \to \pm\infty$, then $y = b$ is a _____ of the graph of f.

In Exercises 1–6, (a) find the domain of the function, (b) complete each table, and (c) discuss the behavior of f near any excluded x-values.

x	$f(x)$
0.5	
0.9	
0.99	
0.999	

x	$f(x)$
1.5	
1.1	
1.01	
1.001	

x	$f(x)$
5	
10	
100	
1000	

x	$f(x)$
-5	
-10	
-100	
-1000	

1. $f(x) = \dfrac{1}{x-1}$

2. $f(x) = \dfrac{5x}{x-1}$

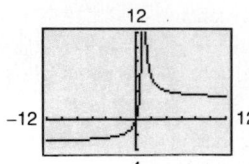

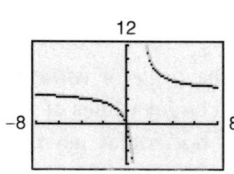

3. $f(x) = \dfrac{3x}{|x-1|}$

4. $f(x) = \dfrac{3}{|x-1|}$

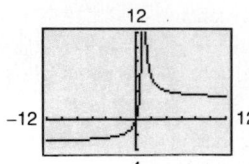

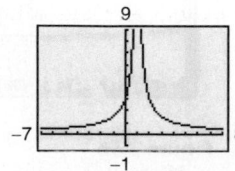

5. $f(x) = \dfrac{3x^2}{x^2-1}$

6. $f(x) = \dfrac{4x}{x^2-1}$

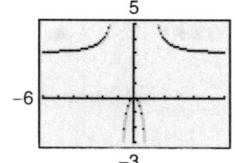

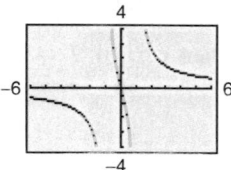

In Exercises 7–12, match the function with its graph. [The graphs are labeled (a), (b), (c), (d), (e), and (f).]

(a)

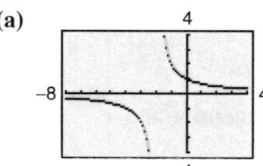

(b)

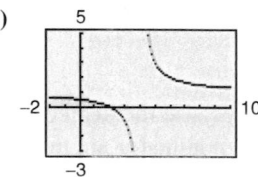

(c)

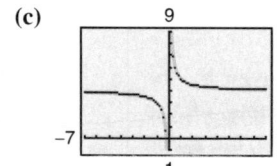

(d)

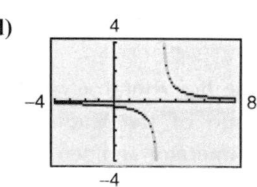

(e)

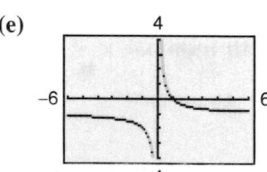

(f)

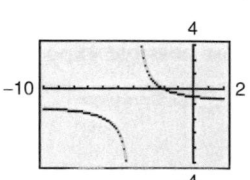

7. $f(x) = \dfrac{2}{x+2}$

8. $f(x) = \dfrac{1}{x-3}$

9. $f(x) = \dfrac{4x+1}{x}$

10. $f(x) = \dfrac{1-x}{x}$

11. $f(x) = \dfrac{x-2}{x-4}$

12. $f(x) = -\dfrac{x+2}{x+4}$

In Exercises 13–18, (a) identify any horizontal and vertical asymptotes and (b) identify any holes in the graph. Verify your answers numerically by creating a table of values.

13. $f(x) = \dfrac{1}{x^2}$

14. $f(x) = \dfrac{3}{(x-2)^3}$

15. $f(x) = \dfrac{x(2+x)}{2x - x^2}$

16. $f(x) = \dfrac{x^2 + 2x + 1}{2x^2 - x - 3}$

17. $f(x) = \dfrac{x^2 - 25}{x^2 + 5x}$

18. $f(x) = \dfrac{3 - 14x - 5x^2}{3 + 7x + 2x^2}$

In Exercises 19–22, (a) find the domain of the function, (b) decide if the function is continuous, and (c) identify any horizontal and vertical asymptotes. Verify your answer to part (a) both graphically by using a graphing utility and numerically by creating a table of values.

19. $f(x) = \dfrac{3x^2 + x - 5}{x^2 + 1}$

20. $f(x) = \dfrac{3x^2 + 1}{x^2 + x + 9}$

21. $f(x) = \dfrac{x - 3}{|x|}$

22. $f(x) = \dfrac{x + 1}{|x| + 1}$

Analytical and Numerical Explanation   **In Exercises 23–26, (a) determine the domains of f and g, (b) simplify f and find any vertical asymptotes of f, (c) identify any holes in the graph of f, (d) complete the table, and (e) explain how the two functions differ.**

23. $f(x) = \dfrac{x^2 - 16}{x - 4}$, $g(x) = x + 4$

x	1	2	3	4	5	6	7
$f(x)$							
$g(x)$							

24. $f(x) = \dfrac{x^2 - 9}{x - 3}$, $g(x) = x + 3$

x	0	1	2	3	4	5	6
$f(x)$							
$g(x)$							

25. $f(x) = \dfrac{x^2 - 1}{x^2 - 2x - 3}$, $g(x) = \dfrac{x - 1}{x - 3}$

x	−2	−1	0	1	2	3	4
$f(x)$							
$g(x)$							

26. $f(x) = \dfrac{x^2 - 4}{x^2 - 3x + 2}$, $g(x) = \dfrac{x + 2}{x - 1}$

x	−3	−2	−1	0	1	2	3
$f(x)$							
$g(x)$							

Exploration   **In Exercises 27–30, determine the value that the function f approaches as the magnitude of x increases. Is $f(x)$ greater than or less than this function value when x is positive and large in magnitude? What about when x is negative and large in magnitude?**

27. $f(x) = 4 - \dfrac{1}{x}$

28. $f(x) = 2 + \dfrac{1}{x - 3}$

29. $f(x) = \dfrac{2x - 1}{x - 3}$

30. $f(x) = \dfrac{2x - 1}{x^2 + 1}$

In Exercises 31–38, find the zeros (if any) of the rational function. Use a graphing utility to verify your answer.

31. $g(x) = \dfrac{x^2 - 4}{x + 3}$

32. $g(x) = \dfrac{x^3 - 8}{x^2 + 4}$

33. $f(x) = 1 - \dfrac{2}{x - 5}$

34. $h(x) = 5 + \dfrac{3}{x^2 + 1}$

35. $g(x) = \dfrac{x^2 - 2x - 3}{x^2 + 1}$

36. $g(x) = \dfrac{x^2 - 5x + 6}{x^2 + 4}$

37. $f(x) = \dfrac{2x^2 - 5x + 2}{2x^2 - 7x + 3}$

38. $f(x) = \dfrac{2x^2 + 3x - 2}{x^2 + x - 2}$

39. *Environment*   The cost C (in millions of dollars) of removing $p\%$ of the industrial and municipal pollutants discharged into a river is given by

$$C = \dfrac{255p}{100 - p}, \quad 0 \le p < 100.$$

(a) Find the cost of removing 10% of the pollutants.

(b) Find the cost of removing 40% of the pollutants.

(c) Find the cost of removing 75% of the pollutants.

(d) Use a graphing utility to graph the cost function. Be sure to choose an appropriate viewing window. Explain why you chose the values that you used in your viewing window.

(e) According to this model, would it be possible to remove 100% of the pollutants? Explain.

40. *Environment*   In a pilot project, a rural township is given recycling bins for separating and storing recyclable products. The cost C (in dollars) for supplying bins to $p\%$ of the population is given by

$$C = \frac{25{,}000p}{100 - p}, \quad 0 \le p < 100.$$

(a) Find the cost of supplying bins to 15% of the population.

(b) Find the cost of supplying bins to 50% of the population.

(c) Find the cost of supplying bins to 90% of the population.

(d) Use a graphing utility to graph the cost function. Be sure to choose an appropriate viewing window. Explain why you chose the values that you used in your viewing window.

(e) According to this model, would it be possible to supply bins to 100% of the residents? Explain.

41. *Data Analysis*   The endpoints of the interval over which distinct vision is possible are called the *near point* and *far point* of the eye (see figure). With increasing age these points normally change. The table shows the approximate near points y (in inches) for various ages x (in years).

Age, x	Near point, y
16	3.0
32	4.7
44	9.8
50	19.7
60	39.4

(a) Find a rational model for the data. Take the reciprocals of the near points to generate the points $(x, 1/y)$. Use the *regression* feature of a graphing utility to find a linear model for the data. The resulting line has the form $1/y = ax + b$. Solve for y.

(b) Use the *table* feature of a graphing utility to create a table showing the predicted near point based on the model for each of the ages in the original table.

(c) Do you think the model can be used to predict the near point for a person who is 70 years old? Explain.

42. *Data Analysis*   Consider a physics laboratory experiment designed to determine an unknown mass. A flexible metal meter stick is clamped to a table with 50 centimeters overhanging the edge (see figure). Known masses M ranging from 200 grams to 2000 grams are attached to the end of the meter stick. For each mass, the meter stick is displaced vertically and then allowed to oscillate. The average time t (in seconds) of one oscillation for each mass is recorded in the table.

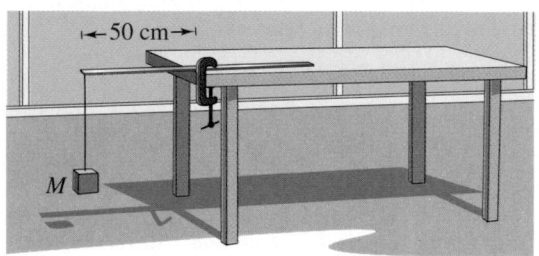

Mass, M	Time, t
200	0.450
400	0.597
600	0.712
800	0.831
1000	0.906
1200	1.003
1400	1.088
1600	1.126
1800	1.218
2000	1.338

A model for the data is given by

$$t = \frac{38M + 16{,}965}{10(M + 5000)}.$$

(a) Use the *table* feature of a graphing utility to create a table showing the estimated time based on the model for each of the masses shown in the table. What can you conclude?

(b) Use the model to approximate the mass of an object when the average time for one oscillation is 1.056 seconds.

43. *Wildlife*   The game commission introduces 100 deer into newly acquired state game lands. The population N of the herd is given by

$$N = \frac{20(5 + 3t)}{1 + 0.04t}, \quad t \geq 0$$

where t is the time in years.

(a) Use a graphing utility to graph the model.

(b) Find the populations when $t = 5$, $t = 10$, and $t = 25$.

(c) What is the limiting size of the herd as time increases? Explain.

44. *Defense*   The table shows the national defense outlays D (in billions of dollars) from 1997 to 2005. The data can be modeled by

$$D = \frac{1.493t^2 - 39.06t + 273.5}{0.0051t^2 - 0.1398t + 1}, \quad 7 \leq t \leq 15$$

where t is the year, with $t = 7$ corresponding to 1997. (Source: U.S. Office of Management and Budget)

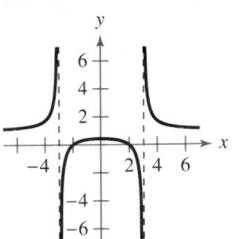

Year	Defense outlays (in billions of dollars)
1997	270.5
1998	268.5
1999	274.9
2000	294.5
2001	305.5
2002	348.6
2003	404.9
2004	455.9
2005	465.9

(a) Use a graphing utility to plot the data and graph the model in the same viewing window. How well does the model represent the data?

(b) Use the model to predict the national defense outlays for the years 2010, 2015, and 2020. Are the predictions reasonable?

(c) Determine the horizontal asymptote of the graph of the model. What does it represent in the context of the situation?

Synthesis

True or False?   In Exercises 45 and 46, determine whether the statement is true or false. Justify your answer.

45. A rational function can have infinitely many vertical asymptotes.

46. A rational function must have at least one vertical asymptote.

Library of Parent Functions   In Exercises 47 and 48, identify the rational function represented by the graph.

47.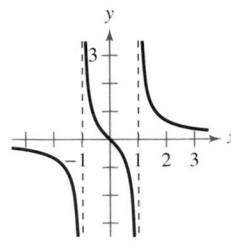

48.

(a) $f(x) = \dfrac{x^2 - 9}{x^2 - 4}$

(b) $f(x) = \dfrac{x^2 - 4}{x^2 - 9}$

(c) $f(x) = \dfrac{x - 4}{x^2 - 9}$

(d) $f(x) = \dfrac{x - 9}{x^2 - 4}$

(a) $f(x) = \dfrac{x^2 - 1}{x^2 + 1}$

(b) $f(x) = \dfrac{x^2 + 1}{x^2 - 1}$

(c) $f(x) = \dfrac{x}{x^2 - 1}$

(d) $f(x) = \dfrac{x}{x^2 + 1}$

Think About It   In Exercises 49–52, write a rational function f that has the specified characteristics. (There are many correct answers.)

49. Vertical asymptote: $x = 2$

Horizontal asymptote: $y = 0$

Zero: $x = 1$

50. Vertical asymptote: $x = -1$

Horizontal asymptote: $y = 0$

Zero: $x = 2$

51. Vertical asymptotes: $x = -2$, $x = 1$

Horizontal asymptote: $y = 2$

Zeros: $x = 3$, $x = -3$

52. Vertical asymptotes: $x = -1$, $x = 2$

Horizontal asymptote: $y = -2$

Zeros: $x = -2$, $x = 3$

Skills Review

In Exercises 53–56, write the general form of the equation of the line that passes through the points.

53. $(3, 2)$, $(0, -1)$

54. $(-6, 1)$, $(4, -5)$

55. $(2, 7)$, $(3, 10)$

56. $(0, 0)$, $(-9, 4)$

In Exercises 57–60, divide using long division.

57. $(x^2 + 5x + 6) \div (x - 4)$

58. $(x^2 - 10x + 15) \div (x - 3)$

59. $(2x^4 + x^2 - 11) \div (x^2 + 5)$

60. $(4x^5 + 3x^3 - 10) \div (2x + 3)$

2.7 Graphs of Rational Functions

The Graph of a Rational Function

To sketch the graph of a rational function, use the following guidelines.

Guidelines for Graphing Rational Functions

Let $f(x) = N(x)/D(x)$, where $N(x)$ and $D(x)$ are polynomials.

1. Simplify f, if possible. Any restrictions on the domain of f not in the simplified function should be listed.

2. Find and plot the y-intercept (if any) by evaluating $f(0)$.

3. Find the zeros of the numerator (if any) by setting the numerator equal to zero. Then plot the corresponding x-intercepts.

4. Find the zeros of the denominator (if any) by setting the denominator equal to zero. Then sketch the corresponding vertical asymptotes using dashed vertical lines and plot the corresponding holes using open circles.

5. Find and sketch any other asymptotes of the graph using dashed lines.

6. Plot at least one point *between* and one point *beyond* each x-intercept and vertical asymptote.

7. Use smooth curves to complete the graph between and beyond the vertical asymptotes, excluding any points where f is not defined.

What you should learn

- Analyze and sketch graphs of rational functions.
- Sketch graphs of rational functions that have slant asymptotes.
- Use rational functions to model and solve real-life problems.

Why you should learn it

The graph of a rational function provides a good indication of the future behavior of a mathematical model. Exercise 86 on page 164 models the population of a herd of elk after their release onto state game lands.

Ed Reschke/Peter Arnold, Inc.

TECHNOLOGY TIP Some graphing utilities have difficulty graphing rational functions that have vertical asymptotes. Often, the utility will connect parts of the graph that are not supposed to be connected. Notice that the graph in Figure 2.57(a) should consist of two *unconnected* portions—one to the left of $x = 2$ and the other to the right of $x = 2$. To eliminate this problem, you can try changing the *mode* of the graphing utility to *dot mode* [see Figure 2.57(b)]. The problem with this mode is that the graph is then represented as a collection of dots rather than as a smooth curve, as shown in Figure 2.57(c). In this text, a blue curve is placed behind the graphing utility's display to indicate where the graph should appear. [See Figure 2.57(c).]

TECHNOLOGY SUPPORT

For instructions on how to use the *connected* mode and the *dot* mode, see Appendix A; for specific keystrokes, go to this textbook's *Online Study Center*.

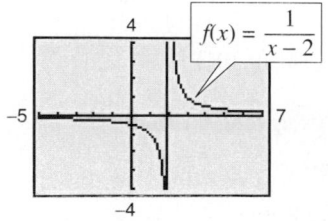

(a) *Connected* mode

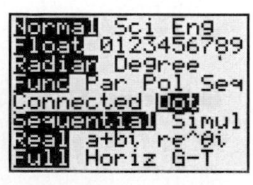

(b) Mode screen

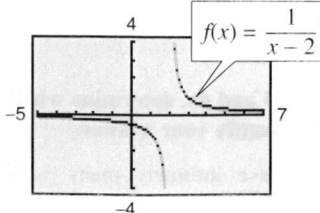

(c) *Dot* mode

Figure 2.57

Example 1   Sketching the Graph of a Rational Function

Sketch the graph of $g(x) = \dfrac{3}{x - 2}$ by hand.

Solution

y-Intercept: $\left(0, -\frac{3}{2}\right)$, because $g(0) = -\frac{3}{2}$

x-Intercept: None, because $3 \neq 0$

Vertical Asymptote: $x = 2$, zero of denominator

Horizontal Asymptote: $y = 0$, because degree of $N(x) <$ degree of $D(x)$

Additional Points:

x	-4	1	2	3	5
$g(x)$	-0.5	-3	Undefined	3	1

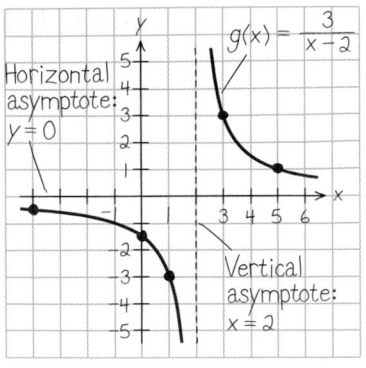

Figure 2.58

By plotting the intercept, asymptotes, and a few additional points, you can obtain the graph shown in Figure 2.58. Confirm this with a graphing utility.

✓CHECKPOINT   Now try Exercise 9.

Note that the graph of g in Example 1 is a vertical stretch and a right shift of the graph of

$$f(x) = \frac{1}{x}$$

because

$$g(x) = \frac{3}{x - 2} = 3\left(\frac{1}{x - 2}\right) = 3f(x - 2).$$

STUDY TIP

Note in the examples in this section that the vertical asymptotes are included in the tables of additional points. This is done to emphasize numerically the behavior of the graph of the function.

Example 2   Sketching the Graph of a Rational Function

Sketch the graph of $f(x) = \dfrac{2x - 1}{x}$ by hand.

Solution

y-Intercept: None, because $x = 0$ is not in the domain

x-Intercept: $\left(\frac{1}{2}, 0\right)$, because $2x - 1 = 0$

Vertical Asymptote: $x = 0$, zero of denominator

Horizontal Asymptote: $y = 2$, because degree of $N(x) =$ degree of $D(x)$

Additional Points:

x	-4	-1	0	$\frac{1}{4}$	4
$f(x)$	2.25	3	Undefined	-2	1.75

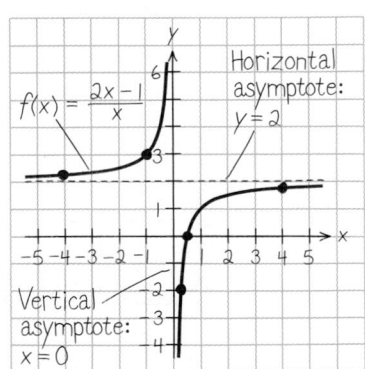

Figure 2.59

By plotting the intercept, asymptotes, and a few additional points, you can obtain the graph shown in Figure 2.59. Confirm this with a graphing utility.

✓CHECKPOINT   Now try Exercise 13.

Example 3   Sketching the Graph of a Rational Function

Sketch the graph of $f(x) = \dfrac{x}{x^2 - x - 2}$.

Solution

Factor the denominator to determine more easily the zeros of the denominator.

$$f(x) = \frac{x}{x^2 - x - 2} = \frac{x}{(x + 1)(x - 2)}.$$

y-Intercept: $(0, 0)$, because $f(0) = 0$

x-Intercept: $(0, 0)$

Vertical Asymptotes: $x = -1$, $x = 2$, zeros of denominator

Horizontal Asymptote: $y = 0$, because degree of $N(x) <$ degree of $D(x)$

Additional Points:

x	-3	-1	-0.5	1	2	3
$f(x)$	-0.3	Undefined	0.4	-0.5	Undefined	0.75

The graph is shown in Figure 2.60.

✓ **CHECKPOINT**   Now try Exercise 21.

Example 4   Sketching the Graph of a Rational Function

Sketch the graph of $f(x) = \dfrac{x^2 - 9}{x^2 - 2x - 3}$.

Solution

By factoring the numerator and denominator, you have

$$f(x) = \frac{x^2 - 9}{x^2 - 2x - 3} = \frac{(x - 3)(x + 3)}{(x - 3)(x + 1)} = \frac{x + 3}{x + 1}, \quad x \neq 3.$$

y-Intercept: $(0, 3)$, because $f(0) = 3$

x-Intercept: $(-3, 0)$

Vertical Asymptote: $x = -1$, zero of (simplified) denominator

Hole: $\left(3, \frac{3}{2}\right)$, f is not defined at $x = 3$

Horizontal Asymptote: $y = 1$, because degree of $N(x) =$ degree of $D(x)$

Additional Points:

x	-5	-2	-1	-0.5	1	3	4
$f(x)$	0.5	-1	Undefined	5	2	Undefined	1.4

The graph is shown in Figure 2.61.

✓ **CHECKPOINT**   Now try Exercise 23.

Exploration

Use a graphing utility to graph

$$f(x) = 1 + \frac{1}{x - \dfrac{1}{x}}.$$

Set the graphing utility to *dot* mode and use a decimal viewing window. Use the *trace* feature to find three "holes" or "breaks" in the graph. Do all three holes represent zeros of the denominator

$$x - \frac{1}{x}?$$

Explain.

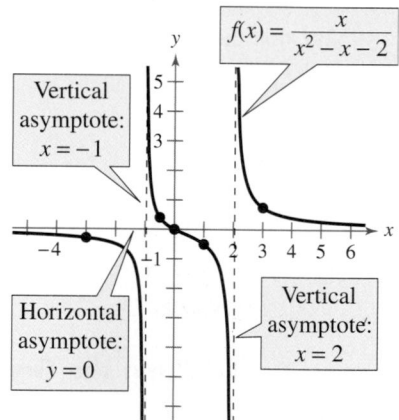

Figure 2.60

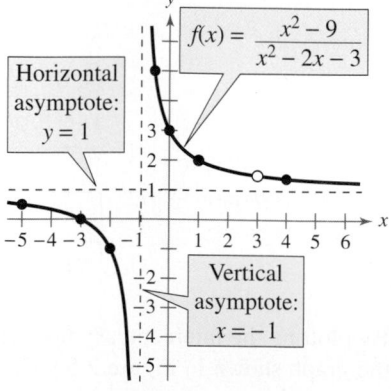

Figure 2.61   *Hole at x = 3*

Slant Asymptotes

Consider a rational function whose denominator is of degree 1 or greater. If the degree of the numerator is exactly *one more* than the degree of the denominator, the graph of the function has a **slant** (or **oblique**) **asymptote.** For example, the graph of

$$f(x) = \frac{x^2 - x}{x + 1}$$

has a slant asymptote, as shown in Figure 2.62. To find the equation of a slant asymptote, use long division. For instance, by dividing $x + 1$ into $x^2 - x$, you have

$$f(x) = \frac{x^2 - x}{x + 1} = \underbrace{x - 2}_{\substack{\text{Slant asymptote} \\ (y = x - 2)}} + \frac{2}{x + 1}.$$

As x increases or decreases without bound, the remainder term $2/(x + 1)$ approaches 0, so the graph of f approaches the line $y = x - 2$, as shown in Figure 2.62.

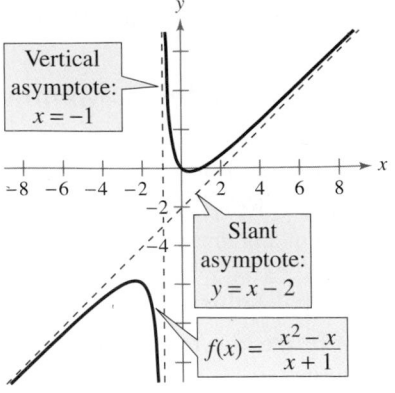

Vertical asymptote: $x = -1$

Slant asymptote: $y = x - 2$

$f(x) = \dfrac{x^2 - x}{x + 1}$

Figure 2.62

Example 5   A Rational Function with a Slant Asymptote

Sketch the graph of $f(x) = \dfrac{x^2 - x - 2}{x - 1}$.

Solution

First write $f(x)$ in two different ways. Factoring the numerator

$$f(x) = \frac{x^2 - x - 2}{x - 1} = \frac{(x - 2)(x + 1)}{x - 1}$$

enables you to recognize the x-intercepts. Long division

$$f(x) = \frac{x^2 - x - 2}{x - 1} = x - \frac{2}{x - 1}$$

enables you to recognize that the line $y = x$ is a slant asymptote of the graph.

y-Intercept:	$(0, 2)$, because $f(0) = 2$
x-Intercepts:	$(-1, 0)$ and $(2, 0)$
Vertical Asymptote:	$x = 1$, zero of denominator
Horizontal Asymptote:	None, because degree of $N(x)$ > degree of $D(x)$
Slant Asymptote:	$y = x$
Additional Points:	

x	-2	0.5	1	1.5	3
$f(x)$	-1.33	4.5	Undefined	-2.5	2

The graph is shown in Figure 2.63.

✓CHECKPOINT   Now try Exercise 45.

Exploration

Do you think it is possible for the graph of a rational function to cross its horizontal asymptote or its slant asymptote? Use the graphs of the following functions to investigate this question. Write a summary of your conclusion. Explain your reasoning.

$$f(x) = \frac{x}{x^2 + 1}$$

$$g(x) = \frac{2x}{3x^2 - 2x + 1}$$

$$h(x) = \frac{x^3}{x^2 + 1}$$

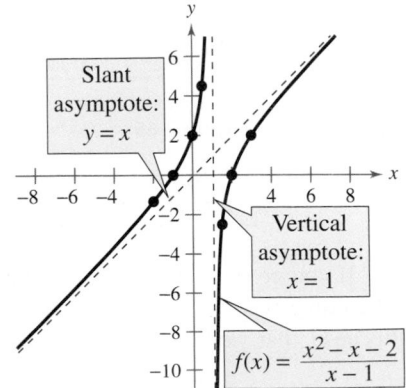

Slant asymptote: $y = x$

Vertical asymptote: $x = 1$

$f(x) = \dfrac{x^2 - x - 2}{x - 1}$

Figure 2.63

Application

Example 6   Finding a Minimum Area

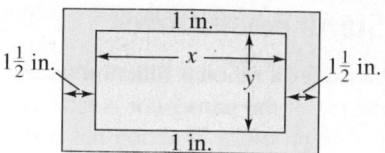

Figure 2.64

A rectangular page is designed to contain 48 square inches of print. The margins on each side of the page are $1\frac{1}{2}$ inches wide. The margins at the top and bottom are each 1 inch deep. What should the dimensions of the page be so that the minimum amount of paper is used?

Graphical Solution

Let A be the area to be minimized. From Figure 2.64, you can write

$$A = (x + 3)(y + 2).$$

The printed area inside the margins is modeled by $48 = xy$ or $y = 48/x$. To find the minimum area, rewrite the equation for A in terms of just one variable by substituting $48/x$ for y.

$$A = (x + 3)\left(\frac{48}{x} + 2\right) = \frac{(x + 3)(48 + 2x)}{x}, \quad x > 0$$

The graph of this rational function is shown in Figure 2.65. Because x represents the width of the printed area, you need consider only the portion of the graph for which x is positive. Using the *minimum* feature or the *zoom* and *trace* features of a graphing utility, you can approximate the minimum value of A to occur when $x \approx 8.5$ inches. The corresponding value of y is $48/8.5 \approx 5.6$ inches. So, the dimensions should be

$$x + 3 \approx 11.5 \text{ inches by } y + 2 \approx 7.6 \text{ inches.}$$

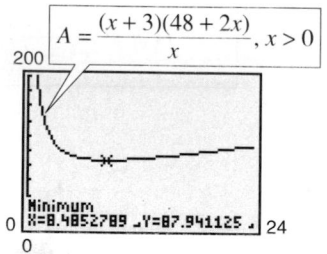

Figure 2.65

✓CHECKPOINT   Now try Exercise 79.

Numerical Solution

Let A be the area to be minimized. From Figure 2.64, you can write

$$A = (x + 3)(y + 2).$$

The printed area inside the margins is modeled by $48 = xy$ or $y = 48/x$. To find the minimum area, rewrite the equation for A in terms of just one variable by substituting $48/x$ for y.

$$A = (x + 3)\left(\frac{48}{x} + 2\right) = \frac{(x + 3)(48 + 2x)}{x}, \quad x > 0$$

Use the *table* feature of a graphing utility to create a table of values for the function

$$y_1 = \frac{(x + 3)(48 + 2x)}{x}$$

beginning at $x = 1$. From the table, you can see that the minimum value of y_1 occurs when x is somewhere between 8 and 9, as shown in Figure 2.66. To approximate the minimum value of y_1 to one decimal place, change the table to begin at $x = 8$ and set the table step to 0.1. The minimum value of y_1 occurs when $x \approx 8.5$, as shown in Figure 2.67. The corresponding value of y is $48/8.5 \approx 5.6$ inches. So, the dimensions should be $x + 3 \approx 11.5$ inches by $y + 2 \approx 7.6$ inches.

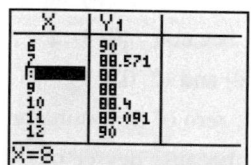

Figure 2.66

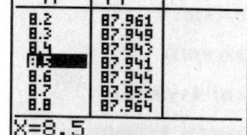

Figure 2.67

If you go on to take a course in calculus, you will learn an analytic technique for finding the exact value of x that produces a minimum area in Example 6. In this case, that value is $x = 6\sqrt{2} \approx 8.485$.

2.7   Exercises

See www.CalcChat.com for worked-out solutions to odd-numbered exercises.

Vocabulary Check

Fill in the blanks.

1. For the rational function $f(x) = N(x)/D(x)$, if the degree of $N(x)$ is exactly one more than the degree of $D(x)$, then the graph of f has a _____ (or oblique) _____ .

2. The graph of $f(x) = 1/x$ has a _____ asymptote at $x = 0$.

In Exercises 1–4, use a graphing utility to graph $f(x) = 2/x$ and the function g in the same viewing window. Describe the relationship between the two graphs.

1. $g(x) = f(x) + 1$

2. $g(x) = f(x - 1)$

3. $g(x) = -f(x)$

4. $g(x) = \frac{1}{2}f(x + 2)$

In Exercises 5–8, use a graphing utility to graph $f(x) = 2/x^2$ and the function g in the same viewing window. Describe the relationship between the two graphs.

5. $g(x) = f(x) - 2$

6. $g(x) = -f(x)$

7. $g(x) = f(x - 2)$

8. $g(x) = \frac{1}{4}f(x)$

In Exercises 9–26, sketch the graph of the rational function by hand. As sketching aids, check for intercepts, vertical asymptotes, horizontal asymptotes, and holes. Use a graphing utility to verify your graph.

9. $f(x) = \dfrac{1}{x + 2}$

10. $f(x) = \dfrac{1}{x - 6}$

11. $C(x) = \dfrac{5 + 2x}{1 + x}$

12. $P(x) = \dfrac{1 - 3x}{1 - x}$

13. $f(t) = \dfrac{1 - 2t}{t}$

14. $g(x) = \dfrac{1}{x + 2} + 2$

15. $f(x) = \dfrac{x^2}{x^2 - 4}$

16. $g(x) = \dfrac{x}{x^2 - 9}$

17. $f(x) = \dfrac{x}{x^2 - 1}$

18. $f(x) = -\dfrac{1}{(x - 2)^2}$

19. $g(x) = \dfrac{4(x + 1)}{x(x - 4)}$

20. $h(x) = \dfrac{2}{x^2(x - 3)}$

21. $f(x) = \dfrac{3x}{x^2 - x - 2}$

22. $f(x) = \dfrac{2x}{x^2 + x - 2}$

23. $f(x) = \dfrac{x^2 + 3x}{x^2 + x - 6}$

24. $g(x) = \dfrac{5(x + 4)}{x^2 + x - 12}$

25. $f(x) = \dfrac{x^2 - 1}{x + 1}$

26. $f(x) = \dfrac{x^2 - 16}{x - 4}$

In Exercises 27–36, use a graphing utility to graph the function. Determine its domain and identify any vertical or horizontal asymptotes.

27. $f(x) = \dfrac{2 + x}{1 - x}$

28. $f(x) = \dfrac{3 - x}{2 - x}$

29. $f(t) = \dfrac{3t + 1}{t}$

30. $h(x) = \dfrac{x - 2}{x - 3}$

31. $h(t) = \dfrac{4}{t^2 + 1}$

32. $g(x) = -\dfrac{x}{(x - 2)^2}$

33. $f(x) = \dfrac{x + 1}{x^2 - x - 6}$

34. $f(x) = \dfrac{x + 4}{x^2 + x - 6}$

35. $f(x) = \dfrac{20x}{x^2 + 1} - \dfrac{1}{x}$

36. $f(x) = 5\left(\dfrac{1}{x - 4} - \dfrac{1}{x + 2}\right)$

Exploration   **In Exercises 37–42, use a graphing utility to graph the function. What do you observe about its asymptotes?**

37. $h(x) = \dfrac{6x}{\sqrt{x^2 + 1}}$

38. $f(x) = -\dfrac{x}{\sqrt{9 + x^2}}$

39. $g(x) = \dfrac{4|x - 2|}{x + 1}$

40. $f(x) = -\dfrac{8|3 + x|}{x - 2}$

41. $f(x) = \dfrac{4(x - 1)^2}{x^2 - 4x + 5}$

42. $g(x) = \dfrac{3x^4 - 5x + 3}{x^4 + 1}$

In Exercises 43–50, sketch the graph of the rational function by hand. As sketching aids, check for intercepts, vertical asymptotes, and slant asymptotes.

43. $f(x) = \dfrac{2x^2 + 1}{x}$

44. $g(x) = \dfrac{1 - x^2}{x}$

45. $h(x) = \dfrac{x^2}{x - 1}$

46. $f(x) = \dfrac{x^3}{x^2 - 1}$

47. $g(x) = \dfrac{x^3}{2x^2 - 8}$

48. $f(x) = \dfrac{x^2 - 1}{x^2 + 4}$

49. $f(x) = \dfrac{x^3 + 2x^2 + 4}{2x^2 + 1}$

50. $f(x) = \dfrac{2x^2 - 5x + 5}{x - 2}$

Graphical Reasoning   In Exercises 51–54, use the graph to estimate any *x*-intercepts of the rational function. Set $y = 0$ and solve the resulting equation to confirm your result.

51. $y = \dfrac{x + 1}{x - 3}$

52. $y = \dfrac{2x}{x - 3}$

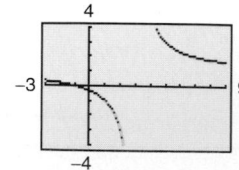

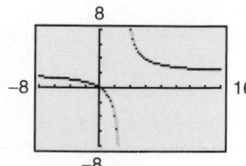

53. $y = \dfrac{1}{x} - x$

54. $y = x - 3 + \dfrac{2}{x}$

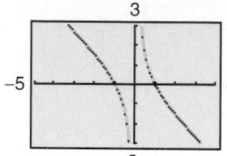

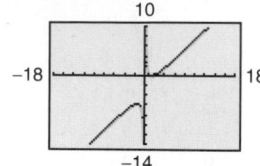

In Exercises 55–58, use a graphing utility to graph the rational function. Determine the domain of the function and identify any asymptotes.

55. $y = \dfrac{2x^2 + x}{x + 1}$

56. $y = \dfrac{x^2 + 5x + 8}{x + 3}$

57. $y = \dfrac{1 + 3x^2 - x^3}{x^2}$

58. $y = \dfrac{12 - 2x - x^2}{2(4 + x)}$

In Exercises 59–64, find all vertical asymptotes, horizontal asymptotes, slant asymptotes, and holes in the graph of the function. Then use a graphing utility to verify your result.

59. $f(x) = \dfrac{x^2 - 5x + 4}{x^2 - 4}$

60. $f(x) = \dfrac{x^2 - 2x - 8}{x^2 - 9}$

61. $f(x) = \dfrac{2x^2 - 5x + 2}{2x^2 - x - 6}$

62. $f(x) = \dfrac{3x^2 - 8x + 4}{2x^2 - 3x - 2}$

63. $f(x) = \dfrac{2x^3 - x^2 - 2x + 1}{x^2 + 3x + 2}$

64. $f(x) = \dfrac{2x^3 + x^2 - 8x - 4}{x^2 - 3x + 2}$

Graphical Reasoning   In Exercises 65–76, use a graphing utility to graph the function and determine any *x*-intercepts. Set $y = 0$ and solve the resulting equation to confirm your result.

65. $y = \dfrac{1}{x + 5} + \dfrac{4}{x}$

66. $y = \dfrac{2}{x + 1} - \dfrac{3}{x}$

67. $y = \dfrac{1}{x + 2} + \dfrac{2}{x + 4}$

68. $y = \dfrac{2}{x + 2} - \dfrac{3}{x - 1}$

69. $y = x - \dfrac{6}{x - 1}$

70. $y = x - \dfrac{9}{x}$

71. $y = x + 2 - \dfrac{1}{x + 1}$

72. $y = 2x - 1 + \dfrac{1}{x - 2}$

73. $y = x + 1 + \dfrac{2}{x - 1}$

74. $y = x + 2 + \dfrac{2}{x + 2}$

75. $y = x + 3 - \dfrac{2}{2x - 1}$

76. $y = x - 1 - \dfrac{2}{2x - 3}$

77. *Concentration of a Mixture*   A 1000-liter tank contains 50 liters of a 25% brine solution. You add *x* liters of a 75% brine solution to the tank.

(a) Show that the concentration *C*, the proportion of brine to the total solution, of the final mixture is given by

$$C = \dfrac{3x + 50}{4(x + 50)}.$$

(b) Determine the domain of the function based on the physical constraints of the problem.

(c) Use a graphing utility to graph the function. As the tank is filled, what happens to the rate at which the concentration of brine increases? What percent does the concentration of brine appear to approach?

78. *Geometry*   A rectangular region of length *x* and width *y* has an area of 500 square meters.

(a) Write the width *y* as a function of *x*.

(b) Determine the domain of the function based on the physical constraints of the problem.

(c) Sketch a graph of the function and determine the width of the rectangle when $x = 30$ meters.

79. **Page Design**   A page that is x inches wide and y inches high contains 30 square inches of print. The margins at the top and bottom are 2 inches deep and the margins on each side are 1 inch wide (see figure).

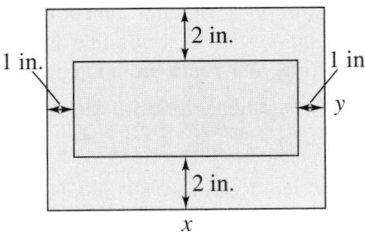

(a) Show that the total area A of the page is given by

$$A = \frac{2x(2x + 11)}{x - 2}.$$

(b) Determine the domain of the function based on the physical constraints of the problem.

(c) Use a graphing utility to graph the area function and approximate the page size such that the minimum amount of paper will be used. Verify your answer numerically using the *table* feature of a graphing utility.

80. **Geometry**   A right triangle is formed in the first quadrant by the x-axis, the y-axis, and a line segment through the point $(3, 2)$ (see figure).

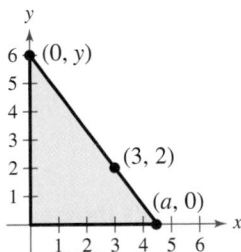

(a) Show that an equation of the line segment is given by

$$y = \frac{2(a - x)}{a - 3}, \quad 0 \le x \le a.$$

(b) Show that the area of the triangle is given by

$$A = \frac{a^2}{a - 3}.$$

(c) Use a graphing utility to graph the area function and estimate the value of a that yields a minimum area. Estimate the minimum area. Verify your answer numerically using the *table* feature of a graphing utility.

81. **Cost**   The ordering and transportation cost C (in thousands of dollars) for the components used in manufacturing a product is given by

$$C = 100\left(\frac{200}{x^2} + \frac{x}{x + 30}\right), \quad x \ge 1$$

where x is the order size (in hundreds). Use a graphing utility to graph the cost function. From the graph, estimate the order size that minimizes cost.

82. **Average Cost**   The cost C of producing x units of a product is given by $C = 0.2x^2 + 10x + 5$, and the average cost per unit is given by

$$\overline{C} = \frac{C}{x} = \frac{0.2x^2 + 10x + 5}{x}, \quad x > 0.$$

Sketch the graph of the average cost function, and estimate the number of units that should be produced to minimize the average cost per unit.

83. **Medicine**   The concentration C of a chemical in the bloodstream t hours after injection into muscle tissue is given by

$$C = \frac{3t^2 + t}{t^3 + 50}, \quad t \ge 0.$$

(a) Determine the horizontal asymptote of the function and interpret its meaning in the context of the problem.

(b) Use a graphing utility to graph the function and approximate the time when the bloodstream concentration is greatest.

(c) Use a graphing utility to determine when the concentration is less than 0.345.

84. **Numerical and Graphical Analysis**   A driver averaged 50 miles per hour on the round trip between Baltimore, Maryland and Philadelphia, Pennsylvania, 100 miles away. The average speeds for going and returning were x and y miles per hour, respectively.

(a) Show that $y = \dfrac{25x}{x - 25}$.

(b) Determine the vertical and horizontal asymptotes of the function.

(c) Use a graphing utility to complete the table. What do you observe?

x	30	35	40	45	50	55	60
y							

(d) Use a graphing utility to graph the function.

(e) Is it possible to average 20 miles per hour in one direction and still average 50 miles per hour on the round trip? Explain.

85. *Comparing Models* The numbers of people A (in thousands) attending women's NCAA Division I college basketball games from 1990 to 2004 are shown in the table. Let t represent the year, with $t = 0$ corresponding to 1990. (Source: NCAA)

Year	Attendance, A (in thousands)
1990	2,777
1991	3,013
1992	3,397
1993	4,193
1994	4,557
1995	4,962
1996	5,234
1997	6,734
1998	7,387
1999	8,010
2000	8,698
2001	8,825
2002	9,533
2003	10,164
2004	10,016

(a) Use the *regression* feature of a graphing utility to find a linear model for the data. Use a graphing utility to plot the data and graph the model in the same viewing window.

(b) Find a rational model for the data. Take the reciprocal of A to generate the points $(t, 1/A)$. Use the *regression* feature of a graphing utility to find a linear model for this data. The resulting line has the form $1/A = at + b$. Solve for A. Use a graphing utility to plot the data and graph the rational model in the same viewing window.

(c) Use the *table* feature of a graphing utility to create a table showing the predicted attendance based on each model for each of the years in the original table. Which model do you prefer? Why?

86. *Elk Population* A herd of elk is released onto state game lands. The expected population P of the herd can be modeled by the equation $P = (10 + 2.7t)/(1 + 0.1t)$, where t is the time in years since the initial number of elk were released.

(a) State the domain of the model. Explain your answer.

(b) Find the initial number of elk in the herd.

(c) Find the populations of elk after 25, 50, and 100 years.

(d) Is there a limit to the size of the herd? If so, what is the expected population?

Use a graphing utility to confirm your results for parts (a) through (d).

Synthesis

True or False? In Exercises 87 and 88, determine whether the statement is true or false. Justify your answer.

87. If the graph of a rational function f has a vertical asymptote at $x = 5$, it is possible to sketch the graph without lifting your pencil from the paper.

88. The graph of a rational function can never cross one of its asymptotes.

Think About It In Exercises 89 and 90, use a graphing utility to graph the function. Explain why there is no vertical asymptote when a superficial examination of the function might indicate that there should be one.

89. $h(x) = \dfrac{6 - 2x}{3 - x}$

90. $g(x) = \dfrac{x^2 + x - 2}{x - 1}$

Think About It In Exercises 91 and 92, write a rational function satisfying the following criteria. (There are many correct answers.)

91. Vertical asymptote: $x = -2$

Slant asymptote: $y = x + 1$

Zero of the function: $x = 2$

92. Vertical asymptote: $x = -4$

Slant asymptote: $y = x - 2$

Zero of the function: $x = 3$

Skills Review

In Exercises 93–96, simplify the expression.

93. $\left(\dfrac{x}{8}\right)^{-3}$

94. $(4x^2)^{-2}$

95. $\dfrac{3^{7/6}}{3^{1/6}}$

96. $\dfrac{(x^{-2})(x^{1/2})}{(x^{-1})(x^{5/2})}$

In Exercises 97–100, use a graphing utility to graph the function and find its domain and range.

97. $f(x) = \sqrt{6 + x^2}$

98. $f(x) = \sqrt{121 - x^2}$

99. $f(x) = -|x + 9|$

100. $f(x) = -x^2 + 9$

101. *Make a Decision* To work an extended application analyzing the total manpower of the Department of Defense, visit this textbook's *Online Study Center*. (Data Source: U.S. Department of Defense)

2.8   Quadratic Models

Classifying Scatter Plots

In real life, many relationships between two variables are parabolic, as in Section 2.1, Example 5. A scatter plot can be used to give you an idea of which type of model will best fit a set of data.

Example 1   Classifying Scatter Plots

Decide whether each set of data could be better modeled by a linear model, $y = ax + b$, or a quadratic model, $y = ax^2 + bx + c$.

a. (0.9, 1.4), (1.3, 1.5), (1.3, 1.9), (1.4, 2.1), (1.6, 2.8), (1.8, 2.9), (2.1, 3.4), (2.1, 3.4), (2.5, 3.6), (2.9, 3.7), (3.2, 4.2), (3.3, 4.3), (3.6, 4.4), (4.0, 4.5), (4.2, 4.8), (4.3, 5.0)

b. (0.9, 2.5), (1.3, 4.03), (1.3, 4.1), (1.4, 4.4), (1.6, 5.1), (1.8, 6.05), (2.1, 7.48), (2.1, 7.6), (2.5, 9.8), (2.9, 12.4), (3.2, 14.3), (3.3, 15.2), (3.6, 18.1), (4.0, 19.9), (4.2, 23.0), (4.3, 23.9)

Solution

Begin by entering the data into a graphing utility, as shown in Figure 2.68.

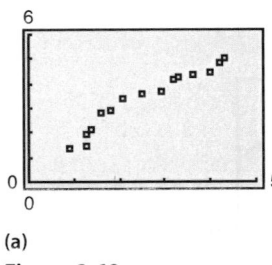

(a)

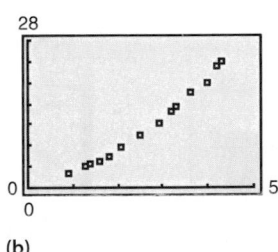

(b)

Figure 2.68

Then display the scatter plots, as shown in Figure 2.69.

(a)

(b)

Figure 2.69

From the scatter plots, it appears that the data in part (a) follow a linear pattern. So, it can be better modeled by a linear function. The data in part (b) follow a parabolic pattern. So, it can be better modeled by a quadratic function.

✓**CHECKPOINT**   Now try Exercise 3.

What you should learn

- Classify scatter plots.
- Use scatter plots and a graphing utility to find quadratic models for data.
- Choose a model that best fits a set of data.

Why you should learn it

Many real-life situations can be modeled by quadratic equations. For instance, in Exercise 15 on page 169, a quadratic equation is used to model the monthly precipitation for San Francisco, California.

Justin Sullivan/Getty Images

Fitting a Quadratic Model to Data

In Section 1.7, you created scatter plots of data and used a graphing utility to find the least squares regression lines for the data. You can use a similar procedure to find a model for nonlinear data. Once you have used a scatter plot to determine the type of model that would best fit a set of data, there are several ways that you can actually find the model. Each method is best used with a computer or calculator, rather than with hand calculations.

Example 2   Fitting a Quadratic Model to Data

A study was done to compare the speed x (in miles per hour) with the mileage y (in miles per gallon) of an automobile. The results are shown in the table. (Source: Federal Highway Administration)

Speed, x	Mileage, y
15	22.3
20	25.5
25	27.5
30	29.0
35	28.8
40	30.0
45	29.9
50	30.2
55	30.4
60	28.8
65	27.4
70	25.3
75	23.3

a. Use a graphing utility to create a scatter plot of the data.

b. Use the *regression* feature of the graphing utility to find a model that best fits the data.

c. Approximate the speed at which the mileage is the greatest.

Solution

a. Begin by entering the data into a graphing utility and displaying the scatter plot, as shown in Figure 2.70. From the scatter plot, you can see that the data appears to follow a parabolic pattern.

b. Using the *regression* feature of a graphing utility, you can find the quadratic model, as shown in Figure 2.71. So, the quadratic equation that best fits the data is given by

$$y = -0.0082x^2 + 0.746x + 13.47. \qquad \text{Quadratic model}$$

c. Graph the data and the model in the same viewing window, as shown in Figure 2.72. Use the *maximum* feature or the *zoom* and *trace* features of the graphing utility to approximate the speed at which the mileage is greatest. You should obtain a maximum of approximately (45, 30), as shown in Figure 2.72. So, the speed at which the mileage is greatest is about 47 miles per hour.

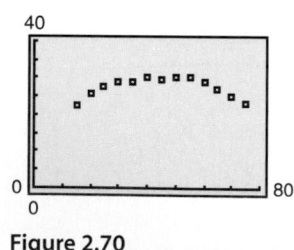

Figure 2.70

Figure 2.71

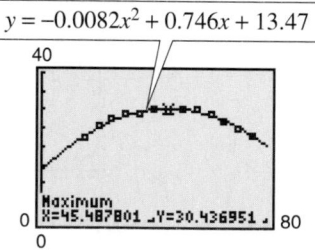

Figure 2.72

✓CHECKPOINT   Now try Exercise 15.

TECHNOLOGY SUPPORT   For instructions on how to use the *regression* feature, see Appendix A; for specific keystrokes, go to this textbook's *Online Study Center*.

Example 3   Fitting a Quadratic Model to Data

A basketball is dropped from a height of about 5.25 feet. The height of the basketball is recorded 23 times at intervals of about 0.02 second.* The results are shown in the table. Use a graphing utility to find a model that best fits the data. Then use the model to predict the time when the basketball will hit the ground.

Solution

Begin by entering the data into a graphing utility and displaying the scatter plot, as shown in Figure 2.73. From the scatter plot, you can see that the data has a parabolic trend. So, using the *regression* feature of the graphing utility, you can find the quadratic model, as shown in Figure 2.74. The quadratic model that best fits the data is given by

$$y = -15.449x^2 - 1.30x + 5.2.$$ Quadratic model

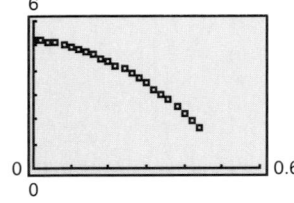

Figure 2.73

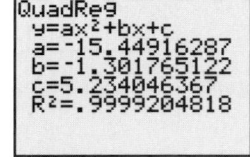

Figure 2.74

Time, x	Height, y
0.0	5.23594
0.02	5.20353
0.04	5.16031
0.06	5.09910
0.08	5.02707
0.099996	4.95146
0.119996	4.85062
0.139992	4.74979
0.159988	4.63096
0.179988	4.50132
0.199984	4.35728
0.219984	4.19523
0.23998	4.02958
0.25993	3.84593
0.27998	3.65507
0.299976	3.44981
0.319972	3.23375
0.339961	3.01048
0.359961	2.76921
0.379951	2.52074
0.399941	2.25786
0.419941	1.98058
0.439941	1.63488

Using this model, you can predict the time when the basketball will hit the ground by substituting 0 for y and solving the resulting equation for x.

$$y = -15.449x^2 - 1.30x + 5.2$$ Write original model.

$$0 = -15.449x^2 - 1.30x + 5.2$$ Substitute 0 for y.

$$x = \frac{-b \pm \sqrt{b^2 - 4ac}}{2a}$$ Quadratic Formula

$$= \frac{-(-1.30) \pm \sqrt{(-1.30)^2 - 4(-15.449)(5.2)}}{2(-15.449)}$$ Substitute for a, b, and c.

$$\approx 0.54$$ Choose positive solution.

So, the solution is about 0.54 second. In other words, the basketball will continue to fall for about $0.54 - 0.44 = 0.1$ second more before hitting the ground.

✓CHECKPOINT   Now try Exercise 17.

Choosing a Model

Sometimes it is not easy to distinguish from a scatter plot which type of model will best fit the data. You should first find several models for the data, using the *Library of Functions,* and then choose the model that best fits the data by comparing the y-values of each model with the actual y-values.

*Data was collected with a Texas Instruments CBL (Calculator-Based Laboratory) System.

Example 4   Choosing a Model

The table shows the amounts y (in billions of dollars) spent on admission to movie theaters in the United States for the years 1997 to 2003. Use the *regression* feature of a graphing utility to find a linear model and a quadratic model for the data. Determine which model better fits the data. (Source: U.S. Bureau of Economic Analysis)

Year	Amount, y
1997	6.3
1998	6.9
1999	7.9
2000	8.6
2001	9.0
2002	9.6
2003	9.9

Solution

Let x represent the year, with $x = 7$ corresponding to 1997. Begin by entering the data into the graphing utility. Then use the *regression* feature to find a linear model (see Figure 2.75) and a quadratic model (see Figure 2.76) for the data.

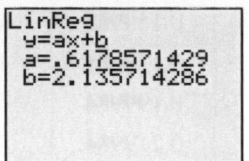

Figure 2.75   *Linear Model*

Figure 2.76   *Quadratic Model*

So, a linear model for the data is given by

$$y = 0.62x + 2.1 \qquad \text{Linear model}$$

and a quadratic model for the data is given by

$$y = -0.049x^2 + 1.59x - 2.6. \qquad \text{Quadratic model}$$

Plot the data and the linear model in the same viewing window, as shown in Figure 2.77. Then plot the data and the quadratic model in the same viewing window, as shown in Figure 2.78. To determine which model fits the data better, compare the y-values given by each model with the actual y-values. The model whose y-values are closest to the actual values is the better fit. In this case, the better-fitting model is the quadratic model.

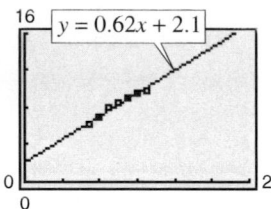

Figure 2.77

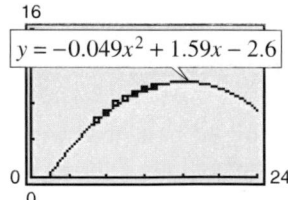

Figure 2.78

✓CHECKPOINT   Now try Exercise 18.

TECHNOLOGY TIP   When you use the regression feature of a graphing utility, the program may output an "r^2-value." This r^2-value is the **coefficient of determination** of the data and gives a measure of how well the model fits the data. The coefficient of determination for the linear model in Example 4 is $r^2 \approx 0.97629$ and the coefficient of determination for the quadratic model is $r^2 \approx 0.99456$. Because the coefficient of determination for the quadratic model is closer to 1, the quadratic model better fits the data.

2.8 Exercises

See www.CalcChat.com for worked-out solutions to odd-numbered exercises.

Vocabulary Check

Fill in the blanks.

1. A scatter plot with either a positive or a negative correlation can be better modeled by a _____ equation.

2. A scatter plot that appears parabolic can be better modeled by a _____ equation.

In Exercises 1–6, determine whether the scatter plot could best be modeled by a linear model, a quadratic model, or neither.

1.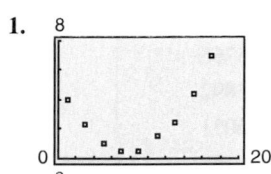

2.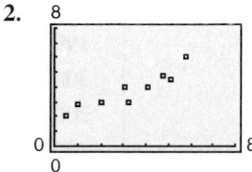

3.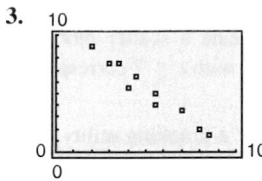

4.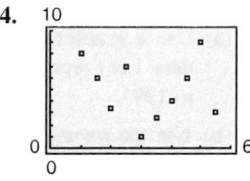

5.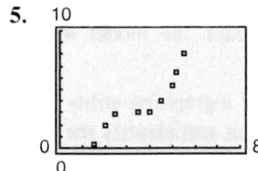

6.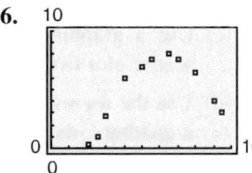

In Exercises 7–10, (a) use a graphing utility to create a scatter plot of the data, (b) determine whether the data could be better modeled by a linear model or a quadratic model, (c) use the *regression* feature of a graphing utility to find a model for the data, (d) use a graphing utility to graph the model with the scatter plot from part (a), and (e) create a table comparing the original data with the data given by the model.

7. $(0, 2.1)$, $(1, 2.4)$, $(2, 2.5)$, $(3, 2.8)$, $(4, 2.9)$, $(5, 3.0)$, $(6, 3.0)$, $(7, 3.2)$, $(8, 3.4)$, $(9, 3.5)$, $(10, 3.6)$

8. $(-2, 11.0)$, $(-1, 10.7)$, $(0, 10.4)$, $(1, 10.3)$, $(2, 10.1)$, $(3, 9.9)$, $(4, 9.6)$, $(5, 9.4)$, $(6, 9.4)$, $(7, 9.2)$, $(8, 9.0)$

9. $(0, 3480)$, $(5, 2235)$, $(10, 1250)$, $(15, 565)$, $(20, 150)$, $(25, 12)$, $(30, 145)$, $(35, 575)$, $(40, 1275)$, $(45, 2225)$, $(50, 3500)$, $(55, 5010)$

10. $(0, 6140)$, $(2, 6815)$, $(4, 7335)$, $(6, 7710)$, $(8, 7915)$, $(10, 7590)$, $(12, 7975)$, $(14, 7700)$, $(16, 7325)$, $(18, 6820)$, $(20, 6125)$, $(22, 5325)$

In Exercises 11–14, (a) use the *regression* feature of a graphing utility to find a linear model and a quadratic model for the data, (b) determine the coefficient of determination for each model, and (c) use the coefficient of determination to determine which model fits the data better.

11. $(1, 4.0)$, $(2, 6.5)$, $(3, 8.8)$, $(4, 10.6)$, $(5, 13.9)$, $(6, 15.0)$, $(7, 17.5)$, $(8, 20.1)$, $(9, 24.0)$, $(10, 27.1)$

12. $(0, 0.1)$, $(1, 2.0)$, $(2, 4.1)$, $(3, 6.3)$, $(4, 8.3)$, $(5, 10.5)$, $(6, 12.6)$, $(7, 14.5)$, $(8, 16.8)$, $(9, 19.0)$

13. $(-6, 10.7)$, $(-4, 9.0)$, $(-2, 7.0)$, $(0, 5.4)$, $(2, 3.5)$, $(4, 1.7)$, $(6, -0.1)$, $(8, -1.8)$, $(10, -3.6)$, $(12, -5.3)$

14. $(-20, 805)$, $(-15, 744)$, $(-10, 704)$, $(-5, 653)$, $(0, 587)$, $(5, 551)$, $(10, 512)$, $(15, 478)$, $(20, 436)$, $(25, 430)$

15. **Meteorology** The table shows the monthly normal precipitation P (in inches) for San Francisco, California. (Source: U.S. National Oceanic and Atmospheric Administration)

Month	Precipitation, P
January	4.45
February	4.01
March	3.26
April	1.17
May	0.38
June	0.11
July	0.03
August	0.07
September	0.20
October	1.40
November	2.49
December	2.89

(a) Use a graphing utility to create a scatter plot of the data. Let t represent the month, with $t = 1$ corresponding to January.

(b) Use the *regression* feature of a graphing utility to find a quadratic model for the data.

(c) Use a graphing utility to graph the model with the scatter plot from part (a).

(d) Use the graph from part (c) to determine in which month the normal precipitation in San Francisco is the least.

16. *Sales*   The table shows the sales S (in millions of dollars) for jogging and running shoes from 1998 to 2004. (Source: National Sporting Goods Association)

Year	Sales, S (in millions of dollars)
1998	1469
1999	1502
2000	1638
2001	1670
2002	1733
2003	1802
2004	1838

(a) Use a graphing utility to create a scatter plot of the data. Let t represent the year, with $t = 8$ corresponding to 1998.

(b) Use the *regression* feature of a graphing utility to find a quadratic model for the data.

(c) Use a graphing utility to graph the model with the scatter plot from part (a).

(d) Use the model to find when sales of jogging and running shoes will exceed 2 billion dollars.

(e) Is this a good model for predicting future sales? Explain.

17. *Sales*   The table shows college textbook sales S (in millions of dollars) in the United States from 2000 to 2005. (Source: Book Industry Study Group, Inc.)

Year	Textbook sales, S (in millions of dollars)
2000	4265.2
2001	4570.7
2002	4899.1
2003	5085.9
2004	5478.6
2005	5703.2

(a) Use a graphing utility to create a scatter plot of the data. Let t represent the year, with $t = 0$ corresponding to 2000.

(b) Use the *regression* feature of a graphing utility to find a quadratic model for the data.

(c) Use a graphing utility to graph the model with the scatter plot from part (a).

(d) Use the model to find when the sales of college textbooks will exceed 10 billion dollars.

(e) Is this a good model for predicting future sales? Explain.

18. *Media*   The table shows the numbers S of FM radio stations in the United States from 1997 to 2003. (Source: Federal Communications Commission)

Year	FM stations, S
1997	5542
1998	5662
1999	5766
2000	5892
2001	6051
2002	6161
2003	6207

(a) Use a graphing utility to create a scatter plot of the data. Let t represent the year, with $t = 7$ corresponding to 1997.

(b) Use the *regression* feature of a graphing utility to find a linear model for the data and identify the coefficient of determination.

(c) Use a graphing utility to graph the model with the scatter plot from part (a).

(d) Use the *regression* feature of a graphing utility to find a quadratic model for the data and identify the coefficient of determination.

(e) Use a graphing utility to graph the quadratic model with the scatter plot from part (a).

(f) Which model is a better fit for the data?

(g) Use each model to find when the number of FM stations will exceed 7000.

19. *Entertainment*   The table shows the amounts A (in dollars) spent per person on the Internet in the United States from 2000 to 2005. (Source: Veronis Suhler Stevenson)

Year	Amount, A (in dollars)
2000	49.64
2001	68.94
2002	84.76
2003	96.35
2004	107.02
2005	117.72

(a) Use a graphing utility to create a scatter plot of the data. Let t represent the year, with $t = 0$ corresponding to 2000.

(b) A cubic model for the data is $S = 0.25444t^3 - 3.0440t^2 + 22.485t + 49.55$ which has an r^2-value of 0.99992. Use a graphing utility to graph this model with the scatter plot from part (a). Is the cubic model a good fit for the data? Explain.

(c) Use the *regression* feature of a graphing utility to find a quadratic model for the data and identify the coefficient of determination.

(d) Use a graphing utility to graph the quadratic model with the scatter plot from part (a). Is the quadratic model a good fit for the data? Explain.

(e) Which model is a better fit for the data? Explain.

(f) The projected amounts $A*$ spent per person on the Internet for the years 2006 to 2008 are shown in the table. Use the models from parts (b) and (c) to predict the amount spent for the same years. Explain why your values may differ from those in the table.

Year	2006	2007	2008
$A*$	127.76	140.15	154.29

20. ***Entertainment*** The table shows the amounts A (in hours) of time per person spent watching television and movies, listening to recorded music, playing video games, and reading books and magazines in the United States from 2000 to 2005. (Source: Veronis Suhler Stevenson)

Year	Amount, A (in hours)
2000	3492
2001	3540
2002	3606
2003	3663
2004	3757
2005	3809

(a) Use a graphing utility to create a scatter plot of the data. Let t represent the year, with $t = 0$ corresponding to 2000.

(b) A cubic model for the data is $A = -1.500t^3 + 13.61t^2 + 33.2t + 3493$ which has an r^2-value of 0.99667. Use a graphing utility to graph this model with the scatter plot from part (a). Is the cubic model a good fit for the data? Explain.

(c) Use the *regression* feature of a graphing utility to find a quadratic model for the data and identify the coefficient of determination.

(d) Use a graphing utility to graph the quadratic model with the scatter plot from part (a). Is the quadratic model a good fit for the data? Explain.

(e) Which model is a better fit for the data? Explain.

(f) The projected amounts $A*$ of time spent per person for the years 2006 to 2008 are shown in the table. Use the models from parts (b) and (c) to predict the number of hours for the same years. Explain why your values may differ from those in the table.

Year	2006	2007	2008
$A*$	3890	3949	4059

Synthesis

True or False? **In Exercises 21 and 22, determine whether the statement is true or false. Justify your answer.**

21. The graph of a quadratic model with a negative leading coefficient will have a maximum value at its vertex.

22. The graph of a quadratic model with a positive leading coefficient will have a minimum value at its vertex.

23. ***Writing*** Explain why the parabola shown in the figure is not a good fit for the data.

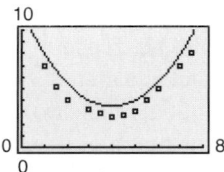

Skills Review

In Exercises 24–27, find (a) $f \circ g$ and (b) $g \circ f$.

24. $f(x) = 2x - 1, \quad g(x) = x^2 + 3$
25. $f(x) = 5x + 8, \quad g(x) = 2x^2 - 1$
26. $f(x) = x^3 - 1, \quad g(x) = \sqrt[3]{x + 1}$
27. $f(x) = \sqrt[3]{x + 5}, \quad g(x) = x^3 - 5$

In Exercises 28–31, determine algebraically whether the function is one-to-one. If it is, find its inverse function. Verify your answer graphically.

28. $f(x) = 2x + 5$

29. $f(x) = \dfrac{x - 4}{5}$

30. $f(x) = x^2 + 5, x \geq 0$

31. $f(x) = 2x^2 - 3, x \geq 0$

In Exercises 32–35, plot the complex number in the complex plane.

32. $1 - 3i$

33. $-2 + 4i$

34. $-5i$

35. $8i$

What Did You Learn?

Key Terms

polynomial function, *p. 92*
linear function, *p. 92*
quadratic function, *p. 92*
continuous, *p. 103*
Leading Coefficient Test, *p. 105*

repeated zeros, *p. 108*
multiplicity, *p. 108*
Intermediate Value Theorem, *p. 111*
synthetic division, *p. 119*
Descartes's Rule of Signs, *p. 124*

upper and lower bounds, *p. 125*
imaginary number, *p. 131*
complex conjugates, *p. 134*
rational function, *p. 146*
slant (oblique) asymptote, *p. 159*

Key Concepts

2.1 ■ Analyze graphs of quadratic functions

The graph of the quadratic function
$f(x) = a(x - h)^2 + k$, $a \neq 0$, is a parabola whose axis
is the vertical line $x = h$ and whose vertex is the point
(h, k). If $a > 0$, the parabola opens upward, and if
$a < 0$, the parabola opens downward.

2.2 ■ Analyze graphs of polynomial functions

1. The graph of the polynomial function $f(x) = a_n x^n + a_{n-1}x^{n-1} + \cdots + a_2 x^2 + a_1 x + a_0$ is smooth
   and continuous, and rises or falls as x moves without
   bound to the left or to the right depending on the
   values of n and a_n.

2. If f is a polynomial function and a is a real number,
   $x = a$ is a zero of the function f, $x = a$ is a solution
   of the polynomial equation $f(x) = 0$, $(x - a)$ is
   a factor of the polynomial $f(x)$, and $(a, 0)$ is an
   x-intercept of the graph of f.

2.3 ■ Rational zeros of polynomial functions

The Rational Zero Test states: If the polynomial
$f(x) = a_n x^n + a_{n-1}x^{n-1} + \cdots + a_2 x^2 + a_1 x + a_0$
has integer coefficients, every rational zero of f has the
form p/q, where p and q have no common factors other
than 1, p is a factor of the constant term a_0, and q is a
factor of the leading coefficient a_n.

2.4 ■ Perform operations with complex numbers and plot complex numbers

1. If a and b are real numbers and $i = \sqrt{-1}$, the
   number $a + bi$ is a complex number written in
   standard form.

2. Add: $(a + bi) + (c + di) = (a + c) + (b + d)i$

   Subtract: $(a + bi) - (c + di) = (a - c) + (b - d)i$

   Multiply: $(a + bi)(c + di) = (ac - bd) + (ad + bc)i$

   Divide: $\dfrac{a + bi}{c + di}\left(\dfrac{c - di}{c - di}\right) = \dfrac{ac + bd}{c^2 + d^2} + \left(\dfrac{bc - ad}{c^2 + d^2}\right)i$

2.5 ■ Real and complex zeros of polynomials

1. The Fundamental Theorem of Algebra states: If $f(x)$
   is a polynomial of degree n, where $n > 0$, then f has
   at least one zero in the complex number system.

2. The Linear Factorization Theorem states: If $f(x)$ is a
   polynomial of degree n, where $n > 0$, f has precisely
   n linear factors $f(x) = a_n(x - c_1)(x - c_2)$
   $\cdots (x - c_n)$, where $c_1, c_2, \ldots, c_n$ are complex
   numbers.

3. Let $f(x)$ be a polynomial function that has real
   coefficients. If $a + bi$, where $b \neq 0$, is a zero of the
   function, the conjugate $a - bi$ is also a zero.

2.6 ■ Domains and asymptotes of rational functions

1. The domain of a rational function of x includes
   all real numbers except x-values that make the
   denominator 0.

2. Let f be the rational function $f(x) = N(x)/D(x)$,
   where $N(x)$ and $D(x)$ have no common factors. The
   graph of f has vertical asymptotes at the zeros of
   $D(x)$. The graph of f has at most one horizontal
   asymptote determined by comparing the degrees
   of $N(x)$ and $D(x)$.

2.7 ■ Sketch the graphs of rational functions

Find and plot the y-intercept. Find the zeros of the
numerator and plot the corresponding x-intercepts. Find
the zeros of the denominator, sketch the corresponding
vertical asymptotes, and plot the corresponding holes.
Find and sketch any other asymptotes. Plot at least one
point between and one point beyond each x-intercept
and vertical asymptote. Use smooth curves to complete
the graph between and beyond the vertical asymptotes.

2.8 ■ Find quadratic models for data

1. Use the *regression* feature of a graphing utility to
   find a quadratic function to model a data set.

2. Compare correlation coefficients to determine
   whether a linear model or a quadratic model is a
   better fit for the data set.

Review Exercises

See www.CalcChat.com for worked-out solutions to odd-numbered exercises.

2.1 In Exercises 1 and 2, use a graphing utility to graph each function in the same viewing window. Describe how the graph of each function is related to the graph of $y = x^2$.

1. (a) $y = 2x^2$ (b) $y = -2x^2$

   (c) $y = x^2 + 2$ (d) $y = (x + 5)^2$

2. (a) $y = x^2 - 3$ (b) $y = 3 - x^2$

   (c) $y = (x - 4)^2$ (d) $y = \frac{1}{2}x^2 + 4$

In Exercises 3–8, sketch the graph of the quadratic function. Identify the vertex and the intercept(s).

3. $f(x) = \left(x + \frac{3}{2}\right)^2 + 1$

4. $f(x) = (x - 4)^2 - 4$

5. $f(x) = \frac{1}{3}(x^2 + 5x - 4)$

6. $f(x) = 3x^2 - 12x + 11$

7. $f(x) = 3 - x^2 - 4x$

8. $f(x) = 30 + 23x + 3x^2$

In Exercises 9–12, write the standard form of the quadratic function that has the indicated vertex and whose graph passes through the given point. Verify your result with a graphing utility.

9. Vertex: $(1, -4)$;   Point: $(2, -3)$

10. Vertex: $(2, 3)$;   Point: $(0, 2)$

11. Vertex: $(-2, -2)$;   Point: $(-1, 0)$

12. Vertex: $\left(-\frac{1}{4}, \frac{3}{2}\right)$;   Point: $(-2, 0)$

13. *Numerical, Graphical, and Analytical Analysis* A rectangle is inscribed in the region bounded by the x-axis, the y-axis, and the graph of $x + 2y - 8 = 0$, as shown in the figure.

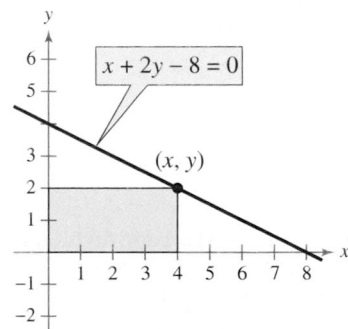

(a) Write the area A of the rectangle as a function of x. Determine the domain of the function in the context of the problem.

(b) Use the *table* feature of a graphing utility to create a table showing possible values of x and the corresponding areas of the rectangle. Use the table to estimate the dimensions that will produce a maximum area.

(c) Use a graphing utility to graph the area function. Use the graph to approximate the dimensions that will produce a maximum area.

(d) Write the area function in standard form to find algebraically the dimensions that will produce a maximum area.

(e) Compare your results from parts (b), (c), and (d).

14. *Cost* A textile manufacturer has daily production costs of

$$C = 10,000 - 110x + 0.45x^2$$

where C is the total cost (in dollars) and x is the number of units produced. Use the *table* feature of a graphing utility to determine how many units should be produced each day to yield a minimum cost.

15. *Gardening* A gardener has 1500 feet of fencing to enclose three adjacent rectangular gardens, as shown in the figure. Determine the dimensions that will produce a maximum enclosed area.

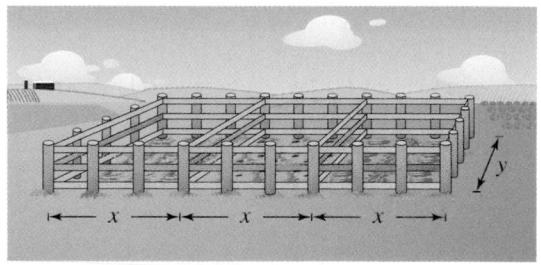

16. *Profit* An online music company sells songs for $1.75 each. The company's cost C per week is given by the model

$$C = 0.0005x^2 + 500$$

where x is the number of songs sold. Therefore, the company's profit P per week is given by the model

$$P = 1.75x - (0.0005x^2 + 500).$$

(a) Use a graphing utility to graph the profit function.

(b) Use the *maximum* feature of the graphing utility to find the number of songs per week that the company needs to sell to maximize their profit.

(c) Confirm your answer to part (b) algebraically.

(d) Determine the company's maximum profit per week.

2.2 In Exercises 17 and 18, sketch the graph of $y = x^n$ and each specified transformation.

17. $y = x^5$

   (a) $f(x) = (x + 4)^5$ (b) $f(x) = x^5 + 1$

   (c) $f(x) = 3 - \frac{1}{2}x^5$ (d) $f(x) = 2(x + 3)^5$

18. $y = x^6$

   (a) $f(x) = x^6 - 2$ (b) $f(x) = -\frac{1}{4}x^6$

   (c) $f(x) = -\frac{1}{2}x^6 - 5$ (d) $f(x) = -(x + 7)^6 + 2$

Graphical Analysis   In Exercises 19 and 20, use a graphing utility to graph the functions f and g in the same viewing window. Zoom out far enough so that the right-hand and left-hand behaviors of f and g appear identical. Show both graphs.

19. $f(x) = \frac{1}{3}x^3 - 2x + 1, \quad g(x) = \frac{1}{3}x^3$

20. $f(x) = -x^4 + 2x^3, \quad g(x) = -x^4$

In Exercises 21–24, use the Leading Coefficient Test to describe the right-hand and left-hand behavior of the graph of the polynomial function.

21. $f(x) = -x^2 + 6x + 9$

22. $f(x) = \frac{1}{3}x^3 + 2x$

23. $g(x) = \frac{3}{4}(x^4 + 3x^2 + 2)$

24. $h(x) = -x^5 - 7x^2 + 10x$

In Exercises 25–30, (a) find the zeros algebraically, (b) use a graphing utility to graph the function, and (c) use the graph to approximate any zeros and compare them with those in part (a).

25. $g(x) = x^4 - x^3 - 2x^2$ **26.** $h(x) = -2x^3 - x^2 + x$

27. $f(t) = t^3 - 3t$ **28.** $f(x) = -(x + 6)^3 - 8$

29. $f(x) = x(x + 3)^2$ **30.** $f(t) = t^4 - 4t^2$

In Exercises 31–34, find a polynomial function that has the given zeros. (There are many correct answers.)

31. $-2, 1, 1, 5$

32. $-3, 0, 1, 4$

33. $3, 2 - \sqrt{3}, 2 + \sqrt{3}$

34. $-7, 4 - \sqrt{6}, 4 + \sqrt{6}$

In Exercises 35 and 36, sketch the graph of the function by (a) applying the Leading Coefficient Test, (b) finding the zeros of the polynomial, (c) plotting sufficient solution points, and (d) drawing a continuous curve through the points.

35. $f(x) = 18 + 27x - 2x^2 - 3x^3$

36. $f(x) = 18 + 27x - 2x^2 - 3x^3$

In Exercises 37–40, (a) use the Intermediate Value Theorem and a graphing utility to find graphically any intervals of length 1 in which the polynomial function is guaranteed to have a zero and, (b) use the *zero* or *root* feature of a graphing utility to approximate the real zeros of the function. Verify your results in part (a) by using the *table* feature of a graphing utility.

37. $f(x) = x^3 + 2x^2 - x - 1$

38. $f(x) = 0.24x^3 - 2.6x - 1.4$

39. $f(x) = x^4 - 6x^2 - 4$

40. $f(x) = 2x^4 + \frac{7}{2}x^3 - 2$

2.3 *Graphical Analysis*   In Exercises 41–44, use a graphing utility to graph the two equations in the same viewing window. Use the graphs to verify that the expressions are equivalent. Verify the results algebraically.

41. $y_1 = \dfrac{x^2}{x - 2}, \quad y_2 = x + 2 + \dfrac{4}{x - 2}$

42. $y_1 = \dfrac{x^2 + 2x - 1}{x + 3}, \quad y_2 = x - 1 + \dfrac{2}{x + 3}$

43. $y_1 = \dfrac{x^4 + 1}{x^2 + 2}, \quad y_2 = x^2 - 2 + \dfrac{5}{x^2 + 2}$

44. $y_1 = \dfrac{x^4 + x^2 - 1}{x^2 + 1}, \quad y_2 = x^2 - \dfrac{1}{x^2 + 1}$

In Exercises 45–52, use long division to divide.

45. $\dfrac{24x^2 - x - 8}{3x - 2}$ **46.** $\dfrac{4x^2 + 7}{3x - 2}$

47. $\dfrac{x^4 - 3x^2 + 2}{x^2 - 1}$

48. $\dfrac{3x^4 + x^2 - 1}{x^2 - 1}$

49. $(5x^3 - 13x^2 - x + 2) \div (x^2 - 3x + 1)$

50. $(x^4 + x^3 - x^2 + 2x) \div (x^2 + 2x)$

51. $\dfrac{6x^4 + 10x^3 + 13x^2 - 5x + 2}{2x^2 - 1}$

52. $\dfrac{x^4 - 3x^3 + 4x^2 - 6x + 3}{x^2 + 2}$

In Exercises 53–58, use synthetic division to divide.

53. $(0.25x^4 - 4x^3) \div (x + 2)$

54. $(0.1x^3 + 0.3x^2 - 0.5) \div (x - 5)$

55. $(6x^4 - 4x^3 - 27x^2 + 18x) \div \left(x - \frac{2}{3}\right)$

56. $(2x^3 + 2x^2 - x + 2) \div \left(x - \frac{1}{2}\right)$

57. $(3x^3 - 10x^2 + 12x - 22) \div (x - 4)$

58. $(2x^3 + 6x^2 - 14x + 9) \div (x - 1)$

In Exercises 59 and 60, use the Remainder Theorem and synthetic division to evaluate the function at each given value. Use a graphing utility to verify your results.

59. $f(x) = x^4 + 10x^3 - 24x^2 + 20x + 44$

    (a) $f(-3)$ (b) $f(-2)$

60. $g(t) = 2t^5 - 5t^4 - 8t + 20$

    (a) $g(-4)$ (b) $g\left(\sqrt{2}\right)$

In Exercises 61–64, (a) verify the given factor(s) of the function f, (b) find the remaining factors of f, (c) use your results to write the complete factorization of f, and (d) list all real zeros of f. Confirm your results by using a graphing utility to graph the function.

	Function	Factor(s)
61.	$f(x) = x^3 + 4x^2 - 25x - 28$	$(x - 4)$
62.	$f(x) = 2x^3 + 11x^2 - 21x - 90$	$(x + 6)$
63.	$f(x) = x^4 - 4x^3 - 7x^2 + 22x + 24$	$(x + 2)$, $(x - 3)$
64.	$f(x) = x^4 - 11x^3 + 41x^2 - 61x + 30$	$(x - 2)$, $(x - 5)$

In Exercises 65 and 66, use the Rational Zero Test to list all possible rational zeros of f. Use a graphing utility to verify that the zeros of f are contained in the list.

65. $f(x) = 4x^3 - 11x^2 + 10x - 3$

66. $f(x) = 10x^3 + 21x^2 - x - 6$

In Exercises 67–70, find all the real zeros of the polynomial function.

67. $f(x) = 6x^3 - 5x^2 + 24x - 20$

68. $f(x) = x^3 - 1.3x^2 - 1.7x + 0.6$

69. $f(x) = 6x^4 - 25x^3 + 14x^2 + 27x - 18$

70. $f(x) = 5x^4 + 126x^2 + 25$

In Exercises 71 and 72, use Descartes's Rule of Signs to determine the possible numbers of positive and negative real zeros of the function.

71. $g(x) = 5x^3 - 6x + 9$

72. $f(x) = 2x^5 - 3x^2 + 2x - 1$

In Exercises 73 and 74, use synthetic division to verify the upper and lower bounds of the real zeros of f.

73. $f(x) = 4x^3 - 3x^2 + 4x - 3$

    Upper bound: $x = 1$; Lower bound: $x = -\frac{1}{4}$

74. $f(x) = 2x^3 - 5x^2 - 14x + 8$

    Upper bound: $x = 8$; Lower bound: $x = -4$

2.4 In Exercises 75–78, write the complex number in standard form.

75. $6 + \sqrt{-25}$

76. $-\sqrt{-12} + 3$

77. $-2i^2 + 7i$

78. $-i^2 - 4i$

In Exercises 79–90, perform the operations and write the result in standard form.

79. $(7 + 5i) + (-4 + 2i)$

80. $\left(\dfrac{\sqrt{2}}{2} - \dfrac{\sqrt{2}}{2}i\right) - \left(\dfrac{\sqrt{2}}{2} + \dfrac{\sqrt{2}}{2}i\right)$

81. $5i(13 - 8i)$

82. $(1 + 6i)(5 - 2i)$

83. $\left(\sqrt{-16} + 3\right)\left(\sqrt{-25} - 2\right)$

84. $\left(5 - \sqrt{-4}\right)\left(5 + \sqrt{-4}\right)$

85. $\sqrt{-9} + 3 + \sqrt{-36}$

86. $7 - \sqrt{-81} + \sqrt{-49}$

87. $(10 - 8i)(2 - 3i)$

88. $i(6 + i)(3 - 2i)$

89. $(3 + 7i)^2 + (3 - 7i)^2$

90. $(4 - i)^2 - (4 + i)^2$

In Exercises 91–94, write the quotient in standard form.

91. $\dfrac{6 + i}{i}$ **92.** $\dfrac{4}{-3i}$

93. $\dfrac{3 + 2i}{5 + i}$ **94.** $\dfrac{1 - 7i}{2 + 3i}$

In Exercises 95 and 96, determine the complex number shown in the complex plane.

95. 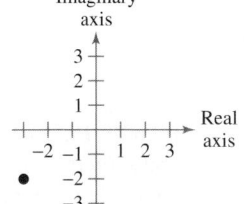   **96.**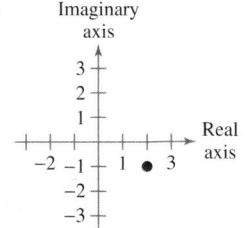

In Exercises 97–102, plot the complex number in the complex plane.

97. $2 - 5i$ **98.** $-1 + 4i$

99. $-6i$ **100.** $7i$

101. 3 **102.** -2

2.5 In Exercises 103 and 104, find all the zeros of the function.

103. $f(x) = 3x(x - 2)^2$

104. $f(x) = (x - 4)(x + 9)^2$

In Exercises 105–110, find all the zeros of the function and write the polynomial as a product of linear factors. Verify your results by using a graphing utility to graph the function.

105. $f(x) = 2x^4 - 5x^3 + 10x - 12$

106. $g(x) = 3x^4 - 4x^3 + 7x^2 + 10x - 4$

107. $h(x) = x^3 - 7x^2 + 18x - 24$

108. $f(x) = 2x^3 - 5x^2 - 9x + 40$

109. $f(x) = x^5 + x^4 + 5x^3 + 5x^2$

110. $f(x) = x^5 - 5x^3 + 4x$

In Exercises 111–116, (a) find all the zeros of the function, (b) write the polynomial as a product of linear factors, and (c) use your factorization to determine the x-intercepts of the graph of the function. Use a graphing utility to verify that the real zeros are the only x-intercepts.

111. $f(x) = x^3 - 4x^2 + 6x - 4$

112. $f(x) = x^3 - 5x^2 - 7x + 51$

113. $f(x) = -3x^3 - 19x^2 - 4x + 12$

114. $f(x) = 2x^3 - 9x^2 + 22x - 30$

115. $f(x) = x^4 + 34x^2 + 225$

116. $f(x) = x^4 + 10x^3 + 26x^2 + 10x + 25$

In Exercises 117–120, find a polynomial function with real coefficients that has the given zeros. (There are many correct answers.)

117. $4, -2, 5i$

118. $2, -2, 2i$

119. $1, -4, -3 + 5i$

120. $-4, -4, 1 + \sqrt{3}i$

In Exercises 121 and 122, write the polynomial (a) as the product of factors that are irreducible over the *rationals*, (b) as the product of linear and quadratic factors that are irreducible over the *reals*, and (c) in completely factored form.

121. $f(x) = x^4 - 2x^3 + 8x^2 - 18x - 9$

(*Hint:* One factor is $x^2 + 9$.)

122. $f(x) = x^4 - 4x^3 + 3x^2 + 8x - 16$

(*Hint:* One factor is $x^2 - x - 4$.)

In Exercises 123 and 124, Use the given zero to find all the zeros of the function.

Function	Zero
124. $f(x) = x^3 + 3x^2 + 4x + 12$	$-2i$
124. $f(x) = 2x^3 - 7x^2 + 14x + 9$	$2 + \sqrt{5}i$

2.6 In Exercises 125–136, (a) find the domain of the function, (b) decide whether the function is continuous, and (c) identify any horizontal and vertical asymptotes.

125. $f(x) = \dfrac{2 - x}{x + 3}$

126. $f(x) = \dfrac{4x}{x - 8}$

127. $f(x) = \dfrac{2}{x^2 - 3x - 18}$

128. $f(x) = \dfrac{2x^2 + 3}{x^2 + x + 3}$

129. $f(x) = \dfrac{7 + x}{7 - x}$

130. $f(x) = \dfrac{6x}{x^2 - 1}$

131. $f(x) = \dfrac{4x^2}{2x^2 - 3}$

132. $f(x) = \dfrac{3x^2 - 11x - 4}{x^2 + 2}$

133. $f(x) = \dfrac{2x - 10}{x^2 - 2x - 15}$

134. $f(x) = \dfrac{x^3 - 4x^2}{x^2 + 3x + 2}$

135. $f(x) = \dfrac{x - 2}{|x| + 2}$

136. $f(x) = \dfrac{2x}{|2x - 1|}$

137. *Seizure of Illegal Drugs*   The cost C (in millions of dollars) for the U.S. government to seize $p\%$ of an illegal drug as it enters the country is given by

$$C = \dfrac{528p}{100 - p}, \quad 0 \le p < 100.$$

(a) Find the costs of seizing 25%, 50%, and 75% of the illegal drug.

(b) Use a graphing utility to graph the function. Be sure to choose an appropriate viewing window. Explain why you chose the values you used in your viewing window.

(c) According to this model, would it be possible to seize 100% of the drug? Explain.

138. *Wildlife* A biology class performs an experiment comparing the quantity of food consumed by a certain kind of moth with the quantity supplied. The model for the experimental data is given by

$$y = \frac{1.568x - 0.001}{6.360x + 1}, \quad x > 0$$

where x is the quantity (in milligrams) of food supplied and y is the quantity (in milligrams) eaten (see figure). At what level of consumption will the moth become satiated?

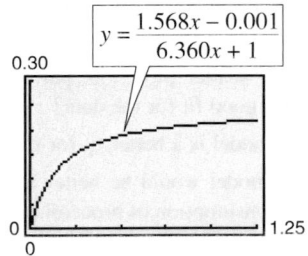

2.7 In Exercises 139–144, find all of the vertical, horizontal, and slant asymptotes, and any holes in the graph of the function. Then use a graphing utility to verify your result.

139. $f(x) = \dfrac{x^2 - 5x + 4}{x^2 - 1}$

140. $f(x) = \dfrac{x^2 - 3x - 8}{x^2 - 4}$

141. $f(x) = \dfrac{2x^2 - 7x + 3}{2x^2 - 3x - 9}$

142. $f(x) = \dfrac{3x^2 + 13x - 10}{2x^2 + 11x + 5}$

143. $f(x) = \dfrac{3x^3 - x^2 - 12x + 4}{x^2 + 3x + 2}$

144. $f(x) = \dfrac{2x^3 + 3x^2 - 2x - 3}{x^2 - 3x + 2}$

In Exercises 145–156, sketch the graph of the rational function by hand. As sketching aids, check for intercepts, vertical asymptotes, horizontal asymptotes, slant asymptotes, and holes.

145. $f(x) = \dfrac{2x - 1}{x - 5}$

146. $f(x) = \dfrac{x - 3}{x - 2}$

147. $f(x) = \dfrac{2x}{x^2 + 4}$

148. $f(x) = \dfrac{2x^2}{x^2 - 4}$

149. $f(x) = \dfrac{x^2}{x^2 + 1}$

150. $f(x) = \dfrac{5x}{x^2 + 1}$

151. $f(x) = \dfrac{2}{(x + 1)^2}$

152. $f(x) = \dfrac{4}{(x - 1)^2}$

153. $f(x) = \dfrac{2x^3}{x^2 + 1}$

154. $f(x) = \dfrac{x^3}{3x^2 - 6}$

155. $f(x) = \dfrac{x^2 - x + 1}{x - 3}$

156. $f(x) = \dfrac{2x^2 + 7x + 3}{x + 1}$

157. *Wildlife* The Parks and Wildlife Commission introduces 80,000 fish into a large human-made lake. The population N of the fish (in thousands) is given by

$$N = \frac{20(4 + 3t)}{1 + 0.05t}, \quad t \geq 0$$

where t is time in years.

(a) Use a graphing utility to graph the function.

(b) Use the graph from part (a) to find the populations when $t = 5$, $t = 10$, and $t = 25$.

(c) What is the maximum number of fish in the lake as time passes? Explain your reasoning.

∫ 158. *Page Design* A page that is x inches wide and y inches high contains 30 square inches of print. The top and bottom margins are 2 inches deep and the margins on each side are 2 inches wide.

(a) Draw a diagram that illustrates the problem.

(b) Show that the total area A of the page is given by

$$A = \frac{2x(2x + 7)}{x - 4}.$$

(c) Determine the domain of the function based on the physical constraints of the problem.

(d) Use a graphing utility to graph the area function and approximate the page size such that the minimum amount of paper will be used. Verify your answer numerically using the *table* feature of a graphing utility.

2.8 In Exercises 159–162, determine whether the scatter plot could best be modeled by a linear model, a quadratic model, or neither.

159.

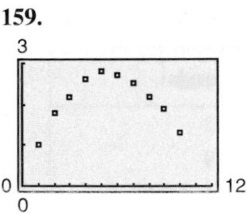

160.

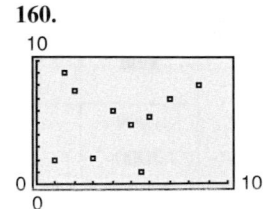

161.

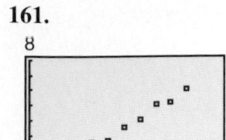

162.

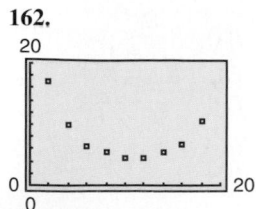

163. *Investment*   The table shows the prices P per fine ounce of gold (in dollars) for the years 1996 to 2004.   (Source: U.S. Geological Survey)

Year	Price per fine ounce, P (in dollars)
1996	389
1997	332
1998	295
1999	280
2000	280
2001	272
2002	311
2003	365
2004	410

(a) Use a graphing utility to create a scatter plot of the data. Let t represent the year, with $t = 6$ corresponding to 1996.

(b) Use the *regression* feature of a graphing utility to find a quadratic model for the data.

(c) Use a graphing utility to graph the model with the scatter plot from part (a). Is the quadratic model a good fit for the data?

(d) Use the model to find when the price per ounce would have exceeded $500.

(e) Do you think the model can be used to predict the price of gold in the future? Explain.

164. *Broccoli*   The table shows the per capita consumptions C (in pounds) of broccoli in the United States for the years 1999 to 2003.   (Source: U.S. Department of Agriculture)

Year	Per capita consumption, C (in pounds)
1999	6.2
2000	5.9
2001	5.4
2002	5.3
2003	5.7

(a) Use a graphing utility to create a scatter plot of the data. Let t represent the year, with $t = 9$ corresponding to 1999.

(b) A cubic model for the data is

$$C = 0.0583t^3 - 1.796t^2 + 17.99t - 52.7.$$

Use a graphing utility to graph the cubic model with the scatter plot from part (a). Is the cubic model a good fit for the data? Explain.

(c) Use the *regression* feature of a graphing utility to find a quadratic model for the data.

(d) Use a graphing utility to graph the quadratic model with the scatter plot from part (a). Is the quadratic model a good fit for the data?

(e) Which model is a better fit for the data? Explain.

(f) Which model would be better for predicting the per capita consumption of broccoli in the future? Explain. Use the model you chose to find the per capita consumption of broccoli in 2010.

Synthesis

True or False?   **In Exercises 165–167, determine whether the statement is true or false. Justify your answer.**

165. The graph of $f(x) = \dfrac{2x^3}{x + 1}$ has a slant asymptote.

166. A fourth-degree polynomial with real coefficients can have -5, $-8i$, $4i$, and 5 as its zeros.

167. The sum of two complex numbers cannot be a real number.

168. *Think About It*   What does it mean for a divisor to divide evenly into a dividend?

169. *Writing*   Write a paragraph discussing whether every rational function has a vertical asymptote.

170. *Error Analysis*   Describe the error.

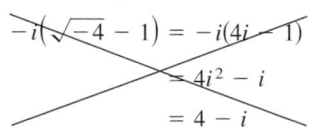

171. *Error Analysis*   Describe the error.

$$-i\left(\sqrt{-4} - 1\right) = -i(4i - 1)$$
$$= -4i^2 - i$$
$$= 4 - i$$

172. Write each of the powers of i as i, 1, or -1.

   (a) i^{40} (b) i^{25} (c) i^{50} (d) i^{67}

2 Chapter Test

See www.CalcChat.com for worked-out solutions to odd-numbered exercises.

Take this test as you would take a test in class. After you are finished, check your work against the answers given in the back of the book.

1. Identify the vertex and intercepts of the graph of $y = x^2 + 4x + 3$.

2. Write an equation of the parabola shown at the right.

3. Find all the real zeros of $f(x) = 4x^3 + 4x^2 + x$. Determine the multiplicity of each zero.

4. Sketch the graph of the function $f(x) = -x^3 + 7x + 6$.

5. Divide using long division: $(3x^3 + 4x - 1) \div (x^2 + 1)$.

6. Divide using synthetic division: $(2x^4 - 5x^2 - 3) \div (x - 2)$.

7. Use synthetic division to evaluate $f(-2)$ for $f(x) = 3x^4 - 6x^2 + 5x - 1$.

Figure for 2

In Exercises 8 and 9, list all the possible rational zeros of the function. Use a graphing utility to graph the function and find all the rational zeros.

8. $g(t) = 2t^4 - 3t^3 + 16t - 24$

9. $h(x) = 3x^5 + 2x^4 - 3x - 2$

10. Find all the zeros of the function $f(x) = x^3 - 7x^2 + 11x + 19$ and write the polynomial as the product of linear factors.

In Exercises 11–14, perform the operations and write the result in standard form.

11. $(-8 - 3i) + (-1 - 15i)$

12. $\left(10 + \sqrt{-20}\right) - \left(4 - \sqrt{-14}\right)$

13. $(2 + i)(6 - i)$

14. $(4 + 3i)^2 - (5 + i)^2$

In Exercises 15–17, write the quotient in standard form.

15. $\dfrac{8 + 5i}{6 - i}$

16. $\dfrac{5i}{2 + i}$

17. $(2i - 1) \div (3i + 2)$

18. Plot the complex number $3 - 2i$ in the complex plane.

In Exercises 19–21, sketch the graph of the rational function. As sketching aids, check for intercepts, vertical asymptotes, horizontal asymptotes, and slant asymptotes.

19. $h(x) = \dfrac{4}{x^2} - 1$

20. $g(x) = \dfrac{x^2 + 2}{x - 1}$

21. $f(x) = \dfrac{2x^2 + 9}{5x^2 + 2}$

22. The table shows the amounts A (in billions of dollars) budgeted for national defense for the years 1998 to 2004. (Source: U.S. Office of Management and Budget)

   (a) Use a graphing utility to create a scatter plot of the data. Let t represent the year, with $t = 8$ corresponding to 1998.

   (b) Use the *regression* feature of a graphing utility to find a quadratic model for the data.

   (c) Use a graphing utility to graph the quadratic model with the scatter plot from part (a). Is the quadratic model a good fit for the data?

   (d) Use the model to estimate the amounts budgeted for the years 2005 and 2010.

   (e) Do you believe the model is useful for predicting the national defense budgets for years beyond 2004? Explain.

Year	Defense budget, A (in billions of dollars)
1998	271.3
1999	292.3
2000	304.1
2001	335.5
2002	362.1
2003	456.2
2004	490.6

Table for 22

Proofs in Mathematics

These two pages contain proofs of four important theorems about polynomial functions. The first two theorems are from Section 2.3, and the second two theorems are from Section 2.5.

The Remainder Theorem   (p. 120)

If a polynomial $f(x)$ is divided by $x - k$, the remainder is

$$r = f(k).$$

Proof

From the Division Algorithm, you have

$$f(x) = (x - k)q(x) + r(x)$$

and because either $r(x) = 0$ or the degree of $r(x)$ is less than the degree of $x - k$, you know that $r(x)$ must be a constant. That is, $r(x) = r$. Now, by evaluating $f(x)$ at $x = k$, you have

$$f(k) = (k - k)q(k) + r$$
$$= (0)q(k) + r = r.$$

To be successful in algebra, it is important that you understand the connection among *factors* of a polynomial, *zeros* of a polynomial function, and *solutions* or *roots* of a polynomial equation. The Factor Theorem is the basis for this connection.

The Factor Theorem   (p. 120)

A polynomial $f(x)$ has a factor $(x - k)$ if and only if $f(k) = 0$.

Proof

Using the Division Algorithm with the factor $(x - k)$, you have

$$f(x) = (x - k)q(x) + r(x).$$

By the Remainder Theorem, $r(x) = r = f(k)$, and you have

$$f(x) = (x - k)q(x) + f(k)$$

where $q(x)$ is a polynomial of lesser degree than $f(x)$. If $f(k) = 0$, then

$$f(x) = (x - k)q(x)$$

and you see that $(x - k)$ is a factor of $f(x)$. Conversely, if $(x - k)$ is a factor of $f(x)$, division of $f(x)$ by $(x - k)$ yields a remainder of 0. So, by the Remainder Theorem, you have $f(k) = 0$.

Linear Factorization Theorem   (p. 139)

If $f(x)$ is a polynomial of degree n, where $n > 0$, then f has precisely n linear factors

$$f(x) = a_n(x - c_1)(x - c_2) \cdots (x - c_n)$$

where $c_1, c_2, \ldots, c_n$ are complex numbers.

Proof

Using the Fundamental Theorem of Algebra, you know that f must have at least one zero, c_1. Consequently, $(x - c_1)$ is a factor of $f(x)$, and you have

$$f(x) = (x - c_1)f_1(x).$$

If the degree of $f_1(x)$ is greater than zero, you again apply the Fundamental Theorem to conclude that f_1 must have a zero c_2, which implies that

$$f(x) = (x - c_1)(x - c_2)f_2(x).$$

It is clear that the degree of $f_1(x)$ is $n - 1$, that the degree of $f_2(x)$ is $n - 2$, and that you can repeatedly apply the Fundamental Theorem n times until you obtain

$$f(x) = a_n(x - c_1)(x - c_2) \cdots (x - c_n)$$

where a_n is the leading coefficient of the polynomial $f(x)$.

Factors of a Polynomial   (p. 141)

Every polynomial of degree $n > 0$ with real coefficients can be written as the product of linear and quadratic factors with real coefficients, where the quadratic factors have no real zeros.

Proof

To begin, you use the Linear Factorization Theorem to conclude that $f(x)$ can be *completely* factored in the form

$$f(x) = d(x - c_1)(x - c_2)(x - c_3) \cdots (x - c_n).$$

If each c_i is real, there is nothing more to prove. If any c_i is complex ($c_i = a + bi$, $b \neq 0$), then, because the coefficients of $f(x)$ are real, you know that the conjugate $c_j = a - bi$ is also a zero. By multiplying the corresponding factors, you obtain

$$(x - c_i)(x - c_j) = [x - (a + bi)][x - (a - bi)]$$
$$= x^2 - 2ax + (a^2 + b^2)$$

where each coefficient is real.

The Fundamental Theorem of Algebra

The Linear Factorization Theorem is closely related to the Fundamental Theorem of Algebra. The Fundamental Theorem of Algebra has a long and interesting history. In the early work with polynomial equations, The Fundamental Theorem of Algebra was thought to have been not true, because imaginary solutions were not considered. In fact, in the very early work by mathematicians such as Abu al-Khwarizmi (c. 800 A.D.), negative solutions were also not considered.

Once imaginary numbers were accepted, several mathematicians attempted to give a general proof of the Fundamental Theorem of Algebra. These included Gottfried von Leibniz (1702), Jean d'Alembert (1746), Leonhard Euler (1749), Joseph-Louis Lagrange (1772), and Pierre Simon Laplace (1795). The mathematician usually credited with the first correct proof of the Fundamental Theorem of Algebra is Carl Friedrich Gauss, who published the proof in his doctoral thesis in 1799.

Progressive Summary (Chapters 1–2)

This chart outlines the topics that have been covered so far in this text.
Progressive Summary charts appear after Chapters 2, 3, 6, 9, and 11. In each
progressive summary, new topics encountered for the first time appear in red.

Algebraic Functions		Transcendental Functions	Other Topics
Polynomial, Rational, Radical			
■ **Rewriting**		■ **Rewriting**	■ **Rewriting**
Polynomial form ↔ Factored form			
Operations with polynomials			
Rationalize denominators			
Simplify rational expressions			
Operations with complex numbers			

■ **Solving**		■ **Solving**	■ **Solving**
Equation	*Strategy*		
Linear	Isolate variable		
Quadratic	Factor, set to zero		
	Extract square roots		
	Complete the square		
	Quadratic Formula		
Polynomial	Factor, set to zero		
	Rational Zero Test		
Rational	Multiply by LCD		
Radical	Isolate, raise to power		
Absolute value	Isolate, form two equations		

■ **Analyzing**		■ **Analyzing**	■ **Analyzing**
Graphically	*Algebraically*		
Intercepts	Domain, Range		
Symmetry	Transformations		
Slope	Composition		
Asymptotes	Standard forms		
End behavior	of equations		
Minimum values	Leading Coefficient		
Maximum values	Test		
	Synthetic division		
	Descartes's Rule of Signs		
Numerically			
Table of values			

Chapter 3

Exponential and Logarithmic Functions

3.1 Exponential Functions and Their Graphs

3.2 Logarithmic Functions and Their Graphs

3.3 Properties of Logarithms

3.4 Solving Exponential and Logarithmic Equations

3.5 Exponential and Logarithmic Models

3.6 Nonlinear Models

Selected Applications

Exponential and logarithmic functions have many real life applications. The applications listed below represent a small sample of the applications in this chapter.

- Radioactive Decay, Exercises 67 and 68, page 194
- Sound Intensity, Exercise 95, page 205
- Home Mortgage, Exercise 96, page 205
- Comparing Models, Exercise 97, page 212
- Forestry, Exercise 138, page 223
- IQ Scores, Exercise 37, page 234
- Newton's Law of Cooling, Exercises 53 and 54, page 236
- Elections, Exercise 27, page 243

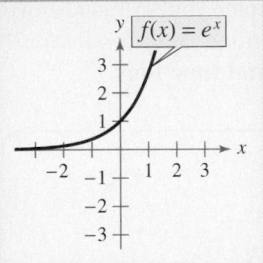
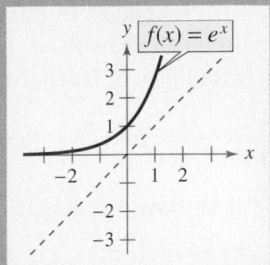
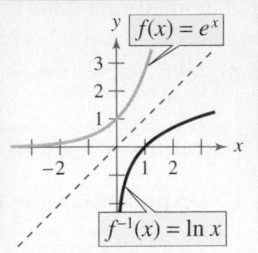

Exponential and logarithmic functions are called transcendental functions because these functions are not algebraic. In Chapter 3, you will learn about the inverse relationship between exponential and logarithmic functions, how to graph these functions, how to solve exponential and logarithmic equations, and how to use these functions in real-life applications.

© Denis O'Regan/Corbis

The relationship between the number of decibels and the intensity of a sound can be modeled by a logarithmic function. A rock concert at a stadium has a decibel rating of 120 decibels. Sounds at this level can cause gradual hearing loss.

3.1 Exponential Functions and Their Graphs

Exponential Functions

So far, this text has dealt mainly with **algebraic functions,** which include polynomial functions and rational functions. In this chapter you will study two types of nonalgebraic functions—*exponential functions* and *logarithmic functions*. These functions are examples of **transcendental functions.**

> **Definition of Exponential Function**
>
> The **exponential function** f **with base** a is denoted by
>
> $$f(x) = a^x$$
>
> where $a > 0$, $a \neq 1$, and x is any real number.

Note that in the definition of an exponential function, the base $a = 1$ is excluded because it yields $f(x) = 1^x = 1$. This is a constant function, not an exponential function.

You have already evaluated a^x for integer and rational values of x. For example, you know that $4^3 = 64$ and $4^{1/2} = 2$. However, to evaluate 4^x for any real number x, you need to interpret forms with *irrational* exponents. For the purposes of this text, it is sufficient to think of

$$a^{\sqrt{2}} \left(\text{where } \sqrt{2} \approx 1.41421356\right)$$

as the number that has the successively closer approximations

$$a^{1.4}, a^{1.41}, a^{1.414}, a^{1.4142}, a^{1.41421}, \ldots .$$

Example 1 shows how to use a calculator to evaluate exponential functions.

Example 1 Evaluating Exponential Functions

Use a calculator to evaluate each function at the indicated value of x.

Function	Value
a. $f(x) = 2^x$	$x = -3.1$
b. $f(x) = 2^{-x}$	$x = \pi$
c. $f(x) = 0.6^x$	$x = \frac{3}{2}$

Solution

Function Value	Graphing Calculator Keystrokes	Display
a. $f(-3.1) = 2^{-3.1}$	2 $\boxed{\wedge}$ $\boxed{(-)}$ 3.1 $\boxed{\text{ENTER}}$	0.1166291
b. $f(\pi) = 2^{-\pi}$	2 $\boxed{\wedge}$ $\boxed{(-)}$ π $\boxed{\text{ENTER}}$	0.1133147
c. $f\left(\frac{3}{2}\right) = (0.6)^{3/2}$	.6 $\boxed{\wedge}$ $\boxed{(}$ 3 $\boxed{\div}$ 2 $\boxed{)}$ $\boxed{\text{ENTER}}$	0.4647580

✓CHECKPOINT   Now try Exercise 3.

Graphs of Exponential Functions

The graphs of all exponential functions have similar characteristics, as shown in Examples 2, 3, and 4.

Example 2   Graphs of $y = a^x$

In the same coordinate plane, sketch the graph of each function by hand.

a. $f(x) = 2^x$ **b.** $g(x) = 4^x$

Solution

The table below lists some values for each function. By plotting these points and connecting them with smooth curves, you obtain the graphs shown in Figure 3.1. Note that both graphs are increasing. Moreover, the graph of $g(x) = 4^x$ is increasing more rapidly than the graph of $f(x) = 2^x$.

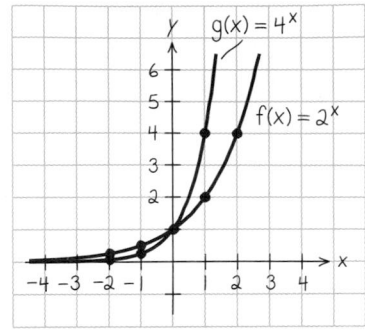

Figure 3.1

x	-2	-1	0	1	2	3
2^x	$\frac{1}{4}$	$\frac{1}{2}$	1	2	4	8
4^x	$\frac{1}{16}$	$\frac{1}{4}$	1	4	16	64

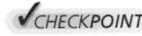 **CHECKPOINT**   Now try Exercise 5.

Example 3   Graphs of $y = a^{-x}$

In the same coordinate plane, sketch the graph of each function by hand.

a. $F(x) = 2^{-x}$ **b.** $G(x) = 4^{-x}$

Solution

The table below lists some values for each function. By plotting these points and connecting them with smooth curves, you obtain the graphs shown in Figure 3.2. Note that both graphs are decreasing. Moreover, the graph of $G(x) = 4^{-x}$ is decreasing more rapidly than the graph of $F(x) = 2^{-x}$.

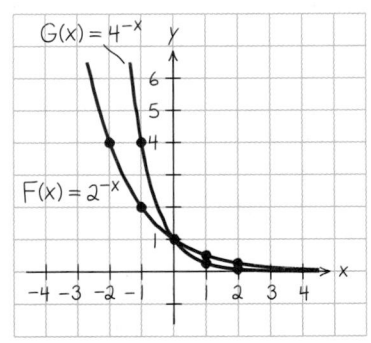

Figure 3.2

x	-3	-2	-1	0	1	2
2^{-x}	8	4	2	1	$\frac{1}{2}$	$\frac{1}{4}$
4^{-x}	64	16	4	1	$\frac{1}{4}$	$\frac{1}{16}$

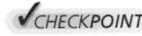 **CHECKPOINT**   Now try Exercise 7.

The properties of exponents can also be applied to real-number exponents. For review, these properties are listed below.

1. $a^x a^y = a^{x+y}$ **2.** $\dfrac{a^x}{a^y} = a^{x-y}$ **3.** $a^{-x} = \dfrac{1}{a^x} = \left(\dfrac{1}{a}\right)^x$ **4.** $a^0 = 1$

5. $(ab)^x = a^x b^x$ **6.** $(a^x)^y = a^{xy}$ **7.** $\left(\dfrac{a}{b}\right)^x = \dfrac{a^x}{b^x}$ **8.** $|a^2| = |a|^2 = a^2$

STUDY TIP

In Example 3, note that the functions $F(x) = 2^{-x}$ and $G(x) = 4^{-x}$ can be rewritten with positive exponents.

$$F(x) = 2^{-x} = \left(\frac{1}{2}\right)^x \quad \text{and}$$

$$G(x) = 4^{-x} = \left(\frac{1}{4}\right)^x$$

Comparing the functions in Examples 2 and 3, observe that

$$F(x) = 2^{-x} = f(-x) \quad \text{and} \quad G(x) = 4^{-x} = g(-x).$$

Consequently, the graph of F is a reflection (in the y-axis) of the graph of f, as shown in Figure 3.3. The graphs of G and g have the same relationship, as shown in Figure 3.4.

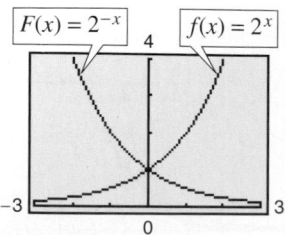

Figure 3.3

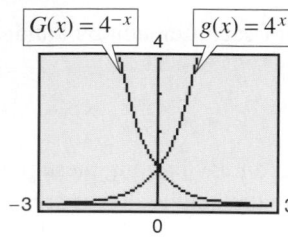

Figure 3.4

The graphs in Figures 3.3 and 3.4 are typical of the graphs of the exponential functions $f(x) = a^x$ and $f(x) = a^{-x}$. They have one y-intercept and one horizontal asymptote (the x-axis), and they are continuous.

Library of Parent Functions: Exponential Function

The *exponential function*

$$f(x) = a^x, \ a > 0, \ a \neq 1$$

is different from all the functions you have studied so far because the variable x is an *exponent*. A distinguishing characteristic of an exponential function is its rapid increase as x increases (for $a > 1$). Many real-life phenomena with patterns of rapid growth (or decline) can be modeled by exponential functions. The basic characteristics of the exponential function are summarized below. A review of exponential functions can be found in the *Study Capsules*.

Graph of $f(x) = a^x$, $a > 1$
Domain: $(-\infty, \infty)$
Range: $(0, \infty)$
Intercept: $(0, 1)$
Increasing on $(-\infty, \infty)$
x-axis is a horizontal asymptote
($a^x \to 0$ as $x \to -\infty$)
Continuous

Graph of $f(x) = a^{-x}$, $a > 1$
Domain: $(-\infty, \infty)$
Range: $(0, \infty)$
Intercept: $(0, 1)$
Decreasing on $(-\infty, \infty)$
x-axis is a horizontal asymptote
($a^{-x} \to 0$ as $x \to \infty$)
Continuous

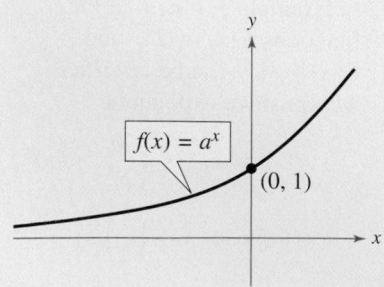

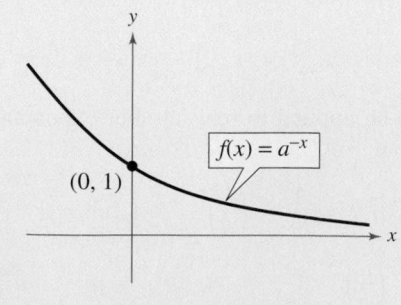

In the following example, the graph of $y = a^x$ is used to graph functions of the form $f(x) = b \pm a^{x+c}$, where b and c are any real numbers.

Example 4   Transformations of Graphs of Exponential Functions

Each of the following graphs is a transformation of the graph of $f(x) = 3^x$.

a. Because $g(x) = 3^{x+1} = f(x + 1)$, the graph of g can be obtained by shifting the graph of f one unit to the *left*, as shown in Figure 3.5.

b. Because $h(x) = 3^x - 2 = f(x) - 2$, the graph of h can be obtained by shifting the graph of f *downward* two units, as shown in Figure 3.6.

c. Because $k(x) = -3^x = -f(x)$, the graph of k can be obtained by *reflecting* the graph of f in the x-axis, as shown in Figure 3.7.

d. Because $j(x) = 3^{-x} = f(-x)$, the graph of j can be obtained by *reflecting* the graph of f in the y-axis, as shown in Figure 3.8.

Prerequisite Skills

If you have difficulty with this example, review shifting and reflecting of graphs in Section 1.4.

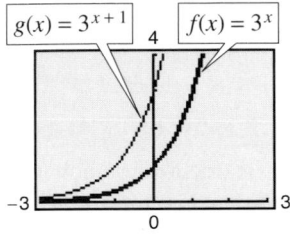

Figure 3.5

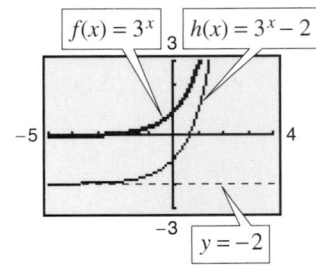

Figure 3.6

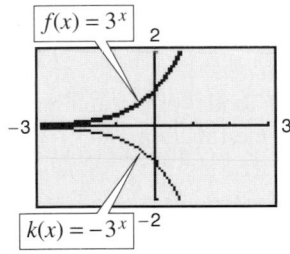

Figure 3.7

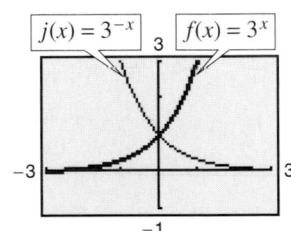

Figure 3.8

Exploration

The following table shows some points on the graphs in Figure 3.5. The functions $f(x)$ and $g(x)$ are represented by Y1 and Y2, respectively. Explain how you can use the table to describe the transformation.

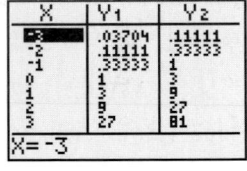

✓CHECKPOINT   Now try Exercise 17.

Notice that the transformations in Figures 3.5, 3.7, and 3.8 keep the x-axis ($y = 0$) as a horizontal asymptote, but the transformation in Figure 3.6 yields a new horizontal asymptote of $y = -2$. Also, be sure to note how the y-intercept is affected by each transformation.

The Natural Base e

For many applications, the convenient choice for a base is the irrational number

$$e = 2.718281828 \ldots .$$

This number is called the **natural base.** The function $f(x) = e^x$ is called the **natural exponential function** and its graph is shown in Figure 3.9. The graph of the exponential function has the same basic characteristics as the graph of the function $f(x) = a^x$ (see page 186). Be sure you see that for the exponential function $f(x) = e^x$, e is the constant $2.718281828 \ldots$, whereas x is the variable.

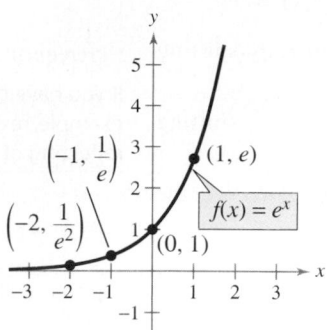

Figure 3.9   *The Natural Exponential Function*

In Example 5, you will see that the number e can be approximated by the expression

$$\left(1 + \frac{1}{x}\right)^x$$ for large values of x.

Example 5   Approximation of the Number e

Evaluate the expression $[1 + (1/x)]^x$ for several large values of x to see that the values approach $e \approx 2.718281828$ as x increases without bound.

Graphical Solution

Use a graphing utility to graph

$$y_1 = [1 + (1/x)]^x \qquad \text{and} \qquad y_2 = e$$

in the same viewing window, as shown in Figure 3.10. Use the *trace* feature of the graphing utility to verify that as x increases, the graph of y_1 gets closer and closer to the line $y_2 = e$.

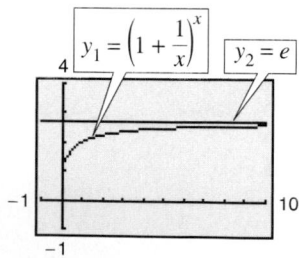

Figure 3.10

✓CHECKPOINT   Now try Exercise 77.

Exploration

Use your graphing utility to graph the functions

$$y_1 = 2^x$$

$$y_2 = e^x$$

$$y_3 = 3^x$$

in the same viewing window. From the relative positions of these graphs, make a guess as to the value of the real number e. Then try to find a number a such that the graphs of $y_2 = e^x$ and $y_4 = a^x$ are as close as possible.

TECHNOLOGY SUPPORT

For instructions on how to use the *trace* feature and the *table* feature, see Appendix A; for specific keystrokes, go to this textbook's *Online Study Center.*

Numerical Solution

Use the *table* feature (in *ask* mode) of a graphing utility to create a table of values for the function $y = [1 + (1/x)]^x$, beginning at $x = 10$ and increasing the x-values as shown in Figure 3.11.

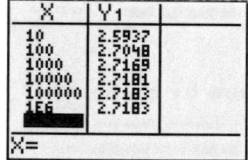

Figure 3.11

From the table, it seems reasonable to conclude that

$$\left(1 + \frac{1}{x}\right)^x \to e \quad \text{as} \quad x \to \infty.$$

Example 6   Evaluating the Natural Exponential Function

Use a calculator to evaluate the function $f(x) = e^x$ at each indicated value of x.

a. $x = -2$ **b.** $x = 0.25$ **c.** $x = -0.4$

Solution

Function Value	Graphing Calculator Keystrokes	Display
a. $f(-2) = e^{-2}$	$\boxed{e^x}$ $\boxed{(-)}$ 2 $\boxed{\text{ENTER}}$	0.1353353
b. $f(0.25) = e^{0.25}$	$\boxed{e^x}$ $.25$ $\boxed{\text{ENTER}}$	1.2840254
c. $f(-0.4) = e^{-0.4}$	$\boxed{e^x}$ $\boxed{(-)}$ $.4$ $\boxed{\text{ENTER}}$	0.6703200

✔CHECKPOINT   Now try Exercise 23.

Exploration

Use a graphing utility to graph $y = (1 + x)^{1/x}$. Describe the behavior of the graph near $x = 0$. Is there a y-intercept? How does the behavior of the graph near $x = 0$ relate to the result of Example 5? Use the *table* feature of a graphing utility to create a table that shows values of y for values of x near $x = 0$, to help you describe the behavior of the graph near this point.

Example 7   Graphing Natural Exponential Functions

Sketch the graph of each natural exponential function.

a. $f(x) = 2e^{0.24x}$ **b.** $g(x) = \frac{1}{2}e^{-0.58x}$

Solution

To sketch these two graphs, you can use a calculator to construct a table of values, as shown below.

x	-3	-2	-1	0	1	2	3
$f(x)$	0.974	1.238	1.573	2.000	2.542	3.232	4.109
$g(x)$	2.849	1.595	0.893	0.500	0.280	0.157	0.088

After constructing the table, plot the points and connect them with smooth curves. Note that the graph in Figure 3.12 is increasing, whereas the graph in Figure 3.13 is decreasing. Use a graphing calculator to verify these graphs.

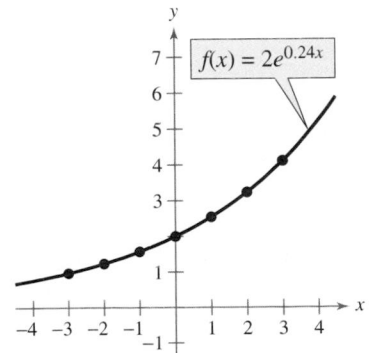

Figure 3.12

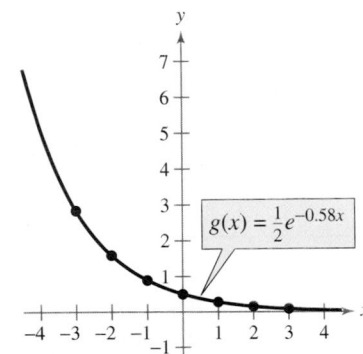

Figure 3.13

✔CHECKPOINT   Now try Exercise 43.

Applications

One of the most familiar examples of exponential growth is that of an investment earning *continuously compounded interest*. Suppose a principal P is invested at an annual interest rate r, compounded once a year. If the interest is added to the principal at the end of the year, the new balance P_1 is $P_1 = P + Pr = P(1 + r)$. This pattern of multiplying the previous principal by $1 + r$ is then repeated each successive year, as shown in the table.

Time in years	Balance after each compounding
0	$P = P$
1	$P_1 = P(1 + r)$
2	$P_2 = P_1(1 + r) = P(1 + r)(1 + r) = P(1 + r)^2$
⋮	⋮
t	$P_t = P(1 + r)^t$

To accommodate more frequent (quarterly, monthly, or daily) compounding of interest, let n be the number of compoundings per year and let t be the number of years. (The product nt represents the total number of times the interest will be compounded.) Then the interest rate per compounding period is r/n, and the account balance after t years is

$$A = P\left(1 + \frac{r}{n}\right)^{nt}.$$ Amount (balance) with n compoundings per year

If you let the number of compoundings n increase without bound, the process approaches what is called **continuous compounding**. In the formula for n compoundings per year, let $m = n/r$. This produces

$$A = P\left(1 + \frac{r}{n}\right)^{nt} = P\left(1 + \frac{1}{m}\right)^{mrt} = P\left[\left(1 + \frac{1}{m}\right)^m\right]^{rt}.$$

As m increases without bound, you know from Example 5 that $[1 + (1/m)]^m$ approaches e. So, for continuous compounding, it follows that

$$P\left[\left(1 + \frac{1}{m}\right)^m\right]^{rt} \rightarrow P[e]^{rt}$$

and you can write $A = Pe^{rt}$. This result is part of the reason that e is the "natural" choice for a base of an exponential function.

Formulas for Compound Interest

After t years, the balance A in an account with principal P and annual interest rate r (in decimal form) is given by the following formulas.

1. For n compoundings per year: $A = P\left(1 + \frac{r}{n}\right)^{nt}$

2. For continuous compounding: $A = Pe^{rt}$

Exploration

Use the formula

$$A = P\left(1 + \frac{r}{n}\right)^{nt}$$

to calculate the amount in an account when $P = \$3000$, $r = 6\%$, $t = 10$ years, and the interest is compounded (a) by the day, (b) by the hour, (c) by the minute, and (d) by the second. Does increasing the number of compoundings per year result in unlimited growth of the amount in the account? Explain.

STUDY TIP

The interest rate r in the formula for compound interest should be written as a decimal. For example, an interest rate of 7% would be written as $r = 0.07$.

Example 8   Finding the Balance for Compound Interest

A total of $9000 is invested at an annual interest rate of 2.5%, compounded annually. Find the balance in the account after 5 years.

Algebraic Solution

In this case,

$$P = 9000, r = 2.5\% = 0.025, n = 1, t = 5.$$

Using the formula for compound interest with n compoundings per year, you have

$$A = P\left(1 + \frac{r}{n}\right)^{nt} \qquad \text{Formula for compound interest}$$

$$= 9000\left(1 + \frac{0.025}{1}\right)^{1(5)} \qquad \text{Substitute for } P, r, n, \text{ and } t.$$

$$= 9000(1.025)^5 \qquad \text{Simplify.}$$

$$\approx \$10,182.67. \qquad \text{Use a calculator.}$$

So, the balance in the account after 5 years will be about $10,182.67.

Graphical Solution

Substitute the values for P, r, and n into the formula for compound interest with n compoundings per year as follows.

$$A = P\left(1 + \frac{r}{n}\right)^{nt} \qquad \text{Formula for compound interest}$$

$$= 9000\left(1 + \frac{0.025}{1}\right)^{(1)t} \qquad \text{Substitute for } P, r, \text{ and } n.$$

$$= 9000(1.025)^t \qquad \text{Simplify.}$$

Use a graphing utility to graph $y = 9000(1.025)^x$. Using the *value* feature or the *zoom* and *trace* features, you can approximate the value of y when $x = 5$ to be about 10,182.67, as shown in Figure 3.14. So, the balance in the account after 5 years will be about $10,182.67.

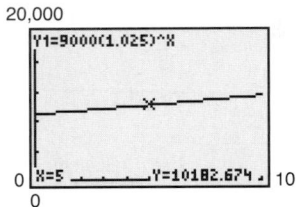

Figure 3.14

✔CHECKPOINT   Now try Exercise 53.

Example 9   Finding Compound Interest

A total of $12,000 is invested at an annual interest rate of 3%. Find the balance after 4 years if the interest is compounded (a) quarterly and (b) continuously.

Solution

a. For quarterly compoundings, $n = 4$. So, after 4 years at 3%, the balance is

$$A = P\left(1 + \frac{r}{n}\right)^{nt} = 12,000\left(1 + \frac{0.03}{4}\right)^{4(4)}$$

$$\approx \$13,523.91.$$

b. For continuous compounding, the balance is

$$A = Pe^{rt} = 12,000e^{0.03(4)}$$

$$\approx \$13,529.96.$$

Note that a continuous-compounding account yields more than a quarterly-compounding account.

✔CHECKPOINT   Now try Exercise 55.

Example 10   Radioactive Decay

Let y represent a mass, in grams, of radioactive strontium (^{90}Sr), whose half-life is 29 years. The quantity of strontium present after t years is $y = 10\left(\frac{1}{2}\right)^{t/29}$.

a. What is the initial mass (when $t = 0$)?

b. How much of the initial mass is present after 80 years?

Algebraic Solution

a. $y = 10\left(\frac{1}{2}\right)^{t/29}$ Write original equation.

$= 10\left(\frac{1}{2}\right)^{0/29}$ Substitute 0 for t.

$= 10$ Simplify.

So, the initial mass is 10 grams.

b. $y = 10\left(\frac{1}{2}\right)^{t/29}$ Write original equation.

$= 10\left(\frac{1}{2}\right)^{80/29}$ Substitute 80 for t.

$\approx 10\left(\frac{1}{2}\right)^{2.759}$ Simplify.

≈ 1.48 Use a calculator.

So, about 1.48 grams is present after 80 years.

✓**CHECKPOINT**   Now try Exercise 67.

Graphical Solution

Use a graphing utility to graph $y = 10\left(\frac{1}{2}\right)^{x/29}$.

a. Use the *value* feature or the *zoom* and *trace* features of the graphing utility to determine that the value of y when $x = 0$ is 10, as shown in Figure 3.15. So, the initial mass is 10 grams.

b. Use the *value* feature or the *zoom* and *trace* features of the graphing utility to determine that the value of y when $x = 80$ is about 1.48, as shown in Figure 3.16. So, about 1.48 grams is present after 80 years.

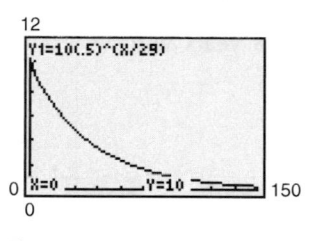

Figure 3.15

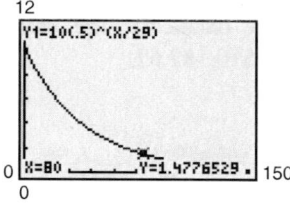

Figure 3.16

Example 11   Population Growth

The approximate number of fruit flies in an experimental population after t hours is given by $Q(t) = 20e^{0.03t}$, where $t \geq 0$.

a. Find the initial number of fruit flies in the population.

b. How large is the population of fruit flies after 72 hours?

c. Graph Q.

Solution

a. To find the initial population, evaluate $Q(t)$ when $t = 0$.

$$Q(0) = 20e^{0.03(0)} = 20e^0 = 20(1) = 20 \text{ flies}$$

b. After 72 hours, the population size is

$$Q(72) = 20e^{0.03(72)} = 20e^{2.16} \approx 173 \text{ flies}.$$

c. The graph of Q is shown in Figure 3.17.

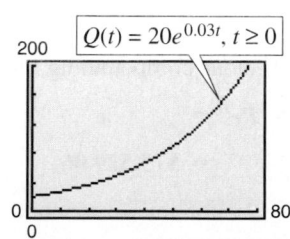

Figure 3.17

✓**CHECKPOINT**   Now try Exercise 69.

3.1 Exercises

See www.CalcChat.com for worked-out solutions to odd-numbered exercises.

Vocabulary Check

Fill in the blanks.

1. Polynomial and rational functions are examples of _____ functions.

2. Exponential and logarithmic functions are examples of nonalgebraic functions, also called _____ functions.

3. The exponential function $f(x) = e^x$ is called the _____ function, and the base e is called the _____ base.

4. To find the amount A in an account after t years with principal P and annual interest rate r compounded n times per year, you can use the formula _____ .

5. To find the amount A in an account after t years with principal P and annual interest rate r compounded continuously, you can use the formula _____ .

In Exercises 1–4, use a calculator to evaluate the function at the indicated value of x. Round your result to three decimal places.

Function	Value
1. $f(x) = 3.4^x$	$x = 6.8$
2. $f(x) = 1.2^x$	$x = \frac{1}{3}$
3. $g(x) = 5^x$	$x = -\pi$
4. $h(x) = 8.6^{-3x}$	$x = -\sqrt{2}$

In Exercises 5–12, graph the exponential function by hand. Identify any asymptotes and intercepts and determine whether the graph of the function is increasing or decreasing.

5. $g(x) = 5^x$

6. $f(x) = \left(\frac{3}{2}\right)^x$

7. $f(x) = 5^{-x}$

8. $h(x) = \left(\frac{3}{2}\right)^{-x}$

9. $h(x) = 5^{x-2}$

10. $g(x) = \left(\frac{3}{2}\right)^{x+2}$

11. $g(x) = 5^{-x} - 3$

12. $f(x) = \left(\frac{3}{2}\right)^{-x} + 2$

Library of Parent Functions   In Exercises 13–16, use the graph of $y = 2^x$ to match the function with its graph. [The graphs are labeled (a), (b), (c), and (d).]

(a)
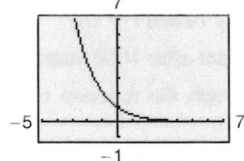

(b)
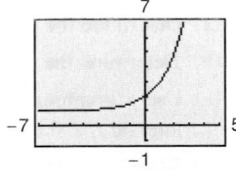

(c)
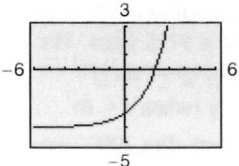

(d)
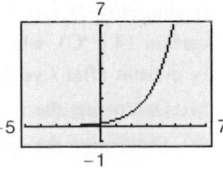

13. $f(x) = 2^{x-2}$

14. $f(x) = 2^{-x}$

15. $f(x) = 2^x - 4$

16. $f(x) = 2^x + 1$

In Exercises 17–22, use the graph of f to describe the transformation that yields the graph of g.

17. $f(x) = 3^x,\ g(x) = 3^{x-5}$

18. $f(x) = -2^x,\ g(x) = 5 - 2^x$

19. $f(x) = \left(\frac{3}{5}\right)^x,\ g(x) = -\left(\frac{3}{5}\right)^{x+4}$

20. $f(x) = 0.3^x,\ g(x) = -0.3^x + 5$

21. $f(x) = 4^x,\ g(x) = 4^{x-2} - 3$

22. $f(x) = \left(\frac{1}{2}\right)^x,\ g(x) = \left(\frac{1}{2}\right)^{-(x+4)}$

In Exercises 23–26, use a calculator to evaluate the function at the indicated value of x. Round your result to the nearest thousandth.

Function	Value
23. $f(x) = e^x$	$x = 9.2$
24. $f(x) = e^{-x}$	$x = -\frac{3}{4}$
25. $g(x) = 50e^{4x}$	$x = 0.02$
26. $h(x) = -5.5e^{-x}$	$x = 200$

In Exercises 27–44, use a graphing utility to construct a table of values for the function. Then sketch the graph of the function. Identify any asymptotes of the graph.

27. $f(x) = \left(\frac{5}{2}\right)^x$

28. $f(x) = \left(\frac{5}{2}\right)^{-x}$

29. $f(x) = 6^x$

30. $f(x) = 2^{x-1}$

31. $f(x) = 3^{x+2}$

32. $f(x) = 4^{x-3} + 3$

33. $y = 2^{-x^2}$

34. $y = 3^{-|x|}$

35. $y = 3^{x-2} + 1$

36. $y = 4^{x+1} - 2$

37. $f(x) = e^{-x}$

38. $s(t) = 3e^{-0.2t}$

39. $f(x) = 3e^{x+4}$

40. $f(x) = 2e^{-0.5x}$

41. $f(x) = 2 + e^{x-5}$

42. $g(x) = 2 - e^{-x}$

43. $s(t) = 2e^{0.12t}$

44. $g(x) = 1 + e^{-x}$

In Exercises 45–48, use a graphing utility to (a) graph the function and (b) find any asymptotes numerically by creating a table of values for the function.

45. $f(x) = \dfrac{8}{1 + e^{-0.5x}}$

46. $g(x) = \dfrac{8}{1 + e^{-0.5/x}}$

47. $f(x) = -\dfrac{6}{2 - e^{0.2x}}$

48. $f(x) = \dfrac{6}{2 - e^{0.2/x}}$

In Exercises 49 and 50, use a graphing utility to find the point(s) of intersection, if any, of the graphs of the functions. Round your result to three decimal places.

49. $y = 20e^{0.05x}$
$y = 1500$

50. $y = 100e^{0.01x}$
$y = 12{,}500$

In Exercises 51 and 52, (a) use a graphing utility to graph the function, (b) use the graph to find the open intervals on which the function is increasing and decreasing, and (c) approximate any relative maximum or minimum values.

51. $f(x) = x^2 e^{-x}$

52. $f(x) = 2x^2 e^{x+1}$

Compound Interest   In Exercises 53–56, complete the table to determine the balance A for P dollars invested at rate r for t years and compounded n times per year.

n	1	2	4	12	365	Continuous
A						

53. $P = \$2500, r = 2.5\%, t = 10$ years

54. $P = \$1000, r = 6\%, t = 10$ years

55. $P = \$2500, r = 4\%, t = 20$ years

56. $P = \$1000, r = 3\%, t = 40$ years

Compound Interest   In Exercises 57–60, complete the table to determine the balance A for $12{,}000$ invested at a rate r for t years, compounded continuously.

t	1	10	20	30	40	50
A						

57. $r = 4\%$

58. $r = 6\%$

59. $r = 3.5\%$

60. $r = 2.5\%$

Annuity   In Exercises 61–64, find the total amount A of an annuity after n months using the annuity formula

$$A = P\left[\dfrac{(1 + r/12)^n - 1}{r/12}\right]$$

where P is the amount deposited every month earning $r\%$ interest, compounded monthly.

61. $P = \$25, r = 12\%, n = 48$ months

62. $P = \$100, r = 9\%, n = 60$ months

63. $P = \$200, r = 6\%, n = 72$ months

64. $P = \$75, r = 3\%, n = 24$ months

65. *Demand*   The demand function for a product is given by

$$p = 5000\left[1 - \dfrac{4}{4 + e^{-0.002x}}\right]$$

where p is the price and x is the number of units.
(a) Use a graphing utility to graph the demand function for $x > 0$ and $p > 0$.
(b) Find the price p for a demand of $x = 500$ units.
(c) Use the graph in part (a) to approximate the highest price that will still yield a demand of at least 600 units.

Verify your answers to parts (b) and (c) numerically by creating a table of values for the function.

66. *Compound Interest*   There are three options for investing $500. The first earns 7% compounded annually, the second earns 7% compounded quarterly, and the third earns 7% compounded continuously.
(a) Find equations that model each investment growth and use a graphing utility to graph each model in the same viewing window over a 20-year period.
(b) Use the graph from part (a) to determine which investment yields the highest return after 20 years. What is the difference in earnings between each investment?

67. *Radioactive Decay*   Let Q represent a mass, in grams, of radioactive radium (^{226}Ra), whose half-life is 1599 years. The quantity of radium present after t years is given by $Q = 25\left(\frac{1}{2}\right)^{t/1599}$.
(a) Determine the initial quantity (when $t = 0$).
(b) Determine the quantity present after 1000 years.
(c) Use a graphing utility to graph the function over the interval $t = 0$ to $t = 5000$.
(d) When will the quantity of radium be 0 grams? Explain.

68. *Radioactive Decay*   Let Q represent a mass, in grams, of carbon 14 (^{14}C), whose half-life is 5715 years. The quantity present after t years is given by $Q = 10\left(\frac{1}{2}\right)^{t/5715}$.
(a) Determine the initial quantity (when $t = 0$).
(b) Determine the quantity present after 2000 years.
(c) Sketch the graph of the function over the interval $t = 0$ to $t = 10{,}000$.

69. *Bacteria Growth* A certain type of bacteria increases according to the model $P(t) = 100e^{0.2197t}$, where t is the time in hours.

(a) Use a graphing utility to graph the model.

(b) Use a graphing utility to approximate $P(0)$, $P(5)$, and $P(10)$.

(c) Verify your answers in part (b) algebraically.

70. *Population Growth* The projected populations of California for the years 2015 to 2030 can be modeled by

$$P = 34.706e^{0.0097t}$$

where P is the population (in millions) and t is the time (in years), with $t = 15$ corresponding to 2015.   (Source: U.S. Census Bureau)

(a) Use a graphing utility to graph the function for the years 2015 through 2030.

(b) Use the *table* feature of a graphing utility to create a table of values for the same time period as in part (a).

(c) According to the model, when will the population of California exceed 50 million?

71. *Inflation* If the annual rate of inflation averages 4% over the next 10 years, the approximate cost C of goods or services during any year in that decade will be modeled by $C(t) = P(1.04)^t$, where t is the time (in years) and P is the present cost. The price of an oil change for your car is presently $23.95.

(a) Use a graphing utility to graph the function.

(b) Use the graph in part (a) to approximate the price of an oil change 10 years from now.

(c) Verify your answer in part (b) algebraically.

72. *Depreciation* In early 2006, a new Jeep Wrangler Sport Edition had a manufacturer's suggested retail price of $23,970. After t years the Jeep's value is given by

$$V(t) = 23{,}970\left(\tfrac{3}{4}\right)^t.$$

(Source: DaimlerChrysler Corporation)

(a) Use a graphing utility to graph the function.

(b) Use a graphing utility to create a table of values that shows the value V for $t = 1$ to $t = 10$ years.

(c) According to the model, when will the Jeep have no value?

Synthesis

True or False? **In Exercises 73 and 74, determine whether the statement is true or false. Justify your answer.**

73. $f(x) = 1^x$ is not an exponential function.

74. $e = \dfrac{271{,}801}{99{,}990}$

75. *Library of Parent Functions* Determine which equation(s) may be represented by the graph shown. (There may be more than one correct answer.)

(a) $y = e^x + 1$

(b) $y = -e^{-x} + 1$

(c) $y = e^{-x} - 1$

(d) $y = e^{-x} + 1$

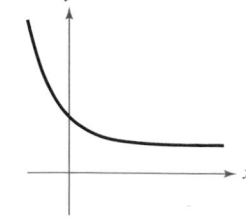

76. *Exploration* Use a graphing utility to graph $y_1 = e^x$ and each of the functions $y_2 = x^2$, $y_3 = x^3$, $y_4 = \sqrt{x}$, and $y_5 = |x|$ in the same viewing window.

(a) Which function increases at the fastest rate for "large" values of x?

(b) Use the result of part (a) to make a conjecture about the rates of growth of $y_1 = e^x$ and $y = x^n$, where n is a natural number and x is "large."

(c) Use the results of parts (a) and (b) to describe what is implied when it is stated that a quantity is growing exponentially.

77. *Graphical Analysis* Use a graphing utility to graph $f(x) = (1 + 0.5/x)^x$ and $g(x) = e^{0.5}$ in the same viewing window. What is the relationship between f and g as x increases without bound?

78. *Think About It* Which functions are exponential? Explain.

(a) $3x$ (b) $3x^2$ (c) 3^x (d) 2^{-x}

Think About It **In Exercises 79–82, place the correct symbol (< or >) between the pair of numbers.**

79. e^{π} _____ π^e

80. 2^{10} _____ 10^2

81. 5^{-3} _____ 3^{-5}

82. $4^{1/2}$ _____ $\left(\tfrac{1}{2}\right)^4$

Skills Review

In Exercises 83–86, determine whether the function has an inverse function. If it does, find f^{-1}.

83. $f(x) = 5x - 7$

84. $f(x) = -\tfrac{2}{3}x + \tfrac{5}{2}$

85. $f(x) = \sqrt[3]{x + 8}$

86. $f(x) = \sqrt{x^2 + 6}$

In Exercises 87 and 88, sketch the graph of the rational function.

87. $f(x) = \dfrac{2x}{x - 7}$

88. $f(x) = \dfrac{x^2 + 3}{x + 1}$

89. *Make a Decision* To work an extended application analyzing the population per square mile in the United States, visit this textbook's *Online Study Center*. (Data Source: U.S. Census Bureau)

3.2   Logarithmic Functions and Their Graphs

Logarithmic Functions

In Section 1.6, you studied the concept of an inverse function. There, you learned that if a function is one-to-one—that is, if the function has the property that no horizontal line intersects its graph more than once—the function must have an inverse function. By looking back at the graphs of the exponential functions introduced in Section 3.1, you will see that every function of the form

$$f(x) = a^x, \qquad a > 0, a \neq 1$$

passes the Horizontal Line Test and therefore must have an inverse function. This inverse function is called the **logarithmic function with base** a.

> **Definition of Logarithmic Function**
>
> For $x > 0$, $a > 0$, and $a \neq 1$,
>
> $$y = \log_a x \qquad \text{if and only if} \qquad x = a^y.$$
>
> The function given by
>
> $$f(x) = \log_a x \qquad \text{Read as "log base } a \text{ of } x."$$
>
> is called the **logarithmic function with base** a.

From the definition above, you can see that every logarithmic equation can be written in an equivalent exponential form and every exponential equation can be written in logarithmic form. The equations $y = \log_a x$ and $x = a^y$ are equivalent.

When evaluating logarithms, remember that *a logarithm is an exponent*. This means that $\log_a x$ is the exponent to which a must be raised to obtain x. For instance, $\log_2 8 = 3$ because 2 must be raised to the third power to get 8.

Example 1   Evaluating Logarithms

Use the definition of logarithmic function to evaluate each logarithm at the indicated value of x.

a. $f(x) = \log_2 x, \quad x = 32$ **b.** $f(x) = \log_3 x, \quad x = 1$
c. $f(x) = \log_4 x, \quad x = 2$ **d.** $f(x) = \log_{10} x, \quad x = \frac{1}{100}$

Solution

a. $f(32) = \log_2 32 = 5$ because $2^5 = 32$.
b. $f(1) = \log_3 1 = 0$ because $3^0 = 1$.
c. $f(2) = \log_4 2 = \frac{1}{2}$ because $4^{1/2} = \sqrt{4} = 2$.
d. $f\left(\frac{1}{100}\right) = \log_{10} \frac{1}{100} = -2$ because $10^{-2} = \frac{1}{10^2} = \frac{1}{100}$.

✓**CHECKPOINT**   Now try Exercise 25.

What you should learn

- Recognize and evaluate logarithmic functions with base a.
- Graph logarithmic functions with base a.
- Recognize, evaluate, and graph natural logarithmic functions.
- Use logarithmic functions to model and solve real-life problems.

Why you should learn it

Logarithmic functions are useful in modeling data that represents quantities that increase or decrease slowly. For instance, Exercises 97 and 98 on page 205 show how to use a logarithmic function to model the minimum required ventilation rates in public school classrooms.

Mark Richards/PhotoEdit

The logarithmic function with base 10 is called the **common logarithmic function.** On most calculators, this function is denoted by $\boxed{\text{LOG}}$. Example 2 shows how to use a calculator to evaluate common logarithmic functions. You will learn how to use a calculator to calculate logarithms to any base in the next section.

Example 2   Evaluating Common Logarithms on a Calculator

Use a calculator to evaluate the function $f(x) = \log_{10} x$ at each value of x.

a. $x = 10$ **b.** $x = 2.5$ **c.** $x = -2$ **d.** $x = \frac{1}{4}$

Solution

Function Value	*Graphing Calculator Keystrokes*	*Display*
a. $f(10) = \log_{10} 10$	$\boxed{\text{LOG}}$ 10 $\boxed{\text{ENTER}}$	1
b. $f(2.5) = \log_{10} 2.5$	$\boxed{\text{LOG}}$ 2.5 $\boxed{\text{ENTER}}$	0.3979400
c. $f(-2) = \log_{10}(-2)$	$\boxed{\text{LOG}}$ $\boxed{(-)}$ 2 $\boxed{\text{ENTER}}$	ERROR
d. $f\left(\frac{1}{4}\right) = \log_{10} \frac{1}{4}$	$\boxed{\text{LOG}}$ $\boxed{(}$ 1 $\boxed{\div}$ 4 $\boxed{)}$ $\boxed{\text{ENTER}}$	-0.6020600

Note that the calculator displays an error message when you try to evaluate $\log_{10}(-2)$. In this case, there is no *real* power to which 10 can be raised to obtain -2.

 *CHECKPOINT*   Now try Exercise 29.

> **TECHNOLOGY TIP**
>
> Some graphing utilities do not give an error message for $\log_{10}(-2)$. Instead, the graphing utility will display a complex number. For the purpose of this text, however, it will be said that the domain of a logarithmic function is the set of positive *real* numbers.

The following properties follow directly from the definition of the logarithmic function with base a.

> **Properties of Logarithms**
>
> **1.** $\log_a 1 = 0$ because $a^0 = 1$.
>
> **2.** $\log_a a = 1$ because $a^1 = a$.
>
> **3.** $\log_a a^x = x$ and $a^{\log_a x} = x$.   Inverse Properties
>
> **4.** If $\log_a x = \log_a y$, then $x = y$.   One-to-One Property

Example 3   Using Properties of Logarithms

a. Solve for x: $\log_2 x = \log_2 3$ **b.** Solve for x: $\log_4 4 = x$

c. Simplify: $\log_5 5^x$ **d.** Simplify: $7^{\log_7 14}$

Solution

a. Using the One-to-One Property (Property 4), you can conclude that $x = 3$.

b. Using Property 2, you can conclude that $x = 1$.

c. Using the Inverse Property (Property 3), it follows that $\log_5 5^x = x$.

d. Using the Inverse Property (Property 3), it follows that $7^{\log_7 14} = 14$.

 *CHECKPOINT*   Now try Exercise 33.

Graphs of Logarithmic Functions

To sketch the graph of $y = \log_a x$, you can use the fact that the graphs of inverse functions are reflections of each other in the line $y = x$.

Example 4   Graphs of Exponential and Logarithmic Functions

In the same coordinate plane, sketch the graph of each function by hand.

a. $f(x) = 2^x$ **b.** $g(x) = \log_2 x$

Solution

a. For $f(x) = 2^x$, construct a table of values. By plotting these points and connecting them with a smooth curve, you obtain the graph of f shown in Figure 3.18.

x	-2	-1	0	1	2	3
$f(x) = 2^x$	$\frac{1}{4}$	$\frac{1}{2}$	1	2	4	8

b. Because $g(x) = \log_2 x$ is the inverse function of $f(x) = 2^x$, the graph of g is obtained by plotting the points $(f(x), x)$ and connecting them with a smooth curve. The graph of g is a reflection of the graph of f in the line $y = x$, as shown in Figure 3.18.

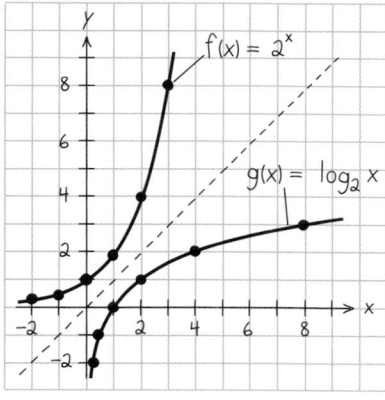

Figure 3.18

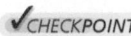 **CHECKPOINT**   Now try Exercise 43.

Before you can confirm the result of Example 4 using a graphing utility, you need to know how to enter $\log_2 x$. You will learn how to do this using the *change-of-base formula* discussed in Section 3.3.

Example 5   Sketching the Graph of a Logarithmic Function

Sketch the graph of the common logarithmic function $f(x) = \log_{10} x$ by hand.

Solution

Begin by constructing a table of values. Note that some of the values can be obtained without a calculator by using the Inverse Property of Logarithms. Others require a calculator. Next, plot the points and connect them with a smooth curve, as shown in Figure 3.19.

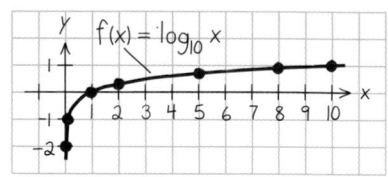

Figure 3.19

	Without calculator				With calculator		
x	$\frac{1}{100}$	$\frac{1}{10}$	1	10	2	5	8
$f(x) = \log_{10} x$	-2	-1	0	1	0.301	0.699	0.903

 **CHECKPOINT**   Now try Exercise 47.

The nature of the graph in Figure 3.19 is typical of functions of the form $f(x) = \log_a x$, $a > 1$. They have one x-intercept and one vertical asymptote. Notice how slowly the graph rises for $x > 1$.

STUDY TIP

In Example 5, you can also sketch the graph of $f(x) = \log_{10} x$ by evaluating the inverse function of f, $g(x) = 10^x$, for several values of x. Plot the points, sketch the graph of g, and then reflect the graph in the line $y = x$ to obtain the graph of f.

Library of Parent Functions: Logarithmic Function

The *logarithmic function*

$$f(x) = \log_a x, \quad a > 0, \ a \neq 1$$

is the inverse function of the exponential function. Its domain is the set of positive real numbers and its range is the set of all real numbers. This is the opposite of the exponential function. Moreover, the logarithmic function has the y-axis as a vertical asymptote, whereas the exponential function has the x-axis as a horizontal asymptote. Many real-life phenomena with a slow rate of growth can be modeled by logarithmic functions. The basic characteristics of the logarithmic function are summarized below. A review of logarithmic functions can be found in the *Study Capsules*.

Graph of $f(x) = \log_a x, \ a > 1$

Domain: $(0, \infty)$

Range: $(-\infty, \infty)$

Intercept: $(1, 0)$

Increasing on $(0, \infty)$

y-axis is a vertical asymptote

$(\log_a x \to -\infty \text{ as } x \to 0^+)$

Continuous

Reflection of graph of $f(x) = a^x$

in the line $y = x$

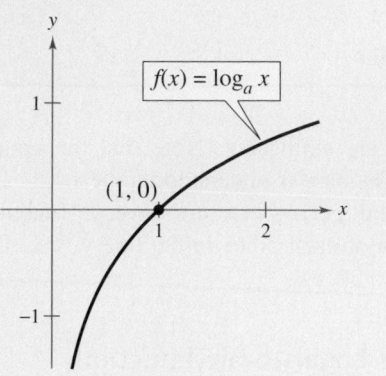

Exploration

Use a graphing utility to graph $y = \log_{10} x$ and $y = 8$ in the same viewing window. Find a viewing window that shows the point of intersection. What is the point of intersection? Use the point of intersection to complete the equation below.

$$\log_{10} \boxed{} = 8$$

Example 6 Transformations of Graphs of Logarithmic Functions

Each of the following functions is a transformation of the graph of $f(x) = \log_{10} x$.

a. Because $g(x) = \log_{10}(x - 1) = f(x - 1)$, the graph of g can be obtained by shifting the graph of f one unit to the *right*, as shown in Figure 3.20.

b. Because $h(x) = 2 + \log_{10} x = 2 + f(x)$, the graph of h can be obtained by shifting the graph of f two units *upward*, as shown in Figure 3.21.

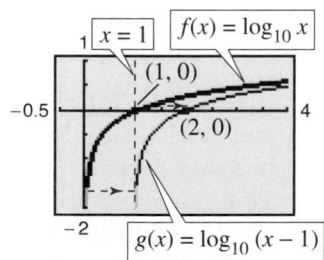

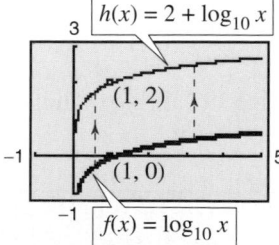

Figure 3.20   **Figure 3.21**

Notice that the transformation in Figure 3.21 keeps the y-axis as a vertical asymptote, but the transformation in Figure 3.20 yields the new vertical asymptote $x = 1$.

 **CHECKPOINT**   Now try Exercise 57.

TECHNOLOGY TIP

When a graphing utility graphs a logarithmic function, it may appear that the graph has an endpoint. This is because some graphing utilities have a limited resolution. So, in this text a blue or light red curve is placed behind the graphing utility's display to indicate where the graph should appear.

The Natural Logarithmic Function

By looking back at the graph of the natural exponential function introduced in Section 3.1, you will see that $f(x) = e^x$ is one-to-one and so has an inverse function. This inverse function is called the **natural logarithmic function** and is denoted by the special symbol $\ln x$, read as "the natural log of x" or "el en of x."

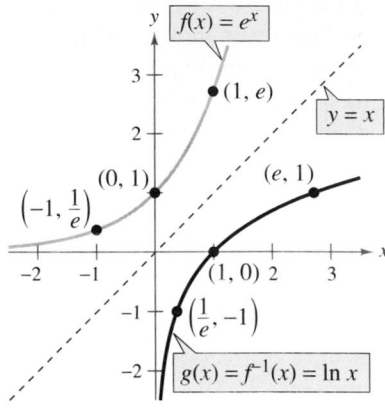

> **The Natural Logarithmic Function**
>
> For $x > 0$,
>
> $\quad y = \ln x$ if and only if $x = e^y$.
>
> The function given by
>
> $\quad f(x) = \log_e x = \ln x$
>
> is called the **natural logarithmic function.**

The equations $y = \ln x$ and $x = e^y$ are equivalent. Note that the natural logarithm $\ln x$ is written without a base. The base is understood to be e.

Because the functions $f(x) = e^x$ and $g(x) = \ln x$ are inverse functions of each other, their graphs are reflections of each other in the line $y = x$. This reflective property is illustrated in Figure 3.22.

Reflection of graph of $f(x) = e^x$ in the line $y = x$

Figure 3.22

Example 7   Evaluating the Natural Logarithmic Function

Use a calculator to evaluate the function $f(x) = \ln x$ at each indicated value of x.

a. $x = 2$ **b.** $x = 0.3$ **c.** $x = -1$

Solution

Function Value	Graphing Calculator Keystrokes	Display
a. $f(2) = \ln 2$	(LN) 2 (ENTER)	0.6931472
b. $f(0.3) = \ln 0.3$	(LN) .3 (ENTER)	-1.2039728
c. $f(-1) = \ln(-1)$	(LN) (−) 1 (ENTER)	ERROR

✓CHECKPOINT   Now try Exercise 63.

The four properties of logarithms listed on page 197 are also valid for natural logarithms.

> **Properties of Natural Logarithms**
>
> **1.** $\ln 1 = 0$ because $e^0 = 1$.
>
> **2.** $\ln e = 1$ because $e^1 = e$.
>
> **3.** $\ln e^x = x$ and $e^{\ln x} = x$.   Inverse Properties
>
> **4.** If $\ln x = \ln y$, then $x = y$.   One-to-One Property

TECHNOLOGY TIP

On most calculators, the natural logarithm is denoted by (LN), as illustrated in Example 7.

STUDY TIP

In Example 7(c), be sure you see that $\ln(-1)$ gives an error message on most calculators. This occurs because the domain of $\ln x$ is the set of *positive* real numbers (see Figure 3.22). So, $\ln(-1)$ is undefined.

Example 8   Using Properties of Natural Logarithms

Use the properties of natural logarithms to rewrite each expression.

a. $\ln \dfrac{1}{e}$ **b.** $e^{\ln 5}$ **c.** $4 \ln 1$ **d.** $2 \ln e$

Solution

a. $\ln \dfrac{1}{e} = \ln e^{-1} = -1$ Inverse Property   **b.** $e^{\ln 5} = 5$ Inverse Property

c. $4 \ln 1 = 4(0) = 0$ Property 1   **d.** $2 \ln e = 2(1) = 2$ Property 2

 **CHECKPOINT**   Now try Exercise 67.

Example 9   Finding the Domains of Logarithmic Functions

Find the domain of each function.

a. $f(x) = \ln(x - 2)$ **b.** $g(x) = \ln(2 - x)$ **c.** $h(x) = \ln x^2$

Algebraic Solution

a. Because $\ln(x - 2)$ is defined only if

$$x - 2 > 0$$

it follows that the domain of f is $(2, \infty)$.

b. Because $\ln(2 - x)$ is defined only if

$$2 - x > 0$$

it follows that the domain of g is $(-\infty, 2)$.

c. Because $\ln x^2$ is defined only if

$$x^2 > 0$$

it follows that the domain of h is all real numbers except $x = 0$.

Graphical Solution

Use a graphing utility to graph each function using an appropriate viewing window. Then use the *trace* feature to determine the domain of each function.

a. From Figure 3.23, you can see that the x-coordinates of the points on the graph appear to extend from the right of 2 to $+\infty$. So, you can estimate the domain to be $(2, \infty)$.

b. From Figure 3.24, you can see that the x-coordinates of the points on the graph appear to extend from $-\infty$ to the left of 2. So, you can estimate the domain to be $(-\infty, 2)$.

c. From Figure 3.25, you can see that the x-coordinates of the points on the graph appear to include all real numbers except $x = 0$. So, you can estimate the domain to be all real numbers except $x = 0$.

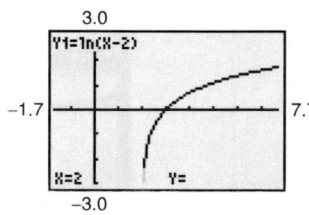

Figure 3.23

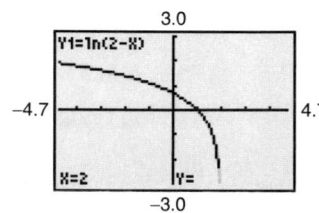

Figure 3.24

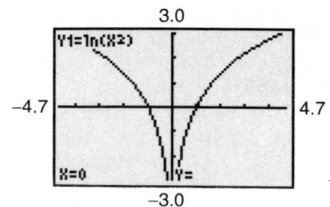

Figure 3.25

 **CHECKPOINT**   Now try Exercise 71.

In Example 9, suppose you had been asked to analyze the function $h(x) = \ln|x - 2|$. How would the domain of this function compare with the domains of the functions given in parts (a) and (b) of the example?

Application

Logarithmic functions are used to model many situations in real life, as shown in the next example.

Example 10 Human Memory Model

Students participating in a psychology experiment attended several lectures on a subject and were given an exam. Every month for a year after the exam, the students were retested to see how much of the material they remembered. The average scores for the group are given by the *human memory model*

$$f(t) = 75 - 6\ln(t + 1), \qquad 0 \le t \le 12$$

where t is the time in months.

a. What was the average score on the original exam $(t = 0)$?

b. What was the average score at the end of $t = 2$ months?

c. What was the average score at the end of $t = 6$ months?

Algebraic Solution

a. The original average score was

$$f(0) = 75 - 6\ln(0 + 1)$$

$$= 75 - 6\ln 1$$

$$= 75 - 6(0)$$

$$= 75.$$

b. After 2 months, the average score was

$$f(2) = 75 - 6\ln(2 + 1)$$

$$= 75 - 6\ln 3$$

$$\approx 75 - 6(1.0986)$$

$$\approx 68.41.$$

c. After 6 months, the average score was

$$f(6) = 75 - 6\ln(6 + 1)$$

$$= 75 - 6\ln 7$$

$$\approx 75 - 6(1.9459)$$

$$\approx 63.32.$$

✓CHECKPOINT   Now try Exercise 91.

Graphical Solution

Use a graphing utility to graph the model $y = 75 - 6\ln(x + 1)$. Then use the *value* or *trace* feature to approximate the following.

a. When $x = 0$, $y = 75$ (see Figure 3.26). So, the original average score was 75.

b. When $x = 2$, $y \approx 68.41$ (see Figure 3.27). So, the average score after 2 months was about 68.41.

c. When $x = 6$, $y \approx 63.32$ (see Figure 3.28). So, the average score after 6 months was about 63.32.

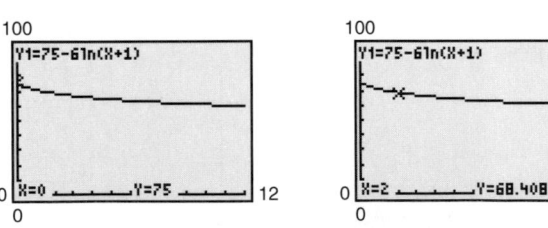

Figure 3.26   Figure 3.27

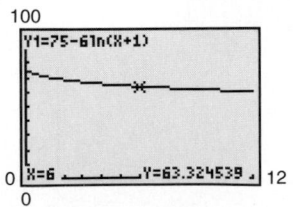

Figure 3.28

3.2 Exercises

See www.CalcChat.com for worked-out solutions to odd-numbered exercises.

Vocabulary Check

Fill in the blanks.

1. The inverse function of the exponential function $f(x) = a^x$ is called the _____ with base a.
2. The common logarithmic function has base _____ .
3. The logarithmic function $f(x) = \ln x$ is called the _____ function.
4. The inverse property of logarithms states that $\log_a a^x = x$ and _____ .
5. The one-to-one property of natural logarithms states that if $\ln x = \ln y$, then _____ .

In Exercises 1–6, write the logarithmic equation in exponential form. For example, the exponential form of $\log_5 25 = 2$ is $5^2 = 25$.

1. $\log_4 64 = 3$
2. $\log_3 81 = 4$
3. $\log_7 \frac{1}{49} = -2$
4. $\log_{10} \frac{1}{1000} = -3$
5. $\log_{32} 4 = \frac{2}{5}$
6. $\log_{16} 8 = \frac{3}{4}$

In Exercises 7–12, write the logarithmic equation in exponential form. For example, the exponential form of $\ln 5 = 1.6094 \ldots$ is $e^{1.6094\cdots} = 5$.

7. $\ln 1 = 0$
8. $\ln 4 = 1.3862 \ldots$
9. $\ln e = 1$
10. $\ln e^3 = 3$
11. $\ln \sqrt{e} = \frac{1}{2}$
12. $\ln \frac{1}{e^2} = -2$

In Exercises 13–18, write the exponential equation in logarithmic form. For example, the logarithmic form of $2^3 = 8$ is $\log_2 8 = 3$.

13. $5^3 = 125$
14. $8^2 = 64$
15. $81^{1/4} = 3$
16. $9^{3/2} = 27$
17. $6^{-2} = \frac{1}{36}$
18. $10^{-3} = 0.001$

In Exercises 19–24, write the exponential equation in logarithmic form. For example, the logarithmic form of $e^2 = 7.3890 \ldots$ is $\ln 7.3890 \ldots = 2$.

19. $e^3 = 20.0855 \ldots$
20. $e^4 = 54.5981 \ldots$
21. $e^{1.3} = 3.6692 \ldots$
22. $e^{2.5} = 12.1824 \ldots$
23. $\sqrt[3]{e} = 1.3956 \ldots$
24. $\frac{1}{e^4} = 0.0183 \ldots$

In Exercises 25–28, evaluate the function at the indicated value of x without using a calculator.

Function	Value
25. $f(x) = \log_2 x$	$x = 16$
26. $f(x) = \log_{16} x$	$x = \frac{1}{4}$
27. $g(x) = \log_{10} x$	$x = \frac{1}{1000}$
28. $g(x) = \log_{10} x$	$x = 10{,}000$

In Exercises 29–32, use a calculator to evaluate the function at the indicated value of x. Round your result to three decimal places.

Function	Value
29. $f(x) = \log_{10} x$	$x = 345$
30. $f(x) = \log_{10} x$	$x = \frac{4}{5}$
31. $h(x) = 6 \log_{10} x$	$x = 14.8$
32. $h(x) = 1.9 \log_{10} x$	$x = 4.3$

In Exercises 33–38, solve the equation for x.

33. $\log_7 x = \log_7 9$
34. $\log_5 5 = x$
35. $\log_6 6^2 = x$
36. $\log_2 2^{-1} = x$
37. $\log_8 x = \log_8 10^{-1}$
38. $\log_4 4^3 = x$

In Exercises 39–42, use the properties of logarithms to rewrite the expression.

39. $\log_4 4^{3x}$
40. $6^{\log_6 36}$
41. $3 \log_2 \frac{1}{2}$
42. $\frac{1}{4} \log_4 16$

In Exercises 43–46, sketch the graph of f. Then use the graph of f to sketch the graph of g.

43. $f(x) = 3^x$
   $g(x) = \log_3 x$
44. $f(x) = 5^x$
   $g(x) = \log_5 x$

45. $f(x) = e^{2x}$

$g(x) = \frac{1}{2} \ln x$

46. $f(x) = 4^x$

$g(x) = \log_4 x$

In Exercises 47–52, find the domain, vertical asymptote, and x-intercept of the logarithmic function, and sketch its graph by hand.

47. $y = \log_2(x + 2)$

48. $y = \log_2(x - 1)$

49. $y = 1 + \log_2 x$

50. $y = 2 - \log_2 x$

51. $y = 1 + \log_2(x - 2)$

52. $y = 2 + \log_2(x + 1)$

Library of Parent Functions In Exercises 53–56, use the graph of $y = \log_3 x$ to match the function with its graph. [The graphs are labeled (a), (b), (c), and (d).]

(a)

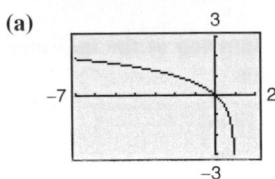

(b)

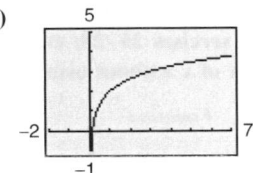

(c)

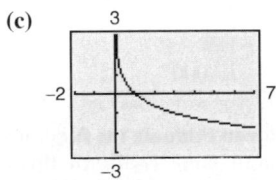

(d)

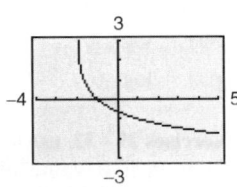

53. $f(x) = \log_3 x + 2$

54. $f(x) = -\log_3 x$

55. $f(x) = -\log_3(x + 2)$

56. $f(x) = \log_3(1 - x)$

In Exercises 57–62, use the graph of f to describe the transformation that yields the graph of g.

57. $f(x) = \log_{10} x,$ $g(x) = -\log_{10} x$

58. $f(x) = \log_{10} x,$ $g(x) = \log_{10}(x + 7)$

59. $f(x) = \log_2 x,$ $g(x) = 4 - \log_2 x$

60. $f(x) = \log_2 x,$ $g(x) = 3 + \log_2 x$

61. $f(x) = \log_8 x,$ $g(x) = -2 + \log_8(x + 3)$

62. $f(x) = \log_8 x,$ $g(x) = 4 + \log_8(x - 1)$

In Exercises 63–66, use a calculator to evaluate the function at the indicated value of x. Round your result to three decimal places.

Function	Value
63. $f(x) = \ln x$	$x = \sqrt{42}$
64. $f(x) = \ln x$	$x = 18.31$
65. $f(x) = -\ln x$	$x = \frac{1}{2}$
66. $f(x) = 3 \ln x$	$x = 0.75$

In Exercises 67–70, use the properties of natural logarithms to rewrite the expression.

67. $\ln e^2$

68. $-\ln e$

69. $e^{\ln 1.8}$

70. $7 \ln e^0$

In Exercises 71–74, find the domain, vertical asymptote, and x-intercept of the logarithmic function, and sketch its graph by hand. Verify using a graphing utility.

71. $f(x) = \ln(x - 1)$

72. $h(x) = \ln(x + 1)$

73. $g(x) = \ln(-x)$

74. $f(x) = \ln(3 - x)$

In Exercises 75–80, use the graph of $f(x) = \ln x$ to describe the transformation that yields the graph of g.

75. $g(x) = \ln(x + 3)$

76. $g(x) = \ln(x - 4)$

77. $g(x) = \ln x - 5$

78. $g(x) = \ln x + 4$

79. $g(x) = \ln(x - 1) + 2$

80. $g(x) = \ln(x + 2) - 5$

In Exercises 81–90, (a) use a graphing utility to graph the function, (b) find the domain, (c) use the graph to find the open intervals on which the function is increasing and decreasing, and (d) approximate any relative maximum or minimum values of the function. Round your result to three decimal places.

81. $f(x) = \frac{x}{2} - \ln \frac{x}{4}$

82. $g(x) = \frac{12 \ln x}{x}$

83. $h(x) = 4x \ln x$

84. $f(x) = \frac{x}{\ln x}$

85. $f(x) = \ln\left(\frac{x + 2}{x - 1}\right)$

86. $f(x) = \ln\left(\frac{2x}{x + 2}\right)$

87. $f(x) = \ln\left(\frac{x^2}{10}\right)$

88. $f(x) = \ln\left(\frac{x}{x^2 + 1}\right)$

89. $f(x) = \sqrt{\ln x}$

90. $f(x) = (\ln x)^2$

91. *Human Memory Model*   Students in a mathematics class were given an exam and then tested monthly with an equivalent exam. The average scores for the class are given by the human memory model

$f(t) = 80 - 17 \log_{10}(t + 1),$ $0 \le t \le 12$

where t is the time in months.

(a) What was the average score on the original exam $(t = 0)$?

(b) What was the average score after 4 months?

(c) What was the average score after 10 months?

Verify your answers in parts (a), (b), and (c) using a graphing utility.

92. *Data Analysis* The table shows the temperatures T (in °F) at which water boils at selected pressures p (in pounds per square inch). (Source: Standard Handbook of Mechanical Engineers)

Pressure, p	Temperature, T
5	162.24°
10	193.21°
14.696 (1 atm)	212.00°
20	227.96°
30	250.33°
40	267.25°
60	292.71°
80	312.03°
100	327.81°

A model that approximates the data is given by

$$T = 87.97 + 34.96 \ln p + 7.91 \sqrt{p}.$$

(a) Use a graphing utility to plot the data and graph the model in the same viewing window. How well does the model fit the data?

(b) Use the graph to estimate the pressure required for the boiling point of water to exceed 300°F.

(c) Calculate T when the pressure is 74 pounds per square inch. Verify your answer graphically.

93. *Compound Interest* A principal P, invested at $5\frac{1}{2}\%$ and compounded continuously, increases to an amount K times the original principal after t years, where

$$t = (\ln K)/0.055.$$

(a) Complete the table and interpret your results.

K	1	2	4	6	8	10	12
t							

(b) Use a graphing utility to graph the function.

94. *Population* The time t in years for the world population to double if it is increasing at a continuous rate of r is given by

$$t = \frac{\ln 2}{r}.$$

(a) Complete the table and interpret your results.

r	0.005	0.010	0.015	0.020	0.025	0.030
t						

(b) Use a graphing utility to graph the function.

95. *Sound Intensity* The relationship between the number of decibels β and the intensity of a sound I in watts per square meter is given by

$$\beta = 10 \log_{10}\left(\frac{I}{10^{-12}}\right).$$

(a) Determine the number of decibels of a sound with an intensity of 1 watt per square meter.

(b) Determine the number of decibels of a sound with an intensity of 10^{-2} watt per square meter.

(c) The intensity of the sound in part (a) is 100 times as great as that in part (b). Is the number of decibels 100 times as great? Explain.

96. *Home Mortgage* The model

$$t = 16.625 \ln\left(\frac{x}{x - 750}\right), \quad x > 750$$

approximates the length of a home mortgage of $150,000 at 6% in terms of the monthly payment. In the model, t is the length of the mortgage in years and x is the monthly payment in dollars.

(a) Use the model to approximate the lengths of a $150,000 mortgage at 6% when the monthly payment is $897.72 and when the monthly payment is $1659.24.

(b) Approximate the total amounts paid over the term of the mortgage with a monthly payment of $897.72 and with a monthly payment of $1659.24. What amount of the total is interest costs for each payment?

Ventilation Rates **In Exercises 97 and 98, use the model**

$$y = 80.4 - 11 \ln x, \quad 100 \le x \le 1500$$

which approximates the minimum required ventilation rate in terms of the air space per child in a public school classroom. In the model, x is the air space per child (in cubic feet) and y is the ventilation rate per child (in cubic feet per minute).

97. Use a graphing utility to graph the function and approximate the required ventilation rate when there is 300 cubic feet of air space per child.

98. A classroom is designed for 30 students. The air-conditioning system in the room has the capacity to move 450 cubic feet of air per minute.

(a) Determine the ventilation rate per child, assuming that the room is filled to capacity.

(b) Use the graph in Exercise 97 to estimate the air space required per child.

(c) Determine the minimum number of square feet of floor space required for the room if the ceiling height is 30 feet.

Synthesis

True or False? **In Exercises 99 and 100, determine whether the statement is true or false. Justify your answer.**

99. You can determine the graph of $f(x) = \log_6 x$ by graphing $g(x) = 6^x$ and reflecting it about the x-axis.

100. The graph of $f(x) = \log_3 x$ contains the point $(27, 3)$.

Think About It **In Exercises 101–104, find the value of the base b so that the graph of $f(x) = \log_b x$ contains the given point.**

101. $(32, 5)$ **102.** $(81, 4)$

103. $\left(\frac{1}{16}, 2\right)$ **104.** $\left(\frac{1}{27}, 3\right)$

Library of Parent Functions **In Exercises 105 and 106, determine which equation(s) may be represented by the graph shown. (There may be more than one correct answer.)**

105. 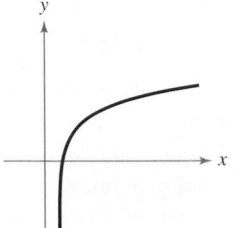   **106.**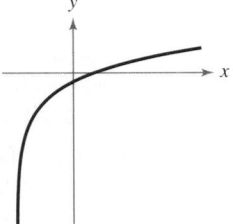

(a) $y = \log_2(x + 1) + 2$ (a) $y = \ln(x - 1) + 2$

(b) $y = \log_2(x - 1) + 2$ (b) $y = \ln(x + 2) - 1$

(c) $y = 2 - \log_2(x - 1)$ (c) $y = 2 - \ln(x - 1)$

(d) $y = \log_2(x + 2) + 1$ (d) $y = \ln(x - 2) + 1$

107. *Writing* Explain why $\log_a x$ is defined only for $0 < a < 1$ and $a > 1$.

108. *Graphical Analysis* Use a graphing utility to graph $f(x) = \ln x$ and $g(x)$ in the same viewing window and determine which is increasing at the greater rate as x approaches $+\infty$. What can you conclude about the rate of growth of the natural logarithmic function?

(a) $g(x) = \sqrt{x}$ (b) $g(x) = \sqrt[4]{x}$

109. *Exploration* The following table of values was obtained by evaluating a function. Determine which of the statements may be true and which must be false.

x	1	2	8
y	0	1	3

(a) y is an exponential function of x.

(b) y is a logarithmic function of x.

(c) x is an exponential function of y.

(d) y is a linear function of x.

110. *Pattern Recognition*

(a) Use a graphing utility to compare the graph of the function $y = \ln x$ with the graph of each function.

$$y_1 = x - 1, \quad y_2 = (x - 1) - \tfrac{1}{2}(x - 1)^2,$$

$$y_3 = (x - 1) - \tfrac{1}{2}(x - 1)^2 + \tfrac{1}{3}(x - 1)^3$$

(b) Identify the pattern of successive polynomials given in part (a). Extend the pattern one more term and compare the graph of the resulting polynomial function with the graph of $y = \ln x$. What do you think the pattern implies?

111. *Numerical and Graphical Analysis*

(a) Use a graphing utility to complete the table for the function

$$f(x) = \frac{\ln x}{x}.$$

x	1	5	10	10^2	10^4	10^6
$f(x)$						

(b) Use the table in part (a) to determine what value $f(x)$ approaches as x increases without bound. Use a graphing utility to confirm the result of part (b).

112. *Writing* Use a graphing utility to determine how many months it would take for the average score in Example 10 to decrease to 60. Explain your method of solving the problem. Describe another way that you can use a graphing utility to determine the answer. Also, make a statement about the general shape of the model. Would a student forget more quickly soon after the test or after some time had passed? Explain your reasoning.

Skills Review

In Exercises 113–120, factor the polynomial.

113. $x^2 + 2x - 3$ **114.** $2x^2 + 3x - 5$

115. $12x^2 + 5x - 3$ **116.** $16x^2 + 16x + 7$

117. $16x^2 - 25$ **118.** $36x^2 - 49$

119. $2x^3 + x^2 - 45x$ **120.** $3x^3 - 5x^2 - 12x$

In Exercises 121–124, evaluate the function for $f(x) = 3x + 2$ and $g(x) = x^3 - 1$.

121. $(f + g)(2)$ **122.** $(f - g)(-1)$

123. $(fg)(6)$ **124.** $\left(\dfrac{f}{g}\right)(0)$

In Exercises 125–128, solve the equation graphically.

125. $5x - 7 = x + 4$ **126.** $-2x + 3 = 8x$

127. $\sqrt{3x - 2} = 9$ **128.** $\sqrt{x - 11} = x + 2$

3.3   Properties of Logarithms

Change of Base

Most calculators have only two types of log keys, one for common logarithms (base 10) and one for natural logarithms (base e). Although common logs and natural logs are the most frequently used, you may occasionally need to evaluate logarithms to other bases. To do this, you can use the following **change-of-base formula.**

Change-of-Base Formula

Let a, b, and x be positive real numbers such that $a \neq 1$ and $b \neq 1$. Then $\log_a x$ can be converted to a different base using any of the following formulas.

Base b	Base 10	Base e
$\log_a x = \dfrac{\log_b x}{\log_b a}$	$\log_a x = \dfrac{\log_{10} x}{\log_{10} a}$	$\log_a x = \dfrac{\ln x}{\ln a}$

One way to look at the change-of-base formula is that logarithms to base a are simply *constant multiples* of logarithms to base b. The constant multiplier is $1/(\log_b a)$.

Example 1   Changing Bases Using Common Logarithms

a. $\log_4 25 = \dfrac{\log_{10} 25}{\log_{10} 4}$ $\qquad\qquad$ $\log_a x = \dfrac{\log_{10} x}{\log_{10} a}$

$\qquad\quad \approx \dfrac{1.39794}{0.60206} \approx 2.32$ $\qquad\qquad$ Use a calculator.

b. $\log_2 12 = \dfrac{\log_{10} 12}{\log_{10} 2} \approx \dfrac{1.07918}{0.30103} \approx 3.58$

✔CHECKPOINT   Now try Exercise 9.

Example 2   Changing Bases Using Natural Logarithms

a. $\log_4 25 = \dfrac{\ln 25}{\ln 4}$ $\qquad\qquad$ $\log_a x = \dfrac{\ln x}{\ln a}$

$\qquad\quad \approx \dfrac{3.21888}{1.38629} \approx 2.32$ $\qquad\qquad$ Use a calculator.

b. $\log_2 12 = \dfrac{\ln 12}{\ln 2} \approx \dfrac{2.48491}{0.69315} \approx 3.58$

✔CHECKPOINT   Now try Exercise 15.

What you should learn

- Rewrite logarithms with different bases.
- Use properties of logarithms to evaluate or rewrite logarithmic expressions.
- Use properties of logarithms to expand or condense logarithmic expressions.
- Use logarithmic functions to model and solve real-life problems.

Why you should learn it

Logarithmic functions can be used to model and solve real-life problems, such as the human memory model in Exercise 96 on page 212.

Gary Conner/PhotoEdit

STUDY TIP

Notice in Examples 1 and 2 that the result is the same whether common logarithms or natural logarithms are used in the change-of-base formula.

Properties of Logarithms

You know from the previous section that the logarithmic function with base a is the *inverse function* of the exponential function with base a. So, it makes sense that the properties of exponents (see Section 3.1) should have corresponding properties involving logarithms. For instance, the exponential property $a^0 = 1$ has the corresponding logarithmic property $\log_a 1 = 0$.

Properties of Logarithms   (See the proof on page 255.)

Let a be a positive number such that $a \neq 1$, and let n be a real number. If u and v are positive real numbers, the following properties are true.

	Logarithm with Base a	*Natural Logarithm*
1. Product Property:	$\log_a(uv) = \log_a u + \log_a v$	$\ln(uv) = \ln u + \ln v$
2. Quotient Property:	$\log_a \dfrac{u}{v} = \log_a u - \log_a v$	$\ln \dfrac{u}{v} = \ln u - \ln v$
3. Power Property:	$\log_a u^n = n \log_a u$	$\ln u^n = n \ln u$

STUDY TIP

There is no general property that can be used to rewrite $\log_a(u \pm v)$. Specifically, $\log_a(x + y)$ is *not* equal to $\log_a x + \log_a y$.

Example 3   Using Properties of Logarithms

Write each logarithm in terms of $\ln 2$ and $\ln 3$.

a. $\ln 6$ **b.** $\ln \dfrac{2}{27}$

Solution

a. $\ln 6 = \ln(2 \cdot 3)$ Rewrite 6 as $2 \cdot 3$.

      $= \ln 2 + \ln 3$ Product Property

b. $\ln \dfrac{2}{27} = \ln 2 - \ln 27$ Quotient Property

         $= \ln 2 - \ln 3^3$ Rewrite 27 as 3^3.

         $= \ln 2 - 3 \ln 3$ Power Property

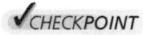 **CHECKPOINT**   Now try Exercise 17.

Example 4   Using Properties of Logarithms

Use the properties of logarithms to verify that $-\log_{10} \frac{1}{100} = \log_{10} 100$.

Solution

     $-\log_{10} \frac{1}{100} = -\log_{10}(100^{-1})$ Rewrite $\frac{1}{100}$ as 100^{-1}.

               $= -(-1) \log_{10} 100$ Power Property

               $= \log_{10} 100$ Simplify.

 **CHECKPOINT**   Now try Exercise 35.

Rewriting Logarithmic Expressions

The properties of logarithms are useful for rewriting logarithmic expressions in forms that simplify the operations of algebra. This is true because they convert complicated products, quotients, and exponential forms into simpler sums, differences, and products, respectively.

Example 5   Expanding Logarithmic Expressions

Use the properties of logarithms to expand each expression.

a. $\log_4 5x^3y$ **b.** $\ln \dfrac{\sqrt{3x-5}}{7}$

Solution

a. $\log_4 5x^3y = \log_4 5 + \log_4 x^3 + \log_4 y$ Product Property

$\qquad\qquad\quad = \log_4 5 + 3 \log_4 x + \log_4 y$ Power Property

b. $\ln \dfrac{\sqrt{3x-5}}{7} = \ln\left[\dfrac{(3x-5)^{1/2}}{7}\right]$ Rewrite radical using rational exponent.

$\qquad\qquad\quad = \ln(3x-5)^{1/2} - \ln 7$ Quotient Property

$\qquad\qquad\quad = \tfrac{1}{2} \ln(3x-5) - \ln 7$ Power Property

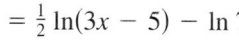**CHECKPOINT**   Now try Exercise 55.

In Example 5, the properties of logarithms were used to *expand* logarithmic expressions. In Example 6, this procedure is reversed and the properties of logarithms are used to *condense* logarithmic expressions.

Exploration

Use a graphing utility to graph the functions

$$y = \ln x - \ln(x - 3)$$

and

$$y = \ln \dfrac{x}{x - 3}$$

in the same viewing window. Does the graphing utility show the functions with the same domain? If so, should it? Explain your reasoning.

Example 6   Condensing Logarithmic Expressions

Use the properties of logarithms to condense each logarithmic expression.

a. $\tfrac{1}{2} \log_{10} x + 3 \log_{10}(x + 1)$ **b.** $2 \ln(x + 2) - \ln x$

c. $\tfrac{1}{3}[\log_2 x + \log_2(x - 4)]$

Solution

a. $\tfrac{1}{2} \log_{10} x + 3 \log_{10}(x + 1) = \log_{10} x^{1/2} + \log_{10}(x + 1)^3$ Power Property

$\qquad\qquad\qquad\qquad\qquad = \log_{10}\left[\sqrt{x}(x + 1)^3\right]$ Product Property

b. $2 \ln(x + 2) - \ln x = \ln(x + 2)^2 - \ln x$ Power Property

$\qquad\qquad\qquad\qquad = \ln \dfrac{(x + 2)^2}{x}$ Quotient Property

c. $\tfrac{1}{3}[\log_2 x + \log_2(x - 4)] = \tfrac{1}{3}\{\log_2[x(x - 4)]\}$ Product Property

$\qquad\qquad\qquad\qquad\quad = \log_2[x(x - 4)]^{1/3}$ Power Property

$\qquad\qquad\qquad\qquad\quad = \log_2 \sqrt[3]{x(x - 4)}$ Rewrite with a radical.

CHECKPOINT   Now try Exercise 71.

Example 7   Finding a Mathematical Model

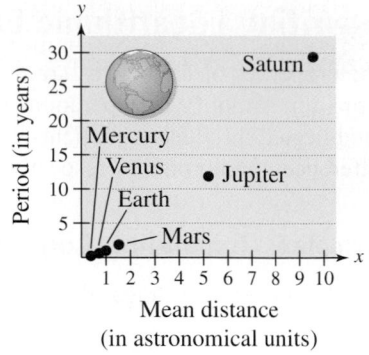

The table shows the mean distance from the sun x and the period y (the time it takes a planet to orbit the sun) for each of the six planets that are closest to the sun. In the table, the mean distance is given in astronomical units (where the Earth's mean distance is defined as 1.0), and the period is given in years. Find an equation that relates y and x.

Figure 3.29

Planet	Mercury	Venus	Earth	Mars	Jupiter	Saturn
Mean distance, x	0.387	0.723	1.000	1.524	5.203	9.555
Period, y	0.241	0.615	1.000	1.881	11.860	29.420

Algebraic Solution

The points in the table are plotted in Figure 3.29. From this figure it is not clear how to find an equation that relates y and x. To solve this problem, take the natural log of each of the x- and y-values in the table. This produces the following results.

Planet	Mercury	Venus	Earth
$\ln x = X$	−0.949	−0.324	0.000
$\ln y = Y$	−1.423	−0.486	0.000

Planet	Mars	Jupiter	Saturn
$\ln x = X$	0.421	1.649	2.257
$\ln y = Y$	0.632	2.473	3.382

Now, by plotting the points in the table, you can see that all six of the points appear to lie in a line, as shown in Figure 3.30. Choose any two points to determine the slope of the line. Using the two points $(0.421, 0.632)$ and $(0, 0)$, you can determine that the slope of the line is

$$m = \frac{0.632 - 0}{0.421 - 0} \approx 1.5 = \frac{3}{2}.$$

By the point-slope form, the equation of the line is $Y = \frac{3}{2}X$, where $Y = \ln y$ and $X = \ln x$. You can therefore conclude that $\ln y = \frac{3}{2} \ln x$.

✓CHECKPOINT   Now try Exercise 97.

Graphical Solution

The points in the table are plotted in Figure 3.29. From this figure it is not clear how to find an equation that relates y and x. To solve this problem, take the natural log of each of the x- and y-values in the table. This produces the following results.

Planet	Mercury	Venus	Earth
$\ln x = X$	−0.949	−0.324	0.000
$\ln y = Y$	−1.423	−0.486	0.000

Planet	Mars	Jupiter	Saturn
$\ln x = X$	0.421	1.649	2.257
$\ln y = Y$	0.632	2.473	3.382

Now, by plotting the points in the table, you can see that all six of the points appear to lie in a line, as shown in Figure 3.30. Using the *linear regression* feature of a graphing utility, you can find a linear model for the data, as shown in Figure 3.31. You can approximate this model to be $Y = 1.5X = \frac{3}{2}X$, where $Y = \ln y$ and $X = \ln x$. From the model, you can see that the slope of the line is $\frac{3}{2}$. So, you can conclude that $\ln y = \frac{3}{2} \ln x$.

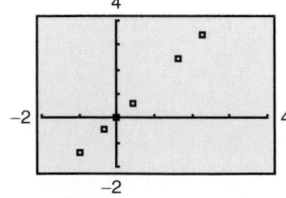

Figure 3.30

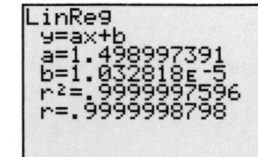

Figure 3.31

In Example 7, try to convert the final equation to $y = f(x)$ form. You will get a function of the form $y = ax^b$, which is called a *power model*.

3.3 Exercises

See www.CalcChat.com for worked-out solutions to odd-numbered exercises.

Vocabulary Check

Fill in the blanks.

1. To evaluate logarithms to any base, you can use the _____ formula.

2. The change-of-base formula for base e is given by $\log_a x = $ _____ .

3. _____ $= n \log_a u$

4. $\ln(uv) = $ _____

In Exercises 1–8, rewrite the logarithm as a ratio of (a) common logarithms and (b) natural logarithms.

1. $\log_5 x$

2. $\log_3 x$

3. $\log_{1/5} x$

4. $\log_{1/3} x$

5. $\log_a \frac{3}{10}$

6. $\log_a \frac{3}{4}$

7. $\log_{2.6} x$

8. $\log_{7.1} x$

In Exercises 9–16, evaluate the logarithm using the change-of-base formula. Round your result to three decimal places.

9. $\log_3 7$

10. $\log_7 4$

11. $\log_{1/2} 4$

12. $\log_{1/8} 64$

13. $\log_9(0.8)$

14. $\log_3(0.015)$

15. $\log_{15} 1460$

16. $\log_{20} 135$

In Exercises 17–20, rewrite the expression in terms of ln 4 and ln 5.

17. $\ln 20$

18. $\ln 500$

19. $\ln \frac{5}{64}$

20. $\ln \frac{2}{5}$

In Exercises 21–24, approximate the logarithm using the properties of logarithms, given that $\log_b 2 \approx 0.3562$, $\log_b 3 \approx 0.5646$, and $\log_b 5 \approx 0.8271$. Round your result to four decimal places.

21. $\log_b 25$

22. $\log_b 30$

23. $\log_b \sqrt{3}$

24. $\log_b\left(\frac{25}{9}\right)$

In Exercises 25–30, use the change-of-base formula $\log_a x = (\ln x)/(\ln a)$ and a graphing utility to graph the function.

25. $f(x) = \log_3(x + 2)$

26. $f(x) = \log_2(x - 1)$

27. $f(x) = \log_{1/2}(x - 2)$

28. $f(x) = \log_{1/3}(x + 1)$

29. $f(x) = \log_{1/4} x^2$

30. $f(x) = \log_{1/2}\left(\frac{x}{2}\right)$

In Exercises 31–34, use the properties of logarithms to rewrite and simplify the logarithmic expression.

31. $\log_4 8$

32. $\log_2(4^2 \cdot 3^4)$

33. $\ln(5e^6)$

34. $\ln \frac{6}{e^2}$

In Exercises 35 and 36, use the properties of logarithms to verify the equation.

35. $\log_5 \frac{1}{250} = -3 - \log_5 2$

36. $-\ln 24 = -(3 \ln 2 + \ln 3)$

In Exercises 37–56, use the properties of logarithms to expand the expression as a sum, difference, and/or constant multiple of logarithms. (Assume all variables are positive.)

37. $\log_{10} 5x$

38. $\log_{10} 10z$

39. $\log_{10} \frac{5}{x}$

40. $\log_{10} \frac{y}{2}$

41. $\log_8 x^4$

42. $\log_6 z^{-3}$

43. $\ln \sqrt{z}$

44. $\ln \sqrt[3]{t}$

45. $\ln xyz$

46. $\ln \frac{xy}{z}$

47. $\log_3 a^2bc^3$

48. $\log_5 x^3y^3z$

49. $\ln\left(a^2\sqrt{a - 1}\right), \quad a > 1$

50. $\ln[z(z - 1)^2], \quad z > 1$

51. $\ln \sqrt[3]{\frac{x}{y}}$

52. $\ln \sqrt{\frac{x^2}{y^3}}$

53. $\ln\left(\frac{x^2 - 1}{x^3}\right), \quad x > 1$

54. $\ln \frac{x}{\sqrt{x^2 + 1}}$

55. $\ln \frac{x^4\sqrt{y}}{z^5}$

56. $\log_b \frac{\sqrt{x}\, y^4}{z^4}$

Graphical Analysis In Exercises 57 and 58, (a) use a graphing utility to graph the two equations in the same viewing window and (b) use the *table* feature of the graphing utility to create a table of values for each equation. (c) What do the graphs and tables suggest? Explain your reasoning.

57. $y_1 = \ln[x^3(x+4)]$, $y_2 = 3\ln x + \ln(x+4)$

58. $y_1 = \ln\left(\dfrac{\sqrt{x}}{x-2}\right)$, $y_2 = \frac{1}{2}\ln x - \ln(x-2)$

In Exercises 59–76, condense the expression to the logarithm of a single quantity.

59. $\ln x + \ln 4$
60. $\ln y + \ln z$
61. $\log_4 z - \log_4 y$
62. $\log_5 8 - \log_5 t$
63. $2\log_2(x+3)$
64. $\frac{5}{2}\log_7(z-4)$
65. $\frac{1}{2}\ln(x^2+4)$
66. $2\ln x + \ln(x+1)$
67. $\ln x - 3\ln(x+1)$
68. $\ln x - 2\ln(x+2)$
69. $\ln(x-2) - \ln(x+2)$
70. $3\ln x + 2\ln y - 4\ln z$
71. $\ln x - 2[\ln(x+2) + \ln(x-2)]$
72. $4[\ln z + \ln(z+5)] - 2\ln(z-5)$
73. $\frac{1}{3}[2\ln(x+3) + \ln x - \ln(x^2-1)]$
74. $2[\ln x - \ln(x+1) - \ln(x-1)]$
75. $\frac{1}{3}[\ln y + 2\ln(y+4)] - \ln(y-1)$
76. $\frac{1}{2}[\ln(x+1) + 2\ln(x-1)] + 3\ln x$

Graphical Analysis In Exercises 77 and 78, (a) use a graphing utility to graph the two equations in the same viewing window and (b) use the *table* feature of the graphing utility to create a table of values for each equation. (c) What do the graphs and tables suggest? Verify your conclusion algebraically.

77. $y_1 = 2[\ln 8 - \ln(x^2+1)]$, $y_2 = \ln\left[\dfrac{64}{(x^2+1)^2}\right]$

78. $y_1 = \ln x + \frac{1}{2}\ln(x+1)$, $y_2 = \ln(x\sqrt{x+1})$

Think About It In Exercises 79 and 80, (a) use a graphing utility to graph the two equations in the same viewing window and (b) use the *table* feature of the graphing utility to create a table of values for each equation. (c) Are the expressions equivalent? Explain.

79. $y_1 = \ln x^2$, $y_2 = 2\ln x$
80. $y_1 = \frac{1}{4}\ln[x^4(x^2+1)]$, $y_2 = \ln x + \frac{1}{4}\ln(x^2+1)$

In Exercises 81–94, find the exact value of the logarithm without using a calculator. If this is not possible, state the reason.

81. $\log_3 9$
82. $\log_6 \sqrt[3]{6}$
83. $\log_4 16^{3.4}$
84. $\log_5\left(\frac{1}{125}\right)$
85. $\log_2(-4)$
86. $\log_4(-16)$
87. $\log_5 75 - \log_5 3$
88. $\log_4 2 + \log_4 32$
89. $\ln e^3 - \ln e^7$
90. $\ln e^6 - 2\ln e^5$
91. $2\ln e^4$
92. $\ln e^{4.5}$
93. $\ln \dfrac{1}{\sqrt{e}}$
94. $\ln \sqrt[5]{e^3}$

95. *Sound Intensity* The relationship between the number of decibels β and the intensity of a sound I in watts per square meter is given by

$$\beta = 10\log_{10}\left(\dfrac{I}{10^{-12}}\right).$$

(a) Use the properties of logarithms to write the formula in a simpler form.

(b) Use a graphing utility to complete the table.

I	10^{-4}	10^{-6}	10^{-8}	10^{-10}	10^{-12}	10^{-14}
β						

(c) Verify your answers in part (b) algebraically.

96. *Human Memory Model* Students participating in a psychology experiment attended several lectures and were given an exam. Every month for the next year, the students were retested to see how much of the material they remembered. The average scores for the group are given by the human memory model

$$f(t) = 90 - 15\log_{10}(t+1), \quad 0 \le t \le 12$$

where t is the time (in months).

(a) Use a graphing utility to graph the function over the specified domain.

(b) What was the average score on the original exam ($t = 0$)?

(c) What was the average score after 6 months?

(d) What was the average score after 12 months?

(e) When did the average score decrease to 75?

97. *Comparing Models* A cup of water at an initial temperature of 78°C is placed in a room at a constant temperature of 21°C. The temperature of the water is measured every 5 minutes during a half-hour period. The results are recorded as ordered pairs of the form (t, T), where t is the time (in minutes) and T is the temperature (in degrees Celsius).

(0, 78.0°), (5, 66.0°), (10, 57.5°), (15, 51.2°),
(20, 46.3°), (25, 42.5°), (30, 39.6°)

(a) The graph of the model for the data should be asymptotic with the graph of the temperature of the room. Subtract the room temperature from each of the temperatures in the ordered pairs. Use a graphing utility to plot the data points (t, T) and $(t, T - 21)$.

(b) An exponential model for the data $(t, T - 21)$ is given by

$$T - 21 = 54.4(0.964)^t.$$

Solve for T and graph the model. Compare the result with the plot of the original data.

(c) Take the natural logarithms of the revised temperatures. Use a graphing utility to plot the points $(t, \ln(T - 21))$ and observe that the points appear linear. Use the *regression* feature of a graphing utility to fit a line to the data. The resulting line has the form

$$\ln(T - 21) = at + b.$$

Use the properties of logarithms to solve for T. Verify that the result is equivalent to the model in part (b).

(d) Fit a rational model to the data. Take the reciprocals of the y-coordinates of the revised data points to generate the points

$$\left(t, \frac{1}{T - 21}\right).$$

Use a graphing utility to plot these points and observe that they appear linear. Use the *regression* feature of a graphing utility to fit a line to the data. The resulting line has the form

$$\frac{1}{T - 21} = at + b.$$

Solve for T, and use a graphing utility to graph the rational function and the original data points.

98. Writing   Write a short paragraph explaining why the transformations of the data in Exercise 97 were necessary to obtain the models. Why did taking the logarithms of the temperatures lead to a linear scatter plot? Why did taking the reciprocals of the temperatures lead to a linear scatter plot?

Synthesis

True or False?   In Exercises 99–106, determine whether the statement is true or false given that $f(x) = \ln x$, where $x > 0$. Justify your answer.

99. $f(ax) = f(a) + f(x), \ a > 0$

100. $f(x - a) = f(x) - f(a), \ x > a$

101. $f\left(\dfrac{x}{a}\right) = \dfrac{f(x)}{f(a)}, \ f(a) \neq 0$

102. $f(x + a) = f(x)f(a), \ a > 0$

103. $\sqrt{f(x)} = \tfrac{1}{2}f(x)$

104. $[f(x)]^n = nf(x)$

105. If $f(x) < 0$, then $0 < x < e$.

106. If $f(x) > 0$, then $x > e$.

107. Proof   Prove that $\dfrac{\log_a x}{\log_{a/b} x} = 1 + \log_a \dfrac{1}{b}$.

108. Think About It   Use a graphing utility to graph

$$f(x) = \ln \frac{x}{2}, \quad g(x) = \frac{\ln x}{\ln 2}, \quad h(x) = \ln x - \ln 2$$

in the same viewing window. Which two functions have identical graphs? Explain why.

In Exercises 109–114, use the change-of-base formula to rewrite the logarithm as a ratio of logarithms. Then use a graphing utility to graph the ratio.

109. $f(x) = \log_2 x$

110. $f(x) = \log_4 x$

111. $f(x) = \log_3 \sqrt{x}$

112. $f(x) = \log_2 \sqrt[3]{x}$

113. $f(x) = \log_5 \dfrac{x}{3}$

114. $f(x) = \log_3 \dfrac{x}{5}$

115. Exploration   For how many integers between 1 and 20 can the natural logarithms be approximated given that $\ln 2 \approx 0.6931$, $\ln 3 \approx 1.0986$,   and $\ln 5 \approx 1.6094$? Approximate these logarithms. (Do not use a calculator.)

Skills Review

In Exercises 116–119, simplify the expression.

116. $\dfrac{24xy^{-2}}{16x^{-3}y}$

117. $\left(\dfrac{2x^2}{3y}\right)^{-3}$

118. $(18x^3y^4)^{-3}(18x^3y^4)^3$

119. $xy(x^{-1} + y^{-1})^{-1}$

In Exercises 120–125, find all solutions of the equation. Be sure to check all your solutions.

120. $x^2 - 6x + 2 = 0$

121. $2x^3 + 20x^2 + 50x = 0$

122. $x^4 - 19x^2 + 48 = 0$

123. $9x^4 - 37x^2 + 4 = 0$

124. $x^3 - 6x^2 - 4x + 24 = 0$

125. $9x^4 - 226x^2 + 25 = 0$

3.4   Solving Exponential and Logarithmic Equations

Introduction

So far in this chapter, you have studied the definitions, graphs, and properties of exponential and logarithmic functions. In this section, you will study procedures for *solving equations* involving exponential and logarithmic functions.

There are two basic strategies for solving exponential or logarithmic equations. The first is based on the One-to-One Properties and the second is based on the Inverse Properties. For $a > 0$ and $a \neq 1$, the following properties are true for all x and y for which $\log_a x$ and $\log_a y$ are defined.

One-to-One Properties

$a^x = a^y$ if and only if $x = y$.

$\log_a x = \log_a y$ if and only if $x = y$.

Inverse Properties

$a^{\log_a x} = x$

$\log_a a^x = x$

Example 1   Solving Simple Exponential and Logarithmic Equations

Original Equation	Rewritten Equation	Solution	Property
a. $2^x = 32$	$2^x = 2^5$	$x = 5$	One-to-One
b. $\ln x - \ln 3 = 0$	$\ln x = \ln 3$	$x = 3$	One-to-One
c. $\left(\frac{1}{3}\right)^x = 9$	$3^{-x} = 3^2$	$x = -2$	One-to-One
d. $e^x = 7$	$\ln e^x = \ln 7$	$x = \ln 7$	Inverse
e. $\ln x = -3$	$e^{\ln x} = e^{-3}$	$x = e^{-3}$	Inverse
f. $\log_{10} x = -1$	$10^{\log_{10} x} = 10^{-1}$	$x = 10^{-1} = \frac{1}{10}$	Inverse

✓**CHECKPOINT**   Now try Exercise 21.

The strategies used in Example 1 are summarized as follows.

Strategies for Solving Exponential and Logarithmic Equations

1. Rewrite the original equation in a form that allows the use of the One-to-One Properties of exponential or logarithmic functions.

2. Rewrite an *exponential* equation in logarithmic form and apply the Inverse Property of logarithmic functions.

3. Rewrite a *logarithmic* equation in exponential form and apply the Inverse Property of exponential functions.

What you should learn

- Solve simple exponential and logarithmic equations.
- Solve more complicated exponential equations.
- Solve more complicated logarithmic equations.
- Use exponential and logarithmic equations to model and solve real-life problems.

Why you should learn it

Exponential and logarithmic equations can be used to model and solve real-life problems. For instance, Exercise 139 on page 223 shows how to use an exponential function to model the average heights of men and women.

Charles Gupton/Corbis

Prerequisite Skills

If you have difficulty with this example, review the properties of logarithms in Section 3.3.

Solving Exponential Equations

Example 2   Solving Exponential Equations

Solve each equation.   **a.** $e^x = 72$ **b.** $3(2^x) = 42$

Algebraic Solution	Graphical Solution

Algebraic Solution

a. $e^x = 72$ Write original equation.

   $\ln e^x = \ln 72$ Take natural log of each side.

   $x = \ln 72 \approx 4.28$ Inverse Property

The solution is $x = \ln 72 \approx 4.28$. Check this in the original equation.

b. $3(2^x) = 42$ Write original equation.

   $2^x = 14$ Divide each side by 3.

   $\log_2 2^x = \log_2 14$ Take log (base 2) of each side.

   $x = \log_2 14$ Inverse Property

   $x = \dfrac{\ln 14}{\ln 2} \approx 3.81$ Change-of-base formula

The solution is $x = \log_2 14 \approx 3.81$. Check this in the original equation.

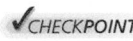 **CHECKPOINT**   Now try Exercise 55.

Graphical Solution

a. Use a graphing utility to graph the left- and right-hand sides of the equation as $y_1 = e^x$ and $y_2 = 72$ in the same viewing window. Use the *intersect* feature or the *zoom* and *trace* features of the graphing utility to approximate the intersection point, as shown in Figure 3.32. So, the approximate solution is $x \approx 4.28$.

b. Use a graphing utility to graph $y_1 = 3(2^x)$ and $y_2 = 42$ in the same viewing window. Use the *intersect* feature or the *zoom* and *trace* features to approximate the intersection point, as shown in Figure 3.33. So, the approximate solution is $x \approx 3.81$.

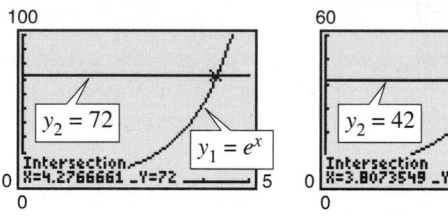

Figure 3.32   **Figure 3.33**

Example 3   Solving an Exponential Equation

Solve $4e^{2x} - 3 = 2$.

Algebraic Solution

   $4e^{2x} - 3 = 2$ Write original equation.

   $4e^{2x} = 5$ Add 3 to each side.

   $e^{2x} = \dfrac{5}{4}$ Divide each side by 4.

   $\ln e^{2x} = \ln \dfrac{5}{4}$ Take natural log of each side.

   $2x = \ln \dfrac{5}{4}$ Inverse Property

   $x = \dfrac{1}{2} \ln \dfrac{5}{4} \approx 0.11$ Divide each side by 2.

The solution is $x = \dfrac{1}{2} \ln \dfrac{5}{4} \approx 0.11$. Check this in the original equation.

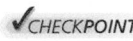 **CHECKPOINT**   Now try Exercise 59.

Graphical Solution

Rather than using the procedure in Example 2, another way to solve the equation graphically is first to rewrite the equation as $4e^{2x} - 5 = 0$, then use a graphing utility to graph $y = 4e^{2x} - 5$. Use the *zero* or *root* feature or the *zoom* and *trace* features of the graphing utility to approximate the value of x for which $y = 0$. From Figure 3.34, you can see that the zero occurs at $x \approx 0.11$. So, the solution is $x \approx 0.11$.

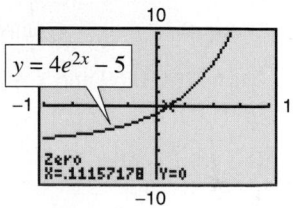

Figure 3.34

Example 4 Solving an Exponential Equation

Solve $2(3^{2t-5}) - 4 = 11$.

Solution

$$2(3^{2t-5}) - 4 = 11 \qquad \text{Write original equation.}$$

$$2(3^{2t-5}) = 15 \qquad \text{Add 4 to each side.}$$

$$3^{2t-5} = \tfrac{15}{2} \qquad \text{Divide each side by 2.}$$

$$\log_3 3^{2t-5} = \log_3 \tfrac{15}{2} \qquad \text{Take log (base 3) of each side.}$$

$$2t - 5 = \log_3 \tfrac{15}{2} \qquad \text{Inverse Property}$$

$$2t = 5 + \log_3 7.5 \qquad \text{Add 5 to each side.}$$

$$t = \tfrac{5}{2} + \tfrac{1}{2}\log_3 7.5 \qquad \text{Divide each side by 2.}$$

$$t \approx 3.42 \qquad \text{Use a calculator.}$$

The solution is $t = \tfrac{5}{2} + \tfrac{1}{2}\log_3 7.5 \approx 3.42$. Check this in the original equation.

✔**CHECKPOINT**　Now try Exercise 49.

> **STUDY TIP**
>
> Remember that to evaluate a logarithm such as $\log_3 7.5$, you need to use the change-of-base formula.
>
> $$\log_3 7.5 = \frac{\ln 7.5}{\ln 3} \approx 1.834$$

When an equation involves two or more exponential expressions, you can still use a procedure similar to that demonstrated in the previous three examples. However, the algebra is a bit more complicated.

Example 5 Solving an Exponential Equation in Quadratic Form

Solve $e^{2x} - 3e^x + 2 = 0$.

Algebraic Solution

$$e^{2x} - 3e^x + 2 = 0 \qquad \text{Write original equation.}$$

$$(e^x)^2 - 3e^x + 2 = 0 \qquad \text{Write in quadratic form.}$$

$$(e^x - 2)(e^x - 1) = 0 \qquad \text{Factor.}$$

$$e^x - 2 = 0 \qquad \text{Set 1st factor equal to 0.}$$

$$e^x = 2 \qquad \text{Add 2 to each side.}$$

$$x = \ln 2 \qquad \text{Solution}$$

$$e^x - 1 = 0 \qquad \text{Set 2nd factor equal to 0.}$$

$$e^x = 1 \qquad \text{Add 1 to each side.}$$

$$x = \ln 1 \qquad \text{Inverse Property}$$

$$x = 0 \qquad \text{Solution}$$

The solutions are $x = \ln 2 \approx 0.69$ and $x = 0$. Check these in the original equation.

✔**CHECKPOINT**　Now try Exercise 61.

Graphical Solution

Use a graphing utility to graph $y = e^{2x} - 3e^x + 2$. Use the *zero* or *root* feature or the *zoom* and *trace* features of the graphing utility to approximate the values of x for which $y = 0$. In Figure 3.35, you can see that the zeros occur at $x = 0$ and at $x \approx 0.69$. So, the solutions are $x = 0$ and $x \approx 0.69$.

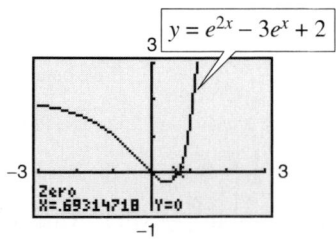

Figure 3.35

Solving Logarithmic Equations

To solve a logarithmic equation, you can write it in exponential form.

$\ln x = 3$	Logarithmic form
$e^{\ln x} = e^3$	Exponentiate each side.
$x = e^3$	Exponential form

This procedure is called *exponentiating* each side of an equation. It is applied after the logarithmic expression has been isolated.

Example 6 Solving Logarithmic Equations

Solve each logarithmic equation.

a. $\ln 3x = 2$ **b.** $\log_3(5x - 1) = \log_3(x + 7)$

Solution

a.

$\ln 3x = 2$	Write original equation.
$e^{\ln 3x} = e^2$	Exponentiate each side.
$3x = e^2$	Inverse Property
$x = \frac{1}{3}e^2 \approx 2.46$	Multiply each side by $\frac{1}{3}$.

The solution is $x = \frac{1}{3}e^2 \approx 2.46$. Check this in the original equation.

b.

$\log_3(5x - 1) = \log_3(x + 7)$	Write original equation.
$5x - 1 = x + 7$	One-to-One Property
$x = 2$	Solve for x.

The solution is $x = 2$. Check this in the original equation.

✔**CHECKPOINT**   Now try Exercise 87.

> **TECHNOLOGY SUPPORT**
>
> For instructions on how to use the *intersect* feature, the *zoom* and *trace* features, and the *zero* or *root* feature, see Appendix A; for specific keystrokes, go to this textbook's *Online Study Center*.

Example 7 Solving a Logarithmic Equation

Solve $5 + 2 \ln x = 4$.

Algebraic Solution

$5 + 2 \ln x = 4$	Write original equation.
$2 \ln x = -1$	Subtract 5 from each side.
$\ln x = -\frac{1}{2}$	Divide each side by 2.
$e^{\ln x} = e^{-1/2}$	Exponentiate each side.
$x = e^{-1/2}$	Inverse Property
$x \approx 0.61$	Use a calculator.

The solution is $x = e^{-1/2} \approx 0.61$. Check this in the original equation.

✔**CHECKPOINT**   Now try Exercise 89.

Graphical Solution

Use a graphing utility to graph $y_1 = 5 + 2 \ln x$ and $y_2 = 4$ in the same viewing window. Use the *intersect* feature or the *zoom* and *trace* features to approximate the intersection point, as shown in Figure 3.36. So, the solution is $x \approx 0.61$.

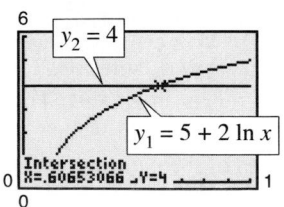

Figure 3.36

Example 8   Solving a Logarithmic Equation

Solve $2 \log_5 3x = 4$.

Solution

$2 \log_5 3x = 4$	Write original equation.
$\log_5 3x = 2$	Divide each side by 2.
$5^{\log_5 3x} = 5^2$	Exponentiate each side (base 5).
$3x = 25$	Inverse Property
$x = \frac{25}{3}$	Divide each side by 3.

The solution is $x = \frac{25}{3}$. Check this in the original equation. Or, perform a graphical check by graphing

$$y_1 = 2 \log_5 3x = 2\left(\frac{\log_{10} 3x}{\log_{10} 5}\right) \quad \text{and} \quad y_2 = 4$$

in the same viewing window. The two graphs should intersect at $x = \frac{25}{3} \approx 8.33$ and $y = 4$, as shown in Figure 3.37.

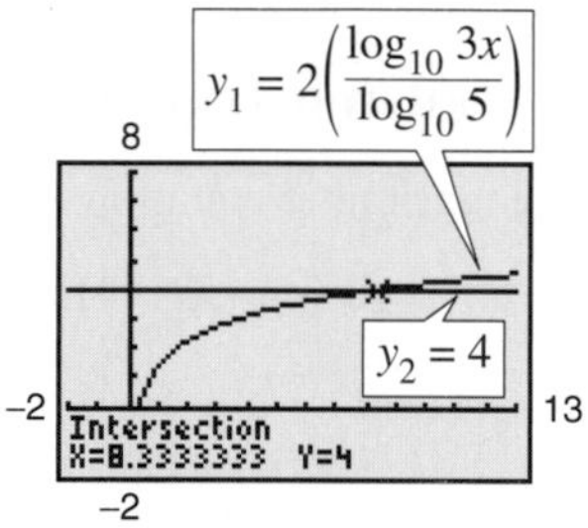

✓CHECKPOINT   Now try Exercise 95.

Figure 3.37

Because the domain of a logarithmic function generally does not include all real numbers, you should be sure to check for extraneous solutions of logarithmic equations, as shown in the next example.

Example 9   Checking for Extraneous Solutions

Solve $\ln(x - 2) + \ln(2x - 3) = 2 \ln x$.

Algebraic Solution

$\ln(x - 2) + \ln(2x - 3) = 2 \ln x$	Write original equation.
$\ln[(x - 2)(2x - 3)] = \ln x^2$	Use properties of logarithms.
$\ln(2x^2 - 7x + 6) = \ln x^2$	Multiply binomials.
$2x^2 - 7x + 6 = x^2$	One-to-One Property
$x^2 - 7x + 6 = 0$	Write in general form.
$(x - 6)(x - 1) = 0$	Factor.
$x - 6 = 0 \implies x = 6$	Set 1st factor equal to 0.
$x - 1 = 0 \implies x = 1$	Set 2nd factor equal to 0.

Finally, by checking these two "solutions" in the original equation, you can conclude that $x = 1$ is not valid. This is because when $x = 1$, $\ln(x - 2) + \ln(2x - 3) = \ln(-1) + \ln(-1)$, which is invalid because -1 is not in the domain of the natural logarithmic function. So, the only solution is $x = 6$.

✓CHECKPOINT   Now try Exercise 103.

Graphical Solution

First rewrite the original equation as $\ln(x - 2) + \ln(2x - 3) - 2 \ln x = 0$. Then use a graphing utility to graph the equation $y = \ln(x - 2) + \ln(2x - 3) - 2 \ln x$. Use the *zero* or *root* feature or the *zoom* and *trace* features of the graphing utility to determine that $x = 6$ is an approximate solution, as shown in Figure 3.38. Verify that 6 is an exact solution algebraically.

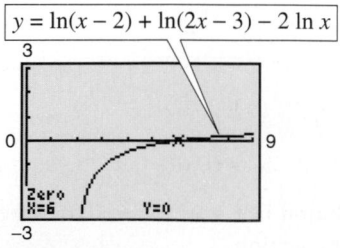

Figure 3.38

Example 10   The Change-of-Base Formula

Prove the change-of-base formula: $\log_a x = \dfrac{\log_b x}{\log_b a}$.

Solution

Begin by letting $y = \log_a x$ and writing the equivalent exponential form $a^y = x$.
Now, taking the logarithms *with base b* of each side produces the following.

$$\log_b a^y = \log_b x$$

$$y \log_b a = \log_b x \qquad \text{Power Property}$$

$$y = \frac{\log_b x}{\log_b a} \qquad \text{Divide each side by } \log_b a.$$

$$\log_a x = \frac{\log_b x}{\log_a a} \qquad \text{Replace } y \text{ with } \log_a x.$$

Equations that involve combinations of algebraic functions, exponential functions, and/or logarithmic functions can be very difficult to solve by algebraic procedures. Here again, you can take advantage of a graphing utility.

Example 11   Approximating the Solution of an Equation

Approximate (to three decimal places) the solution of $\ln x = x^2 - 2$.

Solution

To begin, write the equation so that all terms on one side are equal to 0.

$$\ln x - x^2 + 2 = 0$$

Then use a graphing utility to graph

$$y = -x^2 + 2 + \ln x$$

as shown in Figure 3.39. From this graph, you can see that the equation has two solutions. Next, using the *zero* or *root* feature or the *zoom* and *trace* features, you can approximate the two solutions to be $x \approx 0.138$ and $x \approx 1.564$.

Check

$$\ln x = x^2 - 2 \qquad \text{Write original equation.}$$

$$\ln(0.138) \overset{?}{\approx} (0.138)^2 - 2 \qquad \text{Substitute 0.138 for } x.$$

$$-1.9805 \approx -1.9810 \qquad \text{Solution checks. } \checkmark$$

$$\ln(1.564) \overset{?}{\approx} (1.564)^2 - 2 \qquad \text{Substitute 1.564 for } x.$$

$$0.4472 \approx 0.4461 \qquad \text{Solution checks. } \checkmark$$

So, the two solutions $x \approx 0.138$ and $x \approx 1.564$ seem reasonable.

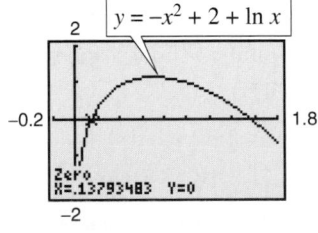

Figure 3.39

✓CHECKPOINT   Now try Exercise 111.

Applications

Example 12 Doubling an Investment

You have deposited $500 in an account that pays 6.75% interest, compounded continuously. How long will it take your money to double?

Solution

Using the formula for continuous compounding, you can find that the balance in the account is

$$A = Pe^{rt} = 500e^{0.0675t}.$$

To find the time required for the balance to double, let $A = 1000$, and solve the resulting equation for t.

$500e^{0.0675t} = 1000$	Substitute 1000 for A.
$e^{0.0675t} = 2$	Divide each side by 500.
$\ln e^{0.0675t} = \ln 2$	Take natural log of each side.
$0.0675t = \ln 2$	Inverse Property
$t = \dfrac{\ln 2}{0.0675} \approx 10.27$	Divide each side by 0.0675.

The balance in the account will double after approximately 10.27 years. This result is demonstrated graphically in Figure 3.40.

 CHECKPOINT Now try Exercise 131.

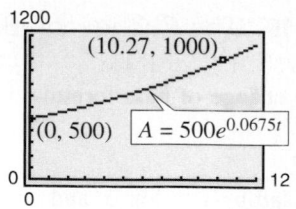

Figure 3.40

Example 13 Average Salary for Public School Teachers

For selected years from 1985 to 2004, the average salary y (in thousands of dollars) for public school teachers for the year t can be modeled by the equation

$$y = -1.562 + 14.584 \ln t, \quad 5 \le t \le 24$$

where $t = 5$ represents 1985 (see Figure 3.41). During which year did the average salary for public school teachers reach $44,000? (Source: National Education Association)

Solution

$-1.562 + 14.584 \ln t = y$	Write original equation.
$-1.562 + 14.584 \ln t = 44.0$	Substitute 44.0 for y.
$14.584 \ln t = 45.562$	Add 1.562 to each side.
$\ln t \approx 3.124$	Divide each side by 14.584.
$e^{\ln t} = e^{3.124}$	Exponentiate each side.
$t \approx 22.74$	Inverse Property

The solution is $t \approx 22.74$ years. Because $t = 5$ represents 1985, it follows that the average salary for public school teachers reached $44,000 in 2002.

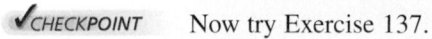 CHECKPOINT Now try Exercise 137.

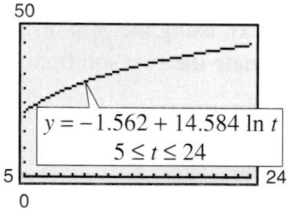

Figure 3.41

3.4   Exercises

See www.CalcChat.com for worked-out solutions to odd-numbered exercises.

Vocabulary Check

Fill in the blanks.

1. To _____ an equation in x means to find all values of x for which the equation is true.

2. To solve exponential and logarithmic equations, you can use the following one-to-one and inverse properties.

   (a) $a^x = a^y$ if and only if _____ .   (b) $\log_a x = \log_a y$ if and only if _____ .

   (c) $a^{\log_a x} =$ _____   (d) $\log_a a^x =$ _____

3. An _____ solution does not satisfy the original equation.

In Exercises 1–8, determine whether each x-value is a solution of the equation.

1. $4^{2x-7} = 64$
   (a) $x = 5$
   (b) $x = 2$

2. $2^{3x+1} = 32$
   (a) $x = -1$
   (b) $x = 2$

3. $3e^{x+2} = 75$
   (a) $x = -2 + e^{25}$
   (b) $x = -2 + \ln 25$
   (c) $x \approx 1.2189$

4. $4e^{x-1} = 60$
   (a) $x = 1 + \ln 15$
   (b) $x \approx 3.7081$
   (c) $x = \ln 16$

5. $\log_4(3x) = 3$
   (a) $x \approx 21.3560$
   (b) $x = -4$
   (c) $x = \frac{64}{3}$

6. $\log_6\left(\frac{5}{3}x\right) = 2$
   (a) $x \approx 20.2882$
   (b) $x = \frac{108}{5}$
   (c) $x = 7.2$

7. $\ln(x-1) = 3.8$
   (a) $x = 1 + e^{3.8}$
   (b) $x \approx 45.7012$
   (c) $x = 1 + \ln 3.8$

8. $\ln(2+x) = 2.5$
   (a) $x = e^{2.5} - 2$
   (b) $x \approx \frac{4073}{400}$
   (c) $x = \frac{1}{2}$

In Exercises 9–16, use a graphing utility to graph f and g in the same viewing window. Approximate the point of intersection of the graphs of f and g. Then solve the equation $f(x) = g(x)$ algebraically.

9. $f(x) = 2^x$
   $g(x) = 8$

10. $f(x) = 27^x$
   $g(x) = 9$

11. $f(x) = 5^{x-2} - 15$
   $g(x) = 10$

12. $f(x) = 2^{-x+1} - 3$
   $g(x) = 13$

13. $f(x) = 4\log_3 x$
   $g(x) = 20$

14. $f(x) = 3\log_5 x$
   $g(x) = 6$

15. $f(x) = \ln e^{x+1}$
   $g(x) = 2x + 5$

16. $f(x) = \ln e^{x-2}$
   $g(x) = 3x + 2$

In Exercises 17–28, solve the exponential equation.

17. $4^x = 16$
18. $3^x = 243$

19. $5^x = \frac{1}{625}$
20. $7^x = \frac{1}{49}$

21. $\left(\frac{1}{8}\right)^x = 64$
22. $\left(\frac{1}{2}\right)^x = 32$

23. $\left(\frac{2}{3}\right)^x = \frac{81}{16}$
24. $\left(\frac{3}{4}\right)^x = \frac{27}{64}$

25. $6(10^x) = 216$
26. $5(8^x) = 325$

27. $2^{x+3} = 256$
28. $3^{x-1} = \frac{1}{81}$

In Exercises 29–38, solve the logarithmic equation.

29. $\ln x - \ln 5 = 0$
30. $\ln x - \ln 2 = 0$

31. $\ln x = -7$
32. $\ln x = -1$

33. $\log_x 625 = 4$
34. $\log_x 25 = 2$

35. $\log_{10} x = -1$
36. $\log_{10} x = -\frac{1}{2}$

37. $\ln(2x-1) = 5$
38. $\ln(3x+5) = 8$

In Exercises 39–44, simplify the expression.

39. $\ln e^{x^2}$
40. $\ln e^{2x-1}$

41. $e^{\ln(5x+2)}$
42. $e^{\ln x^2}$

43. $-1 + \ln e^{2x}$
44. $-8 + e^{\ln x^3}$

In Exercises 45–72, solve the exponential equation algebraically. Round your result to three decimal places. Use a graphing utility to verify your answer.

45. $8^{3x} = 360$
46. $6^{5x} = 3000$

47. $5^{-t/2} = 0.20$
48. $4^{-3t} = 0.10$

49. $5(2^{3-x}) - 13 = 100$

50. $6(8^{-2-x}) + 15 = 2601$

51. $\left(1 + \dfrac{0.10}{12}\right)^{12t} = 2$

52. $\left(16 + \dfrac{0.878}{26}\right)^{3t} = 30$

53. $5000\left[\dfrac{(1+0.005)^x}{0.005}\right] = 250{,}000$

54. $250\left[\dfrac{(1 + 0.01)^x}{0.01}\right] = 150{,}000$

55. $2e^{5x} = 18$

56. $4e^{2x} = 40$

57. $500e^{-x} = 300$

58. $1000e^{-4x} = 75$

59. $7 - 2e^x = 5$

60. $-14 + 3e^x = 11$

61. $e^{2x} - 4e^x - 5 = 0$

62. $e^{2x} - 5e^x + 6 = 0$

63. $250e^{0.02x} = 10{,}000$

64. $100e^{0.005x} = 125{,}000$

65. $e^x = e^{x^2 - 2}$

66. $e^{2x} = e^{x^2 - 8}$

67. $e^{x^2 - 3x} = e^{x - 2}$

68. $e^{-x^2} = e^{x^2 - 2x}$

69. $\dfrac{400}{1 + e^{-x}} = 350$

70. $\dfrac{525}{1 + e^{-x}} = 275$

71. $\dfrac{40}{1 - 5e^{-0.01x}} = 200$

72. $\dfrac{50}{1 - 2e^{-0.001x}} = 1000$

In Exercises 73–76, complete the table to find an interval containing the solution of the equation. Then use a graphing utility to graph both sides of the equation to estimate the solution. Round your result to three decimal places.

73. $e^{3x} = 12$

x	0.6	0.7	0.8	0.9	1.0
e^{3x}					

74. $e^{2x} = 50$

x	1.6	1.7	1.8	1.9	2.0
e^{2x}					

75. $20(100 - e^{x/2}) = 500$

x	5	6	7	8	9
$20(100 - e^{x/2})$					

76. $\dfrac{400}{1 + e^{-x}} = 350$

x	0	1	2	3	4
$\dfrac{400}{1 + e^{-x}}$					

In Exercises 77–80, use the *zero* or *root* feature or the *zoom* and *trace* features of a graphing utility to approximate the solution of the exponential equation accurate to three decimal places.

77. $\left(1 + \dfrac{0.065}{365}\right)^{365t} = 4$

78. $\left(4 - \dfrac{2.471}{40}\right)^{9t} = 21$

79. $\dfrac{3000}{2 + e^{2x}} = 2$

80. $\dfrac{119}{e^{6x} - 14} = 7$

In Exercises 81–84, use a graphing utility to graph the function and approximate its zero accurate to three decimal places.

81. $g(x) = 6e^{1-x} - 25$

82. $f(x) = 3e^{3x/2} - 962$

83. $g(t) = e^{0.09t} - 3$

84. $h(t) = e^{0.125t} - 8$

In Exercises 85–106, solve the logarithmic equation algebraically. Round the result to three decimal places. Verify your answer using a graphing utility.

85. $\ln x = -3$

86. $\ln x = -2$

87. $\ln 4x = 2.1$

88. $\ln 2x = 1.5$

89. $-2 + 2\ln 3x = 17$

90. $3 + 2\ln x = 10$

91. $\log_5(3x + 2) = \log_5(6 - x)$

92. $\log_9(4 + x) = \log_9(2x - 1)$

93. $\log_{10}(z - 3) = 2$

94. $\log_{10} x^2 = 6$

95. $7\log_4(0.6x) = 12$

96. $4\log_{10}(x - 6) = 11$

97. $\ln\sqrt{x + 2} = 1$

98. $\ln\sqrt{x - 8} = 5$

99. $\ln(x + 1)^2 = 2$

100. $\ln(x^2 + 1) = 8$

101. $\log_4 x - \log_4(x - 1) = \tfrac{1}{2}$

102. $\log_3 x + \log_3(x - 8) = 2$

103. $\ln(x + 5) = \ln(x - 1) - \ln(x + 1)$

104. $\ln(x + 1) - \ln(x - 2) = \ln x$

105. $\log_{10} 8x - \log_{10}\left(1 + \sqrt{x}\right) = 2$

106. $\log_{10} 4x - \log_{10}\left(12 + \sqrt{x}\right) = 2$

In Exercises 107–110, complete the table to find an interval containing the solution of the equation. Then use a graphing utility to graph both sides of the equation to estimate the solution. Round your result to three decimal places.

107. $\ln 2x = 2.4$

x	2	3	4	5	6
$\ln 2x$					

108. $3\ln 5x = 10$

x	4	5	6	7	8
$3\ln 5x$					

109. $6\log_3(0.5x) = 11$

x	12	13	14	15	16
$6\log_3(0.5x)$					

110. $5 \log_{10}(x - 2) = 11$

x	150	155	160	165	170
$5 \log_{10}(x - 2)$					

In Exercises 111–116, use the *zero* or *root* feature or the *zoom* and *trace* features of a graphing utility to approximate the solution of the logarithmic equation accurate to three decimal places.

111. $\log_{10} x = x^3 - 3$ **112.** $\log_{10} x^2 = 4$

113. $\ln x + \ln(x - 2) = 1$ **114.** $\ln x + \ln(x + 1) = 2$

115. $\ln(x - 3) + \ln(x + 3) = 1$

116. $\ln x + \ln(x^2 + 4) = 10$

In Exercises 117–122, use a graphing utility to approximate the point of intersection of the graphs. Round your result to three decimal places.

117. $y_1 = 7$
$y_2 = 2^{x-1} - 5$

118. $y_1 = 4$
$y_2 = 3^{x+1} - 2$

119. $y_1 = 80$
$y_2 = 4e^{-0.2x}$

120. $y_1 = 500$
$y_2 = 1500e^{-x/2}$

121. $y_1 = 3.25$
$y_2 = \frac{1}{2}\ln(x + 2)$

122. $y_1 = 1.05$
$y_2 = \ln\sqrt{x - 2}$

$\int$ In Exercises 123–130, solve the equation algebraically. Round the result to three decimal places. Verify your answer using a graphing utility.

123. $2x^2 e^{2x} + 2xe^{2x} = 0$ **124.** $-x^2 e^{-x} + 2xe^{-x} = 0$

125. $-xe^{-x} + e^{-x} = 0$ **126.** $e^{-2x} - 2xe^{-2x} = 0$

127. $2x \ln x + x = 0$ **128.** $\dfrac{1 - \ln x}{x^2} = 0$

129. $\dfrac{1 + \ln x}{2} = 0$ **130.** $2x \ln\left(\dfrac{1}{x}\right) - x = 0$

Compound Interest In Exercises 131–134, find the time required for a $1000 investment to (a) double at interest rate r, compounded continuously, and (b) triple at interest rate r, compounded continuously. Round your results to two decimal places.

131. $r = 7.5\%$ **132.** $r = 6\%$

133. $r = 2.5\%$ **134.** $r = 3.75\%$

135. *Demand* The demand x for a camera is given by

$$p = 500 - 0.5(e^{0.004x})$$

where p is the price in dollars. Find the demands x for prices of (a) $p = \$350$ and (b) $p = \$300$.

136. *Demand* The demand x for a hand-held electronic organizer is given by

$$p = 5000\left(1 - \frac{4}{4 + e^{-0.002x}}\right)$$

where p is the price in dollars. Find the demands x for prices of (a) $p = \$600$ and (b) $p = \$400$.

137. *Medicine* The numbers y of hospitals in the United States from 1995 to 2003 can be modeled by

$$y = 7247 - 596.5 \ln t, \quad 5 \le t \le 13$$

where t represents the year, with $t = 5$ corresponding to 1995. During which year did the number of hospitals fall to 5800? (Source: Health Forum)

138. *Forestry* The yield V (in millions of cubic feet per acre) for a forest at age t years is given by

$$V = 6.7e^{-48.1/t}.$$

(a) Use a graphing utility to graph the function.

(b) Determine the horizontal asymptote of the function. Interpret its meaning in the context of the problem.

(c) Find the time necessary to obtain a yield of 1.3 million cubic feet.

139. *Average Heights* The percent m of American males between the ages of 18 and 24 who are no more than x inches tall is modeled by

$$m(x) = \frac{100}{1 + e^{-0.6114(x - 69.71)}}$$

and the percent f of American females between the ages of 18 and 24 who are no more than x inches tall is modeled by

$$f(x) = \frac{100}{1 + e^{-0.66607(x - 64.51)}}.$$

(Source: U.S. National Center for Health Statistics)

(a) Use a graphing utility to graph the two functions in the same viewing window.

(b) Use the graphs in part (a) to determine the horizontal asymptotes of the functions. Interpret their meanings in the context of the problem.

(c) What is the average height for each sex?

140. *Human Memory Model* In a group project in learning theory, a mathematical model for the proportion P of correct responses after n trials was found to be $P = 0.83/(1 + e^{-0.2n})$.

(a) Use a graphing utility to graph the function.

(b) Use the graph in part (a) to determine any horizontal asymptotes of the function. Interpret the meaning of the upper asymptote in the context of the problem.

(c) After how many trials will 60% of the responses be correct?

141. *Data Analysis* An object at a temperature of 160°C was removed from a furnace and placed in a room at 20°C. The temperature T of the object was measured after each hour h and recorded in the table. A model for the data is given by

$$T = 20[1 + 7(2^{-h})].$$

Hour, h	Temperature
0	160°
1	90°
2	56°
3	38°
4	29°
5	24°

(a) Use a graphing utility to plot the data and graph the model in the same viewing window.

(b) Identify the horizontal asymptote of the graph of the model and interpret the asymptote in the context of the problem.

(c) Approximate the time when the temperature of the object is 100°C.

142. *Finance* The table shows the numbers N of commercial banks in the United States from 1996 to 2005. The data can be modeled by the logarithmic function

$$N = 13,387 - 2190.5 \ln t$$

where t represents the year, with $t = 6$ corresponding to 1996. (Source: Federal Deposit Insurance Corp.)

Year	Number, N
1996	9527
1997	9143
1998	8774
1999	8580
2000	8315
2001	8079
2002	7888
2003	7770
2004	7630
2005	7540

(a) Use the model to determine during which year the number of banks dropped to 7250.

(b) Use a graphing utility to graph the model, and use the graph to verify your answer in part (a).

Synthesis

True or False? **In Exercises 143 and 144, determine whether the statement is true or false. Justify your answer.**

143. An exponential equation must have at least one solution.

144. A logarithmic equation can have at most one extraneous solution.

145. *Writing* Write two or three sentences stating the general guidelines that you follow when (a) solving exponential equations and (b) solving logarithmic equations.

146. *Graphical Analysis* Let $f(x) = \log_a x$ and $g(x) = a^x$, where $a > 1$.

(a) Let $a = 1.2$ and use a graphing utility to graph the two functions in the same viewing window. What do you observe? Approximate any points of intersection of the two graphs.

(b) Determine the value(s) of a for which the two graphs have one point of intersection.

(c) Determine the value(s) of a for which the two graphs have two points of intersection.

147. *Think About It* Is the time required for a continuously compounded investment to quadruple twice as long as the time required for it to double? Give a reason for your answer and verify your answer algebraically.

148. *Writing* Write a paragraph explaining whether or not the time required for a continuously compounded investment to double is dependent on the size of the investment.

Skills Review

In Exercises 149–154, sketch the graph of the function.

149. $f(x) = 3x^3 - 4$

150. $f(x) = -(x + 1)^3 + 2$

151. $f(x) = |x| + 9$

152. $f(x) = |x + 2| - 8$

153. $f(x) = \begin{cases} 2x, & x < 0 \\ -x^2 + 4, & x \geq 0 \end{cases}$

154. $f(x) = \begin{cases} x - 9, & x \leq -1 \\ x^2 + 1, & x > -1 \end{cases}$

3.5 Exponential and Logarithmic Models

Introduction

The five most common types of mathematical models involving exponential functions and logarithmic functions are as follows.

1. **Exponential growth model:** $y = ae^{bx}, \qquad b > 0$

2. **Exponential decay model:** $y = ae^{-bx}, \qquad b > 0$

3. **Gaussian model:** $y = ae^{-(x-b)^2/c}$

4. **Logistic growth model:** $y = \dfrac{a}{1 + be^{-rx}}$

5. **Logarithmic models:** $y = a + b \ln x, \qquad y = a + b \log_{10} x$

The basic shapes of these graphs are shown in Figure 3.42.

What you should learn

- Recognize the five most common types of models involving exponential or logarithmic functions.
- Use exponential growth and decay functions to model and solve real-life problems.
- Use Gaussian functions to model and solve real-life problems.
- Use logistic growth functions to model and solve real-life problems.
- Use logarithmic functions to model and solve real-life problems

Why you should learn it

Exponential growth and decay models are often used to model the population of a country. In Exercise 27 on page 233, you will use such models to predict the population of five countries in 2030.

Kevin Schafer/Peter Arnold, Inc.

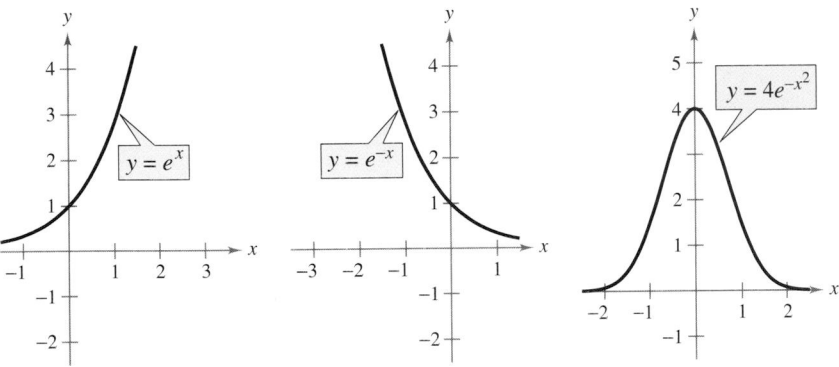

Exponential Growth Model   Exponential Decay Model   Gaussian Model

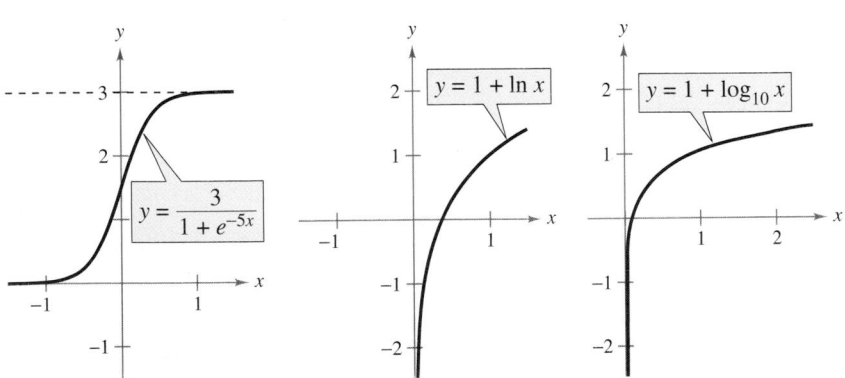

Logistic Growth Model   Natural Logarithmic Model   Common Logarithmic Model
Figure 3.42

You can often gain quite a bit of insight into a situation modeled by an exponential or logarithmic function by identifying and interpreting the function's asymptotes. Use the graphs in Figure 3.42 to identify the asymptotes of each function.

Exponential Growth and Decay

Example 1   Population Growth

Estimates of the world population (in millions) from 1998 through 2007 are shown in the table. A scatter plot of the data is shown in Figure 3.43.   (Source: U.S. Bureau of the Census)

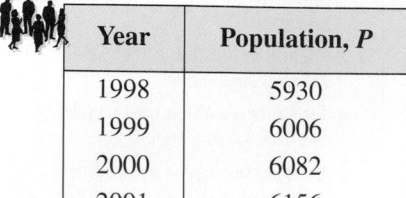

Year	Population, P
1998	5930
1999	6006
2000	6082
2001	6156
2002	6230

Year	Population, P
2003	6303
2004	6377
2005	6451
2006	6525
2007	6600

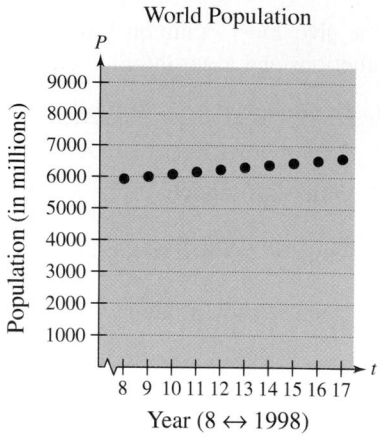

World Population

Figure 3.43

An exponential growth model that approximates this data is given by

$$P = 5400e^{0.011852t}, \qquad 8 \le t \le 17$$

where P is the population (in millions) and $t = 8$ represents 1998. Compare the values given by the model with the estimates shown in the table. According to this model, when will the world population reach 6.8 billion?

Algebraic Solution

The following table compares the two sets of population figures.

Year	1998	1999	2000	2001	2002	2003	2004	2005	2006	2007
Population	5930	6006	6082	6156	6230	6303	6377	6451	6525	6600
Model	5937	6008	6079	6152	6225	6300	6375	6451	6528	6605

To find when the world population will reach 6.8 billion, let $P = 6800$ in the model and solve for t.

$$5400e^{0.011852t} = P \qquad \text{Write original model.}$$

$$5400e^{0.011852t} = 6800 \qquad \text{Substitute 6800 for } P.$$

$$e^{0.011852t} \approx 1.25926 \qquad \text{Divide each side by 5400.}$$

$$\ln e^{0.011852t} \approx \ln 1.25926 \qquad \text{Take natural log of each side.}$$

$$0.011852t \approx 0.23052 \qquad \text{Inverse Property}$$

$$t \approx 19.4 \qquad \text{Divide each side by 0.011852.}$$

According to the model, the world population will reach 6.8 billion in 2009.

 **CHECKPOINT**   Now try Exercise 28.

Graphical Solution

Use a graphing utility to graph the model $y = 5400e^{0.011852x}$ and the data in the same viewing window. You can see in Figure 3.44 that the model appears to closely fit the data.

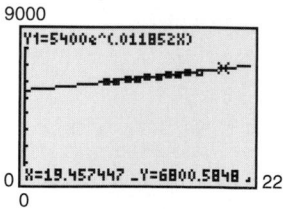

Figure 3.44

Use the *zoom* and *trace* features of the graphing utility to find that the approximate value of x for $y = 6800$ is $x \approx 19.4$. So, according to the model, the world population will reach 6.8 billion in 2009.

An exponential model increases (or decreases) by the same percent each year. What is the annual percent increase for the model in Example 1?

In Example 1, you were given the exponential growth model. Sometimes you must find such a model. One technique for doing this is shown in Example 2.

Example 2   Modeling Population Growth

In a research experiment, a population of fruit flies is increasing according to the law of exponential growth. After 2 days there are 100 flies, and after 4 days there are 300 flies. How many flies will there be after 5 days?

Solution

Let y be the number of flies at time t (in days). From the given information, you know that $y = 100$ when $t = 2$ and $y = 300$ when $t = 4$. Substituting this information into the model $y = ae^{bt}$ produces

$$100 = ae^{2b} \quad \text{and} \quad 300 = ae^{4b}.$$

To solve for b, solve for a in the first equation.

$$100 = ae^{2b} \quad\Longrightarrow\quad a = \frac{100}{e^{2b}} \qquad \text{Solve for } a \text{ in the first equation.}$$

Then substitute the result into the second equation.

$300 = ae^{4b}$	Write second equation.
$300 = \left(\dfrac{100}{e^{2b}}\right)e^{4b}$	Substitute $\dfrac{100}{e^{2b}}$ for a.
$3 = e^{2b}$	Divide each side by 100.
$\ln 3 = \ln e^{2b}$	Take natural log of each side.
$\ln 3 = 2b$	Inverse Property
$\dfrac{1}{2}\ln 3 = b$	Solve for b.

Using $b = \frac{1}{2}\ln 3$ and the equation you found for a, you can determine that

$a = \dfrac{100}{e^{2[(1/2)\ln 3]}}$	Substitute $\frac{1}{2}\ln 3$ for b.
$= \dfrac{100}{e^{\ln 3}}$	Simplify.
$= \dfrac{100}{3} \approx 33.33.$	Inverse Property

So, with $a \approx 33.33$ and $b = \frac{1}{2}\ln 3 \approx 0.5493$, the exponential growth model is

$$y = 33.33e^{0.5493t},$$

as shown in Figure 3.45. This implies that after 5 days, the population will be

$$y = 33.33e^{0.5493(5)} \approx 520 \text{ flies.}$$

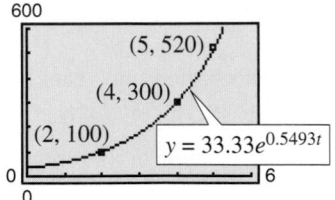

Figure 3.45

✓**CHECKPOINT**   Now try Exercise 29.

In living organic material, the ratio of the content of radioactive carbon isotopes (carbon 14) to the content of nonradioactive carbon isotopes (carbon 12) is about 1 to 10^{12}. When organic material dies, its carbon 12 content remains fixed, whereas its radioactive carbon 14 begins to decay with a half-life of 5715 years. To estimate the age of dead organic material, scientists use the following formula, which denotes the ratio of carbon 14 to carbon 12 present at any time t (in years).

$$R = \frac{1}{10^{12}} e^{-t/8245} \qquad \text{Carbon dating model}$$

The graph of R is shown in Figure 3.46. Note that R decreases as t increases.

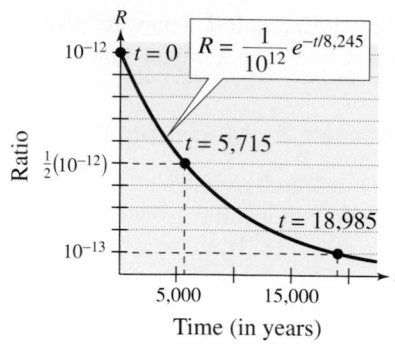

Figure 3.46

Example 3   Carbon Dating

The ratio of carbon 14 to carbon 12 in a newly discovered fossil is

$$R = \frac{1}{10^{13}}.$$

Estimate the age of the fossil.

Algebraic Solution

In the carbon dating model, substitute the given value of R to obtain the following.

$$\frac{1}{10^{12}} e^{-t/8245} = R \qquad \text{Write original model.}$$

$$\frac{e^{-t/8245}}{10^{12}} = \frac{1}{10^{13}} \qquad \text{Substitute } \frac{1}{10^{13}} \text{ for } R.$$

$$e^{-t/8245} = \frac{1}{10} \qquad \text{Multiply each side by } 10^{12}.$$

$$\ln e^{-t/8245} = \ln \frac{1}{10} \qquad \text{Take natural log of each side.}$$

$$-\frac{t}{8245} \approx -2.3026 \qquad \text{Inverse Property}$$

$$t \approx 18{,}985 \qquad \text{Multiply each side by } -8245.$$

So, to the nearest thousand years, you can estimate the age of the fossil to be 19,000 years.

✓CHECKPOINT   Now try Exercise 32.

Graphical Solution

Use a graphing utility to graph the formula for the ratio of carbon 14 to carbon 12 at any time t as

$$y_1 = \frac{1}{10^{12}} e^{-x/8245}.$$

In the same viewing window, graph $y_2 = 1/(10^{13})$. Use the *intersect* feature or the *zoom* and *trace* features of the graphing utility to estimate that $x \approx 18{,}985$ when $y = 1/(10^{13})$, as shown in Figure 3.47.

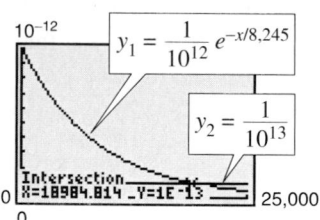

Figure 3.47

So, to the nearest thousand years, you can estimate the age of the fossil to be 19,000 years.

The carbon dating model in Example 3 assumed that the carbon 14 to carbon 12 ratio was one part in 10,000,000,000,000. Suppose an error in measurement occurred and the actual ratio was only one part in 8,000,000,000,000. The fossil age corresponding to the actual ratio would then be approximately 17,000 years. Try checking this result.

Gaussian Models

As mentioned at the beginning of this section, Gaussian models are of the form

$$y = ae^{-(x-b)^2/c}.$$

This type of model is commonly used in probability and statistics to represent populations that are **normally distributed.** For *standard* normal distributions, the model takes the form

$$y = \frac{1}{\sqrt{2\pi}}e^{-x^2/2}.$$

The graph of a Gaussian model is called a **bell-shaped curve.** Try graphing the normal distribution curve with a graphing utility. Can you see why it is called a bell-shaped curve?

   The average value for a population can be found from the bell-shaped curve by observing where the maximum y-value of the function occurs. The x-value corresponding to the maximum y-value of the function represents the average value of the independent variable—in this case, x.

Example 4   SAT Scores

In 2005, the Scholastic Aptitude Test (SAT) mathematics scores for college-bound seniors roughly followed the normal distribution

$$y = 0.0035e^{-(x-520)^2/26,450}, \qquad 200 \le x \le 800$$

where x is the SAT score for mathematics. Use a graphing utility to graph this function and estimate the average SAT score.   (Source: College Board)

Solution

The graph of the function is shown in Figure 3.48. On this bell-shaped curve, the maximum value of the curve represents the average score. Using the *maximum* feature or the *zoom* and *trace* features of the graphing utility, you can see that the average mathematics score for college-bound seniors in 2005 was 520.

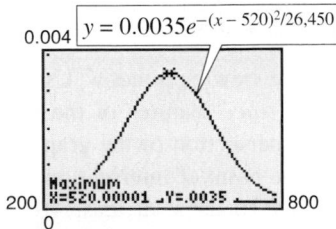

Figure 3.48

✔CHECKPOINT   Now try Exercise 37.

> **TECHNOLOGY SUPPORT**
>
> For instructions on how to use the *maximum* feature, see Appendix A; for specific keystrokes, go to this textbook's *Online Study Center.*

   In Example 4, note that 50% of the seniors who took the test received a score lower than 520.

Logistic Growth Models

Some populations initially have rapid growth, followed by a declining rate of growth, as indicated by the graph in Figure 3.49. One model for describing this type of growth pattern is the **logistic curve** given by the function

$$y = \frac{a}{1 + be^{-rx}}$$

where y is the population size and x is the time. An example is a bacteria culture that is initially allowed to grow under ideal conditions, and then under less favorable conditions that inhibit growth. A logistic growth curve is also called a **sigmoidal curve.**

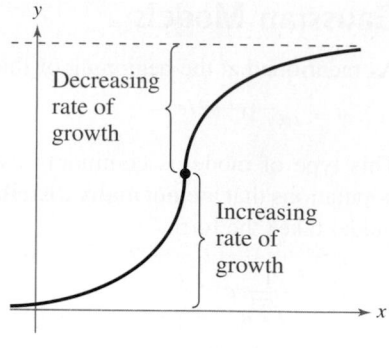

Figure 3.49   Logistic Curve

Example 5   Spread of a Virus

On a college campus of 5000 students, one student returns from vacation with a contagious flu virus. The spread of the virus is modeled by

$$y = \frac{5000}{1 + 4999e^{-0.8t}}, \qquad t \geq 0$$

where y is the total number infected after t days. The college will cancel classes when 40% or more of the students are infected. (a) How many students are infected after 5 days? (b) After how many days will the college cancel classes?

Algebraic Solution

a. After 5 days, the number of students infected is

$$y = \frac{5000}{1 + 4999e^{-0.8(5)}} = \frac{5000}{1 + 4999e^{-4}} \approx 54.$$

b. Classes are cancelled when the number of infected students is $(0.40)(5000) = 2000$.

$$2000 = \frac{5000}{1 + 4999e^{-0.8t}}$$

$$1 + 4999e^{-0.8t} = 2.5$$

$$e^{-0.8t} \approx \frac{1.5}{4999}$$

$$\ln e^{-0.8t} \approx \ln \frac{1.5}{4999}$$

$$-0.8t \approx \ln \frac{1.5}{4999}$$

$$t = -\frac{1}{0.8} \ln \frac{1.5}{4999} \approx 10.14$$

So, after about 10 days, at least 40% of the students will be infected, and classes will be canceled.

Graphical Solution

a. Use a graphing utility to graph $y = \dfrac{5000}{1 + 4999e^{-0.8x}}$.

Use the *value* feature or the *zoom* and *trace* features of the graphing utility to estimate that $y \approx 54$ when $x = 5$. So, after 5 days, about 54 students will be infected.

b. Classes are cancelled when the number of infected students is $(0.40)(5000) = 2000$. Use a graphing utility to graph

$$y_1 = \frac{5000}{1 + 4999e^{-0.8x}} \qquad \text{and} \qquad y_2 = 2000$$

in the same viewing window. Use the *intersect* feature or the *zoom* and *trace* features of the graphing utility to find the point of intersection of the graphs. In Figure 3.50, you can see that the point of intersection occurs near $x \approx 10.14$. So, after about 10 days, at least 40% of the students will be infected, and classes will be canceled.

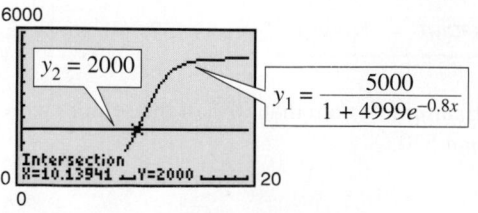

Figure 3.50

✔CHECKPOINT   Now try Exercise 39.

Logarithmic Models

On the Richter scale, the magnitude R of an earthquake of intensity I is given by

$$R = \log_{10} \frac{I}{I_0}$$

where $I_0 = 1$ is the minimum intensity used for comparison. Intensity is a measure of the wave energy of an earthquake.

Example 6   Magnitudes of Earthquakes

On January 22, 2003, an earthquake of magnitude 7.6 in Colima, Mexico killed at least 29 people and left 10,000 people homeless.

In 2001, the coast of Peru experienced an earthquake that measured 8.4 on the Richter scale. In 2003, Colima, Mexico experienced an earthquake that measured 7.6 on the Richter scale. Find the intensity of each earthquake and compare the two intensities.

Solution

Because $I_0 = 1$ and $R = 8.4$, you have

$$8.4 = \log_{10} \frac{I}{1}$$ Substitute 1 for I_0 and 8.4 for R.

$$10^{8.4} = 10^{\log_{10} I}$$ Exponentiate each side.

$$10^{8.4} = I$$ Inverse Property

$$251{,}189{,}000 \approx I.$$ Use a calculator.

For $R = 7.6$, you have

$$7.6 = \log_{10} \frac{I}{1}$$ Substitute 1 for I_0 and 7.6 for R.

$$10^{7.6} = 10^{\log_{10} I}$$ Exponentiate each side.

$$10^{7.6} = I$$ Inverse Property

$$39{,}811{,}000 \approx I.$$ Use a calculator.

Note that an increase of 0.8 unit on the Richter scale (from 7.6 to 8.4) represents an increase in intensity by a factor of

$$\frac{251{,}189{,}000}{39{,}811{,}000} \approx 6.$$

In other words, the 2001 earthquake had an intensity about 6 times as great as that of the 2003 earthquake.

✓CHECKPOINT   Now try Exercise 41.

3.5 Exercises

See www.CalcChat.com for worked-out solutions to odd-numbered exercises.

Vocabulary Check

1. Match the equation with its model.

(a) Exponential growth model

(b) Exponential decay model

(c) Logistic growth model

(d) Logistic decay model

(e) Gaussian model

(f) Natural logarithmic model

(g) Common logarithmic model

(i) $y = ae^{-bx}$, $b > 0$

(ii) $y = a + b \ln x$

(iii) $y = \dfrac{a}{1 + be^{-rx}}$, $r < 0$

(iv) $y = ae^{bx}$, $b > 0$

(v) $y = a + b \log_{10} x$

(vi) $y = \dfrac{1}{1 + be^{-rx}}$, $r > 0$

(vii) $y = ae^{-(x-b)^2/c}$

In Exercises 2–4, fill in the blanks.

2. Gaussian models are commonly used in probability and statistics to represent populations that are _____ distributed.

3. Logistic growth curves are also called _____ curves.

4. The graph of a Gaussian model is called a _____-_____ curve, where the average value or _____ is the x-value corresponding to the maximum y-value of the graph.

Library of Parent Functions In Exercises 1–6, match the function with its graph. [The graphs are labeled (a), (b), (c), (d), (e), and (f).]

(a)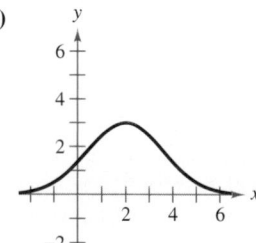

(b)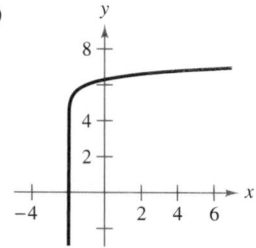

(c)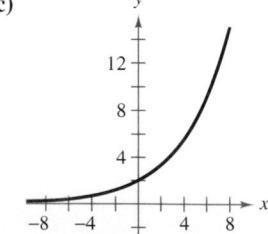

(d)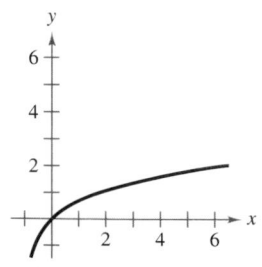

(e)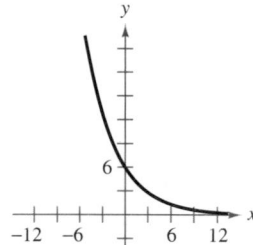

(f)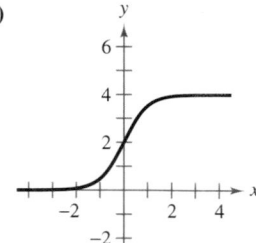

1. $y = 2e^{x/4}$

2. $y = 6e^{-x/4}$

3. $y = 6 + \log_{10}(x + 2)$

4. $y = 3e^{-(x-2)^2/5}$

5. $y = \ln(x + 1)$

6. $y = \dfrac{4}{1 + e^{-2x}}$

Compound Interest In Exercises 7–14, complete the table for a savings account in which interest is compounded continuously.

	Initial Investment	Annual % Rate	Time to Double	Amount After 10 Years
7.	$10,000	3.5%		
8.	$2000	1.5%		
9.	$7500		21 years	

Initial Investment	Annual % Rate	Time to Double	Amount After 10 Years
10. $1000		12 years	
11. $5000			$5665.74
12. $300			$385.21
13.	4.5%		$100,000.00
14.	2%		$2500.00

15. *Compound Interest* Complete the table for the time t necessary for P dollars to triple if interest is compounded continuously at rate r. Create a scatter plot of the data.

r	2%	4%	6%	8%	10%	12%
t						

16. *Compound Interest* Complete the table for the time t necessary for P dollars to triple if interest is compounded annually at rate r. Create a scatter plot of the data.

r	2%	4%	6%	8%	10%	12%
t						

17. *Comparing Investments* If $1 is invested in an account over a 10-year period, the amount A in the account, where t represents the time in years, is given by $A = 1 + 0.075[\![t]\!]$ or $A = e^{0.07t}$ depending on whether the account pays simple interest at $7\frac{1}{2}$% or continuous compound interest at 7%. Use a graphing utility to graph each function in the same viewing window. Which grows at a greater rate? (Remember that $[\![t]\!]$ is the greatest integer function discussed in Section 1.3.)

18. *Comparing Investments* If $1 is invested in an account over a 10-year period, the amount A in the account, where t represents the time in years, is given by

$$A = 1 + 0.06[\![t]\!] \quad \text{or} \quad A = \left(1 + \frac{0.055}{365}\right)^{[\![365t]\!]}$$

depending on whether the account pays simple interest at 6% or compound interest at $5\frac{1}{2}$% compounded daily. Use a graphing utility to graph each function in the same viewing window. Which grows at a greater rate?

Radioactive Decay **In Exercises 19–22, complete the table for the radioactive isotope.**

Isotope	Half-Life (years)	Initial Quantity	Amount After 1000 Years
19. ^{226}Ra	1599	10 g	
20. ^{226}Ra	1599		1.5 g
21. ^{14}C	5715	3 g	
22. ^{239}Pu	24,100		0.4 g

In Exercises 23–26, find the exponential model $y = ae^{bx}$ that fits the points shown in the graph or table.

23.
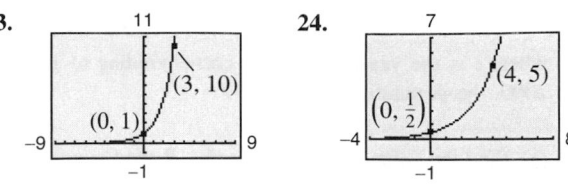

24.

25.

x	0	5
y	4	1

26.

x	0	3
y	1	$\frac{1}{4}$

27. *Population* The table shows the populations (in millions) of five countries in 2000 and the projected populations (in millions) for the year 2010. (Source: U.S. Census Bureau)

Country	2000	2010
Australia	19.2	20.9
Canada	31.3	34.3
Philippines	79.7	95.9
South Africa	44.1	43.3
Turkey	65.7	73.3

(a) Find the exponential growth or decay model, $y = ae^{bt}$ or $y = ae^{-bt}$, for the population of each country by letting $t = 0$ correspond to 2000. Use the model to predict the population of each country in 2030.

(b) You can see that the populations of Australia and Turkey are growing at different rates. What constant in the equation $y = ae^{bt}$ is determined by these different growth rates? Discuss the relationship between the different growth rates and the magnitude of the constant.

(c) You can see that the population of Canada is increasing while the population of South Africa is decreasing. What constant in the equation $y = ae^{bt}$ reflects this difference? Explain.

28. *Population* The populations P (in thousands) of Pittsburgh, Pennsylvania from 1990 to 2004 can be modeled by $P = 372.55e^{-0.01052t}$, where t is the year, with $t = 0$ corresponding to 1990. (Source: U.S. Census Bureau)

(a) According to the model, was the population of Pittsburgh increasing or decreasing from 1990 to 2004? Explain your reasoning.

(b) What were the populations of Pittsburgh in 1990, 2000, and 2004?

(c) According to the model, when will the population be approximately 300,000?

29. Population   The population P (in thousands) of Reno, Nevada can be modeled by

$$P = 134.0e^{kt}$$

where t is the year, with $t = 0$ corresponding to 1990. In 2000, the population was 180,000.   (Source: U.S. Census Bureau)

(a) Find the value of k for the model. Round your result to four decimal places.

(b) Use your model to predict the population in 2010.

30. Population   The population P (in thousands) of Las Vegas, Nevada can be modeled by

$$P = 258.0e^{kt}$$

where t is the year, with $t = 0$ corresponding to 1990. In 2000, the population was 478,000.   (Source: U.S. Census Bureau)

(a) Find the value of k for the model. Round your result to four decimal places.

(b) Use your model to predict the population in 2010.

31. Radioactive Decay   The half-life of radioactive radium (^{226}Ra) is 1599 years. What percent of a present amount of radioactive radium will remain after 100 years?

32. Carbon Dating   Carbon 14 (^{14}C) dating assumes that the carbon dioxide on Earth today has the same radioactive content as it did centuries ago. If this is true, the amount of ^{14}C absorbed by a tree that grew several centuries ago should be the same as the amount of ^{14}C absorbed by a tree growing today. A piece of ancient charcoal contains only 15% as much radioactive carbon as a piece of modern charcoal. How long ago was the tree burned to make the ancient charcoal if the half-life of ^{14}C is 5715 years?

33. Depreciation   A new 2006 SUV that sold for $30,788 has a book value V of $24,000 after 2 years.

(a) Find a linear depreciation model for the SUV.

(b) Find an exponential depreciation model for the SUV. Round the numbers in the model to four decimal places.

(c) Use a graphing utility to graph the two models in the same viewing window.

(d) Which model represents at a greater depreciation rate in the first 2 years?

(e) Explain the advantages and disadvantages of each model to both a buyer and a seller.

34. Depreciation   A new laptop computer that sold for $1150 in 2005 has a book value V of $550 after 2 years.

(a) Find a linear depreciation model for the laptop.

(b) Find an exponential depreciation model for the laptop. Round the numbers in the model to four decimal places.

(c) Use a graphing utility to graph the two models in the same viewing window.

(d) Which model represents at a greater depreciation rate in the first 2 years?

(e) Explain the advantages and disadvantages of each model to both a buyer and a seller.

35. Sales   The sales S (in thousands of units) of a new CD burner after it has been on the market t years are given by $S = 100(1 - e^{kt})$. Fifteen thousand units of the new product were sold the first year.

(a) Complete the model by solving for k.

(b) Use a graphing utility to graph the model.

(c) Use the graph in part (b) to estimate the number of units sold after 5 years.

36. Sales   The sales S (in thousands of units) of a cleaning solution after x hundred dollars is spent on advertising are given by $S = 10(1 - e^{kx})$. When $500 is spent on advertising, 2500 units are sold.

(a) Complete the model by solving for k.

(b) Estimate the number of units that will be sold if advertising expenditures are raised to $700.

37. IQ Scores   The IQ scores for adults roughly follow the normal distribution

$$y = 0.0266e^{-(x-100)^2/450}, \quad 70 \le x \le 115$$

where x is the IQ score.

(a) Use a graphing utility to graph the function.

(b) From the graph in part (a), estimate the average IQ score.

38. Education   The time (in hours per week) a student uses a math lab roughly follows the normal distribution

$$y = 0.7979e^{-(x-5.4)^2/0.5}, \quad 4 \le x \le 7$$

where x is the time spent in the lab.

(a) Use a graphing utility to graph the function.

(b) From the graph in part (a), estimate the average time a student spends per week in the math lab.

39. Wildlife   A conservation organization releases 100 animals of an endangered species into a game preserve. The organization believes that the preserve has a carrying capacity of 1000 animals and that the growth of the herd will follow the logistic curve

$$p(t) = \frac{1000}{1 + 9e^{-0.1656t}}$$

where t is measured in months.

(a) What is the population after 5 months?

(b) After how many months will the population reach 500?

(c) Use a graphing utility to graph the function. Use the graph to determine the values of p at which the horizontal asymptotes occur. Interpret the meaning of the larger asymptote in the context of the problem.

40. *Yeast Growth* The amount Y of yeast in a culture is given by the model

$$Y = \frac{663}{1 + 72e^{-0.547t}}, \quad 0 \le t \le 18$$

where t represents the time (in hours).

(a) Use a graphing utility to graph the model.

(b) Use the model to predict the populations for the 19th hour and the 30th hour.

(c) According to this model, what is the limiting value of the population?

(d) Why do you think the population of yeast follows a logistic growth model instead of an exponential growth model?

Geology In Exercises 41 and 42, use the Richter scale (see page 381) for measuring the magnitudes of earthquakes.

41. Find the intensities I of the following earthquakes measuring R on the Richter scale (let $I_0 = 1$). (Source: U.S. Geological Survey)

(a) Santa Cruz Islands in 2006, $R = 6.1$

(b) Pakistan in 2005, $R = 7.6$

(c) Northern Sumatra in 2004, $R = 9.0$

42. Find the magnitudes R of the following earthquakes of intensity I (let $I_0 = 1$).

(a) $I = 39,811,000$

(b) $I = 12,589,000$

(c) $I = 251,200$

Sound Intensity In Exercises 43–46, use the following information for determining sound intensity. The level of sound β (in decibels) with an intensity I is $\beta = 10 \log_{10}(I/I_0)$, where I_0 is an intensity of 10^{-12} watt per square meter, corresponding roughly to the faintest sound that can be heard by the human ear. In Exercises 43 and 44, find the level of each sound β.

43. (a) $I = 10^{-10}$ watt per m^2 (quiet room)

(b) $I = 10^{-5}$ watt per m^2 (busy street corner)

(c) $I \approx 10^0$ watt per m^2 (threshold of pain)

44. (a) $I = 10^{-4}$ watt per m^2 (door slamming)

(b) $I = 10^{-3}$ watt per m^2 (loud car horn)

(c) $I = 10^{-2}$ watt per m^2 (siren at 30 meters)

45. As a result of the installation of a muffler, the noise level of an engine was reduced from 88 to 72 decibels. Find the percent decrease in the intensity level of the noise due to the installation of the muffler.

46. As a result of the installation of noise suppression materials, the noise level in an auditorium was reduced from 93 to 80 decibels. Find the percent decrease in the intensity level of the noise due to the installation of these materials.

pH Levels In Exercises 47–50, use the acidity model given by pH $= -\log_{10}[\text{H}^+]$, where acidity (pH) is a measure of the hydrogen ion concentration $[\text{H}^+]$ (measured in moles of hydrogen per liter) of a solution.

47. Find the pH if $[\text{H}^+] = 2.3 \times 10^{-5}$.

48. Compute $[\text{H}^+]$ for a solution for which pH $= 5.8$.

49. A grape has a pH of 3.5, and milk of magnesia has a pH of 10.5. The hydrogen ion concentration of the grape is how many times that of the milk of magnesia?

50. The pH of a solution is decreased by one unit. The hydrogen ion concentration is increased by what factor?

51. *Home Mortgage* A \$120,000 home mortgage for 30 years at $7\frac{1}{2}\%$ has a monthly payment of \$839.06. Part of the monthly payment goes toward the interest charge on the unpaid balance, and the remainder of the payment is used to reduce the principal. The amount that goes toward the interest is given by

$$u = M - \left(M - \frac{Pr}{12}\right)\left(1 + \frac{r}{12}\right)^{12t}$$

and the amount that goes toward reduction of the principal is given by

$$v = \left(M - \frac{Pr}{12}\right)\left(1 + \frac{r}{12}\right)^{12t}.$$

In these formulas, P is the size of the mortgage, r is the interest rate, M is the monthly payment, and t is the time (in years).

(a) Use a graphing utility to graph each function in the same viewing window. (The viewing window should show all 30 years of mortgage payments.)

(b) In the early years of the mortgage, the larger part of the monthly payment goes for what purpose? Approximate the time when the monthly payment is evenly divided between interest and principal reduction.

(c) Repeat parts (a) and (b) for a repayment period of 20 years ($M = \$966.71$). What can you conclude?

52. *Home Mortgage* The total interest u paid on a home mortgage of P dollars at interest rate r for t years is given by

$$u = P\left[\frac{rt}{1 - \left(\frac{1}{1 + r/12}\right)^{12t}} - 1\right].$$

Consider a $120,000 home mortgage at $7\frac{1}{2}\%$.

(a) Use a graphing utility to graph the total interest function.

(b) Approximate the length of the mortgage when the total interest paid is the same as the size of the mortgage. Is it possible that a person could pay twice as much in interest charges as the size of his or her mortgage?

53. *Newton's Law of Cooling* At 8:30 A.M., a coroner was called to the home of a person who had died during the night. In order to estimate the time of death, the coroner took the person's temperature twice. At 9:00 A.M. the temperature was 85.7°F, and at 11:00 A.M. the temperature was 82.8°F. From these two temperatures the coroner was able to determine that the time elapsed since death and the body temperature were related by the formula

$$t = -10 \ln \frac{T - 70}{98.6 - 70}$$

where t is the time (in hours elapsed since the person died) and T is the temperature (in degrees Fahrenheit) of the person's body. Assume that the person had a normal body temperature of 98.6°F at death and that the room temperature was a constant 70°F. Use the formula to estimate the time of death of the person. (This formula is derived from a general cooling principle called Newton's Law of Cooling.)

54. *Newton's Law of Cooling* You take a five-pound package of steaks out of a freezer at 11 A.M. and place it in the refrigerator. Will the steaks be thawed in time to be grilled at 6 P.M.? Assume that the refrigerator temperature is 40°F and the freezer temperature is 0°F. Use the formula for Newton's Law of Cooling

$$t = -5.05 \ln \frac{T - 40}{0 - 40}$$

where t is the time in hours (with $t = 0$ corresponding to 11 A.M.) and T is the temperature of the package of steaks (in degrees Fahrenheit).

Synthesis

True or False? **In Exercises 55–58, determine whether the statement is true or false. Justify your answer.**

55. The domain of a logistic growth function cannot be the set of real numbers.

56. The graph of a logistic growth function will always have an x-intercept.

57. The graph of a Gaussian model will never have an x-intercept.

58. The graph of a Gaussian model will always have a maximum point.

Skills Review

Library of Parent Functions **In Exercises 59–62, match the equation with its graph, and identify any intercepts. [The graphs are labeled (a), (b), (c), and (d).]**

(a)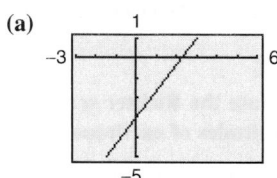

(b)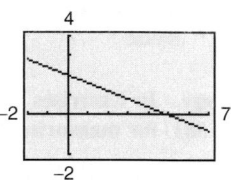

(c)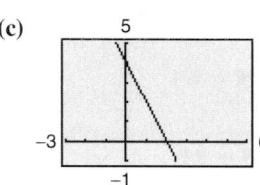

(d)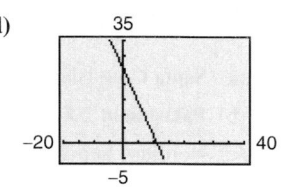

59. $4x - 3y - 9 = 0$

60. $2x + 5y - 10 = 0$

61. $y = 25 - 2.25x$

62. $\frac{x}{2} + \frac{y}{4} = 1$

In Exercises 63–66, use the Leading Coefficient Test to determine the right-hand and left-hand behavior of the graph of the polynomial function.

63. $f(x) = 2x^3 - 3x^2 + x - 1$

64. $f(x) = 5 - x^2 - 4x^4$

65. $g(x) = -1.6x^5 + 4x^2 - 2$

66. $g(x) = 7x^6 + 9.1x^5 - 3.2x^4 + 25x^3$

In Exercises 67 and 68, divide using synthetic division.

67. $(2x^3 - 8x^2 + 3x - 9) \div (x - 4)$

68. $(x^4 - 3x + 1) \div (x + 5)$

69. *Make a Decision* To work an extended application analyzing the net sales for Kohl's Corporation from 1992 to 2005, visit this textbook's *Online Study Center*. (Data Source: Kohl's Illinois, Inc.)

3.6 Nonlinear Models

Classifying Scatter Plots

In Section 1.7, you saw how to fit linear models to data, and in Section 2.8, you saw how to fit quadratic models to data. In real life, many relationships between two variables are represented by different types of growth patterns. A scatter plot can be used to give you an idea of which type of model will best fit a set of data.

Example 1   Classifying Scatter Plots

Decide whether each set of data could best be modeled by an exponential model

$$y = ab^x$$

or a logarithmic model

$$y = a + b \ln x.$$

a. (2, 1), (2.5, 1.2), (3, 1.3), (3.5, 1.5), (4, 1.8), (4.5, 2), (5, 2.4), (5.5, 2.5), (6, 3.1), (6.5, 3.8), (7, 4.5), (7.5, 5), (8, 6.5), (8.5, 7.8), (9, 9), (9.5, 10)

b. (2, 2), (2.5, 3.1), (3, 3.8), (3.5, 4.3), (4, 4.6), (4.5, 5.3), (5, 5.6), (5.5, 5.9), (6, 6.2), (6.5, 6.4), (7, 6.9), (7.5, 7.2), (8, 7.6), (8.5, 7.9), (9, 8), (9.5, 8.2)

Solution

Begin by entering the data into a graphing utility. You should obtain the scatter plots shown in Figure 3.51.

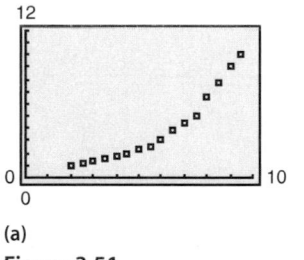

(a)

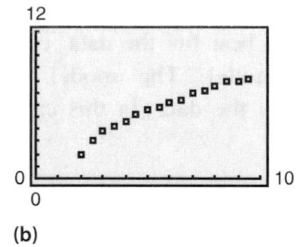

(b)

Figure 3.51

From the scatter plots, it appears that the data in part (a) can be modeled by an exponential function and the data in part (b) can be modeled by a logarithmic function.

✓CHECKPOINT   Now try Exercise 9.

Fitting Nonlinear Models to Data

Once you have used a scatter plot to determine the type of model that would best fit a set of data, there are several ways that you can actually find the model. Each method is best used with a computer or calculator, rather than with hand calculations.

What you should learn

- Classify scatter plots.
- Use scatter plots and a graphing utility to find models for data and choose a model that best fits a set of data.
- Use a graphing utility to find exponential and logistic models for data.

Why you should learn it

Many real-life applications can be modeled by nonlinear equations. For instance, in Exercise 28 on page 243, you are asked to find three different nonlinear models for the price of a half-gallon of ice cream in the United States.

Creatas/PhotoLibrary

TECHNOLOGY SUPPORT

Remember to use the *list editor* of your graphing utility to enter the data in Example 1, as shown below. For instructions on how to use the *list editor*, see Appendix A; for specific keystrokes, go to this textbook's *Online Study Center*.

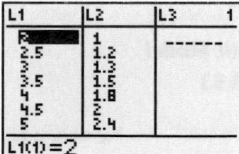

From Example 1(a), you already know that the data can be modeled by an exponential function. In the next example you will determine whether an exponential model best fits the data.

Example 2 Fitting a Model to Data

Fit the following data from Example 1(a) to a quadratic model, an exponential model, and a power model. Identify the coefficient of determination and determine which model best fits the data.

(2, 1), (2.5, 1.2), (3, 1.3), (3.5, 1.5), (4, 1.8), (4.5, 2), (5, 2.4), (5.5, 2.5),
(6, 3.1), (6.5, 3.8), (7, 4.5), (7.5, 5), (8, 6.5), (8.5, 7.8), (9, 9), (9.5, 10)

Solution

Begin by entering the data into a graphing utility. Then use the *regression* feature of the graphing utility to find quadratic, exponential, and power models for the data, as shown in Figure 3.52.

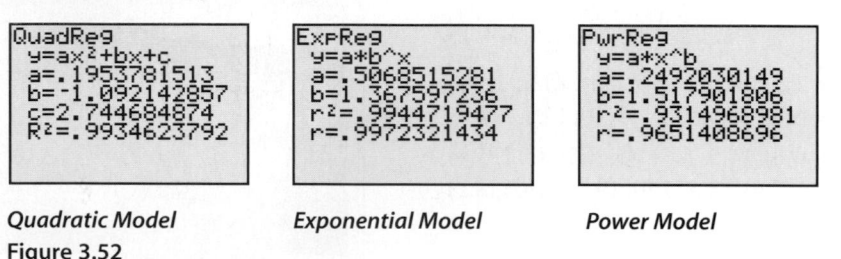

Quadratic Model   **Exponential Model**   **Power Model**
Figure 3.52

TECHNOLOGY SUPPORT

For instructions on how to use the *regression* feature, see Appendix A; for specific keystrokes, go to this textbook's *Online Study Center*.

So, a quadratic model for the data is $y = 0.195x^2 - 1.09x + 2.7$; an exponential model for the data is $y = 0.507(1.368)^x$; and a power model for the data is $y = 0.249x^{1.518}$. Plot the data and each model in the same viewing window, as shown in Figure 3.53. To determine which model best fits the data, compare the coefficients of determination for each model. The model whose r^2-value is closest to 1 is the model that best fits the data. In this case, the best-fitting model is the exponential model.

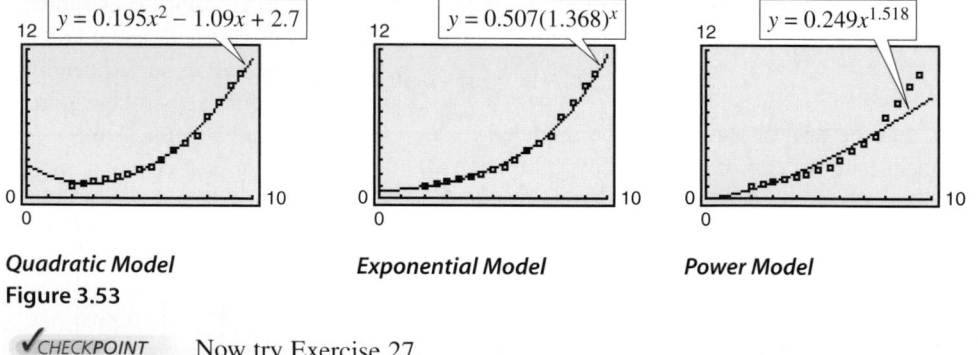

Quadratic Model   **Exponential Model**   **Power Model**
Figure 3.53

✓CHECKPOINT   Now try Exercise 27.

Deciding which model best fits a set of data is a question that is studied in detail in statistics. Recall from Section 1.7 that the model that best fits a set of data is the one whose *sum of squared differences* is the least. In Example 2, the sums of squared differences are 0.89 for the quadratic model, 0.85 for the exponential model, and 14.39 for the power model.

Example 3   Fitting a Model to Data

The table shows the yield y (in milligrams) of a chemical reaction after x minutes. Use a graphing utility to find a logarithmic model and a linear model for the data and identify the coefficient of determination for each model. Determine which model fits the data better.

Minutes, x	Yield, y
1	1.5
2	7.4
3	10.2
4	13.4
5	15.8
6	16.3
7	18.2
8	18.3

Solution

Begin by entering the data into a graphing utility. Then use the *regression* feature of the graphing utility to find logarithmic and linear models for the data, as shown in Figure 3.54.

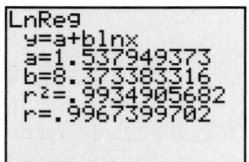

Logarithmic Model

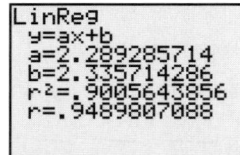

Linear Model

Figure 3.54

So, a logarithmic model for the data is $y = 1.538 + 8.373 \ln x$ and a linear model for the data is $y = 2.29x + 2.3$. Plot the data and each model in the same viewing window, as shown in Figure 3.55. To determine which model fits the data better, compare the coefficients of determination for each model. The model whose coefficient of determination that is closer to 1 is the model that better fits the data. In this case, the better-fitting model is the logarithmic model.

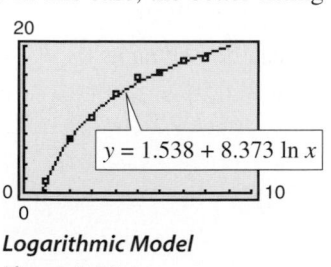

Logarithmic Model

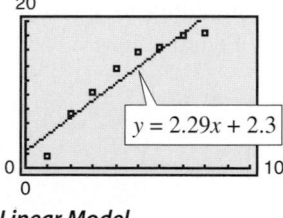

Linear Model

Figure 3.55

✓CHECKPOINT   Now try Exercise 29.

> ### Exploration
>
> Use a graphing utility to find a quadratic model for the data in Example 3. Do you think this model fits the data better than the logarithmic model in Example 3? Explain your reasoning.

In Example 3, the sum of the squared differences for the logarithmic model is 1.55 and the sum of the squared differences for the linear model is 23.86.

Modeling With Exponential and Logistic Functions

Example 4   Fitting an Exponential Model to Data

The table at the right shows the amounts of revenue R (in billions of dollars) collected by the Internal Revenue Service (IRS) for selected years from 1960 to 2005. Use a graphing utility to find a model for the data. Then use the model to estimate the revenue collected in 2010.   (Source: IRS Data Book)

Year	Revenue, R
1960	91.8
1965	114.4
1970	195.7
1975	293.8
1980	519.4
1985	742.9
1990	1056.4
1995	1375.7
2000	2096.9
2005	2268.9

Solution

Let x represent the year, with $x = 0$ corresponding to 1960. Begin by entering the data into a graphing utility and displaying the scatter plot, as shown in Figure 3.56.

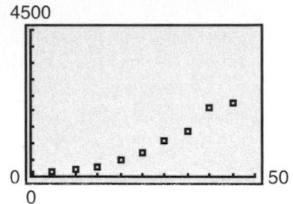

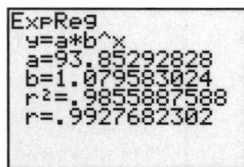

Figure 3.56   **Figure 3.57**

From the scatter plot, it appears that an exponential model is a good fit. Use the *regression* feature of the graphing utility to find the exponential model, as shown in Figure 3.57. Change the model to a natural exponential model, as follows.

$R = 93.85(1.080)^x$ Write original model.

$\approx 93.85e^{(\ln 1.080)x}$ $b = e^{\ln b}$

$\approx 93.85e^{0.077x}$ Simplify.

Graph the data and the model in the same viewing window, as shown in Figure 3.58. From the model, you can see that the revenue collected by the IRS from 1960 to 2005 had an average annual increase of about 8%. From this model, you can estimate the 2010 revenue to be

$R = 93.85e^{0.077x}$ Write original model.

$= 93.85e^{0.077(50)} \approx 4410.3$ billion   Substitute 50 for x.

which is more than twice the amount collected in 2000. You can also use the *value* feature or the *zoom* and *trace* features of a graphing utility to approximate the revenue in 2010 to be $4410.3 billion, as shown in Figure 3.58.

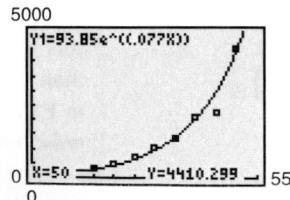

Figure 3.58

✓CHECKPOINT   Now try Exercise 33.

STUDY TIP

You can change an exponential model of the form

$y = ab^x$

to one of the form

$y = ae^{cx}$

by rewriting b in the form

$b = e^{\ln b}$.

For instance,

$y = 3(2^x)$

can be written as

$y = 3(2^x) = 3e^{(\ln 2)x} \approx 3e^{0.693x}$.

The next example demonstrates how to use a graphing utility to fit a logistic model to data.

Example 5   Fitting a Logistic Model to Data

To estimate the amount of defoliation caused by the gypsy moth during a given year, a forester counts the number x of egg masses on $\frac{1}{40}$ of an acre (circle of radius 18.6 feet) in the fall. The percent of defoliation y the next spring is shown in the table.   (Source: USDA, Forest Service)

Egg masses, x	Percent of defoliation, y
0	12
25	44
50	81
75	96
100	99

a. Use the *regression* feature of a graphing utility to find a logistic model for the data.

b. How closely does the model represent the data?

Graphical Solution

a. Enter the data into the graphing utility. Using the *regression* feature of the graphing utility, you can find the logistic model, as shown in Figure 3.59. You can approximate this model to be

$$y = \frac{100}{1 + 7e^{-0.069x}}.$$

b. You can use a graphing utility to graph the actual data and the model in the same viewing window. In Figure 3.60, it appears that the model is a good fit for the actual data.

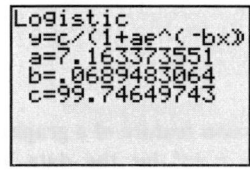

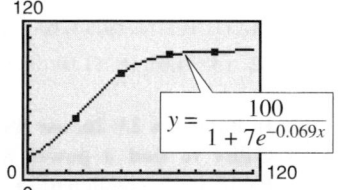

Figure 3.59   **Figure 3.60**

✓CHECKPOINT   Now try Exercise 34.

Numerical Solution

a. Enter the data into the graphing utility. Using the *regression* feature of the graphing utility, you can approximate the logistic model to be

$$y = \frac{100}{1 + 7e^{-0.069x}}.$$

b. You can see how well the model fits the data by comparing the actual values of y with the values of y given by the model, which are labeled y^* in the table below.

x	0	25	50	75	100
y	12	44	81	96	99
y^*	12.5	44.5	81.8	96.2	99.3

In the table, you can see that the model appears to be a good fit for the actual data.

3.6 Exercises

See www.CalcChat.com for worked-out solutions to odd-numbered exercises.

Vocabulary Check

Fill in the blanks.

1. A linear model has the form _____ .

2. A _____ model has the form $y = ax^2 + bx + c$.

3. A power model has the form _____ .

4. One way of determining which model best fits a set of data is to compare the _____ of _____ .

5. An exponential model has the form _____ or _____ .

Library of Functions In Exercises 1–8, determine whether the scatter plot could best be modeled by a linear model, a quadratic model, an exponential model, a logarithmic model, or a logistic model.

1. 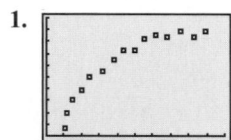   2.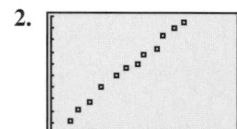

3. 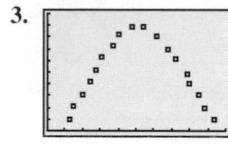   4.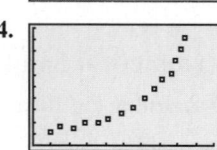

5. 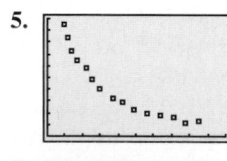   6.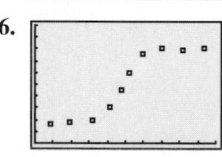

7. 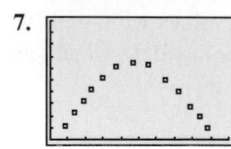   8.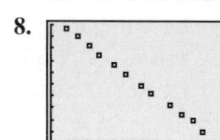

Library of Parent Functions In Exercises 9–14, use a graphing utility to create a scatter plot of the data. Decide whether the data could best be modeled by a linear model, an exponential model, or a logarithmic model.

9. $(1, 2.0), (1.5, 3.5), (2, 4.0), (4, 5.8), (6, 7.0), (8, 7.8)$

10. $(1, 5.8), (1.5, 6.0), (2, 6.5), (4, 7.6), (6, 8.9), (8, 10.0)$

11. $(1, 4.4), (1.5, 4.7), (2, 5.5), (4, 9.9), (6, 18.1), (8, 33.0)$

12. $(1, 11.0), (1.5, 9.6), (2, 8.2), (4, 4.5), (6, 2.5), (8, 1.4)$

13. $(1, 7.5), (1.5, 7.0), (2, 6.8), (4, 5.0), (6, 3.5), (8, 2.0)$

14. $(1, 5.0), (1.5, 6.0), (2, 6.4), (4, 7.8), (6, 8.6), (8, 9.0)$

In Exercises 15–18, use the *regression* feature of a graphing utility to find an exponential model $y = ab^x$ for the data and identify the coefficient of determination. Use the graphing utility to plot the data and graph the model in the same viewing window.

15. $(0, 5), (1, 6), (2, 7), (3, 9), (4, 13)$

16. $(0, 4.0), (2, 6.9), (4, 18.0), (6, 32.3), (8, 59.1),$
    $(10, 118.5)$

17. $(0, 10.0), (1, 6.1), (2, 4.2), (3, 3.8), (4, 3.6)$

18. $(-3, 120.2), (0, 80.5), (3, 64.8), (6, 58.2), (10, 55.0)$

In Exercises 19–22, use the *regression* feature of a graphing utility to find a logarithmic model $y = a + b \ln x$ for the data and identify the coefficient of determination. Use the graphing utility to plot the data and graph the model in the same viewing window.

19. $(1, 2.0), (2, 3.0), (3, 3.5), (4, 4.0), (5, 4.1), (6, 4.2),$
    $(7, 4.5)$

20. $(1, 8.5), (2, 11.4), (4, 12.8), (6, 13.6), (8, 14.2),$
    $(10, 14.6)$

21. $(1, 10), (2, 6), (3, 6), (4, 5), (5, 3), (6, 2)$

22. $(3, 14.6), (6, 11.0), (9, 9.0), (12, 7.6), (15, 6.5)$

In Exercises 23–26, use the *regression* feature of a graphing utility to find a power model $y = ax^b$ for the data and identify the coefficient of determination. Use the graphing utility to plot the data and graph the model in the same viewing window.

23. $(1, 2.0), (2, 3.4), (5, 6.7), (6, 7.3), (10, 12.0)$

24. $(0.5, 1.0), (2, 12.5), (4, 33.2), (6, 65.7), (8, 98.5),$
    $(10, 150.0)$

25. $(1, 10.0), (2, 4.0), (3, 0.7), (4, 0.1)$

26. $(2, 450), (4, 385), (6, 345), (8, 332), (10, 312)$

27. Elections   The table shows the numbers R (in millions) of registered voters in the United States for presidential election years from 1972 to 2004.   (Source: Federal Election Commission)

Year	Number of voters, R
1972	97.3
1976	105.0
1980	113.0
1984	124.2
1988	126.4
1992	133.8
1996	146.2
2000	156.4
2004	174.8

(a) Use the *regression* feature of a graphing utility to find a quadratic model, an exponential model, and a power model for the data. Let t represent the year, with $t = 2$ corresponding to 1972.

(b) Use a graphing utility to graph each model with the original data.

(c) Determine which model best fits the data.

(d) Use the model you chose in part (c) to predict the numbers of registered voters in 2008 and 2012.

28. Consumer Awareness   The table shows the retail prices P (in dollars) of a half-gallon package of ice cream from 1995 to 2004.   (Source: U.S. Bureau of Labor Statistics)

Year	Retail price, P
1995	2.68
1996	2.94
1997	3.02
1998	3.30
1999	3.40
2000	3.66
2001	3.84
2002	3.76
2003	3.90
2004	3.85

(a) Use the *regression* feature of a graphing utility to find a quadratic model, an exponential model, and a power model for the data and to identify the coefficient of determination for each model. Let t represent the year, with $t = 5$ corresponding to 1995.

(b) Use a graphing utility to graph each model with the original data.

(c) Determine which model best fits the data.

(d) Use the model you chose in part (c) to predict the prices of a half-gallon package of ice cream from 2005 through 2010. Are the predictions reasonable? Explain.

29. Population   The populations y (in millions) of the United States for the years 1991 through 2004 are shown in the table, where t represents the year, with $t = 1$ corresponding to 1991.   (Source: U.S. Census Bureau)

Year	Population, P
1991	253.5
1992	256.9
1993	260.3
1994	263.4
1995	266.6
1996	269.7
1997	272.9
1998	276.1
1999	279.3
2000	282.4
2001	285.3
2002	288.2
2003	291.0
2004	293.9

(a) Use the *regression* feature of a graphing utility to find a linear model for the data and to identify the coefficient of determination. Plot the model and the data in the same viewing window.

(b) Use the *regression* feature of a graphing utility to find a power model for the data and to identify the coefficient of determination. Plot the model and the data in the same viewing window.

(c) Use the *regression* feature of a graphing utility to find an exponential model for the data and to identify the coefficient of determination. Plot the model and the data in the same viewing window.

(d) Use the *regression* feature of a graphing utility to find a quadratic model for the data and to identify the coefficient of determination. Plot the model and the data in the same viewing window.

(e) Which model is the best fit for the data? Explain.

(f) Use each model to predict the populations of the United States for the years 2005 through 2010.

(g) Which model is the best choice for predicting the future population of the United States? Explain.

(h) Were your choices of models the same for parts (e) and (g)? If not, explain why your choices were different.

30. *Atmospheric Pressure* The atmospheric pressure decreases with increasing altitude. At sea level, the average air pressure is approximately 1.03323 kilograms per square centimeter, and this pressure is called one atmosphere. Variations in weather conditions cause changes in the atmospheric pressure of up to ±5 percent. The table shows the pressures p (in atmospheres) for various altitudes h (in kilometers).

Altitude, h	Pressure, p
0	1
5	0.55
10	0.25
15	0.12
20	0.06
25	0.02

(a) Use the *regression* feature of a graphing utility to attempt to find the logarithmic model $p = a + b \ln h$ for the data. Explain why the result is an error message.

(b) Use the *regression* feature of a graphing utility to find the logarithmic model $h = a + b \ln p$ for the data.

(c) Use a graphing utility to plot the data and graph the logarithmic model in the same viewing window.

(d) Use the model to estimate the altitude at which the pressure is 0.75 atmosphere.

(e) Use the graph in part (c) to estimate the pressure at an altitude of 13 kilometers.

31. *Data Analysis* A cup of water at an initial temperature of 78°C is placed in a room at a constant temperature of 21°C. The temperature of the water is measured every 5 minutes for a period of $\frac{1}{2}$ hour. The results are recorded in the table, where t is the time (in minutes) and T is the temperature (in degrees Celsius).

Time, t	Temperature, T
0	78.0°
5	66.0°
10	57.5°
15	51.2°
20	46.3°
25	42.5°
30	39.6°

(a) Use the *regression* feature of a graphing utility to find a linear model for the data. Use the graphing utility to plot the data and graph the model in the same viewing window. Does the data appear linear? Explain.

(b) Use the *regression* feature of a graphing utility to find a quadratic model for the data. Use the graphing utility to plot the data and graph the model in the same viewing window. Does the data appear quadratic? Even though the quadratic model appears to be a good fit, explain why it might not be a good model for predicting the temperature of the water when $t = 60$.

(c) The graph of the model should be asymptotic with the graph of the temperature of the room. Subtract the room temperature from each of the temperatures in the table. Use the *regression* feature of a graphing utility to find an exponential model for the revised data. Add the room temperature to this model. Use a graphing utility to plot the original data and graph the model in the same viewing window.

(d) Explain why the procedure in part (c) was necessary for finding the exponential model.

32. *Sales* The table shows the sales S (in millions of dollars) for AutoZone stores from 1995 to 2005. (Source: AutoZone, Inc.)

Year	Sales, S (in millions of dollars)
1995	1808.1
1996	2242.6
1997	2691.4
1998	3242.9
1999	4116.4
2000	4482.7
2001	4818.2
2002	5325.5
2003	5457.1
2004	5637.0
2005	5710.9

(a) Use a graphing utility to create a scatter plot of the data. Let t represent the year, with $t = 5$ corresponding to 1995.

(b) The data can be modeled by the logistic curve

$$S = 1018.4 + \frac{4827.2}{1 + e^{-(t-8.1391)/1.9372}}$$

where t is the year, with $t = 5$ corresponding to 1995, and S is the sales (in millions of dollars). Use the graphing utility to graph the model and the data in the same viewing window. How well does the model fit the data?

(c) Use the model to determine when the sales for AutoZone is expected to reach 5.75 billion dollars.

33. *Vital Statistics*   The table shows the percents P of men who have never been married for different age groups (in years). (Source: U.S. Census Bureau)

Age group	Percent, P
18–19	98.6
20–24	86.4
25–29	56.6
30–34	32.2
35–39	23.4
40–44	17.6
45–54	12.1
55–64	5.9
65–74	4.4
75 and over	3.6

(a) Use the *regression* feature of a graphing utility to find a logistic model for the data. Let x represent the age group, with $x = 1$ corresponding to the 18–19 age group.

(b) Use the graphing utility to graph the model with the original data. How closely does the model represent the data?

34. *Emissions*   The table shows the amounts A (in millions of metric tons) of carbon dioxide emissions from the consumption of fossil fuels in the United States from 1994 to 2003. (Source: U.S. Energy Information Administration)

Year	Carbon dioxide emissions, A (in millions of metric tons)
1994	1418
1995	1442
1996	1481
1997	1512
1998	1521
1999	1541
2000	1586
2001	1563
2002	1574
2003	1582

(a) Use a graphing utility to create a scatter plot of the data. Let t represent the year, with $t = 4$ corresponding to 1994.

(b) Use the *regression* feature of a graphing utility to find a linear model, a quadratic model, a cubic model, a power model, and an exponential model for the data.

(c) Create a table of values for each model. Which model is the best fit for the data? Explain.

(d) Use the best model to predict the emissions in 2015. Is your result reasonable? Explain.

35. *Comparing Models*   The amounts y (in billions of dollars) donated to charity (by individuals, foundations, corporations, and charitable bequests) in the United States from 1995 to 2003 are shown in the table, where t represents the year, with $t = 5$ corresponding to 1995. (Source: AAFRC Trust for Philanthropy)

Year, t	Amount, y
5	124.0
6	138.6
7	159.4
8	177.4
9	201.0
10	227.7
11	229.0
12	234.1
13	240.7

(a) Use the *regression* feature of a graphing utility to find a linear model, a logarithmic model, a quadratic model, an exponential model, and a power model for the data.

(b) Use the graphing utility to graph each model with the original data. Use the graphs to choose the model that you think best fits the data.

(c) For each model, find the sum of the squared differences. Use the results to choose the model that best fits the data.

(d) For each model, find the r^2-value determined by the graphing utility. Use the results to choose the model that best fits the data.

(e) Compare your results from parts (b), (c), and (d).

Synthesis

36. *Writing*   In your own words, explain how to fit a model to a set of data using a graphing utility.

True or False?   In Exercises 37 and 38, determine whether the statement is true or false. Justify your answer.

37. The exponential model $y = ae^{bx}$ represents a growth model if $b > 0$.

38. To change an exponential model of the form $y = ab^x$ to one of the form $y = ae^{cx}$, rewrite b as $b = \ln e^b$.

Skills Review

In Exercises 39–42, find the slope and y-intercept of the equation of the line. Then sketch the line by hand.

39. $2x + 5y = 10$

40. $3x - 2y = 9$

41. $1.2x + 3.5y = 10.5$

42. $0.4x - 2.5y = 12.0$

What Did You Learn?

Key Terms

transcendental function, *p. 184*

exponential function, base *a*, *p. 184*

natural base, *p. 188*

natural exponential function, *p. 188*

continuous compounding, *p. 190*

logarithmic function, base *a*, *p. 196*

common logarithmic function, *p. 197*

natural logarithmic function, *p. 200*

change-of-base formula, *p. 207*

exponential growth model, *p. 225*

exponential decay model, *p. 225*

Gaussian model, *p. 225*

logistic growth model, *p. 225*

logarithmic models, *p. 225*

normally distributed, *p. 229*

bell-shaped curve, *p. 229*

logistic curve, *p. 230*

Key Concepts

3.1 ■ Evaluate and graph exponential functions

1. The exponential function f with base a is denoted by $f(x) = a^x$, where $a > 0$, $a \neq 1$, and x is any real number.

2. The graphs of the exponential functions $f(x) = a^x$ and $f(x) = a^{-x}$ have one y-intercept, one horizontal asymptote (the x-axis), and are continuous.

3. The natural exponential function is $f(x) = e^x$, where e is the constant $2.718281828 \ldots$ Its graph has the same basic characteristics as the graph of $f(x) = a^x$.

3.2 ■ Evaluate and graph logarithmic functions

1. For $x > 0$, $a > 0$, and $a \neq 1$, $y = \log_a x$ if and only if $x = a^y$. The function given by $f(x) = \log_a x$ is called the logarithmic function with base a.

2. The graph of the logarithmic function $f(x) = \log_a x$, where $a > 1$, is the inverse of the graph of $f(x) = a^x$, has one x-intercept, one vertical asymptote (the y-axis), and is continuous.

3. For $x > 0$, $y = \ln x$ if and only if $x = e^y$. The function given by $f(x) = \log_e x = \ln x$ is called the natural logarithmic function. Its graph has the same basic characteristics as the graph of $f(x) = \log_a x$.

3.2 ■ Properties of logarithms

1. $\log_a 1 = 0$ and $\ln 1 = 0$

2. $\log_a a = 1$ and $\ln e = 1$

3. $\log_a a^x = x$, $a^{\log_a x} = x$; $\ln e^x = x$, and $e^{\ln x} = x$

4. If $\log_a x = \log_a y$, then $x = y$. If $\ln x = \ln y$, then $x = y$.

3.3 ■ Change-of-base formulas and properties of logarithms

1. Let a, b, and x be positive real numbers such that $a \neq 1$ and $b \neq 1$. Then $\log_a x$ can be converted to a different base using any of the following formulas.

$$\log_a x = \frac{\log_b x}{\log_b a}, \quad \log_a x = \frac{\log_{10} x}{\log_{10} a}, \quad \log_a x = \frac{\ln x}{\ln a}$$

2. Let a be a positive number such that $a \neq 1$, and let n be a real number. If u and v are positive real numbers, the following properties are true.

Product Property

$$\log_a(uv) = \log_a u + \log_a v \qquad \ln(uv) = \ln u + \ln v$$

Quotient Property

$$\log_a \frac{u}{v} = \log_a u - \log_a v \qquad \ln \frac{u}{v} = \ln u - \ln v$$

Power Property

$$\log_a u^n = n \log_a u \qquad \ln u^n = n \ln u$$

3.4 ■ Solve exponential and logarithmic equations

1. Rewrite the original equation to allow the use of the One-to-One Properties or logarithmic functions.

2. Rewrite an exponential equation in logarithmic form and apply the Inverse Property of logarithmic functions.

3. Rewrite a logarithmic equation in exponential form and apply the Inverse Property of exponential functions.

3.5 ■ Use nonalgebraic models to solve real-life problems

1. Exponential growth model: $y = ae^{bx}$, $b > 0$.

2. Exponential decay model: $y = ae^{-bx}$, $b > 0$.

3. Gaussian model: $y = ae^{-(x-b)^2/c}$.

4. Logistic growth model: $y = a/(1 + be^{-rx})$.

5. Logarithmic models: $y = a + b \ln x$, $y = a + b \log_{10} x$.

3.6 ■ Fit nonlinear models to data

1. Create a scatter plot of the data to determine the type of model (quadratic, exponential, logarithmic, power, or logistic) that would best fit the data.

2. Use a calculator or computer to find the model.

3. The model whose y-values are closest to the actual y-values is the one that fits best.

Review Exercises

See www.CalcChat.com for worked-out solutions to odd-numbered exercises.

3.1 In Exercises 1–8, use a calculator to evaluate the function at the indicated value of x. Round your result to four decimal places.

Function	Value
1. $f(x) = 1.45^x$	$x = 2\pi$
2. $f(x) = 7^x$	$x = -\sqrt{11}$
3. $g(x) = 60^{2x}$	$x = -1.1$
4. $g(x) = 25^{-3x}$	$x = \frac{3}{2}$
5. $f(x) = e^x$	$x = 8$
6. $f(x) = 5e^x$	$x = \sqrt{5}$
7. $f(x) = e^{-x}$	$x = -2.1$
8. $f(x) = -4e^x$	$x = -\frac{3}{5}$

Library of Parent Functions   In Exercises 9–12, match the function with its graph. [The graphs are labeled (a), (b), (c), and (d).]

(a)
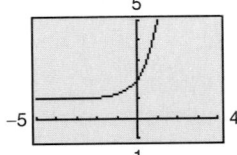

(b)
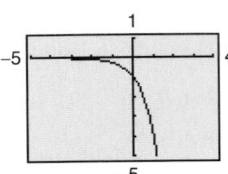

(c)
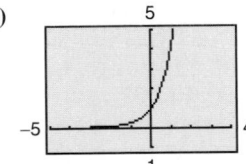

(d)
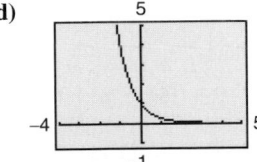

9. $f(x) = 4^x$

10. $f(x) = 4^{-x}$

11. $f(x) = -4^x$

12. $f(x) = 4^x + 1$

In Exercises 13–16, graph the exponential function by hand. Identify any asymptotes and intercepts and determine whether the graph of the function is increasing or decreasing.

13. $f(x) = 6^x$ 14. $f(x) = 0.3^{x+1}$

15. $g(x) = 1 + 6^{-x}$ 16. $g(x) = 0.3^{-x}$

In Exercises 17–22, use a graphing utility to construct a table of values for the function. Then sketch the graph of the function. Identify any asymptotes of the graph.

17. $h(x) = e^{x-1}$ 18. $f(x) = e^{x+2}$

19. $h(x) = -e^x$ 20. $f(x) = 3 - e^{-x}$

21. $f(x) = 4e^{-0.5x}$ 22. $f(x) = 2 + e^{x+3}$

In Exercises 23 and 24, use a graphing utility to (a) graph the exponential function and (b) find any asymptotes numerically by creating a table of values for the function.

23. $f(x) = \dfrac{10}{1 + 2^{-0.05x}}$

24. $f(x) = -\dfrac{12}{1 + 4^{-x}}$

Compound Interest   In Exercises 25 and 26, complete the table to determine the balance A for $10,000 invested at rate r for t years, compounded continuously.

t	1	10	20	30	40	50
A						

25. $r = 8\%$ 26. $r = 3\%$

27. **Depreciation**   After t years, the value of a car that costs $26,000 is modeled by $V(t) = 26,000\left(\frac{3}{4}\right)^t$.

(a) Use a graphing utility to graph the function.

(b) Find the value of the car 2 years after it was purchased.

(c) According to the model, when does the car depreciate most rapidly? Is this realistic? Explain.

28. **Radioactive Decay**   Let Q represent a mass, in grams, of plutonium 241 (^{241}Pu), whose half-life is 14 years. The quantity of plutonium present after t years is given by

$$Q = 100\left(\tfrac{1}{2}\right)^{t/14}.$$

(a) Determine the initial quantity (when $t = 0$).

(b) Determine the quantity present after 10 years.

(c) Use a graphing utility to graph the function over the interval $t = 0$ to $t = 100$.

3.2 In Exercises 29–42, write the logarithmic equation in exponential form or write the exponential equation in logarithmic form.

29. $\log_5 125 = 3$ 30. $\log_6 36 = 2$

31. $\log_{64} 2 = \frac{1}{6}$ 32. $\log_{10}\left(\frac{1}{100}\right) = -2$

33. $\ln e^4 = 4$ 34. $\ln \sqrt{e^3} = \frac{3}{2}$

35. $4^3 = 64$ 36. $3^5 = 243$

37. $25^{3/2} = 125$ 38. $12^{-1} = \frac{1}{12}$

39. $\left(\frac{1}{2}\right)^{-3} = 8$ 40. $\left(\frac{2}{3}\right)^{-2} = \frac{9}{4}$

41. $e^7 = 1096.6331\ldots$ 42. $e^{-3} = 0.0497\ldots$

In Exercises 43–46, evaluate the function at the indicated value of x without using a calculator.

Function   Value

43. $f(x) = \log_6 x$ $x = 216$

44. $f(x) = \log_7 x$ $x = 1$

45. $f(x) = \log_4 x$ $x = \frac{1}{4}$

46. $f(x) = \log_{10} x$ $x = 0.001$

In Exercises 47–50, find the domain, vertical asymptote, and x-intercept of the logarithmic function, and sketch its graph by hand.

47. $g(x) = -\log_2 x + 5$ 48. $g(x) = \log_5(x - 3)$

49. $f(x) = \log_2(x - 1) + 6$ 50. $f(x) = \log_5(x + 2) - 3$

In Exercises 51–54, use a calculator to evaluate the function $f(x) = \ln x$ at the indicated value of x. Round your result to three decimal places, if necessary.

51. $x = 21.5$ 52. $x = 0.98$

53. $x = \sqrt{6}$ 54. $x = \frac{2}{5}$

In Exercises 55–58, solve the equation for x.

55. $\log_5 3 = \log_5 x$ 56. $\log_2 8 = x$

57. $\log_9 x = \log_9 3^{-2}$ 58. $\log_4 4^3 = x$

In Exercises 59–62, use a graphing utility to graph the logarithmic function. Determine the domain and identify any vertical asymptote and x-intercept.

59. $f(x) = \ln x + 3$ 60. $f(x) = \ln(x - 3)$

61. $h(x) = \frac{1}{2} \ln x$ 62. $f(x) = \frac{1}{4} \ln x$

63. *Climb Rate*   The time t (in minutes) for a small plane to climb to an altitude of h feet is given by

$$t = 50 \log_{10}[18{,}000/(18{,}000 - h)]$$

where 18,000 feet is the plane's absolute ceiling.

(a) Determine the domain of the function appropriate for the context of the problem.

(b) Use a graphing utility to graph the function and identify any asymptotes.

(c) As the plane approaches its absolute ceiling, what can be said about the time required to further increase its altitude?

(d) Find the amount of time it will take for the plane to climb to an altitude of 4000 feet.

64. *Home Mortgage*   The model

$$t = 12.542 \ln[x/(x - 1000)], \ x > 1000$$

approximates the length of a home mortgage of $150,000 at 8% in terms of the monthly payment. In the model, t is the length of the mortgage in years and x is the monthly payment in dollars.

(a) Use the model to approximate the length of a $150,000 mortgage at 8% when the monthly payment is $1254.68.

(b) Approximate the total amount paid over the term of the mortgage with a monthly payment of $1254.68. What amount of the total is interest costs?

3.3 In Exercises 65–68, evaluate the logarithm using the change-of-base formula. Do each problem twice, once with common logarithms and once with natural logarithms. Round your results to three decimal places.

65. $\log_4 9$ 66. $\log_{1/2} 5$

67. $\log_{12} 200$ 68. $\log_3 0.28$

In Exercises 69–72, use the change-of-base formula and a graphing utility to graph the function.

69. $f(x) = \log_2(x - 1)$ 70. $f(x) = 2 - \log_3 x$

71. $f(x) = -\log_{1/2}(x + 2)$ 72. $f(x) = \log_{1/3}(x - 1) + 1$

In Exercises 73–76, approximate the logarithm using the properties of logarithms, given that $\log_b 2 \approx 0.3562$, $\log_b 3 \approx 0.5646$, and $\log_b 5 \approx 0.8271$.

73. $\log_b 9$ 74. $\log_b\left(\frac{4}{9}\right)$

75. $\log_b \sqrt{5}$ 76. $\log_b 50$

In Exercises 77–80, use the properties of logarithms to rewrite and simplify the logarithmic expression.

77. $\ln(5e^{-2})$ 78. $\ln \sqrt{e^5}$

79. $\log_{10} 200$ 80. $\log_{10} 0.002$

In Exercises 81–86, use the properties of logarithms to expand the expression as a sum, difference, and/or constant multiple of logarithms. (Assume all variables are positive.)

81. $\log_5 5x^2$ 82. $\log_4 3xy^2$

83. $\log_{10} \dfrac{5\sqrt{y}}{x^2}$ 84. $\ln \dfrac{\sqrt{x}}{4}$

85. $\ln\left(\dfrac{x + 3}{xy}\right)$ 86. $\ln \dfrac{xy^5}{\sqrt{z}}$

In Exercises 87–92, condense the expression to the logarithm of a single quantity.

87. $\log_2 5 + \log_2 x$

88. $\log_6 y - 2 \log_6 z$

89. $\frac{1}{2} \ln(2x - 1) - 2 \ln(x + 1)$

90. $5 \ln(x - 2) - \ln(x + 2) - 3 \ln(x)$

91. $\ln 3 + \frac{1}{3} \ln(4 - x^2) - \ln x$

92. $3[\ln x - 2 \ln(x^2 + 1)] + 2 \ln 5$

93. *Snow Removal* The number of miles s of roads cleared of snow is approximated by the model

$$s = 25 - \frac{13 \ln(h/12)}{\ln 3}, \quad 2 \le h \le 15$$

where h is the depth of the snow (in inches).

(a) Use a graphing utility to graph the function.

(b) Complete the table.

h	4	6	8	10	12	14
s						

(c) Using the graph of the function and the table, what conclusion can you make about the number of miles of roads cleared as the depth of the snow increases?

94. *Human Memory Model* Students in a sociology class were given an exam and then retested monthly with an equivalent exam. The average scores for the class are given by the human memory model $f(t) = 85 - 14 \log_{10}(t + 1)$, where t is the time in months and $0 \le t \le 10$. When will the average score decrease to 71?

3.4 In Exercises 95–108, solve the equation for x without using a calculator.

95. $8^x = 512$ **96.** $3^x = 729$

97. $6^x = \frac{1}{216}$ **98.** $6^{x-2} = 1296$

99. $2^{x+1} = \frac{1}{16}$ **100.** $4^{x/2} = 64$

101. $\log_7 x = 4$ **102.** $\log_x 243 = 5$

103. $\log_2(x - 1) = 3$ **104.** $\log_5(2x + 1) = 2$

105. $\ln x = 4$ **106.** $\ln x = -3$

107. $\ln(x - 1) = 2$ **108.** $\ln(2x + 1) = -4$

In Exercises 109–118, solve the exponential equation algebraically. Round your result to three decimal places.

109. $3e^{-5x} = 132$ **110.** $14e^{3x+2} = 560$

111. $2^x + 13 = 35$ **112.** $6^x - 28 = -8$

113. $-4(5^x) = -68$ **114.** $2(12^x) = 190$

115. $2e^{x-3} - 1 = 4$ **116.** $-e^{x/2} + 1 = \frac{1}{2}$

117. $e^{2x} - 7e^x + 10 = 0$ **118.** $e^{2x} - 6e^x + 8 = 0$

In Exercises 119–128, solve the logarithmic equation algebraically. Round your result to three decimal places.

119. $\ln 3x = 8.2$ **120.** $\ln 5x = 7.2$

121. $\ln x - \ln 3 = 2$ **122.** $\ln x - \ln 5 = 4$

123. $\ln \sqrt{x + 1} = 2$ **124.** $\ln \sqrt{x + 8} = 3$

125. $\log_4(x - 1) = \log_4(x - 2) - \log_4(x + 2)$

126. $\log_5(x + 2) - \log_5 x = \log_5(x + 5)$

127. $\log_{10}(1 - x) = -1$ **128.** $\log_{10}(-x - 4) = 2$

∫ In Exercises 129–132, solve the equation algebraically. Round your result to three decimal places.

129. $xe^x + e^x = 0$ **130.** $2xe^{2x} + e^{2x} = 0$

131. $x \ln x + x = 0$ **132.** $\frac{1 - \ln x}{x^2} = 0$

133. *Compound Interest* You deposit $7550 in an account that pays 7.25% interest, compounded continuously. How long will it take for the money to triple?

134. *Demand* The demand x for a 32-inch television is modeled by $p = 500 - 0.5e^{0.004x}$. Find the demands x for prices of (a) $p = \$450$ and (b) $p = \$400$.

3.5 Library of Parent Functions In Exercises 135–140, match the function with its graph. [The graphs are labeled (a), (b), (c), (d), (e), and (f).]

(a)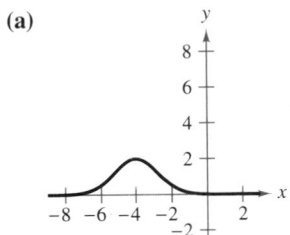

(b)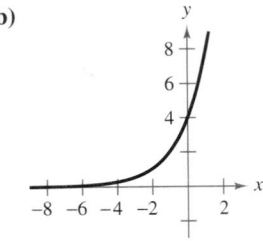

(c)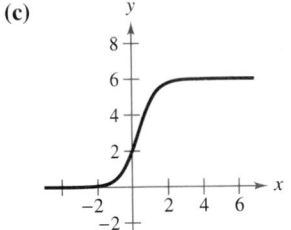

(d)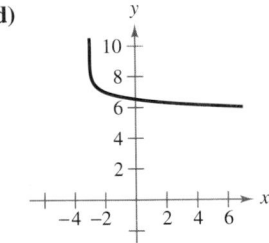

(e)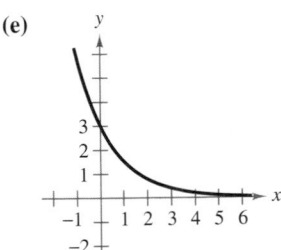

(f)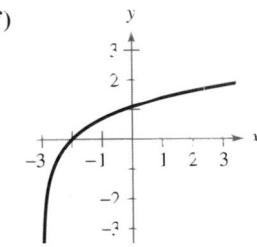

135. $y = 3e^{-2x/3}$ **136.** $y = 4e^{2x/3}$

137. $y = \ln(x + 3)$ **138.** $y = 7 - \log_{10}(x + 3)$

139. $y = 2e^{-(x+4)^2/3}$ **140.** $y = \frac{6}{1 + 2e^{-2x}}$

In Exercises 141–144, find the exponential model $y = ae^{bx}$ **that fits the two points.**

141. $(0, 2)$, $(4, 3)$

142. $(0, 2)$, $(5, 1)$

143. $\left(0, \frac{1}{2}\right)$, $(5, 5)$

144. $(0, 4)$, $\left(5, \frac{1}{2}\right)$

145. ***Population***   The population P (in thousands) of Colorado Springs, Colorado is given by

$$P = 361e^{kt}$$

where t represents the year, with $t = 0$ corresponding to 2000. In 1980, the population was 215,000. Find the value of k and use this result to predict the population in the year 2020.   (Source: U.S. Census Bureau)

146. ***Radioactive Decay***   The half-life of radioactive uranium (^{234}U) is 245,500 years. What percent of the present amount of radioactive uranium will remain after 5000 years?

147. ***Compound Interest***   A deposit of $10,000 is made in a savings account for which the interest is compounded continuously. The balance will double in 12 years.

(a) What is the annual interest rate for this account?

(b) Find the balance after 1 year.

148. ***Test Scores***   The test scores for a biology test follow a normal distribution modeled by

$$y = 0.0499e^{-(x-74)^2/128}, \quad 40 \le x \le 100$$

where x is the test score.

(a) Use a graphing utility to graph the function.

(b) From the graph in part (a), estimate the average test score.

149. ***Typing Speed***   In a typing class, the average number of words per minute N typed after t weeks of lessons was found to be modeled by

$$N = \frac{158}{1 + 5.4e^{-0.12t}}.$$

Find the numbers of weeks necessary to type (a) 50 words per minute and (b) 75 words per minute.

150. ***Geology***   On the Richter scale, the magnitude R of an earthquake of intensity I is modeled by

$$R = \log_{10} \frac{I}{I_0}$$

where $I_0 = 1$ is the minimum intensity used for comparison. Find the intensities I of the following earthquakes measuring R on the Richter scale.

(a) $R = 8.4$ (b) $R = 6.85$ (c) $R = 9.1$

3.6 Library of Parent Functions   In Exercises 151–154, determine whether the scatter plot could best be modeled by a linear model, a quadratic model, an exponential model, a logarithmic model, or a logistic model.

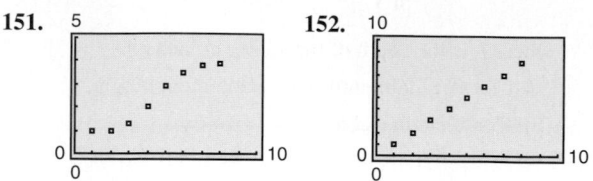

151.

152.

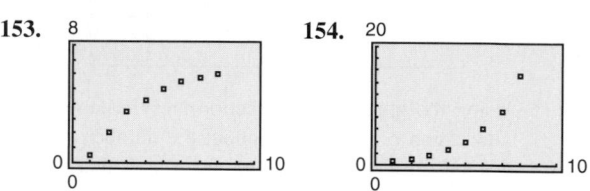

153.

154.

155. ***Fitness***   The table shows the sales S (in millions of dollars) of exercise equipment in the United States from 1998 to 2004.   (Source: National Sporting Goods Association)

Year	Sales, S (in millions of dollars)
1998	3233
1999	3396
2000	3610
2001	3889
2002	4378
2003	4727
2004	4869

(a) Use the *regression* feature of a graphing utility to find a linear model, a quadratic model, an exponential model, a logarithmic model, and a power model for the data and to identify the coefficient of determination for each model. Let t represent the year, with $t = 8$ corresponding to 1998.

(b) Use a graphing utility to graph each model with the original data.

(c) Determine which model best fits the data. Explain.

(d) Use the model you chose in part (c) to predict the sales of exercise equipment in 2010.

(e) Use the model you chose in part (c) to predict the year that sales will reach 5.25 billion dollars.

156. *Sports* The table shows the numbers of female participants P (in thousands) in high school athletic programs from 1991 to 2004. (Source: National Federation of State High School Associations)

Year	Female participants, P (in thousands)
1991	1892
1992	1941
1993	1997
1994	2130
1995	2240
1996	2368
1997	2474
1998	2570
1999	2653
2000	2676
2001	2784
2002	2807
2003	2856
2004	2865

(a) Use the *regression* feature of a graphing utility to find a linear model, a quadratic model, an exponential model, a logarithmic model, and a power model for the data and to identify the coefficient of determination for each model. Let t represent the year, with $t - 1$ corresponding to 1991.

(b) Use a graphing utility to graph each model with the original data.

(c) Determine which model best fits the data. Explain.

(d) Use the model you chose in part (c) to predict the number of participants in 2010.

(e) Use the model you chose in part (c) to predict when the number of participants will exceed 3 million.

157. *Wildlife* A lake is stocked with 500 fish, and the fish population P increases every month. The local fish commission records this increase as shown in the table.

Month, x	Position, P
0	500
6	1488
12	3672
18	6583
24	8650
30	9550
36	9860

(a) Use the *regression* feature of a graphing utility to find a logistic model for the data. Let x represent the month.

(b) Use a graphing utility to graph the model with the original data.

(c) How closely does the model represent the data?

(d) What is the limiting size of the population?

158. *Population* The population P of Italy (in millions) from 1990 to 2005 can be modeled by $P = 56.8e^{0.001603t}$, $0 \le t \le 15$, where t is the year, with $t = 0$ corresponding to 1990. (Source: U.S. Census Bureau)

(a) Use the *table* feature of a graphing utility to create a table of the values of P for $0 \le t \le 15$.

(b) Use the first and last values in your table to create a linear model for the data.

(c) What is the slope of your linear model, and what does it tell you about the situation?

(d) Graph both models in the same viewing window. Explain any differences in the models.

Synthesis

True or False? In Exercises 159–164, determine whether the equation or statement is true or false. Justify your answer.

159. $\log_b b^{2x} = 2x$

160. $e^{x-1} = \dfrac{e^x}{e}$

161. $\ln(x + y) = \ln x + \ln y$

162. $\ln(x + y) = \ln(xy)$

163. The domain of the function $f(x) = \ln x$ is the set of all real numbers.

164. The logarithm of the quotient of two numbers is equal to the difference of the logarithms of the numbers.

165. *Think About It* Without using a calculator, explain why you know that $2^{\sqrt{2}}$ is greater than 2, but less than 4.

166. *Pattern Recognition*

(a) Use a graphing utility to compare the graph of the function $y = e^x$ with the graph of each function below. [$n!$ (read as "n factorial") is defined as $n! = 1 \cdot 2 \cdot 3 \cdot \cdots (n-1) \cdot n$.]

$$y_1 = 1 + \frac{x}{1!}, \quad y_2 = 1 + \frac{x}{1!} + \frac{x^2}{2!},$$

$$y_3 = 1 + \frac{x}{1!} + \frac{x^2}{2!} + \frac{x^3}{3!}$$

(b) Identify the pattern of successive polynomials given in part (a). Extend the pattern one more term and compare the graph of the resulting polynomial function with the graph of $y = e^x$. What do you think this pattern implies?

3 Chapter Test

See www.CalcChat.com for worked-out solutions to odd-numbered exercises.

Take this test as you would take a test in class. After you are finished, check your work against the answers given in the back of the book.

In Exercises 1–3, use a graphing utility to construct a table of values for the function. Then sketch a graph of the function. Identify any asymptotes and intercepts.

1. $f(x) = 10^{-x}$ **2.** $f(x) = -6^{x-2}$ **3.** $f(x) = 1 - e^{2x}$

In Exercises 4–6, evaluate the expression.

4. $\log_7 7^{-0.89}$ **5.** $4.6 \ln e^2$ **6.** $2 - \log_{10} 100$

In Exercises 7–9, use a graphing utility to graph the function. Determine the domain and identify any vertical asymptotes and x-intercepts.

7. $f(x) = -\log_{10} x - 6$ **8.** $f(x) = \ln(x - 4)$ **9.** $f(x) = 1 + \ln(x + 6)$

In Exercises 10–12, evaluate the logarithm using the change-of-base formula. Round your result to three decimal places.

10. $\log_7 44$ **11.** $\log_{2/5} 0.9$ **12.** $\log_{24} 68$

In Exercises 13–15, use the properties of logarithms to expand the expression as a sum, difference, and/or multiple of logarithms.

13. $\log_2 3a^4$ **14.** $\ln \dfrac{5\sqrt{x}}{6}$ **15.** $\ln \dfrac{x\sqrt{x+1}}{2e^4}$

In Exercises 16–18, condense the expression to the logarithm of a single quantity.

16. $\log_3 13 + \log_3 y$ **17.** $4 \ln x - 4 \ln y$ **18.** $\ln x - \ln(x + 2) + \ln(2x - 3)$

In Exercises 19–22, solve the equation for x.

19. $3^x = 81$ **20.** $5^{2x} = 2500$

21. $\log_7 x = 3$ **22.** $\log_{10}(x - 4) = 5$

In Exercises 23–26, solve the equation algebraically. Round your result to three decimal places.

23. $\dfrac{1025}{8 + e^{4x}} = 5$ **24.** $-xe^{-x} + e^{-x} = 0$

25. $\log_{10} x - \log_{10}(8 - 5x) = 2$ **26.** $2x \ln x - x = 0$

27. The half-life of radioactive actinium (^{227}Ac) is 22 years. What percent of a present amount of radioactive actinium will remain after 19 years?

28. The table at the right shows the mail revenues R (in billions of dollars) for the U.S. Postal Service from 1995 to 2004. (Source: U.S. Postal Service)

(a) Use the *regression* feature of a graphing utility to find a quadratic model, an exponential model, and a power model for the data. Let t represent the year, with $t = 5$ corresponding to 1995.

(b) Use a graphing utility to graph each model with the original data.

(c) Determine which model best fits the data. Use the model to predict the mail revenues in 2010.

Year	Revenues, R
1995	52.5
1996	54.5
1997	56.3
1998	58.0
1999	60.4
2000	62.3
2001	63.4
2002	63.8
2003	65.7
2004	65.9

Table for 28

1–3 Cumulative Test

See www.CalcChat.com for worked-out solutions to odd-numbered exercises.

Take this test to review the material in Chapters 1–3. After you are finished, check your work against the answers given in the back of the book.

In Exercises 1–3, (a) write the general form of the equation of the line that satisfies the given conditions and (b) find three additional points through which the line passes.

1. The line contains the points $(-5, 8)$ and $(-1, 4)$.

2. The line contains the point $\left(-\frac{1}{2}, 1\right)$ and has a slope of -2.

3. The line has an undefined slope and contains the point $\left(-\frac{3}{7}, \frac{1}{8}\right)$.

In Exercises 4 and 5, evaluate the function at each value of the independent variable and simplify.

4. $f(x) = \dfrac{x}{x-2}$

   (a) $f(5)$ (b) $f(2)$ (c) $f(5 + 4s)$

5. $f(x) = \begin{cases} 3x - 8, & x < 0 \\ x^2 + 4, & x \geq 0 \end{cases}$

   (a) $f(-8)$ (b) $f(0)$ (c) $f(4)$

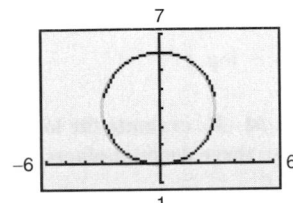

Figure for 6

6. Does the graph at the right represent y as a function of x? Explain.

7. Use a graphing utility to graph the function $f(x) = 2|x - 5| - |x + 5|$. Then determine the open intervals over which the function is increasing, decreasing, or constant.

8. Compare the graphs of each function with the graph of $f(x) = \sqrt[3]{x}$.

   (a) $r(x) = \dfrac{1}{2}\sqrt[3]{x}$ (b) $h(x) = \sqrt[3]{x} + 2$ (c) $g(x) = -\sqrt[3]{x+2}$

In Exercises 9–12, evaluate the indicated function for

$$f(x) = -x^2 + 3x - 10 \quad \text{and} \quad g(x) = 4x + 1.$$

9. $(f + g)(-4)$ 10. $(g - f)\left(\frac{3}{4}\right)$ 11. $(g \circ f)(-2)$ 12. $(fg)(-1)$

13. Determine whether $h(x) = 5x - 2$ has an inverse function. If so, find it.

In Exercises 14–16, sketch the graph of the function. Use a graphing utility to verify the graph.

14. $f(x) = -\frac{1}{2}(x^2 + 4x)$

15. $f(x) = \frac{1}{4}x(x - 2)^2$

16. $f(x) = x^3 + 2x^2 - 9x - 18$

17. Find all the zeros of $f(x) = x^3 + 2x^2 + 4x + 8$.

18. Use a graphing utility to approximate any real zeros of $g(x) = x^3 + 4x^2 - 11$ accurate to three decimal places.

19. Divide $(4x^2 + 14x - 9)$ by $(x + 3)$ using long division.

20. Divide $(2x^3 - 5x^2 + 6x - 20)$ by $(x - 6)$ using synthetic division.

21. Plot the complex number $-5 + 4i$ in the complex plane.

22. Find a polynomial function with real coefficients that has the zeros $0, -3,$ and $1 + \sqrt{5}i$.

In Exercises 23–25, sketch the graph of the rational function. Identify any asymptotes. Use a graphing utility to verify your graph.

23. $f(x) = \dfrac{2x}{x-3}$

24. $f(x) = \dfrac{5x}{x^2+x-6}$

25. $f(x) = \dfrac{x^2-3x+8}{x-2}$

In Exercises 26–29, use a calculator to evaluate the expression. Round your answer to three decimal places.

26. $(1.85)^{3.1}$

27. $58^{\sqrt{5}}$

28. $e^{-20/11}$

29. $4e^{2.56}$

In Exercises 30–33, sketch the graph of the function by hand. Use a graphing utility to verify your graph.

30. $f(x) = -3^{x+4} - 5$

31. $f(x) = -\left(\frac{1}{2}\right)^{-x} - 3$

32. $f(x) = 4 + \log_{10}(x-3)$

33. $f(x) = \ln(4-x)$

In Exercises 34–36, evaluate the logarithm using the change-of-base formula. Round your result to three decimal places.

34. $\log_5 21$

35. $\log_9 6.8$

36. $\log_2\left(\frac{3}{2}\right)$

37. Use the properties of logarithms to expand $\ln\left(\dfrac{x^2-4}{x^2+1}\right)$.

38. Write $2\ln x - \ln(x-1) + \ln(x+1)$ as a logarithm of a single quantity.

In Exercises 39–44, solve the equation algebraically. Round your result to three decimal places and verify your result graphically.

39. $6e^{2x} = 72$

40. $4^{x-5} + 21 = 30$

41. $\log_2 x + \log_2 5 = 6$

42. $250e^{0.05x} = 500{,}000$

43. $2x^2 e^{2x} - 2xe^{2x} = 0$

44. $\ln(2x-5) - \ln x = 1$

45. A rectangular plot of land with a perimeter of 546 feet has a width of x.

(a) Write the area A of the plot as a function of x.

(b) Use a graphing utility to graph the area function. What is the domain of the function?

(c) Approximate the dimensions of the plot when the area is 15,000 square feet.

46. The table shows the average prices y (in dollars) of one gallon of regular gasoline in the United States from 2002 to 2006.   (Source: Energy Information Administration)

(a) Use the *regression* feature of a graphing utility to find a quadratic model, an exponential model, and a power model for the data and identify the coefficient of determination for each. Let x represent the year, with $x = 2$ corresponding to 2002.

(b) Use a graphing utility to graph each model with the original data.

(c) Determine which model best fits the data. Explain.

(d) Use the model you chose in part (c) to predict the average prices of one gallon of gasoline in 2008 and 2010. Are your answers reasonable? Explain.

Year	Average price, y (in dollars)
2002	1.35
2003	1.56
2004	1.85
2005	2.27
2006	2.91

Proofs in Mathematics

Each of the following three properties of logarithms can be proved by using properties of exponential functions.

Properties of Logarithms   (p. 208)

Let a be a positive number such that $a \neq 1$, and let n be a real number. If u and v are positive real numbers, the following properties are true.

	Logarithm with Base a	*Natural Logarithm*
1. Product Property:	$\log_a(uv) = \log_a u + \log_a v$	$\ln(uv) = \ln u + \ln v$
2. Quotient Property:	$\log_a \dfrac{u}{v} = \log_a u - \log_a v$	$\ln \dfrac{u}{v} = \ln u - \ln v$
3. Power Property:	$\log_a u^n = n \log_a u$	$\ln u^n = n \ln u$

Slide Rules

The slide rule was invented by William Oughtred (1574–1660) in 1625. The slide rule is a computational device with a sliding portion and a fixed portion. A slide rule enables you to perform multiplication by using the Product Property of logarithms. There are other slide rules that allow for the calculation of roots and trigonometric functions. Slide rules were used by mathematicians and engineers until the invention of the hand-held calculator in 1972.

Proof

Let

$$x = \log_a u \quad \text{and} \quad y = \log_a v.$$

The corresponding exponential forms of these two equations are

$$a^x = u \quad \text{and} \quad a^y = v.$$

To prove the Product Property, multiply u and v to obtain

$$uv = a^x a^y = a^{x+y}.$$

The corresponding logarithmic form of $uv = a^{x+y}$ is $\log_a(uv) = x + y$. So,

$$\log_a(uv) = \log_a u + \log_a v.$$

To prove the Quotient Property, divide u by v to obtain

$$\frac{u}{v} = \frac{a^x}{a^y} = a^{x-y}.$$

The corresponding logarithmic form of $u/v = a^{x-y}$ is $\log_a(u/v) = x - y$. So,

$$\log_a \frac{u}{v} = \log_a u - \log_a v.$$

To prove the Power Property, substitute a^x for u in the expression $\log_a u^n$, as follows.

$$
\begin{aligned}
\log_a u^n &= \log_a(a^x)^n && \text{Substitute } a^x \text{ for } u. \\
&= \log_a a^{nx} && \text{Property of exponents} \\
&= nx && \text{Inverse Property of logarithms} \\
&= n \log_a u && \text{Substitute } \log_a u \text{ for } x.
\end{aligned}
$$

So, $\log_a u^n = n \log_a u$.

Progressive Summary (Chapters 1–3)

This chart outlines the topics that have been covered so far in this text. Progressive Summary charts appear after Chapters 2, 3, 6, 9, and 11. In each progressive summary, new topics encountered for the first time appear in red.

Algebraic Functions	Transcendental Functions	Other Topics
Polynomial, Rational, Radical	**Exponential, Logarithmic**	

■ Rewriting

Polynomial form ↔ Factored form
Operations with polynomials
Rationalize denominators
Simplify rational expressions
Operations with complex numbers

■ Rewriting (Transcendental)

Exponential form ↔ Logarithmic form
Condense/expand logarithmic
  expressions

■ Rewriting (Other)

■ Solving

Equation	Strategy
Linear	Isolate variable
Quadratic	Factor, set to zero
	Extract square roots
	Complete the square
	Quadratic Formula
Polynomial	Factor, set to zero
	Rational Zero Test
Rational	Multiply by LCD
Radical	Isolate, raise to power
Absolute value	Isolate, form two equations

■ Solving (Transcendental)

Equation	Strategy
Exponential	Take logarithm of each side
Logarithmic	Exponentiate each side

■ Solving (Other)

■ Analyzing

Graphically	Algebraically
Intercepts	Domain, Range
Symmetry	Transformations
Slope	Composition
Asymptotes	Standard forms of equations
End behavior	
Minimum values	Leading Coefficient Test
Maximum values	Synthetic division
	Descartes's Rule of Signs

Numerically
Table of values

■ Analyzing (Transcendental)

Graphically	Algebraically
Intercepts	Domain, Range
Asymptotes	Transformations
	Composition
	Inverse properties

Numerically
Table of values

■ Analyzing (Other)

Chapter 4

Trigonometric Functions

4.1 Radian and Degree Measure

4.2 Trigonometric Functions:
The Unit Circle

4.3 Right Triangle Trigonometry

4.4 Trigonometric Functions of
Any Angle

4.5 Graphs of Sine and Cosine
Functions

4.6 Graphs of Other
Trigonometric Functions

4.7 Inverse Trigonometric
Functions

4.8 Applications and Models

Selected Applications

Trigonometric functions have many real-life applications. The applications listed below represent a small sample of the applications in this chapter.

■ Sports, Exercise 95, page 267

■ Electrical Circuits,
Exercise 73, page 275

■ Machine Shop Calculations,
Exercise 83, page 287

■ Meteorology, Exercise 109,
page 296

■ Sales, Exercise 72, page 306

■ Predator-Prey Model,
Exercise 59, page 317

■ Photography, Exercise 83,
page 329

■ Airplane Ascent, Exercises 29 and
30, page 339

■ Harmonic Motion, Exercises
55–58, page 341

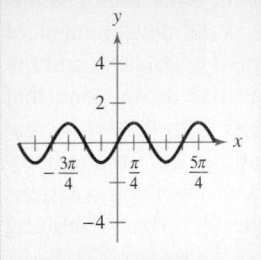
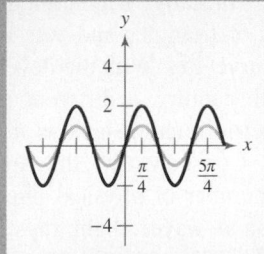
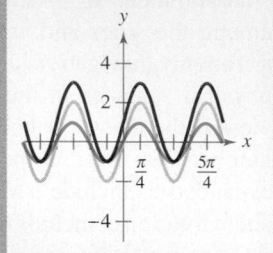

The six trigonometric functions can be defined from a right triangle perspective and as functions of real numbers. In Chapter 4, you will use both perspectives to graph trigonometric functions and solve application problems involving angles and triangles. You will also learn how to graph and evaluate inverse trigonometric functions.

© Winfried Wisniewski/zefa/Corbis

Trigonometric functions are often used to model repeating patterns that occur in real life. For instance, a trigonometric function can be used to model the populations of two species that interact, one of which (the predator) hunts the other (the prey).

4.1   Radian and Degree Measure

Angles

As derived from the Greek language, the word **trigonometry** means "measurement of triangles." Initially, trigonometry dealt with relationships among the sides and angles of triangles and was used in the development of astronomy, navigation, and surveying. With the development of calculus and the physical sciences in the 17th century, a different perspective arose—one that viewed the classic trigonometric relationships as *functions* with the set of real numbers as their domains. Consequently, the applications of trigonometry expanded to include a vast number of physical phenomena involving rotations and vibrations, including sound waves, light rays, planetary orbits, vibrating strings, pendulums, and orbits of atomic particles.

The approach in this text incorporates *both* perspectives, starting with angles and their measure.

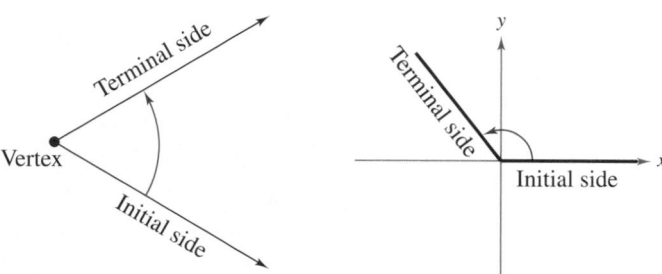

Figure 4.1   **Figure 4.2**

An **angle** is determined by rotating a ray (half-line) about its endpoint. The starting position of the ray is the **initial side** of the angle, and the position after rotation is the **terminal side,** as shown in Figure 4.1. The endpoint of the ray is the **vertex** of the angle. This perception of an angle fits a coordinate system in which the origin is the vertex and the initial side coincides with the positive x-axis. Such an angle is in **standard position,** as shown in Figure 4.2. **Positive angles** are generated by counterclockwise rotation, and **negative angles** by clockwise rotation, as shown in Figure 4.3. Angles are labeled with Greek letters such as α (alpha), β (beta), and θ (theta), as well as uppercase letters such as A, B, and C. In Figure 4.4, note that angles α and β have the same initial and terminal sides. Such angles are **coterminal.**

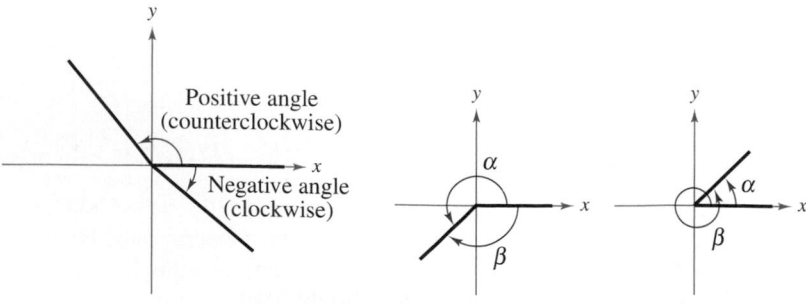

Figure 4.3   **Figure 4.4**

Radian Measure

The **measure of an angle** is determined by the amount of rotation from the initial side to the terminal side. One way to measure angles is in *radians*. This type of measure is especially useful in calculus. To define a radian, you can use a **central angle** of a circle, one whose vertex is the center of the circle, as shown in Figure 4.5.

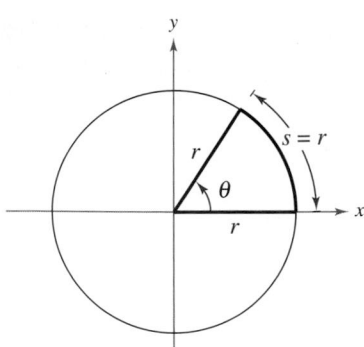

Arc length = radius when θ = 1 radian.

Figure 4.5

> ### Definition of Radian
>
> One **radian** (rad) is the measure of a central angle θ that intercepts an arc s equal in length to the radius r of the circle. See Figure 4.5. Algebraically this means that
>
> $$\theta = \frac{s}{r}$$
>
> where θ is measured in radians.

Because the circumference of a circle is $2\pi r$ units, it follows that a central angle of one full revolution (counterclockwise) corresponds to an arc length of

$$s = 2\pi r.$$

Moreover, because $2\pi \approx 6.28$, there are just over six radius lengths in a full circle, as shown in Figure 4.6. Because the units of measure for s and r are the same, the ratio s/r has no units—it is simply a real number.

Because the radian measure of an angle of one full revolution is 2π, you can obtain the following.

$$\frac{1}{2} \text{ revolution} = \frac{2\pi}{2} = \pi \text{ radians}$$

$$\frac{1}{4} \text{ revolution} = \frac{2\pi}{4} = \frac{\pi}{2} \text{ radians}$$

$$\frac{1}{6} \text{ revolution} = \frac{2\pi}{6} = \frac{\pi}{3} \text{ radians}$$

These and other common angles are shown in Figure 4.7.

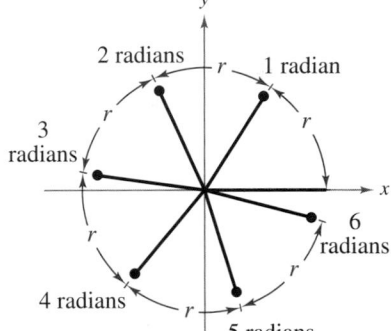

Figure 4.6

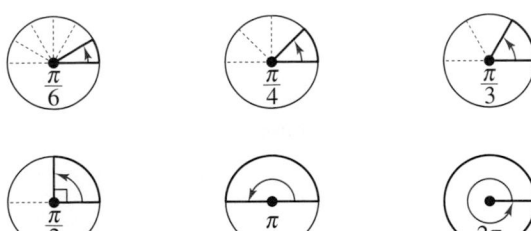

Figure 4.7

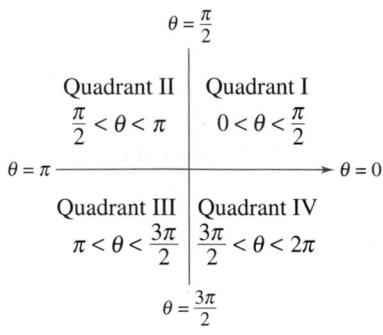

Figure 4.8

Recall that the four quadrants in a coordinate system are numbered I, II, III, and IV. Figure 4.8 shows which angles between 0 and 2π lie in each of the four quadrants. Note that angles between 0 and $\pi/2$ are **acute** and that angles between $\pi/2$ and π are **obtuse**.

Two angles are coterminal if they have the same initial and terminal sides. For instance, the angles 0 and 2π are coterminal, as are the angles $\pi/6$ and $13\pi/6$. You can find an angle that is coterminal to a given angle θ by adding or subtracting 2π (one revolution), as demonstrated in Example 1. A given angle θ has infinitely many coterminal angles. For instance, $\theta = \pi/6$ is coterminal with

$$\frac{\pi}{6} + 2n\pi, \text{ where } n \text{ is an integer.}$$

Example 1   Sketching and Finding Coterminal Angles

a. For the positive angle $\theta = \dfrac{13\pi}{6}$, subtract 2π to obtain a coterminal angle

$$\frac{13\pi}{6} - 2\pi = \frac{\pi}{6}. \qquad \text{See Figure 4.9.}$$

b. For the positive angle $\theta = \dfrac{3\pi}{4}$, subtract 2π to obtain a coterminal angle

$$\frac{3\pi}{4} - 2\pi = -\frac{5\pi}{4}. \qquad \text{See Figure 4.10.}$$

c. For the negative angle $\theta = -\dfrac{2\pi}{3}$, add 2π to obtain a coterminal angle

$$-\frac{2\pi}{3} + 2\pi = \frac{4\pi}{3}. \qquad \text{See Figure 4.11.}$$

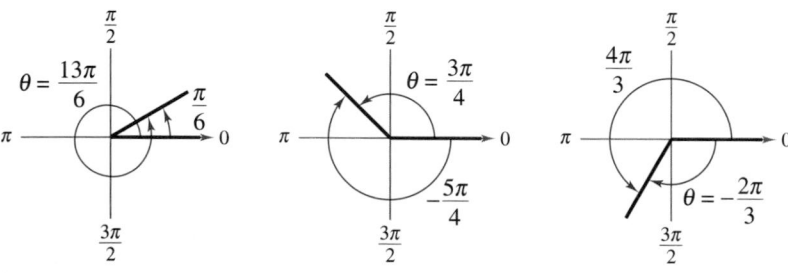

Figure 4.9   Figure 4.10   Figure 4.11

✔CHECKPOINT   Now try Exercise 11.

Two positive angles α and β are **complementary** (complements of each other) if their sum is $\pi/2$. Two positive angles are **supplementary** (supplements of each other) if their sum is π. See Figure 4.12.

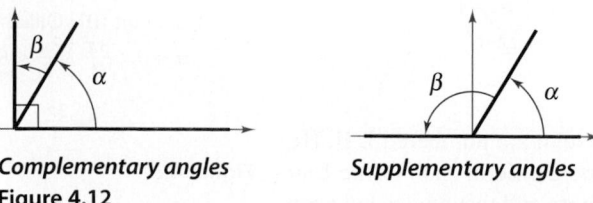

Complementary angles
Figure 4.12   *Supplementary angles*

Example 2   Complementary and Supplementary Angles

If possible, find the complement and the supplement of (a) $\dfrac{2\pi}{5}$ and (b) $\dfrac{4\pi}{5}$.

Solution

a. The complement of $\dfrac{2\pi}{5}$ is

$$\frac{\pi}{2} - \frac{2\pi}{5} = \frac{5\pi}{10} - \frac{4\pi}{10}$$

$$= \frac{\pi}{10}.$$

The supplement of $\dfrac{2\pi}{5}$ is

$$\pi - \frac{2\pi}{5} = \frac{5\pi}{5} - \frac{2\pi}{5}$$

$$= \frac{3\pi}{5}.$$

b. Because $4\pi/5$ is greater than $\pi/2$, it has no complement. (Remember that complements are *positive* angles.) The supplement is

$$\pi - \frac{4\pi}{5} = \frac{5\pi}{5} - \frac{4\pi}{5}$$

$$= \frac{\pi}{5}.$$

✓**CHECKPOINT**   Now try Exercise 15.

Degree Measure

A second way to measure angles is in terms of **degrees,** denoted by the symbol °. A measure of one degree (1°) is equivalent to a rotation of $\frac{1}{360}$ of a complete revolution about the vertex. To measure angles, it is convenient to mark degrees on the circumference of a circle, as shown in Figure 4.13. So, a full revolution (counterclockwise) corresponds to 360°, a half revolution to 180°, a quarter revolution to 90°, and so on.

Because 2π radians corresponds to one complete revolution, degrees and radians are related by the equations

$$360° = 2\pi \text{ rad} \quad \text{and} \quad 180° = \pi \text{ rad}.$$

From the second equation, you obtain

$$1° = \frac{\pi}{180} \text{ rad} \quad \text{and} \quad 1 \text{ rad} = \frac{180°}{\pi}$$

which lead to the conversion rules at the top of the next page.

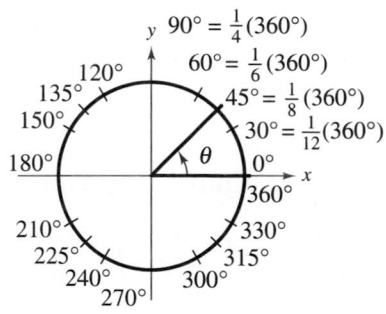

Figure 4.13

> **Conversions Between Degrees and Radians**
>
> 1. To convert degrees to radians, multiply degrees by $\dfrac{\pi \text{ rad}}{180°}$.
>
> 2. To convert radians to degrees, multiply radians by $\dfrac{180°}{\pi \text{ rad}}$.
>
> To apply these two conversion rules, use the basic relationship $\pi \text{ rad} = 180°$. (See Figure 4.14.)

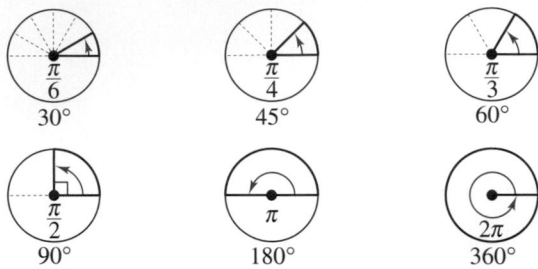

Figure 4.14

When no units of angle measure are specified, *radian measure is implied*. For instance, if you write $\theta = \pi$ or $\theta = 2$, you imply that $\theta = \pi$ radians or $\theta = 2$ radians.

Example 3   Converting from Degrees to Radians

a. $135° = (135 \text{ deg})\left(\dfrac{\pi \text{ rad}}{180 \text{ deg}}\right) = \dfrac{3\pi}{4}$ radians   Multiply by $\dfrac{\pi}{180}$.

b. $540° = (540 \text{ deg})\left(\dfrac{\pi \text{ rad}}{180 \text{ deg}}\right) = 3\pi$ radians   Multiply by $\dfrac{\pi}{180}$.

c. $-270° = (-270 \text{ deg})\left(\dfrac{\pi \text{ rad}}{180 \text{ deg}}\right) = -\dfrac{3\pi}{2}$ radians   Multiply by $\dfrac{\pi}{180}$.

✓**CHECKPOINT**   Now try Exercise 39.

Example 4   Converting from Radians to Degrees

a. $-\dfrac{\pi}{2} \text{ rad} = \left(-\dfrac{\pi}{2} \text{ rad}\right)\left(\dfrac{180 \text{ deg}}{\pi \text{ rad}}\right) = -90°$ Multiply by $\dfrac{180}{\pi}$.

b. $2 \text{ rad} = (2 \text{ rad})\left(\dfrac{180 \text{ deg}}{\pi \text{ rad}}\right) = \dfrac{360}{\pi} \approx 114.59°$ Multiply by $\dfrac{180}{\pi}$.

c. $\dfrac{9\pi}{2} \text{ rad} = \left(\dfrac{9\pi}{2} \text{ rad}\right)\left(\dfrac{180 \text{ deg}}{\pi \text{ rad}}\right) = 810°$ Multiply by $\dfrac{180}{\pi}$.

✓**CHECKPOINT**   Now try Exercise 43.

TECHNOLOGY TIP

With calculators it is convenient to use *decimal* degrees to denote fractional parts of degrees. Historically, however, fractional parts of degrees were expressed in *minutes* and *seconds*, using the prime ($'$) and double prime ($''$) notations, respectively. That is,

$1' = $ one minute $= \frac{1}{60}(1°)$

$1'' = $ one second $= \frac{1}{3600}(1°)$.

Consequently, an angle of 64 degrees, 32 minutes, and 47 seconds was represented by $\theta = 64° \, 32' \, 47''$.

Many calculators have special keys for converting angles in degrees, minutes, and seconds (D° M′ S″) to decimal degree form, and vice versa.

Linear and Angular Speed

The *radian measure* formula $\theta = s/r$ can be used to measure arc length along a circle.

> **Arc Length**
>
> For a circle of radius r, a central angle θ intercepts an arc of length s given by
>
> $$s = r\theta \qquad\qquad \text{Length of circular arc}$$
>
> where θ is measured in radians. Note that if $r = 1$, then $s = \theta$, and the radian measure of θ equals the arc length.

Example 5   Finding Arc Length

A circle has a radius of 4 inches. Find the length of the arc intercepted by a central angle of $240°$, as shown in Figure 4.15.

Solution

To use the formula $s = r\theta$, first convert $240°$ to radian measure.

$$240° = (240 \text{ deg})\left(\frac{\pi \text{ rad}}{180 \text{ deg}}\right) = \frac{4\pi}{3} \text{ radians}$$

Then, using a radius of $r = 4$ inches, you can find the arc length to be

$$s = r\theta = 4\left(\frac{4\pi}{3}\right) = \frac{16\pi}{3} \approx 16.76 \text{ inches}$$

Note that the units for $r\theta$ are determined by the units for r because θ is given in radian measure and therefore has no units.

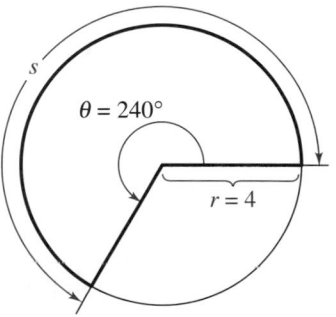

Figure 4.15

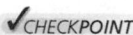CHECKPOINT   Now try Exercise 81.

   The formula for the length of a circular arc can be used to analyze the motion of a particle moving at a *constant speed* along a circular path.

> **Linear and Angular Speed**
>
> Consider a particle moving at a constant speed along a circular arc of radius r. If s is the length of the arc traveled in time t, then the **linear speed** of the particle is
>
> $$\text{Linear speed} = \frac{\text{arc length}}{\text{time}} = \frac{s}{t}.$$
>
> Moreover, if θ is the angle (in radian measure) corresponding to the arc length s, then the **angular speed** of the particle is
>
> $$\text{Angular speed} = \frac{\text{central angle}}{\text{time}} = \frac{\theta}{t}.$$
>
> Linear speed measures how fast the particle moves, and angular speed measures how fast the angle changes.

Example 6 Finding Linear Speed

The second hand of a clock is 10.2 centimeters long, as shown in Figure 4.16. Find the linear speed of the tip of this second hand.

Solution

In one revolution, the arc length traveled is

$$s = 2\pi r$$

$$= 2\pi(10.2) \qquad \text{Substitute for } r.$$

$$= 20.4\pi \text{ centimeters.}$$

The time required for the second hand to travel this distance is

$$t = 1 \text{ minute} = 60 \text{ seconds.}$$

So, the linear speed of the tip of the second hand is

$$\text{Linear speed} = \frac{s}{t}$$

$$= \frac{20.4\pi \text{ centimeters}}{60 \text{ seconds}} \approx 1.07 \text{ centimeters per second.}$$

 Now try Exercise 96.

Figure 4.16

Example 7 Finding Angular and Linear Speed

A 15-inch diameter tire on a car makes 9.3 revolutions per second (see Figure 4.17).

a. Find the angular speed of the tire in radians per second.

b. Find the linear speed of the car.

Solution

a. Because each revolution generates 2π radians, it follows that the tire turns $(9.3)(2\pi) = 18.6\pi$ radians per second. In other words, the angular speed is

$$\text{Angular speed} = \frac{\theta}{t}$$

$$= \frac{18.6\pi \text{ radians}}{1 \text{ second}} = 18.6\pi \text{ radians per second.}$$

b. The linear speed of the tire and car is

$$\text{Linear speed} = \frac{s}{t} = \frac{r\theta}{t}$$

$$= \frac{\left(\frac{1}{2}\right)15(18.6\pi) \text{ inches}}{1 \text{ second}} \approx 438.25 \text{ inches per second.}$$

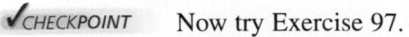 Now try Exercise 97.

Figure 4.17

4.1 Exercises

See www.CalcChat.com for worked-out solutions to odd-numbered exercises.

Vocabulary Check

Fill in the blanks.

1. _____ means "measurement of triangles."

2. An _____ is determined by rotating a ray about its endpoint.

3. An angle with its initial side coinciding with the positive x-axis and the origin as its vertex is said to be in _____ .

4. Two angles that have the same initial and terminal sides are _____ .

5. One _____ is the measure of a central angle that intercepts an arc equal in length to the radius of the circle.

6. Two positive angles that have a sum of $\pi/2$ are _____ angles.

7. Two positive angles that have a sum of π are _____ angles.

8. The angle measure that is equivalent to $\frac{1}{360}$ of a complete revolution about an angle's vertex is one _____ .

9. The _____ speed of a particle is the ratio of the arc length traveled to the time traveled.

10. The _____ speed of a particle is the ratio of the change in the central angle to the time.

In Exercises 1 and 2, estimate the angle to the nearest one-half radian.

1.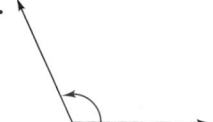

2.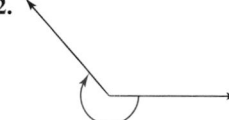

In Exercises 3–6, determine the quadrant in which each angle lies. (The angle measure is given in radians.)

3. (a) $\dfrac{7\pi}{4}$ (b) $\dfrac{11\pi}{4}$

4. (a) $-\dfrac{5\pi}{12}$ (b) $-\dfrac{13\pi}{9}$

5. (a) -1 (b) -2

6. (a) 3.5 (b) 2.25

In Exercises 7–10, sketch each angle in standard position.

7. (a) $\dfrac{3\pi}{4}$ (b) $\dfrac{4\pi}{3}$

8. (a) $-\dfrac{7\pi}{4}$ (b) $-\dfrac{5\pi}{2}$

9. (a) $\dfrac{11\pi}{6}$ (b) $\dfrac{2\pi}{3}$

10. (a) 4 (b) -3

In Exercises 11–14, determine two coterminal angles in radian measure (one positive and one negative) for each angle. (There are many correct answers).

11. (a) 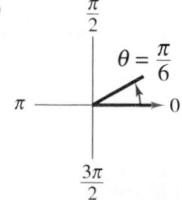   (b)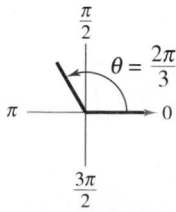

12. (a) 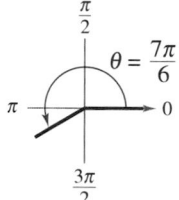   (b)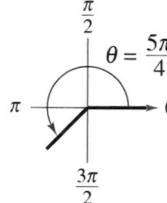

13. (a) $-\dfrac{9\pi}{4}$ (b) $-\dfrac{2\pi}{15}$

14. (a) $\dfrac{7\pi}{8}$ (b) $\dfrac{8\pi}{45}$

In Exercises 15–20, find (if possible) the complement and supplement of the angle.

15. $\dfrac{\pi}{3}$

16. $\dfrac{3\pi}{4}$

17. $\dfrac{\pi}{6}$

18. $\dfrac{2\pi}{3}$

19. 1

20. 2

In Exercises 21 and 22, estimate the number of degrees in the angle

21.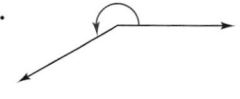

22.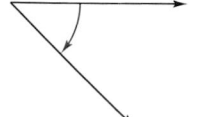

In Exercises 23–26, determine the quadrant in which each angle lies.

23. (a) 150° (b) 282°
24. (a) 87.9° (b) 8.5°
25. (a) −132° 50′ (b) −336° 30′
26. (a) −245.25° (b) −12.35°

In Exercises 27–30, sketch each angle in standard position.

27. (a) 30° (b) 150° 28. (a) −270° (b) −120°
29. (a) 405° (b) 780° 30. (a) −450° (b) −600°

In Exercises 31–34, determine two coterminal angles in degree measure (one positive and one negative) for each angle. (There are many correct answers).

31. (a) 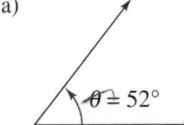   (b)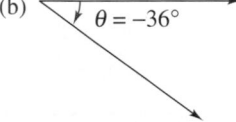

32. (a) 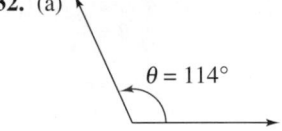   (b)

33. (a) 300° (b) 230°
34. (a) −445° (b) −740°

In Exercises 35–38, find (if possible) the complement and supplement of the angle.

35. 24° 36. 129°
37. 87° 38. 167°

In Exercises 39–42, rewrite each angle in radian measure as a multiple of π. (Do not use a calculator.)

39. (a) 30° (b) 150°
40. (a) 315° (b) 120°
41. (a) −20° (b) −240°
42. (a) −270° (b) 144°

In Exercises 43–46, rewrite each angle in degree measure. (Do not use a calculator.)

43. (a) $\dfrac{3\pi}{2}$ (b) $-\dfrac{7\pi}{6}$

44. (a) -4π (b) 3π

45. (a) $\dfrac{7\pi}{3}$ (b) $-\dfrac{13\pi}{60}$

46. (a) $-\dfrac{15\pi}{6}$ (b) $\dfrac{28\pi}{15}$

In Exercises 47–52, convert the angle measure from degrees to radians. Round your answer to three decimal places.

47. 115° 48. 83.7°
49. −216.35° 50. −46.52°
51. −0.78° 52. 395°

In Exercises 53–58, convert the angle measure from radians to degrees. Round your answer to three decimal places.

53. $\dfrac{\pi}{7}$ 54. $\dfrac{8\pi}{13}$

55. 6.5π 56. -4.2π
57. -2 58. -0.48

In Exercises 59–64, use the angle-conversion capabilities of a graphing utility to convert the angle measure to decimal degree form. Round your answer to three decimal places if necessary.

59. 64° 45′ 60. −124° 30′
61. 85° 18′ 30″ 62. −408° 16′ 25″
63. −125° 36″ 64. 330° 25″

In Exercises 65–70, use the angle-conversion capabilities of a graphing utility to convert the angle measure to D° M′ S″ form.

65. 280.6° 66. −115.8°
67. −345.12° 68. 310.75°
69. −0.355 70. 0.7865

In Exercises 71–74, find the angle in radians.

71. 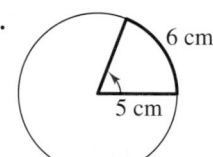   72.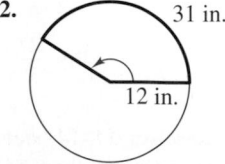

73. 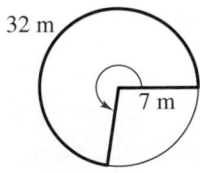   74.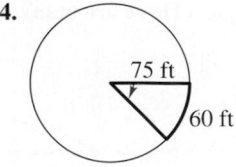

75. Find each angle (in radians) shown on the unit circle.

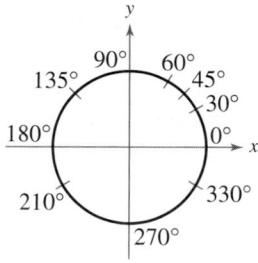

76. Find each angle (in degrees) shown on the unit circle.

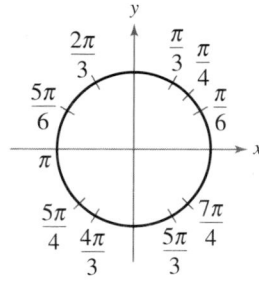

In Exercises 77–80, find the radian measure of the central angle of a circle of radius *r* that intercepts an arc of length *s*.

	Radius *r*	Arc Length *s*
77.	15 inches	8 inches
78.	22 feet	10 feet
79.	14.5 centimeters	35 centimeters
80.	80 kilometers	160 kilometers

In Exercises 81–84, find the length of the arc on a circle of radius *r* intercepted by a central angle *θ*.

	Radius *r*	Central Angle *θ*
81.	14 inches	180°
82.	9 feet	60°
83.	27 meters	$\dfrac{2\pi}{3}$ radians
84.	12 centimeters	$\dfrac{3\pi}{4}$ radians

In Exercises 85–88, find the radius *r* of a circle with an arc length *s* and a central angle *θ*.

	Arc Length *s*	Central Angle *θ*
85.	36 feet	$\dfrac{\pi}{2}$ radians
86.	3 meters	$\dfrac{4\pi}{3}$ radians
87.	82 miles	135°
88.	8 inches	330°

Distance **In Exercises 89 and 90, find the distance between the cities. Assume that Earth is a sphere of radius 4000 miles and the cities are on the same longitude (one city is due north of the other).**

	City	Latitude
89.	Miami	25° 46′ 32″ N
	Erie	42° 7′ 33″ N
90.	Johannesburg, South Africa	26° 8′ S
	Jerusalem, Israel	31° 46′ N

91. *Difference in Latitudes* Assuming that Earth is a sphere of radius 6378 kilometers, what is the difference in the latitudes of Syracuse, New York and Annapolis, Maryland, where Syracuse is 450 kilometers due north of Annapolis?

92. *Difference in Latitudes* Assuming that Earth is a sphere of radius 6378 kilometers, what is the difference in the latitudes of Lynchburg, Virginia and Myrtle Beach, South Carolina, where Lynchburg is 400 kilometers due north of Myrtle Beach?

93. *Instrumentation* A voltmeter's pointer is 6 centimeters in length (see figure). Find the number of degrees through which it rotates when it moves 2.5 centimeters on the scale.

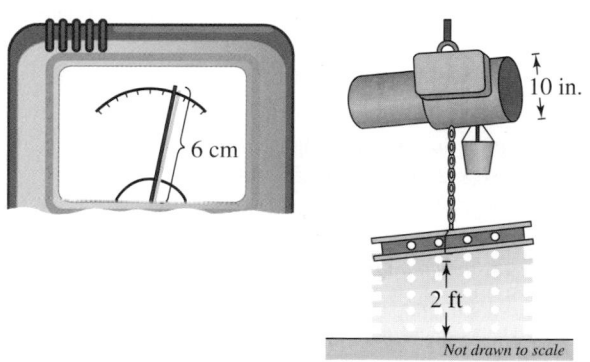

Figure for 93   Figure for 94

94. *Electric Hoist* An electric hoist is used to lift a piece of equipment 2 feet (see figure). The diameter of the drum on the hoist is 10 inches. Find the number of degrees through which the drum must rotate.

95. *Sports* The number of revolutions made by a figure skater for each type of axel jump is given. Determine the measure of the angle generated as the skater performs each jump. Give the answer in both degrees and radians.

(a) Single axel: $1\frac{1}{2}$

(b) Double axel: $2\frac{1}{2}$

(c) Triple axel: $3\frac{1}{2}$

96. Linear Speed A satellite in a circular orbit 1250 kilometers above Earth makes one complete revolution every 110 minutes. What is its linear speed? Assume that Earth is a sphere of radius 6400 kilometers.

97. Construction The circular blade on a saw has a diameter of 7.5 inches (see figure) and rotates at 2400 revolutions per minute.

←——— 7.5 in. ———→

(a) Find the angular speed in radians per second.

(b) Find the linear speed of the saw teeth (in feet per second) as they contact the wood being cut.

98. Construction The circular blade on a saw has a diameter of 7.25 inches and rotates at 4800 revolutions per minute.

(a) Find the angular speed of the blade in radians per second.

(b) Find the linear speed of the saw teeth (in feet per second) as they contact the wood being cut.

99. Angular Speed A computerized spin balance machine rotates a 25-inch diameter tire at 480 revolutions per minute.

(a) Find the road speed (in miles per hour) at which the tire is being balanced.

(b) At what rate should the spin balance machine be set so that the tire is being tested for 70 miles per hour?

100. Angular Speed A DVD is approximately 12 centimeters in diameter. The drive motor of the DVD player is controlled to rotate precisely between 200 and 500 revolutions per minute, depending on what track is being read.

(a) Find an interval for the angular speed of a disc as it rotates.

(b) Find the linear speed of a point on the outermost track as the disc rotates.

Synthesis

True or False? In Exercises 101–103, determine whether the statement is true or false. Justify your answer.

101. A degree is a larger unit of measure than a radian.

102. An angle that measures $-1260°$ lies in Quadrant III.

103. The angles of a triangle can have radian measures $2\pi/3$, $\pi/4$, and $\pi/12$.

104. Writing In your own words, explain the meanings of (a) an angle in standard position, (b) a negative angle, (c) coterminal angles, and (d) an obtuse angle.

105. Writing If the radius of a circle is increasing and the magnitude of a central angle is held constant, how is the length of the intercepted arc changing? Explain your reasoning.

106. Geometry Show that the area of a circular sector of radius r with central angle θ is $A = \frac{1}{2}r^2\theta$, where θ is measured in radians.

Geometry In Exercises 107 and 108, use the result of Exercise 106 to find the area of the sector.

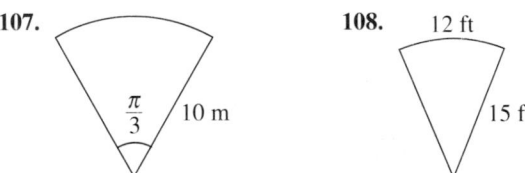

107. $\frac{\pi}{3}$ 10 m

108.   12 ft   15 ft

109. Graphical Reasoning The formulas for the area of a circular sector and arc length are $A = \frac{1}{2}r^2\theta$ and $s = r\theta$, respectively. (r is the radius and θ is the angle measured in radians.)

(a) If $\theta = 0.8$, write the area and arc length as functions of r. What is the domain of each function? Use a graphing utility to graph the functions. Use the graphs to determine which function changes more rapidly as r increases. Explain.

(b) If $r = 10$ centimeters, write the area and arc length as functions of θ. What is the domain of each function? Use a graphing utility to graph and identify the functions.

110. Writing A fan motor turns at a given angular speed. How does the angular speed of the tips of the blades change if a fan of greater diameter is installed on the motor? Explain.

111. Writing In your own words, write a definition for radian.

112. Writing In your own words, explain the difference between 1 radian and 1 degree.

Skills Review

Library of Parent Functions In Exercises 113–118, sketch the graph of $y = x^5$ and the specified transformation.

113. $f(x) = (x - 2)^5$

114. $f(x) = x^5 - 4$

115. $f(x) = 2 - x^5$

116. $f(x) = -(x + 3)^5$

117. $f(x) = (x + 1)^5 - 3$

118. $f(x) = (x - 5)^5 + 1$

4.2 Trigonometric Functions: The Unit Circle

The Unit Circle

The two historical perspectives of trigonometry incorporate different methods for introducing the trigonometric functions. Our first introduction to these functions is based on the unit circle.

Consider the **unit circle** given by

$$x^2 + y^2 = 1 \qquad \text{Unit circle}$$

as shown in Figure 4.18.

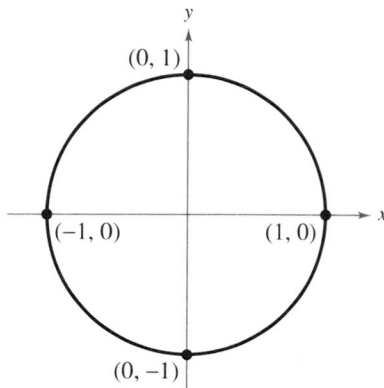

Figure 4.18

Imagine that the real number line is wrapped around this circle, with positive numbers corresponding to a counterclockwise wrapping and negative numbers corresponding to a clockwise wrapping, as shown in Figure 4.19.

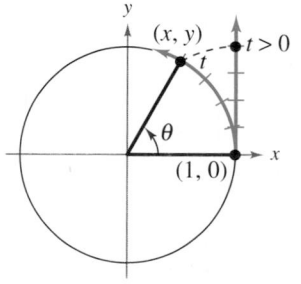 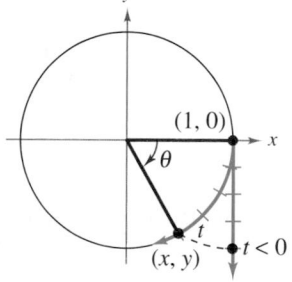

Figure 4.19

As the real number line is wrapped around the unit circle, each real number t corresponds to a point (x, y) on the circle. For example, the real number 0 corresponds to the point $(1, 0)$. Moreover, because the unit circle has a circumference of 2π, the real number 2π also corresponds to the point $(1, 0)$.

In general, each real number t also corresponds to a central angle θ (in standard position) whose radian measure is t. With this interpretation of t, the arc length formula $s = r\theta$ (with $r = 1$) indicates that the real number t is the length of the arc intercepted by the angle θ, given in radians.

The Trigonometric Functions

From the preceding discussion, it follows that the coordinates x and y are two functions of the real variable t. You can use these coordinates to define the six trigonometric functions of t.

sine	**cosecant**
cosine	**secant**
tangent	**cotangent**

These six functions are normally abbreviated sin, csc, cos, sec, tan, and cot, respectively.

Definitions of Trigonometric Functions

Let t be a real number and let (x, y) be the point on the unit circle corresponding to t.

$$\sin t = y \qquad\qquad \csc t = \frac{1}{y}, \quad y \neq 0$$

$$\cos t = x \qquad\qquad \sec t = \frac{1}{x}, \quad x \neq 0$$

$$\tan t = \frac{y}{x}, \quad x \neq 0 \qquad \cot t = \frac{x}{y}, \quad y \neq 0$$

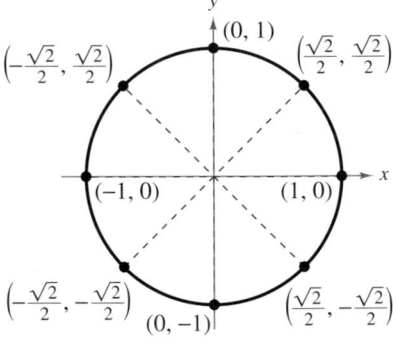

Figure 4.20

Note that the functions in the second column are the *reciprocals* of the corresponding functions in the first column.

In the definitions of the trigonometric functions, note that the tangent and secant are not defined when $x = 0$. For instance, because $t = \pi/2$ corresponds to $(x, y) = (0, 1)$, it follows that $\tan(\pi/2)$ and $\sec(\pi/2)$ are *undefined*. Similarly, the cotangent and cosecant are not defined when $y = 0$. For instance, because $t = 0$ corresponds to $(x, y) = (1, 0)$, cot 0 and csc 0 are *undefined*.

In Figure 4.20, the unit circle has been divided into eight equal arcs, corresponding to t-values of

$$0, \frac{\pi}{4}, \frac{\pi}{2}, \frac{3\pi}{4}, \pi, \frac{5\pi}{4}, \frac{3\pi}{2}, \frac{7\pi}{4}, \text{ and } 2\pi.$$

Similarly, in Figure 4.21, the unit circle has been divided into 12 equal arcs, corresponding to t-values of

$$0, \frac{\pi}{6}, \frac{\pi}{3}, \frac{\pi}{2}, \frac{2\pi}{3}, \frac{5\pi}{6}, \pi, \frac{7\pi}{6}, \frac{4\pi}{3}, \frac{3\pi}{2}, \frac{5\pi}{3}, \frac{11\pi}{6}, \text{ and } 2\pi.$$

Using the (x, y) coordinates in Figures 4.20 and 4.21, you can easily evaluate the exact values of trigonometric functions for common t-values. This procedure is demonstrated in Examples 1 and 2. You should study and learn these exact values for common t-values because they will help you in later sections to perform calculations quickly and easily.

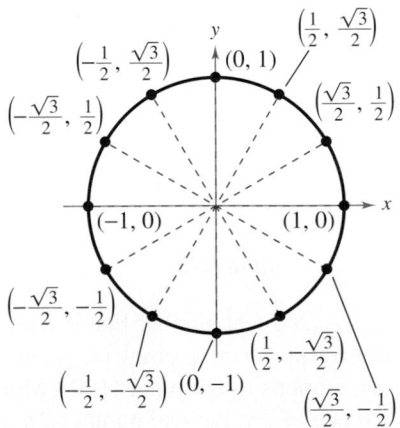

Figure 4.21

Example 1   Evaluating Trigonometric Functions

Evaluate the six trigonometric functions at each real number.

a. $t = \dfrac{\pi}{6}$ **b.** $t = \dfrac{5\pi}{4}$ **c.** $t = 0$ **d.** $t = \pi$

Solution

For each t-value, begin by finding the corresponding point (x, y) on the unit circle. Then use the definitions of trigonometric functions listed on page 270.

a. $t = \pi/6$ corresponds to the point $(x, y) = \left(\sqrt{3}/2,\, 1/2\right)$.

$$\sin\frac{\pi}{6} = y = \frac{1}{2} \qquad\qquad \csc\frac{\pi}{6} = \frac{1}{y} = \frac{1}{1/2} = 2$$

$$\cos\frac{\pi}{6} = x = \frac{\sqrt{3}}{2} \qquad\qquad \sec\frac{\pi}{6} = \frac{1}{x} = \frac{2}{\sqrt{3}} = \frac{2\sqrt{3}}{3}$$

$$\tan\frac{\pi}{6} = \frac{y}{x} = \frac{1/2}{\sqrt{3}/2} = \frac{1}{\sqrt{3}} = \frac{\sqrt{3}}{3} \qquad\qquad \cot\frac{\pi}{6} = \frac{x}{y} = \frac{\sqrt{3}/2}{1/2} = \sqrt{3}$$

b. $t = 5\pi/4$ corresponds to the point $(x, y) = \left(-\sqrt{2}/2,\, -\sqrt{2}/2\right)$.

$$\sin\frac{5\pi}{4} = y = -\frac{\sqrt{2}}{2} \qquad\qquad \csc\frac{5\pi}{4} = \frac{1}{y} = -\frac{2}{\sqrt{2}} = -\sqrt{2}$$

$$\cos\frac{5\pi}{4} = x = -\frac{\sqrt{2}}{2} \qquad\qquad \sec\frac{5\pi}{4} = \frac{1}{x} = -\frac{2}{\sqrt{2}} = -\sqrt{2}$$

$$\tan\frac{5\pi}{4} = \frac{y}{x} = \frac{-\sqrt{2}/2}{-\sqrt{2}/2} = 1 \qquad\qquad \cot\frac{5\pi}{4} = \frac{x}{y} = \frac{-\sqrt{2}/2}{-\sqrt{2}/2} = 1$$

c. $t = 0$ corresponds to the point $(x, y) = (1, 0)$.

$$\sin 0 = y = 0 \qquad\qquad \csc 0 = \frac{1}{y} \text{ is undefined.}$$

$$\cos 0 = x = 1 \qquad\qquad \sec 0 = \frac{1}{x} = \frac{1}{1} = 1$$

$$\tan 0 = \frac{y}{x} = \frac{0}{1} = 0 \qquad\qquad \cot 0 = \frac{x}{y} \text{ is undefined.}$$

d. $t = \pi$ corresponds to the point $(x, y) = (-1, 0)$.

$$\sin \pi = y = 0 \qquad\qquad \csc \pi = \frac{1}{y} \text{ is undefined.}$$

$$\cos \pi = x = -1 \qquad\qquad \sec \pi = \frac{1}{x} = \frac{1}{-1} = -1$$

$$\tan \pi = \frac{y}{x} = \frac{0}{-1} = 0 \qquad\qquad \cot \pi = \frac{x}{y} \text{ is undefined.}$$

✓CHECKPOINT   Now try Exercise 31.

Example 2   Evaluating Trigonometric Functions

Evaluate the six trigonometric functions at $t = -\dfrac{\pi}{3}$.

Solution

Moving *clockwise* around the unit circle, it follows that $t = -\pi/3$ corresponds to the point $(x, y) = \left(1/2, -\sqrt{3}/2\right)$.

$$\sin\left(-\frac{\pi}{3}\right) = -\frac{\sqrt{3}}{2} \qquad\qquad \csc\left(-\frac{\pi}{3}\right) = -\frac{2}{\sqrt{3}} = -\frac{2\sqrt{3}}{3}$$

$$\cos\left(-\frac{\pi}{3}\right) = \frac{1}{2} \qquad\qquad \sec\left(-\frac{\pi}{3}\right) = 2$$

$$\tan\left(-\frac{\pi}{3}\right) = \frac{-\sqrt{3}/2}{1/2} = -\sqrt{3} \quad \cot\left(-\frac{\pi}{3}\right) = \frac{1/2}{-\sqrt{3}/2} = -\frac{1}{\sqrt{3}} = -\frac{\sqrt{3}}{3}$$

✓CHECKPOINT   Now try Exercise 35.

Domain and Period of Sine and Cosine

The *domain* of the sine and cosine functions is the set of all real numbers. To determine the *range* of these two functions, consider the unit circle shown in Figure 4.22. Because $r = 1$, it follows that $\sin t = y$ and $\cos t = x$. Moreover, because (x, y) is on the unit circle, you know that $-1 \le y \le 1$ and $-1 \le x \le 1$. So, the values of sine and cosine also range between -1 and 1.

$$\begin{array}{cc} -1 \le\ y\ \le 1 & -1 \le\ x\ \le 1 \\ & \text{and} \\ -1 \le \sin t \le 1 & -1 \le \cos t \le 1 \end{array}$$

Adding 2π to each value of t in the interval $[0, 2\pi]$ completes a second revolution around the unit circle, as shown in Figure 4.23. The values of $\sin(t + 2\pi)$ and $\cos(t + 2\pi)$ correspond to those of $\sin t$ and $\cos t$. Similar results can be obtained for repeated revolutions (positive or negative) around the unit circle. This leads to the general result

$$\sin(t + 2\pi n) = \sin t$$

and

$$\cos(t + 2\pi n) = \cos t$$

for any integer n and real number t. Functions that behave in such a repetitive (or cyclic) manner are called **periodic.**

Definition of a Periodic Function

A function f is **periodic** if there exists a positive real number c such that

$$f(t + c) = f(t)$$

for all t in the domain of f. The least number c for which f is periodic is called the **period** of f.

Exploration

With your graphing utility in *radian* and *parametric* modes, enter X1T = cos T and Y1T = sin T and use the following settings.

Tmin = 0, Tmax = 6.3,
Tstep = 0.1
Xmin = −1.5, Xmax = 1.5,
Xscl = 1
Ymin = -1, Ymax = 1,
Yscl = 1

1. Graph the entered equations and describe the graph.

2. Use the *trace* feature to move the cursor around the graph. What do the *t*-values represent? What do the *x*- and *y*-values represent?

3. What are the least and greatest values for *x* and *y*?

(0, 1) Unit circle

(−1, 0)   (1, 0) $-1 \le y \le 1$

(0, −1)

$-1 \le x \le 1$

Figure 4.22

$t = \frac{\pi}{2}, \frac{\pi}{2} + 2\pi, \frac{\pi}{2} + 4\pi, \ldots$

$t = \frac{3\pi}{4}, \frac{3\pi}{4} + 2\pi, \ldots$ $t = \frac{\pi}{4}, \frac{\pi}{4} + 2\pi, \ldots$

$t = \pi, 3\pi, \ldots$

$t = 0, 2\pi, \ldots$

$t = \frac{5\pi}{4}, \frac{5\pi}{4} + 2\pi, \ldots$ $t = \frac{7\pi}{4}, \frac{7\pi}{4} + 2\pi, \ldots$

$t = \frac{3\pi}{2}, \frac{3\pi}{2} + 2\pi, \frac{3\pi}{2} + 4\pi, \ldots$

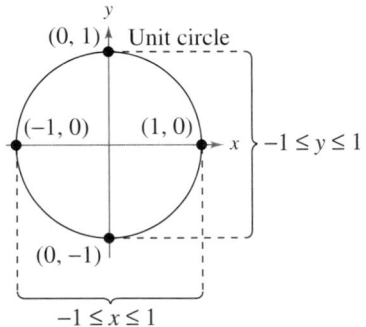

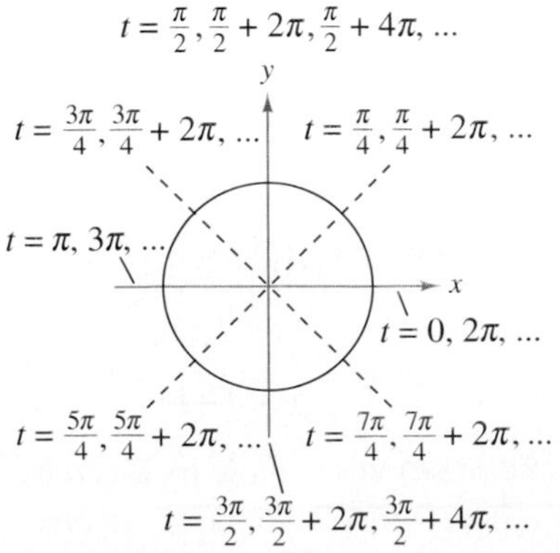

Figure 4.23

Even and Odd Trigonometric Functions

The cosine and secant functions are *even*.

$$\cos(-t) = \cos t \qquad \sec(-t) = \sec t$$

The sine, cosecant, tangent, and cotangent functions are *odd*.

$$\sin(-t) = -\sin t \qquad \csc(-t) = -\csc t$$

$$\tan(-t) = -\tan t \qquad \cot(-t) = -\cot t$$

Prerequisite Skills

To review even and odd functions, see Section 1.3.

Example 3 Using the Period to Evaluate the Sine and Cosine

a. Because $\dfrac{13\pi}{6} = 2\pi + \dfrac{\pi}{6}$, you have

$$\sin\frac{13\pi}{6} = \sin\left(2\pi + \frac{\pi}{6}\right) = \sin\frac{\pi}{6} = \frac{1}{2}.$$

b. Because $-\dfrac{7\pi}{2} = -4\pi + \dfrac{\pi}{2}$, you have

$$\cos\left(-\frac{7\pi}{2}\right) = \cos\left(-4\pi + \frac{\pi}{2}\right) = \cos\frac{\pi}{2} = 0.$$

c. For $\sin t = \dfrac{4}{5}$, $\sin(-t) = -\dfrac{4}{5}$ because the function is odd.

✓**CHECKPOINT**   Now try Exercise 39.

STUDY TIP

It follows from the definition of periodic function that the sine and cosine functions are periodic and have a period of 2π. The other four trigonometric functions are also periodic, and will be discussed further in Section 4.6.

Evaluating Trigonometric Functions with a Calculator

When evaluating a trigonometric function with a calculator, you need to set the calculator to the desired *mode* of measurement (degrees or radians). Most calculators do not have keys for the cosecant, secant, and cotangent functions. To evaluate these functions, you can use the $\boxed{x^{-1}}$ key with their respective reciprocal functions sine, cosine, and tangent. For example, to evaluate $\csc(\pi/8)$, use the fact that

$$\csc\frac{\pi}{8} = \frac{1}{\sin(\pi/8)}$$

and enter the following keystroke sequence in *radian* mode.

$\boxed{(}\ \boxed{\text{SIN}}\ \boxed{(}\ \pi \boxed{\div} 8 \boxed{)}\ \boxed{)}\ \boxed{x^{-1}}\ \boxed{\text{ENTER}}$ Display 2.6131259

Example 4 Using a Calculator

Function	Mode	Graphing Calculator Keystrokes	Display
a. $\sin 2\pi/3$	Radian	$\boxed{\text{SIN}}\ \boxed{(}\ 2\ \boxed{\pi}\ \boxed{\div}\ 3\ \boxed{)}\ \boxed{\text{ENTER}}$	0.8660254
b. $\cot 1.5$	Radian	$\boxed{(}\ \boxed{\text{TAN}}\ \boxed{(}\ 1.5\ \boxed{)}\ \boxed{)}\ \boxed{x^{-1}}\ \boxed{\text{ENTER}}$	0.0709148

✓**CHECKPOINT**   Now try Exercise 55.

TECHNOLOGY TIP

When evaluating trigonometric functions with a calculator, remember to enclose all fractional angle measures in parentheses. For instance, if you want to evaluate $\sin\theta$ for $\theta = \pi/6$, you should enter

$\boxed{\text{SIN}}\ \boxed{(}\ \boxed{\pi}\ \boxed{\div}\ 6\ \boxed{)}\ \boxed{\text{ENTER}}$.

These keystrokes yield the correct value of 0.5.

4.2 Exercises

See www.CalcChat.com for worked-out solutions to odd-numbered exercises.

Vocabulary Check

Fill in the blanks.

1. Each real number t corresponds to a point (x, y) on the _____ .

2. A function f is _____ if there exists a positive real number c such that $f(t + c) = f(t)$ for all t in the domain of f.

3. A function f is _____ if $f(-t) = -f(t)$ and _____ if $f(-t) = f(t)$.

In Exercises 1–4, determine the exact values of the six trigonometric functions of the angle θ.

1.

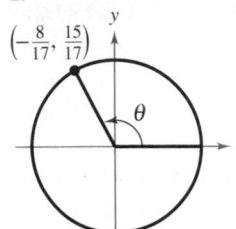

2.

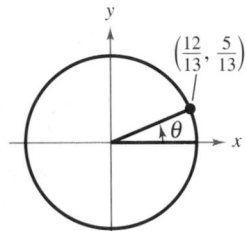

3.

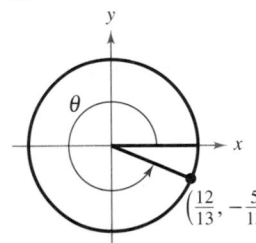

4.

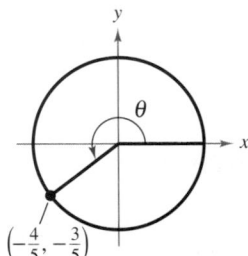

In Exercises 5–16, find the point (x, y) on the unit circle that corresponds to the real number t.

5. $t = \dfrac{\pi}{4}$

6. $t = \dfrac{\pi}{3}$

7. $t = \dfrac{7\pi}{6}$

8. $t = \dfrac{5\pi}{4}$

9. $t = \dfrac{2\pi}{3}$

10. $t = \dfrac{5\pi}{3}$

11. $t = \dfrac{3\pi}{2}$

12. $t = \pi$

13. $t = -\dfrac{7\pi}{4}$

14. $t = -\dfrac{4\pi}{3}$

15. $t = -\dfrac{3\pi}{2}$

16. $t = -2\pi$

In Exercises 17–30, evaluate (if possible) the sine, cosine, and tangent of the real number.

17. $t = \dfrac{\pi}{4}$

18. $t = \dfrac{\pi}{3}$

19. $t = \dfrac{7\pi}{6}$

20. $t = -\dfrac{5\pi}{4}$

21. $t = \dfrac{2\pi}{3}$

22. $t = \dfrac{5\pi}{3}$

23. $t = -\dfrac{5\pi}{3}$

24. $t = \dfrac{11\pi}{6}$

25. $t = -\dfrac{\pi}{6}$

26. $t = -\dfrac{3\pi}{4}$

27. $t = -\dfrac{7\pi}{4}$

28. $t = -\dfrac{4\pi}{3}$

29. $t = -\dfrac{3\pi}{2}$

30. $t = -2\pi$

In Exercises 31–36, evaluate (if possible) the six trigonometric functions of the real number.

31. $t = \dfrac{3\pi}{4}$

32. $t = \dfrac{5\pi}{6}$

33. $t = \dfrac{\pi}{2}$

34. $t = \dfrac{3\pi}{2}$

35. $t = -\dfrac{2\pi}{3}$

36. $t = -\dfrac{7\pi}{4}$

In Exercises 37–44, evaluate the trigonometric function using its period as an aid.

37. $\sin 5\pi$

38. $\cos 7\pi$

39. $\cos \dfrac{8\pi}{3}$

40. $\sin \dfrac{9\pi}{4}$

41. $\cos\left(-\dfrac{13\pi}{6}\right)$

42. $\sin\left(-\dfrac{19\pi}{6}\right)$

43. $\sin\left(-\dfrac{9\pi}{4}\right)$

44. $\cos\left(-\dfrac{8\pi}{3}\right)$

In Exercises 45–50, use the value of the trigonometric function to evaluate the indicated functions.

45. $\sin t = \frac{1}{3}$

  (a) $\sin(-t)$

  (b) $\csc(-t)$

46. $\cos t = -\frac{3}{4}$

  (a) $\cos(-t)$

  (b) $\sec(-t)$

47. $\cos(-t) = -\frac{1}{5}$

  (a) $\cos t$

  (b) $\sec(-t)$

48. $\sin(-t) = \frac{3}{8}$

  (a) $\sin t$

  (b) $\csc t$

49. $\sin t = \frac{4}{5}$

  (a) $\sin(\pi - t)$

  (b) $\sin(t + \pi)$

50. $\cos t = \frac{4}{5}$

  (a) $\cos(\pi - t)$

  (b) $\cos(t + \pi)$

In Exercises 51–68, use a calculator to evaluate the trigonometric expression. Round your answer to four decimal places.

51. $\sin \dfrac{7\pi}{9}$

52. $\tan \dfrac{2\pi}{5}$

53. $\cos \dfrac{11\pi}{5}$

54. $\sin \dfrac{11\pi}{9}$

55. $\csc 1.3$

56. $\cot 3.7$

57. $\cos(-1.7)$

58. $\cos(-2.5)$

59. $\csc 0.8$

60. $\sec 1.8$

61. $\sec 22.8$

62. $\sin(-13.4)$

63. $\cot 2.5$

64. $\tan 1.75$

65. $\csc(-1.5)$

66. $\tan(-2.25)$

67. $\sec(-4.5)$

68. $\csc(-5.2)$

Estimation   **In Exercises 69 and 70, use the figure and a straightedge to approximate the value of each trigonometric function. Check your approximation using a graphing utility. To print an enlarged copy of the graph, go to the website *www.mathgraphs.com*.**

69. (a) $\sin 5$ (b) $\cos 2$

70. (a) $\sin 0.75$ (b) $\cos 2.5$

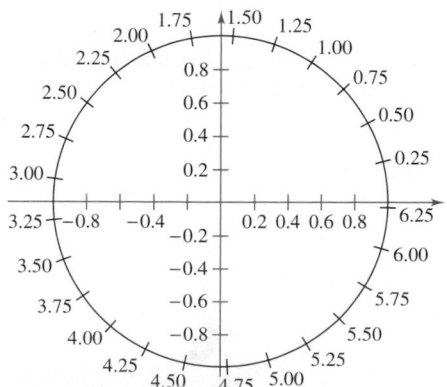

Figure for 69–72

Estimation   **In Exercises 71 and 72, use the figure and a straightedge to approximate the solution of each equation, where $0 \le t < 2\pi$. Check your approximation using a graphing utility. To print an enlarged copy of the graph, go to the website *www.mathgraphs.com*.**

71. (a) $\sin t = 0.25$ (b) $\cos t = -0.25$

72. (a) $\sin t = -0.75$ (b) $\cos t = 0.75$

73. *Electrical Circuits*   The initial current and charge in an electrical circuit are zero. The current when 100 volts is applied to the circuit is given by

$$I = 5e^{-2t} \sin t$$

where the resistance, inductance, and capacitance are 80 ohms, 20 henrys, and 0.01 farad, respectively. Approximate the current (in amperes) $t = 0.7$ second after the voltage is applied.

74. *Electrical Circuits*   Approximate the current (in amperes) in the electrical circuit in Exercise 73 $t = 1.4$ seconds after the voltage is applied.

75. *Harmonic Motion*   The displacement from equilibrium of an oscillating weight suspended by a spring is given by

$$y(t) = \frac{1}{4} \cos 6t$$

where y is the displacement (in feet) and t is the time (in seconds) (see figure). Find the displacement when (a) $t = 0$, (b) $t = \frac{1}{4}$, and (c) $t = \frac{1}{2}$.

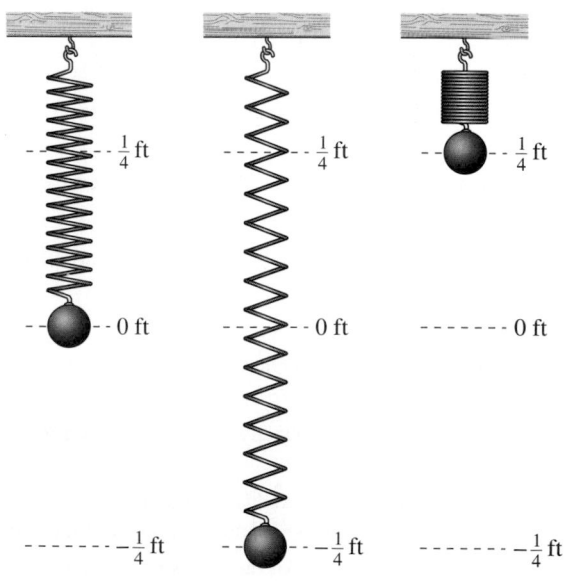

Equilibrium   Maximum negative displacement   Maximum positive displacement

Simple Harmonic Motion

76. *Harmonic Motion* The displacement from equilibrium of an oscillating weight suspended by a spring and subject to the damping effect of friction is given by

$$y(t) = \tfrac{1}{4} e^{-t} \cos 6t$$

where y is the displacement (in feet) and t is the time (in seconds).

(a) What is the initial displacement ($t = 0$)?

(b) Use a graphing utility to complete the table.

t	0.50	1.02	1.54	2.07	2.59
y					

(c) The approximate times when the weight is at its maximum distance from equilibrium are shown in the table in part (b). Explain why the magnitude of the maximum displacement is decreasing. What causes this decrease in maximum displacement in the physical system? What factor in the model measures this decrease?

(d) Find the first two times that the weight is at the equilibrium point ($y = 0$).

Synthesis

True or False? In Exercises 77–80, determine whether the statement is true or false. Justify your answer.

77. Because $\sin(-t) = -\sin t$, it can be said that the sine of a negative angle is a negative number.

78. $\sin a = \sin(a - 6\pi)$

79. The real number 0 corresponds to the point $(0, 1)$ on the unit circle.

80. $\cos\left(-\dfrac{7\pi}{2}\right) = \cos\left(\pi + \dfrac{\pi}{2}\right)$

81. *Exploration* Let (x_1, y_1) and (x_2, y_2) be points on the unit circle corresponding to $t = t_1$ and $t = \pi - t_1$, respectively.

(a) Identify the symmetry of the points (x_1, y_1) and (x_2, y_2).

(b) Make a conjecture about any relationship between $\sin t_1$ and $\sin(\pi - t_1)$.

(c) Make a conjecture about any relationship between $\cos t_1$ and $\cos(\pi - t_1)$.

82. *Exploration* Let (x_1, y_1) and (x_2, y_2) be points on the unit circle corresponding to $t = t_1$ and $t = t_1 + \pi$, respectively.

(a) Identify the symmetry of the points (x_1, y_1) and (x_2, y_2).

(b) Make a conjecture about any relationship between $\sin t_1$ and $\sin(t_1 + \pi)$.

(c) Make a conjecture about any relationship between $\cos t_1$ and $\cos(t_1 + \pi)$.

83. Verify that $\cos 2t \neq 2 \cos t$ by approximating $\cos 1.5$ and $2 \cos 0.75$.

84. Verify that $\sin(t_1 + t_2) \neq \sin t_1 + \sin t_2$ by approximating $\sin 0.25$, $\sin 0.75$, and $\sin 1$.

85. Use the unit circle to verify that the cosine and secant functions are even.

86. Use the unit circle to verify that the sine, cosecant, tangent, and cotangent functions are odd.

87. *Think About It* Because $f(t) = \sin t$ is an odd function and $g(t) = \cos t$ is an even function, what can be said about the function $h(t) = f(t)g(t)$?

88. *Think About It* Because $f(t) = \sin t$ and $g(t) = \tan t$ are odd functions, what can be said about the function $h(t) = f(t)g(t)$?

Skills Review

In Exercises 89–92, find the inverse function f^{-1} of the one-to-one function f. Verify by using a graphing utility to graph both f and f^{-1} in the same viewing window.

89. $f(x) = \tfrac{1}{2}(3x - 2)$

90. $f(x) = \tfrac{1}{4}x^3 + 1$

91. $f(x) = \sqrt{x^2 - 4}, \quad x \geq 2$

92. $f(x) = \dfrac{2x}{x + 1}, \quad x > -1$

In Exercises 93–96, sketch the graph of the rational function by hand. Show all asymptotes. Use a graphing utility to verify your graph.

93. $f(x) = \dfrac{2x}{x - 3}$

94. $f(x) = \dfrac{5x}{x^2 + x - 6}$

95. $f(x) = \dfrac{x^2 + 3x - 10}{2x^2 - 8}$

96. $f(x) = \dfrac{x^3 - 6x^2 + x - 1}{2x^2 - 5x - 8}$

In Exercises 97–100, identify the domain, any intercepts, and any asymptotes of the function.

97. $y = x^2 + 3x - 4$

98. $y = \ln x^4$

99. $f(x) = 3^{x+1} + 2$

100. $f(x) = \dfrac{x - 7}{x^2 + 4x + 4}$

Example 3   Evaluating Trigonometric Functions of 30° and 60°

Use the equilateral triangle shown in Figure 4.27 to find the exact values of sin 60°, cos 60°, sin 30°, and cos 30°.

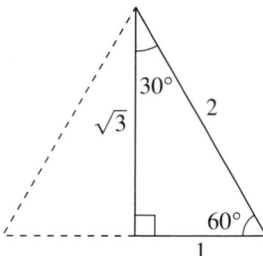

Figure 4.27

Solution

Use the Pythagorean Theorem and the equilateral triangle to verify the lengths of the sides given in Figure 4.27. For $\theta = 60°$, you have adj $= 1$, opp $= \sqrt{3}$, and hyp $= 2$. So,

$$\sin 60° = \frac{\text{opp}}{\text{hyp}} = \frac{\sqrt{3}}{2} \quad \text{and} \quad \cos 60° = \frac{\text{adj}}{\text{hyp}} = \frac{1}{2}.$$

For $\theta = 30°$, adj $= \sqrt{3}$, opp $= 1$, and hyp $= 2$. So,

$$\sin 30° = \frac{\text{opp}}{\text{hyp}} = \frac{1}{2} \quad \text{and} \quad \cos 30° = \frac{\text{adj}}{\text{hyp}} = \frac{\sqrt{3}}{2}.$$

✓CHECKPOINT   Now try Exercise 19.

Sines, Cosines, and Tangents of Special Angles

$$\sin 30° = \sin \frac{\pi}{6} = \frac{1}{2} \qquad \cos 30° = \cos \frac{\pi}{6} = \frac{\sqrt{3}}{2} \qquad \tan 30° = \tan \frac{\pi}{6} = \frac{\sqrt{3}}{3}$$

$$\sin 45° = \sin \frac{\pi}{4} = \frac{\sqrt{2}}{2} \qquad \cos 45° = \cos \frac{\pi}{4} = \frac{\sqrt{2}}{2} \qquad \tan 45° = \tan \frac{\pi}{4} = 1$$

$$\sin 60° = \sin \frac{\pi}{3} = \frac{\sqrt{3}}{2} \qquad \cos 60° = \cos \frac{\pi}{3} = \frac{1}{2} \qquad \tan 60° = \tan \frac{\pi}{3} = \sqrt{3}$$

In the box, note that $\sin 30° = \frac{1}{2} = \cos 60°$. This occurs because 30° and 60° are complementary angles, and, in general, it can be shown from the right triangle definitions that *cofunctions of complementary angles are equal*. That is, if θ is an acute angle, the following relationships are true.

$$\sin(90° - \theta) = \cos \theta \qquad \cos(90° - \theta) = \sin \theta$$

$$\tan(90° - \theta) = \cot \theta \qquad \cot(90° - \theta) = \tan \theta$$

$$\sec(90° - \theta) = \csc \theta \qquad \csc(90° - \theta) = \sec \theta$$

Trigonometric Identities

In trigonometry, a great deal of time is spent studying relationships between trigonometric functions (identities).

Exploration

Select a number t and use your graphing utility to calculate $(\sin t)^2 + (\cos t)^2$. Repeat this experiment for other values of t and explain why the answer is always the same. Is the result true in both *radian* and *degree* modes?

Fundamental Trigonometric Identities

Reciprocal Identities

$$\sin \theta = \frac{1}{\csc \theta} \qquad \cos \theta = \frac{1}{\sec \theta} \qquad \tan \theta = \frac{1}{\cot \theta}$$

$$\csc \theta = \frac{1}{\sin \theta} \qquad \sec \theta = \frac{1}{\cos \theta} \qquad \cot \theta = \frac{1}{\tan \theta}$$

Quotient Identities

$$\tan \theta = \frac{\sin \theta}{\cos \theta} \qquad \cot \theta = \frac{\cos \theta}{\sin \theta}$$

Pythagorean Identities

$$\sin^2 \theta + \cos^2 \theta = 1$$

$$1 + \tan^2 \theta = \sec^2 \theta$$

$$1 + \cot^2 \theta = \csc^2 \theta$$

Note that $\sin^2 \theta$ represents $(\sin \theta)^2$, $\cos^2 \theta$ represents $(\cos \theta)^2$, and so on.

Example 4   Applying Trigonometric Identities

Let θ be an acute angle such that $\cos \theta = 0.8$. Find the values of (a) $\sin \theta$ and (b) $\tan \theta$ using trigonometric identities.

Solution

a. To find the value of $\sin \theta$, use the Pythagorean identity

$$\sin^2 \theta + \cos^2 \theta = 1.$$

So, you have

$$\sin^2 \theta + (0.8)^2 = 1 \qquad \text{Substitute 0.8 for } \cos \theta.$$

$$\sin^2 \theta = 1 - (0.8)^2 = 0.36 \qquad \text{Subtract } (0.8)^2 \text{ from each side.}$$

$$\sin \theta = \sqrt{0.36} = 0.6. \qquad \text{Extract positive square root.}$$

b. Now, knowing the sine and cosine of θ, you can find the tangent of θ to be

$$\tan \theta = \frac{\sin \theta}{\cos \theta} = \frac{0.6}{0.8} = 0.75.$$

Use the definitions of $\cos \theta$ and $\tan \theta$ and the triangle shown in Figure 4.28 to check these results.

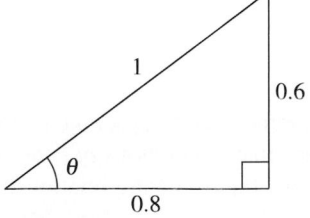

Figure 4.28

✓CHECKPOINT   Now try Exercise 45.

Example 5   Using Trigonometric Identities

Use trigonometric identities to transform one side of the equation into the other $(0 < \theta < \pi/2)$.

a. $\cos \theta \sec \theta = 1$ **b.** $(\sec \theta + \tan \theta)(\sec \theta - \tan \theta) = 1$

Solution

Simplify the expression on the left-hand side of the equation until you obtain the right-hand side.

a. $\cos \theta \sec \theta = \left(\dfrac{1}{\cancel{\sec \theta}} \right) \cancel{\sec} \theta$ Reciprocal identity

$\qquad\qquad = 1$ Divide out common factor.

b. $(\sec \theta + \tan \theta)(\sec \theta - \tan \theta)$

$\qquad = \sec^2 \theta - \sec\theta \tan \theta + \sec \theta \tan \theta - \tan^2 \theta$ Distributive Property

$\qquad = \sec^2\theta - \tan^2\theta$ Simplify.

$\qquad = 1$ Pythagorean identity

✓**CHECKPOINT**   Now try Exercise 55.

Evaluating Trigonometric Functions with a Calculator

To use a calculator to evaluate trigonometric functions of angles measured in degrees, set the calculator to *degree* mode and then proceed as demonstrated in Section 4.2.

Prerequisite Skills

To review evaluating trigonometric functions, see Section 4.2

Example 6   Using a Calculator

Use a calculator to evaluate $\sec(5° \, 40' \, 12'')$.

Solution

Begin by converting to decimal degree form.

$$5° \, 40' \, 12'' = 5° + \left(\frac{40}{60} \right)° + \left(\frac{12}{3600} \right)° = 5.67°$$

Then use a calculator in *degree* mode to evaluate sec 5.67°.

	Function	*Graphing Calculator Keystrokes*	*Display*
$\sec(5°40'12'') = \sec 5.67°$		⟮ COS ⟮ 5.67 ⟯ ⟯ x⁻¹ ENTER	1.0049166

✓**CHECKPOINT**   Now try Exercise 59.

STUDY TIP

Remember that throughout this text, it is assumed that angles are measured in radians unless noted otherwise. For example, sin 1 means the sine of 1 radian and sin 1° means the sine of 1 degree.

Applications Involving Right Triangles

Many applications of trigonometry involve a process called **solving right triangles.** In this type of application, you are usually given one side of a right triangle and one of the acute angles and are asked to find one of the other sides, *or* you are given two sides and are asked to find one of the acute angles.

In Example 7, the angle you are given is the **angle of elevation,** which represents the angle from the horizontal upward to the object. In other applications you may be given the **angle of depression,** which represents the angle from the horizontal downward to the object.

Example 7 Using Trigonometry to Solve a Right Triangle

A surveyor is standing 50 feet from the base of a large tree, as shown in Figure 4.29. The surveyor measures the angle of elevation to the top of the tree as 71.5°. How tall is the tree?

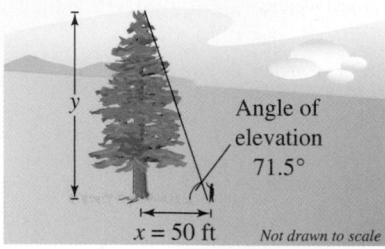

Angle of elevation 71.5°

$x = 50$ ft   *Not drawn to scale*

Figure 4.29

Solution

From Figure 4.29, you can see that

$$\tan 71.5° = \frac{\text{opp}}{\text{adj}} = \frac{y}{x}$$

where $x = 50$ and y is the height of the tree. So, the height of the tree is

$$y = x \tan 71.5°$$

$$\approx 50 \tan 71.5°$$

$$\approx 149.43 \text{ feet.}$$

 **CHECKPOINT** Now try Exercise 77.

Example 8 Using Trigonometry to Solve a Right Triangle

You are 200 yards from a river. Rather than walking directly to the river, you walk 400 yards along a straight path to the river's edge. Find the acute angle θ between this path and the river's edge, as illustrated in Figure 4.30.

Solution

From Figure 4.30, you can see that the sine of the angle θ is

$$\sin \theta = \frac{\text{opp}}{\text{hyp}} = \frac{200}{400} = \frac{1}{2}.$$

Now you should recognize that $\theta = 30°$.

 **CHECKPOINT** Now try Exercise 81.

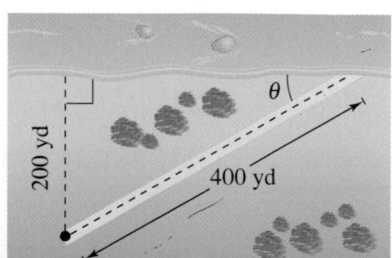

Figure 4.30

In Example 8, you were able to recognize that $\theta = 30°$ is the acute angle that satisfies the equation $\sin \theta = \frac{1}{2}$. Suppose, however, that you were given the equation $\sin \theta = 0.6$ and were asked to find the acute angle θ. Because

$$\sin 30° = \frac{1}{2} = 0.5000$$

and

$$\sin 45° = \frac{1}{\sqrt{2}} \approx 0.7071,$$

you might guess that θ lies somewhere between $30°$ and $45°$. In a later section, you will study a method by which a more precise value of θ can be determined.

TECHNOLOGY TIP

Calculators and graphing utilities have both *degree* and *radian* modes. As you progress through this chapter, be sure you use the correct mode.

Example 9   Solving a Right Triangle

Find the length c of the skateboard ramp shown in Figure 4.31.

Figure 4.31

Solution

From Figure 4.31, you can see that

$$\sin 18.4° = \frac{\text{opp}}{\text{hyp}}$$

$$= \frac{4}{c}.$$

So, the length of the ramp is

$$c = \frac{4}{\sin 18.4°}$$

$$\approx \frac{4}{0.3156}$$

$$\approx 12.67 \text{ feet.}$$

✓CHECKPOINT   Now try Exercise 82.

4.3 Exercises

See www.CalcChat.com for worked-out solutions to odd-numbered exercises.

Vocabulary Check

1. Match the trigonometric function with its right triangle definition.

(a) sine   (b) cosine   (c) tangent

(d) cosecant   (e) secant   (f) cotangent

(i) $\dfrac{\text{hyp}}{\text{adj}}$ (ii) $\dfrac{\text{opp}}{\text{adj}}$ (iii) $\dfrac{\text{opp}}{\text{hyp}}$

(iv) $\dfrac{\text{adj}}{\text{opp}}$ (v) $\dfrac{\text{hyp}}{\text{opp}}$ (vi) $\dfrac{\text{adj}}{\text{hyp}}$

In Exercises 2 and 3, fill in the blanks.

2. Relative to the acute angle θ, the three sides of a right triangle are the _____ , the _____ side, and the _____ side.

3. An angle that measures from the horizontal upward to an object is called the angle of _____ , whereas an angle that measures from the horizontal downward to an object is called the angle of _____ .

In Exercises 1–4, find the exact values of the six trigonometric functions of the angle θ shown in the figure. (Use the Pythagorean Theorem to find the third side of the triangle.)

1.
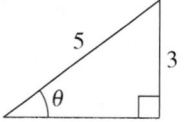

2.
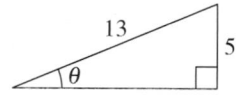

3.
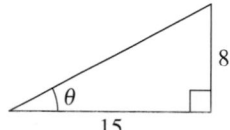

4.
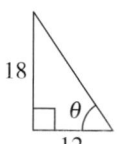

In Exercises 5–8, find the exact values of the six trigonometric functions of the angle θ for each of the triangles. Explain why the function values are the same.

5.
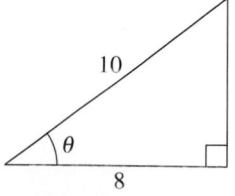

6.
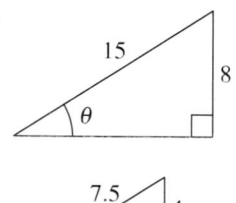

7.
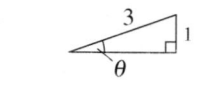

8.
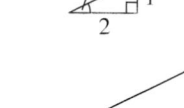

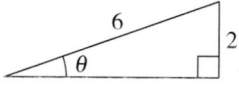

In Exercises 9–16, sketch a right triangle corresponding to the trigonometric function of the acute angle θ. Use the Pythagorean Theorem to determine the third side of the triangle and then find the other five trigonometric functions of θ.

9. $\sin \theta = \frac{5}{6}$

10. $\cot \theta = 5$

11. $\sec \theta = 4$

12. $\cos \theta = \frac{3}{7}$

13. $\tan \theta = 3$

14. $\csc \theta = \frac{17}{4}$

15. $\cot \theta = \frac{9}{4}$

16. $\sin \theta = \frac{3}{8}$

In Exercises 17–26, construct an appropriate triangle to complete the table. ($0 \le \theta \le 90°, 0 \le \theta \le \pi/2$)

Function	θ (deg)	θ (rad)	Function Value
17. sin	30°		
18. cos	45°		
19. tan		$\dfrac{\pi}{3}$	
20. sec		$\dfrac{\pi}{4}$	

Function	θ (deg)	θ (rad)	Function Value
21. cot			$\dfrac{\sqrt{3}}{3}$
22. csc			$\sqrt{2}$
23. cos		$\dfrac{\pi}{6}$	
24. sin		$\dfrac{\pi}{4}$	
25. cot			1
26. tan			$\dfrac{1}{\sqrt{3}}$

In Exercises 27–42, complete the identity.

27. $\sin \theta = \dfrac{1}{\rule{1cm}{0.4pt}}$ 28. $\cos \theta = \dfrac{1}{\rule{1cm}{0.4pt}}$

29. $\tan \theta = \dfrac{1}{\rule{1cm}{0.4pt}}$ 30. $\csc \theta = \dfrac{1}{\rule{1cm}{0.4pt}}$

31. $\sec \theta = \dfrac{1}{\rule{1cm}{0.4pt}}$ 32. $\cot \theta = \dfrac{1}{\rule{1cm}{0.4pt}}$

33. $\tan \theta = \dfrac{\rule{0.8cm}{0.4pt}}{\rule{0.8cm}{0.4pt}}$ 34. $\cot \theta = \dfrac{\rule{0.8cm}{0.4pt}}{\rule{0.8cm}{0.4pt}}$

35. $\sin^2 \theta + \cos^2 \theta =$ 36. $1 + \tan^2 \theta =$
37. $\sin(90° - \theta) =$ 38. $\cos(90° - \theta) =$
39. $\tan(90° - \theta) =$ 40. $\cot(90° - \theta) =$
41. $\sec(90° - \theta) =$ 42. $\csc(90° - \theta) =$

In Exercises 43–48, use the given function value(s) and the trigonometric identities to find the indicated trigonometric functions.

43. $\sin 60° = \dfrac{\sqrt{3}}{2}$, $\cos 60° = \dfrac{1}{2}$
   (a) $\tan 60°$ (b) $\sin 30°$
   (c) $\cos 30°$ (d) $\cot 60°$

44. $\sin 30° = \dfrac{1}{2}$, $\tan 30° = \dfrac{\sqrt{3}}{3}$
   (a) $\csc 30°$ (b) $\cot 60°$
   (c) $\cos 30°$ (d) $\cot 30°$

45. $\csc \theta = 3$, $\sec \theta = \dfrac{3\sqrt{2}}{4}$
   (a) $\sin \theta$ (b) $\cos \theta$
   (c) $\tan \theta$ (d) $\sec(90° - \theta)$

46. $\sec \theta = 5$, $\tan \theta = 2\sqrt{6}$
   (a) $\cos \theta$ (b) $\cot \theta$
   (c) $\cot(90° - \theta)$ (d) $\sin \theta$

47. $\cos \alpha = \dfrac{1}{4}$
   (a) $\sec \alpha$ (b) $\sin \alpha$
   (c) $\cot \alpha$ (d) $\sin(90° - \alpha)$

48. $\tan \beta = 5$
   (a) $\cot \beta$ (b) $\cos \beta$
   (c) $\tan(90° - \beta)$ (d) $\csc \beta$

In Exercises 49–56, use trigonometric identities to transform one side of the equation into the other ($0 < \theta < \pi/2$).

49. $\tan \theta \cot \theta = 1$
50. $\csc \theta \tan \theta = \sec \theta$
51. $\tan \theta \cos \theta = \sin \theta$
52. $\cot \theta \sin \theta = \cos \theta$
53. $(1 + \cos \theta)(1 - \cos \theta) = \sin^2 \theta$
54. $(\csc \theta + \cot \theta)(\csc \theta - \cot \theta) = 1$
55. $\dfrac{\sin \theta}{\cos \theta} + \dfrac{\cos \theta}{\sin \theta} = \csc \theta \sec \theta$
56. $\dfrac{\tan \theta + \cot \theta}{\tan \theta} = \csc^2 \theta$

In Exercises 57–62, use a calculator to evaluate each function. Round your answers to four decimal places. (Be sure the calculator is in the correct angle mode.)

57. (a) $\sin 41°$ (b) $\cos 87°$
58. (a) $\tan 18.5°$ (b) $\cot 71.5°$
59. (a) $\sec 42° 12'$ (b) $\csc 48° 7'$
60. (a) $\cos 8° 50' 25''$ (b) $\sec 8° 50' 25''$
61. (a) $\cot \dfrac{\pi}{16}$ (b) $\tan \dfrac{\pi}{8}$
62. (a) $\sec 1.54$ (b) $\cos 1.25$

In Exercises 63–68, find each value of θ in degrees ($0° < \theta < 90°$) and radians ($0 < \theta < \pi/2$) without using a calculator.

63. (a) $\sin \theta = \dfrac{1}{2}$ (b) $\csc \theta = 2$
64. (a) $\cos \theta = \dfrac{\sqrt{2}}{2}$ (b) $\tan \theta = 1$
65. (a) $\sec \theta = 2$ (b) $\cot \theta = 1$
66. (a) $\tan \theta = \sqrt{3}$ (b) $\cos \theta = \dfrac{1}{2}$
67. (a) $\csc \theta = \dfrac{2\sqrt{3}}{3}$ (b) $\sin \theta = \dfrac{\sqrt{2}}{2}$
68. (a) $\cot \theta = \dfrac{\sqrt{3}}{3}$ (b) $\sec \theta = \sqrt{2}$

In Exercises 69–76, find the exact values of the indicated variables (selected from x, y, and r).

69. Find y and r.

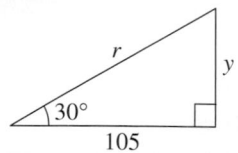

70. Find x and y.

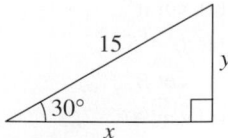

71. Find x and y.

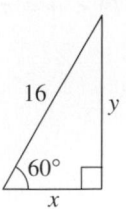

72. Find x and r.

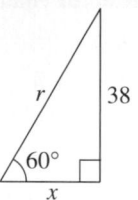

73. Find x and r.

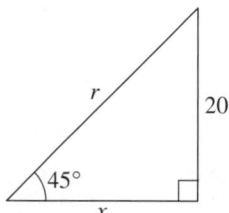

74. Find y and r.

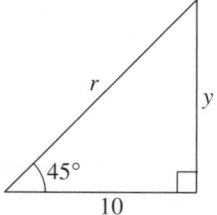

75. Find x and r.

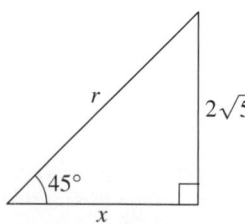

76. Find y and r.

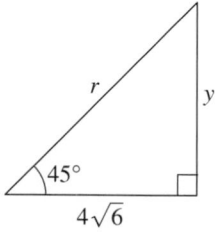

77. _Height_ A six-foot person walks from the base of a streetlight directly toward the tip of the shadow cast by the streetlight. When the person is 16 feet from the streetlight and 5 feet from the tip of the streetlight's shadow, the person's shadow starts to appear beyond the streetlight's shadow.

(a) Draw a right triangle that gives a visual representation of the problem. Show the known quantities and use a variable to indicate the height of the streetlight.

(b) Use a trigonometric function to write an equation involving the unknown quantity.

(c) What is the height of the streetlight?

78. _Height_ A 30-meter line is used to tether a helium-filled balloon. Because of a breeze, the line makes an angle of approximately 75° with the ground.

(a) Draw a right triangle that gives a visual representation of the problem. Show the known quantities and use a variable to indicate the height of the balloon.

(b) Use a trigonometric function to write an equation involving the unknown quantity.

(c) What is the height of the balloon?

79. _Width_ A biologist wants to know the width w of a river (see figure) in order to properly set instruments for studying the pollutants in the water. From point A, the biologist walks downstream 100 feet and sights to point C. From this sighting, it is determined that $\theta = 58°$. How wide is the river? Verify your result numerically.

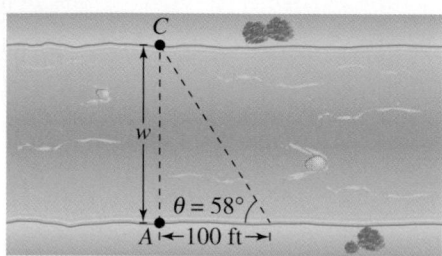

80. _Height of a Mountain_ In traveling across flat land you notice a mountain directly in front of you. Its angle of elevation (to the peak) is 3.5°. After you drive 13 miles closer to the mountain, the angle of elevation is 9° (see figure). Approximate the height of the mountain.

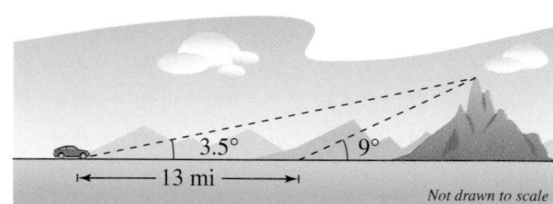

81. _Angle of Elevation_ A zip-line steel cable is being constructed for a reality television competition show. The high end of the zip-line is attached to the top of a 50-foot pole while the lower end is anchored at ground level to a stake 50 feet from the base of the pole.

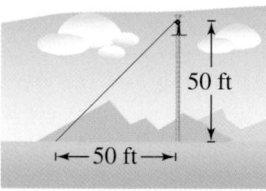

(a) Find the angle of elevation of the zip-line.

(b) Find the number of feet of steel cable needed for the zip-line.

(c) A contestant takes 6 seconds to reach the ground from the top of the zip-line. At what rate is the contestant moving down the line? At what rate is the contestant dropping vertically?

82. **Inclined Plane**   The Johnstown Inclined Plane in Pennsylvania is one of the longest and steepest hoists in the world. The railway cars travel a distance of 896.5 feet at an angle of approximately 35.4°, rising to a height of 1693.5 feet above sea level.

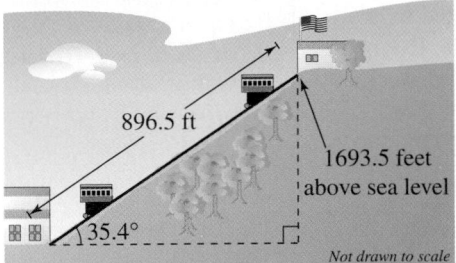

(a) Find the vertical rise of the inclined plane.

(b) Find the elevation of the lower end of the inclined plane.

(c) The cars move up the mountain at a rate of 300 feet per minute. Find the rate at which they rise vertically.

83. **Machine Shop Calculations**   A steel plate has the form of one fourth of a circle with a radius of 60 centimeters. Two 2-centimeter holes are to be drilled in the plate, positioned as shown in the figure. Find the coordinates of the center of each hole.

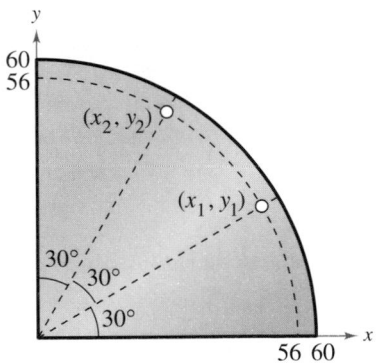

84. **Geometry**   Use a compass to sketch a quarter of a circle of radius 10 centimeters. Using a protractor, construct an angle of 20° in standard position (see figure). Drop a perpendicular from the point of intersection of the terminal side of the angle and the arc of the circle. By actual measurement, calculate the coordinates (x, y) of the point of intersection and use these measurements to approximate the six trigonometric functions of a 20° angle.

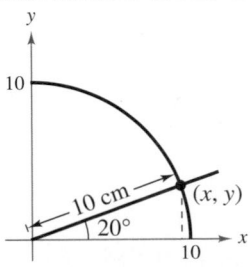

Synthesis

True or False?   In Exercises 85–87, determine whether the statement is true or false. Justify your answer.

85. $\sin 60° \csc 60° = 1$

86. $\sin 45° + \cos 45° = 1$

87. $\cot^2 10° - \csc^2 10° = -1$

88. **Think About It**   You are given the value $\tan \theta$. Is it possible to find the value of $\sec \theta$ without finding the measure of θ? Explain.

89. **Exploration**

(a) Use a graphing utility to complete the table. Round your results to four decimal places.

θ	0°	20°	40°	60°	80°
$\sin \theta$					
$\cos \theta$					
$\tan \theta$					

(b) Classify each of the three trigonometric functions as increasing or decreasing for the table values.

(c) From the values in the table, verify that the tangent function is the quotient of the sine and cosine functions.

90. **Exploration**   Use a graphing utility to complete the table and make a conjecture about the relationship between $\cos \theta$ and $\sin(90° - \theta)$. What are the angles θ and $90° - \theta$ called?

θ	0°	20°	40°	60°	80°
$\cos \theta$					
$\sin(90° - \theta)$					

Skills Review

In Exercises 91–94, use a graphing utility to graph the exponential function.

91. $f(x) = e^{3x}$

92. $f(x) = -e^{3x}$

93. $f(x) = 2 + e^{3x}$

94. $f(x) = -4 + e^{3x}$

In Exercises 95–98, use a graphing utility to graph the logarithmic function.

95. $f(x) = \log_3 x$

96. $f(x) = \log_3 x + 1$

97. $f(x) = -\log_3 x$

98. $f(x) = \log_3(x - 4)$

4.4 Trigonometric Functions of Any Angle

Introduction

In Section 4.3, the definitions of trigonometric functions were restricted to acute angles. In this section, the definitions are extended to cover *any* angle. If θ is an *acute* angle, the definitions here coincide with those given in the preceding section.

Definitions of Trigonometric Functions of Any Angle

Let θ be an angle in standard position with (x, y) a point on the terminal side of θ and $r = \sqrt{x^2 + y^2} \neq 0$.

$$\sin \theta = \frac{y}{r} \qquad\qquad \cos \theta = \frac{x}{r}$$

$$\tan \theta = \frac{y}{x}, \quad x \neq 0 \qquad \cot \theta = \frac{x}{y}, \quad y \neq 0$$

$$\sec \theta = \frac{r}{x}, \quad x \neq 0 \qquad \csc \theta = \frac{r}{y}, \quad y \neq 0$$

Because $r = \sqrt{x^2 + y^2}$ *cannot* be zero, it follows that the sine and cosine functions are defined for any real value of θ. However, if $x = 0$, the tangent and secant of θ are undefined. For example, the tangent of $90°$ is undefined. Similarly, if $y = 0$, the cotangent and cosecant of θ are undefined.

Example 1   Evaluating Trigonometric Functions

Let $(-3, 4)$ be a point on the terminal side of θ. Find the sine, cosine, and tangent of θ.

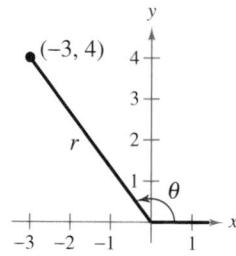

Figure 4.32

Solution

Referring to Figure 4.32, you can see that $x = -3$, $y = 4$, and

$$r = \sqrt{x^2 + y^2} = \sqrt{(-3)^2 + 4^2} = \sqrt{25} = 5.$$

So, you have $\sin \theta = \dfrac{y}{r} = \dfrac{4}{5}$, $\cos \theta = \dfrac{x}{r} = -\dfrac{3}{5}$, and $\tan \theta = \dfrac{y}{x} = -\dfrac{4}{3}$.

✓**CHECKPOINT**   Now try Exercise 1.

The *signs* of the trigonometric functions in the four quadrants can be determined easily from the definitions of the functions. For instance, because $\cos \theta = x/r$, it follows that $\cos \theta$ is positive wherever $x > 0$, which is in Quadrants I and IV. (Remember, r is always positive.) In a similar manner, you can verify the results shown in Figure 4.33.

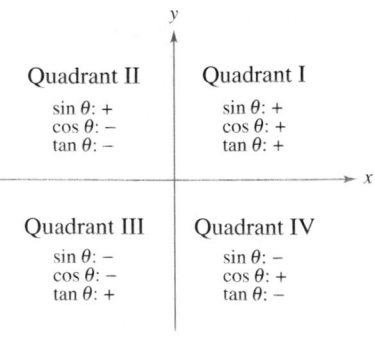

Example 2　Evaluating Trigonometric Functions

Given $\sin \theta = -\frac{2}{3}$ and $\tan \theta > 0$, find $\cos \theta$ and $\cot \theta$.

Solution

Note that θ lies in Quadrant III because that is the only quadrant in which the sine is negative and the tangent is positive. Moreover, using

$$\sin \theta = \frac{y}{r} = -\frac{2}{3}$$

and the fact that y is negative in Quadrant III, you can let $y = -2$ and $r = 3$. Because x is negative in Quadrant III, $x = -\sqrt{9-4} = -\sqrt{5}$, and you have the following.

$$\cos \theta = \frac{x}{r} = \frac{-\sqrt{5}}{3} \qquad \text{Exact value}$$

$$\approx -0.75 \qquad \text{Approximate value}$$

$$\cot \theta = \frac{x}{y} = \frac{-\sqrt{5}}{-2} \qquad \text{Exact value}$$

$$\approx 1.12 \qquad \text{Approximate value}$$

✔CHECKPOINT　Now try Exercise 19.

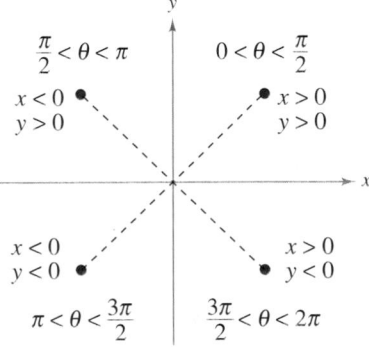

Figure 4.33

Example 3　Trigonometric Functions of Quadrant Angles

Evaluate the sine and cosine functions at the angles 0, $\frac{\pi}{2}$, π, and $\frac{3\pi}{2}$.

Solution

To begin, choose a point on the terminal side of each angle, as shown in Figure 4.34. For each of the four given points, $r = 1$, and you have the following.

$$\sin 0 = \frac{y}{r} = \frac{0}{1} = 0 \qquad \cos 0 = \frac{x}{r} = \frac{1}{1} = 1 \qquad (x, y) = (1, 0)$$

$$\sin \frac{\pi}{2} = \frac{y}{r} = \frac{1}{1} = 1 \qquad \cos \frac{\pi}{2} = \frac{x}{r} = \frac{0}{1} = 0 \qquad (x, y) = (0, 1)$$

$$\sin \pi = \frac{y}{r} = \frac{0}{1} = 0 \qquad \cos \pi = \frac{x}{r} = \frac{-1}{1} = -1 \qquad (x, y) = (-1, 0)$$

$$\sin \frac{3\pi}{2} = \frac{y}{r} = \frac{-1}{1} = -1 \qquad \cos \frac{3\pi}{2} = \frac{x}{r} = \frac{0}{1} = 0 \qquad (x, y) = (0, -1)$$

✔CHECKPOINT　Now try Exercise 29.

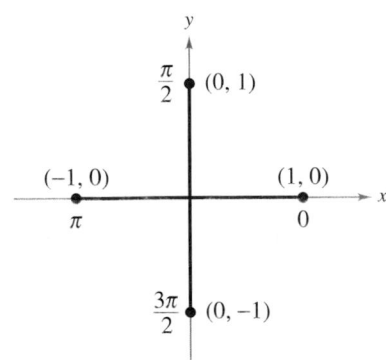

Figure 4.34

Reference Angles

The values of the trigonometric functions of angles greater than 90° (or less than 0°) can be determined from their values at corresponding acute angles called **reference angles.**

> **Definition of Reference Angle**
>
> Let θ be an angle in standard position. Its **reference angle** is the acute angle θ' formed by the terminal side of θ and the horizontal axis.

Figure 4.35 shows the reference angles for θ in Quadrants II, III, and IV.

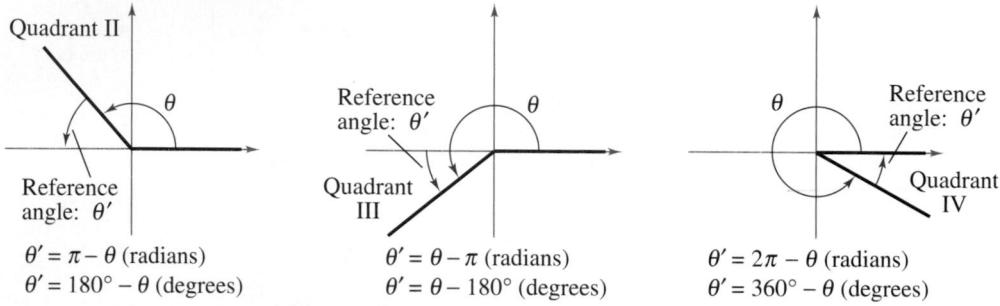

$\theta' = \pi - \theta$ (radians) $\theta' = \theta - \pi$ (radians) $\theta' = 2\pi - \theta$ (radians)
$\theta' = 180° - \theta$ (degrees) $\theta' = \theta - 180°$ (degrees) $\theta' = 360° - \theta$ (degrees)

Figure 4.35

Example 4   Finding Reference Angles

Find the reference angle θ'.

a. $\theta = 300°$ **b.** $\theta = 2.3$ **c.** $\theta = -135°$

Solution

a. Because 300° lies in Quadrant IV, the angle it makes with the x-axis is

$$\theta' = 360° - 300° = 60°. \qquad \text{Degrees}$$

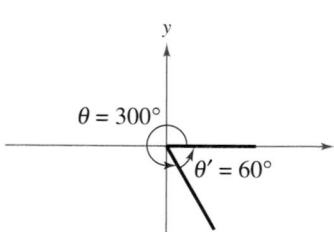

(a)

b. Because 2.3 lies between $\pi/2 \approx 1.5708$ and $\pi \approx 3.1416$, it follows that it is in Quadrant II and its reference angle is

$$\theta' = \pi - 2.3 \approx 0.8416. \qquad \text{Radians}$$

c. First, determine that $-135°$ is coterminal with 225°, which lies in Quadrant III. So, the reference angle is

$$\theta' = 225° - 180° = 45°. \qquad \text{Degrees}$$

Figure 4.36 shows each angle θ and its reference angle θ'.

 CHECKPOINT   Now try Exercise 51.

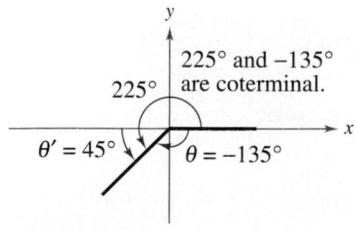

(c)
Figure 4.36

Trigonometric Functions of Real Numbers

To see how a reference angle is used to evaluate a trigonometric function, consider the point (x, y) on the terminal side of θ, as shown in Figure 4.37. By definition, you know that

$$\sin \theta = \frac{y}{r} \quad \text{and} \quad \tan \theta = \frac{y}{x}.$$

For the right triangle with acute angle θ' and sides of lengths $|x|$ and $|y|$, you have

$$\sin \theta' = \frac{\text{opp}}{\text{hyp}} = \frac{|y|}{r}$$

and

$$\tan \theta' = \frac{\text{opp}}{\text{adj}} = \frac{|y|}{|x|}.$$

So, it follows that $\sin \theta$ and $\sin \theta'$ are equal, *except possibly in sign*. The same is true for $\tan \theta$ and $\tan \theta'$ and for the other four trigonometric functions. In all cases, the sign of the function value can be determined by the quadrant in which θ lies.

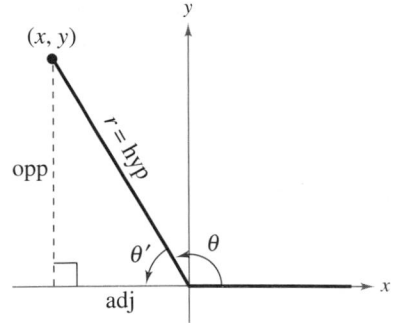

Figure 4.37

Evaluating Trigonometric Functions of Any Angle

To find the value of a trigonometric function of any angle θ:

1. Determine the function value for the associated reference angle θ'.

2. Depending on the quadrant in which θ lies, affix the appropriate sign to the function value.

By using reference angles and the special angles discussed in the previous section, you can greatly extend the scope of *exact* trigonometric values. For instance, knowing the function values of 30° means that you know the function values of all angles for which 30° is a reference angle. For convenience, the following table shows the exact values of the trigonometric functions of special angles and quadrant angles.

STUDY TIP

Learning the table of values at the left is worth the effort because doing so will increase both your efficiency and your confidence. Here is a pattern for the sine function that may help you remember the values.

θ	0°	30°	45°	60°	90°
$\sin \theta$	$\dfrac{\sqrt{0}}{2}$	$\dfrac{\sqrt{1}}{2}$	$\dfrac{\sqrt{2}}{2}$	$\dfrac{\sqrt{3}}{2}$	$\dfrac{\sqrt{4}}{2}$

Reverse the order to get cosine values of the same angles.

Trigonometric Values of Common Angles

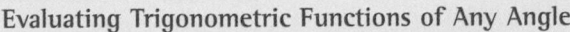

θ (degrees)	0°	30°	45°	60°	90°	180°	270°
θ (radians)	0	$\dfrac{\pi}{6}$	$\dfrac{\pi}{4}$	$\dfrac{\pi}{3}$	$\dfrac{\pi}{2}$	π	$\dfrac{3\pi}{2}$
$\sin \theta$	0	$\dfrac{1}{2}$	$\dfrac{\sqrt{2}}{2}$	$\dfrac{\sqrt{3}}{2}$	1	0	-1
$\cos \theta$	1	$\dfrac{\sqrt{3}}{2}$	$\dfrac{\sqrt{2}}{2}$	$\dfrac{1}{2}$	0	-1	0
$\tan \theta$	0	$\dfrac{\sqrt{3}}{3}$	1	$\sqrt{3}$	Undef.	0	Undef.

Example 5   Trigonometric Functions of Nonacute Angles

Evaluate each trigonometric function.

a. $\cos \dfrac{4\pi}{3}$ **b.** $\tan(-210°)$ **c.** $\csc \dfrac{11\pi}{4}$

Solution

a. Because $\theta = 4\pi/3$ lies in Quadrant III, the reference angle is $\theta' = (4\pi/3) - \pi = \pi/3$, as shown in Figure 4.38. Moreover, the cosine is negative in Quadrant III, so

$$\cos \frac{4\pi}{3} = (-)\cos \frac{\pi}{3} = -\frac{1}{2}.$$

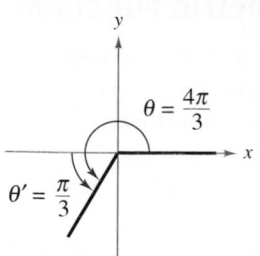

Figure 4.38

b. Because $-210° + 360° = 150°$, it follows that $-210°$ is coterminal with the second-quadrant angle $150°$. Therefore, the reference angle is $\theta' = 180° - 150° = 30°$, as shown in Figure 4.39. Finally, because the tangent is negative in Quadrant II, you have

$$\tan(-210°) = (-)\tan 30° = -\frac{\sqrt{3}}{3}.$$

c. Because $(11\pi/4) - 2\pi = 3\pi/4$, it follows that $11\pi/4$ is coterminal with the second-quadrant angle $3\pi/4$. Therefore, the reference angle is $\theta' = \pi - (3\pi/4) = \pi/4$, as shown in Figure 4.40. Because the cosecant is positive in Quadrant II, you have

$$\csc \frac{11\pi}{4} = (+)\csc \frac{\pi}{4} = \frac{1}{\sin(\pi/4)} = \sqrt{2}.$$

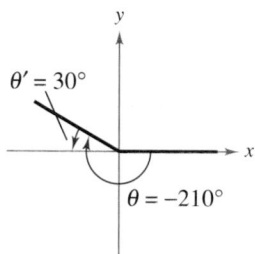

Figure 4.39

 CHECKPOINT   Now try Exercise 63.

The fundamental trigonometric identities listed in the preceding section (for an acute angle θ) are also valid when θ is any angle in the domain of the function.

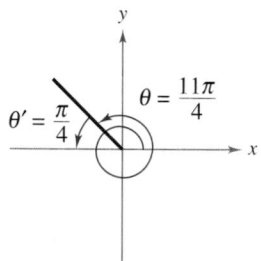

Figure 4.40

Example 6   Using Trigonometric Identities

Let θ be an angle in Quadrant II such that $\sin \theta = \frac{1}{3}$. Find $\cos \theta$ by using trigonometric identities.

Prerequisite Skills

If you have difficulty with this example, review the Fundamental Trigonometric Identities in Section 4.3.

Solution

Using the Pythagorean identity $\sin^2 \theta + \cos^2 \theta = 1$, you obtain

$$\left(\tfrac{1}{3}\right)^2 + \cos^2 \theta = 1$$

$$\cos^2 \theta = 1 - \tfrac{1}{9} = \tfrac{8}{9}.$$

Because $\cos \theta < 0$ in Quadrant II, you can use the negative root to obtain

$$\cos \theta = -\frac{\sqrt{8}}{\sqrt{9}} = -\frac{2\sqrt{2}}{3}.$$

 CHECKPOINT   Now try Exercise 65.

You can use a calculator to evaluate trigonometric functions, as shown in the next example.

Example 7   Using a Calculator

Use a calculator to evaluate each trigonometric function.

a. $\cot 410°$ **b.** $\sin(-7)$ **c.** $\sec \dfrac{\pi}{9}$

Solution

Function	Mode	Graphing Calculator Keystrokes	Display
a. $\cot 410°$	Degree	`( TAN ( 410 ) ) x⁻¹ ENTER`	0.8390996
b. $\sin(-7)$	Radian	`SIN ( (-) 7 ) ENTER`	-0.6569866
c. $\sec \dfrac{\pi}{9}$	Radian	`( COS ( π ÷ 9 ) ) x⁻¹ ENTER`	1.0641778

✓*CHECKPOINT*   Now try Exercise 79.

Exploration

Set your graphing utility to *degree* mode and enter tan 90. What happens? Why? Now set your graphing utility to *radian* mode and enter tan(π/2). Explain the graphing utility's answer.

Library of Parent Functions: Trigonometric Functions

Trigonometric functions are transcendental functions. The six trigonometric functions, sine, cosine, tangent, cosecant, secant, and cotangent, have important uses in construction, surveying, and navigation. Their periodic behavior makes them useful for modeling phenomena such as business cycles, planetary orbits, pendulums, wave motion, and light rays.

The six trigonometric functions can be defined in three different ways.

1. As the ratio of two sides of a right triangle [see Figure 4.41(a)].

2. As coordinates of a point (x, y) in the plane and its distance r from the origin [see Figure 4.41(b)].

3. As functions of any real number, such as time t.

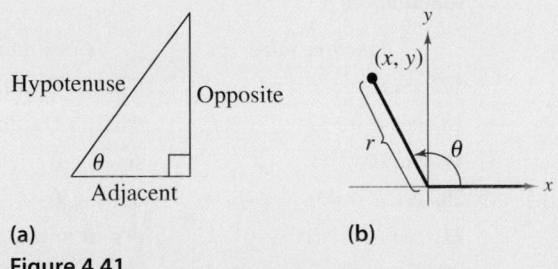

(a)   (b)
Figure 4.41

To be efficient in the use of trigonometric functions, you should learn the trigonometric function values of common angles, such as those listed on page 291. Because pairs of trigonometric functions are related to each other by a variety of identities, it is useful to know the fundamental identities presented in Section 4.3. Finally, trigonometric functions and their identity relationships play a prominent role in calculus. A review of trigonometric functions can be found in the *Study Capsules.*

STUDY TIP

An *algebraic function*, such as a polynomial, can be expressed in terms of variables and constants. *Transcendental functions* are functions which are *not* algebraic, such as exponential functions, logarithmic functions, and trigonometric functions.

4.4 Exercises

See www.CalcChat.com for worked-out solutions to odd-numbered exercises.

Vocabulary Check

In Exercises 1–6, let θ be an angle in standard position with (x, y) a point on the terminal side of θ and $r = \sqrt{x^2 + y^2} \neq 0$. Fill in the blanks.

1. $\sin \theta =$ _____

2. $\dfrac{r}{y} =$ _____

3. $\tan \theta =$ _____

4. $\sec \theta =$ _____

5. $\dfrac{x}{r} =$ _____

6. $\dfrac{x}{y} =$ _____

7. The acute positive angle that is formed by the terminal side of the angle θ and the horizontal axis is called the _____ angle of θ and is denoted by θ'.

Library of Parent Functions In Exercises 1–4, determine the exact values of the six trigonometric functions of the angle θ.

1. (a) 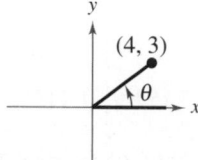   (b)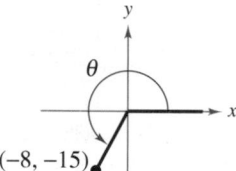

2. (a) 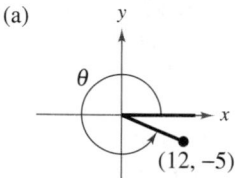   (b)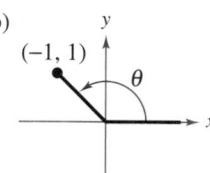

3. (a) 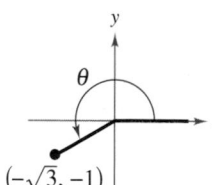   (b)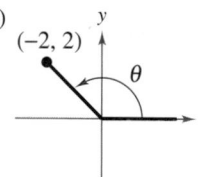

4. (a) 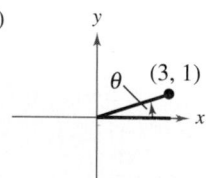   (b)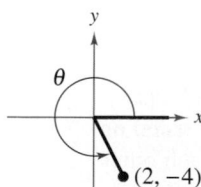

In Exercises 5–12, the point is on the terminal side of an angle in standard position. Determine the exact values of the six trigonometric functions of the angle.

5. $(7, 24)$

6. $(8, 15)$

7. $(5, -12)$

8. $(-24, 10)$

9. $(-4, 10)$

10. $(-5, -6)$

11. $(-10, 8)$

12. $(3, -9)$

In Exercises 13–16, state the quadrant in which θ lies.

13. $\sin \theta < 0$ and $\cos \theta < 0$

14. $\sec \theta > 0$ and $\cot \theta < 0$

15. $\cot \theta > 0$ and $\cos \theta > 0$

16. $\tan \theta > 0$ and $\csc \theta < 0$

In Exercises 17–24, find the values of the six trigonometric functions of θ.

Function Value	Constraint
17. $\sin \theta = \frac{3}{5}$	θ lies in Quadrant II.
18. $\cos \theta = -\frac{4}{5}$	θ lies in Quadrant III.
19. $\tan \theta = -\frac{15}{8}$	$\sin \theta < 0$
20. $\csc \theta = 4$	$\cot \theta < 0$
21. $\sec \theta = -2$	$0 \leq \theta \leq \pi$
22. $\sin \theta = 0$	$\dfrac{\pi}{2} \leq \theta \leq \dfrac{3\pi}{2}$
23. $\cot \theta$ is undefined.	$\dfrac{\pi}{2} \leq \theta \leq \dfrac{3\pi}{2}$
24. $\tan \theta$ is undefined.	$\pi \leq \theta \leq 2\pi$

In Exercises 25–28, the terminal side of θ lies on the given line in the specified quadrant. Find the values of the six trigonometric functions of θ by finding a point on the line.

	Line	*Quadrant*
25.	$y = -x$	II
26.	$y = \frac{1}{3}x$	III
27.	$2x - y = 0$	III
28.	$4x + 3y = 0$	IV

In Exercises 29–36, evaluate the trigonometric function of the quadrant angle.

29. $\sec \pi$ **30.** $\tan \frac{\pi}{2}$

31. $\cot \frac{3\pi}{2}$ **32.** $\csc 0$

33. $\sec 0$ **34.** $\csc \frac{3\pi}{2}$

35. $\cot \pi$ **36.** $\csc \frac{\pi}{2}$

In Exercises 37–44, find the reference angle θ' for the special angle θ. Then sketch θ and θ' in standard position.

37. $\theta = 120°$ **38.** $\theta = 225°$

39. $\theta = -135°$ **40.** $\theta = -330°$

41. $\theta = \frac{5\pi}{3}$ **42.** $\theta = \frac{3\pi}{4}$

43. $\theta = -\frac{5\pi}{6}$ **44.** $\theta = -\frac{2\pi}{3}$

In Exercises 45–52, find the reference angle θ' and sketch θ and θ' in standard position.

45. $\theta = 208°$ **46.** $\theta = 322°$

47. $\theta = -292°$ **48.** $\theta = -165°$

49. $\theta = \frac{11\pi}{5}$ **50.** $\theta = \frac{17\pi}{7}$

51. $\theta = -1.8$ **52.** $\theta = 4.5$

In Exercises 53–64, evaluate the sine, cosine, and tangent of the angle without using a calculator.

53. $225°$ **54.** $300°$

55. $-750°$ **56.** $-495°$

57. $\frac{5\pi}{3}$ **58.** $\frac{3\pi}{4}$

59. $-\frac{\pi}{6}$ **60.** $-\frac{4\pi}{3}$

61. $\frac{11\pi}{4}$ **62.** $\frac{10\pi}{3}$

63. $-\frac{17\pi}{6}$ **64.** $-\frac{20\pi}{3}$

In Exercises 65–70, find the indicated trigonometric value in the specified quadrant.

	Function	*Quadrant*	*Trigonometric Value*
65.	$\sin \theta = -\frac{3}{5}$	IV	$\cos \theta$
66.	$\cot \theta = -3$	II	$\sin \theta$
67.	$\csc \theta = -2$	IV	$\cot \theta$
68.	$\cos \theta = \frac{5}{8}$	I	$\sec \theta$
69.	$\sec \theta = -\frac{9}{4}$	III	$\tan \theta$
70.	$\tan \theta = -\frac{5}{4}$	IV	$\csc \theta$

In Exercises 71–76, use the given value and the trigonometric identities to find the remaining trigonometric functions of the angle.

71. $\sin \theta = \frac{2}{5}$, $\cos \theta < 0$ **72.** $\cos \theta = -\frac{3}{7}$, $\sin \theta < 0$

73. $\tan \theta = -4$, $\cos \theta < 0$ **74.** $\cot \theta = -5$, $\sin \theta > 0$

75. $\csc \theta = -\frac{3}{2}$, $\tan \theta < 0$ **76.** $\sec \theta = -\frac{4}{3}$, $\cot \theta > 0$

In Exercises 77–88, use a calculator to evaluate the trigonometric function. Round your answer to four decimal places. (Be sure the calculator is set in the correct angle mode.)

77. $\sin 10°$ **78.** $\sec 235°$

79. $\tan 245°$ **80.** $\csc 320°$

81. $\cos(-110°)$ **82.** $\cot(-220°)$

83. $\sec(-280°)$ **84.** $\csc 0.33$

85. $\tan \frac{2\pi}{9}$ **86.** $\tan \frac{11\pi}{9}$

87. $\csc\left(-\frac{8\pi}{9}\right)$ **88.** $\cos\left(-\frac{15\pi}{14}\right)$

In Exercises 89–94, find two solutions of the equation. Give your answers in degrees ($0° \le \theta < 360°$) and radians ($0 \le \theta < 2\pi$). Do not use a calculator.

89. (a) $\sin \theta = \frac{1}{2}$ (b) $\sin \theta = -\frac{1}{2}$

90. (a) $\cos \theta = \frac{\sqrt{2}}{2}$ (b) $\cos \theta = -\frac{\sqrt{2}}{2}$

91. (a) $\csc \theta = \frac{2\sqrt{3}}{3}$ (b) $\cot \theta = -1$

92. (a) $\csc \theta = -\sqrt{2}$ (b) $\csc \theta = 2$

93. (a) $\sec \theta = -\frac{2\sqrt{3}}{3}$ (b) $\cos \theta = -\frac{1}{2}$

94. (a) $\cot \theta = -\sqrt{3}$ (b) $\sec \theta = \sqrt{2}$

In Exercises 95–108, find the exact value of each function for the given angle for $f(\theta) = \sin \theta$ and $g(\theta) = \cos \theta$. Do not use a calculator.

(a) $f(\theta) + g(\theta)$ (b) $g(\theta) - f(\theta)$ (c) $[g(\theta)]^2$
(d) $f(\theta)g(\theta)$ (e) $2f(\theta)$ (f) $g(-\theta)$

95. $\theta = 30°$

96. $\theta = 60°$

97. $\theta = 315°$

98. $\theta = 225°$

99. $\theta = 150°$

100. $\theta = 300°$

101. $\theta = \dfrac{7\pi}{6}$

102. $\theta = \dfrac{5\pi}{6}$

103. $\theta = \dfrac{4\pi}{3}$

104. $\theta = \dfrac{5\pi}{3}$

105. $\theta = 270°$

106. $\theta = 180°$

107. $\theta = \dfrac{7\pi}{2}$

108. $\theta = \dfrac{5\pi}{2}$

109. **Meteorology**   The monthly normal temperatures T (in degrees Fahrenheit) for Santa Fe, New Mexico are given by

$$T = 49.5 + 20.5 \cos\left(\frac{\pi t}{6} - \frac{7\pi}{6}\right)$$

where t is the time in months, with $t = 1$ corresponding to January. Find the monthly normal temperature for each month. (Source: National Climatic Data Center)

(a) January   (b) July   (c) December

110. **Sales**   A company that produces water skis, which are seasonal products, forecasts monthly sales over a two-year period to be

$$S = 23.1 + 0.442t + 4.3 \sin \frac{\pi t}{6}$$

where S is measured in thousands of units and t is the time (in months), with $t = 1$ representing January 2006. Estimate sales for each month.

(a) January 2006   (b) February 2007

(c) May 2006   (d) June 2006

111. **Distance**   An airplane flying at an altitude of 6 miles is on a flight path that passes directly over an observer (see figure). If θ is the angle of elevation from the observer to the plane, find the distance from the observer to the plane when (a) $\theta = 30°$, (b) $\theta = 90°$, and (c) $\theta = 120°$.

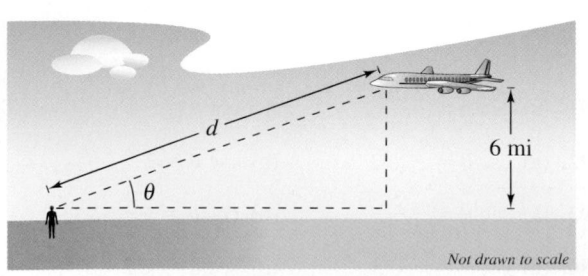

6 mi

Not drawn to scale

112. **Writing**   Consider an angle in standard position with $r = 12$ centimeters, as shown in the figure. Write a short paragraph describing the changes in the magnitudes of x, y, $\sin \theta$, $\cos \theta$, and $\tan \theta$ as θ increases continually from $0°$ to $90°$.

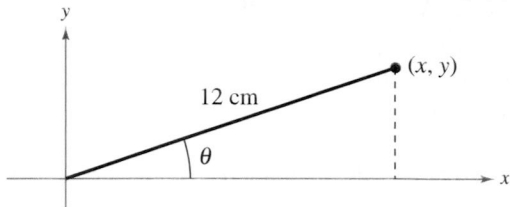

12 cm

(x, y)

Synthesis

True or False?   **In Exercises 113 and 114, determine whether the statement is true or false. Justify your answer.**

113. $\sin 151° = \sin 29°$

114. $-\cot\left(\dfrac{3\pi}{4}\right) = \cot\left(-\dfrac{\pi}{4}\right)$

115. **Conjecture**

(a) Use a graphing utility to complete the table.

θ	$0°$	$20°$	$40°$	$60°$	$80°$
$\sin \theta$					
$\sin(180° - \theta)$					

(b) Make a conjecture about the relationship between $\sin \theta$ and $\sin(180° - \theta)$.

116. **Writing**   Create a table of the six trigonometric functions comparing their domains, ranges, parity (evenness or oddness), periods, and zeros. Then identify, and write a short paragraph describing, any inherent patterns in the trigonometric functions. What can you conclude?

Skills Review

In Exercises 117–126, solve the equation. Round your answer to three decimal places, if necessary.

117. $3x - 7 = 14$

118. $44 - 9x = 61$

119. $x^2 - 2x - 5 = 0$

120. $2x^2 + x - 4 = 0$

121. $\dfrac{3}{x-1} = \dfrac{x+2}{9}$

122. $\dfrac{5}{x} = \dfrac{x+4}{2x}$

123. $4^{3-x} = 726$

124. $\dfrac{4500}{4 + e^{2x}} = 50$

125. $\ln x = -6$

126. $\ln \sqrt{x + 10} = 1$

4.5   Graphs of Sine and Cosine Functions

Basic Sine and Cosine Curves

In this section, you will study techniques for sketching the graphs of the sine and cosine functions. The graph of the sine function is a **sine curve.** In Figure 4.42, the black portion of the graph represents one period of the function and is called **one cycle** of the sine curve. The gray portion of the graph indicates that the basic sine wave repeats indefinitely to the right and left. The graph of the cosine function is shown in Figure 4.43. To produce these graphs with a graphing utility, make sure you set the graphing utility to *radian* mode.

Recall from Section 4.2 that the domain of the sine and cosine functions is the set of all real numbers. Moreover, the range of each function is the interval $[-1, 1]$, and each function has a period of 2π. Do you see how this information is consistent with the basic graphs shown in Figures 4.42 and 4.43?

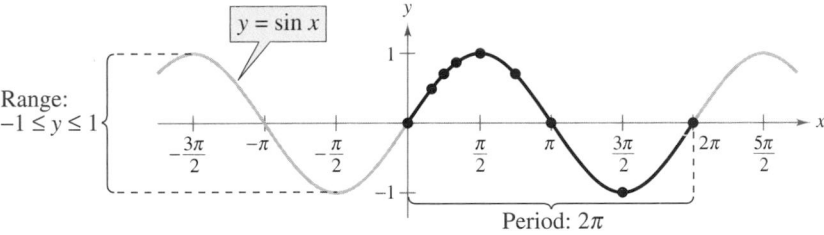

Figure 4.42

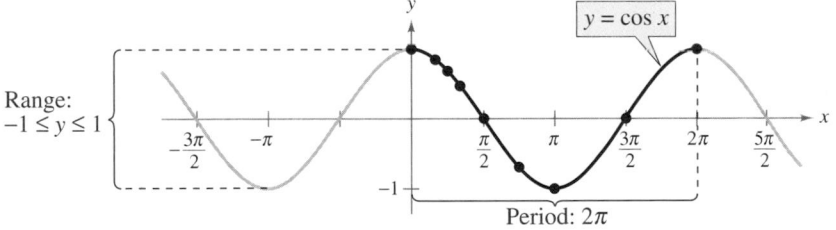

Figure 4.43

What you should learn

- Sketch the graphs of basic sine and cosine functions.
- Use amplitude and period to help sketch the graphs of sine and cosine functions.
- Sketch translations of graphs of sine and cosine functions.
- Use sine and cosine functions to model real-life data.

Why you should learn it

Sine and cosine functions are often used in scientific calculations. For instance, in Exercise 77 on page 306, you can use a trigonometric function to model the percent of the moon's face that is illuminated for any given day in 2006.

Jerry Lodriguss/Photo Researchers, Inc.

The table below lists key points on the graphs of $y = \sin x$ and $y = \cos x$.

x	0	$\dfrac{\pi}{6}$	$\dfrac{\pi}{4}$	$\dfrac{\pi}{3}$	$\dfrac{\pi}{2}$	$\dfrac{3\pi}{4}$	π	$\dfrac{3\pi}{2}$	2π
$\sin x$	0	$\dfrac{1}{2}$	$\dfrac{\sqrt{2}}{2}$	$\dfrac{\sqrt{3}}{2}$	1	$\dfrac{\sqrt{2}}{2}$	0	-1	0
$\cos x$	1	$\dfrac{\sqrt{3}}{2}$	$\dfrac{\sqrt{2}}{2}$	$\dfrac{1}{2}$	0	$-\dfrac{\sqrt{2}}{2}$	-1	0	1

Note in Figures 4.42 and 4.43 that the sine curve is symmetric with respect to the *origin*, whereas the cosine curve is symmetric with respect to the *y-axis*. These properties of symmetry follow from the fact that the sine function is odd whereas the cosine function is even.

To sketch the graphs of the basic sine and cosine functions by hand, it helps to note five *key points* in one period of each graph: the *intercepts*, the *maximum points*, and the *minimum points*. See Figure 4.44.

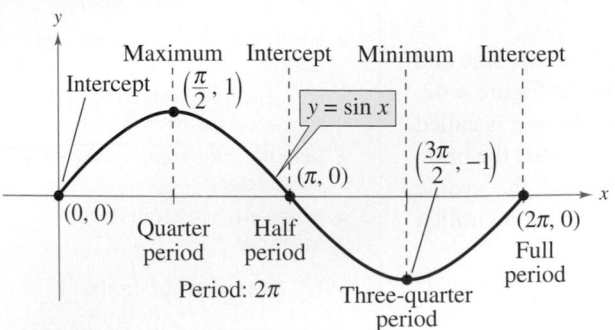 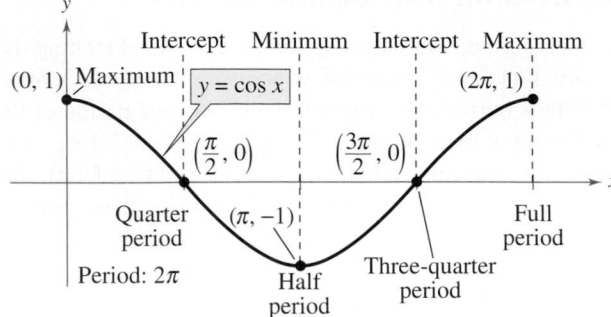

Figure 4.44

Example 1 Using Key Points to Sketch a Sine Curve

Sketch the graph of $y = 2 \sin x$ by hand on the interval $[-\pi, 4\pi]$.

Solution

Note that

$$y = 2 \sin x = 2(\sin x)$$

indicates that the y-values of the key points will have twice the magnitude of those on the graph of $y = \sin x$. Divide the period 2π into four equal parts to get the key points

Intercept	*Maximum*	*Intercept*	*Minimum*		*Intercept*
$(0, 0)$,	$\left(\dfrac{\pi}{2}, 2\right)$,	$(\pi, 0)$,	$\left(\dfrac{3\pi}{2}, -2\right)$,	and	$(2\pi, 0)$.

By connecting these key points with a smooth curve and extending the curve in both directions over the interval $[-\pi, 4\pi]$, you obtain the graph shown in Figure 4.45. Use a graphing utility to confirm this graph. Be sure to set the graphing utility to *radian* mode.

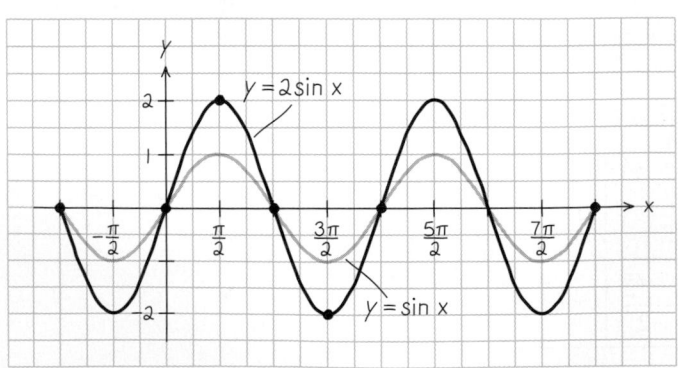

Figure 4.45

✔CHECKPOINT   Now try Exercise 39.

Exploration

Enter the Graphing a Sine Function Program, found at this textbook's *Online Study Center*, into your graphing utility. This program simultaneously draws the unit circle and the corresponding points on the sine curve, as shown below. After the circle and sine curve are drawn, you can connect the points on the unit circle with their corresponding points on the sine curve by pressing [ENTER]. Discuss the relationship that is illustrated.

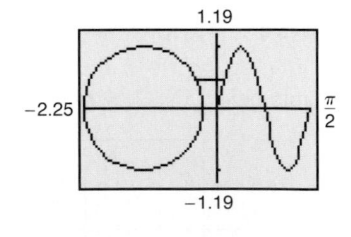

Amplitude and Period of Sine and Cosine Curves

In the rest of this section, you will study the graphic effect of each of the constants a, b, c, and d in equations of the forms

$$y = d + a\sin(bx - c) \quad \text{and} \quad y = d + a\cos(bx - c).$$

The constant factor a in $y = a\sin x$ acts as a *scaling factor*—a *vertical stretch* or *vertical shrink* of the basic sine curve. If $|a| > 1$, the basic sine curve is stretched, and if $|a| < 1$, the basic sine curve is shrunk. The result is that the graph of $y = a\sin x$ ranges between $-a$ and a instead of between -1 and 1. The absolute value of a is the **amplitude** of the function $y = a\sin x$. The range of the function $y = a\sin x$ for $a > 0$ is $-a \le y \le a$.

Definition of Amplitude of Sine and Cosine Curves

The **amplitude** of

$$y = a\sin x \text{ and } y = a\cos x$$

represents half the distance between the maximum and minimum values of the function and is given by

$$\text{Amplitude} = |a|.$$

Example 2   Scaling: Vertical Shrinking and Stretching

On the same set of coordinate axes, sketch the graph of each function by hand.

a. $y = \frac{1}{2}\cos x$ **b.** $y = 3\cos x$

Solution

a. Because the amplitude of $y = \frac{1}{2}\cos x$ is $\frac{1}{2}$, the maximum value is $\frac{1}{2}$ and the minimum value is $-\frac{1}{2}$. Divide one cycle, $0 \le x \le 2\pi$, into four equal parts to get the key points

Maximum	Intercept	Minimum	Intercept	Maximum
$\left(0, \frac{1}{2}\right),$	$\left(\frac{\pi}{2}, 0\right),$	$\left(\pi, -\frac{1}{2}\right),$	$\left(\frac{3\pi}{2}, 0\right),$ and	$\left(2\pi, \frac{1}{2}\right).$

b. A similar analysis shows that the amplitude of $y = 3\cos x$ is 3, and the key points are

Maximum	Intercept	Minimum	Intercept	Maximum
$(0, 3),$	$\left(\frac{\pi}{2}, 0\right),$	$(\pi, -3),$	$\left(\frac{3\pi}{2}, 0\right),$ and	$(2\pi, 3).$

The graphs of these two functions are shown in Figure 4.46. Notice that the graph of $y = \frac{1}{2}\cos x$ is a vertical shrink of the graph of $y = \cos x$ and the graph of $y = 3\cos x$ is a vertical stretch of the graph of $y = \cos x$. Use a graphing utility to confirm these graphs.

✔**CHECKPOINT**   Now try Exercise 40.

TECHNOLOGY TIP

Prerequisite Skills

For a review of transformations of functions, see Section 1.4.

When using a graphing utility to graph trigonometric functions, pay special attention to the viewing window you use. For instance, try graphing $y = [\sin(10x)]/10$ in the standard viewing window in *radian* mode. What do you observe? Use the *zoom* feature to find a viewing window that displays a good view of the graph. For instructions on how to use the *zoom* feature, see Appendix A; for specific keystrokes, go to this textbook's *Online Study Center.*

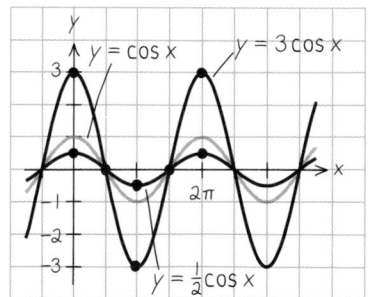

Figure 4.46

You know from Section 1.4 that the graph of $y = -f(x)$ is a *reflection* in the x-axis of the graph of $y = f(x)$. For instance, the graph of $y = -3 \cos x$ is a reflection of the graph of $y = 3 \cos x$, as shown in Figure 4.47.

Because $y = a \sin x$ completes one cycle from $x = 0$ to $x = 2\pi$, it follows that $y = a \sin bx$ completes one cycle from $x = 0$ to $x = 2\pi/b$.

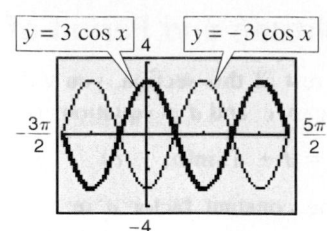

Figure 4.47

> **Period of Sine and Cosine Functions**
>
> Let b be a positive real number. The **period** of $y = a \sin bx$ and $y = a \cos bx$ is given by
>
> $$\text{Period} = \frac{2\pi}{b}.$$

Note that if $0 < b < 1$, the period of $y = a \sin bx$ is greater than 2π and represents a *horizontal stretching* of the graph of $y = a \sin x$. Similarly, if $b > 1$, the period of $y = a \sin bx$ is less than 2π and represents a *horizontal shrinking* of the graph of $y = a \sin x$. If b is negative, the identities $\sin(-x) = -\sin x$ and $\cos(-x) = \cos x$ are used to rewrite the function.

Example 3   Scaling: Horizontal Stretching

Sketch the graph of $y = \sin \dfrac{x}{2}$ by hand.

Solution

The amplitude is 1. Moreover, because $b = \frac{1}{2}$, the period is

$$\frac{2\pi}{b} = \frac{2\pi}{\frac{1}{2}} = 4\pi. \qquad \text{Substitute for } b.$$

Now, divide the period-interval $[0, 4\pi]$ into four equal parts with the values π, 2π, and 3π to obtain the key points on the graph

Intercept	Maximum	Intercept	Minimum		Intercept
$(0, 0)$,	$(\pi, 1)$,	$(2\pi, 0)$,	$(3\pi, -1)$,	and	$(4\pi, 0)$.

The graph is shown in Figure 4.48. Use a graphing utility to confirm this graph.

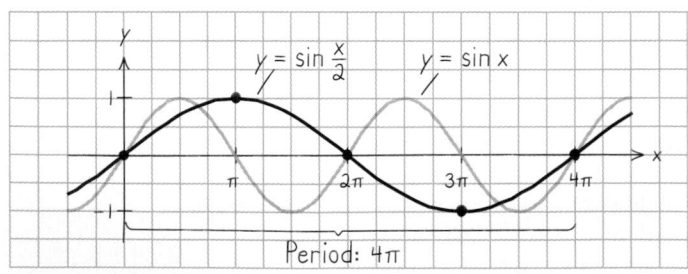

Figure 4.48

✓CHECKPOINT   Now try Exercise 41.

STUDY TIP

In general, to divide a period-interval into four equal parts, successively add "period/4," starting with the left endpoint of the interval. For instance, for the period-interval $[-\pi/6, \pi/2]$ of length $2\pi/3$, you would successively add

$$\frac{2\pi/3}{4} = \frac{\pi}{6}$$

to get $-\pi/6, 0, \pi/6, \pi/3$, and $\pi/2$ as the key points on the graph.

Translations of Sine and Cosine Curves

The constant c in the general equations

$$y = a \sin(bx - c) \qquad \text{and} \qquad y = a \cos(bx - c)$$

creates *horizontal translations* (shifts) of the basic sine and cosine curves. Comparing $y = a \sin bx$ with $y = a \sin(bx - c)$, you find that the graph of $y = a \sin(bx - c)$ completes one cycle from $bx - c = 0$ to $bx - c = 2\pi$. By solving for x, you can find the interval for one cycle to be

Left endpoint   Right endpoint

$$\frac{c}{b} \le x \le \frac{c}{b} + \frac{2\pi}{b}.$$

Period

This implies that the period of $y = a \sin(bx - c)$ is $2\pi/b$, and the graph of $y = a \sin bx$ is shifted by an amount c/b. The number c/b is the **phase shift.**

> **Graphs of Sine and Cosine Functions**
>
> The graphs of $y = a \sin(bx - c)$ and $y = a \cos(bx - c)$ have the following characteristics. (Assume $b > 0$.)
>
> $$\text{Amplitude} = |a| \qquad \text{Period} = 2\pi/b$$
>
> The left and right endpoints of a one-cycle interval can be determined by solving the equations $bx - c = 0$ and $bx - c = 2\pi$.

Example 4   Horizontal Translation

Analyze the graph of $y = \frac{1}{2} \sin\left(x - \frac{\pi}{3}\right)$.

Algebraic Solution

The amplitude is $\frac{1}{2}$ and the period is 2π. By solving the equations

$$x - \frac{\pi}{3} = 0 \qquad \text{and} \qquad x - \frac{\pi}{3} = 2\pi$$

$$x = \frac{\pi}{3} \qquad\qquad\qquad x = \frac{7\pi}{3}$$

you see that the interval $[\pi/3, 7\pi/3]$ corresponds to one cycle of the graph. Dividing this interval into four equal parts produces the following key points.

Intercept	*Maximum*	*Intercept*	*Minimum*	*Intercept*
$\left(\frac{\pi}{3}, 0\right)$,	$\left(\frac{5\pi}{6}, \frac{1}{2}\right)$,	$\left(\frac{4\pi}{3}, 0\right)$,	$\left(\frac{11\pi}{6}, -\frac{1}{2}\right)$,	$\left(\frac{7\pi}{3}, 0\right)$

Graphical Solution

Use a graphing utility set in *radian* mode to graph $y = (1/2) \sin(x - \pi/3)$, as shown in Figure 4.49. Use the *minimum*, *maximum*, and *zero* or *root* features of the graphing utility to approximate the key points $(1.05, 0)$, $(2.62, 0.5)$, $(4.19, 0)$, $(5.76, -0.5)$, and $(7.33, 0)$.

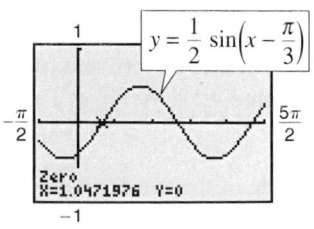

$y = \frac{1}{2} \sin\left(x - \frac{\pi}{3}\right)$

Zero
X=1.0471976 Y=0

Figure 4.49

✓CHECKPOINT   Now try Exercise 43.

Example 5   Horizontal Translation

Use a graphing utility to analyze the graph of $y = -3\cos(2\pi x + 4\pi)$.

Solution

The amplitude is 3 and the period is $2\pi/2\pi = 1$. By solving the equations

$$2\pi x + 4\pi = 0 \qquad \text{and} \qquad 2\pi x + 4\pi = 2\pi$$

$$2\pi x = -4\pi \qquad\qquad 2\pi x = -2\pi$$

$$x = -2 \qquad\qquad\qquad x = -1$$

you see that the interval $[-2, -1]$ corresponds to one cycle of the graph. Dividing this interval into four equal parts produces the key points

Minimum	Intercept	Maximum	Intercept	Minimum
$(-2, -3),$	$(-7/4, 0),$	$(-3/2, 3),$	$(-5/4, 0),$ and	$(-1, -3).$

The graph is shown in Figure 4.50.

✓CHECKPOINT   Now try Exercise 45.

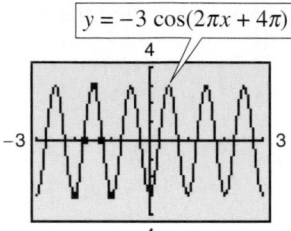

Figure 4.50

The final type of transformation is the *vertical translation* caused by the constant d in the equations

$$y = d + a\sin(bx - c) \qquad \text{and} \qquad y = d + a\cos(bx - c).$$

The shift is d units upward for $d > 0$ and d units downward for $d < 0$. In other words, the graph oscillates about the horizontal line $y = d$ instead of about the x-axis.

Example 6   Vertical Translation

Use a graphing utility to analyze the graph of $y = 2 + 3\cos 2x$.

Solution

The amplitude is 3 and the period is π. The key points over the interval $[0, \pi]$ are

$$(0, 5), \qquad (\pi/4, 2), \qquad (\pi/2, -1), \qquad (3\pi/4, 2), \qquad \text{and} \qquad (\pi, 5).$$

The graph is shown in Figure 4.51. Compared with the graph of $f(x) = 3\cos 2x$, the graph of $y = 2 + 3\cos 2x$ is shifted upward two units.

✓CHECKPOINT   Now try Exercise 49.

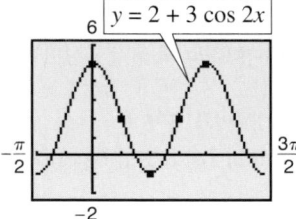

Figure 4.51

Example 7   Finding an Equation for a Graph

Find the amplitude, period, and phase shift for the sine function whose graph is shown in Figure 4.52. Write an equation for this graph.

Solution

The amplitude of this sine curve is 2. The period is 2π, and there is a right phase shift of $\pi/2$. So, you can write $y = 2\sin(x - \pi/2)$.

✓CHECKPOINT   Now try Exercise 65.

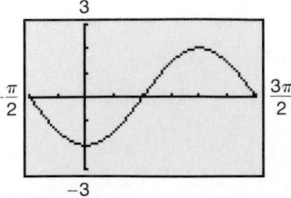

Figure 4.52

Mathematical Modeling

Sine and cosine functions can be used to model many real-life situations, including electric currents, musical tones, radio waves, tides, and weather patterns.

Example 8   Finding a Trigonometric Model

Throughout the day, the depth of the water at the end of a dock in Bangor, Washington varies with the tides. The table shows the depths (in feet) at various times during the morning.   (Source: Nautical Software, Inc.)

Time	Depth, y
Midnight	3.1
2 A.M.	7.8
4 A.M.	11.3
6 A.M.	10.9
8 A.M.	6.6
10 A.M.	1.7
Noon	0.9

a. Use a trigonometric function to model the data.

b. A boat needs at least 10 feet of water to moor at the dock. During what times in the evening can it safely dock?

Solution

a. Begin by graphing the data, as shown in Figure 4.53. You can use either a sine or cosine model. Suppose you use a cosine model of the form

$$y = a \cos(bt - c) + d.$$

The difference between the maximum height and minimum height of the graph is twice the amplitude of the function. So, the amplitude is

$$a = \tfrac{1}{2}[(\text{maximum depth}) - (\text{minimum depth})] = \tfrac{1}{2}(11.3 - 0.9) = 5.2.$$

The cosine function completes one half of a cycle between the times at which the maximum and minimum depths occur. So, the period is

$$p = 2[(\text{time of min. depth}) - (\text{time of max. depth})] = 2(12 - 4) = 16$$

which implies that $b = 2\pi/p \approx 0.393$. Because high tide occurs 4 hours after midnight, consider the left endpoint to be $c/b = 4$, so $c \approx 1.571$. Moreover, because the average depth is $\tfrac{1}{2}(11.3 + 0.9) = 6.1$, it follows that $d = 6.1$. So, you can model the depth with the function

$$y = 5.2 \cos(0.393t - 1.571) + 6.1.$$

b. Using a graphing utility, graph the model with the line $y = 10$. Using the *intersect* feature, you can determine that the depth is at least 10 feet between 6:06 P.M. ($t \approx 18.1$) and 9:48 P.M. ($t \approx 21.8$), as shown in Figure 4.54.

✓CHECKPOINT   Now try Exercise 77.

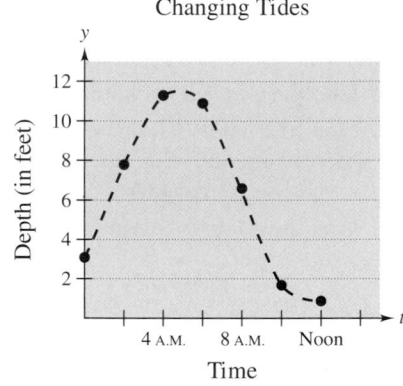

Changing Tides

Figure 4.53

TECHNOLOGY SUPPORT

For instructions on how to use the *intersect* feature, see Appendix A; for specific keystrokes, go to this textbook's *Online Study Center*.

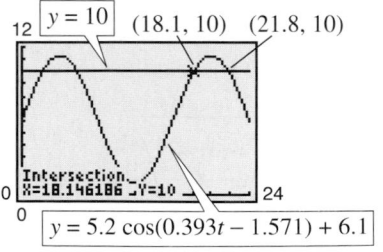

Figure 4.54

4.5 Exercises

See www.CalcChat.com for worked-out solutions to odd-numbered exercises.

Vocabulary Check

Fill in the blanks.

1. The _____ of a sine or cosine curve represents half the distance between the maximum and minimum values of the function.

2. One period of a sine function is called _____ of the sine curve.

3. The period of a sine or cosine function is given by _____ .

4. For the equation $y = a\sin(bx - c)$, $\dfrac{c}{b}$ is the _____ of the graph of the equation.

Library of Parent Functions In Exercises 1 and 2, use the graph of the function to answer the following.

(a) Find the x-intercepts of the graph of $y = f(x)$.

(b) Find the y-intercepts of the graph of $y = f(x)$.

(c) Find the intervals on which the graph $y = f(x)$ is increasing and the intervals on which the graph $y = f(x)$ is decreasing.

(d) Find the relative extrema of the graph of $y = f(x)$.

1. $f(x) = \sin x$

2. $f(x) = \cos x$

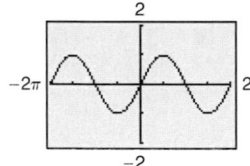

In Exercises 3–14, find the period and amplitude.

3. $y = 3\sin 2x$

4. $y = 2\cos 3x$

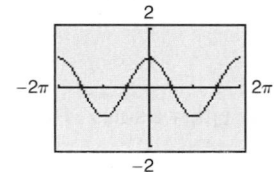

5. $y = \dfrac{5}{2}\cos\dfrac{x}{2}$

6. $y = -3\sin\dfrac{x}{3}$

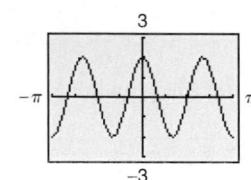

7. $y = \dfrac{2}{3}\sin \pi x$

8. $y = \dfrac{3}{2}\cos\dfrac{\pi x}{2}$

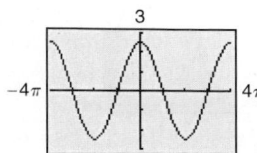

9. $y = -2\sin x$

10. $y = -\cos\dfrac{2x}{5}$

11. $y = \dfrac{1}{4}\cos\dfrac{2x}{3}$

12. $y = \dfrac{5}{2}\cos\dfrac{x}{4}$

13. $y = \dfrac{1}{3}\sin 4\pi x$

14. $y = \dfrac{3}{4}\cos\dfrac{\pi x}{12}$

In Exercises 15–22, describe the relationship between the graphs of f and g. Consider amplitudes, periods, and shifts.

15. $f(x) = \sin x$
    $g(x) = \sin(x - \pi)$

16. $f(x) = \cos x$
    $g(x) = \cos(x + \pi)$

17. $f(x) = \cos 2x$
    $g(x) = -\cos 2x$

18. $f(x) = \sin 3x$
    $g(x) = \sin(-3x)$

19. $f(x) = \cos x$
    $g(x) = -5\cos x$

20. $f(x) = \sin x$
    $g(x) = -\dfrac{1}{2}\sin x$

21. $f(x) = \sin 2x$
    $g(x) = 5 + \sin 2x$

22. $f(x) = \cos 4x$
    $g(x) = -6 + \cos 4x$

In Exercises 23–26, describe the relationship between the graphs of f and g. Consider amplitudes, periods, and shifts.

23.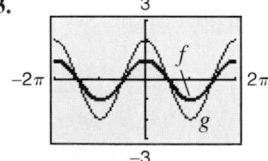

24.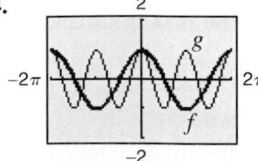

25.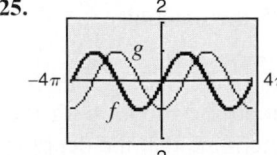

26.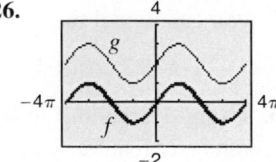

In Exercises 27–34, sketch the graphs of f and g in the same coordinate plane. (Include two full periods.)

27. $f(x) = \sin x$

$g(x) = -4 \sin x$

28. $f(x) = \sin x$

$g(x) = \sin \dfrac{x}{3}$

29. $f(x) = \cos x$

$g(x) = 4 + \cos x$

30. $f(x) = 2 \cos 2x$

$g(x) = -\cos 4x$

31. $f(x) = -\dfrac{1}{2} \sin \dfrac{x}{2}$

$g(x) = 3 - \dfrac{1}{2} \sin \dfrac{x}{2}$

32. $f(x) = 4 \sin \pi x$

$g(x) = 4 \sin \pi x - 2$

33. $f(x) = 2 \cos x$

$g(x) = 2 \cos(x + \pi)$

34. $f(x) = -\cos x$

$g(x) = -\cos\left(x - \dfrac{\pi}{2}\right)$

Conjecture   In Exercises 35–38, use a graphing utility to graph f and g in the same viewing window. (Include two full periods.) Make a conjecture about the functions.

35. $f(x) = \sin x$

$g(x) = \cos\left(x - \dfrac{\pi}{2}\right)$

36. $f(x) = \sin x$

$g(x) = -\cos\left(x + \dfrac{\pi}{2}\right)$

37. $f(x) = \cos x$

$g(x) = -\sin\left(x - \dfrac{\pi}{2}\right)$

38. $f(x) = \cos x$

$g(x) = -\cos(x - \pi)$

In Exercises 39–46, sketch the graph of the function by hand. Use a graphing utility to verify your sketch. (Include two full periods.)

39. $y = 3 \sin x$

40. $y = \frac{1}{4} \cos x$

41. $y = \cos \dfrac{x}{2}$

42. $y = \sin 4x$

43. $y = \sin\left(x - \dfrac{\pi}{4}\right)$

44. $y = \sin(x - \pi)$

45. $y = -8 \cos(x + \pi)$

46. $y = 3 \cos\left(x + \dfrac{\pi}{2}\right)$

In Exercises 47–60, use a graphing utility to graph the function. (Include two full periods.) Identify the amplitude and period of the graph.

47. $y = -2 \sin \dfrac{2\pi x}{3}$

48. $y = -10 \cos \dfrac{\pi x}{6}$

49. $y = -4 + 5 \cos \dfrac{\pi t}{12}$

50. $y = 2 - 2 \sin \dfrac{2\pi x}{3}$

51. $y = \dfrac{2}{3} \cos\left(\dfrac{x}{2} - \dfrac{\pi}{4}\right)$

52. $y = -3 \cos(6x + \pi)$

53. $y = -2 \sin(4x + \pi)$

54. $y = -4 \sin\left(\dfrac{2}{3}x - \dfrac{\pi}{3}\right)$

55. $y = \cos\left(2\pi x - \dfrac{\pi}{2}\right) + 1$

56. $y = 3 \cos\left(\dfrac{\pi x}{2} + \dfrac{\pi}{2}\right) - 3$

57. $y = 5 \sin(\pi - 2x) + 10$

58. $y = 5 \cos(\pi - 2x) + 6$

59. $y = \frac{1}{100} \sin 120\pi t$

60. $y = -\frac{1}{100} \cos 50\pi t$

Graphical Reasoning   In Exercises 61–64, find a and d for the function $f(h) = a \cos x + d$ such that the graph of f matches the figure.

61.

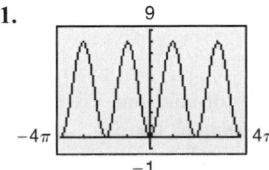

62.

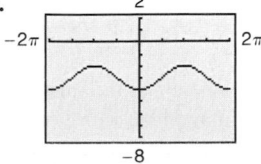

63.

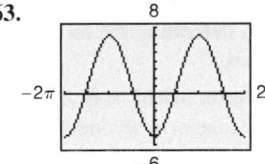

64.

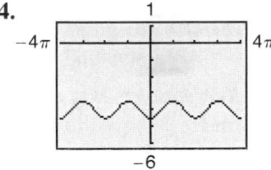

Graphical Reasoning   In Exercises 65–68, find a, b, and c for the function $f(x) = a \sin(bx - c)$ such that the graph of f matches the graph shown.

65.

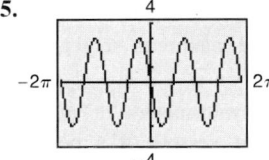

66.

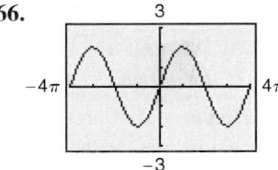

67.

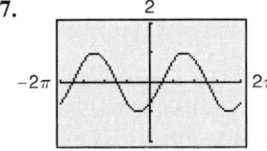

68.

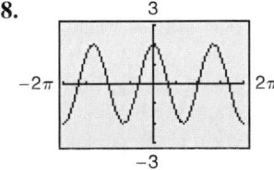

In Exercises 69 and 70, use a graphing utility to graph y_1 and y_2 for all real numbers x in the interval $[-2\pi, 2\pi]$. Use the graphs to find the real numbers x such that $y_1 = y_2$.

69. $y_1 = \sin x$

$y_2 = -\dfrac{1}{2}$

70. $y_1 = \cos x$

$y_2 = -1$

71. *Health* For a person at rest, the velocity v (in liters per second) of air flow during a respiratory cycle (the time from the beginning of one breath to the beginning of the next) is given by $v = 0.85 \sin(\pi t/3)$, where t is the time (in seconds). (Inhalation occurs when $v > 0$, and exhalation occurs when $v < 0$.)

(a) Use a graphing utility to graph v.

(b) Find the time for one full respiratory cycle.

(c) Find the number of cycles per minute.

(d) The model is for a person at rest. How might the model change for a person who is exercising? Explain.

72. *Sales* A company that produces snowboards, which are seasonal products, forecasts monthly sales for 1 year to be

$$S = 74.50 + 43.75 \cos \frac{\pi t}{6}$$

where S is the sales in thousands of units and t is the time in months, with $t = 1$ corresponding to January.

(a) Use a graphing utility to graph the sales function over the one-year period.

(b) Use the graph in part (a) to determine the months of maximum and minimum sales.

73. *Recreation* You are riding a Ferris wheel. Your height h (in feet) above the ground at any time t (in seconds) can be modeled by

$$h = 25 \sin \frac{\pi}{15}(t - 75) + 30.$$

The Ferris wheel turns for 135 seconds before it stops to let the first passengers off.

(a) Use a graphing utility to graph the model.

(b) What are the minimum and maximum heights above the ground?

74. *Health* The pressure P (in millimeters of mercury) against the walls of the blood vessels of a person is modeled by

$$P = 100 - 20 \cos \frac{8\pi}{3}t$$

where t is the time (in seconds). Use a graphing utility to graph the model. One cycle is equivalent to one heartbeat. What is the person's pulse rate in heartbeats per minute?

75. *Fuel Consumption* The daily consumption C (in gallons) of diesel fuel on a farm is modeled by

$$C = 30.3 + 21.6 \sin\left(\frac{2\pi t}{365} + 10.9\right)$$

where t is the time in days, with $t = 1$ corresponding to January 1.

(a) What is the period of the model? Is it what you expected? Explain.

(b) What is the average daily fuel consumption? Which term of the model did you use? Explain.

(c) Use a graphing utility to graph the model. Use the graph to approximate the time of the year when consumption exceeds 40 gallons per day.

76. *Data Analysis* The motion of an oscillating weight suspended from a spring was measured by a motion detector. The data was collected, and the approximate maximum displacements from equilibrium $(y = 2)$ are labeled in the figure. The distance y from the motion detector is measured in centimeters and the time t is measured in seconds.

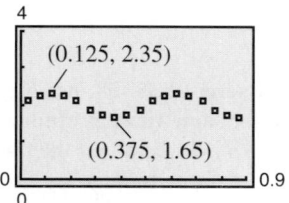

(a) Is y a function of t? Explain.

(b) Approximate the amplitude and period.

(c) Find a model for the data.

(d) Use a graphing utility to graph the model in part (c). Compare the result with the data in the figure.

77. *Data Analysis* The percent y (in decimal form) of the moon's face that is illuminated on day x of the year 2006, where $x = 1$ represents January 1, is shown in the table. (Source: U.S. Naval Observatory)

Day, x	Percent, y
29	0.0
36	0.5
44	1.0
52	0.5
59	0.0
66	0.5

(a) Create a scatter plot of the data.

(b) Find a trigonometric model for the data.

(c) Add the graph of your model in part (b) to the scatter plot. How well does the model fit the data?

(d) What is the period of the model?

(e) Estimate the percent illumination of the moon on June 29, 2007.

78. *Data Analysis* The table shows the average daily high temperatures for Nantucket, Massachusetts N and Athens, Georgia A (in degrees Fahrenheit) for month t, with $t = 1$ corresponding to January.   (Source: U.S. Weather Bureau and the National Weather Service)

Month, t	Nantucket, N	Athens, A
1	40	52
2	41	56
3	42	65
4	53	73
5	62	81
6	71	87
7	78	90
8	76	88
9	70	83
10	59	74
11	48	64
12	40	55

(a) A model for the temperature in Nantucket is given by

$$N(t) = 58 + 19 \sin\left(\frac{2\pi t}{11} - \frac{21\pi}{25}\right).$$

Find a trigonometric model for Athens.

(b) Use a graphing utility to graph the data and the model for the temperatures in Nantucket in the same viewing window. How well does the model fit the data?

(c) Use a graphing utility to graph the data and the model for the temperatures in Athens in the same viewing window. How well does the model fit the data?

(d) Use the models to estimate the average daily high temperature in each city. Which term of the models did you use? Explain.

(e) What is the period of each model? Are the periods what you expected? Explain.

(f) Which city has the greater variability in temperature throughout the year? Which factor of the models determines this variability? Explain.

Synthesis

True or False? In Exercises 79–81, determine whether the statement is true or false. Justify your answer.

79. The graph of $y = 6 - \frac{3}{4}\sin\frac{3x}{10}$ has a period of $\frac{20\pi}{3}$.

80. The function $y = \frac{1}{2}\cos 2x$ has an amplitude that is twice that of the function $y = \cos x$.

81. The graph of $y = -\cos x$ is a reflection of the graph of $y = \sin(x + \pi/2)$ in the x-axis.

82. *Writing* Use a graphing utility to graph the function

$$y = d + a\sin(bx - c)$$

for different values of a, b, c, and d. Write a paragraph describing the changes in the graph corresponding to changes in each constant.

Library of Parent Functions In Exercises 83–86, determine which function is represented by the graph. Do not use a calculator.

83.

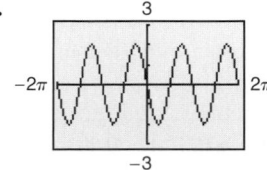

(a) $f(x) = 2\sin 2x$

(b) $f(x) = -2\sin\frac{x}{2}$

(c) $f(x) = -2\cos 2x$

(d) $f(x) = 2\cos\frac{x}{2}$

(e) $f(x) = -2\sin 2x$

84.

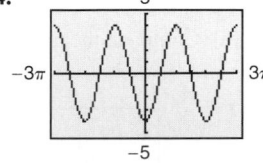

(a) $f(x) = 4\cos(x + \pi)$

(b) $f(x) = 4\cos(4x)$

(c) $f(x) = 4\sin(x - \pi)$

(d) $f(x) = -4\cos(x + \pi)$

(e) $f(x) = 1 - \sin\frac{x}{2}$

85.

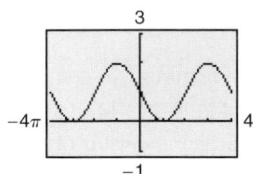

(a) $f(x) = 1 + \sin\frac{x}{2}$

(b) $f(x) = 1 + \cos\frac{x}{2}$

(c) $f(x) = 1 - \sin\frac{x}{2}$

(d) $f(x) = 1 - \cos 2x$

(e) $f(x) = 1 - \sin 2x$

86.

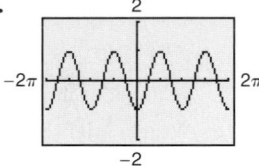

(a) $f(x) = \cos 2x$

(b) $f(x) = \sin\left(\frac{x}{2} - \pi\right)$

(c) $f(x) = \sin(2x + \pi)$

(d) $f(x) = \cos(2x - \pi)$

(e) $f(x) = \sin\frac{x}{2}$

87. *Exploration* In Section 4.2, it was shown that $f(x) = \cos x$ is an even function and $g(x) = \sin x$ is an odd function. Use a graphing utility to graph h and use the graph to determine whether h is even, odd, or neither.

(a) $h(x) = \cos^2 x$ (b) $h(x) = \sin^2 x$

(c) $h(x) = \sin x \cos x$

88. *Conjecture* If f is an even function and g is an odd function, use the results of Exercise 87 to make a conjecture about each of the following.

(a) $h(x) = [f(x)]^2$ (b) $h(x) = [g(x)]^2$

(c) $h(x) = f(x)g(x)$

∫ **89. *Exploration*** Use a graphing utility to explore the ratio $(\sin x)/x$, which appears in calculus.

(a) Complete the table. Round your results to four decimal places.

x	-1	-0.1	-0.01	-0.001
$\dfrac{\sin x}{x}$				

x	0	0.001	0.01	0.1	1
$\dfrac{\sin x}{x}$					

(b) Use a graphing utility to graph the function
$$f(x) = \frac{\sin x}{x}.$$

Use the *zoom* and *trace* features to describe the behavior of the graph as x approaches 0.

(c) Write a brief statement regarding the value of the ratio based on your results in parts (a) and (b).

∫ **90. *Exploration*** Use a graphing utility to explore the ratio $(1 - \cos x)/x$, which appears in calculus.

(a) Complete the table. Round your results to four decimal places.

x	-1	-0.1	-0.01	-0.001
$\dfrac{1 - \cos x}{x}$				

x	0	0.001	0.01	0.1	1
$\dfrac{1 - \cos x}{x}$					

(b) Use a graphing utility to graph the function
$$f(x) = \frac{1 - \cos x}{x}.$$

Use the *zoom* and *trace* features to describe the behavior of the graph as x approaches 0.

(c) Write a brief statement regarding the value of the ratio based on your results in parts (a) and (b).

∫ **91. *Exploration*** Using calculus, it can be shown that the sine and cosine functions can be approximated by the polynomials

$$\sin x \approx x - \frac{x^3}{3!} + \frac{x^5}{5!} \quad \text{and} \quad \cos x \approx 1 - \frac{x^2}{2!} + \frac{x^4}{4!}$$

where x is in radians.

(a) Use a graphing utility to graph the sine function and its polynomial approximation in the same viewing window. How do the graphs compare?

(b) Use a graphing utility to graph the cosine function and its polynomial approximation in the same viewing window. How do the graphs compare?

(c) Study the patterns in the polynomial approximations of the sine and cosine functions and predict the next term in each. Then repeat parts (a) and (b). How did the accuracy of the approximations change when an additional term was added?

∫ **92. *Exploration*** Use the polynomial approximations found in Exercise 91(c) to approximate the following values. Round your answers to four decimal places. Compare the results with those given by a calculator. How does the error in the approximation change as x approaches 0?

(a) $\sin 1$ (b) $\sin \dfrac{1}{2}$ (c) $\sin \dfrac{\pi}{8}$

(d) $\cos(-1)$ (e) $\cos\left(-\dfrac{\pi}{4}\right)$ (f) $\cos\left(-\dfrac{1}{2}\right)$

Skills Review

In Exercises 93 and 94, plot the points and find the slope of the line passing through the points.

93. $(0, 1), (2, 7)$ **94.** $(-1, 4), (3, -2)$

In Exercises 95 and 96, convert the angle measure from radians to degrees. Round your answer to three decimal places.

95. 8.5 **96.** -0.48

97. *Make a Decision* To work an extended application analyzing the normal daily maximum temperature and normal precipitation in Honolulu, Hawaii, visit this textbook's *Online Study Center*. (Data Source: NOAA)

4.6 Graphs of Other Trigonometric Functions

Graph of the Tangent Function

Recall that the tangent function is odd. That is, $\tan(-x) = -\tan x$. Consequently, the graph of $y = \tan x$ is symmetric with respect to the origin. You also know from the identity $\tan x = \sin x / \cos x$ that the tangent function is undefined at values at which $\cos x = 0$. Two such values are $x = \pm \pi/2 \approx \pm 1.5708$.

x	$-\dfrac{\pi}{2}$	-1.57	-1.5	$-\dfrac{\pi}{4}$	0	$\dfrac{\pi}{4}$	1.5	1.57	$\dfrac{\pi}{2}$
$\tan x$	Undef.	-1255.8	-14.1	-1	0	1	14.1	1255.8	Undef.

> $\tan x$ approaches $-\infty$ as x approaches $-\pi/2$ from the right.

> $\tan x$ approaches ∞ as x approaches $\pi/2$ from the left.

As indicated in the table, $\tan x$ increases without bound as x approaches $\pi/2$ from the left, and it decreases without bound as x approaches $-\pi/2$ from the right. So, the graph of $y = \tan x$ has *vertical asymptotes* at $x = \pi/2$ and $x = -\pi/2$, as shown in Figure 4.55. Moreover, because the period of the tangent function is π, vertical asymptotes also occur at $x = \pi/2 + n\pi$, where n is an integer. The domain of the tangent function is the set of all real numbers other than $x = \pi/2 + n\pi$, and the range is the set of all real numbers.

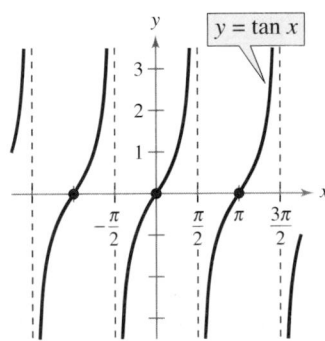

Figure 4.55

Period: π

Domain: all real numbers x,

except $x = \dfrac{\pi}{2} + n\pi$

Range: $(-\infty, \infty)$

Vertical asymptotes: $x = \dfrac{\pi}{2} + n\pi$

Sketching the graph of $y = a \tan(bx - c)$ is similar to sketching the graph of $y = a \sin(bx - c)$ in that you locate key points that identify the intercepts and asymptotes. Two consecutive asymptotes can be found by solving the equations $bx - c = -\pi/2$ and $bx - c = \pi/2$. The midpoint between two consecutive asymptotes is an x-intercept of the graph. The period of the function $y = a \tan(bx - c)$ is the distance between two consecutive asymptotes. The amplitude of a tangent function is not defined. After plotting the asymptotes and the x-intercept, plot a few additional points between the two asymptotes and sketch one cycle. Finally, sketch one or two additional cycles to the left and right.

Example 1   Sketching the Graph of a Tangent Function

Sketch the graph of $y = \tan \dfrac{x}{2}$ by hand.

Solution

By solving the equations $x/2 = -\pi/2$ and $x/2 = \pi/2$, you can see that two consecutive asymptotes occur at $x = -\pi$ and $x = \pi$. Between these two asymptotes, plot a few points, including the x-intercept, as shown in the table. Three cycles of the graph are shown in Figure 4.56. Use a graphing utility to confirm this graph.

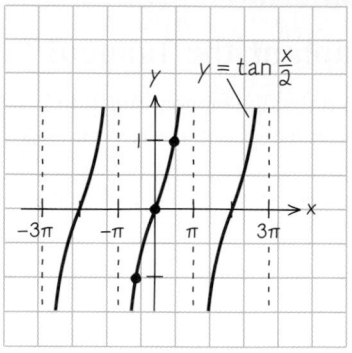

Figure 4.56

x	$-\pi$	$-\dfrac{\pi}{2}$	0	$\dfrac{\pi}{2}$	π
$\tan \dfrac{x}{2}$	Undef.	-1	0	1	Undef.

✔CHECKPOINT   Now try Exercise 5.

Example 2   Sketching the Graph of a Tangent Function

Sketch the graph of $y = -3 \tan 2x$ by hand.

Solution

By solving the equations $2x = -\pi/2$ and $2x = \pi/2$, you can see that two consecutive asymptotes occur at $x = -\pi/4$ and $x = \pi/4$. Between these two asymptotes, plot a few points, including the x-intercept, as shown in the table. Three complete cycles of the graph are shown in Figure 4.57.

x	$-\dfrac{\pi}{4}$	$-\dfrac{\pi}{8}$	0	$\dfrac{\pi}{8}$	$\dfrac{\pi}{4}$
$-3 \tan 2x$	Undef.	3	0	-3	Undef.

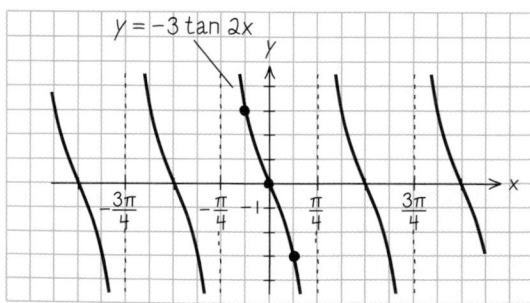

Figure 4.57

✔CHECKPOINT   Now try Exercise 7.

TECHNOLOGY TIP

Your graphing utility may connect parts of the graphs of tangent, cotangent, secant, and cosecant functions that are not supposed to be connected. So, in this text, these functions are graphed on a graphing utility using the *dot* mode. A blue curve is placed behind the graphing utility's display to indicate where the graph should appear. For instructions on how to use the *dot* mode, see Appendix A; for specific keystrokes, go to this textbook's *Online Study Center*.

Graphing utilities are helpful in verifying sketches of trigonometric functions. You can use a graphing utility set in *radian* and *dot* modes to graph the function $y = -3 \tan 2x$ from Example 2, as shown in Figure 4.58. You can use the *zero* or *root* feature or the *zoom* and *trace* features to approximate the key points of the graph.

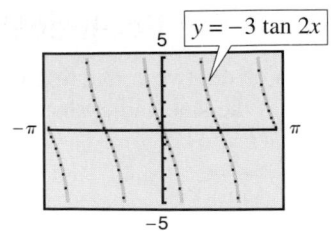

Figure 4.58

By comparing the graphs in Examples 1 and 2, you can see that the graph of $y = a \tan(bx - c)$ increases between consecutive vertical asymptotes when $a > 0$ and decreases between consecutive vertical asymptotes when $a < 0$. In other words, the graph for $a < 0$ is a reflection in the x-axis of the graph for $a > 0$.

Graph of the Cotangent Function

The graph of the cotangent function is similar to the graph of the tangent function. It also has a period of π. However, from the identity

$$y = \cot x = \frac{\cos x}{\sin x}$$

you can see that the cotangent function has vertical asymptotes when $\sin x$ is zero, which occurs at $x = n\pi$, where n is an integer. The graph of the cotangent function is shown in Figure 4.59.

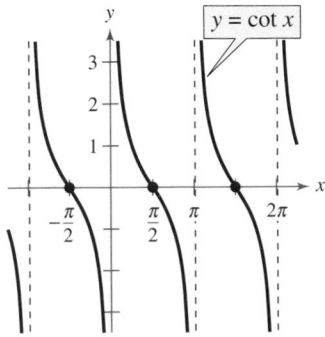

Period: π
Domain: all real numbers x,
      except $x = n\pi$
Range: $(-\infty, \infty)$
Vertical asymptotes: $x = n\pi$
Figure 4.59

Example 3   Sketching the Graph of a Cotangent Function

Sketch the graph of $y = 2 \cot \dfrac{x}{3}$ by hand.

Solution

To locate two consecutive vertical asymptotes of the graph, solve the equations $x/3 = 0$ and $x/3 = \pi$ to see that two consecutive asymptotes occur at $x = 0$ and $x = 3\pi$. Then, between these two asymptotes, plot a few points, including the x-intercept, as shown in the table. Three cycles of the graph are shown in Figure 4.60. Use a graphing utility to confirm this graph. [Enter the function as $y = 2/\tan(x/3)$.] Note that the period is 3π, the distance between consecutive asymptotes.

x	0	$\dfrac{3\pi}{4}$	$\dfrac{3\pi}{2}$	$\dfrac{9\pi}{4}$	3π
$2 \cot \dfrac{x}{3}$	Undef.	2	0	-2	Undef.

✔CHECKPOINT   Now try Exercise 15.

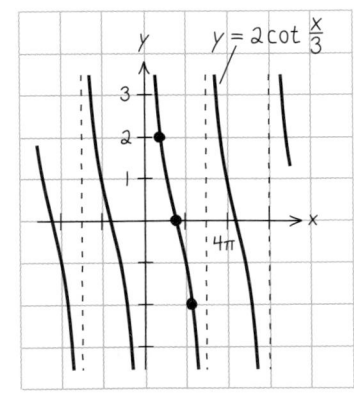

Figure 4.60

Graphs of the Reciprocal Functions

The graphs of the two remaining trigonometric functions can be obtained from the graphs of the sine and cosine functions using the reciprocal identities

$$\csc x = \frac{1}{\sin x} \quad \text{and} \quad \sec x = \frac{1}{\cos x}.$$

For instance, at a given value of x, the y-coordinate for $\sec x$ is the reciprocal of the y-coordinate for $\cos x$. Of course, when $\cos x = 0$, the reciprocal does not exist. Near such values of x, the behavior of the secant function is similar to that of the tangent function. In other words, the graphs of

$$\tan x = \frac{\sin x}{\cos x} \quad \text{and} \quad \sec x = \frac{1}{\cos x}$$

have vertical asymptotes at $x = \pi/2 + n\pi$, where n is an integer (i.e., the values at which the cosine is zero). Similarly,

$$\cot x = \frac{\cos x}{\sin x} \quad \text{and} \quad \csc x = \frac{1}{\sin x}$$

have vertical asymptotes where $\sin x = 0$—that is, at $x = n\pi$.

To sketch the graph of a secant or cosecant function, you should first make a sketch of its reciprocal function. For instance, to sketch the graph of $y = \csc x$, first sketch the graph of $y = \sin x$. Then take the reciprocals of the y-coordinates to obtain points on the graph of $y = \csc x$. You can use this procedure to obtain the graphs shown in Figure 4.61.

Prerequisite Skills

To review the reciprocal identities of trigonometric functions, see Section 4.3.

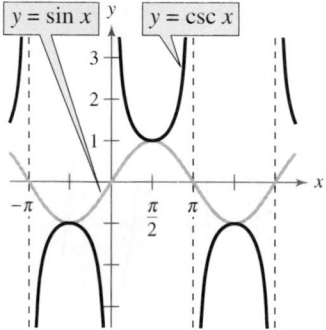

Period: 2π
Domain: all real numbers x,
        except $x = n\pi$

Range: $(-\infty, -1] \cup [1, \infty)$

Vertical asymptotes: $x = n\pi$

Symmetry: origin

Period: 2π
Domain: all real numbers x,
        except $x = \dfrac{\pi}{2} + n\pi$

Range: $(-\infty, -1] \cup [1, \infty)$

Vertical asymptotes: $x = \dfrac{\pi}{2} + n\pi$

Symmetry: y-axis

Figure 4.61

In comparing the graphs of the cosecant and secant functions with those of the sine and cosine functions, note that the "hills" and "valleys" are interchanged. For example, a hill (or maximum point) on the sine curve corresponds to a valley (a local minimum) on the cosecant curve, and a valley (or minimum point) on the sine curve corresponds to a hill (a local maximum) on the cosecant curve, as shown in Figure 4.62. Additionally, *x*-intercepts of the sine and cosine functions become vertical asymptotes of the cosecant and secant functions, respectively (see Figure 4.62).

Example 4 Comparing Trigonometric Graphs

Use a graphing utility to compare the graphs of

$$y = 2\sin\left(x + \frac{\pi}{4}\right) \quad \text{and} \quad y = 2\csc\left(x + \frac{\pi}{4}\right).$$

Solution

The two graphs are shown in Figure 4.63. Note how the hills and valleys of the graphs are related. For the function $y = 2\sin[x + (\pi/4)]$, the amplitude is 2 and the period is 2π. By solving the equations

$$x + \frac{\pi}{4} = 0 \quad \text{and} \quad x + \frac{\pi}{4} = 2\pi$$

you can see that one cycle of the sine function corresponds to the interval from $x = -\pi/4$ to $x = 7\pi/4$. The graph of this sine function is represented by the thick curve in Figure 4.63. Because the sine function is zero at the endpoints of this interval, the corresponding cosecant function

$$y = 2\csc\left(x + \frac{\pi}{4}\right) = 2\left(\frac{1}{\sin[x + (\pi/4)]}\right)$$

has vertical asymptotes at $x = -\pi/4, 3\pi/4, 7\pi/4$, and so on.

✓**CHECKPOINT**   Now try Exercise 25.

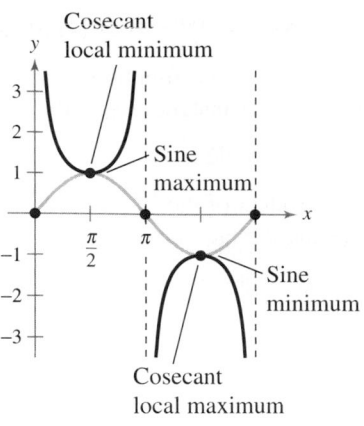

Figure 4.62

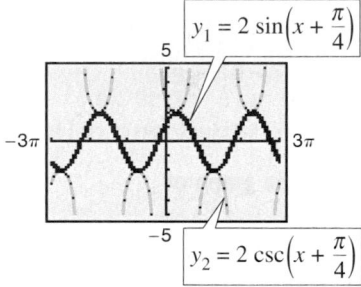

Figure 4.63

Example 5 Comparing Trigonometric Graphs

Use a graphing utility to compare the graphs of $y = \cos 2x$ and $y = \sec 2x$.

Solution

Begin by graphing $y_1 = \cos 2x$ and $y_2 = \sec 2x = 1/\cos 2x$ in the same viewing window, as shown in Figure 4.64. Note that the *x*-intercepts of $y = \cos 2x$

$$\left(-\frac{\pi}{4}, 0\right), \quad \left(\frac{\pi}{4}, 0\right), \quad \left(\frac{3\pi}{4}, 0\right), \ldots$$

correspond to the vertical asymptotes

$$x = -\frac{\pi}{4}, \quad x = \frac{\pi}{4}, \quad x = \frac{3\pi}{4}, \ldots$$

of the graph of $y = \sec 2x$. Moreover, notice that the period of $y = \cos 2x$ and $y = \sec 2x$ is π.

✓**CHECKPOINT**   Now try Exercise 27.

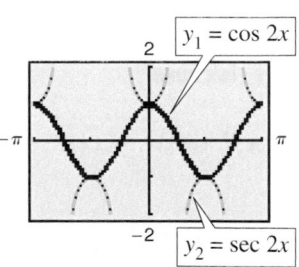

Figure 4.64

Damped Trigonometric Graphs

A *product* of two functions can be graphed using properties of the individual functions. For instance, consider the function

$$f(x) = x \sin x$$

as the product of the functions $y = x$ and $y = \sin x$. Using properties of absolute value and the fact that $|\sin x| \le 1$, you have $0 \le |x| \, |\sin x| \le |x|$. Consequently,

$$-|x| \le x \sin x \le |x|$$

which means that the graph of $f(x) = x \sin x$ lies between the lines $y = -x$ and $y = x$. Furthermore, because

$$f(x) = x \sin x = \pm x \quad \text{at} \quad x = \frac{\pi}{2} + n\pi$$

and

$$f(x) = x \sin x = 0 \quad \text{at} \quad x = n\pi$$

the graph of f touches the line $y = -x$ or the line $y = x$ at $x = \pi/2 + n\pi$ and has x-intercepts at $x = n\pi$. A sketch of f is shown in Figure 4.65. In the function $f(x) = x \sin x$, the factor x is called the **damping factor.**

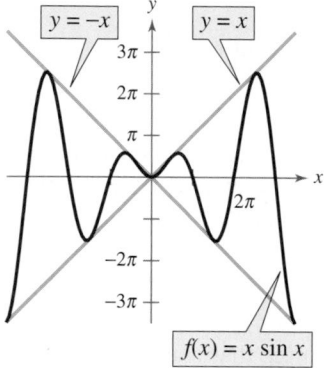

Figure 4.65

Example 6 Analyzing a Damped Sine Curve

Analyze the graph of

$$f(x) = e^{-x} \sin 3x.$$

Solution

Consider $f(x)$ as the product of the two functions

$$y = e^{-x} \quad \text{and} \quad y = \sin 3x$$

each of which has the set of real numbers as its domain. For any real number x, you know that $e^{-x} \ge 0$ and $|\sin 3x| \le 1$. So, $|e^{-x}| \, |\sin 3x| \le e^{-x}$, which means that

$$-e^{-x} \le e^{-x} \sin 3x \le e^{-x}.$$

Furthermore, because

$$f(x) = e^{-x} \sin 3x = \pm e^{-x} \quad \text{at} \quad x = \frac{\pi}{6} + \frac{n\pi}{3}$$

and

$$f(x) = e^{-x} \sin 3x = 0 \quad \text{at} \quad x = \frac{n\pi}{3}$$

the graph of f touches the curves $y = -e^{-x}$ and $y = e^{-x}$ at $x = \pi/6 + n\pi/3$ and has intercepts at $x = n\pi/3$. The graph is shown in Figure 4.66.

✓CHECKPOINT   Now try Exercise 51.

STUDY TIP

Do you see why the graph of $f(x) = x \sin x$ touches the lines $y = \pm x$ at $x = \pi/2 + n\pi$ and why the graph has x-intercepts at $x = n\pi$? Recall that the sine function is equal to ± 1 at $\pi/2, 3\pi/2, 5\pi/2, \ldots$ (odd multiples of $\pi/2$) and is equal to 0 at $\pi, 2\pi, 3\pi, \ldots$ (multiples of π).

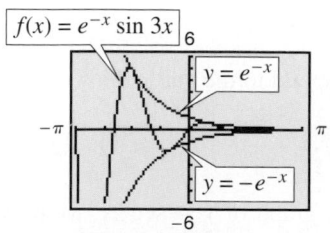

Figure 4.66

Figure 4.67 summarizes the six basic trigonometric functions.

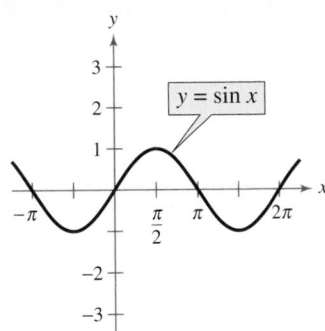

Domain: *all real numbers x*
Range: $[-1, 1]$
Period: 2π

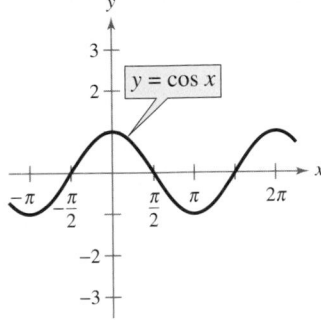

Domain: *all real numbers x*
Range: $[-1, 1]$
Period: 2π

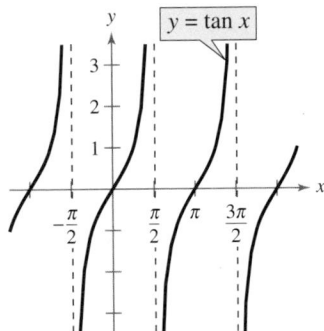

Domain: *all real numbers x,*
$$\text{except } x = \frac{\pi}{2} + n\pi$$
Range: $(-\infty, \infty)$
Period: π

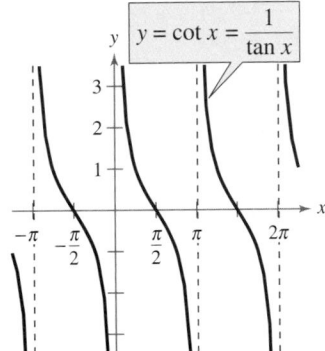

Domain: *all real numbers x,*
$$\text{except } x = n\pi$$
Range: $(-\infty, \infty)$
Period: π

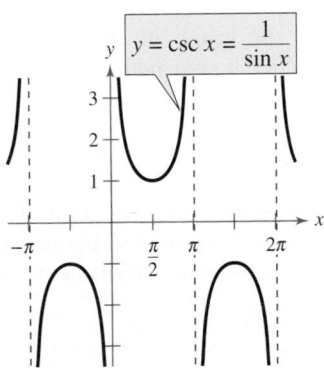

Domain: *all real numbers x,*
$$\text{except } x = n\pi$$
Range: $(-\infty, -1] \cup [1, \infty)$
Period: 2π

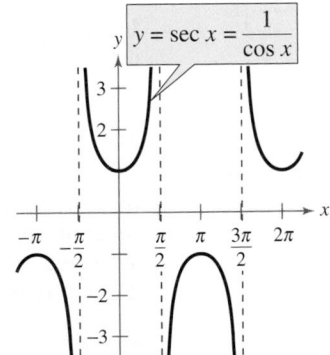

Domain: *all real numbers x,*
$$\text{except } x = \frac{\pi}{2} + n\pi$$
Range: $(-\infty, -1] \cup [1, \infty)$
Period: 2π

Figure 4.67

4.6 Exercises

See www.CalcChat.com for worked-out solutions to odd-numbered exercises.

Vocabulary Check

Fill in the blanks.

1. The graphs of the tangent, cotangent, secant, and cosecant functions have _____ asymptotes.

2. To sketch the graph of a secant or cosecant function, first make a sketch of its _____ function.

3. For the function $f(x) = g(x) \sin x$, $g(x)$ is called the _____ factor of the function.

Library of Parent Functions In Exercises 1–4, use the graph of the function to answer the following.

(a) **Find all x-intercepts of the graph of $y = f(x)$.**

(b) **Find all y-intercepts of the graph of $y = f(x)$.**

(c) **Find the intervals on which the graph $y = f(x)$ is increasing and the intervals on which the graph $y = f(x)$ is decreasing.**

(d) **Find all relative extrema, if any, of the graph of $y = f(x)$.**

(e) **Find all vertical asymptotes, if any, of the graph of $y = f(x)$.**

1. $f(x) = \tan x$

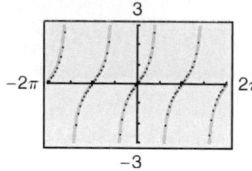

2. $f(x) = \cot x$

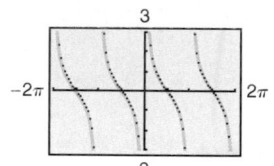

3. $f(x) = \sec x$

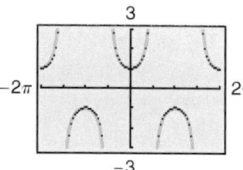

4. $f(x) = \csc x$

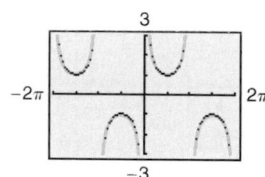

In Exercises 5–24, sketch the graph of the function. (Include two full periods.) Use a graphing utility to verify your result.

5. $y = \frac{1}{2} \tan x$

6. $y = \frac{1}{4} \tan x$

7. $y = -2 \tan 2x$

8. $y = -3 \tan 4x$

9. $y = -\frac{1}{2} \sec x$

10. $y = \frac{1}{4} \sec x$

11. $y = \sec \pi x - 3$

12. $y = -2 \sec 4x + 2$

13. $y = 3 \csc \frac{x}{2}$

14. $y = -\csc \frac{x}{3}$

15. $y = \frac{1}{2} \cot \frac{x}{2}$

16. $y = 3 \cot \pi x$

17. $y = 2 \tan \frac{\pi x}{4}$

18. $y = -\frac{1}{2} \tan \pi x$

19. $y = \frac{1}{2} \sec (2x - \pi)$

20. $y = -\sec(x + \pi)$

21. $y = \csc(\pi - x)$

22. $y = \csc(2x - \pi)$

23. $y = 2 \cot\left(x - \frac{\pi}{2}\right)$

24. $y = \frac{1}{4} \cot(x + \pi)$

In Exercises 25–30, use a graphing utility to graph the function (include two full periods). Graph the corresponding reciprocal function and compare the two graphs. Describe your viewing window.

25. $y = 2 \csc 3x$

26. $y = -\csc(4x - \pi)$

27. $y = -2 \sec 4x$

28. $y = \frac{1}{4} \sec \pi x$

29. $y = \frac{1}{3} \sec\left(\frac{\pi x}{2} + \frac{\pi}{2}\right)$

30. $y = \frac{1}{2} \csc(2x - \pi)$

In Exercises 31–34, use a graph of the function to approximate the solution to the equation on the interval $[-2\pi, 2\pi]$.

31. $\tan x = 1$

32. $\cot x = -\sqrt{3}$

33. $\sec x = -2$

34. $\csc x = \sqrt{2}$

In Exercises 35–38, use the graph of the function to determine whether the function is even, odd, or neither.

35. $f(x) = \sec x$

36. $f(x) = \tan x$

37. $f(x) = \csc 2x$

38. $f(x) = \cot 2x$

In Exercises 39–42, use a graphing utility to graph the two equations in the same viewing window. Use the graphs to determine whether the expressions are equivalent. Verify the results algebraically.

39. $y_1 = \sin x \csc x, \quad y_2 = 1$

40. $y_1 = \sin x \sec x, \quad y_2 = \tan x$

41. $y_1 = \dfrac{\cos x}{\sin x}, \quad y_2 = \cot x$

42. $y_1 = \sec^2 x - 1, \quad y_2 = \tan^2 x$

In Exercises 43–46, match the function with its graph. Describe the behavior of the function as x approaches zero. [The graphs are labeled (a), (b), (c), and (d).]

(a)
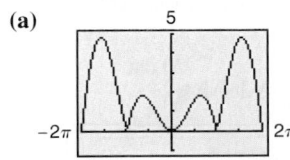

(b)
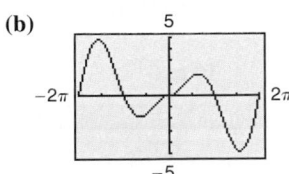

(c)
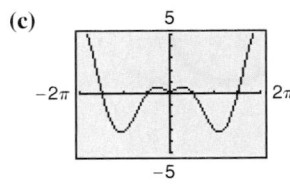

(d)
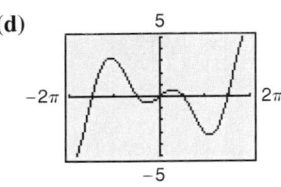

43. $f(x) = x \cos x$

44. $f(x) = |x \sin x|$

45. $g(x) = |x| \sin x$

46. $g(x) = |x| \cos x$

Conjecture In Exercises 47–50, use a graphing utility to graph the functions f and g. Use the graphs to make a conjecture about the relationship between the functions.

47. $f(x) = \sin x + \cos\left(x + \dfrac{\pi}{2}\right)$, $g(x) = 0$

48. $f(x) = \sin x - \cos\left(x + \dfrac{\pi}{2}\right)$, $g(x) = 2 \sin x$

49. $f(x) = \sin^2 x$, $g(x) = \frac{1}{2}(1 - \cos 2x)$

50. $f(x) = \cos^2 \dfrac{\pi x}{2}$, $g(x) = \dfrac{1}{2}(1 + \cos \pi x)$

In Exercises 51–54, use a graphing utility to graph the function and the damping factor of the function in the same viewing window. Describe the behavior of the function as x increases without bound.

51. $f(x) = e^{-x} \cos x$

52. $f(x) = e^{-2x} \sin x$

53. $h(x) = e^{-x^2/4} \cos x$

54. $g(x) = e^{-x^2/2} \sin x$

Exploration In Exercises 55 and 56, use a graphing utility to graph the function. Use the graph to determine the behavior of the function as $x \to c$.

(a) $x \to \dfrac{\pi^+}{2}$ $\left(\text{as } x \text{ approaches } \dfrac{\pi}{2} \text{ from the right}\right)$

(b) $x \to \dfrac{\pi^-}{2}$ $\left(\text{as } x \text{ approaches } \dfrac{\pi}{2} \text{ from the left}\right)$

(c) $x \to -\dfrac{\pi^+}{2}$ $\left(\text{as } x \text{ approaches } -\dfrac{\pi}{2} \text{ from the right}\right)$

(d) $x \to -\dfrac{\pi^-}{2}$ $\left(\text{as } x \text{ approaches } -\dfrac{\pi}{2} \text{ from the left}\right)$

55. $f(x) = \tan x$

56. $f(x) = \sec x$

Exploration In Exercises 57 and 58, use a graphing utility to graph the function. Use the graph to determine the behavior of the function as $x \to c$.

(a) As $x \to 0^+$, the value of $f(x) \to$ ☐ .

(b) As $x \to 0^-$, the value of $f(x) \to$ ☐ .

(c) As $x \to \pi^+$, the value of $f(x) \to$ ☐ .

(d) As $x \to \pi^-$, the value of $f(x) \to$ ☐ .

57. $f(x) = \cot x$

58. $f(x) = \csc x$

59. *Predator-Prey Model* The population P of coyotes (a predator) at time t (in months) in a region is estimated to be

$$P = 10{,}000 + 3000 \sin (\pi t/12)$$

and the population p of rabbits (its prey) is estimated to be

$$p = 15{,}000 + 5000 \cos (\pi t/12).$$

Use the graph of the models to explain the oscillations in the size of each population.

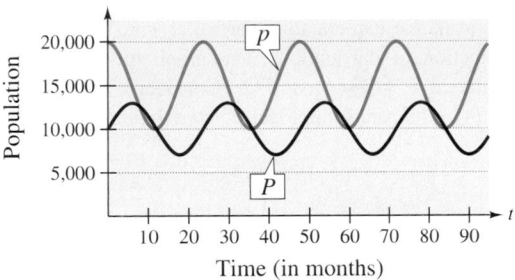

60. *Meteorology* The normal monthly high temperatures H (in degrees Fahrenheit) for Erie, Pennsylvania are approximated by

$$H(t) = 54.33 - 20.38 \cos \dfrac{\pi t}{6} - 15.69 \sin \dfrac{\pi t}{6}$$

and the normal monthly low temperatures L are approximated by

$$L(t) = 39.36 - 15.70 \cos \dfrac{\pi t}{6} - 14.16 \sin \dfrac{\pi t}{6}$$

where t is the time (in months), with $t = 1$ corresponding to January.   (Source: National Oceanic and Atmospheric Association)

(a) Use a graphing utility to graph each function. What is the period of each function?

(b) During what part of the year is the difference between the normal high and normal low temperatures greatest? When is it smallest?

(c) The sun is the farthest north in the sky around June 21, but the graph shows the warmest temperatures at a later date. Approximate the lag time of the temperatures relative to the position of the sun.

61. *Distance* A plane flying at an altitude of 5 miles over level ground will pass directly over a radar antenna (see figure). Let d be the ground distance from the antenna to the point directly under the plane and let x be the angle of elevation to the plane from the antenna. (d is positive as the plane approaches the antenna.) Write d as a function of x and graph the function over the interval $0 < x < \pi$.

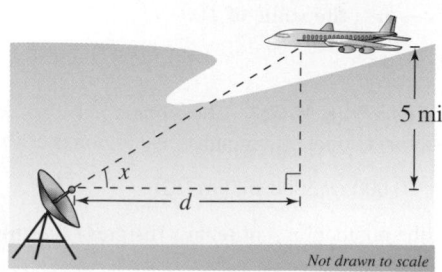

5 mi

Not drawn to scale

62. *Television Coverage* A television camera is on a reviewing platform 36 meters from the street on which a parade will be passing from left to right (see figure). Write the distance d from the camera to a particular unit in the parade as a function of the angle x, and graph the function over the interval $-\pi/2 < x < \pi/2$. (Consider x as negative when a unit in the parade approaches from the left.)

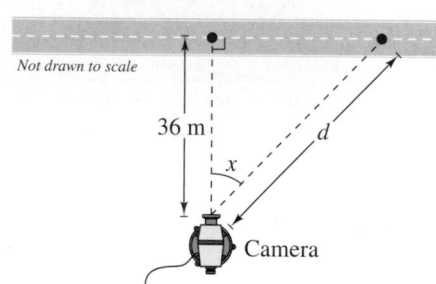

Not drawn to scale

36 m

d

x

Camera

63. *Harmonic Motion* An object weighing W pounds is suspended from a ceiling by a steel spring (see figure). The weight is pulled downward (positive direction) from its equilibrium position and released. The resulting motion of the weight is described by the function $y = \frac{1}{2}e^{-t/4}\cos 4t$, where y is the distance in feet and t is the time in seconds $(t > 0)$.

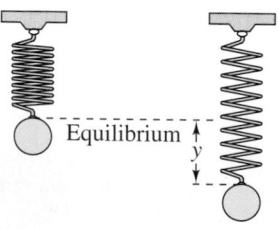

Equilibrium

y

(a) Use a graphing utility to graph the function.

(b) Describe the behavior of the displacement function for increasing values of time t.

64. *Numerical and Graphical Reasoning* A crossed belt connects a 10-centimeter pulley on an electric motor with a 20-centimeter pulley on a saw arbor (see figure). The electric motor runs at 1700 revolutions per minute.

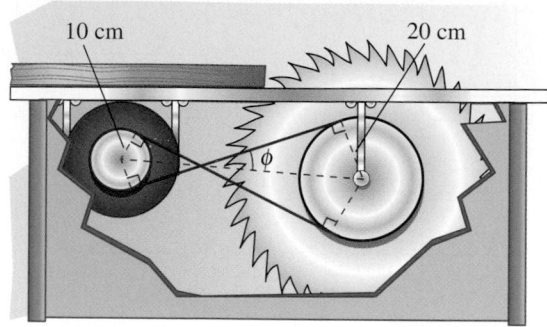

10 cm   20 cm

ϕ

(a) Determine the number of revolutions per minute of the saw.

(b) How does crossing the belt affect the saw in relation to the motor?

(c) Let L be the total length of the belt. Write L as a function of ϕ, where ϕ is measured in radians. What is the domain of the function? (*Hint:* Add the lengths of the straight sections of the belt and the length of belt around each pulley.)

(d) Use a graphing utility to complete the table.

ϕ	0.3	0.6	0.9	1.2	1.5
L					

(e) As ϕ increases, do the lengths of the straight sections of the belt change faster or slower than the lengths of the belts around each pulley?

(f) Use a graphing utility to graph the function over the appropriate domain.

Synthesis

True or False? **In Exercises 65 and 66, determine whether the statement is true or false. Justify your answer.**

65. The graph of $y = -\frac{1}{8}\tan\left(\frac{x}{2} + \pi\right)$ has an asymptote at $x = -3\pi$.

66. For the graph of $y = 2^x \sin x$, as x approaches $-\infty$, y approaches 0.

67. *Graphical Reasoning* Consider the functions $f(x) = 2\sin x$ and $g(x) = \frac{1}{2}\csc x$ on the interval $(0, \pi)$.

(a) Use a graphing utility to graph f and g in the same viewing window.

(b) Approximate the interval in which $f > g$.

(c) Describe the behavior of each of the functions as x approaches π. How is the behavior of g related to the behavior of f as x approaches π?

68. *Pattern Recognition*

(a) Use a graphing utility to graph each function.

$$y_1 = \frac{4}{\pi}\left(\sin \pi x + \frac{1}{3}\sin 3\pi x\right)$$

$$y_2 = \frac{4}{\pi}\left(\sin \pi x + \frac{1}{3}\sin 3\pi x + \frac{1}{5}\sin 5\pi x\right)$$

(b) Identify the pattern in part (a) and find a function y_3 that continues the pattern one more term. Use a graphing utility to graph y_3.

(c) The graphs in parts (a) and (b) approximate the - periodic function in the figure. Find a function y_4 that is a better approximation.

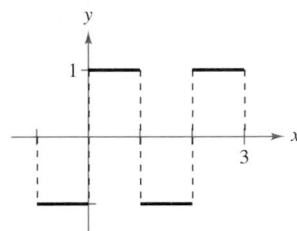

∫ *Exploration*   In Exercises 69 and 70, use a graphing utility to explore the ratio $f(x)$, which appears in calculus.

(a) **Complete the table. Round your results to four decimal places.**

x	-1	-0.1	-0.01	-0.001
$f(x)$				

x	0	0.001	0.01	0.1	1
$f(x)$					

(b) **Use a graphing utility to graph the function $f(x)$. Use the *zoom* and *trace* features to describe the behavior of the graph as x approaches 0.**

(c) **Write a brief statement regarding the value of the ratio based on your results in parts (a) and (b).**

69. $f(x) = \dfrac{\tan x}{x}$

70. $f(x) = \dfrac{\tan 3x}{3x}$

Library of Parent Functions   In Exercises 71 and 72, determine which function is represented by the graph. Do not use a calculator.

71.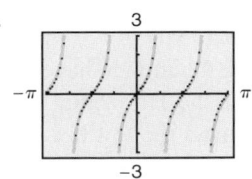

72.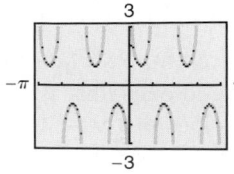

(a) $f(x) = \tan 2x$

(b) $f(x) = \tan \dfrac{x}{2}$

(c) $f(x) = 2\tan x$

(d) $f(x) = -\tan 2x$

(e) $f(x) = -\tan \dfrac{x}{2}$

(a) $f(x) = \sec 4x$

(b) $f(x) = \csc 4x$

(c) $f(x) = \csc \dfrac{x}{4}$

(d) $f(x) = \sec \dfrac{x}{4}$

(e) $f(x) = \csc(4x - \pi)$

∫ 73. *Approximation*   Using calculus, it can be shown that the tangent function can be approximated by the polynomial

$$\tan x \approx x + \frac{2x^3}{3!} + \frac{16x^5}{5!}$$

where x is in radians. Use a graphing utility to graph the tangent function and its polynomial approximation in the same viewing window. How do the graphs compare?

∫ 74. *Approximation*   Using calculus, it can be shown that the secant function can be approximated by the polynomial

$$\sec x \approx 1 + \frac{x^2}{2!} + \frac{5x^4}{4!}$$

where x is in radians. Use a graphing utility to graph the secant function and its polynomial approximation in the same viewing window. How do the graphs compare?

Skills Review

In Exercises 75–78, identify the rule of algebra illustrated by the statement.

75. $5(a - 9) = 5a - 45$

76. $7\left(\frac{1}{7}\right) = 1$

77. $(3 + x) + 0 = 3 + x$

78. $(a + b) + 10 = a + (b + 10)$

In Exercises 79–82, determine whether the function is one-to-one. If it is, find its inverse function.

79. $f(x) = -10$

80. $f(x) = (x - 7)^2 + 3$

81. $f(x) = \sqrt{3x - 14}$

82. $f(x) = \sqrt[3]{x - 5}$

4.7 Inverse Trigonometric Functions

Inverse Sine Function

Recall from Section 1.6 that for a function to have an inverse function, it must be one-to-one—that is, it must pass the Horizontal Line Test. In Figure 4.68 it is obvious that $y = \sin x$ does not pass the test because different values of x yield the same y-value.

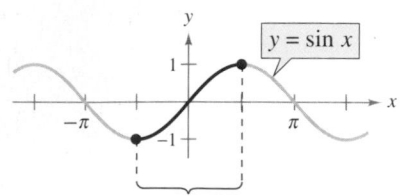

Sin x has an inverse
function on this interval.

Figure 4.68

However, if you restrict the domain to the interval $-\pi/2 \le x \le \pi/2$ (corresponding to the black portion of the graph in Figure 4.68), the following properties hold.

1. On the interval $[-\pi/2, \pi/2]$, the function $y = \sin x$ is increasing.

2. On the interval $[-\pi/2, \pi/2]$, $y = \sin x$ takes on its full range of values, $-1 \le \sin x \le 1$.

3. On the interval $[-\pi/2, \pi/2]$, $y = \sin x$ is one-to-one.

So, on the restricted domain $-\pi/2 \le x \le \pi/2$, $y = \sin x$ has a unique inverse function called the **inverse sine function.** It is denoted by

$$y = \arcsin x \qquad \text{or} \qquad y = \sin^{-1} x.$$

The notation $\sin^{-1} x$ is consistent with the inverse function notation $f^{-1}(x)$. The arcsin x notation (read as "the arcsine of x") comes from the association of a central angle with its intercepted *arc length* on a unit circle. So, arcsin x means the angle (or arc) whose sine is x. Both notations, arcsin x and $\sin^{-1} x$, are commonly used in mathematics, so remember that $\sin^{-1} x$ denotes the *inverse* sine function rather than $1/\sin x$. The values of arcsin x lie in the interval $-\pi/2 \le \arcsin x \le \pi/2$. The graph of $y = \arcsin x$ is shown in Example 2.

Definition of Inverse Sine Function

The **inverse sine function** is defined by

$$y = \arcsin x \qquad \text{if and only if} \qquad \sin y = x$$

where $-1 \le x \le 1$ and $-\pi/2 \le y \le \pi/2$. The domain of $y = \arcsin x$ is $[-1, 1]$ and the range is $[-\pi/2, \pi/2]$.

When evaluating the inverse sine function, it helps to remember the phrase "the arcsine of x is the angle (or number) whose sine is x."

What you should learn

- Evaluate and graph inverse sine functions.
- Evaluate other inverse trigonometric functions.
- Evaluate compositions of trigonometric functions.

Why you should learn it

Inverse trigonometric functions can be useful in exploring how aspects of a real-life problem relate to each other. Exercise 82 on page 329 investigates the relationship between the height of a cone-shaped pile of rock salt, the angle of the cone shape, and the diameter of its base.

Francoise Sauze/Photo Researchers Inc.

Example 1   Evaluating the Inverse Sine Function

If possible, find the exact value.

a. $\arcsin\left(-\dfrac{1}{2}\right)$ **b.** $\sin^{-1}\dfrac{\sqrt{3}}{2}$ **c.** $\sin^{-1} 2$

Solution

a. Because $\sin\left(-\dfrac{\pi}{6}\right) = -\dfrac{1}{2}$, and $-\dfrac{\pi}{6}$ lies in $\left[-\dfrac{\pi}{2}, \dfrac{\pi}{2}\right]$, it follows that

$$\arcsin\left(-\dfrac{1}{2}\right) = -\dfrac{\pi}{6}. \qquad \text{Angle whose sine is } -\tfrac{1}{2}$$

b. Because $\sin\dfrac{\pi}{3} = \dfrac{\sqrt{3}}{2}$, and $\dfrac{\pi}{3}$ lies in $\left[-\dfrac{\pi}{2}, \dfrac{\pi}{2}\right]$, it follows that

$$\sin^{-1}\dfrac{\sqrt{3}}{2} = \dfrac{\pi}{3}. \qquad \text{Angle whose sine is } \sqrt{3}/2$$

c. It is not possible to evaluate $y = \sin^{-1} x$ at $x = 2$ because there is no angle whose sine is 2. Remember that the domain of the inverse sine function is $[-1, 1]$.

✔**CHECKPOINT**   Now try Exercise 1.

Example 2   Graphing the Arcsine Function

Sketch a graph of $y = \arcsin x$ by hand.

Solution

By definition, the equations

$$y = \arcsin x \qquad \text{and} \qquad \sin y = x$$

are equivalent for $-\pi/2 \le y \le \pi/2$. So, their graphs are the same. For the interval $[-\pi/2, \pi/2]$, you can assign values to y in the second equation to make a table of values.

y	$-\dfrac{\pi}{2}$	$-\dfrac{\pi}{4}$	$-\dfrac{\pi}{6}$	0	$\dfrac{\pi}{6}$	$\dfrac{\pi}{4}$	$\dfrac{\pi}{2}$
$x = \sin y$	-1	$-\dfrac{\sqrt{2}}{2}$	$-\dfrac{1}{2}$	0	$\dfrac{1}{2}$	$\dfrac{\sqrt{2}}{2}$	1

Then plot the points and connect them with a smooth curve. The resulting graph of $y = \arcsin x$ is shown in Figure 4.69. Note that it is the reflection (in the line $y = x$) of the black portion of the graph in Figure 4.68. Use a graphing utility to confirm this graph. Be sure you see that Figure 4.69 shows the *entire* graph of the inverse sine function. Remember that the domain of $y = \arcsin x$ is the closed interval $[-1, 1]$ and the range is the closed interval $[-\pi/2, \pi/2]$.

✔**CHECKPOINT**   Now try Exercise 10.

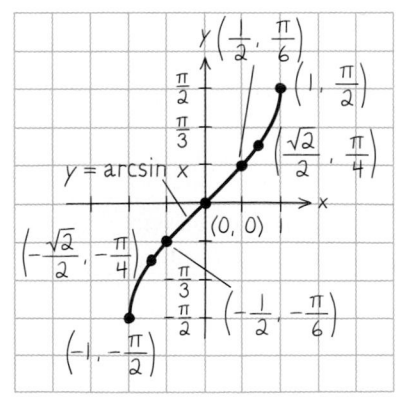

Figure 4.69

Other Inverse Trigonometric Functions

The cosine function is decreasing and one-to-one on the interval $0 \leq x \leq \pi$, as shown in Figure 4.70.

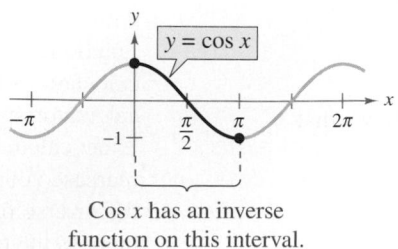

Cos x has an inverse
function on this interval.

Figure 4.70

Consequently, on this interval the cosine function has an inverse function—the **inverse cosine function**—denoted by

$$y = \arccos x \qquad \text{or} \qquad y = \cos^{-1} x.$$

Because $y = \arccos x$ and $x = \cos y$ are equivalent for $0 \leq y \leq \pi$, their graphs are the same, and can be confirmed by the following table of values.

y	0	$\dfrac{\pi}{6}$	$\dfrac{\pi}{3}$	$\dfrac{\pi}{2}$	$\dfrac{2\pi}{3}$	$\dfrac{5\pi}{6}$	π
$x = \cos y$	1	$\dfrac{\sqrt{3}}{2}$	$\dfrac{1}{2}$	0	$-\dfrac{1}{2}$	$-\dfrac{\sqrt{3}}{2}$	-1

Similarly, you can define an **inverse tangent function** by restricting the domain of $y = \tan x$ to the interval $(-\pi/2, \pi/2)$. The following list summarizes the definitions of the three most common inverse trigonometric functions. The remaining three are defined in Exercises 89–91.

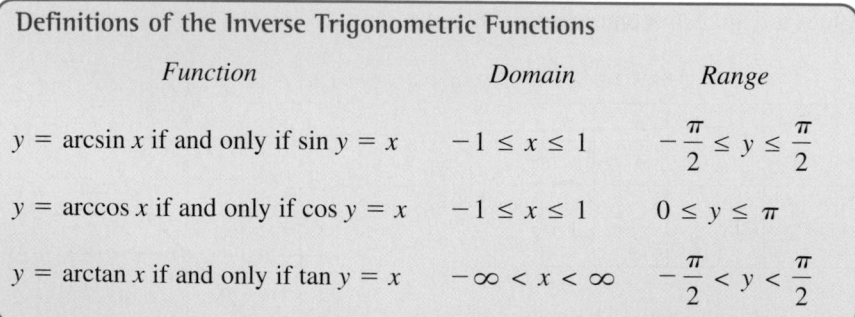

Definitions of the Inverse Trigonometric Functions

Function	Domain	Range
$y = \arcsin x$ if and only if $\sin y = x$	$-1 \leq x \leq 1$	$-\dfrac{\pi}{2} \leq y \leq \dfrac{\pi}{2}$
$y = \arccos x$ if and only if $\cos y = x$	$-1 \leq x \leq 1$	$0 \leq y \leq \pi$
$y = \arctan x$ if and only if $\tan y = x$	$-\infty < x < \infty$	$-\dfrac{\pi}{2} < y < \dfrac{\pi}{2}$

   The graphs of these three inverse trigonometric functions are shown in Figure 4.71.

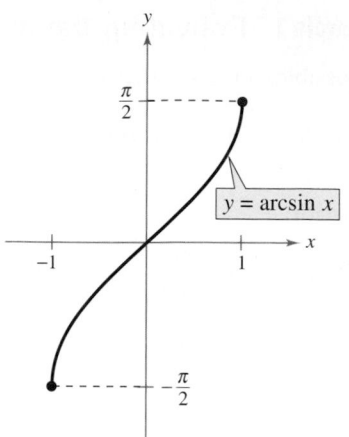

Domain: $[-1, 1]$; *Range:* $\left[-\dfrac{\pi}{2}, \dfrac{\pi}{2}\right]$

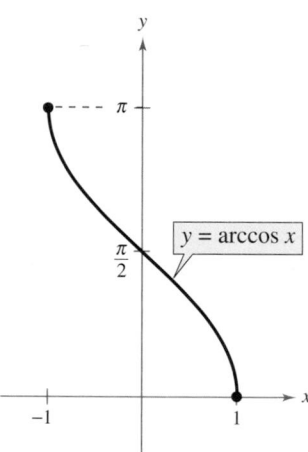

Domain: $[-1, 1]$; *Range:* $[0, \pi]$

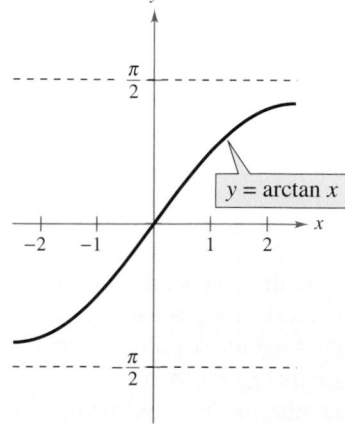

Domain: $(-\infty, \infty)$; *Range:* $\left(-\dfrac{\pi}{2}, \dfrac{\pi}{2}\right)$

Figure 4.71

Example 3   Evaluating Inverse Trigonometric Functions

Find the exact value.

a. $\arccos \dfrac{\sqrt{2}}{2}$ **b.** $\cos^{-1}(-1)$ **c.** $\arctan 0$ **d.** $\tan^{-1}(-1)$

Solution

a. Because $\cos(\pi/4) = \sqrt{2}/2$, and $\pi/4$ lies in $[0, \pi]$, it follows that

$$\arccos \dfrac{\sqrt{2}}{2} = \dfrac{\pi}{4}.$$ Angle whose cosine is $\dfrac{\sqrt{2}}{2}$

b. Because $\cos \pi = -1$, and π lies in $[0, \pi]$, it follows that

$$\cos^{-1}(-1) = \pi.$$ Angle whose cosine is -1

c. Because $\tan 0 = 0$, and 0 lies in $(-\pi/2, \pi/2)$, it follows that

$$\arctan 0 = 0.$$ Angle whose tangent is 0

d. Because $\tan(-\pi/4) = -1$, and $-\pi/4$ lies in $(-\pi/2, \pi/2)$, it follows that

$$\tan^{-1}(-1) = -\dfrac{\pi}{4}.$$ Angle whose tangent is -1

✓CHECKPOINT   Now try Exercise 5.

TECHNOLOGY TIP

You can use the ⌊SIN⁻¹⌋, ⌊COS⁻¹⌋, and ⌊TAN⁻¹⌋ keys on your calculator to approximate values of inverse trigonometric functions. To evaluate the inverse cosecant function, the inverse secant function, or the inverse cotangent function, you can use the inverse sine, inverse cosine, and inverse tangent functions, respectively. For instance, to evaluate $\sec^{-1} 3.4$, enter the expression as shown below.

```
cos⁻¹(1/3.4)
          1.272264126
```

Example 4   Calculators and Inverse Trigonometric Functions

Use a calculator to approximate the value (if possible).

a. $\arctan(-8.45)$ **b.** $\sin^{-1} 0.2447$ **c.** $\arccos 2$

Solution

Function	Mode	Graphing Calculator Keystrokes
a. $\arctan(-8.45)$	Radian	⌊TAN⁻¹⌋ ⌊(⌋ ⌊(−)⌋ 8.45 ⌊)⌋ ⌊ENTER⌋

From the display, it follows that $\arctan(-8.45) \approx -1.4530$.

| **b.** $\sin^{-1} 0.2447$ | Radian | ⌊SIN⁻¹⌋ ⌊(⌋ 0.2447 ⌊)⌋ ⌊ENTER⌋ |

From the display, it follows that $\sin^{-1} 0.2447 \approx 0.2472$.

| **c.** $\arccos 2$ | Radian | ⌊COS⁻¹⌋ ⌊(⌋ 2 ⌊)⌋ ⌊ENTER⌋ |

In *real number* mode, the calculator should display an *error message*, because the domain of the inverse cosine function is $[-1, 1]$.

✓CHECKPOINT   Now try Exercise 15.

TECHNOLOGY TIP   In Example 4, if you had set the calculator to *degree* mode, the display would have been in degrees rather than in radians. This convention is peculiar to calculators. By definition, the values of inverse trigonometric functions are always in *radians*.

Compositions of Functions

Recall from Section 1.6 that for all x in the domains of f and f^{-1}, inverse functions have the properties

$$f(f^{-1}(x)) = x \qquad \text{and} \qquad f^{-1}(f(x)) = x.$$

Inverse Properties

If $-1 \le x \le 1$ and $-\pi/2 \le y \le \pi/2$, then

$$\sin(\arcsin x) = x \qquad \text{and} \qquad \arcsin(\sin y) = y.$$

If $-1 \le x \le 1$ and $0 \le y \le \pi$, then

$$\cos(\arccos x) = x \qquad \text{and} \qquad \arccos(\cos y) = y.$$

If x is a real number and $-\pi/2 < y < \pi/2$, then

$$\tan(\arctan x) = x \qquad \text{and} \qquad \arctan(\tan y) = y.$$

Keep in mind that these inverse properties do not apply for arbitrary values of x and y. For instance,

$$\arcsin\left(\sin\frac{3\pi}{2}\right) = \arcsin(-1) = -\frac{\pi}{2} \ne \frac{3\pi}{2}.$$

In other words, the property $\arcsin(\sin y) = y$ is not valid for values of y outside the interval $[-\pi/2, \pi/2]$.

Example 5 Using Inverse Properties

If possible, find the exact value.

a. $\tan[\arctan(-5)]$ **b.** $\arcsin\left(\sin\dfrac{5\pi}{3}\right)$ **c.** $\cos(\cos^{-1}\pi)$

Solution

a. Because -5 lies in the domain of the arctangent function, the inverse property applies, and you have $\tan[\arctan(-5)] = -5$.

b. In this case, $5\pi/3$ does not lie within the range of the arcsine function, $-\pi/2 \le y \le \pi/2$. However, $5\pi/3$ is coterminal with

$$\frac{5\pi}{3} - 2\pi = -\frac{\pi}{3}$$

which does lie in the range of the arcsine function, and you have

$$\arcsin\left(\sin\frac{5\pi}{3}\right) = \arcsin\left[\sin\left(-\frac{\pi}{3}\right)\right] = -\frac{\pi}{3}.$$

c. The expression $\cos(\cos^{-1}\pi)$ is not defined because $\cos^{-1}\pi$ is not defined. Remember that the domain of the inverse cosine function is $[-1, 1]$.

✓CHECKPOINT   Now try Exercise 31.

Exploration

Use a graphing utility to graph $y = \arcsin(\sin x)$. What are the domain and range of this function? Explain why $\arcsin(\sin 4)$ does not equal 4.

Now graph $y = \sin(\arcsin x)$ and determine the domain and range. Explain why $\sin(\arcsin 4)$ is not defined.

Example 6 shows how to use right triangles to find exact values of compositions of inverse functions.

Example 6   Evaluating Compositions of Functions

Find the exact value.

a. $\tan\left(\arccos\dfrac{2}{3}\right)$ **b.** $\cos\left[\arcsin\left(-\dfrac{3}{5}\right)\right]$

Algebraic Solution

a. If you let $u = \arccos\frac{2}{3}$, then $\cos u = \frac{2}{3}$. Because $\cos u$ is positive, u is a first-quadrant angle. You can sketch and label angle u as shown in Figure 4.72.

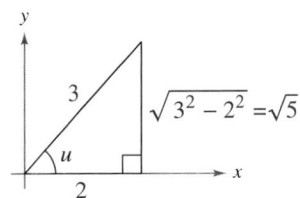

Figure 4.72

Consequently,

$$\tan\left(\arccos\frac{2}{3}\right) = \tan u = \frac{\text{opp}}{\text{adj}} = \frac{\sqrt{5}}{2}.$$

b. If you let $u = \arcsin\left(-\frac{3}{5}\right)$, then $\sin u = -\frac{3}{5}$. Because $\sin u$ is negative, u is a fourth-quadrant angle. You can sketch and label angle u as shown in Figure 4.73.

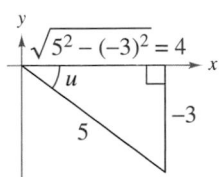

Figure 4.73

Consequently,

$$\cos\left[\arcsin\left(-\frac{3}{5}\right)\right] = \cos u = \frac{\text{adj}}{\text{hyp}} = \frac{4}{5}.$$

Graphical Solution

a. Use a graphing utility set in *radian* mode to graph $y = \tan(\arccos x)$, as shown in Figure 4.74. Use the *value* feature or the *zoom* and *trace* features of the graphing utility to find that the value of the composition of functions when $x = \frac{2}{3} \approx 0.67$ is

$$y = 1.118 \approx \frac{\sqrt{5}}{2}.$$

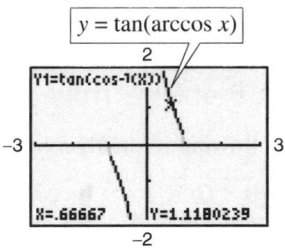

Figure 4.74

b. Use a graphing utility set in *radian* mode to graph $y = \cos(\arcsin x)$, as shown in Figure 4.75. Use the *value* feature or the *zoom* and *trace* features of the graphing utility to find that the value of the composition of functions when $x = -\frac{3}{5} = -0.6$ is

$$y = 0.8 = \frac{4}{5}.$$

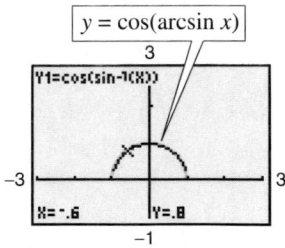

Figure 4.75

✓CHECKPOINT   Now try Exercise 49.

> ## Library of Parent Functions: Inverse Trigonometric Functions
>
> The inverse trigonometric functions are obtained from the trigonometric functions in much the same way that the logarithmic function was developed from the exponential function. However, unlike the exponential function, the trigonometric functions are not one-to-one, and so it is necessary to restrict their domains to intervals on which they pass the Horizontal Line Test. Consequently, the inverse trigonometric functions have restricted domains and ranges, and they are not periodic. A review of inverse trigonometric functions can be found in the *Study Capsules*.
>
> One prominent role played by inverse trigonometric functions is in solving a trigonometric equation in which the argument (angle) of the trigonometric function is the unknown quantity in the equation. You will learn how to solve such equations in the next chapter.
>
> Inverse trigonometric functions play a unique role in calculus. There are two basic operations of calculus. One operation (called *differentiation*) transforms an inverse trigonometric function (a transcendental function) into an algebraic function. The other operation (called *integration*) produces the opposite transformation—from algebraic to transcendental.

Example 7   Some Problems from Calculus

Write each of the following as an algebraic expression in x.

a. $\sin(\arccos 3x)$, $0 \le x \le \frac{1}{3}$ **b.** $\cot(\arccos 3x)$, $0 \le x < \frac{1}{3}$

Solution

If you let $u = \arccos 3x$, then $\cos u = 3x$, where $-1 \le 3x \le 1$. Because

$$\cos u = \text{adj/hyp} = (3x)/1$$

you can sketch a right triangle with acute angle u, as shown in Figure 4.76. From this triangle, you can easily convert each expression to algebraic form.

a. $\sin(\arccos 3x) = \sin u = \dfrac{\text{opp}}{\text{hyp}} = \sqrt{1 - 9x^2}$, $0 \le x \le \dfrac{1}{3}$

b. $\cot(\arccos 3x) = \cot u = \dfrac{\text{adj}}{\text{opp}} = \dfrac{3x}{\sqrt{1 - 9x^2}}$, $0 \le x < \dfrac{1}{3}$

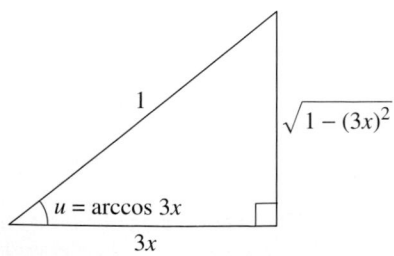

Figure 4.76

A similar argument can be made here for x-values lying in the interval $\left[-\frac{1}{3}, 0\right]$.

✓**CHECKPOINT**   Now try Exercise 55.

4.7   Exercises

See www.CalcChat.com for worked-out solutions to odd-numbered exercises.

Vocabulary Check

Fill in the blanks.

Function	Alternative Notation	Domain	Range
1. $y = \arcsin x$	_____	_____	$-\dfrac{\pi}{2} \le y \le \dfrac{\pi}{2}$
2. _____	$y = \cos^{-1} x$	$-1 \le x \le 1$	_____
3. $y = \arctan x$	_____	_____	_____

Library of Parent Functions   In Exercises 1–9, find the exact value of each expression without using a calculator.

1. (a) $\arcsin \frac{1}{2}$ (b) $\arcsin 0$

2. (a) $\arccos \frac{1}{2}$ (b) $\arccos 0$

3. (a) $\arcsin 1$ (b) $\arccos 1$

4. (a) $\arctan 1$ (b) $\arctan 0$

5. (a) $\arctan \dfrac{\sqrt{3}}{3}$ (b) $\arctan(-1)$

6. (a) $\cos^{-1}\!\left(-\dfrac{\sqrt{2}}{2}\right)$ (b) $\sin^{-1}\!\left(-\dfrac{\sqrt{2}}{2}\right)$

7. (a) $\arctan\left(-\sqrt{3}\right)$ (b) $\arctan \sqrt{3}$

8. (a) $\arccos\!\left(-\dfrac{1}{2}\right)$ (b) $\arcsin \dfrac{\sqrt{2}}{2}$

9. (a) $\sin^{-1} \dfrac{\sqrt{3}}{2}$ (b) $\tan^{-1}\!\left(-\dfrac{\sqrt{3}}{3}\right)$

10. *Numerical and Graphical Analysis*   Consider the function $y = \arcsin x$.

(a) Use a graphing utility to complete the table.

x	-1	-0.8	-0.6	-0.4	-0.2
y					

x	0	0.2	0.4	0.6	0.8	1
y						

(b) Plot the points from the table in part (a) and graph the function. (Do not use a graphing utility.)

(c) Use a graphing utility to graph the inverse sine function and compare the result with your hand-drawn graph in part (b).

(d) Determine any intercepts and symmetry of the graph.

11. *Numerical and Graphical Analysis*   Consider the function $y = \arccos x$.

(a) Use a graphing utility to complete the table.

x	-1	-0.8	-0.6	-0.4	-0.2
y					

x	0	0.2	0.4	0.6	0.8	1
y						

(b) Plot the points from the table in part (a) and graph the function. (Do not use a graphing utility.)

(c) Use a graphing utility to graph the inverse cosine function and compare the result with your hand-drawn graph in part (b).

(d) Determine any intercepts and symmetry of the graph.

12. *Numerical and Graphical Analysis*   Consider the function $y = \arctan x$.

(a) Use a graphing utility to complete the table.

x	-10	-8	-6	-4	-2
y					

x	0	2	4	6	8	10
y						

(b) Plot the points from the table in part (a) and graph the function. (Do not use a graphing utility.)

(c) Use a graphing utility to graph the inverse tangent function and compare the result with your hand-drawn graph in part (b).

(d) Determine the horizontal asymptotes of the graph.

In Exercises 13 and 14, determine the missing coordinates of the points on the graph of the function.

13.

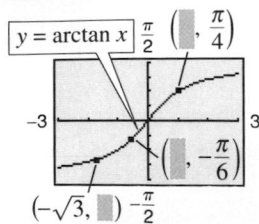

14.

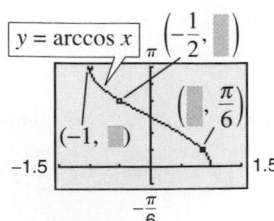

In Exercises 15–20, use a calculator to approximate the value of the expression. Round your answer to the nearest hundredth.

15. $\cos^{-1} 0.75$

16. $\sin^{-1} 0.56$

17. $\arcsin(-0.75)$

18. $\arccos(-0.7)$

19. $\arctan(-6)$

20. $\tan^{-1} 5.9$

In Exercises 21–24, use an inverse trigonometric function to write θ as a function of x.

21.

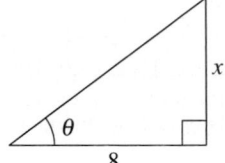

22.

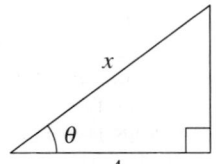

23.

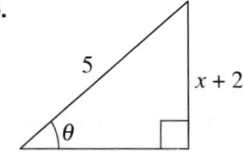

24.

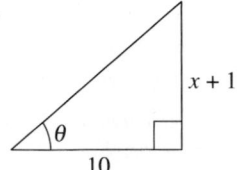

∫ **In Exercises 25–28, find the length of the third side of the triangle in terms of x. Then find θ in terms x for all three inverse trigonometric functions.**

25.

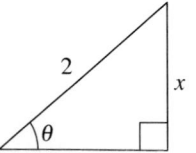

26.

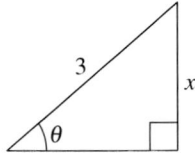

27.

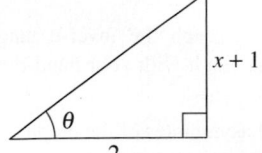

28.

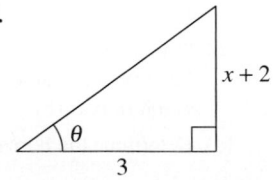

In Exercises 29–46, use the properties of inverse functions to find the exact value of the expression.

29. $\sin(\arcsin 0.7)$

30. $\tan(\arctan 35)$

31. $\cos[\arccos(-0.3)]$

32. $\sin[\arcsin(-0.1)]$

33. $\arcsin(\sin 3\pi)$

34. $\arccos\left(\cos \dfrac{7\pi}{2}\right)$

35. $\tan^{-1}\left(\tan \dfrac{11\pi}{6}\right)$

36. $\sin^{-1}\left(\sin \dfrac{7\pi}{4}\right)$

37. $\sin^{-1}\left(\sin \dfrac{5\pi}{2}\right)$

38. $\cos^{-1}\left(\cos \dfrac{3\pi}{2}\right)$

39. $\sin^{-1}\left(\tan \dfrac{5\pi}{4}\right)$

40. $\cos^{-1}\left(\tan \dfrac{3\pi}{4}\right)$

41. $\tan(\arcsin 0)$

42. $\cos[\arctan(-1)]$

43. $\sin(\arctan 1)$

44. $\sin[\arctan(-1)]$

45. $\arcsin\left[\cos\left(-\dfrac{\pi}{6}\right)\right]$

46. $\arccos\left[\sin\left(-\dfrac{\pi}{6}\right)\right]$

In Exercises 47–54, find the exact value of the expression. Use a graphing utility to verify your result. (*Hint:* Make a sketch of a right triangle.)

47. $\sin\left(\arctan \dfrac{4}{3}\right)$

48. $\sec\left(\arcsin \dfrac{3}{5}\right)$

49. $\cos\left(\arcsin \dfrac{24}{25}\right)$

50. $\csc\left[\arctan\left(-\dfrac{12}{5}\right)\right]$

51. $\sec\left[\arctan\left(-\dfrac{3}{5}\right)\right]$

52. $\tan\left[\arcsin\left(-\dfrac{3}{4}\right)\right]$

53. $\sin\left[\arccos\left(-\dfrac{2}{3}\right)\right]$

54. $\cot\left(\arctan \dfrac{5}{8}\right)$

∫ **In Exercises 55–62, write an algebraic expression that is equivalent to the expression. (*Hint:* Sketch a right triangle, as demonstrated in Example 7.)**

55. $\cot(\arctan x)$

56. $\sin(\arctan x)$

57. $\sin[\arccos(x + 2)]$

58. $\sec[\arcsin(x - 1)]$

59. $\tan\left(\arccos \dfrac{x}{5}\right)$

60. $\cot\left(\arctan \dfrac{4}{x}\right)$

61. $\csc\left(\arctan \dfrac{x}{\sqrt{7}}\right)$

62. $\cos\left(\arcsin \dfrac{x - h}{r}\right)$

In Exercises 63–66, complete the equation.

63. $\arctan \dfrac{14}{x} = \arcsin(\quad), \quad x > 0$

64. $\arcsin \dfrac{\sqrt{36 - x^2}}{6} = \arccos(\quad), \quad 0 \le x \le 6$

65. $\arccos \dfrac{3}{\sqrt{x^2 - 2x + 10}} = \arcsin(\quad)$

66. $\arccos \dfrac{x - 2}{2} = \arctan(\quad), \quad 2 < x < 4$

In Exercises 67–72, use a graphing utility to graph the function.

67. $y = 2 \arccos x$

68. $y = \arcsin \dfrac{x}{2}$

69. $f(x) = \arcsin(x - 2)$

70. $g(t) = \arccos(t + 2)$

71. $f(x) = \arctan 2x$

72. $f(x) = \arccos \dfrac{x}{4}$

In Exercises 73 and 74, write the function in terms of the sine function by using the identity

$$A \cos \omega t + B \sin \omega t \sqrt{A^2 + B^2} \sin\left(\omega t + \arctan \dfrac{A}{B}\right).$$

Use a graphing utility to graph both forms of the function. What does the graph imply?

73. $f(t) = 3 \cos 2t + 3 \sin 2t$

74. $f(t) = 4 \cos \pi t + 3 \sin \pi t$

Exploration **In Exercises 75–80, find the value. If not possible, state the reason.**

75. As $x \to 1^-$, the value of $\arcsin x \to$.

76. As $x \to 1^-$, the value of $\arccos x \to$.

77. As $x \to \infty$, the value of $\arctan x \to$.

78. As $x \to -1^+$, the value of $\arcsin x \to$.

79. As $x \to -1^+$, the value of $\arccos x \to$.

80. As $x \to -\infty$, the value of $\arctan x \to$.

81. *Docking a Boat* A boat is pulled in by means of a winch located on a dock 10 feet above the deck of the boat (see figure). Let θ be the angle of elevation from the boat to the winch and let s be the length of the rope from the winch to the boat.

(a) Write θ as a function of s.

(b) Find θ when $s = 52$ feet and when $s = 26$ feet.

82. *Granular Angle of Repose* Different types of granular substances naturally settle at different angles when stored in cone-shaped piles. This angle θ is called the *angle of repose*. When rock salt is stored in a cone-shaped pile 11 feet high, the diameter of the pile's base is about 34 feet. (Source: Bulk-Store Structures, Inc.)

(a) Draw a diagram that gives a visual representation of the problem. Label all known and unknown quantities.

(b) Find the angle of repose for rock salt.

(c) How tall is a pile of rock salt that has a base diameter of 40 feet?

83. *Photography* A television camera at ground level is filming the lift-off of a space shuttle at a point 750 meters from the launch pad (see figure). Let θ be the angle of elevation to the shuttle and let s be the height of the shuttle.

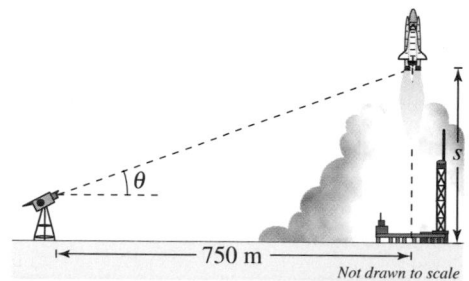

750 m

Not drawn to scale

(a) Write θ as a function of s.

(b) Find θ when $s = 400$ meters and when $s = 1600$ meters.

84. *Photography* A photographer takes a picture of a three-foot painting hanging in an art gallery. The camera lens is 1 foot below the lower edge of the painting (see figure). The angle β subtended by the camera lens x feet from the painting is

$$\beta = \arctan \dfrac{3x}{x^2 + 4}, \quad x > 0.$$

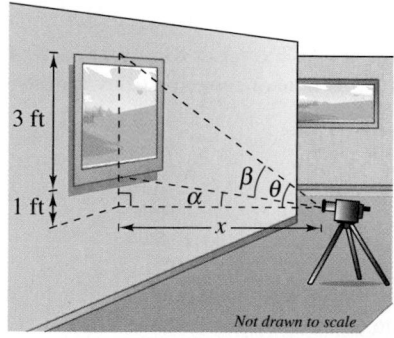

3 ft

1 ft

Not drawn to scale

(a) Use a graphing utility to graph β as a function of x.

(b) Move the cursor along the graph to approximate the distance from the picture when β is maximum.

(c) Identify the asymptote of the graph and discuss its meaning in the context of the problem.

85. *Angle of Elevation*   An airplane flies at an altitude of 6 miles toward a point directly over an observer. Consider θ and x as shown in the figure.

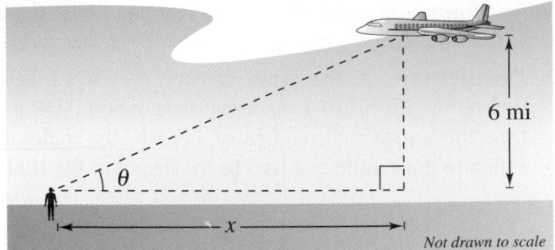

6 mi

θ

x

Not drawn to scale

(a) Write θ as a function of x.

(b) Find θ when $x = 10$ miles and $x = 3$ miles.

86. *Security Patrol*   A security car with its spotlight on is parked 20 meters from a warehouse. Consider θ and x as shown in the figure.

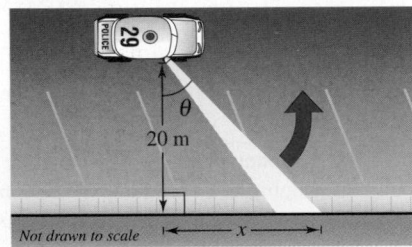

θ

20 m

Not drawn to scale x

(a) Write θ as a function of x.

(b) Find θ when $x = 5$ meters and when $x = 12$ meters.

Synthesis

True or False?   **In Exercises 87 and 88, determine whether the statement is true or false. Justify your answer.**

87. $\sin \dfrac{5\pi}{6} = \dfrac{1}{2} \quad \Longrightarrow \quad \arcsin \dfrac{1}{2} = \dfrac{5\pi}{6}$

88. $\arctan x = \dfrac{\arcsin x}{\arccos x}$

89. Define the inverse cotangent function by restricting the domain of the cotangent function to the interval $(0, \pi)$, and sketch the inverse function's graph.

90. Define the inverse secant function by restricting the domain of the secant function to the intervals $[0, \pi/2)$ and $(\pi/2, \pi]$, and sketch the inverse function's graph.

91. Define the inverse cosecant function by restricting the domain of the cosecant function to the intervals $[-\pi/2, 0)$ and $(0, \pi/2]$, and sketch the inverse function's graph.

92. Use the results of Exercises 89–91 to explain how to graph (a) the inverse cotangent function, (b) the inverse secant function, and (c) the inverse cosecant function on a graphing utility.

In Exercises 93–96, use the results of Exercises 89–91 to evaluate the expression without using a calculator.

93. $\operatorname{arcsec} \sqrt{2}$ **94.** $\operatorname{arcsec} 1$

95. $\operatorname{arccot}\left(-\sqrt{3}\right)$ **96.** $\operatorname{arccsc} 2$

Proof   **In Exercises 97–99, prove the identity.**

97. $\arcsin(-x) = -\arcsin x$

98. $\arctan(-x) = -\arctan x$

99. $\arcsin x + \arccos x = \dfrac{\pi}{2}$

∫ **100.** *Area*   In calculus, it is shown that the area of the region bounded by the graphs of $y = 0$, $y = 1/(x^2 + 1)$, $x = a$, and $x = b$ is given by

$$\text{Area} = \arctan b - \arctan a$$

(see figure). Find the area for each value of a and b.

(a) $a = 0, b = 1$ (b) $a = -1, b = 1$

(c) $a = 0, b = 3$ (d) $a = -1, b = 3$

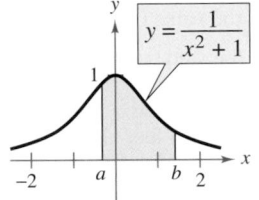

$y = \dfrac{1}{x^2 + 1}$

Skills Review

In Exercises 101–104, simplify the radical expression.

101. $\dfrac{4}{4\sqrt{2}}$ **102.** $\dfrac{2}{\sqrt{3}}$

103. $\dfrac{2\sqrt{3}}{6}$ **104.** $\dfrac{5\sqrt{5}}{2\sqrt{10}}$

In Exercises 105–108, sketch a right triangle corresponding to the trigonometric function of the acute angle θ. Use the Pythagorean Theorem to determine the third side and then find the other five trigonometric functions of θ.

105. $\sin \theta = \dfrac{5}{6}$ **106.** $\tan \theta = 2$

107. $\sin \theta = \dfrac{3}{4}$ **108.** $\sec \theta = 3$

4.8 Applications and Models

Applications Involving Right Triangles

In this section, the three angles of a right triangle are denoted by the letters A, B, and C (where C is the right angle), and the lengths of the sides opposite these angles by the letters a, b, and c (where c is the hypotenuse).

Example 1 Solving a Right Triangle

Solve the right triangle shown in Figure 4.77 for all unknown sides and angles.

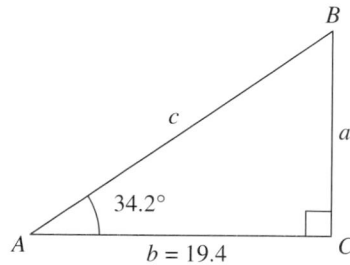

Figure 4.77

Solution

Because $C = 90°$, it follows that $A + B = 90°$ and $B = 90° - 34.2° = 55.8°$. To solve for a, use the fact that

$$\tan A = \frac{\text{opp}}{\text{adj}} = \frac{a}{b} \qquad\Longrightarrow\qquad a = b \tan A.$$

So, $a = 19.4 \tan 34.2° \approx 13.18$. Similarly, to solve for c, use the fact that

$$\cos A = \frac{\text{adj}}{\text{hyp}} = \frac{b}{c} \qquad\Longrightarrow\qquad c = \frac{b}{\cos A}.$$

So, $c = \dfrac{19.4}{\cos 34.2°} \approx 23.46$.

✓CHECKPOINT   Now try Exercise 1.

Recall from Section 4.3 that the term *angle of elevation* denotes the angle from the horizontal upward to an object and that the term *angle of depression* denotes the angle from the horizontal downward to an object. An angle of elevation and an angle of depression are shown in Figure 4.78.

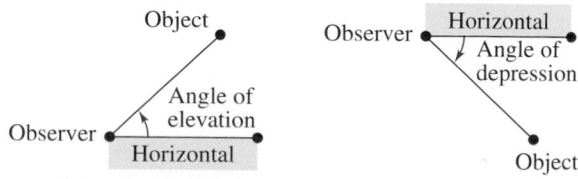

Figure 4.78

What you should learn

- Solve real-life problems involving right triangles.
- Solve real-life problems involving directional bearings.
- Solve real-life problems involving harmonic motion.

Why you should learn it

You can use trigonometric functions to model and solve real-life problems. For instance, Exercise 60 on page 341 shows you how a trigonometric function can be used to model the harmonic motion of a buoy.

Mary Kate Denny/PhotoEdit

Example 2 Finding a Side of a Right Triangle

A safety regulation states that the maximum angle of elevation for a rescue ladder is 72°. A fire department's longest ladder is 110 feet. What is the maximum safe rescue height?

Solution

A sketch is shown in Figure 4.79. From the equation $\sin A = a/c$, it follows that

$$a = c \sin A = 110 \sin 72° \approx 104.62.$$

So, the maximum safe rescue height is about 104.62 feet above the height of the fire truck.

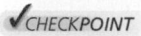

CHECKPOINT Now try Exercise 17.

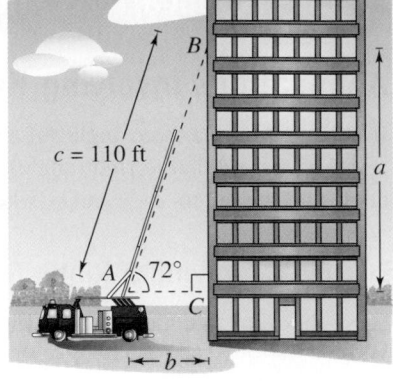

Figure 4.79

Example 3 Finding a Side of a Right Triangle

At a point 200 feet from the base of a building, the angle of elevation to the *bottom* of a smokestack is 35°, and the angle of elevation to the *top* is 53°, as shown in Figure 4.80. Find the height s of the smokestack alone.

Solution

This problem involves two right triangles. For the smaller right triangle, use the fact that $\tan 35° = a/200$ to conclude that the height of the building is

$$a = 200 \tan 35°.$$

Now, for the larger right triangle, use the equation

$$\tan 53° = \frac{a + s}{200}$$

to conclude that $s = 200 \tan 53° - a$. So, the height of the smokestack is

$$s = 200 \tan 53° - a = 200 \tan 53° - 200 \tan 35° \approx 125.37 \text{ feet}.$$

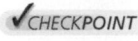

CHECKPOINT Now try Exercise 21.

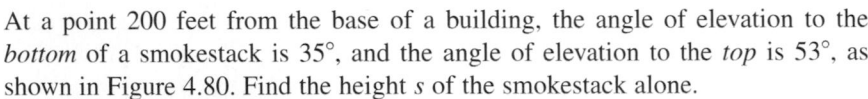

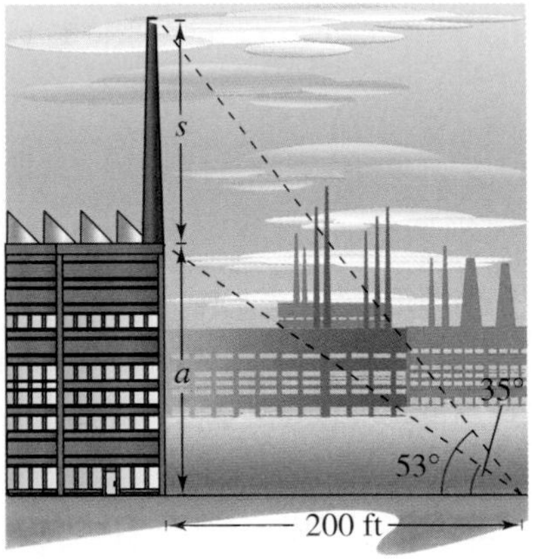

Figure 4.80

Example 4 Finding an Angle of Depression

A swimming pool is 20 meters long and 12 meters wide. The bottom of the pool is slanted such that the water depth is 1.3 meters at the shallow end and 4 meters at the deep end, as shown in Figure 4.81. Find the angle of depression of the bottom of the pool.

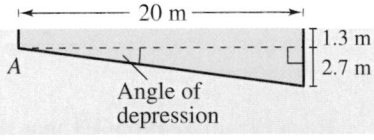

Solution

Using the tangent function, you see that

$$\tan A = \frac{\text{opp}}{\text{adj}} = \frac{2.7}{20} = 0.135.$$

So, the angle of depression is $A = \arctan 0.135 \approx 0.1342$ radian $\approx 7.69°$.

CHECKPOINT Now try Exercise 27.

Figure 4.81

Trigonometry and Bearings

In surveying and navigation, directions are generally given in terms of **bearings.** A bearing measures the acute angle a path or line of sight makes with a fixed north–south line, as shown in Figure 4.82. For instance, the bearing of S 35° E in Figure 4.82(a) means 35 degrees east of south.

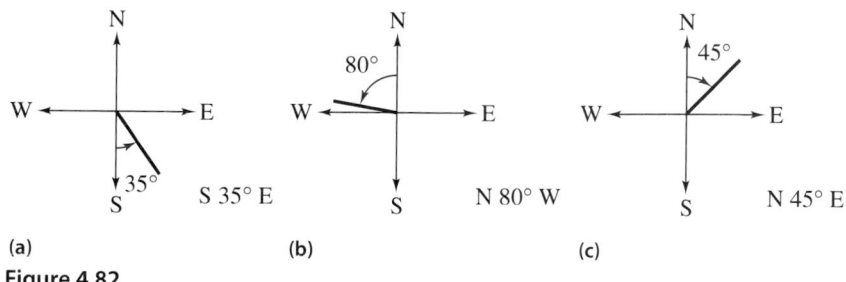

(a)   S 35° E   (b)   N 80° W   (c)   N 45° E

Figure 4.82

Example 5   Finding Directions in Terms of Bearings

A ship leaves port at noon and heads due west at 20 knots, or 20 nautical miles (nm) per hour. At 2 P.M. the ship changes course to N 54° W, as shown in Figure 4.83. Find the ship's bearing and distance from the port of departure at 3 P.M.

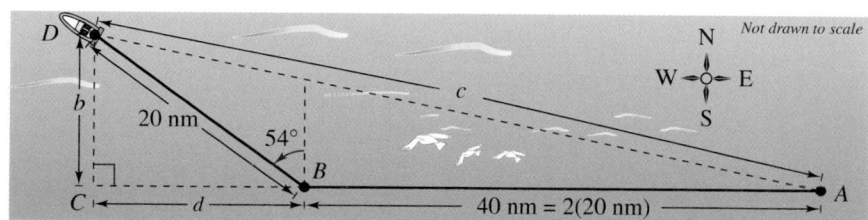

Figure 4.83

Solution

For triangle BCD, you have $B = 90° - 54° = 36°$. The two sides of this triangle can be determined to be

$$b = 20 \sin 36° \quad \text{and} \quad d = 20 \cos 36°.$$

In triangle ACD, you can find angle A as follows.

$$\tan A = \frac{b}{d + 40} = \frac{20 \sin 36°}{20 \cos 36° + 40} \approx 0.2092494$$

$$A \approx \arctan 0.2092494 \approx 0.2062732 \text{ radian} \approx 11.82°$$

The angle with the north-south line is $90° - 11.82° = 78.18°$. So, the bearing of the ship is N 78.18° W. Finally, from triangle ACD, you have $\sin A = b/c$, which yields

$$c = \frac{b}{\sin A} = \frac{20 \sin 36°}{\sin 11.82°} \approx 57.39 \text{ nautical miles.} \qquad \text{Distance from port}$$

✔**CHECKPOINT**   Now try Exercise 33.

STUDY TIP

In *air navigation*, bearings are measured in degrees *clockwise* from north. Examples of air navigation bearings are shown below.

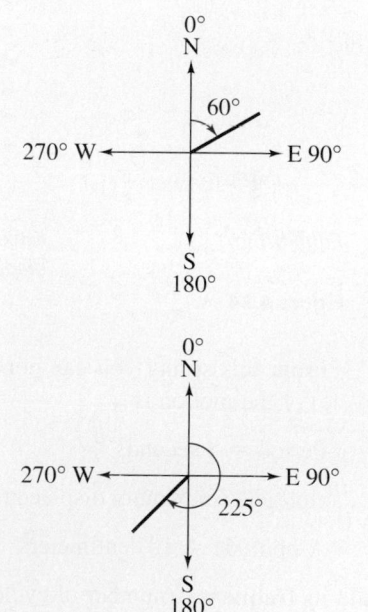

Harmonic Motion

The periodic nature of the trigonometric functions is useful for describing the motion of a point on an object that vibrates, oscillates, rotates, or is moved by wave motion.

For example, consider a ball that is bobbing up and down on the end of a spring, as shown in Figure 4.84. Suppose that 10 centimeters is the maximum distance the ball moves vertically upward or downward from its equilibrium (at-rest) position. Suppose further that the time it takes for the ball to move from its maximum displacement above zero to its maximum displacement below zero and back again is $t = 4$ seconds. Assuming the ideal conditions of perfect elasticity and no friction or air resistance, the ball would continue to move up and down in a uniform and regular manner.

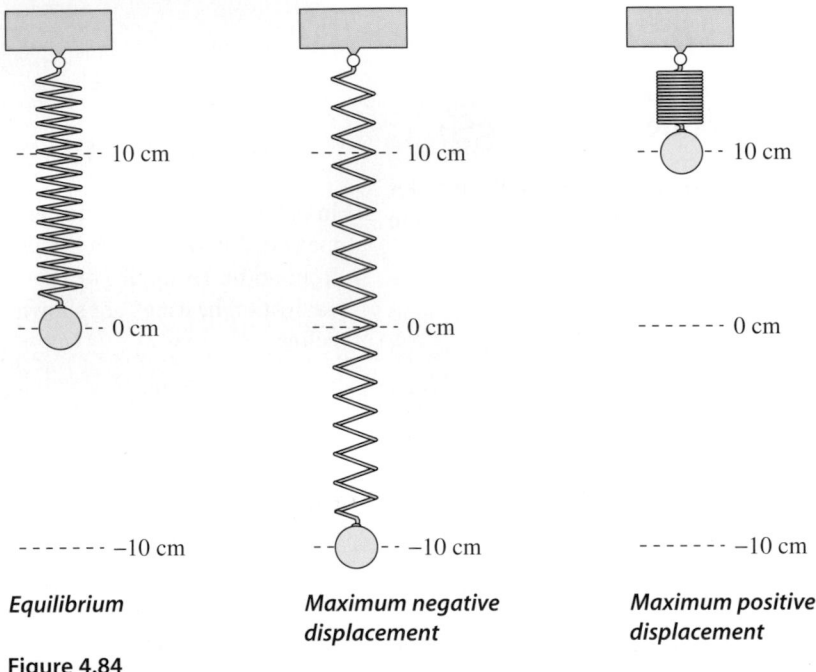

Equilibrium

Maximum negative displacement

Maximum positive displacement

Figure 4.84

From this spring you can conclude that the period (time for one complete cycle) of the motion is

Period = 4 seconds

its amplitude (maximum displacement from equilibrium) is

Amplitude = 10 centimeters

and its **frequency** (number of cycles per second) is

Frequency = $\dfrac{1}{4}$ cycle per second.

Motion of this nature can be described by a sine or cosine function, and is called **simple harmonic motion.**

> **Definition of Simple Harmonic Motion**
>
> A point that moves on a coordinate line is said to be in **simple harmonic motion** if its distance d from the origin at time t is given by either
>
> $$d = a \sin \omega t \quad \text{or} \quad d = a \cos \omega t$$
>
> where a and ω are real numbers such that $\omega > 0$. The motion has amplitude $|a|$, period $2\pi/\omega$, and frequency $\omega/(2\pi)$.

Example 6   Simple Harmonic Motion

Write the equation for the simple harmonic motion of the ball illustrated in Figure 4.84, where the period is 4 seconds. What is the frequency of this motion?

Solution

Because the spring is at equilibrium ($d = 0$) when $t = 0$, you use the equation

$$d = a \sin \omega t.$$

Moreover, because the maximum displacement from zero is 10 and the period is 4, you have the following.

$$\text{Amplitude} = |a| = 10$$

$$\text{Period} = \frac{2\pi}{\omega} = 4 \quad \Longrightarrow \quad \omega = \frac{\pi}{2}$$

Consequently, the equation of motion is

$$d = 10 \sin \frac{\pi}{2} t.$$

Note that the choice of $a = 10$ or $a = -10$ depends on whether the ball initially moves up or down. The frequency is

$$\begin{aligned}
\text{Frequency} &= \frac{\omega}{2\pi} \\
&= \frac{\pi/2}{2\pi} \\
&= \frac{1}{4} \text{ cycle per second.}
\end{aligned}$$

 **CHECKPOINT**   Now try Exercise 51.

Figure 4.85

One illustration of the relationship between sine waves and harmonic motion is the wave motion that results when a stone is dropped into a calm pool of water. The waves move outward in roughly the shape of sine (or cosine) waves, as shown in Figure 4.85. As an example, suppose you are fishing and your fishing bob is attached so that it does not move horizontally. As the waves move outward from the dropped stone, your fishing bob will move up and down in simple harmonic motion, as shown in Figure 4.86.

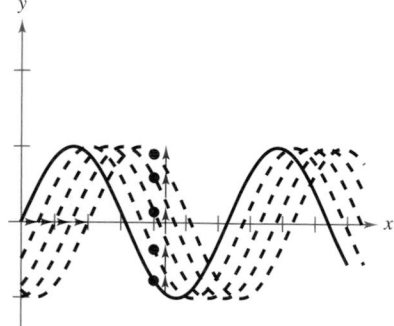

Figure 4.86

Example 7 Simple Harmonic Motion

Given the equation for simple harmonic motion

$$d = 6 \cos \frac{3\pi}{4} t$$

find (a) the maximum displacement, (b) the frequency, (c) the value of d when $t = 4$, and (d) the least positive value of t for which $d = 0$.

Algebraic Solution

The given equation has the form $d = a \cos \omega t$, with $a = 6$ and $\omega = 3\pi/4$.

a. The maximum displacement (from the point of equilibrium) is given by the amplitude. So, the maximum displacement is 6.

b. Frequency $= \dfrac{\omega}{2\pi}$

$$= \frac{3\pi/4}{2\pi}$$

$$= \frac{3}{8} \text{ cycle per unit of time}$$

c. $d = 6 \cos \left[\dfrac{3\pi}{4}(4) \right]$

$$= 6 \cos 3\pi$$

$$= 6(-1)$$

$$= -6$$

d. To find the least positive value of t for which $d = 0$, solve the equation

$$d = 6 \cos \frac{3\pi}{4} t = 0.$$

First divide each side by 6 to obtain

$$\cos \frac{3\pi}{4} t = 0.$$

You know that $\cos t = 0$ when

$$t = \frac{\pi}{2}, \frac{3\pi}{2}, \frac{5\pi}{2}, \ldots$$

Multiply these values by $4/(3\pi)$ to obtain

$$t = \frac{2}{3}, 2, \frac{10}{3}, \ldots$$

So, the least positive value of t is $t = \frac{2}{3}$.

✓CHECKPOINT   Now try Exercise 55.

Graphical Solution

Use a graphing utility set in *radian* mode to graph

$$y = 6 \cos \frac{3\pi}{4} x.$$

a. Use the *maximum* feature of the graphing utility to estimate that the maximum displacement from the point of equilibrium $y = 0$ is 6, as shown in Figure 4.87.

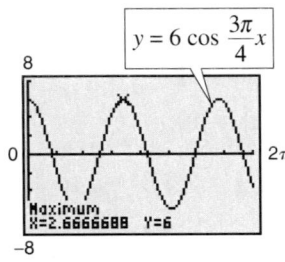

Figure 4.87

b. The period is the time for the graph to complete one cycle, which is $x \approx 2.67$. You can estimate the frequency as follows.

$$\text{Frequency} \approx \frac{1}{2.67} \approx 0.37 \text{ cycle per unit of time}$$

c. Use the *value* or *trace* feature to estimate that the value of y when $x = 4$ is $y = -6$, as shown in Figure 4.88.

d. Use the *zero* or *root* feature to estimate that the least positive value of x for which $y = 0$ is $x \approx 0.67$, as shown in Figure 4.89.

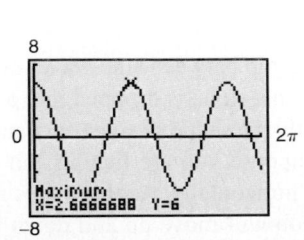

Figure 4.88

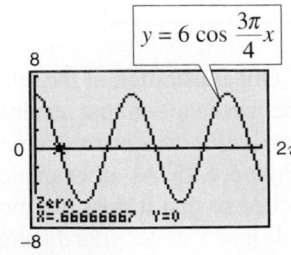

Figure 4.89

4.8 Exercises

See www.CalcChat.com for worked-out solutions to odd-numbered exercises.

Vocabulary Check

Fill in the blanks.

1. An angle that measures from the horizontal upward to an object is called the angle of _____ , whereas an angle that measures from the horizontal downward to an object is called the angle of _____ .

2. A _____ measures the acute angle a path or line of sight makes with a fixed north-south line.

3. A point that moves on a coordinate line is said to be in simple _____ if its distance from the origin at time t is given by either $d = a \sin \omega t$ or $d = a \cos \omega t$.

In Exercises 1–10, solve the right triangle shown in the figure.

1. $A = 30°$, $b = 10$
2. $B = 60°$, $c = 15$
3. $B = 71°$, $b = 14$
4. $A = 7.4°$, $a = 20.5$
5. $a = 6$, $b = 12$
6. $a = 25$, $c = 45$
7. $b = 16$, $c = 54$
8. $b = 1.32$, $c = 18.9$
9. $A = 12° \, 15'$, $c = 430.5$
10. $B = 65° \, 12'$, $a = 145.5$

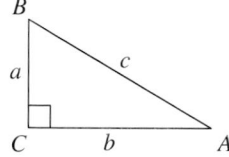

Figure for 1–10

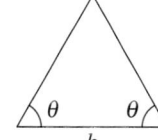

Figure for 11–14

In Exercises 11–14, find the altitude of the isosceles triangle shown in the figure.

11. $\theta = 52°$, $b = 8$ inches
12. $\theta = 18°$, $b = 12$ meters
13. $\theta = 41.6°$, $b = 18.5$ feet
14. $\theta = 72.94°$, $b = 3.26$ centimeters

15. **Length**   A shadow of length L is created by a 60-foot silo when the sun is $\theta°$ above the horizon.

    (a) Draw a right triangle that gives a visual representation of the problem. Label the known and unknown quantities.

    (b) Write L as a function of θ.

    (c) Use a graphing utility to complete the table.

θ	10°	20°	30°	40°	50°
L					

    (d) The angle measure increases in equal increments in the table. Does the length of the shadow change in equal increments? Explain.

16. **Length**   A shadow of length L is created by an 850-foot building when the sun is $\theta°$ above the horizon.

    (a) Draw a right triangle that gives a visual representation of the problem. Label the known and unknown quantities.

    (b) Write L as a function of θ.

    (c) Use a graphing utility to complete the table.

θ	10°	20°	30°	40°	50°
L					

    (d) The angle measure increases in equal increments in the table. Does the length of the shadow change in equal increments? Explain.

17. **Height**   A ladder 20 feet long leans against the side of a house. The angle of elevation of the ladder is 80°. Find the height from the top of the ladder to the ground.

18. **Height**   The angle of elevation from the base to the top of a waterslide is 13°. The slide extends horizontally 58.2 meters. Approximate the height of the waterslide.

19. **Height**   A 100-foot line is attached to a kite. When the kite has pulled the line taut, the angle of elevation to the kite is approximately 50°. Approximate the height of the kite.

20. **Depth**   The sonar of a navy cruiser detects a submarine that is 4000 feet from the cruiser. The angle between the water level and the submarine is 31.5°. How deep is the submarine?

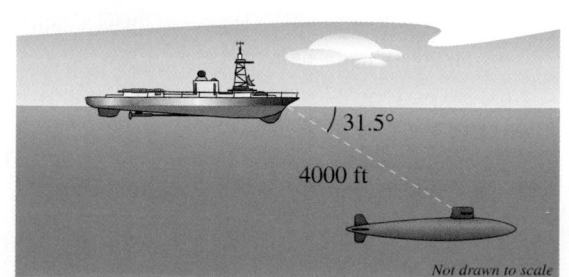

31.5°

4000 ft

Not drawn to scale

21. **Height** From a point 50 feet in front of a church, the angles of elevation to the base of the steeple and the top of the steeple are 35° and 47° 40′, respectively.

    (a) Draw right triangles that give a visual representation of the problem. Label the known and unknown quantities.

    (b) Use a trigonometric function to write an equation involving the unknown quantity.

    (c) Find the height of the steeple.

22. **Height** From a point 100 feet in front of a public library, the angles of elevation to the base of the flagpole and the top of the flagpole are 28° and 39° 45′, respectively. The flagpole is mounted on the front of the library's roof. Find the height of the flagpole.

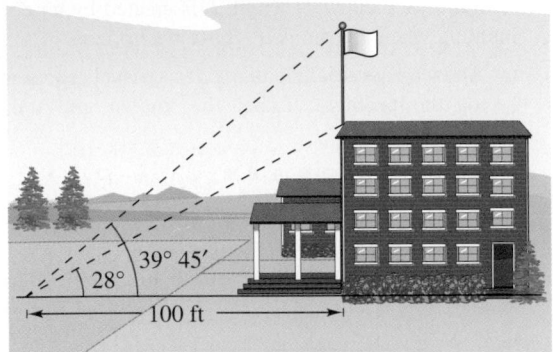

23. **Height** You are holding one of the tethers attached to the top of a giant character balloon in a parade. Before the start of the parade the balloon is upright and the bottom is floating approximately 20 feet above ground level. You are standing approximately 100 feet ahead of the balloon (see figure).

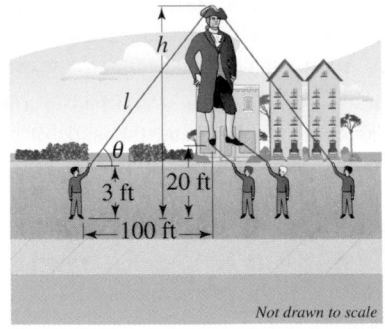

Not drawn to scale

    (a) Find the length ℓ of the tether you will be holding while walking, in terms of h, the height of the balloon.

    (b) Find an expression for the angle of elevation θ from you to the top of the balloon.

    (c) Find the height of the balloon from top to bottom if the angle of elevation to the top of the balloon is 35°.

24. **Height** The designers of a water park are creating a new slide and have sketched some preliminary drawings. The length of the ladder is 30 feet, and its angle of elevation is 60° (see figure).

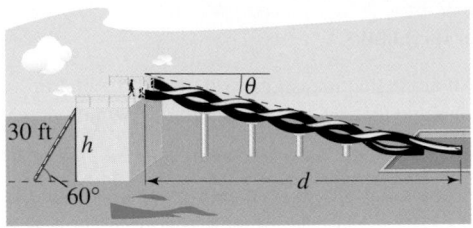

    (a) Find the height h of the slide.

    (b) Find the angle of depression θ from the top of the slide to the end of the slide at the ground in terms of the horizontal distance d the rider travels.

    (c) The angle of depression of the ride is bounded by safety restrictions to be no less than 25° and not more than 30°. Find an interval for how far the rider travels horizontally.

25. **Angle of Elevation** An engineer erects a 75-foot vertical cellular-phone tower. Find the angle of elevation to the top of the tower from a point on level ground 95 feet from its base.

26. **Angle of Elevation** The height of an outdoor basketball backboard is $12\frac{1}{2}$ feet, and the backboard casts a shadow $17\frac{1}{3}$ feet long.

    (a) Draw a right triangle that gives a visual representation of the problem. Label the known and unknown quantities.

    (b) Use a trigonometric function to write an equation involving the unknown quantity.

    (c) Find the angle of elevation of the sun.

27. **Angle of Depression** A Global Positioning System satellite orbits 12,500 miles above Earth's surface (see figure). Find the angle of depression from the satellite to the horizon. Assume the radius of Earth is 4000 miles.

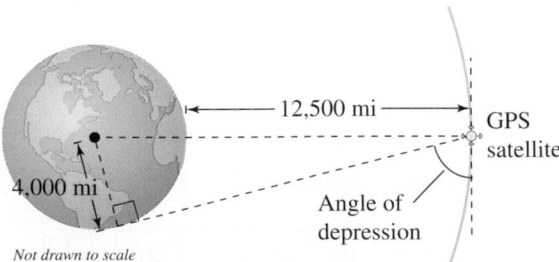

Not drawn to scale

28. *Angle of Depression*   Find the angle of depression from the top of a lighthouse 250 feet above water level to the water line of a ship $2\frac{1}{2}$ miles offshore.

29. *Airplane Ascent*   When an airplane leaves the runway, its angle of climb is 18° and its speed is 275 feet per second. Find the plane's altitude after 1 minute.

30. *Airplane Ascent*   How long will it take the plane in Exercise 29 to climb to an altitude of 10,000 feet? 16,000 feet?

31. *Mountain Descent*   A sign on the roadway at the top of a mountain indicates that for the next 4 miles the grade is 9.5° (see figure). Find the change in elevation for a car descending the mountain.

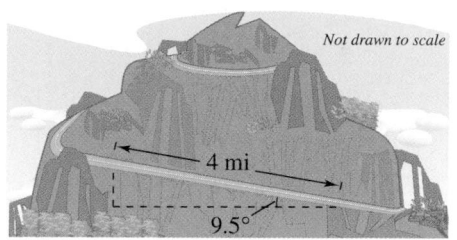

32. *Ski Slope*   A ski slope on a mountain has an angle of elevation of 25.2°. The vertical height of the slope is 1808 feet. How long is the slope?

33. *Navigation*   A ship leaves port at noon and has a bearing of S 29° W. The ship sails at 20 knots. How many nautical miles south and how many nautical miles west will the ship have traveled by 6:00 P.M.?

34. *Navigation*   An airplane flying at 600 miles per hour has a bearing of 52°. After flying for 1.5 hours, how far north and how far east has the plane traveled from its point of departure?

35. *Surveying*   A surveyor wants to find the distance across a pond (see figure). The bearing from A to B is N 32° W. The surveyor walks 50 meters from A, and at the point C the bearing to B is N 68° W. Find (a) the bearing from A to C and (b) the distance from A to B.

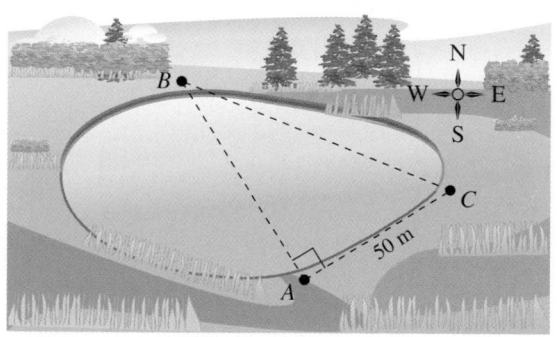

36. *Location of a Fire*   Two fire towers are 30 kilometers apart, where tower A is due west of tower B. A fire is spotted from the towers, and the bearings from A and B are E 14° N and W 34° N, respectively (see figure). Find the distance d of the fire from the line segment AB.

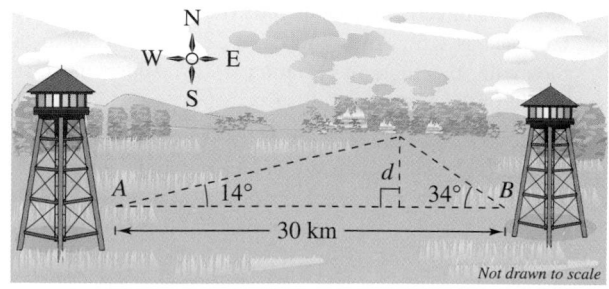

37. *Navigation*   A ship is 45 miles east and 30 miles south of port. The captain wants to sail directly to port. What bearing should be taken?

38. *Navigation*   A plane is 160 miles north and 85 miles east of an airport. The pilot wants to fly directly to the airport. What bearing should be taken?

39. *Distance*   An observer in a lighthouse 350 feet above sea level observes two ships directly offshore. The angles of depression to the ships are 4° and 6.5° (see figure). How far apart are the ships?

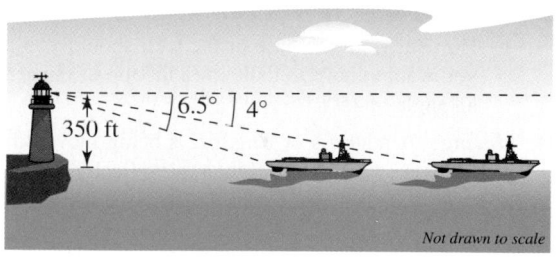

40. *Distance*   A passenger in an airplane flying at an altitude of 10 kilometers sees two towns directly to the east of the plane. The angles of depression to the towns are 28° and 55° (see figure). How far apart are the towns?

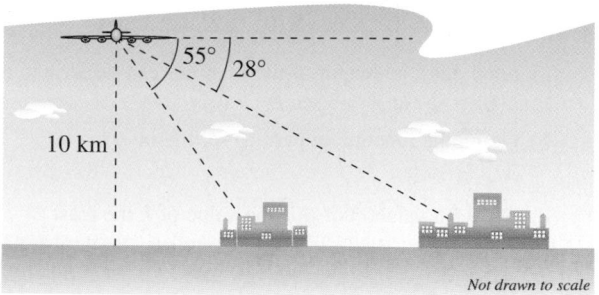

41. *Altitude*   A plane is observed approaching your home and you assume its speed is 550 miles per hour. The angle of elevation to the plane is 16° at one time and 57° one minute later. Approximate the altitude of the plane.

42. *Height*   While traveling across flat land, you notice a mountain directly in front of you. The angle of elevation to the peak is 2.5°. After you drive 18 miles closer to the mountain, the angle of elevation is 10°. Approximate the height of the mountain.

43. *Angle of Elevation*   The top of a drive-in theater screen is 50 feet high and is mounted on a 6-foot-high cement wall. The nearest row of parking is 40 feet from the base of the wall. The furthest row of parking is 150 feet from the base of the wall.

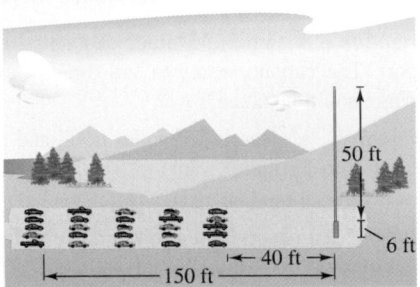

(a) Find the angles of elevation to the *top* of the screen from both the closest row and the furthest row.

(b) How far from the base of the wall should you park if you want to have to look up to the top of the screen at an angle of 45°?

44. *Moving*   A mattress of length L is being moved through two hallways that meet at right angles. Each hallway has a width of three feet (see figure).

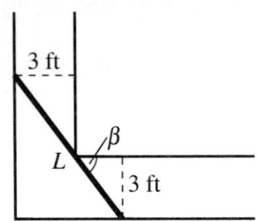

(a) Show that the length of the mattress can be written as $L(\beta) = 3 \csc \beta + 3 \sec \beta$.

(b) Graph the function in part (a) for the interval $0 < \beta < \dfrac{\pi}{2}$.

(c) For what value(s) of β is the value of L the least ?

Geometry   In Exercises 45 and 46, find the angle α between the two nonvertical lines L_1 and L_2. The angle α satisfies the equation

$$\tan \alpha = \left| \frac{m_2 - m_1}{1 + m_2 m_1} \right|$$

where m_1 and m_2 are the slopes of L_1 and L_2, respectively. (Assume $m_1 m_2 \neq -1$.)

45. L_1: $3x - 2y = 5$ **46.** L_1: $2x + y = 8$
  L_2: $x + y = 1$ L_2: $x - 5y = -4$

47. *Geometry*   Determine the angle between the diagonal of a cube and the diagonal of its base, as shown in the figure.

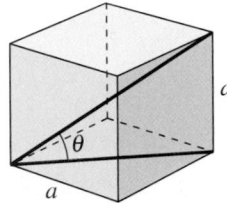   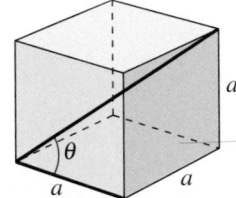

Figure for 47   Figure for 48

48. *Geometry*   Determine the angle between the diagonal of a cube and its edge, as shown in the figure.

49. *Hardware*   Write the distance y across the flat sides of a hexagonal nut as a function of r, as shown in the figure.

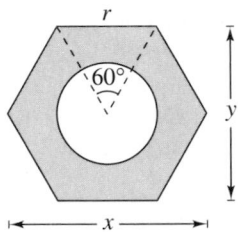

50. *Hardware*   The figure shows a circular piece of sheet metal of diameter 40 centimeters. The sheet contains 12 equally spaced bolt holes. Determine the straight-line distance between the centers of two consecutive bolt holes.

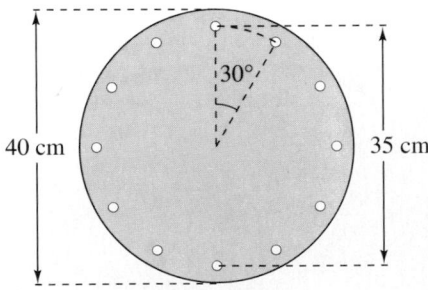

Harmonic Motion **In Exercises 51–54, find a model for simple harmonic motion satisfying the specified conditions.**

Displacement ($t = 0$)	Amplitude	Period
51. 0	8 centimeters	2 seconds
52. 0	3 meters	6 seconds
53. 3 inches	3 inches	1.5 seconds
54. 2 feet	2 feet	10 seconds

Harmonic Motion **In Exercises 55–58, for the simple harmonic motion described by the trigonometric function, find (a) the maximum displacement, (b) the frequency, (c) the value of d when $t = 5$, and (d) the least positive value of t for which $d = 0$. Use a graphing utility to verify your results.**

55. $d = 4 \cos 8\pi t$

56. $d = \frac{1}{2} \cos 20\pi t$

57. $d = \frac{1}{16} \sin 140\pi t$

58. $d = \frac{1}{64} \sin 792\pi t$

59. **Tuning Fork** A point on the end of a tuning fork moves in the simple harmonic motion described by $d = a \sin \omega t$. Find ω given that the tuning fork for middle C has a frequency of 264 vibrations per second.

60. **Wave Motion** A buoy oscillates in simple harmonic motion as waves go past. At a given time it is noted that the buoy moves a total of 3.5 feet from its low point to its high point (see figure), and that it returns to its high point every 10 seconds. Write an equation that describes the motion of the buoy if it is at its high point at time $t = 0$.

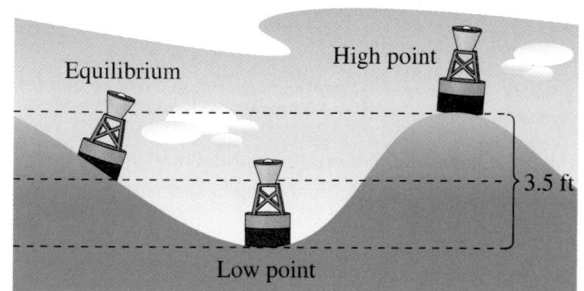

61. **Springs** A ball that is bobbing up and down on the end of a spring has a maximum displacement of 3 inches. Its motion (in ideal conditions) is modeled by

$$y = \frac{1}{4} \cos 16t, \quad t > 0$$

where y is measured in feet and t is the time in seconds.

(a) Use a graphing utility to graph the function.

(b) What is the period of the oscillations?

(c) Determine the first time the ball passes the point of equilibrium ($y = 0$).

62. **Numerical and Graphical Analysis** A two-meter-high fence is 3 meters from the side of a grain storage bin. A grain elevator must reach from ground level outside the fence to the storage bin (see figure). The objective is to determine the shortest elevator that meets the constraints.

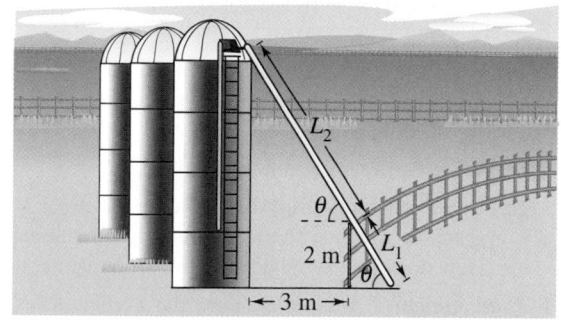

(a) Complete four rows of the table.

θ	L_1	L_2	$L_1 + L_2$
0.1	$\dfrac{2}{\sin 0.1}$	$\dfrac{3}{\cos 0.1}$	23.05
0.2	$\dfrac{2}{\sin 0.2}$	$\dfrac{3}{\cos 0.2}$	13.13

(b) Use the *table* feature of a graphing utility to generate additional rows of the table. Use the table to estimate the minimum length of the elevator.

(c) Write the length $L_1 + L_2$ as a function of θ.

(d) Use a graphing utility to graph the function. Use the graph to estimate the minimum length. How does your estimate compare with that in part (b)?

63. **Numerical and Graphical Analysis** The cross sections of an irrigation canal are isosceles trapezoids, where the lengths of three of the sides are 8 feet (see figure). The objective is to find the angle θ that maximizes the area of the cross sections. [*Hint:* The area of a trapezoid is given by $(h/2)(b_1 + b_2)$.]

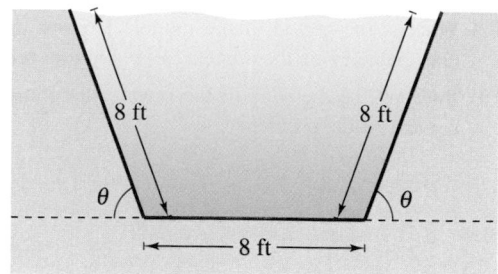

(a) Complete seven rows of the table.

Base 1	Base 2	Altitude	Area
8	$8 + 16\cos 10°$	$8\sin 10°$	22.06
8	$8 + 16\cos 20°$	$8\sin 20°$	42.46

(b) Use the *table* feature of a graphing utility to generate additional rows of the table. Use the table to estimate the maximum cross-sectional area.

(c) Write the area A as a function of θ.

(d) Use a graphing utility to graph the function. Use the graph to estimate the maximum cross-sectional area. How does your estimate compare with that in part (b)?

64. *Data Analysis*   The table shows the average sales S (in millions of dollars) of an outerwear manufacturer for each month t, where $t = 1$ represents January.

Month, t	Sales, S
1	13.46
2	11.15
3	8.00
4	4.85
5	2.54
6	1.70
7	2.54
8	4.85
9	8.00
10	11.15
11	13.46
12	14.30

(a) Create a scatter plot of the data.

(b) Find a trigonometric model that fits the data. Graph the model on your scatter plot. How well does the model fit the data?

(c) What is the period of the model? Do you think it is reasonable given the context? Explain your reasoning.

(d) Interpret the meaning of the model's amplitude in the context of the problem.

65. *Data Analysis*   The times S of sunset (Greenwich Mean Time) at 40° north latitude on the 15th of each month are: 1(16:59), 2(17:35), 3(18:06), 4(18:38), 5(19:08), 6(19:30), 7(19:28), 8(18:57), 9(18:09), 10(17:21), 11(16:44), and 12(16:36). The month is represented by t, with $t = 1$ corresponding to January. A model (in which minutes have been converted to the decimal parts of an hour) for the data is given by

$$S(t) = 18.09 + 1.41 \sin\left(\frac{\pi t}{6} + 4.60\right).$$

(a) Use a graphing utility to graph the data points and the model in the same viewing window.

(b) What is the period of the model? Is it what you expected? Explain.

(c) What is the amplitude of the model? What does it represent in the context of the problem? Explain.

66. *Writing*   Is it true that N 24° E means 24 degrees north of east? Explain.

Synthesis

True or False?   **In Exercises 67 and 68, determine whether the statement is true or false. Justify your answer.**

67. In the right triangle shown below, $a = \dfrac{22.56}{\tan 41.9°}$.

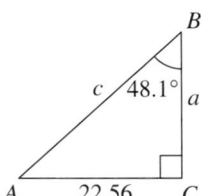

68. For the harmonic motion of a ball bobbing up and down on the end of a spring, one period can be described as the length of one coil of the spring.

Skills Review

In Exercises 69–72, write the standard form of the equation of the line that has the specified characteristics.

69. $m = 4$, passes through $(-1, 2)$

70. $m = -\frac{1}{2}$, passes through $\left(\frac{1}{3}, 0\right)$

71. Passes through $(-2, 6)$ and $(3, 2)$

72. Passes through $\left(\frac{1}{4}, -\frac{2}{3}\right)$ and $\left(-\frac{1}{2}, \frac{1}{3}\right)$

In Exercises 73–76, find the domain of the function.

73. $f(x) = 3x + 8$

74. $f(x) = -x^2 - 1$

75. $g(x) = \sqrt[3]{x + 2}$

76. $g(x) = \sqrt{7 - x}$

What Did You Learn?

Key Terms

initial side of an angle, *p. 258*

terminal side of an angle, *p. 258*

vertex of an angle, *p. 258*

standard position, *p. 258*

positive, negative angles, *p. 258*

coterminal angles, *p. 258*

central angle, *p. 259*

complementary angles, *p. 260*

supplementary angles, *p. 260*

linear speed, *p. 263*

angular speed, *p. 263*

unit circle, *p. 269*

solving right triangles, *p. 281*

angle of elevation, *p. 282*

angle of depression, *p. 282*

reference angle, *p. 290*

phase shift, *p. 301*

damping factor, *p. 314*

bearings, *p. 333*

frequency, *p. 334*

simple harmonic motion, *pp. 334, 335*

Key Concepts

4.1 ■ Convert between degrees and radians

To convert degrees to radians, multiply degrees by $(\pi \text{ rad})/180°$. To convert radians to degrees, multiply radians by $180°/(\pi \text{ rad})$.

4.2 ■ Definitions of trigonometric functions

Let t be a real number and let (x, y) be the point on the unit circle corresponding to t.

$\sin t = y$ $\csc t = 1/y,\ y \neq 0$

$\cos t = x$ $\sec t = 1/x,\ x \neq 0$

$\tan t = y/x,\ x \neq 0$ $\cot t = x/y,\ y \neq 0$

4.3 ■ Trigonometric functions of acute angles

Let θ be an acute angle of a right triangle. Then the six trigonometric functions of the angle θ are defined as:

$\sin \theta = \text{opp/hyp}$ $\cos \theta = \text{adj/hyp}$ $\tan \theta = \text{opp/adj}$

$\csc \theta = \text{hyp/opp}$ $\sec \theta = \text{hyp/adj}$ $\cot \theta = \text{adj/opp}$

4.3 ■ Use the fundamental trigonometric identities

The fundamental trigonometric identities represent relationships between trigonometric functions. (See page 280.)

4.4 ■ Trigonometric functions of any angle

To find the value of a trigonometric function of any angle θ, determine the function value for the associated reference angle θ'. Then, depending on the quadrant in which θ lies, affix the appropriate sign to the function value.

4.5 ■ Graph sine and cosine functions

The graphs of $y = a \sin(bx - c)$ and $y = a \cos(bx - c)$ have the following characteristics. (Assume $b > 0$.) The amplitude is $|a|$. The period is $2\pi/b$. The left and right endpoints of a one-cycle interval can be determined by solving the equations $bx - c = 0$ and $bx - c = 2\pi$. Graph the five key points in one period: the intercepts, the maximum points, and the minimum points.

4.6 ■ Graph other trigonometric functions

1. For tangent and cotangent functions, find the asymptotes, the period, and x-intercepts. Plot additional points between consecutive asymptotes and sketch one cycle, followed by additional cycles to the left and right.

2. For cosecant and secant functions, sketch the reciprocal function (sine or cosine) and take the reciprocals of the y-coordinates to obtain the y-coordinates of the cosecant or secant function. A maximum/minimum point on a sine or cosine function is a local minimum/maximum point on the cosecant or secant function. Also, x-intercepts of the sine and cosine functions become vertical asymptotes of the cosecant and secant functions.

4.7 ■ Evaluate inverse trigonometric functions

1. $y = \arcsin x$ if and only if $\sin y = x$; domain: $-1 \leq x \leq 1$; range: $-\pi/2 \leq y \leq \pi/2$

2. $y = \arccos x$ if and only if $\cos y = x$; domain: $-1 \leq x \leq 1$; range: $0 \leq y \leq \pi$

3. $y = \arctan x$ if and only if $\tan y = x$; domain: $-\infty < x < \infty$; range: $-\pi/2 < y < \pi/2$

4.7 ■ Compositions of trigonometric functions

1. If $-1 \leq x \leq 1$ and $-\pi/2 \leq y \leq \pi/2$, then $\sin(\arcsin x) = x$ and $\arcsin(\sin y) = y$.

2. If $-1 \leq x \leq 1$ and $0 \leq y \leq \pi$, then $\cos(\arccos x) = x$ and $\arccos(\cos y) = y$.

3. If x is a real number and $-\pi/2 < y < \pi/2$, then $\tan(\arctan x) = x$ and $\arctan(\tan y) = y$.

4.8 ■ Solve problems involving harmonic motion

A point that moves on a coordinate line is said to be in simple harmonic motion if its distance d from the origin at time t is given by $d = a \sin \omega t$ or $d = a \cos \omega t$, where a and ω are real numbers such that $\omega > 0$. The motion has amplitude $|a|$, period $2\pi/\omega$, and frequency $\omega/(2\pi)$.

Review Exercises

See www.CalcChat.com for worked-out solutions to odd-numbered exercises.

4.1 In Exercises 1 and 2, estimate the angle to the nearest one-half radian.

1.

2.

In Exercises 3–6, (a) sketch the angle in standard position, (b) determine the quadrant in which the angle lies, and (c) list one positive and one negative coterminal angle.

3. $\dfrac{4\pi}{3}$

4. $\dfrac{11\pi}{6}$

5. $-\dfrac{5\pi}{6}$

6. $-\dfrac{7\pi}{4}$

In Exercises 7–10, find (if possible) the complement and supplement of the angle.

7. $\dfrac{\pi}{8}$

8. $\dfrac{\pi}{12}$

9. $\dfrac{3\pi}{10}$

10. $\dfrac{2\pi}{21}$

In Exercises 11–14, (a) sketch the angle in standard position, (b) determine the quadrant in which the angle lies, and (c) list one positive and one negative coterminal angle.

11. $45°$

12. $210°$

13. $-135°$

14. $-405°$

In Exercises 15–18, find (if possible) the complement and supplement of the angle.

15. $5°$

16. $84°$

17. $171°$

18. $136°$

In Exercises 19–22, use the angle-conversion capabilities of a graphing utility to convert the angle measure to decimal degree form. Round your answer to three decimal places.

19. $135°\,16'\,45''$

20. $-234°\,40''$

21. $5°\,22'\,53''$

22. $280°\,8'\,50''$

In Exercises 23–26, use the angle-conversion capabilities of a graphing utility to convert the angle measure to D°M′S″ form.

23. $135.29°$

24. $25.8°$

25. $-85.36°$

26. $-327.93°$

In Exercises 27–30, convert the angle measure from degrees to radians. Round your answer to three decimal places.

27. $415°$

28. $-355°$

29. $-72°$

30. $94°$

In Exercises 31–34, convert the angle measure from radians to degrees. Round your answer to three decimal places.

31. $\dfrac{5\pi}{7}$

32. $-\dfrac{3\pi}{5}$

33. -3.5

34. 1.55

35. Find the radian measure of the central angle of a circle with a radius of 12 feet that intercepts an arc of length 25 feet.

36. Find the radian measure of the central angle of a circle with a radius of 60 inches that intercepts an arc of length 245 inches.

37. Find the length of the arc on a circle with a radius of 20 meters intercepted by a central angle of $138°$.

38. Find the length of the arc on a circle with a radius of 15 centimeters intercepted by a central angle of $60°$.

39. *Music* The radius of a compact disc is 6 centimeters. Find the linear speed of a point on the circumference of the disc if it is rotating at a speed of 500 revolutions per minute.

40. *Angular Speed* A car is moving at a rate of 28 miles per hour, and the diameter of its wheels is about $2\tfrac{1}{3}$ feet.

(a) Find the number of revolutions per minute the wheels are rotating.

(b) Find the angular speed of the wheels in radians per minute.

4.2 In Exercises 41–48, find the point (x, y) on the unit circle that corresponds to the real number t.

41. $t = \dfrac{7\pi}{4}$

42. $t = \dfrac{3\pi}{4}$

43. $t = \dfrac{5\pi}{6}$

44. $t = \dfrac{4\pi}{3}$

45. $t = -\dfrac{2\pi}{3}$

46. $t = -\dfrac{7\pi}{6}$

47. $t = -\dfrac{5\pi}{4}$

48. $t = -\dfrac{5\pi}{6}$

In Exercises 49–56, evaluate (if possible) the six trigonometric functions of the real number.

49. $t = \dfrac{7\pi}{6}$

50. $t = \dfrac{\pi}{4}$

51. $t = 2\pi$

52. $t = -\pi$

53. $t = -\dfrac{11\pi}{6}$

54. $t = -\dfrac{5\pi}{6}$

55. $t = -\dfrac{\pi}{2}$

56. $t = -\dfrac{\pi}{4}$

In Exercises 57–60, evaluate the trigonometric function using its period as an aid.

57. $\sin \dfrac{11\pi}{4}$

58. $\cos 4\pi$

59. $\sin\left(-\dfrac{17\pi}{6}\right)$

60. $\cos\left(-\dfrac{13\pi}{3}\right)$

In Exercises 61–64, use the value of the trigonometric function to evaluate the indicated functions.

61. $\sin t = \frac{3}{5}$
   (a) $\sin(-t)$
   (b) $\csc(-t)$

62. $\cos t = \frac{5}{13}$
   (a) $\cos(-t)$
   (b) $\sec(-t)$

63. $\sin(-t) = -\frac{2}{3}$
   (a) $\sin t$
   (b) $\csc t$

64. $\cos(-t) = \frac{5}{8}$
   (a) $\cos t$
   (b) $\sec(-t)$

In Exercises 65–68, use a calculator to evaluate the expression. Round your answer to four decimal places.

65. $\cot 2.3$

66. $\sec 4.5$

67. $\cos \dfrac{5\pi}{3}$

68. $\tan\left(-\dfrac{11\pi}{6}\right)$

4.3 In Exercises 69–72, find the exact values of the six trigonometric functions of the angle θ shown in the figure.

69.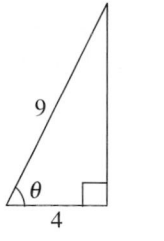

70.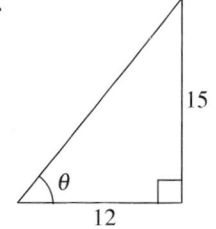

71.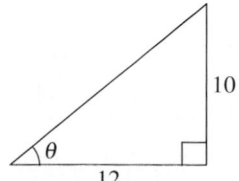

72.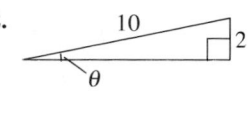

In Exercises 73 and 74, use trigonometric identities to transform one side of the equation into the other.

73. $\csc \theta \tan \theta = \sec \theta$

74. $\dfrac{\cot \theta + \tan \theta}{\cot \theta} = \sec^2 \theta$

In Exercises 75–78, use a calculator to evaluate each function. Round your answers to four decimal places.

75. (a) $\cos 84°$ (b) $\sin 6°$

76. (a) $\csc 52° \, 12'$ (b) $\sec 54° \, 7'$

77. (a) $\cos \dfrac{\pi}{4}$ (b) $\sec \dfrac{\pi}{4}$

78. (a) $\tan \dfrac{3\pi}{20}$ (b) $\cot \dfrac{3\pi}{20}$

79. *Width* An engineer is trying to determine the width of a river (see figure). From point P, the engineer walks downstream 125 feet and sights to point Q. From this sighting, it is determined that $\theta = 62°$. How wide is the river?

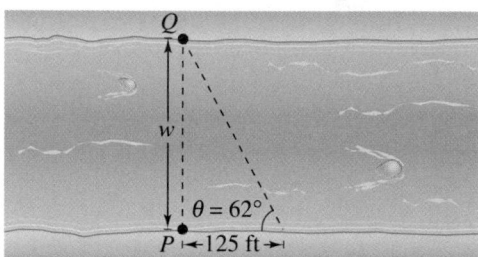

80. *Height* An escalator 152 feet in length rises to a platform and makes a 30° angle with the ground.

(a) Draw a right triangle that gives a visual representation of the problem. Show the known quantities and use a variable to indicate the height of the platform above the ground.

(b) Use a trigonometric function to write an equation involving the unknown quantity.

(c) Find the height of the platform above the ground.

4.4 In Exercises 81–86, the point is on the terminal side of an angle in standard position. Determine the exact values of the six trigonometric functions of the angle.

81. $(12, 16)$

82. $(2, 10)$

83. $(-7, 2)$

84. $(3, -4)$

85. $\left(\frac{2}{3}, \frac{5}{8}\right)$

86. $\left(-\frac{10}{3}, -\frac{2}{3}\right)$

In Exercises 87–90, find the values of the other five trigonometric functions of θ satisfying the given conditions.

87. $\sec \theta = \frac{6}{5}, \quad \tan \theta < 0$

88. $\tan \theta = -\frac{12}{5}, \quad \sin \theta > 0$

89. $\sin \theta = \frac{3}{8}, \quad \cos \theta < 0$

90. $\cos \theta = -\frac{2}{5}, \quad \sin \theta > 0$

In Exercises 91–94, find the reference angle θ' and sketch θ and θ' in standard position.

91. $\theta = 264°$

92. $\theta = 635°$

93. $\theta = -\dfrac{6\pi}{5}$

94. $\theta = \dfrac{17\pi}{3}$

In Exercises 95–102, evaluate the sine, cosine, and tangent of the angle without using a calculator.

95. $240°$ **96.** $315°$

97. $-210°$ **98.** $-315°$

99. $-\dfrac{9\pi}{4}$ **100.** $\dfrac{11\pi}{6}$

101. 4π **102.** $\dfrac{7\pi}{3}$

In Exercises 103–106, use a calculator to evaluate the trigonometric function. Round your answer to four decimal places.

103. $\tan 33°$ **104.** $\csc 105°$

105. $\sec \dfrac{12\pi}{5}$ **106.** $\sin\left(-\dfrac{\pi}{9}\right)$

4.5 In Exercises 107–110, sketch the graph of the function.

107. $f(x) = 3 \sin x$ **108.** $f(x) = 2 \cos x$

109. $f(x) = \frac{1}{4} \cos x$ **110.** $f(x) = \frac{7}{2} \sin x$

In Exercises 111–114, find the period and amplitude.

111.

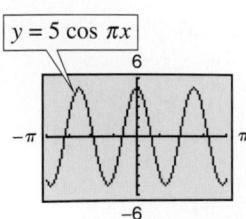

$y = 5 \cos \pi x$

112.

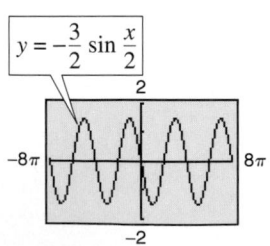

$y = -\dfrac{3}{2} \sin \dfrac{x}{2}$

113.

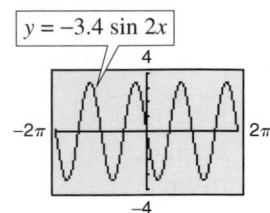

$y = -3.4 \sin 2x$

114.

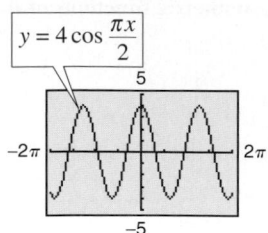

$y = 4 \cos \dfrac{\pi x}{2}$

In Exercises 115–126, sketch the graph of the function. (Include two full periods.)

115. $f(x) = 3 \cos 2\pi x$ **116.** $f(x) = -2 \sin \pi x$

117. $f(x) = 5 \sin \dfrac{2x}{5}$ **118.** $f(x) = 8 \cos\left(-\dfrac{x}{4}\right)$

119. $f(x) = -\dfrac{5}{2} \cos \dfrac{x}{4}$

120. $f(x) = -\dfrac{1}{2} \sin \dfrac{\pi x}{4}$

121. $f(x) = \frac{5}{2} \sin(x - \pi)$

122. $f(x) = 3 \cos(x + \pi)$

123. $f(x) = 2 - \cos \dfrac{\pi x}{2}$

124. $f(x) = \frac{1}{2} \sin \pi x - 3$

125. $f(x) = -3 \cos\left(\dfrac{x}{2} - \dfrac{\pi}{4}\right)$

126. $f(x) = 4 - 2 \cos(4x + \pi)$

Graphical Reasoning In Exercises 127–130, find a, b, and c for the function $f(x) = a \cos(bx - c)$ such that the graph of f matches the graph shown.

127.

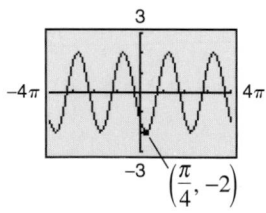

$\left(\dfrac{\pi}{4}, -2\right)$

128.

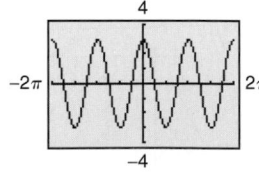

129.

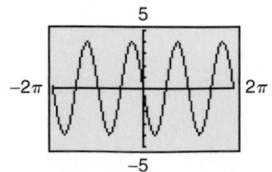

130.

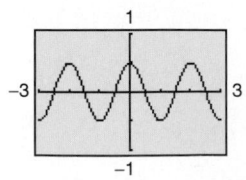

Sales In Exercises 131 and 132, use a graphing utility to graph the sales function over 1 year, where S is the sales (in thousands of units) and t is the time (in months), with $t = 1$ corresponding to January. Determine the months of maximum and minimum sales.

131. $S = 48.4 - 6.1 \cos \dfrac{\pi t}{6}$

132. $S = 56.25 + 9.50 \sin \dfrac{\pi t}{6}$

4.6 In Exercises 133–146, sketch the graph of the function. (Include two full periods.)

133. $f(x) = -\tan \dfrac{\pi x}{4}$ **134.** $f(x) = 4 \tan \pi x$

135. $f(x) = \dfrac{1}{4} \tan\left(x - \dfrac{\pi}{2}\right)$ **136.** $f(x) = 2 + 2 \tan \dfrac{x}{3}$

137. $f(x) = 3 \cot \dfrac{x}{2}$ **138.** $f(x) = \dfrac{1}{2} \cot \dfrac{\pi x}{2}$

139. $f(x) = \dfrac{1}{2} \cot\left(x - \dfrac{\pi}{2}\right)$

140. $f(x) = 4 \cot\left(x + \dfrac{\pi}{4}\right)$

141. $f(x) = \dfrac{1}{4} \sec x$

142. $f(x) = \dfrac{1}{2} \csc x$

143. $f(x) = \dfrac{1}{4} \csc 2x$

144. $f(x) = \dfrac{1}{2} \sec 2\pi x$

145. $f(x) = \sec\left(x - \dfrac{\pi}{4}\right)$

146. $f(x) = \dfrac{1}{2} \csc(2x + \pi)$

In Exercises 147–154, use a graphing utility to graph the function. (Include two full periods.)

147. $f(x) = \dfrac{1}{4} \tan \dfrac{\pi x}{2}$

148. $f(x) = \tan\left(x + \dfrac{\pi}{4}\right)$

149. $f(x) = 4 \cot(2x - \pi)$

150. $f(x) = -2 \cot(4x + \pi)$

151. $f(x) = 2 \sec(x - \pi)$

152. $f(x) = -2 \csc(x - \pi)$

153. $f(x) = \csc\left(3x - \dfrac{\pi}{2}\right)$

154. $f(x) = 3 \csc\left(2x + \dfrac{\pi}{4}\right)$

In Exercises 155–158, use a graphing utility to graph the function and the damping factor of the function in the same viewing window. Describe the behavior of the function as x increases without bound.

155. $f(x) = e^x \sin 2x$

156. $f(x) = e^x \cos x$

157. $f(x) = 2x \cos x$

158. $f(x) = x \sin \pi x$

4.7 In Exercises 159–162, find the exact value of each expression without using a calculator.

159. (a) $\arcsin(-1)$ (b) $\arcsin 4$

160. (a) $\arcsin\left(-\dfrac{1}{2}\right)$ (b) $\arcsin\left(-\dfrac{\sqrt{3}}{2}\right)$

161. (a) $\cos^{-1} \dfrac{\sqrt{2}}{2}$ (b) $\cos^{-1}\left(-\dfrac{\sqrt{3}}{2}\right)$

162. (a) $\tan^{-1}\left(-\sqrt{3}\right)$ (b) $\tan^{-1} 1$

In Exercises 163–170, use a calculator to approximate the value of the expression. **Round your answer to the nearest hundredth.**

163. $\arccos 0.42$ **164.** $\arcsin 0.63$

165. $\sin^{-1}(-0.94)$ **166.** $\cos^{-1}(-0.12)$

167. $\arctan(-12)$ **168.** $\arctan 21$

169. $\tan^{-1} 0.81$ **170.** $\tan^{-1} 6.4$

In Exercises 171 and 172, use an inverse trigonometric function to write θ as a function of x.

171.   **172.**

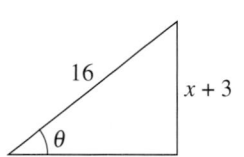   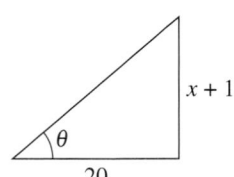

∫ In Exercises 173–176, write an algebraic expression that is equivalent to the expression.

173. $\sec[\arcsin(x - 1)]$

174. $\tan\left(\arccos \dfrac{x}{2}\right)$

175. $\sin\left(\arccos \dfrac{x^2}{4 - x^2}\right)$

176. $\csc(\arcsin 10x)$

4.8

177. Railroad Grade A train travels 3.5 kilometers on a straight track with a grade of 1° 10′ (see figure). What is the vertical rise of the train in that distance?

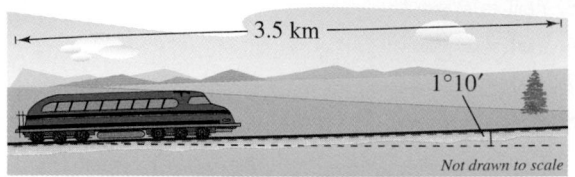

Not drawn to scale

178. Mountain Descent A road sign at the top of a mountain indicates that for the next 4 miles the grade is 12%. Find the angle of the grade and the change in elevation for a car descending the mountain.

179. Distance A passenger in an airplane flying at an altitude of 37,000 feet sees two towns directly to the west of the airplane. The angles of depression to the towns are 32° and 76° (see figure). How far apart are the towns?

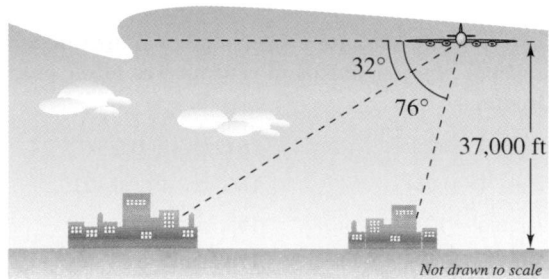

Not drawn to scale

180. Distance From city A to city B, a plane flies 650 miles at a bearing of 48°. From city B to city C, the plane flies 810 miles at a bearing of 115°. Find the distance from A to C and the bearing from A to C.

181. Wave Motion A buoy oscillates in simple harmonic motion as waves go past. At a given time it is noted that the buoy moves a total of 6 feet from its low point to its high point, returning to its high point every 15 seconds (see figure). Write an equation that describes the motion of the buoy if it is at its high point at $t = 0$.

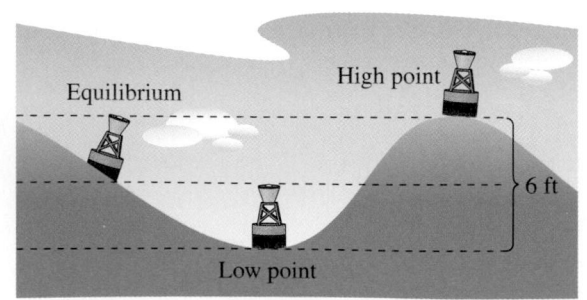

182. Wave Motion Your fishing bobber oscillates in simple harmonic motion from the waves in the lake where you fish. Your bob moves a total of 1.5 inches from its high point to its low point and returns to its high point every 3 seconds. Write an equation modeling the motion of your bob if it is at its high point at $t = 0$.

Synthesis

True or False? **In Exercises 183 and 184, determine whether the statement is true or false. Justify your answer.**

183. $y = \sin \theta$ is not a function because $\sin 30° = \sin 150°$.

184. The tangent function is often useful for modeling simple harmonic motion.

185. Numerical Analysis A 3000-pound automobile is negotiating a circular interchange of radius 300 feet at a speed of s miles per hour (see figure). The relationship between the speed and the angle θ (in degrees) at which the roadway should be banked so that no lateral frictional force is exerted on the tires is $\tan \theta = 0.672s^2/3000$.

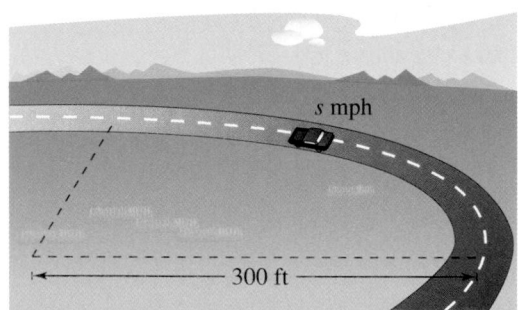

(a) Use a graphing utility to complete the table.

s	10	20	30	40	50	60
θ						

(b) In the table, s is incremented by 10, but θ does not increase by equal increments. Explain.

∫ **186. Approximation** In calculus it can be shown that the arctangent function can be approximated by the polynomial

$$\arctan x \approx x - \frac{x^3}{3} + \frac{x^5}{5} - \frac{x^7}{7}$$

where x is in radians.

(a) Use a graphing utility to graph the arctangent function and its polynomial approximation in the same viewing window. How do the graphs compare?

(b) Study the pattern in the polynomial approximation of the arctangent function and guess the next term. Then repeat part (a). How does the accuracy of the approximation change when additional terms are added?

4 Chapter Test

See www.CalcChat.com for worked-out solutions to odd-numbered exercises.

Take this test as you would take a test in class. After you are finished, check your work against the answers given in the back of the book.

1. Consider an angle that measures $\dfrac{5\pi}{4}$ radians.

   (a) Sketch the angle in standard position.

   (b) Determine two coterminal angles (one positive and one negative).

   (c) Convert the angle to degree measure.

2. A truck is moving at a rate of 90 kilometers per hour, and the diameter of its wheels is 1.25 meters. Find the angular speed of the wheels in radians per minute.

3. Find the exact values of the six trigonometric functions of the angle θ shown in the figure.

4. Given that $\tan \theta = \frac{7}{2}$ and θ is an acute angle, find the other five trigonometric functions of θ.

5. Determine the reference angle θ' of the angle $\theta = 255°$ and sketch θ and θ' in standard position.

6. Determine the quadrant in which θ lies if $\sec \theta < 0$ and $\tan \theta > 0$.

7. Find two exact values of θ in degrees $(0 \le \theta < 360°)$ if $\cos \theta = -\sqrt{2}/2$.

8. Use a calculator to approximate two values of θ in radians $(0 \le \theta < 2\pi)$ if $\csc \theta = 1.030$. Round your answer to two decimal places.

9. Find the five remaining trigonometric functions of θ, given that $\cos \theta = -\frac{3}{5}$ and $\sin \theta > 0$.

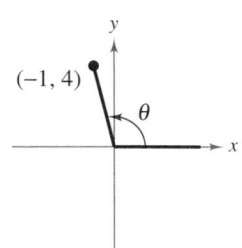

(−1, 4)

θ

y

x

Figure for 3

In Exercises 10–15, sketch the graph of the function. (Include two full periods.)

10. $g(x) = -2 \sin\left(x - \dfrac{\pi}{4}\right)$

11. $f(x) = \dfrac{1}{2} \tan 4x$

12. $f(x) = \frac{1}{2} \sec(x - \pi)$

13. $f(x) = 2 \cos(\pi - 2x) + 3$

14. $f(x) = 2 \csc\left(x + \dfrac{\pi}{2}\right)$

15. $f(x) = 2 \cot\left(x - \dfrac{\pi}{2}\right)$

In Exercises 16 and 17, use a graphing utility to graph the function. If the function is periodic, find its period.

16. $y = \sin 2\pi x + 2 \cos \pi x$

17. $y = 6e^{-0.12t} \cos(0.25t), \quad 0 \le t \le 32$

18. Find a, b, and c for the function $f(x) = a \sin(bx + c)$ such that the graph of f matches the graph at the right.

19. Find the exact value of $\tan\left(\arccos \frac{2}{3}\right)$ without using a calculator.

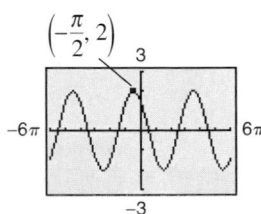

$\left(-\dfrac{\pi}{2}, 2\right)$

3

−6π   6π

−3

Figure for 18

In Exercises 20–22, use a graphing utility to graph the function.

20. $f(x) = 2 \arcsin\left(\dfrac{1}{2}x\right)$

21. $f(x) = 2 \arccos x$

22. $f(x) = \arctan \dfrac{x}{2}$

23. A plane is 160 miles north and 110 miles east of an airport. What bearing should be taken to fly directly to the airport?

Proofs in Mathematics

The Pythagorean Theorem

The Pythagorean Theorem is one of the most famous theorems in mathematics. More than 100 different proofs now exist. James A. Garfield, the twentieth president of the United States, developed a proof of the Pythagorean Theorem in 1876. His proof, shown below, involved the fact that a trapezoid can be formed from two congruent right triangles and an isosceles right triangle.

The Pythagorean Theorem

In a right triangle, the sum of the squares of the lengths of the legs is equal to the square of the length of the hypotenuse, where a and b are the legs and c is the hypotenuse.

$$a^2 + b^2 = c^2$$

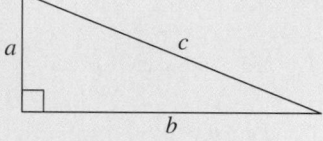

Proof

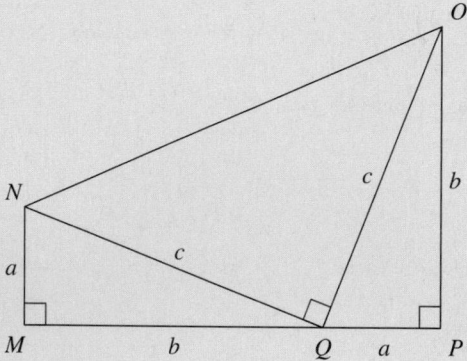

Area of   Area of   Area of   Area of
trapezoid $MNOP$ = $\triangle MNQ$ + $\triangle PQO$ + $\triangle NOQ$

$$\frac{1}{2}(a + b)(a + b) = \frac{1}{2}ab + \frac{1}{2}ab + \frac{1}{2}c^2$$

$$\frac{1}{2}(a + b)(a + b) = ab + \frac{1}{2}c^2$$

$$(a + b)(a + b) = 2ab + c^2$$

$$a^2 + 2ab + b^2 = 2ab + c^2$$

$$a^2 + b^2 = c^2$$

Chapter 5

Analytic Trigonometry

5.1 Using Fundamental Identities

5.2 Verifying Trigonometric Identities

5.3 Solving Trigonometric Equations

5.4 Sum and Difference Formulas

5.5 Multiple-Angle and Product-to-Sum Formulas

Selected Applications

Trigonometric equations and identities have many real-life applications. The applications listed below represent a small sample of the applications in this chapter.

- Friction, Exercise 71, page 367
- Shadow Length, Exercise 72, page 367
- Projectile Motion, Exercise 103, page 378
- Data Analysis: Unemployment Rate, Exercise 105, page 379
- Standing Waves, Exercise 79, page 385
- Harmonic Motion, Exercise 80, page 386
- Railroad Track, Exercise 129, page 397
- Mach Number, Exercise 130, page 398

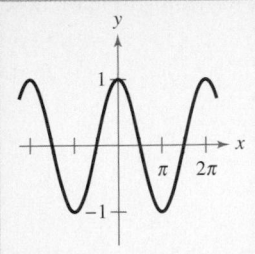 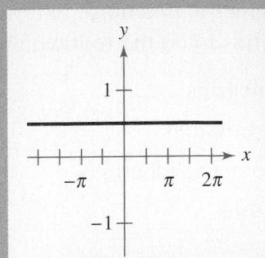 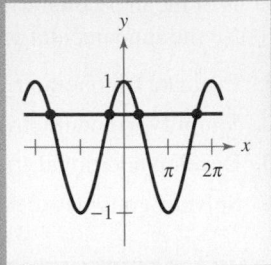

You can use multiple approaches—algebraic, numerical, and graphical—to solve trigonometric equations. In Chapter 5, you will use all three approaches to solve trigonometric equations. You will also use trigonometric identities to evaluate trigonometric functions and simplify trigonometric expressions.

Dan Donovan/MLB Photos/Getty Images

Trigonometry can be used to model projectile motion, such as the flight of a baseball. Given the angle at which the ball leaves the bat and the initial velocity, you can determine the distance the ball will travel.

351

5.1 Using Fundamental Identities

Introduction

In Chapter 4, you studied the basic definitions, properties, graphs, and applications of the individual trigonometric functions. In this chapter, you will learn how to use the fundamental identities to do the following.

1. Evaluate trigonometric functions.
2. Simplify trigonometric expressions.
3. Develop additional trigonometric identities.
4. Solve trigonometric equations.

Fundamental Trigonometric Identities

Reciprocal Identities

$$\sin u = \frac{1}{\csc u} \qquad \cos u = \frac{1}{\sec u} \qquad \tan u = \frac{1}{\cot u}$$

$$\csc u = \frac{1}{\sin u} \qquad \sec u = \frac{1}{\cos u} \qquad \cot u = \frac{1}{\tan u}$$

Quotient Identities

$$\tan u = \frac{\sin u}{\cos u} \qquad \cot u = \frac{\cos u}{\sin u}$$

Pythagorean Identities

$$\sin^2 u + \cos^2 u = 1 \qquad 1 + \tan^2 u = \sec^2 u \qquad 1 + \cot^2 u = \csc^2 u$$

Cofunction Identities

$$\sin\left(\frac{\pi}{2} - u\right) = \cos u \qquad \cos\left(\frac{\pi}{2} - u\right) = \sin u$$

$$\tan\left(\frac{\pi}{2} - u\right) = \cot u \qquad \cot\left(\frac{\pi}{2} - u\right) = \tan u$$

$$\sec\left(\frac{\pi}{2} - u\right) = \csc u \qquad \csc\left(\frac{\pi}{2} - u\right) = \sec u$$

Even/Odd Identities

$$\sin(-u) = -\sin u \qquad \cos(-u) = \cos u \qquad \tan(-u) = -\tan u$$

$$\csc(-u) = -\csc u \qquad \sec(-u) = \sec u \qquad \cot(-u) = -\cot u$$

Using the Fundamental Identities

One common use of trigonometric identities is to use given values of trigonometric functions to evaluate other trigonometric functions.

What you should learn

■ Recognize and write the fundamental trigonometric identities.

■ Use the fundamental trigonometric identities to evaluate trigonometric functions, simplify trigonometric expressions, and rewrite trigonometric expressions.

Why you should learn it

The fundamental trigonometric identities can be used to simplify trigonometric expressions. For instance, Exercise 111 on page 359 shows you how trigonometric identities can be used to simplify an expression for the rate of change of a function, a concept used in calculus.

STUDY TIP

Pythagorean identities are sometimes used in radical form such as

$$\sin u = \pm\sqrt{1 - \cos^2 u}$$

or

$$\tan u = \pm\sqrt{\sec^2 u - 1}$$

where the sign depends on the choice of u.

Example 1   Using Identities to Evaluate a Function

Use the values $\sec u = -\frac{3}{2}$ and $\tan u > 0$ to find the values of all six trigonometric functions.

Solution

Using a reciprocal identity, you have

$$\cos u = \frac{1}{\sec u} = \frac{1}{-3/2} = -\frac{2}{3}.$$

Using a Pythagorean identity, you have

$$\sin^2 u = 1 - \cos^2 u \qquad\qquad \text{Pythagorean identity}$$

$$= 1 - \left(-\frac{2}{3}\right)^2 \qquad\qquad \text{Substitute } -\frac{2}{3} \text{ for } \cos u.$$

$$= 1 - \frac{4}{9} = \frac{5}{9}. \qquad\qquad \text{Simplify.}$$

Because $\sec u < 0$ and $\tan u > 0$, it follows that u lies in Quadrant III. Moreover, because $\sin u$ is negative when u is in Quadrant III, you can choose the negative root and obtain $\sin u = -\sqrt{5}/3$. Now, knowing the values of the sine and cosine, you can find the values of all six trigonometric functions.

$$\sin u = -\frac{\sqrt{5}}{3} \qquad\qquad \csc u = \frac{1}{\sin u} = -\frac{3}{\sqrt{5}} = -\frac{3\sqrt{5}}{5}$$

$$\cos u = -\frac{2}{3} \qquad\qquad \sec u = \frac{1}{\cos u} = -\frac{3}{2}$$

$$\tan u = \frac{\sin u}{\cos u} = \frac{-\sqrt{5}/3}{-2/3} = \frac{\sqrt{5}}{2} \qquad\qquad \cot u = \frac{1}{\tan u} = \frac{2}{\sqrt{5}} = \frac{2\sqrt{5}}{5}$$

✓CHECKPOINT   Now try Exercise 7.

Example 2   Simplifying a Trigonometric Expression

Simplify $\sin x \cos^2 x - \sin x$.

Solution

First factor out a common monomial factor and then use a fundamental identity.

$$\sin x \cos^2 x - \sin x = \sin x(\cos^2 x - 1) \qquad \text{Factor out monomial factor.}$$

$$= -\sin x(1 - \cos^2 x) \qquad \text{Distributive Property}$$

$$= -\sin x(\sin^2 x) \qquad \text{Pythagorean identity}$$

$$= -\sin^3 x \qquad \text{Multiply.}$$

✓CHECKPOINT   Now try Exercise 29.

TECHNOLOGY TIP

You can use a graphing utility to check the result of Example 2. To do this, enter y_1 and y_2 as shown below.

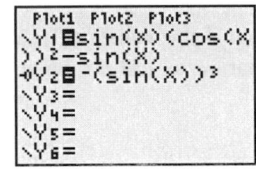

Select the *line* style for y_1 and the *path* style for y_2 (see figure above). The *path* style, denoted by ⟿, traces the leading edge of the graph and draws a path. Now, graph both equations in the same viewing window, as shown below. The two graphs *appear* to coincide, so the expressions *appear* to be equivalent. Remember that in order to be certain that two expressions are equivalent, you need to show their equivalence algebraically, as in Example 2.

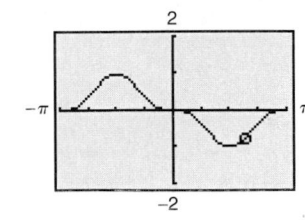

Example 3   Verifying a Trigonometric Identity

Determine whether the equation appears to be an identity.

$$\cos 3x \overset{?}{=} 4\cos^3 x - 3\cos x$$

Numerical Solution

Use the *table* feature of a graphing utility set in *radian* mode to create a table that shows the values of $y_1 = \cos 3x$ and $y_2 = 4\cos^3 x - 3\cos x$ for different values of x, as shown in Figure 5.1. The values of y_1 and y_2 appear to be identical, so $\cos 3x = 4\cos^3 x - 3\cos x$ appears to be an identity.

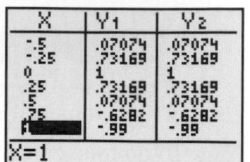

Figure 5.1

Note that if the values of y_1 and y_2 were not identical, then the equation would not be an identity.

✓**CHECKPOINT**   Now try Exercise 39.

Graphical Solution

Use a graphing utility set in *radian* mode to graph $y_1 = \cos 3x$ and $y_2 = 4\cos^3 x - 3\cos x$ in the same viewing window, as shown in Figure 5.2. (Select the *line* style for y_1 and the *path* style for y_2.) Because the graphs appear to coincide, $\cos 3x = 4\cos^3 x - 3\cos x$ appears to be an identity.

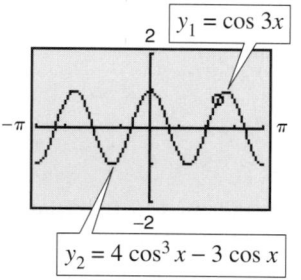

Figure 5.2

Note that if the graphs of y_1 and y_2 did not coincide, then the equation would not be an identity.

Example 4   Verifying a Trigonometric Identity

Verify the identity $\dfrac{\sin \theta}{1 + \cos \theta} + \dfrac{\cos \theta}{\sin \theta} = \csc \theta$.

Algebraic Solution

$$\frac{\sin \theta}{1 + \cos \theta} + \frac{\cos \theta}{\sin \theta} = \frac{(\sin \theta)(\sin \theta) + (\cos \theta)(1 + \cos \theta)}{(1 + \cos \theta)(\sin \theta)}$$

$$= \frac{\sin^2 \theta + \cos^2 \theta + \cos \theta}{(1 + \cos \theta)(\sin \theta)} \qquad \text{Multiply.}$$

$$= \frac{1 + \cos \theta}{(1 + \cos \theta)(\sin \theta)} \qquad \text{Pythagorean identity}$$

$$= \frac{1}{\sin \theta} \qquad \text{Divide out common factor.}$$

$$= \csc \theta \qquad \text{Use reciprocal identity.}$$

Notice how the identity is verified. You start with the left side of the equation (the more complicated side) and use the fundamental trigonometric identities to simplify it until you obtain the right side.

✓**CHECKPOINT**   Now try Exercise 45.

Graphical Solution

Use a graphing utility set in *radian* and *dot* modes to graph y_1 and y_2 in the same viewing window, as shown in Figure 5.3. Because the graphs appear to coincide, this equation appears to be an identity.

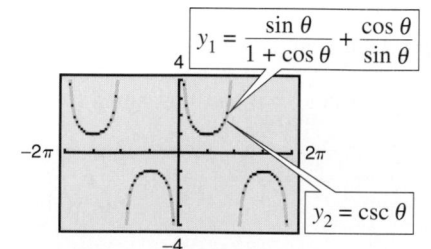

Figure 5.3

When factoring trigonometric expressions, it is helpful to find a polynomial form that fits the expression, as shown in Example 5.

Example 5   Factoring Trigonometric Expressions

Factor (a) $\sec^2 \theta - 1$ and (b) $4 \tan^2 \theta + \tan \theta - 3$.

Solution

a. Here the expression is a difference of two squares, which factors as

$$\sec^2 \theta - 1 = (\sec \theta - 1)(\sec \theta + 1).$$

b. This expression has the polynomial form $ax^2 + bx + c$ and it factors as

$$4 \tan^2 \theta + \tan \theta - 3 = (4 \tan \theta - 3)(\tan \theta + 1).$$

✓CHECKPOINT   Now try Exercise 51.

On occasion, factoring or simplifying can best be done by first rewriting the expression in terms of just *one* trigonometric function or in terms of *sine or cosine alone*. These strategies are illustrated in Examples 6 and 7.

Example 6   Factoring a Trigonometric Expression

Factor $\csc^2 x - \cot x - 3$.

Solution

Use the identity $\csc^2 x = 1 + \cot^2 x$ to rewrite the expression in terms of the cotangent.

$$\csc^2 x - \cot x - 3 = (1 + \cot^2 x) - \cot x - 3 \qquad \text{Pythagorean identity}$$

$$= \cot^2 x - \cot x - 2 \qquad \text{Combine like terms.}$$

$$= (\cot x - 2)(\cot x + 1) \qquad \text{Factor.}$$

✓CHECKPOINT   Now try Exercise 57.

Example 7   Simplifying a Trigonometric Expression

Simplify $\sin t + \cot t \cos t$.

Solution

Begin by rewriting $\cot t$ in terms of sine and cosine.

$$\sin t + \cot t \cos t = \sin t + \left(\frac{\cos t}{\sin t} \right) \cos t \qquad \text{Quotient identity}$$

$$= \frac{\sin^2 t + \cos^2 t}{\sin t} \qquad \text{Add fractions.}$$

$$= \frac{1}{\sin t} = \csc t \qquad \begin{array}{l} \text{Pythagorean identity and} \\ \text{reciprocal identity} \end{array}$$

✓CHECKPOINT   Now try Exercise 67.

The last two examples in this section involve techniques for rewriting expressions into forms that are used in calculus.

Example 8 Rewriting a Trigonometric Expression

Rewrite $\dfrac{1}{1 + \sin x}$ so that it is *not* in fractional form.

Solution

From the Pythagorean identity $\cos^2 x = 1 - \sin^2 x = (1 - \sin x)(1 + \sin x)$, you can see that multiplying both the numerator and the denominator by $(1 - \sin x)$ will produce a monomial denominator.

$$\frac{1}{1 + \sin x} = \frac{1}{1 + \sin x} \cdot \frac{1 - \sin x}{1 - \sin x} \qquad \text{Multiply numerator and denominator by } (1 - \sin x).$$

$$= \frac{1 - \sin x}{1 - \sin^2 x} \qquad \text{Multiply.}$$

$$= \frac{1 - \sin x}{\cos^2 x} \qquad \text{Pythagorean identity}$$

$$= \frac{1}{\cos^2 x} - \frac{\sin x}{\cos^2 x} \qquad \text{Write as separate fractions.}$$

$$= \frac{1}{\cos^2 x} - \frac{\sin x}{\cos x} \cdot \frac{1}{\cos x} \qquad \text{Write as separate fractions.}$$

$$= \sec^2 x - \tan x \sec x \qquad \text{Reciprocal and quotient identities}$$

✓**CHECKPOINT**　　Now try Exercise 69.

Example 9 Trigonometric Substitution

Use the substitution $x = 2 \tan \theta$, $0 < \theta < \pi/2$, to write $\sqrt{4 + x^2}$ as a trigonometric function of θ.

Solution

Begin by letting $x = 2 \tan \theta$. Then you can obtain

$$\sqrt{4 + x^2} = \sqrt{4 + (2 \tan \theta)^2} \qquad \text{Substitute } 2 \tan \theta \text{ for } x.$$

$$= \sqrt{4(1 + \tan^2 \theta)} \qquad \text{Distributive Property}$$

$$= \sqrt{4 \sec^2 \theta} \qquad \text{Pythagorean identity}$$

$$= 2 \sec \theta. \qquad \sec \theta > 0 \text{ for } 0 < \theta < \frac{\pi}{2}$$

✓**CHECKPOINT**　　Now try Exercise 81.

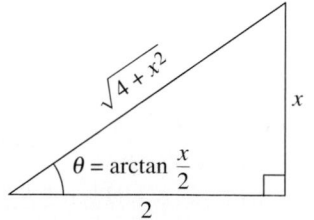

Figure 5.4

Figure 5.4 shows the right triangle illustration of the substitution in Example 9. For $0 < \theta < \pi/2$, you have

$$\text{opp} = x, \text{ adj} = 2, \text{ and hyp} = \sqrt{4 + x^2}.$$

Try using these expressions to obtain the result shown in Example 9.

5.1   Exercises

See www.CalcChat.com for worked-out solutions to odd-numbered exercises.

Vocabulary Check

Fill in the blank to complete the trigonometric identity.

1. $\dfrac{1}{\cos u} =$ _____

2. $\dfrac{1}{\cot u} =$ _____

3. $\dfrac{\cos u}{\sin u} =$ _____

4. $\dfrac{1}{\sin u} =$ _____

5. $1 +$ _____ $= \sec^2 u$

6. $1 + \cot^2 u =$ _____

7. $\cos\left(\dfrac{\pi}{2} - u\right) =$ _____

8. $\csc\left(\dfrac{\pi}{2} - u\right) =$ _____

9. $\tan(-u) =$ _____

10. $\cos(-u) =$ _____

In Exercises 1–14, use the given values to evaluate (if possible) all six trigonometric functions.

1. $\sin x = \dfrac{1}{2}, \quad \cos x = \dfrac{\sqrt{3}}{2}$

2. $\csc \theta = 2, \quad \tan \theta = \dfrac{\sqrt{3}}{3}$

3. $\sec \theta = \sqrt{2}, \quad \sin \theta = -\dfrac{\sqrt{2}}{2}$

4. $\tan x = \dfrac{\sqrt{3}}{3}, \quad \cos x = -\dfrac{\sqrt{3}}{2}$

5. $\tan x = \dfrac{7}{24}, \quad \sec x = -\dfrac{25}{24}$

6. $\cot \phi = -5, \quad \sin \phi = \dfrac{\sqrt{26}}{26}$

7. $\sec \phi = -\dfrac{17}{15}, \quad \sin \phi = \dfrac{8}{17}$

8. $\cos\left(\dfrac{\pi}{2} - x\right) = \dfrac{3}{5}, \quad \cos x = \dfrac{4}{5}$

9. $\sin(-x) = -\dfrac{2}{3}, \quad \tan x = -\dfrac{2\sqrt{5}}{5}$

10. $\csc(-x) = -5, \quad \cos x = \dfrac{\sqrt{24}}{5}$

11. $\tan \theta = 2, \quad \sin \theta < 0$

12. $\sec \theta = -3, \quad \tan \theta < 0$

13. $\csc \theta$ is undefined, $\quad \cos \theta < 0$

14. $\tan \theta$ is undefined, $\quad \sin \theta > 0$

In Exercises 15–20, match the trigonometric expression with one of the following.

(a) $\sec x$ (b) -1 (c) $\cot x$

(d) 1 (e) $-\tan x$ (f) $\sin x$

15. $\sec x \cos x$

16. $\tan x \csc x$

17. $\cot^2 x - \csc^2 x$

18. $(1 - \cos^2 x)(\csc x)$

19. $\dfrac{\sin(-x)}{\cos(-x)}$

20. $\dfrac{\sin[(\pi/2) - x]}{\cos[(\pi/2) - x]}$

In Exercises 21–26, match the trigonometric expression with one of the following.

(a) $\csc x$ (b) $\tan x$ (c) $\sin^2 x$

(d) $\sin x \tan x$ (e) $\sec^2 x$ (f) $\sec^2 x + \tan^2 x$

21. $\sin x \sec x$

22. $\cos^2 x(\sec^2 x - 1)$

23. $\sec^4 x - \tan^4 x$

24. $\cot x \sec x$

25. $\dfrac{\sec^2 x - 1}{\sin^2 x}$

26. $\dfrac{\cos^2[(\pi/2) - x]}{\cos x}$

In Exercises 27–38, use the fundamental identities to simplify the expression. Use the *table* feature of a graphing utility to check your result numerically.

27. $\cot x \sin x$

28. $\cos \beta \tan \beta$

29. $\sin \phi(\csc \phi - \sin \phi)$

30. $\sec^2 x(1 - \sin^2 x)$

31. $\dfrac{\csc x}{\cot x}$

32. $\dfrac{\sec \theta}{\csc \theta}$

33. $\sec \alpha \cdot \dfrac{\sin \alpha}{\tan \alpha}$

34. $\dfrac{\tan^2 \theta}{\sec^2 \theta}$

35. $\sin\left(\dfrac{\pi}{2} - x\right)\csc x$

36. $\cot\left(\dfrac{\pi}{2} - x\right)\cos x$

37. $\dfrac{\cos^2 y}{1 - \sin y}$

38. $\dfrac{1}{\cot^2 x + 1}$

In Exercises 39–44, verify the identity algebraically. Use the *table* feature of a graphing utility to check your result numerically.

39. $\sin\theta + \cos\theta\cot\theta = \csc\theta$

40. $(\sec\theta - \tan\theta)(\csc\theta + 1) = \cot\theta$

41. $\dfrac{\cos\theta}{1 - \sin\theta} = \sec\theta + \tan\theta$

42. $\dfrac{1 + \csc\theta}{\cot\theta + \cos\theta} = \sec\theta$

43. $\dfrac{1 + \cos\theta}{\sin\theta} + \dfrac{\sin\theta}{1 + \cos\theta} = 2\csc\theta$

44. $\dfrac{\sin\theta + \cos\theta}{\sin\theta} - \dfrac{\cos\theta - \sin\theta}{\cos\theta} = \sec\theta\csc\theta$

In Exercises 45–50, verify the identity algebraically. Use a graphing utility to check your result graphically.

45. $\csc\theta\tan\theta = \sec\theta$

46. $\sin\theta\csc\theta - \sin^2\theta = \cos^2\theta$

47. $1 - \dfrac{\sin^2\theta}{1 - \cos\theta} = -\cos\theta$

48. $\dfrac{\tan\theta}{1 + \sec\theta} + \dfrac{1 + \sec\theta}{\tan\theta} = 2\csc\theta$

49. $\dfrac{\cot(-\theta)}{\csc\theta} = -\cos\theta$

50. $\dfrac{\csc\left(\dfrac{\pi}{2} - \theta\right)}{\tan(-\theta)} = -\csc\theta$

In Exercises 51–60, factor the expression and use the fundamental identities to simplify. Use a graphing utility to check your result graphically.

51. $\cot^2 x - \cot^2 x\cos^2 x$

52. $\sec^2 x\tan^2 x + \sec^2 x$

53. $\dfrac{\cos^2 x - 4}{\cos x - 2}$

54. $\dfrac{\csc^2 x - 1}{\csc x - 1}$

55. $\tan^4 x + 2\tan^2 x + 1$

56. $1 - 2\sin^2 x + \sin^4 x$

57. $\sin^4 x - \cos^4 x$

58. $\sec^4 x - \tan^4 x$

59. $\csc^3 x - \csc^2 x - \csc x + 1$

60. $\sec^3 x - \sec^2 x - \sec x + 1$

In Exercises 61–68, perform the indicated operation and use the fundamental identities to simplify.

61. $(\sin x + \cos x)^2$

62. $(\tan x + \sec x)(\tan x - \sec x)$

63. $(\csc x + 1)(\csc x - 1)$

64. $(5 - 5\sin x)(5 + 5\sin x)$

65. $\dfrac{1}{1 + \cos x} + \dfrac{1}{1 - \cos x}$

66. $\dfrac{1}{\sec x + 1} - \dfrac{1}{\sec x - 1}$

67. $\tan x - \dfrac{\sec^2 x}{\tan x}$

68. $\dfrac{\cos x}{1 + \sin x} + \dfrac{1 + \sin x}{\cos x}$

In Exercises 69–72, rewrite the expression so that it is *not* in fractional form.

69. $\dfrac{\sin^2 y}{1 - \cos y}$

70. $\dfrac{5}{\tan x + \sec x}$

71. $\dfrac{3}{\sec x - \tan x}$

72. $\dfrac{\tan^2 x}{\csc x + 1}$

Numerical and Graphical Analysis **In Exercises 73–76, use a graphing utility to complete the table and graph the functions in the same viewing window. Make a conjecture about y_1 and y_2.**

x	0.2	0.4	0.6	0.8	1.0	1.2	1.4
y_1							
y_2							

73. $y_1 = \cos\left(\dfrac{\pi}{2} - x\right),\quad y_2 = \sin x$

74. $y_1 = \cos x + \sin x\tan x,\quad y_2 = \sec x$

75. $y_1 = \dfrac{\cos x}{1 - \sin x},\quad y_2 = \dfrac{1 + \sin x}{\cos x}$

76. $y_1 = \sec^4 x - \sec^2 x,\quad y_2 = \tan^2 x + \tan^4 x$

In Exercises 77–80, use a graphing utility to determine which of the six trigonometric functions is equal to the expression.

77. $\cos x\cot x + \sin x$

78. $\sin x(\cot x + \tan x)$

79. $\sec x - \dfrac{\cos x}{1 + \sin x}$

80. $\dfrac{1}{2}\left(\dfrac{1 + \sin\theta}{\cos\theta} + \dfrac{\cos\theta}{1 + \sin\theta}\right)$

In Exercises 81–92, use the trigonometric substitution to write the algebraic expression as a trigonometric function of θ, where $0 < \theta < \pi/2$.

81. $\sqrt{25 - x^2}, \quad x = 5 \sin \theta$

82. $\sqrt{64 - 16x^2}, \quad x = 2 \cos \theta$

83. $\sqrt{x^2 - 9}, \quad x = 3 \sec \theta$

84. $\sqrt{x^2 + 100}, \quad x = 10 \tan \theta$

85. $\sqrt{9 - x^2}, \quad x = 3 \sin \theta$

86. $\sqrt{4 - x^2}, \quad x = 2 \cos \theta$

87. $\sqrt{4x^2 + 9}, \quad 2x = 3 \tan \theta$

88. $\sqrt{9x^2 + 4}, \quad 3x = 2 \tan \theta$

89. $\sqrt{16x^2 - 9}, \quad 4x = 3 \sec \theta$

90. $\sqrt{9x^2 - 25}, \quad 3x = 5 \sec \theta$

91. $\sqrt{2 - x^2}, \quad x = \sqrt{2} \sin \theta$

92. $\sqrt{5 - x^2}, \quad x = \sqrt{5} \cos \theta$

In Exercises 93–96, use a graphing utility to solve the equation for θ, where $0 \le \theta < 2\pi$.

93. $\sin \theta = \sqrt{1 - \cos^2 \theta}$

94. $\cos \theta = -\sqrt{1 - \sin^2 \theta}$

95. $\sec \theta = \sqrt{1 + \tan^2 \theta}$

96. $\tan \theta = \sqrt{\sec^2 \theta - 1}$

In Exercises 97–100, rewrite the expression as a single logarithm and simplify the result.

97. $\ln|\cos \theta| - \ln|\sin \theta|$

98. $\ln|\csc \theta| + \ln|\tan \theta|$

99. $\ln(1 + \sin x) - \ln|\sec x|$

100. $\ln|\cot t| + \ln(1 + \tan^2 t)$

In Exercises 101–106, show that the identity is *not* true for all values of θ. (There are many correct answers.)

101. $\cos \theta = \sqrt{1 - \sin^2 \theta}$

102. $\tan \theta = \sqrt{\sec^2 \theta - 1}$

103. $\sin \theta = \sqrt{1 - \cos^2 \theta}$

104. $\sec \theta = \sqrt{1 + \tan^2 \theta}$

105. $\csc \theta = \sqrt{1 + \cot^2 \theta}$

106. $\cot \theta = \sqrt{\csc^2 \theta - 1}$

In Exercises 107–110, use the *table* feature of a graphing utility to demonstrate the identity for each value of θ.

107. $\csc^2 \theta - \cot^2 \theta = 1$, (a) $\theta = 132°$ (b) $\theta = \dfrac{2\pi}{7}$

108. $\tan^2 \theta + 1 = \sec^2 \theta$, (a) $\theta = 346°$ (b) $\theta = 3.1$

109. $\cos\left(\dfrac{\pi}{2} - \theta\right) = \sin \theta$, (a) $\theta = 80°$ (b) $\theta = 0.8$

110. $\sin(-\theta) = -\sin \theta$, (a) $\theta = 250°$ (b) $\theta = \frac{1}{2}$

111. **Rate of Change**   The rate of change of the function $f(x) = -\csc x - \sin x$ is given by the expression $\csc x \cot x - \cos x$. Show that this expression can also be written as $\cos x \cot^2 x$.

112. **Rate of Change**   The rate of change of the function $f(x) = \sec x + \cos x$ is given by the expression $\sec x \tan x - \sin x$. Show that this expression can also be written as $\sin x \tan^2 x$.

Synthesis

True or False?   **In Exercises 113 and 114, determine whether the statement is true or false. Justify your answer.**

113. $\sin \theta \csc \theta = 1$

114. $\cos \theta \sec \phi = 1$

In Exercises 115–118, fill in the blanks. (*Note:* $x \to c^+$ indicates that x approaches c from the right, and $x \to c^-$ indicates that x approaches c from the left.)

115. As $x \to \dfrac{\pi^-}{2}$, $\sin x \to$ ▭ and $\csc x \to$ ▭ .

116. As $x \to 0^+$, $\cos x \to$ ▭ and $\sec x \to$ ▭ .

117. As $x \to \dfrac{\pi^-}{2}$, $\tan x \to$ ▭ and $\cot x \to$ ▭ .

118. As $x \to \pi^+$, $\sin x \to$ ▭ and $\csc x \to$ ▭ .

119. Write each of the other trigonometric functions of θ in terms of $\sin \theta$.

120. Write each of the other trigonometric functions of θ in terms of $\cos \theta$.

121. Use the definitions of sine and cosine to derive the Pythagorean identity $\sin^2 \theta + \cos^2 \theta = 1$.

122. **Writing**   Use the Pythagorean identity $\sin^2 \theta + \cos^2 \theta = 1$ to derive the other Pythagorean identities $1 + \tan^2 \theta = \sec^2 \theta$ and $1 + \cot^2 \theta = \csc^2 \theta$. Discuss how to remember these identities and other fundamental identities.

Skills Review

In Exercises 123–126, sketch the graph of the function. (Include two full periods.)

123. $f(x) = \dfrac{1}{2} \sin \pi x$

124. $f(x) = -2 \tan \dfrac{\pi x}{2}$

125. $f(x) = \dfrac{1}{2} \cot\left(x + \dfrac{\pi}{4}\right)$

126. $f(x) = \dfrac{3}{2} \cos(x - \pi) + 3$

5.2 Verifying Trigonometric Identities

Introduction

In this section, you will study techniques for verifying trigonometric identities. In the next section, you will study techniques for solving trigonometric equations. The key to both verifying identities *and* solving equations is your ability to use the fundamental identities and the rules of algebra to rewrite trigonometric expressions.

Remember that a *conditional equation* is an equation that is true for only some of the values in its domain. For example, the conditional equation

$$\sin x = 0 \qquad \text{Conditional equation}$$

is true only for $x = n\pi$, where n is an integer. When you find these values, you are *solving* the equation.

On the other hand, an equation that is true for all real values in the domain of the variable is an *identity*. For example, the familiar equation

$$\sin^2 x = 1 - \cos^2 x \qquad \text{Identity}$$

is true for all real numbers x. So, it is an identity.

Verifying Trigonometric Identities

Verifying that a trigonometric equation is an identity is quite different from solving an equation. There is no well-defined set of rules to follow in verifying trigonometric identities, and the process is best learned by practice.

Guidelines for Verifying Trigonometric Identities

1. Work with one side of the equation at a time. It is often better to work with the more complicated side first.

2. Look for opportunities to factor an expression, add fractions, square a binomial, or create a monomial denominator.

3. Look for opportunities to use the fundamental identities. Note which functions are in the final expression you want. Sines and cosines pair up well, as do secants and tangents, and cosecants and cotangents.

4. If the preceding guidelines do not help, try converting all terms to sines and cosines.

5. Always try *something*. Even making an attempt that leads to a dead end provides insight.

Verifying trigonometric identities is a useful process if you need to convert a trigonometric expression into a form that is more useful algebraically. When you verify an identity, you cannot assume that the two sides of the equation are equal because you are trying to verify that they are equal. As a result, when verifying identities, you cannot use operations such as adding the same quantity to each side of the equation or cross multiplication.

What you should learn

■ Verify trigonometric identities.

Why you should learn it

You can use trigonometric identities to rewrite trigonometric expressions. For instance, Exercise 72 on page 367 shows you how trigonometric identities can be used to simplify an equation that models the length of a shadow cast by a gnomon (a device used to tell time).

BSCHMID/Getty Images

Prerequisite Skills

To review the differences among an identity, an expression, and an equation, see Appendix B.3.

Example 1 Verifying a Trigonometric Identity

Verify the identity $\dfrac{\sec^2\theta - 1}{\sec^2\theta} = \sin^2\theta$.

Solution

Because the left side is more complicated, start with it.

$$\frac{\sec^2\theta - 1}{\sec^2\theta} = \frac{(\tan^2\theta + 1) - 1}{\sec^2\theta} \qquad \text{Pythagorean identity}$$

$$= \frac{\tan^2\theta}{\sec^2\theta} \qquad \text{Simplify.}$$

$$= \tan^2\theta(\cos^2\theta) \qquad \text{Reciprocal identity}$$

$$= \frac{\sin^2\theta}{\cos^2\theta}(\cos^2\theta) \qquad \text{Quotient identity}$$

$$= \sin^2\theta \qquad \text{Simplify.}$$

✓CHECKPOINT   Now try Exercise 5.

> **STUDY TIP**
>
> Remember that an identity is true only for all real values in the domain of the variable. For instance, in Example 1 the identity is not true when $\theta = \pi/2$ because $\sec^2\theta$ is not defined when $\theta = \pi/2$.

There can be more than one way to verify an identity. Here is another way to verify the identity in Example 1.

$$\frac{\sec^2\theta - 1}{\sec^2\theta} = \frac{\sec^2\theta}{\sec^2\theta} - \frac{1}{\sec^2\theta} \qquad \text{Rewrite as the difference of fractions.}$$

$$= 1 - \cos^2\theta \qquad \text{Reciprocal identity}$$

$$= \sin^2\theta \qquad \text{Pythagorean identity}$$

Example 2 Combining Fractions Before Using Identities

Verify the identity $\dfrac{1}{1 - \sin\alpha} + \dfrac{1}{1 + \sin\alpha} = 2\sec^2\alpha$.

Algebraic Solution

$$\frac{1}{1 - \sin\alpha} + \frac{1}{1 + \sin\alpha} = \frac{1 + \sin\alpha + 1 - \sin\alpha}{(1 - \sin\alpha)(1 + \sin\alpha)} \qquad \text{Add fractions.}$$

$$= \frac{2}{1 - \sin^2\alpha} \qquad \text{Simplify.}$$

$$= \frac{2}{\cos^2\alpha} \qquad \text{Pythagorean identity}$$

$$= 2\sec^2\alpha \qquad \text{Reciprocal identity}$$

✓CHECKPOINT   Now try Exercise 31.

Numerical Solution

Use the *table* feature of a graphing utility set in *radian* mode to create a table that shows the values of $y_1 = 1/(1 - \sin x) + 1/(1 + \sin x)$ and $y_2 = 2/\cos^2 x$ for different values of x, as shown in Figure 5.5. From the table, you can see that the values appear to be identical, so $1/(1 - \sin x) + 1/(1 + \sin x) = 2\sec^2 x$ appears to be an identity.

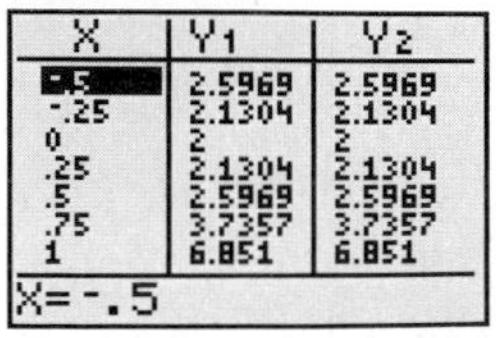

X	Y₁	Y₂
-.5	2.5969	2.5969
-.25	2.1304	2.1304
0	2	2
.25	2.1304	2.1304
.5	2.5969	2.5969
.75	3.7357	3.7357
1	6.851	6.851

X=-.5

Figure 5.5

Example 3 Verifying a Trigonometric Identity

Verify the identity $(\tan^2 x + 1)(\cos^2 x - 1) = -\tan^2 x$.

Algebraic Solution

By applying identities before multiplying, you obtain the following.

$$(\tan^2 x + 1)(\cos^2 x - 1) = (\sec^2 x)(-\sin^2 x) \qquad \text{Pythagorean identities}$$

$$= -\frac{\sin^2 x}{\cos^2 x} \qquad \text{Reciprocal identity}$$

$$= -\left(\frac{\sin x}{\cos x}\right)^2 \qquad \text{Rule of exponents}$$

$$= -\tan^2 x \qquad \text{Quotient identity}$$

Graphical Solution

Use a graphing utility set in *radian* mode to graph the left side of the identity $y_1 = (\tan^2 x + 1)(\cos^2 x - 1)$ and the right side of the identity $y_2 = -\tan^2 x$ in the same viewing window, as shown in Figure 5.6. (Select the *line* style for y_1 and the *path* style for y_2.) Because the graphs appear to coincide, $(\tan^2 x + 1) \cdot (\cos^2 x - 1) = -\tan^2 x$ appears to be an identity.

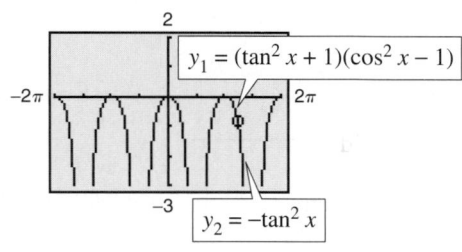

Figure 5.6

✔CHECKPOINT Now try Exercise 39.

Example 4 Converting to Sines and Cosines

Verify the identity $\tan x + \cot x = \sec x \csc x$.

Solution

In this case there appear to be no fractions to add, no products to find, and no opportunities to use the Pythagorean identities. So, try converting the left side to sines and cosines.

$$\tan x + \cot x = \frac{\sin x}{\cos x} + \frac{\cos x}{\sin x} \qquad \text{Quotient identities}$$

$$= \frac{\sin^2 x + \cos^2 x}{\cos x \sin x} \qquad \text{Add fractions.}$$

$$= \frac{1}{\cos x \sin x} \qquad \text{Pythagorean identity}$$

$$= \frac{1}{\cos x} \cdot \frac{1}{\sin x} \qquad \text{Product of fractions}$$

$$= \sec x \csc x \qquad \text{Reciprocal identities}$$

✔CHECKPOINT Now try Exercise 41.

TECHNOLOGY TIP

Although a graphing utility can be useful in helping to verify an identity, you must use algebraic techniques to produce a valid proof. For example, graph the two functions

$$y_1 = \sin 50x$$

$$y_2 = \sin 2x$$

in a trigonometric viewing window. Although their graphs seem identical, $\sin 50x \neq \sin 2x$.

Recall from algebra that *rationalizing the denominator* using conjugates is, on occasion, a powerful simplification technique. A related form of this technique works for simplifying trigonometric expressions as well. For instance, to simplify $1/(1 - \cos x)$, multiply the numerator and the denominator by $1 + \cos x$.

$$\frac{1}{1 - \cos x} = \frac{1}{1 - \cos x}\left(\frac{1 + \cos x}{1 + \cos x}\right)$$

$$= \frac{1 + \cos x}{1 - \cos^2 x}$$

$$= \frac{1 + \cos x}{\sin^2 x}$$

$$= \csc^2 x(1 + \cos x)$$

As shown above, $\csc^2 x(1 + \cos x)$ is considered a simplified form of $1/(1 - \cos x)$ because the expression does not contain any fractions.

Example 5   Verifying a Trigonometric Identity

Verify the identity

$$\sec x + \tan x = \frac{\cos x}{1 - \sin x}.$$

Algebraic Solution

Begin with the *right* side because you can create a monomial denominator by multiplying the numerator and denominator by $(1 + \sin x)$.

$$\frac{\cos x}{1 - \sin x} = \frac{\cos x}{1 - \sin x}\left(\frac{1 + \sin x}{1 + \sin x}\right) \quad \text{Multiply numerator and denominator by } (1 + \sin x).$$

$$= \frac{\cos x + \cos x \sin x}{1 - \sin^2 x} \quad \text{Multiply.}$$

$$= \frac{\cos x + \cos x \sin x}{\cos^2 x} \quad \text{Pythagorean identity}$$

$$= \frac{\cos x}{\cos^2 x} + \frac{\cos x \sin x}{\cos^2 x} \quad \text{Write as separate fractions.}$$

$$= \frac{1}{\cos x} + \frac{\sin x}{\cos x} \quad \text{Simplify.}$$

$$= \sec x + \tan x \quad \text{Identities}$$

✓*CHECKPOINT*   Now try Exercise 47.

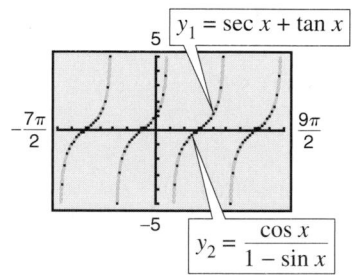

Graphical Solution

Use a graphing utility set in the *radian* and *dot* modes to graph $y_1 = \sec x + \tan x$ and $y_2 = \cos x/(1 - \sin x)$ in the same viewing window, as shown in Figure 5.7. Because the graphs appear to coincide, $\sec x + \tan x = \cos x/(1 - \sin x)$ appears to be an identity.

$y_1 = \sec x + \tan x$

$y_2 = \dfrac{\cos x}{1 - \sin x}$

Figure 5.7

TECHNOLOGY SUPPORT

For instructions on how to use the *radian* and *dot* modes, see Appendix A; for specific keystrokes, go to this textbook's *Online Study Center.*

In Examples 1 through 5, you have been verifying trigonometric identities by working with one side of the equation and converting it to the form given on the other side. On occasion it is practical to work with each side *separately* to obtain one common form equivalent to both sides. This is illustrated in Example 6.

Example 6   Working with Each Side Separately

Verify the identity $\dfrac{\cot^2 \theta}{1 + \csc \theta} = \dfrac{1 - \sin \theta}{\sin \theta}$.

Algebraic Solution

Working with the left side, you have

$$\frac{\cot^2 \theta}{1 + \csc \theta} = \frac{\csc^2 \theta - 1}{1 + \csc \theta} \qquad \text{Pythagorean identity}$$

$$= \frac{(\csc \theta - 1)(\csc \theta + 1)}{1 + \csc \theta} \qquad \text{Factor.}$$

$$= \csc \theta - 1. \qquad \text{Simplify.}$$

Now, simplifying the right side, you have

$$\frac{1 - \sin \theta}{\sin \theta} = \frac{1}{\sin \theta} - \frac{\sin \theta}{\sin \theta} \qquad \text{Write as separate fractions.}$$

$$= \csc \theta - 1. \qquad \text{Reciprocal identity}$$

The identity is verified because both sides are equal to $\csc \theta - 1$.

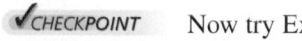 *CHECKPOINT*   Now try Exercise 49.

Numerical Solution

Use the *table* feature of a graphing utility set in *radian* mode to create a table that shows the values of $y_1 = \cot^2 x/(1 + \csc x)$ and $y_2 = (1 - \sin x)/\sin x$ for different values of x, as shown in Figure 5.8. From the table you can see that the values appear to be identical, so $\cot^2 x/(1 + \csc x) = (1 - \sin x)/\sin x$ appears to be an identity.

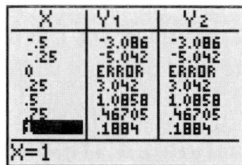

Figure 5.8

In Example 7, powers of trigonometric functions are rewritten as more complicated sums of products of trigonometric functions. This is a common procedure used in calculus.

Example 7   Examples from Calculus

Verify each identity.

a. $\tan^4 x = \tan^2 x \sec^2 x - \tan^2 x$ **b.** $\sin^3 x \cos^4 x = (\cos^4 x - \cos^6 x)\sin x$

Solution

a. $\tan^4 x = (\tan^2 x)(\tan^2 x)$ Write as separate factors.

$\qquad\quad = \tan^2 x(\sec^2 x - 1)$ Pythagorean identity

$\qquad\quad = \tan^2 x \sec^2 x - \tan^2 x$ Multiply.

b. $\sin^3 x \cos^4 x = \sin^2 x \cos^4 x \sin x$ Write as separate factors.

$\qquad\qquad\quad = (1 - \cos^2 x)\cos^4 x \sin x$ Pythagorean identity

$\qquad\qquad\quad = (\cos^4 x - \cos^6 x)\sin x$ Multiply.

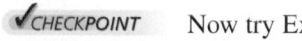 *CHECKPOINT*   Now try Exercise 63.

TECHNOLOGY TIP   Remember that you can use a graphing utility to assist in verifying an identity by creating a table or by graphing.

5.2   Exercises

See www.CalcChat.com for worked-out solutions to odd-numbered exercises.

Vocabulary Check

In Exercises 1 and 2, fill in the blanks.

1. An equation that is true for only some values in its domain is called a _____ equation.

2. An equation that is true for all real values in its domain is called an _____ .

In Exercises 3–10, fill in the blank to complete the trigonometric identity.

3. $\dfrac{1}{\tan u} =$ _____

4. $\dfrac{1}{\csc u} =$ _____

5. $\dfrac{\sin u}{\cos u} =$ _____

6. $\dfrac{1}{\sec u} =$ _____

7. $\sin^2 u +$ _____ $= 1$

8. $\tan\left(\dfrac{\pi}{2} - u\right) =$ _____

9. $\sin(-u) =$ _____

10. $\sec(-u) =$ _____

In Exercises 1–10, verify the identity.

1. $\sin t \csc t = 1$

2. $\sec y \cos y = 1$

3. $\dfrac{\csc^2 x}{\cot x} = \csc x \sec x$

4. $\dfrac{\sin^2 t}{\tan^2 t} = \cos^2 t$

5. $\cos^2 \beta - \sin^2 \beta = 1 - 2\sin^2 \beta$

6. $\cos^2 \beta - \sin^2 \beta = 2\cos^2 \beta - 1$

7. $\tan^2 \theta + 6 = \sec^2 \theta + 5$

8. $2 - \csc^2 z = 1 - \cot^2 z$

9. $(1 + \sin x)(1 - \sin x) = \cos^2 x$

10. $\tan^2 y(\csc^2 y - 1) = 1$

Numerical, Graphical, and Algebraic Analysis   **In Exercises 11–18, use a graphing utility to complete the table and graph the functions in the same viewing window. Use both the table and the graph as evidence that $y_1 = y_2$. Then verify the identity algebraically.**

x	0.2	0.4	0.6	0.8	1.0	1.2	1.4
y_1							
y_2							

11. $y_1 = \dfrac{1}{\sec x \tan x}$, $y_2 = \csc x - \sin x$

12. $y_1 = \dfrac{\csc x - 1}{1 - \sin x}$, $y_2 = \csc x$

13. $y_1 = \csc x - \sin x$, $y_2 = \cos x \cot x$

14. $y_1 = \sec x - \cos x$, $y_2 = \sin x \tan x$

15. $y_1 = \sin x + \cos x \cot x$, $y_2 = \csc x$

16. $y_1 = \cos x + \sin x \tan x$, $y_2 = \sec x$

17. $y_1 = \dfrac{1}{\tan x} + \dfrac{1}{\cot x}$, $y_2 = \tan x + \cot x$

18. $y_1 = \dfrac{1}{\sin x} - \dfrac{1}{\csc x}$, $y_2 = \csc x - \sin x$

Error Analysis   **In Exercises 19 and 20, describe the error.**

19. $(1 + \tan x)[1 + \cot(-x)]$
$= (1 + \tan x)(1 + \cot x)$
$= 1 + \cot x + \tan x + \tan x \cot x$
$= 1 + \cot x + \tan x + 1$
$= 2 + \cot x + \tan x$

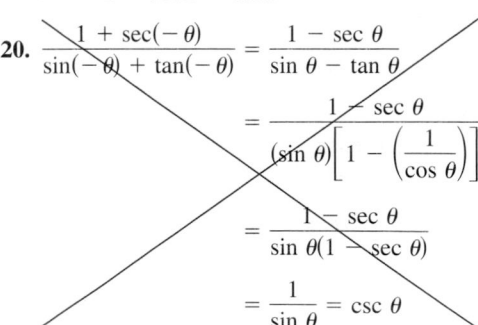

20. $\dfrac{1 + \sec(-\theta)}{\sin(-\theta) + \tan(-\theta)} = \dfrac{1 - \sec \theta}{\sin \theta - \tan \theta}$

$= \dfrac{1 - \sec \theta}{(\sin \theta)\left[1 - \left(\dfrac{1}{\cos \theta}\right)\right]}$

$= \dfrac{1 - \sec \theta}{\sin \theta(1 - \sec \theta)}$

$= \dfrac{1}{\sin \theta} = \csc \theta$

In Exercises 21–30, verify the identity.

21. $\sin^{1/2} x \cos x - \sin^{5/2} x \cos x = \cos^3 x \sqrt{\sin x}$

22. $\sec^6 x(\sec x \tan x) - \sec^4 x(\sec x \tan x) = \sec^5 x \tan^3 x$

23. $\cot\left(\dfrac{\pi}{2} - x\right)\csc x = \sec x$

24. $\dfrac{\sec[(\pi/2) - x]}{\tan[(\pi/2) - x]} = \sec x$

25. $\dfrac{\csc(-x)}{\sec(-x)} = -\cot x$

26. $(1 + \sin y)[1 + \sin(-y)] = \cos^2 y$

27. $\dfrac{\cos(-\theta)}{1 + \sin(-\theta)} = \sec \theta + \tan \theta$

28. $\dfrac{1 + \csc(-\theta)}{\cos(-\theta) + \cot(-\theta)} = \sec \theta$

29. $\dfrac{\sin x \cos y + \cos x \sin y}{\cos x \cos y - \sin x \sin y} = \dfrac{\tan x + \tan y}{1 - \tan x \tan y}$

30. $\dfrac{\tan x + \tan y}{1 - \tan x \tan y} = \dfrac{\cot x + \cot y}{\cot x \cot y - 1}$

In Exercises 31–38, verify the identity algebraically. Use the _table_ feature of a graphing utility to check your result numerically.

31. $\dfrac{\cos x - \cos y}{\sin x + \sin y} + \dfrac{\sin x - \sin y}{\cos x + \cos y} = 0$

32. $\dfrac{\tan x + \cot y}{\tan x \cot y} = \tan y + \cot x$

33. $\sqrt{\dfrac{1 + \sin \theta}{1 - \sin \theta}} = \dfrac{1 + \sin \theta}{|\cos \theta|}$

34. $\sqrt{\dfrac{1 - \cos \theta}{1 + \cos \theta}} = \dfrac{1 - \cos \theta}{|\sin \theta|}$

35. $\sin^2\left(\dfrac{\pi}{2} - x\right) + \sin^2 x = 1$

36. $\sec^2 y - \cot^2\left(\dfrac{\pi}{2} - y\right) = 1$

37. $\sin x \csc\left(\dfrac{\pi}{2} - x\right) = \tan x$

38. $\sec^2\left(\dfrac{\pi}{2} - x\right) - 1 = \cot^2 x$

In Exercises 39–50, verify the identity algebraically. Use a graphing utility to check your result graphically.

39. $2 \sec^2 x - 2 \sec^2 x \sin^2 x - \sin^2 x - \cos^2 x = 1$

40. $\csc x(\csc x - \sin x) + \dfrac{\sin x - \cos x}{\sin x} + \cot x = \csc^2 x$

41. $\dfrac{\cot x \tan x}{\sin x} = \csc x$

42. $\dfrac{1 + \csc \theta}{\sec \theta} - \cot \theta = \cos \theta$

43. $\csc^4 x - 2 \csc^2 x + 1 = \cot^4 x$

44. $\sin x(1 - 2 \cos^2 x + \cos^4 x) = \sin^5 x$

45. $\sec^4 \theta - \tan^4 \theta = 1 + 2 \tan^2 \theta$

46. $\csc^4 \theta - \cot^4 \theta = 2 \csc^2 \theta - 1$

47. $\dfrac{\sin \beta}{1 - \cos \beta} = \dfrac{1 + \cos \beta}{\sin \beta}$

48. $\dfrac{\cot \alpha}{\csc \alpha - 1} = \dfrac{\csc \alpha + 1}{\cot \alpha}$

49. $\dfrac{\tan^3 \alpha - 1}{\tan \alpha - 1} = \tan^2 \alpha + \tan \alpha + 1$

50. $\dfrac{\sin^3 \beta + \cos^3 \beta}{\sin \beta + \cos \beta} = 1 - \sin \beta \cos \beta$

Conjecture **In Exercises 51–54, use a graphing utility to graph the trigonometric function. Use the graph to make a conjecture about a simplification of the expression. Verify the resulting identity algebraically.**

51. $y = \dfrac{1}{\cot x + 1} + \dfrac{1}{\tan x + 1}$

52. $y = \dfrac{\cos x}{1 - \tan x} + \dfrac{\sin x \cos x}{\sin x - \cos x}$

53. $y = \dfrac{1}{\sin x} - \dfrac{\cos^2 x}{\sin x}$

54. $y = \sin t + \dfrac{\cot^2 t}{\csc t}$

In Exercises 55–58, use the properties of logarithms and trigonometric identities to verify the identity.

55. $\ln|\cot \theta| = \ln|\cos \theta| - \ln|\sin \theta|$

56. $\ln|\sec \theta| = -\ln|\cos \theta|$

57. $-\ln(1 + \cos \theta) = \ln(1 - \cos \theta) - 2 \ln|\sin \theta|$

58. $-\ln|\csc \theta + \cot \theta| = \ln|\csc \theta - \cot \theta|$

In Exercises 59–62, use the cofunction identities to evaluate the expression without using a calculator.

59. $\sin^2 35° + \sin^2 55°$

60. $\cos^2 14° + \cos^2 76°$

61. $\cos^2 20° + \cos^2 52° + \cos^2 38° + \cos^2 70°$

62. $\sin^2 18° + \sin^2 40° + \sin^2 50° + \sin^2 72°$

∫ **In Exercises 63–66, powers of trigonometric functions are rewritten to be useful in calculus. Verify the identity.**

63. $\tan^5 x = \tan^3 x \sec^2 x - \tan^3 x$

64. $\sec^4 x \tan^2 x = (\tan^2 x + \tan^4 x)\sec^2 x$

65. $\cos^3 x \sin^2 x = (\sin^2 x - \sin^4 x)\cos x$

66. $\sin^4 x + \cos^4 x = 1 - 2\cos^2 x + 2\cos^4 x$

In Exercises 67–70, verify the identity.

67. $\tan(\sin^{-1} x) = \dfrac{x}{\sqrt{1 - x^2}}$ **68.** $\cos(\sin^{-1} x) = \sqrt{1 - x^2}$

69. $\tan\left(\sin^{-1} \dfrac{x - 1}{4}\right) = \dfrac{x - 1}{\sqrt{16 - (x - 1)^2}}$

70. $\tan\left(\cos^{-1} \dfrac{x + 1}{2}\right) = \dfrac{\sqrt{4 - (x + 1)^2}}{x + 1}$

71. *Friction*   The forces acting on an object weighing W units on an inclined plane positioned at an angle of θ with the horizontal (see figure) are modeled by

$$\mu W \cos \theta = W \sin \theta$$

where μ is the coefficient of friction. Solve the equation for μ and simplify the result.

72. *Shadow Length*   The length s of the shadow cast by a vertical *gnomon* (a device used to tell time) of height h when the angle of the sun above the horizon is θ can be modeled by the equation

$$s = \frac{h \sin(90° - \theta)}{\sin \theta}.$$

Show that the equation is equivalent to $s = h \cot \theta$.

Synthesis

True or False?   **In Exercises 73–76, determine whether the statement is true or false. Justify your answer.**

73. There can be more than one way to verify a trigonometric identity.

74. Of the six trigonometric functions, two are even.

75. The equation $\sin^2 \theta + \cos^2 \theta = 1 + \tan^2 \theta$ is an identity, because $\sin^2(0) + \cos^2(0) = 1$ and $1 + \tan^2(0) = 1$.

76. $\sin(x^2) = \sin^2(x)$

In Exercises 77–80, (a) verify the identity and (b) determine if the identity is true for the given value of x. Explain.

77. $\dfrac{\sin x}{1 + \cos x} = \dfrac{1 - \cos x}{\sin x}$, $x = 0$

78. $\dfrac{\sec x}{\tan x} = \dfrac{\tan x}{\sec x - \cos x}$, $x = \pi$

79. $\csc x - \cot x = \dfrac{\sin x}{1 + \cos x}$, $x = \dfrac{\pi}{2}$

80. $\dfrac{\cot x - 1}{\cot x + 1} = \dfrac{1 - \tan x}{1 + \tan x}$, $x = \dfrac{\pi}{4}$

∫ **In Exercises 81–84, use the trigonometric substitution to write the algebraic expression as a trigonometric function of θ, where $0 < \theta < \pi/2$. Assume $a > 0$.**

81. $\sqrt{a^2 - u^2}$, $u = a \sin \theta$

82. $\sqrt{a^2 - u^2}$, $u = a \cos \theta$

83. $\sqrt{a^2 + u^2}$, $u = a \tan \theta$

84. $\sqrt{u^2 - a^2}$, $u = a \sec \theta$

Think About It   **In Exercises 85 and 86, explain why the equation is not an identity and find one value of the variable for which the equation is not true.**

85. $\sqrt{\tan^2 x} = \tan x$ **86.** $\sin \theta = \sqrt{1 - \cos^2 \theta}$

87. Verify that for all integers n, $\cos\left[\dfrac{(2n + 1)\pi}{2}\right] = 0$.

88. Verify that for all integers n, $\sin\left[\dfrac{(12n + 1)\pi}{6}\right] = \dfrac{1}{2}$.

Skills Review

In Exercises 89–92, find a polynomial function with real coefficients that has the given zeros. (There are many correct answers.)

89. $1, 8i, -8i$ **90.** $i, -i, 4i, -4i$

91. $4, 6 + i, 6 - i$ **92.** $0, 0, 2, 1 - i$

In Exercises 93–96, sketch the graph of the function by hand.

93. $f(x) = 2^x + 3$ **94.** $f(x) = -2^{x-3}$

95. $f(x) = 2^{-x} + 1$ **96.** $f(x) = 2^{x-1} + 3$

In Exercises 97–100, state the quadrant in which θ lies.

97. $\csc \theta > 0$ and $\tan \theta < 0$ **98.** $\cot \theta > 0$ and $\cos \theta < 0$

99. $\sec \theta > 0$ and $\sin \theta < 0$ **100.** $\cot \theta > 0$ and $\sec \theta < 0$

5.3 Solving Trigonometric Equations

Introduction

To solve a trigonometric equation, use standard algebraic techniques such as collecting like terms and factoring. Your preliminary goal is to isolate the trigonometric function involved in the equation.

Example 1   Solving a Trigonometric Equation

$$2 \sin x - 1 = 0 \qquad \text{Original equation}$$

$$2 \sin x = 1 \qquad \text{Add 1 to each side.}$$

$$\sin x = \tfrac{1}{2} \qquad \text{Divide each side by 2.}$$

To solve for x, note in Figure 5.9 that the equation $\sin x = \tfrac{1}{2}$ has solutions $x = \pi/6$ and $x = 5\pi/6$ in the interval $[0, 2\pi)$. Moreover, because $\sin x$ has a period of 2π, there are infinitely many other solutions, which can be written as

$$x = \frac{\pi}{6} + 2n\pi \qquad \text{and} \qquad x = \frac{5\pi}{6} + 2n\pi \qquad \text{General solution}$$

where n is an integer, as shown in Figure 5.9.

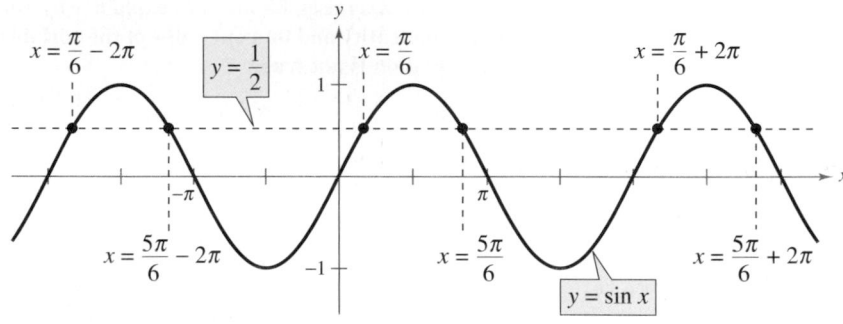

Figure 5.9

✓CHECKPOINT   Now try Exercise 25.

Figure 5.10 verifies that the equation $\sin x = \tfrac{1}{2}$ has infinitely many solutions. Any angles that are coterminal with $\pi/6$ or $5\pi/6$ are also solutions of the equation.

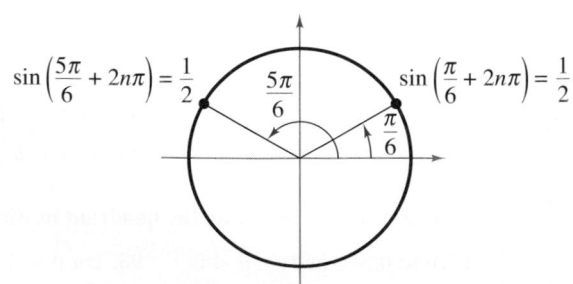

Figure 5.10

What you should learn

- Use standard algebraic techniques to solve trigonometric equations.
- Solve trigonometric equations of quadratic type.
- Solve trigonometric equations involving multiple angles.
- Use inverse trigonometric functions to solve trigonometric equations.

Why you should learn it

You can use trigonometric equations to solve a variety of real-life problems. For instance, Exercise 100 on page 378 shows you how solving a trigonometric equation can help answer questions about the position of the sun in Cheyenne, Wyoming.

SuperStock

Prerequisite Skills

If you have trouble finding coterminal angles, review Section 4.1.

Example 2   Collecting Like Terms

Find all solutions of $\sin x + \sqrt{2} = -\sin x$ in the interval $[0, 2\pi)$.

Algebraic Solution

Rewrite the equation so that $\sin x$ is isolated on one side of the equation.

$$\sin x + \sqrt{2} = -\sin x \qquad \text{Write original equation.}$$

$$\sin x + \sin x = -\sqrt{2} \qquad \begin{array}{l}\text{Add } \sin x \text{ to and subtract}\\ \sqrt{2} \text{ from each side.}\end{array}$$

$$2 \sin x = -\sqrt{2} \qquad \text{Combine like terms.}$$

$$\sin x = -\frac{\sqrt{2}}{2} \qquad \text{Divide each side by 2.}$$

The solutions in the interval $[0, 2\pi)$ are

$$x = \frac{5\pi}{4} \qquad \text{and} \qquad x = \frac{7\pi}{4}.$$

 **CHECKPOINT**   Now try Exercise 35.

Numerical Solution

Use the *table* feature of a graphing utility set in *radian* mode to create a table that shows the values of $y_1 = \sin x + \sqrt{2}$ and $y_2 = -\sin x$ for different values of x. Your table should go from $x = 0$ to $x = 2\pi$ using increments of $\pi/8$, as shown in Figure 5.11. From the table, you can see that the values of y_1 and y_2 appear to be identical when $x \approx 3.927 \approx 5\pi/4$ and $x \approx 5.4978 \approx 7\pi/4$. These values are the approximate solutions of $\sin x + \sqrt{2} = -\sin x$.

X	Y1	Y2
3.1416	1.4142	0
3.5343	1.0315	.38268
3.927	.70711	.70711
4.3197	.49033	.92388
4.7124	.41421	1
5.1051	.49033	.92388
5.4978	.70711	.70711

X=5.497787143782

Figure 5.11

Example 3   Extracting Square Roots

Solve $3 \tan^2 x - 1 = 0$.

Solution

Rewrite the equation so that $\tan x$ is isolated on one side of the equation.

$$3 \tan^2 x = 1 \qquad \text{Add 1 to each side.}$$

$$\tan^2 x = \frac{1}{3} \qquad \text{Divide each side by 3.}$$

$$\tan x = \pm\frac{1}{\sqrt{3}} \qquad \text{Extract square roots.}$$

Because $\tan x$ has a period of π, first find all solutions in the interval $[0, \pi)$. These are $x = \pi/6$ and $x = 5\pi/6$. Finally, add multiples of π to each of these solutions to get the general form

$$x = \frac{\pi}{6} + n\pi \qquad \text{and} \qquad x = \frac{5\pi}{6} + n\pi \qquad \text{General solution}$$

where n is an integer. The graph of $y = 3 \tan^2 x - 1$, shown in Figure 5.12, confirms this result.

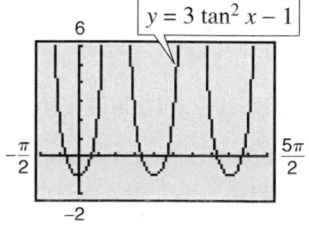

Figure 5.12

> **TECHNOLOGY SUPPORT**
>
> For instructions on how to use the *table* feature, see Appendix A; for specific keystrokes, go to this textbook's *Online Study Center*.

CHECKPOINT   Now try Exercise 37.

Recall that the solutions of an equation correspond to the x-intercepts of the graph of the equation. For instance, the graph in Figure 5.12 has x-intercepts at $\pi/6, 5\pi/6, 7\pi/6$, and so on.

The equations in Examples 1, 2, and 3 involved only one trigonometric function. When two or more functions occur in the same equation, collect all terms on one side and try to separate the functions by factoring or by using appropriate identities. This may produce factors that yield no solutions, as illustrated in Example 4.

Example 4   Factoring

Solve $\cot x \cos^2 x = 2 \cot x$.

Solution

Begin by rewriting the equation so that all terms are collected on one side of the equation.

$$\cot x \cos^2 x = 2 \cot x \qquad \text{Write original equation.}$$

$$\cot x \cos^2 x - 2 \cot x = 0 \qquad \text{Subtract 2 cot } x \text{ from each side.}$$

$$\cot x(\cos^2 x - 2) = 0 \qquad \text{Factor.}$$

By setting each of these factors equal to zero, you obtain the following.

$$\cot x = 0 \qquad \text{and} \qquad \cos^2 x - 2 = 0$$
$$\cos^2 x = 2$$
$$\cos x = \pm\sqrt{2}$$

The equation $\cot x = 0$ has the solution $x = \pi/2$ [in the interval $(0, \pi)$]. No solution is obtained for $\cos x = \pm\sqrt{2}$ because $\pm\sqrt{2}$ are outside the range of the cosine function. Because $\cot x$ has a period of π, the general form of the solution is obtained by adding multiples of π to $x = \pi/2$, to get

$$x = \frac{\pi}{2} + n\pi \qquad \text{General solution}$$

where n is an integer. The graph of $y = \cot x \cos^2 x - 2 \cot x$ (in *dot* mode), shown in Figure 5.13, confirms this result.

 **CHECKPOINT**   Now try Exercise 39.

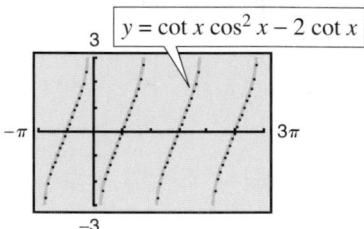
$y = \cot x \cos^2 x - 2 \cot x$

Figure 5.13

Exploration

Using the equation in Example 4, explain what would happen if you divided each side of the equation by cot x. Why is this an incorrect method to use when solving an equation?

Equations of Quadratic Type

Many trigonometric equations are of quadratic type $ax^2 + bx + c = 0$. Here are a few examples.

Quadratic in sin x	*Quadratic in sec x*
$2\sin^2 x - \sin x - 1 = 0$	$\sec^2 x - 3\sec x - 2 = 0$
$2(\sin x)^2 - \sin x - 1 = 0$	$(\sec x)^2 - 3(\sec x) - 2 = 0$

To solve equations of this type, factor the quadratic or, if factoring is not possible, use the Quadratic Formula.

Example 5   Factoring an Equation of Quadratic Type

Find all solutions of $2 \sin^2 x - \sin x - 1 = 0$ in the interval $[0, 2\pi)$.

Algebraic Solution

Treating the equation as a quadratic in $\sin x$ and factoring produces the following.

$$2 \sin^2 x - \sin x - 1 = 0 \qquad \text{Write original equation.}$$

$$(2 \sin x + 1)(\sin x - 1) = 0 \qquad \text{Factor.}$$

Setting each factor equal to zero, you obtain the following solutions in the interval $[0, 2\pi)$.

$$2 \sin x + 1 = 0 \qquad \text{and} \quad \sin x - 1 = 0$$

$$\sin x = -\frac{1}{2} \qquad\qquad \sin x = 1$$

$$x = \frac{7\pi}{6}, \frac{11\pi}{6} \qquad\qquad x = \frac{\pi}{2}$$

✓CHECKPOINT   Now try Exercise 49.

Graphical Solution

Use a graphing utility set in *radian* mode to graph $y = 2 \sin^2 x - \sin x - 1$ for $0 \leq x < 2\pi$, as shown in Figure 5.14. Use the *zero* or *root* feature or the *zoom* and *trace* features to approximate the x-intercepts to be

$$x \approx 1.571 \approx \frac{\pi}{2}, \ x \approx 3.665 \approx \frac{7\pi}{6}, \ \text{and} \ x \approx 5.760 \approx \frac{11\pi}{6}.$$

These values are the approximate solutions of $2 \sin^2 x - \sin x - 1 = 0$ in the interval $[0, 2\pi)$.

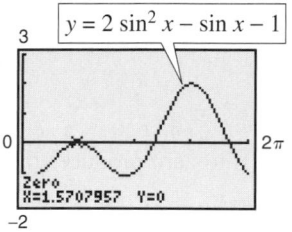

Figure 5.14

When working with an equation of quadratic type, be sure that the equation involves a *single* trigonometric function, as shown in the next example.

Example 6   Rewriting with a Single Trigonometric Function

Solve $2 \sin^2 x + 3 \cos x - 3 = 0$.

Solution

Begin by rewriting the equation so that it has only cosine functions.

$$2 \sin^2 x + 3 \cos x - 3 = 0 \qquad \text{Write original equation.}$$

$$2(1 - \cos^2 x) + 3 \cos x - 3 = 0 \qquad \text{Pythagorean identity}$$

$$2 \cos^2 x - 3 \cos x + 1 = 0 \qquad \begin{array}{l}\text{Combine like terms and}\\ \text{multiply each side by } -1.\end{array}$$

$$(2 \cos x - 1)(\cos x - 1) = 0 \qquad \text{Factor.}$$

By setting each factor equal to zero, you can find the solutions in the interval $[0, 2\pi)$ to be $x = 0$, $x = \pi/3$, and $x = 5\pi/3$. Because $\cos x$ has a period of 2π, the general solution is

$$x = 2n\pi, \qquad x = \frac{\pi}{3} + 2n\pi, \qquad x = \frac{5\pi}{3} + 2n\pi \qquad \text{General solution}$$

where n is an integer. The graph of $y = 2 \sin^2 x + 3 \cos x - 3$, shown in Figure 5.15, confirms this result.

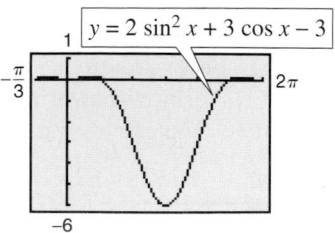

Figure 5.15

✓CHECKPOINT   Now try Exercise 51.

Sometimes you must square each side of an equation to obtain a quadratic. Because this procedure can introduce extraneous solutions, you should check any solutions in the original equation to see whether they are valid or extraneous.

Example 7   Squaring and Converting to Quadratic Type

Find all solutions of $\cos x + 1 = \sin x$ in the interval $[0, 2\pi)$.

Solution

It is not clear how to rewrite this equation in terms of a single trigonometric function. Notice what happens when you square each side of the equation.

$\cos x + 1 = \sin x$	Write original equation.
$\cos^2 x + 2\cos x + 1 = \sin^2 x$	Square each side.
$\cos^2 x + 2\cos x + 1 = 1 - \cos^2 x$	Pythagorean identity
$2\cos^2 x + 2\cos x = 0$	Combine like terms.
$2\cos x(\cos x + 1) = 0$	Factor.

Setting each factor equal to zero produces the following.

$$2\cos x = 0 \qquad \text{and} \qquad \cos x + 1 = 0$$

$$\cos x = 0 \qquad\qquad\qquad \cos x = -1$$

$$x = \frac{\pi}{2}, \frac{3\pi}{2} \qquad\qquad\qquad x = \pi$$

Because you squared the original equation, check for extraneous solutions.

Check

$\cos \dfrac{\pi}{2} + 1 \overset{?}{=} \sin \dfrac{\pi}{2}$	Substitute $\pi/2$ for x.
$0 + 1 = 1$	Solution checks. ✓
$\cos \dfrac{3\pi}{2} + 1 \overset{?}{=} \sin \dfrac{3\pi}{2}$	Substitute $3\pi/2$ for x.
$0 + 1 \neq -1$	Solution does not check.
$\cos \pi + 1 \overset{?}{=} \sin \pi$	Substitute π for x.
$-1 + 1 = 0$	Solution checks. ✓

Of the three possible solutions, $x = 3\pi/2$ is extraneous. So, in the interval $[0, 2\pi)$, the only solutions are $x = \pi/2$ and $x = \pi$. The graph of $y = \cos x + 1 - \sin x$, shown in Figure 5.16, confirms this result because the graph has two x-intercepts (at $x = \pi/2$ and $x = \pi$) in the interval $[0, 2\pi)$.

✓CHECKPOINT   Now try Exercise 53.

Exploration

Use a graphing utility to confirm the solutions found in Example 7 in two different ways. Do both methods produce the same x-values? Which method do you prefer? Why?

1. Graph both sides of the equation and find the x-coordinates of the points at which the graphs intersect.

   Left side: $y = \cos x + 1$

   Right side: $y = \sin x$

2. Graph the equation $y = \cos x + 1 - \sin x$ and find the x-intercepts of the graph.

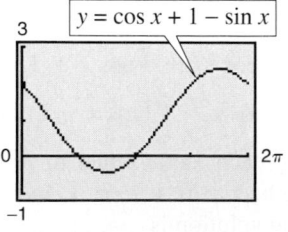

Figure 5.16

Functions Involving Multiple Angles

The next two examples involve trigonometric functions of multiple angles of the forms $\sin ku$ and $\cos ku$. To solve equations of these forms, first solve the equation for ku, then divide your result by k.

Example 8   Functions of Multiple Angles

Solve $2 \cos 3t - 1 = 0$.

Solution

$2 \cos 3t - 1 = 0$	Write original equation.
$2 \cos 3t = 1$	Add 1 to each side.
$\cos 3t = \dfrac{1}{2}$	Divide each side by 2.

In the interval $[0, 2\pi)$, you know that $3t = \pi/3$ and $3t = 5\pi/3$ are the only solutions. So in general, you have $3t = \pi/3 + 2n\pi$ and $3t = 5\pi/3 + 2n\pi$. Dividing this result by 3, you obtain the general solution

$$t = \frac{\pi}{9} + \frac{2n\pi}{3} \qquad \text{and} \qquad t = \frac{5\pi}{9} + \frac{2n\pi}{3} \qquad \text{General solution}$$

where n is an integer. This solution is confirmed graphically in Figure 5.17.

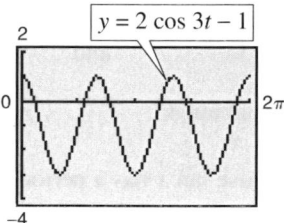

Figure 5.17

✓CHECKPOINT   Now try Exercise 65.

Example 9   Functions of Multiple Angles

Solve $3 \tan \dfrac{x}{2} + 3 = 0$.

Solution

$3 \tan \dfrac{x}{2} + 3 = 0$	Write original equation.
$3 \tan \dfrac{x}{2} = -3$	Subtract 3 from each side.
$\tan \dfrac{x}{2} = -1$	Divide each side by 3.

In the interval $[0, \pi)$, you know that $x/2 = 3\pi/4$ is the only solution. So in general, you have $x/2 = 3\pi/4 + n\pi$. Multiplying this result by 2, you obtain the general solution

$$x = \frac{3\pi}{2} + 2n\pi \qquad \text{General solution}$$

where n is an integer. This solution is confirmed graphically in Figure 5.18.

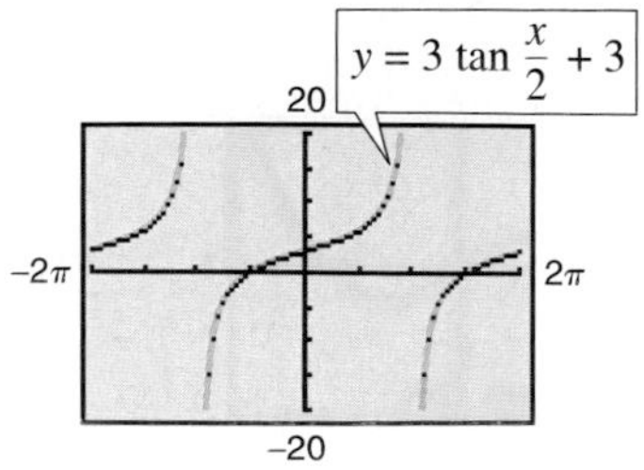

Figure 5.18

✓CHECKPOINT   Now try Exercise 71.

Using Inverse Functions

Example 10   Using Inverse Functions

Find all solutions of $\sec^2 x - 2 \tan x = 4$.

Solution

$\sec^2 x - 2 \tan x = 4$	Write original equation.
$1 + \tan^2 x - 2 \tan x - 4 = 0$	Pythagorean identity
$\tan^2 x - 2 \tan x - 3 = 0$	Combine like terms.
$(\tan x - 3)(\tan x + 1) = 0$	Factor.

Setting each factor equal to zero, you obtain two solutions in the interval $(-\pi/2, \pi/2)$. [Recall that the range of the inverse tangent function is $(-\pi/2, \pi/2)$.]

$$\tan x = 3 \qquad \text{and} \qquad \tan x = -1$$

$$x = \arctan 3 \qquad\qquad x = \arctan(-1) = -\frac{\pi}{4}$$

Finally, because $\tan x$ has a period of π, add multiples of π to obtain

$$x = \arctan 3 + n\pi \qquad \text{and} \qquad x = -\frac{\pi}{4} + n\pi \qquad \text{General solution}$$

where n is an integer. This solution is confirmed graphically in Figure 5.19.

✓CHECKPOINT   Now try Exercise 47.

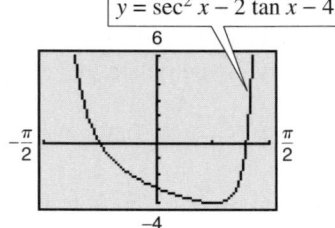

Figure 5.19

With some trigonometric equations, there is no reasonable way to find the solutions algebraically. In such cases, you can still use a graphing utility to approximate the solutions.

Example 11   Approximating Solutions

Approximate the solutions of $x = 2 \sin x$ in the interval $[-\pi, \pi]$.

Solution

Use a graphing utility to graph $y = x - 2 \sin x$ in the interval $[-\pi, \pi]$. Using the *zero* or *root* feature or the *zoom* and *trace* features, you can see that the solutions are $x \approx -1.8955$, $x = 0$, and $x \approx 1.8955$. (See Figure 5.20.)

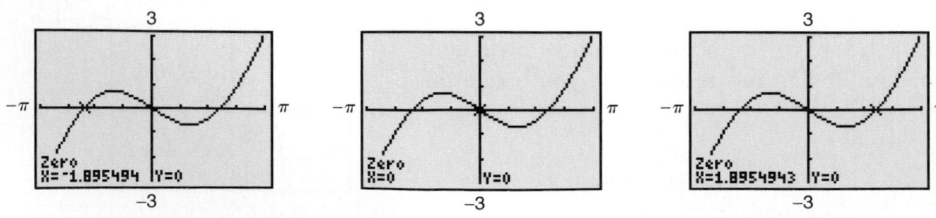

Figure 5.20 $y = x - 2 \sin x$

✓CHECKPOINT   Now try Exercise 85.

Example 12   Surface Area of a Honeycomb

The surface area of a honeycomb is given by the equation

$$S = 6hs + \frac{3}{2}s^2\left(\frac{\sqrt{3} - \cos\theta}{\sin\theta}\right), \qquad 0 < \theta \le 90°$$

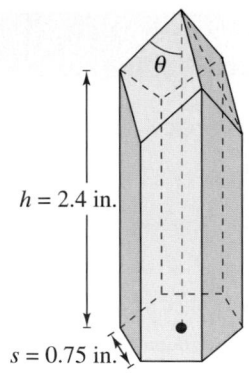

where $h = 2.4$ inches, $s = 0.75$ inch, and θ is the angle indicated in Figure 5.21.

$h = 2.4$ in.

a. What value of θ gives a surface area of 12 square inches?

b. What value of θ gives the minimum surface area?

Solution

$s = 0.75$ in.

a. Let $h = 2.4$, $s = 0.75$, and $S = 12$.

Figure 5.21

$$S = 6hs + \frac{3}{2}s^2\left(\frac{\sqrt{3} - \cos\theta}{\sin\theta}\right)$$

$$12 = 6(2.4)(0.75) + \frac{3}{2}(0.75)^2\left(\frac{\sqrt{3} - \cos\theta}{\sin\theta}\right)$$

$$12 = 10.8 + 0.84375\left(\frac{\sqrt{3} - \cos\theta}{\sin\theta}\right)$$

TECHNOLOGY SUPPORT

For instructions on how to use the *degree* mode and the *minimum* feature, see Appendix A; for specific keystrokes, go to this textbook's *Online Study Center.*

$$0 = 0.84375\left(\frac{\sqrt{3} - \cos\theta}{\sin\theta}\right) - 1.2$$

Using a graphing utility set to *degree* mode, you can graph the function

$$y = 0.84375\left(\frac{\sqrt{3} - \cos x}{\sin x}\right) - 1.2.$$

Using the *zero* or *root* feature or the *zoom* and *trace* features, you can determine that $\theta \approx 49.9°$ and $\theta \approx 59.9°$. (See Figure 5.22.)

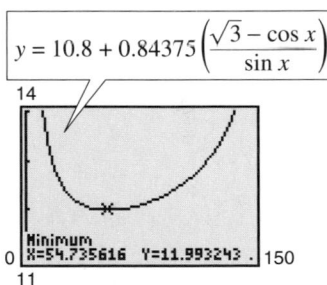

$$y = 10.8 + 0.84375\left(\frac{\sqrt{3} - \cos x}{\sin x}\right)$$

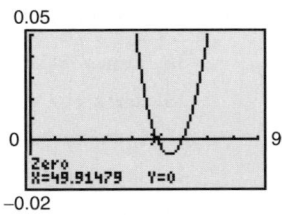   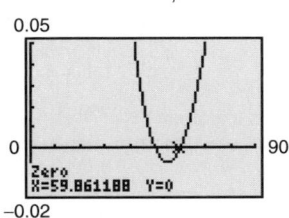

Figure 5.23

Figure 5.22 $y = 0.84375\left(\dfrac{\sqrt{3} - \cos x}{\sin x}\right) - 1.2$

b. From part (a), let $h = 2.4$ and $s = 0.75$ to obtain

$$S = 10.8 + 0.84375\left(\frac{\sqrt{3} - \cos\theta}{\sin\theta}\right).$$

Graph this function using a graphing utility set to *degree* mode. Use the *minimum* feature or the *zoom* and *trace* features to approximate the minimum point on the graph, which occurs at $\theta \approx 54.7°$, as shown in Figure 5.23.

✓CHECKPOINT   Now try Exercise 93.

STUDY TIP

By using calculus, it can be shown that the exact minimum value is

$$\theta = \arccos\left(\frac{1}{\sqrt{3}}\right) \approx 54.7356°.$$

5.3 Exercises

See www.CalcChat.com for worked-out solutions to odd-numbered exercises.

Vocabulary Check

Fill in the blanks.

1. The equation $2 \cos x - 1 = 0$ has the solutions $x = \dfrac{\pi}{3} + 2n\pi$ and $x = \dfrac{5\pi}{3} + 2n\pi$, which are called _____ solutions.

2. The equation $\tan^2 x - 5 \tan x + 6 = 0$ is an equation of _____ type.

3. A solution of an equation that does not satisfy the original equation is called an _____ solution.

In Exercises 1–6, verify that each x-value is a solution of the equation.

1. $2 \cos x - 1 = 0$

   (a) $x = \dfrac{\pi}{3}$ (b) $x = \dfrac{5\pi}{3}$

2. $\sec x - 2 = 0$

   (a) $x = \dfrac{\pi}{3}$ (b) $x = \dfrac{5\pi}{3}$

3. $3 \tan^2 2x - 1 = 0$

   (a) $x = \dfrac{\pi}{12}$ (b) $x = \dfrac{5\pi}{12}$

4. $4 \cos^2 2x - 2 = 0$

   (a) $x = \dfrac{\pi}{8}$ (b) $x = \dfrac{7\pi}{8}$

5. $2 \sin^2 x - \sin x - 1 = 0$

   (a) $x = \dfrac{\pi}{2}$ (b) $x = \dfrac{7\pi}{6}$

6. $\sec^4 x - 3 \sec^2 x - 4 = 0$

   (a) $x = \dfrac{2\pi}{3}$ (b) $x = \dfrac{5\pi}{3}$

In Exercises 7–12, find all solutions of the equation in the interval $[0°, 360°)$.

7. $\sin x = \dfrac{1}{2}$ 8. $\cos x = \dfrac{\sqrt{3}}{2}$

9. $\cos x = -\dfrac{1}{2}$ 10. $\sin x = -\dfrac{\sqrt{2}}{2}$

11. $\tan x = 1$ 12. $\tan x = -\sqrt{3}$

In Exercises 13–24, find all solutions of the equation in the interval $[0, 2\pi)$.

13. $\cos x = -\dfrac{\sqrt{3}}{2}$ 14. $\sin x = -\dfrac{1}{2}$

15. $\cot x = -1$ 16. $\sin x = \dfrac{\sqrt{3}}{2}$

17. $\tan x = -\dfrac{\sqrt{3}}{3}$ 18. $\cos x = \dfrac{\sqrt{2}}{2}$

19. $\csc x = -2$ 20. $\sec x = \sqrt{2}$

21. $\cot x = \sqrt{3}$ 22. $\sec x = 2$

23. $\tan x = -1$ 24. $\csc x = -\sqrt{2}$

In Exercises 25–34, solve the equation.

25. $2 \cos x + 1 = 0$ 26. $\sqrt{2} \sin x + 1 = 0$

27. $\sqrt{3} \sec x - 2 = 0$ 28. $\cot x + 1 = 0$

29. $3 \csc^2 x - 4 = 0$ 30. $3 \cot^2 x - 1 = 0$

31. $4 \cos^2 x - 1 = 0$ 32. $\cos x(\cos x - 1) = 0$

33. $\sin^2 x = 3 \cos^2 x$

34. $(3 \tan^2 x - 1)(\tan^2 x - 3) = 0$

In Exercises 35–48, find all solutions of the equation in the interval $[0, 2\pi)$ algebraically. Use the *table* feature of a graphing utility to check your answers numerically.

35. $\tan x + \sqrt{3} = 0$ 36. $2 \sin x + 1 = 0$

37. $\csc^2 x - 2 = 0$ 38. $\tan^2 x - 1 = 0$

39. $3 \tan^3 x = \tan x$ 40. $2 \sin^2 x = 2 + \cos x$

41. $\sec^2 x - \sec x = 2$ 42. $\sec x \csc x = 2 \csc x$

43. $2 \sin x + \csc x = 0$ 44. $\sec x + \tan x = 1$

45. $\cos x + \sin x \tan x = 2$ 46. $\sin^2 x + \cos x + 1 = 0$

47. $\sec^2 x + \tan x = 3$ 48. $2 \cos^2 x + \cos x - 1 = 0$

In Exercises 49–56, use a graphing utility to approximate the solutions of the equation in the interval $[0, 2\pi)$ by setting the equation equal to 0, graphing the new equation, and using the *zero* or *root* feature to approximate the x-intercepts of the graph.

49. $2 \sin^2 x + 3 \sin x + 1 = 0$

50. $2 \sec^2 x + \tan^2 x - 3 = 0$

51. $4 \sin^2 x = 2 \cos x + 1$

52. $\csc^2 x = 3 \csc x + 4$

53. $\csc x + \cot x = 1$

54. $4 \sin x = \cos x - 2$

55. $\dfrac{\cos x \cot x}{1 - \sin x} = 3$

56. $\dfrac{1 + \sin x}{\cos x} + \dfrac{\cos x}{1 + \sin x} = 4$

In Exercises 57–60, (a) use a graphing utility to graph each function in the interval $[0, 2\pi)$, (b) write an equation whose solutions are the points of intersection of the graphs, and (c) use the *intersect* feature of the graphing utility to find the points of intersection (to four decimal places).

57. $y = \sin 2x, \quad y = x^2 - 2x$

58. $y = \cos x, \quad y = x + x^2$

59. $y = \sin^2 x, \quad y = e^x - 4x$

60. $y = \cos^2 x, \quad y = e^{-x} + x - 1$

In Exercises 61–72, solve the multiple-angle equation.

61. $\cos \dfrac{x}{4} = 0$

62. $\sin \dfrac{x}{2} = 0$

63. $\sin 4x = 1$

64. $\cos 2x = -1$

65. $\sin 2x = -\dfrac{\sqrt{3}}{2}$

66. $\sec 4x = 2$

67. $2 \sin^2 2x = 1$

68. $\tan^2 3x = 3$

69. $\tan 3x(\tan x - 1) = 0$

70. $\cos 2x(2 \cos x + 1) = 0$

71. $\cos \dfrac{x}{2} = \dfrac{\sqrt{2}}{2}$

72. $\tan \dfrac{x}{3} = 1$

In Exercises 73–76, approximate the *x*-intercepts of the graph. Use a graphing utility to check your solutions.

73. $y = \sin \dfrac{\pi x}{2} + 1$

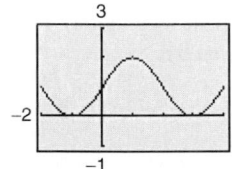

74. $y = \sin \pi x + \cos \pi x$

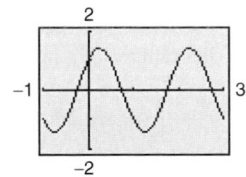

75. $y = \tan^2 \left(\dfrac{\pi x}{6} \right) - 3$

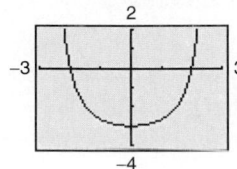

76. $y = \sec^4 \left(\dfrac{\pi x}{8} \right) - 4$

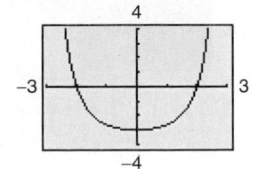

In Exercises 77–84, use a graphing utility to approximate the solutions of the equation in the interval $[0, 2\pi)$.

77. $2 \cos x - \sin x = 0$

78. $2 \sin x + \cos x = 0$

79. $x \tan x - 1 = 0$

80. $2x \sin x - 2 = 0$

81. $\sec^2 x + 0.5 \tan x = 1$

82. $\csc^2 x + 0.5 \cot x = 5$

83. $12 \sin^2 x - 13 \sin x + 3 = 0$

84. $3 \tan^2 x + 4 \tan x - 4 = 0$

In Exercises 85–88, use a graphing utility to approximate the solutions (to three decimal places) of the equation in the given interval.

85. $3 \tan^2 x + 5 \tan x - 4 = 0, \quad \left[-\dfrac{\pi}{2}, \dfrac{\pi}{2} \right]$

86. $\cos^2 x - 2 \cos x - 1 = 0, \quad [0, \pi]$

87. $4 \cos^2 x - 2 \sin x + 1 = 0, \quad \left[-\dfrac{\pi}{2}, \dfrac{\pi}{2} \right]$

88. $2 \sec^2 x + \tan x - 6 = 0, \quad \left[-\dfrac{\pi}{2}, \dfrac{\pi}{2} \right]$

∫ **In Exercises 89–94, (a) use a graphing utility to graph the function and approximate the maximum and minimum points (to four decimal places) of the graph in the interval $[0, 2\pi]$, and (b) solve the trigonometric equation and verify that the *x*-coordinates of the maximum and minimum points of f are among its solutions (the trigonometric equation is found using calculus).**

Function	*Trigonometric Equation*
89. $f(x) = \sin 2x$	$2 \cos 2x = 0$
90. $f(x) = \cos 2x$	$-2 \sin 2x = 0$
91. $f(x) = \sin^2 x + \cos x$	$2 \sin x \cos x - \sin x = 0$
92. $f(x) = \cos^2 x - \sin x$	$-2 \sin x \cos x - \cos x = 0$
93. $f(x) = \sin x + \cos x$	$\cos x - \sin x = 0$
94. $f(x) = 2 \sin x + \cos 2x$	$2 \cos x - 4 \sin x \cos x = 0$

***Fixed Point*   In Exercises 95 and 96, find the smallest positive fixed point of the function f. [A *fixed point* of a function f is a real number c such that $f(c) = c$.]**

95. $f(x) = \tan \dfrac{\pi x}{4}$

96. $f(x) = \cos x$

97. Graphical Reasoning   Consider the function

$$f(x) = \cos\frac{1}{x}$$

and its graph shown in the figure.

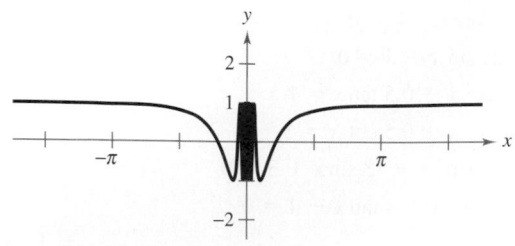

(a) What is the domain of the function?

(b) Identify any symmetry or asymptotes of the graph.

(c) Describe the behavior of the function as $x \to 0$.

(d) How many solutions does the equation $\cos(1/x) = 0$ have in the interval $[-1, 1]$? Find the solutions.

(e) Does the equation $\cos(1/x) = 0$ have a greatest solution? If so, approximate the solution. If not, explain.

98. Graphical Reasoning   Consider the function

$$f(x) = \frac{\sin x}{x}$$

and its graph shown in the figure.

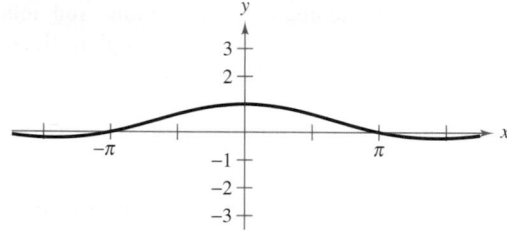

(a) What is the domain of the function?

(b) Identify any symmetry or asymptotes of the graph.

(c) Describe the behavior of the function as $x \to 0$.

(d) How many solutions does the equation $(\sin x)/x = 0$ have in the interval $[-8, 8]$? Find the solutions.

99. Sales   The monthly sales S (in thousands of units) of lawn mowers are approximated by

$$S = 74.50 - 43.75 \cos\frac{\pi t}{6}$$

where t is the time (in months), with $t = 1$ corresponding to January. Determine the months during which sales exceed 100,000 units.

100. Position of the Sun   Cheyenne, Wyoming has a latitude of 41° N. At this latitude, the position of the sun at sunrise can be modeled by

$$D = 31 \sin\left(\frac{2\pi}{365}t - 1.4\right)$$

where t is the time (in days) and $t = 1$ represents January 1. In this model, D represents the number of degrees north or south of due east at which the sun rises. Use a graphing utility to determine the days on which the sun is more than 20° north of due east at sunrise.

101. Harmonic Motion   A weight is oscillating on the end of a spring (see figure). The position of the weight relative to the point of equilibrium is given by

$$y = \tfrac{1}{12}(\cos 8t - 3\sin 8t)$$

where y is the displacement (in meters) and t is the time (in seconds). Find the times at which the weight is at the point of equilibrium $(y = 0)$ for $0 \le t \le 1$.

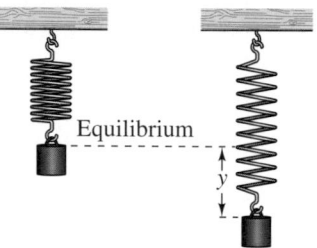

Equilibrium

102. Damped Harmonic Motion   The displacement from equilibrium of a weight oscillating on the end of a spring is given by $y = 1.56e^{-0.22t}\cos 4.9t$, where y is the displacement (in feet) and t is the time (in seconds). Use a graphing utility to graph the displacement function for $0 \le t \le 10$. Find the time beyond which the displacement does not exceed 1 foot from equilibrium.

103. Projectile Motion   A batted baseball leaves the bat at an angle of θ with the horizontal and an initial velocity of $v_0 = 100$ feet per second. The ball is caught by an outfielder 300 feet from home plate (see figure). Find θ if the range r of a projectile is given by

$$r = \frac{1}{32}v_0^2 \sin 2\theta.$$

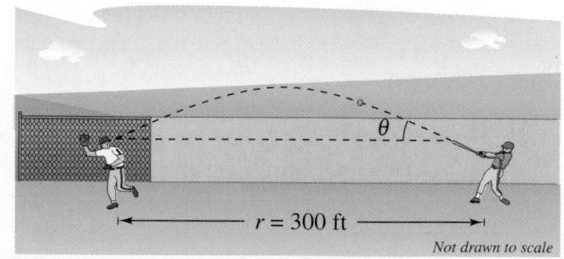

$r = 300$ ft

Not drawn to scale

104. *Area* The area of a rectangle inscribed in one arc of the graph of $y = \cos x$ (see figure) is given by

$$A = 2x \cos x, \quad 0 \le x \le \frac{\pi}{2}.$$

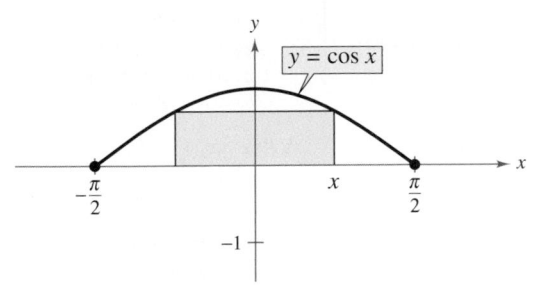

(a) Use a graphing utility to graph the area function, and approximate the area of the largest inscribed rectangle.

(b) Determine the values of x for which $A \ge 1$.

105. *Data Analysis: Unemployment Rate* The table shows the unemployment rates r in the United States for selected years from 1990 through 2004. The time t is measured in years, with $t = 0$ corresponding to 1990.   (Source: U.S. Bureau of Labor Statistics)

Time, t	Rate, r	Time, t	Rate, r
0	5.6	8	4.5
2	7.5	10	4.0
4	6.1	12	5.8
6	5.4	14	5.5

(a) Use a graphing utility to create a scatter plot of the data.

(b) Which of the following models best represents the data? Explain your reasoning.

   (1) $r = 1.24 \sin(0.47t + 0.40) + 5.45$

   (2) $r = 1.24 \sin(0.47t - 0.01) + 5.45$

   (3) $r = \sin(0.10t + 5.61) + 4.80$

   (4) $r = 896 \sin(0.57t - 2.05) + 6.48$

(c) What term in the model gives the average unemployment rate? What is the rate?

(d) Economists study the lengths of business cycles, such as unemployment rates. Based on this short span of time, use the model to determine the length of this cycle.

(e) Use the model to estimate the next time the unemployment rate will be 5% or less.

106. *Quadratic Approximation* Consider the function

$$f(x) = 3 \sin(0.6x - 2).$$

(a) Approximate the zero of the function in the interval $[0, 6]$.

(b) A quadratic approximation agreeing with f at $x = 5$ is

$$g(x) = -0.45x^2 + 5.52x - 13.70.$$

Use a graphing utility to graph f and g in the same viewing window. Describe the result.

(c) Use the Quadratic Formula to find the zeros of g. Compare the zero in the interval $[0, 6]$ with the result of part (a).

Synthesis

True or False? **In Exercises 107–109, determine whether the statement is true or false. Justify your answer.**

107. All trigonometric equations have either an infinite number of solutions or no solution.

108. The solutions of any trigonometric equation can always be found from its solutions in the interval $[0, 2\pi)$.

109. If you correctly solve a trigonometric equation down to the statement $\sin x = 3.4$, then you can finish solving the equation by using an inverse trigonometric function.

110. *Writing* Describe the difference between verifying an identity and solving an equation.

Skills Review

In Exercises 111–114, convert the angle measure from degrees to radians. Round your answer to three decimal places.

111. $124°$

112. $486°$

113. $-0.41°$

114. $-210.55°$

In Exercises 115 and 116, solve for x.

115.

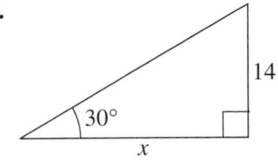

116.

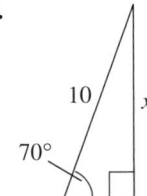

117. *Distance* From the 100-foot roof of a condominium on the coast, a tourist sights a cruise ship. The angle of depression is 2.5°. How far is the ship from the shoreline?

118. *Make a Decision* To work an extended application analyzing the normal daily high temperatures in Phoenix and in Seattle, visit this textbook's *Online Study Center*. (Data Source: NOAA)

5.4 Sum and Difference Formulas

Using Sum and Difference Formulas

In this section and the following section, you will study the uses of several trigonometric identities and formulas.

Sum and Difference Formulas   (See the proofs on page 404.)

$$\sin(u + v) = \sin u \cos v + \cos u \sin v \qquad \tan(u + v) = \frac{\tan u + \tan v}{1 - \tan u \tan v}$$

$$\sin(u - v) = \sin u \cos v - \cos u \sin v$$

$$\cos(u + v) = \cos u \cos v - \sin u \sin v \qquad \tan(u - v) = \frac{\tan u - \tan v}{1 + \tan u \tan v}$$

$$\cos(u - v) = \cos u \cos v + \sin u \sin v$$

Exploration

Use a graphing utility to graph $y_1 = \cos(x + 2)$ and $y_2 = \cos x + \cos 2$ in the same viewing window. What can you conclude about the graphs? Is it true that $\cos(x + 2) = \cos x + \cos 2$?

Use a graphing utility to graph $y_1 = \sin(x + 4)$ and $y_2 = \sin x + \sin 4$ in the same viewing window. What can you conclude about the graphs? Is it true that $\sin(x + 4) = \sin x + \sin 4$?

Examples 1 and 2 show how **sum and difference formulas** can be used to find exact values of trigonometric functions involving sums or differences of special angles.

Example 1 Evaluating a Trigonometric Function

Find the exact value of $\cos 75°$.

Solution

To find the exact value of $\cos 75°$, use the fact that $75° = 30° + 45°$. Consequently, the formula for $\cos(u + v)$ yields

$$\cos 75° = \cos(30° + 45°)$$

$$= \cos 30° \cos 45° - \sin 30° \sin 45°$$

$$= \frac{\sqrt{3}}{2}\left(\frac{\sqrt{2}}{2}\right) - \frac{1}{2}\left(\frac{\sqrt{2}}{2}\right)$$

$$= \frac{\sqrt{6} - \sqrt{2}}{4}.$$

Try checking this result on your calculator. You will find that $\cos 75° \approx 0.259$.

✓**CHECKPOINT**   Now try Exercise 1.

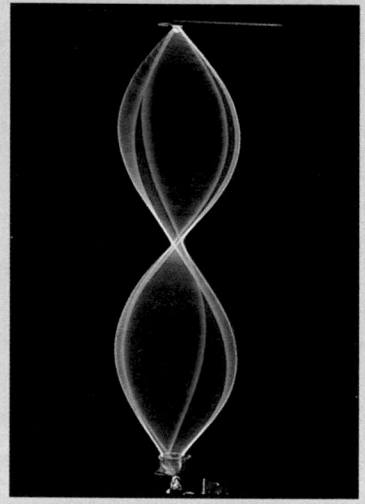
Prerequisite Skills

To review sines, cosines, and tangents of special angles, see Section 4.3.

Example 2   Evaluating a Trigonometric Function

Find the exact value of $\sin \dfrac{\pi}{12}$.

Solution

Using the fact that $\pi/12 = \pi/3 - \pi/4$ together with the formula for $\sin(u - v)$, you obtain

$$\sin \frac{\pi}{12} = \sin\left(\frac{\pi}{3} - \frac{\pi}{4}\right) = \sin \frac{\pi}{3} \cos \frac{\pi}{4} - \cos \frac{\pi}{3} \sin \frac{\pi}{4}$$

$$= \frac{\sqrt{3}}{2}\left(\frac{\sqrt{2}}{2}\right) - \frac{1}{2}\left(\frac{\sqrt{2}}{2}\right) = \frac{\sqrt{6} - \sqrt{2}}{4}.$$

✓**CHECKPOINT**   Now try Exercise 3.

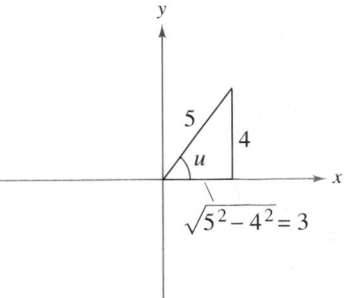

Figure 5.24

Example 3   Evaluating a Trigonometric Expression

Find the exact value of $\sin(u + v)$ given

$$\sin u = \frac{4}{5}, \text{ where } 0 < u < \frac{\pi}{2} \quad \text{and} \quad \cos v = -\frac{12}{13}, \text{ where } \frac{\pi}{2} < v < \pi.$$

Solution

Because $\sin u = 4/5$ and u is in Quadrant I, $\cos u = 3/5$, as shown in Figure 5.24. Because $\cos v = -12/13$ and v is in Quadrant II, $\sin v = 5/13$, as shown in Figure 5.25. You can find $\sin(u + v)$ as follows.

$$\sin(u + v) = \sin u \cos v + \cos u \sin v$$

$$= \left(\frac{4}{5}\right)\left(-\frac{12}{13}\right) + \left(\frac{3}{5}\right)\left(\frac{5}{13}\right) = -\frac{48}{65} + \frac{15}{65} = -\frac{33}{65}$$

✓**CHECKPOINT**   Now try Exercise 35.

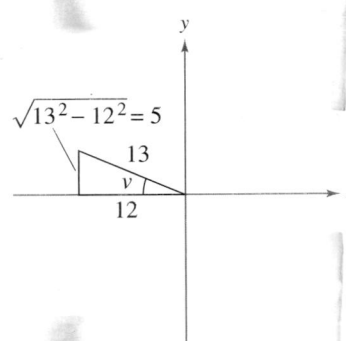

Figure 5.25

Example 4   An Application of a Sum Formula

Write $\cos(\arctan 1 + \arccos x)$ as an algebraic expression.

Solution

This expression fits the formula for $\cos(u + v)$. Angles $u = \arctan 1$ and $v = \arccos x$ are shown in Figure 5.26.

$$\cos(u + v) = \cos(\arctan 1)\cos(\arccos x) - \sin(\arctan 1)\sin(\arccos x)$$

$$= \frac{1}{\sqrt{2}} \cdot x - \frac{1}{\sqrt{2}} \cdot \sqrt{1 - x^2} = \frac{x - \sqrt{1 - x^2}}{\sqrt{2}}.$$

✓**CHECKPOINT**   Now try Exercise 43.

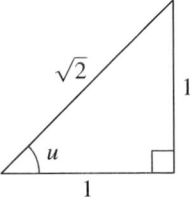

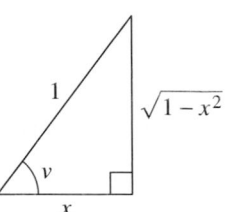

Figure 5.26

Example 5   Proving a Cofunction Identity

Prove the cofunction identity $\cos\left(\dfrac{\pi}{2} - x\right) = \sin x$.

Solution

Using the formula for $\cos(u - v)$, you have

$$\cos\left(\frac{\pi}{2} - x\right) = \cos\frac{\pi}{2}\cos x + \sin\frac{\pi}{2}\sin x$$

$$= (0)(\cos x) + (1)(\sin x)$$

$$= \sin x.$$

✓CHECKPOINT   Now try Exercise 63.

Sum and difference formulas can be used to derive **reduction formulas** involving expressions such as

$$\sin\left(\theta + \frac{n\pi}{2}\right) \quad \text{and} \quad \cos\left(\theta + \frac{n\pi}{2}\right), \text{ where } n \text{ is an integer.}$$

Example 6   Deriving Reduction Formulas

Simplify each expression.

a. $\cos\left(\theta - \dfrac{3\pi}{2}\right)$ **b.** $\tan(\theta + 3\pi)$

Solution

a. Using the formula for $\cos(u - v)$, you have

$$\cos\left(\theta - \frac{3\pi}{2}\right) = \cos\theta\cos\frac{3\pi}{2} + \sin\theta\sin\frac{3\pi}{2}$$

$$= (\cos\theta)(0) + (\sin\theta)(-1)$$

$$= -\sin\theta.$$

b. Using the formula for $\tan(u + v)$, you have

$$\tan(\theta + 3\pi) = \frac{\tan\theta + \tan 3\pi}{1 - \tan\theta\tan 3\pi}$$

$$= \frac{\tan\theta + 0}{1 - (\tan\theta)(0)}$$

$$= \tan\theta.$$

Note that the period of $\tan\theta$ is π, so the period of $\tan(\theta + 3\pi)$ is the same as the period of $\tan\theta$.

✓CHECKPOINT   Now try Exercise 67.

Example 7   Solving a Trigonometric Equation

Find all solutions of $\sin\left(x + \dfrac{\pi}{4}\right) + \sin\left(x - \dfrac{\pi}{4}\right) = -1$ in the interval $[0, 2\pi)$.

Algebraic Solution

Using sum and difference formulas, rewrite the equation as

$$\sin x \cos \frac{\pi}{4} + \cos x \sin \frac{\pi}{4} + \sin x \cos \frac{\pi}{4} - \cos x \sin \frac{\pi}{4} = -1$$

$$2 \sin x \cos \frac{\pi}{4} = -1$$

$$2(\sin x)\left(\frac{\sqrt{2}}{2}\right) = -1$$

$$\sin x = -\frac{1}{\sqrt{2}}$$

$$\sin x = -\frac{\sqrt{2}}{2}.$$

So, the only solutions in the interval $[0, 2\pi)$ are

$$x = \frac{5\pi}{4} \quad \text{and} \quad x = \frac{7\pi}{4}.$$

✓**CHECKPOINT**   Now try Exercise 71.

Graphical Solution

Use a graphing utility set in *radian* mode to graph $\quad y = \sin\left(x + \dfrac{\pi}{4}\right) + \sin\left(x - \dfrac{\pi}{4}\right) + 1,$

as shown in Figure 5.27. Use the *zero* or *root* feature or the *zoom* and *trace* features to approximate the x-intercepts in the interval $[0, 2\pi)$ to be

$$x \approx 3.927 \approx \frac{5\pi}{4} \text{ and } x \approx 5.498 \approx \frac{7\pi}{4}.$$

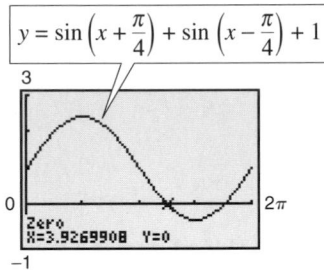

Figure 5.27

The next example was taken from calculus. It is used to derive the formula for the derivative of the cosine function.

Example 8   An Application from Calculus

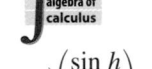

Verify $\dfrac{\cos(x + h) - \cos x}{h} = (\cos x)\left(\dfrac{\cos h - 1}{h}\right) - (\sin x)\left(\dfrac{\sin h}{h}\right), h \neq 0.$

Solution

Using the formula for $\cos(u + v)$, you have

$$\frac{\cos(x + h) - \cos x}{h} = \frac{\cos x \cos h - \sin x \sin h - \cos x}{h}$$

$$= \frac{\cos x(\cos h - 1) - \sin x \sin h}{h}$$

$$= (\cos x)\left(\frac{\cos h - 1}{h}\right) - (\sin x)\left(\frac{\sin h}{h}\right).$$

✓**CHECKPOINT**   Now try Exercise 93.

TECHNOLOGY SUPPORT

For instructions on how to use the *zero* or *root* feature and the *zoom* and *trace* features, see Appendix A; for specific keystrokes, go to this textbook's *Online Study Center*.

5.4 Exercises

See www.CalcChat.com for worked-out solutions to odd-numbered exercises.

Vocabulary Check

Fill in the blank to complete the trigonometric formula.

1. $\sin(u - v) = $ _____
2. $\cos(u + v) = $ _____
3. $\tan(u + v) = $ _____
4. $\sin(u + v) = $ _____
5. $\cos(u - v) = $ _____
6. $\tan(u - v) = $ _____

In Exercises 1–6, find the exact value of each expression.

1. (a) $\cos(240° - 0°)$ (b) $\cos 240° - \cos 0°$
2. (a) $\sin(405° + 120°)$ (b) $\sin 405° + \sin 120°$
3. (a) $\cos\left(\dfrac{\pi}{4} + \dfrac{\pi}{3}\right)$ (b) $\cos\dfrac{\pi}{4} + \cos\dfrac{\pi}{3}$
4. (a) $\sin\left(\dfrac{2\pi}{3} + \dfrac{5\pi}{6}\right)$ (b) $\sin\dfrac{2\pi}{3} + \sin\dfrac{5\pi}{6}$
5. (a) $\sin(315° - 60°)$ (b) $\sin 315° - \sin 60°$
6. (a) $\sin\left(\dfrac{7\pi}{6} - \dfrac{\pi}{3}\right)$ (b) $\sin\dfrac{7\pi}{6} - \sin\dfrac{\pi}{3}$

In Exercises 7–22, find the exact values of the sine, cosine, and tangent of the angle.

7. $105° = 60° + 45°$
8. $165° = 135° + 30°$
9. $195° = 225° - 30°$
10. $255° = 300° - 45°$
11. $\dfrac{11\pi}{12} = \dfrac{3\pi}{4} + \dfrac{\pi}{6}$
12. $\dfrac{17\pi}{12} = \dfrac{7\pi}{6} + \dfrac{\pi}{4}$
13. $-\dfrac{\pi}{12} = \dfrac{\pi}{6} - \dfrac{\pi}{4}$
14. $-\dfrac{19\pi}{12} = \dfrac{2\pi}{3} - \dfrac{9\pi}{4}$
15. $75°$
16. $15°$
17. $-225°$
18. $-165°$
19. $\dfrac{13\pi}{12}$
20. $\dfrac{5\pi}{12}$
21. $-\dfrac{7\pi}{12}$
22. $-\dfrac{13\pi}{12}$

In Exercises 23–30, write the expression as the sine, cosine, or tangent of an angle.

23. $\cos 60° \cos 20° - \sin 60° \sin 20°$
24. $\sin 110° \cos 80° + \cos 110° \sin 80°$
25. $\dfrac{\tan 325° - \tan 86°}{1 + \tan 325° \tan 86°}$
26. $\dfrac{\tan 154° - \tan 49°}{1 + \tan 154° \tan 49°}$
27. $\sin 3.5 \cos 1.2 - \cos 3.5 \sin 1.2$
28. $\cos 0.96 \cos 0.42 + \sin 0.96 \sin 0.42$
29. $\cos\dfrac{\pi}{9} \cos\dfrac{\pi}{7} - \sin\dfrac{\pi}{9} \sin\dfrac{\pi}{7}$
30. $\sin\dfrac{4\pi}{9} \cos\dfrac{\pi}{8} + \cos\dfrac{4\pi}{9} \sin\dfrac{\pi}{8}$

Numerical, Graphical, and Algebraic Analysis **In Exercises 31–34, use a graphing utility to complete the table and graph the two functions in the same viewing window. Use both the table and the graph as evidence that $y_1 = y_2$. Then verify the identity algebraically.**

x	0.2	0.4	0.6	0.8	1.0	1.2	1.4
y_1							
y_2							

31. $y_1 = \sin\left(\dfrac{\pi}{6} + x\right)$, $y_2 = \dfrac{1}{2}(\cos x + \sqrt{3} \sin x)$
32. $y_1 = \cos\left(\dfrac{5\pi}{4} - x\right)$, $y_2 = -\dfrac{\sqrt{2}}{2}(\cos x + \sin x)$
33. $y_1 = \cos(x + \pi) \cos(x - \pi)$, $y_2 = \cos^2 x$
34. $y_1 = \sin(x + \pi) \sin(x - \pi)$, $y_2 = \sin^2 x$

In Exercises 35–38, find the exact value of the trigonometric function given that $\sin u = \dfrac{5}{13}$ and $\cos v = -\dfrac{3}{5}$. (Both u and v are in Quadrant II.)

35. $\sin(u + v)$
36. $\cos(v - u)$
37. $\tan(u + v)$
38. $\sin(u - v)$

In Exercises 39–42, find the exact value of the trigonometric function given that $\sin u = -\dfrac{8}{17}$ and $\cos v = -\dfrac{4}{5}$. (Both u and v are in Quadrant III.)

39. $\cos(u + v)$
40. $\tan(u + v)$
41. $\sin(v - u)$
42. $\cos(u - v)$

In Exercises 43–46, write the trigonometric expression as an algebraic expression.

43. $\sin(\arcsin x + \arccos x)$ 44. $\cos(\arccos x - \arcsin x)$

45. $\sin(\arctan 2x - \arccos x)$ 46. $\cos(\arcsin x - \arctan 2x)$

In Exercises 47–54, find the value of the expression without using a calculator.

47. $\sin(\sin^{-1} 1 + \cos^{-1} 1)$

48. $\cos[\sin^{-1}(-1) + \cos^{-1} 0]$

49. $\sin[\sin^{-1} 1 - \cos^{-1}(-1)]$

50. $\cos[\cos^{-1}(-1) - \cos^{-1} 1]$

51. $\sin\left(\sin^{-1}\dfrac{1}{2} - \cos^{-1}\dfrac{1}{2}\right)$

52. $\cos\left[\cos^{-1}\left(-\dfrac{1}{2}\right) + \sin^{-1} 1\right]$

53. $\tan\left(\sin^{-1} 0 + \sin^{-1}\dfrac{1}{2}\right)$

54. $\tan\left(\cos^{-1}\dfrac{\sqrt{2}}{2} - \sin^{-1} 0\right)$

In Exercises 55–58, evaluate the trigonometric function without using a calculator.

55. $\sin\left[\dfrac{\pi}{2} + \sin^{-1}(-1)\right]$ 56. $\sin[\cos^{-1}(-1) + \pi]$

57. $\cos(\pi + \sin^{-1} 1)$ 58. $\cos[\pi - \cos^{-1}(-1)]$

In Exercises 59–62, use right triangles to evaluate the expression.

59. $\sin\left(\cos^{-1}\dfrac{3}{5} - \sin^{-1}\dfrac{5}{13}\right)$

60. $\cos\left(\sin^{-1}\dfrac{12}{13} + \cos^{-1}\dfrac{8}{17}\right)$

61. $\sin\left(\tan^{-1}\dfrac{3}{4} + \sin^{-1}\dfrac{3}{5}\right)$

62. $\tan\left(\sin^{-1}\dfrac{4}{5} - \cos^{-1}\dfrac{5}{13}\right)$

In Exercises 63–70, verify the identity.

63. $\sin\left(\dfrac{\pi}{2} + x\right) = \cos x$ 64. $\sin(3\pi - x) = \sin x$

65. $\tan(x + \pi) - \tan(\pi - x) = 2\tan x$

66. $\tan\left(\dfrac{\pi}{4} - \theta\right) = \dfrac{1 - \tan\theta}{1 + \tan\theta}$

67. $\sin(x + y) + \sin(x - y) = 2\sin x \cos y$

68. $\cos(x + y) + \cos(x - y) = 2\cos x \cos y$

69. $\cos(x + y)\cos(x - y) = \cos^2 x - \sin^2 y$

70. $\sin(x + y)\sin(x - y) = \sin^2 x - \sin^2 y$

In Exercises 71–74, find the solution(s) of the equation in the interval $[0, 2\pi)$. Use a graphing utility to verify your results.

71. $\sin\left(x + \dfrac{\pi}{3}\right) + \sin\left(x - \dfrac{\pi}{3}\right) = 1$

72. $\cos\left(x + \dfrac{\pi}{6}\right) - \cos\left(x - \dfrac{\pi}{6}\right) = 1$

73. $\tan(x + \pi) + 2\sin(x + \pi) = 0$

74. $2\sin\left(x + \dfrac{\pi}{2}\right) + 3\tan(\pi - x) = 0$

In Exercises 75–78, use a graphing utility to approximate the solutions of the equation in the interval $[0, 2\pi)$.

75. $\cos\left(x + \dfrac{\pi}{4}\right) + \cos\left(x - \dfrac{\pi}{4}\right) = 1$

76. $\sin\left(x + \dfrac{\pi}{2}\right) - \cos\left(x + \dfrac{3\pi}{2}\right) = 0$

77. $\tan(x + \pi) - \cos\left(x + \dfrac{\pi}{2}\right) = 0$

78. $\tan(\pi - x) + 2\cos\left(x + \dfrac{3\pi}{2}\right) = 0$

79. *Standing Waves*   The equation of a standing wave is obtained by adding the displacements of two waves traveling in opposite directions (see figure). Assume that each of the waves has amplitude A, period T, and wavelength λ. If the models for these waves are

$$y_1 = A\cos 2\pi\left(\dfrac{t}{T} - \dfrac{x}{\lambda}\right) \text{ and } y_2 = A\cos 2\pi\left(\dfrac{t}{T} + \dfrac{x}{\lambda}\right)$$

show that

$$y_1 + y_2 = 2A\cos\dfrac{2\pi t}{T}\cos\dfrac{2\pi x}{\lambda}.$$

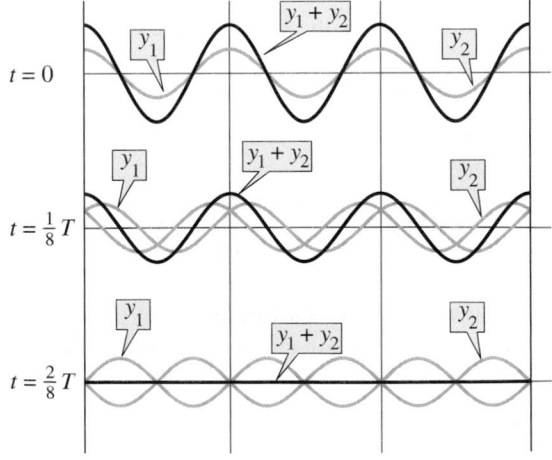

80. Harmonic Motion   A weight is attached to a spring suspended vertically from a ceiling. When a driving force is applied to the system, the weight moves vertically from its equilibrium position, and this motion is modeled by

$$y = \frac{1}{3}\sin 2t + \frac{1}{4}\cos 2t$$

where y is the distance from equilibrium (in feet) and t is the time (in seconds).

(a) Use a graphing utility to graph the model.

(b) Use the identity

$$a \sin B\theta + b \cos B\theta = \sqrt{a^2 + b^2}\, \sin(B\theta + C)$$

where $C = \arctan(b/a)$, $a > 0$, to write the model in the form $y = \sqrt{a^2 + b^2}\, \sin(Bt + C)$. Use a graphing utility to verify your result.

(c) Find the amplitude of the oscillations of the weight.

(d) Find the frequency of the oscillations of the weight.

Synthesis

True or False?   In Exercises 81 and 82, determine whether the statement is true or false. Justify your answer.

81. $\cos(u \pm v) = \cos u \pm \cos v$

82. $\sin\left(x - \dfrac{11\pi}{2}\right) = \cos x$

In Exercises 83–86, verify the identity.

83. $\cos(n\pi + \theta) = (-1)^n \cos \theta,$ n is an integer.

84. $\sin(n\pi + \theta) = (-1)^n \sin \theta,$ n is an integer.

85. $a \sin B\theta + b \cos B\theta = \sqrt{a^2 + b^2}\, \sin(B\theta + C)$, where $C = \arctan(b/a)$ and $a > 0$.

86. $a \sin B\theta + b \cos B\theta = \sqrt{a^2 + b^2}\, \cos(B\theta - C)$, where $C = \arctan(a/b)$ and $b > 0$.

In Exercises 87–90, use the formulas given in Exercises 85 and 86 to write the expression in the following forms. Use a graphing utility to verify your results.

(a) $\sqrt{a^2 + b^2}\, \sin(B\theta + C)$

(b) $\sqrt{a^2 + b^2}\, \cos(B\theta - C)$

87. $\sin \theta + \cos \theta$

88. $3 \sin 2\theta + 4 \cos 2\theta$

89. $12 \sin 3\theta + 5 \cos 3\theta$

90. $\sin 2\theta - \cos 2\theta$

In Exercises 91 and 92, use the formulas given in Exercises 85 and 86 to write the trigonometric expression in the form $a \sin B\theta + b \cos B\theta$.

91. $2 \sin\left(\theta + \dfrac{\pi}{2}\right)$

92. $5 \cos\left(\theta + \dfrac{\pi}{4}\right)$

∫ **93.** Verify the following identity used in calculus.

$$\frac{\sin(x+h) - \sin x}{h} = \frac{\cos x \sin h}{h} - \frac{\sin x(1 - \cos h)}{h}$$

∫ **94. Exploration**   Let $x = \pi/3$ in the identity in Exercise 93 and define the functions f and g as follows.

$$f(h) = \frac{\sin(\pi/3 + h) - \sin(\pi/3)}{h}$$

$$g(h) = \cos\frac{\pi}{3}\left(\frac{\sin h}{h}\right) - \sin\frac{\pi}{3}\left(\frac{1 - \cos h}{h}\right)$$

(a) What are the domains of the functions f and g?

(b) Use a graphing utility to complete the table.

h	0.01	0.02	0.05	0.1	0.2	0.5
$f(h)$						
$g(h)$						

(c) Use a graphing utility to graph the functions f and g.

(d) Use the table and graph to make a conjecture about the values of the functions f and g as $h \to 0$.

95. Conjecture   Three squares of side s are placed side by side (see figure). Make a conjecture about the relationship between the sum $u + v$ and w. Prove your conjecture by using the identity for the tangent of the sum of two angles.

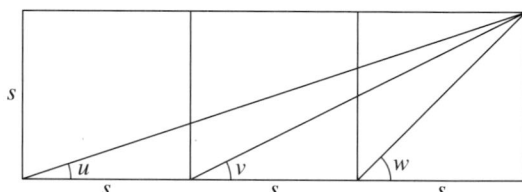

96. (a) Write a sum formula for $\sin(u + v + w)$.

(b) Write a sum formula for $\tan(u + v + w)$.

Skills Review

In Exercises 97–100, find the *x*- and *y*-intercepts of the graph of the equation. Use a graphing utility to verify your results.

97. $y = -\frac{1}{2}(x - 10) + 14$

98. $y = x^2 - 3x - 40$

99. $y = |2x - 9| - 5$

100. $y = 2x\sqrt{x + 7}$

In Exercises 101–104, evaluate the expression without using a calculator.

101. $\arccos\left(\dfrac{\sqrt{3}}{2}\right)$

102. $\arctan\left(-\sqrt{3}\right)$

103. $\sin^{-1} 1$

104. $\tan^{-1} 0$

5.5 Multiple-Angle and Product-to-Sum Formulas

Multiple–Angle Formulas

In this section, you will study four additional categories of trigonometric identities.

1. The first category involves *functions of multiple angles* such as $\sin ku$ and $\cos ku$.
2. The second category involves *squares of trigonometric functions* such as $\sin^2 u$.
3. The third category involves *functions of half-angles* such as $\sin(u/2)$.
4. The fourth category involves *products of trigonometric functions* such as $\sin u \cos v$.

You should learn the **double-angle formulas** below because they are used most often.

> **Double-Angle Formulas**   (See the proofs on page 405.)
>
> $$\sin 2u = 2 \sin u \cos u \qquad \cos 2u = \cos^2 u - \sin^2 u$$
>
> $$= 2 \cos^2 u - 1$$
>
> $$\tan 2u = \frac{2 \tan u}{1 - \tan^2 u} \qquad = 1 - 2 \sin^2 u$$

Example 1 Solving a Multiple-Angle Equation

Solve $2 \cos x + \sin 2x = 0$.

Solution

Begin by rewriting the equation so that it involves functions of x (rather than $2x$). Then factor and solve as usual.

$$2 \cos x + \sin 2x = 0 \qquad \text{Write original equation.}$$

$$2 \cos x + 2 \sin x \cos x = 0 \qquad \text{Double-angle formula}$$

$$2 \cos x(1 + \sin x) = 0 \qquad \text{Factor.}$$

$$\cos x = 0 \qquad\qquad 1 + \sin x = 0 \qquad \text{Set factors equal to zero.}$$

$$x = \frac{\pi}{2}, \frac{3\pi}{2} \qquad\qquad x = \frac{3\pi}{2} \qquad \text{Solutions in } [0, 2\pi)$$

So, the general solution is

$$x = \frac{\pi}{2} + 2n\pi \quad \text{and} \quad x = \frac{3\pi}{2} + 2n\pi \qquad \text{General solution}$$

where n is an integer. Try verifying this solution graphically.

✓CHECKPOINT   Now try Exercise 3.

What you should learn

- Use multiple-angle formulas to rewrite and evaluate trigonometric functions.
- Use power-reducing formulas to rewrite and evaluate trigonometric functions.
- Use half-angle formulas to rewrite and evaluate trigonometric functions.
- Use product-to-sum and sum-to-product formulas to rewrite and evaluate trigonometric functions.

Why you should learn it

You can use a variety of trigonometric formulas to rewrite trigonometric functions in more convenient forms. For instance, Exercise 130 on page 398 shows you how to use a half-angle formula to determine the apex angle of a sound wave cone caused by the speed of an airplane.

NASA-Liaison/Getty Images

Example 2 Using Double–Angle Formulas to Analyze Graphs

Analyze the graph of $y = 4\cos^2 x - 2$ in the interval $[0, 2\pi]$.

Solution

Using a double-angle formula, you can rewrite the original function as

$$y = 4\cos^2 x - 2$$
$$= 2(2\cos^2 x - 1)$$
$$= 2\cos 2x.$$

Using the techniques discussed in Section 4.5, you can recognize that the graph of this function has an amplitude of 2 and a period of π. The key points in the interval $[0, \pi]$ are as follows.

Maximum	Intercept	Minimum	Intercept	Maximum
$(0, 2)$	$\left(\dfrac{\pi}{4}, 0\right)$	$\left(\dfrac{\pi}{2}, -2\right)$	$\left(\dfrac{3\pi}{4}, 0\right)$	$(\pi, 2)$

Two cycles of the graph are shown in Figure 5.28.

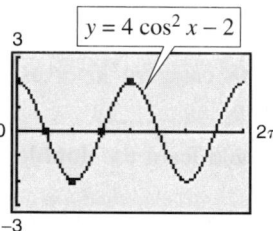

Figure 5.28

✓**CHECKPOINT**   Now try Exercise 7.

Example 3 Evaluating Functions Involving Double Angles

Use the following to find $\sin 2\theta$, $\cos 2\theta$, and $\tan 2\theta$.

$$\cos\theta = \frac{5}{13}, \qquad \frac{3\pi}{2} < \theta < 2\pi$$

Solution

In Figure 5.29, you can see that $\sin\theta = y/r = -12/13$. Consequently, using each of the double-angle formulas, you can write the double angles as follows.

$$\sin 2\theta = 2\sin\theta\cos\theta = 2\left(-\frac{12}{13}\right)\left(\frac{5}{13}\right) = -\frac{120}{169}$$

$$\cos 2\theta = 2\cos^2\theta - 1 = 2\left(\frac{25}{169}\right) - 1 = -\frac{119}{169}$$

$$\tan 2\theta = \frac{2\tan\theta}{1 - \tan^2\theta} = \frac{2(-12/5)}{1 - (-12/5)^2} = \frac{120}{119}$$

✓**CHECKPOINT**   Now try Exercise 13.

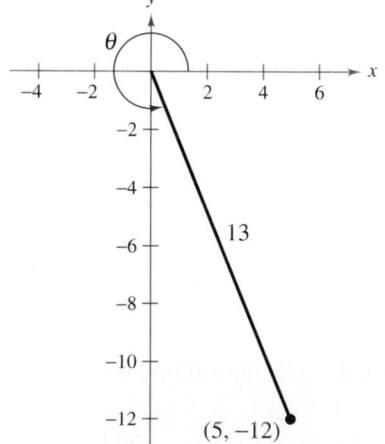

Figure 5.29

The double-angle formulas are not restricted to the angles 2θ and θ. Other *double* combinations, such as 4θ and 2θ or 6θ and 3θ, are also valid. Here are two examples.

$$\sin 4\theta = 2\sin 2\theta\cos 2\theta \qquad \text{and} \qquad \cos 6\theta = \cos^2 3\theta - \sin^2 3\theta$$

By using double-angle formulas together with the sum formulas derived in the preceding section, you can form other multiple-angle formulas.

Example 4   Deriving a Triple-Angle Formula

$$
\begin{aligned}
\sin 3x &= \sin(2x + x) && \text{Rewrite as a sum.}\\
&= \sin 2x \cos x + \cos 2x \sin x && \text{Sum formula}\\
&= 2 \sin x \cos x \cos x + (1 - 2 \sin^2 x) \sin x && \text{Double-angle formula}\\
&= 2 \sin x \cos^2 x + \sin x - 2 \sin^3 x && \text{Multiply.}\\
&= 2 \sin x(1 - \sin^2 x) + \sin x - 2 \sin^3 x && \text{Pythagorean identity}\\
&= 2 \sin x - 2 \sin^3 x + \sin x - 2 \sin^3 x && \text{Multiply.}\\
&= 3 \sin x - 4 \sin^3 x && \text{Simplify.}
\end{aligned}
$$

✓CHECKPOINT   Now try Exercise 19.

Power-Reducing Formulas

The double-angle formulas can be used to obtain the following **power-reducing formulas.**

Power-Reducing Formulas   (See the proofs on page 405.)

$$
\sin^2 u = \frac{1 - \cos 2u}{2} \qquad \cos^2 u = \frac{1 + \cos 2u}{2} \qquad \tan^2 u = \frac{1 - \cos 2u}{1 + \cos 2u}
$$

Example 5   Reducing a Power

Rewrite $\sin^4 x$ as a sum of first powers of the cosines of multiple angles.

Solution

$$
\begin{aligned}
\sin^4 x &= (\sin^2 x)^2 && \text{Property of exponents}\\
&= \left(\frac{1 - \cos 2x}{2}\right)^2 && \text{Power-reducing formula}\\
&= \frac{1}{4}(1 - 2 \cos 2x + \cos^2 2x) && \text{Expand binomial.}\\
&= \frac{1}{4}\left(1 - 2 \cos 2x + \frac{1 + \cos 4x}{2}\right) && \text{Power-reducing formula}\\
&= \frac{1}{4} - \frac{1}{2} \cos 2x + \frac{1}{8} + \frac{1}{8} \cos 4x && \text{Distributive Property}\\
&= \frac{3}{8} - \frac{1}{2} \cos 2x + \frac{1}{8} \cos 4x && \text{Simplify.}\\
&= \frac{1}{8}(3 - 4 \cos 2x + \cos 4x) && \text{Factor.}
\end{aligned}
$$

✓CHECKPOINT   Now try Exercise 23.

STUDY TIP

Power-reducing formulas are often used in calculus. Example 5 shows a typical power reduction that is used in calculus. Note the repeated use of power-reducing formulas.

Half–Angle Formulas

You can derive some useful alternative forms of the power-reducing formulas by replacing u with $u/2$. The results are called **half-angle formulas.**

Half-Angle Formulas

$$\sin\frac{u}{2} = \pm\sqrt{\frac{1-\cos u}{2}} \qquad\qquad \cos\frac{u}{2} = \pm\sqrt{\frac{1+\cos u}{2}}$$

$$\tan\frac{u}{2} = \frac{1-\cos u}{\sin u} = \frac{\sin u}{1+\cos u}$$

The signs of $\sin\dfrac{u}{2}$ and $\cos\dfrac{u}{2}$ depend on the quadrant in which $\dfrac{u}{2}$ lies.

Example 6   Using a Half–Angle Formula

Find the exact value of $\sin 105°$.

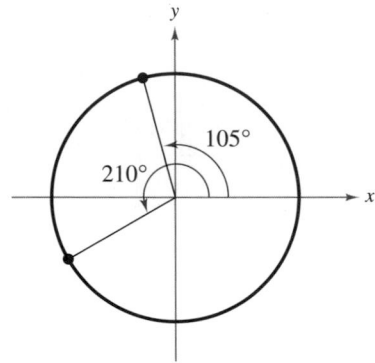

Figure 5.30

STUDY TIP

To find the exact value of a trigonometric function with an angle in D°M′S″ form using a half-angle formula, first convert the angle measure to decimal degree form. Then multiply the angle measure by 2.

Solution

Begin by noting that $105°$ is half of $210°$. Then, using the half-angle formula for $\sin(u/2)$ and the fact that $105°$ lies in Quadrant II (see Figure 5.30), you have

$$\sin 105° = \sqrt{\frac{1-\cos 210°}{2}}$$

$$= \sqrt{\frac{1-(-\cos 30°)}{2}}$$

$$= \sqrt{\frac{1+\left(\sqrt{3}/2\right)}{2}} = \frac{\sqrt{2+\sqrt{3}}}{2}.$$

The positive square root is chosen because $\sin\theta$ is positive in Quadrant II.

✓*CHECKPOINT*   Now try Exercise 39.

TECHNOLOGY TIP

Use your calculator to verify the result obtained in Example 6. That is, evaluate $\sin 105°$ and $\left(\sqrt{2+\sqrt{3}}\right)/2$. You will notice that both expressions yield the same result.

Example 7   Solving a Trigonometric Equation

Find all solutions of $2 - \sin^2 x = 2 \cos^2 \dfrac{x}{2}$ in the interval $[0, 2\pi)$.

Algebraic Solution

$$2 - \sin^2 x = 2 \cos^2 \frac{x}{2} \qquad \text{Write original equation.}$$

$$2 - \sin^2 x = 2\left(\pm\sqrt{\frac{1 + \cos x}{2}}\right)^2 \qquad \text{Half-angle formula}$$

$$2 - \sin^2 x = 2\left(\frac{1 + \cos x}{2}\right) \qquad \text{Simplify.}$$

$$2 - \sin^2 x = 1 + \cos x \qquad \text{Simplify.}$$

$$2 - (1 - \cos^2 x) = 1 + \cos x \qquad \text{Pythagorean identity}$$

$$\cos^2 x - \cos x = 0 \qquad \text{Simplify.}$$

$$\cos x(\cos x - 1) = 0 \qquad \text{Factor.}$$

By setting the factors $\cos x$ and $(\cos x - 1)$ equal to zero, you find that the solutions in the interval $[0, 2\pi)$ are

$$x = \frac{\pi}{2}, \quad x = \frac{3\pi}{2}, \quad \text{and} \quad x = 0.$$

 CHECKPOINT   Now try Exercise 57.

Graphical Solution

Use a graphing utility set in *radian* mode to graph $y = 2 - \sin^2 x - 2 \cos^2(x/2)$, as shown in Figure 5.31. Use the *zero* or *root* feature or the *zoom* and *trace* features to approximate the x-intercepts in the interval $[0, 2\pi)$ to be

$$x = 0, \quad x \approx 1.5708 \approx \frac{\pi}{2}, \quad \text{and} \quad x \approx 4.7124 \approx \frac{3\pi}{2}.$$

These values are the approximate solutions of $2 - \sin^2 x = 2 \cos^2 \dfrac{x}{2}$ in the interval $[0, 2\pi)$.

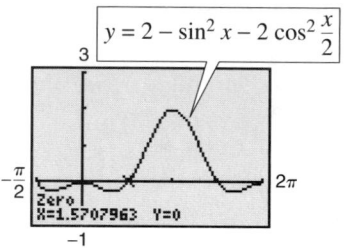

Figure 5.31

Product-to-Sum Formulas

Each of the following **product-to-sum formulas** is easily verified using the sum and difference formulas discussed in the preceding section.

Product-to-Sum Formulas

$$\sin u \sin v = \frac{1}{2}[\cos(u - v) - \cos(u + v)]$$

$$\cos u \cos v = \frac{1}{2}[\cos(u - v) + \cos(u + v)]$$

$$\sin u \cos v = \frac{1}{2}[\sin(u + v) + \sin(u - v)]$$

$$\cos u \sin v = \frac{1}{2}[\sin(u + v) - \sin(u - v)]$$

Product-to-sum formulas are used in calculus to evaluate integrals involving the products of sines and cosines of two different angles.

Example 8   Writing Products as Sums

Rewrite the product as a sum or difference.

$$\cos 5x \sin 4x$$

Solution

$$\cos 5x \sin 4x = \frac{1}{2}[\sin(5x + 4x) - \sin(5x - 4x)]$$

$$= \frac{1}{2}\sin 9x - \frac{1}{2}\sin x$$

✓CHECKPOINT   Now try Exercise 63.

TECHNOLOGY TIP

You can use a graphing utility to verify the solution in Example 8. Graph $y_1 = \cos 5x \sin 4x$ and $y_2 = \frac{1}{2}\sin 9x - \frac{1}{2}\sin x$ in the same viewing window. Notice that the graphs coincide. So, you can conclude that the two expressions are equivalent.

Occasionally, it is useful to reverse the procedure and write a sum of trigonometric functions as a product. This can be accomplished with the following **sum-to-product formulas.**

Sum-to-Product Formulas   (See the proof on page 406.)

$$\sin u + \sin v = 2 \sin\left(\frac{u + v}{2}\right)\cos\left(\frac{u - v}{2}\right)$$

$$\sin u - \sin v = 2 \cos\left(\frac{u + v}{2}\right)\sin\left(\frac{u - v}{2}\right)$$

$$\cos u + \cos v = 2 \cos\left(\frac{u + v}{2}\right)\cos\left(\frac{u - v}{2}\right)$$

$$\cos u - \cos v = -2 \sin\left(\frac{u + v}{2}\right)\sin\left(\frac{u - v}{2}\right)$$

Example 9   Using a Sum-to-Product Formula

Find the exact value of $\cos 195° + \cos 105°$.

Solution

Using the appropriate sum-to-product formula, you obtain

$$\cos 195° + \cos 105° = 2 \cos\left(\frac{195° + 105°}{2}\right)\cos\left(\frac{195° - 105°}{2}\right)$$

$$= 2 \cos 150° \cos 45°$$

$$= 2\left(-\frac{\sqrt{3}}{2}\right)\left(\frac{\sqrt{2}}{2}\right)$$

$$= -\frac{\sqrt{6}}{2}.$$

✓CHECKPOINT   Now try Exercise 81.

Example 10   Solving a Trigonometric Equation

Find all solutions of $\sin 5x + \sin 3x = 0$ in the interval $[0, 2\pi)$.

Algebraic Solution

$$\sin 5x + \sin 3x = 0 \qquad \text{Write original equation.}$$

$$2 \sin\left(\frac{5x + 3x}{2}\right) \cos\left(\frac{5x - 3x}{2}\right) = 0 \qquad \text{Sum-to-product formula}$$

$$2 \sin 4x \cos x = 0 \qquad \text{Simplify.}$$

By setting the factor $\sin 4x$ equal to zero, you can find that the solutions in the interval $[0, 2\pi)$ are

$$x = 0, \frac{\pi}{4}, \frac{\pi}{2}, \frac{3\pi}{4}, \pi, \frac{5\pi}{4}, \frac{3\pi}{2}, \frac{7\pi}{4}.$$

Moreover, the equation $\cos x = 0$ yields no additional solutions. Note that the general solution is

$$x = \frac{n\pi}{4}$$

where n is an integer.

✔CHECKPOINT   Now try Exercise 85.

Graphical Solution

Use a graphing utility set in *radian* mode to graph $y = \sin 5x + \sin 3x$, as shown in Figure 5.32. Use the *zero* or *root* feature or the *zoom* and *trace* features to approximate the x-intercepts in the interval $[0, 2\pi)$ to be

$$x \approx 0, \ x \approx 0.7854 \approx \frac{\pi}{4}, x \approx 1.5708 \approx \frac{\pi}{2},$$

$$x \approx 2.3562 \approx \frac{3\pi}{4}, \ x \approx 3.1416 \approx \pi, x \approx 3.9270 \approx \frac{5\pi}{4},$$

$$x \approx 4.7124 \approx \frac{3\pi}{2}, \ x \approx 5.4978 \approx \frac{7\pi}{4}.$$

These values are the approximate solutions of $\sin 5x + \sin 3x = 0$ in the interval $[0, 2\pi)$.

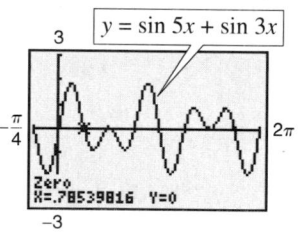

Figure 5.32

Example 11   Verifying a Trigonometric Identity

Verify the identity $\dfrac{\sin t + \sin 3t}{\cos t + \cos 3t} = \tan 2t$.

Algebraic Solution

Using appropriate sum-to-product formulas, you have

$$\frac{\sin t + \sin 3t}{\cos t + \cos 3t} = \frac{2 \sin 2t \cos(-t)}{2 \cos 2t \cos(-t)}$$

$$= \frac{\sin 2t}{\cos 2t}$$

$$= \tan 2t.$$

✔CHECKPOINT   Now try Exercise 105.

Numerical Solution

Use the *table* feature of a graphing utility set in *radian* mode to create a table that shows the values of $y_1 = (\sin x + \sin 3x)/(\cos x + \cos 3x)$ and $y_2 = \tan 2x$ for different values of x, as shown in Figure 5.33. In the table, you can see that the values appear to be identical, so $(\sin x + \sin 3x)/(\cos x + \cos 3x) = \tan 2x$ appears to be an identity.

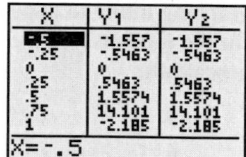

Figure 5.33

5.5   Exercises

See www.CalcChat.com for worked-out solutions to odd-numbered exercises.

Vocabulary Check

Fill in the blank to complete the trigonometric formula.

1. $\sin 2u = $ _____

2. $\cos^2 u = $ _____

3. _____ $= 1 - 2 \sin^2 u$

4. _____ $= \dfrac{\sin u}{1 + \cos u}$

5. $\tan 2u = $ _____

6. $\cos u \cos v = $ _____

7. _____ $= \dfrac{1 - \cos 2u}{2}$

8. _____ $= \pm \sqrt{\dfrac{1 + \cos u}{2}}$

9. $\sin u \cos v = $ _____

10. $\sin u + \sin v = $ _____

In Exercises 1 and 2, use the figure to find the exact value of each trigonometric function.

1.

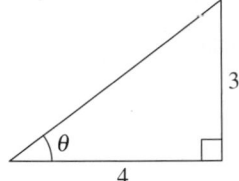

(a) $\sin \theta$ (b) $\cos \theta$

(c) $\cos 2\theta$ (d) $\sin 2\theta$

(e) $\tan 2\theta$ (f) $\sec 2\theta$

(g) $\csc 2\theta$ (h) $\cot 2\theta$

2.

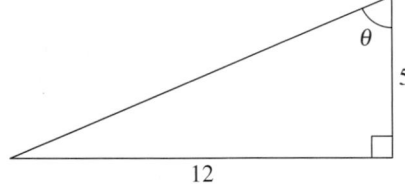

(a) $\sin \theta$ (b) $\cos \theta$

(c) $\sin 2\theta$ (d) $\cos 2\theta$

(e) $\tan 2\theta$ (f) $\cot 2\theta$

(g) $\sec 2\theta$ (h) $\csc 2\theta$

In Exercises 3–12, use a graphing utility to approximate the solutions of the equation in the interval $[0, 2\pi)$. If possible, find the exact solutions algebraically.

3. $\sin 2x - \sin x = 0$

4. $\sin 2x + \cos x = 0$

5. $4 \sin x \cos x = 1$

6. $\sin 2x \sin x = \cos x$

7. $\cos 2x - \cos x = 0$

8. $\tan 2x - \cot x = 0$

9. $\sin 4x = -2 \sin 2x$

10. $(\sin 2x + \cos 2x)^2 = 1$

11. $\cos 2x + \sin x = 0$

12. $\tan 2x - 2 \cos x = 0$

In Exercises 13–18, find the exact values of $\sin 2u$, $\cos 2u$, and $\tan 2u$ using the double-angle formulas.

13. $\sin u = \frac{3}{5}$, $0 < u < \pi/2$

14. $\cos u = -\frac{2}{7}$, $\pi/2 < u < \pi$

15. $\tan u = \frac{1}{2}$, $\pi < u < 3\pi/2$

16. $\cot u = -6$, $3\pi/2 < u < 2\pi$

17. $\sec u = -\frac{5}{2}$, $\pi/2 < u < \pi$

18. $\csc u = 3$, $\pi/2 < u < \pi$

In Exercises 19–22, use a double-angle formula to rewrite the expression. Use a graphing utility to graph both expressions to verify that both forms are the same.

19. $8 \sin x \cos x$

20. $4 \sin x \cos x + 1$

21. $6 - 12 \sin^2 x$

22. $(\cos x + \sin x)(\cos x - \sin x)$

∫ **In Exercises 23–36, rewrite the expression in terms of the first power of the cosine. Use a graphing utility to graph both expressions to verify that both forms are the same.**

23. $\cos^4 x$

24. $\sin^4 x$

25. $\sin^2 x \cos^2 x$

26. $\cos^6 x$

27. $\sin^2 x \cos^4 x$

28. $\sin^4 x \cos^2 x$

29. $\sin^2 2x$

30. $\cos^2 2x$

31. $\cos^2 \dfrac{x}{2}$

32. $\sin^2 \dfrac{x}{2}$

33. $\sin^2 2x \cos^2 2x$

34. $\sin^2 \dfrac{x}{2} \cos^2 \dfrac{x}{2}$

35. $\sin^4 \dfrac{x}{2}$

36. $\cos^4 \dfrac{x}{2}$

In Exercises 37 and 38, use the figure to find the exact value of each trigonometric function.

37.

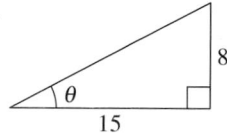

(a) $\cos \dfrac{\theta}{2}$ (b) $\sin \dfrac{\theta}{2}$

(c) $\tan \dfrac{\theta}{2}$ (d) $\sec \dfrac{\theta}{2}$

(e) $\csc \dfrac{\theta}{2}$ (f) $\cot \dfrac{\theta}{2}$

(g) $2 \sin \dfrac{\theta}{2} \cos \dfrac{\theta}{2}$ (h) $2 \cos \dfrac{\theta}{2} \tan \dfrac{\theta}{2}$

38.

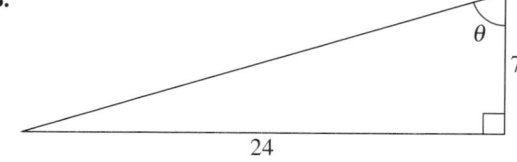

(a) $\sin \dfrac{\theta}{2}$ (b) $\cos \dfrac{\theta}{2}$

(c) $\tan \dfrac{\theta}{2}$ (d) $\cot \dfrac{\theta}{2}$

(e) $\sec \dfrac{\theta}{2}$ (f) $\csc \dfrac{\theta}{2}$

(g) $2 \sin \dfrac{\theta}{2} \cos \dfrac{\theta}{2}$ (h) $\cos 2\theta$

In Exercises 39–46, use the half-angle formulas to determine the exact values of the sine, cosine, and tangent of the angle.

39. $15°$ **40.** $165°$

41. $112° \ 30'$ **42.** $157° \ 30'$

43. $\dfrac{\pi}{8}$ **44.** $\dfrac{\pi}{12}$

45. $\dfrac{3\pi}{8}$ **46.** $\dfrac{7\pi}{12}$

In Exercises 47–52, find the exact values of $\sin(u/2)$, $\cos(u/2)$, and $\tan(u/2)$ using the half-angle formulas.

47. $\sin u = \frac{5}{13}, \quad \pi/2 < u < \pi$
48. $\cos u = \frac{7}{25}, \quad 0 < u < \pi/2$

49. $\tan u = -\frac{8}{5}, \quad 3\pi/2 < u < 2\pi$
50. $\cot u = 7, \quad \pi < u < 3\pi/2$
51. $\csc u = -\frac{5}{3}, \quad \pi < u < 3\pi/2$
52. $\sec u = -\frac{7}{2}, \quad \pi/2 < u < \pi$

In Exercises 53–56, use the half-angle formulas to simplify the expression.

53. $\sqrt{\dfrac{1 - \cos 6x}{2}}$

54. $\sqrt{\dfrac{1 + \cos 4x}{2}}$

55. $-\sqrt{\dfrac{1 - \cos 8x}{1 + \cos 8x}}$

56. $-\sqrt{\dfrac{1 - \cos(x - 1)}{2}}$

In Exercises 57–60, find the solutions of the equation in the interval $[0, 2\pi)$. Use a graphing utility to verify your answers.

57. $\sin \dfrac{x}{2} - \cos x = 0$

58. $\sin \dfrac{x}{2} + \cos x - 1 = 0$

59. $\cos \dfrac{x}{2} - \sin x = 0$

60. $\tan \dfrac{x}{2} - \sin x = 0$

In Exercises 61–72, use the product-to-sum formulas to write the product as a sum or difference.

61. $6 \sin \dfrac{\pi}{3} \cos \dfrac{\pi}{3}$

62. $4 \sin \dfrac{\pi}{3} \cos \dfrac{5\pi}{6}$

63. $\sin 5\theta \cos 3\theta$

64. $5 \sin 3\alpha \sin 4\alpha$

65. $10 \cos 75° \cos 15°$

66. $6 \sin 45° \cos 15°$

67. $5 \cos(-5\beta) \cos 3\beta$

68. $\cos 2\theta \cos 4\theta$

69. $\sin(x + y) \sin(x - y)$

70. $\sin(x + y) \cos(x - y)$

71. $\cos(\theta - \pi) \sin(\theta + \pi)$

72. $\sin(\theta + \pi) \sin(\theta - \pi)$

In Exercises 73–80, use the sum-to-product formulas to write the sum or difference as a product.

73. $\sin 5\theta - \sin \theta$

74. $\sin 3\theta + \sin \theta$

75. $\cos 6x + \cos 2x$

76. $\sin x + \sin 7x$

77. $\sin(\alpha + \beta) - \sin(\alpha - \beta)$

78. $\cos(\phi + 2\pi) + \cos \phi$

79. $\cos\left(\theta + \dfrac{\pi}{2}\right) - \cos\left(\theta - \dfrac{\pi}{2}\right)$

80. $\sin\left(x + \dfrac{\pi}{2}\right) + \sin\left(x - \dfrac{\pi}{2}\right)$

In Exercises 81–84, use the sum-to-product formulas to find the exact value of the expression.

81. $\sin 195° + \sin 105°$

82. $\cos 165° - \cos 75°$

83. $\cos \dfrac{5\pi}{12} + \cos \dfrac{\pi}{12}$

84. $\sin \dfrac{11\pi}{12} - \sin \dfrac{7\pi}{12}$

In Exercises 85–88, find the solutions of the equation in the interval $[0, 2\pi)$. Use a graphing utility to verify your answers.

85. $\sin 6x + \sin 2x = 0$

86. $\cos 2x - \cos 6x = 0$

87. $\dfrac{\cos 2x}{\sin 3x - \sin x} - 1 = 0$

88. $\sin^2 3x - \sin^2 x = 0$

In Exercises 89–92, use the figure and trigonometric identities to find the exact value of the trigonometric function in two ways.

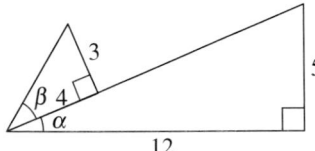

89. $\sin^2 \alpha$

90. $\cos^2 \alpha$

91. $\sin \alpha \cos \beta$

92. $\cos \alpha \sin \beta$

In Exercises 93–110, verify the identity algebraically. Use a graphing utility to check your result graphically.

93. $\csc 2\theta = \dfrac{\csc \theta}{2 \cos \theta}$

94. $\sec 2\theta = \dfrac{\sec^2 \theta}{2 - \sec^2 \theta}$

95. $\cos^2 2\alpha - \sin^2 2\alpha = \cos 4\alpha$

96. $\cos^4 x - \sin^4 x = \cos 2x$

97. $(\sin x + \cos x)^2 = 1 + \sin 2x$

98. $\sin \dfrac{\alpha}{3} \cos \dfrac{\alpha}{3} = \dfrac{1}{2} \sin \dfrac{2\alpha}{3}$

99. $1 + \cos 10y = 2 \cos^2 5y$

100. $\dfrac{\cos 3\beta}{\cos \beta} = 1 - 4 \sin^2 \beta$

101. $\sec \dfrac{u}{2} = \pm \sqrt{\dfrac{2 \tan u}{\tan u + \sin u}}$

102. $\tan \dfrac{u}{2} = \csc u - \cot u$

103. $\cos 3\beta = \cos^3 \beta - 3 \sin^2 \beta \cos \beta$

104. $\sin 4\beta = 4 \sin \beta \cos \beta (1 - 2 \sin^2 \beta)$

105. $\dfrac{\cos 4x - \cos 2x}{2 \sin 3x} = -\sin x$

106. $\dfrac{\cos 3x - \cos x}{\sin 3x - \sin x} = -\tan 2x$

107. $\dfrac{\cos 4x + \cos 2x}{\sin 4x + \sin 2x} = \cot 3x$

108. $\dfrac{\cos t + \cos 3t}{\sin 3t - \sin t} = \cot t$

109. $\sin\left(\dfrac{\pi}{6} + x\right) + \sin\left(\dfrac{\pi}{6} - x\right) = \cos x$

110. $\cos\left(\dfrac{\pi}{3} + x\right) + \cos\left(\dfrac{\pi}{3} - x\right) = \cos x$

In Exercises 111–114, rewrite the function using the power-reducing formulas. Then use a graphing utility to graph the function.

111. $f(x) = \sin^2 x$

112. $f(x) = \cos^2 x$

113. $f(x) = \cos^4 x$

114. $f(x) = \sin^3 x$

In Exercises 115–120, write the trigonometric expression as an algebraic expression.

115. $\sin(2 \arcsin x)$

116. $\cos(2 \arccos x)$

117. $\cos(2 \arcsin x)$

118. $\sin(2 \arccos x)$

119. $\cos(2 \arctan x)$

120. $\sin(2 \arctan x)$

∫ **In Exercises 121–124, (a) use a graphing utility to graph the function and approximate the maximum and minimum points of the graph in the interval $[0, 2\pi]$, and (b) solve the trigonometric equation and verify that the x-coordinates of the maximum and minimum points of f are among its solutions (calculus is required to find the trigonometric equation).**

Function	*Trigonometric Equation*
121. $f(x) = 4 \sin \dfrac{x}{2} + \cos x$	$2 \cos \dfrac{x}{2} - \sin x = 0$
122. $f(x) = \cos 2x - 2 \sin x$	$-2 \cos x (2 \sin x + 1) = 0$
123. $f(x) = 2 \cos \dfrac{x}{2} + \sin 2x$	$2 \cos 2x - \sin \dfrac{x}{2} = 0$
124. $f(x) = 2 \sin \dfrac{x}{2} -$	$10 \sin\left(2x - \dfrac{\pi}{4}\right) + \cos \dfrac{x}{2} = 0$
$\quad\quad 5 \cos\left(2x - \dfrac{\pi}{4}\right)$	

∫ **In Exercises 125 and 126, the graph of a function f is shown over the interval $[0, 2\pi]$. (a) Find the x-intercepts of the graph of f algebraically. Verify your solutions by using the *zero* or *root* feature of a graphing utility. (b) The x-coordinates of the extrema or turning points of the graph of f are solutions of the trigonometric equation (calculus is required to find the trigonometric equation). Find the solutions of the equation algebraically. Verify the solutions using the *maximum* and *minimum* features of a graphing utility.**

125. *Function:* $f(x) = \sin 2x - \sin x$

    Trigonometric equation: $2 \cos 2x - \cos x = 0$

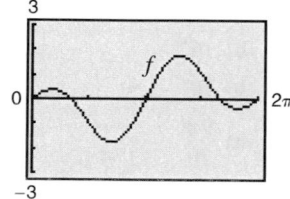

126. *Function:* $f(x) = \cos 2x + \sin x$

    Trigonometric equation: $-2 \sin 2x + \cos x = 0$

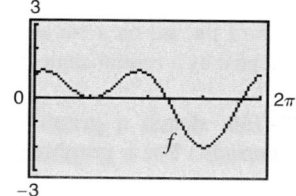

127. *Projectile Motion*   The range of a projectile fired at an angle θ with the horizontal and with an initial velocity of v_0 feet per second is given by

$$r = \frac{1}{32} v_0^2 \sin 2\theta$$

where r is measured in feet.

(a) Rewrite the expression for the range in terms of θ.

(b) Find the range r if the initial velocity of a projectile is 80 feet per second at an angle of $\theta = 42°$.

(c) Find the initial velocity required to fire a projectile 300 feet at an angle of $\theta = 40°$.

(d) For a given initial velocity, what angle of elevation yields a maximum range? Explain.

128. *Geometry*   The length of each of the two equal sides of an isosceles triangle is 10 meters (see figure). The angle between the two sides is θ.

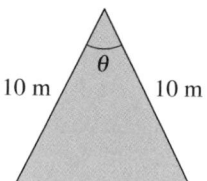

(a) Write the area of the triangle as a function of $\theta/2$.

(b) Write the area of the triangle as a function of θ and determine the value of θ such that the area is a maximum.

129. *Railroad Track*   When two railroad tracks merge, the overlapping portions of the tracks are in the shape of a circular arc (see figure). The radius of each arc r (in feet) and the angle θ are related by

$$\frac{x}{2} = 2r \sin^2 \frac{\theta}{2}.$$

Write a formula for x in terms of $\cos \theta$.

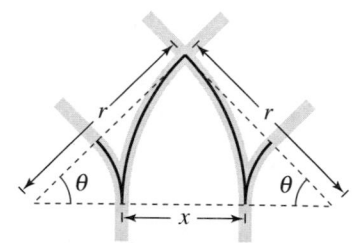

130. *Mach Number* The mach number M of an airplane is the ratio of its speed to the speed of sound. When an airplane travels faster than the speed of sound, the sound waves form a cone behind the airplane (see figure). The mach number is related to the apex angle θ of the cone by

$$\sin \frac{\theta}{2} = \frac{1}{M}.$$

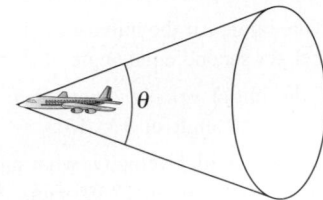

(a) Find the angle θ that corresponds to a mach number of 1.

(b) Find the angle θ that corresponds to a mach number of 4.5.

(c) The speed of sound is about 760 miles per hour. Determine the speed of an object having the mach numbers in parts (a) and (b).

(d) Rewrite the equation as a trigonometric function of θ.

Synthesis

True or False? **In Exercises 131 and 132, determine whether the statement is true or false. Justify your answer.**

131. $\sin \dfrac{x}{2} = -\sqrt{\dfrac{1 - \cos x}{2}}, \quad \pi \le x \le 2\pi$

132. The graph of $y = 4 - 8 \sin^2 x$ has a maximum at $(\pi, 4)$.

133. *Conjecture* Consider the function

$$f(x) = 2 \sin x \left(2 \cos^2 \frac{x}{2} - 1 \right).$$

(a) Use a graphing utility to graph the function.

(b) Make a conjecture about the function that is an identity with f.

(c) Verify your conjecture algebraically.

134. *Exploration* Consider the function

$$f(x) = \sin^4 x + \cos^4 x.$$

(a) Use the power-reducing formulas to write the function in terms of cosine to the first power.

(b) Determine another way of rewriting the function. Use a graphing utility to rule out incorrectly rewritten functions.

(c) Add a trigonometric term to the function so that it becomes a perfect square trinomial. Rewrite the function as a perfect square trinomial minus the term that you added. Use a graphing utility to rule out incorrectly rewritten functions.

(d) Rewrite the result of part (c) in terms of the sine of a double angle. Use a graphing utility to rule out incorrectly rewritten functions.

(e) When you rewrite a trigonometric expression, the result may not be the same as a friend's. Does this mean that one of you is wrong? Explain.

135. *Writing* Describe how you can use a double-angle formula or a half-angle formula to derive a formula for the area of an isosceles triangle. Use a labeled sketch to illustrate your derivation. Then write two examples that show how your formula can be used.

136. (a) Write a formula for $\cos 3\theta$.

(b) Write a formula for $\cos 4\theta$.

Skills Review

In Exercises 137–140, (a) plot the points, (b) find the distance between the points, and (c) find the midpoint of the line segment connecting the points.

137. $(5, 2), (-1, 4)$

138. $(-4, -3), (6, 10)$

139. $\left(0, \frac{1}{2}\right), \left(\frac{4}{3}, \frac{5}{2}\right)$

140. $\left(\frac{1}{3}, \frac{2}{3}\right), \left(-1, -\frac{3}{2}\right)$

In Exercises 141–144, find (if possible) the complement and supplement of each angle.

141. (a) $55°$ (b) $162°$

142. (a) $109°$ (b) $78°$

143. (a) $\dfrac{\pi}{18}$ (b) $\dfrac{9\pi}{20}$

144. (a) 0.95 (b) 2.76

145. Find the radian measure of the central angle of a circle with a radius of 15 inches that intercepts an arc of length 7 inches.

146. Find the length of the arc on a circle of radius 21 centimeters intercepted by a central angle of $35°$.

In Exercises 147–150, sketch a graph of the function. (Include two full periods.) Use a graphing utility to verify your graph.

147. $f(x) = \dfrac{3}{2} \cos 2x$

148. $f(x) = \dfrac{5}{2} \sin \dfrac{1}{2}x$

149. $f(x) = \dfrac{1}{2} \tan 2\pi x$

150. $f(x) = \dfrac{1}{4} \sec \dfrac{\pi x}{2}$

What Did You Learn?

Key Terms

sum and difference formulas, *p. 380*
reduction formulas, *p. 382*
double-angle formulas, *p. 387*

power-reducing formulas, *p. 389*
half-angle formulas, *p. 390*

product-to-sum formulas, *p. 391*
sum-to-product formulas, *p. 392*

Key Concepts

5.1 ■ Use the fundamental trigonometric identities

The fundamental trigonometric identities can be used to evaluate trigonometric functions, simplify trigonometric expressions, develop additional trigonometric identities, and solve trigonometric equations.

5.2 ■ Verify trigonometric identities

1. Work with one side of the equation at a time. It is often better to work with the more complicated side first.

2. Look for opportunities to factor an expression, add fractions, square a binomial, or create a monomial denominator.

3. Look for opportunities to use the fundamental identities. Note which functions are in the final expression you want. Sines and cosines pair up well, as do secants and tangents, and cosecants and cotangents.

4. Try converting all terms to sines and cosines.

5. Always try something.

5.3 ■ Solve trigonometric equations

1. Use algebraic techniques, such as collecting like terms, extracting square roots, factoring, and the Quadratic Formula to isolate the trigonometric function involved in the equation.

2. If there is no reasonable way to find the solution(s) of a trigonometric equation algebraically, use a graphing utility to approximate the solution(s).

5.4 ■ Use sum and difference formulas to evaluate trigonometric functions, verify identities, and solve trigonometric equations

$$\sin(u \pm v) = \sin u \cos v \pm \cos u \sin v$$

$$\cos(u \pm v) = \cos u \cos v \mp \sin u \sin v$$

$$\tan(u \pm v) = \frac{\tan u \pm \tan v}{1 \mp \tan u \tan v}$$

5.5 ■ Use multiple-angle formulas, power-reducing formulas, half-angle formulas, product-to-sum formulas, and sum-to-product formulas

Double-Angle Formulas:

$$\sin 2u = 2 \sin u \cos u \qquad \cos 2u = \cos^2 u - \sin^2 u$$

$$\tan 2u = \frac{2 \tan u}{1 - \tan^2 u} \qquad\qquad = 2 \cos^2 u - 1$$

$$= 1 - 2 \sin^2 u$$

Power-Reducing Formulas:

$$\sin^2 u = \frac{1 - \cos 2u}{2}$$

$$\cos^2 u = \frac{1 + \cos 2u}{2}$$

$$\tan^2 u = \frac{1 - \cos 2u}{1 + \cos 2u}$$

Half-Angle Formulas:

$$\sin \frac{u}{2} = \pm \sqrt{\frac{1 - \cos u}{2}} \qquad \cos \frac{u}{2} = \pm \sqrt{\frac{1 + \cos u}{2}}$$

$$\tan \frac{u}{2} = \frac{1 - \cos u}{\sin u} = \frac{\sin u}{1 + \cos u}$$

Product-to-Sum Formulas:

$$\sin u \sin v = \tfrac{1}{2}[\cos(u - v) - \cos(u + v)]$$

$$\cos u \cos v = \tfrac{1}{2}[\cos(u - v) + \cos(u + v)]$$

$$\sin u \cos v = \tfrac{1}{2}[\sin(u + v) + \sin(u - v)]$$

$$\cos u \sin v = \tfrac{1}{2}[\sin(u + v) - \sin(u - v)]$$

Sum-to-Product Formulas:

$$\sin u + \sin v = 2 \sin\left(\frac{u + v}{2}\right) \cos\left(\frac{u - v}{2}\right)$$

$$\sin u - \sin v = 2 \cos\left(\frac{u + v}{2}\right) \sin\left(\frac{u - v}{2}\right)$$

$$\cos u + \cos v = 2 \cos\left(\frac{u + v}{2}\right) \cos\left(\frac{u - v}{2}\right)$$

$$\cos u - \cos v = -2 \sin\left(\frac{u + v}{2}\right) \sin\left(\frac{u - v}{2}\right)$$

Review Exercises

See www.CalcChat.com for worked-out solutions to odd-numbered exercises.

5.1 In Exercises 1–10, name the trigonometric function that is equivalent to the expression.

1. $\dfrac{1}{\cos x}$ **2.** $\dfrac{1}{\sin x}$

3. $\dfrac{1}{\sec x}$ **4.** $\dfrac{1}{\tan x}$

5. $\sqrt{1 - \cos^2 x}$ **6.** $\sqrt{1 + \tan^2 x}$

7. $\csc\left(\dfrac{\pi}{2} - x\right)$ **8.** $\cot\left(\dfrac{\pi}{2} - x\right)$

9. $\sec(-x)$ **10.** $\tan(-x)$

In Exercises 11–14, use the given values to evaluate (if possible) the remaining trigonometric functions of the angle.

11. $\sin x = \dfrac{4}{5}, \quad \cos x = \dfrac{3}{5}$

12. $\tan \theta = \dfrac{2}{3}, \quad \sec \theta = \dfrac{\sqrt{13}}{3}$

13. $\sin\left(\dfrac{\pi}{2} - x\right) = \dfrac{1}{\sqrt{2}}, \quad \sin x = -\dfrac{1}{\sqrt{2}}$

14. $\csc\left(\dfrac{\pi}{2} - \theta\right) = 3, \quad \sin \theta = \dfrac{2\sqrt{2}}{3}$

In Exercises 15–22, use the fundamental identities to simplify the expression. Use the *table* feature of a graphing utility to check your result numerically.

15. $\dfrac{1}{\tan^2 x + 1}$ **16.** $\dfrac{\sec^2 x - 1}{\sec x - 1}$

17. $\dfrac{\sin^2 \alpha - \cos^2 \alpha}{\sin^2 \alpha - \sin \alpha \cos \alpha}$ **18.** $\dfrac{\sin^3 \beta + \cos^3 \beta}{\sin \beta + \cos \beta}$

19. $\tan^2 \theta(\csc^2 \theta - 1)$ **20.** $\csc^2 x(1 - \cos^2 x)$

21. $\tan\left(\dfrac{\pi}{2} - x\right)\sec x$ **22.** $\dfrac{\sin(-x)\cot x}{\sin\left(\dfrac{\pi}{2} - x\right)}$

∫ **23.** *Rate of Change* The rate of change of the function $f(x) = 2\sqrt{\sin x}$ is given by the expression $\sin^{-1/2} x \cos x$. Show that this expression can also be written as $\cot x\sqrt{\sin x}$.

∫ **24.** *Rate of Change* The rate of change of the function $f(x) = \csc x - \cot x$ is given by the expression $\csc^2 x - \csc x \cot x$. Show that this expression can also be written as $(1 - \cos x)/\sin^2 x$.

5.2 In Exercises 25–36, verify the identity.

25. $\cos x(\tan^2 x + 1) = \sec x$

26. $\sec^2 x \cot x - \cot x = \tan x$

27. $\sin^3 \theta + \sin \theta \cos^2 \theta = \sin \theta$

28. $\cot^2 x - \cos^2 x = \cot^2 x \cos^2 x$

29. $\sin^5 x \cos^2 x = (\cos^2 x - 2\cos^4 x + \cos^6 x)\sin x$

30. $\cos^3 x \sin^2 x = (\sin^2 x - \sin^4 x)\cos x$

31. $\sqrt{\dfrac{1 - \sin \theta}{1 + \sin \theta}} = \dfrac{1 - \sin \theta}{|\cos \theta|}$

32. $\sqrt{1 - \cos x} = \dfrac{|\sin x|}{\sqrt{1 + \cos x}}$

33. $\dfrac{\csc(-x)}{\sec(-x)} = -\cot x$

34. $\dfrac{1 + \sec(-x)}{\sin(-x) + \tan(-x)} = -\csc x$

35. $\csc^2\left(\dfrac{\pi}{2} - x\right) - 1 = \tan^2 x$

36. $\tan\left(\dfrac{\pi}{2} - x\right)\sec x = \csc x$

5.3 In Exercises 37–48, solve the equation.

37. $2\sin x - 1 = 0$ **38.** $\tan x + 1 = 0$

39. $\sin x = \sqrt{3} - \sin x$ **40.** $4\cos x = 1 + 2\cos x$

41. $3\sqrt{3}\tan x = 3$ **42.** $\dfrac{1}{2}\sec x - 1 = 0$

43. $3\csc^2 x = 4$ **44.** $4\tan^2 x - 1 = \tan^2 x$

45. $4\cos^2 x - 3 = 0$ **46.** $\sin x(\sin x + 1) = 0$

47. $\sin x - \tan x = 0$ **48.** $\csc x - 2\cot x = 0$

In Exercises 49–52, find all solutions of the equation in the interval $[0, 2\pi)$. Use a graphing utility to check your answers.

49. $2\cos^2 x - \cos x = 1$

50. $2\sin^2 x - 3\sin x = -1$

51. $\cos^2 x + \sin x = 1$

52. $\sin^2 x + 2\cos x = 2$

In Exercises 53–58, find all solutions of the multiple-angle equation in the interval $[0, 2\pi)$.

53. $2\sin 2x - \sqrt{2} = 0$ **54.** $\sqrt{3}\tan 3x = 0$

55. $\cos 4x(\cos x - 1) = 0$ **56.** $3\csc^2 5x = -4$

57. $\cos 4x - 7\cos 2x = 8$ **58.** $\sin 4x - \sin 2x = 0$

In Exercises 59–62, solve the equation.

59. $2 \sin 2x - 1 = 0$
60. $2 \cos 4x + \sqrt{3} = 0$
61. $2 \sin^2 3x - 1 = 0$
62. $4 \cos^2 2x - 3 = 0$

In Exercises 63–66, use the inverse functions where necessary to find all solutions of the equation in the interval $[0, 2\pi)$.

63. $\sin^2 x - 2 \sin x = 0$
64. $3 \cos^2 x + 5 \cos x = 0$
65. $\tan^2 \theta + 3 \tan \theta - 10 = 0$
66. $\sec^2 x + 6 \tan x + 4 = 0$

5.4 In Exercises 67–70, find the exact values of the sine, cosine, and tangent of the angle.

67. $285° = 315° - 30°$
68. $345° = 300° + 45°$
69. $\dfrac{31\pi}{12} = \dfrac{11\pi}{6} + \dfrac{3\pi}{4}$
70. $\dfrac{13\pi}{12} = \dfrac{11\pi}{6} - \dfrac{3\pi}{4}$

In Exercises 71–74, write the expression as the sine, cosine, or tangent of an angle.

71. $\sin 130° \cos 50° + \cos 130° \sin 50°$
72. $\cos 45° \cos 120° - \sin 45° \sin 120°$
73. $\dfrac{\tan 25° + \tan 10°}{1 - \tan 25° \tan 10°}$
74. $\dfrac{\tan 63° - \tan 118°}{1 + \tan 63° \tan 118°}$

In Exercises 75–80, find the exact value of the trigonometric function given that $\sin u = \frac{3}{5}$ and $\cos v = -\frac{7}{25}$. (Both u and v are in Quadrant II.)

75. $\sin(u + v)$
76. $\tan(u + v)$
77. $\tan(u - v)$
78. $\sin(u - v)$
79. $\cos(u + v)$
80. $\cos(u - v)$

In Exercises 81–84, find the value of the expression without using a calculator.

81. $\sin[\sin^{-1} 0 + \cos^{-1}(-1)]$
82. $\cos(\cos^{-1} 1 + \sin^{-1} 0)$
83. $\cos[\cos^{-1} 1 - \sin^{-1}(-1)]$
84. $\tan[\cos^{-1}(-1) - \cos^{-1} 1]$

In Exercises 85–90, verify the identity.

85. $\cos\left(x + \dfrac{\pi}{2}\right) = -\sin x$
86. $\sin\left(x - \dfrac{3\pi}{2}\right) = \cos x$
87. $\cot\left(\dfrac{\pi}{2} - x\right) = \tan x$
88. $\sin(\pi - x) = \sin x$
89. $\cos 3x = 4 \cos^3 x - 3 \cos x$
90. $\dfrac{\sin(\alpha + \beta)}{\cos \alpha \cos \beta} = \tan \alpha + \tan \beta$

In Exercises 91 and 92, find the solutions of the equation in the interval $[0, 2\pi)$.

91. $\sin\left(x + \dfrac{\pi}{2}\right) - \sin\left(x - \dfrac{\pi}{2}\right) = \sqrt{2}$
92. $\cos\left(x + \dfrac{\pi}{4}\right) - \cos\left(x - \dfrac{\pi}{4}\right) = 1$

5.5 In Exercises 93–96, find the exact values of $\sin 2u$, $\cos 2u$, and $\tan 2u$ using the double-angle formulas.

93. $\sin u = -\dfrac{5}{7}, \quad \pi < u < \dfrac{3\pi}{2}$
94. $\cos u = \dfrac{4}{5}, \quad \dfrac{3\pi}{2} < u < 2\pi$
95. $\tan u = -\dfrac{2}{9}, \quad \dfrac{\pi}{2} < u < \pi$
96. $\cos u = -\dfrac{2}{\sqrt{5}}, \quad \dfrac{\pi}{2} < u < \pi$

In Exercises 97–100, use double-angle formulas to verify the identity algebraically. Use a graphing utility to check your result graphically.

97. $6 \sin x \cos x = 3 \sin 2x$
98. $4 \sin x \cos x + 2 = 2 \sin 2x + 2$
99. $1 - 4 \sin^2 x \cos^2 x = \cos^2 2x$
100. $\sin 4x = 8 \cos^3 x \sin x - 4 \cos x \sin x$

101. *Projectile Motion* A baseball leaves the hand of the first baseman at an angle of θ with the horizontal and with an initial velocity of $v_0 = 80$ feet per second. The ball is caught by the second baseman 100 feet away. Find θ if the range r of a projectile is given by $r = \frac{1}{32} v_0^2 \sin 2\theta$.

102. *Projectile Motion* Use the equation in Exercise 101 to find θ when a golf ball is hit with an initial velocity of $v_0 = 50$ feet per second and lands 77 feet away.

In Exercises 103–106, use the power-reducing formulas to rewrite the expression in terms of the first power of the cosine.

103. $\sin^6 x$
104. $\cos^4 x \sin^4 x$
105. $\cos^4 2x$
106. $\sin^4 2x$

In Exercises 107–110, use the half-angle formulas to determine the exact values of the sine, cosine, and tangent of the angle.

107. $105°$
108. $112° 30'$
109. $\dfrac{7\pi}{8}$
110. $\dfrac{11\pi}{12}$

In Exercises 111–114, find the exact values of $\sin(u/2)$, $\cos(u/2)$, and $\tan(u/2)$ using the half-angle formulas.

111. $\sin u = \dfrac{3}{5}, \qquad 0 < u < \dfrac{\pi}{2}$

112. $\tan u = \dfrac{21}{20}, \qquad \pi < u < \dfrac{3\pi}{2}$

113. $\cos u = -\dfrac{2}{7}, \qquad \dfrac{\pi}{2} < u < \pi$

114. $\sec u = -6, \qquad \dfrac{\pi}{2} < u < \pi$

In Exercises 115 and 116, use the half-angle formulas to simplify the expression.

115. $-\sqrt{\dfrac{1 + \cos 8x}{2}}$

116. $\dfrac{\sin 10x}{1 + \cos 10x}$

Geometry In Exercises 117 and 118, a trough for feeding cattle is 4 meters long and its cross sections are isosceles triangles with two equal sides of $\frac{1}{2}$ meter (see figure). The angle between the equal sides is θ.

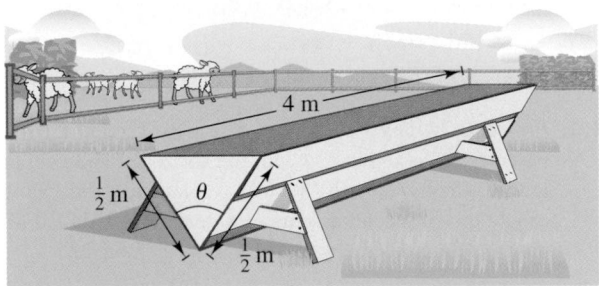

117. Write the trough's volume as a function of $\theta/2$.

118. Write the volume of the trough as a function of θ and determine the value of θ such that the volume is maximum.

In Exercises 119–122, use the product-to-sum formulas to write the product as a sum or difference.

119. $6 \sin \dfrac{\pi}{4} \cos \dfrac{\pi}{4}$

120. $4 \sin 15° \sin 45°$

121. $\sin 5\alpha \sin 4\alpha$

122. $\cos 6\theta \sin 8\theta$

In Exercises 123–126, use the sum-to-product formulas to write the sum or difference as a product.

123. $\cos 5\theta + \cos 4\theta$

124. $\sin 3\theta + \sin 2\theta$

125. $\sin\left(x + \dfrac{\pi}{4}\right) - \sin\left(x - \dfrac{\pi}{4}\right)$

126. $\cos\left(x + \dfrac{\pi}{6}\right) - \cos\left(x - \dfrac{\pi}{6}\right)$

Harmonic Motion In Exercises 127–130, a weight is attached to a spring suspended vertically from a ceiling. When a driving force is applied to the system, the weight moves vertically from its equilibrium position. This motion is described by the model

$$y = 1.5 \sin 8t - 0.5 \cos 8t$$

where y is the distance from equilibrium in feet and t is the time in seconds.

127. Write the model in the form

$$y = \sqrt{a^2 + b^2} \sin(Bt + C).$$

128. Use a graphing utility to graph the model.

129. Find the amplitude of the oscillations of the weight.

130. Find the frequency of the oscillations of the weight.

Synthesis

True or False? In Exercises 131–134, determine whether the statement is true or false. Justify your answer.

131. If $\dfrac{\pi}{2} < \theta < \pi$, then $\cos \dfrac{\theta}{2} < 0$.

132. $\sin(x + y) = \sin x + \sin y$

133. $4 \sin(-x) \cos(-x) = -2 \sin 2x$

134. $4 \sin 45° \cos 15° = 1 + \sqrt{3}$

135. List the reciprocal identities, quotient identities, and Pythagorean identities from memory.

136. Is $\cos \theta = \sqrt{1 - \sin^2 \theta}$ an identity? Explain.

In Exercises 137 and 138, use the graphs of y_1 and y_2 to determine how to change y_2 to a new function y_3 such that $y_1 = y_3$.

137. $y_1 = \sec^2\left(\dfrac{\pi}{2} - x\right)$

$y_2 = \cot^2 x$

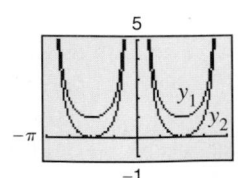

138. $y_1 = \dfrac{\cos 3x}{\cos x}$

$y_2 = (2 \sin x)^2$

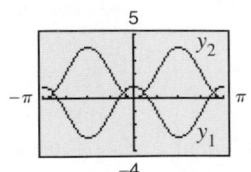

5 Chapter Test

See www.CalcChat.com for worked-out solutions to odd-numbered exercises.

Take this test as you would take a test in class. After you are finished, check your work against the answers given in the back of the book.

1. If $\tan \theta = \frac{3}{2}$ and $\cos \theta < 0$, use the fundamental identities to evaluate the other five trigonometric functions of θ.

2. Use the fundamental identities to simplify $\csc^2 \beta (1 - \cos^2 \beta)$.

3. Factor and simplify $\dfrac{\sec^4 x - \tan^4 x}{\sec^2 x + \tan^2 x}$.

4. Add and simplify $\dfrac{\cos \theta}{\sin \theta} + \dfrac{\sin \theta}{\cos \theta}$.

5. Determine the values of θ, $0 \le \theta < 2\pi$, for which $\tan \theta = -\sqrt{\sec^2 \theta - 1}$ is true.

6. Use a graphing utility to graph the functions $y_1 = \sin x + \cos x \cot x$ and $y_2 = \csc x$. Make a conjecture about y_1 and y_2. Verify your result algebraically.

In Exercises 7–12, verify the identity.

7. $\sin \theta \sec \theta = \tan \theta$

8. $\sec^2 x \tan^2 x + \sec^2 x = \sec^4 x$

9. $\dfrac{\csc \alpha + \sec \alpha}{\sin \alpha + \cos \alpha} = \cot \alpha + \tan \alpha$

10. $\cos\left(x + \dfrac{\pi}{2}\right) = -\sin x$

11. $\sin(n\pi + \theta) = (-1)^n \sin \theta$, n is an integer.

12. $(\sin x + \cos x)^2 = 1 + \sin 2x$

13. Find the exact value of $\tan 105°$.

14. Rewrite $\sin^4 x \tan^2 x$ in terms of the first power of the cosine.

15. Use a half-angle formula to simplify the expression $\dfrac{\sin 4\theta}{1 + \cos 4\theta}$.

16. Write $4 \cos 2\theta \sin 4\theta$ as a sum or difference.

17. Write $\sin 3\theta - \sin 4\theta$ as a product.

In Exercises 18–21, find all solutions of the equation in the interval $[0, 2\pi)$.

18. $\tan^2 x + \tan x = 0$

19. $\sin 2\alpha - \cos \alpha = 0$

20. $4 \cos^2 x - 3 = 0$

21. $\csc^2 x - \csc x - 2 = 0$

22. Use a graphing utility to approximate the solutions of the equation $3 \cos x - x = 0$ accurate to three decimal places.

23. Use the figure to find the exact values of $\sin 2u$, $\cos 2u$, and $\tan 2u$.

24. The *index of refraction* n of a transparent material is the ratio of the speed of light in a vacuum to the speed of light in the material. For the triangular glass prism in the figure, $n = 1.5$ and $\alpha = 60°$. Find the angle θ for the glass prism if

$$n = \frac{\sin(\theta/2 + \alpha/2)}{\sin(\theta/2)}.$$

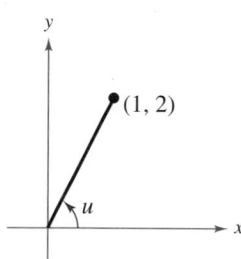

Figure for 23

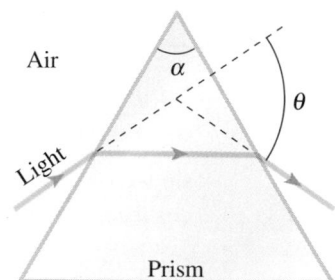

Figure for 24

Proofs in Mathematics

Sum and Difference Formulas   (p. 380)

$\sin(u + v) = \sin u \cos v + \cos u \sin v$

$\sin(u - v) = \sin u \cos v - \cos u \sin v$

$\cos(u + v) = \cos u \cos v - \sin u \sin v$

$\cos(u - v) = \cos u \cos v + \sin u \sin v$

$\tan(u + v) = \dfrac{\tan u + \tan v}{1 - \tan u \tan v}$

$\tan(u - v) = \dfrac{\tan u - \tan v}{1 + \tan u \tan v}$

Proof

You can use the figures at the right for the proofs of the formulas for $\cos(u \pm v)$. In the top figure, let A be the point $(1, 0)$ and then use u and v to locate the points $B = (x_1, y_1)$, $C = (x_2, y_2)$, and $D = (x_3, y_3)$ on the unit circle. So, $x_i^2 + y_i^2 = 1$ for $i = 1, 2,$ and 3. For convenience, assume that $0 < v < u < 2\pi$. In the bottom figure, note that arcs AC and BD have the same length. So, line segments AC and BD are also equal in length, which implies that

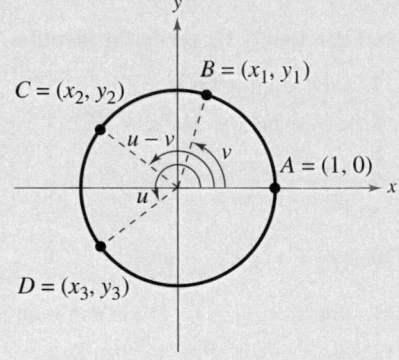

$$\sqrt{(x_2 - 1)^2 + (y_2 - 0)^2} = \sqrt{(x_3 - x_1)^2 + (y_3 - y_1)^2}$$

$$x_2^2 - 2x_2 + 1 + y_2^2 = x_3^2 - 2x_1x_3 + x_1^2 + y_3^2 - 2y_1y_3 + y_1^2$$

$$(x_2^2 + y_2^2) + 1 - 2x_2 = (x_3^2 + y_3^2) + (x_1^2 + y_1^2) - 2x_1x_3 - 2y_1y_3$$

$$1 + 1 - 2x_2 = 1 + 1 - 2x_1x_3 - 2y_1y_3$$

$$x_2 = x_3x_1 + y_3y_1.$$

Finally, by substituting the values $x_2 = \cos(u - v)$, $x_3 = \cos u$, $x_1 = \cos v$, $y_3 = \sin u$, and $y_1 = \sin v$, you obtain $\cos(u - v) = \cos u \cos v + \sin u \sin v$. The formula for $\cos(u + v)$ can be established by considering $u + v = u - (-v)$ and using the formula just derived to obtain

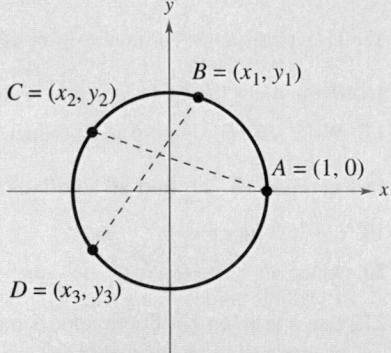

$$\cos(u + v) = \cos[u - (-v)] = \cos u \cos(-v) + \sin u \sin(-v)$$

$$= \cos u \cos v - \sin u \sin v.$$

You can use the sum and difference formulas for sine and cosine to prove the formulas for $\tan(u \pm v)$.

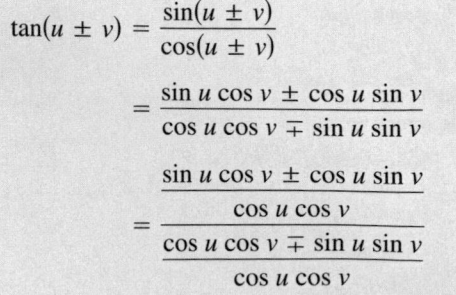

$\tan(u \pm v) = \dfrac{\sin(u \pm v)}{\cos(u \pm v)}$ Quotient identity

$= \dfrac{\sin u \cos v \pm \cos u \sin v}{\cos u \cos v \mp \sin u \sin v}$ Sum and difference formulas

$= \dfrac{\dfrac{\sin u \cos v \pm \cos u \sin v}{\cos u \cos v}}{\dfrac{\cos u \cos v \mp \sin u \sin v}{\cos u \cos v}}$ Divide numerator and denominator by $\cos u \cos v$.

$$= \dfrac{\dfrac{\sin u \cos v}{\cos u \cos v} \pm \dfrac{\cos u \sin v}{\cos u \cos v}}{\dfrac{\cos u \cos v}{\cos u \cos v} \mp \dfrac{\sin u \sin v}{\cos u \cos v}}$$

Write as separate fractions.

$$= \dfrac{\dfrac{\sin u}{\cos u} \pm \dfrac{\sin v}{\cos v}}{1 \mp \dfrac{\sin u}{\cos u} \cdot \dfrac{\sin v}{\cos v}}$$

Product of fractions

$$= \dfrac{\tan u \pm \tan v}{1 \mp \tan u \tan v}$$

Quotient identity

Double-Angle Formulas   (p. 387)

$$\sin 2u = 2 \sin u \cos u \qquad \cos 2u = \cos^2 u - \sin^2 u$$

$$\tan 2u = \dfrac{2 \tan u}{1 - \tan^2 u} \qquad\qquad = 2 \cos^2 u - 1$$

$$= 1 - 2 \sin^2 u$$

Proof

To prove all three formulas, let $v = u$ in the corresponding sum formulas.

$$\sin 2u = \sin(u + u) = \sin u \cos u + \cos u \sin u = 2 \sin u \cos u$$

$$\cos 2u = \cos(u + u) = \cos u \cos u - \sin u \sin u = \cos^2 u - \sin^2 u$$

$$\tan 2u = \tan(u + u) = \dfrac{\tan u + \tan u}{1 - \tan u \tan u} = \dfrac{2 \tan u}{1 - \tan^2 u}$$

Power-Reducing Formulas   (p. 389)

$$\sin^2 u = \dfrac{1 - \cos 2u}{2} \qquad \cos^2 u = \dfrac{1 + \cos 2u}{2} \qquad \tan^2 u = \dfrac{1 - \cos 2u}{1 + \cos 2u}$$

Proof

To prove the first formula, solve for $\sin^2 u$ in the double-angle formula $\cos 2u = 1 - 2 \sin^2 u$, as follows.

$$\cos 2u = 1 - 2 \sin^2 u \qquad\qquad \text{Write double-angle formula.}$$

$$2 \sin^2 u = 1 - \cos 2u \qquad\qquad \text{Subtract } \cos 2u \text{ from and add } 2 \sin^2 u \text{ to each side.}$$

$$\sin^2 u = \dfrac{1 - \cos 2u}{2} \qquad\qquad \text{Divide each side by 2.}$$

Trigonometry and Astronomy

Trigonometry was used by early astronomers to calculate measurements in the universe. Trigonometry was used to calculate the circumference of Earth and the distance from Earth to the moon. Another major accomplishment in astronomy using trigonometry was computing distances to stars.

In a similar way you can prove the second formula, by solving for $\cos^2 u$ in the double-angle formula

$$\cos 2u = 2\cos^2 u - 1.$$

To prove the third formula, use a quotient identity, as follows.

$$\tan^2 u = \frac{\sin^2 u}{\cos^2 u}$$

$$= \frac{\dfrac{1 - \cos 2u}{2}}{\dfrac{1 + \cos 2u}{2}}$$

$$= \frac{1 - \cos 2u}{1 + \cos 2u}$$

Sum-to-Product Formulas   (p. 392)

$$\sin u + \sin v = 2 \sin\left(\frac{u+v}{2}\right)\cos\left(\frac{u-v}{2}\right)$$

$$\sin u - \sin v = 2 \cos\left(\frac{u+v}{2}\right)\sin\left(\frac{u-v}{2}\right)$$

$$\cos u + \cos v = 2 \cos\left(\frac{u+v}{2}\right)\cos\left(\frac{u-v}{2}\right)$$

$$\cos u - \cos v = -2 \sin\left(\frac{u+v}{2}\right)\sin\left(\frac{u-v}{2}\right)$$

Proof

To prove the first formula, let $x = u + v$ and $y = u - v$. Then substitute $u = (x + y)/2$ and $v = (x - y)/2$ in the product-to-sum formula.

$$\sin u \cos v = \frac{1}{2}[\sin(u + v) + \sin(u - v)]$$

$$\sin\left(\frac{x+y}{2}\right)\cos\left(\frac{x-y}{2}\right) = \frac{1}{2}(\sin x + \sin y)$$

$$2 \sin\left(\frac{x+y}{2}\right)\cos\left(\frac{x-y}{2}\right) = \sin x + \sin y$$

The other sum-to-product formulas can be proved in a similar manner.

Chapter 6

6.1 Law of Sines
6.2 Law of Cosines
6.3 Vectors in the Plane
6.4 Vectors and Dot Products
6.5 Trigonometric Form of a Complex Number

Selected Applications

Triangles and vectors have many real-life applications. The applications listed below represent a small sample of the applications in this chapter.

- Flight Path, Exercise 27, page 415
- Bridge Design, Exercise 28, page 415
- Surveying, Exercise 38, page 422
- Landau Building, Exercise 45, page 422
- Velocity, Exercises 83 and 84, page 435
- Navigation, Exercises 89 and 90, page 436
- Revenue, Exercises 59 and 60, page 446
- Work, Exercise 63, page 447

Additional Topics in Trigonometry

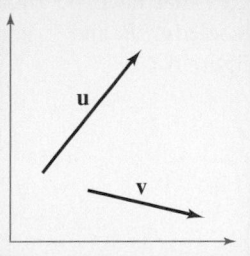 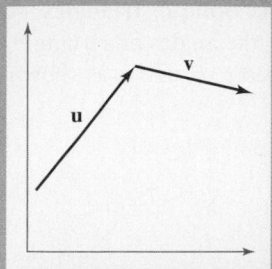 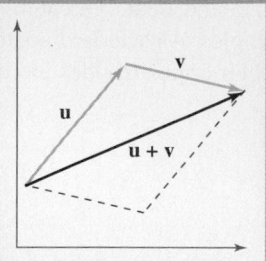

Vectors indicate quantities that involve both magnitude and direction. In Chapter 6, you will study the operations of vectors in the plane and you will learn how to represent vector operations geometrically. You will also learn to solve oblique triangles and write complex numbers in trigonometric form.

Lester Lefkowitz/Getty Images

Vectors can be used to find the airspeed and direction of an airplane that will allow the airplane to maintain its groundspeed and direction.

6.1 Law of Sines

Introduction

In Chapter 4 you looked at techniques for solving right triangles. In this section and the next, you will solve **oblique triangles**—triangles that have no right angles. As standard notation, the angles of a triangle are labeled A, B, and C, and their opposite sides are labeled a, b, and c, as shown in Figure 6.1.

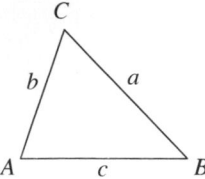

Figure 6.1

To solve an oblique triangle, you need to know the measure of at least one side and the measures of any two other parts of the triangle—two sides, two angles, or one angle and one side. This breaks down into the following four cases.

1. Two angles and any side (AAS or ASA)
2. Two sides and an angle opposite one of them (SSA)
3. Three sides (SSS)
4. Two sides and their included angle (SAS)

The first two cases can be solved using the **Law of Sines**, whereas the last two cases require the Law of Cosines (see Section 6.2).

Law of Sines   (See the proof on page 468.)

If ABC is a triangle with sides a, b, and c, then

$$\frac{a}{\sin A} = \frac{b}{\sin B} = \frac{c}{\sin C}.$$

Oblique Triangles

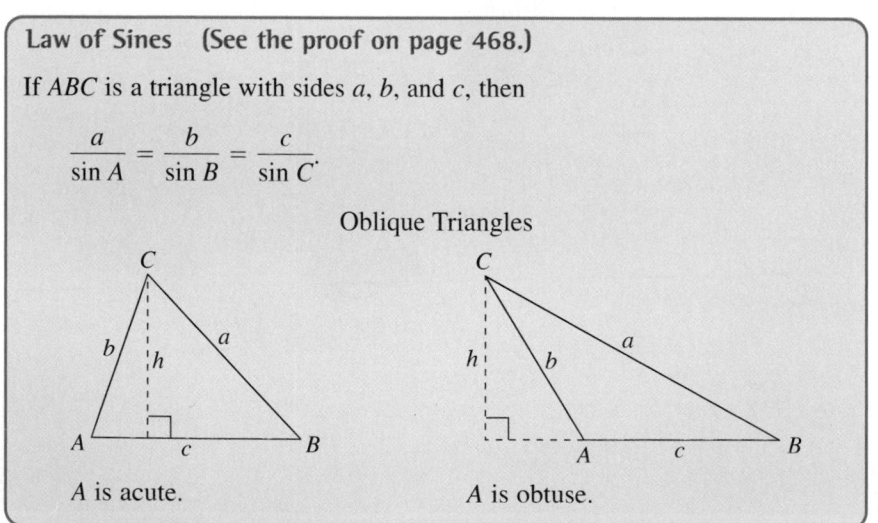

A is acute. A is obtuse.

The Law of Sines can also be written in the reciprocal form

$$\frac{\sin A}{a} = \frac{\sin B}{b} = \frac{\sin C}{c}.$$

SOH CAH TOA

What you should learn

- Use the Law of Sines to solve oblique triangles (AAS or ASA).
- Use the Law of Sines to solve oblique triangles (SSA).
- Find areas of oblique triangles.
- Use the Law of Sines to model and solve real-life problems.

Why you should learn it

You can use the Law of Sines to solve real-life problems involving oblique triangles. For instance, Exercise 32 on page 415 shows how the Law of Sines can be used to help determine the distance from a boat to the shoreline.

© Owen Franken/Corbis

STUDY TIP

Notice in Figure 6.1 that angle A is the included angle between sides b and c, angle B is the included angle between sides a and c, and angle C is the included angle between sides a and b.

Example 1   Given Two Angles and One Side—AAS

For the triangle in Figure 6.2, $C = 102.3°$, $B = 28.7°$, and $b = 27.4$ feet. Find the remaining angle and sides.

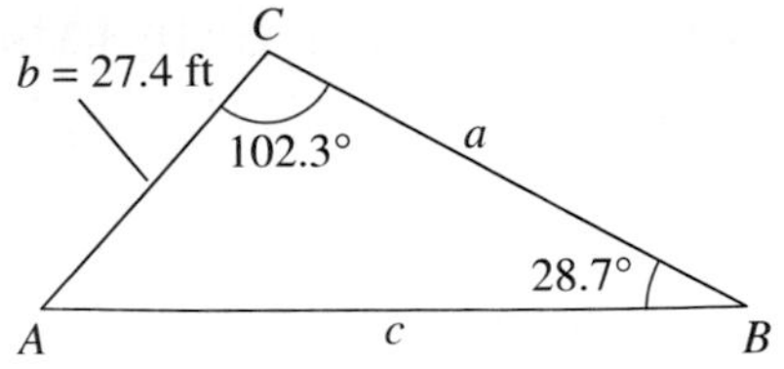

Solution

The third angle of the triangle is

$$A = 180° - B - C$$

$$= 180° - 28.7° - 102.3°$$

$$= 49.0°.$$

By the Law of Sines, you have

$$\frac{a}{\sin A} = \frac{b}{\sin B} = \frac{c}{\sin C}.$$

Using $b = 27.4$ produces

$$a = \frac{b}{\sin B}(\sin A) = \frac{27.4}{\sin 28.7°}(\sin 49.0°) \approx 43.06 \text{ feet}$$

and

$$c = \frac{b}{\sin B}(\sin C) = \frac{27.4}{\sin 28.7°}(\sin 102.3°) \approx 55.75 \text{ feet}.$$

✓CHECKPOINT   Now try Exercise 3.

C
b = 27.4 ft
102.3°
a
28.7°
A   *c*   *B*

Figure 6.2

Example 2   Given Two Angles and One Side—ASA

A pole tilts *toward* the sun at an 8° angle from the vertical, and it casts a 22-foot shadow. The angle of elevation from the tip of the shadow to the top of the pole is 43°. How tall is the pole?

Solution

In Figure 6.3, $A = 43°$ and $B = 90° + 8° = 98°$. So, the third angle is

$$C = 180° - A - B = 180° - 43° - 98° = 39°.$$

By the Law of Sines, you have

$$\frac{a}{\sin A} = \frac{c}{\sin C}.$$

Because $c = 22$ feet, the length of the pole is

$$a = \frac{c}{\sin C}(\sin A) = \frac{22}{\sin 39°}(\sin 43°) \approx 23.84 \text{ feet}.$$

✓CHECKPOINT   Now try Exercise 25.

For practice, try reworking Example 2 for a pole that tilts *away from* the sun under the same conditions.

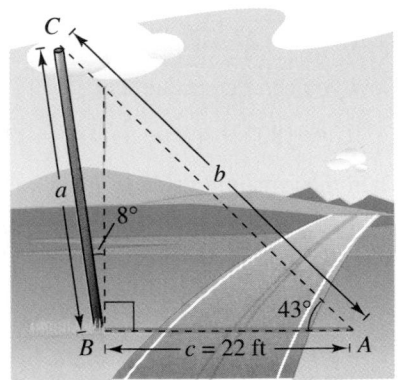

Figure 6.3

The Ambiguous Case (SSA)

In Examples 1 and 2 you saw that two angles and one side determine a unique triangle. However, if two sides and one opposite angle are given, three possible situations can occur: (1) no such triangle exists, (2) one such triangle exists, or (3) two distinct triangles satisfy the conditions.

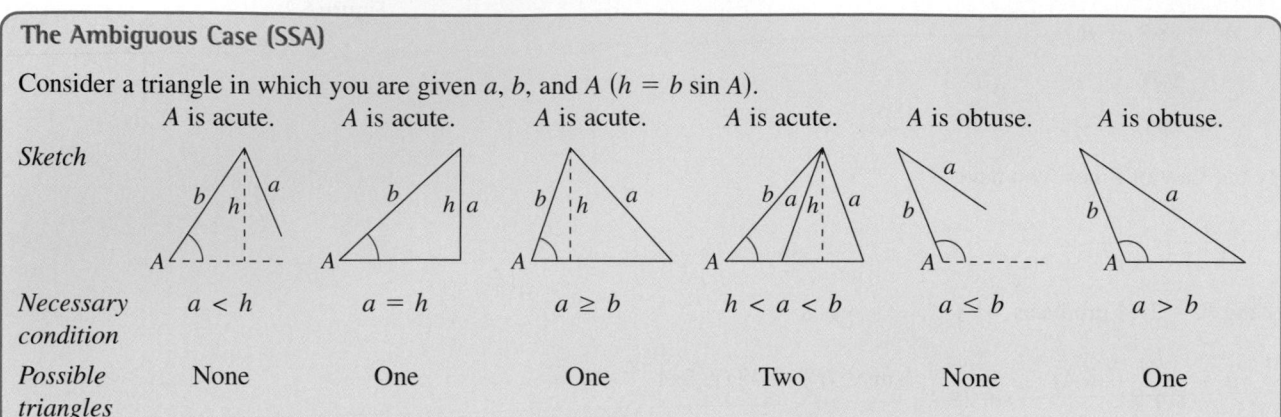

The Ambiguous Case (SSA)

Consider a triangle in which you are given a, b, and A ($h = b \sin A$).

	A is acute.	A is acute.	A is acute.	A is acute.	A is obtuse.	A is obtuse.
Sketch						
Necessary condition	$a < h$	$a = h$	$a \geq b$	$h < a < b$	$a \leq b$	$a > b$
Possible triangles	None	One	One	Two	None	One

Example 3   Single-Solution Case—SSA

For the triangle in Figure 6.4, $a = 22$ inches, $b = 12$ inches, and $A = 42°$. Find the remaining side and angles.

Solution

By the Law of Sines, you have

$$\frac{\sin B}{b} = \frac{\sin A}{a} \qquad \text{Reciprocal form}$$

$$\sin B = b\left(\frac{\sin A}{a}\right) \qquad \text{Multiply each side by } b.$$

$$\sin B = 12\left(\frac{\sin 42°}{22}\right) \qquad \text{Substitute for } A, a, \text{ and } b.$$

$$B \approx 21.41°. \qquad B \text{ is acute.}$$

Now you can determine that

$$C \approx 180° - 42° - 21.41° = 116.59°.$$

Then the remaining side is given by

$$\frac{c}{\sin C} = \frac{a}{\sin A}$$

$$c = \frac{a}{\sin A}(\sin C) = \frac{22}{\sin 42°}(\sin 116.59°) \approx 29.40 \text{ inches.}$$

✓**CHECKPOINT**   Now try Exercise 13.

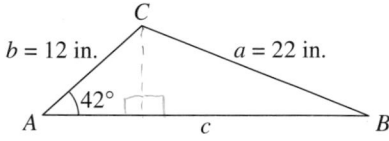

Figure 6.4   One solution: a > b

Example 1   Given Two Angles and One Side—AAS

For the triangle in Figure 6.2, $C = 102.3°$, $B = 28.7°$, and $b = 27.4$ feet. Find the remaining angle and sides.

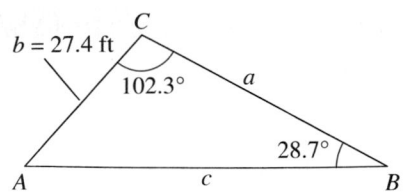

Figure 6.2

Solution

The third angle of the triangle is

$$A = 180° - B - C$$
$$= 180° - 28.7° - 102.3°$$
$$= 49.0°.$$

By the Law of Sines, you have

$$\frac{a}{\sin A} = \frac{b}{\sin B} = \frac{c}{\sin C}.$$

Using $b = 27.4$ produces

$$a = \frac{b}{\sin B}(\sin A) = \frac{27.4}{\sin 28.7°}(\sin 49.0°) \approx 43.06 \text{ feet}$$

and

$$c = \frac{b}{\sin B}(\sin C) = \frac{27.4}{\sin 28.7°}(\sin 102.3°) \approx 55.75 \text{ feet}.$$

 *CHECKPOINT*   Now try Exercise 3.

Example 2   Given Two Angles and One Side—ASA

A pole tilts *toward* the sun at an 8° angle from the vertical, and it casts a 22-foot shadow. The angle of elevation from the tip of the shadow to the top of the pole is 43°. How tall is the pole?

Solution

In Figure 6.3, $A = 43°$ and $B = 90° + 8° = 98°$. So, the third angle is

$$C = 180° - A - B = 180° - 43° - 98° = 39°.$$

By the Law of Sines, you have

$$\frac{a}{\sin A} = \frac{c}{\sin C}.$$

Because $c = 22$ feet, the length of the pole is

$$a = \frac{c}{\sin C}(\sin A) = \frac{22}{\sin 39°}(\sin 43°) \approx 23.84 \text{ feet}.$$

 *CHECKPOINT*   Now try Exercise 25.

For practice, try reworking Example 2 for a pole that tilts *away from* the sun under the same conditions.

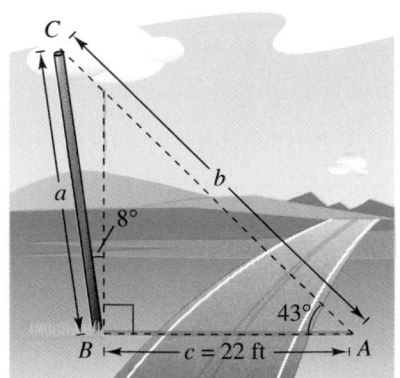

Figure 6.3

STUDY TIP

When you are solving triangles, a careful sketch is useful as a quick test for the feasibility of an answer. Remember that the longest side lies opposite the largest angle, and the shortest side lies opposite the smallest angle.

The Ambiguous Case (SSA)

In Examples 1 and 2 you saw that two angles and one side determine a unique triangle. However, if two sides and one opposite angle are given, three possible situations can occur: (1) no such triangle exists, (2) one such triangle exists, or (3) two distinct triangles satisfy the conditions.

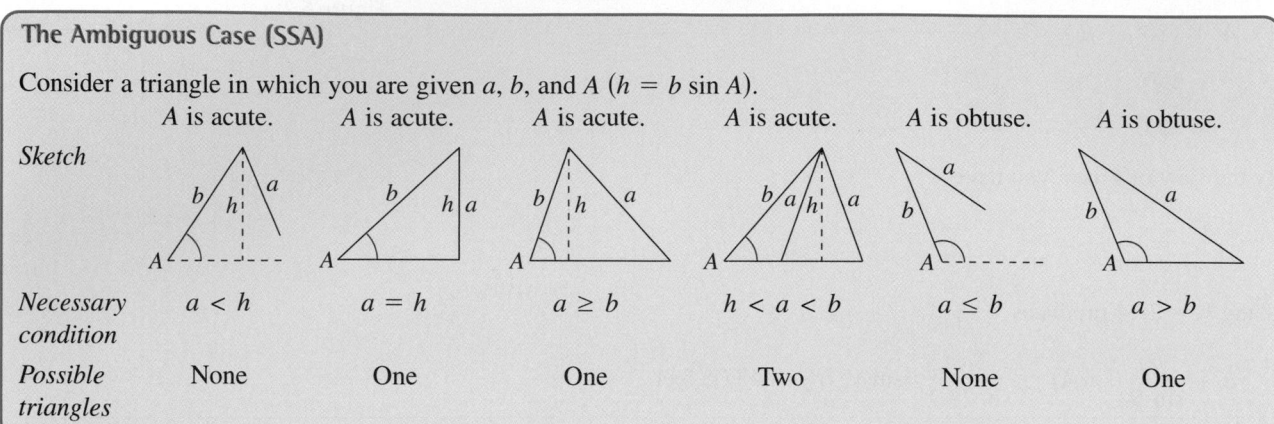

The Ambiguous Case (SSA)

Consider a triangle in which you are given a, b, and A ($h = b \sin A$).

	A is acute.	A is acute.	A is acute.	A is acute.	A is obtuse.	A is obtuse.
Necessary condition	$a < h$	$a = h$	$a \geq b$	$h < a < b$	$a \leq b$	$a > b$
Possible triangles	None	One	One	Two	None	One

Example 3   Single-Solution Case—SSA

For the triangle in Figure 6.4, $a = 22$ inches, $b = 12$ inches, and $A = 42°$. Find the remaining side and angles.

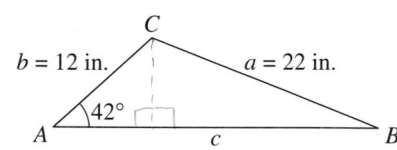

Figure 6.4   One solution: $a > b$

Solution

By the Law of Sines, you have

$$\frac{\sin B}{b} = \frac{\sin A}{a} \qquad \text{Reciprocal form}$$

$$\sin B = b\left(\frac{\sin A}{a}\right) \qquad \text{Multiply each side by } b.$$

$$\sin B = 12\left(\frac{\sin 42°}{22}\right) \qquad \text{Substitute for } A, a, \text{ and } b.$$

$$B \approx 21.41°. \qquad B \text{ is acute.}$$

Now you can determine that

$$C \approx 180° - 42° - 21.41° = 116.59°.$$

Then the remaining side is given by

$$\frac{c}{\sin C} = \frac{a}{\sin A}$$

$$c = \frac{a}{\sin A}(\sin C) = \frac{22}{\sin 42°}(\sin 116.59°) \approx 29.40 \text{ inches.}$$

✓**CHECKPOINT**   Now try Exercise 13.

Example 4   No-Solution Case—SSA

Show that there is no triangle for which $a = 15$, $b = 25$, and $A = 85°$.

Solution

Begin by making the sketch shown in Figure 6.5. From this figure it appears that no triangle is formed. You can verify this by using the Law of Sines.

$$\frac{\sin B}{b} = \frac{\sin A}{a}$$ Reciprocal form

$$\sin B = b\left(\frac{\sin A}{a}\right)$$ Multiply each side by b.

$$\sin B = 25\left(\frac{\sin 85°}{15}\right) \approx 1.6603 > 1$$

This contradicts the fact that $|\sin B| \le 1$. So, no triangle can be formed having sides $a = 15$ and $b = 25$ and an angle of $A = 85°$.

 Now try Exercise 15.

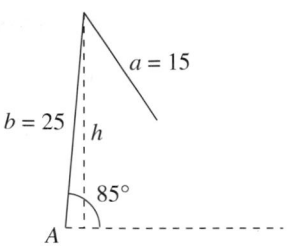

Figure 6.5   No solution: $a < h$

Example 5   Two-Solution Case—SSA

Find two triangles for which $a = 12$ meters, $b = 31$ meters, and $A = 20.5°$.

Solution

Because $h = b \sin A = 31(\sin 20.5°) \approx 10.86$ meters, you can conclude that there are two possible triangles (because $h < a < b$). By the Law of Sines, you have

$$\frac{\sin B}{b} = \frac{\sin A}{a}$$ Reciprocal form

$$\sin B = b\left(\frac{\sin A}{a}\right) = 31\left(\frac{\sin 20.5°}{12}\right) \approx 0.9047.$$

There are two angles $B_1 \approx 64.8°$ and $B_2 \approx 180° - 64.8° = 115.2°$ between $0°$ and $180°$ whose sine is 0.9047. For $B_1 \approx 64.8°$, you obtain

$$C \approx 180° - 20.5° - 64.8° = 94.7°$$

$$c = \frac{a}{\sin A}(\sin C) = \frac{12}{\sin 20.5°}(\sin 94.7°) \approx 34.15 \text{ meters.}$$

For $B_2 \approx 115.2°$, you obtain

$$C \approx 180° - 20.5° - 115.2° = 44.3°$$

$$c = \frac{a}{\sin A}(\sin C) = \frac{12}{\sin 20.5°}(\sin 44.3°) \approx 23.93 \text{ meters.}$$

The resulting triangles are shown in Figure 6.6.

 Now try Exercise 17.

STUDY TIP

In Example 5, the height h of the triangle can be found using the formula

$$\sin A = \frac{h}{b}$$

or

$$h = b \sin A.$$

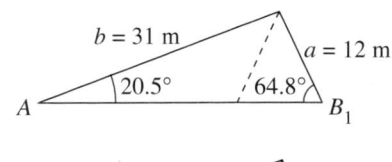

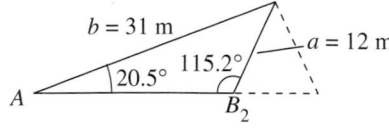

Figure 6.6   Two solutions: $h < a < b$

Area of an Oblique Triangle

The procedure used to prove the Law of Sines leads to a simple formula for the area of an oblique triangle. Referring to Figure 6.7, note that each triangle has a height of $h = b \sin A$. To see this when A is obtuse, substitute the reference angle $180° - A$ for A. Now the height of the triangle is given by

$$h = b \sin(180° - A).$$

Using the difference formula for sine, the height is given by

$$h = b(\sin 180° \cos A - \cos 180° \sin A) \qquad \sin(u - v) = \sin u \cos v - \cos u \sin v$$

$$= b[0 \cdot \cos A - (-1) \cdot \sin A]$$

$$= b \sin A.$$

Consequently, the area of each triangle is given by

$$\text{Area} = \frac{1}{2}(\text{base})(\text{height}) = \frac{1}{2}(c)(b \sin A) = \frac{1}{2}bc \sin A.$$

By similar arguments, you can develop the formulas

$$\text{Area} = \frac{1}{2}ab \sin C = \frac{1}{2}ac \sin B.$$

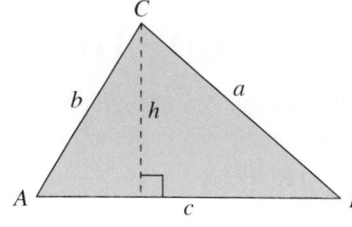

A is acute.

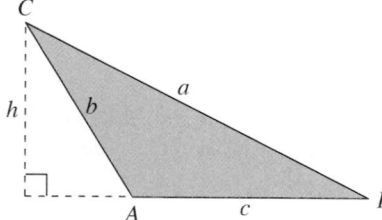

A is obtuse.

Figure 6.7

Area of an Oblique Triangle

The area of any triangle is one-half the product of the lengths of two sides times the sine of their included angle. That is,

$$\text{Area} = \frac{1}{2}bc \sin A = \frac{1}{2}ab \sin C = \frac{1}{2}ac \sin B.$$

Note that if angle A is 90°, the formula gives the area of a right triangle as

$$\text{Area} = \frac{1}{2}bc = \frac{1}{2}(\text{base})(\text{height}).$$

Similar results are obtained for angles C and B equal to 90°.

Example 6   Finding the Area of an Oblique Triangle

Find the area of a triangular lot having two sides of lengths 90 meters and 52 meters and an included angle of 102°.

Solution

Consider $a = 90$ meters, $b = 52$ meters, and $C = 102°$, as shown in Figure 6.8. Then the area of the triangle is

$$\text{Area} = \frac{1}{2}ab \sin C = \frac{1}{2}(90)(52)(\sin 102°) \approx 2288.87 \text{ square meters.}$$

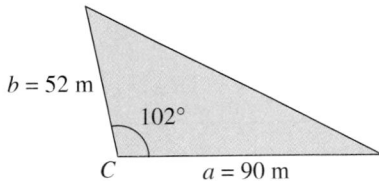

Figure 6.8

✓*CHECKPOINT*   Now try Exercise 19.

Example 7   An Application of the Law of Sines

The course for a boat race starts at point A and proceeds in the direction S 52° W to point B, then in the direction S 40° E to point C, and finally back to A, as shown in Figure 6.9. Point C lies 8 kilometers directly south of point A. Approximate the total distance of the race course.

Solution

Because lines BD and AC are parallel, it follows that $\angle BCA \cong \angle DBC$. Consequently, triangle ABC has the measures shown in Figure 6.10. For angle B, you have $B = 180° - 52° - 40° = 88°$. Using the Law of Sines

$$\frac{a}{\sin 52°} = \frac{b}{\sin 88°} = \frac{c}{\sin 40°}$$

you can let $b = 8$ and obtain

$$a = \frac{8}{\sin 88°}(\sin 52°) \approx 6.31$$

and

$$c = \frac{8}{\sin 88°}(\sin 40°) \approx 5.15.$$

The total length of the course is approximately

$$\text{Length} \approx 8 + 6.31 + 5.15 = 19.46 \text{ kilometers.}$$

✓*CHECKPOINT*   Now try Exercise 27.

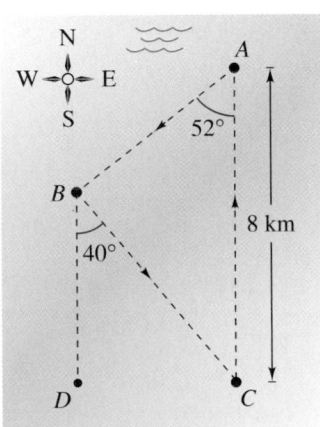

Figure 6.9

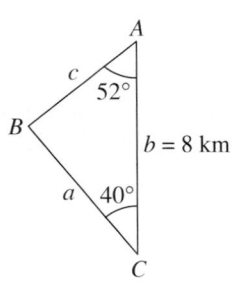

Figure 6.10

6.1 Exercises

See www.CalcChat.com for worked-out solutions to odd-numbered exercises.

Vocabulary Check

Fill in the blanks.

1. An _____ triangle is one that has no right angles.

2. Law of Sines: $\dfrac{a}{\sin A} = $ _____ $= \dfrac{c}{\sin C}$

3. The Law of Sines can be used to solve a triangle for cases

   (a) _____ angle(s) and _____ side(s), which can be denoted _____ or _____ ,

   (b) _____ side(s) and _____ angle(s), which can be denoted _____ .

4. To find the area of any triangle, use one of the following three formulas: Area = _____ , _____ , or _____ .

In Exercises 1–18, use the Law of Sines to solve the triangle. If two solutions exist, find both.

1.
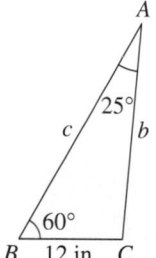

2.
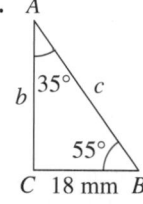

3.
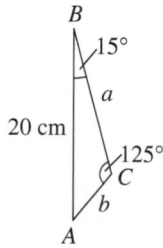

4.
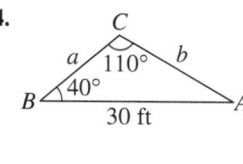

5.
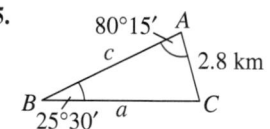

6.
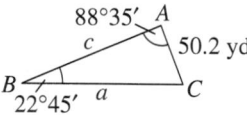

7. $A = 36°$, $a = 8$, $b = 5$

8. $A = 60°$, $a = 9$, $c = 10$

9. $A = 102.4°$, $C = 16.7°$, $a = 21.6$

10. $A = 24.3°$, $C = 54.6°$, $c = 2.68$

11. $A = 110° \, 15'$, $a = 48$, $b = 16$

12. $B = 2° \, 45'$, $b = 6.2$, $c = 5.8$

13. $A = 110°$, $a = 125$, $b = 100$

14. $A = 110°$, $a = 125$, $b = 200$

15. $A = 76°$, $a = 18$, $b = 20$

16. $A = 76°$, $a = 34$, $b = 21$

17. $A = 58°$, $a = 11.4$, $b = 12.8$

18. $A = 58°$, $a = 4.5$, $b = 12.8$

In Exercises 19–24, find the area of the triangle having the indicated angle and sides.

19. $C = 110°$, $a = 6$, $b = 10$

20. $B = 130°$, $a = 92$, $c = 30$

21. $A = 38° \, 45'$, $b = 67$, $c = 85$

22. $A = 5° \, 15'$, $b = 4.5$, $c = 22$

23. $B = 75° \, 15'$, $a = 103$, $c = 58$

24. $C = 85° \, 45'$, $a = 16$, $b = 20$

25. **Height** A flagpole at a right angle to the horizontal is located on a slope that makes an angle of 14° with the horizontal. The flagpole casts a 16-meter shadow up the slope when the angle of elevation from the tip of the shadow to the sun is 20°.

    (a) Draw a triangle that represents the problem. Show the known quantities on the triangle and use a variable to indicate the height of the flagpole.

    (b) Write an equation involving the unknown quantity.

    (c) Find the height of the flagpole.

26. **Height** You are standing 40 meters from the base of a tree that is leaning 8° from the vertical away from you. The angle of elevation from your feet to the top of the tree is 20° 50′.

    (a) Draw a triangle that represents the problem. Show the known quantities on the triangle and use a variable to indicate the height of the tree.

    (b) Write an equation involving the unknown height of the tree.

    (c) Find the height of the tree.

27. Flight Path A plane flies 500 kilometers with a bearing of 316° (clockwise from north) from Naples to Elgin (see figure). The plane then flies 720 kilometers from Elgin to Canton. Find the bearing of the flight from Elgin to Canton.

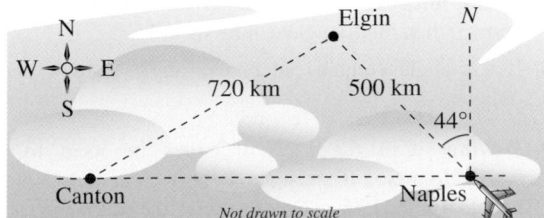

Not drawn to scale

28. Bridge Design A bridge is to be built across a small lake from a gazebo to a dock (see figure). The bearing from the gazebo to the dock is S 41° W. From a tree 100 meters from the gazebo, the bearings to the gazebo and the dock are S 74° E and S 28° E, respectively. Find the distance from the gazebo to the dock.

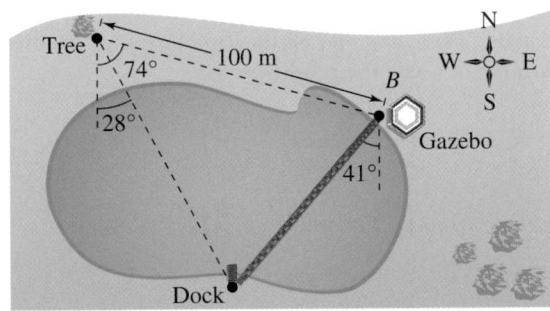

29. Railroad Track Design The circular arc of a railroad curve has a chord of length 3000 feet and a central angle of 40°.

(a) Draw a diagram that visually represents the problem. Show the known quantities on the diagram and use the variables r and s to represent the radius of the arc and the length of the arc, respectively.

(b) Find the radius r of the circular arc.

(c) Find the length s of the circular arc.

30. Glide Path A pilot has just started on the glide path for landing at an airport with a runway of length 9000 feet. The angles of depression from the plane to the ends of the runway are 17.5° and 18.8°.

(a) Draw a diagram that visually represents the problem.

(b) Find the air distance the plane must travel until touching down on the near end of the runway.

(c) Find the ground distance the plane must travel until touching down.

(d) Find the altitude of the plane when the pilot begins the descent.

31. Locating a Fire The bearing from the Pine Knob fire tower to the Colt Station fire tower is N 65° E, and the two towers are 30 kilometers apart. A fire spotted by rangers in each tower has a bearing of N 80° E from Pine Knob and S 70° E from Colt Station. Find the distance of the fire from each tower.

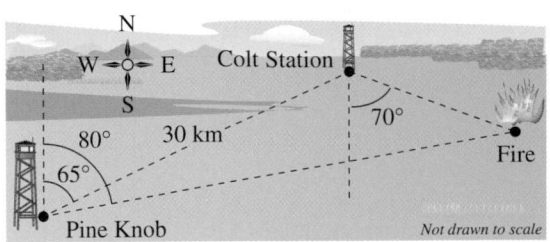

Not drawn to scale

32. Distance A boat is sailing due east parallel to the shoreline at a speed of 10 miles per hour. At a given time the bearing to a lighthouse is S 70° E, and 15 minutes later the bearing is S 63° E (see figure). The lighthouse is located at the shoreline. Find the distance from the boat to the shoreline.

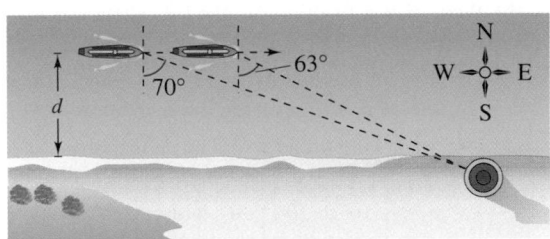

33. Angle of Elevation A 10-meter telephone pole casts a 17-meter shadow directly down a slope when the angle of elevation of the sun is 42° (see figure). Find θ, the angle of elevation of the ground.

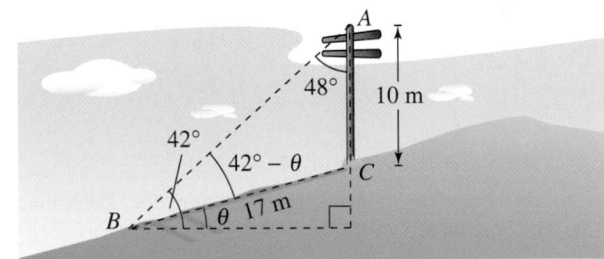

34. Distance The angles of elevation θ and ϕ to an airplane are being continuously monitored at two observation points A and B, respectively, which are 2 miles apart, and the airplane is east of both points in the same vertical plane.

(a) Draw a diagram that illustrates the problem.

(b) Write an equation giving the distance d between the plane and point B in terms of θ and ϕ.

35. *Shadow Length* The Leaning Tower of Pisa in Italy leans because it was built on unstable soil—a mixture of clay, sand, and water. The tower is approximately 58.36 meters tall from its foundation (see figure). The top of the tower leans about 5.45 meters off center.

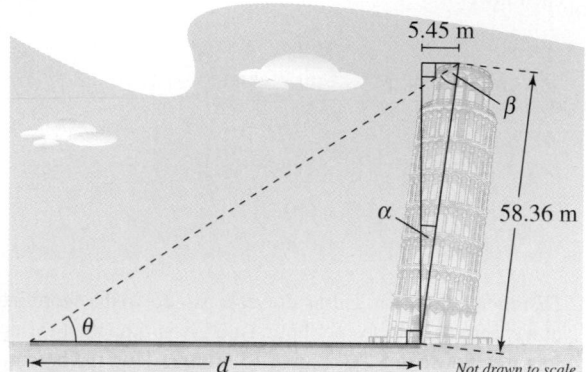

5.45 m

β

α

58.36 m

θ

d

Not drawn to scale

(a) Find the angle of lean α of the tower.

(b) Write β as a function of d and θ, where θ is the angle of elevation to the sun.

(c) Use the Law of Sines to write an equation for the length d of the shadow cast by the tower in terms of θ.

(d) Use a graphing utility to complete the table.

θ	10°	20°	30°	40°	50°	60°
d						

36. *Graphical and Numerical Analysis* In the figure, α and β are positive angles.

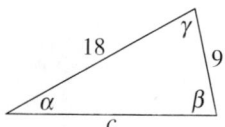

18

γ

9

α

c

β

(a) Write α as a function of β.

(b) Use a graphing utility to graph the function. Determine its domain and range.

(c) Use the result of part (b) to write c as a function of β.

(d) Use a graphing utility to graph the function in part (c). Determine its domain and range.

(e) Use a graphing utility to complete the table. What can you conclude?

β	0.4	0.8	1.2	1.6	2.0	2.4	2.8
α							
c							

Synthesis

True or False? **In Exercises 37 and 38, determine whether the statement is true or false. Justify your answer.**

37. If any three sides or angles of an oblique triangle are known, then the triangle can be solved.

38. If a triangle contains an obtuse angle, then it must be oblique.

39. *Writing* Can the Law of Sines be used to solve a right triangle? If so, write a short paragraph explaining how to use the Law of Sines to solve the following triangle. Is there an easier way to solve the triangle? Explain.

$$B = 50°, C = 90°, a = 10$$

40. *Think About It* Given $A = 36°$ and $a = 5$, find values of b such that the triangle has (a) one solution, (b) two solutions, and (c) no solution.

Mollweide's Formula **In Exercises 41 and 42, solve the triangle. Then use one of the two forms of Mollweide's Formula to verify the results.**

$$(a + b) \sin\left(\frac{C}{2}\right) = c \cos\left(\frac{A - B}{2}\right) \qquad \text{Form 1}$$

$$(a - b) \cos\left(\frac{C}{2}\right) = c \sin\left(\frac{A - B}{2}\right) \qquad \text{Form 2}$$

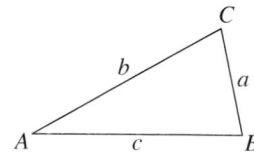

C

b

a

A

c

B

41. Solve the triangle for $A = 45°$, $B = 52°$, and $a = 16$. Then use form 1 of Mollweide's Formula to verify your solution.

42. Solve the triangle for $A = 42°$, $B = 60°$, and $a = 24$. Then use form 2 of Mollweide's Formula to verify your solution.

Skills Review

In Exercises 43 and 44, use the given values to find (if possible) the values of the remaining four trigonometric functions of θ.

43. $\cos \theta = \frac{5}{13}$, $\sin \theta = -\frac{12}{13}$

44. $\tan \theta = \frac{2}{9}$, $\csc \theta = -\frac{\sqrt{85}}{2}$

In Exercises 45–48, write the product as a sum or difference.

45. $6 \sin 8\theta \cos 3\theta$

46. $2 \cos 2\theta \cos 5\theta$

47. $3 \cos \frac{\pi}{6} \sin \frac{5\pi}{3}$

48. $\frac{5}{2} \sin \frac{3\pi}{4} \sin \frac{5\pi}{6}$

6.2   Law of Cosines

Introduction

Two cases remain in the list of conditions needed to solve an oblique triangle—SSS and SAS. To use the Law of Sines, you must know at least one side and its opposite angle. If you are given three sides (SSS), or two sides and their included angle (SAS), none of the ratios in the Law of Sines would be complete. In such cases you can use the **Law of Cosines.**

Law of Cosines   (See the proof on page 469.)

| *Standard Form* | *Alternative Form* |

$$a^2 = b^2 + c^2 - 2bc \cos A$$ $$\cos A = \frac{b^2 + c^2 - a^2}{2bc}$$

$$b^2 = a^2 + c^2 - 2ac \cos B$$ $$\cos B = \frac{a^2 + c^2 - b^2}{2ac}$$

$$c^2 = a^2 + b^2 - 2ab \cos C$$ $$\cos C = \frac{a^2 + b^2 - c^2}{2ab}$$

Example 1   Three Sides of a Triangle—SSS

Find the three angles of the triangle shown in Figure 6.11.

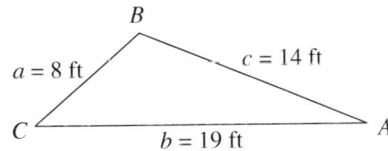

Figure 6.11

Solution

It is a good idea first to find the angle opposite the longest side—side b in this case. Using the alternative form of the Law of Cosines, you find that

$$\cos B = \frac{a^2 + c^2 - b^2}{2ac} = \frac{8^2 + 14^2 - 19^2}{2(8)(14)} \approx -0.45089.$$

Because $\cos B$ is negative, you know that B is an *obtuse* angle given by $B \approx 116.80°$. At this point it is simpler to use the Law of Sines to determine A.

$$\sin A = a\left(\frac{\sin B}{b}\right) \approx 8\left(\frac{\sin 116.80°}{19}\right) \approx 0.37583$$

Because B is obtuse, A must be acute, because a triangle can have at most one obtuse angle. So, $A \approx 22.08°$ and $C \approx 180° - 22.08° - 116.80° = 41.12°$.

✓CHECKPOINT   Now try Exercise 1.

What you should learn

■ Use the Law of Cosines to solve oblique triangles (SSS or SAS).

■ Use the Law of Cosines to model and solve real-life problems.

■ Use Heron's Area Formula to find areas of triangles.

Why you should learn it

You can use the Law of Cosines to solve real-life problems involving oblique triangles. For instance, Exercise 42 on page 422 shows you how the Law of Cosines can be used to determine the length of the guy wires that anchor a tower.

© Gregor Schuster/zefa/Corbis

Do you see why it was wise to find the largest angle *first* in Example 1? Knowing the cosine of an angle, you can determine whether the angle is acute or obtuse. That is,

$$\cos \theta > 0 \quad \text{for} \quad 0° < \theta < 90° \qquad \text{Acute}$$

$$\cos \theta < 0 \quad \text{for} \quad 90° < \theta < 180°. \qquad \text{Obtuse}$$

So, in Example 1, once you found that angle B was obtuse, you knew that angles A and C were both acute. Furthermore, if the largest angle is acute, the remaining two angles are also acute.

Example 2 Two Sides and the Included Angle—SAS

Find the remaining angles and side of the triangle shown in Figure 6.12.

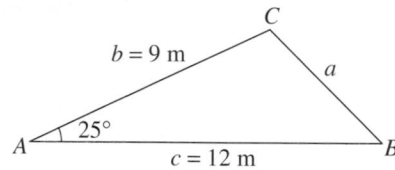

Figure 6.12

Solution

Use the Law of Cosines to find the unknown side a in the figure.

$$a^2 = b^2 + c^2 - 2bc \cos A$$

$$a^2 = 9^2 + 12^2 - 2(9)(12) \cos 25° \approx 29.2375$$

$$a \approx 5.4072$$

Because $a \approx 5.4072$ meters, you now know the ratio $\sin A/a$ and you can use the reciprocal form of the Law of Sines to solve for B.

$$\frac{\sin B}{b} = \frac{\sin A}{a}$$

$$\sin B = b\left(\frac{\sin A}{a}\right) = 9\left(\frac{\sin 25°}{5.4072}\right) \approx 0.7034$$

There are two angles between $0°$ and $180°$ whose sine is 0.7034, $B_1 \approx 44.7°$ and $B_2 \approx 180° - 44.7° = 135.3°$. For $B_1 \approx 44.7°$,

$$C_1 \approx 180° - 25° - 44.7° = 110.3°.$$

For $B_2 \approx 135.3°$,

$$C_2 \approx 180° - 25° - 135.3° = 19.7°.$$

Because side c is the longest side of the triangle, C must be the largest angle of the triangle. So, $B \approx 44.7°$ and $C \approx 110.3°$.

✓**CHECKPOINT**   Now try Exercise 5.

Exploration

What familiar formula do you obtain when you use the third form of the Law of Cosines

$$c^2 = a^2 + b^2 - 2ab \cos C$$

and you let $C = 90°$? What is the relationship between the Law of Cosines and this formula?

STUDY TIP

When solving an oblique triangle given three sides, you use the alternative form of the Law of Cosines to solve for an angle. When solving an oblique triangle given two sides and their included angle, you use the standard form of the Law of Cosines to solve for an unknown side.

Exploration

In Example 2, suppose $A = 115°$. After solving for a, which angle would you solve for next, B or C? Are there two possible solutions for that angle? If so, how can you determine which angle is the correct solution?

Applications

Example 3   An Application of the Law of Cosines

The pitcher's mound on a women's softball field is 43 feet from home plate and the distance between the bases is 60 feet, as shown in Figure 6.13. (The pitcher's mound is *not* halfway between home plate and second base.) How far is the pitcher's mound from first base?

Solution

In triangle *HPF*, $H = 45°$ (line *HP* bisects the right angle at *H*), $f = 43$, and $p = 60$. Using the Law of Cosines for this SAS case, you have

$$h^2 = f^2 + p^2 - 2fp \cos H$$

$$= 43^2 + 60^2 - 2(43)(60) \cos 45°$$

$$\approx 1800.33.$$

So, the approximate distance from the pitcher's mound to first base is

$$h \approx \sqrt{1800.33} \approx 42.43 \text{ feet.}$$

✓**CHECKPOINT**   Now try Exercise 37.

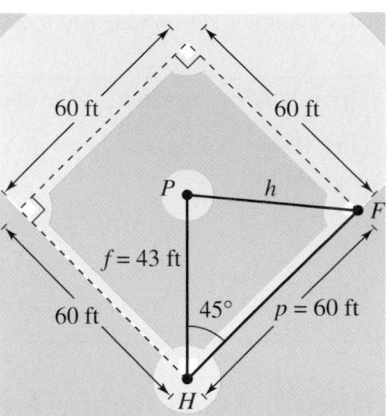

Figure 6.13

Example 4   An Application of the Law of Cosines

A ship travels 60 miles due east, then adjusts its course northward, as shown in Figure 6.14. After traveling 80 miles in the new direction, the ship is 139 miles from its point of departure. Describe the bearing from point *B* to point *C*.

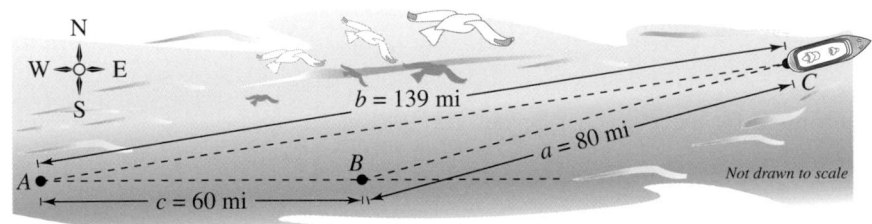

Figure 6.14

Solution

You have $a = 80$, $b = 139$, and $c = 60$; so, using the alternative form of the Law of Cosines, you have

$$\cos B = \frac{a^2 + c^2 - b^2}{2ac} = \frac{80^2 + 60^2 - 139^2}{2(80)(60)} \approx -0.97094.$$

So, $B \approx \arccos(-0.97094) \approx 166.15°$. Therefore, the bearing measured from due north from point *B* to point *C* is $166.15° - 90° = 76.15°$, or N 76.15° E.

✓**CHECKPOINT**   Now try Exercise 39.

Heron's Area Formula

The Law of Cosines can be used to establish the following formula for the area of a triangle. This formula is called **Heron's Area Formula** after the Greek mathematician Heron (ca. 100 B.C.).

Heron's Area Formula   (See the proof on page 470.)

Given any triangle with sides of lengths a, b, and c, the area of the triangle is given by

$$\text{Area} = \sqrt{s(s-a)(s-b)(s-c)}$$

where $s = \dfrac{a+b+c}{2}$.

Example 5   Using Heron's Area Formula

Find the area of a triangle having sides of lengths $a = 43$ meters, $b = 53$ meters, and $c = 72$ meters.

Solution

Because $s = (a + b + c)/2 = 168/2 = 84$, Heron's Area Formula yields

$$\text{Area} = \sqrt{s(s-a)(s-b)(s-c)}$$
$$= \sqrt{84(84-43)(84-53)(84-72)}$$
$$= \sqrt{84(41)(31)(12)}$$
$$\approx 1131.89 \text{ square meters.}$$

✓**CHECKPOINT**   Now try Exercise 45.

You have now studied three different formulas for the area of a triangle.

Formulas for Area of a Triangle

1. **Standard Formula:** $\text{Area} = \frac{1}{2}bh$

2. **Oblique Triangle:** $\text{Area} = \frac{1}{2}bc \sin A = \frac{1}{2}ab \sin C = \frac{1}{2}ac \sin B$

3. **Heron's Area Formula:** $\text{Area} = \sqrt{s(s-a)(s-b)(s-c)}$

Exploration

Can the formulas above be used to find the area of any type of triangle? Explain the advantages and disadvantages of using one formula over another.

6.2 Exercises

See www.CalcChat.com for worked-out solutions to odd-numbered exercises.

Vocabulary Check

Fill in the blanks.

1. The standard form of the Law of Cosines for $\cos C = \dfrac{a^2 + b^2 - c^2}{2ab}$ is _____ .

2. _____ Formula is established by using the Law of Cosines.

3. Three different formulas for the area of a triangle are given by Area = _____ ,
   Area = $\frac{1}{2}bc \sin A = \frac{1}{2}ab \sin C = \frac{1}{2}ac \sin B$, and Area = _____ .

In Exercises 1–20, use the Law of Cosines to solve the triangle.

1.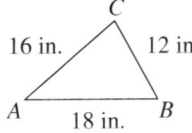

2.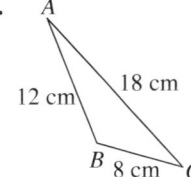

3.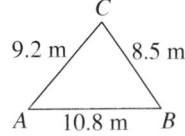

4.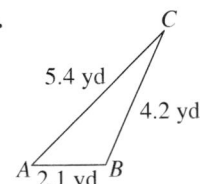

5.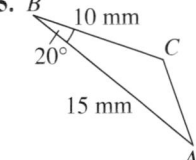

6.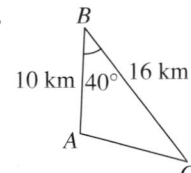

7.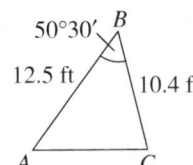

8.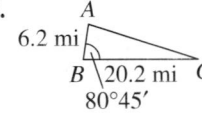

9. $a = 6$, $b = 8$, $c = 12$

10. $a = 9$, $b = 3$, $c = 11$

11. $A = 50°$, $b = 15$, $c = 30$

12. $C = 108°$, $a = 10$, $b = 7$

13. $a = 9$, $b = 12$, $c = 15$

14. $a = 45$, $b = 30$, $c = 72$

15. $a = 75.4$, $b = 48$, $c = 48$

16. $a = 1.42$, $b = 0.75$, $c = 1.25$

17. $B = 8° \, 15'$, $a = 26$, $c = 18$

18. $B = 10° \, 35'$, $a = 40$, $c = 30$

19. $B = 75° \, 20'$, $a = 6.2$, $c = 9.5$

20. $C = 15° \, 15'$, $a = 6.25$, $b = 2.15$

In Exercises 21–26, complete the table by solving the parallelogram shown in the figure. (The lengths of the diagonals are given by c and d.)

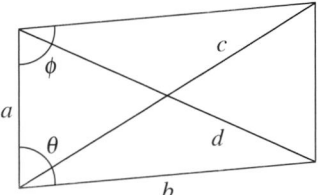

	a	b	c	d	θ	ϕ
21.	4	8			30°	
22.	25	35				120°
23.	10	14	20			
24.	40	60		80		
25.	15		25	20		
26.		25	50	35		

In Exercises 27–36, use Heron's Area Formula to find the area of the triangle.

27.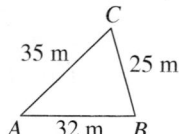

28.

29. $a = 5$, $b = 8$, $c = 10$

30. $a = 14$, $b = 17$, $c = 7$

31.

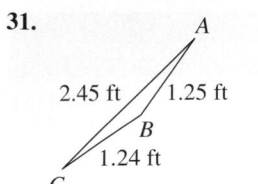

32.

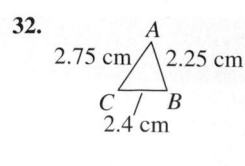

33. $a = 3.5$, $b = 10.2$, $c = 9$

34. $a = 75.4$, $b = 52$, $c = 52$

35. $a = 10.59$, $b = 6.65$, $c = 12.31$

36. $a = 4.45$, $b = 1.85$, $c = 3.00$

37. *Navigation* A plane flies 810 miles from Franklin to Centerville with a bearing of 75° (clockwise from north). Then it flies 648 miles from Centerville to Rosemont with a bearing of 32°. Draw a diagram that visually represents the problem, and find the straight-line distance and bearing from Rosemont to Franklin.

38. *Surveying* To approximate the length of a marsh, a surveyor walks 380 meters from point A to point B. Then the surveyor turns 80° and walks 240 meters to point C (see figure). Approximate the length AC of the marsh.

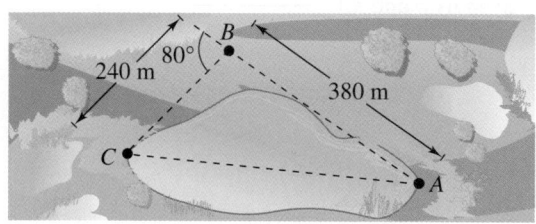

39. *Navigation* A boat race runs along a triangular course marked by buoys A, B, and C. The race starts with the boats headed west for 3600 meters. The other two sides of the course lie to the north of the first side, and their lengths are 1500 meters and 2800 meters. Draw a diagram that visually represents the problem, and find the bearings for the last two legs of the race.

40. *Streetlight Design* Determine the angle θ in the design of the streetlight shown in the figure.

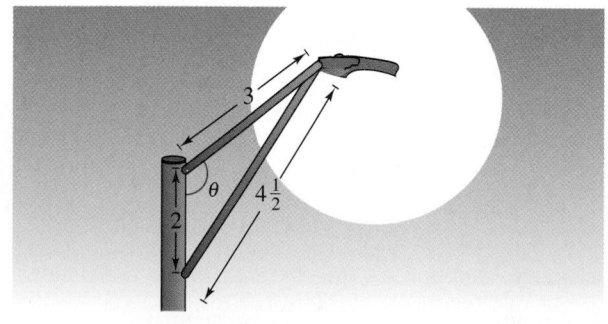

41. *Distance* Two ships leave a port at 9 A.M. One travels at a bearing of N 53° W at 12 miles per hour, and the other travels at a bearing of S 67° W at 16 miles per hour. Approximate how far apart the ships are at noon that day.

42. *Length* A 100-foot vertical tower is to be erected on the side of a hill that makes a 6° angle with the horizontal (see figure). Find the length of each of the two guy wires that will be anchored 75 feet uphill and downhill from the base of the tower.

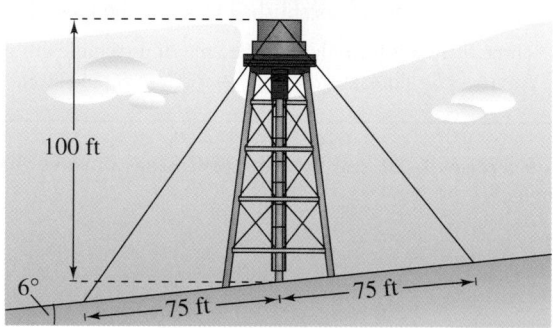

43. *Trusses* Q is the midpoint of the line segment $\overline{PR}$ in the truss rafter shown in the figure. What are the lengths of the line segments $\overline{PQ}$, $\overline{QS}$, and $\overline{RS}$?

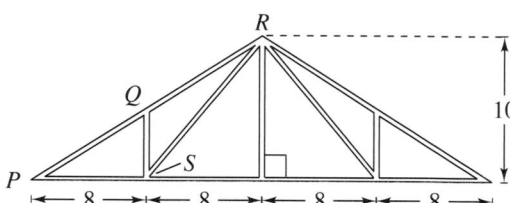

44. *Awning Design* A retractable awning above a patio lowers at an angle of 50° from the exterior wall at a height of 10 feet above the ground (see figure). No direct sunlight is to enter the door when the angle of elevation of the sun is greater than 70°. What is the length x of the awning?

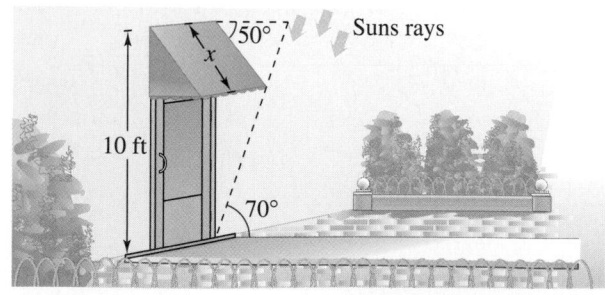

45. *Landau Building* The Landau Building in Cambridge, Massachusetts has a triangular-shaped base. The lengths of the sides of the triangular base are 145 feet, 257 feet, and 290 feet. Find the area of the base of the building.

46. *Geometry* A parking lot has the shape of a parallelogram (see figure). The lengths of two adjacent sides are 70 meters and 100 meters. The angle between the two sides is 70°. What is the area of the parking lot?

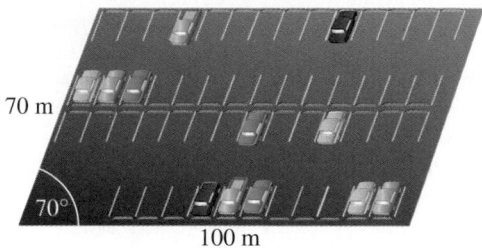

70 m

70°

100 m

47. *Engine Design* An engine has a seven-inch connecting rod fastened to a crank (see figure).

(a) Use the Law of Cosines to write an equation giving the relationship between x and θ.

(b) Write x as a function of θ. (Select the sign that yields positive values of x.)

(c) Use a graphing utility to graph the function in part (b).

(d) Use the graph in part (c) to determine the total distance the piston moves in one cycle.

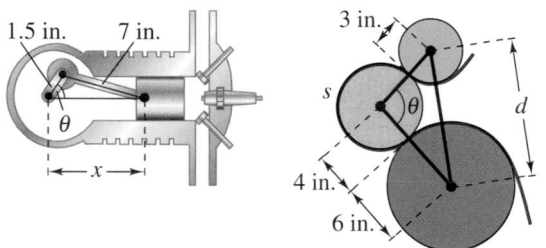

1.5 in.   7 in.   3 in.

θ s θ d

x   4 in.

                                6 in.

Figure for 47   Figure for 48

48. *Manufacturing* In a process with continuous paper, the paper passes across three rollers of radii 3 inches, 4 inches, and 6 inches (see figure). The centers of the three-inch and six-inch rollers are d inches apart, and the length of the arc in contact with the paper on the four-inch roller is s inches.

(a) Use the Law of Cosines to write an equation giving the relationship between d and θ.

(b) Write θ as a function of d.

(c) Write s as a function of θ.

(d) Complete the table.

d (inches)	9	10	12	13	14	15	16
θ (degrees)							
s (inches)							

Synthesis

True or False?   **In Exercises 49–51, determine whether the statement is true or false. Justify your answer.**

49. A triangle with side lengths of 10 feet, 16 feet, and 5 feet can be solved using the Law of Cosines.

50. Two sides and their included angle determine a unique triangle.

51. In Heron's Area Formula, s is the average of the lengths of the three sides of the triangle.

Proofs   **In Exercises 52–54, use the Law of Cosines to prove each of the following.**

52. $\dfrac{1}{2} bc(1 + \cos A) = \left(\dfrac{a + b + c}{2}\right)\left(\dfrac{-a + b + c}{2}\right)$

53. $\dfrac{1}{2} bc(1 - \cos A) = \left(\dfrac{a - b + c}{2}\right)\left(\dfrac{a + b - c}{2}\right)$

54. $\dfrac{\cos A}{a} + \dfrac{\cos B}{b} + \dfrac{\cos C}{c} = \dfrac{a^2 + b^2 + c^2}{2abc}$

55. *Proof*   Use a half-angle formula and the Law of Cosines to show that, for any triangle,

$$\cos\left(\dfrac{C}{2}\right) = \sqrt{\dfrac{s(s - c)}{ab}}$$

where $s = \frac{1}{2}(a + b + c)$.

56. *Proof*   Use a half-angle formula and the Law of Cosines to show that, for any triangle,

$$\sin\left(\dfrac{C}{2}\right) = \sqrt{\dfrac{(s - a)(s - b)}{ab}}$$

where $s = \frac{1}{2}(a + b + c)$.

57. *Writing*   Describe how the Law of Cosines can be used to solve the ambiguous case of the oblique triangle ABC, where $a = 12$ feet, $b = 30$ feet, and $A = 20°$. Is the result the same as when the Law of Sines is used to solve the triangle? Describe the advantages and the disadvantages of each method.

58. *Writing*   In Exercise 57, the Law of Cosines was used to solve a triangle in the two-solution case of SSA. Can the Law of Cosines be used to solve the no-solution and single-solution cases of SSA? Explain.

Skills Review

In Exercises 59–62, evaluate the expression without using a calculator.

59. $\arcsin(-1)$

60. $\cos^{-1} 0$

61. $\tan^{-1} \sqrt{3}$

62. $\arcsin\left(-\dfrac{\sqrt{3}}{2}\right)$

6.3 Vectors in the Plane

Introduction

Many quantities in geometry and physics, such as area, time, and temperature, can be represented by a single real number. Other quantities, such as force and velocity, involve both *magnitude* and *direction* and cannot be completely characterized by a single real number. To represent such a quantity, you can use a **directed line segment,** as shown in Figure 6.15. The directed line segment $\overrightarrow{PQ}$ has **initial point** P and **terminal point** Q. Its **magnitude,** or **length,** is denoted by $\|\overrightarrow{PQ}\|$ and can be found by using the Distance Formula.

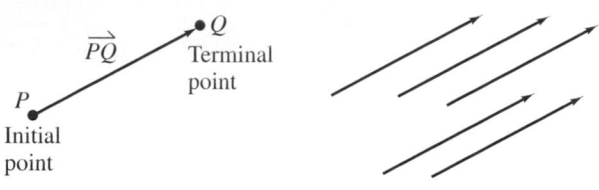

Figure 6.15   Figure 6.16

Two directed line segments that have the same magnitude and direction are *equivalent*. For example, the directed line segments in Figure 6.16 are all equivalent. The set of all directed line segments that are equivalent to a given directed line segment $\overrightarrow{PQ}$ is a **vector v in the plane,** written $\mathbf{v} = \overrightarrow{PQ}$. Vectors are denoted by lowercase, boldface letters such as $\mathbf{u}$, $\mathbf{v}$, and $\mathbf{w}$.

Example 1 Equivalent Directed Line Segments

Let $\mathbf{u}$ be represented by the directed line segment from $P = (0, 0)$ to $Q = (3, 2)$, and let $\mathbf{v}$ be represented by the directed line segment from $R = (1, 2)$ to $S = (4, 4)$, as shown in Figure 6.17. Show that $\mathbf{u} = \mathbf{v}$.

Solution

From the Distance Formula, it follows that $\overrightarrow{PQ}$ and $\overrightarrow{RS}$ have the *same magnitude*.

$$\|\overrightarrow{PQ}\| = \sqrt{(3 - 0)^2 + (2 - 0)^2} = \sqrt{13}$$

$$\|\overrightarrow{RS}\| = \sqrt{(4 - 1)^2 + (4 - 2)^2} = \sqrt{13}$$

Moreover, both line segments have the *same direction*, because they are both directed toward the upper right on lines having the same slope.

Slope of $\overrightarrow{PQ} = \dfrac{2 - 0}{3 - 0} = \dfrac{2}{3}$

Slope of $\overrightarrow{RS} = \dfrac{4 - 2}{4 - 1} = \dfrac{2}{3}$

So, $\overrightarrow{PQ}$ and $\overrightarrow{RS}$ have the same magnitude and direction, and it follows that $\mathbf{u} = \mathbf{v}$.

 **CHECKPOINT** Now try Exercise 1.

Sandra Baker/Getty Images

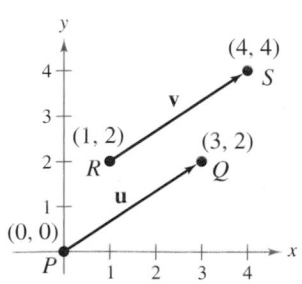

Figure 6.17

Component Form of a Vector

The directed line segment whose initial point is the origin is often the most convenient representative of a set of equivalent directed line segments. This representative of the vector **v** is in **standard position.**

A vector whose initial point is at the origin $(0, 0)$ can be uniquely represented by the coordinates of its terminal point (v_1, v_2). This is the **component form of a vector v,** written as

$$\mathbf{v} = \langle v_1, v_2 \rangle.$$

The coordinates v_1 and v_2 are the *components* of **v**. If both the initial point and the terminal point lie at the origin, **v** is the **zero vector** and is denoted by $\mathbf{0} = \langle 0, 0 \rangle$.

Component Form of a Vector

The component form of the vector with initial point $P = (p_1, p_2)$ and terminal point $Q = (q_1, q_2)$ is given by

$$\overrightarrow{PQ} = \langle q_1 - p_1, q_2 - p_2 \rangle = \langle v_1, v_2 \rangle = \mathbf{v}.$$

The **magnitude** (or length) of **v** is given by

$$\|\mathbf{v}\| = \sqrt{(q_1 - p_1)^2 + (q_2 - p_2)^2} = \sqrt{v_1^2 + v_2^2}.$$

If $\|\mathbf{v}\| = 1$, **v** is a **unit vector.** Moreover, $\|\mathbf{v}\| = 0$ if and only if **v** is the zero vector **0**.

Two vectors $\mathbf{u} = \langle u_1, u_2 \rangle$ and $\mathbf{v} = \langle v_1, v_2 \rangle$ are *equal* if and only if $u_1 = v_1$ and $u_2 = v_2$. For instance, in Example 1, the vector **u** from $P = (0, 0)$ to $Q = (3, 2)$ is

$$\mathbf{u} = \overrightarrow{PQ} = \langle 3 - 0, 2 - 0 \rangle = \langle 3, 2 \rangle$$

and the vector **v** from $R = (1, 2)$ to $S = (4, 4)$ is

$$\mathbf{v} = \overrightarrow{RS} = \langle 4 - 1, 4 - 2 \rangle = \langle 3, 2 \rangle.$$

> **TECHNOLOGY TIP**
>
> You can graph vectors with a graphing utility by graphing directed line segments. Consult the user's guide for your graphing utility for specific instructions.

Example 2   Finding the Component Form of a Vector

Find the component form and magnitude of the vector **v** that has initial point $(4, -7)$ and terminal point $(-1, 5)$.

Solution

Let $P = (4, -7) = (p_1, p_2)$ and $Q = (-1, 5) = (q_1, q_2)$, as shown in Figure 6.18. Then, the components of $\mathbf{v} = \langle v_1, v_2 \rangle$ are

$$v_1 = q_1 - p_1 = -1 - 4 = -5$$

$$v_2 = q_2 - p_2 = 5 - (-7) = 12.$$

So, $\mathbf{v} = \langle -5, 12 \rangle$ and the magnitude of **v** is

$$\|\mathbf{v}\| = \sqrt{(-5)^2 + 12^2} = \sqrt{169} = 13.$$

✓**CHECKPOINT**   Now try Exercise 5.

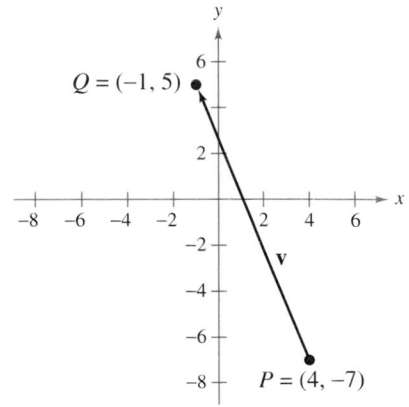

Figure 6.18

Vector Operations

The two basic vector operations are **scalar multiplication** and **vector addition.** Geometrically, the product of a vector **v** and a scalar k is the vector that is $|k|$ times as long as **v**. If k is positive, $k\mathbf{v}$ has the same direction as **v**, and if k is negative, $k\mathbf{v}$ has the opposite direction of **v**, as shown in Figure 6.19.

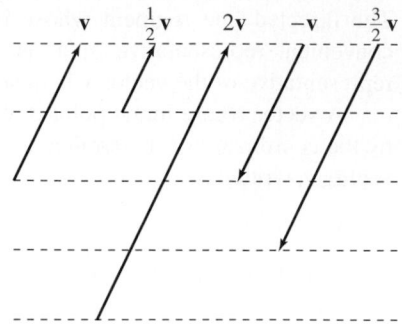

To add two vectors **u** and **v** geometrically, first position them (without changing their lengths or directions) so that the initial point of the second vector **v** coincides with the terminal point of the first vector **u.** The sum $\mathbf{u} + \mathbf{v}$ is the vector formed by joining the initial point of the first vector **u** with the terminal point of the second vector **v**, as shown in Figure 6.20. This technique is called the **parallelogram law** for vector addition because the vector $\mathbf{u} + \mathbf{v}$, often called the **resultant** of vector addition, is the diagonal of a parallelogram having **u** and **v** as its adjacent sides.

Figure 6.19

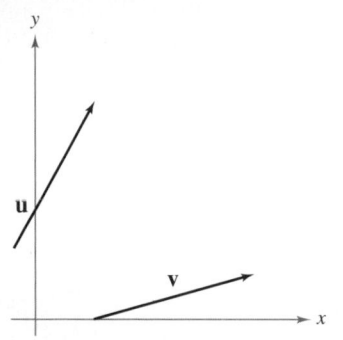
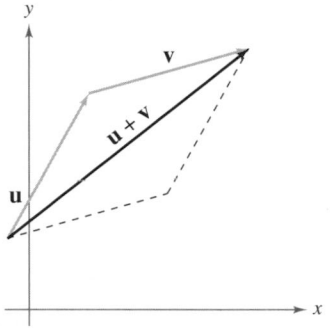

Figure 6.20

Definition of Vector Addition and Scalar Multiplication

Let $\mathbf{u} = \langle u_1, u_2 \rangle$ and $\mathbf{v} = \langle v_1, v_2 \rangle$ be vectors and let k be a scalar (a real number). Then the **sum** of **u** and **v** is the vector

$$\mathbf{u} + \mathbf{v} = \langle u_1 + v_1, u_2 + v_2 \rangle \qquad \text{Sum}$$

and the **scalar multiple** of k times **u** is the vector

$$k\mathbf{u} = k\langle u_1, u_2 \rangle = \langle ku_1, ku_2 \rangle. \qquad \text{Scalar multiple}$$

The **negative** of $\mathbf{v} = \langle v_1, v_2 \rangle$ is

$$-\mathbf{v} = (-1)\mathbf{v}$$

$$= \langle -v_1, -v_2 \rangle \qquad \text{Negative}$$

and the **difference** of **u** and **v** is

$$\mathbf{u} - \mathbf{v} = \mathbf{u} + (-\mathbf{v}) \qquad \text{Add } (-\mathbf{v}). \text{ See Figure 6.21.}$$

$$= \langle u_1 - v_1, u_2 - v_2 \rangle. \qquad \text{Difference}$$

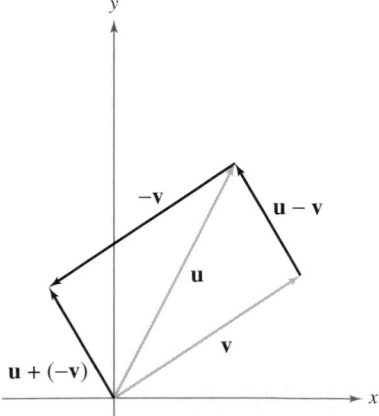

Figure 6.21

To represent $\mathbf{u} - \mathbf{v}$ geometrically, you can use directed line segments with the *same* initial point. The difference $\mathbf{u} - \mathbf{v}$ is the vector from the terminal point of **v** to the terminal point of **u**, which is equal to $\mathbf{u} + (-\mathbf{v})$, as shown in Figure 6.21.

The component definitions of vector addition and scalar multiplication are illustrated in Example 3. In this example, notice that each of the vector operations can be interpreted geometrically.

Example 3 Vector Operations

Let $\mathbf{v} = \langle -2, 5 \rangle$ and $\mathbf{w} = \langle 3, 4 \rangle$, and find each of the following vectors.

a. $2\mathbf{v}$ **b.** $\mathbf{w} - \mathbf{v}$ **c.** $\mathbf{v} + 2\mathbf{w}$ **d.** $2\mathbf{v} - 3\mathbf{w}$

Solution

a. Because $\mathbf{v} = \langle -2, 5 \rangle$, you have

$$2\mathbf{v} = 2\langle -2, 5 \rangle$$

$$= \langle 2(-2), 2(5) \rangle$$

$$= \langle -4, 10 \rangle.$$

A sketch of $2\mathbf{v}$ is shown in Figure 6.22.

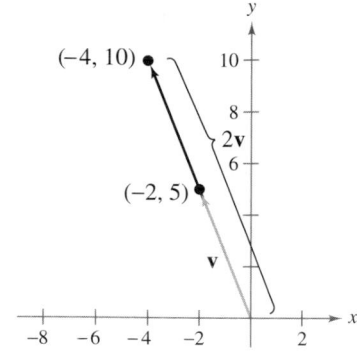

Figure 6.22

b. The difference of $\mathbf{w}$ and $\mathbf{v}$ is

$$\mathbf{w} - \mathbf{v} = \langle 3 - (-2), 4 - 5 \rangle$$

$$= \langle 5, -1 \rangle.$$

A sketch of $\mathbf{w} - \mathbf{v}$ is shown in Figure 6.23. Note that the figure shows the vector difference $\mathbf{w} - \mathbf{v}$ as the sum $\mathbf{w} + (-\mathbf{v})$.

c. The sum of $\mathbf{v}$ and $2\mathbf{w}$ is

$$\mathbf{v} + 2\mathbf{w} = \langle -2, 5 \rangle + 2\langle 3, 4 \rangle$$

$$= \langle -2, 5 \rangle + \langle 2(3), 2(4) \rangle$$

$$= \langle -2, 5 \rangle + \langle 6, 8 \rangle$$

$$= \langle -2 + 6, 5 + 8 \rangle$$

$$= \langle 4, 13 \rangle.$$

A sketch of $\mathbf{v} + 2\mathbf{w}$ is shown in Figure 6.24.

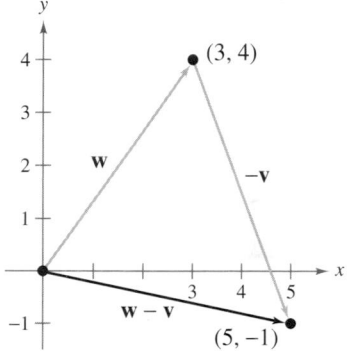

Figure 6.23

d. The difference of $2\mathbf{v}$ and $3\mathbf{w}$ is

$$2\mathbf{v} - 3\mathbf{w} = 2\langle -2, 5 \rangle - 3\langle 3, 4 \rangle$$

$$= \langle 2(-2), 2(5) \rangle - \langle 3(3), 3(4) \rangle$$

$$= \langle -4, 10 \rangle - \langle 9, 12 \rangle$$

$$= \langle -4 - 9, 10 - 12 \rangle$$

$$= \langle -13, -2 \rangle.$$

A sketch of $2\mathbf{v} - 3\mathbf{w}$ is shown in Figure 6.25. Note that the figure shows the vector difference $2\mathbf{v} - 3\mathbf{w}$ as the sum $2\mathbf{v} + (-3\mathbf{w})$.

✔CHECKPOINT   Now try Exercise 25.

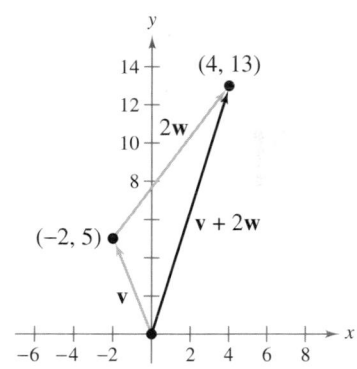

Figure 6.24

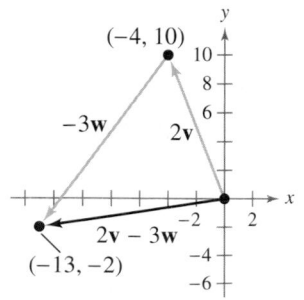

Figure 6.25

Vector addition and scalar multiplication share many of the properties of ordinary arithmetic.

Properties of Vector Addition and Scalar Multiplication

Let $\mathbf{u}$, $\mathbf{v}$, and $\mathbf{w}$ be vectors and let c and d be scalars. Then the following properties are true.

1. $\mathbf{u} + \mathbf{v} = \mathbf{v} + \mathbf{u}$
2. $(\mathbf{u} + \mathbf{v}) + \mathbf{w} = \mathbf{u} + (\mathbf{v} + \mathbf{w})$
3. $\mathbf{u} + \mathbf{0} = \mathbf{u}$
4. $\mathbf{u} + (-\mathbf{u}) = \mathbf{0}$
5. $c(d\mathbf{u}) = (cd)\mathbf{u}$
6. $(c + d)\mathbf{u} = c\mathbf{u} + d\mathbf{u}$
7. $c(\mathbf{u} + \mathbf{v}) = c\mathbf{u} + c\mathbf{v}$
8. $1(\mathbf{u}) = \mathbf{u}, \ 0(\mathbf{u}) = \mathbf{0}$
9. $\|c\mathbf{v}\| = |c|\,\|\mathbf{v}\|$

STUDY TIP

Property 9 can be stated as follows: The magnitude of the vector $c\mathbf{v}$ is the absolute value of c times the magnitude of $\mathbf{v}$.

Unit Vectors

In many applications of vectors, it is useful to find a unit vector that has the same direction as a given nonzero vector $\mathbf{v}$. To do this, you can divide $\mathbf{v}$ by its length to obtain

$$\mathbf{u} = \text{unit vector} = \frac{\mathbf{v}}{\|\mathbf{v}\|} = \left(\frac{1}{\|\mathbf{v}\|}\right)\mathbf{v}. \qquad \text{Unit vector in direction of } \mathbf{v}$$

Note that $\mathbf{u}$ is a scalar multiple of $\mathbf{v}$. The vector $\mathbf{u}$ has a magnitude of 1 and the same direction as $\mathbf{v}$. The vector $\mathbf{u}$ is called a **unit vector in the direction of v.**

Example 4   Finding a Unit Vector

Find a unit vector in the direction of $\mathbf{v} = \langle -2, 5 \rangle$ and verify that the result has a magnitude of 1.

Solution

The unit vector in the direction of $\mathbf{v}$ is

$$\frac{\mathbf{v}}{\|\mathbf{v}\|} = \frac{\langle -2, 5 \rangle}{\sqrt{(-2)^2 + (5)^2}}$$

$$= \frac{1}{\sqrt{29}}\langle -2, 5 \rangle$$

$$= \left\langle \frac{-2}{\sqrt{29}}, \frac{5}{\sqrt{29}} \right\rangle = \left\langle \frac{-2\sqrt{29}}{29}, \frac{5\sqrt{29}}{29} \right\rangle.$$

This vector has a magnitude of 1 because

$$\sqrt{\left(\frac{-2\sqrt{29}}{29}\right)^2 + \left(\frac{5\sqrt{29}}{29}\right)^2} = \sqrt{\frac{116}{841} + \frac{725}{841}} = \sqrt{\frac{841}{841}} = 1.$$

✓**CHECKPOINT**   Now try Exercise 37.

The unit vectors $\langle 1, 0 \rangle$ and $\langle 0, 1 \rangle$ are called the **standard unit vectors** and are denoted by

$$\mathbf{i} = \langle 1, 0 \rangle \qquad \text{and} \qquad \mathbf{j} = \langle 0, 1 \rangle$$

as shown in Figure 6.26. (Note that the lowercase letter $\mathbf{i}$ is written in boldface to distinguish it from the imaginary number $i = \sqrt{-1}$.) These vectors can be used to represent any vector $\mathbf{v} = \langle v_1, v_2 \rangle$ as follows.

$$\mathbf{v} = \langle v_1, v_2 \rangle$$
$$= v_1 \langle 1, 0 \rangle + v_2 \langle 0, 1 \rangle$$
$$= v_1 \mathbf{i} + v_2 \mathbf{j}$$

The scalars v_1 and v_2 are called the **horizontal and vertical components of v,** respectively. The vector sum

$$v_1 \mathbf{i} + v_2 \mathbf{j}$$

is called a **linear combination** of the vectors $\mathbf{i}$ and $\mathbf{j}$. Any vector in the plane can be written as a linear combination of the standard unit vectors $\mathbf{i}$ and $\mathbf{j}$.

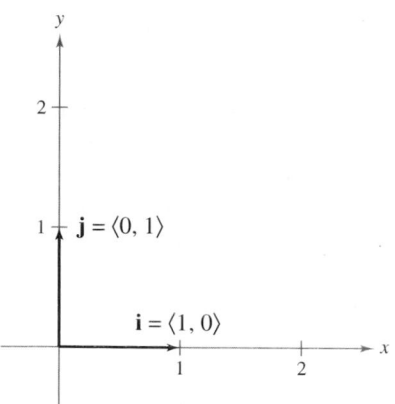

Figure 6.26

Example 5   Writing a Linear Combination of Unit Vectors

Let $\mathbf{u}$ be the vector with initial point $(2, -5)$ and terminal point $(-1, 3)$. Write $\mathbf{u}$ as a linear combination of the standard unit vectors $\mathbf{i}$ and $\mathbf{j}$.

Solution

Begin by writing the component form of the vector $\mathbf{u}$.

$$\mathbf{u} = \langle -1 - 2, 3 - (-5) \rangle$$
$$= \langle -3, 8 \rangle$$
$$= -3\mathbf{i} + 8\mathbf{j}$$

This result is shown graphically in Figure 6.27.

 Now try Exercise 51.

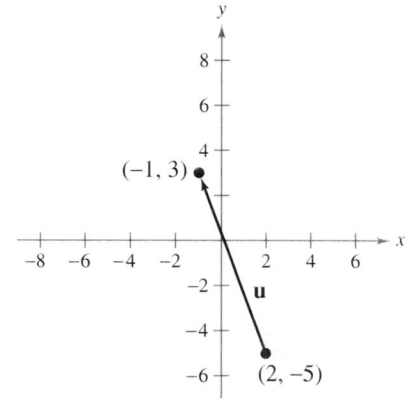

Figure 6.27

Example 6   Vector Operations

Let $\mathbf{u} = -3\mathbf{i} + 8\mathbf{j}$ and $\mathbf{v} = 2\mathbf{i} - \mathbf{j}$. Find $2\mathbf{u} - 3\mathbf{v}$.

Solution

You could solve this problem by converting $\mathbf{u}$ and $\mathbf{v}$ to component form. This, however, is not necessary. It is just as easy to perform the operations in unit vector form.

$$2\mathbf{u} - 3\mathbf{v} = 2(-3\mathbf{i} + 8\mathbf{j}) - 3(2\mathbf{i} - \mathbf{j})$$
$$= -6\mathbf{i} + 16\mathbf{j} - 6\mathbf{i} + 3\mathbf{j}$$
$$= -12\mathbf{i} + 19\mathbf{j}$$

✓CHECKPOINT   Now try Exercise 57.

Direction Angles

If **u** is a *unit vector* such that θ is the angle (measured counterclockwise) from the positive x-axis to **u**, the terminal point of **u** lies on the unit circle and you have

$$\mathbf{u} = \langle x, y \rangle = \langle \cos\theta, \sin\theta \rangle = (\cos\theta)\mathbf{i} + (\sin\theta)\mathbf{j}$$

as shown in Figure 6.28. The angle θ is the **direction angle** of the vector **u**.

Suppose that **u** is a unit vector with direction angle θ. If $\mathbf{v} = a\mathbf{i} + b\mathbf{j}$ is any vector that makes an angle θ with the positive x-axis, then it has the same direction as **u** and you can write

$$\mathbf{v} = \|\mathbf{v}\| \langle \cos\theta, \sin\theta \rangle$$
$$= \|\mathbf{v}\| (\cos\theta)\mathbf{i} + \|\mathbf{v}\| (\sin\theta)\mathbf{j}.$$

Because $\mathbf{v} = a\mathbf{i} + b\mathbf{j} = \|\mathbf{v}\| (\cos\theta)\mathbf{i} + \|\mathbf{v}\| (\sin\theta)\mathbf{j}$, it follows that the direction angle θ for **v** is determined from

$$\tan\theta = \frac{\sin\theta}{\cos\theta} \qquad \text{Quotient identity}$$

$$= \frac{\|\mathbf{v}\| \sin\theta}{\|\mathbf{v}\| \cos\theta} \qquad \text{Multiply numerator and denominator by } \|\mathbf{v}\|.$$

$$= \frac{b}{a}. \qquad \text{Simplify.}$$

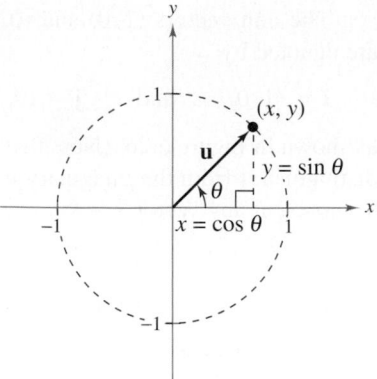

Figure 6.28

Example 7 Finding Direction Angles of Vectors

Find the direction angle of each vector.

a. $\mathbf{u} = 3\mathbf{i} + 3\mathbf{j}$ **b.** $\mathbf{v} = 3\mathbf{i} - 4\mathbf{j}$

Solution

a. The direction angle is

$$\tan\theta = \frac{b}{a} = \frac{3}{3} = 1.$$

So, $\theta = 45°$, as shown in Figure 6.29.

b. The direction angle is

$$\tan\theta = \frac{b}{a} = \frac{-4}{3}.$$

Moreover, because $\mathbf{v} = 3\mathbf{i} - 4\mathbf{j}$ lies in Quadrant IV, θ lies in Quadrant IV and its reference angle is

$$\theta' = \left| \arctan\left(-\frac{4}{3}\right) \right| \approx |-53.13°| = 53.13°.$$

So, it follows that $\theta \approx 360° - 53.13° = 306.87°$, as shown in Figure 6.30.

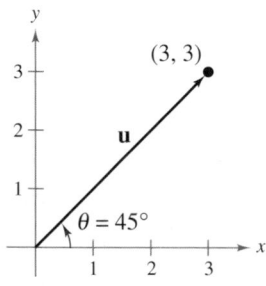

Figure 6.29

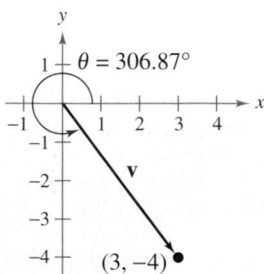

Figure 6.30

CHECKPOINT   Now try Exercise 65.

Applications of Vectors

Example 8   Finding the Component Form of a Vector

Find the component form of the vector that represents the velocity of an airplane descending at a speed of 100 miles per hour at an angle of $30°$ below the horizontal, as shown in Figure 6.31.

Solution

The velocity vector $\mathbf{v}$ has a magnitude of 100 and a direction angle of $\theta = 210°$.

$$\mathbf{v} = \|\mathbf{v}\| (\cos \theta)\mathbf{i} + \|\mathbf{v}\| (\sin \theta)\mathbf{j}$$

$$= 100(\cos 210°)\mathbf{i} + 100(\sin 210°)\mathbf{j}$$

$$= 100\left(-\frac{\sqrt{3}}{2}\right)\mathbf{i} + 100\left(-\frac{1}{2}\right)\mathbf{j}$$

$$= -50\sqrt{3}\,\mathbf{i} - 50\mathbf{j} = \langle -50\sqrt{3},\, -50 \rangle$$

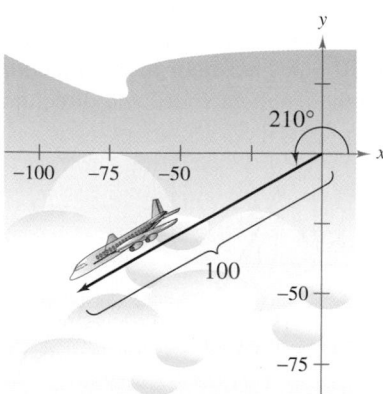

Figure 6.31

You can check that $\mathbf{v}$ has a magnitude of 100 as follows.

$$\|\mathbf{v}\| = \sqrt{\left(-50\sqrt{3}\right)^2 + (-50)^2}$$

$$= \sqrt{7500 + 2500} = \sqrt{10,000} = 100 \qquad \text{Solution checks. } \checkmark$$

 **Now try Exercise 83.**

Example 9   Using Vectors to Determine Weight

A force of 600 pounds is required to pull a boat and trailer up a ramp inclined at $15°$ from the horizontal. Find the combined weight of the boat and trailer.

Solution

Based on Figure 6.32, you can make the following observations.

$\|\overrightarrow{BA}\|$ = force of gravity = combined weight of boat and trailer

$\|\overrightarrow{BC}\|$ = force against ramp

$\|\overrightarrow{AC}\|$ = force required to move boat up ramp = 600 pounds

By construction, triangles BWD and ABC are similar. So, angle ABC is $15°$. In triangle ABC you have

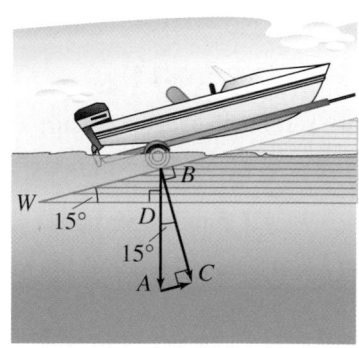

Figure 6.32

$$\sin 15° = \frac{\|\overrightarrow{AC}\|}{\|\overrightarrow{BA}\|} = \frac{600}{\|\overrightarrow{BA}\|}$$

$$\|\overrightarrow{BA}\| = \frac{600}{\sin 15°} \approx 2318.$$

So, the combined weight is approximately 2318 pounds. (In Figure 6.32, note that $\overrightarrow{AC}$ is parallel to the ramp.)

 **Now try Exercise 85.**

Example 10 Using Vectors to Find Speed and Direction

An airplane is traveling at a speed of 500 miles per hour with a bearing of 330° at a fixed altitude with a negligible wind velocity, as shown in Figure 6.33(a). As the airplane reaches a certain point, it encounters a wind blowing with a velocity of 70 miles per hour in the direction N 45° E, as shown in Figure 6.33(b). What are the resultant speed and direction of the airplane?

STUDY TIP

Recall from Section 4.8 that in air navigation, bearings are measured in degrees clockwise from north.

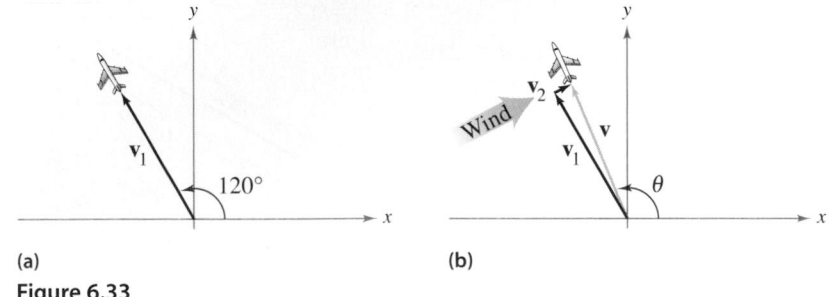

(a)   (b)

Figure 6.33

Solution

Using Figure 6.33, the velocity of the airplane (alone) is

$$\mathbf{v}_1 = 500\langle \cos 120°, \sin 120° \rangle$$
$$= \langle -250, 250\sqrt{3} \rangle$$

and the velocity of the wind is

$$\mathbf{v}_2 = 70\langle \cos 45°, \sin 45° \rangle$$
$$= \langle 35\sqrt{2}, 35\sqrt{2} \rangle.$$

So, the velocity of the airplane (in the wind) is

$$\mathbf{v} = \mathbf{v}_1 + \mathbf{v}_2$$
$$= \langle -250 + 35\sqrt{2}, 250\sqrt{3} + 35\sqrt{2} \rangle$$
$$\approx \langle -200.5, 482.5 \rangle$$

and the resultant speed of the airplane is

$$\|\mathbf{v}\| = \sqrt{(-200.5)^2 + (482.5)^2}$$
$$\approx 522.5 \text{ miles per hour.}$$

Finally, if θ is the direction angle of the flight path, you have

$$\tan \theta = \frac{482.5}{-200.5} \approx -2.4065$$

which implies that

$$\theta \approx 180° + \arctan(-2.4065) \approx 180° - 67.4° = 112.6°.$$

So, the true direction of the airplane is 337.4°.

✔**CHECKPOINT**   Now try Exercise 89.

6.3 Exercises

Vocabulary Check

Fill in the blanks.

1. A _____ can be used to represent a quantity that involves both magnitude and direction.

2. The directed line segment $\overrightarrow{PQ}$ has _____ point P and _____ point Q.

3. The _____ of the directed line segment $\overrightarrow{PQ}$ is denoted by $\|\overrightarrow{PQ}\|$.

4. The set of all directed line segments that are equivalent to a given directed line segment $\overrightarrow{PQ}$ is a _____ **v** in the plane.

5. The directed line segment whose initial point is the origin is said to be in _____ .

6. A vector that has a magnitude of 1 is called a _____ .

7. The two basic vector operations are scalar _____ and vector _____ .

8. The vector **u** + **v** is called the _____ of vector addition.

9. The vector sum $v_1\mathbf{i} + v_2\mathbf{j}$ is called a _____ of the vectors **i** and **j**, and the scalars v_1 and v_2 are called the _____ and _____ components of **v**, respectively.

In Exercises 1 and 2, show that u = v.

1.
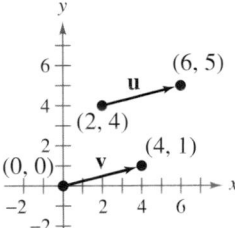

2.
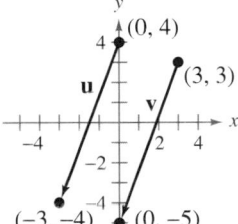

In Exercises 3–12, find the component form and the magnitude of the vector v.

3.
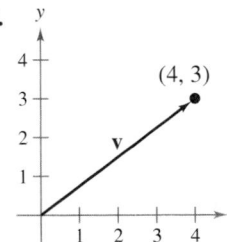

4.
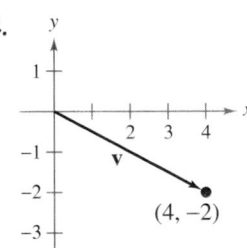

5.
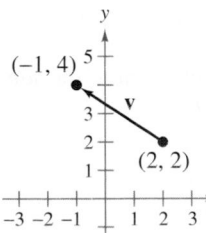

6.
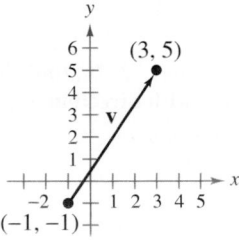

7.
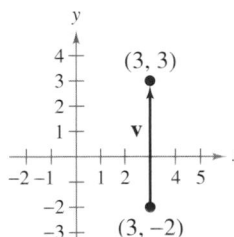

8.
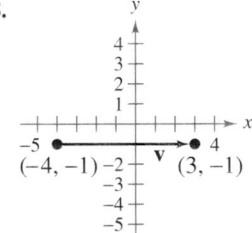

	Initial Point	Terminal Point
9.	$\left(\frac{2}{5}, 1\right)$	$\left(1, \frac{2}{5}\right)$
10.	$\left(\frac{7}{2}, 0\right)$	$\left(0, -\frac{7}{2}\right)$
11.	$\left(-\frac{2}{3}, -1\right)$	$\left(\frac{1}{2}, \frac{4}{5}\right)$
12.	$\left(\frac{5}{2}, -2\right)$	$\left(1, \frac{2}{5}\right)$

In Exercises 13–18, use the figure to sketch a graph of the specified vector. To print an enlarged copy of the graph, go to the website *www.mathgraphs.com*.

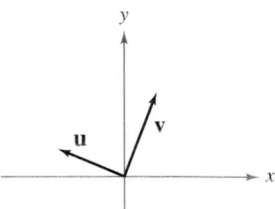

13. $-\mathbf{v}$

14. $3\mathbf{u}$

15. $\mathbf{u} + \mathbf{v}$

16. $\mathbf{u} - \mathbf{v}$

17. $\mathbf{u} + 2\mathbf{v}$

18. $\mathbf{v} - \frac{1}{2}\mathbf{u}$

In Exercises 19–24, use the figure to sketch a graph of the specified vector. To print an enlarged copy of the graph, go to the website *www.mathgraphs.com*.

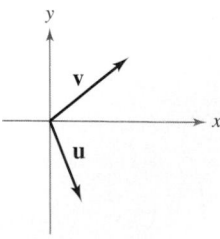

19. $2\mathbf{u}$ 20. $-3\mathbf{v}$

21. $\mathbf{u} + 2\mathbf{v}$ 22. $\frac{1}{2}\mathbf{v}$

23. $\mathbf{v} - \frac{1}{2}\mathbf{u}$ 24. $2\mathbf{u} + 3\mathbf{v}$

In Exercises 25–30, find (a) $\mathbf{u} + \mathbf{v}$, (b) $\mathbf{u} - \mathbf{v}$, (c) $2\mathbf{u} - 3\mathbf{v}$, and (d) $\mathbf{v} + 4\mathbf{u}$.

25. $\mathbf{u} = \langle 4, 2 \rangle$, $\mathbf{v} = \langle 7, 1 \rangle$ 26. $\mathbf{u} = \langle 5, 3 \rangle$, $\mathbf{v} = \langle -4, 0 \rangle$

27. $\mathbf{u} = \langle -6, -8 \rangle$, $\mathbf{v} = \langle 2, 4 \rangle$

28. $\mathbf{u} = \langle 0, -5 \rangle$, $\mathbf{v} = \langle -3, 9 \rangle$

29. $\mathbf{u} = \mathbf{i} + \mathbf{j}$, $\mathbf{v} = 2\mathbf{i} - 3\mathbf{j}$ 30. $\mathbf{u} = 2\mathbf{i} - \mathbf{j}$, $\mathbf{v} = -\mathbf{i} + \mathbf{j}$

In Exercises 31–34, use the figure and write the vector in terms of the other two vectors.

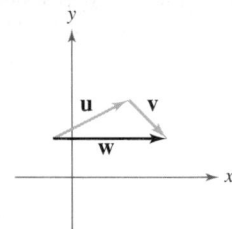

31. $\mathbf{w}$ 32. $\mathbf{v}$

33. $\mathbf{u}$ 34. $2\mathbf{v}$

In Exercises 35–44, find a unit vector in the direction of the given vector.

35. $\mathbf{u} = \langle 6, 0 \rangle$ 36. $\mathbf{u} = \langle 0, -2 \rangle$

37. $\mathbf{v} = \langle -1, 1 \rangle$ 38. $\mathbf{v} = \langle 3, -4 \rangle$

39. $\mathbf{v} = \langle -24, -7 \rangle$ 40. $\mathbf{v} = \langle 8, -20 \rangle$

41. $\mathbf{v} = 4\mathbf{i} - 3\mathbf{j}$ 42. $\mathbf{w} = \mathbf{i} - 2\mathbf{j}$

43. $\mathbf{w} = 2\mathbf{j}$ 44. $\mathbf{w} = -3\mathbf{i}$

In Exercises 45–50, find the vector v with the given magnitude and the same direction as u.

Magnitude	Direction
45. $\|\mathbf{v}\| = 8$	$\mathbf{u} = \langle 5, 6 \rangle$
46. $\|\mathbf{v}\| = 3$	$\mathbf{u} = \langle 4, -4 \rangle$
47. $\|\mathbf{v}\| = 7$	$\mathbf{u} = 3\mathbf{i} + 4\mathbf{j}$
48. $\|\mathbf{v}\| = 10$	$\mathbf{u} = 2\mathbf{i} - 3\mathbf{j}$
49. $\|\mathbf{v}\| = 8$	$\mathbf{u} = -2\mathbf{i}$
50. $\|\mathbf{v}\| = 4$	$\mathbf{u} = 5\mathbf{j}$

In Exercises 51–54, the initial and terminal points of a vector are given. Write the vector as a linear combination of the standard unit vectors i and j.

Initial Point	Terminal Point
51. $(-3, 1)$	$(4, 5)$
52. $(0, -2)$	$(3, 6)$
53. $(-1, -5)$	$(2, 3)$
54. $(-6, 4)$	$(0, 1)$

In Exercises 55–60, find the component form of v and sketch the specified vector operations geometrically, where $\mathbf{u} = 2\mathbf{i} - \mathbf{j}$ and $\mathbf{w} = \mathbf{i} + 2\mathbf{j}$.

55. $\mathbf{v} = \frac{3}{2}\mathbf{u}$ 56. $\mathbf{v} = \frac{2}{3}\mathbf{w}$

57. $\mathbf{v} = \mathbf{u} + 2\mathbf{w}$ 58. $\mathbf{v} = -\mathbf{u} + \mathbf{w}$

59. $\mathbf{v} = \frac{1}{2}(3\mathbf{u} + \mathbf{w})$ 60. $\mathbf{v} = 2\mathbf{u} - 2\mathbf{w}$

In Exercises 61–66, find the magnitude and direction angle of the vector v.

61. $\mathbf{v} = 5(\cos 30°\mathbf{i} + \sin 30°\mathbf{j})$

62. $\mathbf{v} = 8(\cos 135°\mathbf{i} + \sin 135°\mathbf{j})$

63. $\mathbf{v} = 6\mathbf{i} - 6\mathbf{j}$ 64. $\mathbf{v} = -4\mathbf{i} - 7\mathbf{j}$

65. $\mathbf{v} = -2\mathbf{i} + 5\mathbf{j}$ 66. $\mathbf{v} = 12\mathbf{i} + 15\mathbf{j}$

In Exercises 67–72, find the component form of v given its magnitude and the angle it makes with the positive *x*-axis. Sketch v.

Magnitude	Angle
67. $\|\mathbf{v}\| = 3$	$\theta = 0°$
68. $\|\mathbf{v}\| = 1$	$\theta = 45°$
69. $\|\mathbf{v}\| = 3\sqrt{2}$	$\theta = 150°$
70. $\|\mathbf{v}\| = 4\sqrt{3}$	$\theta = 90°$
71. $\|\mathbf{v}\| = 2$	v in the direction $\mathbf{i} + 3\mathbf{j}$
72. $\|\mathbf{v}\| = 3$	v in the direction $3\mathbf{i} + 4\mathbf{j}$

In Exercises 73–76, find the component form of the sum of u and v with direction angles $\theta_{\mathbf{u}}$ and $\theta_{\mathbf{v}}$.

Magnitude	Angle
73. $\|\mathbf{u}\| = 5$	$\theta_{\mathbf{u}} = 60°$
$\|\mathbf{v}\| = 5$	$\theta_{\mathbf{v}} = 90°$

Magnitude	Angle
74. $\|\mathbf{u}\| = 2$	$\theta_{\mathbf{u}} = 30°$
$\|\mathbf{v}\| = 2$	$\theta_{\mathbf{v}} = 90°$
75. $\|\mathbf{u}\| = 20$	$\theta_{\mathbf{u}} = 45°$
$\|\mathbf{v}\| = 50$	$\theta_{\mathbf{v}} = 150°$
76. $\|\mathbf{u}\| = 35$	$\theta_{\mathbf{u}} = 25°$
$\|\mathbf{v}\| = 50$	$\theta_{\mathbf{v}} = 120°$

In Exercises 77 and 78, use the Law of Cosines to find the angle α between the vectors. (Assume $0° \le \alpha \le 180°$.)

77. $\mathbf{v} = \mathbf{i} + \mathbf{j}, \quad \mathbf{w} = 2(\mathbf{i} - \mathbf{j})$

78. $\mathbf{v} = 3\mathbf{i} + \mathbf{j}, \quad \mathbf{w} = 2\mathbf{i} - \mathbf{j}$

In Exercises 79 and 80, graph the vectors and the resultant of the vectors. Find the magnitude and direction of the resultant.

79. **80.**

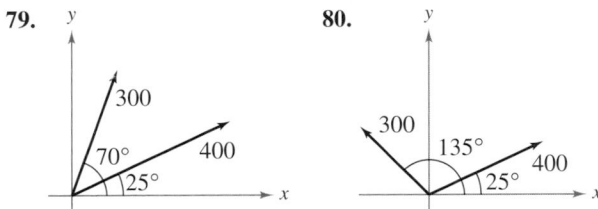

Resultant Force   In Exercises 81 and 82, find the angle between the forces given the magnitude of their resultant. (*Hint:* Write force 1 as a vector in the direction of the positive *x*-axis and force 2 as a vector at an angle θ with the positive *x*-axis.)

Force 1	Force 2	Resultant Force
81. 45 pounds	60 pounds	90 pounds
82. 3000 pounds	1000 pounds	3750 pounds

83. *Velocity*   A ball is thrown with an initial velocity of 70 feet per second, at an angle of 40° with the horizontal (see figure). Find the vertical and horizontal components of the velocity.

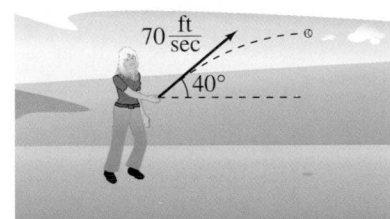

84. *Velocity*   A gun with a muzzle velocity of 1200 feet per second is fired at an angle of 4° with the horizontal. Find the vertical and horizontal components of the velocity.

85. *Tension*   Use the figure to determine the tension in each cable supporting the load.

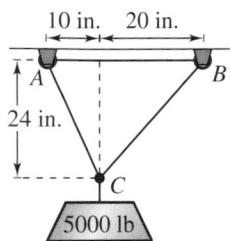

86. *Tension*   The cranes shown in the figure are lifting an object that weighs 20,240 pounds. Find the tension in the cable of each crane.

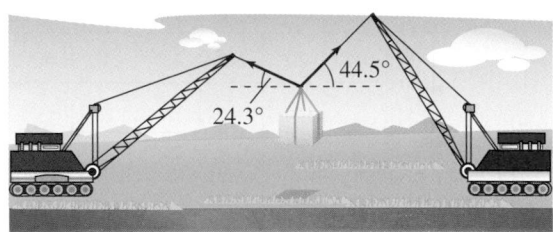

87. *Numerical and Graphical Analysis*   A loaded barge is being towed by two tugboats, and the magnitude of the resultant is 6000 pounds directed along the axis of the barge (see figure). Each tow line makes an angle of θ degrees with the axis of the barge.

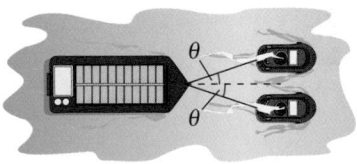

(a) Write the resultant tension T of each tow line as a function of θ. Determine the domain of the function.

(b) Use a graphing utility to complete the table.

θ	10°	20°	30°	40°	50°	60°
T						

(c) Use a graphing utility to graph the tension function.

(d) Explain why the tension increases as θ increases.

88. *Numerical and Graphical Analysis* To carry a 100-pound cylindrical weight, two people lift on the ends of short ropes that are tied to an eyelet on the top center of the cylinder. Each rope makes an angle of θ degrees with the vertical (see figure).

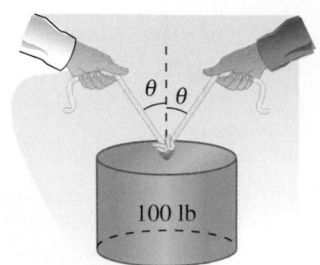

100 lb

(a) Write the tension T of each rope as a function of θ. Determine the domain of the function.

(b) Use a graphing utility to complete the table.

θ	10°	20°	30°	40°	50°	60°
T						

(c) Use a graphing utility to graph the tension function.

(d) Explain why the tension increases as θ increases.

89. *Navigation* An airplane is flying in the direction 148° with an airspeed of 860 kilometers per hour. Because of the wind, its groundspeed and direction are, respectively, 800 kilometers per hour and 140°. Find the direction and speed of the wind.

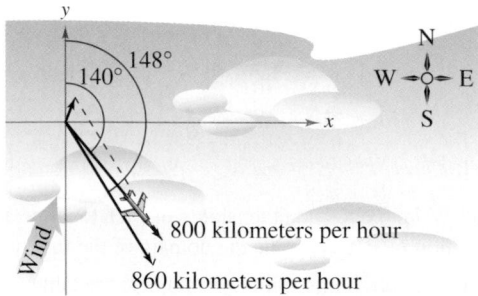

90. *Navigation* A commercial jet is flying from Miami to Seattle. The jet's velocity with respect to the air is 580 miles per hour, and its bearing is 332°. The wind, at the altitude of the plane, is blowing from the southwest with a velocity of 60 miles per hour.

(a) Draw a figure that gives a visual representation of the problem.

(b) Write the velocity of the wind as a vector in component form.

(c) Write the velocity of the jet relative to the air as a vector in component form.

(d) What is the speed of the jet with respect to the ground?

(e) What is the true direction of the jet?

91. *Numerical and Graphical Analysis* Forces with magnitudes of 150 newtons and 220 newtons act on a hook (see figure).

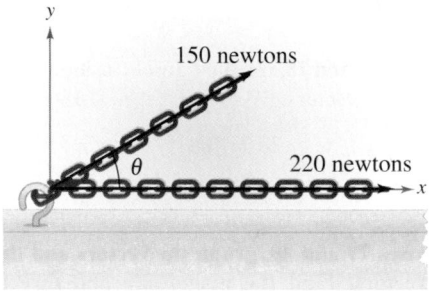

(a) Find the direction and magnitude of the resultant of the forces when $\theta = 30°$.

(b) Write the magnitude M of the resultant and the direction α of the resultant as functions of θ, where $0° \leq \theta \leq 180°$.

(c) Use a graphing utility to complete the table.

θ	0°	30°	60°	90°	120°	150°	180°
M							
α							

(d) Use a graphing utility to graph the two functions.

(e) Explain why one function decreases for increasing θ, whereas the other doesn't.

92. *Numerical and Graphical Analysis* A tetherball weighing 1 pound is pulled outward from the pole by a horizontal force **u** until the rope makes an angle of θ degrees with the pole (see figure).

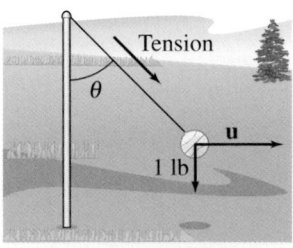

(a) Write the tension T in the rope and the magnitude of **u** as functions of θ. Determine the domains of the functions.

(b) Use a graphing utility to complete the table.

θ	0°	10°	20°	30°	40°	50°	60°
T							
$\|\mathbf{u}\|$							

(c) Use a graphing utility to graph the two functions for $0° \le \theta \le 60°$.

(d) Compare T and $\|\mathbf{u}\|$ as θ increases.

Synthesis

True or False?   **In Exercises 93–96, determine whether the statement is true or false. Justify your answer.**

93. If $\mathbf{u}$ and $\mathbf{v}$ have the same magnitude and direction, then $\mathbf{u} = \mathbf{v}$.

94. If $\mathbf{u}$ is a unit vector in the direction of $\mathbf{v}$, then $\mathbf{v} = \|\mathbf{v}\|\,\mathbf{u}$.

95. If $\mathbf{v} = a\mathbf{i} + b\mathbf{j} = \mathbf{0}$, then $a = -b$.

96. If $\mathbf{u} = a\mathbf{i} + b\mathbf{j}$ is a unit vector, then $a^2 + b^2 = 1$.

True or False?   **In Exercises 97–104, use the figure to determine whether the statement is true or false. Justify your answer.**

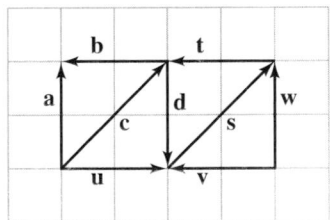

97. $\mathbf{a} = -\mathbf{d}$

98. $\mathbf{c} = \mathbf{s}$

99. $\mathbf{a} + \mathbf{u} = \mathbf{c}$

100. $\mathbf{v} + \mathbf{w} = -\mathbf{s}$

101. $\mathbf{a} + \mathbf{w} = -2\mathbf{d}$

102. $\mathbf{a} + \mathbf{d} = \mathbf{0}$

103. $\mathbf{u} - \mathbf{v} = -2(\mathbf{b} + \mathbf{t})$

104. $\mathbf{t} - \mathbf{w} = \mathbf{b} - \mathbf{a}$

105. ***Think About It***   Consider two forces of equal magnitude acting on a point.

(a) If the magnitude of the resultant is the sum of the magnitudes of the two forces, make a conjecture about the angle between the forces.

(b) If the resultant of the forces is $\mathbf{0}$, make a conjecture about the angle between the forces.

(c) Can the magnitude of the resultant be greater than the sum of the magnitudes of the two forces? Explain.

106. ***Graphical Reasoning***   Consider two forces

$$\mathbf{F}_1 = \langle 10, 0\rangle \quad \text{and} \quad \mathbf{F}_2 = 5\langle \cos\theta, \sin\theta\rangle.$$

(a) Find $\|\mathbf{F}_1 + \mathbf{F}_2\|$ as a function of θ.

(b) Use a graphing utility to graph the function for $0 \le \theta < 2\pi$.

(c) Use the graph in part (b) to determine the range of the function. What is its maximum, and for what value of θ does it occur? What is its minimum, and for what value of θ does it occur?

(d) Explain why the magnitude of the resultant is never 0.

107. ***Proof***   Prove that $(\cos\theta)\mathbf{i} + (\sin\theta)\mathbf{j}$ is a unit vector for any value of θ.

108. ***Technology***   Write a program for your graphing utility that graphs two vectors and their difference given the vectors in component form.

In Exercises 109 and 110, use the program in Exercise 108 to find the difference of the vectors shown in the graph.

109.   **110.**

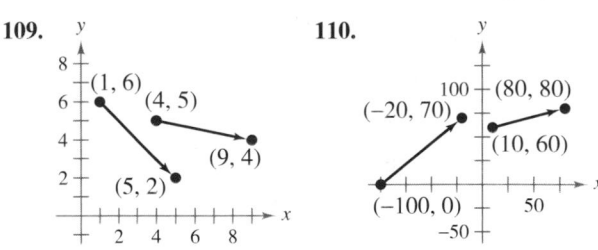

Skills Review

In Exercises 111–116, simplify the expression.

111. $\left(\dfrac{6x^4}{7y^{-2}}\right)(14x^{-1}y^5)$

112. $(5s^5t^{-5})\left(\dfrac{3s^{-2}}{50t^{-1}}\right)$

113. $(18x)^0(4xy)^2(3x^{-1})$

114. $(5ab^2)(a^{-3}b^0)(2a^0b)^{-2}$

115. $(2.1 \times 10^9)(3.4 \times 10^{-4})$

116. $(6.5 \times 10^6)(3.8 \times 10^4)$

∫ **In Exercises 117–120, use the trigonometric substitution to write the algebraic expression as a trigonometric function of θ, where $0 < \theta < \pi/2$.**

117. $\sqrt{49 - x^2}$, $x = 7\sin\theta$

118. $\sqrt{x^2 - 49}$, $x = 7\sec\theta$

119. $\sqrt{x^2 + 100}$, $x = 10\cot\theta$

120. $\sqrt{x^2 - 4}$, $x = 2\csc\theta$

In Exercises 121–124, solve the equation.

121. $\cos x(\cos x + 1) = 0$

122. $\sin x(2\sin x + \sqrt{2}) = 0$

123. $3\sec x + 4 = 10$

124. $\cos x\cot x - \cos x = 0$

6.4 Vectors and Dot Products

The Dot Product of Two Vectors

So far you have studied two vector operations—vector addition and multiplication by a scalar—each of which yields another vector. In this section, you will study a third vector operation, the **dot product.** This product yields a scalar, rather than a vector.

Definition of Dot Product

The **dot product** of $\mathbf{u} = \langle u_1, u_2 \rangle$ and $\mathbf{v} = \langle v_1, v_2 \rangle$ is given by

$$\mathbf{u} \cdot \mathbf{v} = u_1v_1 + u_2v_2.$$

Properties of the Dot Product (See the proofs on page 471.)

Let $\mathbf{u}$, $\mathbf{v}$, and $\mathbf{w}$ be vectors in the plane or in space and let c be a scalar.

1. $\mathbf{u} \cdot \mathbf{v} = \mathbf{v} \cdot \mathbf{u}$
2. $\mathbf{0} \cdot \mathbf{v} = 0$
3. $\mathbf{u} \cdot (\mathbf{v} + \mathbf{w}) = \mathbf{u} \cdot \mathbf{v} + \mathbf{u} \cdot \mathbf{w}$
4. $\mathbf{v} \cdot \mathbf{v} = \|\mathbf{v}\|^2$
5. $c(\mathbf{u} \cdot \mathbf{v}) = c\mathbf{u} \cdot \mathbf{v} = \mathbf{u} \cdot c\mathbf{v}$

Example 1 Finding Dot Products

Find each dot product.
a. $\langle 4, 5 \rangle \cdot \langle 2, 3 \rangle$
b. $\langle 2, -1 \rangle \cdot \langle 1, 2 \rangle$
c. $\langle 0, 3 \rangle \cdot \langle 4, -2 \rangle$

Solution

a. $\langle 4, 5 \rangle \cdot \langle 2, 3 \rangle = 4(2) + 5(3) = 8 + 15 = 23$
b. $\langle 2, -1 \rangle \cdot \langle 1, 2 \rangle = 2(1) + (-1)(2) = 2 - 2 = 0$
c. $\langle 0, 3 \rangle \cdot \langle 4, -2 \rangle = 0(4) + 3(-2) = 0 - 6 = -6$

✓CHECKPOINT Now try Exercise 1.

In Example 1, be sure you see that the dot product of two vectors is a scalar (a real number), not a vector. Moreover, notice that the dot product can be positive, zero, or negative.

What you should learn

- Find the dot product of two vectors and use properties of the dot product.
- Find angles between vectors and determine whether two vectors are orthogonal.
- Write vectors as sums of two vector components.
- Use vectors to find the work done by a force.

Why you should learn it

You can use the dot product of two vectors to solve real-life problems involving two vector quantities. For instance, Exercise 61 on page 446 shows you how the dot product can be used to find the force necessary to keep a truck from rolling down a hill.

Alan Thornton/Getty Images

Example 2   Using Properties of Dot Products

Let $\mathbf{u} = \langle -1, 3 \rangle$, $\mathbf{v} = \langle 2, -4 \rangle$, and $\mathbf{w} = \langle 1, -2 \rangle$. Find each dot product.

a. $(\mathbf{u} \cdot \mathbf{v})\mathbf{w}$ **b.** $\mathbf{u} \cdot 2\mathbf{v}$

Solution

Begin by finding the dot product of $\mathbf{u}$ and $\mathbf{v}$.

$$\begin{aligned}
\mathbf{u} \cdot \mathbf{v} &= \langle -1, 3 \rangle \cdot \langle 2, -4 \rangle \\
&= (-1)(2) + 3(-4) \\
&= -14
\end{aligned}$$

a. $\begin{aligned}[t] (\mathbf{u} \cdot \mathbf{v})\mathbf{w} &= -14\langle 1, -2 \rangle \\ &= \langle -14, 28 \rangle \end{aligned}$

b. $\begin{aligned}[t] \mathbf{u} \cdot 2\mathbf{v} &= 2(\mathbf{u} \cdot \mathbf{v}) \\ &= 2(-14) \\ &= -28 \end{aligned}$

Notice that the product in part (a) is a vector, whereas the product in part (b) is a scalar. Can you see why?

✔*CHECKPOINT*   Now try Exercise 9.

Example 3   Dot Product and Magnitude

The dot product of $\mathbf{u}$ with itself is 5. What is the magnitude of $\mathbf{u}$?

Solution

Because $\|\mathbf{u}\|^2 = \mathbf{u} \cdot \mathbf{u} = 5$, it follows that

$$\begin{aligned}
\|\mathbf{u}\| &= \sqrt{\mathbf{u} \cdot \mathbf{u}} \\
&= \sqrt{5}.
\end{aligned}$$

✔*CHECKPOINT*   Now try Exercise 11.

The Angle Between Two Vectors

The **angle between two nonzero vectors** is the angle θ, $0 \leq \theta \leq \pi$, between their respective standard position vectors, as shown in Figure 6.34. This angle can be found using the dot product. (Note that the angle between the zero vector and another vector is not defined.)

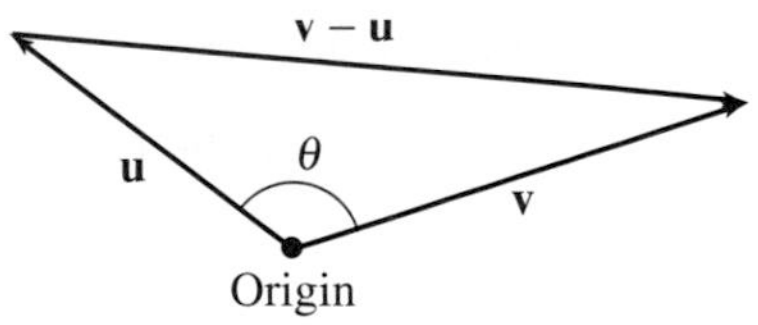

Figure 6.34

> **Angle Between Two Vectors**   (See the proof on page 471.)
>
> If θ is the angle between two nonzero vectors $\mathbf{u}$ and $\mathbf{v}$, then
>
> $$\cos \theta = \frac{\mathbf{u} \cdot \mathbf{v}}{\|\mathbf{u}\| \, \|\mathbf{v}\|}.$$

Example 4   Finding the Angle Between Two Vectors

Find the angle between $\mathbf{u} = \langle 4, 3 \rangle$ and $\mathbf{v} = \langle 3, 5 \rangle$.

Solution

$$\cos \theta = \frac{\mathbf{u} \cdot \mathbf{v}}{\|\mathbf{u}\| \, \|\mathbf{v}\|}$$

$$= \frac{\langle 4, 3 \rangle \cdot \langle 3, 5 \rangle}{\|\langle 4, 3 \rangle\| \, \|\langle 3, 5 \rangle\|}$$

$$= \frac{27}{5\sqrt{34}}$$

This implies that the angle between the two vectors is

$$\theta = \arccos \frac{27}{5\sqrt{34}} \approx 22.2°$$

as shown in Figure 6.35.

✓ **CHECKPOINT**   Now try Exercise 17.

Rewriting the expression for the angle between two vectors in the form

$$\mathbf{u} \cdot \mathbf{v} = \|\mathbf{u}\| \, \|\mathbf{v}\| \cos \theta \qquad \text{Alternative form of dot product}$$

produces an alternative way to calculate the dot product. From this form, you can see that because $\|\mathbf{u}\|$ and $\|\mathbf{v}\|$ are always positive, $\mathbf{u} \cdot \mathbf{v}$ and $\cos \theta$ will always have the same sign. Figure 6.36 shows the five possible orientations of two vectors.

TECHNOLOGY TIP

The graphing utility program Finding the Angle Between Two Vectors, found at this textbook's *Online Study Center*, graphs two vectors $\mathbf{u} = \langle a, b \rangle$ and $\mathbf{v} = \langle c, d \rangle$ in standard position and finds the measure of the angle between them. Use the program to verify Example 4.

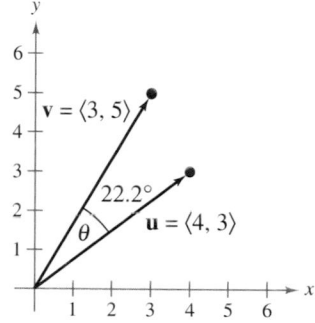

Figure 6.35

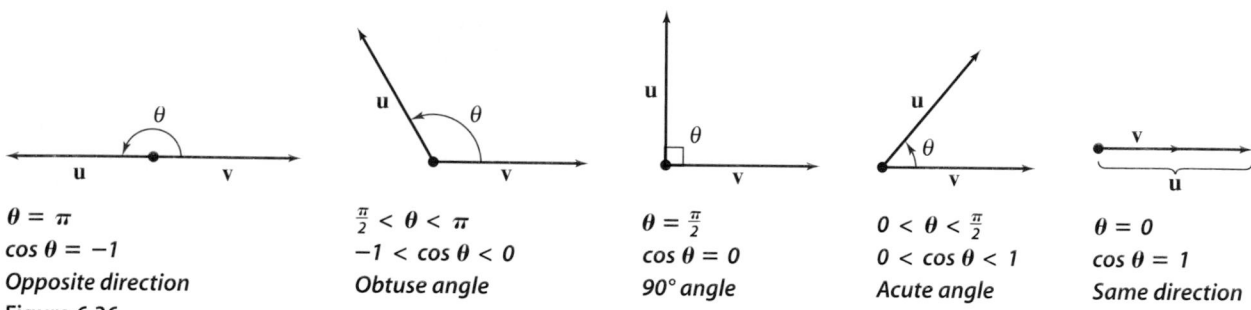

$\theta = \pi$	$\frac{\pi}{2} < \theta < \pi$	$\theta = \frac{\pi}{2}$	$0 < \theta < \frac{\pi}{2}$	$\theta = 0$
$\cos \theta = -1$	$-1 < \cos \theta < 0$	$\cos \theta = 0$	$0 < \cos \theta < 1$	$\cos \theta = 1$
Opposite direction	*Obtuse angle*	*90° angle*	*Acute angle*	*Same direction*

Figure 6.36

Definition of Orthogonal Vectors

The vectors $\mathbf{u}$ and $\mathbf{v}$ are **orthogonal** if $\mathbf{u} \cdot \mathbf{v} = 0$.

The terms *orthogonal* and *perpendicular* mean essentially the same thing—meeting at right angles. Even though the angle between the zero vector and another vector is not defined, it is convenient to extend the definition of orthogonality to include the zero vector. In other words, the zero vector is orthogonal to every vector $\mathbf{u}$ because $\mathbf{0} \cdot \mathbf{u} = 0$.

Example 5 Determining Orthogonal Vectors

Are the vectors $\mathbf{u} = \langle 2, -3 \rangle$ and $\mathbf{v} = \langle 6, 4 \rangle$ orthogonal?

Solution

Begin by finding the dot product of the two vectors.

$$\mathbf{u} \cdot \mathbf{v} = \langle 2, -3 \rangle \cdot \langle 6, 4 \rangle = 2(6) + (-3)(4) = 0$$

Because the dot product is 0, the two vectors are orthogonal, as shown in Figure 6.37.

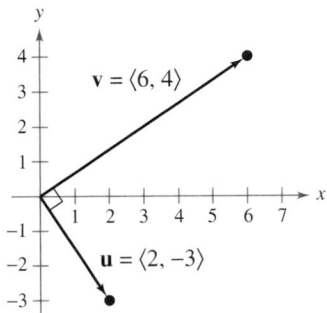

Figure 6.37

✓CHECKPOINT   Now try Exercise 35.

Finding Vector Components

You have already seen applications in which two vectors are added to produce a resultant vector. Many applications in physics and engineering pose the reverse problem—decomposing a given vector into the sum of two **vector components.**

Consider a boat on an inclined ramp, as shown in Figure 6.38. The force $\mathbf{F}$ due to gravity pulls the boat *down* the ramp and *against* the ramp. These two orthogonal forces, $\mathbf{w}_1$ and $\mathbf{w}_2$, are vector components of $\mathbf{F}$. That is,

$$\mathbf{F} = \mathbf{w}_1 + \mathbf{w}_2. \qquad \text{Vector components of } \mathbf{F}$$

The negative of component $\mathbf{w}_1$ represents the force needed to keep the boat from rolling down the ramp, and $\mathbf{w}_2$ represents the force that the tires must withstand against the ramp. A procedure for finding $\mathbf{w}_1$ and $\mathbf{w}_2$ is shown on the next page.

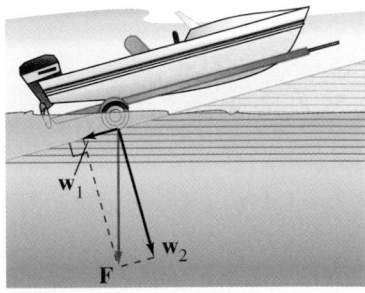

Figure 6.38

Definition of Vector Components

Let $\mathbf{u}$ and $\mathbf{v}$ be nonzero vectors such that

$$\mathbf{u} = \mathbf{w}_1 + \mathbf{w}_2$$

where $\mathbf{w}_1$ and $\mathbf{w}_2$ are orthogonal and $\mathbf{w}_1$ is parallel to (or a scalar multiple of) $\mathbf{v}$, as shown in Figure 6.39. The vectors $\mathbf{w}_1$ and $\mathbf{w}_2$ are called **vector components** of $\mathbf{u}$. The vector $\mathbf{w}_1$ is the **projection** of $\mathbf{u}$ onto $\mathbf{v}$ and is denoted by

$$\mathbf{w}_1 = \text{proj}_{\mathbf{v}}\mathbf{u}.$$

The vector $\mathbf{w}_2$ is given by $\mathbf{w}_2 = \mathbf{u} - \mathbf{w}_1$.

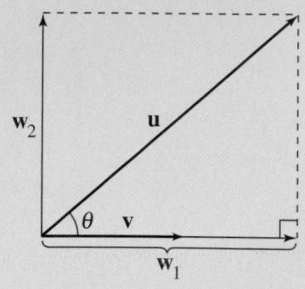
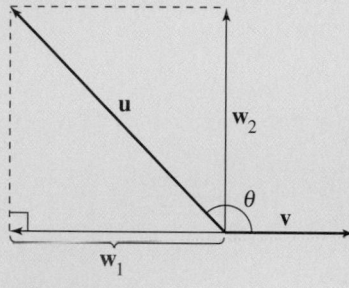

θ *is acute.* θ *is obtuse*

Figure 6.39

From the definition of vector components, you can see that it is easy to find the component $\mathbf{w}_2$ once you have found the projection of $\mathbf{u}$ onto $\mathbf{v}$. To find the projection, you can use the dot product, as follows.

$$\mathbf{u} = \mathbf{w}_1 + \mathbf{w}_2 = c\mathbf{v} + \mathbf{w}_2 \qquad \text{$\mathbf{w}_1$ is a scalar multiple of $\mathbf{v}$.}$$

$$\mathbf{u} \cdot \mathbf{v} = (c\mathbf{v} + \mathbf{w}_2) \cdot \mathbf{v} \qquad \text{Take dot product of each side with $\mathbf{v}$.}$$

$$= c\mathbf{v} \cdot \mathbf{v} + \mathbf{w}_2 \cdot \mathbf{v}$$

$$= c\|\mathbf{v}\|^2 + 0 \qquad \text{$\mathbf{w}_2$ and $\mathbf{v}$ are orthogonal.}$$

So,

$$c = \frac{\mathbf{u} \cdot \mathbf{v}}{\|\mathbf{v}\|^2}$$

and

$$\mathbf{w}_1 = \text{proj}_{\mathbf{v}}\mathbf{u} = c\mathbf{v} = \frac{\mathbf{u} \cdot \mathbf{v}}{\|\mathbf{v}\|^2}\mathbf{v}.$$

Projection of $\mathbf{u}$ onto $\mathbf{v}$

Let $\mathbf{u}$ and $\mathbf{v}$ be nonzero vectors. The projection of $\mathbf{u}$ onto $\mathbf{v}$ is given by

$$\text{proj}_{\mathbf{v}}\mathbf{u} = \left(\frac{\mathbf{u} \cdot \mathbf{v}}{\|\mathbf{v}\|^2}\right)\mathbf{v}.$$

Example 6   Decomposing a Vector into Components

Find the projection of $\mathbf{u} = \langle 3, -5 \rangle$ onto $\mathbf{v} = \langle 6, 2 \rangle$. Then write $\mathbf{u}$ as the sum of two orthogonal vectors, one of which is $\text{proj}_\mathbf{v}\mathbf{u}$.

Solution

The projection of $\mathbf{u}$ onto $\mathbf{v}$ is

$$\mathbf{w}_1 = \text{proj}_\mathbf{v}\mathbf{u} = \left(\frac{\mathbf{u} \cdot \mathbf{v}}{\|\mathbf{v}\|^2}\right)\mathbf{v} = \left(\frac{8}{40}\right)\langle 6, 2 \rangle = \left\langle \frac{6}{5}, \frac{2}{5} \right\rangle$$

as shown in Figure 6.40. The other component, $\mathbf{w}_2$, is

$$\mathbf{w}_2 = \mathbf{u} - \mathbf{w}_1 = \langle 3, -5 \rangle - \left\langle \frac{6}{5}, \frac{2}{5} \right\rangle = \left\langle \frac{9}{5}, -\frac{27}{5} \right\rangle.$$

So, $\mathbf{u} = \mathbf{w}_1 + \mathbf{w}_2 = \left\langle \frac{6}{5}, \frac{2}{5} \right\rangle + \left\langle \frac{9}{5}, -\frac{27}{5} \right\rangle = \langle 3, -5 \rangle.$

CHECKPOINT   Now try Exercise 45.

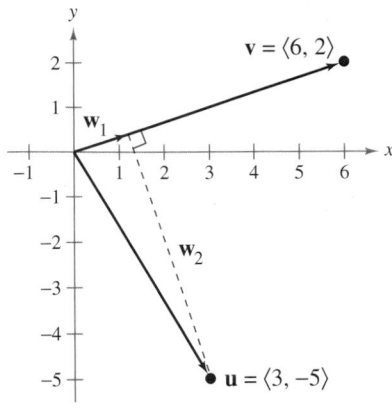

Figure 6.40

Example 7   Finding a Force

A 200-pound cart sits on a ramp inclined at $30°$, as shown in Figure 6.41. What force is required to keep the cart from rolling down the ramp?

Solution

Because the force due to gravity is vertical and downward, you can represent the gravitational force by the vector

$$\mathbf{F} = -200\mathbf{j}. \qquad \text{\small Force due to gravity}$$

To find the force required to keep the cart from rolling down the ramp, project $\mathbf{F}$ onto a unit vector $\mathbf{v}$ in the direction of the ramp, as follows.

$$\mathbf{v} = (\cos 30°)\mathbf{i} + (\sin 30°)\mathbf{j} = \frac{\sqrt{3}}{2}\mathbf{i} + \frac{1}{2}\mathbf{j} \qquad \text{\small Unit vector along ramp}$$

Figure 6.41

Therefore, the projection of $\mathbf{F}$ onto $\mathbf{v}$ is

$$\mathbf{w}_1 = \text{proj}_\mathbf{v}\mathbf{F} = \left(\frac{\mathbf{F} \cdot \mathbf{v}}{\|\mathbf{v}\|^2}\right)\mathbf{v}$$

$$= (\mathbf{F} \cdot \mathbf{v})\mathbf{v}$$

$$= (-200)\left(\frac{1}{2}\right)\mathbf{v}$$

$$= -100\left(\frac{\sqrt{3}}{2}\mathbf{i} + \frac{1}{2}\mathbf{j}\right).$$

The magnitude of this force is 100, and therefore a force of 100 pounds is required to keep the cart from rolling down the ramp.

CHECKPOINT   Now try Exercise 61.

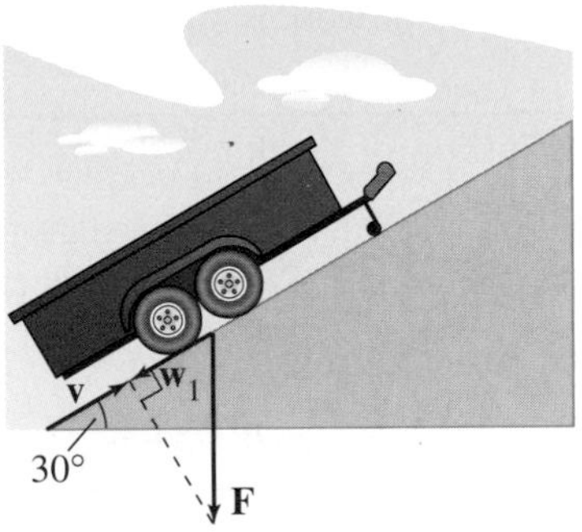

Work

The work W done by a constant force $\mathbf{F}$ acting along the line of motion of an object is given by

$$W = (\text{magnitude of force})(\text{distance}) = \|\mathbf{F}\|\|\overrightarrow{PQ}\|$$

as shown in Figure 6.42. If the constant force $\mathbf{F}$ is not directed along the line of motion (see Figure 6.43), the work W done by the force is given by

$$W = \|\text{proj}_{\overrightarrow{PQ}} \mathbf{F}\|\|\overrightarrow{PQ}\| \qquad \text{Projection form for work}$$

$$= (\cos \theta)\|\mathbf{F}\|\|\overrightarrow{PQ}\| \qquad \|\text{proj}_{\overrightarrow{PQ}} \mathbf{F}\| = (\cos \theta)\|\mathbf{F}\|$$

$$= \mathbf{F} \cdot \overrightarrow{PQ}. \qquad \text{Dot product form for work}$$

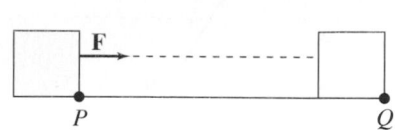

Force acts along the line of motion.

Figure 6.42

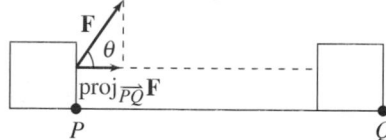

Force acts at angle θ with the line of motion.

Figure 6.43

This notion of work is summarized in the following definition.

Definition of Work

The **work** W done by a constant force $\mathbf{F}$ as its point of application moves along the vector $\overrightarrow{PQ}$ is given by either of the following.

1. $W = \|\text{proj}_{\overrightarrow{PQ}} \mathbf{F}\|\|\overrightarrow{PQ}\|$ Projection form

2. $W = \mathbf{F} \cdot \overrightarrow{PQ}$ Dot product form

---

Example 8 Finding Work

To close a barn's sliding door, a person pulls on a rope with a constant force of 50 pounds at a constant angle of 60°, as shown in Figure 6.44. Find the work done in moving the door 12 feet to its closed position.

Solution

Using a projection, you can calculate the work as follows.

$$W = \|\text{proj}_{\overrightarrow{PQ}} \mathbf{F}\|\|\overrightarrow{PQ}\| \qquad \text{Projection form for work}$$

$$= (\cos 60°)\|\mathbf{F}\|\|\overrightarrow{PQ}\|$$

$$= \frac{1}{2}(50)(12) = 300 \text{ foot-pounds}$$

So, the work done is 300 foot-pounds. You can verify this result by finding the vectors $\mathbf{F}$ and $\overrightarrow{PQ}$ and calculating their dot product.

Figure 6.44

✓**CHECKPOINT**   Now try Exercise 63.

6.4 Exercises

See www.CalcChat.com for worked-out solutions to odd-numbered exercises.

Vocabulary Check

Fill in the blanks.

1. The _____ of two vectors yields a scalar, rather than a vector.

2. If θ is the angle between two nonzero vectors $\mathbf{u}$ and $\mathbf{v}$, then $\cos\theta =$ _____ .

3. The vectors $\mathbf{u}$ and $\mathbf{v}$ are _____ if $\mathbf{u} \cdot \mathbf{v} = 0$.

4. The projection of $\mathbf{u}$ onto $\mathbf{v}$ is given by $\text{proj}_{\mathbf{v}}\mathbf{u} =$ _____ .

5. The work W done by a constant force $\mathbf{F}$ as its point of application moves along the vector $\overrightarrow{PQ}$ is given by either $W =$ _____ or $W =$ _____ .

In Exercises 1–4, find the dot product of u and v.

1. $\mathbf{u} = \langle 6, 3 \rangle$
   $\mathbf{v} = \langle 2, -4 \rangle$

2. $\mathbf{u} = \langle -4, 1 \rangle$
   $\mathbf{v} = \langle 2, -3 \rangle$

3. $\mathbf{u} = 5\mathbf{i} + \mathbf{j}$
   $\mathbf{v} = 3\mathbf{i} - \mathbf{j}$

4. $\mathbf{u} = 3\mathbf{i} + 2\mathbf{j}$
   $\mathbf{v} = -2\mathbf{i} + \mathbf{j}$

In Exercises 5–10, use the vectors $\mathbf{u} = \langle 2, 2 \rangle$, $\mathbf{v} = \langle -3, 4 \rangle$, and $\mathbf{w} = \langle 1, -4 \rangle$ to find the indicated quantity. State whether the result is a vector or a scalar.

5. $\mathbf{u} \cdot \mathbf{u}$

6. $\mathbf{v} \cdot \mathbf{w}$

7. $\mathbf{u} \cdot 2\mathbf{v}$

8. $4\mathbf{u} \cdot \mathbf{v}$

9. $(3\mathbf{w} \cdot \mathbf{v})\mathbf{u}$

10. $(\mathbf{u} \cdot 2\mathbf{v})\mathbf{w}$

In Exercises 11–16, use the dot product to find the magnitude of u.

11. $\mathbf{u} = \langle -5, 12 \rangle$

12. $\mathbf{u} = \langle 2, -4 \rangle$

13. $\mathbf{u} = 20\mathbf{i} + 25\mathbf{j}$

14. $\mathbf{u} = 6\mathbf{i} - 10\mathbf{j}$

15. $\mathbf{u} = -4\mathbf{j}$

16. $\mathbf{u} = 9\mathbf{i}$

In Exercises 17–24, find the angle θ between the vectors.

17. $\mathbf{u} = \langle -1, 0 \rangle$
    $\mathbf{v} = \langle 0, 2 \rangle$

18. $\mathbf{u} = \langle 4, 4 \rangle$
    $\mathbf{v} = \langle -2, 0 \rangle$

19. $\mathbf{u} = 3\mathbf{i} + 4\mathbf{j}$
    $\mathbf{v} = -2\mathbf{i} + 3\mathbf{j}$

20. $\mathbf{u} = 2\mathbf{i} - 3\mathbf{j}$
    $\mathbf{v} = \mathbf{i} - 2\mathbf{j}$

21. $\mathbf{u} = 2\mathbf{i}$
    $\mathbf{v} = -3\mathbf{j}$

22. $\mathbf{u} = 4\mathbf{j}$
    $\mathbf{v} = -3\mathbf{i}$

23. $\mathbf{u} = \cos\left(\dfrac{\pi}{3}\right)\mathbf{i} + \sin\left(\dfrac{\pi}{3}\right)\mathbf{j}$

    $\mathbf{v} = \cos\left(\dfrac{3\pi}{4}\right)\mathbf{i} + \sin\left(\dfrac{3\pi}{4}\right)\mathbf{j}$

24. $\mathbf{u} = \cos\left(\dfrac{\pi}{4}\right)\mathbf{i} + \sin\left(\dfrac{\pi}{4}\right)\mathbf{j}$

    $\mathbf{v} = \cos\left(\dfrac{2\pi}{3}\right)\mathbf{i} + \sin\left(\dfrac{2\pi}{3}\right)\mathbf{j}$

In Exercises 25–28, graph the vectors and find the degree measure of the angle between the vectors.

25. $\mathbf{u} = 2\mathbf{i} - 4\mathbf{j}$
    $\mathbf{v} = 3\mathbf{i} - 5\mathbf{j}$

26. $\mathbf{u} = -6\mathbf{i} - 3\mathbf{j}$
    $\mathbf{v} = -8\mathbf{i} + 4\mathbf{j}$

27. $\mathbf{u} = 6\mathbf{i} - 2\mathbf{j}$
    $\mathbf{v} = 8\mathbf{i} - 5\mathbf{j}$

28. $\mathbf{u} = 2\mathbf{i} - 3\mathbf{j}$
    $\mathbf{v} = 4\mathbf{i} + 3\mathbf{j}$

In Exercises 29 and 30, use vectors to find the interior angles of the triangle with the given vertices.

29. $(1, 2), (3, 4), (2, 5)$

30. $(-3, 0), (2, 2), (0, 6)$

In Exercises 31 and 32, find $\mathbf{u} \cdot \mathbf{v}$, where θ is the angle between u and v.

31. $\|\mathbf{u}\| = 9, \|\mathbf{v}\| = 36, \theta = \dfrac{3\pi}{4}$

32. $\|\mathbf{u}\| = 4, \|\mathbf{v}\| = 12, \ \theta = \dfrac{\pi}{3}$

In Exercises 33–38, determine whether u and v are orthogonal, parallel, or neither.

33. $\mathbf{u} = \langle -12, 30 \rangle$
    $\mathbf{v} = \langle \frac{1}{2}, -\frac{5}{4} \rangle$

34. $\mathbf{u} = \langle 15, 45 \rangle$
    $\mathbf{v} = \langle -5, 12 \rangle$

35. $\mathbf{u} = \frac{1}{4}(3\mathbf{i} - \mathbf{j})$
    $\mathbf{v} = 5\mathbf{i} + 6\mathbf{j}$

36. $\mathbf{u} = \mathbf{j}$
    $\mathbf{v} = \mathbf{i} - 2\mathbf{j}$

37. $\mathbf{u} = 2\mathbf{i} - 2\mathbf{j}$
     $\mathbf{v} = -\mathbf{i} - \mathbf{j}$

38. $\mathbf{u} = 8\mathbf{i} + 4\mathbf{j}$
     $\mathbf{v} = -2\mathbf{i} - \mathbf{j}$

In Exercises 39–44, find the value of k so that the vectors u and v are orthogonal.

39. $\mathbf{u} = 2\mathbf{i} - k\mathbf{j}$
     $\mathbf{v} = 3\mathbf{i} + 2\mathbf{j}$

40. $\mathbf{u} = 3\mathbf{i} + 2\mathbf{j}$
     $\mathbf{v} = 2\mathbf{i} - k\mathbf{j}$

41. $\mathbf{u} = \mathbf{i} + 4\mathbf{j}$
     $\mathbf{v} = 2k\mathbf{i} - 5\mathbf{j}$

42. $\mathbf{u} = -3k\mathbf{i} + 5\mathbf{j}$
     $\mathbf{v} = 2\mathbf{i} - 4\mathbf{j}$

43. $\mathbf{u} = -3k\mathbf{i} + 2\mathbf{j}$
     $\mathbf{v} = -6\mathbf{i}$

44. $\mathbf{u} = 4\mathbf{i} - 4k\mathbf{j}$
     $\mathbf{v} = 3\mathbf{j}$

In Exercises 45–48, find the projection of u onto v. Then write u as the sum of two orthogonal vectors, one of which is $\text{proj}_{\mathbf{v}}\,\mathbf{u}$.

45. $\mathbf{u} = \langle 3, 4 \rangle$
     $\mathbf{v} = \langle 8, 2 \rangle$

46. $\mathbf{u} = \langle 4, 2 \rangle$
     $\mathbf{v} = \langle 1, -2 \rangle$

47. $\mathbf{u} = \langle 0, 3 \rangle$
     $\mathbf{v} = \langle 2, 15 \rangle$

48. $\mathbf{u} = \langle -5, -1 \rangle$
     $\mathbf{v} = \langle -1, 1 \rangle$

In Exercises 49–52, use the graph to determine mentally the projection of u onto v. (The coordinates of the terminal points of the vectors in standard position are given.) Use the formula for the projection of u onto v to verify your result.

49.
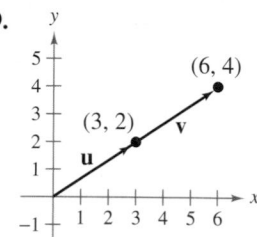

50.
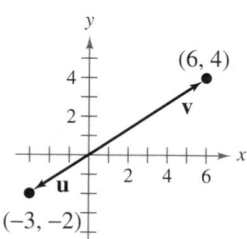

51.
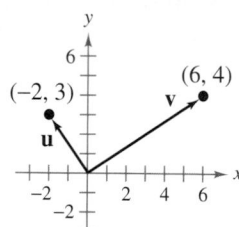

52.
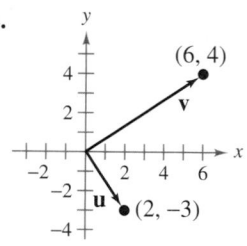

In Exercises 53–56, find two vectors in opposite directions that are orthogonal to the vector u. (There are many correct answers.)

53. $\mathbf{u} = \langle 2, 6 \rangle$

54. $\mathbf{u} = \langle -7, 5 \rangle$

55. $\mathbf{u} = \frac{1}{2}\mathbf{i} - \frac{3}{4}\mathbf{j}$

56. $\mathbf{u} = -\frac{5}{2}\mathbf{i} - 3\mathbf{j}$

Work   **In Exercises 57 and 58, find the work done in moving a particle from P to Q if the magnitude and direction of the force are given by v.**

57. $P = (0, 0)$, $Q = (4, 7)$, $\mathbf{v} = \langle 1, 4 \rangle$

58. $P = (1, 3)$, $Q = (-3, 5)$, $\mathbf{v} = -2\mathbf{i} + 3\mathbf{j}$

59. *Revenue*   The vector $\mathbf{u} = \langle 1245, 2600 \rangle$ gives the numbers of units of two types of picture frames produced by a company. The vector $\mathbf{v} = \langle 12.20, 8.50 \rangle$ gives the price (in dollars) of each frame, respectively. (a) Find the dot product $\mathbf{u} \cdot \mathbf{v}$ and explain its meaning in the context of the problem. (b) Identify the vector operation used to increase prices by 2 percent.

60. *Revenue*   The vector $\mathbf{u} = \langle 3240, 2450 \rangle$ gives the numbers of hamburgers and hot dogs, respectively, sold at a fast food stand in one week. The vector $\mathbf{v} = \langle 1.75, 1.25 \rangle$ gives the prices in dollars for the food items. (a) Find the dot product $\mathbf{u} \cdot \mathbf{v}$ and explain its meaning in the context of the problem. (b) Identify the vector operation used to increase prices by $2\frac{1}{2}$ percent.

61. *Braking Load*   A truck with a gross weight of 30,000 pounds is parked on a slope of $d°$ (see figure). Assume that the only force to overcome is the force of gravity.

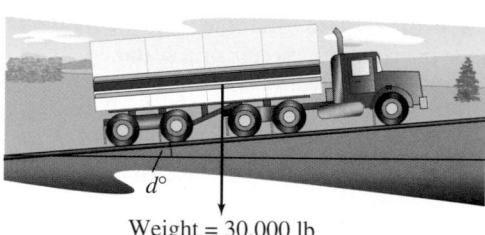

Weight = 30,000 lb

(a) Find the force required to keep the truck from rolling down the hill in terms of the slope d.

(b) Use a graphing utility to complete the table.

d	0°	1°	2°	3°	4°	5°
Force						

d	6°	7°	8°	9°	10°
Force					

(c) Find the force perpendicular to the hill when $d = 5°$.

62. *Braking Load*   A sport utility vehicle with a gross weight of 5400 pounds is parked on a slope of 10°. Assume that the only force to overcome is the force of gravity. Find the force required to keep the vehicle from rolling down the hill. Find the force perpendicular to the hill.

63. *Work*   A tractor pulls a log d meters and the tension in the cable connecting the tractor and log is approximately 1600 kilograms (15,691 newtons). The direction of the force is 30° above the horizontal (see figure).

(a) Find the work done in terms of the distance d.

(b) Use a graphing utility to complete the table.

d	0	200	400	800
Work				

64. *Work*   A force of 45 pounds in the direction of 30° above the horizontal is required to slide a table across a floor. Find the work done if the table is dragged 20 feet.

65. *Work*   One of the events in a local strongman contest is to pull a cement block 100 feet. If a force of 250 pounds was used to pull the block at an angle of 30° with the horizontal, find the work done in pulling the block.

66. *Work*   A toy wagon is pulled by exerting a force of 25 pounds on a handle that makes a 20° angle with the horizontal. Find the work done in pulling the wagon 50 feet.

67. *Work*   A toy wagon is pulled by exerting a force of 20 pounds on a handle that makes a 25° angle with the horizontal. Find the work done in pulling the wagon 40 feet.

68. *Work*   A mover exerts a horizontal force of 25 pounds on a crate as it is pushed up a ramp that is 12 feet long and inclined at an angle of 20° above the horizontal. Find the work done in pushing the crate up the ramp.

Synthesis

True or False?   **In Exercises 69 and 70, determine whether the statement is true or false. Justify your answer.**

69. The vectors $\mathbf{u} = \langle 0, 0 \rangle$ and $\mathbf{v} = \langle -12, 6 \rangle$ are orthogonal.

70. The work W done by a constant force $\mathbf{F}$ acting along the line of motion of an object is represented by a vector.

71. If $\mathbf{u} = \langle \cos \theta, \sin \theta \rangle$ and $\mathbf{v} = \langle \sin \theta, -\cos \theta \rangle$, are $\mathbf{u}$ and $\mathbf{v}$ orthogonal, parallel, or neither? Explain.

72. *Think About It*   What is known about θ, the angle between two nonzero vectors $\mathbf{u}$ and $\mathbf{v}$, if each of the following is true?

(a) $\mathbf{u} \cdot \mathbf{v} = 0$ (b) $\mathbf{u} \cdot \mathbf{v} > 0$ (c) $\mathbf{u} \cdot \mathbf{v} < 0$

73. *Think About It*   What can be said about the vectors $\mathbf{u}$ and $\mathbf{v}$ under each condition?

(a) The projection of $\mathbf{u}$ onto $\mathbf{v}$ equals $\mathbf{u}$.

(b) The projection of $\mathbf{u}$ onto $\mathbf{v}$ equals $\mathbf{0}$.

74. *Proof*   Use vectors to prove that the diagonals of a rhombus are perpendicular.

75. *Proof*   Prove the following.

$$\|\mathbf{u} - \mathbf{v}\|^2 = \|\mathbf{u}\|^2 + \|\mathbf{v}\|^2 - 2\mathbf{u} \cdot \mathbf{v}$$

76. *Proof*   Prove that if $\mathbf{u}$ is orthogonal to $\mathbf{v}$ and $\mathbf{w}$, then $\mathbf{u}$ is orthogonal to $c\mathbf{v} + d\mathbf{w}$ for any scalars c and d.

77. *Proof*   Prove that if $\mathbf{u}$ is a unit vector and θ is the angle between $\mathbf{u}$ and $\mathbf{i}$, then $\mathbf{u} = \cos \theta \mathbf{i} + \sin \theta \mathbf{j}$.

78. *Proof*   Prove that if $\mathbf{u}$ is a unit vector and θ is the angle between $\mathbf{u}$ and $\mathbf{j}$, then

$$\mathbf{u} = \cos\left(\frac{\pi}{2} - \theta\right)\mathbf{i} + \sin\left(\frac{\pi}{2} - \theta\right)\mathbf{j}.$$

Skills Review

In Exercises 79–82, describe how the graph of g is related to the graph of f.

79. $g(x) = f(x - 4)$ **80.** $g(x) = -f(x)$

81. $g(x) = f(x) + 6$ **82.** $g(x) = f(2x)$

In Exercises 83–90, perform the operation and write the result in standard form.

83. $\sqrt{-4} - 1$ **84.** $\sqrt{-8} + 5$

85. $3i(4 - 5i)$ **86.** $-2i(1 + 6i)$

87. $(1 + 3i)(1 - 3i)$ **88.** $(7 - 4i)(7 + 4i)$

89. $\dfrac{3}{1 + i} + \dfrac{2}{2 - 3i}$ **90.** $\dfrac{6}{4 - i} - \dfrac{3}{1 + i}$

In Exercises 91–94, plot the complex number in the complex plane.

91. $-2i$ **92.** $3i$

93. $1 + 8i$ **94.** $9 - 7i$

6.5 Trigonometric Form of a Complex Number

The Complex Plane

Recall from Section 2.4 that you can represent a complex number $z = a + bi$ as the point (a, b) in a coordinate plane (the complex plane). The horizontal axis is called the real axis and the vertical axis is called the imaginary axis, as shown in Figure 6.45.

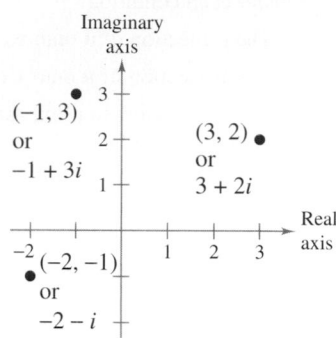

Figure 6.45

The **absolute value of a complex number** $a + bi$ is defined as the distance between the origin $(0, 0)$ and the point (a, b).

> **Definition of the Absolute Value of a Complex Number**
>
> The **absolute value** of the complex number $z = a + bi$ is given by
>
> $$|a + bi| = \sqrt{a^2 + b^2}.$$

If the complex number $a + bi$ is a real number (that is, if $b = 0$), then this definition agrees with that given for the absolute value of a real number

$$|a + 0i| = \sqrt{a^2 + 0^2} = |a|.$$

Example 1   Finding the Absolute Value of a Complex Number

Plot $z = -2 + 5i$ and find its absolute value.

Solution

The number is plotted in Figure 6.46. It has an absolute value of

$$|z| = \sqrt{(-2)^2 + 5^2}$$
$$= \sqrt{29}.$$

✔*CHECKPOINT*   Now try Exercise 5.

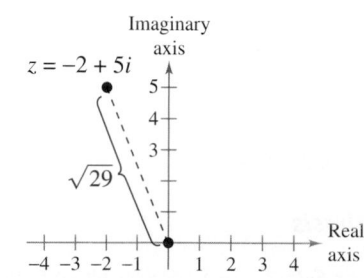

Figure 6.46

What you should learn

- Find absolute values of complex numbers.
- Write trigonometric forms of complex numbers.
- Multiply and divide complex numbers written in trigonometric form.
- Use DeMoivre's Theorem to find powers of complex numbers.
- Find *n*th roots of complex numbers.

Why you should learn it

You can use the trigonometric form of a complex number to perform operations with complex numbers. For instance, in Exercises 141–148 on page 459, you can use the trigonometric form of a complex number to help you solve polynomial equations.

Prerequisite Skills

To review complex numbers and the complex plane, see Section 2.4.

Trigonometric Form of a Complex Number

In Section 2.4 you learned how to add, subtract, multiply, and divide complex numbers. To work effectively with *powers* and *roots* of complex numbers, it is helpful to write complex numbers in trigonometric form. In Figure 6.47, consider the nonzero complex number $a + bi$. By letting θ be the angle from the positive real axis (measured counterclockwise) to the line segment connecting the origin and the point (a, b), you can write

$$a = r \cos \theta \qquad \text{and} \qquad b = r \sin \theta$$

where $r = \sqrt{a^2 + b^2}$. Consequently, you have

$$a + bi = (r \cos \theta) + (r \sin \theta)i$$

from which you can obtain the **trigonometric form of a complex number.**

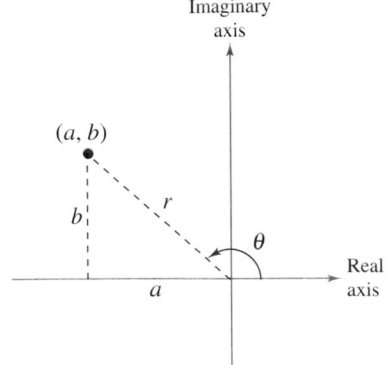

Figure 6.47

> **Trigonometric Form of a Complex Number**
>
> The **trigonometric form** of the complex number $z = a + bi$ is given by
>
> $$z = r(\cos \theta + i \sin \theta)$$
>
> where $a = r \cos \theta$, $b = r \sin \theta$, $r = \sqrt{a^2 + b^2}$, and $\tan \theta = b/a$.
> The number r is the **modulus** of z, and θ is called an **argument** of z.

The trigonometric form of a complex number is also called the *polar form.* Because there are infinitely many choices for θ, the trigonometric form of a complex number is not unique. Normally, θ is restricted to the interval $0 \leq \theta < 2\pi$, although on occasion it is convenient to use $\theta < 0$.

Example 2   Writing a Complex Number in Trigonometric Form

Write the complex number $z = -2 - 2\sqrt{3}i$ in trigonometric form.

Solution

The absolute value of z is

$$r = \left| -2 - 2\sqrt{3}i \right| = \sqrt{(-2)^2 + \left(-2\sqrt{3}\right)^2} = \sqrt{16} = 4$$

and the angle θ is given by

$$\tan \theta = \frac{b}{a} = \frac{-2\sqrt{3}}{-2} = \sqrt{3}.$$

Because $\tan(\pi/3) = \sqrt{3}$ and $z = -2 - 2\sqrt{3}i$ lies in Quadrant III, choose θ to be $\theta = \pi + \pi/3 = 4\pi/3$. So, the trigonometric form is

$$z = r(\cos \theta + i \sin \theta) = 4\left(\cos \frac{4\pi}{3} + i \sin \frac{4\pi}{3} \right).$$

See Figure 6.48.

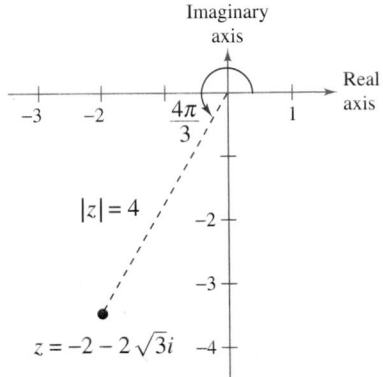

Figure 6.48

✓CHECKPOINT   Now try Exercise 19.

Example 3   Writing a Complex Number in Standard Form

Write the complex number in standard form $a + bi$.

$$z = \sqrt{8}\left[\cos\left(-\frac{\pi}{3}\right) + i \sin\left(-\frac{\pi}{3}\right)\right]$$

Solution

Because $\cos(-\pi/3) = 1/2$ and $\sin(-\pi/3) = -\sqrt{3}/2$, you can write

$$z = \sqrt{8}\left[\cos\left(-\frac{\pi}{3}\right) + i \sin\left(-\frac{\pi}{3}\right)\right]$$

$$= \sqrt{8}\left[\frac{1}{2} - \frac{\sqrt{3}}{2}i\right]$$

$$= 2\sqrt{2}\left[\frac{1}{2} - \frac{\sqrt{3}}{2}i\right] = \sqrt{2} - \sqrt{6}i.$$

✓CHECKPOINT   Now try Exercise 37.

TECHNOLOGY TIP

A graphing utility can be used to convert a complex number in trigonometric form to standard form. For instance, enter the complex number $\sqrt{2}(\cos \pi/4 + i \sin \pi/4)$ in your graphing utility and press ENTER. You should obtain the standard form $1 + i$, as shown below.

```
√(2)(cos(π/4)+is
in(π/4))
              1+1i
```

Multiplication and Division of Complex Numbers

The trigonometric form adapts nicely to multiplication and division of complex numbers. Suppose you are given two complex numbers

$$z_1 = r_1(\cos \theta_1 + i \sin \theta_1) \qquad \text{and} \qquad z_2 = r_2(\cos \theta_2 + i \sin \theta_2).$$

The product of z_1 and z_2 is

$$z_1 z_2 = r_1 r_2(\cos \theta_1 + i \sin \theta_1)(\cos \theta_2 + i \sin \theta_2)$$

$$= r_1 r_2[(\cos \theta_1 \cos \theta_2 - \sin \theta_1 \sin \theta_2) + i(\sin \theta_1 \cos \theta_2 + \cos \theta_1 \sin \theta_2)].$$

Using the sum and difference formulas for cosine and sine, you can rewrite this equation as

$$z_1 z_2 = r_1 r_2[\cos(\theta_1 + \theta_2) + i \sin(\theta_1 + \theta_2)].$$

This establishes the first part of the following rule. The second part is left for you to verify (see Exercise 158).

Product and Quotient of Two Complex Numbers

Let $z_1 = r_1(\cos \theta_1 + i \sin \theta_1)$ and $z_2 = r_2(\cos \theta_2 + i \sin \theta_2)$ be complex numbers.

$$z_1 z_2 = r_1 r_2[\cos(\theta_1 + \theta_2) + i \sin(\theta_1 + \theta_2)] \qquad \text{Product}$$

$$\frac{z_1}{z_2} = \frac{r_1}{r_2}[\cos(\theta_1 - \theta_2) + i \sin(\theta_1 - \theta_2)], \qquad z_2 \neq 0 \qquad \text{Quotient}$$

Note that this rule says that to *multiply* two complex numbers you multiply moduli and add arguments, whereas to *divide* two complex numbers you divide moduli and subtract arguments.

Example 4   **Multiplying Complex Numbers in Trigonometric Form**

Find the product $z_1 z_2$ of the complex numbers.

$$z_1 = 2\left(\cos\frac{2\pi}{3} + i\sin\frac{2\pi}{3}\right) \qquad z_2 = 8\left(\cos\frac{11\pi}{6} + i\sin\frac{11\pi}{6}\right)$$

Solution

$$z_1 z_2 = 2\left(\cos\frac{2\pi}{3} + i\sin\frac{2\pi}{3}\right) \cdot 8\left(\cos\frac{11\pi}{6} + i\sin\frac{11\pi}{6}\right)$$

$$= 16\left[\cos\left(\frac{2\pi}{3} + \frac{11\pi}{6}\right) + i\sin\left(\frac{2\pi}{3} + \frac{11\pi}{6}\right)\right]$$

$$= 16\left(\cos\frac{5\pi}{2} + i\sin\frac{5\pi}{2}\right)$$

$$= 16\left(\cos\frac{\pi}{2} + i\sin\frac{\pi}{2}\right)$$

$$= 16[0 + i(1)] = 16i$$

You can check this result by first converting to the standard forms $z_1 = -1 + \sqrt{3}i$ and $z_2 = 4\sqrt{3} - 4i$ and then multiplying algebraically, as in Section 2.4.

$$z_1 z_2 = \left(-1 + \sqrt{3}i\right)\left(4\sqrt{3} - 4i\right)$$

$$= -4\sqrt{3} + 4i + 12i + 4\sqrt{3} = 16i$$

✓*CHECKPOINT*   Now try Exercise 55.

Example 5   **Dividing Complex Numbers in Trigonometric Form**

Find the quotient z_1/z_2 of the complex numbers.

$$z_1 = 24(\cos 300° + i\sin 300°) \qquad z_2 = 8(\cos 75° + i\sin 75°)$$

Solution

$$\frac{z_1}{z_2} = \frac{24(\cos 300° + i\sin 300°)}{8(\cos 75° + i\sin 75°)}$$

$$= \frac{24}{8}[\cos(300° - 75°) + i\sin(300° - 75°)]$$

$$= 3(\cos 225° + i\sin 225°)$$

$$= 3\left[\left(-\frac{\sqrt{2}}{2}\right) + i\left(-\frac{\sqrt{2}}{2}\right)\right] = -\frac{3\sqrt{2}}{2} - \frac{3\sqrt{2}}{2}i$$

✓*CHECKPOINT*   Now try Exercise 61.

Powers of Complex Numbers

The trigonometric form of a complex number is used to raise a complex number to a power. To accomplish this, consider repeated use of the multiplication rule.

$$z = r(\cos \theta + i \sin \theta)$$

$$z^2 = r(\cos \theta + i \sin \theta)r(\cos \theta + i \sin \theta) = r^2(\cos 2\theta + i \sin 2\theta)$$

$$z^3 = r^2(\cos 2\theta + i \sin 2\theta)r(\cos \theta + i \sin \theta) = r^3(\cos 3\theta + i \sin 3\theta)$$

$$z^4 = r^4(\cos 4\theta + i \sin 4\theta)$$

$$z^5 = r^5(\cos 5\theta + i \sin 5\theta)$$

$$\vdots$$

This pattern leads to **DeMoivre's Theorem,** which is named after the French mathematician Abraham DeMoivre (1667–1754).

DeMoivre's Theorem

If $z = r(\cos \theta + i \sin \theta)$ is a complex number and n is a positive integer, then

$$z^n = [r(\cos \theta + i \sin \theta)]^n$$

$$= r^n(\cos n\theta + i \sin n\theta).$$

Exploration

Plot the numbers i, i^2, i^3, i^4, and i^5 in the complex plane. Write each number in trigonometric form and describe what happens to the angle θ as you form higher powers of i^n.

Example 6   Finding Powers of a Complex Number

Use DeMoivre's Theorem to find $\left(-1 + \sqrt{3}i\right)^{12}$.

Solution

First convert the complex number to trigonometric form using

$$r = \sqrt{(-1)^2 + \left(\sqrt{3}\right)^2} = 2 \quad \text{and} \quad \theta = \arctan \frac{\sqrt{3}}{-1} = \frac{2\pi}{3}.$$

So, the trigonometric form is

$$-1 + \sqrt{3}i = 2\left(\cos \frac{2\pi}{3} + i \sin \frac{2\pi}{3}\right).$$

Then, by DeMoivre's Theorem, you have

$$\left(-1 + \sqrt{3}i\right)^{12} = \left[2\left(\cos \frac{2\pi}{3} + i \sin \frac{2\pi}{3}\right)\right]^{12}$$

$$= 2^{12}\left[\cos\left(12 \cdot \frac{2\pi}{3}\right) + i \sin\left(12 \cdot \frac{2\pi}{3}\right)\right]$$

$$= 4096(\cos 8\pi + i \sin 8\pi)$$

$$= 4096(1 + 0) = 4096.$$

✓**CHECKPOINT**   Now try Exercise 91.

Roots of Complex Numbers

Recall that a consequence of the Fundamental Theorem of Algebra is that a polynomial equation of degree n has n solutions in the complex number system. So, an equation such as $x^6 = 1$ has six solutions, and in this particular case you can find the six solutions by factoring and using the Quadratic Formula.

$$x^6 - 1 = 0$$

$$(x^3 - 1)(x^3 + 1) = 0$$

$$(x - 1)(x^2 + x + 1)(x + 1)(x^2 - x + 1) = 0$$

Consequently, the solutions are

$$x = \pm 1, \qquad x = \frac{-1 \pm \sqrt{3}i}{2}, \qquad \text{and} \qquad x = \frac{1 \pm \sqrt{3}i}{2}.$$

Each of these numbers is a sixth root of 1. In general, the **nth root of a complex number** is defined as follows.

Definition of an nth Root of a Complex Number

The complex number $u = a + bi$ is an **nth root** of the complex number z if

$$z = u^n = (a + bi)^n.$$

To find a formula for an nth root of a complex number, let u be an nth root of z, where

$$u = s(\cos \beta + i \sin \beta) \qquad \text{and} \qquad z = r(\cos \theta + i \sin \theta).$$

By DeMoivre's Theorem and the fact that $u^n = z$, you have

$$s^n (\cos n\beta + i \sin n\beta) = r(\cos \theta + i \sin \theta).$$

Taking the absolute value of each side of this equation, it follows that $s^n = r$. Substituting back into the previous equation and dividing by r, you get

$$\cos n\beta + i \sin n\beta = \cos \theta + i \sin \theta.$$

So, it follows that

$$\cos n\beta = \cos \theta \qquad \text{and} \qquad \sin n\beta = \sin \theta.$$

Because both sine and cosine have a period of 2π, these last two equations have solutions if and only if the angles differ by a multiple of 2π. Consequently, there must exist an integer k such that

$$n\beta = \theta + 2\pi k$$

$$\beta = \frac{\theta + 2\pi k}{n}.$$

By substituting this value of β into the trigonometric form of u, you get the result stated in the theorem on the following page.

Exploration

The nth roots of a complex number are useful for solving some polynomial equations. For instance, explain how you can use DeMoivre's Theorem to solve the polynomial equation

$$x^4 + 16 = 0.$$

[*Hint:* Write -16 as

$$16(\cos \pi + i \sin \pi).]$$

*n*th Roots of a Complex Number

For a positive integer n, the complex number $z = r(\cos \theta + i \sin \theta)$ has exactly n distinct nth roots given by

$$\sqrt[n]{r}\left(\cos \frac{\theta + 2\pi k}{n} + i \sin \frac{\theta + 2\pi k}{n}\right)$$

where $k = 0, 1, 2, \ldots, n - 1$.

When $k > n - 1$ the roots begin to repeat. For instance, if $k = n$, the angle

$$\frac{\theta + 2\pi n}{n} = \frac{\theta}{n} + 2\pi$$

is coterminal with θ/n, which is also obtained when $k = 0$.

The formula for the nth roots of a complex number z has a nice geometrical interpretation, as shown in Figure 6.49. Note that because the nth roots of z all have the same magnitude $\sqrt[n]{r}$, they all lie on a circle of radius $\sqrt[n]{r}$ with center at the origin. Furthermore, because successive nth roots have arguments that differ by $2\pi/n$, the n roots are equally spaced around the circle.

You have already found the sixth roots of 1 by factoring and by using the Quadratic Formula. Example 7 shows how you can solve the same problem with the formula for nth roots.

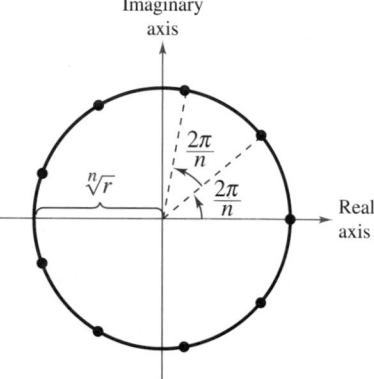

Figure 6.49

Example 7 Finding the *n*th Roots of a Real Number

Find all the sixth roots of 1.

Solution

First write 1 in the trigonometric form $1 = 1(\cos 0 + i \sin 0)$. Then, by the nth root formula with $n = 6$ and $r = 1$, the roots have the form

$$\sqrt[6]{1}\left(\cos \frac{0 + 2\pi k}{6} + i \sin \frac{0 + 2\pi k}{6}\right) = \cos \frac{\pi k}{3} + i \sin \frac{\pi k}{3}.$$

So, for $k = 0, 1, 2, 3, 4$, and 5, the sixth roots are as follows. (See Figure 6.50.)

$$\cos 0 + i \sin 0 = 1$$

$$\cos \frac{\pi}{3} + i \sin \frac{\pi}{3} = \frac{1}{2} + \frac{\sqrt{3}}{2} i \qquad \text{Incremented by } \frac{2\pi}{n} = \frac{2\pi}{6} = \frac{\pi}{3}$$

$$\cos \frac{2\pi}{3} + i \sin \frac{2\pi}{3} = -\frac{1}{2} + \frac{\sqrt{3}}{2} i$$

$$\cos \pi + i \sin \pi = -1$$

$$\cos \frac{4\pi}{3} + i \sin \frac{4\pi}{3} = -\frac{1}{2} - \frac{\sqrt{3}}{2} i$$

$$\cos \frac{5\pi}{3} + i \sin \frac{5\pi}{3} = \frac{1}{2} - \frac{\sqrt{3}}{2} i$$

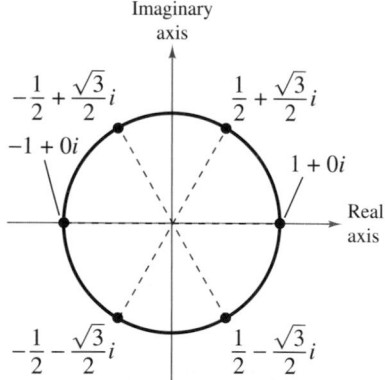

Figure 6.50

✓CHECKPOINT   Now try Exercise 135.

In Figure 6.50, notice that the roots obtained in Example 7 all have a magnitude of 1 and are equally spaced around the unit circle. Also notice that the complex roots occur in conjugate pairs, as discussed in Section 2.5. The *n* distinct *n*th roots of 1 are called the **nth roots of unity.**

Example 8 Finding the *n*th Roots of a Complex Number

Find the three cube roots of $z = -2 + 2i$.

Solution

The absolute value of z is

$$r = |-2 + 2i| = \sqrt{(-2)^2 + 2^2} = \sqrt{8}$$

and the angle θ is given by

$$\tan \theta = \frac{b}{a} = \frac{2}{-2} = -1.$$

Because z lies in Quadrant II, the trigonometric form of z is

$$z = -2 + 2i$$
$$= \sqrt{8}(\cos 135° + i \sin 135°).$$

By the formula for *n*th roots, the cube roots have the form

$$\sqrt[6]{8}\left(\cos \frac{135° + 360°k}{3} + i \sin \frac{135° + 360°k}{3}\right).$$

Finally, for $k = 0$, 1, and 2, you obtain the roots

$$\sqrt[6]{8}\left(\cos \frac{135° + 360°(0)}{3} + i \sin \frac{135° + 360°(0)}{3}\right)$$
$$= \sqrt{2}(\cos 45° + i \sin 45°)$$
$$= 1 + i$$

$$\sqrt[6]{8}\left(\cos \frac{135° + 360°(1)}{3} + i \sin \frac{135° + 360°(1)}{3}\right)$$
$$= \sqrt{2}(\cos 165° + i \sin 165°)$$
$$\approx -1.3660 + 0.3660i$$

$$\sqrt[6]{8}\left(\cos \frac{135° + 360°(2)}{3} + i \sin \frac{135° + 360°(2)}{3}\right)$$
$$= \sqrt{2}(\cos 285° + i \sin 285°)$$
$$\approx 0.3660 - 1.3660i.$$

See Figure 6.51.

✓**CHECKPOINT**   Now try Exercise 139.

Exploration

Use a graphing utility set in *parametric* and *radian* modes to display the graphs of

$$X1T = \cos T$$

and

$$Y1T = \sin T.$$

Set the viewing window so that $-1.5 \le X \le 1.5$ and $-1 \le Y \le 1$. Then, using $0 \le T \le 2\pi$, set the "Tstep" to $2\pi/n$ for various values of *n*. Explain how the graphing utility can be used to obtain the *n*th roots of unity.

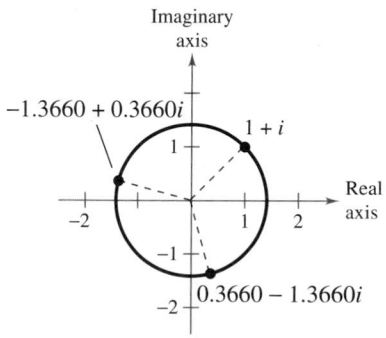

Figure 6.51

6.5 Exercises

See www.CalcChat.com for worked-out solutions to odd-numbered exercises.

Vocabulary Check

Fill in the blanks.

1. The _____ of a complex number $a + bi$ is the distance between the origin $(0, 0)$ and the point (a, b).

2. The _____ of a complex number $z = a + bi$ is given by $z = r(\cos \theta + i \sin \theta)$, where r is the _____ of z and θ is the _____ of z.

3. _____ Theorem states that if $z = r(\cos \theta + i \sin \theta)$ is a complex number and n is a positive integer, then $z^n = r^n(\cos n\theta + i \sin n\theta)$.

4. The complex number $u = a + bi$ is an _____ of the complex number z if $z = u^n = (a + bi)^n$.

In Exercises 1–8, plot the complex number and find its absolute value.

1. $6i$

2. $-2i$

3. -4

4. 7

5. $-4 + 4i$

6. $-5 - 12i$

7. $3 + 6i$

8. $10 - 3i$

In Exercises 9–16, write the complex number in trigonometric form without using a calculator.

9.
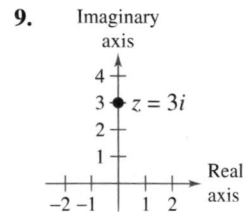

10.
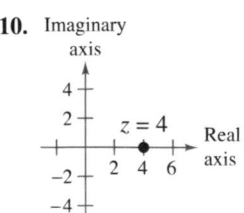

11.
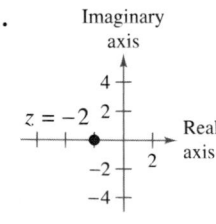

12.

13.
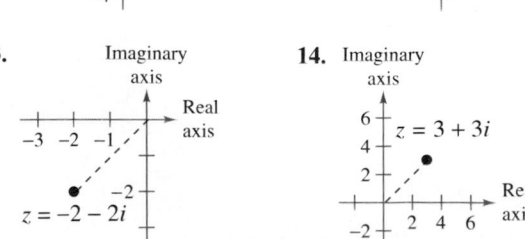

14.

15.
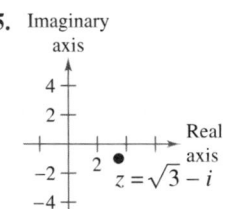

16.
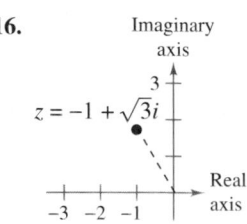

In Exercises 17–36, represent the complex number graphically, and find the trigonometric form of the number.

17. $5 - 5i$ 18. $2 + 2i$

19. $\sqrt{3} + i$ 20. $-1 - \sqrt{3}i$

21. $-2(1 + \sqrt{3}i)$ 22. $\frac{5}{2}(\sqrt{3} - i)$

23. $-8i$ 24. $4i$

25. $-7 + 4i$ 26. $5 - i$

27. 3 28. 6

29. $3 + \sqrt{3}i$ 30. $2\sqrt{2} - i$

31. $-1 - 2i$ 32. $1 + 3i$

33. $5 + 2i$ 34. $-3 + i$

35. $3\sqrt{2} - 7i$ 36. $-8 - 5\sqrt{3}i$

In Exercises 37–48, represent the complex number graphically, and find the standard form of the number.

37. $2(\cos 120° + i \sin 120°)$

38. $5(\cos 135° + i \sin 135°)$

39. $\frac{3}{2}(\cos 330° + i \sin 330°)$

40. $\frac{3}{4}(\cos 315° + i \sin 315°)$

41. $3.75\left(\cos \dfrac{3\pi}{4} + i \sin \dfrac{3\pi}{4}\right)$

42. $1.5\left(\cos\dfrac{\pi}{2} + i\sin\dfrac{\pi}{2}\right)$

43. $6\left(\cos\dfrac{\pi}{3} + i\sin\dfrac{\pi}{3}\right)$

44. $8\left(\cos\dfrac{5\pi}{6} + i\sin\dfrac{5\pi}{6}\right)$

45. $4\left(\cos\dfrac{3\pi}{2} + i\sin\dfrac{3\pi}{2}\right)$

46. $9(\cos 0 + i\sin 0)$

47. $3[\cos(18°\,45') + i\sin(18°\,45')]$

48. $6[\cos(230°\,30') + i\sin(230°\,30')]$

In Exercises 49–52, use a graphing utility to represent the complex number in standard form.

49. $5\left(\cos\dfrac{\pi}{9} + i\sin\dfrac{\pi}{9}\right)$

50. $12\left(\cos\dfrac{3\pi}{5} + i\sin\dfrac{3\pi}{5}\right)$

51. $9(\cos 58° + i\sin 58°)$

52. $4(\cos 216.5° + i\sin 216.5°)$

In Exercises 53 and 54, represent the powers z, z^2, z^3, and z^4 graphically. Describe the pattern.

53. $z = \dfrac{\sqrt{2}}{2}(1 + i)$

54. $z = \dfrac{1}{2}(1 + \sqrt{3}i)$

In Exercises 55–66, perform the operation and leave the result in trigonometric form.

55. $\left[3\left(\cos\dfrac{\pi}{3} + i\sin\dfrac{\pi}{3}\right)\right]\left[4\left(\cos\dfrac{\pi}{6} + i\sin\dfrac{\pi}{6}\right)\right]$

56. $\left[\dfrac{3}{2}\left(\cos\dfrac{\pi}{6} + i\sin\dfrac{\pi}{6}\right)\right]\left[6\left(\cos\dfrac{\pi}{4} + i\sin\dfrac{\pi}{4}\right)\right]$

57. $\left[\frac{5}{3}(\cos 140° + i\sin 140°)\right]\left[\frac{2}{3}(\cos 60° + i\sin 60°)\right]$

58. $\left[\frac{1}{2}(\cos 115° + i\sin 115°)\right]\left[\frac{4}{5}(\cos 300° + i\sin 300°)\right]$

59. $\left[\frac{11}{20}(\cos 290° + i\sin 290°)\right]\left[\frac{2}{5}(\cos 200° + i\sin 200°)\right]$

60. $(\cos 5° + i\sin 5°)(\cos 20° + i\sin 20°)$

61. $\dfrac{\cos 50° + i\sin 50°}{\cos 20° + i\sin 20°}$

62. $\dfrac{5(\cos 4.3 + i\sin 4.3)}{4(\cos 2.1 + i\sin 2.1)}$

63. $\dfrac{2(\cos 120° + i\sin 120°)}{4(\cos 40° + i\sin 40°)}$

64. $\dfrac{\cos\left(\dfrac{7\pi}{4}\right) + i\sin\left(\dfrac{7\pi}{4}\right)}{\cos \pi + i\sin \pi}$

65. $\dfrac{18(\cos 54° + i\sin 54°)}{3(\cos 102° + i\sin 102°)}$

66. $\dfrac{9(\cos 20° + i\sin 20°)}{5(\cos 75° + i\sin 75°)}$

In Exercises 67–82, (a) write the trigonometric forms of the complex numbers, (b) perform the indicated operation using the trigonometric forms, and (c) perform the indicated operation using the standard forms and check your result with that of part (b).

67. $(2 - 2i)(1 + i)$

68. $(3 - 3i)(1 - i)$

69. $(2 + 2i)(1 - i)$

70. $(\sqrt{3} + i)(1 + i)$

71. $-2i(1 + i)$

72. $3i(1 + i)$

73. $-2i(\sqrt{3} - i)$

74. $-i(1 + \sqrt{3}i)$

75. $2(1 - i)$

76. $-4(1 + i)$

77. $\dfrac{3 + 3i}{1 - \sqrt{3}i}$

78. $\dfrac{2 + 2i}{1 + \sqrt{3}i}$

79. $\dfrac{5}{2 + 2i}$

80. $\dfrac{2}{\sqrt{3} - i}$

81. $\dfrac{4i}{-1 + i}$

82. $\dfrac{2i}{1 - \sqrt{3}i}$

In Exercises 83–90, sketch the graph of all complex numbers z satisfying the given condition.

83. $|z| = 2$

84. $|z| = 5$

85. $|z| = 4$

86. $|z| = 6$

87. $\theta = \dfrac{\pi}{6}$

88. $\theta = \dfrac{\pi}{4}$

89. $\theta = \dfrac{5\pi}{6}$

90. $\theta = \dfrac{2\pi}{3}$

In Exercises 91–110, use DeMoivre's Theorem to find the indicated power of the complex number. Write the result in standard form.

91. $(1 + i)^3$

92. $(2 + 2i)^6$

93. $(-1 + i)^{10}$

94. $(1 - i)^8$

95. $2(\sqrt{3} + i)^5$

96. $4(1 - \sqrt{3}i)^3$

97. $[5(\cos 20° + i \sin 20°)]^3$

98. $[3(\cos 150° + i \sin 150°)]^4$

99. $\left(\cos \dfrac{5\pi}{4} + i \sin \dfrac{5\pi}{4}\right)^{10}$

100. $\left[2\left(\cos \dfrac{\pi}{2} + i \sin \dfrac{\pi}{2}\right)\right]^{12}$

101. $[2(\cos 1.25 + i \sin 1.25)]^4$

102. $[4(\cos 2.8 + i \sin 2.8)]^5$

103. $[2(\cos \pi + i \sin \pi)]^8$

104. $(\cos 0 + i \sin 0)^{20}$

105. $(3 - 2i)^5$

106. $(\sqrt{5} - 4i)^4$

107. $[4(\cos 10° + i \sin 10°)]^6$

108. $[3(\cos 15° + i \sin 15°)]^4$

109. $\left[3\left(\cos \dfrac{\pi}{8} + i \sin \dfrac{\pi}{8}\right)\right]^2$

110. $\left[2\left(\cos \dfrac{\pi}{10} + i \sin \dfrac{\pi}{10}\right)\right]^5$

111. Show that $-\dfrac{1}{2}(1 + \sqrt{3}i)$ is a sixth root of 1.

112. Show that $2^{-1/4}(1 - i)$ is a fourth root of -2.

Graphical Reasoning **In Exercises 113–116, use the graph of the roots of a complex number. (a) Write each of the roots in trigonometric form. (b) Identify the complex number whose roots are given. (c) Use a graphing utility to verify the results of part (b).**

113.
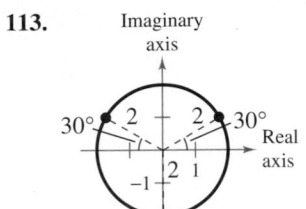

114.
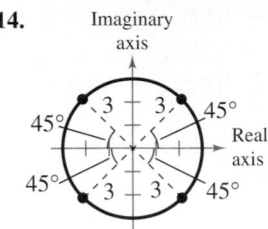

115.
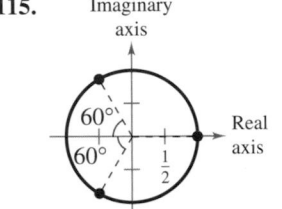

116.
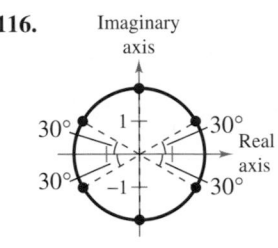

In Exercises 117–124, find the square roots of the complex number.

117. $2i$

118. $5i$

119. $-3i$

120. $-6i$

121. $2 - 2i$

122. $2 + 2i$

123. $1 + \sqrt{3}i$

124. $1 - \sqrt{3}i$

In Exercises 125–140, (a) use the theorem on page 454 to find the indicated roots of the complex number, (b) represent each of the roots graphically, and (c) write each of the roots in standard form.

125. Square roots of $5(\cos 120° + i \sin 120°)$

126. Square roots of $16(\cos 60° + i \sin 60°)$

127. Fourth roots of $16\left(\cos \dfrac{4\pi}{3} + i \sin \dfrac{4\pi}{3}\right)$

128. Fifth roots of $32\left(\cos \dfrac{5\pi}{6} + i \sin \dfrac{5\pi}{6}\right)$

129. Cube roots of $-27i$

130. Fourth roots of $625i$

131. Cube roots of $-\dfrac{125}{2}(1 + \sqrt{3}i)$

132. Cube roots of $-4\sqrt{2}(1-i)$

133. Cube roots of $64i$

134. Fourth roots of i

135. Fifth roots of 1

136. Cube roots of 1000

137. Cube roots of -125

138. Fourth roots of -4

139. Fifth roots of $128(-1+i)$

140. Sixth roots of $729i$

In Exercises 141–148, use the theorem on page 454 to find all the solutions of the equation, and represent the solutions graphically.

141. $x^4 - i = 0$

142. $x^3 + 27 = 0$

143. $x^5 + 243 = 0$

144. $x^4 - 81 = 0$

145. $x^4 + 16i = 0$

146. $x^6 - 64i = 0$

147. $x^3 - (1-i) = 0$

148. $x^4 + (1+i) = 0$

Electrical Engineering   **In Exercises 149–154, use the formula to find the missing quantity for the given conditions. The formula**

$$E = I \cdot Z$$

where E represents voltage, I represents current, and Z represents impedance (a measure of opposition to a sinusoidal electric current), is used in electrical engineering. Each variable is a complex number.

149. $I = 10 + 2i$
    $Z = 4 + 3i$

150. $I = 12 + 2i$
    $Z = 3 + 5i$

151. $I = 2 + 4i$
    $E = 5 + 5i$

152. $I = 10 + 2i$
    $E = 4 + 5i$

153. $E = 12 + 24i$
    $Z = 12 + 20i$

154. $E = 15 + 12i$
    $Z = 25 + 24i$

Synthesis

True or False?   **In Exercises 155–157, determine whether the statement is true or false. Justify your answer.**

155. $\frac{1}{2}(1 - \sqrt{3}i)$ is a ninth root of -1.

156. $\sqrt{3} + i$ is a solution of the equation $x^2 - 8i = 0$.

157. Geometrically, the nth roots of any complex number z are all equally spaced around the unit circle centered at the origin.

158. Given two complex numbers $z_1 = r_1(\cos\theta_1 + i\sin\theta_1)$ and $z_2 = r_2(\cos\theta_2 + i\sin\theta_2)$, $z_2 \neq 0$, show that

$$\frac{z_1}{z_2} = \frac{r_1}{r_2}[\cos(\theta_1 - \theta_2) + i\sin(\theta_1 - \theta_2)].$$

159. Show that $\bar{z} = r[\cos(-\theta) + i\sin(-\theta)]$ is the complex conjugate of $z = r(\cos\theta + i\sin\theta)$.

160. Use the trigonometric forms of z and $\bar{z}$ in Exercise 159 to find (a) $z\bar{z}$ and (b) $z/\bar{z}$, $\bar{z} \neq 0$.

161. Show that the negative of $z = r(\cos\theta + i\sin\theta)$ is $-z = r[\cos(\theta + \pi) + i\sin(\theta + \pi)]$.

162. *Writing*   The famous formula

$$e^{a+bi} = e^a(\cos b + i\sin b)$$

is called Euler's Formula, after the Swiss mathematician Leonhard Euler (1707–1783). This formula gives rise to the equation

$$e^{\pi i} + 1 = 0.$$

This equation relates the five most famous numbers in mathematics—0, 1, π, e, and i—in a single equation. Show how Euler's Formula can be used to derive this equation. Write a short paragraph summarizing your work.

Skills Review

Harmonic Motion   **In Exercises 163–166, for the simple harmonic motion described by the trigonometric function, find the maximum displacement from equilibrium and the lowest possible positive value of t for which $d = 0$.**

163. $d = 16\cos\dfrac{\pi}{4}t$

164. $d = \dfrac{1}{16}\sin\dfrac{5\pi}{4}t$

165. $d = \frac{1}{8}\cos 12\pi t$

166. $d = \frac{1}{12}\sin 60\pi t$

In Exercises 167–170, find all solutions of the equation in the interval $[0, 2\pi)$. Use a graphing utility to verify your answers.

167. $2\cos(x + \pi) + 2\cos(x - \pi) = 0$

168. $\sin\left(x + \dfrac{3\pi}{2}\right) - \sin\left(x - \dfrac{3\pi}{2}\right) = 0$

169. $\sin\left(x - \dfrac{\pi}{3}\right) - \sin\left(x + \dfrac{\pi}{3}\right) = \dfrac{3}{2}$

170. $\tan(x + \pi) - \cos\left(x + \dfrac{5\pi}{2}\right) = 0$

What Did You Learn?

Key Terms

oblique triangle, p. 408
directed line segment, p. 424
vector $\mathbf{v}$ in the plane, p. 424
standard position of a vector, p. 425
zero vector, p. 425
parallelogram law, p. 426
resultant, p. 426

standard unit vector, p. 429
horizontal and vertical components of
  $\mathbf{v}$, p. 429
linear combination, p. 429
direction angle, p. 430
vector components, p. 441
absolute value of a complex number,
  p. 448

trigonometric form of a complex
  number, p. 449
modulus, p. 449
argument, p. 449
DeMoivre's Theorem, p. 452
nth root of a complex number, p. 453
nth roots of unity, p. 455

Key Concepts

6.1 ■ Use the Law of Sines to solve oblique triangles

If ABC is a triangle with sides a, b, and c, then the Law of Sines is as follows.

$$\frac{a}{\sin A} = \frac{b}{\sin B} = \frac{c}{\sin C}$$

6.1 ■ Find areas of oblique triangles

To find the area of a triangle given two sides and their included angle, use one of the following formulas.

$$\text{Area} = \frac{1}{2}bc\sin A = \frac{1}{2}ab\sin C = \frac{1}{2}ac\sin B$$

6.2 ■ Use the Law of Cosines to solve oblique triangles

1. $a^2 = b^2 + c^2 - 2bc\cos A$ or $\cos A = \dfrac{b^2 + c^2 - a^2}{2bc}$

2. $b^2 = a^2 + c^2 - 2ac\cos B$ or $\cos B = \dfrac{a^2 + c^2 - b^2}{2ac}$

3. $c^2 = a^2 + b^2 - 2ab\cos C$ or $\cos C = \dfrac{a^2 + b^2 - c^2}{2ab}$

6.2 ■ Use Heron's Area Formula to find areas of triangles

Given any triangle with sides of lengths a, b, and c, the area of the triangle is given by

$$\sqrt{s(s - a)(s - b)(s - c)}$$

where $s = (a + b + c)/2$.

6.3 ■ Use vectors in the plane

1. The component form of a vector $\mathbf{v}$ with initial point $P = (p_1, p_2)$ and terminal point $Q = (q_1, q_2)$ is given by $\overrightarrow{PQ} = \langle q_1 - p_1, q_2 - p_2 \rangle = \langle v_1, v_2 \rangle = \mathbf{v}$.

2. The magnitude (or length) of $\mathbf{v}$ is given by
$$\|\mathbf{v}\| = \sqrt{(q_1 - p_1)^2 + (q_2 - p_2)^2} = \sqrt{v_1^2 + v_2^2}.$$

3. The sum of $\mathbf{u} = \langle u_1, u_2 \rangle$ and $\mathbf{v} = \langle v_1, v_2 \rangle$ is
$$\mathbf{u} + \mathbf{v} = \langle u_1 + v_1, u_2 + v_2 \rangle.$$

4. The scalar multiple of k times $\mathbf{u}$ is $k\mathbf{u} = \langle ku_1, ku_2 \rangle$.

5. A unit vector $\mathbf{u}$ in the direction of $\mathbf{v}$ is $\mathbf{u} = \left(\dfrac{1}{\|\mathbf{v}\|}\right)\mathbf{v}$.

6.4 ■ Use the dot product of two vectors

1. The dot product of $\mathbf{u}$ and $\mathbf{v}$ is $\mathbf{u} \cdot \mathbf{v} = u_1 v_1 + u_2 v_2$.

2. If θ is the angle between two nonzero vectors $\mathbf{u}$ and $\mathbf{v}$, then $\cos\theta = \dfrac{\mathbf{u} \cdot \mathbf{v}}{\|\mathbf{u}\|\,\|\mathbf{v}\|}$.

3. The vectors $\mathbf{u}$ and $\mathbf{v}$ are orthogonal if $\mathbf{u} \cdot \mathbf{v} = 0$.

4. The projection of $\mathbf{u}$ onto $\mathbf{v}$ is given by
$$\text{proj}_{\mathbf{v}}\mathbf{u} = \left(\frac{\mathbf{u} \cdot \mathbf{v}}{\|\mathbf{v}\|^2}\right)\mathbf{v}.$$

5. The work W done by a constant force $\mathbf{F}$ as its point of application moves along the vector $\overrightarrow{PQ}$ is given by either $W = \|\text{proj}_{\overrightarrow{PQ}}\,\mathbf{F}\|\,\|\overrightarrow{PQ}\|$ or $W = \mathbf{F} \cdot \overrightarrow{PQ}$.

6.5 ■ Find nth roots of complex numbers

For a positive integer n, the complex number

$$z = r(\cos\theta + i\sin\theta)$$

has exactly n distinct nth roots given by

$$\sqrt[n]{r}\left(\cos\frac{\theta + 2\pi k}{n} + i\sin\frac{\theta + 2\pi k}{n}\right)$$

where $k = 0, 1, 2, \ldots, n - 1$.

See www.CalcChat.com for worked-out solutions to odd-numbered exercises.

6.1 In Exercises 1–12, use the Law of Sines to solve the triangle. If two solutions exist, find both.

1. $A = 32°$, $B = 50°$, $a = 16$
2. $A = 38°$, $B = 58°$, $a = 12$
3. $B = 25°$, $C = 105°$, $c = 25$
4. $B = 20°$, $C = 115°$, $c = 30$
5. $A = 60°15'$, $B = 45°30'$, $b = 4.8$
6. $A = 82°45'$, $B = 28°45'$, $b = 40.2$
7. $A = 75°$, $a = 2.5$, $b = 16.5$
8. $A = 15°$, $a = 5$, $b = 10$
9. $B = 115°$, $a = 9$, $b = 14.5$
10. $B = 150°$, $a = 64$, $b = 10$
11. $C = 50°$, $a = 25$, $c = 22$
12. $B = 25°$, $a = 6.2$, $b = 4$

In Exercises 13–16, find the area of the triangle having the indicated angle and sides.

13. $A = 27°$, $b = 5$, $c = 8$
14. $B = 80°$, $a = 4$, $c = 8$
15. $C = 122°$, $b = 18$, $a = 29$
16. $C = 100°$, $a = 120$, $b = 74$

17. *Height* From a distance of 50 meters, the angle of elevation to the top of a building is 17°. Approximate the height of the building.

18. *Distance* A family is traveling due west on a road that passes a famous landmark. At a given time the bearing to the landmark is N 62° W, and after the family travels 5 miles farther, the bearing is N 38° W. What is the closest the family will come to the landmark while on the road?

19. *Height* A tree stands on a hillside of slope 28° from the horizontal. From a point 75 feet down the hill, the angle of elevation to the top of the tree is 45° (see figure). Find the height of the tree.

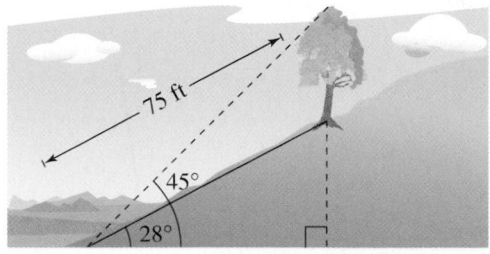

20. *Width* A surveyor finds that a tree on the opposite bank of a river has a bearing of N 22° 30′ E from a certain point and a bearing of N 15° W from a point 400 feet downstream. Find the width of the river.

6.2 In Exercises 21–32, use the Law of Cosines to solve the triangle.

21. $a = 18$, $b = 12$, $c = 15$
22. $a = 10$, $b = 12$, $c = 16$
23. $a = 9$, $b = 12$, $c = 20$
24. $a = 7$, $b = 15$, $c = 19$
25. $a = 6.5$, $b = 10.2$, $c = 16$
26. $a = 6.2$, $b = 6.4$, $c = 2.1$
27. $C = 65°$, $a = 25$, $b = 12$
28. $B = 48°$, $a = 18$, $c = 12$
29. $B = 110°$, $a = 4$, $c = 4$
30. $B = 150°$, $a = 10$, $c = 20$
31. $B = 55°30'$, $a = 12.4$, $c = 18.5$
32. $B = 85°15'$, $a = 24.2$, $c = 28.2$

33. *Geometry* The lengths of the diagonals of a parallelogram are 10 feet and 16 feet. Find the lengths of the sides of the parallelogram if the diagonals intersect at an angle of 28°.

34. *Geometry* The lengths of the diagonals of a parallelogram are 30 meters and 40 meters. Find the lengths of the sides of the parallelogram if the diagonals intersect at an angle of 34°.

35. *Navigation* Two planes leave Washington, D.C.'s Dulles International Airport at approximately the same time. One is flying at 425 miles per hour at a bearing of 355°, and the other is flying at 530 miles per hour at a bearing of 67° (see figure). Determine the distance between the planes after they have flown for 2 hours.

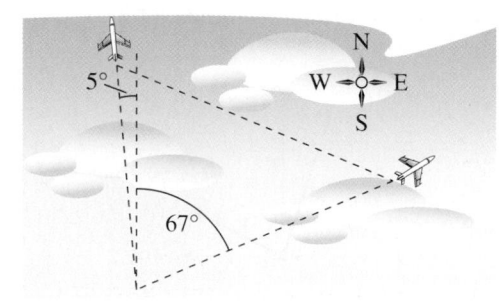

36. *Surveying* To approximate the length of a marsh, a surveyor walks 425 meters from point A to point B. The surveyor then turns 65° and walks 300 meters to point C (see figure). Approximate the length AC of the marsh.

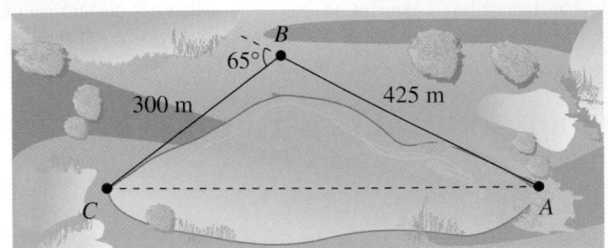

In Exercises 37–40, use Heron's Area Formula to find the area of the triangle with the given side lengths.

37. $a = 4$, $b = 5$, $c = 7$

38. $a = 15$, $b = 8$, $c = 10$

39. $a = 64.8$, $b = 49.2$, $c = 24.1$

40. $a = 8.55$, $b = 5.14$, $c = 12.73$

6.3 In Exercises 41–46, find the component form of the vector v satisfying the given conditions.

41.

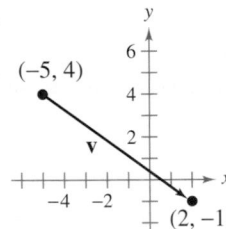

42.

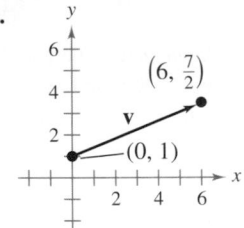

43. Initial point: $(0, 10)$; terminal point: $(7, 3)$

44. Initial point: $(1, 5)$; terminal point: $(15, 9)$

45. $\|\mathbf{v}\| = 8$, $\theta = 120°$ **46.** $\|\mathbf{v}\| = \frac{1}{2}$, $\theta = 225°$

In Exercises 47–52, use the figure to sketch a graph of the specified vector. To print an enlarged copy of the graph, go to the website *www.mathgraphs.com*.

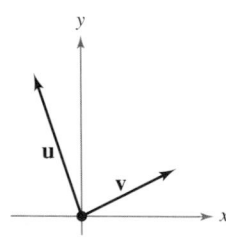

47. $2\mathbf{u}$

48. $-\frac{1}{2}\mathbf{v}$

49. $2\mathbf{u} + \mathbf{v}$

50. $\mathbf{u} + 2\mathbf{v}$

51. $\mathbf{u} - 2\mathbf{v}$

52. $\mathbf{v} - 2\mathbf{u}$

In Exercises 53–60, find (a) u + v, (b) u − v, (c) 3u, and (d) 2v + 5u.

53. $\mathbf{u} = \langle -1, -3 \rangle$, $\mathbf{v} = \langle -3, 6 \rangle$

54. $\mathbf{u} = \langle 4, 5 \rangle$, $\mathbf{v} = \langle 0, -1 \rangle$

55. $\mathbf{u} = \langle -5, 2 \rangle$, $\mathbf{v} = \langle 4, 4 \rangle$

56. $\mathbf{u} = \langle 1, -8 \rangle$, $\mathbf{v} = \langle 3, -2 \rangle$

57. $\mathbf{u} = 2\mathbf{i} - \mathbf{j}$, $\mathbf{v} = 5\mathbf{i} + 3\mathbf{j}$

58. $\mathbf{u} = -6\mathbf{j}$, $\mathbf{v} = \mathbf{i} + \mathbf{j}$

59. $\mathbf{u} = 4\mathbf{i}$, $\mathbf{v} = -\mathbf{i} + 6\mathbf{j}$

60. $\mathbf{u} = -7\mathbf{i} - 3\mathbf{j}$, $\mathbf{v} = 4\mathbf{i} - \mathbf{j}$

In Exercises 61–64, find the component form of w and sketch the specified vector operations geometrically, where $\mathbf{u} = 6\mathbf{i} - 5\mathbf{j}$ and $\mathbf{v} = 10\mathbf{i} + 3\mathbf{j}$.

61. $\mathbf{w} = 3\mathbf{v}$ **62.** $\mathbf{w} = \frac{1}{2}\mathbf{v}$

63. $\mathbf{w} = 4\mathbf{u} + 5\mathbf{v}$ **64.** $\mathbf{w} = 3\mathbf{v} - 2\mathbf{u}$

In Exercises 65–68, find a unit vector in the direction of the given vector.

65. $\mathbf{u} = \langle 0, -6 \rangle$ **66.** $\mathbf{v} = \langle -12, -5 \rangle$

67. $\mathbf{v} = 5\mathbf{i} - 2\mathbf{j}$ **68.** $\mathbf{w} = -7\mathbf{i}$

In Exercises 69 and 70, write a linear combination of the standard unit vectors i and j for the given initial and terminal points.

69. Initial point: $(-8, 3)$

  Terminal point: $(1, -5)$

70. Initial point: $(2, -3.2)$

  Terminal point: $(-6.4, 10.8)$

In Exercises 71 and 72, write the vector v in the form $\|\mathbf{v}\|[(\cos \theta)\mathbf{i} + (\sin \theta)\mathbf{j}]$.

71. $\mathbf{v} = -10\mathbf{i} + 10\mathbf{j}$ **72.** $\mathbf{v} = 4\mathbf{i} - \mathbf{j}$

In Exercises 73 and 74, graph the vectors and the resultant of the vectors. Find the magnitude and direction of the resultant.

73.

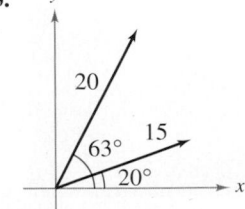

74.

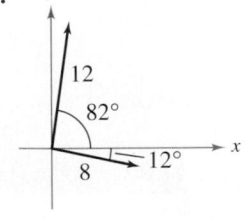

75. *Resultant Force* Three forces with magnitudes of 250 pounds, 100 pounds, and 200 pounds act on an object at angles of $60°$, $150°$, and $-90°$, respectively, with the positive x-axis. Find the direction and magnitude of the resultant of these forces.

76. *Resultant Force* Forces with magnitudes of 85 pounds and 50 pounds act on a single point. The angle between the forces is $15°$. Describe the resultant force.

77. *Tension* A 180-pound weight is supported by two ropes, as shown in the figure. Find the tension in each rope.

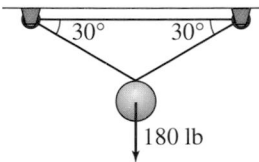

78. *Cable Tension* In a manufacturing process, an electric hoist lifts 200-pound ingots. Find the tension in the supporting cables (see figure).

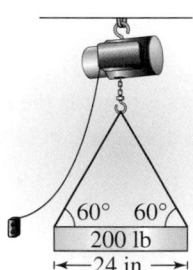

79. *Navigation* An airplane has an airspeed of 430 miles per hour at a bearing of $135°$. The wind velocity is 35 miles per hour in the direction N $30°$ E. Find the resultant speed and direction of the plane.

80. *Navigation* An airplane has an airspeed of 724 kilometers per hour at a bearing of $30°$. The wind velocity is from the west at 32 kilometers per hour. Find the resultant speed and direction of the plane.

6.4 **In Exercises 81–84, find the dot product of u and v.**

81. $\mathbf{u} = \langle 0, -2 \rangle$
  $\mathbf{v} = \langle 1, 10 \rangle$

82. $\mathbf{u} = \langle -4, 5 \rangle$
  $\mathbf{v} = \langle 3, -1 \rangle$

83. $\mathbf{u} = 6\mathbf{i} - \mathbf{j}$
  $\mathbf{v} = 2\mathbf{i} + 5\mathbf{j}$

84. $\mathbf{u} = 8\mathbf{i} - 7\mathbf{j}$
  $\mathbf{v} = 3\mathbf{i} - 4\mathbf{j}$

In Exercises 85–88, use the vectors $\mathbf{u} = \langle -3, -4 \rangle$ and $\mathbf{v} = \langle 2, 1 \rangle$ to find the indicated quantity.

85. $\mathbf{u} \cdot \mathbf{u}$

86. $\|\mathbf{v}\| - 3$

87. $4\mathbf{u} \cdot \mathbf{v}$

88. $(\mathbf{u} \cdot \mathbf{v})\mathbf{u}$

In Exercises 89–92, find the angle θ between the vectors.

89. $\mathbf{u} = \langle 2\sqrt{2}, -4 \rangle$, $\mathbf{v} = \langle -\sqrt{2}, 1 \rangle$

90. $\mathbf{u} = \langle 3, 1 \rangle$, $\mathbf{v} = \langle 4, 5 \rangle$

91. $\mathbf{u} = \cos\dfrac{7\pi}{4}\mathbf{i} + \sin\dfrac{7\pi}{4}\mathbf{j}$, $\mathbf{v} = \cos\dfrac{5\pi}{6}\mathbf{i} + \sin\dfrac{5\pi}{6}\mathbf{j}$

92. $\mathbf{u} = \cos 45°\mathbf{i} + \sin 45°\mathbf{j}$
  $\mathbf{v} = \cos 300°\mathbf{i} + \sin 300°\mathbf{j}$

In Exercises 93–96, graph the vectors and find the degree measure of the angle between the vectors.

93. $\mathbf{u} = 4\mathbf{i} + \mathbf{j}$
  $\mathbf{v} = \mathbf{i} - 4\mathbf{j}$

94. $\mathbf{u} = 6\mathbf{i} + 2\mathbf{j}$
  $\mathbf{v} = -3\mathbf{i} - \mathbf{j}$

95. $\mathbf{u} = 7\mathbf{i} - 5\mathbf{j}$
  $\mathbf{v} = 10\mathbf{i} + 3\mathbf{j}$

96. $\mathbf{u} = -5.3\mathbf{i} + 2.8\mathbf{j}$
  $\mathbf{v} = -8.1\mathbf{i} - 4\mathbf{j}$

In Exercises 97–100, determine whether u and v are orthogonal, parallel, or neither.

97. $\mathbf{u} = \langle 39, -12 \rangle$
  $\mathbf{v} = \langle -26, 8 \rangle$

98. $\mathbf{u} = \langle 8, -4 \rangle$
  $\mathbf{v} = \langle 5, 10 \rangle$

99. $\mathbf{u} = \langle 8, 5 \rangle$
  $\mathbf{v} = \langle -2, 4 \rangle$

100. $\mathbf{u} = \langle -15, 51 \rangle$
  $\mathbf{v} = \langle 20, -68 \rangle$

In Exercises 101–104, find the value of k so that the vectors u and v are orthogonal.

101. $\mathbf{u} = \mathbf{i} - k\mathbf{j}$
  $\mathbf{v} = \mathbf{i} + 2\mathbf{j}$

102. $\mathbf{u} = 2\mathbf{i} + \mathbf{j}$
  $\mathbf{v} = -\mathbf{i} - k\mathbf{j}$

103. $\mathbf{u} = k\mathbf{i} - \mathbf{j}$
  $\mathbf{v} = 2\mathbf{i} - 2\mathbf{j}$

104. $\mathbf{u} = k\mathbf{i} - 2\mathbf{j}$
  $\mathbf{v} = \mathbf{i} + 4\mathbf{j}$

In Exercises 105–108, find the projection of u onto v. Then write u as the sum of two orthogonal vectors, one of which is $\text{proj}_\mathbf{v}\,\mathbf{u}$.

105. $\mathbf{u} = \langle -4, 3 \rangle$, $\mathbf{v} = \langle -8, -2 \rangle$

106. $\mathbf{u} = \langle 5, 6 \rangle$, $\mathbf{v} = \langle 10, 0 \rangle$

107. $\mathbf{u} = \langle 2, 7 \rangle$, $\mathbf{v} = \langle 1, -1 \rangle$

108. $\mathbf{u} = \langle -3, 5 \rangle$, $\mathbf{v} = \langle -5, 2 \rangle$

109. *Work* Determine the work done by a crane lifting an 18,000-pound truck 48 inches.

110. *Braking Force* A 500-pound motorcycle is headed up a hill inclined at $12°$. What force is required to keep the motorcycle from rolling back down the hill when stopped at a red light?

6.5 In Exercises 111–114, plot the complex number and find its absolute value.

111. $-i$

112. $5i$

113. $7 - 5i$

114. $-3 + 9i$

In Exercises 115–120, write the complex number in trigonometric form without using a calculator.

115. $2 - 2i$

116. $-2 + 2i$

117. $-\sqrt{3} - i$

118. $-\sqrt{3} + i$

119. $-2i$

120. $4i$

In Exercises 121–124, perform the operation and leave the result in trigonometric form.

121. $\left[\dfrac{5}{2}\left(\cos \dfrac{\pi}{2} + i \sin \dfrac{\pi}{2}\right)\right]\left[4\left(\cos \dfrac{\pi}{4} + i \sin \dfrac{\pi}{4}\right)\right]$

122. $\left[2\left(\cos \dfrac{2\pi}{3} + i \sin \dfrac{2\pi}{3}\right)\right]\left[3\left(\cos \dfrac{\pi}{6} + i \sin \dfrac{\pi}{6}\right)\right]$

123. $\dfrac{20(\cos 320° + i \sin 320°)}{5(\cos 80° + i \sin 80°)}$

124. $\dfrac{3(\cos 230° + i \sin 230°)}{9(\cos 95° + i \sin 95°)}$

In Exercises 125–130, (a) write the trigonometric forms of the complex numbers, (b) perform the indicated operation using the trigonometric forms, and (c) perform the indicated operation using the standard forms and check your result with that of part (b).

125. $(2 - 2i)(3 + 3i)$

126. $(4 + 4i)(-1 - i)$

127. $-i(2 + 2i)$

128. $4i(1 - i)$

129. $\dfrac{3 - 3i}{2 + 2i}$

130. $\dfrac{-1 - i}{-2 - 2i}$

In Exercises 131–134, use DeMoivre's Theorem to find the indicated power of the complex number. Write the result in standard form.

131. $\left[5\left(\cos \dfrac{\pi}{12} + i \sin \dfrac{\pi}{12}\right)\right]^4$

132. $\left[2\left(\cos \dfrac{4\pi}{15} + i \sin \dfrac{4\pi}{15}\right)\right]^5$

133. $(2 + 3i)^6$

134. $(1 - i)^8$

In Exercises 135–140, find the square roots of the complex number.

135. $-\sqrt{3} + i$

136. $\sqrt{3} - i$

137. $-2i$

138. $-5i$

139. $-2 - 2i$

140. $-2 + 2i$

In Exercises 141–144, (a) use the theorem on page 454 to find the indicated roots of the complex number, (b) represent each of the roots graphically, and (c) write each of the roots in standard form.

141. Sixth roots of $-729i$

142. Fourth roots of $256i$

143. Cube roots of 8

144. Fifth roots of -1024

In Exercises 145–148, use the theorem on page 454 to find all solutions of the equation, and represent the solutions graphically.

145. $x^4 + 256 = 0$

146. $x^5 - 32i = 0$

147. $x^3 + 8i = 0$

148. $x^4 + 81 = 0$

Synthesis

In Exercises 149 and 150, determine whether the statement is true or false. Justify your answer.

149. The Law of Sines is true if one of the angles in the triangle is a right angle.

150. When the Law of Sines is used, the solution is always unique.

151. What characterizes a vector in the plane?

152. Which vectors in the figure appear to be equivalent?

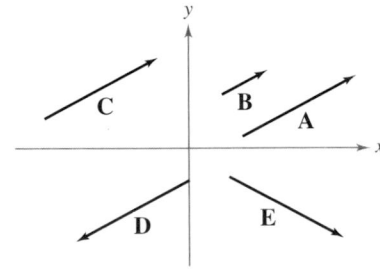

153. The figure shows z_1 and z_2. Describe $z_1 z_2$ and z_1/z_2.

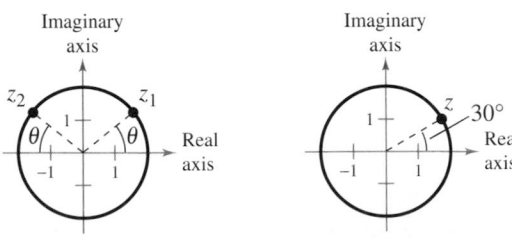

Figure for 153   Figure for 154

154. One of the fourth roots of a complex number z is shown in the graph.

(a) How many roots are not shown?

(b) Describe the other roots.

6 Chapter Test

See www.CalcChat.com for worked-out solutions to odd-numbered exercises.

Take this test as you would take a test in class. After you are finished, check your work against the answers given in the back of the book.

In Exercises 1–6, use the given information to solve the triangle. If two solutions exist, find both solutions.

1. $A = 36°$, $B = 98°$, $c = 16$

2. $a = 4$, $b = 8$, $c = 10$

3. $A = 35°$, $b = 8$, $c = 12$

4. $A = 25°$, $b = 28$, $a = 18$

5. $B = 130°$, $c = 10.1$, $b = 5.2$

6. $A = 150°$, $b = 4.8$, $a = 9.4$

7. Find the length of the pond shown at the right.

8. A triangular parcel of land has borders of lengths 55 meters, 85 meters, and 100 meters. Find the area of the parcel of land.

9. Find the component form and magnitude of the vector **w** that has initial point $(-8, -12)$ and terminal point $(4, 1)$.

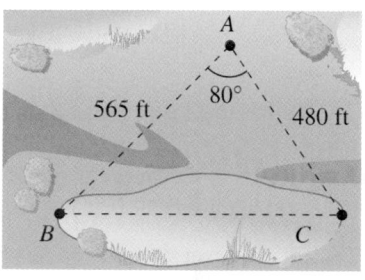

Figure for 7

In Exercises 10–13, find (a) 2v + u, (b) u − 3v, and (c) 5u − v.

10. $\mathbf{u} = \langle 0, -4 \rangle$, $\mathbf{v} = \langle -2, -8 \rangle$

11. $\mathbf{u} = \langle -2, -3 \rangle$, $\mathbf{v} = \langle -1, -10 \rangle$

12. $\mathbf{u} = \mathbf{i} - \mathbf{j}$, $\mathbf{v} = 6\mathbf{i} + 9\mathbf{j}$

13. $\mathbf{u} = 2\mathbf{i} + 3\mathbf{j}$, $\mathbf{v} = -\mathbf{i} - 2\mathbf{j}$

14. Find a unit vector in the direction of $\mathbf{v} = 7\mathbf{i} + 4\mathbf{j}$.

15. Find the component form of the vector **v** with $\|\mathbf{v}\| = 12$, in the same direction as $\mathbf{u} = \langle 3, -5 \rangle$.

16. Forces with magnitudes of 250 pounds and 130 pounds act on an object at angles of $45°$ and $-60°$, respectively, with the positive x-axis. Find the direction and magnitude of the resultant of these forces.

17. Find the dot product of $\mathbf{u} = \langle -9, 4 \rangle$ and $\mathbf{v} = \langle 1, 3 \rangle$.

18. Find the angle between the vectors $\mathbf{u} = 7\mathbf{i} + 2\mathbf{j}$ and $\mathbf{v} = -4\mathbf{j}$.

19. Are the vectors $\mathbf{u} = \langle 6, -4 \rangle$ and $\mathbf{v} = \langle 2, -3 \rangle$ orthogonal? Explain.

20. Find the projection of $\mathbf{u} = \langle 6, 7 \rangle$ onto $\mathbf{v} = \langle -5, -1 \rangle$. Then write **u** as the sum of two orthogonal vectors.

21. Write the complex number $z = -2 + 2i$ in trigonometric form.

22. Write the complex number $100(\cos 240° + i \sin 240°)$ in standard form.

In Exercises 23 and 24, use DeMoivre's Theorem to find the indicated power of the complex number. Write the result in standard form.

23. $\left[3\left(\cos \dfrac{5\pi}{6} + i \sin \dfrac{5\pi}{6} \right) \right]^8$

24. $(3 - 3i)^6$

25. Find the fourth roots of $128(1 + \sqrt{3}\,i)$.

26. Find all solutions of the equation $x^4 - 625i = 0$ and represent the solutions graphically.

4–6 Cumulative Test

See www.CalcChat.com for worked-out solutions to odd-numbered exercises.

Take this test to review the material in Chapters 4–6. After you are finished, check your work against the answers given in the back of the book.

1. Consider the angle $\theta = -150°$.

   (a) Sketch the angle in standard position.

   (b) Determine a coterminal angle in the interval $[0°, 360°)$.

   (c) Convert the angle to radian measure.

   (d) Find the reference angle θ'.

   (e) Find the exact values of the six trigonometric functions of θ.

2. Convert the angle $\theta = 2.55$ radians to degrees. Round your answer to one decimal place.

3. Find $\cos \theta$ if $\tan \theta = -\frac{12}{5}$ and $\sin \theta > 0$.

In Exercises 4–6, sketch the graph of the function by hand. (Include two full periods.) Use a graphing utility to verify your graph.

4. $f(x) = 3 - 2 \sin \pi x$ 5. $f(x) = \tan 3x$ 6. $f(x) = \frac{1}{2} \sec(x + \pi)$

7. Find a, b, and c such that the graph of the function $h(x) = a \cos(bx + c)$ matches the graph in the figure at the right.

In Exercises 8 and 9, find the exact value of the expression without using a calculator.

8. $\sin\left(\arctan \frac{3}{4}\right)$

9. $\tan\left[\arcsin\left(-\frac{1}{2}\right)\right]$

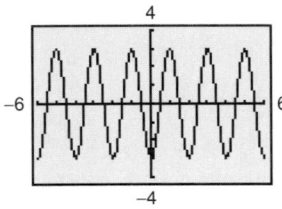

10. Write an algebraic expression equivalent to $\sin(\arctan 2x)$.

11. Subtract and simplify: $\dfrac{\sin \theta - 1}{\cos \theta} - \dfrac{\cos \theta}{\sin \theta - 1}$.

Figure for 7

In Exercises 12–14, verify the identity.

12. $\cot^2 \alpha(\sec^2 \alpha - 1) = 1$

13. $\sin(x + y) \sin(x - y) = \sin^2 x - \sin^2 y$

14. $\sin^2 x \cos^2 x = \frac{1}{8}(1 - \cos 4x)$

In Exercises 15 and 16, solve the equation.

15. $\sin^2 x + 2 \sin x + 1 = 0$

16. $3 \tan \theta - \cot \theta = 0$

17. Approximate the solutions to the equation $\cos^2 x - 5 \cos x - 1 = 0$ in the interval $[0, 2\pi)$.

In Exercises 18 and 19, use a graphing utility to graph the function and approximate its zeros in the interval $[0, 2\pi)$. If possible, find the exact values of the zeros algebraically.

18. $y = \dfrac{1 + \sin x}{\cos x} + \dfrac{\cos x}{1 + \sin x} - 4$

19. $y = \tan^3 x - \tan^2 x + 3 \tan x - 3$

20. Given that $\sin u = \frac{12}{13}$, $\cos v = \frac{3}{5}$, and angles u and v are both in Quadrant I, find $\tan(u - v)$.

21. If $\tan \theta = \frac{1}{2}$, find the exact value of $\tan 2\theta$, $0 < \theta < \frac{\pi}{2}$.

22. If $\tan \theta = \frac{4}{3}$, find the exact value of $\sin \frac{\theta}{2}$, $\pi < \theta < \frac{3\pi}{2}$.

23. Write $\cos 8x + \cos 4x$ as a product.

In Exercises 24–27, verify the identity.

24. $\tan x(1 - \sin^2 x) = \frac{1}{2} \sin 2x$

25. $\sin 3\theta \sin \theta = \frac{1}{2}(\cos 2\theta - \cos 4\theta)$

26. $\sin 3x \cos 2x = \frac{1}{2}(\sin 5x + \sin x)$

27. $\dfrac{2 \cos 3x}{\sin 4x - \sin 2x} = \csc x$

In Exercises 28–31, use the information to solve the triangle shown at the right.

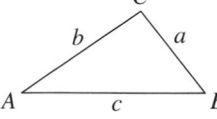

28. $A = 46°$, $a = 14$, $b = 5$

29. $A = 32°$, $b = 8$, $c = 10$

30. $A = 24°$, $C = 101°$, $a = 10$

31. $a = 24$, $b = 30$, $c = 47$

Figure for 28–31

32. Two sides of a triangle have lengths 14 inches and 19 inches. Their included angle measures 82°. Find the area of the triangle.

33. Find the area of a triangle with sides of lengths 12 inches, 16 inches, and 18 inches.

34. Write the vector $\mathbf{u} = \langle 3, 5 \rangle$ as a linear combination of the standard unit vectors $\mathbf{i}$ and $\mathbf{j}$.

35. Find a unit vector in the direction of $\mathbf{v} = \mathbf{i} - 2\mathbf{j}$.

36. Find $\mathbf{u} \cdot \mathbf{v}$ for $\mathbf{u} = 3\mathbf{i} + 4\mathbf{j}$ and $\mathbf{v} = \mathbf{i} - 2\mathbf{j}$.

37. Find k so that $\mathbf{u} = \mathbf{i} + 2k\mathbf{j}$ and $\mathbf{v} = 2\mathbf{i} - \mathbf{j}$ are orthogonal.

38. Find the projection of $\mathbf{u} = \langle 8, -2 \rangle$ onto $\mathbf{v} = \langle 1, 5 \rangle$. Then write $\mathbf{u}$ as the sum of two orthogonal vectors.

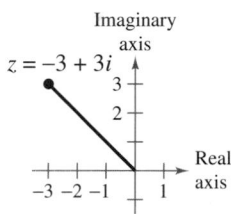

39. Find the trigonometric form of the complex number shown at the right.

Figure for 39

40. Write the complex number $8\left(\cos \frac{5\pi}{6} + i \sin \frac{5\pi}{6} \right)$ in standard form.

41. Find the product $[4(\cos 30° + i \sin 30°)][6(\cos 120° + i \sin 120°)]$. Write the answer in standard form.

42. Find the square roots of $2 + i$.

43. Find the three cube roots of 1.

44. Write all the solutions of the equation $x^5 + 243 = 0$.

45. From a point 200 feet from a flagpole, the angles of elevation to the bottom and top of the flag are $16° \, 45'$ and $18°$, respectively. Approximate the height of the flag to the nearest foot.

46. Write a model for a particle in simple harmonic motion with a displacement of 4 inches and a period of 8 seconds.

47. An airplane's velocity with respect to the air is 500 kilometers per hour, with a bearing of 30°. The wind at the altitude of the plane has a velocity of 50 kilometers per hour with a bearing of N 60° E. What is the true direction of the plane, and what is its speed relative to the ground?

48. Forces of 60 pounds and 100 pounds have a resultant force of 125 pounds. Find the angle between the two forces.

Proofs in Mathematics

Law of Sines (p. 408)

If ABC is a triangle with sides a, b, and c, then

$$\frac{a}{\sin A} = \frac{b}{\sin B} = \frac{c}{\sin C}.$$

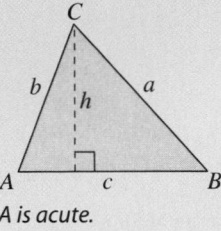

A is acute.

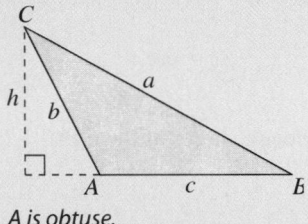

A is obtuse.

Law of Tangents

Besides the Law of Sines and the Law of Cosines, there is also a Law of Tangents, which was developed by Francois Viète (1540–1603). The Law of Tangents follows from the Law of Sines and the sum-to-product formulas for sine and is defined as follows.

$$\frac{a + b}{a - b} = \frac{\tan[(A + B)/2]}{\tan[(A - B)/2]}$$

The Law of Tangents can be used to solve a triangle when two sides and the included angle are given (SAS). Before calculators were invented, the Law of Tangents was used to solve the SAS case instead of the Law of Cosines, because computation with a table of tangent values was easier.

Proof

Let h be the altitude of either triangle shown in the figure above. Then you have

$$\sin A = \frac{h}{b} \quad \text{or} \quad h = b \sin A$$

$$\sin B = \frac{h}{a} \quad \text{or} \quad h = a \sin B.$$

Equating these two values of h, you have

$$a \sin B = b \sin A \quad \text{or} \quad \frac{a}{\sin A} = \frac{b}{\sin B}.$$

Note that $\sin A \neq 0$ and $\sin B \neq 0$ because no angle of a triangle can have a measure of $0°$ or $180°$. In a similar manner, construct an altitude from vertex B to side AC (extended in the obtuse triangle), as shown at the right. Then you have

$$\sin A = \frac{h}{c} \quad \text{or} \quad h = c \sin A$$

$$\sin C = \frac{h}{a} \quad \text{or} \quad h = a \sin C.$$

Equating these two values of h, you have

$$a \sin C = c \sin A \quad \text{or} \quad \frac{a}{\sin A} = \frac{c}{\sin C}.$$

By the Transitive Property of Equality, you know that

$$\frac{a}{\sin A} = \frac{b}{\sin B} = \frac{c}{\sin C}.$$

So, the Law of Sines is established.

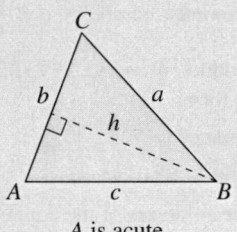

A is acute.

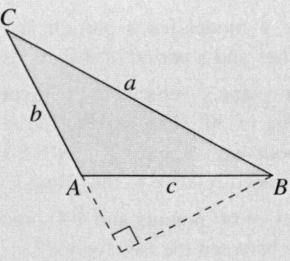

A is obtuse.

Law of Cosines (p. 417)

Standard Form	*Alternative Form*

$$a^2 = b^2 + c^2 - 2bc \cos A \qquad \cos A = \frac{b^2 + c^2 - a^2}{2bc}$$

$$b^2 = a^2 + c^2 - 2ac \cos B \qquad \cos B = \frac{a^2 + c^2 - b^2}{2ac}$$

$$c^2 = a^2 + b^2 - 2ab \cos C \qquad \cos C = \frac{a^2 + b^2 - c^2}{2ab}$$

Proof

To prove the first formula, consider the top triangle at the right, which has three acute angles. Note that vertex B has coordinates $(c, 0)$. Furthermore, C has coordinates (x, y), where $x = b \cos A$ and $y = b \sin A$. Because a is the distance from vertex C to vertex B, it follows that

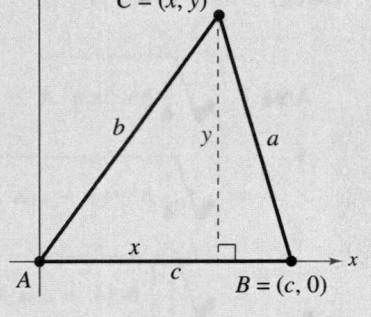

$a = \sqrt{(x - c)^2 + (y - 0)^2}$	Distance Formula
$a^2 = (x - c)^2 + (y - 0)^2$	Square each side.
$a^2 = (b \cos A - c)^2 + (b \sin A)^2$	Substitute for x and y.
$a^2 = b^2 \cos^2 A - 2bc \cos A + c^2 + b^2 \sin^2 A$	Expand.
$a^2 = b^2(\sin^2 A + \cos^2 A) + c^2 - 2bc \cos A$	Factor out b^2.
$a^2 = b^2 + c^2 - 2bc \cos A.$	$\sin^2 A + \cos^2 A = 1$

To prove the second formula, consider the bottom triangle at the right, which also has three acute angles. Note that vertex A has coordinates $(c, 0)$. Furthermore, C has coordinates (x, y), where $x = a \cos B$ and $y = a \sin B$. Because b is the distance from vertex C to vertex A, it follows that

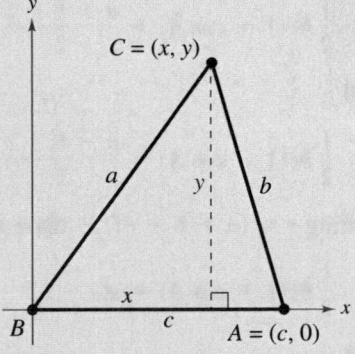

$b = \sqrt{(x - c)^2 + (y - 0)^2}$	Distance Formula
$b^2 = (x - c)^2 + (y - 0)^2$	Square each side.
$b^2 = (a \cos B - c)^2 + (a \sin B)^2$	Substitute for x and y.
$b^2 = a^2 \cos^2 B - 2ac \cos B + c^2 + a^2 \sin^2 B$	Expand.
$b^2 = a^2(\sin^2 B + \cos^2 B) + c^2 - 2ac \cos B$	Factor out a^2.
$b^2 = a^2 + c^2 - 2ac \cos B.$	$\sin^2 B + \cos^2 B = 1$

A similar argument is used to establish the third formula.

Heron's Area Formula   (p. 420)

Given any triangle with sides of lengths a, b, and c, the area of the triangle is

$$\text{Area} = \sqrt{s(s - a)(s - b)(s - c)}$$

where $s = \dfrac{a + b + c}{2}$.

Proof

From Section 6.1, you know that

$$\text{Area} = \frac{1}{2} bc \sin A \qquad \text{Formula for the area of an oblique triangle}$$

$$(\text{Area})^2 = \frac{1}{4} b^2 c^2 \sin^2 A \qquad \text{Square each side.}$$

$$\text{Area} = \sqrt{\frac{1}{4} b^2 c^2 \sin^2 A} \qquad \text{Take the square root of each side.}$$

$$= \sqrt{\frac{1}{4} b^2 c^2 (1 - \cos^2 A)} \qquad \text{Pythagorean Identity}$$

$$= \sqrt{\left[\frac{1}{2} bc(1 + \cos A)\right]\left[\frac{1}{2} bc(1 - \cos A)\right]}. \qquad \text{Factor.}$$

Using the Law of Cosines, you can show that

$$\frac{1}{2} bc(1 + \cos A) = \frac{a + b + c}{2} \cdot \frac{-a + b + c}{2}$$

and

$$\frac{1}{2} bc(1 - \cos A) = \frac{a - b + c}{2} \cdot \frac{a + b - c}{2}.$$

Letting $s = (a + b + c)/2$, these two equations can be rewritten as

$$\frac{1}{2} bc(1 + \cos A) = s(s - a)$$

and

$$\frac{1}{2} bc(1 - \cos A) = (s - b)(s - c).$$

By substituting into the last formula for area, you can conclude that

$$\text{Area} = \sqrt{s(s - a)(s - b)(s - c)}.$$

Properties of the Dot Product   (p. 438)

Let $\mathbf{u}$, $\mathbf{v}$, and $\mathbf{w}$ be vectors in the plane or in space and let c be a scalar.

1. $\mathbf{u} \cdot \mathbf{v} = \mathbf{v} \cdot \mathbf{u}$ 2. $\mathbf{0} \cdot \mathbf{v} = 0$

3. $\mathbf{u} \cdot (\mathbf{v} + \mathbf{w}) = \mathbf{u} \cdot \mathbf{v} + \mathbf{u} \cdot \mathbf{w}$ 4. $\mathbf{v} \cdot \mathbf{v} = \|\mathbf{v}\|^2$

5. $c(\mathbf{u} \cdot \mathbf{v}) = c\mathbf{u} \cdot \mathbf{v} = \mathbf{u} \cdot c\mathbf{v}$

Proof

Let $\mathbf{u} = \langle u_1, u_2 \rangle$, $\mathbf{v} = \langle v_1, v_2 \rangle$, $\mathbf{w} = \langle w_1, w_2 \rangle$, $\mathbf{0} = \langle 0, 0 \rangle$, and let c be a scalar.

1. $\mathbf{u} \cdot \mathbf{v} = u_1 v_1 + u_2 v_2 = v_1 u_1 + v_2 u_2 = \mathbf{v} \cdot \mathbf{u}$

2. $\mathbf{0} \cdot \mathbf{v} = 0 \cdot v_1 + 0 \cdot v_2 = 0$

3. $\mathbf{u} \cdot (\mathbf{v} + \mathbf{w}) = \mathbf{u} \cdot \langle v_1 + w_1, v_2 + w_2 \rangle$

$$= u_1(v_1 + w_1) + u_2(v_2 + w_2)$$

$$= u_1 v_1 + u_1 w_1 + u_2 v_2 + u_2 w_2$$

$$= (u_1 v_1 + u_2 v_2) + (u_1 w_1 + u_2 w_2) = \mathbf{u} \cdot \mathbf{v} + \mathbf{u} \cdot \mathbf{w}$$

4. $\mathbf{v} \cdot \mathbf{v} = v_1^2 + v_2^2 = \left(\sqrt{v_1^2 + v_2^2}\right)^2 = \|\mathbf{v}\|^2$

5. $c(\mathbf{u} \cdot \mathbf{v}) = c(\langle u_1, u_2 \rangle \cdot \langle v_1, v_2 \rangle)$

$$= c(u_1 v_1 + u_2 v_2)$$

$$= (cu_1)v_1 + (cu_2)v_2$$

$$= \langle cu_1, cu_2 \rangle \cdot \langle v_1, v_2 \rangle$$

$$= c\mathbf{u} \cdot \mathbf{v}$$

Angle Between Two Vectors   (p. 439)

If θ is the angle between two nonzero vectors $\mathbf{u}$ and $\mathbf{v}$, then $\cos \theta = \dfrac{\mathbf{u} \cdot \mathbf{v}}{\|\mathbf{u}\| \, \|\mathbf{v}\|}$.

Proof

Consider the triangle determined by vectors $\mathbf{u}$, $\mathbf{v}$, and $\mathbf{v} - \mathbf{u}$, as shown in the figure. By the Law of Cosines, you can write

$$\|\mathbf{v} - \mathbf{u}\|^2 = \|\mathbf{u}\|^2 + \|\mathbf{v}\|^2 - 2\|\mathbf{u}\| \, \|\mathbf{v}\| \cos \theta$$

$$(\mathbf{v} - \mathbf{u}) \cdot (\mathbf{v} - \mathbf{u}) = \|\mathbf{u}\|^2 + \|\mathbf{v}\|^2 - 2\|\mathbf{u}\| \, \|\mathbf{v}\| \cos \theta$$

$$(\mathbf{v} - \mathbf{u}) \cdot \mathbf{v} - (\mathbf{v} - \mathbf{u}) \cdot \mathbf{u} = \|\mathbf{u}\|^2 + \|\mathbf{v}\|^2 - 2\|\mathbf{u}\| \, \|\mathbf{v}\| \cos \theta$$

$$\mathbf{v} \cdot \mathbf{v} - \mathbf{u} \cdot \mathbf{v} - \mathbf{v} \cdot \mathbf{u} + \mathbf{u} \cdot \mathbf{u} = \|\mathbf{u}\|^2 + \|\mathbf{v}\|^2 - 2\|\mathbf{u}\| \, \|\mathbf{v}\| \cos \theta$$

$$\|\mathbf{v}\|^2 - 2\mathbf{u} \cdot \mathbf{v} + \|\mathbf{u}\|^2 = \|\mathbf{u}\|^2 + \|\mathbf{v}\|^2 - 2\|\mathbf{u}\| \, \|\mathbf{v}\| \cos \theta$$

$$\cos \theta = \frac{\mathbf{u} \cdot \mathbf{v}}{\|\mathbf{u}\| \, \|\mathbf{v}\|}.$$

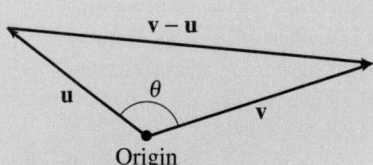

Progressive Summary (Chapters 1–6)

This chart outlines the topics that have been covered so far in this text. Progressive Summary charts appear after Chapters 2, 3, 6, 9 and 11. In each progressive summary, new topics encountered for the first time appear in red.

Algebraic Functions	Transcendental Functions	Other Topics
Polynomial, Rational, Radical	**Exponential, Logarithmic Trigonometric, Inverse Trigonometric**	
■ **Rewriting**	■ **Rewriting**	■ **Rewriting**
Polynomial form ↔ Factored form Operations with polynomials Rationalize denominators Simplify rational expressions Operations with complex numbers	Exponential form ↔ Logarithmic form Condense/expand logarithmic expressions Simplify trigonometric expressions Prove trigonometric identities Use conversion formulas Operations with vectors Powers and roots of complex numbers	
■ **Solving**	■ **Solving**	■ **Solving**

Equation	*Strategy*	*Equation*	*Strategy*
Linear	Isolate variable	Exponential	Take logarithm of each side
Quadratic	Factor, set to zero Extract square roots Complete the square Quadratic Formula	Logarithmic	Exponentiate each side

Polynomial	Factor, set to zero
	Rational Zero Test
		Trigonometric	Isolate function
			  factor, use inverse
			  function

Rational	Multiply by LCD	Multiple angle	Use trigonometric identities
Radical	Isolate, raise to power	or high powers	
Absolute Value	Isolate, form two equations		

■ **Analyzing**		■ **Analyzing**		■ **Analyzing**
Graphically	*Algebraically*	*Graphically*	*Algebraically*	
Intercepts	Domain, Range	Intercepts	Domain, Range	
Symmetry	Transformations	Asymptotes	Transformations	
Slope	Composition	Minimum values	Composition	
Asymptotes	Standard forms	Maximum values	Inverse properties	
End behavior	of equations		Amplitude, period	
Minimum values	Leading Coefficient		Reference angles	
Maximum values	Test	*Numerically*		
	Synthetic division	Table of values		
	Descartes's Rule of Signs			
Numerically				
Table of values				

Chapter 7

Linear Systems and Matrices

7.1 Solving Systems of Equations

7.2 Systems of Linear Equations in Two Variables

7.3 Multivariable Linear Systems

7.4 Matrices and Systems of Equations

7.5 Operations with Matrices

7.6 The Inverse of a Square Matrix

7.7 The Determinant of a Square Matrix

7.8 Applications of Matrices and Determinants

Selected Applications

Linear systems and matrices have many real life applications. The applications listed below represent a small sample of the applications in this chapter.

- Break-Even Analysis, Exercises 71 and 72, page 483
- Airplane Speed, Exercises 71 and 72, page 492
- Vertical Motion, Exercises 83–86, page 507
- Data Analysis, Exercise 81, page 524
- Voting Preference, Exercise 83, page 539
- Investment Portfolio, Exercises 65–68, page 549
- Sports, Exercise 27, page 567
- Supply and Demand, Exercises 35 and 36, page 570

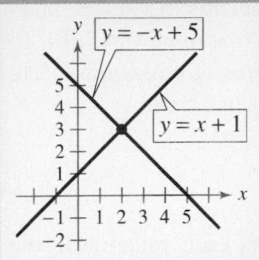 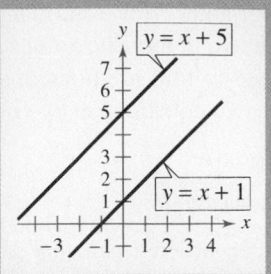 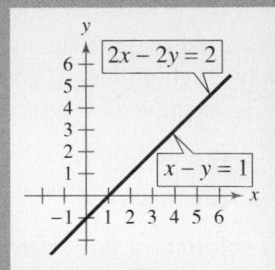

Until now, you have been working with single equations. In Chapter 7, you will solve systems of two or more equations in two or more variables. You can solve systems of equations algebraically, or graphically by finding the point of intersection of the graphs. You will also use matrices to represent data and to solve systems of linear equations.

Joseph Pobereskin/Getty Images

Systems of equations can be used to determine when a company can expect to earn a profit, incur a loss, or break even.

7.1 Solving Systems of Equations

The Methods of Substitution and Graphing

So far in this text, most problems have involved either a function of one variable or a single equation in two variables. However, many problems in science, business, and engineering involve two or more equations in two or more variables. To solve such problems, you need to find solutions of **systems of equations.** Here is an example of a system of two equations in two unknowns, x and y.

$$\begin{cases} 2x + y = 5 & \text{Equation 1} \\ 3x - 2y = 4 & \text{Equation 2} \end{cases}$$

A **solution** of this system is an ordered pair that satisfies each equation in the system. Finding the set of all such solutions is called **solving the system of equations.** For instance, the ordered pair $(2, 1)$ is a solution of this system. To check this, you can substitute 2 for x and 1 for y in *each* equation.

In this section, you will study two ways to solve systems of equations, beginning with the **method of substitution.**

The Method of Substitution

To use the **method of substitution** to solve a system of two equations in x and y, perform the following steps.

1. Solve one of the equations for one variable in terms of the other.

2. Substitute the expression found in Step 1 into the other equation to obtain an equation in one variable.

3. Solve the equation obtained in Step 2.

4. Back-substitute the value(s) obtained in Step 3 into the expression obtained in Step 1 to find the value(s) of the other variable.

5. Check that each solution satisfies *both* of the original equations.

When using the **method of graphing,** note that the solution of the system corresponds to the **point(s) of intersection** of the graphs.

The Method of Graphing

To use the **method of graphing** to solve a system of two equations in x and y, perform the following steps.

1. Solve both equations for y in terms of x.

2. Use a graphing utility to graph both equations in the same viewing window.

3. Use the *intersect* feature or the *zoom* and *trace* features of the graphing utility to approximate the point(s) of intersection of the graphs.

4. Check that each solution satisfies *both* of the original equations.

What you should learn

- Use the methods of substitution and graphing to solve systems of equations in two variables.
- Use systems of equations to model and solve real-life problems.

Why you should learn it

You can use systems of equations in situations in which the variables must satisfy two or more conditions. For instance, Exercise 76 on page 483 shows how to use a system of equations to compare two models for estimating the number of board feet in a 16-foot log.

Bruce Hands/Getty Images

STUDY TIP

When using the method of substitution, it does not matter which variable you choose to solve for first. Whether you solve for y first or x first, you will obtain the same solution. When making your choice, you should choose the variable and equation that are easier to work with.

Example 1   Solving a System of Equations

Solve the system of equations.

$$\begin{cases} x + y = 4 & \text{Equation 1} \\ x - y = 2 & \text{Equation 2} \end{cases}$$

Algebraic Solution

Begin by solving for y in Equation 1.

$y = 4 - x$ Solve for y in Equation 1.

Next, substitute this expression for y into Equation 2 and solve the resulting single-variable equation for x.

$x - y = 2$	Write Equation 2.
$x - (4 - x) = 2$	Substitute $4 - x$ for y.
$x - 4 + x = 2$	Distributive Property
$2x = 6$	Combine like terms.
$x = 3$	Divide each side by 2.

Finally, you can solve for y by *back-substituting* $x = 3$ into the equation $y = 4 - x$ to obtain

$y = 4 - x$	Write revised Equation 1.
$y = 4 - 3$	Substitute 3 for x.
$y = 1.$	Solve for y.

The solution is the ordered pair $(3, 1)$. Check this as follows.

Check $(3, 1)$ *in Equation 1:*

$x + y = 4$	Write Equation 1.
$3 + 1 \overset{?}{=} 4$	Substitute for x and y.
$4 = 4$	Solution checks in Equation 1. ✓

Check $(3, 1)$ *in Equation 2:*

$x - y = 2$	Write Equation 2.
$3 - 1 \overset{?}{=} 2$	Substitute for x and y.
$2 = 2$	Solution checks in Equation 2. ✓

✓CHECKPOINT   Now try Exercise 13.

Graphical Solution

Begin by solving both equations for y. Then use a graphing utility to graph the equations $y_1 = 4 - x$ and $y_2 = x - 2$ in the same viewing window. Use the *intersect* feature (see Figure 7.1) or the *zoom* and *trace* features of the graphing utility to approximate the point of intersection of the graphs.

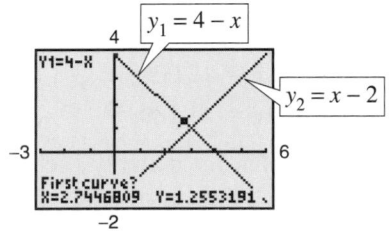

Figure 7.1

The point of intersection is $(3, 1)$, as shown in Figure 7.2.

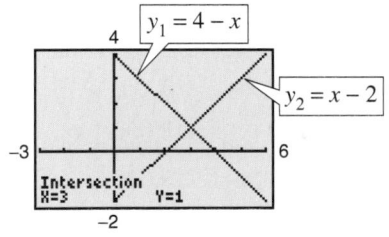

Figure 7.2

Check that $(3, 1)$ is the exact solution as follows.

Check $(3, 1)$ *in Equation 1:*

$3 + 1 \overset{?}{=} 4$	Substitute for x and y in Equation 1.
$4 = 4$	Solution checks in Equation 1. ✓

Check $(3, 1)$ *in Equation 2:*

$3 - 1 \overset{?}{=} 2$	Substitute for x and y in Equation 2.
$2 = 2$	Solution checks in Equation 2. ✓

In the algebraic solution of Example 1, note that the term *back-substitution* implies that you work *backwards*. First you solve for one of the variables, and then you substitute that value *back* into one of the equations in the system to find the value of the other variable.

Example 2   Solving a System by Substitution

A total of $12,000 is invested in two funds paying 9% and 11% simple interest. The yearly interest is $1180. How much is invested at each rate?

Solution

Verbal Model:
$$\boxed{\begin{array}{c}9\% \\ \text{fund}\end{array}} + \boxed{\begin{array}{c}11\% \\ \text{fund}\end{array}} = \boxed{\begin{array}{c}\text{Total} \\ \text{investment}\end{array}}$$

$$\boxed{\begin{array}{c}9\% \\ \text{interest}\end{array}} + \boxed{\begin{array}{c}11\% \\ \text{interest}\end{array}} = \boxed{\begin{array}{c}\text{Total} \\ \text{interest}\end{array}}$$

Labels: Amount in 9% fund $= x$ Amount in 11% fund $= y$ (dollars)
        Interest for 9% fund $= 0.09x$ Interest for 11% fund $= 0.11y$ (dollars)
        Total investment $= \$12,000$ Total interest $= \$1180$ (dollars)

System:
$$\begin{cases} x + y = 12{,}000 & \text{Equation 1} \\ 0.09x + 0.11y = 1{,}180 & \text{Equation 2} \end{cases}$$

To begin, it is convenient to multiply each side of Equation 2 by 100. This eliminates the need to work with decimals.

$9x + 11y = 118{,}000$ Revised Equation 2

To solve this system, you can solve for x in Equation 1.

$x = 12{,}000 - y$ Revised Equation 1

Next, substitute this expression for x into revised Equation 2 and solve the resulting equation for y.

$9x + 11y = 118{,}000$ Write revised Equation 2.

$9(12{,}000 - y) + 11y = 118{,}000$ Substitute $12{,}000 - y$ for x.

$108{,}000 - 9y + 11y = 118{,}000$ Distributive Property

$2y = 10{,}000$ Combine like terms.

$y = 5000$ Divide each side by 2.

Finally, back-substitute the value $y = 5000$ to solve for x.

$x = 12{,}000 - y$ Write revised Equation 1.

$x = 12{,}000 - 5000$ Substitute 5000 for y.

$x = 7000$ Simplify.

The solution is $(7000, 5000)$. So, $7000 is invested at 9% and $5000 is invested at 11% to yield yearly interest of $1180. Check this in the original system.

✓CHECKPOINT   Now try Exercise 75.

The equations in Examples 1 and 2 are linear. Substitution and graphing can also be used to solve systems in which one or both of the equations are nonlinear.

TECHNOLOGY TIP

Remember that a good way to check the answers you obtain in this section is to use a graphing utility. For instance, enter the two equations in Example 2

$$y_1 = 12{,}000 - x$$

$$y_2 = \frac{1180 - 0.09x}{0.11}$$

and find an appropriate viewing window that shows where the lines intersect. Then use the *intersect* feature or the *zoom* and *trace* features to find the point of intersection.

Example 3   Substitution: Two-Solution Case

Solve the system of equations: $\begin{cases} x^2 + 4x - y = 7 & \text{Equation 1} \\ 2x - y = -1 & \text{Equation 2} \end{cases}$.

Algebraic Solution

Begin by solving for y in Equation 2 to obtain $y = 2x + 1$. Next, substitute this expression for y into Equation 1 and solve for x.

$x^2 + 4x - y = 7$	Write Equation 1.
$x^2 + 4x - (2x + 1) = 7$	Substitute $2x + 1$ for y.
$x^2 + 4x - 2x - 1 = 7$	Distributive Property
$x^2 + 2x - 8 = 0$	Write in general form.
$(x + 4)(x - 2) = 0$	Factor.
$x + 4 = 0 \implies x = -4$	Set 1st factor equal to 0.
$x - 2 = 0 \implies x = 2$	Set 2nd factor equal to 0.

Back-substituting these values of x into revised Equation 2 produces

$$y = 2(-4) + 1 = -7 \quad \text{and} \quad y = 2(2) + 1 = 5.$$

So, the solutions are $(-4, -7)$ and $(2, 5)$. Check these in the original system.

✓CHECKPOINT   Now try Exercise 23.

Graphical Solution

To graph each equation, first solve both equations for y. Then use a graphing utility to graph the equations in the same viewing window. Use the *intersect* feature or the *zoom* and *trace* features to approximate the points of intersection of the graphs. The points of intersection are $(-4, -7)$ and $(2, 5)$, as shown in Figure 7.3. Check that $(-4, -7)$ and $(2, 5)$ are the exact solutions by substituting *both* ordered pairs into *both* equations.

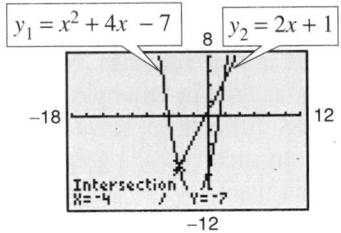

Figure 7.3

Example 4   Substitution: No-Solution Case

Solve the system of equations.

$$\begin{cases} -x + y = 4 & \text{Equation 1} \\ x^2 + y = 3 & \text{Equation 2} \end{cases}$$

Solution

Begin by solving for y in Equation 1 to obtain $y = x + 4$. Next, substitute this expression for y into Equation 2 and solve for x.

$x^2 + y = 3$	Write Equation 2.
$x^2 + (x + 4) = 3$	Substitute $x + 4$ for y.
$x^2 + x + 1 = 0$	Simplify.
$x = \dfrac{-1 \pm \sqrt{3}\,i}{2}$	Quadratic Formula

Because this yields two complex values, the equation $x^2 + x + 1 = 0$ has no *real* solution. So, the original system of equations has no *real* solution.

✓CHECKPOINT   Now try Exercise 25.

STUDY TIP

When using substitution, solve for the variable that is not raised to a power in either equation. For instance, in Example 4 it would not be practical to solve for x in Equation 2. Can you see why?

Exploration

Graph the system of equations in Example 4. Do the graphs of the equations intersect? Why or why not?

From Examples 1, 3, and 4, you can see that a system of two equations in two unknowns can have exactly one solution, more than one solution, or no solution. For instance, in Figure 7.4, the two equations in Example 1 graph as two lines with a *single point* of intersection. The two equations in Example 3 graph as a parabola and a line with *two points* of intersection, as shown in Figure 7.5. The two equations in Example 4 graph as a line and a parabola that have *no points* of intersection, as shown in Figure 7.6.

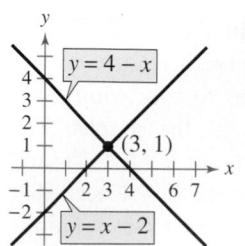

One Intersection Point
Figure 7.4

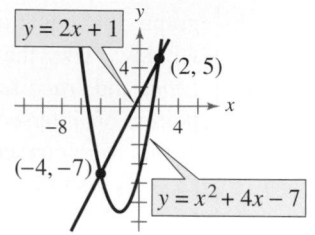

Two Intersection Points
Figure 7.5

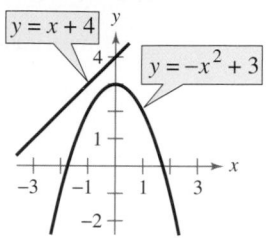

No Intersection Points
Figure 7.6

Example 5 shows the value of a graphical approach to solving systems of equations in two variables. Notice what would happen if you tried only the substitution method in Example 5. You would obtain the equation $x + \ln x = 1$. It would be difficult to solve this equation for x using standard algebraic techniques. In such cases, a graphical approach to solving systems of equations is more convenient.

Example 5 Solving a System of Equations Graphically

Solve the system of equations.

$$\begin{cases} y = \ln x & \text{Equation 1} \\ x + y = 1 & \text{Equation 2} \end{cases}$$

Solution

From the graphs of these equations, it is clear that there is only one point of intersection. Use the *intersect* feature or the *zoom* and *trace* features of a graphing utility to approximate the solution point as $(1, 0)$, as shown in Figure 7.7. You can confirm this by substituting $(1, 0)$ into *both* equations.

Check $(1, 0)$ *in Equation 1:*

$y = \ln x$ Write Equation 1.

$0 = \ln 1$ Equation 1 checks. ✓

Check $(1, 0)$ *in Equation 2:*

$x + y = 1$ Write Equation 2.

$1 + 0 = 1$ Equation 2 checks. ✓

✓CHECKPOINT   Now try Exercise 45.

TECHNOLOGY SUPPORT

For instructions on how to use the *intersect* feature and the *zoom* and *trace* features, see Appendix A; for specific keystrokes, go to this textbook's *Online Study Center*.

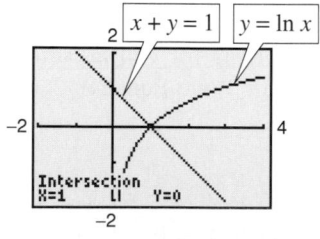

Figure 7.7

Points of Intersection and Applications

The total cost C of producing x units of a product typically has two components: the initial cost and the cost per unit. When enough units have been sold that the total revenue R equals the total cost C, the sales are said to have reached the **break-even point.** You will find that the break-even point corresponds to the point of intersection of the cost and revenue curves.

Example 6   Break-Even Analysis

A small business invests \$10,000 in equipment to produce a new soft drink. Each bottle of the soft drink costs \$0.65 to produce and is sold for \$1.20. How many bottles must be sold before the business breaks even?

Solution

The total cost of producing x bottles is

Total cost	=	Cost per bottle	·	Number of bottles	+	Initial cost

$$C = 0.65x + 10,000. \qquad \text{Equation 1}$$

The revenue obtained by selling x bottles is

Total revenue	=	Price per bottle	·	Number of bottles

$$R = 1.20x. \qquad \text{Equation 2}$$

Because the break-even point occurs when $R = C$, you have $C = 1.20x$, and the system of equations to solve is

$$\begin{cases} C = 0.65x + 10,000 \\ C = 1.20x \end{cases}.$$

Now you can solve by substitution.

$1.20x = 0.65x + 10,000$ Substitute $1.20x$ for C in Equation 1.

$0.55x = 10,000$ Subtract $0.65x$ from each side.

$x = \dfrac{10,000}{0.55} \approx 18,182$ bottles.   Divide each side by 0.55.

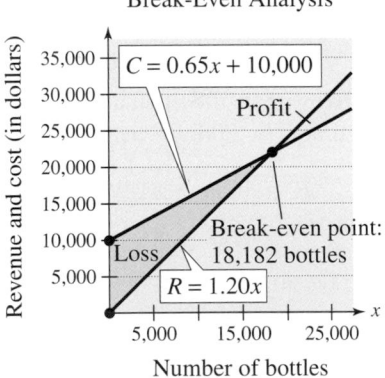

Break-Even Analysis

Figure 7.8

Note in Figure 7.8 that revenue less than the break-even point corresponds to an overall loss, whereas revenue greater than the break-even point corresponds to a profit. Verify the break-even point using the *intersect* feature or the *zoom* and *trace* features of a graphing utility.

✓CHECKPOINT   Now try Exercise 71.

Another way to view the solution in Example 6 is to consider the profit function $P = R - C$. The break-even point occurs when the profit is 0, which is the same as saying that $R = C$.

Example 7   State Populations

From 1998 to 2004, the population of Colorado increased more rapidly than the population of Alabama. Two models that approximate the populations P (in thousands) are

$$\begin{cases} P = 3488 + 81.9t & \text{Colorado} \\ P = 4248 + 19.9t & \text{Alabama} \end{cases}$$

where t represents the year, with $t = 8$ corresponding to 1998.   (Source: U.S. Census Bureau)

a. According to these two models, when would you expect the population of Colorado to have exceeded the population of Alabama?

b. Use the two models to estimate the populations of both states in 2010.

TECHNOLOGY SUPPORT

For instructions on how to use the *value* feature, see Appendix A; for specific keystrokes, go to this textbook's *Online Study Center.*

Algebraic Solution

a. Because the first equation has already been solved for P in terms of t, you can substitute this value into the second equation and solve for t, as follows.

$$3488 + 81.9t = 4248 + 19.9t$$

$$81.9t - 19.9t = 4248 - 3488$$

$$62.0t = 760$$

$$t \approx 12.26$$

So, from the given models, you would expect that the population of Colorado exceeded the population of Alabama after $t \approx 12.26$ years, which was sometime during 2002.

b. To estimate the populations of both states in 2010, substitute $t = 20$ into each model and evaluate, as follows.

$$P = 3488 + 81.9t \qquad \text{Model for Colorado}$$

$$= 3488 + 81.9(20) \qquad \text{Substitute 20 for } t.$$

$$= 5126 \qquad \text{Simplify.}$$

$$P = 4248 + 19.9t \qquad \text{Model for Alabama}$$

$$= 4248 + 19.9(20) \qquad \text{Substitute 20 for } t.$$

$$= 4646 \qquad \text{Simplify.}$$

So, according to the models, Colorado's population in 2010 will be 5,126,000 and Alabama's population in 2010 will be 4,646,000.

✓CHECKPOINT   Now try Exercise 77.

Graphical Solution

a. Use a graphing utility to graph $y_1 = 3488 + 81.9x$ and $y_2 = 4248 + 19.9x$ in the same viewing window. Use the *intersect* feature or the *zoom* and *trace* features of the graphing utility to approximate the point of intersection of the graphs. The point of intersection occurs at $x \approx 12.26$, as shown in Figure 7.9. So, it appears that the population of Colorado exceeded the population of Alabama sometime during 2002.

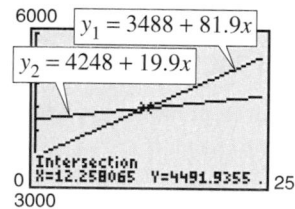

Figure 7.9

b. To estimate the populations of both states in 2010, use the *value* feature or the *zoom* and *trace* features of the graphing utility to find the value of y when $x = 20$. (Be sure to adjust your viewing window.) So, from Figure 7.10, you can see that Colorado's population in 2010 will be 5126 thousand, or 5,126,000, and Alabama's population in 2010 will be 4646 thousand, or 4,646,000.

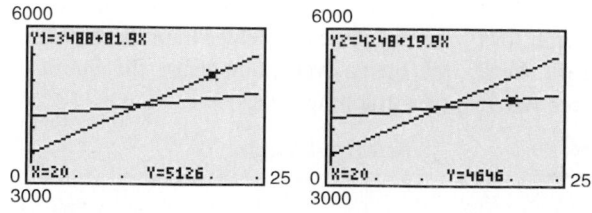

Figure 7.10

7.1 Exercises

See www.CalcChat.com for worked-out solutions to odd-numbered exercises.

Vocabulary Check

Fill in the blanks.

1. A set of two or more equations in two or more unknowns is called a _____ of _____ .

2. A _____ of a system of equations is an ordered pair that satisfies each equation in the system.

3. The first step in solving a system of equations by the _____ of _____ is to solve one of the equations for one variable in terms of the other variable.

4. Graphically, the solution to a system of equations is called the _____ of _____ .

5. In business applications, the _____ occurs when revenue equals cost.

In Exercises 1–4, determine whether each ordered pair is a solution of the system of equations.

1. $\begin{cases} 4x - y = 1 \\ 6x + y = -6 \end{cases}$
   (a) $(0, -3)$ (b) $(-1, -5)$
   (c) $\left(-\frac{3}{2}, 3\right)$ (d) $\left(-\frac{1}{2}, -3\right)$

2. $\begin{cases} 4x^2 + y = 3 \\ -x - y = 11 \end{cases}$
   (a) $(2, -13)$ (b) $(-2, -9)$
   (c) $\left(-\frac{3}{2}, 6\right)$ (d) $\left(-\frac{7}{4}, -\frac{37}{4}\right)$

3. $\begin{cases} y = -2e^x \\ 3x - y = 2 \end{cases}$
   (a) $(-2, 0)$ (b) $(0, -2)$
   (c) $(0, -3)$ (d) $(-1, -5)$

4. $\begin{cases} -\log_{10} x + 3 = y \\ \frac{1}{9}x + y = \frac{28}{9} \end{cases}$
   (a) $(100, 1)$ (b) $(10, 2)$
   (c) $(1, 3)$ (d) $(1, 1)$

In Exercises 5–12, solve the system by the method of substitution. Check your solution graphically.

5. $\begin{cases} 2x + y = 6 \\ -x + y = 0 \end{cases}$

6. $\begin{cases} x - y = -4 \\ x + 2y = 5 \end{cases}$

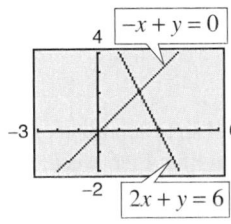
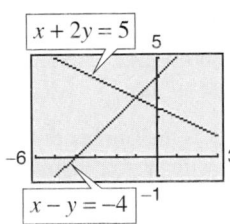

7. $\begin{cases} x - y = -4 \\ x^2 - y = -2 \end{cases}$

8. $\begin{cases} -2x + y = -5 \\ x^2 + y^2 = 25 \end{cases}$

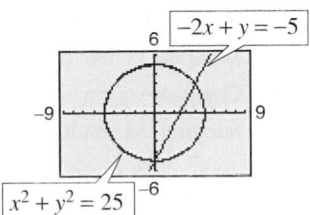

9. $\begin{cases} 3x + y = 2 \\ x^3 - 2 + y = 0 \end{cases}$

10. $\begin{cases} x + y = 0 \\ x^3 - 5x - y = 0 \end{cases}$

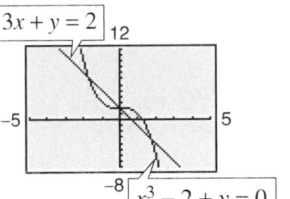
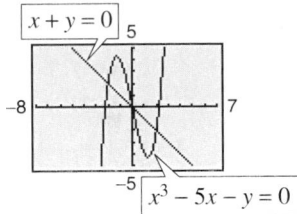

11. $\begin{cases} -\frac{7}{2}x - y = -18 \\ 8x^2 - 2y^3 = 0 \end{cases}$

12. $\begin{cases} y = x^3 - 3x^2 + 4 \\ y = -2x + 4 \end{cases}$

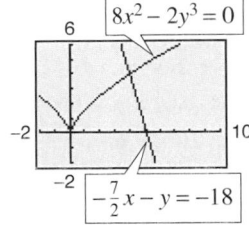
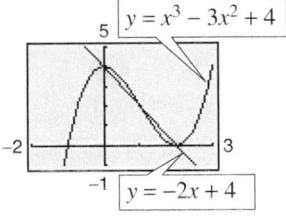

In Exercises 13–28, solve the system by the method of substitution. Use a graphing utility to verify your results.

13. $\begin{cases} x - y = 0 \\ 5x - 3y = 10 \end{cases}$

14. $\begin{cases} x + 2y = 1 \\ 5x - 4y = -23 \end{cases}$

15. $\begin{cases} 2x - y + 2 = 0 \\ 4x + y - 5 = 0 \end{cases}$

16. $\begin{cases} 6x - 3y - 4 = 0 \\ x + 2y - 4 = 0 \end{cases}$

17. $\begin{cases} 1.5x + 0.8y = 2.3 \\ 0.3x - 0.2y = 0.1 \end{cases}$

18. $\begin{cases} 0.5x + 3.2y = 9.0 \\ 0.2x - 1.6y = -3.6 \end{cases}$

19. $\begin{cases} \frac{1}{5}x + \frac{1}{2}y = 8 \\ x + y = 20 \end{cases}$

20. $\begin{cases} \frac{1}{2}x + \frac{3}{4}y = 10 \\ \frac{3}{4}x - y = 4 \end{cases}$

21. $\begin{cases} -\frac{5}{3}x + y = 5 \\ -5x + 3y = 6 \end{cases}$ **22.** $\begin{cases} -\frac{2}{3}x + y = 2 \\ 2x - 3y = 6 \end{cases}$

23. $\begin{cases} x^2 - 2x + y = 8 \\ x - y = -2 \end{cases}$ **24.** $\begin{cases} 2x^2 - 2x - y = 14 \\ 2x - y = -2 \end{cases}$

25. $\begin{cases} 2x^2 - y = 1 \\ x - y = 2 \end{cases}$ **26.** $\begin{cases} 2x^2 + y = 3 \\ x + y = 4 \end{cases}$

27. $\begin{cases} x^3 - y = 0 \\ x - y = 0 \end{cases}$ **28.** $\begin{cases} y = -x \\ y = x^3 + 3x^2 + 2x \end{cases}$

In Exercises 29–36, solve the system graphically. Verify your solutions algebraically.

29. $\begin{cases} -x + 2y = 2 \\ 3x + y = 15 \end{cases}$ **30.** $\begin{cases} x + y = 0 \\ 3x - 2y = 10 \end{cases}$

31. $\begin{cases} x - 3y = -2 \\ 5x + 3y = 17 \end{cases}$ **32.** $\begin{cases} -x + 2y = 1 \\ x - y = 2 \end{cases}$

33. $\begin{cases} x^2 + y = 1 \\ x + y = 2 \end{cases}$ **34.** $\begin{cases} x^2 - y = 4 \\ x - y = 2 \end{cases}$

35. $\begin{cases} -x + y = 3 \\ x^2 + y^2 - 6x - 27 = 0 \end{cases}$

36. $\begin{cases} y^2 - 4x + 11 = 0 \\ -\frac{1}{2}x + y = -\frac{1}{2} \end{cases}$

In Exercises 37–50, use a graphing utility to approximate all points of intersection of the graph of the system of equations. Round your results to three decimal places. Verify your solutions by checking them in the original system.

37. $\begin{cases} 7x + 8y = 24 \\ x - 8y = 8 \end{cases}$ **38.** $\begin{cases} x - y = 0 \\ 5x - 2y = 6 \end{cases}$

39. $\begin{cases} x - y^2 = -1 \\ x - y = 5 \end{cases}$ **40.** $\begin{cases} x - y^2 = -2 \\ x - 2y = 6 \end{cases}$

41. $\begin{cases} x^2 + y^2 = 8 \\ y = x^2 \end{cases}$ **42.** $\begin{cases} x^2 + y^2 = 25 \\ (x - 8)^2 + y^2 = 41 \end{cases}$

43. $\begin{cases} y = e^x \\ x - y + 1 = 0 \end{cases}$ **44.** $\begin{cases} y = -4e^{-x} \\ y + 3x + 8 = 0 \end{cases}$

45. $\begin{cases} x + 2y = 8 \\ y = 2 + \ln x \end{cases}$

46. $\begin{cases} y = -2 + \ln(x - 1) \\ 3y + 2x = 9 \end{cases}$

47. $\begin{cases} y = \sqrt{x} + 4 \\ y = 2x + 1 \end{cases}$ **48.** $\begin{cases} x - y = 3 \\ \sqrt{x} - y = 1 \end{cases}$

49. $\begin{cases} x^2 + y^2 = 169 \\ x^2 - 8y = 104 \end{cases}$ **50.** $\begin{cases} x^2 + y^2 = 4 \\ 2x^2 - y = 2 \end{cases}$

In Exercises 51–64, solve the system graphically or algebraically. Explain your choice of method.

51. $\begin{cases} 2x - y = 0 \\ x^2 - y = -1 \end{cases}$ **52.** $\begin{cases} x + y = 4 \\ x^2 + y = 2 \end{cases}$

53. $\begin{cases} 3x - 7y = -6 \\ x^2 - y^2 = 4 \end{cases}$ **54.** $\begin{cases} x^2 + y^2 = 25 \\ 2x + y = 10 \end{cases}$

55. $\begin{cases} x^2 + y^2 = 1 \\ x + y = 4 \end{cases}$ **56.** $\begin{cases} x^2 + y^2 = 4 \\ x - y = 5 \end{cases}$

57. $\begin{cases} y = 2x + 1 \\ y = \sqrt{x + 2} \end{cases}$ **58.** $\begin{cases} y = 2x - 1 \\ y = \sqrt{x + 1} \end{cases}$

59. $\begin{cases} y - e^{-x} = 1 \\ y - \ln x = 3 \end{cases}$ **60.** $\begin{cases} 2 \ln x + y = 4 \\ e^x - y = 0 \end{cases}$

61. $\begin{cases} y = x^3 - 2x^2 + 1 \\ y = 1 - x^2 \end{cases}$ **62.** $\begin{cases} y = x^3 - 2x^2 + x - 1 \\ y = -x^2 + 3x - 1 \end{cases}$

63. $\begin{cases} xy - 1 = 0 \\ 2x - 4y + 7 = 0 \end{cases}$ **64.** $\begin{cases} xy - 2 = 0 \\ 3x - 2y + 4 = 0 \end{cases}$

Break-Even Analysis **In Exercises 65–68, use a graphing utility to graph the cost and revenue functions in the same viewing window. Find the sales x necessary to break even $(R = C)$ and the corresponding revenue R obtained by selling x units. (Round to the nearest whole unit.)**

	Cost	*Revenue*
65.	$C = 8650x + 250{,}000$	$R = 9950x$
66.	$C = 2.65x + 350{,}000$	$R = 4.15x$
67.	$C = 5.5\sqrt{x} + 10{,}000$	$R = 3.29x$
68.	$C = 7.8\sqrt{x} + 18{,}500$	$R = 12.84x$

69. ***DVD Rentals***   The daily DVD rentals of a newly released animated film and a newly released horror film from a movie rental store can be modeled by the equations

$$\begin{cases} N = 360 - 24x & \text{Animated film} \\ N = 24 + 18x & \text{Horror film} \end{cases}$$

where N is the number of DVDs rented and x represents the week, with $x = 1$ corresponding to the first week of release.

(a) Use the *table* feature of a graphing utility to find the numbers of rentals of each movie for each of the first 12 weeks of release.

(b) Use the results of part (a) to determine the solution to the system of equations.

(c) Solve the system of equations algebraically.

(d) Compare your results from parts (b) and (c).

(e) Interpret the results in the context of the situation.

70. *Sports* The points scored during each of the first 12 games by two players on a girl's high school basketball team can be modeled by the equations

$$\begin{cases} P_S = 24 - 2x & \text{Sofia} \\ P_P = 12 + 2x & \text{Paige} \end{cases}$$

where P represents the points scored by each player and x represents the number of games played, with $x = 1$ corresponding to the first game.

(a) Use the *table* feature of a graphing utility to find the numbers of points scored by each player for each of the first 12 games.

(b) Use the results of part (a) to determine the solution to the system of equations.

(c) Solve the system of equations algebraically.

(d) Compare your results from parts (b) and (c).

(e) Interpret the results in the context of the situation.

71. *Break-Even Analysis* A small software company invests $16,000 to produce a software package that will sell for $55.95. Each unit can be produced for $35.45.

(a) Write the cost and revenue functions for x units produced and sold.

(b) Use a graphing utility to graph the cost and revenue functions in the same viewing window. Use the graph to approximate the number of units that must be sold to break even, and verify the result algebraically.

72. *Break-Even Analysis* A small fast food restaurant invests $5000 to produce a new food item that will sell for $3.49. Each item can be produced for $2.16.

(a) Write the cost and revenue functions for x items produced and sold.

(b) Use a graphing utility to graph the cost and revenue functions in the same viewing window. Use the graph to approximate the number of items that must be sold to break even, and verify the result algebraically.

73. *Choice of Two Jobs* You are offered two different jobs selling dental supplies. One company offers a straight commission of 6% of sales. The other company offers a salary of $350 per week plus 3% of sales. How much would you have to sell in a week in order to make the straight commission offer the better offer?

74. *Choice of Two Jobs* You are offered two jobs selling college textbooks. One company offers an annual salary of $25,000 plus a year-end bonus of 1% of your total sales. The other company offers an annual salary of $20,000 plus a year-end bonus of 2% of your total sales. How much would you have to sell in a year to make the second offer the better offer?

75. *Investment* A total of $20,000 is invested in two funds paying 6.5% and 8.5% simple interest. The 6.5% investment has a lower risk. The investor wants a yearly interest check of $1600 from the investments.

(a) Write a system of equations in which one equation represents the total amount invested and the other equation represents the $1600 required in interest. Let x and y represent the amounts invested at 6.5% and 8.5%, respectively.

(b) Use a graphing utility to graph the two equations in the same viewing window. As the amount invested at 6.5% increases, how does the amount invested at 8.5% change? How does the amount of interest change? Explain.

(c) What amount should be invested at 6.5% to meet the requirement of $1600 per year in interest?

76. *Log Volume* You are offered two different rules for estimating the number of board feet in a 16-foot log. (A board foot is a unit of measure for lumber equal to a board 1 foot square and 1 inch thick.) One rule is the *Doyle Log Rule* and is modeled by

$$V = (D - 4)^2, \quad 5 \le D \le 40$$

and the other rule is the *Scribner Log Rule* and is modeled by

$$V = 0.79D^2 - 2D - 4, \quad 5 \le D \le 40$$

where D is the diameter (in inches) of the log and V is its volume in (board feet).

(a) Use a graphing utility to graph the two log rules in the same viewing window.

(b) For what diameter do the two rules agree?

(c) You are selling large logs by the board foot. Which rule would you use? Explain your reasoning.

77. *Population* The populations (in thousands) of Missouri M and Tennessee T from 1990 to 2004 can by modeled by the system

$$\begin{cases} M = 47.4t + 5104 & \text{Missouri} \\ T = 76.5t + 4875 & \text{Tennessee} \end{cases}$$

where t is the year, with $t = 0$ corresponding to 1990. (Source: U.S. Census Bureau)

(a) Record in a table the populations of the two states for the years 1990, 1994, 1998, 2002, 2006, and 2010.

(b) According to the table, over what period of time does the population of Tennessee exceed that of Missouri?

(c) Use a graphing utility to graph the models in the same viewing window. Estimate the point of intersection of the models.

(d) Find the point of intersection algebraically.

(e) Summarize your findings of parts (b) through (d).

78. *Tuition*   The table shows the average costs (in dollars) of one year's tuition for public and private universities in the United States from 2000 to 2004.   (Source: U.S, National Center for Education Statistics)

Year	Public universities	Private universities
2000	2506	14,081
2001	2562	15,000
2002	2700	15,742
2003	2903	16,383
2004	3313	17,442

(a) Use the *regression* feature of a graphing utility to find a quadratic model T_{public} for tuition at public universities and a linear model $T_{private}$ for tuition at private universities. Let x represent the year, with $x = 0$ corresponding to 2000.

(b) Use a graphing utility to graph the models with the original data in the same viewing window.

(c) Use the graph in part (b) to determine the year after 2004 in which tuition at public universities will exceed tuition at private universities.

(d) Algebraically determine the year in which tuition at public universities will exceed tuition at private universities.

(e) Compare your results from parts (c) and (d).

Geometry   **In Exercises 79 and 80, find the dimensions of the rectangle meeting the specified conditions.**

79. The perimeter is 30 meters and the length is 3 meters greater than the width.

80. The perimeter is 280 centimeters and the width is 20 centimeters less than the length.

81. *Geometry*   What are the dimensions of a rectangular tract of land if its perimeter is 40 miles and its area is 96 square miles?

82. *Geometry*   What are the dimensions of an isosceles right triangle with a two-inch hypotenuse and an area of 1 square inch?

Synthesis

True or False?   **In Exercises 83 and 84, determine whether the statement is true or false. Justify your answer.**

83. In order to solve a system of equations by substitution, you must always solve for y in one of the two equations and then back-substitute.

84. If a system consists of a parabola and a circle, then it can have at most two solutions.

85. *Think About It*   When solving a system of equations by substitution, how do you recognize that the system has no solution?

86. *Writing*   Write a brief paragraph describing any advantages of substitution over the graphical method of solving a system of equations.

87. *Exploration*   Find the equations of lines whose graphs intersect the graph of the parabola $y = x^2$ at (a) two points, (b) one point, and (c) no points. (There are many correct answers.)

88. *Exploration*   Create systems of two linear equations in two variables that have (a) no solution, (b) one distinct solution, and (c) infinitely many solutions. (There are many correct answers.)

89. *Exploration*   Create a system of linear equations in two variables that has the solution $(2, -1)$ as its only solution. (There are many correct answers.)

90. *Conjecture*   Consider the system of equations.

$$\begin{cases} y = b^x \\ y = x^b \end{cases}$$

(a) Use a graphing utility to graph the system of equations for $b = 2$ and $b = 4$.

(b) For a fixed value of $b > 1$, make a conjecture about the number of points of intersection of the graphs in part (a).

Skills Review

In Exercises 91–96, find the general form of the equation of the line passing through the two points.

91. $(-2, 7), (5, 5)$

92. $(3, 4), (10, 6)$

93. $(6, 3), (10, 3)$

94. $(4, -2), (4, 5)$

95. $\left(\frac{3}{5}, 0\right), (4, 6)$

96. $\left(-\frac{7}{3}, 8\right), \left(\frac{5}{2}, \frac{1}{2}\right)$

In Exercises 97–102, find the domain of the function and identify any horizontal or vertical asymptotes.

97. $f(x) = \dfrac{5}{x - 6}$

98. $f(x) = \dfrac{2x - 7}{3x + 2}$

99. $f(x) = \dfrac{x^2 + 2}{x^2 - 16}$

100. $f(x) = 3 - \dfrac{2}{x^2}$

101. $f(x) = \dfrac{x + 1}{x^2 + 1}$

102. $f(x) = \dfrac{x - 4}{x^2 + 16}$

7.2 Systems of Linear Equations in Two Variables

The Method of Elimination

In Section 7.1, you studied two methods for solving a system of equations: substitution and graphing. Now you will study the **method of elimination** to solve a system of linear equations in two variables. The key step in this method is to obtain, for one of the variables, coefficients that differ only in sign so that *adding* the equations eliminates the variable.

$$\begin{aligned} 3x + 5y &= 7 \qquad && \text{Equation 1} \\ -3x - 2y &= -1 \qquad && \text{Equation 2} \\ \hline 3y &= 6 \qquad && \text{Add equations.} \end{aligned}$$

Note that by adding the two equations, you eliminate the x-terms and obtain a single equation in y. Solving this equation for y produces $y = 2$, which you can then back-substitute into one of the original equations to solve for x.

Example 1 Solving a System by Elimination

Solve the system of linear equations.

$$\begin{cases} 3x + 2y = 4 \qquad & \text{Equation 1} \\ 5x - 2y = 8 \qquad & \text{Equation 2} \end{cases}$$

Solution

You can eliminate the y-terms by adding the two equations.

$$\begin{aligned} 3x + 2y &= 4 \qquad && \text{Write Equation 1.} \\ 5x - 2y &= 8 \qquad && \text{Write Equation 2.} \\ \hline 8x &= 12 \qquad && \text{Add equations.} \end{aligned}$$

So, $x = \frac{3}{2}$. By back-substituting into Equation 1, you can solve for y.

$$\begin{aligned} 3x + 2y &= 4 \qquad && \text{Write Equation 1.} \\ 3\left(\tfrac{3}{2}\right) + 2y &= 4 \qquad && \text{Substitute } \tfrac{3}{2} \text{ for } x. \\ y &= -\tfrac{1}{4} \qquad && \text{Solve for } y. \end{aligned}$$

The solution is $\left(\frac{3}{2}, -\frac{1}{4}\right)$. You can check the solution *algebraically* by substituting into the original system, or graphically as shown in Section 7.1.

Check

$$\begin{aligned} 3\left(\tfrac{3}{2}\right) + 2\left(-\tfrac{1}{4}\right) &\overset{?}{=} 4 \qquad && \text{Substitute into Equation 1.} \\ \tfrac{9}{2} - \tfrac{1}{2} &= 4 \qquad && \text{Equation 1 checks. } \checkmark \\ 5\left(\tfrac{3}{2}\right) - 2\left(-\tfrac{1}{4}\right) &\overset{?}{=} 8 \qquad && \text{Substitute into Equation 2.} \\ \tfrac{15}{2} + \tfrac{1}{2} &= 8 \qquad && \text{Equation 2 checks. } \checkmark \end{aligned}$$

✓CHECKPOINT   Now try Exercise 7.

What you should learn

- Use the method of elimination to solve systems of linear equations in two variables.
- Graphically interpret the number of solutions of a system of linear equations in two variables.
- Use systems of linear equations in two variables to model and solve real-life problems.

Why you should learn it

You can use systems of linear equations to model many business applications. For instance, Exercise 76 on page 493 shows how to use a system of linear equations to compare sales of two competing companies.

Spencer Platt/Getty Images

Exploration

Use the method of substitution to solve the system given in Example 1. Which method is easier?

> **The Method of Elimination**
>
> To use the **method of elimination** to solve a system of two linear equations in x and y, perform the following steps.
>
> 1. Obtain coefficients for x (or y) that differ only in sign by multiplying all terms of one or both equations by suitably chosen constants.
>
> 2. Add the equations to eliminate one variable; solve the resulting equation.
>
> 3. Back-substitute the value obtained in Step 2 into either of the original equations and solve for the other variable.
>
> 4. Check your solution in both of the original equations.

Example 2 Solving a System by Elimination

Solve the system of linear equations.

$$\begin{cases} 5x + 3y = 9 & \text{Equation 1} \\ 2x - 4y = 14 & \text{Equation 2} \end{cases}$$

Algebraic Solution

You can obtain coefficients of y that differ only in sign by multiplying Equation 1 by 4 and multiplying Equation 2 by 3.

$$5x + 3y = 9 \implies 20x + 12y = 36 \quad \text{Multiply Equation 1 by 4.}$$
$$2x - 4y = 14 \implies \underline{6x - 12y = 42} \quad \text{Multiply Equation 2 by 3.}$$
$$26x \qquad = 78 \quad \text{Add equations.}$$

From this equation, you can see that $x = 3$. By back-substituting this value of x into Equation 2, you can solve for y.

$$2x - 4y = 14 \qquad \text{Write Equation 2.}$$
$$2(3) - 4y = 14 \qquad \text{Substitute 3 for } x.$$
$$-4y = 8 \qquad \text{Combine like terms.}$$
$$y = -2 \qquad \text{Solve for } y.$$

The solution is $(3, -2)$. You can check the solution algebraically by substituting into the original system.

Graphical Solution

Solve each equation for y. Then use a graphing utility to graph $y_1 = 3 - \frac{5}{3}x$ and $y_2 = -\frac{7}{2} + \frac{1}{2}x$ in the same viewing window. Use the *intersect* feature or the *zoom* and *trace* features to approximate the point of intersection of the graphs. The point of intersection is $(3, -2)$, as shown in Figure 7.11. You can determine that this is the exact solution by checking $(3, -2)$ in both equations.

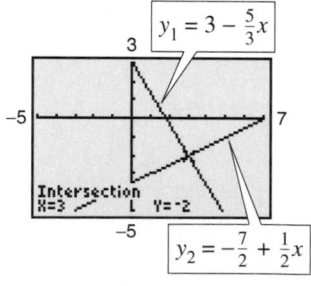

Figure 7.11

✓**CHECKPOINT**   Now try Exercise 9.

In Example 2, the original system and the system obtained by multiplying by constants are called **equivalent systems** because they have precisely the same solution set. The operations that can be performed on a system of linear equations to produce an equivalent system are (1) interchanging any two equations, (2) multiplying an equation by a nonzero constant, and (3) adding a multiple of one equation to any other equation in the system.

Graphical Interpretation of Two-Variable Systems

It is possible for any system of equations to have exactly one solution, two or more solutions, or no solution. If a system of *linear* equations has two different solutions, it must have an *infinite* number of solutions. To see why this is true, consider the following graphical interpretations of a system of two linear equations in two variables.

Graphical Interpretation of Solutions

For a system of two linear equations in two variables, the number of solutions is one of the following.

Number of Solutions	*Graphical Interpretation*
1. Exactly one solution	The two lines intersect at one point.
2. Infinitely many solutions	The two lines are coincident (identical).
3. No solution	The two lines are parallel.

> **Exploration**
>
> Rewrite each system of equations in slope-intercept form and use a graphing utility to graph each system. What is the relationship between the slopes of the two lines and the number of points of intersection?
>
> **a.** $\begin{cases} y = 5x + 1 \\ y - x = -5 \end{cases}$
>
> **b.** $\begin{cases} 3y = 4x - 1 \\ -8x + 2 = -6y \end{cases}$
>
> **c.** $\begin{cases} 2y = -x + 3 \\ -4 = y + \frac{1}{2}x \end{cases}$

A system of linear equations is **consistent** if it has at least one solution. It is **inconsistent** if it has no solution.

Example 3   Recognizing Graphs of Linear Systems

Match each system of linear equations (a, b, c) with its graph (i, ii, iii) in Figure 7.12. Describe the number of solutions. Then state whether the system is consistent or inconsistent.

a. $\begin{cases} 2x - 3y = 3 \\ -4x + 6y = 6 \end{cases}$ **b.** $\begin{cases} 2x - 3y = 3 \\ x + 2y = 5 \end{cases}$ **c.** $\begin{cases} 2x - 3y = 3 \\ -4x + 6y = -6 \end{cases}$

> **Prerequisite Skills**
>
> If you have difficulty with this example, review graphing of linear equations in Appendix B.3.

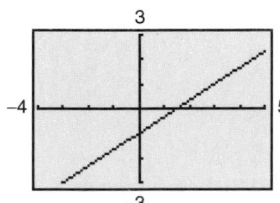

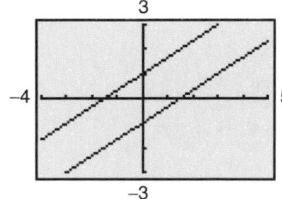

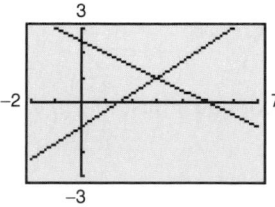

i.   ii.   iii.

Figure 7.12

Solution

a. The graph is a pair of parallel lines (ii). The lines have no point of intersection, so the system has no solution. The system is inconsistent.

b. The graph is a pair of intersecting lines (iii). The lines have one point of intersection, so the system has exactly one solution. The system is consistent.

c. The graph is a pair of lines that coincide (i). The lines have infinitely many points of intersection, so the system has infinitely many solutions. The system is consistent.

✓CHECKPOINT   Now try Exercises 17–20.

In Examples 4 and 5, note how you can use the method of elimination to determine that a system of linear equations has no solution or infinitely many solutions.

Example 4　The Method of Elimination: No–Solution Case

Solve the system of linear equations.

$$\begin{cases} x - 2y = 3 & \text{Equation 1} \\ -2x + 4y = 1 & \text{Equation 2} \end{cases}$$

Algebraic Solution

To obtain coefficients that differ only in sign, multiply Equation 1 by 2.

$$x - 2y = 3 \quad \Longrightarrow \quad 2x - 4y = 6$$
$$\underline{-2x + 4y = 1} \quad \Longrightarrow \quad \underline{-2x + 4y = 1}$$
$$0 = 7$$

By adding the equations, you obtain $0 = 7$. Because there are no values of x and y for which $0 = 7$, this is a false statement. So, you can conclude that the system is inconsistent and has no solution.

Graphical Solution

Solving each equation for y yields $y_1 = -\frac{3}{2} + \frac{1}{2}x$ and $y_2 = \frac{1}{4} + \frac{1}{2}x$. Notice that the lines have the same slope and different y-intercepts, so they are parallel. You can use a graphing utility to verify this by graphing both equations in the same viewing window, as shown in Figure 7.13. Then try using the *intersect* feature to find a point of intersection. Because the graphing utility cannot find a point of intersection, you will get an error message. Therefore, the system has no solution.

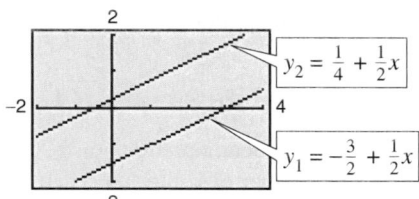

Figure 7.13

✓CHECKPOINT　Now try Exercise 23.

Example 5　The Method of Elimination: Infinitely Many Solutions Case

Solve the system of linear equations: $\begin{cases} 2x - y = 1 & \text{Equation 1} \\ 4x - 2y = 2 & \text{Equation 2} \end{cases}$.

Solution

To obtain coefficients that differ only in sign, multiply Equation 1 by -2.

$$2x - y = 1 \quad \Longrightarrow \quad -4x + 2y = -2 \qquad \text{Multiply Equation 1 by } -2.$$
$$\underline{4x - 2y = 2} \quad \Longrightarrow \quad \underline{4x - 2y = 2} \qquad \text{Write Equation 2.}$$
$$0 = 0 \qquad \text{Add equations.}$$

Because $0 = 0$ for all values of x and y, the two equations turn out to be equivalent (have the same solution set). You can conclude that the system has infinitely many solutions. The solution set consists of all points (x, y) lying on the line $2x - y = 1$, as shown in Figure 7.14.

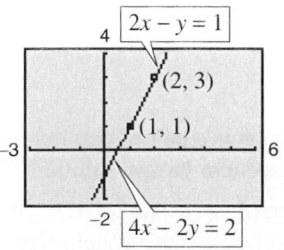

Figure 7.14

✓CHECKPOINT　Now try Exercise 25.

In Example 4, note that the occurrence of the false statement $0 = 7$ indicates that the system has no solution. In Example 5, note that the occurrence of a statement that is true for all values of the variables—in this case, $0 = 0$—indicates that the system has infinitely many solutions.

Example 6 illustrates a strategy for solving a system of linear equations that has decimal coefficients.

Example 6 A Linear System Having Decimal Coefficients

Solve the system of linear equations.

$$\begin{cases} 0.02x - 0.05y = -0.38 & \text{Equation 1} \\ 0.03x + 0.04y = 1.04 & \text{Equation 2} \end{cases}$$

Solution

Because the coefficients in this system have two decimal places, you can begin by multiplying each equation by 100 to produce a system with integer coefficients.

$$\begin{cases} 2x - 5y = -38 & \text{Revised Equation 1} \\ 3x + 4y = 104 & \text{Revised Equation 2} \end{cases}$$

Now, to obtain coefficients that differ only in sign, multiply revised Equation 1 by 3 and multiply revised Equation 2 by -2.

$2x - 5y = -38$ ⟹	$6x - 15y = -114$	Multiply revised Equation 1 by 3.
$3x + 4y = 104$ ⟹	$-6x - 8y = -208$	Multiply revised Equation 2 by -2.
	$-23y = -322$	Add equations.

So, you can conclude that $y = \dfrac{-322}{-23} = 14$. Back-substituting this value into revised Equation 2 produces the following.

$3x + 4y = 104$	Write revised Equation 2.
$3x + 4(14) = 104$	Substitute 14 for y.
$3x = 48$	Combine like terms.
$x = 16$	Solve for x.

The solution is $(16, 14)$. Check this as follows in the original system.

Check $(16, 14)$ in Equation 1:

$0.02x - 0.05y = -0.38$	Write Equation 1.
$0.02(16) - 0.05(14) \overset{?}{=} -0.38$	Substitute for x and y.
$-0.38 = 0.38$	Solution checks in Equation 1.

Check $(16, 14)$ in Equation 2:

$0.03x + 0.04y = 1.04$	Write Equation 2.
$0.03(16) + 0.04(14) \overset{?}{=} 1.04$	Substitute for x and y.
$1.04 = 1.04$	Solution checks in Equation 2.

✓CHECKPOINT   Now try Exercise 33.

TECHNOLOGY SUPPORT

The general solution of the linear system

$$\begin{cases} ax + by = c \\ dx + ey = f \end{cases}$$

is $x = (ce - bf)/(ae - bd)$ and $y = (af - cd)/(ae - bd)$. If $ae - bd = 0$, the system does not have a unique solution. A program (called Systems of Linear Equations) for solving such a system is available at this textbook's *Online Study Center.* Try using this program to check the solution of the system in Example 6.

Application

At this point, you may be asking the question "How can I tell which application problems can be solved using a system of linear equations?" The answer comes from the following considerations.

1. Does the problem involve more than one unknown quantity?

2. Are there two (or more) equations or conditions to be satisfied?

If one or both of these conditions are met, the appropriate mathematical model for the problem may be a system of linear equations.

Example 7   An Application of a Linear System

An airplane flying into a headwind travels the 2000-mile flying distance between Cleveland, Ohio and Fresno, California in 4 hours and 24 minutes. On the return flight, the same distance is traveled in 4 hours. Find the airspeed of the plane and the speed of the wind, assuming that both remain constant.

Solution

The two unknown quantities are the speeds of the wind and the plane. If r_1 is the speed of the plane and r_2 is the speed of the wind, then

$r_1 - r_2 =$ speed of the plane *against* the wind

$r_1 + r_2 =$ speed of the plane *with* the wind

as shown in Figure 7.15. Using the formula distance = (rate)(time) for these two speeds, you obtain the following equations.

$$2000 = (r_1 - r_2)\left(4 + \frac{24}{60}\right)$$

$$2000 = (r_1 + r_2)(4)$$

Original flight

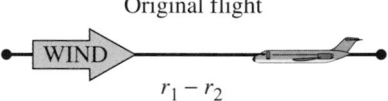

These two equations simplify as follows.

$$\begin{cases} 5000 = 11r_1 - 11r_2 & \text{Equation 1} \\ 500 = r_1 + r_2 & \text{Equation 2} \end{cases}$$

Return flight

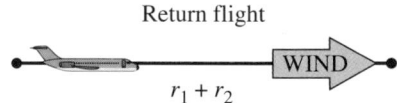

Figure 7.15

To solve this system by elimination, multiply Equation 2 by 11.

$$5000 = 11r_1 - 11r_2 \implies 5000 = 11r_1 - 11r_2 \quad \text{Write Equation 1.}$$

$$\underline{500 = r_1 + r_2} \implies \underline{5500 = 11r_1 + 11r_2} \quad \text{Multiply Equation 2 by 11.}$$

$$10,500 = 22r_1 \quad \text{Add equations.}$$

So,

$$r_1 = \frac{10,500}{22} = \frac{5250}{11} \approx 477.27 \text{ miles per hour} \qquad \text{Speed of plane}$$

$$r_2 = 500 - \frac{5250}{11} = \frac{250}{11} \approx 22.73 \text{ miles per hour.} \qquad \text{Speed of wind}$$

Check this solution in the original statement of the problem.

✓**CHECKPOINT**   Now try Exercise 71.

7.2 Exercises

See www.CalcChat.com for worked-out solutions to odd-numbered exercises.

Vocabulary Check

Fill in the blanks.

1. The first step in solving a system of equations by the _____ of _____ is to obtain coefficients for x (or y) that differ only in sign.

2. Two systems of equations that have the same solution set are called _____ systems.

3. A system of linear equations that has at least one solution is called _____ , whereas a system of linear equations that has no solution is called _____ .

In Exercises 1–6, solve the system by the method of elimination. Label each line with its equation.

1. $\begin{cases} 2x + y = 5 \\ x - y = 1 \end{cases}$

2. $\begin{cases} x + 3y = 1 \\ -x + 2y = 4 \end{cases}$

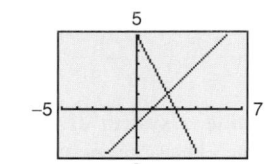 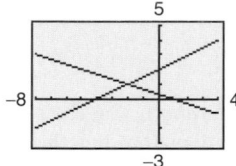

3. $\begin{cases} x + y = 0 \\ 3x + 2y = 1 \end{cases}$

4. $\begin{cases} 2x - y = 3 \\ 4x + 3y = 21 \end{cases}$

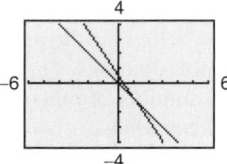 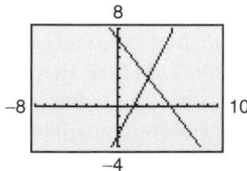

5. $\begin{cases} x - y = 2 \\ -2x + 2y = 5 \end{cases}$

6. $\begin{cases} 3x - 2y = 5 \\ -6x + 4y = -10 \end{cases}$

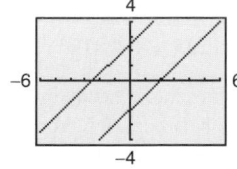 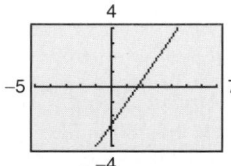

In Exercises 7–16, solve the system by the method of elimination and check any solutions algebraically.

7. $\begin{cases} x + 2y = 4 \\ x - 2y = 1 \end{cases}$

8. $\begin{cases} 3x - 2y = 5 \\ x + 2y = 7 \end{cases}$

9. $\begin{cases} 2x + 3y = 18 \\ 5x - y = 11 \end{cases}$

10. $\begin{cases} x + 7y = 12 \\ 3x - 5y = 10 \end{cases}$

11. $\begin{cases} 3r + 2s = 10 \\ 2r + 5s = 3 \end{cases}$

12. $\begin{cases} 2r + 4s = 5 \\ 16r + 50s = 55 \end{cases}$

13. $\begin{cases} 5u + 6v = 24 \\ 3u + 5v = 18 \end{cases}$

14. $\begin{cases} 3u + 11v = 4 \\ -2u - 5v = 9 \end{cases}$

15. $\begin{cases} 1.8x + 1.2y = 4 \\ 9x + 6y = 3 \end{cases}$

16. $\begin{cases} 3.1x - 2.9y = -10.2 \\ 31x - 12y = 34 \end{cases}$

In Exercises 17–20, match the system of linear equations with its graph. State the number of solutions. Then state whether the system is consistent or inconsistent. [The graphs are labeled (a), (b), (c), and (d).]

(a)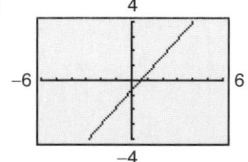

(b)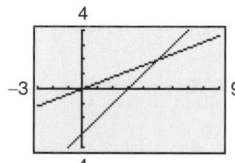

(c)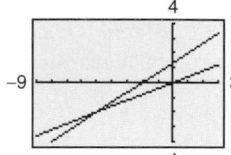

(d)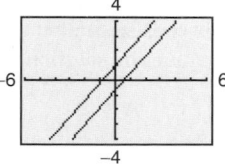

17. $\begin{cases} 2x - 5y = 0 \\ x - y = 3 \end{cases}$

18. $\begin{cases} -7x + 6y = -4 \\ 14x - 12y = 8 \end{cases}$

19. $\begin{cases} 2x - 5y = 0 \\ 2x - 3y = -4 \end{cases}$

20. $\begin{cases} 7x - 6y = -6 \\ -7x + 6y = -4 \end{cases}$

In Exercises 21–40, solve the system by the method of elimination and check any solutions using a graphing utility.

21. $\begin{cases} 4x + 3y = 3 \\ 3x + 11y = 13 \end{cases}$

22. $\begin{cases} 2x + 5y = 8 \\ 5x + 8y = 10 \end{cases}$

23. $\begin{cases} \frac{2}{5}x - \frac{3}{2}y = 4 \\ \frac{1}{5}x - \frac{3}{4}y = -2 \end{cases}$ **24.** $\begin{cases} \frac{2}{3}x + \frac{1}{6}y = \frac{2}{3} \\ 4x + y = 4 \end{cases}$

25. $\begin{cases} \frac{3}{4}x + y = \frac{1}{8} \\ \frac{9}{4}x + 3y = \frac{3}{8} \end{cases}$ **26.** $\begin{cases} \frac{1}{4}x + \frac{1}{6}y = 1 \\ -3x - 2y = 0 \end{cases}$

27. $\begin{cases} \frac{x+3}{4} + \frac{y-1}{3} = 1 \\ 2x - y = 12 \end{cases}$ **28.** $\begin{cases} \frac{x+2}{4} + \frac{y-1}{4} = 1 \\ x - y = 4 \end{cases}$

29. $\begin{cases} \frac{x-1}{2} + \frac{y+2}{3} = 4 \\ x - 2y = 5 \end{cases}$ **30.** $\begin{cases} \frac{x-1}{2} + \frac{y-2}{2} = 1 \\ x - y = 2 \end{cases}$

31. $\begin{cases} 2.5x - 3y = 1.5 \\ 2x - 2.4y = 1.2 \end{cases}$ **32.** $\begin{cases} 6.3x + 7.2y = 5.4 \\ 5.6x + 6.4y = 4.8 \end{cases}$

33. $\begin{cases} 0.2x - 0.5y = -27.8 \\ 0.3x + 0.4y = 68.7 \end{cases}$ **34.** $\begin{cases} 0.2x + 0.6y = -1 \\ x - 0.5y = 2 \end{cases}$

35. $\begin{cases} 0.05x - 0.03y = 0.21 \\ 0.07x + 0.02y = 0.16 \end{cases}$ **36.** $\begin{cases} 0.2x + 0.4y = -0.2 \\ x + 0.5y = -2.5 \end{cases}$

37. $\begin{cases} \frac{1}{x} + \frac{3}{y} = 2 \\ \frac{4}{x} - \frac{1}{y} = -5 \end{cases}$ **38.** $\begin{cases} \frac{2}{x} - \frac{1}{y} = 0 \\ \frac{4}{x} - \frac{3}{y} = -1 \end{cases}$

39. $\begin{cases} \frac{1}{x} + \frac{2}{y} = 5 \\ \frac{3}{x} - \frac{4}{y} = -5 \end{cases}$ **40.** $\begin{cases} \frac{2}{x} - \frac{1}{y} = 5 \\ \frac{6}{x} + \frac{1}{y} = 11 \end{cases}$

In Exercises 41–46, use a graphing utility to graph the lines in the system. Use the graphs to determine whether the system is consistent or inconsistent. If the system is consistent, determine the solution. Verify your results algebraically.

41. $\begin{cases} 2x - 5y = 0 \\ x - y = 3 \end{cases}$ **42.** $\begin{cases} 2x + y = 5 \\ x - 2y = -1 \end{cases}$

43. $\begin{cases} \frac{3}{5}x - y = 3 \\ -3x + 5y = 9 \end{cases}$ **44.** $\begin{cases} 4x - 6y = 9 \\ \frac{16}{3}x - 8y = 12 \end{cases}$

45. $\begin{cases} 8x - 14y = 5 \\ 2x - 3.5y = 1.25 \end{cases}$ **46.** $\begin{cases} -x + 7y = 3 \\ -\frac{1}{7}x + y = 5 \end{cases}$

In Exercises 47–54, use a graphing utility to graph the two equations. Use the graphs to approximate the solution of the system. Round your results to three decimal places.

47. $\begin{cases} 6y = 42 \\ 6x - y = 16 \end{cases}$ **48.** $\begin{cases} 4y = -8 \\ 7x - 2y = 25 \end{cases}$

49. $\begin{cases} \frac{3}{2}x - \frac{1}{5}y = 8 \\ -2x + 3y = 3 \end{cases}$ **50.** $\begin{cases} \frac{3}{4}x - \frac{5}{2}y = -9 \\ -x + 6y = 28 \end{cases}$

51. $\begin{cases} \frac{1}{3}x + y = -\frac{1}{3} \\ 5x - 3y = 7 \end{cases}$ **52.** $\begin{cases} 5x - y = -4 \\ 2x + \frac{3}{5}y = \frac{2}{5} \end{cases}$

53. $\begin{cases} 0.5x + 2.2y = 9 \\ 6x + 0.4y = -22 \end{cases}$ **54.** $\begin{cases} 2.4x + 3.8y = -17.6 \\ 4x - 0.2y = -3.2 \end{cases}$

In Exercises 55–62, use any method to solve the system.

55. $\begin{cases} 3x - 5y = 7 \\ 2x + y = 9 \end{cases}$ **56.** $\begin{cases} -x + 3y = 17 \\ 4x + 3y = 7 \end{cases}$

57. $\begin{cases} y = 4x + 3 \\ y = -5x - 12 \end{cases}$ **58.** $\begin{cases} 7x + 3y = 16 \\ y = x + 1 \end{cases}$

59. $\begin{cases} x - 5y = 21 \\ 6x + 5y = 21 \end{cases}$ **60.** $\begin{cases} y = -3x - 8 \\ y = 15 - 2x \end{cases}$

61. $\begin{cases} -2x + 8y = 19 \\ y = x - 3 \end{cases}$ **62.** $\begin{cases} 4x - 3y = 6 \\ -5x + 7y = -1 \end{cases}$

Exploration **In Exercises 63–66, find a system of linear equations that has the given solution. (There are many correct answers.)**

63. $(0, 8)$ **64.** $(3, -4)$

65. $\left(3, \frac{5}{2}\right)$ **66.** $\left(-\frac{2}{3}, -10\right)$

Supply and Demand **In Exercises 67–70, find the *point of equilibrium* of the demand and supply equations. The point of equilibrium is the price p and the number of units x that satisfy both the demand and supply equations.**

Demand	Supply
67. $p = 50 - 0.5x$	$p = 0.125x$
68. $p = 100 - 0.05x$	$p = 25 + 0.1x$
69. $p = 140 - 0.00002x$	$p = 80 + 0.00001x$
70. $p = 400 - 0.0002x$	$p = 225 + 0.0005x$

71. *Airplane Speed* An airplane flying into a headwind travels the 1800-mile flying distance between New York City and Albuquerque, New Mexico in 3 hours and 36 minutes. On the return flight, the same distance is traveled in 3 hours. Find the airspeed of the plane and the speed of the wind, assuming that both remain constant.

72. *Airplane Speed* Two planes start from Boston's Logan International Airport and fly in opposite directions. The second plane starts $\frac{1}{2}$ hour after the first plane, but its speed is 80 kilometers per hour faster. Find the airspeed of each plane if 2 hours after the first plane departs, the planes are 3200 kilometers apart.

73. *Ticket Sales* A minor league baseball team had a total attendance one evening of 1175. The tickets for adults and children sold for $5.00 and $3.50, respectively. The ticket revenue that night was $5087.50.

(a) Create a system of linear equations to find the numbers of adults A and children C that attended the game.

(b) Solve your system of equations by elimination or by substitution. Explain your choice.

(c) Use the *intersect* feature or the *zoom* and *trace* features of a graphing utility to solve your system.

74. *Consumerism* One family purchases five cold drinks and three snow cones for $8.50. A second family purchases six cold drinks and four snow cones for $10.50.

(a) Create a system of linear equations to find the prices of a cold drink C and a snow cone S.

(b) Solve your system of equations by elimination or by substitution. Explain your choice.

(c) Use the *intersect* feature or the *zoom* and *trace* features of a graphing utility to solve your system.

75. *Produce* A grocer sells oranges for $0.95 each and grapefruits for $1.05 each. You purchase a mix of 16 oranges and grapefruits and pay $15.90. How many of each type of fruit did you buy?

76. *Sales* The sales S (in millions of dollars) for Family Dollar Stores, Inc. and Dollar General Corporation stores from 1995 to 2005 can be modeled by

$$\begin{cases} S - 440.36t = -1023.0 & \text{Family Dollar} \\ S - 691.48t = -2122.8 & \text{Dollar General} \end{cases}$$

where t is the year, with $t = 5$ corresponding to 1995. (Source: Family Dollar Stores, Inc.; Dollar General Corporation)

(a) Solve the system of equations using the method of your choice. Explain why you chose that method.

(b) Interpret the meaning of the solution in the context of the problem.

77. *Revenues* Revenues for a video rental store on a particular Friday evening are $867.50 for 310 rentals. The rental fee for movies is $3.00 each and the rental fee for video game cartridges is $2.50 each. Determine the number of each type that are rented that evening.

78. *Sales* On Saturday night, the manager of a shoe store evaluates the receipts of the previous week's sales. Two hundred fifty pairs of two different styles of running shoes were sold. One style sold for $75.50 and the other sold for $89.95. The receipts totaled $20,031. The cash register that was supposed to record the number of each type of shoe sold malfunctioned. Can you recover the information? If so, how many shoes of each type were sold?

Fitting a Line to Data **In Exercises 79–82, find the least squares regression line** $y = ax + b$ **for the points** $(x_1, y_1), (x_2, y_2), \ldots, (x_n, y_n)$ **by solving the system for** a **and** b**. Then use the regression feature of a graphing utility to confirm your result. (For an explanation of how the coefficients of** a **and** b **in the system are obtained, see Appendix B.)**

79. $\begin{cases} 5b + 10a = 20.2 \\ 10b + 30a = 50.1 \end{cases}$ **80.** $\begin{cases} 5b + 10a = 11.7 \\ 10b + 30a = 25.6 \end{cases}$

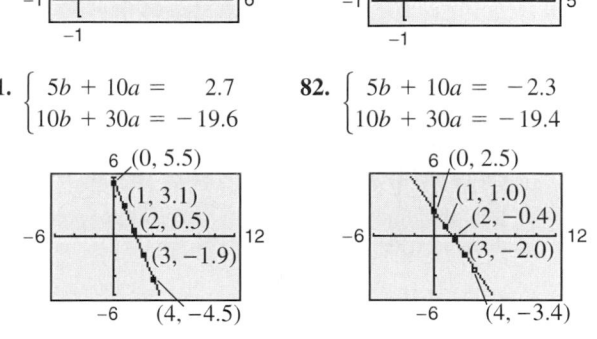

81. $\begin{cases} 5b + 10a = 2.7 \\ 10b + 30a = -19.6 \end{cases}$ **82.** $\begin{cases} 5b + 10a = -2.3 \\ 10b + 30a = -19.4 \end{cases}$

83. *Data Analysis* A farmer used four test plots to determine the relationship between wheat yield (in bushels per acre) and the amount of fertilizer applied (in hundreds of pounds per acre). The results are shown in the table.

Fertilizer, x	Yield, y
1.0	32
1.5	41
2.0	48
2.5	53

(a) Find the least squares regression line $y = ax + b$ for the data by solving the system for a and b.

$$\begin{cases} 4b + 7.0a = 174 \\ 7b + 13.5a = 322 \end{cases}$$

(b) Use the *regression* feature of a graphing utility to confirm the result in part (a).

(c) Use a graphing utility to plot the data and graph the linear model from part (a) in the same viewing window.

(d) Use the linear model from part (a) to predict the yield for a fertilizer application of 160 pounds per acre.

84. *Data Analysis* A candy store manager wants to know the demand for a candy bar as a function of the price. The daily sales for different prices of the product are shown in the table.

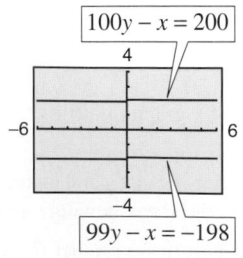

Price, x	Demand, y
$1.00	45
$1.20	37
$1.50	23

(a) Find the least squares regression line $y = ax + b$ for the data by solving the system for a and b.

$$\begin{cases} 3.00b + 3.70a = 105.00 \\ 3.70b + 4.69a = 123.90 \end{cases}$$

(b) Use the *regression* feature of a graphing utility to confirm the result in part (a).

(c) Use a graphing utility to plot the data and graph the linear model from part (a) in the same viewing window.

(d) Use the linear model from part (a) to predict the demand when the price is $1.75.

Synthesis

True or False? In Exercises 85 and 86, determine whether the statement is true or false. Justify your answer.

85. If a system of linear equations has two distinct solutions, then it has an infinite number of solutions.

86. If a system of linear equations has no solution, then the lines must be parallel.

Think About It In Exercises 87 and 88, the graphs of the two equations appear to be parallel. Yet, when the system is solved algebraically, it is found that the system does have a solution. Find the solution and explain why it does not appear on the portion of the graph that is shown.

87. $\begin{cases} 100y - x = 200 \\ 99y - x = -198 \end{cases}$

88. $\begin{cases} 21x - 20y = 0 \\ 13x - 12y = 120 \end{cases}$

$100y - x = 200$
4
−6 ⟷ 6
−4
$99y - x = -198$

$21x - 20y = 0$
12
−6 ⟷ 6
−12
$13x - 12y = 120$

89. *Writing* Briefly explain whether or not it is possible for a consistent system of linear equations to have exactly two solutions.

90. *Think About It* Give examples of (a) a system of linear equations that has no solution and (b) a system of linear equations that has an infinite number of solutions. (There are many correct answers.)

In Exercises 91 and 92, find the value of k such that the system of equations is inconsistent.

91. $\begin{cases} 4x - 8y = -3 \\ 2x + ky = 16 \end{cases}$

92. $\begin{cases} 15x + 3y = 6 \\ -10x + ky = 9 \end{cases}$

∫ *Advanced Applications* In Exercises 93 and 94, solve the system of equations for u and v. While solving for these variables, consider the transcendental functions as constants. (Systems of this type are found in a course in differential equations.)

93. $\begin{cases} u \sin x + v \cos x = 0 \\ u \cos x - v \sin x = \sec x \end{cases}$

94. $\begin{cases} u \cos 2x + v \sin 2x = 0 \\ u(-2 \sin 2x) + v(2 \cos 2x) = \csc 2x \end{cases}$

Skills Review

In Exercises 95–100, solve the inequality and graph the solution on a real number line.

95. $-11 - 6x \geq 33$

96. $-6 \leq 3x - 10 < 6$

97. $|x - 8| < 10$

98. $|x + 10| \geq -3$

99. $2x^2 + 3x - 35 < 0$

100. $3x^2 + 12x > 0$

In Exercises 101–106, write the expression as the logarithm of a single quantity.

101. $\ln x + \ln 6$

102. $\ln x - 5 \ln(x + 3)$

103. $\log_9 12 - \log_9 x$

104. $\frac{1}{4}\log_6 3 + \frac{1}{4}\log_6 x$

105. $2 \ln x - \ln(x + 2)$

106. $\frac{1}{2}\ln(x^2 + 4) - \ln x$

107. *Make a Decision* To work an extended application analyzing the average undergraduate tuition, room, and board charges at private colleges in the United States from 1985 to 2003, visit this textbook's *Online Study Center*. (Data Source: U.S. Census Bureau)

7.3   Multivariable Linear Systems

Row–Echelon Form and Back–Substitution

The method of elimination can be applied to a system of linear equations in more than two variables. When elimination is used to solve a system of linear equations, the goal is to rewrite the system in a form to which back-substitution can be applied. To see how this works, consider the following two systems of linear equations.

System of Three Linear Equations in Three Variables (See Example 2):

$$\begin{cases} x - 2y + 3z = 9 \\ -x + 3y + z = -2 \\ 2x - 5y + 5z = 17 \end{cases}$$

Equivalent System in Row-Echelon Form (See Example 1):

$$\begin{cases} x - 2y + 3z = 9 \\ y + 4z = 7 \\ z = 2 \end{cases}$$

The second system is said to be in **row-echelon form,** which means that it has a "stair-step" pattern with leading coefficients of 1. After comparing the two systems, it should be clear that it is easier to solve the system in row-echelon form, using back-substitution.

Example 1   Using Back–Substitution in Row–Echelon Form

Solve the system of linear equations.

$$\begin{cases} x - 2y + 3z = 9 & \text{Equation 1} \\ y + 4z = 7 & \text{Equation 2} \\ z = 2 & \text{Equation 3} \end{cases}$$

Solution

From Equation 3, you know the value of z. To solve for y, substitute $z = 2$ into Equation 2 to obtain

$$y + 4(2) = 7 \qquad \text{Substitute 2 for } z.$$
$$y = -1. \qquad \text{Solve for } y.$$

Finally, substitute $y = -1$ and $z = 2$ into Equation 1 to obtain

$$x - 2(-1) + 3(2) = 9 \qquad \text{Substitute } -1 \text{ for } y \text{ and 2 for } z.$$
$$x = 1. \qquad \text{Solve for } x.$$

The solution is $x = 1$, $y = -1$, and $z = 2$, which can be written as the **ordered triple** $(1, -1, 2)$. Check this in the original system of equations.

✓CHECKPOINT   Now try Exercise 5.

Gaussian Elimination

Two systems of equations are *equivalent* if they have the same solution set. To solve a system that is not in row-echelon form, first convert it to an *equivalent* system that *is* in row-echelon form by using one or more of the elementary row operations shown below. This process is called **Gaussian elimination,** after the German mathematician Carl Friedrich Gauss (1777–1855).

Elementary Row Operations for Systems of Equations

1. Interchange two equations.

2. Multiply one of the equations by a nonzero constant.

3. Add a multiple of one equation to another equation.

Example 2 Using Gaussian Elimination to Solve a System

Solve the system of linear equations.

$$\begin{cases} x - 2y + 3z = 9 & \text{Equation 1} \\ -x + 3y + z = -2 & \text{Equation 2} \\ 2x - 5y + 5z = 17 & \text{Equation 3} \end{cases}$$

Solution

Because the leading coefficient of the first equation is 1, you can begin by saving the x at the upper left and eliminating the other x-terms from the first column.

$$\begin{cases} x - 2y + 3z = 9 \\ y + 4z = 7 \\ 2x - 5y + 5z = 17 \end{cases}$$

Adding the first equation to the second equation produces a new second equation.

$$\begin{cases} x - 2y + 3z = 9 \\ y + 4z = 7 \\ -y - z = -1 \end{cases}$$

Adding -2 times the first equation to the third equation produces a new third equation.

Now that all but the first x have been eliminated from the first column, go to work on the second column. (You need to eliminate y from the third equation.)

$$\begin{cases} x - 2y + 3z = 9 \\ y + 4z = 7 \\ 3z = 6 \end{cases}$$

Adding the second equation to the third equation produces a new third equation.

Finally, you need a coefficient of 1 for z in the third equation.

$$\begin{cases} x - 2y + 3z = 9 \\ y + 4z = 7 \\ z = 2 \end{cases}$$

Multiplying the third equation by $\frac{1}{3}$ produces a new third equation.

This is the same system that was solved in Example 1. As in that example, you can conclude that the solution is $x = 1$, $y = -1$, and $z = 2$, written as $(1, -1, 2)$.

✔**CHECKPOINT** Now try Exercise 15.

The goal of Gaussian elimination is to use elementary row operations on a system in order to isolate one variable. You can then solve for the value of the variable and use back-substitution to find the values of the remaining variables.

The next example involves an inconsistent system—one that has no solution. The key to recognizing an inconsistent system is that at some stage in the elimination process, you obtain a false statement such as $0 = -2$.

Example 3   An Inconsistent System

Solve the system of linear equations.

$$\begin{cases} x - 3y + z = 1 & \text{Equation 1} \\ 2x - y - 2z = 2 & \text{Equation 2} \\ x + 2y - 3z = -1 & \text{Equation 3} \end{cases}$$

Solution

$$\begin{cases} x - 3y + z = 1 \\ \quad 5y - 4z = 0 \\ x + 2y - 3z = -1 \end{cases}$$

Adding -2 times the first equation to the second equation produces a new second equation.

$$\begin{cases} x - 3y + z = 1 \\ \quad 5y - 4z = 0 \\ \quad 5y - 4z = -2 \end{cases}$$

Adding -1 times the first equation to the third equation produces a new third equation.

$$\begin{cases} x - 3y + z = 1 \\ \quad 5y - 4z = 0 \\ \quad\quad\quad 0 = -2 \end{cases}$$

Adding -1 times the second equation to the third equation produces a new third equation.

Because $0 = -2$ is a false statement, you can conclude that this system is inconsistent and so has no solution. Moreover, because this system is equivalent to the original system, you can conclude that the original system also has no solution.

✓CHECKPOINT   Now try Exercise 21.

As with a system of linear equations in two variables, the number of solutions of a system of linear equations in more than two variables must fall into one of three categories.

The Number of Solutions of a Linear System

For a system of linear equations, exactly one of the following is true.

1. There is exactly one solution.

2. There are infinitely many solutions.

3. There is no solution.

A system of linear equations is called *consistent* if it has at least one solution. A consistent system with exactly one solution is **independent.** A consistent system with infinitely many solutions is **dependent.** A system of linear equations is called *inconsistent* if it has no solution.

Example 4   A System with Infinitely Many Solutions

Solve the system of linear equations.

$$\begin{cases} x + y - 3z = -1 & \text{Equation 1} \\ \phantom{x + {}} y - z = 0 & \text{Equation 2} \\ -x + 2y \phantom{{} - 3z} = 1 & \text{Equation 3} \end{cases}$$

Solution

$$\begin{cases} x + y - 3z = -1 \\ \phantom{x + {}} y - z = 0 \\ \phantom{x + {}} 3y - 3z = 0 \end{cases}$$

> Adding the first equation to the third equation produces a new third equation.

$$\begin{cases} x + y - 3z = -1 \\ \phantom{x + {}} y - z = 0 \\ 0 = 0 \end{cases}$$

> Adding -3 times the second equation to the third equation produces a new third equation.

This result means that Equation 3 depends on Equations 1 and 2 in the sense that it gives us no additional information about the variables. So, the original system is equivalent to the system

$$\begin{cases} x + y - 3z = -1 \\ \phantom{x + {}} y - z = 0 \end{cases}.$$

In the last equation, solve for y in terms of z to obtain $y = z$. Back-substituting for y in the previous equation produces $x = 2z - 1$. Finally, letting $z = a$, where a is a real number, the solutions to the original system are all of the form

$$x = 2a - 1, \qquad y = a, \qquad \text{and} \qquad z = a.$$

So, every ordered triple of the form

$$(2a - 1, a, a), \qquad a \text{ is a real number}$$

is a solution of the system.

✓**CHECKPOINT**   Now try Exercise 25.

STUDY TIP

There are an infinite number of solutions to Example 4, but they are all of a specific form. By selecting, for example, a-values of 0, 1, and 3, you can verify that $(-1, 0, 0)$, $(1, 1, 1)$, and $(5, 3, 3)$ are specific solutions. It is incorrect to say simply that the solution to Example 4 is "infinite." You must also specify the form of the solutions.

In Example 4, there are other ways to write the same infinite set of solutions. For instance, the solutions could have been written as

$$\left(b, \tfrac{1}{2}(b + 1), \tfrac{1}{2}(b + 1)\right), \qquad b \text{ is a real number.}$$

This description produces the same set of solutions, as shown below.

Substitution	Solution	
$a = 0$	$(2(0) - 1, 0, 0) = (-1, 0, 0)$	Same solution
$b = -1$	$\left(-1, \tfrac{1}{2}(-1 + 1), \tfrac{1}{2}(-1 + 1)\right) = (-1, 0, 0)$	
$a = 1$	$(2(1) - 1, 1, 1) = (1, 1, 1)$	Same solution
$b = 1$	$\left(1, \tfrac{1}{2}(1 + 1), \tfrac{1}{2}(1 + 1)\right) = (1, 1, 1)$	
$a = 2$	$(2(2) - 1, 2, 2) = (3, 2, 2)$	Same solution
$b = 3$	$\left(3, \tfrac{1}{2}(3 + 1), \tfrac{1}{2}(3 + 1)\right) = (3, 2, 2)$	

Nonsquare Systems

So far, each system of linear equations you have looked at has been *square*, which means that the number of equations is equal to the number of variables. In a **nonsquare system of equations,** the number of equations differs from the number of variables. A system of linear equations cannot have a unique solution unless there are at least as many equations as there are variables in the system.

Example 5 A System with Fewer Equations than Variables

Solve the system of linear equations.

$$\begin{cases} x - 2y + z = 2 & \text{Equation 1} \\ 2x - y - z = 1 & \text{Equation 2} \end{cases}$$

Solution

Begin by rewriting the system in row-echelon form.

$$\begin{cases} x - 2y + z = 2 \\ 3y - 3z = -3 \end{cases}$$

Adding -2 times the first equation to the second equation produces a new second equation.

$$\begin{cases} x - 2y + z = 2 \\ y - z = -1 \end{cases}$$

Multiplying the second equation by $\frac{1}{3}$ produces a new second equation.

Solve for y in terms of z to obtain $y = z - 1$. By back-substituting into Equation 1, you can solve for x as follows.

$$x - 2(z - 1) + z = 2 \qquad \text{Substitute for } y \text{ in Equation 1.}$$
$$x - 2z + 2 + z = 2 \qquad \text{Distributive Property}$$
$$x = z \qquad \text{Solve for } x.$$

Finally, by letting $z = a$ where a is a real number, you have the solution $x = a$, $y = a - 1$, and $z = a$. So, every ordered triple of the form

$$(a, a - 1, a), \qquad a \text{ is a real number}$$

is a solution of the system.

✓**CHECKPOINT**   Now try Exercise 37.

In Example 5, try choosing some values of a to obtain different solutions of the system, such as $(1, 0, 1)$, $(2, 1, 2)$, and $(3, 2, 3)$. Then check each of the solutions in the original system, as follows.

Check: $(1, 0, 1)$

$$1 - 2(0) + 1 \overset{?}{=} 2$$
$$2 = 2 \checkmark$$

$$2(1) - 0 - 1 \overset{?}{=} 1$$
$$1 = 1 \checkmark$$

Check: $(2, 1, 2)$

$$2 - 2(1) + 2 \overset{?}{=} 2$$
$$2 = 2 \checkmark$$

$$2(2) - 1 - 2 \overset{?}{=} 1$$
$$1 = 1 \checkmark$$

Check: $(3, 2, 3)$

$$3 - 2(2) + 3 \overset{?}{=} 2$$
$$2 = 2 \checkmark$$

$$2(3) - 2 - 3 \overset{?}{=} 1$$
$$1 = 1 \checkmark$$

Graphical Interpretation of Three-Variable Systems

Solutions of equations in three variables can be pictured using a **three-dimensional coordinate system.** To construct such a system, begin with the xy-coordinate plane in a horizontal position. Then draw the z-axis as a vertical line through the origin.

Every ordered triple (x, y, z) corresponds to a point on the three-dimensional coordinate system. For instance, the points corresponding to

$$(-2, 5, 4), \qquad (2, -5, 3), \qquad \text{and} \qquad (3, 3, -2)$$

are shown in Figure 7.16.

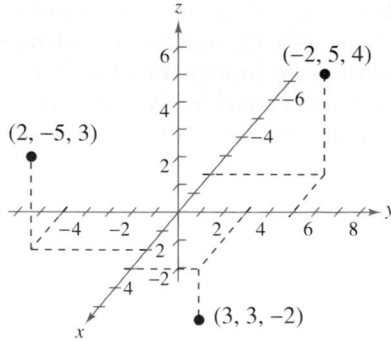

Figure 7.16

The **graph of an equation in three variables** consists of all points (x, y, z) that are solutions of the equation. The graph of a linear equation in three variables is a *plane*. Sketching graphs on a three-dimensional coordinate system is difficult because the sketch itself is only two-dimensional.

One technique for sketching a plane is to find the three points at which the plane intersects the axes. For instance, the plane

$$3x + 2y + 4z = 12$$

intersects the x-axis at the point $(4, 0, 0)$, the y-axis at the point $(0, 6, 0)$, and the z-axis at the point $(0, 0, 3)$. By plotting these three points, connecting them with line segments, and shading the resulting triangular region, you can sketch a portion of the graph, as shown in Figure 7.17.

The graph of a system of three linear equations in three variables consists of *three* planes. When these planes intersect in a single point, the system has exactly one solution (see Figure 7.18). When the three planes have no point in common, the system has no solution (see Figures 7.19 and 7.20). When the three planes intersect in a line or a plane, the system has infinitely many solutions (see Figures 7.21 and 7.22).

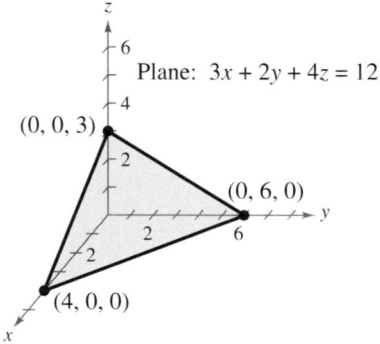

Figure 7.17

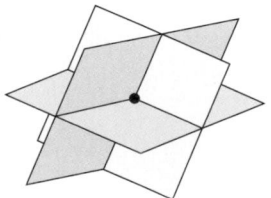

Solution: One point
Figure 7.18

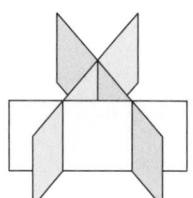

Solution: None
Figure 7.19

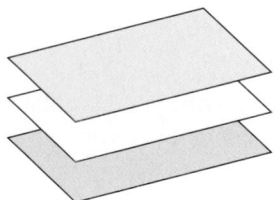

Solution: None
Figure 7.20

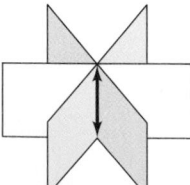

Solution: One line
Figure 7.21

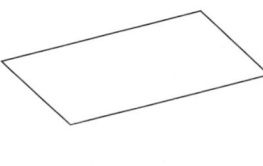

Solution: One plane
Figure 7.22

TECHNOLOGY TIP

Three-dimensional graphing utilities and computer algebra systems, such as *Derive* and *Mathematica*, are very efficient in producing three-dimensional graphs. They are good tools to use while studying calculus. If you have access to such a utility, try reproducing the plane shown in Figure 7.17.

Partial Fraction Decomposition and Other Applications

A rational expression can often be written as the sum of two or more simpler rational expressions. For example, the rational expression

$$\frac{x + 7}{x^2 - x - 6}$$

can be written as the sum of two fractions with linear denominators. That is,

$$\frac{x + 7}{x^2 - x - 6} = \underbrace{\frac{2}{x - 3}}_{\substack{\text{Partial} \\ \text{fraction}}} + \underbrace{\frac{-1}{x + 2}}_{\substack{\text{Partial} \\ \text{fraction}}}.$$

Each fraction on the right side of the equation is a **partial fraction,** and together they make up the **partial fraction decomposition** of the left side.

Decomposition of $N(x)/D(x)$ into Partial Fractions

1. *Divide if improper:* If $N(x)/D(x)$ is an improper fraction [degree of $N(x) \geq$ degree of $D(x)$], divide the denominator into the numerator to obtain

$$\frac{N(x)}{D(x)} = (\text{polynomial}) + \frac{N_1(x)}{D(x)}$$

   and apply Steps 2, 3, and 4 (below) to the proper rational expression $N_1(x)/D(x)$.

2. *Factor denominator:* Completely factor the denominator into factors of the form

$$(px + q)^m \qquad \text{and} \qquad (ax^2 + bx + c)^n$$

   where $(ax^2 + bx + c)$ is irreducible over the reals.

3. *Linear factors:* For *each* factor of the form $(px + q)^m$, the partial fraction decomposition must include the following sum of m fractions.

$$\frac{A_1}{(px + q)} + \frac{A_2}{(px + q)^2} + \cdots + \frac{A_m}{(px + q)^m}$$

4. *Quadratic factors:* For *each* factor of the form $(ax^2 + bx + c)^n$, the partial fraction decomposition must include the following sum of n fractions.

$$\frac{B_1 x + C_1}{ax^2 + bx + c} + \frac{B_2 x + C_2}{(ax^2 + bx + c)^2} + \cdots + \frac{B_n x + C_n}{(ax^2 + bx + c)^n}$$

One of the most important applications of partial fractions is in calculus. If you go on to take a course in calculus, you will learn how partial fractions can be used in a calculus operation called antidifferentiation.

Example 6   Partial Fraction Decomposition: Distinct Linear Factors

Write the partial fraction decomposition of

$$\frac{x+7}{x^2 - x - 6}.$$

Solution

Because $x^2 - x - 6 = (x - 3)(x + 2)$, you should include one partial fraction with a constant numerator for each linear factor of the denominator and write

$$\frac{x+7}{x^2 - x - 6} = \frac{A}{x - 3} + \frac{B}{x + 2}.$$

Multiplying each side of this equation by the least common denominator, $(x - 3)(x + 2)$, leads to the **basic equation**

$x + 7 = A(x + 2) + B(x - 3)$	Basic equation
$= Ax + 2A + Bx - 3B$	Distributive Property
$= (A + B)x + 2A - 3B.$	Write in polynomial form.

By equating coefficients of like terms on opposite sides of the equation, you obtain the following system of linear equations.

$$\begin{cases} A + B = 1 \\ 2A - 3B = 7 \end{cases}$$

Equation 1

Equation 2

You can solve the system of linear equations as follows.

$A + B = 1$	$\Rightarrow$	$3A + 3B = 3$	Multiply Equation 1 by 3.
$2A - 3B = 7$	$\Rightarrow$	$2A - 3B = 7$	Write Equation 2.
		$5A = 10$	Add equations.

From this equation, you can see that $A = 2$. By back-substituting this value of A into Equation 1, you can determine that $B = -1$. So, the partial fraction decomposition is

$$\frac{x+7}{x^2 - x - 6} = \frac{2}{x - 3} - \frac{1}{x + 2}.$$

Check this result by combining the two partial fractions on the right side of the equation.

✓CHECKPOINT   Now try Exercise 69.

TECHNOLOGY TIP   You can graphically check the decomposition found in Example 6. To do this, use a graphing utility to graph

$$y_1 = \frac{x+7}{x^2 - x - 6} \quad \text{and} \quad y_2 = \frac{2}{x - 3} - \frac{1}{x + 2}$$

in the same viewing window. The graphs should be identical, as shown in Figure 7.23.

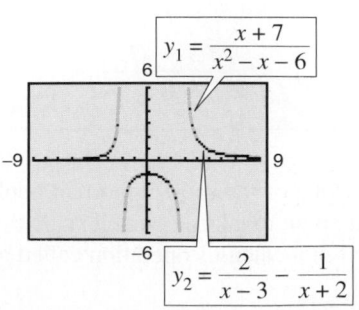

$$y_1 = \frac{x+7}{x^2 - x - 6}$$

$$y_2 = \frac{2}{x - 3} - \frac{1}{x + 2}$$

Figure 7.23

The next example shows how to find the partial fraction decomposition for a rational function whose denominator has a repeated linear factor.

Example 7 Partial Fraction Decomposition: Repeated Linear Factors

Write the partial fraction decomposition of $\dfrac{5x^2 + 20x + 6}{x^3 + 2x^2 + x}$.

Solution

Because the denominator factors as

$$x^3 + 2x^2 + x = x(x^2 + 2x + 1)$$
$$= x(x + 1)^2$$

you should include one partial fraction with a constant numerator for each power of x and $(x + 1)$ and write

$$\frac{5x^2 + 20x + 6}{x^3 + 2x^2 + x} = \frac{A}{x} + \frac{B}{x + 1} + \frac{C}{(x + 1)^2}.$$

Multiplying by the LCD, $x(x + 1)^2$, leads to the basic equation

$$5x^2 + 20x + 6 = A(x + 1)^2 + Bx(x + 1) + Cx \quad \text{Basic equation}$$
$$= Ax^2 + 2Ax + A + Bx^2 + Bx + Cx \quad \text{Expand.}$$
$$= (A + B)x^2 + (2A + B + C)x + A. \quad \text{Polynomial form}$$

By equating coefficients of like terms on opposite sides of the equation, you obtain the following system of linear equations.

$$\begin{cases} A + B & = 5 \\ 2A + B + C & = 20 \\ A & = 6 \end{cases}$$

Substituting 6 for A in the first equation produces

$$6 + B = 5$$
$$B = -1.$$

Substituting 6 for A and -1 for B in the second equation produces

$$2(6) + (-1) + C = 20$$
$$C = 9.$$

So, the partial fraction decomposition is

$$\frac{5x^2 + 20x + 6}{x^3 + 2x^2 + x} = \frac{6}{x} - \frac{1}{x + 1} + \frac{9}{(x + 1)^2}.$$

Check this result by combining the three partial fractions on the right side of the equation.

✓**CHECKPOINT**   Now try Exercise 73.

Exploration

Partial fraction decomposition is practical only for rational functions whose denominators factor "nicely." For example, the factorization of the expression $x^2 - x - 5$ is

$$\left(x - \frac{1 - \sqrt{21}}{2}\right)\left(x - \frac{1 + \sqrt{21}}{2}\right).$$

Write the basic equation and try to complete the decomposition for

$$\frac{x + 7}{x^2 - x - 5}.$$

What problems do you encounter?

Example 8 Vertical Motion

The height at time t of an object that is moving in a (vertical) line with constant acceleration a is given by the *position equation* $s = \frac{1}{2}at^2 + v_0t + s_0$. The height s is measured in feet, t is measured in seconds, v_0 is the initial velocity (in feet per second) at $t = 0$, and s_0 is the initial height. Find the values of a, v_0, and s_0 if $s = 52$ at $t = 1$, $s = 52$ at $t = 2$, and $s = 20$ at $t = 3$, as shown in Figure 7.24.

Solution

You can obtain three linear equations in a, v_0, and s_0 as follows.

When $t = 1$: $\frac{1}{2}a(1)^2 + v_0(1) + s_0 = 52$ ⟹ $a + 2v_0 + 2s_0 = 104$

When $t = 2$: $\frac{1}{2}a(2)^2 + v_0(2) + s_0 = 52$ ⟹ $2a + 2v_0 + s_0 = 52$

When $t = 3$: $\frac{1}{2}a(3)^2 + v_0(3) + s_0 = 20$ ⟹ $9a + 6v_0 + 2s_0 = 40$

Solving this system yields $a = -32$, $v_0 = 48$, and $s_0 = 20$.

✓CHECKPOINT   Now try Exercise 83.

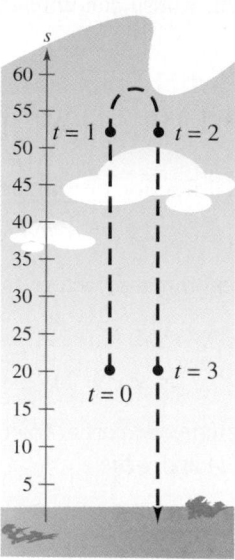

Figure 7.24

Example 9 Data Analysis: Curve-Fitting

Find a quadratic equation $y = ax^2 + bx + c$ whose graph passes through the points $(-1, 3)$, $(1, 1)$, and $(2, 6)$.

Solution

Because the graph of $y = ax^2 + bx + c$ passes through the points $(-1, 3)$, $(1, 1)$, and $(2, 6)$, you can write the following.

When $x = -1$, $y = 3$: $a(-1)^2 + b(-1) + c = 3$

When $x = 1$, $y = 1$: $a(1)^2 + b(1) + c = 1$

When $x = 2$, $y = 6$: $a(2)^2 + b(2) + c = 6$

This produces the following system of linear equations.

$$\begin{cases} a - b + c = 3 & \text{Equation 1} \\ a + b + c = 1 & \text{Equation 2} \\ 4a + 2b + c = 6 & \text{Equation 3} \end{cases}$$

The solution of this system is $a = 2$, $b = -1$, and $c = 0$. So, the equation of the parabola is $y = 2x^2 - x$, and its graph is shown in Figure 7.25.

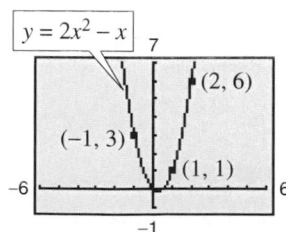

Figure 7.25

✓CHECKPOINT   Now try Exercise 87.

STUDY TIP

When you use a system of linear equations to solve an application problem, it is wise to interpret your solution in the context of the problem to see if it makes sense. For instance, in Example 8 the solution results in the position equation

$$s = -16t^2 + 48t + 20$$

which implies that the object was thrown upward at a velocity of 48 feet per second from a height of 20 feet. The object undergoes a constant downward acceleration of 32 feet per second squared. (In physics, this is the value of the acceleration due to gravity.)

7.3 Exercises

See www.CalcChat.com for worked-out solutions to odd-numbered exercises.

Vocabulary Check

Fill in the blanks.

1. A system of equations that is in _____ form has a "stair-step" pattern with leading coefficients of 1.

2. A solution to a system of three linear equations in three unknowns can be written as an _____ , which has the form (x, y, z).

3. The process used to write a system of equations in row-echelon form is called _____ elimination.

4. A system of linear equations that has exactly one solution is called _____ , whereas a system of linear equations that has infinitely many solutions is called _____ .

5. A system of equations is called _____ if the number of equations differs from the number of variables in the system.

6. Solutions of equations in three variables can be pictured using a _____ coordinate system.

7. The process of writing a rational expression as the sum of two or more simpler rational expressions is called _____ .

In Exercises 1–4, determine whether each ordered triple is a solution of the system of equations.

1. $\begin{cases} 3x - y + z = 1 \\ 2x \quad\ - 3z = -14 \\ \quad\ 5y + 2z = 8 \end{cases}$

   (a) $(3, 5, -3)$ (b) $(-1, 0, 4)$
   (c) $(-4, 1, 2)$ (d) $(1, 0, 4)$

2. $\begin{cases} 3x + 4y - z = 17 \\ 5x - y + 2z = -2 \\ 2x - 3y + 7z = -21 \end{cases}$

   (a) $(1, 5, 6)$ (b) $(-2, -4, 2)$
   (c) $(1, 3, -2)$ (d) $(0, 7, 0)$

3. $\begin{cases} 4x + y - z = 0 \\ -8x - 6y + z = -\frac{7}{4} \\ 3x - y = -\frac{9}{4} \end{cases}$

   (a) $(0, 1, 1)$ (b) $\left(-\frac{3}{2}, \frac{5}{4}, -\frac{5}{4}\right)$
   (c) $\left(-\frac{1}{2}, \frac{3}{4}, -\frac{5}{4}\right)$ (d) $\left(-\frac{1}{2}, 2, 0\right)$

4. $\begin{cases} -4x - y - 8z = -6 \\ y + z = 0 \\ 4x - 7y = 6 \end{cases}$

   (a) $(-2, -2, 2)$ (b) $\left(-\frac{33}{2}, -10, 10\right)$
   (c) $\left(\frac{1}{8}, -\frac{1}{2}, \frac{1}{2}\right)$ (d) $\left(-\frac{11}{2}, -4, 4\right)$

In Exercises 5–10, use back-substitution to solve the system of linear equations.

5. $\begin{cases} 2x - y + 5z = 16 \\ y + 2z = 2 \\ z = 2 \end{cases}$

6. $\begin{cases} 4x - 3y - 2z = -17 \\ 6y - 5z = -12 \\ z = -2 \end{cases}$

7. $\begin{cases} 2x + y - 3z = 10 \\ y + z = 12 \\ z = 2 \end{cases}$

8. $\begin{cases} x - y + 2z = 22 \\ 3y - 8z = -9 \\ z = -3 \end{cases}$

9. $\begin{cases} 4x - 2y + z = 8 \\ -y + z = 4 \\ z = 2 \end{cases}$

10. $\begin{cases} 5x - 8z = 22 \\ 3y - 5z = 10 \\ z = -4 \end{cases}$

In Exercises 11–14, perform the row operation and write the equivalent system.

11. Add Equation 1 to Equation 2.

$\begin{cases} x - 2y + 3z = 5 & \text{Equation 1} \\ -x + 3y - 5z = 4 & \text{Equation 2} \\ 2x \qquad - 3z = 0 & \text{Equation 3} \end{cases}$

What did this operation accomplish?

12. Add -2 times Equation 1 to Equation 3.

$\begin{cases} x - 2y + 3z = 5 & \text{Equation 1} \\ -x + 3y - 5z = 4 & \text{Equation 2} \\ 2x \qquad - 3z = 0 & \text{Equation 3} \end{cases}$

What did this operation accomplish?

13. Add -2 times Equation 1 to Equation 2.

$\begin{cases} x - 2y + z = 1 & \text{Equation 1} \\ 2x - y + 3z = 0 & \text{Equation 2} \\ 3x - y - 4z = 1 & \text{Equation 3} \end{cases}$

What did this operation accomplish?

14. Add -3 times Equation 1 to Equation 3.

$$\begin{cases} x + 2y - 4z = 2 & \text{Equation 1} \\ -x + 4y - z = -2 & \text{Equation 2} \\ 3x - 4y + z = 1 & \text{Equation 3} \end{cases}$$

What did this operation accomplish?

In Exercises 15–48, solve the system of linear equations and check any solution algebraically.

15. $\begin{cases} x + y + z = 6 \\ 2x - y + z = 3 \\ 3x - z = 0 \end{cases}$

16. $\begin{cases} x + y + z = 2 \\ -x + 3y + 2z = 8 \\ 4x + y = 4 \end{cases}$

17. $\begin{cases} 2x + 2z = 2 \\ 5x + 3y = 4 \\ 3y - 4z = 4 \end{cases}$

18. $\begin{cases} 6y + 4z = -18 \\ 3x + 3y = 9 \\ 2x - 3z = 12 \end{cases}$

19. $\begin{cases} 4x + y - 3z = 11 \\ 2x - 3y + 2z = 9 \\ x + y + z = -3 \end{cases}$

20. $\begin{cases} 2x + 4y + z = -4 \\ 2x - 4y + 6z = 13 \\ 4x - 2y + z = 6 \end{cases}$

21. $\begin{cases} 3x - 2y + 4z = 1 \\ x + y - 2z = 3 \\ 2x - 3y + 6z = 8 \end{cases}$

22. $\begin{cases} 5x - 3y + 2z = 3 \\ 2x + 4y - z = 7 \\ x - 11y + 4z = 3 \end{cases}$

23. $\begin{cases} 3x + 3y + 5z = 1 \\ 3x + 5y + 9z = 0 \\ 5x + 9y + 17z = 0 \end{cases}$

24. $\begin{cases} 2x + y + 3z = 1 \\ 2x + 6y + 8z = 3 \\ 6x + 8y + 18z = 5 \end{cases}$

25. $\begin{cases} x + 2y - 7z = -4 \\ 2x + y + z = 13 \\ 3x + 9y - 36z = -33 \end{cases}$

26. $\begin{cases} 2x + y - 3z = 4 \\ 4x + 2z = 10 \\ -2x + 3y - 13z = -8 \end{cases}$

27. $\begin{cases} x - y + 2z = 6 \\ 2x + y + z = 3 \\ x + y + z = 2 \end{cases}$

28. $\begin{cases} x + y - z = 3 \\ 2x - y - z = 1 \\ x + y - 2z = 5 \end{cases}$

29. $\begin{cases} 3x - 3y + 6z = 6 \\ x + 2y - z = 5 \\ 5x - 8y + 13z = 7 \end{cases}$

30. $\begin{cases} x + 4z = 13 \\ 4x - 2y + z = 7 \\ 2x - 2y - 7z = -19 \end{cases}$

31. $\begin{cases} x - 2y + 3z = 4 \\ 3x - y + 2z = 0 \\ x + 3y - 4z = -2 \end{cases}$

32. $\begin{cases} -x + 3y + z = 4 \\ 4x - 2y - 5z = -7 \\ 2x + 4y - 3z = 12 \end{cases}$

33. $\begin{cases} x + 4z = 1 \\ x + y + 10z = 10 \\ 2x - y + 2z = -5 \end{cases}$

34. $\begin{cases} 3x - 2y - 6z = -4 \\ -3x + 2y + 6z = 1 \\ x - y - 5z = -3 \end{cases}$

35. $\begin{cases} x + 2y + z = 1 \\ x - 2y + 3z = -3 \\ 2x + y + z = -1 \end{cases}$

36. $\begin{cases} x - 2y + z = 2 \\ 2x + 2y - 3z = -4 \\ 5x + z = 1 \end{cases}$

37. $\begin{cases} x - 2y + 5z = 2 \\ 4x - z = 0 \end{cases}$

38. $\begin{cases} 12x + 5y + z = 0 \\ 23x + 4y - z = 0 \end{cases}$

39. $\begin{cases} 2x - 3y + z = -2 \\ -4x + 9y = 7 \end{cases}$

40. $\begin{cases} 10x - 3y + 2z = 0 \\ 19x - 5y - z = 0 \end{cases}$

41. $\begin{cases} x - 3y + 2z = 18 \\ 5x - 13y + 12z = 80 \end{cases}$

42. $\begin{cases} 2x + 3y + 3z = 7 \\ 4x + 18y + 15z = 44 \end{cases}$

43. $\begin{cases} x - y + 2z - w = 0 \\ 2x + y + z - w = 0 \\ x + y - w = -1 \\ x + y - 2z + w = 1 \end{cases}$

44. $\begin{cases} x - 2y - z + 2w = 6 \\ 2x - 3y + z + w = 3 \\ x - 2y + z + w = 2 \\ y - z + w = 3 \end{cases}$

45. $\begin{cases} \dfrac{1}{x} + \dfrac{2}{y} - \dfrac{3}{z} = 3 \\ \dfrac{1}{x} - \dfrac{2}{y} + \dfrac{1}{z} = 1 \\ \dfrac{2}{x} + \dfrac{2}{y} - \dfrac{3}{z} = 4 \end{cases}$

46. $\begin{cases} \dfrac{4}{x} - \dfrac{2}{y} + \dfrac{1}{z} = -3 \\ \dfrac{1}{x} + \dfrac{2}{y} - \dfrac{2}{z} = -1 \\ \dfrac{2}{x} + \dfrac{1}{y} - \dfrac{3}{z} = 0 \end{cases}$

47. $\begin{cases} \dfrac{2}{x} - \dfrac{1}{y} + \dfrac{2}{z} = 4 \\ \dfrac{1}{x} + \dfrac{2}{y} - \dfrac{2}{z} = -2 \\ \dfrac{3}{x} + \dfrac{3}{y} + \dfrac{4}{z} = 2 \end{cases}$

48. $\begin{cases} \dfrac{1}{x} + \dfrac{1}{y} + \dfrac{2}{z} = 1 \\ \dfrac{2}{x} + \dfrac{1}{y} - \dfrac{2}{z} = 0 \\ \dfrac{3}{x} - \dfrac{1}{y} + \dfrac{4}{z} = 6 \end{cases}$

Exploration **In Exercises 49–52, find a system of linear equations that has the given solution. (There are many correct answers.)**

49. $(4, -1, 2)$

50. $(-5, -2, 1)$

51. $\left(3, -\dfrac{1}{2}, \dfrac{7}{4}\right)$

52. $\left(-\dfrac{3}{2}, 4, -7\right)$

Three-Dimensional Graphics **In Exercises 53–56, sketch the plane represented by the linear equation. Then list four points that lie in the plane.**

53. $2x + 3y + 4z = 12$

54. $x + y + z = 6$

55. $2x + y + z = 4$

56. $x + 2y + 2z = 6$

In Exercises 57–62, write the form of the partial fraction decomposition of the rational expression. Do not solve for the constants.

57. $\dfrac{7}{x^2 - 14x}$

58. $\dfrac{x - 2}{x^2 + 4x + 3}$

59. $\dfrac{12}{x^3 - 10x^2}$

60. $\dfrac{x^2 - 3x + 2}{4x^3 + 11x^2}$

61. $\dfrac{4x^2 + 3}{(x - 5)^3}$

62. $\dfrac{6x + 5}{(x + 2)^4}$

In Exercises 63–80, write the partial fraction decomposition for the rational expression. Check your result algebraically by combining fractions, and check your result graphically by using a graphing utility to graph the rational expression and the partial fractions in the same viewing window.

63. $\dfrac{1}{x^2 - 1}$

64. $\dfrac{1}{4x^2 - 9}$

65. $\dfrac{1}{x^2 + x}$

66. $\dfrac{3}{x^2 - 3x}$

67. $\dfrac{1}{2x^2 + x}$

68. $\dfrac{5}{x^2 + x - 6}$

69. $\dfrac{5 - x}{2x^2 + x - 1}$

70. $\dfrac{x - 2}{x^2 + 4x + 3}$

71. $\dfrac{x^2 + 12x + 12}{x^3 - 4x}$

72. $\dfrac{x^2 + 12x - 9}{x^3 - 9x}$

73. $\dfrac{4x^2 + 2x - 1}{x^2(x + 1)}$

74. $\dfrac{2x - 3}{(x - 1)^2}$

75. $\dfrac{27 - 7x}{x(x - 3)^2}$

76. $\dfrac{x^2 - x + 2}{x(x - 1)^2}$

77. $\dfrac{2x^3 - x^2 + x + 5}{x^2 + 3x + 2}$

78. $\dfrac{x^3 + 2x^2 - x + 1}{x^2 + 3x - 4}$

79. $\dfrac{x^4}{(x - 1)^3}$

80. $\dfrac{4x^4}{(2x - 1)^3}$

Graphical Analysis In Exercises 81 and 82, write the partial fraction decomposition for the rational function. Identify the graph of the rational function and the graph of each term of its decomposition. State any relationship between the vertical asymptotes of the rational function and the vertical asymptotes of the terms of the decomposition.

81. $y = \dfrac{x - 12}{x(x - 4)}$

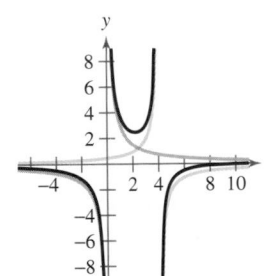

82. $y = \dfrac{2(4x - 3)}{x^2 - 9}$

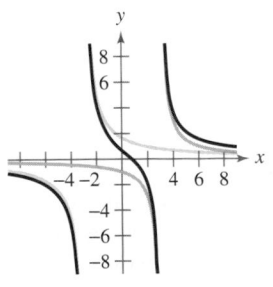

Vertical Motion In Exercises 83–86, an object moving vertically is at the given heights at the specified times. Find the position equation $s = \frac{1}{2}at^2 + v_0 t + s_0$ for the object.

83. At $t = 1$ second, $s = 128$ feet.
    At $t = 2$ seconds, $s = 80$ feet.
    At $t = 3$ seconds, $s = 0$ feet.

84. At $t = 1$ second, $s = 48$ feet.
    At $t = 2$ seconds, $s = 64$ feet.
    At $t = 3$ seconds, $s = 48$ feet.

85. At $t = 1$ second, $s = 452$ feet.
    At $t = 2$ seconds, $s = 372$ feet.
    At $t = 3$ seconds, $s = 260$ feet.

86. At $t = 1$ second, $s = 132$ feet.
    At $t = 2$ seconds, $s = 100$ feet.
    At $t = 3$ seconds, $s = 36$ feet.

In Exercises 87–90, find the equation of the parabola

$$y = ax^2 + bx + c$$

that passes through the points. To verify your result, use a graphing utility to plot the points and graph the parabola.

87. $(0, 0), (2, -2), (4, 0)$

88. $(0, 3), (1, 4), (2, 3)$

89. $(2, 0), (3, -1), (4, 0)$

90. $(1, 3), (2, 2), (3, -3)$

In Exercises 91–94, find the equation of the circle

$$x^2 + y^2 + Dx + Ey + F = 0$$

that passes through the points. To verify your result, use a graphing utility to plot the points and graph the circle.

91. $(0, 0), (2, 2), (4, 0)$

92. $(0, 0), (0, 6), (3, 3)$

93. $(-3, -1), (2, 4), (-6, 8)$

94. $(-6, -1), (-4, 3), (2, -5)$

95. *Borrowing* A small corporation borrowed $775,000 to expand its software line. Some of the money was borrowed at 8%, some at 9%, and some at 10%. How much was borrowed at each rate if the annual interest was $67,000 and the amount borrowed at 8% was four times the amount borrowed at 10%?

96. *Borrowing* A small corporation borrowed $1,000,000 to expand its line of toys. Some of the money was borrowed at 8%, some at 10%, and some at 12%. How much was borrowed at each rate if the annual interest was $97,200 and the amount borrowed at 8% was two times the amount borrowed at 10%?

Investment Portfolio In Exercises 97 and 98, consider an investor with a portfolio totaling $500,000 that is invested in certificates of deposit, municipal bonds, blue-chip stocks, and growth or speculative stocks. How much is invested in each type of investment?

97. The certificates of deposit pay 8% annually, and the municipal bonds pay 9% annually. Over a five-year period, the investor expects the blue-chip stocks to return 12% annually and the growth stocks to return 15% annually. The investor wants a combined annual return of 10% and also wants to have only one-fourth of the portfolio invested in municipal bonds.

98. The certificates of deposit pay 9% annually, and the municipal bonds pay 5% annually. Over a five-year period, the investor expects the blue-chip stocks to return 12% annually and the growth stocks to return 14% annually. The investor wants a combined annual return of 10% and also wants to have only one-fourth of the portfolio invested in stocks.

99. **Sports** In the 2005 Women's NCAA Championship basketball game, Baylor University defeated Michigan State University by a score of 84 to 62. Baylor won by scoring a combination of two-point field goals, three-point field goals, and one-point free throws. The number of two-point field goals was six more than the number of free throws, and four times the number of three-point field goals. Find the combination of scores that won the National Championship for Baylor. (Source: NCAA)

100. **Sports** In the 2005 Men's NCAA Championship basketball game, the University of North Carolina defeated the University of Illinois by a score of 75 to 70. North Carolina won by scoring a combination of two-point field goals, three-point field goals, and one-point free throws. The number of free throws was three more than the number of three-point field goals, and six less than the number of two-point field goals. Find the combination of scores that won the National Championship for North Carolina. (Source: NCAA)

101. **Sports** On February 5, 2006, in Super Bowl XL, the Pittsburgh Steelers beat the Seattle Seahawks by a score of 21 to 10. The scoring in that game was a combination of touchdowns, extra-point kicks, and field goals, worth 6 points, 1 point, and 3 points, respectively. There were a total of nine scoring plays by both teams. The number of touchdowns scored was four times the number of field goals scored. The number of extra-point kicks scored was equal to the number of touchdowns. How many touchdowns, extra-point kicks, and field goals were scored during the game? (Source: SuperBowl.com)

102. **Sports** The University of Georgia and Florida State University scored a total of 39 points during the 2003 Sugar Bowl. The points came from a total of 11 different scoring plays, which were a combination of touchdowns, extra-point kicks, and field goals, worth 6, 1, and 3 points, respectively. The same numbers of touchdowns and field goals were scored. How many touchdowns, extra-point kicks, and field goals were scored during the game? (Source: espn.com)

103. **Electrical Networks** When Kirchhoff's Laws are applied to the electrical network in the figure, the currents I_1, I_2, and I_3 are the solution of the system

$$\begin{cases} I_1 - I_2 + I_3 = 0 \\ 3I_1 + 2I_2 \qquad = 7. \\ \qquad 2I_2 + 4I_3 = 8 \end{cases}$$

Find the currents.

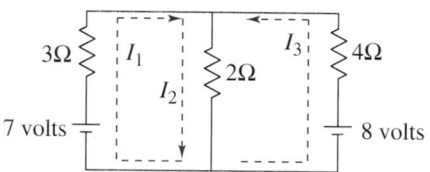

104. **Pulley System** A system of pulleys is loaded with 128-pound and 32-pound weights (see figure). The tensions t_1 and t_2 in the ropes and the acceleration a of the 32-pound weight are modeled by the system

$$\begin{cases} t_1 - 2t_2 \qquad = 0 \\ t_1 \qquad - 2a = 128 \\ \quad t_2 + a = 32 \end{cases}$$

where t_1 and t_2 are measured in pounds and a is in feet per second squared. Solve the system.

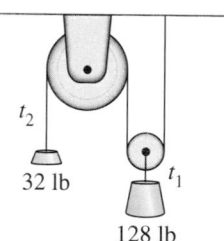

105. **Sports** The Augusta National Golf Club in Augusta, Georgia, is an 18-hole course that consists of par-3 holes, par-4 holes, and par-5 holes. A golfer who shoots par has a total of 72 strokes for the entire course. There are two more par-4 holes than twice the number of par-5 holes, and the number of par-3 holes is equal to the number of par-5 holes. Find the numbers of par-3, par-4, and par-5 holes on the course. (Source: Augusta National, Inc.)

106. *Sports* St Andrews Golf Course in St Andrews, Scotland is one of the oldest golf courses in the world. It is an 18-hole course that consists of par-3 holes, par-4 holes, and par-5 holes. A golfer who shoots par at The Old Course at St Andrews has a total of 72 strokes for the entire course. There are seven times as many par-4 holes as par-5 holes, and the sum of the numbers of par-3 and par-5 holes is four. Find the numbers of par-3, par-4, and par-5 holes on the course. (Source: St Andrews Links Trust)

Fitting a Parabola In Exercises 107–110, find the least squares regression parabola $y = ax^2 + bx + c$ for the points $(x_1, y_1), (x_2, y_2), \ldots, (x_n, y_n)$ by solving the following system of linear equations for a, b, and c. Then use the *regression* feature of a graphing utility to confirm your result. (For an explanation of how the coefficients of a, b, and c in the system are obtained, see Appendix C.)

107.
$$\begin{cases} 4c & + 40a = & 19 \\ & 40b & = -12 \\ 40c & + 544a = & 160 \end{cases}$$

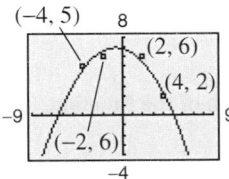

108.
$$\begin{cases} 5c & + 10a = & 8 \\ & 10b & = 12 \\ 10c & + 34a = & 22 \end{cases}$$

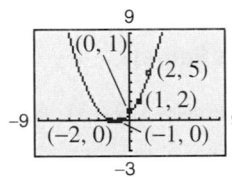

109.
$$\begin{cases} 4c + 9b + 29a = 20 \\ 9c + 29b + 99a = 70 \\ 29c + 99b + 353a = 254 \end{cases}$$

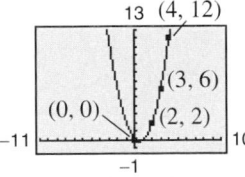

110.
$$\begin{cases} 4c + 6b + 14a = 25 \\ 6c + 14b + 36a = 21 \\ 14c + 36b + 98a = 33 \end{cases}$$

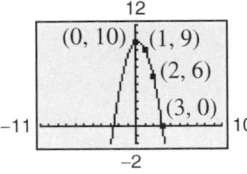

111. *Data Analysis* During the testing of a new automobile braking system, the speeds x (in miles per hour) and the stopping distances y (in feet) were recorded in the table.

Speed, x	Stopping distance, y
30	55
40	105
50	188

(a) Use the data to create a system of linear equations. Then find the least squares regression parabola for the data by solving the system.

(b) Use a graphing utility to graph the parabola and the data in the same viewing window.

(c) Use the model to estimate the stopping distance for a speed of 70 miles per hour.

112. *Data Analysis* A wildlife management team studied the reproduction rates of deer in three five-acre tracts of a wildlife preserve. In each tract, the number of females x and the percent of females y that had offspring the following year were recorded. The results are shown in the table.

Number, x	Percent, y
120	68
140	55
160	30

(a) Use the data to create a system of linear equations. Then find the least squares regression parabola for the data by solving the system.

(b) Use a graphing utility to graph the parabola and the data in the same viewing window.

(c) Use the model to predict the percent of females that had offspring when there were 170 females.

113. *Thermodynamics* The magnitude of the range R of exhaust temperatures (in degrees Fahrenheit) in an experimental diesel engine is approximated by the model

$$R = \frac{2000(4 - 3x)}{(11 - 7x)(7 - 4x)}, \quad 0 \le x \le 1$$

where x is the relative load (in foot-pounds).

(a) Write the partial fraction decomposition for the rational function.

(b) The decomposition in part (a) is the difference of two fractions. The absolute values of the terms give the expected maximum and minimum temperatures of the exhaust gases. Use a graphing utility to graph each term.

114. *Environment* The predicted cost C (in thousands of dollars) for a company to remove $p\%$ of a chemical from its waste water is given by the model

$$C = \frac{120p}{10,000 - p^2}, \quad 0 \le p < 100.$$

Write the partial fraction decomposition for the rational function. Verify your result by using the *table* feature of a graphing utility to create a table comparing the original function with the partial fractions.

Synthesis

True or False?   **In Exercises 115–117, determine whether the statement is true or false. Justify your answer.**

115. The system

$$\begin{cases} x + 4y - 5z = 8 \\ \quad\;\; 2y + \;\; z = 5 \\ \qquad\qquad z = 1 \end{cases}$$

is in row-echelon form.

116. If a system of three linear equations is inconsistent, then its graph has no points common to all three equations.

117. For the rational expression

$$\frac{x}{(x + 10)(x - 10)^2}$$

the partial fraction decomposition is of the form

$$\frac{A}{x + 10} + \frac{B}{(x - 10)^2}.$$

118. *Error Analysis*   You are tutoring a student in algebra. In trying to find a partial fraction decomposition, your student writes the following.

$$\frac{x^2 + 1}{x(x - 1)} = \frac{A}{x} + \frac{B}{x - 1}$$

$$x^2 + 1 = A(x - 1) + Bx \qquad \text{Basic equation}$$

$$x^2 + 1 = (A + B)x - A$$

Your student then forms the following system of linear equations.

$$\begin{cases} A + B = 0 \\ -A \qquad = 1 \end{cases}$$

Solve the system and check the partial fraction decomposition it yields. Has your student worked the problem correctly? If not, what went wrong?

In Exercises 119–122, write the partial fraction decomposition for the rational expression. Check your result algebraically. Then assign a value to the constant a and check the result graphically.

119. $\dfrac{1}{a^2 - x^2}$

120. $\dfrac{1}{(x + 1)(a - x)}$

121. $\dfrac{1}{y(a - y)}$

122. $\dfrac{1}{x(x + a)}$

123. *Think About It*   Are the two systems of equations equivalent? Give reasons for your answer.

$$\begin{cases} x + 3y - \;\; z = 6 \\ 2x - \;\; y + 2z = 1 \\ 3x + 2y - \;\; z = 2 \end{cases} \qquad \begin{cases} x + 3y - \;\; z = \quad 6 \\ \quad\;\; -7y + 4z = \quad 1 \\ \quad\;\; -7y - 4z = -16 \end{cases}$$

124. *Writing*   When using Gaussian elimination to solve a system of linear equations, explain how you can recognize that the system has no solution. Give an example that illustrates your answer.

∫ *Advanced Applications*   **In Exercises 125–128, find values of x, y, and λ that satisfy the system. These systems arise in certain optimization problems in calculus. (λ is called a *Lagrange multiplier*.)**

125. $\begin{cases} y + \lambda = 0 \\ x + \lambda = 0 \\ x + y - 10 = 0 \end{cases}$
   126. $\begin{cases} 2x + \lambda = 0 \\ 2y + \lambda = 0 \\ x + y - 4 = 0 \end{cases}$

127. $\begin{cases} 2x - 2x\lambda = 0 \\ -2y + \lambda = 0 \\ y - x^2 = 0 \end{cases}$
   128. $\begin{cases} 2 + 2x + 2\lambda = 0 \\ 2x + 1 + \lambda = 0 \\ 2x + y - 100 = 0 \end{cases}$

Skills Review

In Exercises 129–134, sketch the graph of the function.

129. $f(x) = -3x + 7$ **130.** $f(x) = 6 - x$

131. $f(x) = -2x^2$ **132.** $f(x) = \frac{1}{4}x^2 + 1$

133. $f(x) = -x^2(x - 3)$ **134.** $f(x) = \frac{1}{2}x^3 - 1$

In Exercises 135–138, (a) determine the real zeros of f and (b) sketch the graph of f.

135. $f(x) = x^3 + x^2 - 12x$

136. $f(x) = -8x^4 + 32x^2$

137. $f(x) = 2x^3 + 5x^2 - 21x - 36$

138. $f(x) = 6x^3 - 29x^2 - 6x + 5$

In Exercises 139–142, use a graphing utility to create a table of values for the function. Then sketch the graph of the function by hand.

139. $y = \left(\frac{1}{2}\right)^{x-2}$ **140.** $y = \left(\frac{1}{3}\right)^{x} + 1$

141. $y = 2^{x+1} - 1$ **142.** $y = 3^{x-1} + 2$

143. *Make a Decision*   To work an extended application analyzing the earnings per share for Wal-Mart Stores, Inc. from 1988 to 2005, visit this textbook's *Online Study Center*.   (Data Source: Wal-Mart Stores, Inc.)

7.4   Matrices and Systems of Equations

Matrices

In this section, you will study a streamlined technique for solving systems of linear equations. This technique involves the use of a rectangular array of real numbers called a **matrix.** The plural of matrix is *matrices.*

Definition of Matrix

If m and n are positive integers, an $m \times n$ (read "m by n") matrix is a rectangular array

$$
\begin{array}{ccccc}
 & \text{Column 1} & \text{Column 2} & \text{Column 3} & \cdots & \text{Column } n \\
\text{Row 1} & \begin{bmatrix} a_{11} & a_{12} & a_{13} & \cdots & a_{1n} \\ a_{21} & a_{22} & a_{23} & \cdots & a_{2n} \\ a_{31} & a_{32} & a_{33} & \cdots & a_{3n} \\ \vdots & \vdots & \vdots & & \vdots \\ a_{m1} & a_{m2} & a_{m3} & \cdots & a_{mn} \end{bmatrix}
\end{array}
$$

in which each **entry** a_{ij} of the matrix is a real number. An $m \times n$ matrix has m rows and n columns.

The entry in the ith row and jth column is denoted by the *double subscript* notation a_{ij}. For instance, the entry a_{23} is the entry in the second row and third column. A matrix having m rows and n columns is said to be of **order** $m \times n$. If $m = n$, the matrix is **square** of order n. For a square matrix, the entries $a_{11}, a_{22}, a_{33}, \ldots$ are the **main diagonal** entries.

Example 1   Order of Matrices

Determine the order of each matrix.

a. $[2]$ **b.** $\begin{bmatrix} 1 & -3 & 0 & \frac{1}{2} \end{bmatrix}$ **c.** $\begin{bmatrix} 0 & 0 \\ 0 & 0 \end{bmatrix}$ **d.** $\begin{bmatrix} 5 & 0 \\ 2 & -2 \\ -7 & 4 \end{bmatrix}$

Solution

a. This matrix has *one* row and *one* column. The order of the matrix is 1×1.

b. This matrix has *one* row and *four* columns. The order of the matrix is 1×4.

c. This matrix has *two* rows and *two* columns. The order of the matrix is 2×2.

d. This matrix has *three* rows and *two* columns. The order of the matrix is 3×2.

✓CHECKPOINT   Now try Exercise 3.

A matrix that has only one row [such as the matrix in Example 1(b)] is called a **row matrix,** and a matrix that has only one column is called a **column matrix.**

A matrix derived from a system of linear equations (each written in standard form with the constant term on the right) is the **augmented matrix** of the system. Moreover, the matrix derived from the coefficients of the system (but not including the constant terms) is the **coefficient matrix** of the system.

System
$$\begin{cases} x - 4y + 3z = 5 \\ -x + 3y - z = -3 \\ 2x \qquad - 4z = 6 \end{cases}$$

Augmented Matrix
$$\begin{bmatrix} 1 & -4 & 3 & \vdots & 5 \\ -1 & 3 & -1 & \vdots & -3 \\ 2 & 0 & -4 & \vdots & 6 \end{bmatrix}$$

Coefficient Matrix
$$\begin{bmatrix} 1 & -4 & 3 \\ -1 & 3 & -1 \\ 2 & 0 & -4 \end{bmatrix}$$

Note the use of 0 for the missing coefficient of the y-variable in the third equation, and also note the fourth column (of constant terms) in the augmented matrix. The optional dotted line in the augmented matrix helps to separate the coefficients of the linear system from the constant terms.

When forming either the coefficient matrix or the augmented matrix of a system, you should begin by vertically aligning the variables in the equations and using 0's for any missing coefficients of variables.

Example 2 Writing an Augmented Matrix

Write the augmented matrix for the system of linear equations.

$$\begin{cases} x + 3y = 9 \\ -y + 4z = -2 \\ x - 5z = 0 \end{cases}$$

Solution

Begin by writing the linear system and aligning the variables.

$$\begin{cases} x + 3y \qquad = 9 \\ -y + 4z = -2 \\ x \qquad - 5z = 0 \end{cases}$$

Next, use the coefficients and constant terms as the matrix entries. Include zeros for any missing coefficients.

$$\begin{matrix} R_1 \\ R_2 \\ R_3 \end{matrix} \begin{bmatrix} 1 & 3 & 0 & \vdots & 9 \\ 0 & -1 & 4 & \vdots & -2 \\ 1 & 0 & -5 & \vdots & 0 \end{bmatrix}$$

The notation R_n is used to designate each row in the matrix. For example, Row 1 is represented by R_1.

✓CHECKPOINT   Now try Exercise 9.

Elementary Row Operations

In Section 7.3, you studied three operations that can be used on a system of linear equations to produce an equivalent system. These operations are: interchange two equations, multiply an equation by a nonzero constant, and add a multiple of an equation to another equation. In matrix terminology, these three operations correspond to **elementary row operations.** An elementary row operation on an augmented matrix of a given system of linear equations produces a new augmented matrix corresponding to a new (but equivalent) system of linear equations. Two matrices are **row-equivalent** if one can be obtained from the other by a sequence of elementary row operations.

Elementary Row Operations for Matrices

   1. Interchange two rows.

   2. Multiply a row by a nonzero constant.

   3. Add a multiple of a row to another row.

Although elementary row operations are simple to perform, they involve a lot of arithmetic. Because it is easy to make a mistake, you should get in the habit of noting the elementary row operations performed in each step so that you can go back and check your work.

Example 3 demonstrates the elementary row operations described above.

Example 3   Elementary Row Operations

a. Interchange the first and second rows of the original matrix.

Original Matrix

$$\begin{bmatrix} 0 & 1 & 3 & 4 \\ -1 & 2 & 0 & 3 \\ 2 & -3 & 4 & 1 \end{bmatrix}$$

New Row-Equivalent Matrix

$$\begin{matrix} R_2 \\ R_1 \end{matrix} \begin{bmatrix} -1 & 2 & 0 & 3 \\ 0 & 1 & 3 & 4 \\ 2 & -3 & 4 & 1 \end{bmatrix}$$

b. Multiply the first row of the original matrix by $\frac{1}{2}$.

Original Matrix

$$\begin{bmatrix} 2 & -4 & 6 & -2 \\ 1 & 3 & -3 & 0 \\ 5 & -2 & 1 & 2 \end{bmatrix}$$

New Row-Equivalent Matrix

$$\frac{1}{2}R_1 \rightarrow \begin{bmatrix} 1 & -2 & 3 & -1 \\ 1 & 3 & -3 & 0 \\ 5 & -2 & 1 & 2 \end{bmatrix}$$

c. Add -2 times the first row of the original matrix to the third row.

Original Matrix

$$\begin{bmatrix} 1 & 2 & -4 & 3 \\ 0 & 3 & -2 & -1 \\ 2 & 1 & 5 & -2 \end{bmatrix}$$

New Row-Equivalent Matrix

$$\begin{matrix} \\ \\ -2R_1 + R_3 \rightarrow \end{matrix} \begin{bmatrix} 1 & 2 & -4 & 3 \\ 0 & 3 & -2 & -1 \\ 0 & -3 & 13 & -8 \end{bmatrix}$$

Note that the elementary row operation is written beside the row that is *changed*.

✓CHECKPOINT   Now try Exercise 21.

TECHNOLOGY TIP

Most graphing utilities can perform elementary row operations on matrices. The top screen below shows how one graphing utility displays the original matrix in Example 3(a). The bottom screen below shows the new row-equivalent matrix in Example 3(a). The new row-equivalent matrix is obtained by using the *row swap* feature of the graphing utility. For instructions on how to use the *matrix* feature and the *row swap* feature (and other elementary row operations features) of a graphing utility, see Appendix A; for specific keystrokes, go to this textbook's *Online Study Center*.

```
[A]
   [[0  1  3  4]
    [-1 2  0  3]
    [2 -3  4  1]]
```

```
rowSwap([A],1,2)
   [[-1 2  0  3]
    [0  1  3  4]
    [2 -3  4  1]]
```

Gaussian Elimination with Back–Substitution

In Example 2 of Section 7.3, you used Gaussian elimination with back-substitution to solve a system of linear equations. The next example demonstrates the matrix version of Gaussian elimination. The basic difference between the two methods is that with matrices you do not need to keep writing the variables.

Example 4   Comparing Linear Systems and Matrix Operations

Linear System   *Associated Augmented Matrix*

$$\begin{cases} x - 2y + 3z = 9 \\ -x + 3y + z = -2 \\ 2x - 5y + 5z = 17 \end{cases}$$
$$\begin{bmatrix} 1 & -2 & 3 & \vdots & 9 \\ -1 & 3 & 1 & \vdots & -2 \\ 2 & -5 & 5 & \vdots & 17 \end{bmatrix}$$

Add the first equation to the second equation.

Add the first row to the second row $(R_1 + R_2)$.

$$\begin{cases} x - 2y + 3z = 9 \\ y + 4z = 7 \\ 2x - 5y + 5z = 17 \end{cases}$$
$$R_1 + R_2 \rightarrow \begin{bmatrix} 1 & -2 & 3 & \vdots & 9 \\ 0 & 1 & 4 & \vdots & 7 \\ 2 & -5 & 5 & \vdots & 17 \end{bmatrix}$$

Add -2 times the first equation to the third equation.

Add -2 times the first row to the third row $(-2R_1 + R_3)$.

$$\begin{cases} x - 2y + 3z = 9 \\ y + 4z = 7 \\ -y - z = -1 \end{cases}$$
$$-2R_1 + R_3 \rightarrow \begin{bmatrix} 1 & -2 & 3 & \vdots & 9 \\ 0 & 1 & 4 & \vdots & 7 \\ 0 & -1 & -1 & \vdots & -1 \end{bmatrix}$$

Add the second equation to the third equation.

Add the second row to the third row $(R_2 + R_3)$.

$$\begin{cases} x - 2y + 3z = 9 \\ y + 4z = 7 \\ 3z = 6 \end{cases}$$
$$R_2 + R_3 \rightarrow \begin{bmatrix} 1 & -2 & 3 & \vdots & 9 \\ 0 & 1 & 4 & \vdots & 7 \\ 0 & 0 & 3 & \vdots & 6 \end{bmatrix}$$

Multiply the third equation by $\frac{1}{3}$.

Multiply the third row by $\frac{1}{3} \left(\frac{1}{3}R_3\right)$.

$$\begin{cases} x - 2y + 3z = 9 \\ y + 4z = 7 \\ z = 2 \end{cases}$$
$$\frac{1}{3}R_3 \rightarrow \begin{bmatrix} 1 & -2 & 3 & \vdots & 9 \\ 0 & 1 & 4 & \vdots & 7 \\ 0 & 0 & 1 & \vdots & 2 \end{bmatrix}$$

At this point, you can use back-substitution to find that the solution is $x = 1$, $y = -1$, and $z = 2$, as was done in Example 2 of Section 7.3.

✓**CHECKPOINT**   Now try Exercise 29.

Remember that you should check a solution by substituting the values of x, y, and z into each equation in the original system.

The last matrix in Example 4 is in **row-echelon form.** The term *echelon* refers to the stair-step pattern formed by the nonzero elements of the matrix. To be in this form, a matrix must have the properties listed on the next page.

Row-Echelon Form and Reduced Row-Echelon Form

A matrix in **row-echelon form** has the following properties.

1. Any rows consisting entirely of zeros occur at the bottom of the matrix.

2. For each row that does not consist entirely of zeros, the first nonzero entry is 1 (called a **leading 1**).

3. For two successive (nonzero) rows, the leading 1 in the higher row is farther to the left than the leading 1 in the lower row.

A matrix in *row-echelon form* is in **reduced row-echelon form** if every column that has a leading 1 has zeros in every position above and below its leading 1.

It is worth mentioning that the row-echelon form of a matrix is not unique. That is, two different sequences of elementary row operations may yield different row-echelon forms.

Example 5   Row–Echelon Form

Determine whether each matrix is in row-echelon form. If it is, determine whether the matrix is in reduced row-echelon form.

a. $\begin{bmatrix} 1 & 2 & -1 & 4 \\ 0 & 1 & 0 & 3 \\ 0 & 0 & 1 & -2 \end{bmatrix}$ **b.** $\begin{bmatrix} 1 & 2 & -1 & 2 \\ 0 & 0 & 0 & 0 \\ 0 & 1 & 2 & -4 \end{bmatrix}$

c. $\begin{bmatrix} 1 & -5 & 2 & -1 & 3 \\ 0 & 0 & 1 & 3 & -2 \\ 0 & 0 & 0 & 1 & 4 \\ 0 & 0 & 0 & 0 & 1 \end{bmatrix}$ **d.** $\begin{bmatrix} 1 & 0 & 0 & -1 \\ 0 & 1 & 0 & 2 \\ 0 & 0 & 1 & 3 \\ 0 & 0 & 0 & 0 \end{bmatrix}$

e. $\begin{bmatrix} 1 & 2 & -3 & 4 \\ 0 & 2 & 1 & -1 \\ 0 & 0 & 1 & -3 \end{bmatrix}$ **f.** $\begin{bmatrix} 0 & 1 & 0 & 5 \\ 0 & 0 & 1 & 3 \\ 0 & 0 & 0 & 0 \end{bmatrix}$

Solution

The matrices in (a), (c), (d), and (f) are in row-echelon form. The matrices in (d) and (f) are in *reduced* row-echelon form because every column that has a leading 1 has zeros in every position above and below its leading 1. The matrix in (b) is not in row-echelon form because the row of all zeros does not occur at the bottom of the matrix. The matrix in (e) is not in row-echelon form because the first nonzero entry in row 2 is not a leading 1.

✓CHECKPOINT   Now try Exercise 23.

Every matrix is row-equivalent to a matrix in row-echelon form. For instance, in Example 5, you can change the matrix in part (e) to row-echelon form by multiplying its second row by $\frac{1}{2}$. What elementary row operation could you perform on the matrix in part (b) so that it would be in row-echelon form?

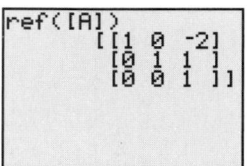

Gaussian elimination with back-substitution works well for solving systems of linear equations by hand or with a computer. For this algorithm, the order in which the elementary row operations are performed is important. You should operate *from left to right by columns*, using elementary row operations to obtain zeros in all entries directly below the leading 1's.

Example 6 Gaussian Elimination with Back–Substitution

Solve the system $\begin{cases} y + z - 2w = -3 \\ x + 2y - z \qquad = 2 \\ 2x + 4y + z - 3w = -2 \\ x - 4y - 7z - w = -19 \end{cases}$.

Solution

$$\begin{bmatrix} 0 & 1 & 1 & -2 & \vdots & -3 \\ 1 & 2 & -1 & 0 & \vdots & 2 \\ 2 & 4 & 1 & -3 & \vdots & -2 \\ 1 & -4 & -7 & -1 & \vdots & -19 \end{bmatrix}$$

Write augmented matrix.

$$\begin{matrix} R_2 \\ R_1 \end{matrix} \begin{bmatrix} 1 & 2 & -1 & 0 & \vdots & 2 \\ 0 & 1 & 1 & -2 & \vdots & -3 \\ 2 & 4 & 1 & -3 & \vdots & -2 \\ 1 & -4 & -7 & -1 & \vdots & -19 \end{bmatrix}$$

Interchange R_1 and R_2 so first column has leading 1 in upper left corner.

$$\begin{matrix} \\ \\ -2R_1 + R_3 \rightarrow \\ -R_1 + R_4 \rightarrow \end{matrix} \begin{bmatrix} 1 & 2 & -1 & 0 & \vdots & 2 \\ 0 & 1 & 1 & -2 & \vdots & -3 \\ 0 & 0 & 3 & -3 & \vdots & -6 \\ 0 & -6 & -6 & -1 & \vdots & -21 \end{bmatrix}$$

Perform operations on R_3 and R_4 so first column has zeros below its leading 1.

$$\begin{matrix} \\ \\ \\ 6R_2 + R_4 \rightarrow \end{matrix} \begin{bmatrix} 1 & 2 & -1 & 0 & \vdots & 2 \\ 0 & 1 & 1 & -2 & \vdots & -3 \\ 0 & 0 & 3 & -3 & \vdots & -6 \\ 0 & 0 & 0 & -13 & \vdots & -39 \end{bmatrix}$$

Perform operations on R_4 so second column has zeros below its leading 1.

$$\begin{matrix} \\ \\ \frac{1}{3}R_3 \rightarrow \\ -\frac{1}{13}R_4 \rightarrow \end{matrix} \begin{bmatrix} 1 & 2 & -1 & 0 & \vdots & 2 \\ 0 & 1 & 1 & -2 & \vdots & -3 \\ 0 & 0 & 1 & -1 & \vdots & -2 \\ 0 & 0 & 0 & 1 & \vdots & 3 \end{bmatrix}$$

Perform operations on R_3 and R_4 so third and fourth columns have leading 1's.

The matrix is now in row-echelon form, and the corresponding system is

$$\begin{cases} x + 2y - z \qquad = 2 \\ y + z - 2w = -3 \\ z - w = -2 \\ w = 3 \end{cases}.$$

Using back-substitution, you can determine that the solution is $x = -1$, $y = 2$, $z = 1$, and $w = 3$. Check this in the original system of equations.

✓**CHECKPOINT** Now try Exercise 53.

The following steps summarize the procedure used in Example 6.

> **Gaussian Elimination with Back–Substitution**
>
> 1. Write the augmented matrix of the system of linear equations.
>
> 2. Use elementary row operations to rewrite the augmented matrix in row-echelon form.
>
> 3. Write the system of linear equations corresponding to the matrix in row-echelon form and use back-substitution to find the solution.

Remember that it is possible for a system to have no solution. If, in the elimination process, you obtain a row with zeros except for the last entry, you can conclude that the system is inconsistent.

Example 7   A System with No Solution

Solve the system $\begin{cases} x - y + 2z = 4 \\ x \quad\;\; + \;\; z = 6 \\ 2x - 3y + 5z = 4 \\ 3x + 2y - \;\; z = 1 \end{cases}$.

Solution

$$\begin{bmatrix} 1 & -1 & 2 & \vdots & 4 \\ 1 & 0 & 1 & \vdots & 6 \\ 2 & -3 & 5 & \vdots & 4 \\ 3 & 2 & -1 & \vdots & 1 \end{bmatrix}$$ Write augmented matrix.

$$\begin{matrix} \\ -R_1 + R_2 \rightarrow \\ -2R_1 + R_3 \rightarrow \\ -3R_1 + R_4 \rightarrow \end{matrix} \begin{bmatrix} 1 & -1 & 2 & \vdots & 4 \\ 0 & 1 & -1 & \vdots & 2 \\ 0 & -1 & 1 & \vdots & -4 \\ 0 & 5 & -7 & \vdots & -11 \end{bmatrix}$$ Perform row operations.

$$\begin{matrix} \\ \\ R_2 + R_3 \rightarrow \\ \\ \end{matrix} \begin{bmatrix} 1 & -1 & 2 & \vdots & 4 \\ 0 & 1 & -1 & \vdots & 2 \\ 0 & 0 & 0 & \vdots & -2 \\ 0 & 5 & -7 & \vdots & -11 \end{bmatrix}$$ Perform row operations.

Note that the third row of this matrix consists of zeros except for the last entry. This means that the original system of linear equations is *inconsistent.* You can see why this is true by converting back to a system of linear equations. Because the third equation is not possible, the system has no solution.

$$\begin{cases} x - y + 2z = \;\;\; 4 \\ y - \;\; z = \;\;\; 2 \\ \quad\quad\;\; 0 = -2 \\ 5y - 7z = -11 \end{cases}$$

✔**CHECKPOINT**   Now try Exercise 51.

Gauss–Jordan Elimination

With Gaussian elimination, elementary row operations are applied to a matrix to obtain a (row-equivalent) row-echelon form of the matrix. A second method of elimination, called **Gauss-Jordan elimination** after Carl Friedrich Gauss (1777–1855) and Wilhelm Jordan (1842–1899), continues the reduction process until a *reduced* row-echelon form is obtained. This procedure is demonstrated in Example 8.

Example 8   Gauss–Jordan Elimination

Use Gauss-Jordan elimination to solve the system.

$$\begin{cases} x - 2y + 3z = 9 \\ -x + 3y + z = -2 \\ 2x - 5y + 5z = 17 \end{cases}$$

Solution

In Example 4, Gaussian elimination was used to obtain the row-echelon form

$$\begin{bmatrix} 1 & -2 & 3 & \vdots & 9 \\ 0 & 1 & 4 & \vdots & 7 \\ 0 & 0 & 1 & \vdots & 2 \end{bmatrix}.$$

Now, rather than using back-substitution, apply additional elementary row operations until you obtain a matrix in *reduced* row-echelon form. To do this, you must produce zeros above each of the leading 1's, as follows.

$$2R_2 + R_1 \rightarrow \begin{bmatrix} 1 & 0 & 11 & \vdots & 23 \\ 0 & 1 & 4 & \vdots & 7 \\ 0 & 0 & 1 & \vdots & 2 \end{bmatrix}$$

Perform operations on R_1 so second column has a zero above its leading 1.

$$\begin{matrix} -11R_3 + R_1 \rightarrow \\ -4R_3 + R_2 \rightarrow \end{matrix} \begin{bmatrix} 1 & 0 & 0 & \vdots & 1 \\ 0 & 1 & 0 & \vdots & -1 \\ 0 & 0 & 1 & \vdots & 2 \end{bmatrix}$$

Perform operations on R_1 and R_2 so third column has zeros above its leading 1.

The matrix is now in reduced row-echelon form. Converting back to a system of linear equations, you have

$$\begin{cases} x = 1 \\ y = -1 \\ z = 2 \end{cases}$$

which is the same solution that was obtained using Gaussian elimination.

✓CHECKPOINT   Now try Exercise 55.

The beauty of Gauss-Jordan elimination is that, from the reduced row-echelon form, you can simply read the solution without the need for back-substitution.

TECHNOLOGY SUPPORT

For a demonstration of a graphical approach to Gauss-Jordan elimination on a 2×3 matrix, see the Visualizing Row Operations Program, available for several models of graphing calculators at this textbook's *Online Study Center.*

The elimination procedures described in this section employ an algorithmic approach that is easily adapted to computer programs. However, the procedure makes no effort to avoid fractional coefficients. For instance, in the elimination procedure for the system

$$\begin{cases} 2x - 5y + 5z = 17 \\ 3x - 2y + 3z = 11 \\ -3x + 3y \quad\quad = -6 \end{cases}$$

you may be inclined to multiply the first row by $\frac{1}{2}$ to produce a leading 1, which will result in working with fractional coefficients. For hand computations, you can sometimes avoid fractions by judiciously choosing the order in which you apply elementary row operations.

Example 9   A System with an Infinite Number of Solutions

Solve the system $\begin{cases} 2x + 4y - 2z = 0 \\ 3x + 5y \quad\quad = 1 \end{cases}$.

Solution

$$\begin{bmatrix} 2 & 4 & -2 & \vdots & 0 \\ 3 & 5 & 0 & \vdots & 1 \end{bmatrix} \quad \frac{1}{2}R_1 \rightarrow \begin{bmatrix} 1 & 2 & -1 & \vdots & 0 \\ 3 & 5 & 0 & \vdots & 1 \end{bmatrix}$$

$$-3R_1 + R_2 \rightarrow \begin{bmatrix} 1 & 2 & -1 & \vdots & 0 \\ 0 & -1 & 3 & \vdots & 1 \end{bmatrix}$$

$$-R_2 \rightarrow \begin{bmatrix} 1 & 2 & -1 & \vdots & 0 \\ 0 & 1 & -3 & \vdots & -1 \end{bmatrix}$$

$$-2R_2 + R_1 \rightarrow \begin{bmatrix} 1 & 0 & 5 & \vdots & 2 \\ 0 & 1 & -3 & \vdots & -1 \end{bmatrix}$$

The corresponding system of equations is

$$\begin{cases} x + 5z = 2 \\ y - 3z = -1 \end{cases}.$$

Solving for x and y in terms of z, you have $x = -5z + 2$ and $y = 3z - 1$. To write a solution of the system that does not use any of the three variables of the system, let a represent any real number and let $z = a$. Now substitute a for z in the equations for x and y.

$$x = -5z + 2 = -5a + 2$$

$$y = 3z - 1 = 3a - 1$$

So, the solution set has the form

$$(-5a + 2, 3a - 1, a).$$

Recall from Section 5.3 that a solution set of this form represents an infinite number of solutions. Try substituting values for a to obtain a few solutions. Then check each solution in the original system of equations.

✓CHECKPOINT   Now try Exercise 57.

Example 10   Analysis of a Network

Set up a system of linear equations representing the network shown in Figure 7.26. In a network, it is assumed that the total flow into a junction (blue circle) is equal to the total flow out of the junction.

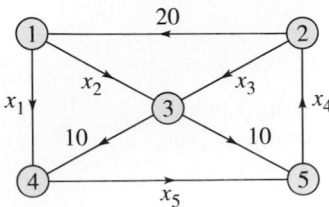

Figure 7.26

Solution

Because Junction 1 in Figure 7.26 has 20 units flowing into it, there must be 20 units flowing out of it. This is represented by the linear equation $x_1 + x_2 = 20$. Because Junction 2 has 20 units flowing out of it, there must be 20 units flowing into it. This is represented by $x_4 - x_3 = 20$ or $-x_3 + x_4 = 20$. A linear equation can be written for each of the network's five junctions, so the network is modeled by the following system.

$$
\begin{cases}
x_1 + x_2 & = 20 \qquad &\text{Junction 1}\\
-x_3 + x_4 & = 20 \qquad &\text{Junction 2}\\
x_2 + x_3 & = 20 \qquad &\text{Junction 3}\\
-x_1 \qquad + x_5 & = 10 \qquad &\text{Junction 4}\\
x_4 - x_5 & = 10 \qquad &\text{Junction 5}
\end{cases}
$$

Using Gauss-Jordan elimination on the augmented matrix produces the matrix in reduced row-echelon form.

Augmented Matrix

$$
\left[
\begin{array}{ccccc:c}
1 & 1 & 0 & 0 & 0 & 20\\
0 & 0 & -1 & 1 & 0 & 20\\
0 & 1 & 1 & 0 & 0 & 20\\
-1 & 0 & 0 & 0 & 1 & 10\\
0 & 0 & 0 & 1 & -1 & 10
\end{array}
\right]
$$

Matrix in Reduced Row-Echelon Form

$$
\left[
\begin{array}{ccccc:c}
1 & 0 & 0 & 0 & -1 & -10\\
0 & 1 & 0 & 0 & 1 & 30\\
0 & 0 & 1 & 0 & -1 & -10\\
0 & 0 & 0 & 1 & -1 & 10\\
0 & 0 & 0 & 0 & 0 & 0
\end{array}
\right]
$$

Letting $x_5 = t$, where t is a real number, you have $x_1 = t - 10$, $x_2 = -t + 30$, $x_3 = t - 10$, and $x_4 = t + 10$. So, this system has an infinite number of solutions.

✓CHECKPOINT   Now try Exercise 85.

7.4 Exercises

See www.CalcChat.com for worked-out solutions to odd-numbered exercises.

Vocabulary Check

Fill in the blanks.

1. A rectangular array of real numbers that can be used to solve a system of linear equations is called a _____ .

2. A matrix is _____ if the number of rows equals the number of columns.

3. A matrix with only one row is called a _____ and a matrix with only one column is called a _____ .

4. The matrix derived from a system of linear equations is called the _____ of the system.

5. The matrix derived from the coefficients of a system of linear equations is called the _____ of the system.

6. Two matrices are called _____ if one of the matrices can be obtained from the other by a sequence of elementary row operations.

7. A matrix in row-echelon form is in _____ if every column that has a leading 1 has zeros in every position above and below its leading 1.

8. The process of using row operations to write a matrix in reduced row-echelon form is called _____ .

In Exercises 1–6, determine the order of the matrix.

1. $\begin{bmatrix} 7 & 0 \end{bmatrix}$

2. $\begin{bmatrix} 6 & -3 & 10 & 8 \end{bmatrix}$

3. $\begin{bmatrix} 4 \\ 32 \\ 3 \end{bmatrix}$

4. $\begin{bmatrix} -3 & 7 & 15 & 0 \\ 0 & 0 & 3 & 3 \\ 1 & 1 & 6 & 7 \end{bmatrix}$

5. $\begin{bmatrix} 33 & 45 \\ -9 & 20 \end{bmatrix}$

6. $\begin{bmatrix} 3 & -1 & 4 \\ 6 & 0 & -5 \end{bmatrix}$

In Exercises 7–10, write the augmented matrix for the system of linear equations.

7. $\begin{cases} 6x - 7y = 11 \\ -2x + 5y = -1 \end{cases}$

8. $\begin{cases} 7x + 4y = 22 \\ 5x - 9y = 15 \end{cases}$

9. $\begin{cases} x + 10y - 2z = 2 \\ 5x - 3y + 4z = 0 \\ 2x + y = 6 \end{cases}$

10. $\begin{cases} x - 3y + z = 1 \\ 4y = 0 \\ 7z = -5 \end{cases}$

In Exercises 11–14, write the system of linear equations represented by the augmented matrix. (Use the variables x, y, z, and w, if applicable.)

11. $\begin{bmatrix} 3 & 4 & \vdots & 9 \\ 1 & -1 & \vdots & -3 \end{bmatrix}$

12. $\begin{bmatrix} 7 & -5 & \vdots & 0 \\ 8 & 3 & \vdots & -2 \end{bmatrix}$

13. $\begin{bmatrix} 9 & 12 & 3 & \vdots & 0 \\ -2 & 18 & 5 & \vdots & 10 \\ 1 & 7 & -8 & \vdots & -4 \end{bmatrix}$

14. $\begin{bmatrix} 6 & 2 & -1 & -5 & \vdots & -25 \\ -1 & 0 & 7 & 3 & \vdots & 7 \\ 4 & -1 & -10 & 6 & \vdots & 23 \\ 0 & 8 & 1 & -11 & \vdots & -21 \end{bmatrix}$

In Exercises 15–18, fill in the blanks using elementary row operations to form a row-equivalent matrix.

15. $\begin{bmatrix} 1 & 4 & 3 \\ 2 & 10 & 5 \end{bmatrix}$

$\begin{bmatrix} 1 & 4 & 3 \\ 0 & & -1 \end{bmatrix}$

16. $\begin{bmatrix} 3 & 6 & 8 \\ 4 & -3 & 6 \end{bmatrix}$

$\begin{bmatrix} 1 & & \frac{8}{3} \\ 4 & -3 & 6 \end{bmatrix}$

17. $\begin{bmatrix} 1 & 1 & 4 & -1 \\ 3 & 8 & 10 & 3 \\ -2 & 1 & 12 & 6 \end{bmatrix}$

$\begin{bmatrix} 1 & 1 & 4 & -1 \\ 0 & 5 & & \\ 0 & 3 & & \end{bmatrix}$

$\begin{bmatrix} 1 & 1 & 4 & -1 \\ 0 & 1 & -\frac{2}{5} & \frac{6}{5} \\ 0 & 3 & & \end{bmatrix}$

18. $\begin{bmatrix} 2 & 4 & 8 & 3 \\ 1 & -1 & -3 & 2 \\ 2 & 6 & 4 & 9 \end{bmatrix}$

$\begin{bmatrix} 1 & & & \\ 1 & -1 & -3 & 2 \\ 2 & 6 & 4 & 9 \end{bmatrix}$

$\begin{bmatrix} 1 & 2 & 4 & \frac{3}{2} \\ 0 & & -7 & \frac{1}{2} \\ 0 & 2 & & \end{bmatrix}$

In Exercises 19–22, identify the elementary row operation performed to obtain the new row-equivalent matrix.

Original Matrix	New Row-Equivalent Matrix

19. $\begin{bmatrix} -3 & 6 & 0 \\ 5 & 2 & -2 \end{bmatrix}$ $\begin{bmatrix} -18 & 0 & 6 \\ 5 & 2 & -2 \end{bmatrix}$

20. $\begin{bmatrix} 3 & -1 & -4 \\ -4 & 3 & 7 \end{bmatrix}$ $\begin{bmatrix} 3 & -1 & -4 \\ 5 & 0 & -5 \end{bmatrix}$

21. $\begin{bmatrix} 0 & -1 & -5 & 5 \\ -1 & 3 & -7 & 6 \\ 4 & -5 & 1 & 3 \end{bmatrix}$ $\begin{bmatrix} -1 & 3 & -7 & 6 \\ 0 & -1 & -5 & 5 \\ 4 & -5 & 1 & 3 \end{bmatrix}$

Original Matrix   *New Row-Equivalent Matrix*

22. $\begin{bmatrix} -1 & -2 & 3 & -2 \\ 2 & -5 & 1 & -7 \\ 5 & 4 & -7 & 6 \end{bmatrix}$ $\begin{bmatrix} -1 & -2 & 3 & -2 \\ 2 & -5 & 1 & -7 \\ 0 & -6 & 8 & -4 \end{bmatrix}$

In Exercises 23–28, determine whether the matrix is in row-echelon form. If it is, determine if it is also in reduced row-echelon form.

23. $\begin{bmatrix} 1 & 0 & 0 & 0 \\ 0 & 1 & 1 & 5 \\ 0 & 0 & 0 & 0 \end{bmatrix}$ **24.** $\begin{bmatrix} 1 & 3 & 0 & 0 \\ 0 & 0 & 1 & 8 \\ 0 & 0 & 0 & 0 \end{bmatrix}$

25. $\begin{bmatrix} 3 & 0 & 3 & 7 \\ 0 & -2 & 0 & 4 \\ 0 & 0 & 1 & 5 \end{bmatrix}$ **26.** $\begin{bmatrix} 1 & 0 & 2 & 1 \\ 0 & 1 & -3 & 10 \\ 0 & 0 & 1 & 0 \end{bmatrix}$

27. $\begin{bmatrix} 1 & 0 & 0 & 1 \\ 0 & 1 & 0 & -1 \\ 0 & 0 & 0 & 2 \end{bmatrix}$ **28.** $\begin{bmatrix} 1 & 0 & 1 & 0 \\ 0 & 1 & 0 & 2 \\ 0 & 0 & 1 & 0 \end{bmatrix}$

29. Perform the sequence of row operations on the matrix. What did the operations accomplish?

$$\begin{bmatrix} 1 & 2 & 3 \\ 2 & -1 & -4 \\ 3 & 1 & -1 \end{bmatrix}$$

   (a) Add -2 times R_1 to R_2.

   (b) Add -3 times R_1 to R_3.

   (c) Add -1 times R_2 to R_3.

   (d) Multiply R_2 by $-\frac{1}{5}$.

   (e) Add -2 times R_2 to R_1.

30. Perform the sequence of row operations on the matrix. What did the operations accomplish?

$$\begin{bmatrix} 7 & 1 \\ 0 & 2 \\ -3 & 4 \\ 4 & 1 \end{bmatrix}$$

   (a) Add R_3 to R_4.

   (b) Interchange R_1 and R_4.

   (c) Add 3 times R_1 to R_3.

   (d) Add -7 times R_1 to R_4.

   (e) Multiply R_2 by $\frac{1}{2}$.

   (f) Add the appropriate multiples of R_2 to R_1, R_3, and R_4.

31. Repeat steps (a) through (e) in Exercise 29 using a graphing utility.

32. Repeat steps (a) through (f) in Exercise 30 using a graphing utility.

In Exercises 33–36, write the matrix in row-echelon form. Remember that the row-echelon form of a matrix is not unique.

33. $\begin{bmatrix} 1 & 2 & 3 & 0 \\ -1 & 4 & 0 & -5 \\ 2 & 6 & 3 & 10 \end{bmatrix}$

34. $\begin{bmatrix} 1 & 2 & -1 & 3 \\ 3 & 7 & -5 & 14 \\ -2 & -1 & -3 & 8 \end{bmatrix}$

35. $\begin{bmatrix} 1 & -1 & -1 & 1 \\ 5 & -4 & 1 & 8 \\ -6 & 8 & 18 & 0 \end{bmatrix}$

36. $\begin{bmatrix} 1 & -3 & 0 & -7 \\ -3 & 10 & 1 & 23 \\ 4 & -10 & 2 & -24 \end{bmatrix}$

In Exercises 37–40, use the matrix capabilities of a graphing utility to write the matrix in reduced row-echelon form.

37. $\begin{bmatrix} 3 & 3 & 3 \\ -1 & 0 & -4 \\ 2 & 4 & -2 \end{bmatrix}$ **38.** $\begin{bmatrix} 1 & 3 & 2 \\ 5 & 15 & 9 \\ 2 & 6 & 10 \end{bmatrix}$

39. $\begin{bmatrix} -4 & 1 & 0 & 6 \\ 1 & -2 & 3 & -4 \end{bmatrix}$

40. $\begin{bmatrix} 5 & 1 & 2 & 4 \\ -1 & 5 & 10 & -32 \end{bmatrix}$

In Exercises 41–44, write the system of linear equations represented by the augmented matrix. Then use back-substitution to find the solution. (Use the variables x, y, and z, if applicable.)

41. $\begin{bmatrix} 1 & -2 & \vdots & 4 \\ 0 & 1 & \vdots & -3 \end{bmatrix}$

42. $\begin{bmatrix} 1 & 8 & \vdots & 12 \\ 0 & 1 & \vdots & 3 \end{bmatrix}$

43. $\begin{bmatrix} 1 & -1 & 2 & \vdots & 4 \\ 0 & 1 & -1 & \vdots & 2 \\ 0 & 0 & 1 & \vdots & -2 \end{bmatrix}$

44. $\begin{bmatrix} 1 & 2 & -2 & \vdots & -1 \\ 0 & 1 & 1 & \vdots & 9 \\ 0 & 0 & 1 & \vdots & -3 \end{bmatrix}$

In Exercises 45–48, an augmented matrix that represents a system of linear equations (in the variables x and y or x, y, and z) has been reduced using Gauss-Jordan elimination. Write the solution represented by the augmented matrix.

45. $\begin{bmatrix} 1 & 0 & \vdots & 7 \\ 0 & 1 & \vdots & -5 \end{bmatrix}$ **46.** $\begin{bmatrix} 1 & 0 & \vdots & -2 \\ 0 & 1 & \vdots & 4 \end{bmatrix}$

47. $\begin{bmatrix} 1 & 0 & 0 & \vdots & -4 \\ 0 & 1 & 0 & \vdots & -8 \\ 0 & 0 & 1 & \vdots & 2 \end{bmatrix}$

48. $\begin{bmatrix} 1 & 0 & 0 & \vdots & 3 \\ 0 & 1 & 0 & \vdots & -1 \\ 0 & 0 & 1 & \vdots & 0 \end{bmatrix}$

In Exercises 49–54, use matrices to solve the system of equations, if possible. Use Gaussian elimination with back-substitution.

49. $\begin{cases} x + 2y = 7 \\ 2x + y = 8 \end{cases}$

50. $\begin{cases} 2x + 6y = 16 \\ 2x + 3y = 7 \end{cases}$

51. $\begin{cases} -x + y = -22 \\ 3x + 4y = 4 \\ 4x - 8y = 32 \end{cases}$

52. $\begin{cases} x + 2y = 0 \\ x + y = 6 \\ 3x - 2y = 8 \end{cases}$

53. $\begin{cases} 3x + 2y - z + w = 0 \\ x - y + 4z + 2w = 25 \\ -2x + y + 2z - w = 2 \\ x + y + z + w = 6 \end{cases}$

54. $\begin{cases} x - 4y + 3z - 2w = 9 \\ 3x - 2y + z - 4w = -13 \\ -4x + 3y - 2z + w = -4 \\ -2x + y - 4z + 3w = -10 \end{cases}$

In Exercises 55–60, use matrices to solve the system of equations, if possible. Use Gauss-Jordan elimination.

55. $\begin{cases} x - 3z = -2 \\ 3x + y - 2z = 5 \\ 2x + 2y + z = 4 \end{cases}$

56. $\begin{cases} 2x - y + 3z = 24 \\ 2y - z = 14 \\ 7x - 5y = 6 \end{cases}$

57. $\begin{cases} x + y - 5z = 3 \\ x - 2z = 1 \\ 2x - y - z = 0 \end{cases}$

58. $\begin{cases} 2x + 3z = 3 \\ 4x - 3y + 7z = 5 \\ 8x - 9y + 15z = 9 \end{cases}$

59. $\begin{cases} -x + y - z = -14 \\ 2x - y + z = 21 \\ 3x + 2y + z = 19 \end{cases}$

60. $\begin{cases} 2x + 2y - z = 2 \\ x - 3y + z = 28 \\ -x + y = 14 \end{cases}$

In Exercises 61–64, use the matrix capabilities of a graphing utility to reduce the augmented matrix corresponding to the system of equations, and solve the system.

61. $\begin{cases} 3x + 3y + 12z = 6 \\ x + y + 4z = 2 \\ 2x + 5y + 20z = 10 \\ -x + 2y + 8z = 4 \end{cases}$

62. $\begin{cases} x + y + z = 0 \\ 2x + 3y + z = 0 \\ 3x + 5y + z = 0 \end{cases}$

63. $\begin{cases} 2x + 10y + 2z = 6 \\ x + 5y + 2z = 6 \\ x + 5y + z = 3 \\ -3x - 15y - 3z = -9 \end{cases}$

64. $\begin{cases} 2x + y - z + 2w = -6 \\ 3x + 4y + w = 1 \\ x + 5y + 2z + 6w = -3 \\ 5x + 2y - z - w = 3 \end{cases}$

In Exercises 65–68, determine whether the two systems of linear equations yield the same solution. If so, find the solution.

65. (a) $\begin{cases} x - 2y + z = -6 \\ y - 5z = 16 \\ z = -3 \end{cases}$

(b) $\begin{cases} x + y - 2z = 6 \\ y + 3z = -8 \\ z = -3 \end{cases}$

66. (a) $\begin{cases} x - 3y + 4z = -11 \\ y - z = -4 \\ z = 2 \end{cases}$

(b) $\begin{cases} x + 4y = -11 \\ y + 3z = 4 \\ z = 2 \end{cases}$

67. (a) $\begin{cases} x - 4y + 5z = 27 \\ y - 7z = -54 \\ z = 8 \end{cases}$

(b) $\begin{cases} x - 6y + z = 15 \\ y + 5z = 42 \\ z = 8 \end{cases}$

68. (a) $\begin{cases} x + 3y - z = 19 \\ y + 6z = -18 \\ z = -4 \end{cases}$

(b) $\begin{cases} x - y + 3z = -15 \\ y - 2z = 14 \\ z = -4 \end{cases}$

In Exercises 69–72, use a system of equations to find the equation of the parabola $y = ax^2 + bx + c$ that passes through the points. Solve the system using matrices. Use a graphing utility to verify your result.

69.

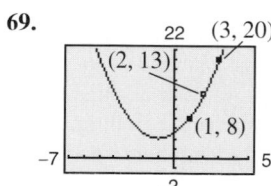

70.

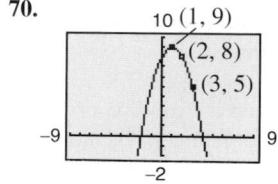

71.

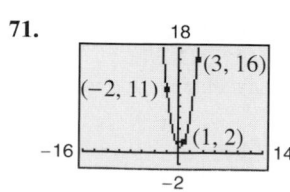

72.

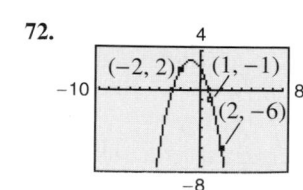

In Exercises 73 and 74, use a system of equations to find the quadratic function $f(x) = ax^2 + bx + c$ that satisfies the equations. Solve the system using matrices.

73. $f(-2) = -15$
$f(-1) = 7$
$f(1) = -3$

74. $f(-2) = -3$
$f(1) = -3$
$f(2) = -11$

In Exercises 75 and 76, use a system of equations to find the cubic function $f(x) = ax^3 + bx^2 + cx + d$ that satisfies the equations. Solve the system using matrices.

75. $f(-2) = -7$
$f(-1) = 2$
$f(1) = -4$
$f(2) = -7$

76. $f(-2) = -17$
$f(-1) = -5$
$f(1) = 1$
$f(2) = 7$

77. Borrowing Money A small corporation borrowed $1,500,000 to expand its line of shoes. Some of the money was borrowed at 7%, some at 8%, and some at 10%. Use a system of equations to determine how much was borrowed at each rate if the annual interest was $130,500 and the amount borrowed at 10% was four times the amount borrowed at 7%. Solve the system using matrices.

78. Borrowing Money A small corporation borrowed $500,000 to build a new office building. Some of the money was borrowed at 9%, some at 10%, and some at 12%. Use a system of equations to determine how much was borrowed at each rate if the annual interest was $52,000 and the amount borrowed at 10% was $2\frac{1}{2}$ times the amount borrowed at 9%. Solve the system using matrices.

79. Electrical Network The currents in an electrical network are given by the solution of the system

$$\begin{cases} I_1 - I_2 + I_3 = 0 \\ 2I_1 + 2I_2 \qquad = 7 \\ \qquad 2I_2 + 4I_3 = 8 \end{cases}$$

where I_1, I_2, and I_3 are measured in amperes. Solve the system of equations using matrices.

80. Mathematical Modeling A videotape of the path of a ball thrown by a baseball player was analyzed with a grid covering the TV screen. The tape was paused three times, and the position of the ball was measured each time. The coordinates obtained are shown in the table (x and y are measured in feet).

Horizontal distance, x	Height, y
0	5.0
15	9.6
30	12.4

(a) Use a system of equations to find the equation of the parabola $y = ax^2 + bx + c$ that passes through the points. Solve the system using matrices.

(b) Use a graphing utility to graph the parabola.

(c) Graphically approximate the maximum height of the ball and the point at which the ball strikes the ground.

(d) Algebraically approximate the maximum height of the ball and the point at which the ball strikes the ground.

81. Data Analysis The table shows the average retail prices y (in dollars) of prescriptions from 2002 to 2004. (Source: National Association of Chain Drug Stores)

Year	Price, y (in dollars)
2002	55.37
2003	59.52
2004	63.59

(a) Use a system of equations to find the equation of the parabola $y = at^2 + bt + c$ that passes through the points. Let t represent the year, with $t = 2$ corresponding to 2002. Solve the system using matrices.

(b) Use a graphing utility to graph the parabola and plot the data points.

(c) Use the equation in part (a) to estimate the average retail prices in 2005, 2010, and 2015.

(d) Are your estimates in part (c) reasonable? Explain.

82. Data Analysis The table shows the average annual salaries y (in thousands of dollars) for public school classroom teachers in the United States from 2002 to 2004. (Source: Educational Research Service)

Year	Annual salary, y (in thousands of dollars)
2002	43.8
2003	45.0
2005	45.6

(a) Use a system of equations to find the equation of the parabola $y = at^2 + bt + c$ that passes through the points. Let t represent the year, with $t = 2$ corresponding to 2002. Solve the system using matrices.

(b) Use a graphing utility to graph the parabola and plot the data points.

(c) Use the equation in part (a) to estimate the average annual salaries in 2005, 2010, and 2015.

(d) Are your estimates in part (c) reasonable? Explain.

83. *Paper*   A wholesale paper company sells a 100-pound package of computer paper that consists of three grades, glossy, semi-gloss, and matte, for printing photographs. Glossy costs $5.50 per pound, semi-gloss costs $4.25 per pound, and matte costs $3.75 per pound. One half of the 100-pound package consists of the two less expensive grades. The cost of the 100-pound package is $480. Set up and solve a system of equations, using matrices, to find the number of pounds of each grade of paper in a 100-pound package.

84. *Tickets*   The theater department of a high school has collected the receipts for a production of *Phantom of the Opera*, which total $1030. The ticket prices were $3.50 for students, $5.00 for adults, and $2.50 for children under 12 years of age. Twice as many adults attended as children, and the number of students that attended was 20 more than one-half the number of adults. Set up and solve a system of equations, using matrices, to find the numbers of tickets sold to adults, students, and children.

Network Analysis   **In Exercises 85 and 86, answer the questions about the specified network.**

85.   Water flowing through a network of pipes (in thousands of cubic meters per hour) is shown below.

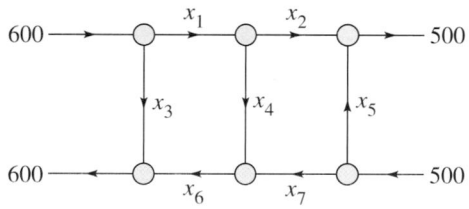

(a) Use matrices to solve this system for the water flow represented by x_i, $i = 1, 2, 3, 4, 5, 6,$ and 7.

(b) Find the network flow pattern when $x_6 = 0$ and $x_7 = 0$.

(c) Find the network flow pattern when $x_5 = 1000$ and $x_6 = 0$.

86.   The flow of traffic (in vehicles per hour) through a network of streets is shown below.

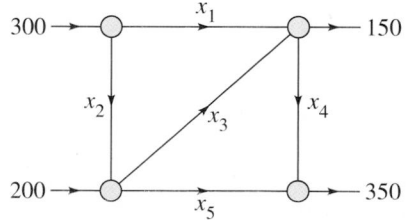

(a) Use matrices to solve this system for the traffic flow represented by x_i, $i = 1, 2, 3, 4,$ and 5.

(b) Find the traffic flow when $x_2 = 200$ and $x_3 = 50$.

(c) Find the traffic flow when $x_2 = 150$ and $x_3 = 0$.

Synthesis

True or False?   **In Exercises 87 and 88, determine whether the statement is true or false. Justify your answer.**

87. $\begin{bmatrix} 6 & 0 & -3 & 10 \\ -2 & 5 & -6 & 2 \end{bmatrix}$ is a 4×2 matrix.

88.   Gaussian elimination reduces a matrix until a reduced row-echelon form is obtained.

89. *Think About It*   The augmented matrix represents a system of linear equations (in the variables x, y, and z) that has been reduced using Gauss-Jordan elimination. Write a system of equations with *nonzero* coefficients that is represented by the reduced matrix. (There are many correct answers.)

$$\begin{bmatrix} 1 & 0 & 3 & \vdots & -2 \\ 0 & 1 & 4 & \vdots & 1 \\ 0 & 0 & 0 & \vdots & 0 \end{bmatrix}$$

90. *Think About It*

(a) Describe the row-echelon form of an augmented matrix that corresponds to a system of linear equations that is inconsistent.

(b) Describe the row-echelon form of an augmented matrix that corresponds to a system of linear equations that has an infinite number of solutions.

91. *Error Analysis*   One of your classmates has submitted the following steps for the solution of a system by Gauss-Jordan elimination. Find the error(s) in the solution. Write a short paragraph explaining the error(s) to your classmate.

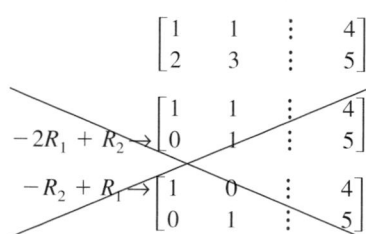

92. *Writing*   In your own words, describe the difference between a matrix in row-echelon form and a matrix in reduced row-echelon form.

Skills Review

In Exercises 93–96, sketch the graph of the function. Identify any asymptotes.

93. $f(x) = \dfrac{7}{-x - 1}$

94. $f(x) = \dfrac{4x}{5x^2 + 2}$

95. $f(x) = \dfrac{x^2 - 2x - 3}{x - 4}$

96. $f(x) = \dfrac{x^2 - 36}{x + 1}$

7.5 Operations with Matrices

Equality of Matrices

In Section 7.4, you used matrices to solve systems of linear equations. There is a rich mathematical theory of matrices, and its applications are numerous. This section and the next two introduce some fundamentals of matrix theory. It is standard mathematical convention to represent matrices in any of the following three ways.

Representation of Matrices

1. A matrix can be denoted by an uppercase letter such as A, B, or C.

2. A matrix can be denoted by a representative element enclosed in brackets, such as $[a_{ij}]$, $[b_{ij}]$, or $[c_{ij}]$.

3. A matrix can be denoted by a rectangular array of numbers such as

$$A = [a_{ij}] = \begin{bmatrix} a_{11} & a_{12} & a_{13} & \cdots & a_{1n} \\ a_{21} & a_{22} & a_{23} & \cdots & a_{2n} \\ a_{31} & a_{32} & a_{33} & \cdots & a_{3n} \\ \vdots & \vdots & \vdots & & \vdots \\ a_{m1} & a_{m2} & a_{m3} & \cdots & a_{mn} \end{bmatrix}.$$

Two matrices $A = [a_{ij}]$ and $B = [b_{ij}]$ are **equal** if they have the same order $(m \times n)$ and all of their corresponding entries are equal. For instance, using the matrix equation

$$\begin{bmatrix} a_{11} & a_{12} \\ a_{21} & a_{22} \end{bmatrix} = \begin{bmatrix} 2 & -1 \\ -3 & 0 \end{bmatrix}$$

you can conclude that $a_{11} = 2$, $a_{12} = -1$, $a_{21} = -3$, and $a_{22} = 0$.

Matrix Addition and Scalar Multiplication

You can add two matrices (of the same order) by adding their corresponding entries.

Definition of Matrix Addition

If $A = [a_{ij}]$ and $B = [b_{ij}]$ are matrices of order $m \times n$, their sum is the $m \times n$ matrix given by

$$A + B = [a_{ij} + b_{ij}].$$

The sum of two matrices of different orders is undefined.

Example 1 Addition of Matrices

a. $\begin{bmatrix} -1 & 2 \\ 0 & 1 \end{bmatrix} + \begin{bmatrix} 1 & 3 \\ -1 & 2 \end{bmatrix} = \begin{bmatrix} -1 + 1 & 2 + 3 \\ 0 + (-1) & 1 + 2 \end{bmatrix} = \begin{bmatrix} 0 & 5 \\ -1 & 3 \end{bmatrix}$

b. $\begin{bmatrix} 1 \\ -3 \\ -2 \end{bmatrix} + \begin{bmatrix} -1 \\ 3 \\ 2 \end{bmatrix} = \begin{bmatrix} 0 \\ 0 \\ 0 \end{bmatrix}$

c. The sum of

$$A = \begin{bmatrix} 2 & 1 & 0 \\ 4 & 0 & -1 \end{bmatrix} \quad \text{and} \quad B = \begin{bmatrix} 0 & 1 \\ -1 & 3 \end{bmatrix}$$

is undefined because A is of order 2×3 and B is of order 2×2.

✓**CHECKPOINT**   Now try Exercise 7(a).

> **TECHNOLOGY TIP**
>
> Try using a graphing utility to find the sum of the two matrices in Example 1(c). Your graphing utility should display an error message similar to the one shown below.
>
>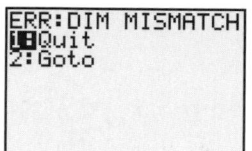

Most graphing utilities can perform matrix operations. Example 2 shows how a graphing utility can be used to add two matrices.

Example 2 Addition of Matrices

Use a graphing utility to find the sum of

$$A = \begin{bmatrix} 0.5 & 1.3 & -2.6 \\ 1.1 & 2.3 & 3.4 \end{bmatrix} \quad \text{and} \quad B = \begin{bmatrix} -3.2 & 4.8 & 9.6 \\ -4.5 & 3.2 & -1.7 \end{bmatrix}.$$

Solution

Use the *matrix editor* to enter A and B in the graphing utility (see Figure 7.27). Then, find the sum, as shown in Figure 7.28.

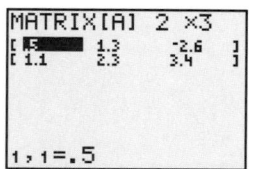
Matrix A
Figure 7.27

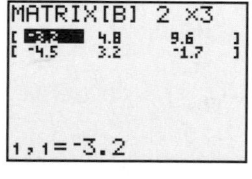
Matrix B

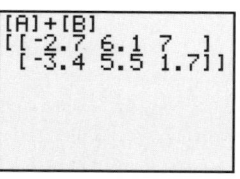
A + B
Figure 7.28

> **TECHNOLOGY SUPPORT**
>
> For instructions on how to use the *matrix editor*, see Appendix A; for specific keystrokes, go to this textbook's *Online Study Center.*

✓**CHECKPOINT**   Now try Exercise 17.

In operations with matrices, numbers are usually referred to as **scalars.** In this text, scalars will always be real numbers. You can multiply a matrix A by a scalar c by multiplying each entry in A by c.

Definition of Scalar Multiplication

If $A = [a_{ij}]$ is an $m \times n$ matrix and c is a scalar, the **scalar multiple** of A by c is the $m \times n$ matrix given by

$$cA = [ca_{ij}].$$

The symbol $-A$ represents the negation of A, which is the scalar product $(-1)A$. Moreover, if A and B are of the same order, then $A - B$ represents the sum of A and $(-1)B$. That is,

$$A - B = A + (-1)B.$$ Subtraction of matrices

Example 3   Scalar Multiplication and Matrix Subtraction

For the following matrices, find (a) $3A$, (b) $-B$, and (c) $3A - B$.

$$A = \begin{bmatrix} 2 & 2 & 4 \\ -3 & 0 & -1 \\ 2 & 1 & 2 \end{bmatrix} \quad \text{and} \quad B = \begin{bmatrix} 2 & 0 & 0 \\ 1 & -4 & 3 \\ -1 & 3 & 2 \end{bmatrix}$$

Solution

a. $3A = 3\begin{bmatrix} 2 & 2 & 4 \\ -3 & 0 & -1 \\ 2 & 1 & 2 \end{bmatrix}$ Scalar multiplication

$$= \begin{bmatrix} 3(2) & 3(2) & 3(4) \\ 3(-3) & 3(0) & 3(-1) \\ 3(2) & 3(1) & 3(2) \end{bmatrix}$$ Multiply each entry by 3.

$$= \begin{bmatrix} 6 & 6 & 12 \\ -9 & 0 & -3 \\ 6 & 3 & 6 \end{bmatrix}$$ Simplify.

b. $-B = (-1)\begin{bmatrix} 2 & 0 & 0 \\ 1 & -4 & 3 \\ -1 & 3 & 2 \end{bmatrix}$ Definition of negation

$$= \begin{bmatrix} -2 & 0 & 0 \\ -1 & 4 & -3 \\ 1 & -3 & -2 \end{bmatrix}$$ Multiply each entry by -1.

c. $3A - B = \begin{bmatrix} 6 & 6 & 12 \\ -9 & 0 & -3 \\ 6 & 3 & 6 \end{bmatrix} - \begin{bmatrix} 2 & 0 & 0 \\ 1 & -4 & 3 \\ -1 & 3 & 2 \end{bmatrix}$ Matrix subtraction

$$= \begin{bmatrix} 4 & 6 & 12 \\ -10 & 4 & -6 \\ 7 & 0 & 4 \end{bmatrix}$$ Subtract corresponding entries.

✓CHECKPOINT   Now try Exercise 7(b), (c), and (d).

It is often convenient to rewrite the scalar multiple cA by factoring c out of every entry in the matrix. For instance, in the following example, the scalar $\frac{1}{2}$ has been factored out of the matrix.

$$\begin{bmatrix} \frac{1}{2} & -\frac{3}{2} \\ \frac{5}{2} & \frac{1}{2} \end{bmatrix} = \begin{bmatrix} \frac{1}{2}(1) & \frac{1}{2}(-3) \\ \frac{1}{2}(5) & \frac{1}{2}(1) \end{bmatrix} = \frac{1}{2}\begin{bmatrix} 1 & -3 \\ 5 & 1 \end{bmatrix}$$

Example 4 Scalar Multiplication and Matrix Subtraction

For the following matrices, use a graphing utility to find $\frac{1}{2}A - \frac{1}{4}B$.

$$A = \begin{bmatrix} -1 & 8 \\ 6 & 2 \end{bmatrix} \quad \text{and} \quad B = \begin{bmatrix} 0 & 4 \\ 5 & -3 \end{bmatrix}$$

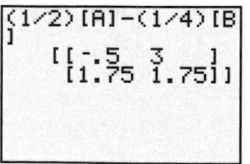

Solution

Use the *matrix editor* to enter A and B into the graphing utility. Then, find $\frac{1}{2}A - \frac{1}{4}B$, as shown in Figure 7.29

Figure 7.29

✓ CHECKPOINT   Now try Exercise 19.

The properties of matrix addition and scalar multiplication are similar to those of addition and multiplication of real numbers. One important property of addition of real numbers is that the number 0 is the additive identity. That is, $c + 0 = c$ for any real number c. For matrices, a similar property holds. That is, if A is an $m \times n$ matrix and O is the $m \times n$ **zero matrix** consisting entirely of zeros, then $A + O = A$.

In other words, O is the **additive identity** for the set of all $m \times n$ matrices. For example, the following matrices are the additive identities for the sets of all 2×3 and 2×2 matrices.

$$O = \begin{bmatrix} 0 & 0 & 0 \\ 0 & 0 & 0 \end{bmatrix} \quad \text{and} \quad O = \begin{bmatrix} 0 & 0 \\ 0 & 0 \end{bmatrix}$$

2×3 zero matrix 2×2 zero matrix

Properties of Matrix Addition and Scalar Multiplication

Let A, B, and C be $m \times n$ matrices and let c and d be scalars.

1. $A + B = B + A$ Commutative Property of Matrix Addition
2. $A + (B + C) = (A + B) + C$ Associative Property of Matrix Addition
3. $(cd)A = c(dA)$ Associative Property of Scalar Multiplication
4. $1A = A$ Scalar Identity
5. $A + O = A$ Additive Identity
6. $c(A + B) = cA + cB$ Distributive Property
7. $(c + d)A = cA + dA$ Distributive Property

STUDY TIP

Note that the Associative Property of Matrix Addition allows you to write expressions such as $A + B + C$ without ambiguity because the same sum occurs no matter how the matrices are grouped. This same reasoning applies to sums of four or more matrices.

Example 5 Addition of More than Two Matrices

By adding corresponding entries, you obtain the following sum of four matrices.

$$\begin{bmatrix} 1 \\ 2 \\ -3 \end{bmatrix} + \begin{bmatrix} -1 \\ -1 \\ 2 \end{bmatrix} + \begin{bmatrix} 0 \\ 1 \\ 4 \end{bmatrix} + \begin{bmatrix} 2 \\ -3 \\ -2 \end{bmatrix} = \begin{bmatrix} 2 \\ -1 \\ 1 \end{bmatrix}$$

✓ CHECKPOINT   Now try Exercise 13.

Example 6   Using the Distributive Property

$$3\left(\begin{bmatrix} -2 & 0 \\ 4 & 1 \end{bmatrix} + \begin{bmatrix} 4 & -2 \\ 3 & 7 \end{bmatrix}\right) = 3\begin{bmatrix} -2 & 0 \\ 4 & 1 \end{bmatrix} + 3\begin{bmatrix} 4 & -2 \\ 3 & 7 \end{bmatrix}$$

$$= \begin{bmatrix} -6 & 0 \\ 12 & 3 \end{bmatrix} + \begin{bmatrix} 12 & -6 \\ 9 & 21 \end{bmatrix} = \begin{bmatrix} 6 & -6 \\ 21 & 24 \end{bmatrix}$$

✓ CHECKPOINT   Now try Exercise 15.

> **STUDY TIP**
>
> In Example 6, you could add the two matrices first and then multiply the resulting matrix by 3. The result would be the same.

The algebra of real numbers and the algebra of matrices have many similarities. For example, compare the following solutions.

Real Numbers (Solve for x.)	$m \times n$ Matrices (Solve for X.)
$x + a = b$	$X + A = B$
$x + a + (-a) = b + (-a)$	$X + A + (-A) = B + (-A)$
$x + 0 = b - a$	$X + O = B - A$
$x = b - a$	$X = B - A$

The algebra of real numbers and the algebra of matrices also have important differences, which will be discussed later.

Example 7   Solving a Matrix Equation

Solve for X in the equation $3X + A = B$, where

$$A = \begin{bmatrix} 1 & -2 \\ 0 & 3 \end{bmatrix} \quad \text{and} \quad B = \begin{bmatrix} -3 & 4 \\ 2 & 1 \end{bmatrix}.$$

Solution

Begin by solving the equation for X to obtain

$$3X = B - A$$

$$X = \frac{1}{3}(B - A).$$

Now, using the matrices A and B, you have

$$X = \frac{1}{3}\left(\begin{bmatrix} -3 & 4 \\ 2 & 1 \end{bmatrix} - \begin{bmatrix} 1 & -2 \\ 0 & 3 \end{bmatrix}\right) \qquad \text{Substitute the matrices.}$$

$$= \frac{1}{3}\begin{bmatrix} -4 & 6 \\ 2 & -2 \end{bmatrix} \qquad \text{Subtract matrix } A \text{ from matrix } B.$$

$$= \begin{bmatrix} -\frac{4}{3} & 2 \\ \frac{2}{3} & -\frac{2}{3} \end{bmatrix}. \qquad \text{Multiply the resulting matrix by } \frac{1}{3}.$$

✓ CHECKPOINT   Now try Exercise 23.

Matrix Multiplication

Another basic matrix operation is **matrix multiplication.** At first glance, the following definition may seem unusual. You will see later, however, that this definition of the product of two matrices has many practical applications.

Definition of Matrix Multiplication

If $A = [a_{ij}]$ is an $m \times n$ matrix and $B = [b_{ij}]$ is an $n \times p$ matrix, the product AB is an $m \times p$ matrix given by

$$AB = [c_{ij}]$$

where $c_{ij} = a_{i1}b_{1j} + a_{i2}b_{2j} + a_{i3}b_{3j} + \cdots + a_{in}b_{nj}.$

The definition of matrix multiplication indicates a *row-by-column* multiplication, where the entry in the ith row and jth column of the product AB is obtained by multiplying the entries in the ith row of A by the corresponding entries in the jth column of B and then adding the results. The general pattern for matrix multiplication is as follows.

$$
\begin{bmatrix}
a_{11} & a_{12} & a_{13} & \cdots & a_{1n} \\
a_{21} & a_{22} & a_{23} & \cdots & a_{2n} \\
a_{31} & a_{32} & a_{33} & \cdots & a_{3n} \\
\vdots & \vdots & \vdots & & \vdots \\
a_{i1} & a_{i2} & a_{i3} & \cdots & a_{in} \\
\vdots & \vdots & \vdots & & \vdots \\
a_{m1} & a_{m2} & a_{m3} & \cdots & a_{mn}
\end{bmatrix}
\begin{bmatrix}
b_{11} & b_{12} & \cdots & b_{1j} & \cdots & b_{1p} \\
b_{21} & b_{22} & \cdots & b_{2j} & \cdots & b_{2p} \\
b_{31} & b_{32} & \cdots & b_{3j} & \cdots & b_{3p} \\
\vdots & \vdots & & \vdots & & \vdots \\
b_{n1} & b_{n2} & \cdots & b_{nj} & \cdots & b_{np}
\end{bmatrix}
=
\begin{bmatrix}
c_{11} & c_{12} & \cdots & c_{1j} & \cdots & c_{1p} \\
c_{21} & c_{22} & \cdots & c_{2j} & \cdots & c_{2p} \\
\vdots & \vdots & & \vdots & & \vdots \\
c_{i1} & c_{i2} & \cdots & c_{ij} & \cdots & c_{ip} \\
\vdots & \vdots & & \vdots & & \vdots \\
c_{m1} & c_{m2} & \cdots & c_{mj} & \cdots & c_{mp}
\end{bmatrix}
$$

$$a_{i1}b_{1j} + a_{i2}b_{2j} + a_{i3}b_{3j} + \cdots + a_{in}b_{nj} = c_{ij}$$

Example 8   Finding the Product of Two Matrices

Find the product AB using $A = \begin{bmatrix} -1 & 3 \\ 4 & -2 \\ 5 & 0 \end{bmatrix}$ and $B = \begin{bmatrix} -3 & 2 \\ -4 & 1 \end{bmatrix}.$

Solution

First, note that the product AB is defined because the number of columns of A is equal to the number of rows of B. Moreover, the product AB has order 3×2. To find the entries of the product, multiply each row of A by each column of B.

$$AB = \begin{bmatrix} -1 & 3 \\ 4 & -2 \\ 5 & 0 \end{bmatrix} \begin{bmatrix} -3 & 2 \\ -4 & 1 \end{bmatrix}$$

$$= \begin{bmatrix} (-1)(-3) + (3)(-4) & (-1)(2) + (3)(1) \\ (4)(-3) + (-2)(-4) & (4)(2) + (-2)(1) \\ (5)(-3) + (0)(-4) & (5)(2) + (0)(1) \end{bmatrix} = \begin{bmatrix} -9 & 1 \\ -4 & 6 \\ -15 & 10 \end{bmatrix}$$

✓CHECKPOINT   Now try Exercise 27.

Be sure you understand that for the product of two matrices to be defined, the number of *columns* of the first matrix must equal the number of *rows* of the second matrix. That is, the middle two indices must be the same. The outside two indices give the order of the product, as shown in the following diagram.

$$A \quad \times \quad B \quad = \quad AB$$
$$m \times n \qquad n \times p \qquad m \times p$$

Equal
Order of AB

Example 9 Matrix Multiplication

a. $\begin{bmatrix} 1 & 0 & 3 \\ 2 & -1 & -2 \end{bmatrix} \begin{bmatrix} -2 & 4 & 2 \\ 1 & 0 & 0 \\ -1 & 1 & -1 \end{bmatrix} = \begin{bmatrix} -5 & 7 & -1 \\ -3 & 6 & 6 \end{bmatrix}$

$\qquad 2 \times 3 \qquad\qquad 3 \times 3 \qquad\qquad 2 \times 3$

b. $\begin{bmatrix} 3 & 4 \\ -2 & 5 \end{bmatrix} \begin{bmatrix} 1 & 0 \\ 0 & 1 \end{bmatrix} = \begin{bmatrix} 3 & 4 \\ -2 & 5 \end{bmatrix}$

$\qquad 2 \times 2 \qquad 2 \times 2 \qquad 2 \times 2$

c. $\begin{bmatrix} 1 & 2 \\ 1 & 1 \end{bmatrix} \begin{bmatrix} -1 & 2 \\ 1 & -1 \end{bmatrix} = \begin{bmatrix} 1 & 0 \\ 0 & 1 \end{bmatrix}$

$\qquad 2 \times 2 \qquad 2 \times 2 \qquad 2 \times 2$

d. $\begin{bmatrix} 1 & -2 & -3 \end{bmatrix} \begin{bmatrix} 2 \\ -1 \\ 1 \end{bmatrix} = \begin{bmatrix} 1 \end{bmatrix}$

$\qquad 1 \times 3 \qquad 3 \times 1 \quad 1 \times 1$

e. $\begin{bmatrix} 2 \\ -1 \\ 1 \end{bmatrix} \begin{bmatrix} 1 & -2 & -3 \end{bmatrix} = \begin{bmatrix} 2 & -4 & -6 \\ -1 & 2 & 3 \\ 1 & -2 & -3 \end{bmatrix}$

$\quad 3 \times 1 \qquad 1 \times 3 \qquad\qquad 3 \times 3$

f. The product AB for the following matrices is not defined.

$$A = \begin{bmatrix} -2 & 1 \\ 1 & -3 \\ 1 & 4 \end{bmatrix} \quad \text{and} \quad B = \begin{bmatrix} -2 & 3 & 1 & 4 \\ 0 & 1 & -1 & 2 \\ 2 & -1 & 0 & 1 \end{bmatrix}$$

$\qquad\qquad 3 \times 2 \qquad\qquad\qquad 3 \times 4$

✓**CHECKPOINT** Now try Exercise 29.

Exploration

Use the following matrices to find AB, BA, $(AB)C$, and $A(BC)$. What do your results tell you about matrix multiplication and commutativity and associativity?

$$A = \begin{bmatrix} 1 & 2 \\ 3 & 4 \end{bmatrix}$$

$$B = \begin{bmatrix} 0 & 1 \\ 2 & 3 \end{bmatrix}$$

$$C = \begin{bmatrix} 3 & 0 \\ 0 & 1 \end{bmatrix}$$

In parts (d) and (e) of Example 9, note that the two products are different. Matrix multiplication is not, in general, commutative. That is, for most matrices, $AB \neq BA$. This is one way in which the algebra of real numbers and the algebra of matrices differ.

Example 10   Matrix Multiplication

Use a graphing utility to find the product AB using

$$A = \begin{bmatrix} 1 & 2 & 3 \\ 2 & -5 & 1 \end{bmatrix} \quad \text{and} \quad B = \begin{bmatrix} -3 & 2 & 1 \\ 4 & -2 & 0 \\ 1 & 2 & 3 \end{bmatrix}.$$

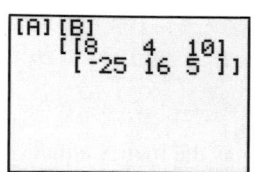

Solution

Note that the order of A is 2×3 and the order of B is 3×3. So, the product will have order 2×3. Use the *matrix editor* to enter A and B into the graphing utility. Then, find the product, as shown in Figure 7.30.

Figure 7.30

✓CHECKPOINT   Now try Exercise 39.

Properties of Matrix Multiplication

Let A, B, and C be matrices and let c be a scalar.

1. $A(BC) = (AB)C$ Associative Property of Matrix Multiplication

2. $A(B + C) = AB + AC$ Left Distributive Property

3. $(A + B)C = AC + BC$ Right Distributive Property

4. $c(AB) = (cA)B = A(cB)$ Associative Property of Scalar Multiplication

---

Definition of Identity Matrix

The $n \times n$ matrix that consists of 1's on its main diagonal and 0's elsewhere is called the **identity matrix of order n** and is denoted by

$$I_n = \begin{bmatrix} 1 & 0 & 0 & \ldots & 0 \\ 0 & 1 & 0 & \ldots & 0 \\ 0 & 0 & 1 & \ldots & 0 \\ \vdots & \vdots & \vdots & & \vdots \\ 0 & 0 & 0 & \ldots & 1 \end{bmatrix}. \quad \text{Identity matrix}$$

Note that an identity matrix must be *square*. When the order is understood to be n, you can denote I_n simply by I.

If A is an $n \times n$ matrix, the identity matrix has the property that $AI_n = A$ and $I_nA = A$. For example,

$$\begin{bmatrix} 3 & -2 & 5 \\ 1 & 0 & 4 \\ -1 & 2 & -3 \end{bmatrix}\begin{bmatrix} 1 & 0 & 0 \\ 0 & 1 & 0 \\ 0 & 0 & 1 \end{bmatrix} = \begin{bmatrix} 3 & -2 & 5 \\ 1 & 0 & 4 \\ -1 & 2 & -3 \end{bmatrix} \quad AI = A$$

and

$$\begin{bmatrix} 1 & 0 & 0 \\ 0 & 1 & 0 \\ 0 & 0 & 1 \end{bmatrix}\begin{bmatrix} 3 & -2 & 5 \\ 1 & 0 & 4 \\ -1 & 2 & -3 \end{bmatrix} = \begin{bmatrix} 3 & -2 & 5 \\ 1 & 0 & 4 \\ -1 & 2 & -3 \end{bmatrix}. \quad IA = A$$

Applications

Matrix multiplication can be used to represent a system of linear equations. Note how the system

$$\begin{cases} a_{11}x_1 + a_{12}x_2 + a_{13}x_3 = b_1 \\ a_{21}x_1 + a_{22}x_2 + a_{23}x_3 = b_2 \\ a_{31}x_1 + a_{32}x_2 + a_{33}x_3 = b_3 \end{cases}$$

can be written as the matrix equation $AX = B$, where A is the *coefficient matrix* of the system and X and B are column matrices.

$$\begin{bmatrix} a_{11} & a_{12} & a_{13} \\ a_{21} & a_{22} & a_{23} \\ a_{31} & a_{32} & a_{33} \end{bmatrix} \begin{bmatrix} x_1 \\ x_2 \\ x_3 \end{bmatrix} = \begin{bmatrix} b_1 \\ b_2 \\ b_3 \end{bmatrix}$$
$$A \quad \times \quad X \quad = \quad B$$

STUDY TIP

The column matrix B is also called a *constant* matrix. Its entries are the constant terms in the system of equations.

Example 11 Solving a System of Linear Equations

Consider the system of linear equations $\begin{cases} x_1 - 2x_2 + x_3 = -4 \\ x_2 + 2x_3 = 4. \\ 2x_1 + 3x_2 - 2x_3 = 2 \end{cases}$

a. Write this system as a matrix equation $AX = B$.

b. Use Gauss-Jordan elimination on $[A \vdots B]$ to solve for the matrix X.

Solution

a. In matrix form $AX = B$, the system can be written as follows.

$$\begin{bmatrix} 1 & -2 & 1 \\ 0 & 1 & 2 \\ 2 & 3 & -2 \end{bmatrix} \begin{bmatrix} x_1 \\ x_2 \\ x_3 \end{bmatrix} = \begin{bmatrix} -4 \\ 4 \\ 2 \end{bmatrix}$$

b. The augmented matrix is

$$[A \vdots B] = \begin{bmatrix} 1 & -2 & 1 & \vdots & -4 \\ 0 & 1 & 2 & \vdots & 4 \\ 2 & 3 & -2 & \vdots & 2 \end{bmatrix}.$$

Using Gauss-Jordan elimination, you can rewrite this equation as

$$[I \vdots X] = \begin{bmatrix} 1 & 0 & 0 & \vdots & -1 \\ 0 & 1 & 0 & \vdots & 2 \\ 0 & 0 & 1 & \vdots & 1 \end{bmatrix}.$$

So, the solution of the system of linear equations is $x_1 = -1$, $x_2 = 2$, and $x_3 = 1$. The solution of the matrix equation is

$$X = \begin{bmatrix} x_1 \\ x_2 \\ x_3 \end{bmatrix} = \begin{bmatrix} -1 \\ 2 \\ 1 \end{bmatrix}.$$

✓CHECKPOINT   Now try Exercise 57.

TECHNOLOGY TIP

Most graphing utilities can be used to obtain the reduced row-echelon form of a matrix. The screen below shows how one graphing utility displays the reduced row-echelon form of the augmented matrix in Example 11.

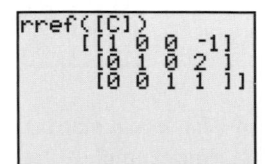

Example 12 Health Care

A company offers three types of health care plans with two levels of coverage to its employees. The current annual costs for these plans are represented by the matrix A. If the annual costs are expected to increase by 4% next year, what will be the annual costs for each plan next year?

$$
A = \begin{array}{c} \overbrace{\hspace{6cm}}^{\text{Plan}} \end{array}
$$

$$
\begin{array}{ccc}
\text{Premium} & \text{HMO} & \text{HMO Plus}
\end{array}
$$

$$
A = \begin{bmatrix} 694 & 451 & 489 \\ 1725 & 1187 & 1248 \end{bmatrix} \begin{array}{l} \text{Single} \\ \text{Family} \end{array} \left.\right\} \begin{array}{l} \text{Coverage} \\ \text{level} \end{array}
$$

Solution

Because an increase of 4% corresponds to 100% + 4%, multiply A by 104% or 1.04. So, the annual costs for each health care plan next year are as follows.

$$
1.04A = 1.04 \begin{bmatrix} 694 & 451 & 489 \\ 1725 & 1187 & 1248 \end{bmatrix} = \begin{bmatrix} 722 & 469 & 509 \\ 1794 & 1234 & 1298 \end{bmatrix} \begin{array}{l} \text{Single} \\ \text{Family} \end{array} \left.\right\} \begin{array}{l} \text{Coverage} \\ \text{level} \end{array}
$$

with column headers: Premium, HMO, HMO Plus under "Plan"

✓CHECKPOINT   Now try Exercise 77.

Example 13 Softball Team Expenses

Two softball teams submit equipment lists to their sponsors, as shown in the table at the right. Each bat costs $80, each ball costs $6, and each glove costs $60. Use matrices to find the total cost of equipment for each team.

Equipment	Women's Team	Men's Team
Bats	12	15
Balls	45	38
Gloves	15	17

Solution

The equipment lists E and the costs per item C can be written in matrix form as

$$
E = \begin{bmatrix} 12 & 15 \\ 45 & 38 \\ 15 & 17 \end{bmatrix} \quad \text{and} \quad C = \begin{bmatrix} 80 & 6 & 60 \end{bmatrix}.
$$

You can find the total cost of the equipment for each team using the product CE because the number of columns of C (3 columns) equals the number of rows of E (3 rows). Therefore, the total cost of equipment for each team is given by

$$
CE = \begin{bmatrix} 80 & 6 & 60 \end{bmatrix} \begin{bmatrix} 12 & 15 \\ 45 & 38 \\ 15 & 17 \end{bmatrix}
$$

$$
= \begin{bmatrix} 80(12) + 6(45) + 60(15) & 80(15) + 6(38) + 60(17) \end{bmatrix}
$$

$$
= \begin{bmatrix} 2130 & 2448 \end{bmatrix}.
$$

So, the total cost of equipment for the women's team is $2130, and the total cost of equipment for the men's team is $2448.

✓CHECKPOINT   Now try Exercise 79.

STUDY TIP

Notice in Example 13 that you cannot find the total cost using the product EC because EC is not defined. That is, the number of columns of E (2 columns) does not equal the number of rows of C (1 row).

7.5 Exercises

See www.CalcChat.com for worked-out solutions to odd-numbered exercises.

Vocabulary Check

In Exercises 1–4, fill in the blanks.

1. Two matrices are _____ if all of their corresponding entries are equal.

2. When working with matrices, real numbers are often referred to as _____ .

3. A matrix consisting entirely of zeros is called a _____ matrix and is denoted by _____ .

4. The $n \times n$ matrix consisting of 1's on its main diagonal and 0's elsewhere is called the _____ matrix of order n.

In Exercises 5 and 6, match the matrix property with the correct form. A, B, and C are matrices, and c and d are scalars.

5. (a) $(cd)A = c(dA)$
  (b) $A + B = B + A$
  (c) $1A = A$
  (d) $c(A + B) = cA + cB$
  (e) $A + (B + C) = (A + B) + C$

  (i) Commutative Property of Matrix Addition
  (ii) Associative Property of Matrix Addition
  (iii) Associative Property of Scalar Multiplication
  (iv) Scalar Identity
  (v) Distributive Property

6. (a) $A(B + C) = AB + AC$
  (b) $c(AB) = (cA)B = A(cB)$
  (c) $A(BC) = (AB)C$
  (d) $(A + B)C = AC + BC$

  (i) Associative Property of Matrix Multiplication
  (ii) Left Distributive Property
  (iii) Right Distributive Property
  (iv) Associative Property of Scalar Multiplication

In Exercises 1–4, find x and y or x, y, and z.

1. $\begin{bmatrix} x & -2 \\ 7 & y \end{bmatrix} = \begin{bmatrix} -4 & -2 \\ 7 & 22 \end{bmatrix}$

2. $\begin{bmatrix} -5 & x \\ y & 8 \end{bmatrix} = \begin{bmatrix} -5 & 13 \\ 12 & 8 \end{bmatrix}$

3. $\begin{bmatrix} 16 & 4 & 5 & 4 \\ -3 & 13 & 15 & 12 \\ 0 & 2 & 4 & 0 \end{bmatrix} = \begin{bmatrix} 16 & 4 & 2x+7 & 4 \\ -3 & 13 & & 15 & 3y \\ 0 & 2 & 3z-14 & 0 \end{bmatrix}$

4. $\begin{bmatrix} x+4 & 8 & -3 \\ 1 & 22 & 2y \\ 7 & -2 & z+2 \end{bmatrix} = \begin{bmatrix} 2x+9 & 8 & -3 \\ 1 & 22 & -8 \\ 7 & -2 & 11 \end{bmatrix}$

In Exercises 5–12, find, if possible, (a) $A + B$, (b) $A - B$, (c) $3A$, and (d) $3A - 2B$. Use the matrix capabilities of a graphing utility to verify your results.

5. $A = \begin{bmatrix} 1 & -1 \\ 2 & -1 \end{bmatrix}$, $B = \begin{bmatrix} 2 & -1 \\ -1 & 8 \end{bmatrix}$

6. $A = \begin{bmatrix} 1 & 2 \\ 2 & 1 \end{bmatrix}$, $B = \begin{bmatrix} -3 & -2 \\ 4 & 2 \end{bmatrix}$

7. $A = \begin{bmatrix} 8 & -1 \\ 2 & 3 \\ -4 & 5 \end{bmatrix}$, $B = \begin{bmatrix} 1 & 6 \\ -1 & -5 \\ 1 & 10 \end{bmatrix}$

8. $A = \begin{bmatrix} 1 & -1 & 3 \\ 0 & 6 & 9 \end{bmatrix}$, $B = \begin{bmatrix} -2 & 0 & -5 \\ -3 & 4 & -7 \end{bmatrix}$

9. $A = \begin{bmatrix} 4 & 5 & -1 & 3 & 4 \\ 1 & 2 & -2 & -1 & 0 \end{bmatrix}$,
$B = \begin{bmatrix} 1 & 0 & -1 & 1 & 0 \\ -6 & 8 & 2 & -3 & -7 \end{bmatrix}$

10. $A = \begin{bmatrix} -1 & 4 & 0 \\ 3 & -2 & 2 \\ 5 & 4 & -1 \\ 0 & 8 & -6 \\ -4 & -1 & 0 \end{bmatrix}$, $B = \begin{bmatrix} -3 & 5 & 1 \\ 2 & -4 & -7 \\ 10 & -9 & -1 \\ 3 & 2 & -4 \\ 0 & 1 & -2 \end{bmatrix}$

11. $A = \begin{bmatrix} 6 & 0 & 3 \\ -1 & -4 & 0 \end{bmatrix}$, $B = \begin{bmatrix} 8 & -1 \\ 4 & -3 \end{bmatrix}$

12. $A = \begin{bmatrix} 3 \\ 2 \\ -1 \end{bmatrix}$, $B = \begin{bmatrix} -4 & 6 & 2 \end{bmatrix}$

In Exercises 13–16, evaluate the expression.

13. $\begin{bmatrix} -5 & 0 \\ 3 & -6 \end{bmatrix} + \begin{bmatrix} 7 & 1 \\ -2 & -1 \end{bmatrix} + \begin{bmatrix} -10 & -8 \\ 14 & 6 \end{bmatrix}$

14. $\begin{bmatrix} 6 & 9 \\ -1 & 0 \\ 7 & 1 \end{bmatrix} + \begin{bmatrix} 0 & 5 \\ -2 & -1 \\ 3 & -6 \end{bmatrix} + \begin{bmatrix} -13 & -7 \\ 4 & -1 \\ -6 & 0 \end{bmatrix}$

15. $4\left(\begin{bmatrix} -4 & 0 & 1 \\ 0 & 2 & 3 \end{bmatrix} - \begin{bmatrix} 2 & 1 & -2 \\ 3 & -6 & 0 \end{bmatrix} \right)$

16. $\frac{1}{2}([5 \quad -2 \quad 4 \quad 0] + [14 \quad 6 \quad -18 \quad 9])$

In Exercises 17–20, use the matrix capabilities of a graphing utility to evaluate the expression. Round your results to the nearest thousandths, if necessary.

17. $\begin{bmatrix} 2 & 5 \\ -1 & -4 \end{bmatrix} + \begin{bmatrix} -3 & 0 \\ 2 & 2 \end{bmatrix}$

18. $\begin{bmatrix} 14 & -11 \\ -22 & 19 \end{bmatrix} + \begin{bmatrix} -22 & 20 \\ 13 & 6 \end{bmatrix}$

19. $-\frac{1}{2}\begin{bmatrix} 3.211 & 6.829 \\ -1.004 & 4.914 \\ 0.055 & -3.889 \end{bmatrix} - 8\begin{bmatrix} 1.630 & -3.090 \\ 5.256 & 8.335 \\ -9.768 & 4.251 \end{bmatrix}$

20. $-1\begin{bmatrix} 4 & 11 \\ -2 & -1 \\ 9 & 3 \end{bmatrix} + \frac{1}{6}\left(\begin{bmatrix} -5 & -1 \\ 3 & 4 \\ 0 & 13 \end{bmatrix} + \begin{bmatrix} 7 & 5 \\ -9 & -1 \\ 6 & -1 \end{bmatrix} \right)$

In Exercises 21–24, solve for X when

$$A = \begin{bmatrix} -2 & -1 \\ 1 & 0 \\ 3 & -4 \end{bmatrix} \quad \text{and} \quad B = \begin{bmatrix} 0 & 3 \\ 2 & 0 \\ -4 & -1 \end{bmatrix}.$$

21. $X = 3A - 2B$

22. $2X = 2A - B$

23. $2X + 3A = B$

24. $2A + 4B = -2X$

In Exercises 25–32, find AB, if possible.

25. $A = \begin{bmatrix} 2 & 1 \\ -3 & 4 \\ -1 & 6 \end{bmatrix}, \quad B = \begin{bmatrix} 0 & -3 & 0 \\ 4 & 0 & 2 \\ 8 & -2 & 7 \end{bmatrix}$

26. $A = \begin{bmatrix} 0 & -1 & 2 \\ 6 & 0 & 3 \\ 7 & -1 & 8 \end{bmatrix}, \quad B = \begin{bmatrix} 2 & -1 \\ 4 & -5 \\ 1 & 6 \end{bmatrix}$

27. $A = \begin{bmatrix} -1 & 6 \\ -4 & 5 \\ 0 & 3 \end{bmatrix}, \quad B = \begin{bmatrix} 2 & 3 \\ 0 & 9 \end{bmatrix}$

28. $A = \begin{bmatrix} 1 & 0 & 0 \\ 0 & 4 & 0 \\ 0 & 0 & -2 \end{bmatrix}, \quad B = \begin{bmatrix} 3 & 0 & 0 \\ 0 & -1 & 0 \\ 0 & 0 & 5 \end{bmatrix}$

29. $A = \begin{bmatrix} 5 & 0 & 0 \\ 0 & -8 & 0 \\ 0 & 0 & 7 \end{bmatrix}, \quad B = \begin{bmatrix} \frac{1}{5} & 0 & 0 \\ 0 & -\frac{1}{8} & 0 \\ 0 & 0 & \frac{1}{2} \end{bmatrix}$

30. $A = \begin{bmatrix} 0 & 0 & 5 \\ 0 & 0 & -3 \\ 0 & 0 & 4 \end{bmatrix}, \quad B = \begin{bmatrix} 6 & -11 & 4 \\ 8 & 16 & 4 \\ 0 & 0 & 0 \end{bmatrix}$

31. $A = \begin{bmatrix} 5 \\ 6 \end{bmatrix}, \quad B = [-3 \quad -1 \quad -5 \quad -9]$

32. $A = \begin{bmatrix} 1 & 0 & 3 & -2 \\ 6 & 13 & 8 & -17 \end{bmatrix}, \quad B = \begin{bmatrix} 1 & 6 \\ 4 & 2 \end{bmatrix}$

In Exercises 33–38, find, if possible, (a) AB, (b) BA, and (c) A^2. (*Note:* $A^2 = AA$.) Use the matrix capabilities of a graphing utility to verify your results.

33. $A = \begin{bmatrix} 1 & 2 \\ 5 & 2 \end{bmatrix}, \quad B = \begin{bmatrix} 2 & -1 \\ -1 & 8 \end{bmatrix}$

34. $A = \begin{bmatrix} 6 & 3 \\ -2 & -4 \end{bmatrix}, \quad B = \begin{bmatrix} -2 & 0 \\ 2 & 4 \end{bmatrix}$

35. $A = \begin{bmatrix} 3 & -1 \\ 1 & 3 \end{bmatrix}, \quad B = \begin{bmatrix} 1 & -3 \\ 3 & 1 \end{bmatrix}$

36. $A = \begin{bmatrix} 1 & -1 \\ 1 & 1 \end{bmatrix}, \quad B = \begin{bmatrix} 1 & 3 \\ -3 & 1 \end{bmatrix}$

37. $A = \begin{bmatrix} 7 \\ 8 \\ -1 \end{bmatrix}, \quad B = [1 \quad 1 \quad 2]$

38. $A = [3 \quad 2 \quad 1], \quad B = \begin{bmatrix} 2 \\ 3 \\ 0 \end{bmatrix}$

In Exercises 39–44, use the matrix capabilities of a graphing utility to find AB, if possible.

39. $A = \begin{bmatrix} 7 & 5 & -4 \\ -2 & 5 & 1 \\ 10 & -4 & -7 \end{bmatrix}, \quad B = \begin{bmatrix} 2 & -2 & 3 \\ 8 & 1 & 4 \\ -4 & 2 & -8 \end{bmatrix}$

40. $A = \begin{bmatrix} 1 & -12 & 4 \\ 14 & 10 & 12 \\ 6 & -15 & 3 \end{bmatrix}, \quad B = \begin{bmatrix} 12 & 10 \\ -6 & 12 \\ 10 & 16 \end{bmatrix}$

41. $A = \begin{bmatrix} -3 & 8 & -6 & 8 \\ -12 & 15 & 9 & 6 \\ 5 & -1 & 1 & 5 \end{bmatrix}, \quad B = \begin{bmatrix} 3 & 1 & 6 \\ 24 & 15 & 14 \\ 16 & 10 & 21 \\ 8 & -4 & 10 \end{bmatrix}$

42. $A = \begin{bmatrix} -2 & 6 & 12 \\ 21 & -5 & 6 \\ 13 & -2 & 9 \end{bmatrix}, \quad B = \begin{bmatrix} 3 & 0 \\ -7 & 18 \\ 34 & 14 \\ 0.5 & 1.4 \end{bmatrix}$

43. $A = \begin{bmatrix} 9 & 10 & -38 & 18 \\ 100 & -50 & 250 & 75 \end{bmatrix}$,

$B = \begin{bmatrix} 52 & -85 & 27 & 45 \\ 40 & -35 & 60 & 82 \end{bmatrix}$

44. $A = \begin{bmatrix} 16 & -18 \\ -4 & 13 \\ -9 & 21 \end{bmatrix}$, $B = \begin{bmatrix} -7 & 20 & -1 \\ 7 & 15 & 26 \end{bmatrix}$

In Exercises 45–48, use the matrix capabilities of a graphing utility to evaluate the expression.

45. $\begin{bmatrix} 3 & 1 \\ 0 & -2 \end{bmatrix} \begin{bmatrix} 1 & 0 \\ -2 & 2 \end{bmatrix} \begin{bmatrix} 1 & 0 \\ 2 & 4 \end{bmatrix}$

46. $-3\left(\begin{bmatrix} 6 & 5 & -1 \\ 1 & -2 & 0 \end{bmatrix} \begin{bmatrix} 0 & 3 \\ -1 & -3 \\ 4 & 1 \end{bmatrix} \right)$

47. $\begin{bmatrix} 0 & 2 & -2 \\ 4 & 1 & 2 \end{bmatrix} \left(\begin{bmatrix} 4 & 0 \\ 0 & -1 \\ -1 & 2 \end{bmatrix} + \begin{bmatrix} -2 & 3 \\ -3 & 5 \\ 0 & -3 \end{bmatrix} \right)$

48. $\begin{bmatrix} 3 \\ -1 \\ 5 \\ 7 \end{bmatrix} ([5 \quad -6] + [7 \quad -1] + [-8 \quad 9])$

In Exercises 49–52, use matrix multiplication to determine whether each matrix is a solution of the system of equations. Use a graphing utility to verify your results.

49. $\begin{cases} x + 2y = 4 \\ 3x + 2y = 0 \end{cases}$

(a) $\begin{bmatrix} 2 \\ 1 \end{bmatrix}$ (b) $\begin{bmatrix} -2 \\ 3 \end{bmatrix}$

(c) $\begin{bmatrix} -4 \\ 4 \end{bmatrix}$ (d) $\begin{bmatrix} 2 \\ -3 \end{bmatrix}$

50. $\begin{cases} 6x + 2y = 0 \\ -x + 5y = 16 \end{cases}$

(a) $\begin{bmatrix} -1 \\ 3 \end{bmatrix}$ (b) $\begin{bmatrix} 2 \\ -6 \end{bmatrix}$

(c) $\begin{bmatrix} 3 \\ -9 \end{bmatrix}$ (d) $\begin{bmatrix} -3 \\ 9 \end{bmatrix}$

51. $\begin{cases} -2x - 3y = -6 \\ 4x + 2y = 20 \end{cases}$

(a) $\begin{bmatrix} 3 \\ 0 \end{bmatrix}$ (b) $\begin{bmatrix} 6 \\ -2 \end{bmatrix}$

(c) $\begin{bmatrix} -6 \\ 6 \end{bmatrix}$ (d) $\begin{bmatrix} 4 \\ 2 \end{bmatrix}$

52. $\begin{cases} 5x - 7y = -15 \\ 3x + y = 17 \end{cases}$

(a) $\begin{bmatrix} 4 \\ 5 \end{bmatrix}$ (b) $\begin{bmatrix} 5 \\ 2 \end{bmatrix}$

(c) $\begin{bmatrix} -4 \\ -5 \end{bmatrix}$ (d) $\begin{bmatrix} 2 \\ 11 \end{bmatrix}$

In Exercises 53–60, (a) write the system of equations as a matrix equation $AX = B$ and (b) use Gauss-Jordan elimination on the augmented matrix $[A \vdots B]$ to solve for the matrix X. Use a graphing utility to check your solution.

53. $\begin{cases} -x_1 + x_2 = 4 \\ -2x_1 + x_2 = 0 \end{cases}$

54. $\begin{cases} 2x_1 + 3x_2 = 5 \\ x_1 + 4x_2 = 10 \end{cases}$

55. $\begin{cases} -2x_1 - 3x_2 = -4 \\ 6x_1 + x_2 = -36 \end{cases}$

56. $\begin{cases} -4x_1 + 9x_2 = -13 \\ x_1 - 3x_2 = 12 \end{cases}$

57. $\begin{cases} x_1 - 2x_2 + 3x_3 = 9 \\ -x_1 + 3x_2 - x_3 = -6 \\ 2x_1 - 5x_2 + 5x_3 = 17 \end{cases}$

58. $\begin{cases} x_1 + x_2 - 3x_3 = -1 \\ -x_1 + 2x_2 = 1 \\ x_1 - x_2 + x_3 = 2 \end{cases}$

59. $\begin{cases} x_1 - 5x_2 + 2x_3 = -20 \\ -3x_1 + x_2 - x_3 = 8 \\ -2x_2 + 5x_3 = -16 \end{cases}$

60. $\begin{cases} x_1 - x_2 + 4x_3 = 17 \\ x_1 + 3x_2 = -11 \\ -6x_2 + 5x_3 = 40 \end{cases}$

In Exercises 61–66, use a graphing utility to perform the operations for the matrices A, B, and C, and the scalar c. Write a brief statement comparing the results of parts (a) and (b).

$A = \begin{bmatrix} 1 & 2 & -1 \\ 0 & -2 & 3 \\ 4 & -3 & 2 \end{bmatrix}$, $B = \begin{bmatrix} 2 & 3 & 0 \\ 4 & 1 & -2 \\ -1 & 2 & 0 \end{bmatrix}$,

$C = \begin{bmatrix} 3 & -2 & 1 \\ -4 & 0 & 3 \\ -1 & 3 & -2 \end{bmatrix}$, and $c = 3$

61. (a) $A(B + C)$ (b) $AB + AC$

62. (a) $(B + C)A$ (b) $BA + CA$

63. (a) $(A + B)^2$ (b) $A^2 + AB + BA + B^2$

64. (a) $(A - B)^2$ (b) $A^2 - AB - BA + B^2$

65. (a) $A(BC)$ (b) $(AB)C$

66. (a) $c(AB)$ (b) $(cA)B$

In Exercises 67–74, perform the operations (a) using a graphing utility and (b) by hand algebraically. If it is not possible to perform the operation(s), state the reason.

$A = \begin{bmatrix} 1 & 2 & -2 \\ -1 & 1 & 0 \end{bmatrix}$, $B = \begin{bmatrix} -1 & 4 & -1 \\ -2 & -1 & 0 \end{bmatrix}$,

$C = \begin{bmatrix} 1 & 2 \\ -2 & 3 \\ 1 & 0 \end{bmatrix}$, $c = 2$, and $d = -3$

67. $A + cB$

68. $A(B + C)$

69. $c(AB)$

70. $B + dA$

71. $CA - BC$

72. dAB^2

73. cdA

74. $cA + dB$

In Exercises 75 and 76, use the matrix capabilities of a graphing utility to find $f(A) = a_0 I_n + a_1 A + a_2 A^2$.

75. $f(x) = x^2 - 5x + 2, \quad A = \begin{bmatrix} 2 & 0 \\ 4 & 5 \end{bmatrix}$

76. $f(x) = x^2 - 7x + 6, \quad A = \begin{bmatrix} 5 & 4 \\ 1 & 2 \end{bmatrix}$

77. *Manufacturing* A corporation has three factories, each of which manufactures acoustic guitars and electric guitars. The number of units of guitars produced at factory j in one day is represented by a_{ij} in the matrix

$$A = \begin{bmatrix} 70 & 50 & 25 \\ 35 & 100 & 70 \end{bmatrix}.$$

Find the production levels if production is increased by 20%.

78. *Manufacturing* A corporation has four factories, each of which manufactures sport utility vehicles and pickup trucks. The number of units of vehicle i produced at factory j in one day is represented by a_{ij} in the matrix

$$A = \begin{bmatrix} 100 & 90 & 70 & 30 \\ 40 & 20 & 60 & 60 \end{bmatrix}.$$

Find the production levels if production is increased by 10%.

79. *Agriculture* A fruit grower raises two crops, apples and peaches. Each of these crops is shipped to three different outlets. The number of units of crop i that are shipped to outlet j is represented by a_{ij} in the matrix

$$A = \begin{bmatrix} 125 & 100 & 75 \\ 100 & 175 & 125 \end{bmatrix}.$$

The profit per unit is represented by the matrix

$$B = [\$3.50 \quad \$6.00].$$

Find the product BA and state what each entry of the product represents.

80. *Revenue* A manufacturer produces three models of portable CD players, which are shipped to two warehouses. The number of units of model i that are shipped to warehouse j is represented by a_{ij} in the matrix

$$A = \begin{bmatrix} 5{,}000 & 4{,}000 \\ 6{,}000 & 10{,}000 \\ 8{,}000 & 5{,}000 \end{bmatrix}.$$

The price per unit is represented by the matrix

$$B = [\$39.50 \quad \$44.50 \quad \$56.50].$$

Compute BA and state what each entry of the product represents.

81. *Inventory* A company sells five models of computers through three retail outlets. The inventories are given by S. The wholesale and retail prices are given by T. Compute ST and interpret the result.

$$S = \begin{array}{c} \\ \\ \\ \end{array} \overset{\text{Model}}{\overset{\text{A B C D E}}{\begin{bmatrix} 3 & 2 & 2 & 3 & 0 \\ 0 & 2 & 3 & 4 & 3 \\ 4 & 2 & 1 & 3 & 2 \end{bmatrix}}} \begin{array}{l} 1 \\ 2 \\ 3 \end{array} \text{Outlet}$$

$$T = \overset{\text{Price}}{\overset{\text{Wholesale \quad Retail}}{\begin{bmatrix} \$840 & \$1100 \\ \$1200 & \$1350 \\ \$1450 & \$1650 \\ \$2650 & \$3000 \\ \$3050 & \$3200 \end{bmatrix}}} \begin{array}{l} A \\ B \\ C \\ D \\ E \end{array} \text{Model}$$

82. *Labor/Wage Requirements* A company that manufactures boats has the following labor-hour and wage requirements. Compute ST and interpret the result.

Labor per Boat

$$S = \overset{\text{Department}}{\overset{\text{Cutting \quad Assembly \quad Packaging}}{\begin{bmatrix} 1.0 \text{ hr} & 0.5 \text{ hr} & 0.2 \text{ hr} \\ 1.6 \text{ hr} & 1.0 \text{ hr} & 0.2 \text{ hr} \\ 2.5 \text{ hr} & 2.0 \text{ hr} & 0.4 \text{ hr} \end{bmatrix}}} \begin{array}{l} \text{Small} \\ \text{Medium} \\ \text{Large} \end{array} \begin{array}{l} \text{Boat} \\ \text{Size} \end{array}$$

Wages per Hour

$$T = \overset{\text{Plant}}{\overset{\text{A \quad B}}{\begin{bmatrix} \$12 & \$10 \\ \$9 & \$8 \\ \$6 & \$5 \end{bmatrix}}} \begin{array}{l} \text{Cutting} \\ \text{Assembly} \\ \text{Packaging} \end{array} \text{Department}$$

83. *Voting Preference* The matrix

$$P = \overset{\text{From}}{\overset{\text{R \quad D \quad I}}{\begin{bmatrix} 0.6 & 0.1 & 0.1 \\ 0.2 & 0.7 & 0.1 \\ 0.2 & 0.2 & 0.8 \end{bmatrix}}} \begin{array}{l} \text{R} \\ \text{D} \\ \text{I} \end{array} \text{To}$$

is called a *stochastic matrix*. Each entry p_{ij} $(i \neq j)$ represents the proportion of the voting population that changes from party i to party j, and p_{ii} represents the proportion that remains loyal to the party from one election to the next. Compute and interpret P^2.

84. *Voting Preference* Use a graphing utility to find P^3, P^4, P^5, P^6, P^7, and P^8 for the matrix given in Exercise 83. Can you detect a pattern as P is raised to higher powers?

Synthesis

True or False? **In Exercises 85 and 86, determine whether the statement is true or false. Justify your answer.**

85. Two matrices can be added only if they have the same order.

86. Matrix multiplication is commutative.

Think About It **In Exercises 87–94, let matrices A, B, C, and D be of orders 2×3, 2×3, 3×2, and 2×2, respectively. Determine whether the matrices are of proper order to perform the operation(s). If so, give the order of the answer.**

87. $A + 2C$ **88.** $B - 3C$

89. AB **90.** BC

91. $BC - D$ **92.** $CB - D$

93. $D(A - 3B)$ **94.** $(BC - D)A$

Think About It **In Exercises 95–98, use the matrices**

$$A = \begin{bmatrix} 2 & -1 \\ 1 & 3 \end{bmatrix} \text{ and } B = \begin{bmatrix} -1 & 1 \\ 0 & -2 \end{bmatrix}.$$

95. Show that $(A + B)^2 \neq A^2 + 2AB + B^2$.

96. Show that $(A - B)^2 \neq A^2 - 2AB + B^2$.

97. Show that $(A + B)(A - B) \neq A^2 - B^2$.

98. Show that $(A + B)^2 = A^2 + AB + BA + B^2$.

99. *Think About It* If a, b, and c are real numbers such that $c \neq 0$ and $ac = bc$, then $a = b$. However, if A, B, and C are nonzero matrices such that $AC = BC$, then A is *not necessarily* equal to B. Illustrate this using the following matrices.

$$A = \begin{bmatrix} 0 & 1 \\ 0 & 1 \end{bmatrix}, \quad B = \begin{bmatrix} 1 & 0 \\ 1 & 0 \end{bmatrix}, \quad C = \begin{bmatrix} 2 & 3 \\ 2 & 3 \end{bmatrix}$$

100. *Think About It* If a and b are real numbers such that $ab = 0$, then $a = 0$ or $b = 0$. However, if A and B are matrices such that $AB = O$, it is *not necessarily* true that $A = O$ or $B = O$. Illustrate this using the following matrices.

$$A = \begin{bmatrix} 3 & 3 \\ 4 & 4 \end{bmatrix}, \quad B = \begin{bmatrix} 1 & -1 \\ -1 & 1 \end{bmatrix}$$

101. *Exploration* Let $i = \sqrt{-1}$ and let

$$A = \begin{bmatrix} i & 0 \\ 0 & i \end{bmatrix} \quad \text{and} \quad B = \begin{bmatrix} 0 & -i \\ i & 0 \end{bmatrix}.$$

(a) Find A^2, A^3, and A^4. Identify any similarities with i^2, i^3, and i^4.

(b) Find and identify B^2.

102. *Conjecture* Let A and B be unequal diagonal matrices of the same order. (A *diagonal matrix* is a square matrix in which each entry not on the main diagonal is zero.) Determine the products AB for several pairs of such matrices. Make a conjecture about a quick rule for such products.

103. *Exploration* Consider matrices of the form

$$A = \begin{bmatrix} 0 & a_{12} & a_{13} & a_{14} & \cdots & a_{1n} \\ 0 & 0 & a_{23} & a_{24} & \cdots & a_{2n} \\ 0 & 0 & 0 & a_{34} & \cdots & a_{3n} \\ \vdots & \vdots & \vdots & \vdots & \cdots & \vdots \\ 0 & 0 & 0 & 0 & \cdots & a_{(n-1)n} \\ 0 & 0 & 0 & 0 & \cdots & 0 \end{bmatrix}.$$

(a) Write a 2×2 matrix and a 3×3 matrix in the form of A.

(b) Use a graphing utility to raise each of the matrices to higher powers. Describe the result.

(c) Use the result of part (b) to make a conjecture about powers of A if A is a 4×4 matrix. Use a graphing utility to test your conjecture.

(d) Use the results of parts (b) and (c) to make a conjecture about powers of an $n \times n$ matrix A.

104. *Writing* Two competing companies offer cable television to a city with 100,000 households. Gold Cable Company has 25,000 subscribers and Galaxy Cable Company has 30,000 subscribers. (The other 45,000 households do not subscribe.) The percent changes in cable subscriptions each year are shown below. Write a short paragraph explaining how matrix multiplication can be used to find the number of subscribers each company will have in 1 year.

		Percent Changes		
		From Gold	From Galaxy	From Nonsubscriber
Percent Changes	To Gold	0.70	0.15	0.15
	To Galaxy	0.20	0.80	0.15
	To Nonsubscriber	0.10	0.05	0.70

Skills Review

In Exercises 105–108, condense the expression to the logarithm of a single quantity.

105. $3 \ln 4 - \frac{1}{3} \ln(x^2 + 3)$

106. $\ln x - 3[\ln(x + 6) + \ln(x - 6)]$

107. $\frac{1}{2}[2 \ln(x + 5) + \ln x - \ln(x - 8)]$

108. $\frac{3}{2} \ln 7t^4 - \frac{3}{5} \ln t^5$

7.6 The Inverse of a Square Matrix

The Inverse of a Matrix

This section further develops the algebra of matrices. To begin, consider the real number equation $ax = b$. To solve this equation for x, multiply each side of the equation by a^{-1} (provided that $a \neq 0$).

$$ax = b$$
$$(a^{-1}a)x = a^{-1}b$$
$$(1)x = a^{-1}b$$
$$x = a^{-1}b$$

The number a^{-1} is called the *multiplicative inverse of a* because $a^{-1}a = 1$. The definition of the multiplicative **inverse of a matrix** is similar.

Definition of the Inverse of a Square Matrix

Let A be an $n \times n$ matrix and let I_n be the $n \times n$ identity matrix. If there exists a matrix A^{-1} such that

$$AA^{-1} = I_n = A^{-1}A$$

then A^{-1} is called the **inverse** of A. The symbol A^{-1} is read "A inverse."

Example 1   The Inverse of a Matrix

Show that B is the inverse of A, where $A = \begin{bmatrix} -1 & 2 \\ -1 & 1 \end{bmatrix}$ and $B = \begin{bmatrix} 1 & -2 \\ 1 & -1 \end{bmatrix}$.

Solution

To show that B is the inverse of A, show that $AB = I = BA$, as follows.

$$AB = \begin{bmatrix} -1 & 2 \\ -1 & 1 \end{bmatrix}\begin{bmatrix} 1 & -2 \\ 1 & -1 \end{bmatrix} = \begin{bmatrix} -1+2 & 2-2 \\ -1+1 & 2-1 \end{bmatrix} = \begin{bmatrix} 1 & 0 \\ 0 & 1 \end{bmatrix}$$

$$BA = \begin{bmatrix} 1 & -2 \\ 1 & -1 \end{bmatrix}\begin{bmatrix} -1 & 2 \\ -1 & 1 \end{bmatrix} = \begin{bmatrix} -1+2 & 2-2 \\ -1+1 & 2-1 \end{bmatrix} = \begin{bmatrix} 1 & 0 \\ 0 & 1 \end{bmatrix}$$

As you can see, $AB = I = BA$. This is an example of a square matrix that has an inverse. Note that not all square matrices have inverses.

✓**CHECKPOINT**   Now try Exercise 3.

Recall that it is not always true that $AB = BA$, even if both products are defined. However, if A and B are both square matrices and $AB = I_n$, it can be shown that $BA = I_n$. So, in Example 1, you need only check that $AB = I_2$.

What you should learn

- Verify that two matrices are inverses of each other.
- Use Gauss-Jordan elimination to find inverses of matrices.
- Use a formula to find inverses of 2×2 matrices.
- Use inverse matrices to solve systems of linear equations.

Why you should learn it

A system of equations can be solved using the inverse of the coefficient matrix. This method is particularly useful when the coefficients are the same for several systems, but the constants are different. Exercise 77 on page 550 shows how to use an inverse matrix to find a model for the number of people participating in snowboarding.

Chris Noble/Getty Images

Finding Inverse Matrices

If a matrix A has an inverse, A is called **invertible** (or **nonsingular**); otherwise, A is called **singular.** A nonsquare matrix cannot have an inverse. To see this, note that if A is of order $m \times n$ and B is of order $n \times m$ (where $m \neq n$), the products AB and BA are of different orders and so cannot be equal to each other. Not all square matrices have inverses, as you will see at the bottom of page 544. If, however, a matrix does have an inverse, that inverse is unique. Example 2 shows how to use systems of equations to find the inverse of a matrix.

Example 2   Finding the Inverse of a Matrix

Find the inverse of

$$A = \begin{bmatrix} 1 & 4 \\ -1 & -3 \end{bmatrix}.$$

Solution

To find the inverse of A, try to solve the matrix equation $AX = I$ for X.

$$\begin{bmatrix} 1 & 4 \\ -1 & -3 \end{bmatrix}\begin{bmatrix} x_{11} & x_{12} \\ x_{21} & x_{22} \end{bmatrix} = \begin{bmatrix} 1 & 0 \\ 0 & 1 \end{bmatrix}$$

$$\begin{bmatrix} x_{11} + 4x_{21} & x_{12} + 4x_{22} \\ -x_{11} - 3x_{21} & -x_{12} - 3x_{22} \end{bmatrix} = \begin{bmatrix} 1 & 0 \\ 0 & 1 \end{bmatrix}$$

Equating corresponding entries, you obtain the following two systems of linear equations.

$$\begin{cases} x_{11} + 4x_{21} = 1 \\ -x_{11} - 3x_{21} = 0 \end{cases} \qquad \begin{cases} x_{12} + 4x_{22} = 0 \\ -x_{12} - 3x_{22} = 1 \end{cases}$$

Solve the first system using elementary row operations to determine that $x_{11} = -3$ and $x_{21} = 1$. From the second system you can determine that $x_{12} = -4$ and $x_{22} = 1$. Therefore, the inverse of A is

$$X = A^{-1}$$
$$= \begin{bmatrix} -3 & -4 \\ 1 & 1 \end{bmatrix}.$$

You can use matrix multiplication to check this result.

Check

$$AA^{-1} = \begin{bmatrix} 1 & 4 \\ -1 & -3 \end{bmatrix}\begin{bmatrix} -3 & -4 \\ 1 & 1 \end{bmatrix} = \begin{bmatrix} 1 & 0 \\ 0 & 1 \end{bmatrix} ✓$$

$$A^{-1}A = \begin{bmatrix} -3 & -4 \\ 1 & 1 \end{bmatrix}\begin{bmatrix} 1 & 4 \\ -1 & -3 \end{bmatrix} = \begin{bmatrix} 1 & 0 \\ 0 & 1 \end{bmatrix} ✓$$

✓**CHECKPOINT**   Now try Exercise 11.

Exploration

Most graphing utilities are capable of finding the inverse of a square matrix. Try using a graphing utility to find the inverse of the matrix

$$A = \begin{bmatrix} 2 & -3 & 1 \\ -1 & 2 & -1 \\ -2 & 0 & 1 \end{bmatrix}.$$

After you find A^{-1}, store it as $[B]$ and use the graphing utility to find $[A] \times [B]$ and $[B] \times [A]$. What can you conclude?

In Example 2, note that the two systems of linear equations have the *same coefficient matrix A*. Rather than solve the two systems represented by

$$\begin{bmatrix} 1 & 4 & \vdots & 1 \\ -1 & -3 & \vdots & 0 \end{bmatrix}$$

and

$$\begin{bmatrix} 1 & 4 & \vdots & 0 \\ -1 & -3 & \vdots & 1 \end{bmatrix}$$

separately, you can solve them *simultaneously* by *adjoining* the identity matrix to the coefficient matrix to obtain

$$\overset{A}{} \qquad \overset{I}{}$$

$$\begin{bmatrix} 1 & 4 & \vdots & 1 & 0 \\ -1 & -3 & \vdots & 0 & 1 \end{bmatrix}.$$

This "doubly augmented" matrix can be represented as $[A \,\vdots\, I]$. By applying Gauss-Jordan elimination to this matrix, you can solve *both* systems with a single elimination process.

$$\begin{bmatrix} 1 & 4 & \vdots & 1 & 0 \\ -1 & -3 & \vdots & 0 & 1 \end{bmatrix}$$

$$R_1 + R_2 \rightarrow \begin{bmatrix} 1 & 4 & \vdots & 1 & 0 \\ 0 & 1 & \vdots & 1 & 1 \end{bmatrix}$$

$$-4R_2 + R_1 \rightarrow \begin{bmatrix} 1 & 0 & \vdots & -3 & -4 \\ 0 & 1 & \vdots & 1 & 1 \end{bmatrix}$$

So, from the "doubly augmented" matrix $[A \,\vdots\, I]$, you obtained the matrix $[I \,\vdots\, A^{-1}]$.

$$\overset{A}{} \qquad \overset{I}{} \qquad\qquad \overset{I}{} \qquad \overset{A^{-1}}{}$$

$$\begin{bmatrix} 1 & 4 & \vdots & 1 & 0 \\ -1 & -3 & \vdots & 0 & 1 \end{bmatrix} \implies \begin{bmatrix} 1 & 0 & \vdots & -3 & -4 \\ 0 & 1 & \vdots & 1 & 1 \end{bmatrix}$$

This procedure (or algorithm) works for any square matrix that has an inverse.

Finding an Inverse Matrix

Let A be a square matrix of order n.

1. Write the $n \times 2n$ matrix that consists of the given matrix A on the left and the $n \times n$ identity matrix I on the right to obtain $[A \,\vdots\, I]$.

2. If possible, row reduce A to I using elementary row operations on the *entire* matrix $[A \,\vdots\, I]$. The result will be the matrix $[I \,\vdots\, A^{-1}]$. If this is not possible, A is not invertible.

3. Check your work by multiplying to see that $AA^{-1} = I = A^{-1}A$.

Example 3 Finding the Inverse of a Matrix

Find the inverse of $A = \begin{bmatrix} 1 & -1 & 0 \\ 1 & 0 & -1 \\ 6 & -2 & -3 \end{bmatrix}$.

Solution

Begin by adjoining the identity matrix to A to form the matrix

$$[A \ \vdots \ I] = \begin{bmatrix} 1 & -1 & 0 & \vdots & 1 & 0 & 0 \\ 1 & 0 & -1 & \vdots & 0 & 1 & 0 \\ 6 & -2 & -3 & \vdots & 0 & 0 & 1 \end{bmatrix}.$$

Use elementary row operations to obtain the form $[I \ \vdots \ A^{-1}]$, as follows.

$$\begin{bmatrix} 1 & 0 & 0 & \vdots & -2 & -3 & 1 \\ 0 & 1 & 0 & \vdots & -3 & -3 & 1 \\ 0 & 0 & 1 & \vdots & -2 & -4 & 1 \end{bmatrix}$$

Therefore, the matrix A is invertible and its inverse is

$$A^{-1} = \begin{bmatrix} -2 & -3 & 1 \\ -3 & -3 & 1 \\ -2 & -4 & 1 \end{bmatrix}.$$

Try using a graphing utility to confirm this result by multiplying A by A^{-1} to obtain I.

✓CHECKPOINT   Now try Exercise 17.

TECHNOLOGY TIP

Most graphing utilities can find the inverse of a matrix. A graphing utility can be used to check matrix operations. This saves valuable time otherwise spent doing minor arithmetic calculations.

The algorithm shown in Example 3 applies to any $n \times n$ matrix A. When using this algorithm, if the matrix A does not reduce to the identity matrix, then A does not have an inverse. For instance, the following matrix has no inverse.

$$A = \begin{bmatrix} 1 & 2 & 0 \\ 3 & -1 & 2 \\ -2 & 3 & -2 \end{bmatrix}$$

To see why matrix A above has no inverse, begin by adjoining the identity matrix to A to form

$$[A \ \vdots \ I] = \begin{bmatrix} 1 & 2 & 0 & \vdots & 1 & 0 & 0 \\ 3 & -1 & 2 & \vdots & 0 & 1 & 0 \\ -2 & 3 & -2 & \vdots & 0 & 0 & 1 \end{bmatrix}.$$

Then use elementary row operations to obtain

$$\begin{bmatrix} 1 & 2 & 0 & \vdots & 1 & 0 & 0 \\ 0 & -7 & 2 & \vdots & -3 & 1 & 0 \\ 0 & 0 & 0 & \vdots & -1 & 1 & 1 \end{bmatrix}.$$

At this point in the elimination process you can see that it is impossible to obtain the identity matrix I on the left. Therefore, A is not invertible.

Example 4   Finding the Inverse of a Matrix

Use a graphing utility to find the inverse of $A = \begin{bmatrix} 1 & 2 & -2 \\ 1 & -1 & 0 \\ 0 & -1 & 4 \end{bmatrix}$.

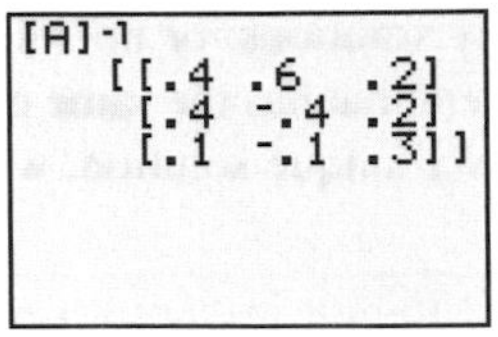

```
[A]-1
   [[.4 .6  .2]
    [.4 -.4 .2]
    [.1 -.1 .3]]
```

Figure 7.31

Solution

Use the *matrix editor* to enter A into the graphing utility. Use the inverse key $\boxed{x^{-1}}$ to find the inverse of the matrix, as shown in Figure 7.31. Check this result algebraically by multiplying A by A^{-1} to obtain I.

✔**CHECKPOINT**   Now try Exercise 23.

The Inverse of a 2 × 2 Matrix

Using Gauss-Jordan elimination to find the inverse of a matrix works well (even as a computer technique) for matrices of order 3×3 or greater. For 2×2 matrices, however, many people prefer to use a formula for the inverse rather than Gauss-Jordan elimination. This simple formula, which works *only* for 2×2 matrices, is explained as follows. If A is the 2×2 matrix given by

$$A = \begin{bmatrix} a & b \\ c & d \end{bmatrix}$$

then A is invertible if and only if $ad - bc \neq 0$. If $ad - bc \neq 0$, the inverse is given by

$$A^{-1} = \frac{1}{ad - bc}\begin{bmatrix} d & -b \\ -c & a \end{bmatrix}. \qquad \text{Formula for inverse of matrix } A$$

The denominator $ad - bc$ is called the *determinant* of the 2×2 matrix A. You will study determinants in the next section.

Example 5   Finding the Inverse of a 2 × 2 Matrix

If possible, find the inverse of $A = \begin{bmatrix} 3 & -1 \\ -2 & 2 \end{bmatrix}$.

Solution

Apply the formula for the inverse of a 2×2 matrix to obtain

$$ad - bc = (3)(2) - (-1)(-2) = 4.$$

Because this quantity is not zero, the inverse is formed by interchanging the entries on the main diagonal, changing the signs of the other two entries, and multiplying by the scalar $\frac{1}{4}$, as follows.

$$A^{-1} = \frac{1}{4}\begin{bmatrix} 2 & 1 \\ 2 & 3 \end{bmatrix} = \begin{bmatrix} \frac{1}{2} & \frac{1}{4} \\ \frac{1}{2} & \frac{3}{4} \end{bmatrix}$$

✔**CHECKPOINT**   Now try Exercise 29.

> **Exploration**
>
> Use a graphing utility to find the inverse of the matrix
>
> $$A = \begin{bmatrix} 1 & -3 \\ -2 & 6 \end{bmatrix}.$$
>
> What message appears on the screen? Why does the graphing utility display this message?

Systems of Linear Equations

You know that a system of linear equations can have exactly one solution, infinitely many solutions, or no solution. If the coefficient matrix A of a *square* system (a system that has the same number of equations as variables) is invertible, the system has a unique solution, which is defined as follows.

> ### A System of Equations with a Unique Solution
>
> If A is an invertible matrix, the system of linear equations represented by $AX = B$ has a unique solution given by
>
> $$X = A^{-1}B.$$

The formula $X = A^{-1}B$ is used on most graphing utilities to solve linear systems that have invertible coefficient matrices. That is, you enter the $n \times n$ coefficient matrix $[A]$ and the $n \times 1$ column matrix $[B]$. The solution X is given by $[A]^{-1}[B]$.

Example 6   Solving a System of Equations Using an Inverse

Use an inverse matrix to solve the system.

$$\begin{cases} 2x + 3y + z = -1 \\ 3x + 3y + z = 1 \\ 2x + 4y + z = -2 \end{cases}$$

Solution

Begin by writing the system as $AX = B$.

$$\begin{bmatrix} 2 & 3 & 1 \\ 3 & 3 & 1 \\ 2 & 4 & 1 \end{bmatrix} \begin{bmatrix} x \\ y \\ z \end{bmatrix} = \begin{bmatrix} -1 \\ 1 \\ -2 \end{bmatrix}$$

Then, use Gauss-Jordan elimination to find A^{-1}.

$$A^{-1} = \begin{bmatrix} -1 & 1 & 0 \\ -1 & 0 & 1 \\ 6 & -2 & -3 \end{bmatrix}$$

Finally, multiply B by A^{-1} on the left to obtain the solution.

$$X = A^{-1}B$$
$$= \begin{bmatrix} -1 & 1 & 0 \\ -1 & 0 & 1 \\ 6 & -2 & -3 \end{bmatrix} \begin{bmatrix} -1 \\ 1 \\ -2 \end{bmatrix} = \begin{bmatrix} 2 \\ -1 \\ -2 \end{bmatrix}$$

So, the solution is $x = 2$, $y = -1$, and $z = -2$. Use a graphing utility to verify A^{-1} for the system of equations.

✓**CHECKPOINT**   Now try Exercise 55.

STUDY TIP

Remember that matrix multiplication is not commutative. So, you must multiply matrices in the correct order. For instance, in Example 6, you must multiply B by A^{-1} on the left.

7.6 Exercises

See www.CalcChat.com for worked-out solutions to odd-numbered exercises.

Vocabulary Check

Fill in the blanks.

1. In a _____ matrix, the number of rows equals the number of columns.

2. If there exists an $n \times n$ matrix A^{-1} such that $AA^{-1} = I_n = A^{-1}A$, then A^{-1} is called the _____ of A.

3. If a matrix A has an inverse, it is called invertible or _____ ; if it does not have an inverse, it is called _____ .

In Exercises 1–6, show that B is the inverse of A.

1. $A = \begin{bmatrix} 2 & 1 \\ 5 & 3 \end{bmatrix}$, $B = \begin{bmatrix} 3 & -1 \\ -5 & 2 \end{bmatrix}$

2. $A = \begin{bmatrix} 1 & -1 \\ -1 & 2 \end{bmatrix}$, $B = \begin{bmatrix} 2 & 1 \\ 1 & 1 \end{bmatrix}$

3. $A = \begin{bmatrix} 1 & 2 \\ 3 & 4 \end{bmatrix}$, $B = \begin{bmatrix} -2 & 1 \\ \frac{3}{2} & -\frac{1}{2} \end{bmatrix}$

4. $A = \begin{bmatrix} 1 & -1 \\ 2 & 3 \end{bmatrix}$, $B = \begin{bmatrix} \frac{3}{5} & \frac{1}{5} \\ -\frac{2}{5} & \frac{1}{5} \end{bmatrix}$

5. $A = \begin{bmatrix} 2 & -17 & 11 \\ -1 & 11 & -7 \\ 0 & 3 & -2 \end{bmatrix}$, $B = \begin{bmatrix} 1 & 1 & 2 \\ 2 & 4 & -3 \\ 3 & 6 & -5 \end{bmatrix}$

6. $A = \begin{bmatrix} 1 & 0 & -1 \\ -1 & 1 & 0 \\ 1 & 2 & 0 \end{bmatrix}$, $B = \frac{1}{3}\begin{bmatrix} 0 & -2 & 1 \\ 0 & 1 & 1 \\ -3 & -2 & 1 \end{bmatrix}$

In Exercises 7–10, use the matrix capabilities of a graphing utility to show that B is the inverse of A.

7. $A = \begin{bmatrix} -1 & -4 \\ 1 & 2 \end{bmatrix}$, $B = \begin{bmatrix} 1 & 2 \\ -\frac{1}{2} & -\frac{1}{2} \end{bmatrix}$

8. $A = \begin{bmatrix} 11 & -12 \\ 2 & -2 \end{bmatrix}$, $B = \begin{bmatrix} -1 & 6 \\ -1 & \frac{11}{2} \end{bmatrix}$

9. $A = \begin{bmatrix} 1.6 & 2 \\ -3.5 & -4.5 \end{bmatrix}$, $B = \begin{bmatrix} 22.5 & 10 \\ -17.5 & -8 \end{bmatrix}$

10. $A = \begin{bmatrix} 4 & 0 & -2 \\ 1 & 2 & -4 \\ 0 & 3 & 1 \end{bmatrix}$, $B = \begin{bmatrix} 0.28 & -0.12 & 0.08 \\ -0.02 & 0.08 & 0.28 \\ 0.06 & -0.24 & 0.16 \end{bmatrix}$

In Exercises 11–20, find the inverse of the matrix (if it exists).

11. $\begin{bmatrix} 2 & 0 \\ 0 & 3 \end{bmatrix}$

12. $\begin{bmatrix} 1 & 2 \\ 3 & 7 \end{bmatrix}$

13. $\begin{bmatrix} 1 & -2 \\ 2 & -3 \end{bmatrix}$

14. $\begin{bmatrix} -7 & 33 \\ 4 & -19 \end{bmatrix}$

15. $\begin{bmatrix} 2 & 7 & 1 \\ -3 & -9 & 2 \end{bmatrix}$

16. $\begin{bmatrix} -2 & 5 \\ 6 & -15 \\ 0 & 1 \end{bmatrix}$

17. $\begin{bmatrix} 1 & 1 & 1 \\ 3 & 5 & 4 \\ 3 & 6 & 5 \end{bmatrix}$

18. $\begin{bmatrix} 1 & 2 & 2 \\ 3 & 7 & 9 \\ -1 & -4 & -7 \end{bmatrix}$

19. $\begin{bmatrix} -5 & 0 & 0 \\ 2 & 0 & 0 \\ -1 & 5 & 7 \end{bmatrix}$

20. $\begin{bmatrix} 1 & 0 & 0 \\ 3 & 5 & 0 \\ 2 & 5 & 0 \end{bmatrix}$

In Exercises 21–28, use the matrix capabilities of a graphing utility to find the inverse of the matrix (if it exists).

21. $\begin{bmatrix} 1 & 0 & -2 \\ 4 & 0 & 0 \\ -2 & 0 & 3 \end{bmatrix}$

22. $\begin{bmatrix} 1 & 2 & -1 \\ 3 & 7 & -10 \\ -5 & -7 & -15 \end{bmatrix}$

23. $\begin{bmatrix} -\frac{1}{2} & \frac{3}{4} & \frac{1}{4} \\ 1 & 0 & -\frac{3}{2} \\ 0 & -1 & \frac{1}{2} \end{bmatrix}$

24. $\begin{bmatrix} -\frac{5}{6} & \frac{1}{3} & \frac{11}{6} \\ 0 & \frac{2}{3} & 2 \\ 1 & -\frac{1}{2} & -\frac{5}{2} \end{bmatrix}$

25. $\begin{bmatrix} 0.1 & 0.2 & 0.3 \\ -0.3 & 0.2 & 0.2 \\ 0.5 & 0.4 & 0.4 \end{bmatrix}$

26. $\begin{bmatrix} 0.6 & 0 & -0.3 \\ 0.7 & -1 & 0.2 \\ 1 & 0 & -0.9 \end{bmatrix}$

27. $\begin{bmatrix} -1 & 0 & 1 & 0 \\ 0 & 2 & 0 & -1 \\ 2 & 0 & -1 & 0 \\ 0 & -1 & 0 & 1 \end{bmatrix}$

28. $\begin{bmatrix} 1 & -2 & -1 & -2 \\ 3 & -5 & -2 & -3 \\ 2 & -5 & -2 & -5 \\ -1 & 4 & 4 & 11 \end{bmatrix}$

In Exercises 29–36, use the formula on page 545 to find the inverse of the 2×2 matrix.

29. $\begin{bmatrix} 5 & 1 \\ -2 & -2 \end{bmatrix}$

30. $\begin{bmatrix} -8 & 0 \\ 11 & -10 \end{bmatrix}$

31. $\begin{bmatrix} \frac{7}{2} & -\frac{3}{4} \\ \frac{1}{5} & \frac{4}{5} \end{bmatrix}$

32. $\begin{bmatrix} -\frac{1}{4} & -\frac{2}{3} \\ \frac{1}{3} & \frac{8}{9} \end{bmatrix}$

33. $\begin{bmatrix} 2 & 3 \\ -1 & 5 \end{bmatrix}$ **34.** $\begin{bmatrix} 1 & -2 \\ -3 & 2 \end{bmatrix}$

35. $\begin{bmatrix} -1 & 0 \\ 3 & -2 \end{bmatrix}$ **36.** $\begin{bmatrix} 2 & -5 \\ 3 & 1 \end{bmatrix}$

In Exercises 37–40, find the value of the constant k such that $B = A^{-1}$.

37. $A = \begin{bmatrix} 1 & 2 \\ -2 & 0 \end{bmatrix}$, $B = \begin{bmatrix} k & -\frac{1}{2} \\ \frac{1}{2} & \frac{1}{4} \end{bmatrix}$

38. $A = \begin{bmatrix} -1 & 1 \\ 2 & 1 \end{bmatrix}$, $B = \begin{bmatrix} -\frac{1}{3} & \frac{1}{3} \\ k & \frac{1}{3} \end{bmatrix}$

39. $A = \begin{bmatrix} -1 & 2 \\ -3 & 1 \end{bmatrix}$, $B = \begin{bmatrix} \frac{1}{5} & -\frac{2}{5} \\ k & -\frac{1}{5} \end{bmatrix}$

40. $A = \begin{bmatrix} -1 & -2 \\ 0 & 2 \end{bmatrix}$, $B = \begin{bmatrix} -1 & -1 \\ 0 & k \end{bmatrix}$

In Exercises 41–44, use the inverse matrix found in Exercise 13 to solve the system of linear equations.

41. $\begin{cases} x - 2y = 5 \\ 2x - 3y = 10 \end{cases}$ **42.** $\begin{cases} x - 2y = 0 \\ 2x - 3y = 3 \end{cases}$

43. $\begin{cases} x - 2y = 4 \\ 2x - 3y = 2 \end{cases}$ **44.** $\begin{cases} x - 2y = 1 \\ 2x - 3y = -2 \end{cases}$

In Exercises 45 and 46, use the inverse matrix found in Exercise 17 to solve the system of linear equations.

45. $\begin{cases} x + y + z = 0 \\ 3x + 5y + 4z = 5 \\ 3x + 6y + 5z = 2 \end{cases}$ **46.** $\begin{cases} x + y + z = -1 \\ 3x + 5y + 4z = 2 \\ 3x + 6y + 5z = 0 \end{cases}$

In Exercises 47 and 48, use the inverse matrix found in Exercise 28 and the matrix capabilities of a graphing utility to solve the system of linear equations.

47. $\begin{cases} x_1 - 2x_2 - x_3 - 2x_4 = 0 \\ 3x_1 - 5x_2 - 2x_3 - 3x_4 = 1 \\ 2x_1 - 5x_2 - 2x_3 - 5x_4 = -1 \\ -x_1 + 4x_2 + 4x_3 + 11x_4 = 2 \end{cases}$

48. $\begin{cases} x_1 - 2x_2 - x_3 - 2x_4 = 1 \\ 3x_1 - 5x_2 - 2x_3 - 3x_4 = -2 \\ 2x_1 - 5x_2 - 2x_3 - 5x_4 = 0 \\ -x_1 + 4x_2 + 4x_3 + 11x_4 = -3 \end{cases}$

In Exercises 49–56, use an inverse matrix to solve (if possible) the system of linear equations.

49. $\begin{cases} 3x + 4y = -2 \\ 5x + 3y = 4 \end{cases}$ **50.** $\begin{cases} 18x + 12y = 13 \\ 30x + 24y = 23 \end{cases}$

51. $\begin{cases} -0.4x + 0.8y = 1.6 \\ 2x - 4y = 5 \end{cases}$ **52.** $\begin{cases} 0.2x - 0.6y = 2.4 \\ -x + 1.4y = -8.8 \end{cases}$

53. $\begin{cases} -\frac{1}{4}x + \frac{3}{8}y = -2 \\ \frac{3}{2}x + \frac{3}{4}y = -12 \end{cases}$ **54.** $\begin{cases} \frac{5}{6}x - y = -20 \\ \frac{4}{3}x - \frac{7}{2}y = -51 \end{cases}$

55. $\begin{cases} 4x - y + z = -5 \\ 2x + 2y + 3z = 10 \\ 5x - 2y + 6z = 1 \end{cases}$ **56.** $\begin{cases} 4x - 2y + 3z = -2 \\ 2x + 2y + 5z = 16 \\ 8x - 5y - 2z = 4 \end{cases}$

In Exercises 57–60, use the matrix capabilities of a graphing utility to solve (if possible) the system of linear equations.

57. $\begin{cases} 5x - 3y + 2z = 2 \\ 2x + 2y - 3z = 3 \\ -x + 7y - 8z = 4 \end{cases}$ **58.** $\begin{cases} 2x + 3y + 5z = 4 \\ 3x + 5y - 9z = 7 \\ 5x + 9y + 17z = 13 \end{cases}$

59. $\begin{cases} 7x - 3y + 2w = 41 \\ -2x + y - w = -13 \\ 4x + z - 2w = 12 \\ -x + y - w = -8 \end{cases}$

60. $\begin{cases} 2x + 5y + w = 11 \\ x + 4y + 2z - 2w = -7 \\ 2x - 2y + 5z + w = 3 \\ x - 3w = -1 \end{cases}$

Computer Graphics **In Exercises 61–64, the matrix product AX performs the translation of the point (x, y) to the point $(x + h, y + k)$, when**

$$A = \begin{bmatrix} 1 & 0 & h \\ 0 & 1 & k \\ 0 & 0 & 1 \end{bmatrix} \text{ and } X = \begin{bmatrix} x \\ y \\ 1 \end{bmatrix}.$$

(a) What are the coordinates of the point (x, y)?

(b) Predict the coordinates of the translated point.

(c) Find $B = AX$ and compare your result with part (b).

(d) Find A^{-1}.

(e) Find $A^{-1}B$. What does $A^{-1}B$ represent?

61. $A = \begin{bmatrix} 1 & 0 & 2 \\ 0 & 1 & -1 \\ 0 & 0 & 1 \end{bmatrix}$, $X = \begin{bmatrix} -3 \\ 2 \\ 1 \end{bmatrix}$

62. $A = \begin{bmatrix} 1 & 0 & -2 \\ 0 & 1 & 3 \\ 0 & 0 & 1 \end{bmatrix}$, $X = \begin{bmatrix} 1 \\ -2 \\ 1 \end{bmatrix}$

63. $A = \begin{bmatrix} 1 & 0 & -3 \\ 0 & 1 & 4 \\ 0 & 0 & 1 \end{bmatrix}$, $X = \begin{bmatrix} 2 \\ -4 \\ 1 \end{bmatrix}$

64. $A = \begin{bmatrix} 1 & 0 & 3 \\ 0 & 1 & -2 \\ 0 & 0 & 1 \end{bmatrix}$, $X = \begin{bmatrix} 0 \\ -3 \\ 1 \end{bmatrix}$

Investment Portfolio   In Exercises 65–68, consider a person who invests in AAA-rated bonds, A-rated bonds, and B-rated bonds. The average yields are 6.5% on AAA bonds, 7% on A bonds, and 9% on B bonds. The person invests twice as much in B bonds as in A bonds. Let x, y, and z represent the amounts invested in AAA, A, and B bonds, respectively.

$$\begin{cases} x + y + z = \text{(total investment)} \\ 0.065x + 0.07y + 0.09z = \text{(annual return)} \\ 2y - z = 0 \end{cases}$$

Use the inverse of the coefficient matrix of this system to find the amount invested in each type of bond.

	Total Investment	Annual Return
65.	$25,000	$1900
66.	$10,000	$760
67.	$65,000	$5050
68.	$500,000	$38,000

Circuit Analysis   In Exercises 69 and 70, consider the circuit in the figure. The currents I_1, I_2, and I_3, in amperes, are given by the solution of the system of linear equations

$$\begin{cases} 2I_1 + 4I_3 = E_1 \\ I_2 + 4I_3 = E_2 \\ I_1 + I_2 - I_3 = 0 \end{cases}$$

where E_1 and E_2 are voltages. Use the inverse of the coefficient matrix of this system to find the unknown currents for the voltages.

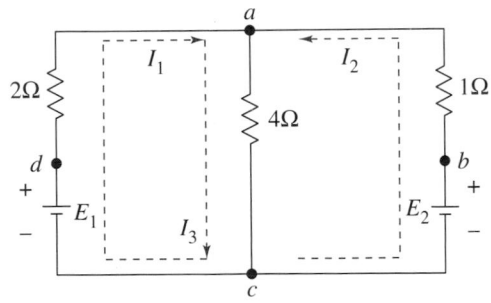

69. $E_1 = 14$ volts, $E_2 = 28$ volts

70. $E_1 = 10$ volts, $E_2 = 10$ volts

Production   In Exercises 71–74, a small home business specializes in gourmet-baked goods, muffins, bones, and cookies, for dogs. In addition to other ingredients, each muffin requires 2 units of beef, 3 units of chicken, and 2 units of liver. Each bone requires 1 unit of beef, 1 unit of chicken, and 1 unit of liver. Each cookie requires 2 units of beef, 1 unit of chicken, and 1.5 units of liver. Find the number of muffins, bones, and cookies that the company can create with the given amounts of ingredients.

71. 700 units of beef
500 units of chicken
600 units of liver

72. 525 units of beef
480 units of chicken
500 units of liver

73. 800 units of beef
750 units of chicken
725 units of liver

74. 1000 units of beef
950 units of chicken
900 units of liver

75. *Coffee*   A coffee manufacturer sells a 10-pound package of coffee for $26 that contains three flavors of coffee. French vanilla coffee costs $2 per pound, hazelnut flavored coffee costs $2.50 per pound, and Swiss chocolate flavored coffee costs $3 per pound. The package contains the same amount of hazelnut as Swiss chocolate. Let f represent the number of pounds of French vanilla, h represent the number of pounds of hazelnut, and s represent the number of pounds of Swiss chocolate.

(a) Write a system of linear equations that represents the situation.

(b) Write a matrix equation that corresponds to your system.

(c) Solve your system of linear equations using an inverse matrix. Find the number of pounds of each flavor of coffee in the 10-pound package.

76. *Flowers*   A florist is creating 10 centerpieces for the tables at a wedding reception. Roses cost $2.50 each, lilies cost $4 each, and irises cost $2 each. The customer has a budget of $300 allocated for the centerpieces and wants each centerpiece to contain 12 flowers, with twice as many roses as the number of irises and lilies combined.

(a) Write a system of linear equations that represents the situation.

(b) Write a matrix equation that corresponds to your system.

(c) Solve your system of linear equations using an inverse matrix. Find the number of flowers of each type that the florist can use to create the 10 centerpieces.

77. *Data Analysis*   The table shows the numbers of people y (in thousands) who participated in snowboarding from 2002 to 2004.   (Source: National Sporting Goods Association)

Year	Snowboarders, y (in thousands)
2002	5343
2003	5589
2004	6309

(a) The data can be approximated by a parabola. Create a system of linear equations for the data. Let t represent the year, with $t = 2$ corresponding to 2002.

(b) Use the matrix capabilities of a graphing utility to find an inverse matrix to solve the system in part (a) and find the least squares regression parabola $y = at^2 + bt + c$.

(c) Use a graphing utility to graph the parabola with the data points.

(d) Use the result of part (b) to estimate the numbers of snowboarders in 2005, 2010, and 2015.

(e) Are your estimates from part (d) reasonable? Explain.

78. *Data Analysis*   The table shows the numbers of international travelers y (in millions) to the United States from Europe from 2002 to 2004.   (Source: U.S. Department of Commerce)

Year	Travelers, y (in millions)
2002	8603
2003	8639
2004	9686

(a) The data can be approximated by a parabola. Create a system of linear equations for the data. Let t represent the year, with $t = 2$ corresponding to 2002.

(b) Use the matrix capabilities of a graphing utility to find an inverse matrix to solve the system in part (a) and find the least squares regression parabola $y = at^2 + bt + c$.

(c) Use a graphing utility to graph the parabola with the data points.

(d) Use the result of part (b) to estimate the numbers of international travelers to the United States from Europe in 2005, 2010, and 2015.

(e) Are your estimates from part (d) reasonable? Explain.

Synthesis

True or False?   **In Exercises 79 and 80, determine whether the statement is true or false. Justify your answer.**

79. Multiplication of an invertible matrix and its inverse is commutative.

80. No nonsquare matrices have inverses.

81. If A is a 2×2 matrix given by $A = \begin{bmatrix} a & b \\ c & d \end{bmatrix}$, then A is invertible if and only if $ad - bc \neq 0$. If $ad - bc \neq 0$, verify that the inverse is $A^{-1} = \dfrac{1}{ad - bc} \begin{bmatrix} d & -b \\ -c & a \end{bmatrix}$.

82. *Exploration*   Consider the matrices of the form

$$A = \begin{bmatrix} a_{11} & 0 & 0 & 0 & \dots & 0 \\ 0 & a_{22} & 0 & 0 & \dots & 0 \\ 0 & 0 & a_{33} & 0 & \dots & 0 \\ \vdots & \vdots & \vdots & \vdots & \dots & \vdots \\ 0 & 0 & 0 & 0 & \dots & a_{nn} \end{bmatrix}.$$

(a) Write a 2×2 matrix and a 3×3 matrix in the form of A. Find the inverse of each.

(b) Use the result of part (a) to make a conjecture about the inverse of a matrix in the form of A.

Skills Review

In Exercises 83–86, simplify the complex fraction.

83. $\dfrac{\left(\dfrac{9}{x}\right)}{\left(\dfrac{6}{x} + 2\right)}$

84. $\dfrac{\left(1 + \dfrac{2}{x}\right)}{\left(1 - \dfrac{4}{x}\right)}$

85. $\dfrac{\left(\dfrac{4}{x^2 - 9} + \dfrac{2}{x - 2}\right)}{\left(\dfrac{1}{x + 3} + \dfrac{1}{x - 3}\right)}$

86. $\dfrac{\left(\dfrac{1}{x + 1} + \dfrac{1}{2}\right)}{\left(\dfrac{3}{2x^2 + 4x + 2}\right)}$

In Exercises 87–90, solve the equation algebraically. Round your result to three decimal places.

87. $e^{2x} + 2e^x - 15 = 0$

88. $e^{2x} - 10e^x + 24 = 0$

89. $7 \ln 3x = 12$

90. $\ln(x + 9) = 2$

91. *Make a Decision*   To work an extended application analyzing the number of U.S. households with color televisions from 1985 to 2005, visit this textbook's *Online Study Center.*   (Data Source: Nielsen Media Research)

7.7   The Determinant of a Square Matrix

The Determinant of a 2 × 2 Matrix

Every *square* matrix can be associated with a real number called its **determinant.** Determinants have many uses, and several will be discussed in this and the next section. Historically, the use of determinants arose from special number patterns that occur when systems of linear equations are solved. For instance, the system

$$\begin{cases} a_1x + b_1y = c_1 \\ a_2x + b_2y = c_2 \end{cases}$$

has a solution

$$x = \frac{c_1b_2 - c_2b_1}{a_1b_2 - a_2b_1}$$

and

$$y = \frac{a_1c_2 - a_2c_1}{a_1b_2 - a_2b_1}$$

provided that $a_1b_2 - a_2b_1 \neq 0$. Note that the denominator of each fraction is the same. This denominator is called the *determinant* of the coefficient matrix of the system.

Coefficient Matrix   *Determinant*

$$A = \begin{bmatrix} a_1 & b_1 \\ a_2 & b_2 \end{bmatrix} \qquad \det(A) = a_1b_2 - a_2b_1$$

The determinant of the matrix A can also be denoted by vertical bars on both sides of the matrix, as indicated in the following definition.

Definition of the Determinant of a 2 × 2 Matrix

The **determinant** of the matrix

$$A = \begin{bmatrix} a_1 & b_1 \\ a_2 & b_2 \end{bmatrix}$$

is given by

$$\det(A) = |A| = \begin{vmatrix} a_1 & b_1 \\ a_2 & b_2 \end{vmatrix}$$

$$= a_1b_2 - a_2b_1.$$

In this text, $\det(A)$ and $|A|$ are used interchangeably to represent the determinant of A. Although vertical bars are also used to denote the absolute value of a real number, the context will show which use is intended.

What you should learn

- Find the determinants of 2 × 2 matrices.
- Find minors and cofactors of square matrices.
- Find the determinants of square matrices.
- Find the determinants of triangular matrices.

Why you should learn it

Determinants are often used in other branches of mathematics. For instance, Exercises 61–66 on page 558 show some types of determinants that are useful in calculus.

A convenient method for remembering the formula for the determinant of a 2×2 matrix is shown in the following diagram.

$$\det(A) = \begin{vmatrix} a_1 & b_1 \\ a_2 & b_2 \end{vmatrix} = a_1 b_2 - a_2 b_1$$

Note that the determinant is the difference of the products of the two diagonals of the matrix.

Example 1   The Determinant of a 2×2 Matrix

Find the determinant of each matrix.

a. $A = \begin{bmatrix} 2 & -3 \\ 1 & 2 \end{bmatrix}$ **b.** $B = \begin{bmatrix} 2 & 1 \\ 4 & 2 \end{bmatrix}$ **c.** $C = \begin{bmatrix} 0 & \frac{3}{2} \\ 2 & 4 \end{bmatrix}$

Solution

a. $\det(A) = \begin{vmatrix} 2 & -3 \\ 1 & 2 \end{vmatrix} = 2(2) - 1(-3)$

$= 4 + 3 = 7$

b. $\det(B) = \begin{vmatrix} 2 & 1 \\ 4 & 2 \end{vmatrix} = 2(2) - 4(1)$

$= 4 - 4 = 0$

c. $\det(C) = \begin{vmatrix} 0 & \frac{3}{2} \\ 2 & 4 \end{vmatrix} = 0(4) - 2\left(\frac{3}{2}\right)$

$= 0 - 3 = -3$

✓CHECKPOINT   Now try Exercise 5.

> **Exploration**
>
> Try using a graphing utility to find the determinant of
> $$A = \begin{bmatrix} 3 & -1 & 1 \\ 0 & 2 & 1 \end{bmatrix}.$$
> What message appears on the screen? Why does the graphing utility display this message?

Notice in Example 1 that the determinant of a matrix can be positive, zero, or negative.

The determinant of a matrix of order 1×1 is defined simply as the entry of the matrix. For instance, if $A = \begin{bmatrix} -2 \end{bmatrix}$, then $\det(A) = -2$.

TECHNOLOGY TIP   Most graphing utilities can evaluate the determinant of a matrix. For instance, you can evaluate the determinant of the matrix A in Example 1(a) by entering the matrix as $[A]$ (see Figure 7.32) and then choosing the *determinant* feature. The result should be 7, as in Example 1(a) (see Figure 7.33).

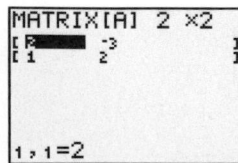

Figure 7.32

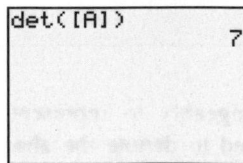

Figure 7.33

> **TECHNOLOGY SUPPORT**
>
> For instructions on how to use the *determinant* feature, see Appendix A; for specific keystrokes, go to this textbook's *Online Study Center*.

Minors and Cofactors

To define the determinant of a square matrix of order 3×3 or higher, it is helpful to introduce the concepts of **minors** and **cofactors.**

Minors and Cofactors of a Square Matrix

If A is a square matrix, the **minor** M_{ij} of the entry a_{ij} is the determinant of the matrix obtained by deleting the ith row and jth column of A. The **cofactor** C_{ij} of the entry a_{ij} is given by

$$C_{ij} = (-1)^{i+j}M_{ij}.$$

Sign Patterns for Cofactors

$$\begin{bmatrix} + & - & + \\ - & + & - \\ + & - & + \end{bmatrix}$$

3×3 matrix

$$\begin{bmatrix} + & - & + & - \\ - & + & - & + \\ + & - & + & - \\ - & + & - & + \end{bmatrix}$$

4×4 matrix

$$\begin{bmatrix} + & - & + & - & + & \cdots \\ - & + & - & + & - & \cdots \\ + & - & + & - & + & \cdots \\ - & + & - & + & - & \cdots \\ + & - & + & - & + & \cdots \\ \vdots & \vdots & \vdots & \vdots & \vdots & \end{bmatrix}$$

$n \times n$ matrix

Example 2   Finding the Minors and Cofactors of a Matrix

Find all the minors and cofactors of

$$A = \begin{bmatrix} 0 & 2 & 1 \\ 3 & -1 & 2 \\ 4 & 0 & 1 \end{bmatrix}.$$

Solution

To find the minor M_{11}, delete the first row and first column of A and evaluate the determinant of the resulting matrix.

$$\begin{bmatrix} 0 & 2 & 1 \\ 3 & -1 & 2 \\ 4 & 0 & 1 \end{bmatrix}, \quad M_{11} = \begin{vmatrix} -1 & 2 \\ 0 & 1 \end{vmatrix} = -1(1) - 0(2) = -1$$

Similarly, to find M_{12}, delete the first row and second column.

$$\begin{bmatrix} 0 & 2 & 1 \\ 3 & -1 & 2 \\ 4 & 0 & 1 \end{bmatrix}, \quad M_{12} = \begin{vmatrix} 3 & 2 \\ 4 & 1 \end{vmatrix} = 3(1) - 4(2) = -5$$

Continuing this pattern, you obtain all the minors.

$$M_{11} = -1 \qquad M_{12} = -5 \qquad M_{13} = 4$$
$$M_{21} = 2 \qquad M_{22} = -4 \qquad M_{23} = -8$$
$$M_{31} = 5 \qquad M_{32} = -3 \qquad M_{33} = -6$$

Now, to find the cofactors, combine these minors with the checkerboard pattern of signs for a 3×3 matrix shown at the upper right.

$$C_{11} = -1 \qquad C_{12} = 5 \qquad C_{13} = 4$$
$$C_{21} = -2 \qquad C_{22} = -4 \qquad C_{23} = 8$$
$$C_{31} = 5 \qquad C_{32} = 3 \qquad C_{33} = -6$$

✔**CHECKPOINT**   Now try Exercise 17.

STUDY TIP

In the sign patterns for cofactors above, notice that *odd* positions (where $i + j$ is odd) have negative signs and *even* positions (where $i + j$ is even) have positive signs.

The Determinant of a Square Matrix

The following definition is called *inductive* because it uses determinants of matrices of order $n - 1$ to define determinants of matrices of order n.

> **Determinant of a Square Matrix**
>
> If A is a square matrix (of order 2×2 or greater), the determinant of A is the sum of the entries in any row (or column) of A multiplied by their respective cofactors. For instance, expanding along the first row yields
>
> $$|A| = a_{11}C_{11} + a_{12}C_{12} + \cdots + a_{1n}C_{1n}.$$
>
> Applying this definition to find a determinant is called **expanding by cofactors.**

Try checking that for a 2×2 matrix

$$A = \begin{bmatrix} a_1 & b_1 \\ a_2 & b_2 \end{bmatrix}$$

the definition of the determinant above yields

$$|A| = a_1 b_2 - a_2 b_1$$

as previously defined.

Example 3 The Determinant of a Matrix of Order 3×3

Find the determinant of $A = \begin{bmatrix} 0 & 2 & 1 \\ 3 & -1 & 2 \\ 4 & 0 & 1 \end{bmatrix}$.

Solution

Note that this is the same matrix that was in Example 2. There you found the cofactors of the entries in the first row to be

$$C_{11} = -1, \qquad C_{12} = 5, \qquad \text{and} \qquad C_{13} = 4.$$

So, by the definition of the determinant of a square matrix, you have

$$|A| = a_{11}C_{11} + a_{12}C_{12} + a_{13}C_{13} \qquad \text{First-row expansion}$$

$$= 0(-1) + 2(5) + 1(4)$$

$$= 14.$$

✓CHECKPOINT Now try Exercise 23.

In Example 3, the determinant was found by expanding by the cofactors in the first row. You could have used any row or column. For instance, you could have expanded along the second row to obtain

$$|A| = a_{21}C_{21} + a_{22}C_{22} + a_{23}C_{23} \qquad \text{Second-row expansion}$$

$$= 3(-2) + (-1)(-4) + 2(8)$$

$$= 14.$$

When expanding by cofactors, you do not need to find cofactors of zero entries, because zero times its cofactor is zero.

$$a_{ij}C_{ij} = (0)C_{ij} = 0$$

So, the row (or column) containing the most zeros is usually the best choice for expansion by cofactors.

Triangular Matrices

Evaluating determinants of matrices of order 4×4 or higher can be tedious. There is, however, an important exception: the determinant of a **triangular** matrix. A triangular matrix is a square matrix with all zero entries either below or above its main diagonal. A square matrix is **upper triangular** if it has all zero entries below its main diagonal and **lower triangular** if it has all zero entries above its main diagonal. A matrix that is both upper and lower triangular is called **diagonal.** That is, a diagonal matrix is a square matrix in which all entries above and below the main diagonal are zero.

Exploration

The formula for the determinant of a triangular matrix (discussed at the left) is only one of many properties of matrices. You can use a computer or calculator to discover other properties. For instance, how is $|cA|$ related to $|A|$? How are $|A|$ and $|B|$ related to $|AB|$?

Upper Triangular Matrix

$$\begin{bmatrix} a_{11} & a_{12} & a_{13} & \cdots & a_{1n} \\ 0 & a_{22} & a_{23} & \cdots & a_{2n} \\ 0 & 0 & a_{33} & \cdots & a_{3n} \\ \vdots & \vdots & \vdots & & \vdots \\ 0 & 0 & 0 & \cdots & a_{nn} \end{bmatrix}$$

Lower Triangular Matrix

$$\begin{bmatrix} a_{11} & 0 & 0 & \cdots & 0 \\ a_{21} & a_{22} & 0 & \cdots & 0 \\ a_{31} & a_{32} & a_{33} & \cdots & 0 \\ \vdots & \vdots & \vdots & & \vdots \\ a_{n1} & a_{n2} & a_{n3} & \cdots & a_{nn} \end{bmatrix}$$

Diagonal Matrix

$$\begin{bmatrix} a_{11} & 0 & 0 & \cdots & 0 \\ 0 & a_{22} & 0 & \cdots & 0 \\ 0 & 0 & a_{33} & \cdots & 0 \\ \vdots & \vdots & \vdots & & \vdots \\ 0 & 0 & 0 & \cdots & a_{nn} \end{bmatrix}$$

To find the determinant of a triangular matrix of any order, simply form the product of the entries on the main diagonal.

Example 4   The Determinant of a Triangular Matrix

a. $\begin{vmatrix} 2 & 0 & 0 & 0 \\ 4 & -2 & 0 & 0 \\ -5 & 6 & 1 & 0 \\ 1 & 5 & 3 & 3 \end{vmatrix} = (2)(-2)(1)(3) = -12$

b. $\begin{vmatrix} -1 & 0 & 0 & 0 & 0 \\ 0 & 3 & 0 & 0 & 0 \\ 0 & 0 & 2 & 0 & 0 \\ 0 & 0 & 0 & 4 & 0 \\ 0 & 0 & 0 & 0 & -2 \end{vmatrix} = (-1)(3)(2)(4)(-2) = 48$

✓*CHECKPOINT*   Now try Exercise 29.

7.7 Exercises

See www.CalcChat.com for worked-out solutions to odd-numbered exercises.

Vocabulary Check

Fill in the blanks.

1. Both $\det(A)$ and $|A|$ represent the _____ of the matrix A.

2. The _____ M_{ij} of the entry a_{ij} is the determinant of the matrix obtained by deleting the ith row and jth column of the square matrix A.

3. The _____ C_{ij} of the entry a_{ij} is given by $(-1)^{i+j}M_{ij}$.

4. One way of finding the determinant of a matrix of order 2×2 or greater is _____ .

5. A square matrix with all zero entries either above or below its main diagonal is called a _____ matrix.

6. A matrix that is both upper and lower triangular is called a _____ matrix.

In Exercises 1–12, find the determinant of the matrix.

1. $[4]$

2. $[-10]$

3. $\begin{bmatrix} 8 & 4 \\ 2 & 3 \end{bmatrix}$

4. $\begin{bmatrix} -9 & 0 \\ 6 & 2 \end{bmatrix}$

5. $\begin{bmatrix} 6 & 2 \\ -5 & 3 \end{bmatrix}$

6. $\begin{bmatrix} 3 & -3 \\ 4 & -8 \end{bmatrix}$

7. $\begin{bmatrix} -7 & 6 \\ \frac{1}{2} & 3 \end{bmatrix}$

8. $\begin{bmatrix} 4 & -3 \\ 0 & 0 \end{bmatrix}$

9. $\begin{bmatrix} 2 & -1 & 0 \\ 4 & 2 & 1 \\ 4 & 2 & 1 \end{bmatrix}$

10. $\begin{bmatrix} -2 & 2 & 3 \\ 1 & -1 & 0 \\ 0 & 1 & 4 \end{bmatrix}$

11. $\begin{bmatrix} -1 & 2 & -5 \\ 0 & 3 & 4 \\ 0 & 0 & 3 \end{bmatrix}$

12. $\begin{bmatrix} 1 & 0 & 0 \\ -4 & -1 & 0 \\ 5 & 1 & 5 \end{bmatrix}$

In Exercises 13 and 14, use the matrix capabilities of a graphing utility to find the determinant of the matrix.

13. $\begin{bmatrix} 0.3 & 0.2 & 0.2 \\ 0.2 & 0.2 & 0.2 \\ -0.4 & 0.4 & 0.3 \end{bmatrix}$

14. $\begin{bmatrix} 0.1 & 0.2 & 0.3 \\ -0.3 & 0.2 & 0.2 \\ 0.5 & 0.4 & 0.4 \end{bmatrix}$

In Exercises 15–18, find all (a) minors and (b) cofactors of the matrix.

15. $\begin{bmatrix} 3 & 4 \\ 2 & -5 \end{bmatrix}$

16. $\begin{bmatrix} 11 & 0 \\ -3 & 2 \end{bmatrix}$

17. $\begin{bmatrix} -4 & 6 & 3 \\ 7 & -2 & 8 \\ 1 & 0 & -5 \end{bmatrix}$

18. $\begin{bmatrix} -2 & 9 & 4 \\ 7 & -6 & 0 \\ 6 & 7 & -6 \end{bmatrix}$

In Exercises 19–22, find the determinant of the matrix by the method of expansion by cofactors. Expand using the indicated row or column.

19. $\begin{bmatrix} -3 & 2 & 1 \\ 4 & 5 & 6 \\ 2 & -3 & 1 \end{bmatrix}$

   (a) Row 1
   (b) Column 2

20. $\begin{bmatrix} -3 & 4 & 2 \\ 6 & 3 & 1 \\ 4 & -7 & -8 \end{bmatrix}$

   (a) Row 2
   (b) Column 3

21. $\begin{bmatrix} 6 & 0 & -3 & 5 \\ 4 & 13 & 6 & -8 \\ -1 & 0 & 7 & 4 \\ 8 & 6 & 0 & 2 \end{bmatrix}$

   (a) Row 2
   (b) Column 2

22. $\begin{bmatrix} 10 & 8 & 3 & -7 \\ 4 & 0 & 5 & -6 \\ 0 & 3 & 2 & 7 \\ 1 & 0 & -3 & 2 \end{bmatrix}$

   (a) Row 3
   (b) Column 1

In Exercises 23–28, find the determinant of the matrix. Expand by cofactors on the row or column that appears to make the computations easiest.

23. $\begin{bmatrix} 1 & 4 & -2 \\ 3 & 2 & 0 \\ -1 & 4 & 3 \end{bmatrix}$

24. $\begin{bmatrix} -3 & 0 & 0 \\ 7 & 11 & 0 \\ 1 & 2 & 2 \end{bmatrix}$

25. $\begin{bmatrix} 2 & 6 & 6 & 2 \\ 2 & 7 & 3 & 6 \\ 1 & 5 & 0 & 1 \\ 3 & 7 & 0 & 7 \end{bmatrix}$

26. $\begin{bmatrix} 3 & 6 & -5 & 4 \\ -2 & 0 & 6 & 0 \\ 1 & 1 & 2 & 2 \\ 0 & 3 & -1 & -1 \end{bmatrix}$

27. $\begin{bmatrix} 3 & 2 & 4 & -1 & 5 \\ -2 & 0 & 1 & 3 & 2 \\ 1 & 0 & 0 & 4 & 0 \\ 6 & 0 & 2 & -1 & 0 \\ 3 & 0 & 5 & 1 & 0 \end{bmatrix}$

28.
$$\begin{bmatrix} 5 & 2 & 0 & 0 & -2 \\ 0 & 1 & 4 & 3 & 2 \\ 0 & 0 & 2 & 6 & 3 \\ 0 & 0 & 3 & 4 & 1 \\ 0 & 0 & 0 & 0 & 2 \end{bmatrix}$$

In Exercises 29–32, evaluate the determinant. Do not use a graphing utility.

29.
$$\begin{vmatrix} 4 & 0 & 0 & 0 \\ 6 & -5 & 0 & 0 \\ 1 & 3 & 1 & 0 \\ 1 & -2 & 7 & 3 \end{vmatrix}$$

30.
$$\begin{vmatrix} 5 & -10 & 1 & 1 \\ 0 & 6 & 3 & 4 \\ 0 & 0 & -2 & -1 \\ 0 & 0 & 0 & -1 \end{vmatrix}$$

31.
$$\begin{vmatrix} -6 & 7 & 2 & 0 & 5 \\ 0 & -1 & 3 & 4 & -3 \\ 0 & 0 & -7 & 0 & 4 \\ 0 & 0 & 0 & -2 & 1 \\ 0 & 0 & 0 & 0 & -2 \end{vmatrix}$$

32.
$$\begin{vmatrix} -2 & 0 & 0 & 0 & 0 \\ -1 & 4 & 0 & 0 & 0 \\ 3 & 5 & 1 & 0 & 0 \\ 6 & -11 & 8 & 10 & 0 \\ 0 & 13 & -9 & 0 & -3 \end{vmatrix}$$

In Exercises 33–36, use the matrix capabilities of a graphing utility to evaluate the determinant.

33.
$$\begin{vmatrix} 1 & -1 & 8 & 4 \\ 2 & 6 & 0 & -4 \\ 2 & 0 & 2 & 6 \\ 0 & 2 & 8 & 0 \end{vmatrix}$$

34.
$$\begin{vmatrix} 0 & -3 & 8 & 2 \\ 8 & 1 & -1 & 6 \\ -4 & 6 & 0 & 9 \\ -7 & 0 & 0 & 14 \end{vmatrix}$$

35.
$$\begin{vmatrix} 3 & -2 & 4 & 3 & 1 \\ -1 & 0 & 2 & 1 & 0 \\ 5 & -1 & 0 & 3 & 2 \\ 4 & 7 & -8 & 0 & 0 \\ 1 & 2 & 3 & 0 & 2 \end{vmatrix}$$

36.
$$\begin{vmatrix} -2 & 0 & 0 & 0 & 0 \\ 0 & 3 & 0 & 0 & 0 \\ 0 & 0 & -1 & 0 & 0 \\ 0 & 0 & 0 & 2 & 0 \\ 0 & 0 & 0 & 0 & -4 \end{vmatrix}$$

In Exercises 37–40, find (a) $|A|$, (b) $|B|$, (c) AB, and (d) $|AB|$.

37. $A = \begin{bmatrix} -1 & 0 \\ 0 & 3 \end{bmatrix}$, $B = \begin{bmatrix} 2 & 0 \\ 0 & -1 \end{bmatrix}$

38. $A = \begin{bmatrix} 4 & 0 \\ 3 & -2 \end{bmatrix}$, $B = \begin{bmatrix} -1 & 1 \\ -2 & 2 \end{bmatrix}$

39. $A = \begin{bmatrix} -1 & 2 & 1 \\ 1 & 0 & 1 \\ 0 & 1 & 0 \end{bmatrix}$, $B = \begin{bmatrix} -1 & 0 & 0 \\ 0 & 2 & 0 \\ 0 & 0 & 3 \end{bmatrix}$

40. $A = \begin{bmatrix} 2 & 0 & 1 \\ 1 & -1 & 2 \\ 3 & 1 & 0 \end{bmatrix}$, $B = \begin{bmatrix} 2 & -1 & 4 \\ 0 & 1 & 3 \\ 3 & -2 & 1 \end{bmatrix}$

In Exercises 41 and 42, use the matrix capabilities of a graphing utility to find (a) $|A|$, (b) $|B|$, (c) AB, and (d) $|AB|$.

41. $A = \begin{bmatrix} 6 & 4 & 0 & 1 \\ 2 & -3 & -2 & -4 \\ 0 & 1 & 5 & 0 \\ -1 & 0 & -1 & 1 \end{bmatrix}$,

$B = \begin{bmatrix} 0 & -5 & 0 & -2 \\ -2 & 4 & -1 & -4 \\ 3 & 0 & 1 & 0 \\ 1 & -2 & 3 & 0 \end{bmatrix}$

42. $A = \begin{bmatrix} -1 & 5 & 2 & 0 \\ 0 & 0 & 1 & 1 \\ 3 & -3 & -1 & 0 \\ 4 & 2 & 4 & -1 \end{bmatrix}$,

$B = \begin{bmatrix} 1 & 5 & 0 & 0 \\ 10 & -1 & 2 & 4 \\ 2 & 0 & 0 & 1 \\ -3 & 2 & 5 & 0 \end{bmatrix}$

In Exercises 43–48, evaluate the determinants to verify the equation.

43. $\begin{vmatrix} w & x \\ y & z \end{vmatrix} = - \begin{vmatrix} y & z \\ w & x \end{vmatrix}$

44. $\begin{vmatrix} w & cx \\ y & cz \end{vmatrix} = c \begin{vmatrix} w & x \\ y & z \end{vmatrix}$

45. $\begin{vmatrix} w & x \\ y & z \end{vmatrix} = \begin{vmatrix} w & x + cw \\ y & z + cy \end{vmatrix}$

46. $\begin{vmatrix} w & x \\ cw & cx \end{vmatrix} = 0$

47. $\begin{vmatrix} 1 & x & x^2 \\ 1 & y & y^2 \\ 1 & z & z^2 \end{vmatrix} = (y - x)(z - x)(z - y)$

48. $\begin{vmatrix} a + b & a & a \\ a & a + b & a \\ a & a & a + b \end{vmatrix} = b^2(3a + b)$

In Exercises 49–60, solve for x.

49. $\begin{vmatrix} x & 2 \\ 1 & x \end{vmatrix} = 2$

50. $\begin{vmatrix} x & 4 \\ -1 & x \end{vmatrix} = 20$

51. $\begin{vmatrix} 2x & -3 \\ -2 & 2x \end{vmatrix} = 3$ **52.** $\begin{vmatrix} x & 2 \\ 4 & 9x \end{vmatrix} = 8$

53. $\begin{vmatrix} x & 1 \\ 2 & x-2 \end{vmatrix} = -1$ **54.** $\begin{vmatrix} x+1 & 2 \\ -1 & x \end{vmatrix} = 4$

55. $\begin{vmatrix} x+3 & 2 \\ 1 & x+2 \end{vmatrix} = 0$ **56.** $\begin{vmatrix} x-2 & -1 \\ -3 & x \end{vmatrix} = 0$

57. $\begin{vmatrix} 2x & 1 \\ -1 & x-1 \end{vmatrix} = x$ **58.** $\begin{vmatrix} x-1 & x \\ x+1 & 2 \end{vmatrix} = -8$

59. $\begin{vmatrix} 1 & 2 & x \\ -1 & 3 & 2 \\ 3 & -2 & 1 \end{vmatrix} = 0$ **60.** $\begin{vmatrix} 1 & x & -2 \\ 1 & 3 & 3 \\ 0 & 2 & -2 \end{vmatrix} = 0$

$\int$ **In Exercises 61–66, evaluate the determinant, in which the entries are functions. Determinants of this type occur when changes of variables are made in calculus.**

61. $\begin{vmatrix} 4u & -1 \\ -1 & 2v \end{vmatrix}$ **62.** $\begin{vmatrix} 3x^2 & -3y^2 \\ 1 & 1 \end{vmatrix}$

63. $\begin{vmatrix} e^{2x} & e^{3x} \\ 2e^{2x} & 3e^{3x} \end{vmatrix}$ **64.** $\begin{vmatrix} e^{-x} & xe^{-x} \\ -e^{-x} & (1-x)e^{-x} \end{vmatrix}$

65. $\begin{vmatrix} x & \ln x \\ 1 & 1/x \end{vmatrix}$ **66.** $\begin{vmatrix} x & x\ln x \\ 1 & 1+\ln x \end{vmatrix}$

Synthesis

True or False? **In Exercises 67 and 68, determine whether the statement is true or false. Justify your answer.**

67. If a square matrix has an entire row of zeros, the determinant will always be zero.

68. If two columns of a square matrix are the same, the determinant of the matrix will be zero.

69. *Exploration* Find square matrices A and B to demonstrate that $|A + B| \neq |A| + |B|$.

70. *Conjecture* Consider square matrices in which the entries are consecutive integers. An example of such a matrix is

$$\begin{bmatrix} 4 & 5 & 6 \\ 7 & 8 & 9 \\ 10 & 11 & 12 \end{bmatrix}.$$

Use a graphing utility to evaluate four determinants of this type. Make a conjecture based on the results. Then verify your conjecture.

In Exercises 71–74, (a) find the determinant of A, (b) find A^{-1}, (c) find $\det(A^{-1})$, and (d) compare your results from parts (a) and (c). Make a conjecture with regard to your results.

71. $A = \begin{bmatrix} 1 & 2 \\ -2 & 2 \end{bmatrix}$ **72.** $A = \begin{bmatrix} 5 & -1 \\ 2 & -1 \end{bmatrix}$

73. $A = \begin{bmatrix} 1 & -3 & -2 \\ -1 & 3 & 1 \\ 0 & 2 & -2 \end{bmatrix}$ **74.** $A = \begin{bmatrix} -1 & 3 & 2 \\ 1 & 3 & -1 \\ 1 & 1 & -2 \end{bmatrix}$

In Exercises 75–77, a property of determinants is given (A and B are square matrices). State how the property has been applied to the given determinants and use a graphing utility to verify the results.

75. If B is obtained from A by interchanging two rows of A or by interchanging two columns of A, then $|B| = -|A|$.

(a) $\begin{vmatrix} 1 & 3 & 4 \\ -7 & 2 & -5 \\ 6 & 1 & 2 \end{vmatrix} = -\begin{vmatrix} 1 & 4 & 3 \\ -7 & -5 & 2 \\ 6 & 2 & 1 \end{vmatrix}$

(b) $\begin{vmatrix} 1 & 3 & 4 \\ -2 & 2 & 0 \\ 1 & 6 & 2 \end{vmatrix} = -\begin{vmatrix} 1 & 6 & 2 \\ -2 & 2 & 0 \\ 1 & 3 & 4 \end{vmatrix}$

76. If B is obtained from A by adding a multiple of a row of A to another row of A or by adding a multiple of a column of A to another column of A, then $|B| = |A|$.

(a) $\begin{vmatrix} 1 & -3 \\ 5 & 2 \end{vmatrix} = \begin{vmatrix} 1 & -3 \\ 0 & 17 \end{vmatrix}$

(b) $\begin{vmatrix} 5 & 4 & 2 \\ 2 & -3 & 4 \\ 7 & 6 & 3 \end{vmatrix} = \begin{vmatrix} 1 & 10 & -6 \\ 2 & -3 & 4 \\ 7 & 6 & 3 \end{vmatrix}$

77. If B is obtained from A by multiplying a row of A by a nonzero constant c or by multiplying a column of A by a nonzero constant c, then $|B| = c|A|$.

(a) $\begin{vmatrix} 5 & 10 \\ 2 & -3 \end{vmatrix} = 5\begin{vmatrix} 1 & 2 \\ 2 & -3 \end{vmatrix}$

(b) $\begin{vmatrix} 1 & 8 & -3 \\ 3 & -12 & 6 \\ 7 & 4 & 9 \end{vmatrix} = 12\begin{vmatrix} 1 & 2 & -1 \\ 3 & -3 & 2 \\ 7 & 1 & 3 \end{vmatrix}$

78. *Writing* Write an argument that explains why the determinant of a 3×3 triangular matrix is the product of its main diagonal entries.

Skills Review

In Exercises 79–82, factor the expression.

79. $x^2 - 3x + 2$ **80.** $x^2 + 5x + 6$

81. $4y^2 - 12y + 9$ **82.** $4y^2 - 28y + 49$

In Exercises 83 and 84, solve the system of equations using the method of substitution or the method of elimination.

83. $\begin{cases} 3x - 10y = 46 \\ x + y = -2 \end{cases}$ **84.** $\begin{cases} 5x + 7y = 23 \\ -4x - 2y = -4 \end{cases}$

7.8 Applications of Matrices and Determinants

Area of a Triangle

In this section, you will study some additional applications of matrices and determinants. The first involves a formula for finding the area of a triangle whose vertices are given by three points on a rectangular coordinate system.

> ### Area of a Triangle
>
> The area of a triangle with vertices (x_1, y_1), (x_2, y_2), and (x_3, y_3) is
>
> $$\text{Area} = \pm \frac{1}{2} \begin{vmatrix} x_1 & y_1 & 1 \\ x_2 & y_2 & 1 \\ x_3 & y_3 & 1 \end{vmatrix}$$
>
> where the symbol $(\pm)$ indicates that the appropriate sign should be chosen to yield a positive area.

What you should learn

- Use determinants to find areas of triangles.
- Use determinants to decide whether points are collinear.
- Use Cramer's Rule to solve systems of linear equations.
- Use matrices to encode and decode messages.

Why you should learn it

Determinants and Cramer's Rule can be used to find the least squares regression parabola that models lawn care retail sales, as shown in Exercise 28 on page 568.

ThinkStock/Index Stock Imagery

Example 1 Finding the Area of a Triangle

Find the area of the triangle whose vertices are $(1, 0)$, $(2, 2)$, and $(4, 3)$, as shown in Figure 7.34.

Solution

Let $(x_1, y_1) = (1, 0)$, $(x_2, y_2) = (2, 2)$, and $(x_3, y_3) = (4, 3)$. Then, to find the area of the triangle, evaluate the determinant by expanding along row 1.

$$\begin{vmatrix} x_1 & y_1 & 1 \\ x_2 & y_2 & 1 \\ x_3 & y_3 & 1 \end{vmatrix} = \begin{vmatrix} 1 & 0 & 1 \\ 2 & 2 & 1 \\ 4 & 3 & 1 \end{vmatrix}$$

$$= 1(-1)^2 \begin{vmatrix} 2 & 1 \\ 3 & 1 \end{vmatrix} + 0(-1)^3 \begin{vmatrix} 2 & 1 \\ 4 & 1 \end{vmatrix} + 1(-1)^4 \begin{vmatrix} 2 & 2 \\ 4 & 3 \end{vmatrix}$$

$$= 1(-1) + 0 + 1(-2)$$

$$= -3$$

Using this value, you can conclude that the area of the triangle is

$$\text{Area} = -\frac{1}{2} \begin{vmatrix} 1 & 0 & 1 \\ 2 & 2 & 1 \\ 4 & 3 & 1 \end{vmatrix}$$

$$= -\frac{1}{2}(-3)$$

$$= \frac{3}{2} \text{ square units.}$$

✓CHECKPOINT   Now try Exercise 1.

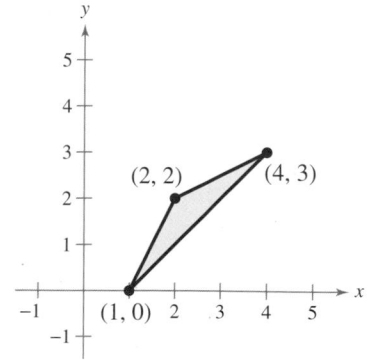

Figure 7.34

Collinear Points

What if the three points in Example 1 had been on the same line? What would have happened had the area formula been applied to three such points? The answer is that the determinant would have been zero. Consider, for instance, the three collinear points $(0, 1)$, $(2, 2)$, and $(4, 3)$, as shown in Figure 7.35. The area of the "triangle" that has these three points as vertices is

$$\frac{1}{2}\begin{vmatrix} 0 & 1 & 1 \\ 2 & 2 & 1 \\ 4 & 3 & 1 \end{vmatrix} = \frac{1}{2}\left[0(-1)^2\begin{vmatrix} 2 & 1 \\ 3 & 1 \end{vmatrix} + 1(-1)^3\begin{vmatrix} 2 & 1 \\ 4 & 1 \end{vmatrix} + 1(-1)^4\begin{vmatrix} 2 & 2 \\ 4 & 3 \end{vmatrix} \right]$$

$$= \frac{1}{2}[0 - 1(-2) + 1(-2)]$$

$$= 0$$

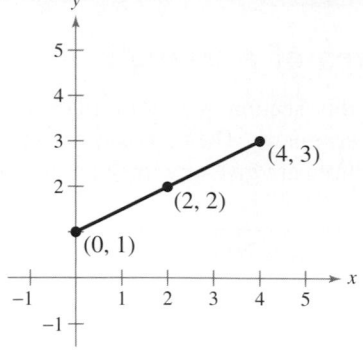

Figure 7.35

This result is generalized as follows.

Test for Collinear Points

Three points (x_1, y_1), (x_2, y_2), and (x_3, y_3) are **collinear** (lie on the same line) if and only if

$$\begin{vmatrix} x_1 & y_1 & 1 \\ x_2 & y_2 & 1 \\ x_3 & y_3 & 1 \end{vmatrix} = 0.$$

Example 2 Testing for Collinear Points

Determine whether the points $(-2, -2)$, $(1, 1)$, and $(7, 5)$ are collinear. (See Figure 7.36.)

Solution

Letting $(x_1, y_1) = (-2, -2)$, $(x_2, y_2) = (1, 1)$, and $(x_3, y_3) = (7, 5)$ and expanding along row 1, you have

$$\begin{vmatrix} x_1 & y_1 & 1 \\ x_2 & y_2 & 1 \\ x_3 & y_3 & 1 \end{vmatrix} = \begin{vmatrix} -2 & -2 & 1 \\ 1 & 1 & 1 \\ 7 & 5 & 1 \end{vmatrix}$$

$$= -2(-1)^2\begin{vmatrix} 1 & 1 \\ 5 & 1 \end{vmatrix} + (-2)(-1)^3\begin{vmatrix} 1 & 1 \\ 7 & 1 \end{vmatrix} + 1(-1)^4\begin{vmatrix} 1 & 1 \\ 7 & 5 \end{vmatrix}$$

$$= -2(-4) + 2(-6) + 1(-2)$$

$$= -6.$$

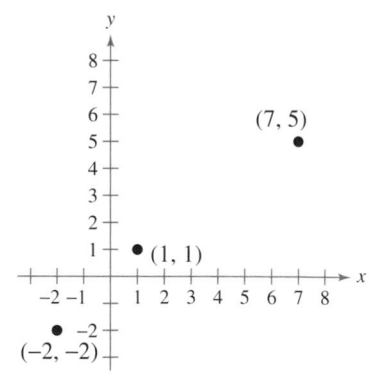

Figure 7.36

Because the value of this determinant is *not* zero, you can conclude that the three points are not collinear.

✓CHECKPOINT   Now try Exercise 9.

Cramer's Rule

So far, you have studied three methods for solving a system of linear equations: substitution, elimination with equations, and elimination with matrices. You will now study one more method, **Cramer's Rule,** named after Gabriel Cramer (1704–1752). This rule uses determinants to write the solution of a system of linear equations. To see how Cramer's Rule works, take another look at the solution described at the beginning of Section 7.7. There, it was pointed out that the system

$$\begin{cases} a_1x + b_1y = c_1 \\ a_2x + b_2y = c_2 \end{cases}$$

has a solution

$$x = \frac{c_1b_2 - c_2b_1}{a_1b_2 - a_2b_1} \quad \text{and} \quad y = \frac{a_1c_2 - a_2c_1}{a_1b_2 - a_2b_1}$$

provided that $a_1b_2 - a_2b_1 \neq 0$. Each numerator and denominator in this solution can be expressed as a determinant, as follows.

$$x = \frac{c_1b_2 - c_2b_1}{a_1b_2 - a_2b_1} = \frac{\begin{vmatrix} c_1 & b_1 \\ c_2 & b_2 \end{vmatrix}}{\begin{vmatrix} a_1 & b_1 \\ a_2 & b_2 \end{vmatrix}}$$

$$y = \frac{a_1c_2 - a_2c_1}{a_1b_2 - a_2b_1} = \frac{\begin{vmatrix} a_1 & c_1 \\ a_2 & c_2 \end{vmatrix}}{\begin{vmatrix} a_1 & b_1 \\ a_2 & b_2 \end{vmatrix}}$$

Relative to the original system, the denominators of x and y are simply the determinant of the *coefficient* matrix of the system. This determinant is denoted by D. The numerators of x and y are denoted by D_x and D_y, respectively. They are formed by using the column of constants as replacements for the coefficients of x and y, as follows.

Coefficient Matrix	D	D_x	D_y
$\begin{bmatrix} a_1 & b_1 \\ a_2 & b_2 \end{bmatrix}$	$\begin{vmatrix} a_1 & b_1 \\ a_2 & b_2 \end{vmatrix}$	$\begin{vmatrix} c_1 & b_1 \\ c_2 & b_2 \end{vmatrix}$	$\begin{vmatrix} a_1 & c_1 \\ a_2 & c_2 \end{vmatrix}$

For example, given the system

$$\begin{cases} 2x - 5y = 3 \\ -4x + 3y = 8 \end{cases}$$

the coefficient matrix, D, D_x, and D_y are as follows.

Coefficient Matrix	D	D_x	D_y
$\begin{bmatrix} 2 & -5 \\ -4 & 3 \end{bmatrix}$	$\begin{vmatrix} 2 & -5 \\ -4 & 3 \end{vmatrix}$	$\begin{vmatrix} 3 & -5 \\ 8 & 3 \end{vmatrix}$	$\begin{vmatrix} 2 & 3 \\ -4 & 8 \end{vmatrix}$

Cramer's Rule generalizes easily to systems of n equations in n variables. The value of each variable is given as the quotient of two determinants. The denominator is the determinant of the coefficient matrix, and the numerator is the determinant of the matrix formed by replacing the column corresponding to the variable being solved for with the column representing the constants. For instance, the solution for x_3 in the following system is shown.

$$\begin{cases} a_{11}x_1 + a_{12}x_2 + a_{13}x_3 = b_1 \\ a_{21}x_1 + a_{22}x_2 + a_{23}x_3 = b_2 \\ a_{31}x_1 + a_{32}x_2 + a_{33}x_3 = b_3 \end{cases} \qquad x_3 = \frac{|A_3|}{|A|} = \frac{\begin{vmatrix} a_{11} & a_{12} & b_1 \\ a_{21} & a_{22} & b_2 \\ a_{31} & a_{32} & b_3 \end{vmatrix}}{\begin{vmatrix} a_{11} & a_{12} & a_{13} \\ a_{21} & a_{22} & a_{23} \\ a_{31} & a_{32} & a_{33} \end{vmatrix}}$$

Cramer's Rule

If a system of n linear equations in n variables has a coefficient matrix A with a *nonzero* determinant $|A|$, the solution of the system is

$$x_1 = \frac{|A_1|}{|A|}, \quad x_2 = \frac{|A_2|}{|A|}, \quad \ldots, \quad x_n = \frac{|A_n|}{|A|}$$

where the ith column of A_i is the column of constants in the system of equations. If the determinant of the coefficient matrix is zero, the system has either no solution or infinitely many solutions.

STUDY TIP

Cramer's Rule does not apply when the determinant of the coefficient matrix is zero. This would create division by zero, which is undefined.

Example 3 Using Cramer's Rule for a 2 × 2 System

Use Cramer's Rule to solve the system $\begin{cases} 4x - 2y = 10 \\ 3x - 5y = 11 \end{cases}$.

Solution

To begin, find the determinant of the coefficient matrix.

$$D = \begin{vmatrix} 4 & -2 \\ 3 & -5 \end{vmatrix} = -20 - (-6) = -14$$

Because this determinant is not zero, apply Cramer's Rule.

$$x = \frac{D_x}{D} = \frac{\begin{vmatrix} 10 & -2 \\ 11 & -5 \end{vmatrix}}{-14} = \frac{(-50) - (-22)}{-14} = \frac{-28}{-14} = 2$$

$$y = \frac{D_y}{D} = \frac{\begin{vmatrix} 4 & 10 \\ 3 & 11 \end{vmatrix}}{-14} = \frac{44 - 30}{-14} = \frac{14}{-14} = -1$$

So, the solution is $x = 2$ and $y = -1$. Check this in the original system.

✓**CHECKPOINT** Now try Exercise 15.

Example 4 Using Cramer's Rule for a 3×3 System

Use Cramer's Rule and a graphing utility, if possible, to solve the system of linear equations.

$$\begin{cases} -x \quad + \; z = \quad 4 \\ 2x - y + \; z = -3 \\ \quad\quad y - 3z = \quad 1 \end{cases}$$

Solution

Using a graphing utility to evaluate the determinant of the coefficient matrix A, you find that Cramer's Rule cannot be applied because $|A| = 0$.

✓CHECKPOINT   Now try Exercise 17.

TECHNOLOGY TIP

Try using a graphing utility to evaluate D_x/D from Example 4. You should obtain the error message shown below.

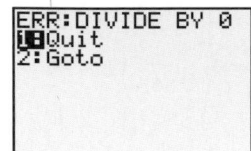

Example 5 Using Cramer's Rule for a 3×3 System

Use Cramer's Rule, if possible, to solve the system of linear equations.

$$\begin{cases} -x + 2y - 3z = 1 \\ 2x \quad\quad + \; z = 0 \\ 3x - 4y + 4z = 2 \end{cases}$$

Coefficient Matrix

$$\implies \begin{bmatrix} -1 & 2 & -3 \\ 2 & 0 & 1 \\ 3 & -4 & 4 \end{bmatrix}$$

Solution

The coefficient matrix above can be expanded along the second row, as follows.

$$D = 2(-1)^3 \begin{vmatrix} 2 & -3 \\ -4 & 4 \end{vmatrix} + 0(-1)^4 \begin{vmatrix} -1 & -3 \\ 3 & 4 \end{vmatrix} + 1(-1)^5 \begin{vmatrix} -1 & 2 \\ 3 & -4 \end{vmatrix}$$

$$= -2(-4) + 0 - 1(-2) = 10$$

Because this determinant is not zero, you can apply Cramer's Rule.

$$x = \frac{D_x}{D} = \frac{\begin{vmatrix} 1 & 2 & -3 \\ 0 & 0 & 1 \\ 2 & -4 & 4 \end{vmatrix}}{10} = \frac{8}{10} = \frac{4}{5}$$

$$y = \frac{D_y}{D} = \frac{\begin{vmatrix} -1 & 1 & -3 \\ 2 & 0 & 1 \\ 3 & 2 & 4 \end{vmatrix}}{10} = \frac{-15}{10} = -\frac{3}{2}$$

$$z = \frac{D_z}{D} = \frac{\begin{vmatrix} -1 & 2 & 1 \\ 2 & 0 & 0 \\ 3 & -4 & 2 \end{vmatrix}}{10} = \frac{-16}{10} = -\frac{8}{5}$$

The solution is $\left(\frac{4}{5}, -\frac{3}{2}, -\frac{8}{5}\right)$. Check this in the original system.

✓CHECKPOINT   Now try Exercise 21.

Cryptography

A **cryptogram** is a message written according to a secret code. (The Greek word *kryptos* means "hidden.") Matrix multiplication can be used to encode and decode messages. To begin, you need to assign a number to each letter in the alphabet (with 0 assigned to a blank space), as follows.

0 = _	9 = I	18 = R
1 = A	10 = J	19 = S
2 = B	11 = K	20 = T
3 = C	12 = L	21 = U
4 = D	13 = M	22 = V
5 = E	14 = N	23 = W
6 = F	15 = O	24 = X
7 = G	16 = P	25 = Y
8 = H	17 = Q	26 = Z

Then the message is converted to numbers and partitioned into **uncoded row matrices,** each having n entries, as demonstrated in Example 6.

Example 6   Forming Uncoded Row Matrices

Write the uncoded row matrices of order 1×3 for the message

MEET ME MONDAY.

Solution

Partitioning the message (including blank spaces, but ignoring punctuation) into groups of three produces the following uncoded row matrices.

$$[13 \quad 5 \quad 5] \quad [20 \quad 0 \quad 13] \quad [5 \quad 0 \quad 13] \quad [15 \quad 14 \quad 4] \quad [1 \quad 25 \quad 0]$$

$$\text{M} \quad \text{E} \quad \text{E} \quad \text{T} \quad \quad \text{M} \quad \text{E} \quad \quad \text{M} \quad \text{O} \quad \text{N} \quad \text{D} \quad \text{A} \quad \text{Y}$$

Note that a blank space is used to fill out the last uncoded row matrix.

✓CHECKPOINT   Now try Exercise 29.

To encode a message, choose an $n \times n$ invertible matrix A by using the techniques demonstrated in Section 7.6 and multiply the uncoded row matrices by A (on the right) to obtain **coded row matrices.** Here is an example.

Uncoded Matrix   Encoding Matrix A   Coded Matrix

$$[13 \quad 5 \quad 5] \begin{bmatrix} 1 & -2 & 2 \\ -1 & 1 & 3 \\ 1 & -1 & -4 \end{bmatrix} = [13 \quad -26 \quad 21]$$

This technique is further illustrated in Example 7.

Example 7   Encoding a Message

Use the following matrix to encode the message MEET ME MONDAY.

$$A = \begin{bmatrix} 1 & -2 & 2 \\ -1 & 1 & 3 \\ 1 & -1 & -4 \end{bmatrix}$$

Solution

The coded row matrices are obtained by multiplying each of the uncoded row matrices found in Example 6 by the matrix A, as follows.

Uncoded Matrix   *Encoding Matrix A*   *Coded Matrix*

$$[13 \quad 5 \quad 5] \begin{bmatrix} 1 & -2 & 2 \\ -1 & 1 & 3 \\ 1 & -1 & -4 \end{bmatrix} = [13 \ -26 \quad 21]$$

$$[20 \quad 0 \quad 13] \begin{bmatrix} 1 & -2 & 2 \\ -1 & 1 & 3 \\ 1 & -1 & -4 \end{bmatrix} = [33 \ -53 \ -12]$$

$$[5 \quad 0 \quad 13] \begin{bmatrix} 1 & -2 & 2 \\ -1 & 1 & 3 \\ 1 & -1 & -4 \end{bmatrix} = [18 \ -23 \ -42]$$

$$[15 \quad 14 \quad 4] \begin{bmatrix} 1 & -2 & 2 \\ -1 & 1 & 3 \\ 1 & -1 & -4 \end{bmatrix} = [5 \ -20 \quad 56]$$

$$[1 \quad 25 \quad 0] \begin{bmatrix} 1 & -2 & 2 \\ -1 & 1 & 3 \\ 1 & -1 & -4 \end{bmatrix} = [-24 \quad 23 \quad 77]$$

So, the sequence of coded row matrices is

$$[13 \ -26 \ 21][33 \ -53 \ -12][18 \ -23 \ -42][5 \ -20 \ 56][-24 \ 23 \ 77].$$

Finally, removing the matrix notation produces the following cryptogram.

$$13 \ -26 \ 21 \ 33 \ -53 \ -12 \ 18 \ -23 \ -42 \ 5 \ -20 \ 56 \ -24 \ 23 \ 77$$

✓CHECKPOINT   Now try Exercise 31.

> **TECHNOLOGY TIP**
>
> An efficient method for encoding the message at the left with your graphing utility is to enter A as a 3×3 matrix. Let B be the 5×3 matrix whose rows are the uncoded row matrices
>
> $$B = \begin{bmatrix} 13 & 5 & 5 \\ 20 & 0 & 13 \\ 5 & 0 & 13 \\ 15 & 14 & 4 \\ 1 & 25 & 0 \end{bmatrix}.$$
>
> The product BA gives the coded row matrices.

For those who do not know the encoding matrix A, decoding the cryptogram found in Example 7 is difficult. But for an authorized receiver who knows the encoding matrix A, decoding is simple. The receiver need only multiply the coded row matrices by A^{-1} (on the right) to retrieve the uncoded row matrices. Here is an example.

$$\underbrace{[13 \ -26 \quad 21]}_{\text{Coded}} \underbrace{\begin{bmatrix} -1 & -10 & -8 \\ -1 & -6 & -5 \\ 0 & -1 & -1 \end{bmatrix}}_{A^{-1}} = \underbrace{[13 \quad 5 \quad 5]}_{\text{Uncoded}}$$

Example 8   Decoding a Message

Use the inverse of the matrix

$$A = \begin{bmatrix} 1 & -2 & 2 \\ -1 & 1 & 3 \\ 1 & -1 & -4 \end{bmatrix}$$

to decode the cryptogram

$$13 \ -26 \ 21 \ 33 \ -53 \ -12 \ 18 \ -23 \ -42 \ 5 \ -20 \ 56 \ -24 \ 23 \ 77$$

Solution

First find A^{-1} by using the techniques demonstrated in Section 7.6. A^{-1} is the decoding matrix. Next partition the message into groups of three to form the coded row matrices. Then multiply each coded row matrix by A^{-1} (on the right).

Coded Matrix   Decoding Matrix A^{-1} Decoded Matrix

$$\begin{bmatrix} 13 & -26 & 21 \end{bmatrix} \begin{bmatrix} -1 & -10 & -8 \\ -1 & -6 & -5 \\ 0 & -1 & -1 \end{bmatrix} = \begin{bmatrix} 13 & 5 & 5 \end{bmatrix}$$

$$\begin{bmatrix} 33 & -53 & -12 \end{bmatrix} \begin{bmatrix} -1 & -10 & -8 \\ -1 & -6 & -5 \\ 0 & -1 & -1 \end{bmatrix} = \begin{bmatrix} 20 & 0 & 13 \end{bmatrix}$$

$$\begin{bmatrix} 18 & -23 & -42 \end{bmatrix} \begin{bmatrix} -1 & -10 & -8 \\ -1 & -6 & -5 \\ 0 & -1 & -1 \end{bmatrix} = \begin{bmatrix} 5 & 0 & 13 \end{bmatrix}$$

$$\begin{bmatrix} 5 & -20 & 56 \end{bmatrix} \begin{bmatrix} -1 & -10 & -8 \\ -1 & -6 & -5 \\ 0 & -1 & -1 \end{bmatrix} = \begin{bmatrix} 15 & 14 & 4 \end{bmatrix}$$

$$\begin{bmatrix} -24 & 23 & 77 \end{bmatrix} \begin{bmatrix} -1 & -10 & -8 \\ -1 & -6 & -5 \\ 0 & -1 & -1 \end{bmatrix} = \begin{bmatrix} 1 & 25 & 0 \end{bmatrix}$$

So, the message is as follows.

$$\begin{bmatrix} 13 & 5 & 5 \end{bmatrix} \ \begin{bmatrix} 20 & 0 & 13 \end{bmatrix} \ \begin{bmatrix} 5 & 0 & 13 \end{bmatrix} \ \begin{bmatrix} 15 & 14 & 4 \end{bmatrix} \ \begin{bmatrix} 1 & 25 & 0 \end{bmatrix}$$

$$\text{M} \quad \text{E} \quad \text{E} \quad \text{T} \quad \text{M} \quad \text{E} \quad \text{M} \quad \text{O} \quad \text{N} \quad \text{D} \quad \text{A} \quad \text{Y}$$

✓CHECKPOINT   Now try Exercise 35.

TECHNOLOGY TIP   An efficient method for decoding the cryptogram in Example 8 with your graphing utility is to enter A as a 3×3 matrix and then find A^{-1}. Let B be the 5×3 matrix whose rows are the coded row matrices, as shown at the right. The product BA^{-1} gives the decoded row matrices.

$$B = \begin{bmatrix} 13 & -26 & 21 \\ 33 & -53 & -12 \\ 18 & -23 & -42 \\ 5 & -20 & 56 \\ -24 & 23 & 77 \end{bmatrix}$$

7.8   Exercises

See www.CalcChat.com for worked-out solutions to odd-numbered exercises.

Vocabulary Check

Fill in the blanks.

1. Three points are _____ if they lie on the same line.

2. The method of using determinants to solve a system of linear equations is called _____ .

3. A message written according to a secret code is called a _____ .

4. To encode a message, choose an invertible matrix A and multiply the _____ row matrices by A (on the right) to obtain _____ row matrices.

In Exercises 1–6, use a determinant to find the area of the figure with the given vertices.

1. $(-2, 4), (2, 3), (-1, 5)$
2. $(-3, 5), (2, 6), (3, -5)$
3. $\left(0, \frac{1}{2}\right), \left(\frac{5}{2}, 0\right), (4, 3)$
4. $\left(\frac{9}{2}, 0\right), (2, 6), \left(0, -\frac{3}{2}\right)$

5.
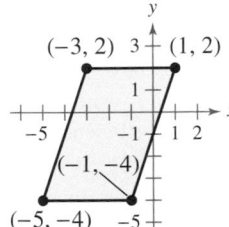

6.
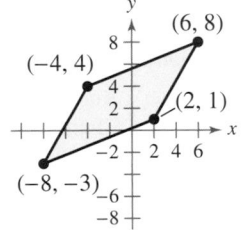

In Exercises 7 and 8, find x such that the triangle has an area of 4 square units.

7. $(-1, 5), (-2, 0), (x, 2)$
8. $(-4, 2), (-3, 5), (-1, x)$

In Exercises 9–12, use a determinant to determine whether the points are collinear.

9. $(3, -1), (0, -3), (12, 5)$
10. $(3, -5), (6, 1), (4, 2)$
11. $\left(2, -\frac{1}{2}\right), (-4, 4), (6, -3)$
12. $\left(0, \frac{1}{2}\right), (2, -1), \left(-4, \frac{7}{2}\right)$

In Exercises 13 and 14, find x such that the points are collinear.

13. $(1, -2), (x, 2), (5, 6)$
14. $(-6, 2), (-5, x), (-3, 5)$

In Exercises 15–22, use Cramer's Rule to solve (if possible) the system of equations.

15. $\begin{cases} -7x + 11y = -1 \\ 3x - 9y = 9 \end{cases}$

16. $\begin{cases} 4x - 3y = -10 \\ 6x + 9y = 12 \end{cases}$

17. $\begin{cases} 3x + 2y = -2 \\ 6x + 4y = 4 \end{cases}$

18. $\begin{cases} 6x - 5y = 17 \\ -13x + 3y = -76 \end{cases}$

19. $\begin{cases} -0.4x + 0.8y = 1.6 \\ 0.2x + 0.3y = 2.2 \end{cases}$

20. $\begin{cases} 2.4x - 0.8y = 10.8 \\ 4.6x + 1.2y = 24.8 \end{cases}$

21. $\begin{cases} 4x - y + z = -5 \\ 2x + 2y + 3z = 10 \\ 5x - 2y + 6z = 1 \end{cases}$

22. $\begin{cases} 4x - 2y + 3z = -2 \\ 2x + 2y + 5z = 16 \\ 8x - 5y - 2z = 4 \end{cases}$

In Exercises 23–26, solve the system of equations using (a) Gaussian elimination and (b) Cramer's Rule. Which method do you prefer, and why?

23. $\begin{cases} 3x + 3y + 5z = 1 \\ 3x + 5y + 9z = 2 \\ 5x + 9y + 17z = 4 \end{cases}$

24. $\begin{cases} 2x + 3y - 5z = 1 \\ 3x + 5y + 9z = -16 \\ 5x + 9y + 17z = -30 \end{cases}$

25. $\begin{cases} 2x - y + z = 5 \\ x - 2y - z = 1 \\ 3x + y + z = 4 \end{cases}$

26. $\begin{cases} 3x - y - 3z = 1 \\ 2x + y + 2z = -4 \\ x + y - z = 5 \end{cases}$

27. **Sports** The average salaries (in thousands of dollars) for football players in the National Football League from 2000 to 2004 are shown in the table. (Source: National Football League Players Association)

Year	Annual salary (in thousands of dollars)
2000	787
2001	986
2002	1180
2003	1259
2004	1331

The coefficients of the least squares regression parabola $y = at^2 + bt + c$, where y represents the average salary (in thousands of dollars) and t represents the year, with $t = 0$ corresponding to 2000, can be found by solving the system

$$\begin{cases} 5c + 10b + 30a = 5,543 \\ 10c + 30b + 100a = 12,447. \\ 30c + 100b + 354a = 38,333 \end{cases}$$

(a) Use Cramer's Rule to solve the system and write the least squares regression parabola for the data.

(b) Use a graphing utility to graph the parabola with the data.

(c) Do you believe the model can be used to predict the average salaries for future years? Explain.

28. *Retail Sales*   The retail sales (in millions of dollars) for lawn care in the United States from 2000 to 2004 are shown in the table.   (Source: The National Gardening Association)

Year	Retail Sales (in millions of dollars)
2000	9,794
2001	12,672
2002	11,963
2003	10,413
2004	8,887

The coefficients of the least squares regression parabola $y = at^2 + bt + c$, where y represents the retail sales (in millions of dollars) and t represents the year, with $t = 0$ corresponding to 2000, can be found by solving the system

$$\begin{cases} 5c + 10b + 30a = 53{,}729 \\ 10c + 30b + 100a = 103{,}385. \\ 30c + 100b + 354a = 296{,}433 \end{cases}$$

(a) Use Cramer's Rule to solve the system and write the least squares regression parabola for the data.

(b) Use a graphing utility to graph the parabola with the data.

(c) Use the graph to estimate when the retail sales decreased to $7,500,000,000.

In Exercises 29 and 30, write the uncoded 1 × 3 row matrices for the message. Then encode the message using the encoding matrix.

Message   *Encoding Matrix*

29. CALL ME TOMORROW
$$\begin{bmatrix} 1 & -1 & 0 \\ 1 & 0 & -1 \\ -6 & 2 & 3 \end{bmatrix}$$

30. PLEASE SEND MONEY
$$\begin{bmatrix} 4 & 2 & 1 \\ -3 & -3 & -1 \\ 3 & 2 & 1 \end{bmatrix}$$

In Exercises 31 and 32, write a cryptogram for the message using the matrix A.

$$A = \begin{bmatrix} 1 & 2 & 2 \\ 3 & 7 & 9 \\ -1 & -4 & -7 \end{bmatrix}$$

31. GONE FISHING   **32.** HAPPY BIRTHDAY

In Exercises 33–35, use A^{-1} to decode the cryptogram.

33. $A = \begin{bmatrix} 1 & 2 \\ 3 & 5 \end{bmatrix}$ 11 21 64 112 25 50 29 53
23 46 40 75 55 92

34. $A = \begin{bmatrix} 2 & 3 \\ 3 & 4 \end{bmatrix}$ 85 120 6 8 10 15 84 117 42
56 90 125 60 80 30 45 19 26

35. $A = \begin{bmatrix} 1 & 2 & 1 \\ -1 & 0 & 2 \\ 1 & -1 & -2 \end{bmatrix}$ 38 36 −1 11 17 11 42 15
−27 −5 18 37 26 28 17 8
24 20 32 20 −7 23 −1 −19

36. The following cryptogram was encoded with a 2×2 matrix.

8 21 −15 −10 −13 −13 5 10 5 25 5 19 −1 6 20 40 −18 −18 1 16

The last word of the message is _RON. What is the message?

Synthesis

True or False?   **In Exercises 37 and 38, determine whether the statement is true or false. Justify your answer.**

37. Cramer's Rule cannot be used to solve a system of linear equations if the determinant of the coefficient matrix is zero.

38. In a system of linear equations, if the determinant of the coefficient matrix is zero, the system has no solution.

39. *Writing*   At this point in the book, you have learned several methods for solving a system of linear equations. Briefly describe which method(s) you find easiest to use and which method(s) you find most difficult to use.

40. *Writing*   Use your school's library, the Internet, or some other reference source to research a few current real-life uses of cryptography. Write a short summary of these uses. Include a description of how messages are encoded and decoded in each case.

Skills Review

In Exercises 41–44, find the general form of the equation of the line that passes through the two points.

41. $(-1, 5), (7, 3)$ **42.** $(0, -6), (-2, 10)$

43. $(3, -3), (10, -1)$ **44.** $(-4, 12), (4, 2)$

In Exercises 45 and 46, sketch the graph of the rational function. Identify any asymptotes.

45. $f(x) = \dfrac{2x^2}{x^2 + 4}$ **46.** $f(x) = \dfrac{2x}{x^2 + 3x - 18}$

What Did You Learn?

Key Terms

systems of equations, *p. 474*
solution of a system, *p. 474*
equivalent systems, *p. 486*
consistent, *p. 487*
inconsistent, *p. 487*
independent, dependent, *p. 497*
partial fraction decomposition, *p. 501*
row-equivalent matrices, *p. 513*

row-echelon form, *p. 515*
reduced row-echelon form, *p. 515*
Gauss-Jordan elimination, *p. 518*
scalar, scalar multiple, *p. 527*
zero matrix, *p. 529*
properties of matrix addition, *p. 529*
properties of matrix multiplication, *p. 533*

identity matrix of order *n*, *p. 533*
determinant, *p. 551*
minor, cofactor, *p. 553*
triangular matrix, *p. 555*
diagonal matrix, *p. 555*
area of a triangle, *p. 559*
test for collinear points, *p. 560*
cryptogram, *p. 564*

Key Concepts

7.1 ■ Use the method of substitution

Solve one of the equations for one variable. Substitute the expression found in Step 1 into the other equation to obtain an equation in one variable. Solve the equation obtained in Step 2. Back-substitute the value(s) obtained in Step 3 into the expression obtained in Step 1 to find the value(s) of the other variable. Check the solution(s).

7.1 ■ Use the method of graphing

Solve both equations for *y* in terms of *x*. Use a graphing utility to graph both equations in the same viewing window. Use the *intersect* feature or the *zoom* and *trace* features of the graphing utility to approximate the point(s) of intersection of the graphs. Check the solution(s).

7.2 ■ Use the method of elimination

Obtain coefficients for *x* (or *y*) that differ only in sign by multiplying all terms of one or both equations by suitably chosen constants. Add the equations to eliminate one variable and solve the resulting equation. Back-substitute the value obtained in Step 2 into either of the original equations and solve for the other variable. Check the solution(s).

7.3 ■ Use Gaussian elimination to solve systems

Use elementary row operations to convert a system of linear equations to row-echelon form. (1) Interchange two equations. (2) Multiply one of the equations by a nonzero constant. (3) Add a multiple of one equation to another equation.

7.4 ■ Use matrices and Gaussian elimination

Write the augmented matrix of the system. Use elementary row operations to rewrite the augmented matrix in row-echelon form. Write the system of linear equations corresponding to the matrix in row-echelon form and use back-substitution to find the solution.

7.5 ■ Perform matrix operations

1. Let $A = [a_{ij}]$ and $B = [b_{ij}]$ be matrices of order $m \times n$ and let *c* be a scalar.

$$A + B = [a_{ij} + b_{ij}] \qquad cA = [ca_{ij}]$$

2. Let $A = [a_{ij}]$ be an $m \times n$ matrix and let $B = [b_{ij}]$ be an $n \times p$ matrix. The product AB is an $m \times p$ matrix given by $AB = [c_{ij}]$, where
$c_{ij} = a_{i1}b_{1j} + a_{i2}b_{2j} + a_{i3}b_{3j} + \cdots + a_{in}b_{nj}$.

7.6 ■ Find inverse matrices

1. Use Gauss-Jordan elimination. Write the $n \times 2n$ matrix $[A \vdots I]$. Row reduce *A* to *I* using elementary row operations. The result will be the matrix $[I \vdots A^{-1}]$.

2. If $A = \begin{bmatrix} a & b \\ c & d \end{bmatrix}$ and $ad - bc \neq 0$, then

$$A^{-1} = \frac{1}{ad - bc} \begin{bmatrix} d & -b \\ -c & a \end{bmatrix}.$$

7.7 ■ Find the determinants of square matrices

1. $\det(A) = |A| = \begin{vmatrix} a_1 & b_1 \\ a_2 & b_2 \end{vmatrix} = a_1 b_2 - a_2 b_1$

2. The determinant of a square matrix *A* is the sum of the entries in any row or column of *A* multiplied by their respective cofactors.

7.8 ■ Use Cramer's Rule to solve linear systems

If a system of *n* linear equations in *n* variables has a coefficient matrix *A* with a nonzero determinant $|A|$, the solution of the system is

$$x_1 = \frac{|A_1|}{A}, x_2 = \frac{|A_2|}{A}, \ldots x_n = \frac{|A_n|}{A}$$

where the *i*th column of A_i is the column of constants in the system of equations.

Review Exercises

See www.CalcChat.com for worked-out solutions to odd-numbered exercises.

7.1 In Exercises 1–6, solve the system by the method of substitution.

1. $\begin{cases} x + y = 2 \\ x - y = 0 \end{cases}$

2. $\begin{cases} 2x - 3y = 3 \\ x - y = 0 \end{cases}$

3. $\begin{cases} x^2 - y^2 = 9 \\ x - y = 1 \end{cases}$

4. $\begin{cases} x^2 + y^2 = 169 \\ 3x + 2y = 39 \end{cases}$

5. $\begin{cases} y = 2x^2 \\ y = x^4 - 2x^2 \end{cases}$

6. $\begin{cases} x = y + 3 \\ x = y^2 + 1 \end{cases}$

In Exercises 7–16, use a graphing utility to approximate all points of intersection of the graphs of the equations in the system. Verify your solutions by checking them in the original system.

7. $\begin{cases} 5x + 6y = 7 \\ -x - 4y = 0 \end{cases}$

8. $\begin{cases} 8x - 3y = -3 \\ 2x + 5y = 28 \end{cases}$

9. $\begin{cases} y^2 - 4x = 0 \\ x + y = 0 \end{cases}$

10. $\begin{cases} y^2 - x = -1 \\ y + 2x = 5 \end{cases}$

11. $\begin{cases} y = 3 - x^2 \\ y = 2x^2 + x + 1 \end{cases}$

12. $\begin{cases} y = 2x^2 - 4x + 1 \\ y = x^2 - 4x + 3 \end{cases}$

13. $\begin{cases} y = 2(6 - x) \\ y = 2^{x-2} \end{cases}$

14. $\begin{cases} 3x + y = 16 \\ y = 1 + 3^x \end{cases}$

15. $\begin{cases} y = \ln(x + 2) + 1 \\ x + y = 0 \end{cases}$

16. $\begin{cases} y = \ln(x - 1) + 3 \\ y = 4 - \frac{1}{2}x \end{cases}$

17. *Break-Even Analysis*   You set up a business and make an initial investment of $10,000. The unit cost of the product is $2.85 and the selling price is $4.95. How many units must you sell to break even?

18. *Choice of Two Jobs*   You are offered two sales jobs. One company offers an annual salary of $22,500 plus a year-end bonus of 1.5% of your total sales. The other company offers an annual salary of $20,000 plus a year-end bonus of 2% of your total sales. How much would you have to sell in a year to make the second offer the better offer?

19. *Geometry*   The perimeter of a rectangle is 480 meters and its length is 1.5 times its width. Find the dimensions of the rectangle.

20. *Geometry*   The perimeter of a rectangle is 68 feet and its width is $\frac{8}{9}$ times its length. Find the dimensions of the rectangle.

7.2 In Exercises 21–28, solve the system by the method of elimination.

21. $\begin{cases} 2x - y = 2 \\ 6x + 8y = 39 \end{cases}$

22. $\begin{cases} 40x + 30y = 24 \\ 20x - 50y = -14 \end{cases}$

23. $\begin{cases} \frac{1}{5}x + \frac{3}{10}y = \frac{7}{50} \\ \frac{2}{5}x + \frac{1}{2}y = \frac{1}{5} \end{cases}$

24. $\begin{cases} \frac{5}{12}x - \frac{3}{4}y = \frac{25}{4} \\ -x + \frac{7}{8}y = -\frac{38}{5} \end{cases}$

25. $\begin{cases} 3x - 2y = 0 \\ 3x + 2(y + 5) = 10 \end{cases}$

26. $\begin{cases} 7x + 12y = 63 \\ 2x + 3y = 15 \end{cases}$

27. $\begin{cases} 1.25x - 2y = 3.5 \\ 5x - 8y = 14 \end{cases}$

28. $\begin{cases} 1.5x + 2.5y = 8.5 \\ 6x + 10y = 24 \end{cases}$

In Exercises 29–34, use a graphing utility to graph the lines in the system. Use the graphs to determine whether the system is consistent or inconsistent. If the system is consistent, determine the solution. Verify your results algebraically.

29. $\begin{cases} 3x + 2y = 0 \\ x - y = 4 \end{cases}$

30. $\begin{cases} x + y = 6 \\ -2x - 2y = -12 \end{cases}$

31. $\begin{cases} \frac{1}{4}x - \frac{1}{5}y = 2 \\ -5x + 4y = 8 \end{cases}$

32. $\begin{cases} \frac{7}{2}x - 7y = -1 \\ -x + 2y = 4 \end{cases}$

33. $\begin{cases} 2x - 2y = 8 \\ 4x - 1.5y = -5.5 \end{cases}$

34. $\begin{cases} -x + 3.2y = 10.4 \\ -2x - 9.6y = 6.4 \end{cases}$

Supply and Demand   **In Exercises 35 and 36, find the point of equilibrium of the demand and supply equations.**

	Demand	Supply
35.	$p = 37 - 0.0002x$	$p = 22 + 0.00001x$
36.	$p = 120 - 0.0001x$	$p = 45 + 0.0002x$

37. *Airplane Speed*   Two planes leave Pittsburgh and Philadelphia at the same time, each going to the other city. One plane flies 25 miles per hour faster than the other. Find the airspeed of each plane if the cities are 275 miles apart and the planes pass each other after 40 minutes of flying time.

38. *Investment Portfolio*   A total of $46,000 is invested in two corporate bonds that pay 6.75% and 7.25% simple interest. The investor wants an annual interest income of $3245 from the investments. What is the most that can be invested in the 6.75% bond?

7.3 In Exercises 39 and 40, use back-substitution to solve the system of linear equations.

39. $\begin{cases} x - 4y + 3z = -14 \\ -y + z = -5 \\ z = -2 \end{cases}$

40. $\begin{cases} x - 7y + 8z = -14 \\ y - 9z = 26 \\ z = -3 \end{cases}$

In Exercises 41–48, solve the system of linear equations and check any solution algebraically.

41. $\begin{cases} x + 3y - z = 13 \\ 2x \quad\quad - 5z = 23 \\ 4x - y - 2z = 14 \end{cases}$

42. $\begin{cases} x + 2y + 6z = 4 \\ 3x - 2y + z = 4 \\ 4x + \quad\quad 2z = 0 \end{cases}$

43. $\begin{cases} x - 2y + z = -6 \\ 2x - 3y \quad\quad = -7 \\ -x + 3y - 3z = 11 \end{cases}$

44. $\begin{cases} 2x + \quad\quad 6z = -9 \\ 3x - 2y + 11z = -16 \\ 3x - y + 7z = -11 \end{cases}$

45. $\begin{cases} x - 2y + 3z = -5 \\ 2x + 4y + 5z = 1 \\ x + 2y + z = 0 \end{cases}$

46. $\begin{cases} x - 2y + z = 5 \\ 2x + 3y + z = 5 \\ x + y + 2z = 3 \end{cases}$

47. $\begin{cases} 5x - 12y + 7z = 16 \\ 3x - 7y + 4z = 9 \end{cases}$

48. $\begin{cases} 2x + 5y - 19z = 34 \\ 3x + 8y - 31z = 54 \end{cases}$

In Exercises 49 and 50, sketch the plane represented by the linear equation. Then list four points that lie in the plane.

49. $2x - 4y + z = 8$

50. $3x + 3y - z = 9$

In Exercises 51–56, write the partial fraction decomposition for the rational expression. Check your result algebraically by combining the fractions, and check your result graphically by using a graphing utility to graph the rational expression and the partial fractions in the same viewing window.

51. $\dfrac{4 - x}{x^2 + 6x + 8}$

52. $\dfrac{-x}{x^2 + 3x + 2}$

53. $\dfrac{x^2 + 2x}{x^3 - x^2 + x - 1}$

54. $\dfrac{3x^3 + 4x}{x^4 + 2x^2 + 1}$

55. $\dfrac{x^2 + 3x - 3}{x^3 + 2x^2 + x + 2}$

56. $\dfrac{2x^2 - x + 7}{x^4 + 8x^2 + 16}$

In Exercises 57 and 58, find the equation of the parabola $y = ax^2 + bx + c$ that passes through the points. To verify your result, use a graphing utility to plot the points and graph the parabola.

57. $(-1, -4), (1, -2), (2, 5)$ 58. $(-1, 0), (1, 4), (2, 3)$

59. *Agriculture* A mixture of 6 gallons of chemical A, 8 gallons of chemical B, and 13 gallons of chemical C is required to kill a destructive crop insect. Commercial spray X contains 1, 2, and 2 parts, respectively, of these chemicals. Commercial spray Y contains only chemical C. Commercial spray Z contains chemicals A, B, and C in equal amounts. How much of each type of commercial spray is needed to obtain the desired mixture?

60. *Investment Portfolio* An inheritance of $20,000 is divided among three investments yielding $1780 in interest per year. The interest rates for the three investments are 7%, 9%, and 11%. Find the amount of each investment if the second and third are $3000 and $1000 less than the first, respectively.

7.4 In Exercises 61–64, determine the order of the matrix.

61. $\begin{bmatrix} -3 \\ 1 \\ 10 \end{bmatrix}$

62. $\begin{bmatrix} 3 & -1 & 0 & 6 \\ -2 & 7 & 1 & 4 \end{bmatrix}$

63. $\begin{bmatrix} 14 \end{bmatrix}$

64. $\begin{bmatrix} 6 & 7 & -5 & 0 & -8 \end{bmatrix}$

In Exercises 65–68, write the augmented matrix for the system of linear equations.

65. $\begin{cases} 3x - 10y = 15 \\ 5x + 4y = 22 \end{cases}$

66. $\begin{cases} -x + y = 12 \\ 10x - 4y = -90 \end{cases}$

67. $\begin{cases} 8x - 7y + 4z = 12 \\ 3x - 5y + 2z = 20 \\ 5x + 3y - 3z = 26 \end{cases}$

68. $\begin{cases} 3x - 5y + z = 25 \\ -4x \quad\quad - 2z = -14 \\ 6x + y \quad\quad = 15 \end{cases}$

In Exercises 69 and 70, write the system of linear equations represented by the augmented matrix. (Use the variables x, y, z, and w, if applicable.)

69. $\begin{bmatrix} 5 & 1 & 7 & \vdots & -9 \\ 4 & 2 & 0 & \vdots & 10 \\ 9 & 4 & 2 & \vdots & 3 \end{bmatrix}$

70. $\begin{bmatrix} 13 & 16 & 7 & 3 & \vdots & 2 \\ 1 & 21 & 8 & 5 & \vdots & 12 \\ 4 & 10 & -4 & 3 & \vdots & -1 \end{bmatrix}$

In Exercises 71 and 72, write the matrix in row-echelon form. Remember that the row-echelon form of a matrix is not unique.

71. $\begin{bmatrix} 0 & 1 & 1 \\ 1 & 2 & 3 \\ 2 & 2 & 2 \end{bmatrix}$

72. $\begin{bmatrix} 3 & 5 & 2 \\ 1 & -2 & 4 \\ -2 & 0 & 5 \end{bmatrix}$

In Exercises 73–76, use the matrix capabilities of a graphing utility to write the matrix in reduced row-echelon form.

73. $\begin{bmatrix} 3 & -2 & 1 & 0 \\ 4 & -3 & 0 & 1 \end{bmatrix}$

74. $\begin{bmatrix} 1 & 1 & 2 & 1 & 0 & 0 \\ -1 & 0 & 3 & 0 & 1 & 0 \\ 1 & 2 & 8 & 0 & 0 & 1 \end{bmatrix}$

75. $\begin{bmatrix} 1.5 & 3.6 & 4.2 \\ 0.2 & 1.4 & 1.8 \\ 2.0 & 4.4 & 6.4 \end{bmatrix}$

76. $\begin{bmatrix} 4.1 & 8.3 & 1.6 \\ 3.2 & -1.7 & 2.4 \\ -2.3 & 1.0 & 1.2 \end{bmatrix}$

In Exercises 77–84, use matrices to solve the system of equations, if possible. Use Gaussian elimination with back-substitution.

77. $\begin{cases} 5x + 4y = 2 \\ -x + y = -22 \end{cases}$

78. $\begin{cases} 2x - 5y = 2 \\ 3x - 7y = 1 \end{cases}$

79. $\begin{cases} 2x + y = 0.3 \\ 3x - y = -1.3 \end{cases}$

80. $\begin{cases} 0.2x - 0.1y = 0.07 \\ 0.4x - 0.5y = -0.01 \end{cases}$

81. $\begin{cases} 2x + 3y + 3z = 3 \\ 6x + 6y + 12z = 13 \\ 12x + 9y - z = 2 \end{cases}$

82. $\begin{cases} x + 2y + 6z = 1 \\ 2x + 5y + 15z = 4 \\ 3x + y + 3z = -6 \end{cases}$

83. $\begin{cases} x + 2y - z = 1 \\ y + z = 0 \end{cases}$

84. $\begin{cases} x - y + 4z - w = 4 \\ x + 3y - 2z + w = -4 \\ y - z + w = -3 \\ 2x + z + w = 0 \end{cases}$

In Exercises 85–88, use matrices to solve the system of equations, if possible. Use Gauss-Jordan elimination.

85. $\begin{cases} -x + y + 2z = 1 \\ 2x + 3y + z = -2 \\ 5x + 4y + 2z = 4 \end{cases}$

86. $\begin{cases} 4x + 4y + 4z = 5 \\ 4x - 2y - 8z = 1 \\ 5x + 3y + 8z = 6 \end{cases}$

87. $\begin{cases} x + y + 2z = 4 \\ x - y + 4z = 1 \\ 2x - y + 2z = 1 \end{cases}$

88. $\begin{cases} x + y + 4z = 0 \\ 2x + y + 2z = 0 \\ -x + y - 2z = -1 \end{cases}$

In Exercises 89–92, use the matrix capabilities of a graphing utility to reduce the augmented matrix corresponding to the system of equations, and solve the system.

89. $\begin{cases} x + 2y - z = 7 \\ -y - z = 4 \\ 4x - z = 16 \end{cases}$

90. $\begin{cases} 3x + 6z = 0 \\ -2x + y = 5 \\ y + 2z = 3 \end{cases}$

91. $\begin{cases} 3x - y + 5z - 2w = -44 \\ x + 6y + 4z - w = 1 \\ 5x - y + z + 3w = -15 \\ 4y - z - 8w = 58 \end{cases}$

92. $\begin{cases} 4x + 12y + 2z = 20 \\ x + 6y + 4z = 12 \\ x + 6y + z = 8 \\ -2x - 10y - 2z = -10 \end{cases}$

7.5 **In Exercises 93–96, find x and y.**

93. $\begin{bmatrix} -1 & x \\ y & 9 \end{bmatrix} = \begin{bmatrix} -1 & 12 \\ -7 & 9 \end{bmatrix}$

94. $\begin{bmatrix} -1 & 0 \\ x & 5 \\ -4 & y \end{bmatrix} = \begin{bmatrix} -1 & 0 \\ 8 & 5 \\ -4 & 0 \end{bmatrix}$

95. $\begin{bmatrix} x + 3 & 4 & -4y \\ 0 & -3 & 2 \\ -2 & y + 5 & 6x \end{bmatrix} = \begin{bmatrix} 5x - 1 & 4 & -44 \\ 0 & -3 & 2 \\ -2 & 16 & 6 \end{bmatrix}$

96. $\begin{bmatrix} -9 & 4 & 2 & -5 \\ 0 & -3 & 7 & -4 \\ 6 & -1 & 1 & 0 \end{bmatrix} = \begin{bmatrix} -9 & 4 & x - 10 & -5 \\ 0 & -3 & 7 & 2y \\ \frac{1}{2}x & -1 & 1 & 0 \end{bmatrix}$

In Exercises 97–100, find, if possible, (a) $A + B$, (b) $A - B$, (c) $4A$, and (d) $A + 3B$.

97. $A = \begin{bmatrix} 7 & 3 \\ -1 & 5 \end{bmatrix}$, $B = \begin{bmatrix} 10 & -20 \\ 14 & -3 \end{bmatrix}$

98. $A = \begin{bmatrix} -11 & 16 & 19 \\ -7 & -2 & 1 \end{bmatrix}$, $B = \begin{bmatrix} 6 & 0 \\ 8 & -4 \\ -2 & 10 \end{bmatrix}$

99. $A = \begin{bmatrix} 6 & 0 & 7 \\ 5 & -1 & 2 \\ 3 & 2 & 3 \end{bmatrix}$, $B = \begin{bmatrix} 0 & 5 & 1 \\ -4 & 8 & 6 \\ 2 & -1 & 1 \end{bmatrix}$

100. $A = \begin{bmatrix} 2 & -3 & 6 \\ 0 & 4 & 1 \end{bmatrix}$, $B = \begin{bmatrix} -3 & 5 & 5 \\ 1 & 1 & 1 \end{bmatrix}$

In Exercises 101–104, evaluate the expression. If it is not possible, explain why.

101. $\begin{bmatrix} 2 & 1 & 0 \\ 0 & 5 & -4 \end{bmatrix} - 3\begin{bmatrix} 5 & 3 & -6 \\ 0 & -2 & 5 \end{bmatrix}$

102. $-2\begin{bmatrix} 1 & 2 \\ 5 & -4 \\ 6 & 0 \end{bmatrix} + 8\begin{bmatrix} 7 & 1 \\ 1 & 2 \\ 1 & 4 \end{bmatrix}$

103. $-\begin{bmatrix} 8 & -1 \\ -2 & 4 \end{bmatrix} - 5\begin{bmatrix} -2 & 0 \\ 3 & -1 \end{bmatrix} + \begin{bmatrix} 7 & -8 \\ 4 & 3 \end{bmatrix}$

104. $6\left(\begin{bmatrix} -4 & -1 & -3 & 4 \\ 2 & -5 & 7 & -10 \end{bmatrix} + \begin{bmatrix} -1 & 1 & 13 & -7 \\ 14 & -3 & 8 & -1 \end{bmatrix} \right)$

In Exercises 105 and 106, use the matrix capabilities of a graphing utility to evaluate the expression.

105. $3\begin{bmatrix} \frac{8}{3} & -2 & \frac{5}{6} \\ 1 & \frac{4}{3} & -1 \end{bmatrix} + 6\begin{bmatrix} 4 & -\frac{5}{12} & -\frac{2}{3} \\ \frac{1}{2} & 7 & 6 \end{bmatrix}$

106. $-5\begin{bmatrix} 2.7 & 0.2 \\ 7.3 & -2.9 \\ 8.6 & 2.1 \end{bmatrix} + 4\begin{bmatrix} 4.4 & -2.3 \\ 6.6 & 11.6 \\ -1.5 & 3.9 \end{bmatrix}$

In Exercises 107–110, solve for X when

$$A = \begin{bmatrix} -4 & 0 \\ 1 & -5 \\ -3 & 2 \end{bmatrix} \quad \text{and} \quad B = \begin{bmatrix} 1 & 2 \\ -2 & 1 \\ 4 & 4 \end{bmatrix}.$$

107. $X = 3A - 2B$

108. $6X = 4A + 3B$

109. $3X + 2A = B$

110. $2A - 5B = 3X$

In Exercises 111–114, find AB, if possible.

111. $A = \begin{bmatrix} 1 & 2 \\ 5 & -4 \\ 6 & 0 \end{bmatrix}$, $B = \begin{bmatrix} 6 & -2 & 8 \\ 4 & 0 & 0 \end{bmatrix}$

112. $A = \begin{bmatrix} 1 & 5 & 6 \\ 2 & -4 & 0 \end{bmatrix}$, $B = \begin{bmatrix} 7 & 5 & 2 \\ 0 & 1 & 0 \end{bmatrix}$

113. $A = \begin{bmatrix} 3 & -2 & 0 \\ 1 & 4 & 9 \end{bmatrix}$, $B = \begin{bmatrix} 7 & 0 \\ 5 & 3 \\ -1 & 3 \end{bmatrix}$

114. $A = \begin{bmatrix} 1 & 3 & 2 \\ 0 & 2 & -4 \\ 1 & -1 & 3 \end{bmatrix}$, $B = \begin{bmatrix} 4 & -3 & 2 \\ 0 & 3 & -1 \\ 0 & 6 & 2 \end{bmatrix}$

In Exercises 115–118, use the matrix capabilities of a graphing utility to evaluate the expression.

115. $\begin{bmatrix} 4 & 1 \\ 11 & -7 \\ 12 & 3 \end{bmatrix} \begin{bmatrix} 3 & -5 & 6 \\ 2 & -2 & -2 \end{bmatrix}$

116. $\begin{bmatrix} -2 & 3 & 10 \\ 4 & -2 & 2 \end{bmatrix} \begin{bmatrix} 1 & 1 \\ -5 & 2 \\ 3 & 2 \end{bmatrix}$

117. $\begin{bmatrix} 2 & 1 \\ 6 & 0 \end{bmatrix} \left(\begin{bmatrix} 4 & 2 \\ -3 & 1 \end{bmatrix} + \begin{bmatrix} -2 & 4 \\ 0 & 4 \end{bmatrix} \right)$

118. $\begin{bmatrix} 1 & -1 \\ 4 & 2 \end{bmatrix} \left(\begin{bmatrix} 0 & 3 \\ 1 & 2 \end{bmatrix} \begin{bmatrix} 1 & 0 \\ 5 & -3 \end{bmatrix} \right)$

119. *Sales* At a dairy mart, the numbers of gallons of skim, 2%, and whole milk sold on Friday, Saturday, and Sunday of a particular week are given by the following matrix.

	Skim milk	2% milk	Whole milk	
$A =$	40	64	52	Friday
	60	82	76	Saturday
	76	96	84	Sunday

A second matrix gives the selling price per gallon and the profit per gallon for each of the three types of milk sold by the dairy mart.

	Selling price per gallon	Profit per gallon	
$B =$	2.65	0.25	Skim milk
	2.81	0.30	2% milk
	2.93	0.35	Whole milk

(a) Find AB. What is the meaning of AB in the context of the situation?

(b) Find the dairy mart's profit for Friday through Sunday.

120. *Exercise* The numbers of calories burned by individuals of different weights performing different types of aerobic exercises for 20-minute time periods are shown in the matrix.

	120-lb person	150-lb person	
$B =$	109	136	Bicycling
	127	159	Jogging
	64	79	Walking

(a) A 120-pound person and a 150-pound person bicycle for 40 minutes, jog for 10 minutes, and walk for 60 minutes. Organize a matrix A for the time spent exercising in units of 20-minute intervals.

(b) Find the product AB.

(c) Explain the meaning of the product AB in the context of the situation.

7.6 In Exercises 121 and 122, show that B is the inverse of A.

121. $A = \begin{bmatrix} -4 & -1 \\ 7 & 2 \end{bmatrix}$, $B = \begin{bmatrix} -2 & -1 \\ 7 & 4 \end{bmatrix}$

122. $A = \begin{bmatrix} 1 & 1 & 0 \\ 1 & 0 & 1 \\ 6 & 2 & 3 \end{bmatrix}$, $B = \begin{bmatrix} -2 & -3 & 1 \\ 3 & 3 & -1 \\ 2 & 4 & -1 \end{bmatrix}$

In Exercises 123–126, find the inverse of the matrix (if it exists).

123. $\begin{bmatrix} -6 & 5 \\ -5 & 4 \end{bmatrix}$

124. $\begin{bmatrix} -3 & -5 \\ 2 & 3 \end{bmatrix}$

125. $\begin{bmatrix} -1 & -2 & -2 \\ 3 & 7 & 9 \\ 1 & 4 & 7 \end{bmatrix}$

126. $\begin{bmatrix} 0 & -2 & 1 \\ -5 & -2 & -3 \\ 7 & 3 & 4 \end{bmatrix}$

In Exercises 127–130, use the matrix capabilities of a graphing utility to find the inverse of the matrix (if it exists).

127. $\begin{bmatrix} 2 & 6 \\ 3 & -6 \end{bmatrix}$ **128.** $\begin{bmatrix} 3 & -10 \\ 4 & 2 \end{bmatrix}$

129. $\begin{bmatrix} 1 & 2 & 0 \\ -1 & 1 & 1 \\ 0 & -1 & 0 \end{bmatrix}$ **130.** $\begin{bmatrix} 1 & -1 & -2 \\ 0 & 1 & -2 \\ 1 & 2 & -4 \end{bmatrix}$

In Exercises 131–134, use the formula on page 545 to find the inverse of the 2×2 matrix.

131. $\begin{bmatrix} -7 & 2 \\ -8 & 2 \end{bmatrix}$ **132.** $\begin{bmatrix} 10 & 4 \\ 7 & 3 \end{bmatrix}$

133. $\begin{bmatrix} -1 & 10 \\ 2 & 20 \end{bmatrix}$ **134.** $\begin{bmatrix} -6 & -5 \\ 3 & 3 \end{bmatrix}$

In Exercises 135–140, use an inverse matrix to solve (if possible) the system of linear equations.

135. $\begin{cases} x + 5y = -1 \\ 3x - 5y = 5 \end{cases}$ **136.** $\begin{cases} 2x + 3y = -10 \\ 4x - y = 1 \end{cases}$

137. $\begin{cases} 3x + 2y - z = 6 \\ x - y + 2z = -1 \\ 5x + y + z = 7 \end{cases}$ **138.** $\begin{cases} -x + 4y - 2z = 12 \\ 2x - 9y + 5z = -25 \\ -x + 5y - 4z = 10 \end{cases}$

139. $\begin{cases} x + 2y + z - w = -2 \\ 2x + y + z + w = 1 \\ x - y - 3z = 0 \\ z + w = 1 \end{cases}$

140. $\begin{cases} x + y + z + w = 1 \\ x - y + 2z + w = -3 \\ y + w = 2 \\ x + w = 2 \end{cases}$

In Exercises 141–144, use the matrix capabilities of a graphing utility to solve (if possible) the system of linear equations.

141. $\begin{cases} x + 2y = -1 \\ 3x + 4y = -5 \end{cases}$ **142.** $\begin{cases} x + 3y = 23 \\ -6x + 2y = -18 \end{cases}$

143. $\begin{cases} -3x - 3y - 4z = 2 \\ y + z = -1 \\ 4x + 3y + 4z = -1 \end{cases}$ **144.** $\begin{cases} 2x + 3y - 4z = 1 \\ x - y + 2z = -4 \\ 3x + 7y - 10z = 0 \end{cases}$

7.7 In Exercises 145–148, find the determinant of the matrix.

145. $\begin{bmatrix} 8 & 5 \\ 2 & -4 \end{bmatrix}$ **146.** $\begin{bmatrix} -9 & 11 \\ 7 & -4 \end{bmatrix}$

147. $\begin{bmatrix} 50 & -30 \\ 10 & 5 \end{bmatrix}$ **148.** $\begin{bmatrix} 14 & -24 \\ 12 & -15 \end{bmatrix}$

In Exercises 149–152, find all (a) minors and (b) cofactors of the matrix.

149. $\begin{bmatrix} 2 & -1 \\ 7 & 4 \end{bmatrix}$ **150.** $\begin{bmatrix} 3 & 6 \\ 5 & -4 \end{bmatrix}$

151. $\begin{bmatrix} 3 & 2 & -1 \\ -2 & 5 & 0 \\ 1 & 8 & 6 \end{bmatrix}$ **152.** $\begin{bmatrix} 8 & 3 & 4 \\ 6 & 5 & -9 \\ -4 & 1 & 2 \end{bmatrix}$

In Exercises 153–158, find the determinant of the matrix. Expand by cofactors on the row or column that appears to make the computations easiest.

153. $\begin{bmatrix} -2 & 4 & 1 \\ -6 & 0 & 2 \\ 5 & 3 & 4 \end{bmatrix}$ **154.** $\begin{bmatrix} 4 & 7 & -1 \\ 2 & -3 & 4 \\ -5 & 1 & -1 \end{bmatrix}$

155. $\begin{bmatrix} 1 & 0 & -2 \\ 0 & 1 & 0 \\ -2 & 0 & 1 \end{bmatrix}$ **156.** $\begin{bmatrix} 0 & 3 & 1 \\ 5 & -2 & 1 \\ 1 & 6 & 1 \end{bmatrix}$

157. $\begin{bmatrix} 3 & 0 & -4 & 0 \\ 0 & 8 & 1 & 2 \\ 6 & 1 & 8 & 2 \\ 0 & 3 & -4 & 1 \end{bmatrix}$ **158.** $\begin{bmatrix} -5 & 6 & 0 & 0 \\ 0 & 1 & -1 & 2 \\ -3 & 4 & -5 & 1 \\ 1 & 6 & 0 & 3 \end{bmatrix}$

In Exercises 159 and 160, evaluate the determinant. Do not use a graphing utility.

159. $\begin{vmatrix} 8 & 6 & 0 & 2 \\ 0 & -1 & 1 & -4 \\ 0 & 0 & 4 & 5 \\ 0 & 0 & 0 & 3 \end{vmatrix}$ **160.** $\begin{vmatrix} -5 & 0 & 0 & 0 \\ 7 & -2 & 0 & 0 \\ 11 & 21 & 2 & 0 \\ -6 & 9 & 12 & 14 \end{vmatrix}$

7.8 In Exercises 161–168, use a determinant to find the area of the figure with the given vertices.

161. $(1, 0), (5, 0), (5, 8)$

162. $(-4, 0), (4, 0), (0, 6)$

163. $\left(\frac{1}{2}, 1\right), \left(2, -\frac{5}{2}\right), \left(\frac{3}{2}, 1\right)$

164. $\left(\frac{3}{2}, 1\right), \left(4, -\frac{1}{2}\right), (4, 2)$

165. $(2, 4), (5, 6), (4, 1)$

166. $(-3, 2), (2, -3), (-4, -4)$

167. $(-2, -1), (4, 9), (-2, -9), (4, 1)$

168. $(-4, 8), (4, 0), (-4, 0), (4, -8)$

In Exercises 169 and 170, use a determinant to determine whether the points are collinear.

169. $(-1, 7), (2, 5), (4, 1)$

170. $(0, -5), (2, 1), (4, 7)$

In Exercises 171–178, use Cramer's Rule to solve (if possible) the system of equations.

171. $\begin{cases} x + 2y = 5 \\ -x + y = 1 \end{cases}$ **172.** $\begin{cases} 2x - y = -10 \\ 3x + 2y = -1 \end{cases}$

173. $\begin{cases} 5x - 2y = 6 \\ -11x + 3y = -23 \end{cases}$ **174.** $\begin{cases} 3x + 8y = -7 \\ 9x - 5y = 37 \end{cases}$

175. $\begin{cases} -2x + 3y - 5z = -11 \\ 4x - y + z = -3 \\ -x - 4y + 6z = 15 \end{cases}$

176. $\begin{cases} 5x - 2y + z = 15 \\ 3x - 3y - z = -7 \\ 2x - y - 7z = -3 \end{cases}$

177. $\begin{cases} x - 3y + 2z = 2 \\ 2x + 2y - 3z = 3 \\ x - 7y + 8z = -4 \end{cases}$ **178.** $\begin{cases} 14x - 21y - 7z = 10 \\ -4x + 2y - 2z = 4 \\ 56x - 21y + 7z = 5 \end{cases}$

In Exercises 179 and 180, solve the system of equations using (a) Gaussian elimination and (b) Cramer's Rule. Which method do you prefer, and why?

179. $\begin{cases} x - 3y + 2z = 5 \\ 2x + y - 4z = -1 \\ 2x + 4y + 2z = 3 \end{cases}$ **180.** $\begin{cases} x + 2y - z = -3 \\ 2x - y + z = -1 \\ 4x - 2y - z = 5 \end{cases}$

In Exercises 181 and 182, write the uncoded 1×3 row matrices for the message. Then encode the message using the encoding matrix.

Message	*Encoding Matrix*
181. I HAVE A DREAM	$\begin{bmatrix} 2 & -2 & 0 \\ 3 & 0 & -3 \\ -6 & 2 & 3 \end{bmatrix}$
182. JUST DO IT	$\begin{bmatrix} 2 & 1 & 0 \\ -6 & -6 & -2 \\ 3 & 2 & 1 \end{bmatrix}$

In Exercises 183–186, use A^{-1} to decode the cryptogram.

183. $A = \begin{bmatrix} 1 & 0 & -1 \\ -1 & -2 & 0 \\ 1 & -2 & 2 \end{bmatrix}$ 32 −46 37 9 −48 15
3 −14 10 −1 −6 2
−8 −22 −3

184. $A = \begin{bmatrix} 1 & 2 & 0 \\ -1 & 1 & 2 \\ 2 & -1 & 2 \end{bmatrix}$ 30 −7 30 5 10 80 37
34 16 40 −7 38 −3 8
36 16 −1 58 23 46 0

185. $A = \begin{bmatrix} 1 & -1 & 0 \\ 0 & 1 & 2 \\ 1 & 1 & -2 \end{bmatrix}$ 21 −11 14 29 −11 −18
32 −6 −26 31 −19
−12 10 6 26 13 −11
−2 37 28 −8 5 13 36

186. $A = \begin{bmatrix} 1 & 1 & 0 \\ -1 & 2 & -2 \\ 1 & -1 & -2 \end{bmatrix}$ 9 15 −54 13 32 −26 8
−6 −14 −4 26 −70
−1 56 −38 28 27 −46
−13 27 −30 26 23 −48
25 4 −26 −11 31 −58
13 39 −34

187. *Population* The populations (in millions) of Florida for selected years from 1998 to 2004 are shown in the table. (Source: U.S. Census Bureau)

Year	Population (in millions)
1998	14.9
2000	16.0
2002	16.7
2004	17.4

A system of linear equations that can be used to find the least squares regression line $y = at + b$, where y is the population (in millions) and t is the year, with $t = 8$ corresponding to 1998, is

$$\begin{cases} 4b + 44a = 65 \\ 44b + 504a = 723.2 \end{cases}$$

(a) Use Cramer's Rule to solve the system and find the least squares regression line.

(b) Use a graphing utility to graph the line from part (a).

(c) Use the graph from part (b) to estimate when the population of Florida will exceed 20 million.

(d) Use your regression equation to find algebraically when the population will exceed 20 million.

Synthesis

True or False? In Exercises 188 and 189, determine whether the statement is true or false. Justify your answer.

188. Solving a system of equations graphically will always give an exact solution.

189. $\begin{vmatrix} a_{11} & a_{12} & a_{13} \\ a_{21} & a_{22} & a_{23} \\ a_{31} + c_1 & a_{32} + c_2 & a_{33} + c_3 \end{vmatrix} =$

$\begin{vmatrix} a_{11} & a_{12} & a_{13} \\ a_{21} & a_{22} & a_{23} \\ a_{31} & a_{32} & a_{33} \end{vmatrix} + \begin{vmatrix} a_{11} & a_{12} & a_{13} \\ a_{21} & a_{22} & a_{23} \\ c_1 & c_2 & c_3 \end{vmatrix}$

190. What is the relationship between the three elementary row operations performed on an augmented matrix and the operations that lead to equivalent systems of equations?

191. Under what conditions does a matrix have an inverse?

7 Chapter Test

See www.CalcChat.com for worked-out solutions to odd-numbered exercises.

Take this test as you would take a test in class. After you are finished, check your work against the answers given in the back of the book.

In Exercises 1–3, solve the system by the method of substitution. Check your solution graphically.

1. $\begin{cases} x - y = 6 \\ 3x + 5y = 2 \end{cases}$

2. $\begin{cases} y = x - 1 \\ y = (x - 1)^3 \end{cases}$

3. $\begin{cases} 4x - y^2 = 7 \\ x - y = 3 \end{cases}$

In Exercises 4–6, solve the system by the method of elimination.

4. $\begin{cases} 2x + 5y = -11 \\ 5x - y = 19 \end{cases}$

5. $\begin{cases} 3x - 2y + z = 0 \\ 6x + 2y + 3z = -2 \\ 3x - 4y + 5z = 5 \end{cases}$

6. $\begin{cases} x - 4y - z = 3 \\ 2x - 5y + z = 0 \\ 3x - 3y + 2z = -1 \end{cases}$

7. Find the equation of the parabola $y = ax^2 + bx + c$ that passes through the points $(0, 6)$, $(-2, 2)$, and $\left(3, \frac{9}{2}\right)$.

In Exercises 8 and 9, write the partial fraction decomposition for the rational expression.

8. $\dfrac{5x - 2}{(x - 1)^2}$

9. $\dfrac{x^3 + x^2 + x + 2}{x^4 + x^2}$

$\begin{cases} -2x + 2y + 3z = 7 \\ x - y = -5 \\ y + 4z = -1 \end{cases}$

System for 13

In Exercises 10 and 11, use matrices to solve the system of equations, if possible.

10. $\begin{cases} 2x + y + 2z = 4 \\ 2x + 2y = 5 \\ 2x - y + 6z = 2 \end{cases}$

11. $\begin{cases} 2x + 3y + z = 10 \\ 2x - 3y - 3z = 22 \\ 4x - 2y + 3z = -2 \end{cases}$

12. If possible, find (a) $A - B$, (b) $3A$, (c) $3A - 2B$, and (d) AB.

$$A = \begin{bmatrix} 5 & 4 & 4 \\ -4 & -4 & 0 \\ 1 & 2 & 0 \end{bmatrix}, \quad B = \begin{bmatrix} 4 & 4 & 0 \\ 3 & 2 & 1 \\ 1 & -2 & 0 \end{bmatrix}$$

13. Find A^{-1} for $A = \begin{bmatrix} -2 & 2 & 3 \\ 1 & -1 & 0 \\ 0 & 1 & 4 \end{bmatrix}$ and use A^{-1} to solve the system at the right.

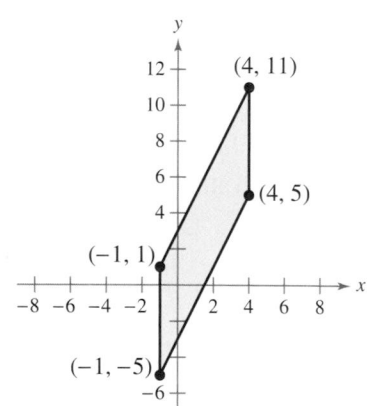

Figure for 16

In Exercises 14 and 15, find the determinant of the matrix.

14. $\begin{bmatrix} -25 & 18 \\ 6 & -7 \end{bmatrix}$

15. $\begin{bmatrix} 4 & 0 & 3 \\ 1 & -8 & 2 \\ 3 & 2 & 2 \end{bmatrix}$

16. Use determinants to find the area of the parallelogram shown at the right.

17. Use Cramer's Rule to solve (if possible) $\begin{cases} 2x - 2y = 3 \\ x + 4y = -1 \end{cases}$.

18. The flow of traffic (in vehicles per hour) through a network of streets is shown at the right. Solve the system for the traffic flow represented by x_i, $i = 1, 2, 3, 4$, and 5.

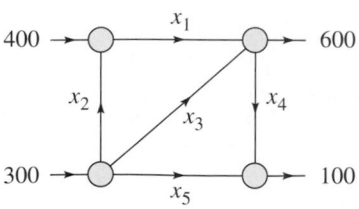

Figure for 18

Proofs in Mathematics

An **indirect proof** can be useful in proving statements of the form "p implies q." Recall that the conditional statement $p \rightarrow q$ is false only when p is true and q is false. To prove a conditional statement indirectly, assume that p is true and q is false. If this assumption leads to an impossibility, then you have proved that the conditional statement is true. An indirect proof is also called a **proof by contradiction.**

You can use an indirect proof to prove the following conditional statement,

"If a is a positive integer and a^2 is divisible by 2, then a is divisible by 2,"

as follows. First, assume that p, "a is a positive integer and a^2 is divisible by 2" is true and q, "a is divisible by 2," is false. This means that a is not divisible by 2. If so, a is odd and can be written as $a = 2n + 1$, where n is an integer.

$a = 2n + 1$ Definition of an odd integer

$a^2 = 4n^2 + 4n + 1$ Square each side.

$a^2 = 2(2n^2 + 2n) + 1$ Distributive Property

So, by the definition of an odd integer, a^2 is odd. This contradicts the assumption, and you can conclude that a is divisible by 2.

Example   Using an Indirect Proof

Use an indirect proof to prove that $\sqrt{2}$ is an irrational number.

Solution

Begin by assuming that $\sqrt{2}$ is *not* an irrational number. Then $\sqrt{2}$ can be written as the quotient of two integers a and $b (b \neq 0)$ that have no common factors.

$\sqrt{2} = \dfrac{a}{b}$ Assume that $\sqrt{2}$ is a rational number.

$2 = \dfrac{a^2}{b^2}$ Square each side.

$2b^2 = a^2$ Multiply each side by b^2.

This implies that 2 is a factor of a^2. So, 2 is also a factor of a, and a can be written as $2c$, where c is an integer.

$2b^2 = (2c)^2$ Substitute $2c$ for a.

$2b^2 = 4c^2$ Simplify.

$b^2 = 2c^2$ Divide each side by 2.

This implies that 2 is a factor of b^2 and also a factor of b. So, 2 is a factor of both a and b. This contradicts the assumption that a and b have no common factors. So, you can conclude that $\sqrt{2}$ is an irrational number.

Proofs without words are pictures or diagrams that give a visual understanding of why a theorem or statement is true. They can also provide a starting point for writing a formal proof. The following proof shows that a 2×2 determinant is the area of a parallelogram.

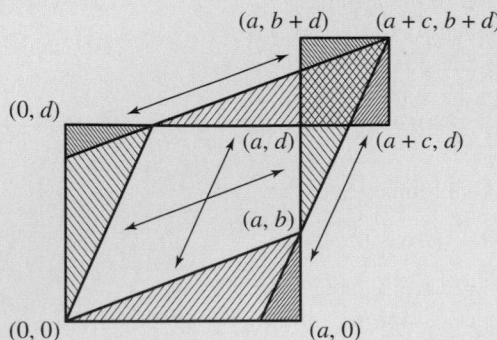

$$\begin{vmatrix} a & b \\ c & d \end{vmatrix} = ad - bc = \|\square\| - \|\,\square\,\| = \|\square\|$$

The following is a color-coded version of the proof along with a brief explanation of why this proof works.

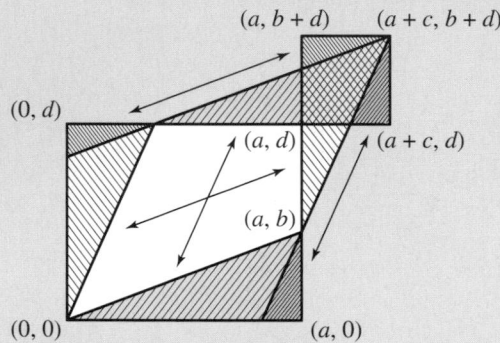

$$\begin{vmatrix} a & b \\ c & d \end{vmatrix} = ad - bc = \|\square\| - \|\,\square\,\| = \|\square\|$$

Area of $\square$ = Area of orange $\triangle$ + Area of yellow $\triangle$ + Area of blue $\triangle$ + Area of pink $\triangle$ + Area of white quadrilateral

Area of $\square$ = Area of orange $\triangle$ + Area of pink $\triangle$ + Area of green quadrilateral

Area of $\square$ = Area of white quadrilateral + Area of blue $\triangle$ + Area of yellow $\triangle$ − Area of green quadrilateral
= Area of $\square$ − Area of $\square$

From "Proof Without Words" by Solomon W. Golomb, *Mathematics Magazine*, March 1985. Vol. 58, No. 2, pg. 107. Reprinted with permission.

Chapter 8

Sequences, Series, and Probability

8.1 Sequences and Series
8.2 Arithmetic Sequences and Partial Sums
8.3 Geometric Sequences and Series
8.4 Mathematical Induction
8.5 The Binomial Theorem
8.6 Counting Principles
8.7 Probability

Selected Applications

Sequences, series, and probability have many real-life applications. The applications listed below represent a small sample of the applications in this chapter.

- Average Wages, Exercise 119, page 590
- Falling Object, Exercise 84, page 599
- Multiplier Effect, Exercises 91–96, page 609
- Tower of Hanoi, Exercise 56, page 618
- Health, Exercise 109, page 626
- PIN Codes, Exercise 18, page 634
- Data Analysis, Exercise 35, page 647
- Course Schedule, Exercise 113, page 653

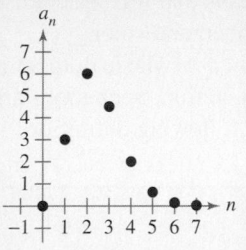 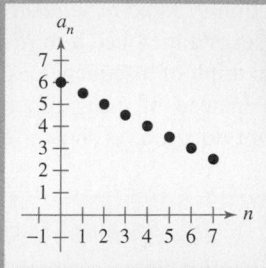 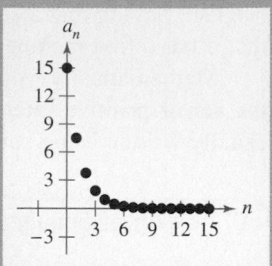

Sequences and series describe algebraic patterns. Graphs of sequences allow you to obtain a graphical perspective of the algebraic pattern described. In Chapter 8, you will study sequences and series extensively. You will also learn how to use mathematical induction to prove formulas and how to use the Binomial Theorem to calculate binomial coefficients, and you will study probability theory.

Bill Lai/Index Stock

Personal identification numbers, or PINs, are numerical passcodes for accessing such things as automatic teller machines, online accounts, and entrance to secure buildings. PINs are randomly generated, or consumers can create their own PIN.

8.1 Sequences and Series

Sequences

In mathematics, the word *sequence* is used in much the same way as in ordinary English. Saying that a collection is listed *in sequence* means that it is ordered so that it has a first member, a second member, a third member, and so on.

Mathematically, you can think of a sequence as a *function* whose domain is the set of positive integers. Instead of using function notation, sequences are usually written using subscript notation, as shown in the following definition.

> ### Definition of Sequence
>
> An **infinite sequence** is a function whose domain is the set of positive integers. The function values
>
> $$a_1, a_2, a_3, a_4, \ldots, a_n, \ldots$$
>
> are the **terms** of the sequence. If the domain of a function consists of the first n positive integers only, the sequence is a **finite sequence.**

On occasion, it is convenient to begin subscripting a sequence with 0 instead of 1 so that the terms of the sequence become $a_0, a_1, a_2, a_3, \ldots$. The domain of the function is the set of nonnegative integers.

Example 1 Writing the Terms of a Sequence

Write the first four terms of each sequence.

a. $a_n = 3n - 2$ **b.** $a_n = 3 + (-1)^n$

Solution

a. The first four terms of the sequence given by $a_n = 3n - 2$ are

$$a_1 = 3(1) - 2 = 1 \qquad \text{1st term}$$
$$a_2 = 3(2) - 2 = 4 \qquad \text{2nd term}$$
$$a_3 = 3(3) - 2 = 7 \qquad \text{3rd term}$$
$$a_4 = 3(4) - 2 = 10. \qquad \text{4th term}$$

b. The first four terms of the sequence given by $a_n = 3 + (-1)^n$ are

$$a_1 = 3 + (-1)^1 = 3 - 1 = 2 \qquad \text{1st term}$$
$$a_2 = 3 + (-1)^2 = 3 + 1 = 4 \qquad \text{2nd term}$$
$$a_3 = 3 + (-1)^3 = 3 - 1 = 2 \qquad \text{3rd term}$$
$$a_4 = 3 + (-1)^4 = 3 + 1 = 4. \qquad \text{4th term}$$

✔**CHECKPOINT** Now try Exercise 1.

What you should learn

- Use sequence notation to write the terms of sequences.
- Use factorial notation.
- Use summation notation to write sums.
- Find sums of infinite series.
- Use sequences and series to model and solve real-life problems.

Why you should learn it

Sequences and series are useful in modeling sets of values in order to identify patterns. For instance, Exercise 121 on page 590 shows how a sequence can be used to model the revenue of a pizza franchise from 1999 to 2006.

Santi Visalli/Tips Images

TECHNOLOGY TIP

To graph a sequence using a graphing utility, set the mode to *dot* and *sequence* and enter the sequence. Try graphing the sequences in Example 1 and using the *value* or *trace* feature to identify the terms. For instructions on how to use the *dot* mode, *sequence* mode, *value* feature, and *trace* feature, see Appendix A; for specific keystrokes, go to this textbook's *Online Study Center*.

Example 2   Writing the Terms of a Sequence

Write the first five terms of the sequence given by $a_n = \dfrac{(-1)^n}{2n-1}$.

Algebraic Solution

The first five terms of the sequence are as follows.

$$a_1 = \frac{(-1)^1}{2(1)-1} = \frac{-1}{2-1} = -1 \qquad \text{1st term}$$

$$a_2 = \frac{(-1)^2}{2(2)-1} = \frac{1}{4-1} = \frac{1}{3} \qquad \text{2nd term}$$

$$a_3 = \frac{(-1)^3}{2(3)-1} = \frac{-1}{6-1} = -\frac{1}{5} \qquad \text{3rd term}$$

$$a_4 = \frac{(-1)^4}{2(4)-1} = \frac{1}{8-1} = \frac{1}{7} \qquad \text{4th term}$$

$$a_5 = \frac{(-1)^5}{2(5)-1} = \frac{-1}{10-1} = -\frac{1}{9} \qquad \text{5th term}$$

✓*CHECKPOINT*   Now try Exercise 11.

Numerical Solution

Set your graphing utility to *sequence* mode. Enter the sequence into your graphing utility, as shown in Figure 8.1. Use the *table* feature (in *ask* mode) to create a table showing the terms of the sequence u_n for $n = 1, 2, 3, 4,$ and 5. From Figure 8.2, you can estimate the first five terms of the sequence as follows.

$$u_1 = -1, \qquad u_2 = 0.33333 \approx \tfrac{1}{3}, \qquad u_3 = -0.2 = -\tfrac{1}{5},$$

$$u_4 = 0.14286 \approx \tfrac{1}{7}, \qquad \text{and} \qquad u_5 = -0.1111 \approx -\tfrac{1}{9}$$

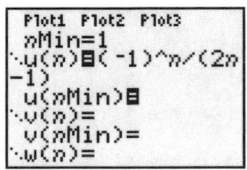

Figure 8.1

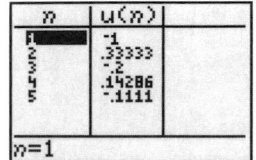

Figure 8.2

Simply listing the first few terms is not sufficient to define a unique sequence—the *n*th term *must be given*. To see this, consider the following sequences, both of which have the same first three terms.

$$\frac{1}{2}, \frac{1}{4}, \frac{1}{8}, \frac{1}{16}, \ldots, \frac{1}{2^n}, \ldots$$

$$\frac{1}{2}, \frac{1}{4}, \frac{1}{8}, \frac{1}{15}, \ldots, \frac{6}{(n+1)(n^2-n+6)}, \ldots$$

TECHNOLOGY SUPPORT

For instructions on how to use the *table* feature, see Appendix A; for specific keystrokes, go to this textbook's *Online Study Center*.

Example 3   Finding the *n*th Term of a Sequence

Write an expression for the apparent *n*th term (a_n) of each sequence.

a. 1, 3, 5, 7, . . .   **b.** 2, 5, 10, 17, . . .

Solution

a.   *n:* 1 2 3 4 . . . *n*

Terms: 1 3 5 7 . . . a_n

Apparent Pattern: Each term is 1 less than twice *n*. So, the apparent *n*th term is $a_n = 2n - 1$.

b.   *n:* 1 2   3   4 . . . *n*

Terms: 2 5 10 17 . . . a_n

Apparent Pattern: Each term is 1 more than the square of *n*. So, the apparent *n*th term is $a_n = n^2 + 1$.

✓*CHECKPOINT*   Now try Exercise 43.

Some sequences are defined **recursively.** To define a sequence recursively, you need to be given one or more of the first few terms. All other terms of the sequence are then defined using previous terms. A well-known example is the Fibonacci sequence, shown in Example 4.

Example 4   The Fibonacci Sequence: A Recursive Sequence

The Fibonacci sequence is defined recursively as follows.

$$a_0 = 1, \ a_1 = 1, \ a_k = a_{k-2} + a_{k-1}, \qquad \text{where } k \geq 2$$

Write the first six terms of this sequence.

Solution

$a_0 = 1$ 0th term is given.

$a_1 = 1$ 1st term is given.

$a_2 = a_{2-2} + a_{2-1} = a_0 + a_1 = 1 + 1 = 2$ Use recursion formula.

$a_3 = a_{3-2} + a_{3-1} = a_1 + a_2 = 1 + 2 = 3$ Use recursion formula.

$a_4 = a_{4-2} + a_{4-1} = a_2 + a_3 = 2 + 3 = 5$ Use recursion formula.

$a_5 = a_{5-2} + a_{5-1} = a_3 + a_4 = 3 + 5 = 8$ Use recursion formula.

✓**CHECKPOINT**   Now try Exercise 57.

Factorial Notation

Some very important sequences in mathematics involve terms that are defined with special types of products called **factorials.**

Definition of Factorial

If n is a positive integer, n **factorial** is defined as

$$n! = 1 \cdot 2 \cdot 3 \cdot 4 \ \cdots \ (n - 1) \cdot n.$$

As a special case, zero factorial is defined as $0! = 1$.

Here are some values of $n!$ for the first few nonnegative integers. Notice that $0! = 1$ by definition.

$0! = 1$

$1! = 1$

$2! = 1 \cdot 2 = 2$

$3! = 1 \cdot 2 \cdot 3 = 6$

$4! = 1 \cdot 2 \cdot 3 \cdot 4 = 24$

$5! = 1 \cdot 2 \cdot 3 \cdot 4 \cdot 5 = 120$

The value of n does not have to be very large before the value of $n!$ becomes huge. For instance, $10! = 3,628,800$.

Example 2   Writing the Terms of a Sequence

Write the first five terms of the sequence given by $a_n = \dfrac{(-1)^n}{2n - 1}$.

Algebraic Solution

The first five terms of the sequence are as follows.

$$a_1 = \frac{(-1)^1}{2(1) - 1} = \frac{-1}{2 - 1} = -1 \qquad \text{1st term}$$

$$a_2 = \frac{(-1)^2}{2(2) - 1} = \frac{1}{4 - 1} = \frac{1}{3} \qquad \text{2nd term}$$

$$a_3 = \frac{(-1)^3}{2(3) - 1} = \frac{-1}{6 - 1} = -\frac{1}{5} \qquad \text{3rd term}$$

$$a_4 = \frac{(-1)^4}{2(4) - 1} = \frac{1}{8 - 1} = \frac{1}{7} \qquad \text{4th term}$$

$$a_5 = \frac{(-1)^5}{2(5) - 1} = \frac{-1}{10 - 1} = -\frac{1}{9} \qquad \text{5th term}$$

✔**CHECKPOINT**   Now try Exercise 11.

Numerical Solution

Set your graphing utility to *sequence* mode. Enter the sequence into your graphing utility, as shown in Figure 8.1. Use the *table* feature (in *ask* mode) to create a table showing the terms of the sequence u_n for $n = 1, 2, 3, 4,$ and 5. From Figure 8.2, you can estimate the first five terms of the sequence as follows.

$$u_1 = -1, \qquad u_2 = 0.33333 \approx \tfrac{1}{3}, \qquad u_3 = -0.2 = -\tfrac{1}{5},$$

$$u_4 = 0.14286 \approx \tfrac{1}{7}, \qquad \text{and} \qquad u_5 = -0.1111 \approx -\tfrac{1}{9}$$

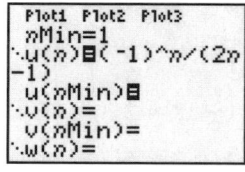

Figure 8.1

Figure 8.2

Simply listing the first few terms is not sufficient to define a unique sequence—the nth term *must be given*. To see this, consider the following sequences, both of which have the same first three terms.

$$\frac{1}{2}, \frac{1}{4}, \frac{1}{8}, \frac{1}{16}, \ldots, \frac{1}{2^n}, \ldots$$

$$\frac{1}{2}, \frac{1}{4}, \frac{1}{8}, \frac{1}{15}, \ldots, \frac{6}{(n + 1)(n^2 - n + 6)}, \ldots$$

TECHNOLOGY SUPPORT

For instructions on how to use the *table* feature, see Appendix A; for specific keystrokes, go to this textbook's *Online Study Center*.

Example 3   Finding the nth Term of a Sequence

Write an expression for the apparent nth term (a_n) of each sequence.

a. $1, 3, 5, 7, \ldots$ **b.** $2, 5, 10, 17, \ldots$

Solution

a.   *n:* 1 2 3 4 . . . n

   Terms: 1 3 5 7 . . . a_n

   Apparent Pattern: Each term is 1 less than twice n. So, the apparent nth term is $a_n = 2n - 1$.

b.   *n:* 1 2 3 4 . . . n

   Terms: 2 5 10 17 . . . a_n

   Apparent Pattern: Each term is 1 more than the square of n. So, the apparent nth term is $a_n = n^2 + 1$.

✔**CHECKPOINT**   Now try Exercise 43.

Some sequences are defined **recursively.** To define a sequence recursively, you need to be given one or more of the first few terms. All other terms of the sequence are then defined using previous terms. A well-known example is the Fibonacci sequence, shown in Example 4.

Example 4   The Fibonacci Sequence: A Recursive Sequence

The Fibonacci sequence is defined recursively as follows.

$$a_0 = 1, \ a_1 = 1, \ a_k = a_{k-2} + a_{k-1}, \qquad \text{where } k \geq 2$$

Write the first six terms of this sequence.

Solution

$a_0 = 1$	0th term is given.
$a_1 = 1$	1st term is given.
$a_2 = a_{2-2} + a_{2-1} = a_0 + a_1 = 1 + 1 = 2$	Use recursion formula.
$a_3 = a_{3-2} + a_{3-1} = a_1 + a_2 = 1 + 2 = 3$	Use recursion formula.
$a_4 = a_{4-2} + a_{4-1} = a_2 + a_3 = 2 + 3 = 5$	Use recursion formula.
$a_5 = a_{5-2} + a_{5-1} = a_3 + a_4 = 3 + 5 = 8$	Use recursion formula.

✔CHECKPOINT   Now try Exercise 57.

Factorial Notation

Some very important sequences in mathematics involve terms that are defined with special types of products called **factorials.**

Definition of Factorial

If n is a positive integer, **n factorial** is defined as

$$n! = 1 \cdot 2 \cdot 3 \cdot 4 \cdots (n - 1) \cdot n.$$

As a special case, zero factorial is defined as $0! = 1$.

Here are some values of $n!$ for the first few nonnegative integers. Notice that $0! = 1$ by definition.

$$0! = 1$$

$$1! = 1$$

$$2! = 1 \cdot 2 = 2$$

$$3! = 1 \cdot 2 \cdot 3 = 6$$

$$4! = 1 \cdot 2 \cdot 3 \cdot 4 = 24$$

$$5! = 1 \cdot 2 \cdot 3 \cdot 4 \cdot 5 = 120$$

The value of n does not have to be very large before the value of $n!$ becomes huge. For instance, $10! = 3,628,800$.

Exploration

Most graphing utilities have the capability to compute $n!$. Use your graphing utility to compare $3 \cdot 5!$ and $(3 \cdot 5)!$. How do they differ? How large a value of $n!$ will your graphing utility allow you to compute?

Factorials follow the same conventions for order of operations as do exponents. For instance,

$$2n! = 2(n!) = 2(1 \cdot 2 \cdot 3 \cdot 4 \cdots n)$$

whereas $(2n)! = 1 \cdot 2 \cdot 3 \cdot 4 \cdots 2n$.

Example 5 Writing the Terms of a Sequence Involving Factorials

Write the first five terms of the sequence given by $a_n = \dfrac{2^n}{n!}$. Begin with $n = 0$.

Algebraic Solution

$$a_0 = \frac{2^0}{0!} = \frac{1}{1} = 1 \qquad \text{0th term}$$

$$a_1 = \frac{2^1}{1!} = \frac{2}{1} = 2 \qquad \text{1st term}$$

$$a_2 = \frac{2^2}{2!} = \frac{4}{2} = 2 \qquad \text{2nd term}$$

$$a_3 = \frac{2^3}{3!} = \frac{8}{6} = \frac{4}{3} \qquad \text{3rd term}$$

$$a_4 = \frac{2^4}{4!} = \frac{16}{24} = \frac{2}{3} \qquad \text{4th term}$$

✓CHECKPOINT   Now try Exercise 65.

Graphical Solution

Using a graphing utility set to *dot* and *sequence* modes, enter the sequence $u_n = 2^n/n!$, as shown in Figure 8.3. Set the viewing window to $0 \le n \le 4$, $0 \le x \le 6$, and $0 \le y \le 4$. Then graph the sequence, as shown in Figure 8.4. Use the *value* or *trace* feature to approximate the first five terms as follows.

$$u_0 = 1, \quad u_1 = 2, \quad u_2 = 2, \quad u_3 \approx 1.333 \approx \tfrac{4}{3}, \quad u_4 \approx 0.667 \approx \tfrac{2}{3}$$

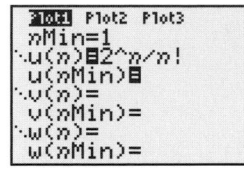

Figure 8.3

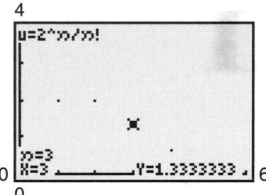

Figure 8.4

When working with fractions involving factorials, you will often find that the fractions can be reduced to simplify the computations.

Example 6 Evaluating Factorial Expressions

Simplify each factorial expression.

a. $\dfrac{8!}{2! \cdot 6!}$ **b.** $\dfrac{2! \cdot 6!}{3! \cdot 5!}$ **c.** $\dfrac{n!}{(n-1)!}$

Solution

a. $\dfrac{8!}{2! \cdot 6!} = \dfrac{1 \cdot 2 \cdot 3 \cdot 4 \cdot 5 \cdot 6 \cdot 7 \cdot 8}{1 \cdot 2 \cdot 1 \cdot 2 \cdot 3 \cdot 4 \cdot 5 \cdot 6} = \dfrac{7 \cdot 8}{2} = 28$

b. $\dfrac{2! \cdot 6!}{3! \cdot 5!} = \dfrac{1 \cdot 2 \cdot 1 \cdot 2 \cdot 3 \cdot 4 \cdot 5 \cdot 6}{1 \cdot 2 \cdot 3 \cdot 1 \cdot 2 \cdot 3 \cdot 4 \cdot 5} = \dfrac{6}{3} = 2$

c. $\dfrac{n!}{(n-1)!} = \dfrac{1 \cdot 2 \cdot 3 \cdots (n-1) \cdot n}{1 \cdot 2 \cdot 3 \cdots (n-1)} = n$

✓CHECKPOINT   Now try Exercise 75.

STUDY TIP

Note in Example 6(a) that you can also simplify the computation as follows.

$$\frac{8!}{2! \cdot 6!} = \frac{8 \cdot 7 \cdot 6!}{2! \cdot 6!}$$

$$= \frac{8 \cdot 7}{2 \cdot 1} = 28$$

Summation Notation

There is a convenient notation for the sum of the terms of a finite sequence. It is called **summation notation** or **sigma notation** because it involves the use of the uppercase Greek letter sigma, written as Σ.

Definition of Summation Notation

The sum of the first n terms of a sequence is represented by

$$\sum_{i=1}^{n} a_i = a_1 + a_2 + a_3 + a_4 + \cdots + a_n$$

where i is called the **index of summation**, n is the **upper limit of summation**, and 1 is the **lower limit of summation**.

STUDY TIP

Summation notation is an instruction to add the terms of a sequence. From the definition at the left, the upper limit of summation tells you where to end the sum. Summation notation helps you generate the appropriate terms of the sequence prior to finding the actual sum, which may be unclear.

Example 7 Sigma Notation for Sums

a. $\displaystyle\sum_{i=1}^{5} 3i = 3(1) + 3(2) + 3(3) + 3(4) + 3(5)$

$\qquad\qquad = 3(1 + 2 + 3 + 4 + 5)$

$\qquad\qquad = 3(15)$

$\qquad\qquad = 45$

b. $\displaystyle\sum_{k=3}^{6} (1 + k^2) = (1 + 3^2) + (1 + 4^2) + (1 + 5^2) + (1 + 6^2)$

$\qquad\qquad\qquad = 10 + 17 + 26 + 37 = 90$

c. $\displaystyle\sum_{n=0}^{8} \frac{1}{n!} = \frac{1}{0!} + \frac{1}{1!} + \frac{1}{2!} + \frac{1}{3!} + \frac{1}{4!} + \frac{1}{5!} + \frac{1}{6!} + \frac{1}{7!} + \frac{1}{8!}$

$\qquad\qquad = 1 + 1 + \frac{1}{2} + \frac{1}{6} + \frac{1}{24} + \frac{1}{120} + \frac{1}{720} + \frac{1}{5040} + \frac{1}{40{,}320}$

$\qquad\qquad \approx 2.71828$

For the summation in part (c), note that the sum is very close to the irrational number $e \approx 2.718281828$. It can be shown that as more terms of the sequence whose nth term is $1/n!$ are added, the sum becomes closer and closer to e.

✔CHECKPOINT   Now try Exercise 79.

In Example 7, note that the lower limit of a summation does not have to be 1. Also note that the index of summation does not have to be the letter i. For instance, in part (b), the letter k is the index of summation.

TECHNOLOGY SUPPORT

For instructions on how to use the *sum sequence* feature, see Appendix A; for specific keystrokes, go to this textbook's *Online Study Center.*

TECHNOLOGY TIP   Most graphing utilities are able to sum the first n terms of a sequence. Figure 8.5 shows an example of how one graphing utility displays the sum of the terms of the sequence below using the *sum sequence* feature.

$$a_n = \frac{1}{n!} \quad \text{from} \quad n = 0 \quad \text{to} \quad n = 8$$

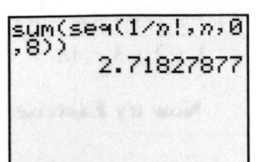

Figure 8.5

Properties of Sums (See the proofs on page 656.)

1. $\displaystyle\sum_{i=1}^{n} c = cn$, c is a constant. 2. $\displaystyle\sum_{i=1}^{n} ca_i = c\sum_{i=1}^{n} a_i$, c is a constant.

3. $\displaystyle\sum_{i=1}^{n} (a_i + b_i) = \sum_{i=1}^{n} a_i + \sum_{i=1}^{n} b_i$ 4. $\displaystyle\sum_{i=1}^{n} (a_i - b_i) = \sum_{i=1}^{n} a_i - \sum_{i=1}^{n} b_i$

Series

Many applications involve the sum of the terms of a finite or an infinite sequence. Such a sum is called a **series.**

Definition of a Series

Consider the infinite sequence $a_1, a_2, a_3, \ldots, a_i, \ldots$

1. The sum of the first n terms of the sequence is called a **finite series** or the **partial sum** of the sequence and is denoted by

$$a_1 + a_2 + a_3 + \cdots + a_n = \sum_{i=1}^{n} a_i.$$

2. The sum of all the terms of the infinite sequence is called an **infinite series** and is denoted by

$$a_1 + a_2 + a_3 + \cdots + a_i + \cdots = \sum_{i=1}^{\infty} a_i.$$

Example 8 Finding the Sum of a Series

For the series $\displaystyle\sum_{i=1}^{\infty} \frac{3}{10^i}$, find (a) the third partial sum and (b) the sum.

Solution

a. The third partial sum is

$$\sum_{i=1}^{3} \frac{3}{10^i} = \frac{3}{10^1} + \frac{3}{10^2} + \frac{3}{10^3} = 0.3 + 0.03 + 0.003 = 0.333.$$

b. The sum of the series is

$$\sum_{i=1}^{\infty} \frac{3}{10^i} = \frac{3}{10^1} + \frac{3}{10^2} + \frac{3}{10^3} + \frac{3}{10^4} + \frac{3}{10^5} + \cdots$$

$$= 0.3 + 0.03 + 0.003 + 0.0003 + 0.00003 + \cdots$$

$$= 0.33333\ldots = \frac{1}{3}.$$

✓**CHECKPOINT** Now try Exercise 109.

Notice in Example 8(b) that the sum of an infinite series can be a finite number.

STUDY TIP

Variations in the upper and lower limits of summation can produce quite different-looking summation notations for *the same sum*. For example, the following two sums have identical terms.

$$\sum_{i=1}^{3} 3(2^i) = 3(2^1 + 2^2 + 2^3)$$

$$\sum_{i=0}^{2} 3(2^{i+1}) = 3(2^1 + 2^2 + 2^3)$$

Application

Sequences have many applications in situations that involve recognizable patterns. One such model is illustrated in Example 9.

Example 9   Population of the United States

From 1970 to 2004, the resident population of the United States can be approximated by the model

$$a_n = 205.5 + 1.82n + 0.024n^2, \qquad n = 0, 1, \ldots, 34$$

where a_n is the population (in millions) and n represents the year, with $n = 0$ corresponding to 1970. Find the last five terms of this finite sequence.   (Source: U.S. Census Bureau)

Algebraic Solution

The last five terms of this finite sequence are as follows.

$a_{30} = 205.5 + 1.82(30) + 0.024(30)^2$

$\quad = 281.7$ 2000 population

$a_{31} = 205.5 + 1.82(31) + 0.024(31)^2$

$\quad \approx 285.0$ 2001 population

$a_{32} = 205.5 + 1.82(32) + 0.024(32)^2$

$\quad \approx 288.3$ 2002 population

$a_{33} = 205.5 + 1.82(33) + 0.024(33)^2$

$\quad \approx 291.7$ 2003 population

$a_{34} = 205.5 + 1.82(34) + 0.024(34)^2$

$\quad \approx 295.1$ 2004 population

✔CHECKPOINT   Now try Exercise 113.

Graphical Solution

Using a graphing utility set to *dot* and *sequence* modes, enter the sequence

$$u_n = 205.5 + 1.82n + 0.024n^2.$$

Set the viewing window to $0 \le n \le 35, 0 \le x \le 35$, and $200 \le y \le 300$. Then graph the sequence. Use the *value* or *trace* feature to approximate the last five terms, as shown in Figure 8.6.

$a_{30} = 281.7,$

$a_{31} \approx 285.0,$

$a_{32} \approx 288.3,$

$a_{33} \approx 291.7,$

$a_{34} \approx 295.1$

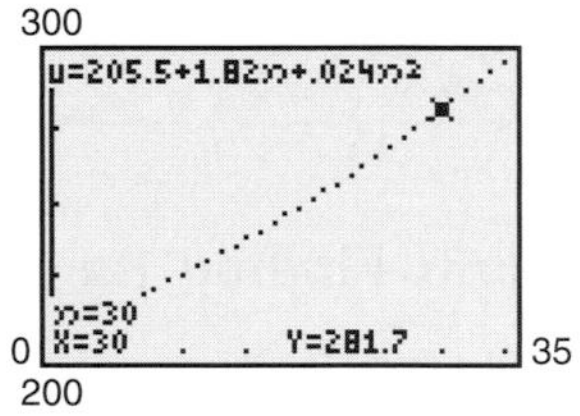

```
300
u=205.5+1.82n+.024n2
                        ⊠
 ⎸
0  n=30
   X=30        Y=281.7      35
200
```

Figure 8.6

Exploration

A $3 \times 3 \times 3$ cube is created using 27 unit cubes (a unit cube has a length, width, and height of 1 unit) and only the faces of each cube that are visible are painted blue (see Figure 8.7). Complete the table below to determine how many unit cubes of the $3 \times 3 \times 3$ cube have no blue faces, one blue face, two blue faces, and three blue faces. Do the same for a $4 \times 4 \times 4$ cube, a $5 \times 5 \times 5$ cube, and a $6 \times 6 \times 6$ cube, and add your results to the table below. What type of pattern do you observe in the table? Write a formula you could use to determine the column values for an $n \times n \times n$ cube.

Cube \ Number of blue faces	0	1	2	3
$3 \times 3 \times 3$				

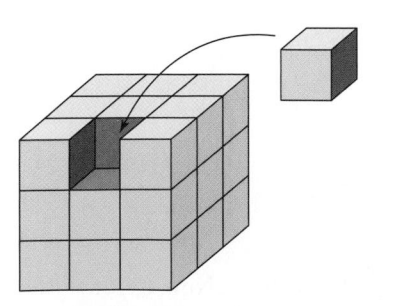

Figure 8.7

8.1 Exercises

See www.CalcChat.com for worked-out solutions to odd-numbered exercises.

Vocabulary Check

Fill in the blanks.

1. An _____ is a function whose domain is the set of positive integers.

2. The function values $a_1, a_2, a_3, a_4, \ldots, a_n, \ldots$ are called the _____ of a sequence.

3. A sequence is a _____ sequence if the domain of the function consists of the first n positive integers.

4. If you are given one or more of the first few terms of a sequence, and all other terms of the sequence are defined using previous terms, then the sequence is defined _____ .

5. If n is a positive integer, n _____ is defined as $n! = 1 \cdot 2 \cdot 3 \cdot 4 \cdots (n-1) \cdot n$.

6. The notation used to represent the sum of the terms of a finite sequence is _____ or sigma notation.

7. For the sum $\displaystyle\sum_{i=1}^{n} a_i$, i is called the _____ of summation, n is the _____ of summation, and 1 is the _____ of summation.

8. The sum of the terms of a finite or an infinite sequence is called a _____

9. The _____ of a sequence is the sum of the first n terms of the sequence.

In Exercises 1–20, write the first five terms of the sequence. (Assume n begins with 1.) Use the *table* feature of a graphing utility to verify your results.

1. $a_n = 2n + 5$

2. $a_n = 4n - 7$

3. $a_n = 2^n$

4. $a_n = \left(\frac{1}{2}\right)^n$

5. $a_n = \left(-\frac{1}{2}\right)^n$

6. $a_n = (-2)^n$

7. $a_n = \dfrac{n+1}{n}$

8. $a_n = \dfrac{n}{n+1}$

9. $a_n = \dfrac{n}{n^2 + 1}$

10. $a_n = \dfrac{2n}{n+1}$

11. $a_n = \dfrac{1 + (-1)^n}{n}$

12. $a_n = \dfrac{1 + (-1)^n}{2n}$

13. $a_n = 1 - \dfrac{1}{2^n}$

14. $a_n = \dfrac{3^n}{4^n}$

15. $a_n = \dfrac{1}{n^{3/2}}$

16. $a_n = \dfrac{1}{\sqrt{n}}$

17. $a_n = \dfrac{(-1)^n}{n^2}$

18. $a_n = (-1)^n \left(\dfrac{n}{n+1}\right)$

19. $a_n = (2n - 1)(2n + 1)$

20. $a_n = n(n-1)(n-2)$

In Exercises 21–26, find the indicated term of the sequence.

21. $a_n = (-1)^n(3n - 2)$

$a_{25} = $ ▩

22. $a_n = (-1)^{n-1}[n(n-1)]$

$a_{16} = $ ▩

23. $a_n = \dfrac{n^2}{n^2 + 1}$

$a_{10} = $ ▩

24. $a_n = \dfrac{n^2}{2n + 1}$

$a_5 = $ ▩

25. $a_n = \dfrac{2^n}{2^n + 1}$

$a_6 = $ ▩

26. $a_n = \dfrac{2^{n+1}}{2^n + 1}$

$a_7 = $ ▩

In Exercises 27–32, use a graphing utility to graph the first 10 terms of the sequence. (Assume n begins with 1.)

27. $a_n = \dfrac{2}{3}n$

28. $a_n = 2 - \dfrac{4}{n}$

29. $a_n = 16(-0.5)^{n-1}$

30. $a_n = 8(0.75)^{n-1}$

31. $a_n = \dfrac{2n}{n+1}$

32. $a_n = \dfrac{3n^2}{n^2 + 1}$

In Exercises 33–38, use the *table* feature of a graphing utility to find the first 10 terms of the sequence. (Assume n begins with 1.)

33. $a_n = 2(3n - 1) + 5$

34. $a_n = 2n(n + 1)(n + 2)$

35. $a_n = 1 + \dfrac{n+1}{n}$

36. $a_n = \dfrac{4n^2}{n+2}$

37. $a_n = (-1)^n + 1$

38. $a_n = (-1)^{n+1} + 1$

In Exercises 39–42, match the sequence with its graph. [The graphs are labeled (a), (b), (c), and (d).]

(a)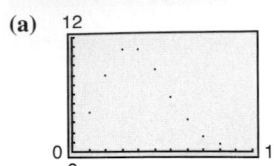

(b)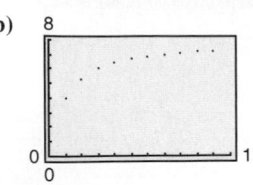

(c)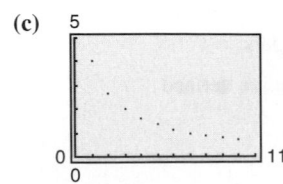

(d)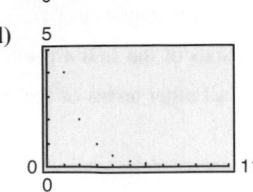

39. $a_n = \dfrac{8}{n+1}$

40. $a_n = \dfrac{8n}{n+1}$

41. $a_n = 4(0.5)^{n-1}$

42. $a_n = \dfrac{4^n}{n!}$

In Exercises 43–56, write an expression for the *apparent n*th term of the sequence. (Assume *n* begins with 1.)

43. $1, 4, 7, 10, 13, \ldots$

44. $3, 7, 11, 15, 19, \ldots$

45. $0, 3, 8, 15, 24, \ldots$

46. $1, \frac{1}{4}, \frac{1}{9}, \frac{1}{16}, \frac{1}{25}, \ldots$

47. $\frac{2}{3}, \frac{3}{4}, \frac{4}{5}, \frac{5}{6}, \frac{6}{7}, \ldots$

48. $\frac{2}{1}, \frac{3}{3}, \frac{4}{5}, \frac{5}{7}, \frac{6}{9}, \ldots$

49. $\frac{1}{2}, \frac{-1}{4}, \frac{1}{8}, \frac{-1}{16}, \ldots$

50. $\frac{1}{3}, -\frac{2}{9}, \frac{4}{27}, -\frac{8}{81}, \ldots$

51. $1 + \frac{1}{1}, 1 + \frac{1}{2}, 1 + \frac{1}{3}, 1 + \frac{1}{4}, 1 + \frac{1}{5}, \ldots$

52. $1 + \frac{1}{2}, 1 + \frac{3}{4}, 1 + \frac{7}{8}, 1 + \frac{15}{16}, 1 + \frac{31}{32}, \ldots$

53. $1, \dfrac{1}{2}, \dfrac{1}{6}, \dfrac{1}{24}, \dfrac{1}{120}, \ldots$

54. $1, 2, \dfrac{2^2}{2}, \dfrac{2^3}{6}, \dfrac{2^4}{24}, \dfrac{2^5}{120}, \ldots$

55. $1, 3, 1, 3, 1, \ldots$

56. $1, -1, 1, -1, 1, \ldots$

In Exercises 57–60, write the first five terms of the sequence defined recursively.

57. $a_1 = 28, \quad a_{k+1} = a_k - 4$

58. $a_1 = 15, \quad a_{k+1} = a_k + 3$

59. $a_1 = 3, \quad a_{k+1} = 2(a_k - 1)$

60. $a_1 = 32, \quad a_{k+1} = \frac{1}{2}a_k$

In Exercises 61–64, write the first five terms of the sequence defined recursively. Use the pattern to write the *n*th term of the sequence as a function of *n*. (Assume *n* begins with 1.)

61. $a_1 = 6, \quad a_{k+1} = a_k + 2$

62. $a_1 = 25, \quad a_{k+1} = a_k - 5$

63. $a_1 = 81, \quad a_{k+1} = \frac{1}{3}a_k$

64. $a_1 = 14, \quad a_{k+1} = -2a_k$

In Exercises 65–70, write the first five terms of the sequence. (Assume *n* begins with 0.) Use the *table* feature of a graphing utility to verify your results.

65. $a_n = \dfrac{1}{n!}$

66. $a_n = \dfrac{1}{(n+1)!}$

67. $a_n = \dfrac{n!}{2n+1}$

68. $a_n = \dfrac{n^2}{(n+1)!}$

69. $a_n = \dfrac{(-1)^{2n}}{(2n)!}$

70. $a_n = \dfrac{(-1)^{2n+1}}{(2n+1)!}$

In Exercises 71–78, simplify the factorial expression.

71. $\dfrac{2!}{4!}$

72. $\dfrac{5!}{7!}$

73. $\dfrac{12!}{4! \cdot 8!}$

74. $\dfrac{10! \cdot 3!}{4! \cdot 6!}$

75. $\dfrac{(n+1)!}{n!}$

76. $\dfrac{(n+2)!}{n!}$

77. $\dfrac{(2n-1)!}{(2n+1)!}$

78. $\dfrac{(2n+2)!}{(2n)!}$

In Exercises 79–90, find the sum.

79. $\displaystyle\sum_{i=1}^{5}(2i+1)$

80. $\displaystyle\sum_{i=1}^{6}(3i-1)$

81. $\displaystyle\sum_{k=1}^{4}10$

82. $\displaystyle\sum_{k=1}^{5}6$

83. $\displaystyle\sum_{i=0}^{4}i^2$

84. $\displaystyle\sum_{i=0}^{5}3i^2$

85. $\displaystyle\sum_{k=0}^{3}\dfrac{1}{k^2+1}$

86. $\displaystyle\sum_{j=3}^{5}\dfrac{1}{j}$

87. $\displaystyle\sum_{i=1}^{4}\left[(i-1)^2+(i+1)^3\right]$

88. $\displaystyle\sum_{k=2}^{5}(k+1)(k-3)$

89. $\displaystyle\sum_{i=1}^{4}2^i$

90. $\displaystyle\sum_{j=0}^{4}(-2)^j$

In Exercises 91–94, use a graphing utility to find the sum.

91. $\displaystyle\sum_{j=1}^{6}(24-3j)$

92. $\displaystyle\sum_{j=1}^{10}\dfrac{3}{j+1}$

93. $\displaystyle\sum_{k=0}^{4}\dfrac{(-1)^k}{k+1}$

94. $\displaystyle\sum_{k=0}^{4}\dfrac{(-1)^k}{k!}$

In Exercises 95–104, use sigma notation to write the sum. Then use a graphing utility to find the sum.

95. $\dfrac{1}{3(1)} + \dfrac{1}{3(2)} + \dfrac{1}{3(3)} + \cdots + \dfrac{1}{3(9)}$

96. $\dfrac{5}{1+1} + \dfrac{5}{1+2} + \dfrac{5}{1+3} + \cdots + \dfrac{5}{1+15}$

97. $\left[2\left(\tfrac{1}{8}\right) + 3\right] + \left[2\left(\tfrac{2}{8}\right) + 3\right] + \cdots + \left[2\left(\tfrac{8}{8}\right) + 3\right]$

98. $\left[1 - \left(\tfrac{1}{6}\right)^2\right] + \left[1 - \left(\tfrac{2}{6}\right)^2\right] + \cdots + \left[1 - \left(\tfrac{6}{6}\right)^2\right]$

99. $3 - 9 + 27 - 81 + 243 - 729$

100. $1 - \tfrac{1}{2} + \tfrac{1}{4} - \tfrac{1}{8} + \cdots - \tfrac{1}{128}$

101. $\dfrac{1}{1^2} - \dfrac{1}{2^2} + \dfrac{1}{3^2} - \dfrac{1}{4^2} + \cdots - \dfrac{1}{20^2}$

102. $\dfrac{1}{1 \cdot 3} + \dfrac{1}{2 \cdot 4} + \dfrac{1}{3 \cdot 5} + \cdots + \dfrac{1}{10 \cdot 12}$

103. $\tfrac{1}{4} + \tfrac{3}{8} + \tfrac{7}{16} + \tfrac{15}{32} + \tfrac{31}{64}$

104. $\tfrac{1}{2} + \tfrac{2}{4} + \tfrac{6}{8} + \tfrac{24}{16} + \tfrac{120}{32} + \tfrac{720}{64}$

In Exercises 105–108, find the indicated partial sum of the series.

105. $\displaystyle\sum_{i=1}^{\infty} 5\left(\tfrac{1}{2}\right)^i$ **106.** $\displaystyle\sum_{i=1}^{\infty} 2\left(\tfrac{1}{3}\right)^i$

Fourth partial sum   Fifth partial sum

107. $\displaystyle\sum_{n=1}^{\infty} 4\left(-\tfrac{1}{2}\right)^n$ **108.** $\displaystyle\sum_{n=1}^{\infty} 8\left(-\tfrac{1}{4}\right)^n$

Third partial sum   Fourth partial sum

In Exercises 109–112, find (a) the fourth partial sum and (b) the sum of the infinite series.

109. $\displaystyle\sum_{i=1}^{\infty} 6\left(\tfrac{1}{10}\right)^i$ **110.** $\displaystyle\sum_{k=1}^{\infty} 4\left(\tfrac{1}{10}\right)^k$

111. $\displaystyle\sum_{k=1}^{\infty} \left(\tfrac{1}{10}\right)^k$ **112.** $\displaystyle\sum_{i=1}^{\infty} 2\left(\tfrac{1}{10}\right)^i$

113. *Compound Interest*   A deposit of $5000 is made in an account that earns 3% interest compounded quarterly. The balance in the account after n quarters is given by

$$A_n = 5000\left(1 + \frac{0.03}{4}\right)^n, \quad n = 1, 2, 3, \ldots.$$

(a) Compute the first eight terms of this sequence.

(b) Find the balance in this account after 10 years by computing the 40th term of the sequence.

114. *Compound Interest*   A deposit of $100 is made *each month* in an account that earns 12% interest compounded monthly. The balance in the account after n months is given by

$$A_n = 100(101)[(1.01)^n - 1], \quad n = 1, 2, 3, \ldots.$$

(a) Compute the first six terms of this sequence.

(b) Find the balance in this account after 5 years by computing the 60th term of the sequence.

(c) Find the balance in this account after 20 years by computing the 240th term of the sequence.

115. *Fish*   A landlocked lake has been selected to be stocked in the year 2008 with 5500 trout, and to be restocked each year thereafter with 500 trout. Each year the fish population declines 25% due to harvesting and other natural causes.

(a) Write a recursive sequence that gives the population p_n of trout in the lake in terms of the year n since stocking began.

(b) Use the recursion formula from part (a) to find the numbers of trout in the lake in the years 2009, 2010, and 2011.

(c) Use a graphing utility to find the number of trout in the lake as time passes infinitely. Explain your result.

116. *Tree Farm*   A tree farm in the year 2010 has 10,000 Douglas fir trees on its property. Each year thereafter 10% of the fir trees are harvested and 750 new fir saplings are then planted in their place.

(a) Write a recursive sequence that gives the current number t_n of fir trees on the farm in the year n, with $n = 0$ corresponding to 2010.

(b) Use the recursion formula from part (a) to find the numbers of fir trees for $n = 1, 2, 3$, and 4. Interpret these values in the context of the situation.

(c) Use a graphing utility to find the number of fir trees as time passes infinitely. Explain your result.

117. *Investment*   You decide to place $50 at the beginning of each month into a Roth IRA for your education. The account earns 6% compounded monthly.

(a) Find a recursive sequence that yields the total a_n in the account, where n is the number of months since you began depositing the $50.

(b) Use the recursion formula from part (a) to find the amount in the IRA after 12 deposits.

(c) Use a graphing utility to find the amount in the IRA after 50 deposits.

Use the *table* feature of a graphing utility to verify your answers in parts (b) and (c).

118. *Mortgage Payments*   You borrow $150,000 at 9% interest compounded monthly for 30 years to buy a new home. The monthly mortgage payment has been determined to be $1206.94.

(a) Find a recursive sequence that gives the balance b_n of the mortgage remaining after each monthly payment n has been made.

(b) Use the *table* feature of a graphing utility to find the balance remaining for every five years where $0 \le n \le 360$.

(c) What is the total amount paid for a $150,000 loan under these conditions? Explain your answer.

(d) How much interest will be paid over the life of the loan?

119. *Average Wages* The average hourly wage rates r_n (in dollars) for instructional teacher aides in the United States from 1998 to 2005 are shown in the table. (Source: Educational Research Service)

Year	Average wage rate, r_n (in dollars)
1998	9.46
1999	9.80
2000	10.00
2001	10.41
2002	10.68
2003	10.93
2004	11.22
2005	11.35

A sequence that models the data is

$r_n = -0.0092n^2 + 0.491n + 6.11$

where n is the year, with $n = 8$ corresponding to 1998.

(a) Use a graphing utility to plot the data and graph the model in the same viewing window.

(b) Use the model to find the average hourly wage rates for instructional teacher aides in 2010 and 2015.

(c) Are your results in part (b) reasonable? Explain.

(d) Use the model to find when the average hourly wage rate will reach $12.

120. *Education* The preprimary enrollments s_n (in thousands) in the United States from 1998 to 2003 are shown in the table. (Source: U.S. Census Bureau)

Year	Number of students, s_n (in thousands)
1998	7788
1999	7844
2000	7592
2001	7441
2002	7504
2003	7921

A sequence that models the data is

$s_n = 33.787n^3 - 1009.56n^2 + 9840.6n - 23{,}613$

where n is the year, with $n = 8$ corresponding to 1998.

(a) Use a graphing utility to plot the data and graph the model in the same viewing window.

(b) Use the model to find the numbers of students enrolled in a preprimary school in 2005, 2010, and 2015.

(c) Are your results in part (c) reasonable? Explain.

121. *Revenue* The revenues R_n (in millions of dollars) for California Pizza Kitchen, Inc. from 1999 to 2006 are shown in the table. (Source: California Pizza Kitchen, Inc.)

Year	Revenue, R_n (in millions of dollars)
1999	179.2
2000	210.8
2001	249.3
2002	306.3
2003	359.9
2004	422.5
2005	480.0
2006	560.0

(a) Use a graphing utility to plot the data. Let n represent the year, with $n = 9$ corresponding to 1999.

(b) Use the *regression* feature of a graphing utility to find a linear sequence and a quadratic sequence that model the data. Identify the coefficient of determination for each model.

(c) Separately graph each model in the same viewing window as the data.

(d) Decide which of the models is a better fit for the data. Explain.

(e) Use the model you chose in part (d) to predict the revenues for the years 2010 and 2015.

(f) Use your model from part (d) to find when the revenues will reach one billion dollars.

122. *Sales* The sales S_n (in billions of dollars) for Anheuser-Busch Companies, Inc. from 1995 to 2006 are shown in the table. (Source: Anheuser-Busch Companies, Inc.)

Year	Sales, S_n (in billions of dollars)
1995	10.3
1996	10.9
1997	11.1
1998	11.2
1999	11.7
2000	12.3
2001	12.9
2002	13.6
2003	14.1
2004	14.9
2005	15.1
2006	15.5

(a) Use a graphing utility to plot the data. Let n represent the year, with $n = 5$ corresponding to 1995.

(b) Use the *regression* feature of a graphing utility to find a linear sequence and a quadratic sequence that model the data. Identify the coefficient of determination for each model.

(c) Separately graph each model in the same viewing window as the data.

(d) Decide which of the models is a better fit for the data. Explain.

(e) Use the model you chose in part (d) to predict the sales for Anheuser-Busch for the years 2010 and 2015.

(f) Use your model from part (d) to find when the sales will reach 20 billion dollars.

Synthesis

True or False? **In Exercises 123 and 124, determine whether the statement is true or false. Justify your answer.**

123. $\displaystyle\sum_{i=1}^{4} (i^2 + 2i) = \sum_{i=1}^{4} i^2 + 2\sum_{i=1}^{4} i$

124. $\displaystyle\sum_{j=1}^{4} 2^j = \sum_{j=3}^{6} 2^{j-2}$

Fibonacci Sequence **In Exercises 125 and 126, use the Fibonacci sequence. (See Example 4.)**

125. Write the first 12 terms of the Fibonacci sequence a_n and the first 10 terms of the sequence given by

$$b_n = \frac{a_{n+1}}{a_n}, \quad n > 0.$$

126. Using the definition of b_n given in Exercise 125, show that b_n can be defined recursively by

$$b_n = 1 + \frac{1}{b_{n-1}}.$$

Exploration **In Exercises 127–130, let**

$$a_n = \frac{\left(1 + \sqrt{5}\right)^n - \left(1 - \sqrt{5}\right)^n}{2^n \sqrt{5}}$$

be a sequence with nth term a_n.

127. Use the *table* feature of a graphing utility to find the first five terms of the sequence.

128. Do you recognize the terms of the sequence in Exercise 127? What sequence is it?

129. Find an expression for a_{n+1} and a_{n+2} in terms of n.

130. Use the result from Exercise 129 to show that $a_{n+2} = a_{n+1} + a_n$. Is this result the same as your answer to Exercise 127? Explain.

FOR FURTHER INFORMATION: Use the Internet or your library to read more about the Fibonacci sequence in the publication called the *Fibonacci Quarterly*, a journal dedicated to this famous result in mathematics.

∫ **In Exercises 131–140, write the first five terms of the sequence.**

131. $a_n = \dfrac{x^n}{n!}$

132. $a_n = \dfrac{x^2}{n^2}$

133. $a_n = \dfrac{(-1)^n x^{2n+1}}{2n+1}$

134. $a_n = \dfrac{(-1)^n x^{n+1}}{n+1}$

135. $a_n = \dfrac{(-1)^n x^{2n}}{(2n)!}$

136. $a_n = \dfrac{(-1)^n x^{2n+1}}{(2n+1)!}$

137. $a_n = \dfrac{(-1)^n x^n}{n!}$

138. $a_n = \dfrac{(-1)^n x^{n+1}}{(n+1)!}$

139. $a_n = \dfrac{(-1)^{n+1}(x+1)^n}{n!}$

140. $a_n = \dfrac{(-1)^n (x-1)^n}{(n+1)!}$

In Exercises 141–146, write the first five terms of the sequence. Then find an expression for the nth partial sum.

141. $a_n = \dfrac{1}{2n} - \dfrac{1}{2n+2}$

142. $a_n = \dfrac{1}{n} - \dfrac{1}{n+1}$

143. $a_n = \dfrac{1}{n+1} - \dfrac{1}{n+2}$

144. $a_n = \dfrac{1}{n} - \dfrac{1}{n+2}$

145. $a_n = \ln n$

146. $a_n = 1 - \ln(n+1)$

Skills Review

In Exercises 147–150, find, if possible, (a) $A - B$, (b) $2B - 3A$, (c) AB, and (d) BA.

147. $A = \begin{bmatrix} 6 & 5 \\ 3 & 4 \end{bmatrix}$, $B = \begin{bmatrix} -2 & 4 \\ 6 & -3 \end{bmatrix}$

148. $A = \begin{bmatrix} 10 & 7 \\ -4 & 6 \end{bmatrix}$, $B = \begin{bmatrix} 0 & -12 \\ 8 & 11 \end{bmatrix}$

149. $A = \begin{bmatrix} -2 & -3 & 6 \\ 4 & 5 & 7 \\ 1 & 7 & 4 \end{bmatrix}$, $B = \begin{bmatrix} 1 & 4 & 2 \\ 0 & 1 & 6 \\ 0 & 3 & 1 \end{bmatrix}$

150. $A = \begin{bmatrix} -1 & 4 & 0 \\ 5 & 1 & 2 \\ 0 & -1 & 3 \end{bmatrix}$, $B = \begin{bmatrix} 0 & 4 & 0 \\ 3 & 1 & -2 \\ -1 & 0 & 2 \end{bmatrix}$

8.2 Arithmetic Sequences and Partial Sums

Arithmetic Sequences

A sequence whose consecutive terms have a common difference is called an **arithmetic sequence.**

Definition of Arithmetic Sequence

A sequence is **arithmetic** if the differences between consecutive terms are the same. So, the sequence

$$a_1, a_2, a_3, a_4, \ldots, a_n, \ldots$$

is arithmetic if there is a number d such that

$$a_2 - a_1 = a_3 - a_2 = a_4 - a_3 = \cdots = d.$$

The number d is the **common difference** of the arithmetic sequence.

Example 1 Examples of Arithmetic Sequences

a. The sequence whose nth term is $4n + 3$ is arithmetic. For this sequence, the common difference between consecutive terms is 4.

$$\underbrace{7, 11}, 15, 19, \ldots, 4n + 3, \ldots \qquad \text{Begin with } n = 1.$$
$$11 - 7 = 4$$

b. The sequence whose nth term is $7 - 5n$ is arithmetic. For this sequence, the common difference between consecutive terms is -5.

$$\underbrace{2, -3}, -8, -13, \ldots, 7 - 5n, \ldots \qquad \text{Begin with } n = 1.$$
$$-3 - 2 = -5$$

c. The sequence whose nth term is $\frac{1}{4}(n + 3)$ is arithmetic. For this sequence, the common difference between consecutive terms is $\frac{1}{4}$.

$$\underbrace{1, \frac{5}{4}}, \frac{3}{2}, \frac{7}{4}, \ldots, \frac{n + 3}{4}, \ldots \qquad \text{Begin with } n = 1.$$
$$\frac{5}{4} - 1 = \frac{1}{4}$$

✓**CHECKPOINT**   Now try Exercise 9.

The sequence $1, 4, 9, 16, \ldots$, whose nth term is n^2, is *not* arithmetic. The difference between the first two terms is

$$a_2 - a_1 = 4 - 1 = 3$$

but the difference between the second and third terms is

$$a_3 - a_2 = 9 - 4 = 5.$$

In Example 1, notice that each of the arithmetic sequences has an nth term that is of the form $dn + c$, where the common difference of the sequence is d.

The nth Term of an Arithmetic Sequence

The nth term of an arithmetic sequence has the form

$$a_n = dn + c$$

where d is the common difference between consecutive terms of the sequence and $c = a_1 - d$.

An arithmetic sequence $a_n = dn + c$ can be thought of as "counting by d's" after a shift of c units from d. For instance, the sequence

$$2, 6, 10, 14, 18, \ldots$$

has a common difference of 4, so you are counting by 4's after a shift of two units below 4 (beginning with $a_1 = 2$). So, the nth term is $4n - 2$. Similarly, the nth term of the sequence

$$6, 11, 16, 21, \ldots$$

is $5n + 1$ because you are counting by 5's after a shift of one unit above 5 (beginning with $a_1 = 6$).

Example 2   Finding the nth Term of an Arithmetic Sequence

Find a formula for the nth term of the arithmetic sequence whose common difference is 3 and whose first term is 2.

Solution

Because the sequence is arithmetic, you know that the formula for the nth term is of the form $a_n = dn + c$. Moreover, because the common difference is $d = 3$, the formula must have the form $a_n = 3n + c$. Because $a_1 = 2$, it follows that

$$c = a_1 - d = 2 - 3 = -1.$$

So, the formula for the nth term is $a_n = 3n - 1$. The sequence therefore has the following form.

$$2, 5, 8, 11, 14, \ldots, 3n - 1, \ldots$$

A graph of the first 15 terms of the sequence is shown in Figure 8.8. Notice that the points lie on a line. This makes sense because a_n is a linear function of n. In other words, the terms "arithmetic" and "linear" are closely connected.

✓**CHECKPOINT**   Now try Exercise 17.

Another way to find a formula for the nth term of the sequence in Example 2 is to begin by writing the terms of the sequence.

a_1	a_2	a_3	a_4	a_5	a_6	a_7	
2	$2 + 3$	$5 + 3$	$8 + 3$	$11 + 3$	$14 + 3$	$17 + 3$	$\ldots$
2	5	8	11	14	17	20	$\ldots$

So, you can reason that the nth term is of the form

$$a_n = dn + c = 3n - 1.$$

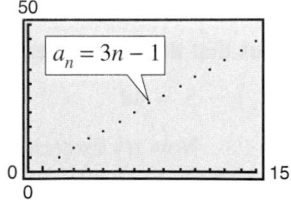

Figure 8.8

Example 3 Writing the Terms of an Arithmetic Sequence

The fourth term of an arithmetic sequence is 20, and the 13th term is 65. Write the first several terms of this sequence.

Solution

The fourth and 13th terms of the sequence are related by

$$a_{13} = a_4 + 9d.$$

Using $a_4 = 20$ and $a_{13} = 65$, you have $65 = 20 + 9d$. So, you can conclude that $d = 5$, which implies that the sequence is as follows.

a_1	a_2	a_3	a_4	a_5	a_6	a_7	a_8	a_9	a_{10}	a_{11}	a_{12}	a_{13}	. . .
5,	10,	15,	20,	25,	30,	35,	40,	45,	50,	55,	60,	65,	. . .

✔CHECKPOINT   Now try Exercise 31.

> **STUDY TIP**
>
> In Example 3, the relationship between the fourth and 13th terms can be found by subtracting the equation for the fourth term, $a_4 = 4d + c$, from the equation for the 13th term, $a_{13} = 13d + c$. The result, $a_{13} - a_4 = 9d$, can be rewritten as $a_{13} = a_4 + 9d$.

If you know the nth term of an arithmetic sequence *and* you know the common difference of the sequence, you can find the $(n + 1)$th term by using the *recursion formula*

$$a_{n+1} = a_n + d. \qquad \text{Recursion formula}$$

With this formula, you can find any term of an arithmetic sequence, *provided* that you know the preceding term. For instance, if you know the first term, you can find the second term. Then, knowing the second term, you can find the third term, and so on.

Example 4 Using a Recursion Formula

Find the seventh term of the arithmetic sequence whose first two terms are 2 and 9.

Solution

For this sequence, the common difference is $d = 9 - 2 = 7$. Next find a formula for the nth term. Because the first term is 2, it follows that

$$c = a_1 - d = 2 - 7 = -5.$$

Therefore, a formula for the nth term is

$$a_n = dn + c = 7n - 5.$$

which implies that the seventh term is

$$a_7 = 7(7) - 5 = 44.$$

✔CHECKPOINT   Now try Exercise 39.

> **STUDY TIP**
>
> Another way to find the seventh term in Example 4 is to determine the common difference, $d = 7$, and then simply write out the first seven terms (by repeatedly adding 7).
>
> 2, 9, 16, 23, 30, 37, 44
>
> As you can see, the seventh term is 44.

If you substitute $a_1 - d$ for c in the formula $a_n = dn + c$, the nth term of an arithmetic sequence has the alternative recursion formula

$$a_n = a_1 + (n - 1)d. \qquad \text{Alternative recursion formula}$$

Use this formula to solve Example 4. You should obtain the same answer.

The Sum of a Finite Arithmetic Sequence

There is a simple formula for the *sum* of a finite arithmetic sequence.

> **The Sum of a Finite Arithmetic Sequence**   (See the proof on page 657.)
>
> The sum of a finite arithmetic sequence with n terms is given by
>
> $$S_n = \frac{n}{2}(a_1 + a_n).$$

Be sure you see that this formula works only for *arithmetic* sequences. Using this formula reduces the amount of time it takes to find the sum of an arithmetic sequence, as you will see in the following example.

Example 5   Finding the Sum of a Finite Arithmetic Sequence

Find each sum.

a. $1 + 3 + 5 + 7 + 9 + 11 + 13 + 15 + 17 + 19$

b. Sum of the integers from 1 to 100

Solution

a. To begin, notice that the sequence is arithmetic (with a common difference of 2). Moreover, the sequence has 10 terms. So, the sum of the sequence is

$$S_n = 1 + 3 + 5 + 7 + 9 + 11 + 13 + 15 + 17 + 19$$

$$= \frac{n}{2}(a_1 + a_n) \qquad \text{Formula for sum of an arithmetic sequence}$$

$$= \frac{10}{2}(1 + 19) \qquad \text{Substitute 10 for } n, \text{ 1 for } a_1, \text{ and 19 for } a_n.$$

$$= 5(20) = 100. \qquad \text{Simplify.}$$

b. The integers from 1 to 100 form an arithmetic sequence that has 100 terms. So, you can use the formula for the sum of an arithmetic sequence, as follows.

$$S_n = 1 + 2 + 3 + 4 + 5 + 6 + \cdots + 99 + 100$$

$$= \frac{n}{2}(a_1 + a_n) \qquad \text{Formula for sum of an arithmetic sequence}$$

$$= \frac{100}{2}(1 + 100) \qquad \text{Substitute 100 for } n, \text{ 1 for } a_1, \text{ and 100 for } a_n.$$

$$= 50(101) = 5050 \qquad \text{Simplify.}$$

✓**CHECKPOINT**   Now try Exercise 53.

The sum of the first n terms of an infinite sequence is called the **nth partial sum.** The nth partial sum of an arithmetic sequence can be found by using the formula for the sum of a finite arithmetic sequence.

Example 6   Finding a Partial Sum of an Arithmetic Sequence

Find the 150th partial sum of the arithmetic sequence 5, 16, 27, 38, 49, ...

Solution

For this arithmetic sequence, you have $a_1 = 5$ and $d = 16 - 5 = 11$. So,

$$c = a_1 - d = 5 - 11 = -6$$

and the nth term is $a_n = 11n - 6$. Therefore, $a_{150} = 11(150) - 6 = 1644$, and the sum of the first 150 terms is

$$S_n = \frac{n}{2}(a_1 + a_n) \qquad \text{nth partial sum formula}$$

$$= \frac{150}{2}(5 + 1644) \qquad \text{Substitute 150 for n, 5 for a_1, and 1644 for a_n.}$$

$$= 75(1649) = 123{,}675. \qquad \text{Simplify.}$$

✓**CHECKPOINT**   Now try Exercise 61.

Applications

Example 7   Seating Capacity

An auditorium has 20 rows of seats. There are 20 seats in the first row, 21 seats in the second row, 22 seats in the third row, and so on (see Figure 8.9). How many seats are there in all 20 rows?

Solution

The numbers of seats in the 20 rows form an arithmetic sequence for which the common difference is $d = 1$. Because

$$c = a_1 - d = 20 - 1 = 19$$

you can determine that the formula for the nth term of the sequence is $a_n = n + 19$. So, the 20th term of the sequence is $a_{20} = 20 + 19 = 39$, and the total number of seats is

$$S_n = 20 + 21 + 22 + \cdots + 39$$

$$= \frac{20}{2}(20 + 39) \qquad \text{Substitute 20 for n, 20 for a_1, and 39 for a_n.}$$

$$= 10(59) = 590. \qquad \text{Simplify.}$$

✓**CHECKPOINT**   Now try Exercise 81.

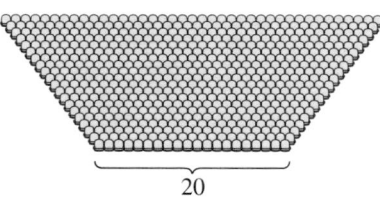

Figure 8.9

Example 8 Total Sales

A small business sells $10,000 worth of sports memorabilia during its first year. The owner of the business has set a goal of increasing annual sales by $7500 each year for 19 years. Assuming that this goal is met, find the total sales during the first 20 years this business is in operation.

Algebraic Solution

The annual sales form an arithmetic sequence in which $a_1 = 10{,}000$ and $d = 7500$. So,

$$c = a_1 - d$$

$$= 10{,}000 - 7500$$

$$= 2500$$

and the nth term of the sequence is

$$a_n = 7500n + 2500.$$

This implies that the 20th term of the sequence is

$$a_{20} = 7500(20) + 2500$$

$$= 152{,}500.$$

The sum of the first 20 terms of the sequence is

$$S_n = \frac{n}{2}(a_1 + a_n) \qquad \text{nth partial sum formula}$$

$$= \frac{20}{2}(10{,}000 + 152{,}500) \qquad \begin{array}{l}\text{Substitute 20 for } n, 10{,}000 \\ \text{for } a_1, \text{ and } 152{,}500 \text{ for } a_n.\end{array}$$

$$= 10(162{,}500) \qquad \text{Simplify.}$$

$$= 1{,}625{,}000. \qquad \text{Simplify.}$$

So, the total sales for the first 20 years are $1,625,000.

Numerical Solution

The annual sales form an arithmetic sequence in which $a_1 = 10{,}000$ and $d = 7500$. So,

$$c = a_1 - d$$

$$= 10{,}000 - 7500$$

$$= 2500.$$

The nth term of the sequence is given by

$$u_n = 7500n + 2500.$$

You can use the *list editor* of a graphing utility to create a table that shows the sales for each of the 20 years. First, enter the numbers 1 through 20 in L_1. Then enter $7500*L_1 + 2500$ for L_2. You should obtain a table like the one shown in Figure 8.10. Finally, use the *sum* feature of the graphing utility to find the sum of the data in L_2, as shown in Figure 8.11. So, the total sales for the first 20 years are $1,625,000.

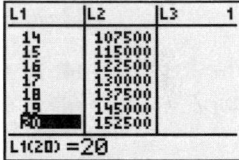

Figure 8.10

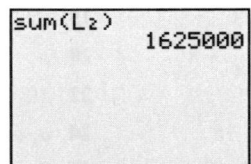

Figure 8.11

✓CHECKPOINT   Now try Exercise 83.

If you go on to take a course in calculus, you will study sequences and series in detail. You will learn that sequences and series play a major role in the study of calculus.

TECHNOLOGY SUPPORT

For instructions on how to use the *list editor* and *sum* features, see Appendix A; for specific keystrokes, go to this textbook's *Online Study Center.*

8.2 Exercises

See www.CalcChat.com for worked-out solutions to odd-numbered exercises.

Vocabulary Check

Fill in the blanks.

1. A sequence is called an _____ sequence if the differences between consecutive terms are the same. This difference is called the _____ difference.

2. The nth term of an arithmetic sequence has the form _____ .

3. The formula $S_n = \dfrac{n}{2}(a_1 + a_n)$ can be used to find the sum of the first n terms of an arithmetic sequence, called the _____ .

In Exercises 1–8, determine whether or not the sequence is arithmetic. If it is, find the common difference.

1. $10, 8, 6, 4, 2, \ldots$
2. $4, 9, 14, 19, 24, \ldots$
3. $3, \frac{5}{2}, 2, \frac{3}{2}, 1, \ldots$
4. $\frac{1}{3}, \frac{2}{3}, \frac{4}{3}, \frac{8}{3}, \frac{16}{3}, \ldots$
5. $-24, -16, -8, 0, 8, \ldots$
6. $\ln 1, \ln 2, \ln 3, \ln 4, \ln 5, \ldots$
7. $3.7, 4.3, 4.9, 5.5, 6.1, \ldots$
8. $1^2, 2^2, 3^2, 4^2, 5^2, \ldots$

In Exercises 9–16, write the first five terms of the sequence. Determine whether or not the sequence is arithmetic. If it is, find the common difference. (Assume n begins with 1.)

9. $a_n = 8 + 13n$
10. $a_n = 2^n + n$
11. $a_n = \dfrac{1}{n+1}$
12. $a_n = 1 + (n-1)4$
13. $a_n = 150 - 7n$
14. $a_n = 2^{n-1}$
15. $a_n = 3 + 2(-1)^n$
16. $a_n = 3 - 4(n+6)$

In Exercises 17–26, find a formula for a_n for the arithmetic sequence.

17. $a_1 = 1, d = 3$
18. $a_1 = 15, d = 4$
19. $a_1 = 100, d = -8$
20. $a_1 = 0, d = -\frac{2}{3}$
21. $4, \frac{3}{2}, -1, -\frac{7}{2}, \ldots$
22. $10, 5, 0, -5, -10, \ldots$
23. $a_1 = 5, a_4 = 15$
24. $a_1 = -4, a_5 = 16$
25. $a_3 = 94, a_6 = 85$
26. $a_5 = 190, a_{10} = 115$

In Exercises 27–34, write the first five terms of the arithmetic sequence. Use the _table_ feature of a graphing utility to verify your results.

27. $a_1 = 5, d = 6$
28. $a_1 = 5, d = -\frac{3}{4}$
29. $a_1 = -10, d = -12$
30. $a_4 = 16, a_{10} = 46$
31. $a_8 = 26, a_{12} = 42$
32. $a_6 = -38, a_{11} = -73$
33. $a_3 = 19, a_{15} = -1.7$
34. $a_5 = 16, a_{14} = 38.5$

In Exercises 35–38, write the first five terms of the arithmetic sequence. Find the common difference and write the nth term of the sequence as a function of n.

35. $a_1 = 15, \quad a_{k+1} = a_k + 4$
36. $a_1 = 200, \quad a_{k+1} = a_k - 10$
37. $a_1 = \frac{3}{5}, \quad a_{k+1} = -\frac{1}{10} + a_k$
38. $a_1 = 1.5, \quad a_{k+1} = a_k - 2.5$

In Exercises 39–42, the first two terms of the arithmetic sequence are given. Find the missing term. Use the _table_ feature of a graphing utility to verify your results.

39. $a_1 = 5, \quad a_2 = 11, \quad a_{10} = $ ▢
40. $a_1 = 3, \quad a_2 = 13, \quad a_9 = $ ▢
41. $a_1 = 4.2, \quad a_2 = 6.6, \quad a_7 = $ ▢
42. $a_1 = -0.7, \quad a_2 = -13.8, \quad a_8 = $ ▢

In Exercises 43–46, use a graphing utility to graph the first 10 terms of the sequence. (Assume n begins with 1.)

43. $a_n = 15 - \frac{3}{2}n$
44. $a_n = -5 + 2n$
45. $a_n = 0.5n + 4$
46. $a_n = -0.9n + 2$

In Exercises 47–52, use the _table_ feature of a graphing utility to find the first 10 terms of the sequence. (Assume n begins with 1.)

47. $a_n = 4n - 5$
48. $a_n = 17 + 3n$
49. $a_n = 20 - \frac{3}{4}n$
50. $a_n = \frac{4}{5}n + 12$
51. $a_n = 1.5 + 0.05n$
52. $a_n = 8 - 12.5n$

In Exercises 53–60, find the sum of the finite arithmetic sequence.

53. $2 + 4 + 6 + 8 + 10 + 12 + 14 + 16 + 18 + 20$
54. $1 + 4 + 7 + 10 + 13 + 16 + 19$
55. $-1 + (-3) + (-5) + (-7) + (-9)$

56. $-5 + (-3) + (-1) + 1 + 3 + 5$

57. Sum of the first 50 positive even integers

58. Sum of the first 100 positive odd integers

59. Sum of the integers from -100 to 30

60. Sum of the integers from -10 to 50

In Exercises 61–66, find the indicated *n*th partial sum of the arithmetic sequence.

61. $8, 20, 32, 44, \ldots, \quad n = 10$

62. $-6, -2, 2, 6, \ldots, \quad n = 50$

63. $0.5, 1.3, 2.1, 2.9, \ldots, \quad n = 10$

64. $4.2, 3.7, 3.2, 2.7, \ldots, \quad n = 12$

65. $a_1 = 100, \ a_{25} = 220, \quad n = 25$

66. $a_1 = 15, \ a_{100} = 307, \quad n = 100$

In Exercises 67–74, find the partial sum without using a graphing utility.

67. $\displaystyle\sum_{n=1}^{50} n$

68. $\displaystyle\sum_{n=1}^{100} 2n$

69. $\displaystyle\sum_{n=1}^{100} 5n$

70. $\displaystyle\sum_{n=51}^{100} 7n$

71. $\displaystyle\sum_{n=11}^{30} n - \sum_{n=1}^{10} n$

72. $\displaystyle\sum_{n=51}^{100} n - \sum_{n=1}^{50} n$

73. $\displaystyle\sum_{n=1}^{500} (n + 8)$

74. $\displaystyle\sum_{n=1}^{250} (1000 - n)$

In Exercises 75–80, use a graphing utility to find the partial sum.

75. $\displaystyle\sum_{n=1}^{20} (2n + 1)$

76. $\displaystyle\sum_{n=0}^{50} (50 - 2n)$

77. $\displaystyle\sum_{n=1}^{100} \frac{n + 1}{2}$

78. $\displaystyle\sum_{n=0}^{100} \frac{4 - n}{4}$

79. $\displaystyle\sum_{i=1}^{60} \left(250 - \tfrac{2}{5}i\right)$

80. $\displaystyle\sum_{j=1}^{200} (10.5 + 0.025j)$

81. *Brick Pattern*   A brick patio has the approximate shape of a trapezoid, as shown in the figure. The patio has 18 rows of bricks. The first row has 14 bricks and the 18th row has 31 bricks. How many bricks are in the patio?

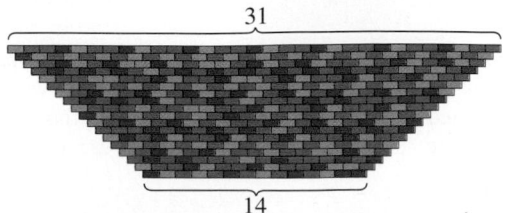

82. *Number of Logs*   Logs are stacked in a pile, as shown in the figure. The top row has 15 logs and the bottom row has 24 logs. How many logs are in the pile?

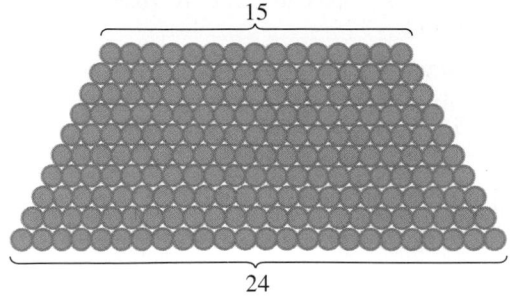

83. *Sales*   A small hardware store makes a profit of $20,000 during its first year. The store owner sets a goal of increasing profits by $5000 each year for 4 years. Assuming that this goal is met, find the total profit during the first 5 years of business.

84. *Falling Object*   An object with negligible air resistance is dropped from an airplane. During the first second of fall, the object falls 4.9 meters; during the second second, it falls 14.7 meters; during the third second, it falls 24.5 meters; and during the fourth second, it falls 34.3 meters. If this arithmetic pattern continues, how many meters will the object fall in 10 seconds?

85. *Sales*   The table shows the sales a_n (in billions of dollars) for Coca-Cola Enterprises, Inc. from 1997 to 2004. (Source: Coca-Cola Enterprises, Inc.)

Year	Sales, a_n (in billions of dollars)
1997	11.3
1998	13.4
1999	14.4
2000	14.8
2001	15.7
2002	16.9
2003	17.3
2004	18.2

(a) Use the *regression* feature of a graphing utility to find an arithmetic sequence for the data. Let n represent the year, with $n = 7$ corresponding to 1997.

(b) Use the sequence from part (a) to approximate the annual sales for Coca-Cola Enterprises, Inc. for the years 1997 to 2004. How well does the model fit the data?

(c) Use the sequence to find the total annual sales for Coca Cola for the years from 1997 to 2004.

(d) Use the sequence to predict the total annual sales for the years 2005 to 2012. Is your total reasonable? Explain.

86. Education The table shows the numbers a_n (in thousands) of master's degrees conferred in the United States from 1995 to 2003. (Source: U.S. National Center for Education Statistics)

Year	Master's degrees conferred, a_n (in thousands)
1995	398
1996	406
1997	419
1998	430
1999	440
2000	457
2001	468
2002	482
2003	512

(a) Use the *regression* feature of a graphing utility to find an arithmetic sequence for the data. Let n represent the year, with $n = 5$ corresponding to 1995.

(b) Use the sequence from part (a) to approximate the numbers of master's degrees conferred for the years 1995 to 2003. How well does the model fit the data?

(c) Use the sequence to find the total number of master's degrees conferred over the period from 1995 to 2003.

(d) Use the sequence to predict the total number of master's degrees conferred over the period from 2004 to 2014. Is your total reasonable? Explain.

Synthesis

True or False? **In Exercises 87 and 88, determine whether the statement is true or false. Justify your answer.**

87. Given an arithmetic sequence for which only the first and second terms are known, it is possible to find the nth term.

88. If the only known information about a finite arithmetic sequence is its first term and its last term, then it is possible to find the sum of the sequence.

In Exercises 89 and 90, find the first 10 terms of the sequence.

89. $a_1 = x, d = 2x$

90. $a_1 = -y, d = 5y$

91. *Think About It* The sum of the first 20 terms of an arithmetic sequence with a common difference of 3 is 650. Find the first term.

92. *Think About It* The sum of the first n terms of an arithmetic sequence with first term a_1 and common difference d is S_n. Determine the sum if each term is increased by 5. Explain.

93. *Think About It* Decide whether it is possible to fill in the blanks in each of the sequences such that the resulting sequence is arithmetic. If so, find a recursion formula for the sequence. Write a short paragraph explaining how you made your decisions.

(a) -7, ___, ___, ___, ___, ___, 11

(b) 17, ___, ___, ___, ___, ___, ___, 59

(c) $2, 6$, ___, ___, 162

(d) $4, 7.5$, ___, ___, ___, ___, ___, 28.5

(e) $8, 12$, ___, ___, ___, 60.75

94. *Gauss* Carl Friedrich Gauss, a famous nineteenth century mathematician, was a child prodigy. It was said that when Gauss was 10 he was asked by his teacher to add the numbers from 1 to 100. Almost immediately, Gauss found the answer by mentally finding the summation. Write an explanation of how he arrived at his conclusion, and then find the formula for the sum of the first n natural numbers.

In Exercises 95–98, find the sum using the method from Exercise 94.

95. The first 200 natural numbers

96. The first 100 even natural numbers from 2 to 200, inclusive

97. The first 51 odd natural numbers from 1 to 101, inclusive

98. The first 100 multiples of 4 from 4 to 400, inclusive

Skills Review

In Exercises 99 and 100, use Gauss-Jordan elimination to solve the system of equations.

99. $\begin{cases} 2x - y + 7z = -10 \\ 3x + 2y - 4z = 17 \\ 6x - 5y + z = -20 \end{cases}$

100. $\begin{cases} -x + 4y + 10z = 4 \\ 5x - 3y + z = 31 \\ 8x + 2y - 3z = -5 \end{cases}$

In Exercises 101 and 102, use a determinant to find the area of the triangle with the given vertices.

101. $(0, 0), (4, -3), (2, 6)$

102. $(-1, 2), (5, 1), (3, 8)$

103. *Make a Decision* To work an extended application analyzing the amount of municipal waste recovered in the United States from 1983 to 2003, visit this textbook's *Online Study Center.* (*Data Source: U.S. Census Bureau*)

8.3 Geometric Sequences and Series

Geometric Sequences

In Section 8.2, you learned that a sequence whose consecutive terms have a common *difference* is an arithmetic sequence. In this section, you will study another important type of sequence called a **geometric sequence.** Consecutive terms of a geometric sequence have a common *ratio*.

> **Definition of Geometric Sequence**
>
> A sequence is **geometric** if the ratios of consecutive terms are the same. So, the sequence $a_1, a_2, a_3, a_4, \ldots, a_n, \ldots$ is geometric if there is a number r such that
>
> $$\frac{a_2}{a_1} = \frac{a_3}{a_2} = \frac{a_4}{a_3} = \cdots = r, \qquad r \neq 0.$$
>
> The number r is the **common ratio** of the sequence.

Example 1 Examples of Geometric Sequences

a. The sequence whose nth term is 2^n is geometric. For this sequence, the common ratio between consecutive terms is 2.

$$2, 4, 8, 16, \ldots, 2^n, \ldots \qquad \text{Begin with } n = 1.$$

$$\frac{4}{2} = 2$$

b. The sequence whose nth term is $4(3^n)$ is geometric. For this sequence, the common ratio between consecutive terms is 3.

$$12, 36, 108, 324, \ldots, 4(3^n), \ldots \qquad \text{Begin with } n = 1.$$

$$\frac{36}{12} = 3$$

c. The sequence whose nth term is $\left(-\frac{1}{3}\right)^n$ is geometric. For this sequence, the common ratio between consecutive terms is $-\frac{1}{3}$.

$$-\frac{1}{3}, \frac{1}{9}, -\frac{1}{27}, \frac{1}{81}, \ldots, \left(-\frac{1}{3}\right)^n, \ldots \qquad \text{Begin with } n = 1.$$

$$\frac{1/9}{-1/3} = -\frac{1}{3}$$

✓**CHECKPOINT**   Now try Exercise 1.

The sequence $1, 4, 9, 16, \ldots$, whose nth term is n^2, is *not* geometric. The ratio of the second term to the first term is

$$\frac{a_2}{a_1} = \frac{4}{1} = 4$$

but the ratio of the third term to the second term is $\dfrac{a_3}{a_2} = \dfrac{9}{4}$.

What you should learn

- Recognize, write, and find the nth terms of geometric sequences.
- Find nth partial sums of geometric sequences.
- Find sums of infinite geometric series.
- Use geometric sequences to model and solve real-life problems.

Why you should learn it

Geometric sequences can reduce the amount of time it takes to find the sum of a sequence of numbers with a common ratio. For instance, Exercise 99 on page 610 shows how to use a geometric sequence to estimate the distance a bungee jumper travels after jumping off a bridge.

Brand X Pictures/age fotostock

STUDY TIP

In Example 1, notice that each of the geometric sequences has an nth term of the form ar^n, where r is the common ratio of the sequence.

The nth Term of a Geometric Sequence

The nth term of a geometric sequence has the form

$$a_n = a_1 r^{n-1}$$

where r is the common ratio of consecutive terms of the sequence. So, every geometric sequence can be written in the following form.

$$a_1, \quad a_2, \quad a_3, \quad a_4, \quad a_5, \quad \ldots, \quad a_n, \quad \ldots$$

$$a_1, a_1 r, a_1 r^2, a_1 r^3, a_1 r^4, \ldots, a_1 r^{n-1}, \ldots$$

If you know the nth term of a geometric sequence, you can find the $(n + 1)$th term by multiplying by r. That is, $a_{n+1} = a_n r$.

Example 2   Finding the Terms of a Geometric Sequence

Write the first five terms of the geometric sequence whose first term is $a_1 = 3$ and whose common ratio is $r = 2$.

Solution

Starting with 3, repeatedly multiply by 2 to obtain the following.

$a_1 = 3$	1st term		$a_4 = 3(2^3) = 24$	4th term
$a_2 = 3(2^1) = 6$	2nd term		$a_5 = 3(2^4) = 48$	5th term
$a_3 = 3(2^2) = 12$	3rd term			

✓CHECKPOINT   Now try Exercise 11.

Example 3   Finding a Term of a Geometric Sequence

Find the 15th term of the geometric sequence whose first term is 20 and whose common ratio is 1.05.

Algebraic Solution

$$a_n = a_1 r^{n-1} \qquad \text{Formula for a geometric sequence}$$

$$a_{15} = 20(1.05)^{15-1} \qquad \text{Substitute 20 for } a_1, 1.05 \text{ for } r, \text{ and 15 for } n.$$

$$\approx 39.60 \qquad \text{Use a calculator.}$$

Numerical Solution

For this sequence, $r = 1.05$ and $a_1 = 20$. So, $a_n = 20(1.05)^{n-1}$. Use the *table* feature of a graphing utility to create a table that shows the values of $u_n = 20(1.05)^{n-1}$ for $n = 1$ through $n = 15$. From Figure 8.12, the number in the 15th row is approximately 39.60, so the 15th term of the geometric sequence is about 39.60.

n	$u(n)$
9	29.549
10	31.027
11	32.578
12	34.207
13	35.917
14	37.713
15	39.599

$u(n)=39.59863199$

Figure 8.12

✓CHECKPOINT   Now try Exercise 25.

Example 4   Finding a Term of a Geometric Sequence

Find a formula for the nth term of the following geometric sequence. What is the ninth term of the sequence?

$$5, 15, 45, \ldots$$

Solution

The common ratio of this sequence is

$$r = \frac{15}{5} = 3.$$

Because the first term is $a_1 = 5$, the formula must have the form

$$a_n = a_1 r^{n-1} = 5(3)^{n-1}.$$

You can determine the ninth term $(n = 9)$ to be

$$a_9 = 5(3)^{9-1} \qquad \text{Substitute 9 for } n.$$

$$= 5(6561) = 32,805. \qquad \text{Simplify.}$$

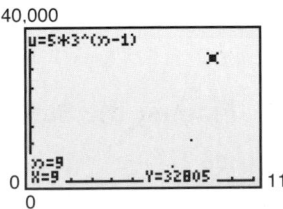

Figure 8.13

A graph of the first nine terms of the sequence is shown in Figure 8.13. Notice that the points lie on an exponential curve. This makes sense because a_n is an exponential function of n.

✓CHECKPOINT   Now try Exercise 33.

If you know *any* two terms of a geometric sequence, you can use that information to find a formula for the nth term of the sequence.

Example 5   Finding a Term of a Geometric Sequence

The fourth term of a geometric sequence is 125, and the 10th term is $125/64$. Find the 14th term. (Assume that the terms of the sequence are positive.)

Solution

The 10th term is related to the fourth term by the equation

$$a_{10} = a_4 r^6. \qquad \text{Multiply 4th term by } r^{10-4}.$$

Because $a_{10} = 125/64$ and $a_4 = 125$, you can solve for r as follows.

$$\frac{125}{64} = 125r^6$$

$$\frac{1}{64} = r^6 \quad \Longrightarrow \quad \frac{1}{2} = r$$

You can obtain the 14th term by multiplying the 10th term by r^4.

$$a_{14} = a_{10}r^4 = \frac{125}{64}\left(\frac{1}{2}\right)^4 = \frac{125}{1024}$$

✓CHECKPOINT   Now try Exercise 31.

STUDY TIP

Remember that r is the common ratio of consecutive terms of a geometric sequence. So, in Example 5,

$$a_{10} = a_1 r^9$$

$$= a_1 \cdot r \cdot r \cdot r \cdot r^6$$

$$= a_1 \cdot \frac{a_2}{a_1} \cdot \frac{a_3}{a_2} \cdot \frac{a_4}{a_3} \cdot r^6$$

$$= a_4 r^6.$$

The Sum of a Finite Geometric Sequence

The formula for the sum of a *finite* geometric sequence is as follows.

The Sum of a Finite Geometric Sequence   (See the proof on page 657.)

The sum of the finite geometric sequence

$$a_1, \; a_1r, \; a_1r^2, \; a_1r^3, \; a_1r^4, \ldots, \; a_1r^{n-1}$$

with common ratio $r \neq 1$ is given by

$$S_n = \sum_{i=1}^{n} a_1 r^{i-1} = a_1\left(\frac{1 - r^n}{1 - r}\right).$$

Example 6   Finding the Sum of a Finite Geometric Sequence

Find the sum $\displaystyle\sum_{n=1}^{12} 4(0.3)^n$.

Solution

By writing out a few terms, you have

$$\sum_{n=1}^{12} 4(0.3)^n = 4(0.3)^1 + 4(0.3)^2 + 4(0.3)^3 + \cdots + 4(0.3)^{12}.$$

Now, because $a_1 = 4(0.3)$, $r = 0.3$, and $n = 12$, you can apply the formula for the sum of a finite geometric sequence to obtain

$$\sum_{n=1}^{12} 4(0.3)^n = a_1\left(\frac{1 - r^n}{1 - r}\right) \qquad \text{Formula for sum of a finite geometric sequence}$$

$$= 4(0.3)\left[\frac{1 - (0.3)^{12}}{1 - 0.3}\right] \qquad \text{Substitute } 4(0.3) \text{ for } a_1, \, 0.3 \text{ for } r, \text{ and } 12 \text{ for } n.$$

$$\approx 1.71. \qquad \text{Use a calculator.}$$

✓**CHECKPOINT**   Now try Exercise 45.

TECHNOLOGY TIP

Using the *sum sequence* feature of a graphing utility, you can calculate the sum of the sequence in Example 6 to be about 1.7142848, as shown below.

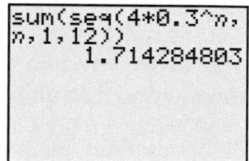

Calculate the sum beginning at $n = 0$. You should obtain a sum of 5.7142848.

When using the formula for the sum of a geometric sequence, be careful to check that the index begins at $i = 1$. If the index begins at $i = 0$, you must adjust the formula for the nth partial sum. For instance, if the index in Example 6 had begun with $n = 0$, the sum would have been

$$\sum_{n=0}^{12} 4(0.3)^n = 4(0.3)^0 + \sum_{n=1}^{12} 4(0.3)^n$$

$$= 4 + \sum_{n=1}^{12} 4(0.3)^n$$

$$\approx 4 + 1.71$$

$$= 5.71.$$

Geometric Series

The sum of the terms of an infinite geometric sequence is called an **infinite geometric series** or simply a **geometric series.**

The formula for the sum of a *finite geometric sequence* can, depending on the value of r, be extended to produce a formula for the sum of an *infinite geometric series.* Specifically, if the common ratio r has the property that $|r| < 1$, it can be shown that r^n becomes arbitrarily close to zero as n increases without bound. Consequently,

$$a_1 \left(\frac{1 - r^n}{1 - r} \right) \longrightarrow a_1 \left(\frac{1 - 0}{1 - r} \right) \quad \text{as} \quad n \longrightarrow \infty.$$

This result is summarized as follows.

The Sum of an Infinite Geometric Series

If $|r| < 1$, then the infinite geometric series

$$a_1 + a_1 r + a_1 r^2 + a_1 r^3 + \cdots + a_1 r^{n-1} + \cdots$$

has the sum

$$S = \sum_{i=0}^{\infty} a_1 r^i = \frac{a_1}{1 - r}.$$

Note that if $|r| \geq 1$, the series does not have a sum.

Example 7   Finding the Sum of an Infinite Geometric Series

Use a graphing utility to find the first six partial sums of the series. Then find the sum of the series.

$$\sum_{n=1}^{\infty} 4(0.6)^{n-1}$$

Solution

You can use the *cumulative sum* feature to find the first six partial sums of the series, as shown in Figure 8.14. By scrolling to the right, you can determine that the first six partial sums are as follows.

$$4, 6.4, 7.84, 8.704, 9.2224, 9.53344$$

Use the formula for the sum of an infinite geometric series to find the sum.

$$\sum_{n=1}^{\infty} 4(0.6)^{n-1} = 4(1) + 4(0.6) + 4(0.6)^2 + 4(0.6)^3 + \cdots + 4(0.6)^{n-1} + \cdots$$

$$= \frac{4}{1 - 0.6} = 10 \qquad \frac{a_1}{1 - r}$$

✓**CHECKPOINT**   Now try Exercise 65.

> ### Exploration
>
> Notice that the formula for the sum of an infinite geometric series requires that $|r| < 1$. What happens if $r = 1$ or $r = -1$? Give examples of infinite geometric series for which $|r| > 1$ and convince yourself that they do not have finite sums.

TECHNOLOGY SUPPORT

For instructions on how to use the *cumulative sum* feature, see Appendix A; for specific keystrokes, go to this textbook's *Online Study Center.*

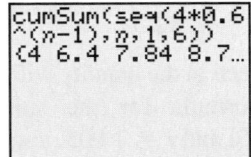

Figure 8.14

Example 8   Finding the Sum of an Infinite Geometric Series

Find the sum $3 + 0.3 + 0.03 + 0.003 + \cdots$.

Solution

$$3 + 0.3 + 0.03 + 0.003 + \cdots = 3 + 3(0.1) + 3(0.1)^2 + 3(0.1)^3 + \cdots$$

$$= \frac{3}{1 - 0.1} \qquad \frac{a_1}{1 - r}$$

$$= \frac{10}{3}$$

$$\approx 3.33$$

✓CHECKPOINT   Now try Exercise 69.

Exploration

Notice in Example 7 that when using a graphing utility to find the sum of a series, you cannot enter ∞ as the upper limit of summation. Can you still find the sum using a graphing utility? If so, which partial sum will result in 10, the exact sum of the series?

Application

Example 9   Increasing Annuity

A deposit of $50 is made on the first day of each month in a savings account that pays 6% compounded monthly. What is the balance at the end of 2 years? (This type of savings plan is called an **increasing annuity.**)

Solution

The first deposit will gain interest for 24 months, and its balance will be

$$A_{24} = 50\left(1 + \frac{0.06}{12}\right)^{24} = 50(1.005)^{24}.$$

The second deposit will gain interest for 23 months, and its balance will be

$$A_{23} = 50\left(1 + \frac{0.06}{12}\right)^{23} = 50(1.005)^{23}.$$

The last deposit will gain interest for only 1 month, and its balance will be

$$A_1 = 50\left(1 + \frac{0.06}{12}\right)^1 = 50(1.005).$$

The total balance in the annuity will be the sum of the balances of the 24 deposits. Using the formula for the sum of a finite geometric sequence, with $A_1 = 50(1.005)$ and $r = 1.005$, you have

$$S_n = a_1\left(\frac{1 - r^n}{1 - r}\right) \qquad \begin{array}{l}\text{Formula for sum of a finite}\\ \text{geometric sequence}\end{array}$$

$$S_{24} = 50(1.005)\left[\frac{1 - (1.005)^{24}}{1 - 1.005}\right] \qquad \begin{array}{l}\text{Substitute } 50(1.005) \text{ for } a_1,\\ 1.005 \text{ for } r, \text{ and } 24 \text{ for } n.\end{array}$$

$$\approx \$1277.96. \qquad \text{Simplify.}$$

✓CHECKPOINT   Now try Exercise 85.

STUDY TIP

Recall from Section 3.1 that the compound interest formula is

$$A = P\left(1 + \frac{r}{n}\right)^{nt}.$$

So, in Example 9, $50 is the principal, 0.06 is the interest rate, 12 is the number of compoundings per year, and 2 is the time in years. If you substitute these values, you obtain

$$A = 50\left(1 + \frac{0.06}{12}\right)^{12(2)}$$

$$= 50\left(1 + \frac{0.06}{12}\right)^{24}$$

8.3 Exercises

See www.CalcChat.com for worked-out solutions to odd-numbered exercises.

Vocabulary Check

Fill in the blanks.

1. A sequence is called a _____ sequence if the ratios of consecutive terms are the same. This ratio is called the _____ ratio.

2. The nth term of a geometric sequence has the form _____ .

3. The formula for the sum of a finite geometric sequence is given by _____ .

4. The sum of the terms of an infinite geometric sequence is called a _____ .

5. The formula for the sum of an infinite geometric series is given by _____ .

In Exercises 1–10, determine whether or not the sequence is geometric. If it is, find the common ratio.

1. $5, 15, 45, 135, \ldots$
2. $3, 12, 48, 192, \ldots$
3. $6, 18, 30, 42, \ldots$
4. $1, -2, 4, -8, \ldots$
5. $1, -\frac{1}{2}, \frac{1}{4}, -\frac{1}{8}, \ldots$
6. $5, 1, 0.2, 0.04, \ldots$
7. $\frac{1}{8}, \frac{1}{4}, \frac{1}{2}, 1, \ldots$
8. $9, -6, 4, -\frac{8}{3}, \ldots$
9. $1, \frac{1}{2}, \frac{1}{3}, \frac{1}{4}, \ldots$
10. $\frac{1}{5}, \frac{2}{7}, \frac{3}{9}, \frac{4}{11}, \ldots$

In Exercises 11–18, write the first five terms of the geometric sequence.

11. $a_1 = 6, \quad r = 3$
12. $a_1 = 4, \quad r = 2$
13. $a_1 = 1, \quad r = \frac{1}{2}$
14. $a_1 = 2, \quad r = \frac{1}{3}$
15. $a_1 = 5, \quad r = -\frac{1}{10}$
16. $a_1 = 6, \quad r = -\frac{1}{4}$
17. $a_i = 1, \quad r = e$
18. $a_1 = 4, \quad r = \sqrt{3}$

In Exercises 19–24, write the first five terms of the geometric sequence. Find the common ratio and write the nth term of the sequence as a function of n.

19. $a_1 = 64, \quad a_{k+1} = \frac{1}{2}a_k$
20. $a_1 = 81, \quad a_{k+1} = \frac{1}{3}a_k$
21. $a_1 = 9, \quad a_{k+1} = 2a_k$
22. $a_1 = 5, \quad a_{k+1} = -3a_k$
23. $a_1 = 6, \quad a_{k+1} = -\frac{3}{2}a_k$
24. $a_1 = 30, \quad a_{k+1} = -\frac{2}{3}a_k$

In Exercises 25–32, find the nth term of the geometric sequence. Use the _table_ feature of a graphing utility to verify your answer numerically.

25. $a_1 = 4, \quad a_4 = \frac{1}{2}, \quad n = 10$
26. $a_1 = 5, \quad a_3 = \frac{45}{4}, \quad n = 8$
27. $a_1 = 6, \quad r = -\frac{1}{3}, \quad n = 12$
28. $a_1 = 8, \quad r = -\frac{3}{4}, \quad n = 9$
29. $a_1 = 500, \quad r = 1.02, \quad n = 14$
30. $a_1 = 1000, \quad r = 1.005, \quad n = 11$
31. $a_2 = -18, \quad a_5 = \frac{2}{3}, \quad n = 6$
32. $a_3 = \frac{16}{3}, \quad a_5 = \frac{64}{27}, \quad n = 7$

In Exercises 33–36, find a formula for the nth term of the geometric sequence. Then find the indicated nth term of the geometric sequence.

33. 9th term: $7, 21, 63, \ldots$
34. 7th term: $3, 36, 432, \ldots$
35. 10th term: $5, 30, 180, \ldots$
36. 22nd term: $4, 8, 16, \ldots$

In Exercises 37–40, use a graphing utility to graph the first 10 terms of the sequence.

37. $a_n = 12(-0.75)^{n-1}$
38. $a_n = 20(1.25)^{n-1}$
39. $a_n = 2(1.3)^{n-1}$
40. $a_n = 10(-1.2)^{n-1}$

In Exercises 41 and 42, find the first four terms of the sequence of partial sums of the geometric series. In a sequence of partial sums, the term S_n is the sum of the first n terms of the sequence. For instance, S_2 is the sum of the first two terms.

41. $8, -4, 2, -1, \frac{1}{2}, \ldots$
42. $8, 12, 18, 27, \frac{81}{2}, \ldots$

In Exercises 43 and 44, use a graphing utility to create a table showing the sequence of partial sums for the first 10 terms of the series.

43. $\displaystyle\sum_{n=1}^{\infty} 16\left(\frac{1}{2}\right)^{n-1}$
44. $\displaystyle\sum_{n=1}^{\infty} 4(0.2)^{n-1}$

In Exercises 45–54, find the sum. Use a graphing utility to verify your result.

45. $\displaystyle\sum_{n=1}^{9} 2^{n-1}$
46. $\displaystyle\sum_{n=1}^{9} (-2)^{n-1}$

47. $\sum_{i=1}^{7} 64\left(-\frac{1}{2}\right)^{i-1}$

48. $\sum_{i=1}^{6} 32\left(\frac{1}{4}\right)^{i-1}$

49. $\sum_{n=0}^{20} 3\left(\frac{3}{2}\right)^{n}$

50. $\sum_{n=0}^{15} 2\left(\frac{4}{3}\right)^{n}$

51. $\sum_{i=1}^{10} 8\left(-\frac{1}{4}\right)^{i-1}$

52. $\sum_{i=1}^{10} 5\left(-\frac{1}{3}\right)^{i-1}$

53. $\sum_{n=0}^{5} 300(1.06)^{n}$

54. $\sum_{n=0}^{6} 500(1.04)^{n}$

In Exercises 55–58, use summation notation to write the sum.

55. $5 + 15 + 45 + \cdots + 3645$

56. $7 + 14 + 28 + \cdots + 896$

57. $2 - \frac{1}{2} + \frac{1}{8} - \cdots + \frac{1}{2048}$

58. $15 - 3 + \frac{3}{5} - \cdots - \frac{3}{625}$

In Exercises 59–72, find the sum of the infinite geometric series, if possible. If not possible, explain why.

59. $\sum_{n=0}^{\infty} 10\left(\frac{4}{5}\right)^{n}$

60. $\sum_{n=0}^{\infty} 6\left(\frac{2}{3}\right)^{n}$

61. $\sum_{n=0}^{\infty} 5\left(-\frac{1}{2}\right)^{n}$

62. $\sum_{n=0}^{\infty} 9\left(-\frac{2}{3}\right)^{n}$

63. $\sum_{n=1}^{\infty} 2\left(\frac{7}{3}\right)^{n-1}$

64. $\sum_{n=1}^{\infty} 8\left(\frac{5}{3}\right)^{n-1}$

65. $\sum_{n=0}^{\infty} 10(0.11)^{n}$

66. $\sum_{n=0}^{\infty} 5(0.45)^{n}$

67. $\sum_{n=0}^{\infty} -3(-0.9)^{n}$

68. $\sum_{n=0}^{\infty} -10(-0.2)^{n}$

69. $8 + 6 + \frac{9}{2} + \frac{27}{8} + \cdots$

70. $9 + 6 + 4 + \frac{8}{3} + \cdots$

71. $3 - 1 + \frac{1}{3} - \frac{1}{9} + \cdots$

72. $-6 + 5 - \frac{25}{6} + \frac{125}{36} - \cdots$

In Exercises 73–76, find the rational number representation of the repeating decimal.

73. $0.\overline{36}$

74. $0.\overline{297}$

75. $1.2\overline{5}$

76. $1.3\overline{8}$

77. Compound Interest   A principal of $1000 is invested at 3% interest. Find the amount after 10 years if the interest is compounded (a) annually, (b) semiannually, (c) quarterly, (d) monthly, and (e) daily.

78. Compound Interest   A principal of $2500 is invested at 4% interest. Find the amount after 20 years if the interest is compounded (a) annually, (b) semiannually, (c) quarterly, (d) monthly, and (e) daily.

79. Annuity   A deposit of $100 is made at the beginning of each month in an account that pays 3% interest, compounded monthly. The balance A in the account at the end of 5 years is given by

$$A = 100\left(1 + \frac{0.03}{12}\right)^{1} + \cdots + 100\left(1 + \frac{0.03}{12}\right)^{60}.$$

Find A.

80. Annuity   A deposit of $50 is made at the beginning of each month in an account that pays 2% interest, compounded monthly. The balance A in the account at the end of 5 years is given by

$$A = 50\left(1 + \frac{0.02}{12}\right)^{1} + \cdots + 50\left(1 + \frac{0.02}{12}\right)^{60}.$$

Find A.

81. Annuity   A deposit of P dollars is made at the beginning of each month in an account earning an annual interest rate r, compounded monthly. The balance A after t years is given by

$$A = P\left(1 + \frac{r}{12}\right) + P\left(1 + \frac{r}{12}\right)^{2} + \cdots + P\left(1 + \frac{r}{12}\right)^{12t}.$$

Show that the balance is given by

$$A = P\left[\left(1 + \frac{r}{12}\right)^{12t} - 1\right]\left(1 + \frac{12}{r}\right).$$

82. Annuity   A deposit of P dollars is made at the beginning of each month in an account earning an annual interest rate r, compounded continuously. The balance A after t years is given by $A = Pe^{r/12} + Pe^{2r/12} + \cdots + Pe^{12tr/12}$. Show that the balance is given by

$$A = \frac{Pe^{r/12}(e^{rt} - 1)}{e^{r/12} - 1}.$$

Annuities   **In Exercises 83–86, consider making monthly deposits of P dollars in a savings account earning an annual interest rate r. Use the results of Exercises 81 and 82 to find the balances A after t years if the interest is compounded (a) monthly and (b) continuously.**

83. $P = \$50$, $r = 7\%$, $t = 20$ years

84. $P = \$75$, $r = 4\%$, $t = 25$ years

85. $P = \$100$, $r = 5\%$, $t = 40$ years

86. $P = \$20$, $r = 6\%$, $t = 50$ years

87. Geometry   The sides of a square are 16 inches in length. A new square is formed by connecting the midpoints of the sides of the original square, and two of the resulting triangles are shaded (see figure on the next page). If this process is repeated five more times, determine the total area of the shaded region.

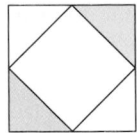 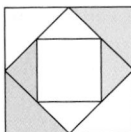 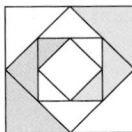 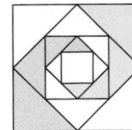

Figure for 87

88. Geometry   The sides of a square are 27 inches in length. New squares are formed by dividing the original square into nine squares. The center square is then shaded (see figure). If this process is repeated three more times, determine the total area of the shaded region.

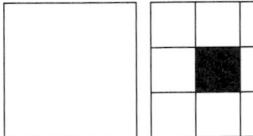 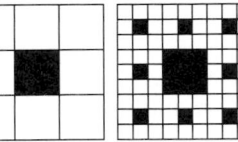 

89. Temperature   The temperature of water in an ice cube tray is 70°F when it is placed in the freezer. Its temperature n hours after being placed in the freezer is 20% less than 1 hour earlier.

(a) Find a formula for the nth term of the geometric sequence that gives the temperature of the water n hours after it is placed in the freezer.

(b) Find the temperatures of the water 6 hours and 12 hours after it is placed in the freezer.

(c) Use a graphing utility to graph the sequence to approximate the time required for the water to freeze.

90. Sphereflake   The sphereflake shown is a computer-generated fractal that was created by Eric Haines. The radius of the large sphere is 1. Attached to the large sphere are nine spheres of radius $\frac{1}{3}$. Attached to each of the smaller spheres are nine spheres of radius $\frac{1}{9}$. This process is continued infinitely.

Eric Haines

(a) Write a formula in series notation that gives the surface area of the sphereflake.

(b) Write a formula in series notation that gives the volume of the sphereflake.

(c) Determine if either the surface area or the volume of the sphereflake is finite or infinite. If either is finite, find the value.

Multiplier Effect   In Exercises 91–96, use the following information. A tax rebate is given to property owners by the state government with the anticipation that each property owner will spend approximately $p\%$ of the rebate, and in turn each recipient of this amount will spend $p\%$ of what they receive, and so on. Economists refer to this exchange of money and its circulation within the economy as the "multiplier effect." The multiplier effect operates on the principle that one individual's expenditure is another individual's income. Find the total amount put back into the state's economy, if this effect continues without end.

	Tax rebate	$p\%$
91.	$400	75%
92.	$500	70%
93.	$250	80%
94.	$350	75%
95.	$600	72.5%
96.	$450	77.5%

97. Salary Options   You have been hired at a company and the administration offers you two salary options.

Option 1: a starting salary of $30,000 for the first year with salary increases of 2.5% per year for four years and then a reevaluation of performance

Option 2: a starting salary of $32,500 for the first year with salary increases of 2% per year for four years and then a reevaluation of performance

(a) Which option do you choose if you want to make the greater cumulative amount for the five-year period? Explain your reasoning.

(b) Which option do you choose if you want to make the greater amount the year prior to reevaluation? Explain your reasoning.

98. Manufacturing   An electronics game manufacturer producing a new product estimates the annual sales to be 8000 units. Each year, 10% of the units that have been sold will become inoperative. So, 8000 units will be in use after 1 year, $[8000 + 0.9(8000)]$ units will be in use after 2 years, and so on.

(a) Write a formula in series notation for the number of units that will be operative after n years.

(b) Find the numbers of units that will be operative after 10 years, 20 years, and 50 years.

(c) If this trend continues indefinitely, will the number of units that will be operative be finite? If so, how many? If not, explain your reasoning.

99. *Distance* A bungee jumper is jumping off the New River Gorge Bridge in West Virginia, which has a height of 876 feet. The cord stretches 850 feet and the jumper rebounds 75% of the distance fallen.

(a) After jumping and rebounding 10 times, how far has the jumper traveled downward? How far has the jumper traveled upward? What is the total distance traveled downward and upward?

(b) Approximate the total distance both downward and upward, that the jumper travels before coming to rest.

∫ **100.** *Distance* A ball is dropped from a height of 16 feet. Each time it drops h feet, it rebounds $0.81h$ feet.

(a) Find the total vertical distance traveled by the ball.

(b) The ball takes the following times (in seconds) for each fall.

$$s_1 = -16t^2 + 16, \qquad s_1 = 0 \text{ if } t = 1$$
$$s_2 = -16t^2 + 16(0.81), \qquad s_2 = 0 \text{ if } t = 0.9$$
$$s_3 = -16t^2 + 16(0.81)^2, \qquad s_3 = 0 \text{ if } t = (0.9)^2$$
$$s_4 = -16t^2 + 16(0.81)^3, \qquad s_4 = 0 \text{ if } t = (0.9)^3$$
$$\vdots \qquad\qquad \vdots$$
$$s_n = -16t^2 + 16(0.81)^{n-1}, \quad s_n = 0 \text{ if } t = (0.9)^{n-1}$$

Beginning with s_2, the ball takes the same amount of time to bounce up as it does to fall, and so the total time elapsed before it comes to rest is

$$t = 1 + 2\sum_{n=1}^{\infty} (0.9)^n.$$

Find this total time.

Synthesis

True or False? In Exercises 101 and 102, determine whether the statement is true or false. Justify your answer.

101. A sequence is geometric if the ratios of consecutive differences of consecutive terms are the same.

102. You can find the nth term of a geometric sequence by multiplying its common ratio by the first term of the sequence raised to the $(n - 1)$th power.

In Exercises 103 and 104, write the first five terms of the geometric sequence.

103. $a_1 = 3, r = \dfrac{x}{2}$

104. $a_1 = \dfrac{1}{2}, r = 7x$

In Exercises 105 and 106, find the nth term of the geometric sequence.

105. $a_1 = 100, r = e^x, n = 9$

106. $a_1 = 4, r = \dfrac{4x}{3}, n = 6$

107. *Graphical Reasoning* Use a graphing utility to graph each function. Identify the horizontal asymptote of the graph and determine its relationship to the sum.

(a) $f(x) = 6\left[\dfrac{1 - (0.5)^x}{1 - (0.5)}\right], \quad \displaystyle\sum_{n=0}^{\infty} 6\left(\dfrac{1}{2}\right)^n$

(b) $f(x) = 2\left[\dfrac{1 - (0.8)^x}{1 - (0.8)}\right], \quad \displaystyle\sum_{n=0}^{\infty} 2\left(\dfrac{4}{5}\right)^n$

108. *Writing* Write a brief paragraph explaining why the terms of a geometric sequence decrease in magnitude when $-1 < r < 1$.

109. *Writing* Write a brief paragraph explaining how to use the first two terms of a geometric sequence to find the nth term.

110. *Exploration* The terms of a geometric sequence can be written as

$$a_1, \ a_2 = a_1 r, \ a_3 = a_2 r, \ a_4 = a_3 r, \ \ldots$$

Write each term of the sequence in terms of a_1 and r. Then based on the pattern, write the nth term of the geometric sequence.

Skills Review

111. *Average Speed* A truck traveled at an average speed of 50 miles per hour on a 200-mile trip. On the return trip, the average speed was 42 miles per hour. Find the average speed for the round trip.

112. *Work Rate* Your friend can mow a lawn in 4 hours and you can mow it in 6 hours. How long will it take both of you to mow the lawn working together?

In Exercises 113 and 114, find the determinant of the matrix.

113. $\begin{bmatrix} -1 & 3 & 4 \\ -2 & 8 & 0 \\ 2 & 5 & -1 \end{bmatrix}$ **114.** $\begin{bmatrix} -1 & 0 & 4 \\ -4 & 3 & 5 \\ 0 & 2 & -3 \end{bmatrix}$

115. *Make a Decision* To work an extended application analyzing the monthly profits of a clothing manufacturer over a period of 36 months, visit this textbook's *Online Study Center*.

8.4 Mathematical Induction

Introduction

In this section, you will study a form of mathematical proof called **mathematical induction.** It is important that you clearly see the logical need for it, so let's take a closer look at the problem discussed in Example 5(a) on page 595.

$$S_1 = 1 = 1^2$$

$$S_2 = 1 + 3 = 2^2$$

$$S_3 = 1 + 3 + 5 = 3^2$$

$$S_4 = 1 + 3 + 5 + 7 = 4^2$$

$$S_5 = 1 + 3 + 5 + 7 + 9 = 5^2$$

Judging from the pattern formed by these first five sums, it appears that the sum of the first n odd integers is

$$S_n = 1 + 3 + 5 + 7 + 9 + \cdots + (2n - 1) = n^2.$$

Although this particular formula is valid, it is important for you to see that recognizing a pattern and then simply *jumping to the conclusion* that the pattern must be true for all values of n is *not* a logically valid method of proof. There are many examples in which a pattern appears to be developing for small values of n but then fails at some point. One of the most famous cases of this is the conjecture by the French mathematician Pierre de Fermat (1601–1665), who speculated that all numbers of the form

$$F_n = 2^{2^n} + 1, \quad n = 0, 1, 2, \ldots$$

are prime. For $n = 0, 1, 2, 3,$ and 4, the conjecture is true.

$$F_0 = 3$$

$$F_1 = 5$$

$$F_2 = 17$$

$$F_3 = 257$$

$$F_4 = 65{,}537$$

The size of the next *Fermat number* ($F_5 = 4{,}294{,}967{,}297$) is so great that it was difficult for Fermat to determine whether or not it was prime. However, another well-known mathematician, Leonhard Euler (1707–1783), later found a factorization

$$F_5 = 4{,}294{,}967{,}297$$

$$= 641(6{,}700{,}417)$$

which proved that F_5 is not prime and therefore Fermat's conjecture was false.

Just because a rule, pattern, or formula seems to work for several values of n, you cannot simply decide that it is valid for *all* values of n without going through a *legitimate proof*. Mathematical induction is one method of proof.

What you should learn

- Use mathematical induction to prove statements involving a positive integer n.
- Find the sums of powers of integers.
- Find finite differences of sequences.

Why you should learn it

Finite differences can be used to determine what type of model can be used to represent a sequence. For instance, in Exercise 55 on page 618, you will use finite differences to find a model that represents the number of sides of the nth Koch snowflake.

Courtesy of Stefan Steinhaus

> **The Principle of Mathematical Induction**
>
> Let P_n be a statement involving the positive integer n. If
>
> 1. P_1 is true, and
>
> 2. the truth of P_k implies the truth of P_{k+1} for every positive integer k,
>
> then P_n must be true for all positive integers n.

STUDY TIP

It is important to recognize that in order to prove a statement by induction, *both* parts of the Principle of Mathematical Induction are necessary.

To apply the Principle of Mathematical Induction, you need to be able to determine the statement P_{k+1} for a given statement P_k. To determine P_{k+1}, substitute $k + 1$ for k in the statement P_k.

Example 1   A Preliminary Example

Find P_{k+1} for each P_k.

a. $P_k : S_k = \dfrac{k^2(k + 1)^2}{4}$

b. $P_k : S_k = 1 + 5 + 9 + \cdots + [4(k - 1) - 3] + (4k - 3)$

c. $P_k : k + 3 < 5k^2$

d. $P_k : 3^k \geq 2k + 1$

Solution

a. $P_{k+1} : S_{k+1} = \dfrac{(k + 1)^2(k + 1 + 1)^2}{4}$ Replace k by $k + 1$.

$\qquad = \dfrac{(k + 1)^2(k + 2)^2}{4}$ Simplify.

b. $P_{k+1} : S_{k+1} = 1 + 5 + 9 + \cdots + \{4[(k + 1) - 1] - 3\} + [4(k + 1) - 3]$

$\qquad = 1 + 5 + 9 + \cdots + (4k - 3) + (4k + 1)$

c. $P_{k+1} : (k + 1) + 3 < 5(k + 1)^2$

$\qquad k + 4 < 5(k^2 + 2k + 1)$

d. $P_{k+1} : 3^{k+1} \geq 2(k + 1) + 1$

$\qquad 3^{k+1} \geq 2k + 3$

✓**CHECKPOINT**   Now try Exercise 5.

A well-known illustration used to explain why the Principle of Mathematical Induction works is the unending line of dominoes represented by Figure 8.15. If the line actually contains infinitely many dominoes, it is clear that you could not knock down the entire line by knocking down only *one domino* at a time. However, suppose it were true that each domino would knock down the next one as it fell. Then you could knock them all down simply by pushing the first one and starting a chain reaction. Mathematical induction works in the same way. If the truth of P_k implies the truth of P_{k+1} and if P_1 is true, the chain reaction proceeds as follows: P_1 implies P_2, P_2 implies P_3, P_3 implies P_4, and so on.

Figure 8.15

When using mathematical induction to prove a *summation* formula (such as the one in Example 2), it is helpful to think of S_{k+1} as

$$S_{k+1} = S_k + a_{k+1}$$

where a_{k+1} is the $(k+1)$th term of the original sum.

Example 2   Using Mathematical Induction

Use mathematical induction to prove the following formula.

$$S_n = 1 + 3 + 5 + 7 + \cdots + (2n - 1) = n^2$$

Solution

Mathematical induction consists of two distinct parts. First, you must show that the formula is true when $n = 1$.

1. When $n = 1$, the formula is valid because

$$S_1 = 1 = 1^2.$$

The second part of mathematical induction has two steps. The first step is to assume that the formula is valid for *some* integer k. The second step is to use this assumption to prove that the formula is valid for the next integer, $k + 1$.

2. Assuming that the formula

$$S_k = 1 + 3 + 5 + 7 + \cdots + (2k - 1) = k^2$$

is true, you must show that the formula $S_{k+1} = (k + 1)^2$ is true.

$$
\begin{aligned}
S_{k+1} &= 1 + 3 + 5 + 7 + \cdots + (2k - 1) + [2(k + 1) - 1] \\
&= [1 + 3 + 5 + 7 + \cdots + (2k - 1)] + (2k + 2 - 1) \\
&= S_k + (2k + 1) \qquad \text{Group terms to form } S_k. \\
&= k^2 + 2k + 1 \qquad \text{Replace } S_k \text{ by } k^2. \\
&= (k + 1)^2
\end{aligned}
$$

Combining the results of parts (1) and (2), you can conclude by mathematical induction that the formula is valid for all positive integer values of n.

✓**CHECKPOINT**   Now try Exercise 7.

It occasionally happens that a statement involving natural numbers is *not* true for the first $k - 1$ positive integers but *is* true for all values of $n \geq k$. In these instances, you use a slight variation of the Principle of Mathematical Induction in which you verify P_k rather than P_1. This variation is called the *Extended Principle of Mathematical Induction*. To see the validity of this principle, note from Figure 8.15 that all but the first $k - 1$ dominoes can be knocked down by knocking over the kth domino. This suggests that you can prove a statement P_n to be true for $n \geq k$ by showing that P_k is true and that P_k implies P_{k+1}. In Exercises 25–30 in this section, you are asked to apply this extension of mathematical induction.

Example 3 Using Mathematical Induction

Use mathematical induction to prove the formula

$$S_n = 1^2 + 2^2 + 3^2 + 4^2 + \cdots + n^2 = \frac{n(n+1)(2n+1)}{6}$$

for all integers $n \geq 1$.

Solution

1. When $n = 1$, the formula is valid because

$$S_1 = 1^2 = \frac{1(1+1)(2 \cdot 1 + 1)}{6} = \frac{1(2)(3)}{6}.$$

2. Assuming that

$$S_k = 1^2 + 2^2 + 3^2 + 4^2 + \cdots + k^2 = \frac{k(k+1)(2k+1)}{6}$$

you must show that

$$S_{k+1} = \frac{(k+1)(k+1+1)[2(k+1)+1]}{6} = \frac{(k+1)(k+2)(2k+3)}{6}.$$

To do this, write the following.

$$
\begin{aligned}
S_{k+1} &= S_k + a_{k+1} \\
&= (1^2 + 2^2 + 3^2 + 4^2 + \cdots + k^2) + (k+1)^2 \\
&= \frac{k(k+1)(2k+1)}{6} + (k+1)^2 \qquad \text{By assumption} \\
&= \frac{k(k+1)(2k+1) + 6(k+1)^2}{6} \\
&= \frac{(k+1)[k(2k+1) + 6(k+1)]}{6} \\
&= \frac{(k+1)(2k^2 + 7k + 6)}{6} \\
&= \frac{(k+1)(k+2)(2k+3)}{6}
\end{aligned}
$$

Combining the results of parts (1) and (2), you can conclude by mathematical induction that the formula is valid for *all* integers $n \geq 1$.

✓*CHECKPOINT*   Now try Exercise 13.

When proving a formula by mathematical induction, the only statement that you *need* to verify is P_1. As a check, it is a good idea to try verifying some of the other statements. For instance, in Example 3, try verifying P_2 and P_3.

Sums of Powers of Integers

The formula in Example 3 is one of a collection of useful summation formulas. This and other formulas dealing with the sums of various powers of the first n positive integers are summarized below.

Sums of Powers of Integers

1. $\displaystyle\sum_{i=1}^{n} i = 1 + 2 + 3 + 4 + \cdots + n = \frac{n(n+1)}{2}$

2. $\displaystyle\sum_{i=1}^{n} i^2 = 1^2 + 2^2 + 3^2 + 4^2 + \cdots + n^2 = \frac{n(n+1)(2n+1)}{6}$

3. $\displaystyle\sum_{i=1}^{n} i^3 = 1^3 + 2^3 + 3^3 + 4^3 + \cdots + n^3 = \frac{n^2(n+1)^2}{4}$

4. $\displaystyle\sum_{i=1}^{n} i^4 = 1^4 + 2^4 + 3^4 + 4^4 + \cdots + n^4 = \frac{n(n+1)(2n+1)(3n^2+3n-1)}{30}$

5. $\displaystyle\sum_{i=1}^{n} i^5 = 1^5 + 2^5 + 3^5 + 4^5 + \cdots + n^5 = \frac{n^2(n+1)^2(2n^2+2n-1)}{12}$

Each of these formulas for sums can be proven by mathematical induction. (See Exercises 13–16 in this section.)

Example 4 Proving an Inequality by Mathematical Induction

Prove that $n < 2^n$ for all integers $n \geq 1$.

Solution

1. For $n = 1$ and $n = 2$, the formula is true because

$$1 < 2^1 \text{ and } 2 < 2^2.$$

2. Assuming that

$$k < 2^k$$

you need to show that $k + 1 < 2^{k+1}$. Multiply each side of $k < 2^k$ by 2.

$$2(k) < 2(2^k) = 2^{k+1}$$

Because $k + 1 < k + k = 2k$ for all $k > 1$, it follows that

$$k + 1 < 2k < 2^{k+1}$$

or

$$k + 1 < 2^{k+1}.$$

Combining the results of parts (1) and (2), you can conclude by mathematical induction that $n < 2^n$ for all integers $n \geq 1$.

✓CHECKPOINT   Now try Exercise 25.

Finite Differences

The **first differences** of a sequence are found by subtracting consecutive terms. The **second differences** are found by subtracting consecutive first differences. The first and second differences of the sequence $3, 5, 8, 12, 17, 23, \ldots$ are as follows.

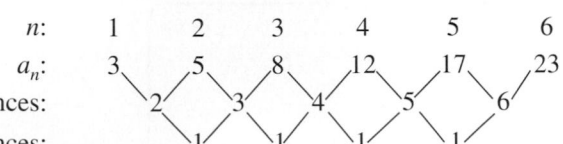

> **STUDY TIP**
>
> For a linear model, the *first* differences are the same nonzero number. For a quadratic model, the *second* differences are the same nonzero number.

For this sequence, the second differences are all the same. When this happens, and the second differences are nonzero, the sequence has a perfect *quadratic* model. If the first differences are all the same nonzero number, the sequence has a *linear* model—that is, it is arithmetic.

Example 5 Finding a Quadratic Model

Find the quadratic model for the sequence $3, 5, 8, 12, 17, 23, \ldots$.

Solution

You know from the second differences shown above that the model is quadratic and has the form

$$a_n = an^2 + bn + c.$$

By substituting 1, 2, and 3 for n, you can obtain a system of three linear equations in three variables.

> **Prerequisite Skills**
>
> If you have difficulty in solving the system of equations in this example, review Gaussian elimination in Section 7.3.

$a_1 = a(1)^2 + b(1) + c = 3$ Substitute 1 for n.

$a_2 = a(2)^2 + b(2) + c = 5$ Substitute 2 for n.

$a_3 = a(3)^2 + b(3) + c = 8$ Substitute 3 for n.

You now have a system of three equations in a, b, and c.

$$\begin{cases} a + b + c = 3 & \text{Equation 1} \\ 4a + 2b + c = 5 & \text{Equation 2} \\ 9a + 3b + c = 8 & \text{Equation 3} \end{cases}$$

Solving this system of equations using the techniques discussed in Chapter 7, you can find the solution to be $a = \frac{1}{2}$, $b = \frac{1}{2}$, and $c = 2$. So, the quadratic model is

$$a_n = \tfrac{1}{2}n^2 + \tfrac{1}{2}n + 2.$$

Check the values of a_1, a_2, and a_3 as follows.

Check

$a_1 = \frac{1}{2}(1)^2 + \frac{1}{2}(1) + 2 = 3$ Solution checks. ✓

$a_2 = \frac{1}{2}(2)^2 + \frac{1}{2}(2) + 2 = 5$ Solution checks. ✓

$a_3 = \frac{1}{2}(3)^2 + \frac{1}{2}(3) + 2 = 8$ Solution checks. ✓

✓**CHECKPOINT**   Now try Exercise 51.

8.4 Exercises

See www.CalcChat.com for worked-out solutions to odd-numbered exercises.

Vocabulary Check

Fill in the blanks.

1. The first step in proving a formula by _____ is to show that the formula is true when $n = 1$.

2. The _____ differences of a sequence are found by subtracting consecutive terms.

3. A sequence is an _____ sequence if the first differences are all the same nonzero number.

4. If the _____ differences of a sequence are all the same nonzero number, then the sequence has a perfect quadratic model.

In Exercises 1–6, find P_{k+1} for the given P_k.

1. $P_k = \dfrac{5}{k(k + 1)}$

2. $P_k = \dfrac{4}{(k + 2)(k + 3)}$

3. $P_k = \dfrac{2^k}{(k + 1)!}$

4. $P_k = \dfrac{2^{k-1}}{k!}$

5. $P_k = 1 + 6 + 11 + \cdots + [5(k - 1) - 4] + (5k - 4)$

6. $P_k = 7 + 13 + 19 + \cdots + [6(k - 1) + 1] + (6k + 1)$

In Exercises 7–20, use mathematical induction to prove the formula for every positive integer n.

7. $2 + 4 + 6 + 8 + \cdots + 2n = n(n + 1)$

8. $3 + 11 + 19 + 27 + \cdots + (8n - 5) = n(4n - 1)$

9. $3 + 8 + 13 + 18 + \cdots + (5n - 2) = \dfrac{n}{2}(5n + 1)$

10. $1 + 4 + 7 + 10 + \cdots + (3n - 2) = \dfrac{n}{2}(3n - 1)$

11. $1 + 2 + 2^2 + 2^3 + \cdots + 2^{n-1} = 2^n - 1$

12. $2(1 + 3 + 3^2 + 3^3 + \cdots + 3^{n-1}) = 3^n - 1$

13. $\displaystyle\sum_{i=1}^{n} i = \dfrac{n(n + 1)}{2}$

14. $\displaystyle\sum_{i=1}^{n} i^3 = \dfrac{n^2(n + 1)^2}{4}$

15. $\displaystyle\sum_{i=1}^{n} i^4 = \dfrac{n(n + 1)(2n + 1)(3n^2 + 3n - 1)}{30}$

16. $\displaystyle\sum_{i=1}^{n} i^5 = \dfrac{n^2(n + 1)^2(2n^2 + 2n - 1)}{12}$

17. $\displaystyle\sum_{i=1}^{n} i(i + 1) = \dfrac{n(n + 1)(n + 2)}{3}$

18. $\displaystyle\sum_{i=1}^{n} \dfrac{1}{(2i - 1)(2i + 1)} = \dfrac{n}{2n + 1}$

19. $\displaystyle\sum_{i=1}^{n} \dfrac{1}{i(i + 1)} = \dfrac{n}{n + 1}$

20. $\displaystyle\sum_{i=1}^{n} \dfrac{1}{i(i + 1)(i + 2)} = \dfrac{n(n + 3)}{4(n + 1)(n + 2)}$

In Exercises 21–24, find the sum using the formulas for the sums of powers of integers.

21. $\displaystyle\sum_{n=1}^{50} n^3$

22. $\displaystyle\sum_{n=1}^{10} n^4$

23. $\displaystyle\sum_{n=1}^{12} (n^2 - n)$

24. $\displaystyle\sum_{n=1}^{40} (n^3 - n)$

In Exercises 25–30, prove the inequality for the indicated integer values of n.

25. $n! > 2^n$, $n \geq 4$

26. $\left(\dfrac{4}{3}\right)^n > n$, $n \geq 7$

27. $\dfrac{1}{\sqrt{1}} + \dfrac{1}{\sqrt{2}} + \dfrac{1}{\sqrt{3}} + \cdots + \dfrac{1}{\sqrt{n}} > \sqrt{n}$, $n \geq 2$

28. $\left(\dfrac{x}{y}\right)^{n+1} < \left(\dfrac{x}{y}\right)^n$, $n \geq 1$ and $0 < x < y$

29. $(1 + a)^n \geq na$, $n \geq 1$ and $a > 1$

30. $3^n > n\,2^n$, $n \geq 1$

In Exercises 31–42, use mathematical induction to prove the property for all positive integers n.

31. $(ab)^n = a^n b^n$

32. $\left(\dfrac{a}{b}\right)^n = \dfrac{a^n}{b^n}$

33. If $x_1 \neq 0$, $x_2 \neq 0, \ldots, x_n \neq 0$, then
$(x_1 x_2 x_3 \cdots x_n)^{-1} = x_1^{-1} x_2^{-1} x_3^{-1} \cdots x_n^{-1}$.

34. If $x_1 > 0$, $x_2 > 0, \ldots, x_n > 0$, then
$\ln(x_1 x_2 \cdots x_n) = \ln x_1 + \ln x_2 + \cdots + \ln x_n$.

35. Generalized Distributive Law:
$x(y_1 + y_2 + \cdots + y_n) = xy_1 + xy_2 + \cdots + xy_n$

36. $(a + bi)^n$ and $(a - bi)^n$ are complex conjugates for all $n \geq 1$.

37. A factor of $(n^3 + 3n^2 + 2n)$ is 3.

38. A factor of $(n^3 + 5n + 6)$ is 3.

39. A factor of $(n^3 - n + 3)$ is 3.

40. A factor of $(n^4 - n + 4)$ is 2.

41. A factor of $(2^{2n+1} + 1)$ is 3.

42. A factor of $(2^{4n-2} + 1)$ is 5.

In Exercises 43–50, write the first five terms of the sequence beginning with the given term. Then calculate the first and second differences of the sequence. Does the sequence have a linear model, a quadratic model, or neither?

43. $a_1 = 0$

$a_n = a_{n-1} + 3$

44. $a_1 = 2$

$a_n = n - a_{n-1}$

45. $a_1 = 3$

$a_n = a_{n-1} - n$

46. $a_2 = -3$

$a_n = -2a_{n-1}$

47. $a_0 = 0$

$a_n = a_{n-1} + n$

48. $a_0 = 2$

$a_n = (a_{n-1})^2$

49. $a_1 = 2$

$a_n = a_{n-1} + 2$

50. $a_1 = 0$

$a_n = a_{n-1} + 2n$

In Exercises 51–54, find a quadratic model for the sequence with the indicated terms.

51. 3, 3, 5, 9, 15, 23, . . .

52. 7, 6, 7, 10, 15, 22, . . .

53. $a_0 = -3$, $a_2 = 1$, $a_4 = 9$

54. $a_0 = 3$, $a_2 = 0$, $a_6 = 36$

55. **Koch Snowflake**   A *Koch snowflake* is created by starting with an equilateral triangle with sides one unit in length. Then, on each side of the triangle, a new equilateral triangle is created on the middle third of that side. This process is repeated continuously, as shown in the figure.

(a) Determine a formula for the number of sides of the nth Koch snowflake. Use mathematical induction to prove your answer.

(b) Determine a formula for the area of the nth Koch snowflake. Recall that the area A of an equilateral triangle with side s is $A = (\sqrt{3}/4)s^2$.

(c) Determine a formula for the perimeter of the nth Koch snowflake.

56. **Tower of Hanoi**   The *Tower of Hanoi* puzzle is a game in which three pegs are attached to a board and one of the pegs has n disks sitting on it, as shown in the figure. Each disk on that peg must sit on a larger disk. The strategy of the game is to move the entire pile of disks, one at a time, to another peg. At no time may a disk sit on a smaller disk.

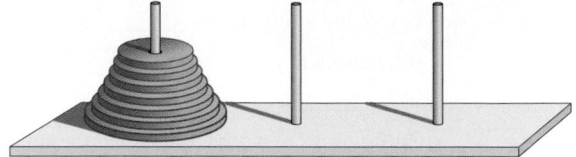

(a) Find the number of moves if there are three disks.

(b) Find the number of moves if there are four disks.

(c) Use your results from parts (a) and (b) to find a formula for the number of moves if there are n disks.

(d) Use mathematical induction to prove the formula you found in part (c).

Synthesis

True or False?   **In Exercises 57–59, determine whether the statement is true or false. Justify your answer.**

57. If the statement P_k is true and P_k implies P_{k+1}, then P_1 is also true.

58. If a sequence is arithmetic, then the first differences of the sequence are all zero.

59. A sequence with n terms has $n - 1$ second differences.

60. **Think About It**   What conclusion can be drawn from the given information about the sequence of statements P_n?

(a) P_3 is true and P_k implies P_{k+1}.

(b) $P_1, P_2, P_3, \ldots, P_{50}$ are all true.

(c) P_1, P_2, and P_3 are all true, but the truth of P_k does not imply that P_{k+1} is true.

(d) P_2 is true and P_{2k} implies P_{2k+2}.

Skills Review

In Exercises 61–64, find the product.

61. $(2x^2 - 1)^2$

62. $(2x - y)^2$

63. $(5 - 4x)^3$

64. $(2x - 4y)^3$

In Exercises 65–68, simplify the expression.

65. $3\sqrt{-27} - \sqrt{-12}$

66. $\sqrt[3]{125} + 4\sqrt[3]{-8} - 2\sqrt[3]{-54}$

67. $10\left(\sqrt[3]{64} - 2\sqrt[3]{-16}\right)$

68. $\left(-5 + \sqrt{-9}\right)^2$

8.5 The Binomial Theorem

Binomial Coefficients

Recall that a *binomial* is a polynomial that has two terms. In this section, you will study a formula that provides a quick method of raising a binomial to a power. To begin, look at the expansion of

$$(x + y)^n$$

for several values of n.

$$(x + y)^0 = 1$$
$$(x + y)^1 = x + y$$
$$(x + y)^2 = x^2 + 2xy + y^2$$
$$(x + y)^3 = x^3 + 3x^2y + 3xy^2 + y^3$$
$$(x + y)^4 = x^4 + 4x^3y + 6x^2y^2 + 4xy^3 + y^4$$
$$(x + y)^5 = x^5 + 5x^4y + 10x^3y^2 + 10x^2y^3 + 5xy^4 + y^5$$

There are several observations you can make about these expansions.

1. In each expansion, there are $n + 1$ terms.
2. In each expansion, x and y have symmetric roles. The powers of x decrease by 1 in successive terms, whereas the powers of y increase by 1.
3. The sum of the powers of each term is n. For instance, in the expansion of $(x + y)^5$, the sum of the powers of each term is 5.

$$4 + 1 = 5 \quad 3 + 2 = 5$$

$$(x + y)^5 = x^5 + 5x^4y^1 + 10x^3y^2 + 10x^2y^3 + 5x^1y^4 + y^5$$

4. The coefficients increase and then decrease in a symmetric pattern.

The coefficients of a binomial expansion are called **binomial coefficients.** To find them, you can use the **Binomial Theorem.**

> **The Binomial Theorem** (See the proof on page 658.)
>
> In the expansion of $(x + y)^n$
>
> $$(x + y)^n = x^n + nx^{n-1}y + \cdots + {}_nC_r\, x^{n-r}y^r + \cdots + nxy^{n-1} + y^n$$
>
> the coefficient of $x^{n-r}y^r$ is
>
> $${}_nC_r = \frac{n!}{(n-r)!r!}.$$
>
> The symbol $\binom{n}{r}$ is often used in place of ${}_nC_r$ to denote binomial coefficients.

Prerequisite Skills

Review the definition of factorial, $n!$, in Section 8.1.

Example 1 Finding Binomial Coefficients

Find each binomial coefficient.

a. $_8C_2$ **b.** $\binom{10}{3}$ **c.** $_7C_0$ **d.** $\binom{8}{8}$

Solution

a. $_8C_2 = \dfrac{8!}{6! \cdot 2!} = \dfrac{(8 \cdot 7) \cdot 6!}{6! \cdot 2!} = \dfrac{8 \cdot 7}{2 \cdot 1} = 28$

b. $\binom{10}{3} = \dfrac{10!}{7! \cdot 3!} = \dfrac{(10 \cdot 9 \cdot 8) \cdot 7!}{7! \cdot 3!} = \dfrac{10 \cdot 9 \cdot 8}{3 \cdot 2 \cdot 1} = 120$

c. $_7C_0 = \dfrac{7!}{7! \cdot 0!} = 1$

d. $\binom{8}{8} = \dfrac{8!}{0! \cdot 8!} = 1$

✓CHECKPOINT   Now try Exercise 1.

When $r \neq 0$ and $r \neq n$, as in parts (a) and (b) of Example 1, there is a simple pattern for evaluating binomial coefficients that works because there will always be factorial terms that divide out from the expression.

2 factors

$_8C_2 = \dfrac{8 \cdot 7}{2 \cdot 1}$ and $\binom{10}{3} = \dfrac{10 \cdot 9 \cdot 8}{3 \cdot 2 \cdot 1}$

2 factorial   3 factorial

3 factors

Example 2 Finding Binomial Coefficients

Find each binomial coefficient using the pattern shown above.

a. $_7C_3$ **b.** $_7C_4$ **c.** $_{12}C_1$ **d.** $_{12}C_{11}$

Solution

a. $_7C_3 = \dfrac{7 \cdot 6 \cdot 5}{3 \cdot 2 \cdot 1} = 35$ **b.** $_7C_4 = \dfrac{7 \cdot 6 \cdot 5 \cdot 4}{4 \cdot 3 \cdot 2 \cdot 1} = 35$

c. $_{12}C_1 = \dfrac{12}{1} = 12$

d. $_{12}C_{11} = \dfrac{12!}{1! \cdot 11!} = \dfrac{(12) \cdot 11!}{1! \cdot 11!} = \dfrac{12}{1} = 12$

✓CHECKPOINT   Now try Exercise 5.

It is not a coincidence that the results in parts (a) and (b) of Example 2 are the same and that the results in parts (c) and (d) are the same. In general, it is true that

$_nC_r = {}_nC_{n-r}.$

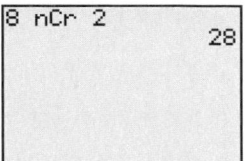
Exploration

Find each pair of binomial coefficients.

a. $_7C_0, {}_7C_7$ **d.** $_7C_1, {}_7C_6$
b. $_8C_0, {}_8C_8$ **e.** $_8C_1, {}_8C_7$
c. $_{10}C_0, {}_{10}C_{10}$ **f.** $_{10}C_1, {}_{10}C_9$

What do you observe about the pairs in (a), (b), and (c)? What do you observe about the pairs in (d), (e), and (f)? Write two conjectures from your observations. Develop a convincing argument for your two conjectures.

Binomial Expansions

As mentioned at the beginning of this section, when you write out the coefficients for a binomial that is raised to a power, you are **expanding a binomial.** The formulas for binomial coefficients give you an easy way to expand binomials, as demonstrated in the next four examples.

Example 3   Expanding a Binomial

Write the expansion of the expression $(x + 1)^3$.

Solution

The binomial coefficients are

$$_3C_0 = 1, \ _3C_1 = 3, \ _3C_2 = 3, \ \text{and} \ _3C_3 = 1.$$

Therefore, the expansion is as follows.

$$(x + 1)^3 = (1)x^3 + (3)x^2(1) + (3)x(1^2) + (1)(1^3)$$

$$= x^3 + 3x^2 + 3x + 1$$

✓**CHECKPOINT**   Now try Exercise 17.

To expand binomials representing *differences*, rather than sums, you alternate signs. Here is an example.

$$(x - 1)^3 = [x + (-1)]^3$$

$$= (1)x^3 + (3)x^2(-1) + (3)x(-1)^2 + (1)(-1)^3$$

$$= x^3 - 3x^2 + 3x - 1$$

Example 4   Expanding Binomial Expressions

Write the expansion of each expression.

a. $(2x - 3)^4$

b. $(x - 2y)^4$

Solution

The binomial coefficients are

$$_4C_0 = 1, \ _4C_1 = 4, \ _4C_2 = 6, \ _4C_3 = 4, \ \text{and} \ _4C_4 = 1.$$

Therefore, the expansions are as follows.

a. $(2x - 3)^4 = (1)(2x)^4 - (4)(2x)^3(3) + (6)(2x)^2(3^2) - (4)(2x)(3^3) + (1)(3^4)$

$$= 16x^4 - 96x^3 + 216x^2 - 216x + 81$$

b. $(x - 2y)^4 = (1)x^4 - (4)x^3(2y) + (6)x^2(2y)^2 - (4)x(2y)^3 + (1)(2y)^4$

$$= x^4 - 8x^3y + 24x^2y^2 - 32xy^3 + 16y^4$$

✓**CHECKPOINT**   Now try Exercise 29.

TECHNOLOGY TIP

You can use a graphing utility to check the expansion in Example 4(a) by graphing the original binomial expression and the expansion in the same viewing window. The graphs should coincide, as shown below.

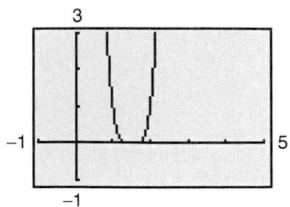

Example 5   Expanding a Binomial

Write the expansion of the expression $(x^2 + 4)^3$.

Solution

Expand using the binomial coefficients from Example 3.

$$(x^2 + 4)^3 = (1)(x^2)^3 + (3)(x^2)^2(4) + (3)x^2(4^2) + (1)(4^3)$$

$$= x^6 + 12x^4 + 48x^2 + 64$$

✓CHECKPOINT   Now try Exercise 31.

Sometimes you will need to find a specific term in a binomial expansion. Instead of writing out the entire expansion, you can use the fact that, from the Binomial Theorem, the $(r + 1)$st term is

$$_nC_r x^{n-r} y^r.$$

For example, if you wanted to find the third term of the expression in Example 5, you could use the formula above with $n = 3$ and $r = 2$ to obtain

$$_3C_2(x^2)^{3-2} \cdot 4^2 = 3(x^2) \cdot 16$$

$$= 48x^2.$$

Example 6   Finding a Term or Coefficient in a Binomial Expansion

a. Find the sixth term of $(a + 2b)^8$.

b. Find the coefficient of the term a^6b^5 in the expansion of $(2a - 5b)^{11}$.

Solution

a. To find the sixth term in this binomial expansion, use $n = 8$ and $r = 5$ [the formula is for the $(r + 1)$st term, so r is one less than the number of the term that you are looking for] to get

$$_8C_5 a^{8-5}(2b)^5 = 56 \cdot a^3 \cdot (2b)^5$$

$$= 56(2^5)a^3b^5$$

$$= 1792a^3b^5.$$

b. In this case, $n = 11$, $r = 5$, $x = 2a$, and $y = -5b$. Substitute these values to obtain

$$_nC_r x^{n-r} y^r = {}_{11}C_5(2a)^6(-5b)^5$$

$$= (462)(64a^6)(-3125b^5)$$

$$= -92,400,000a^6b^5.$$

So, the coefficient is $-92,400,000$.

✓CHECKPOINT   Now try Exercises 49 and 61.

Pascal's Triangle

There is a convenient way to remember the pattern for binomial coefficients. By arranging the coefficients in a triangular pattern, you obtain the following array, which is called **Pascal's Triangle.** This triangle is named after the famous French mathematician Blaise Pascal (1623–1662).

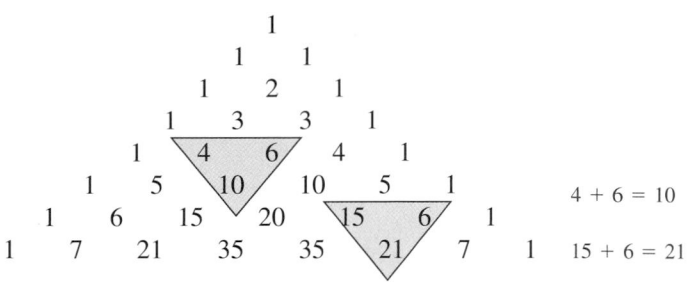

$$4 + 6 = 10$$

$$15 + 6 = 21$$

Exploration

Complete the table and describe the result.

n	r	$_nC_r$	$_nC_{n-r}$
9	5		
7	1		
12	4		
6	0		
10	7		

What characteristics of Pascal's Triangle are illustrated by this table?

The first and last number in each row of Pascal's Triangle is 1. Every other number in each row is formed by adding the two numbers immediately above the number. Pascal noticed that the numbers in this triangle are precisely the same numbers as the coefficients of binomial expansions, as follows.

$$(x + y)^0 = 1 \qquad \text{0th row}$$
$$(x + y)^1 = 1x + 1y \qquad \text{1st row}$$
$$(x + y)^2 = 1x^2 + 2xy + 1y^2 \qquad \text{2nd row}$$
$$(x + y)^3 = 1x^3 + 3x^2y + 3xy^2 + 1y^3 \qquad \text{3rd row}$$
$$(x + y)^4 = 1x^4 + 4x^3y + 6x^2y^2 + 4xy^3 + 1y^4 \qquad \vdots$$
$$(x + y)^5 = 1x^5 + 5x^4y + 10x^3y^2 + 10x^2y^3 + 5xy^4 + 1y^5$$
$$(x + y)^6 = 1x^6 + 6x^5y + 15x^4y^2 + 20x^3y^3 + 15x^2y^4 + 6xy^5 + 1y^6$$
$$(x + y)^7 = 1x^7 + 7x^6y + 21x^5y^2 + 35x^4y^3 + 35x^3y^4 + 21x^2y^5 + 7xy^6 + 1y^7$$

The top row of Pascal's Triangle is called the *zeroth row* because it corresponds to the binomial expansion $(x + y)^0 = 1$. Similarly, the next row is called the *first row* because it corresponds to the binomial expansion $(x + y)^1 = 1(x) + 1(y)$. In general, the *nth row* of Pascal's Triangle gives the coefficients of $(x + y)^n$.

Example 7   Using Pascal's Triangle

Use the seventh row of Pascal's Triangle to find the binomial coefficients.

$$_8C_0,\ _8C_1,\ _8C_2,\ _8C_3,\ _8C_4,\ _8C_5,\ _8C_6,\ _8C_7,\ _8C_8$$

Solution

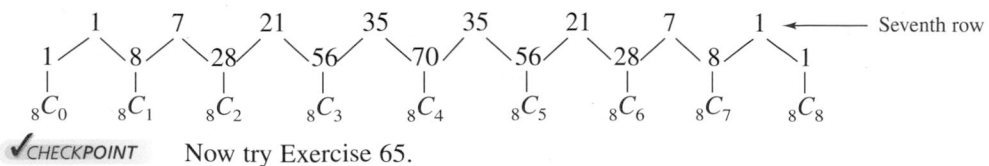

Seventh row

✓CHECKPOINT   Now try Exercise 65.

8.5   Exercises

See www.CalcChat.com for worked-out solutions to odd-numbered exercises.

Vocabulary Check

Fill in the blanks.

1. The coefficients of a binomial expansion are called _____ .

2. To find binomial coefficients you can use the _____ or _____ .

3. The notation used to denote a binomial coefficient is _____ or _____ .

4. When you write out the coefficients for a binomial that is raised to a power, you are _____ a _____ .

In Exercises 1–10, find the binomial coefficient.

1. $_7C_5$

2. $_9C_6$

3. $\binom{12}{0}$

4. $\binom{20}{20}$

5. $_{20}C_{15}$

6. $_{12}C_3$

7. $_{14}C_1$

8. $_{18}C_{17}$

9. $\binom{100}{98}$

10. $\binom{10}{7}$

In Exercises 11–16, use a graphing utility to find $_nC_r$.

11. $_{41}C_{36}$

12. $_{34}C_4$

13. $_{100}C_{98}$

14. $_{500}C_{498}$

15. $_{250}C_2$

16. $_{1000}C_2$

In Exercises 17–48, use the Binomial Theorem to expand and simplify the expression.

17. $(x + 2)^4$

18. $(x + 1)^6$

19. $(a + 3)^3$

20. $(a + 2)^4$

21. $(y - 2)^4$

22. $(y - 2)^5$

23. $(x + y)^5$

24. $(x + y)^6$

25. $(3r + 2s)^6$

26. $(4x + 3y)^4$

27. $(x - y)^5$

28. $(2x - y)^5$

29. $(1 - 4x)^3$

30. $(5 - 2y)^3$

31. $(x^2 + 2)^4$

32. $(3 - y^2)^3$

33. $(x^2 - 5)^5$

34. $(y^2 + 1)^6$

35. $(x^2 + y^2)^4$

36. $(x^2 + y^2)^6$

37. $(x^3 - y)^6$

38. $(2x^3 - y)^5$

39. $\left(\dfrac{1}{x} + y\right)^5$

40. $\left(\dfrac{1}{x} + 2y\right)^6$

41. $\left(\dfrac{2}{x} - y\right)^4$

42. $\left(\dfrac{2}{x} - 3y\right)^5$

43. $(4x - 1)^3 - 2(4x - 1)^4$

44. $(x + 3)^5 - 4(x + 3)^4$

45. $2(x - 3)^4 + 5(x - 3)^2$

46. $3(x + 1)^5 + 4(x + 1)^3$

47. $-3(x - 2)^3 - 4(x + 1)^6$

48. $5(x + 2)^5 - 2(x - 1)^2$

In Exercises 49–56, find the specified nth term in the expansion of the binomial.

49. $(x + 8)^{10}$, $n = 4$

50. $(x - 5)^6$, $n = 7$

51. $(x - 6y)^5$, $n = 3$

52. $(x - 10z)^7$, $n = 4$

53. $(4x + 3y)^9$, $n = 8$

54. $(5a + 6b)^5$, $n = 5$

55. $(10x - 3y)^{12}$, $n = 9$

56. $(7x + 2y)^{15}$, $n = 8$

In Exercises 57–64, find the coefficient a of the given term in the expansion of the binomial.

	Binomial	Term
57.	$(x + 3)^{12}$	ax^4
58.	$(x + 4)^{12}$	ax^5
59.	$(x - 2y)^{10}$	ax^8y^2
60.	$(4x - y)^{10}$	ax^2y^8
61.	$(3x - 2y)^9$	ax^6y^3
62.	$(2x - 3y)^8$	ax^4y^4
63.	$(x^2 + y)^{10}$	ax^8y^6
64.	$(z^2 - 1)^{12}$	az^6

In Exercises 65–68, use Pascal's Triangle to find the binomial coefficient.

65. $_7C_5$

66. $_6C_3$

67. $_6C_5$

68. $_5C_2$

In Exercises 69–72, expand the binomial by using Pascal's Triangle to determine the coefficients.

69. $(3t - 2v)^4$

70. $(5v - 2z)^4$

71. $(2x - 3y)^5$

72. $(5y + 2)^5$

In Exercises 73–76, use the Binomial Theorem to expand and simplify the expression.

73. $\left(\sqrt{x} + 5\right)^4$

74. $\left(4\sqrt{t} - 1\right)^3$

75. $(x^{2/3} - y^{1/3})^3$

76. $(u^{3/5} + v^{1/5})^5$

$\int$ **In Exercises 77–82, expand the expression in the difference quotient and simplify.**

$$\frac{f(x + h) - f(x)}{h}, \quad h \neq 0$$

77. $f(x) = x^3$

78. $f(x) = x^4$

79. $f(x) = x^6$

80. $f(x) = x^8$

81. $f(x) = \sqrt{x}$

82. $f(x) = \dfrac{1}{x}$

In Exercises 83–96, use the Binomial Theorem to expand the complex number. Simplify your result. (Remember that $i = \sqrt{-1}$.

83. $(1 + i)^4$

84. $(4 - i)^5$

85. $(4 + i)^4$

86. $(2 - i)^5$

87. $(2 - 3i)^6$

88. $(3 - 2i)^6$

89. $\left(5 + \sqrt{-16}\right)^3$

90. $\left(5 + \sqrt{-9}\right)^3$

91. $\left(4 + \sqrt{3}i\right)^4$

92. $\left(5 - \sqrt{3}i\right)^4$

93. $\left(-\dfrac{1}{2} + \dfrac{\sqrt{3}}{2}i\right)^3$

94. $\left(\dfrac{1}{2} - \dfrac{\sqrt{3}}{2}i\right)^3$

95. $\left(\dfrac{1}{4} - \dfrac{\sqrt{3}}{4}i\right)^3$

96. $\left(\dfrac{1}{3} - \dfrac{\sqrt{3}}{3}i\right)^3$

Approximation **In Exercises 97–100, use the Binomial Theorem to approximate the quantity accurate to three decimal places. For example, in Exercise 97, use the expansion**

$$(1.02)^8 = (1 + 0.02)^8 = 1 + 8(0.02) + 28(0.02)^2 + \cdots.$$

97. $(1.02)^8$

98. $(2.005)^{10}$

99. $(2.99)^{12}$

100. $(1.98)^9$

Graphical Reasoning **In Exercises 101 and 102, use a graphing utility to graph f and g in the same viewing window. What is the relationship between the two graphs? Use the Binomial Theorem to write the polynomial function g in standard form.**

101. $f(x) = x^3 - 4x, \quad g(x) = f(x + 3)$

102. $f(x) = -x^4 + 4x^2 - 1, \quad g(x) = f(x - 5)$

Graphical Reasoning **In Exercises 103 and 104, use a graphing utility to graph the functions in the given order and in the same viewing window. Compare the graphs. Which two functions have identical graphs, and why?**

103. (a) $f(x) = (1 - x)^3$

  (b) $g(x) = 1 - 3x$

  (c) $h(x) = 1 - 3x + 3x^2$

  (d) $p(x) = 1 - 3x + 3x^2 - x^3$

104. (a) $f(x) = \left(1 - \frac{1}{2}x\right)^4$

  (b) $g(x) = 1 - 2x + \frac{3}{2}x^2$

  (c) $h(x) = 1 - 2x + \frac{3}{2}x^2 - \frac{1}{2}x^3$

  (d) $p(x) = 1 - 2x + \frac{3}{2}x^2 - \frac{1}{2}x^3 + \frac{1}{16}x^4$

Probability **In Exercises 105–108, consider n independent trials of an experiment in which each trial has two possible outcomes, success or failure. The probability of a success on each trial is p and the probability of a failure is $q = 1 - p$. In this context, the term $_nC_k\, p^k q^{n-k}$ in the expansion of $(p + q)^n$ gives the probability of k successes in the n trials of the experiment.**

105. A fair coin is tossed seven times. To find the probability of obtaining four heads, evaluate the term

$$_7C_4\left(\tfrac{1}{2}\right)^4\left(\tfrac{1}{2}\right)^3$$

in the expansion of $\left(\tfrac{1}{2} + \tfrac{1}{2}\right)^7$.

106. The probability of a baseball player getting a hit during any given time at bat is $\frac{1}{4}$. To find the probability that the player gets three hits during the next 10 times at bat, evaluate the term

$$_{10}C_3\left(\tfrac{1}{4}\right)^3\left(\tfrac{3}{4}\right)^7$$

in the expansion of $\left(\tfrac{1}{4} + \tfrac{3}{4}\right)^{10}$.

107. The probability of a sales representative making a sale with any one customer is $\frac{1}{3}$. The sales representative makes eight contacts a day. To find the probability of making four sales, evaluate the term

$$_8C_4\left(\tfrac{1}{3}\right)^4\left(\tfrac{2}{3}\right)^4$$

in the expansion of $\left(\tfrac{1}{3} + \tfrac{2}{3}\right)^8$.

108. To find the probability that the sales representative in Exercise 107 makes four sales if the probability of a sale with any one customer is $\frac{1}{2}$, evaluate the term

$$_8C_4\left(\tfrac{1}{2}\right)^4\left(\tfrac{1}{2}\right)^4$$

in the expansion of $\left(\tfrac{1}{2} + \tfrac{1}{2}\right)^8$.

109. Health The death rates f (per 100,000 people) attributed to heart disease in the United States from 1985 to 2003 can be modeled by the equation

$$f(t) = 0.064t^2 - 9.30t + 416.5, \quad 5 \le t \le 23$$

where t represents the year, with $t = 5$ corresponding to 1985 (see figure). (Source: U.S. National Center for Health Statistics)

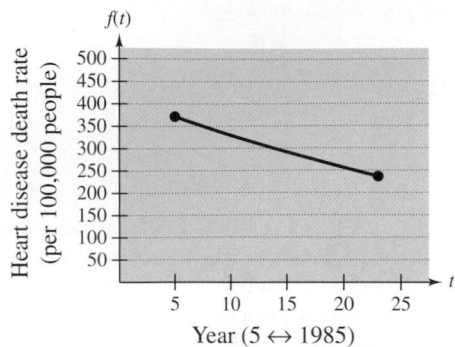

(a) Adjust the model so that $t = 0$ corresponds to 2000 rather than $t = 5$ corresponding to 1985. To do this, shift the graph of f 20 units to the left and obtain $g(t) = f(t + 20)$. Write $g(t)$ in standard form.

(b) Use a graphing utility to graph f and g in the same viewing window.

110. Education The average tuition, room, and board costs f (in dollars) for undergraduates at public institutions from 1985 through 2005 can be approximated by the model

$$f(t) = 6.22t^2 + 115.2t + 2730, \quad 5 \le t \le 25$$

where t represents the year, with $t = 5$ corresponding to 1985 (see figure). (Source: National Center for Education Statistics)

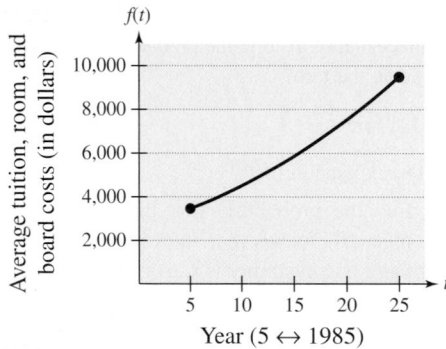

(a) Adjust the model so that $t = 0$ corresponds to 2000 rather than $t = 5$ corresponding to 1985. To do this, shift the graph of f 20 units to the left and obtain $g(t) = f(t + 20)$. Write $g(t)$ in standard form.

(b) Use a graphing utility to graph f and g in the same viewing window.

Synthesis

True or False? In Exercises 111 and 112, determine whether the statement is true or false. Justify your answer.

111. One of the terms in the expansion of $(x - 2y)^{12}$ is $7920x^4y^8$.

112. The x^{10}-term and the x^{14}-term in the expansion of $(x^2 + 3)^{12}$ have identical coefficients.

113. Writing In your own words, explain how to form the rows of Pascal's Triangle.

114. Form rows 8–10 of Pascal's Triangle.

115. Think About It How do the expansions of $(x + y)^n$ and $(x - y)^n$ differ?

116. Error Analysis You are a math instructor and receive the following solutions from one of your students on a quiz. Find the error(s) in each solution and write a short paragraph discussing ways that your student could avoid the error(s) in the future.

(a) Find the second term in the expansion of $(2x - 3y)^5$.

$$\cancel{5(2x)^4(3y)^2 = 720x^4y^2}$$

(b) Find the fourth term in the expansion of $\left(\frac{1}{2}x + 7y\right)^6$.

$$\cancel{{}_6C_4\left(\tfrac{1}{2}x\right)^2(7y)^4 = 9003.75x^2y^4}$$

Proof In Exercises 117–120, prove the property for all integers r and n, where $0 \le r \le n$.

117. ${}_nC_r = {}_nC_{n-r}$

118. ${}_nC_0 - {}_nC_1 + {}_nC_2 - \cdots \pm {}_nC_n = 0$

119. ${}_{n+1}C_r = {}_nC_r + {}_nC_{r-1}$

120. The sum of the numbers in the nth row of Pascal's Triangle is 2^n.

Skills Review

In Exercises 121–124, describe the relationship between the graphs of f and g.

121. $g(x) = f(x) + 8$

122. $g(x) = f(x - 3)$

123. $g(x) = f(-x)$

124. $g(x) = -f(x)$

In Exercises 125 and 126, find the inverse of the matrix.

125. $\begin{bmatrix} -6 & 5 \\ -5 & 4 \end{bmatrix}$

126. $\begin{bmatrix} 1.2 & -2.3 \\ -2 & 4 \end{bmatrix}$

8.6   Counting Principles

Simple Counting Problems

The last two sections of this chapter present a brief introduction to some of the basic counting principles and their application to probability. In the next section, you will see that much of probability has to do with counting the number of ways an event can occur.

Example 1   Selecting Pairs of Numbers at Random

Eight pieces of paper are numbered from 1 to 8 and placed in a box. One piece of paper is drawn from the box, its number is written down, and the piece of paper is *returned to the box*. Then, a second piece of paper is drawn from the box, and its number is written down. Finally, the two numbers are added together. In how many different ways can a sum of 12 be obtained?

Solution

To solve this problem, count the number of different ways that a sum of 12 can be obtained using two numbers from 1 to 8.

First number	4	5	6	7	8
Second number	8	7	6	5	4

From this list, you can see that a sum of 12 can occur in five different ways.

✔CHECKPOINT   Now try Exercise 7.

Example 2   Selecting Pairs of Numbers at Random

Eight pieces of paper are numbered from 1 to 8 and placed in a box. Two pieces of paper are drawn from the box *at the same time*, and the numbers on the pieces of paper are written down and totaled. In how many different ways can a sum of 12 be obtained?

Solution

To solve this problem, count the number of different ways that a sum of 12 can be obtained using two *different* numbers from 1 to 8.

First number	4	5	7	8
Second number	8	7	5	4

So, a sum of 12 can be obtained in four different ways.

✔CHECKPOINT   Now try Exercise 8.

STUDY TIP

The difference between the counting problems in Examples 1 and 2 can be described by saying that the random selection in Example 1 occurs **with replacement,** whereas the random selection in Example 2 occurs **without replacement,** which eliminates the possibility of choosing two 6's.

The Fundamental Counting Principle

Examples 1 and 2 describe simple counting problems in which you can *list* each possible way that an event can occur. When it is possible, this is always the best way to solve a counting problem. However, some events can occur in so many different ways that it is not feasible to write out the entire list. In such cases, you must rely on formulas and counting principles. The most important of these is the **Fundamental Counting Principle.**

> **Fundamental Counting Principle**
>
> Let E_1 and E_2 be two events. The first event E_1 can occur in m_1 different ways. After E_1 has occurred, E_2 can occur in m_2 different ways. The number of ways that the two events can occur is $m_1 \cdot m_2$.

Example 3   Using the Fundamental Counting Principle

How many different pairs of letters from the English alphabet are possible?

Solution

There are two events in this situation. The first event is the choice of the first letter, and the second event is the choice of the second letter. Because the English alphabet contains 26 letters, it follows that the number of two-letter pairs is

$$26 \cdot 26 = 676.$$

   Now try Exercise 9.

Example 4   Using the Fundamental Counting Principle

Telephone numbers in the United States currently have 10 digits. The first three are the *area code* and the next seven are the *local telephone number*. How many different telephone numbers are possible within each area code? (Note that at this time, a local telephone number cannot begin with 0 or 1.)

Solution

Because the first digit cannot be 0 or 1, there are only eight choices for the first digit. For each of the other six digits, there are 10 choices.

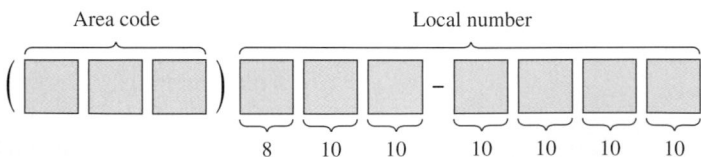

So, the number of local telephone numbers that are possible within each area code is $8 \cdot 10 \cdot 10 \cdot 10 \cdot 10 \cdot 10 \cdot 10 = 8,000,000$.

   Now try Exercise 15.

Permutations

One important application of the Fundamental Counting Principle is in determining the number of ways that n elements can be arranged (in order). An ordering of n elements is called a **permutation** of the elements.

> **Definition of Permutation**
>
> A **permutation** of n different elements is an ordering of the elements such that one element is first, one is second, one is third, and so on.

Example 5   Finding the Number of Permutations of n Elements

How many permutations are possible of the letters A, B, C, D, E, and F?

Solution

Consider the following reasoning.

First position:	Any of the *six* letters
Second position:	Any of the remaining *five* letters
Third position:	Any of the remaining *four* letters
Fourth position:	Any of the remaining *three* letters
Fifth position:	Any of the remaining *two* letters
Sixth position:	The *one* remaining letter

So, the numbers of choices for the six positions are as follows.

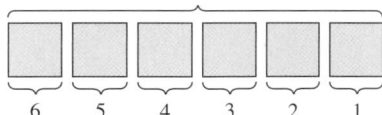

Permutations of six letters

6   5   4   3   2   1

The total number of permutations of the six letters is

$$6! = 6 \cdot 5 \cdot 4 \cdot 3 \cdot 2 \cdot 1 = 720.$$

✓CHECKPOINT   Now try Exercise 33.

> **Number of Permutations of n Elements**
>
> The number of permutations of n elements is given by
>
> $$n \cdot (n - 1) \cdots 4 \cdot 3 \cdot 2 \cdot 1 = n!.$$
>
> In other words, there are $n!$ different ways that n elements can be ordered.

It is useful, on occasion, to order a *subset* of a collection of elements rather than the entire collection. For example, you might want to choose and order r elements out of a collection of n elements. Such an ordering is called a **permutation of n elements taken r at a time.**

Example 6   Counting Horse Race Finishes

Eight horses are running in a race. In how many different ways can these horses come in first, second, and third? (Assume that there are no ties.)

Solution

Here are the different possibilities.

Win (first position):	*Eight* choices
Place (second position):	*Seven* choices
Show (third position):	*Six* choices

The numbers of choices for the three positions are as follows.

Different orders of horses

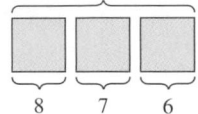

8   7   6

So, using the Fundamental Counting Principle, you can determine that there are

$$8 \cdot 7 \cdot 6 = 336$$

different ways in which the eight horses can come in first, second, and third.

✓CHECKPOINT   Now try Exercise 37.

Permutations of *n* Elements Taken *r* at a Time

The number of **permutations of *n* elements taken *r* at a time** is given by

$$_nP_r = \frac{n!}{(n-r)!}$$

$$= n(n-1)(n-2)\cdots(n-r+1).$$

Using this formula, you can rework Example 6 to find that the number of permutations of eight horses taken three at a time is

$$_8P_3 = \frac{8!}{5!}$$

$$= \frac{8 \cdot 7 \cdot 6 \cdot \cancel{5!}}{\cancel{5!}}$$

$$= 336$$

which is the same answer obtained in the example.

TECHNOLOGY TIP   Most graphing utilities are programmed to evaluate $_nP_r$. Figure 8.16 shows how one graphing utility evaluates the permutation $_8P_3$. For instructions on how to use the $_nP_r$ feature, see Appendix A; for specific keystrokes, go to this textbook's *Online Study Center*.

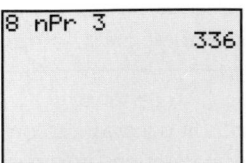

Figure 8.16

Remember that for permutations, order is important. So, if you are looking at the possible permutations of the letters A, B, C, and D taken three at a time, the permutations (A, B, D) and (B, A, D) would be counted as different because the *order* of the elements is different.

Suppose, however, that you are asked to find the possible permutations of the letters A, A, B, and C. The total number of permutations of the four letters would be $_4P_4 = 4!$. However, not all of these arrangements would be *distinguishable* because there are two A's in the list. To find the number of distinguishable permutations, you can use the following formula.

Distinguishable Permutations

Suppose a set of n objects has n_1 of one kind of object, n_2 of a second kind, n_3 of a third kind, and so on, with

$$n = n_1 + n_2 + n_3 + \cdots + n_k.$$

The number of **distinguishable permutations** of the n objects is given by

$$\frac{n!}{n_1! \cdot n_2! \cdot n_3! \cdots n_k!}.$$

Example 7   Distinguishable Permutations

In how many distinguishable ways can the letters in BANANA be written?

Solution

This word has six letters, of which three are A's, two are N's, and one is a B. So, the number of distinguishable ways in which the letters can be written is

$$\frac{6!}{3! \cdot 2! \cdot 1!} = \frac{6 \cdot 5 \cdot 4 \cdot 3!}{3! \cdot 2!} = 60.$$

The 60 different distinguishable permutations are as follows.

AAABNN	AAANBN	AAANNB	AABANN
AABNAN	AABNNA	AANABN	AANANB
AANBAN	AANBNA	AANNAB	AANNBA
ABAANN	ABANAN	ABANNA	ABNAAN
ABNANA	ABNNAA	ANAABN	ANAANB
ANABAN	ANABNA	ANANAB	ANANBA
ANBAAN	ANBANA	ANBNAA	ANNAAB
ANNABA	ANNBAA	BAAANN	BAANAN
BAANNA	BANAAN	BANANA	BANNAA
BNAAAN	BNAANA	BNANAA	BNNAAA
NAAABN	NAAANB	NAABAN	NAABNA
NAANAB	NAANBA	NABAAN	NABANA
NABNAA	NANAAB	NANABA	NANBAA
NBAAAN	NBAANA	NBANAA	NBNAAA
NNAAAB	NNAABA	NNABAA	NNBAAA

✓CHECKPOINT   Now try Exercise 45.

Combinations

When you count the number of possible permutations of a set of elements, order is important. As a final topic in this section, you will look at a method for selecting subsets of a larger set in which order *is not* important. Such subsets are called **combinations of n elements taken r at a time.** For instance, the combinations

    {A, B, C}   and   {B, A, C}

are equivalent because both sets contain the same three elements, and the order in which the elements are listed is not important. So, you would count only one of the two sets. A common example of a combination is a card game in which the player is free to reorder the cards after they have been dealt.

Example 8   Combinations of n Elements Taken r at a Time

In how many different ways can three letters be chosen from the letters A, B, C, D, and E? (The order of the three letters is not important.)

Solution

The following subsets represent the different combinations of three letters that can be chosen from five letters.

    {A, B, C}   {A, B, D}

    {A, B, E}   {A, C, D}

    {A, C, E}   {A, D, E}

    {B, C, D}   {B, C, E}

    {B, D, E}   {C, D, E}

From this list, you can conclude that there are 10 different ways in which three letters can be chosen from five letters.

✓CHECKPOINT   Now try Exercise 57.

> **Combinations of n Elements Taken r at a Time**
>
> The number of **combinations of n elements taken r at a time** is given by
>
> $$_nC_r = \frac{n!}{(n-r)!r!}.$$

Note that the formula for $_nC_r$ is the same one given for binomial coefficients. To see how this formula is used, solve the counting problem in Example 8. In that problem, you are asked to find the number of combinations of five elements taken three at a time. So, $n = 5$, $r = 3$, and the number of combinations is

$$_5C_3 = \frac{5!}{2!3!} = \frac{5 \cdot \overset{2}{\cancel{4}} \cdot \cancel{3!}}{\cancel{2} \cdot 1 \cdot \cancel{3!}} = 10$$

which is the same answer obtained in Example 8.

STUDY TIP

When solving problems involving counting principles, you need to be able to distinguish among the various counting principles in order to determine which is necessary to solve the problem correctly. To do this, ask yourself the following questions.

1. Is the order of the elements important? *Permutation*

2. Are the chosen elements a subset of a larger set in which order is not important? *Combination*

3. Does the problem involve two or more separate events? *Fundamental Counting Principle*

Example 9   Counting Card Hands

A standard poker hand consists of five cards dealt from a deck of 52. How many different poker hands are possible? (After the cards are dealt, the player may reorder them, so order is not important.)

Solution

You can find the number of different poker hands by using the formula for the number of combinations of 52 elements taken five at a time, as follows.

$$_{52}C_5 = \frac{52!}{47!5!}$$

$$= \frac{52 \cdot 51 \cdot 50 \cdot 49 \cdot 48 \cdot 47!}{5 \cdot 4 \cdot 3 \cdot 2 \cdot 1 \cdot 47!}$$

$$= 2,598,960$$

✔CHECKPOINT   Now try Exercise 59.

Example 10   Forming a Team

You are forming a 12-member swim team from 10 girls and 15 boys. The team must consist of five girls and seven boys. How many different 12-member teams are possible?

Solution

There are $_{10}C_5$ ways of choosing five girls. There are $_{15}C_7$ ways of choosing seven boys. By the Fundamental Counting Principle, there are $_{10}C_5 \cdot {_{15}C_7}$ ways of choosing five girls and seven boys.

$$_{10}C_5 \cdot {_{15}C_7} = \frac{10!}{5! \cdot 5!} \cdot \frac{15!}{8! \cdot 7!}$$

$$= 252 \cdot 6435$$

$$= 1,621,620$$

So, there are 1,621,620 12-member swim teams possible. You can verify this by using the $_nC_r$ feature of a graphing utility, as shown in Figure 8.17.

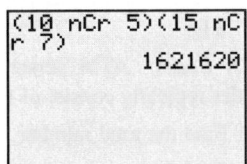

Figure 8.17

✔CHECKPOINT   Now try Exercise 69.

8.6   Exercises

See www.CalcChat.com for worked-out solutions to odd-numbered exercises.

Vocabulary Check

Fill in the blanks.

1. The _____ states that if there are m_1 ways for one event to occur and m_2 ways for a second event to occur, then there are $m_1 \cdot m_2$ ways for both events to occur.

2. An ordering of n elements is called a _____ of the elements.

3. The number of permutations of n elements taken r at a time is given by the formula _____ .

4. The number of _____ of n objects is given by $\dfrac{n!}{n_1! \cdot n_2! \cdot n_3! \cdot \cdots \cdot n_k!}$.

5. When selecting subsets of a larger set in which order is not important, you are finding the number of _____ of n elements taken r at a time.

Random Selection   In Exercises 1–8, determine the number of ways in which a computer can randomly generate one or more such integers, or pairs of integers, from 1 through 12.

1. An odd integer

2. An even integer

3. A prime integer

4. An integer that is greater than 6

5. An integer that is divisible by 4

6. An integer that is divisible by 3

7. A pair of integers whose sum is 8

8. A pair of distinct integers whose sum is 8

9. **Consumer Awareness**   A customer can choose one of four amplifiers, one of six compact disc players, and one of five speaker models for an entertainment system. Determine the number of possible system configurations.

10. **Course Schedule**   A college student is preparing a course schedule for the next semester. The student must select one of two mathematics courses, one of three science courses, and one of five courses from the social sciences and humanities. How many schedules are possible?

11. **True-False Exam**   In how many ways can a 10-question true-false exam be answered? (Assume that no questions are omitted.)

12. **Attaché Case**   An attaché case has two locks, each of which is a three-digit number sequence where digits may be repeated. Find the total number of combinations of the two locks in order to open the attaché case.

13. **Three-Digit Numbers**   How many three-digit numbers can be formed under each condition?

    (a) The leading digit cannot be a 0.

    (b) The leading digit cannot be a 0 and no repetition of digits is allowed.

14. **Four-Digit Numbers**   How many four-digit numbers can be formed under each condition?

    (a) The leading digit cannot be a 0 and the number must be less than 5000.

    (b) The leading digit cannot be a 0 and the number must be even.

15. **Telephone Numbers**   In 2006, the state of Nevada had two area codes. Using the information about telephone numbers given in Example 4, how many telephone numbers could Nevada's phone system have accommodated?

16. **Telephone Numbers**   In 2006, the state of Kansas had four area codes. Using the information about telephone numbers given in Example 4, how many telephone numbers could Kansas's phone system have accommodated?

17. **Radio Stations**   Typically radio stations are identified by four "call letters." Radio stations east of the Mississippi River have call letters that start with the letter W and radio stations west of the Mississippi River have call letters that start with the letter K.

    (a) Find the number of different sets of radio station call letters that are possible in the United States.

    (b) Find the number of different sets of radio station call letters that are possible if the call letters must include a Q.

18. **PIN Codes**   ATM personal identification number (PIN) codes typically consist of four-digit sequences of numbers.

    (a) Find the total number of ATM codes possible.

    (b) Find the total number of ATM codes possible if the first digit is not a 0.

19. **ZIP Codes**   In 1963, the United States Postal Service launched the Zoning Improvement Plan (ZIP) Code to streamline the mail-delivery system. A ZIP code consists of a five-digit sequence of numbers.

(a) Find the number of ZIP codes consisting of five digits.

(b) Find the number of ZIP codes consisting of five digits, if the first digit is a 1 or a 2.

20. *ZIP Codes* In 1983, in order to identify small geographic segments, such as city blocks or a single building, within a delivery code, the post office began to use an expanded ZIP code called ZIP+4, which is composed of the original five-digit code plus a four-digit add-on code.

(a) Find the number of ZIP codes consisting of five digits followed by the four additional digits.

(b) Find the number of ZIP codes consisting of five digits followed by the four additional digits, if the first number of the five-digit code is a 1 or a 2.

21. *Entertainment* Three couples have reserved seats in a row for a concert. In how many different ways can they be seated if

(a) there are no seating restrictions?

(b) the two members of each couple wish to sit together?

22. *Single File* In how many orders can five girls and three boys walk through a doorway single file if

(a) there are no restrictions?

(b) the girls walk through before the boys?

In Exercises 23–28, evaluate $_nP_r$ using the formula from this section.

23. $_4P_4$

24. $_5P_5$

25. $_8P_3$

26. $_{20}P_2$

27. $_5P_4$

28. $_7P_4$

In Exercises 29–32, evaluate $_nP_r$ using a graphing utility.

29. $_{20}P_6$

30. $_{10}P_8$

31. $_{120}P_4$

32. $_{100}P_5$

33. *Posing for a Photograph* In how many ways can five children posing for a photograph line up in a row?

34. *Riding in a Car* In how many ways can four people sit in a four-passenger car?

35. *U.S. Supreme Court* The nine justices of the U.S. Supreme Court pose for a photograph while standing in a straight line, as opposed to the typical pose of two rows. How many different orders of the justices are possible for this photograph?

36. *Manufacturing* Four processes are involved in assembling a product, and they can be performed in any order. The management wants to test each order to determine which is the least time-consuming. How many different orders will have to be tested?

37. *Choosing Officers* From a pool of 12 candidates, the offices of president, vice-president, secretary, and treasurer will be filled. In how many ways can the offices be filled?

38. *Batting Order* How many different batting orders can a baseball coach create from a team of 15 players, if there are nine positions to fill?

39. *School Locker* A school locker has a dial lock on which there are 37 numbers from 0 to 36. Find the total number of possible lock combinations, if the lock requires a three-digit sequence of left-right-left and the numbers can be repeated.

40. *Sports* Eight sprinters have qualified for the finals in the 100-meter dash at the NCAA national track meet. How many different orders of the top three finishes are possible? (Assume there are no ties.)

In Exercises 41 and 42, use the letters A, B, C, and D.

41. Write all permutations of the letters.

42. Write all permutations of the letters if the letters B and C must remain between the letters A and D.

In Exercises 43–46, find the number of distinguishable permutations of the group of letters.

43. A, A, G, E, E, E, M

44. B, B, B, T, T, T, T, T

45. A, L, G, E, B, R, A

46. M, I, S, S, I, S, S, I, P, P, I

In Exercises 47–52, evaluate $_nC_r$ using the formula from this section.

47. $_5C_2$

48. $_6C_3$

49. $_4C_1$

50. $_5C_1$

51. $_{25}C_0$

52. $_{20}C_0$

In Exercises 53–56, evaluate $_nC_r$ using a graphing utility.

53. $_{20}C_4$

54. $_{10}C_7$

55. $_{42}C_5$

56. $_{50}C_6$

In Exercises 57 and 58, use the letters A, B, C, D, E, and F.

57. Write all possible selections of two letters that can be formed from the letters. (The order of the two letters is not important.)

58. Write all possible selections of three letters that can be formed from the letters. (The order of the three letters is not important.)

59. *Forming a Committee* As of June 2006, the U.S. Senate Committee on Indian Affairs had 14 members. If party affiliation is not a factor in selection, how many different committees are possible from the 100 U.S. senators?

60. *Exam Questions* You can answer any 12 questions from a total of 14 questions on an exam. In how many different ways can you select the questions?

61. *Lottery* In Washington's Lotto game, a player chooses six distinct numbers from 1 to 49. In how many ways can a player select the six numbers?

62. *Lottery* Powerball is played with 55 white balls, numbered 1 through 55, and 42 red balls, numbered 1 through 42. Five white balls and one red ball, the Powerball, are drawn. In how many ways can a player select the six numbers?

63. *Geometry* Three points that are not collinear determine three lines. How many lines are determined by nine points, no three of which are collinear?

64. *Defective Units* A shipment of 25 television sets contains three defective units. In how many ways can a vending company purchase four of these units and receive (a) all good units, (b) two good units, and (c) at least two good units?

65. *Poker Hand* You are dealt five cards from an ordinary deck of 52 playing cards. In how many ways can you get a full house? (A full house consists of three of one kind and two of another. For example, 8-8-8-5-5 and K-K-K-10-10 are full houses.)

66. *Card Hand* Five cards are chosen from a standard deck of 52 cards. How many five-card combinations contain two jacks and three aces?

67. *Job Applicants* A clothing manufacturer interviews 12 people for four openings in the human resources department of the company. Five of the 12 people are women. If all 12 are qualified, in how many ways can the employer fill the four positions if (a) the selection is random and (b) exactly two women are selected?

68. *Job Applicants* A law office interviews paralegals for 10 openings. There are 13 paralegals with two years of experience and 20 paralegals with one year of experience. How many combinations of seven paralegals with two years of experience and three paralegals with one year of experience are possible?

69. *Forming a Committee* A six-member research committee is to be formed having one administrator, three faculty members, and two students. There are seven administrators, 12 faculty members, and 20 students in contention for the committee. How many six-member committees are possible?

70. *Interpersonal Relationships* The number of possible interpersonal relationships increases dramatically as the size of a group increases. Determine the number of different two-person relationships that are possible in a group of people of size (a) 3, (b) 8, (c) 12, and (d) 20.

Geometry **In Exercises 71–74, find the number of diagonals of the polygon. (A line segment connecting any two nonadjacent vertices is called a *diagonal* of a polygon.)**

71. Pentagon

72. Hexagon

73. Octagon

74. Decagon

In Exercises 75–82, solve for n.

75. $14 \cdot {}_nP_3 = {}_{n+2}P_4$

76. ${}_nP_5 = 18 \cdot {}_{n-2}P_4$

77. ${}_nP_4 = 10 \cdot {}_{n-1}P_3$

78. ${}_nP_6 = 12 \cdot {}_{n-1}P_5$

79. ${}_{n+1}P_3 = 4 \cdot {}_nP_2$

80. ${}_{n+2}P_3 = 6 \cdot {}_{n+2}P_1$

81. $4 \cdot {}_{n+1}P_2 = {}_{n+2}P_3$

82. $5 \cdot {}_{n-1}P_1 = {}_nP_2$

Synthesis

True or False? **In Exercises 83 and 84, determine whether the statement is true or false. Justify your answer.**

83. The number of pairs of letters that can be formed from any of the first 13 letters in the alphabet (A–M), where repetitions are allowed, is an example of a permutation.

84. The number of permutations of n elements can be derived by using the Fundamental Counting Principle.

85. *Think About It* Can your calculator evaluate ${}_{100}P_{80}$? If not, explain why.

86. *Writing* Explain in your own words the meaning of ${}_nP_r$.

87. What is the relationship between ${}_nC_r$ and ${}_nC_{n-r}$?

88. Without calculating the numbers, determine which of the following is greater. Explain.

(a) The number of combinations of 10 elements taken six at a time

(b) The number of permutations of 10 elements taken six at a time

Proof **In Exercises 89–92, prove the identity.**

89. ${}_nP_{n-1} = {}_nP_n$

90. ${}_nC_n = {}_nC_0$

91. ${}_nC_{n-1} = {}_nC_1$

92. ${}_nC_r = \dfrac{{}_nP_r}{r!}$

Skills Review

In Exercises 93–96, solve the equation. Round your answer to three decimal places, if necessary.

93. $\sqrt{x-3} = x - 6$

94. $\dfrac{4}{t} + \dfrac{3}{2t} = 1$

95. $\log_2(x-3) = 5$

96. $e^{x/3} = 16$

In Exercises 97–100, use Cramer's Rule to solve the system of equations.

97. $\begin{cases} -5x + 3y = -14 \\ 7x - 2y = 2 \end{cases}$

98. $\begin{cases} 8x + y = 35 \\ 6x + 2y = 10 \end{cases}$

99. $\begin{cases} -3x - 4y = -1 \\ 9x + 5y = -4 \end{cases}$

100. $\begin{cases} 10x - 11y = -74 \\ -8x - 4y = 8 \end{cases}$

8.7 Probability

The Probability of an Event

Any happening whose result is uncertain is called an **experiment.** The possible results of the experiment are **outcomes,** the set of all possible outcomes of the experiment is the **sample space** of the experiment, and any subcollection of a sample space is an **event.**

For instance, when a six-sided die is tossed, the sample space can be represented by the numbers 1 through 6. For the experiment to be fair, each of the outcomes is *equally likely*.

To describe a sample space in such a way that each outcome is equally likely, you must sometimes distinguish between or among various outcomes in ways that appear artificial. Example 1 illustrates such a situation.

Example 1   Finding the Sample Space

Find the sample space for each of the following.

a. One coin is tossed.

b. Two coins are tossed.

c. Three coins are tossed.

Solution

a. Because the coin will land either heads up (denoted by H) or tails up (denoted by T), the sample space S is

$$S = \{H, T\}.$$

b. Because either coin can land heads up or tails up, the possible outcomes are as follows.

HH = heads up on both coins

HT = heads up on first coin and tails up on second coin

TH = tails up on first coin and heads up on second coin

TT = tails up on both coins

So, the sample space is

$$S = \{HH, HT, TH, TT\}.$$

Note that this list distinguishes between the two cases HT and TH, even though these two outcomes appear to be similar.

c. Following the notation of part (b), the sample space is

$$S = \{HHH, HHT, HTH, HTT, THH, THT, TTH, TTT\}.$$

Note that this list distinguishes among the cases HHT, HTH, and THH, and among the cases HTT, THT, and TTH.

✓CHECKPOINT   Now try Exercise 1.

To calculate the probability of an event, count the number of outcomes in the event and in the sample space. The *number of equally likely outcomes* in event E is denoted by $n(E)$, and the number of equally likely outcomes in the sample space S is denoted by $n(S)$. The probability that event E will occur is given by $n(E)/n(S)$.

The Probability of an Event

If an event E has $n(E)$ equally likely outcomes and its sample space S has $n(S)$ equally likely outcomes, the **probability** of event E is given by

$$P(E) = \frac{n(E)}{n(S)}.$$

Because the number of outcomes in an event must be less than or equal to the number of outcomes in the sample space, the probability of an event must be a number from 0 to 1, inclusive. That is,

$$0 \leq P(E) \leq 1$$

as indicated in Figure 8.18. If $P(E) = 0$, event E *cannot occur*, and E is called an **impossible event.** If $P(E) = 1$, event E *must occur*, and E is called a **certain event.**

Example 2 Finding the Probability of an Event

a. Two coins are tossed. What is the probability that both land heads up?

b. A card is drawn from a standard deck of playing cards. What is the probability that it is an ace?

Solution

a. Following the procedure in Example 1(b), let

$$E = \{HH\}$$

and

$$S = \{HH, HT, TH, TT\}.$$

The probability of getting two heads is

$$P(E) = \frac{n(E)}{n(S)} = \frac{1}{4}.$$

b. Because there are 52 cards in a standard deck of playing cards and there are four aces (one of each suit), the probability of drawing an ace is

$$P(E) = \frac{n(E)}{n(S)}$$

$$= \frac{4}{52} = \frac{1}{13}.$$

✓**CHECKPOINT**   Now try Exercise 7.

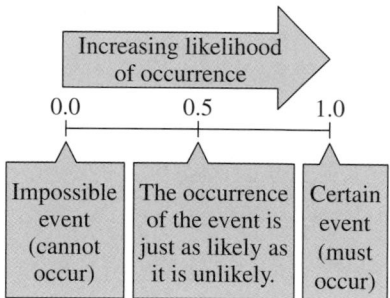

Figure 8.18

STUDY TIP

You can write a probability as a fraction, a decimal, or a percent. For instance, in Example 2(a), the probability of getting two heads can be written as $\frac{1}{4}$, 0.25, or 25%.

Example 3   Finding the Probability of an Event

Two six-sided dice are tossed. What is the probability that a total of 7 is rolled? (See Figure 8.19.)

Solution

Because there are six possible outcomes on each die, you can use the Fundamental Counting Principle to conclude that there are $6 \cdot 6 = 36$ different outcomes when two dice are tossed. To find the probability of rolling a total of 7, you must first count the number of ways this can occur.

Figure 8.19

First die	1	2	3	4	5	6
Second die	6	5	4	3	2	1

So, a total of 7 can be rolled in six ways, which means that the probability of rolling a 7 is

$$P(E) = \frac{n(E)}{n(S)} = \frac{6}{36} = \frac{1}{6}.$$

✓CHECKPOINT   Now try Exercise 15.

You could have written out each sample space in Examples 2 and 3 and simply counted the outcomes in the desired events. For larger sample spaces, however, using the counting principles discussed in Section 8.6 should save you time.

Example 4   Finding the Probability of an Event

Twelve-sided dice, as shown in Figure 8.20, can be constructed (in the shape of regular dodecahedrons) such that each of the numbers from 1 to 6 appears twice on each die. Show that these dice can be used in any game requiring ordinary six-sided dice without changing the probabilities of different outcomes.

Solution

For an ordinary six-sided die, each of the numbers 1, 2, 3, 4, 5, and 6 occurs only once, so the probability of any particular number coming up is

$$P(E) = \frac{n(E)}{n(S)} = \frac{1}{6}.$$

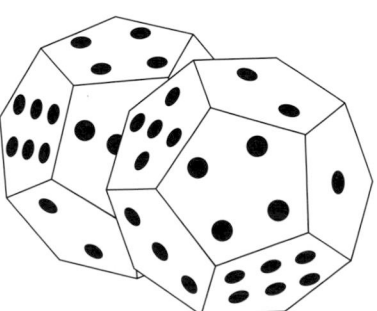

Figure 8.20

For a 12-sided die, each number occurs twice, so the probability of any particular number coming up is

$$P(E) = \frac{n(E)}{n(S)} = \frac{2}{12} = \frac{1}{6}.$$

✓CHECKPOINT   Now try Exercise 17.

Example 5   The Probability of Winning a Lottery

In Delaware's Multi-Win Lotto game, a player chooses six different numbers from 1 to 35. If these six numbers match the six numbers drawn (in any order) by the lottery commission, the player wins (or shares) the top prize. What is the probability of winning the top prize if the player buys one ticket?

Solution

To find the number of elements in the sample space, use the formula for the number of combinations of 35 elements taken six at a time.

$$n(S) = {}_{35}C_6$$

$$= \frac{35 \cdot 34 \cdot 33 \cdot 32 \cdot 31 \cdot 30}{6 \cdot 5 \cdot 4 \cdot 3 \cdot 2 \cdot 1} = 1{,}623{,}160$$

> **Prerequisite Skills**
>
> Review combinations of *n* elements taken *r* at a time in Section 8.6, if you have difficulty with this example.

If a person buys only one ticket, the probability of winning is

$$P(E) = \frac{n(E)}{n(S)} = \frac{1}{1{,}623{,}160}.$$

✓ *CHECKPOINT*   Now try Exercise 19.

Example 6   Random Selection

The numbers of colleges and universities in various regions of the United States in 2004 are shown in Figure 8.21. One institution is selected at random. What is the probability that the institution is in one of the three southern regions? (Source: U.S. National Center for Education Statistics)

Solution

From the figure, the total number of colleges and universities is 4231. Because there are $398 + 287 + 720 = 1405$ colleges and universities in the three southern regions, the probability that the institution is in one of these regions is

$$P(E) = \frac{n(E)}{n(S)} = \frac{1405}{4231} \approx 0.33.$$

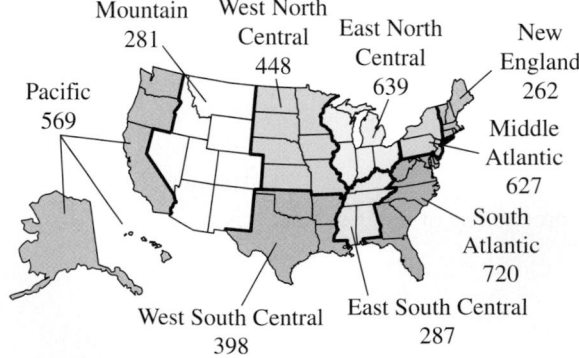

Mountain
281

West North
Central
448

East North
Central
639

New
England
262

Pacific
569

Middle
Atlantic
627

South
Atlantic
720

West South Central
398

East South Central
287

Figure 8.21

✓ *CHECKPOINT*   Now try Exercise 31.

Mutually Exclusive Events

Two events A and B (from the same sample space) are **mutually exclusive** if A and B have no outcomes in common. In the terminology of sets, the intersection of A and B is the empty set, which is expressed as

$$P(A \cap B) = 0.$$

For instance, if two dice are tossed, the event A of rolling a total of 6 and the event B of rolling a total of 9 are mutually exclusive. To find the probability that one or the other of two mutually exclusive events will occur, you can *add* their individual probabilities.

Probability of the Union of Two Events

If A and B are events in the same sample space, the probability of A *or* B occurring is given by

$$P(A \cup B) = P(A) + P(B) - P(A \cap B).$$

If A and B are mutually exclusive, then

$$P(A \cup B) = P(A) + P(B).$$

Example 7   The Probability of a Union

One card is selected from a standard deck of 52 playing cards. What is the probability that the card is either a heart or a face card?

Solution

Because the deck has 13 hearts, the probability of selecting a heart (event A) is

$$P(A) = \frac{13}{52}.$$

Similarly, because the deck has 12 face cards, the probability of selecting a face card (event B) is

$$P(B) = \frac{12}{52}.$$

Because three of the cards are hearts and face cards (see Figure 8.22), it follows that

$$P(A \cap B) = \frac{3}{52}.$$

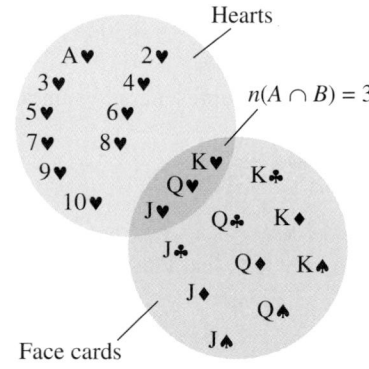

Figure 8.22

Finally, applying the formula for the probability of the union of two events, you can conclude that the probability of selecting a heart or a face card is

$$P(A \cup B) = P(A) + P(B) - P(A \cap B)$$

$$= \frac{13}{52} + \frac{12}{52} - \frac{3}{52} = \frac{22}{52} \approx 0.42.$$

 *CHECKPOINT*   Now try Exercise 49.

Example 8   Probability of Mutually Exclusive Events

The personnel department of a company has compiled data on the numbers of employees who have been with the company for various periods of time. The results are shown in the table.

Years of service	Numbers of employees
0–4	157
5–9	89
10–14	74
15–19	63
20–24	42
25–29	38
30–34	37
35–39	21
40–44	8

If an employee is chosen at random, what is the probability that the employee has (a) 4 or fewer years of service and (b) 9 or fewer years of service?

Solution

a. To begin, add the number of employees and find that the total is 529. Next, let event A represent choosing an employee with 0 to 4 years of service. Then the probability of choosing an employee who has 4 or fewer years of service is

$$P(A) = \frac{157}{529} \approx 0.30.$$

b. Let event B represent choosing an employee with 5 to 9 years of service. Then

$$P(B) = \frac{89}{529}.$$

Because event A from part (a) and event B have no outcomes in common, you can conclude that these two events are mutually exclusive and that

$$P(A \cup B) = P(A) + P(B)$$

$$= \frac{157}{529} + \frac{89}{529}$$

$$= \frac{246}{529}$$

$$\approx 0.47.$$

So, the probability of choosing an employee who has 9 or fewer years of service is about 0.47.

✓CHECKPOINT   Now try Exercise 51.

Independent Events

Two events are **independent** if the occurrence of one has no effect on the occurrence of the other. For instance, rolling a total of 12 with two six-sided dice has no effect on the outcome of future rolls of the dice. To find the probability that two independent events will occur, *multiply* the probabilities of each.

Probability of Independent Events

If A and B are **independent events,** the probability that both A and B will occur is given by

$$P(A \text{ and } B) = P(A) \cdot P(B).$$

Example 9   Probability of Independent Events

A random number generator on a computer selects three integers from 1 to 20. What is the probability that all three numbers are less than or equal to 5?

Solution

The probability of selecting a number from 1 to 5 is

$$P(A) = \frac{5}{20} = \frac{1}{4}.$$

So, the probability that all three numbers are less than or equal to 5 is

$$P(A) \cdot P(A) \cdot P(A) = \left(\frac{1}{4}\right)\left(\frac{1}{4}\right)\left(\frac{1}{4}\right) = \frac{1}{64}.$$

✔**CHECKPOINT**   Now try Exercise 52.

Example 10   Probability of Independent Events

In 2004, approximately 65% of the population of the United States was 25 years old or older. In a survey, 10 people were chosen at random from the population. What is the probability that all 10 were 25 years old or older?   (Source: U.S. Census Bureau)

Solution

Let A represent choosing a person who was 25 years old or older. The probability of choosing a person who was 25 years old or older is 0.65, the probability of choosing a second person who was 25 years old or older is 0.65, and so on. Because these events are independent, you can conclude that the probability that all 10 people were 25 years old or older is

$$[P(A)]^{10} = (0.65)^{10} \approx 0.01.$$

✔**CHECKPOINT**   Now try Exercise 53.

The Complement of an Event

The **complement of an event** A is the collection of all outcomes in the sample space that are *not* in A. The complement of event A is denoted by A'. Because $P(A \text{ or } A') = 1$ and because A and A' are mutually exclusive, it follows that $P(A) + P(A') = 1$. So, the probability of A' is given by

$$P(A') = 1 - P(A).$$

For instance, if the probability of *winning* a game is

$$P(A) = \frac{1}{4}$$

then the probability of *losing* the game is

$$P(A') = 1 - \frac{1}{4}$$

$$= \frac{3}{4}.$$

Exploration

You are in a class with 22 other people. What is the probability that at least two out of the 23 people will have a birthday on the same day of the year? What if you know the probability of everyone having the same birthday? Do you think this information would help you to find the answer?

> **Probability of a Complement**
>
> Let A be an event and let A' be its complement. If the probability of A is $P(A)$, then the probability of the complement is given by
>
> $$P(A') = 1 - P(A).$$

Example 11   Finding the Probability of a Complement

A manufacturer has determined that a machine averages one faulty unit for every 1000 it produces. What is the probability that an order of 200 units will have one or more faulty units?

Solution

To solve this problem as stated, you would need to find the probabilities of having exactly one faulty unit, exactly two faulty units, exactly three faulty units, and so on. However, using complements, you can simply find the probability that all units are perfect and then subtract this value from 1. Because the probability that any given unit is perfect is 999/1000, the probability that all 200 units are perfect is

$$P(A) = \left(\frac{999}{1000}\right)^{200}$$

$$\approx 0.82.$$

So, the probability that at least one unit is faulty is

$$P(A') = 1 - P(A)$$

$$\approx 0.18.$$

✓CHECKPOINT   Now try Exercise 55.

8.7 Exercises

See www.CalcChat.com for worked-out solutions to odd-numbered exercises.

Vocabulary Check

In Exercises 1–7, fill in the blanks.

1. An _____ is an event whose result is uncertain, and the possible results of the event are called _____ .

2. The set of all possible outcomes of an experiment is called the _____ .

3. To determine the _____ of an event, you can use the formula $P(E) = \dfrac{n(E)}{n(S)}$, where $n(E)$ is the number of outcomes in the event and $n(S)$ is the number of outcomes in the sample space.

4. If $P(E) = 0$, then E is an _____ event, and if $P(E) = 1$, then E is a _____ event.

5. If two events from the same sample space have no outcomes in common, then the two events are _____ .

6. If the occurrence of one event has no effect on the occurrence of a second event, then the events are _____ .

7. The _____ of an event A is the collection of all outcomes in the sample space that are not in A.

8. Match the probability formula with the correct probability name.

   (a) Probability of the union of two events   (i) $P(A \cup B) = P(A) + P(B)$

   (b) Probability of mutually exclusive events   (ii) $P(A') = 1 - P(A)$

   (c) Probability of independent events   (iii) $P(A \cup B) = P(A) + P(B) - P(A \cap B)$

   (d) Probability of a complement   (iv) $P(A \text{ and } B) = P(A) \cdot P(B)$

In Exercises 1–6, determine the sample space for the experiment.

1. A coin and a six-sided die are tossed.

2. A six-sided die is tossed twice and the sum of the results is recorded.

3. A taste tester has to rank three varieties of orange juice, A, B, and C, according to preference.

4. Two marbles are selected (without replacement) from a sack containing two red marbles, two blue marbles, and one yellow marble. The color of each marble is recorded.

5. Two county supervisors are selected from five supervisors, A, B, C, D, and E, to study a recycling plan.

6. A sales representative makes presentations of a product in three homes per day. In each home there may be a sale (denote by S) or there may be no sale (denote by F).

Tossing a Coin **In Exercises 7–10, find the probability for the experiment of tossing a coin three times. Use the sample space $S = \{HHH, HHT, HTH, HTT, THH, THT, TTH, TTT\}$.**

7. The probability of getting exactly two tails

8. The probability of getting a head on the first toss

9. The probability of getting at least one head

10. The probability of getting at least two heads

Drawing a Card **In Exercises 11–14, find the probability for the experiment of selecting one card from a standard deck of 52 playing cards.**

11. The card is a face card.

12. The card is not a black face card.

13. The card is a face card or an ace.

14. The card is a 9 or lower. (Aces are low.)

Tossing a Die **In Exercises 15–18, find the probability for the experiment of tossing a six-sided die twice.**

15. The sum is 6.

16. The sum is at least 8.

17. The sum is less than 11.

18. The sum is odd or prime.

Drawing Marbles **In Exercises 19–22, find the probability for the experiment of drawing two marbles (without replacement) from a bag containing one green, two yellow, and three red marbles.**

19. Both marbles are red.

20. Both marbles are yellow.

21. Neither marble is yellow.

22. The marbles are of different colors.

In Exercises 23–26, you are given the probability that an event *will* happen. Find the probability that the event *will not* happen.

23. $P(E) = 0.75$ 24. $P(E) = 0.\overline{2}$

25. $P(E) = \frac{2}{3}$ 26. $P(E) = \frac{7}{8}$

In Exercises 27–30, you are given the probability that an event *will not* happen. Find the probability that the event *will* happen.

27. $P(E') = 0.12$ 28. $P(E') = 0.84$

29. $P(E') = \frac{13}{20}$ 30. $P(E') = \frac{61}{100}$

31. ***Graphical Reasoning***   In 2004, there were approximately 8.15 million unemployed workers in the United States. The circle graph shows the age profile of these unemployed workers.   (Source: U.S. Bureau of Labor Statistics)

Ages of Unemployed Workers

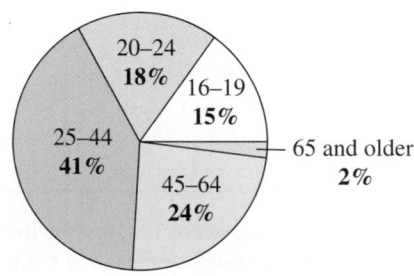

(a) Estimate the number of unemployed workers in the 16–19 age group.

(b) What is the probability that a person selected at random from the population of unemployed workers is in the 25–44 age group?

(c) What is the probability that a person selected at random from the population of unemployed workers is in the 45–64 age group?

(d) What is the probability that a person selected at random from the population of unemployed workers is 45 or older?

32. ***Graphical Reasoning***   The circle graph shows the numbers of children of the 42 U.S. presidents as of 2006.   (Source: infoplease.com)

Children of U.S. Presidents

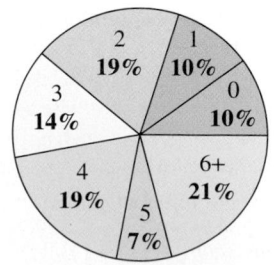

(a) Determine the number of presidents who had no children.

(b) Determine the number of presidents who had four children.

(c) What is the probability that a president selected at random had five or more children?

(d) What is the probability that a president selected at random had three children?

33. ***Graphical Reasoning***   The total population of the United States in 2004 was approximately 293.66 million. The circle graph shows the race profile of the population. (Source: U.S. Census Bureau)

United States Population by Race

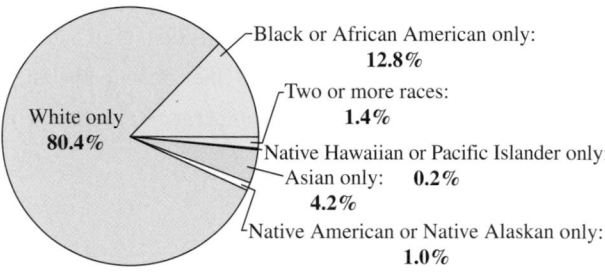

(a) Estimate the population of African Americans.

(b) A person is selected at random. Find the probability that this person is Native American or Native Alaskan.

(c) A person is selected at random. Find the probability that this person is Native American, Native Alaskan, Native Hawaiian, or Pacific Islander.

34. ***Graphical Reasoning***   The educational attainment of the United States population age 25 years or older in 2004 is shown in the circle graph. Use the fact that the population of people 25 years or older was 186.88 million in 2004. (Source: U.S. Census Bureau)

Educational Attainment

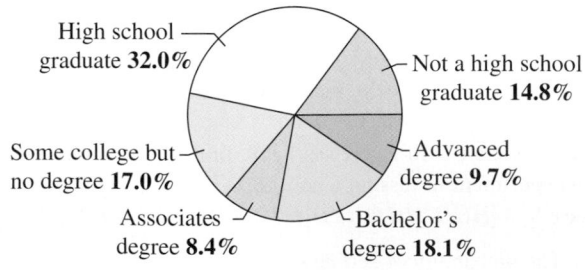

(a) Estimate the number of people 25 or older who have high school diplomas.

(b) Estimate the number of people 25 or older who have advanced degrees.

(c) Find the probability that a person 25 or older selected at random has earned a Bachelor's degree or higher.

(d) Find the probability that a person 25 or older selected at random has earned a high school diploma or gone on to post-secondary education.

(e) Find the probability that a person 25 or older selected at random has earned an Associate's degree or higher.

35. **Data Analysis** One hundred college students were interviewed to determine their political party affiliations and whether they favored a balanced budget amendment to the Constitution. The results of the study are listed in the table, where D represents Democrat and R represents Republican.

	D	R	Total
Favor	23	32	55
Oppose	25	9	34
Unsure	7	4	11
Total	55	45	100

A person is selected at random from the sample. Find the probability that the person selected is (a) a person who doesn't favor the amendment, (b) a Republican, (c) a Democrat who favors the amendment.

36. **Data Analysis** A study of the effectiveness of a flu vaccine was conducted with a sample of 500 people. Some participants in the study were given no vaccine, some were given one injection, and some were given two injections. The results of the study are shown in the table.

	Flu	No flu	Total
No vaccine	7	149	156
One injection	2	52	54
Two Injections	13	277	290
Total	22	478	500

A person is selected at random from the sample. Find the probability that the person selected (a) had two injections, (b) did not get the flu, and (c) got the flu and had one injection.

37. **Alumni Association** A college sends a survey to selected members of the class of 2006. Of the 1254 people who graduated that year, 672 are women, of whom 124 went on to graduate school. Of the 582 male graduates, 198 went on to graduate school. An alumni member is selected at random. What is the probability that the person is (a) female, (b) male, and (c) female and did not attend graduate school?

38. **Education** In a high school graduating class of 128 students, 52 are on the honor roll. Of these, 48 are going on to college; of the other 76 students, 56 are going on to college. A student is selected at random from the class. What is the probability that the person chosen is (a) going to college, (b) not going to college, and (c) not going to college and on the honor roll?

39. **Election** Taylor, Moore, and Perez are candidates for public office. It is estimated that Moore and Perez have about the same probability of winning, and Taylor is believed to be twice as likely to win as either of the others. Find the probability of each candidate's winning the election.

40. **Payroll Error** The employees of a company work in six departments: 31 are in sales, 54 are in research, 42 are in marketing, 20 are in engineering, 47 are in finance, and 58 are in production. One employee's paycheck is lost. What is the probability that the employee works in the research department?

In Exercises 41–52, the sample spaces are large and you should use the counting principles discussed in Section 6.6.

41. **Preparing for a Test** A class is given a list of 20 study problems from which 10 will be chosen as part of an upcoming exam. A given student knows how to solve 15 of the problems. Find the probability that the student will be able to answer (a) all 10 questions on the exam, (b) exactly 8 questions on the exam, and (c) at least 9 questions on the exam.

42. **Payroll Mix-Up** Five paychecks and envelopes are addressed to five different people. The paychecks are randomly inserted into the envelopes. What is the probability that (a) exactly one paycheck is inserted in the correct envelope and (b) at least one paycheck is inserted in the correct envelope?

43. **Game Show** On a game show you are given five digits to arrange in the proper order to form the price of a car. If you are correct, you win the car. What is the probability of winning if you (a) guess the position of each digit and (b) know the first digit and guess the others?

44. **Card Game** The deck of a card game is made up of 108 cards. Twenty-five each are red, yellow, blue, and green, and eight are wild cards. Each player is randomly dealt a seven-card hand. What is the probability that a hand will contain (a) exactly two wild cards, and (b) two wild cards, two red cards, and three blue cards?

45. **Radio Stations** Typically radio stations are identified by four "call letters." Radio stations east of the Mississippi River have call letters that start with the letter W and radio stations west of the Mississippi River have call letters that start with the letter K. Assuming the station call letters are equally distributed, what is the probability that a radio station selected at random has call letters that contain (a) a Q and a Y, and (b) a Q, a Y, and an X?

46. *PIN Codes* ATM personal identification number (PIN) codes typically consist of four-digit sequences of numbers. Find the probability that if you forget your PIN, you can guess the correct sequence (a) at random and (b) if you recall the first two digits.

47. *Lottery* Powerball is played with 55 white balls, numbered 1 through 55, and 42 red balls, numbered 1 through 42. Five white balls and one red ball, the Powerball, are drawn to determine the winning ticket(s). Find the probability that you purchase a winning ticket if you purchase (a) 100 tickets and (b) 1000 tickets with different combinations of numbers.

48. *ZIP Codes* The U.S. Postal Service is to deliver a letter to a certain postal ZIP+4 code. Find the probability that the ZIP+4 code is correct if the sender (a) randomly chooses the code, (b) knows the five-digit code but must randomly choose the last four digits, and (c) knows the five-digit code and the first two digits of the plus 4 code.

49. *Drawing a Card* One card is selected at random from a standard deck of 52 playing cards. Find the probability that (a) the card is an even-numbered card, (b) the card is a heart or a diamond, and (c) the card is a nine or a face card.

50. *Poker Hand* Five cards are drawn from an ordinary deck of 52 playing cards. What is the probability of getting a full house? (A full house consists of three of one kind and two of another kind.)

51. *Defective Units* A shipment of 12 microwave ovens contains three defective units. A vending company has ordered four of these units, and because all are packaged identically, the selection will be random. What is the probability that (a) all four units are good, (b) exactly two units are good, and (c) at least two units are good?

52. *Random Number Generator* Two integers from 1 through 40 are chosen by a random number generator. What is the probability that (a) the numbers are both even, (b) one number is even and one is odd, (c) both numbers are less than 30, and (d) the same number is chosen twice?

53. *Consumerism* Suppose that the methods used by shoppers to pay for merchandise are as shown in the circle graph. Two shoppers are chosen at random. What is the probability that both shoppers paid for their purchases only in cash?

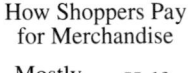

How Shoppers Pay
for Merchandise

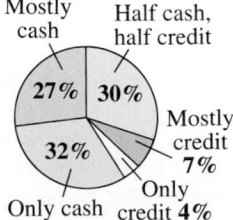

54. *Flexible Work Hours* In a survey, people were asked if they would prefer to work flexible hours—even if it meant slower career advancement—so they could spend more time with their families. The results of the survey are shown in the circle graph. Three people from the survey are chosen at random. What is the probability that all three people would prefer flexible work hours?

Flexible Work Hours

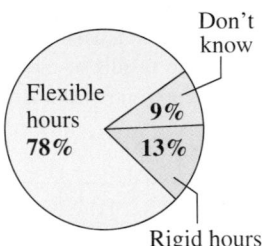

55. *Backup System* A space vehicle has an independent backup system for one of its communication networks. The probability that either system will function satisfactorily for the duration of a flight is 0.985. What is the probability that during a given flight (a) both systems function satisfactorily, (b) at least one system functions satisfactorily, and (c) both systems fail?

56. *Backup Vehicle* A fire company keeps two rescue vehicles to serve the community. Because of the demand on the vehicles and the chance of mechanical failure, the probability that a specific vehicle is available when needed is 90%. The availability of one vehicle is *independent* of the other. Find the probability that (a) both vehicles are available at a given time, (b) neither vehicle is available at a given time, and (c) at least one vehicle is available at a given time.

57. *Making a Sale* A sales representative makes sales on approximately one-fifth of all calls. On a given day, the representative contacts six potential clients. What is the probability that a sale will be made with (a) all six contacts, (b) none of the contacts, and (c) at least one contact?

58. *A Boy or a Girl?* Assume that the probability of the birth of a child of a particular sex is 50%. In a family with four children, what is the probability that (a) all the children are boys, (b) all the children are the same sex, and (c) there is at least one boy?

59. *Estimating* π A coin of diameter d is dropped onto a paper that contains a grid of squares d units on a side (see figure on the next page).

(a) Find the probability that the coin covers a vertex of one of the squares on the grid.

(b) Perform the experiment 100 times and use the results to approximate π.

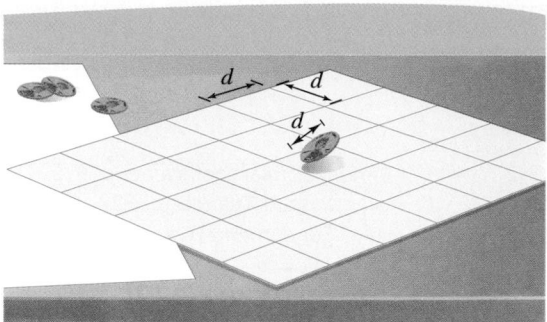

Figure for 59

60. *Geometry*   You and a friend agree to meet at your favorite fast food restaurant between 5:00 and 6:00 P.M. The one who arrives first will wait 15 minutes for the other, after which the first person will leave (see figure). What is the probability that the two of you will actually meet, assuming that your arrival times are random within the hour?

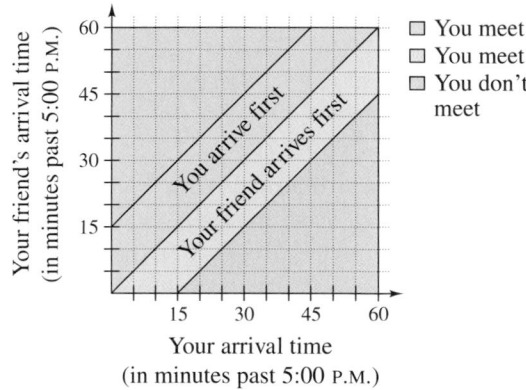

☐ You meet
☐ You meet
☐ You don't meet

Your arrival time
(in minutes past 5:00 P.M.)

Synthesis

True or False?   In Exercises 61 and 62, determine whether the statement is true or false. Justify your answer.

61. If the probability of an outcome in a sample space is 1, then the probability of the other outcomes in the sample space is 0.

62. Rolling a number less than 3 on a normal six-sided die has a probability of $\frac{1}{3}$. The complement of this event is to roll a number greater than 3, and its probability is $\frac{1}{2}$.

63. *Pattern Recognition and Exploration*   Consider a group of n people.

(a) Explain why the following pattern gives the probability that the n people have distinct birthdays.

$$n = 2: \quad \frac{365}{365} \cdot \frac{364}{365} = \frac{365 \cdot 364}{365^2}$$

$$n = 3: \quad \frac{365}{365} \cdot \frac{364}{365} \cdot \frac{363}{365} = \frac{365 \cdot 364 \cdot 363}{365^3}$$

(b) Use the pattern in part (a) to write an expression for the probability that four people ($n = 4$) have distinct birthdays.

(c) Let P_n be the probability that the n people have distinct birthdays. Verify that this probability can be obtained recursively by

$$P_1 = 1 \quad \text{and} \quad P_n = \frac{365 - (n-1)}{365} P_{n-1}.$$

(d) Explain why $Q_n = 1 - P_n$ gives the probability that at least two people in a group of n people have the same birthday.

(e) Use the results of parts (c) and (d) to complete the table.

n	10	15	20	23	30	40	50
P_n							
Q_n							

(f) How many people must be in a group so that the probability of at least two of them having the same birthday is greater than $\frac{1}{2}$? Explain.

64. *Think About It*   The weather forecast indicates that the probability of rain is 40%. Explain what this means.

Skills Review

In Exercises 65–68, solve the rational equation.

65. $\dfrac{2}{x-5} = 4$

66. $\dfrac{3}{2x+3} - 4 = \dfrac{-1}{2x+3}$

67. $\dfrac{3}{x-2} + \dfrac{x}{x+2} = 1$

68. $\dfrac{2}{x} - \dfrac{5}{x-2} = \dfrac{-13}{x^2-2x}$

In Exercises 69–72, solve the equation algebraically. Round your result to three decimal places.

69. $e^x + 7 = 35$

70. $200e^{-x} = 75$

71. $4 \ln 6x = 16$

72. $5 \ln 2x - 4 = 11$

In Exercises 73–76, evaluate $_nP_r$. Verify your result using a graphing utility.

73. $_5P_3$

74. $_{10}P_4$

75. $_{11}P_8$

76. $_9P_2$

In Exercises 77–80, evaluate $_nC_r$. Verify your result using a graphing utility.

77. $_6C_2$

78. $_9C_5$

79. $_{11}C_8$

80. $_{16}C_{13}$

What Did You Learn?

Key Terms

infinite sequence, *p. 580*
finite sequence, *p. 580*
recursively defined sequence, *p. 582*
n factorial, *p. 582*
summation notation, *p. 584*
series, *p. 585*

arithmetic sequence, *p. 592*
common difference, *p. 592*
geometric sequence, *p. 601*
common ratio, *p. 601*
first differences, *p. 616*
second differences, *p. 616*

binomial coefficients, *p. 619*
Pascal's Triangle, *p. 623*
Fundamental Counting Principle,
   p. 628
sample space, *p. 637*
mutually exclusive events, *p. 641*

Key Concepts

8.1 ■ Find the sum of an infinite sequence

Consider the infinite sequence $a_1, a_2, a_3, \ldots a_i, \ldots$.

1. The sum of the first n terms of the sequence is the finite series or partial sum of the sequence and is denoted by

$$a_1 + a_2 + a_3 + \cdots + a_n = \sum_{i=1}^{n} a_i$$

   where i is the index of summation, n is the upper limit of summation, and 1 is the lower limit of summation.

2. The sum of all terms of the infinite sequence is called an infinite series and is denoted by

$$a_1 + a_2 + a_3 + \cdots + a_i + \cdots = \sum_{i=1}^{\infty} a_i.$$

8.2 ■ Find the *n*th term and the *n*th partial sum of an arithmetic sequence

1. The nth term of an arithmetic sequence is $a_n = dn + c$, where d is the common difference between consecutive terms and $c = a_1 - d$.

2. The sum of a finite arithmetic sequence with n terms is given by $S_n = (n/2)(a_1 + a_n)$.

8.3 ■ Find the *n*th term and the *n*th partial sum of a geometric sequence

1. The nth term of a geometric sequence is $a_n = a_1 r^{n-1}$, where r is the common ratio of consecutive terms.

2. The sum of a finite geometric sequence
$a_1, a_1 r, a_1 r^2, a_1 r^3, a_1 r^4, \ldots, a_1 r^{n-1}$ with common

   ratio $r \neq 1$ is $S_n = \sum_{i=1}^{n} a_1 r^{i-1} = a_1 \left(\dfrac{1 - r^n}{1 - r} \right)$.

8.3 ■ Find the sum of an infinite geometric series

If $|r| < 1$, then the infinite geometric series $a_1 + a_1 r + a_1 r^2 + a_1 r^3 + \cdots + a_1 r^{n-1} + \cdots$ has the sum

$$S = \sum_{i=0}^{\infty} a_1 r^i = \frac{a_1}{1 - r}.$$

8.4 ■ Use mathematical induction

Let P_n be a statement with the positive integer n. If P_1 is true, and the truth of P_k implies the truth of P_{k+1} for every positive integer k, then P_n must be true for all positive integers n.

8.5 ■ Use the Binomial Theorem to expand a binomial

In the expansion of $(x + y)^n = x^n + nx^{n-1}y + \cdots + {}_nC_r x^{n-r}y^r + \cdots + nxy^{n-1} + y^n$, the coefficient of $x^{n-r}y^r$ is ${}_nC_r = n!/[(n - r)!r!]$.

8.6 ■ Solve counting problems

1. If one event can occur in m_1 different ways and a second event can occur in m_2 different ways, then the number of ways that the two events can occur is $m_1 \cdot m_2$.

2. The number of permutations of n elements is $n!$.

3. The number of permutations of n elements taken r at a time is given by ${}_nP_r = n!/(n - r)!$

4. The number of distinguishable permutations of n objects is given by $n!/(n_1! \cdot n_2! \cdot n_3! \cdots n_k!)$.

5. The number of combinations of n elements taken r at a time is given by ${}_nC_r = n!/[(n - r)!r!]$.

8.7 ■ Find probabilities of events

1. If an event E has $n(E)$ equally likely outcomes and its sample space S has $n(S)$ equally likely outcomes, the probability of event E is $P(E) = n(E)/n(S)$.

2. If A and B are events in the same sample space, the probability of A or B occurring is given by $P(A \cup B) = P(A) + P(B) - P(A \cap B)$. If A and B are mutually exclusive, then $P(A \cup B) = P(A) + P(B)$.

3. If A and B are independent events, the probability that A and B will occur is $P(A \text{ and } B) = P(A) \cdot P(B)$.

4. Let A be an event and let A' be its complement. If the probability of A is $P(A)$, then the probability of the complement is $P(A') = 1 - P(A)$.

Review Exercises

See www.CalcChat.com for worked-out solutions to odd-numbered exercises.

8.1 In Exercises 1–4, write the first five terms of the sequence. (Assume n begins with 1.)

1. $a_n = \dfrac{2^n}{2^n + 1}$

2. $a_n = \dfrac{1}{n} - \dfrac{1}{n + 1}$

3. $a_n = \dfrac{(-1)^n}{n!}$

4. $a_n = \dfrac{(-1)^n}{(2n + 1)!}$

In Exercises 5–8, write an expression for the *apparent n*th term of the sequence. (Assume n begins with 1.)

5. $5, 10, 15, 20, 25, \ldots$

6. $50, 48, 46, 44, 42, \ldots$

7. $2, \frac{2}{3}, \frac{2}{5}, \frac{2}{7}, \frac{2}{9}, \ldots$

8. $\frac{3}{2}, \frac{4}{3}, \frac{5}{4}, \frac{6}{5}, \frac{7}{6}, \ldots$

In Exercises 9 and 10, write the first five terms of the sequence defined recursively.

9. $a_1 = 9, a_{k+1} = a_k - 4$

10. $a_1 = 49, a_{k+1} = a_k + 6$

In Exercises 11–14, simplify the factorial expression.

11. $\dfrac{18!}{20!}$

12. $\dfrac{10!}{8!}$

13. $\dfrac{(n + 1)!}{(n - 1)!}$

14. $\dfrac{2n!}{(n + 1)!}$

In Exercises 15–22, find the sum.

15. $\displaystyle\sum_{i=1}^{6} 5$

16. $\displaystyle\sum_{k=2}^{5} 4k$

17. $\displaystyle\sum_{j=1}^{4} \dfrac{6}{j^2}$

18. $\displaystyle\sum_{i=1}^{8} \dfrac{i}{i + 1}$

19. $\displaystyle\sum_{k=1}^{100} 2k^3$

20. $\displaystyle\sum_{j=0}^{40} (j^2 + 1)$

21. $\displaystyle\sum_{n=0}^{50} (n^2 + 3)$

22. $\displaystyle\sum_{n=1}^{100} \left(\dfrac{1}{n} - \dfrac{1}{n + 1} \right)$

In Exercises 23–26, use sigma notation to write the sum. Then use a graphing utility to find the sum.

23. $\dfrac{1}{2(1)} + \dfrac{1}{2(2)} + \dfrac{1}{2(3)} + \cdots + \dfrac{1}{2(20)}$

24. $2(1^2) + 2(2^2) + 2(3^2) + \cdots + 2(9^2)$

25. $\dfrac{1}{2} + \dfrac{2}{3} + \dfrac{3}{4} + \cdots + \dfrac{9}{10}$

26. $1 - \dfrac{1}{3} + \dfrac{1}{9} - \dfrac{1}{27} + \cdots$

In Exercises 27–30, find (a) the fourth partial sum and (b) the sum of the infinite series.

27. $\displaystyle\sum_{k=1}^{\infty} \dfrac{5}{10^k}$

28. $\displaystyle\sum_{k=1}^{\infty} \dfrac{3}{2^k}$

29. $\displaystyle\sum_{k=1}^{\infty} 2(0.5)^k$

30. $\displaystyle\sum_{k=1}^{\infty} 4(0.25)^k$

31. *Compound Interest* A deposit of \$2500 is made in an account that earns 2% interest compounded quarterly. The balance in the account after n quarters is given by

$$a_n = 2500\left(1 + \dfrac{0.02}{4}\right)^n, \qquad n = 1, 2, 3, \ldots.$$

(a) Compute the first eight terms of this sequence.

(b) Find the balance in this account after 10 years by computing the 40th term of the sequence.

32. *Education* The numbers a_n of full-time faculty (in thousands) employed in institutions of higher education in the United States from 1991 to 2003 can be approximated by the model

$$a_n = 0.41n^2 + 2.7n + 532, \quad n = 1, 2, 3, \ldots, 13$$

where n is the year, with $n = 1$ corresponding to 1991. (Source: U.S. National Center for Education Statistics)

(a) Find the terms of this finite sequence for the given values of n.

(b) Use a graphing utility to graph the sequence for the given values of n.

(c) Use a graphing utility to construct a bar graph of the sequence for the given values of n.

(d) Use the sequence to predict the numbers of full-time faculty for the years 2004 to 2010. Do your results seem reasonable? Explain.

8.2 In Exercises 33–36, determine whether or not the sequence is arithmetic. If it is, find the common difference.

33. $5, 3, 1, -1, -3, \ldots$

34. $0, 1, 3, 6, 10, \ldots$

35. $\frac{1}{2}, 1, \frac{3}{2}, 2, \frac{5}{2}, \ldots$

36. $\frac{9}{9}, \frac{8}{9}, \frac{7}{9}, \frac{6}{9}, \frac{5}{9}, \ldots$

In Exercises 37–40, write the first five terms of the arithmetic sequence.

37. $a_1 = 3, \ d = 4$

38. $a_1 = 8, \ d = -2$

39. $a_4 = 10, \ a_{10} = 28$

40. $a_2 = 14, \ a_6 = 22$

In Exercises 41–44, write the first five terms of the arithmetic sequence. Find the common difference and write the nth term of the sequence as a function of n.

41. $a_1 = 35$, $a_{k+1} = a_k - 3$

42. $a_1 = 15$, $a_{k+1} = a_k + \frac{5}{2}$

43. $a_1 = 9$, $a_{k+1} = a_k + 7$

44. $a_1 = 100$, $a_{k+1} = a_k - 5$

In Exercises 45 and 46, find a formula for a_n for the arithmetic sequence and find the sum of the first 20 terms of the sequence.

45. $a_1 = 100$, $d = -3$ **46.** $a_1 = 10$, $a_3 = 28$

In Exercises 47–50, find the partial sum. Use a graphing utility to verify your result.

47. $\displaystyle\sum_{j=1}^{10} (2j - 3)$ **48.** $\displaystyle\sum_{j=1}^{8} (20 - 3j)$

49. $\displaystyle\sum_{k=1}^{11} \left(\frac{2}{3}k + 4\right)$ **50.** $\displaystyle\sum_{k=1}^{25} \left(\frac{3k + 1}{4}\right)$

51. Find the sum of the first 100 positive multiples of 5.

52. Find the sum of the integers from 20 to 80 (inclusive).

53. *Job Offer*   The starting salary for an accountant is $34,000 with a guaranteed salary increase of $2250 per year for the first 4 years of employment. Determine (a) the salary during the fifth year and (b) the total compensation through 5 full years of employment.

54. *Baling Hay*   In his first trip baling hay around a field, a farmer makes 123 bales. In his second trip he makes 11 fewer bales. Because each trip is shorter than the preceding trip, the farmer estimates that the same pattern will continue. Estimate the total number of bales made if there are another six trips around the field.

8.3 In Exercises 55–58, determine whether or not the sequence is geometric. If it is, find the common ratio.

55. 5, 10, 20, 40, . . .   **56.** $\frac{1}{2}$, $\frac{2}{3}$, $\frac{3}{4}$, $\frac{4}{5}$, . . .

57. 54, -18, 6, -2, . . .   **58.** $\frac{1}{3}$, $-\frac{2}{3}$, $\frac{4}{3}$, $-\frac{8}{3}$, . . .

In Exercises 59–62, write the first five terms of the geometric sequence.

59. $a_1 = 4$, $r = -\frac{1}{4}$ **60.** $a_1 = 2$, $r = \frac{3}{2}$

61. $a_1 = 9$, $a_3 = 4$ **62.** $a_1 = 2$, $a_3 = 12$

In Exercises 63–66, write the first five terms of the geometric sequence. Find the common ratio and write the nth term of the sequence as a function of n.

63. $a_1 = 120$, $a_{k+1} = \frac{1}{3}a_k$

64. $a_1 = 200$, $a_{k+1} = 0.1a_k$

65. $a_1 = 25$, $a_{k+1} = -\frac{3}{5}a_k$

66. $a_1 = 18$, $a_{k+1} = \frac{5}{3}a_k$

In Exercises 67–70, find the nth term of the geometric sequence and find the sum of the first 20 terms of the sequence.

67. $a_1 = 16$, $a_2 = -8$ **68.** $a_3 = 6$, $a_4 = 1$

69. $a_1 = 100$, $r = 1.05$ **70.** $a_1 = 5$, $r = 0.2$

In Exercises 71–78, find the sum. Use a graphing utility to verify your result.

71. $\displaystyle\sum_{i=1}^{7} 2^{i-1}$ **72.** $\displaystyle\sum_{i=1}^{5} 3^{i-1}$

73. $\displaystyle\sum_{n=1}^{7} (-4)^{n-1}$ **74.** $\displaystyle\sum_{n=1}^{4} 12\left(-\frac{1}{2}\right)^{n-1}$

75. $\displaystyle\sum_{n=0}^{4} 250(1.02)^n$ **76.** $\displaystyle\sum_{n=0}^{5} 400(1.08)^n$

77. $\displaystyle\sum_{i=1}^{10} 10\left(\frac{3}{5}\right)^{i-1}$ **78.** $\displaystyle\sum_{i=1}^{15} 20(0.2)^{i-1}$

In Exercises 79–82, find the sum of the infinite geometric series.

79. $\displaystyle\sum_{i=1}^{\infty} 4\left(\frac{7}{8}\right)^{i-1}$ **80.** $\displaystyle\sum_{i=1}^{\infty} 6\left(\frac{1}{3}\right)^{i-1}$

81. $\displaystyle\sum_{k=1}^{\infty} 4\left(\frac{2}{3}\right)^{k-1}$ **82.** $\displaystyle\sum_{k=1}^{\infty} 1.3\left(\frac{1}{10}\right)^{k-1}$

83. *Depreciation*   A company buys a fleet of six vans for $120,000. During the next 5 years, the fleet will depreciate at a rate of 30% per year. (That is, at the end of each year, the depreciated value will be 70% of the value at the beginning of the year.)

(a) Find the formula for the nth term of a geometric sequence that gives the value of the fleet t full years after it was purchased.

(b) Find the depreciated value of the fleet at the end of 5 full years.

84. *Annuity*   A deposit of $75 is made at the beginning of each month in an account that pays 4% interest, compounded monthly. The balance A in the account at the end of 4 years is given by

$$A = 75\left(1 + \frac{0.04}{12}\right)^1 + \cdots + 75\left(1 + \frac{0.04}{12}\right)^{48}.$$

Find A.

8.4 In Exercises 85–88, use mathematical induction to prove the formula for every positive integer n.

85. $2 + 7 + \cdots + (5n - 3) = \dfrac{n}{2}(5n - 1)$

86. $1 + \dfrac{3}{2} + 2 + \dfrac{5}{2} + \cdots + \dfrac{1}{2}(n + 1) = \dfrac{n}{4}(n + 3)$

87. $\displaystyle\sum_{i=0}^{n-1} ar^i = \dfrac{a(1 - r^n)}{1 - r}$

88. $\displaystyle\sum_{k=0}^{n-1} (a + kd) = \dfrac{n}{2}[2a + (n - 1)d]$

In Exercises 89–92, find the sum using the formulas for the sums of powers of integers.

89. $\displaystyle\sum_{n=1}^{30} n$

90. $\displaystyle\sum_{n=1}^{10} n^2$

91. $\displaystyle\sum_{n=1}^{7} (n^4 - n)$

92. $\displaystyle\sum_{n=1}^{6} (n^5 - n^2)$

In Exercises 93–96, write the first five terms of the sequence beginning with a_1. Then calculate the first and second differences of the sequence. Does the sequence have a linear model, a quadratic model, or neither?

93. $a_1 = 5$
$a_n = a_{n-1} + 5$

94. $a_1 = -3$
$a_n = a_{n-1} - 2n$

95. $a_1 = 16$
$a_n = a_{n-1} - 1$

96. $a_1 = 1$
$a_n = n - a_{n-1}$

8.5 In Exercises 97–100, find the binomial coefficient. Use a graphing utility to verify your result.

97. $_{10}C_8$

98. $_{12}C_5$

99. $\dbinom{9}{4}$

100. $\dbinom{14}{12}$

In Exercises 101–104, use Pascal's Triangle to find the binomial coefficient.

101. $_6C_3$

102. $_9C_7$

103. $\dbinom{8}{4}$

104. $\dbinom{10}{5}$

In Exercises 105–110, use the Binomial Theorem to expand and simplify the expression. $\left(\text{Recall that } i = \sqrt{-1}.\right)$

105. $(x + 5)^4$

106. $(y - 3)^3$

107. $(a - 4b)^5$

108. $(3x + y)^7$

109. $(7 + 2i)^4$

110. $(4 - 5i)^3$

8.6
111. *Numbers in a Hat* Slips of paper numbered 1 through 14 are placed in a hat. In how many ways can two numbers be drawn so that the sum of the numbers is 12? Assume the random selection is without replacement.

112. *Aircraft Boarding* Eight people are boarding an aircraft. Two have tickets for first class and board before those in economy class. In how many ways can the eight people board the aircraft?

113. *Course Schedule* A college student is preparing a course schedule of four classes for the next semester. The student can choose from the open sections shown in the table.

Course	Sections
Math 100	001–004
Economics 110	001–003
English 105	001–006
Humanities 101	001–003

(a) Find the number of possible schedules that the student can create from the offerings.

(b) Find the number of possible schedules that the student can create from the offerings if two of the Math 100 sections are closed.

(c) Find the number of possible schedules that the student can create from the offerings if two of the Math 100 sections and four of the English 105 sections are closed.

114. *Telemarketing* A telemarketing firm is making calls to prospective customers by randomly dialing a seven-digit phone number within an area code.

(a) Find the number of possible calls that the telemarketer can make.

(b) If the telemarketing firm is calling only within an exchange that begins with a "7" or a "6", how many different calls are possible?

(c) If the telemarketing firm is calling only within an exchange that does not begin with a "0" or a "1," how many calls are possible?

In Exercises 115–122, evaluate the expression. Use a graphing utility to verify your result.

115. $_{10}C_8$

116. $_8C_6$

117. $_{12}P_{10}$

118. $_6P_4$

119. $_{100}C_{98}$

120. $_{50}C_{48}$

121. $_{1000}P_2$

122. $_{500}P_2$

In Exercises 123 and 124, find the number of distinguishable permutations of the group of letters.

123. C, A, L, C, U, L, U, S

124. I, N, T, E, G, R, A, T, E

125. *Sports* There are 10 bicyclists entered in a race. In how many different orders could the 10 bicyclists finish? (Assume there are no ties.)

126. *Sports* From a pool of seven juniors and eleven seniors, four co-captains will be chosen for the football team. How many different combinations are possible if two juniors and two seniors are to be chosen?

127. *Exam Questions* A student can answer any 15 questions from a total of 20 questions on an exam. In how many different ways can the student select the questions?

128. *Lottery* In the Lotto Texas game, a player chooses six distinct numbers from 1 to 54. In how many ways can a player select the six numbers?

In Exercises 129 and 130, solve for *n*.

129. $_{n+1}P_2 = 4 \cdot {_nP_1}$ **130.** $8 \cdot {_nP_2} = {_{n+1}P_3}$

8.7

131. *Apparel* A man has five pairs of socks (no two pairs are the same color). He randomly selects two socks from a drawer. What is the probability that he gets a matched pair?

132. *Bookshelf Order* A child returns a five-volume set of books to a bookshelf. The child is not able to read, and so cannot distinguish one volume from another. What is the probability that the books are shelved in the correct order?

133. *Data Analysis* A sample of college students, faculty members, and administrators were asked whether they favored a proposed increase in the annual activity fee to enhance student life on campus. The results of the study are shown in the table.

	Favor	Oppose	Total
Students	237	163	400
Faculty	37	38	75
Admin.	18	7	25
Total	292	208	500

A person is selected at random from the sample. Find each probability.

(a) The person is not in favor of the proposal.

(b) The person is a student.

(c) The person is a faculty member and is in favor of the proposal.

134. *Tossing a Die* A six-sided die is rolled six times. What is the probability that each side appears exactly once?

135. *Poker Hand* Five cards are drawn from an ordinary deck of 52 playing cards. Find the probability of getting two pairs. (For example, the hand could be A-A-5-5-Q or 4-4-7-7-K.)

136. *Drawing a Card* You randomly select a card from a 52-card deck. What is the probability that the card is *not* a club?

Synthesis

True or False? **In Exercises 137 and 138, determine whether the statement is true or false. Justify your answer.**

137. $\dfrac{(n+2)!}{n!} = (n+2)(n+1)$ **138.** $\displaystyle\sum_{k=1}^{8} 3k = 3\sum_{k=1}^{8} k$

139. *Writing* In your own words, explain what makes a sequence (a) arithmetic and (b) geometric.

140. *Think About It* How do the two sequences differ?

(a) $a_n = \dfrac{(-1)^n}{n}$ (b) $a_n = \dfrac{(-1)^{n+1}}{n}$

141. *Graphical Reasoning* The graphs of two sequences are shown below. Identify each sequence as arithmetic or geometric. Explain your reasoning.

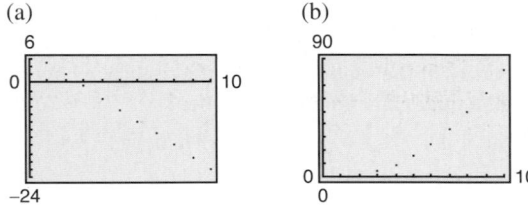

142. *Population Growth* Consider an idealized population with the characteristic that each member of the population produces one offspring at the end of every time period. If each member has a life span of three time periods and the population begins with 10 newborn members, then the following table shows the populations during the first five time periods.

Age Bracket	Time Period				
	1	**2**	**3**	**4**	**5**
0–1	10	10	20	40	70
1–2		10	10	20	40
2–3			10	10	20
Total	10	20	40	70	130

The sequence for the total populations has the property that

$$S_n = S_{n-1} + S_{n-2} + S_{n-3}, \quad n > 3.$$

Find the total populations during the next five time periods.

143. *Writing* Explain what a recursion formula is.

144. *Writing* Explain why the terms of a geometric sequence of positive terms decrease when $0 < r < 1$.

145. *Think About It* How do the expansions of $(x - y)^n$ and $(-x + y)^n$ differ?

146. The probability of an event must be a real number in what interval? Is the interval open or closed?

8 Chapter Test

Take this test as you would take a test in class. After you are finished, check your work against the answers given in the back of the book.

In Exercises 1–4, write the first five terms of the sequence.

1. $a_n = \left(-\frac{2}{3}\right)^{n-1}$ (Begin with $n = 1$.)   **2.** $a_1 = 12$ and $a_{k+1} = a_k + 4$

3. $b_n = \dfrac{(-1)^n x^n}{n}$ (Begin with $n = 1$.)   **4.** $b_n = \dfrac{(-1)^{2n+1} x^{2n+1}}{(2n+1)!}$ (Begin with $n = 1$.)

5. Simplify $\dfrac{11! \cdot 4!}{4! \cdot 7!}$.   **6.** Simplify $\dfrac{n!}{(n+1)!}$.   **7.** Simplify $\dfrac{2n!}{(n-1)!}$.

8. Write an expression for the *apparent* nth term of the sequence $2, 5, 10, 17, 26, \ldots$. (Assume n begins with 1).

In Exercises 9 and 10, find a formula for the nth term of the sequence.

9. Arithmetic: $a_1 = 5000$, $d = -100$ **10.** Geometric: $a_1 = 4$, $a_{k+1} = \frac{1}{2} a_k$

11. Use sigma notation to write $\dfrac{2}{3(1)+1} + \dfrac{2}{3(2)+1} + \cdots + \dfrac{2}{3(12)+1}$.

12. Use sigma notation to write $2 + \frac{1}{2} + \frac{1}{8} + \frac{1}{32} + \frac{1}{128} + \cdots$.

In Exercises 13–15, find the sum.

13. $\displaystyle\sum_{n=1}^{7} (8n - 5)$ **14.** $\displaystyle\sum_{n=1}^{8} 24\left(\frac{1}{6}\right)^{n-1}$ **15.** $5 - 2 + \frac{4}{5} - \frac{8}{25} + \frac{16}{125} - \cdots$

16. Use mathematical induction to prove the formula

$$3 + 6 + 9 + \cdots + 3n = \frac{3n(n+1)}{2}.$$

17. Use the Binomial Theorem to expand and simplify $(2a - 5b)^4$.

In Exercises 18–21, evaluate the expression.

18. $_9C_3$ **19.** $_{20}C_3$ **20.** $_9P_2$ **21.** $_{70}P_3$

22. Solve for n in $4 \cdot {}_nP_3 = {}_{n+1}P_4$.

23. How many distinct license plates can be issued consisting of one letter followed by a three-digit number?

24. Four students are randomly selected from a class of 25 to answer questions from a reading assignment. In how many ways can the four be selected?

25. A card is drawn from a standard deck of 52 playing cards. Find the probability that it is a red face card.

26. In 2006, six of the eleven men's basketball teams in the Big Ten Conference were to participate in the NCAA Men's Basketball Championship Tournament. If six of the eleven schools are selected at random, what is the probability that the six teams chosen were the actual six teams selected to play?

27. Two integers from 1 to 60 are chosen by a random number generator. What is the probability that (a) both numbers are odd, (b) both numbers are less than 12, and (c) the same number is chosen twice?

28. A weather forecast indicates that the probability of snow is 75%. What is the probability that it will not snow?

Proofs in Mathematics

Properties of Sums   (p. 585)

1. $\displaystyle\sum_{i=1}^{n} c = cn,$ c is a constant.

2. $\displaystyle\sum_{i=1}^{n} ca_i = c\sum_{i=1}^{n} a_i,$ c is a constant.

3. $\displaystyle\sum_{i=1}^{n} (a_i + b_i) = \sum_{i=1}^{n} a_i + \sum_{i=1}^{n} b_i$

4. $\displaystyle\sum_{i=1}^{n} (a_i - b_i) = \sum_{i=1}^{n} a_i - \sum_{i=1}^{n} b_i$

Proof

Each of these properties follows directly from the properties of real numbers.

1. $\displaystyle\sum_{i=1}^{n} c = c + c + c + \cdots + c = cn$ n terms

The Distributive Property is used in the proof of Property 2.

2. $\displaystyle\sum_{i=1}^{n} ca_i = ca_1 + ca_2 + ca_3 + \cdots + ca_n$

$= c(a_1 + a_2 + a_3 + \cdots + a_n) = c\sum_{i=1}^{n} a_i$

The proof of Property 3 uses the Commutative and Associative Properties of Addition.

3.
$\displaystyle\sum_{i=1}^{n} (a_i + b_i) = (a_1 + b_1) + (a_2 + b_2) + (a_3 + b_3) + \cdots + (a_n + b_n)$

$= (a_1 + a_2 + a_3 + \cdots + a_n) + (b_1 + b_2 + b_3 + \cdots + b_n)$

$= \sum_{i=1}^{n} a_i + \sum_{i=1}^{n} b_i$

The proof of Property 4 uses the Commutative and Associative Properties of Addition and the Distributive Property.

4.
$\displaystyle\sum_{i=1}^{n} (a_i - b_i) = (a_1 - b_1) + (a_2 - b_2) + (a_3 - b_3) + \cdots + (a_n - b_n)$

$= (a_1 + a_2 + a_3 + \cdots + a_n) + (-b_1 - b_2 - b_3 - \cdots - b_n)$

$= (a_1 + a_2 + a_3 + \cdots + a_n) - (b_1 + b_2 + b_3 + \cdots + b_n)$

$= \sum_{i=1}^{n} a_i - \sum_{i=1}^{n} b_i$

Infinite Series

The study of infinite series was considered a novelty in the fourteenth century. Logician Richard Suiseth, whose nickname was Calculator, solved this problem.

If throughout the first half of a given time interval a variation continues at a certain intensity; throughout the next quarter of the interval at double the intensity; throughout the following eighth at triple the intensity and so ad infinitum; The average intensity for the whole interval will be the intensity of the variation during the second subinterval (or double the intensity).

This is the same as saying that the sum of the infinite series

$\dfrac{1}{2} + \dfrac{2}{4} + \dfrac{3}{8} + \cdots + \dfrac{n}{2^n} + \cdots$

is 2.

The Sum of a Finite Arithmetic Sequence   (p. 595)

The sum of a finite arithmetic sequence with n terms is given by

$$S_n = \frac{n}{2}(a_1 + a_n).$$

Proof

Begin by generating the terms of the arithmetic sequence in two ways. In the first way, repeatedly add d to the first term to obtain

$$S_n = a_1 + a_2 + a_3 + \cdots + a_{n-2} + a_{n-1} + a_n$$
$$= a_1 + [a_1 + d] + [a_1 + 2d] + \cdots + [a_1 + (n-1)d].$$

In the second way, repeatedly subtract d from the nth term to obtain

$$S_n = a_n + a_{n-1} + a_{n-2} + \cdots + a_3 + a_2 + a_1$$
$$= a_n + [a_n - d] + [a_n - 2d] + \cdots + [a_n - (n-1)d].$$

If you add these two versions of S_n, the multiples of d subtract out and you obtain

$$2S_n = (a_1 + a_n) + (a_1 + a_n) + (a_1 + a_n) + \cdots + (a_1 + a_n) \quad n \text{ terms}$$
$$2S_n = n(a_1 + a_n)$$
$$S_n = \frac{n}{2}(a_1 + a_n).$$

The Sum of a Finite Geometric Sequence   (p. 604)

The sum of the finite geometric sequence

$$a_1, \; a_1r, \; a_1r^2, \; a_1r^3, \; a_1r^4, \; \ldots, \; a_1r^{n-1}$$

with common ratio $r \neq 1$ is given by $S_n = \sum_{i=1}^{n} a_1 r^{i-1} = a_1\left(\dfrac{1 - r^n}{1 - r}\right).$

Proof

$$S_n = a_1 + a_1r + a_1r^2 + \cdots + a_1r^{n-2} + a_1r^{n-1}$$
$$rS_n = a_1r + a_1r^2 + a_1r^3 + \cdots + a_1r^{n-1} + a_1r^n \qquad \text{Multiply by } r.$$

Subtracting the second equation from the first yields

$$S_n - rS_n = a_1 - a_1r^n.$$

So, $S_n(1 - r) = a_1(1 - r^n)$, and, because $r \neq 1$, you have $S_n = a_1\left(\dfrac{1 - r^n}{1 - r}\right).$

The Binomial Theorem   (p. 619)

In the expansion of $(x + y)^n$

$$(x + y)^n = x^n + nx^{n-1}y + \cdots + {}_nC_r\, x^{n-r}y^r + \cdots + nxy^{n-1} + y^n$$

the coefficient of $x^{n-r}y^r$ is

$${}_nC_r = \frac{n!}{(n - r)!r!}.$$

Proof

The Binomial Theorem can be proved quite nicely using mathematical induction. The steps are straightforward but look a little messy, so only an outline of the proof is presented.

1. If $n = 1$, you have $(x + y)^1 = x^1 + y^1 = {}_1C_0x + {}_1C_1y$, and the formula is valid.

2. Assuming that the formula is true for $n = k$, the coefficient of $x^{k-r}y^r$ is

$${}_kC_r = \frac{k!}{(k - r)!r!} = \frac{k(k - 1)(k - 2) \cdots (k - r + 1)}{r!}.$$

To show that the formula is true for $n = k + 1$, look at the coefficient of $x^{k+1-r}y^r$ in the expansion of

$$(x + y)^{k+1} = (x + y)^k(x + y).$$

From the right-hand side, you can determine that the term involving $x^{k+1-r}y^r$ is the sum of two products.

$$({}_kC_r x^{k-r}y^r)(x) + ({}_kC_{r-1}x^{k+1-r}y^{r-1})(y)$$

$$= \left[\frac{k!}{(k - r)!r!} + \frac{k!}{(k + 1 - r)!(r - 1)!}\right]x^{k+1-r}y^r$$

$$= \left[\frac{(k + 1 - r)k!}{(k + 1 - r)!r!} + \frac{k!r}{(k + 1 - r)!r!}\right]x^{k+1-r}y^r$$

$$= \left[\frac{k!(k + 1 - r + r)}{(k + 1 - r)!r!}\right]x^{k+1-r}y^r$$

$$= \left[\frac{(k + 1)!}{(k + 1 - r)!r!}\right]x^{k+1-r}y^r$$

$$= {}_{k+1}C_r x^{k+1-r}y^r$$

So, by mathematical induction, the Binomial Theorem is valid for all positive integers n.

Chapter 9

9.1 Circles and Parabolas

9.2 Ellipses

9.3 Hyperbolas

9.4 Rotation and Systems of Quadratic Equations

9.5 Parametric Equations

9.6 Polar Coordinates

9.7 Graphs of Polar Equations

9.8 Polar Equations of Conics

Selected Applications

Analytic geometry concepts have many real-life applications. The applications listed below represent a small sample of the applications in this chapter.

■ Earthquake, Exercise 35, page 667

■ Suspension Bridge, Exercise 93, page 669

■ Architecture, Exercises 47–49, page 678

■ Satellite Orbit, Exercise 54, page 679

■ Navigation, Exercise 46, page 688

■ Projectile Motion, Exercises 55–58, page 706

■ Planetary Motion, Exercises 49–55, page 727

■ Sports, Exercises 95–98, page 732

Topics in Analytic Geometry

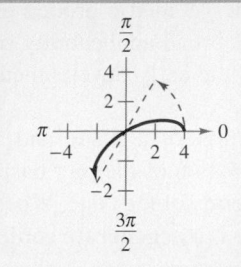
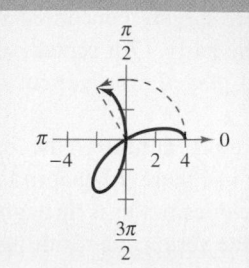
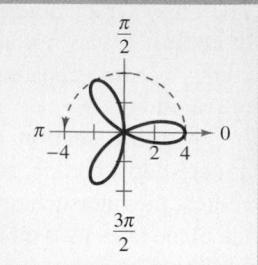

Conics are used to represent many real-life phenomena such as reflectors used in flashlights, orbits of planets, and navigation. In Chapter 9, you will learn how to write and graph equations of conics in rectangular and polar coordinates. You will also learn how to graph other polar equations and curves represented by parametric equations.

AP/Wide World Photos

Satellites are used to monitor weather patterns, collect scientific data, and assist in navigation. Satellites orbit Earth in elliptical paths.

9.1 Circles and Parabolas

Conics

Conic sections were discovered during the classical Greek period, 600 to 300 B.C. The early Greek studies were largely concerned with the geometric properties of conics. It was not until the early 17th century that the broad applicability of conics became apparent and played a prominent role in the early development of calculus.

A **conic section** (or simply **conic**) is the intersection of a plane and a double-napped cone. Notice in Figure 9.1 that in the formation of the four basic conics, the intersecting plane does not pass through the vertex of the cone. When the plane does pass through the vertex, the resulting figure is a **degenerate conic,** as shown in Figure 9.2.

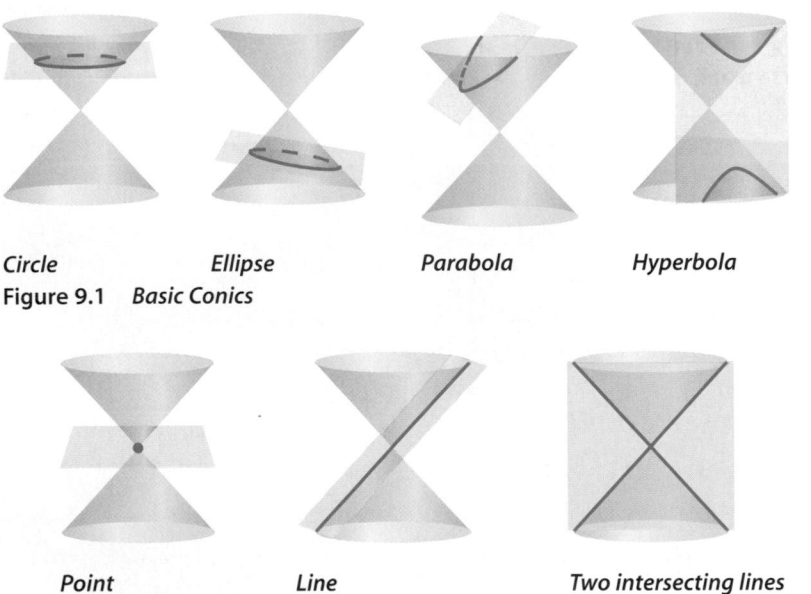

Circle   Ellipse   Parabola   Hyperbola
Figure 9.1   Basic Conics

Point   Line   Two intersecting lines
Figure 9.2   Degenerate Conics

There are several ways to approach the study of conics. You could begin by defining conics in terms of the intersections of planes and cones, as the Greeks did, or you could define them algebraically, in terms of the general second-degree equation

$$Ax^2 + Bxy + Cy^2 + Dx + Ey + F = 0.$$

However, you will study a third approach, in which each of the conics is defined as a **locus** (collection) of points satisfying a certain geometric property. For example, the definition of a circle as *the collection of all points (x, y) that are equidistant from a fixed point (h, k)* leads to the standard equation of a circle

$$(x - h)^2 + (y - k)^2 = r^2. \qquad \text{Equation of circle}$$

Circles

The definition of a circle as a locus of points is a more general definition of a circle as it applies to conics.

Definition of a Circle

A **circle** is the set of all points (x, y) in a plane that are equidistant from a fixed point (h, k), called the **center** of the circle. (See Figure 9.3.) The distance r between the center and any point (x, y) on the circle is the **radius.**

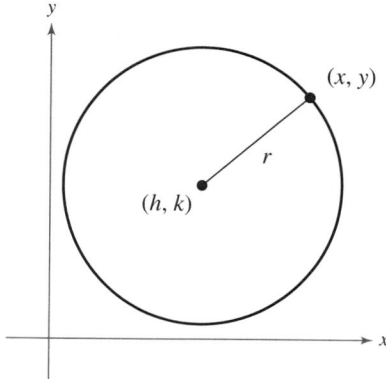

Figure 9.3

The Distance Formula can be used to obtain an equation of a circle whose center is (h, k) and whose radius is r.

$\sqrt{(x - h)^2 + (y - k)^2} = r$ Distance Formula

$(x - h)^2 + (y - k)^2 = r^2$ Square each side.

Standard Form of the Equation of a Circle

The **standard form of the equation of a circle** is

$$(x - h)^2 + (y - k)^2 = r^2.$$

The point (h, k) is the center of the circle, and the positive number r is the radius of the circle. The standard form of the equation of a circle whose center is the origin, $(h, k) = (0, 0)$, is

$$x^2 + y^2 = r^2.$$

Example 1   Finding the Standard Equation of a Circle

The point $(1, 4)$ is on a circle whose center is at $(-2, -3)$, as shown in Figure 9.4. Write the standard form of the equation of the circle.

Solution

The radius of the circle is the distance between $(-2, -3)$ and $(1, 4)$.

$r = \sqrt{[1 - (-2)]^2 + [4 - (-3)]^2}$ Use Distance Formula.

$\quad = \sqrt{3^2 + 7^2}$ Simplify.

$\quad = \sqrt{58}$ Simplify.

The equation of the circle with center $(h, k) = (-2, -3)$ and radius $r = \sqrt{58}$ is

$(x - h)^2 + (y - k)^2 = r^2$ Standard form

$[x - (-2)]^2 + [y - (-3)]^2 = \left(\sqrt{58}\right)^2$ Substitute for h, k, and r.

$(x + 2)^2 + (y + 3)^2 = 58.$ Simplify.

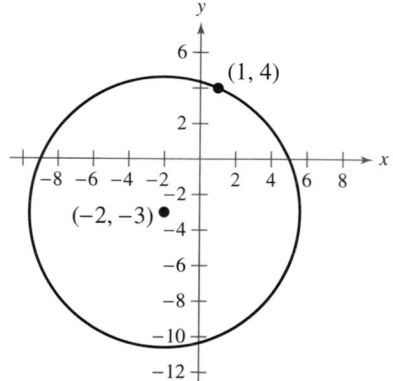

Figure 9.4

✓**CHECKPOINT**   Now try Exercise 3.

Example 2 Sketching a Circle

Sketch the circle given by the equation

$$x^2 - 6x + y^2 - 2y + 6 = 0$$

and identify its center and radius.

Solution

Begin by writing the equation in standard form.

$$x^2 - 6x + y^2 - 2y + 6 = 0 \qquad \text{Write original equation.}$$

$$(x^2 - 6x + 9) + (y^2 - 2y + 1) = -6 + 9 + 1 \qquad \text{Complete the squares.}$$

$$(x - 3)^2 + (y - 1)^2 = 4 \qquad \text{Write in standard form.}$$

In this form, you can see that the graph is a circle whose center is the point $(3, 1)$ and whose radius is $r = \sqrt{4} = 2$. Plot several points that are two units from the center. The points $(5, 1)$, $(3, 3)$, $(1, 1)$, and $(3, -1)$ are convenient. Draw a circle that passes through the four points, as shown in Figure 9.5.

✓CHECKPOINT   Now try Exercise 23.

Prerequisite Skills

To use a graphing utility to graph a circle, review Appendix B.2.

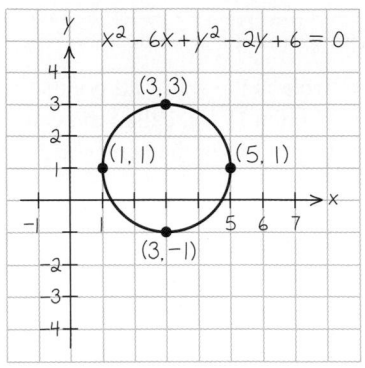

Figure 9.5

Example 3 Finding the Intercepts of a Circle

Find the x- and y-intercepts of the graph of the circle given by the equation

$$(x - 4)^2 + (y - 2)^2 = 16.$$

Solution

To find any x-intercepts, let $y = 0$. To find any y-intercepts, let $x = 0$.

x-intercepts:

$$(x - 4)^2 + (0 - 2)^2 = 16 \qquad \text{Substitute 0 for } y.$$

$$(x - 4)^2 = 12 \qquad \text{Simplify.}$$

$$x - 4 = \pm\sqrt{12} \qquad \text{Take square root of each side.}$$

$$x = 4 \pm 2\sqrt{3} \qquad \text{Add 4 to each side.}$$

y-intercepts:

$$(0 - 4)^2 + (y - 2)^2 = 16 \qquad \text{Substitute 0 for } x.$$

$$(y - 2)^2 = 0 \qquad \text{Simplify.}$$

$$y - 2 = 0 \qquad \text{Take square root of each side.}$$

$$y = 2 \qquad \text{Add 2 to each side.}$$

So the x-intercepts are $\left(4 + 2\sqrt{3}, 0\right)$ and $\left(4 - 2\sqrt{3}, 0\right)$, and the y-intercept is $(0, 2)$, as shown in Figure 9.6.

✓CHECKPOINT   Now try Exercise 29.

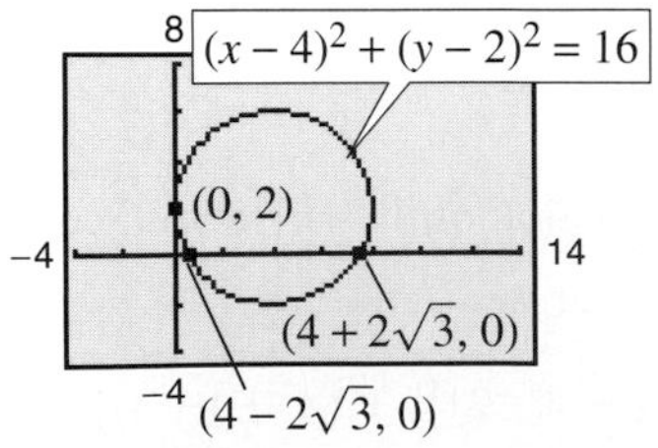

Figure 9.6

Parabolas

In Section 2.1, you learned that the graph of the quadratic function

$$f(x) = ax^2 + bx + c$$

is a parabola that opens upward or downward. The following definition of a parabola is more general in the sense that it is independent of the orientation of the parabola.

Definition of a Parabola

A **parabola** is the set of all points (x, y) in a plane that are equidistant from a fixed line, the **directrix,** and a fixed point, the **focus,** not on the line. (See Figure 9.7.) The midpoint between the focus and the directrix is the **vertex,** and the line passing through the focus and the vertex is the **axis** of the parabola.

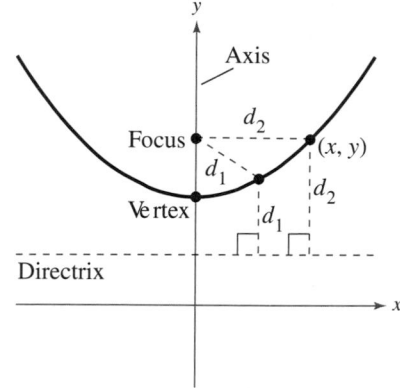

Figure 9.7

Note in Figure 9.7 that a parabola is symmetric with respect to its axis. Using the definition of a parabola, you can derive the following **standard form of the equation of a parabola** whose directrix is parallel to the x-axis or to the y-axis.

Standard Equation of a Parabola   (See the proof on page 737.)

The **standard form of the equation of a parabola** with vertex at (h, k) is as follows.

$$(x - h)^2 = 4p(y - k), \ p \neq 0 \qquad \text{Vertical axis; directrix: } y = k - p$$

$$(y - k)^2 = 4p(x - h), \ p \neq 0 \qquad \text{Horizontal axis; directrix: } x = h - p$$

The focus lies on the axis p units (*directed distance*) from the vertex. If the vertex is at the origin $(0, 0)$, the equation takes one of the following forms.

$$x^2 = 4py \qquad \text{Vertical axis}$$

$$y^2 = 4px \qquad \text{Horizontal axis}$$

See Figure 9.8.

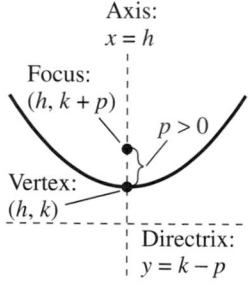

$(x - h)^2 = 4p(y - k)$

(a) Vertical axis: $p > 0$

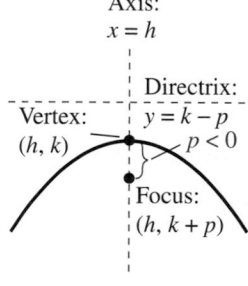

$(x - h)^2 = 4p(y - k)$

(b) Vertical axis: $p < 0$

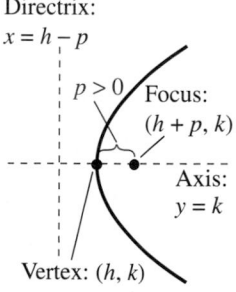

$(y - k)^2 = 4p(x - h)$

(c) Horizontal axis: $p > 0$

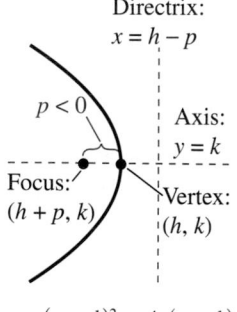

$(y - k)^2 = 4p(x - h)$

(d) Horizontal axis: $p < 0$

Figure 9.8

Example 4 Finding the Standard Equation of a Parabola

Find the standard form of the equation of the parabola with vertex at the origin and focus $(0, 4)$.

Solution

Because the axis of the parabola is vertical, passing through $(0, 0)$ and $(0, 4)$, consider the equation

$$x^2 = 4py.$$

Because the focus is $p = 4$ units from the vertex, the equation is

$$x^2 = 4(4)y$$

$$x^2 = 16y.$$

You can obtain the more common quadratic form as follows.

$$x^2 = 16y \qquad \text{Write original equation.}$$

$$\tfrac{1}{16}x^2 = y \qquad \text{Multiply each side by } \tfrac{1}{16}.$$

Use a graphing utility to confirm that the graph is a parabola, as shown in Figure 9.9.

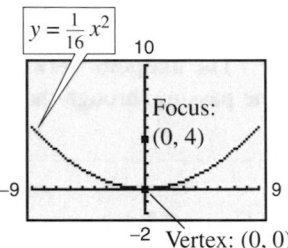

Figure 9.9

✓CHECKPOINT Now try Exercise 45.

Example 5 Finding the Focus of a Parabola

Find the focus of the parabola given by $y = -\tfrac{1}{2}x^2 - x + \tfrac{1}{2}$.

Solution

To find the focus, convert to standard form by completing the square.

$$y = -\tfrac{1}{2}x^2 - x + \tfrac{1}{2} \qquad \text{Write original equation.}$$

$$-2y = x^2 + 2x - 1 \qquad \text{Multiply each side by } -2.$$

$$1 - 2y = x^2 + 2x \qquad \text{Add 1 to each side.}$$

$$1 + 1 - 2y = x^2 + 2x + 1 \qquad \text{Complete the square.}$$

$$2 - 2y = x^2 + 2x + 1 \qquad \text{Combine like terms.}$$

$$-2(y - 1) = (x + 1)^2 \qquad \text{Write in standard form.}$$

Comparing this equation with

$$(x - h)^2 = 4p(y - k)$$

you can conclude that $h = -1$, $k = 1$, and $p = -\tfrac{1}{2}$. Because p is negative, the parabola opens downward, as shown in Figure 9.10. Therefore, the focus of the parabola is

$$(h, k + p) = \left(-1, \tfrac{1}{2}\right). \qquad \text{Focus}$$

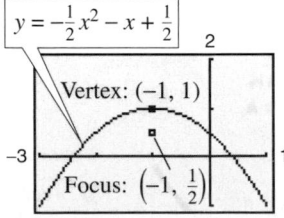

Figure 9.10

✓CHECKPOINT Now try Exercise 63.

Example 6   Finding the Standard Equation of a Parabola

Find the standard form of the equation of the parabola with vertex $(1, 0)$ and focus at $(2, 0)$.

Solution

Because the axis of the parabola is horizontal, passing through $(1, 0)$ and $(2, 0)$, consider the equation

$$(y - k)^2 = 4p(x - h)$$

where $h = 1$, $k = 0$, and $p = 2 - 1 = 1$. So, the standard form is

$$(y - 0)^2 = 4(1)(x - 1) \quad \Longrightarrow \quad y^2 = 4(x - 1).$$

The parabola is shown in Figure 9.11.

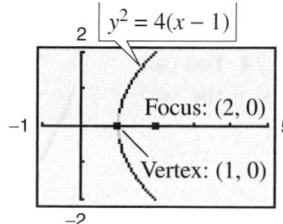

Figure 9.11

✓CHECKPOINT   Now try Exercise 77.

TECHNOLOGY TIP   Use a graphing utility to confirm the equation found in Example 6. To do this, it helps to graph the equation using two separate equations: $y_1 = \sqrt{4(x - 1)}$ (upper part) and $y_2 = -\sqrt{4(x - 1)}$ (lower part). Note that when you graph conics using two separate equations, your graphing utility may not connect the two parts. This is because some graphing utilities are limited in their resolution. So, in this text, a blue curve is placed behind the graphing utility's display to indicate where the graph should appear.

Reflective Property of Parabolas

A line segment that passes through the focus of a parabola and has endpoints on the parabola is called a **focal chord.** The specific focal chord perpendicular to the axis of the parabola is called the **latus rectum.**

Parabolas occur in a wide variety of applications. For instance, a parabolic reflector can be formed by revolving a parabola about its axis. The resulting surface has the property that all incoming rays parallel to the axis are reflected through the focus of the parabola. This is the principle behind the construction of the parabolic mirrors used in reflecting telescopes. Conversely, the light rays emanating from the focus of a parabolic reflector used in a flashlight are all parallel to one another, as shown in Figure 9.12.

A line is **tangent** to a parabola at a point on the parabola if the line intersects, but does not cross, the parabola at the point. Tangent lines to parabolas have special properties related to the use of parabolas in constructing reflective surfaces.

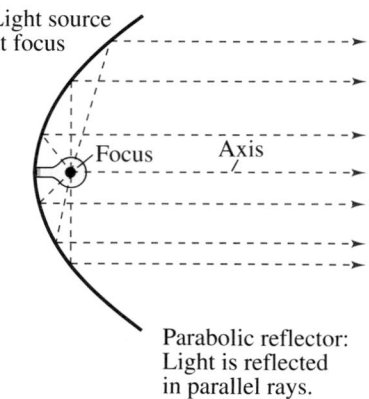

Parabolic reflector:
Light is reflected
in parallel rays.

Figure 9.12

> **Reflective Property of a Parabola**
>
> The tangent line to a parabola at a point P makes equal angles with the following two lines (see Figure 9.13).
>
> **1.** The line passing through P and the focus
>
> **2.** The axis of the parabola

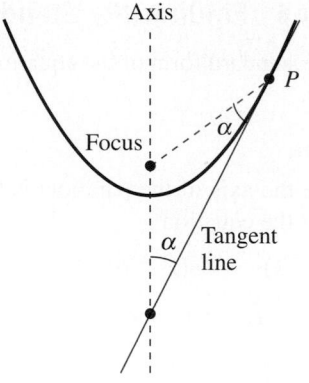

Figure 9.13

Example 7 Finding the Tangent Line at a Point on a Parabola

Find the equation of the tangent line to the parabola given by $y = x^2$ at the point $(1, 1)$.

Solution

For this parabola, $p = \frac{1}{4}$ and the focus is $\left(0, \frac{1}{4}\right)$, as shown in Figure 9.14. You can find the y-intercept $(0, b)$ of the tangent line by equating the lengths of the two sides of the isosceles triangle shown in Figure 9.14:

$$d_1 = \frac{1}{4} - b$$

and

$$d_2 = \sqrt{(1 - 0)^2 + \left(1 - \frac{1}{4}\right)^2} = \frac{5}{4}.$$

Note that $d_1 = \frac{1}{4} - b$ rather than $b - \frac{1}{4}$. The order of subtraction for the distance is important because the distance must be positive. Setting $d_1 = d_2$ produces

$$\frac{1}{4} - b = \frac{5}{4}$$

$$b = -1.$$

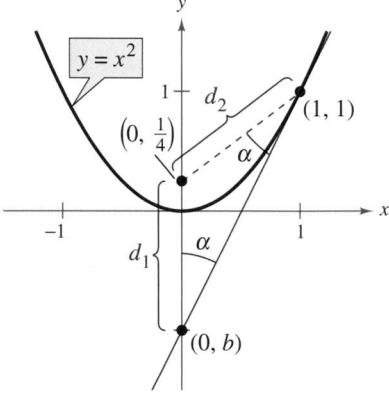

Figure 9.14

So, the slope of the tangent line is

$$m = \frac{1 - (-1)}{1 - 0} = 2$$

and the equation of the tangent line in slope-intercept form is

$$y = 2x - 1.$$

✓CHECKPOINT Now try Exercise 85.

TECHNOLOGY TIP Try using a graphing utility to confirm the result of Example 7. By graphing

$$y_1 = x^2 \quad \text{and} \quad y_2 = 2x - 1$$

in the same viewing window, you should be able to see that the line touches the parabola at the point $(1, 1)$.

9.1   Exercises

See www.CalcChat.com for worked-out solutions to odd-numbered exercises.

Vocabulary Check

Fill in the blanks.

1. A _____ is the intersection of a plane and a double-napped cone.

2. A collection of points satisfying a geometric property can also be referred to as a _____ of points.

3. A _____ is the set of all points (x, y) in a plane that are equidistant from a fixed point, called the _____ .

4. A _____ is the set of all points (x, y) in a plane that are equidistant from a fixed line, called the _____ , and a fixed point, called the _____ , not on the line.

5. The _____ of a parabola is the midpoint between the focus and the directrix.

6. The line that passes through the focus and vertex of a parabola is called the _____ of the parabola.

7. A line is _____ to a parabola at a point on the parabola if the line intersects, but does not cross, the parabola at the point.

In Exercises 1–6, find the standard form of the equation of the circle with the given characteristics.

1. Center at origin;   radius: $\sqrt{18}$

2. Center at origin;   radius: $4\sqrt{2}$

3. Center: $(3, 7)$;   point on circle: $(1, 0)$

4. Center: $(6, -3)$;   point on circle: $(-2, 4)$

5. Center: $(-3, -1)$;   diameter: $2\sqrt{7}$

6. Center: $(5, -6)$;   diameter: $4\sqrt{3}$

In Exercises 7–12, identify the center and radius of the circle.

7. $x^2 + y^2 = 49$ 8. $x^2 + y^2 = 1$

9. $(x + 2)^2 + (y - 7)^2 = 16$

10. $(x + 9)^2 + (y + 1)^2 = 36$

11. $(x - 1)^2 + y^2 = 15$ 12. $x^2 + (y + 12)^2 = 24$

In Exercises 13–20, write the equation of the circle in standard form. Then identify its center and radius.

13. $\frac{1}{4}x^2 + \frac{1}{4}y^2 = 1$ 14. $\frac{1}{9}x^2 + \frac{1}{9}y^2 = 1$

15. $\frac{4}{3}x^2 + \frac{4}{3}y^2 = 1$ 16. $\frac{9}{2}x^2 + \frac{9}{2}y^2 = 1$

17. $x^2 + y^2 - 2x + 6y + 9 = 0$

18. $x^2 + y^2 - 10x - 6y + 25 = 0$

19. $4x^2 + 4y^2 + 12x - 24y + 41 = 0$

20. $9x^2 + 9y^2 + 54x - 36y + 17 = 0$

In Exercises 21–28, sketch the circle. Identify its center and radius.

21. $x^2 = 16 - y^2$ 22. $y^2 = 81 - x^2$

23. $x^2 + 4x + y^2 + 4y - 1 = 0$

24. $x^2 - 6x + y^2 + 6y + 14 = 0$

25. $x^2 - 14x + y^2 + 8y + 40 = 0$

26. $x^2 + 6x + y^2 - 12y + 41 = 0$

27. $x^2 + 2x + y^2 - 35 = 0$ 28. $x^2 + y^2 + 10y + 9 = 0$

In Exercises 29–34, find the x- and y-intercepts of the graph of the circle.

29. $(x - 2)^2 + (y + 3)^2 = 9$

30. $(x + 5)^2 + (y - 4)^2 = 25$

31. $x^2 - 2x + y^2 - 6y - 27 = 0$

32. $x^2 + 8x + y^2 + 2y + 9 = 0$

33. $(x - 6)^2 + (y + 3)^2 = 16$

34. $(x + 7)^2 + (y - 8)^2 = 4$

35. *Earthquake*   An earthquake was felt up to 81 miles from its epicenter. You were located 60 miles west and 45 miles south of the epicenter.

(a) Let the epicenter be at the point $(0, 0)$. Find the standard equation that describes the outer boundary of the earthquake.

(b) Would you have felt the earthquake?

(c) Verify your answer to part (b) by graphing the equation of the outer boundary of the earthquake and plotting your location. How far were you from the outer boundary of the earthquake?

36. *Landscaper*   A landscaper has installed a circular sprinkler system that covers an area of 1800 square feet.

(a) Find the radius of the region covered by the sprinkler system. Round your answer to three decimal places.

(b) If the landscaper wants to cover an area of 2400 square feet, how much longer does the radius need to be?

In Exercises 37–42, match the equation with its graph. [The graphs are labeled (a), (b), (c), (d), (e), and (f).]

(a)
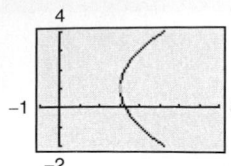

(b)
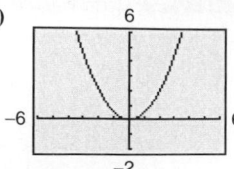

(c)
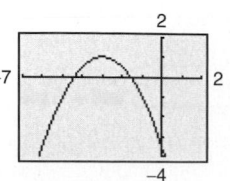

(d)
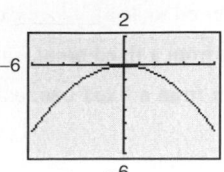

(e)
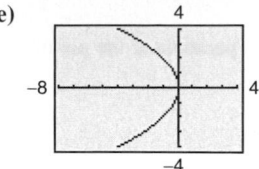

(f)
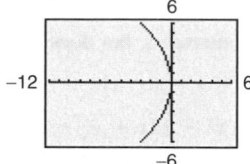

37. $y^2 = -4x$ **38.** $x^2 = 2y$

39. $x^2 = -8y$ **40.** $y^2 = -12x$

41. $(y - 1)^2 = 4(x - 3)$ **42.** $(x + 3)^2 = -2(y - 1)$

In Exercises 43–54, find the standard form of the equation of the parabola with the given characteristic(s) and vertex at the origin.

43.
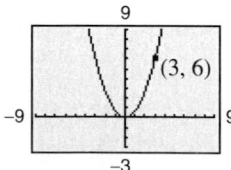

44.
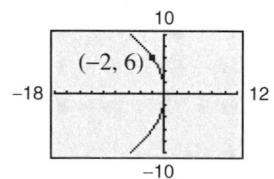

45. Focus: $\left(0, -\frac{3}{2}\right)$ **46.** Focus: $\left(\frac{5}{2}, 0\right)$

47. Focus: $(-2, 0)$ **48.** Focus: $(0, 1)$

49. Directrix: $y = -1$ **50.** Directrix: $y = 3$

51. Directrix: $x = 2$ **52.** Directrix: $x = -3$

53. Horizontal axis and passes through the point $(4, 6)$

54. Vertical axis and passes through the point $(-3, -3)$

In Exercises 55–72, find the vertex, focus, and directrix of the parabola and sketch its graph.

55. $y = \frac{1}{2}x^2$ **56.** $y = -4x^2$

57. $y^2 = -6x$ **58.** $y^2 = 3x$

59. $x^2 + 8y = 0$ **60.** $x + y^2 = 0$

61. $(x + 1)^2 + 8(y + 3) = 0$

62. $(x - 5) + (y + 4)^2 = 0$

63. $y^2 + 6y + 8x + 25 = 0$

64. $y^2 - 4y - 4x = 0$

65. $\left(x + \frac{3}{2}\right)^2 = 4(y - 2)$

66. $\left(x + \frac{1}{2}\right)^2 = 4(y - 1)$

67. $y = \frac{1}{4}(x^2 - 2x + 5)$

68. $x = \frac{1}{4}(y^2 + 2y + 33)$

69. $x^2 + 4x + 6y - 2 = 0$

70. $x^2 - 2x + 8y + 9 = 0$

71. $y^2 + x + y = 0$

72. $y^2 - 4x - 4 = 0$

In Exercises 73–82, find the standard form of the equation of the parabola with the given characteristics.

73.
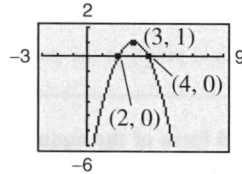

74.
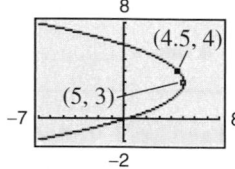

75. Vertex: $(-2, 0)$; focus: $\left(-\frac{3}{2}, 0\right)$

76. Vertex: $(3, -3)$; focus: $\left(3, -\frac{9}{4}\right)$

77. Vertex: $(5, 2)$; focus: $(3, 2)$

78. Vertex: $(-1, 2)$; focus: $(-1, 0)$

79. Vertex: $(0, 4)$; directrix: $y = 2$

80. Vertex: $(-2, 1)$; directrix: $x = 1$

81. Focus: $(2, 2)$; directrix: $x = -2$

82. Focus: $(0, 0)$; directrix: $y = 4$

In Exercises 83 and 84, the equations of a parabola and a tangent line to the parabola are given. Use a graphing utility to graph both in the same viewing window. Determine the coordinates of the point of tangency.

Parabola	*Tangent Line*
83. $y^2 - 8x = 0$	$x - y + 2 = 0$
84. $x^2 + 12y = 0$	$x + y - 3 = 0$

In Exercises 85–88, find an equation of the tangent line to the parabola at the given point and find the x-intercept of the line.

85. $x^2 = 2y$, $(4, 8)$

86. $x^2 = 2y$, $\left(-3, \frac{9}{2}\right)$

87. $y = -2x^2$, $(-1, -2)$

88. $y = -2x^2$, $(2, -8)$

89. *Revenue*   The revenue R (in dollars) generated by the sale of x 32-inch televisions is modeled by $R = 375x - \frac{3}{2}x^2$. Use a graphing utility to graph the function and approximate the sales that will maximize revenue.

90. *Beam Deflection* A simply supported beam is 64 feet long and has a load at the center (see figure). The deflection (bending) of the beam at its center is 1 inch. The shape of the deflected beam is parabolic.

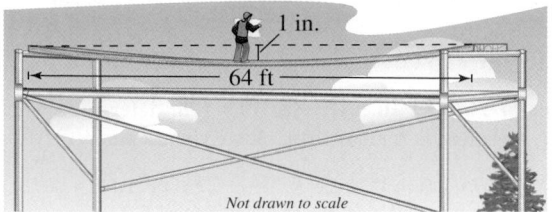

1 in.

64 ft

Not drawn to scale

(a) Find an equation of the parabola. (Assume that the origin is at the center of the beam.)

(b) How far from the center of the beam is the deflection equal to $\frac{1}{2}$ inch?

91. *Automobile Headlight* The filament of an automobile headlight is at the focus of a parabolic reflector, which sends light out in a straight beam (see figure).

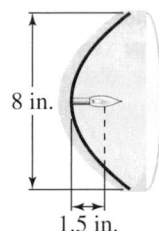

8 in.

1.5 in.

(a) The filament of the headlight is 1.5 inches from the vertex. Find an equation for the cross section of the reflector.

(b) The reflector is 8 inches wide. Find the depth of the reflector.

92. *Solar Cooker* You want to make a solar hot dog cooker using aluminum foil-lined cardboard, shaped as a parabolic trough. The figure shows how to suspend the hot dog with a wire through the foci of the ends of the parabolic trough. The parabolic end pieces are 12 inches wide and 4 inches deep. How far from the bottom of the trough should the wire be inserted?

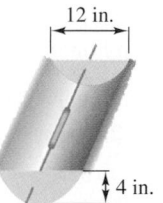

12 in.

4 in.

93. *Suspension Bridge* Each cable of the Golden Gate Bridge is suspended (in the shape of a parabola) between two towers that are 1280 meters apart. The top of each tower is 152 meters above the roadway. The cables touch the roadway midway between the towers.

(a) Draw a sketch of the bridge. Locate the origin of a rectangular coordinate system at the center of the roadway. Label the coordinates of the known points.

(b) Write an equation that models the cables.

(c) Complete the table by finding the height y of the suspension cables over the roadway at a distance of x meters from the center of the bridge.

x	0	200	400	500	600
y					

94. *Road Design* Roads are often designed with parabolic surfaces to allow rain to drain off. A particular road that is 32 feet wide is 0.4 foot higher in the center than it is on the sides (see figure).

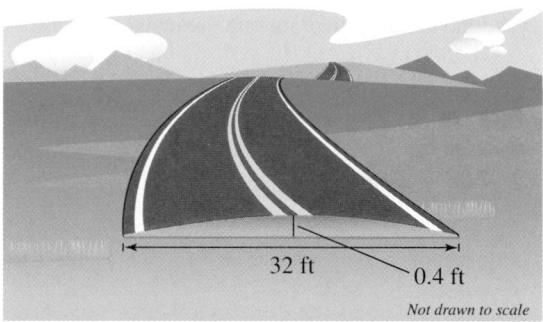

32 ft

0.4 ft

Not drawn to scale

(a) Find an equation of the parabola that models the road surface. (Assume that the origin is at the center of the road.)

(b) How far from the center of the road is the road surface 0.1 foot lower than in the middle?

95. *Highway Design* Highway engineers design a parabolic curve for an entrance ramp from a straight street to an interstate highway (see figure). Find an equation of the parabola.

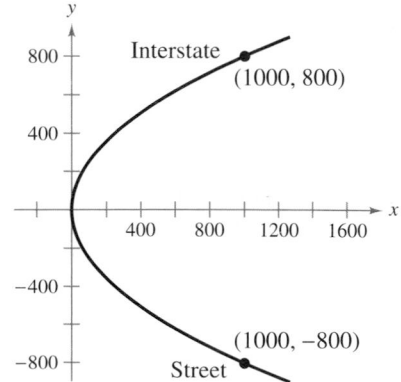

Interstate

(1000, 800)

Street

(1000, −800)

96. *Satellite Orbit* A satellite in a 100-mile-high circular orbit around Earth has a velocity of approximately 17,500 miles per hour. If this velocity is multiplied by $\sqrt{2}$, the satellite will have the minimum velocity necessary to escape Earth's gravity, and it will follow a parabolic path with the center of Earth as the focus (see figure).

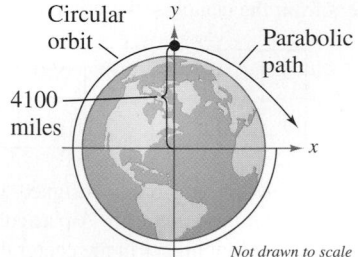

Not drawn to scale

(a) Find the escape velocity of the satellite.

(b) Find an equation of its path (assume the radius of Earth is 4000 miles).

97. *Path of a Projectile* The path of a softball is modeled by

$$-12.5(y - 7.125) = (x - 6.25)^2.$$

The coordinates x and y are measured in feet, with $x = 0$ corresponding to the position from which the ball was thrown.

(a) Use a graphing utility to graph the trajectory of the softball.

(b) Use the *zoom* and *trace* features of the graphing utility to approximate the highest point the ball reaches and the distance the ball travels.

98. *Projectile Motion* Consider the path of a projectile projected horizontally with a velocity of v feet per second at a height of s feet, where the model for the path is $x^2 = -\frac{1}{16}v^2(y - s)$. In this model, air resistance is disregarded, y is the height (in feet) of the projectile, and x is the horizontal distance (in feet) the projectile travels. A ball is thrown from the top of a 75-foot tower with a velocity of 32 feet per second.

(a) Find the equation of the parabolic path.

(b) How far does the ball travel horizontally before striking the ground?

In Exercises 99–102, find an equation of the tangent line to the circle at the indicated point. Recall from geometry that the tangent line to a circle is perpendicular to the radius of the circle at the point of tangency.

	Circle	*Point*
99.	$x^2 + y^2 = 25$	$(3, -4)$
100.	$x^2 + y^2 = 169$	$(-5, 12)$
101.	$x^2 + y^2 = 12$	$(2, -2\sqrt{2})$
102.	$x^2 + y^2 = 24$	$(-2\sqrt{5}, 2)$

Synthesis

True or False? **In Exercises 103–108, determine whether the statement is true or false. Justify your answer.**

103. The equation $x^2 + (y + 5)^2 = 25$ represents a circle with its center at the origin and a radius of 5.

104. The graph of the equation $x^2 + y^2 = r^2$ will have x-intercepts $(\pm r, 0)$ and y-intercepts $(0, \pm r)$.

105. A circle is a degenerate conic.

106. It is possible for a parabola to intersect its directrix.

107. The point which lies on the graph of a parabola closest to its focus is the vertex of the parabola.

108. The directrix of the parabola $x^2 = y$ intersects, or is tangent to, the graph of the parabola at its vertex, $(0, 0)$.

109. *Writing* Cross sections of television antenna dishes are parabolic in shape (see figure). Write a paragraph describing why these dishes are parabolic. Include a graphical representation of your description.

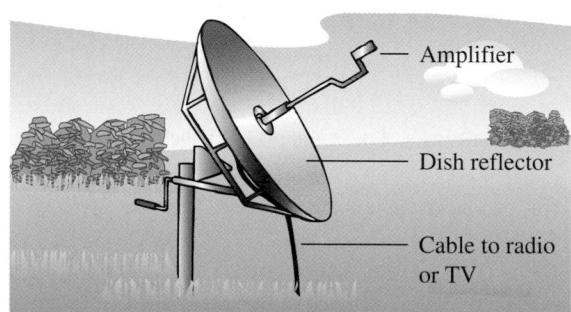

110. *Think About It* The equation $x^2 + y^2 = 0$ is a degenerate conic. Sketch the graph of this equation and identify the degenerate conic. Describe the intersection of the plane with the double-napped cone for this particular conic.

Think About It **In Exercises 111 and 112, change the equation so that its graph matches the description.**

111. $(y - 3)^2 = 6(x + 1)$; upper half of parabola

112. $(y + 1)^2 = 2(x - 2)$; lower half of parabola

Skills Review

In Exercises 113–116, use a graphing utility to approximate any relative minimum or maximum values of the function.

113. $f(x) = 3x^3 - 4x + 2$

114. $f(x) = 2x^2 + 3x$

115. $f(x) = x^4 + 2x + 2$

116. $f(x) = x^5 - 3x - 1$

9.2   Ellipses

Introduction

The third type of conic is called an **ellipse.** It is defined as follows.

> ### Definition of an Ellipse
>
> An **ellipse** is the set of all points (x, y) in a plane, the sum of whose distances from two distinct fixed points (**foci**) is constant. [See Figure 9.15(a).]

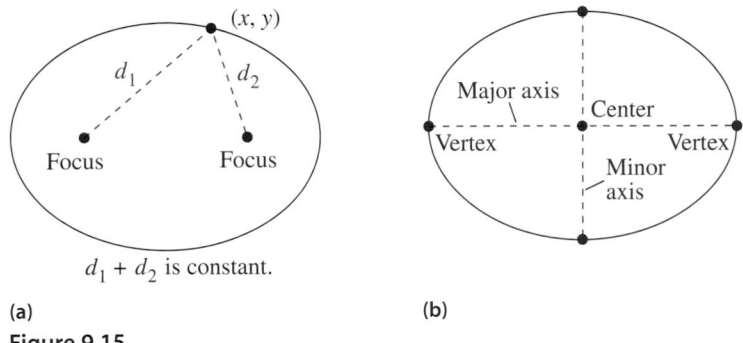

$d_1 + d_2$ is constant.

(a)   (b)

Figure 9.15

The line through the foci intersects the ellipse at two points called **vertices.** The chord joining the vertices is the **major axis,** and its midpoint is the **center** of the ellipse. The chord perpendicular to the major axis at the center is the **minor axis.** [See Figure 9.15(b).]

You can visualize the definition of an ellipse by imagining two thumbtacks placed at the foci, as shown in Figure 9.16. If the ends of a fixed length of string are fastened to the thumbtacks and the string is drawn taut with a pencil, the path traced by the pencil will be an ellipse.

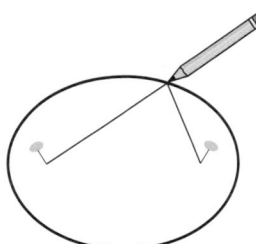

Figure 9.16

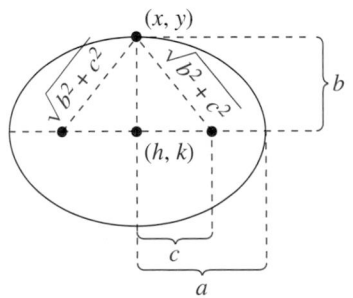

$$2\sqrt{b^2 + c^2} = 2a$$
$$b^2 + c^2 = a^2$$

Figure 9.17

To derive the standard form of the equation of an ellipse, consider the ellipse in Figure 9.17 with the following points: center, (h, k); vertices, $(h \pm a, k)$; foci, $(h \pm c, k)$. Note that the center is the midpoint of the segment joining the foci.

The sum of the distances from any point on the ellipse to the two foci is constant. Using a vertex point, this constant sum is

$$(a + c) + (a - c) = 2a \qquad \text{Length of major axis}$$

or simply the length of the major axis.

Now, if you let (x, y) be *any* point on the ellipse, the sum of the distances between (x, y) and the two foci must also be $2a$. That is,

$$\sqrt{[x - (h - c)]^2 + (y - k)^2} + \sqrt{[x - (h + c)]^2 + (y - k)^2} = 2a.$$

Finally, in Figure 9.17, you can see that $b^2 = a^2 - c^2$, which implies that the equation of the ellipse is

$$b^2(x - h)^2 + a^2(y - k)^2 = a^2 b^2$$

$$\frac{(x - h)^2}{a^2} + \frac{(y - k)^2}{b^2} = 1.$$

You would obtain a similar equation in the derivation by starting with a vertical major axis. Both results are summarized as follows.

Standard Equation of an Ellipse

The **standard form of the equation of an ellipse** with center (h, k) and major and minor axes of lengths $2a$ and $2b$, respectively, where $0 < b < a$, is

$$\frac{(x - h)^2}{a^2} + \frac{(y - k)^2}{b^2} = 1 \qquad \text{Major axis is horizontal.}$$

$$\frac{(x - h)^2}{b^2} + \frac{(y - k)^2}{a^2} = 1. \qquad \text{Major axis is vertical.}$$

The foci lie on the major axis, c units from the center, with $c^2 = a^2 - b^2$. If the center is at the origin $(0, 0)$, the equation takes one of the following forms.

$$\frac{x^2}{a^2} + \frac{y^2}{b^2} = 1 \qquad \text{Major axis is horizontal.}$$

$$\frac{x^2}{b^2} + \frac{y^2}{a^2} = 1 \qquad \text{Major axis is vertical.}$$

Exploration

On page 671 it was noted that an ellipse can be drawn using two thumbtacks, a string of fixed length (greater than the distance between the two tacks), and a pencil. Try doing this. Vary the length of the string and the distance between the thumbtacks. Explain how to obtain ellipses that are almost circular. Explain how to obtain ellipses that are long and narrow.

Figure 9.18 shows both the vertical and horizontal orientations for an ellipse.

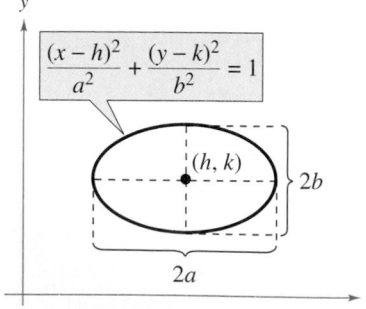

Major axis is horizontal.
Figure 9.18

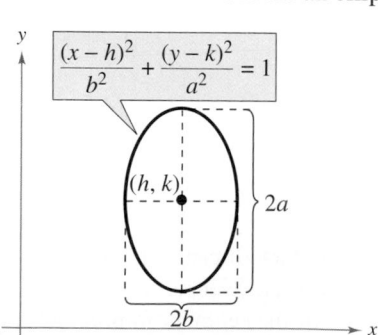

Major axis is vertical.

Example 1 Finding the Standard Equation of an Ellipse

Find the standard form of the equation of the ellipse having foci at $(0, 1)$ and $(4, 1)$ and a major axis of length 6, as shown in Figure 9.19.

Solution

By the Midpoint Formula, the center of the ellipse is $(2, 1)$ and the distance from the center to one of the foci is $c = 2$. Because $2a = 6$, you know that $a = 3$. Now, from $c^2 = a^2 - b^2$, you have

$$b = \sqrt{a^2 - c^2} = \sqrt{9 - 4} = \sqrt{5}.$$

Because the major axis is horizontal, the standard equation is

$$\frac{(x - 2)^2}{3^2} + \frac{(y - 1)^2}{(\sqrt{5})^2} = 1.$$

✓**CHECKPOINT**   Now try Exercise 35.

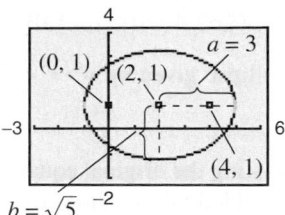

Figure 9.19

TECHNOLOGY SUPPORT

For instructions on how to use the *zoom* and *trace* features, see Appendix A; for specific keystrokes, go to this textbook's *Online Study Center*.

Example 2 Sketching an Ellipse

Sketch the ellipse given by $4x^2 + y^2 = 36$ and identify the center and vertices.

Algebraic Solution

$$4x^2 + y^2 = 36 \qquad \text{Write original equation.}$$

$$\frac{4x^2}{36} + \frac{y^2}{36} = \frac{36}{36} \qquad \text{Divide each side by 36.}$$

$$\frac{x^2}{3^2} + \frac{y^2}{6^2} = 1 \qquad \text{Write in standard form.}$$

The center of the ellipse is $(0, 0)$. Because the denominator of the y^2-term is larger than the denominator of the x^2-term, you can conclude that the major axis is vertical. Moreover, because $a = 6$, the vertices are $(0, -6)$ and $(0, 6)$. Finally, because $b = 3$, the endpoints of the minor axis are $(-3, 0)$ and $(3, 0)$, as shown in Figure 9.20.

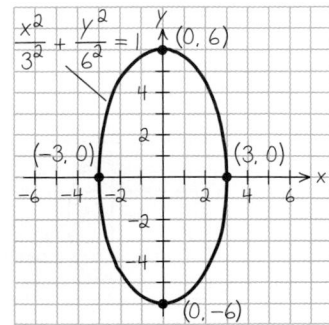

Figure 9.20

✓**CHECKPOINT**   Now try Exercise 13.

Graphical Solution

Solve the equation of the ellipse for y as follows.

$$4x^2 + y^2 = 36$$

$$y^2 = 36 - 4x^2$$

$$y = \pm\sqrt{36 - 4x^2}$$

Then use a graphing utility to graph $y_1 = \sqrt{36 - 4x^2}$ and $y_2 = -\sqrt{36 - 4x^2}$ in the same viewing window. Be sure to use a square setting. From the graph in Figure 9.21, you can see that the major axis is vertical and its center is at the point $(0, 0)$. You can use the *zoom* and *trace* features to approximate the vertices to be $(0, 6)$ and $(0, -6)$.

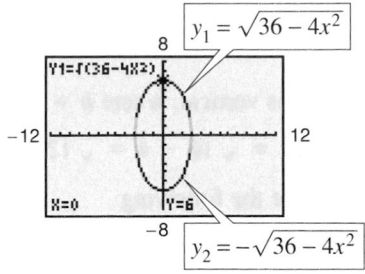

Figure 9.21

Example 3   Graphing an Ellipse

Graph the ellipse given by $x^2 + 4y^2 + 6x - 8y + 9 = 0$.

Solution

Begin by writing the original equation in standard form. In the third step, note that 9 and 4 are added to *both* sides of the equation when completing the squares.

$$x^2 + 4y^2 + 6x - 8y + 9 = 0 \qquad \text{Write original equation.}$$

$$\left(x^2 + 6x + \quad\right) + 4\left(y^2 - 2y + \quad\right) = -9 \qquad \begin{array}{l}\text{Group terms and factor 4}\\ \text{out of } y\text{-terms.}\end{array}$$

$$(x^2 + 6x + 9) + 4(y^2 - 2y + 1) = -9 + 9 + 4(1)$$

$$(x + 3)^2 + 4(y - 1)^2 = 4 \qquad \begin{array}{l}\text{Write in completed}\\ \text{square form.}\end{array}$$

$$\frac{(x + 3)^2}{2^2} + \frac{(y - 1)^2}{1^2} = 1 \qquad \text{Write in standard form.}$$

Now you see that the center is $(h, k) = (-3, 1)$. Because the denominator of the x-term is $a^2 = 2^2$, the endpoints of the major axis lie two units to the right and left of the center. Similarly, because the denominator of the y-term is $b^2 = 1^2$, the endpoints of the minor axis lie one unit up and down from the center. The graph of this ellipse is shown in Figure 9.22.

✔CHECKPOINT   Now try Exercise 15.

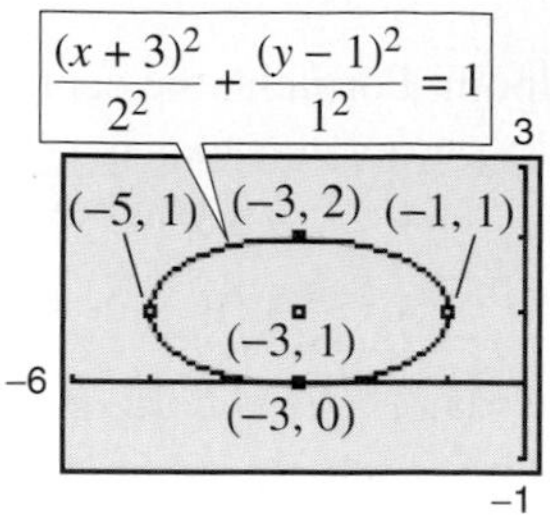

Figure 9.22

$$\frac{(x + 3)^2}{2^2} + \frac{(y - 1)^2}{1^2} = 1$$

(points shown: $(-5, 1)$, $(-3, 2)$, $(-1, 1)$, $(-3, 1)$, $(-3, 0)$)

Example 4   Analyzing an Ellipse

Find the center, vertices, and foci of the ellipse $4x^2 + y^2 - 8x + 4y - 8 = 0$.

Solution

By completing the square, you can write the original equation in standard form.

$$4x^2 + y^2 - 8x + 4y - 8 = 0 \qquad \text{Write original equation.}$$

$$4\left(x^2 - 2x + \quad\right) + \left(y^2 + 4y + \quad\right) = 8 \qquad \begin{array}{l}\text{Group terms and factor 4}\\ \text{out of } x\text{-terms.}\end{array}$$

$$4(x^2 - 2x + 1) + (y^2 + 4y + 4) = 8 + 4(1) + 4$$

$$4(x - 1)^2 + (y + 2)^2 = 16 \qquad \begin{array}{l}\text{Write in completed}\\ \text{square form.}\end{array}$$

$$\frac{(x - 1)^2}{2^2} + \frac{(y + 2)^2}{4^2} = 1 \qquad \text{Write in standard form.}$$

So, the major axis is vertical, where $h = 1$, $k = -2$, $a = 4$, $b = 2$, and

$$c = \sqrt{a^2 - b^2} = \sqrt{16 - 4} = \sqrt{12} = 2\sqrt{3}.$$

Therefore, you have the following.

Center: $(1, -2)$ Vertices: $(1, -6)$ Foci: $\left(1, -2 - 2\sqrt{3}\right)$

$\qquad\qquad\qquad\qquad\quad (1, 2)$ $\qquad\qquad\quad \left(1, -2 + 2\sqrt{3}\right)$

The graph of the ellipse is shown in Figure 9.23.

✔CHECKPOINT   Now try Exercise 17.

TECHNOLOGY TIP

You can use a graphing utility to graph an ellipse by graphing the upper and lower portions in the same viewing window. For instance, to graph the ellipse in Example 3, first solve for y to obtain

$$y_1 = 1 + \sqrt{1 - \frac{(x + 3)^2}{4}}$$

and

$$y_2 = 1 - \sqrt{1 - \frac{(x + 3)^2}{4}}.$$

Use a viewing window in which $-6 \le x \le 0$ and $-1 \le y \le 3$. You should obtain the graph shown in Figure 9.22.

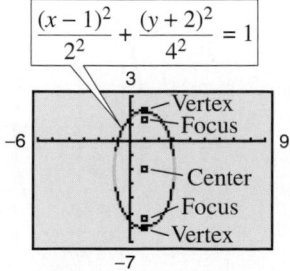

$$\frac{(x - 1)^2}{2^2} + \frac{(y + 2)^2}{4^2} = 1$$

Figure 9.23

Application

Ellipses have many practical and aesthetic uses. For instance, machine gears, supporting arches, and acoustic designs often involve elliptical shapes. The orbits of satellites and planets are also ellipses. Example 5 investigates the elliptical orbit of the moon about Earth.

Example 5   An Application Involving an Elliptical Orbit

The moon travels about Earth in an elliptical orbit with Earth at one focus, as shown in Figure 9.24. The major and minor axes of the orbit have lengths of 768,800 kilometers and 767,640 kilometers, respectively. Find the greatest and smallest distances (the *apogee* and *perigee*) from Earth's center to the moon's center.

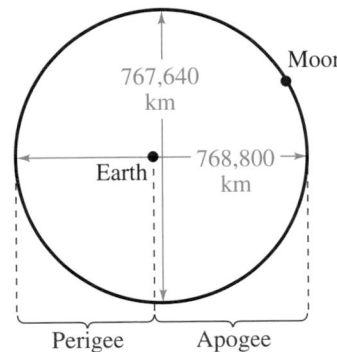

Figure 9.24

> **STUDY TIP**
>
> Note in Example 5 and Figure 9.24 that Earth *is not* the center of the moon's orbit.

Solution

Because $2a = 768,800$ and $2b = 767,640$, you have

$$a = 384,400 \quad \text{and} \quad b = 383,820$$

which implies that

$$c = \sqrt{a^2 - b^2}$$
$$= \sqrt{384,400^2 - 383,820^2}$$
$$\approx 21,108.$$

So, the greatest distance between the center of Earth and the center of the moon is

$$a + c \approx 384,400 + 21,108$$
$$= 405,508 \text{ kilometers}$$

and the smallest distance is

$$a - c \approx 384,400 - 21,108$$
$$= 363,292 \text{ kilometers}.$$

✓CHECKPOINT   Now try Exercise 53.

Eccentricity

One of the reasons it was difficult for early astronomers to detect that the orbits of the planets are ellipses is that the foci of the planetary orbits are relatively close to their centers, and so the orbits are nearly circular. To measure the ovalness of an ellipse, you can use the concept of **eccentricity.**

Definition of Eccentricity

The **eccentricity** e of an ellipse is given by the ratio $e = \dfrac{c}{a}$.

Note that $0 < e < 1$ for *every* ellipse.

To see how this ratio is used to describe the shape of an ellipse, note that because the foci of an ellipse are located along the major axis between the vertices and the center, it follows that

$$0 < c < a.$$

For an ellipse that is nearly circular, the foci are close to the center and the ratio c/a is small [see Figure 9.25(a)]. On the other hand, for an elongated ellipse, the foci are close to the vertices and the ratio c/a is close to 1 [see Figure 9.25(b)].

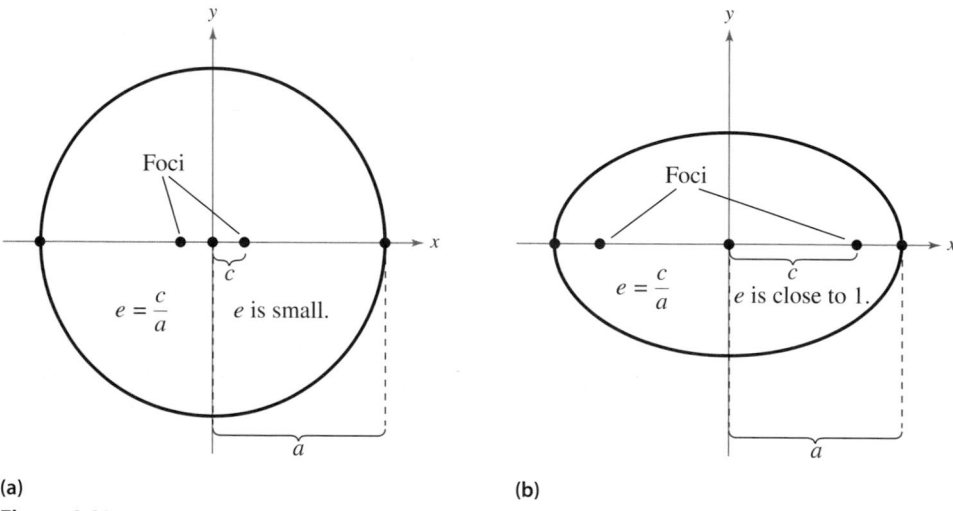

(a)

(b)

Figure 9.25

The orbit of the moon has an eccentricity of $e \approx 0.0549$, and the eccentricities of the eight planetary orbits are as follows.

Mercury:	$e \approx 0.2056$	Jupiter:	$e \approx 0.0484$
Venus:	$e \approx 0.0068$	Saturn:	$e \approx 0.0542$
Earth:	$e \approx 0.0167$	Uranus:	$e \approx 0.0472$
Mars:	$e \approx 0.0934$	Neptune:	$e \approx 0.0086$

9.2 Exercises

See www.CalcChat.com for worked-out solutions to odd-numbered exercises.

Vocabulary Check

Fill in the blanks.

1. An _____ is the set of all points (x, y) in a plane, the sum of whose distances from two distinct fixed points is constant.

2. The chord joining the vertices of an ellipse is called the _____ , and its midpoint is the _____ of the ellipse.

3. The chord perpendicular to the major axis at the center of an ellipse is called the _____ of the ellipse.

4. You can use the concept of _____ to measure the ovalness of an ellipse.

In Exercises 1–6, match the equation with its graph. [The graphs are labeled (a), (b), (c), (d), (e), and (f).]

(a)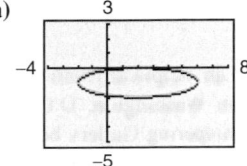

(b)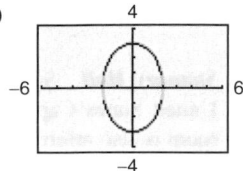

(c)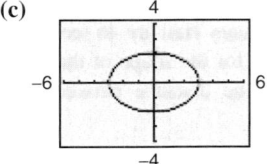

(d)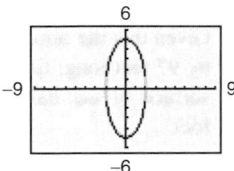

(e)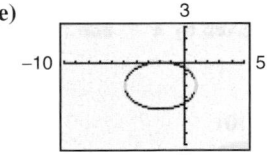

(f)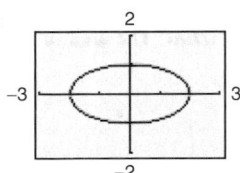

1. $\dfrac{x^2}{4} + \dfrac{y^2}{9} = 1$

2. $\dfrac{x^2}{9} + \dfrac{y^2}{4} = 1$

3. $\dfrac{x^2}{4} + \dfrac{y^2}{25} = 1$

4. $\dfrac{x^2}{4} + y^2 = 1$

5. $\dfrac{(x-2)^2}{16} + (y+1)^2 = 1$

6. $\dfrac{(x+2)^2}{9} + \dfrac{(y+2)^2}{4} = 1$

In Exercises 7–12, find the center, vertices, foci, and eccentricity of the ellipse, and sketch its graph. Use a graphing utility to verify your graph.

7. $\dfrac{x^2}{64} + \dfrac{y^2}{9} = 1$

8. $\dfrac{x^2}{16} + \dfrac{y^2}{81} = 1$

9. $\dfrac{(x-4)^2}{16} + \dfrac{(y+1)^2}{25} = 1$

10. $\dfrac{(x+3)^2}{12} + \dfrac{(y-2)^2}{16} = 1$

11. $\dfrac{(x+5)^2}{\frac{9}{4}} + (y-1)^2 = 1$ 12. $(x+2)^2 + \dfrac{(y+4)^2}{\frac{1}{4}} = 1$

In Exercises 13–22, (a) find the standard form of the equation of the ellipse, (b) find the center, vertices, foci, and eccentricity of the ellipse, (c) sketch the ellipse, and use a graphing utility to verify your graph.

13. $x^2 + 9y^2 = 36$ 14. $16x^2 + y^2 = 16$

15. $9x^2 + 4y^2 + 36x - 24y + 36 = 0$

16. $9x^2 + 4y^2 - 54x + 40y + 37 = 0$

17. $6x^2 + 2y^2 + 18x - 10y + 2 = 0$

18. $x^2 + 4y^2 - 6x + 20y - 2 = 0$

19. $16x^2 + 25y^2 - 32x + 50y + 16 = 0$

20. $9x^2 + 25y^2 - 36x - 50y + 61 = 0$

21. $12x^2 + 20y^2 - 12x + 40y - 37 = 0$

22. $36x^2 + 9y^2 + 48x - 36y + 43 = 0$

In Exercises 23–30, find the standard form of the equation of the ellipse with the given characteristics and center at the origin.

23.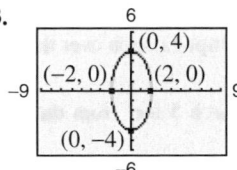

24.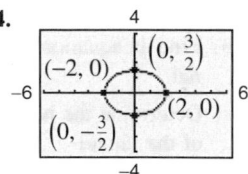

25. Vertices: $(\pm 3, 0)$; foci: $(\pm 2, 0)$

26. Vertices: $(0, \pm 8)$; foci: $(0, \pm 4)$

27. Foci: $(\pm 3, 0)$; major axis of length 8

28. Foci: $(\pm 2, 0)$; major axis of length 12

29. Vertices: $(0, \pm 5)$; passes through the point $(4, 2)$

30. Vertical major axis; passes through points $(0, 4)$ and $(2, 0)$

In Exercises 31–40, find the standard form of the equation of the ellipse with the given characteristics.

31.

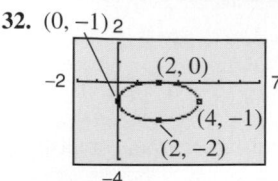

32. $(0, -1)$

33. Vertices: $(0, 2)$, $(8, 2)$; minor axis of length 2

34. Foci: $(0, 0)$, $(4, 0)$; major axis of length 6

35. Foci: $(0, 0)$, $(0, 8)$; major axis of length 36

36. Center: $(2, -1)$; vertex: $\left(2, \frac{1}{2}\right)$; minor axis of length 2

37. Vertices: $(3, 1)$, $(3, 9)$; minor axis of length 6

38. Center: $(3, 2)$; $a = 3c$; foci: $(1, 2)$, $(5, 2)$

39. Center: $(0, 4)$; $a = 2c$; vertices: $(-4, 4)$, $(4, 4)$

40. Vertices: $(5, 0)$, $(5, 12)$; endpoints of the minor axis: $(0, 6)$, $(10, 6)$

In Exercises 41–44, find the eccentricity of the ellipse.

41. $\dfrac{x^2}{4} + \dfrac{y^2}{9} = 1$ 42. $\dfrac{x^2}{25} + \dfrac{y^2}{36} = 1$

43. $x^2 + 9y^2 - 10x + 36y + 52 = 0$

44. $4x^2 + 3y^2 - 8x + 18y + 19 = 0$

45. Find an equation of the ellipse with vertices $(\pm 5, 0)$ and eccentricity $e = \frac{4}{5}$.

46. Find an equation of the ellipse with vertices $(0, \pm 8)$ and eccentricity $e = \frac{1}{2}$.

47. *Architecture*   A semielliptical arch over a tunnel for a road through a mountain has a major axis of 100 feet and a height at the center of 40 feet.

    (a) Draw a rectangular coordinate system on a sketch of the tunnel with the center of the road entering the tunnel at the origin. Identify the coordinates of the known points.

    (b) Find an equation of the semielliptical arch over the tunnel.

    (c) Determine the height of the arch 5 feet from the edge of the tunnel.

48. *Architecture*   A semielliptical arch through a railroad underpass has a major axis of 32 feet and a height at the center of 12 feet.

    (a) Draw a rectangular coordinate system on a sketch of the underpass with the center of the road entering the underpass at the origin. Identify the coordinates of the known points.

    (b) Find an equation of the semielliptical arch over the underpass.

    (c) Will a truck that is 10 feet wide and 9 feet tall be able to drive through the underpass without crossing the center line? Explain your reasoning.

49. *Architecture*   A fireplace arch is to be constructed in the shape of a semiellipse. The opening is to have a height of 2 feet at the center and a width of 6 feet along the base (see figure). The contractor draws the outline of the ellipse on the wall by the method discussed on page 671. Give the required positions of the tacks and the length of the string.

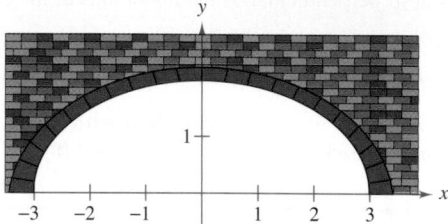

50. *Statuary Hall*   Statuary Hall is an elliptical room in the United States Capitol Building in Washington, D.C. The room is also referred to as the Whispering Gallery because a person standing at one focus of the room can hear even a whisper spoken by a person standing at the other focus. Given that the dimensions of Statuary Hall are 46 feet wide by 97 feet long, find an equation for the shape of the floor surface of the hall. Determine the distance between the foci.

51. *Geometry*   The area of the ellipse in the figure is twice the area of the circle. What is the length of the major axis? (*Hint:* The area of an ellipse is given by $A = \pi ab$.)

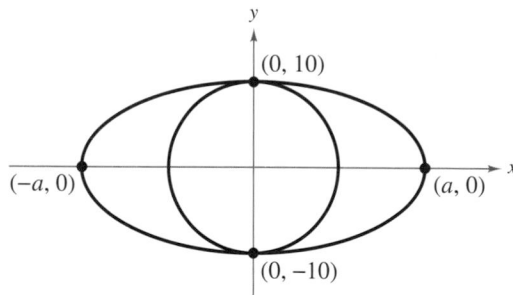

52. *Astronomy*   Halley's comet has an elliptical orbit with the sun at one focus. The eccentricity of the orbit is approximately 0.97. The length of the major axis of the orbit is about 35.88 astronomical units. (An astronomical unit is about 93 million miles.) Find the standard form of the equation of the orbit. Place the center of the orbit at the origin and place the major axis on the x-axis.

53. *Astronomy*   The comet Encke has an elliptical orbit with the sun at one focus. Encke's orbit ranges from 0.34 to 4.08 astronomical units from the sun. Find the standard form of the equation of the orbit. Place the center of the orbit at the origin and place the major axis on the x-axis.

54. *Satellite Orbit* The first artificial satellite to orbit Earth was Sputnik I (launched by the former Soviet Union in 1957). Its highest point above Earth's surface was 947 kilometers, and its lowest point was 228 kilometers. The center of Earth was a focus of the elliptical orbit, and the radius of Earth is 6378 kilometers (see figure). Find the eccentricity of the orbit.

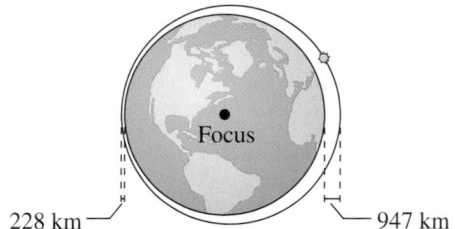

55. *Geometry* A line segment through a focus with endpoints on an ellipse, perpendicular to the major axis, is called a **latus rectum** of the ellipse. Therefore, an ellipse has two latera recta. Knowing the length of the latera recta is helpful in sketching an ellipse because this information yields other points on the curve (see figure). Show that the length of each latus rectum is $2b^2/a$.

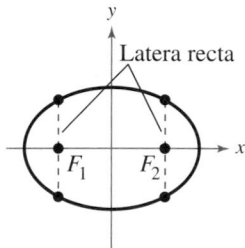

In Exercises 56–59, sketch the ellipse using the latera recta (see Exercise 55).

56. $\dfrac{x^2}{4} + \dfrac{y^2}{1} = 1$

57. $\dfrac{x^2}{9} + \dfrac{y^2}{16} = 1$

58. $9x^2 + 4y^2 = 36$

59. $5x^2 + 3y^2 = 15$

60. *Writing* Write an equation of an ellipse in standard form and graph it on paper. Do not write the equation on your graph. Exchange graphs with another student. Use the graph you receive to reconstruct the equation of the ellipse it represents and find its eccentricity. Compare your results and write a short paragraph discussing your findings.

Synthesis

True or False? **In Exercises 61 and 62, determine whether the statement is true or false. Justify your answer.**

61. It is easier to distinguish the graph of an ellipse from the graph of a circle if the eccentricity of the ellipse is large (close to 1).

62. The area of a circle with diameter $d = 2r = 8$ is greater than the area of an ellipse with major axis $2a = 8$.

63. *Think About It* At the beginning of this section it was noted that an ellipse can be drawn using two thumbtacks, a string of fixed length (greater than the distance between the two tacks), and a pencil (see Figure 9.16). If the ends of the string are fastened at the tacks and the string is drawn taut with a pencil, the path traced by the pencil is an ellipse.

(a) What is the length of the string in terms of a?

(b) Explain why the path is an ellipse.

64. *Exploration* Consider the ellipse

$$\frac{x^2}{a^2} + \frac{y^2}{b^2} = 1, \quad a + b = 20.$$

(a) The area of the ellipse is given by $A = \pi ab$. Write the area of the ellipse as a function of a.

(b) Find the equation of an ellipse with an area of 264 square centimeters.

(c) Complete the table using your equation from part (a) and make a conjecture about the shape of the ellipse with a maximum area.

a	8	9	10	11	12	13
A						

(d) Use a graphing utility to graph the area function to support your conjecture in part (c).

65. *Think About It* Find the equation of an ellipse such that for any point on the ellipse, the sum of the distances from the point $(2, 2)$ and $(10, 2)$ is 36.

66. *Proof* Show that $a^2 = b^2 + c^2$ for the ellipse

$$\frac{x^2}{a^2} + \frac{y^2}{b^2} = 1$$

where $a > 0$, $b > 0$, and the distance from the center of the ellipse $(0, 0)$ to a focus is c.

Skills Review

In Exercises 67–70, determine whether the sequence is arithmetic, geometric, or neither.

67. 66, 55, 44, 33, 22, . . .

68. 80, 40, 20, 10, 5, . . .

69. $\frac{1}{4}, \frac{1}{2}, 1, 2, 4, . . .$

70. $-\frac{1}{2}, \frac{1}{2}, \frac{3}{2}, \frac{5}{2}, \frac{7}{2}, . . .$

In Exercises 71–74, find the sum.

71. $\displaystyle\sum_{n=0}^{6} 3^n$

72. $\displaystyle\sum_{n=0}^{6} (-3)^n$

73. $\displaystyle\sum_{n=1}^{10} 4\left(\frac{3}{4}\right)^{n-1}$

74. $\displaystyle\sum_{n=0}^{10} 5\left(\frac{4}{3}\right)^n$

9.3 Hyperbolas

Introduction

The definition of a **hyperbola** is similar to that of an ellipse. The difference is that for an ellipse, the *sum* of the distances between the foci and a point on the ellipse is constant; whereas for a hyperbola, the *difference* of the distances between the foci and a point on the hyperbola is constant.

Definition of a Hyperbola

A **hyperbola** is the set of all points (x, y) in a plane, the difference of whose distances from two distinct fixed points, the **foci,** is a positive constant. [See Figure 9.26(a).]

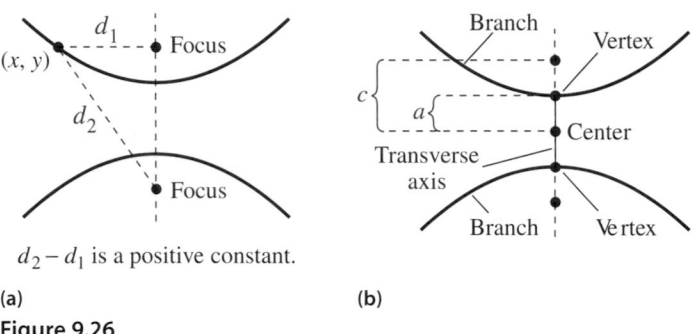

$d_2 - d_1$ is a positive constant.

(a)   (b)
Figure 9.26

The graph of a hyperbola has two disconnected parts called the **branches.** The line through the two foci intersects the hyperbola at two points called the **vertices.** The line segment connecting the vertices is the **transverse axis,** and the midpoint of the transverse axis is the **center** of the hyperbola [see Figure 9.26(b)]. The development of the **standard form of the equation of a hyperbola** is similar to that of an ellipse. Note that a, b, and c are related differently for hyperbolas than for ellipses. For a hyperbola, the distance between the foci and the center is greater than the distance between the vertices and the center.

Standard Equation of a Hyperbola

The **standard form of the equation of a hyperbola** with center at (h, k) is

$$\frac{(x - h)^2}{a^2} - \frac{(y - k)^2}{b^2} = 1 \qquad \text{Transverse axis is horizontal.}$$

$$\frac{(y - k)^2}{a^2} - \frac{(x - h)^2}{b^2} = 1. \qquad \text{Transverse axis is vertical.}$$

The vertices are a units from the center, and the foci are c units from the center. Moreover, $c^2 = a^2 + b^2$. If the center of the hyperbola is at the origin $(0, 0)$, the equation takes one of the following forms.

$$\frac{x^2}{a^2} - \frac{y^2}{b^2} = 1 \quad \text{\small Transverse axis is horizontal.} \qquad \frac{y^2}{a^2} - \frac{x^2}{b^2} = 1 \quad \text{\small Transverse axis is vertical.}$$

What you should learn

- Write equations of hyperbolas in standard form.
- Find asymptotes of and graph hyperbolas.
- Use properties of hyperbolas to solve real-life problems.
- Classify conics from their general equations.

Why you should learn it

Hyperbolas can be used to model and solve many types of real-life problems. For instance, in Exercise 44 on page 688, hyperbolas are used to locate the position of an explosion that was recorded by three listening stations.

James Foote/Photo Researchers, Inc.

Figure 9.27 shows both the horizontal and vertical orientations for a hyperbola.

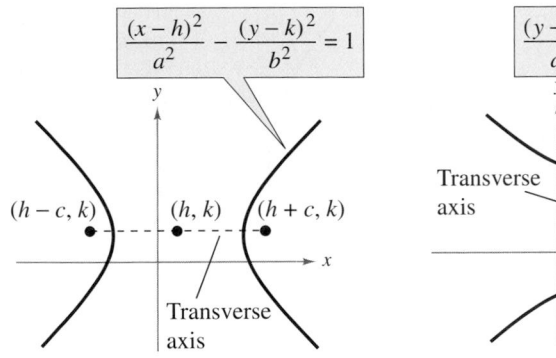

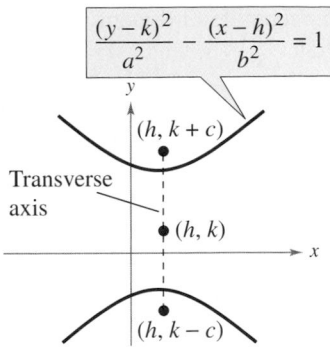

Transverse axis is horizontal.
Figure 9.27

Transverse axis is vertical.

Example 1 Finding the Standard Equation of a Hyperbola

Find the standard form of the equation of the hyperbola with foci $(-1, 2)$ and $(5, 2)$ and vertices $(0, 2)$ and $(4, 2)$.

Solution

By the Midpoint Formula, the center of the hyperbola occurs at the point $(2, 2)$. Furthermore, $c = 3$ and $a = 2$, and it follows that

$$b = \sqrt{c^2 - a^2}$$
$$= \sqrt{3^2 - 2^2}$$
$$= \sqrt{9 - 4}$$
$$= \sqrt{5}.$$

So, the hyperbola has a horizontal transverse axis and the standard form of the equation of the hyperbola is

$$\frac{(x - 2)^2}{2^2} - \frac{(y - 2)^2}{\left(\sqrt{5}\right)^2} = 1.$$

Figure 9.28 shows the hyperbola.

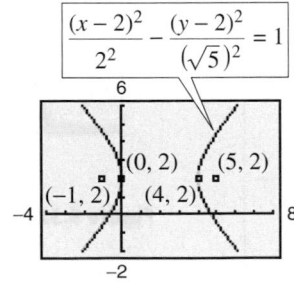

Figure 9.28

✔CHECKPOINT   Now try Exercise 33.

TECHNOLOGY TIP

When using a graphing utility to graph an equation, you must solve the equation for y before entering it into the graphing utility. When graphing equations of conics, it can be difficult to solve for y, which is why it is very important to know the algebra used to solve equations for y.

Asymptotes of a Hyperbola

Each hyperbola has two **asymptotes** that intersect at the center of the hyperbola. The asymptotes pass through the corners of a rectangle of dimensions $2a$ by $2b$, with its center at (h, k), as shown in Figure 9.29.

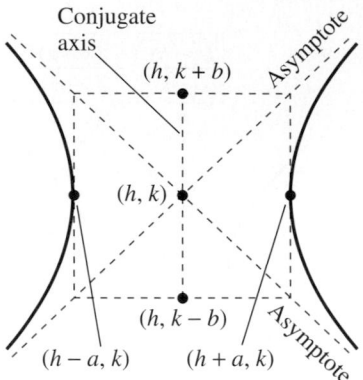

Asymptotes of a Hyperbola

$$y = k \pm \frac{b}{a}(x - h)$$ Asymptotes for horizontal transverse axis

$$y = k \pm \frac{a}{b}(x - h)$$ Asymptotes for vertical transverse axis

The **conjugate axis** of a hyperbola is the line segment of length $2b$ joining $(h, k + b)$ and $(h, k - b)$ if the transverse axis is horizontal, and the line segment of length $2b$ joining $(h + b, k)$ and $(h - b, k)$ if the transverse axis is vertical.

Figure 9.29

Example 2   Sketching a Hyperbola

Sketch the hyperbola whose equation is $4x^2 - y^2 = 16$.

Algebraic Solution

$$4x^2 - y^2 = 16$$ Write original equation.

$$\frac{4x^2}{16} - \frac{y^2}{16} = \frac{16}{16}$$ Divide each side by 16.

$$\frac{x^2}{2^2} - \frac{y^2}{4^2} = 1$$ Write in standard form.

Because the x^2-term is positive, you can conclude that the transverse axis is horizontal. So, the vertices occur at $(-2, 0)$ and $(2, 0)$, the endpoints of the conjugate axis occur at $(0, -4)$ and $(0, 4)$, and you can sketch the rectangle shown in Figure 9.30. Finally, by drawing the asymptotes, $y = 2x$ and $y = -2x$, through the corners of this rectangle, you can complete the sketch, as shown in Figure 9.31.

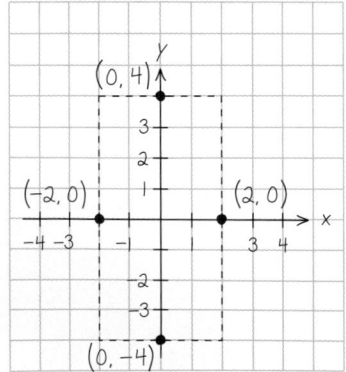

Figure 9.30

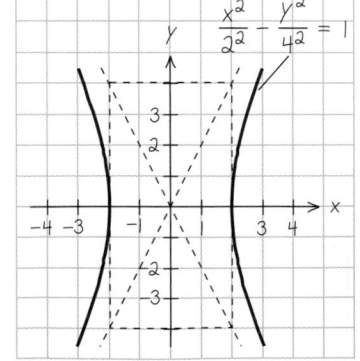

Figure 9.31

✔CHECKPOINT   Now try Exercise 15.

Graphical Solution

Solve the equation of the hyperbola for y as follows.

$$4x^2 - y^2 = 16$$

$$4x^2 - 16 = y^2$$

$$\pm\sqrt{4x^2 - 16} = y$$

Then use a graphing utility to graph $y_1 = \sqrt{4x^2 - 16}$ and $y_2 = -\sqrt{4x^2 - 16}$ in the same viewing window. Be sure to use a square setting. From the graph in Figure 9.32, you can see that the transverse axis is horizontal. You can use the *zoom* and *trace* features to approximate the vertices to be $(-2, 0)$ and $(2, 0)$.

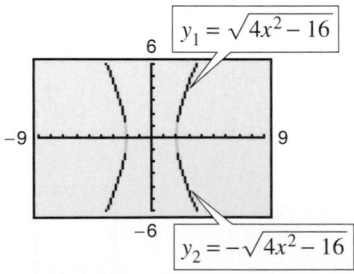

Figure 9.32

Example 3   Finding the Asymptotes of a Hyperbola

Sketch the hyperbola given by

$$4x^2 - 3y^2 + 8x + 16 = 0$$

and find the equations of its asymptotes.

Solution

$4x^2 - 3y^2 + 8x + 16 = 0$	Write original equation.
$4(x^2 + 2x) - 3y^2 = -16$	Subtract 16 from each side and factor.
$4(x^2 + 2x + 1) - 3y^2 = -16 + 4(1)$	Complete the square.
$4(x + 1)^2 - 3y^2 = -12$	Write in completed square form.
$\dfrac{y^2}{2^2} - \dfrac{(x + 1)^2}{\left(\sqrt{3}\right)^2} = 1$	Write in standard form.

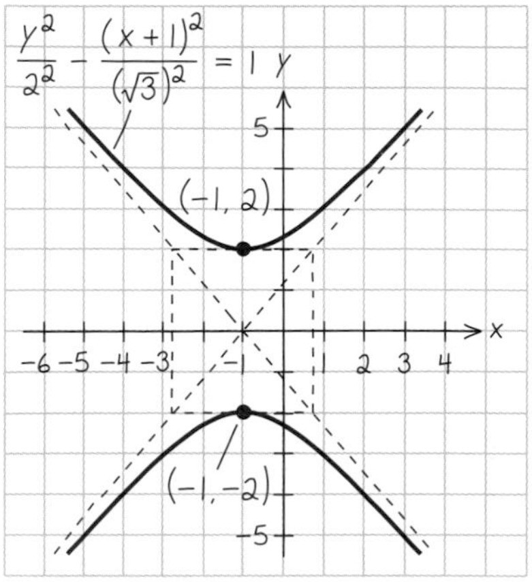

From this equation you can conclude that the hyperbola has a vertical transverse axis, is centered at $(-1, 0)$, has vertices $(-1, 2)$ and $(-1, -2)$, and has a conjugate axis with endpoints $\left(-1 - \sqrt{3}, 0\right)$ and $\left(-1 + \sqrt{3}, 0\right)$. To sketch the hyperbola, draw a rectangle through these four points. The asymptotes are the lines passing through the corners of the rectangle, as shown in Figure 9.33. Finally, using $a = 2$ and $b = \sqrt{3}$, you can conclude that the equations of the asymptotes are

$$y = \frac{2}{\sqrt{3}}(x + 1) \qquad \text{and} \qquad y = -\frac{2}{\sqrt{3}}(x + 1).$$

✓CHECKPOINT   Now try Exercise 19.

Figure 9.33

TECHNOLOGY TIP   You can use a graphing utility to graph a hyperbola by graphing the upper and lower portions in the same viewing window. For instance, to graph the hyperbola in Example 3, first solve for y to obtain

$$y_1 = 2\sqrt{1 + \frac{(x + 1)^2}{3}} \quad \text{and} \quad y_2 = -2\sqrt{1 + \frac{(x + 1)^2}{3}}.$$

Use a viewing window in which $-9 \le x \le 9$ and $-6 \le y \le 6$. You should obtain the graph shown in Figure 9.34. Notice that the graphing utility does not draw the asymptotes. However, if you trace along the branches, you will see that the values of the hyperbola approach the asymptotes.

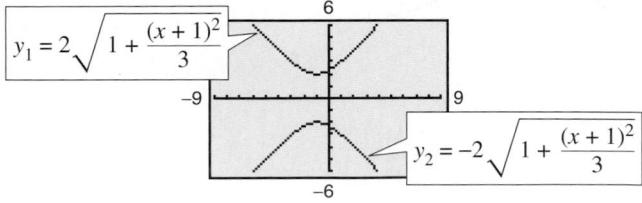

Figure 9.34

Example 4 Using Asymptotes to Find the Standard Equation

Find the standard form of the equation of the hyperbola having vertices $(3, -5)$ and $(3, 1)$ and having asymptotes

$$y = 2x - 8 \qquad \text{and} \qquad y = -2x + 4$$

as shown in Figure 9.35.

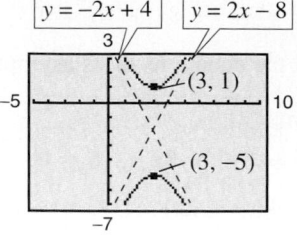

Figure 9.35

Solution

By the Midpoint Formula, the center of the hyperbola is $(3, -2)$. Furthermore, the hyperbola has a vertical transverse axis with $a = 3$. From the original equations, you can determine the slopes of the asymptotes to be

$$m_1 = 2 = \frac{a}{b} \qquad \text{and} \qquad m_2 = -2 = -\frac{a}{b}$$

and because $a = 3$, you can conclude that $b = \frac{3}{2}$. So, the standard form of the equation is

$$\frac{(y + 2)^2}{3^2} - \frac{(x - 3)^2}{\left(\dfrac{3}{2}\right)^2} = 1.$$

✓**CHECKPOINT** Now try Exercise 39.

As with ellipses, the *eccentricity* of a hyperbola is

$$e = \frac{c}{a} \qquad \text{Eccentricity}$$

and because $c > a$ it follows that $e > 1$. If the eccentricity is large, the branches of the hyperbola are nearly flat, as shown in Figure 9.36(a). If the eccentricity is close to 1, the branches of the hyperbola are more pointed, as shown in Figure 9.36(b).

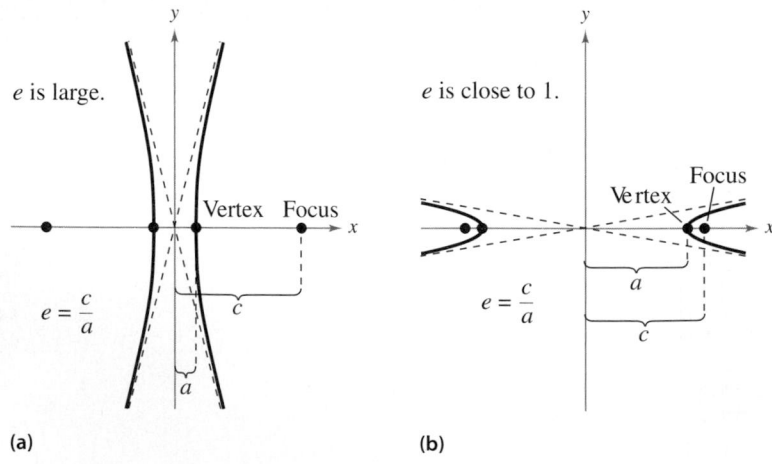

(a)

(b)

Figure 9.36

Applications

The following application was developed during World War II. It shows how the properties of hyperbolas can be used in radar and other detection systems.

Example 5   An Application Involving Hyperbolas

Two microphones, 1 mile apart, record an explosion. Microphone A receives the sound 2 seconds before microphone B. Where did the explosion occur?

Solution

Assuming sound travels at 1100 feet per second, you know that the explosion took place 2200 feet farther from B than from A, as shown in Figure 9.37. The locus of all points that are 2200 feet closer to A than to B is one branch of the hyperbola

$$\frac{x^2}{a^2} - \frac{y^2}{b^2} = 1$$

where

$$c = \frac{5280}{2} = 2640 \quad \text{and} \quad a = \frac{2200}{2} = 1100.$$

So, $b^2 = c^2 - a^2 = 2640^2 - 1100^2 = 5{,}759{,}600$, and you can conclude that the explosion occurred somewhere on the right branch of the hyperbola

$$\frac{x^2}{1{,}210{,}000} - \frac{y^2}{5{,}759{,}600} = 1.$$

✓CHECKPOINT   Now try Exercise 43.

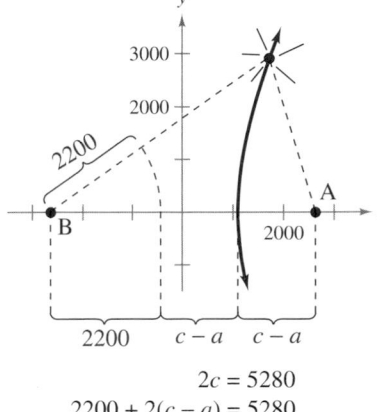

$$2c = 5280$$
$$2200 + 2(c - a) = 5280$$

Figure 9.37

Another interesting application of conic sections involves the orbits of comets in our solar system. Of the 610 comets identified prior to 1970, 245 have elliptical orbits, 295 have parabolic orbits, and 70 have hyperbolic orbits. The center of the sun is a focus of each of these orbits, and each orbit has a vertex at the point where the comet is closest to the sun, as shown in Figure 9.38. Undoubtedly, there are many comets with parabolic or hyperbolic orbits that have not been identified. You get to see such comets only *once*. Comets with elliptical orbits, such as Halley's comet, are the only ones that remain in our solar system.

If p is the distance between the vertex and the focus in meters, and v is the velocity of the comet at the vertex in meters per second, then the type of orbit is determined as follows.

1. Ellipse: $v < \sqrt{2GM/p}$

2. Parabola: $v = \sqrt{2GM/p}$

3. Hyperbola: $v > \sqrt{2GM/p}$

In each of these equations, $M \approx 1.989 \times 10^{30}$ kilograms (the mass of the sun) and $G \approx 6.67 \times 10^{-11}$ cubic meter per kilogram-second squared (the universal gravitational constant).

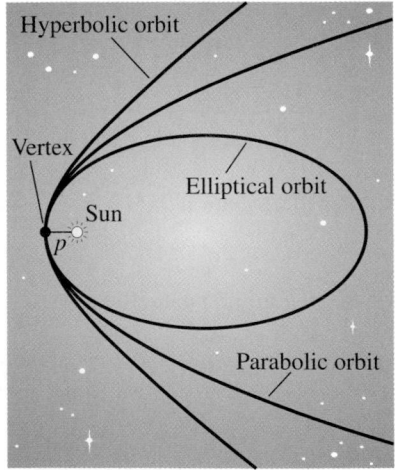

Figure 9.38

General Equations of Conics

> **Classifying a Conic from Its General Equation**
>
> The graph of $Ax^2 + Bxy + Cy^2 + Dx + Ey + F = 0$ is one of the following.
>
> 1. Circle: $A = C$ $A \neq 0$
> 2. Parabola: $AC = 0$ $A = 0$ or $C = 0$, but not both.
> 3. Ellipse: $AC > 0$ A and C have like signs.
> 4. Hyperbola: $AC < 0$ A and C have unlike signs.

The test above is valid *if* the graph is a conic. The test does not apply to equations such as $x^2 + y^2 = -1$, whose graphs are not conics.

Example 6 Classifying Conics from General Equations

Classify the graph of each equation.

a. $4x^2 - 9x + y - 5 = 0$

b. $4x^2 - y^2 + 8x - 6y + 4 = 0$

c. $2x^2 + 4y^2 - 4x + 12y = 0$

d. $2x^2 + 2y^2 - 8x + 12y + 2 = 0$

Solution

a. For the equation $4x^2 - 9x + y - 5 = 0$, you have

$$AC = 4(0) = 0. \qquad \text{Parabola}$$

So, the graph is a parabola.

b. For the equation $4x^2 - y^2 + 8x - 6y + 4 = 0$, you have

$$AC = 4(-1) < 0. \qquad \text{Hyperbola}$$

So, the graph is a hyperbola.

c. For the equation $2x^2 + 4y^2 - 4x + 12y = 0$, you have

$$AC = 2(4) > 0. \qquad \text{Ellipse}$$

So, the graph is an ellipse.

d. For the equation $2x^2 + 2y^2 - 8x + 12y + 2 = 0$, you have

$$A = C = 2. \qquad \text{Circle}$$

So, the graph is a circle.

✓**CHECKPOINT**   Now try Exercise 49.

STUDY TIP

Notice in Example 6(a) that there is no y^2-term in the equation. Therefore, $C = 0$.

9.3 Exercises

See www.CalcChat.com for worked-out solutions to odd-numbered exercises.

Vocabulary Check

Fill in the blanks.

1. A _____ is the set of all points (x, y) in a plane, the difference of whose distances from two distinct fixed points is a positive constant.

2. The graph of a hyperbola has two disconnected parts called _____ .

3. The line segment connecting the vertices of a hyperbola is called the _____ , and the midpoint of the line segment is the _____ of the hyperbola.

4. Each hyperbola has two _____ that intersect at the center of the hyperbola.

5. The general form of the equation of a conic is given by _____ .

In Exercises 1–4, match the equation with its graph. [The graphs are labeled (a), (b), (c), and (d).]

(a)
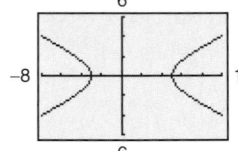

(b)
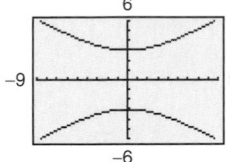

(c)
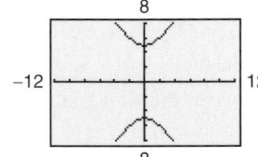

(d)
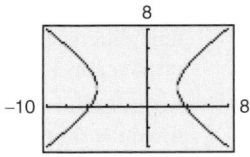

1. $\dfrac{y^2}{9} - \dfrac{x^2}{25} = 1$

2. $\dfrac{y^2}{25} - \dfrac{x^2}{9} = 1$

3. $\dfrac{(x-1)^2}{16} - \dfrac{y^2}{4} = 1$

4. $\dfrac{(x+1)^2}{16} - \dfrac{(y-2)^2}{9} = 1$

In Exercises 5–14, find the center, vertices, foci, and asymptotes of the hyperbola, and sketch its graph using the asymptotes as an aid. Use a graphing utility to verify your graph.

5. $x^2 - y^2 = 1$

6. $\dfrac{x^2}{9} - \dfrac{y^2}{25} = 1$

7. $\dfrac{y^2}{1} - \dfrac{x^2}{4} = 1$

8. $\dfrac{y^2}{9} - \dfrac{x^2}{1} = 1$

9. $\dfrac{y^2}{25} - \dfrac{x^2}{81} = 1$

10. $\dfrac{x^2}{36} - \dfrac{y^2}{4} = 1$

11. $\dfrac{(x-1)^2}{4} - \dfrac{(y+2)^2}{1} = 1$

12. $\dfrac{(x+3)^2}{144} - \dfrac{(y-2)^2}{25} = 1$

13. $\dfrac{(y+5)^2}{\frac{1}{9}} - \dfrac{(x-1)^2}{\frac{1}{4}} = 1$

14. $\dfrac{(y-1)^2}{\frac{1}{4}} - \dfrac{(x+3)^2}{\frac{1}{16}} = 1$

In Exercises 15–24, (a) find the standard form of the equation of the hyperbola, (b) find the center, vertices, foci, and asymptotes of the hyperbola, (c) sketch the hyperbola, and use a graphing utility to verify your graph.

15. $4x^2 - 9y^2 = 36$

16. $25x^2 - 4y^2 = 100$

17. $2x^2 - 3y^2 = 6$

18. $6y^2 - 3x^2 = 18$

19. $9x^2 - y^2 - 36x - 6y + 18 = 0$

20. $x^2 - 9y^2 + 36y - 72 = 0$

21. $x^2 - 9y^2 + 2x - 54y - 80 = 0$

22. $16y^2 - x^2 + 2x + 64y + 63 = 0$

23. $9y^2 - x^2 + 2x + 54y + 62 = 0$

24. $9x^2 - y^2 + 54x + 10y + 55 = 0$

In Exercises 25–30, find the standard form of the equation of the hyperbola with the given characteristics and center at the origin.

25. Vertices: $(0, \pm2)$; foci: $(0, \pm4)$

26. Vertices: $(\pm3, 0)$; foci: $(\pm6, 0)$

27. Vertices: $(\pm1, 0)$; asymptotes: $y = \pm5x$

28. Vertices: $(0, \pm3)$; asymptotes: $y = \pm3x$

29. Foci: $(0, \pm8)$; asymptotes: $y = \pm4x$

30. Foci: $(\pm10, 0)$; asymptotes: $y = \pm\frac{3}{4}x$

In Exercises 31–42, find the standard form of the equation of the hyperbola with the given characteristics.

31. Vertices: $(2, 0)$, $(6, 0)$; foci: $(0, 0)$, $(8, 0)$

32. Vertices: $(2, 3)$, $(2, -3)$; foci: $(2, 5)$, $(2, -5)$

33. Vertices: $(4, 1)$, $(4, 9)$; foci: $(4, 0)$, $(4, 10)$

34. Vertices: $(-2, 1)$, $(2, 1)$; foci: $(-3, 1)$, $(3, 1)$

35. Vertices: $(2, 3)$, $(2, -3)$;
      passes through the point $(0, 5)$

36. Vertices: $(-2, 1)$, $(2, 1)$;
      passes through the point $(5, 4)$

37. Vertices: $(0, 4)$, $(0, 0)$;
      passes through the point $\left(\sqrt{5}, -1 \right)$

38. Vertices: $(1, 2)$, $(1, -2)$;
      passes through the point $\left(0, \sqrt{5} \right)$

39. Vertices: $(1, 2)$, $(3, 2)$;
      asymptotes: $y = x$, $y = 4 - x$

40. Vertices: $(3, 0)$, $(3, -6)$;
      asymptotes: $y = x - 6$, $y = -x$

41. Vertices: $(0, 2)$, $(6, 2)$;
      asymptotes: $y = \frac{2}{3}x$, $y = 4 - \frac{2}{3}x$

42. Vertices: $(3, 0)$, $(3, 4)$;
      asymptotes: $y = \frac{2}{3}x$, $y = 4 - \frac{2}{3}x$

43. *Sound Location* You and a friend live 4 miles apart (on the same "east-west" street) and are talking on the phone. You hear a clap of thunder from lightning in a storm, and 18 seconds later your friend hears the thunder. Find an equation that gives the possible places where the lightning could have occurred. (Assume that the coordinate system is measured in feet and that sound travels at 1100 feet per second.)

44. *Sound Location* Three listening stations located at $(3300, 0)$, $(3300, 1100)$, and $(-3300, 0)$ monitor an explosion. The last two stations detect the explosion 1 second and 4 seconds after the first, respectively. Determine the coordinates of the explosion. (Assume that the coordinate system is measured in feet and that sound travels at 1100 feet per second.)

45. *Pendulum* The base for a pendulum of a clock has the shape of a hyperbola (see figure).

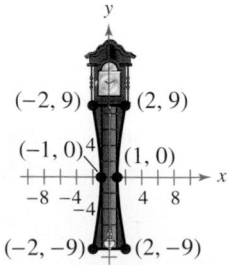

(a) Write an equation of the cross section of the base.

(b) Each unit in the coordinate plane represents $\frac{1}{2}$ foot. Find the width of the base of the pendulum 4 inches from the bottom.

46. *Navigation* Long distance radio navigation for aircraft and ships uses synchronized pulses transmitted by widely separated transmitting stations. These pulses travel at the speed of light (186,000 miles per second). The difference in the times of arrival of these pulses at an aircraft or ship is constant on a hyperbola having the transmitting stations as foci. Assume that two stations, 300 miles apart, are positioned on a rectangular coordinate system at coordinates $(-150, 0)$ and $(150, 0)$, and that a ship is traveling on a hyperbolic path with coordinates $(x, 75)$ (see figure).

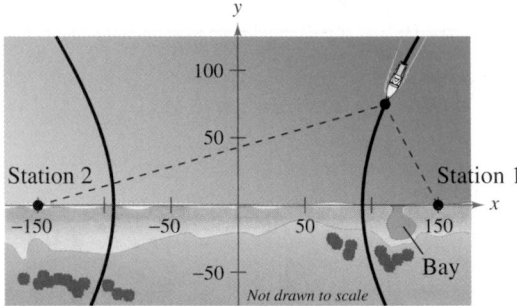

(a) Find the x-coordinate of the position of the ship if the time difference between the pulses from the transmitting stations is 1000 microseconds (0.001 second).

(b) Determine the distance between the ship and station 1 when the ship reaches the shore.

(c) The captain of the ship wants to enter a bay located between the two stations. The bay is 30 miles from station 1. What should be the time difference between the pulses?

(d) The ship is 60 miles offshore when the time difference in part (c) is obtained. What is the position of the ship?

47. *Hyperbolic Mirror* A hyperbolic mirror (used in some telescopes) has the property that a light ray directed at a focus will be reflected to the other focus. The focus of a hyperbolic mirror (see figure) has coordinates $(24, 0)$. Find the vertex of the mirror if the mount at the top edge of the mirror has coordinates $(24, 24)$.

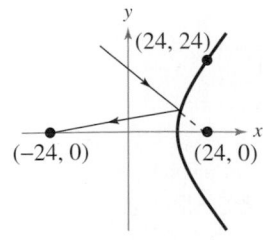

48. *Panoramic Photo*   A panoramic photo can be taken using a hyperbolic mirror. The camera is pointed toward the vertex of the mirror and the camera's optical center is positioned at one focus of the mirror (see figure). An equation for the cross-section of the mirror is

$$\frac{y^2}{25} - \frac{x^2}{16} = 1.$$

Find the distance from the camera's optical center to the mirror.

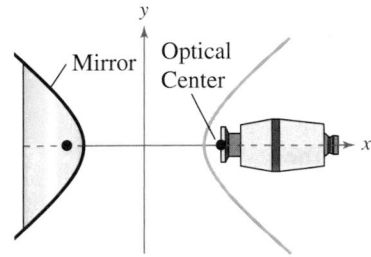

In Exercises 49–58, classify the graph of the equation as a circle, a parabola, an ellipse, or a hyperbola.

49. $9x^2 + 4y^2 - 18x + 16y - 119 = 0$

50. $x^2 + y^2 - 4x - 6y - 23 = 0$

51. $16x^2 - 9y^2 + 32x + 54y - 209 = 0$

52. $x^2 + 4x - 8y + 20 = 0$

53. $y^2 + 12x + 4y + 28 = 0$

54. $4x^2 + 25y^2 + 16x + 250y + 541 = 0$

55. $x^2 + y^2 + 2x - 6y = 0$

56. $y^2 - x^2 + 2x - 6y - 8 = 0$

57. $x^2 - 6x - 2y + 7 = 0$

58. $9x^2 + 4y^2 - 90x + 8y + 228 = 0$

Synthesis

True or False?   **In Exercises 59–62, determine whether the statement is true or false. Justify your answer.**

59. In the standard form of the equation of a hyperbola, the larger the ratio of b to a, the larger the eccentricity of the hyperbola.

60. In the standard form of the equation of a hyperbola, the trivial solution of two intersecting lines occurs when $b = 0$.

61. If $D \neq 0$ and $E \neq 0$, then the graph of $x^2 - y^2 + Dx + Ey = 0$ is a hyperbola.

62. If the asymptotes of the hyperbola $x^2/a^2 - y^2/b^2 = 1$, where $a, b > 0$, intersect at right angles, then $a = b$.

63. *Think About It*   Consider a hyperbola centered at the origin with a horizontal transverse axis. Use the definition of a hyperbola to derive its standard form.

64. *Writing*   Explain how the central rectangle of a hyperbola can be used to sketch its asymptotes.

65. Use the figure to show that $|d_2 - d_1| = 2a$.

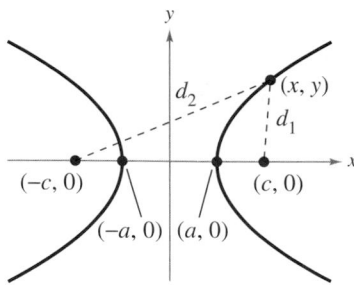

66. *Think About It*   Find the equation of the hyperbola for any point on which, the difference between its distances from the points $(2, 2)$ and $(10, 2)$ is 6.

67. *Proof*   Show that $c^2 = a^2 + b^2$ for the equation of the hyperbola

$$\frac{x^2}{a^2} - \frac{y^2}{b^2} = 1$$

where the distance from the center of the hyperbola $(0, 0)$ to a focus is c.

68. *Proof*   Prove that the graph of the equation

$$Ax^2 + Cy^2 + Dx + Ey + F = 0$$

is one of the following (except in degenerate cases).

Conic	Condition
(a) Circle	$A = C$
(b) Parabola	$A = 0$ or $C = 0$ (but not both)
(c) Ellipse	$AC > 0$
(d) Hyperbola	$AC < 0$

Skills Review

In Exercises 69–72, perform the indicated operation.

69. Subtract: $(x^3 - 3x^2) - (6 - 2x - 4x^2)$

70. Multiply: $\left(3x - \frac{1}{2}\right)(x + 4)$

71. Divide: $\dfrac{x^3 - 3x + 4}{x + 2}$

72. Expand: $[(x + y) + 3]^2$

In Exercises 73–78, factor the polynomial completely.

73. $x^3 - 16x$ **74.** $x^2 + 14x + 49$

75. $2x^3 - 24x^2 + 72x$

76. $6x^3 - 11x^2 - 10x$

77. $16x^3 + 54$

78. $4 - x + 4x^2 - x^3$

9.4   Rotation and Systems of Quadratic Equations

Rotation

In the preceding section, you learned that the equation of a conic with axes parallel to the coordinate axes has a standard form that can be written in the general form

$$Ax^2 + Cy^2 + Dx + Ey + F = 0. \qquad \text{Horizontal or vertical axes}$$

In this section, you will study the equations of conics whose axes are rotated so that they are not parallel to either the x-axis or the y-axis. The general equation for such conics contains an xy-term.

$$Ax^2 + Bxy + Cy^2 + Dx + Ey + F = 0 \qquad \text{Equation in } xy\text{-plane}$$

To eliminate this xy-term, you can use a procedure called **rotation of axes.** The objective is to rotate the x- and y-axes until they are parallel to the axes of the conic. The rotated axes are denoted as the x'-axis and the y'-axis, as shown in Figure 9.39.

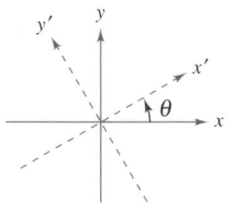

Figure 9.39

After the rotation, the equation of the conic in the new $x'y'$-plane will have the form

$$A'(x')^2 + C'(y')^2 + D'x' + E'y' + F' = 0. \qquad \text{Equation in } x'y'\text{-plane}$$

Because this equation has no xy-term, you can obtain a standard form by completing the square. The following theorem identifies how much to rotate the axes to eliminate the xy-term and also the equations for determining the new coefficients A', C', D', E', and F'.

> **Rotation of Axes to Eliminate an xy-Term (See the proof on page 738.)**
>
> The general second-degree equation
>
> $$Ax^2 + Bxy + Cy^2 + Dx + Ey + F = 0$$
>
> can be rewritten as
>
> $$A'(x')^2 + C'(y')^2 + D'x' + E'y' + F' = 0$$
>
> by rotating the coordinate axes through an angle θ, where $\cot 2\theta = \dfrac{A - C}{B}$.
>
> The coefficients of the new equation are obtained by making the substitutions
>
> $$x = x' \cos \theta - y' \sin \theta \qquad \text{and} \qquad y = x' \sin \theta + y' \cos \theta.$$

What you should learn

- Rotate the coordinate axes to eliminate the xy-term in equations of conics.
- Use the discriminant to classify conics.
- Solve systems of quadratic equations.

Why you should learn it

As illustrated in Exercises 3–14 on page 697, rotation of the coordinate axes can help you identify the graph of a general second-degree equation.

Example 1   Rotation of Axes for a Hyperbola

Rotate the axes to eliminate the xy-term in the equation $xy - 1 = 0$. Then write the equation in standard form and sketch its graph.

Solution

Because $A = 0$, $B = 1$, and $C = 0$, you have

$$\cot 2\theta = \frac{A - C}{B} = 0 \quad \Longrightarrow \quad 2\theta = \frac{\pi}{2} \quad \Longrightarrow \quad \theta = \frac{\pi}{4}$$

which implies that

$$x = x' \cos \frac{\pi}{4} - y' \sin \frac{\pi}{4}$$

$$= x'\left(\frac{1}{\sqrt{2}}\right) - y'\left(\frac{1}{\sqrt{2}}\right)$$

$$= \frac{x' - y'}{\sqrt{2}}$$

and

$$y = x' \sin \frac{\pi}{4} + y' \cos \frac{\pi}{4}$$

$$= x'\left(\frac{1}{\sqrt{2}}\right) + y'\left(\frac{1}{\sqrt{2}}\right)$$

$$= \frac{x' + y'}{\sqrt{2}}.$$

The equation in the $x'y'$-system is obtained by substituting these expressions into the equation $xy - 1 = 0$.

$$\left(\frac{x' - y'}{\sqrt{2}}\right)\left(\frac{x' + y'}{\sqrt{2}}\right) - 1 = 0$$

$$\frac{(x')^2 - (y')^2}{2} - 1 = 0$$

$$\frac{(x')^2}{(\sqrt{2})^2} - \frac{(y')^2}{(\sqrt{2})^2} = 1 \qquad \text{Write in standard form.}$$

In the $x'y'$-system, this is a hyperbola centered at the origin with vertices at $(\pm\sqrt{2}, 0)$, as shown in Figure 9.40. To find the coordinates of the vertices in the xy-system, substitute the coordinates $(\pm\sqrt{2}, 0)$ into the equations

$$x = \frac{x' - y'}{\sqrt{2}} \qquad \text{and} \qquad y = \frac{x' + y'}{\sqrt{2}}.$$

This substitution yields the vertices $(1, 1)$ and $(-1, -1)$ in the xy-system. Note also that the asymptotes of the hyperbola have equations $y' = \pm x'$, which correspond to the original x- and y-axes.

✔CHECKPOINT   Now try Exercise 3.

STUDY TIP

Remember that the substitutions

$$x = x' \cos \theta - y' \sin \theta$$

and

$$y = x' \sin \theta + y' \cos \theta$$

were developed to eliminate the $x'y'$-term in the rotated system. You can use this as a check on your work. In other words, if your final equation contains an $x'y'$-term, you know that you made a mistake.

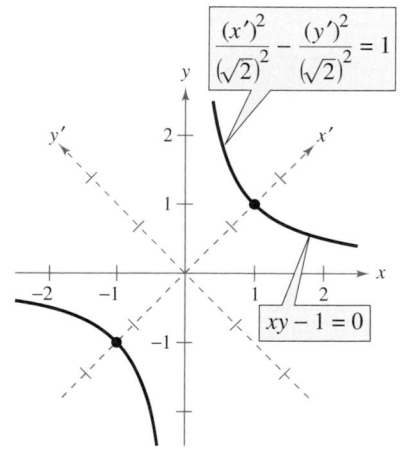

Vertices:
In $x'y'$-system: $(\sqrt{2}, 0), (-\sqrt{2}, 0)$
In xy-system: $(1, 1), (-1, -1)$
Figure 9.40

Example 2 Rotation of Axes for an Ellipse

Rotate the axes to eliminate the xy-term in the equation

$$7x^2 - 6\sqrt{3}\,xy + 13y^2 - 16 = 0.$$

Then write the equation in standard form and sketch its graph.

Prerequisite Skills

To review conics, see Sections 9.1–9.3.

Solution

Because $A = 7$, $B = -6\sqrt{3}$, and $C = 13$, you have

$$\cot 2\theta = \frac{A - C}{B} = \frac{7 - 13}{-6\sqrt{3}} = \frac{1}{\sqrt{3}}$$

which implies that $\theta = \pi/6$. The equation in the $x'y'$-system is obtained by making the substitutions

$$x = x' \cos \frac{\pi}{6} - y' \sin \frac{\pi}{6}$$

$$= x'\left(\frac{\sqrt{3}}{2}\right) - y'\left(\frac{1}{2}\right)$$

$$= \frac{\sqrt{3}\,x' - y'}{2}$$

and

$$y = x' \sin \frac{\pi}{6} + y' \cos \frac{\pi}{6}$$

$$= x'\left(\frac{1}{2}\right) + y'\left(\frac{\sqrt{3}}{2}\right)$$

$$= \frac{x' + \sqrt{3}\,y'}{2}$$

into the original equation. So, you have

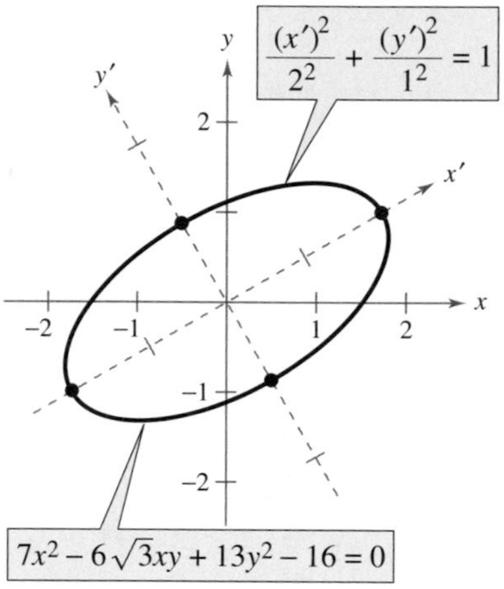

Figure 9.41

Vertices:
In $x'y'$-system: $(\pm 2, 0)$, $(0, \pm 1)$
In xy-system: $\left(\sqrt{3}, 1\right)$, $\left(-\sqrt{3}, 1\right)$,
$\left(\dfrac{1}{2}, \dfrac{\sqrt{3}}{2}\right)$, $\left(-\dfrac{1}{2}, \dfrac{\sqrt{3}}{2}\right)$

$$7x^2 - 6\sqrt{3}\,xy + 13y^2 - 16 = 0$$

$$7\left(\frac{\sqrt{3}\,x' - y'}{2}\right)^2 - 6\sqrt{3}\left(\frac{\sqrt{3}\,x' - y'}{2}\right)\left(\frac{x' + \sqrt{3}\,y'}{2}\right) + 13\left(\frac{x' + \sqrt{3}\,y'}{2}\right)^2 - 16 = 0$$

which simplifies to

$$4(x')^2 + 16(y')^2 - 16 = 0$$

$$4(x')^2 + 16(y')^2 = 16$$

$$\frac{(x')^2}{2^2} + \frac{(y')^2}{1^2} = 1. \qquad \text{Write in standard form.}$$

This is the equation of an ellipse centered at the origin with vertices $(\pm 2, 0)$ in the $x'y'$-system, as shown in Figure 9.41.

✓CHECKPOINT   Now try Exercise 11.

Example 3   Rotation of Axes for a Parabola

Rotate the axes to eliminate the xy-term in the equation

$$x^2 - 4xy + 4y^2 + 5\sqrt{5}\,y + 1 = 0.$$

Then write the equation in standard form and sketch its graph.

Solution

Because $A = 1$, $B = -4$, and $C = 4$, you have

$$\cot 2\theta = \frac{A - C}{B} = \frac{1 - 4}{-4} = \frac{3}{4}.$$

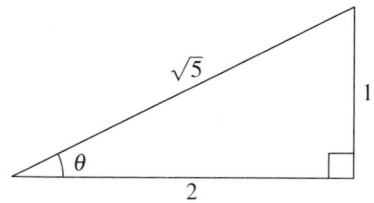

Figure 9.42

Using the identity $\cot 2\theta = (\cot^2 \theta - 1)/(2 \cot \theta)$ produces

$$\cot 2\theta = \frac{3}{4} = \frac{\cot^2 \theta - 1}{2 \cot \theta}$$

from which you obtain the equation

$$4 \cot^2 \theta - 4 = 6 \cot \theta$$

$$4 \cot^2 \theta - 6 \cot \theta - 4 = 0$$

$$(2 \cot \theta - 4)(2 \cot \theta + 1) = 0.$$

Considering $0 < \theta < \pi/2$, you have $2 \cot \theta = 4$. So,

$$\cot \theta = 2 \quad \Longrightarrow \quad \theta \approx 26.6°.$$

From the triangle in Figure 9.42, you obtain $\sin \theta = 1/\sqrt{5}$ and $\cos \theta = 2/\sqrt{5}$.
So, you use the substitutions

$$x = x' \cos \theta - y' \sin \theta = x'\left(\frac{2}{\sqrt{5}}\right) - y'\left(\frac{1}{\sqrt{5}}\right) = \frac{2x' - y'}{\sqrt{5}}$$

$$y = x' \sin \theta + y' \cos \theta = x'\left(\frac{1}{\sqrt{5}}\right) + y'\left(\frac{2}{\sqrt{5}}\right) = \frac{x' + 2y'}{\sqrt{5}}.$$

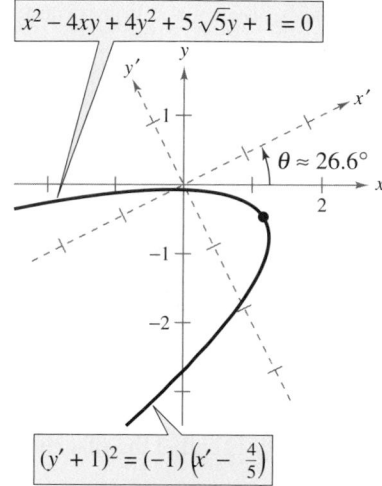

$x^2 - 4xy + 4y^2 + 5\sqrt{5}\,y + 1 = 0$

$(y' + 1)^2 = (-1)\left(x' - \frac{4}{5}\right)$

Vertex:

In $x'y'$-system: $\left(\frac{4}{5}, -1\right)$

In xy-system: $\left(\dfrac{13}{5\sqrt{5}}, -\dfrac{6}{5\sqrt{5}}\right)$

Figure 9.43

Substituting these expressions into the original equation, you have

$$\left(\frac{2x' - y'}{\sqrt{5}}\right)^2 - 4\left(\frac{2x' - y'}{\sqrt{5}}\right)\left(\frac{x' + 2y'}{\sqrt{5}}\right) + 4\left(\frac{x' + 2y'}{\sqrt{5}}\right)^2 + 5\sqrt{5}\left(\frac{x' + 2y'}{\sqrt{5}}\right) + 1 = 0$$

which simplifies as follows.

$$5(y')^2 + 5x' + 10y' + 1 = 0$$

$$5[(y')^2 + 2y'] = -5x' - 1 \qquad \text{Group terms.}$$

$$5(y' + 1)^2 = -5x' + 4 \qquad \text{Write in completed square form.}$$

$$(y' + 1)^2 = (-1)\left(x' - \tfrac{4}{5}\right) \qquad \text{Write in standard form.}$$

The graph of this equation is a parabola with vertex at $\left(\frac{4}{5}, -1\right)$ in the $x'y'$-plane.
Its axis is parallel to the x'-axis in the $x'y'$-system, as shown in Figure 9.43.

 **CHECKPOINT**   Now try Exercise 13.

Invariants Under Rotation

In the rotation of axes theorem listed at the beginning of this section, note that the constant term is the same in both equations—that is, $F' = F$. Such quantities are **invariant under rotation.** The next theorem lists some other rotation invariants.

Rotation Invariants

The rotation of the coordinate axes through an angle θ that transforms the equation $Ax^2 + Bxy + Cy^2 + Dx + Ey + F = 0$ into the form

$$A'(x')^2 + C'(y')^2 + D'x' + E'y' + F' = 0$$

has the following rotation invariants.

1. $F = F'$

2. $A + C = A' + C'$

3. $B^2 - 4AC = (B')^2 - 4A'C'$

You can use the results of this theorem to classify the graph of a second-degree equation *with* an xy-term in much the same way that you classify the graph of a second-degree equation *without* an xy-term. Note that because $B' = 0$, the invariant $B^2 - 4AC$ reduces to

$$B^2 - 4AC = -4A'C'. \qquad \text{Discriminant}$$

This quantity is called the **discriminant** of the equation

$$Ax^2 + Bxy + Cy^2 + Dx + Ey + F = 0.$$

Now, from the classification procedure given in Section 9.3, you know that the sign of $A'C'$ determines the type of graph for the equation

$$A'(x')^2 + C'(y')^2 + D'x' + E'y' + F' = 0.$$

Consequently, the sign of $B^2 - 4AC$ will determine the type of graph for the original equation, as shown in the following classification.

Classification of Conics by the Discriminant

The graph of the equation $Ax^2 + Bxy + Cy^2 + Dx + Ey + F = 0$ is, except in degenerate cases, determined by its discriminant as follows.

1. Ellipse or circle: $B^2 - 4AC < 0$

2. Parabola: $B^2 - 4AC = 0$

3. Hyperbola: $B^2 - 4AC > 0$

For example, in the general equation $3x^2 + 7xy + 5y^2 - 6x - 7y + 15 = 0$, you have $A = 3$, $B = 7$, and $C = 5$. So, the discriminant is

$$B^2 - 4AC = 7^2 - 4(3)(5) = 49 - 60 = -11.$$

Because $-11 < 0$, the graph of the equation is an ellipse or a circle.

Example 4   Rotations and Graphing Utilities

For each equation, classify the graph of the equation, use the Quadratic Formula to solve for y, and then use a graphing utility to graph the equation.

a. $2x^2 - 3xy + 2y^2 - 2x = 0$ **b.** $x^2 - 6xy + 9y^2 - 2y + 1 = 0$

c. $3x^2 + 8xy + 4y^2 - 7 = 0$

Solution

a. Because $B^2 - 4AC = 9 - 16 < 0$, the graph is a circle or an ellipse. Solve for y as follows.

$$2x^2 - 3xy + 2y^2 - 2x = 0 \qquad \text{Write original equation.}$$

$$2y^2 - 3xy + (2x^2 - 2x) = 0 \qquad \text{Quadratic form } ay^2 + by + c = 0$$

$$y = \frac{-(-3x) \pm \sqrt{(-3x)^2 - 4(2)(2x^2 - 2x)}}{2(2)}$$

$$y = \frac{3x \pm \sqrt{x(16 - 7x)}}{4}$$

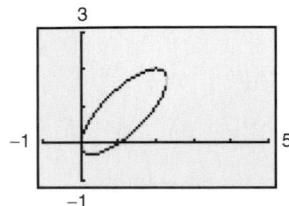

Figure 9.44

Graph both of the equations to obtain the ellipse shown in Figure 9.44.

$$y_1 = \frac{3x + \sqrt{x(16 - 7x)}}{4} \qquad \text{Top half of ellipse}$$

$$y_2 = \frac{3x - \sqrt{x(16 - 7x)}}{4} \qquad \text{Bottom half of ellipse}$$

b. Because $B^2 - 4AC = 36 - 36 = 0$, the graph is a parabola.

$$x^2 - 6xy + 9y^2 - 2y + 1 = 0 \qquad \text{Write original equation.}$$

$$9y^2 - (6x + 2)y + (x^2 + 1) = 0 \qquad \text{Quadratic form } ay^2 + by + c = 0$$

$$y = \frac{(6x + 2) \pm \sqrt{(6x + 2)^2 - 4(9)(x^2 + 1)}}{2(9)}$$

$$y = \frac{3x + 1 \pm \sqrt{2(3x - 4)}}{9}$$

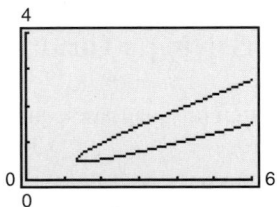

Figure 9.45

Graph both of the equations to obtain the parabola shown in Figure 9.45.

c. Because $B^2 - 4AC = 64 - 48 > 0$, the graph is a hyperbola.

$$3x^2 + 8xy + 4y^2 - 7 = 0 \qquad \text{Write original equation.}$$

$$4y^2 + 8xy + (3x^2 - 7) = 0 \qquad \text{Quadratic form } ay^2 + by + c = 0$$

$$y = \frac{-8x \pm \sqrt{(8x)^2 - 4(4)(3x^2 - 7)}}{2(4)}$$

$$y = \frac{-2x \pm \sqrt{x^2 + 7}}{2}$$

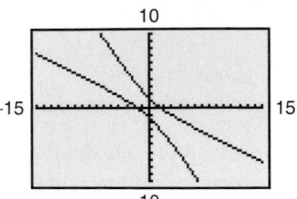

Figure 9.46

Graph both of the equations to obtain the hyperbola shown in Figure 9.46.

✓**CHECKPOINT**   Now try Exercise 27.

Systems of Quadratic Equations

To find the points of intersection of two conics, you can use elimination or substitution, as demonstrated in Examples 5 and 6.

TECHNOLOGY SUPPORT

For instructions on how to use the *intersect* feature, see Appendix A; for specific keystrokes, go to this textbook's *Online Study Center*.

Example 5   Solving a Quadratic System by Elimination

Solve the system of quadratic equations
$$\begin{cases} x^2 + y^2 - 16x + 39 = 0 & \text{Equation 1} \\ x^2 - y^2 \qquad\quad - 9 = 0 & \text{Equation 2} \end{cases}$$

Algebraic Solution

You can eliminate the y^2-term by adding the two equations. The resulting equation can then be solved for x.

$$2x^2 - 16x + 30 = 0$$

$$2(x - 3)(x - 5) = 0$$

There are two real solutions: $x = 3$ and $x = 5$. The corresponding y-values are $y = 0$ and $y = \pm 4$. So, the solutions of the system are $(3, 0)$, $(5, 4)$, and $(5, -4)$.

✔CHECKPOINT   Now try Exercise 41.

Graphical Solution

Begin by solving each equation for y as follows.

$$y = \pm\sqrt{-x^2 + 16x - 39} \qquad y = \pm\sqrt{x^2 - 9}$$

Use a graphing utility to graph all four equations $y_1 = \sqrt{-x^2 + 16x - 39}$, $y_2 = -\sqrt{-x^2 + 16x - 39}$, $y_3 = \sqrt{x^2 - 9}$, and $y_4 = -\sqrt{x^2 - 9}$ in the same viewing window. Use the *intersect* feature of the graphing utility to approximate the points of intersection to be $(3, 0)$, $(5, 4)$, and $(5, -4)$, as shown in Figure 9.47.

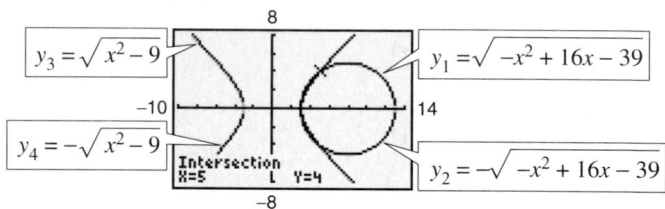

Figure 9.47

Example 6   Solving a Quadratic System by Substitution

Solve the system of quadratic equations
$$\begin{cases} x^2 + 4y^2 - 4x - 8y + 4 = 0 & \text{Equation 1} \\ x^2 + 4y \qquad\quad - 4 = 0 & \text{Equation 2} \end{cases}$$

Solution

Because Equation 2 has no y^2-term, solve the equation for y to obtain $y = 1 - \frac{1}{4}x^2$. Next, substitute this into Equation 1 and solve for x.

$$x^2 + 4\left(1 - \tfrac{1}{4}x^2\right)^2 - 4x - 8\left(1 - \tfrac{1}{4}x^2\right) + 4 = 0$$

$$x^2 + 4 - 2x^2 + \tfrac{1}{4}x^4 - 4x - 8 + 2x^2 + 4 = 0$$

$$x^4 + 4x^2 - 16x = 0$$

$$x(x - 2)(x^2 + 2x + 8) = 0$$

In factored form, you can see that the equation has two real solutions: $x = 0$ and $x = 2$. The corresponding values of y are $y = 1$ and $y = 0$. This implies that the solutions of the system of equations are $(0, 1)$ and $(2, 0)$, as shown in Figure 9.48.

✔CHECKPOINT   Now try Exercise 47.

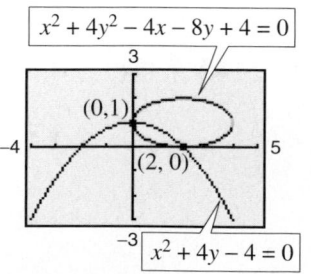

Figure 9.48

9.4 Exercises

See www.CalcChat.com for worked-out solutions to odd-numbered exercises.

Vocabulary Check

Fill in the blanks.

1. The procedure used to eliminate the xy-term in a general second-degree equation is called _____ of _____ .

2. Quantities that are equal in both the original equation of a conic and the equation of the rotated conic are _____ .

3. The quantity $B^2 - 4AC$ is called the _____ of the equation $Ax^2 + Bxy + Cy^2 + Dx + Ey + F = 0$.

In Exercises 1 and 2, the $x'y'$-coordinate system has been rotated θ degrees from the xy-coordinate system. The coordinates of a point in the xy-coordinate system are given. Find the coordinates of the point in the rotated coordinate system.

1. $\theta = 90°$, $(0, 3)$

2. $\theta = 45°$, $(3, 3)$

In Exercises 3–14, rotate the axes to eliminate the xy-term in the equation. Then write the equation in standard form. Sketch the graph of the resulting equation, showing both sets of axes.

3. $xy + 1 = 0$

4. $xy - 2 = 0$

5. $x^2 - 4xy + y^2 + 1 = 0$

6. $xy + x - 2y + 3 = 0$

7. $xy - 2y - 4x = 0$

8. $2x^2 - 3xy - 2y^2 + 10 = 0$

9. $5x^2 - 6xy + 5y^2 - 12 = 0$

10. $13x^2 + 6\sqrt{3}xy + 7y^2 - 16 = 0$

11. $3x^2 - 2\sqrt{3}xy + y^2 + 2x + 2\sqrt{3}y = 0$

12. $16x^2 - 24xy + 9y^2 - 60x - 80y + 100 = 0$

13. $9x^2 + 24xy + 16y^2 + 90x - 130y = 0$

14. $9x^2 + 24xy + 16y^2 + 80x - 60y = 0$

In Exercises 15–20, use a graphing utility to graph the conic. Determine the angle θ through which the axes are rotated. Explain how you used the graphing utility to obtain the graph.

15. $x^2 + 3xy + y^2 = 20$

16. $x^2 - 4xy + 2y^2 = 8$

17. $17x^2 + 32xy - 7y^2 = 75$

18. $40x^2 + 36xy + 25y^2 = 52$

19. $32x^2 + 48xy + 8y^2 = 50$

20. $4x^2 - 12xy + 9y^2 + (4\sqrt{13} - 12)x$
    $- (6\sqrt{13} + 8)y = 91$

In Exercises 21–26, match the graph with its equation. [The graphs are labeled (a), (b), (c), (d), (e), and (f).]

(a)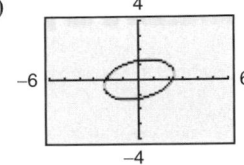

(b)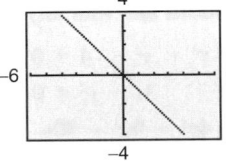

(c)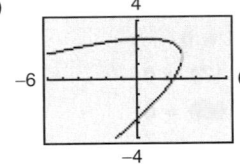

(d)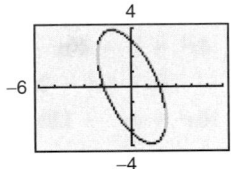

(e)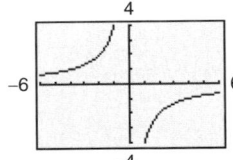

(f)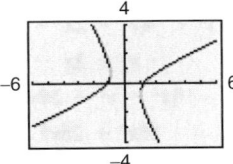

21. $xy + 4 = 0$

22. $x^2 + 2xy + y^2 = 0$

23. $-2x^2 + 3xy + 2y^2 + 3 = 0$

24. $x^2 - xy + 3y^2 - 5 = 0$

25. $3x^2 + 2xy + y^2 - 10 = 0$

26. $x^2 - 4xy + 4y^2 + 10x - 30 = 0$

In Exercises 27–34, (a) use the discriminant to classify the graph of the equation, (b) use the Quadratic Formula to solve for y, and (c) use a graphing utility to graph the equation.

27. $16x^2 - 24xy + 9y^2 - 30x - 40y = 0$

28. $x^2 - 4xy - 2y^2 - 6 = 0$

29. $15x^2 - 8xy + 7y^2 - 45 = 0$

30. $2x^2 + 4xy + 5y^2 + 3x - 4y - 20 = 0$

31. $x^2 - 6xy - 5y^2 + 4x - 22 = 0$

32. $36x^2 - 60xy + 25y^2 + 9y = 0$

33. $x^2 + 4xy + 4y^2 - 5x - y - 3 = 0$

34. $x^2 + xy + 4y^2 + x + y - 4 = 0$

In Exercises 35–38, sketch (if possible) the graph of the degenerate conic.

35. $y^2 - 16x^2 = 0$

36. $x^2 + y^2 - 2x + 6y + 10 = 0$

37. $x^2 + 2xy + y^2 - 1 = 0$

38. $x^2 - 10xy + y^2 = 0$

In Exercises 39–46, solve the system of quadratic equations algebraically by the method of elimination. Then verify your results by using a graphing utility to graph the equations and find any points of intersection of the graphs.

39. $\begin{cases} x^2 + y^2 - 4 = 0 \\ 3x - y^2 = 0 \end{cases}$

40. $\begin{cases} 4x^2 + 9y^2 - 36y = 0 \\ x^2 + y^2 - 27 = 0 \end{cases}$

41. $\begin{cases} -4x^2 - y^2 - 16x + 24y - 16 = 0 \\ 4x^2 + y^2 + 40x - 24y + 208 = 0 \end{cases}$

42. $\begin{cases} x^2 - 4y^2 - 20x - 64y - 172 = 0 \\ 16x^2 + 4y^2 - 320x + 64y + 1600 = 0 \end{cases}$

43. $\begin{cases} x^2 - y^2 - 12x + 16y - 64 = 0 \\ x^2 + y^2 - 12x - 16y + 64 = 0 \end{cases}$

44. $\begin{cases} x^2 + 4y^2 - 2x - 8y + 1 = 0 \\ -x^2 + 2x - 4y - 1 = 0 \end{cases}$

45. $\begin{cases} -16x^2 - y^2 + 24y - 80 = 0 \\ 16x^2 + 25y^2 - 400 = 0 \end{cases}$

46. $\begin{cases} 16x^2 - y^2 + 16y - 128 = 0 \\ y^2 - 48x - 16y - 32 = 0 \end{cases}$

In Exercises 47–52, solve the system of quadratic equations algebraically by the method of substitution. Then verify your results by using a graphing utility to graph the equations and find any points of intersection of the graphs.

47. $\begin{cases} 2x^2 - y^2 + 6 = 0 \\ 2x + y = 0 \end{cases}$ **48.** $\begin{cases} 6x^2 + 3y^2 - 12 = 0 \\ x + y - 2 = 0 \end{cases}$

49. $\begin{cases} 10x^2 - 25y^2 - 100x + 160 = 0 \\ y^2 - 2x + 16 = 0 \end{cases}$

50. $\begin{cases} 4x^2 - y^2 - 8x + 6y - 9 = 0 \\ 2x^2 - 3y^2 + 4x + 18y - 43 = 0 \end{cases}$

51. $\begin{cases} xy + x - 2y + 3 = 0 \\ x^2 + 4y^2 - 9 = 0 \end{cases}$

52. $\begin{cases} 5x^2 - 2xy + 5y^2 - 12 = 0 \\ x + y - 1 = 0 \end{cases}$

Synthesis

True or False?   **In Exercises 53 and 54, determine whether the statement is true or false. Justify your answer.**

53. The graph of $x^2 + xy + ky^2 + 6x + 10 = 0$, where k is any constant less than $\frac{1}{4}$, is a hyperbola.

54. After using a rotation of axes to eliminate the xy-term from an equation of the form

$$Ax^2 + Bxy + Cy^2 + Dx + Ey + F = 0$$

the coefficients of the x^2- and y^2-terms remain A and C, respectively.

Skills Review

In Exercises 55–58, sketch the graph of the rational function. Identify all intercepts and asymptotes.

55. $g(x) = \dfrac{2}{2 - x}$ **56.** $f(x) = \dfrac{2x}{2 - x}$

57. $h(t) = \dfrac{t^2}{2 - t}$ **58.** $g(s) = \dfrac{2}{4 - s^2}$

In Exercises 59–62, if possible, find (a) AB, (b) BA, and (c) A^2.

59. $A = \begin{bmatrix} 1 & -3 \\ 2 & 5 \end{bmatrix}$, $B = \begin{bmatrix} 0 & 6 \\ 5 & -1 \end{bmatrix}$

60. $A = \begin{bmatrix} 1 & 5 \\ 0 & -2 \end{bmatrix}$, $B = \begin{bmatrix} 3 & 2 \\ -3 & 8 \end{bmatrix}$

61. $A = \begin{bmatrix} 4 & -2 & 5 \end{bmatrix}$, $B = \begin{bmatrix} 3 \\ -4 \\ 5 \end{bmatrix}$

62. $A = \begin{bmatrix} 0 & -2 & 0 \\ 1 & 1 & 5 \\ 3 & 4 & 0 \end{bmatrix}$, $B = \begin{bmatrix} 1 & 0 & -3 \\ -4 & 5 & -1 \\ 6 & 3 & 2 \end{bmatrix}$

In Exercises 63–70, graph the function.

63. $f(x) = |x + 3|$ **64.** $f(x) = |x - 4| + 1$

65. $g(x) = \sqrt{4 - x^2}$ **66.** $g(x) = \sqrt{3x - 2}$

67. $h(t) = -(t - 2)^3 + 3$ **68.** $h(t) = \frac{1}{2}(t + 4)^3$

69. $f(t) = [\![t - 5]\!] + 1$ **70.** $f(t) = -2[\![t]\!] + 3$

In Exercises 71–74, find the area of the triangle.

71. $C = 110°$, $a = 8$, $b = 12$

72. $B = 70°$, $a = 25$, $c = 16$

73. $a = 11$, $b = 18$, $c = 10$

74. $a = 23$, $b = 35$, $c = 27$

9.5   Parametric Equations

Plane Curves

Up to this point, you have been representing a graph by a single equation involving *two* variables such as x and y. In this section, you will study situations in which it is useful to introduce a *third* variable to represent a curve in the plane.

To see the usefulness of this procedure, consider the path of an object that is propelled into the air at an angle of 45°. If the initial velocity of the object is 48 feet per second, it can be shown that the object follows the parabolic path

$$y = -\frac{x^2}{72} + x$$ Rectangular equation

as shown in Figure 9.49. However, this equation does not tell the whole story. Although it does tell you *where* the object has been, it doesn't tell you *when* the object was at a given point (x, y) on the path. To determine this time, you can introduce a third variable t, called a **parameter.** It is possible to write both x and y as functions of t to obtain the **parametric equations**

$$x = 24\sqrt{2}t$$ Parametric equation for x

$$y = -16t^2 + 24\sqrt{2}t.$$ Parametric equation for y

From this set of equations you can determine that at time $t = 0$, the object is at the point $(0, 0)$. Similarly, at time $t = 1$, the object is at the point $\left(24\sqrt{2}, 24\sqrt{2} - 16\right)$, and so on.

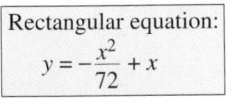

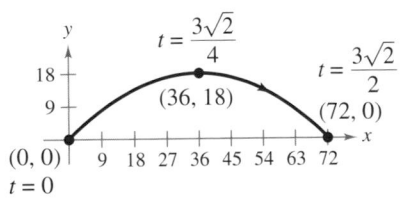

Rectangular equation:
$$y = -\frac{x^2}{72} + x$$

Parametric equations:
$$x = 24\sqrt{2}t$$
$$y = -16t^2 + 24\sqrt{2}t$$

Curvilinear motion: two variables for position, one variable for time
Figure 9.49

For this particular motion problem, x and y are continuous functions of t, and the resulting path is a **plane curve.** (Recall that a *continuous function* is one whose graph can be traced without lifting the pencil from the paper.)

Definition of a Plane Curve

If f and g are continuous functions of t on an interval l, the set of ordered pairs $(f(t), g(t))$ is a **plane curve** C. The equations given by

$$x = f(t) \qquad \text{and} \qquad y = g(t)$$

are **parametric equations** for C, and t is the **parameter.**

Sketching a Plane Curve

One way to sketch a curve represented by a pair of parametric equations is to plot points in the xy-plane. Each set of coordinates (x, y) is determined from a value chosen for the parameter t. By plotting the resulting points in the order of *increasing* values of t, you trace the curve in a specific direction. This is called the **orientation** of the curve.

Example 1   Sketching a Plane Curve

Sketch the curve given by the parametric equations

$$x = t^2 - 4 \quad \text{and} \quad y = \frac{t}{2}, \qquad -2 \le t \le 3.$$

Describe the orientation of the curve.

Solution

Using values of t in the interval, the parametric equations yield the points (x, y) shown in the table.

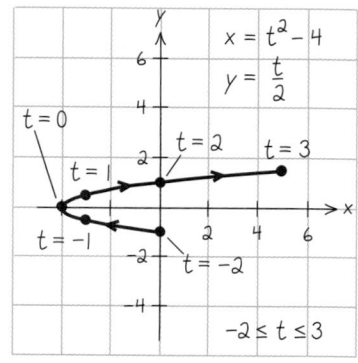

t	-2	-1	0	1	2	3
x	0	-3	-4	-3	0	5
y	-1	$-\frac{1}{2}$	0	$\frac{1}{2}$	1	$\frac{3}{2}$

By plotting these points in the order of increasing t, you obtain the curve shown in Figure 9.50. The arrows on the curve indicate its orientation as t increases from -2 to 3. So, if a particle were moving on this curve, it would start at $(0, -1)$ and then move along the curve to the point $\left(5, \frac{3}{2}\right)$.

Figure 9.50

✓**CHECKPOINT**   Now try Exercises 7(a) and (b).

Note that the graph shown in Figure 9.50 does not define y as a function of x. This points out one benefit of parametric equations—they can be used to represent graphs that are more general than graphs of functions.

Two different sets of parametric equations can have the same graph. For example, the set of parametric equations

$$x = 4t^2 - 4 \quad \text{and} \quad y = t, \qquad -1 \le t \le \frac{3}{2}$$

has the same graph as the set given in Example 1. However, by comparing the values of t in Figures 9.50 and 9.51, you can see that this second graph is traced out more *rapidly* (considering t as time) than the first graph. So, in applications, different parametric representations can be used to represent various *speeds* at which objects travel along a given path.

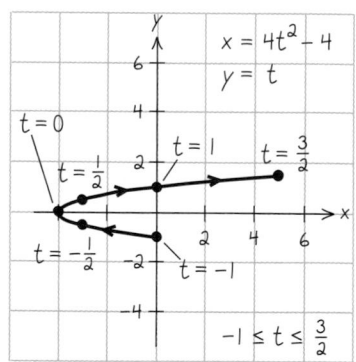

Figure 9.51

> **TECHNOLOGY TIP**   Most graphing utilities have a *parametric* mode. So, another way to display a curve represented by a pair of parametric equations is to use a graphing utility, as shown in Example 2. For instructions on how to use the *parametric* mode, see Appendix A; for specifice keystrokes, go to this textbook's *Online Study Center*.

Example 2   Using a Graphing Utility in Parametric Mode

Use a graphing utility to graph the curves represented by the parametric equations. Using the graph and the Vertical Line Test, for which curve is y a function of x?

a. $x = t^2, y = t^3$ **b.** $x = t, y = t^3$ **c.** $x = t^2, y = t$

Prerequisite Skills

See Section 1.3 to review the Vertical Line Test.

Solution

Begin by setting the graphing utility to *parametric* mode. When choosing a viewing window, you must set not only minimum and maximum values of x and y, but also minimum and maximum values of t.

a. Enter the parametric equations for x and y, as shown in Figure 9.52. Use the viewing window shown in Figure 9.53. The curve is shown in Figure 9.54. From the graph, you can see that y *is not* a function of x.

Figure 9.52

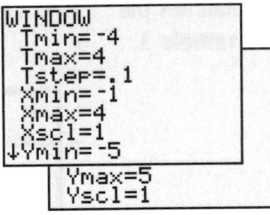

Figure 9.53

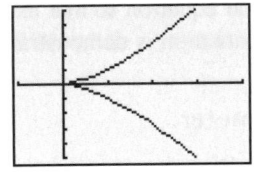

Figure 9.54

b. Enter the parametric equations for x and y, as shown in Figure 9.55. Use the viewing window shown in Figure 9.56. The curve is shown in Figure 9.57. From the graph, you can see that y *is* a function of x.

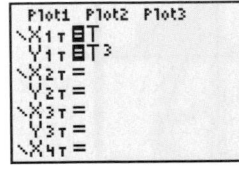

Figure 9.55

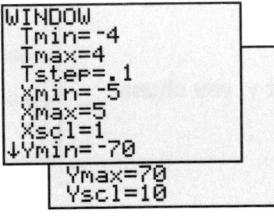

Figure 9.56

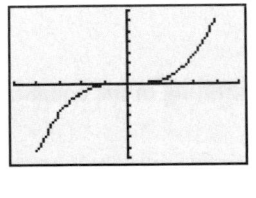

Figure 9.57

c. Enter the parametric equations for x and y, as shown in Figure 9.58. Use the viewing window shown in Figure 9.59. The curve is shown in Figure 9.60. From the graph, you can see that y *is not* a function of x.

Figure 9.58

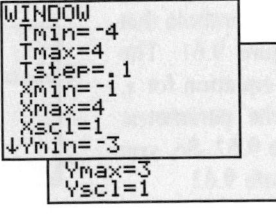

Figure 9.59

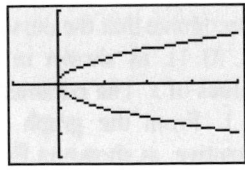

Figure 9.60

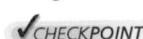CHECKPOINT   Now try Exercise 7(c).

Exploration

Use a graphing utility set in *parametric* mode to graph the curve

$$x = t \quad \text{and} \quad y = 1 - t^2$$

Set the viewing window so that $-4 \le x \le 4$ and $-12 \le y \le 2$. Now, graph the curve with various settings for t. Use the following.

a. $0 \le t \le 3$

b. $-3 \le t \le 0$

c. $-3 \le t \le 3$

Compare the curves given by the different t settings. Repeat this experiment using $x = -t$. How does this change the results?

TECHNOLOGY TIP

Notice in Example 2 that in order to set the viewing windows of parametric graphs, you have to scroll down to enter the Ymax and Yscl values.

Eliminating the Parameter

Many curves that are represented by sets of parametric equations have graphs that can also be represented by rectangular equations (in x and y). The process of finding the rectangular equation is called **eliminating the parameter.**

Parametric equations	$\Rightarrow$	Solve for t in one equation.	$\Rightarrow$	Substitute in second equation.	$\Rightarrow$	Rectangular equation

$$x = t^2 - 4 \qquad\qquad t = 2y \qquad\qquad\qquad x = (2y)^2 - 4 \qquad x = 4y^2 - 4$$
$$y = \tfrac{1}{2}t$$

Now you can recognize that the equation $x = 4y^2 - 4$ represents a parabola with a horizontal axis and vertex at $(-4, 0)$.

When converting equations from parametric to rectangular form, you may need to alter the domain of the rectangular equation so that its graph matches the graph of the parametric equations. This situation is demonstrated in Example 3.

Example 3   Eliminating the Parameter

Identify the curve represented by the equations

$$x = \frac{1}{\sqrt{t + 1}} \qquad \text{and} \qquad y = \frac{t}{t + 1}.$$

Solution

Solving for t in the equation for x produces

$$x^2 = \frac{1}{t + 1} \qquad \text{or} \qquad \frac{1}{x^2} = t + 1$$

which implies that $t = (1/x^2) - 1$. Substituting in the equation for y, you obtain the rectangular equation

$$y = \frac{t}{t + 1}$$

$$= \frac{\left(\dfrac{1}{x^2}\right) - 1}{\left(\dfrac{1}{x^2}\right) - 1 + 1} = \frac{\dfrac{1 - x^2}{x^2}}{\left(\dfrac{1}{x^2}\right)} = \frac{1 - x^2}{x^2} \cdot \frac{x^2}{x^2} = 1 - x^2.$$

From the rectangular equation, you can recognize that the curve is a parabola that opens downward and has its vertex at $(0, 1)$, as shown in Figure 9.61. The rectangular equation is defined for all values of x. The parametric equation for x, however, is defined only when $t > -1$. From the graph of the parametric equations, you can see that x is always positive, as shown in Figure 9.62. So, you should restrict the domain of x to positive values, as shown in Figure 9.63.

✓**CHECKPOINT**   Now try Exercise 7(d).

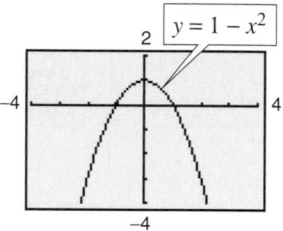

Figure 9.61

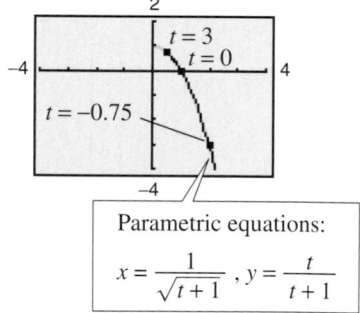

Figure 9.62

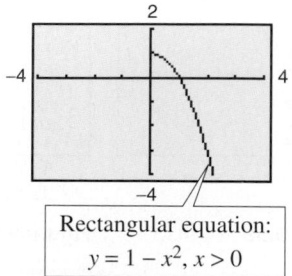

Figure 9.63

Example 4   Eliminating the Parameter

Sketch the curve represented by $x = 3 \cos \theta$ and $y = 4 \sin \theta$, $0 \le \theta \le 2\pi$, by eliminating the parameter.

Solution

Begin by solving for $\cos \theta$ and $\sin \theta$ in the equations.

$$\cos \theta = \frac{x}{3} \quad \text{and} \quad \sin \theta = \frac{y}{4} \qquad \text{Solve for } \cos \theta \text{ and } \sin \theta.$$

Use the identity $\sin^2 \theta + \cos^2 \theta = 1$ to form an equation involving only x and y.

$$\cos^2 \theta + \sin^2 \theta = 1 \qquad \text{Pythagorean identity}$$

$$\left(\frac{x}{3}\right)^2 + \left(\frac{y}{4}\right)^2 = 1 \qquad \text{Substitute } \frac{x}{3} \text{ for } \cos \theta \text{ and } \frac{y}{4} \text{ for } \sin \theta.$$

$$\frac{x^2}{9} + \frac{y^2}{16} = 1 \qquad \text{Rectangular equation}$$

From this rectangular equation, you can see that the graph is an ellipse centered at $(0, 0)$, with vertices $(0, 4)$ and $(0, -4)$, and minor axis of length $2b = 6$, as shown in Figure 9.64. Note that the elliptic curve is traced out *counterclockwise*.

✓*CHECKPOINT*   Now try Exercise 23.

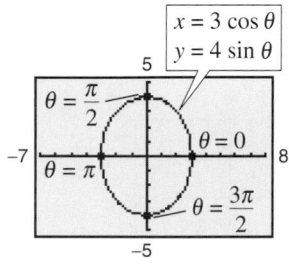

Figure 9.64

Finding Parametric Equations for a Graph

How can you determine a set of parametric equations for a given graph or a given physical description? From the discussion following Example 1, you know that such a representation is not unique. This is further demonstrated in Example 5.

Example 5   Finding Parametric Equations for a Given Graph

Find a set of parametric equations to represent the graph of $y = 1 - x^2$ using the parameters (a) $t = x$ and (b) $t = 1 - x$.

Solution

a. Letting $t = x$, you obtain the following parametric equations.

$$x = t \qquad \text{Parametric equation for } x$$

$$y = 1 - t^2 \qquad \text{Parametric equation for } y$$

The graph of these equations is shown in Figure 9.65.

b. Letting $t = 1 - x$, you obtain the following parametric equations.

$$x = 1 - t \qquad \text{Parametric equation for } x$$

$$y = 1 - (1 - t)^2 = 2t - t^2 \qquad \text{Parametric equation for } y$$

The graph of these equations is shown in Figure 9.66. Note that the graphs in Figures 9.65 and 9.66 have opposite orientations.

✓*CHECKPOINT*   Now try Exercise 45.

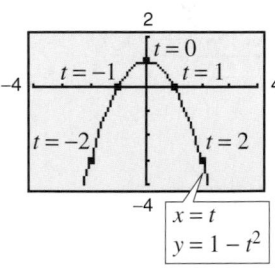

Figure 9.65

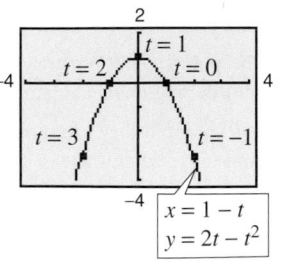

Figure 9.66

9.5 Exercises

See www.CalcChat.com for worked-out solutions to odd-numbered exercises.

Vocabulary Check

Fill in the blanks.

1. If f and g are continuous functions of t on an interval I, the set of ordered pairs $(f(t), g(t))$ is a _____ C. The equations given by $x = f(t)$ and $y = g(t)$ are _____ for C, and t is the _____ .

2. The _____ of a curve is the direction in which the curve is traced out for increasing values of the parameter.

3. The process of converting a set of parametric equations to rectangular form is called _____ the _____ .

In Exercises 1–6, match the set of parametric equations with its graph. [The graphs are labeled (a), (b), (c), (d), (e), and (f).]

(a)
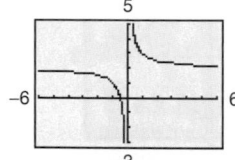

(b)
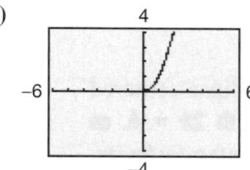

(c)
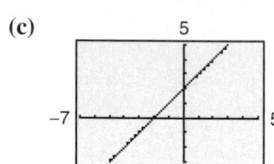

(d)
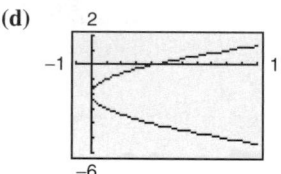

(e)
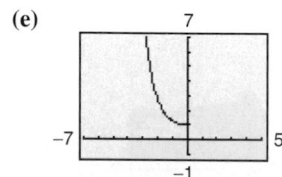

(f)
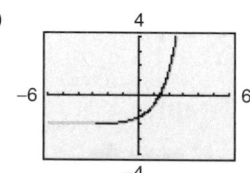

1. $x = t,\ y = t + 2$

2. $x = t^2,\ y = t - 2$

3. $x = \sqrt{t},\ y = t$

4. $x = \dfrac{1}{t},\ y = t + 2$

5. $x = \ln t,\ y = \frac{1}{2}t - 2$

6. $x = -2\sqrt{t},\ y = e^t$

7. Consider the parametric equations $x = \sqrt{t}$ and $y = 2 - t$.

   (a) Create a table of x- and y-values using $t = 0, 1, 2, 3,$ and 4.

   (b) Plot the points (x, y) generated in part (a) and sketch a graph of the parametric equations.

   (c) Use a graphing utility to graph the curve represented by the parametric equations.

   (d) Find the rectangular equation by eliminating the parameter. Sketch its graph. How does the graph differ from those in parts (b) and (c)?

8. Consider the parametric equations $x = 4\cos^2\theta$ and $y = 4\sin\theta$.

   (a) Create a table of x- and y-values using $\theta = -\pi/2,\ -\pi/4, 0, \pi/4,$ and $\pi/2$.

   (b) Plot the points (x, y) generated in part (a) and sketch a graph of the parametric equations.

   (c) Use a graphing utility to graph the curve represented by the parametric equations.

   (d) Find the rectangular equation by eliminating the parameter. Sketch its graph. How does the graph differ from those in parts (b) and (c)?

Library of Parent Functions   In Exercises 9 and 10, determine the plane curve whose graph is shown.

9. (a) $x = t^2$
      $y = 2t + 1$

   (b) $x = 2t + 1$
      $y = t^2$

   (c) $x = 2t - 1$
      $y = t^2$

   (d) $x = -2t + 1$
      $y = t^2$

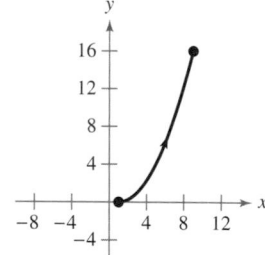

10. (a) $x = 2 - \cos\theta$
       $y = 3 - \sin\theta$

    (b) $x = 3 - \cos\theta$
       $y = 2 + \sin\theta$

    (c) $x = 2 + \cos\theta$
       $y = 3 - \sin\theta$

    (d) $x = 3 + \cos\theta$
       $y = 2 - \sin\theta$

In Exercises 11–26, sketch the curve represented by the parametric equations (indicate the orientation of the curve). Use a graphing utility to confirm your result. Then eliminate the parameter and write the corresponding rectangular equation whose graph represents the curve. Adjust the domain of the resulting rectangular equation, if necessary.

11. $x = t, y = -4t$

12. $x = t, y = \frac{1}{2}t$

13. $x = 3t - 3, y = 2t + 1$

14. $x = 3 - 2t, y = 2 + 3t$

15. $x = \frac{1}{4}t, y = t^2$

16. $x = t, y = t^3$

17. $x = t + 2, y = t^2$

18. $x = \sqrt{t}, y = 1 - t$

19. $x = 2t, y = |t - 2|$

20. $x = |t - 1|, y = t + 2$

21. $x = 2\cos\theta, y = 3\sin\theta$

22. $x = \cos\theta, y = 4\sin\theta$

23. $x = e^{-t}, y = e^{3t}$

24. $x = e^{2t}, y = e^t$

25. $x = t^3, y = 3\ln t$

26. $x = \ln 2t, y = 2t^2$

In Exercises 27–32, use a graphing utility to graph the curve represented by the parametric equations.

27. $x = 4 + 3\cos\theta$
   $y = -2 + \sin\theta$

28. $x = 4 + 3\cos\theta$
   $y = -2 + 2\sin\theta$

29. $x = 4\sec\theta$
   $y = 2\tan\theta$

30. $x = \sec\theta$
   $y = \tan\theta$

31. $x = t/2$
   $y = \ln(t^2 + 1)$

32. $x = 10 - 0.01e^t$
   $y = 0.4t^2$

In Exercises 33 and 34, determine how the plane curves differ from each other.

33. (a) $x = t$
       $y = 2t + 1$
   (b) $x = \cos\theta$
       $y = 2\cos\theta + 1$
   (c) $x = e^{-t}$
       $y = 2e^{-t} + 1$
   (d) $x = e^t$
       $y = 2e^t + 1$

34. (a) $x = 2\sqrt{t}$
       $y = 4 - \sqrt{t}$
   (b) $x = 2\sqrt[3]{t}$
       $y = 4 - \sqrt[3]{t}$
   (c) $x = 2(t + 1)$
       $y = 3 - t$
   (d) $x = -2t^2$
       $y = 4 + t^2$

In Exercises 35–38, eliminate the parameter and obtain the standard form of the rectangular equation.

35. Line through (x_1, y_1) and (x_2, y_2):
   $x = x_1 + t(x_2 - x_1)$
   $y = y_1 + t(y_2 - y_1)$

36. Circle: $x = h + r\cos\theta, y = k + r\sin\theta$

37. Ellipse: $x = h + a\cos\theta, y = k + b\sin\theta$

38. Hyperbola: $x = h + a\sec\theta, y = k + b\tan\theta$

In Exercises 39–42, use the results of Exercises 37–40 to find a set of parametric equations for the line or conic.

39. Line: passes through $(1, 4)$ and $(6, -3)$

40. Circle: center: $(2, 5)$; radius: 4

41. Ellipse: vertices: $(\pm 5, 0)$; foci: $(\pm 4, 0)$

42. Hyperbola: vertices: $(0, \pm 1)$; foci: $(0, \pm 2)$

In Exercises 43–48, find two different sets of parametric equations for the given rectangular equation. (There are many correct answers.)

43. $y = 5x - 3$

44. $y = 4 - 7x$

45. $y = \dfrac{1}{x}$

46. $y = \dfrac{1}{2x}$

47. $y = 6x^2 - 5$

48. $y = x^3 + 2x$

In Exercises 49 and 50, use a graphing utility to graph the curve represented by the parametric equations.

49. Witch of Agnesi: $x = 2\cot\theta, y = 2\sin^2\theta$

50. Folium of Descartes: $x = \dfrac{3t}{1 + t^3}, y = \dfrac{3t^2}{1 + t^3}$

In Exercises 51–54, match the parametric equations with the correct graph. [The graphs are labeled (a), (b), (c), and (d).]

(a)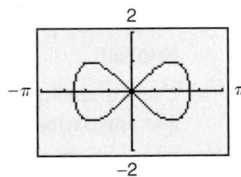

(b)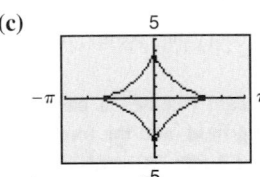

(c)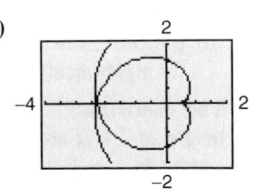

(d)

51. Lissajous curve: $x = 2\cos\theta, y = \sin 2\theta$

52. Evolute of ellipse: $x = 2\cos^3\theta, y = 4\sin^3\theta$

53. Involute of circle: $x = \frac{1}{2}(\cos\theta + \theta\sin\theta)$
   $y = \frac{1}{2}(\sin\theta - \theta\cos\theta)$

54. Serpentine curve: $x = \frac{1}{2}\cot\theta, y = 4\sin\theta\cos\theta$

Projectile Motion In Exercises 55–58, consider a projectile launched at a height of *h* feet above the ground at an angle of θ with the horizontal. The initial velocity is v_0 feet per second and the path of the projectile is modeled by the parametric equations

$$x = (v_0 \cos \theta)t \text{ and } y = h + (v_0 \sin \theta)t - 16t^2.$$

55. The center field fence in a ballpark is 10 feet high and 400 feet from home plate. A baseball is hit at a point 3 feet above the ground. It leaves the bat at an angle of θ degrees with the horizontal at a speed of 100 miles per hour (see figure).

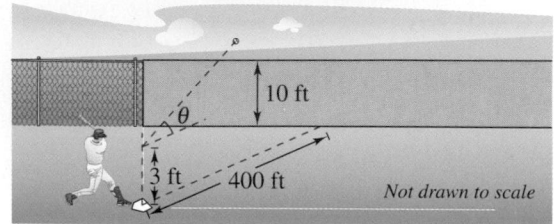

10 ft

θ

3 ft   400 ft

Not drawn to scale

(a) Write a set of parametric equations for the path of the baseball.

(b) Use a graphing utility to graph the path of the baseball for $\theta = 15°$. Is the hit a home run?

(c) Use a graphing utility to graph the path of the baseball for $\theta = 23°$. Is the hit a home run?

(d) Find the minimum angle required for the hit to be a home run.

56. The right field fence in a ballpark is 10 feet high and 314 feet from home plate. A baseball is hit at a point 2.5 feet above the ground. It leaves the bat at an angle of $\theta = 40°$ with the horizontal at a speed of 105 feet per second.

(a) Write a set of parametric equations for the path of the baseball.

(b) Use a graphing utility to graph the path of the baseball and approximate its maximum height.

(c) Use a graphing utility to find the horizontal distance that the baseball travels. Is the hit a home run?

(d) Explain how you could find the result in part (c) algebraically.

57. The quarterback of a football team releases a pass at a height of 7 feet above the playing field, and the football is caught by a receiver at a height of 4 feet, 30 yards directly downfield. The pass is released at an angle of 35° with the horizontal.

(a) Write a set of parametric equations for the path of the football.

(b) Find the speed of the football when it is released.

(c) Use a graphing utility to graph the path of the football and approximate its maximum height.

(d) Find the time the receiver has to position himself after the quarterback releases the football.

58. To begin a football game, a kicker kicks off from his team's 35-yard line. The football is kicked at an angle of 50° with the horizontal at an initial velocity of 85 feet per second.

(a) Write a set of parametric equations for the path of the kick.

(b) Use a graphing utility to graph the path of the kick and approximate its maximum height.

(c) Use a graphing utility to find the horizontal distance that the kick travels.

(d) Explain how you could find the result in part (c) algebraically.

Synthesis

True or False? In Exercises 59–62, determine whether the statement is true or false. Justify your answer.

59. The two sets of parametric equations $x = t$, $y = t^2 + 1$ and $x = 3t$, $y = 9t^2 + 1$ correspond to the same rectangular equation.

60. Because the graph of the parametric equations $x = t^2$, $y = t^2$ and $x = t$, $y = t$ both represent the line $y = x$, they are the same plane curve.

61. If y is a function of t and x is a function of t, then y must be a function of x.

62. The parametric equations $x = at + h$ and $y = bt + k$, where $a \neq 0$ and $b \neq 0$, represent a circle centered at (h, k), if $a = b$.

63. As θ increases, the ellipse given by the parametric equations $x = \cos \theta$ and $y = 2 \sin \theta$ is traced out *counterclockwise*. Find a parametric representation for which the same ellipse is traced out *clockwise*.

64. *Think About It* The graph of the parametric equations $x = t^3$ and $y = t - 1$ is shown below. Would the graph change for the equations $x = (-t)^3$ and $y = -t - 1$? If so, how would it change?

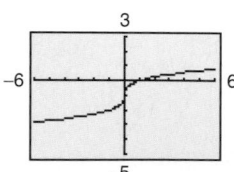

3

−6   6

−5

Skills Review

In Exercises 65–68, check for symmetry with respect to both axes and the origin. Then determine whether the function is even, odd, or neither.

65. $f(x) = \dfrac{4x^2}{x^2 + 1}$ **66.** $f(x) = \sqrt{x}$

67. $y = e^x$ **68.** $(x - 2)^2 = y + 4$

9.6   Polar Coordinates

Introduction

So far, you have been representing graphs of equations as collections of points (x, y) in the rectangular coordinate system, where x and y represent the directed distances from the coordinate axes to the point (x, y). In this section, you will study a second coordinate system called the **polar coordinate system.**

To form the polar coordinate system in the plane, fix a point O, called the **pole** (or **origin**), and construct from O an initial ray called the **polar axis,** as shown in Figure 9.67. Then each point P in the plane can be assigned **polar coordinates** (r, θ) as follows.

1. $r =$ *directed distance* from O to P

2. $\theta =$ *directed angle*, counterclockwise from the polar axis to segment $\overline{OP}$

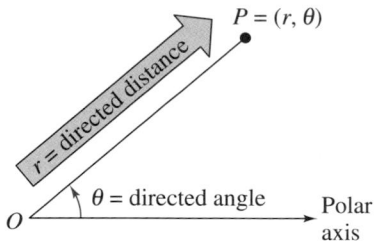

Figure 9.67

Example 1   Plotting Points in the Polar Coordinate System

a. The point $(r, \theta) = (2, \pi/3)$ lies two units from the pole on the terminal side of the angle $\theta = \pi/3$, as shown in Figure 9.68.

b. The point $(r, \theta) = (3, -\pi/6)$ lies three units from the pole on the terminal side of the angle $\theta = -\pi/6$, as shown in Figure 9.69.

c. The point $(r, \theta) = (3, 11\pi/6)$ coincides with the point $(3, -\pi/6)$, as shown in Figure 9.70.

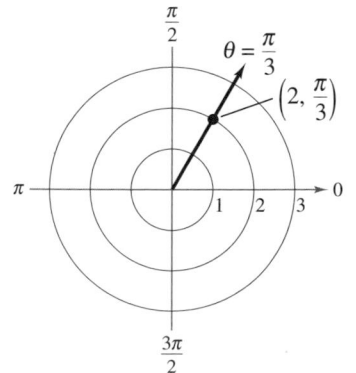

Figure 9.68

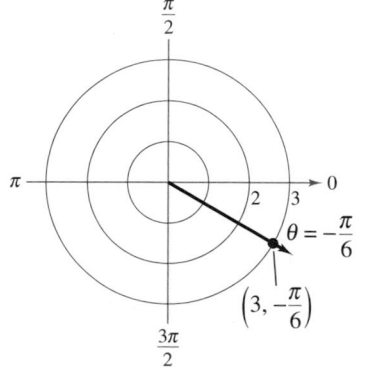

Figure 9.69

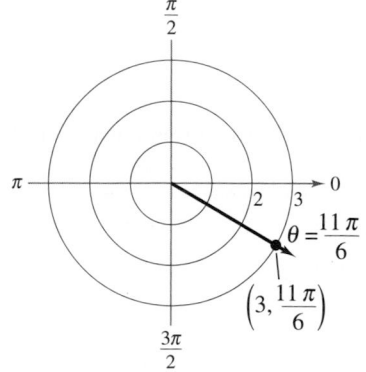

Figure 9.70

✓CHECKPOINT   Now try Exercise 5.

In rectangular coordinates, each point (x, y) has a unique representation. This is not true for polar coordinates. For instance, the coordinates (r, θ) and $(r, \theta + 2\pi)$ represent the same point, as illustrated in Example 1. Another way to obtain multiple representations of a point is to use negative values for r. Because r is a *directed distance*, the coordinates (r, θ) and $(-r, \theta + \pi)$ represent the same point. In general, the point (r, θ) can be represented as

$$(r, \theta) = (r, \theta \pm 2n\pi) \qquad \text{or} \qquad (r, \theta) = (-r, \theta \pm (2n + 1)\pi)$$

where n is any integer. Moreover, the pole is represented by $(0, \theta)$, where θ is any angle.

Example 2   Multiple Representations of Points

Plot the point $(3, -3\pi/4)$ and find three additional polar representations of this point, using $-2\pi < \theta < 2\pi$.

Solution

The point is shown in Figure 9.71. Three other representations are as follows.

$$\left(3, -\frac{3\pi}{4} + 2\pi\right) = \left(3, \frac{5\pi}{4}\right) \qquad\qquad \text{Add } 2\pi \text{ to } \theta.$$

$$\left(-3, -\frac{3\pi}{4} - \pi\right) = \left(-3, -\frac{7\pi}{4}\right) \qquad\qquad \text{Replace } r \text{ by } -r; \text{ subtract } \pi \text{ from } \theta.$$

$$\left(-3, -\frac{3\pi}{4} + \pi\right) = \left(-3, \frac{\pi}{4}\right) \qquad\qquad \text{Replace } r \text{ by } -r; \text{ add } \pi \text{ to } \theta.$$

 **CHECKPOINT**   Now try Exercise 7.

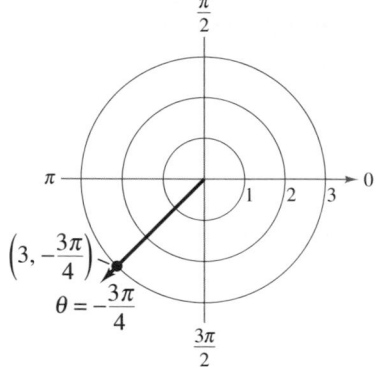

$$\left(3, -\tfrac{3\pi}{4}\right) = \left(3, \tfrac{5\pi}{4}\right) = \left(-3, -\tfrac{7\pi}{4}\right) = \left(-3, \tfrac{\pi}{4}\right) = \cdots$$

Figure 9.71

Coordinate Conversion

To establish the relationship between polar and rectangular coordinates, let the polar axis coincide with the positive x-axis and the pole with the origin, as shown in Figure 9.72. Because (x, y) lies on a circle of radius r, it follows that $r^2 = x^2 + y^2$. Moreover, for $r > 0$, the definitions of the trigonometric functions imply that

$$\tan \theta = \frac{y}{x}, \qquad \cos \theta = \frac{x}{r}, \qquad \text{and} \qquad \sin \theta = \frac{y}{r}.$$

You can show that the same relationships hold for $r < 0$.

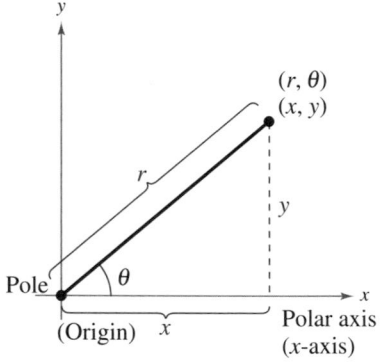

Figure 9.72

Coordinate Conversion

The polar coordinates (r, θ) are related to the rectangular coordinates (x, y) as follows.

$$x = r \cos \theta \qquad \text{and} \qquad \tan \theta = \frac{y}{x}$$

$$y = r \sin \theta \qquad\qquad\qquad r^2 = x^2 + y^2$$

Example 3   Polar-to-Rectangular Conversion

Convert each point to rectangular coordinates.

a. $(2, \pi)$ **b.** $\left(\sqrt{3}, \dfrac{\pi}{6}\right)$

Solution

a. For the point $(r, \theta) = (2, \pi)$, you have the following.

$$x = r \cos \theta = 2 \cos \pi = -2$$

$$y = r \sin \theta = 2 \sin \pi = 0$$

The rectangular coordinates are $(x, y) = (-2, 0)$. (See Figure 9.73.)

b. For the point $(r, \theta) = \left(\sqrt{3}, \pi/6\right)$, you have the following.

$$x = r \cos \theta = \sqrt{3} \cos \frac{\pi}{6} = \sqrt{3}\left(\frac{\sqrt{3}}{2}\right) = \frac{3}{2}$$

$$y = r \sin \theta = \sqrt{3} \sin \frac{\pi}{6} = \sqrt{3}\left(\frac{1}{2}\right) = \frac{\sqrt{3}}{2}$$

The rectangular coordinates are $(x, y) = \left(3/2, \sqrt{3}/2\right)$. (See Figure 9.73.)

✓**CHECKPOINT**   Now try Exercise 13.

Example 4   Rectangular-to-Polar Conversion

Convert each point to polar coordinates.

a. $(-1, 1)$ **b.** $(0, 2)$

Solution

a. For the second-quadrant point $(x, y) = (-1, 1)$, you have

$$\tan \theta = \frac{y}{x} = \frac{1}{-1} = -1$$

$$\theta = \frac{3\pi}{4}.$$

Because θ lies in the same quadrant as (x, y), use positive r.

$$r = \sqrt{x^2 + y^2} = \sqrt{(-1)^2 + (1)^2} = \sqrt{2}$$

So, *one* set of polar coordinates is $(r, \theta) = \left(\sqrt{2}, 3\pi/4\right)$, as shown in Figure 9.74.

b. Because the point $(x, y) = (0, 2)$ lies on the positive y-axis, choose

$$\theta = \pi/2 \quad \text{and} \quad r = 2.$$

This implies that one set of polar coordinates is $(r, \theta) = (2, \pi/2)$, as shown in Figure 9.75.

✓**CHECKPOINT**   Now try Exercise 29.

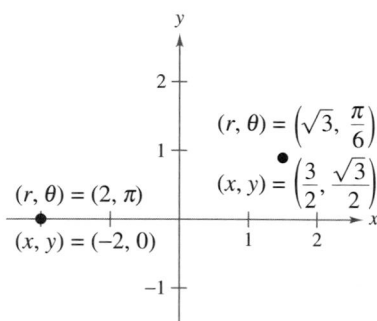

Figure 9.73

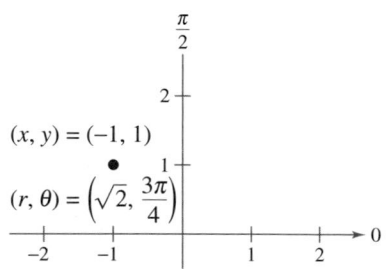

Figure 9.74

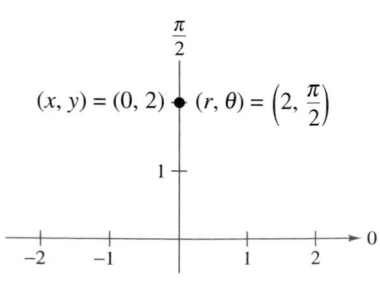

Figure 9.75

Equation Conversion

By comparing Examples 3 and 4, you see that point conversion from the polar to the rectangular system is straightforward, whereas point conversion from the rectangular to the polar system is more involved. For equations, the opposite is true. To convert a rectangular equation to polar form, you simply replace x by $r \cos \theta$ and y by $r \sin \theta$. For instance, the rectangular equation $y = x^2$ can be written in polar form as follows.

$$y = x^2 \qquad \text{Rectangular equation}$$

$$r \sin \theta = (r \cos \theta)^2 \qquad \text{Polar equation}$$

$$r = \sec \theta \tan \theta \qquad \text{Simplest form}$$

On the other hand, converting a polar equation to rectangular form requires considerable ingenuity.

Example 5 demonstrates several polar-to-rectangular conversions that enable you to sketch the graphs of some polar equations.

Example 5   Converting Polar Equations to Rectangular Form

Describe the graph of each polar equation and find the corresponding rectangular equation.

a. $r = 2$ **b.** $\theta = \dfrac{\pi}{3}$ **c.** $r = \sec \theta$

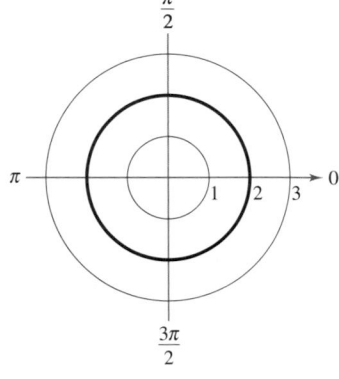

Figure 9.76

Solution

a. The graph of the polar equation $r = 2$ consists of all points that are two units from the pole. In other words, this graph is a circle centered at the origin with a radius of 2, as shown in Figure 9.76. You can confirm this by converting to rectangular form, using the relationship $r^2 = x^2 + y^2$.

$$\underbrace{r = 2}_{\text{Polar equation}} \quad \Longrightarrow \quad r^2 = 2^2 \quad \Longrightarrow \quad \underbrace{x^2 + y^2 = 2^2}_{\text{Rectangular equation}}$$

b. The graph of the polar equation $\theta = \pi/3$ consists of all points on the line that makes an angle of $\pi/3$ with the positive x-axis, as shown in Figure 9.77. To convert to rectangular form, you make use of the relationship $\tan \theta = y/x$.

$$\underbrace{\theta = \frac{\pi}{3}}_{\text{Polar equation}} \quad \Longrightarrow \quad \tan \theta = \sqrt{3} \quad \Longrightarrow \quad \underbrace{y = \sqrt{3}x}_{\text{Rectangular equation}}$$

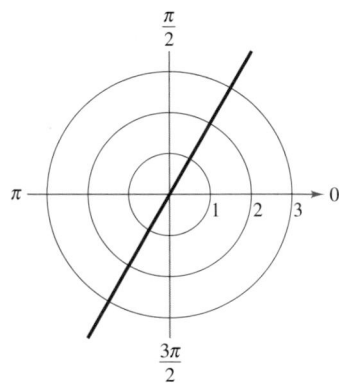

Figure 9.77

c. The graph of the polar equation $r = \sec \theta$ is not evident by simple inspection, so you convert to rectangular form by using the relationship $r \cos \theta = x$.

$$\underbrace{r = \sec \theta}_{\text{Polar equation}} \quad \Longrightarrow \quad r \cos \theta = 1 \quad \Longrightarrow \quad \underbrace{x = 1}_{\text{Rectangular equation}}$$

Now you can see that the graph is a vertical line, as shown in Figure 9.78.

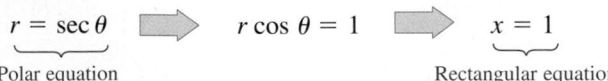 **CHECKPOINT**   Now try Exercise 83.

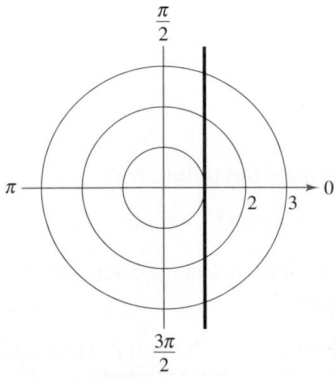

Figure 9.78

9.6 Exercises

See www.CalcChat.com for worked-out solutions to odd-numbered exercises.

Vocabulary Check

Fill in the blanks.

1. The origin of the polar coordinate system is called the _____ .
2. For the point (r, θ), r is the _____ from O to P and θ is the _____ counterclockwise from the polar axis to segment $\overline{OP}$.
3. To graph the point (r, θ), you use the _____ coordinate system.

In Exercises 1–4, a point in polar coordinates is given. Find the corresponding rectangular coordinates for the point.

1. $\left(4, \dfrac{\pi}{2}\right)$

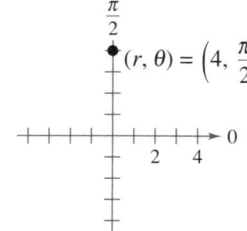

2. $\left(4, \dfrac{3\pi}{2}\right)$

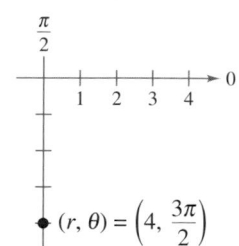

3. $\left(-1, \dfrac{5\pi}{4}\right)$

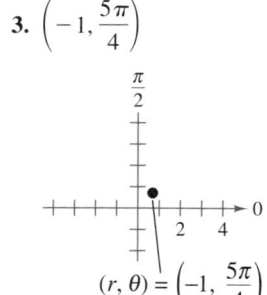

4. $\left(2, -\dfrac{\pi}{4}\right)$

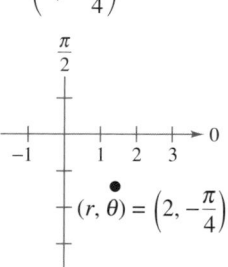

In Exercises 5–12, plot the point given in polar coordinates and find three additional polar representations of the point, using $-2\pi < \theta < 2\pi$.

5. $\left(3, \dfrac{5\pi}{6}\right)$

6. $\left(2, \dfrac{3\pi}{4}\right)$

7. $\left(-1, -\dfrac{\pi}{3}\right)$

8. $\left(-3, -\dfrac{7\pi}{6}\right)$

9. $\left(\sqrt{3}, \dfrac{5\pi}{6}\right)$

10. $\left(5\sqrt{2}, -\dfrac{11\pi}{6}\right)$

11. $\left(\dfrac{3}{2}, -\dfrac{3\pi}{2}\right)$

12. $\left(0, -\dfrac{\pi}{4}\right)$

In Exercises 13–20, plot the point given in polar coordinates and find the corresponding rectangular coordinates for the point.

13. $\left(4, -\dfrac{\pi}{3}\right)$

14. $\left(2, \dfrac{7\pi}{6}\right)$

15. $\left(-1, -\dfrac{3\pi}{4}\right)$

16. $\left(-3, -\dfrac{2\pi}{3}\right)$

17. $\left(0, -\dfrac{7\pi}{6}\right)$

18. $\left(0, \dfrac{5\pi}{4}\right)$

19. $\left(\sqrt{2}, 2.36\right)$

20. $(-3, -1.57)$

In Exercises 21–28, use a graphing utility to find the rectangular coordinates of the point given in polar coordinates. Round your results to two decimal places.

21. $\left(2, \dfrac{2\pi}{9}\right)$

22. $\left(4, \dfrac{11\pi}{9}\right)$

23. $(-4.5, 1.3)$

24. $(8.25, 3.5)$

25. $(2.5, 1.58)$

26. $(5.4, 2.85)$

27. $(-4.1, -0.5)$

28. $(8.2, -3.2)$

In Exercises 29–36, plot the point given in rectangular coordinates and find *two* sets of polar coordinates for the point for $0 \le \theta < 2\pi$.

29. $(-7, 0)$

30. $(0, -5)$

31. $(1, 1)$

32. $(-3, -3)$

33. $\left(-\sqrt{3}, -\sqrt{3}\right)$

34. $\left(\sqrt{3}, -1\right)$

35. $(6, 9)$

36. $(5, 12)$

In Exercises 37–42, use a graphing utility to find one set of polar coordinates for the point given in rectangular coordinates. (There are many correct answers.)

37. $(3, -2)$

38. $(-5, 2)$

39. $\left(\sqrt{3}, 2\right)$

40. $\left(3\sqrt{2}, 3\sqrt{2}\right)$

41. $\left(\dfrac{5}{2}, \dfrac{4}{3}\right)$

42. $\left(\dfrac{7}{4}, \dfrac{3}{2}\right)$

In Exercises 43–60, convert the rectangular equation to polar form. Assume $a > 0$.

43. $x^2 + y^2 = 9$

44. $x^2 + y^2 = 16$

45. $y = 4$

46. $y = x$

47. $x = 8$

48. $x = a$

49. $3x - 6y + 2 = 0$

50. $4x + 7y - 2 = 0$

51. $xy = 4$

52. $2xy = 1$

53. $(x^2 + y^2)^2 = 9(x^2 - y^2)$

54. $y^2 - 8x - 16 = 0$

55. $x^2 + y^2 - 6x = 0$

56. $x^2 + y^2 - 8y = 0$

57. $x^2 + y^2 - 2ax = 0$

58. $x^2 + y^2 - 2ay = 0$

59. $y^2 = x^3$

60. $x^2 = y^3$

In Exercises 61–80, convert the polar equation to rectangular form.

61. $r = 6 \sin \theta$

62. $r = 2 \cos \theta$

63. $\theta = \dfrac{4\pi}{3}$

64. $\theta = \dfrac{5\pi}{3}$

65. $\theta = \dfrac{5\pi}{6}$

66. $\theta = \dfrac{11\pi}{6}$

67. $\theta = \dfrac{\pi}{2}$

68. $\theta = \pi$

69. $r = 4$

70. $r = 10$

71. $r = -3 \csc \theta$

72. $r = 2 \sec \theta$

73. $r^2 = \cos \theta$

74. $r^2 = \sin 2\theta$

75. $r = 2 \sin 3\theta$

76. $r = 3 \cos 2\theta$

77. $r = \dfrac{1}{1 - \cos \theta}$

78. $r = \dfrac{2}{1 + \sin \theta}$

79. $r = \dfrac{6}{2 - 3 \sin \theta}$

80. $r = \dfrac{6}{2 \cos \theta - 3 \sin \theta}$

In Exercises 81–86, describe the graph of the polar equation and find the corresponding rectangular equation. Sketch its graph.

81. $r = 7$

82. $r = 8$

83. $\theta = \dfrac{\pi}{4}$

84. $\theta = \dfrac{7\pi}{6}$

85. $r = 3 \sec \theta$

86. $r = 2 \csc \theta$

Synthesis

True or False?   **In Exercises 87 and 88, determine whether the statement is true or false. Justify your answer.**

87. If (r_1, θ_1) and (r_2, θ_2) represent the same point in the polar coordinate system, then $|r_1| = |r_2|$.

88. If (r, θ_1) and (r, θ_2) represent the same point in the polar coordinate system, then $\theta_1 = \theta_2 + 2\pi n$ for some integer n.

89. *Think About It*

(a) Show that the distance between the points (r_1, θ_1) and (r_2, θ_2) is

$$\sqrt{r_1^2 + r_2^2 - 2r_1 r_2 \cos(\theta_1 - \theta_2)}.$$

(b) Describe the positions of the points relative to each other for $\theta_1 = \theta_2$. Simplify the Distance Formula for this case. Is the simplification what you expected? Explain.

(c) Simplify the Distance Formula for $\theta_1 - \theta_2 = 90°$. Is the simplification what you expected? Explain.

(d) Choose two points in the polar coordinate system and find the distance between them. Then choose different polar representations of the same two points and apply the Distance Formula again. Discuss the result.

90. *Writing* Write a short paragraph explaining the differences between the rectangular coordinate system and the polar coordinate system.

Skills Review

In Exercises 91–96, use the Law of Sines or the Law of Cosines to solve the triangle.

91. $a = 13, b = 19, c = 25$

92. $A = 24°, a = 10, b = 6$

93. $A = 56°, C = 38°, c = 12$

94. $B = 71°, a = 21, c = 29$

95. $C = 35°, a = 8, b = 4$

96. $B = 64°, b = 52, c = 44$

In Exercises 97–102, use any method to solve the system of equations.

97. $\begin{cases} 5x - 7y = -11 \\ -3x + y = -3 \end{cases}$

98. $\begin{cases} 3x + 5y = 10 \\ 4x - 2y = -5 \end{cases}$

99. $\begin{cases} 3a - 2b + c = 0 \\ 2a + b - 3c = 0 \\ a - 3b + 9c = 0 \end{cases}$

100. $\begin{cases} 5u + 7v + 9w = 15 \\ u - 2v - 3w = 7 \\ 8u - 2v + w = 0 \end{cases}$

101. $\begin{cases} -x + y + 2z = 1 \\ 2x + 3y + z = -2 \\ 5x + 4y + 2z = 4 \end{cases}$

102. $\begin{cases} 2x_1 + x_2 + 2x_3 = 4 \\ 2x_1 + 2x_2 = 5 \\ 2x_1 - x_2 + 6x_3 = 2 \end{cases}$

In Exercises 103–106, use a determinant to determine whether the points are collinear.

103. $(4, -3), (6, -7), (-2, -1)$

104. $(-2, 4), (0, 1), (4, -5)$

105. $(-6, -4), (-1, -3), (1.5, -2.5)$

106. $(-2.3, 5), (-0.5, 0), (1.5, -3)$

9.7   Graphs of Polar Equations

Introduction

In previous chapters you sketched graphs in rectangular coordinate systems. You began with the basic point-plotting method. Then you used sketching aids such as a graphing utility, symmetry, intercepts, asymptotes, periods, and shifts to further investigate the nature of the graph. This section approaches curve sketching in the polar coordinate system similarly.

Example 1   Graphing a Polar Equation by Point Plotting

Sketch the graph of the polar equation $r = 4 \sin \theta$ by hand.

Solution

The sine function is periodic, so you can get a full range of r-values by considering values of θ in the interval $0 \le \theta \le 2\pi$, as shown in the table.

θ	0	$\dfrac{\pi}{6}$	$\dfrac{\pi}{3}$	$\dfrac{\pi}{2}$	$\dfrac{2\pi}{3}$	$\dfrac{5\pi}{6}$	π	$\dfrac{7\pi}{6}$	$\dfrac{3\pi}{2}$	$\dfrac{11\pi}{6}$	2π
r	0	2	$2\sqrt{3}$	4	$2\sqrt{3}$	2	0	-2	-4	-2	0

By plotting these points as shown in Figure 9.79, it appears that the graph is a circle of radius 2 whose center is the point $(x, y) = (0, 2)$.

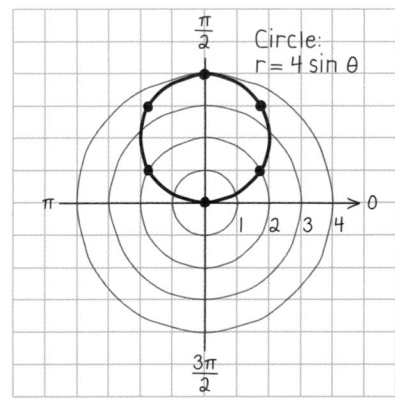

Figure 9.79

✔CHECKPOINT   Now try Exercise 25.

You can confirm the graph found in Example 1 in three ways.

1. *Convert to Rectangular Form*   Multiply each side of the polar equation by r and convert the result to rectangular form.

2. *Use a Polar Coordinate Mode*   Set your graphing utility to *polar* mode and graph the polar equation. (Use $0 \le \theta \le \pi$, $-6 \le x \le 6$, and $-4 \le y \le 4$.)

3. *Use a Parametric Mode*   Set your graphing utility to *parametric* mode and graph $x = (4 \sin t) \cos t$ and $y = (4 \sin t) \sin t$.

What you should learn

- Graph polar equations by point plotting.
- Use symmetry as a sketching aid.
- Use zeros and maximum r-values as sketching aids.
- Recognize special polar graphs.

Why you should learn it

Several common figures, such as the circle in Exercise 4 on page 720, are easier to graph in the polar coordinate system than in the rectangular coordinate system.

Prerequisite Skills

If you have trouble finding the sines of the angles in Example 1, review Trigonometric Functions of Any Angle in Section 4.4.

Most graphing utilities have a *polar* graphing mode. If yours doesn't, you can rewrite the polar equation $r = f(\theta)$ in parametric form, using t as a parameter, as follows.

$$x = f(t) \cos t \quad \text{and} \quad y = f(t) \sin t$$

Symmetry

In Figure 9.79, note that as θ increases from 0 to 2π the graph is traced out twice. Moreover, note that the graph is *symmetric with respect to the line* $\theta = \pi/2$. Had you known about this symmetry and retracing ahead of time, you could have used fewer points. The three important types of symmetry to consider in polar curve sketching are shown in Figure 9.80.

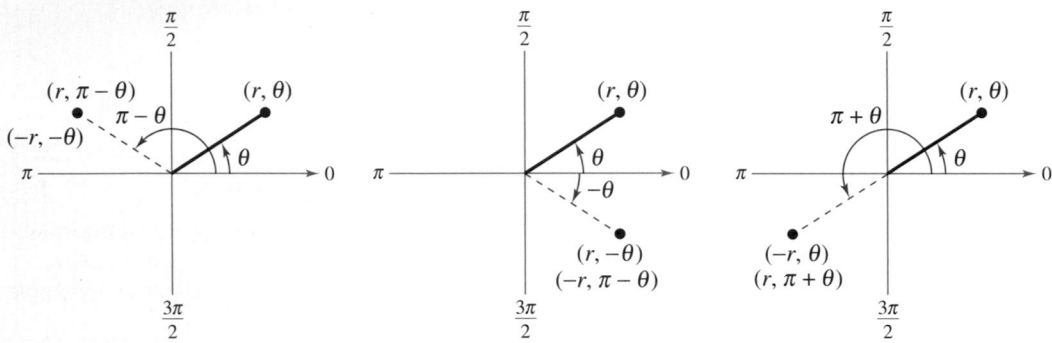

Symmetry with Respect to the Line $\theta = \dfrac{\pi}{2}$

Symmetry with Respect to the Polar Axis

Symmetry with Respect to the Pole

Figure 9.80

Testing for Symmetry in Polar Coordinates

The graph of a polar equation is symmetric with respect to the following if the given substitution yields an equivalent equation.

1. The line $\theta = \dfrac{\pi}{2}$:   Replace (r, θ) by $(r, \pi - \theta)$ or $(-r, -\theta)$.

2. The polar axis: Replace (r, θ) by $(r, -\theta)$ or $(-r, \pi - \theta)$.

3. The pole: Replace (r, θ) by $(r, \pi + \theta)$ or $(-r, \theta)$.

You can determine the symmetry of the graph of $r = 4 \sin \theta$ (see Example 1) as follows.

1. Replace (r, θ) by $(-r, -\theta)$:

$$-r = 4 \sin(-\theta) \quad \Longrightarrow \quad r = -4 \sin(-\theta) = 4 \sin \theta$$

2. Replace (r, θ) by $(r, -\theta)$: $r = 4 \sin(-\theta) = -4 \sin \theta$

3. Replace (r, θ) by $(-r, \theta)$: $-r = 4 \sin \theta \quad \Longrightarrow \quad r = -4 \sin \theta$

So, the graph of $r = 4 \sin \theta$ is symmetric with respect to the line $\theta = \pi/2$.

STUDY TIP

Recall from Section 4.2 that the sine function is odd. That is, $\sin(-\theta) = -\sin \theta$.

Example 2   Using Symmetry to Sketch a Polar Graph

Use symmetry to sketch the graph of $r = 3 + 2 \cos \theta$ by hand.

Solution

Replacing (r, θ) by $(r, -\theta)$ produces

$$r = 3 + 2 \cos(-\theta) = 3 + 2 \cos \theta. \qquad \cos(-u) = \cos u$$

So, by using the even trigonometric identity, you can conclude that the curve is symmetric with respect to the polar axis. Plotting the points in the table and using polar axis symmetry, you obtain the graph shown in Figure 9.81. This graph is called a **limaçon.**

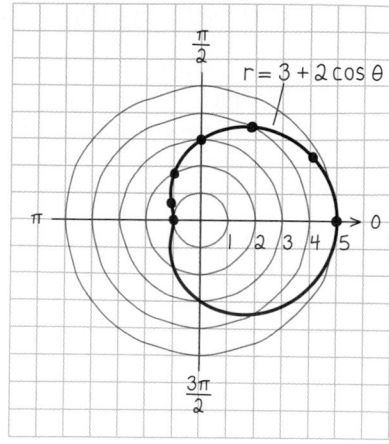

θ	0	$\dfrac{\pi}{6}$	$\dfrac{\pi}{3}$	$\dfrac{\pi}{2}$	$\dfrac{2\pi}{3}$	$\dfrac{5\pi}{6}$	π
r	5	$3 + \sqrt{3}$	4	3	2	$3 - \sqrt{3}$	1

Use a graphing utility to confirm this graph.

✓CHECKPOINT   Now try Exercise 29.

Figure 9.81

The three tests for symmetry in polar coordinates on page 714 are sufficient to guarantee symmetry, but they are not necessary. For instance, Figure 9.82 shows the graph of

$$r = \theta + 2\pi. \qquad \text{Spiral of Archimedes}$$

From the figure, you can see that the graph is symmetric with respect to the line $\theta = \pi/2$. Yet the tests on page 714 fail to indicate symmetry because neither of the following replacements yields an equivalent equation.

Original Equation	*Replacement*	*New Equation*
$r = \theta + 2\pi$	(r, θ) by $(-r, -\theta)$	$-r = -\theta + 2\pi$
$r = \theta + 2\pi$	(r, θ) by $(r, \pi - \theta)$	$r = -\theta + 3\pi$

The equations discussed in Examples 1 and 2 are of the form

$$r = 4 \sin \theta = f(\sin \theta) \qquad \text{and} \qquad r = 3 + 2 \cos \theta = g(\cos \theta).$$

The graph of the first equation is symmetric with respect to the line $\theta = \pi/2$, and the graph of the second equation is symmetric with respect to the polar axis. This observation can be generalized to yield the following *quick tests for symmetry.*

> **Quick Tests for Symmetry in Polar Coordinates**
>
> **1.** The graph of $r = f(\sin \theta)$ is symmetric with respect to the line $\theta = \dfrac{\pi}{2}$.
>
> **2.** The graph of $r = g(\cos \theta)$ is symmetric with respect to the polar axis.

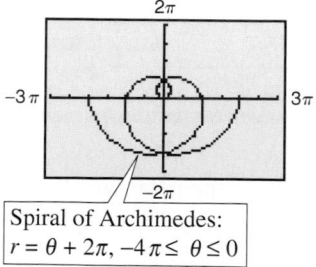

Spiral of Archimedes:
$r = \theta + 2\pi, \ -4\pi \leq \theta \leq 0$

Figure 9.82

Zeros and Maximum *r*-Values

Two additional aids to sketching graphs of polar equations involve knowing the θ-values for which $|r|$ is maximum and knowing the θ-values for which $r = 0$. In Example 1, the maximum value of $|r|$ for $r = 4 \sin \theta$ is $|r| = 4$, and this occurs when $\theta = \pi/2$ (see Figure 9.79). Moreover, $r = 0$ when $\theta = 0$.

Example 3   Finding Maximum *r*-Values of a Polar Graph

Find the maximum value of r for the graph of $r = 1 - 2 \cos \theta$.

Graphical Solution

Because the polar equation is of the form

$$r = 1 - 2 \cos \theta = g(\cos \theta)$$

you know the graph is symmetric with respect to the polar axis. You can confirm this by graphing the polar equation. Set your graphing utility to *polar* mode and enter the equation, as shown in Figure 9.83. (In the graph, θ varies from 0 to 2π.) To find the maximum *r*-value for the graph, use your graphing utility's *trace* feature and you should find that the graph has a maximum *r*-value of 3, as shown in Figure 9.84. This value of r occurs when $\theta = \pi$. In the graph, note that the point $(3, \pi)$ is farthest from the pole.

Numerical Solution

To approximate the maximum value of r for the graph of $r = 1 - 2 \cos \theta$, use the *table* feature of a graphing utility to create a table that begins at $\theta = 0$ and increments by $\pi/12$, as shown in Figure 9.85. From the table, the maximum value of r appears to be 3 when $\theta = 3.1416 \approx \pi$. If a second table that begins at $\theta = \pi/2$ and increments by $\pi/24$ is created, as shown in Figure 9.86, the maximum value of r still appears to be 3 when $\theta = 3.1416 \approx \pi$.

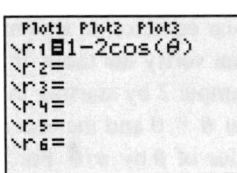

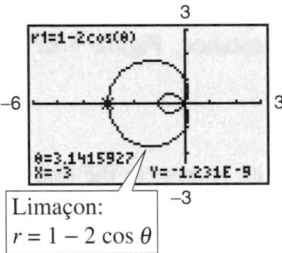

Limaçon:
$r = 1 - 2 \cos \theta$

Figure 9.83   **Figure 9.84**

Note how the negative *r*-values determine the *inner loop* of the graph in Figure 9.84. This type of graph is a limaçon.

✓ CHECKPOINT   Now try Exercise 19.

Figure 9.85

Figure 9.86

Exploration

The graph of the polar equation $r = e^{\cos \theta} - 2 \cos 4\theta + \sin^5(\theta/12)$ is called *the butterfly curve*, as shown in Figure 9.87.

a. The graph in Figure 9.87 was produced using $0 \le \theta \le 2\pi$. Does this show the entire graph? Explain your reasoning.

b. Use the *trace* feature of your graphing utility to approximate the maximum *r*-value of the graph. Does this value change if you use $0 \le \theta \le 4\pi$ instead of $0 \le \theta \le 2\pi$? Explain.

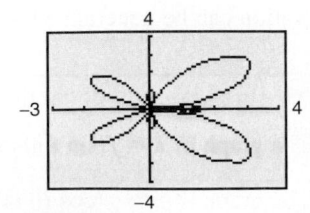

Figure 9.87

Some curves reach their zeros and maximum *r*-values at more than one point, as shown in Example 4.

Example 4   Analyzing a Polar Graph

Analyze the graph of $r = 2 \cos 3\theta$.

Solution

Symmetry:	With respect to the polar axis
Maximum value of $\lvert r \rvert$:	$\lvert r \rvert = 2$ when $3\theta = 0, \pi, 2\pi, 3\pi$
	or $\theta = 0, \pi/3, 2\pi/3, \pi$
Zeros of r:	$r = 0$ when $3\theta = \pi/2, 3\pi/2, 5\pi/2$
	or $\theta = \pi/6, \pi/2, 5\pi/6$

θ	0	$\dfrac{\pi}{12}$	$\dfrac{\pi}{6}$	$\dfrac{\pi}{4}$	$\dfrac{\pi}{3}$	$\dfrac{5\pi}{12}$	$\dfrac{\pi}{2}$
r	2	$\sqrt{2}$	0	$-\sqrt{2}$	-2	$-\sqrt{2}$	0

By plotting these points and using the specified symmetry, zeros, and maximum values, you can obtain the graph shown in Figure 9.88. This graph is called a **rose curve**, and each loop on the graph is called a *petal*. Note how the entire curve is generated as θ increases from 0 to π.

 $0 \le \theta \le \dfrac{\pi}{6}$

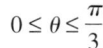 $0 \le \theta \le \dfrac{\pi}{3}$

$0 \le \theta \le \dfrac{\pi}{2}$

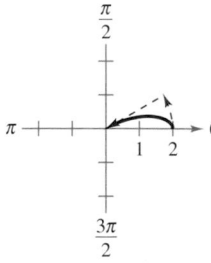

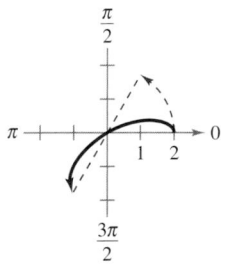

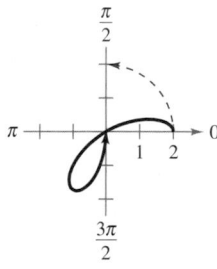

$0 \le \theta \le \dfrac{2\pi}{3}$

$0 \le \theta \le \dfrac{5\pi}{6}$

$0 \le \theta \le \pi$

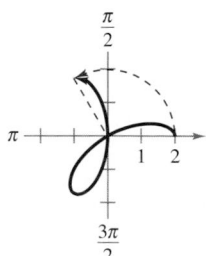

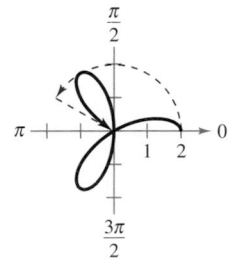

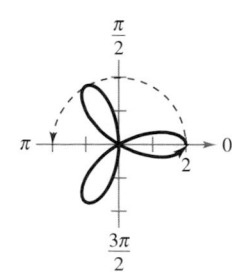

Figure 9.88

✓ *CHECKPOINT*   Now try Exercise 33.

Exploration

Notice that the rose curve in Example 4 has three petals. How many petals do the rose curves $r = 2 \cos 4\theta$ and $r = 2 \sin 3\theta$ have? Determine the numbers of petals for the curves $r = 2 \cos n\theta$ and $r = 2 \sin n\theta$, where n is a positive integer.

Special Polar Graphs

Several important types of graphs have equations that are simpler in polar form than in rectangular form. For example, the circle

$$r = 4 \sin \theta$$

in Example 1 has the more complicated rectangular equation

$$x^2 + (y - 2)^2 = 4.$$

Several other types of graphs that have simple polar equations are shown below.

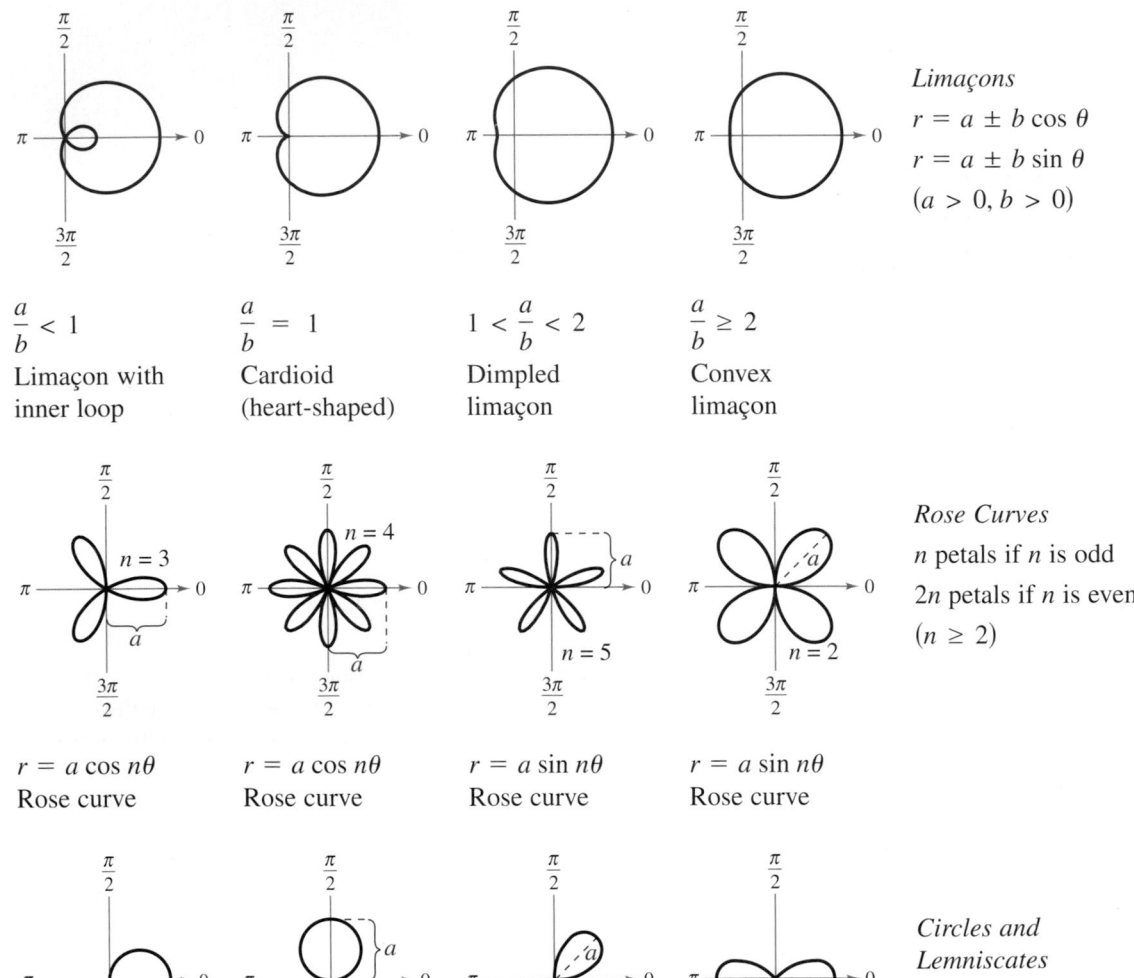

Limaçons
$r = a \pm b \cos \theta$
$r = a \pm b \sin \theta$
$(a > 0, b > 0)$

$\dfrac{a}{b} < 1$
Limaçon with inner loop

$\dfrac{a}{b} = 1$
Cardioid (heart-shaped)

$1 < \dfrac{a}{b} < 2$
Dimpled limaçon

$\dfrac{a}{b} \geq 2$
Convex limaçon

Rose Curves
n petals if n is odd
$2n$ petals if n is even
$(n \geq 2)$

$r = a \cos n\theta$
Rose curve

$r = a \cos n\theta$
Rose curve

$r = a \sin n\theta$
Rose curve

$r = a \sin n\theta$
Rose curve

Circles and Lemniscates

$r = a \cos \theta$
Circle

$r = a \sin \theta$
Circle

$r^2 = a^2 \sin 2\theta$
Lemniscate

$r^2 = a^2 \cos 2\theta$
Lemniscate

Example 5   Analyzing a Rose Curve

Analyze the graph of $r = 3 \cos 2\theta$.

Solution

Type of curve:	Rose curve with $2n = 4$ petals
Symmetry:	With respect to the polar axis, the line $\theta = \pi/2$, and the pole
Maximum value of $\|r\|$:	$\|r\| = 3$ when $\theta = 0, \pi/2, \pi, 3\pi/2$
Zeros of r:	$r = 0$ when $\theta = \pi/4, 3\pi/4$

Using a graphing utility, enter the equation, as shown in Figure 9.89 (with $0 \le \theta \le 2\pi$). You should obtain the graph shown in Figure 9.90.

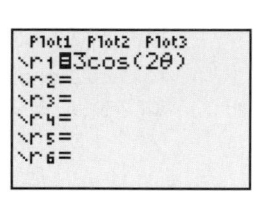

Figure 9.89

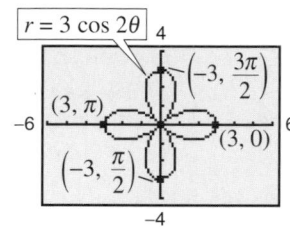

Figure 9.90

✓CHECKPOINT   Now try Exercise 37.

Example 6   Analyzing a Lemniscate

Analyze the graph of $r^2 = 9 \sin 2\theta$.

Solution

Type of curve:	Lemniscate
Symmetry:	With respect to the pole
Maximum value of $\|r\|$:	$\|r\| = 3$ when $\theta = \pi/4$
Zeros of r:	$r = 0$ when $\theta = 0, \pi/2$

Using a graphing utility, enter the equation, as shown in Figure 9.91 (with $0 \le \theta \le 2\pi$). You should obtain the graph shown in Figure 9.92.

Figure 9.91

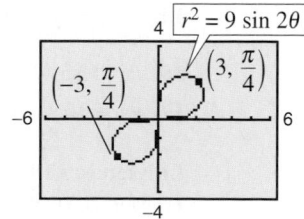

Figure 9.92

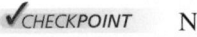CHECKPOINT   Now try Exercise 45.

9.7 Exercises

See www.CalcChat.com for worked-out solutions to odd-numbered exercises.

Vocabulary Check

Fill in the blanks.

1. The graph of $r = f(\sin \theta)$ is symmetric with respect to the line _____ .
2. The graph of $r = g(\cos \theta)$ is symmetric with respect to the _____ .
3. The equation $r = 2 + \cos \theta$ represents a _____ .
4. The equation $r = 2 \cos \theta$ represents a _____ .
5. The equation $r^2 = 4 \sin 2\theta$ represents a _____ .
6. The equation $r = 1 + \sin \theta$ represents a _____ .

In Exercises 1–6, identify the type of polar graph.

1.
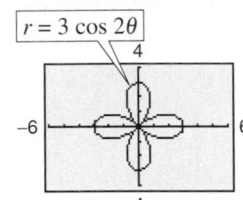
$r = 3 \cos 2\theta$

2.
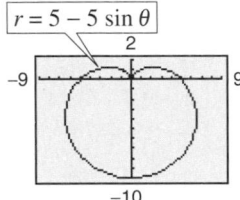
$r = 5 - 5 \sin \theta$

3.
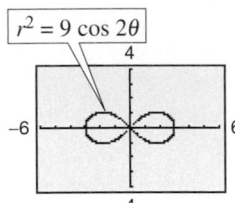
$r^2 = 9 \cos 2\theta$

4.
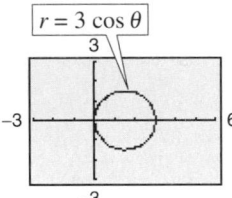
$r = 3 \cos \theta$

5.
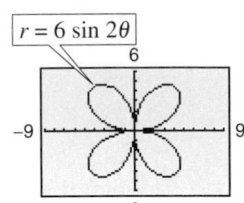
$r = 6 \sin 2\theta$

6.
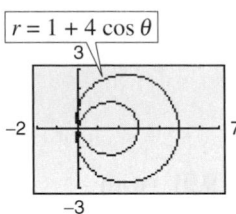
$r = 1 + 4 \cos \theta$

Library of Parent Functions   In Exercises 7–10, determine the equation of the polar curve whose graph is shown.

7.
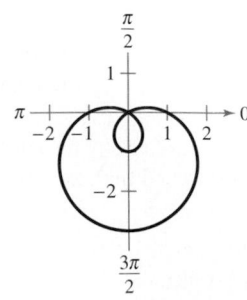

8.
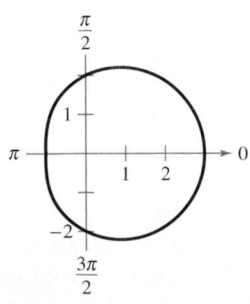

(a) $r = 1 - 2 \sin \theta$
(b) $r = 1 + 2 \sin \theta$
(c) $r = 1 + 2 \cos \theta$
(d) $r = 1 - 2 \cos \theta$

(a) $r = 2 - \cos \theta$
(b) $r = 2 - \sin \theta$
(c) $r = 2 + \cos \theta$
(d) $r = 2 + \sin \theta$

9.
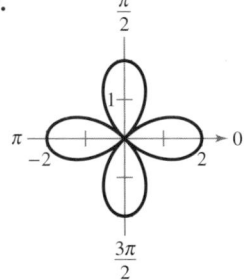

10.
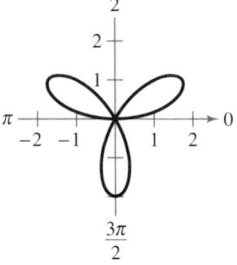

(a) $r = 2 \cos 4\theta$

(b) $r = \cos 4\theta$

(c) $r = 2 \cos 2\theta$

(d) $r = 2 \cos \dfrac{\theta}{2}$

(a) $r = 2 \sin 6\theta$

(b) $r = 2 \cos \left(\dfrac{3\theta}{2} \right)$

(c) $r = 2 \sin \left(\dfrac{3\theta}{2} \right)$

(d) $r = 2 \sin 3\theta$

In Exercises 11–18, test for symmetry with respect to $\theta = \pi/2$, the polar axis, and the pole.

11. $r = 14 + 4 \cos \theta$
12. $r = 12 \cos 3\theta$
13. $r = \dfrac{4}{1 + \sin \theta}$
14. $r = \dfrac{2}{1 - \cos \theta}$
15. $r = 6 \sin \theta$
16. $r = 4 \csc \theta \cos \theta$
17. $r^2 = 16 \sin 2\theta$
18. $r^2 = 25 \cos 4\theta$

In Exercises 19–22, find the maximum value of $|r|$ and any zeros of r. Verify your answers numerically.

19. $r = 10 - 10 \sin \theta$
20. $r = 6 + 12 \cos \theta$
21. $r = 4 \cos 3\theta$
22. $r = \sin 2\theta$

In Exercises 23–36, sketch the graph of the polar equation. Use a graphing utility to verify your graph.

23. $r = 5$

24. $\theta = -\dfrac{5\pi}{3}$

25. $r = 3\sin\theta$

26. $r = 2\cos\theta$

27. $r = 3(1 - \cos\theta)$

28. $r = 4(1 + \sin\theta)$

29. $r = 3 - 4\cos\theta$

30. $r = 1 - 2\sin\theta$

31. $r = 4 + 5\sin\theta$

32. $r = 3 + 6\cos\theta$

33. $r = 5\cos 3\theta$

34. $r = -\sin 5\theta$

35. $r = 7\sin 2\theta$

36. $r = 3\cos 5\theta$

In Exercises 37–52, use a graphing utility to graph the polar equation. Describe your viewing window.

37. $r = 8\cos 2\theta$

38. $r = \cos 2\theta$

39. $r = 2(5 - \sin\theta)$

40. $r = 6 - 4\sin\theta$

41. $r = 3 - 6\cos\theta$

42. $r = 3 - 2\sin\theta$

43. $r = \dfrac{3}{\sin\theta - 2\cos\theta}$

44. $r = \dfrac{6}{2\sin\theta - 3\cos\theta}$

45. $r^2 = 4\cos 2\theta$

46. $r^2 = 9\sin\theta$

47. $r = 4\sin\theta\cos^2\theta$

48. $r = 2\cos(3\theta - 2)$

49. $r = 2\csc\theta + 6$

50. $r = 4 - \sec\theta$

51. $r = e^{2\theta}$

52. $r = e^{\theta/2}$

In Exercises 53–58, use a graphing utility to graph the polar equation. Find an interval for θ for which the graph is traced *only once*.

53. $r = 3 - 2\cos\theta$

54. $r = 2(1 - 2\sin\theta)$

55. $r = 2\cos\left(\dfrac{3\theta}{2}\right)$

56. $r = 3\sin\left(\dfrac{5\theta}{2}\right)$

57. $r^2 = \sin 2\theta$

58. $r^2 = \dfrac{1}{\theta}$

In Exercises 59–62, use a graphing utility to graph the polar equation and show that the indicated line is an asymptote of the graph.

	Name of Graph	Polar Equation	Asymptote
59.	Conchoid	$r = 2 - \sec\theta$	$x = -1$
60.	Conchoid	$r = 2 + \csc\theta$	$y = 1$
61.	Hyperbolic spiral	$r = 2/\theta$	$y = 2$
62.	Strophoid	$r = 2\cos 2\theta\sec\theta$	$x = -2$

Synthesis

True or False?   **In Exercises 63 and 64, determine whether the statement is true or false. Justify your answer.**

63. The graph of $r = 10\sin 5\theta$ is a rose curve with five petals.

64. A rose curve will always have symmetry with respect to the line $\theta = \pi/2$.

65. *Writing*   Use a graphing utility to graph the polar equation

$$r = \cos 5\theta + n\cos\theta, \quad 0 \le \theta < \pi$$

for the integers $n = -5$ to $n = 5$. As you graph these equations, you should see the graph change shape from a heart to a bell. Write a short paragraph explaining which values of n produce the heart-shaped curves and which values of n produce the bell-shaped curves.

66. The graph of $r = f(\theta)$ is rotated about the pole through an angle ϕ. Show that the equation of the rotated graph is $r = f(\theta - \phi)$.

67. Consider the graph of $r = f(\sin\theta)$.

(a) Show that if the graph is rotated counterclockwise $\pi/2$ radians about the pole, the equation of the rotated graph is $r = f(-\cos\theta)$.

(b) Show that if the graph is rotated counterclockwise π radians about the pole, the equation of the rotated graph is $r = f(-\sin\theta)$.

(c) Show that if the graph is rotated counterclockwise $3\pi/2$ radians about the pole, the equation of the rotated graph is $r = f(\cos\theta)$.

In Exercises 68–70, use the results of Exercises 66 and 67.

68. Write an equation for the limaçon $r = 2 - \sin\theta$ after it has been rotated through each given angle.

(a) $\dfrac{\pi}{4}$ (b) $\dfrac{\pi}{2}$ (c) π (d) $\dfrac{3\pi}{2}$

69. Write an equation for the rose curve $r = 2\sin 2\theta$ after it has been rotated through each given angle.

(a) $\dfrac{\pi}{6}$ (b) $\dfrac{\pi}{2}$ (c) $\dfrac{2\pi}{3}$ (d) π

70. Sketch the graph of each equation.

(a) $r = 1 - \sin\theta$ (b) $r = 1 - \sin\left(\theta - \dfrac{\pi}{4}\right)$

71. *Exploration*   Use a graphing utility to graph the polar equation $r = 2 + k\cos\theta$ for $k = 0$, $k = 1$, $k = 2$, and $k = 3$. Identify each graph.

72. *Exploration*   Consider the polar equation $r = 3\sin k\theta$.

(a) Use a graphing utility to graph the equation for $k = 1.5$. Find the interval for θ for which the graph is traced only once.

(b) Use a graphing utility to graph the equation for $k = 2.5$. Find the interval for θ for which the graph is traced only once.

(c) Is it possible to find an interval for θ for which the graph is traced only once for any rational number k? Explain.

9.8 Polar Equations of Conics

Alternative Definition of Conics

In Sections 9.2 and 9.3, you learned that the rectangular equations of ellipses and hyperbolas take simple forms when the origin lies at the *center*. As it happens, there are many important applications of conics in which it is more convenient to use one of the *foci* as the origin. In this section, you will learn that polar equations of conics take simple forms if one of the foci lies at the pole.

To begin, consider the following alternative definition of a conic that uses the concept of eccentricity (a measure of the flatness of the conic).

> **Alternative Definition of a Conic**
>
> The locus of a point in the plane that moves such that its distance from a fixed point (focus) is in a constant ratio to its distance from a fixed line (directrix) is a **conic**. The constant ratio is the **eccentricity** of the conic and is denoted by e. Moreover, the conic is an **ellipse** if $e < 1$, a **parabola** if $e = 1$, and a **hyperbola** if $e > 1$. (See Figure 9.93.)

In Figure 9.93, note that for each type of conic, the focus is at the pole.

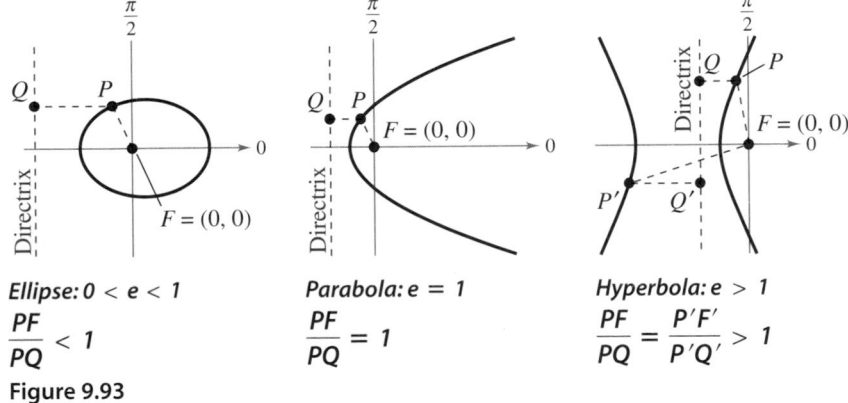

Ellipse: $0 < e < 1$
$$\frac{PF}{PQ} < 1$$

Parabola: $e = 1$
$$\frac{PF}{PQ} = 1$$

Hyperbola: $e > 1$
$$\frac{PF}{PQ} = \frac{P'F'}{P'Q'} > 1$$

Figure 9.93

Polar Equations of Conics

The benefit of locating a focus of a conic at the pole is that the equation of the conic becomes simpler.

> **Polar Equations of Conics** (See the proof on page 739.)
>
> The graph of a polar equation of the form
>
> **1.** $r = \dfrac{ep}{1 \pm e \cos \theta}$ or   **2.** $r = \dfrac{ep}{1 \pm e \sin \theta}$
>
> is a conic, where $e > 0$ is the eccentricity and $|p|$ is the distance between the focus (pole) and the directrix.

What you should learn

- Define conics in terms of eccentricities.
- Write and graph equations of conics in polar form.
- Use equations of conics in polar form to model real-life problems.

Why you should learn it

The orbits of planets and satellites can be modeled by polar equations. For instance, in Exercise 55 on page 727, you will use polar equations to model the orbits of Neptune and Pluto.

Kevin Kelley/Getty Images

Prerequisite Skills

To review the characteristics of conics, see Sections 9.1–9.3.

An equation of the form

$$r = \frac{ep}{1 \pm e \cos \theta}$$ Vertical directrix

corresponds to a conic with a vertical directrix and symmetry with respect to the polar axis. An equation of the form

$$r = \frac{ep}{1 \pm e \sin \theta}$$ Horizontal directrix

corresponds to a conic with a horizontal directrix and symmetry with respect to $\theta = \pi/2$. Moreover, the converse is also true—that is, any conic with a focus at the pole and having a horizontal or vertical directrix can be represented by one of the given equations.

Example 1   Identifying a Conic from Its Equation

Identify the type of conic represented by the equation $r = \dfrac{15}{3 - 2 \cos \theta}$.

Algebraic Solution

To identify the type of conic, rewrite the equation in the form $r = ep/(1 \pm e \cos \theta)$.

$$r = \frac{15}{3 - 2 \cos \theta}$$

$$= \frac{5}{1 - (2/3) \cos \theta}$$ Divide numerator and denominator by 3.

Because $e = \frac{2}{3} < 1$, you can conclude that the graph is an ellipse.

Graphical Solution

Use a graphing utility in *polar* mode to graph $r = \dfrac{15}{3 - 2 \cos \theta}$.

Be sure to use a square setting. From the graph in Figure 9.94, you can see that the conic appears to be an ellipse.

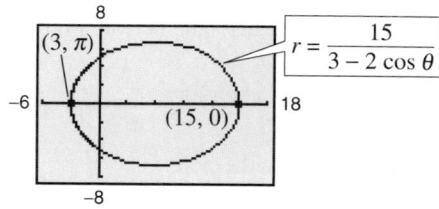

Figure 9.94

✔**CHECKPOINT**   Now try Exercise 11.

For the ellipse in Figure 9.94, the major axis is horizontal and the vertices lie at $(r, \theta) = (15, 0)$ and $(r, \theta) = (3, \pi)$. So, the length of the *major* axis is $2a = 18$. To find the length of the *minor* axis, you can use the equations $e = c/a$ and $b^2 = a^2 - c^2$ to conclude that

$$b^2 = a^2 - c^2$$

$$= a^2 - (ea)^2$$

$$= a^2(1 - e^2).$$ Ellipse

Because $e = \frac{2}{3}$, you have $b^2 = 9^2\left[1 - \left(\frac{2}{3}\right)^2\right] = 45$, which implies that $b = \sqrt{45} = 3\sqrt{5}$. So, the length of the minor axis is $2b = 6\sqrt{5}$. A similar analysis for hyperbolas yields

$$b^2 = c^2 - a^2$$

$$= (ea)^2 - a^2$$

$$= a^2(e^2 - 1).$$ Hyperbola

Example 2   Analyzing the Graph of a Polar Equation

Analyze the graph of the polar equation

$$r = \frac{32}{3 + 5\sin\theta}.$$

Solution

Dividing the numerator and denominator by 3 produces

$$r = \frac{32/3}{1 + (5/3)\sin\theta}.$$

Because $e = \frac{5}{3} > 1$, the graph is a hyperbola. The transverse axis of the hyperbola lies on the line $\theta = \pi/2$ and the vertices occur at $(r, \theta) = (4, \pi/2)$ and $(r, \theta) = (-16, 3\pi/2)$. Because the length of the transverse axis is 12, you can see that $a = 6$. To find b, write

$$b^2 = a^2(e^2 - 1) = 6^2\left[\left(\frac{5}{3}\right)^2 - 1\right] = 64.$$

So, $b = 8$. You can use a and b to determine that the asymptotes are $y = 10 \pm \frac{3}{4}x$, as shown in Figure 9.95.

✓**CHECKPOINT**   Now try Exercise 23.

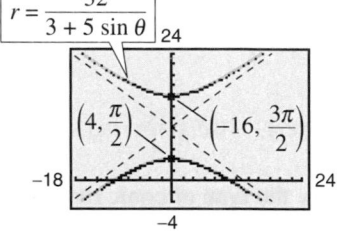

Figure 9.95

In the next example, you are asked to find a polar equation for a specified conic. To do this, let p be the distance between the pole and the directrix.

1. Horizontal directrix above the pole: $r = \dfrac{ep}{1 + e\sin\theta}$

2. Horizontal directrix below the pole: $r = \dfrac{ep}{1 - e\sin\theta}$

3. Vertical directrix to the right of the pole: $r = \dfrac{ep}{1 + e\cos\theta}$

4. Vertical directrix to the left of the pole: $r = \dfrac{ep}{1 - e\cos\theta}$

Example 3   Finding the Polar Equation of a Conic

Find the polar equation of the parabola whose focus is the pole and whose directrix is the line $y = 3$.

Solution

From Figure 9.96, you can see that the directrix is horizontal and above the pole. Moreover, because the eccentricity of a parabola is $e = 1$ and the distance between the pole and the directrix is $p = 3$, you have the equation

$$r = \frac{ep}{1 + e\sin\theta} = \frac{3}{1 + \sin\theta}.$$

✓**CHECKPOINT**   Now try Exercise 33.

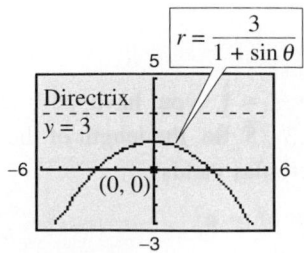

Figure 9.96

Application

Kepler's Laws (listed below), named after the German astronomer Johannes Kepler (1571–1630), can be used to describe the orbits of the planets about the sun.

1. Each planet moves in an elliptical orbit with the sun as a focus.

2. A ray from the sun to the planet sweeps out equal areas of the ellipse in equal times.

3. The square of the period (the time it takes for a planet to orbit the sun) is proportional to the cube of the mean distance between the planet and the sun.

Although Kepler simply stated these laws on the basis of observation, they were later validated by Isaac Newton (1642–1727). In fact, Newton was able to show that each law can be deduced from a set of universal laws of motion and gravitation that govern the movement of all heavenly bodies, including comets and satellites. This is illustrated in the next example, which involves the comet named after the English mathematician and physicist Edmund Halley (1656–1742).

If you use Earth as a reference with a period of 1 year and a distance of 1 astronomical unit (an *astronomical unit* is defined as the mean distance between Earth and the sun, or about 93 million miles), the proportionality constant in Kepler's third law is 1. For example, because Mars has a mean distance to the sun of $d \approx 1.524$ astronomical units, its period P is given by $d^3 = P^2$. So, the period of Mars is $P \approx 1.88$ years.

Example 4   Halley's Comet

Halley's comet has an elliptical orbit with an eccentricity of $e \approx 0.967$. The length of the major axis of the orbit is approximately 35.88 astronomical units. Find a polar equation for the orbit. How close does Halley's comet come to the sun?

Solution

Using a vertical axis, as shown in Figure 9.97, choose an equation of the form $r = ep/(1 + e \sin \theta)$. Because the vertices of the ellipse occur at $\theta = \pi/2$ and $\theta = 3\pi/2$, you can determine the length of the major axis to be the sum of the r-values of the vertices. That is,

$$2a = \frac{0.967p}{1 + 0.967} + \frac{0.967p}{1 - 0.967} \approx 29.79p \approx 35.88.$$

So, $p \approx 1.204$ and $ep \approx (0.967)(1.204) \approx 1.164$. Using this value of ep in the equation, you have

$$r = \frac{1.164}{1 + 0.967 \sin \theta}$$

where r is measured in astronomical units. To find the closest point to the sun (the focus), substitute $\theta = \pi/2$ into this equation to obtain

$$r = \frac{1.164}{1 + 0.967 \sin(\pi/2)} \approx 0.59 \text{ astronomical units} \approx 55{,}000{,}000 \text{ miles.}$$

✓**CHECKPOINT**   Now try Exercise 51.

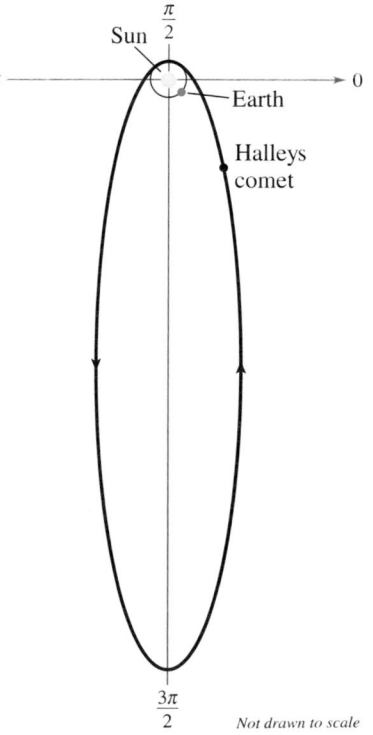

Figure 9.97

9.8 Exercises

See www.CalcChat.com for worked-out solutions to odd-numbered exercises.

Vocabulary Check

In Exercises 1 and 2, fill in the blanks.

1. The locus of a point in the plane that moves such that its distance from a fixed point (focus) is in a constant ratio to its distance from a fixed line (directrix) is a _____ .

2. The constant ratio is the _____ of the conic and is denoted by _____ .

3. Match the conic with its eccentricity.

    (a) $e < 1$ (i) ellipse

    (b) $e = 1$ (ii) hyperbola

    (c) $e > 1$ (iii) parabola

Graphical Reasoning **In Exercises 1–4, use a graphing utility to graph the polar equation for (a) $e = 1$, (b) $e = 0.5$, and (c) $e = 1.5$. Identify the conic for each equation.**

1. $r = \dfrac{2e}{1 + e\cos\theta}$ 2. $r = \dfrac{2e}{1 - e\cos\theta}$

3. $r = \dfrac{2e}{1 - e\sin\theta}$ 4. $r = \dfrac{2e}{1 + e\sin\theta}$

In Exercises 5–10, match the polar equation with its graph. [The graphs are labeled (a), (b), (c), (d), (e), and (f).]

(a)

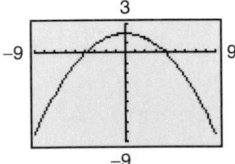

(b)

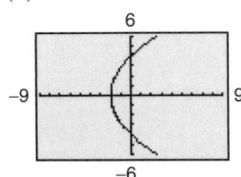

(c)

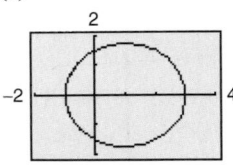

(d)

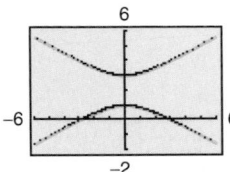

(e)

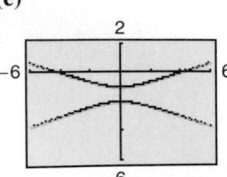

(f)

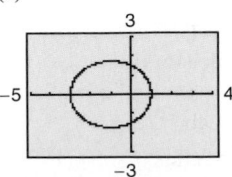

5. $r = \dfrac{4}{1 - \cos\theta}$ 6. $r = \dfrac{3}{2 - \cos\theta}$

7. $r = \dfrac{3}{2 + \cos\theta}$ 8. $r = \dfrac{4}{1 - 3\sin\theta}$

9. $r = \dfrac{3}{1 + 2\sin\theta}$ 10. $r = \dfrac{4}{1 + \sin\theta}$

In Exercises 11–20, identify the conic represented by the equation algebraically. Use a graphing utility to confirm your result.

11. $r = \dfrac{2}{1 - \cos\theta}$ 12. $r = \dfrac{2}{1 + \sin\theta}$

13. $r = \dfrac{4}{4 - \cos\theta}$ 14. $r = \dfrac{7}{7 + \sin\theta}$

15. $r = \dfrac{8}{4 + 3\sin\theta}$ 16. $r = \dfrac{6}{3 - 2\cos\theta}$

17. $r = \dfrac{6}{2 + \sin\theta}$ 18. $r = \dfrac{5}{-1 + 2\cos\theta}$

19. $r = \dfrac{3}{4 - 8\cos\theta}$ 20. $r = \dfrac{10}{3 + 9\sin\theta}$

In Exercises 21–26, use a graphing utility to graph the polar equation. Identify the graph.

21. $r = \dfrac{-5}{1 - \sin\theta}$ 22. $r = \dfrac{-1}{2 + 4\sin\theta}$

23. $r = \dfrac{14}{14 + 17\sin\theta}$ 24. $r = \dfrac{12}{2 - \cos\theta}$

25. $r = \dfrac{3}{-4 + 2\cos\theta}$ 26. $r = \dfrac{4}{1 - 2\cos\theta}$

In Exercises 27–32, use a graphing utility to graph the rotated conic.

27. $r = \dfrac{2}{1 - \cos(\theta - \pi/4)}$ (See Exercise 11.)

28. $r = \dfrac{7}{7 + \sin(\theta - \pi/3)}$ (See Exercise 14.)

29. $r = \dfrac{4}{4 - \cos(\theta + 3\pi/4)}$ (See Exercise 13.)

30. $r = \dfrac{6}{3 - 2\cos(\theta + \pi/2)}$ (See Exercise 16.)

31. $r = \dfrac{8}{4 + 3\sin(\theta + \pi/6)}$ (See Exercise 15.)

32. $r = \dfrac{5}{-1 + 2\cos(\theta + 2\pi/3)}$ (See Exercise 18.)

In Exercises 33–48, find a polar equation of the conic with its focus at the pole.

Conic	Eccentricity	Directrix
33. Parabola	$e = 1$	$x = -1$
34. Parabola	$e = 1$	$y = -4$
35. Ellipse	$e = \frac{1}{2}$	$y = 1$
36. Ellipse	$e = \frac{3}{4}$	$y = -4$
37. Hyperbola	$e = 2$	$x = 1$
38. Hyperbola	$e = \frac{3}{2}$	$x = -1$

Conic	Vertex or Vertices
39. Parabola	$\left(1, -\dfrac{\pi}{2}\right)$
40. Parabola	$(8, 0)$
41. Parabola	$(5, \pi)$
42. Parabola	$\left(10, \dfrac{\pi}{2}\right)$
43. Ellipse	$(2, 0), (10, \pi)$
44. Ellipse	$\left(2, \dfrac{\pi}{2}\right), \left(4, \dfrac{3\pi}{2}\right)$
45. Ellipse	$(20, 0), (4, \pi)$
46. Hyperbola	$\left(1, \dfrac{3\pi}{2}\right), \left(9, \dfrac{3\pi}{2}\right)$
47. Hyperbola	$\left(4, \dfrac{\pi}{2}\right), \left(-1, \dfrac{3\pi}{2}\right)$
48. Hyperbola	$\left(4, \dfrac{\pi}{2}\right), \left(1, \dfrac{\pi}{2}\right)$

49. **Planetary Motion**   The planets travel in elliptical orbits with the sun at one focus. Assume that the focus is at the pole, the major axis lies on the polar axis, and the length of the major axis is $2a$ (see figure). Show that the polar equation of the orbit of a planet is

$$r = \frac{(1 - e^2)a}{1 - e\cos\theta}$$

where e is the eccentricity.

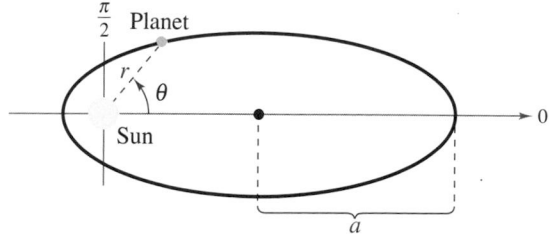

50. **Planetary Motion**   Use the result of Exercise 49 to show that the minimum distance (*perihelion*) from the sun to a planet is $r = a(1 - e)$ and that the maximum distance (*aphelion*) is $r = a(1 + e)$.

Planetary Motion **In Exercises 51–54, use the results of Exercises 49 and 50 to find the polar equation of the orbit of the planet and the perihelion and aphelion distances.**

51. Earth $a = 92.956 \times 10^6$ miles
               $e = 0.0167$

52. Mercury $a = 35.983 \times 10^6$ miles
               $e = 0.2056$

53. Jupiter $a = 77.841 \times 10^7$ kilometers
               $e = 0.0484$

54. Saturn $a = 142.673 \times 10^7$ kilometers
               $e = 0.0542$

55. **Planetary Motion**   Use the results of Exercises 49 and 50, where for the planet Neptune, $a = 4.498 \times 10^9$ kilometers and $e = 0.0086$ and for the dwarf planet Pluto, $a = 5.906 \times 10^9$ kilometers and $e = 0.2488$.

    (a) Find the polar equation of the orbit of each planet.

    (b) Find the perihelion and aphelion distances for each planet.

    (c) Use a graphing utility to graph both Neptune's and Pluto's equations of orbit in the same viewing window.

    (d) Is Pluto ever closer to the sun than Neptune? Until recently, Pluto was considered the ninth planet. Why was Pluto called the ninth planet and Neptune the eighth planet?

    (e) Do the orbits of Neptune and Pluto intersect? Will Neptune and Pluto ever collide? Why or why not?

56. *Explorer 18* On November 27, 1963, the United States launched a satellite named *Explorer 18*. Its low and high points above the surface of Earth were 119 miles and 122,800 miles, respectively (see figure). The center of Earth is at one focus of the orbit.

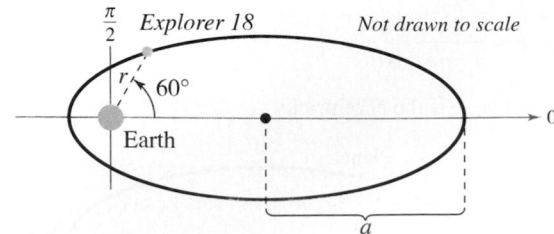

(a) Find the polar equation for the orbit (assume the radius of Earth is 4000 miles).

(b) Find the distance between the surface of Earth and the satellite when $\theta = 60°$.

(c) Find the distance between the surface of Earth and the satellite when $\theta = 30°$.

Synthesis

True or False? **In Exercises 57 and 58, determine whether the statement is true or false. Justify your answer.**

57. The graph of $r = 4/(-3 - 3\sin\theta)$ has a horizontal directrix above the pole.

58. The conic represented by the following equation is an ellipse.

$$r^2 = \frac{16}{9 - 4\cos\left(\theta + \frac{\pi}{4}\right)}$$

59. Show that the polar equation for the ellipse

$$\frac{x^2}{a^2} + \frac{y^2}{b^2} = 1 \quad \text{is} \quad r^2 = \frac{b^2}{1 - e^2\cos^2\theta}.$$

60. Show that the polar equation for the hyperbola

$$\frac{x^2}{a^2} - \frac{y^2}{b^2} = 1 \quad \text{is} \quad r^2 = \frac{-b^2}{1 - e^2\cos^2\theta}.$$

In Exercises 61–66, use the results of Exercises 59 and 60 to write the polar form of the equation of the conic.

61. $\dfrac{x^2}{169} + \dfrac{y^2}{144} = 1$ **62.** $\dfrac{x^2}{9} - \dfrac{y^2}{16} = 1$

63. $\dfrac{x^2}{25} + \dfrac{y^2}{16} = 1$ **64.** $\dfrac{x^2}{36} - \dfrac{y^2}{4} = 1$

65. Hyperbola   One focus: $(5, \pi/2)$

Vertices: $(4, \pi/2), (4, -\pi/2)$

66. Ellipse   One focus: $(4, 0)$

Vertices: $(5, 0), (5, \pi)$

67. *Exploration* Consider the polar equation

$$r = \frac{4}{1 - 0.4\cos\theta}.$$

(a) Identify the conic without graphing the equation.

(b) Without graphing the following polar equations, describe how each differs from the given polar equation. Use a graphing utility to verify your results.

$$r = \frac{4}{1 + 0.4\cos\theta}, \quad r = \frac{4}{1 - 0.4\sin\theta}$$

68. *Exploration* The equation

$$r = \frac{ep}{1 \pm e\sin\theta}$$

is the equation of an ellipse with $e < 1$. What happens to the lengths of both the major axis and the minor axis when the value of e remains fixed and the value of p changes? Use an example to explain your reasoning.

69. *Writing* In your own words, define the term *eccentricity* and explain how it can be used to classify conics.

70. What conic does the polar equation given by $r = a\sin\theta + b\cos\theta$ represent?

Skills Review

In Exercises 71–76, solve the equation.

71. $4\sqrt{3}\tan\theta - 3 = 1$

72. $6\cos x - 2 = 1$

73. $12\sin^2\theta = 9$

74. $9\csc^2 x - 10 = 2$

75. $2\cot x = 5\cos\dfrac{\pi}{2}$

76. $\sqrt{2}\sec\theta = 2\csc\dfrac{\pi}{4}$

In Exercises 77–80 find the value of the trigonometric function given that u and v are in Quadrant IV and $\sin u = -\frac{3}{5}$ and $\cos v = 1/\sqrt{2}$.

77. $\cos(u + v)$

78. $\sin(u + v)$

79. $\sin(u - v)$

80. $\cos(u - v)$

In Exercises 81–84, evaluate the expression. Do not use a calculator.

81. $_{12}C_9$ **82.** $_{18}C_{16}$

83. $_{10}P_3$ **84.** $_{29}P_2$

What Did You Learn?

Key Terms

conic (conic section), *p. 660*

degenerate conic, *p. 660*

circle, *p. 661*

center (of a circle), *p. 661*

radius, *p. 661*

parabola, *p. 663*

directrix, *p. 663*

focus (of a parabola), *p. 663*

vertex (of a parabola), *p. 663*

axis (of a parabola), *p. 663*

ellipse, *p. 671*

foci (of an ellipse), *p. 671*

vertices (of an ellipse), *p. 671*

major axis (of an ellipse), *p. 671*

center (of an ellipse), *p. 671*

minor axis (of an ellipse), *p. 671*

hyperbola, *p. 680*

foci (of a hyperbola), *p. 680*

branches, *p. 680*

vertices (of a hyperbola), *p. 680*

transverse axis, *p. 680*

center (of a hyperbola), *p. 680*

conjugate axis, *p. 682*

asymptotes (of a hyperbola), *p. 682*

rotation of axes, *p. 690*

invariant under rotation, *p. 694*

discriminant, *p. 694*

parameter, *p. 699*

parametric equations, *p. 699*

plane curve, *p. 699*

orientation, *p. 700*

eliminating the parameter, *p. 702*

polar coordinate system, *p. 707*

pole, *p. 707*

polar axis, *p. 707*

polar coordinates, *p. 707*

limaçon, *p. 715*

rose curve, *p. 717*

eccentricity, *p. 722*

Key Concepts

9.1–9.3 ■ Write and graph translations of conics

1. Circle with center (h, k) and radius r:

$$(x - h)^2 + (y - k)^2 = r^2$$

2. Parabola with vertex (h, k):

$(x - h)^2 = 4p(y - k)$ Vertical axis

$(y - k)^2 = 4p(x - h)$ Horizontal axis

3. Ellipse with center (h, k) and $0 < b < a$:

$\dfrac{(x - h)^2}{a^2} + \dfrac{(y - k)^2}{b^2} = 1$ Horizontal major axis

$\dfrac{(x - h)^2}{b^2} + \dfrac{(y - k)^2}{a^2} = 1$ Vertical major axis

4. Hyperbola with center (h, k):

$\dfrac{(x - h)^2}{a^2} - \dfrac{(y - k)^2}{b^2} = 1$ Horizontal transverse axis

$\dfrac{(y - k)^2}{a^2} - \dfrac{(x - h)^2}{b^2} = 1$ Vertical transverse axis

9.4 ■ Rotation and systems of quadratic equations

1. Rotate the coordinate axis to eliminate the xy-term.

2. Use the discriminant to classify conics.

9.5 ■ Graph curves that are represented by sets of parametric equations

1. Sketch a curve represented by a pair of parametric equations by plotting points in the order of increasing values of t in the xy-plane.

2. Use the *parametric* mode of a graphing utility to graph a set of parametric equations.

9.6 ■ Convert points from rectangular to polar form and vice versa

1. Polar-to-Rectangular: $x = r \cos \theta$, $y = r \sin \theta$

2. Rectangular-to-Polar: $\tan \theta = \dfrac{x}{y}$, $r^2 = x^2 + y^2$

9.7 ■ Use symmetry to aid in sketching graphs of polar equations

1. Symmetry with respect to the line $\theta = \dfrac{\pi}{2}$: Replace (r, θ) by $(r, \pi - \theta)$ or $(-r, -\theta)$.

2. Symmetry with respect to the polar axis: Replace (r, θ) by $(r, -\theta)$ or $(-r, \pi - \theta)$.

3. Symmetry with respect to the pole: Replace (r, θ) by $(r, \pi + \theta)$ or $(-r, \theta)$.

4. The graph of $r = f(\sin \theta)$ is symmetric with respect to the line $\theta = \dfrac{\pi}{2}$.

5. The graph of $r = g(\cos \theta)$ is symmetric with respect to the polar axis.

9.8 ■ Write and graph equations of conics in polar form

The graph of a polar equation of the form

$$r = \frac{ep}{1 \pm e \cos \theta} \quad \text{or} \quad r = \frac{ep}{1 \pm e \sin \theta}$$

is a conic, where $e > 0$ is the eccentricity and $|p|$ is the distance between the focus (pole) and the directrix.

Review Exercises

See www.CalcChat.com for worked-out solutions to odd-numbered exercises.

9.1 In Exercises 1–4, find the standard form of the equation of the circle with the given characteristics.

1. Center at origin; point on the circle: $(-3, -4)$

2. Center at origin; point on the circle: $(8, -15)$

3. Endpoints of a diameter: $(-1, 2)$ and $(5, 6)$

4. Endpoints of a diameter: $(-2, 3)$ and $(6, -5)$

In Exercises 5–8, write the equation of the circle in standard form. Then identify its center and radius.

5. $\frac{1}{2}x^2 + \frac{1}{2}y^2 = 18$

6. $\frac{3}{4}x^2 + \frac{3}{4}y^2 = 1$

7. $16x^2 + 16y^2 - 16x + 24y - 3 = 0$

8. $4x^2 + 4y^2 + 32x - 24y + 51 = 0$

In Exercises 9 and 10, sketch the circle. Identify its center and radius.

9. $x^2 + y^2 + 4x + 6y - 3 = 0$

10. $x^2 + y^2 + 8x - 10y - 8 = 0$

In Exercises 11 and 12, find the x- and y-intercepts of the graph of the circle.

11. $(x - 3)^2 + (y + 1)^2 = 7$

12. $(x + 5)^2 + (y - 6)^2 = 27$

In Exercises 13–16, find the vertex, focus, and directrix of the parabola, and sketch its graph. Use a graphing utility to verify your graph.

13. $4x - y^2 = 0$

14. $y = -\frac{1}{8}x^2$

15. $\frac{1}{2}y^2 + 18x = 0$

16. $\frac{1}{4}y - 8x^2 = 0$

In Exercises 17–20, find the standard form of the equation of the parabola with the given characteristics.

17. Vertex: $(0, 0)$

Focus: $(-6, 0)$

18. Vertex: $(4, 2)$

Focus: $(4, 0)$

19.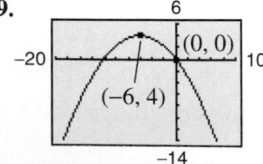

20.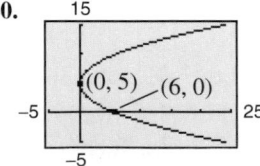

In Exercises 21 and 22, find an equation of the tangent line to the parabola at the given point and find the x-intercept of the line.

21. $x^2 = -2y$, $(2, -2)$

22. $y^2 = -2x$, $(-8, -4)$

23. *Architecture* A parabolic archway (see figure) is 12 meters high at the vertex. At a height of 10 meters, the width of the archway is 8 meters. How wide is the archway at ground level?

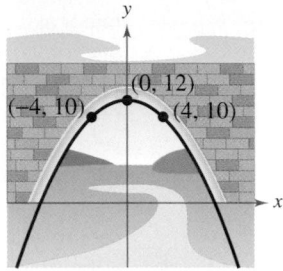

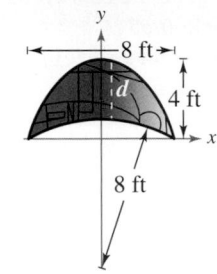

Figure for 23   Figure for 24

24. *Architecture* A church window (see figure) is bounded on top by a parabola and below by the arc of a circle.

(a) Find equations of the parabola and the circle.

(b) Use a graphing utility to create a table showing the vertical distances d between the circle and the parabola for various values of x.

x	0	1	2	3	4
d					

9.2 In Exercises 25–28, find the center, vertices, foci, and eccentricity of the ellipse and sketch its graph. Use a graphing utility to verify your graph.

25. $\frac{x^2}{4} + \frac{y^2}{16} = 1$

26. $\frac{x^2}{9} + \frac{y^2}{8} = 1$

27. $\frac{(x - 4)^2}{6} + \frac{(y + 4)^2}{9} = 1$

28. $\frac{(x + 1)^2}{16} + \frac{(y - 3)^2}{6} = 1$

In Exercises 29–32, (a) find the standard form of the equation of the ellipse, (b) find the center, vertices, foci, and eccentricity of the ellipse, and (c) sketch the ellipse. Use a graphing utility to verify your graph.

29. $16x^2 + 9y^2 - 32x + 72y + 16 = 0$

30. $4x^2 + 25y^2 + 16x - 150y + 141 = 0$

31. $3x^2 + 8y^2 + 12x - 112y + 403 = 0$

32. $x^2 + 20y^2 - 5x + 120y + 185 = 0$

In Exercises 33–36, find the standard form of the equation of the ellipse with the given characteristics.

33. Vertices: $(\pm 5, 0)$; foci: $(\pm 4, 0)$

34. Vertices: $(0, \pm 6)$; passes through the point $(2, 2)$

35. Vertices: $(-3, 0)$, $(7, 0)$; foci: $(0, 0)$, $(4, 0)$

36. Vertices: $(2, 0)$, $(2, 4)$; foci: $(2, 1)$, $(2, 3)$

37. *Architecture*   A semielliptical archway is to be formed over the entrance to an estate. The arch is to be set on pillars that are 10 feet apart and is to have a height (atop the pillars) of 4 feet. Where should the foci be placed in order to sketch the arch?

38. *Wading Pool*   You are building a wading pool that is in the shape of an ellipse. Your plans give an equation for the elliptical shape of the pool measured in feet as

$$\frac{x^2}{324} + \frac{y^2}{196} = 1.$$

Find the longest distance across the pool, the shortest distance, and the distance between the foci.

39. *Planetary Motion*   Saturn moves in an elliptical orbit with the sun at one focus. The smallest distance and the greatest distance of the planet from the sun are 1.3495×10^9 and 1.5045×10^9 kilometers, respectively. Find the eccentricity of the orbit, defined by $e = c/a$.

40. *Planetary Motion*   Mercury moves in an elliptical orbit with the sun at one focus. The eccentricity of Mercury's orbit is $e = 0.2056$. The length of the major axis is 72 million miles. Find the standard equation of Mercury's orbit. Place the center of the orbit at the origin and the major axis on the x-axis.

9.3 In Exercises 41–46, (a) find the standard form of the equation of the hyperbola, (b) find the center, vertices, foci, and eccentricity of the hyperbola, and (c) sketch the hyperbola. Use a graphing utility to verify your graph in part (c).

41. $5y^2 - 4x^2 = 20$ **42.** $x^2 - y^2 = \frac{9}{4}$

43. $9x^2 - 16y^2 - 18x - 32y - 151 = 0$

44. $-4x^2 + 25y^2 - 8x + 150y + 121 = 0$

45. $y^2 - 4x^2 - 2y - 48x + 59 = 0$

46. $9x^2 - y^2 - 72x + 8y + 119 = 0$

In Exercises 47–50, find the standard form of the equation of the hyperbola with the given characteristics.

47. Vertices: $(\pm 4, 0)$; foci: $(\pm 6, 0)$

48. Vertices: $(0, \pm 1)$; foci: $(0, \pm 3)$

49. Foci: $(0, 0)$, $(8, 0)$; asymptotes: $y = \pm 2(x - 4)$

50. Foci: $(3, \pm 2)$; asymptotes: $y = \pm 2(x - 3)$

51. *Navigation*   Radio transmitting station A is located 200 miles east of transmitting station B. A ship is in an area to the north and 40 miles west of station A. Synchronized radio pulses transmitted at 186,000 miles per second by the two stations are received 0.0005 second sooner from station A than from station B. How far north is the ship?

52. *Sound Location*   Two of your friends live 4 miles apart on the same "east-west" street, and you live halfway between them. You are having a three-way phone conversation when you hear an explosion. Six seconds later your friend to the east hears the explosion, and your friend to the west hears it 8 seconds after you do. Find equations of two hyperbolas that would locate the explosion. (Assume that the coordinate system is measured in feet and that sound travels at 1100 feet per second.)

In Exercises 53–56, classify the graph of the equation as a circle, a parabola, an ellipse, or a hyperbola.

53. $3x^2 + 2y^2 - 12x + 12y + 29 = 0$

54. $4x^2 + 4y^2 - 4x + 8y - 11 = 0$

55. $5x^2 - 2y^2 + 10x - 4y + 17 = 0$

56. $-4y^2 + 5x + 3y + 7 = 0$

9.4 In Exercises 57–60, rotate the axes to eliminate the xy-term in the equation. Then write the equation in standard form. Sketch the graph of the resulting equation, showing both sets of axes.

57. $xy - 4 = 0$ **58.** $x^2 - 10xy + y^2 + 1 = 0$

59. $5x^2 - 2xy + 5y^2 - 12 = 0$

60. $4x^2 + 8xy + 4y^2 + 7\sqrt{2}x + 9\sqrt{2}y = 0$

In Exercises 61–64, (a) use the discriminant to classify the graph of the equation, (b) use the Quadratic Formula to solve for y, and (c) use a graphing utility to graph the equation.

61. $16x^2 - 8xy + y^2 - 10x + 5y = 0$

62. $13x^2 - 8xy + 7y^2 - 45 = 0$

63. $x^2 + y^2 + 2xy + 2\sqrt{2}x - 2\sqrt{2}y + 2 = 0$

64. $x^2 - 10xy + y^2 + 1 = 0$

In Exercises 65 and 66, use any method to solve the system of quadratic equations algebraically. Then verify your results by using a graphing utility to graph the equations and find any points of intersection of the graphs.

65. $\begin{cases} -4x^2 - y^2 - 32x + 24y - 64 = 0 \\ 4x^2 + y^2 + 56x - 24y + 304 = 0 \end{cases}$

66. $\begin{cases} x^2 + y^2 - 25 = 0 \\ 9x - 4y^2 = 0 \end{cases}$

9.5 In Exercises 67 and 68, complete the table for the set of parametric equations. Plot the points (x, y) and sketch a graph of the parametric equations.

67. $x = 3t - 2$ and $y = 7 - 4t$

t	-2	-1	0	1	2	3
x						
y						

68. $x = \sqrt{t}$ and $y = 8 - t$

t	0	1	2	3	4
x					
y					

In Exercises 69–74, sketch the curve represented by the parametric equations (indicate the orientation of the curve). Then eliminate the parameter and write the corresponding rectangular equation whose graph represents the curve. Adjust the domain of the resulting rectangular equation, if necessary.

69. $x = 5t - 1$
$y = 2t + 5$

70. $x = 4t + 1$
$y = 8 - 3t$

71. $x = t^2 + 2$
$y = 4t^2 - 3$

72. $x = \ln 4t$
$y = t^2$

73. $x = t^3$
$y = \dfrac{1}{2}t^2$

74. $x = \dfrac{4}{t}$
$y = t^2 - 1$

In Exercises 75–86, use a graphing utility to graph the curve represented by the parametric equations.

75. $x = \sqrt[3]{t},\ y = t$

76. $x = t,\ y = \sqrt[3]{t}$

77. $x = \dfrac{1}{t},\ y = t$

78. $x = t,\ y = \dfrac{1}{t}$

79. $x = 2t,\ y = 4t$

80. $x = t^2,\ y = \sqrt{t}$

81. $x = 1 + 4t,\ y = 2 - 3t$

82. $x = t + 4,\ y = t^2$

83. $x = 3,\ y = t$

84. $x = t,\ y = 2$

85. $x = 6 \cos \theta$
$y = 6 \sin \theta$

86. $x = 3 + 3 \cos \theta$
$y = 2 + 5 \sin \theta$

In Exercises 87–90, find two different sets of parametric equations for the given rectangular equation. (There are many correct answers.)

87. $y = 6x + 2$

88. $y = 10 - x$

89. $y = x^2 + 2$

90. $y = 2x^3 + 5x$

In Exercises 91–94, find a set of parametric equations for the line that passes through the given points. (There are many correct answers.)

91. $(3, 5),\ (8, 5)$

92. $(2, -1),\ (2, 4)$

93. $(-1, 6),\ (10, 0)$

94. $(0, 0),\ \left(\tfrac{5}{2}, 6\right)$

Sports   In Exercises 95–98, the quarterback of a football team releases a pass at a height of 7 feet above the playing field, and the football is caught at a height of 4 feet, 30 yards directly downfield. The pass is released at an angle of 35° with the horizontal. The parametric equations for the path of the football are given by $x = 0.82v_0 t$ and $y = 7 + 0.57v_0 t - 16t^2$ where v_0 is the speed of the football (in feet per second) when it is released.

95. Find the speed of the football when it is released.

96. Write a set of parametric equations for the path of the ball.

97. Use a graphing utility to graph the path of the ball and approximate its maximum height.

98. Find the time the receiver has to position himself after the quarterback releases the ball.

9.6 In Exercises 99–104, plot the point given in polar coordinates and find three additional polar representations of the point, using $-2\pi < \theta < 2\pi$.

99. $\left(1, \dfrac{\pi}{4}\right)$

100. $\left(-5, -\dfrac{\pi}{3}\right)$

101. $\left(-2, -\dfrac{11\pi}{6}\right)$

102. $\left(1, \dfrac{5\pi}{6}\right)$

103. $\left(\sqrt{5}, -\dfrac{4\pi}{3}\right)$

104. $\left(\sqrt{10}, \dfrac{3\pi}{4}\right)$

In Exercises 105–110, plot the point given in polar coordinates and find the corresponding rectangular coordinates for the point.

105. $\left(5, -\dfrac{7\pi}{6}\right)$

106. $\left(-4, \dfrac{2\pi}{3}\right)$

107. $\left(2, -\dfrac{5\pi}{3}\right)$

108. $\left(-1, \dfrac{11\pi}{6}\right)$

109. $\left(3, \dfrac{3\pi}{4}\right)$

110. $\left(0, \dfrac{\pi}{2}\right)$

In Exercises 111–114, plot the point given in rectangular coordinates and find two sets of polar coordinates for the point for $0 \le \theta < 2\pi$.

111. $(0, -9)$

112. $(-3, 4)$

113. $(5, -5)$

114. $\left(-3, -\sqrt{3}\right)$

In Exercises 115–122, convert the rectangular equation to polar form.

115. $x^2 + y^2 = 9$

116. $x^2 + y^2 = 20$

117. $x^2 + y^2 - 4x = 0$

118. $x^2 + y^2 - 6y = 0$

119. $xy = 5$

120. $xy = -2$

121. $4x^2 + y^2 = 1$

122. $2x^2 + 3y^2 = 1$

In Exercises 123–130, convert the polar equation to rectangular form.

123. $r = 5$

124. $r = 12$

125. $r = 3 \cos \theta$

126. $r = 8 \sin \theta$

127. $r^2 = \cos 2\theta$

128. $r^2 = \sin \theta$

129. $\theta = \dfrac{5\pi}{6}$

130. $\theta = \dfrac{4\pi}{3}$

9.7 In Exercises 131–136, sketch the graph of the polar equation by hand. Then use a graphing utility to verify your graph.

131. $r = 5$

132. $r = 3$

133. $\theta = \dfrac{\pi}{2}$

134. $\theta = -\dfrac{5\pi}{6}$

135. $r = 5 \cos \theta$

136. $r = 2 \sin \theta$

In Exercises 137–144, identify and then sketch the graph of the polar equation. Identify any symmetry, maximum r-values, and zeros of r. Use a graphing utility to verify your graph.

137. $r = 5 + 4 \cos \theta$

138. $r = 1 + 4 \sin \theta$

139. $r = 3 - 5 \sin \theta$

140. $r = 2 - 6 \cos \theta$

141. $r = -3 \cos 2\theta$

142. $r = \cos 5\theta$

143. $r^2 = 5 \sin 2\theta$

144. $r^2 = \cos 2\theta$

9.8 In Exercises 145–150, identify the conic represented by the equation algebraically. Then use a graphing utility to graph the polar equation.

145. $r = \dfrac{2}{1 - \sin \theta}$

146. $r = \dfrac{1}{1 + 2 \sin \theta}$

147. $r = \dfrac{4}{5 - 3 \cos \theta}$

148. $r = \dfrac{6}{-1 + 4 \cos \theta}$

149. $r = \dfrac{5}{6 + 2 \sin \theta}$

150. $r = \dfrac{3}{4 - 4 \cos \theta}$

In Exercises 151–154, find a polar equation of the conic with its focus at the pole.

151. Parabola, vertex: $(2, \pi)$

152. Parabola, vertex: $(2, \pi/2)$

153. Ellipse, vertices: $(5, 0), (1, \pi)$

154. Hyperbola, vertices: $(1, 0), (7, 0)$

155. *Planetary Motion*   The planet Mars has an elliptical orbit with an eccentricity of $e \approx 0.093$. The length of the major axis of the orbit is approximately 3.05 astronomical units. Find a polar equation for the orbit and its perihelion and aphelion distances.

156. *Astronomy*   An asteroid takes a parabolic path with Earth as its focus. It is about 6,000,000 miles from Earth at its closest approach. Write the polar equation of the path of the asteroid with its vertex at $\theta = -\pi/2$. Find the distance between the asteroid and Earth when $\theta = -\pi/3$.

Synthesis

True or False?   In Exercises 157 and 158, determine whether the statement is true or false. Justify your answer.

157. The graph of $\frac{1}{4}x^2 - y^4 = 1$ represents the equation of a hyperbola.

158. There is only one set of parametric equations that represents the line $y = 3 - 2x$.

Writing   In Exercises 159 and 160, an equation and four variations are given. In your own words, describe how the graph of each of the variations differs from the graph of the original equation.

159. $y^2 = 8x$

    (a) $(y - 2)^2 = 8x$ (b) $y^2 = 8(x + 1)$

    (c) $y^2 = -8x$ (d) $y^2 = 4x$

160. $\dfrac{x^2}{4} + \dfrac{y^2}{9} = 1$

    (a) $\dfrac{x^2}{9} + \dfrac{y^2}{4} = 1$ (b) $\dfrac{x^2}{4} + \dfrac{y^2}{4} = 1$

    (c) $\dfrac{x^2}{4} + \dfrac{y^2}{25} = 1$ (d) $\dfrac{(x - 3)^2}{4} + \dfrac{y^2}{9} = 1$

161. Consider an ellipse whose major axis is horizontal and 10 units in length. The number b in the standard form of the equation of the ellipse must be less than what real number? Describe the change in the shape of the ellipse as b approaches this number.

162. The graph of the parametric equations $x = 2 \sec t$ and $y = 3 \tan t$ is shown in the figure. Would the graph change for the equations $x = 2 \sec(-t)$ and $y = 3 \tan(-t)$? If so, how would it change?

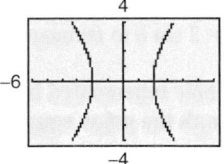

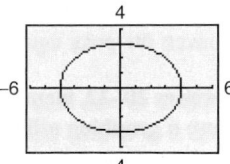

Figure for 162   Figure for 163

163. The path of a moving object is modeled by the parametric equations $x = 4 \cos t$ and $y = 3 \sin t$, where t is time (see figure). How would the path change for each of the following?

    (a) $x = 4 \cos 2t, y = 3 \sin 2t$

    (b) $x = 5 \cos t, y = 3 \sin t$

9 Chapter Test

See www.CalcChat.com for worked-out solutions to odd-numbered exercises.

Take this test as you would take a test in class. After you are finished, check your work against the answers given in the back of the book.

In Exercises 1–3, graph the conic and identify any vertices and foci.

1. $y^2 - 8x = 0$ **2.** $y^2 - 4x + 4 = 0$ **3.** $x^2 - 4y^2 - 4x = 0$

4. Find the standard form of the equation of the parabola with focus $(8, -2)$ and directrix $x = 4$, and sketch the parabola.

5. Find the standard form of the equation of the ellipse shown at the right.

6. Find the standard form of the equation of the hyperbola with vertices $(0, \pm 3)$ and asymptotes $y = \pm \frac{3}{2}x$.

7. Use a graphing utility to graph the conic $x^2 - \dfrac{y^2}{4} = 1$. Describe your viewing window.

8. (a) Determine the number of degrees the axis must be rotated to eliminate the xy-term of the conic $x^2 + 6xy + y^2 - 6 = 0$.

 (b) Graph the conic in part (a) and use a graphing utility to confirm your result.

9. Solve the system of equations at the right algebraically by the method of substitution. Then verify your results by using a graphing utility.

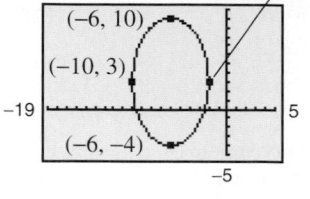

Figure for 5

$\begin{cases} x^2 + 2y^2 - 4x + 6y - 5 = 0 \\ \qquad\quad x + \ y + 5 = 0 \end{cases}$

System for 9

In Exercises 10–12, sketch the curve represented by the parametric equations. Then eliminate the parameter and write the corresponding rectangular equation whose graph represents the curve.

10. $x = t^2 - 6$
$y = \dfrac{1}{2}t - 1$

11. $x = \sqrt{t^2 + 2}$
$y = \dfrac{t}{4}$

12. $x = 2 + 3 \cos \theta$
$y = 2 \sin \theta$

In Exercises 13–15, find two different sets of parametric equations for the given rectangular equation. (There are many correct answers.)

13. $y = x^2 + 10$ **14.** $x + y^2 = 4$ **15.** $x^2 + 4y^2 - 16 = 0$

16. Convert the polar coordinate $\left(-14, \dfrac{5\pi}{3}\right)$ to rectangular form.

17. Convert the rectangular coordinate $(2, -2)$ to polar form and find two additional polar representations of this point. (There are many correct answers.)

18. Convert the rectangular equation $x^2 + y^2 - 12y = 0$ to polar form.

19. Convert the polar equation $r = 2 \sin \theta$ to rectangular form.

In Exercises 20–22, identify the conic represented by the polar equation algebraically. Then use a graphing utility to graph the polar equation.

20. $r = 2 + 3 \sin \theta$ **21.** $r = \dfrac{1}{1 - \cos \theta}$ **22.** $r = \dfrac{4}{2 + 3 \sin \theta}$

23. Find a polar equation of an ellipse with its focus at the pole, an eccentricity of $e = \frac{1}{4}$, and directrix at $y = 4$.

24. Find a polar equation of a hyperbola with its focus at the pole, an eccentricity of $e = \frac{5}{4}$, and directrix at $y = 2$.

25. For the polar equation $r = 8 \cos 3\theta$, find the maximum value of $|r|$ and any zeros of r. Verify your answers numerically.

7–9 Cumulative Test

See www.CalcChat.com for worked-out solutions to odd-numbered exercises.

Take this test to review the material in Chapters 7–9. After you are finished, check your work against the answers given in the back of the book.

In Exercises 1–4, use any method to solve the system of equations.

1. $\begin{cases} -x - 3y = 5 \\ 4x + 2y = 10 \end{cases}$

2. $\begin{cases} 2x - y^2 = 0 \\ x - y = 4 \end{cases}$

3. $\begin{cases} 2x - 3y + z = 13 \\ -4x + y - 2z = -6 \\ x - 3y + 3z = 12 \end{cases}$

4. $\begin{cases} x - 4y + 3z = 5 \\ 5x + 2y - z = 1 \\ -2x - 8y = 30 \end{cases}$

In Exercises 5–8, perform the matrix operations given

$$A = \begin{bmatrix} -3 & 0 & -4 \\ 2 & 4 & 5 \\ -4 & 8 & 1 \end{bmatrix} \quad \text{and} \quad B = \begin{bmatrix} -1 & 5 & 2 \\ 6 & -3 & 3 \\ 0 & 4 & -2 \end{bmatrix}.$$

5. $3A - 2B$ **6.** $5A + 3B$ **7.** AB **8.** BA

9. Find (a) the inverse of A (if it exists) and (b) the determinant of A.

$$A = \begin{bmatrix} 1 & 2 & -1 \\ 3 & 7 & -10 \\ -5 & -7 & -15 \end{bmatrix}$$

10. Use a determinant to find the area of the triangle with vertices $(0, 0)$, $(6, 2)$, and $(8, 10)$.

11. Write the first five terms of each sequence a_n. (Assume that n begins with 1.)

(a) $a_n = \dfrac{(-1)^{n+1}}{2n + 3}$ (b) $a_n = 3(2)^{n-1}$

In Exercises 12–15, find the sum. Use a graphing utility to verify your result.

12. $\displaystyle\sum_{k=1}^{6} (7k - 2)$ **13.** $\displaystyle\sum_{k=1}^{4} \dfrac{2}{k^2 + 4}$ **14.** $\displaystyle\sum_{n=0}^{10} 9\left(\dfrac{3}{4}\right)^n$ **15.** $\displaystyle\sum_{n=0}^{50} 100\left(-\dfrac{1}{2}\right)^n$

In Exercises 16–18, find the sum of the inifinite geometric series.

16. $\displaystyle\sum_{n=0}^{\infty} 3\left(-\dfrac{3}{5}\right)^n$ **17.** $\displaystyle\sum_{n=1}^{\infty} 5(-0.02)^n$ **18.** $4 - 2 + 1 - \dfrac{1}{2} + \dfrac{1}{4} - \cdots$

19. Use mathematical induction to prove the formula
$$3 + 7 + 11 + 15 + \cdots + (4n - 1) = n(2n + 1).$$

In Exercises 20–23, use the Binomial Theorem to expand and simplify the expression.

20. $(x + 3)^4$ **21.** $(2x + y^2)^5$ **22.** $(x - 2y)^6$ **23.** $(3a - 4b)^8$

In Exercises 24–27, find the number of distinguishable permutations of the group of letters.

24. M, I, A, M, I   **25.** B, U, B, B, L, E

26. B, A, S, K, E, T, B, A, L, L   **27.** A, N, T, A, R, C, T, I, C, A

In Exercises 28–31, identify the conic and sketch its graph.

28. $\dfrac{(y + 3)^2}{36} - \dfrac{(x - 5)^2}{121} = 1$

29. $\dfrac{(x - 2)^2}{4} + \dfrac{(y + 1)^2}{9} = 1$

30. $y^2 - x^2 = 16$

31. $x^2 + y^2 - 2x - 4y + 1 = 0$

In Exercises 32–34, find the standard form of the equation of the conic.

32.
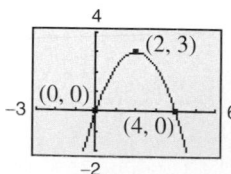

33.
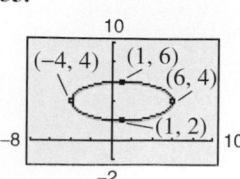

34.
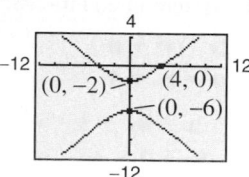

35. Use a graphing utility to graph $x^2 - 4xy + 2y^2 = 6$. Determine the angle θ through which the axes are rotated.

In Exercises 36–38, (a) sketch the curve represented by the parametric equations, (b) use a graphing utility to verify your graph, and (c) eliminate the parameter and write the corresponding rectangular equation whose graph represents the curve. Adjust the domain of the resulting rectangular equation, if necessary.

36. $x = 2t + 1$
$\quad y = t^2$

37. $x = \cos \theta$
$\quad y = 2 \sin^2 \theta$

38. $x = 4 \ln t$
$\quad y = \frac{1}{2}t^2$

In Exercises 39–42, find two different sets of parametric equations for the given rectangular equation. (There are many correct answers.)

39. $y = 3x - 2$

40. $x^2 - y^2 = 16$

41. $y = \dfrac{2}{x}$

42. $y = \dfrac{e^{2x}}{e^{2x} + 1}$

In Exercises 43–46, plot the point given in polar coordinates and find three additional polar representations for $-2\pi < \theta < 2\pi$.

43. $\left(8, \dfrac{5\pi}{6}\right)$

44. $\left(5, -\dfrac{3\pi}{4}\right)$

45. $\left(-2, \dfrac{5\pi}{4}\right)$

46. $\left(-3, -\dfrac{11\pi}{6}\right)$

47. Convert the rectangular equation $4x + 4y + 1 = 0$ to polar form.

48. Convert the polar equation $r = 2 \cos \theta$ to rectangular form.

49. Convert the polar equation $r = \dfrac{2}{4 - 5 \cos \theta}$ to rectangular form.

In Exercises 50–52, identify the graph represented by the polar equation algebraically. Then use a graphing utility to graph the polar equation.

50. $r = -\dfrac{\pi}{6}$

51. $r = 3 - 2 \sin \theta$

52. $r = 2 + 5 \cos \theta$

53. The salary for the first year of a job is $32,500. During the next 14 years, the salary increases by 5% each year. Determine the total compensation over the 15-year period.

54. On a game show, the digits 3, 4, and 5 must be arranged in the proper order to form the price of an appliance. If they are arranged correctly, the contestant wins the appliance. What is the probability of winning if the contestant knows that the price is at least $400?

55. A parabolic archway is 16 meters high at the vertex. At a height of 14 meters, the width of the archway is 12 meters, as shown in the figure at the right. How wide is the archway at ground level?

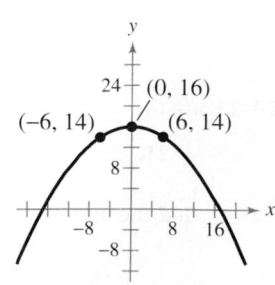

Figure for 55

Proofs in Mathematics

Standard Equation of a Parabola   (p. 663)

The standard form of the equation of a parabola with vertex at (h, k) is as follows.

$(x - h)^2 = 4p(y - k), \quad p \neq 0$ Vertical axis, directrix: $y = k - p$

$(y - k)^2 = 4p(x - h), \quad p \neq 0$ Horizontal axis, directrix: $x = h - p$

The focus lies on the axis p units (*directed distance*) from the vertex. If the vertex is at the origin $(0, 0)$, the equation takes one of the following forms.

$x^2 = 4py$ Vertical axis

$y^2 = 4px$ Horizontal axis

Parabolic Paths

There are many natural occurrences of parabolas in real life. For instance, the famous astronomer Galileo discovered in the 17th century that an object that is projected upward and obliquely to the pull of gravity travels in a parabolic path. Examples of this are the center of gravity of a jumping dolphin and the path of water molecules in a drinking fountain.

Proof

For the case in which the directrix is parallel to the x-axis and the focus lies above the vertex, as shown in the top figure, if (x, y) is any point on the parabola, then, by definition, it is equidistant from the focus $(h, k + p)$ and the directrix $y = k - p$. So, you have

$$\sqrt{(x - h)^2 + [y - (k + p)]^2} = y - (k - p)$$

$$(x - h)^2 + [y - (k + p)]^2 = [y - (k - p)]^2$$

$$(x - h)^2 + y^2 - 2y(k + p) + (k + p)^2 = y^2 - 2y(k - p) + (k - p)^2$$

$$(x - h)^2 + y^2 - 2ky - 2py + k^2 + 2pk + p^2 = y^2 - 2ky + 2py + k^2 - 2pk + p^2$$

$$(x - h)^2 - 2py + 2pk = 2py - 2pk$$

$$(x - h)^2 = 4p(y - k).$$

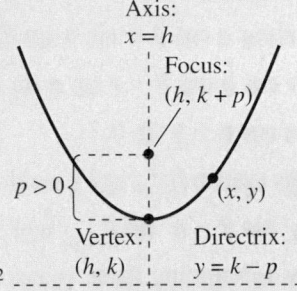

Parabola with vertical axis

For the case in which the directrix is parallel to the y-axis and the focus lies to the right of the vertex, as shown in the bottom figure, if (x, y) is any point on the parabola, then, by definition, it is equidistant from the focus $(h + p, k)$ and the directrix $x = h - p$. So, you have

$$\sqrt{[x - (h + p)]^2 + (y - k)^2} = x - (h - p)$$

$$[x - (h + p)]^2 + (y - k)^2 = [x - (h - p)]^2$$

$$x^2 - 2x(h + p) + (h + p)^2 + (y - k)^2 = x^2 - 2x(h - p) + (h - p)^2$$

$$x^2 - 2hx - 2px + h^2 + 2ph + p^2 + (y - k)^2 = x^2 - 2hx + 2px + h^2 - 2ph + p^2$$

$$-2px + 2ph + (y - k)^2 = 2px - 2ph$$

$$(y - k)^2 = 4p(x - h).$$

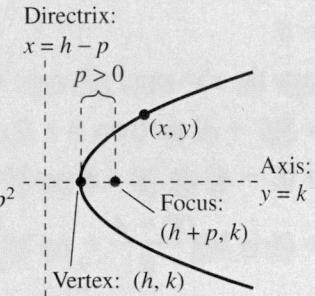

Parabola with horizontal axis

Note that if a parabola is centered at the origin, then the two equations above would simplify to $x^2 = 4py$ and $y^2 = 4px$, respectively.

> **Rotation of Axes to Eliminate an xy-Term**   (p. 690)
>
> The general second-degree equation $Ax^2 + Bxy + Cy^2 + Dx + Ey + F = 0$ can be rewritten as
>
> $$A'(x')^2 + C'(y')^2 + D'x' + E'y' + F' = 0$$
>
> by rotating the coordinate axes through an angle θ, where
>
> $$\cot 2\theta = \frac{A - C}{B}.$$
>
> The coefficients of the new equation are obtained by making the substitutions $x = x' \cos \theta - y' \sin \theta$ and $y = x' \sin \theta + y' \cos \theta$.

Proof

You need to discover how the coordinates in the xy-system are related to the coordinates in the $x'y'$-system. To do this, choose a point $P = (x, y)$ in the original system and attempt to find its coordinates (x', y') in the rotated system. In either system, the distance r between the point P and the origin is the same. So, the equations for x, y, x', and y' are those given in the figure. Using the formulas for the sine and cosine of the difference of two angles, you have the following.

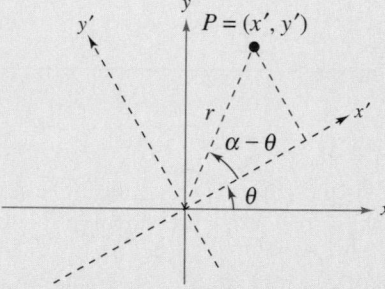

Rotated: $x' = r \cos(\alpha - \theta)$
$y' = r \sin(\alpha - \theta)$

$$
\begin{aligned}
x' &= r \cos(\alpha - \theta) & \qquad y' &= r \sin(\alpha - \theta) \\
&= r(\cos \alpha \cos \theta + \sin \alpha \sin \theta) & &= r(\sin \alpha \cos \theta - \cos \alpha \sin \theta) \\
&= r \cos \alpha \cos \theta + r \sin \alpha \sin \theta & &= r \sin \alpha \cos \theta - r \cos \alpha \sin \theta \\
&= x \cos \theta + y \sin \theta & &= y \cos \theta - x \sin \theta
\end{aligned}
$$

Solving this system for x and y yields

$$x = x' \cos \theta - y' \sin \theta \qquad \text{and} \qquad y = x' \sin \theta + y' \cos \theta.$$

Finally, by substituting these values for x and y into the original equation and collecting terms, you obtain

$$A' = A \cos^2 \theta + B \cos \theta \sin \theta + C \sin^2 \theta$$

$$C' = A \sin^2 \theta - B \cos \theta \sin \theta + C \cos^2 \theta$$

$$D' = D \cos \theta + E \sin \theta$$

$$E' = -D \sin \theta + E \cos \theta$$

$$F' = F.$$

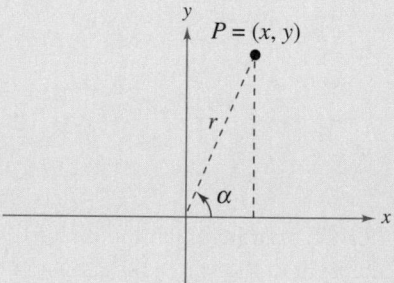

Original: $x = r \cos \alpha$
$y = r \sin \alpha$

To eliminate the $x'y'$-term, you must select θ such that $B' = 0$.

$$
\begin{aligned}
B' &= 2(C - A) \sin \theta \cos \theta + B(\cos^2 \theta - \sin^2 \theta) \\
&= (C - A) \sin 2\theta + B \cos 2\theta \\
&= B(\sin 2\theta)\left(\frac{C - A}{B} + \cot 2\theta\right) = 0, \quad \sin 2\theta \neq 0
\end{aligned}
$$

If $B = 0$, no rotation is necessary because the xy-term is not present in the original equation. If $B \neq 0$, the only way to make $B' = 0$ is to let

$$\cot 2\theta = \frac{A - C}{B}, \quad B \neq 0.$$

So, you have established the desired results.

> **Polar Equations of Conics   (p. 722)**
>
> The graph of a polar equation of the form
>
> **1.** $r = \dfrac{ep}{1 \pm e \cos \theta}$ or   **2.** $r = \dfrac{ep}{1 \pm e \sin \theta}$
>
> is a conic, where $e > 0$ is the eccentricity and $|p|$ is the distance between the focus (pole) and the directrix.

Proof

A proof for $r = ep/(1 + e \cos \theta)$ with $p > 0$ is shown here. The proofs of the other cases are similar. In the figure, consider a vertical directrix, p units to the right of the focus $F = (0, 0)$. If $P = (r, \theta)$ is a point on the graph of

$$r = \frac{ep}{1 + e \cos \theta}$$

the distance between P and the directrix is

$$PQ = |p - x|$$

$$= |p - r \cos \theta|$$

$$= \left| p - \left(\frac{ep}{1 + e \cos \theta} \right) \cos \theta \right|$$

$$= \left| p \left(1 - \frac{e \cos \theta}{1 + e \cos \theta} \right) \right|$$

$$= \left| \frac{p}{1 + e \cos \theta} \right|$$

$$= \left| \frac{r}{e} \right|.$$

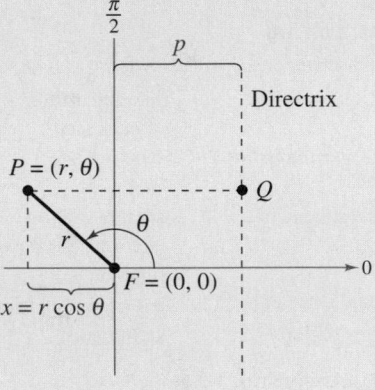

Moreover, because the distance between P and the pole is simply $PF = |r|$, the ratio of PF to PQ is

$$\frac{PF}{PQ} = \frac{|r|}{\left| \dfrac{r}{e} \right|} = |e| = e$$

and, by definition, the graph of the equation must be a conic.

Progressive Summary (Chapters 3–9)

This chart outlines the topics that have been covered so far in this text. Progressive Summary charts appear after Chapters 2, 3, 6, 9, and 11. In each progressive summary, new topics encountered for the first time appear in red.

Transcendental Functions	**Systems and Series**	**Other Topics**
Exponential, Logarithmic, Trigonometric, Inverse Trigonometric	Systems, Sequences, Series	Conics, Parametric and Polar Equations,

■ Rewriting

Transcendental Functions	Systems and Series	Other Topics
Exponential form ↔ Logarithmic form	Row operations for systems of equations	Standard forms of conics
Condense/expand logarithmic expressions	Partial fraction decomposition	Eliminate parameters
Simplify trigonometric expressions	Operations with matrices	Rectangular form ↔ Parametric form
Prove trigonometric identities	Matrix form of a system of equations	Rectangular form ↔ Polar form
Use conversion formulas	nth term of a sequence	
Operations with vectors	Summation form of a series	
Powers and roots of complex numbers		

■ Solving

Transcendental Functions

Equation	Strategy
Exponential	Take logarithm of each side
Logarithmic	Exponentiate each side
Trigonometric	Isolate function factor, use inverse function
Multiple angle or high powers	Use trigonometric identities

Systems and Series

Equation	Strategy
System of linear equations	Substitution
	Elimination
	Gaussian
	Gauss-Jordan
	Inverse matrices
	Cramer's Rule

Other Topics

Equation	Strategy
Conics	Convert to standard form
	Convert to polar form

■ Analyzing

Transcendental Functions

Graphically	Algebraically
Intercepts	Domain, Range
Asymptotes	Transformations
Minimum values	Composition
Maximum values	Inverse properties
	Amplitude, period
	Reference angles

Numerically

Table of values

Systems and Series

Systems:
  Intersecting, parallel, and coincident lines, determinants
Sequences:
  Graphing utility in *dot* mode, nth term, partial sums, summation formulas

Other Topics

Conics:
  Table of values, vertices, foci, axes, symmetry, asymptotes, translations, eccentricity
Parametric forms:
  Point plotting, eliminate parameters
Polar forms:
  Point plotting, special equations, symmetry, zeros, eccentricity, maximum r-values, directrix

Chapter 10

10.1 The Three-Dimensional Coordinate System

10.2 Vectors in Space

10.3 The Cross Product of Two Vectors

10.4 Lines and Planes in Space

Selected Applications

Three-dimensional analytic geometry concepts have many real-life applications. The applications listed below represent a small sample of the applications in this chapter.

- Crystals, Exercises 75 and 76, pages 748 and 749
- Geography, Exercise 78, page 749
- Tension, Exercises 71 and 72, page 756
- Torque, Exercises 57 and 58, page 763
- Machine Design, Exercise 59, page 772
- Mechanical Design, Exercise 60, page 772

Analytic Geometry in Three Dimensions

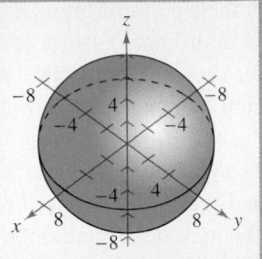
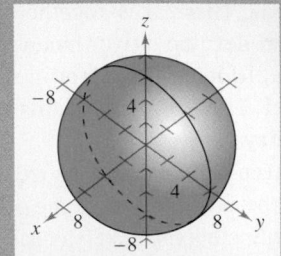
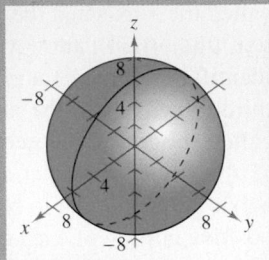

Until now, you have been working mainly with two-dimensional coordinate systems. In Chapter 10, you will learn how to plot points, find distances between points, and represent vectors in three-dimensional coordinate systems. You will also write equations of spheres and graph traces of surfaces in space, as shown above.

Arnold Fisher/Photo Researchers, Inc.

The three-dimensional coordinate system is used in chemistry to help understand the structures of crystals. For example, isometric crystals are shaped like cubes.

10.1 The Three–Dimensional Coordinate System

The Three–Dimensional Coordinate System

Recall that the Cartesian plane is determined by two perpendicular number lines called the x-axis and the y-axis. These axes, together with their point of intersection (the origin), allow you to develop a two-dimensional coordinate system for identifying points in a plane. To identify a point in space, you must introduce a third dimension to the model. The geometry of this three-dimensional model is called **solid analytic geometry.**

You can construct a **three-dimensional coordinate system** by passing a z-axis perpendicular to both the x- and y-axes at the origin. Figure 10.1 shows the positive portion of each coordinate axis. Taken as pairs, the axes determine three **coordinate planes:** the **xy-plane,** the **xz-plane,** and the **yz-plane.** These three coordinate planes separate the three-dimensional coordinate system into eight **octants.** The first octant is the one in which all three coordinates are positive. In this three-dimensional system, a point P in space is determined by an ordered triple (x, y, z), where x, y, and z are as follows.

$x = $ directed distance from yz-plane to P

$y = $ directed distance from xz-plane to P

$z = $ directed distance from xy-plane to P

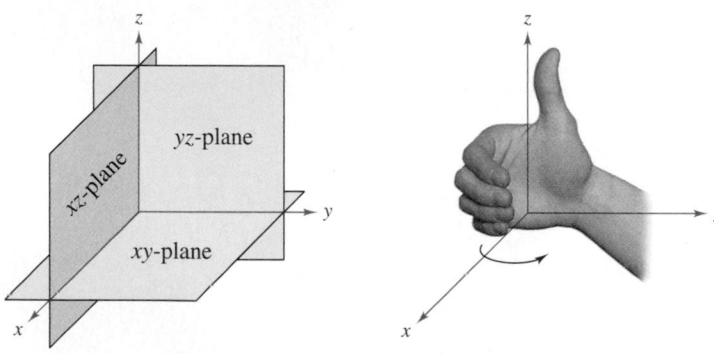

Figure 10.1   **Figure 10.2**

A three-dimensional coordinate system can have either a **left-handed** or a **right-handed** orientation. In this text, you will work exclusively with right-handed systems, as illustrated in Figure 10.2. In a right-handed system, Octants II, III, and IV are found by rotating counterclockwise around the positive z-axis. Octant V is vertically below Octant I. Octants VI, VII, and VIII are then found by rotating counterclockwise around the negative z-axis. See Figure 10.3.

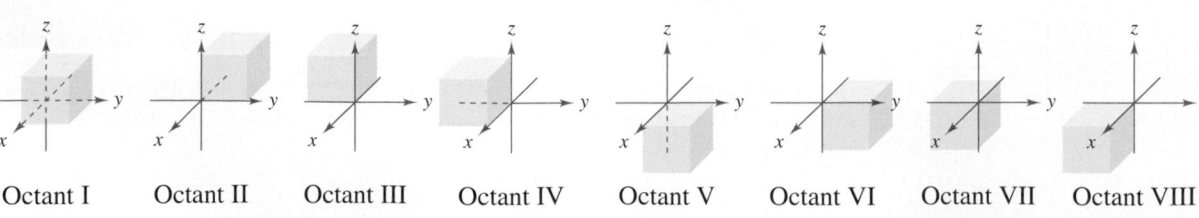

Octant I   Octant II   Octant III   Octant IV   Octant V   Octant VI   Octant VII   Octant VIII

Figure 10.3

What you should learn

- Plot points in the three-dimensional coordinate system.
- Find distances between points in space and find midpoints of line segments joining points in space.
- Write equations of spheres in standard form and find traces of surfaces in space.

Why you should learn it

The three-dimensional coordinate system can be used to graph equations that model surfaces in space, such as the spherical shape of Earth, as shown in Exercise 78 on page 749.

NASA

Prerequisite Skills

To review the Cartesian Plane, see Appendix B.1.

Example 1   Plotting Points in Space

Plot each point in space.

a. $(2, \; 3, 3)$ **b.** $(-2, 6, 2)$ **c.** $(1, 4, 0)$ **d.** $(2, 2, -3)$

Solution

To plot the point $(2, -3, 3)$, notice that $x = 2$, $y = -3$, and $z = 3$. To help visualize the point, locate the point $(2, -3)$ in the xy-plane (denoted by a cross in Figure 10.4). The point $(2, -3, 3)$ lies three units above the cross. The other three points are also shown in Figure 10.4.

✓**CHECKPOINT**   Now try Exercise 7.

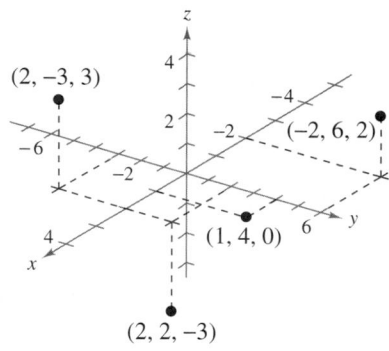

Figure 10.4

The Distance and Midpoint Formulas

Many of the formulas established for the two-dimensional coordinate system can be extended to three dimensions. For example, to find the distance between two points in space, you can use the Pythagorean Theorem twice, as shown in Figure 10.5.

Distance Formula in Space

The distance between the points (x_1, y_1, z_1) and (x_2, y_2, z_2) given by the **Distance Formula in Space** is

$$d = \sqrt{(x_2 - x_1)^2 + (y_2 - y_1)^2 + (z_2 - z_1)^2}.$$

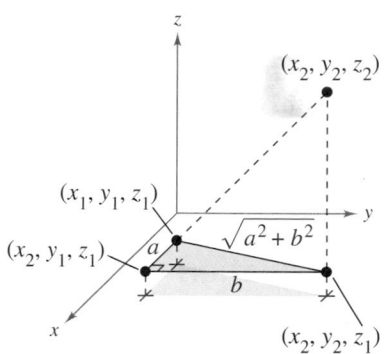

Example 2   Finding the Distance Between Two Points in Space

Find the distance between $(0, 1, 3)$ and $(1, 4, -2)$.

Solution

$$\begin{aligned}
d &= \sqrt{(x_2 - x_1)^2 + (y_2 - y_1)^2 + (z_2 - z_1)^2} && \text{Distance Formula in Space} \\
&= \sqrt{(1 - 0)^2 + (4 - 1)^2 + (-2 - 3)^2} && \text{Substitute.} \\
&= \sqrt{1 + 9 + 25} = \sqrt{35} && \text{Simplify.}
\end{aligned}$$

✓**CHECKPOINT**   Now try Exercise 21.

Notice the similarity between the Distance Formulas in the plane and in space. The Midpoint Formulas in the plane and in space are also similar.

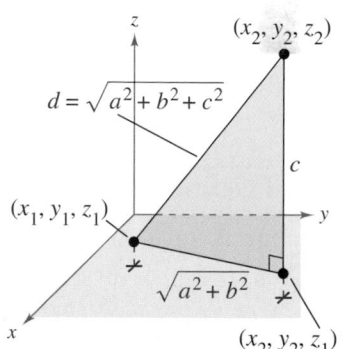

Figure 10.5

Midpoint Formula in Space

The midpoint of the line segment joining the points (x_1, y_1, z_1) and (x_2, y_2, z_2) given by the **Midpoint Formula in Space** is

$$\left(\frac{x_1 + x_2}{2}, \frac{y_1 + y_2}{2}, \frac{z_1 + z_2}{2} \right).$$

Example 3   Using the Midpoint Formula in Space

Find the midpoint of the line segment joining $(5, -2, 3)$ and $(0, 4, 4)$.

Solution

Using the Midpoint Formula in Space, the midpoint is

$$\left(\frac{5 + 0}{2}, \frac{-2 + 4}{2}, \frac{3 + 4}{2}\right) = \left(\frac{5}{2}, 1, \frac{7}{2}\right)$$

as shown in Figure 10.6.

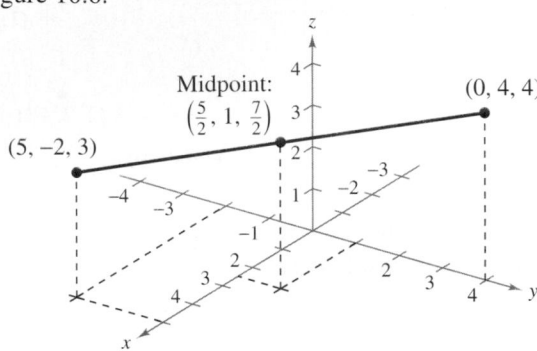

Figure 10.6

✓CHECKPOINT   Now try Exercise 35.

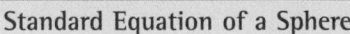

The Equation of a Sphere

A **sphere** with center (h, k, j) and radius r is defined as the set of all points (x, y, z) such that the distance between (x, y, z) and (h, k, j) is r, as shown in Figure 10.7. Using the Distance Formula, this condition can be written as

$$\sqrt{(x - h)^2 + (y - k)^2 + (z - j)^2} = r.$$

By squaring each side of this equation, you obtain the standard equation of a sphere.

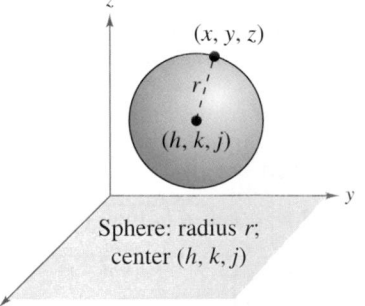

Sphere: radius r; center (h, k, j)

Figure 10.7

Standard Equation of a Sphere

The **standard equation of a sphere** with center (h, k, j) and radius r is given by

$$(x - h)^2 + (y - k)^2 + (z - j)^2 = r^2.$$

Notice the similarity of this formula to the equation of a circle in the plane.

$(x - h)^2 + (y - k)^2 + (z - j)^2 = r^2$ Equation of sphere in space

$(x - h)^2 + (y - k)^2 = r^2$ Equation of circle in the plane

As is true with the equation of a circle, the equation of a sphere is simplified when the center lies at the origin. In this case, the equation is

$x^2 + y^2 + z^2 = r^2.$ Sphere with center at origin

Example 4 Finding the Equation of a Sphere

Find the standard equation of the sphere with center $(2, 4, 3)$ and radius 3. Does this sphere intersect the xy-plane?

Solution

$$(x - h)^2 + (y - k)^2 + (z - j)^2 = r^2 \qquad \text{Standard equation}$$

$$(x - 2)^2 + (y - 4)^2 + (z - 3)^2 = 3^2 \qquad \text{Substitute.}$$

From the graph shown in Figure 10.8, you can see that the center of the sphere lies three units above the xy-plane. Because the sphere has a radius of 3, you can conclude that it does intersect the xy-plane—at the point $(2, 4, 0)$.

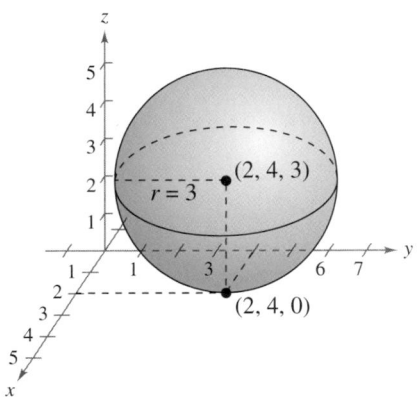

Figure 10.8

✔CHECKPOINT   Now try Exercise 45.

> **Exploration**
>
> Find the equation of the sphere that has the points $(3, -2, 6)$ and $(-1, 4, 2)$ as endpoints of a diameter. Explain how this problem gives you a chance to use all three formulas discussed so far in this section: the Distance Formula in Space, the Midpoint Formula in Space, and the standard equation of a sphere.

Example 5 Finding the Center and Radius of a Sphere

Find the center and radius of the sphere given by

$$x^2 + y^2 + z^2 - 2x + 4y - 6z + 8 = 0.$$

Solution

To obtain the standard equation of this sphere, complete the square as follows.

$$x^2 + y^2 + z^2 - 2x + 4y - 6z + 8 = 0$$

$$\left(x^2 - 2x + \boxed{}\right) + \left(y^2 + 4y + \boxed{}\right) + \left(z^2 - 6z + \boxed{}\right) = -8$$

$$(x^2 - 2x + 1) + (y^2 + 4y + 4) + (z^2 - 6z + 9) = -8 + 1 + 4 + 9$$

$$(x - 1)^2 + (y + 2)^2 + (z - 3)^2 = \left(\sqrt{6}\right)^2$$

So, the center of the sphere is $(1, -2, 3)$, and its radius is $\sqrt{6}$. See Figure 10.9.

✔CHECKPOINT   Now try Exercise 55.

> Sphere:
> $$(x - 1)^2 + (y + 2)^2 + (z - 3)^2 = \left(\sqrt{6}\right)^2$$

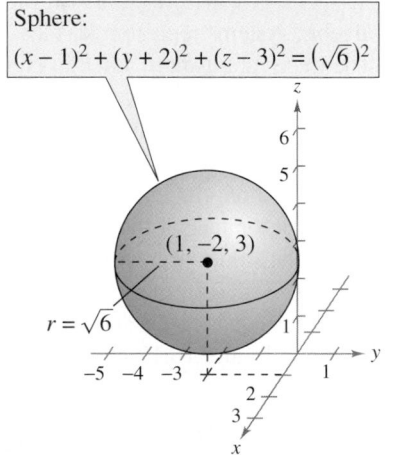

Figure 10.9

Note in Example 5 that the points satisfying the equation of the sphere are "surface points," not "interior points." In general, the collection of points satisfying an equation involving x, y, and z is called a **surface in space.**

Finding the intersection of a surface with one of the three coordinate planes (or with a plane parallel to one of the three coordinate planes) helps one visualize the surface. Such an intersection is called a **trace** of the surface. For example, the *xy*-trace of a surface consists of all points that are common to both the surface *and* the *xy*-plane. Similarly, the *xz*-trace of a surface consists of all points that are common to both the surface and the *xz*-plane.

Example 6   Finding a Trace of a Surface

Sketch the *xy*-trace of the sphere given by $(x - 3)^2 + (y - 2)^2 + (z + 4)^2 = 5^2$.

Solution

To find the *xy*-trace of this surface, use the fact that every point in the *xy*-plane has a *z*-coordinate of zero. This means that if you substitute $z = 0$ into the original equation, the resulting equation will represent the intersection of the surface with the *xy*-plane.

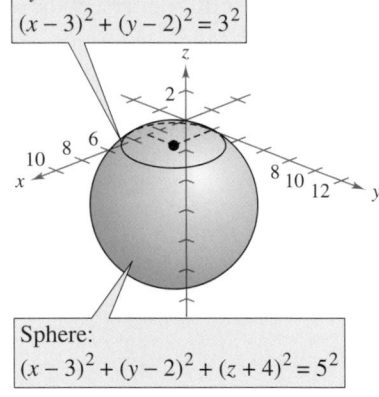

$$(x - 3)^2 + (y - 2)^2 + (z + 4)^2 = 5^2 \qquad \text{Write original equation.}$$

$$(x - 3)^2 + (y - 2)^2 + (0 + 4)^2 = 5^2 \qquad \text{Substitute 0 for } z.$$

$$(x - 3)^2 + (y - 2)^2 + 16 = 25 \qquad \text{Simplify.}$$

$$(x - 3)^2 + (y - 2)^2 = 9 \qquad \text{Subtract 16 from each side.}$$

$$(x - 3)^2 + (y - 2)^2 = 3^2 \qquad \text{Equation of circle}$$

From this form, you can see that the *xy*-trace is a circle of radius 3, as shown in Figure 10.10.

xy-trace:
$(x - 3)^2 + (y - 2)^2 = 3^2$

Sphere:
$(x - 3)^2 + (y - 2)^2 + (z + 4)^2 = 5^2$

Figure 10.10

✓CHECKPOINT   Now try Exercise 65.

TECHNOLOGY TIP   Most three-dimensional graphing utilities and computer algebra systems represent surfaces by sketching several traces of the surface. The traces are usually taken in equally spaced parallel planes. To graph an equation involving *x*, *y*, and *z* with a three-dimensional "function grapher," you must first set the graphing mode to *three-dimensional* and solve the equation for *z*. After entering the equation, you need to specify a rectangular viewing cube (the three-dimensional analog of a viewing window). For instance, to graph the top half of the sphere from Example 6, solve the equation for *z* to obtain the solutions $z = -4 \pm \sqrt{25 - (x - 3)^2 - (y - 2)^2}$. The equation $z = -4 + \sqrt{25 - (x - 3)^2 - (y - 2)^2}$ represents the top half of the sphere. Enter this equation, as shown in Figure 10.11. Next, use the viewing cube shown in Figure 10.12. Finally, you can display the graph, as shown in Figure 10.13.

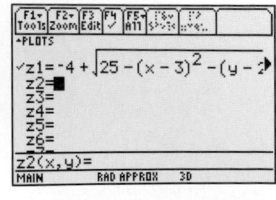

Figure 10.11

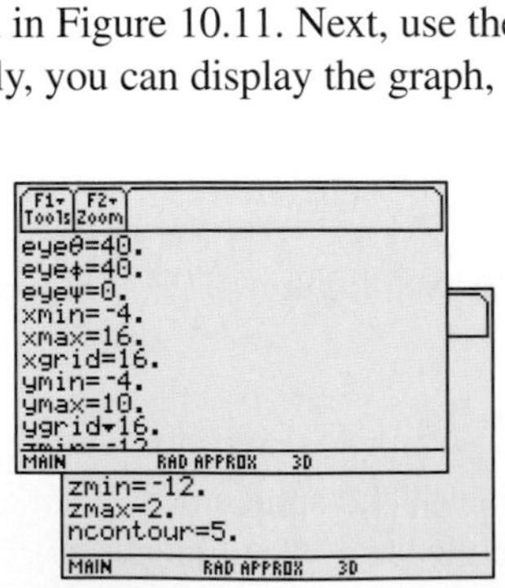

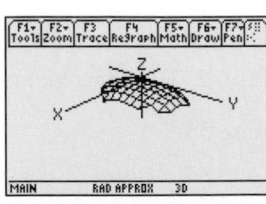

Figure 10.12

Figure 10.13

10.1 Exercises

See www.CalcChat.com for worked-out solutions to odd-numbered exercises.

Vocabulary Check

Fill in the blanks.

1. A _____ coordinate system can be formed by passing a z-axis perpendicular to both the x-axis and the y-axis at the origin.

2. The three coordinate planes of a three-dimensional coordinate system are the _____ , the _____ , and the _____ .

3. The coordinate planes of a three-dimensional coordinate system separate the coordinate system into eight _____ .

4. The distance between the points (x_1, y_1, z_1) and (x_2, y_2, z_2) can be found using the _____ in Space.

5. The midpoint of the line segment joining the points (x_1, y_1, z_1) and (x_2, y_2, z_2) given by the Midpoint Formula in Space is _____ .

6. A _____ is the set of all points (x, y, z) such that the distance between (x, y, z) and a fixed point (h, k, j) is r.

7. A _____ in _____ is the collection of points satisfying an equation involving x, y, and z.

8. The intersection of a surface with one of the three coordinate planes is called a _____ of the surface.

In Exercises 1–4, approximate the coordinates of the points.

1.
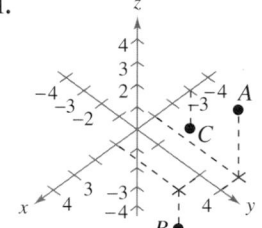

2.
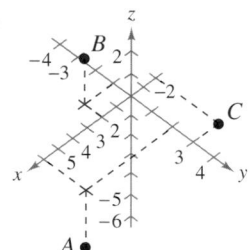

3.
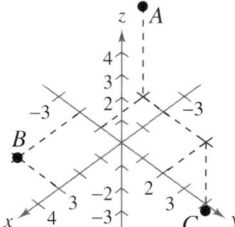

4.
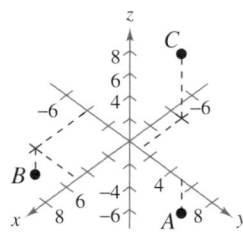

In Exercises 5–10, plot each point in the same three-dimensional coordinate system.

5. (a) $(2, 1, 3)$
   (b) $(-1, 2, 1)$
7. (a) $(3, -1, 0)$
   (b) $(-4, 2, 2)$
9. (a) $(3, -2, 5)$
   (b) $\left(\frac{3}{2}, 4, -2\right)$

6. (a) $(3, 0, 0)$
   (b) $(-3, -2, -1)$
8. (a) $(0, 4, -3)$
   (b) $(4, 0, 4)$
10. (a) $(5, -2, 2)$
    (b) $(5, -2, -2)$

In Exercises 11–14, find the coordinates of the point.

11. The point is located three units behind the yz-plane, three units to the right of the xz-plane, and four units above the xy-plane.

12. The point is located six units in front of the yz-plane, one unit to the left of the xz-plane, and one unit below the xy-plane.

13. The point is located on the x-axis, 10 units in front of the yz-plane.

14. The point is located in the yz-plane, two units to the right of the xz-plane, and eight units above the xy-plane.

In Exercises 15–20, determine the octant(s) in which (x, y, z) is located so that the condition(s) is (are) satisfied.

15. $x > 0, y < 0, z > 0$ 16. $x < 0, y > 0, z < 0$

17. $z > 0$ 18. $y < 0$

19. $xy < 0$ 20. $yz > 0$

In Exercises 21–26, find the distance between the points.

21.
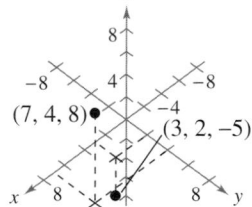

22.
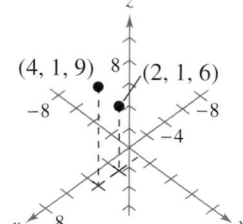

23. $(-1, 4, -2), (6, 0, -9)$
24. $(1, 1, -7), (-2, -3, -7)$
25. $(0, -3, 0), (1, 0, -10)$
26. $(2, -4, 0), (0, 6, -3)$

In Exercises 27–30, find the lengths of the sides of the right triangle. Show that these lengths satisfy the Pythagorean Theorem.

27.

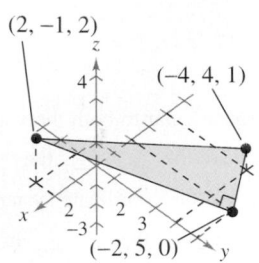

28. $(2, -1, 2)$

29. $(0, 0, 0)$, $(2, 2, 1)$, $(2, -4, 4)$

30. $(1, 0, 1)$, $(1, 3, 1)$, $(1, 0, 3)$

In Exercises 31–34, find the lengths of the sides of the triangle with the indicated vertices, and determine whether the triangle is a right triangle, an isosceles triangle, or neither.

31. $(1, -3, -2)$, $(5, -1, 2)$, $(-1, 1, 2)$

32. $(5, 3, 4)$, $(7, 1, 3)$, $(3, 5, 3)$

33. $(4, -1, -2)$, $(8, 1, 2)$, $(2, 3, 2)$

34. $(1, -2, -1)$, $(3, 0, 0)$, $(3, -6, 3)$

In Exercises 35–40, find the midpoint of the line segment joining the points.

35. $(3, -6, 10)$, $(-3, 4, 4)$

36. $(-1, 5, -3)$, $(3, 7, -1)$

37. $(6, -2, 5)$, $(-4, 2, 6)$

38. $(-3, 5, 5)$, $(-6, 4, 8)$

39. $(-2, 8, 10)$, $(7, -4, 2)$

40. $(9, -5, 1)$, $(9, -2, -4)$

In Exercises 41–50, find the standard form of the equation of the sphere with the given characteristic.

41.

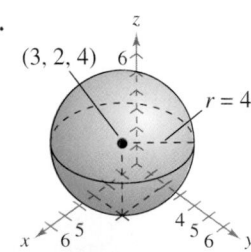

42.

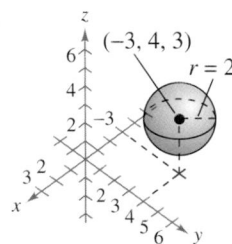

43.

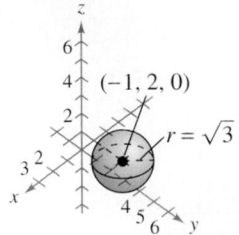

44.

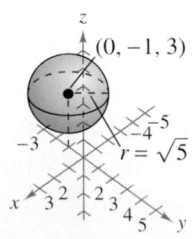

45. Center: $(0, 4, 3)$; radius: 3

46. Center: $(2, -1, 8)$; radius: 6

47. Center: $(-3, 7, 5)$; diameter: 10

48. Center: $(0, 5, -9)$; diameter: 8

49. Endpoints of a diameter: $(3, 0, 0)$, $(0, 0, 6)$

50. Endpoints of a diameter: $(2, -2, 2)$, $(-1, 4, 6)$

In Exercises 51–64, find the center and radius of the sphere.

51. $x^2 + y^2 + z^2 - 5x = 0$

52. $x^2 + y^2 + z^2 - 8y = 0$

53. $x^2 + y^2 + z^2 - 4x + 2y = 0$

54. $x^2 + y^2 + z^2 - x - y - z = 0$

55. $x^2 + y^2 + z^2 - 4x + 2y - 6z + 10 = 0$

56. $x^2 + y^2 + z^2 - 6x + 4y + 9 = 0$

57. $x^2 + y^2 + z^2 + 4x - 8z + 19 = 0$

58. $x^2 + y^2 + z^2 - 8y - 6z + 13 = 0$

59. $9x^2 + 9y^2 + 9z^2 - 18x - 6y - 72z + 73 = 0$

60. $2x^2 + 2y^2 + 2z^2 - 2x - 6y - 4z + 5 = 0$

61. $4x^2 + 4y^2 + 4z^2 - 8x + 16y - 1 = 0$

62. $9x^2 + 9y^2 + 9z^2 - 18x + 36y + 54 - 126 = 0$

63. $9x^2 + 9y^2 + 9z^2 - 6x + 18y + 1 = 0$

64. $4x^2 + 4y^2 + 4z^2 - 4x - 32y + 8z + 33 = 0$

In Exercises 65–70, sketch the graph of the equation and sketch the specified trace.

65. $(x - 1)^2 + y^2 + z^2 = 36$; xz-trace

66. $x^2 + (y + 3)^2 + z^2 = 25$; yz-trace

67. $(x + 2)^2 + (y - 3)^2 + z^2 = 9$; yz-trace

68. $x^2 + (y - 1)^2 + (z + 1)^2 = 4$; xy-trace

69. $x^2 + y^2 + z^2 - 2x - 4z + 1 = 0$; yz-trace

70. $x^2 + y^2 + z^2 - 4y - 6z - 12 = 0$; xz-trace

In Exercises 71–74, use a three-dimensional graphing utility to graph the sphere.

71. $x^2 + y^2 + z^2 - 6x - 8y - 10z + 46 = 0$

72. $x^2 + y^2 + z^2 + 6y - 8z + 21 = 0$

73. $4x^2 + 4y^2 + 4z^2 - 8x - 16y + 8z - 25 = 0$

74. $9x^2 + 9y^2 + 9z^2 + 18x - 18y + 36z + 35 = 0$

75. *Crystals* Crystals are classified according to their symmetry. Crystals shaped like cubes are classified as isometric. The vertices of an isometric crystal mapped onto a three-dimensional coordinate system are shown in the figure on the next page. Determine (x, y, z).

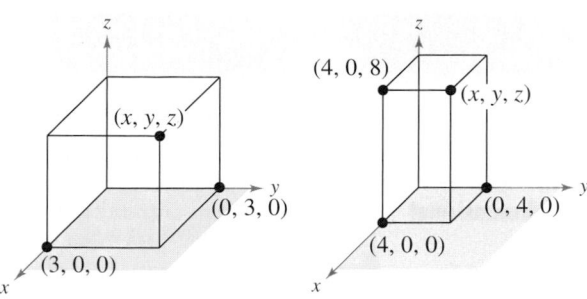

Figure for 75   Figure for 76

76. Crystals Crystals shaped like rectangular prisms are classified as tetragonal. The vertices of a tetragonal crystal mapped onto a three-dimensional coordinate system are shown in the figure. Determine (x, y, z).

77. Architecture A spherical building has a diameter of 165 feet. The center of the building is placed at the origin of a three-dimensional coordinate system. What is the equation of the sphere?

78. Geography Assume that Earth is a sphere with a radius of 3963 miles. The center of Earth is placed at the origin of a three-dimensional coordinate system.

(a) What is the equation of the sphere?

(b) Lines of longitude that run north-south could be represented by what trace(s)? What shape would each of these traces form?

(c) Lines of latitude that run east-west could be represented by what trace(s)? What shape would each of these traces form?

Synthesis

True or False? **In Exercises 79 and 80, determine whether the statement is true or false. Justify your answer.**

79. In the ordered triple (x, y, z) that represents point P in space, x is the directed distance from the xy-plane to P.

80. The surface consisting of all points (x, y, z) in space that are the same distance r from the point (h, k, j) has a circle as its xy-trace.

81. Think About It What is the z-coordinate of any point in the xy-plane? What is the y-coordinate of any point in the xz-plane? What is the x-coordinate of any point in the yz-plane?

82. Writing In two-dimensional coordinate geometry, the graph of the equation $ax + by + c = 0$ is a line. In three-dimensional coordinate geometry, what is the graph of the equation $ax + by + cz = 0$? Is it a line? Explain your reasoning.

83. A sphere intersects the yz-plane. Describe the trace.

84. A plane intersects the xy-plane. Describe the trace.

85. A line segment has (x_1, y_1, z_1) as one endpoint and (x_m, y_m, z_m) as its midpoint. Find the other endpoint (x_2, y_2, z_2) of the line segment in terms of $x_1, y_1, z_1, x_m, y_m,$ and z_m.

86. Use the result of Exercise 85 to find the coordinates of one endpoint of a line segment if the coordinates of the other endpoint and the midpoint are $(3, 0, 2)$ and $(5, 8, 7)$, respectively.

Skills Review

In Exercises 87–92, solve the quadratic equation by completing the square.

87. $v^2 + 3v - 2 = 0$ **88.** $z^2 - 7z - 19 = 0$

89. $x^2 - 5x + 5 = 0$ **90.** $x^2 + 3x - 1 = 0$

91. $4y^2 + 4y - 9 = 0$ **92.** $2x^2 + 5x - 8 = 0$

In Exercises 93–96, find the magnitude and direction angle of the vector v.

93. $\mathbf{v} = 3\mathbf{i} - 3\mathbf{j}$ **94.** $\mathbf{v} = -\mathbf{i} + 2\mathbf{j}$

95. $\mathbf{v} = 4\mathbf{i} + 5\mathbf{j}$ **96.** $\mathbf{v} = 10\mathbf{i} - 7\mathbf{j}$

In Exercises 97 and 98, find the dot product of u and v.

97. $\mathbf{u} = \langle -4, 1 \rangle$ **98.** $\mathbf{u} = \langle -1, 0 \rangle$

   $\mathbf{v} = \langle 3, 5 \rangle$ $\mathbf{v} = \langle -2, -6 \rangle$

In Exercises 99–102, write the first five terms of the sequence beginning with the given term. Then calculate the first and second differences of the sequence. State whether the sequence has a linear model, a quadratic model, or neither.

99. $a_0 = 1$ **100.** $a_0 = 0$

   $a_n = a_{n-1} + n^2$ $a_n = a_{n-1} - 1$

101. $a_1 = -1$ **102.** $a_1 = 4$

   $a_n = a_{n-1} + 3$ $a_n = a_{n-1} - 2n$

In Exercises 103–110, find the standard form of the equation of the conic with the given characteristics.

103. Circle: center: $(-5, 1)$; radius: 7

104. Circle: center: $(3, -6)$; radius: 9

105. Parabola: vertex: $(4, 1)$; focus: $(1, 1)$

106. Parabola: vertex: $(-2, 5)$; focus: $(-2, 0)$

107. Ellipse: vertices: $(0, 3)$, $(6, 3)$;
   minor axis of length 4

108. Ellipse: foci: $(0, 0)$, $(0, 6)$; major axis of length 9

109. Hyperbola: vertices: $(4, 0)$, $(8, 0)$;
   foci: $(0, 0)$, $(12, 0)$

110. Hyperbola: vertices: $(3, 1)$, $(3, 9)$;
   foci: $(3, 0)$, $(3, 10)$

10.2 Vectors in Space

Vectors in Space

Physical forces and velocities are not confined to the plane, so it is natural to extend the concept of vectors from two-dimensional space to three-dimensional space. In space, vectors are denoted by ordered triples

$$\mathbf{v} = \langle v_1, v_2, v_3 \rangle. \qquad \text{Component form}$$

The **zero vector** is denoted by $\mathbf{0} = \langle 0, 0, 0 \rangle$. Using the unit vectors $\mathbf{i} = \langle 1, 0, 0 \rangle$, $\mathbf{j} = \langle 0, 1, 0 \rangle$, and $\mathbf{k} = \langle 0, 0, 1 \rangle$ in the direction of the positive z-axis, the **standard unit vector notation** for $\mathbf{v}$ is

$$\mathbf{v} = v_1\mathbf{i} + v_2\mathbf{j} + v_3\mathbf{k} \qquad \text{Unit vector form}$$

as shown in Figure 10.14. If $\mathbf{v}$ is represented by the directed line segment from $P(p_1, p_2, p_3)$ to $Q(q_1, q_2, q_3)$, as shown in Figure 10.15, the **component form** of $\mathbf{v}$ is produced by subtracting the coordinates of the initial point from the coordinates of the terminal point

$$\mathbf{v} = \langle v_1, v_2, v_3 \rangle = \langle q_1 - p_1, q_2 - p_2, q_3 - p_3 \rangle.$$

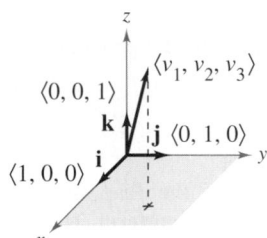

Figure 10.14

Figure 10.15

Vectors in space can be used to represent many physical forces, such as tension in the wires used to support auditorium lights, as shown in Exercise 71 on page 756.

SuperStock

What you should learn
- Find the component forms of, the unit vectors in the same direction of, the magnitudes of, the dot products of, and the angles between vectors in space.
- Determine whether vectors in space are parallel or orthogonal.
- Use vectors in space to solve real-life problems.

Why you should learn it

Vectors in Space

1. Two vectors are **equal** if and only if their corresponding components are equal.

2. The **magnitude** (or **length**) of $\mathbf{u} = \langle u_1, u_2, u_3 \rangle$ is
$$\|\mathbf{u}\| = \sqrt{u_1^2 + u_2^2 + u_3^2}.$$

3. A **unit vector** $\mathbf{u}$ in the direction of $\mathbf{v}$ is $\mathbf{u} = \dfrac{\mathbf{v}}{\|\mathbf{v}\|},\ \mathbf{v} \neq \mathbf{0}$.

4. The **sum** of $\mathbf{u} = \langle u_1, u_2, u_3 \rangle$ and $\mathbf{v} = \langle v_1, v_2, v_3 \rangle$ is
$$\mathbf{u} + \mathbf{v} = \langle u_1 + v_1, u_2 + v_2, u_3 + v_3 \rangle. \quad \text{Vector addition}$$

5. The **scalar multiple** of the real number c and $\mathbf{u} = \langle u_1, u_2, u_3 \rangle$ is
$$c\mathbf{u} = \langle cu_1, cu_2, cu_3 \rangle. \quad \text{Scalar multiplication}$$

6. The **dot product** of $\mathbf{u} = \langle u_1, u_2, u_3 \rangle$ and $\mathbf{v} = \langle v_1, v_2, v_3 \rangle$ is
$$\mathbf{u} \cdot \mathbf{v} = u_1v_1 + u_2v_2 + u_3v_3. \quad \text{Dot product}$$

Prerequisite Skills

To review definitions of vectors in the plane, see Section 6.3.

Example 1   Finding the Component Form of a Vector

Find the component form and magnitude of the vector **v** having initial point $(3, 4, 2)$ and terminal point $(3, 6, 4)$. Then find a unit vector in the direction of **v**.

Solution

The component form of **v** is

$$\mathbf{v} = \langle 3 - 3, 6 - 4, 4 - 2 \rangle = \langle 0, 2, 2 \rangle$$

which implies that its magnitude is

$$\|\mathbf{v}\| = \sqrt{0^2 + 2^2 + 2^2} = \sqrt{8} = 2\sqrt{2}.$$

The unit vector in the direction of **v** is

$$\mathbf{u} = \frac{\mathbf{v}}{\|\mathbf{v}\|} = \frac{1}{2\sqrt{2}}\langle 0, 2, 2 \rangle = \left\langle 0, \frac{1}{\sqrt{2}}, \frac{1}{\sqrt{2}} \right\rangle = \left\langle 0, \frac{\sqrt{2}}{2}, \frac{\sqrt{2}}{2} \right\rangle.$$

✔CHECKPOINT   Now try Exercise 5.

TECHNOLOGY TIP

Some graphing utilities have the capability to perform vector operations, such as the dot product. Consult the user's guide for your graphing utility for specific instructions.

Example 2   Finding the Dot Product of Two Vectors

Find the dot product of $\langle 4, 0, 1 \rangle$ and $\langle -1, 3, 2 \rangle$.

Solution

$$\langle 4, 0, 1 \rangle \cdot \langle -1, 3, 2 \rangle = 4(-1) + 0(3) + 1(2)$$

$$= -4 + 0 + 2 = -2$$

Note that the dot product of two vectors is a real number, not a vector.

✔CHECKPOINT   Now try Exercise 39.

As was discussed in Section 6.4, the **angle between two nonzero vectors** is the angle θ, $0 \le \theta \le \pi$, between their respective standard position vectors, as shown in Figure 10.16. This angle can be found using the dot product. (Note that the angle between the zero vector and another vector is not defined.)

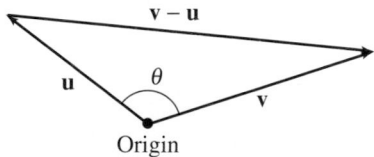

Figure 10.16

Angle Between Two Vectors

If θ is the angle between two nonzero vectors **u** and **v**, then $\cos \theta = \dfrac{\mathbf{u} \cdot \mathbf{v}}{\|\mathbf{u}\| \|\mathbf{v}\|}$.

If the dot product of two nonzero vectors is zero, the angle between the vectors is 90°. Such vectors are called **orthogonal.** For instance, the standard unit vectors **i**, **j**, and **k** are orthogonal to each other.

Example 3   Finding the Angle Between Two Vectors

Find the angle between $\mathbf{u} = \langle 1, 0, 2 \rangle$ and $\mathbf{v} = \langle 3, 1, 0 \rangle$.

Solution

$$\cos \theta = \frac{\mathbf{u} \cdot \mathbf{v}}{\|\mathbf{u}\| \, \|\mathbf{v}\|} = \frac{\langle 1, 0, 2 \rangle \cdot \langle 3, 1, 0 \rangle}{\|\langle 1, 0, 2 \rangle\| \, \|\langle 3, 1, 0 \rangle\|} = \frac{3}{\sqrt{50}}$$

This implies that the angle between the two vectors is

$$\theta = \arccos \frac{3}{\sqrt{50}} \approx 64.9°$$

as shown in Figure 10.17.

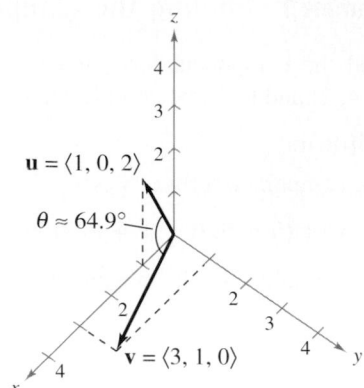

Figure 10.17

✓**CHECKPOINT**   Now try Exercise 43.

Parallel Vectors

Recall from the definition of scalar multiplication that positive scalar multiples of a nonzero vector $\mathbf{v}$ have the same direction as $\mathbf{v}$, whereas negative multiples have the direction opposite that of $\mathbf{v}$. In general, two nonzero vectors $\mathbf{u}$ and $\mathbf{v}$ are **parallel** if there is some scalar c such that $\mathbf{u} = c\mathbf{v}$. For example, in Figure 10.18, the vectors $\mathbf{u}$, $\mathbf{v}$, and $\mathbf{w}$ are parallel because $\mathbf{u} = 2\mathbf{v}$ and $\mathbf{w} = -\mathbf{v}$.

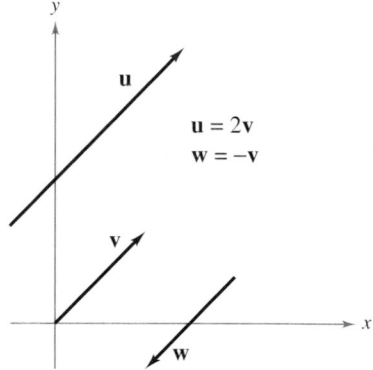

Figure 10.18

Example 4   Parallel Vectors

Vector $\mathbf{w}$ has initial point $(1, -2, 0)$ and terminal point $(3, 2, 1)$. Which of the following vectors is parallel to $\mathbf{w}$?

a. $\mathbf{u} = \langle 4, 8, 2 \rangle$ **b.** $\mathbf{v} = \langle 4, 8, 4 \rangle$

Solution

Begin by writing $\mathbf{w}$ in component form.

$$\mathbf{w} = \langle 3 - 1, 2 - (-2), 1 - 0 \rangle = \langle 2, 4, 1 \rangle$$

a. Because

$$\mathbf{u} = \langle 4, 8, 2 \rangle$$

$$= 2\langle 2, 4, 1 \rangle$$

$$= 2\mathbf{w}$$

you can conclude that $\mathbf{u}$ *is* parallel to $\mathbf{w}$.

b. In this case, you need to find a scalar c such that

$$\langle 4, 8, 4 \rangle = c\langle 2, 4, 1 \rangle.$$

However, equating corresponding components produces $c = 2$ for the first two components and $c = 4$ for the third. So, the equation has no solution, and the vectors $\mathbf{v}$ and $\mathbf{w}$ are *not* parallel.

✓**CHECKPOINT**   Now try Exercise 47.

You can use vectors to determine whether three points are collinear (lie on the same line). The points P, Q, and R are **collinear** if and only if the vectors $\overrightarrow{PQ}$ and $\overrightarrow{PR}$ are parallel.

Example 5   Using Vectors to Determine Collinear Points

Determine whether the points $P(2, -1, 4)$, $Q(5, 4, 6)$, and $R(-4, -11, 0)$ are collinear.

Solution

The component forms of $\overrightarrow{PQ}$ and $\overrightarrow{PR}$ are

$$\overrightarrow{PQ} = \langle 5 - 2, 4 - (-1), 6 - 4 \rangle = \langle 3, 5, 2 \rangle$$

and

$$\overrightarrow{PR} = \langle -4 - 2, -11 - (-1), 0 - 4 \rangle = \langle -6, -10, -4 \rangle.$$

Because $\overrightarrow{PR} = -2\overrightarrow{PQ}$, you can conclude that they are parallel. Therefore, the points P, Q, and R lie on the same line, as shown in Figure 10.19.

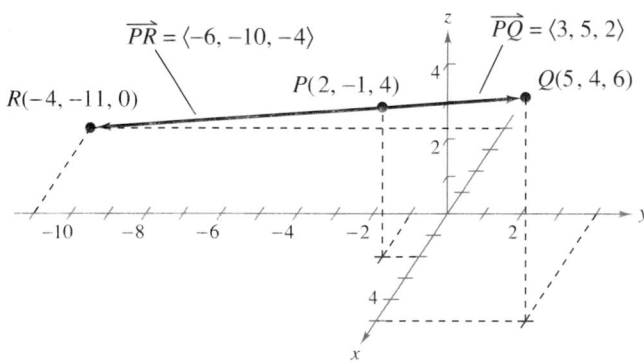

Figure 10.19

✓CHECKPOINT   Now try Exercise 57.

Example 6   Finding the Terminal Point of a Vector

The initial point of the vector $\mathbf{v} = \langle 4, 2, -1 \rangle$ is $P(3, -1, 6)$. What is the terminal point of this vector?

Solution

Using the component form of the vector whose initial point is $P(3, -1, 6)$ and whose terminal point is $Q(q_1, q_2, q_3)$, you can write

$$\overrightarrow{PQ} = \langle q_1 - p_1, q_2 - p_2, q_3 - p_3 \rangle$$

$$= \langle q_1 - 3, q_2 + 1, q_3 - 6 \rangle = \langle 4, 2, -1 \rangle.$$

This implies that $q_1 - 3 = 4$, $q_2 + 1 = 2$, and $q_3 - 6 = -1$. The solutions of these three equations are $q_1 = 7$, $q_2 = 1$, and $q_3 = 5$. So, the terminal point is $Q(7, 1, 5)$.

✓CHECKPOINT   Now try Exercise 63.

Application

In Section 6.3, you saw how to use vectors to solve an equilibrium problem in a plane. The next example shows how to use vectors to solve an equilibrium problem in space.

Example 7 Solving an Equilibrium Problem

A weight of 480 pounds is supported by three ropes. As shown in Figure 10.20, the weight is located at $S(0, 2, -1)$. The ropes are tied to the points $P(2, 0, 0)$, $Q(0, 4, 0)$, and $R(-2, 0, 0)$. Find the force (or tension) on each rope.

Solution

The (downward) force of the weight is represented by the vector

$$\mathbf{w} = \langle 0, 0, -480 \rangle.$$

The force vectors corresponding to the ropes are as follows.

$$\mathbf{u} = \|\mathbf{u}\| \frac{\overrightarrow{SP}}{\|\overrightarrow{SP}\|} = \|\mathbf{u}\| \frac{\langle 2 - 0, 0 - 2, 0 - (-1) \rangle}{3} = \|\mathbf{u}\| \left\langle \frac{2}{3}, -\frac{2}{3}, \frac{1}{3} \right\rangle$$

$$\mathbf{v} = \|\mathbf{v}\| \frac{\overrightarrow{SQ}}{\|\overrightarrow{SQ}\|} = \|\mathbf{v}\| \frac{\langle 0 - 0, 4 - 2, 0 - (-1) \rangle}{\sqrt{5}} = \|\mathbf{v}\| \left\langle 0, \frac{2}{\sqrt{5}}, \frac{1}{\sqrt{5}} \right\rangle$$

$$\mathbf{z} = \|\mathbf{z}\| \frac{\overrightarrow{SR}}{\|\overrightarrow{SR}\|} = \|\mathbf{z}\| \frac{\langle -2 - 0, 0 - 2, 0 - (-1) \rangle}{3} = \|\mathbf{z}\| \left\langle -\frac{2}{3}, -\frac{2}{3}, \frac{1}{3} \right\rangle$$

For the system to be in equilibrium, it must be true that

$$\mathbf{u} + \mathbf{v} + \mathbf{z} + \mathbf{w} = \mathbf{0} \quad \text{or} \quad \mathbf{u} + \mathbf{v} + \mathbf{z} = -\mathbf{w}.$$

This yields the following system of linear equations.

$$\begin{cases} \dfrac{2}{3}\|\mathbf{u}\| \quad\quad\quad\quad -\dfrac{2}{3}\|\mathbf{z}\| = \quad 0 \\[2mm] -\dfrac{2}{3}\|\mathbf{u}\| + \dfrac{2}{\sqrt{5}}\|\mathbf{v}\| - \dfrac{2}{3}\|\mathbf{z}\| = \quad 0 \\[2mm] \dfrac{1}{3}\|\mathbf{u}\| + \dfrac{1}{\sqrt{5}}\|\mathbf{v}\| + \dfrac{1}{3}\|\mathbf{z}\| = 480 \end{cases}$$

Using the techniques demonstrated in Chapter 7, you can find the solution of the system to be

$$\|\mathbf{u}\| = 360.0$$

$$\|\mathbf{v}\| \approx 536.7$$

$$\|\mathbf{z}\| = 360.0.$$

So, the rope attached at point P has 360 pounds of tension, the rope attached at point Q has about 536.7 pounds of tension, and the rope attached at point R has 360 pounds of tension.

✓CHECKPOINT   Now try Exercise 71.

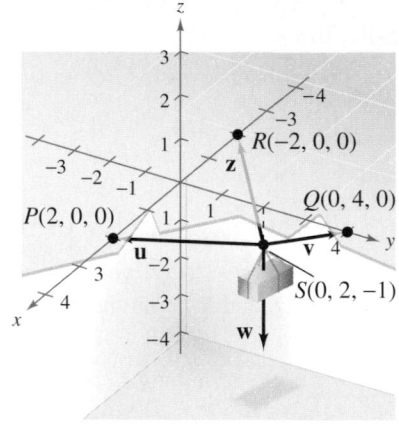

Figure 10.20

10.2 Exercises

See www.CalcChat.com for worked-out solutions to odd-numbered exercises.

Vocabulary Check

Fill in the blanks.

1. The _____ vector is denoted by $\mathbf{0} = \langle 0, 0, 0 \rangle$.

2. The standard unit vector notation for a vector $\mathbf{v}$ in space is given by _____ .

3. The _____ of a vector $\mathbf{v}$ is produced by subtracting the coordinates of the initial point from the corresponding coordinates of the terminal point.

4. If the dot product of two nonzero vectors is zero, the angle between the vectors is 90° and the vectors are called _____ .

5. Two nonzero vectors $\mathbf{u}$ and $\mathbf{v}$ are _____ if there is some scalar c such that $\mathbf{u} = c\mathbf{v}$.

In Exercises 1–4, (a) find the component form of the vector v and (b) sketch the vector with its initial point at the origin.

1.
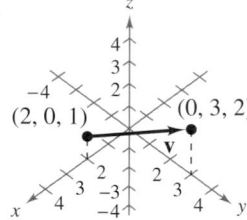

2.
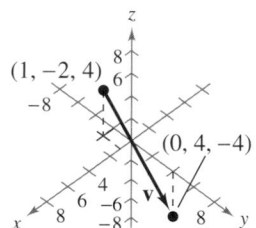

3.
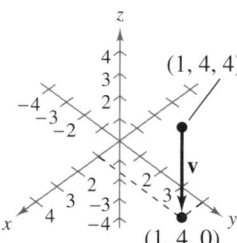

4.
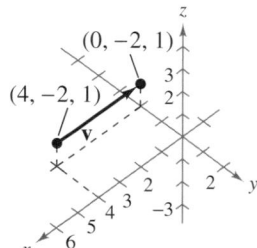

In Exercises 5–8, (a) write the component form of the vector v, (b) find the magnitude of v, and (c) find a unit vector in the direction of v.

Initial point	Terminal point
5. $(-6, 4, -2)$	$(1, -1, 3)$
6. $(-7, 3, 5)$	$(0, 0, 2)$
7. $(-1, 2, -4)$	$(1, 4, -4)$
8. $(0, -1, 0)$	$(0, 2, 1)$

In Exercises 9–12, sketch each scalar multiple of v.

9. $\mathbf{v} = \langle 1, 1, 3 \rangle$

   (a) $2\mathbf{v}$ (b) $-\mathbf{v}$ (c) $\frac{3}{2}\mathbf{v}$ (d) $0\mathbf{v}$

10. $\mathbf{v} = \langle -1, 2, 2 \rangle$

   (a) $-\mathbf{v}$ (b) $2\mathbf{v}$ (c) $\frac{1}{2}\mathbf{v}$ (d) $\frac{5}{2}\mathbf{v}$

11. $\mathbf{v} = 2\mathbf{i} + 2\mathbf{j} - \mathbf{k}$

   (a) $2\mathbf{v}$ (b) $-\mathbf{v}$ (c) $\frac{5}{2}\mathbf{v}$ (d) $0\mathbf{v}$

12. $\mathbf{v} = \mathbf{i} - 2\mathbf{j} + \mathbf{k}$

   (a) $4\mathbf{v}$ (b) $-2\mathbf{v}$ (c) $\frac{1}{2}\mathbf{v}$ (d) $0\mathbf{v}$

In Exercises 13–20, find the vector z, given $\mathbf{u} = \langle -1, 3, 2 \rangle$, $\mathbf{v} = \langle 1, -2, -2 \rangle$, and $\mathbf{w} = \langle 5, 0, -5 \rangle$. Use a graphing utility to verify your answer.

13. $\mathbf{z} = \mathbf{u} - 2\mathbf{v}$ 14. $\mathbf{z} = 7\mathbf{u} + \mathbf{v} - \frac{1}{5}\mathbf{w}$

15. $2\mathbf{z} - 4\mathbf{u} = \mathbf{w}$ 16. $\mathbf{u} + \mathbf{v} + \mathbf{z} = \mathbf{0}$

17. $\mathbf{z} = 2\mathbf{u} - 3\mathbf{v} + \frac{1}{2}\mathbf{w}$ 18. $\mathbf{z} = 3\mathbf{w} - 2\mathbf{v} + \mathbf{u}$

19. $4\mathbf{z} = 4\mathbf{w} - \mathbf{u} + \mathbf{v}$ 20. $\mathbf{u} + 2\mathbf{v} + \mathbf{z} = \mathbf{w}$

In Exercises 21–30, find the magnitude of v.

21. $\mathbf{v} = \langle 7, 8, 7 \rangle$ 22. $\mathbf{v} = \langle -2, 0, -5 \rangle$

23. $\mathbf{v} = \langle 1, -2, 4 \rangle$ 24. $\mathbf{v} = \langle -1, 0, 3 \rangle$

25. $\mathbf{v} = 2\mathbf{i} - 4\mathbf{j} + \mathbf{k}$ 26. $\mathbf{v} = \mathbf{i} + 3\mathbf{j} - \mathbf{k}$

27. $\mathbf{v} = 4\mathbf{i} - 3\mathbf{j} - 7\mathbf{k}$ 28. $\mathbf{v} = 2\mathbf{i} - \mathbf{j} + 6\mathbf{k}$

29. Initial point: $(1, -3, 4)$; terminal point: $(1, 0, -1)$

30. Initial point: $(0, -1, 0)$; terminal point: $(1, 2, -2)$

In Exercises 31–34, find a unit vector (a) in the direction of u and (b) in the direction opposite of u.

31. $\mathbf{u} = 5\mathbf{i} - 12\mathbf{k}$ 32. $\mathbf{u} = 3\mathbf{i} - 4\mathbf{k}$

33. $\mathbf{u} = 8\mathbf{i} + 3\mathbf{j} - \mathbf{k}$ 34. $\mathbf{u} = -3\mathbf{i} + 5\mathbf{j} + 10\mathbf{k}$

In Exercises 35–38, use a graphing utility to determine the specified quantity where $\mathbf{u} = \langle -1, 3, 4 \rangle$ and $\mathbf{v} = \langle 5, 4.5, -6 \rangle$.

35. $6\mathbf{u} - 4\mathbf{v}$ 36. $2\mathbf{u} + \frac{5}{2}\mathbf{v}$

37. $\| \mathbf{u} + \mathbf{v} \|$ 38. $\dfrac{\mathbf{v}}{\| \mathbf{v} \|}$

In Exercises 39–42, find the dot product of u and v.

39. $u = \langle 4, 4, -1 \rangle$
$v = \langle 2, -5, -8 \rangle$

40. $u = \langle 3, -1, 6 \rangle$
$v = \langle 4, -10, 1 \rangle$

41. $u = 2i - 5j + 3k$
$v = 9i + 3j - k$

42. $u = 3j - 6k$
$v = 6i - 4j - 2k$

In Exercises 43–46, find the angle θ between the vectors.

43. $u = \langle 0, 2, 2 \rangle$
$v = \langle 3, 0, -4 \rangle$

44. $u = \langle -1, 3, 0 \rangle$
$v = \langle 1, 2, -1 \rangle$

45. $u = 10i + 40j$
$v = -3j + 8k$

46. $u = 8j - 20k$
$v = 10i - 5k$

In Exercises 47–54, determine whether u and v are orthogonal, parallel, or neither.

47. $u = \langle -12, 6, 15 \rangle$
$v = \langle 8, -4, -10 \rangle$

48. $u = \langle -1, 3, -1 \rangle$
$v = \langle 2, -1, 5 \rangle$

49. $u = \frac{3}{4}i - \frac{1}{2}j + 2k$
$v = 4i + 10j + k$

50. $u = -i + \frac{1}{2}j - k$
$v = 8i - 4j + 8k$

51. $u = j + 6k$
$v = i - 2j - k$

52. $u = 4i - k$
$v = i$

53. $u = -2i + 3j - k$
$v = 2i + j - k$

54. $u = 2i - 3j + k$
$v = -i - j - k$

In Exercises 55–58, use vectors to determine whether the points are collinear.

55. $(5, 4, 1), (7, 3, -1), (4, 5, 3)$

56. $(-2, 7, 4), (-4, 8, 1), (0, 6, 7)$

57. $(1, 3, 2), (-1, 2, 5), (3, 4, -1)$

58. $(0, 4, 4), (-1, 5, 6), (-2, 6, 7)$

In Exercises 59–62, the vertices of a triangle are given. Determine whether the traingle is an acute triangle, an obtuse triangle, or a right triangle. Explain your reasoning.

59. $(1, 2, 0), (0, 0, 0), (-2, 1, 0)$

60. $(-3, 0, 0), (0, 0, 0), (1, 2, 3)$

61. $(2, -3, 4), (0, 1, 2), (-1, 2, 0)$

62. $(2, -7, 3), (-1, 5, 8), (4, 6, -1)$

In Exercises 63–66, the vector v and its initial point are given. Find the terminal point.

63. $v = \langle 2, -4, 7 \rangle$
Initial point: $(1, 5, 0)$

64. $v = \langle 4, -1, -1 \rangle$
Initial point: $(6, -4, 3)$

65. $v = \langle 4, \frac{3}{2}, -\frac{1}{4} \rangle$
Initial point: $\left(2, 1, -\frac{3}{2}\right)$

66. $v = \langle \frac{5}{2}, -\frac{1}{2}, 4 \rangle$
Initial point: $\left(3, 2, -\frac{1}{2}\right)$

67. Determine the values of c such that $\|cu\| = 3$, where $u = i + 2j + 3k$.

68. Determine the values of c such that $\|cu\| = 12$, where $u = -2i + 2j - 4k$.

In Exercises 69 and 70, write the component form of v.

69. v lies in the yz-plane, has magnitude 4, and makes an angle of $45°$ with the positive y-axis.

70. v lies in the xz-plane, has magnitude 10, and makes an angle of $60°$ with the positive z-axis.

71. *Tension* The lights in an auditorium are 30-pound disks of radius 24 inches. Each disk is supported by three equally spaced 60-inch wires attached to the ceiling (see figure). Find the tension in each wire.

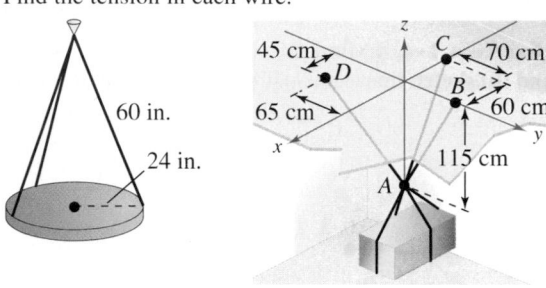

Figure for 71   Figure for 72

72. *Tension* The weight of a crate is 500 newtons. Find the tension in each of the supporting cables shown in the figure.

Synthesis

True or False? **In Exercises 73 and 74, determine whether the statement is true or false. Justify your answer.**

73. If the dot product of two nonzero vectors is zero, then the angle between the vectors is a right angle.

74. If AB and AC are parallel vectors, then points A, B, and C are collinear.

75. *Exploration* Let $u = i + j$, $v = j + k$,   and $w = au + bv$.

   (a) Sketch u and v.

   (b) If $w = 0$, show that a and b must both be zero.

   (c) Find a and b such that $w = i + 2j + k$.

   (d) Show that no choice of a and b yields $w = i + 2j + 3k$.

76. *Think About It* The initial and terminal points of v are (x_1, y_1, z_1) and (x, y, z), respectively. Describe the set of all points (x, y, z) such that $\|v\| = 4$.

77. What is known about the nonzero vectors u and v if $u \cdot v < 0$? Explain.

78. *Writing* Consider the two nonzero vectors u and v. Describe the geometric figure generated by the terminal points of the vectors tv, $u + tv$, and $su + tv$, where s and t represent real numbers.

10.3 The Cross Product of Two Vectors

The Cross Product

Many applications in physics, engineering, and geometry involve finding a vector in space that is orthogonal to two given vectors. In this section, you will study a product that will yield such a vector. It is called the **cross product,** and it is most conveniently defined and calculated using the standard unit vector form.

Definition of Cross Product of Two Vectors in Space

Let

$$\mathbf{u} = u_1\mathbf{i} + u_2\mathbf{j} + u_3\mathbf{k} \qquad \text{and} \qquad \mathbf{v} = v_1\mathbf{i} + v_2\mathbf{j} + v_3\mathbf{k}$$

be vectors in space. The **cross product** of $\mathbf{u}$ and $\mathbf{v}$ is the vector

$$\mathbf{u} \times \mathbf{v} = (u_2v_3 - u_3v_2)\mathbf{i} - (u_1v_3 - u_3v_1)\mathbf{j} + (u_1v_2 - u_2v_1)\mathbf{k}.$$

It is important to note that this definition applies only to three-dimensional vectors. The cross product is not defined for two-dimensional vectors.

A convenient way to calculate $\mathbf{u} \times \mathbf{v}$ is to use the following *determinant form* with cofactor expansion. (This 3×3 determinant form is used simply to help remember the formula for the cross product—it is technically not a determinant because the entries of the corresponding matrix are not all real numbers.)

$$\mathbf{u} \times \mathbf{v} = \begin{vmatrix} \mathbf{i} & \mathbf{j} & \mathbf{k} \\ u_1 & u_2 & u_3 \\ v_1 & v_2 & v_3 \end{vmatrix}$$

⇐ Put $\mathbf{u}$ in Row 2.
⇐ Put $\mathbf{v}$ in Row 3.

$$= \begin{vmatrix} u_2 & u_3 \\ v_2 & v_3 \end{vmatrix}\mathbf{i} - \begin{vmatrix} u_1 & u_3 \\ v_1 & v_3 \end{vmatrix}\mathbf{j} + \begin{vmatrix} u_1 & u_2 \\ v_1 & v_2 \end{vmatrix}\mathbf{k}$$

$$= (u_2v_3 - u_3v_2)\mathbf{i} - (u_1v_3 - u_3v_1)\mathbf{j} + (u_1v_2 - u_2v_1)\mathbf{k}$$

Note the minus sign in front of the $\mathbf{j}$-component. Recall from Section 7.7 that each of the three 2×2 determinants can be evaluated by using the following pattern.

$$\begin{vmatrix} a_1 & b_1 \\ a_2 & b_2 \end{vmatrix} = a_1b_2 - a_2b_1$$

What you should learn

■ Find cross products of vectors in space.
■ Use geometric properties of cross products of vectors in space.
■ Use triple scalar products to find volumes of parallelepipeds.

Why you should learn it

The cross product of two vectors in space has many applications in physics and engineering. For instance, in Exercise 57 on page 763, the cross product is used to find the torque on the crank of a bicycle's brake.

Alamy

Exploration

Find each cross product. What can you conclude?

a. $\mathbf{i} \times \mathbf{j}$ **b.** $\mathbf{i} \times \mathbf{k}$ **c.** $\mathbf{j} \times \mathbf{k}$

Example 1 Finding Cross Products

Given $\mathbf{u} = \mathbf{i} + 2\mathbf{j} + \mathbf{k}$ and $\mathbf{v} = 3\mathbf{i} + \mathbf{j} + 2\mathbf{k}$, find each cross product.

a. $\mathbf{u} \times \mathbf{v}$ **b.** $\mathbf{v} \times \mathbf{u}$ **c.** $\mathbf{v} \times \mathbf{v}$

Solution

a. $\mathbf{u} \times \mathbf{v} = \begin{vmatrix} \mathbf{i} & \mathbf{j} & \mathbf{k} \\ 1 & 2 & 1 \\ 3 & 1 & 2 \end{vmatrix}$

$\qquad = \begin{vmatrix} 2 & 1 \\ 1 & 2 \end{vmatrix} \mathbf{i} - \begin{vmatrix} 1 & 1 \\ 3 & 2 \end{vmatrix} \mathbf{j} + \begin{vmatrix} 1 & 2 \\ 3 & 1 \end{vmatrix} \mathbf{k}$

$\qquad = (4 - 1)\mathbf{i} - (2 - 3)\mathbf{j} + (1 - 6)\mathbf{k}$

$\qquad = 3\mathbf{i} + \mathbf{j} - 5\mathbf{k}$

b. $\mathbf{v} \times \mathbf{u} = \begin{vmatrix} \mathbf{i} & \mathbf{j} & \mathbf{k} \\ 3 & 1 & 2 \\ 1 & 2 & 1 \end{vmatrix}$

$\qquad = \begin{vmatrix} 1 & 2 \\ 2 & 1 \end{vmatrix} \mathbf{i} - \begin{vmatrix} 3 & 2 \\ 1 & 1 \end{vmatrix} \mathbf{j} + \begin{vmatrix} 3 & 1 \\ 1 & 2 \end{vmatrix} \mathbf{k}$

$\qquad = (1 - 4)\mathbf{i} - (3 - 2)\mathbf{j} + (6 - 1)\mathbf{k}$

$\qquad = -3\mathbf{i} - \mathbf{j} + 5\mathbf{k}$

Note that this result is the negative of that in part (a).

c. $\mathbf{v} \times \mathbf{v} = \begin{vmatrix} \mathbf{i} & \mathbf{j} & \mathbf{k} \\ 3 & 1 & 2 \\ 3 & 1 & 2 \end{vmatrix} = \mathbf{0}$

✓**CHECKPOINT**   Now try Exercise 1.

TECHNOLOGY TIP

Some graphing utilities have the capability to perform vector operations, such as the cross product. Consult the user's guide for your graphing utility for specific instructions.

Exploration

Calculate $\mathbf{u} \times \mathbf{v}$ and $-(\mathbf{v} \times \mathbf{u})$ for several values of $\mathbf{u}$ and $\mathbf{v}$. What do your results imply? Interpret your results geometrically.

The results obtained in Example 1 suggest some interesting algebraic properties of the cross product. For instance,

$$\mathbf{u} \times \mathbf{v} = -(\mathbf{v} \times \mathbf{u}) \qquad \text{and} \qquad \mathbf{v} \times \mathbf{v} = \mathbf{0}.$$

These properties, and several others, are summarized in the following list.

Algebraic Properties of the Cross Product   (See the proof on page 777.)

Let $\mathbf{u}$, $\mathbf{v}$, and $\mathbf{w}$ be vectors in space and let c be a scalar.

1. $\mathbf{u} \times \mathbf{v} = -(\mathbf{v} \times \mathbf{u})$
2. $\mathbf{u} \times (\mathbf{v} + \mathbf{w}) = (\mathbf{u} \times \mathbf{v}) + (\mathbf{u} \times \mathbf{w})$
3. $c(\mathbf{u} \times \mathbf{v}) = (c\mathbf{u}) \times \mathbf{v} = \mathbf{u} \times (c\mathbf{v})$
4. $\mathbf{u} \times \mathbf{0} = \mathbf{0} \times \mathbf{u} = \mathbf{0}$
5. $\mathbf{u} \times \mathbf{u} = \mathbf{0}$
6. $\mathbf{u} \cdot (\mathbf{v} \times \mathbf{w}) = (\mathbf{u} \times \mathbf{v}) \cdot \mathbf{w}$

Geometric Properties of the Cross Product

The first property listed on the preceding page indicates that the cross product is *not commutative*. In particular, this property indicates that the vectors $\mathbf{u} \times \mathbf{v}$ and $\mathbf{v} \times \mathbf{u}$ have equal lengths but opposite directions. The following list gives some other *geometric* properties of the cross product of two vectors.

Geometric Properties of the Cross Product (See the proof on page 778.)

Let $\mathbf{u}$ and $\mathbf{v}$ be nonzero vectors in space, and let θ be the angle between $\mathbf{u}$ and $\mathbf{v}$.

1. $\mathbf{u} \times \mathbf{v}$ is orthogonal to both $\mathbf{u}$ and $\mathbf{v}$.
2. $\|\mathbf{u} \times \mathbf{v}\| = \|\mathbf{u}\| \, \|\mathbf{v}\| \sin \theta$
3. $\mathbf{u} \times \mathbf{v} = \mathbf{0}$ if and only if $\mathbf{u}$ and $\mathbf{v}$ are scalar multiples of each other.
4. $\|\mathbf{u} \times \mathbf{v}\|$ = area of parallelogram having $\mathbf{u}$ and $\mathbf{v}$ as adjacent sides.

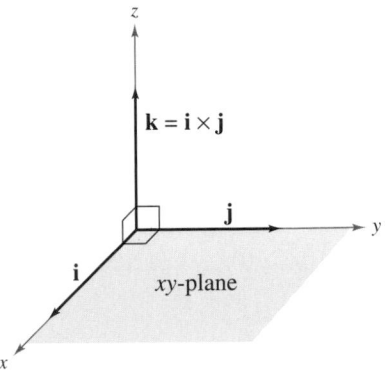

Figure 10.21

Both $\mathbf{u} \times \mathbf{v}$ and $\mathbf{v} \times \mathbf{u}$ are perpendicular to the plane determined by $\mathbf{u}$ and $\mathbf{v}$. One way to remember the orientations of the vectors $\mathbf{u}$, $\mathbf{v}$, and $\mathbf{u} \times \mathbf{v}$ is to compare them with the unit vectors $\mathbf{i}$, $\mathbf{j}$, and $\mathbf{k} = \mathbf{i} \times \mathbf{j}$, as shown in Figure 10.21. The three vectors $\mathbf{u}$, $\mathbf{v}$, and $\mathbf{u} \times \mathbf{v}$ form a *right-handed system*.

Example 2   Using the Cross Product

Find a unit vector that is orthogonal to both

$$\mathbf{u} = 3\mathbf{i} - 4\mathbf{j} + \mathbf{k} \qquad \text{and} \qquad \mathbf{v} = -3\mathbf{i} + 6\mathbf{j}.$$

Solution

The cross product $\mathbf{u} \times \mathbf{v}$, as shown in Figure 10.22, is orthogonal to both $\mathbf{u}$ and $\mathbf{v}$.

$$\mathbf{u} \times \mathbf{v} = \begin{vmatrix} \mathbf{i} & \mathbf{j} & \mathbf{k} \\ 3 & -4 & 1 \\ -3 & 6 & 0 \end{vmatrix}$$

$$= -6\mathbf{i} - 3\mathbf{j} + 6\mathbf{k}$$

Because

$$\|\mathbf{u} \times \mathbf{v}\| = \sqrt{(-6)^2 + (-3)^2 + 6^2}$$

$$= \sqrt{81}$$

$$= 9$$

a unit vector orthogonal to both $\mathbf{u}$ and $\mathbf{v}$ is

$$\frac{\mathbf{u} \times \mathbf{v}}{\|\mathbf{u} \times \mathbf{v}\|} = -\frac{2}{3}\mathbf{i} - \frac{1}{3}\mathbf{j} + \frac{2}{3}\mathbf{k}.$$

✓**CHECKPOINT**   Now try Exercise 29.

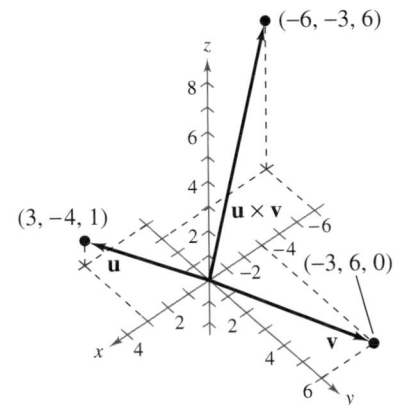

Figure 10.22

In Example 2, note that you could have used the cross product $\mathbf{v} \times \mathbf{u}$ to form a unit vector that is orthogonal to both $\mathbf{u}$ and $\mathbf{v}$. With that choice, you would have obtained the *negative* of the unit vector found in the example.

The fourth geometric property of the cross product states that $\|\mathbf{u} \times \mathbf{v}\|$ is the area of the parallelogram that has $\mathbf{u}$ and $\mathbf{v}$ as adjacent sides. A simple example of this is given by the unit square with adjacent sides of $\mathbf{i}$ and $\mathbf{j}$. Because

$$\mathbf{i} \times \mathbf{j} = \mathbf{k}$$

and $\|\mathbf{k}\| = 1$, it follows that the square has an area of 1. This geometric property of the cross product is illustrated further in the next example.

Example 3   Geometric Application of the Cross Product

Show that the quadrilateral with vertices at the following points is a parallelogram. Then find the area of the parallelogram. Is the parallelogram a rectangle?

$$A(5, 2, 0), \qquad B(2, 6, 1), \qquad C(2, 4, 7), \qquad D(5, 0, 6)$$

Solution

From Figure 10.23 you can see that the sides of the quadrilateral correspond to the following four vectors.

$$\overrightarrow{AB} = -3\mathbf{i} + 4\mathbf{j} + \mathbf{k}$$

$$\overrightarrow{CD} = 3\mathbf{i} - 4\mathbf{j} - \mathbf{k} = -\overrightarrow{AB}$$

$$\overrightarrow{AD} = 0\mathbf{i} - 2\mathbf{j} + 6\mathbf{k}$$

$$\overrightarrow{CB} = 0\mathbf{i} + 2\mathbf{j} - 6\mathbf{k} = -\overrightarrow{AD}$$

Because $\overrightarrow{CD} = -\overrightarrow{AB}$ and $\overrightarrow{CB} = -\overrightarrow{AD}$, you can conclude that $\overrightarrow{AB}$ is parallel to $\overrightarrow{CD}$ and $\overrightarrow{AD}$ is parallel to $\overrightarrow{CB}$. It follows that the quadrilateral is a parallelogram with $\overrightarrow{AB}$ and $\overrightarrow{AD}$ as adjacent sides. Moreover, because

$$\overrightarrow{AB} \times \overrightarrow{AD} = \begin{vmatrix} \mathbf{i} & \mathbf{j} & \mathbf{k} \\ -3 & 4 & 1 \\ 0 & -2 & 6 \end{vmatrix} = 26\mathbf{i} + 18\mathbf{j} + 6\mathbf{k}$$

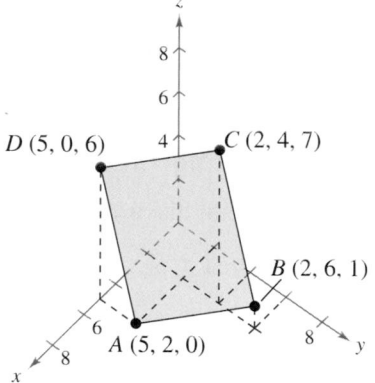

Figure 10.23

the area of the parallelogram is

$$\|\overrightarrow{AB} \times \overrightarrow{AD}\| = \sqrt{26^2 + 18^2 + 6^2} = \sqrt{1036} \approx 32.19.$$

You can tell whether the parallelogram is a rectangle by finding the angle between the vectors $\overrightarrow{AB}$ and $\overrightarrow{AD}$.

$$\sin \theta = \frac{\|\overrightarrow{AB} \times \overrightarrow{AD}\|}{\|\overrightarrow{AB}\| \, \|\overrightarrow{AD}\|}$$

$$= \frac{\sqrt{1036}}{\sqrt{26}\sqrt{40}} \approx 0.998$$

$$\theta = \arcsin 0.998$$

$$\theta \approx 86.4°$$

Because $\theta \neq 90°$, the parallelogram is not a rectangle.

✓CHECKPOINT   Now try Exercise 41.

Exploration

If you connect the terminal points of two vectors $\mathbf{u}$ and $\mathbf{v}$ that have the same initial points, a triangle is formed. Is it possible to use the cross product $\mathbf{u} \times \mathbf{v}$ to determine the area of the triangle? Explain. Verify your conclusion using two vectors from Example 3.

The Triple Scalar Product

For the vectors $\mathbf{u}$, $\mathbf{v}$, and $\mathbf{w}$ in space, the dot product of $\mathbf{u}$ and $\mathbf{v} \times \mathbf{w}$ is called the **triple scalar product** of $\mathbf{u}$, $\mathbf{v}$, and $\mathbf{w}$.

> **The Triple Scalar Product**
>
> For $\mathbf{u} = u_1\mathbf{i} + u_2\mathbf{j} + u_3\mathbf{k}$, $\mathbf{v} = v_1\mathbf{i} + v_2\mathbf{j} + v_3\mathbf{k}$, and $\mathbf{w} = w_1\mathbf{i} + w_2\mathbf{j} + w_3\mathbf{k}$, the **triple scalar product** is given by
>
> $$\mathbf{u} \cdot (\mathbf{v} \times \mathbf{w}) = \begin{vmatrix} u_1 & u_2 & u_3 \\ v_1 & v_2 & v_3 \\ w_1 & w_2 & w_3 \end{vmatrix}.$$

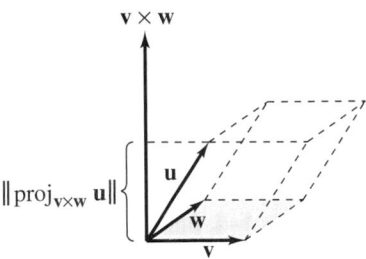

If the vectors $\mathbf{u}$, $\mathbf{v}$, and $\mathbf{w}$ do not lie in the same plane, the triple scalar product $\mathbf{u} \cdot (\mathbf{v} \times \mathbf{w})$ can be used to determine the volume of the parallelepiped (a polyhedron, all of whose faces are parallelograms) with $\mathbf{u}$, $\mathbf{v}$, and $\mathbf{w}$ as adjacent edges, as shown in Figure 10.24.

Area of base $= \|\mathbf{v} \times \mathbf{w}\|$
Volume of
parallelepiped $= |\mathbf{u} \cdot (\mathbf{v} \times \mathbf{w})|$
Figure 10.24

> **Geometric Property of Triple Scalar Product**
>
> The volume V of a parallelepiped with vectors $\mathbf{u}$, $\mathbf{v}$, and $\mathbf{w}$ as adjacent edges is given by
>
> $$V = \left| \mathbf{u} \cdot (\mathbf{v} \times \mathbf{w}) \right|.$$

Example 4   Volume by the Triple Scalar Product

Find the volume of the parallelepiped having

$$\mathbf{u} = 3\mathbf{i} - 5\mathbf{j} + \mathbf{k}, \qquad \mathbf{v} = 2\mathbf{j} - 2\mathbf{k}, \qquad \text{and} \qquad \mathbf{w} = 3\mathbf{i} + \mathbf{j} + \mathbf{k}$$

as adjacent edges, as shown in Figure 10.25.

Solution

The value of the triple scalar product is

$$\mathbf{u} \cdot (\mathbf{v} \times \mathbf{w}) = \begin{vmatrix} 3 & -5 & 1 \\ 0 & 2 & -2 \\ 3 & 1 & 1 \end{vmatrix}$$

$$= 3\begin{vmatrix} 2 & -2 \\ 1 & 1 \end{vmatrix} - (-5)\begin{vmatrix} 0 & -2 \\ 3 & 1 \end{vmatrix} + 1\begin{vmatrix} 0 & 2 \\ 3 & 1 \end{vmatrix}$$

$$= 3(4) + 5(6) + 1(-6)$$

$$= 36.$$

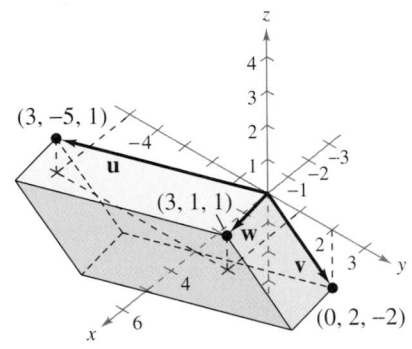

Figure 10.25

So, the volume of the parallelepiped is

$$\left| \mathbf{u} \cdot (\mathbf{v} \times \mathbf{w}) \right| = |36| = 36.$$

✓**CHECKPOINT**   Now try Exercise 53.

10.3 Exercises

See www.CalcChat.com for worked-out solutions to odd-numbered exercises.

Vocabulary Check

Fill in the blanks.

1. To find a vector in space that is orthogonal to two given vectors, find the _____ of the two vectors.

2. $\mathbf{u} \times \mathbf{u} =$ _____

3. $\|\mathbf{u} \times \mathbf{v}\| =$ _____

4. The dot product of $\mathbf{u}$ and $\mathbf{v} \times \mathbf{w}$ is called the _____ of $\mathbf{u}$, $\mathbf{v}$, and $\mathbf{w}$.

In Exercises 1–4, find the cross product of the unit vectors and sketch the result.

1. $\mathbf{j} \times \mathbf{i}$
2. $\mathbf{k} \times \mathbf{j}$
3. $\mathbf{i} \times \mathbf{k}$
4. $\mathbf{k} \times \mathbf{i}$

In Exercises 5–20, find $\mathbf{u} \times \mathbf{v}$ and show that it is orthogonal to both u and v.

5. $\mathbf{u} = \langle 1, -1, 0 \rangle$
   $\mathbf{v} = \langle 0, 1, -1 \rangle$

6. $\mathbf{u} = \langle -1, 1, 0 \rangle$
   $\mathbf{v} = \langle 1, 0, -1 \rangle$

7. $\mathbf{u} = \langle 3, -2, 5 \rangle$
   $\mathbf{v} = \langle 0, -1, 1 \rangle$

8. $\mathbf{u} = \langle 2, -3, 1 \rangle$
   $\mathbf{v} = \langle 1, -2, 1 \rangle$

9. $\mathbf{u} = \langle -10, 0, 6 \rangle$
   $\mathbf{v} = \langle 7, 0, 0 \rangle$

10. $\mathbf{u} = \langle -5, 5, 11 \rangle$
    $\mathbf{v} = \langle 2, 2, 3 \rangle$

11. $\mathbf{u} = 6\mathbf{i} + 2\mathbf{j} + \mathbf{k}$
    $\mathbf{v} = \mathbf{i} + 3\mathbf{j} - 2\mathbf{k}$

12. $\mathbf{u} = \mathbf{i} + \frac{3}{2}\mathbf{j} - \frac{5}{2}\mathbf{k}$
    $\mathbf{v} = \frac{1}{2}\mathbf{i} - \frac{3}{4}\mathbf{j} + \frac{1}{4}\mathbf{k}$

13. $\mathbf{u} = 2\mathbf{i} + 4\mathbf{j} + 3\mathbf{k}$
    $\mathbf{v} = -\mathbf{i} + 3\mathbf{j} - 2\mathbf{k}$

14. $\mathbf{u} = 3\mathbf{i} - 2\mathbf{j} + \mathbf{k}$
    $\mathbf{v} = -2\mathbf{i} + \mathbf{j} + 2\mathbf{k}$

15. $\mathbf{u} = \frac{1}{2}\mathbf{i} - \frac{2}{3}\mathbf{j} + \mathbf{k}$
    $\mathbf{v} = -\frac{3}{4}\mathbf{i} + \mathbf{j} + \frac{1}{4}\mathbf{k}$

16. $\mathbf{u} = \frac{2}{5}\mathbf{i} - \frac{1}{4}\mathbf{j} + \frac{1}{2}\mathbf{k}$
    $\mathbf{v} = -\frac{3}{5}\mathbf{i} + \mathbf{j} + \frac{1}{5}\mathbf{k}$

17. $\mathbf{u} = 6\mathbf{k}$
    $\mathbf{v} = -\mathbf{i} + 3\mathbf{j} + \mathbf{k}$

18. $\mathbf{u} = \frac{2}{3}\mathbf{i}$
    $\mathbf{v} = \frac{1}{3}\mathbf{j} - 3\mathbf{k}$

19. $\mathbf{u} = -\mathbf{i} + \mathbf{k}$
    $\mathbf{v} = \mathbf{j} - 2\mathbf{k}$

20. $\mathbf{u} = \mathbf{i} - 2\mathbf{k}$
    $\mathbf{v} = -\mathbf{j} + \mathbf{k}$

In Exercises 21–26, use a graphing utility to find $\mathbf{u} \times \mathbf{v}$.

21. $\mathbf{u} = \langle 2, 4, 3 \rangle$
    $\mathbf{v} = \langle 0, -2, 1 \rangle$

22. $\mathbf{u} = \langle 4, -2, 6 \rangle$
    $\mathbf{v} = \langle -1, 5, 7 \rangle$

23. $\mathbf{u} = \mathbf{i} - 2\mathbf{j} + 4\mathbf{k}$
    $\mathbf{v} = -4\mathbf{i} + 2\mathbf{j} - \mathbf{k}$

24. $\mathbf{u} = 2\mathbf{i} - \mathbf{j} + 3\mathbf{k}$
    $\mathbf{v} = -\mathbf{i} + \mathbf{j} - 4\mathbf{k}$

25. $\mathbf{u} = 6\mathbf{i} - 5\mathbf{j} + \mathbf{k}$
    $\mathbf{v} = \frac{1}{2}\mathbf{i} - \frac{3}{4}\mathbf{j} + \frac{2}{10}\mathbf{k}$

26. $\mathbf{u} = 8\mathbf{i} - 4\mathbf{j} + 2\mathbf{k}$
    $\mathbf{v} = \frac{1}{2}\mathbf{i} + \frac{3}{4}\mathbf{j} - \frac{1}{4}\mathbf{k}$

In Exercises 27–34, find a unit vector orthogonal to u and v.

27. $\mathbf{u} = \langle 1, 2, 3 \rangle$
    $\mathbf{v} = \langle 2, -3, 0 \rangle$

28. $\mathbf{u} = \langle 2, -1, 3 \rangle$
    $\mathbf{v} = \langle 1, 0, -2 \rangle$

29. $\mathbf{u} = 3\mathbf{i} + \mathbf{j}$
    $\mathbf{v} = \mathbf{j} + \mathbf{k}$

30. $\mathbf{u} = \mathbf{i} + 2\mathbf{j}$
    $\mathbf{v} = \mathbf{i} - 3\mathbf{k}$

31. $\mathbf{u} = -3\mathbf{i} + 2\mathbf{j} - 5\mathbf{k}$
    $\mathbf{v} = \frac{1}{2}\mathbf{i} - \frac{3}{4}\mathbf{j} + \frac{1}{10}\mathbf{k}$

32. $\mathbf{u} = 7\mathbf{i} - 14\mathbf{j} + 5\mathbf{k}$
    $\mathbf{v} = 14\mathbf{i} + 28\mathbf{j} - 15\mathbf{k}$

33. $\mathbf{u} = \mathbf{i} + \mathbf{j} - \mathbf{k}$
    $\mathbf{v} = \mathbf{i} + \mathbf{j} + \mathbf{k}$

34. $\mathbf{u} = \mathbf{i} - 2\mathbf{j} + 2\mathbf{k}$
    $\mathbf{v} = 2\mathbf{i} - \mathbf{j} - 2\mathbf{k}$

In Exercises 35–40, find the area of the parallelogram that has the vectors as adjacent sides.

35. $\mathbf{u} = \mathbf{k}$
    $\mathbf{v} = \mathbf{i} + \mathbf{k}$

36. $\mathbf{u} = \mathbf{i} + 2\mathbf{j} + 2\mathbf{k}$
    $\mathbf{v} = \mathbf{i} + \mathbf{k}$

37. $\mathbf{u} = 3\mathbf{i} + 4\mathbf{j} + 6\mathbf{k}$
    $\mathbf{v} = 2\mathbf{i} - \mathbf{j} + 5\mathbf{k}$

38. $\mathbf{u} = -2\mathbf{i} + 3\mathbf{j} + 2\mathbf{k}$
    $\mathbf{v} = \mathbf{i} + 2\mathbf{j} + 4\mathbf{k}$

39. $\mathbf{u} = \langle 2, 2, -3 \rangle$
    $\mathbf{v} = \langle 0, 2, 3 \rangle$

40. $\mathbf{u} = \langle 4, -3, 2 \rangle$
    $\mathbf{v} = \langle 5, 0, 1 \rangle$

In Exercises 41 and 42, (a) verify that the points are the vertices of a parallelogram, (b) find its area, and (c) determine whether the parallelogram is a rectangle.

41. $A(2, -1, 4)$, $B(3, 1, 2)$, $C(0, 5, 6)$, $D(-1, 3, 8)$

42. $A(1, 1, 1)$, $B(2, 3, 4)$, $C(6, 5, 2)$, $D(7, 7, 5)$

In Exercises 43–46, find the area of the triangle with the given vertices. (The area A of the triangle having u and v as adjacent sides is given by $A = \frac{1}{2}\|\mathbf{u} \times \mathbf{v}\|$.)

43. $(0, 0, 0)$, $(1, 2, 3)$, $(-3, 0, 0)$

44. $(1, -4, 3)$, $(2, 0, 2)$, $(-2, 2, 0)$

45. $(2, 3, -5)$, $(-2, -2, 0)$, $(3, 0, 6)$

46. $(2, 4, 0)$, $(-2, -4, 0)$, $(0, 0, 4)$

In Exercises 47–50, find the triple scalar product.

47. $\mathbf{u} = \langle 2, 3, 3 \rangle$, $\mathbf{v} = \langle 4, 4, 0 \rangle$, $\mathbf{w} = \langle 0, 0, 4 \rangle$

48. $\mathbf{u} = \langle 2, 0, 1 \rangle$, $\mathbf{v} = \langle 0, 3, 0 \rangle$, $\mathbf{w} = \langle 0, 0, 1 \rangle$

49. $\mathbf{u} = 2\mathbf{i} + 3\mathbf{j} + \mathbf{k}$, $\mathbf{v} = \mathbf{i} - \mathbf{j}$, $\mathbf{w} = 4\mathbf{i} + 3\mathbf{j} + \mathbf{k}$

50. $\mathbf{u} = \mathbf{i} + 4\mathbf{j} - 7\mathbf{k}$, $\mathbf{v} = 2\mathbf{i} + 4\mathbf{k}$, $\mathbf{w} = -3\mathbf{j} + 6\mathbf{k}$

In Exercises 51–54, use the triple scalar product to find the volume of the parallelepiped having adjacent edges u, v, and w.

51. $\mathbf{u} = \mathbf{i} + \mathbf{j}$
 $\mathbf{v} = \mathbf{j} + \mathbf{k}$
 $\mathbf{w} = \mathbf{i} + \mathbf{k}$

52. $\mathbf{u} = \mathbf{i} + \mathbf{j} + 3\mathbf{k}$
 $\mathbf{v} = 3\mathbf{j} + 3\mathbf{k}$
 $\mathbf{w} = 3\mathbf{i} + 3\mathbf{k}$

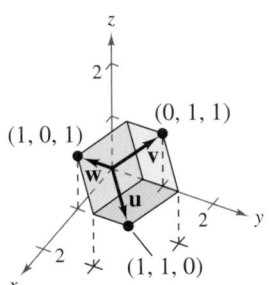

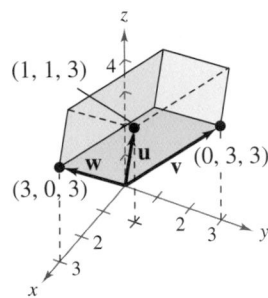

53. $\mathbf{u} = \langle 0, 2, 2 \rangle$
 $\mathbf{v} = \langle 0, 0, -2 \rangle$
 $\mathbf{w} = \langle 3, 0, 2 \rangle$

54. $\mathbf{u} = \langle 1, 2, -1 \rangle$
 $\mathbf{v} = \langle -1, 2, 2 \rangle$
 $\mathbf{w} = \langle 2, 0, 1 \rangle$

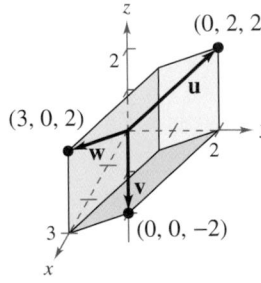

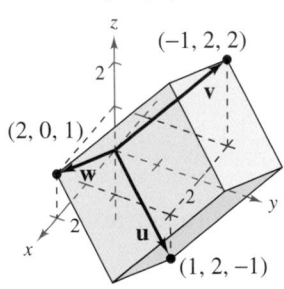

In Exercises 55 and 56, find the volume of the parallelepiped with the given vertices.

55. $A(0, 0, 0)$, $B(4, 0, 0)$, $C(4, -2, 3)$, $D(0, -2, 3)$,
 $E(4, 5, 3)$, $F(0, 5, 3)$, $G(0, 3, 6)$, $H(4, 3, 6)$

56. $A(0, 0, 0)$, $B(1, 1, 0)$, $C(1, 0, 2)$, $D(0, 1, 1)$,
 $E(2, 1, 2)$, $F(1, 1, 3)$, $G(1, 2, 1)$, $H(2, 2, 3)$

57. *Torque*   The brakes on a bicycle are applied by using a downward force of p pounds on the pedal when the six-inch crank makes a $40°$ angle with the horizontal (see figure). Vectors representing the position of the crank and the force are $\mathbf{V} = \frac{1}{2}(\cos 40° \mathbf{j} + \sin 40° \mathbf{k})$ and $\mathbf{F} = -p\mathbf{k}$, respectively.

(a) The magnitude of the torque on the crank is given by $\|\mathbf{V} \times \mathbf{F}\|$. Using the given information, write the torque T on the crank as a function of p.

(b) Use the function from part (a) to complete the table.

p	15	20	25	30	35	40	45
T							

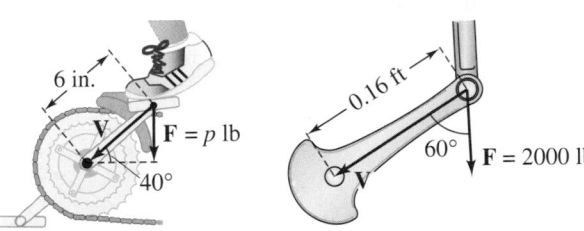

Figure for 57   Figure for 58

58. *Torque*   Both the magnitude and direction of the force on a crankshaft change as the crankshaft rotates. Use the technique given in Exercise 57 to find the magnitude of the torque on the crankshaft using the position and data shown in the figure.

Synthesis

True or False?   **In Exercises 59 and 60, determine whether the statement is true or false. Justify your answer.**

59. The cross product is not defined for vectors in the plane.

60. If $\mathbf{u}$ and $\mathbf{v}$ are vectors in space that are nonzero and not parallel, then $\mathbf{u} \times \mathbf{v} = \mathbf{v} \times \mathbf{u}$.

61. *Proof*   Prove that $\|\mathbf{u} \times \mathbf{v}\| = \|\mathbf{u}\| \, \|\mathbf{v}\|$ if $\mathbf{u}$ and $\mathbf{v}$ are orthogonal.

62. *Proof*   Prove that $\mathbf{u} \times (\mathbf{v} \times \mathbf{w}) = (\mathbf{u} \cdot \mathbf{w})\mathbf{v} - (\mathbf{u} \cdot \mathbf{v})\mathbf{w}$.

63. *Proof*   Prove that the triple scalar product of $\mathbf{u}$, $\mathbf{v}$, and $\mathbf{w}$ is given by

$$\mathbf{u} \cdot (\mathbf{v} \times \mathbf{w}) = \begin{vmatrix} u_1 & u_2 & u_3 \\ v_1 & v_2 & v_3 \\ w_1 & w_2 & w_3 \end{vmatrix}.$$

64. *Proof*   Consider the vectors $\mathbf{u} = \langle \cos \alpha, \sin \alpha, 0 \rangle$ and $\mathbf{v} = \langle \cos \beta, \sin \beta, 0 \rangle$, where $\alpha > \beta$. Find the cross product of the vectors and use the result to prove the identity $\sin(\alpha - \beta) = \sin \alpha \cos \beta - \cos \alpha \sin \beta$.

Skills Review

In Exercises 65–68, evaluate the expression without using a calculator.

65. $\cos 480°$

66. $\tan 300°$

67. $\sin \dfrac{19\pi}{6}$

68. $\cos \dfrac{17\pi}{6}$

10.4   Lines and Planes in Space

Lines in Space

In the plane, *slope* is used to determine an equation of a line. In space, it is more convenient to use *vectors* to determine the equation of a line.

In Figure 10.26, consider the line L through the point $P(x_1, y_1, z_1)$ and parallel to the vector

$$\mathbf{v} = \langle a, b, c \rangle. \qquad \text{Direction vector for } L$$

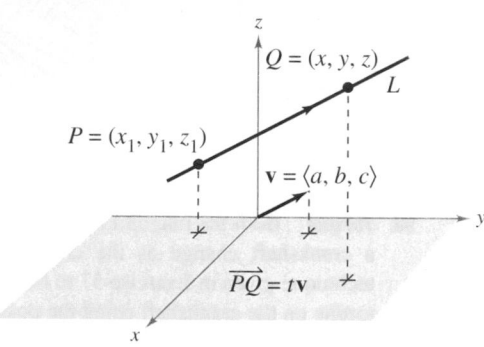

Figure 10.26

The vector $\mathbf{v}$ is the **direction vector** for the line L, and a, b, and c are the **direction numbers.** One way of describing the line L is to say that it consists of all points $Q(x, y, z)$ for which the vector $\overrightarrow{PQ}$ is parallel to $\mathbf{v}$. This means that $\overrightarrow{PQ}$ is a scalar multiple of $\mathbf{v}$, and you can write $\overrightarrow{PQ} = t\mathbf{v}$, where t is a scalar.

$$\overrightarrow{PQ} = \langle x - x_1, y - y_1, z - z_1 \rangle$$

$$= \langle at, bt, ct \rangle$$

$$= t\mathbf{v}$$

By equating corresponding components, you can obtain the **parametric equations of a line in space.**

Parametric Equations of a Line in Space

A line L parallel to the nonzero vector $\mathbf{v} = \langle a, b, c \rangle$ and passing through the point $P(x_1, y_1, z_1)$ is represented by the parametric equations

$$x = x_1 + at, \qquad y = y_1 + bt, \qquad \text{and} \qquad z = z_1 + ct.$$

If the direction numbers a, b, and c are all nonzero, you can eliminate the parameter t to obtain the **symmetric equations** of a line.

$$\frac{x - x_1}{a} = \frac{y - y_1}{b} = \frac{z - z_1}{c} \qquad \text{Symmetric equations}$$

Example 1   Finding Parametric and Symmetric Equations

Find parametric and symmetric equations of the line L that passes through the point $(1, -2, 4)$ and is parallel to $\mathbf{v} = \langle 2, 4, -4 \rangle$.

Solution

To find a set of parametric equations of the line, use the coordinates $x_1 = 1$, $y_1 = -2$, and $z_1 = 4$ and direction numbers $a = 2$, $b = 4$, and $c = -4$ (see Figure 10.27).

$$x = 1 + 2t, \qquad y = -2 + 4t, \qquad z = 4 - 4t \qquad \text{Parametric equations}$$

Because a, b, and c are all nonzero, a set of symmetric equations is

$$\frac{x - 1}{2} = \frac{y + 2}{4} = \frac{z - 4}{-4}. \qquad \text{Symmetric equations}$$

✓**CHECKPOINT**   Now try Exercise 1.

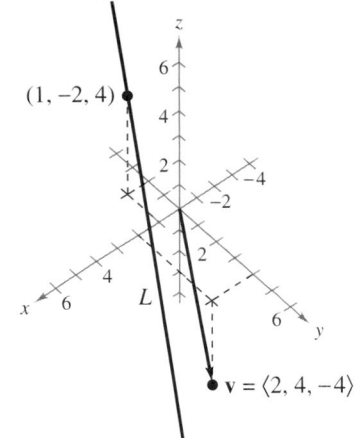

Figure 10.27

Neither the parametric equations nor the symmetric equations of a given line are unique. For instance, in Example 1, by letting $t = 1$ in the parametric equations you would obtain the point $(3, 2, 0)$. Using this point with the direction numbers $a = 2$, $b = 4$, and $c = -4$ produces the parametric equations

$$x = 3 + 2t, \qquad y = 2 + 4t, \quad \text{and} \quad z = -4t.$$

Example 2   Parametric and Symmetric Equations of a Line Through Two Points

Find a set of parametric and symmetric equations of the line that passes through the points $(-2, 1, 0)$ and $(1, 3, 5)$.

Solution

Begin by letting $P = (-2, 1, 0)$ and $Q = (1, 3, 5)$. Then a direction vector for the line passing through P and Q is

$$\begin{aligned} \mathbf{v} &= \overrightarrow{PQ} \\ &= \langle 1 - (-2), 3 - 1, 5 - 0 \rangle \\ &= \langle 3, 2, 5 \rangle \\ &= \langle a, b, c \rangle. \end{aligned}$$

Using the direction numbers $a = 3$, $b = 2$, and $c = 5$ with the point $P(-2, 1, 0)$, you can obtain the parametric equations

$$x = -2 + 3t, \qquad y = 1 + 2t, \quad \text{and} \quad z = 5t. \qquad \text{Parametric equations}$$

Because a, b, and c are all nonzero, a set of symmetric equations is

$$\frac{x + 2}{3} = \frac{y - 1}{2} = \frac{z}{5}. \qquad \text{Symmetric equations}$$

✓**CHECKPOINT**   Now try Exercise 7.

STUDY TIP

To check the answer to Example 2, verify that the two original points lie on the line. To see this, substitute $t = 0$ and $t = 1$ into the parametric equations as follows.

$t = 0$:

$$x = -2 + 3(0) = -2$$
$$y = 1 + 2(0) = 1$$
$$z = 5(0) = 0$$

$t = 1$:

$$x = -2 + 3(1) = 1$$
$$y = 1 + 2(1) = 3$$
$$z = 5(1) = 5$$

Planes in Space

You have seen how an equation of a line in space can be obtained from a point on the line and a vector *parallel* to it. You will now see that an equation of a plane in space can be obtained from a point in the plane and a vector *normal* (perpendicular) to the plane.

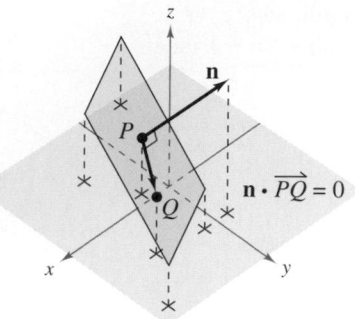

Figure 10.28

Consider the plane containing the point $P(x_1, y_1, z_1)$ having a nonzero normal vector $\mathbf{n} = \langle a, b, c \rangle$, as shown in Figure 10.28. This plane consists of all points $Q(x, y, z)$ for which the vector $\overrightarrow{PQ}$ is orthogonal to $\mathbf{n}$. Using the dot product, you can write the following.

$$\mathbf{n} \cdot \overrightarrow{PQ} = 0$$

$$\langle a, b, c \rangle \cdot \langle x - x_1, y - y_1, z - z_1 \rangle = 0$$

$$a(x - x_1) + b(y - y_1) + c(z - z_1) = 0$$

The third equation of the plane is said to be in **standard form.**

Standard Equation of a Plane in Space

The plane containing the point (x_1, y_1, z_1) and having nonzero normal vector $\mathbf{n} = \langle a, b, c \rangle$ can be represented by the **standard form of the equation of a plane** $a(x - x_1) + b(y - y_1) + c(z - z_1) = 0$.

Regrouping terms yields the **general form of the equation of a plane** in space

$$ax + by + cz + d = 0. \qquad \text{General form of equation of plane}$$

Given the general form of the equation of a plane, it is easy to find a normal vector to the plane. Use the coefficients of x, y, and z to write $\mathbf{n} = \langle a, b, c \rangle$.

Exploration

Consider the following four planes.

$$2x + 3y - z = 2 \qquad\qquad 4x + 6y - 2z = 5$$

$$-2x - 3y + z = -2 \qquad\qquad -6x - 9y + 3z = 11$$

What are the normal vectors for each plane? What can you say about the relative positions of these planes in space?

Example 3   Finding an Equation of a Plane in Three-Space

Find the general equation of the plane passing through the points $(2, 1, 1)$, $(0, 4, 1)$, and $(-2, 1, 4)$.

Solution

To find the equation of the plane, you need a point in the plane and a vector that is normal to the plane. There are three choices for the point, but no normal vector is given. To obtain a normal vector, use the cross product of vectors $\mathbf{u}$ and $\mathbf{v}$ extending from the point $(2, 1, 1)$ to the points $(0, 4, 1)$ and $(-2, 1, 4)$, as shown in Figure 10.29. The component forms of $\mathbf{u}$ and $\mathbf{v}$ are

$$\mathbf{u} = \langle 0 - 2, 4 - 1, 1 - 1 \rangle = \langle -2, 3, 0 \rangle$$

$$\mathbf{v} = \langle -2 - 2, 1 - 1, 4 - 1 \rangle = \langle -4, 0, 3 \rangle$$

and it follows that

$$\mathbf{n} = \mathbf{u} \times \mathbf{v} = \begin{vmatrix} \mathbf{i} & \mathbf{j} & \mathbf{k} \\ -2 & 3 & 0 \\ -4 & 0 & 3 \end{vmatrix}$$

$$= 9\mathbf{i} + 6\mathbf{j} + 12\mathbf{k}$$

$$= \langle a, b, c \rangle$$

is normal to the given plane. Using the direction numbers for $\mathbf{n}$ and the point $(x_1, y_1, z_1) = (2, 1, 1)$, you can determine an equation of the plane to be

$$a(x - x_1) + b(y - y_1) + c(z - z_1) = 0$$

$$9(x - 2) + 6(y - 1) + 12(z - 1) = 0 \qquad \text{Standard form}$$

$$9x + 6y + 12z - 36 = 0$$

$$3x + 2y + 4z - 12 = 0. \qquad \text{General form}$$

Check that each of the three points satisfies the equation $3x + 2y + 4z - 12 = 0$.

✓ *CHECKPOINT*   Now try Exercise 25.

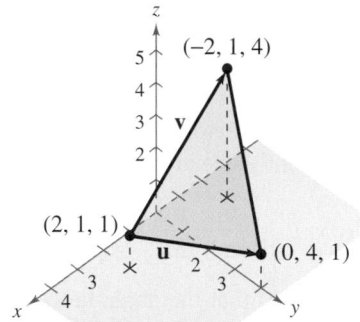

Figure 10.29

Two distinct planes in three-space either are parallel or intersect in a line. If they intersect, you can determine the angle θ $(0 \leq \theta \leq 90°)$ between them from the angle between their normal vectors, as shown in Figure 10.30. Specifically, if vectors $\mathbf{n}_1$ and $\mathbf{n}_2$ are normal to two intersecting planes, the angle θ between the normal vectors is equal to the **angle between the two planes** and is given by

$$\cos \theta = \frac{|\mathbf{n}_1 \cdot \mathbf{n}_2|}{\|\mathbf{n}_1\| \, \|\mathbf{n}_2\|}. \qquad \text{Angle between two planes}$$

Consequently, two planes with normal vectors $\mathbf{n}_1$ and $\mathbf{n}_2$ are

1. *perpendicular* if $\mathbf{n}_1 \cdot \mathbf{n}_2 = 0$.

2. *parallel* if $\mathbf{n}_1$ is a scalar multiple of $\mathbf{n}_2$.

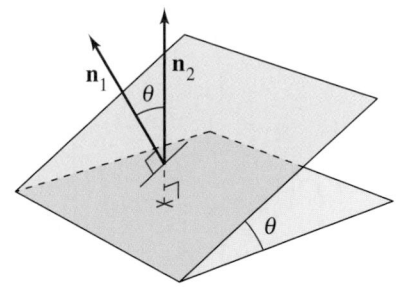

Figure 10.30

Example 4   Finding the Line of Intersection of Two Planes

Find the angle between the two planes given by

$$x - 2y + z = 0 \qquad \text{Equation for plane 1}$$
$$2x + 3y - 2z = 0 \qquad \text{Equation for plane 2}$$

and find parametric equations of their line of intersection (see Figure 10.31).

Solution

The normal vectors for the planes are $\mathbf{n}_1 = \langle 1, -2, 1 \rangle$ and $\mathbf{n}_2 = \langle 2, 3, -2 \rangle$. Consequently, the angle between the two planes is determined as follows.

$$\cos \theta = \frac{|\mathbf{n}_1 \cdot \mathbf{n}_2|}{\|\mathbf{n}_1\| \, \|\mathbf{n}_2\|} = \frac{|-6|}{\sqrt{6}\sqrt{17}} = \frac{6}{\sqrt{102}} \approx 0.59409.$$

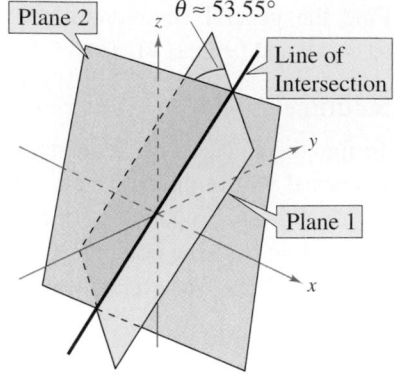

Figure 10.31

This implies that the angle between the two planes is $\theta \approx 53.55°$. You can find the line of intersection of the two planes by simultaneously solving the two linear equations representing the planes. One way to do this is to multiply the first equation by -2 and add the result to the second equation.

$$x - 2y + z = 0 \implies -2x + 4y - 2z = 0$$
$$\underline{2x + 3y - 2z = 0} \qquad \underline{2x + 3y - 2z = 0}$$
$$7y - 4z = 0 \implies y = \frac{4z}{7}$$

Substituting $y = 4z/7$ back into one of the original equations, you can determine that $x = z/7$. Finally, by letting $t = z/7$, you obtain the parametric equations

$$x = t = x_1 + at, \qquad y = 4t = y_1 + bt, \qquad z = 7t = z_1 + ct.$$

Because $(x_1, y_1, z_1) = (0, 0, 0)$ lies in both planes, you can substitute for $x_1, y_1,$ and z_1 in these parametric equations, which indicates that $a = 1, b = 4,$ and $c = 7$ are direction numbers for the line of intersection.

✓**CHECKPOINT**   Now try Exercise 45.

Note that the direction numbers in Example 4 can be obtained from the cross product of the two normal vectors as follows.

$$\mathbf{n}_1 \times \mathbf{n}_2 = \begin{vmatrix} \mathbf{i} & \mathbf{j} & \mathbf{k} \\ 1 & -2 & 1 \\ 2 & 3 & -2 \end{vmatrix}$$

$$= \begin{vmatrix} -2 & 1 \\ 3 & -2 \end{vmatrix} \mathbf{i} - \begin{vmatrix} 1 & 1 \\ 2 & -2 \end{vmatrix} \mathbf{j} + \begin{vmatrix} 1 & -2 \\ 2 & 3 \end{vmatrix} \mathbf{k}$$

$$= \mathbf{i} + 4\mathbf{j} + 7\mathbf{k}$$

This means that the *line of intersection of the two planes is parallel to the cross product of their normal vectors.*

Sketching Planes in Space

As discussed in Section 10.1, if a plane in space intersects one of the coordinate planes, the line of intersection is called the *trace* of the given plane in the coordinate plane. To sketch a plane in space, it is helpful to find its points of intersection with the coordinate axes and its traces in the coordinate planes. For example, consider the plane

$$3x + 2y + 4z = 12. \qquad \text{Equation of plane}$$

You can find the xy-trace by letting $z = 0$ and sketching the line

$$3x + 2y = 12 \qquad xy\text{-trace}$$

in the xy-plane. This line intersects the x-axis at $(4, 0, 0)$ and the y-axis at $(0, 6, 0)$. In Figure 10.32, this process is continued by finding the yz-trace and the xz-trace and then shading the triangular region lying in the first octant.

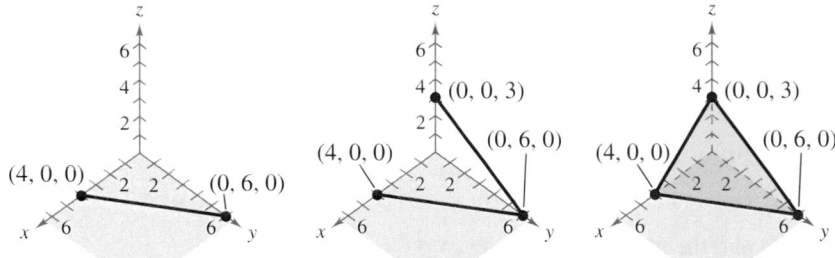

(a) *xy-trace* $(z = 0)$:
    $3x + 2y = 12$

(b) *yz-trace* $(x = 0)$:
    $2y + 4z = 12$

(c) *xz-trace* $(y = 0)$:
    $3x + 4z = 12$

Figure 10.32

If the equation of a plane has a missing variable, such as $2x + z = 1$, the plane must be *parallel to the axis* represented by the missing variable, as shown in Figure 10.33. If two variables are missing from the equation of a plane, then it is *parallel to the coordinate plane* represented by the missing variables, as shown in Figure 10.34.

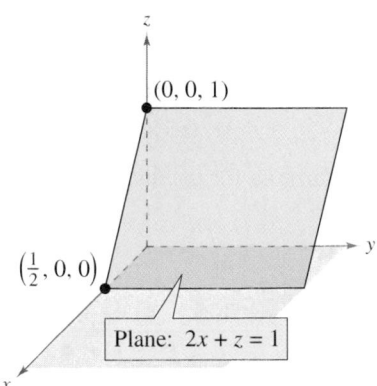

Figure 10.33   *Plane is parallel to y-axis.*

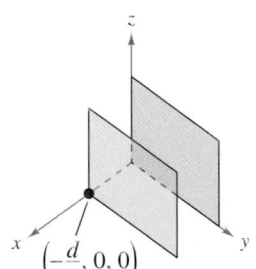 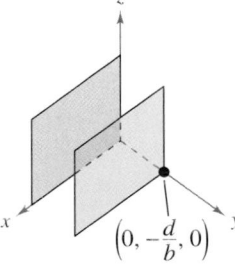 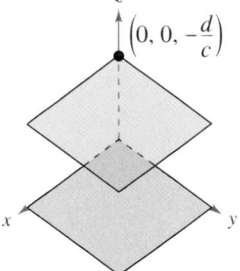

(a) *Plane $ax + d = 0$
    is parallel to yz-plane.*

(b) *Plane $by + d = 0$
    is parallel to xz-plane.*

(c) *Plane $cz + d = 0$
    is parallel to xy-plane.*

Figure 10.34

Distance Between a Point and a Plane

The distance D between a point Q and a plane is the length of the shortest line segment connecting Q to the plane, as shown in Figure 10.35. If P is *any* point in the plane, you can find this distance by projecting the vector $\overrightarrow{PQ}$ onto the normal vector $\mathbf{n}$. The length of this projection is the desired distance.

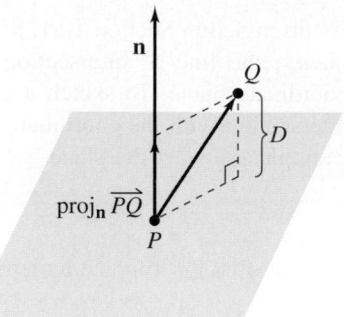

Distance Between a Point and a Plane

The **distance between a plane and a point Q** (not in the plane) is

$$D = \|\text{proj}_{\mathbf{n}}\overrightarrow{PQ}\| = \frac{|\overrightarrow{PQ} \cdot \mathbf{n}|}{\|\mathbf{n}\|}$$

where P is a point in the plane and $\mathbf{n}$ is normal to the plane.

$D = \|\text{proj}_{\mathbf{n}}\overrightarrow{PQ}\|$

Figure 10.35

To find a point in the plane given by $ax + by + cz + d = 0$, where $a \neq 0$, let $y = 0$ and $z = 0$. Then, from the equation $ax + d = 0$, you can conclude that the point $(-d/a, 0, 0)$ lies in the plane.

Example 5 Finding the Distance Between a Point and a Plane

Find the distance between the point $Q(1, 5, -4)$ and the plane $3x - y + 2z = 6$.

Solution

You know that $\mathbf{n} = \langle 3, -1, 2 \rangle$ is normal to the given plane. To find a point in the plane, let $y = 0$ and $z = 0$, and obtain the point $P(2, 0, 0)$. The vector from P to Q is

$$\overrightarrow{PQ} = \langle 1 - 2, 5 - 0, -4 - 0 \rangle$$
$$= \langle -1, 5, -4 \rangle.$$

The formula for the distance between a point and a plane produces

$$D = \frac{|\overrightarrow{PQ} \cdot \mathbf{n}|}{\|\mathbf{n}\|}$$

$$= \frac{|\langle -1, 5, -4 \rangle \cdot \langle 3, -1, 2 \rangle|}{\sqrt{9 + 1 + 4}}$$

$$= \frac{|-3 - 5 - 8|}{\sqrt{14}}$$

$$= \frac{16}{\sqrt{14}}.$$

✓**CHECKPOINT** Now try Exercise 57.

The choice of the point P in Example 5 is arbitrary. Try choosing a different point to verify that you obtain the same distance.

10.4 Exercises

See www.CalcChat.com for worked-out solutions to odd-numbered exercises.

Vocabulary Check

Fill in the blanks.

1. The _____ vector for a line L is given by $\mathbf{v} =$ _____ .

2. The _____ of a line in space are given by $x = x_1 + at$, $y = y_1 + bt$, and $z = z_1 + ct$.

3. If the direction numbers a, b, and c of the vector $\mathbf{v} = \langle a, b, c \rangle$ are all nonzero, you can eliminate the parameter to obtain the _____ of a line.

4. A vector that is perpendicular to a plane is called _____ .

5. The standard form of the equation of a plane is given by _____ .

In Exercises 1–6, find (a) a set of parametric equations and (b) if possible, a set of symmetric equations for the line through the point and parallel to the specified vector or line. For each line, write the direction numbers as integers. (There are many correct answers.)

Point	Parallel to
1. $(0, 0, 0)$	$\mathbf{v} = \langle 1, 2, 3 \rangle$
2. $(3, -5, 1)$	$\mathbf{v} = \langle 3, -7, -10 \rangle$
3. $(-4, 1, 0)$	$\mathbf{v} = \frac{1}{2}\mathbf{i} + \frac{4}{3}\mathbf{j} - \mathbf{k}$
4. $(5, 0, 10)$	$\mathbf{v} = 4\mathbf{i} + 3\mathbf{k}$
5. $(2, -3, 5)$	$x = 5 + 2t$, $y = 7 - 3t$, $z = -2 + t$
6. $(1, 0, 1)$	$x = 3 + 3t$, $y = 5 - 2t$, $z = -7 + t$

In Exercises 7–14, find (a) a set of parametric equations and (b) if possible, a set of symmetric equations of the line that passes through the given points. For each line, write the direction numbers as integers. (There are many correct answers.)

7. $(2, 0, 2), (1, 4, -3)$

8. $(2, 3, 0), (10, 8, 12)$

9. $(-3, 8, 15), (1, -2, 16)$

10. $(2, 3, -1), (1, -5, 3)$

11. $(3, 1, 2), (-1, 1, 5)$

12. $(2, -1, 5), (2, 1, -3)$

13. $\left(-\frac{1}{2}, 2, \frac{1}{2}\right), \left(1, -\frac{1}{2}, 0\right)$

14. $\left(-\frac{3}{2}, \frac{3}{2}, 2\right), (3, -5, -4)$

In Exercises 15 and 16, sketch a graph of the line.

15. $x = 2t$, $y = 2 + t$, $z = 1 + \frac{1}{2}t$

16. $x = 5 - 2t$, $y = 1 + t$, $z = 5 - \frac{1}{2}t$

In Exercises 17–22, find the general form of the equation of the plane passing through the point and normal to the specified vector or line.

Point	Perpendicular to
17. $(2, 1, 2)$	$\mathbf{n} = \mathbf{i}$
18. $(1, 0, -3)$	$\mathbf{n} = \mathbf{k}$
19. $(5, 6, 3)$	$\mathbf{n} = -2\mathbf{i} + \mathbf{j} - 2\mathbf{k}$
20. $(0, 0, 0)$	$\mathbf{n} = -3\mathbf{j} + 5\mathbf{k}$
21. $(2, 0, 0)$	$x = 3 - t$, $y = 2 - 2t$, $z = 4 + t$
22. $(0, 0, 6)$	$x = 1 - t$, $y = 2 + t$, $z = 4 - 2t$

In Exercises 23–26, find the general form of the equation of the plane passing through the three points.

23. $(0, 0, 0), (1, 2, 3), (-2, 3, 3)$

24. $(4, -1, 3), (2, 5, 1), (-1, 2, 1)$

25. $(2, 3, -2), (3, 4, 2), (1, -1, 0)$

26. $(5, -1, 4), (1, -1, 2), (2, 1, -3)$

In Exercises 27–32, find the general form of the equation of the plane with the given characteristics.

27. Passes through $(2, 5, 3)$ and is parallel to xz-plane

28. Passes through $(1, 2, 3)$ and is parallel to yz-plane

29. Passes through $(0, 2, 4)$ and $(-1, -2, 0)$ and is perpendicular to yz-plane

30. Passes through $(1, -2, 4)$ and $(4, 0, -1)$ and is perpendicular to xz-plane

31. Passes through $(2, 2, 1)$ and $(-1, 1, -1)$ and is perpendicular to $2x - 3y + z = 3$

32. Passes through $(1, 2, 0)$ and $(-1, -1, 2)$ and is perpendicular to $2x - 3y + z = 6$

In Exercises 33–40, find a set of parametric equations of the line. (There are many correct answers.)

33. Passes through $(2, 3, 4)$ and is parallel to xz-plane and the yz-plane

34. Passes through $(-4, 5, 2)$ and is parallel to xy-plane and the yz-plane

35. Passes through $(2, 3, 4)$ and is perpendicular to the plane given by $3x + 2y - z = 6$

36. Passes through $(-4, 5, 2)$ and is perpendicular to the plane given by $-x + 2y + z = 5$

37. Passes through $(5, -3, -4)$ and is parallel to $\mathbf{v} = \langle 2, -1, 3 \rangle$

38. Passes through $(-1, 4, -3)$ and is parallel to $\mathbf{v} = 5\mathbf{i} - \mathbf{j}$

39. Passes through $(2, 1, 2)$ and is parallel to $x = -t$, $y = 1 + t, z = -2 + t$

40. Passes through $(-6, 0, 8)$ and is parallel to $x = 5 - 2t$, $y = -4 + 2t, z = 0$

In Exercises 41–44, determine whether the planes are parallel, orthogonal, or neither. If they are neither parallel nor orthogonal, find the angle of intersection.

41. $5x - 3y + z = 4$
$x + 4y + 7z = 1$

42. $3x + y - 4z = 3$
$-9x - 3y + 12z = 4$

43. $2x - z = 1$
$4x + y + 8z = 10$

44. $x - 5y - z = 1$
$5x - 25y - 5z = -3$

In Exercises 45–48, (a) find the angle between the two planes and (b) find parametric equations of their line of intersection.

45. $3x - 4y + 5z = 6$
$x + y - z = 2$

46. $x - 3y + z = -2$
$2x + 5z + 3 = 0$

47. $x + y - z = 0$
$2x - 5y - z = 1$

48. $2x + 4y - 2z = 1$
$-3x - 6y + 3z = 10$

In Exercises 49–54, plot the intercepts and sketch a graph of the plane.

49. $x + 2y + 3z = 6$

50. $2x - y + 4z = 4$

51. $x + 2y = 4$

52. $y + z = 5$

53. $3x + 2y - z = 6$

54. $x - 3z = 6$

In Exercises 55–58, find the distance between the point and the plane.

55. $(0, 0, 0)$
$8x - 4y + z = 8$

56. $(3, 2, 1)$
$x - y + 2z = 4$

57. $(4, -2, -2)$
$2x - y + z = 4$

58. $(-1, 2, 5)$
$2x + 3y + z = 12$

59. *Machine Design*   A tractor fuel tank has the shape and dimensions shown in the figure. In fabricating the tank, it is necessary to know the angle between two adjacent sides. Find this angle.

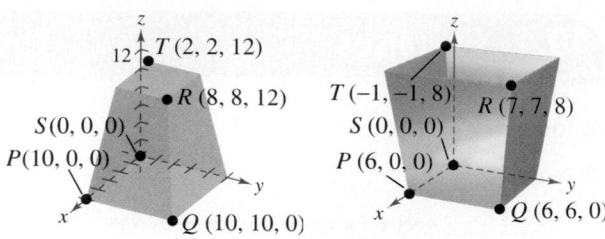

Figure for 59   Figure for 60

60. *Mechanical Design*   A chute at the top of a grain elevator of a combine funnels the grain into a bin, as shown in the figure. Find the angle between two adjacent sides.

Synthesis

True or False?   **In Exercises 61 and 62, determine whether the statement is true or false. Justify your answer.**

61. Every two lines in space are either intersecting or parallel.

62. Two nonparallel planes in space will always intersect.

63. The direction numbers of two distinct lines in space are 10, -18, 20, and -15, 27, -30. What is the relationship between the lines? Explain.

64. *Exploration*

(a) Describe and find an equation for the surface generated by all points (x, y, z) that are two units from the point $(4, -1, 1)$.

(b) Describe and find an equation for the surface generated by all points (x, y, z) that are two units from the plane $4x - 3y + z = 10$.

Skills Review

In Exercises 65–68, convert the polar equation to rectangular form.

65. $r = 10$

66. $\theta = \dfrac{3\pi}{4}$

67. $r = 3 \cos \theta$

68. $r = \dfrac{1}{2 - \cos \theta}$

In Exercises 69–72, convert the rectangular equation to polar form.

69. $x^2 + y^2 = 49$

70. $x^2 + y^2 - 4x = 0$

71. $y = 5$

72. $2x - y + 1 = 0$

What Did You Learn?

Key Terms

solid analytic geometry, p. 742
three-dimensional coordinate system, p. 742
xy-plane, xz-plane, and yz-plane, p. 742
octants, p. 742
left-handed or right-handed orientation, p. 742
surface in space, p. 745

trace, p. 746
zero vector, p. 750
standard unit vector notation, p. 750
component form, p. 750
orthogonal, p. 751
parallel, p. 752
collinear, p. 753
triple scalar product, p. 761

direction vector, p. 764
direction numbers, p. 764
parametric equations, p. 764
symmetric equations, p. 764
general form of the equation of a plane, p. 766
angle between two planes, p. 767

Key Concepts

10.1 ■ Find the distance between two points in space

The distance between the points (x_1, y_1, z_1) and (x_2, y_2, z_2) given by the Distance Formula in Space is

$$d = \sqrt{(x_2 - x_1)^2 + (y_2 - y_1)^2 + (z_2 - z_1)^2}.$$

10.1 ■ Find the midpoint of a line segment joining two points in space

The midpoint of the line segment joining the points (x_1, y_1, z_1) and (x_2, y_2, z_2) given by the Midpoint Formula in Space is

$$\left(\frac{x_1 + x_2}{2}, \frac{y_1 + y_2}{2}, \frac{z_2 + z_2}{2} \right)$$

10.1 ■ Write the standard equation of a sphere

The standard equation of a sphere with center (h, k, j) and radius r is given by

$$(x - h)^2 + (y - k)^2 + (z - j)^2 = r^2.$$

10.2 ■ Use vectors in space

1. Two vectors are equal if and only if their corresponding components are equal.

2. The magnitude (or length) of $\mathbf{u} = \langle u_1, u_2, u_3 \rangle$ is

$$\|\mathbf{u}\| = \sqrt{u_1^2 + u_2^2 + u_3^2}.$$

3. A unit vector $\mathbf{u}$ in the direction of $\mathbf{v}$ is

$$\mathbf{u} = \frac{\mathbf{v}}{\|\mathbf{v}\|}, \quad \mathbf{v} \neq 0.$$

4. The sum of $\mathbf{u} = \langle u_1, u_2, u_3 \rangle$ and $\mathbf{v} = \langle v_1, v_2, v_3 \rangle$ is

$$\mathbf{u} + \mathbf{v} = \langle u_1 + v_1, u_2 + v_2, u_3 + v_3 \rangle.$$

5. The scalar multiple of the real number c and $\mathbf{u} = \langle u_1, u_2, u_3 \rangle$ is $c\mathbf{u} = \langle cu_1, cu_2, cu_3 \rangle$.

6. The dot product of $\mathbf{u} = \langle u_1, u_2, u_3 \rangle$ and $\mathbf{v} = \langle v_1, v_2, v_3 \rangle$ is $\mathbf{u} \cdot \mathbf{v} = u_1 v_1 + u_2 v_2 + u_3 v_3$.

10.2 ■ Find the angle between two vectors

If θ is the angle between two nonzero vectors $\mathbf{u}$ and $\mathbf{v}$, then

$$\cos \theta = \frac{\mathbf{u} \cdot \mathbf{v}}{\|\mathbf{u}\| \, \|\mathbf{v}\|}.$$

10.3 ■ Use the cross product of two vectors

Let $\mathbf{u} = u_1 \mathbf{i} + u_2 \mathbf{j} + u_3 \mathbf{k}$ and $\mathbf{v} = v_1 \mathbf{i} + v_2 \mathbf{j} + v_3 \mathbf{k}$ be vectors in space. The cross product of $\mathbf{u}$ and $\mathbf{v}$ is

$$\mathbf{u} \times \mathbf{v} = (u_2 v_3 - u_3 v_2)\mathbf{i} - (u_1 v_3 - u_3 v_1)\mathbf{j} + (u_1 v_2 - u_2 v_1)\mathbf{k}.$$

Let $\mathbf{u}$, $\mathbf{v}$, and $\mathbf{w}$ be vectors in space and let c be a scalar.

1. $\mathbf{u} \times \mathbf{v} = -(\mathbf{v} \times \mathbf{u})$

2. $\mathbf{u} \times (\mathbf{v} + \mathbf{w}) = (\mathbf{u} \times \mathbf{v}) + (\mathbf{u} \times \mathbf{w})$

3. $c(\mathbf{u} \times \mathbf{v}) = (c\mathbf{u}) \times \mathbf{v} = \mathbf{u} \times (c\mathbf{v})$

4. $\mathbf{u} \times \mathbf{0} = \mathbf{0} \times \mathbf{u} = \mathbf{0}$

5. $\mathbf{u} \times \mathbf{u} = \mathbf{0}$

6. $\mathbf{u} \cdot (\mathbf{v} \times \mathbf{w}) = (\mathbf{u} \times \mathbf{v}) \cdot \mathbf{w}$

10.4 ■ Write parametric equations of a line in space

A line L parallel to nonzero the vector $\mathbf{v} = \langle a, b, c \rangle$ and passing through the point $P(x_1, y_1, z_1)$ is represented by the parametric equations $x = x_1 + at$, $y = y_1 + bt$, and $z = z_1 + ct$.

10.4 ■ Write the standard equation of a plane in space

The plane containing the point (x_1, y_1, z_1) and having nonzero normal vector $\mathbf{n} = \langle a, b, c \rangle$ can be represented by the standard form of the equation of a plane

$$a(x - x_1) + b(y - y_1) + c(z - z_1) = 0.$$

Review Exercises

See www.CalcChat.com for worked-out solutions to odd-numbered exercises.

10.1 In Exercises 1 and 2, plot each point in the same three-dimensional coordinate system.

1. (a) $(5, -1, 2)$

(b) $(-3, 3, 0)$

2. (a) $(2, 4, -3)$

(b) $(0, 0, 5)$

3. Find the coordinates of the point in the xy-plane four units to the right of the xz-plane and five units behind the yz-plane.

4. Find the coordinates of the point located on the y-axis and seven units to the left of the xz-plane.

In Exercises 5 and 6, find the distance between the points.

5. $(4, 0, 7), (5, 2, 1)$

6. $(2, 3, -4), (-1, -3, 0)$

In Exercises 7 and 8, find the lengths of the sides of the right triangle. Show that these lengths satisfy the Pythagorean Theorem.

7.

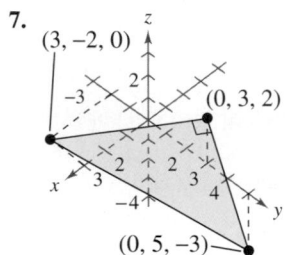

8.

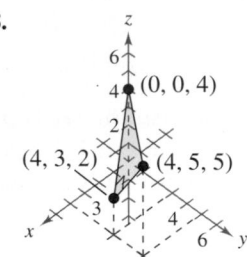

In Exercises 9–12, find the midpoint of the line segment joining the points.

9. $(-2, 3, 2), (2, -5, -2)$

10. $(7, 1, -4), (1, -1, 2)$

11. $(10, 6, -12), (-8, -2, -6)$

12. $(-5, -3, 1), (-7, -9, -5)$

In Exercises 13–16, find the standard form of the equation of the sphere with the given characteristics.

13. Center: $(2, 3, 5)$; radius: 1

14. Center: $(3, -2, 4)$; radius: 4

15. Center: $(1, 5, 2)$; diameter: 12

16. Center: $(0, 4, -1)$; diameter: 15

In Exercises 17 and 18, find the center and radius of the sphere.

17. $x^2 + y^2 + z^2 - 4x - 6y + 4 = 0$

18. $x^2 + y^2 + z^2 - 10x + 6y - 4z + 34 = 0$

In Exercises 19 and 20, sketch the graph of the equation and sketch the specified trace.

19. $x^2 + (y - 3)^2 + z^2 = 16$

(a) xz-trace

(b) yz-trace

20. $(x + 2)^2 + (y - 1)^2 + z^2 = 9$

(a) xy-trace

(b) yz-trace

10.2 In Exercises 21–24, (a) write the component form of the vector v, (b) find the magnitude of v, and (c) find a unit vector in the direction of v.

	Initial point	*Terminal point*
21.	$(2, -1, 4)$	$(3, 3, 0)$
22.	$(2, -1, 2)$	$(-3, 2, 3)$
23.	$(7, -4, 3)$	$(-3, 2, 10)$
24.	$(0, 3, -1)$	$(5, -8, 6)$

In Exercises 25–28, find the dot product of u and v.

25. $\mathbf{u} = \langle -1, 4, 3 \rangle$

$\mathbf{v} = \langle 0, -6, 5 \rangle$

26. $\mathbf{u} = \langle 8, -4, 2 \rangle$

$\mathbf{v} = \langle 2, 5, 2 \rangle$

27. $\mathbf{u} = 2\mathbf{i} - \mathbf{j} + \mathbf{k}$

$\mathbf{v} = \mathbf{i} - \mathbf{k}$

28. $\mathbf{u} = 2\mathbf{i} + \mathbf{j} - 2\mathbf{k}$

$\mathbf{v} = \mathbf{i} - 3\mathbf{j} + 2\mathbf{k}$

In Exercises 29–32, find the angle θ between the vectors.

29. $\mathbf{u} = \langle 4, -1, 5 \rangle$

$\mathbf{v} = \langle 3, 2, -2 \rangle$

30. $\mathbf{u} = \langle 10, -5, 15 \rangle$

$\mathbf{v} = \langle -2, 1, -3 \rangle$

31. $\mathbf{u} = \langle 2\sqrt{2}, -4, 4 \rangle$

$\mathbf{v} = \langle -\sqrt{2}, 1, 2 \rangle$

32. $\mathbf{u} = \langle 3, 1, -1 \rangle$

$\mathbf{v} = \langle 4, 5, 2 \rangle$

In Exercises 33–36, determine whether u and v are orthogonal, parallel, or neither.

33. $\mathbf{u} = \langle 7, -2, 3 \rangle$

$\mathbf{v} = \langle -1, 4, 5 \rangle$

34. $\mathbf{u} = \langle -4, 3, -6 \rangle$

$\mathbf{v} = \langle 16, -12, 24 \rangle$

35. $\mathbf{u} = \langle 39, -12, 21 \rangle$

$\mathbf{v} = \langle -26, 8, -14 \rangle$

36. $\mathbf{u} = \langle 8, 5, -8 \rangle$

$\mathbf{v} = \langle -2, 4, \frac{1}{2} \rangle$

In Exercises 37–40, use vectors to determine whether the points are collinear.

37. $(5, 2, 0), (2, 6, 1), (2, 4, 7)$

38. $(6, 3, -1), (5, 8, 3), (7, -2, -5)$

39. $(3, 4, -1), (-1, 6, 9), (5, 3, -6)$

40. $(5, -4. 7), (8, -5, 5), (11, 6, 3)$

41. *Tension* A load of 300 pounds is supported by three cables, as shown in the figure. Find the tension in each of the supporting cables.

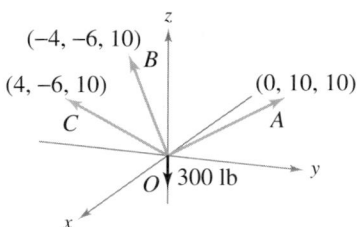

42. *Tension* Determine the tension in each of the supporting cables in Exercise 41 if the load is 200 pounds.

10.3 **In Exercises 43 and 44, find u × v.**

43. $\mathbf{u} = \langle -2, 8, 2 \rangle$
$\mathbf{v} = \langle 1, 1, -1 \rangle$

44. $\mathbf{u} = \langle 10, 15, 5 \rangle$
$\mathbf{v} = \langle 5, -3, 0 \rangle$

In Exercises 45 and 46, find a unit vector orthogonal to u and v.

45. $\mathbf{u} = -3\mathbf{i} + 2\mathbf{j} - 5\mathbf{k}$
$\mathbf{v} = 10\mathbf{i} - 15\mathbf{j} + 2\mathbf{k}$

46. $\mathbf{u} = 4\mathbf{k}$
$\mathbf{v} = \mathbf{i} + 12\mathbf{k}$

In Exercises 47 and 48, verify that the points are the vertices of a parallelogram and find its area.

47. $A(2, -1, 1), B(5, 1, 4), C(0, 1, 1), D(3, 3, 4)$

48. $A(0, 4, 0), B(1, 4, 1), C(0, 6, 0), D(1, 6, 1)$

In Exercises 49 and 50, find the volume of the parallelepiped with the given vertices.

49. $A(0, 0, 0), B(3, 0, 0), C(0, 5, 1), D(3, 5, 1),$
$E(2, 0, 5), F(5, 0, 5), G(2, 5, 6), H(5, 5, 6)$

50. $A(0, 0, 0), B(2, 0, 0), C(2, 4, 0), D(0, 4, 0),$
$E(0, 0, 6), F(2, 0, 6), G(2, 4, 6), H(0, 4, 6)$

10.4 **In Exercises 51 and 52, find sets of (a) parametric equations and (b) symmetric equations for the line that passes through the given points. For each line, write the direction numbers as intergers. (There are many correct answers.)**

51. $(3, 0, 2), (9, 11, 6)$ **52.** $(-1, 4, 3), (8, 10, 5)$

In Exercises 53–56, find a set of (a) parametric equations and (b) symmetric equations for the specified line. (There are many correct answers.)

53. Passes through $(-1, 3, 5)$ and $(3, 6, -1)$

54. Passes through $(0, -10, 3)$ and $(5, 10, 0)$

55. Passes through $(0, 0, 0)$ and is parallel to $\mathbf{v} = \left\langle -2, \frac{5}{2}, 1 \right\rangle$

56. Passes through $(3, 2, 1)$ and is parallel to $x = y = z$

In Exercises 57–60, find the general form of the equation of the plane with the given characteristics.

57. Passes through $(0, 0, 0), (5, 0, 2),$ and $(2, 3, 8)$

58. Passes through $(-1, 3, 4), (4, -2, 2),$ and $(2, 8, 6)$

59. Passes through $(5, 3, 2)$ and is parallel to the xy-plane

60. Passes through $(0, 0, 6)$ and is perpendicular to $x = 1 - t,$ $y = 2 + t,$ and $z = 4 - 2t$

In Exercises 61–64, plot the intercepts and sketch a graph of the plane.

61. $3x - 2y + 3z = 6$

62. $5x - y - 5z = 5$

63. $2x - 3z = 6$

64. $4y - 3z = 12$

In Exercises 65–68, find the distance between the point and the plane.

65. $(2, 3, 10)$
$2x - 20y + 6z = 6$

66. $(1, 2, 3)$
$2x - y + z = 4$

67. $(0, 0, 0)$
$x - 10y + 3z = 2$

68. $(0, 0, 0)$
$2x + 3y + z = 12$

Synthesis

True or False? **In Exercises 69 and 70, determine whether the statement is true or false. Justify your answer.**

69. The cross product is commutative.

70. The triple scalar product of three vectors in space is a scalar.

In Exercises 71–74, let $\mathbf{u} = \langle 3, -2, 1 \rangle$, $\mathbf{v} = \langle 2, -4, -3 \rangle$, and $\mathbf{w} = \langle -1, 2, 2 \rangle$.

71. Show that $\mathbf{u} \cdot \mathbf{u} = \|\mathbf{u}\|^2$.

72. Show that $\mathbf{u} \times \mathbf{v} = -(\mathbf{v} \times \mathbf{u})$.

73. Show that $\mathbf{u} \cdot (\mathbf{v} + \mathbf{w}) = \mathbf{u} \cdot \mathbf{v} + \mathbf{u} \cdot \mathbf{w}$.

74. Show that $\mathbf{u} \times (\mathbf{v} + \mathbf{w}) = (\mathbf{u} \times \mathbf{v}) + (\mathbf{u} \times \mathbf{w})$.

75. *Writing* Define the cross product of vectors $\mathbf{u}$ and $\mathbf{v}$.

76. *Writing* State the geometric properties of the cross product.

77. *Writing* If the magnitudes of two vectors are doubled, how will the magnitude of the cross product of vectors change?

78. *Writing* The vertices of a triangle in space are (x_1, y_1, z_1), (x_2, y_2, z_2), and (x_3, y_3, z_3). Explain how to find a vector perpendicular to the triangle.

10   Chapter Test   See www.CalcChat.com for worked-out solutions to odd-numbered exercises.

Take this test as you would take a test in class. After you are finished, check your work against the answers given in the back of the book.

1. Plot each point in the same three-dimensional coordinate system.

   (a) $(5, -2, 3)$ (b) $(-2, -2, 3)$ (c) $(-1, 2, 1)$

In Exercises 2–4, use the points $A(8, -2, 5)$, $B(6, 4, -1)$, and $C(-4, 3, 0)$ to solve the problem.

2. Consider the triangle with vertices A, B, and C. Is it a right triangle? Explain.

3. Find the coordinates of the midpoint of the line segment joining points A and B.

4. Find the standard form of the equation of the sphere for which A and B are the end-points of a diameter. Sketch the sphere and its xz-trace.

In Exercises 5 and 6, find the component form and the magnitude of the vector v.

5. Initial point: $(2, -1, 3)$
   Terminal point: $(4, 4, -7)$

6. Initial point: $(6, 2, 0)$
   Terminal point: $(3, -3, 8)$

In Exercises 7–10, let u and v be the vectors from $A(8, -2, 5)$ to $B(6, 4, -1)$ and from A to $C(-4, 3, 0)$ respectively.

7. Write **u** and **v** in component form.

8. Find (a) $\|\mathbf{v}\|$, (b) $\mathbf{u} \cdot \mathbf{v}$, and (c) $\mathbf{u} \times \mathbf{v}$.

9. Find the angle between **u** and **v**.

10. Find a set of (a) parametric equations and (b) symmetric equations for the line through points A and B.

In Exercises 11 and 12, determine whether u and v are orthogonal, parallel, or neither.

11. $\mathbf{u} = \mathbf{i} - 2\mathbf{j} - \mathbf{k}$
    $\mathbf{v} = \mathbf{j} + 6\mathbf{k}$

12. $\mathbf{u} = 2\mathbf{i} - 3\mathbf{j} + \mathbf{k}$
    $\mathbf{v} = -\mathbf{i} - \mathbf{j} - \mathbf{k}$

13. Verify that the points $A(2. -3, 1)$, $B(6, 5, -1)$, $C(3, -6, 4)$, and $D(7, 2, 2)$ are the vertices of a parallelogram, and find its area.

14. Find the general form of the equation of the plane passing through the points $(-3, -4, 2)$, $(-3, 4, 1)$, and $(1, 1, -2)$.

15. Find the volume of the parallelepiped at the right with the given vertices.
    $A(0, 0, 5)$, $B(0, 10, 5)$, $C(4, 10, 5)$, $D(4, 0, 5)$,
    $E(0, 1, 0)$, $F(0, 11, 0)$, $G(4, 11, 0)$, $H(4, 1, 0)$

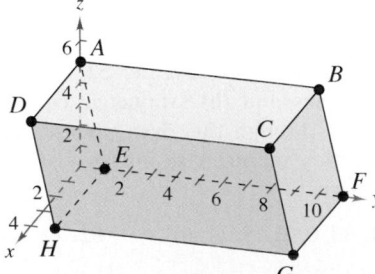

Figure for 15

In Exercises 16 and 17, label the intercepts and sketch a graph of the plane.

16. $2x + 3y + 4z = 12$

17. $5x - y - 2z = 10$

18. Find the distance between the point $(2, -1, 6)$ and the plane $3x - 2y + z = 6$.

Proofs in Mathematics

Algebraic Properties of the Cross Product   (p. 758)

Let $\mathbf{u}$, $\mathbf{v}$, and $\mathbf{w}$ be vectors in space and let c be a scalar.

1. $\mathbf{u} \times \mathbf{v} = -(\mathbf{v} \times \mathbf{u})$

2. $\mathbf{u} \times (\mathbf{v} + \mathbf{w}) = (\mathbf{u} \times \mathbf{v}) + (\mathbf{u} \times \mathbf{w})$

3. $c(\mathbf{u} \times \mathbf{v}) = (c\mathbf{u}) \times \mathbf{v} = \mathbf{u} \times (c\mathbf{v})$

4. $\mathbf{u} \times \mathbf{0} = \mathbf{0} \times \mathbf{u} = \mathbf{0}$

5. $\mathbf{u} \times \mathbf{u} = \mathbf{0}$

6. $\mathbf{u} \cdot (\mathbf{v} \times \mathbf{w}) = (\mathbf{u} \times \mathbf{v}) \cdot \mathbf{w}$

Notation for Dot and Cross Products

The notation for the dot products and cross products of vectors was first introduced by the American physicist Josiah Willard Gibbs (1839–1903). In the early 1880s, Gibbs built a system to represent physical quantities called *vector analysis*. The system was a departure from William Hamilton's theory of quaternions.

Proof

Let $\mathbf{u} = u_1\mathbf{i} + u_2\mathbf{j} + u_3\mathbf{k}$, $\mathbf{v} = v_1\mathbf{i} + v_2\mathbf{j} + v_3\mathbf{k}$, $\mathbf{w} = w_1\mathbf{i} + w_2\mathbf{j} + w_3\mathbf{k}$, $\mathbf{0} = 0\mathbf{i} + 0\mathbf{j} + 0\mathbf{k}$, and let c be a scalar.

1. $\mathbf{u} \times \mathbf{v} = (u_2v_3 - u_3v_2)\mathbf{i} - (u_1v_3 - u_3v_1)\mathbf{j} + (u_1v_2 - u_2v_1)\mathbf{k}$

$\mathbf{v} \times \mathbf{u} = (v_2u_3 - v_3u_2)\mathbf{i} - (v_1u_3 - v_3u_1)\mathbf{j} + (v_1u_2 - v_2u_1)\mathbf{k}$

So, this implies $\mathbf{u} \times \mathbf{v} = -(\mathbf{v} \times \mathbf{u})$.

2. $\mathbf{u} \times (\mathbf{v} + \mathbf{w}) = [u_2(v_3 + w_3) - u_3(v_2 + w_2)]\mathbf{i} - [u_1(v_3 + w_3) - u_3(v_1 + w_1)]\mathbf{j} + [u_1(v_2 + w_2) - u_2(v_1 + w_1)]\mathbf{k}$

$= (u_2v_3 - u_3v_2)\mathbf{i} - (u_1v_3 - u_3v_1)\mathbf{j} + (u_1v_2 - u_2v_1)\mathbf{k} +$
$(u_2w_3 - u_3w_2)\mathbf{i} - (u_1w_3 - u_3w_1)\mathbf{j} + (u_1w_2 - u_2w_1)\mathbf{k}$

$= (\mathbf{u} \times \mathbf{v}) + (\mathbf{u} \times \mathbf{w})$

3. $(c\mathbf{u}) \times \mathbf{v} = (cu_2v_3 - cu_3v_2)\mathbf{i} - (cu_1v_3 - cu_3v_1)\mathbf{j} + (cu_1v_2 - cu_2v_1)\mathbf{k}$

$= c[(u_2v_3 - u_3v_2)\mathbf{i} - (u_1v_3 - u_3v_1)\mathbf{j} + (u_1v_2 - u_2v_1)\mathbf{k}] = c(\mathbf{u} \times \mathbf{v})$

4. $\mathbf{u} \times \mathbf{0} = (u_2 \cdot 0 - u_3 \cdot 0)\mathbf{i} - (u_1 \cdot 0 - u_3 \cdot 0)\mathbf{j} + (u_1 \cdot 0 - u_2 \cdot 0)\mathbf{k}$

$= 0\mathbf{i} + 0\mathbf{j} + 0\mathbf{k} = \mathbf{0}$

$\mathbf{0} \times \mathbf{u} = (0 \cdot u_3 - 0 \cdot u_2)\mathbf{i} - (0 \cdot u_3 - 0 \cdot u_1)\mathbf{j} + (0 \cdot u_2 - 0 \cdot u_1)\mathbf{k}$

$= 0\mathbf{i} + 0\mathbf{j} + 0\mathbf{k} = \mathbf{0}$

So, this implies $\mathbf{u} \times \mathbf{0} = \mathbf{0} \times \mathbf{u} = \mathbf{0}$.

5. $\mathbf{u} \times \mathbf{u} = (u_2u_3 - u_3u_2)\mathbf{i} - (u_1u_3 - u_3u_1)\mathbf{j} + (u_1u_2 - u_2u_1)\mathbf{k} = \mathbf{0}$

6. $\mathbf{u} \cdot (\mathbf{v} \times \mathbf{w}) = \begin{vmatrix} u_1 & u_2 & u_3 \\ v_1 & v_2 & v_3 \\ w_1 & w_2 & w_3 \end{vmatrix}$ and $(\mathbf{u} \times \mathbf{v}) \cdot \mathbf{w} = \mathbf{w} \cdot (\mathbf{u} \times \mathbf{v}) = \begin{vmatrix} w_1 & w_2 & w_3 \\ u_1 & u_2 & u_3 \\ v_1 & v_2 & v_3 \end{vmatrix}$

$\mathbf{u} \cdot (\mathbf{v} \times \mathbf{w}) = u_1(v_2w_3 - w_2v_3) - u_2(v_1w_3 - w_1v_3) + u_3(v_1w_2 - w_1v_2)$

$= u_1v_2w_3 - u_1w_2v_3 - u_2v_1w_3 + u_2w_1v_3 + u_3v_1w_2 - u_3w_1v_2$

$= u_2w_1v_3 - u_3w_1v_2 - u_1w_2v_3 + u_3v_1w_2 + u_1v_2w_3 - u_2v_1w_3$

$= w_1(u_2v_3 - v_2u_3) - w_2(u_1v_3 - v_1u_3) + w_3(u_1v_2 - v_1u_2)$

$= (\mathbf{u} \times \mathbf{v}) \cdot \mathbf{w}$

Geometric Properties of the Cross Product (p. 759)

Let **u** and **v** be nonzero vectors in space, and let θ be the angle between **u** and **v**.

1. $\mathbf{u} \times \mathbf{v}$ is orthogonal to both **u** and **v**.

2. $\|\mathbf{u} \times \mathbf{v}\| = \|\mathbf{u}\|\,\|\mathbf{v}\|\sin\theta$

3. $\mathbf{u} \times \mathbf{v} = \mathbf{0}$ if and only if **u** and **v** are scalar multiples of each other.

4. $\|\mathbf{u} \times \mathbf{v}\| =$ area of parallelogram having **u** and **v** as adjacent sides.

Proof

Let $\mathbf{u} = u_1\mathbf{i} + u_2\mathbf{j} + u_3\mathbf{k}$, $\mathbf{v} = v_1\mathbf{i} + v_2\mathbf{j} + v_3\mathbf{k}$, and $\mathbf{0} = 0\mathbf{i} + 0\mathbf{j} + 0\mathbf{k}$.

1. $\mathbf{u} \times \mathbf{v} = (u_2v_3 - u_3v_2)\mathbf{i} - (u_1v_3 - u_3v_1)\mathbf{j} + (u_1v_2 - u_2v_1)\mathbf{k}$

$(\mathbf{u} \times \mathbf{v}) \cdot \mathbf{u} = (u_2v_3 - u_3v_2)u_1 - (u_1v_3 - u_3v_1)u_2 + (u_1v_2 - u_2v_1)u_3$

$\qquad = u_1u_2v_3 - u_1u_3v_2 - u_1u_2v_3 + u_2u_3v_1 + u_1u_3v_2 - u_2u_3v_1 = 0$

$(\mathbf{u} \times \mathbf{v}) \cdot \mathbf{v} = (u_2v_3 - u_3v_2)v_1 - (u_1v_3 - u_3v_1)v_2 + (u_1v_2 - u_2v_1)v_3$

$\qquad = u_2v_1v_3 - u_3v_1v_2 - u_1v_2v_3 + u_3v_1v_2 + u_1v_2v_3 - u_2v_1v_3 = 0$

Because two vectors are orthogonal if their dot product is zero, it follows that $\mathbf{u} \times \mathbf{v}$ is orthogonal to both **u** and **v**.

2. Note that $\cos\theta = \dfrac{\mathbf{u} \cdot \mathbf{v}}{\|\mathbf{u}\|\,\|\mathbf{v}\|}$. Therefore,

$\|\mathbf{u}\|\,\|\mathbf{v}\|\sin\theta = \|\mathbf{u}\|\,\|\mathbf{v}\|\sqrt{1 - \cos^2\theta}$

$\qquad = \|\mathbf{u}\|\,\|\mathbf{v}\|\sqrt{1 - \dfrac{(\mathbf{u} \cdot \mathbf{v})^2}{\|\mathbf{u}\|^2\,\|\mathbf{v}\|^2}}$

$\qquad = \sqrt{\|\mathbf{u}\|^2\,\|\mathbf{v}\|^2 - (\mathbf{u} \cdot \mathbf{v})^2}$

$\qquad = \sqrt{(u_1^2 + u_2^2 + u_3^2)(v_1^2 + v_2^2 + v_3^2) - (u_1v_1 + u_2v_2 + u_3v_3)^2}$

$\qquad = \sqrt{(u_2v_3 - u_3v_2)^2 + (u_1v_3 - u_3v_1)^2 + (u_1v_2 - u_2v_1)^2} = \|\mathbf{u} \times \mathbf{v}\|.$

3. If **u** and **v** are scalar multiples of each other, then $\mathbf{u} = c\mathbf{v}$ for some scalar c.

$\mathbf{u} \times \mathbf{v} = (c\mathbf{v}) \times \mathbf{v} = c(\mathbf{v} \times \mathbf{v}) = c(\mathbf{0}) = \mathbf{0}$

If $\mathbf{u} \times \mathbf{v} = \mathbf{0}$, then $\|\mathbf{u}\|\,\|\mathbf{v}\|\sin\theta = 0$. (Assume $\mathbf{u} \neq \mathbf{0}$ and $\mathbf{v} \neq \mathbf{0}$.) So, $\sin\theta = 0$, and $\theta = 0$ or $\theta = \pi$. In either case, because θ is the angle between the vectors, **u** and **v** are parallel. Therefore, $\mathbf{u} = c\mathbf{v}$ for some scalar c.

4. The figure at the left is a parallelogram having **v** and **u** as adjacent sides. Because the height of the parallelogram is $\|\mathbf{v}\|\sin\theta$, the area is

$\qquad$ Area = (base)(height)

$\qquad\qquad = \|\mathbf{u}\|\,\|\mathbf{v}\|\sin\theta$

$\qquad\qquad = \|\mathbf{u} \times \mathbf{v}\|.$

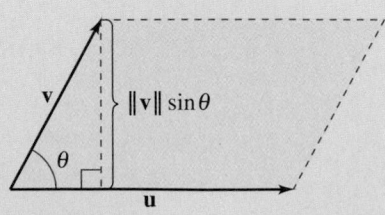

Chapter 11

Limits and an Introduction to Calculus

11.1 Introduction to Limits

11.2 Techniques for Evaluating Limits

11.3 The Tangent Line Problem

11.4 Limits at Infinity and Limits of Sequences

11.5 The Area Problem

Selected Applications

Limit concepts have many real-life applications. The applications listed below represent a small sample of the applications in this chapter.

- Free-Falling Object, Exercises 77 and 78, page 799
- Communications, Exercise 80, page 800
- Market Research, Exercise 64, page 809
- Rate of Change, Exercise 65, page 809
- Average Cost, Exercise 54, page 818
- School Enrollment, Exercise 55, page 818
- Geometry, Exercise 45, page 827

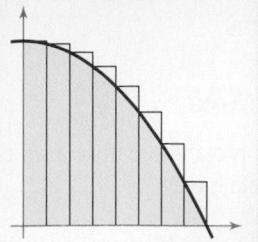
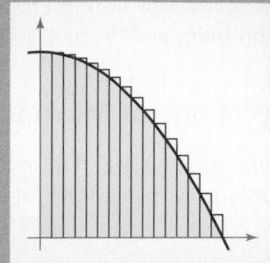
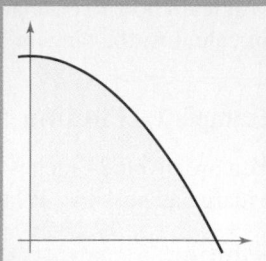

The limit process is a fundamental concept of calculus. In Chapter 11, you will learn many properties of limits and how the limit process can be used to find areas of regions bounded by the graphs of functions. You will also learn how the limit process can be used to find slopes of tangent lines to graphs.

Inga Spence/Index Stock

Americans produce over 200 million pounds of waste each year. Many residents and businesses recycle about 28% of the waste produced. Limits can be used to determine the average cost of recycling material as the amount of material increases infinitely.

11.1 Introduction to Limits

The Limit Concept

The notion of a limit is a *fundamental* concept of calculus. In this chapter, you will learn how to evaluate limits and how they are used in the two basic problems of calculus: the tangent line problem and the area problem.

Example 1 Finding a Rectangle of Maximum Area

You are given 24 inches of wire and are asked to form a rectangle whose area is as large as possible. What dimensions should the rectangle have?

Solution

Let w represent the width of the rectangle and let l represent the length of the rectangle. Because

$$2w + 2l = 24 \qquad \text{Perimeter is 24.}$$

it follows that $l = 12 - w$, as shown in Figure 11.1. So, the area of the rectangle is

$$A = lw \qquad\qquad \text{Formula for area}$$
$$= (12 - w)w \qquad \text{Substitute } 12 - w \text{ for } l.$$
$$= 12w - w^2. \qquad \text{Simplify.}$$

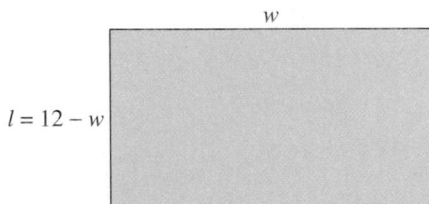

Figure 11.1

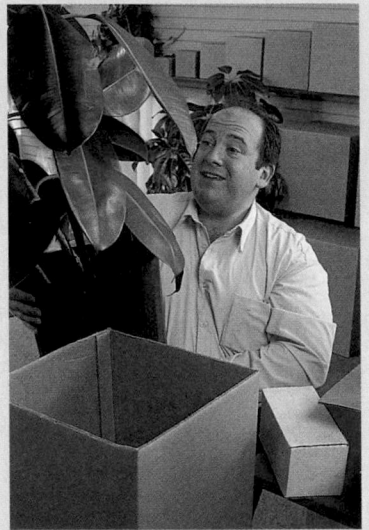

Dick Luria/Getty Images

Using this model for area, you can experiment with different values of w to see how to obtain the maximum area. After trying several values, it appears that the maximum area occurs when $w = 6$, as shown in the table.

Width, w	5.0	5.5	5.9	6.0	6.1	6.5	7.0
Area, A	35.00	35.75	35.99	36.00	35.99	35.75	35.00

In limit terminology, you can say that "the limit of A as w approaches 6 is 36." This is written as

$$\lim_{w \to 6} A = \lim_{w \to 6} (12w - w^2) = 36.$$

✓**CHECKPOINT**   Now try Exercise 1.

Definition of Limit

> **Definition of Limit**
>
> If $f(x)$ becomes arbitrarily close to a unique number L as x approaches c from either side, the **limit** of $f(x)$ as x approaches c is L. This is written as
>
> $$\lim_{x \to c} f(x) = L.$$

Example 2   Estimating a Limit Numerically

Use a table to estimate numerically the limit: $\lim_{x \to 2} (3x - 2)$.

Solution

Let $f(x) = 3x - 2$. Then construct a table that shows values of $f(x)$ for two sets of x-values—one set that approaches 2 from the left and one that approaches 2 from the right.

x	1.9	1.99	1.999	2.0	2.001	2.01	2.1
$f(x)$	3.700	3.970	3.997	?	4.003	4.030	4.300

From the table, it appears that the closer x gets to 2, the closer $f(x)$ gets to 4. So, you can estimate the limit to be 4. Figure 11.2 adds further support to this conclusion.

✓CHECKPOINT   Now try Exercise 3.

Figure 11.2

In Figure 11.2, note that the graph of $f(x) = 3x - 2$ is continuous. For graphs that are not continuous, finding a limit can be more difficult.

Example 3   Estimating a Limit Numerically

Use a table to estimate numerically the limit: $\lim_{x \to 0} \dfrac{x}{\sqrt{x + 1} - 1}$.

Solution

Let $f(x) = x/(\sqrt{x + 1} - 1)$. Then construct a table that shows values of $f(x)$ for two sets of x-values—one set that approaches 0 from the left and one that approaches 0 from the right.

x	-0.01	-0.001	-0.0001	0	0.0001	0.001	0.01
$f(x)$	1.99499	1.99949	1.99995	?	2.00005	2.00050	2.00499

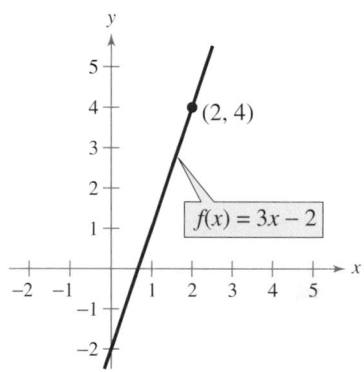

$\lim_{x \to 0} f(x) = 2$

$(0, 2)$

$f(x) = \dfrac{x}{\sqrt{x + 1} - 1}$

f is undefined at $x = 0$.

Figure 11.3

From the table, it appears that the limit is 2. The graph shown in Figure 11.3 verifies that the limit is 2.

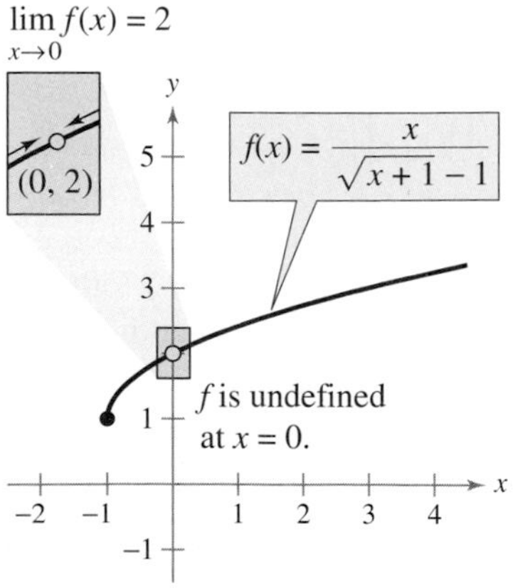

✓CHECKPOINT   Now try Exercise 5.

In Example 3, note that $f(x)$ has a limit when $x \to 0$ even though the function is not defined when $x = 0$. This often happens, and it is important to realize that *the existence or nonexistence of $f(x)$ at $x = c$ has no bearing on the existence of the limit of $f(x)$ as x approaches c.*

Example 4   Using a Graphing Utility to Estimate a Limit

Estimate the limit: $\lim\limits_{x \to 1} \dfrac{x^3 - x^2 + x - 1}{x - 1}$.

Numerical Solution

Let $f(x) = (x^3 - x^2 + x - 1)/(x - 1)$. Because you are finding the limit of $f(x)$ as x approaches 1, use the *table* feature of a graphing utility to create a table that shows the value of the function for x starting at $x = 0.9$ and setting the table step to 0.01, as shown in Figure 11.4(a). Then change the table so that x starts at 0.99 and set the table step to 0.001, as shown in Figure 11.4(b). From the tables, you can estimate the limit to be 2.

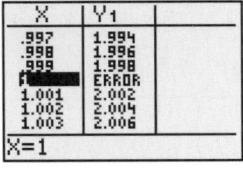

(a)   (b)

Figure 11.4

✓CHECKPOINT   Now try Exercise 11.

Graphical Solution

Use a graphing utility to graph $y = (x^3 - x^2 + x - 1)/(x - 1)$ using a *decimal* setting. Then use the *zoom* and *trace* features to determine that as x gets closer and closer to 1, y gets closer and closer to 2 from the left and from the right, as shown in Figure 11.5. Using the *trace* feature, notice that there is no value given for y when $x = 1$, and that there is a hole or break in the graph when $x = 1$.

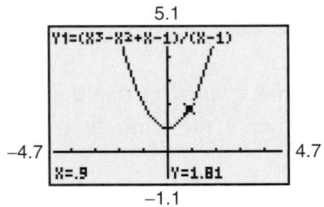

Figure 11.5

Example 5   Using a Graph to Find a Limit

Find the limit of $f(x)$ as x approaches 3, where f is defined as

$$f(x) = \begin{cases} 2, & x \neq 3 \\ 0, & x = 3 \end{cases}.$$

Solution

Because $f(x) = 2$ for all x other than $x = 3$ and because the value of $f(3)$ is immaterial, it follows that the limit is 2 (see Figure 11.6). So, you can write

$$\lim\limits_{x \to 3} f(x) = 2.$$

The fact that $f(3) = 0$ has no bearing on the existence or value of the limit as x approaches 3. For instance, if the function were defined as

$$f(x) = \begin{cases} 2, & x \neq 3 \\ 4, & x = 3 \end{cases}$$

the limit as x approaches 3 would be the same.

✓CHECKPOINT   Now try Exercise 25.

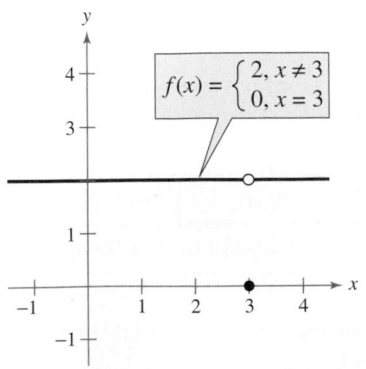

Figure 11.6

Limits That Fail to Exist

Next, you will examine some functions for which limits do not exist.

Example 6 Comparing Left and Right Behavior

Show that the limit does not exist.

$$\lim_{x \to 0} \frac{|x|}{x}$$

Solution

Consider the graph of the function given by $f(x) = |x|/x$. In Figure 11.7, you can see that for positive x-values

$$\frac{|x|}{x} = 1, \qquad x > 0$$

and for negative x-values

$$\frac{|x|}{x} = -1, \qquad x < 0.$$

This means that no matter how close x gets to 0, there will be both positive and negative x-values that yield $f(x) = 1$ and $f(x) = -1$. This implies that the limit does not exist.

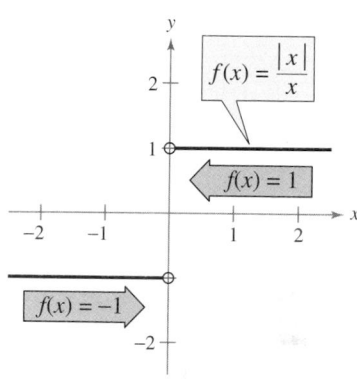

Figure 11.7

 *CHECKPOINT*   Now try Exercise 31.

Example 7 Unbounded Behavior

Discuss the existence of the limit.

$$\lim_{x \to 0} \frac{1}{x^2}$$

Solution

Let $f(x) = 1/x^2$. In Figure 11.8, note that as x approaches 0 from either the right or the left, $f(x)$ increases without bound. This means that by choosing x close enough to 0, you can force $f(x)$ to be as large as you want. For instance, $f(x)$ will be larger than 100 if you choose x that is within $\frac{1}{10}$ of 0. That is,

$$0 < |x| < \frac{1}{10} \quad \Longrightarrow \quad f(x) = \frac{1}{x^2} > 100.$$

Similarly, you can force $f(x)$ to be larger than 1,000,000, as follows.

$$0 < |x| < \frac{1}{1000} \quad \Longrightarrow \quad f(x) = \frac{1}{x^2} > 1,000,000$$

Because $f(x)$ is not approaching a unique real number L as x approaches 0, you can conclude that the limit does not exist.

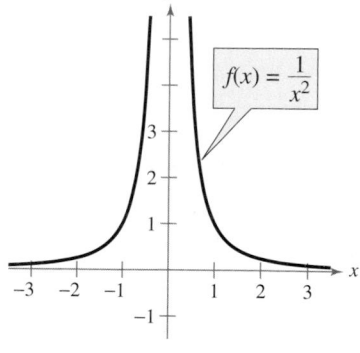

Figure 11.8

CHECKPOINT   Now try Exercise 32.

Example 8 Oscillating Behavior

Discuss the existence of the limit.

$$\lim_{x \to 0} \sin\left(\frac{1}{x}\right)$$

Solution

Let $f(x) = \sin(1/x)$. In Figure 11.9, you can see that as x approaches 0, $f(x)$ oscillates between -1 and 1. Therefore, the limit does not exist because no matter how close you are to 0, it is possible to choose values of x_1 and x_2 such that $\sin(1/x_1) = 1$ and $\sin(1/x_2) = -1$, as indicated in the table.

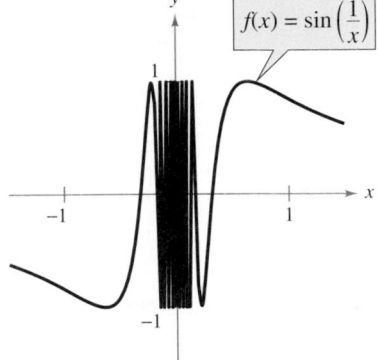

Figure 11.9

x	$\frac{2}{\pi}$	$\frac{2}{3\pi}$	$\frac{2}{5\pi}$	$\frac{2}{7\pi}$	$\frac{2}{9\pi}$	$\frac{2}{11\pi}$	$x \to 0$
$\sin\left(\frac{1}{x}\right)$	1	-1	1	-1	1	-1	Limit does not exist.

✔CHECKPOINT   Now try Exercise 33.

Examples 6, 7, and 8 show three of the most common types of behavior associated with the *nonexistence* of a limit.

Conditions Under Which Limits Do Not Exist

The limit of $f(x)$ as $x \to c$ does not exist if any of the following conditions is true.

1. $f(x)$ approaches a different number from the right side of c than it approaches from the left side of c.   Example 6

2. $f(x)$ increases or decreases without bound as x approaches c.   Example 7

3. $f(x)$ oscillates between two fixed values as x approaches c.   Example 8

TECHNOLOGY TIP   A graphing utility can help you discover the behavior of a function near the x-value at which you are trying to evaluate a limit. When you do this, however, you should realize that you can't always trust the graphs that graphing utilities display. For instance, if you use a graphing utility to graph the function in Example 8 over an interval containing 0, you will most likely obtain an incorrect graph, as shown in Figure 11.10. The reason that a graphing utility can't show the correct graph is that the graph has infinitely many oscillations over any interval that contains 0.

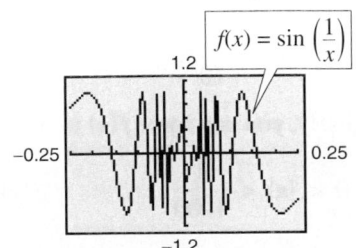

Figure 11.10

Properties of Limits and Direct Substitution

You have seen that sometimes the limit of $f(x)$ as $x \to c$ is simply $f(c)$. In such cases, it is said that the limit can be evaluated by **direct substitution.** That is,

$$\lim_{x \to c} f(x) = f(c). \qquad \text{Substitute } c \text{ for } x.$$

There are many "well-behaved" functions, such as polynomial functions and rational functions with nonzero denominators, that have this property. Some of the basic ones are included in the following list.

Exploration

Use a graphing utility to graph the tangent function. What are $\lim_{x \to 0} \tan x$ and $\lim_{x \to \pi/4} \tan x$? What can you say about the existence of the limit $\lim_{x \to \pi/2} \tan x$?

Basic Limits

Let b and c be real numbers and let n be a positive integer.

1. $\lim_{x \to c} b = b$

2. $\lim_{x \to c} x = c$

3. $\lim_{x \to c} x^n = c^n$ (See the proof on page 835.)

4. $\lim_{x \to c} \sqrt[n]{x} = \sqrt[n]{c}$,   for n even and $c > 0$

Trigonometric functions can also be included in this list. For instance,

$$\lim_{x \to \pi} \sin x = \sin \pi$$
$$= 0$$

and

$$\lim_{x \to 0} \cos x = \cos 0$$
$$= 1.$$

By combining the basic limits with the following operations, you can find limits for a wide variety of functions.

Properties of Limits

Let b and c be real numbers, let n be a positive integer, and let f and g be functions with the following limits.

$$\lim_{x \to c} f(x) = L \qquad \text{and} \qquad \lim_{x \to c} g(x) = K$$

1. Scalar multiple: $\lim_{x \to c} [b f(x)] = bL$

2. Sum or difference: $\lim_{x \to c} [f(x) \pm g(x)] = L \pm K$

3. Product: $\lim_{x \to c} [f(x) g(x)] = LK$

4. Quotient: $\lim_{x \to c} \dfrac{f(x)}{g(x)} = \dfrac{L}{K}$,   provided $K \neq 0$

5. Power: $\lim_{x \to c} [f(x)]^n = L^n$

Example 9   Direct Substitution and Properties of Limits

Find each limit.

a. $\lim_{x \to 4} x^2$ **b.** $\lim_{x \to 4} 5x$ **c.** $\lim_{x \to \pi} \dfrac{\tan x}{x}$

d. $\lim_{x \to 9} \sqrt{x}$ **e.** $\lim_{x \to \pi} (x \cos x)$ **f.** $\lim_{x \to 3} (x + 4)^2$

Solution

You can use the properties of limits and direct substitution to evaluate each limit.

a. $\lim_{x \to 4} x^2 = (4)^2$ Direct Substitution

$\qquad\qquad = 16$

b. $\lim_{x \to 4} 5x = 5 \lim_{x \to 4} x$ Scalar Multiple Property

$\qquad\quad = 5(4)$

$\qquad\quad = 20$

c. $\lim_{x \to \pi} \dfrac{\tan x}{x} = \dfrac{\lim_{x \to \pi} \tan x}{\lim_{x \to \pi} x}$ Quotient Property

$\qquad\qquad = \dfrac{0}{\pi} = 0$

d. $\lim_{x \to 9} \sqrt{x} = \sqrt{9}$

$\qquad\qquad = 3$

e. $\lim_{x \to \pi} (x \cos x) = \left(\lim_{x \to \pi} x\right)\left(\lim_{x \to \pi} \cos x\right)$ Product Property

$\qquad\qquad\quad = \pi(\cos \pi)$

$\qquad\qquad\quad = -\pi$

f. $\lim_{x \to 3} (x + 4)^2 = \left[\left(\lim_{x \to 3} x\right) + \left(\lim_{x \to 3} 4\right)\right]^2$ Sum and Power Properties

$\qquad\qquad\quad = (3 + 4)^2$

$\qquad\qquad\quad = 7^2 = 49$

✓**CHECKPOINT**   Now try Exercise 49.

TECHNOLOGY TIP

When evaluating limits, remember that there are several ways to solve most problems. Often, a problem can be solved *numerically*, *graphically*, or *algebraically*. The limits in Example 9 were found algebraically. You can verify these solutions numerically and/or graphically. For instance, to verify the limit in Example 9(a) numerically, use the *table* feature of a graphing utility to create a table, as shown in Figure 11.11. From the table, you can see that the limit as x approaches 4 is 16. Now, to verify the limit graphically, use a graphing utility to graph $y = x^2$. Using the *zoom* and *trace* features, you can determine that the limit as x approaches 4 is 16, as shown in Figure 11.12.

Exploration

Sketch the graph of each function. Then find the limits of each function as x approaches 1 and as x approaches 2. What conclusions can you make?

a. $f(x) = x + 1$

b. $g(x) = \dfrac{x^2 - 1}{x - 1}$

c. $h(x) = \dfrac{x^3 - 2x^2 - x + 2}{x^2 - 3x + 2}$

Use a graphing utility to graph each function above. Does the graphing utility distinguish among the three graphs? Write a short explanation of your findings.

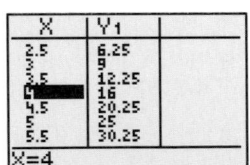

Figure 11.11

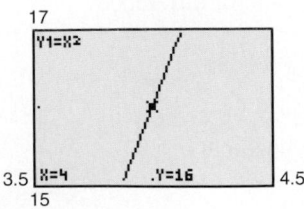

Figure 11.12

The results of using direct substitution to evaluate limits of polynomial and rational functions are summarized as follows.

Limits of Polynomial and Rational Functions

1. If p is a polynomial function and c is a real number, then

$$\lim_{x \to c} p(x) = p(c). \quad \text{(See the proof on page 835.)}$$

2. If r is a rational function given by $r(x) = p(x)/q(x)$, and c is a real number such that $q(c) \neq 0$, then

$$\lim_{x \to c} r(x) = r(c) = \frac{p(c)}{q(c)}.$$

Example 10   Evaluating Limits by Direct Substitution

Find each limit.

a. $\displaystyle\lim_{x \to -1} (x^2 + x - 6)$ **b.** $\displaystyle\lim_{x \to -1} \frac{x^2 + x - 6}{x + 3}$

Solution

The first function is a polynomial function and the second is a rational function (with a nonzero denominator at $x = -1$). So, you can evaluate the limits by direct substitution.

a. $\displaystyle\lim_{x \to -1} (x^2 + x - 6) = (-1)^2 + (-1) - 6$

$$= -6$$

b. $\displaystyle\lim_{x \to -1} \frac{x^2 + x - 6}{x + 3} = \frac{(-1)^2 + (-1) - 6}{-1 + 3}$

$$= -\frac{6}{2}$$

$$= -3$$

✓*CHECKPOINT*   Now try Exercise 53.

Exploration

Use a graphing utility to graph the function

$$f(x) = \frac{x^2 - 3x - 10}{x - 5}.$$

Use the *trace* feature to approximate $\lim_{x \to 4} f(x)$. What do you think $\lim_{x \to 5} f(x)$ equals? Is f defined at $x = 5$? Does this affect the existence of the limit as x approaches 5?

11.1 Exercises

See www.CalcChat.com for worked-out solutions to odd-numbered exercises.

Vocabulary Check

Fill in the blanks.

1. If $f(x)$ becomes arbitrarily close to a unique number L as x approaches c from either side, the _____ of $f(x)$ as x approaches c is L.

2. The limit of $f(x)$ as $x \to c$ does not exist if $f(x)$ _____ between two fixed values.

3. To evaluate the limit of a polynomial function, use _____ .

---

1. **Geometry** You create an open box from a square piece of material, 24 centimeters on a side. You cut equal squares from the corners and turn up the sides.

   (a) Draw and label a diagram that represents the box.

   (b) Verify that the volume of the box is given by

   $$V = 4x(12 - x)^2.$$

   (c) The box has a maximum volume when $x = 4$. Use a graphing utility to complete the table and observe the behavior of the function as x approaches 4. Use the table to find $\lim\limits_{x \to 4} V$.

x	3	3.5	4	4.1	4.5	5
V						

   (d) Use a graphing utility to graph the volume function. Verify that the volume is maximum when $x = 4$.

2. **Geometry** You are given wire and are asked to form a right triangle with a hypotenuse of $\sqrt{18}$ inches whose area is as large as possible.

   (a) Draw and label a diagram that shows the base x and height y of the triangle.

   (b) Verify that the area of the triangle is given by

   $$A = \tfrac{1}{2}x\sqrt{18 - x^2}.$$

   (c) The triangle has a maximum area when $x = 3$ inches. Use a graphing utility to complete the table and observe the behavior of the function as x approaches 3. Use the table to find $\lim\limits_{x \to 3} A$.

x	2	2.5	2.9	3	3.1	3.5	4
A							

   (d) Use a graphing utility to graph the area function. Verify that the area is maximum when $x = 3$ inches.

In Exercises 3–10, complete the table and use the result to estimate the limit numerically. Determine whether or not the limit can be reached.

3. $\lim\limits_{x \to 2} (5x + 4)$

x	1.9	1.99	1.999	2	2.001	2.01	2.1
$f(x)$				?			

4. $\lim\limits_{x \to 1} (2x^2 + x - 4)$

x	0.9	0.99	0.999	1	1.001	1.01	1.1
$f(x)$				?			

5. $\lim\limits_{x \to 3} \dfrac{x - 3}{x^2 - 9}$

x	2.9	2.99	2.999	3	3.001	3.01	3.1
$f(x)$				?			

6. $\lim\limits_{x \to -1} \dfrac{x + 1}{x^2 - x - 2}$

x	−1.1	−1.01	−1.001	−1	−0.999
$f(x)$				?	

x	−0.99	−0.9
$f(x)$		

7. $\lim\limits_{x \to 0} \dfrac{\sin 2x}{x}$

x	−0.1	−0.01	−0.001	0	0.001
$f(x)$				?	

x	0.01	0.1
$f(x)$		

8. $\lim\limits_{x \to 0} \dfrac{\tan x}{2x}$

x	-0.1	-0.01	-0.001	0	0.001
$f(x)$				?	

x	0.01	0.1
$f(x)$		

9. $\lim\limits_{x \to 0} \dfrac{e^{2x} - 1}{x}$

x	-0.1	-0.01	-0.001	0	0.001
$f(x)$				?	

x	0.01	0.1
$f(x)$		

10. $\lim\limits_{x \to 1} \dfrac{\ln x}{x - 1}$

x	0.9	0.99	0.999	1	1.001	1.01	1.1
$f(x)$				?			

In Exercises 11–24, use the *table* feature of a graphing utility to create a table for the function and use the result to estimate the limit numerically. Use the graphing utility to graph the corresponding function to confirm your result graphically.

11. $\lim\limits_{x \to 1} \dfrac{x - 1}{x^2 + 2x - 3}$

12. $\lim\limits_{x \to -2} \dfrac{x + 2}{x^2 + 5x + 6}$

13. $\lim\limits_{x \to 0} \dfrac{\sqrt{x + 5} - \sqrt{5}}{x}$

14. $\lim\limits_{x \to -3} \dfrac{\sqrt{1 - x} - 2}{x + 3}$

15. $\lim\limits_{x \to -4} \dfrac{\dfrac{x}{x + 2} - 2}{x + 4}$

16. $\lim\limits_{x \to 2} \dfrac{\dfrac{1}{x + 2} - \dfrac{1}{4}}{x - 2}$

17. $\lim\limits_{x \to 0} \dfrac{\sin x}{x}$

18. $\lim\limits_{x \to 0} \dfrac{\cos x - 1}{x}$

19. $\lim\limits_{x \to 0} \dfrac{\sin^2 x}{x}$

20. $\lim\limits_{x \to 0} \dfrac{2x}{\tan 4x}$

21. $\lim\limits_{x \to 0} \dfrac{e^{2x} - 1}{2x}$

22. $\lim\limits_{x \to 0} \dfrac{1 - e^{-4x}}{x}$

23. $\lim\limits_{x \to 1} \dfrac{\ln(2x - 1)}{x - 1}$

24. $\lim\limits_{x \to 1} \dfrac{\ln(x^2)}{x - 1}$

In Exercises 25–28, graph the function and find the limit (if it exists) as x approaches 2.

25. $f(x) = \begin{cases} 2x + 1, & x < 2 \\ x + 3, & x \geq 2 \end{cases}$

26. $f(x) = \begin{cases} 8 - x^2, & x \leq 2 \\ x + 2, & x > 2 \end{cases}$

27. $f(x) = \begin{cases} 2x + 1, & x \leq 2 \\ x + 4, & x > 2 \end{cases}$

28. $f(x) = \begin{cases} -2x, & x \leq 2 \\ x^2 - 4x + 1, & x > 2 \end{cases}$

In Exercises 29–36, use the graph to find the limit (if it exists). If the limit does not exist, explain why.

29. $\lim\limits_{x \to -4} (x^2 - 3)$

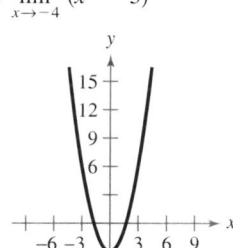

30. $\lim\limits_{x \to 2} \dfrac{3x^2 - 12}{x - 2}$

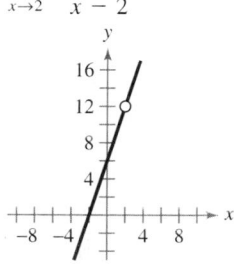

31. $\lim\limits_{x \to -2} \dfrac{|x + 2|}{x + 2}$

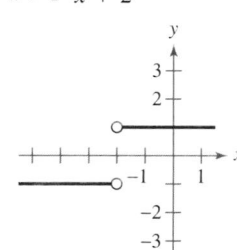

32. $\lim\limits_{x \to 1} \dfrac{1}{x - 1}$

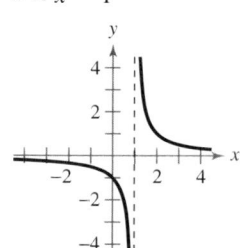

33. $\lim\limits_{x \to 0} 2 \cos \dfrac{\pi}{x}$

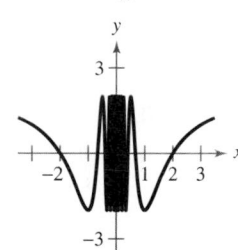

34. $\lim\limits_{x \to -1} \sin \dfrac{\pi x}{2}$

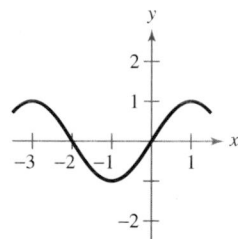

35. $\lim\limits_{x \to \pi/2} \tan x$

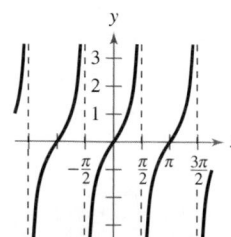

36. $\lim\limits_{x \to \pi/2} \sec x$

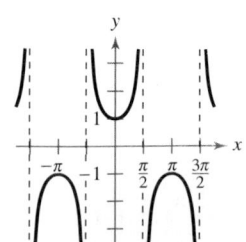

In Exercises 37–46, use a graphing utility to graph the function and use the graph to determine whether or not the limit exists. If the limit does not exist, explain why.

37. $f(x) = \dfrac{5}{2 + e^{1/x}}, \quad \lim\limits_{x \to 0} f(x)$

38. $f(x) = \dfrac{e^x - 1}{x}, \quad \lim\limits_{x \to 0} f(x)$

39. $f(x) = \cos \dfrac{1}{x}, \quad \lim\limits_{x \to 0} f(x)$

40. $f(x) = \sin \pi x, \quad \lim\limits_{x \to -1} f(x)$

41. $f(x) = \dfrac{\sqrt{x + 3} - 1}{x - 4}, \quad \lim\limits_{x \to 4} f(x)$

42. $f(x) = \dfrac{\sqrt{x + 5} - 4}{x - 2}, \quad \lim\limits_{x \to 2} f(x)$

43. $f(x) = \dfrac{x - 1}{x^2 - 4x + 3}, \quad \lim\limits_{x \to 1} f(x)$

44. $f(x) = \dfrac{7}{x - 3}, \quad \lim\limits_{x \to 3} f(x)$

45. $f(x) = \ln(x + 3), \quad \lim\limits_{x \to 4} f(x)$

46. $f(x) = \ln(7 - x), \quad \lim\limits_{x \to -1} f(x)$

In Exercises 47 and 48, use the given information to evaluate each limit.

47. $\lim\limits_{x \to c} f(x) = 3, \quad \lim\limits_{x \to c} g(x) = 6$

(a) $\lim\limits_{x \to c} [-2g(x)]$

(b) $\lim\limits_{x \to c} [f(x) + g(x)]$

(c) $\lim\limits_{x \to c} \dfrac{f(x)}{g(x)}$

(d) $\lim\limits_{x \to c} \sqrt{f(x)}$

48. $\lim\limits_{x \to c} f(x) = 5, \quad \lim\limits_{x \to c} g(x) = -2$

(a) $\lim\limits_{x \to c} [f(x) + g(x)]^2$

(b) $\lim\limits_{x \to c} [6f(x)g(x)]$

(c) $\lim\limits_{x \to c} \dfrac{5g(x)}{4f(x)}$

(d) $\lim\limits_{x \to c} \dfrac{1}{\sqrt{f(x)}}$

In Exercises 49 and 50, find (a) $\lim\limits_{x \to 2} f(x)$, (b) $\lim\limits_{x \to 2} g(x)$, (c) $\lim\limits_{x \to 2} [f(x)g(x)]$. and (d) $\lim\limits_{x \to 2} [g(x) - f(x)]$.

49. $f(x) = x^3, \quad g(x) = \dfrac{\sqrt{x^2 + 5}}{2x^2}$

50. $f(x) = \dfrac{x}{3 - x}, \quad g(x) = \sin \pi x$

In Exercises 51–70, find the limit by direct substitution.

51. $\lim\limits_{x \to 5} (10 - x^2)$

52. $\lim\limits_{x \to -2} \left(\tfrac{1}{2}x^3 - 5x\right)$

53. $\lim\limits_{x \to -3} (2x^2 + 4x + 1)$

54. $\lim\limits_{x \to -2} (x^3 - 6x + 5)$

55. $\lim\limits_{x \to 3} \left(-\dfrac{9}{x}\right)$

56. $\lim\limits_{x \to -5} \dfrac{6}{x + 2}$

57. $\lim\limits_{x \to -3} \dfrac{3x}{x^2 + 1}$

58. $\lim\limits_{x \to 4} \dfrac{x - 1}{x^2 + 2x + 3}$

59. $\lim\limits_{x \to -2} \dfrac{5x + 3}{2x - 9}$

60. $\lim\limits_{x \to 3} \dfrac{x^2 + 1}{x}$

61. $\lim\limits_{x \to -1} \sqrt{x + 2}$

62. $\lim\limits_{x \to 3} \sqrt[3]{x^2 - 1}$

63. $\lim\limits_{x \to 7} \dfrac{5x}{\sqrt{x + 2}}$

64. $\lim\limits_{x \to 8} \dfrac{\sqrt{x + 1}}{x - 4}$

65. $\lim\limits_{x \to 3} e^x$

66. $\lim\limits_{x \to e} \ln x$

67. $\lim\limits_{x \to \pi} \sin 2x$

68. $\lim\limits_{x \to \pi} \tan x$

69. $\lim\limits_{x \to 1/2} \arcsin x$

70. $\lim\limits_{x \to 1} \arccos \dfrac{x}{2}$

Synthesis

True or False?   In Exercises 71 and 72, determine whether the statement is true or false. Justify your answer.

71. The limit of a function as x approaches c does not exist if the function approaches -3 from the left of c and 3 from the right of c.

72. The limit of the product of two functions is equal to the product of the limits of the two functions.

73. *Think About It*   From Exercises 3 to 10, select a limit that can be reached and one that cannot be reached.

(a) Use a graphing utility to graph the corresponding functions using a standard viewing window. Do the graphs reveal whether or not the limit can be reached? Explain.

(b) Use a graphing utility to graph the corresponding functions using a *decimal* setting. Do the graphs reveal whether or not the limit can be reached? Explain.

74. *Think About It*   Use the results of Exercise 73 to draw a conclusion as to whether or not you can use the graph generated by a graphing utility to determine reliably if a limit can be reached.

75. *Think About It*

(a) If $f(2) = 4$, can you conclude anything about $\lim\limits_{x \to 2} f(x)$? Explain your reasoning.

(b) If $\lim\limits_{x \to 2} f(x) = 4$, can you conclude anything about $f(2)$? Explain your reasoning.

76. *Writing*   Write a brief description of the meaning of the notation $\lim\limits_{x \to 5} f(x) = 12$.

Skills Review

In Exercises 77–82, simplify the rational expression.

77. $\dfrac{5 - x}{3x - 15}$

78. $\dfrac{x^2 - 81}{9 - x}$

79. $\dfrac{15x^2 + 7x - 4}{15x^2 + x - 2}$

80. $\dfrac{x^2 - 12x + 36}{x^2 - 7x + 6}$

81. $\dfrac{x^3 + 27}{x^2 + x - 6}$

82. $\dfrac{x^3 - 8}{x^2 - 4}$

11.2 Techniques for Evaluating Limits

Dividing Out Technique

In Section 11.1, you studied several types of functions whose limits can be evaluated by direct substitution. In this section, you will study several techniques for evaluating limits of functions for which direct substitution fails.

Suppose you were asked to find the following limit.

$$\lim_{x \to -3} \frac{x^2 + x - 6}{x + 3}$$

Direct substitution fails because -3 is a zero of the denominator. By using a table, however, it appears that the limit of the function as $x \to -3$ is -5.

x	-3.01	-3.001	-3.0001	-3	-2.9999	-2.999	-2.99
$\dfrac{x^2 + x - 6}{x + 3}$	-5.01	-5.001	-5.0001	?	-4.9999	-4.999	-4.99

Another way to find the limit of this function is shown in Example 1.

Example 1   Dividing Out Technique

Find the limit: $\displaystyle \lim_{x \to -3} \frac{x^2 + x - 6}{x + 3}$.

Solution

Begin by factoring the numerator and dividing out any common factors.

$$\lim_{x \to -3} \frac{x^2 + x - 6}{x + 3} = \lim_{x \to -3} \frac{(x - 2)(x + 3)}{x + 3} \qquad \text{Factor numerator.}$$

$$= \lim_{x \to -3} \frac{(x - 2)(x + 3)}{x + 3} \qquad \text{Divide out common factor.}$$

$$= \lim_{x \to -3} (x - 2) \qquad \text{Simplify.}$$

$$= -3 - 2 \qquad \text{Direct substitution}$$

$$= -5 \qquad \text{Simplify.}$$

✓ CHECKPOINT   Now try Exercise 7.

This procedure for evaluating a limit is called the **dividing out technique.** The validity of this technique stems from the fact that if two functions agree at all but a single number c, they must have identical limit behavior at $x = c$. In Example 1, the functions given by

$$f(x) = \frac{x^2 + x - 6}{x + 3} \qquad \text{and} \qquad g(x) = x - 2$$

agree at all values of x other than $x = -3$. So, you can use $g(x)$ to find the limit of $f(x)$.

What you should learn

- Use the dividing out technique to evaluate limits of functions.
- Use the rationalizing technique to evaluate limits of functions.
- Approximate limits of functions graphically and numerically.
- Evaluate one-sided limits of functions.
- Evaluate limits of difference quotients from calculus.

Why you should learn it

Many definitions in calculus involve the limit of a function. For instance, in Exercises 77 and 78 on page 799, the definition of the velocity of a free-falling object at any instant in time involves finding the limit of a position function.

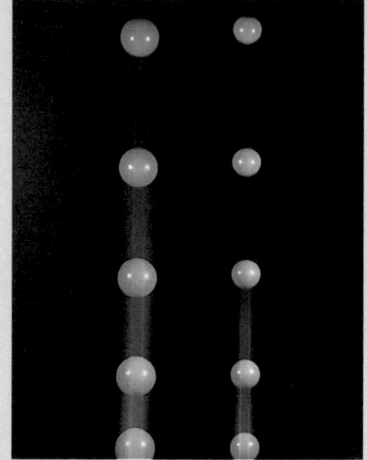

Peticolas Megna/Fundamental Photographs

Prerequisite Skills

To review factoring techniques, see Appendix G, Study Capsule 1.

The dividing out technique should be applied only when direct substitution produces 0 in both the numerator *and* the denominator. The resulting fraction, $\frac{0}{0}$, has no meaning as a real number. It is called an **indeterminate form** because you cannot, from the form alone, determine the limit. When you try to evaluate a limit of a rational function by direct substitution and encounter this form, you can conclude that the numerator and denominator must have a common factor. After factoring and dividing out, you should try direct substitution again.

Example 2 Dividing Out Technique

Find the limit.

$$\lim_{x \to 1} \frac{x - 1}{x^3 - x^2 + x - 1}$$

Solution

Begin by substituting $x = 1$ into the numerator and denominator.

$$1 - 1 = 0 \qquad\qquad \text{Numerator is 0 when } x = 1.$$

$$1^3 - 1^2 + 1 - 1 = 0 \qquad\qquad \text{Denominator is 0 when } x = 1.$$

Because both the numerator and denominator are zero when $x = 1$, direct substitution will not yield the limit. To find the limit, you should factor the numerator and denominator, divide out any common factors, and then try direct substitution again.

$$\lim_{x \to 1} \frac{x - 1}{x^3 - x^2 + x - 1} = \lim_{x \to 1} \frac{x - 1}{(x - 1)(x^2 + 1)} \qquad \text{Factor denominator.}$$

$$= \lim_{x \to 1} \frac{\cancel{x - 1}}{\cancel{(x - 1)}(x^2 + 1)} \qquad \text{Divide out common factor.}$$

$$= \lim_{x \to 1} \frac{1}{x^2 + 1} \qquad \text{Simplify.}$$

$$= \frac{1}{1^2 + 1} \qquad \text{Direct substitution}$$

$$= \frac{1}{2} \qquad \text{Simplify.}$$

This result is shown graphically in Figure 11.13.

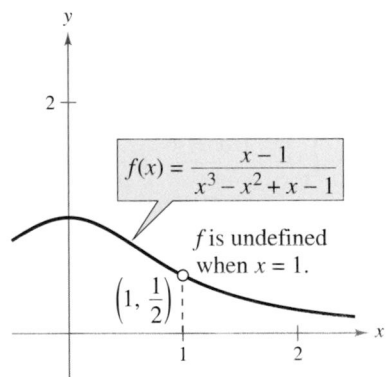

Figure 11.13

✔CHECKPOINT   Now try Exercise 9.

In Example 2, the factorization of the denominator can be obtained by dividing by $(x - 1)$ or by grouping as follows.

$$x^3 - x^2 + x - 1 = x^2(x - 1) + (x - 1)$$

$$= (x - 1)(x^2 + 1)$$

Rationalizing Technique

Another way to find the limits of some functions is first to rationalize the numerator of the function. This is called the **rationalizing technique.** Recall that rationalizing the numerator means multiplying the numerator and denominator by the conjugate of the numerator. For instance, the conjugate of $\sqrt{x} + 4$ is $\sqrt{x} - 4$.

Example 3   Rationalizing Technique

Find the limit: $\displaystyle\lim_{x \to 0} \frac{\sqrt{x+1} - 1}{x}$.

Solution

By direct substitution, you obtain the indeterminate form $\frac{0}{0}$.

$$\lim_{x \to 0} \frac{\sqrt{x+1} - 1}{x} = \frac{\sqrt{0+1} - 1}{0} = \frac{0}{0} \qquad \text{Indeterminate form}$$

In this case, you can rewrite the fraction by rationalizing the numerator.

$$\frac{\sqrt{x+1} - 1}{x} = \left(\frac{\sqrt{x+1} - 1}{x}\right)\left(\frac{\sqrt{x+1} + 1}{\sqrt{x+1} + 1}\right)$$

$$= \frac{(x+1) - 1}{x(\sqrt{x+1} + 1)} \qquad \text{Multiply.}$$

$$= \frac{x}{x(\sqrt{x+1} + 1)} \qquad \text{Simplify.}$$

$$= \frac{\cancel{x}}{\cancel{x}(\sqrt{x+1} + 1)} \qquad \text{Divide out common factor.}$$

$$= \frac{1}{\sqrt{x+1} + 1}, \qquad x \neq 0 \qquad \text{Simplify.}$$

Now you can evaluate the limit by direct substitution.

$$\lim_{x \to 0} \frac{\sqrt{x+1} - 1}{x} = \lim_{x \to 0} \frac{1}{\sqrt{x+1} + 1} = \frac{1}{\sqrt{0+1} + 1} = \frac{1}{1+1} = \frac{1}{2}$$

You can reinforce your conclusion that the limit is $\frac{1}{2}$ by constructing a table, as shown below, or by sketching a graph, as shown in Figure 11.14.

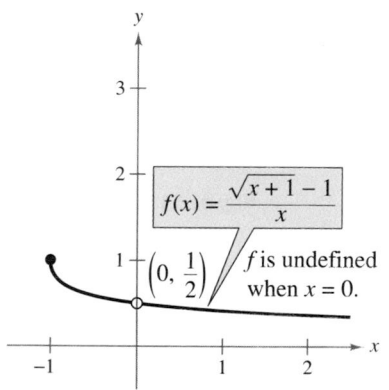

Figure 11.14

x	-0.1	-0.01	-0.001	0	0.001	0.01	0.1
$f(x)$	0.5132	0.5013	0.5001	?	0.4999	0.4988	0.4881

✓CHECKPOINT   Now try Exercise 17.

The rationalizing technique for evaluating limits is based on multiplication by a convenient form of 1. In Example 3, the convenient form is

$$1 = \frac{\sqrt{x+1} + 1}{\sqrt{x+1} + 1}.$$

Using Technology

The dividing out and rationalizing techniques may not work well for finding limits of nonalgebraic functions. You often need to use more sophisticated analytic techniques to find limits of these types of functions.

Example 4   Approximating a Limit

Approximate the limit: $\lim\limits_{x \to 0} (1 + x)^{1/x}$.

Numerical Solution

Let $f(x) = (1 + x)^{1/x}$. Because you are finding the limit when $x = 0$, use the *table* feature of a graphing utility to create a table that shows the values of f for x starting at $x = -0.01$ and setting the table step to 0.001, as shown in Figure 11.15. Because 0 is halfway between -0.001 and 0.001, use the average of the values of f at these two x-coordinates to estimate the limit as follows.

$$\lim\limits_{x \to 0} (1 + x)^{1/x} \approx \frac{2.7196 + 2.7169}{2} = 2.71825$$

The actual limit can be found algebraically to be $e \approx 2.71828$.

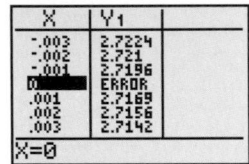

Figure 11.15

Graphical Solution

To approximate the limit graphically, graph the function $y = (1 + x)^{1/x}$, as shown in Figure 11.16. Using the *zoom* and *trace* features of the graphing utility, choose two points on the graph of f, such as

$$(-0.00017, 2.7185) \quad \text{and} \quad (0.00017, 2.7181)$$

as shown in Figure 11.17. Because the x-coordinates of these two points are equidistant from 0, you can approximate the limit to be the average of the y-coordinates. That is,

$$\lim\limits_{x \to 0} (1 + x)^{1/x} \approx \frac{2.7185 + 2.7181}{2} = 2.7183.$$

The actual limit can be found algebraically to be $e \approx 2.71828$.

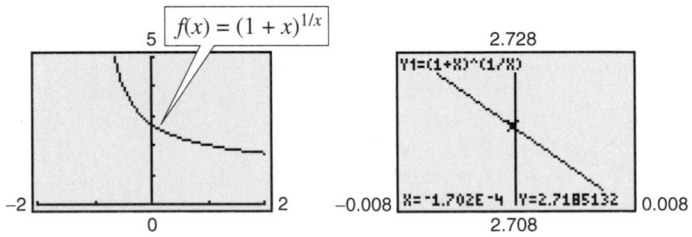

Figure 11.16   **Figure 11.17**

Example 5   Approximating a Limit Graphically

Approximate the limit: $\lim\limits_{x \to 0} \dfrac{\sin x}{x}$.

Solution

Direct substitution produces the indeterminate form $\frac{0}{0}$. To approximate the limit, begin by using a graphing utility to graph $f(x) = (\sin x)/x$, as shown in Figure 11.18. Then use the *zoom* and *trace* features of the graphing utility to choose a point on each side of 0, such as $(-0.0012467, 0.9999997)$ and $(0.0012467, 0.9999997)$. Finally, approximate the limit as the average of the y-coordinates of these two points, $\lim\limits_{x \to 0} (\sin x)/x \approx 0.9999997$. It can be shown algebraically that this limit is exactly 1.

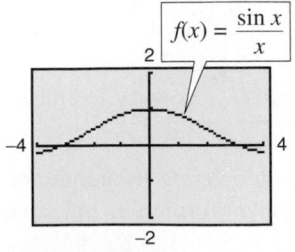

Figure 11.18

TECHNOLOGY TIP   The graphs shown in Figures 11.16 and 11.18 appear to be continuous at $x = 0$. But when you try to use the *trace* or the *value* feature of a graphing utility to determine the value of y when $x = 0$, there is no value given. Some graphing utilities can show breaks or holes in a graph when an appropriate viewing window is used. Because the holes in the graphs in Figures 11.16 and 11.18 occur on the y-axis, the holes are not visible.

TECHNOLOGY SUPPORT

For instructions on how to use the *zoom* and *trace* features and the *value* feature, see Appendix A; for specific keystrokes, go to this textbook's *Online Study Center*.

One-Sided Limits

In Section 11.1, you saw that one way in which a limit can fail to exist is when a function approaches a different value from the left side of c than it approaches from the right side of c. This type of behavior can be described more concisely with the concept of a **one-sided limit.**

$$\lim_{x \to c^-} f(x) = L_1 \text{ or } f(x) \to L_1 \text{ as } x \to c^- \qquad \text{Limit from the left}$$

$$\lim_{x \to c^+} f(x) = L_2 \text{ or } f(x) \to L_2 \text{ as } x \to c^+ \qquad \text{Limit from the right}$$

Example 6   Evaluating One-Sided Limits

Find the limit as $x \to 0$ from the left and the limit as $x \to 0$ from the right for

$$f(x) = \frac{|2x|}{x}.$$

Solution

From the graph of f, shown in Figure 11.19, you can see that $f(x) = -2$ for all $x < 0$. Therefore, the limit from the left is

$$\lim_{x \to 0^-} \frac{|2x|}{x} = -2. \qquad \text{Limit from the left}$$

Because $f(x) = 2$ for all $x > 0$, the limit from the right is

$$\lim_{x \to 0^+} \frac{|2x|}{x} = 2. \qquad \text{Limit from the right}$$

✓**CHECKPOINT**   Now try Exercise 53.

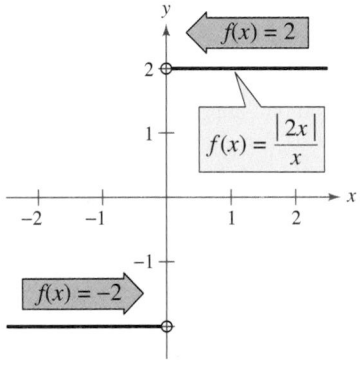

Figure 11.19

In Example 6, note that the function approaches different limits from the left and from the right. In such cases, the limit of $f(x)$ as $x \to c$ does not exist. For the limit of a function to exist as $x \to c$, it must be true that both one-sided limits exist and are equal.

> **Existence of a Limit**
>
> If f is a function and c and L are real numbers, then
>
> $$\lim_{x \to c} f(x) = L$$
>
> if and only if both the left and right limits *exist* and are *equal* to L.

Example 7   Finding One-Sided Limits

Find the limit of $f(x)$ as x approaches 1.

$$f(x) = \begin{cases} 4 - x, & x < 1 \\ 4x - x^2, & x > 1 \end{cases}$$

Prerequisite Skills

For a review of piecewise-defined functions, see Section 1.2.

Solution

Remember that you are concerned about the value of f *near* $x = 1$ rather than *at* $x = 1$. So, for $x < 1$, $f(x)$ is given by $4 - x$, and you can use direct substitution to obtain

$$\lim_{x \to 1^-} f(x) = \lim_{x \to 1^-} (4 - x) = 4 - 1 = 3.$$

For $x > 1$, $f(x)$ is given by $4x - x^2$, and you can use direct substitution to obtain

$$\lim_{x \to 1^+} f(x) = \lim_{x \to 1^+} (4x - x^2) = 4(1) - 1^2 = 3.$$

Because the one-sided limits both exist and are equal to 3, it follows that

$$\lim_{x \to 1} f(x) = 3.$$

The graph in Figure 11.20 confirms this conclusion.

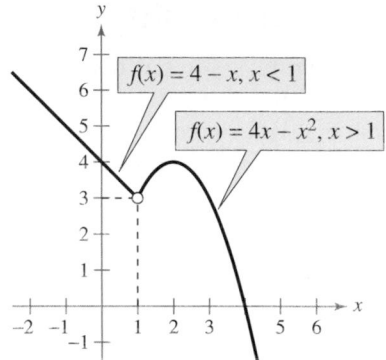

Figure 11.20

✔CHECKPOINT   Now try Exercise 57.

Example 8   Comparing Limits from the Left and Right

To ship a package overnight, a delivery service charges \$17.80 for the first pound and \$1.40 for each additional pound or portion of a pound. Let x represent the weight of a package and let $f(x)$ represent the shipping cost. Show that the limit of $f(x)$ as $x \to 2$ does not exist.

$$f(x) = \begin{cases} 17.80, & 0 < x \le 1 \\ 19.20, & 1 < x \le 2 \\ 20.60, & 2 < x \le 3 \end{cases}$$

Solution

The graph of f is shown in Figure 11.21. The limit of $f(x)$ as x approaches 2 from the left is

$$\lim_{x \to 2^-} f(x) = 19.20$$

whereas the limit of $f(x)$ as x approaches 2 from the right is

$$\lim_{x \to 2^+} f(x) = 20.60.$$

Because these one-sided limits are not equal, the limit of $f(x)$ as $x \to 2$ does not exist.

✔CHECKPOINT   Now try Exercise 81.

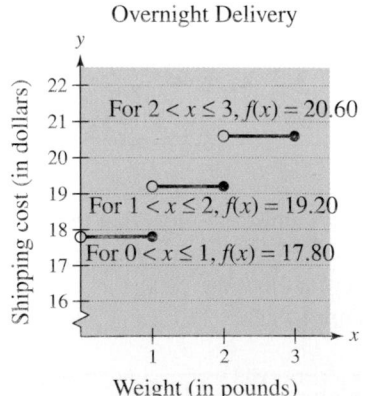

Figure 11.21

A Limit from Calculus

In the next section, you will study an important type of limit from calculus—the limit of a *difference quotient*.

Example 9   Evaluating a Limit from Calculus

For the function given by $f(x) = x^2 - 1$, find

$$\lim_{h \to 0} \frac{f(3 + h) - f(3)}{h}.$$

Solution

Direct substitution produces an indeterminate form.

$$\lim_{h \to 0} \frac{f(3 + h) - f(3)}{h} = \lim_{h \to 0} \frac{[(3 + h)^2 - 1] - [(3)^2 - 1]}{h}$$

$$= \lim_{h \to 0} \frac{9 + 6h + h^2 - 1 - 9 + 1}{h}$$

$$= \lim_{h \to 0} \frac{6h + h^2}{h}$$

$$= \frac{0}{0}$$

By factoring and dividing out, you obtain the following.

$$\lim_{h \to 0} \frac{f(3 + h) - f(3)}{h} = \lim_{h \to 0} \frac{6h + h^2}{h} = \lim_{h \to 0} \frac{h(6 + h)}{h}$$

$$= \lim_{h \to 0} (6 + h)$$

$$= 6 + 0$$

$$= 6$$

So, the limit is 6.

✓CHECKPOINT   Now try Exercise 73.

Note that for any *x*-value, the limit of a difference quotient is an expression of the form

$$\lim_{h \to 0} \frac{f(x + h) - f(x)}{h}.$$

Direct substitution into the difference quotient always produces the indeterminate form $\frac{0}{0}$. For instance,

$$\lim_{h \to 0} \frac{f(x + h) - f(x)}{h} = \frac{f(x + 0) - f(x)}{0} = \frac{f(x) - f(x)}{0} = \frac{0}{0}.$$

Prerequisite Skills

For a review of evaluating difference quotients, refer to Section 1.2.

11.2 Exercises

See www.CalcChat.com for worked-out solutions to odd-numbered exercises.

Vocabulary Check

Fill in the blanks.

1. To evaluate the limit of a rational function that has common factors in its numerator and denominator, use the _____ .

2. The fraction $\frac{0}{0}$ has no meaning as a real number and therefore is called an _____ .

3. The limit $\lim\limits_{x \to c^-} f(x) = L$ is an example of a _____ .

4. The limit of a _____ is an expression of the form $\lim\limits_{h \to 0} \dfrac{f(x + h) - f(x)}{h}$.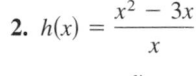

In Exercises 1–4, use the graph to determine each limit (if it exists). Then identify another function that agrees with the given function at all but one point.

1. $g(x) = \dfrac{-2x^2 + x}{x}$

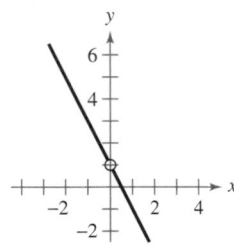

(a) $\lim\limits_{x \to 0} g(x)$

(b) $\lim\limits_{x \to -1} g(x)$

(c) $\lim\limits_{x \to -2} g(x)$

2. $h(x) = \dfrac{x^2 - 3x}{x}$

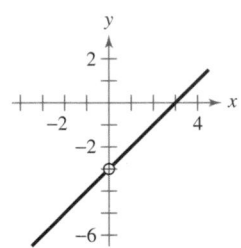

(a) $\lim\limits_{x \to -2} h(x)$

(b) $\lim\limits_{x \to 0} h(x)$

(c) $\lim\limits_{x \to 3} h(x)$

3. $g(x) = \dfrac{x^3 - x}{x - 1}$

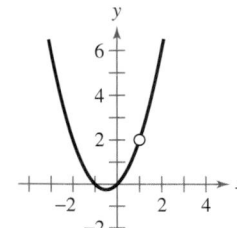

(a) $\lim\limits_{x \to 1} g(x)$

(b) $\lim\limits_{x \to -1} g(x)$

(c) $\lim\limits_{x \to 0} g(x)$

4. $f(x) = \dfrac{x^2 - 1}{x + 1}$

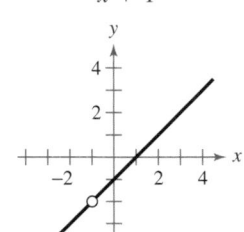

(a) $\lim\limits_{x \to 1} f(x)$

(b) $\lim\limits_{x \to 2} f(x)$

(c) $\lim\limits_{x \to -1} f(x)$

In Exercises 5–28, find the limit (if it exists). Use a graphing utility to verify your result graphically.

5. $\lim\limits_{x \to 6} \dfrac{x - 6}{x^2 - 36}$

6. $\lim\limits_{x \to 9} \dfrac{9 - x}{x^2 - 81}$

7. $\lim\limits_{x \to -1} \dfrac{1 - 2x - 3x^2}{1 + x}$

8. $\lim\limits_{x \to -4} \dfrac{2x^2 + 7x - 4}{x + 4}$

9. $\lim\limits_{t \to 2} \dfrac{t^3 - 8}{t - 2}$

10. $\lim\limits_{a \to -4} \dfrac{a^3 + 64}{a + 4}$

11. $\lim\limits_{x \to 1} \dfrac{x^4 - 1}{x^4 - 3x^2 - 4}$

12. $\lim\limits_{x \to 2} \dfrac{x^4 - 2x^2 - 8}{x^4 - 6x^2 + 8}$

13. $\lim\limits_{x \to -1} \dfrac{x^3 + 2x^2 - x - 2}{x^3 + 4x^2 - x - 4}$

14. $\lim\limits_{x \to -3} \dfrac{x^3 + 2x^2 - 9x - 18}{x^3 + x^2 - 9x - 9}$

15. $\lim\limits_{x \to 2} \dfrac{x^3 + 2x^2 - 5x - 6}{x^3 - 7x + 6}$

16. $\lim\limits_{x \to 3} \dfrac{x^3 - 4x^2 - 3x + 18}{x^3 - 4x^2 + x + 6}$

17. $\lim\limits_{y \to 0} \dfrac{\sqrt{5 + y} - \sqrt{5}}{y}$

18. $\lim\limits_{z \to 0} \dfrac{\sqrt{7 - z} - \sqrt{7}}{z}$

19. $\lim\limits_{x \to -3} \dfrac{\sqrt{x + 7} - 2}{x + 3}$

20. $\lim\limits_{x \to 2} \dfrac{4 - \sqrt{18 - x}}{x - 2}$

21. $\lim\limits_{x \to 0} \dfrac{\dfrac{1}{x + 1} - 1}{x}$

22. $\lim\limits_{x \to 0} \dfrac{\dfrac{1}{x - 8} + \dfrac{1}{8}}{x}$

23. $\lim\limits_{x \to 0} \dfrac{\sec x}{\tan x}$

24. $\lim\limits_{x \to \pi/2} \dfrac{1 - \sin x}{\cos x}$

25. $\lim\limits_{x \to 0} \dfrac{\cos 2x}{\cot 2x}$

26. $\lim\limits_{x \to 0} \dfrac{\sin x - x}{\sin x}$

27. $\lim\limits_{x \to \pi/2} \dfrac{\sin x - 1}{x}$

28. $\lim\limits_{x \to \pi} \dfrac{1 + \cos x}{x}$

In Exercises 29–36, use a graphing utility to graph the function and approximate the limit.

29. $\lim\limits_{x \to 0} \dfrac{\sqrt{x + 3} - \sqrt{3}}{x}$

30. $\lim\limits_{x \to 0} \dfrac{\sqrt{x + 4} - 2}{x}$

31. $\lim\limits_{x \to 0} \dfrac{\sqrt{2x+1}-1}{x}$

32. $\lim\limits_{x \to 9} \dfrac{3-\sqrt{x}}{x-9}$

33. $\lim\limits_{x \to 2} \dfrac{x^5-32}{x-2}$

34. $\lim\limits_{x \to 1} \dfrac{x^4-1}{x-1}$

35. $\lim\limits_{x \to 0} \dfrac{\dfrac{1}{x+4}-\dfrac{1}{4}}{x}$

36. $\lim\limits_{x \to 0} \dfrac{\dfrac{1}{2+x}-\dfrac{1}{2}}{x}$

In Exercises 37–48, use a graphing utility to graph the function and approximate the limit. Write an approximation that is accurate to three decimal places.

37. $\lim\limits_{x \to 0} \dfrac{e^{2x}-1}{x}$

38. $\lim\limits_{x \to 0} \dfrac{1-e^{-x}}{x}$

39. $\lim\limits_{x \to 0^+} (x \ln x)$

40. $\lim\limits_{x \to 0^+} (x^2 \ln x)$

41. $\lim\limits_{x \to 0} \dfrac{\sin 2x}{x}$

42. $\lim\limits_{x \to 0} \dfrac{\sin 3x}{x}$

43. $\lim\limits_{x \to 0} \dfrac{\tan x}{x}$

44. $\lim\limits_{x \to 0} \dfrac{1-\cos 2x}{x}$

45. $\lim\limits_{x \to 1} \dfrac{1-\sqrt[3]{x}}{1-x}$

46. $\lim\limits_{x \to 1} \dfrac{\sqrt[3]{x}-x}{x-1}$

47. $\lim\limits_{x \to 0} (1-x)^{2/x}$

48. $\lim\limits_{x \to 0} (1+2x)^{1/x}$

Graphical, Numerical, and Algebraic Analysis   **In Exercises 49–52, (a) graphically approximate the limit (if it exists) by using a graphing utility to graph the function, (b) numerically approximate the limit (if it exists) by using the *table* feature of a graphing utility to create a table, and (c) algebraically evaluate the limit (if it exists) by the appropriate technique(s).**

49. $\lim\limits_{x \to 1^-} \dfrac{x-1}{x^2-1}$

50. $\lim\limits_{x \to 5^+} \dfrac{5-x}{25-x^2}$

51. $\lim\limits_{x \to 16^+} \dfrac{4-\sqrt{x}}{x-16}$

52. $\lim\limits_{x \to 0^-} \dfrac{\sqrt{x+2}-\sqrt{2}}{x}$

In Exercises 53–60, graph the function. Determine the limit (if it exists) by evaluating the corresponding one-sided limits.

53. $\lim\limits_{x \to 6} \dfrac{|x-6|}{x-6}$

54. $\lim\limits_{x \to 2} \dfrac{|x-2|}{x-2}$

55. $\lim\limits_{x \to 1} \dfrac{1}{x^2+1}$

56. $\lim\limits_{x \to 1} \dfrac{1}{x^2-1}$

57. $\lim\limits_{x \to 2} f(x)$ where $f(x) = \begin{cases} x-1, & x \le 2 \\ 2x-3, & x > 2 \end{cases}$

58. $\lim\limits_{x \to 1} f(x)$ where $f(x) = \begin{cases} 2x+1, & x < 1 \\ 4-x^2, & x \ge 1 \end{cases}$

59. $\lim\limits_{x \to 1} f(x)$ where $f(x) = \begin{cases} 4-x^2, & x \le 1 \\ 3-x, & x > 1 \end{cases}$

60. $\lim\limits_{x \to 0} f(x)$ where $f(x) = \begin{cases} 4-x^2, & x \le 0 \\ x+4, & x > 0 \end{cases}$

In Exercises 61–66, use a graphing utility to graph the function and the equations $y = x$ and $y = -x$ in the same viewing window. Use the graph to find $\lim\limits_{x \to 0} f(x)$.

61. $f(x) = x \cos x$

62. $f(x) = |x \sin x|$

63. $f(x) = |x| \sin x$

64. $f(x) = |x| \cos x$

65. $f(x) = x \sin \dfrac{1}{x}$

66. $f(x) = x \cos \dfrac{1}{x}$

In Exercises 67 and 68, state which limit can be evaluated by using direct substitution. Then evaluate or approximate each limit.

67. (a) $\lim\limits_{x \to 0} x^2 \sin x^2$ (b) $\lim\limits_{x \to 0} \dfrac{\sin x^2}{x^2}$

68. (a) $\lim\limits_{x \to 0} \dfrac{x}{\cos x}$ (b) $\lim\limits_{x \to 0} \dfrac{1-\cos x}{x}$

∫ **In Exercises 69–76, find $\lim\limits_{h \to 0} \dfrac{f(x+h)-f(x)}{h}$.**

69. $f(x) = 3x-1$

70. $f(x) = 5-6x$

71. $f(x) = \sqrt{x}$

72. $f(x) = \sqrt{x-2}$

73. $f(x) = x^2-3x$

74. $f(x) = 4-2x-x^2$

75. $f(x) = \dfrac{1}{x+2}$

76. $f(x) = \dfrac{1}{x-1}$

Free-Falling Object   **In Exercises 77 and 78, use the position function $s(t) = -16t^2 + 128$, which gives the height (in feet) of a free-falling object. The velocity at time $t = a$ seconds is given by**

$$\lim\limits_{t \to a} \dfrac{s(a)-s(t)}{a-t}.$$

77. Find the velocity when $t = 1$ second.

78. Find the velocity when $t = 2$ seconds.

79. *Communications*   The cost of a cellular phone call within your calling area is $1.00 for the first minute and $0.25 for each additional minute or portion of a minute. A model for the cost C is given by $C(t) = 1.00 - 0.25[\![-(t-1)]\!]$, where t is the time in minutes. (Recall from Section 1.3 that $f(x) = [\![x]\!]$ = the greatest integer less than or equal to x.)

(a) Sketch the graph of C for $0 < t \le 5$.

(b) Complete the table and observe the behavior of C as t approaches 3.5. Use the graph from part (a) and the table to find $\lim\limits_{t \to 3.5} C(t)$.

t	3	3.3	3.4	3.5	3.6	3.7	4
C				?			

(c) Complete the table and observe the behavior of C as t approaches 3. Does the limit of $C(t)$ as t approaches 3 exist? Explain.

t	2	2.5	2.9	3	3.1	3.5	4
C				?			

80. *Communications*   The cost of a cellular phone call within your calling area is \$1.25 for the first minute and \$0.15 for each additional minute or portion of a minute. A model for the cost C is given by $C(t) = 1.25 - 0.15[\![-(t-1)]\!]$, where t is the time in minutes. (Recall from Section 1.3 that $f(x) = [\![x]\!] =$ the greatest integer less than or equal to x.)

(a) Sketch the graph of C for $0 < t \le 5$.

(b) Complete the table and observe the behavior of C as t approaches 3.5. Use the graph from part (a) and the table to find $\lim_{t \to 3.5} C(t)$.

t	3	3.3	3.4	3.5	3.6	3.7	4
C				?			

(c) Complete the table and observe the behavior of C as t approaches 3. Does the limit of $C(t)$ as t approaches 3 exist? Explain.

t	2	2.5	2.9	3	3.1	3.5	4
C				?			

81. *Salary Contract*   A union contract guarantees a 20% salary increase yearly for 3 years. For a current salary of \$32,500, the salary $f(t)$ (in thousands of dollars) for the next 3 years is given by

$$f(t) = \begin{cases} 32.50, & 0 < t \le 1 \\ 39.00, & 1 < t \le 2 \\ 46.80, & 2 < t \le 3 \end{cases}$$

where t represents the time in years. Show that the limit of f as $t \to 2$ does not exist.

82. *Consumer Awareness*   The cost of sending a package overnight is \$14.40 for the first pound and \$3.90 for each additional pound or portion of a pound. A plastic mailing bag can hold up to 3 pounds. The cost $f(x)$ of sending a package in a plastic mailing bag is given by

$$f(x) = \begin{cases} 14.40, & 0 < x \le 1 \\ 18.30, & 1 < x \le 2 \\ 22.20, & 2 < x \le 3 \end{cases}$$

where x represents the weight of the package (in pounds). Show that the limit of f as $x \to 1$ does not exist.

Synthesis

True or False?   **In Exercises 83 and 84, determine whether the statement is true or false. Justify your answer.**

83. When your attempt to find the limit of a rational function yields the indeterminate form $\frac{0}{0}$, the rational function's numerator and denominator have a common factor.

84. If $f(c) = L$, then $\lim_{x \to c} f(x) = L$.

85. *Think About It*

(a) Sketch the graph of a function for which $f(2)$ is defined but for which the limit of $f(x)$ as x approaches 2 does not exist.

(b) Sketch the graph of a function for which the limit of $f(x)$ as x approaches 1 is 4 but for which $f(1) \ne 4$.

86. *Writing*   Consider the limit of the rational function $p(x)/q(x)$. What conclusion can you make if direct substitution produces each expression? Write a short paragraph explaining your reasoning.

(a) $\lim_{x \to c} \dfrac{p(x)}{q(x)} = \dfrac{0}{1}$ 　　 (b) $\lim_{x \to c} \dfrac{p(x)}{q(x)} = \dfrac{1}{1}$

(c) $\lim_{x \to c} \dfrac{p(x)}{q(x)} = \dfrac{1}{0}$ 　　 (d) $\lim_{x \to c} \dfrac{p(x)}{q(x)} = \dfrac{0}{0}$

Skills Review

87. Write an equation of the line that passes through $(6, -10)$ and is perpendicular to the line that passes through $(4, -6)$ and $(3, -4)$.

88. Write an equation of the line that passes through $(1, -1)$ and is parallel to the line that passes through $(3, -3)$ and $(5, -2)$.

In Exercises 89–94, identify the type of conic algebraically. Then use a graphing utility to graph the conic.

89. $r = \dfrac{3}{1 + \cos \theta}$ 　　 90. $r = \dfrac{12}{3 + 2 \sin \theta}$

91. $r = \dfrac{9}{2 + 3 \cos \theta}$ 　　 92. $r = \dfrac{4}{4 + \cos \theta}$

93. $r = \dfrac{5}{1 - \sin \theta}$ 　　 94. $r = \dfrac{6}{3 - 4 \sin \theta}$

In Exercises 95–98, determine whether the vectors are orthogonal, parallel, or neither.

95. $\langle 7, -2, 3 \rangle, \langle -1, 4, 5 \rangle$

96. $\langle 5, 5, 0 \rangle, \langle 0, 5, 1 \rangle$

97. $\langle -4, 3, -6 \rangle, \langle 12, -9, 18 \rangle$

98. $\langle 2, -3, 1 \rangle, \langle -2, 2, 2 \rangle$

11.3   The Tangent Line Problem

Tangent Line to a Graph

Calculus is a branch of mathematics that studies rates of change of functions. If you go on to take a course in calculus, you will learn that rates of change have many applications in real life.

Earlier in the text, you learned how the slope of a line indicates the rate at which a line rises or falls. For a line, this rate (or slope) is the same at every point on the line. For graphs other than lines, the rate at which the graph rises or falls changes from point to point. For instance, in Figure 11.22, the parabola is rising more quickly at the point (x_1, y_1) than it is at the point (x_2, y_2). At the vertex (x_3, y_3), the graph levels off, and at the point (x_4, y_4), the graph is falling.

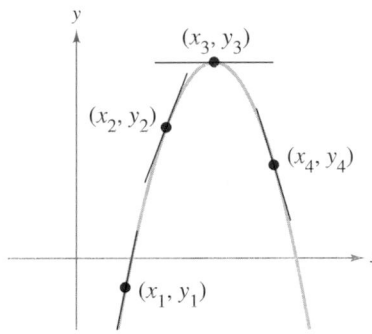

Figure 11.22

To determine the rate at which a graph rises or falls at a *single point*, you can find the slope of the tangent line at that point. In simple terms, the **tangent line** to the graph of a function f at a point $P(x_1, y_1)$ is the line that best approximates the slope of the graph at the point. Figure 11.23 shows other examples of tangent lines.

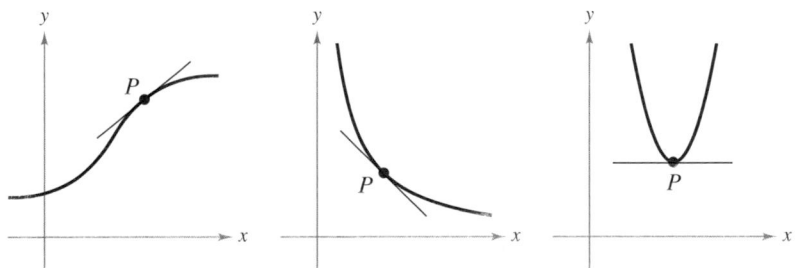

Figure 11.23

From geometry, you know that a line is tangent to a circle if the line intersects the circle at only one point. Tangent lines to noncircular graphs, however, can intersect the graph at more than one point. For instance, in the first graph in Figure 11.23, if the tangent line were extended, it would intersect the graph at a point other than the point of tangency.

What you should learn

- Use a tangent line to approximate the slope of a graph at a point.
- Use the limit definition of slope to find exact slopes of graphs.
- Find derivatives of functions and use derivatives to find slopes of graphs.

Why you should learn it

The derivative, or the slope of the tangent line to the graph of a function at a point, can be used to analyze rates of change. For instance, in Exercise 65 on page 809, the derivative is used to analyze the rate of change of the volume of a spherical balloon.

© Richard Hutchings/Corbis

Prerequisite Skills

For a review of the slopes of lines, see Section 1.1.

Slope of a Graph

Because a tangent line approximates the slope of a graph at a point, the problem of finding the slope of a graph at a point is the same as finding the slope of the tangent line at the point.

Example 1   Visually Approximating the Slope of a Graph

Use the graph in Figure 11.24 to approximate the slope of the graph of $f(x) = x^2$ at the point $(1, 1)$.

Solution

From the graph of $f(x) = x^2$, you can see that the tangent line at $(1, 1)$ rises approximately two units for each unit change in x. So, you can estimate the slope of the tangent line at $(1, 1)$ to be

$$\text{Slope} = \frac{\text{change in } y}{\text{change in } x}$$

$$\approx \frac{2}{1}$$

$$= 2.$$

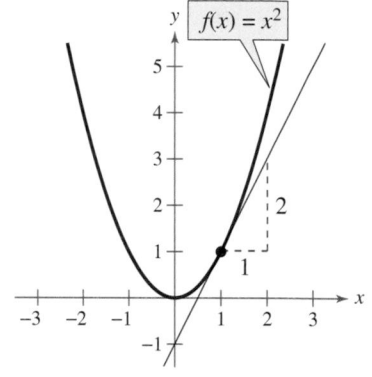

Figure 11.24

Because the tangent line at the point $(1, 1)$ has a slope of about 2, you can conclude that the graph of f has a slope of about 2 at the point $(1, 1)$.

 Now try Exercise 1.

When you are visually approximating the slope of a graph, remember that the scales on the horizontal and vertical axes may differ. When this happens (as it frequently does in applications), the slope of the tangent line is distorted, and you must be careful to account for the difference in scales.

Example 2   Approximating the Slope of a Graph

Figure 11.25 graphically depicts the monthly normal temperatures (in degrees Fahrenheit) for Dallas, Texas. Approximate the slope of this graph at the indicated point and give a physical interpretation of the result.   (Source: National Climatic Data Center)

Solution

From the graph, you can see that the tangent line at the given point falls approximately 16 units for each two-unit change in x. So, you can estimate the slope at the given point to be

$$\text{Slope} = \frac{\text{change in } y}{\text{change in } x} \approx \frac{-16}{2} = -8 \text{ degrees per month.}$$

This means that you can expect the monthly normal temperature in November to be about 8 degrees lower than the normal temperature in October.

 Now try Exercise 3.

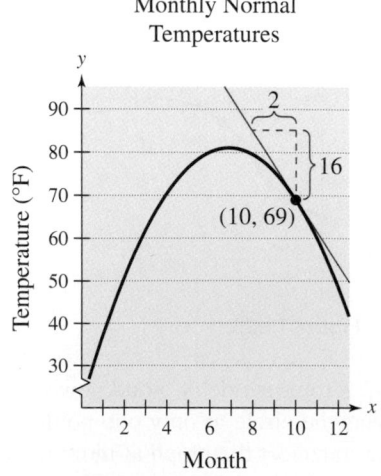

Figure 11.25

Slope and the Limit Process

In Examples 1 and 2, you approximated the slope of a graph at a point by creating a graph and then "eyeballing" the tangent line at the point of tangency. A more systematic method of approximating tangent lines makes use of a **secant line** through the point of tangency and a second point on the graph, as shown in Figure 11.26. If $(x, f(x))$ is the point of tangency and $(x + h, f(x + h))$ is a second point on the graph of f, the slope of the secant line through the two points is given by

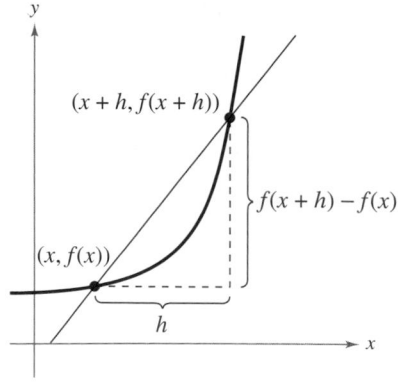

Figure 11.26

$$m_{sec} = \frac{f(x + h) - f(x)}{h}. \qquad \text{Slope of secant line}$$

The right side of this equation is called the **difference quotient.** The denominator h is the *change in x*, and the numerator is the *change in y*. The beauty of this procedure is that you obtain more and more accurate approximations of the slope of the tangent line by choosing points closer and closer to the point of tangency, as shown in Figure 11.27.

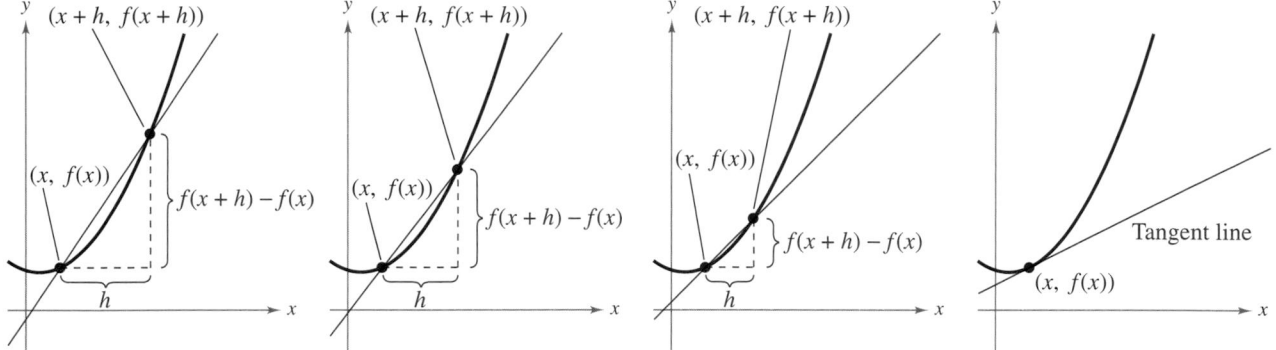

As h approaches 0, the secant line approaches the tangent line.
Figure 11.27

Using the limit process, you can find the *exact* slope of the tangent line at $(x, f(x))$.

Definition of the Slope of a Graph

The **slope** m of the graph of f at the point $(x, f(x))$ is equal to the slope of its tangent line at $(x, f(x))$, and is given by

$$m = \lim_{h \to 0} m_{sec}$$

$$= \lim_{h \to 0} \frac{f(x + h) - f(x)}{h}$$

provided this limit exists.

From the above definition and from Section 11.2, you can see that the difference quotient is used frequently in calculus. Using the difference quotient to find the slope of a tangent line to a graph is a major concept of calculus.

Example 3   Finding the Slope of a Graph

Find the slope of the graph of $f(x) = x^2$ at the point $(-2, 4)$.

Solution

Find an expression that represents the slope of a secant line at $(-2, 4)$.

$$m_{\text{sec}} = \frac{f(-2 + h) - f(-2)}{h}$$ Set up difference quotient.

$$= \frac{(-2 + h)^2 - (-2)^2}{h}$$ Substitute into $f(x) = x^2$.

$$= \frac{4 - 4h + h^2 - 4}{h}$$ Expand terms.

$$= \frac{-4h + h^2}{h}$$ Simplify.

$$= \frac{\cancel{h}(-4 + h)}{\cancel{h}}$$ Factor and divide out.

$$= -4 + h, \quad h \neq 0$$ Simplify.

Next, take the limit of m_{sec} as h approaches 0.

$$m = \lim_{h \to 0} m_{\text{sec}} = \lim_{h \to 0} (-4 + h) = -4 + 0 = -4$$

The graph has a slope of -4 at the point $(-2, 4)$, as shown in Figure 11.28.

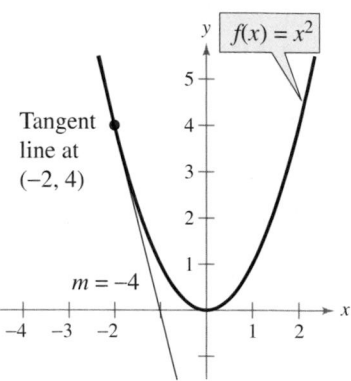

Figure 11.28

✓CHECKPOINT   Now try Exercise 5.

Example 4   Finding the Slope of a Graph

Find the slope of $f(x) = -2x + 4$.

Solution

$$m = \lim_{h \to 0} \frac{f(x + h) - f(x)}{h}$$ Set up difference quotient.

$$= \lim_{h \to 0} \frac{[-2(x + h) + 4] - (-2x + 4)}{h}$$ Substitute into $f(x) = -2x + 4$.

$$= \lim_{h \to 0} \frac{-2x - 2h + 4 + 2x - 4}{h}$$ Expand terms.

$$= \lim_{h \to 0} \frac{-2\cancel{h}}{\cancel{h}}$$ Divide out.

$$= -2$$ Simplify.

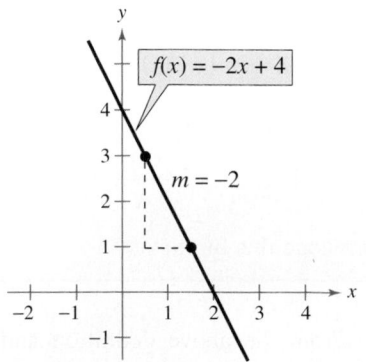

Figure 11.29

You know from your study of linear functions that the line given by $f(x) = -2x + 4$ has a slope of -2, as shown in Figure 11.29. This conclusion is consistent with that obtained by the limit definition of slope, as shown above.

✓CHECKPOINT   Now try Exercise 7.

It is important that you see the difference between the ways the difference quotients were set up in Examples 3 and 4. In Example 3, you were finding the slope of a graph at a specific point $(c, f(c))$. To find the slope in such a case, you can use the following form of the difference quotient.

$$m = \lim_{h \to 0} \frac{f(c + h) - f(c)}{h}$$ Slope at specific point

In Example 4, however, you were finding a *formula* for the slope at *any* point on the graph. In such cases, you should use x, rather than c, in the difference quotient.

$$m = \lim_{h \to 0} \frac{f(x + h) - f(x)}{h}$$ Formula for slope

Except for linear functions, this form will always produce a function of x, which can then be evaluated to find the slope at any desired point.

Example 5   Finding a Formula for the Slope of a Graph

Find a formula for the slope of the graph of $f(x) = x^2 + 1$. What are the slopes at the points $(-1, 2)$ and $(2, 5)$?

Solution

$$m_{\text{sec}} = \frac{f(x + h) - f(x)}{h}$$ Set up difference quotient.

$$= \frac{[(x + h)^2 + 1] - (x^2 + 1)}{h}$$ Substitute into $f(x) = x^2 + 1$.

$$= \frac{x^2 + 2xh + h^2 + 1 - x^2 - 1}{h}$$ Expand terms.

$$= \frac{2xh + h^2}{h}$$ Simplify.

$$= \frac{\cancel{h}(2x + h)}{\cancel{h}}$$ Factor and divide out.

$$= 2x + h, \quad h \neq 0$$ Simplify.

Next, take the limit of m_{sec} as h approaches 0.

$$m = \lim_{h \to 0} m_{\text{sec}} = \lim_{h \to 0} (2x + h) = 2x + 0 = 2x$$

Using the formula $m = 2x$ for the slope at $(x, f(x))$, you can find the slope at the specified points. At $(-1, 2)$, the slope is

$$m = 2(-1) = -2$$

and at $(2, 5)$, the slope is

$$m = 2(2) = 4.$$

The graph of f is shown in Figure 11.30.

✔**CHECKPOINT**   Now try Exercise 13.

Try verifying the result in Example 5 by using a graphing utility to graph the function and the tangent lines at $(-1, 2)$ and $(2, 5)$ as

$$y_1 = x^2 + 1$$

$$y_2 = -2x$$

$$y_3 = 4x - 3$$

in the same viewing window. Some graphing utilities even have a *tangent* feature that automatically graphs the tangent line to a curve at a given point. If you have such a graphing utility, try verifying the solution of Example 5 using this feature. For instructions on how to use the *tangent* feature, see Appendix A; for specific keystrokes, go to this textbook's *Online Study Center*.

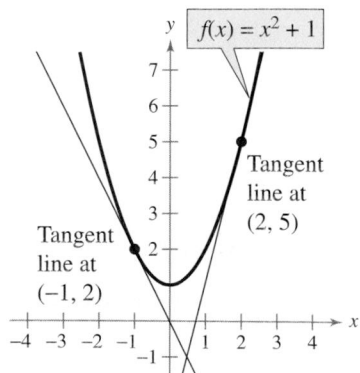

Figure 11.30

The Derivative of a Function

In Example 5, you started with the function $f(x) = x^2 + 1$ and used the limit process to derive another function, $m = 2x$, that represents the slope of the graph of f at the point $(x, f(x))$. This derived function is called the **derivative** of f at x. It is denoted by $f'(x)$, which is read as "f prime of x."

Definition of the Derivative

The **derivative** of f at x is given by

$$f'(x) = \lim_{h \to 0} \frac{f(x + h) - f(x)}{h}$$

provided this limit exists.

Remember that the derivative $f'(x)$ is a formula for the slope of the tangent line to the graph of f at the point $(x, f(x))$.

Example 6   Finding a Derivative   algebra of calculus

Find the derivative of $f(x) = 3x^2 - 2x$.

Solution

$$f'(x) = \lim_{h \to 0} \frac{f(x + h) - f(x)}{h}$$

$$= \lim_{h \to 0} \frac{[3(x + h)^2 - 2(x + h)] - (3x^2 - 2x)}{h}$$

$$= \lim_{h \to 0} \frac{3x^2 + 6xh + 3h^2 - 2x - 2h - 3x^2 + 2x}{h}$$

$$= \lim_{h \to 0} \frac{6xh + 3h^2 - 2h}{h}$$

$$= \lim_{h \to 0} \frac{\cancel{h}(6x + 3h - 2)}{\cancel{h}}$$

$$= \lim_{h \to 0} (6x + 3h - 2)$$

$$= 6x + 3(0) - 2$$

$$= 6x - 2$$

So, the derivative of $f(x) = 3x^2 - 2x$ is $f'(x) = 6x - 2$.

✓**CHECKPOINT**   Now try Exercise 29.

Note that in addition to $f'(x)$, other notations can be used to denote the derivative of $y = f(x)$. The most common are

$$\frac{dy}{dx}, \quad y', \quad \frac{d}{dx}[f(x)], \quad \text{and} \quad D_x[y].$$

Example 7   Using the Derivative

Find $f'(x)$ for $f(x) = \sqrt{x}$. Then find the slopes of the graph of f at the points $(1, 1)$ and $(4, 2)$ and equations of the tangent lines to the graph at the points.

Solution

$$f'(x) = \lim_{h \to 0} \frac{f(x + h) - f(x)}{h}$$

$$= \lim_{h \to 0} \frac{\sqrt{x + h} - \sqrt{x}}{h}$$

Because direct substitution yields the indeterminate form $\frac{0}{0}$, you should use the rationalizing technique discussed in Section 11.2 to find the limit.

$$f'(x) = \lim_{h \to 0} \left(\frac{\sqrt{x + h} - \sqrt{x}}{h} \right) \left(\frac{\sqrt{x + h} + \sqrt{x}}{\sqrt{x + h} + \sqrt{x}} \right)$$

$$= \lim_{h \to 0} \frac{(x + h) - x}{h(\sqrt{x + h} + \sqrt{x})}$$

$$= \lim_{h \to 0} \frac{\cancel{h}}{\cancel{h}(\sqrt{x + h} + \sqrt{x})}$$

$$= \lim_{h \to 0} \frac{1}{\sqrt{x + h} + \sqrt{x}}$$

$$= \frac{1}{\sqrt{x + 0} + \sqrt{x}} = \frac{1}{2\sqrt{x}}$$

At the point $(1, 1)$, the slope is

$$f'(1) = \frac{1}{2\sqrt{1}} = \frac{1}{2}.$$

An equation of the tangent line at the point $(1, 1)$ is

$y - y_1 = m(x - x_1)$	Point-slope form
$y - 1 = \frac{1}{2}(x - 1)$	Substitute $\frac{1}{2}$ for m, 1 for x_1, and 1 for y_1.
$y = \frac{1}{2}x + \frac{1}{2}.$	Tangent line

At the point $(4, 2)$, the slope is

$$f'(4) = \frac{1}{2\sqrt{4}} = \frac{1}{4}.$$

An equation of the tangent line at the point $(4, 2)$ is

$y - y_1 = m(x - x_1)$	Point-slope form
$y - 2 = \frac{1}{4}(x - 4)$	Substitute $\frac{1}{4}$ for m, 4 for x_1, and 2 for y_1.
$y = \frac{1}{4}x + 1.$	Tangent line

The graphs of f and the tangent lines at the points $(1, 1)$ and $(4, 2)$ are shown in Figure 11.31.

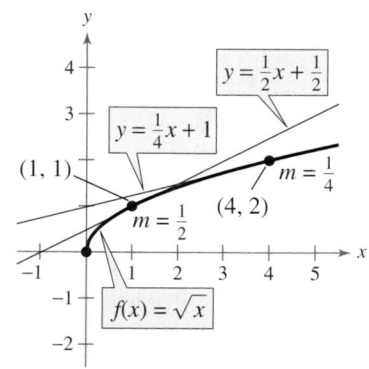

Figure 11.31

✓CHECKPOINT   Now try Exercise 39.

11.3 Exercises

See www.CalcChat.com for worked-out solutions to odd-numbered exercises.

Vocabulary Check

Fill in the blanks.

1. _____ is the study of the rates of change of functions.

2. The _____ to the graph of a function at a point is the line that best approximates the slope of the graph at the point.

3. A _____ is a line through the point of tangency and a second point on the graph.

4. The slope of the secant line is represented by the _____ $m_{\sec} = \dfrac{f(x + h) - f(x)}{h}$.

5. The _____ of a function f at x represents the slope of the graph of f at the point $(x, f(x))$.

In Exercises 1–4, use the figure to approximate the slope of the curve at the point (x, y).

1.
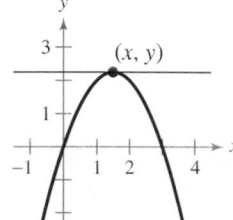

2.
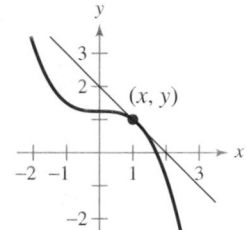

3.
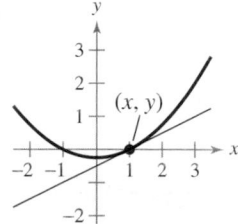

4.
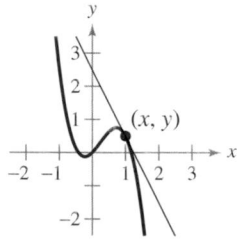

In Exercises 5–12, use the limit process to find the slope of the graph of the function at the specified point. Use a graphing utility to confirm your result.

5. $g(x) = x^2 - 4x$, $(3, -3)$

6. $f(x) = 10x - 2x^2$, $(3, 12)$

7. $g(x) = 5 - 2x$, $(1, 3)$

8. $h(x) = 2x + 5$, $(-1, -3)$

9. $g(x) = \dfrac{4}{x}$, $(2, 2)$

10. $g(x) = \dfrac{1}{x - 2}$, $\left(4, \dfrac{1}{2}\right)$

11. $h(x) = \sqrt{x}$, $(9, 3)$

12. $h(x) = \sqrt{x + 10}$, $(-1, 3)$

In Exercises 13–18, find a formula for the slope of the graph of f at the point $(x, f(x))$. Then use it to find the slopes at the two specified points.

13. $g(x) = 4 - x^2$

(a) $(0, 4)$

(b) $(-1, 3)$

14. $g(x) = x^3$

(a) $(1, 1)$

(b) $(-2, -8)$

15. $g(x) = \dfrac{1}{x + 4}$

(a) $\left(0, \dfrac{1}{4}\right)$

(b) $\left(-2, \dfrac{1}{2}\right)$

16. $f(x) = \dfrac{1}{x + 2}$

(a) $\left(0, \dfrac{1}{2}\right)$

(b) $(-1, 1)$

17. $g(x) = \sqrt{x - 1}$

(a) $(5, 2)$

(b) $(10, 3)$

18. $f(x) = \sqrt{x - 4}$

(a) $(5, 1)$

(b) $(8, 2)$

In Exercises 19–24, use a graphing utility to graph the function and the tangent line at the point $(1, f(1))$. Use the graph to approximate the slope of the tangent line.

19. $f(x) = x^2 - 2$

20. $f(x) = x^2 - 2x + 1$

21. $f(x) = \sqrt{2 - x}$

22. $f(x) = \sqrt{x + 3}$

23. $f(x) = \dfrac{4}{x + 1}$

24. $f(x) = \dfrac{3}{2 - x}$

∫ **In Exercises 25–38, find the derivative of the function.**

25. $f(x) = 5$

26. $f(x) = -1$

27. $f(x) = 9 - \dfrac{1}{3}x$

28. $f(x) = -5x + 2$

29. $f(x) = 4 - 3x^2$

30. $f(x) = x^2 - 3x + 4$

31. $f(x) = \dfrac{1}{x^2}$

32. $f(x) = \dfrac{1}{x^3}$

33. $f(x) = \sqrt{x - 4}$

34. $f(x) = \sqrt{x + 8}$

35. $f(x) = \dfrac{1}{x + 2}$

36. $f(x) = \dfrac{1}{x - 5}$

37. $f(x) = \dfrac{1}{\sqrt{x - 9}}$

38. $f(x) = \dfrac{1}{\sqrt{x + 1}}$

∫ **In Exercises 39–46, (a) find the slope of the graph of f at the given point, (b) then find an equation of the tangent line to the graph at the point, and (c) graph the function and the tangent line.**

39. $f(x) = x^2 - 1$, $(2, 3)$

40. $f(x) = 4 - x^2$, $(1, 3)$

41. $f(x) = x^3 - 2x, (1, -1)$ **42.** $f(x) = x^3 - x, (2, 6)$

43. $f(x) = \sqrt{x + 1}, (3, 2)$ **44.** $f(x) = \sqrt{x - 2}, (3, 1)$

45. $f(x) = \dfrac{1}{x + 5}, (-4, 1)$ **46.** $f(x) = \dfrac{1}{x - 3}, (4, 1)$

In Exercises 47–50, use a graphing utility to graph f over the interval $[-2, 2]$ and complete the table. Compare the value of the first derivative with a visual approximation of the slope of the graph.

x	-2	-1.5	-1	-0.5	0	0.5	1	1.5	2
$f(x)$									
$f'(x)$									

47. $f(x) = \frac{1}{2}x^2$ **48.** $f(x) = \frac{1}{4}x^3$

49. $f(x) = \sqrt{x + 3}$ **50.** $f(x) = \dfrac{x^2 - 4}{x + 4}$

∫ **In Exercises 51–54, find the derivative of f. Use the derivative to determine any points on the graph of f at which the tangent line is horizontal. Use a graphing utility to verify your results.**

51. $f(x) = x^2 - 4x + 3$

52. $f(x) = x^2 - 6x + 4$

53. $f(x) = 3x^3 - 9x$

54. $f(x) = x^3 + 3x$

∫ **In Exercises 55–62, use the function and its derivative to determine any points on the graph of f at which the tangent line is horizontal. Use a graphing utility to verify your results.**

55. $f(x) = x^4 - 2x^2, f'(x) = 4x^3 - 4x$

56. $f(x) = 3x^4 + 4x^3, f'(x) = 12x^3 + 12x^2$

57. $f(x) = 2 \cos x + x, f'(x) = -2 \sin x + 1$, over the interval $(0, 2\pi)$

58. $f(x) = x - 2 \sin x, f'(x) = 1 - 2 \cos x$, over the interval $(0, 2\pi)$

59. $f(x) = x^2 e^x, f'(x) = x^2 e^x + 2xe^x$

60. $f(x) = xe^{-x}, f'(x) = e^{-x} - xe^{-x}$

61. $f(x) = x \ln x, f'(x) = \ln x + 1$

62. $f(x) = \dfrac{\ln x}{x}, f'(x) = \dfrac{1 - \ln x}{x^2}$

63. *Population*   The projected populations y (in thousands) of New Jersey for selected years from 2010 to 2025 are shown in the table. (Source: U.S. Census Bureau)

Year	Population (in thousands)
2010	9018
2015	9256
2020	9462
2025	9637

Table for 63

(a) Use the *regression* feature of a graphing utility to find a quadratic model for the data. Let t represent the year, with $t = 10$ corresponding to 2010.

(b) Use a graphing utility to graph the model found in part (a). Estimate the slope of the graph when $t = 20$, and interpret the result.

(c) Find the derivative of the model in part (a). Then evaluate the derivative for $t = 20$.

(d) Write a brief statement regarding your results for parts (a) through (c).

64. *Market Research*   The data in the table shows the number N (in thousands) of books sold when the price per book is p (in dollars).

Price, p	Number of books, N
$10	900
$15	630
$20	396
$25	227
$30	102
$35	36

(a) Use the *regression* feature of a graphing utility to find a quadratic model for the data.

(b) Use a graphing utility to graph the model found in part (a). Estimate the slopes of the graph when $p = \$15$ and $p = \$30$.

(c) Use a graphing utility to graph the tangent lines to the model when $p = \$15$ and $p = \$30$. Compare the slopes given by the graphing utility with your estimates in part (b).

(d) The slopes of the tangent lines at $p = \$15$ and $p = \$30$ are not the same. Explain what this means to the company selling the books.

∫ **65.** *Rate of Change*   A spherical balloon is inflated. The volume V is approximated by the formula $V(r) = \frac{4}{3}\pi r^3$, where r is the radius.

(a) Find the derivative of V with respect to r.

(b) Evaluate the derivative when the radius is 4 inches.

(c) What type of unit would be applied to your answer in part (b)? Explain.

∫ **66. Rate of Change** An approximately spherical benign tumor is reducing in size. The surface area S is given by the formula $S(r) = 4\pi r^2$, where r is the radius.

(a) Find the derivative of S with respect to r.

(b) Evaluate the derivative when the radius is 2 millimeters.

(c) What type of unit would be applied to your answer in part (b)? Explain.

∫ **67. Vertical Motion** A water balloon is thrown upward from the top of an 80-foot building with a velocity of 64 feet per second. The height or displacement s (in feet) of the balloon can be modeled by the position function $s(t) = -16t^2 + 64t + 80$, where t is the time in seconds from when it was thrown.

(a) Find a formula for the instantaneous rate of change of the balloon.

(b) Find the average rate of change of the balloon after the first three seconds of flight. Explain your results.

(c) Find the time at which the balloon reaches its maximum height. Explain your method.

(d) Velocity is given by the derivative of the position function. Find the velocity of the balloon as it impacts the ground.

(e) Use a graphing utility to graph the model and verify your results for parts (a)–(d).

∫ **68. Vertical Motion** A Sacajawea dollar is dropped from the top of a 120-foot building. The height or displacement s (in feet) of the coin can be modeled by the position function $s(t) = -16t^2 + 120$, where t is the time in seconds from when it was dropped.

(a) Find a formula for the instantaneous rate of change of the coin.

(b) Find the average rate of change of the coin after the first two seconds of free fall. Explain your results.

(c) Velocity is given by the derivative of the position function. Find the velocity of the coin as it impacts the ground.

(d) Find the time when the coin's velocity is -60 feet per second.

(e) Use a graphing utility to graph the model and verify your results for parts (a)–(d).

Synthesis

True or False? **In Exercises 69 and 70, determine whether the statement is true or false. Justify your answer.**

69. The slope of the graph of $y = x^2$ is different at every point on the graph of f.

70. A tangent line to a graph can intersect the graph only at the point of tangency.

Library of Parent Functions **In Exercises 71–74, match the function with the graph of its *derivative*. It is not necessary to find the derivative of the function. [The graphs are labeled (a), (b), (c), and (d).]**

(a)
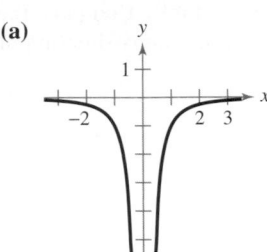

(b)
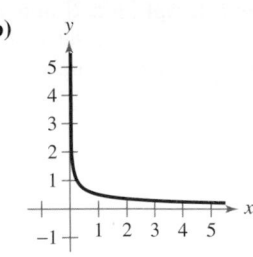

(c)
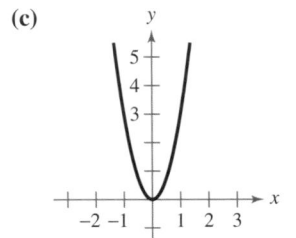

(d)
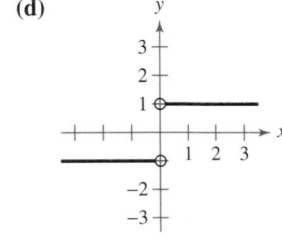

71. $f(x) = \sqrt{x}$

72. $f(x) = \dfrac{1}{x}$

73. $f(x) = |x|$

74. $f(x) = x^3$

75. *Think About It* Sketch the graph of a function whose derivative is always positive.

76. *Think About It* Sketch the graph of a function whose derivative is always negative.

Skills Review

In Exercises 77–80, sketch the graph of the rational function. As sketching aids, check for intercepts, vertical asymptotes, horizontal asymptotes, and slant asymptotes. Use a graphing utility to verify your graph.

77. $f(x) = \dfrac{1}{x^2 - x - 2}$

78. $f(x) = \dfrac{x - 2}{x^2 - 4x + 3}$

79. $f(x) = \dfrac{x^2 - x - 2}{x - 2}$

80. $f(x) = \dfrac{x^2 - 16}{x + 4}$

In Exercises 81–84, find the cross product of the vectors.

81. $\langle 1, 1, 1 \rangle, \langle 2, 1, -1 \rangle$

82. $\langle -10, 0, 6 \rangle, \langle 7, 0, 0 \rangle$

83. $\langle -4, 10, 0 \rangle, \langle 4, -1, 0 \rangle$

84. $\langle 8, -7, 14 \rangle, \langle -1, 8, 4 \rangle$

11.4 Limits at Infinity and Limits of Sequences

Limits at Infinity and Horizontal Asymptotes

As pointed out at the beginning of this chapter, there are two basic problems in calculus: finding **tangent lines** and finding the **area** of a region. In Section 11.3, you saw how limits can be used to solve the tangent line problem. In this section and the next, you will see how a different type of limit, a *limit at infinity*, can be used to solve the area problem. To get an idea of what is meant by a limit at infinity, consider the function

$$f(x) = \frac{x+1}{2x}.$$

The graph of f is shown in Figure 11.32. From earlier work, you know that $y = \frac{1}{2}$ is a horizontal asymptote of the graph of this function. Using limit notation, this can be written as follows.

$$\lim_{x \to -\infty} f(x) = \frac{1}{2} \qquad \text{Horizontal asymptote to the left}$$

$$\lim_{x \to \infty} f(x) = \frac{1}{2} \qquad \text{Horizontal asymptote to the right}$$

These limits mean that the value of $f(x)$ gets arbitrarily close to $\frac{1}{2}$ as x decreases or increases without bound.

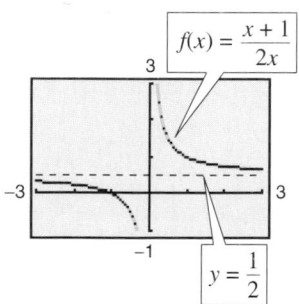

Figure 11.32

Definition of Limits at Infinity

If f is a function and L_1 and L_2 are real numbers, the statements

$$\lim_{x \to -\infty} f(x) = L_1 \qquad \text{Limit as } x \text{ approaches } -\infty$$

and

$$\lim_{x \to \infty} f(x) = L_2 \qquad \text{Limit as } x \text{ approaches } \infty$$

denote the **limits at infinity.** The first statement is read *"the limit of $f(x)$ as x approaches $-\infty$ is L_1,"* and the second is read *"the limit of $f(x)$ as x approaches ∞ is L_2."*

What you should learn

- Evaluate limits of functions at infinity.
- Find limits of sequences.

Why you should learn it

Finding limits at infinity is useful in highway safety applications. For instance, in Exercise 56 on page 819, you are asked to find a limit at infinity to predict the number of injuries due to motor vehicle accidents in the United States.

AP Photos

TECHNOLOGY TIP

Recall from Section 2.7 that some graphing utilities have difficulty graphing rational functions. In this text, rational functions are graphed using the *dot* mode of a graphing utility, and a blue curve is placed behind the graphing utility's display to indicate where the graph should appear.

To help evaluate limits at infinity, you can use the following definition.

> **Limits at Infinity**
>
> If r is a positive real number, then
>
> $$\lim_{x \to \infty} \frac{1}{x^r} = 0. \qquad \text{Limit toward the right}$$
>
> Furthermore, if x^r is defined when $x < 0$, then
>
> $$\lim_{x \to -\infty} \frac{1}{x^r} = 0. \qquad \text{Limit toward the left}$$

Limits at infinity share many of the properties of limits listed in Section 11.1. Some of these properties are demonstrated in the next example.

Example 1   Evaluating a Limit at Infinity

Find the limit.

$$\lim_{x \to \infty} \left(4 - \frac{3}{x^2} \right)$$

Algebraic Solution

Use the properties of limits listed in Section 11.1.

$$\lim_{x \to \infty} \left(4 - \frac{3}{x^2} \right) = \lim_{x \to \infty} 4 - \lim_{x \to \infty} \frac{3}{x^2}$$

$$= \lim_{x \to \infty} 4 - 3 \left(\lim_{x \to \infty} \frac{1}{x^2} \right)$$

$$= 4 - 3(0)$$

$$= 4$$

So, the limit of $f(x) = 4 - \dfrac{3}{x^2}$ as x approaches ∞ is 4.

✓CHECKPOINT   Now try Exercise 9.

Graphical Solution

Use a graphing utility to graph $y = 4 - 3/x^2$. Then use the *trace* feature to determine that as x gets larger and larger, y gets closer and closer to 4, as shown in Figure 11.33. Note that the line $y = 4$ is a horizontal asymptote to the right.

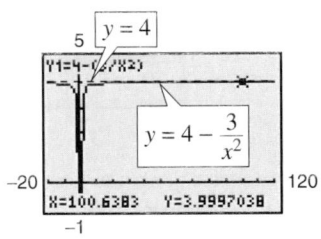

Figure 11.33

In Figure 11.33, it appears that the line $y = 4$ is also a horizontal asymptote *to the left*. You can verify this by showing that

$$\lim_{x \to -\infty} \left(4 - \frac{3}{x^2} \right) = 4.$$

The graph of a rational function need not have a horizontal asymptote. If it does, however, its left and right asymptotes must be the same.

When evaluating limits at infinity for more complicated rational functions, divide the numerator and denominator by the *highest-powered term* in the denominator. This enables you to evaluate each limit using the limits at infinity at the top of this page.

Example 2   Comparing Limits at Infinity

Find the limit as x approaches ∞ for each function.

a. $f(x) = \dfrac{-2x + 3}{3x^2 + 1}$ **b.** $f(x) = \dfrac{-2x^2 + 3}{3x^2 + 1}$ **c.** $f(x) = \dfrac{-2x^3 + 3}{3x^2 + 1}$

Solution

In each case, begin by dividing both the numerator and denominator by x^2, the highest-powered term in the denominator.

a. $\displaystyle\lim_{x\to\infty} \frac{-2x + 3}{3x^2 + 1} = \lim_{x\to\infty} \frac{-\dfrac{2}{x} + \dfrac{3}{x^2}}{3 + \dfrac{1}{x^2}}$

$$= \frac{-0 + 0}{3 + 0}$$

$$= 0$$

b. $\displaystyle\lim_{x\to\infty} \frac{-2x^2 + 3}{3x^2 + 1} = \lim_{x\to\infty} \frac{-2 + \dfrac{3}{x^2}}{3 + \dfrac{1}{x^2}}$

$$= \frac{-2 + 0}{3 + 0}$$

$$= -\frac{2}{3}$$

c. $\displaystyle\lim_{x\to\infty} \frac{-2x^3 + 3}{3x^2 + 1} = \lim_{x\to\infty} \frac{-2x + \dfrac{3}{x^2}}{3 + \dfrac{1}{x^2}}$

As $3/x^2$ in the numerator approaches 0, the numerator approaches $-\infty$. In this case, you can conclude that the limit does not exist because the numerator decreases without bound as the denominator approaches 3.

✓CHECKPOINT   Now try Exercise 15.

Exploration

Use a graphing utility to complete the table below to verify that

$$\lim_{x\to\infty} \frac{1}{x} = 0.$$

x	10^0	10^1	10^2
$\dfrac{1}{x}$			

x	10^3	10^4	10^5
$\dfrac{1}{x}$			

Make a conjecture about

$$\lim_{x\to 0} \frac{1}{x}.$$

In Example 2, observe that when the degree of the numerator is less than the degree of the denominator, as in part (a), the limit is 0. When the degrees of the numerator and denominator are equal, as in part (b), the limit is the ratio of the coefficients of the highest-powered terms. When the degree of the numerator is greater than the degree of the denominator, as in part (c), the limit does not exist.

This result seems reasonable when you realize that for large values of x, the highest-powered term of a polynomial is the most "influential" term. That is, a polynomial tends to behave as its highest-powered term behaves as x approaches positive or negative infinity.

> **Limits at Infinity for Rational Functions**
>
> Consider the rational function $f(x) = N(x)/D(x)$, where
>
> $$N(x) = a_n x^n + \cdots + a_0 \qquad \text{and} \qquad D(x) = b_m x^m + \cdots + b_0.$$
>
> The limit of $f(x)$ as x approaches positive or negative infinity is as follows.
>
> $$\lim_{x \to \pm\infty} f(x) = \begin{cases} 0, & n < m \\ \dfrac{a_n}{b_m}, & n = m \end{cases}$$
>
> If $n > m$, the limit does not exist.

Example 3   Finding the Average Cost

You are manufacturing greeting cards that cost \$0.50 per card to produce. Your initial investment is \$5000, which implies that the total cost C of producing x cards is given by $C = 0.50x + 5000$. The average cost $\overline{C}$ per card is given by

$$\overline{C} = \frac{C}{x} = \frac{0.50x + 5000}{x}.$$

Find the average cost per card when (a) $x = 1000$, (b) $x = 10{,}000$, and (c) $x = 100{,}000$. (d) What is the limit of $\overline{C}$ as x approaches infinity?

Solution

a. When $x = 1000$, the average cost per card is

$$\overline{C} = \frac{0.50(1000) + 5000}{1000} \qquad x = 1000$$

$$= \$5.50.$$

b. When $x = 10{,}000$, the average cost per card is

$$\overline{C} = \frac{0.50(10{,}000) + 5000}{10{,}000} \qquad x = 10{,}000$$

$$= \$1.00.$$

c. When $x = 100{,}000$, the average cost per card is

$$\overline{C} = \frac{0.50(100{,}000) + 5000}{100{,}000} \qquad x = 100{,}000$$

$$= \$0.55.$$

d. As x approaches infinity, the limit of $\overline{C}$ is

$$\lim_{x \to \infty} \frac{0.50x + 5000}{x} = \$0.50. \qquad x \to \infty$$

The graph of $\overline{C}$ is shown in Figure 11.34.

✓**CHECKPOINT**   Now try Exercise 53.

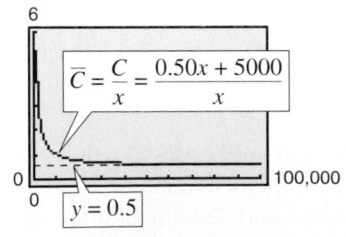

As $x \to \infty$, the average cost per card approaches \$0.50.

Figure 11.34

Limits of Sequences

Limits of sequences have many of the same properties as limits of functions. For instance, consider the sequence whose nth term is $a_n = 1/2^n$.

$$\frac{1}{2}, \frac{1}{4}, \frac{1}{8}, \frac{1}{16}, \frac{1}{32}, \ldots$$

As n increases without bound, the terms of this sequence get closer and closer to 0, and the sequence is said to **converge** to 0. Using limit notation, you can write

$$\lim_{n \to \infty} \frac{1}{2^n} = 0.$$

The following relationship shows how limits of functions of x can be used to evaluate the limit of a sequence.

> **Limit of a Sequence**
>
> Let f be a function of a real variable such that
> $$\lim_{x \to \infty} f(x) = L.$$
> If $\{a_n\}$ is a sequence such that $f(n) = a_n$ for every positive integer n, then
> $$\lim_{n \to \infty} a_n = L.$$

A sequence that does not converge is said to **diverge.** For instance, the sequence $1, -1, 1, -1, 1, \ldots$ diverges because it does not approach a unique number.

Example 4 Finding the Limit of a Sequence

Find the limit of each sequence. (Assume n begins with 1.)

a. $a_n = \dfrac{2n + 1}{n + 4}$ **b.** $b_n = \dfrac{2n + 1}{n^2 + 4}$ **c.** $c_n = \dfrac{2n^2 + 1}{4n^2}$

Solution

a. $\lim\limits_{n \to \infty} \dfrac{2n + 1}{n + 4} = 2$ $\qquad \dfrac{3}{5}, \dfrac{5}{6}, \dfrac{7}{7}, \dfrac{9}{8}, \dfrac{11}{9}, \dfrac{13}{10}, \ldots \to 2$

b. $\lim\limits_{n \to \infty} \dfrac{2n + 1}{n^2 + 4} = 0$ $\qquad \dfrac{3}{5}, \dfrac{5}{8}, \dfrac{7}{13}, \dfrac{9}{20}, \dfrac{11}{29}, \dfrac{13}{40}, \ldots \to 0$

c. $\lim\limits_{n \to \infty} \dfrac{2n^2 + 1}{4n^2} = \dfrac{1}{2}$ $\qquad \dfrac{3}{4}, \dfrac{9}{16}, \dfrac{19}{36}, \dfrac{33}{64}, \dfrac{51}{100}, \dfrac{73}{144}, \ldots \to \dfrac{1}{2}$

✔**CHECKPOINT**   Now try Exercise 39.

TECHNOLOGY TIP

There are a number of ways to use a graphing utility to generate the terms of a sequence. For instance, you can display the first 10 terms of the sequence

$$a_n = \frac{1}{2^n}$$

using the *sequence* feature or the *table* feature. For instructions on how to use the *sequence* feature and the *table* feature, see Appendix A; for specific keystrokes, go to this textbook's *Online Study Center*.

STUDY TIP

You can use the definition of limits at infinity for rational functions on page 814 to verify the limits of the sequences in Example 4.

In the next section, you will encounter limits of sequences such as that shown in Example 5. A strategy for evaluating such limits is to begin by writing the nth term in standard rational function form. Then you can determine the limit by comparing the degrees of the numerator and denominator, as shown on page 814.

Example 5   Finding the Limit of a Sequence

Find the limit of the sequence whose nth term is

$$a_n = \frac{8}{n^3}\left[\frac{n(n+1)(2n+1)}{6}\right].$$

Algebraic Solution

Begin by writing the nth term in standard rational function form— as the ratio of two polynomials.

$$a_n = \frac{8}{n^3}\left[\frac{n(n+1)(2n+1)}{6}\right] \qquad \text{Write original } n\text{th term.}$$

$$= \frac{8(n)(n+1)(2n+1)}{6n^3} \qquad \text{Multiply fractions.}$$

$$= \frac{8n^3 + 12n^2 + 4n}{3n^3} \qquad \text{Write in standard rational form.}$$

From this form, you can see that the degree of the numerator is equal to the degree of the denominator. So, the limit of the sequence is the ratio of the coefficients of the highest-powered terms.

$$\lim_{n\to\infty} \frac{8n^3 + 12n^2 + 4n}{3n^3} = \frac{8}{3}$$

✓CHECKPOINT   Now try Exercise 49.

Numerical Solution

Enter the sequence into a graphing utility. Be sure the graphing utility is set to *sequence* mode. Then use the *table* feature (in *ask* mode) to create a table that shows the values of a_n as n becomes larger and larger, as shown in Figure 11.35.

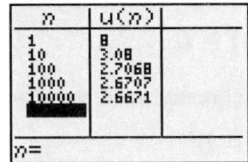

n	u(n)
1	8
10	3.08
100	2.7068
1000	2.6707
10000	2.6671

n=

Figure 11.35

From the table, you can estimate that as n approaches ∞, a_n gets closer and closer to $2.667 \approx \frac{8}{3}$.

Exploration

In the table in Example 5 above, the value of a_n approaches its limit of $\frac{8}{3}$ rather slowly. (The first term to be accurate to three decimal places is $a_{4801} \approx 2.667$.) Each of the following sequences converges to 0. Which converges the quickest? Which converges the slowest? Why? Write a short paragraph discussing your conclusions.

a. $a_n = \dfrac{1}{n}$ **b.** $b_n = \dfrac{1}{n^2}$ **c.** $c_n = \dfrac{1}{2^n}$

d. $d_n = \dfrac{1}{n!}$ **e.** $h_n = \dfrac{2^n}{n!}$

11.4 Exercises

Vocabulary Check

Fill in the blanks.

1. A _____ at _____ can be used to solve the area problem in calculus.

2. A sequence that has a limit is said to _____ .

3. A sequence that does not have a limit is said to _____ .

In Exercises 1–8, match the function with its graph, using horizontal asymptotes as aids. [The graphs are labeled (a), (b), (c), (d), (e), (f), (g), and (h).]

(a)
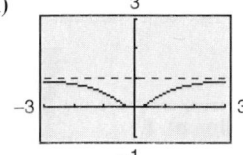

(b)
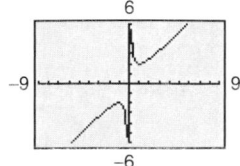

(c)
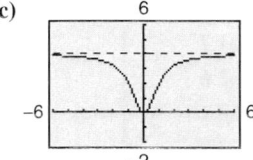

(d)
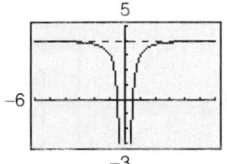

(e)
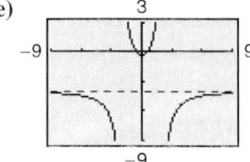

(f)
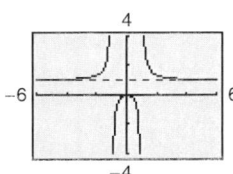

(g)
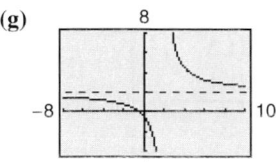

(h)
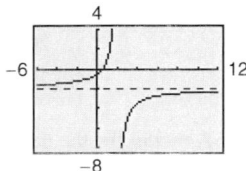

1. $f(x) = \dfrac{4x^2}{x^2 + 1}$

2. $f(x) = \dfrac{x^2}{x^2 + 1}$

3. $f(x) = 4 - \dfrac{1}{x^2}$

4. $f(x) = x + \dfrac{1}{x}$

5. $f(x) = \dfrac{x^2}{x^2 - 1}$

6. $f(x) = \dfrac{2x + 1}{x - 2}$

7. $f(x) = \dfrac{1 - 2x}{x - 2}$

8. $f(x) = \dfrac{1 - 4x^2}{x^2 - 4}$

In Exercises 9–28, find the limit (if it exists). If the limit does not exist, explain why. Use a graphing utility to verify your result graphically.

9. $\lim\limits_{x \to \infty} \dfrac{3}{x^2}$

10. $\lim\limits_{x \to \infty} \dfrac{5}{2x}$

11. $\lim\limits_{x \to \infty} \dfrac{3 + x}{3 - x}$

12. $\lim\limits_{x \to \infty} \dfrac{2 - 7x}{2 + 3x}$

13. $\lim\limits_{x \to -\infty} \dfrac{5x - 2}{6x + 1}$

14. $\lim\limits_{x \to -\infty} \dfrac{5 - 3x}{x + 4}$

15. $\lim\limits_{x \to -\infty} \dfrac{4x^2 - 3}{2 - x^2}$

16. $\lim\limits_{x \to -\infty} \dfrac{x^2 + 3}{5x^2 - 4}$

17. $\lim\limits_{t \to \infty} \dfrac{t^2}{t + 3}$

18. $\lim\limits_{y \to \infty} \dfrac{4y^4}{y^2 + 3}$

19. $\lim\limits_{t \to \infty} \dfrac{4t^2 + 3t - 1}{3t^2 + 2t - 5}$

20. $\lim\limits_{x \to \infty} \dfrac{5 - 6x - 3x^2}{2x^2 + x + 4}$

21. $\lim\limits_{y \to -\infty} \dfrac{3 + 8y - 4y^2}{3 - y - 2y^2}$

22. $\lim\limits_{t \to -\infty} \dfrac{t^2 + 9t - 10}{2 + 4t - 3t^2}$

23. $\lim\limits_{x \to -\infty} \dfrac{-(x^2 + 3)}{(2 - x)^2}$

24. $\lim\limits_{x \to \infty} \dfrac{2x^2 - 6}{(x - 1)^2}$

25. $\lim\limits_{x \to -\infty} \left[\dfrac{x}{(x + 1)^2} - 4 \right]$

26. $\lim\limits_{x \to \infty} \left[7 + \dfrac{2x^2}{(x + 3)^2} \right]$

27. $\lim\limits_{t \to \infty} \left(\dfrac{1}{3t^2} - \dfrac{5t}{t + 2} \right)$

28. $\lim\limits_{x \to \infty} \left[\dfrac{x}{2x + 1} + \dfrac{3x^2}{(x - 3)^2} \right]$

In Exercises 29–34, use a graphing utility to graph the function and verify that the horizontal asymptote corresponds to the limit at infinity.

29. $y = \dfrac{3x}{1 - x}$

30. $y = \dfrac{x^2}{x^2 + 4}$

31. $y = \dfrac{2x}{1 - x^2}$

32. $y = \dfrac{2x + 1}{x^2 - 1}$

33. $y = 1 - \dfrac{3}{x^2}$

34. $y = 2 + \dfrac{1}{x}$

Numerical and Graphical Analysis In Exercises 35–38, (a) complete the table and numerically estimate the limit as *x* approaches infinity and (b) use a graphing utility to graph the function and estimate the limit graphically.

n	10^0	10^1	10^2	10^3	10^4	10^5	10^6
$f(x)$							

35. $f(x) = x - \sqrt{x^2 + 2}$ **36.** $f(x) = 3x - \sqrt{9x^2 + 1}$

37. $f(x) = 3\left(2x - \sqrt{4x^2 + x}\right)$

38. $f(x) = 4\left(4x - \sqrt{16x^2 - x}\right)$

In Exercises 39–48, write the first five terms of the sequence and find the limit of the sequence (if it exists). If the limit does not exist, explain why. Assume *n* begins with 1.

39. $a_n = \dfrac{n + 1}{n^2 + 1}$ **40.** $a_n = \dfrac{n}{n^2 + 1}$

41. $a_n = \dfrac{n}{2n + 1}$ **42.** $a_n = \dfrac{4n - 1}{n + 3}$

43. $a_n = \dfrac{n^2}{3n + 2}$ **44.** $a_n = \dfrac{4n^2 + 1}{2n}$

45. $a_n = \dfrac{(n + 1)!}{n!}$ **46.** $a_n = \dfrac{(3n - 1)!}{(3n + 1)!}$

47. $a_n = \dfrac{(-1)^n}{n}$ **48.** $a_n = \dfrac{(-1)^{n+1}}{n^2}$

In Exercises 49–52, use a graphing utility to complete the table and estimate the limit of the sequence as *n* approaches infinity. Then find the limit algebraically.

n	10^0	10^1	10^2	10^3	10^4	10^5	10^6
a_n							

49. $a_n = \dfrac{1}{n}\left(n + \dfrac{1}{n}\left[\dfrac{n(n + 1)}{2}\right]\right)$

50. $a_n = \dfrac{4}{n}\left(n + \dfrac{4}{n}\left[\dfrac{n(n + 1)}{2}\right]\right)$

51. $a_n = \dfrac{16}{n^3}\left[\dfrac{n(n + 1)(2n + 1)}{6}\right]$

52. $a_n = \dfrac{n(n + 1)}{n^2} - \dfrac{1}{n^4}\left[\dfrac{n(n + 1)}{2}\right]^2$

53. *Average Cost* The cost function for a certain model of a personal digital assistant (PDA) is given by $C = 13.50x + 45{,}750$, where C is the cost (in dollars) and x is the number of PDAs produced.

(a) Write a model for the average cost per unit produced.

(b) Find the average costs per unit when $x = 100$ and $x = 1000$.

(c) Determine the limit of the average cost function as *x* approaches infinity. Explain the meaning of the limit in the context of the problem.

54. *Average Cost* The cost function for a company to recycle *x* tons of material is given by $C = 1.25x + 10{,}500$, where C is the cost (in dollars).

(a) Write a model for the average cost per ton of material recycled.

(b) Find the average costs of recycling 100 tons of material and 1000 tons of material.

(c) Determine the limit of the average cost function as *x* approaches infinity. Explain the meaning of the limit in the context of the problem.

55. *School Enrollment* The table shows the school enrollments *E* (in millions) in the United States for the years 1990 through 2003. (Source: U.S. National Center for Education Statistics)

Year	Enrollment, E (in millions)
1990	60.3
1991	61.7
1992	62.6
1993	63.1
1994	63.9
1995	64.8
1996	65.7
1997	66.5
1998	67.0
1999	67.7
2000	68.7
2001	69.9
2002	71.2
2003	71.4

A model for the data is given by

$$E(t) = \frac{0.702t^2 + 61.49}{0.009t^2 + 1}, \quad 0 \le t \le 13$$

where *t* represents the year, with $t = 0$ corresponding to 1990.

(a) Use a graphing utility to create a scatter plot of the data and graph the model in the same viewing window. How do they compare?

(b) Use the model to predict the enrollments in 2004 and 2008.

(c) Find the limit of the model as $t \to \infty$ and interpret its meaning in the context of the situation.

(d) Can the model be used to predict the enrollments for future years? Explain.

56. *Highway Safety*   The table shows the numbers of injuries N (in thousands) from motor vehicle accidents in the United States for the years 1991 through 2004.   (Source: U.S. National Highway Safety Administration)

Year	Injuries, N (in thousands)
1991	3097
1992	3070
1993	3149
1994	3266
1995	3465
1996	3483
1997	3348
1998	3192
1999	3236
2000	3189
2001	3033
2002	2926
2003	2889
2004	2788

A model for the data is given by

$$N(t) = \frac{40.8189t^2 - 500.059t + 2950.8}{0.0157t^2 - 0.192t + 1}, \quad 1 \le t \le 14$$

where t represents the year, with $t = 1$ corresponding to 1991.

(a) Use a graphing utility to create a scatter plot of the data and graph the model in the same viewing window. How do they compare?

(b) Use the model to predict the numbers of injuries in 2006 and 2010.

(c) Find the limit of the model as $t \to \infty$ and interpret its meaning in the context of the situation.

(d) Can the model be used to predict the numbers of injuries for future years? Explain.

Synthesis

True or False?   **In Exercises 57–60, determine whether the statement is true or false. Justify your answer.**

57. Every rational function has a horizontal asymptote.

58. If $f(x)$ increases without bound as x approaches c, then the limit of $f(x)$ exists.

59. If a sequence converges, then it has a limit.

60. When the degrees of the numerator and denominator of a rational function are equal, the limit as x goes to infinity does not exist.

61. *Think About It*   Find the functions f and g such that both $f(x)$ and $g(x)$ increase without bound as x approaches c, but $\lim\limits_{x \to c} [f(x) - g(x)] \ne \infty$.

62. *Think About It*   Use a graphing utility to graph the function $f(x) = x/\sqrt{x^2 + 1}$. How many horizontal asymptotes does the function appear to have? What are the horizontal asymptotes?

Exploration   **In Exercises 63–66, use a graphing utility to create a scatter plot of the terms of the sequence. Determine whether the sequence converges or diverges. If it converges, estimate its limit.**

63. $a_n = 4\left(\frac{2}{3}\right)^n$ **64.** $a_n = 3\left(\frac{3}{2}\right)^n$

65. $a_n = \dfrac{3[1 - (1.5)^n]}{1 - 1.5}$ **66.** $a_n = \dfrac{3[1 - (0.5)^n]}{1 - 0.5}$

Skills Review

In Exercises 67 and 68, sketch the graphs of y and each transformation on the same rectangular coordinate system.

67. $y = x^4$

    (a) $f(x) = (x + 3)^4$ (b) $f(x) = x^4 - 1$

    (c) $f(x) = -2 + x^4$ (d) $f(x) = \frac{1}{2}(x - 4)^4$

68. $y = x^3$

    (a) $f(x) = (x + 2)^3$ (b) $f(x) = 3 + x^3$

    (c) $f(x) = 2 - \frac{1}{4}x^3$ (d) $f(x) = 3(x + 1)^3$

In Exercises 69–72, divide using long division.

69. $(x^4 + 2x^3 - 3x^2 - 8x - 4) \div (x^2 - 4)$

70. $(2x^5 - 8x^3 + 4x - 1) \div (x^2 - 2x + 1)$

71. $(3x^4 + 17x^3 + 10x^2 - 9x - 8) \div (3x + 2)$

72. $(10x^3 + 51x^2 + 48x - 28) \div (5x - 2)$

In Exercises 73–76, find all the real zeros of the polynomial function. Use a graphing utility to graph the function and verify that the real zeros are the x-intercepts of the graph of the function.

73. $f(x) = x^4 - x^3 - 20x^2$ **74.** $f(x) = x^5 + x^3 - 6x$

75. $f(x) = x^3 - 3x^2 + 2x - 6$

76. $f(x) = x^3 - 4x^2 - 25x + 100$

In Exercises 77–80, find the sum.

77. $\displaystyle\sum_{i=1}^{6} (2i + 3)$ **78.** $\displaystyle\sum_{i=0}^{4} 5i^2$

79. $\displaystyle\sum_{k=1}^{10} 15$ **80.** $\displaystyle\sum_{k=0}^{8} \frac{3}{k^2 + 1}$

11.5   The Area Problem

Limits of Summations

Earlier in the text, you used the concept of a limit to obtain a formula for the sum S of an infinite geometric series

$$S = a_1 + a_1 r + a_1 r^2 + \cdots = \sum_{i=1}^{\infty} a_1 r^{i-1} = \frac{a_1}{1-r}, \quad |r| < 1.$$

Using limit notation, this sum can be written as

$$S = \lim_{n \to \infty} \sum_{i=1}^{n} a_1 r^{i-1} = \lim_{n \to \infty} \frac{a_1(1-r^n)}{1-r} = \frac{a_1}{1-r}.$$

The following summation formulas and properties are used to evaluate finite and infinite summations.

Summation Formulas and Properties

1. $\sum_{i=1}^{n} c = cn$, c is a constant.

2. $\sum_{i=1}^{n} i = \dfrac{n(n+1)}{2}$

3. $\sum_{i=1}^{n} i^2 = \dfrac{n(n+1)(2n+1)}{6}$

4. $\sum_{i=1}^{n} i^3 = \dfrac{n^2(n+1)^2}{4}$

5. $\sum_{i=1}^{n} (a_i \pm b_i) = \sum_{i=1}^{n} a_i \pm \sum_{i=1}^{n} b_i$

6. $\sum_{i=1}^{n} ka_i = k \sum_{i=1}^{n} a_i$, k is a constant.

What you should learn

■ Find limits of summations.

■ Use rectangles to approximate areas of plane regions.

■ Use limits of summations to find areas of plane regions.

Why you should learn it

Limits of summations are useful in determining areas of plane regions. For instance, in Exercise 46 on page 827, you are asked to find the limit of a summation to determine the area of a parcel of land bounded by a stream and two roads.

Adam Woolfitt/Corbis

Example 1   Evaluating a Summation

Evaluate the summation.

$$\sum_{i=1}^{200} i = 1 + 2 + 3 + 4 + \cdots + 200$$

Solution

Using Formula 2 with $n = 200$, you can write

$$\sum_{i=1}^{n} i = \frac{n(n+1)}{2}$$

$$\sum_{i=1}^{200} i = \frac{200(200+1)}{2}$$

$$= \frac{40{,}200}{2}$$

$$= 20{,}100.$$

✔**CHECKPOINT**   Now try Exercise 1.

Prerequisite Skills

Recall that the sum of a finite geometric sequence is given by

$$\sum_{i=1}^{n} a_1 r^{i-1} = a_1 \left(\frac{1-r^n}{1-r} \right).$$

Furthermore, if $0 < |r| < 1$, then $r^n \to 0$ as $n \to \infty$. To review sums of geometric sequences, see Section 8.3.

Example 2   Evaluating a Summation

Evaluate the summation

$$S = \sum_{i=1}^{n} \frac{i+2}{n^2} = \frac{3}{n^2} + \frac{4}{n^2} + \frac{5}{n^2} + \cdots + \frac{n+2}{n^2}$$

for $n = 10, 100, 1000,$ and $10,000$.

Solution

Begin by applying summation formulas and properties to simplify S. In the second line of the solution, note that $1/n^2$ can be factored out of the sum because n is considered to be constant. You could not factor i out of the summation because i is the (variable) index of summation.

$$S = \sum_{i=1}^{n} \frac{i+2}{n^2}$$ Write original form of summation.

$$= \frac{1}{n^2} \sum_{i=1}^{n} (i+2)$$ Factor constant $1/n^2$ out of sum.

$$= \frac{1}{n^2} \left(\sum_{i=1}^{n} i + \sum_{i=1}^{n} 2 \right)$$ Write as two sums.

$$= \frac{1}{n^2} \left[\frac{n(n+1)}{2} + 2n \right]$$ Apply Formulas 1 and 2.

$$= \frac{1}{n^2} \left(\frac{n^2 + 5n}{2} \right)$$ Add fractions.

$$= \frac{n+5}{2n}$$ Simplify.

Now you can evaluate the sum by substituting the appropriate values of n, as shown in the following table.

n	10	100	1000	10,000
$\sum_{i=1}^{n} \frac{i+2}{n^2} = \frac{n+5}{2n}$	0.75	0.525	0.5025	0.50025

✓CHECKPOINT   Now try Exercise 11.

In Example 2, note that the sum appears to approach a limit as n increases. To find the limit of $(n+5)/2n$ as n approaches infinity, you can use the techniques from Section 11.4 to write

$$\lim_{n \to \infty} \frac{n+5}{2n} = \frac{1}{2}.$$

TECHNOLOGY TIP

Some graphing utilities have a *sum sequence* feature that is useful for computing summations. For instructions on how to use the *sum sequence* feature, see Appendix A; for specific keystrokes, go to this textbook's *Online Study Center*.

Be sure you notice the strategy used in Example 2. Rather than separately evaluating the sums

$$\sum_{i=1}^{10} \frac{i+2}{n^2}, \qquad \sum_{i=1}^{100} \frac{i+2}{n^2}, \qquad \sum_{i=1}^{1000} \frac{i+2}{n^2}, \qquad \sum_{i=1}^{10,000} \frac{i+2}{n^2}$$

it was more efficient first to convert to rational form using the summation formulas and properties listed on page 820.

$$S = \underbrace{\sum_{i=1}^{n} \frac{i+2}{n^2}}_{\substack{\text{Summation} \\ \text{form}}} = \underbrace{\frac{n+5}{2n}}_{\text{Rational form}}$$

With this rational form, each sum can be evaluated by simply substituting appropriate values of n.

Example 3   Finding the Limit of a Summation

Find the limit of $S(n)$ as $n \to \infty$.

$$S(n) = \sum_{i=1}^{n} \left(1 + \frac{i}{n}\right)^2 \left(\frac{1}{n}\right)$$

Solution

Begin by rewriting the summation in rational form.

$$S(n) = \sum_{i=1}^{n} \left(1 + \frac{i}{n}\right)^2 \left(\frac{1}{n}\right) \qquad \text{Write original form of summation.}$$

$$= \sum_{i=1}^{n} \left(\frac{n^2 + 2ni + i^2}{n^2}\right)\left(\frac{1}{n}\right) \qquad \text{Square } (1 + i/n) \text{ and write as a single fraction.}$$

$$= \frac{1}{n^3} \sum_{i=1}^{n} (n^2 + 2ni + i^2) \qquad \text{Factor constant } 1/n^3 \text{ out of the sum.}$$

$$= \frac{1}{n^3} \left(\sum_{i=1}^{n} n^2 + \sum_{i=1}^{n} 2ni + \sum_{i=1}^{n} i^2\right) \qquad \text{Write as three sums.}$$

$$= \frac{1}{n^3} \left\{ n^3 + 2n\left[\frac{n(n+1)}{2}\right] + \frac{n(n+1)(2n+1)}{6}\right\} \qquad \text{Use summation formulas.}$$

$$= \frac{14n^3 + 9n^2 + n}{6n^3} \qquad \text{Simplify.}$$

In this rational form, you can now find the limit as $n \to \infty$.

$$\lim_{n \to \infty} S(n) = \lim_{n \to \infty} \frac{14n^3 + 9n^2 + n}{6n^3}$$

$$= \frac{14}{6} = \frac{7}{3}$$

✓CHECKPOINT   Now try Exercise 13.

STUDY TIP

As you can see from Example 3, there is a lot of algebra involved in rewriting a summation in rational form. You may want to review simplifying rational expressions if you are having difficulty with this procedure.

Prerequisite Skills

To review limits at infinity, see Section 11.4.

The Area Problem

You now have the tools needed to solve the second basic problem of calculus: the area problem. The problem is to find the *area* of the region R bounded by the graph of a nonnegative, continuous function f, the x-axis, and the vertical lines $x = a$ and $x = b$, as shown in Figure 11.36.

   If the region R is a square, a triangle, a trapezoid, or a semicircle, you can find its area by using a geometric formula. For more general regions, however, you must use a different approach—one that involves the limit of a summation. The basic strategy is to use a collection of rectangles of equal width that approximates the region R, as illustrated in Example 4.

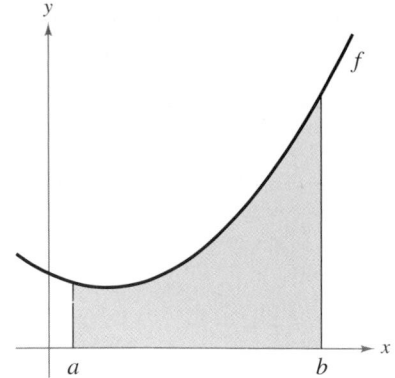

Figure 11.36

Example 4 Approximating the Area of a Region

Use the five rectangles in Figure 11.37 to approximate the area of the region bounded by the graph of $f(x) = 6 - x^2$, the x-axis, and the lines $x = 0$ and $x = 2$.

Solution

Because the length of the interval along the x-axis is 2 and there are five rectangles, the width of each rectangle is $\frac{2}{5}$. The height of each rectangle can be obtained by evaluating f at the right endpoint of each interval. The five intervals are as follows.

$$\left[0, \frac{2}{5}\right], \quad \left[\frac{2}{5}, \frac{4}{5}\right], \quad \left[\frac{4}{5}, \frac{6}{5}\right], \quad \left[\frac{6}{5}, \frac{8}{5}\right], \quad \left[\frac{8}{5}, \frac{10}{5}\right]$$

Notice that the right endpoint of each interval is $\frac{2}{5}i$ for $i = 1, 2, 3, 4$, and 5. The sum of the areas of the five rectangles is

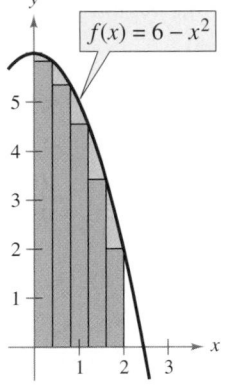

Figure 11.37

$$\sum_{i=1}^{5} f\left(\frac{2i}{5}\right)\left(\frac{2}{5}\right) = \sum_{i=1}^{5}\left[6 - \left(\frac{2i}{5}\right)^2\right]\left(\frac{2}{5}\right)$$

$$= \frac{2}{5}\left(\sum_{i=1}^{5} 6 - \frac{4}{25}\sum_{i=1}^{5} i^2\right)$$

$$= \frac{2}{5}\left(30 - \frac{44}{5}\right) = \frac{212}{25} = 8.48.$$

(Height, Width labels over the first summation)

So, you can approximate the area of R as 8.48 square units.

✓CHECKPOINT   Now try Exercise 19.

   By increasing the number of rectangles used in Example 4, you can obtain closer and closer approximations of the area of the region. For instance, using 25 rectangles of width $\frac{2}{25}$ each, you can approximate the area to be $A \approx 9.17$ square units. The following table shows even better approximations.

n	5	25	100	1000	5000
Approximate area	8.48	9.17	9.29	9.33	9.33

Based on the procedure illustrated in Example 4, the *exact* **area of a plane region** *R* is given by the limit of the sum of *n* rectangles as *n* approaches ∞.

Area of a Plane Region

Let f be continuous and nonnegative on the interval $[a, b]$. The **area** A of the region bounded by the graph of f, the x-axis, and the vertical lines $x = a$ and $x = b$ is given by

$$A = \lim_{n \to \infty} \sum_{i=1}^{n} \underbrace{f\left(a + \frac{(b-a)i}{n}\right)}_{\text{Height}} \underbrace{\left(\frac{b-a}{n}\right)}_{\text{Width}}.$$

Example 5   Finding the Area of a Region

Find the area of the region bounded by the graph of $f(x) = x^2$ and the x-axis between $x = 0$ and $x = 1$, as shown in Figure 11.38.

Solution

Begin by finding the dimensions of the rectangles.

Width: $\dfrac{b-a}{n} = \dfrac{1-0}{n} = \dfrac{1}{n}$

Height: $f\left(a + \dfrac{(b-a)i}{n}\right) = f\left(0 + \dfrac{(1-0)i}{n}\right) = f\left(\dfrac{i}{n}\right) = \dfrac{i^2}{n^2}$

Next, approximate the area as the sum of the areas of n rectangles.

$$A \approx \sum_{i=1}^{n} f\left(a + \frac{(b-a)i}{n}\right)\left(\frac{b-a}{n}\right)$$

$$= \sum_{i=1}^{n} \left(\frac{i^2}{n^2}\right)\left(\frac{1}{n}\right)$$

$$= \sum_{i=1}^{n} \frac{i^2}{n^3}$$

$$= \frac{1}{n^3} \sum_{i=1}^{n} i^2$$

$$= \frac{1}{n^3}\left[\frac{n(n+1)(2n+1)}{6}\right]$$

$$= \frac{2n^3 + 3n^2 + n}{6n^3}$$

Finally, find the exact area by taking the limit as n approaches ∞.

$$A = \lim_{n \to \infty} \frac{2n^3 + 3n^2 + n}{6n^3} = \frac{1}{3}$$

✓CHECKPOINT   Now try Exercise 33.

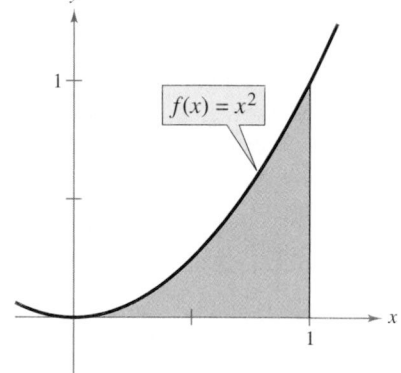

$f(x) = x^2$

Figure 11.38

Example 6   Finding the Area of a Region

Find the area of the region bounded by the graph of $f(x) = 3x - x^2$ and the x-axis between $x = 1$ and $x = 2$, as shown in Figure 11.39.

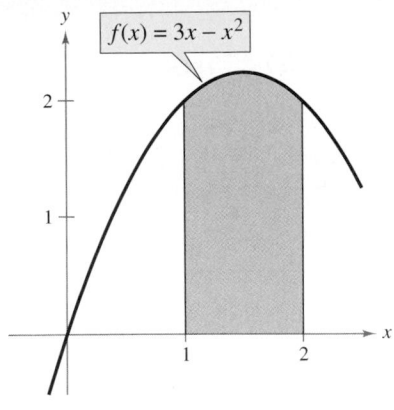

Solution

Begin by finding the dimensions of the rectangles.

Width: $\dfrac{b - a}{n} = \dfrac{2 - 1}{n} = \dfrac{1}{n}$

Height: $f\left(a + \dfrac{(b - a)i}{n}\right) = f\left(1 + \dfrac{i}{n}\right)$

$$= 3\left(1 + \dfrac{i}{n}\right) - \left(1 + \dfrac{i}{n}\right)^2$$

$$= 3 + \dfrac{3i}{n} - \left(1 + \dfrac{2i}{n} + \dfrac{i^2}{n^2}\right)$$

$$= 2 + \dfrac{i}{n} - \dfrac{i^2}{n^2}$$

Figure 11.39

Next, approximate the area as the sum of the areas of n rectangles.

$$A \approx \sum_{i=1}^{n} f\left(a + \dfrac{(b - a)i}{n}\right)\left(\dfrac{b - a}{n}\right)$$

$$= \sum_{i=1}^{n}\left(2 + \dfrac{i}{n} - \dfrac{i^2}{n^2}\right)\left(\dfrac{1}{n}\right)$$

$$= \dfrac{1}{n}\sum_{i=1}^{n} 2 + \dfrac{1}{n^2}\sum_{i=1}^{n} i - \dfrac{1}{n^3}\sum_{i=1}^{n} i^2$$

$$= \dfrac{1}{n}(2n) + \dfrac{1}{n^2}\left[\dfrac{n(n + 1)}{2}\right] - \dfrac{1}{n^3}\left[\dfrac{n(n + 1)(2n + 1)}{6}\right]$$

$$= 2 + \dfrac{n^2 + n}{2n^2} - \dfrac{2n^3 + 3n^2 + n}{6n^3}$$

$$= 2 + \dfrac{1}{2} + \dfrac{1}{2n} - \dfrac{1}{3} - \dfrac{1}{2n} - \dfrac{1}{6n^2}$$

$$= \dfrac{13}{6} - \dfrac{1}{6n^2}$$

Finally, find the exact area by taking the limit as n approaches ∞.

$$A = \lim_{n \to \infty}\left(\dfrac{13}{6} - \dfrac{1}{6n^2}\right)$$

$$= \dfrac{13}{6}$$

✓**CHECKPOINT**   Now try Exercise 39.

11.5   Exercises

See www.CalcChat.com for worked-out solutions to odd-numbered exercises.

Vocabulary Check

Fill in the blanks.

1. $\displaystyle\sum_{i=1}^{n} i =$ _____   **2.** $\displaystyle\sum_{i=1}^{n} i^3 =$ _____

3. The exact _____ of a plane region R is given by the limit of the sum of n rectangles as n approaches ∞.

In Exercises 1–8, evaluate the sum using the summation formulas and properties.

1. $\displaystyle\sum_{i=1}^{60} 7$

2. $\displaystyle\sum_{i=1}^{45} 3$

3. $\displaystyle\sum_{i=1}^{20} i^3$

4. $\displaystyle\sum_{i=1}^{30} i^2$

5. $\displaystyle\sum_{k=1}^{20} (k^3 + 2)$

6. $\displaystyle\sum_{k=1}^{50} (2k + 1)$

7. $\displaystyle\sum_{j=1}^{25} (j^2 + j)$

8. $\displaystyle\sum_{j=1}^{10} (j^3 - 3j^2)$

In Exercises 9–16, (a) rewrite the sum as a rational function $S(n)$. (b) Use $S(n)$ to complete the table. (c) Find $\displaystyle\lim_{n\to\infty} S(n)$.

n	10^0	10^1	10^2	10^3	10^4
$S(n)$					

9. $\displaystyle\sum_{i=1}^{n} \frac{i^3}{n^4}$

10. $\displaystyle\sum_{i=1}^{n} \frac{i}{n^2}$

11. $\displaystyle\sum_{i=1}^{n} \frac{3}{n^3}(1 + i^2)$

12. $\displaystyle\sum_{i=1}^{n} \frac{2i + 3}{n^2}$

13. $\displaystyle\sum_{i=1}^{n} \left(\frac{i^2}{n^3} + \frac{2}{n}\right)\left(\frac{1}{n}\right)$

14. $\displaystyle\sum_{i=1}^{n} \left[3 - 2\left(\frac{i}{n}\right)\right]\left(\frac{1}{n}\right)$

15. $\displaystyle\sum_{i=1}^{n} \left[1 - \left(\frac{i}{n}\right)^2\right]\left(\frac{1}{n}\right)$

16. $\displaystyle\sum_{i=1}^{n} \left[\frac{4}{n} + \left(\frac{2i}{n^2}\right)\right]\left(\frac{2i}{n}\right)$

In Exercises 17–20, approximate the area of the region using the indicated number of rectangles of equal width.

17. $f(x) = x + 4$

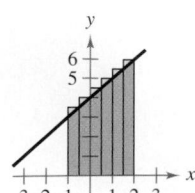

18. $f(x) = 2 - x^2$

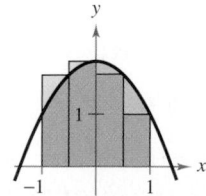

19. $f(x) = \frac{1}{4}x^3$

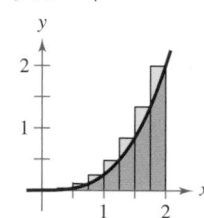

20. $f(x) = \frac{1}{2}(x - 1)^3$

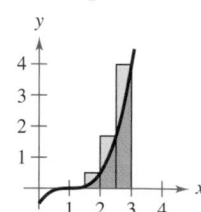

In Exercises 21–24, complete the table showing the approximate area of the region in the graph using n rectangles of equal width.

n	4	8	20	50
Approximate area				

21. $f(x) = -\frac{1}{3}x + 4$

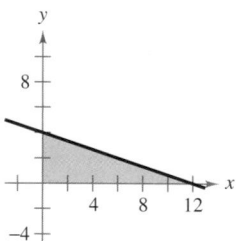

22. $f(x) = 9 - x^2$

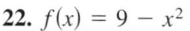
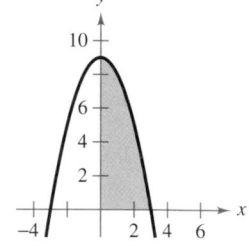

23. $f(x) = \frac{1}{9}x^3$

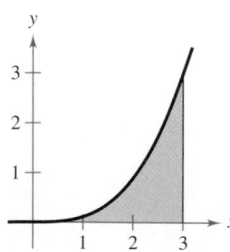

24. $f(x) = 3 - \frac{1}{4}x^3$

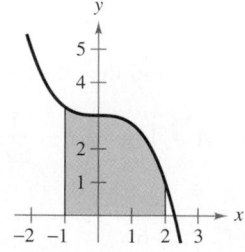

In Exercises 25–32, complete the table using the function $f(x)$**, over the specified interval** $[a. b]$**, to approximate the area of the region bounded by the graph of** $y = f(x)$**, the** x**-axis, and the vertical lines** $x = a$ **and** $x = b$ **using the indicated number of rectangles. Then find the exact area as** $n \to \infty$**.**

n	4	8	20	50	100	∞
Area						

	Function	Interval
25.	$f(x) = 2x + 5$	$[0, 4]$
26.	$f(x) = 3x + 1$	$[0, 4]$
27.	$f(x) = 16 - 2x$	$[1, 5]$
28.	$f(x) = 20 - 2x$	$[2, 6]$
29.	$f(x) = 9 - x^2$	$[0, 2]$
30.	$f(x) = x^2 + 1$	$[4, 6]$
31.	$f(x) = \frac{1}{2}x + 4$	$[-1, 3]$
32.	$f(x) = \frac{1}{2}x + 1$	$[-2, 2]$

In Exercises 33–44, use the limit process to find the area of the region between the graph of the function and the x**-axis over the specified interval.**

	Function	Interval
33.	$f(x) = 4x + 1$	$[0, 1]$
34.	$f(x) = 3x + 2$	$[0, 2]$
35.	$f(x) = -2x + 3$	$[0, 1]$
36.	$f(x) = 3x - 4$	$[2, 5]$
37.	$f(x) = 2 - x^2$	$[-1, 1]$
38.	$f(x) = x^2 + 2$	$[0, 1]$
39.	$g(x) = 8 - x^3$	$[1, 2]$
40.	$g(x) = 64 - x^3$	$[1, 4]$
41.	$g(x) = 2x - x^3$	$[0, 1]$
42.	$g(x) = 4x - x^3$	$[0, 2]$
43.	$f(x) = \frac{1}{4}(x^2 + 4x)$	$[1, 4]$
44.	$f(x) = x^2 - x^3$	$[-1, 1]$

45. *Geometry* The boundaries of a parcel of land are two edges modeled by the coordinate axes and a stream modeled by the equation $y = (-3.0 \times 10^{-6})x^3 + 0.002x^2 - 1.05x + 400$. Use a graphing utility to graph the equation. Find the area of the property. Assume all distances are measured in feet.

46. *Geometry* The table shows the measurements (in feet) of a lot bounded by a stream and two straight roads that meet at right angles (see figure).

x	0	50	100	150	200	250	300
y	450	362	305	268	245	156	0

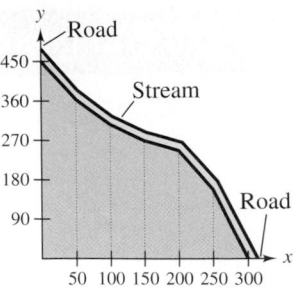

Figure for 46

(a) Use the *regression* feature of a graphing utility to find a model of the form $y = ax^3 + bx^2 + cx + d$.

(b) Use a graphing utility to plot the data and graph the model in the same viewing window.

(c) Use the model in part (a) to estimate the area of the lot.

Synthesis

True or False? **In Exercises 47 and 48, determine whether the statement is true or false. Justify your answer.**

47. The sum of the first n positive integers is $n(n + 1)/2$.

48. The exact area of a region is given by the limit of the sum of n rectangles as n approaches 0.

49. *Writing* Describe the process of finding the area of a region bounded by the graph of a nonnegative, continuous function f, the x-axis, and the vertical lines $x = a$ and $x = b$.

50. *Think About It* Determine which value best approximates the area of the region shown in the graph. (Make your selection on the basis of the sketch of the region and *not* by performing any calculations.)

(a) -2 (b) 1   (c) 4   (d) 6   (e) 9

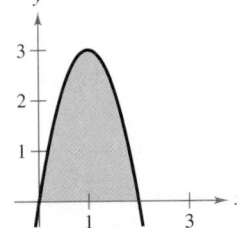

Skills Review

In Exercises 51 and 52, solve the equation.

51. $2 \tan x = \tan 2x$ **52.** $\cos 2x - 3 \sin x = 2$

In Exercises 53–56, use the vectors $\mathbf{u} = \langle 4, -5 \rangle$ **and** $\mathbf{v} = \langle -1, -2 \rangle$ **to find the indicated quantity.**

53. $(\mathbf{u} \cdot \mathbf{v})\mathbf{u}$ **54.** $3\mathbf{u} \cdot \mathbf{v}$

55. $\|\mathbf{v}\| - 2$ **56.** $\|\mathbf{u}\|^2 - \|\mathbf{v}\|^2$

What Did You Learn?

Key Terms

limit, *p. 781*
direct substitution, *p. 785*
dividing out technique, *p. 791*
indeterminate form, *p. 792*
rationalizing technique, *p. 793*

one-sided limit, *p. 795*
tangent line, *p. 801*
secant line, *p. 803*
difference quotient, *p. 803*

slope of a graph, *p. 803*
limits at infinity, *p. 811*
converge, *p. 815*
diverge, *p. 815*

Key Concepts

11.1 ■ Conditions under which limits do not exist

The limit of $f(x)$ as $x \to c$ does not exist if any of the following conditions is true.

1. $f(x)$ approaches a different number from the right side of c than it approaches from the left side of c.

2. $f(x)$ increases or decreases without bound as $x \to c$.

3. $f(x)$ oscillates between two fixed values as $x \to c$.

11.1 ■ Use basic limits and properties of limits

Let b and c be real numbers, let n be a positive integer, and let f and g be functions with the limits $\lim_{x \to c} f(x) = L$ and $\lim_{x \to c} g(x) = K$.

1. $\lim_{x \to c} b = b$ 2. $\lim_{x \to c} x = c$ 3. $\lim_{x \to c} x^n = c^n$

4. $\lim_{x \to c} \sqrt[n]{x} = \sqrt[n]{c}$, n even, $c > 0$ 5. $\lim_{x \to c} [bf(x)] = bL$

6. $\lim_{x \to c} [f(x) \pm g(x)] = L \pm K$ 7. $\lim_{x \to c} [f(x)g(x)] = LK$

8. $\lim_{x \to c} \dfrac{f(x)}{g(x)} = \dfrac{L}{K},\ \ K \neq 0$ 9. $\lim_{x \to c} [f(x)]^n = L^n$

11.1 ■ Limits of polynomial and rational functions

1. If p is a polynomial function and c is a real number, then $\lim_{x \to c} p(x) = p(c)$.

2. If r is a rational function given by $r(x) = p(x)/q(x)$, and c is a real number such that $q(c) \neq 0$, then

$$\lim_{x \to c} r(x) = r(c) = \frac{p(c)}{q(c)}.$$

11.2 ■ Determine the existence of limits

If f is a function and c and L are real numbers, then $\lim_{x \to c} f(x) = L$ if and only if both the left and right limits exist and are equal to L.

11.3 ■ Find derivatives of functions

The derivative of f at x is given by

$$f'(x) = \lim_{h \to 0} \frac{f(x + h) - f(x)}{h}$$

provided this limit exists.

11.4 ■ Evaluate limits at infinity

1. If r is a positive real number, then $\lim_{x \to \infty} (1/x^r) = 0$.

   If x^r is defined when $x < 0$, then $\lim_{x \to -\infty} (1/x^r) = 0$.

2. Consider the rational function $f(x) = N(x)/D(x)$, where $N(x) = a_n x^n + \cdots + a_0$ and $D(x) = b_m x^m + \cdots + b_0$. The limit of $f(x)$ as x approaches positive or negative infinity is as follows. Note that if $n > m$, the limit does not exist.

$$\lim_{x \to \pm\infty} f(x) = \begin{cases} 0, & n < m \\ \dfrac{a_n}{b_m}, & n = m \end{cases}$$

11.4 ■ Find limits of sequences

Let f be a function of a real variable such that $\lim_{x \to \infty} f(x) = L$. If $\{a_n\}$ is a sequence such that $f(n) = a_n$ for every positive integer n, then $\lim_{n \to \infty} a_n = L$.

11.5 ■ Use summation formulas and properties

1. $\displaystyle\sum_{i=1}^{n} c = cn,\ \ c$ is a constant.   2. $\displaystyle\sum_{i=1}^{n} i = \frac{n(n + 1)}{2}$

3. $\displaystyle\sum_{i=1}^{n} i^2 = \frac{n(n + 1)(2n + 1)}{6}$ 4. $\displaystyle\sum_{i=1}^{n} i^3 = \frac{n^2(n + 1)^2}{4}$

5. $\displaystyle\sum_{i=1}^{n} (a_i \pm b_i) = \sum_{i=1}^{n} a_i \pm \sum_{i=1}^{n} b_i$

6. $\displaystyle\sum_{i=1}^{n} ka_i = k\sum_{i=1}^{n} a_i,\ \ k$ is a constant.

11.5 ■ Find the area of a plane region

Let f be continuous and nonnegative on the interval $[a, b]$. The area A of the region bounded by the graph of f, the x-axis, and the vertical lines $x = a$ and $x = b$ is given by

$$A = \lim_{n \to \infty} \sum_{i=1}^{n} \underbrace{f\left(a + \frac{(b - a)i}{n}\right)}_{\text{Height}} \underbrace{\left(\frac{b - a}{n}\right)}_{\text{Width}}.$$

Review Exercises

See www.CalcChat.com for worked-out solutions to odd-numbered exercises.

11.1 In Exercises 1–4, complete the table and use the result to estimate the limit numerically. Determine whether or not the limit can be reached.

1. $\lim\limits_{x \to 3} (6x - 1)$

x	2.9	2.99	2.999	3	3.001	3.01	3.1
$f(x)$				?			

2. $\lim\limits_{x \to 2} \dfrac{x - 2}{3x^2 - 4x - 4}$

x	1.9	1.99	1.999	2	2.001	2.01	2.1
$f(x)$				?			

3. $\lim\limits_{x \to 0} \dfrac{1 - e^{-x}}{x}$

x	-0.1	-0.01	-0.001	0
$f(x)$				?

x	0.001	0.01	0.1
$f(x)$			

4. $\lim\limits_{x \to 0} \dfrac{\ln(1 - x)}{x}$

x	-0.1	-0.01	-0.001	0
$f(x)$				?

x	0.001	0.01	0.1
$f(x)$			

In Exercises 5–8, use the graph to find the limit (if it exists). If the limit does not exist, explain why.

5. $\lim\limits_{x \to 1} (3 - x)$

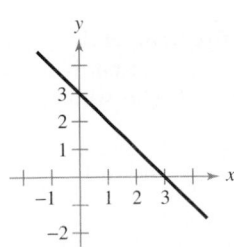

6. $\lim\limits_{x \to 2} \dfrac{1}{x - 2}$

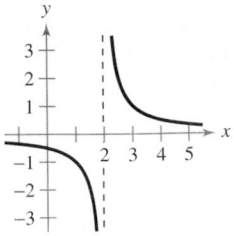

7. $\lim\limits_{x \to 1} \dfrac{x^2 - 1}{x - 1}$

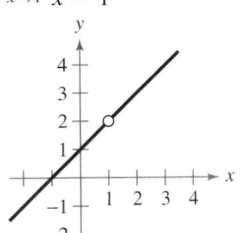

8. $\lim\limits_{x \to -1} (2x^2 + 1)$

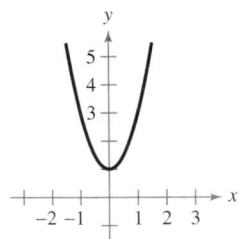

In Exercises 9 and 10, use the given information to evaluate each limit.

9. $\lim\limits_{x \to c} f(x) = 4, \ \lim\limits_{x \to c} g(x) = 5$

(a) $\lim\limits_{x \to c} [f(x)]^3$

(b) $\lim\limits_{x \to c} [3f(x) - g(x)]$

(c) $\lim\limits_{x \to c} [f(x)g(x)]$

(d) $\lim\limits_{x \to c} \dfrac{f(x)}{g(x)}$

10. $\lim\limits_{x \to c} f(x) = 27, \ \lim\limits_{x \to c} g(x) = 12$

(a) $\lim\limits_{x \to c} \sqrt[3]{f(x)}$

(b) $\lim\limits_{x \to c} \dfrac{f(x)}{18}$

(c) $\lim\limits_{x \to c} [f(x)g(x)]$

(d) $\lim\limits_{x \to c} [f(x) - 2g(x)]$

In Exercises 11–24, find the limit by direct substitution.

11. $\lim\limits_{x \to 4} \left(\tfrac{1}{2}x + 3\right)$

12. $\lim\limits_{x \to 3} (5x - 4)$

13. $\lim\limits_{x \to 2} (5x - 3)(3x + 5)$

14. $\lim\limits_{x \to -2} (5 - 2x - x^2)$

15. $\lim\limits_{t \to 3} \dfrac{t^2 + 1}{t}$

16. $\lim\limits_{x \to 2} \dfrac{3x + 5}{5x - 3}$

17. $\lim\limits_{x \to -2} \sqrt[3]{4x}$

18. $\lim\limits_{x \to -1} \sqrt{5 - x}$

19. $\lim\limits_{x \to \pi} \sin 3x$

20. $\lim\limits_{x \to 0} \tan x$

21. $\lim\limits_{x \to -1} 2e^x$

22. $\lim\limits_{x \to 4} \ln x$

23. $\lim\limits_{x \to -1/2} \arcsin x$

24. $\lim\limits_{x \to 0} \arctan x$

11.2 In Exercises 25–36, find the limit (if it exists). Use a graphing utility to verify your result graphically.

25. $\lim\limits_{t \to -2} \dfrac{t + 2}{t^2 - 4}$

26. $\lim\limits_{t \to 3} \dfrac{t^2 - 9}{t - 3}$

27. $\lim\limits_{x \to 5} \dfrac{x - 5}{x^2 + 5x - 50}$

28. $\lim\limits_{x \to -1} \dfrac{x + 1}{x^2 - 5x - 6}$

29. $\lim\limits_{x \to -2} \dfrac{x^2 - 4}{x^3 + 8}$

30. $\lim\limits_{x \to 4} \dfrac{x^3 - 64}{x^2 - 16}$

31. $\lim\limits_{x \to -1} \dfrac{\dfrac{1}{x + 2} - 1}{x + 1}$

32. $\lim\limits_{x \to 0} \dfrac{\dfrac{1}{x + 1} - 1}{x}$

33. $\lim\limits_{u \to 0} \dfrac{\sqrt{4 + u} - 2}{u}$

34. $\lim\limits_{v \to 0} \dfrac{\sqrt{v + 9} - 3}{v}$

35. $\lim\limits_{x \to 5} \dfrac{\sqrt{x - 1} - 2}{x - 5}$

36. $\lim\limits_{x \to 1} \dfrac{\sqrt{3} - \sqrt{x + 2}}{1 - x}$

Graphical and Numerical Analysis In Exercises 37–44, (a) graphically approximate the limit (if it exists) by using a graphing utility to graph the function and (b) numerically approximate the limit (if it exists) by using the *table* feature of a graphing utility to create a table.

37. $\lim\limits_{x \to 3} \dfrac{x - 3}{x^2 - 9}$

38. $\lim\limits_{x \to 4} \dfrac{4 - x}{16 - x^2}$

39. $\lim\limits_{x \to 0} e^{-2/x}$

40. $\lim\limits_{x \to 0} e^{-4/x^2}$

41. $\lim\limits_{x \to 0} \dfrac{\sin 4x}{2x}$

42. $\lim\limits_{x \to 0} \dfrac{\tan 2x}{x}$

43. $\lim\limits_{x \to 1^+} \dfrac{\sqrt{2x + 1} - \sqrt{3}}{x - 1}$

44. $\lim\limits_{x \to 1^+} \dfrac{1 - \sqrt{x}}{x - 1}$

In Exercises 45–52, graph the function. Determine the limit (if it exists) by evaluating the corresponding one-sided limits.

45. $\lim\limits_{x \to 3} \dfrac{|x - 3|}{x - 3}$

46. $\lim\limits_{x \to 8} \dfrac{|8 - x|}{8 - x}$

47. $\lim\limits_{x \to 2} \dfrac{2}{x^2 - 4}$

48. $\lim\limits_{x \to -3} \dfrac{1}{x^2 + 9}$

49. $\lim\limits_{x \to 5} \dfrac{|x - 5|}{x - 5}$

50. $\lim\limits_{x \to -2} \dfrac{|x + 2|}{x + 2}$

51. $\lim\limits_{x \to 2} f(x)$ where $f(x) = \begin{cases} 5 - x, & x \le 2 \\ x^2 - 3, & x > 2 \end{cases}$

52. $\lim\limits_{x \to 0} f(x)$ where $f(x) = \begin{cases} x - 6, & x \ge 0 \\ x^2 - 4, & x < 0 \end{cases}$

∫ In Exercises 53 and 54, find $\lim\limits_{h \to 0} \dfrac{f(x + h) - f(x)}{h}$.

53. $f(x) = 3x - x^2$

54. $f(x) = x^2 - 5x - 2$

11.3 In Exercises 55 and 56, approximate the slope of the tangent line to the graph at the point (x, y).

55.

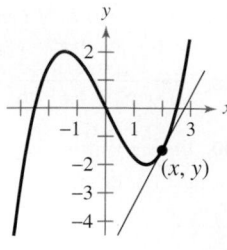

56.

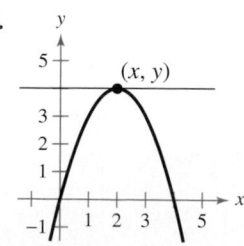

In Exercises 57–62, use a graphing utility to graph the function and the tangent line at the point $(2, f(2))$. Use the graph to approximate the slope of the tangent line.

57. $f(x) = x^2 - 2x$

58. $f(x) = 6 - x^2$

59. $f(x) = \sqrt{x + 2}$

60. $f(x) = \sqrt{x^2 + 5}$

61. $f(x) = \dfrac{6}{x - 4}$

62. $f(x) = \dfrac{1}{3 - x}$

In Exercises 63–66, find a formula for the slope of the graph of f at the point $(x, f(x))$. Then use it to find the slopes at the two specified points.

63. $f(x) = x^2 - 4x$

    (a) $(0, 0)$

    (b) $(5, 5)$

64. $f(x) = \tfrac{1}{4}x^4$

    (a) $(-2, 4)$

    (b) $\left(1, \tfrac{1}{4}\right)$

65. $f(x) = \dfrac{4}{x - 6}$

    (a) $(7, 4)$

    (b) $(8, 2)$

66. $f(x) = \sqrt{x}$

    (a) $(1, 1)$

    (b) $(4, 2)$

∫ In Exercises 67–78, find the derivative of the function.

67. $f(x) = 5$

68. $g(x) = -3$

69. $h(x) = 5 - \tfrac{1}{2}x$

70. $f(x) = 3x$

71. $g(x) = 2x^2 - 1$

72. $f(x) = -x^3 + 4x$

73. $f(t) = \sqrt{t + 5}$

74. $g(t) = \sqrt{t - 3}$

75. $g(s) = \dfrac{4}{s + 5}$

76. $g(t) = \dfrac{6}{5 - t}$

77. $g(x) = \dfrac{1}{\sqrt{x + 4}}$

78. $f(x) = \dfrac{1}{\sqrt{12 - x}}$

11.4 In Exercises 79–86, find the limit (if it exists). If the limit does not exist, explain why. Use a graphing utility to verify your result graphically.

79. $\lim\limits_{x \to \infty} \dfrac{4x}{2x - 3}$

80. $\lim\limits_{x \to \infty} \dfrac{7x}{14x + 2}$

81. $\lim\limits_{x \to -\infty} \dfrac{2x}{x^2 - 25}$

82. $\lim\limits_{x \to -\infty} \dfrac{3x}{(1 - x)^3}$

83. $\lim\limits_{x \to \infty} \dfrac{x^2}{2x + 3}$

84. $\lim\limits_{y \to \infty} \dfrac{3y^4}{y^2 + 1}$

85. $\lim\limits_{x \to \infty} \left[\dfrac{x}{(x - 2)^2} + 3 \right]$

86. $\lim\limits_{x \to \infty} \left[2 - \dfrac{2x^2}{(x + 1)^2} \right]$

In Exercises 87–92, write the first five terms of the sequence and find the limit of the sequence (if it exists). If the limit does not exist, explain why. Assume n begins with 1.

87. $a_n = \dfrac{2n - 3}{5n + 4}$

88. $a_n = \dfrac{2n}{n^2 + 1}$

89. $a_n = \dfrac{(-1)^n}{n^3}$ **90.** $a_n = \dfrac{(-1)^{n+1}}{n}$

91. $a_n = \dfrac{1}{2n^2}[3 - 2n(n + 1)]$

92. $a_n = \left(\dfrac{2}{n}\right)\left\{n + \dfrac{2}{n}\left[\dfrac{n(n-1)}{2} - n\right]\right\}$

11.5 In Exercises 93 and 94, (a) use the summation formulas and properties to rewrite the sum as a rational function $S(n)$. (b) Use $S(n)$ to complete the table. (c) Find $\displaystyle\lim_{n\to\infty} S(n)$.

n	10^0	10^1	10^2	10^3	10^4
$S(n)$					

93. $\displaystyle\sum_{i=1}^{n}\left(\dfrac{4i^2}{n^2} - \dfrac{i}{n}\right)\left(\dfrac{1}{n}\right)$ **94.** $\displaystyle\sum_{i=1}^{n}\left[4 - \left(\dfrac{3i}{n}\right)^2\right]\left(\dfrac{3i}{n^2}\right)$

In Exercises 95 and 96, approximate the area of the region using the indicated number of rectangles of equal width.

95. $f(x) = 4 - x$ **96.** $f(x) = 4 - x^2$

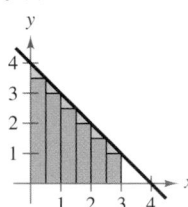   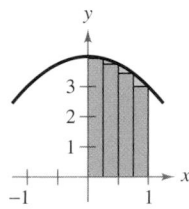

In Exercises 97 and 98, complete the table showing the approximate area of the region in the graph using n rectangles of equal width.

n	4	8	20	50
Approximate area				

97. $f(x) = \frac{1}{4}x^2$ **98.** $f(x) = 4x - x^2$

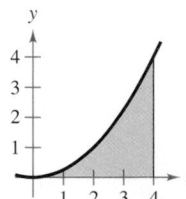   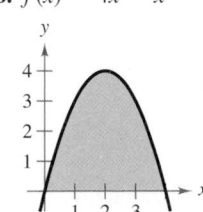

In Exercises 99–106, use the limit process to find the area of the region between the graph of the function and the x-axis over the specified interval.

	Function	Interval
99.	$f(x) = 10 - x$	$[0, 10]$
100.	$f(x) = 2x - 6$	$[3, 6]$
101.	$f(x) = x^2 + 4$	$[-1, 2]$
102.	$f(x) = 8(x - x^2)$	$[0, 1]$
103.	$f(x) = x^3 + 1$	$[0, 2]$
104.	$f(x) = 1 - x^3$	$[-3, -1]$
105.	$f(x) = 3(x^3 - x^2)$	$[1, 3]$
106.	$f(x) = 5 - (x + 2)^2$	$[-2, 0]$

107. *Geometry*   The table shows the measurements (in feet) of a lot bounded by a stream and two straight roads that meet at right angles (see figure).

x	0	100	200	300	400	500
y	125	125	120	112	90	90

x	600	700	800	900	1000
y	95	88	75	35	0

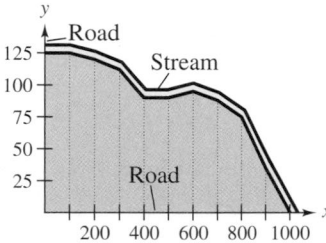

(a) Use the *regression* feature of a graphing utility to find a model of the form $y = ax^3 + bx^2 + cx + d$.

(b) Use a graphing utility to plot the data and graph the model in the same viewing window.

(c) Use the model in part (a) to estimate the area of the lot.

Synthesis

True or False?   In Exercises 108 and 109, determine whether the statement is true or false. Justify your answer.

108. The limit of the sum of two functions is the sum of the limits of the two functions.

109. If the degree of the numerator $N(x)$ of a rational function $f(x) = N(x)/D(x)$ is greater than the degree of its denominator $D(x)$, then the limit of the rational function as x approaches ∞ is 0.

110. *Writing*   Write a short paragraph explaining several reasons why the limit of a function may not exist.

11   Chapter Test

See www.CalcChat.com for worked-out solutions to odd-numbered exercises.

Take this test as you would take a test in class. After you are finished, check your work against the answers given in the back of the book.

In Exercises 1–3, use a graphing utility to graph the function and approximate the limit (if it exists). Then find the limit (if it exists) algebraically by using appropriate techniques.

1. $\lim\limits_{x \to -2} \dfrac{x^2 - 1}{2x}$ **2.** $\lim\limits_{x \to 1} \dfrac{-x^2 + 5x - 3}{1 - x}$ **3.** $\lim\limits_{x \to 5} \dfrac{\sqrt{x} - 2}{x - 5}$

In Exercises 4 and 5, use a graphing utility to graph the function and approximate the limit. Write an approximation that is accurate to four decimal places. Then create a table to verify your limit numerically.

4. $\lim\limits_{x \to 0} \dfrac{\sin 3x}{x}$ **5.** $\lim\limits_{x \to 0} \dfrac{e^{2x} - 1}{x}$

6. Find a formula for the slope of the graph of f at the point $(x, f(x))$. Then use it to find the slope at the specified point.

(a) $f(x) = 3x^2 - 5x - 2$, $(2, 0)$ (b) $f(x) = 2x^3 + 6x$, $(-1, -8)$

In Exercises 7–9, find the derivative of the function.

7. $f(x) = 5 - \dfrac{2}{5}x$ **8.** $f(x) = 2x^2 + 4x - 1$ **9.** $f(x) = \dfrac{1}{x + 3}$

In Exercises 10–12, find the limit (if it exists). If the limit does not exist, explain why. Use a graphing utility to verify your result graphically.

10. $\lim\limits_{x \to \infty} \dfrac{6}{5x - 1}$ **11.** $\lim\limits_{x \to \infty} \dfrac{1 - 3x^2}{x^2 - 5}$ **12.** $\lim\limits_{x \to -\infty} \dfrac{x^2}{3x + 2}$

In Exercises 13 and 14, write the first five terms of the sequence and find the limit of the sequence (if it exists). If the limit does not exist, explain why. Assume n begins with 1.

13. $a_n = \dfrac{n^2 + 3n - 4}{2n^2 + n - 2}$ **14.** $a_n = \dfrac{1 + (-1)^n}{n}$

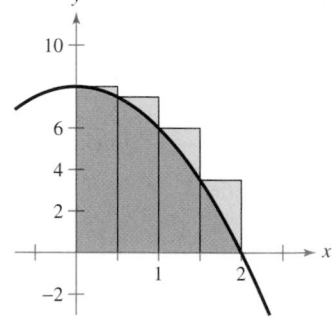

Figure for 15

15. Approximate the area of the region bounded by the graph of $f(x) = 8 - 2x^2$ shown at the right using the indicated number of rectangles of equal width.

In Exercises 16 and 17, use the limit process to find the area of the region between the graph of the function and the x-axis over the specified interval.

16. $f(x) = x + 2$; interval: $[-2, 2]$ **17.** $f(x) = 3 - x^2$; interval: $[-1, 1]$

18. The table shows the height of a space shuttle during its first 5 seconds of motion.
(a) Use the *regression* feature of a graphing utility to find a quadratic model $y = ax^2 + bx + c$ for the data.
(b) The value of the derivative of the model is the rate of change of height with respect to time, or the velocity, at that instant. Find the velocity of the shuttle after 5 seconds.

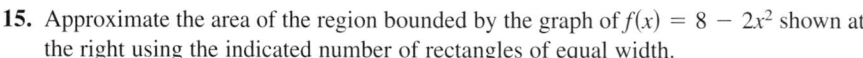

Time (seconds), x	Height (feet), y
0	0
1	1
2	23
3	60
4	115
5	188

Table for 18

10–11 Cumulative Test

See www.CalcChat.com for worked-out solutions to odd-numbered exercises.

Take this test to review the material in Chapters 10 and 11. After you are finished, check your work against the answers given in the back of the book.

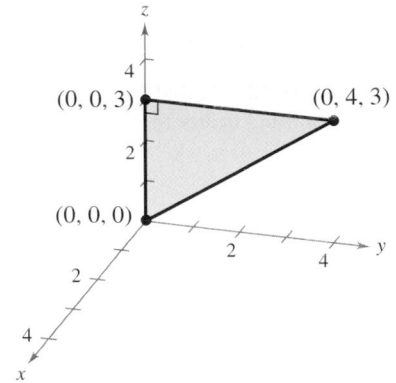

Figure for 4

In Exercises 1 and 2, find the coordinates of the point.

1. The point is located six units behind the yz-plane, one unit to the right of the xz-plane, and three units above the xy-plane.

2. The point is located on the y-axis, four units to the left of the xz-plane.

3. Find the distance between the points $(-2, 3, -6)$ and $(4, -5, 1)$.

4. Find the lengths of the sides of the right triangle at the right. Show that these lengths satisfy the Pythagorean Theorem.

5. Find the coordinates of the midpoint of the line segment joining $(3, 4, -1)$ and $(-5, 0, 2)$.

6. Find an equation of the sphere for which the endpoints of a diameter are $(0, 0, 0)$ and $(4, 4, 8)$.

7. Sketch the graph of the equation $(x - 2)^2 + (y + 1)^2 + z^2 = 4$, and then sketch the xy-trace and the yz-trace.

8. For the vectors $\mathbf{u} = \langle 2, -6, 0 \rangle$ and $\mathbf{v} = \langle -4, 5, 3 \rangle$, find $\mathbf{u} \cdot \mathbf{v}$ and $\mathbf{u} \times \mathbf{v}$.

In Exercises 9–11, determine whether u and v are orthogonal, parallel, or neither.

9. $\mathbf{u} = \langle 4, 4, 0 \rangle$
   $\mathbf{v} = \langle 0, -8, 6 \rangle$

10. $\mathbf{u} = \langle 4, -2, 10 \rangle$
    $\mathbf{v} = \langle -2, 6, 2 \rangle$

11. $\mathbf{u} = \langle -1, 6, -3 \rangle$
    $\mathbf{v} = \langle 3, -18, 9 \rangle$

12. Find the volume of the parallelepiped with the vertices $A(1, 3, 2)$, $B(3, 4, 2)$, $C(3, 2, 2)$, $D(1, 1, 2)$, $E(1, 3, 5)$, $F(3, 4, 5)$, $G(3, 2, 5)$, and $H(1, 1, 5)$.

13. Find sets of (a) parametric equations and (b) symmetric equations for the line passing through the points $(-2, 3, 0)$ and $(5, 8, 25)$.

14. Find the parametric form of the equation of the line passing through the point $(-1, 2, 0)$ and perpendicular to $2x - 4y + z = 8$.

15. Find an equation of the plane passing through the points $(0, 0, 0)$, $(-2, 3, 0)$, and $(5, 8, 25)$.

16. Label the intercepts and sketch the graph of the plane given by $3x - 6y - 12z = 24$.

17. Find the distance between the point $(0, 0, 25)$ and the plane $2x - 5y + z = 10$.

18. A plastic wastebasket has the shape and dimensions shown in the figure. In fabricating a mold for making the wastebasket, it is necessary to know the angle between two adjacent sides. Find the angle.

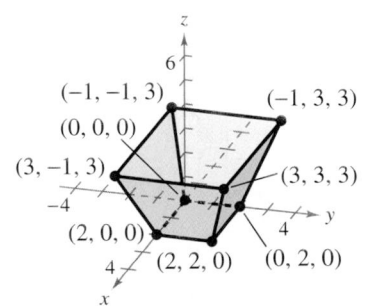

Figure for 18

In Exercises 19–27, find the limit (if it exists). If the limit does not exist, explain why. Use a graphing utility to verify your result graphically.

19. $\displaystyle\lim_{x \to 4} (5x - x^2)$

20. $\displaystyle\lim_{x \to -2^+} \frac{x + 2}{x^2 + x - 2}$

21. $\displaystyle\lim_{x \to 7} \frac{x - 7}{x^2 - 49}$

22. $\displaystyle\lim_{x \to 0} \frac{\sqrt{x + 4} - 2}{x}$

23. $\displaystyle\lim_{x \to 4^-} \frac{|x - 4|}{x - 4}$

24. $\displaystyle\lim_{x \to 0} \sin\left(\frac{\pi}{x}\right)$

25. $\displaystyle\lim_{x \to 0} \frac{\dfrac{1}{x + 2} - \dfrac{1}{2}}{x}$

26. $\displaystyle\lim_{x \to 0} \frac{\sqrt{x + 1} - 1}{x}$

27. $\displaystyle\lim_{x \to 2^-} \frac{x - 2}{x^2 - 4}$

In Exercises 28–31, find a formula for the slope of f at the point $(x, f(x))$. Then use it to find the slope at the specified point.

28. $f(x) = 4 - x^2$, $(0, 4)$

29. $f(x) = \sqrt{x + 3}$, $(-2, 1)$

30. $f(x) = \dfrac{1}{x + 3}$, $\left(1, \dfrac{1}{4}\right)$

31. $f(x) = x^2 - x$, $(1, 0)$

In Exercises 32–37, find the limit (if it exists). If the limit does not exist, explain why. Use a graphing utility to verify your result graphically.

32. $\displaystyle\lim_{x \to \infty} \dfrac{x^3}{x^2 - 9}$

33. $\displaystyle\lim_{x \to \infty} \dfrac{3 - 7x}{x + 4}$

34. $\displaystyle\lim_{x \to \infty} \dfrac{3x^2 + 1}{x^2 + 4}$

35. $\displaystyle\lim_{x \to \infty} \dfrac{2x}{x^2 + 3x - 2}$

36. $\displaystyle\lim_{x \to \infty} \dfrac{3 - x}{x^2 + 1}$

37. $\displaystyle\lim_{x \to \infty} \dfrac{3 + 4x - x^3}{2x^2 + 3}$

In Exercises 38–40, evaluate the sum using the summation formulas and properties.

38. $\displaystyle\sum_{i=1}^{50} (1 - i^2)$

39. $\displaystyle\sum_{k=1}^{20} (3k^2 - 2k)$

40. $\displaystyle\sum_{i=1}^{40} (12 + i^3)$

In Exercises 41–44, approximate the area of the region using the indicated number of rectangles of equal width.

41.

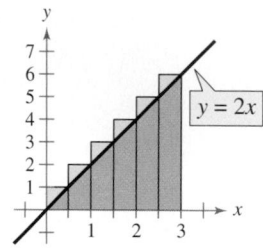

42.

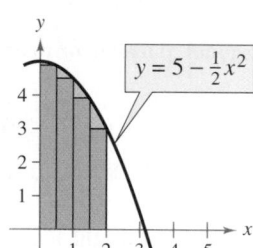

43.

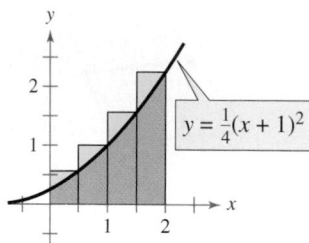

44.

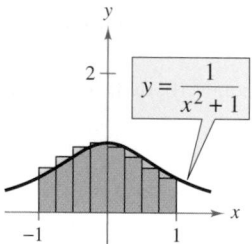

In Exercises 45–50, use the limit process to find the area of the region between the graph of the function and the x-axis over the specified interval.

45. $f(x) = x + 2$

Interval: $[0, 1]$

46. $f(x) = 8 - 2x$

Interval: $[-4, 4]$

47. $f(x) = 2x + 5$

Interval: $[-1, 3]$

48. $f(x) = x^2 + 1$

Interval: $[0, 4]$

49. $f(x) = 4 - x^2$

Interval: $[0, 2]$

50. $f(x) = 1 - x^3$

Interval: $[0, 1]$

Proofs in Mathematics

Many of the proofs of the definitions and properties presented in this chapter are beyond the scope of this text. Included below are simple proofs for the limit of a power function and the limit of a polynomial function.

Limit of a Power Function (*p. 785*)

$\lim\limits_{x \to c} x^n = c^n$, c is a real number and n is a positive integer.

Proof

$$\lim_{x \to c} x^n = \lim_{x \to c} \underbrace{(x \cdot x \cdot x \cdot \cdots \cdot x)}_{n \text{ factors}}$$

$$= \underbrace{\lim_{x \to c} x \cdot \lim_{x \to c} x \cdot \lim_{x \to c} x \cdot \cdots \cdot \lim_{x \to c} x}_{n \text{ factors}} \qquad \text{Product Property of Limits}$$

$$= \underbrace{c \cdot c \cdot c \cdot \cdots \cdot c}_{n \text{ factors}} \qquad \text{Limit of the identity function}$$

$$= c^n \qquad \text{Exponential form}$$

Limit of a Polynomial Function (*p. 787*)

If p is a polynomial function and c is a real number, then

$$\lim_{x \to c} p(x) = p(c).$$

Proof

Let p be a polynomial function such that

$$p(x) = a_n x^n + a_{n-1} x^{n-1} + \cdots + a_2 x^2 + a_1 x + a_0.$$

Because a polynomial function is the sum of monomial functions, you can write the following.

$$\lim_{x \to c} p(x) = \lim_{x \to c} \left(a_n x^n + a_{n-1} x^{n-1} + \cdots + a_2 x^2 + a_1 x + a_0 \right)$$

$$= \lim_{x \to c} a_n x^n + \lim_{x \to c} a_{n-1} x^{n-1} + \cdots + \lim_{x \to c} a_2 x^2 + \lim_{x \to c} a_1 x + \lim_{x \to c} a_0$$

$$= a_n c^n + a_{n-1} c^{n-1} + \cdots + a_2 c^2 + a_1 c + a_0 \qquad \text{Scalar Multiple Property of Limits and limit of a power function}$$

$$= p(c) \qquad \text{p evaluated at c}$$

Proving Limits

To prove most of the definitions and properties from this chapter, you must use the *formal definition of a limit*. This definition is called the *epsilon-delta definition* and was first introduced by Karl Weierstrass (1815–1897). If you go on to take a course in calculus, you will use this definition of a limit extensively.

Progressive Summary (Chapters 3–11)

This chart outlines the topics that have been covered so far in this text. Progressive summary charts appear after Chapters 2, 3, 6, 9, and 11. In each progressive summary, new topics encountered for the first time appear in red.

Transcendental Functions	Systems and Series	Other Topics
Exponential, Logarithmic, Trigonometric, Inverse Trigonometric	Systems, Sequences, Series	Conics, Parametric and Polar Equations, 3-space, Limits

■ Rewriting

Transcendental Functions	Systems and Series	Other Topics
Exponential form $\leftrightarrow$ Logarithmic form	Row operations for systems of equations	Standard forms of conics
Condense/expand logarithmic functions	Partial fraction decomposition	Eliminate parameters
Simplify trigonometric expressions	Operations with matrices	Rectangular form $\leftrightarrow$ Parametric form
Prove trigonometric identities	Matrix form of a system of equations	Rectangular form $\leftrightarrow$ Polar form
Use conversion formulas	nth term of a sequence	Rationalize numerator
Operations with vectors	Summation form of a series	Difference quotient
Powers of roots and complex numbers		

■ Solving

Transcendental Functions

Equation	Strategy
Exponential	Take logarithm of each side
Logarithmic	Exponentiate each side
Trigonometric	Isolate function or factor. Use inverse function
Multiple angle or high powers	Use trigonometric identities

Systems and Series

Equation	Strategy
System of linear equations	Substitution / Elimination / Gaussian / Gauss-Jordan / Inverse matrices / Cramer's Rule

Other Topics

Equation	Strategy
Conics	Convert to standard form. Convert to polar form
3-space	Equation of plane / Equation of line / Cross-product of vectors
Limits	Direct substitution / One-sided limits / Limits at infinity

■ Analyzing

Transcendental Functions

Graphically	Algebraically
Intercepts	Domain, Range
Asymptotes	Transformations
Minimum values	Composition
Maximum values	Inverse properties
	Amplitude, period
	Reference angles

Numerically
Table of values

Systems and Series

Systems:
  Intersecting, parallel, and coincident lines, determinants
Sequences:
  Graphing utility in *dot* mode, nth term, partial sums, summation formulas

Other Topics

Conics:
  Table of values, vertices, foci, axes, symmetry, asymptotes, translations, eccentricity
Parametric forms:
  Point plotting, eliminate parameters
Polar forms:
  Point plotting, special equations, symmetry, zeros, eccentricity, maximum r-values, directrix
3-space:
  Point plotting, intercepts, traces, vectors

Appendix A: Technology Support

Introduction

Graphing utilities such as graphing calculators and computers with graphing software are very valuable tools for visualizing mathematical principles, verifying solutions to equations, exploring mathematical ideas, and developing mathematical models. Although graphing utilities are extremely helpful in learning mathematics, their use does not mean that learning algebra is any less important. In fact, the combination of knowledge of mathematics and the use of graphing utilities enables you to explore mathematics more easily and to a greater depth. If you are using a graphing utility in this course, it is up to you to learn its capabilities and to practice using this tool to enhance your mathematical learning.

In this text, there are many opportunities to use a graphing utility, some of which are described below.

Uses of a Graphing Utility

  1. Check or validate answers to problems obtained using algebraic methods.

  2. Discover and explore algebraic properties, rules, and concepts.

  3. Graph functions, and approximate solutions to equations involving functions.

  4. Efficiently perform complicated mathematical procedures such as those found in many real-life applications.

  5. Find mathematical models for sets of data.

In this appendix, the features of graphing utilities are discussed from a generic perspective and are listed in alphabetical order. To learn how to use the features of a specific graphing utility, consult your user's manual or go to this textbook's *Online Study Center*. Additional keystroke guides are available for most graphing utilities, and your college library may have a videotape on how to use your graphing utility.

Many graphing utilities are designed to act as "function graphers." In this course, functions and their graphs are studied in detail. You may recall from previous courses that a function can be thought of as a rule that describes the relationship between two variables. These rules are frequently written in terms of x and y. For example, the equation

$$y = 3x + 5$$

represents y as a function of x.

Many graphing utilities have an *equation editor* feature that requires that an equation be written in "$y =$" form in order to be entered, as shown in Figure A.1. (You should note that your *equation editor* screen may not look like the screen shown in Figure A.1.)

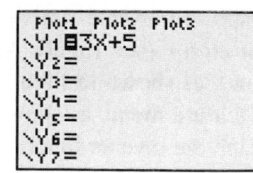

Figure A.1

Cumulative Sum Feature

The *cumulative sum* feature finds partial sums of a series. For example, to find the first four partial sums of the series

$$\sum_{k=1}^{4} 2(0.1)^k$$

choose the *cumulative sum* feature, which is found in the *operations* menu of the *list* feature (see Figure A.2). To use this feature, you will also have to use the *sequence* feature (see Figure A.2 and page A15). You must enter an expression for the sequence, a variable, the lower limit of summation, and the upper limit of summation, as shown in Figure A.3. After pressing ENTER, you can see that the first four partial sums are 0.2, 0.22, 0.222, and 0.2222. You may have to scroll to the right in order to see all the partial sums.

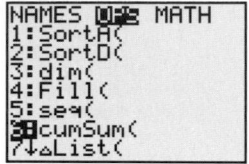

Figure A.2

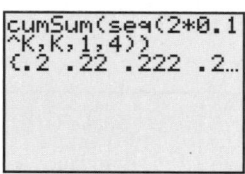

Figure A.3

Determinant Feature

The *determinant* feature evaluates the determinant of a square matrix. For example, to evaluate the determinant of the matrix shown at the right, enter the 3×3 matrix in the graphing utility using the *matrix editor*, as shown in Figure A.4. Then choose the *determinant* feature from the *math* menu of the *matrix* feature, as shown in Figure A.5. Once you choose the matrix name, A, press ENTER and you should obtain a determinant of -50, as shown in Figure A.6.

$$A = \begin{bmatrix} 7 & -1 & 0 \\ 2 & 2 & 3 \\ -6 & 4 & 1 \end{bmatrix}$$

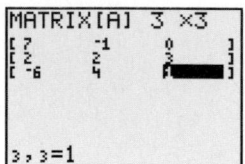

Figure A.4

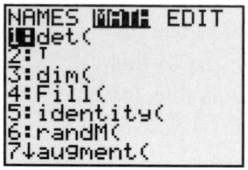

Figure A.5

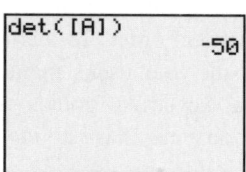

Figure A.6

Draw Inverse Feature

The *draw inverse* feature graphs the inverse function of a one-to-one function. For instance, to graph the inverse function of $f(x) = x^3 + 4$, first enter the function in the *equation editor* (see Figure A.7) and graph the function (using a square viewing window), as shown in Figure A.8. Then choose the *draw inverse* feature from the *draw* feature menu, as shown in Figure A.9. You must enter the function you want to graph the inverse function of, as shown in Figure A.10. Finally, press ENTER to obtain the graph of the inverse function of $f(x) = x^3 + 4$, as shown in Figure A.11. This feature can be used only when the graphing utility is in *function* mode.

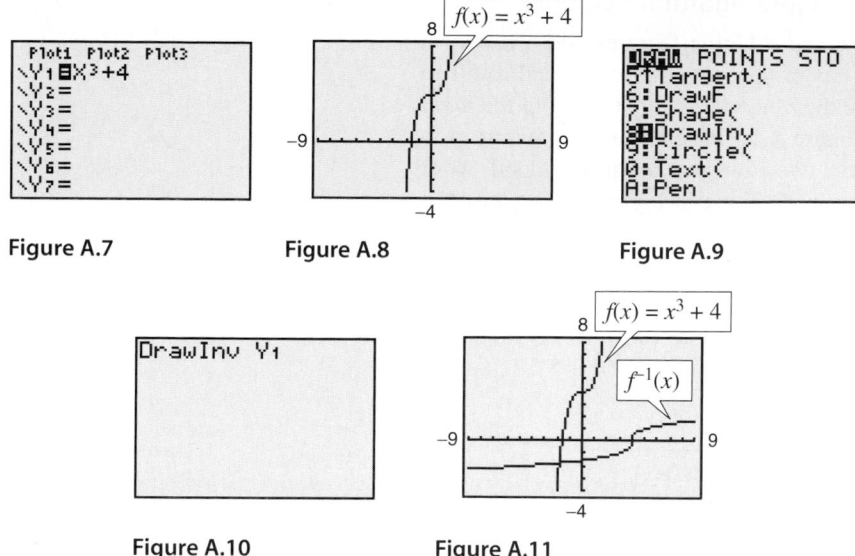

Figure A.7   Figure A.8   Figure A.9

Figure A.10   Figure A.11

Elementary Row Operations Features

Most graphing utilities can perform elementary row operations on matrices.

Row Swap Feature

The *row swap* feature interchanges two rows of a matrix. To interchange rows 1 and 3 of the matrix shown at the right, first enter the matrix in the graphing utility using the *matrix editor*, as shown in Figure A.12. Then choose the *row swap* feature from the *math* menu of the *matrix* feature, as shown in Figure A.13. When using this feature, you must enter the name of the matrix and the two rows that are to be interchanged. After pressing ENTER, you should obtain the matrix shown in Figure A.14. Because the resulting matrix will be used to demonstrate the other elementary row operation features, use the *store* feature to copy the resulting matrix to [A], as shown in Figure A.15.

$$A = \begin{bmatrix} -1 & -2 & 1 & 2 \\ 2 & -4 & 6 & -2 \\ 1 & 3 & -3 & 0 \end{bmatrix}$$

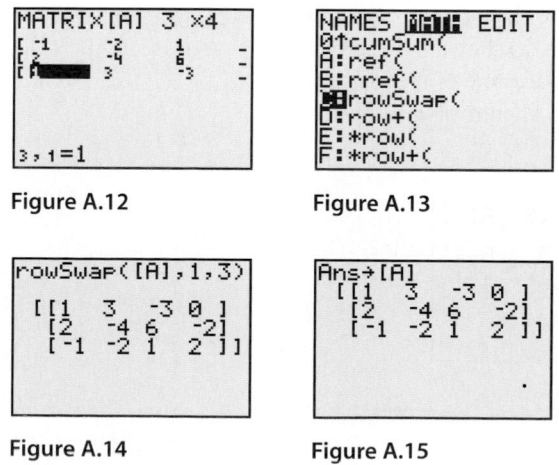

Figure A.12   Figure A.13

Figure A.14   Figure A.15

TECHNOLOGY TIP

The *store* feature of a graphing utility is used to store a value in a variable or to copy one matrix to another matrix. For instance, as shown at the left, after performing a row operation on a matrix, you can copy the answer to another matrix (see Figure A.15). You can then perform another row operation on the copied matrix. If you want to continue performing row operations to obtain a matrix in row-echelon form or reduced row-echelon form, you must copy the resulting matrix to a new matrix before each operation.

Row Addition and Row Multiplication and Addition Features

The *row addition* and *row multiplication and addition* features add a row or a multiple of a row of a matrix to another row of the same matrix. To add row 1 to row 3 of the matrix stored in [A], choose the *row addition* feature from the *math* menu of the *matrix* feature, as shown in Figure A.16. When using this feature, you must enter the name of the matrix and the two rows that are to be added. After pressing (ENTER), you should obtain the matrix shown in Figure A.17. Copy the resulting matrix to [A].

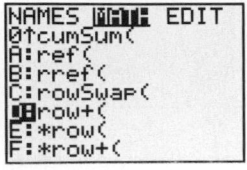

Figure A.16   **Figure A.17**

To add -2 times row 1 to row 2 of the matrix stored in [A], choose the *row multiplication and addition* feature from the *math* menu of the *matrix* feature, as shown in Figure A.18. When using this feature, you must enter the constant, the name of the matrix, the row the constant is multiplied by, and the row to be added to. After pressing (ENTER), you should obtain the matrix shown in Figure A.19. Copy the resulting matrix to [A].

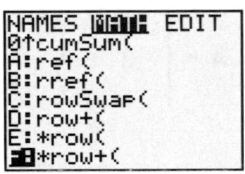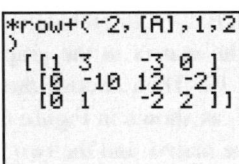

Figure A.18   **Figure A.19**

Row Multiplication Feature

The *row multiplication* feature multiplies a row of a matrix by a nonzero constant. To multiply row 2 of the matrix stored in [A] by $-\frac{1}{10}$, choose the *row multiplication* feature from the *math* menu of the *matrix* feature, as shown in Figure A.20. When using this feature, you must enter the constant, the name of the matrix, and the row to be multiplied. After pressing (ENTER), you should obtain the matrix shown in Figure A.21.

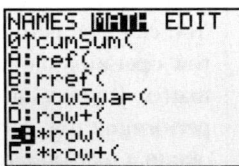

Figure A.20   **Figure A.21**

Intersect Feature

The *intersect* feature finds the point(s) of intersection of two graphs. The *intersect* feature is found in the *calculate* menu (see Figure A.22). To find the point(s) of intersection of the graphs of $y_1 = -x + 2$ and $y_2 = x + 4$, first enter the equations in the *equation editor*, as shown in Figure A.23. Then graph the equations, as shown in Figure A.24. Next, use the *intersect* feature to find the point of intersection. Trace the cursor along the graph of y_1 near the intersection and press (ENTER) (see Figure A.25). Then trace the cursor along the graph of y_2 near the intersection and press (ENTER) (see Figure A.26). Marks are then placed on the graph at these points (see Figure A.27). Finally, move the cursor near the point of intersection and press (ENTER). In Figure A.28, you can see that the coordinates of the point of intersection are displayed at the bottom of the window. So, the point of intersection is $(-1, 3)$.

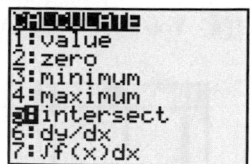

Figure A.22

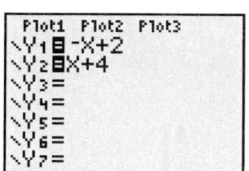

Figure A.23

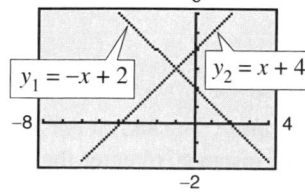

Figure A.24

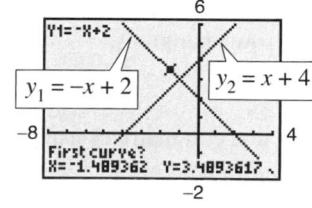

Figure A.25

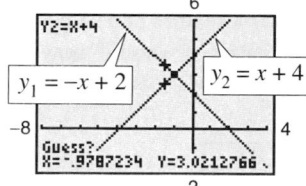

Figure A.26

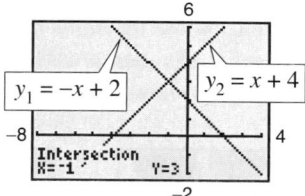

Figure A.27

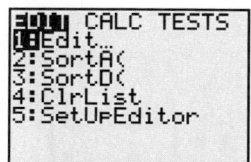

Figure A.28

List Editor

Most graphing utilities can store data in lists. The *list editor* can be used to create tables and to store statistical data. The *list editor* can be found in the *edit* menu of the *statistics* feature, as shown in Figure A.29. To enter the numbers 1 through 10 in a list, first choose a list (L_1) and then begin entering the data into each row, as shown in Figure A.30.

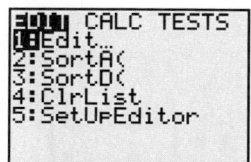

Figure A.29

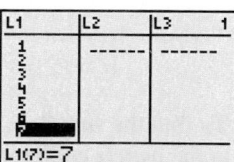

Figure A.30

You can also attach a formula to a list. For instance, you can multiply each of the data values in L_1 by 3. First, display the *list editor* and move the cursor to the top line. Then move the cursor onto the list to which you want to attach the formula (L_2). Finally, enter the formula 3* L_1 (see Figure A.31) and then press ENTER. You should obtain the list shown in Figure A.32.

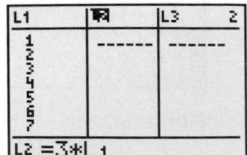

Figure A.31   Figure A.32

Matrix Feature

The *matrix* feature of a graphing utility has many uses, such as evaluating a determinant and performing row operations.

Matrix Editor

You can define, display, and edit matrices using the *matrix editor*. The *matrix editor* can be found in the *edit* menu of the *matrix* feature. For instance, to enter the matrix shown at the right, first choose the matrix name [A], as shown in Figure A.33. Then enter the dimension of the matrix (in this case, the dimension is 2×3) and enter the entries of the matrix, as shown in Figure A.34. To display the matrix on the home screen, choose the *name* menu of the *matrix* feature and select the matrix [A] (see Figure A.35), then press ENTER. The matrix A should now appear on the home screen, as shown in Figure A.36.

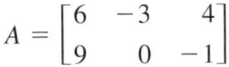

$$A = \begin{bmatrix} 6 & -3 & 4 \\ 9 & 0 & -1 \end{bmatrix}$$

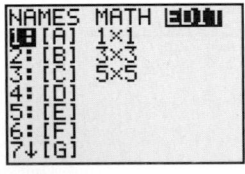

Figure A.33   Figure A.34

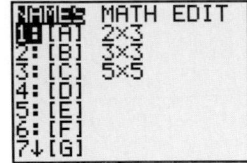 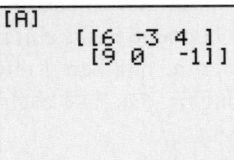

Figure A.35   Figure A.36

Matrix Operations

Most graphing utilities can perform matrix operations. To find the sum $A + B$ of the matrices shown at the right, first enter the matrices in the *matrix editor* as [A] and [B]. Then find the sum, as shown in Figure A.37. Scalar multiplication can be performed in a similar manner. For example, you can evaluate $7A$, where A is the matrix at the right, as shown in Figure A.38. To find the product AB of the matrices A and B at the right, first be sure that the product is defined. Because the number of columns of A (2 columns) equals the number of rows of B (2 rows), you can find the product AB, as shown in Figure A.39.

$$A = \begin{bmatrix} -3 & 5 \\ 0 & 4 \end{bmatrix}$$

$$B = \begin{bmatrix} 7 & -2 \\ -1 & 2 \end{bmatrix}$$

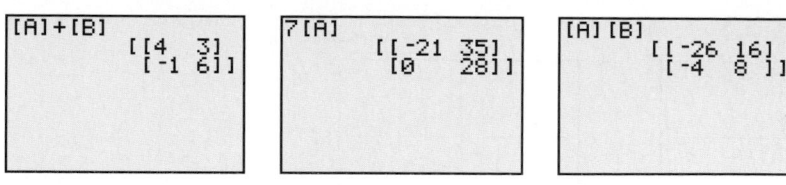

Figure A.37   **Figure A.38**   **Figure A.39**

Inverse Matrix

Some graphing utilities may not have an *inverse matrix* feature. However, you can find the inverse of a square matrix by using the inverse key $\boxed{x^{-1}}$. To find the inverse of the matrix shown at the right, enter the matrix in the *matrix editor* as [A]. Then find the inverse, as shown in Figure A.40.

$$A = \begin{bmatrix} 1 & -2 & 1 \\ -1 & 3 & 0 \\ 2 & 4 & 5 \end{bmatrix}$$

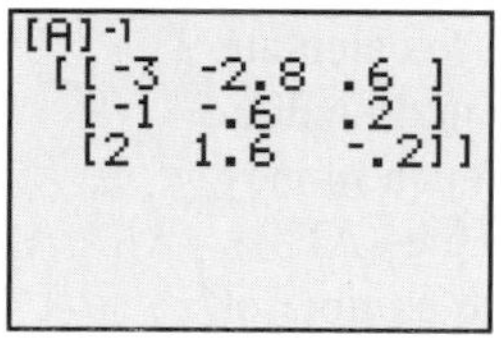

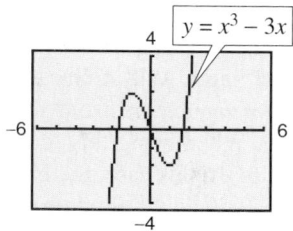

Figure A.40

Maximum and Minimum Features

The *maximum* and *minimum* features find relative extrema of a function. For instance, the graph of $y = x^3 - 3x$ is shown in Figure A.41. In the figure, the graph appears to have a relative maximum at $x = -1$ and a relative minimum at $x = 1$. To find the exact values of the relative extrema, you can use the *maximum* and *minimum* features found in the *calculate* menu (see Figure A.42). First, to find the relative maximum, choose the *maximum* feature and trace the cursor along the graph to a point left of the maximum and press $\boxed{\text{ENTER}}$ (see Figure A.43). Then trace the cursor along the graph to a point right of the maximum and press $\boxed{\text{ENTER}}$ (see Figure A.44). Note the two arrows near the top of the display marking the left and right bounds, as shown in Figure A.45. Next, trace the cursor along the graph between the two bounds and as close to the maximum as you can (see Figure A.45) and press $\boxed{\text{ENTER}}$. In Figure A.46, you can see that the coordinates of the maximum point are displayed at the bottom of the window. So, the relative maximum is $(-1, 2)$.

Figure A.41

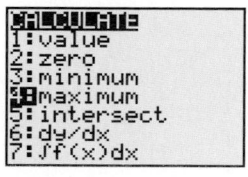

Figure A.42

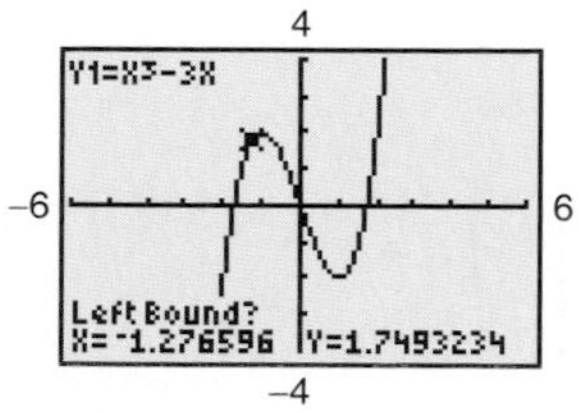

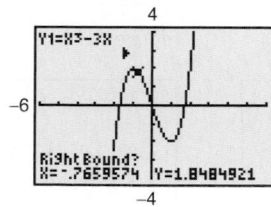

Figure A.43   **Figure A.44**

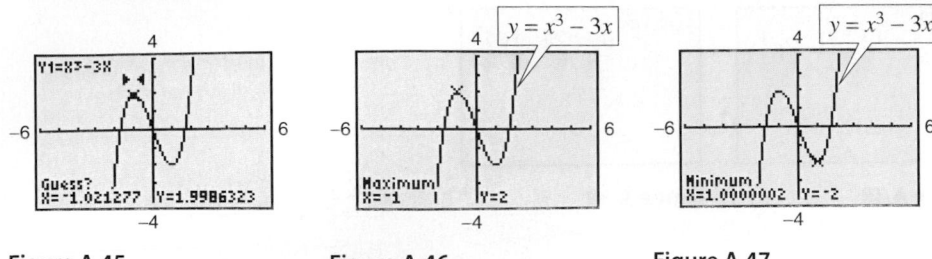

Figure A.45   **Figure A.46**   **Figure A.47**

You can find the relative minimum in a similar manner. In Figure A.47, you can see that the relative minimum is $(1, -2)$.

Mean and Median Features

In real-life applications, you often encounter large data sets and want to calculate statistical values. The *mean* and *median* features calculate the mean and median of a data set. For instance, in a survey, 100 people were asked how much money (in dollars) per week they withdraw from an automatic teller machine (ATM). The results are shown in the table below. The frequency represents the number of responses.

Amount	10	20	30	40	50	60	70	80	90	100
Frequency	3	8	10	19	24	13	13	7	2	1

To find the mean and median of the data set, first enter the data in the *list editor*, as shown in Figure A.48. Enter the amount in L_1 and the frequency in L_2. Then choose the *mean* feature from the *math* menu of the *list* feature, as shown in Figure A.49. When using this feature, you must enter a list and a frequency list (if applicable). In this case, the list is L_1 and the frequency list is L_2. After pressing ⟨ENTER⟩, you should obtain a mean of $49.80, as shown in Figure A.50. You can follow the same steps (except choose the *median* feature) to find the median of the data. You should obtain a median of $50, as shown in Figure A.51.

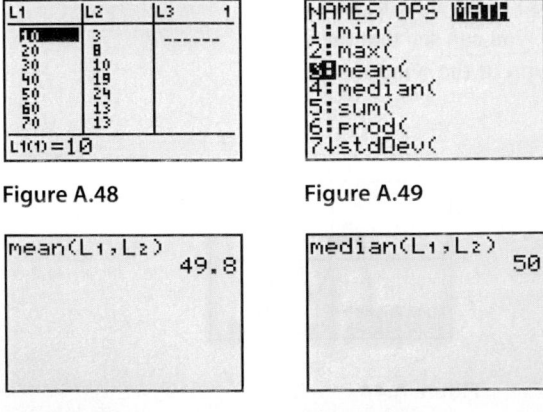

Figure A.48   **Figure A.49**

Figure A.50   **Figure A.51**

Mode Settings

Mode settings of a graphing utility control how the utility displays and interprets numbers and graphs. The default mode settings are shown in Figure A.52.

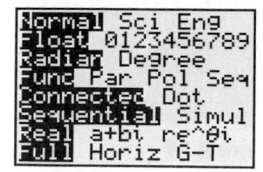

Figure A.52

Radian and Degree Modes

The trigonometric functions can be applied to angles measured in either radians or degrees. When your graphing utility is in *radian* mode, it interprets angle values as radians and displays answers in radians. When your graphing utility is in *degree* mode, it interprets angle values as degrees and displays answers in degrees. For instance, to calculate $\sin(\pi/6)$, make sure the calculator is in *radian* mode. You should obtain an answer of 0.5, as shown in Figure A.53. To calculate $\sin 45°$, make sure the calculator is in *degree* mode, as shown in Figure A.54. You should obtain an approximate answer of 0.7071, as shown in Figure A.55. If you did not change the mode of the calculator before evaluating $\sin 45°$, you would obtain an answer of approximately 0.8509, which is the sine of 45 radians.

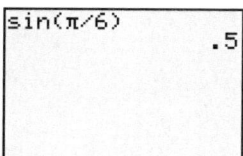
Figure A.53

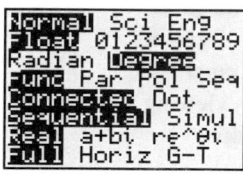
Figure A.54

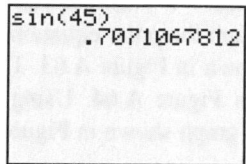
Figure A.55

Function, Parametric, Polar, and Sequence Modes

Most graphing utilities can graph using four different modes.

Function Mode   The *function* mode is used to graph standard algebraic and trigonometric functions. For instance, to graph $y = 2x^2$, use the *function* mode, as shown in Figure A.52. Then enter the equation in the *equation editor*, as shown in Figure A.56. Using a standard viewing window (see Figure A.57), you obtain the graph shown in Figure A.58.

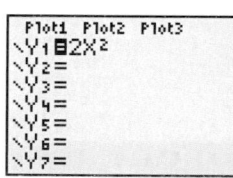
Figure A.56

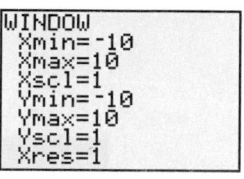
Figure A.57

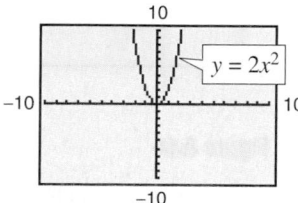
Figure A.58

Parametric Mode   To graph parametric equations such as $x = t + 1$ and $y = t^2$, use the *parametric* mode, as shown in Figure A.59. Then enter the equations in the *equation editor*, as shown in Figure A.60. Using the viewing window shown in Figure A.61, you obtain the graph shown in Figure A.62.

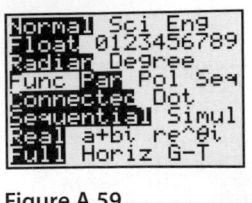

Figure A.59

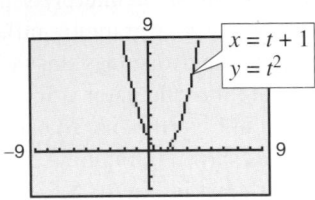

Figure A.60

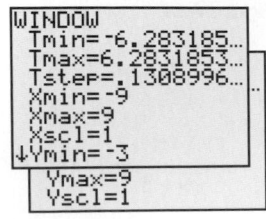

Figure A.61

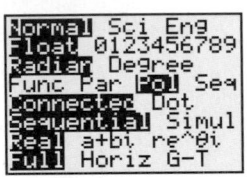

Figure A.62

Polar Mode   To graph polar equations of the form $r = f(\theta)$, you can use the *polar* mode of a graphing utility. For instance, to graph the polar equation $r = 2 \cos \theta$, use the *polar* mode (and *radian* mode), as shown in Figure A.63. Then enter the equation in the *equation editor*, as shown in Figure A.64. Using the viewing window shown in Figure A.65, you obtain the graph shown in Figure A.66.

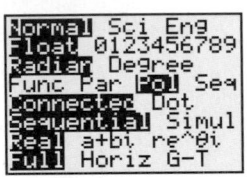

Figure A.63

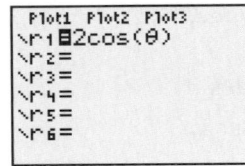

Figure A.64

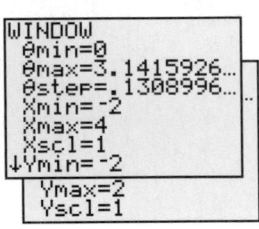

Figure A.65

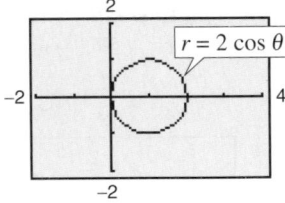

Figure A.66

Sequence Mode   To graph the first five terms of a sequence such as $a_n = 4n - 5$, use the *sequence* mode, as shown in Figure A.67. Then enter the sequence in the *equation editor*, as shown in Figure A.68 (assume that n begins with 1). Using the viewing window shown in Figure A.69, you obtain the graph shown in Figure A.70.

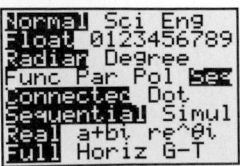

Figure A.67

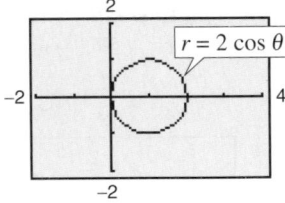

Figure A.68

TECHNOLOGY TIP

Note that when using the different graphing modes of a graphing utility, the utility uses different variables. When the utility is in *function* mode, it uses the variables x and y. In *parametric* mode, the utility uses the variables x, y, and t. In *polar* mode, the utility uses the variables r and θ. In *sequence* mode, the utility uses the variables u (instead of a) and n.

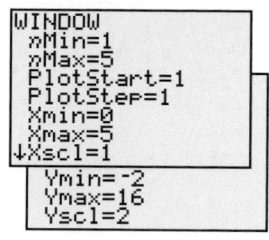

Figure A.69

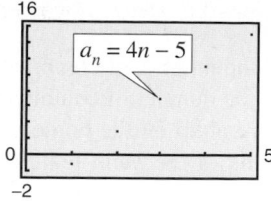

Figure A.70

Connected and Dot Modes

Graphing utilities use the point-plotting method to graph functions. When a graphing utility is in *connected* mode, the utility connects the points that are plotted. When the utility is in *dot* mode, it does not connect the points that are plotted. For example, the graph of $y = x^3$ in *connected* mode is shown in Figure A.71. To graph this function using *dot* mode, first change the mode to *dot* mode (see Figure A.72) and then graph the equation, as shown in Figure A.73. As you can see in Figure A.73, the graph is a collection of dots.

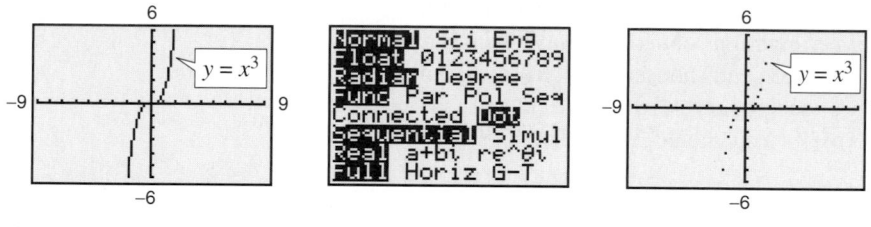

Figure A.71   Figure A.72   Figure A.73

A problem arises in some graphing utilities when the *connected* mode is used. Graphs with vertical asymptotes, such as rational functions and tangent functions, appear to be connected. For instance, the graph of

$$y = \frac{1}{x + 3}$$

is shown in Figure A.74. Notice how the two portions of the graph appear to be connected with a vertical line at $x = -3$. From your study of rational functions, you know that the graph has a vertical asymptote at $x = -3$ and therefore is undefined when $x = -3$. When using a graphing utility to graph rational functions and other functions that have vertical asymptotes, you should use the *dot* mode to eliminate extraneous vertical lines. Because the *dot* mode of a graphing utility displays a graph as a collection of dots rather than as a smooth curve, in this text, a blue or light red curve is placed behind the graphing utility's display to indicate where the graph should appear, as shown in Figure A.75.

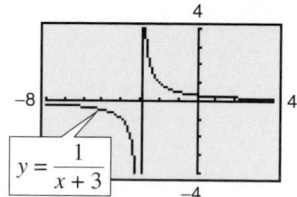

Figure A.74

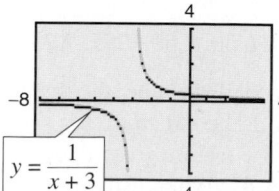

Figure A.75

$_nC_r$ Feature

The $_nC_r$ feature calculates binomial coefficients and the number of combinations of n elements taken r at a time. For example, to find the number of combinations of eight elements taken five at a time, enter 8 (the n-value) on the home screen and choose the $_nC_r$ feature from the *probability* menu of the *math* feature (see Figure A.76). Next, enter 5 (the r-value) on the home screen and press ENTER. You should obtain 56, as shown in Figure A.77.

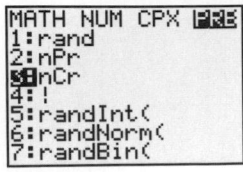

Figure A.76

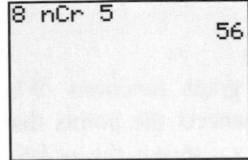

Figure A.77

$_nP_r$ Feature

The $_nP_r$ feature calculates the number of permutations of n elements taken r at a time. For example, to find the number of permutations of six elements taken four at a time, enter 6 (the n-value) on the home screen and choose the $_nP_r$ feature from the *probability* menu of the *math* feature (see Figure A.78). Next enter 4 (the r-value) on the home screen and press ENTER. You should obtain 360, as shown in Figure A.79.

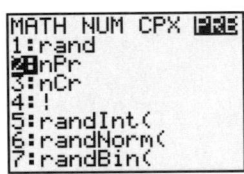

Figure A.78

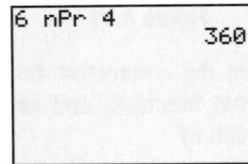

Figure A.79

One-Variable Statistics Feature

Graphing utilities are useful in calculating statistical values for a set of data. The *one-variable statistics* feature analyzes data with one measured variable. This feature outputs the mean of the data, the sum of the data, the sum of the data squared, the sample standard deviation of the data, the population standard deviation of the data, the number of data points, the minimum data value, the maximum data value, the first quartile of the data, the median of the data, and the third quartile of the data. Consider the following data, which shows the hourly earnings (in dollars) for 12 retail sales associates.

$$5.95,\ 8.15,\ 6.35,\ 7.05,\ 6.80,\ 6.10,\ 7.15,\ 8.20,\ 6.50,\ 7.50,\ 7.95,\ 9.25$$

You can use the *one-variable statistics* feature to determine the mean and standard deviation of the data. First, enter the data in the *list editor*, as shown in Figure A.80. Then choose the *one-variable statistics* feature from the *calculate* menu of the *statistics* feature, as shown in Figure A.81. When using this feature, you must enter a list. In this case, the list is L_1. In Figure A.82, you can see that the mean of the data is $\bar{x} \approx 7.25$ and the standard deviation of the data is $\sigma x \approx 0.95$.

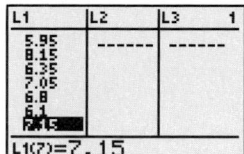

Figure A.80

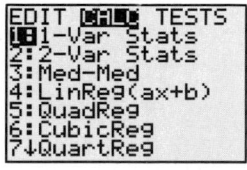

Figure A.81

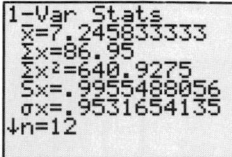

Figure A.82

Regression Feature

Throughout the text, you are asked to use the *regression* feature of a graphing utility to find models for sets of data. Most graphing utilities have built-in regression programs for the following.

Regression	*Form of Model*
Linear	$y = ax + b$ or $y = a + bx$
Quadratic	$y = ax^2 + bx + c$
Cubic	$y = ax^3 + bx^2 + cx + d$
Quartic	$y = ax^4 + bx^3 + cx^2 + dx + e$
Logarithmic	$y = a + b \ln(x)$
Exponential	$y = ab^x$
Power	$y = ax^b$
Logistic	$y = \dfrac{c}{1 + ae^{-bx}}$
Sine	$y = a \sin(bx + c) + d$

For example, you can find a linear model for the number y of television sets (in millions) in U.S. households in the years 1996 through 2005, shown in the table.   (Source: Television Bureau of Advertising, Inc.)

Year	Number, y
1996	222.8
1997	228.7
1998	235.0
1999	240.3
2000	245.0
2001	248.2
2002	254.4
2003	260.2
2004	268.3
2005	287.0

First, let x represent the year, with $x = 6$ corresponding to 1996. Then enter the data in the *list editor*, as shown in Figure A.83. Note that L_1 contains the years and L_2 contains the numbers of television sets that correspond to the years. Now choose the *linear regression* feature from the *calculate* menu of the *statistics* feature, as shown in Figure A.84. In Figure A.85, you can see that a linear model for the data is given by $y = 6.22x + 183.7$.

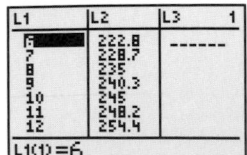

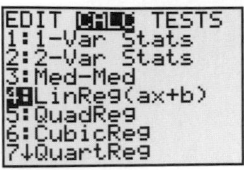

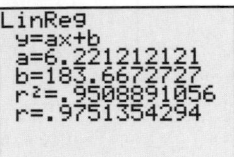

Figure A.83   **Figure A.84**   **Figure A.85**

When you use the *regression* feature of a graphing utility, you will notice that the program may also output an "*r*-value." (For some calculators, make sure you select the *diagnostics on* feature before you use the *regression* feature. Otherwise, the calculator will not output an *r*-value.) The *r*-value or *correlation coefficient* measures how well the linear model fits the data. The closer the value of $|r|$ is to 1, the better the fit. For the data above, $r \approx 0.97514$, which implies that the model is a good fit for the data.

Row-Echelon and Reduced Row-Echelon Features

Some graphing utilities have features that can automatically transform a matrix to row-echelon form and reduced row-echelon form. These features can be used to check your solutions to systems of equations.

Row-Echelon Feature

Consider the system of equations and the corresponding augmented matrix shown below.

$$\begin{array}{cc} \textit{Linear System} & \textit{Augmented Matrix} \\ \begin{cases} 2x + 5y - 3z = 4 \\ 4x + y \quad\quad = 2 \\ -x + 3y - 2z = -1 \end{cases} & \begin{bmatrix} 2 & 5 & -3 & \vdots & 4 \\ 4 & 1 & 0 & \vdots & 2 \\ -1 & 3 & -2 & \vdots & -1 \end{bmatrix} \end{array}$$

You can use the *row-echelon* feature of a graphing utility to write the augmented matrix in row-echelon form. First, enter the matrix in the graphing utility using the *matrix editor*, as shown in Figure A.86. Next, choose the *row-echelon* feature from the *math* menu of the *matrix* feature, as shown in Figure A.87. When using this feature, you must enter the name of the matrix. In this case, the name of the matrix is [A]. You should obtain the matrix shown in Figure A.88. You may have to scroll to the right in order to see all the entries of the matrix.

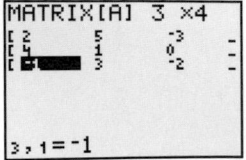

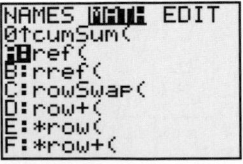

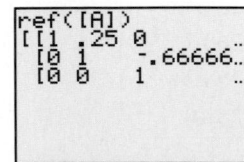

Figure A.86   **Figure A.87**   **Figure A.88**

Reduced Row-Echelon Feature

To write the augmented matrix in reduced row-echelon form, follow the same steps used to write a matrix in row-echelon form except choose the *reduced row-echelon* feature, as shown in Figure A.89. You should obtain the matrix shown in Figure A.90. From Figure A.90, you can conclude that the solution to the system is $x = 3$, $y = -10$, and $z = -16$.

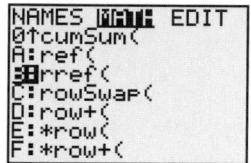

Figure A.89

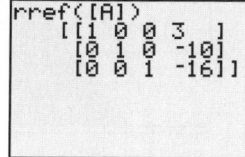

Figure A.90

Sequence Feature

The *sequence* feature is used to display the terms of sequences. For instance, to determine the first five terms of the arithmetic sequence

$a_n = 3n + 5$ Assume n begins with 1.

set the graphing utility to *sequence* mode. Then choose the *sequence* feature from the *operations* menu of the *list* feature, as shown in Figure A.91. When using this feature, you must enter the sequence, the variable (in this case n), the beginning value (in this case 1), and the end value (in this case 5). The first five terms of the sequence are 8, 11, 14, 17, and 20, as shown in Figure A.92. You may have to scroll to the right in order to see all the terms of the sequence.

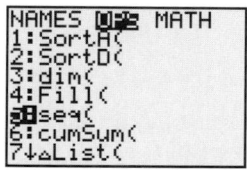

Figure A.91

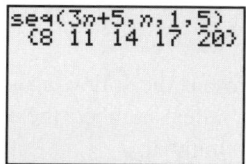

Figure A.92

Shade Feature

Most graphing utilities have a *shade* feature that can be used to graph inequalities. For instance, to graph the inequality $y \leq 2x - 3$, first enter the equation $y = 2x - 3$ in the *equation editor*, as shown in Figure A.93. Next, using a standard viewing window (see Figure A.94), graph the equation, as shown in Figure A.95.

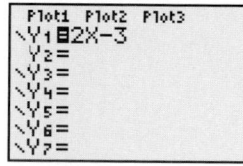

Figure A.93

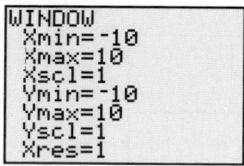

Figure A.94

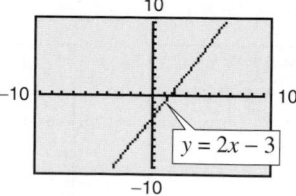

Figure A.95

Because the inequality sign is ≤, you want to shade the region below the line $y = 2x - 3$. Choose the *shade* feature from the *draw* feature menu, as shown in Figure A.96. You must enter a lower function and an upper function. In this case, the lower function is -10 (this is the smallest y-value in the viewing window) and the upper function is Y_1 ($y = 2x - 3$), as shown in Figure A.97. Then press [ENTER] to obtain the graph shown in Figure A.98.

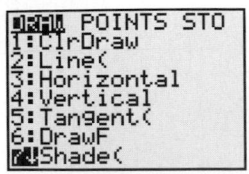

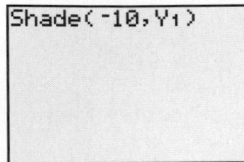

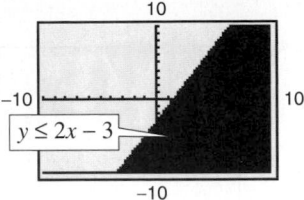

Figure A.96   Figure A.97   Figure A.98

If you wanted to graph the inequality $y \geq 2x - 3$ (using a standard viewing window), you would enter the lower function as Y_1 ($y = 2x - 3$) and the upper function as 10 (the largest y-value in the viewing window).

Statistical Plotting Feature

The *statistical plotting* feature plots data that is stored in lists. Most graphing utilities can display the following types of plots.

Plot Type	Variables
Scatter plot	x-list, y-list
xy line graph	x-list, y-list
Histogram	x-list, frequency
Box-and-whisker plot	x-list, frequency
Normal probability plot	data list, data axis

For example, use a box-and-whisker plot to represent the following set of data. Then use the graphing utility plot to find the smallest number, the lower quartile, the median, the upper quartile, and the largest number.

17, 19, 21, 27, 29, 30, 37, 27, 15, 23, 19, 16

First, use the *list editor* to enter the values in a list, as shown in Figure A.99. Then go to the *statistical plotting editor*. In this editor you will turn the plot on, select the box-and-whisker plot, select the list you entered in the *list editor*, and enter the frequency of each item in the list, as shown in Figure A.100. Now use the *zoom* feature and choose the *zoom stat* option to set the viewing window and plot the graph, as shown in Figure A.101. Use the *trace* feature to find that the smallest number is 15, the lower quartile is 18, the median is 22, the upper quartile is 28, and the largest number is 37.

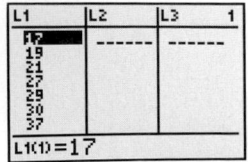

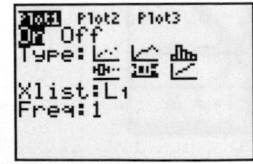

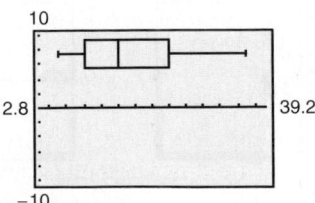

Figure A.99   Figure A.100   Figure A.101

Sum Feature

The *sum* feature finds the sum of a list of data. For instance, the data below represents a student's quiz scores on 10 quizzes throughout an algebra course.

22, 23, 19, 24, 20, 15, 25, 21, 18, 24

To find the total quiz points the student earned, enter the data in the *list editor*, as shown in Figure A.102. To find the sum, choose the *sum* feature from the *math* menu of the *list* feature, as shown in Figure A.103. You must enter a list. In this case the list is L_1. You should obtain a sum of 211, as shown in Figure A.104.

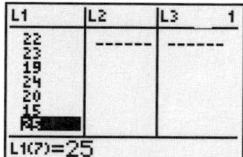
Figure A.102

Figure A.103

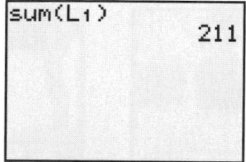
Figure A.104

Sum Sequence Feature

The *sum* feature and the *sequence* feature can be combined to find the sum of a sequence or series. For example, to find the sum

$$\sum_{k=0}^{10} 5^{k+1}$$

first choose the *sum* feature from the *math* menu of the *list* feature, as shown in Figure A.105. Then choose the *sequence* feature from the *operations* menu of the *list* feature, as shown in Figure A.106. You must enter an expression for the sequence, a variable, the lower limit of summation, and the upper limit of summation. After pressing $\boxed{\text{ENTER}}$, you should obtain the sum 61,035,155, as shown in Figure A.107.

Figure A.105

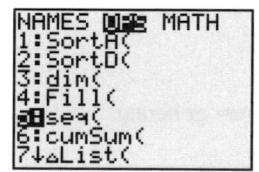
Figure A.106

Figure A.107

Table Feature

Most graphing utilities are capable of displaying a table of values with x-values and one or more corresponding y-values. These tables can be used to check solutions of an equation and to generate ordered pairs to assist in graphing an equation by hand.

To use the *table* feature, enter an equation in the *equation editor*. The table may have a setup screen, which allows you to select the starting x-value and the table step or x-increment. You may then have the option of automatically generating values for x and y or building your own table using the *ask* mode (see Figure A.108).

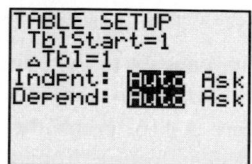
Figure A.108

For example, enter the equation

$$y = \frac{3x}{x + 2}$$

in the *equation editor*, as shown in Figure A.109. In the table setup screen, set the table to start at $x = -4$ and set the table step to 1, as shown in Figure A.110. When you view the table, notice that the first x-value is -4 and that each value after it increases by 1. Also notice that the Y_1 column gives the resulting y-value for each x-value, as shown in Figure A.111. The table shows that the y-value for $x = -2$ is ERROR. This means that the equation is undefined when $x = -2$.

Figure A.109

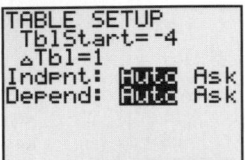

Figure A.110

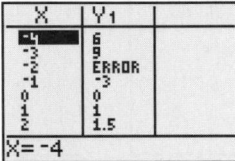

Figure A.111

With the same equation in the *equation editor*, set the independent variable in the table to *ask* mode, as shown in Figure A.112. In this mode, you do not need to set the starting x-value or the table step because you are entering any value you choose for x. You may enter any real value for x—integers, fractions, decimals, irrational numbers, and so forth. If you enter $x = 1 + \sqrt{3}$, the graphing utility may rewrite the number as a decimal approximation, as shown in Figure A.113. You can continue to build your own table by entering additional x-values in order to generate y-values, as shown in Figure A.114.

Figure A.112

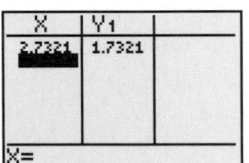

Figure A.113

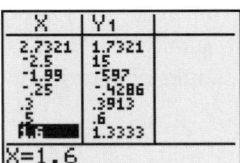

Figure A.114

If you have several equations in the *equation editor*, the table may generate y-values for each equation.

Tangent Feature

Some graphing utilities have the capability of drawing a tangent line to a graph at a given point. For instance, consider the equation

$$y = -x^3 + x + 2.$$

To draw the line tangent to the point $(1, 2)$ on the graph of y, enter the equation in the *equation editor*, as shown in Figure A.115. Using the viewing window shown in Figure A.116, graph the equation, as shown in Figure A.117. Next, choose the *tangent* feature from the *draw* feature menu, as shown in Figure A.118. You can either move the cursor to select a point or enter the x-value at which you want the tangent line to be drawn. Because you want the tangent line to the point $(1, 2)$, enter 1 (see Figure A.119) and then press ENTER. The x-value you entered and the equation of the tangent line are displayed at the bottom of the window, as shown in Figure A.120.

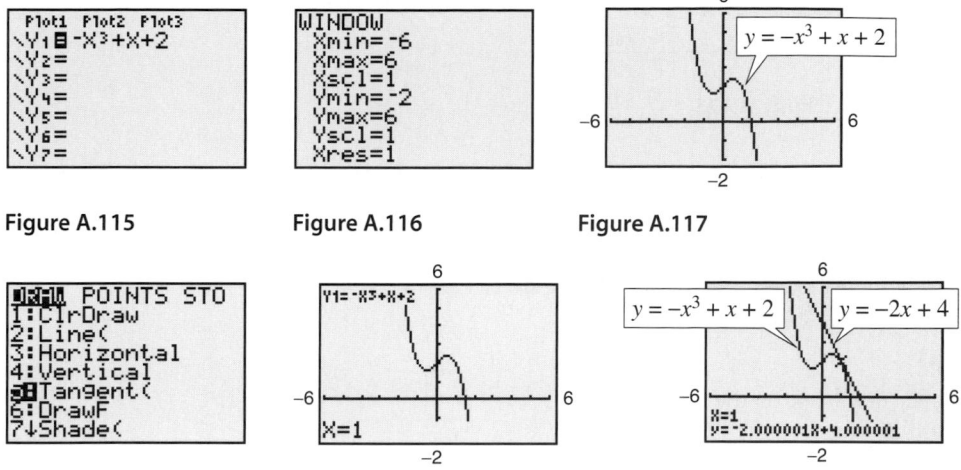

Figure A.115   Figure A.116   Figure A.117

Figure A.118   Figure A.119   Figure A.120

Trace Feature

For instructions on how to use the *trace* feature, see Zoom and Trace Features on page A23.

Value Feature

The *value* feature finds the value of a function y for a given x-value. To find the value of a function such as $f(x) = 0.5x^2 - 1.5x$ at $x = 1.8$, first enter the function in the *equation editor* (see Figure A.121) and then graph the function (using a standard viewing window), as shown in Figure A.122. Next, choose the *value* feature from the *calculate* menu, as shown in Figure A.123. You will see "X= " displayed at the bottom of the window. Enter the x-value, in this case $x = 1.8$, as shown in Figure A.124. When entering an x-value, be sure it is between the Xmin and Xmax values you entered for the viewing window. Then press ENTER. In Figure A.125, you can see that when $x = 1.8$, $y = -1.08$.

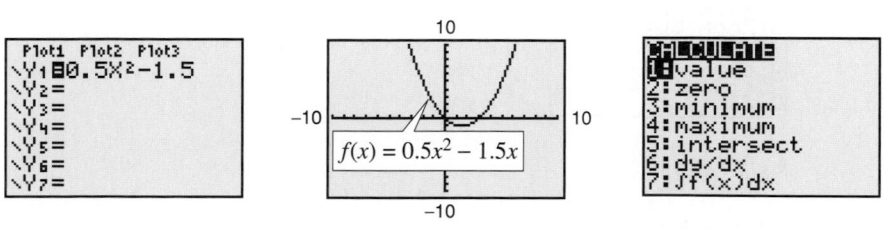

Figure A.121   Figure A.122   Figure A.123

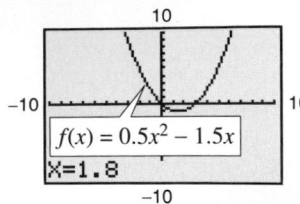

Figure A.124

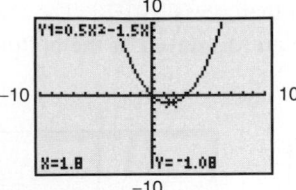

Figure A.125

Viewing Window

A viewing window for a graph is a rectangular portion of the coordinate plane. A viewing window is determined by the following six values (see Figure A.126).

Xmin = the smallest value of x
Xmax = the largest value of x
Xscl = the number of units per tick mark on the x-axis
Ymin = the smallest value of y
Ymax = the largest value of y
Yscl = the number of units per tick mark on the y-axis

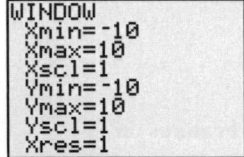

Figure A.126

When you enter these six values in a graphing utility, you are setting the viewing window. On some graphing utilities there is a seventh value for the viewing window labeled Xres. This sets the pixel resolution (1 through 8). For instance, when Xres = 1, functions are evaluated and graphed at each pixel on the x-axis. Some graphing utilities have a standard viewing window, as shown in Figure A.127. To initialize the standard viewing window quickly, choose the *standard viewing window* feature from the *zoom* feature menu (see page A23), as shown in Figure A.128.

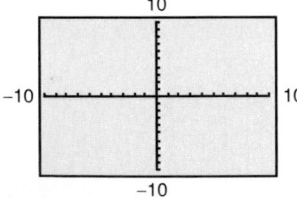

Figure A.127

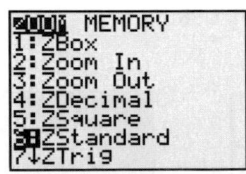

Figure A.128

By choosing different viewing windows for a graph, it is possible to obtain different impressions of the graph's shape. For instance, Figure A.129 shows four different viewing windows for the graph of $y = 0.1x^4 - x^3 + 2x^2$. Of these viewing windows, the one shown in part (a) is the most complete.

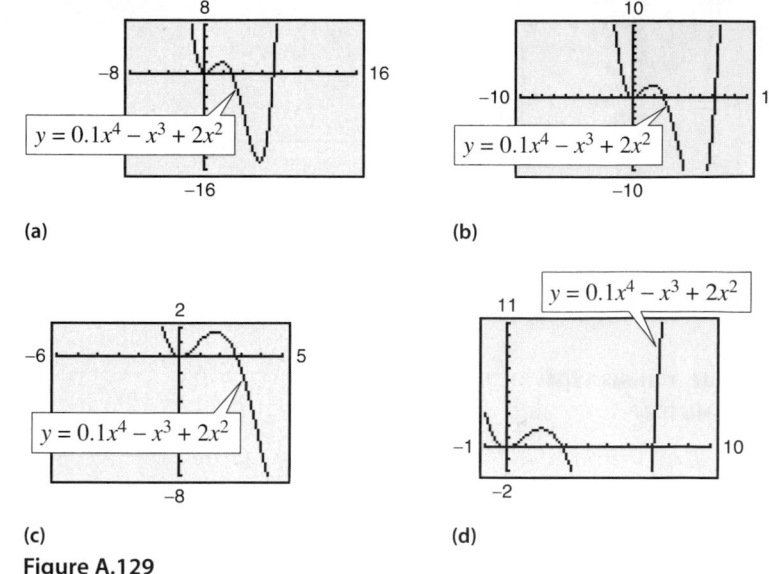

(a)

(b)

(c)

(d)

Figure A.129

On most graphing utilities, the display screen is two-thirds as high as it is wide. On such screens, you can obtain a graph with a true geometric perspective by using a square setting—one in which

$$\frac{Y\text{max} - Y\text{min}}{X\text{max} - X\text{min}} = \frac{2}{3}.$$

One such setting is shown in Figure A.130. Notice that the x and y tick marks are equally spaced on a square setting, but not on a standard setting (see Figure A.127). To initialize the square viewing window quickly, choose the *square viewing window* feature from the *zoom* feature menu (see page A23), as shown in Figure A.131.

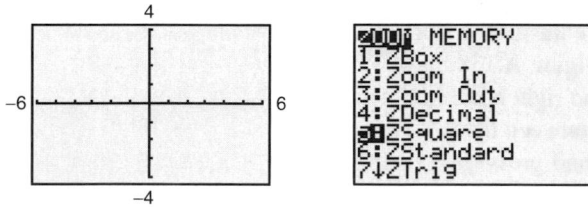

Figure A.130   **Figure A.131**

To see how the viewing window affects the geometric perspective, graph the semicircles $y_1 = \sqrt{9 - x^2}$ and $y_2 = -\sqrt{9 - x^2}$ using a standard viewing window, as shown in Figure A.132. Notice how the circle appears elliptical rather than circular. Now graph y_1 and y_2 using a square viewing window, as shown in Figure A.133. Notice how the circle appears circular. (Note that when you graph the two semicircles, your graphing utility may not connect them. This is because some graphing utilities are limited in their resolution. So, in this text, a blue or light red curve is placed behind the graphing utility's display to indicate where the graph should appear.)

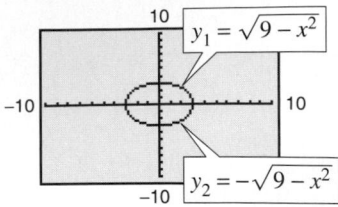

Figure A.132

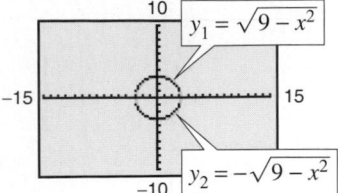

Figure A.133

Zero or Root Feature

The *zero* or *root* feature finds the real zeros of the various types of functions studied in this text. To find the zeros of a function such as

$$f(x) = 2x^3 - 4x$$

first enter the function in the *equation editor*, as shown in Figure A.134. Now graph the equation (using a standard viewing window), as shown in Figure A.135. From the graph you can see that the graph of the function crosses the x-axis three times, so the function has three zeros.

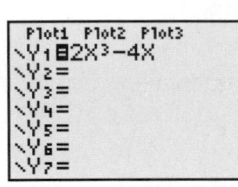

Figure A.134

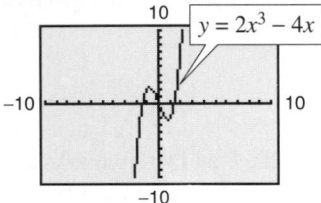

Figure A.135

To find these zeros, choose the *zero* feature found in the *calculate* menu (see Figure A.136). Next, trace the cursor along the graph to a point left of one of the zeros and press (ENTER) (see Figure A.137). Then trace the cursor along the graph to a point right of the zero and press (ENTER) (see Figure A.138). Note the two arrows near the top of the display marking the left and right bounds, as shown in Figure A.139. Now trace the cursor along the graph between the two bounds and as close to the zero as you can (see Figure A.140) and press (ENTER). In Figure A.141, you can see that one zero of the function is $x \approx -1.414214$.

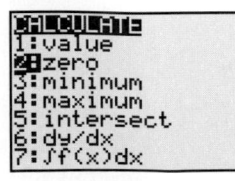

Figure A.136

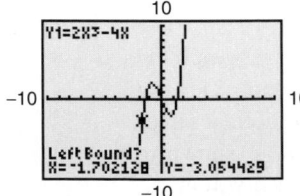

Figure A.137

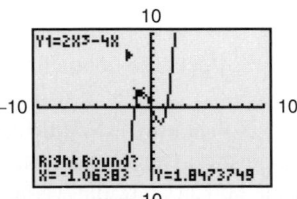

Figure A.138

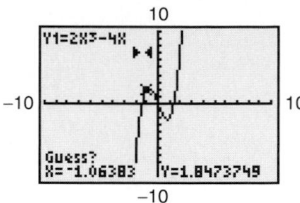

Figure A.139

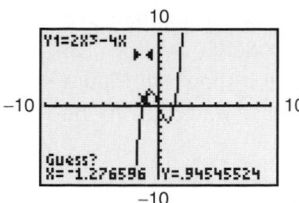

Figure A.140

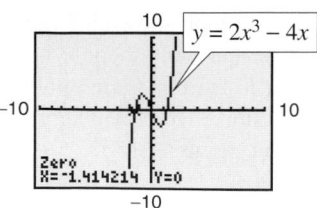

Figure A.141

Repeat this process to determine that the other two zeros of the function are $x = 0$ (see Figure A.142) and $x \approx 1.414214$ (see Figure A.143).

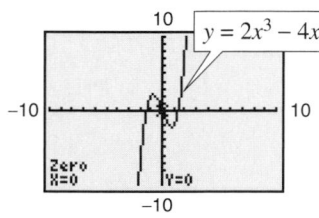

Figure A.142

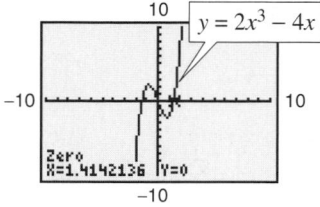

Figure A.143

Zoom and Trace Features

The *zoom* feature enables you to adjust the viewing window of a graph quickly (see Figure A.144). For example, the *zoom box* feature allows you to create a new viewing window by drawing a box around any part of the graph.

Figure A.144

The *trace* feature moves from point to point along a graph. For instance, enter the equation $y = 2x^3 - 3x + 2$ in the *equation editor* (see Figure A.145) and graph the equation, as shown in Figure A.146. To activate the *trace* feature, press (TRACE); then use the arrow keys to move the cursor along the graph. As you trace the graph, the coordinates of each point are displayed, as shown in Figure A.147.

Figure A.145

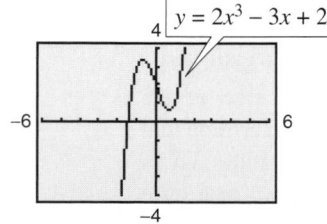

Figure A.146

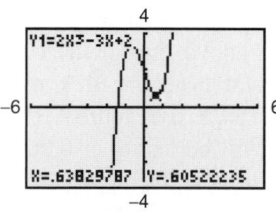

Figure A.147

The *trace* feature combined with the *zoom* feature enables you to obtain better and better approximations of desired points on a graph. For instance, you can use the *zoom* feature to approximate the *x*-intercept of the graph of $y = 2x^3 - 3x + 2$. From the viewing window shown in Figure A.146, the graph appears to have only one *x*-intercept. This intercept lies between -2 and -1. To zoom in on the *x*-intercept, choose the *zoom-in* feature from the *zoom* feature menu, as shown in Figure A.148. Next, trace the cursor to the point you want to zoom in on, in this case the *x*-intercept (see Figure A.149). Then press

ENTER. You should obtain the graph shown in Figure A.150. Now, using the *trace* feature, you can approximate the *x*-intercept to be $x \approx -1.468085$, as shown in Figure A.151. Use the *zoom-in* feature again to obtain the graph shown in Figure A.152. Using the *trace* feature, you can approximate the *x*-intercept to be $x \approx -1.476064$, as shown in Figure A.153.

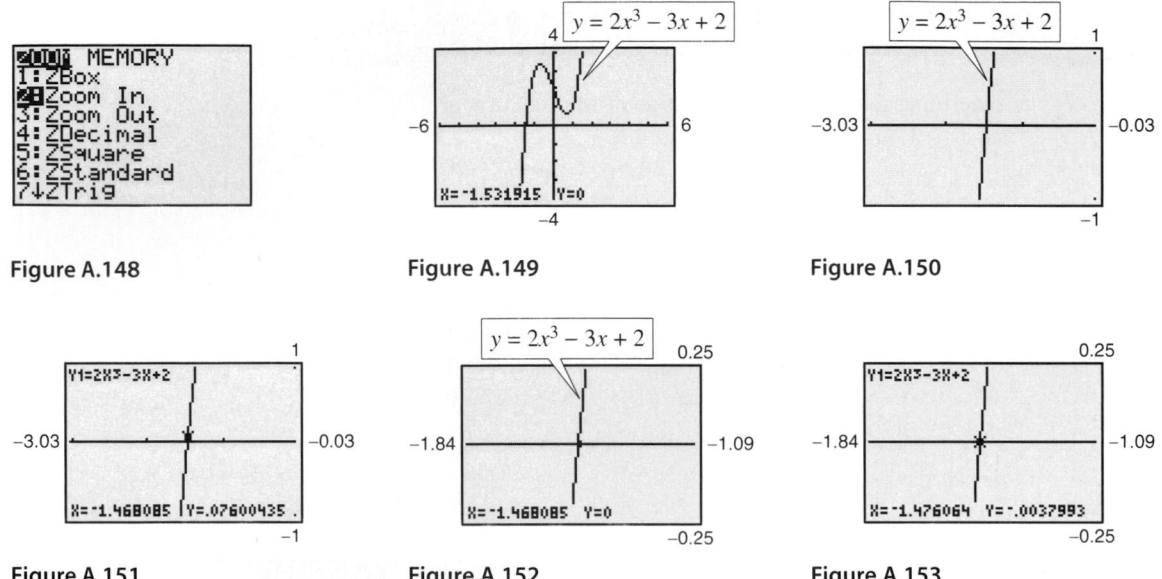

Figure A.148   Figure A.149   Figure A.150

Figure A.151   Figure A.152   Figure A.153

Here are some suggestions for using the *zoom* feature.

1. With each successive zoom-in, adjust the scale so that the viewing window shows at least one tick mark on each side of the *x*-intercept.

2. The error in your approximation will be less than the distance between two scale marks.

3. The *trace* feature can usually be used to add one more decimal place of accuracy without changing the viewing window.

You can adjust the scale in Figure A.153 to obtain a better approximation of the *x*-intercept. Using the suggestions above, change the viewing window settings so that the viewing window shows at least one tick mark on each side of the *x*-intercept, as shown in Figure A.154. From Figure A.154, you can determine that the error in your approximation will be less than 0.001 (the Xscl value). Then, using the *trace* feature, you can improve the approximation, as shown in Figure A.155. To three decimal places, the *x*-intercept is $x \approx -1.476$.

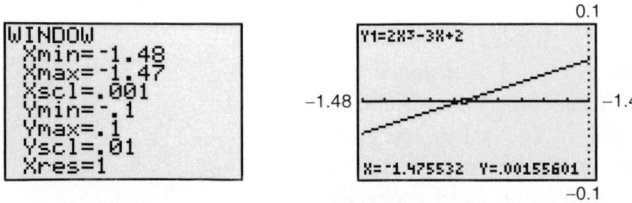

Figure A.154   Figure A.155

Appendix B: Review of Graphs, Equations, and Inequalities

B.1 The Cartesian Plane

The Cartesian Plane

Just as you can represent real numbers by points on a real number line, you can represent ordered pairs of real numbers by points in a plane called the **rectangular coordinate system,** or the **Cartesian plane,** after the French mathematician René Descartes (1596–1650).

The Cartesian plane is formed by using two real number lines intersecting at right angles, as shown in Figure B.1. The horizontal real number line is usually called the **x-axis,** and the vertical real number line is usually called the **y-axis.** The point of intersection of these two axes is the **origin,** and the two axes divide the plane into four parts called **quadrants.**

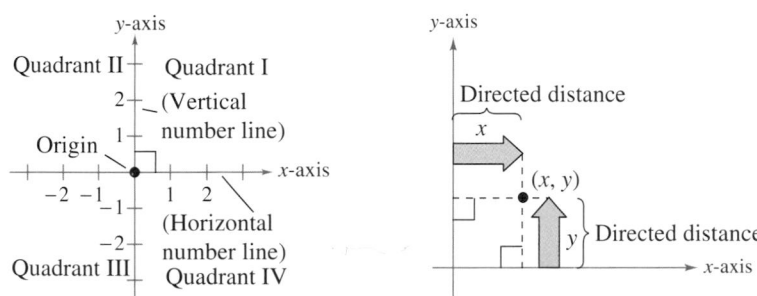

Figure B.1   *The Cartesian Plane*   **Figure B.2   *Ordered Pair (x, y)***

Each point in the plane corresponds to an **ordered pair** (x, y) of real numbers x and y, called **coordinates** of the point. The **x-coordinate** represents the directed distance from the y-axis to the point, and the **y-coordinate** represents the directed distance from the x-axis to the point, as shown in Figure B.2.

The notation (x, y) denotes both a point in the plane and an open interval on the real number line. The context will tell you which meaning is intended.

Example 1   Plotting Points in the Cartesian Plane

Plot the points $(-1, 2)$, $(3, 4)$, $(0, 0)$, $(3, 0)$, and $(-2, -3)$.

Solution

To plot the point $(-1, 2)$, imagine a vertical line through -1 on the x-axis and a horizontal line through 2 on the y-axis. The intersection of these two lines is the point $(-1, 2)$. This point is one unit to the left of the y-axis and two units up from the x-axis. The other four points can be plotted in a similar way (see Figure B.3).

✓CHECKPOINT   Now try Exercise 3.

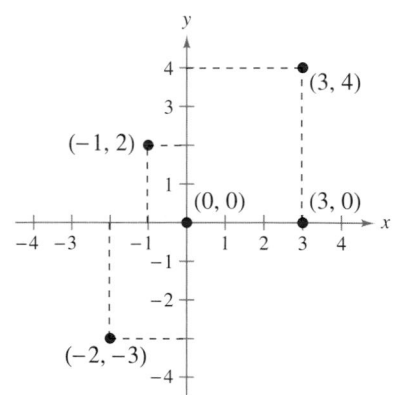

Figure B.3

What you should learn

- Plot points in the Cartesian plane and sketch scatter plots.
- Use the Distance Formula to find the distance between two points.
- Use the Midpoint Formula to find the midpoint of a line segment.
- Find the equation of a circle.
- Translate points in the plane.

Why you should learn it

The Cartesian plane can be used to represent relationships between two variables. For instance, Exercise 85 on page A34 shows how to graphically represent the number of recording artists inducted to the Rock and Roll hall of Fame from 1986 to 2006.

The beauty of a rectangular coordinate system is that it enables you to see relationships between two variables. It would be difficult to overestimate the importance of Descartes's introduction of coordinates to the plane. Today, his ideas are in common use in virtually every scientific and business-related field.

In the next example, data is represented graphically by points plotted on a rectangular coordinate system. This type of graph is called a **scatter plot.**

Example 2 Sketching a Scatter Plot

The amounts A (in millions of dollars) spent on archery equipment in the United States from 1999 to 2004 are shown in the table, where t represents the year. Sketch a scatter plot of the data by hand. (Source: National Sporting Goods Association)

Year, t	Amount, A
1999	262
2000	259
2001	276
2002	279
2003	281
2004	282

Solution

Before you sketch the scatter plot, it is helpful to represent each pair of values by an ordered pair (t, A), as follows.

(1999, 262), (2000, 259), (2001, 276), (2002, 279), (2003, 281), (2004, 282)

To sketch a scatter plot of the data shown in the table, first draw a vertical axis to represent the amount (in millions of dollars) and a horizontal axis to represent the year. Then plot the resulting points, as shown in Figure B.4. Note that the break in the t-axis indicates that the numbers 0 through 1998 have been omitted.

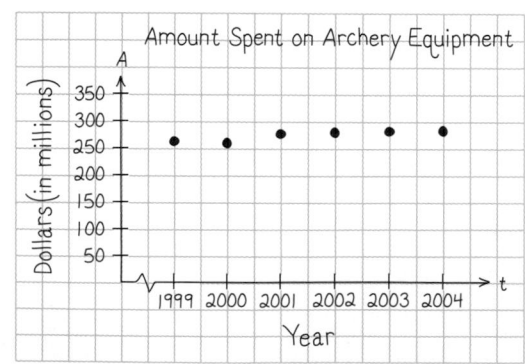

Figure B.4

✓CHECKPOINT   Now try Exercise 21.

TECHNOLOGY TIP You can use a graphing utility to graph the scatter plot in Example 2. First, enter the data into the graphing utility's *list editor* as shown in Figure B.5. Then use the *statistical plotting* feature to set up the scatter plot, as shown in Figure B.6. Finally, display the scatter plot (use a viewing window in which $1998 \le x \le 2005$ and $0 \le y \le 300$), as shown in Figure B.7.

TECHNOLOGY SUPPORT

For instructions on how to use the *list editor*, see Appendix A; for specific keystrokes, go to this textbook's *Online Study Center*.

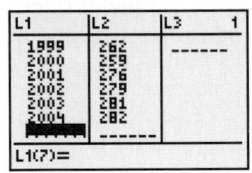

Figure B.5

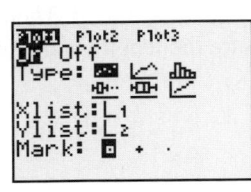

Figure B.6

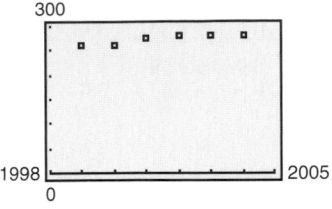

Figure B.7

Some graphing utilities have a *ZoomStat* feature, as shown in Figure B.8. This feature automatically selects an appropriate viewing window that displays all the data in the list editor, as shown in Figure B.9.

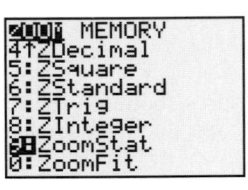

Figure B.8

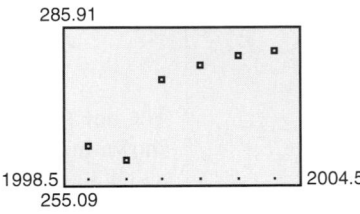

Figure B.9

The Distance Formula

Recall from the Pythagorean Theorem that, for a right triangle with hypotenuse of length c and sides of lengths a and b, you have $a^2 + b^2 = c^2$, as shown in Figure B.10. (The converse is also true. That is, if $a^2 + b^2 = c^2$, then the triangle is a right triangle.)

Suppose you want to determine the distance d between two points (x_1, y_1) and (x_2, y_2) in the plane. With these two points, a right triangle can be formed, as shown in Figure B.11. The length of the vertical side of the triangle is $|y_2 - y_1|$, and the length of the horizontal side is $|x_2 - x_1|$. By the Pythagorean Theorem,

$$d^2 = |x_2 - x_1|^2 + |y_2 - y_1|^2$$
$$d = \sqrt{|x_2 - x_1|^2 + |y_2 - y_1|^2}$$
$$d = \sqrt{(x_2 - x_1)^2 + (y_2 - y_1)^2}.$$

This result is called the **Distance Formula.**

The Distance Formula

The distance d between the points (x_1, y_1) and (x_2, y_2) in the plane is

$$d = \sqrt{(x_2 - x_1)^2 + (y_2 - y_1)^2}.$$

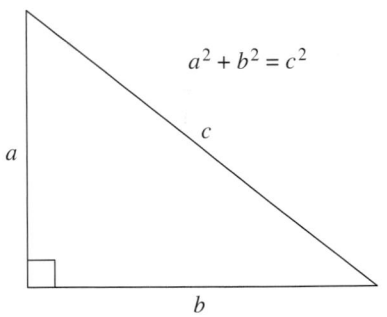

Figure B.10

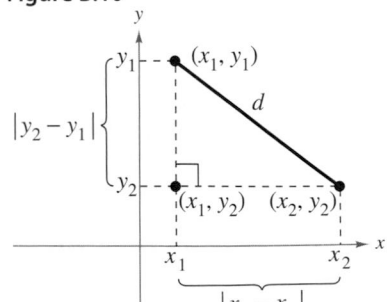

Figure B.11

Example 3   Finding a Distance

Find the distance between the points $(-2, 1)$ and $(3, 4)$.

Algebraic Solution

Let $(x_1, y_1) = (-2, 1)$ and $(x_2, y_2) = (3, 4)$. Then apply the Distance Formula as follows.

$$d = \sqrt{(x_2 - x_1)^2 + (y_2 - y_1)^2}$$ Distance Formula

$$= \sqrt{[3 - (-2)]^2 + (4 - 1)^2}$$ Substitute for $x_1, y_1, x_2,$ and y_2.

$$= \sqrt{(5)^2 + (3)^2}$$ Simplify.

$$= \sqrt{34} \approx 5.83$$ Simplify.

So, the distance between the points is about 5.83 units.

     You can use the Pythagorean Theorem to check that the distance is correct.

$$d^2 \overset{?}{=} 3^2 + 5^2$$ Pythagorean Theorem

$$\left(\sqrt{34}\right)^2 \overset{?}{=} 3^2 + 5^2$$ Substitute for d.

$$34 = 34$$ Distance checks. ✓

Graphical Solution

Use centimeter graph paper to plot the points $A(-2, 1)$ and $B(3, 4)$. Carefully sketch the line segment from A to B. Then use a centimeter ruler to measure the length of the segment.

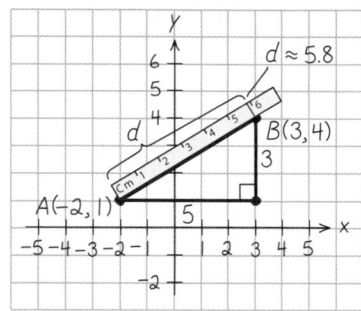

Figure B.12

The line segment measures about 5.8 centimeters, as shown in Figure B.12. So, the distance between the points is about 5.8 units.

✓CHECKPOINT   Now try Exercise 23.

Example 4   Verifying a Right Triangle

Show that the points $(2, 1)$, $(4, 0)$, and $(5, 7)$ are the vertices of a right triangle.

Solution

The three points are plotted in Figure B.13. Using the Distance Formula, you can find the lengths of the three sides as follows.

$$d_1 = \sqrt{(5 - 2)^2 + (7 - 1)^2} = \sqrt{9 + 36} = \sqrt{45}$$

$$d_2 = \sqrt{(4 - 2)^2 + (0 - 1)^2} = \sqrt{4 + 1} = \sqrt{5}$$

$$d_3 = \sqrt{(5 - 4)^2 + (7 - 0)^2} = \sqrt{1 + 49} = \sqrt{50}$$

Because $(d_1)^2 + (d_2)^2 = 45 + 5 = 50 = (d_3)^2$, you can conclude that the triangle must be a right triangle.

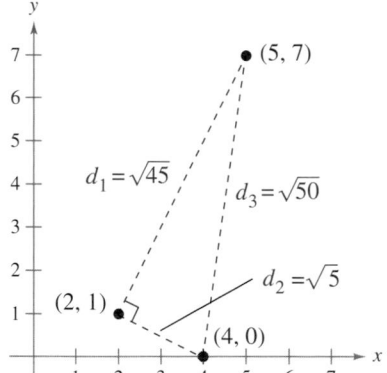

Figure B.13

✓CHECKPOINT   Now try Exercise 37.

The Midpoint Formula

To find the **midpoint** of the line segment that joins two points in a coordinate plane, find the average values of the respective coordinates of the two endpoints using the **Midpoint Formula.**

> **The Midpoint Formula**
>
> The midpoint of the line segment joining the points (x_1, y_1) and (x_2, y_2) is given by the Midpoint Formula
>
> $$\text{Midpoint} = \left(\frac{x_1 + x_2}{2}, \frac{y_1 + y_2}{2} \right).$$

Example 5   Finding a Line Segment's Midpoint

Find the midpoint of the line segment joining the points $(-5, -3)$ and $(9, 3)$.

Solution

Let $(x_1, y_1) = (-5, -3)$ and $(x_2, y_2) = (9, 3)$.

$$\text{Midpoint} = \left(\frac{x_1 + x_2}{2}, \frac{y_1 + y_2}{2} \right) \qquad \text{Midpoint Formula}$$

$$= \left(\frac{-5 + 9}{2}, \frac{-3 + 3}{2} \right) \qquad \text{Substitute for } x_1, y_1, x_2, \text{ and } y_2.$$

$$= (2, 0) \qquad \text{Simplify.}$$

The midpoint of the line segment is $(2, 0)$, as shown in Figure B.14.

Figure B.14

✔**CHECKPOINT**   Now try Exercise 49.

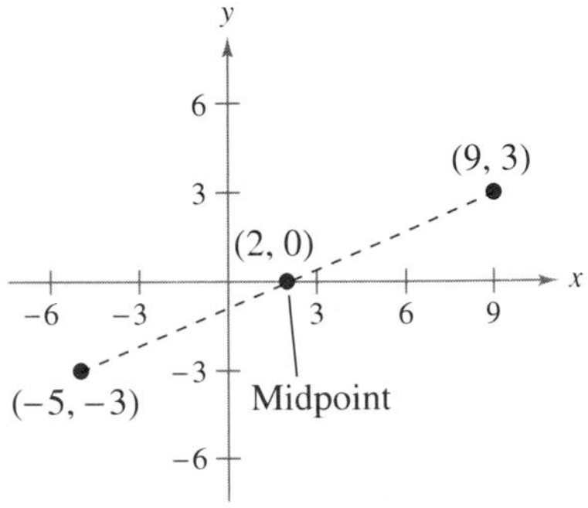

Example 6   Estimating Annual Sales

Kraft Foods Inc. had annual sales of $29.71 billion in 2002 and $32.17 billion in 2004. Without knowing any additional information, what would you estimate the 2003 sales to have been?   (Source: Kraft Foods Inc.)

Solution

One solution to the problem is to assume that sales followed a *linear* pattern. With this assumption, you can estimate the 2003 sales by finding the midpoint of the line segment connecting the points $(2002, 29.71)$ and $(2004, 32.17)$.

$$\text{Midpoint} = \left(\frac{2002 + 2004}{2}, \frac{29.71 + 32.17}{2} \right)$$

$$= (2003, 30.94)$$

So, you would estimate the 2003 sales to have been about $30.94 billion, as shown in Figure B.15. (The actual 2003 sales were $31.01 billion.)

Figure B.15

✔**CHECKPOINT**   Now try Exercise 55.

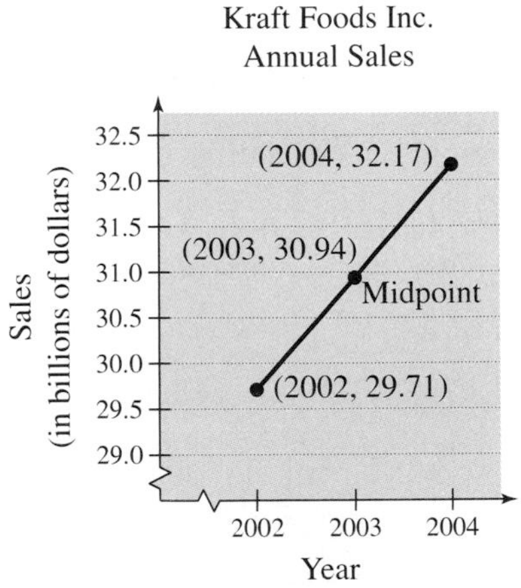

The Equation of a Circle

The Distance Formula provides a convenient way to define circles. A **circle of radius r** with center at the point (h, k) is shown in Figure B.16. The point (x, y) is on this circle if and only if its distance from the center (h, k) is r. This means that a **circle** in the plane consists of all points (x, y) that are a given positive distance r from a fixed point (h, k). Using the Distance Formula, you can express this relationship by saying that the point (x, y) lies on the circle if and only if

$$\sqrt{(x - h)^2 + (y - k)^2} = r.$$

By squaring each side of this equation, you obtain the **standard form of the equation of a circle.**

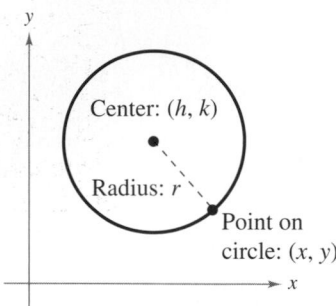

Figure B.16

> **Standard Form of the Equation of a Circle**
>
> The **standard form of the equation of a circle** is
>
> $$(x - h)^2 + (y - k)^2 = r^2.$$
>
> The point (h, k) is the **center** of the circle, and the positive number r is the **radius** of the circle. The standard form of the equation of a circle whose center is the origin, $(h, k) = (0, 0)$, is $x^2 + y^2 = r^2$.

Example 7   Writing the Equation of a Circle

The point $(3, 4)$ lies on a circle whose center is at $(-1, 2)$, as shown in Figure B.17. Write the standard form of the equation of this circle.

Solution

The radius r of the circle is the distance between $(-1, 2)$ and $(3, 4)$.

$r = \sqrt{[3 - (-1)]^2 + (4 - 2)^2}$ Substitute for x, y, h, and k.

$\quad = \sqrt{16 + 4}$ Simplify.

$\quad = \sqrt{20}$ Radius

Using $(h, k) = (-1, 2)$ and $r = \sqrt{20}$, the equation of the circle is

$(x - h)^2 + (y - k)^2 = r^2$ Equation of circle

$[x - (-1)]^2 + (y - 2)^2 = \left(\sqrt{20}\right)^2$ Substitute for h, k, and r.

$(x + 1)^2 + (y - 2)^2 = 20.$ Standard form

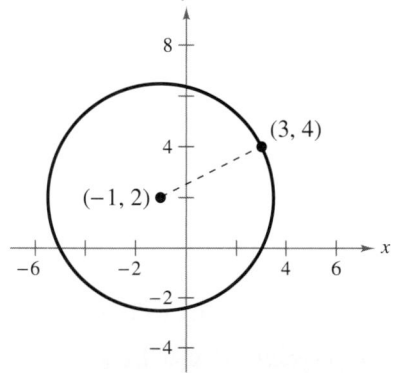

Figure B.17

✓*CHECKPOINT*   Now try Exercise 61.

Example 8   Translating Points in the Plane

The triangle in Figure B.18 has vertices at the points $(-1, 2)$, $(1, -4)$, and $(2, 3)$. Shift the triangle three units to the right and two units upward and find the vertices of the shifted triangle, as shown in Figure B.19.

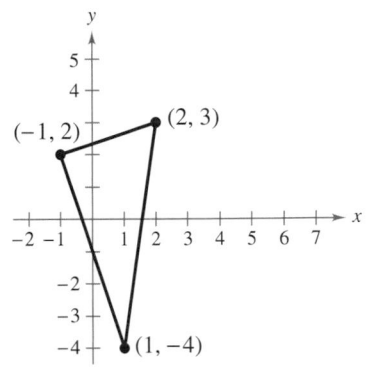

Figure B.18   **Figure B.19**

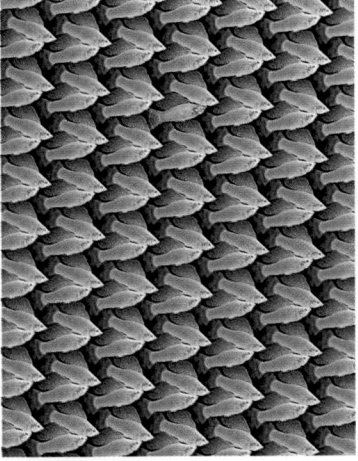

Paul Morrell

Much of computer graphics, including this computer-generated goldfish tessellation, consists of transformations of points in a coordinate plane. One type of transformation, a translation, is illustrated in Example 8. Other types of transformations include reflections, rotations, and stretches.

Solution

To shift the vertices three units to the right, add 3 to each of the x-coordinates. To shift the vertices two units upward, add 2 to each of the y-coordinates.

Original Point	Translated Point
$(-1, 2)$	$(-1 + 3, 2 + 2) = (2, 4)$
$(1, -4)$	$(1 + 3, -4 + 2) = (4, -2)$
$(2, 3)$	$(2 + 3, 3 + 2) = (5, 5)$

Plotting the translated points and sketching the line segments between them produces the shifted triangle shown in Figure B.19.

✓CHECKPOINT   Now try Exercise 79.

Example 8 shows how to translate points in a coordinate plane. The following transformed points are related to the original points as follows.

Original Point	Transformed Point	
(x, y)	$(-x, y)$	$(-x, y)$ is a reflection of the original point in the y-axis.
(x, y)	$(x, -y)$	$(x, -y)$ is a reflection of the original point in the x-axis.
(x, y)	$(-x, -y)$	$(-x, -y)$ is a reflection of the original point through the origin.

The figures provided with Example 8 were not really essential to the solution. Nevertheless, it is strongly recommended that you develop the habit of including sketches with your solutions, even if they are not required, because they serve as useful problem-solving tools.

B.1 Exercises

See www.CalcChat.com for worked-out solutions to odd-numbered exercises.

Vocabulary Check

1. Match each term with its definition.

(a) x-axis

(b) y-axis

(c) origin

(d) quadrants

(e) x-coordinate

(f) y-coordinate

(i) point of intersection of vertical axis and horizontal axis

(ii) directed distance from the x-axis

(iii) horizontal real number line

(iv) four regions of the coordinate plane

(v) directed distance from the y-axis

(vi) vertical real number line

In Exercises 2–5, fill in the blanks.

2. An ordered pair of real numbers can be represented in a plane called the rectangular coordinate system or the _____ plane.

3. The _____ is a result derived from the Pythagorean Theorem.

4. Finding the average values of the respective coordinates of the two endpoints of a line segment in a coordinate plane is also known as using the _____ .

5. The standard form of the equation of a circle is _____ , where the point (h, k) is the _____ of the circle and the positive number r is the _____ of the circle.

In Exercises 1 and 2, approximate the coordinates of the points.

1.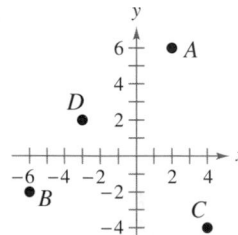

2.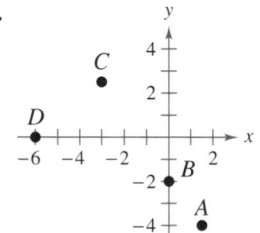

In Exercises 3–6, plot the points in the Cartesian plane.

3. $(-4, 2), (-3, -6), (0, 5), (1, -4)$

4. $(4, -2), (0, 0), (-4, 0), (-5, -5)$

5. $(3, 8), (0.5, -1), (5, -6), (-2, -2.5)$

6. $\left(1, -\frac{1}{2}\right), \left(-\frac{3}{4}, 2\right), (3, -3), \left(\frac{3}{2}, \frac{4}{3}\right)$

In Exercises 7–10, find the coordinates of the point.

7. The point is located five units to the left of the y-axis and four units above the x-axis.

8. The point is located three units below the x-axis and two units to the right of the y-axis.

9. The point is located six units below the x-axis and the coordinates of the point are equal.

10. The point is on the x-axis and 10 units to the left of the y-axis.

In Exercises 11–20, determine the quadrant(s) in which (x, y) is located so that the condition(s) is (are) satisfied.

11. $x > 0$ and $y < 0$

12. $x < 0$ and $y < 0$

13. $x = -4$ and $y > 0$

14. $x > 2$ and $y = 3$

15. $y < -5$

16. $x > 4$

17. $x < 0$ and $-y > 0$

18. $-x > 0$ and $y < 0$

19. $xy > 0$

20. $xy < 0$

In Exercises 21 and 22, sketch a scatter plot of the data shown in the table.

21. *Sales*   The table shows the sales y (in millions of dollars) for Apple Computer, Inc. for the years 1997–2006. (Source: Value Line)

Year	Sales, y (in millions of dollars)
1997	7,081
1998	5,941
1999	6,134
2000	7,983
2001	5,363
2002	5,742
2003	6,207
2004	8,279
2005	13,900
2006	16,600

22. *Meteorology*   The table shows the lowest temperature on record *y* (in degrees Fahrenheit) in Duluth, Minnesota for each month *x*, where *x* = 1 represents January.   (Source: NOAA)

Month, *x*	Temperature, *y*
1	−39
2	−39
3	−29
4	−5
5	17
6	27
7	35
8	32
9	22
10	8
11	−23
12	−34

In Exercises 23–32, find the distance between the points algebraically and verify graphically by using centimeter graph paper and a centimeter ruler.

23. $(6, -3), (6, 5)$

24. $(1, 4), (8, 4)$

25. $(-3, -1), (2, -1)$

26. $(-3, -4), (-3, 6)$

27. $(-2, 6), (3, -6)$

28. $(8, 5), (0, 20)$

29. $\left(\frac{1}{2}, \frac{4}{3}\right), (2, -1)$

30. $\left(-\frac{2}{3}, 3\right), \left(-1, \frac{5}{4}\right)$

31. $(-4.2, 3.1), (-12.5, 4.8)$

32. $(9.5, -2.6), (-3.9, 8.2)$

In Exercises 33–36, (a) find the length of each side of the right triangle and (b) show that these lengths satisfy the Pythagorean Theorem.

33.

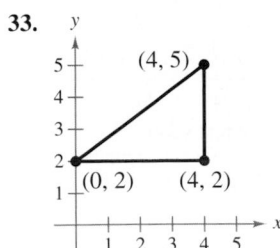

34.

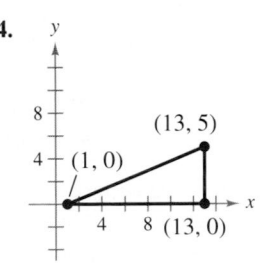

35.

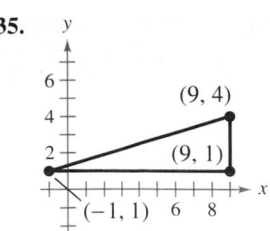

36.

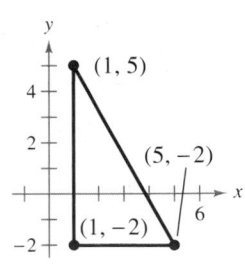

In Exercises 37–44, show that the points form the vertices of the polygon.

37. Right triangle: $(4, 0), (2, 1), (-1, -5)$

38. Right triangle: $(-1, 3), (3, 5), (5, 1)$

39. Isosceles triangle: $(1, -3), (3, 2), (-2, 4)$

40. Isosceles triangle: $(2, 3), (4, 9), (-2, 7)$

41. Parallelogram: $(2, 5), (0, 9), (-2, 0), (0, -4)$

42. Parallelogram: $(0, 1), (3, 7), (4, 4), (1, -2)$

43. Rectangle: $(-5, 6), (0, 8), (-3, 1), (2, 3)$ (*Hint:* Show that the diagonals are of equal length.)

44. Rectangle: $(2, 4), (3, 1), (1, 2), (4, 3)$ (*Hint:* Show that the diagonals are of equal length.)

In Exercises 45–54, (a) plot the points, (b) find the distance between the points, and (c) find the midpoint of the line segment joining the points.

45. $(1, 1), (9, 7)$ **46.** $(1, 12), (6, 0)$

47. $(-4, 10), (4, -5)$

48. $(-7, -4), (2, 8)$

49. $(-1, 2), (5, 4)$

50. $(2, 10), (10, 2)$

51. $\left(\frac{1}{2}, 1\right), \left(-\frac{5}{2}, \frac{4}{3}\right)$

52. $\left(-\frac{1}{3}, -\frac{1}{3}\right), \left(-\frac{1}{6}, -\frac{1}{2}\right)$

53. $(6.2, 5.4), (-3.7, 1.8)$

54. $(-16.8, 12.3), (5.6, 4.9)$

Revenue   **In Exercises 55 and 56, use the Midpoint Formula to estimate the annual revenues (in millions of dollars) for Wendy's Intl., Inc. and Papa John's Intl. in 2003. The revenues for the two companies in 2000 and 2006 are shown in the tables. Assume that the revenues followed a linear pattern.**   (Source: Value Line)

55. Wendy's Intl., Inc.

Year	Annual revenue (in millions of dollars)
2000	2237
2006	3950

56. Papa John's Intl.

Year	Annual revenue (in millions of dollars)
2000	945
2006	1005

57. *Exploration*   A line segment has (x_1, y_1) as one endpoint and (x_m, y_m) as its midpoint. Find the other endpoint (x_2, y_2) of the line segment in terms of x_1, y_1, x_m, and y_m. Use the result to find the coordinates of the endpoint of a line segment if the coordinates of the other endpoint and midpoint are, respectively,

(a) $(1, -2), (4, -1)$

(b) $(-5, 11), (2, 4)$

58. *Exploration*   Use the Midpoint Formula three times to find the three points that divide the line segment joining (x_1, y_1) and (x_2, y_2) into four parts. Use the result to find the points that divide the line segment joining the given points into four equal parts.

(a) $(1, -2), (4, -1)$

(b) $(-2, -3), (0, 0)$

In Exercises 59–72, write the standard form of the equation of the specified circle.

59. Center: $(0, 0)$; radius: 3

60. Center: $(0, 0)$; radius: 6

61. Center: $(2, -1)$; radius: 4

62. Center: $\left(0, \frac{1}{3}\right)$; radius: $\frac{1}{3}$

63. Center: $(-1, 2)$; solution point: $(0, 0)$

64. Center: $(3, -2)$; solution point: $(-1, 1)$

65. Endpoints of a diameter: $(0, 0), (6, 8)$

66. Endpoints of a diameter: $(-4, -1), (4, 1)$

67. Center: $(-2, 1)$; tangent to the x-axis

68. Center: $(3, -2)$; tangent to the y-axis

69. The circle inscribed in the square with vertices $(7, -2)$, $(-1, -2), (-1, -10)$, and $(7, -10)$

70. The circle inscribed in the square with vertices $(-12, 10)$, $(8, 10), (8, -10)$, and $(-12, -10)$

71.

72.

In Exercises 73–78, find the center and radius, and sketch the circle.

73. $x^2 + y^2 = 25$

74. $x^2 + y^2 = 16$

75. $(x - 1)^2 + (y + 3)^2 = 4$

76. $x^2 + (y - 1)^2 = 49$

77. $\left(x - \frac{1}{2}\right)^2 + \left(y - \frac{1}{2}\right)^2 = \frac{9}{4}$

78. $\left(x - \frac{2}{3}\right)^2 + \left(y + \frac{1}{4}\right)^2 = \frac{25}{9}$

In Exercises 79–82, the polygon is shifted to a new position in the plane. Find the coordinates of the vertices of the polygon in the new position.

79.   **80.**

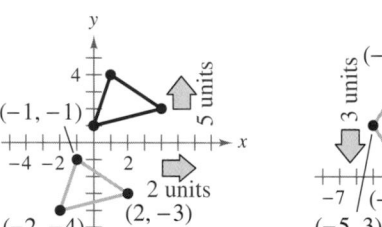

81. Original coordinates of vertices:

$(0, 2), (3, 5), (5, 2), (2, -1)$

Shift: three units upward, one unit to the left

82. Original coordinates of vertices:

$(1, -1), (3, 2), (1, -2)$

Shift: two units downward, three units to the left

Analyzing Data   **In Exercises 83 and 84, refer to the scatter plot, which shows the mathematics entrance test scores x and the final examination scores y in an algebra course for a sample of 10 students.**

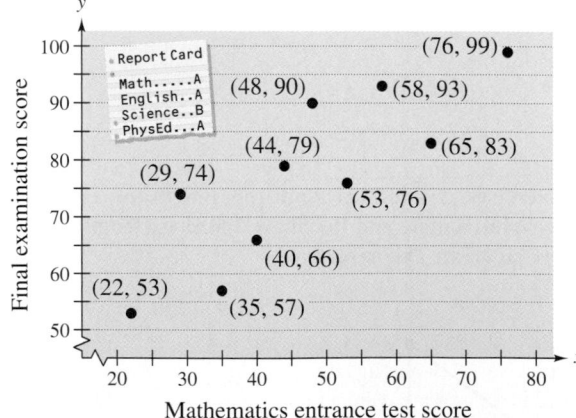

83. Find the entrance exam score of any student with a final exam score in the 80s.

84. Does a higher entrance exam score necessarily imply a higher final exam score? Explain.

85. *Rock and Roll Hall of Fame* The graph shows the numbers of recording artists inducted into the Rock and Roll Hall of Fame from 1986 to 2006.

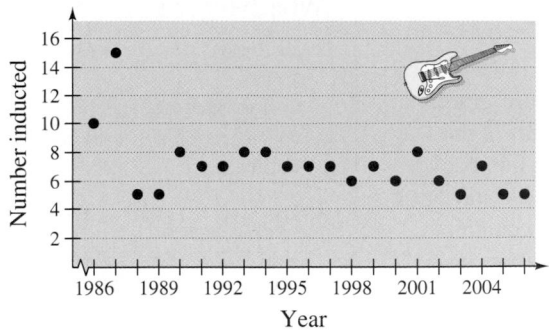

(a) Describe any trends in the data. From these trends, predict the number of artists that will be elected in 2007.

(b) Why do you think the numbers elected in 1986 and 1987 were greater than in other years?

86. *Flying Distance* A jet plane flies from Naples, Italy in a straight line to Rome, Italy, which is 120 kilometers north and 150 kilometers west of Naples. How far does the plane fly?

87. *Sports* In a football game, a quarterback throws a pass from the 15-yard line, 10 yards from the sideline, as shown in the figure. The pass is caught on the 40-yard line, 45 yards from the same sideline. How long is the pass?

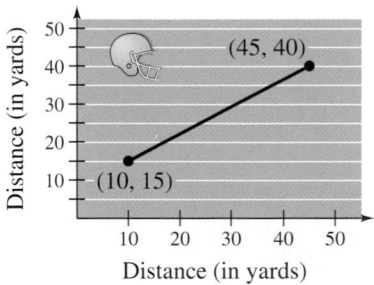

88. *Sports* A major league baseball diamond is a square with 90-foot sides. Place a coordinate system over the baseball diamond so that home plate is at the origin and the first base line lies on the positive x-axis (see figure). Let one unit in the coordinate plane represent one foot. The right fielder fields the ball at the point $(300, 25)$. How far does the right fielder have to throw the ball to get a runner out at home plate? How far does the right fielder have to throw the ball to get a runner out at third base? (Round your answers to one decimal place.)

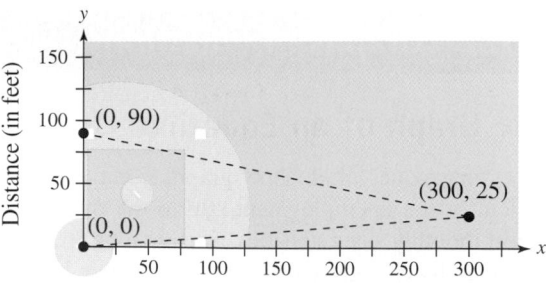

Figure for 88

89. *Boating* A yacht named Beach Lover leaves port at noon and travels due north at 16 miles per hour. At the same time another yacht, The Fisherman, leaves the same port and travels west at 12 miles per hour.

(a) Using graph paper, plot the coordinates of each yacht at 2 P.M. and 4 P.M. Let the port be at the origin of your coordinate system.

(b) Find the distance between the yachts at 2 P.M. and 4 P.M. Are the yachts twice as far from each other at 4 P.M. as they were at 2 P.M.?

90. *Make a Conjecture* Plot the points $(2, 1)$, $(-3, 5)$, and $(7, -3)$ on a rectangular coordinate system. Then change the signs of the indicated coordinate(s) of each point and plot the three new points on the same rectangular coordinate system. Make a conjecture about the location of a point when each of the following occurs.

(a) The sign of the x-coordinate is changed.

(b) The sign of the y-coordinate is changed.

(c) The signs of both the x- and y-coordinates are changed.

91. Show that the coordinates $(2, 6)$, $\left(2 + 2\sqrt{3}, 0\right)$, and $\left(2 - 2\sqrt{3}, 0\right)$ form the vertices of an equilateral triangle.

92. Show that the coordinates $(-2, -1)$, $(4, 7)$, and $(2, -4)$ form the vertices of a right triangle.

Synthesis

True or False? **In Exercises 93–95, determine whether the statement is true or false. Justify your answer.**

93. In order to divide a line segment into 16 equal parts, you would have to use the Midpoint Formula 16 times.

94. The points $(-8, 4)$, $(2, 11)$ and $(-5, 1)$ represent the vertices of an isosceles triangle.

95. If four points represent the vertices of a polygon, and the four sides are equal, then the polygon must be a square.

96. *Think About It* What is the y-coordinate of any point on the x-axis? What is the x-coordinate of any point on the y-axis?

97. *Think About It* When plotting points on the rectangular coordinate system, is it true that the scales on the x- and y-axes must be the same? Explain.

B.2   Graphs of Equations

The Graph of an Equation

News magazines often show graphs comparing the rate of inflation, the federal deficit, or the unemployment rate to the time of year. Businesses use graphs to report monthly sales statistics. Such graphs provide geometric pictures of the way one quantity changes with respect to another. Frequently, the relationship between two quantities is expressed as an **equation.** This section introduces the basic procedure for determining the geometric picture associated with an equation.

For an equation in the variables x and y, a point (a, b) is a **solution point** if substitution of a for x and b for y satisfies the equation. Most equations have *infinitely many* solution points. For example, the equation $3x + y = 5$ has solution points $(0, 5)$, $(1, 2)$, $(2, -1)$, $(3, -4)$, and so on. The set of all solution points of an equation is the **graph of the equation.**

Example 1   Determining Solution Points

Determine whether (a) $(2, 13)$ and (b) $(-1, -3)$ lie on the graph of $y = 10x - 7$.

Solution

a.

$y = 10x - 7$	Write original equation.
$13 \overset{?}{=} 10(2) - 7$	Substitute 2 for x and 13 for y.
$13 = 13$	$(2, 13)$ is a solution. ✓

The point $(2, 13)$ *does* lie on the graph of $y = 10x - 7$ because it is a solution point of the equation.

b.

$y = 10x - 7$	Write original equation.
$-3 \overset{?}{=} 10(-1) - 7$	Substitute -1 for x and -3 for y.
$-3 \neq -17$	$(-1, -3)$ is not a solution.

The point $(-1, -3)$ *does not* lie on the graph of $y = 10x - 7$ because it is not a solution point of the equation.

✓CHECKPOINT   Now try Exercise 3.

The basic technique used for sketching the graph of an equation is the point-plotting method.

Sketching the Graph of an Equation by Point Plotting

1. If possible, rewrite the equation so that one of the variables is isolated on one side of the equation.
2. Make a table of values showing several solution points.
3. Plot these points on a rectangular coordinate system.
4. Connect the points with a smooth curve or line.

Example 2   Sketching a Graph by Point Plotting

Use point plotting and graph paper to sketch the graph of $3x + y = 6$.

Solution

In this case you can isolate the variable y.

$y = 6 - 3x$ Solve equation for y.

Using negative, zero, and positive values for x, you can obtain the following table of values (solution points).

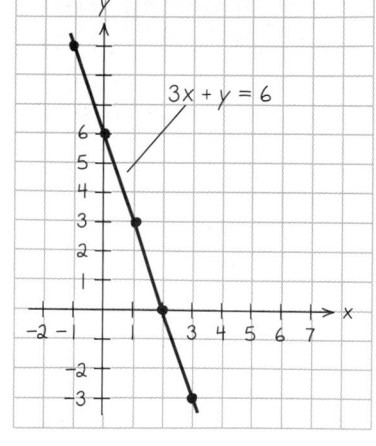

Figure B.20

x	-1	0	1	2	3
$y = 6 - 3x$	9	6	3	0	-3
Solution point	$(-1, 9)$	$(0, 6)$	$(1, 3)$	$(2, 0)$	$(3, -3)$

Next, plot these points and connect them, as shown in Figure B.20. It appears that the graph is a straight line. You will study lines extensively in Section 1.1.

✔CHECKPOINT   Now try Exercise 7.

The points at which a graph touches or crosses an axis are called the **intercepts** of the graph. For instance, in Example 2 the point $(0, 6)$ is the y-intercept of the graph because the graph crosses the y-axis at that point. The point $(2, 0)$ is the x-intercept of the graph because the graph crosses the x-axis at that point.

Example 3   Sketching a Graph by Point Plotting

Use point plotting and graph paper to sketch the graph of $y = x^2 - 2$.

Solution

Because the equation is already solved for y, make a table of values by choosing several convenient values of x and calculating the corresponding values of y.

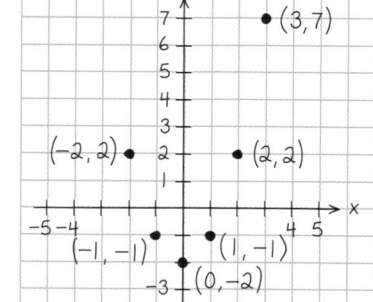

(a)

x	-2	-1	0	1	2	3
$y = x^2 - 2$	2	-1	-2	-1	2	7
Solution point	$(-2, 2)$	$(-1, -1)$	$(0, -2)$	$(1, -1)$	$(2, 2)$	$(3, 7)$

Next, plot the corresponding solution points, as shown in Figure B.21(a). Finally, connect the points with a smooth curve, as shown in Figure B.21(b). This graph is called a *parabola*. You will study parabolas in Section 2.1.

✔CHECKPOINT   Now try Exercise 8.

In this text, you will study two basic ways to create graphs: *by hand* and *using a graphing utility*. For instance, the graphs in Figures B.20 and B.21 were sketched by hand and the graph in Figure B.25 was created using a graphing utility.

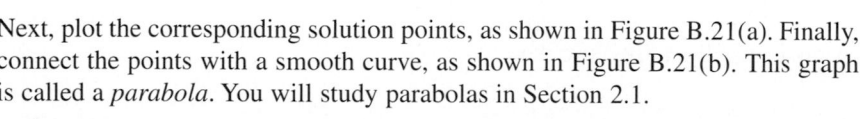

(b)

Figure B.21

Using a Graphing Utility

One of the disadvantages of the point-plotting method is that to get a good idea about the shape of a graph, you need to plot *many* points. With only a few points, you could misrepresent the graph of an equation. For instance, consider the equation

$$y = \frac{1}{30}x(x^4 - 10x^2 + 39).$$

Suppose you plotted only five points: $(-3, -3)$, $(-1, -1)$, $(0, 0)$, $(1, 1)$, and $(3, 3)$, as shown in Figure B.22(a). From these five points, you might assume that the graph of the equation is a line. That, however, is not correct. By plotting several more points and connecting the points with a smooth curve, you can see that the actual graph is not a line at all, as shown in Figure B.22(b).

TECHNOLOGY SUPPORT

This section presents a brief overview of how to use a graphing utility to graph an equation. For more extensive coverage of this topic, see Appendix A and the *Graphing Technology Guide* at this textbook's *Online Study Center*.

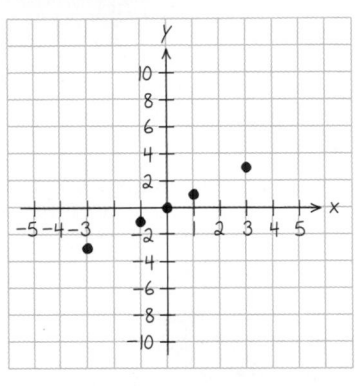

(a)

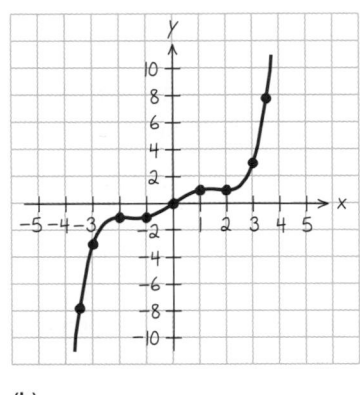

(b)

Figure B.22

From this, you can see that the point-plotting method leaves you with a dilemma. This method can be very inaccurate if only a few points are plotted, and it is very time-consuming to plot a dozen (or more) points. Technology can help solve this dilemma. Plotting several (even several hundred) points on a rectangular coordinate system is something that a computer or calculator can do easily.

TECHNOLOGY TIP   The point-plotting method is the method used by *all* graphing utilities. Each computer or calculator screen is made up of a grid of hundreds or thousands of small areas called *pixels*. Screens that have many pixels per square inch are said to have a higher *resolution* than screens with fewer pixels.

Using a Graphing Utility to Graph an Equation

To graph an equation involving x and y on a graphing utility, use the following procedure.

1. Rewrite the equation so that y is isolated on the left side.
2. Enter the equation in the graphing utility.
3. Determine a *viewing window* that shows all important features of the graph.
4. Graph the equation.

Example 4   Using a Graphing Utility to Graph an Equation

Use a graphing utility to graph $2y + x^3 = 4x$.

Solution

To begin, solve the equation for y in terms of x.

$$2y + x^3 = 4x \qquad\qquad \text{Write original equation.}$$

$$2y = -x^3 + 4x \qquad\qquad \text{Subtract } x^3 \text{ from each side.}$$

$$y = -\frac{x^3}{2} + 2x \qquad\qquad \text{Divide each side by 2.}$$

Enter this equation in a graphing utility (see Figure B.23). Using a standard viewing window (see Figure B.24), you can obtain the graph shown in Figure B.25.

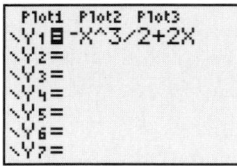

Figure B.23

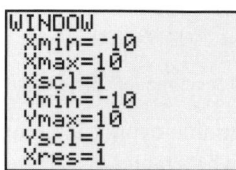

Figure B.24

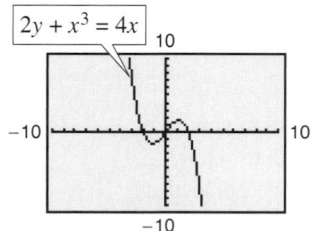

Figure B.25

> **TECHNOLOGY TIP**
>
> Many graphing utilities are capable of creating a table of values such as the following, which shows some points of the graph in Figure B.25. For instructions on how to use the *table* feature, see Appendix A; for specific keystrokes, go to this textbook's *Online Study Center*.
>
>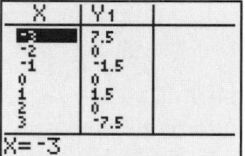

✓CHECKPOINT   Now try Exercise 41.

TECHNOLOGY TIP   By choosing different viewing windows for a graph, it is possible to obtain very different impressions of the graph's shape. For instance, Figure B.26 shows three different viewing windows for the graph of the equation in Example 4. However, none of these views shows *all* of the important features of the graph as does Figure B.25. For instructions on how to set up a viewing window, see Appendix A; for specific keystrokes, go to this textbook's *Online Study Center*.

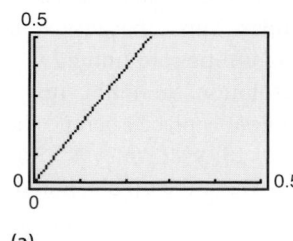

(a)

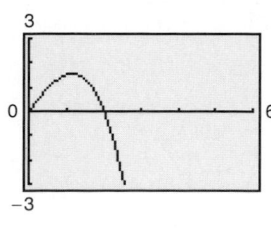

(b)

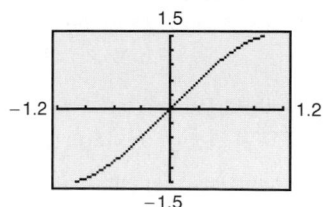

(c)

Figure B.26

TECHNOLOGY TIP   The standard viewing window on many graphing utilities does not give a true geometric perspective because the screen is rectangular, which distorts the image. That is, perpendicular lines will not appear to be perpendicular and circles will not appear to be circular. To overcome this, you can use a *square setting*, as demonstrated in Example 5.

Example 5   Using a Graphing Utility to Graph a Circle

Use a graphing utility to graph $x^2 + y^2 = 9$.

Solution

The graph of $x^2 + y^2 = 9$ is a circle whose center is the origin and whose radius is 3. To graph the equation, begin by solving the equation for y.

$$x^2 + y^2 = 9 \qquad \text{Write original equation.}$$

$$y^2 = 9 - x^2 \qquad \text{Subtract } x^2 \text{ from each side.}$$

$$y = \pm\sqrt{9 - x^2} \qquad \text{Take the square root of each side.}$$

Remember that when you take the square root of a variable expression, you must account for both the positive and negative solutions. The graph of

$$y = \sqrt{9 - x^2} \qquad \text{Upper semicircle}$$

is the upper semicircle. The graph of

$$y = -\sqrt{9 - x^2} \qquad \text{Lower semicircle}$$

is the lower semicircle. Enter *both* equations in your graphing utility and generate the resulting graphs. In Figure B.27, note that if you use a standard viewing window, the two graphs do not appear to form a circle. You can overcome this problem by using a *square setting*, in which the horizontal and vertical tick marks have equal spacing, as shown in Figure B.28. On many graphing utilities, a square setting can be obtained by using a y to x ratio of 2 to 3. For instance, in Figure B.28, the y to x ratio is

$$\frac{Y_{\max} - Y_{\min}}{X_{\max} - X_{\min}} = \frac{4 - (-4)}{6 - (-6)} = \frac{8}{12} = \frac{2}{3}.$$

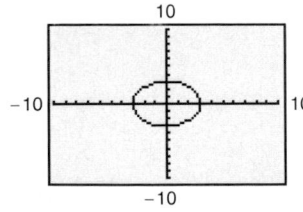

Figure B.27

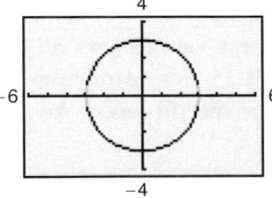

Figure B.28

✔CHECKPOINT   Now try Exercise 63.

Prerequisite Skills

To review the equation of a circle, see Section B.1.

TECHNOLOGY TIP

Notice that when you graph a circle by graphing two separate equations for y, your graphing utility may not connect the two semicircles. This is because some graphing utilities are limited in their resolution. So, in this text, a blue curve is placed behind the graphing utility's display to indicate where the graph should appear.

Applications

Throughout this course, you will learn that there are many ways to approach a problem. Two of the three common approaches are illustrated in Example 6.

An Algebraic Approach: Use the rules of algebra.

A Graphical Approach: Draw and use a graph.

A Numerical Approach: Construct and use a table.

You should develop the habit of using at least two approaches to solve every problem in order to build your intuition and to check that your answer is reasonable.

The following two applications show how to develop mathematical models to represent real-world situations. You will see that both a graphing utility and algebra can be used to understand and solve the problems posed.

Example 6   Running a Marathon

A runner runs at a constant rate of 4.9 miles per hour. The verbal model and algebraic equation relating distance run and elapsed time are as follows.

Verbal Model: Distance = Rate · Time   *Equation:* $d = 4.9t$

a. Determine how far the runner can run in 3.1 hours.

b. Determine how long it will take to run a 26.2-mile marathon.

TECHNOLOGY SUPPORT

For instructions on how to use the *value* feature, the *zoom* and *trace* features, and the *table* feature of a graphing utility, see Appendix A; for specific keystrokes, go to this textbook's *Online Study Center*.

Algebraic Solution

a. To begin, find how far the runner can run in 3.1 hours by substituting 3.1 for t in the equation.

$d = 4.9t$	Write original equation.
$= 4.9(3.1)$	Substitute 3.1 for t.
≈ 15.2	Use a calculator.

So, the runner can run about 15.2 miles in 3.1 hours. Use estimation to check your answer. Because 4.9 is about 5 and 3.1 is about 3, the distance is about $5(3) = 15$. So, 15.2 is reasonable.

b. You can find how long it will take to run a 26.2-mile marathon as follows. (For help with solving linear equations, see Appendix E.)

$d = 4.9t$	Write original equation.
$26.2 = 4.9t$	Substitute 26.2 for d.
$\dfrac{26.2}{4.9} = t$	Divide each side by 4.9.
$5.3 \approx t$	Use a calculator.

So, it will take about 5.3 hours to run 26.2 miles.

✔*CHECKPOINT*   Now try Exercise 71.

Graphical Solution

a. Use a graphing utility to graph the equation $d = 4.9t$. (Represent d by y and t by x.) Be sure to use a viewing window that shows the graph at $x = 3.1$. Then use the *value* feature or the *zoom* and *trace* features of the graphing utility to estimate that when $x = 3.1$, the distance is $y \approx 15.2$ miles, as shown in Figure B.29(a).

b. Adjust the viewing window so that it shows the graph at $y = 26.2$. Use the *zoom* and *trace* features to estimate that when $y \approx 26.2$, the time is $x \approx 5.3$ hours, as shown in Figure B.29(b).

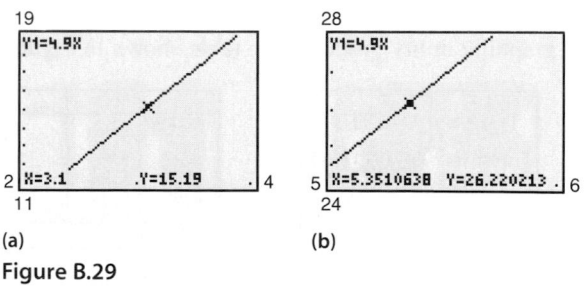

(a)   (b)

Figure B.29

Note that the viewing window on your graphing utility may differ slightly from those shown in Figure B.29.

Example 7 Monthly Wage

You receive a monthly salary of $2000 plus a commission of 10% of sales. The verbal model and algebraic equation relating the wages, the salary, and the commission are as follows.

Verbal Model: Wages = Salary + Commission on sales

Equation: $y = 2000 + 0.1x$

a. Sales are $1480 in August. What are your wages for that month?

b. You receive $2225 for September. What are your sales for that month?

Numerical Solution

a. To find the wages in August, evaluate the equation when $x = 1480$.

$$y = 2000 + 0.1x \qquad \text{Write original equation.}$$

$$= 2000 + 0.1(1480) \qquad \text{Substitute 1480 for } x.$$

$$= 2148 \qquad \text{Simplify.}$$

So, your wages in August are $2148.

b. You can use the *table* feature of a graphing utility to create a table that shows the wages for different sales amounts. First enter the equation in the graphing utility. Then set up a table, as shown in Figure B.30. The graphing utility produces the table shown in Figure B.31.

Figure B.30   **Figure B.31**

From the table, you can see that wages of $2225 result from sales between $2200 and $2300. You can improve this estimate by setting up the table shown in Figure B.32. The graphing utility produces the table shown in Figure B.33.

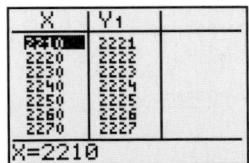

Figure B.32   **Figure B.33**

From the table, you can see that wages of $2225 result from sales of $2250.

CHECKPOINT   Now try Exercise 73.

Graphical Solution

a. You can use a graphing utility to graph $y = 2000 + 0.1x$ and then estimate the wages when $x = 1480$. Be sure to use a viewing window that shows the graph for $x \geq 0$ and $y > 2000$. Then, by using the *value* feature or the *zoom* and *trace* features near $x = 1480$, you can estimate that the wages are about $2148, as shown in Figure B.34(a).

b. Use the graphing utility to find the value along the x-axis (sales) that corresponds to a y-value of 2225 (wages). Using the *zoom* and *trace* features, you can estimate the sales to be about $2250, as shown in Figure B.34(b).

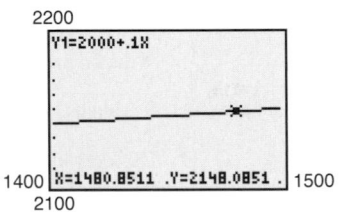

(a) Zoom near $x = 1480$

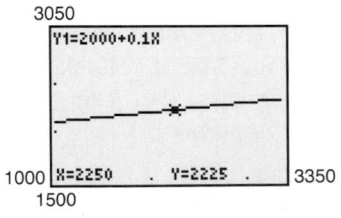

(b) Zoom near $y = 2225$

Figure B.34

B.2 Exercises

See www.CalcChat.com for worked-out solutions to odd-numbered exercises.

Vocabulary Check

Fill in the blanks.

1. For an equation in x and y, if substitution of a for x and b for y satisfies the equation, then the point (a, b) is a _____ .

2. The set of all solution points of an equation is the _____ of the equation.

3. The points at which a graph touches or crosses an axis are called the _____ of the graph.

In Exercises 1–6, determine whether each point lies on the graph of the equation.

Equation		Points	
1. $y = \sqrt{x + 4}$		(a) $(0, 2)$	(b) $(5, 3)$
2. $y = x^2 - 3x + 2$		(a) $(2, 0)$	(b) $(-2, 8)$
3. $y = 4 - |x - 2|$		(a) $(1, 5)$	(b) $(1.2, 3.2)$
4. $2x - y - 3 = 0$		(a) $(1, 2)$	(b) $(1, -1)$
5. $x^2 + y^2 = 20$		(a) $(3, -2)$	(b) $(-4, 2)$
6. $y = \frac{1}{3}x^3 - 2x^2$		(a) $\left(2, -\frac{16}{3}\right)$	(b) $(-3, 9)$

In Exercises 7 and 8, complete the table. Use the resulting solution points to sketch the graph of the equation. Use a graphing utility to verify the graph.

7. $3x - 2y = 2$

x	-2	0	$\frac{2}{3}$	1	2
y					
Solution point					

8. $2x + y = x^2$

x	-1	0	1	2	3
y					
Solution point					

9. *Exploration*

(a) Complete the table for the equation $y = \frac{1}{4}x - 3$.

x	-2	-1	0	1	2
y					

(b) Use the solution points to sketch the graph. Then use a graphing utility to verify the graph.

(c) Repeat parts (a) and (b) for the equation $y = -\frac{1}{4}x - 3$. Describe any differences between the graphs.

10. *Exploration*

(a) Complete the table for the equation

$$y = \frac{6x}{x^2 + 1}.$$

x	-2	-1	0	1	2
y					

(b) Use the solution points to sketch the graph. Then use a graphing utility to verify the graph.

(c) Continue the table in part (a) for x-values of 5, 10, 20, and 40. What is the value of y approaching? Can y be negative for positive values of x? Explain.

In Exercises 11–16, match the equation with its graph. [The graphs are labeled (a), (b), (c), (d), (e), and (f).]

(a)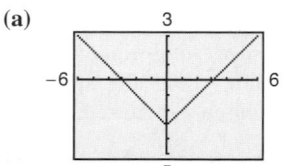

(b)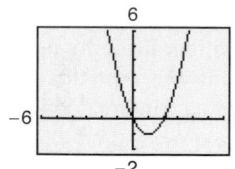

(c)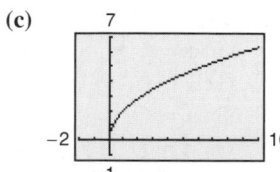

(d)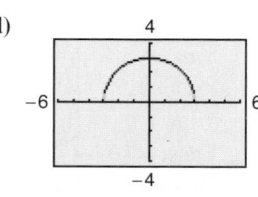

(e)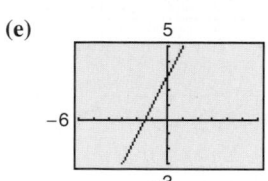

(f)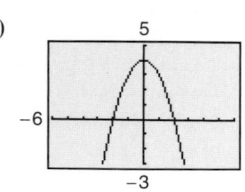

11. $y = 2x + 3$

12. $y = 4 - x^2$

13. $y = x^2 - 2x$

14. $y = \sqrt{9 - x^2}$

15. $y = 2\sqrt{x}$

16. $y = |x| - 3$

In Exercises 17–30, sketch the graph of the equation.

17. $y = -4x + 1$ 18. $y = 2x - 3$

19. $y = 2 - x^2$ 20. $y = x^2 - 1$

21. $y = x^2 - 3x$ 22. $y = -x^2 - 4x$

23. $y = x^3 + 2$ 24. $y = x^3 - 3$

25. $y = \sqrt{x - 3}$ 26. $y = \sqrt{1 - x}$

27. $y = |x - 2|$ 28. $y = 5 - |x|$

29. $x = y^2 - 1$ 30. $x = y^2 + 4$

In Exercises 31–44, use a graphing utility to graph the equation. Use a standard viewing window. Approximate any x- or y-intercepts of the graph.

31. $y = x - 7$ 32. $y = x + 1$

33. $y = 3 - \frac{1}{2}x$ 34. $y = \frac{2}{3}x - 1$

35. $y = \dfrac{2x}{x - 1}$ 36. $y = \dfrac{4}{x}$

37. $y = x\sqrt{x + 3}$

38. $y = (6 - x)\sqrt{x}$

39. $y = \sqrt[3]{x - 8}$

40. $y = \sqrt[3]{x + 1}$

41. $x^2 - y = 4x - 3$

42. $2y - x^2 + 8 = 2x$

43. $y - 4x = x^2(x - 4)$

44. $x^3 + y = 1$

In Exercises 45–48, use a graphing utility to graph the equation. Begin by using a standard viewing window. Then graph the equation a second time using the specified viewing window. Which viewing window is better? Explain.

45. $y = \frac{5}{2}x + 5$ 46. $y = -3x + 50$

Xmin = 0
Xmax = 6
Xscl = 1
Ymin = 0
Ymax = 10
Yscl = 1

Xmin = -1
Xmax = 4
Xscl = 1
Ymin = -5
Ymax = 60
Yscl = 5

47. $y = -x^2 + 10x - 5$ 48. $y = 4(x + 5)\sqrt{4 - x}$

Xmin = -1
Xmax = 11
Xscl = 1
Ymin = -5
Ymax = 25
Yscl = 5

Xmin = -6
Xmax = 6
Xscl = 1
Ymin = -5
Ymax = 50
Yscl = 5

In Exercises 49–54, describe the viewing window of the graph shown.

49. $y = -10x + 50$ 50. $y = 4x^2 - 25$

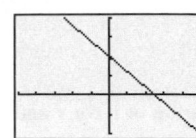   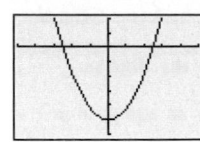

51. $y = \sqrt{x + 2} - 1$ 52. $y = x^3 - 3x^2 + 4$

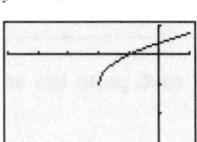   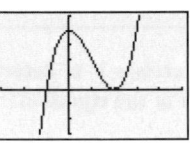

53. $y = |x| + |x - 10|$ 54. $y = 8\sqrt[3]{x - 6}$

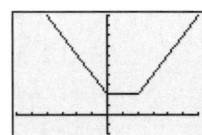   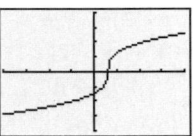

In Exercises 55–58, explain how to use a graphing utility to verify that $y_1 = y_2$. Identify the rule of algebra that is illustrated.

55. $y_1 = \frac{1}{4}(x^2 - 8)$
    $y_2 = \frac{1}{4}x^2 - 2$

56. $y_1 = \frac{1}{2}x + (x + 1)$
    $y_2 = \frac{3}{2}x + 1$

57. $y_1 = \dfrac{1}{5}[10(x^2 - 1)]$
    $y_2 = 2(x^2 - 1)$

58. $y_1 = (x - 3) \cdot \dfrac{1}{x - 3}$
    $y_2 = 1$

In Exercises 59–62, use a graphing utility to graph the equation. Use the *trace* feature of the graphing utility to approximate the unknown coordinate of each solution point accurate to two decimal places. (*Hint:* You may need to use the *zoom* feature of the graphing utility to obtain the required accuracy.)

59. $y = \sqrt{5 - x}$ 60. $y = x^3(x - 3)$
    (a) $(2, y)$ (a) $(2.25, y)$
    (b) $(x, 3)$ (b) $(x, 20)$

61. $y = x^5 - 5x$ 62. $y = |x^2 - 6x + 5|$
    (a) $(-0.5, y)$ (a) $(2, y)$
    (b) $(x, -4)$ (b) $(x, 1.5)$

In Exercises 63–66, solve for y and use a graphing utility to graph each of the resulting equations in the same viewing window. (Adjust the viewing window so that the circle appears circular.)

63. $x^2 + y^2 = 16$

64. $x^2 + y^2 = 36$

65. $(x - 1)^2 + (y - 2)^2 = 4$

66. $(x - 3)^2 + (y - 1)^2 = 25$

In Exercises 67 and 68, determine which equation is the best choice for the graph of the circle shown.

67.

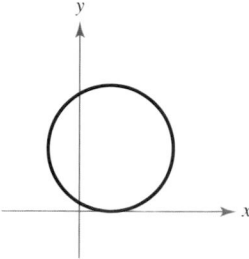

(a) $(x - 1)^2 + (y - 2)^2 = 4$

(b) $(x + 1)^2 + (y - 2)^2 = 4$

(c) $(x - 1)^2 + (y - 2)^2 = 16$

(d) $(x + 1)^2 + (y + 2)^2 = 4$

68.

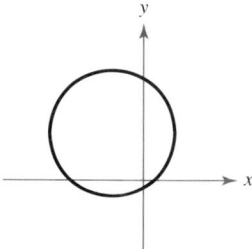

(a) $(x - 2)^2 + (y - 3)^2 = 4$

(b) $(x - 2)^2 + (y - 3)^2 = 16$

(c) $(x + 2)^2 + (y - 3)^2 = 16$

(d) $(x + 2)^2 + (y - 3)^2 = 4$

In Exercises 69 and 70, determine whether each point lies on the graph of the circle. (There may be more than one correct answer.)

69. $(x - 1)^2 + (y - 2)^2 = 25$

(a) $(1, 2)$ (b) $(-2, 6)$

(c) $(5, -1)$ (d) $\left(0, 2 + 2\sqrt{6}\right)$

70. $(x + 2)^2 + (y - 3)^2 = 25$

(a) $(-2, 3)$ (b) $(0, 0)$

(c) $(1, -1)$ (d) $\left(-1, 3 - 2\sqrt{6}\right)$

71. *Depreciation*   A manufacturing plant purchases a new molding machine for $225,000. The depreciated value (decreased value) y after t years is $y = 225,000 - 20,000t$, for $0 \le t \le 8$.

(a) Use the constraints of the model to graph the equation using an appropriate viewing window.

(b) Use the *value* feature or the *zoom* and *trace* features of a graphing utility to determine the value of y when $t = 5.8$. Verify your answer algebraically.

(c) Use the *value* feature or the *zoom* and *trace* features of a graphing utility to determine the value of y when $t = 2.35$. Verify your answer algebraically.

72. *Consumerism*   You buy a personal watercraft for $8100. The depreciated value y after t years is $y = 8100 - 929t$, for $0 \le t \le 6$.

(a) Use the constraints of the model to graph the equation using an appropriate viewing window.

(b) Use the *zoom* and *trace* features of a graphing utility to determine the value of t when $y = 5545.25$. Verify your answer algebraically.

(c) Use the *value* feature or the *zoom* and *trace* features of a graphing utility to determine the value of y when $t = 5.5$. Verify your answer algebraically.

73. *Data Analysis*   The table shows the median (middle) sales prices (in thousands of dollars) of new one-family homes in the southern United States from 1995 to 2004.   (Sources: U.S. Census Bureau and U.S. Department of Housing and Urban Development)

Year	Median sales price, y
1995	124.5
1996	126.2
1997	129.6
1998	135.8
1999	145.9
2000	148.0
2001	155.4
2002	163.4
2003	168.1
2004	181.1

A model for the median sales price during this period is given by

$$y = -0.0049t^3 + 0.443t^2 - 0.75t + 116.7, \quad 5 \le t \le 14$$

where y represents the sales price and t represents the year, with $t = 5$ corresponding to 1995.

(a) Use the model and the *table* feature of a graphing utility to find the median sales prices from 1995 to 2004. How well does the model fit the data? Explain.

(b) Use a graphing utility to graph the data from the table and the model in the same viewing window. How well does the model fit the data? Explain.

(c) Use the model to estimate the median sales prices in 2008 and 2010. Do the values seem reasonable? Explain.

(d) Use the *zoom* and *trace* features of a graphing utility to determine during which year(s) the median sales price was approximately $150,000.

74. *Population Statistics* The table shows the life expectancies of a child (at birth) in the United States for selected years from 1930 to 2000. (Source: U.S. National Center for Health Statistics)

Year	Life expectancy, y
1930	59.7
1940	62.9
1950	68.2
1960	69.7
1970	70.8
1980	73.7
1990	75.4
2000	77.0

A model for the life expectancy during this period is given by

$$y = \frac{59.617 + 1.18t}{1 + 0.012t}, \quad 0 \le t \le 70$$

where y represents the life expectancy and t is the time in years, with $t = 0$ corresponding to 1930.

(a) Use a graphing utility to graph the data from the table above and the model in the same viewing window. How well does the model fit the data? Explain.

(b) What does the y-intercept of the graph of the model represent?

(c) Use the *zoom* and *trace* features of a graphing utility to determine the year when the life expectancy was 73.2. Verify your answer algebraically.

(d) Determine the life expectancy in 1948 both graphically and algebraically.

(e) Use the model to estimate the life expectancy of a child born in 2010.

75. *Geometry* A rectangle of length x and width w has a perimeter of 12 meters.

(a) Draw a diagram that represents the rectangle. Use the specified variables to label its sides.

(b) Show that the width of the rectangle is $w = 6 - x$ and that its area is $A = x(6 - x)$.

(c) Use a graphing utility to graph the area equation.

(d) Use the *zoom* and *trace* features of a graphing utility to determine the value of A when $w = 4.9$ meters. Verify your answer algebraically.

(e) From the graph in part (c), estimate the dimensions of the rectangle that yield a maximum area.

76. Find the standard form of the equation of the circle for which the endpoints of a diameter are $(0, 0)$ and $(4, -6)$.

Synthesis

True or False? **In Exercises 77 and 78, determine whether the statement is true or false. Justify your answer.**

77. A parabola can have only one x-intercept.

78. The graph of a linear equation can have either no x-intercepts or only one x-intercept.

79. *Writing* Explain how to find an appropriate viewing window for the graph of an equation.

80. *Writing* Your employer offers you a choice of wage scales: a monthly salary of $3000 plus commission of 7% of sales or a salary of $3400 plus a 5% commission. Write a short paragraph discussing how you would choose your option. At what sales level would the options yield the same salary?

81. *Writing* Given the equation $y = 250x + 1000$, write a possible explanation of what the equation could represent in real life.

82. *Writing* Given the equation $y = -0.1x + 10$, write a possible explanation of what the equation could represent in real life.

B.3   Solving Equations Algebraically and Graphically

Equations and Solutions of Equations

An **equation** in x is a statement that two algebraic expressions are equal. For example, $3x - 5 = 7$, $x^2 - x - 6 = 0$, and $\sqrt{2x} = 4$ are equations. To **solve** an equation in x means to find all values of x for which the equation is true. Such values are **solutions.** For instance, $x = 4$ is a solution of the equation $3x - 5 = 7$ because $3(4) - 5 = 7$ is a true statement.

The solutions of an equation depend on the kinds of numbers being considered. For instance, in the set of rational numbers, $x^2 = 10$ has no solution because there is no rational number whose square is 10. However, in the set of real numbers, the equation has the two solutions $\sqrt{10}$ and $-\sqrt{10}$.

An equation that is true for *every* real number in the domain (the domain is the set of all real numbers for which the equation is defined) of the variable is called an **identity.** For example, $x^2 - 9 = (x + 3)(x - 3)$ is an identity because it is a true statement for any real value of x, and $x/(3x^2) = 1/(3x)$, where $x \neq 0$, is an identity because it is true for any nonzero real value of x.

An equation that is true for just *some* (or even none) of the real numbers in the domain of the variable is called a **conditional equation.** The equation $x^2 - 9 = 0$ is conditional because $x = 3$ and $x = -3$ are the only values in the domain that satisfy the equation. The equation $2x + 1 = 2x - 3$ is also conditional because there are no real values of x for which the equation is true.

A **linear equation in one variable x** is an equation that can be written in the standard form $ax + b = 0$, where a and b are real numbers, with $a \neq 0$. For a review of solving one- and two-step linear equations, see Appendix E.

To solve an equation involving fractional expressions, find the least common denominator (LCD) of all terms in the equation and multiply every term by this LCD. This procedure clears the equation of fractions, as demonstrated in Example 1.

What you should learn

- Solve linear equations.
- Find x- and y-intercepts of graphs of equations.
- Find solutions of equations graphically.
- Find the points of intersection of two graphs.
- Solve polynomial equations.
- Solve equations involving radicals, fractions, or absolute values.

Why you should learn it

Knowing how to solve equations algebraically and graphically can help you solve real-life problems. For instance, in Exercise 195 on page A62, you will find the point of intersection of the graphs of two population models both algebraically and graphically.

Example 1   Solving an Equation Involving Fractions

Solve $\dfrac{x}{3} + \dfrac{3x}{4} = 2$.

Solution

$$\frac{x}{3} + \frac{3x}{4} = 2 \qquad \text{Write original equation.}$$

$$(12)\frac{x}{3} + (12)\frac{3x}{4} = (12)2 \qquad \text{Multiply each term by the LCD of 12.}$$

$$4x + 9x = 24 \qquad \text{Divide out and multiply.}$$

$$13x = 24 \qquad \text{Combine like terms.}$$

$$x = \frac{24}{13} \qquad \text{Divide each side by 13.}$$

✓CHECKPOINT   Now try Exercise 23.

STUDY TIP

After solving an equation, you should check each solution in the original equation. For instance, you can check the solution to Example 1 as follows.

$$\frac{x}{3} + \frac{3x}{4} = 2$$

$$\frac{\frac{24}{13}}{3} + \frac{3\left(\frac{24}{13}\right)}{4} \overset{?}{=} 2$$

$$\frac{8}{13} + \frac{18}{13} \overset{?}{=} 2$$

$$2 = 2 \checkmark$$

When multiplying or dividing an equation by a *variable* expression, it is possible to introduce an **extraneous solution**—one that does not satisfy the original equation. The next example demonstrates the importance of checking your solution when you have multiplied or divided by a variable expression.

Example 2   An Equation with an Extraneous Solution

Solve $\dfrac{1}{x-2} = \dfrac{3}{x+2} - \dfrac{6x}{x^2-4}$.

Algebraic Solution

The LCD is

$$x^2 - 4 = (x+2)(x-2).$$

Multiplying each term by the LCD and simplifying produces the following.

$$\frac{1}{x-2}(x+2)(x-2)$$

$$= \frac{3}{x+2}(x+2)(x-2) - \frac{6x}{x^2-4}(x+2)(x-2)$$

$$x + 2 = 3(x-2) - 6x, \quad x \neq \pm 2$$

$$x + 2 = 3x - 6 - 6x$$

$$4x = -8$$

$$x = -2 \qquad \text{Extraneous solution}$$

A check of $x = -2$ in the original equation shows that it yields a denominator of zero. So, $x = -2$ is an extraneous solution, and the original equation has *no solution*.

✓CHECKPOINT   Now try Exercise 39.

Graphical Solution

Use a graphing utility (in *dot* mode) to graph the left and right sides of the equation,

$$y_1 = \frac{1}{x-2} \quad \text{and} \quad y_2 = \frac{3}{x+2} - \frac{6x}{x^2-4}$$

in the same viewing window, as shown in Figure B.35. The graphs of the equations do not appear to intersect. This means that there is no point for which the left side of the equation y_1 is equal to the right side of the equation y_2. So, the equation appears to have *no solution*.

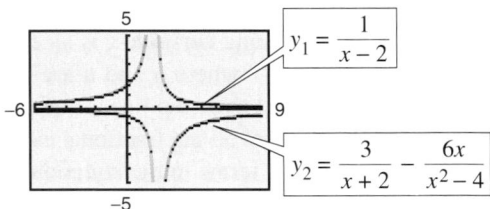

Figure B.35

Intercepts and Solutions

In Section B.2, you learned that the intercepts of a graph are the points at which the graph intersects the *x*- or *y*-axis.

Definition of Intercepts

1. The point $(a, 0)$ is called an **x-intercept** of the graph of an equation if it is a solution point of the equation. To find the *x*-intercept(s), set *y* equal to 0 and solve the equation for *x*.

2. The point $(0, b)$ is called a **y-intercept** of the graph of an equation if it is a solution point of the equation. To find the *y*-intercept(s), set *x* equal to 0 and solve the equation for *y*.

STUDY TIP

Recall that the least common denominator of several rational expressions consists of the product of all prime factors in the denominators, with each factor given the highest power of its occurrence in any denominator.

Sometimes it is convenient to denote the x-intercept as simply the x-coordinate of the point $(a, 0)$ rather than the point itself. Unless it is necessary to make a distinction, "intercept" will be used to mean either the point or the coordinate.

It is possible for a graph to have no intercepts, one intercept, or several intercepts. For instance, consider the three graphs shown in Figure B.36.

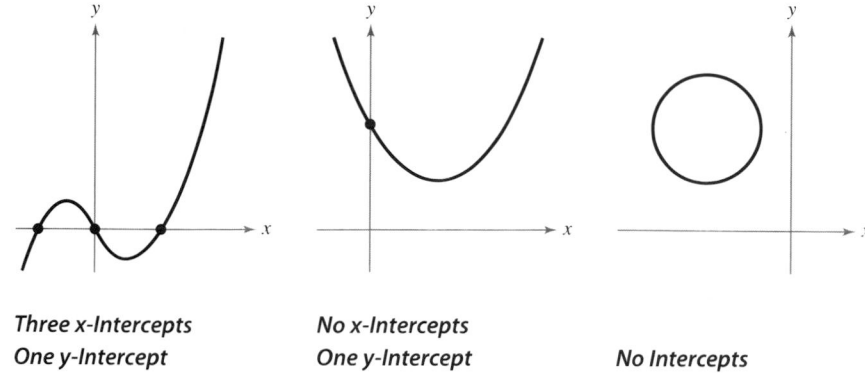

Three x-Intercepts
One y-Intercept
Figure B.36

No x-Intercepts
One y-Intercept

No Intercepts

Example 3   Finding x- and y-Intercepts

Find the x- and y-intercepts of the graph of $2x + 3y = 5$.

Solution

To find the x-intercept, let $y = 0$ and solve for x. This produces

$$2x = 5 \quad \Longrightarrow \quad x = \tfrac{5}{2}$$

which implies that the graph has one x-intercept: $\left(\tfrac{5}{2}, 0\right)$. To find the y-intercept, let $x = 0$ and solve for y. This produces

$$3y = 5 \quad \Longrightarrow \quad y = \tfrac{5}{3}$$

which implies that the graph has one y-intercept: $\left(0, \tfrac{5}{3}\right)$. See Figure B.37.

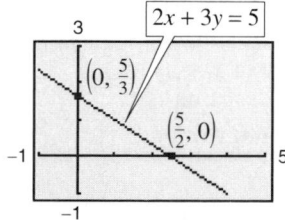

Figure B.37

✓CHECKPOINT   Now try Exercise 41.

The concepts of x-intercepts and solutions of equations are closely related. In fact, the following statements are equivalent.

1. The point $(a, 0)$ is an *x-intercept* of the graph of an equation.
2. The number a is a *solution* of the equation $y = 0$.

The close connection between x-intercepts and solutions is crucial to your study of algebra. You can take advantage of this connection in two ways. Use your algebraic equation-solving skills to find the x-intercepts of a graph, and use your graphing skills to approximate the solutions of an equation.

Finding Solutions Graphically

Polynomial equations of degree 1 or 2 can be solved in relatively straightforward ways. Solving polynomial equations of higher degree can, however, be quite difficult, especially if you rely only on algebraic techniques. For such equations, a graphing utility can be very helpful.

Graphical Approximations of Solutions of an Equation

1. Write the equation in *general form*, $y = 0$, with the nonzero terms on one side of the equation and zero on the other side.

2. Use a graphing utility to graph the equation. Be sure the viewing window shows all the relevant features of the graph.

3. Use the *zero* or *root* feature or the *zoom* and *trace* features of the graphing utility to approximate the x-intercepts of the graph.

Chapter 2 shows techniques for determining the number of solutions of a polynomial equation. For now, you should know that a polynomial equation of degree n cannot have more than n different solutions.

Example 4   Finding Solutions of an Equation Graphically

Use a graphing utility to approximate the solutions of $2x^3 - 3x + 2 = 0$.

Solution

Graph the function $y = 2x^3 - 3x + 2$. You can see from the graph that there is one x-intercept. It lies between -2 and -1 and is approximately -1.5. By using the *zero* or *root* feature of a graphing utility, you can improve the approximation. Choose a left bound of $x = -2$ (see Figure B.38) and a right bound of $x = -1$ (see Figure B.39). To two-decimal-place accuracy, the solution is $x \approx -1.48$, as shown in Figure B.40. Check this approximation on your calculator. You will find that the value of y is $y = 2(-1.48)^3 - 3(-1.48) + 2 \approx -0.04$.

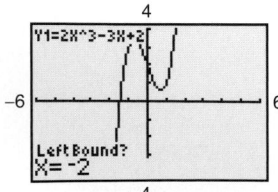

Figure B.38

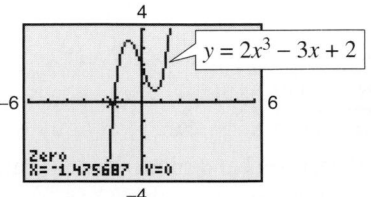

Figure B.39

Figure B.40

TECHNOLOGY SUPPORT

For instructions on how to use the *zero* or *root* feature, see Appendix A; for specific keystrokes, go to this textbook's *Online Study Center*.

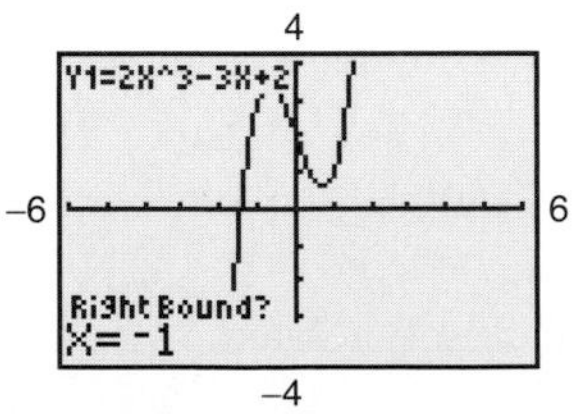

✓CHECKPOINT   Now try Exercise 53.

TECHNOLOGY TIP   You can also use a graphing calculator's *zoom* and *trace* features to approximate the solution(s) of an equation. Here are some suggestions for using the *zoom-in* feature of a graphing utility.

1. With each successive zoom-in, adjust the *x*-scale (if necessary) so that the resulting viewing window shows at least the two scale marks between which the solution lies.

2. The accuracy of the approximation will always be such that the error is less than the distance between two scale marks.

3. If you have a *trace* feature on your graphing utility, you can generally add one more decimal place of accuracy without changing the viewing window.

Unless stated otherwise, this book will approximate all real solutions with an error of *at most* 0.01.

Example 5   Approximating Solutions of an Equation Graphically

Use a graphing utility to approximate the solutions of $x^2 + 3 = 5x$.

Solution

In general form, this equation is

$$x^2 - 5x + 3 = 0.$$ Equation in general form

So, you can begin by graphing

$$y = x^2 - 5x + 3$$ Function to be graphed

as shown in Figure B.41. This graph has two *x*-intercepts, and by using the *zoom* and *trace* features you can approximate the corresponding solutions to be $x \approx 0.70$ and $x \approx 4.30$, as shown in Figures B.42 and B.43.

TECHNOLOGY TIP

Remember that the more decimal places in the solution, the more accurate the solution is. You can reach the desired accuracy when zooming in as follows.

• To approximate the zero to the nearest hundredth, set the *x*-scale to 0.01.

• To approximate the zero to the nearest thousandth, set the *x*-scale to 0.001.

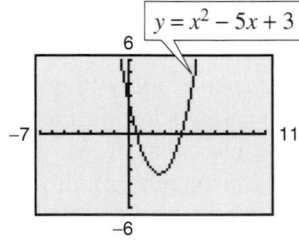

Figure B.41

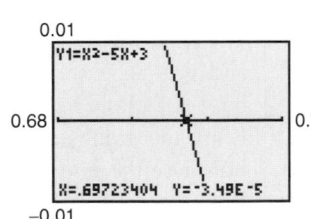

Figure B.42

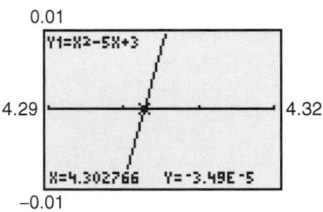

Figure B.43

✓*CHECKPOINT*   Now try Exercise 55.

TECHNOLOGY TIP   Remember from Example 4 that the built-in *zero* or *root* features of a graphing utility will approximate solutions of equations or *x*-intercepts of graphs. If your graphing utility has such features, try using them to approximate the solutions in Example 5.

Points of Intersection of Two Graphs

An ordered pair that is a solution of two different equations is called a **point of intersection** of the graphs of the two equations. For instance, in Figure B.44 you can see that the graphs of the following equations have two points of intersection.

$y = x + 2$ Equation 1

$y = x^2 - 2x - 2$ Equation 2

The point $(-1, 1)$ is a solution of both equations, and the point $(4, 6)$ is a solution of both equations. To check this algebraically, substitute $x = -1$ and $x = 4$ into each equation.

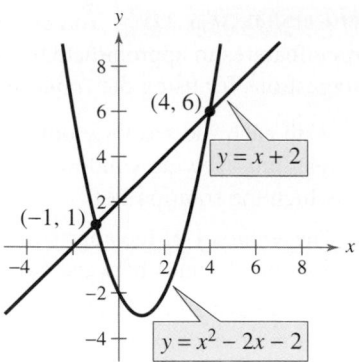

Figure B.44

Check that $(-1, 1)$ is a solution.

Equation 1: $y = -1 + 2 = 1$ Solution checks. ✓

Equation 2: $y = (-1)^2 - 2(-1) - 2 = 1$ Solution checks. ✓

Check that $(4, 6)$ is a solution.

Equation 1: $y = 4 + 2 = 6$ Solution checks. ✓

Equation 2: $y = (4)^2 - 2(4) - 2 = 6$ Solution checks. ✓

To find the points of intersection of the graphs of two equations, solve each equation for y (or x) and set the two results equal to each other. The resulting equation will be an equation in one variable that can be solved using standard procedures, as shown in Example 6.

> **TECHNOLOGY SUPPORT**
>
> For instructions on how to use the *intersect* feature, see Appendix A; for specific keystrokes, go to this textbook's *Online Study Center.*

Example 6   Finding Points of Intersection

Find the points of intersection of the graphs of $2x - 3y = -2$ and $4x - y = 6$.

Algebraic Solution

To begin, solve each equation for y to obtain

$$y = \frac{2}{3}x + \frac{2}{3} \quad \text{and} \quad y = 4x - 6.$$

Next, set the two expressions for y equal to each other and solve the resulting equation for x, as follows.

$\frac{2}{3}x + \frac{2}{3} = 4x - 6$ Equate expressions for y.

$2x + 2 = 12x - 18$ Multiply each side by 3.

$-10x = -20$ Subtract $12x$ and 2 from each side.

$x = 2$ Divide each side by -10.

When $x = 2$, the y-value of each of the original equations is 2. So, the point of intersection is $(2, 2)$.

✓CHECKPOINT   Now try Exercise 93.

Graphical Solution

To begin, solve each equation for y to obtain $y_1 = \frac{2}{3}x + \frac{2}{3}$ and $y_2 = 4x - 6$. Then use a graphing utility to graph both equations in the same viewing window. In Figure B.45, the graphs appear to have one point of intersection. Use the *intersect* feature of the graphing utility to approximate the point of intersection to be $(2, 2)$.

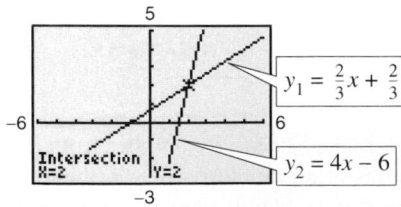

Figure B.45

> **TECHNOLOGY TIP**   Another way to approximate the points of intersection of two graphs is to graph both equations with a graphing utility and use the *zoom* and *trace* features to find the point or points at which the two graphs intersect.

Example 7   Approximating Points of Intersection Graphically

Approximate the point(s) of intersection of the graphs of the following equations.

$$y = x^2 - 3x - 4 \qquad \text{Equation 1 (quadratic function)}$$

$$y = x^3 + 3x^2 - 2x - 1 \qquad \text{Equation 2 (cubic function)}$$

Solution

Begin by using a graphing utility to graph both equations, as shown in Figure B.46. From this display, you can see that the two graphs have only one point of intersection. Then, using the *zoom* and *trace* features, approximate the point of intersection to be $(-2.17, 7.25)$, as shown in Figure B.47.

To test the reasonableness of this approximation, you can evaluate both equations at $x = -2.17$.

Quadratic Equation:

$$y = (-2.17)^2 - 3(-2.17) - 4 \approx 7.22$$

Cubic Equation:

$$y = (-2.17)^3 + 3(-2.17)^2 - 2(-2.17) - 1 \approx 7.25$$

Because both equations yield approximately the same y-value, you can approximate the coordinates of the point of intersection to be $x \approx -2.17$ and $y \approx 7.25$.

✓CHECKPOINT   Now try Exercise 97.

> **TECHNOLOGY TIP**   If you choose to use the *intersect* feature of your graphing utility to find the point of intersection of the graphs in Example 7, you will see that it yields the same result.

The method shown in Example 7 gives a nice graphical picture of the points of intersection of two graphs. However, for actual approximation purposes, it is better to use the algebraic procedure described in Example 6. That is, the point of intersection of $y = x^2 - 3x - 4$ and $y = x^3 + 3x^2 - 2x - 1$ coincides with the solution of the equation

$$x^3 + 3x^2 - 2x - 1 = x^2 - 3x - 4 \qquad \text{Equate } y\text{-values.}$$

$$x^3 + 2x^2 + x + 3 = 0. \qquad \text{Write in general form.}$$

By graphing $y = x^3 + 2x^2 + x + 3$ with a graphing utility and using the *zoom* and *trace* features (or the *zero* or *root* feature), you can approximate the solution of this equation to be $x \approx -2.17$. The corresponding y-value for *both* of the functions given in Example 7 is $y \approx 7.25$.

> **TECHNOLOGY TIP**
>
> The table shows some points on the graphs of the equations in Example 6. You can find the points of intersection of the graphs by finding the value(s) of x for which y_1 and y_2 are equal.

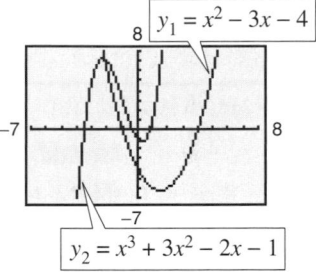

Figure B.46

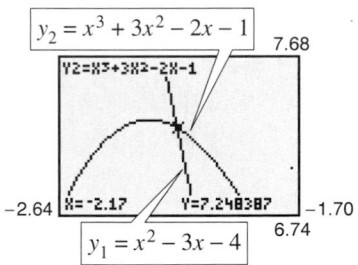

Figure B.47

Solving Polynomial Equations Algebraically

Polynomial equations can be classified by their *degree*. The degree of a polynomial equation is the highest degree of its terms. For polynomials in more than one variable, the degree of the *term* is the sum of the exponents of the variables in the term. For instance, the degree of the polynomial $-2x^3y^6 + 4xy - x^7y^4$ is 11 because the sum of the exponents in the last term is the greatest.

Degree	*Name*	*Example*
First	Linear equation	$6x + 2 = 4$
Second	Quadratic equation	$2x^2 - 5x + 3 = 0$
Third	Cubic equation	$x^3 - x = 0$
Fourth	Quartic equation	$x^4 - 3x^2 + 2 = 0$
Fifth	Quintic equation	$x^5 - 12x^2 + 7x + 4 = 0$

In general, the higher the degree, the more difficult it is to solve the equation—either algebraically or graphically.

You should be familiar with the following four methods for solving quadratic equations *algebraically*.

Solving a Quadratic Equation

Method	*Example*

Factoring: If $ab = 0$, then $a = 0$ or $b = 0$.

$$x^2 - x - 6 = 0$$
$$(x - 3)(x + 2) = 0$$
$$x - 3 = 0 \implies x = 3$$
$$x + 2 = 0 \implies x = -2$$

Extracting Square Roots: If $u^2 = c$, where $c > 0$, then $u = \pm\sqrt{c}$.

$$(x + 3)^2 = 16$$
$$x + 3 = \pm 4$$
$$x = -3 \pm 4$$
$$x = 1 \quad \text{or} \quad x = -7$$

Completing the Square: If $x^2 + bx = c$, then

$$x^2 + bx + \left(\frac{b}{2}\right)^2 = c + \left(\frac{b}{2}\right)^2$$

$$\left(x + \frac{b}{2}\right)^2 = c + \frac{b^2}{4}.$$

$$x^2 + 6x = 5$$
$$x^2 + 6x + 3^2 = 5 + 3^2$$
$$(x + 3)^2 = 14$$
$$x + 3 = \pm\sqrt{14}$$
$$x = -3 \pm \sqrt{14}$$

Quadratic Formula: If $ax^2 + bx + c = 0$, then

$$x = \frac{-b \pm \sqrt{b^2 - 4ac}}{2a}.$$

$$2x^2 + 3x - 1 = 0$$
$$x = \frac{-3 \pm \sqrt{3^2 - 4(2)(-1)}}{2(2)}$$
$$= \frac{-3 \pm \sqrt{17}}{4}$$

The methods used to solve quadratic equations can sometimes be extended to polynomial equations of higher degree, as shown in the next two examples.

Example 8   Solving a Polynomial Equation by Factoring

Solve $2x^3 - 6x^2 - 6x + 18 = 0$.

Solution

This equation has a common factor of 2. You can simplify the equation by first dividing each side of the equation by 2.

$$2x^3 - 6x^2 - 6x + 18 = 0 \qquad \text{Write original equation.}$$

$$x^3 - 3x^2 - 3x + 9 = 0 \qquad \text{Divide each side by 2.}$$

$$x^2(x - 3) - 3(x - 3) = 0 \qquad \text{Group terms.}$$

$$(x - 3)(x^2 - 3) = 0 \qquad \text{Factor by grouping.}$$

$$x - 3 = 0 \implies x = 3 \qquad \text{Set 1st factor equal to 0.}$$

$$x^2 - 3 = 0 \implies x = \pm\sqrt{3} \qquad \text{Set 2nd factor equal to 0.}$$

The equation has three solutions: $x = 3$, $x = \sqrt{3}$, and $x = -\sqrt{3}$. Check these solutions in the original equation. Figure B.48 verifies the solutions graphically.

✔**CHECKPOINT**   Now try Exercise 157.

> **STUDY TIP**
>
> Many cubic polynomial equations can be solved using factoring by grouping, as illustrated in Example 8.

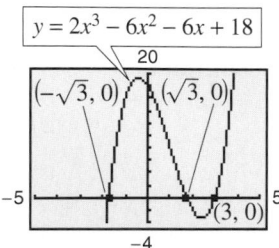

Figure B.48

Occasionally, mathematical models involve equations that are of **quadratic type.** In general, an equation is of quadratic type if it can be written in the form

$$au^2 + bu + c = 0$$

where $a \neq 0$ and u is an algebraic expression.

Example 9   Solving an Equation of Quadratic Type

Solve $x^4 - 3x^2 + 2 = 0$.

Solution

This equation is of quadratic type with $u = x^2$. To solve this equation, you can use the Quadratic Formula.

$$x^4 - 3x^2 + 2 = 0 \qquad \text{Write original equation.}$$

$$(x^2)^2 - 3(x^2) + 2 = 0 \qquad \text{Write in quadratic form.}$$

$$x^2 = \frac{-(-3) \pm \sqrt{(-3)^2 - 4(1)(2)}}{2(1)} \qquad \text{Quadratic Formula}$$

$$x^2 = \frac{3 \pm 1}{2} \qquad \text{Simplify.}$$

$$x^2 = 2 \implies x = \pm\sqrt{2} \qquad \text{Solutions}$$

$$x^2 = 1 \implies x = \pm 1 \qquad \text{Solutions}$$

The equation has four solutions: $x = -1$, $x = 1$, $x = \sqrt{2}$, and $x = -\sqrt{2}$. Check these solutions in the original equation. Figure B.49 verifies the solutions graphically.

✔**CHECKPOINT**   Now try Exercise 155.

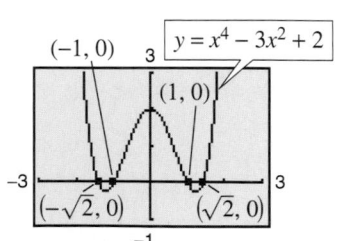

Figure B.49

Other Types of Equations

An equation involving a radical expression can often be cleared of radicals by raising each side of the equation to an appropriate power. When using this procedure, it is crucial to check for extraneous solutions because of the restricted domain of a radical equation.

Example 10   Solving an Equation Involving a Radical

Solve $\sqrt{2x + 7} - x = 2$.

Algebraic Solution

$$\sqrt{2x + 7} - x = 2 \qquad \text{Write original equation.}$$

$$\sqrt{2x + 7} = x + 2 \qquad \text{Isolate radical.}$$

$$2x + 7 = x^2 + 4x + 4 \qquad \text{Square each side.}$$

$$x^2 + 2x - 3 = 0 \qquad \text{Write in general form.}$$

$$(x + 3)(x - 1) = 0 \qquad \text{Factor.}$$

$$x + 3 = 0 \quad \Longrightarrow \quad x = -3 \qquad \text{Set 1st factor equal to 0.}$$

$$x - 1 = 0 \quad \Longrightarrow \quad x = 1 \qquad \text{Set 2nd factor equal to 0.}$$

By substituting into the original equation, you can determine that $x = -3$ is extraneous, whereas $x = 1$ is valid. So, the equation has only one real solution: $x = 1$.

✔CHECKPOINT   Now try Exercise 169.

Graphical Solution

First rewrite the equation as $\sqrt{2x + 7} - x - 2 = 0$. Then use a graphing utility to graph $y = \sqrt{2x + 7} - x - 2$, as shown in Figure B.50. Notice that the domain is $x \geq -\frac{7}{2}$ because the expression under the radical cannot be negative. There appears to be one solution near $x = 1$. Use the *zoom* and *trace* features, as shown in Figure B.51, to approximate the only solution to be $x = 1$.

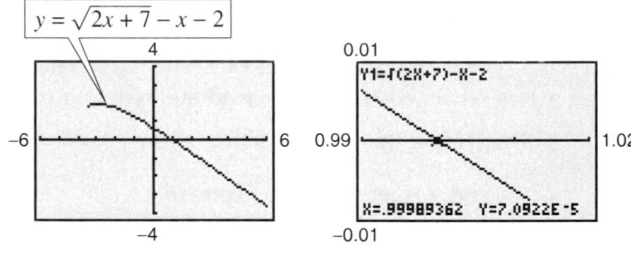

Figure B.50   Figure B.51

Example 11   Solving an Equation Involving Two Radicals

$$\sqrt{2x + 6} - \sqrt{x + 4} = 1 \qquad \text{Original equation}$$

$$\sqrt{2x + 6} = 1 + \sqrt{x + 4} \qquad \text{Isolate } \sqrt{2x + 6}.$$

$$2x + 6 = 1 + 2\sqrt{x + 4} + (x + 4) \qquad \text{Square each side.}$$

$$x + 1 = 2\sqrt{x + 4} \qquad \text{Isolate } 2\sqrt{x + 4}.$$

$$x^2 + 2x + 1 = 4(x + 4) \qquad \text{Square each side.}$$

$$x^2 - 2x - 15 = 0 \qquad \text{Write in general form.}$$

$$(x - 5)(x + 3) = 0 \qquad \text{Factor.}$$

$$x - 5 = 0 \quad \Longrightarrow \quad x = 5 \qquad \text{Set 1st factor equal to 0.}$$

$$x + 3 = 0 \quad \Longrightarrow \quad x = -3 \qquad \text{Set 2nd factor equal to 0.}$$

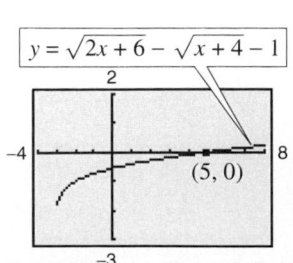

Figure B.52

By substituting into the original equation, you can determine that $x = -3$ is extraneous, whereas $x = 5$ is valid. Figure B.52 verifies that $x = 5$ is the only solution.

✔CHECKPOINT   Now try Exercise 173.

Example 12   Solving an Equation with Rational Exponents

Solve $(x + 1)^{2/3} = 4$.

Algebraic Solution

$(x + 1)^{2/3} = 4$	Write original equation.
$\sqrt[3]{(x + 1)^2} = 4$	Rewrite in radical form.
$(x + 1)^2 = 64$	Cube each side.
$x + 1 = \pm 8$	Take square root of each side.
$x = 7, x = -9$	Subtract 1 from each side.

Substitute $x = 7$ and $x = -9$ into the original equation to determine that both are valid solutions.

✔CHECKPOINT   Now try Exercise 175.

Graphical Solution

Use a graphing utility to graph $y_1 = \sqrt[3]{(x + 1)^2}$ and $y_2 = 4$ in the same viewing window. Use the *intersect* feature of the graphing utility to approximate the solutions to be $x = -9$ and $x = 7$, as shown in Figure B.53.

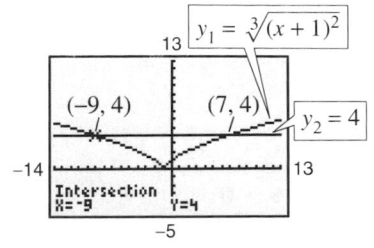

Figure B.53

As demonstrated in Example 1, you can solve an equation involving fractions algebraically by multiplying each side of the equation by the least common denominator of all terms in the equation to clear the equation of fractions.

Example 13   Solving an Equation Involving Fractions

Solve $\dfrac{2}{x} = \dfrac{3}{x - 2} - 1$.

Solution

For this equation, the least common denominator of the three terms is $x(x - 2)$, so you can begin by multiplying each term of the equation by this expression.

$\dfrac{2}{x} = \dfrac{3}{x - 2} - 1$	Write original equation.
$x(x - 2)\dfrac{2}{x} = x(x - 2)\dfrac{3}{x - 2} - x(x - 2)(1)$	Multiply each term by the LCD.
$2(x - 2) = 3x - x(x - 2), \quad x \neq 0, 2$	Simplify.
$x^2 - 3x - 4 = 0$	Write in general form.
$(x - 4)(x + 1) = 0$	Factor.
$x - 4 = 0 \implies x = 4$	Set 1st factor equal to 0.
$x + 1 = 0 \implies x = -1$	Set 2nd factor equal to 0.

> **TECHNOLOGY TIP**
>
> Graphs of functions involving variable denominators can be tricky because graphing utilities skip over points at which the denominator is zero. Graphs of such functions are introduced in Sections 2.6 and 2.7.

The equation has two solutions: $x = 4$ and $x = -1$. Check these solutions in the original equation. Use a graphing utility to verify these solutions graphically.

✔CHECKPOINT   Now try Exercise 179.

Example 14   Solving an Equation Involving Absolute Value

Solve $|x^2 - 3x| = -4x + 6$.

Solution

Begin by writing the equation as $|x^2 - 3x| + 4x - 6 = 0$. From the graph of $y = |x^2 - 3x| + 4x - 6$ in Figure B.54, you can estimate the solutions to be $x = -3$ and $x = 1$. These solutions can be verified by substitution into the equation. To solve an equation involving absolute value *algebraically*, you must consider the fact that the expression inside the absolute value symbols can be positive or negative. This results in *two* separate equations, each of which must be solved.

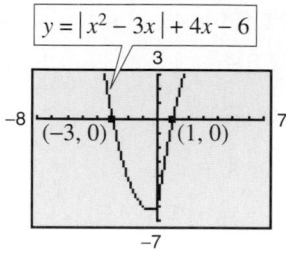

Figure B.54

First Equation:

$$x^2 - 3x = -4x + 6$$ Use positive expression.

$$x^2 + x - 6 = 0$$ Write in general form.

$$(x + 3)(x - 2) = 0$$ Factor.

$$x + 3 = 0 \implies x = -3$$ Set 1st factor equal to 0.

$$x - 2 = 0 \implies x = 2$$ Set 2nd factor equal to 0.

Second Equation:

$$-(x^2 - 3x) = -4x + 6$$ Use negative expression.

$$x^2 - 7x + 6 = 0$$ Write in general form.

$$(x - 1)(x - 6) = 0$$ Factor.

$$x - 1 = 0 \implies x = 1$$ Set 1st factor equal to 0.

$$x - 6 = 0 \implies x = 6$$ Set 2nd factor equal to 0.

Check

$$|(-3)^2 - 3(-3)| \overset{?}{=} -4(-3) + 6$$ Substitute -3 for x.

$$18 = 18$$ -3 checks. ✓

$$|2^2 - 3(2)| \overset{?}{=} -4(2) + 6$$ Substitute 2 for x.

$$2 \neq -2$$ 2 does not check.

$$|1^2 - 3(1)| \overset{?}{=} -4(1) + 6$$ Substitute 1 for x.

$$2 = 2$$ 1 checks. ✓

$$|6^2 - 3(6)| \overset{?}{=} -4(6) + 6$$ Substitute 6 for x.

$$18 \neq -18$$ 6 does not check.

The equation has only two solutions: $x = -3$ and $x = 1$, just as you obtained by graphing.

✓**CHECKPOINT**   Now try Exercise 185.

B.3   Exercises

See www.CalcChat.com for worked-out solutions to odd-numbered exercises.

Vocabulary Check

Fill in the blanks.

1. An _____ is a statement that equates two algebraic expressions.

2. To find all values that satisfy an equation is to _____ the equation.

3. When solving an equation, it is possible to introduce an _____ solution, which is a value that does not satisfy the original equation.

4. The points $(a, 0)$ and $(0, b)$ are called the _____ and _____ respectively, of the graph of an equation.

5. An ordered pair that is a solution of two different equations is called a _____ of the graphs of the two equations.

In Exercises 1–6, determine whether each value of x is a solution of the equation.

$$\qquad Equation \qquad\qquad\qquad Values$$

1. $\dfrac{5}{2x} - \dfrac{4}{x} = 3$

    (a) $x = -\frac{1}{2}$ (b) $x = 4$

    (c) $x = 0$ (d) $x = \frac{1}{4}$

2. $\dfrac{x}{2} + \dfrac{6x}{7} = \dfrac{19}{14}$

    (a) $x = -2$ (b) $x = 1$

    (c) $x = \frac{1}{2}$ (d) $x = 7$

3. $3 + \dfrac{1}{x + 2} = 4$

    (a) $x = -1$ (b) $x = -2$

    (c) $x = 0$ (d) $x = 5$

4. $\dfrac{(x + 5)(x - 3)}{2} = 24$

    (a) $x = -3$ (b) $x = -2$

    (c) $x = 7$ (d) $x = 9$

5. $\dfrac{\sqrt{x + 4}}{6} + 3 = 4$

    (a) $x = -3$ (b) $x = 0$

    (c) $x = 21$ (d) $x = 32$

6. $\dfrac{\sqrt[3]{x - 8}}{3} = -\dfrac{2}{3}$

    (a) $x = -16$ (b) $x = 0$

    (c) $x = 9$ (d) $x = 16$

In Exercises 7–12, determine whether the equation is an identity or a conditional equation.

7. $2(x - 1) = 2x - 2$

8. $-7(x - 3) + 4x = 3(7 - x)$

9. $x^2 - 8x + 5 = (x - 4)^2 - 11$

10. $x^2 + 2(3x - 2) = x^2 + 6x - 4$

11. $3 + \dfrac{1}{x + 1} = \dfrac{4x}{x + 1}$ 12. $\dfrac{5}{x} + \dfrac{3}{x} = 24$

In Exercises 13–16, solve the equation using two methods. Then decide which method is easier and explain why.

13. $\dfrac{3x}{8} - \dfrac{4x}{3} = 4$ 14. $\dfrac{3z}{8} - \dfrac{z}{10} = 6$

15. $\dfrac{2x}{5} + 5x = \dfrac{4}{3}$ 16. $\dfrac{4y}{3} - 2y = \dfrac{16}{5}$

In Exercises 17–40, solve the equation (if possible). Use a graphing utility to verify your solution.

17. $3x - 5 = 2x + 7$ 18. $5x + 3 = 6 - 2x$

19. $4y + 2 - 5y = 7 - 6y$ 20. $5y + 1 = 8y - 5 + 6y$

21. $3(y - 5) = 3 + 5y$ 22. $5(z - 4) + 4z = 5 - 6z$

23. $\dfrac{x}{5} - \dfrac{x}{2} = 3$ 24. $\dfrac{5x}{4} + \dfrac{1}{2} = x - \dfrac{1}{2}$

25. $\dfrac{3}{2}(z + 5) - \dfrac{1}{4}(z + 24) = 0$ 26. $\dfrac{3x}{2} + \dfrac{1}{4}(x - 2) = 10$

27. $\dfrac{2(z - 4)}{5} + 5 = 10z$ 28. $\dfrac{5}{3} + 2(y + 1) = \dfrac{10}{3}$

29. $\dfrac{100 - 4u}{3} = \dfrac{5u + 6}{4} + 6$ 30. $\dfrac{17 + y}{y} + \dfrac{32 + y}{y} = 100$

31. $\dfrac{5x - 4}{5x + 4} = \dfrac{2}{3}$ 32. $\dfrac{10x + 3}{5x + 6} = \dfrac{1}{2}$

33. $\dfrac{1}{x - 3} + \dfrac{1}{x + 3} = \dfrac{10}{x^2 - 9}$

34. $\dfrac{1}{x - 2} + \dfrac{3}{x + 3} = \dfrac{4}{x^2 + x - 6}$

35. $\dfrac{7}{2x + 1} - \dfrac{8x}{2x - 1} = -4$

36. $\dfrac{x}{x + 4} + \dfrac{4}{x + 4} + 2 = 0$

37. $\dfrac{1}{x} + \dfrac{2}{x-5} = 0$

38. $3 = 2 + \dfrac{2}{z+2}$

39. $\dfrac{3}{x^2-3x} + \dfrac{4}{x} = \dfrac{1}{x-3}$

40. $\dfrac{6}{x} - \dfrac{2}{x+3} = \dfrac{3(x+5)}{x(x+3)}$

In Exercises 41–52, use a graphing utility to find the x- and y-intercepts of the graph of the equation.

41. $y = x - 5$

42. $y = -\frac{3}{4}x - 3$

43. $y = x^2 + x - 2$

44. $y = 4 - x^2$

45. $y = x\sqrt{x+2}$

46. $y = -\frac{1}{2}x\sqrt{x+3} + 1$

47. $y = \dfrac{4}{x}$

48. $y = \dfrac{3x-1}{4x}$

49. $y = |x-2| - 4$

50. $y = 3 - \frac{1}{2}|x+1|$

51. $xy - 2y - x + 1 = 0$

52. $xy - x + 4y = 0$

Graphical Analysis **In Exercises 53–56, use a graphing utility to graph the equation and approximate any x-intercepts. Set $y = 0$ and solve the resulting equation. Compare the results with the x-intercepts of the graph.**

53. $y = 2(x-1) - 4$

54. $y = 4(x+3) - 2$

55. $y = 20 - (3x - 10)$

56. $y = 10 + 2(x - 2)$

In Exercises 57–62, the solution(s) of the equation are given. Verify the solution(s) both algebraically and graphically.

	Equation	*Solution(s)*
57.	$y = 5(4 - x)$	$x = 4$
58.	$y = 3(x - 5) + 9$	$x = 2$
59.	$y = x^3 - 6x^2 + 5x$	$x = 0, 5, 1$
60.	$y = x^3 - 9x^2 + 18x$	$x = 0, 3, 6$
61.	$y = \dfrac{x+2}{3} - \dfrac{x-1}{5} - 1$	$x = 1$
62.	$y = x - 3 - \dfrac{10}{x}$	$x = -2, 5$

In Exercises 63–86, solve the equation algebraically. Then verify your algebraic solution by writing the equation in the form $y = 0$ and using a graphing utility to graph the equation.

63. $2.7x - 0.4x = 1.2$

64. $3.5x - 8 = 0.5x$

65. $25(x - 3) = 12(x + 2) - 10$

66. $1200 = 300 + 2(x - 500)$

67. $\dfrac{3x}{2} + \dfrac{1}{4}(x - 2) = 10$

68. $\dfrac{2x}{3} + \dfrac{1}{2}(x - 5) = 6$

69. $0.60x + 0.40(100 - x) = 1.2$

70. $0.75x + 0.2(80 - x) = 20$

71. $\dfrac{2x}{3} = 10 - \dfrac{24}{x}$

72. $\dfrac{x-3}{25} = \dfrac{x-5}{12}$

73. $\dfrac{3}{x+2} - \dfrac{4}{x-2} = 5$

74. $\dfrac{6}{x} + \dfrac{8}{x+5} = 3$

75. $(x+2)^2 = x^2 - 6x + 1$

76. $(x+1)^2 + 2(x-2) = (x+1)(x-2)$

77. $2x^3 - x^2 - 18x + 9 = 0$

78. $4x^3 + 12x^2 - 8x - 24 = 0$

79. $x^4 = 2x^2 - 1$

80. $5 = 3x^{1/3} + 2x^{2/3}$

81. $\dfrac{2}{x+2} = 3$

82. $\dfrac{5}{x} = 1 + \dfrac{3}{x+2}$

83. $|x - 3| = 4$

84. $|x + 1| = 6$

85. $\sqrt{x-2} = 3$

86. $\sqrt{x-4} = 8$

In Exercises 87–92, determine any point(s) of intersection of the equations algebraically. Then use a graphing utility to verify your results.

87. $y = 2 - x$
      $y = 2x - 1$

88. $x - y = -4$
      $x + 2y = 5$

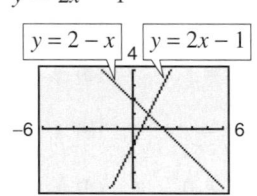
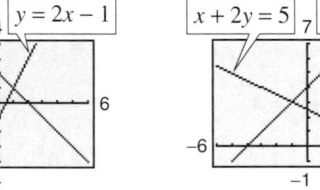

89. $x - y = -4$
      $x^2 - y = -2$

90. $3x + y = 2$
      $x^3 + y = 0$

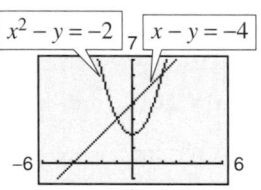

91. $y = x^2 - x + 1$
      $y = x^2 + 2x + 4$

92. $y = -x^2 + 3x + 1$
      $y = -x^2 - 2x - 4$

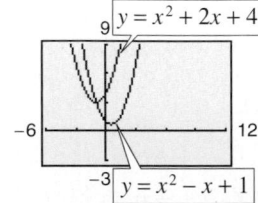
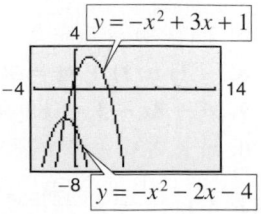

In Exercises 93–98, use a graphing utility to approximate any points of intersection (accurate to three decimal places) of the graphs of the equations. Verify your results algebraically.

93. $y = 9 - 2x$
$y = x - 3$

94. $y = \frac{1}{3}x + 2$
$y = \frac{5}{2}x - 11$

95. $y = 4 - x^2$
$y = 2x - 1$

96. $y = x^3 - 3$
$y = 5 - 2x$

97. $y = 2x^2$
$y = x^4 - 2x^2$

98. $y = -x$
$y = 2x - x^2$

In Exercises 99–108, solve the quadratic equation by factoring. Check your solutions in the original equation.

99. $6x^2 + 3x = 0$

100. $9x^2 - 1 = 0$

101. $x^2 - 2x - 8 = 0$

102. $x^2 - 10x + 9 = 0$

103. $3 + 5x - 2x^2 = 0$

104. $2x^2 = 19x + 33$

105. $x^2 + 4x = 12$

106. $-x^2 + 8x = 12$

107. $(x + a)^2 - b^2 = 0$

108. $x^2 + 2ax + a^2 = 0$

In Exercises 109–118, solve the equation by extracting square roots. List both the exact solutions and the decimal solutions rounded to two decimal places.

109. $x^2 = 49$

110. $x^2 = 144$

111. $(x - 12)^2 = 16$

112. $(x - 5)^2 = 25$

113. $(3x - 1)^2 + 6 = 0$

114. $(2x + 3)^2 + 25 = 0$

115. $(2x - 1)^2 = 12$

116. $(4x + 7)^2 = 44$

117. $(x - 7)^2 = (x + 3)^2$

118. $(x + 5)^2 = (x + 4)^2$

In Exercises 119–128, solve the quadratic equation by completing the square. Verify your answer graphically.

119. $x^2 + 4x - 32 = 0$

120. $x^2 - 2x - 3 = 0$

121. $x^2 + 6x + 2 = 0$

122. $x^2 + 8x + 14 = 0$

123. $9x^2 - 18x + 3 = 0$

124. $4x^2 - 4x - 99 = 0$

125. $-6 + 2x - x^2 = 0$

126. $-x^2 + x - 1 = 0$

127. $2x^2 + 5x - 8 = 0$

128. $9x^2 - 12x - 14 = 0$

In Exercises 129–138, use the Quadratic Formula to solve the equation. Use a graphing utility to verify your solutions graphically.

129. $2 + 2x - x^2 = 0$

130. $x^2 - 10x + 22 = 0$

131. $x^2 + 8x - 4 = 0$

132. $4x^2 - 4x - 4 = 0$

133. $x^2 + 3x = -8$

134. $x^2 + 16 = -5x$

135. $28x - 49x^2 = 4$

136. $9x^2 + 24x + 16 = 0$

137. $3x^2 + 16x + 17 = 0$

138. $9x^2 - 6x - 37 = 0$

In Exercises 139–148, solve the equation using any convenient method. Use a graphing utility to verify your solutions graphically.

139. $x^2 - 2x - 1 = 0$

140. $11x^2 + 33x = 0$

141. $(x + 3)^2 = 81$

142. $(x - 1)^2 = -1$

143. $x^2 - 14x + 49 = 0$

144. $x^2 - 2x + \frac{13}{4} = 0$

145. $x^2 - x - \frac{11}{4} = 0$

146. $x^2 + 3x - \frac{3}{4} = 0$

147. $(x + 1)^2 = x^2$

148. $a^2x^2 - b^2 = 0, \ a \neq 0$

In Exercises 149–166, find all solutions of the equation algebraically. Use a graphing utility to verify the solutions graphically.

149. $4x^4 - 16x^2 = 0$

150. $8x^4 - 18x^2 = 0$

151. $5x^3 + 30x^2 + 45x = 0$

152. $9x^4 - 24x^3 + 16x^2 = 0$

153. $4x^4 - 18x^2 = 0$

154. $20x^3 - 125x = 0$

155. $x^4 - 4x^2 + 3 = 0$

156. $x^4 + 5x^2 - 36 = 0$

157. $x^3 - 3x^2 - x + 3 = 0$

158. $x^4 + 2x^3 - 8x - 16 = 0$

159. $4x^4 - 65x^2 + 16 = 0$

160. $36t^4 + 29t^2 - 7 = 0$

161. $\frac{1}{t^2} + \frac{8}{t} + 15 = 0$

162. $6 - \frac{1}{x} - \frac{1}{x^2} = 0$

163. $6\left(\frac{s}{s+1}\right)^2 + 5\left(\frac{s}{s+1}\right) - 6 = 0$

164. $8\left(\frac{t}{t-1}\right)^2 - 2\left(\frac{t}{t-1}\right) - 3 = 0$

165. $2x + 9\sqrt{x} - 5 = 0$

166. $6x - 7\sqrt{x} - 3 = 0$

In Exercises 167–186, find all solutions of the equation algebraically. Check your solutions both algebraically and graphically.

167. $\sqrt{x - 10} - 4 = 0$

168. $\sqrt{2x + 5} + 3 = 0$

169. $\sqrt{x + 1} - 3x = 1$

170. $\sqrt{x + 5} - 2x = 3$

171. $\sqrt[3]{2x + 1} + 8 = 0$

172. $\sqrt[3]{4x - 3} + 2 = 0$

173. $\sqrt{x} - \sqrt{x - 5} = 1$

174. $\sqrt{x} + \sqrt{x - 20} = 10$

175. $(x - 5)^{2/3} = 16$

176. $(x^2 - x - 22)^{4/3} = 16$

177. $3x(x - 1)^{1/2} + 2(x - 1)^{3/2} = 0$

178. $4x^2(x - 1)^{1/3} + 6x(x - 1)^{4/3} = 0$

179. $\frac{1}{x} - \frac{1}{x + 1} = 3$

180. $\frac{x}{x^2 - 4} + \frac{1}{x + 2} = 3$

181. $x = \frac{3}{x} + \frac{1}{2}$

182. $4x + 1 = \frac{3}{x}$

183. $|2x - 1| = 5$

184. $|3x + 2| = 7$

185. $|x| = x^2 + x - 3$

186. $|x - 10| = x^2 - 10x$

Graphical Analysis **In Exercises 187–194, (a) use a graphing utility to graph the equation, (b) use the graph to approximate any x-intercepts of the graph, (c) set y = 0 and solve the resulting equation, and (d) compare the result of part (c) with the x-intercepts of the graph.**

187. $y = x^3 - 2x^2 - 3x$

188. $y = x^4 - 10x^2 + 9$

189. $y = \sqrt{11x - 30} - x$

190. $y = 2x - \sqrt{15 - 4x}$

191. $y = \dfrac{1}{x} - \dfrac{4}{x - 1} - 1$

192. $y = x + \dfrac{9}{x + 1} - 5$

193. $y = |x + 1| - 2$

194. $y = |x - 2| - 3$

195. *State Populations* The populations (in thousands) of South Carolina S, and Arizona A, from 1980 to 2004 can be modeled by

$$S = 45.2t + 3087, \quad 0 \le t \le 24$$

$$A = 128.2t + 2533, \quad 0 \le t \le 24$$

where t represents the year, with $t = 0$ corresponding to 1980. (Source: U.S. Census Bureau)

(a) Use a graphing utility to graph each model in the same viewing window over the appropriate domain. Approximate the point of intersection. Round your result to one decimal place. Explain the meaning of the coordinates of the point.

(b) Find the point of intersection algebraically. Round your result to one decimal place. What does the point of intersection represent?

(c) Explain the meaning of the slopes of both models and what it tells you about the population growth rates.

(d) Use the models to estimate the population of each state in 2010. Do the values seem reasonable? Explain.

196. *Medical Costs* The average retail prescription prices P (in dollars) from 1997 through 2004 can be approximated by the model

$$P = 0.1220t^2 + 1.529t + 18.72, \quad 7 \le t \le 14$$

where t represents the year, with $t = 7$ corresponding to 1997. (Source: National Association of Chain Drug Stores)

(a) Determine algebraically when the average retail price was $40 and $50.

(b) Verify your answer to part (a) by creating a table of values for the model.

(c) Use a graphing utility to graph the model.

(d) According to the model, when will the average retail price reach $75?

(e) Do you believe the model could be used to predict the average retail prices for years beyond 2004? Explain your reasoning.

197. *Biology* The metabolic rate of an ectothermic organism increases with increasing temperature within a certain range. Experimental data for oxygen consumption C (in microliters per gram per hour) of a beetle at certain temperatures yielded the model

$$C = 0.45x^2 - 1.65x + 50.75, \quad 10 \le x \le 25$$

where x is the air temperature in degrees Celsius.

(a) Use a graphing utility to graph the consumption model over the specified domain.

(b) Use the graph to approximate the air temperature resulting in oxygen consumption of 150 microliters per gram per hour.

(c) The temperature is increased from 10°C to 20°C. The oxygen consumption is increased by approximately what factor?

198. *Saturated Steam* The temperature T (in degrees Fahrenheit) of saturated steam increases as pressure increases. This relationship is approximated by

$$T = 75.82 - 2.11x + 43.51\sqrt{x}, \quad 5 \le x \le 40$$

where x is the absolute pressure in pounds per square inch.

(a) Use a graphing utility to graph the model over the specified domain.

(b) The temperature of steam at sea level ($x = 14.696$) is 212°F. Evaluate the model at this pressure and verify the result graphically.

(c) Use the model to approximate the pressure for a steam temperature of 240°F.

Synthesis

True or False? **In Exercises 199 and 200, determine whether the statement is true or false. Justify your answer.**

199. Two linear equations can have either one point of intersection or no points of intersection.

200. An equation can never have more than one extraneous solution.

201. *Think About It* Find c such that $x = 3$ is a solution to the linear equation $2x - 5c = 10 + 3c - 3x$.

202. *Think About It* Find c such that $x = 2$ is a solution to the linear equation $5x + 2c = 12 + 4x - 2c$.

203. *Exploration* Given that a and b are nonzero real numbers, determine the solutions of the equations.

(a) $ax^2 + bx = 0$ (b) $ax^2 - ax = 0$

B.4 Solving Inequalities Algebraically and Graphically

Properties of Inequalities

The inequality symbols $<$, $\leq$, $>$, and $\geq$ were used to compare two numbers and to denote subsets of real numbers. For instance, the simple inequality $x \geq 3$ denotes all real numbers x that are greater than or equal to 3.

In this section, you will study inequalities that contain more involved statements such as

$$5x - 7 > 3x + 9 \qquad \text{and} \qquad -3 \leq 6x - 1 < 3.$$

As with an equation, you **solve an inequality** in the variable x by finding all values of x for which the inequality is true. These values are **solutions** of the inequality and are said to **satisfy** the inequality. For instance, the number 9 is a solution of the first inequality listed above because

$$5(9) - 7 > 3(9) + 9$$

$$38 > 36.$$

On the other hand, the number 7 is not a solution because

$$5(7) - 7 \not> 3(7) + 9$$

$$28 \not> 30.$$

The set of all real numbers that are solutions of an inequality is the **solution set** of the inequality.

The set of all points on the real number line that represent the solution set is the **graph of the inequality.** Graphs of many types of inequalities consist of intervals on the real number line.

The procedures for solving linear inequalities in one variable are much like those for solving linear equations. To isolate the variable, you can make use of the **properties of inequalities.** These properties are similar to the properties of equality, but there are two important exceptions. When each side of an inequality is multiplied or divided by a negative number, *the direction of the inequality symbol must be reversed* in order to maintain a true statement. Here is an example.

$$-2 < 5 \qquad \text{Original inequality}$$

$$(-3)(-2) > (-3)(5) \qquad \text{Multiply each side by } -3 \text{ and reverse the inequality.}$$

$$6 > -15 \qquad \text{Simplify.}$$

Two inequalities that have the same solution set are **equivalent inequalities.** For instance, the inequalities

$$x + 2 < 5 \qquad \text{and} \qquad x < 3$$

are equivalent. To obtain the second inequality from the first, you can subtract 2 from each side of the inequality. The properties listed at the top of the next page describe operations that can be used to create equivalent inequalities.

What you should learn

- Use properties of inequalities to solve linear inequalities.
- Solve inequalities involving absolute values.
- Solve polynomial inequalities.
- Solve rational inequalities.
- Use inequalities to model and solve real-life problems.

Why you should learn it

An inequality can be used to determine when a real-life quantity exceeds a given level. For instance, Exercises 85–88 on page A74 show how to use linear inequalities to determine when the number of hours per person spent playing video games exceeded the number of hours per person spent reading newspapers.

Prerequisite Skills

To review techniques for solving linear inequalities, see Appendix E.

Properties of Inequalities

Let a, b, c, and d be real numbers.

1. *Transitive Property*

$$a < b \text{ and } b < c \implies a < c$$

2. *Addition of Inequalities*

$$a < b \text{ and } c < d \implies a + c < b + d$$

3. *Addition of a Constant*

$$a < b \implies a + c < b + c$$

4. *Multiplying by a Constant*

For $c > 0$, $a < b \implies ac < bc$

For $c < 0$, $a < b \implies ac > bc$

Exploration

Use a graphing utility to graph $f(x) = 5x - 7$ and $g(x) = 3x + 9$ in the same viewing window. (Use $-1 \le x \le 15$ and $-5 \le y \le 50$.) For which values of x does the graph of f lie above the graph of g? Explain how the answer to this question can be used to solve the inequality in Example 1.

Each of the properties above is true if the symbol $<$ is replaced by $\le$ and $>$ is replaced by $\ge$. For instance, another form of Property 3 is as follows.

$$a \le b \implies a + c \le b + c$$

Solving a Linear Inequality

The simplest type of inequality to solve is a **linear inequality** in one variable, such as $2x + 3 > 4$. (See Appendix E for help with solving one-step linear inequalities.)

Example 1   Solving a Linear Inequality

Solve $5x - 7 > 3x + 9$.

Solution

$5x - 7 > 3x + 9$	Write original inequality.
$2x - 7 > 9$	Subtract $3x$ from each side.
$2x > 16$	Add 7 to each side.
$x > 8$	Divide each side by 2.

So, the solution set is all real numbers that are greater than 8. The interval notation for this solution set is $(8, \infty)$. The number line graph of this solution set is shown in Figure B.55. Note that a parenthesis at 8 on the number line indicates that 8 *is not* part of the solution set.

✔CHECKPOINT   Now try Exercise 13.

Note that the four inequalities forming the solution steps of Example 1 are all *equivalent* in the sense that each has the same solution set.

STUDY TIP

Checking the solution set of an inequality is not as simple as checking the solution(s) of an equation because there are simply too many x-values to substitute into the original inequality. However, you can get an indication of the validity of the solution set by substituting a few convenient values of x. For instance, in Example 1, try substituting $x = 5$ and $x = 10$ into the original inequality.

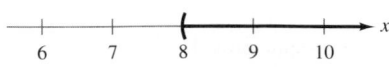

Figure B.55   *Solution Interval:* $(8, \infty)$

Example 2 Solving an Inequality

Solve $1 - \frac{3}{2}x \geq x - 4$.

Algebraic Solution

$1 - \frac{3}{2}x \geq x - 4$	Write original inequality.
$2 - 3x \geq 2x - 8$	Multiply each side by the LCD.
$2 - 5x \geq -8$	Subtract $2x$ from each side.
$-5x \geq -10$	Subtract 2 from each side.
$x \leq 2$	Divide each side by -5 and reverse the inequality.

The solution set is all real numbers that are less than or equal to 2. The interval notation for this solution set is $(-\infty, 2]$. The number line graph of this solution set is shown in Figure B.56. Note that a bracket at 2 on the number line indicates that 2 *is* part of the solution set.

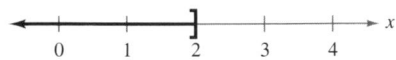

Figure B.56 *Solution Interval:* $(-\infty, 2]$

✔CHECKPOINT   Now try Exercise 15.

Graphical Solution

Use a graphing utility to graph $y_1 = 1 - \frac{3}{2}x$ and $y_2 = x - 4$ in the same viewing window. In Figure B.57, you can see that the graphs appear to intersect at the point $(2, -2)$. Use the *intersect* feature of the graphing utility to confirm this. The graph of y_1 lies above the graph of y_2 to the left of their point of intersection, which implies that $y_1 \geq y_2$ for all $x \leq 2$.

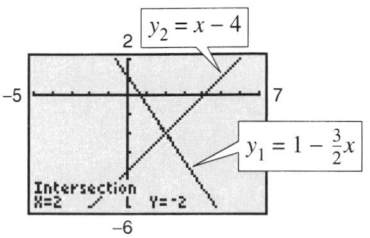

Figure B.57

Sometimes it is possible to write two inequalities as a **double inequality,** as demonstrated in Example 3.

Example 3 Solving a Double Inequality

Solve $-3 \leq 6x - 1$ and $6x - 1 < 3$.

Algebraic Solution

$-3 \leq 6x - 1 < 3$	Write as a double inequality.
$-2 \leq 6x < 4$	Add 1 to each part.
$-\frac{1}{3} \leq x < \frac{2}{3}$	Divide by 6 and simplify.

The solution set is all real numbers that are greater than or equal to $-\frac{1}{3}$ *and* less than $\frac{2}{3}$. The interval notation for this solution set is $\left[-\frac{1}{3}, \frac{2}{3}\right)$. The number line graph of this solution set is shown in Figure B.58.

Figure B.58 *Solution Interval:* $\left[-\frac{1}{3}, \frac{2}{3}\right)$

✔CHECKPOINT   Now try Exercise 17.

Graphical Solution

Use a graphing utility to graph $y_1 = 6x - 1$, $y_2 = -3$, and $y_3 = 3$ in the same viewing window. In Figure B.59, you can see that the graphs appear to intersect at the points $\left(-\frac{1}{3}, -3\right)$ and $\left(\frac{2}{3}, 3\right)$. Use the *intersect* feature of the graphing utility to confirm this. The graph of y_1 lies above the graph of y_2 to the right of $\left(-\frac{1}{3}, -3\right)$ *and* the graph of y_1 lies below the graph of y_3 to the left of $\left(\frac{2}{3}, 3\right)$. This implies that $y_2 \leq y_1 < y_3$ when $-\frac{1}{3} \leq x < \frac{2}{3}$.

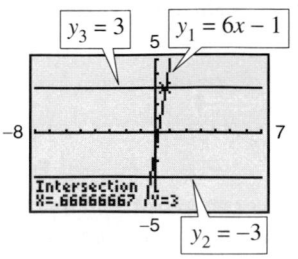

Figure B.59

Inequalities Involving Absolute Values

> **Solving an Absolute Value Inequality**
>
> Let x be a variable or an algebraic expression and let a be a real number such that $a \geq 0$.
>
> **1.** The solutions of $|x| < a$ are all values of x that lie between $-a$ and a.
>
> $$|x| < a \quad \text{if and only if} \quad -a < x < a. \qquad \text{Double inequality}$$
>
> **2.** The solutions of $|x| > a$ are all values of x that are less than $-a$ or greater than a.
>
> $$|x| > a \quad \text{if and only if} \quad x < -a \quad \text{or} \quad x > a. \qquad \text{Compound inequality}$$
>
> These rules are also valid if $<$ is replaced by $\leq$ and $>$ is replaced by $\geq$.

Example 4   Solving Absolute Value Inequalities

Solve each inequality.

a. $|x - 5| < 2$ **b.** $|x - 5| > 2$

Algebraic Solution

a.
$$|x - 5| < 2 \qquad \text{Write original inequality.}$$
$$-2 < x - 5 < 2 \qquad \text{Write double inequality.}$$
$$3 < x < 7 \qquad \text{Add 5 to each part.}$$

The solution set is all real numbers that are greater than 3 *and* less than 7. The interval notation for this solution set is $(3, 7)$. The number line graph of this solution set is shown in Figure B.60.

b. The absolute value inequality $|x - 5| > 2$ is equivalent to the following compound inequality: $x - 5 < -2$ *or* $x - 5 > 2$.

Solve first inequality: $x - 5 < -2$ Write first inequality.

$\qquad\qquad\qquad\qquad x < 3$ Add 5 to each side.

Solve second inequality: $x - 5 > 2$ Write second inequality.

$\qquad\qquad\qquad\qquad x > 7$ Add 5 to each side.

The solution set is all real numbers that are less than 3 *or* greater than 7. The interval notation for this solution set is $(-\infty, 3) \cup (7, \infty)$. The symbol $\cup$ is called a *union* symbol and is used to denote the combining of two sets. The number line graph of this solution set is shown in Figure B.61.

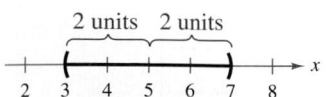

Figure B.60

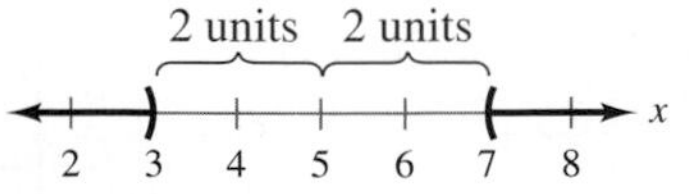

Figure B.61

✓CHECKPOINT   Now try Exercise 31.

Graphical Solution

a. Use a graphing utility to graph $y_1 = |x - 5|$ and $y_2 = 2$ in the same viewing window. In Figure B.62, you can see that the graphs appear to intersect at the points $(3, 2)$ and $(7, 2)$. Use the *intersect* feature of the graphing utility to confirm this. The graph of y_1 lies below the graph of y_2 when $3 < x < 7$. So, you can approximate the solution set to be all real numbers greater than 3 *and* less than 7.

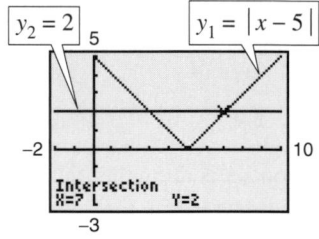

Figure B.62

b. In Figure B.62, you can see that the graph of y_1 lies above the graph of y_2 when $x < 3$ *or* when $x > 7$. So, you can approximate the solution set to be all real numbers that are less than 3 *or* greater than 7.

Polynomial Inequalities

To solve a polynomial inequality such as $x^2 - 2x - 3 < 0$, use the fact that a polynomial can change signs only at its zeros (the x-values that make the polynomial equal to zero). Between two consecutive zeros, a polynomial must be entirely positive or entirely negative. This means that when the real zeros of a polynomial are put in order, they divide the real number line into intervals in which the polynomial has no sign changes. These zeros are the **critical numbers** of the inequality, and the resulting open intervals are the **test intervals** for the inequality. For instance, the polynomial above factors as

$$x^2 - 2x - 3 = (x + 1)(x - 3)$$

and has two zeros, $x = -1$ and $x = 3$, which divide the real number line into three test intervals: $(-\infty, -1)$, $(-1, 3)$, and $(3, \infty)$. To solve the inequality $x^2 - 2x - 3 < 0$, you need to test only one value in each test interval.

> **Finding Test Intervals for a Polynomial**
>
> To determine the intervals on which the values of a polynomial are entirely negative or entirely positive, use the following steps.
>
> 1. Find all real zeros of the polynomial, and arrange the zeros in increasing order. The zeros of a polynomial are its critical numbers.
>
> 2. Use the critical numbers to determine the test intervals.
>
> 3. Choose one representative x-value in each test interval and evaluate the polynomial at that value. If the value of the polynomial is negative, the polynomial will have negative values for *every* x-value in the interval. If the value of the polynomial is positive, the polynomial will have positive values for *every* x-value in the interval.

TECHNOLOGY TIP

Some graphing utilities will produce graphs of inequalities. For instance, you can graph $2x^2 + 5x > 12$ by setting the graphing utility to *dot* mode and entering $y = 2x^2 + 5x > 12$. Using the settings $-10 \le x \le 10$ and $-4 \le y \le 4$, your graph should look like the graph shown below. Solve the problem algebraically to verify that the solution is $(-\infty, -4) \cup \left(\frac{3}{2}, \infty\right)$.

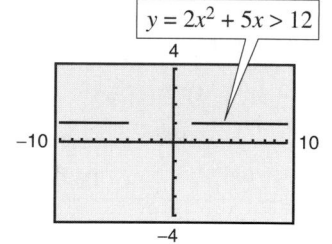

Example 5   Investigating Polynomial Behavior

To determine the intervals on which $x^2 - 3$ is entirely negative and those on which it is entirely positive, factor the quadratic as $x^2 - 3 = \left(x + \sqrt{3}\right)\left(x - \sqrt{3}\right)$. The critical numbers occur at $x = -\sqrt{3}$ and $x = \sqrt{3}$. So, the test intervals for the quadratic are $\left(-\infty, -\sqrt{3}\right)$, $\left(-\sqrt{3}, \sqrt{3}\right)$, and $\left(\sqrt{3}, \infty\right)$. In each test interval, choose a representative x-value and evaluate the polynomial, as shown in the table.

Interval	x-Value	Value of Polynomial	Sign of Polynomial
$\left(-\infty, -\sqrt{3}\right)$	$x = -3$	$(-3)^2 - 3 = 6$	Positive
$\left(-\sqrt{3}, \sqrt{3}\right)$	$x = 0$	$(0)^2 - 3 = -3$	Negative
$\left(\sqrt{3}, \infty\right)$	$x = 5$	$(5)^2 - 3 = 22$	Positive

The polynomial has negative values for every x in the interval $\left(-\sqrt{3}, \sqrt{3}\right)$ and positive values for every x in the intervals $\left(-\infty, -\sqrt{3}\right)$ and $\left(\sqrt{3}, \infty\right)$. This result is shown graphically in Figure B.63.

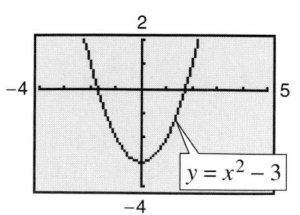

Figure B.63

✓CHECKPOINT   Now try Exercise 49.

To determine the test intervals for a polynomial inequality, the inequality must first be written in general form with the polynomial on one side.

Example 6 Solving a Polynomial Inequality

Solve $2x^2 + 5x > 12$.

Algebraic Solution

$$2x^2 + 5x - 12 > 0 \qquad \text{Write inequality in general form.}$$

$$(x + 4)(2x - 3) > 0 \qquad \text{Factor.}$$

Critical Numbers: $x = -4$, $x = \frac{3}{2}$

Test Intervals: $(-\infty, -4), \left(-4, \frac{3}{2}\right), \left(\frac{3}{2}, \infty\right)$

Test: Is $(x + 4)(2x - 3) > 0$?

After testing these intervals, you can see that the polynomial $2x^2 + 5x - 12$ is positive on the open intervals $(-\infty, -4)$ and $\left(\frac{3}{2}, \infty\right)$. Therefore, the solution set of the inequality is

$$(-\infty, -4) \cup \left(\frac{3}{2}, \infty\right).$$

✔CHECKPOINT   Now try Exercise 55.

Graphical Solution

First write the polynomial inequality $2x^2 + 5x > 12$ as $2x^2 + 5x - 12 > 0$. Then use a graphing utility to graph $y = 2x^2 + 5x - 12$. In Figure B.64, you can see that the graph is *above* the x-axis when x is less than -4 *or* when x is greater than $\frac{3}{2}$. So, you can graphically approximate the solution set to be $(-\infty, -4) \cup \left(\frac{3}{2}, \infty\right)$.

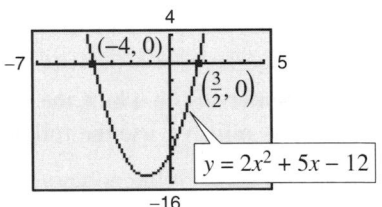

Figure B.64

Example 7 Solving a Polynomial Inequality

Solve $2x^3 - 3x^2 - 32x > -48$.

Solution

$$2x^3 - 3x^2 - 32x + 48 > 0 \qquad \text{Write inequality in general form.}$$

$$x^2(2x - 3) - 16(2x - 3) > 0 \qquad \text{Factor by grouping.}$$

$$(x^2 - 16)(2x - 3) > 0 \qquad \text{Distributive Property}$$

$$(x - 4)(x + 4)(2x - 3) > 0 \qquad \text{Factor difference of two squares.}$$

The critical numbers are $x = -4$, $x = \frac{3}{2}$, and $x = 4$; and the test intervals are $(-\infty, -4), \left(-4, \frac{3}{2}\right), \left(\frac{3}{2}, 4\right)$, and $(4, \infty)$.

Interval	x-Value	Polynomial Value	Conclusion
$(-\infty, -4)$	$x = -5$	$2(-5)^3 - 3(-5)^2 - 32(-5) + 48 = -117$	Negative
$\left(-4, \frac{3}{2}\right)$	$x = 0$	$2(0)^3 - 3(0)^2 - 32(0) + 48 = 48$	Positive
$\left(\frac{3}{2}, 4\right)$	$x = 2$	$2(2)^3 - 3(2)^2 - 32(2) + 48 = -12$	Negative
$(4, \infty)$	$x = 5$	$2(5)^3 - 3(5)^2 - 32(5) + 48 = 63$	Positive

From this you can conclude that the polynomial is positive on the open intervals $\left(-4, \frac{3}{2}\right)$ and $(4, \infty)$. So, the solution set is $\left(-4, \frac{3}{2}\right) \cup (4, \infty)$.

✔CHECKPOINT   Now try Exercise 61.

STUDY TIP

When solving a quadratic inequality, be sure you have accounted for the particular type of inequality symbol given in the inequality. For instance, in Example 7, note that the original inequality contained a "greater than" symbol and the solution consisted of two open intervals. If the original inequality had been

$$2x^3 - 3x^2 - 32x \geq -48$$

the solution would have consisted of the closed interval $\left[-4, \frac{3}{2}\right]$ and the interval $[4, \infty)$.

Example 8 Unusual Solution Sets

a. The solution set of

$$x^2 + 2x + 4 > 0$$

consists of the entire set of real numbers, $(-\infty, \infty)$. In other words, the value of the quadratic $x^2 + 2x + 4$ is positive for every real value of x, as indicated in Figure B.65(a). (Note that this quadratic inequality has *no* critical numbers. In such a case, there is only one test interval—the entire real number line.)

b. The solution set of

$$x^2 + 2x + 1 \le 0$$

consists of the single real number $\{-1\}$, because the quadratic $x^2 + 2x + 1$ has one critical number, $x = -1$, and it is the only value that satisfies the inequality, as indicated in Figure B.65(b).

c. The solution set of

$$x^2 + 3x + 5 < 0$$

is empty. In other words, the quadratic $x^2 + 3x + 5$ is not less than zero for any value of x, as indicated in Figure B.65(c).

d. The solution set of

$$x^2 - 4x + 4 > 0$$

consists of all real numbers *except* the number 2. In interval notation, this solution set can be written as $(-\infty, 2) \cup (2, \infty)$. The graph of $x^2 - 4x + 4$ lies above the x-axis except at $x = 2$, where it touches it, as indicated in Figure B.65(d).

TECHNOLOGY TIP

One of the advantages of technology is that you can solve complicated polynomial inequalities that might be difficult, or even impossible, to factor. For instance, you could use a graphing utility to approximate the solution to the inequality

$$x^3 - 0.2x^2 - 3.16x + 1.4 < 0.$$

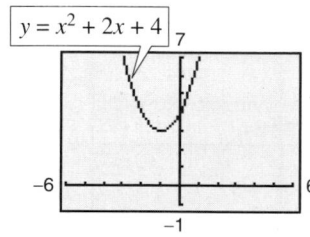

(a)

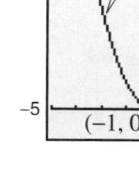

(b)

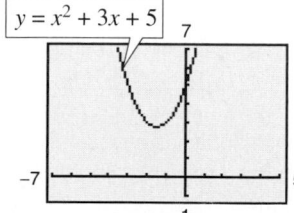

(c)

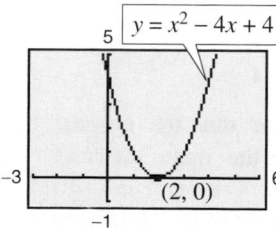

(d)

Figure B.65

✓CHECKPOINT   Now try Exercise 59.

Rational Inequalities

The concepts of critical numbers and test intervals can be extended to inequalities involving rational expressions. To do this, use the fact that the value of a rational expression can change sign only at its *zeros* (the x-values for which its numerator is zero) and its *undefined values* (the x-values for which its denominator is zero). These two types of numbers make up the *critical numbers* of a rational inequality.

Example 9 Solving a Rational Inequality

Solve $\dfrac{2x - 7}{x - 5} \le 3$.

Algebraic Solution

$$\frac{2x - 7}{x - 5} \le 3 \qquad \text{Write original inequality.}$$

$$\frac{2x - 7}{x - 5} - 3 \le 0 \qquad \text{Write in general form.}$$

$$\frac{2x - 7 - 3x + 15}{x - 5} \le 0 \qquad \text{Write as single fraction.}$$

$$\frac{-x + 8}{x - 5} \le 0 \qquad \text{Simplify.}$$

Now, in standard form you can see that the critical numbers are $x = 5$ and $x = 8$, and you can proceed as follows.

Critical Numbers: $x = 5, x = 8$
Test Intervals: $(-\infty, 5), (5, 8), (8, \infty)$
Test: Is $\dfrac{-x + 8}{x - 5} \le 0$?

Interval	x-Value	Polynomial Value	Conclusion
$(-\infty, 5)$	$x = 0$	$\dfrac{-0 + 8}{0 - 5} = -\dfrac{8}{5}$	Negative
$(5, 8)$	$x = 6$	$\dfrac{-6 + 8}{6 - 5} = 2$	Positive
$(8, \infty)$	$x = 9$	$\dfrac{-9 + 8}{9 - 5} = -\dfrac{1}{4}$	Negative

By testing these intervals, you can determine that the rational expression $(-x + 8)/(x - 5)$ is negative in the open intervals $(-\infty, 5)$ and $(8, \infty)$. Moreover, because $(-x + 8)/(x - 5) = 0$ when $x = 8$, you can conclude that the solution set of the inequality is $(-\infty, 5) \cup [8, \infty)$.

✔**CHECKPOINT**   Now try Exercise 69.

Graphical Solution

Use a graphing utility to graph

$$y_1 = \frac{2x - 7}{x - 5} \quad \text{and} \quad y_2 = 3$$

in the same viewing window. In Figure B.66, you can see that the graphs appear to intersect at the point $(8, 3)$. Use the *intersect* feature of the graphing utility to confirm this. The graph of y_1 lies below the graph of y_2 in the intervals $(-\infty, 5)$ and $[8, \infty)$. So, you can graphically approximate the solution set to be all real numbers less than 5 *or* greater than or equal to 8.

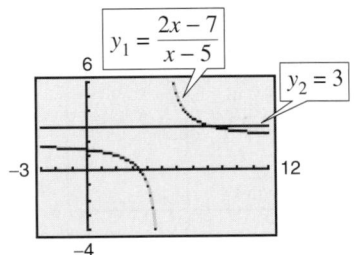

Figure B.66

Note in Example 9 that $x = 5$ is not included in the solution set because the inequality is undefined when $x = 5$.

Application

The *implied domain* of a function is the set of all *x*-values for which the function is defined. A common type of implied domain is used to avoid even roots of negative numbers, as shown in Example 10.

Example 10   Finding the Domain of an Expression

Find the domain of $\sqrt{64 - 4x^2}$.

Solution

Because $\sqrt{64 - 4x^2}$ is defined only if $64 - 4x^2$ is nonnegative, the domain is given by $64 - 4x^2 \geq 0$.

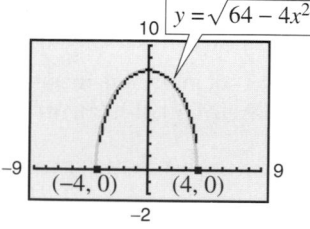

$64 - 4x^2 \geq 0$	Write in general form.
$16 - x^2 \geq 0$	Divide each side by 4.
$(4 - x)(4 + x) \geq 0$	Factor.

The inequality has two critical numbers: $x = -4$ and $x = 4$. A test shows that $64 - 4x^2 \geq 0$ in the *closed interval* $[-4, 4]$. The graph of $y = \sqrt{64 - 4x^2}$, shown in Figure B.67, confirms that the domain is $[-4, 4]$.

Figure B.67

✔CHECKPOINT   Now try Exercise 77.

Example 11   Height of a Projectile

A projectile is fired straight upward from ground level with an initial velocity of 384 feet per second. During what time period will its height exceed 2000 feet?

Solution

The position of an object moving vertically can be modeled by the *position equation*

$$s = -16t^2 + v_0 t + s_0$$

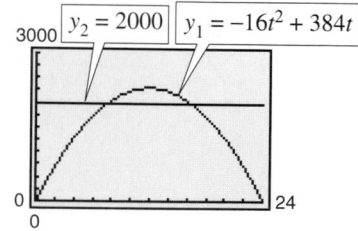

where *s* is the height in feet and *t* is the time in seconds. In this case, $s_0 = 0$ and $v_0 = 384$. So, you need to solve the inequality $-16t^2 + 384t > 2000$. Using a graphing utility, graph $y_1 = -16t^2 + 384t$ and $y_2 = 2000$, as shown in Figure B.68. From the graph, you can determine that $-16t^2 + 384t > 2000$ for *t* between approximately 7.6 and 16.4. You can verify this result algebraically.

Figure B.68

$-16t^2 + 384t > 2000$	Write original inequality.
$t^2 - 24t < -125$	Divide by -16 and reverse inequality.
$t^2 - 24t + 125 < 0$	Write in general form.

By the Quadratic Formula the critical numbers are $t = 12 - \sqrt{19}$ and $t = 12 + \sqrt{19}$, or approximately 7.64 and 16.36. A test will verify that the height of the projectile will exceed 2000 feet when $7.64 < t < 16.36$; that is, during the time interval $(7.64, 16.36)$ seconds.

✔CHECKPOINT   Now try Exercise 81.

B.4 Exercises

See www.CalcChat.com for worked-out solutions to odd-numbered exercises.

Vocabulary Check

Fill in the blanks.

1. To solve a linear inequality in one variable, you can use the properties of inequalities, which are identical to those used to solve an equation, with the exception of multiplying or dividing each side by a _____ constant.

2. It is sometimes possible to write two inequalities as one inequality, called a _____ inequality.

3. The solutions to $|x| \le a$ are those values of x such that _____ .

4. The solutions to $|x| \ge a$ are those values of x such that _____ or _____ .

5. The critical numbers of a rational expression are its _____ and its _____ .

In Exercises 1–6, match the inequality with its graph. [The graphs are labeled (a), (b), (c), (d), (e), and (f).]

(a)

(b)

(c)

(d)

(e)

(f)

1. $x < 3$
2. $x \ge 5$
3. $-3 < x \le 4$
4. $0 \le x \le \frac{9}{2}$
5. $-1 \le x \le \frac{5}{2}$
6. $-1 < x < \frac{5}{2}$

In Exercises 7–10, determine whether each value of x is a solution of the inequality.

Inequality		Values	
7. $5x - 12 > 0$	(a) $x = 3$	(b) $x = -3$	
	(c) $x = \frac{5}{2}$	(d) $x = \frac{3}{2}$	
8. $-5 < 2x - 1 \le 1$	(a) $x = -\frac{1}{2}$	(b) $x = -\frac{5}{2}$	
	(c) $x = \frac{4}{3}$	(d) $x = 0$	
9. $-1 < \dfrac{3-x}{2} \le 1$	(a) $x = 0$	(b) $x = \sqrt{5}$	
	(c) $x = 1$	(d) $x = 5$	
10. $|x - 10| \ge 3$	(a) $x = 13$	(b) $x = -1$	
	(c) $x = 14$	(d) $x = 9$	

In Exercises 11–20, solve the inequality and sketch the solution on the real number line. Use a graphing utility to verify your solution graphically.

11. $-10x < 40$

12. $6x > 15$

13. $4(x + 1) < 2x + 3$

14. $2x + 7 < 3(x - 4)$

15. $\frac{3}{4}x - 6 \le x - 7$

16. $3 + \frac{2}{7}x > x - 2$

17. $-8 \le 1 - 3(x - 2) < 13$

18. $0 \le 2 - 3(x + 1) < 20$

19. $-4 < \dfrac{2x - 3}{3} < 4$

20. $0 \le \dfrac{x + 3}{2} < 5$

Graphical Analysis **In Exercises 21–24, use a graphing utility to approximate the solution.**

21. $5 - 2x \ge 1$

22. $20 < 6x - 1$

23. $3(x + 1) < x + 7$

24. $4(x - 3) \le 8 - x$

In Exercises 25–28, use a graphing utility to graph the equation and graphically approximate the values of x that satisfy the specified inequalities. Then solve each inequality algebraically.

Equation	Inequalities	
25. $y = 2x - 3$	(a) $y \ge 1$	(b) $y \le 0$
26. $y = -3x + 8$	(a) $-1 \le y \le 3$	(b) $y \le 0$
27. $y = -\frac{1}{2}x + 2$	(a) $0 \le y \le 3$	(b) $y \ge 0$
28. $y = \frac{2}{3}x + 1$	(a) $y \le 5$	(b) $y \ge 0$

In Exercises 29–36, solve the inequality and sketch the solution on the real number line. Use a graphing utility to verify your solutions graphically.

29. $|5x| > 10$ **30.** $\left|\frac{x}{2}\right| \leq 1$

31. $|x - 7| < 6$ **32.** $|x - 20| \geq 4$

33. $|x + 14| + 3 > 17$ **34.** $\left|\frac{x - 3}{2}\right| \geq 5$

35. $10|1 - 2x| < 5$ **36.** $3|4 - 5x| \leq 9$

In Exercises 37 and 38, use a graphing utility to graph the equation and graphically approximate the values of x that satisfy the specified inequalities. Then solve each inequality algebraically.

Equation	*Inequalities*	
37. $y = \lvert x - 3\rvert$	(a) $y \leq 2$	(b) $y \geq 4$
38. $y = \lvert \frac{1}{2}x + 1\rvert$	(a) $y \leq 4$	(b) $y \geq 1$

In Exercises 39–46, use absolute value notation to define the interval (or pair of intervals) on the real number line.

39.

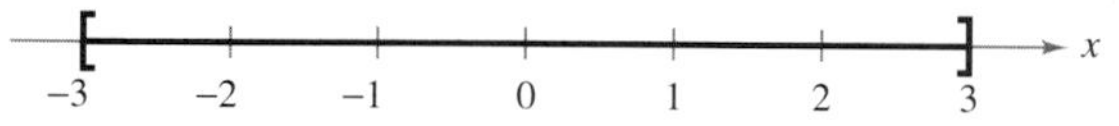

40.

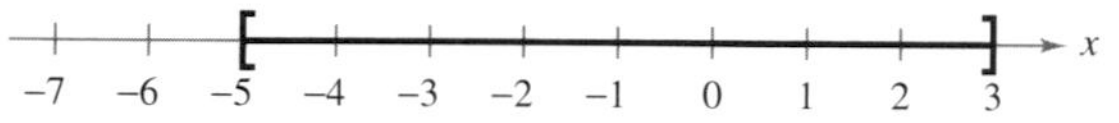

41.

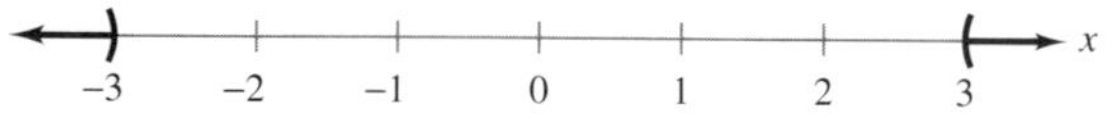

42.

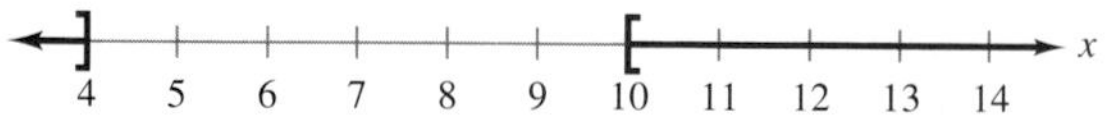

43. All real numbers within 10 units of 7

44. All real numbers no more than 8 units from -5

45. All real numbers at least 5 units from 3

46. All real numbers more than 3 units from -1

In Exercises 47–52, determine the intervals on which the polynomial is entirely negative and those on which it is entirely positive.

47. $x^2 - 4x - 5$ **48.** $x^2 - 3x - 4$

49. $2x^2 - 4x - 3$ **50.** $2x^2 - x - 5$

51. $x^2 - 4x + 5$ **52.** $-x^2 + 6x - 10$

In Exercises 53–62, solve the inequality and graph the solution on the real number line. Use a graphing utility to verify your solution graphically.

53. $(x + 2)^2 < 25$ **54.** $(x - 3)^2 \geq 1$

55. $x^2 + 4x + 4 \geq 9$ **56.** $x^2 - 6x + 9 < 16$

57. $x^3 - 4x \geq 0$ **58.** $x^4(x - 3) \leq 0$

59. $3x^2 - 11x + 16 \leq 0$ **60.** $4x^2 + 12x + 9 \leq 0$

61. $2x^3 + 5x^2 > 6x + 9$ **62.** $2x^3 + 3x^2 < 11x + 6$

In Exercises 63 and 64, use the graph of the function to solve the equation or inequality.

(a) $f(x) = g(x)$ **(b)** $f(x) \geq g(x)$ **(c)** $f(x) > g(x)$

63.

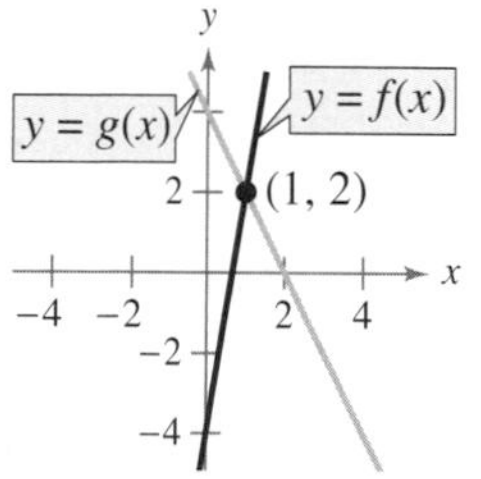

64.

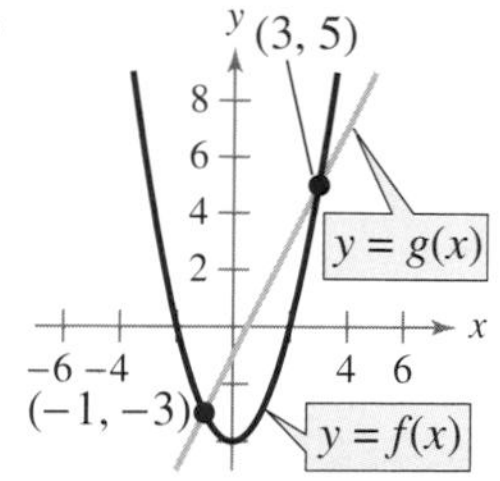

In Exercises 65 and 66, use a graphing utility to graph the equation and graphically approximate the values of x that satisfy the specified inequalities. Then solve each inequality algebraically.

Equation	*Inequalities*	
65. $y = -x^2 + 2x + 3$	(a) $y \leq 0$	(b) $y \geq 3$
66. $y = x^3 - x^2 - 16x + 16$	(a) $y \leq 0$	(b) $y \geq 36$

In Exercises 67–70, solve the inequality and graph the solution on the real number line. Use a graphing utility to verify your solution graphically.

67. $\frac{1}{x} - x > 0$ **68.** $\frac{1}{x} - 4 < 0$

69. $\frac{x + 6}{x + 1} - 2 < 0$ **70.** $\frac{x + 12}{x + 2} - 3 \geq 0$

In Exercises 71 and 72, use a graphing utility to graph the equation and graphically approximate the values of x that satisfy the specified inequalities. Then solve each inequality algebraically.

Equation	*Inequalities*	
71. $y = \frac{3x}{x - 2}$	(a) $y \leq 0$	(b) $y \geq 6$
72. $y = \frac{5x}{x^2 + 4}$	(a) $y \geq 1$	(b) $y \leq 0$

In Exercises 73–78, find the domain of x in the expression.

73. $\sqrt{x - 5}$ **74.** $\sqrt[4]{6x + 15}$

75. $\sqrt[3]{6 - x}$ **76.** $\sqrt[3]{2x^2 - 8}$

77. $\sqrt{x^2 - 4}$ **78.** $\sqrt[4]{4 - x^2}$

79. *Population* The graph models the population P (in thousands) of Las Vegas, Nevada from 1990 to 2004, where t is the year, with $t = 0$ corresponding to 1990. Also shown is the line $y = 1000$. Use the graphs of the model and the horizontal line to write an equation or an inequality that could be solved to answer the question. Then answer the question.   (Source: U.S. Census Bureau)

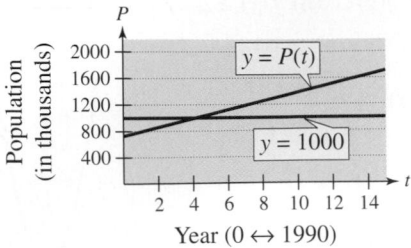

Year ($0 \leftrightarrow 1990$)

(a) In what year does the population of Las Vegas reach one million?

(b) Over what time period is the population of Las Vegas less than one million? greater than one million?

80. *Population* The graph models the population P (in thousands) of Pittsburgh, Pennsylvania from 1993 to 2004, where t is the year, with $t = 3$ corresponding to 1993. Also shown is the line $y = 2450$. Use the graphs of the model and the horizontal line to write an equation or an inequality that could be solved to answer the question. Then answer the question.   (Source: U.S. Census Bureau)

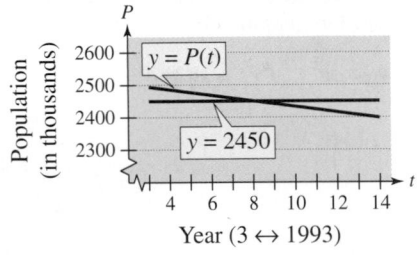

Year ($3 \leftrightarrow 1993$)

(a) In what year did the population of Pittsburgh equal 2.45 million?

(b) Over what time period is the population of Pittsburgh less than 2.45 million? greater than 2.45 million?

81. *Height of a Projectile* A projectile is fired straight upward from ground level with an initial velocity of 160 feet per second.

(a) At what instant will it be back at ground level?

(b) When will the height exceed 384 feet?

82. *Height of a Projectile* A projectile is fired straight upward from ground level with an initial velocity of 128 feet per second.

(a) At what instant will it be back at ground level?

(b) When will the height be less than 128 feet?

83. *Education*. The numbers D of doctorate degrees (in thousands) awarded to female students from 1990 to 2003 in the United States can be approximated by the model $D = -0.0165t^2 + 0.755t + 14.06, 0 \leq t \leq 13$, where t is the year, with $t = 0$ corresponding to 1990.   (Source: U.S. National Center for Education Statistics)

(a) Use a graphing utility to graph the model.

(b) Use the *zoom* and *trace* features to find when the number of degrees was between 15 and 20 thousand.

(c) Algebraically verify your results from part (b).

(d) According to the model, will the number of degrees exceed 30 thousand? If so, when? If not, explain.

84. *Data Analysis* You want to determine whether there is a relationship between an athlete's weight x (in pounds) and the athlete's maximum bench-press weight y (in pounds). Sample data from 12 athletes is shown below.

(165, 170), (184, 185), (150, 200), (210, 255), (196, 205), (240, 295), (202, 190), (170, 175), (185, 195), (190, 185), (230, 250), (160, 150)

(a) Use a graphing utility to plot the data.

(b) A model for this data is $y = 1.3x - 36$. Use a graphing utility to graph the equation in the same viewing window used in part (a).

(c) Use the graph to estimate the value of x that predict a maximum bench-press weight of at least 200 pounds.

(d) Use the graph to write a statement about the accuracy of the model. If you think the graph indicates that an athlete's weight is not a good indicator of the athlete's maximum bench-press weight, list other factors that might influence an individual's maximum bench-press weight.

Leisure Time In Exercises 85–88, use the models below which approximate the annual numbers of hours per person spent reading daily newspapers N and playing video games V for the years 2000 to 2005, where t is the year, with $t = 0$ corresponding to 2000.   (Source: Veronis Suhler Stevenson)

Daily Newspapers: $N = -2.51t + 179.6, \quad 0 \leq t \leq 5$

Video Games: $V = 3.37t + 57.9, \quad 0 \leq t \leq 5$

85. Solve the inequality $V(t) \geq 65$. Explain what the solution of the inequality represents.

86. Solve the inequality $N(t) \leq 175$. Explain what the solution of the inequality represents.

87. Solve the equation $V(t) = N(t)$. Explain what the solution of the equation represents.

88. Solve the inequality $V(t) > N(t)$. Explain what the solution of the inequality represents.

Music   In Exercises 89–92, use the following information. Michael Kasha of Florida State University used physics and mathematics to design a new classical guitar. He used the model for the frequency of the vibrations on a circular plate

$$v = \frac{2.6t}{d^2} \sqrt{\frac{E}{\rho}}$$

where v is the frequency (in vibrations per second), t is the plate thickness (in millimeters), d is the diameter of the plate, E is the elasticity of the plate material, and ρ is the density of the plate material. For fixed values of d, E, and ρ, the graph of the equation is a line, as shown in the figure.

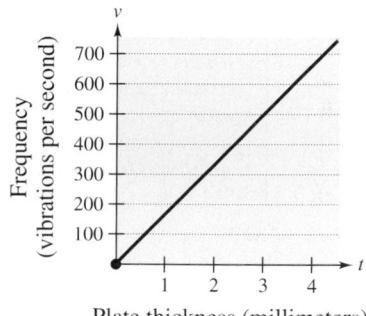

Plate thickness (millimeters)

89. Estimate the frequency when the plate thickness is 2 millimeters.

90. Estimate the plate thickness when the frequency is 600 vibrations per second.

91. Approximate the interval for the plate thickness when the frequency is between 200 and 400 vibrations per second.

92. Approximate the interval for the frequency when the plate thickness is less than 3 millimeters.

In Exercises 93 and 94, (a) write equations that represent each option, (b) use a graphing utility to graph the options in the same viewing window, (c) determine when each option is the better choice, and (d) explain which option you would choose.

93. *Cellular Phones*   You are trying to decide between two different cellular telephone contracts, option A and option B. Option A has a monthly fee of $12 plus $0.15 per minute. Option B has no monthly fee but charges $0.20 per minute. All other monthly charges are identical.

94. *Moving*   You are moving from your home to your dorm room, and the moving company has offered you two options. The charges for gasoline, insurance, and all other incidental fees are equal.

Option A: $200 plus $18 per hour to move all of your belongings from your home to your dorm room.

Option B: $24 per hour to move all of your belongings from your home to your dorm room.

Synthesis

True or False?   In Exercises 95 and 96, determine whether the statement is true or false. Justify your answer.

95. If $-10 \le x \le 8$, then $-10 \ge -x$ and $-x \ge -8$.

96. The solution set of the inequality $\frac{3}{2}x^2 + 3x + 6 \ge 0$ is the entire set of real numbers.

In Exercises 97 and 98, consider the polynomial $(x - a)(x - b)$ and the real number line (see figure).

97. Identify the points on the line where the polynomial is zero.

98. In each of the three subintervals of the line, write the sign of each factor and the sign of the product. For which x-values does the polynomial possibly change signs?

99. *Proof*   The arithmetic mean of a and b is given by $(a + b)/2$. Order the statements of the proof to show that if $a < b$, then $a < (a + b)/2 < b$.

   i. $a < \dfrac{a + b}{2} < b$

   ii. $2a < 2b$

   iii. $2a < a + b < 2b$

   iv. $a < b$

100. *Proof*   The geometric mean of a and b is given by $\sqrt{ab}$. Order the statements of the proof to show that if $0 < a < b$, then $a < \sqrt{ab} < b$.

   i. $a^2 < ab < b^2$

   ii. $0 < a < b$

   iii. $a < \sqrt{ab} < b$

B.5 Representing Data Graphically

Line Plots

Statistics is the branch of mathematics that studies techniques for collecting, organizing, and interpreting data. In this section, you will study several ways to organize data. The first is a **line plot,** which uses a portion of a real number line to order numbers. Line plots are especially useful for ordering small sets of numbers (about 50 or less) by hand.

Many statistical measures can be obtained from a line plot. Two such measures are the *frequency* and *range* of the data. The **frequency** measures the number of times a value occurs in a data set. The **range** is the difference between the greatest and smallest data values. For example, consider the data values

20, 21, 21, 25, 32.

The frequency of 21 in the data set is 2 because 21 occurs twice. The range is 12 because the difference between the greatest and smallest data values is $32 - 20 = 12$.

Example 1   Constructing a Line Plot

Use a line plot to organize the following test scores. Which score occurs with the greatest frequency? What is the range of scores?

93, 70, 76, 67, 86, 93, 82, 78, 83, 86, 64, 78, 76, 66, 83
83, 96, 74, 69, 76, 64, 74, 79, 76, 88, 76, 81, 82, 74, 70

Solution

Begin by scanning the data to find the smallest and largest numbers. For the data, the smallest number is 64 and the largest is 96. Next, draw a portion of a real number line that includes the interval [64, 96]. To create the line plot, start with the first number, 93, and enter an $\times$ above 93 on the number line. Continue recording $\times$'s for each number in the list until you obtain the line plot shown in Figure B.69. From the line plot, you can see that 76 occurs with the greatest frequency. Because the range is the difference between the greatest and smallest data values, the range of scores is $96 - 64 = 32$.

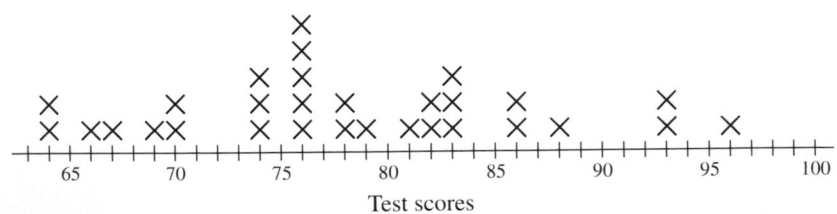

Test scores

Figure B.69

✓CHECKPOINT   Now try Exercise 1.

Histograms and Frequency Distributions

When you want to organize large sets of data, it is useful to group the data into intervals and plot the frequency of the data in each interval. A **frequency distribution** can be used to construct a **histogram.** A histogram uses a portion of a real number line as its horizontal axis. The bars of a histogram are not separated by spaces.

Example 2   Constructing a Histogram

The table at the right shows the percent of the resident population of each state and the District of Columbia that was at least 65 years old in 2004. Construct a frequency distribution and a histogram for the data.   (Source: U.S. Census Bureau)

Solution

To begin constructing a frequency distribution, you must first decide on the number of intervals. There are several ways to group the data. However, because the smallest number is 6.4 and the largest is 16.8, it seems that six intervals would be appropriate. The first would be the interval $[6, 8)$, the second would be $[8, 10)$, and so on. By tallying the data into the six intervals, you obtain the frequency distribution shown below. You can construct the histogram by drawing a vertical axis to represent the number of states and a horizontal axis to represent the percent of the population 65 and older. Then, for each interval, draw a vertical bar whose height is the total tally, as shown in Figure B.70.

Interval	Tally
$[6, 8)$	|
$[8, 10)$	||||
$[10, 12)$	⩵⩵ ||
$[12, 14)$	⩵⩵ ⩵⩵ ⩵⩵ ⩵⩵ ⩵⩵ ⩵⩵ ||
$[14, 16)$	⩵⩵ |
$[16, 18)$	|

AK	6.4	MT	13.7
AL	13.2	NC	12.1
AR	13.8	ND	14.7
AZ	12.7	NE	13.3
CA	10.7	NH	12.1
CO	9.8	NJ	12.9
CT	13.5	NM	12.1
DC	12.1	NV	11.2
DE	13.1	NY	13.0
FL	16.8	OH	13.3
GA	9.6	OK	13.2
HI	13.6	OR	12.8
IA	14.7	PA	15.3
ID	11.4	RI	13.9
IL	12.0	SC	12.4
IN	12.4	SD	14.2
KS	13.0	TN	12.5
KY	12.5	TX	9.9
LA	11.7	UT	8.7
MA	13.3	VA	11.4
MD	11.4	VT	13.0
ME	14.4	WA	11.3
MI	12.3	WI	13.0
MN	12.1	WV	15.3
MO	13.3	WY	12.1
MS	12.2		

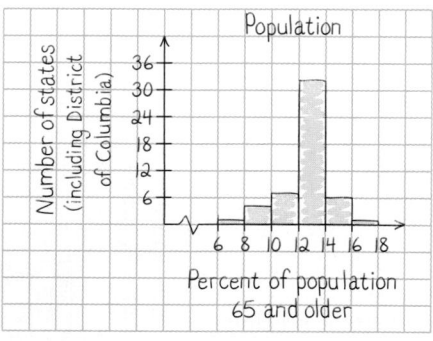

Figure B.70

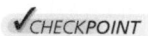 CHECKPOINT   Now try Exercise 5.

Example 3　Constructing a Histogram

A company has 48 sales representatives who sold the following numbers of units during the first quarter of 2008. Construct a frequency distribution for the data.

107	162	184	170	177	102	145	141
105	193	167	149	195	127	193	191
150	153	164	167	171	163	141	129
109	171	150	138	100	164	147	153
171	163	118	142	107	144	100	132
153	107	124	162	192	134	187	177

Interval	Tally
100–109	﷼ |||
110–119		
120–129	|||
130–139	|||
140–149	﷼ ||
150–159	﷼
160–169	﷼ |||
170–179	﷼	
180–189	||
190–199	﷼

Solution

To begin constructing a frequency distribution, you must first decide on the number of intervals. There are several ways to group the data. However, because the smallest number is 100 and the largest is 195, it seems that 10 intervals would be appropriate. The first interval would be 100–109, the second would be 110–119, and so on. By tallying the data into the 10 intervals, you obtain the distribution shown at the right above. A histogram for the distribution is shown in Figure B.71.

✓CHECKPOINT　Now try Exercise 6.

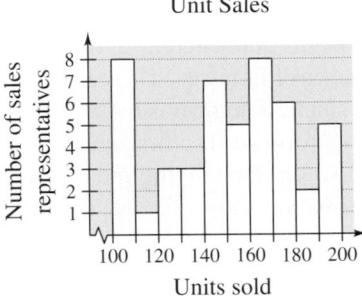

Figure B.71

Bar Graphs

A **bar graph** is similar to a histogram, except that the bars can be either horizontal or vertical and the labels of the bars are not necessarily numbers. Another difference between a bar graph and a histogram is that the bars in a bar graph are usually separated by spaces.

Example 4　Constructing a Bar Graph

The data below show the monthly normal precipitation (in inches) in Houston, Texas. Construct a bar graph for the data. What can you conclude?　(Source: National Climatic Data Center)

January	3.7	February	3.0	March	3.4
April	3.6	May	5.2	June	5.4
July	3.2	August	3.8	September	4.3
October	4.5	November	4.2	December	3.7

Solution

To create a bar graph, begin by drawing a vertical axis to represent the precipitation and a horizontal axis to represent the month. The bar graph is shown in Figure B.72. From the graph, you can see that Houston receives a fairly consistent amount of rain throughout the year—the driest month tends to be February and the wettest month tends to be June.

✓CHECKPOINT　Now try Exercise 7.

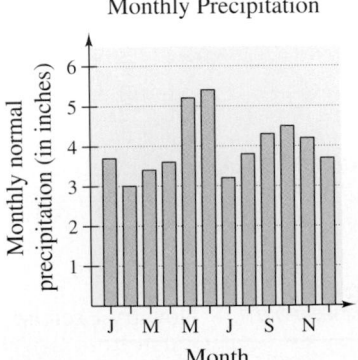

Figure B.72

Example 5 Constructing a Double Bar Graph

The table shows the percents of associate degrees awarded to males and females for selected fields of study in the United States in 2003. Construct a double bar graph for the data.   (Source: U.S. National Center for Education Statistics)

Field of Study	% Female	% Male
Agriculture and Natural Resources	36.4	63.6
Biological Sciences/ Life Sciences	70.4	29.6
Business and Management	66.8	33.2
Education	80.5	19.5
Engineering	16.5	83.5
Law and Legal Studies	89.6	10.4
Liberal/General Sciences	63.1	36.9
Mathematics	36.5	63.5
Physical Sciences	44.7	55.3
Social Sciences	65.3	34.7

Solution

For the data, a horizontal bar graph seems to be appropriate. This makes it easier to label and read the bars. Such a graph is shown in Figure B.73.

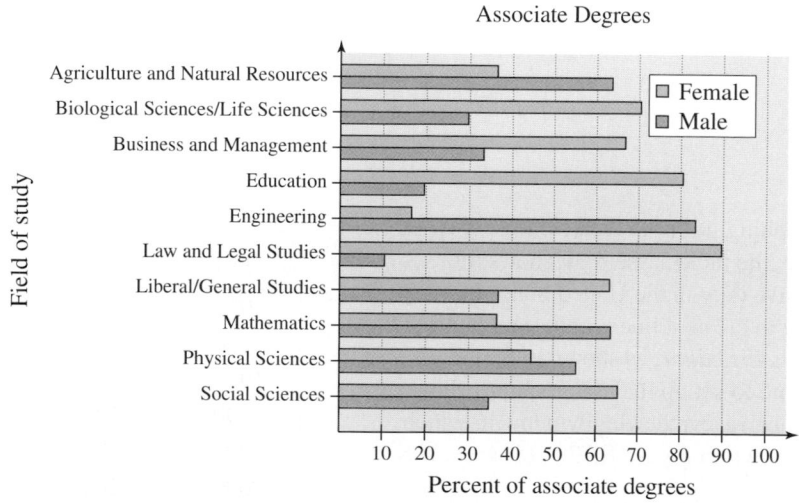

Figure B.73

✓CHECKPOINT   Now try Exercise 11.

Line Graphs

A **line graph** is similar to a standard coordinate graph. Line graphs are usually used to show trends over periods of time.

Example 6   Constructing a Line Graph

The table at the right shows the number of immigrants (in thousands) entering the United States for each decade from 1901 to 2000. Construct a line graph for the data. What can you conclude?   (Source: U.S. Immigration and Naturalization Service)

Decade	Number
1901–1910	8795
1911–1920	5736
1921–1930	4107
1931–1940	528
1941–1950	1035
1951–1960	2515
1961–1970	3322
1971–1980	4493
1981–1990	7338
1991–2000	9095

Solution

Begin by drawing a vertical axis to represent the number of immigrants in thousands. Then label the horizontal axis with decades and plot the points shown in the table. Finally, connect the points with line segments, as shown in Figure B.74. From the line graph, you can see that the number of immigrants hit a low point during the depression of the 1930s. Since then the number has steadily increased.

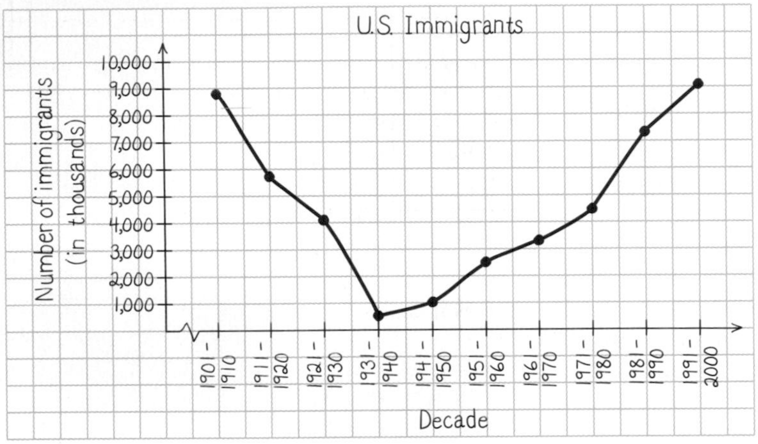

Figure B.74

✓CHECKPOINT   Now try Exercise 17.

TECHNOLOGY TIP   You can use a graphing utility to create different types of graphs, such as line graphs. For instance, the table at the right shows the numbers N (in thousands) of women on active duty in the United States military for selected years. To use a graphing utility to create a line graph of the data, first enter the data into the graphing utility's *list editor*, as shown in Figure B.75. Then use the *statistical plotting* feature to set up the line graph, as shown in Figure B.76. Finally, display the line graph (use a viewing window in which $1970 \le x \le 2010$ and $0 \le y \le 250$), as shown in Figure B.77.   (Source: U.S. Department of Defense)

Year	Number
1975	97
1980	171
1985	212
1990	227
1995	196
2000	203
2005	203

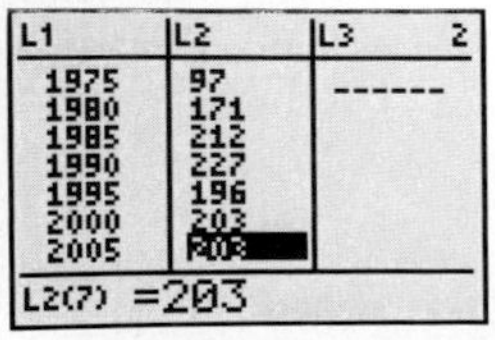

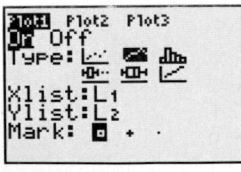

Figure B.75   **Figure B.76**

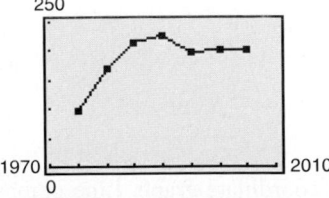

Figure B.77

B.5   Exercises

See www.CalcChat.com for worked-out solutions to odd-numbered exercises.

Vocabulary Check

Fill in the blanks.

1. _____ is the branch of mathematics that studies techniques for collecting, organizing, and interpreting data.

2. _____ are useful for ordering small sets of numbers by hand.

3. A _____ uses a portion of a real number line as its horizontal axis, and the bars are not separated by spaces.

4. You use a _____ to construct a histogram.

5. The bars in a _____ can be either vertical or horizontal.

6. _____ show trends over periods of time.

---

1. **Consumer Awareness** The line plot shows a sample of prices of unleaded regular gasoline in 25 different cities.

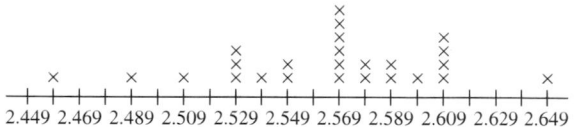

(a) What price occurred with the greatest frequency?

(b) What is the range of prices?

2. **Agriculture** The line plot shows the weights (to the nearest hundred pounds) of 30 head of cattle sold by a rancher.

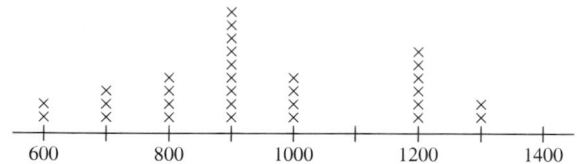

(a) What weight occurred with the greatest frequency?

(b) What is the range of weights?

Quiz and Exam Scores In Exercises 3 and 4, use the following scores from an algebra class of 30 students. The scores are for one 25-point quiz and one 100-point exam.

Quiz 20, 15, 14, 20, 16, 19, 10, 21, 24, 15, 15, 14, 15, 21, 19, 15, 20, 18, 18, 22, 18, 16, 18, 19, 21, 19, 16, 20, 14, 12

Exam 77, 100, 77, 70, 83, 89, 87, 85, 81, 84, 81, 78, 89, 78, 88, 85, 90, 92, 75, 81, 85, 100, 98, 81, 78, 75, 85, 89, 82, 75

3. Construct a line plot for the quiz. Which score(s) occurred with the greatest frequency?

4. Construct a line plot for the exam. Which score(s) occurred with the greatest frequency?

5. **Agriculture** The list shows the numbers of farms (in thousands) in the 50 states in 2004. Use a frequency distribution and a histogram to organize the data. (Source: U.S. Department of Agriculture)

AK 1	AL 44	AR 48	AZ 10
CA 77	CO 31	CT 4	DE 2
FL 43	GA 49	HI 6	IA 90
ID 25	IL 73	IN 59	KS 65
KY 85	LA 27	MA 6	MD 12
ME 7	MI 53	MN 80	MO 106
MS 42	MT 28	NC 52	ND 30
NE 48	NH 3	NJ 10	NM 18
NV 3	NY 36	OH 77	OK 84
OR 40	PA 58	RI 1	SC 24
SD 32	TN 85	TX 229	UT 15
VA 48	VT 6	WA 35	WI 77
WV 21	WY 9		

6. **Schools** The list shows the numbers of public high school graduates (in thousands) in the 50 states and the District of Columbia in 2004. Use a frequency distribution and a histogram to organize the data. (Source: U.S. National Center for Education Statistics)

AK 7.1	AL 37.6	AR 26.9	AZ 57.0
CA 342.6	CO 42.9	CT 34.4	DC 3.2
DE 6.8	FL 129.0	GA 69.7	HI 10.3
IA 33.8	ID 15.5	IL 121.3	IN 57.6
KS 30.0	KY 36.2	LA 36.2	MA 57.9
MD 53.0	ME 13.4	MI 106.3	MN 59.8
MO 57.0	MS 23.6	MT 10.5	NC 71.4
ND 7.8	NE 20.0	NH 13.3	NJ 88.3
NM 18.1	NV 16.2	NY 150.9	OH 116.3
OK 36.7	OR 32.5	PA 121.6	RI 9.3
SC 32.1	SD 9.1	TN 43.6	TX 236.7
UT 29.9	VA 71.7	VT 7.0	WA 60.4
WI 62.3	WV 17.1	WY 5.7	

7. Business The table shows the numbers of Wal-Mart stores from 1995 to 2006. Construct a bar graph for the data. Write a brief statement regarding the number of Wal-Mart stores over time.   (Source: Value Line)

Year	Number of stores
1995	2943
1996	3054
1997	3406
1998	3599
1999	3985
2000	4189
2001	4414
2002	4688
2003	4906
2004	5289
2005	5650
2006	6050

8. Business The table shows the revenues (in billions of dollars) for Costco Wholesale from 1995 to 2006. Construct a bar graph for the data. Write a brief statement regarding the revenue of Costco Wholesale stores over time.   (Source: Value Line)

Year	Revenue (in billions of dollars)
1995	18.247
1996	19.566
1997	21.874
1998	24.270
1999	27.456
2000	32.164
2001	34.797
2002	38.762
2003	42.546
2004	48.107
2005	52.935
2006	58.600

Tuition In Exercises 9 and 10, the double bar graph shows the mean tuitions (in dollars) charged by public and private institutions of higher education in the United States from 1999 to 2004.   (Source: U.S. National Center for Education Statistics)

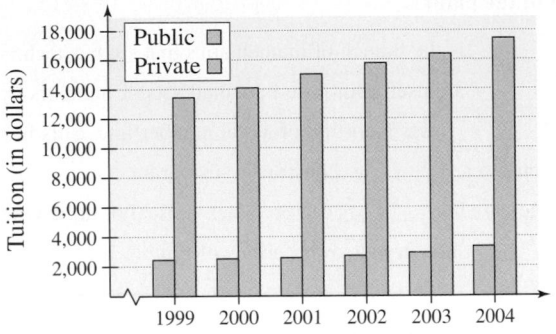

9. Approximate the difference in tuition charges for public and private schools for each year.

10. Approximate the increase or decrease in tuition charges for each type of institution from year to year.

11. College Enrollment The table shows the total college enrollments (in thousands) for women and men in the United States from 1997 to 2003. Construct a double bar graph for the data.   (Source: U.S. National Center for Education Statistics)

Year	Women (in thousands)	Men (in thousands)
1997	8106.3	6396.0
1998	8137.7	6369.3
1999	8300.6	6490.6
2000	8590.5	6721.8
2001	8967.2	6960.8
2002	9410.0	7202.0
2003	9652.0	7259.0

12. Population The table shows the populations (in millions) in the coastal regions of the United States in 1970 and 2003. Construct a double bar graph for the data.   (Source: U.S. Census Bureau)

Region	1970 population (in millions)	2003 population (in millions)
Atlantic	52.1	67.1
Gulf of Mexico	10.0	18.9
Great Lakes	26.0	27.5
Pacific	22.8	39.4

Advertising   **In Exercises 13 and 14, use the line graph, which shows the costs of a 30-second television spot (in thousands of dollars) during the Super Bowl from 1995 to 2005.**   (Source: The Associated Press)

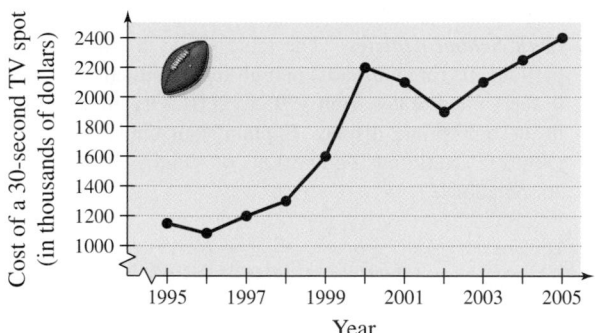

13. Approximate the percent increase in the cost of a 30-second spot from Super Bowl XXX in 1996 to Super Bowl XXXIX in 2005.

14. Estimate the increase or decrease in the cost of a 30-second spot from (a) Super Bowl XXIX in 1995 to Super Bowl XXXIII in 1999, and (b) Super Bowl XXXIV in 2000 to Super Bowl XXXIX in 2005.

Retail Price   **In Exercises 15 and 16, use the line graph, which shows the average retail price (in dollars) of one pound of 100% ground beef in the United States for each month in 2004.**   (Source: U.S. Bureau of Labor Statistics)

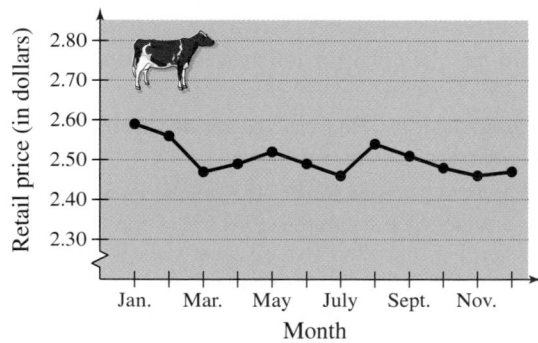

15. What is the highest price of one pound of 100% ground beef shown in the graph? When did this price occur?

16. What was the difference between the highest price and the lowest price of one pound of 100% ground beef in 2004?

17. *Labor*   The table shows the total numbers of women in the work force (in thousands) in the United States from 1995 to 2004. Construct a line graph for the data. Write a brief statement describing what the graph reveals. (Source: U.S. Bureau of Labor Statistics)

Year	Women in the work force (in thousands)
1995	60,944
1996	61,857
1997	63,036
1998	63,714
1999	64,855
2000	66,303
2001	66,848
2002	67,363
2003	68,272
2004	68,421

18. *SAT Scores*   The table shows the average Scholastic Aptitude Test (SAT) Math Exam scores for college-bound seniors in the United States for selected years from 1970 to 2005. Construct a line graph for the data. Write a brief statement describing what the graph reveals.   (Source: The College Entrance Examination Board)

Year	SAT scores
1970	512
1975	498
1980	492
1985	500
1990	501
1995	506
2000	514
2005	520

19. *Hourly Earnings*   The table on page A84 shows the average hourly earnings (in dollars) of production workers in the United States from 1994 to 2005. Use a graphing utility to construct a line graph for the data.   (Source: U.S. Bureau of Labor Statistics)

Year	Hourly earnings (in dollars)
1994	11.19
1995	11.47
1996	11.84
1997	12.27
1998	12.77
1999	13.25
2000	13.73
2001	14.27
2002	14.73
2003	15.19
2004	15.48
2005	15.90

Table for 19

20. *Internet Access* The list shows the percent of households in each of the 50 states and the District of Columbia with Internet access in 2003. Use a graphing utility to organize the data in the graph of your choice. Explain your choice of graph. (Source: U.S. Department of Commerce)

AK 67.6	AL 45.7	AR 42.4	AZ 55.2
CA 59.6	CO 63.0	CT 62.9	DC 53.2
DE 56.8	FL 55.6	GA 53.5	HI 55.0
IA 57.1	ID 56.4	IL 51.1	IN 51.0
KS 54.3	KY 49.6	LA 44.1	MA 58.1
MD 59.2	ME 57.9	MI 52.0	MN 61.6
MO 53.0	MS 38.9	MT 50.4	NC 51.1
ND 53.2	NE 55.4	NH 65.2	NJ 60.5
NM 44.5	NV 55.2	NY 53.3	OH 52.5
OK 48.4	OR 61.0	PA 54.7	RI 55.7
SC 45.6	SD 53.6	TN 48.9	TX 51.8
UT 62.6	VA 60.3	VT 58.1	WA 62.3
WI 57.4	WV 47.6	WY 57.7	

Cellular Phones **In Exercises 21 and 22, use the table, which shows the average monthly cellular telephone bills (in dollars) in the United States from 1999 to 2004.** (Source: Telecommunications & Internet Association)

Year	Average monthly bill (in dollars)
1999	41.24
2000	45.27
2001	47.37
2002	48.40
2003	49.91
2004	50.64

21. Organize the data in an appropriate display. Explain your choice of graph.

22. The average monthly bills in 1990 and 1995 were $80.90 and $51.00, respectively. How would you explain the trend(s) in the data?

23. *High School Athletes* The table shows the numbers of participants (in thousands) in high school athletic programs in the United States from 1995 to 2004. Organize the data in an appropriate display. Explain your choice of graph. (Source: National Federation of State High School Associations)

Year	Female athletes (in thousands)	Male athletes (in thousands)
1995	2240	3536
1996	2368	3634
1997	2474	3706
1998	2570	3763
1999	2653	3832
2000	2676	3862
2001	2784	3921
2002	2807	3961
2003	2856	3989
2004	2865	4038

Synthesis

24. *Writing* Describe the differences between a bar graph and a histogram.

25. *Think About It* How can you decide which type of graph to use when you are organizing data?

26. *Graphical Interpretation* The graphs shown below represent the same data points. Which of the two graphs is misleading, and why? Discuss other ways in which graphs can be misleading. Try to find another example of a misleading graph in a newspaper or magazine. Why is it misleading? Why would it be beneficial for someone to use a misleading graph?

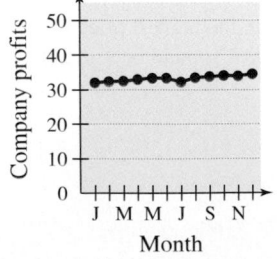
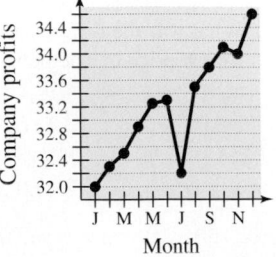

Appendix C: Concepts in Statistics

C.1 Measures of Central Tendency and Dispersion

Mean, Median, and Mode

In many real-life situations, it is helpful to describe data by a single number that is most representative of the entire collection of numbers. Such a number is called a **measure of central tendency.** The most commonly used measures are as follows.

1. The **mean,** or **average,** of n numbers is the sum of the numbers divided by n.
2. The **median** of n numbers is the middle number when the numbers are written in numerical order. If n is even, the median is the average of the two middle numbers.
3. The **mode** of n numbers is the number that occurs most frequently. If two numbers tie for most frequent occurrence, the collection has two modes and is called **bimodal.**

What you should learn

- Find and interpret the mean, median, and mode of a set of data.
- Determine the measure of central tendency that best represents a set of data.
- Find the standard deviation of a set of data.
- Use box-and-whisker plots.

Why you should learn it

Measures of central tendency and dispersion provide a convenient way to describe and compare sets of data. For instance, in Exercise 34 on page A93, the mean and standard deviation are used to analyze the prices of gold for the years 1982 through 2005.

Example 1   Comparing Measures of Central Tendency

On an interview for a job, the interviewer tells you that the average annual income of the company's 25 employees is $60,849. The actual annual incomes of the 25 employees are shown below. What are the mean, median, and mode of the incomes?

$17,305	$478,320	$45,678	$18,980	$17,408
$25,676	$28,906	$12,500	$24,540	$33,450
$12,500	$33,855	$37,450	$20,432	$28,956
$34,983	$36,540	$250,921	$36,853	$16,430
$32,654	$98,213	$48,980	$94,024	$35,671

Solution

The mean of the incomes is

$$\text{Mean} = \frac{17,305 + 478,320 + 45,678 + 18,980 + \cdots + 35,671}{25}$$

$$= \frac{1,521,225}{25} = \$60,849.$$

To find the median, order the incomes as follows.

$12,500	$12,500	$16,430	$17,305	$17,408
$18,980	$20,432	$24,540	$25,676	$28,906
$28,956	$32,654	$33,450	$33,855	$34,983
$35,671	$36,540	$36,853	$37,450	$45,678
$48,980	$94,024	$98,213	$250,921	$478,320

From this list, you can see that the median income is $33,450. You can also see that $12,500 is the only income that occurs more than once. So, the mode is $12,500.

✓*CHECKPOINT*   Now try Exercise 1.

In Example 1, was the interviewer telling you the truth about the annual incomes? Technically, the person was telling the truth because the average is (generally) defined to be the mean. However, of the three measures of central tendency—*mean:* $60,849, *median:* $33,450, *mode:* $12,500—it seems clear that the median is most representative. The mean is inflated by the two highest salaries.

Choosing a Measure of Central Tendency

Which of the three measures of central tendency is most representative of a particular data set? The answer is that it depends on the distribution of the data *and* the way in which you plan to use the data.

For instance, in Example 1, the mean salary of $60,849 does not seem very representative to a potential employee. To a city income tax collector who wants to estimate 1% of the total income of the 25 employees, however, the mean is precisely the right measure.

Example 2   Choosing a Measure of Central Tendency

Which measure of central tendency is most representative of the data given in each frequency distribution?

a.

Number	1	2	3	4	5	6	7	8	9
Frequency	7	20	15	11	8	3	2	0	15

b.

Number	1	2	3	4	5	6	7	8	9
Frequency	9	8	7	6	5	6	7	8	9

c.

Number	1	2	3	4	5	6	7	8	9
Frequency	6	1	2	3	5	5	8	3	0

Solution

a. For this data set, the mean is 4.23, the median is 3, and the mode is 2. Of these, the median or mode is probably the most representative measure.

b. For this data set, the mean and median are each 5 and the modes are 1 and 9 (the distribution is bimodal). Of these, the mean or median is the most representative measure.

c. For this data set, the mean is 4.59, the median is 5, and the mode is 7. Of these, the mean or median is the most representative measure.

✔CHECKPOINT   Now try Exercise 15.

Variance and Standard Deviation

Very different sets of numbers can have the same mean. You will now study two **measures of dispersion,** which give you an idea of how much the numbers in a set differ from the mean of the set. These two measures are called the *variance* of the set and the *standard deviation* of the set.

TECHNOLOGY TIP

Calculating the mean and median of a large data set can become time consuming. Most graphing utilities have *mean* and *median* features that can be used to find the means and medians of data sets. Enter the data from Example 2(a) in the *list editor* of a graphing utility. Then use the *mean* and *median* features to verify the solution to Example 2(a), as shown below.

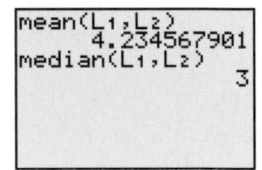

For instructions on how to use the *list* feature, the *mean* feature, and the *median* feature, see Appendix A; for specific keystrokes, go to this textbook's *Online Study Center.*

Definitions of Variance and Standard Deviation

Consider a set of numbers $\{x_1, x_2, \ldots, x_n\}$ with a mean of $\bar{x}$. The **variance** of the set is

$$v = \frac{(x_1 - \bar{x})^2 + (x_2 - \bar{x})^2 + \cdots + (x_n - \bar{x})^2}{n}$$

and the **standard deviation** of the set is $\sigma = \sqrt{v}$ (σ is the lowercase Greek letter *sigma*).

The standard deviation of a set is a measure of how much a typical number in the set differs from the mean. The greater the standard deviation, the more the numbers in the set vary from the mean. For instance, each of the following sets has a mean of 5.

$$\{5, 5, 5, 5\}, \qquad \{4, 4, 6, 6\}, \qquad \text{and} \qquad \{3, 3, 7, 7\}$$

The standard deviations of the sets are 0, 1, and 2.

$$\sigma_1 = \sqrt{\frac{(5 - 5)^2 + (5 - 5)^2 + (5 - 5)^2 + (5 - 5)^2}{4}} = 0$$

$$\sigma_2 = \sqrt{\frac{(4 - 5)^2 + (4 - 5)^2 + (6 - 5)^2 + (6 - 5)^2}{4}} = 1$$

$$\sigma_3 = \sqrt{\frac{(3 - 5)^2 + (3 - 5)^2 + (7 - 5)^2 + (7 - 5)^2}{4}} = 2$$

Example 3   Estimations of Standard Deviation

Consider the three frequency distributions represented by the bar graphs in Figure C.1. Which set has the smallest standard deviation? Which has the largest?

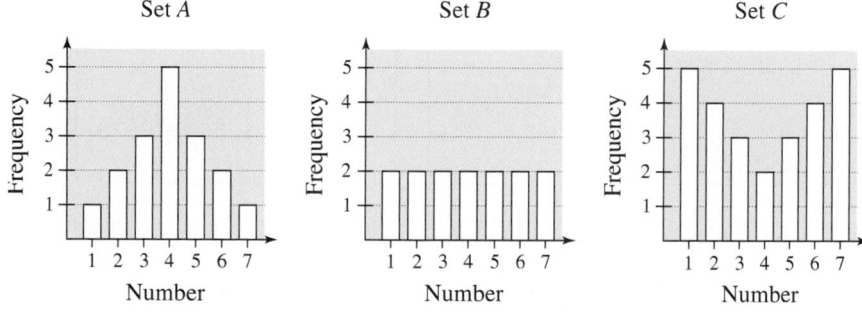

Figure C.1

Solution

Of the three sets, the numbers in set *A* are grouped most closely to the center and the numbers in set *C* are the most dispersed. So, set *A* has the smallest standard deviation and set *C* has the largest standard deviation.

✔CHECKPOINT   Now try Exercise 17.

STUDY TIP

In Example 3, you may find it helpful to write each set numerically. For instance, set *A* is

$$\{1, 2, 2, 3, 3, 3, 4, 4, 4,$$
$$4, 4, 5, 5, 5, 6, 6, 7\}.$$

Example 4   Finding a Standard Deviation

Find the standard deviation of each set shown in Example 3.

Solution

Because of the symmetry of each bar graph, you can conclude that each has a mean of $\bar{x} = 4$. The standard deviation of set A is

$$\sigma = \sqrt{\frac{(-3)^2 + 2(-2)^2 + 3(-1)^2 + 5(0)^2 + 3(1)^2 + 2(2)^2 + (3)^2}{17}}$$

$$\approx 1.53.$$

The standard deviation of set B is

$$\sigma = \sqrt{\frac{2(-3)^2 + 2(-2)^2 + 2(-1)^2 + 2(0)^2 + 2(1)^2 + 2(2)^2 + 2(3)^2}{14}}$$

$$= 2.$$

The standard deviation of set C is

$$\sigma = \sqrt{\frac{5(-3)^2 + 4(-2)^2 + 3(-1)^2 + 2(0)^2 + 3(1)^2 + 4(2)^2 + 5(3)^2}{26}}$$

$$\approx 2.22.$$

These values confirm the results of Example 3. That is, set A has the smallest standard deviation and set C has the largest.

✔CHECKPOINT   Now try Exercise 21.

The following alternative formula provides a more efficient way to compute the standard deviation.

Alternative Formula for Standard Deviation

The standard deviation of $\{x_1, x_2, \ldots, x_n\}$ is given by

$$\sigma = \sqrt{\frac{x_1^2 + x_2^2 + \cdots + x_n^2}{n} - \bar{x}^2}.$$

Because of lengthy computations, this formula is difficult to verify. Conceptually, however, the process is straightforward. It consists of showing that the expressions

$$\sqrt{\frac{(x_1 - \bar{x})^2 + (x_2 - \bar{x})^2 + \cdots + (x_n - \bar{x})^2}{n}}$$

and

$$\sqrt{\frac{x_1^2 + x_2^2 + \cdots + x_n^2}{n} - \bar{x}^2}$$

are equivalent. Try verifying this equivalence for the set $\{x_1, x_2, x_3\}$ with $\bar{x} = (x_1 + x_2 + x_3)/3$.

TECHNOLOGY TIP

Calculating the standard deviation of a large data set can become time-consuming. Most graphing utilities have *statistical* features that can be used to find different statistical values of data sets. Enter the data from set A in Example 3 in the *list editor* of a graphing utility. Then use the *one-variable statistics* feature to verify the solution to Example 4, as shown below.

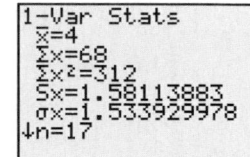

In the figure above, the standard deviation is represented as σx, which is about 1.53. For instructions on how to use the *one-variable statistics* feature, see Appendix A; for specific keystrokes, go to this textbook's *Online Study Center*.

Example 5   Using the Alternative Formula

Use the alternative formula for standard deviation to find the standard deviation of the following set of numbers.

5, 6, 6, 7, 7, 8, 8, 8, 9, 10

Solution

Begin by finding the mean of the set, which is 7.4. So, the standard deviation is

$$\sigma = \sqrt{\frac{5^2 + 2(6^2) + 2(7^2) + 3(8^2) + 9^2 + 10^2}{10} - (7.4)^2}$$

$$= \sqrt{\frac{568}{10} - 54.76} = \sqrt{2.04} \approx 1.43.$$

You can use the *one-variable statistics* feature of a graphing utility to check this result.

✓**CHECKPOINT**   Now try Exercise 27.

A well-known theorem in statistics, called *Chebychev's Theorem*, states that at least

$$1 - \frac{1}{k^2}$$

of the numbers in a distribution must lie within k standard deviations of the mean. So, at least 75% of the numbers in a collection must lie within two standard deviations of the mean, and at least 88.9% of the numbers must lie within three standard deviations of the mean. For most distributions, these percents are low. For instance, in all three distributions shown in Example 3, 100% of the numbers lie within two standard deviations of the mean.

Example 6   Describing a Distribution

The table at the right shows the number of outpatient visits to hospitals (in millions) in each state and the District of Columbia in 2003. Find the mean and standard deviation of the numbers. What percent of the numbers lie within two standard deviations of the mean?   (Source: Health Forum)

Solution

Begin by entering the numbers in a graphing utility. Then use the *one-variable statistics* feature to obtain $\bar{x} \approx 11.12$ and $\sigma \approx 11.10$. The interval that contains all numbers that lie within two standard deviations of the mean is

$$[11.12 - 2(11.10), \ 11.12 + 2(11.10)] \quad \text{or} \quad [-11.08, 33.32].$$

From the table you can see that all but two of the numbers (96%) lie in this interval—all but the numbers that correspond to the numbers of outpatient visits to hospitals in California and New York.

✓**CHECKPOINT**   Now try Exercise 35.

AK	1	MT	3
AL	9	NC	15
AR	5	ND	2
AZ	7	NE	4
CA	48	NH	3
CO	7	NJ	15
CT	7	NM	5
DC	2	NV	2
DE	2	NY	48
FL	22	OH	30
GA	13	OK	6
HI	2	OR	8
IA	10	PA	33
ID	3	RI	2
IL	27	SC	7
IN	15	SD	2
KS	6	TN	10
KY	9	TX	32
LA	11	UT	5
MA	20	VA	11
MD	7	VT	2
ME	4	WA	10
MI	27	WI	12
MN	9	WV	6
MO	16	WY	1
MS	4		

Box-and-Whisker Plots

Standard deviation is the measure of dispersion that is associated with the mean. **Quartiles** measure dispersion associated with the median.

> **Definition of Quartiles**
>
> Consider an ordered set of numbers whose median is m. The **lower quartile** is the median of the numbers that occur on or before m. The **upper quartile** is the median of the numbers that occur on or after m.

Example 7 Finding Quartiles of a Set

Find the lower and upper quartiles of the following set.

34, 14, 24, 16, 12, 18, 20, 24, 16, 26, 13, 27

Solution

Begin by ordering the set.

12, 13, 14,   16, 16, 18,   20, 24, 24,   26, 27, 34
  1st 25%   2nd 25%   3rd 25%   4th 25%

The median of the entire set is 19. The median of the six numbers that are less than 19 is 15. So, the lower quartile is 15. The median of the six numbers that are greater than 19 is 25. So, the upper quartile is 25.

✓*CHECKPOINT*   Now try Exercise 37(a).

Quartiles are represented graphically by a **box-and-whisker plot,** as shown in Figure C.2. In the plot, notice that five numbers are listed: the smallest number, the lower quartile, the median, the upper quartile, and the largest number. Also notice that the numbers are spaced proportionally, as though they were on a real number line.

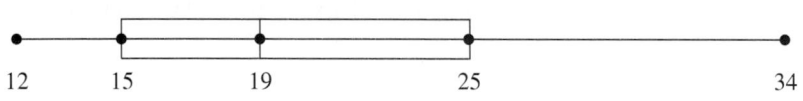

12   15   19   25   34

Figure C.2

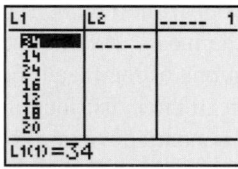

Figure C.3

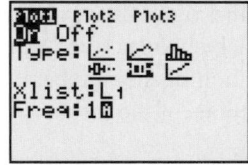

Figure C.4

TECHNOLOGY TIP   You can use a graphing utility to graph the box-and-whisker plot in Figure C.2. First enter the data in the graphing utility's *list editor*, as shown in Figure C.3. Then use the *statistical plotting* feature to set up the box-and-whisker plot, as shown in Figure C.4. Finally, display the box-and-whisker plot (using the *ZoomStat* feature), as shown in Figure C.5. For instructions on how to use the *list editor* and the *statistical plotting* features, see Appendix A; for specific keystrokes, go to this textbook's *Online Study Center.*

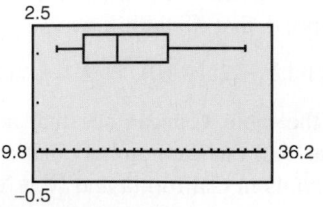

Figure C.5

The next example shows how to find quartiles when the number of elements in a set is not divisible by 4.

Example 8 Sketching Box-and-Whisker Plots

Sketch a box-and-whisker plot for each data set.

a. 82, 82, 83, 85, 87, 89, 90, 94, 95, 95, 96, 98, 99

b. 11, 13, 13, 15, 17, 17, 20, 24, 24, 27

Solution

a. This set has 13 numbers. The median is 90 (the seventh number). The lower quartile is 84 (the median of the first six numbers). The upper quartile is 95.5 (the median of the last six numbers). See Figure C.6.

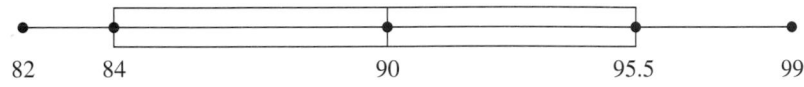

82
84
90
95.5
99

Figure C.6

b. This set has 10 numbers. The median is 17 (the average of the fifth and sixth numbers). The lower quartile is 13 (the median of the first five numbers). The upper quartile is 24 (the median of the last five numbers). See Figure C.7.

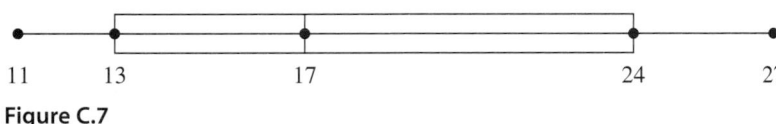

11
13
17
24
27

Figure C.7

CHECKPOINT Now try Exercise 37(b).

C.1 Exercises

See www.CalcChat.com for worked-out solutions to odd-numbered exercises.

Vocabulary Check

Fill in the blanks.

1. A single number that is the most representative of a data set is called a _______ of _______ .

2. If two numbers are tied for the most frequent occurrence, the collection has two _______ and is called _______ .

3. Two measures of dispersion are called the ______ and the ______ of a data set.

4. _______ measure dispersion associated with the median.

In Exercises 1–6, find the mean, median, and mode of the data set.

1. 5, 12, 7, 14, 8, 9, 7

2. 30, 37, 32, 39, 33, 34, 32

3. 5, 12, 7, 24, 8, 9, 7

4. 20, 37, 32, 39, 33, 34, 32

5. 5, 12, 7, 14, 9, 7

6. 30, 37, 32, 39, 34, 32

7. *Reasoning*

(a) Compare your answers in Exercises 1 and 3 with those in Exercises 2 and 4. Which of the measures of central tendency is sensitive to extreme measurements? Explain your reasoning.

(b) Add 6 to each measurement in Exercise 1 and calculate the mean, median, and mode of the revised measurements. How are the measures of central tendency changed?

(c) If a constant k is added to each measurement in a set of data, how will the measures of central tendency change?

8. *Consumer Awareness*   A person had the following monthly bills for electricity. What are the mean and median of the collection of bills?

January	$67.92	February	$59.84
March	$52.00	April	$52.50
May	$57.99	June	$65.35
July	$81.76	August	$74.98
September	$87.82	October	$83.18
November	$65.35	December	$57.00

9. *Car Rental*   A car rental company kept the following record of the numbers of miles a rental car was driven. What are the mean, median, and mode of the data?

Monday	410	Tuesday	260
Wednesday	320	Thursday	320
Friday	460	Saturday	150

10. *Families*   A study was done on families having six children. The table shows the numbers of families in the study with the indicated numbers of girls. Determine the mean, median, and mode of the data.

Number of girls	0	1	2	3	4	5	6
Frequency	1	24	45	54	50	19	7

11. *Bowling Scores*   The table shows the bowling scores for a three-game series of a three-member team.

Team member	Game 1	Game 2	Game 3
Jay	181	222	196
Hank	199	195	205
Buck	202	251	235

(a) Find the mean for each team member.

(b) Find the mean for the entire team for the three-game series.

(c) Find the median for the entire team for the three-game series.

12. *Selling Price*   The selling prices of 12 new homes built in one subdivision are listed.

$525,000	$375,000	$425,000	$550,000
$385,000	$500,000	$550,000	$425,000
$475,000	$500,000	$350,000	$450,000

(a) Find the mean, mode, and median of the selling prices.

(b) Which measure of central tendency best describes the prices? Explain.

13. *Think About It*   Construct a collection of numbers that has the following properties. If this is not possible, explain why.

Mean = 6, median = 4, mode = 4

14. *Think About It*   Construct a collection of numbers that has the following properties. If this is not possible, explain why.

Mean = 6, median = 6, mode = 4

15. *Test Scores*   An English professor records the following scores for a 100-point exam.

99, 64, 80, 77, 59, 72, 87, 79, 92, 88, 90, 42, 20, 89, 42, 100, 98, 84, 78, 91

Which measure of central tendency best describes these test scores?

16. *Shoe Sales*   A salesman sold eight pairs of men's brown dress shoes. The sizes of the eight pairs were as follows: $10\frac{1}{2}$, 8, 12, $10\frac{1}{2}$, 10, $9\frac{1}{2}$, 11, and $10\frac{1}{2}$. Which measure (or measures) of central tendency best describes (describe) the typical shoe size for this data?

In Exercises 17 and 18, line plots of data sets are given. Determine the mean and standard deviation of each set.

17. (a)
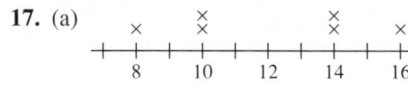

(b)
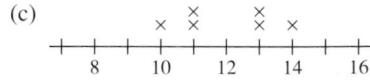

(c)
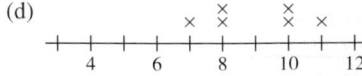

(d)
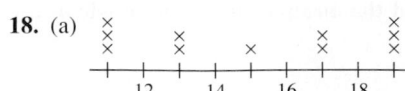

18. (a)

(b)

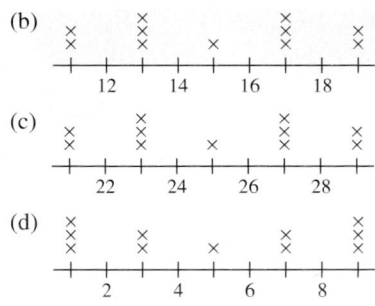

(c)

(d)

In Exercises 19–26, find the mean ($\bar{x}$), variance (v), and standard deviation (σ) of the set.

19. 4, 10, 8, 2

20. 3, 15, 6, 9, 2

21. 0, 1, 1, 2, 2, 2, 3, 3, 4

22. 2, 2, 2, 2, 2, 2

23. 1, 2, 3, 4, 5, 6, 7

24. 1, 1, 1, 5, 5, 5

25. 49, 62, 40, 29, 32, 70

26. 1.5, 0.4, 2.1, 0.7, 0.8

In Exercises 27–30, use the alternative formula to find the standard deviation of the set.

27. 2, 4, 6, 6, 13, 5

28. 246, 336, 473, 167, 219, 359

29. 8.1, 6.9, 3.7, 4.2, 6.1

30. 9.0, 7.5, 3.3, 7.4, 6.0

31. *Reasoning* Without calculating the standard deviation, explain why the set {4, 4, 20, 20} has a standard deviation of 8.

32. *Reasoning* If the standard deviation of a set of numbers is 0, what does this imply about the set?

33. *Test Scores* An instructor adds five points to each student's exam score. Will this change the mean or standard deviation of the exam scores? Explain.

34. *Price of Gold* The following data represents the average prices of gold (in dollars per fine ounce) for the years 1982 to 2005. Use a computer or graphing utility to find the mean, variance, and standard deviation of the data. What percent of the data lies within two standard deviations of the mean? (Source: National Mining Association)

376,   424,   361,   317,   368,   447,

437,   381,   384,   362,   344,   360,

384,   384,   388,   331,   294,   279,

279,   271,   310,   363,   410,   445

35. *Test Scores* The scores on a mathematics exam given to 600 science and engineering students at a college had a mean and standard deviation of 235 and 28, respectively. Use Chebychev's Theorem to determine the intervals containing at least $\frac{3}{4}$ and at least $\frac{8}{9}$ of the scores. How would the intervals change if the standard deviation were 16?

36. *Think About It* The histograms represent the test scores of two classes of a college course in mathematics. Which histogram has the smaller standard deviation?

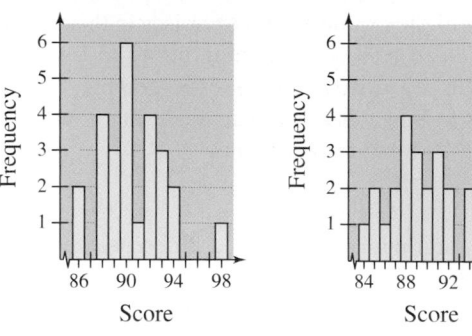

In Exercises 37–40, (a) find the lower and upper quartiles of the data and (b) sketch a box-and-whisker plot for the data without using a graphing utility.

37. 23, 15, 14, 23, 13, 14, 13, 20, 12

38. 11, 10, 11, 14, 17, 16, 14, 11, 8, 14, 20

39. 46, 48, 48, 50, 52, 47, 51, 47, 49, 53

40. 25, 20, 22, 28, 24, 28, 25, 19, 27, 29, 28, 21

In Exercises 41–44, use a graphing utility to create a box-and-whisker plot for the data.

41. 19, 12, 14, 9, 14, 15, 17, 13, 19, 11, 10, 19

42. 9, 5, 5, 5, 6, 5, 4, 12, 7, 10, 7, 11, 8, 9, 9

43. 20.1, 43.4, 34.9, 23.9, 33.5, 24.1, 22.5, 42.4, 25.7, 17.4, 23.8, 33.3, 17.3, 36.4, 21.8

44. 78.4, 76.3, 107.5, 78.5, 93.2, 90.3, 77.8, 37.1, 97.1, 75.5, 58.8, 65.6

45. *Product Lifetime* A company has redesigned a product in an attempt to increase the lifetime of the product. The two sets of data list the lifetimes (in months) of 20 units with the original design and 20 units with the new design. Create a box-and-whisker plot for each set of data, and then comment on the differences between the plots.

Original Design

15.1	78.3	56.3	68.9	30.6
27.2	12.5	42.7	72.7	20.2
53.0	13.5	11.0	18.4	85.2
10.8	38.3	85.1	10.0	12.6

New Design

55.8	71.5	25.6	19.0	23.1
37.2	60.0	35.3	18.9	80.5
46.7	31.1	67.9	23.5	99.5
54.0	23.2	45.5	24.8	87.8

C.2 Least Squares Regression

In many of the examples and exercises in this text, you have been asked to use the *regression* feature of a graphing utility to find mathematical models for sets of data. The *regression* feature of a graphing utility uses the **method of least squares** to find a mathematical model for a set of data. As a measure of how well a model fits a set of data points

$$\{(x_1, y_1), (x_2, y_2), (x_3, y_3), \ldots, (x_n, y_n)\}$$

you can add the squares of the differences between the actual y-values and the values given by the model to obtain the **sum of the squared differences.** For instance, the table shows the heights x (in feet) and the diameters y (in inches) of eight trees. The table also shows the values of a linear model $y^* = 0.54x - 29.5$ for each x-value. The sum of squared differences for the model is 51.7.

x	70	72	75	76	85	78	77	80
y	8.3	10.5	11.0	11.4	12.9	14.0	16.3	18.0
y^*	8.3	9.38	11.0	11.54	16.4	12.62	12.08	13.7
$(y - y^*)^2$	0	1.2544	0	0.0196	12.25	1.9044	17.8084	18.49

The model that has the *least* sum of squared differences is the **least squares regression** line for the data. The least squares regression line for the data in the table is $y \approx 0.43x - 20.3$. The sum of squared differences is 43.3.

To find the least squares regression line $y = ax + b$ for the points $\{(x_1, y_1), (x_2, y_2), (x_3, y_3), \ldots, (x_n, y_n)\}$ algebraically, you need to solve the following system for a and b.

$$\begin{cases} nb + \left(\sum_{i=1}^n x_i\right)a = \sum_{i=1}^n y_i \\ \left(\sum_{i=1}^n x_i\right)b + \left(\sum_{i=1}^n x_i^2\right)a = \sum_{i=1}^n x_i y_i \end{cases}$$

In the system,

$$\sum_{i=1}^n x_i = x_1 + x_2 + \cdots + x_n$$

$$\sum_{i=1}^n y_i = y_1 + y_2 + \cdots + y_n$$

$$\sum_{i=1}^n x_i^2 = x_1^2 + x_2^2 + \cdots + x_n^2$$

$$\sum_{i=1}^n x_i y_i = x_1 y_1 + x_2 y_2 + \cdots + x_n y_n.$$

What you should learn

- Use the sum of squared differences to determine a least squares regression line.
- Find a least squares regression line for a set of data.
- Find a least squares regression parabola for a set of data.

Why you should learn it

The method of least squares provides a way of creating a mathematical model for a set of data, which can then be analyzed.

TECHNOLOGY SUPPORT

For instructions on how to use the *regression* feature, see Appendix A; for specific keystrokes, go to this textbook's *Online Study Center.*

TECHNOLOGY TIP Recall from Section 1.7 that when you use the *regression* feature of a graphing utility, the program may output a correlation coefficient, r. When $|r|$ is close to 1, the model is a good fit for the data.

Example 1   Finding a Least Squares Regression Line

Find the least squares regression line for $(-3, 0)$, $(-1, 1)$, $(0, 2)$, and $(2, 3)$.

Solution

Begin by constructing a table, as shown below.

x	y	xy	x^2
-3	0	0	9
-1	1	-1	1
0	2	0	0
2	3	6	4
$\sum\limits_{i=1}^{n} x_i = -2$	$\sum\limits_{i=1}^{n} y_i = 6$	$\sum\limits_{i=1}^{n} x_i y_i = 5$	$\sum\limits_{i=1}^{n} x_i^2 = 14$

Applying the system for the least squares regression line with $n = 4$ produces

$$\begin{cases} nb + \left(\sum\limits_{i=1}^{n} x_i \right) a = \sum\limits_{i=1}^{n} y_i \\ \left(\sum\limits_{i=1}^{n} x_i \right) b + \left(\sum\limits_{i=1}^{n} x_i^2 \right) a = \sum\limits_{i=1}^{n} x_i y_i \end{cases} \implies \begin{cases} 4b - 2a = 6 \\ -2b + 14a = 5 \end{cases}.$$

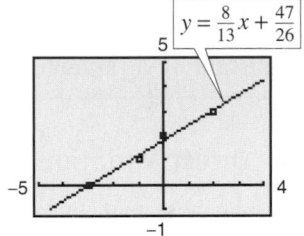

Solving this system of equations produces $a = \frac{8}{13}$ and $b = \frac{47}{26}$. So, the least squares regression line is $y = \frac{8}{13}x + \frac{47}{26}$, as shown in Figure C.8.

Figure C.8

✓CHECKPOINT   Now try Exercise 1.

The least squares regression parabola $y = ax^2 + bx + c$ for the points

$$\{(x_1, y_1), (x_2, y_2), (x_3, y_3), \ldots, (x_n, y_n)\}$$

is obtained in a similar manner by solving the following system of three equations in three unknowns for a, b, and c.

$$\begin{cases} nc + \left(\sum\limits_{i=1}^{n} x_i \right) b + \left(\sum\limits_{i=1}^{n} x_i^2 \right) a = \sum\limits_{i=1}^{n} y_i \\ \left(\sum\limits_{i=1}^{n} x_i \right) c + \left(\sum\limits_{i=1}^{n} x_i^2 \right) b + \left(\sum\limits_{i=1}^{n} x_i^3 \right) a = \sum\limits_{i=1}^{n} x_i y_i \\ \left(\sum\limits_{i=1}^{n} x_i^2 \right) c + \left(\sum\limits_{i=1}^{n} x_i^3 \right) b + \left(\sum\limits_{i=1}^{n} x_i^4 \right) a = \sum\limits_{i=1}^{n} x_i^2 y_i \end{cases}$$

C.2   Exercises

See www.CalcChat.com for worked-out solutions to odd-numbered exercises.

In Exercises 1–4, find the least squares regression line for the points. Verify your answer with a graphing utility.

1. $(-4, 1)$, $(-3, 3)$, $(-2, 4)$, $(-1, 6)$

2. $(0, -1)$, $(2, 0)$, $(4, 3)$, $(6, 5)$

3. $(-3, 1)$, $(-1, 2)$, $(1, 2)$, $(4, 3)$

4. $(0, -1)$, $(2, 1)$, $(3, 2)$, $(5, 3)$

Appendix D: Variation

Direct Variation

There are two basic types of linear models. The more general model has a *y*-intercept that is nonzero.

$$y = mx + b, \quad b \neq 0$$

The simpler model $y = kx$ has a *y*-intercept that is zero. In the simpler model, *y* is said to **vary directly** as *x*, or to be **directly proportional** to *x*.

> **Direct Variation**
>
> The following statements are equivalent.
>
> **1.** *y* **varies directly** as *x*.
>
> **2.** *y* is **directly proportional** to *x*.
>
> **3.** $y = kx$ for some nonzero constant *k*.
>
> *k* is the **constant of variation** or the **constant of proportionality**.

What you should learn

- Write mathematical models for direct variation.
- Write mathematical models for direct variation as an *n*th power.
- Write mathematical models for inverse variation.
- Write mathematical models for joint variation.

Why you should learn it

You can use functions as models to represent a wide variety of real-life data sets. For instance, in Exercise 55 on page A102, a variation model can be used to model the water temperatures of the ocean at various depths.

Example 1   Direct Variation

In Pennsylvania, the state income tax is directly proportional to *gross income*. You are working in Pennsylvania and your state income tax deduction is $46.05 for a gross monthly income of $1500. Find a mathematical model that gives the Pennsylvania state income tax in terms of gross income.

Solution

Verbal Model: State income tax $= k \cdot$ Gross income

Labels:
State income tax $= y$ (dollars)
Gross income $= x$ (dollars)
Income tax rate $= k$ (percent in decimal form)

Equation: $y = kx$

To solve for *k*, substitute the given information in the equation $y = kx$, and then solve for *k*.

$$y = kx \qquad \text{Write direct variation model.}$$
$$46.05 = k(1500) \qquad \text{Substitute } y = 46.05 \text{ and } x = 1500.$$
$$0.0307 = k \qquad \text{Simplify.}$$

So, the equation (or model) for state income tax in Pennsylvania is

$$y = 0.0307x.$$

In other words, Pennsylvania has a state income tax rate of 3.07% of gross income. The graph of this equation is shown in Figure D.1.

✓**CHECKPOINT**   Now try Exercise 7.

Pennsylvania Taxes

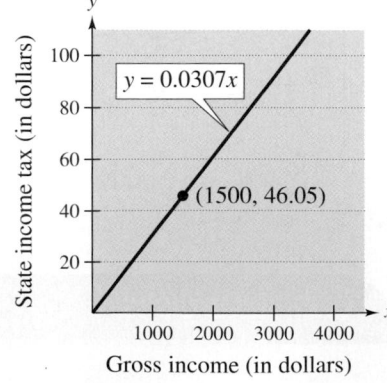

Figure D.1

Direct Variation as an *n*th Power

Another type of direct variation relates one variable to a *power* of another variable. For example, in the formula for the area of a circle

$$A = \pi r^2$$

the area A is directly proportional to the square of the radius r. Note that for this formula, π is the constant of proportionality.

Direct Variation as an *n*th Power

The following statements are equivalent.

1. y **varies directly as the *n*th power** of x.

2. y is **directly proportional to the *n*th power** of x.

3. $y = kx^n$ for some constant k.

STUDY TIP

Note that the direct variation model $y = kx$ is a special case of $y = kx^n$ with $n = 1$.

Example 2   **Direct Variation as an *n*th Power**

The distance a ball rolls down an inclined plane is directly proportional to the square of the time it rolls. During the first second, the ball rolls 8 feet. (See Figure D.2.)

a. Write an equation relating the distance traveled to the time.

b. How far will the ball roll during the first 3 seconds?

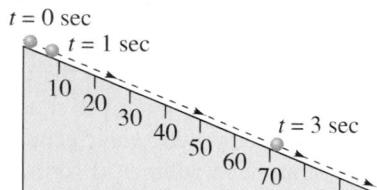

Figure D.2

Solution

a. Letting d be the distance (in feet) the ball rolls and letting t be the time (in seconds), you have

$$d = kt^2.$$

Now, because $d = 8$ when $t = 1$, you can see that $k = 8$, as follows.

$$d = kt^2$$
$$8 = k(1)^2$$
$$8 = k$$

So, the equation relating distance to time is

$$d = 8t^2.$$

b. When $t = 3$, the distance traveled is $d = 8(3)^2 = 8(9) = 72$ feet.

✓CHECKPOINT   Now try Exercise 15.

In Examples 1 and 2, the direct variations are such that an *increase* in one variable corresponds to an *increase* in the other variable. This is also true in the model $d = \frac{1}{5}F$, $F > 0$, where an increase in F results in an increase in d. You should not, however, assume that this always occurs with direct variation. For example, in the model $y = -3x$, an increase in x results in a *decrease* in y, and yet y is said to vary directly as x.

Inverse Variation

> **Inverse Variation**
>
> The following statements are equivalent.
>
> **1.** y **varies inversely** as x.
>
> **2.** y is **inversely proportional** to x.
>
> **3.** $y = \dfrac{k}{x}$ for some constant k.

If x and y are related by an equation of the form $y = k/x^n$, then y varies inversely as the nth power of x (or y is inversely proportional to the nth power of x).

Some applications of variation involve problems with *both* direct and inverse variation in the same model. These types of models are said to have **combined variation.**

Example 3 Direct and Inverse Variation

A gas law states that the volume of an enclosed gas varies directly as the temperature *and* inversely as the pressure, as shown in Figure D.3. The pressure of a gas is 0.75 kilogram per square centimeter when the temperature is 294 K and the volume is 8000 cubic centimeters. (a) Write an equation relating pressure, temperature, and volume. (b) Find the pressure when the temperature is 300 K and the volume is 7000 cubic centimeters.

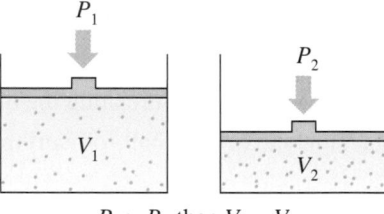

$$P_2 > P_1 \text{ then } V_2 < V_1$$

Figure D.3 *If the temperature is held constant and pressure increases, volume decreases.*

Solution

a. Let V be volume (in cubic centimeters), let P be pressure (in kilograms per square centimeter), and let T be temperature (in Kelvin). Because V varies directly as T and inversely as P, you have

$$V = \frac{kT}{P}.$$

Now, because $P = 0.75$ when $T = 294$ and $V = 8000$, you have

$$8000 = \frac{k(294)}{0.75}$$

$$k = \frac{6000}{294} = \frac{1000}{49}.$$

So, the equation relating pressure, temperature, and volume is

$$V = \frac{1000}{49}\left(\frac{T}{P}\right).$$

b. When $T = 300$ and $V = 7000$, the pressure is

$$P = \frac{1000}{49}\left(\frac{300}{7000}\right) = \frac{300}{343} \approx 0.87 \text{ kilogram per square centimeter.}$$

 ✓**CHECKPOINT** Now try Exercise 49.

Joint Variation

In Example 3, note that when a direct variation and an inverse variation occur in the same statement, they are coupled with the word "and." To describe two different *direct* variations in the same statement, the word **jointly** is used.

Joint Variation

The following statements are equivalent.

1. z **varies jointly** as x and y.

2. z is **jointly proportional** to x and y.

3. $z = kxy$ for some constant k.

If x, y, and z are related by an equation of the form

$$z = kx^n y^m$$

then z varies jointly as the nth power of x and the mth power of y.

Example 4 Joint Variation

The *simple* interest for a certain savings account is jointly proportional to the time and the principal. After one quarter (3 months), the interest on a principal of $5000 is $43.75.

a. Write an equation relating the interest, principal, and time.

b. Find the interest after three quarters.

Solution

a. Let I = interest (in dollars), P = principal (in dollars), and t = time (in years). Because I is jointly proportional to P and t, you have

$$I = kPt.$$

For $I = 43.75$, $P = 5000$, and $t = \frac{1}{4}$, you have

$$43.75 = k(5000)\left(\frac{1}{4}\right)$$

which implies that $k = 4(43.75)/5000 = 0.035$. So, the equation relating interest, principal, and time is

$$I = 0.035Pt$$

which is the familiar equation for simple interest where the constant of proportionality, 0.035, represents an annual interest rate of 3.5%.

b. When $P = \$5000$ and $t = \frac{3}{4}$, the interest is

$$I = (0.035)(5000)\left(\frac{3}{4}\right)$$

$$= \$131.25.$$

✓CHECKPOINT   Now try Exercise 51.

D Exercises

See www.CalcChat.com for worked-out solutions to odd-numbered exercises.

Vocabulary Check

Fill in the blanks.

1. Direct variation models can be described as "y varies directly as x," or "y is _____ _____ to x."

2. In direct variation models of the form $y = kx$, k is called the _____ of _____.

3. The direct variation model $y = kx^n$ can be described as "y varies directly as the nth power of x," or "y is _____ _____ to the nth power of x."

4. The mathematical model $y = \dfrac{k}{x}$ is an example of _____ variation.

5. Mathematical models that involve both direct and inverse variation are said to have _____ variation.

6. The joint variation model $z = kxy$ can be described as "z varies jointly as x and y," or "z is _____ _____ to x and y."

In Exercises 1–4, assume that y is directly proportional to x. Use the given x-value and y-value to find a linear model that relates y and x.

1. $x = 5$, $y = 12$
2. $x = 2$, $y = 14$
3. $x = 10$, $y = 2050$
4. $x = 6$, $y = 580$

5. *Measurement* On a yardstick with scales in inches and centimeters, you notice that 13 inches is approximately the same length as 33 centimeters. Use this information to find a mathematical model that relates centimeters to inches. Then use the model to find the numbers of centimeters in 10 inches and 20 inches.

6. *Measurement* When buying gasoline, you notice that 14 gallons of gasoline is approximately the same amount of gasoline as 53 liters. Use this information to find a linear model that relates gallons to liters. Then use the model to find the numbers of liters in 5 gallons and 25 gallons.

7. *Taxes* Property tax is based on the assessed value of a property. A house that has an assessed value of $150,000 has a property tax of $5520. Find a mathematical model that gives the amount of property tax y in terms of the assessed value x of the property. Use the model to find the property tax on a house that has an assessed value of $200,000.

8. *Taxes* State sales tax is based on retail price. An item that sells for $145.99 has a sales tax of $10.22. Find a mathematical model that gives the amount of sales tax y in terms of the retail price x. Use the model to find the sales tax on a $540.50 purchase.

Hooke's Law In Exercises 9 and 10, use Hooke's Law for springs, which states that the distance a spring is stretched (or compressed) varies directly as the force on the spring.

9. A force of 265 newtons stretches a spring 0.15 meter (see figure).

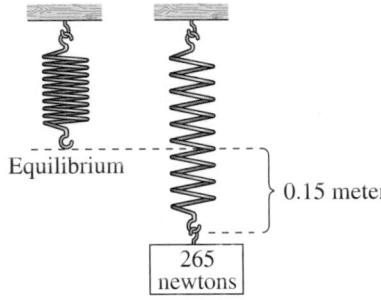

Equilibrium

0.15 meter

265 newtons

Figure for 9

(a) How far will a force of 90 newtons stretch the spring?

(b) What force is required to stretch the spring 0.1 meter?

10. A force of 220 newtons stretches a spring 0.12 meter. What force is required to stretch the spring 0.16 meter?

In Exercises 11–14, use the given value of k to complete the table for the direct variation model $y = kx^2$. Plot the points on a rectangular coordinate system.

x	2	4	6	8	10
$y = kx^2$					

11. $k = 1$
12. $k = 2$
13. $k = \frac{1}{2}$
14. $k = \frac{1}{4}$

Ecology In Exercises 15 and 16, use the fact that the diameter of the largest particle that can be moved by a stream varies approximately directly as the square of the velocity of the stream.

15. A stream with a velocity of $\frac{1}{4}$ mile per hour can move coarse sand particles about 0.02 inch in diameter. Approximate the velocity required to carry particles 0.12 inch in diameter.

16. A stream of velocity v can move particles of diameter d or less. By what factor does d increase when the velocity is doubled?

In Exercises 17–20, use the given value of k to complete the table for the inverse variation model $y = k/x^2$. Plot the points on a rectangular coordinate system.

x	2	4	6	8	10
$y = \dfrac{k}{x^2}$					

17. $k = 2$ **18.** $k = 5$

19. $k = 10$ **20.** $k = 20$

In Exercises 21–24, determine whether the variation model is of the form $y = kx$ or $y = k/x$, and find k.

21.

x	5	10	15	20	25
y	1	$\frac{1}{2}$	$\frac{1}{3}$	$\frac{1}{4}$	$\frac{1}{5}$

22.

x	5	10	15	20	25
y	2	4	6	8	10

23.

x	5	10	15	20	25
y	-3.5	-7	-10.5	-14	-17.5

24.

x	5	10	15	20	25
y	24	12	8	6	$\frac{24}{5}$

In Exercises 25–34, find a mathematical model for the verbal statement.

25. A varies directly as the square of r.

26. V varies directly as the cube of e.

27. y varies inversely as the square of x.

28. h varies inversely as the square root of s.

29. F varies directly as g and inversely as r^2.

30. z is jointly proportional to the square of x and the cube of y.

31. *Boyle's Law:* For a constant temperature, the pressure P of a gas is inversely proportional to the volume V of the gas.

32. *Logistic Growth:* The rate of growth R of a population is jointly proportional to the size S of the population and the difference between S and the maximum population size L that the environment can support.

33. *Newton's Law of Cooling:* The rate of change R of the temperature of an object is proportional to the difference between the temperature T of the object and the temperature T_e of the environment in which the object is placed.

34. *Newton's Law of Universal Gravitation:* The gravitational attraction F between two objects of masses m_1 and m_2 is proportional to the product of the masses and inversely proportional to the square of the distance r between the objects.

In Exercises 35–40, write a sentence using the variation terminology of this section to describe the formula.

35. *Area of a triangle:* $A = \frac{1}{2}bh$

36. *Surface area of a sphere:* $S = 4\pi r^2$

37. *Volume of a sphere:* $V = \frac{4}{3}\pi r^3$

38. *Volume of a right circular cylinder:* $V = \pi r^2 h$

39. *Average speed:* $r = \dfrac{d}{t}$

40. *Free vibrations:* $\omega = \sqrt{\dfrac{kg}{W}}$

In Exercises 41–48, find a mathematical model representing the statement. (In each case, determine the constant of proportionality.)

41. A varies directly as r^2. ($A = 9\pi$ when $r = 3$.)

42. y varies inversely as x. ($y = 3$ when $x = 25$.)

43. y is inversely proportional to x. ($y = 7$ when $x = 4$.)

44. z varies jointly as x and y. ($z = 64$ when $x = 4$ and $y = 8$.)

45. F is jointly proportional to r and the third power of s. ($F = 4158$ when $r = 11$ and $s = 3$.)

46. P varies directly as x and inversely as the square of y. ($P = \frac{28}{3}$ when $x = 42$ and $y = 9$.)

47. z varies directly as the square of x and inversely as y. ($z = 6$ when $x = 6$ and $y = 4$.)

48. v varies jointly as p and q and inversely as the square of s. ($v = 1.5$ when $p = 4.1$, $q = 6.3$, and $s = 1.2$.)

Resistance **In Exercises 49 and 50, use the fact that the resistance of a wire carrying an electrical current is directly proportional to its length and inversely proportional to its cross-sectional area.**

49. If #28 copper wire (which has a diameter of 0.0126 inch) has a resistance of 66.17 ohms per thousand feet, what length of #28 copper wire will produce a resistance of 33.5 ohms?

50. A 14-foot piece of copper wire produces a resistance of 0.05 ohm. Use the constant of proportionality from Exercise 49 to find the diameter of the wire.

51. Work The work W (in joules) done when an object is lifted varies jointly with the mass m (in kilograms) of the object and the height h (in meters) that the object is lifted. The work done when a 120-kilogram object is lifted 1.8 meters is 2116.8 joules. How much work is done when a 100-kilogram object is lifted 1.5 meters?

52. Spending The prices of three sizes of pizza at a pizza shop are as follows.

9-inch: $8.78, 12-inch: $11.78, 15-inch: $14.18

You would expect that the price of a certain size of pizza would be directly proportional to its surface area. Is that the case for this pizza shop? If not, which size of pizza is the best buy?

53. Fluid Flow The velocity v of a fluid flowing in a conduit is inversely proportional to the cross-sectional area of the conduit. (Assume that the volume of the flow per unit of time is held constant.) Determine the change in the velocity of water flowing from a hose when a person places a finger over the end of the hose to decrease its cross-sectional area by 25%.

54. Beam Load The maximum load that can be safely supported by a horizontal beam varies jointly as the width of the beam and the square of its depth, and inversely as the length of the beam. Determine the changes in the maximum safe load under the following conditions.

(a) The width and length of the beam are doubled.

(b) The width and depth of the beam are doubled.

(c) All three of the dimensions are doubled.

(d) The depth of the beam is halved.

55. Ocean Temperatures An oceanographer took readings of the water temperatures C (in degrees Celsius) at several depths d (in meters). The data collected is shown in the table.

Depth, d	Temperature, C
1000	4.2°
2000	1.9°
3000	1.4°
4000	1.2°
5000	0.9°

(a) Sketch a scatter plot of the data.

(b) Does it appear that the data can be modeled by the inverse variation model $C = k/d$? If so, find k for each pair of coordinates.

(c) Determine the mean value of k from part (b) to find the inverse variation model $C = k/d$.

(d) Use a graphing utility to plot the data points and the inverse model in part (c).

(e) Use the model to approximate the depth at which the water temperature is 3°C.

56. Physics Experiment An experiment in a physics lab requires a student to measure the compressed lengths y (in centimeters) of a spring when various forces of F pounds are applied. The data is shown in the table.

Force, F	Length, y
0	0
2	1.15
4	2.3
6	3.45
8	4.6
10	5.75
12	6.9

(a) Sketch a scatter plot of the data.

(b) Does it appear that the data can be modeled by Hooke's Law? If so, estimate k. (See Exercises 9 and 10.)

(c) Use the model in part (b) to approximate the force required to compress the spring 9 centimeters.

Synthesis

True or False? In Exercises 57 and 58, decide whether the statement is true or false. Justify your answer.

57. If y varies directly as x, then if x increases, y will increase as well.

58. In the equation for kinetic energy, $E = \frac{1}{2}mv^2$, the amount of kinetic energy E is directly proportional to the mass m of an object and the square of its velocity v.

Think About It In Exercises 59 and 60, use the graph to determine whether y varies directly as some power of x or inversely as some power of x. Explain.

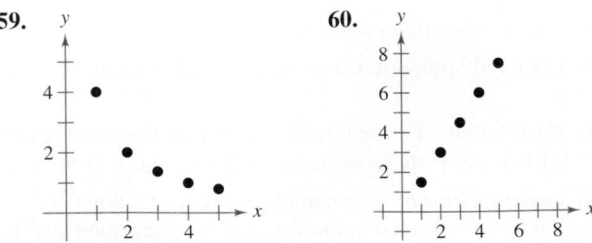

Appendix E: Solving Linear Equations and Inequalities

Linear Equations

A *linear equation* in one variable x is an equation that can be written in the standard form $ax + b = 0$, where a and b are real numbers with $a \neq 0$.

A linear equation in one variable, written in standard form, has exactly one solution. To see this, consider the following steps. (Remember that $a \neq 0$.)

$$ax + b = 0 \qquad \text{Original equation}$$

$$ax = -b \qquad \text{Subtract } b \text{ from each side.}$$

$$x = -\frac{b}{a} \qquad \text{Divide each side by } a.$$

To solve a linear equation in x, isolate x on one side of the equation by creating a sequence of *equivalent* (and usually simpler) equations, each having the same solution(s) as the original equation. The operations that yield equivalent equations come from the Substitution Principle and the Properties of Equality.

Generating Equivalent Equations

An equation can be transformed into an *equivalent equation* by one or more of the following steps.

		Original Equation	*Equivalent Equation*
1.	Remove symbols of grouping, combine like terms, or simplify fractions on one or both sides of the equation.	$2x - x = 4$	$x = 4$
2.	Add (or subtract) the same quantity to (from) *each* side of the equation.	$x + 1 = 6$	$x = 5$
3.	Multiply (or divide) *each* side of the equation by the same *nonzero* quantity.	$2x = 6$	$x = 3$
4.	Interchange the two sides of the equation.	$2 = x$	$x = 2$

After solving an equation, check each solution in the original equation. For example, you can check the solution to the equation in Step 2 above as follows.

$$x + 1 = 6 \qquad \text{Write original equation.}$$

$$5 + 1 \overset{?}{=} 6 \qquad \text{Substitute 5 for } x.$$

$$6 = 6 \qquad \text{Solution checks.} \quad \checkmark$$

What you should learn

- Solve linear equations in one variable.
- Solve linear inequalities in one variable.

Why you should learn it

The method of solving linear equations is used to determine the intercepts of the graph of a linear function. The method of solving linear inequalities is used to determine the domains of different functions.

Example 1   Solving Linear Equations

a.

$3x - 6 = 0$	Original equation
$3x - 6 + 6 = 0 + 6$	Add 6 to each side.
$3x = 6$	Simplify.
$\dfrac{3x}{3} = \dfrac{6}{3}$	Divide each side by 3.
$x = 2$	Simplify.

b.

$4(2x + 3) = 6$	Original equation
$8x + 12 = 6$	Distributive Property
$8x + 12 - 12 = 6 - 12$	Subtract 12 from each side.
$8x = -6$	Simplify.
$x = -\dfrac{3}{4}$	Divide each side by 8 and simplify.

✓CHECKPOINT   Now try Exercise 15.

Linear Inequalities

Solving a linear inequality in one variable is much like solving a linear equation in one variable. To solve the inequality, you isolate the variable on one side using transformations that produce *equivalent inequalities*, which have the same solution(s) as the original inequality.

Generating Equivalent Inequalities

An inequality can be transformed into an *equivalent inequality* by one or more of the following steps.

	Original Inequality	*Equivalent Inequality*
1. Remove symbols of grouping, combine like terms, or simplify fractions on one or both sides of the inequality.	$4x + x \geq 2$	$5x \geq 2$
2. Add (or subtract) the same number to (from) *each* side of the inequality.	$x - 3 < 5$	$x < 8$
3. Multiply (or divide) each side of the inequality by the same *positive* number.	$\tfrac{1}{2}x > 3$	$x > 6$
4. Multiply (or divide) each side of the inequality by the same *negative* number and *reverse* the inequality symbol.	$-2x \leq 6$	$x \geq -3$

Example 2   Solving Linear Inequalities

a.

$x + 5 \geq 3$	Original inequality
$x + 5 - 5 \geq 3 - 5$	Subtract 5 from each side.
$x \geq -2$	Simplify.

The solution is all real numbers greater than or equal to -2, which is denoted by $[-2, \infty)$. Check several numbers that are greater than or equal to -2 in the original inequality.

b.

$-4.2m > 6.3$	Original inequality
$\dfrac{-4.2m}{-4.2} < \dfrac{6.3}{-4.2}$	Divide each side by -4.2 and reverse inequality symbol.
$m < -1.5$	Simplify.

The solution is all real numbers less than -1.5, which is denoted by $(-\infty, -1.5)$. Check several numbers that are less than -1.5 in the original inequality.

✓**CHECKPOINT**   Now try Exercise 29.

STUDY TIP

Remember that when you multiply or divide by a negative number, you must *reverse the inequality symbol*, as shown in Example 2(b).

E   Exercises

See www.CalcChat.com for worked-out solutions to odd-numbered exercises.

Vocabulary Check

Fill in the blanks.

1. A _____ equation in one variable x is an equation that can be written in the standard form $ax + b = 0$.

2. To solve a linear inequality, isolate the variable on one side using transformations that produce _____ .

In Exercises 1–22, solve the equation and check your solution.

1. $x + 11 = 15$
2. $x + 3 = 9$
3. $x - 2 = 5$
4. $x - 5 = 1$
5. $3x = 12$
6. $2x = 6$
7. $\dfrac{x}{5} = 4$
8. $\dfrac{x}{4} = 5$
9. $8x + 7 = 39$
10. $12x - 5 = 43$
11. $24 - 7x = 3$
12. $13 + 6x = 61$
13. $8x - 5 = 3x + 20$
14. $7x + 3 = 3x - 17$
15. $-2(x + 5) = 10$
16. $4(3 - x) = 9$
17. $2x + 3 = 2x - 2$
18. $8(x - 2) = 4(2x - 4)$
19. $\frac{3}{2}(x + 5) - \frac{1}{4}(x + 24) = 0$
20. $\frac{3}{2}x + \frac{1}{4}(x - 2) = 10$
21. $0.25x + 0.75(10 - x) = 3$
22. $0.60x + 0.40(100 - x) = 50$

In Exercises 23–44, solve the inequality and check your solution.

23. $x + 6 < 8$
24. $3 + x > -10$
25. $-x - 8 > -17$
26. $-3 + x < 19$
27. $6 + x \leq -8$
28. $x - 10 \geq -6$
29. $\frac{4}{5}x > 8$
30. $\frac{2}{3}x < -4$
31. $-\frac{3}{4}x > -3$
32. $-\frac{1}{6}x < -2$
33. $4x < 12$
34. $10x > -40$
35. $-11x \leq -22$
36. $-7x \geq 21$
37. $x - 3(x + 1) \geq 7$
38. $2(4x - 5) - 3x \leq -15$
39. $7x - 12 < 4x + 6$
40. $11 - 6x \leq 2x + 7$
41. $\frac{3}{4}x - 6 \leq x - 7$
42. $3 + \frac{2}{7}x > x - 2$
43. $3.6x + 11 \geq -3.4$
44. $15.6 - 1.3x < -5.2$

Appendix F: Systems of Inequalities

F.1 Solving Systems of Inequalities

The Graph of an Inequality

The statements $3x - 2y < 6$ and $2x^2 + 3y^2 \geq 6$ are inequalities in two variables. An ordered pair (a, b) is a **solution of an inequality** in x and y if the inequality is true when a and b are substituted for x and y, respectively. The **graph of an inequality** is the collection of all solutions of the inequality. To sketch the graph of an inequality, begin by sketching the graph of the *corresponding equation*. The graph of the equation will normally separate the plane into two or more regions. In each such region, one of the following must be true.

1. *All* points in the region are solutions of the inequality.

2. *No* point in the region is a solution of the inequality.

So, you can determine whether the points in an entire region satisfy the inequality by simply testing *one* point in the region.

Sketching the Graph of an Inequality in Two Variables

1. Replace the inequality sign with an equal sign and sketch the graph of the corresponding equation. Use a dashed line for $<$ or $>$ and a solid line for $\leq$ or $\geq$. (A dashed line means that all points on the line or curve *are not* solutions of the inequality. A solid line means that all points on the line or curve *are* solutions of the inequality.)

2. Test one point in each of the regions formed by the graph in Step 1. If the point satisfies the inequality, shade the entire region to denote that every point in the region satisfies the inequality.

Example 1   Sketching the Graph of an Inequality

Sketch the graph of $y \geq x^2 - 1$ by hand.

Solution

Begin by graphing the corresponding *equation* $y = x^2 - 1$, which is a parabola, as shown in Figure F.1. By testing a point *above* the parabola $(0, 0)$ and a point *below* the parabola $(0, -2)$, you can see that $(0, 0)$ satisfies the inequality because $0 \geq 0^2 - 1$ and that $(0, -2)$ does not satisfy the inequality because $-2 \not\geq 0^2 - 1$. So, the points that satisfy the inequality are those lying above and those lying on the parabola.

✓**CHECKPOINT**   Now try Exercise 9.

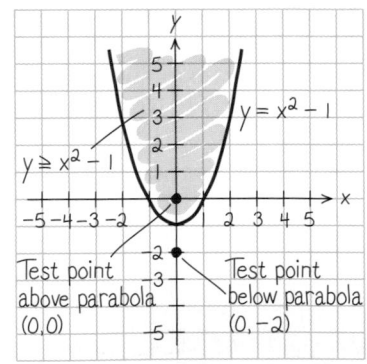

Figure F.1

The inequality in Example 1 is a nonlinear inequality in two variables. Most of the following examples involve **linear inequalities** such as $ax + by < c$ (a and b are not both zero). The graph of a linear inequality is a half-plane lying on one side of the line $ax + by = c$.

Example 2   Sketching the Graphs of Linear Inequalities

Sketch the graph of each linear inequality.

a. $x > -2$ **b.** $y \le 3$

Solution

a. The graph of the corresponding equation $x = -2$ is a vertical line. The points that satisfy the inequality $x > -2$ are those lying to the right of (but not on) this line, as shown in Figure F.2.

b. The graph of the corresponding equation $y = 3$ is a horizontal line. The points that satisfy the inequality $y \le 3$ are those lying below (or on) this line, as shown in Figure F.3.

TECHNOLOGY TIP

A graphing utility can be used to graph an inequality. For instance, to graph $y \ge x - 2$, enter $y = x - 2$ and use the *shade* feature of the graphing utility to shade the correct part of the graph. You should obtain the graph shown below.

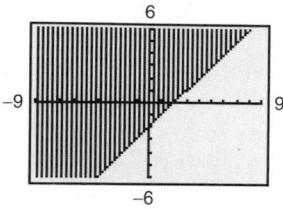

For instructions on how to use the *shade* feature, see Appendix A; for specific keystrokes, go to this textbook's *Online Study Center*.

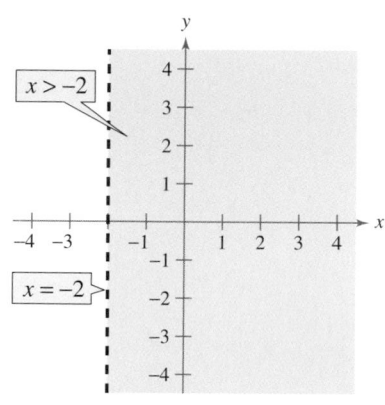

Figure F.2

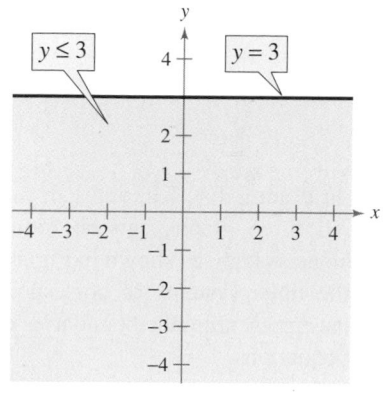

Figure F.3

 **CHECKPOINT**   Now try Exercise 15.

Example 3   Sketching the Graph of a Linear Inequality

Sketch the graph of $x - y < 2$.

Solution

The graph of the corresponding equation $x - y = 2$ is a line, as shown in Figure F.4. Because the origin $(0, 0)$ satisfies the inequality, the graph consists of the half-plane lying above the line. (Try checking a point below the line. Regardless of which point below the line you choose, you will see that it does not satisfy the inequality.)

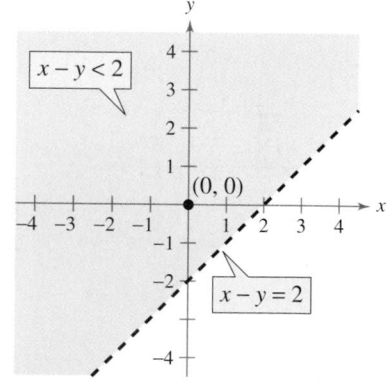

Figure F.4

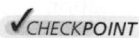 **CHECKPOINT**   Now try Exercise 17.

To graph a linear inequality, it can help to write the inequality in slope-intercept form. For instance, by writing $x - y < 2$ in Example 3 in the form

$$y > x - 2$$

you can see that the solution points lie *above* the line $y = x - 2$ (or $x - y = 2$), as shown in Figure F.4.

Systems of Inequalities

Many practical problems in business, science, and engineering involve systems of linear inequalities. A **solution of a system of inequalities** in x and y is a point (x, y) that satisfies each inequality in the system.

To sketch the graph of a system of inequalities in two variables, first sketch the graph of each individual inequality (on the same coordinate system) and then find the region that is *common* to every graph in the system. For systems of *linear* inequalities, it is helpful to find the vertices of the solution region.

Example 4   Solving a System of Inequalities

Sketch the graph (and label the vertices) of the solution set of the system.

$$\begin{cases} x - y < 2 & \text{Inequality 1} \\ x > -2 & \text{Inequality 2} \\ y \le 3 & \text{Inequality 3} \end{cases}$$

Solution

The graphs of these inequalities are shown in Figures F.4, F.2, and F.3, respectively. The triangular region common to all three graphs can be found by superimposing the graphs on the same coordinate system, as shown in Figure F.5. To find the vertices of the region, solve the three systems of corresponding equations obtained by taking pairs of equations representing the boundaries of the individual regions and solving these pairs of equations.

Vertex A: $(-2, -4)$ *Vertex B:* $(5, 3)$ *Vertex C:* $(-2, 3)$

$$\begin{cases} x - y = 2 \\ x = -2 \end{cases} \qquad \begin{cases} x - y = 2 \\ y = 3 \end{cases} \qquad \begin{cases} x = -2 \\ y = 3 \end{cases}$$

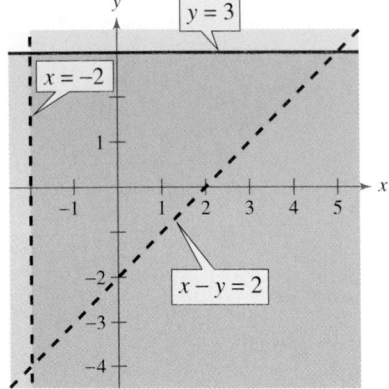

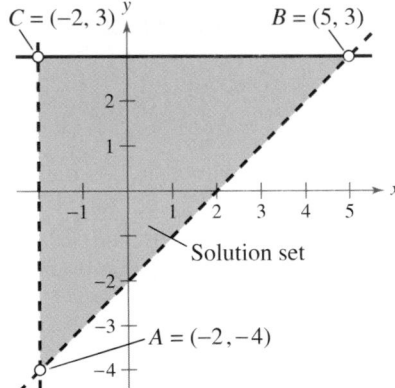

Figure F.5

> ## STUDY TIP
>
> Using different colored pencils to shade the solution of each inequality in a system makes identifying the solution of the system of inequalities easier. The region common to every graph in the system is where all shaded regions overlap. This region represents the solution set of the system.

Note in Figure F.5 that the vertices of the region are represented by open dots. This means that the vertices *are not* solutions of the system of inequalities.

✓*CHECKPOINT*   Now try Exercise 47.

For the triangular region shown in Figure F.5, each point of intersection of a pair of boundary lines corresponds to a vertex. With more complicated regions, two border lines can sometimes intersect at a point that is not a vertex of the region, as shown in Figure F.6. To keep track of which points of intersection are actually vertices of the region, you should sketch the region and refer to your sketch as you find each point of intersection.

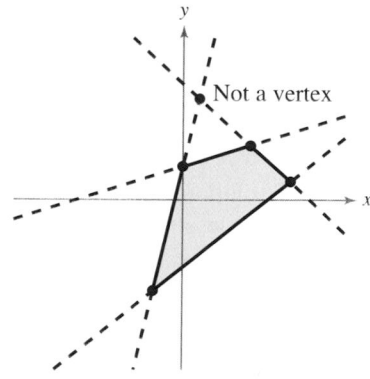

Figure F.6

Example 5   Solving a System of Inequalities

Sketch the region containing all points that satisfy the system of inequalities.

$$\begin{cases} x^2 - y \le 1 & \text{Inequality 1} \\ -x + y \le 1 & \text{Inequality 2} \end{cases}$$

Solution

As shown in Figure F.7, the points that satisfy the inequality $x^2 - y \le 1$ are the points lying above (or on) the parabola given by

$$y = x^2 - 1. \qquad \text{Parabola}$$

The points that satisfy the inequality $-x + y \le 1$ are the points lying below (or on) the line given by

$$y = x + 1. \qquad \text{Line}$$

To find the points of intersection of the parabola and the line, solve the system of corresponding equations.

$$\begin{cases} x^2 - y = 1 \\ -x + y = 1 \end{cases}$$

Using the method of substitution, you can find the solutions to be $(-1, 0)$ and $(2, 3)$. So, the region containing all points that satisfy the system is indicated by the purple shaded region in Figure F.7.

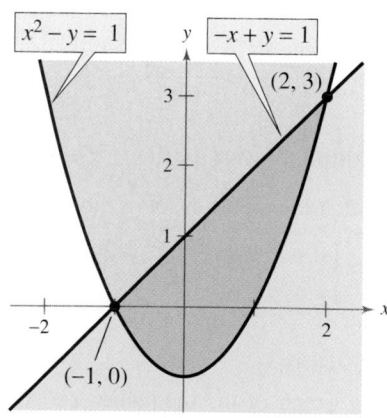

Figure F.7

**CHECKPOINT**   Now try Exercise 55.

When solving a system of inequalities, you should be aware that the system might have no solution, or it might be represented by an unbounded region in the plane. These two possibilities are shown in Examples 6 and 7.

Example 6   A System with No Solution

Sketch the solution set of the system of inequalities.

$$\begin{cases} x + y > 3 & \text{Inequality 1} \\ x + y < -1 & \text{Inequality 2} \end{cases}$$

Solution

From the way the system is written, it is clear that the system has no solution, because the quantity $(x + y)$ cannot be both less than -1 and greater than 3. Graphically, the inequality $x + y > 3$ is represented by the half-plane lying above the line $x + y = 3$, and the inequality $x + y < -1$ is represented by the half-plane lying below the line $x + y = -1$, as shown in Figure F.8. These two half-planes have no points in common. So the system of inequalities has no solution.

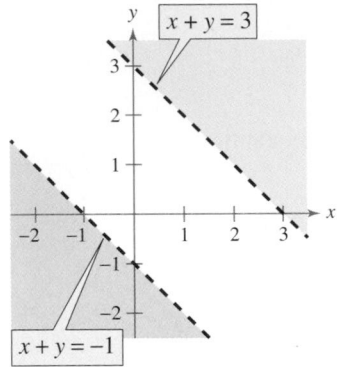

Figure F.8   No Solution

✓*CHECKPOINT*   Now try Exercise 51.

Example 7   An Unbounded Solution Set

Sketch the solution set of the system of inequalities.

$$\begin{cases} x + y < 3 & \text{Inequality 1} \\ x + 2y > 3 & \text{Inequality 2} \end{cases}$$

Solution

The graph of the inequality $x + y < 3$ is the half-plane that lies below the line $x + y = 3$, as shown in Figure F.9. The graph of the inequality $x + 2y > 3$ is the half-plane that lies above the line $x + 2y = 3$. The intersection of these two half-planes is an *infinite wedge* that has a vertex at $(3, 0)$. This unbounded region represents the solution set.

✓*CHECKPOINT*   Now try Exercise 53.

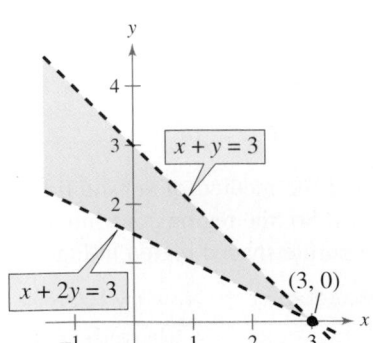

Figure F.9   Unbounded Region

Applications

The next example discusses two concepts that economists call *consumer surplus* and *producer surplus*. As shown in Figure F.10, the *point of equilibrium* is defined by the price p and the number of units x that satisfy both the demand and supply equations. Consumer surplus is defined as the area of the region that lies *below* the demand curve, *above* the horizontal line passing through the equilibrium point, and to the right of the p-axis. Similarly, the producer surplus is defined as the area of the region that lies *above* the supply curve, *below* the horizontal line passing through the equilibrium point, and to the right of the p-axis. The consumer surplus is a measure of the amount that consumers would have been willing to pay *above what they actually paid*, whereas the producer surplus is a measure of the amount that producers would have been willing to receive *below what they actually received*.

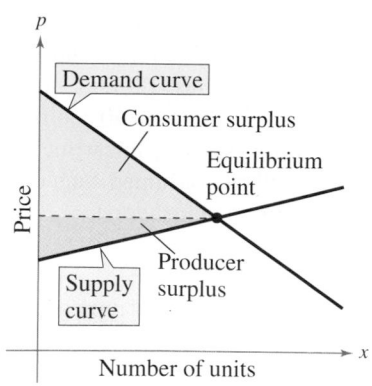

Figure F.10

Example 8   Consumer Surplus and Producer Surplus

The demand and supply functions for a new type of calculator are given by

$$\begin{cases} p = 150 - 0.00001x & \text{Demand equation} \\ p = 60 + 0.00002x & \text{Supply equation} \end{cases}$$

where p is the price (in dollars) and x represents the number of units. Find the consumer surplus and producer surplus for these two equations.

Solution

Begin by finding the point of equilibrium by setting the two equations equal to each other and solving for x.

$$60 + 0.00002x = 150 - 0.00001x \qquad \text{Set equations equal to each other.}$$
$$0.00003x = 90 \qquad \text{Combine like terms.}$$
$$x = 3,000,000 \qquad \text{Solve for } x.$$

So, the solution is $x = 3,000,000$, which corresponds to an equilibrium price of $p = \$120$. So, the consumer surplus and producer surplus are the areas of the following triangular regions.

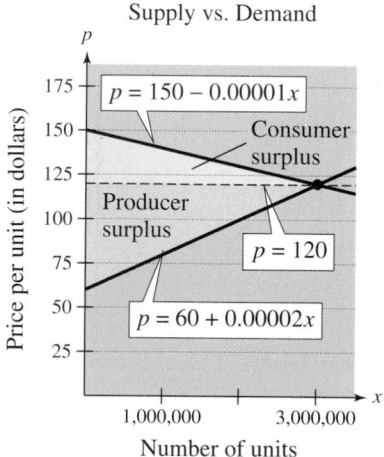

Supply vs. Demand

Figure F.11

Consumer Surplus	Producer Surplus
$\begin{cases} p \le 150 - 0.00001x \\ p \ge 120 \\ x \ge 0 \end{cases}$	$\begin{cases} p \ge 60 + 0.00002x \\ p \le 120 \\ x \ge 0 \end{cases}$

In Figure F.11, you can see that the consumer and producer surpluses are defined as the areas of the shaded triangles.

$$\text{Consumer surplus} = \tfrac{1}{2}(\text{base})(\text{height}) = \tfrac{1}{2}(3,000,000)(30) = \$45,000,000$$

$$\text{Producer surplus} = \tfrac{1}{2}(\text{base})(\text{height}) = \tfrac{1}{2}(3,000,000)(60) = \$90,000,000$$

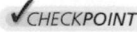 CHECKPOINT   Now try Exercise 75.

Example 9 Nutrition

The minimum daily requirements from the liquid portion of a diet are 300 calories, 36 units of vitamin A, and 90 units of vitamin C. A cup of dietary drink X provides 60 calories, 12 units of vitamin A, and 10 units of vitamin C. A cup of dietary drink Y provides 60 calories, 6 units of vitamin A, and 30 units of vitamin C. Set up a system of linear inequalities that describes how many cups of each drink should be consumed each day to meet the minimum daily requirements for calories and vitamins.

Solution

Begin by letting x and y represent the following.

x = number of cups of dietary drink X

y = number of cups of dietary drink Y

To meet the minimum daily requirements, the following inequalities must be satisfied.

$$\begin{cases} 60x + 60y \geq 300 & \text{Calories} \\ 12x + 6y \geq 36 & \text{Vitamin A} \\ 10x + 30y \geq 90 & \text{Vitamin C} \\ x \geq 0 \\ y \geq 0 \end{cases}$$

The last two inequalities are included because x and y cannot be negative. The graph of this system of inequalities is shown in Figure F.12. (More is said about this application in Example 6 in Section F.2.)

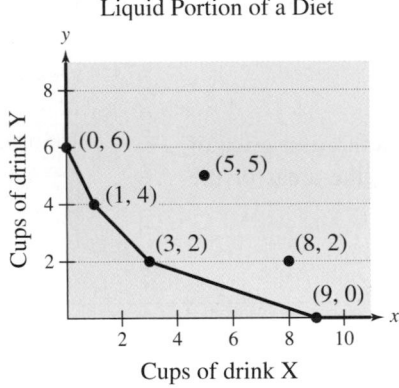

Liquid Portion of a Diet

Figure F.12

From the graph, you can see that two solutions (other than the vertices) that will meet the minimum daily requirements for calories and vitamins are $(5, 5)$ and $(8, 2)$. There are many other solutions.

✓CHECKPOINT   Now try Exercise 81.

STUDY TIP

When using a system of inequalities to represent a real-life application in which the variables cannot be negative, remember to include inequalities for this constraint. For instance, in Example 9, x and y cannot be negative, so the inequalities $x \geq 0$ and $y \geq 0$ must be included in the system.

F.1   Exercises

See www.CalcChat.com for worked-out solutions to odd-numbered exercises.

Vocabulary Check

Fill in the blanks.

1. An ordered pair (a, b) is a _____ of an inequality in x and y if the inequality is true when a and b are substituted for x and y, respectively.

2. The _____ of an inequality is the collection of all solutions of the inequality.

3. The graph of a _____ inequality is a half-plane lying on one side of the line $ax + by = c$.

4. The _____ of _____ is defined by the price p and the number of units x that satisfy both the demand and supply equations.

In Exercises 1–8, match the inequality with its graph. [The graphs are labeled (a), (b), (c), (d), (e), (f), (g), and (h).]

1. $x < 2$

2. $y \geq 3$

3. $2x + 3y \geq 6$

4. $2x - y \leq -2$

5. $x^2 + y^2 < 9$

6. $(x - 2)^2 + (y - 3)^2 > 9$

7. $xy > 1$

8. $y \leq 1 - x^2$

(a)
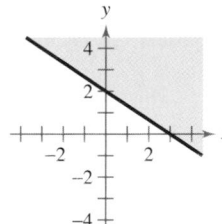

(b)
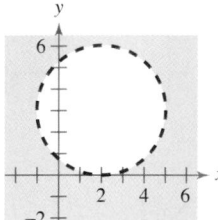

(c)
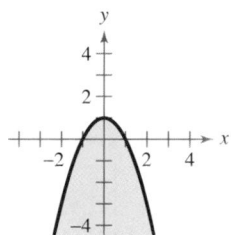

(d)
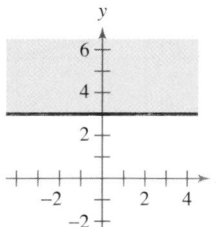

(e)
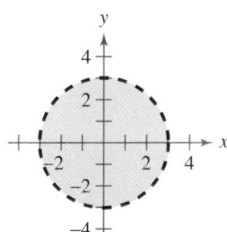

(f)
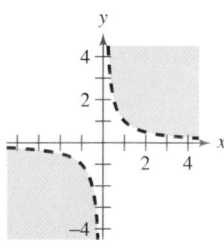

(g)
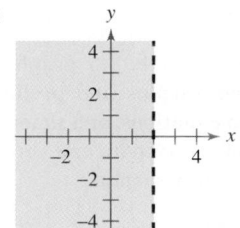

(h)
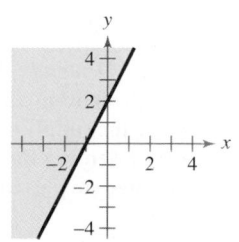

In Exercises 9–28, sketch the graph of the inequality.

9. $y < 2 - x^2$

10. $y - 4 \leq x^2$

11. $y^2 + 1 \geq x$

12. $y^2 - x < 0$

13. $x \geq 4$

14. $x \leq -5$

15. $y \geq -1$

16. $y \leq 3$

17. $2y - x \geq 4$

18. $5x + 3y \geq -15$

19. $2x + 3y < 6$

20. $5x - 2y > 10$

21. $4x - 3y \leq 24$

22. $2x + 7y \leq 28$

23. $y > 3x^2 + 1$

24. $y + 9 \geq x^2$

25. $2x - y^2 > 0$

26. $4x + y^2 > 1$

27. $(x + 1)^2 + y^2 < 9$

28. $(x - 1)^2 + (y - 4)^2 > 9$

In Exercises 29–40, use a graphing utility to graph the inequality. Use the *shade* feature to shade the region representing the solution.

29. $y \geq \frac{2}{3}x - 1$

30. $y \leq 6 - \frac{3}{2}x$

31. $y < -3.8x + 1.1$

32. $y \geq -20.74 + 2.66x$

33. $x^2 + 5y - 10 \leq 0$

34. $2x^2 - y - 3 > 0$

35. $y \leq \dfrac{1}{1 + x^2}$

36. $y > \dfrac{-10}{x^2 + x + 4}$

37. $y < \ln x$

38. $y \geq 4 - \ln(x + 5)$

39. $y > 3^{-x-4}$

40. $y \leq 2^{2x-1} - 3$

In Exercises 41–44, write an inequality for the shaded region shown in the graph.

41.

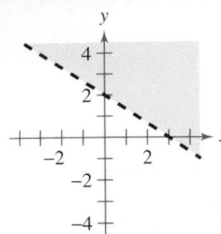

42.

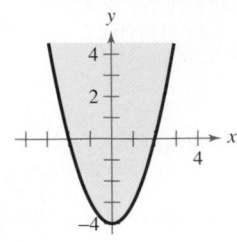

43.

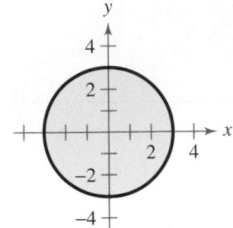

44.

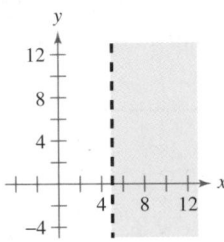

In Exercises 45 and 46, determine whether each ordered pair is a solution of the system of inequalities.

45. $\begin{cases} -2x + 5y \geq 3 \\ \quad\quad\quad y < 4 \\ -4x + 2y < 7 \end{cases}$ (a) $(0, 2)$ (b) $(-6, 4)$

(c) $(-8, -2)$ (d) $(-3, 2)$

46. $\begin{cases} x^2 + y^2 \geq 36 \\ -3x + y \leq 10 \\ \frac{2}{3}x - y \geq 5 \end{cases}$ (a) $(-1, 7)$ (b) $(-5, 1)$

(c) $(6, 0)$ (d) $(4, -8)$

In Exercises 47–64, sketch the graph of the solution of the system of inequalities.

47. $\begin{cases} x + y \leq 1 \\ -x + y \leq 1 \\ \quad\quad y \geq 0 \end{cases}$ **48.** $\begin{cases} 3x + 2y < 6 \\ x \quad\quad > 0 \\ \quad\quad y > 0 \end{cases}$

49. $\begin{cases} -3x + 2y < 6 \\ \quad x - 4y > -2 \\ \quad 2x + y < 3 \end{cases}$ **50.** $\begin{cases} x - 7y > -36 \\ 5x + 2y > 5 \\ 6x - 5y > 6 \end{cases}$

51. $\begin{cases} 3x + y \leq y^2 \\ x - y > 0 \end{cases}$ **52.** $\begin{cases} y^2 - 3x \geq 9 \\ x + y \geq -3 \end{cases}$

53. $\begin{cases} 2x + y < 2 \\ x + 3y > 2 \end{cases}$ **54.** $\begin{cases} x - 2y < -6 \\ 2x - 4y > -9 \end{cases}$

55. $\begin{cases} x < y^2 \\ x > y + 2 \end{cases}$ **56.** $\begin{cases} x - y^2 > 0 \\ x - y < 2 \end{cases}$

57. $\begin{cases} x^2 + y^2 \leq 9 \\ x^2 + y^2 \geq 1 \end{cases}$ **58.** $\begin{cases} x^2 + y^2 \leq 25 \\ 4x - 3y \leq 0 \end{cases}$

59. $\begin{cases} y \leq \sqrt{3x} + 1 \\ y \geq x^2 + 1 \end{cases}$ **60.** $\begin{cases} y < -x^2 + 2x + 3 \\ y > x^2 - 4x + 3 \end{cases}$

61. $\begin{cases} y < x^3 - 2x + 1 \\ y > -2x \\ x \leq 1 \end{cases}$ **62.** $\begin{cases} y \geq x^4 - 2x^2 + 1 \\ y \leq 1 - x^2 \end{cases}$

63. $\begin{cases} x^2y \geq 1 \\ 0 < x \leq 4 \\ y \leq 4 \end{cases}$ **64.** $\begin{cases} y \leq e^{-x^2/2} \\ y \geq 0 \\ -2 \leq x \leq 2 \end{cases}$

In Exercises 65–74, find a set of inequalities to describe the region.

65.

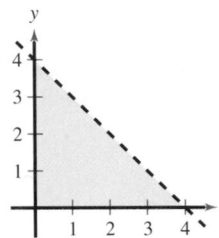

66.

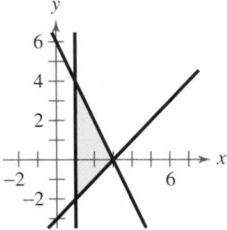

67.

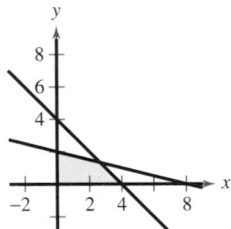

68.

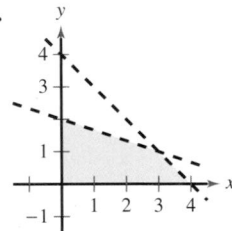

69.

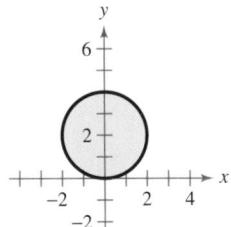

70.

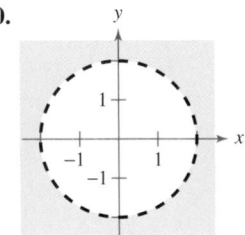

71. Rectangle: Vertices at $(2, 1), (5, 1), (5, 7), (2, 7)$

72. Parallelogram: Vertices at $(0, 0), (4, 0), (1, 4), (5, 4)$

73. Triangle: Vertices at $(0, 0), (5, 0), (2, 3)$

74. Triangle: Vertices at $(-1, 0), (1, 0), (0, 1)$

Supply and Demand **In Exercises 75–78, (a) graph the system representing the consumer surplus and producer surplus for the supply and demand equations, and shade the region representing the solution of the system, and (b) find the consumer surplus and the producer surplus.**

Demand	*Supply*
75. $p = 50 - 0.5x$	$p = 0.125x$

Demand	Supply
76. $p = 100 - 0.05x$	$p = 25 + 0.1x$
77. $p = 300 - 0.0002x$	$p = 225 + 0.0005x$
78. $p = 140 - 0.00002x$	$p = 80 + 0.00001x$

In Exercises 79–82, (a) find a system of inequalities that models the problem and (b) graph the system, shading the region that represents the solution of the system.

79. *Investment Analysis* A person plans to invest some or all of $30,000 in two different interest-bearing accounts. Each account is to contain at least $7500, and one account should have at least twice the amount that is in the other account.

80. *Ticket Sales* For a summer concert event, one type of ticket costs $20 and another costs $35. The promoter of the concert must sell at least 20,000 tickets, including at least 10,000 of the $20 tickets and at least 5000 of the $35 tickets, and the gross receipts must total at least $300,000 in order for the concert to be held.

81. *Nutrition* A dietitian is asked to design a special dietary supplement using two different foods. The minimum daily requirements of the new supplement are 280 units of calcium, 160 units of iron, and 180 units of vitamin B. Each ounce of food X contains 20 units of calcium, 15 units of iron, and 10 units of vitamin B. Each ounce of food Y contains 10 units of calcium, 10 units of iron, and 20 units of vitamin B.

82. *Inventory* A store sells two models of computers. Because of the demand, the store stocks at least twice as many units of model A as units of model B. The costs to the store for models A and B are $800 and $1200, respectively. The management does not want more than $20,000 in computer inventory at any one time, and it wants at least four model A computers and two model B computers in inventory at all times.

83. *Construction* You design an exercise facility that has an indoor running track with an exercise floor inside the track (see figure). The track must be at least 125 meters long, and the exercise floor must have an area of at least 500 square meters.

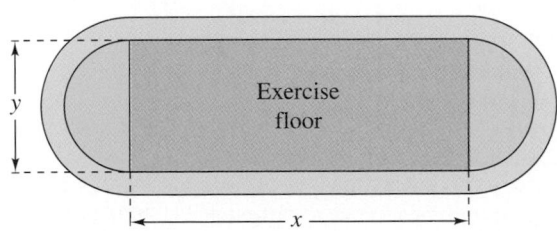

Exercise floor

(a) Find a system of inequalities describing the requirements of the facility.

(b) Sketch the graph of the system in part (a).

84. *Graphical Reasoning* Two concentric circles have radii of x and y meters, where $y > x$ (see figure). The area between the boundaries of the circles must be at least 10 square meters.

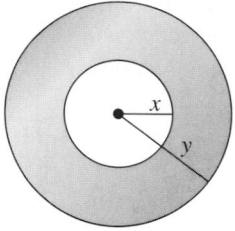

(a) Find a system of inequalities describing the constraints on the circles.

(b) Graph the inequality in part (a).

(c) Identify the graph of the line $y = x$ in relation to the boundary of the inequality. Explain its meaning in the context of the problem.

Synthesis

True or False?   **In Exercises 85 and 86, determine whether the statement is true or false. Justify your answer.**

85. The area of the figure defined by the system below is 99 square units.

$$\begin{cases} x \geq -3 \\ x \leq 6 \\ y \leq 5 \\ y \geq -6 \end{cases}$$

86. The graph below shows the solution of the system

$$\begin{cases} y \leq 6 \\ -4x - 9y > 6. \\ 3x + y^2 \geq 2 \end{cases}$$

87. *Think About It* After graphing the boundary of an inequality in x and y, how do you decide on which side of the boundary the solution set of the inequality lies?

88. *Writing* Describe the difference between the solution set of a system of equations and the solution set of a system of inequalities.

F.2   Linear Programming

Linear Programming: A Graphical Approach

Many applications in business and economics involve a process called **optimization,** in which you are asked to find the minimum or maximum value of a quantity. In this section, you will study an optimization strategy called **linear programming.**

A two-dimensional linear programming problem consists of a linear **objective function** and a system of linear inequalities called **constraints.** The objective function gives the quantity that is to be maximized (or minimized), and the constraints determine the set of **feasible solutions.** For example, suppose you are asked to maximize the value of

$$z = ax + by \qquad \text{Objective function}$$

subject to a set of constraints that determines the region in Figure F.13. Because every point in the shaded region satisfies each constraint, it is not clear how you should find the point that yields a maximum value of z. Fortunately, it can be shown that if there is an optimal solution, it must occur at one of the vertices. So, *you can find the maximum value of z by testing z at each of the vertices.*

Optimal Solution of a Linear Programming Problem

If a linear programming problem has a solution, it must occur at a vertex of the set of feasible solutions. If there is more than one solution, at least one of them must occur at such a vertex. In either case, the value of the objective function is unique.

Here are some guidelines for solving a linear programming problem in two variables in which an objective function is to be maximized or minimized.

Solving a Linear Programming Problem

1. Sketch the region corresponding to the system of constraints. (The points inside or on the boundary of the region are *feasible solutions.*)

2. Find the vertices of the region.

3. Test the objective function at each of the vertices and select the values of the variables that optimize the objective function. For a bounded region, both a minimum and a maximum value will exist. (For an unbounded region, *if* an optimal solution exists, it will occur at a vertex.)

Figure F.13

Example 1   Solving a Linear Programming Problem

Find the maximum value of

$$z = 3x + 2y \qquad \text{Objective function}$$

subject to the following constraints.

$$\left. \begin{array}{r} x \geq 0 \\ y \geq 0 \\ x + 2y \leq 4 \\ x - y \leq 1 \end{array} \right\} \qquad \text{Constraints}$$

Solution

The constraints form the region shown in Figure F.14. At the four vertices of this region, the objective function has the following values.

At $(0, 0)$: $z = 3(0) + 2(0) = 0$

At $(1, 0)$: $z = 3(1) + 2(0) = 3$

At $(2, 1)$: $z = 3(2) + 2(1) = 8$ \qquad Maximum value of z

At $(0, 2)$: $z = 3(0) + 2(2) = 4$

So, the maximum value of z is 8, and this value occurs when $x = 2$ and $y = 1$.

 *CHECKPOINT*   Now try Exercise 13.

In Example 1, try testing some of the *interior* points in the region. You will see that the corresponding values of z are less than 8. Here are some examples.

At $(1, 1)$: $z = 3(1) + 2(1) = 5$

At $\left(1, \frac{1}{2}\right)$: $z = 3(1) + 2\left(\frac{1}{2}\right) = 4$

At $\left(\frac{1}{2}, \frac{3}{2}\right)$: $z = 3\left(\frac{1}{2}\right) + 2\left(\frac{3}{2}\right) = \frac{9}{2}$

To see why the maximum value of the objective function in Example 1 must occur at a vertex, consider writing the objective function in the form

$$y = -\frac{3}{2}x + \frac{z}{2} \qquad \text{Family of lines}$$

where $z/2$ is the y-intercept of the objective function. This equation represents a family of lines, each of slope $-\frac{3}{2}$. Of these infinitely many lines, you want the one that has the largest z-value while still intersecting the region determined by the constraints. In other words, of all the lines with a slope of $-\frac{3}{2}$, you want the one that has the largest y-intercept *and* intersects the given region, as shown in Figure F.15. It should be clear that such a line will pass through one (or more) of the vertices of the region.

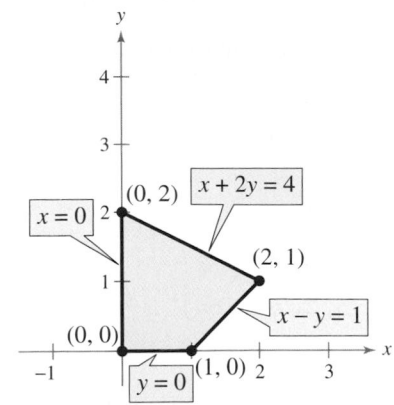

Figure F.14

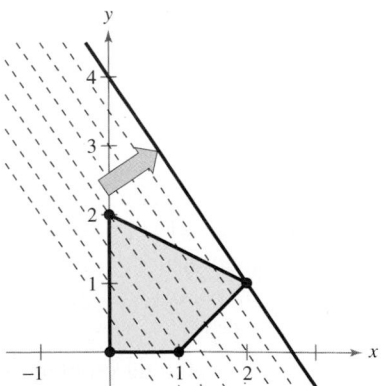

Figure F.15

The next example shows that the same basic procedure can be used to solve a problem in which the objective function is to be *minimized*.

Example 2   Solving a Linear Programming Problem

Find the minimum value of

$$z = 5x + 7y \qquad \text{Objective function}$$

where $x \geq 0$ and $y \geq 0$, subject to the following constraints.

$$\left. \begin{aligned} 2x + 3y &\geq 6 \\ 3x - y &\leq 15 \\ -x + y &\leq 4 \\ 2x + 5y &\leq 27 \end{aligned} \right\} \qquad \text{Constraints}$$

Solution

The region bounded by the constraints is shown in Figure F.16. By testing the objective function at each vertex, you obtain the following.

At $(0, 2)$: $z = 5(0) + 7(2) = 14$ Minimum value of z
At $(0, 4)$: $z = 5(0) + 7(4) = 28$
At $(1, 5)$: $z = 5(1) + 7(5) = 40$
At $(6, 3)$: $z = 5(6) + 7(3) = 51$
At $(5, 0)$: $z = 5(5) + 7(0) = 25$
At $(3, 0)$: $z = 5(3) + 7(0) = 15$

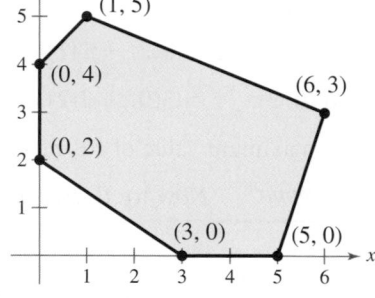

Figure F.16

So, the minimum value of z is 14, and this value occurs when $x = 0$ and $y = 2$.

✓CHECKPOINT   Now try Exercise 15.

Example 3   Solving a Linear Programming Problem

Find the maximum value of

$$z = 5x + 7y \qquad \text{Objective function}$$

where $x \geq 0$ and $y \geq 0$, subject to the following constraints.

$$\left. \begin{aligned} 2x + 3y &\geq 6 \\ 3x - y &\leq 15 \\ -x + y &\leq 4 \\ 2x + 5y &\leq 27 \end{aligned} \right\} \qquad \text{Constraints}$$

Solution

This linear programming problem is identical to that given in Example 2 above, *except* that the objective function is *maximized* instead of minimized. Using the values of z at the vertices shown in Example 2, you can conclude that the maximum value of z is 51, and that this value occurs when $x = 6$ and $y = 3$.

✓CHECKPOINT   Now try Exercise 21.

It is possible for the maximum (or minimum) value in a linear programming problem to occur at *two* different vertices. For instance, at the vertices of the region shown in Figure F.17, the objective function

$z = 2x + 2y$ Objective function

has the following values.

At $(0, 0)$: $z = 2(0) + 2(0) = 0$
At $(0, 4)$: $z = 2(0) + 2(4) = 8$
At $(2, 4)$: $z = 2(2) + 2(4) = 12$ Maximum value of z
At $(5, 1)$: $z = 2(5) + 2(1) = 12$ Maximum value of z
At $(5, 0)$: $z = 2(5) + 2(0) = 10$

In this case, you can conclude that the objective function has a maximum value (of 12) not only at the vertices $(2, 4)$ and $(5, 1)$, but also at *any point on the line segment connecting these two vertices*, as shown in Figure F.17. Note that by rewriting the objective function as

$$y = -x + \frac{1}{2}z$$

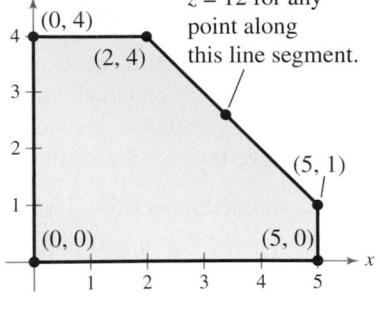

Figure F.17

you can see that its graph has the same slope as the line through the vertices $(2, 4)$ and $(5, 1)$.

Some linear programming problems have no optimal solutions. This can occur if the region determined by the constraints is *unbounded*.

Example 4   An Unbounded Region

Find the maximum value of

$z = 4x + 2y$ Objective function

where $x \geq 0$ and $y \geq 0$, subject to the following constraints.

$$\left.\begin{array}{r} x + 2y \geq 4 \\ 3x + y \geq 7 \\ -x + 2y \leq 7 \end{array}\right\}$$ Constraints

Solution

The region determined by the constraints is shown in Figure F.18. For this unbounded region, there is no maximum value of z. To see this, note that the point $(x, 0)$ lies in the region for all values of $x \geq 4$. By choosing large values of x, you can obtain values of $z = 4(x) + 2(0) = 4x$ that are as large as you want. So, there is no maximum value of z. For the vertices of the region, the objective function has the following values. So, there *is* a minimum value of z, $z = 10$, which occurs at the vertex $(2, 1)$.

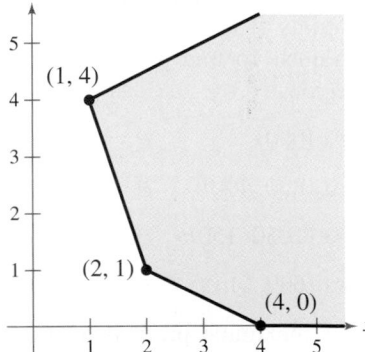

Figure F.18

At $(1, 4)$: $z = 4(1) + 2(4) = 12$
At $(2, 1)$: $z = 4(2) + 2(1) = 10$ Minimum value of z
At $(4, 0)$: $z = 4(4) + 2(0) = 16$

✔CHECKPOINT   Now try Exercise 31.

Applications

Example 5 shows how linear programming can be used to find the maximum profit in a business application.

Example 5   Optimizing Profit

A manufacturer wants to maximize the profit from selling two types of boxed chocolates. A box of chocolate covered creams yields a profit of $1.50, and a box of chocolate covered cherries yields a profit of $2.00. Market tests and available resources have indicated the following constraints.

1. The combined production level should not exceed 1200 boxes per month.
2. The demand for a box of chocolate covered cherries is no more than half the demand for a box of chocolate covered creams.
3. The production level of a box of chocolate covered creams is less than or equal to 600 boxes plus three times the production level of a box of chocolate covered cherries.

Solution

Let x be the number of boxes of chocolate covered creams and y be the number of boxes of chocolate covered cherries. The objective function (for the combined profit) is given by

$$P = 1.5x + 2y. \qquad \text{Objective function}$$

The three constraints translate into the following linear inequalities.

1. $x + y \le 1200$ ⟹ $x + y \le 1200$

2. $y \le \frac{1}{2}x$ ⟹ $-x + 2y \le 0$

3. $x \le 3y + 600$ ⟹ $x - 3y \le 600$

Because neither x nor y can be negative, you also have the two additional constraints of $x \ge 0$ and $y \ge 0$. Figure F.19 shows the region determined by the constraints. To find the maximum profit, test the value of P at each vertex of the region.

At $(0, 0)$: $P = 1.5(0) \quad + 2(0) = 0$

At $(800, 400)$: $P = 1.5(800) \ + 2(400) = 2000$ Maximum profit

At $(1050, 150)$: $P = 1.5(1050) + 2(150) = 1875$

At $(600, 0)$: $P = 1.5(600) \ + 2(0) = 900$

So, the maximum profit is $2000, and it occurs when the monthly production consists of 800 boxes of chocolate covered creams and 400 boxes of chocolate covered cherries.

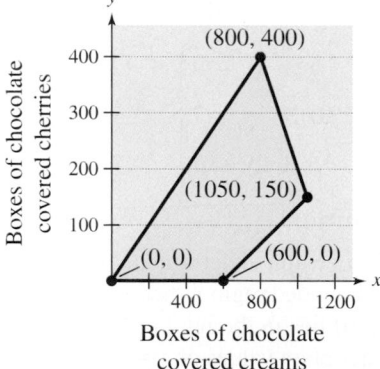

Figure F.19

✓CHECKPOINT   Now try Exercise 35.

In Example 5, suppose the manufacturer improves the production of chocolate covered creams so that a profit of $2.50 per box is obtained. The maximum profit can now be found using the objective function $P = 2.5x + 2y$. By testing the values of P at the vertices of the region, you find that the maximum profit is now $2925, which occurs when $x = 1050$ and $y = 150$.

Example 6 Optimizing Cost

The minimum daily requirements from the liquid portion of a diet are 300 calories, 36 units of vitamin A, and 90 units of vitamin C. A cup of dietary drink X costs $0.12 and provides 60 calories, 12 units of vitamin A, and 10 units of vitamin C. A cup of dietary drink Y costs $0.15 and provides 60 calories, 6 units of vitamin A, and 30 units of vitamin C. How many cups of each drink should be consumed each day to minimize the cost and still meet the daily requirements?

Solution

As in Example 9 on page A112, let x be the number of cups of dietary drink X and let y be the number of cups of dietary drink Y.

$$\left.\begin{array}{lll} \textit{For Calories:} & 60x + 60y \geq 300 \\ \textit{For Vitamin A:} & 12x + 6y \geq 36 \\ \textit{For Vitamin C:} & 10x + 30y \geq 90 \\ & x \geq 0 \\ & y \geq 0 \end{array}\right\} \quad \text{Constraints}$$

The cost C is given by

$$C = 0.12x + 0.15y. \qquad \text{Objective function}$$

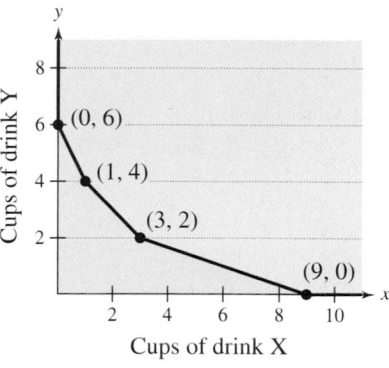

Liquid Portion of a Diet

Figure F.20

The graph of the region determined by the constraints is shown in Figure F.20. To determine the minimum cost, test C at each vertex of the region.

At $(0, 6)$: $C = 0.12(0) + 0.15(6) = 0.90$

At $(1, 4)$: $C = 0.12(1) + 0.15(4) = 0.72$

At $(3, 2)$: $C = 0.12(3) + 0.15(2) = 0.66$ Minimum value of C

At $(9, 0)$: $C = 0.12(9) + 0.15(0) = 1.08$

So, the minimum cost is $0.66 per day, and this cost occurs when three cups of drink X and two cups of drink Y are consumed each day.

✓CHECKPOINT   Now try Exercise 37.

TECHNOLOGY TIP   You can check the points of the vertices of the constraints by using a graphing utility to graph the equations that represent the boundaries of the inequalities. Then use the *intersect* feature to confirm the vertices.

F.2   Exercises

See www.CalcChat.com for worked-out solutions to odd-numbered exercises.

Vocabulary Check

Fill in the blanks.

1. In the process called _____ , you are asked to find the minimum or maximum value of a quantity.

2. The _____ of a linear programming problem gives the quantity that is to be maximized or minimized.

3. The _____ of a linear programming problem determine the set of _____ .

In Exercises 1–12, find the minimum and maximum values of the objective function and where they occur, subject to the indicated constraints. (For each exercise, the graph of the region determined by the constraints is provided.)

1. Objective function:

   $z = 3x + 5y$

   Constraints:

   $$x \geq 0$$
   $$y \geq 0$$
   $$x + y \leq 6$$

2. Objective function:

   $z = 2x + 8y$

   Constraints:

   $$x \geq 0$$
   $$y \geq 0$$
   $$2x + y \leq 4$$

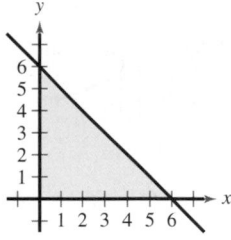

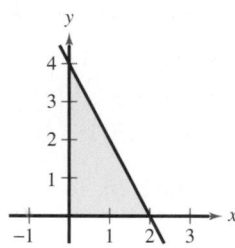

3. Objective function:

   $z = 10x + 7y$

   Constraints:
   See Exercise 1.

4. Objective function:

   $z = 7x + 3y$

   Constraints:
   See Exercise 2.

5. Objective function:

   $z = 3x + 2y$

   Constraints:

   $$x \geq 0$$
   $$y \geq 0$$
   $$x + 3y \leq 15$$
   $$4x + y \leq 16$$

6. Objective function:

   $z = 4x + 3y$

   Constraints:

   $$x \geq 0$$
   $$2x + 3y \geq 6$$
   $$3x - 2y \leq 9$$
   $$x + 5y \leq 20$$

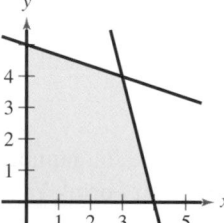

Figure for 5

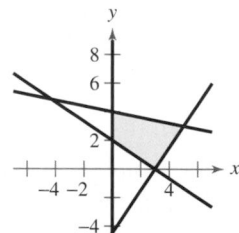

Figure for 6

7. Objective function:

   $z = 5x + 0.5y$

   Constraints:
   See Exercise 5.

8. Objective function:

   $z = x + 6y$

   Constraints:
   See Exercise 6.

9. Objective function:

   $z = 10x + 7y$

   Constraints:

   $$0 \leq x \leq 60$$
   $$0 \leq y \leq 45$$
   $$5x + 6y \leq 420$$

10. Objective function:

    $z = 50x + 35y$

    Constraints:

    $$x \geq 0$$
    $$y \geq 0$$
    $$8x + 9y \leq 7200$$
    $$8x + 9y \geq 5400$$

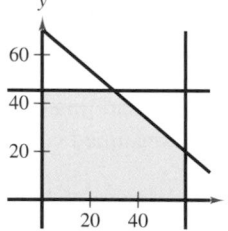

11. Objective function:

    $z = 25x + 30y$

    Constraints:
    See Exercise 9.

12. Objective function:

    $z = 15x + 20y$

    Constraints:
    See Exercise 10.

In Exercises 13–26, sketch the region determined by the constraints. Then find the minimum and maximum values of the objective function and where they occur, subject to the indicated constraints.

13. Objective function:

$z = 6x + 10y$

Constraints:

$$x \geq 0$$
$$y \geq 0$$
$$2x + 5y \leq 10$$

14. Objective function:

$z = 7x + 8y$

Constraints:

$$x \geq 0$$
$$y \geq 0$$
$$x + \tfrac{1}{2}y \leq 4$$

15. Objective function:

$z = 3x + 4y$

Constraints:

$$x \geq 0$$
$$y \geq 0$$
$$2x + 5y \leq 50$$
$$4x + y \leq 28$$

16. Objective function:

$z = 4x + 5y$

Constraints:

$$x \geq 0$$
$$y \geq 0$$
$$2x + 2y \leq 10$$
$$x + 2y \leq 6$$

17. Objective function:

$z = x + 2y$

Constraints:
See Exercise 15.

18. Objective function:

$z = 2x + 4y$

Constraints:
See Exercise 16.

19. Objective function:

$z = 2x$

Constraints:
See Exercise 15.

20. Objective function:

$z = 3y$

Constraints:
See Exercise 16.

21. Objective function:

$z = 4x + y$

Constraints:

$$x \geq 0$$
$$y \geq 0$$
$$x + 2y \leq 40$$
$$2x + 3y \geq 72$$

22. Objective function:

$z = x$

Constraints:

$$x \geq 0$$
$$y \geq 0$$
$$2x + 3y \leq 60$$
$$2x + y \leq 28$$
$$4x + y \leq 48$$

23. Objective function:

$z = x + 4y$

Constraints:
See Exercise 21.

24. Objective function:

$z = y$

Constraints:
See Exercise 22.

25. Objective function:

$z = 2x + 3y$

Constraints:
See Exercise 21.

26. Objective function:

$z = 3x + 2y$

Constraints:
See Exercise 22.

Exploration   **In Exercises 27–30, perform the following.**

(a) Graph the region bounded by the following constraints.

$$3x + y \leq 15$$
$$4x + 3y \leq 30$$
$$x \geq 0$$
$$y \geq 0$$

(b) Graph the objective function for the given maximum value of z on the same set of coordinate axes as the graph of the constraints.

(c) Use the graph to determine the feasible point or points that yield the maximum. Explain how you arrived at your answer.

Objective Function	Maximum
27. $z = 2x + y$	$z = 12$
28. $z = 5x + y$	$z = 25$
29. $z = x + y$	$z = 10$
30. $z = 3x + y$	$z = 15$

In Exercises 31–34, the linear programming problem has an unusual characteristic. Sketch a graph of the solution region for the problem and describe the unusual characteristic. The objective function is to be maximized in each case.

31. Objective function:

$z = x + y$

Constraints:

$$x \geq 0$$
$$y \geq 0$$
$$-x + y \leq 1$$
$$-x + 2y \leq 4$$

32. Objective function:

$z = 2.5x + y$

Constraints:

$$x \geq 0$$
$$y \geq 0$$
$$3x + 5y \leq 15$$
$$5x + 2y \leq 10$$

33. Objective function:

$z = x + y$

Constraints:

$$x \geq 0$$
$$y \geq 0$$
$$-x + y \leq 0$$
$$-3x + y \geq 3$$

34. Objective function:

$z = -x + 2y$

Constraints:

$$x \geq 0$$
$$y \geq 0$$
$$x \leq 10$$
$$x + y \leq 7$$

35. *Optimizing Revenue*   An accounting firm has 800 hours of staff time and 96 hours of reviewing time available each week. The firm charges $2000 for an audit and $300 for a tax return. Each audit requires 100 hours of staff time and 8 hours of review time. Each tax return requires 12.5 hours of staff time and 2 hours of review time. (a) What numbers of audits and tax returns will yield the maximum revenue? (b) What is the maximum revenue?

36. *Optimizing Profit*   A manufacturer produces two models of snowboards. The amounts of time (in hours) required for assembling, painting, and packaging the two models are as follows.

	Model A	Model B
Assembling	2.5	3
Painting	2	1
Packaging	0.75	1.25

The total amounts of time available for assembling, painting, and packaging are 4000 hours, 2500 hours, and 1500 hours, respectively. The profits per unit are $50 for model A and $52 for model B.

(a) How many of each model should be produced to maximize profit?

(b) What is the maximum profit?

37. *Optimizing Cost*   A farming cooperative mixes two brands of cattle feed. Brand X costs $25 per bag and contains two units of nutritional element A, two units of nutritional element B, and two units of nutritional element C. Brand Y costs $20 per bag and contains one unit of nutritional element A, nine units of nutritional element B, and three units of nutritional element C. The minimum requirements for nutritional elements A, B, and C are 12 units, 36 units, and 24 units, respectively.

(a) Find the number of bags of each brand that should be mixed to produce a mixture having a minimum cost per bag.

(b) What is the minimum cost?

38. *Optimizing Cost*   A pet supply company mixes two brands of dry dog food. Brand X costs $15 per bag and contains eight units of nutritional element A, one unit of nutritional element B, and two units of nutritional element C. Brand Y costs $30 per bag and contains two units of nutritional element A, one unit of nutritional element B, and seven units of nutritional element C. Each bag of mixed dog food must contain at least 16 units, 5 units, and 20 units of nutritional elements A, B, and C, respectively.

(a) Find the numbers of bags of brands X and Y that should be mixed to produce a mixture meeting the minimum nutritional requirements and having a minimum cost per bag.

(b) What is the minimum cost?

Synthesis

True or False?   **In Exercises 39 and 40, determine whether the statement is true or false. Justify your answer.**

39. If an objective function has a maximum value at the adjacent vertices (4, 7) and (8, 3), you can conclude that it also has a maximum value at the points (4.5, 6.5) and (7.8, 3.2).

40. When solving a linear programming problem, if the objective function has a maximum value at two adjacent vertices, you can assume that there are an infinite number of points that will produce the maximum value.

Think About It   **In Exercises 41–44, find an objective function that has a maximum or minimum value at the indicated vertex of the constraint region shown below. (There are many correct answers.)**

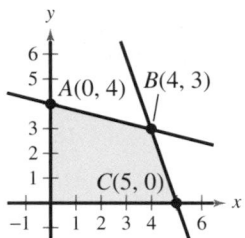

41. The maximum occurs at vertex A.

42. The maximum occurs at vertex B.

43. The maximum occurs at vertex C.

44. The minimum occurs at vertex C.

In Exercises 45 and 46, determine values of t such that the objective function has a maximum value at each indicated vertex.

45. Objective function:

$z = 3x + ty$

Constraints:

$$x \geq 0$$
$$y \geq 0$$
$$x + 3y \leq 15$$
$$4x + y \leq 16$$

(a) (0, 5)

(b) (3, 4)

46. Objective function:

$z = 3x + ty$

Constraints:

$$x \geq 0$$
$$y \geq 0$$
$$x + 2y \leq 4$$
$$x - y \leq 1$$

(a) (2, 1)

(b) (0, 2)

Appendix G Study Capsules

Study Capsule 1 Algebraic Expressions and Functions

Properties

Exponents and Radicals

Properties of Exponents

1. $a^m \cdot a^n = a^{m+n}$ 2. $\dfrac{a^m}{a^n} = a^{m-n}$ 3. $(a^m)^n = a^{mn}$ 4. $a^{-n} = \dfrac{1}{a^n};\ \dfrac{1}{a^{-n}} = a^n$ 5. $a^0 = 1,\ a \neq 0$

Properties of Radicals

1. $\sqrt{a \cdot b} = \sqrt{a} \cdot \sqrt{b}$ 2. $\sqrt{\dfrac{a}{b}} = \dfrac{\sqrt{a}}{\sqrt{b}}$ 3. $\sqrt{a^2} = |a|$ 4. $\sqrt[n]{a} = a^{1/n}$ 5. $\sqrt[n]{a^m} = a^{m/n} = \left(\sqrt[n]{a}\right)^m,\ a > 0$

Methods

Polynomials and Factoring

Factoring Quadratics

1. $x^2 + bx + c = (x + \boxed{})(x + \boxed{})$

   Fill blanks with factors of c that add up to b.

2. $ax^2 + bx + c = (\boxed{} x + \boxed{})(\boxed{} x + \boxed{})$

   Fill blanks with factors of a and of c, so that the binomial product has a middle factor of bx.

Factoring Polynomials

Factor a polynomial $ax^3 + bx^2 + cx + d$ by grouping.

Examples

$x^2 - 7x + 12 = (x + \boxed{})(x + \boxed{})$ Factor 12 as $(-3)(-4)$.

$ = (x - 3)(x - 4)$

Factors of 4

$4x^2 + 4x - 15 = (\boxed{} x + \boxed{})(\boxed{} x + \boxed{})$

Factors of -15

$ = (2x - 3)(2x + 5)$ Factor 4 as $(2)(2)$. Factor -15 as $(-3)(5)$.

$4x^3 + 12x^2 - x - 3$

$= (4x^3 + 12x^2) - (x + 3)$ Group by pairs.

$= 4x^2(x + 3) - (x + 3)$ Factor out monomial.

$= (x + 3)(4x^2 - 1)$ Factor out binomial.

$= (x + 3)(2x + 1)(2x - 1)$ Difference of squares

Simplifying Expressions

Fractional Expressions

1. Factor completely and simplify.

$\dfrac{2x^3 - 4x^2 - 6x}{2x^2 - 18} = \dfrac{2x(x^2 - 2x - 3)}{2(x^2 - 9)}$ Factor out monomials.

$= \dfrac{2x(x - 3)(x + 1)}{2(x + 3)(x - 3)}$ Factor quadratics.

$= \dfrac{x(x + 1)}{x + 3},\ x \neq 3$ Divide out common factors.

2. Rationalize denominator. (*Note:* Radicals in the numerator can be rationalized in a similar manner.)

$\dfrac{3x}{\sqrt{x - 5} + 2} = \dfrac{3x}{\sqrt{x - 5} + 2} \cdot \dfrac{\sqrt{x - 5} - 2}{\sqrt{x - 5} - 2}$ Multiply by conjugate.

$= \dfrac{3x\left(\sqrt{x - 5} - 2\right)}{(x - 5) - 4}$ Difference of squares

$= \dfrac{3x\left(\sqrt{x - 5} - 2\right)}{x - 9}$ Simplify.

Study Capsule 1   Algebraic Expressions and Functions (continued)

Equations and Graphs

Equations

Slope of a Line Passing Through (x_1, y_1) and (x_2, y_2):

$$m = \frac{y_2 - y_1}{x_2 - x_1}$$

$m_1 = m_2$	Parallel lines
$m_1 = -\dfrac{1}{m_2}$	Perpendicular lines

Equations of Lines

$y = mx + b$	Slope-intercept form
$y - y_1 = m(x - x_1)$	Point-slope form
$Ax + By + C = 0$	General form
$x = a,\ y = b$	Vertical and horizontal lines

Distance Between Points (x_1, y_1) and (x_2, y_2):

$$d = \sqrt{(x_2 - x_1)^2 + (y_2 - y_1)^2}$$

Graphs

Graphing Equations by Point Plotting
1. Write equation in the form $y = .\ .\ .\ .$
2. Make a table of values.
3. Find intercepts.
4. Use symmetry.
5. Plot points and connect with smooth curve.

Graphing Equations with a Graphing Utility
1. Enter the equation in the form $y = .\ .\ .\ .$
2. Identify domain and range.
3. Set an appropriate viewing window.

Functions and Graphs

Functions

Definition: f is a function if to each element x in the domain of f there corresponds exactly one element y in the range of f.

Notation: $y = f(x)$

f is the name of the function.

y is the dependent variable, or the output value.

x is the independent variable, or the input value.

$f(x)$ is the value of the function at x.

Examples

Polynomial Function: $f(x) = 2x^3 - 3x^2 - 4x + 6$

Piecewise-Defined Function: $f(x) = \begin{cases} 2 - 3x, & x > 1 \\ x^2 + 2x, & x \le 1 \end{cases}$

Inverse Functions f and f^{-1}: Their graphs are reflections of each other in the line $y = x$.

$$f(f^{-1}(x)) = x \quad \text{and} \quad f^{-1}(f(x)) = x$$

To find the inverse function of $y = f(x)$, if it exists, interchange x and y, then solve for y. The result is $f^{-1}(x)$.

Transformations and Compositions

Transformations

Transformations of the Graph of $y = f(x)$

Vertical shifts: $h(x) = f(x) + c$	Upward c units
$h(x) = f(x) - c$	Downward c units
Horizontal shifts: $h(x) = f(x - c)$	Right shift c units
$h(x) = f(x + c)$	Left shift c units
Reflections: $h(x) = -f(x)$	Reflection in x-axis
$h(x) = f(-x)$	Reflection in y-axis
Stretches/Shrinks: $h(x) = cf(x)$	Vertical stretch, $c > 1$
	Vertical shrink, $c < 1$
$h(x) = f(cx)$	Horizontal stretch, $0 < c < 1$
	Horizontal shrink, $c > 1$

Compositions

Compositions of Functions

$$(f \circ g)(x) = f(g(x))$$
$$(g \circ f)(x) = g(f(x))$$

Examples

$f(x) = x^2, \quad g(x) = 2x - 1$

$$\begin{aligned}(f \circ g)(x) &= f(g(x)) \\ &= f(2x - 1) \\ &= (2x - 1)^2 \\ &= 4x^2 - 4x + 1\end{aligned}$$

$$\begin{aligned}(g \circ f)(x) &= g(f(x)) \\ &= g(x^2) \\ &= 2(x^2) - 1 \\ &= 2x^2 - 1\end{aligned}$$

Study Capsule 2   Graphing Algebraic Functions

Graphical Analysis	**Examples**

Linear Functions

Graph of $f(x) = mx + b$ is a line.

1. m = slope of a line
2. y-intercept: $(0, b)$
3. x-intercept: $(k, 0)$, where k is solution to $0 = mx + b$

To graph the linear equation
$3x - 4y - 8 = 0$, solve for y
to get $y = \frac{3}{4}x - 2$.
So, $m = \frac{3}{4}$, the y-intercept is
$(0, -2)$, and the x-intercept is
$\left(\frac{8}{3}, 0\right)$.

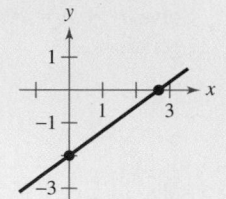

Quadratic Functions

Graph of $y = ax^2 + bx + c$ is a parabola (U-shaped).

1. Opens upward if $a > 0$.
   Opens downward if $a < 0$.
2. Vertex: $\left(-\dfrac{b}{2a}, f\left(-\dfrac{b}{2a}\right)\right)$
3. Vertex is minimum if $a > 0$.
   Vertex is maximum if $a < 0$.
4. Axis of symmetry: $x = -\dfrac{b}{2a}$

$y = -x^2 + 5x - 4$ opens
downward because $a = -1$.
Vertex:
$\left(\dfrac{-5}{2(-1)}, f\left(\dfrac{-5}{2(-1)}\right)\right) = \left(\dfrac{5}{2}, \dfrac{9}{4}\right)$
Vertex is a maximum.
Axis of symmetry is $x = \frac{5}{2}$.

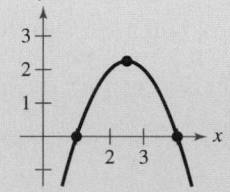

Polynomial Functions

Graph of $f(x) = a_n x^n + a_{n-1}x^{n-1} + \cdots + a_1 x + a_0$
has the following characteristics.

1. x-intercepts occur at zeros of f.
   y-intercept is $(0, a_0)$.
2. Right-hand and left-hand behaviors:

	$a_n > 0$	$a_n < 0$
n is odd	Falls to left, rises to right	Rises to left, falls to right
n is even	Rises to left and right	Falls to left and right

$y = -x^3 + 3x^2 + 4x - 12$
$\quad = (x - 3)(4 - x^2)$
x-intercepts: $(\pm 2, 0), (3, 0)$
y-intercept: $(0, -12)$
End behavior: Up to left and
down to right because $a_n < 0$
and n is odd.

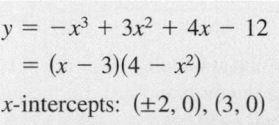

Rational Functions

Graph of
$$f(x) = \frac{N(x)}{D(x)} = \frac{a_n x^n + a_{n-1}x^{n-1} + \cdots + a_1 x + a_0}{b_m x^m + b_{m-1}x^{m-1} + \cdots + b_1 x + b_0}$$
where N and D have no common factors, has the following characteristics.

1. x-intercepts occur at zeros of $N(x)$.
2. Vertical asymptotes occur at zeros of $D(x)$.
3. Horizontal asymptote occurs at $y = 0$ when $n < m$, and at $y = a_n/b_m$ when $n = m$.

$y = \dfrac{x^2 - 1}{x^2 - 2x}$
$\quad = \dfrac{(x + 1)(x - 1)}{x(x - 2)}$
x-intercepts: $(\pm 1, 0)$
Vertical asymptotes:
$\quad x = 0, x = 2$
Horizontal asymptote:
$\quad y = \frac{1}{1} = 1$

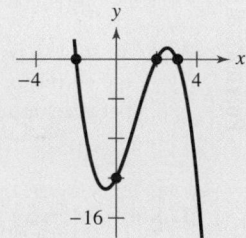

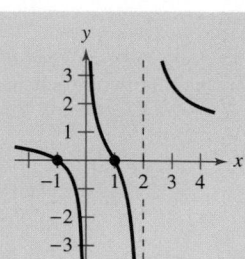

Study Capsule 3   Zeros of Algebraic Functions

Solution Strategy	Examples

Linear Functions

Solve $ax + b = c$ by isolating x using *inverse* operations.

$$-3x + 5 = 8 \qquad \text{Original equation}$$
$$-3x = 3 \qquad \text{First, subtract 5 from each side.}$$
$$x = \tfrac{3}{-3} = -1 \qquad \text{Then, divide each side by } -3.$$

Quadratic Functions

Solve for $ax^2 + bx + c = 0$ using one of the following methods.

1. Factor.

2. Use the Quadratic Formula $x = \dfrac{-b \pm \sqrt{b^2 - 4ac}}{2a}$.

3. Complete the square and/or extract square roots.

$$2x^2 + 5x - 3 = 0$$

1. $(2x - 1)(x + 3) = 0$ Factor.

$$2x - 1 = 0 \qquad x + 3 = 0 \qquad \text{Set factors equal to 0.}$$
$$x = \tfrac{1}{2} \text{ or} \qquad x = -3 \qquad \text{Solve for } x.$$

2. $x = \dfrac{-5 \pm \sqrt{5^2 - 4(2)(-3)}}{2(2)}$ $a = 2, b = 5, c = -3$

$$x = \frac{-5 \pm \sqrt{49}}{4} = \frac{-5 \pm 7}{4} \implies x = \tfrac{1}{2}, -3$$

Polynomial Functions

Solve $a_n x^n + a_{n-1} x^{n-1} + \cdots + a_1 x + a_0 = 0$ by using the Rational Zero Test in combination with synthetic division.

Possible rational zeros $= \dfrac{\pm \text{ factors of } a_0}{\pm \text{ factors of } a_n}$

Note: To solve a polynomial inequality, find the zeros of the corresponding equation and test the inequality between and beyond each zero.

$x^3 + x^2 - 4x - 4 = 0$, where $a_n = 1$ and $a_0 = -4$.

Possible rational zeros are $\pm 1, \pm 2,$ and ± 4, so try
$x = -1$: $f(-1) = (-1)^3 + (-1)^2 - 4(-1) - 4 = 0$
Synthetic division using the zero $x = -1$

$$\begin{array}{r|rrrr}
-1 & 1 & 1 & -4 & -4 \\
   &   & -1 & 0 & 4 \\
\hline
   & 1 & 0 & -4 & 0
\end{array} \implies x^2 - 4 = 0$$

So, the zeros are -1 and ± 2.

Other Functions

1. Solve an equation involving *radicals* (or fractional powers) by isolating the radical and then raising each side to the appropriate power to obtain a polynomial equation.

2. Solve an equation involving *fractions* by multiplying each side by the LCD of the fraction to obtain a polynomial equation.

3. Solve an *absolute value* equation, $|f(x)| = g(x)$, by solving for x in the two equations $f(x) = g(x)$ and $-f(x) = g(x)$.

4. To solve $|f(x)| \le c$, isolate x in $-c \le f(x) \le c$.

5. To solve $|f(x)| \ge c$, isolate x in both $f(x) \ge c$ and $f(x) \le -c$.

1. $2\sqrt{x + 3} - x = 4$ Original equation
$$2\sqrt{x + 3} = x + 4 \qquad \text{Isolate radical term.}$$
$$4(x + 3) = (x + 4)^2 \qquad \text{Raise each side to 2nd power.}$$
$$0 = x^2 + 4x + 4 \qquad \text{Standard form}$$
$$0 = (x + 2)^2 \qquad x = -2 \text{ is repeated zero.}$$

2. $6 + \dfrac{2}{x + 3} = \dfrac{6x + 1}{3}$ Original equation
$$6(3)(x + 3) + 2(3) = (6x + 1)(x + 3) \qquad \text{Multiply by LCD.}$$
$$0 = 6x^2 + x - 57 \qquad \text{Standard form}$$
$$0 = (6x + 19)(x - 3) \qquad \text{Factor.}$$

3. $x^2 - 5x = |x - 5|$ Isolate absolute value.
$$x^2 - 5x = (x - 5) \quad \text{or} \quad x^2 - 5x = -(x - 5)$$
$$x^2 - 6x + 5 = 0 \qquad\qquad x^2 - 4x - 5 = 0$$
$$(x - 5)(x - 1) = 0 \qquad (x - 5)(x + 1) = 0$$
$$x = 5, \ x = 1 \qquad\qquad x = 5, \ x = -1$$

Note: The only solutions are $x = 5$ and $x = -1$.

Study Capsule 4   Exponential and Logarithmic Functions

Exponential Functions	**Logarithmic Functions**

Definitions and Graphs

Definition: The exponential function f with base a is denoted by $f(x) = a^x$, where $a > 0$, $a \neq 1$, and x is any real number.

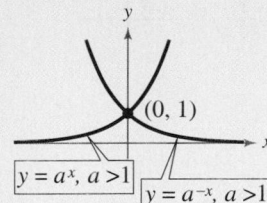

Domain: $(-\infty, \infty)$

Range: $(0, \infty)$

Intercept: $(0, 1)$

Horizontal asymptote: $y = 0$

Definition: For $x > 0$, $a > 0$, and $a \neq 1$, the logarithmic function f with base a is $f(x) = \log_a x$, where $y = \log_a x$ if and only if $x = a^y$.

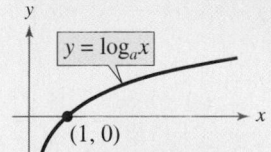

Domain: $(0, \infty)$

Range: $(-\infty, \infty)$

Intercept: $(1, 0)$

Vertical asymptote: $x = 0$

Properties

One-to-One: $a^x = a^y \implies x = y$

Inverse: $a^{\log_a x} = x$

Product: $a^u \cdot a^v = a^{u+v}$

Quotient: $\dfrac{a^u}{a^v} = a^{u-v}$

Power: $(a^u)^v = a^{u \cdot v}$

Others: $a^0 = 1$

Note: The same properties hold for the natural base e, where e is the constant $2.718281828\ldots$.

One-to-One: $\log_a x = \log_a y \implies x = y$

Inverse: $\log_a a^x = x$ **Power:** $\log_a u^v = v \log_a u$

Product: $\log_a(u \cdot v) = \log_a u + \log_a v$

Quotient: $\log_a\left(\dfrac{u}{v}\right) = \log_a u - \log_a v$

Others: $\log_a a = 1$; $\log_a 1 = 0$; $\log_a 0$ is undefined.

Change of Base: $\log_a x = \dfrac{\log x}{\log a} = \dfrac{\ln x}{\ln a}$, where $\log x$ and $\ln e$ denote bases 10 and e.

Note: The same properties apply for bases 10 and e.

Solving Equations

Solve an exponential equation by isolating the exponential term and taking the logarithm of each side.

$2^x - 5 = 0$	Original equation
$2^x = 5$	Isolate exponential term.
$\log_2 2^x = \log_2 5$	Take log of each side.
$x = \log_2 5$	Inverse Property
$x = \dfrac{\ln 5}{\ln 2}$	Change-of-base formula
$x \approx 2.32$	Use a calculator.

Some exponential equations can be solved by using the Inverse Property.

$3e^{2x} - 2 = 5$	Original equation
$3e^{2x} = 7$	Add 2 to each side.
$e^{2x} = \frac{7}{3}$	Isolate exponential term.
$\ln e^{2x} = \ln \frac{7}{3}$	Take natural log of each side.
$2x = \ln \frac{7}{3}$	Inverse Property
$x = \frac{1}{2} \ln \frac{7}{3}$	Multiply each side by $\frac{1}{2}$.
$x \approx 0.42$	Use a calculator.

Solve a logarithmic equation by isolating the logarithmic term and exponentiating each side.

$6 + 2\log_{10} x = 3$	Original equation
$\log_{10} x = -\frac{3}{2}$	Isolate logarithmic term.
$10^{\log_{10} x} = 10^{-3/2}$	Exponentiate using base 10.
$x = 10^{-3/2}$	Inverse Property
$x \approx 0.0316$	Use a calculator.

Properties of logarithms are useful in rewriting equations in forms that are easier to solve.

$\ln(x + 4) - \ln(x - 2) = \ln x$	Original equation
$\ln\left(\dfrac{x + 4}{x - 2}\right) = \ln x$	Quotient Property
$\dfrac{x + 4}{x - 2} = x$	One-to-One Property
$x + 4 = x^2 - 2x$	
$0 = x^2 - 3x - 4$	Standard form
$0 = (x - 4)(x + 1)$	Factor.

$x = 4$ is a valid solution. $x = -1$ is not in the domain of $\ln(x - 2)$.

Study Capsule 5   Trigonometric Functions

Definitions

Right Triangle

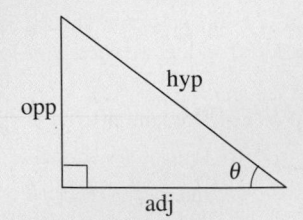

$$\sin \theta = \frac{\text{opp}}{\text{hyp}}, \quad \csc \theta = \frac{\text{hyp}}{\text{opp}}$$

$$\cos \theta = \frac{\text{adj}}{\text{hyp}}, \quad \sec \theta = \frac{\text{hyp}}{\text{adj}}$$

$$\tan \theta = \frac{\text{opp}}{\text{adj}}, \quad \cot \theta = \frac{\text{adj}}{\text{opp}}$$

Coordinate Plane

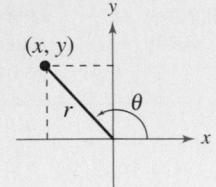

$$x^2 + y^2 = r^2$$

$$\sin \theta = \frac{y}{r}, \quad \csc \theta = \frac{r}{y}$$

$$\cos \theta = \frac{x}{r}, \quad \sec \theta = \frac{r}{x}$$

$$\tan \theta = \frac{y}{x}, \quad \cot \theta = \frac{x}{y}$$

Unit Circle

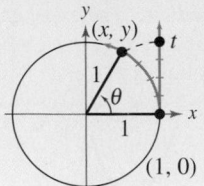

$$x^2 + y^2 = 1, (x, y) = (\cos t, \sin t)$$

$$\sin t = y, \quad \csc t = \frac{1}{y}$$

$$\cos t = x \quad \sec t = \frac{1}{x}$$

$$\tan t = \frac{y}{x}, \quad \cot t = \frac{x}{y}$$

Identities

Fundamental Identities

Reciprocal: $\csc \theta = \dfrac{1}{\sin \theta}, \ \sec \theta = \dfrac{1}{\cos \theta}, \ \cot \theta = \dfrac{1}{\tan \theta}$

Quotient: $\tan \theta = \dfrac{\sin \theta}{\cos \theta}, \ \cot \theta = \dfrac{\cos \theta}{\sin \theta}$

Pythagorean: $\sin^2 \theta + \cos^2 \theta = 1$

$1 + \tan^2 \theta = \sec^2 \theta$

$1 + \cot^2 \theta = \csc^2 \theta$

Cofunction: $\sin\left(\dfrac{\pi}{2} - \theta\right) = \cos \theta, \ \cos\left(\dfrac{\pi}{2} - \theta\right) = \sin \theta$

$\tan\left(\dfrac{\pi}{2} - \theta\right) = \cot \theta, \ \cot\left(\dfrac{\pi}{2} - \theta\right) = \tan \theta$

$\sec\left(\dfrac{\pi}{2} - \theta\right) = \csc \theta, \ \cos\left(\dfrac{\pi}{2} - \theta\right) = \sin \theta$

Even/Odd: $\sin(-t) = -\sin t, \ \cos(-t) = \cos t$

$\tan(-t) = -\tan t, \ \csc(-t) = -\csc t$

$\sec(-t) = \sec t, \quad \cot(-t) = -\cot t$

Evaluations

Table of Values

θ	sin	cos	tan
0°	0	1	0
30°	1/2	$\sqrt{3}/2$	$\sqrt{3}/3$
45°	$\sqrt{2}/2$	$\sqrt{2}/2$	1
60°	$\sqrt{3}/2$	1/2	$\sqrt{3}$
90°	1	0	undef.
180°	0	-1	0
270°	-1	0	undef.

Reference Angles

Evaluate functions of $\theta > 90°$ using the reference angle θ', which is the angle between the x-axis and the terminal side of θ.

Example: The third-quadrant angle $\theta = 240°$ has a reference angle $\theta' = 240° - 180° = 60°$. So,

$$\sin 240° = -\sin 60° = -\frac{\sqrt{3}}{2}.$$

Graphing Utility

To evaluate functions of θ, follow this sequence:

1. Choose mode (degree or radian).
2. Enter function (sin, cos, tan).
3. Enter angle.

Study Capsule 5   Trigonometric Functions (continued)

Graphs

Sine

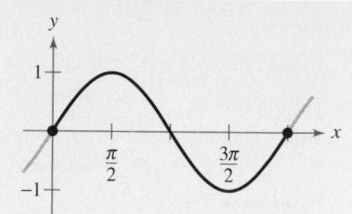

For $y = \sin \theta$, period $= 2\pi$ and amplitude $= 1$.

For $y = a \sin(bx - c)$, period $= 2\pi/b$, amplitude $= a$, and horizontal shift $= c/b$.

Cosine

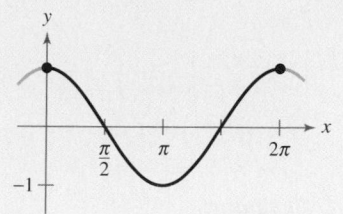

For $y = \cos \theta$, period $= 2\pi$ and amplitude $= 1$.

For $y = a \cos(bx - c)$, period $= 2\pi/b$, amplitude $= a$, and horizontal shift $= c/b$.

Tangent

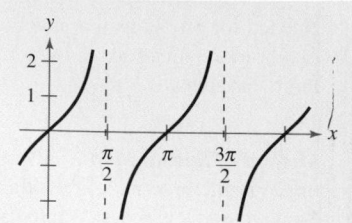

For $y = \tan \theta$, period $= \pi$ and vertical asymptotes are $x = \pi/2$ and $x = 3\pi/2$.

Identities and Equations

Proving Identities

Use known identities to rewrite just one side of the given statement in the form of the other side.

Solving Equations

Linear forms: Isolate the function and use the inverse function to determine the angle θ.

Quadratic forms: Extract roots, or factor and solve the resulting linear equations.

Inverse Functions

• $y = \arcsin x$ if and only if $\sin y = x$, where $-1 \le x \le 1$ and $-\pi/2 \le y \le \pi/2$.

• $y = \arccos x$ if and only if $\cos y = x$, where $-1 \le x \le 1$ and $0 \le y \le \pi$.

• $y = \arctan x$ if and only if $\tan y = x$, where $-\infty \le x \le \infty$ and $-\pi/2 \le y \le \pi/2$.

Definitions and Formulas

Laws of Sine and Cosine

Laws of Sines: For $\triangle ABC$ with sides, a, b, and c

$$\frac{a}{\sin A} = \frac{b}{\sin B} = \frac{c}{\sin C}.$$

Laws of Cosines:

$a^2 = b^2 + c^2 - 2bc \cos A$

$b^2 = a^2 + c^2 - 2ac \cos B$

$c^2 = a^2 + b^2 - 2ab \cos C$

Note: Use the Law of Sines for cases AAS, ASA, and SSA. Use the Law of Cosines for cases SSS and SAS.

Vectors

• A *vector* from point $P(x_1, y_1)$ to $Q(x_2, y_2)$ is $\mathbf{v} = \langle x_2 - x_1, y_2 - y_1 \rangle$ $= \langle v_1, v_2 \rangle$. The equivalent unit vector form is $\mathbf{v} = v_1\mathbf{i} + v_2\mathbf{j}$.

• The *magnitude* of $\mathbf{v}$ is $\|\mathbf{v}\| = \sqrt{v_1^2 + v_2^2}$.

• $\mathbf{u} = \mathbf{v}/\|\mathbf{v}\|$ is a *unit vector* in the direction of $\mathbf{v}$.

• $\mathbf{u} = \langle \cos \theta, \sin \theta \rangle$, where θ is the positive angle from the x-axis to the unit vector $\mathbf{u}$.

• The *dot product* of $\mathbf{u} = \langle u_1, u_2 \rangle$ and $\mathbf{v} = \langle v_1, v_2 \rangle$ is $\mathbf{u} \cdot \mathbf{v} = u_1v_1 + u_2v_2$.

Powers and nth Roots

Trigonometric form of a complex number:

$z = a + bi = r(\cos \theta + i \sin \theta)$, where $r = \sqrt{a^2 + b^2}$, $a = r \cos \theta$, and $b = r \sin \theta$.

DeMoivre's Theorem: For a positive integer n and complex number $z = r(\cos \theta + i \sin \theta)$, then

$z^n = [r(\cos \theta + i \sin \theta)]^n$ $= r^n(\cos n\theta + i \sin n\theta)$.

nth Root: The nth root of $z = r(\cos \theta + i \sin \theta)$ is $\sqrt[n]{r}\{\cos[(\theta + 2\pi k)/n] + i \sin[(\theta + 2\pi k)/n]\}$, where $k = 0, 1, \ldots, n - 1$.

Study Capsule 6   Linear Systems and Matrices

Systems of Equations

Substitution

Needed for problems that involve two or more equations in two or more variables.

Method of Substitution: Solve for one variable in terms of the other. Substitute this expression into the other equation and solve this one-variable equation. Back-substitute to find the value of the other variable.

Examples

Linear:

$$\begin{cases} 2x + y = 1 \\ x - y = 5 \end{cases} \Rightarrow y = x - 5$$

$$2x + (x - 5) = 1 \Rightarrow x = 2$$

$$2 - y = 5 \Rightarrow y = -3$$

Nonlinear:

$$\begin{cases} x - y^2 = 1 \\ x - y = 3 \end{cases} \Rightarrow y = x - 3$$

$$x - (x - 3)^2 = 1$$

$$x^2 - 7x + 10 = 0$$

$$(x - 5)(x - 2) = 0$$

$$x = 5, y = 2 \text{ and } x = 2, y = -1$$

Graphical Interpretation

Linear:

Lines intersect $\Rightarrow$ one solution

Lines parallel $\Rightarrow$ no solution

Lines coincide $\Rightarrow$ infinite number of solutions

Nonlinear:

Graphs intersect at one point.

Graphs intersect at multiple points.

Graphs do not intersect.

Algebraic Methods

Elimination

Method of Elimination: Obtain coefficients for x (or y) that differ only in sign by multiplying one or both equations by appropriate constants. Then add the equations to eliminate one variable. Solve the remaining one-variable equation. Back-substitute into one of the original equations to find the value of the other variable.

$$\begin{cases} 2x + y = 3 \\ 4x + 3y = -1 \end{cases} \Rightarrow \begin{array}{r} -6x - 3y = -9 \\ 4x + 3y = -1 \\ \hline -2x \qquad = -10 \\ x \qquad = 5, y = -7 \end{array}$$

Gaussian Elimination: For systems of linear equations in more than two variables, use elementary row operations to rewrite the system in row-echelon form. Back-substitute into one of the original equations to find the value of each remaining variable.

$$\begin{cases} x + 2y + z = 3 \\ 2x + 5y - z = -4 \\ 3x - 2y - z = 5 \end{cases} \begin{array}{l} \\ -2R_1 + R_2 \to \\ -3R_1 + R_3 \to \end{array} \begin{cases} x + 2y + z = 3 \\ y - 3z = -10 \\ -8y - 4z = -4 \end{cases}$$

Using $8R_2 + R_3$ for row 3, the row-echelon form is

$$\begin{cases} x + 2y + z = 3 \\ y - 3z = -10 \\ z = 3 \end{cases}$$

Back-substitution yields $y = -1$ and $x = 2$.

Types of Systems:

Consistent and independent, if one solution

Consistent and dependent, if infinitely many solutions

Inconsistent, if no solution

Matrix Methods

Gauss-Jordan Elimination: Form the augmented matrix for a system of equations and apply elementary row operations until a reduced row-echelon matrix is obtained.

Augmented Matrix

$$\left[\begin{array}{ccc:c} 1 & 1 & 1 & -1 \\ 3 & 5 & 4 & 2 \\ 3 & 6 & 5 & 0 \end{array} \right]$$

Reduced Row-Echelon

$$\left[\begin{array}{ccc:c} 1 & 0 & 0 & 1 \\ 0 & 1 & 0 & 7 \\ 0 & 0 & 1 & -9 \end{array} \right]$$

Solution: $(1, 7, -9)$

Solve a Matrix Equation: Solve the matrix equation $AX = B$, using the inverse A^{-1} to obtain $X = A^{-1}B$. The inverse of A is found by converting the matrix $[A \vdots I]$ into the form $[I \vdots A^{-1}]$, where I is the identity matrix.

$$AX = B \qquad \begin{bmatrix} 1 & 1 & 1 \\ 3 & 5 & 4 \\ 3 & 6 & 5 \end{bmatrix} \begin{bmatrix} x \\ y \\ z \end{bmatrix} = \begin{bmatrix} -1 \\ 2 \\ 0 \end{bmatrix}$$

$$X = A^{-1}B \qquad \begin{bmatrix} x \\ y \\ z \end{bmatrix} = \begin{bmatrix} 1 & 1 & -1 \\ -3 & 2 & -1 \\ 3 & -3 & 2 \end{bmatrix} \begin{bmatrix} -1 \\ 2 \\ 0 \end{bmatrix} = \begin{bmatrix} 1 \\ 7 \\ -9 \end{bmatrix}$$

Cramer's Rule: $x = \dfrac{|A_1|}{|A|}$, $y = \dfrac{|A_2|}{|A|}$, $z = \dfrac{|A_3|}{|A|}$, where

$$|A| = \begin{vmatrix} 1 & 1 & 1 \\ 3 & 5 & 4 \\ 3 & 6 & 5 \end{vmatrix}, \quad |A_1| = \begin{vmatrix} -1 & 1 & 1 \\ 2 & 5 & 4 \\ 0 & 6 & 5 \end{vmatrix},$$

$$|A_2| = \begin{vmatrix} 1 & -1 & 1 \\ 3 & 2 & 4 \\ 3 & 0 & 5 \end{vmatrix}, \quad |A_3| = \begin{vmatrix} 1 & 1 & -1 \\ 3 & 5 & 2 \\ 3 & 6 & 0 \end{vmatrix}$$

Study Capsule 7   Sequences, Series, and Probability

General

Sequences

Definition: An infinite sequence $\{a_n\}$ has function values $a_1, a_2, a_3, \ldots, a_n, \ldots$ called the *terms* of the sequence.

Skills: Use or find the form of the nth term

1. Given the form of a_n, write the first five terms.
2. Given the first several terms, find a_n.

Arithmetic

Definition: $\{a_n\}$ is *arithmetic* if the *difference* between consecutive terms is a common value d.

Skills: Given d and a_1, or given two specific terms

1. Find the first five terms of $\{a_n\}$.
2. Find the form of a_n.

In general, $a_n = a_1 + (n - 1)d$.

Geometric

Definition: $\{a_n\}$ is *geometric* if the *ratio* of two consecutive terms is a common value r.

Skills: Given r and a_1, or given two specific terms

1. Find the first five terms of $\{a_n\}$.
2. Find the form of a_n.

In general, $a_n = a_1 r^{n-1}$.

Sums and Series

Summation Notation: $\sum_{i=1}^{n} a_i$

There is no general formula for calculating the nth partial sum or the sum of an infinite series.

nth Partial Sum:

$$S_n = \sum_{i=1}^{n} a_i = \frac{n}{2}(a_1 + a_n)$$

where $a_n = a_1 + (n - 1)d$

Infinite Series:

$$S = \sum_{n=1}^{\infty} a_n = [\text{sum is not finite}]$$

nth Partial Sum:

$$S_n = \sum_{i=1}^{n} a_i = a_1\left(\frac{1 - r^n}{1 - r}\right), \ r \neq 1$$

Infinite Series:

$$S = \sum_{n=1}^{\infty} a_n = \frac{a_1}{1 - r}, \ |r| < 1$$

Binomial Theorem, Counting Principles, and Probability

Binomial Theorem

Binomial Theorem:

$(x + y)^n = x^n + nx^{n-1}y + \cdots + {}_nC_r x^{n-r}y^r + \cdots + nxy^{n-1} + y^n$.

Skills:

1. Calculate the binomial coefficients using the formula

$${}_nC_r = \binom{n}{r} = \frac{n!}{(n - r)! \, r!}$$

or by using Pascal's Triangle.

2. Expand a binomial.

Example: Expand $(3x - 2y)^4$.

Using Pascal's Triangle for $n = 4$, the coefficients are 1, 4, 6, 4, 1.

Using the theorem pattern, decrease powers of $3x$ and increase powers of $2y$. The expansion is $(3x - 2y)^4 =$
$(1)(3x)^4 - (4)(3x)^3(2y) + (6)(3x)^2(2y)^2 - 4(3x)(2y)^3 + (1)(2y)^4$.

Counting Principles

Fundamental Counting Principle:

If event E_1 can occur in m_1 different ways and following E_1, event E_2 can occur in m_2 different ways, then the number of ways the two events can occur is $m_1 \cdot m_2$.

Permutations: (order is important)

The number of permutations (orderings) of n elements is ${}_nP_n = n! = n(n - 1)(n - 2) \cdots 3 \cdot 2 \cdot 1$.

The number of permutations of n elements taken r at a time is

${}_nP_r = n!/[(n - r)!] = n(n - 1)(n - 2) \cdots (n - r + 1)$.

Combinations: (order is not important)

The number of combinations of n elements taken r at a time is

${}_nC_r = n!/[(n - r)!r!]$.

Probability

Probability of Event E

$$P(E) = \frac{n(E)}{n(S)}$$

where event E has $n(E)$ equally likely outcomes and sample space S has $n(S)$ equally likely outcomes.

Probability Formulas

$P(A \text{ or } B) = P(A \cup B)$

$P(A \text{ and } B) = P(A \cap B)$

$P(A \cup B) = P(A) + P(B) - P(A \cap B)$

$P(A \cup B) = P(A) + P(B)$, if A and B have no outcomes in common.

$P(A \cap B) = P(A) \cdot P(B)$, if A and B are independent events.

$P(\text{complement of } A) = P(A')$
$= 1 - P(A)$

Study Capsule 8   Conics, Parametric and Polar Equations

Conics

Definitions

Circle: Locus of points equidistant from a fixed point, the center

Parabola: Locus of points equidistant from a fixed point (its focus) and a fixed line (the directrix)

Ellipse: Locus of points, so that the *sum* of their distances from two fixed points (foci) is constant

Hyperbola: Locus of points, so that the *difference* of their distances from two fixed points (foci) is constant

Eccentricity: $e = \dfrac{c}{a}$

parabola: $e = 1$

ellipse $0 < e < 1$

hyperbola $e > 1$

Standard Equations

Circle: [center (h, k), radius $= r$]
$$(x - h)^2 + (y - k)^2 = r^2$$

Parabola: [vertex (h, k), and $p = $ distance from the vertex to the focus]
$$(x - h)^2 = 4p(y - k)$$
$$(y - k)^2 = 4p(x - h)$$

Ellipse: [major axis $2a$, minor axis $2b$, distance from the center to focus is c, and $c^2 = a^2 - b^2$]
$$(x - h)^2/a^2 + (y - k)^2/b^2 = 1$$
$$(y - h)^2/a^2 - (y - k)^2/b^2 = 1$$

Hyperbola: $c^2 = a^2 + b^2$
$$(x - h)^2/a^2 - (y - k)^2/b^2 = 1$$
$$(y - k)^2/a^2 - (x - h)^2/b^2 = 1$$

Basic Problem Types

1. Given information needed to find the parts (center, radius, vertices, foci, etc.), write the standard equation of the conic.

2. Given the general equation of a conic, complete the square and find the required parts of the conic. Then sketch its graph.

3. Given a general quadratic equation, use the coefficients A and C to classify the conic, as stated on page 686.

4. Given a general equation with an xy-term, use the discriminant $B^2 - 4AC$ to classify the conic, as stated on page 694.

Parametric Equations

Definitions

Definition: Parametric equations are used where the coordinates x and y are each a function of a third variable, called a *parameter.* Common parameters are time t and angle θ.

Standard Equations

Plane Curve C: If f and g are continuous functions of t on an interval I, the set of ordered pairs $(x(t), y(t))$ is a *plane curve C.* The equations $x = f(t)$ and $y = g(t)$ are *parametric equations* for C.

Basic Problem Types

1. Given the parametric equations for a plane curve C, construct a three-row table of values using input for t. Plot the resulting (x, y) points and sketch curve C. Then identify the orientation of the curve.

2. Given a set of parametric equations, eliminate the parameter and write the corresponding rectangular equation.

3. Given a rectangular equation, find a corresponding set of parametric equations using an appropriate parameter.

Polar Equations

Definitions

Definition: Point P in a polar coordinate system is denoted by $P(r, \theta)$, where r is the directed distance from the origin O to the point P and θ is the counterclockwise angle from the polar axis to $\overrightarrow{OP}$.

Multiple Representation of Points:
$$(r, \theta) = (r, \theta \pm 2n\pi)$$
$$= (-r, \theta \pm (2n + 1)\pi)$$

Symmetry: The tests for symmetry of a polar equation, $r = f(\theta)$, are given on page 714.

Standard Equations

Conversion Equations: Polar and rectangular coordinates are related by the equations $x = r \cos \theta$, $y = r \sin \theta$, and $\tan \theta = y/x$, $r^2 = x^2 + y^2$.

Polar Equations of Concis: The focus is located at the pole and p is the distance from the focus to the directrix.
$$r = ep/(1 \pm e \cos \theta)$$
$$r = ep/(1 \pm e \sin \theta)$$

Basic Problem Types

1. Convert polar coordinates or equations to rectangular form or visa versa.

2. Given the polar equation of a conic, analyze its graph.

3. Given information needed to find values of e and p, write the polar equation of the conic.

4. Use point plotting, symmetry, zeros, and maximum r-values to sketch graphs of special polar curves.

Study Capsule 9   Lines, Planes, and Vectors in 3–Space

Definitions, Formulas, and Techniques

Points in 3-space: points $P(x_1, y_1, z_1)$ and $Q(x_2, y_2, z_2)$

$d = \sqrt{(x_2 - x_1)^2 + (y_2 - y_1)^2 + (z_2 - z_1)^2}$ Distance between P and Q

$\left(\dfrac{x_1 + x_2}{2}, \dfrac{y_1 + y_2}{2}, \dfrac{z_1 + z_2}{2}\right)$ Midpoint between P and Q

Equation of a sphere: center (h, k, j) and radius r

$(x - h)^2 + (y - k)^2 + (z - j)^2 = r^2$ Standard equation

Vectors in 3-space: vectors $\mathbf{u} = \langle u_1, u_2, u_3 \rangle$ and $\mathbf{v} = \langle v_1, v_2, v_3 \rangle$

$\|\mathbf{u}\| = \sqrt{u_1^2 + u_2^2 + u_3^2}$ Magnitude $\dfrac{\mathbf{v}}{\|\mathbf{v}\|}$, $\mathbf{v} \neq \mathbf{0}$ Unit vector

Operations with vectors u and v:

$\mathbf{u} + \mathbf{v} = \langle u_1 + v_1, u_2 + v_2, u_3 + v_3 \rangle$ Sum

$c\mathbf{u} = \langle cu_1, cu_2, cu_3 \rangle$ Scalar product

$\mathbf{u} \cdot \mathbf{v} = u_1 v_1 + u_2 v_2 + u_3 v_3$ Dot product

$\cos \theta = \dfrac{\mathbf{u} \cdot \mathbf{v}}{\|\mathbf{u}\| \, \|\mathbf{v}\|}$ θ is the angle between $\mathbf{u}$ and $\mathbf{v}$.

$\mathbf{u} \times \mathbf{v} = \begin{vmatrix} \mathbf{i} & \mathbf{j} & \mathbf{k} \\ u_1 & u_2 & u_3 \\ v_1 & v_2 & v_3 \end{vmatrix}$ Cross product

$\mathbf{u} \cdot (\mathbf{v} \times \mathbf{w}) = \begin{vmatrix} u_1 & u_2 & u_3 \\ v_1 & v_2 & v_3 \\ w_1 & w_2 & w_3 \end{vmatrix}$ Triple scalar product of $\mathbf{u}$, $\mathbf{v}$, and $\mathbf{w}$

$|\mathbf{u} \cdot (\mathbf{v} \times \mathbf{w})|$ = volume of parallelepiped with vectors $\mathbf{u}$, $\mathbf{v}$, and $\mathbf{w}$ as adjacent edges.

Properties of cross products: See pages 758 and 759.

Line in 3-space: Line L is parallel to vector $\mathbf{v} = a\mathbf{i} + b\mathbf{j} + c\mathbf{k}$ and passes through point (x_1, y_1, z_1).

$x = x_1 + at, \quad y = y_1 + bt, \quad z = z_1 + ct$ Parametric equations for I.

$\dfrac{x - x_1}{a} = \dfrac{y - y_1}{b} = \dfrac{z - z_1}{c}$ Symmetric equations for I.

Plane in 3-space: containing point (x_1, y_1, z_1) with normal vector $\mathbf{n} = \langle a, b, c \rangle$

$a(x - x_1) + b(y - y_1) + c(z - z_1) = 0$ Standard equation

$ax + by + cz + d = 0$ General equation

$D = \|\text{proj}_{\mathbf{n}} \overrightarrow{PQ}\| = \dfrac{\overrightarrow{PQ} \cdot \mathbf{n}}{\|\mathbf{n}\|}$ Distance between a plane and point Q (not in the plane), where P is a point in the plane and $\mathbf{n}$ is normal to the plane.

Basic Problem Types

Given three points $A(-1, 3, 4)$, $B(4, -2, 2)$, and $C(2, 8, 6)$:

1. Plot each point in a 3-D coordinate system.
2. Find the distance between points A and B.
3. Find the midpoint of the line segment $\overline{AC}$.
4. Find the standard equation of a sphere with points A and C as endpoints of a diameter.
5. Find the component forms of vectors $\mathbf{u} = \overrightarrow{AC}$ and $\mathbf{v} = \overrightarrow{AB}$.
6. Find $\|\mathbf{u}\|$, $\mathbf{u} \cdot \mathbf{v}$, and $\mathbf{u} \times \mathbf{v}$.
7. Find the angle θ between $\mathbf{u}$ and $\mathbf{v}$.
8. Find a set of parametric equations and the corresponding symmetric forms for the lines through points A and B.
9. Determine whether $\mathbf{u}$ and $\mathbf{v}$ are orthogonal, parallel, or neither.
10. Find the cross product of $\mathbf{u}$, $\mathbf{v}$, and $\mathbf{w}$.
11. Find the general equation of the plane containing points A, B, and C.
12. Find the distance between the point $Q(1, 2, 2)$ and the plane containing A, B, and C.

Given the general equation of a sphere:

1. Find the center (h, k, j) and radius r.
2. Sketch the sphere and show its trace in any one of the coordinate planes.

Given the parametric equations of a line L in space:

1. Find two points on the line L and sketch its graph.
2. Find a vector $\mathbf{u}$ parallel to line L.

Given the general equation of a plane in space:

1. Find the intercepts and sketch the plane.
2. Find a unit vector perpendicular (normal) to the plane.
3. Find the distance between a point Q, not in the plane, and the plane.

Study Capsule 10   Limits and an Introduction to Calculus

Definitions, Notations, and Techniques

Limit notation:

$\lim\limits_{x \to c} f(x) = L$ means that $f(x)$ approaches L as x approaches c from either side. Equivalently, $f(x) \to L$.

$\lim\limits_{x \to c^+} f(x) = $ limit from the *right* of c.

$\lim\limits_{x \to c^-} f(x) = $ limit from the *left* of c.

$\lim\limits_{x \to \pm\infty} f(x) = $ limit *at infinity* or *at negative infinity*.

Existence or nonexistence of limits:

$\lim\limits_{x \to c} f(x)$ exists if the limit from the left and right exist and are equal.

$\lim\limits_{x \to c} f(x)$ fails to exist if:

1. limits from the left and right are not equal.
2. $f(x)$ increases or decreases without bound as $x \to c$.
3. $f(x)$ oscillates between two fixed values as $x \to c$.

Evaluating limits:

By direct substitution of c into $f(x)$—that is, $\lim\limits_{x \to c} f(x) = f(c)$

If $f(x) = \dfrac{N(x)}{D(x)}$ and direct substitution yields $\dfrac{0}{0}$, either:

1. Divide out the common factor in $N(x)$ and $D(x)$, or
2. Rationalize either $N(x)$ or $D(x)$.

For limits at infinity, where $N(x)$ has leading coefficient a_n and degree n, and $D(x)$ has leading coefficient b_m and degree m:

$$\lim\limits_{x \to \pm\infty} \frac{N(x)}{D(x)} = \begin{cases} 0, & n < m \\ \dfrac{a_n}{b_m}, & n = m \\ \text{No limit}, & n > m \end{cases}$$

Σ-notation formulas

$$\sum_{i=1}^{n} c = cn \qquad\qquad \sum_{i=1}^{n} i = \frac{n(n+1)}{2}$$

$$\sum_{i=1}^{n} i^2 = \frac{n(n+1)(2n+1)}{6} \qquad \sum_{i=1}^{n} i^3 = \frac{n^2(n+1)^2}{4}$$

Basic Problem Types

Evaluate $\lim\limits_{x \to a} f(x)$:

1. *Numerically*, using a graphing utility and a table of values
2. *Graphically*, using the *zoom* and *trace* features of a graphing utility on the graph
3. By *direct substitution*: $\lim\limits_{x \to c} f(x) = f(c)$
4. By evaluating the corresponding one-sided limits

Evaluate the limit (as $h \to 0$) of the difference quotient

$$\frac{f(x+h) - f(x)}{h}:$$

1. To find the *slope* of the *tangent line* to the graph of f at a point (x_1, y_1)
2. To find the derivative, $f'(x)$, of the function f

Given the form of the nth term, a_n, of an infinite sequence:

1. Find the first several terms of the sequence.
2. Find the limit, if it exists, of the sequence as $n \to \infty$.

Use Σ-notation and limits to:

1. Find the sum of the first n terms of a sequence.
2. Find the area of a region bounded by the graph of f, the x-axis, and the vertical lines $x = a$ and $x = b$. Use the formula

$$A = \lim_{n \to \infty} \sum_{i=1}^{n} f\left(a + \frac{(b-a)i}{n} \right)\left(\frac{b-a}{n} \right).$$

Answers to Odd-Numbered Exercises and Tests

Chapter 1

Section 1.1   (page 11)

Vocabulary Check   (page 11)

1. (a) iii   (b) i   (c) v   (d) ii   (e) iv

2. slope   **3.** parallel   **4.** perpendicular

5. linear extrapolation

1. (a) L_2 (b) L_3 (c) L_1

3. 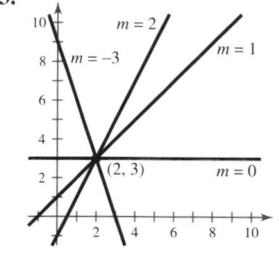   **5.** $\frac{3}{2}$

7. 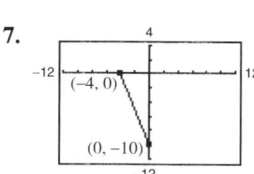   **9.**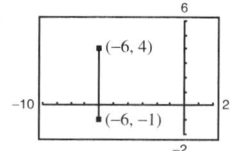

$m = -\frac{5}{2}$ m is undefined.

11. $(0, 1), (3, 1), (-1, 1)$ **13.** $(1, 4), (1, 6), (1, 9)$

15. $(-1, -7), (-2, -5), (-5, 1)$

17. $(3, -4), (5, -3), (9, -1)$

19. (a) $m = 5$;
    y-intercept: $(0, 3)$
(b)

21. (a) Slope is undefined;
    there is no y-intercept.
(b)

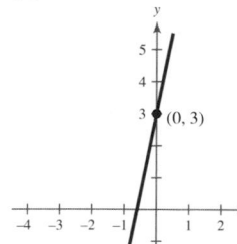   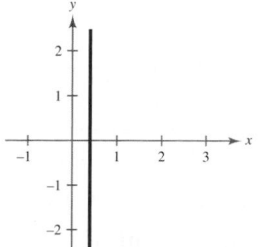

23. (a) $m = 0$; y-intercept: $\left(0, -\frac{5}{3}\right)$

(b)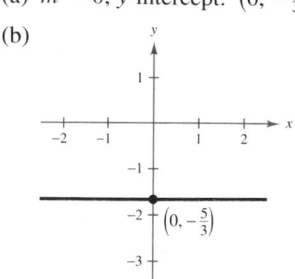

25. $3x - y - 2 = 0$ **27.** $y + \frac{1}{2}x + 2 = 0$

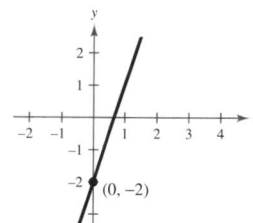   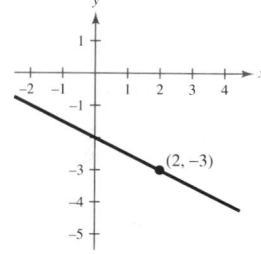

29. $x - 6 = 0$ **31.** $y - \frac{3}{2} = 0$

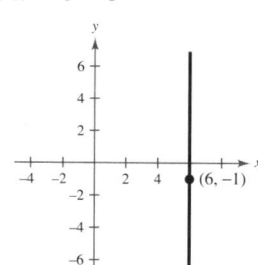   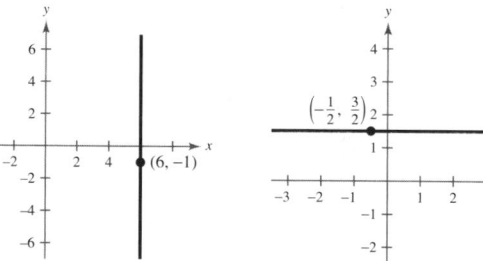

33. $y = -\frac{3}{5}x + 2$ **35.** $x + 8 = 0$

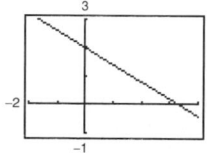   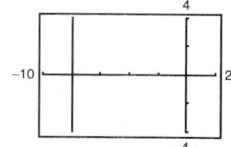

37. $y = -\frac{1}{2}x + \frac{3}{2}$ **39.** $y = -\frac{6}{5}x - \frac{18}{25}$

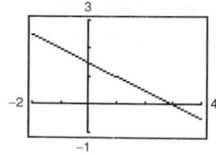   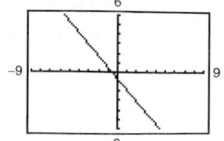

CHAPTER 1

41. $y = \frac{2}{5}x + \frac{1}{5}$ **43.** $y = 2x - 5$

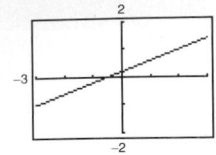

45. $37,300

47. $m = \frac{1}{2}$; y-intercept: $(0, -2)$; a line that rises from left to right

49. Slope is undefined; no y-intercept; a vertical line at $x = -6$

51.

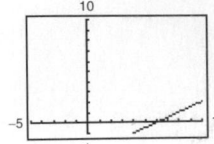   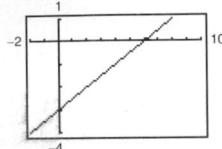

The second setting gives a more complete graph, with a view of both intercepts.

53. Perpendicular   **55.** Parallel

57. (a) $y = 2x - 3$ (b) $y = -\frac{1}{2}x + 2$

59. (a) $y = -\frac{3}{4}x + \frac{3}{8}$ (b) $y = \frac{4}{3}x + \frac{127}{72}$

61. (a) $x = 3$ (b) $y = -2$

63. $y = 2x + 1$ **65.** $y = -\frac{1}{2}x + 1$

67. The lines $y = \frac{1}{2}x$ and $y = -2x$ are perpendicular.

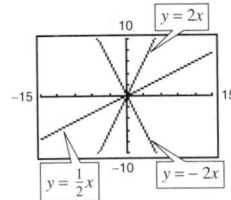

69. The lines $y = -\frac{1}{2}x$ and $y = -\frac{1}{2}x + 3$ are parallel. Both are perpendicular to $y = 2x - 4$.

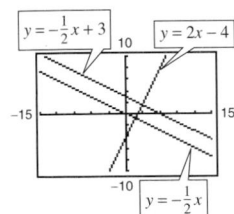

71. (a) The greatest increase $(+0.86)$ was from 1998 to 1999 and the greatest decrease (-0.78) was from 1999 to 2000.

(b) $y = -0.067t + 1.24$, $t = 5$ corresponds to 1995.

(c) There is a decrease of about $0.067 per year.

(d) -0.1, Answers will vary.

73. 12 feet   **75.** $V = 125t + 1790$

77. $V = -2000t + 32,400$

79. b; slope $= -10$; the amount owed decreases by $10 per week.

80. c; slope $= 1.5$; 1.50; the hourly wage increases by $1.50 per unit produced.

81. a; slope $= 0.35$; expenses increase by $0.35 per mile.

82. d; slope $= -100$; the value depreciates $100 per year.

83. (a) $V = 25,000 - 2300t$

(b)

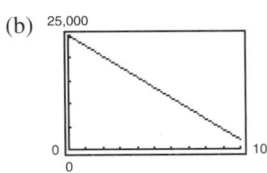

t	0	1	2	3	4
V	25,000	22,700	20,400	18,100	15,800

t	5	6	7	8	9	10
V	13,500	11,200	8900	6600	4300	2000

85. (a) $C = 16.75t + 36,500$ (b) $R = 27t$

(c) $P = 10.25t - 36,500$ (d) $t \approx 3561$ hours

87. (a) Increase of about 341 students per year

(b) 72,962; 77,395; 78,418

(c) $y = 341x + 75,008$, where $x = 1$ corresponds to 1991; $m = 341$; the slope determines the average increase in enrollment.

89. False. The slopes $\left(\frac{2}{7}$ and $-\frac{11}{7}\right)$ are not equal.

91.

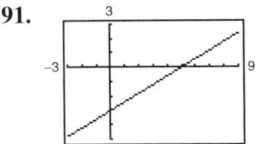 a and b represent the x- and y-intercepts.

93.

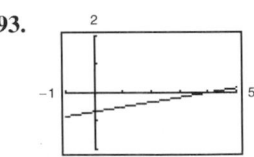 a and b represent the x- and y-intercepts.

95. $3x + 2y - 6 = 0$ **97.** $12x + 3y + 2 = 0$

99. a   **101.** c

103. No. Answers will vary. Sample answer: The line, $y = 2$, does not have an x-intercept.

105. Yes. Once a parallel line is established to the given line, there are an infinite number of distances away from that line, creating an infinite number of parallel lines.

107. Yes. $x + 20$ **109.** No   **111.** No

113. $(x - 9)(x + 3)$ **115.** $(2x - 5)(x + 8)$

117. Answers will vary.

Section 1.2   (page 24)

1. Yes. Each element of the domain is assigned to exactly one element of the range.

3. No. The National League, an element in the domain, is assigned to three items in the range, the Cubs, the Pirates, and the Dodgers; the American League, an element in the domain, is also assigned to three items in the range, the Orioles, the Yankees, and the Twins.

5. Yes. Each input value is matched with one output value.

7. No. The inputs 7 and 10 are both matched with two different outputs.

9. (a) Function

(b) Not a function because the element 1 in A corresponds to two elements, -2 and 1, in B.

(c) Function

(d) Not a function because the element 2 in A corresponds to no element in B.

11. Each is a function of the year. To each year there corresponds one and only one circulation.

13. Not a function   **15.** Function   **17.** Function

19. Not a function   **21.** Function   **23.** Not a function

25. (a) $\dfrac{1}{5}$ (b) 1   (c) $\dfrac{1}{4t + 1}$ (d) $\dfrac{1}{x + c + 1}$

27. (a) 7   (b) -11 (c) $3t + 7$

29. (a) 0   (b) -0.75 (c) $x^2 + 2x$

31. (a) 1   (b) 2.5   (c) $3 - 2|x|$

33. (a) $-\dfrac{1}{9}$ (b) Undefined   (c) $\dfrac{1}{y^2 + 6y}$

35. (a) 1   (b) -1 (c) $\dfrac{|t|}{t}$

37. (a) -1 (b) 2   (c) 6

39. (a) 6   (b) 3   (c) 10

41. (a) 0   (b) 4   (c) 17

43.

t	-5	-4	-3	-2	-1
$h(t)$	1	$\frac{1}{2}$	0	$\frac{1}{2}$	1

45.

x	-2	-1	0	1	2
$f(x)$	5	$\frac{9}{2}$	4	1	0

47. 5   **49.** $\frac{4}{3}$ **51.** 2, -1 **53.** All real numbers x

55. All real numbers t except $t = 0$ **57.** All real numbers x

59. All real numbers x except $x = 0, -2$

61. All real numbers y such that $y > 10$

63.

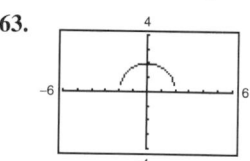

Domain: $[-2, 2]$; range: $[0, 2]$

65.

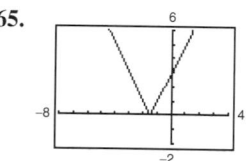

Domain: $(-\infty, \infty)$; range: $[0, \infty)$

67. $\{(-2, 4), (-1, 1), (0, 0), (1, 1), (2, 4)\}$

69. $\{(-2, 4), (-1, 3), (0, 2), (1, 3), (2, 4)\}$

71. $A = \dfrac{C^2}{4\pi}$

73. (a) \$3375

(b)

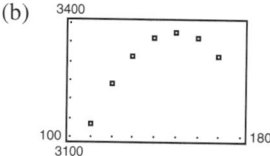

Yes, it is a function.

(c) $P(x) = \begin{cases} 30x, & x \le 100 \\ 45x - 0.15x^2, & x > 100 \end{cases}$

75. $A = \dfrac{x^2}{2(x - 2)}, \; x > 2$

77. (a) $V = 108x^2 - 4x^3$

Domain: all real numbers x such that $0 < x < 27$

(b)
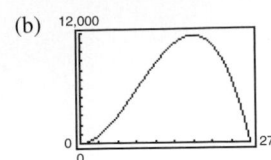

$x = 18$ inches, $y = 36$ inches

79. $7 \le x \le 12$, $1 \le x \le 6$; Answers will vary.

81. 4.63; $4630 in monthly revenue in November

83.

t	0	1	2	3	4	5	6
$n(t)$	577	647	704	749	782	803	811

t	7	8	9	10	11	12	13
$n(t)$	846	871	896	921	946	971	996

85. (a)

y	5	10	20
$F(y)$	26,474	149,760	847,170

y	30	40
$F(y)$	2,334,527	4,792,320

Each time the depth is doubled, the force increases by more than 5.7 times.

(b)
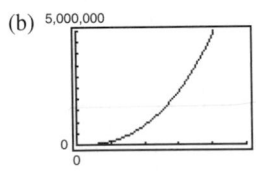

```
Xmin = 0
Xmax = 50
Xscl = 10
Ymin = 0
Ymax = 5,000,000
Yscl = 500,000
```

(c) Depth $\approx$ 21.37 feet
   Use the *trace* and *zoom* features on a graphing utility.

87. $2, c \ne 0$ **89.** $3 + h, h \ne 0$ **91.** $-\dfrac{1}{t}, t \ne 1$

93. False. The range is $[-1, \infty)$.

95. $f(x) = \begin{cases} x + 4, & x \le 0 \\ 4 - x^2, & x > 0 \end{cases}$

97. $f(x) = \begin{cases} 2 - x, & x \le -2 \\ 4, & -2 < x < 3 \\ x + 1, & x \ge 3 \end{cases}$

99. The domain is the set of input values of a function. The range is the set of output values.

101. $\dfrac{12x + 20}{x + 2}$ **103.** $\dfrac{(x + 6)(x + 10)}{5(x + 3)}, x \ne 0, \dfrac{1}{2}$

Section 1.3   (page 38)

1. Domain: $(-\infty, \infty)$; range: $(-\infty, 1]$; $f(0) = 1$

3. Domain: $[-4, 4]$; range: $[0, 4]$; $f(0) = 4$

5.
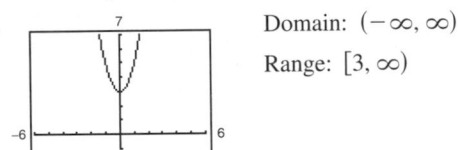

Domain: $(-\infty, \infty)$

Range: $[3, \infty)$

7.
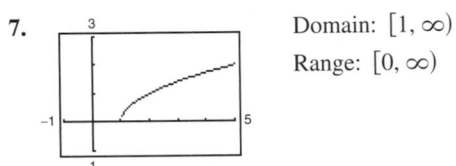

Domain: $[1, \infty)$

Range: $[0, \infty)$

9.
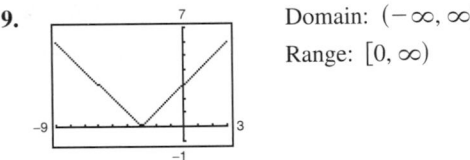

Domain: $(-\infty, \infty)$

Range: $[0, \infty)$

11. (a) $(-\infty, \infty)$ (b) $-2, 3$ (c) x-intercepts
(d) -6 (e) y-intercept   (f) $-6; (1, -6)$
(g) $-4; (-1, -4)$ (h) 6

13. (a) $(-\infty, \infty)$ (b) $-1, 3$ (c) x-intercepts
(d) -1 (e) y-intercept   (f) $-2; (1, -2)$
(g) $0; (-1, 0)$ (h) 2

15. Function. Graph the given function over the window shown in the figure.

17. Not a function. Solve for y and graph the two resulting functions.

19. Increasing on $(-\infty, \infty)$

21. Increasing on $(-\infty, 0), (2, \infty)$
   Decreasing on $(0, 2)$

23. (a)
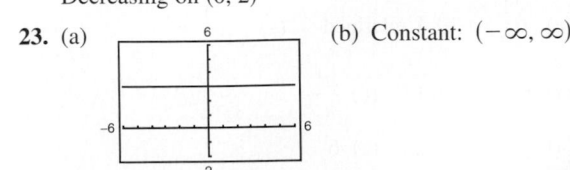

(b) Constant: $(-\infty, \infty)$

25. (a)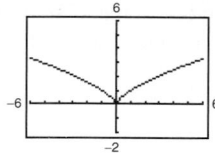

(b) Decreasing on $(-\infty, 0)$
Increasing on $(0, \infty)$

27. (a)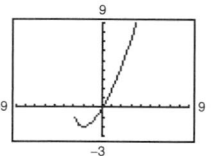

(b) Increasing on $(-2, \infty)$
Decreasing on $(-3, -2)$

29. (a)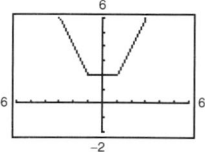

(b) Decreasing on $(-\infty, -1)$
Constant on $(-1, 1)$
Increasing on $(1, \infty)$

31.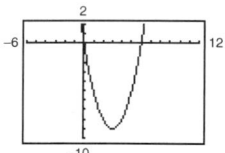

Relative minimum: $(3, -9)$

33.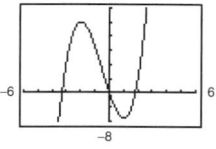

Relative minimum: $(1, -7)$
Relative maximum: $(-2, 20)$

35.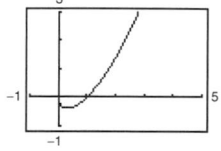

Minimum: $(0.33, -0.38)$

37. (a)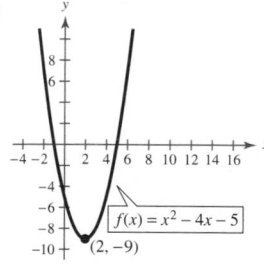

(b) Relative minimum at $(2, -9)$

(c) Answers are the same.

39. (a)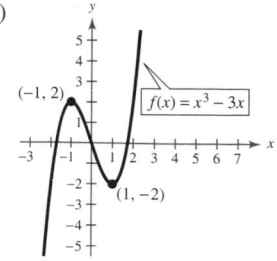

(b) Relative minimum at $(1, -2)$
Relative maximum at $(-1, 2)$

(c) Answers are the same.

41. (a)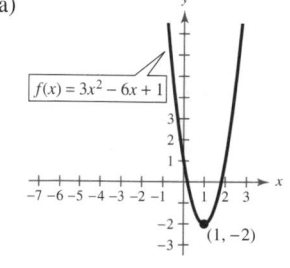

(b) Relative minimum at $(1, -2)$

(c) Answers are the same.

43.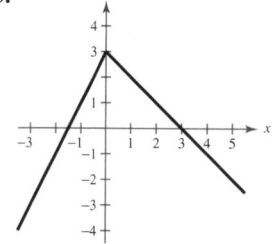

45.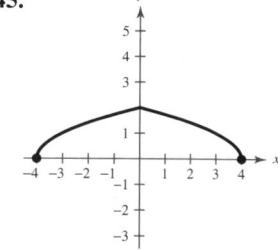

47.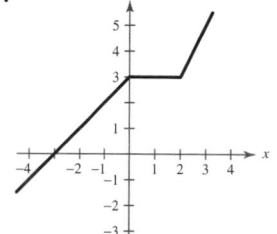

49.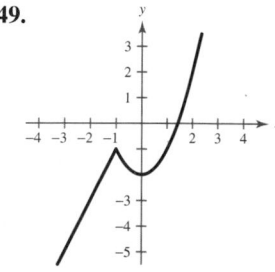

51.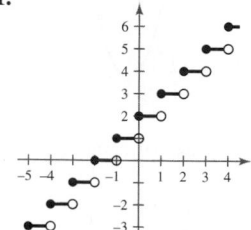

53.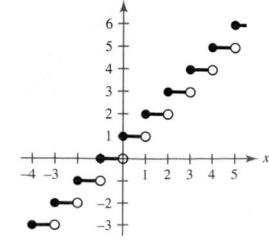

55.

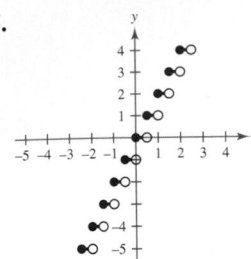

57.

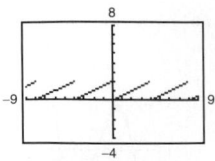

Domain: $(-\infty, \infty)$

Range: $[0, 2)$

Sawtooth pattern

59. Neither even nor odd   **61.** Odd function

63. Odd function   **65.** Even function

67. (a) $\left(\frac{3}{2}, 4\right)$ (b) $\left(\frac{3}{2}, -4\right)$

69. (a) $(-4, 9)$ (b) $(-4, -9)$

71. (a) $(-x, -y)$ (b) $(-x, y)$

73. Even function   **75.** Neither even nor odd

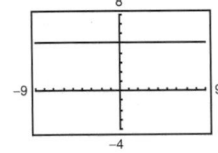
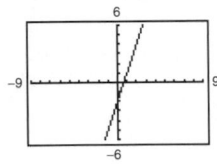

77. Even function   **79.** Neither even nor odd

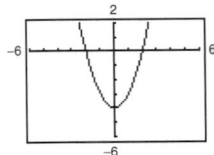
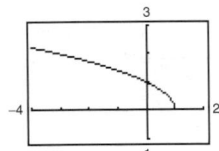

81. Neither even nor odd

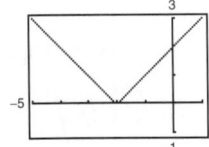

83. $(-\infty, 4]$ **85.** $(-\infty, -3], [3, \infty)$

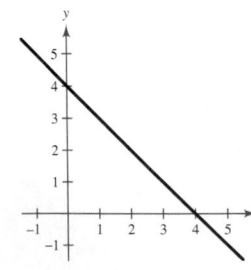
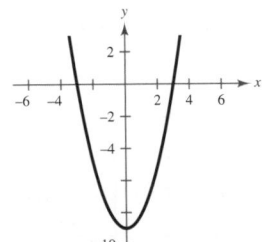

87. (a) C_2 is the appropriate model. The cost of the first minute is $1.05 and the cost increases $0.38 when the next minute begins, and so on.

(b)

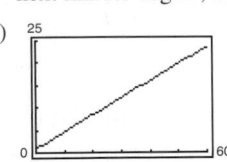

$7.89

89. $h = -x^2 + 4x - 3$, $1 \le x \le 3$

91. (a)

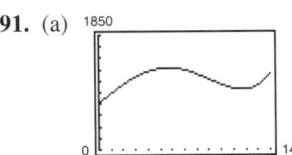

(b) Increasing from 1990 to 1995 and from 2001 to 2004; decreasing from 1995 to 2001

(c) About 1,821,00.

93. False. Counterexample: $f(x) = \sqrt{1 + x^2}$

95. c   **96.** d   **97.** b   **98.** e

99. a   **100.** f

101. Proof

103. (a) Even. g is a reflection in the x-axis.

(b) Even. g is a reflection in the y-axis.

(c) Even. g is a vertical shift downward.

(d) Neither even nor odd. g is shifted to the right and reflected in the x-axis.

105. No. x is not a function of y because horizontal lines can be drawn to intersect the graph twice. Therefore, each y-value corresponds to two distinct x-values when $-5 < y < 5$.

107. Terms: $-2x^2, 8x$; coefficients: $-2, 8$

109. Terms: $\frac{x}{3}, -5x^2, x^3$; coefficients: $\frac{1}{3}, -5, 1$

111. (a) $d = 4\sqrt{5}$ (b) Midpoint: $(2, 5)$

113. (a) $d = \sqrt{41}$ (b) Midpoint: $\left(\frac{1}{2}, \frac{3}{2}\right)$

115. (a) 29   (b) -6 (c) $5x - 16$

117. (a) 0   (b) 36   (c) $6\sqrt{3}$

119. $h + 4, h \ne 0$

Section 1.4   (page 48)

Vocabulary Check   (page 48)

1. quadratic function   **2.** absolute value function

3. rigid transformations   **4.** $-f(x), f(-x)$

5. $c > 1, 0 < c < 1$

6. (a) ii   (b) iv   (c) iii   (d) i

1.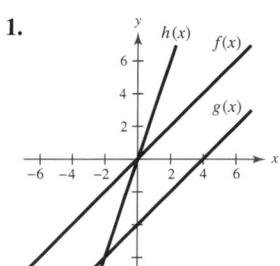

3.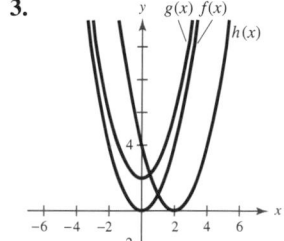

5.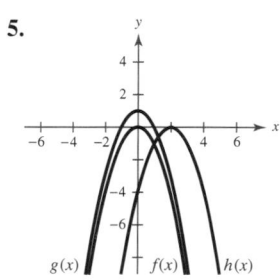

7.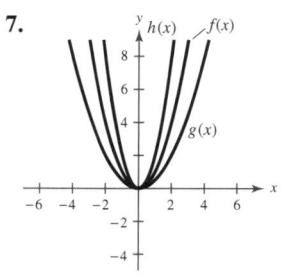

9.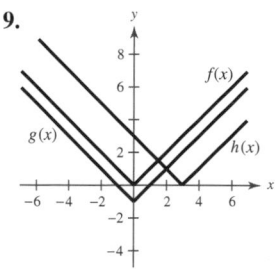

11.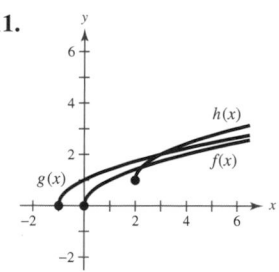

13. (a) 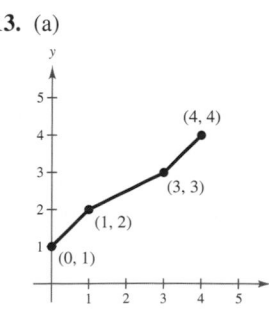   (b)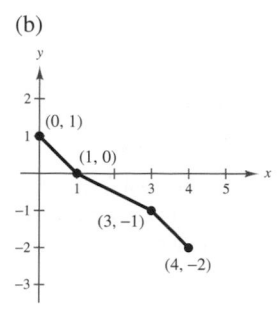

(c) 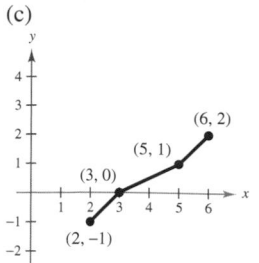   (d)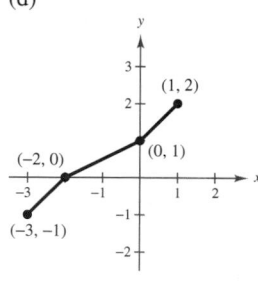

(e) 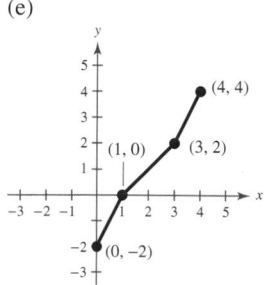   (f)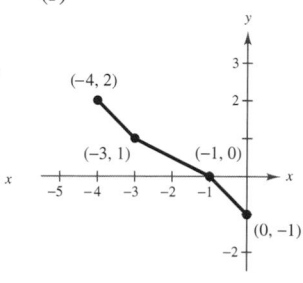

(g)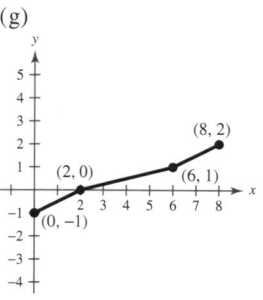

15. Vertical shift of $y = x$; $y = x + 3$

17. Vertical shift of $y = x^2$; $y = x^2 - 1$

19. Reflection in the x-axis and a vertical shift one unit upward of $y = \sqrt{x}$; $y = 1 - \sqrt{x}$

21. Reflection in the x-axis and vertical shift one unit downward

23. Horizontal shift two units to the right

25. Vertical stretch

27. Horizontal shift five units to the left

29. Reflection in the x-axis   **31.** Vertical stretch

33. Reflection in the x-axis and vertical shift four units upward

35. Horizontal shift two units to the left and vertical shrink

37. Horizontal stretch and vertical shift two units upward

39.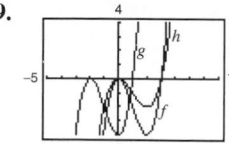

g is a horizontal shift and h is a vertical shrink.

41.

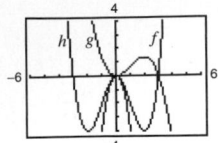

g is a vertical shrink and a reflection in the x-axis and h is a reflection in the y-axis.

43. (a) $f(x) = x^2$

(b) Horizontal shift five units to the left, reflection in the x-axis, and vertical shift two units upward

(c)

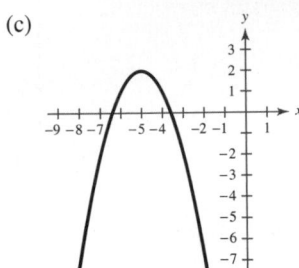

(d) $g(x) = 2 - f(x + 5)$

45. (a) $f(x) = x^2$

(b) Horizontal shift four units to the right, vertical stretch, and vertical shift three units upward

(c)

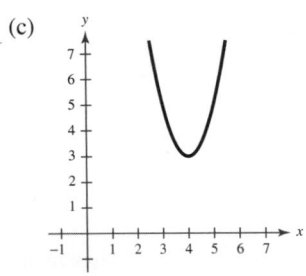

(d) $g(x) = 3 + 2f(x - 4)$

47. (a) $f(x) = x^3$

(b) Horizontal shift two units to the right and vertical stretch

(c)

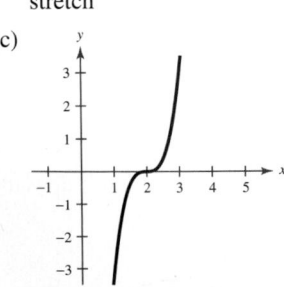

(d) $g(x) = 3f(x - 2)$

49. (a) $f(x) = x^3$

(b) Horizontal shift one unit to the right and vertical shift two units upward

(c)

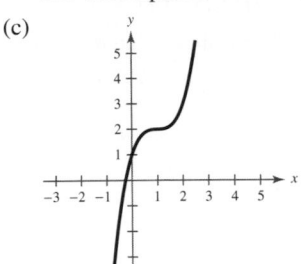

(d) $g(x) = f(x - 1) + 2$

51. (a) $f(x) = |x|$

(b) Horizontal shift four units to the left and vertical shift eight units upward

(c)

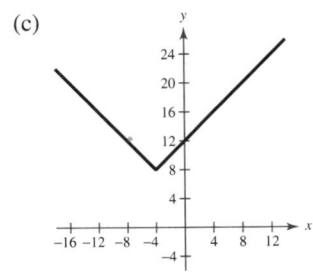

(d) $g(x) = f(x + 4) + 8$

53. (a) $f(x) = |x|$

(b) Horizontal shift one unit to the right, reflection in the x-axis, vertical stretch, and vertical shift four units downward

(c)

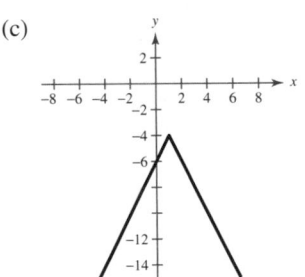

(d) $g(x) = -2f(x - 1) - 4$

55. (a) $f(x) = \sqrt{x}$

(b) Horizontal shift three units to the left, reflection in the x-axis, vertical shrink, and vertical shift one unit downward

(c)

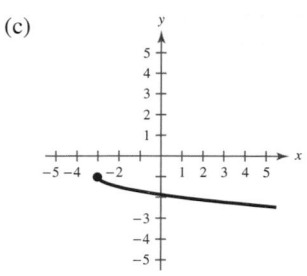

(d) $g(x) = -\frac{1}{2}f(x + 3) - 1$

57. (a) Vertical stretch and vertical shift

(b)

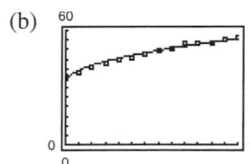

(c) $G(t) = 33.0 + 6.2\sqrt{t + 13}$, where $t = 0$ corresponds to 2003. Answers will vary.

59. False. When $f(x) = x^2$, $f(-x) = (-x)^2 = x^2$. Because $f(x) = f(-x)$ in this case, $y = f(-x)$ is not a reflection of $y = f(x)$ across the x-axis in all cases.

61. (a) $x = -2$ and $x = 3$

(b) $x = -3$ and $x = 2$

(c) $x = -3$ and $x = 2$

(d) Cannot be determined

(e) $x = 0$ and $x = 5$

63. (a) Increasing on the interval $(-\infty, -2)$ and decreasing on the interval $(-2, \infty)$

(b) Increasing on the interval $(2, \infty)$ and decreasing on the interval $(-\infty, 2)$

(c) Increasing on the interval $(-\infty, 2)$ and decreasing on the interval $(2, \infty)$

(d) Increasing on the interval $(-\infty, 2)$ and decreasing on the interval $(2, \infty)$

(e) Increasing on the interval $(-\infty, 1)$ and decreasing on the interval $(1, \infty)$

65. c   **67.** c   **69.** Neither

71. All real numbers x except $x = 9$

73. All real numbers x such that $-10 \le x \le 10$

Section 1.5   (page 58)

1.

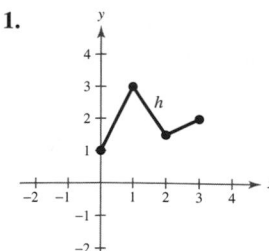

3.

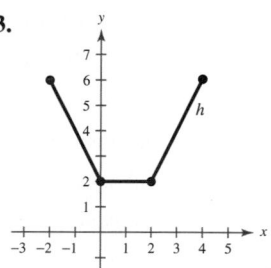

5. (a) $2x$ (b) 6   (c) $x^2 - 9$

(d) $\dfrac{x + 3}{x - 3}$, All real numbers x, except $x = 3$.

7. (a) $x^2 - x + 1$ (b) $x^2 + x - 1$ (c) $x^2 - x^3$

(d) $\dfrac{x^2}{1 - x}$, All real numbers x, except $x = 1$.

9. (a) $x^2 + 5 + \sqrt{1 - x}$ (b) $x^2 + 5 - \sqrt{1 - x}$

(c) $(x^2 + 5)\sqrt{1 - x}$ (d) $\dfrac{x^2 + 5}{\sqrt{1 - x}}$, $x < 1$

11. (a) $\dfrac{x + 1}{x^2}$ (b) $\dfrac{x - 1}{x^2}$ (c) $\dfrac{1}{x^3}$ (d) x, $x \ne 0$

13. 9   **15.** 1   **17.** 30   **19.** $-\frac{24}{7}$

21. $4t^2 - 2t + 1$ **23.** $-125t^3 - 50t^2 + 5t + 2$

25. $\dfrac{t^2 - 1}{-t - 2}$

27.

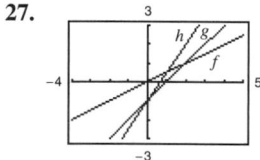

29.

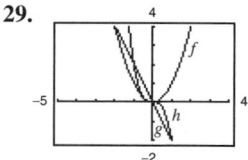

31.

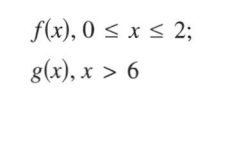

$f(x), 0 \le x \le 2;$

$g(x), x > 6$

33.

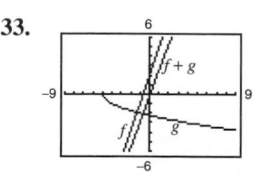

$f(x), 0 \le x \le 2;$

$f(x), x > 6$

35. (a) $(x - 1)^2$ (b) $x^2 - 1$ (c) 1

37. (a) $20 - 3x$ (b) $-3x$ (c) 20

39. (a) All real numbers x such that $x \ge -4$

(b) All real numbers x (c) All real numbers x

41. (a) All real numbers x

   (b) All real numbers x such that $x \geq 0$

   (c) All real numbers x such that $x \geq 0$

43. (a) All real numbers x except $x = 0$

   (b) All real numbers x

   (c) All real numbers x except $x = -3$

45. (a) All real numbers x (b) All real numbers x

   (c) All real numbers x

47. (a) All real numbers x

   (b) All real numbers x except $x = \pm 2$

   (c) All real numbers x except $x = \pm 2$

49. (a) $(f \circ g)(x) = \sqrt{x^2 + 4}$

   $(g \circ f)(x) = x + 4,\ x \geq -4;$

   Domain of $f \circ g$: all real numbers x

   (b)

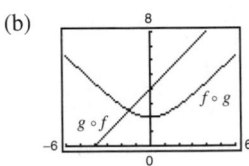

   $f \circ g \neq g \circ f$

51. (a) $(f \circ g)(x) = x;\ (g \circ f)(x) = x;$

   Domain of $f \circ g$: all real numbers x

   (b)

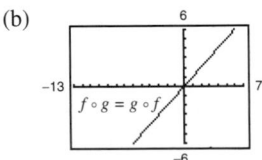

   $f \circ g = g \circ f$

53. (a) $(f \circ g)(x) = x^4;\ (g \circ f)(x) = x^4;$

   Domain of $f \circ g$: all real numbers x

   (b)

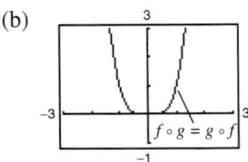

   $f \circ g = g \circ f$

55. (a) $(f \circ g)(x) = 24 - 5x;\ (g \circ f)(x) = -5x$

   (b) $24 - 5x \neq -5x$

(c)

x	0	1	2	3
$g(x)$	4	3	2	1
$(f \circ g)(x)$	24	19	14	9

x	0	1	2	3
$f(x)$	4	9	14	19
$(g \circ f)(x)$	0	-5	-10	-15

57. (a) $(f \circ g)(x) = \sqrt{x^2 + 1};\ (g \circ f)(x) = x + 1,\ x \geq -6$

   (b) $x + 1 \neq \sqrt{x^2 + 1}$

(c)

x	0	1	2	3
$g(x)$	-5	-4	-1	4
$(f \circ g)(x)$	1	$\sqrt{2}$	$\sqrt{5}$	$\sqrt{10}$

x	0	1	2	3
$f(x)$	$\sqrt{6}$	$\sqrt{7}$	$\sqrt{8}$	3
$(g \circ f)(x)$	1	2	3	4

59. (a) $(f \circ g)(x) = |2x + 2|;\ (g \circ f)(x) = 2|x + 3| - 1$

   (b) $(f \circ g)(x) = \begin{cases} 2x + 2, & x \geq -1 \\ -2x - 2, & x < -1 \end{cases}$

   $(g \circ f)(x) = \begin{cases} 2x + 5, & x \geq -3 \\ -2x - 7, & x < -3 \end{cases}$

   $(f \circ g)(x) \neq (g \circ f)(x)$

(c)

x	-2	-1	0	1	2
$g(x)$	-16	-2	0	2	16
$(f \circ g)(x)$	-16	2	0	2	16

x	-2	-1	0	1	2
$f(x)$	2	1	0	1	2
$(g \circ f)(x)$	-16	2	0	2	16

61. (a) 3   (b) 0   **63.** (a) 0   (b) 4

65. $f(x) = x^2,\ g(x) = 2x + 1$

67. $f(x) = \sqrt[3]{x},\ g(x) = x^2 - 4$

69. $f(x) = \dfrac{1}{x},\ g(x) = x + 2$

71. $f(x) = x^2 + 2x$, $g(x) = x + 4$

73. (a) $T = \frac{3}{4}x + \frac{1}{15}x^2$

(b)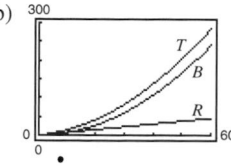

(c) B. For example, $B(60) = 240$, whereas $R(60)$ is only 45.

75.

Year	1995	1996	1997	1998
y_1	140	151.4	162.8	174.2
y_2	325.8	342.8	364.4	390.6
y_3	458.8	475.3	497.9	526.5

Year	1999	2000	2001	2002
y_1	185.6	197	208.4	219.8
y_2	421.5	457	497.1	541.8
y_3	561.2	602	648.8	701.7

Year	2003	2004	2005
y_1	231.2	242.6	254
y_2	591.2	645.2	703.8
y_3	760.7	825.7	896.8

Answers will vary.

77. $(A \circ r)(t) = 0.36\pi t^2$

$(A \circ r)(t)$ represents the area of the circle at time t.

79. (a) $C(x(t)) = 3000t + 750$; $C(x(t))$ represents the production cost after t production hours.

(b) 200 units

(c)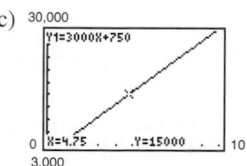

$t = 4.75$, or 4 hours 45 minutes

81. (a) $N(T(t))$ or $(N \circ T)(t) = 40t^2 + 590$; $N(T(t))$ or $(N \circ T)(t)$ represents the number of bacteria after t hours outside the refrigerator.

(b) $(N \circ T)(6) = 2030$; There are 2030 bacteria in a refrigerated food product after 6 hours outside the refrigerator.

(c) About 2.3 hours

83. $g(f(x))$ represents 3% of an amount over \$500,000.

85. False. $(f \circ g)(x) = 6x + 1 \neq 6x + 6 = (g \circ f)(x)$

87. (a) $O(M(Y)) = 2\left(6 + \frac{1}{2}Y\right) = 12 + Y$

(b) Middle child is 8 years old, youngest child is 4 years old.

89 and 91. Proof

93. $(0, -5), (1, -5), (2, -7)$

95. $\left(0, 2\sqrt{6}\right), \left(1, \sqrt{23}\right), \left(2, 2\sqrt{5}\right)$

97. $10x - y + 38 = 0$ **99.** $30x + 11y - 34 = 0$

Section 1.6   (page 69)

Vocabulary Check   (page 69)

1. inverse, f^{-1} **2.** range, domain   **3.** $y = x$

4. one-to-one   **5.** Horizontal

1. $f^{-1}(x) = \dfrac{x}{6}$ **3.** $f^{-1}(x) = x - 7$

5. $f^{-1}(x) = \frac{1}{2}(x - 1)$ **7.** $f^{-1}(x) = x^3$

9. (a) $f(g(x)) = f\left(-\dfrac{2x + 6}{7}\right)$

$= -\dfrac{7}{2}\left(-\dfrac{2x + 6}{7}\right) - 3 = x$

$g(f(x)) = g\left(-\dfrac{7}{2}x - 3\right)$

$= -\dfrac{2\left(-\frac{7}{2}x - 3\right) + 6}{7} = x$

(b)

x	0	2	-2	6
$f(x)$	-3	-10	4	-24

x	-3	-10	4	-24
$g(x)$	0	2	-2	6

11. (a) $f(g(x)) = f\left(\sqrt[3]{x - 5}\right) = \left(\sqrt[3]{x - 5}\right)^3 + 5 = x$

$g(f(x)) = g(x^3 + 5) = \sqrt[3]{(x^3 + 5) - 5} = x$

(b)

x	0	1	−1	−2	4
f(x)	5	6	4	−3	69

x	5	6	4	−3	69
g(x)	0	1	−1	−2	4

13. (a) $f(g(x)) = f(8 + x^2)$
$$= -\sqrt{(8 + x^2) - 8}$$
$$= -\sqrt{x^2} = -(-x) = x, \quad x \le 0$$
$g(f(x)) = g(-\sqrt{x - 8})$
$$= 8 + (-\sqrt{x - 8})^2$$
$$= 8 + (x - 8) = x, \quad x \ge 8$$

(b)

x	8	9	12	15
f(x)	0	−1	−2	$-\sqrt{7}$

x	0	−1	−2	$-\sqrt{7}$
g(x)	8	9	12	15

15. $f(g(x)) = f(\sqrt[3]{x}) = (\sqrt[3]{x})^3 = x$
$g(f(x)) = g(x^3) = \sqrt[3]{x^3} = x$

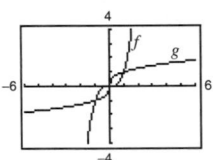

Reflections in the line $y = x$

17. $f(g(x)) = f(x^2 + 4), \quad x \ge 0$
$$= \sqrt{(x^2 + 4) - 4} = x$$
$g(f(x)) = g(\sqrt{x - 4})$
$$= (\sqrt{x - 4})^2 + 4 = x$$

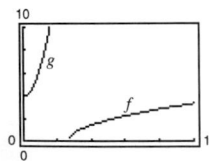

Reflections in the line $y = x$

19. $f(g(x)) = f(\sqrt[3]{1 - x}) = 1 - (\sqrt[3]{1 - x})^3 = x$
$g(f(x)) = g(1 - x^3) = \sqrt[3]{1 - (1 - x^3)} = x$

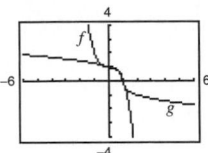

Reflections in the line $y = x$

21. c   **22.** b   **23.** a   **24.** d

25. (a)

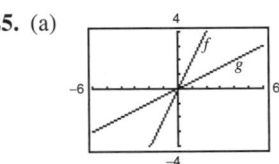

(b)

x	−4	−2	0	2	4
f(x)	−8	−4	0	4	8

x	−8	−4	0	4	8
g(x)	−4	−2	0	2	4

27. (a)

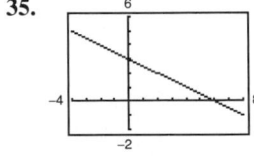

(b)

x	−3	−2	−1	0	2	3	4
f(x)	−2	−1	$-\frac{1}{2}$	$-\frac{1}{5}$	$\frac{1}{7}$	$\frac{1}{4}$	$\frac{1}{3}$

x	−2	−1	$-\frac{1}{2}$	$-\frac{1}{5}$	$\frac{1}{7}$	$\frac{1}{4}$	$\frac{1}{3}$
g(x)	−3	−2	−1	0	2	3	4

29. Not a function   **31.** Function; one-to-one

33. Function; one-to-one

35.

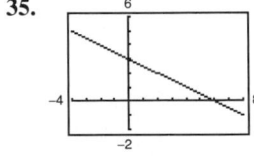

One-to-one

37.

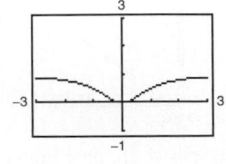

Not one-to-one

39.

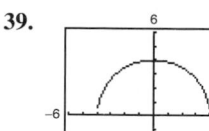

Not one-to-one

41.

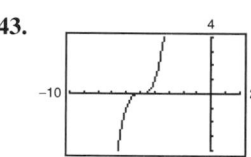

Not one-to-one

43.

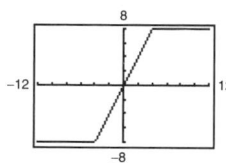

One-to-one

45.

Not one-to-one

47. Not one-to-one

49. $f^{-1}(x) = \dfrac{5x - 4}{3}$

51. Not one-to-one

53. $f^{-1}(x) = \sqrt{x} - 3,\ x \ge 0$

55. $f^{-1}(x) = \dfrac{x^2 - 3}{2},\ x \ge 0$

57. $f^{-1}(x) = 2 - x,\ x \ge 0$

59. $f^{-1}(x) = \dfrac{x + 3}{2}$

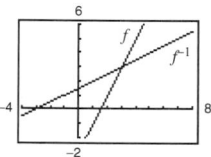

Reflections in the line $y = x$

61. $f^{-1}(x) = \sqrt[5]{x}$

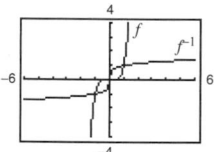

Reflections in the line $y = x$

63. $f^{-1}(x) = x^{5/3}$

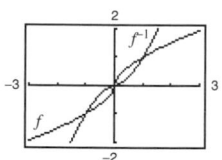

Reflections in the line $y = x$

65. $f^{-1}(x) = \sqrt{4 - x^2},\ 0 \le x \le 2$

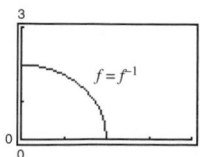

The graphs are the same.

67. $f^{-1}(x) = \dfrac{4}{x}$

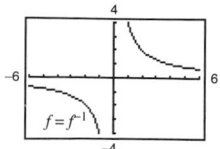

The graphs are the same.

69. $f^{-1}(x) = \sqrt{x} + 2$

Domain of f: all real numbers x such that $x \ge 2$
Range of f: all real numbers y such that $y \ge 0$
Domain of f^{-1}: all real numbers x such that $x \ge 0$
Range of f^{-1}: all real numbers y such that $y \ge 2$

71. $f^{-1}(x) = x - 2$

Domain of f: all real numbers x such that $x \ge -2$
Range of f: all real numbers y such that $y \ge 0$
Domain of f^{-1}: all real numbers x such that $x \ge 0$
Range of f^{-1}: all real numbers y such that $y \ge -2$

73. $f^{-1}(x) = \sqrt{x} - 3$

Domain of f: all real numbers x such that $x \ge -3$
Range of f: all real numbers y such that $y \ge 0$
Domain of f^{-1}: all real numbers x such that $x \ge 0$
Range of f^{-1}: all real numbers y such that $y \ge -3$

75. $f^{-1}(x) = \dfrac{\sqrt{-2(x - 5)}}{2}$

Domain of f: all real numbers x such that $x \ge 0$
Range of f: all real numbers y such that $y \le 5$
Domain of f^{-1}: all real numbers x such that $x \le 5$
Range of f^{-1}: all real numbers y such that $y \ge 0$

77. $f^{-1}(x) = x + 3$

Domain of f: all real numbers x such that $x \ge 4$
Range of f: all real numbers y such that $y \ge 1$
Domain of f^{-1}: all real numbers x such that $x \ge 1$
Range of f^{-1}: all real numbers y such that $y \ge 4$

79.

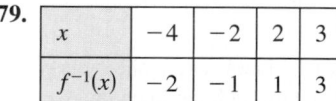

x	-4	-2	2	3
$f^{-1}(x)$	-2	-1	1	3

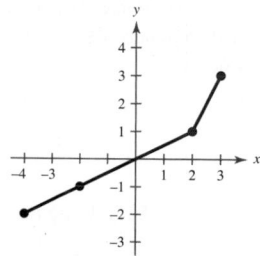

81. $\frac{1}{2}$ **83.** -2 **85.** 0   **87.** 2

89. (a) and (b)

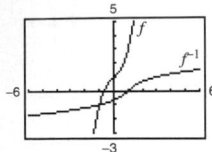

(c) Inverse function because it satisfies the Vertical Line Test

91. (a) and (b)

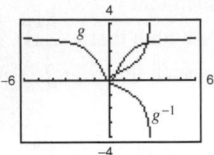

(c) Not an inverse function because it does not satisfy the Vertical Line Test

93. 32   **95.** 600   **97.** $2\sqrt[3]{x+3}$

99. $\dfrac{x+1}{2}$ **101.** $\dfrac{x+1}{2}$

103. (a) f is one-to-one because no two elements in the domain (men's U.S. shoe sizes) correspond to the same element in the range (men's European shoe sizes).

(b) 45   (c) 10   (d) 41   (e) 13

105. (a) Yes

(b) $f^{-1}(t)$ represents the year new car sales totaled $\$t$ billion.

(c) 10 or 2000

(d) No. The inverse would not be a function because f would not be one-to-one.

107. False. For example, $y = x^2$ is even, but does not have an inverse.

109. Proof

111. f and g are not inverses of each other because one is not the graph of the other when reflected through the line $y = x$.

113. f and g are inverses of each other because one is the graph of the other when reflected through the line $y = x$.

115. This situation could be represented by a one-to-one function. The inverse function would represent the number of miles completed in terms of time in hours.

117. This function could not be represented by a one-to-one function because it oscillates.

119. $9x$, $x \neq 0$ **121.** $-(x+6)$, $x \neq 6$

123. Function   **125.** Not a function   **127.** Function

Section 1.7   (page 78)

1. (a)

(b) Answers will vary.

3. Negative correlation   **5.** No correlation

7. (a)

(b) $y = 0.46x + 1.62$; 0.95095

(c)

(d) The models appear valid.

9. (a)

(b) $y = 0.95x + 0.92$; 0.90978

(c)

(d) The models appear valid.

11. (a)

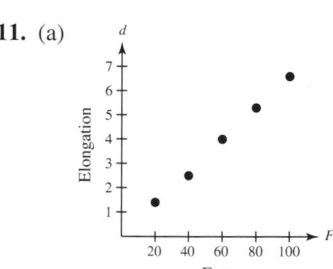

(b) $d = 0.07F - 0.3$

(c) $d = 0.066F$

(d) 3.63 centimeters

13. (a)

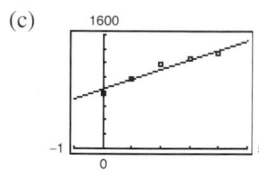

(b) $S = 136.1t + 836$

(c)

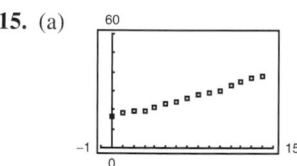

The model fits the data.

(d) 2005: \$1,516,500; 2010: \$2,197,000;
Answers will vary.

(e) 136.1; The slope represents the average annual
increase in salaries (in thousands of dollars).

15. (a)

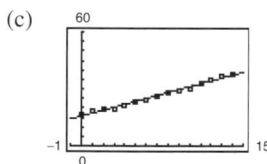

(b) $C = 1.552t + 15.70; 0.99544$

(c)

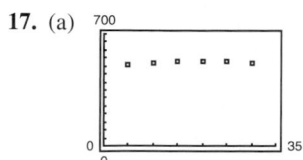

(d) Yes; answers will vary.

(e) 2005: \$38.98; 2010: \$46.74

(f) Answers will vary.

17. (a)

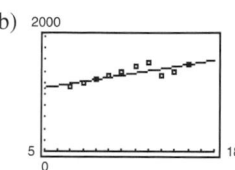

(b) $P = 0.6t + 512$

(c)

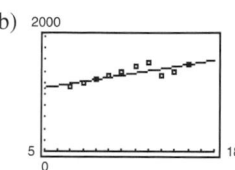

The model is not the best fit for the data.

(d) 542,000 people; answers will vary.

19. (a) $T = 36.7t + 926; 0.79495$

(b)

(c) The slope represents an increase of about 37 Target
stores annually.

(d) 2013

(e)

Year	1997	1998	1999	2000	2001
Actual T-values (in thousands)	1130	1182	1243	1307	1381
T-values from model (in thousands)	1183	1220	1256	1293	1330

Year	2002	2003	2004	2005	2006
Actual T-values (in thousands)	1475	1553	1308	1400	1505
T-values from model (in thousands)	1366	1403	1440	1477	1513

Answers will vary.

21. True. To have positive correlation, the y-values tend to
increase as x increases.

23. Answers will vary.   **25.** (a) 10   (b) $2w^2 + 5w + 7$

27. (a) 5   (b) 1   **29.** $-\frac{3}{5}$ **31.** $-\frac{1}{4}, \frac{3}{2}$

33. $\dfrac{7 \pm \sqrt{17}}{4}$

CHAPTER 1

Review Exercises   (page 83)

1.

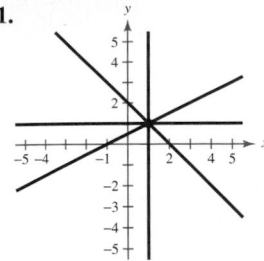

3.

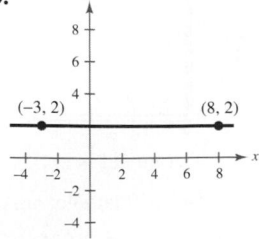

$m = 0$

5.

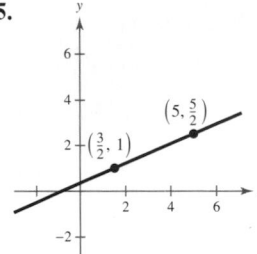

$m = \frac{3}{7}$

7.

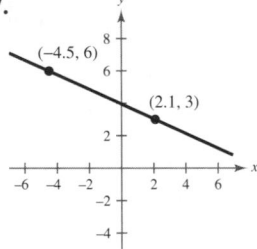

$m = -\frac{5}{11}$

9. $x - 4y - 6 = 0$; $(6, 0)$, $(10, 1)$, $(-2, -2)$

11. $3x - 2y - 10 = 0$; $(4, 1)$, $(2, -2)$, $(-2, -8)$

13. $5x + 5y + 24 = 0$; $\left(-5, \frac{1}{5}\right)$, $\left(-4, -\frac{4}{5}\right)$, $\left(-6, \frac{6}{5}\right)$

15. $y - 6 = 0$; $(0, 6)$, $(1, 6)$, $(-1, 6)$

17. $x - 10 = 0$; $(10, 1)$, $(10, 3)$, $(10, -2)$

19. $y = -1$

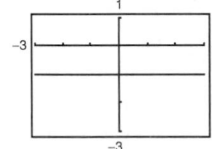

21. $y = \frac{2}{7}x + \frac{2}{7}$

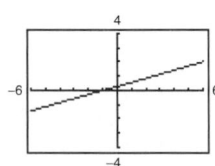

23. $V = 850t + 5700$ **25.** $V = 42.70t + 283.9$

27. \$210,000

29. (a) $y = \frac{5}{4}x - \frac{23}{4}$ **31.** (a) $x = -6$

     (b) $y = -\frac{4}{5}x + \frac{2}{5}$ (b) $y = 2$

33. (a) Not a function because element 20 in A corresponds to two elements, 4 and 6, in B.

     (b) Function

     (c) Function

     (d) Not a function because 30 in A corresponds to no element in B.

35. Not a function   **37.** Function

39. (a) 2   (b) 10   (c) $b^6 + 1$ (d) $x^2 - 2x + 2$

41. (a) -3 (b) -1 (c) 2   (d) 6

43. All real numbers x except $x = -2$

45. All real numbers x such that $-5 \le x \le 5$.

47. All real numbers s except $s = 3$

49. (a) $C = 5.35x + 16,000$ (b) $P = 2.85x - 16,000$

51. $2h + 4x + 3$, $h \ne 0$

53.

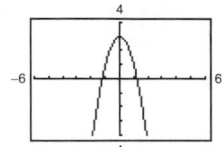

55.

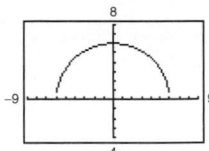

Domain: all real numbers x Domain: $[-6, 6]$

Range: $(-\infty, 3]$ Range: $[0, 6]$

57. (a)

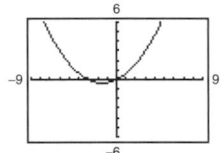

59. (a)

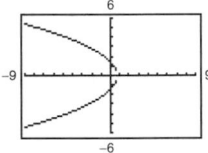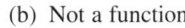

     (b) Function   (b) Not a function

61. (a)

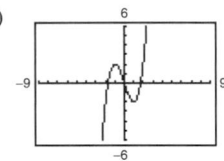

     (b) Increasing on $(-\infty, -1)$, $(1, \infty)$

         Decreasing on $(-1, 1)$

63. (a)

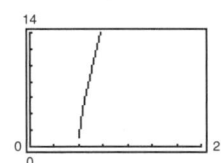

     (b) Increasing on $(6, \infty)$

65. Relative maximum: $(0, 16)$

     Relative minima: $(-2, 0)$, $(2, 0)$

67. Relative maximum: $(3, 27)$

69.

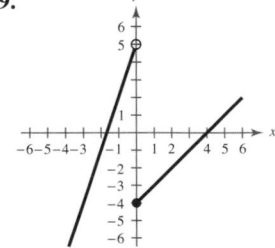

71.

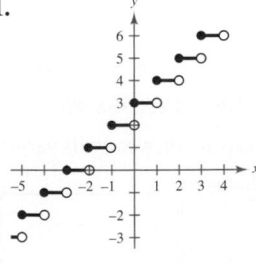

73. Even   **75.** Even   **77.** Neither

79. Constant function $f(x) = c$

Vertical shift two units downward

$g(x) = -2$

81. Quadratic function $f(x) = x^2$

Vertical shift one unit upward, horizontal shift two units right

$g(x) = (x - 2)^2 + 1$

83. Absolute value function $f(x) = |x|$

Vertical shift three units upward

$g(x) = |x| + 3$

85.
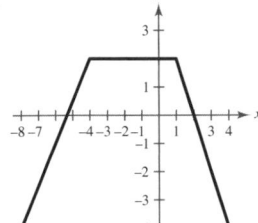

87.
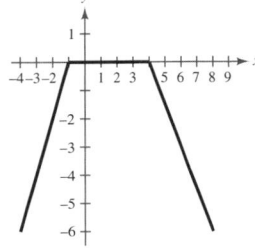

89. (a) Quadratic function $f(x) = x^2$

(b) Vertical shift six units downward

(c)
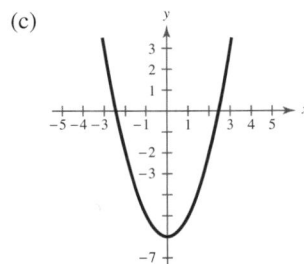

(d) $h(x) = f(x) - 6$

91. (a) Cubic function $f(x) = x^3$

(b) Horizontal shift two units right, vertical shift five units upward

(c)
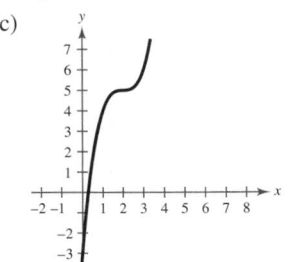

(d) $h(x) = f(x - 2) + 5$

93. (a) Quadratic function $f(x) = x^2$

(b) Horizontal shift two units to the right, reflection in the x-axis, vertical shift eight units downward

(c)
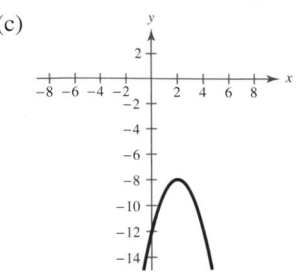

(d) $h(x) = -f(x - 2) - 8$

95. (a) Square root function $f(x) = \sqrt{x}$

(b) Reflection in the x-axis, vertical shift five units upward

(c)
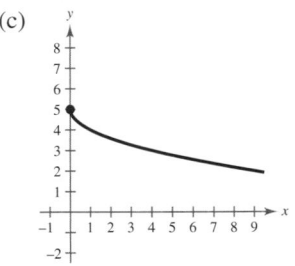

(d) $h(x) = -f(x) + 5$

97. (a) Square root function $f(x) = \sqrt{x}$

(b) Horizontal shift one unit to the right, vertical shift, three units upward

(c)
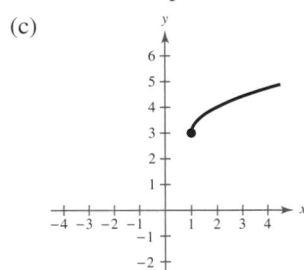

(d) $h(x) = f(x - 1) + 3$

99. (a) Absolute value function $f(x) = |x|$

(b) Reflection in the x-axis, vertical shrink by a factor of $\frac{1}{2}$, vertical shift nine units upward

(c)
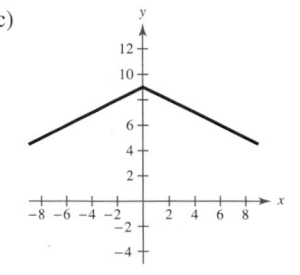

(d) $h(x) = -\frac{1}{2}f(x) + 9$

CHAPTER 1

101. -7 **103.** -42 **105.** 5   **107.** 23

109. -97 **111.** $f(x) = x^2,\ g(x) = x + 3$

113. $f(x) = \sqrt{x},\ g(x) = 4x + 2$

115. $f(x) = \dfrac{4}{x},\ g(x) = x + 2$

117. 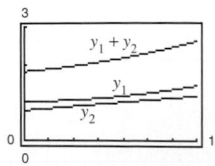   **119.** $f^{-1}(x) = \dfrac{x}{6}$

121. $f^{-1}(x) = 2x - 6$

123. Answers will vary.

125. 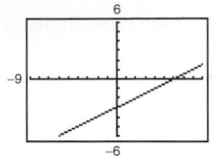   **127.**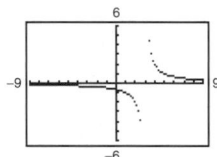

One-to-one   One-to-one

129. $f^{-1}(x) = 2x + 10$ **131.** $f^{-1}(x) = \sqrt[3]{\dfrac{x+3}{4}}$

133. $f^{-1}(x) = x^2 - 10,\ x \geq 0$

135. Negative correlation

137. (a)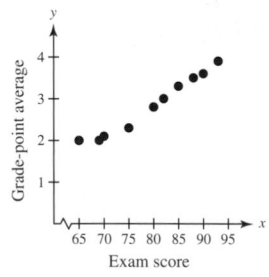

(b) Yes. Answers will vary.

139. (a)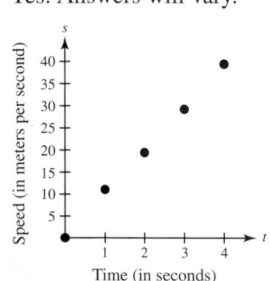

(b) Answers will vary.

Sample answer: $s = 10t - 0.4$

(c) $s = 9.7t + 0.4$; 0.99933; This model fits the data better.

(d) 24.65 meters per second

141. $y = 95.174x - 458.42$

143. The model doesn't fit the data very well.

145. False. The point $(-1, 28)$ does not lie on the graph of the function $g(x) = -(x - 6)^2 - 3$.

147. False. For example, $f(x) = 4 - x = f^{-1}(x)$.

Chapter Test   (page 88)

1. (a) $5x + 2y - 8 = 0$ (b) $2x - 5y + 20 = 0$

2. $y = -x + 1$

3. No. To some x there correspond more than one value of y.

4. (a) -9 (b) 1   (c) $|t - 4| - 15$

5. $(-\infty, 3]$ **6.** $C = 5.60x + 24{,}000$

$$P = 93.9x - 24{,}000$$

7. Odd   **8.** Even

9. Increasing: $(-2, 0),\ (2, \infty)$

Decreasing: $(-\infty, -2),\ (0, 2)$

10. Increasing: $(-2, 2)$

Constant: $(-\infty, -2),\ (2, \infty)$

11. 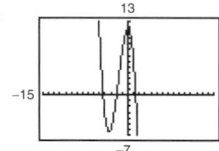   Relative minimum: $(-3.33, -6.52)$

Relative maximum: $(0, 12)$

12. 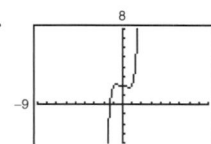   Relative minimum: $(0.77, 1.81)$

Relative maximum: $(-0.77, 2.19)$

13. (a) Cubic function $f(x) = x^3$

(b) Horizontal shift five units to the right, reflection in the x-axis, vertical stretch, and vertical shift three units upward

(c)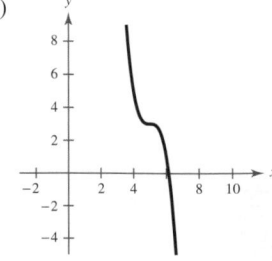

14. (a) Square root function $f(x) = \sqrt{x}$

(b) Reflection in the y-axis and horizontal shift seven units to the left

(c)

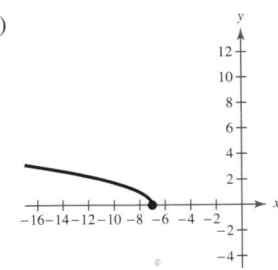

15. (a) Absolute value function $f(x) = |x|$

(b) Reflection in the y-axis (no effect), vertical stretch, and vertical shift seven units downward

(c)

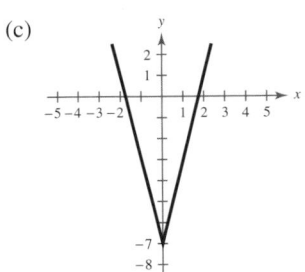

16. (a) $x^2 - \sqrt{2 - x}$, $(-\infty, 2]$　(b) $\dfrac{x^2}{\sqrt{2 - x}}$, $(-\infty, 2)$

(c) $2 - x$, $(-\infty, 2]$　(d) $\sqrt{2 - x^2}$, $\left[-\sqrt{2}, \sqrt{2}\right]$

17. $f^{-1}(x) = \sqrt[3]{x - 8}$　　**18.** No inverse

19. $f^{-1}(x) = \left(\frac{8}{3}x\right)^{2/3}$, $x \geq 0$

20. $S = 18.3t - 76.2$; 2005

Chapter 2

Section 2.1　(page 99)

1. c　　**2.** d　　**3.** b　　**4.** a

5.

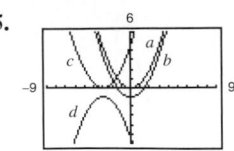

(a) Vertical shrink

(b) Vertical shrink and vertical shift one unit downward

(c) Vertical shrink and a horizontal shift three units to the left

(d) Vertical shrink, reflection in the x-axis, a horizontal shift three units to the left, and a vertical shift one unit downward

7.

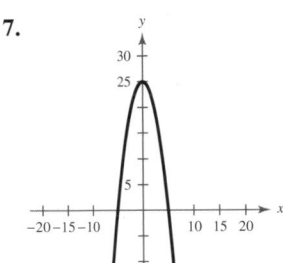

Vertex: $(0, 25)$

x-intercepts: $(\pm 5, 0)$

9.

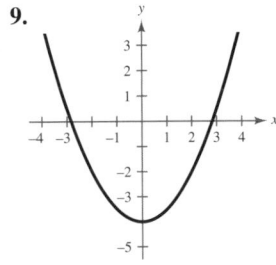

Vertex: $(0, -4)$

x-intercepts: $\left(\pm 2\sqrt{2}, 0\right)$

11.

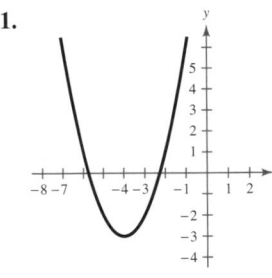

Vertex: $(-4, -3)$

x-intercepts:

$\left(\pm\sqrt{3} - 4, 0\right)$

13.

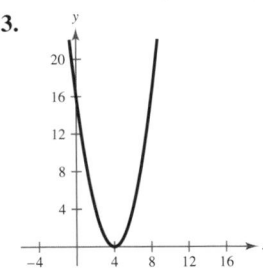

Vertex: $(4, 0)$

x-intercept: $(4, 0)$

15.

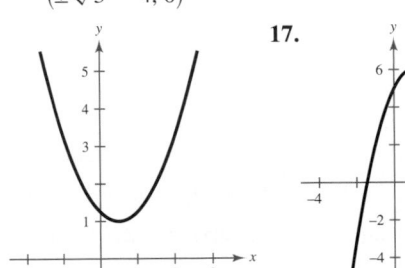

Vertex: $\left(\frac{1}{2}, 1\right)$

x-intercept: None

17.

Vertex: $(1, 6)$

x-intercepts:

$\left(1 \pm \sqrt{6}, 0\right)$

CHAPTER 2

19.

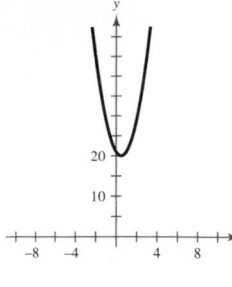

Vertex: $\left(\frac{1}{2}, 20\right)$

x-intercept: None

21.

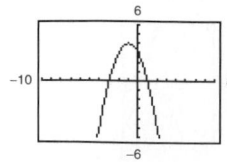

Vertex: $(-1, 4)$

x-intercepts:

$(1, 0), (-3, 0)$

23.

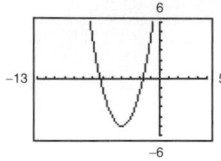

Vertex: $(-4, -5)$

x-intercepts:

$\left(-4 \pm \sqrt{5}, 0\right)$

25.

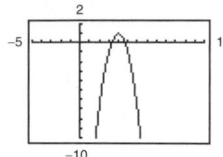

Vertex: $(4, 1)$

x-intercepts:

$\left(4 \pm \frac{1}{2}\sqrt{2}, 0\right)$

27. $y = -(x + 1)^2 + 4$ **29.** $f(x) = (x + 2)^2 + 5$

31. $y = 4(x - 1)^2 - 2$ **33.** $y = -\frac{104}{125}\left(x - \frac{1}{2}\right)^2 + 1$

35. $(5, 0), (-1, 0)$; They are the same.

37. $(-4, 0)$; They are the same.

39.

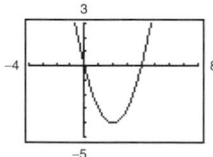

$(0, 0), (4, 0)$; They are the same.

41.

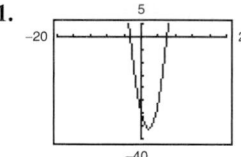

$\left(-\frac{5}{2}, 0\right), (6, 0)$;

They are the same.

43.

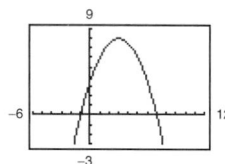

$(7, 0), (-1, 0)$;

They are the same.

45. $f(x) = x^2 - 2x - 3$

$g(x) = -x^2 + 2x + 3$

47. $f(x) = 2x^2 + 7x + 3$

$g(x) = -2x^2 - 7x - 3$

49. $55, 55$ **51.** $12, 6$

53. (a)

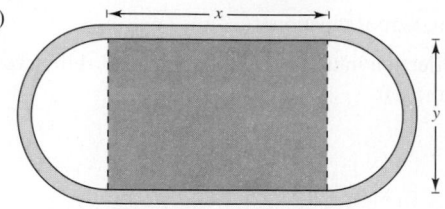

(b) $r = \frac{1}{2}y$; $d = y\pi$ (c) $y = \dfrac{200 - 2x}{\pi}$

(d) $A = x\left(\dfrac{200 - 2x}{\pi}\right)$

(e)

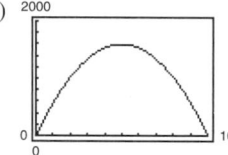

$x = 50$ meters, $y = \dfrac{100}{\pi}$ meters

55. (a)

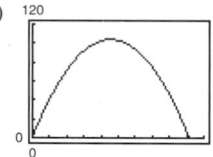

(b) $\frac{3}{2}$ feet

(c) About 104 feet

(d) About 228.6 feet

57. 20 fixtures

59. (a) \$14,000,000; \$14,375,000; \$13,500,000

(b) \$24   (c) \$14,400,000   (d) Answers will vary.

61. (a)

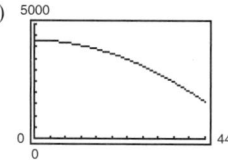

(b) 1960; 4306 cigarettes per person per year; Answers will vary.

(c) 8909 cigarettes per smoker per year; 24 cigarettes per smoker per day

63. True. The vertex is $(0, -1)$ and the parabola opens down.

65. c, d   **67.** $b = \pm 20$ **69.** $b = \pm 8$

71. Model (a). The profits are positive and rising.

73. $(1.2, 6.8)$ **75.** $(2, 5), (-3, 0)$

77. Answers will vary.

Section 2.2 (page 112)

Vocabulary Check (page 112)

1. continuous   **2.** Leading Coefficient Test

3. $n, n - 1$, relative extrema

4. solution, $(x - a)$, x-intercept   **5.** touches, crosses

6. Intermediate Value

1. f   **2.** h   **3.** c   **4.** a   **5.** e   **6.** d

7. g   **8.** b

9. (a)   (b)

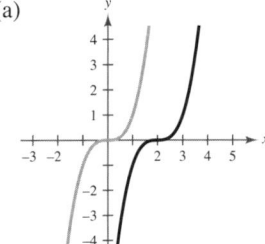

(c)   (d)

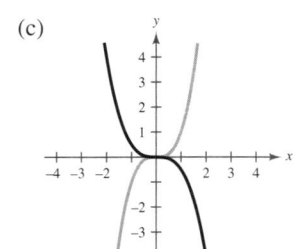

11.   **13.**

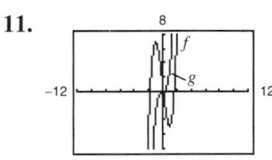

15. Rises to the left,   **17.** Falls to the left,
rises to the right   falls to the right

19. Falls to the left,   **21.** Falls to the left,
rises to the right   falls to the right

23. ± 5 (multiplicity 1)   **25.** 3 (multiplicity 2)

27. $1, -2$ (multiplicity 1)

29. 2 (multiplicity 2), 0 (multiplicity 1)

31. $\dfrac{-5 \pm \sqrt{37}}{2}$ (multiplicity 1)

33. (a) $\left(2 \pm \sqrt{3}, 0\right)$

(b)

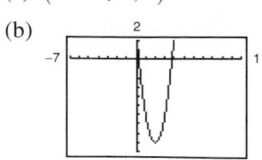

(c) $(0.27, 0)$, $(3.73, 0)$; answers are approximately the same.

35. (a) $(-1, 0)$, $(1, 0)$

(b)

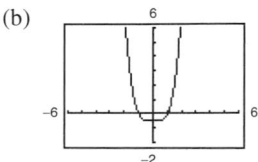

(c) $(-1, 0)$, $(1, 0)$; answers are approximately the same.

37. (a) $(0, 0)$, $\left(\pm\sqrt{2}, 0\right)$

(b)

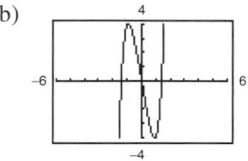

(c) $(-1.41, 0)$, $(0, 0)$, $(1.41, 0)$; answers are
approximately the same.

39. (a) $\left(\pm\sqrt{5}, 0\right)$

(b)

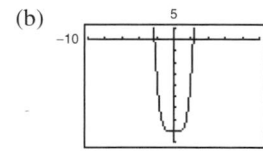

(c) $(-2.236, 0)$, $(2.236, 0)$

41. (a) $(4, 0)$, $(\pm 5, 0)$

(b)

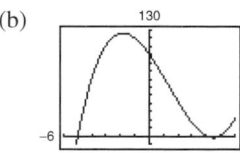

(c) $(-5, 0)$, $(4, 0)$, $(5, 0)$

43. (a) $(0, 0)$, $\left(\tfrac{5}{2}, 0\right)$

(b)

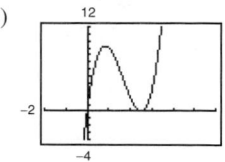

(c) $(0, 0)$, $\left(\tfrac{5}{2}, 0\right)$

CHAPTER 2

45.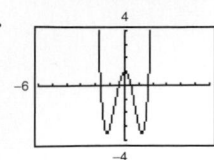

Zeros: $\pm 1.680, \pm 0.421$

Relative maximum: $(0, 1)$

Relative minima:

$(\pm 1.225, -3.500)$

47.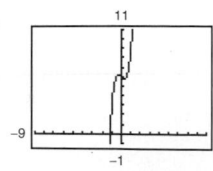

Zero: -1.178

Relative maximum:

$(-0.324, 6.218)$

Relative minimum:

$(0.324, 5.782)$

49. $f(x) = x^2 - 4x$ **51.** $f(x) = x^3 + 5x^2 + 6x$

53. $f(x) = x^4 - 4x^3 - 9x^2 + 36x$

55. $f(x) = x^2 - 2x - 2$

57. $f(x) = x^3 - 10x^2 + 27x - 22$

59. $f(x) = x^3 + 5x^2 + 8x + 4$

61. $f(x) = x^4 + 2x^3 - 23x^2 - 24x + 144$

63. $f(x) = -x^3 - 4x^2 - 5x - 2$

65. Not possible; odd-degree polynomials must have an odd number of real solutions.

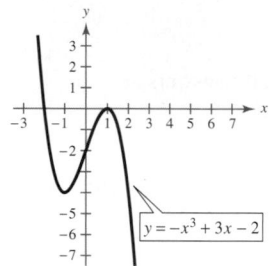

$y = -x^3 + 3x - 2$

67.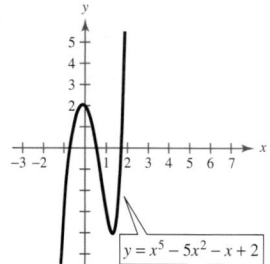

$y = x^5 - 5x^2 - x + 2$

69. (a) Falls to the left, rises to the right

(b) $(0, 0), (3, 0), (-3, 0)$

(c) Answers will vary.

(d)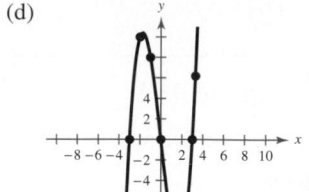

71. (a) Falls to the left, rises to the right

(b) $(0, 0), (3, 0)$ (c) Answers will vary.

(d)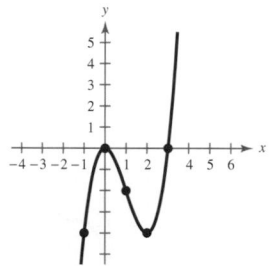

73. (a) Falls to the left, falls to the right

(b) $(\pm 2, 0), (\pm \sqrt{5}, 0)$ (c) Answers will vary.

(d)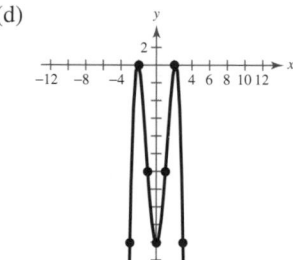

75. (a) Falls to the left, rises to the right

(b) $(\pm 3, 0)$ (c) Answers will vary.

(d)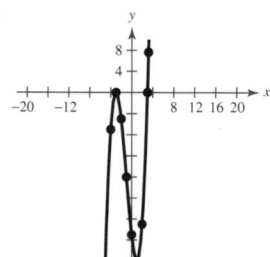

77. (a) Falls to the left, falls to the right

(b) $(-2, 0), (2, 0)$

(c) Answers will vary.

(d)

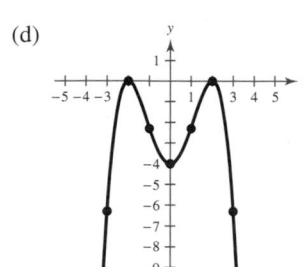

79. (a)

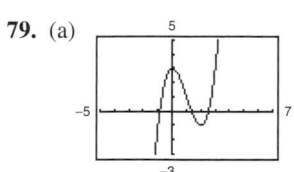

$(-1, 0), (1, 2), (2, 3)$

(b) $-0.879, 1.347, 2.532$

81. (a)

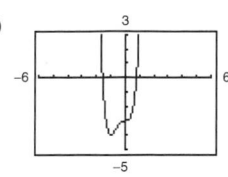

$(-2, -1), (0, 1)$

(b) $-1.585, 0.779$

83.

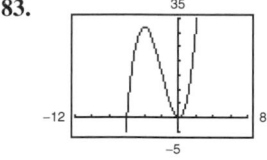

Two x-intercepts

85.

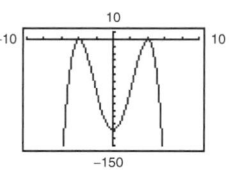

y-axis symmetry
Two x-intercepts

87.

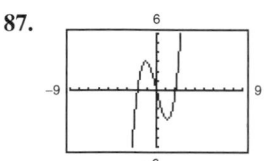

Origin symmetry
Three x-intercepts

89.

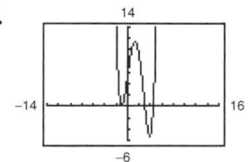

Three x-intercepts

91. (a) Answers will vary.

(b) Domain: $0 < x < 18$

(c)

Height, x	Volume, V
1	1156
2	2048
3	2700
4	3136
5	3380
6	3456
7	3388

$5 < x < 7$

(d)

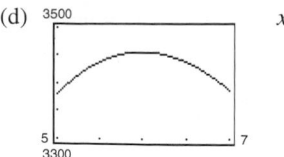

$x = 6$

93. $(200, 160)$

95.

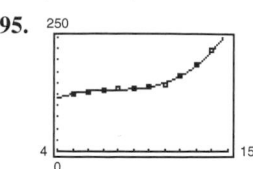

The model fits the data.

97. Northeast: $730,200; South: $285,000; Answers will vary.

99. True; $y = x^6$

101. False; the graph touches the x-axis at $x = 1$.

103. True   **105.** b   **107.** a   **109.** 33

111. $-\frac{4}{3} \approx -1.3$ **113.** 72

115. $x \le -\frac{1}{2}, x \ge 1$ **117.** $x \le -24, x \ge 8$

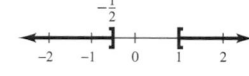

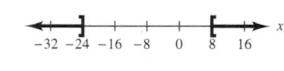

Section 2.3   (page 127)

1. $2x + 4, \ x \ne -3$ **3.** $x^3 + 3x^2 - 1, \ x \ne -2$

5. $x^2 - 3x + 1, \ x \ne -\frac{5}{4}$ **7.** $7x^2 - 14x + 28 - \dfrac{53}{x + 2}$

9. $3x + 5 - \dfrac{2x - 3}{2x^2 + 1}$ **11.** $x - \dfrac{x + 9}{x^2 + 1}$

13. $2x - \dfrac{17x - 5}{x^2 - 2x + 1}$ **15.** $3x^2 - 2x + 5, \ x \ne 5$

17. $6x^2 + 25x + 74 + \dfrac{248}{x - 3}$ **19.** $9x^2 - 16, \ x \ne 2$

CHAPTER 2

21. $x^2 - 8x + 64$, $x \neq -8$

23. $4x^2 + 14x - 30$, $x \neq -\frac{1}{2}$

25. 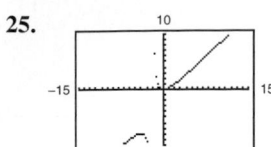   **27.**

29. $f(x) = (x - 4)(x^2 + 3x - 2) + 3$, $f(4) = 3$

31. $f(x) = \left(x - \sqrt{2}\right)\left[x^2 + \left(3 + \sqrt{2}\right)x + 3\sqrt{2}\right] - 8$,
$f\left(\sqrt{2}\right) = -8$

33. $f(x) = \left(x - 1 + \sqrt{3}\right)\left[4x^2 - \left(2 + 4\sqrt{3}\right)x - \left(2 + 2\sqrt{3}\right)\right]$,
$f\left(1 - \sqrt{3}\right) = 0$

35. (a) -2 (b) 1 (c) $-\frac{1}{4}$ (d) 5

37. (a) -35 (b) -22 (c) -10 (d) -211

39. $(x - 2)(x + 3)(x - 1)$ **41.** $(2x - 1)(x - 5)(x - 2)$
Zeros: $2, -3, 1$ Zeros: $\frac{1}{2}, 5, 2$

43. (a) Answers will vary.   (b) $(2x - 1)$, $(x - 1)$
(c) $(x + 2)(x - 1)(2x - 1)$ (d) $-2, 1, \frac{1}{2}$

45. (a) Answers will vary.   (b) $(x - 1)$, $(x - 2)$
(c) $(x - 5)(x + 4)(x - 1)(x - 2)$ (d) $-4, 1, 2, 5$

47. (a) Answers will vary.   (b) $(x + 7)$, $(3x - 2)$
(c) $(2x + 1)(3x - 2)(x + 7)$ (d) $-7, -\frac{1}{2}, \frac{2}{3}$

49. $\pm 1, -3$

51. $\pm 1, \pm 3, \pm 5, \pm 9, \pm 15, \pm 45, \pm\frac{1}{2}, \pm\frac{3}{2}, \pm\frac{5}{2}, \pm\frac{9}{2}, \pm\frac{15}{2}, \pm\frac{45}{2}$;
$-1, \frac{3}{2}, 3, 5$

53. $-1, 2$ **55.** $-6, \frac{1}{2}, 1$ **57.** $-3, -\frac{3}{2}, \frac{1}{2}, 4$

59. $-\frac{5}{2}, -2, \pm 1, \frac{3}{2}$

61. (a) $-2, 0.268, 3.732$ (b) -2
(c) $h(t) = (t + 2)\left(t - 2 + \sqrt{3}\right)\left(t - 2 - \sqrt{3}\right)$

63. (a) $0, 3, 4, -1.414, 1.414$ (b) $0, 3, 4$
(c) $h(x) = x(x - 3)(x - 4)\left(x + \sqrt{2}\right)\left(x - \sqrt{2}\right)$

65. $4, 2$, or 0 positive real zeros, no negative real zeros

67. 2 or 0 positive real zeros, 1 negative real zero

69. (a) 1 positive real zero, 2 or 0 negative real zeros
(b) $\pm 1, \pm 2, \pm 4$
(c) 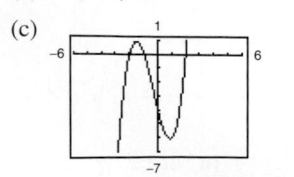   (d) $-2, -1, 2$

71. (a) 3 or 1 positive real zeros, 1 negative real zero
(b) $\pm 1, \pm 2, \pm 4, \pm 8, \pm\frac{1}{2}$
(c) 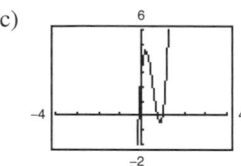   (d) $-\frac{1}{2}, 1, 2, 4$

73. (a) 2 or 0 positive real zeros, 1 negative real zero
(b) $\pm 1, \pm 3, \pm\frac{1}{2}, \pm\frac{3}{2}, \pm\frac{1}{4}, \pm\frac{3}{4}, \pm\frac{1}{8}, \pm\frac{3}{8}, \pm\frac{1}{16}, \pm\frac{3}{16}, \pm\frac{1}{32}, \pm\frac{3}{32}$
(c) 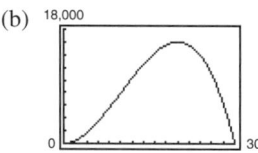   (d) $-\frac{1}{8}, \frac{3}{4}, 1$

75. Answers will vary; $1.937, 3.705$

77. Answers will vary; ± 2 **79.** $\pm 2, \pm\frac{3}{2}$ **81.** $\pm 1, \frac{1}{4}$

83. d   **84.** a   **85.** b   **86.** c   **87.** $-\frac{1}{2}, 2 \pm \sqrt{3}, 1$

89. $-1, \frac{3}{2}, 4 \pm \sqrt{17}$

91. (a)   (b) The model fits the data.
(c) 307.8; Answers will vary.

93. (a) Answers will vary.
(b) $20 \times 20 \times 40$

(c) $15, \dfrac{15 \pm 15\sqrt{5}}{2}$;

$\dfrac{15 - 15\sqrt{5}}{2}$ represents a negative volume.

95. False. If $(7x + 4)$ is a factor of f, then $-\frac{4}{7}$ is a zero of f.

97. $-2(x - 1)^2(x + 2)$

99. $-(x - 2)(x + 2)(x + 1)(x - 1)$ **101.** 7

103. (a) $x + 1$, $x \neq 1$
(b) $x^2 + x + 1$, $x \neq 1$
(c) $x^3 + x^2 + x + 1$, $x \neq 1$

$\dfrac{x^n - 1}{x - 1} = x^{n-1} + x^{n-2} + \cdots + x^2 + x + 1$, $x \neq 1$

105. $\pm\dfrac{5}{3}$ **107.** $\dfrac{-3 \pm \sqrt{3}}{2}$ **109.** $f(x) = x^2 + 12x$

111. $f(x) = x^4 - 6x^3 + 3x^2 + 10x$

Section 2.4   (page 137)

> ### Vocabulary Check   (page 137)
>
> **1.** (a) ii   (b) iii   (c) i   **2.** $\sqrt{-1}, -1$
>
> **3.** complex, $a + bi$ **4.** real, imaginary
>
> **5.** Mandelbrot Set

1. $a = -9, b = 4$ **3.** $a = 6, b = 5$ **5.** $5 + 4i$

7. -6 **9.** $-1 - 5i$ **11.** -75 **13.** $0.3i$

15. $-3 + 3i$ **17.** $7 - 3\sqrt{2}\,i$ **19.** $-14 + 20i$

21. $\frac{19}{6} + \frac{37}{6}i$ **23.** $-4.2 + 7.5i$ **25.** $-2\sqrt{3}$

27. -10 **29.** $5 + i$ **31.** $-20 + 32i$ **33.** 24

35. $80i$ **37.** $4 - 3i; 25$ **39.** $-6 + \sqrt{5}\,i; 41$

41. $-\sqrt{20}\,i; 20$ **43.** $3 + \sqrt{-2}; 11$

45. $-6i$ **47.** $\frac{8}{41} + \frac{10}{41}i$ **49.** $\frac{3}{5} + \frac{4}{5}i$

51. $-\frac{40}{1681} - \frac{9}{1681}i$ **53.** $-\frac{1}{2} - \frac{5}{2}i$ **55.** $\frac{62}{949} + \frac{297}{949}i$

57. $-1 + 6i$ **59.** $-375\sqrt{3}\,i$ **61.** i

63. (a) 8   (b) 8   (c) 8; Answers will vary.

65. $4 + 3i$ **67.** $5i$ **69.** 2

71. 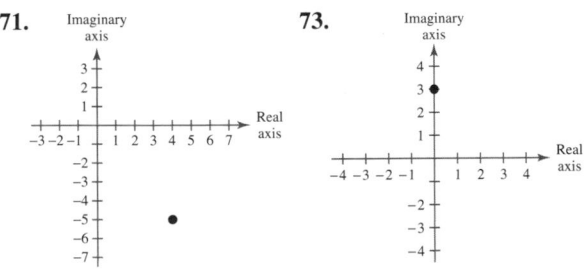   **73.**

75.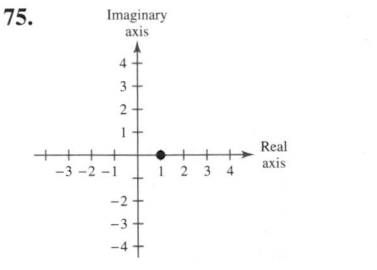

77. $0.5i, -0.25 + 0.5i, -0.1875 + 0.25i, -0.0273 + 0.4063i,$
$-0.1643 + 0.4778i, -0.2013 + 0.3430i$; Yes, bounded

79. $3.12 - 0.97i$

81. False. Any real number is equal to its conjugate.

83. False. Example: $(1 + i) + (1 - i) = 2$, which is not an imaginary number.

85. True   **87.** $16x^2 - 25$ **89.** $3x^2 + \frac{23}{2}x - 2$

Section 2.5   (page 144)

> ### Vocabulary Check   (page 144)
>
> **1.** Fundamental Theorem, Algebra
>
> **2.** Linear Factorization Theorem
>
> **3.** irreducible, reals   **4.** complex conjugate

1. $-3, 0, 0$ **3.** $-9, \pm 4i$

5. Zeros: $4, -i, i$. One real zero; they are the same.

7. Zeros: $\sqrt{2}\,i, \sqrt{2}\,i, -\sqrt{2}\,i, -\sqrt{2}\,i$. No real zeros; they are the same.

9. $2 \pm \sqrt{3}$
$\left(x - 2 - \sqrt{3}\right)\left(x - 2 + \sqrt{3}\right)$

11. $6 \pm \sqrt{10}$
$\left(x - 6 - \sqrt{10}\right)\left(x - 6 + \sqrt{10}\right)$

13. $\pm 5i$
$(x + 5i)(x - 5i)$

15. $\pm\frac{3}{2}, \pm\frac{3}{2}i$
$(2x - 3)(2x + 3)(2x - 3i)(2x + 3i)$

17. $\dfrac{1 \pm \sqrt{223}i}{2}$
$\left(z - \dfrac{1 - \sqrt{223}i}{2}\right)\left(z - \dfrac{1 + \sqrt{223}i}{2}\right)$

19. $\pm i, \pm 3i$
$(x + i)(x - i)(x + 3i)(x - 3i)$

21. $\frac{5}{3}, \pm 4i$
$(3x - 5)(x - 4i)(x + 4i)$

23. $-5, 4 \pm 3i$
$(t + 5)(t - 4 + 3i)(t - 4 - 3i)$

25. $1 \pm \sqrt{5}\,i, -\frac{1}{5}$
$(5x + 1)\left(x - 1 + \sqrt{5}\,i\right)\left(x - 1 - \sqrt{5}\,i\right)$

27. $2, 2, \pm 2i$
$(x - 2)^2(x + 2i)(x - 2i)$

29. (a) $7 \pm \sqrt{3}$ (b) $\left(x - 7 - \sqrt{3}\right)\left(x - 7 + \sqrt{3}\right)$
(c) $\left(7 \pm \sqrt{3}, 0\right)$

31. (a) $\frac{3}{2}, \pm 2i$ (b) $(2x - 3)(x - 2i)(x + 2i)$ (c) $\left(\frac{3}{2}, 0\right)$

CHAPTER 2

33. (a) $-6, 3 \pm 4i$ (b) $(x + 6)(x - 3 - 4i)(x - 3 + 4i)$

(c) $(-6, 0)$

35. (a) $\pm 4i, \pm 3i$ (b) $(x + 4i)(x - 4i)(x + 3i)(x - 3i)$

(c) None

37. $f(x) = x^3 - 2x^2 + x - 2$

39. $f(x) = x^4 - 12x^3 + 53x^2 - 100x + 68$

41. $f(x) = x^4 + 3x^3 - 7x^2 + 15x$

43. (a) $-(x - 1)(x + 2)(x - 2i)(x + 2i)$

(b) $f(x) = -(x^4 + x^3 + 2x^2 + 4x - 8)$

45. (a) $-2(x + 1)\left(x - 2 + \sqrt{5}i\right)\left(x - 2 - \sqrt{5}i\right)$

(b) $f(x) = -2x^3 + 6x^2 - 10x - 18$

47. (a) $(x^2 + 1)(x^2 - 7)$

(b) $(x^2 + 1)\left(x + \sqrt{7}\right)\left(x - \sqrt{7}\right)$

(c) $(x + i)(x - i)\left(x + \sqrt{7}\right)\left(x - \sqrt{7}\right)$

49. (a) $(x^2 - 6)(x^2 - 2x + 3)$

(b) $\left(x + \sqrt{6}\right)\left(x - \sqrt{6}\right)(x^2 - 2x + 3)$

(c) $\left(x + \sqrt{6}\right)\left(x - \sqrt{6}\right)\left(x - 1 - \sqrt{2}i\right)\left(x - 1 + \sqrt{2}i\right)$

51. $-\frac{3}{2}, \pm 5i$ **53.** $-3, 5 \pm 2i$ **55.** $-\frac{2}{3}, 1 \pm \sqrt{3}i$

57. $\frac{3}{4}, \frac{1}{2}\left(1 \pm \sqrt{5}i\right)$ **59.** (a) $1.000, 2.000$ (b) $-3 \pm \sqrt{2}i$

61. (a) 0.750 (b) $\dfrac{1}{2} \pm \dfrac{\sqrt{5}}{2}i$

63. No. Setting $h = 64$ and solving the resulting equation yields imaginary roots.

65. False. A third-degree polynomial must have at least one real zero.

67. (a) $k = 4$ (b) $k < 0$

69.

Vertex: $\left(\frac{7}{2}, -\frac{81}{4}\right)$

Intercepts:

$(-1, 0), (8, 0), (0, -8)$

71.

Vertex: $\left(-\frac{15}{12}, -\frac{169}{24}\right)$

Intercepts:

$\left(-\frac{3}{2}, 0\right), \left(\frac{2}{3}, 0\right), (0, -6)$

Section 2.6   (page 152)

Vocabulary Check   (page 152)

1. rational functions   **2.** vertical asymptote

3. horizontal asymptote

1. (a) Domain: all real numbers x except $x = 1$

(b)

x	$f(x)$	x	$f(x)$
0.5	-2	1.5	2
0.9	-10	1.1	10
0.99	-100	1.01	100
0.999	-1000	1.001	1000

x	$f(x)$	x	$f(x)$
5	0.25	-5	$-0.\overline{16}$
10	$0.\overline{1}$	-10	$-0.\overline{09}$
100	$0.\overline{01}$	-100	$-0.\overline{0099}$
1000	$0.\overline{001}$	-1000	$-0.\overline{000999}$

(c) f approaches $-\infty$ from the left and ∞ from the right of $x = 1$.

3. (a) Domain: all real numbers x except $x = 1$

(b)

x	$f(x)$	x	$f(x)$
0.5	3	1.5	9
0.9	27	1.1	33
0.99	297	1.01	303
0.999	2997	1.001	3003

x	$f(x)$	x	$f(x)$
5	3.75	-5	-2.5
10	$3.\overline{33}$	-10	-2.727
100	$3.\overline{03}$	-100	-2.97
1000	$3.\overline{003}$	-1000	-2.997

(c) f approaches ∞ from both the left and the right of $x = 1$.

5. (a) Domain: all real numbers x except $x = \pm 1$

(b)

x	$f(x)$	x	$f(x)$
0.5	-1	1.5	5.4
0.9	-12.79	1.1	17.29
0.99	-147.8	1.01	152.3
0.999	-1498	1.001	1502.3

x	$f(x)$	x	$f(x)$
5	3.125	-5	3.125
10	$3.\overline{03}$	-10	$3.\overline{03}$
100	$3.\overline{0003}$	-100	$3.\overline{0003}$
1000	3	-1000	3.000003

(c) f approaches ∞ from the left and $-\infty$ from the right of $x = -1$. f approaches $-\infty$ from the left and ∞ from the right of $x = 1$.

7. a   **8.** d   **9.** c   **10.** e   **11.** b   **12.** f

13. (a) Vertical asymptote: $x = 0$

Horizontal asymptote: $y = 0$

(b) Holes: none

15. (a) Vertical asymptote: $x = 2$

Horizontal asymptote: $y = -1$

(b) Hole at $x = 0$

17. (a) Vertical asymptote: $x = 0$

Horizontal asymptote: $y = 1$

(b) Hole at $x = -5$

19. (a) Domain: all real numbers x

(b) Continuous

(c) Vertical asymptote: none

Horizontal asymptote: $y = 3$

21. (a) Domain: all real numbers x except $x = 0$

(b) Not continuous

(c) Vertical asymptote: $x = 0$

Horizontal asymptotes: $y = \pm 1$

23. (a) Domain of f: all real numbers x except $x = 4$; domain of g: all real numbers x

(b) $x + 4, x \neq 4$; Vertical asymptote: none   (c) $x = 4$

(d)

x	1	2	3	4	5	6	7
$f(x)$	5	6	7	Undef.	9	10	11
$g(x)$	5	6	7	8	9	10	11

(e) Values differ only where f is undefined.

25. (a) Domain of f: all real numbers x except $x = -1, 3$; domain of g: all real numbers x except $x = 3$

(b) $\dfrac{x-1}{x-3}, x \neq -1$; Vertical asymptote: $x = 3$

(c) $x = -1$

(d)

x	-2	-1	0	1	2	3	4
$f(x)$	$\frac{3}{5}$	Undef.	$\frac{1}{3}$	0	-1	Undef.	3
$g(x)$	$\frac{3}{5}$	$\frac{1}{2}$	$\frac{1}{3}$	0	-1	Undef.	3

(e) Values differ only where f is undefined and g is defined.

27. 4; less than; greater than   **29.** 2; greater than; less than

31. ± 2 **33.** 7   **35.** $-1, 3$ **37.** 2

39. (a) \$28.33 million   (b) \$170 million

(c) \$765 million

(d)

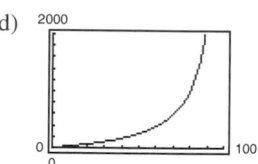

Answers will vary.

(e) No. The function is undefined at the 100% level.

41. (a) $y = \dfrac{1}{0.445 - 0.007x}$

(b)

Age, x	16	32	44	50	60
Near point, y	3.0	4.5	7.3	10.5	40

(c) No; the function is negative for $x = 70$.

43. (a)

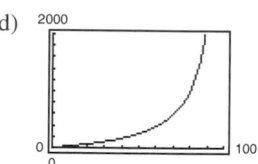

(b) 333 deer, 500 deer, 800 deer

(c) 1500. Because the degrees of the numerator and the denominator are equal, the limiting size is the ratio of the leading coefficients, $60/0.04 = 1500$.

45. False. The degree of the denominator gives the maximum possible number of vertical asymptotes, and the degree is finite.

47. b   **49.** $f(x) = \dfrac{x-1}{x^3 - 8}$ **51.** $f(x) = \dfrac{2x^2 - 18}{x^2 + x - 2}$

53. $x - y - 1 = 0$ **55.** $3x - y + 1 = 0$

57. $x + 9 + \dfrac{42}{x - 4}$ **59.** $2x^2 - 9 + \dfrac{34}{x^2 + 5}$

Section 2.7   (page 161)

1.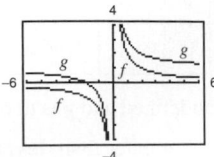

Vertical shift

3.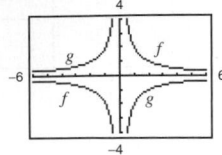

Reflection in the x-axis

5.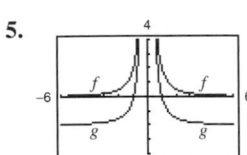

Vertical shift

7.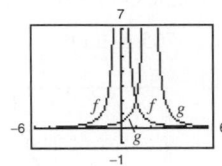

Horizontal shift

9.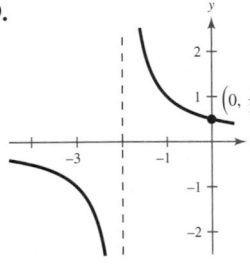

11.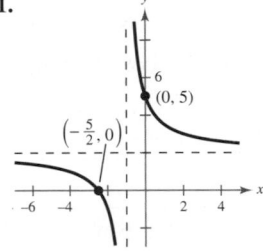

13.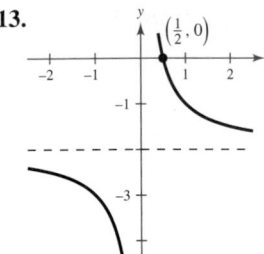

15.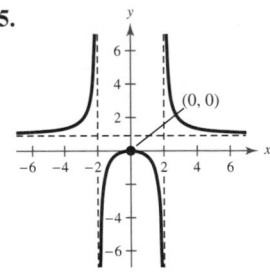

17.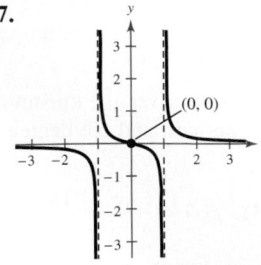

19.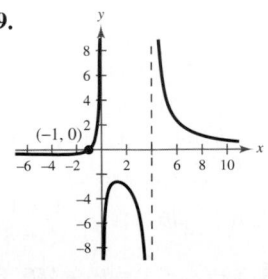

21.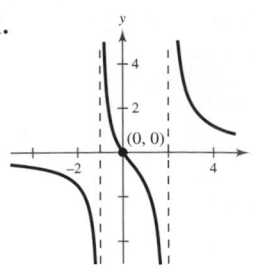

23.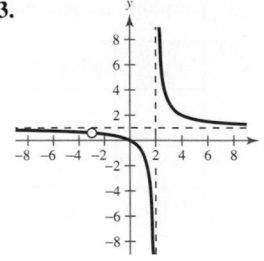

There is a hole at $x = -3$.

25.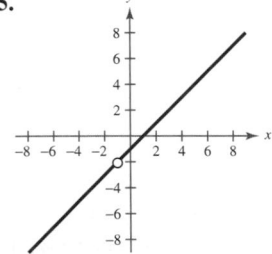

There is a hole at $x = -1$.

27.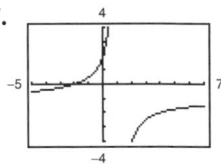

Domain: $(-\infty, 1), (1, \infty)$

Vertical asymptote: $x = 1$

Horizontal asymptote:
$y = -1$

29.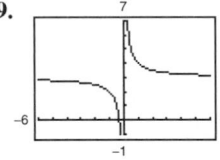

Domain: $(-\infty, 0), (0, \infty)$

Vertical asymptote: $t = 0$

Horizontal asymptote: $y = 3$

31.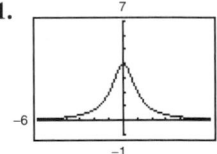

Domain: $(-\infty, \infty)$

Horizontal asymptote: $y = 0$

33.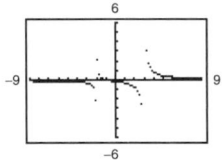

Domain: $(-\infty, -2)$,
$(-2, 3), (3, \infty)$

Vertical asymptotes:
$x = -2, x = 3$

Horizontal asymptote: $y = 0$

35.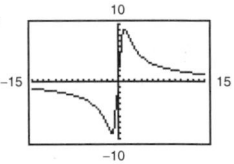

Domain: $(-\infty, 0), (0, \infty)$

Vertical asymptote: $x = 0$

Horizontal asymptote:
$y = 0$

37.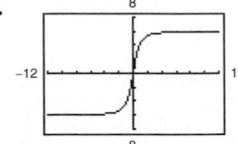

There are two horizontal asymptotes, $y = \pm 6$.

39.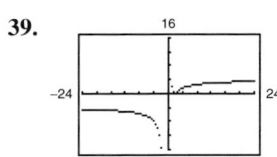

There are two horizontal asymptotes, $y = \pm 4$, and one vertical asymptote, $x = -1$.

41.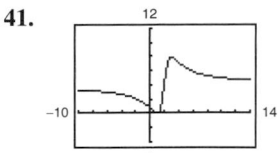

The graph crosses the horizontal asymptote, $y = 4$.

43. 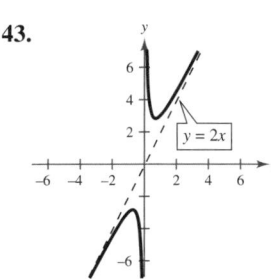   **45.**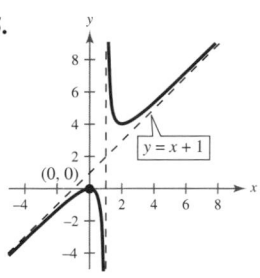

47. 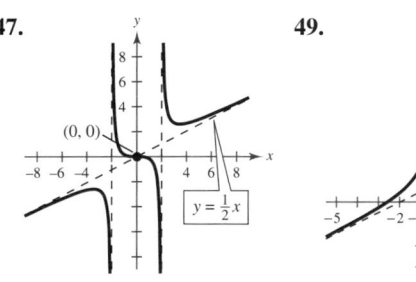   **49.**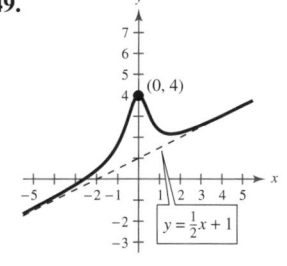

51. $(-1, 0)$ **53.** $(1, 0), (-1, 0)$

55. 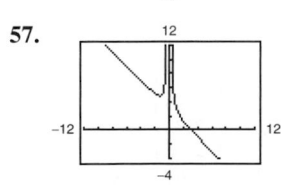   Domain: $(-\infty, -1), (-1, \infty)$
Vertical asymptote: $x = -1$
Slant asymptote: $y = 2x - 1$

57. Domain: $(-\infty, 0), (0, \infty)$
Vertical asymptote: $x = 0$
Slant asymptote: $y = -x + 3$

59. Vertical asymptotes: $x = \pm 2$; horizontal asymptote: $y = 1$; slant asymptote: none; holes: none

61. Vertical asymptote: $x = -\frac{3}{2}$; horizontal asymptote: $y = 1$; slant asymptote: none; hole at $x = 2$

63. Vertical asymptote: $x = -2$; horizontal asymptote: none; slant asymptote: $y = 2x - 7$; hole at $x = -1$

65. 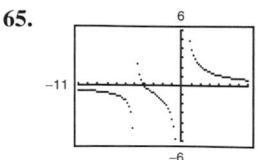   **67.**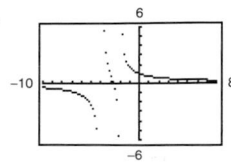

$(-4, 0)$ $\left(-\frac{8}{3}, 0\right)$

69. 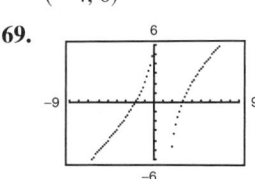   **71.**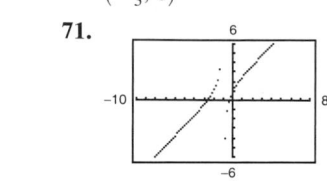

$(3, 0), (-2, 0)$ $\left(\dfrac{-3 \pm \sqrt{5}}{2}, 0\right)$

73. 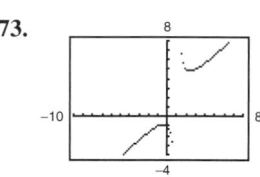   **75.**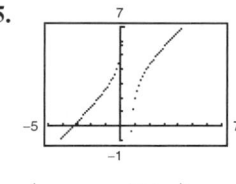

None $\left(\dfrac{-5 \pm \sqrt{65}}{4}, 0\right)$

77. (a) Answers will vary.   (b) $[0, 950]$

(c)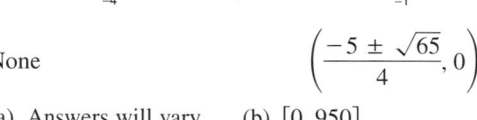

The concentration increases more slowly; the concentration approaches 75%.

79. (a) Answers will vary.   (b) $(2, \infty)$

(c)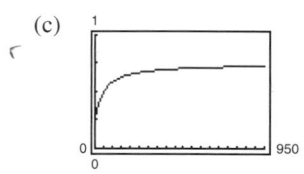

5.9 inches $\times$ 11.8 inches

81.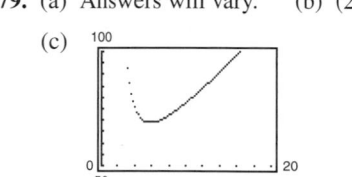

$x \approx 40$

83. (a) $C = 0$. The chemical will eventually dissipate.

(b) 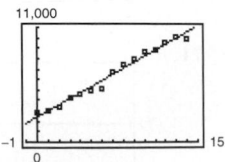 $t \approx 4.5$ hours

(c) Before ≈ 2.6 hours and after ≈ 8.3 hours

85. (a) $A = 583.8t + 2414$

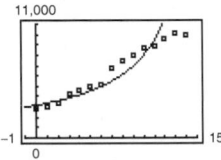

(b) $A = \dfrac{1}{-0.0000185t + 0.000315}$

$= \dfrac{-54054.05}{t - 17.03}$

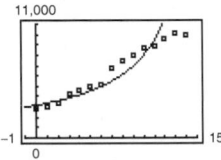

(c)

Year, x	1990	1991	1992	1993
Original data, y	2777	3013	3397	4193
Model from (a), y	2414	2998	3582	4165
Model from (b), y	3174	3372	3596	3853

Year, x	1994	1995	1996	1997
Original data, y	4557	4962	5234	6734
Model from (a), y	4749	5333	5917	6501
Model from (b), y	4148	4493	4901	5389

Year, x	1998	1999	2000	2001
Original data, y	7387	8010	8698	8825
Model from (a), y	7084	7668	8252	8836
Model from (b), y	5986	6732	7689	8964

Year, x	2002	2003	2004
Original data, y	9533	10,164	10,016
Model from (a), y	9420	10,003	10,587
Model from (b), y	10,746	13,413	17,840

Answers will vary.

87. False. A graph with a vertical asymptote is not continuous.

89.

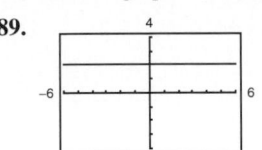

The denominator is a factor of the numerator.

91. $f(x) = \dfrac{x^2 + 3x - 10}{x + 2}$ **93.** $\dfrac{512}{x^3}$ **95.** 3

97.

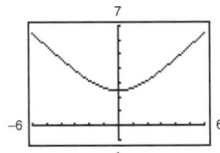

Domain: $(-\infty, \infty)$
Range: $\left[\sqrt{6}, \infty\right)$

99.

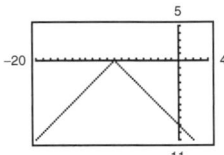

Domain: $(-\infty, \infty)$
Range: $(-\infty, 0]$

101. Answers will vary.

Section 2.8   (page 169)

> **Vocabulary Check**   (page 169)
>
> **1.** linear   **2.** quadratic

1. Quadratic   **3.** Linear   **5.** Neither

7. (a)

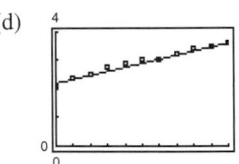

(b) Linear

(c) $y = 0.14x + 2.2$

(d)

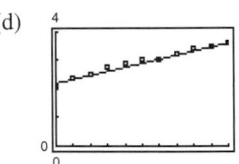

(e)

x	0	1	2	3	4	5
Actual, y	2.1	2.4	2.5	2.8	2.9	3.0
Model, y	2.2	2.3	2.5	2.6	2.8	2.9

x	6	7	8	9	10
Actual, y	3.0	3.2	3.4	3.5	3.6
Model, y	3.0	3.2	3.3	3.5	3.6

9. (a) 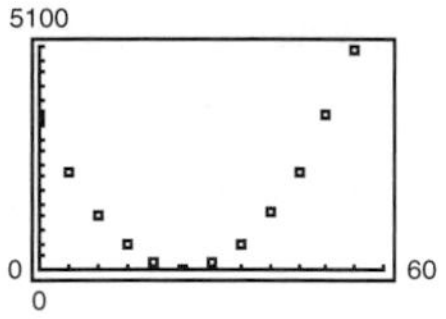

(b) Quadratic

(c) $y = 5.55x^2 - 277.5x + 3478$

(d) 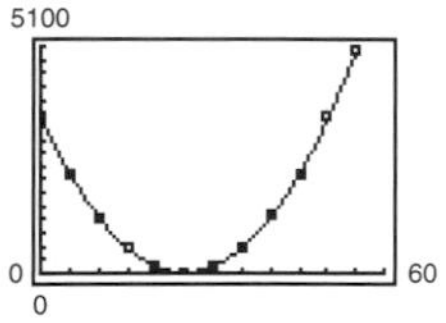

(e)

x	0	5	10	15	20
Actual, y	3480	2235	1250	565	150
Model, y	3478	2229	1258	564	148

x	25	30	35	40
Actual, y	12	145	575	1275
Model, y	9	148	564	1258

x	45	50	55
Actual, y	2225	3500	5010
Model, y	2229	3478	5004

11. (a) $y = 2.48x + 1.1$; $y = 0.071x^2 + 1.69x + 2.7$

(b) 0.98995; 0.99519 (c) Quadratic

13. (a) $y = -0.89x + 5.3$; $y = 0.001x^2 - 0.90x + 5.3$

(b) 0.99982; 0.99987 (c) Quadratic

15. (a) 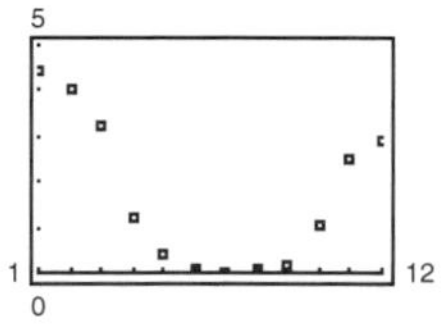

(b) $P = 0.1322t^2 - 1.901t + 6.87$

(c) 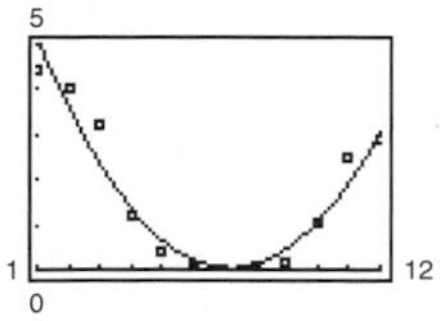

(d) July

17. (a) 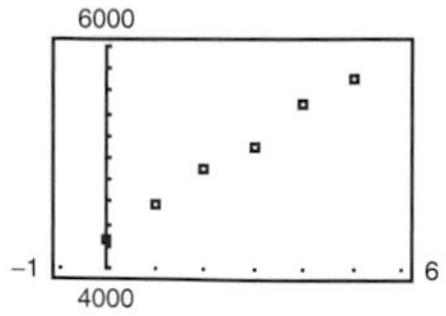

(b) $S = -2.630t^2 + 301.74t + 4270.2$

(c) 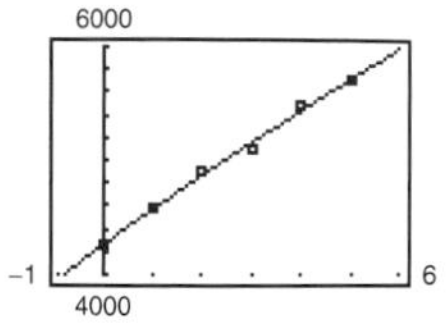

(d) 2024 (e) Answers will vary.

19. (a) 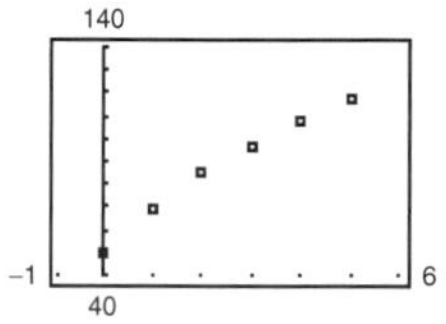

(b) 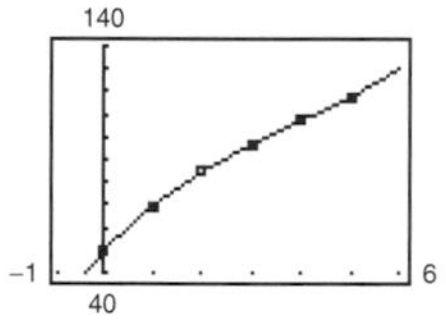

Answers will vary.

(c) $A = -1.1357t^2 + 18.999t + 50.30$; 0.99859

(d) 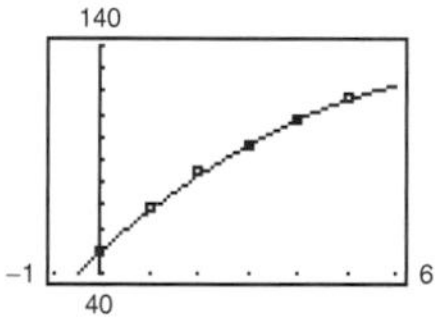

Answers will vary.

(e) Cubic model; the coefficient of determination is closer to 1.

(f)

Year	2006	2007	2008
A^*	127.76	140.15	154.29
Cubic model	129.91	145.13	164.96
Quadratic model	123.40	127.64	129.60

Explanations will vary.

21. True **23.** The model is consistently above the data.

25. (a) $(f \circ g)(x) = 10x^2 + 3$

(b) $(g \circ f)(x) = 50x^2 + 160x + 127$

27. (a) $(f \circ g)(x) = x$ (b) $(g \circ f)(x) = x$

29. $f^{-1}(x) = 5x + 4$ **31.** $f^{-1}(x) = \dfrac{\sqrt{2x + 6}}{2}$

33. 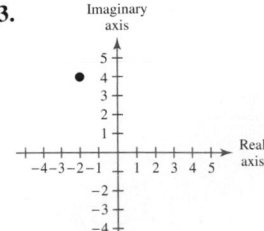   **35.**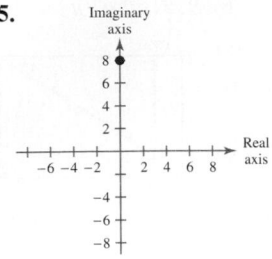

Review Exercises   (page 173)

1.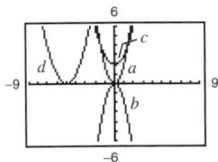

(a) Vertical stretch

(b) Vertical stretch, reflection in the x-axis

(c) Vertical shift

(d) Horizontal shift

3.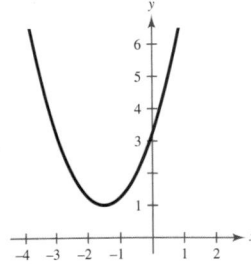

Vertex: $\left(-\dfrac{3}{2}, 1\right)$

Intercept: $\left(0, \dfrac{13}{4}\right)$

5.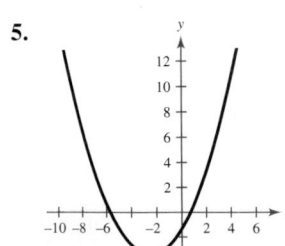

Vertex: $\left(-\dfrac{5}{2}, -\dfrac{41}{12}\right)$

Intercepts:

$\left(0, -\dfrac{4}{3}\right), \left(\dfrac{-5 \pm \sqrt{41}}{2}, 0\right)$

7.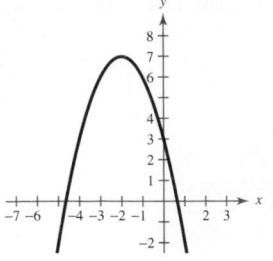

Vertex: $(-2, 7)$

Intercepts:

$(0, 3), \left(-2 \pm \sqrt{7}, 0\right)$

9. $f(x) = (x - 1)^2 - 4$ **11.** $f(x) = 2(x + 2)^2 - 2$

13. (a) $A = x\left(\dfrac{8 - x}{2}\right), 0 < x < 8$

(b)

x	1	2	3	4	5	6
Area	$\frac{7}{2}$	6	$\frac{15}{2}$	8	$\frac{15}{2}$	6

$x = 4, y = 2$

(c)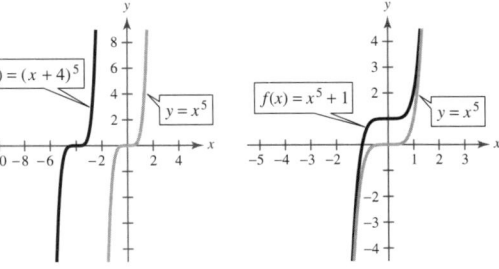

$x = 4, y = 2$

(d) $A = -\dfrac{1}{2}(x - 4)^2 + 8$

(e) Answers will vary.

15. $x = 125$ feet, $y = 187.5$ feet

17. (a) 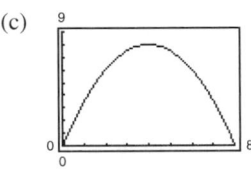   (b)

(c) 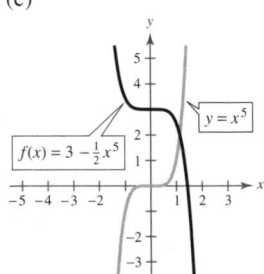   (d)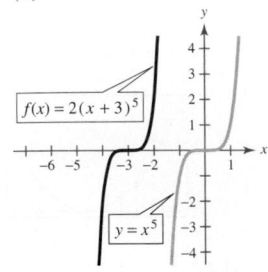

19.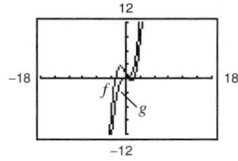

21. Falls to the left, falls to the right

23. Rises to the left, rises to the right

25. (a) $x = -1, 0, 0, 2$

(b)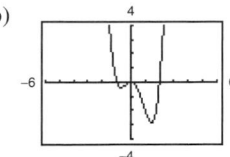

(c) $x = -1, 0, 0, 2$; They are the same.

27. (a) $t = 0, \pm\sqrt{3}$

(b)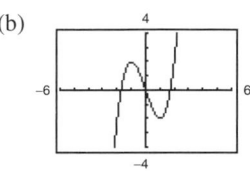

(c) $t = 0, \pm 1.73$; They are the same.

29. (a) $x = -3, -3, 0$

(b)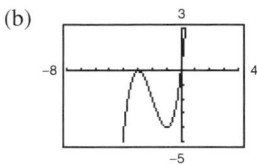

(c) $x = -3, -3, 0$; They are the same.

31. $f(x) = x^4 - 5x^3 - 3x^2 + 17x - 10$

33. $f(x) = x^3 - 7x^2 + 13x - 3$

35. (a) Rises to the left, rises to the right

(b) $x = -3, -1, 3, 3$

(c) and (d)

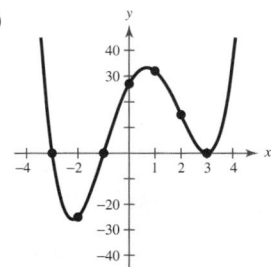

37. (a) $(-3, -2), (-1, 0), (0, 1)$

(b) $x = -2.25, -0.56, 0.80$

39. (a) $(-3, -2), (2, 3)$ (b) $x = -2.57, 2.57$

41. 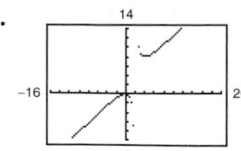   **43.**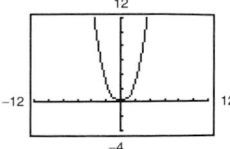

45. $8x + 5 + \dfrac{2}{3x - 2}$ **47.** $x^2 - 2, \ x \neq \pm 1$

49. $5x + 2, \ x \neq \dfrac{3 \pm \sqrt{5}}{2}$ **51.** $3x^2 + 5x + 8 + \dfrac{10}{2x^2 - 1}$

53. $0.25x^3 - 4.5x^2 + 9x - 18 + \dfrac{36}{x + 2}$

55. $6x^3 - 27x, \ x \neq \dfrac{2}{3}$ **57.** $3x^2 + 2x + 20 + \dfrac{58}{x - 4}$

59. (a) -421 (b) -156

61. (a) Answers will vary.

(b) $(x + 1)(x + 7)$

(c) $f(x) = (x - 4)(x + 1)(x + 7)$

(d) $x = 4, -1, -7$

63. (a) Answers will vary.

(b) $(x + 1)(x - 4)$

(c) $f(x) = (x + 2)(x - 3)(x + 1)(x - 4)$

(d) $x = -2, 3, -1, 4$

65. $\pm 1, \pm 3, \pm\frac{3}{2}, \pm\frac{3}{4}, \pm\frac{1}{2}, \pm\frac{1}{4}$ **67.** $x = \frac{5}{6}$

69. $x = -1, \frac{3}{2}, 3, \frac{2}{3}$

71. 2 or 0 positive real zeros

1 negative real zero

73. Answers will vary.   **75.** $6 + 5i$ **77.** $2 + 7i$

79. $3 + 7i$ **81.** $40 + 65i$ **83.** $-26 + 7i$

85. $3 + 9i$ **87.** $-4 - 46i$ **89.** -80

91. $1 - 6i$ **93.** $\frac{17}{26} + \frac{7}{26}i$ **95.** $-3 - 2i$

97.

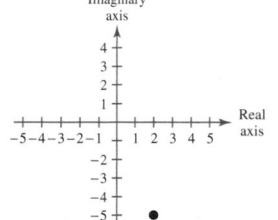

99.

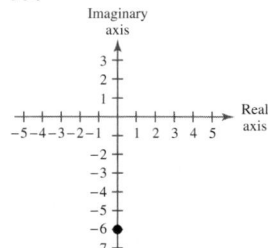

101.

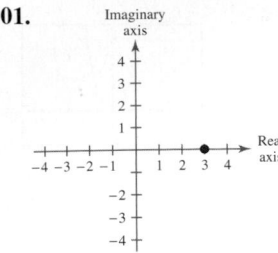

103. $x = 0, 2, 2$

105. $x = 2, -\frac{3}{2}, 1 \pm i$;
$(x - 2)(2x + 3)(x - 1 + i)(x - 1 - i)$

107. $x = 4, \dfrac{3 \pm \sqrt{15}i}{2}$;

$(x - 4)\left(x - \dfrac{3 + \sqrt{15}i}{2}\right)\left(x - \dfrac{3 - \sqrt{15}i}{2}\right)$

109. $x = 0, -1, \pm\sqrt{5}i$; $x^2(x + 1)\left(x + \sqrt{5}i\right)\left(x - \sqrt{5}i\right)$

111. (a) $x = 2, 1 \pm i$

(b) $(x - 2)(x - 1 - i)(x - 1 + i)$ (c) $(2, 0)$

113. (a) $x = -6, -1, \frac{2}{3}$

(b) $-(x + 1)(x + 6)(3x - 2)$

(c) $(-6, 0), (-1, 0), \left(\frac{2}{3}, 0\right)$

115. (a) $\pm 3i, \pm 5i$

(b) $(x - 3i)(x + 3i)(x - 5i)(x + 5i)$ (c) None

117. $f(x) = x^4 - 2x^3 + 17x^2 - 50x - 200$

119. $f(x) = x^4 + 9x^3 + 48x^2 + 78x - 136$

121. (a) $(x^2 + 9)(x^2 - 2x - 1)$

(b) $(x^2 + 9)\left(x - 1 + \sqrt{2}\right)\left(x - 1 - \sqrt{2}\right)$

(c) $(x + 3i)(x - 3i)\left(x - 1 + \sqrt{2}\right)\left(x - 1 - \sqrt{2}\right)$

123. $x = -3, \pm 2i$

125. (a) Domain: all real numbers x except $x = -3$

(b) Not continuous

(c) Vertical asymptote: $x = -3$

Horizontal asymptote: $y = -1$

127. (a) Domain: all real numbers x except $x = 6, -3$

(b) Not continuous

(c) Vertical asymptotes: $x = 6, x = -3$

Horizontal asymptote: $y = 0$

129. (a) Domain: all real numbers x except $x = 7$

(b) Not continuous

(c) Vertical asymptote: $x = 7$

Horizontal asymptote: $y = -1$

131. (a) Domain: all real numbers x except $x = \pm\dfrac{\sqrt{6}}{2}$

(b) Not continuous

(c) Vertical asymptotes: $x = \pm\dfrac{\sqrt{6}}{2}$

Horizontal asymptote: $y = 2$

133. (a) Domain: all real numbers x except $x = 5, -3$

(b) Not continuous

(c) Vertical asymptote: $x = -3$

Horizontal asymptote: $y = 0$

135. (a) Domain: all real numbers x

(b) Continuous

(c) Horizontal asymptotes: $y = \pm 1$

137. (a) \$176 million; \$528 million; \$1584 million

(b)

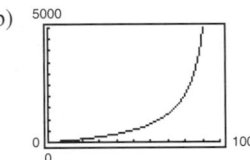

Answers will vary.

(c) No. As $p \to 100$, the cost approaches ∞.

139. Vertical asymptote: $x = -1$

Horizontal asymptote: $y = 1$

Hole at $x = 1$

141. Vertical asymptote: $x = -\frac{3}{2}$

Horizontal asymptote: $y = 1$

Hole at $x = 3$

143. Vertical asymptote: $x = -1$

Slant asymptote: $y = 3x - 10$

Hole at $x = -2$

145.

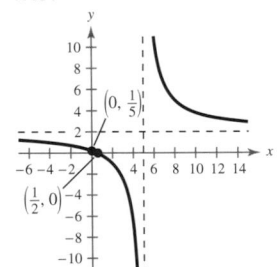

147.

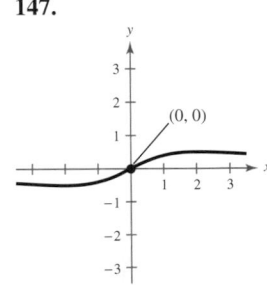

149.

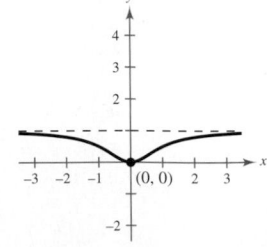

151.

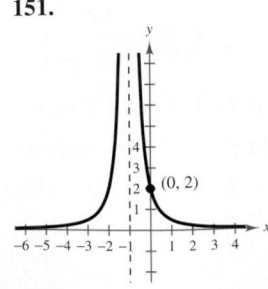

153.

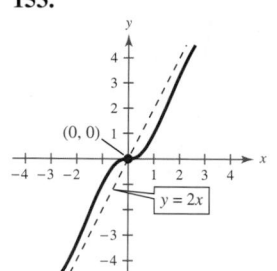

155.

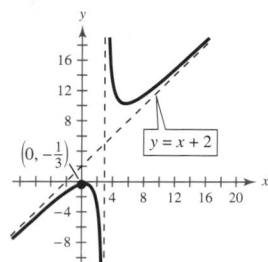

157. (a)

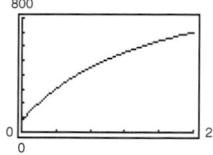

(b) 304,000; 453,333; 702,222

(c) 1,200,000, because N has a horizontal asymptote at $y = 1200$.

159. Quadratic   **161.** Linear

163. (a)

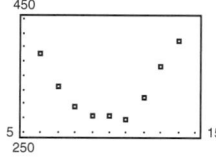

(b) $y = 8.03x^2 - 157.1x + 1041$

(c)

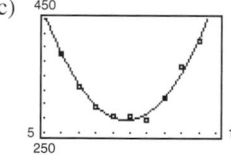

Answers will vary.

(d) 2005

(e) Answers will vary.

165. False. For the graph of a rational function to have a slant asymptote, the degree of its numerator must be exactly one more than the degree of its denominator.

167. False. Example: $(1 + 2i) + (1 - 2i) = 2$.

169. Answers will vary.

171. The first step is completed incorrectly: $\sqrt{-4} = 2i \neq 4i$

Chapter Test   (page 179)

1. Vertex: $(-2, -1)$

Intercepts: $(0, 3), (-3, 0), (-1, 0)$

2. $y = (x - 3)^2 - 6$

3. 0, multiplicity 1; $-\frac{1}{2}$, multiplicity 2

4.

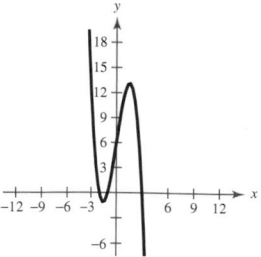

5. $3x + \dfrac{x - 1}{x^2 + 1}$ **6.** $2x^3 + 4x^2 + 3x + 6 + \dfrac{9}{x - 2}$

7. 13

8. $\pm 1, \pm 2, \pm 3, \pm 4, \pm 6, \pm 8, \pm 12, \pm 24, \pm\frac{1}{2}, \pm\frac{3}{2}$

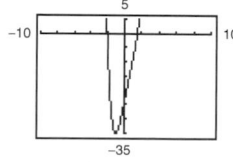

$t = -2, \frac{3}{2}$

9. $\pm 1, \pm 2, \pm\frac{1}{3}, \pm\frac{2}{3}$

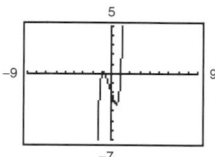

$x = \pm 1, -\frac{2}{3}$

10. $x = -1, 4 \pm \sqrt{3}\,i$

$(x + 1)(x - 4 + \sqrt{3}\,i)(x - 4 - \sqrt{3}\,i)$

11. $-9 - 18i$ **12.** $6 + (2\sqrt{5} + \sqrt{14})i$ **13.** $13 + 4i$

14. $-17 + 14i$ **15.** $\frac{43}{37} + \frac{38}{37}i$ **16.** $1 + 2i$

17. $\frac{4}{13} + \frac{7}{13}i$

18.

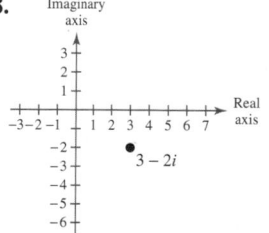

19.

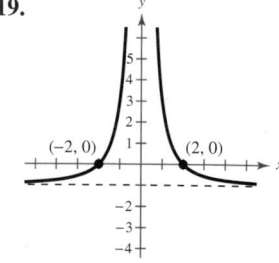

CHAPTER 2

20.

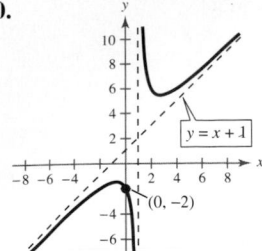

21.

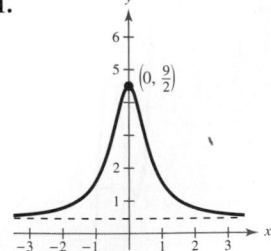

22. (a)

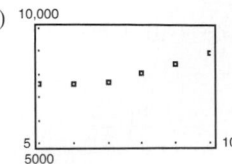

(b) $y = 5.582x^2 - 85.53x + 602.0$

(c)

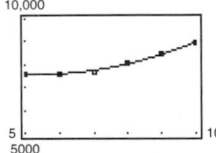

Answers will vary.

(d) $574.96 billion; $1124.17 billion

(e) Answers will vary.

Chapter 3

Section 3.1   (page 193)

Vocabulary Check   (page 193)

1. algebraic   **2.** transcendental

3. natural exponential, natural

4. $A = P\left(1 + \dfrac{r}{n}\right)^{nt}$ **5.** $A = Pe^{rt}$

1. 4112.033   **3.** 0.006

5.

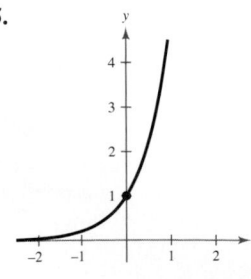

$y = 0, (0, 1)$, increasing

7.

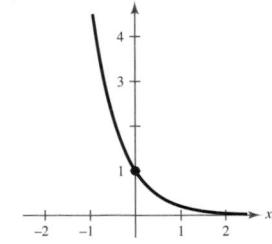

$y = 0, (0, 1)$, decreasing

9.

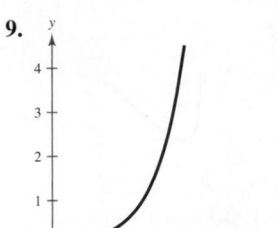

$y = 0, \left(0, \frac{1}{25}\right)$, increasing

11.

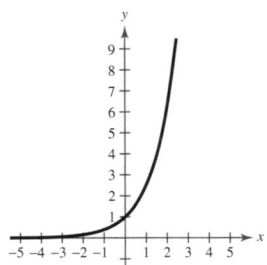

$y = -3, (0, -2)$, $(-0.683, 0)$, decreasing

13. d   **14.** a   **15.** c   **16.** b

17. Right shift of five units

19. Left shift of four units and reflection in the x-axis

21. Right shift of two units and downward shift of three units

23. 9897.129   **25.** 54.164

27.

x	-2	-1	0	1	2
$f(x)$	0.16	0.4	1	2.5	6.25

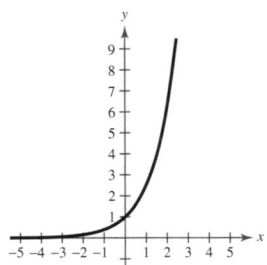

Asymptote: $y = 0$

29.

x	-2	-1	0	1	2
$f(x)$	0.03	0.17	1	6	36

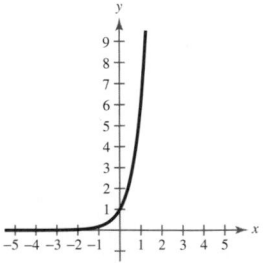

Asymptote: $y = 0$

31.

x	-3	-2	-1	0	1
$f(x)$	0.33	1	3	9	27

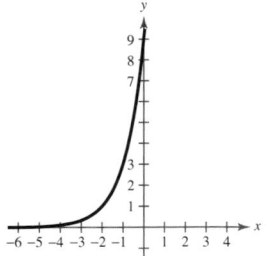

Asymptote: $y = 0$

33.

x	-2	-1	0	1	2
y	0.06	0.5	1	0.5	0.06

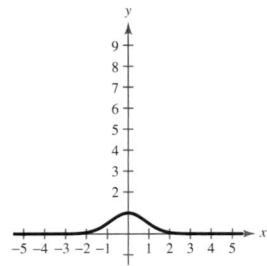

Asymptote: $y = 0$

35.

x	-1	0	1	2	3	4
y	1.04	1.11	1.33	2	4	10

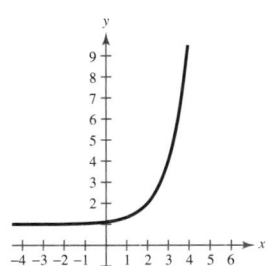

Asymptote: $y = 1$

37.

x	-2	-1	0	1	2
$f(x)$	7.39	2.72	1	0.37	0.14

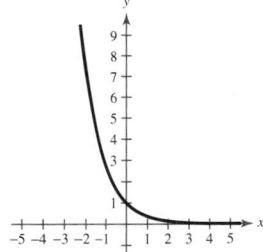

Asymptote: $y = 0$

39.

x	-6	-5	-4	-3	-2
$f(x)$	0.41	1.10	3	8.15	22.17

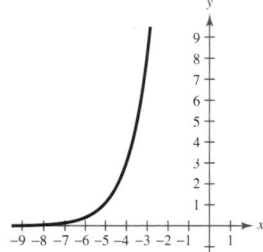

Asymptote: $y = 0$

41.

x	3	4	5	6	7
$f(x)$	2.14	2.37	3	4.72	9.39

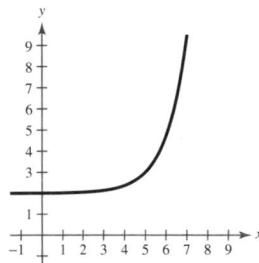

Asymptote: $y = 2$

CHAPTER 3

43.

t	-2	-1	0	1	2
$s(t)$	1.57	1.77	2	2.26	2.54

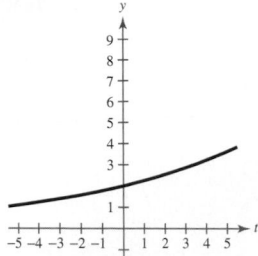

Asymptote: $y = 0$

45. (a)

(b) $y = 0, y = 8$

47. (a)

(b) $y = -3, y = 0, x \approx 3.47$

49. $(86.350, 1500)$

51. (a)

(b) Decreasing on $(-\infty, 0), (2, \infty)$

     Increasing on $(0, 2)$

(c) Relative minimum: $(0, 0)$

     Relative maximum: $(2, 0.54)$

53.

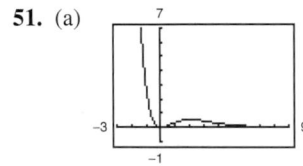

n	1	2	4	12
A	\$3200.21	\$3205.09	\$3207.57	\$3209.23

n	365	Continuous
A	\$3210.04	\$3210.06

55.

n	1	2	4	12
A	\$5477.81	\$5520.10	\$5541.79	\$5556.46

n	365	Continuous
A	\$5563.61	\$5563.85

57.

t	1	10	20
A	\$12,489.73	\$17,901.90	\$26,706.49

t	30	40	50
A	\$39,841.40	\$59,436.39	\$88,668.67

59.

t	1	10	20
A	\$12,427.44	\$17,028.81	\$24,165.03

t	30	40	50
A	\$34,291.81	\$48,662.40	\$69,055.23

61. \$1530.57   **63.** \$17,281.77

65. (a)

(b) \$421.12

(c) \$350.13

67. (a) 25 grams   (b) 16.21 grams

(c)

(d) Never. The graph has a horizontal asymptote at $Q = 0$.

69. (a)

(b) 100; 300; 900

71. (a)

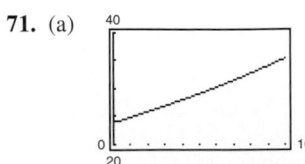

(b) and (c) \$35.45

73. True. The definition of an exponential function is $f(x) = a^x, a > 0, a \neq 1$.

75. d

77.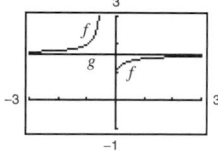

 $f(x)$ approaches $g(x) \approx 1.6487$.

79. >   **81.** >   **83.** $f^{-1}(x) = \dfrac{x+7}{5}$

85. $f^{-1}(x) = x^3 - 8$

87.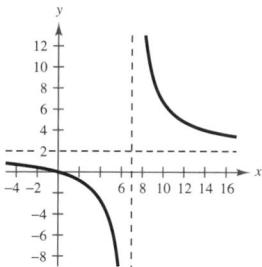

89. Answers will vary.

Section 3.2   (page 203)

Vocabulary Check   (page 203)

1. logarithmic function   **2.** 10

3. natural logarithmic   **4.** $a^{\log_a x} = x$ **5.** $x = y$

1. $4^3 = 64$ **3.** $7^{-2} = \frac{1}{49}$ **5.** $32^{2/5} = 4$ **7.** $e^0 = 1$

9. $e^1 = e$ **11.** $e^{1/2} = \sqrt{e}$ **13.** $\log_5 125 = 3$

15. $\log_{81} 3 = \frac{1}{4}$ **17.** $\log_6 \frac{1}{36} = -2$

19. $\ln 20.0855\ldots = 3$ **21.** $\ln 3.6692\ldots = 1.3$

23. $\ln 1.3596\ldots = \frac{1}{3}$ **25.** 4   **27.** -3

29. 2.538   **31.** 7.022   **33.** 9   **35.** 2   **37.** $\frac{1}{10}$

39. $3x$ **41.** -3

43. 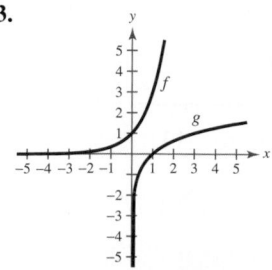   **45.**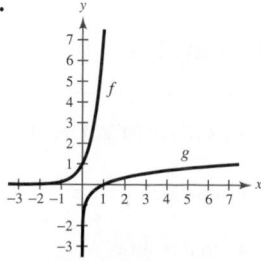

47. 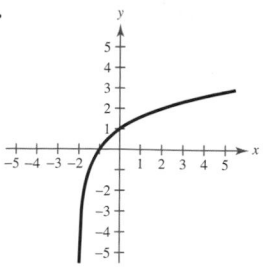   **49.**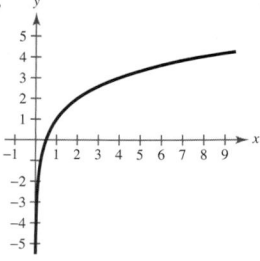

Domain: $(-2, \infty)$

Vertical asymptote: $x = -2$

x-intercept: $(-1, 0)$

Domain: $(0, \infty)$

Vertical asymptote: $x = 0$

x-intercept: $\left(\frac{1}{2}, 0\right)$

51.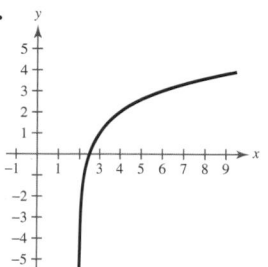

Domain: $(2, \infty)$

Vertical asymptote: $x = 2$

x-intercept: $\left(\frac{5}{2}, 0\right)$

53. b   **54.** c   **55.** d   **56.** a

57. Reflection in the x-axis

59. Reflection in the x-axis, vertical shift four units upward

61. Horizontal shift three units to the left and vertical shift two units downward

63. 1.869   **65.** 0.693   **67.** 2   **69.** 1.8

71. 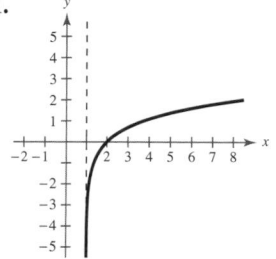   **73.**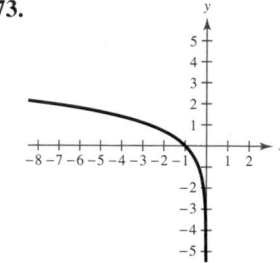

Domain: $(1, \infty)$

Vertical asymptote: $x = 1$

x-intercept: $(2, 0)$

Domain: $(-\infty, 0)$

Vertical asymptote: $x = 0$

x-intercept: $(-1, 0)$

75. Horizontal shift three units to the left

77. Vertical shift five units downward

79. Horizontal shift one unit to the right and vertical shift two units upward

81. (a) 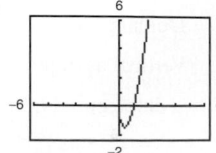   (b) Domain: $(0, \infty)$

(c) Decreasing on $(0, 2)$; increasing on $(2, \infty)$

(d) Relative minimum: $(2, 1.693)$

83. (a) 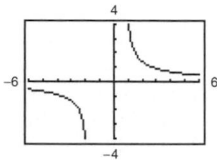   (b) Domain: $(0, \infty)$

(c) Decreasing on $(0, 0.368)$; increasing on $(0.368, \infty)$

(d) Relative minimum: $(0.37, -1.47), (0.368, -1.472)$

85. (a)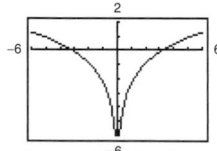

(b) Domain: $(-\infty, -2), (1, \infty)$

(c) Decreasing on $(-\infty, -2); (1, \infty)$

(d) No relative maxima or minima

87. (a)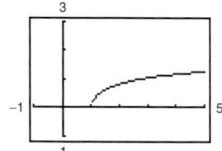

(b) Domain: $(-\infty, 0), (0, \infty)$

(c) Decreasing on $(-\infty, 0)$; increasing on $(0, \infty)$

(d) No relative maxima or minima

89. (a)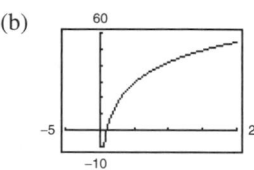

(b) Domain: $[1, \infty)$

(c) Increasing on $(1, \infty)$

(d) Relative minimum: $(1, 0)$

91. (a) 80   (b) 68.12   (c) 62.30

93. (a)

K	1	2	4	6
t	0	12.60	25.21	32.57

K	8	10	12
t	37.81	41.87	45.18

It takes 12.60 years for the principal to double.

(b)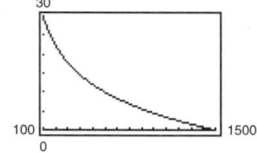

95. (a) 120 decibels   (b) 100 decibels

(c) No. The function is not linear.

97.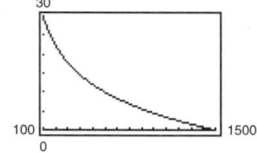

17.66 cubic feet per minute

99. False. Reflect $g(x)$ about the line $y = x$.

101. 2   **103.** $\frac{1}{4}$ **105.** b

107. $\log_a x$ is the inverse of a^x only if $0 < a < 1$ and $a > 1$, so $\log_a x$ is defined only for $0 < a < 1$ and $a > 1$.

109. (a) False   (b) True   (c) True   (d) False

111. (a)

x	1	5	10	10^2
$f(x)$	0	0.32	0.23	0.046

x	10^4	10^6
$f(x)$	0.00092	0.0000138

(b) 0

113. $(x + 3)(x - 1)$ **115.** $(4x + 3)(3x - 1)$

117. $(4x + 5)(4x - 5)$ **119.** $x(2x - 9)(x + 5)$

121. 15   **123.** 4300   **125.** 2.75   **127.** 27.67

Section 3.3   (page 211)

Vocabulary Check   (page 211)

1. change-of-base   **2.** $\dfrac{\ln x}{\ln a}$ **3.** $\log_a u^n$

4. $\ln u + \ln v$

1. (a) $\dfrac{\log_{10} x}{\log_{10} 5}$ (b) $\dfrac{\ln x}{\ln 5}$ **3.** (a) $\dfrac{\log_{10} x}{\log_{10} \frac{1}{5}}$ (b) $\dfrac{\ln x}{\ln \frac{1}{5}}$

5. (a) $\dfrac{\log_{10} \frac{3}{10}}{\log_{10} a}$ (b) $\dfrac{\ln \frac{3}{10}}{\ln a}$

7. (a) $\dfrac{\log_{10} x}{\log_{10} 2.6}$ (b) $\dfrac{\ln x}{\ln 2.6}$ **9.** 1.771 **11.** -2

13. -0.102 **15.** 2.691 **17.** $\ln 5 + \ln 4$

19. $\ln 5 - 3 \ln 4$ **21.** 1.6542 **23.** 0.2823

25. 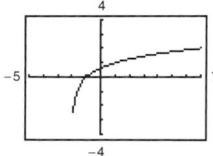   **27.**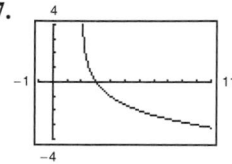

29.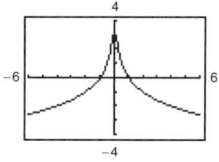

31. $\frac{3}{2}$ **33.** $6 + \ln 5$ **35.** Answers will vary.

37. $\log_{10} 5 + \log_{10} x$ **39.** $\log_{10} 5 - \log_{10} x$

41. $4 \log_8 x$ **43.** $\frac{1}{2} \ln z$ **45.** $\ln x + \ln y + \ln z$

47. $2 \log_3 a + \log_3 b + 3 \log_3 c$

49. $2 \ln a + \frac{1}{2} \ln(a - 1)$ **51.** $\frac{1}{3} \ln x - \frac{1}{3} \ln y$

53. $\ln(x + 1) + \ln(x - 1) - 3 \ln x, \; x > 1$

55. $4 \ln x + \frac{1}{2} \ln y - 5 \ln z$

57. (a)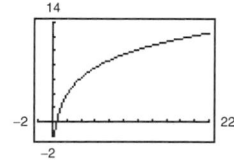

(b)

x	1	2	3	4	5	6
y_1	1.61	3.87	5.24	6.24	7.03	7.68
y_2	1.61	3.87	5.24	6.24	7.03	7.68

x	7	8	9	10	11
y_1	8.24	8.72	9.16	9.55	9.90
y_2	8.24	8.72	9.16	9.55	9.90

(c) $y_1 = y_2$ for positive values of x.

59. $\ln 4x$ **61.** $\log_4 \dfrac{z}{y}$ **63.** $\log_2(x + 3)^2$

65. $\ln \sqrt{x^2 + 4}$ **67.** $\ln \dfrac{x}{(x + 1)^3}$ **69.** $\ln \dfrac{x - 2}{x + 2}$

71. $\ln \dfrac{x}{(x^2 - 4)^2}$ **73.** $\ln \sqrt[3]{\dfrac{x(x + 3)^2}{x^2 - 1}}$

75. $\ln \dfrac{\sqrt[3]{y(y + 4)^2}}{y - 1}$

77. (a)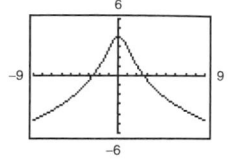

(b)

x	-5	-4	-3	-2	-1
y_1	-2.36	-1.51	-0.45	0.94	2.77
y_2	-2.36	-1.51	-0.45	0.94	2.77

x	0	1	2	3	4	5
y_1	4.16	2.77	0.94	-0.45	-1.51	-2.36
y_2	4.16	2.77	0.94	-0.45	-1.51	-2.36

(c) $y_1 = y_2$

79. (a)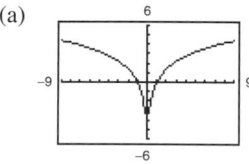

(b)

x	-5	-4	-3	-2	-1
y_1	3.22	2.77	2.20	1.39	0
y_2	Error	Error	Error	Error	Error

x	0	1	2	3	4	5
y_1	Error	0	1.39	2.20	2.77	3.22
y_2	Error	0	1.39	2.20	2.77	3.22

(c) No. The domains differ.

81. 2 **83.** 6.8

85. Not possible; -4 is not in the domain of $\log_2 x$.

87. 2 **89.** -4 **91.** 8 **93.** $-\frac{1}{2}$

95. (a) $\beta = 120 + 10 \log_{10} I$

CHAPTER 3

(b)

I	10^{-4}	10^{-6}	10^{-8}	10^{-10}	10^{-12}	10^{-14}
β	80	60	40	20	0	-20

97. (a)

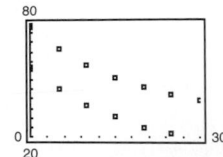

(b) $T = 54.438(0.964)^t + 21$

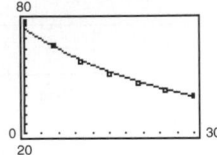

The model fits the data.

(c)

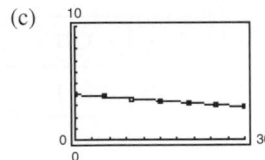

$\ln(T - 21) = -0.037t + 3.997$

$T = e^{(-0.037t + 3.997)} + 21$

(d)

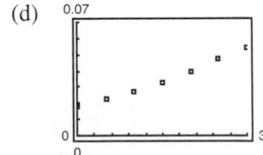

$\dfrac{1}{T - 21} = 0.0012t + 0.0162$

$T = \dfrac{1}{0.0012t + 0.0162} + 21$

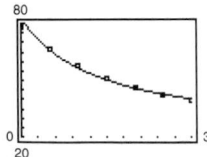

99. True   **101.** False. $f\left(\dfrac{x}{a}\right) = f(x) - f(a)$, not $\dfrac{f(x)}{f(a)}$.

103. False. $f(\sqrt{x}) = \tfrac{1}{2}f(x)$

105. True. When $f(x) = 0$, $x = 1 < e$.   **107.** Proof

109. $f(x) = \dfrac{\log_{10} x}{\log_{10} 2}$

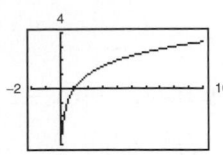

111. $f(x) = \dfrac{\log_{10} \sqrt{x}}{\log_{10} 3}$

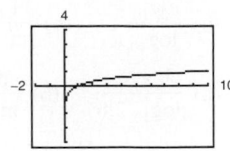

113. $f(x) = \dfrac{\log_{10}(x/3)}{\log_{10} 5}$

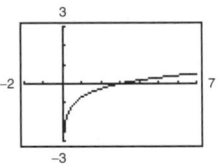

115. $\ln 1 = 0$ $\ln 9 \approx 2.1972$

$\ln 2 \approx 0.6931$ $\ln 10 \approx 2.3025$

$\ln 3 \approx 1.0986$ $\ln 12 \approx 2.4848$

$\ln 4 \approx 1.3862$ $\ln 15 \approx 2.7080$

$\ln 5 \approx 1.6094$ $\ln 16 \approx 2.7724$

$\ln 6 \approx 1.7917$ $\ln 18 \approx 2.8903$

$\ln 8 \approx 2.0793$ $\ln 20 \approx 2.9956$

117. $\dfrac{27y^3}{8x^6}$ **119.** $\dfrac{x^2y^2}{x + y}$ **121.** $0, -5$

123. $\pm\tfrac{1}{3}, \pm 2$ **125.** $\pm\tfrac{1}{3}, \pm 5$

Section 3.4   (page 221)

Vocabulary Check   (page 221)

1. solve

2. (a) $x = y$ (b) $x = y$ (c) x (d) x

3. extraneous

1. (a) Yes   (b) No

3. (a) No   (b) Yes   (c) Yes, approximate

5. (a) Yes, approximate   (b) No   (c) Yes

7. (a) Yes   (b) Yes, approximate   (c) No

9.

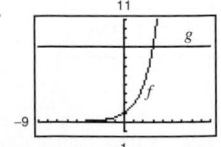

11.

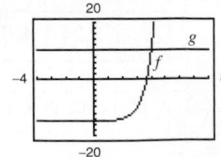

(3, 8); 3   (4, 10); 4

13.

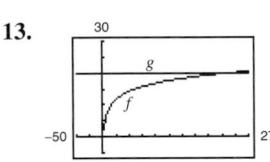

(243, 20); 243

15.

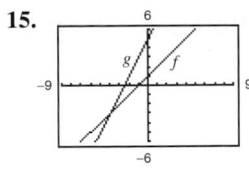

(−4, −3); −4

17. 2 **19.** −4 **21.** −2 **23.** −4 **25.** $\log_{10} 36$

27. 5 **29.** 5 **31.** e^{-7} **33.** 5 **35.** 0.1

37. $\dfrac{e^5 + 1}{2}$ **39.** x^2 **41.** $5x + 2$

43. $2x - 1$ **45.** 0.944 **47.** 2 **49.** −1.498

51. 6.960 **53.** −277.951 **55.** 0.439 **57.** 0.511

59. 0 **61.** 1.609 **63.** 184.444 **65.** −1, 2

67. 0.586, 3.414 **69.** 1.946 **71.** 183.258

73.

x	0.6	0.7	0.8	0.9	1.0
e^{3x}	6.05	8.17	11.02	14.88	20.09

(0.8, 0.9)

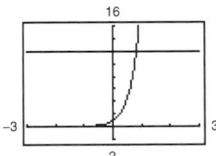

0.828

75.

x	5	6	7	8	9
$20(100 - e^{x/2})$	1756	1598	1338	908	200

(8, 9)

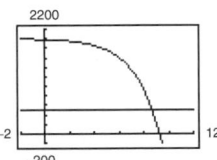

8.635

77. 21.330 **79.** 3.656

81.

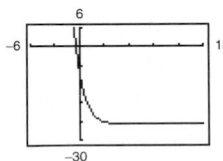

$x = -0.427$

83.

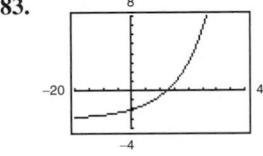

$x = 12.207$

85. 0.050 **87.** 2.042 **89.** 4453.242 **91.** 1

93. 103 **95.** 17.945 **97.** 5.389

99. 1.718, −3.718 **101.** 2 **103.** No real solution

105. 180.384

107.

x	2	3	4	5	6
$\ln 2x$	1.39	1.79	2.08	2.30	2.48

(5, 6)

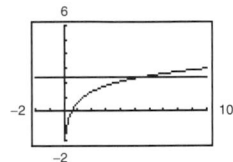

5.512

109.

x	12	13	14	15	16
$6 \log_3(0.5x)$	9.79	10.22	10.63	11.00	11.36

(14, 15)

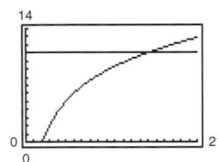

14.988

111. 1.469, 0.001 **113.** 2.928 **115.** 3.423

117.

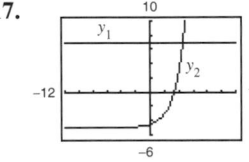

(4.585, 7)

119.

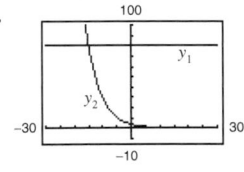

(−14.979, 80)

121.

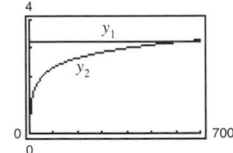

(663.142, 3.25)

123. −1, 0 **125.** 1 **127.** $e^{-1/2} \approx 0.607$

129. $e^{-1} \approx 0.368$ **131.** (a) 9.24 years (b) 14.65 years

133. (a) 27.73 years (b) 43.94 years

135. (a) 1426 units (b) 1498 units **137.** 2001

139. (a)

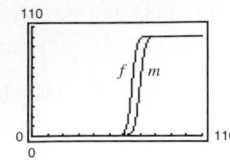

(b) $y = 100$ and $y = 0$; The percentages cannot exceed 100% or be less than 0%.

(c) Males: 69.71 inches; females: 64.51 inches

141. (a)

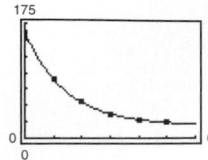

(b) $y = 20$. The object's temperature cannot cool below the room's temperature.

(c) 0.81 hour

143. False; $e^x = 0$ has no solutions.

145. (a) Isolate the exponential term by grouping, factoring, etc. Take the logarithms of both sides and solve for the variable.

(b) Isolate the logarithmic term by grouping, collapsing, etc. Exponentiate both sides and solve for the variable.

147. Yes. The investment will double every $\dfrac{\ln 2}{r}$ years.

149.

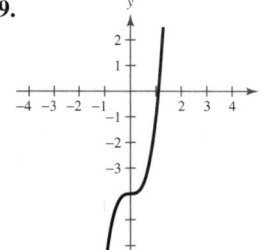

151.

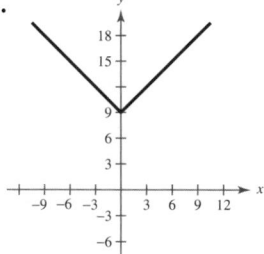

153.

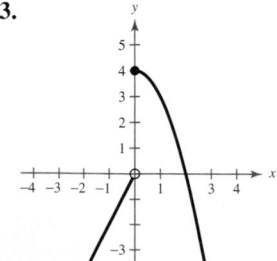

Section 3.5 (page 232)

Vocabulary Check (page 232)

1. (a) iv (b) i (c) vi (d) iii
  (e) vii (f) ii (g) v

2. Normally **3.** Sigmoidal

4. Bell-shaped, mean

1. c **2.** e **3.** b **4.** a **5.** d **6.** f

	Initial Investment	Annual % Rate	Time to Double	Amount After 10 Years
7.	$10,000	3.5%	19.8 years	$14,190.68
9.	$7500	3.30%	21 years	$10,432.26
11.	$5000	1.25%	55.45 years	$5,665.74
13.	$63,762.82	4.5%	15.40 years	$100,000.00

15.

r	2%	4%	6%	8%	10%	12%
t	54.93	27.47	18.31	13.73	10.99	9.16

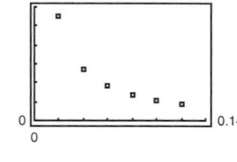

17.

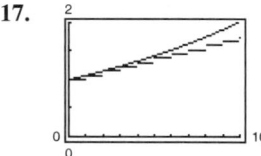

Continuous compounding

Isotope	Half-Life (years)	Initial Quantity	Amount After 1000 Years
19. ^{226}Ra	1599	10 grams	6.48 grams
21. ^{14}C	5715	3 grams	2.66 grams

23. $y = e^{0.768x}$ **25.** $y = 4e^{-0.2773x}$

27. (a) Australia: $y = 19.2e^{0.00848t}$; 24.76 million

Canada: $y = 31.3e^{0.00915t}$; 41.19 million

Phillipines: $y = 79.7e^{0.0185t}$; 138.83 million

South Africa: $y = 44.1e^{-0.00183t}$; 41.74 million

Turkey: $y = 65.7e^{0.01095t}$; 91.25 million

(b) b; Population changes at a faster rate for a greater magnitude of b.

(c) b; b is positive when the population is increasing and negative when the population is decreasing.

29. (a) 0.0295 (b) About 241,734 people **31.** 95.8%

33. (a) $V = -3394t + 30,788$ (b) $V = 30,788e^{-0.1245t}$

(c)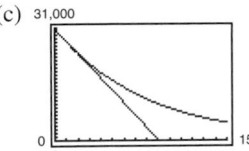

(d) Exponential model (e) Answers will vary.

35. (a) $S(t) = 100(1 - e^{-0.1625t})$

(b) 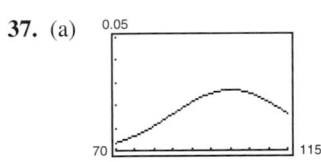 (c) 55,625 units

37. (a) 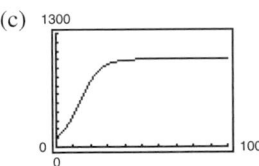 (b) 100

39. (a) ≈ 203 animals (b) ≈ 13 months

(c)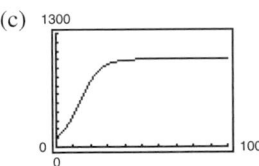

The horizontal asymptotes occur at $p = 1000$ and $p = 0$. The asymptote at $p = 1000$ means there will not be more than 1000 animals in the preserve.

41. (a) 1,258,925 (b) 39,810,717

(c) 1,000,000,000

43. (a) 20 decibels (b) 70 decibels (c) 120 decibels

45. 97.49% **47.** 4.64 **49.** 10,000,000 times

51. (a)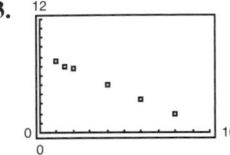

(b) Interest; $t \approx 20.7$ years

(c) 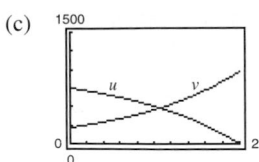 Interest; $t \approx 10.73$ years
Answers will vary.

53. 3:00 A.M.

55. False. The domain can be all real numbers.

57. True. Any x-value in the Gaussian model will give a positive y-value.

59. a; $(0, -3)$, $\left(\frac{9}{4}, 0\right)$ **60.** b; $(0, 2)$, $(5, 0)$

61. d; $(0, 25)$, $\left(\frac{100}{9}, 0\right)$ **62.** c; $(0, 4)$, $(2, 0)$

63. Falls to the left and rises to the right

65. Rises to the left, falls to the right

67. $2x^2 + 3 + \dfrac{3}{x - 4}$

69. Answers will vary.

Section 3.6 (page 242)

Vocabulary Check (page 242)

1. $y = ax + b$ **2.** quadratic **3.** $y = ax^b$

4. sum, squared differences **5.** $y = ab^x$, ae^{cx}

1. Logarithmic model **3.** Quadratic model

5. Exponential model **7.** Quadratic model

9.   **11.**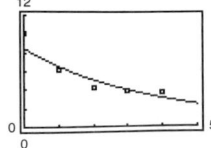

Logarithmic model Exponential model

13.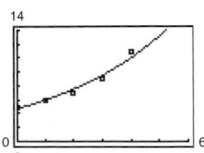

Linear model

15. $y = 4.752(1.2607)^x$; **17.** $y = 8.463(0.7775)^x$;
0.96773 0.86639

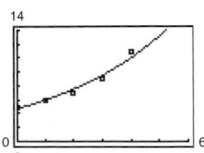 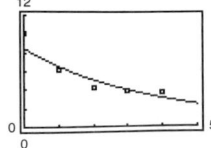

19. $y = 2.083 + 1.257 \ln x$; 0.98672

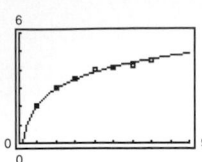

21. $y = 9.826 - 4.097 \ln x$; 0.93704

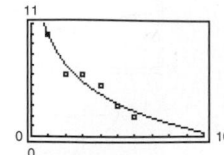

23. $y = 1.985x^{0.760}$; 0.99686

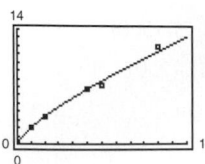

25. $y = 16.103x^{-3.174}$; 0.88161

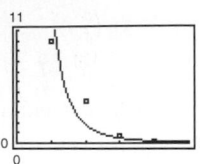

27. (a) Quadratic model: $y = 0.031t^2 + 1.13t + 97.1$

Exponential model: $y = 94.435(1.0174)^t$

Power model: $y = 77.837t^{0.1918}$

(b) Quadratic model: Exponential model:

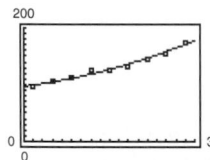   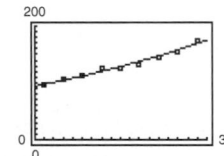

Power model:

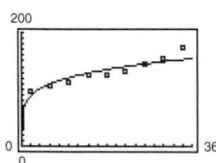

(c) Exponential model

(d) 2008: 181.89 million; 2012: 194.89 million

29. (a) Linear model: $P = 3.11t + 250.9$; 0.99942

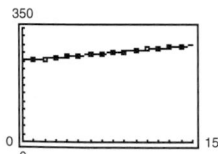

(b) Power model: $P = 246.52t^{0.0587}$; 0.90955

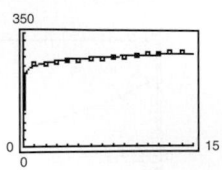

(c) Exponential model: $P = 251.57(1.0114)^t$; 0.99811

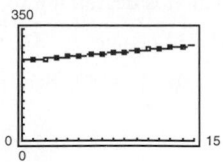

(d) Quadratic model: $P = -0.020t^2 + 3.41t + 250.1$; 0.99994

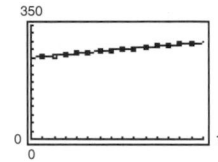

(e) Quadratic model; the coefficient of determination for the quadratic model is closest to 1, so the quadratic model best fits the data.

(f) Linear model:

Year	2005	2006	2007	2008	2009	2010
Population (in millions)	297.6	300.7	303.8	306.9	310.0	313.1

Power model:

Year	2005	2006	2007	2008	2009	2010
Population (in millions)	289.0	290.1	291.1	292.1	293.0	293.9

Exponential model:

Year	2005	2006	2007	2008	2009	2010
Population (in millions)	298.2	301.6	305.0	308.5	312.0	315.6

Quadratic model:

Year	2005	2006	2007	2008	2009	2010
Population (in millions)	296.8	299.5	302.3	305.0	307.7	310.3

(g) and (h) Answers will vary.

31. (a) Linear model: $T = -1.24t + 73.0$

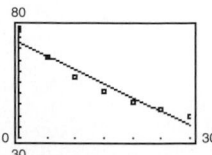

The data does not appear to be linear. Answers will vary.

(b) Quadratic model: $T = 0.034t^2 - 2.26t + 77.3$

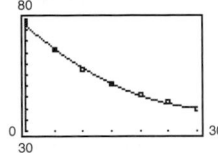

The data appears to be quadratic. When $t = 60$, the temperature of the water should decrease, not increase as shown in this model.

(c) Exponential model: $T = 54.438(0.964)^t + 21$

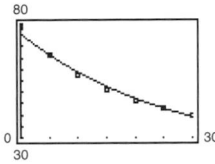

(d) Answers will vary.

33. (a) $P = \dfrac{162.4}{1 + 0.34e^{0.5609x}}$

(b)

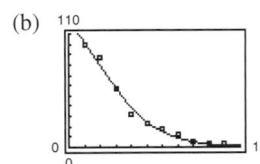

The model closely represents the actual data.

35. (a) Linear model: $y = 15.71t + 51.0$

Logarithmic model: $y = 134.67 \ln t - 97.5$

Quadratic model: $y = -1.292t^2 + 38.96t - 45$

Exponential model: $y = 85.97(1.091)^t$

Power model: $y = 37.274t^{0.7506}$

(b) Linear model: Logarithmic model:

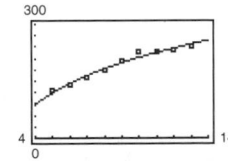

Quadratic model: Exponential model:

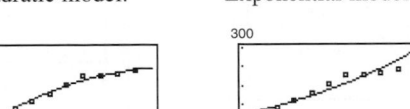
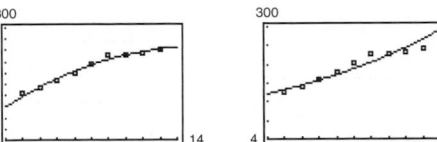

Power model:

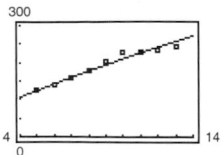

The quadratic model best fits the data.

(c) Linear model: 803.9, logarithmic model: 411.7, quadratic model: 289.8, exponential model: 1611.4, and power model: 667.1. The quadratic model best fits the data.

(d) Linear model: 0.9485, logarithmic model: 0.9736, quadratic model: 0.9814, exponential model: 0.9274, and power model: 0.9720. The quadratic model best fits the data.

(e) The quadratic model is the best-fitting model.

37. True

39. Slope: $-\frac{2}{5}$; y-intercept: $(0, 2)$

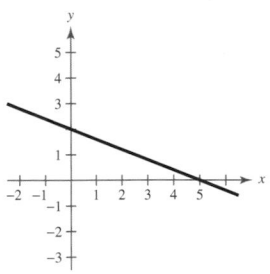

41. Slope: $-\frac{12}{35}$; y-intercept: $(0, 3)$

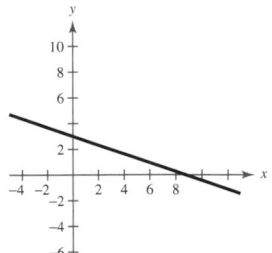

Review Exercises   (page 247)

1. 10.3254   **3.** 0.0001   **5.** 2980.9580

7. 8.1662   **9.** c   **10.** d   **11.** b   **12.** a

13.

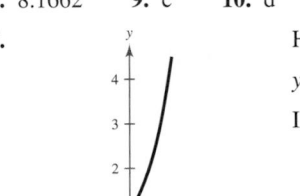

Horizontal asymptote: $y = 0$

y-intercept: $(0, 1)$

Increasing on $(-\infty, \infty)$

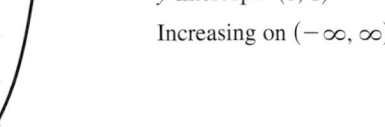

CHAPTER 3

15.

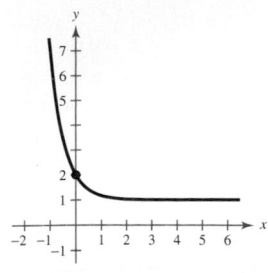

Horizontal asymptote: $y = 1$

y-intercept: $(0, 2)$

Decreasing on $(-\infty, \infty)$

17.

x	0	1	2	3	4
$h(x)$	0.37	1	2.72	7.39	20.09

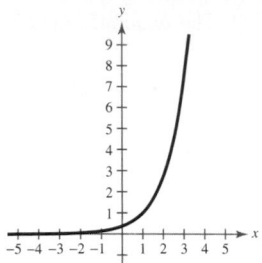

Horizontal asymptote: $y = 0$

19.

x	-2	-1	0	1	2
$h(x)$	-0.14	-0.37	-1	-2.72	-7.39

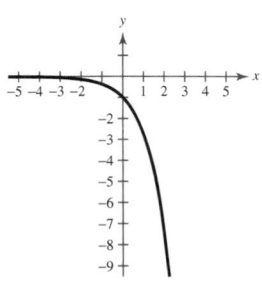

Horizontal asymptote: $y = 0$

21.

x	-1	0	1	2	3	4
$f(x)$	6.59	4	2.43	1.47	0.89	0.54

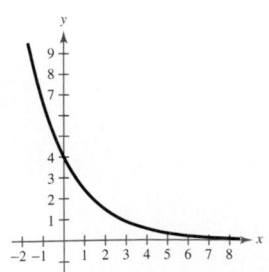

Horizontal asymptote: $y = 0$

23. (a)

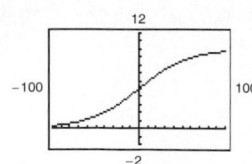

(b) Horizontal asymptotes:
$y = 0$ and $y = 10$

25.

t	1	10	20
A	\$10,832.87	\$22,255.41	\$49,530.32

t	30	40	50
A	\$110,231.76	\$245,325.30	\$545,981.50

27. (a)

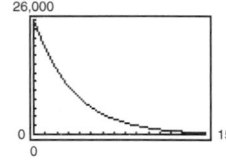

(b) \$14,625

(c) When it is first sold; Yes; Answers will vary.

29. $5^3 = 125$ **31.** $64^{1/6} = 2$ **33.** $e^4 = e^4$

35. $\log_4 64 = 3$ **37.** $\log_{25} 125 = \frac{3}{2}$

39. $\log_{1/2} 8 = -3$ **41.** $\ln 1096.6331 \ldots = 7$

43. 3   **45.** -1

47. Domain: $(0, \infty)$ **49.** Domain: $(1, \infty)$

Vertical asymptote: Vertical asymptote:

$x = 0$ $x = 1$

x-intercept: $(32, 0)$ x-intercept: $(1.016, 0)$

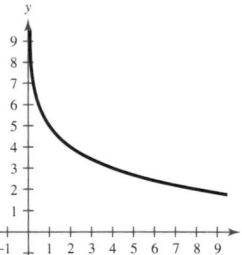

51. 3.068   **53.** 0.896   **55.** 3   **57.** $\frac{1}{9}$

59.   **61.**

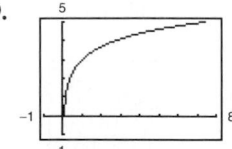   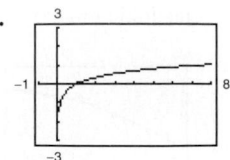

Domain: $(0, \infty)$ Domain: $(0, \infty)$

Vertical asymptote: $x = 0$ Vertical asymptote: $x = 0$

x-intercept: $(0.05, 0)$ x-intercept: $(1, 0)$

63. (a) $0 \le h < 18,000$

(b)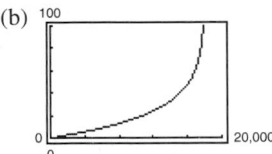

Asymptote: $h = 18,000$

(c) The time required to increase its altitude further increases.

(d) 5.46 minutes

65. 1.585   **67.** 2.132

69. 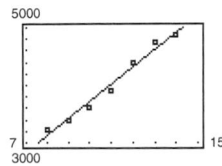   **71.**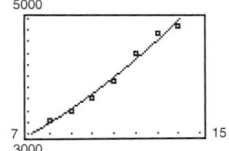

73. 1.13   **75.** 0.41   **77.** $\ln 5 - 2$ **79.** $2 + \log_{10} 2$

81. $1 + 2 \log_5 x$ **83.** $\log_{10} 5 + \frac{1}{2} \log_{10} y - 2 \log_{10} x$

85. $\ln(x + 3) - \ln x - \ln y$ **87.** $\log_2 5x$

89. $\ln \dfrac{\sqrt{2x - 1}}{(x + 1)^2}$ **91.** $\ln \dfrac{3 \sqrt[3]{4 - x^2}}{x}$

93. (a)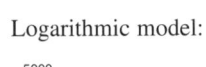

(b)

h	4	6	8	10	12	14
s	38	33	30	27	25	23

(c) The decrease in productivity starts to level off.

95. 3   **97.** -3 **99.** -5 **101.** 2401   **103.** 9

105. e^4 **107.** $e^2 + 1$ **109.** -0.757 **111.** 4.459

113. 1.760   **115.** 3.916   **117.** 1.609, 0.693

119. 1213.650   **121.** 22.167   **123.** 53.598

125. No solution   **127.** 0.9   **129.** -1 **131.** 0.368

133. 15.15 years   **135.** e   **136.** b   **137.** f

138. d   **139.** a   **140.** c   **141.** $y = 2e^{0.1014x}$

143. $y = \frac{1}{2} e^{0.4605x}$ **145.** $k \approx 0.0259$; 606,000

147. (a) 5.78%   (b) \$10,595.03

149. (a) 7.64 weeks   (b) 13.21 weeks

151. Logistic model   **153.** Logarithmic model

155. (a) Linear model: $S = 297.8t + 739$; 0.97653

Quadratic model: $S = 11.79t^2 + 38.5t + 2118$; 0.98112

Exponential model: $S = 1751.5(1.077)^t$; 0.98225
Logarithmic model: $S = 3169.8 \ln t - 3532$; 0.95779

Power model: $S = 598.1t^{0.7950}$; 0.97118

(b) Linear model: Quadratic model:

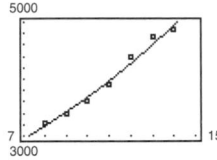

Exponential model: Logarithmic model:

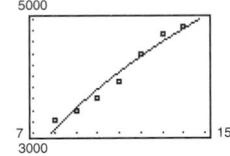

Power model:

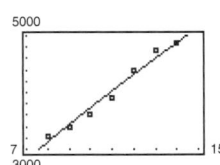

(c) Exponential model; the coefficient of determination for the exponential model is closest to 1.

(d) \$7722 million   (e) 2004

157. (a) $P = \dfrac{9999.887}{1 + 19.0e^{-0.2x}}$

(b)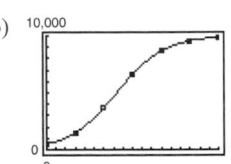

(c) The model fits the data well.   (d) 10,000 fish

159. True   **161.** False. $\ln(xy) = \ln x + \ln y$

163. False. $x > 0$

165. Because $1 < \sqrt{2} < 2$, then $2^1 < 2^{\sqrt{2}} < 2^2$.

Chapter Test   (page 252)

1.

x	−2	−1	0	1	2
f(x)	100	10	1	0.1	0.01

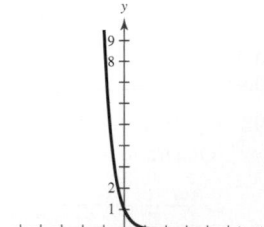

Horizontal asymptote: $y = 0$

y-intercept: $(0, 1)$

2.

x	0	2	3	4
f(x)	−0.03	−1	−6	−36

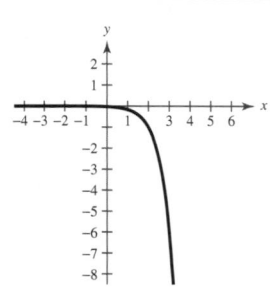

Horizontal asymptote: $y = 0$

y-intercept: $\left(0, -\frac{1}{36}\right)$

3.

x	−2	−1	0	1	2
f(x)	0.9817	0.8647	0	−6.3891	−53.5982

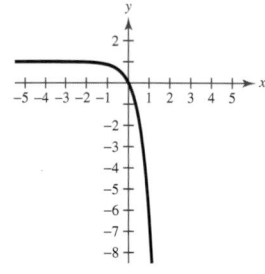

Horizontal asymptote: $y = 1$

Intercept: $(0, 0)$

4. −0.89   **5.** 9.2   **6.** 0

7.

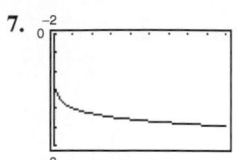

Domain: $(0, \infty)$

Vertical asymptote:
$x = 0$

Intercept: $(0, 0)$

8.

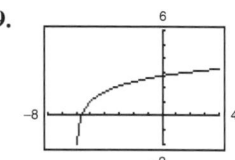

Domain: $(4, \infty)$

Vertical asymptote:
$x = 4$

x-intercept: $(5, 0)$

9.

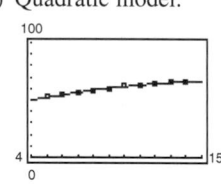

Domain: $(-6, \infty)$

Vertical asymptote: $x = -6$

x-intercept: $(-5.63, 0)$

10. 1.945   **11.** 0.115   **12.** 1.328

13. $\log_2 3 + 4 \log_2 a$ **14.** $\ln 5 + \frac{1}{2} \ln x - \ln 6$

15. $\ln x + \frac{1}{2} \ln(x + 1) - \ln 2 - 4$ **16.** $\log_3 13y$

17. $\ln\left(\dfrac{x^4}{y^4}\right)$ **18.** $\ln\left[\dfrac{x(2x - 3)}{x + 2}\right]$ **19.** 4

20. 2.431   **21.** 343   **22.** 100,004   **23.** 1.321

24. 1   **25.** 1.597   **26.** 1.649   **27.** 54.96%

28. (a) Quadratic model: $R = -0.092t^2 + 3.29t + 38.1$

Exponential model: $R = 46.99(1.026)^t$

Power model: $R = 36.00t^{0.233}$

(b) Quadratic model: Exponential model:

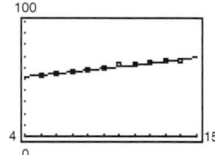

Power model:

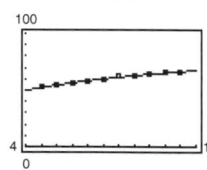

(c) Quadratic model, 67.1

Cumulative Test for Chapters 1–3
(page 253)

1. (a) $x + y - 3 = 0$

   (b) Answers will vary. Sample answer: $(0, 3), (1, 2), (2, 1)$

2. (a) $2x + y = 0$

   (b) Answers will vary. Sample answer:
   $(0, 0), (1, -2), (2, -4)$

3. (a) $7x + 3 = 0$

   (b) Answers will vary. Sample answer:
   $\left(-\frac{3}{7}, 0\right), \left(-\frac{3}{7}, 1\right), \left(-\frac{3}{7}, -3\right)$

4. (a) $\dfrac{5}{3}$ (b) Undefined   (c) $\dfrac{5 + 4s}{3 + 4s}$

5. (a) -32 (b) 4 (c) 20

6. No. It doesn't pass the Vertical Line Test.

7. 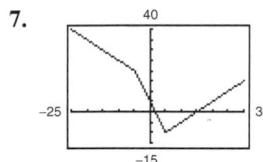   Decreasing on $(-\infty, 5)$
   Increasing on $(5, \infty)$

8. (a) Vertical shrink   (b) Vertical shift

   (c) Horizontal shift and reflection in the x-axis

9. -53 10. $\dfrac{197}{16}$ 11. -79 12. 42

13. $h^{-1}(x) = \dfrac{x + 2}{5}$

14. 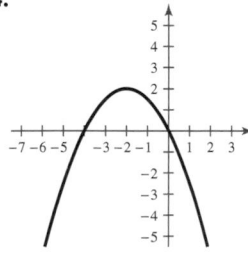   15.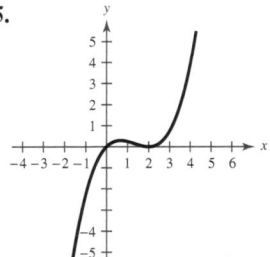

16.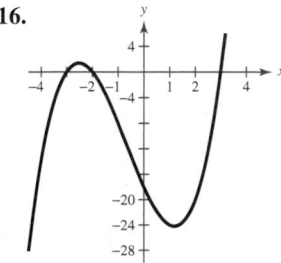

17. $x = -2, \pm 2i$ 18. 1.424

19. $4x + 2 - \dfrac{15}{x + 3}$ 20. $2x^2 + 7x + 48 + \dfrac{268}{x - 6}$

21.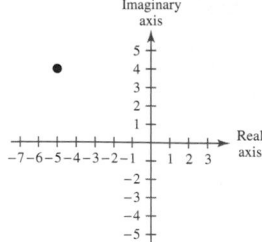

22. Answers will vary. Sample answer: $f(x) = x^4 + x^3 + 18x$

23. 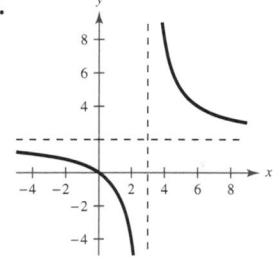   24.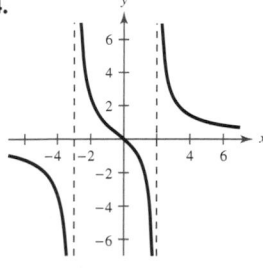

   Asymptotes: Asymptotes:
   $x = 3, y = 2$ $x = -3, x = 2, y = 0$

25.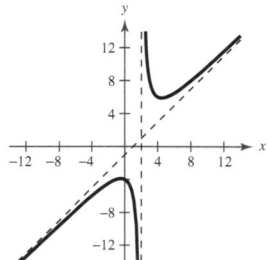

   Asymptotes: $x = 2, y = x - 1$

26. 6.733 27. 8772.934 28. 0.162 29. 51.743

30. 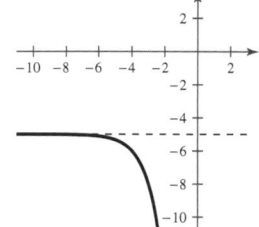   31.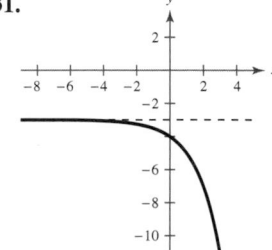

32. 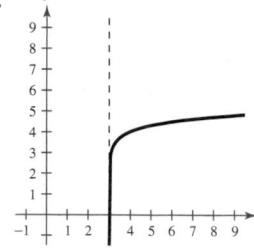   33.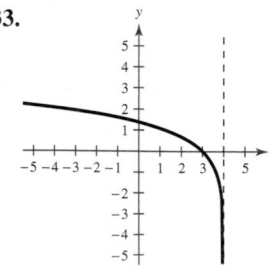

CHAPTER 3

34. 1.892   **35.** 0.872   **36.** 0.585

37. $\ln(x+2) + \ln(x-2) - \ln(x^2+1)$

38. $\ln\left(\dfrac{x^2(x+1)}{x-1}\right)$ **39.** 1.242   **40.** 6.585   **41.** 12.8

42. 152.018   **43.** 0, 1   **44.** No solution

45. (a) $A = x(273 - x)$

   (b) 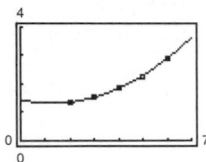 $0 < x < 273$

   (c) 76.23 feet × 196.77 feet

46. (a) Quadratic model:
    $y = 0.0707x^2 - 0.183x + 1.45$; 0.99871
    Exponential model: $y = 0.89(1.21)^x$; 0.98862
    Power model: $y = 0.79(x)^{0.68}$; 0.93130

   (b) Quadratic model: Exponential model:

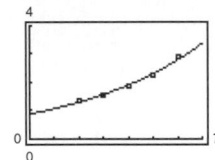

   Power model:

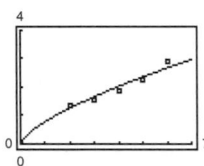

   (c) Quadratic model; the coefficient of determination for the quadratic model is closest to 1.

   (d) $4.51, $6.69; Answers will vary.

Chapter 4

Section 4.1   (page 265)

Vocabulary Check   (page 265)

1. Trigonometry   **2.** angle

3. standard position   **4.** coterminal   **5.** radian

6. complementary   **7.** supplementary

8. degree   **9.** linear   **10.** angular

1. 2   **3.** (a) Quadrant IV   (b) Quadrant II

5. (a) Quadrant IV   (b) Quadrant III

7. (a) 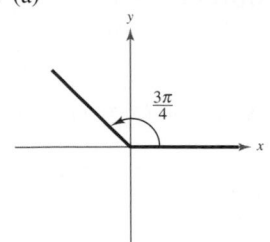   (b)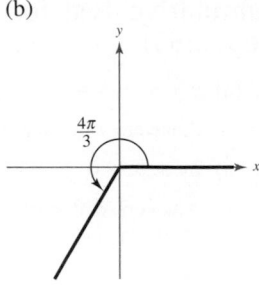

9. (a) 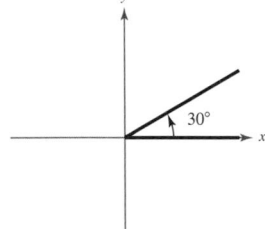   (b)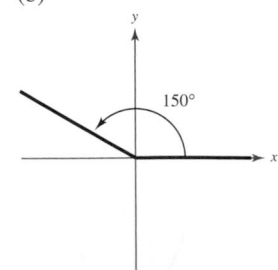

11. (a) $\dfrac{13\pi}{6}, -\dfrac{11\pi}{6}$ (b) $\dfrac{8\pi}{3}, -\dfrac{4\pi}{3}$

13. (a) $\dfrac{7\pi}{4}, -\dfrac{\pi}{4}$ (b) $\dfrac{28\pi}{15}, -\dfrac{32\pi}{15}$

15. Complement: $\dfrac{\pi}{6}$; supplement: $\dfrac{2\pi}{3}$

17. Complement: $\dfrac{\pi}{3}$; supplement: $\dfrac{5\pi}{6}$

19. Complement: 0.571; supplement: 2.142

21. 210°   **23.** (a) Quadrant II   (b) Quadrant IV

25. (a) Quadrant III   (b) Quadrant I

27. (a) 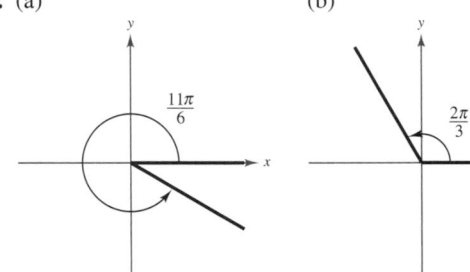   (b)

29. (a) 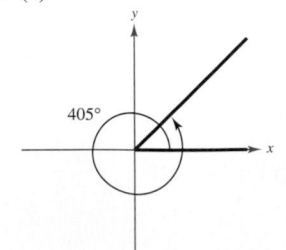   (b)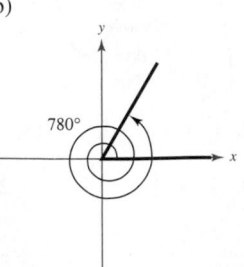

31. (a) $412°, -308°$ (b) $324°, -396°$

33. (a) $660°, -60°$ (b) $590°, -130°$

35. Complement: $66°$; supplement: $156°$

37. Complement: $3°$; supplement: $93°$

39. (a) $\dfrac{\pi}{6}$ (b) $\dfrac{5\pi}{6}$ **41.** (a) $-\dfrac{\pi}{9}$ (b) $-\dfrac{4\pi}{3}$

43. (a) $270°$ (b) $-210°$ **45.** (a) $420°$ (b) $-39°$

47. 2.007 **49.** -3.776 **51.** -0.014 **53.** $25.714°$

55. $1170°$ **57.** $-114.592°$ **59.** $64.75°$ **61.** $85.308°$

63. $-125.01°$ **65.** $280°\,36'$ **67.** $-345°\,7'\,12''$

69. $-20°\,20'\,24''$ **71.** $\dfrac{6}{5}$ rad   **73.** $\dfrac{32}{7}$ rad

75. $0, \dfrac{\pi}{6}, \dfrac{\pi}{4}, \dfrac{\pi}{3}, \dfrac{\pi}{2}, \dfrac{3\pi}{4}, \pi, \dfrac{7\pi}{6}, \dfrac{3\pi}{2}, \dfrac{11\pi}{6}$ **77.** $\dfrac{8}{15}$ rad

79. $\dfrac{70}{29}$ rad   **81.** 14π inches   **83.** 18π meters

85. 22.92 feet   **87.** 34.80 miles

89. 1141.81 miles   **91.** $4°\,2'\,33''$ **93.** $23.87°$

95. (a) $540° = 3\pi$ rad   (b) $900° = 5\pi$ rad

(c) $1260° = 7\pi$ rad

97. (a) 80π radians per second   (b) 25π feet per second

99. (a) 35.70 miles per hour

(b) 941.18 revolutions per minute

101. False. A radian is larger: 1 rad $\approx 57.3°$.

103. True. The sum of the angles of a triangle must equal $180° = \pi$ radians, and $\dfrac{2\pi}{3} + \dfrac{\pi}{4} + \dfrac{\pi}{12} = \pi$.

105. The length of the intercepted arc is increasing. Answers will vary.

107. $\dfrac{50\pi}{3}$ square meters

109. (a) $A = 0.4r^2,\ r > 0;\ s = 0.8r,\ r > 0$

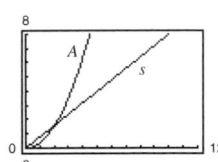

The area function changes more rapidly for $r > 1$ because it is quadratic and the arc length function is linear.

(b) $A = 50\theta,\ 0 < \theta < 2\pi;\ s = 10\theta,\ 0 < \theta < 2\pi$

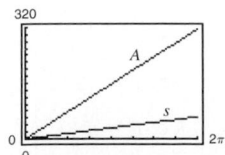

111. Answers will vary.

113.

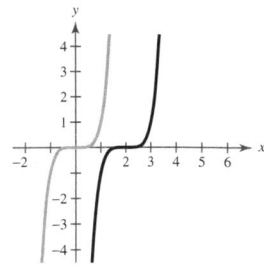

115.

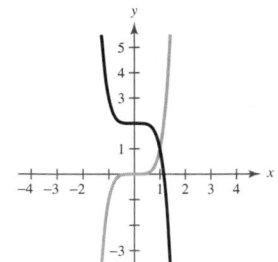

117.

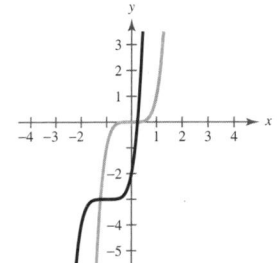

Section 4.2   (page 274)

1. $\sin\theta = \dfrac{15}{17}$

$\cos\theta = -\dfrac{8}{17}$

$\tan\theta = -\dfrac{15}{8}$

$\csc\theta = \dfrac{17}{15}$

$\sec\theta = -\dfrac{17}{8}$

$\cot\theta = -\dfrac{8}{15}$

3. $\sin\theta = -\dfrac{5}{13}$

$\cos\theta = \dfrac{12}{13}$

$\tan\theta = -\dfrac{5}{12}$

$\csc\theta = -\dfrac{13}{5}$

$\sec\theta = \dfrac{13}{12}$

$\cot\theta = -\dfrac{12}{5}$

5. $\left(\dfrac{\sqrt{2}}{2}, \dfrac{\sqrt{2}}{2}\right)$ **7.** $\left(-\dfrac{\sqrt{3}}{2}, -\dfrac{1}{2}\right)$ **9.** $\left(-\dfrac{1}{2}, \dfrac{\sqrt{3}}{2}\right)$

11. $(0, -1)$ **13.** $\left(\dfrac{\sqrt{2}}{2}, \dfrac{\sqrt{2}}{2}\right)$ **15.** $(0, 1)$

17. $\sin\dfrac{\pi}{4} = \dfrac{\sqrt{2}}{2}$

$\cos\dfrac{\pi}{4} = \dfrac{\sqrt{2}}{2}$

$\tan\dfrac{\pi}{4} = 1$

19. $\sin\dfrac{7\pi}{6} = -\dfrac{1}{2}$

$\cos\dfrac{7\pi}{6} = -\dfrac{\sqrt{3}}{2}$

$\tan\dfrac{7\pi}{6} = \dfrac{\sqrt{3}}{3}$

21. $\sin \dfrac{2\pi}{3} = \dfrac{\sqrt{3}}{2}$

$\cos \dfrac{2\pi}{3} = -\dfrac{1}{2}$

$\tan \dfrac{2\pi}{3} = -\sqrt{3}$

23. $\sin\left(-\dfrac{5\pi}{3}\right) = \dfrac{\sqrt{3}}{2}$

$\cos\left(-\dfrac{5\pi}{3}\right) = \dfrac{1}{2}$

$\tan\left(-\dfrac{5\pi}{3}\right) = \sqrt{3}$

25. $\sin\left(-\dfrac{\pi}{6}\right) = -\dfrac{1}{2}$

$\cos\left(-\dfrac{\pi}{6}\right) = \dfrac{\sqrt{3}}{2}$

$\tan\left(-\dfrac{\pi}{6}\right) = -\dfrac{\sqrt{3}}{3}$

27. $\sin\left(-\dfrac{7\pi}{4}\right) = \dfrac{\sqrt{2}}{2}$

$\cos\left(-\dfrac{7\pi}{4}\right) = \dfrac{\sqrt{2}}{2}$

$\tan\left(-\dfrac{7\pi}{4}\right) = 1$

29. $\sin\left(-\dfrac{3\pi}{2}\right) = 1$

$\cos\left(-\dfrac{3\pi}{2}\right) = 0$

$\tan\left(-\dfrac{3\pi}{2}\right)$ is undefined.

31. $\sin \dfrac{3\pi}{4} = \dfrac{\sqrt{2}}{2}$

$\cos \dfrac{3\pi}{4} = -\dfrac{\sqrt{2}}{2}$

$\tan \dfrac{3\pi}{4} = -1$

$\csc \dfrac{3\pi}{4} = \sqrt{2}$

$\sec \dfrac{3\pi}{4} = -\sqrt{2}$

$\cot \dfrac{3\pi}{4} = -1$

33. $\sin \dfrac{\pi}{2} = 1$

$\cos \dfrac{\pi}{2} = 0$

$\tan \dfrac{\pi}{2}$ is undefined.

$\csc \dfrac{\pi}{2} = 1$

$\sec \dfrac{\pi}{2}$ is undefined.

$\cot \dfrac{\pi}{2} = 0$

35. $\sin\left(-\dfrac{2\pi}{3}\right) = -\dfrac{\sqrt{3}}{2}$

$\cos\left(-\dfrac{2\pi}{3}\right) = -\dfrac{1}{2}$

$\tan\left(-\dfrac{2\pi}{3}\right) = \sqrt{3}$

$\csc\left(-\dfrac{2\pi}{3}\right) = -\dfrac{2\sqrt{3}}{3}$

$\sec\left(-\dfrac{2\pi}{3}\right) = -2$

$\cot\left(-\dfrac{2\pi}{3}\right) = \dfrac{\sqrt{3}}{3}$

37. 0 **39.** $-\dfrac{1}{2}$ **41.** $\dfrac{\sqrt{3}}{2}$ **43.** $-\dfrac{\sqrt{2}}{2}$

45. (a) $-\dfrac{1}{3}$ (b) -3 **47.** (a) $-\dfrac{1}{5}$ (b) -5

49. (a) $\dfrac{4}{5}$ (b) $-\dfrac{4}{5}$ **51.** 0.6428 **53.** 0.8090

55. 1.0378 **57.** -0.1288 **59.** 1.3940

61. -1.4486 **63.** -1.3386 **65.** -1.0025

67. -4.7439 **69.** (a) -0.9 (b) -0.4

71. (a) $0.25, 2.89$ (b) $1.82, 4.46$ **73.** 0.79 ampere

75. (a) 0.25 foot   (b) 0.02 foot   (c) -0.25 foot

77. False. $\sin(-t) = -\sin t$ means that the function is odd, not that the sine of a negative angle is a negative number.

79. False. The real number 0 corresponds to the point $(1, 0)$.

81. (a) y-axis   (b) $\sin t_1 = \sin(\pi - t_1)$

(c) $\cos(\pi - t_1) = -\cos t_1$

83. Answers will vary.   **85.** Answers will vary.

87. It is an odd function.

89. $f^{-1}(x) = \dfrac{2}{3}(x + 1)$ **91.** $f^{-1}(x) = \sqrt{x^2 + 4},\, x \ge 0$

93.

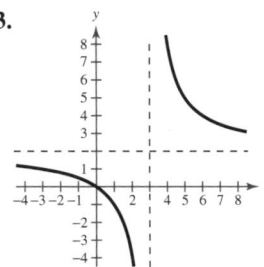

95.

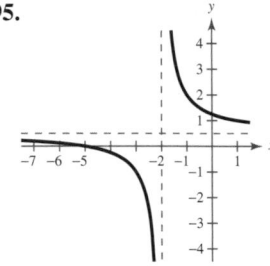

97. Domain: all real numbers x

Intercepts: $(-4, 0), (1, 0), (0, -4)$

No asymptotes

99. Domain: all real numbers x

Intercept: $(0, 5)$

Asymptote: $y = 2$

Section 4.3   (page 284)

Vocabulary Check   (page 284)

1. (a) iii   (b) vi   (c) ii   (d) v   (e) i   (f) iv

2. hypotenuse, opposite, adjacent

3. elevation, depression

1. $\sin \theta = \dfrac{3}{5}$

$\cos \theta = \dfrac{4}{5}$

$\tan \theta = \dfrac{3}{4}$

$\csc \theta = \dfrac{5}{3}$

$\sec \theta = \dfrac{5}{4}$

$\cot \theta = \dfrac{4}{3}$

3. $\sin \theta = \dfrac{8}{17}$

$\cos \theta = \dfrac{15}{17}$

$\tan \theta = \dfrac{8}{15}$

$\csc \theta = \dfrac{17}{8}$

$\sec \theta = \dfrac{17}{15}$

$\cot \theta = \dfrac{15}{8}$

5. $\sin \theta = \dfrac{3}{5}$

$\cos \theta = \dfrac{4}{5}$

$\tan \theta = \dfrac{3}{4}$

$\csc \theta = \dfrac{5}{3}$

$\sec \theta = \dfrac{5}{4}$

$\cot \theta = \dfrac{4}{3}$

The triangles are similar and corresponding sides are proportional.

7. $\sin \theta = \dfrac{1}{3}$

$\cos \theta = \dfrac{2\sqrt{2}}{3}$

$\tan \theta = \dfrac{\sqrt{2}}{4}$

$\csc \theta = 3$

$\sec \theta = \dfrac{3\sqrt{2}}{4}$

$\cot \theta = 2\sqrt{2}$

The triangles are similar and corresponding sides are proportional.

9. $\cos \theta = \dfrac{\sqrt{11}}{6}$

$\tan \theta = \dfrac{5\sqrt{11}}{11}$

$\csc \theta = \dfrac{6}{5}$

$\sec \theta = \dfrac{6\sqrt{11}}{11}$

$\cot \theta = \dfrac{\sqrt{11}}{5}$

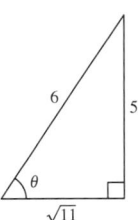

11. $\sin \theta = \dfrac{\sqrt{15}}{4}$

$\cos \theta = \dfrac{1}{4}$

$\tan \theta = \sqrt{15}$

$\csc \theta = \dfrac{4\sqrt{15}}{15}$

$\cot \theta = \dfrac{\sqrt{15}}{15}$

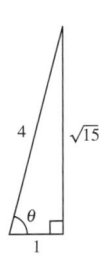

13. $\sin \theta = \dfrac{3\sqrt{10}}{10}$

$\cos \theta = \dfrac{\sqrt{10}}{10}$

$\csc \theta = \dfrac{\sqrt{10}}{3}$

$\sec \theta = \sqrt{10}$

$\cot \theta = \dfrac{1}{3}$

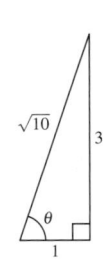

15. $\sin \theta = \dfrac{4\sqrt{97}}{97}$

$\cos \theta = \dfrac{9\sqrt{97}}{97}$

$\tan \theta = \dfrac{4}{9}$

$\csc \theta = \dfrac{\sqrt{97}}{4}$

$\sec \theta = \dfrac{\sqrt{97}}{9}$

17. $\dfrac{\pi}{6}, \dfrac{1}{2}$ **19.** $60°, \sqrt{3}$ **21.** $60°, \dfrac{\pi}{3}$ **23.** $30°, \dfrac{\sqrt{3}}{2}$

25. $45°, \dfrac{\pi}{4}$ **27.** $\csc \theta$ **29.** $\cot \theta$ **31.** $\cos \theta$

33. $\dfrac{\sin \theta}{\cos \theta}$ **35.** 1   **37.** $\cos \theta$ **39.** $\cot \theta$

41. $\csc \theta$ **43.** (a) $\sqrt{3}$ (b) $\dfrac{1}{2}$ (c) $\dfrac{\sqrt{3}}{2}$ (d) $\dfrac{\sqrt{3}}{3}$

45. (a) $\dfrac{1}{3}$ (b) $\dfrac{2\sqrt{2}}{3}$ (c) $\dfrac{\sqrt{2}}{4}$ (d) 3

47. (a) 4   (b) $\pm\dfrac{\sqrt{15}}{4}$ (c) $\pm\dfrac{\sqrt{15}}{15}$ (d) $\dfrac{1}{4}$

49–55. Answers will vary.   **57.** (a) 0.6561   (b) 0.0523

59. (a) 1.3499   (b) 1.3432

61. (a) 5.0273   (b) 0.4142

63. (a) $30° = \dfrac{\pi}{6}$ (b) $30° = \dfrac{\pi}{6}$

65. (a) $60° = \dfrac{\pi}{3}$ (b) $45° = \dfrac{\pi}{4}$

67. (a) $60° = \dfrac{\pi}{3}$ (b) $45° = \dfrac{\pi}{4}$

69. $y = 35\sqrt{3}, r = 70\sqrt{3}$ **71.** $x = 8, y = 8\sqrt{3}$

73. $x = 20, r = 20\sqrt{2}$ **75.** $x = 2\sqrt{5}, r = 2\sqrt{10}$

77. (a) 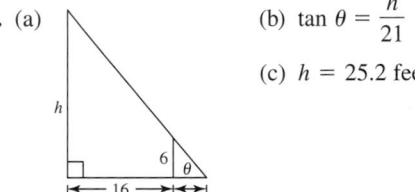   (b) $\tan \theta = \dfrac{h}{21}$

(c) $h = 25.2$ feet

79. 160 feet

81. (a) $45°$

(b) $50\sqrt{2}$ feet

(c) $\dfrac{25\sqrt{2}}{3}$ feet per second; $\dfrac{25}{3}$ feet per second

CHAPTER 4

83. $(x_1, y_1) = \left(28\sqrt{3}, 28\right)$
$(x_2, y_2) = \left(28, 28\sqrt{3}\right)$

85. True. $\csc x = \dfrac{1}{\sin x}$ **87.** True. $1 + \cot^2 x = \csc^2 x$

89. (a)

θ	0°	20°	40°	60°	80°
$\sin \theta$	0	0.3420	0.6428	0.8660	0.9848
$\cos \theta$	1	0.9397	0.7660	0.5	0.1736
$\tan \theta$	0	0.3640	0.8391	1.7321	5.6713

(b) Sine: increasing

Cosine: decreasing

Tangent: increasing

(c) Answers will vary.

91. 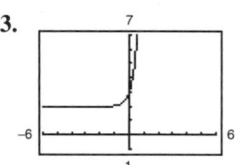   **93.**

95. 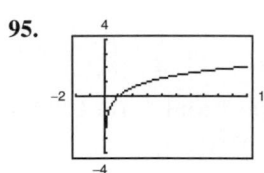   **97.**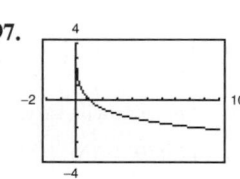

Section 4.4   (page 294)

Vocabulary Check   (page 294)

1. $\dfrac{y}{r}$ **2.** $\csc \theta$ **3.** $\dfrac{y}{x}$ **4.** $\dfrac{r}{x}$

5. $\cos \theta$ **6.** $\cot \theta$ **7.** reference

---

1. (a) $\sin \theta = \dfrac{3}{5}$ (b) $\sin \theta = -\dfrac{15}{17}$

$\cos \theta = \dfrac{4}{5}$ $\cos \theta = -\dfrac{8}{17}$

$\tan \theta = \dfrac{3}{4}$ $\tan \theta = \dfrac{15}{8}$

$\csc \theta = \dfrac{5}{3}$ $\csc \theta = -\dfrac{17}{15}$

$\sec \theta = \dfrac{5}{4}$ $\sec \theta = -\dfrac{17}{8}$

$\cot \theta = \dfrac{4}{3}$ $\cot \theta = \dfrac{8}{15}$

3. (a) $\sin \theta = -\dfrac{1}{2}$ (b) $\sin \theta = \dfrac{\sqrt{2}}{2}$

$\cos \theta = -\dfrac{\sqrt{3}}{2}$ $\cos \theta = -\dfrac{\sqrt{2}}{2}$

$\tan \theta = \dfrac{\sqrt{3}}{3}$ $\tan \theta = -1$

$\csc \theta = -2$ $\csc \theta = \sqrt{2}$

$\sec \theta = -\dfrac{2\sqrt{3}}{3}$ $\sec \theta = -\sqrt{2}$

$\cot \theta = \sqrt{3}$ $\cot \theta = -1$

5. $\sin \theta = \frac{24}{25}$ **7.** $\sin \theta = -\frac{12}{13}$

$\cos \theta = \frac{7}{25}$ $\cos \theta = \frac{5}{13}$

$\tan \theta = \frac{24}{7}$ $\tan \theta = -\frac{12}{5}$

$\csc \theta = \frac{25}{24}$ $\csc \theta = -\frac{13}{12}$

$\sec \theta = \frac{25}{7}$ $\sec \theta = \frac{13}{5}$

$\cot \theta = \frac{7}{24}$ $\cot \theta = -\frac{5}{12}$

9. $\sin \theta = \dfrac{5\sqrt{29}}{29}$ **11.** $\sin \theta = \dfrac{4\sqrt{41}}{41}$

$\cos \theta = -\dfrac{2\sqrt{29}}{29}$ $\cos \theta = -\dfrac{5\sqrt{41}}{41}$

$\tan \theta = -\dfrac{5}{2}$ $\tan \theta = -\dfrac{4}{5}$

$\csc \theta = \dfrac{\sqrt{29}}{5}$ $\csc \theta = \dfrac{\sqrt{41}}{4}$

$\sec \theta = -\dfrac{\sqrt{29}}{2}$ $\sec \theta = -\dfrac{\sqrt{41}}{5}$

$\cot \theta = -\dfrac{2}{5}$ $\cot \theta = -\dfrac{5}{4}$

13. Quadrant III   **15.** Quadrant I

17. $\sin \theta = \frac{3}{5}$ **19.** $\sin \theta = -\frac{15}{17}$

$\cos \theta = -\frac{4}{5}$ $\cos \theta = \frac{8}{17}$

$\tan \theta = -\frac{3}{4}$ $\tan \theta = -\frac{15}{8}$

$\csc \theta = \frac{5}{3}$ $\csc \theta = -\frac{17}{15}$

$\sec \theta = -\frac{5}{4}$ $\sec \theta = \frac{17}{8}$

$\cot \theta = -\frac{4}{3}$ $\cot \theta = -\frac{8}{15}$

21. $\sin \theta = \dfrac{\sqrt{3}}{2}$

$\cos \theta = -\dfrac{1}{2}$

$\tan \theta = -\sqrt{3}$

$\csc \theta = \dfrac{2\sqrt{3}}{3}$

$\sec \theta = -2$

$\cot \theta = -\dfrac{\sqrt{3}}{3}$

23. $\sin \theta = 0$

$\cos \theta = -1$

$\tan \theta = 0$

$\csc \theta$ is undefined.

$\sec \theta = -1$

$\cot \theta$ is undefined.

25. $\sin \theta = \dfrac{\sqrt{2}}{2}$

$\cos \theta = -\dfrac{\sqrt{2}}{2}$

$\tan \theta = -1$

$\csc \theta = \sqrt{2}$

$\sec \theta = -\sqrt{2}$

$\cot \theta = -1$

27. $\sin \theta = -\dfrac{2\sqrt{5}}{5}$

$\cos \theta = -\dfrac{\sqrt{5}}{5}$

$\tan \theta = 2$

$\csc \theta = -\dfrac{\sqrt{5}}{2}$

$\sec \theta = -\sqrt{5}$

$\cot \theta = \dfrac{1}{2}$

29. -1 **31.** 0 **33.** 1 **35.** Undefined

37. $\theta' = 60°$

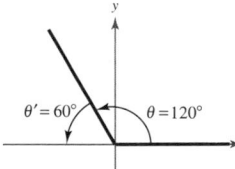

39. $\theta' = 45°$

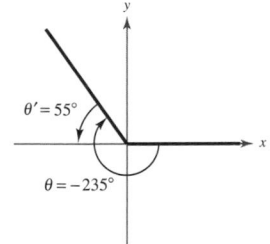

41. $\theta' = \dfrac{\pi}{3}$

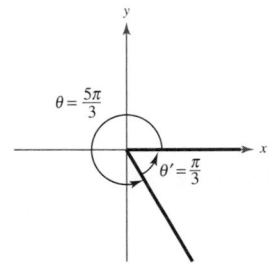

43. $\theta' = \dfrac{\pi}{6}$

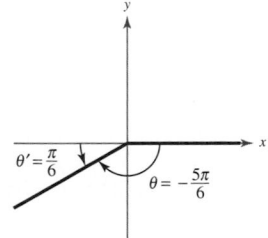

45. $\theta' = 28°$

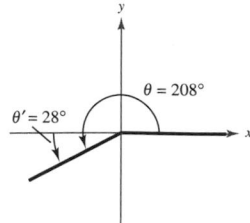

47. $\theta' = 68°$

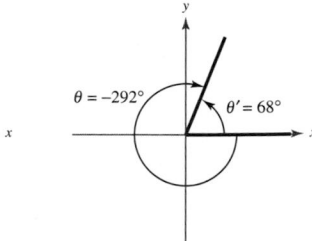

49. $\theta' = \dfrac{\pi}{5}$

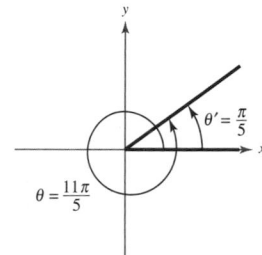

51. $\theta' = 1.342$

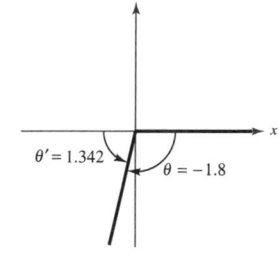

53. $\sin 225° = -\dfrac{\sqrt{2}}{2}$

$\cos 225° = -\dfrac{\sqrt{2}}{2}$

$\tan 225° = 1$

55. $\sin(-750°) = -\dfrac{1}{2}$

$\cos(-750°) = \dfrac{\sqrt{3}}{2}$

$\tan(-750°) = -\dfrac{\sqrt{3}}{3}$

57. $\sin \dfrac{5\pi}{3} = -\dfrac{\sqrt{3}}{2}$

$\cos \dfrac{5\pi}{3} = \dfrac{1}{2}$

$\tan \dfrac{5\pi}{3} = -\sqrt{3}$

59. $\sin\left(-\dfrac{\pi}{6}\right) = -\dfrac{1}{2}$

$\cos\left(-\dfrac{\pi}{6}\right) = \dfrac{\sqrt{3}}{2}$

$\tan\left(-\dfrac{\pi}{6}\right) = -\dfrac{\sqrt{3}}{3}$

61. $\sin \dfrac{11\pi}{4} = \dfrac{\sqrt{2}}{2}$

$\cos \dfrac{11\pi}{4} = -\dfrac{\sqrt{2}}{2}$

$\tan \dfrac{11\pi}{4} = -1$

63. $\sin\left(-\dfrac{17\pi}{6}\right) = -\dfrac{1}{2}$

$\cos\left(-\dfrac{17\pi}{6}\right) = -\dfrac{\sqrt{3}}{2}$

$\tan\left(-\dfrac{17\pi}{6}\right) = \dfrac{\sqrt{3}}{3}$

65. $\dfrac{4}{5}$ **67.** $-\sqrt{3}$ **69.** $\dfrac{\sqrt{65}}{4}$

CHAPTER 4

71. $\cos \theta = -\dfrac{\sqrt{21}}{5}$ **73.** $\sin \theta = \dfrac{4\sqrt{17}}{17}$

$\tan \theta = -\dfrac{2\sqrt{21}}{21}$ $\cos \theta = -\dfrac{\sqrt{17}}{17}$

$\csc \theta = \dfrac{5}{2}$ $\csc \theta = \dfrac{\sqrt{17}}{4}$

$\sec \theta = -\dfrac{5\sqrt{21}}{21}$ $\sec \theta = -\sqrt{17}$

$\cot \theta = -\dfrac{\sqrt{21}}{2}$ $\cot \theta = -\dfrac{1}{4}$

75. $\sin \theta = -\dfrac{2}{3}$

$\cos \theta = \dfrac{\sqrt{5}}{3}$

$\tan \theta = -\dfrac{2\sqrt{5}}{5}$

$\sec \theta = \dfrac{3\sqrt{5}}{5}$

$\cot \theta = -\dfrac{\sqrt{5}}{2}$

77. 0.1736   **79.** 2.1445   **81.** -0.3420

83. 5.7588   **85.** 0.8391   **87.** -2.9238

89. (a) $30° = \dfrac{\pi}{6}, 150° = \dfrac{5\pi}{6}$ (b) $210° = \dfrac{7\pi}{6}, 330° = \dfrac{11\pi}{6}$

91. (a) $60° = \dfrac{\pi}{3}, 120° = \dfrac{2\pi}{3}$ (b) $135° = \dfrac{3\pi}{4}, 315° = \dfrac{7\pi}{4}$

93. (a) $150° = \dfrac{5\pi}{6}, 210° = \dfrac{7\pi}{6}$

(b) $120° = \dfrac{2\pi}{3}, 240° = \dfrac{4\pi}{3}$

95. (a) $\dfrac{1 + \sqrt{3}}{2}$ (b) $\dfrac{\sqrt{3} - 1}{2}$ (c) $\dfrac{3}{4}$

(d) $\dfrac{\sqrt{3}}{4}$ (e) 1   (f) $\dfrac{\sqrt{3}}{2}$

97. (a) 0   (b) $\sqrt{2}$ (c) $\dfrac{1}{2}$

(d) $-\dfrac{1}{2}$ (e) $-\sqrt{2}$ (f) $\dfrac{\sqrt{2}}{2}$

99. (a) $\dfrac{1 - \sqrt{3}}{2}$ (b) $-\dfrac{1 + \sqrt{3}}{2}$ (c) $\dfrac{3}{4}$

(d) $-\dfrac{\sqrt{3}}{4}$ (e) 1   (f) $-\dfrac{\sqrt{3}}{2}$

101. (a) $-\dfrac{1 + \sqrt{3}}{2}$ (b) $\dfrac{1 - \sqrt{3}}{2}$ (c) $\dfrac{3}{4}$

(d) $\dfrac{\sqrt{3}}{4}$ (e) -1 (f) $-\dfrac{\sqrt{3}}{2}$

103. (a) $-\dfrac{1 + \sqrt{3}}{2}$ (b) $\dfrac{-1 + \sqrt{3}}{2}$ (c) $\dfrac{1}{4}$

(d) $\dfrac{\sqrt{3}}{4}$ (e) $-\sqrt{3}$ (f) $-\dfrac{1}{2}$

105. (a) -1 (b) 1   (c) 0   (d) 0   (e) -2 (f) 0
107. (a) -1 (b) 1   (c) 0   (d) 0   (e) -2 (f) 0
109. (a) 29° F   (b) 70° F   (c) 31.75° F
111. (a) 12 miles   (b) 6 miles   (c) 6.93 miles
113. True. The angles have the same reference angle.
115. (a)

θ	0°	20°	40°
$\sin \theta$	0	0.3420	0.6428
$\sin(180° - \theta)$	0	0.3420	0.6428

θ	60°	80°
$\sin \theta$	0.8660	0.9848
$\sin(180° - \theta)$	0.8660	0.9848

(b) $\sin \theta = \sin(180° - \theta)$

117. 7   **119.** $3.449, -1.449$ **121.** $4.908, -5.908$
123. -1.752 **125.** 0.002

Section 4.5   (page 304)

Vocabulary Check   (page 304)

1. amplitude   **2.** one cycle   **3.** $\dfrac{2\pi}{b}$

4. phase shift

---

1. (a) $x = -2\pi, -\pi, 0, \pi, 2\pi$

(b) $y = 0$

(c) Increasing: $\left(-2\pi, -\dfrac{3\pi}{2}\right), \left(-\dfrac{\pi}{2}, \dfrac{\pi}{2}\right), \left(\dfrac{3\pi}{2}, 2\pi\right)$

Decreasing: $\left(-\dfrac{3\pi}{2}, -\dfrac{\pi}{2}\right), \left(\dfrac{\pi}{2}, -\dfrac{3\pi}{2}\right)$

(d) Relative maxima: $\left(-\dfrac{3\pi}{2}, 1\right), \left(\dfrac{\pi}{2}, 1\right)$

Relative minima: $\left(-\dfrac{\pi}{2}, -1\right), \left(\dfrac{3\pi}{2}, -1\right)$

3. Period: π **5.** Period: 4π **7.** Period: 2
Amplitude: 3   Amplitude: $\frac{5}{2}$ Amplitude: $\frac{2}{3}$

9. Period: 2π **11.** Period: 3π
Amplitude: 2   Amplitude: $\frac{1}{4}$

13. Period: $\frac{1}{2}$ **15.** g is a shift of f π units to the right.
Amplitude: $\frac{1}{3}$

17. g is a reflection of f in the x-axis.

19. g is a reflection of f in the x-axis and has five times the amplitude of f.

21. g is a shift of f five units upward.

23. g has twice the amplitude of f.

25. g is a horizontal shift of f π units to the right.

27. 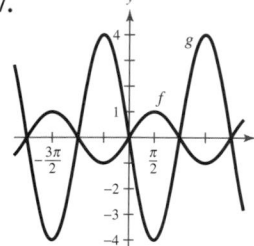   **29.**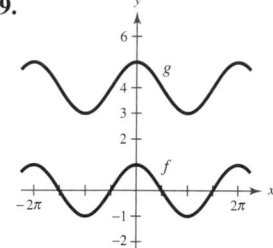

31. 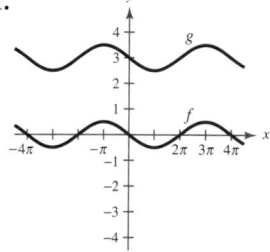   **33.**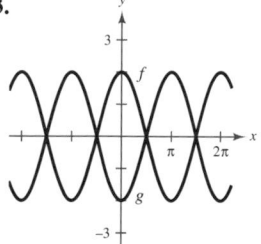

35. 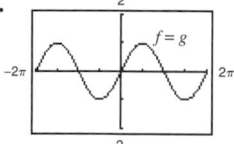   **37.**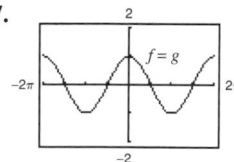

39. 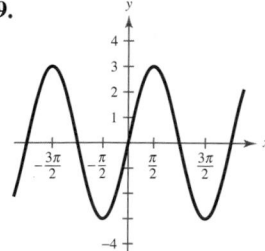   **41.**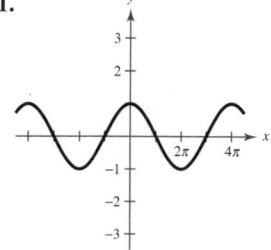

43. 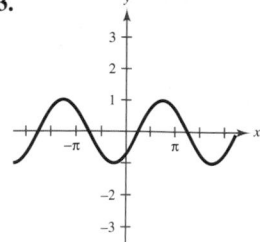   **45.**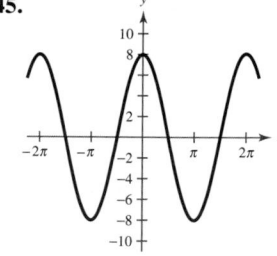

47. 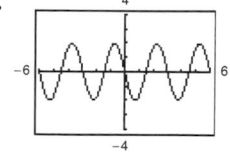   **49.**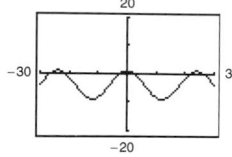
Amplitude: 2   Amplitude: 5
Period: 3   Period: 24

51. 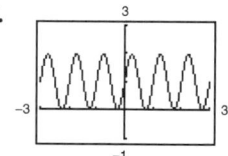   **53.**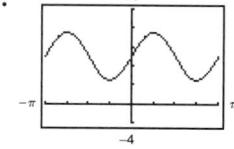
Amplitude: $\frac{2}{3}$ Amplitude: 2
Period: 4π Period: $\frac{\pi}{2}$

55. 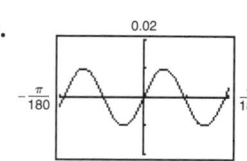   **57.**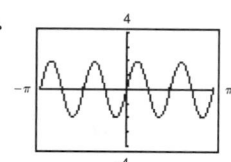
Amplitude: 1   Amplitude: 5
Period: 1   Period: π

59.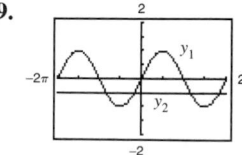
Amplitude: $\frac{1}{100}$
Period: $\frac{1}{60}$

61. $a = -4, d = 4$ **63.** $a = -6, d = 1$

65. $a = -3, b = 2, c = 0$ **67.** $a = 1, b = 1, c = \frac{\pi}{4}$

69.
$x = -\frac{\pi}{6}, -\frac{5\pi}{6}, \frac{7\pi}{6}, \frac{11\pi}{6}$

CHAPTER 4

71. (a)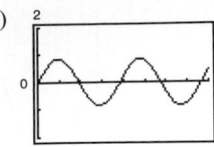

(b) 6 seconds

(c) 10 cycles per minute

(d) The period of the model would decrease because the time for a respiratory cycle would decrease.

73. (a)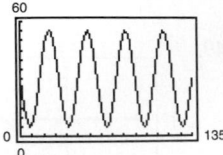

(b) Minimum height: 5 feet

Maximum height: 55 feet

75. (a) 365 days. The cycle is 1 year.

(b) 30.3 gallons per day. The average is the constant term of the model.

(c)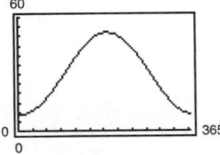

Consumption exceeds 40 gallons per day from the beginning of May through part of September.

77. (a)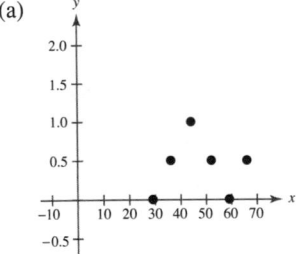

(b) $y = 0.506 \sin(0.209x - 1.336) + 0.526$

(c)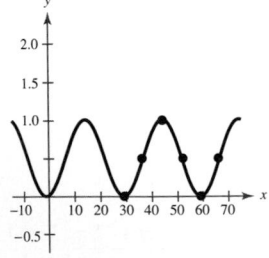

Answers will vary.

(d) 30.06

(e) 27.09%

79. True. $2\pi \cdot \dfrac{10}{3} = \dfrac{20\pi}{3}$ **81.** True   **83.** e   **85.** c

87. (a) 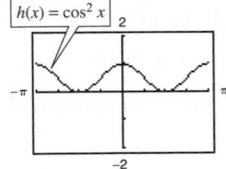   (b)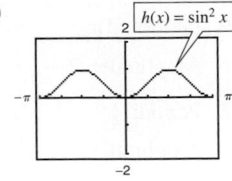

Even   Even

(c)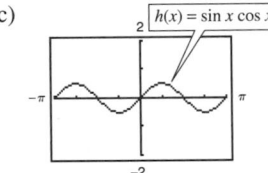

Odd

89. (a)

x	-1	-0.1	-0.01
$\dfrac{\sin x}{x}$	0.8415	0.9983	1.0000

x	-0.001	0	0.001
$\dfrac{\sin x}{x}$	1.0000	Undefined	1.0000

x	0.01	0.1	1
$\dfrac{\sin x}{x}$	1.0000	0.9983	0.8415

(b)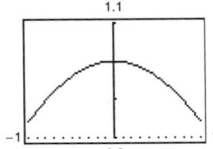

$f \to 1$ as $x \to 0$

(c) The ratio approaches 1 as x approaches 0.

91. (a)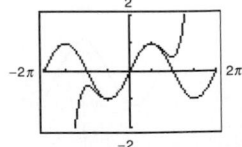

The polynomial function is a good approximation of the sine function when x is close to 0.

(b)

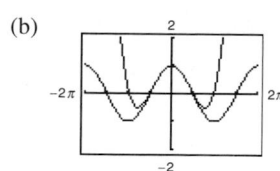

The polynomial function is a good approximation of the cosine function when x is close to 0.

(c) $\sin x \approx x - \dfrac{x^3}{3!} + \dfrac{x^5}{5!} - \dfrac{x^7}{7!}$

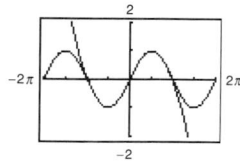

$\cos x \approx 1 - \dfrac{x^2}{2!} + \dfrac{x^4}{4!} - \dfrac{x^6}{6!}$

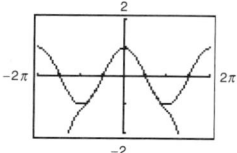

The accuracy increased.

93.

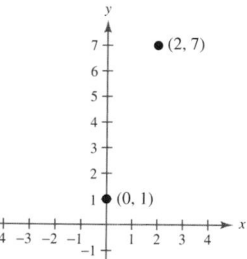

$m = 3$

95. $487.014°$ **97.** Answers will vary.

Section 4.6   (page 316)

Vocabulary Check   (page 316)

1. vertical   **2.** reciprocal   **3.** damping

1. (a) $x = -2\pi, -\pi, 0, \pi, 2\pi$ (b) $y = 0$

(c) Increasing on $\left(-2\pi, -\dfrac{3\pi}{2}\right), \left(-\dfrac{3\pi}{2}, -\dfrac{\pi}{2}\right), \left(-\dfrac{\pi}{2}, \dfrac{\pi}{2}\right),$

$\left(\dfrac{\pi}{2}, \dfrac{3\pi}{2}\right), \left(\dfrac{3\pi}{2}, 2\pi\right)$

(d) No relative extrema

(e) $x = -\dfrac{3\pi}{2}, -\dfrac{\pi}{2}, \dfrac{\pi}{2}, \dfrac{3\pi}{2}$

3. (a) No x-intercepts

(b) $y = 0$

(c) Increasing on $\left(-2\pi, -\dfrac{3\pi}{2}\right), \left(-\dfrac{3\pi}{2}, -\pi\right),$

$\left(0, \dfrac{\pi}{2}\right), \left(\dfrac{\pi}{2}, \pi\right)$

Decreasing on $\left(-\pi, -\dfrac{\pi}{2}\right), \left(-\dfrac{\pi}{2}, 0\right),$

$\left(\pi, \dfrac{3\pi}{2}\right), \left(\dfrac{3\pi}{2}, 2\pi\right)$

(d) Relative minima: $(-2\pi, 1), (0, 1), (2\pi, 1)$

Relative maxima: $(-\pi, -1), (\pi, -1)$

(e) $x = -\dfrac{3\pi}{2}, -\dfrac{\pi}{2}, \dfrac{\pi}{2}, \dfrac{3\pi}{2}$

5.

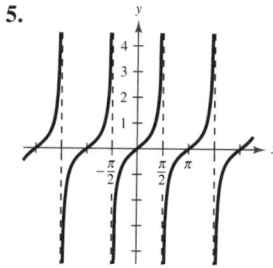

7.

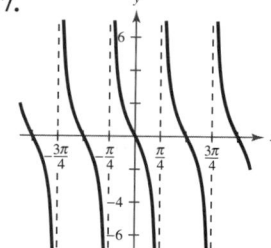

9.

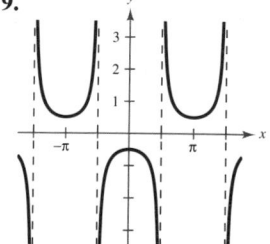

11.

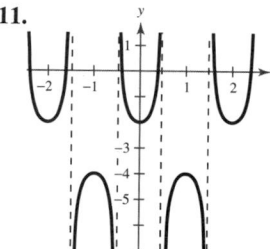

13.

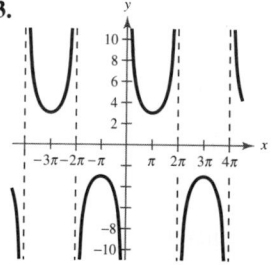

15.

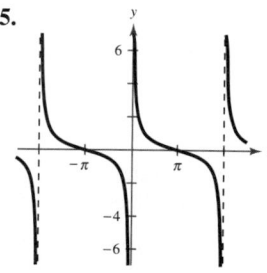

17.

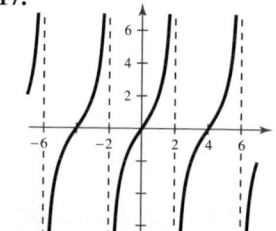

19.

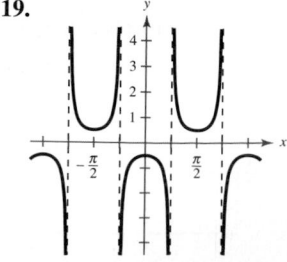

CHAPTER 4

21. 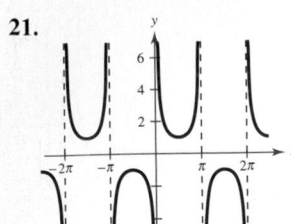 **23.**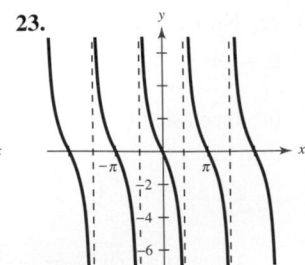

25.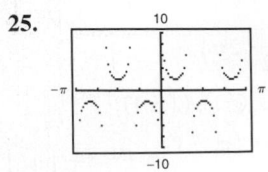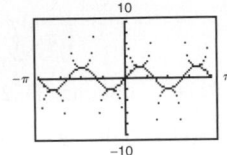

Answers will vary.

27.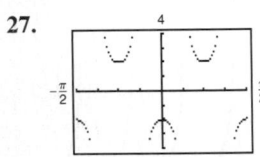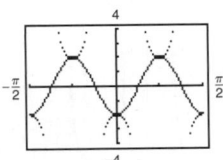

Answers will vary.

29.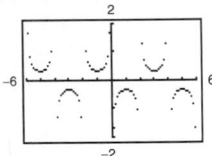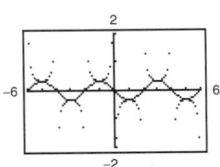

Answers will vary.

31. $-5.498, -2.356, 0.785, 3.927$

33. $-4.189, -2.094, 2.094, 4.189$

35. Even   **37.** Odd

39. 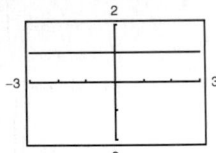   **41.**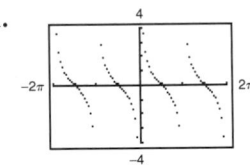

Not equivalent;   Equivalent
y_1 is undefined at $x = 0$.

43. d; as x approaches 0, $f(x)$ approaches 0.

44. a; as x approaches 0, $f(x)$ approaches 0.

45. b; as x approaches 0, $g(x)$ approaches 0.

46. c; as x approaches 0, $g(x)$ approaches 0.

47. 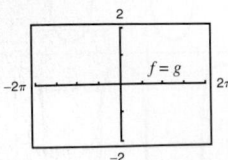   **49.**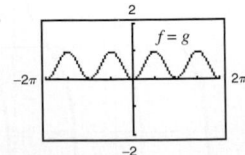

51. 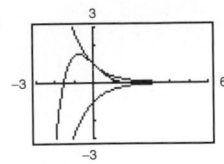   **53.**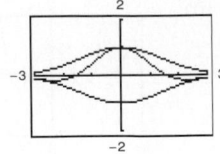

$f \to 0$ as $x \to \infty$ $h \to 0$ as $x \to \infty$

55. (a) $f \to -\infty$ (b) $f \to \infty$ (c) $f \to -\infty$ (d) $f \to \infty$

57. (a) $f \to \infty$ (b) $f \to -\infty$ (c) $f \to \infty$ (d) $f \to -\infty$

59. As the predator population increases, the number of prey decreases. When the number of prey is small, the number of predators decreases.

61. $d = 5 \cot x$

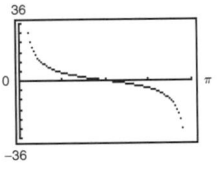

63. (a)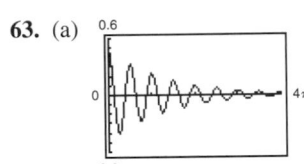

(b) Not periodic and damped; approaches 0 as t increases.

65. True

67. (a)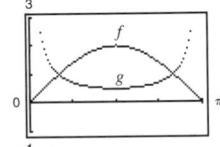

(b) $0.524 < x < 2.618$

(c) f approaches 0 and g approaches ∞, because g is the reciprocal of f.

69. (a)

x	-1	-0.1	-0.01
$\dfrac{\tan x}{x}$	1.5574	1.0033	1.0000

x	-0.001	0	0.001
$\dfrac{\tan x}{x}$	1.0000	Undefined	1.0000

x	0.01	0.1	1
$\dfrac{\tan x}{x}$	1.0000	1.0033	1.5574

(b)

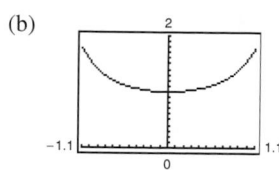

$f \to 1$ as $x \to 0$

(c) The ratio approaches 1 as x approaches 0.

71. a

73.

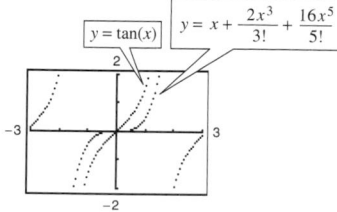

The polynomial function is a good approximation of the tangent function when x is close to 0.

75. Distributive Property   **77.** Additive Identity Property

79. Not one-to-one

81. One-to-one. $f^{-1}(x) = \dfrac{x^2 + 14}{3}, \ x \geq 0$

Section 4.7   (page 327)

Vocabulary Check   (page 327)

1. $y = \sin^{-1} x, \ -1 \leq x \leq 1$

2. $y = \arccos x, \ 0 \leq y \leq \pi$

3. $y = \tan^{-1} x, \ -\infty < x < \infty, \ -\dfrac{\pi}{2} < y < \dfrac{\pi}{2}$

1. (a) $\dfrac{\pi}{6}$ (b) 0   **3.** (a) $\dfrac{\pi}{2}$ (b) 0

5. (a) $\dfrac{\pi}{6}$ (b) $-\dfrac{\pi}{4}$ **7.** (a) $-\dfrac{\pi}{3}$ (b) $\dfrac{\pi}{3}$

9. (a) $\dfrac{\pi}{3}$ (b) $-\dfrac{\pi}{6}$

11. (a)

x	-1	-0.8	-0.6	-0.4	-0.2	
y	3.142	2.498	2.214	1.982	1.772	

x	0	0.2	0.4	0.6	0.8	1
y	1.571	1.369	1.159	0.927	0.644	0

(b)

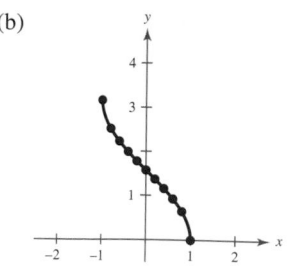

(c)

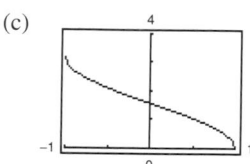

They are the same.

(d) Intercepts: $\left(0, \dfrac{\pi}{2}\right), (1, 0)$

13. $\left(-\sqrt{3}, -\dfrac{\pi}{3}\right), \left(-\dfrac{\sqrt{3}}{3}, -\dfrac{\pi}{6}\right), \left(1, \dfrac{\pi}{4}\right)$ **15.** 0.72

17. -0.85 **19.** -1.41 **21.** $\theta = \arctan \dfrac{x}{8}$

23. $\theta = \arcsin \dfrac{x + 2}{5}$

25. $\sqrt{4 - x^2}; \ \theta = \arcsin \dfrac{x}{2}, \ \theta = \arccos \dfrac{\sqrt{4 - x^2}}{2},$

$\theta = \arctan \dfrac{x}{\sqrt{4 - x^2}}$

27. $\sqrt{x^2 + 2x + 5}; \ \theta = \arcsin \dfrac{x + 1}{\sqrt{x^2 + 2x + 5}},$

$\theta = \arccos \dfrac{2}{\sqrt{x^2 + 2x + 5}}, \ \theta = \arctan \dfrac{x + 1}{2}$

29. 0.7   **31.** -0.3 **33.** 0   **35.** $-\dfrac{\pi}{6}$ **37.** $\dfrac{\pi}{2}$

39. $\dfrac{\pi}{2}$ **41.** 0   **43.** $\dfrac{\sqrt{2}}{2}$ **45.** $\dfrac{\pi}{3}$ **47.** $\dfrac{4}{5}$ **49.** $\dfrac{7}{25}$

51. $\dfrac{\sqrt{34}}{5}$ **53.** $\dfrac{\sqrt{5}}{3}$ **55.** $\dfrac{1}{x}$ **57.** $\sqrt{-x^2 - 4x - 3}$

59. $\dfrac{\sqrt{25 - x^2}}{x}$ **61.** $\dfrac{\sqrt{x^2 + 7}}{x}$

63. $\dfrac{14}{\sqrt{x^2 + 196}}$ **65.** $\dfrac{|x - 1|}{\sqrt{x^2 - 2x + 10}}$

67.

69.

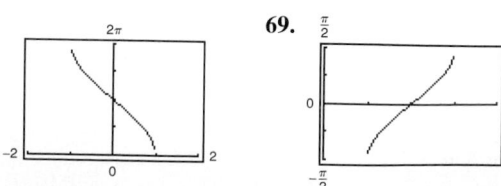

CHAPTER 4

71.

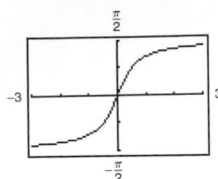

73. $3\sqrt{2}\sin\left(2t + \dfrac{\pi}{4}\right)$

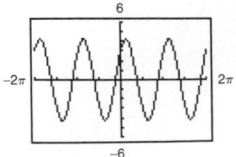

The two forms are equivalent.

75. $\dfrac{\pi}{2}$ **77.** $\dfrac{\pi}{2}$ **79.** π

81. (a) $\theta = \arcsin\dfrac{10}{s}$ (b) 0.19 rad, 0.39 rad

83. (a) $\theta = \arctan\dfrac{s}{750}$ (b) 0.49 rad, 1.13 rad

85. (a) $\theta = \arctan\dfrac{6}{x}$ (b) 0.54 rad, 1.11 rad

87. False. $5\pi/6$ is not in the range of the arcsine function.

89.   **91.**

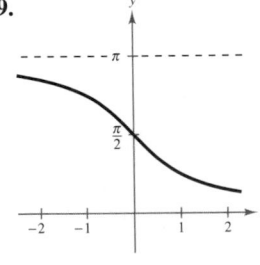   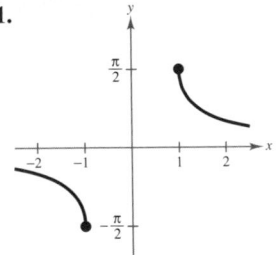

93. $\dfrac{\pi}{4}$ **95.** $\dfrac{5\pi}{6}$ **97.** Proof   **99.** Proof

101. $\dfrac{\sqrt{2}}{2}$ **103.** $\dfrac{\sqrt{3}}{3}$

105. $\cos\theta = \dfrac{\sqrt{11}}{6}$

$\tan\theta = \dfrac{5\sqrt{11}}{11}$

$\csc\theta = \dfrac{6}{5}$

$\sec\theta = \dfrac{6\sqrt{11}}{11}$

$\cot\theta = \dfrac{\sqrt{11}}{5}$

107. $\cos\theta = \dfrac{\sqrt{7}}{4}$

$\tan\theta = \dfrac{3\sqrt{7}}{7}$

$\csc\theta = \dfrac{4}{3}$

$\sec\theta = \dfrac{4\sqrt{7}}{7}$

$\cot\theta = \dfrac{\sqrt{7}}{3}$

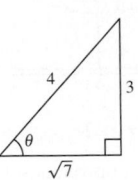

Section 4.8   (page 337)

Vocabulary Check   (page 337)

1. elevation, depression   2. bearing

3. harmonic motion

1. $B = 60°$ **3.** $A = 19°$ **5.** $A \approx 26.57°$
   $a \approx 5.77$ $a \approx 4.82$ $B \approx 63.43°$
   $c \approx 11.55$ $c \approx 14.81$ $c \approx 13.42$

7. $A \approx 72.76°$ **9.** $B = 77°45'$
   $B \approx 17.24°$ $a \approx 91.34$
   $a \approx 51.58$ $b \approx 420.70$

11. 5.12 inches   **13.** 8.21 feet

15. (a)

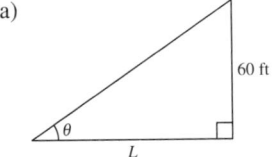

(b) $L = 60\cot\theta$

(c)

θ	10°	20°	30°	40°	50°
L	340.28	164.85	103.92	71.51	50.35

(d) No. The cotangent is not a linear function.

17. 19.70 feet   **19.** 76.60 feet

21. (a)

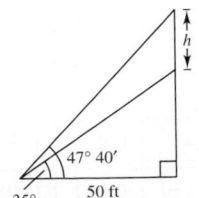

(b) $h = 50(\tan 47°40' - \tan 35°)$ (c) 19.87 feet

23. (a) $l = \sqrt{h^2 + 34h + 10{,}289}$

(b) $\theta = \arccos \dfrac{100}{l}$

(c) 53 feet

25. $38.29°$ **27.** $75.97°$

29. 5099 feet   **31.** 0.66 mile

33. 104.95 nautical miles south, 58.18 nautical miles west

35. (a) N 58° E   (b) 68.82 meters

37. N 56.31° W   **39.** 1933.32 feet   **41.** 3.23 miles

43. (a) $54.46°$; $20.47°$ (b) 56 feet

45. $78.69°$ **47.** $35.26°$ **49.** $y = \sqrt{3}\,r$

51. $d = 8 \sin \pi t$ **53.** $d = 3 \cos\left(\dfrac{4\pi}{3}t\right)$

55. (a) 4   (b) 4   (c) 4   (d) $\frac{1}{16}$

57. (a) $\frac{1}{16}$ (b) 70   (c) 0   (d) $\frac{1}{140}$ **59.** $\omega = 528\pi$

61. (a)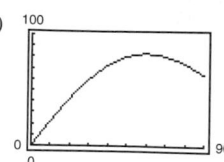

(b) $\dfrac{\pi}{8}$ second

(c) $\dfrac{\pi}{32}$ second

63. (a)

Base 1	Base 2	Altitude	Area
8	8 + 16 cos 10°	8 sin 10°	22.06
8	8 + 16 cos 20°	8 sin 20°	42.46
8	8 + 16 cos 30°	8 sin 30°	59.71
8	8 + 16 cos 40°	8 sin 40°	72.65
8	8 + 16 cos 50°	8 sin 50°	80.54
8	8 + 16 cos 60°	8 sin 60°	83.14
8	8 + 16 cos 70°	8 sin 70°	80.71

(b)

Base 1	Base 2	Altitude	Area
8	8 + 16 cos 56°	8 sin 56°	82.73
8	8 + 16 cos 58°	8 sin 58°	83.04
8	8 + 16 cos 59°	8 sin 59°	83.11
8	8 + 16 cos 60°	8 sin 60°	83.14
8	8 + 16 cos 61°	8 sin 61°	83.11
8	8 + 16 cos 62°	8 sin 62°	83.04

83.14 square feet

(c) $A = 64(1 + \cos \theta)(\sin \theta)$

(d)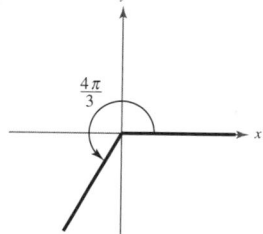

83.14 square feet; They are the same.

65. (a)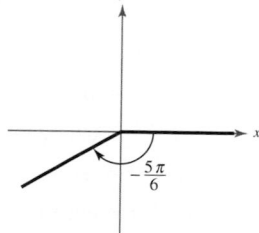

(b) 12 months. Yes; one period is 1 year.

(c) 1.41 hours; maximum displacement of 1.41 hours from the average sunset time of 18.09

67. False. $a = \dfrac{22.56}{\tan 48.1°}$ **69.** $4x - y + 6 = 0$

71. $4x + 5y - 22 = 0$ **73.** All real numbers x

75. All real numbers x

Review Exercises   (page 344)

1. 1 rad

3. (a)

(b) Quadrant III

(c) $\dfrac{10\pi}{3}$, $-\dfrac{2\pi}{3}$

5. (a)

(b) Quadrant III

(c) $\dfrac{7\pi}{6}$, $-\dfrac{17\pi}{6}$

7. Complement: $\dfrac{3\pi}{8}$; supplement: $\dfrac{7\pi}{8}$

CHAPTER 4

9. Complement: $\frac{\pi}{5}$; supplement: $\frac{7\pi}{10}$

11. (a)

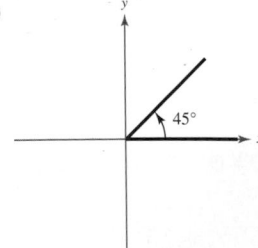

(b) Quadrant I

(c) $405°, -315°$

13. (a)

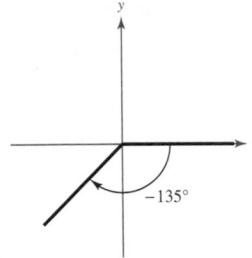

(b) Quadrant III

(c) $225°, -495°$

15. Complement: $85°$; supplement: $175°$

17. Complement: none; supplement: $9°$ **19.** $135.279°$

21. $5.381°$ **23.** $135°\,17'24''$ **25.** $-85°\,21'36''$

27. 7.243 **29.** -1.257 **31.** $128.571°$

33. $-200.535°$ **35.** $\frac{25}{12}$ rad   **37.** $\frac{46\pi}{3}$ meters

39. $6000\,\pi$ centimeters per minute

41. $\left(\frac{\sqrt{2}}{2}, -\frac{\sqrt{2}}{2}\right)$ **43.** $\left(-\frac{\sqrt{3}}{2}, \frac{1}{2}\right)$

45. $\left(-\frac{1}{2}, -\frac{\sqrt{3}}{2}\right)$ **47.** $\left(-\frac{\sqrt{2}}{2}, \frac{\sqrt{2}}{2}\right)$

49. $\sin\frac{7\pi}{6} = -\frac{1}{2}$

$\cos\frac{7\pi}{6} = -\frac{\sqrt{3}}{2}$

$\tan\frac{7\pi}{6} = \frac{\sqrt{3}}{3}$

$\csc\frac{7\pi}{6} = -2$

$\sec\frac{7\pi}{6} = -\frac{2\sqrt{3}}{3}$

$\cot\frac{7\pi}{6} = \sqrt{3}$

51. $\sin 2\pi = 0$

$\cos 2\pi = 1$

$\tan 2\pi = 0$

$\csc 2\pi$ is undefined.

$\sec 2\pi = 1$

$\cot 2\pi$ is undefined.

53. $\sin\left(-\frac{11\pi}{6}\right) = \frac{1}{2}$

$\cos\left(-\frac{11\pi}{6}\right) = \frac{\sqrt{3}}{2}$

$\tan\left(-\frac{11\pi}{6}\right) = \frac{\sqrt{3}}{3}$

$\csc\left(-\frac{11\pi}{6}\right) = 2$

$\sec\left(-\frac{11\pi}{6}\right) = \frac{2\sqrt{3}}{3}$

$\cot\left(-\frac{11\pi}{6}\right) = \sqrt{3}$

55. $\sin\left(-\frac{\pi}{2}\right) = -1$

$\cos\left(-\frac{\pi}{2}\right) = 0$

$\tan\left(-\frac{\pi}{2}\right)$ is undefined.

$\csc\left(-\frac{\pi}{2}\right) = -1$

$\sec\left(-\frac{\pi}{2}\right)$ is undefined.

$\cot\left(-\frac{\pi}{2}\right) = 0$

57. $\frac{\sqrt{2}}{2}$ **59.** $-\frac{1}{2}$ **61.** (a) $-\frac{3}{5}$ (b) $-\frac{5}{3}$

63. (a) $\frac{2}{3}$ (b) $\frac{3}{2}$ **65.** -0.8935 **67.** 0.5

69. $\sin\theta = \frac{\sqrt{65}}{9}$

$\cos\theta = \frac{4}{9}$

$\tan\theta = \frac{\sqrt{65}}{4}$

$\csc\theta = \frac{9\sqrt{65}}{65}$

$\sec\theta = \frac{9}{4}$

$\cot\theta = \frac{4\sqrt{65}}{65}$

71. $\sin\theta = \frac{5\sqrt{61}}{61}$

$\cos\theta = \frac{6\sqrt{61}}{61}$

$\tan\theta = \frac{5}{6}$

$\csc\theta = \frac{\sqrt{61}}{5}$

$\sec\theta = \frac{\sqrt{61}}{6}$

$\cot\theta = \frac{6}{5}$

73. Answers will vary.   **75.** (a) 0.1045 (b) 0.1045

77. (a) 0.7071 (b) 1.4142 **79.** 235 feet

81. $\sin\theta = \frac{4}{5}$

$\cos\theta = \frac{3}{5}$

$\tan\theta = \frac{4}{3}$

$\csc\theta = \frac{5}{4}$

$\sec\theta = \frac{5}{3}$

$\cot\theta = \frac{3}{4}$

83. $\sin\theta = \frac{2\sqrt{53}}{53}$

$\cos\theta = -\frac{7\sqrt{53}}{53}$

$\tan\theta = -\frac{2}{7}$

$\csc\theta = \frac{\sqrt{53}}{2}$

$\sec\theta = -\frac{\sqrt{53}}{7}$

$\cot\theta = -\frac{7}{2}$

85. $\sin\theta = \dfrac{15\sqrt{481}}{481}$

$\cos\theta = \dfrac{16\sqrt{481}}{481}$

$\tan\theta = \dfrac{15}{16}$

$\csc\theta = \dfrac{\sqrt{481}}{15}$

$\sec\theta = \dfrac{\sqrt{481}}{16}$

$\cot\theta = \dfrac{16}{15}$

89. $\cos\theta = -\dfrac{\sqrt{55}}{8}$

$\tan\theta = -\dfrac{3\sqrt{55}}{55}$

$\csc\theta = \dfrac{8}{3}$

$\sec\theta = -\dfrac{8\sqrt{55}}{55}$

$\cot\theta = -\dfrac{\sqrt{55}}{3}$

87. $\sin\theta = -\dfrac{\sqrt{11}}{6}$

$\cos\theta = \dfrac{5}{6}$

$\tan\theta = -\dfrac{\sqrt{11}}{5}$

$\csc\theta = -\dfrac{6\sqrt{11}}{11}$

$\cot\theta = -\dfrac{5\sqrt{11}}{11}$

91. $\theta' = 84°$ **93.** $\theta' = \dfrac{\pi}{5}$

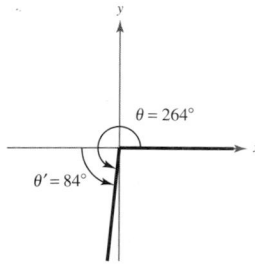 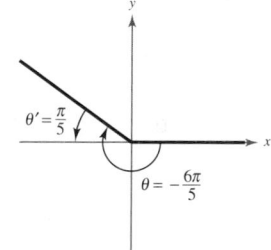

95. $\sin 240° = -\dfrac{\sqrt{3}}{2}$

$\cos 240° = -\dfrac{1}{2}$

$\tan 240° = \sqrt{3}$

97. $\sin(-210°) = \dfrac{1}{2}$

$\cos(-210°) = -\dfrac{\sqrt{3}}{2}$

$\tan(-210°) = -\dfrac{\sqrt{3}}{3}$

99. $\sin\left(-\dfrac{9\pi}{4}\right) = -\dfrac{\sqrt{2}}{2}$

$\cos\left(-\dfrac{9\pi}{4}\right) = \dfrac{\sqrt{2}}{2}$

$\tan\left(-\dfrac{9\pi}{4}\right) = -1$

101. $\sin 4\pi = 0$

$\cos 4\pi = 1$

$\tan 4\pi = 0$

103. 0.6494   **105.** 3.2361

107. 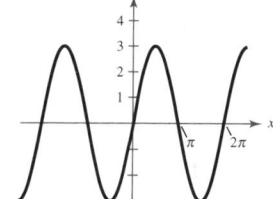   **109.**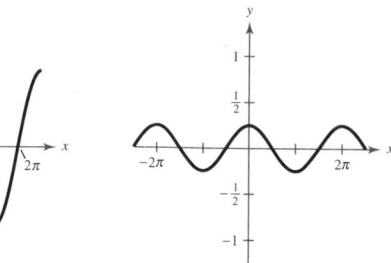

111. Period: 2; amplitude: 5

113. Period: π; amplitude: 3.4

115. 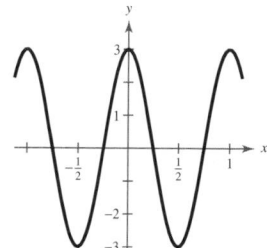   **117.**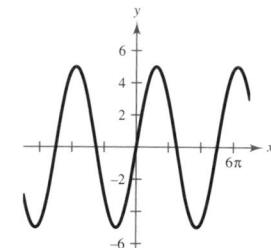

119. 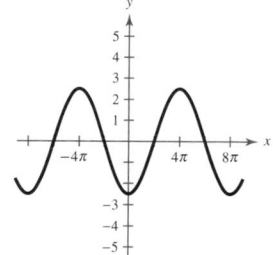   **121.**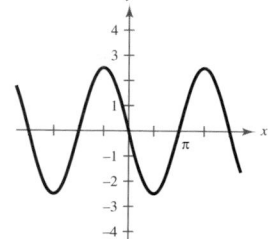

123. 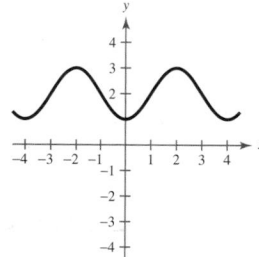   **125.**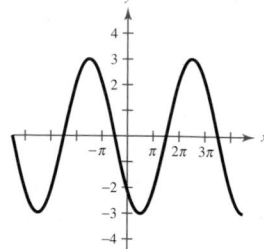

127. $a = -2, b = 1, c = \dfrac{\pi}{4}$ **129.** $a = -4, b = 2, c = \dfrac{\pi}{2}$

CHAPTER 4

131.

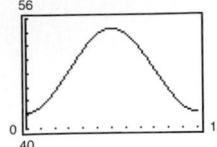

Maximum sales: June
Minimum sales: December

133.

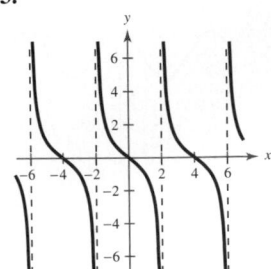

135.

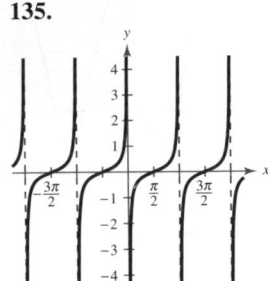

137.

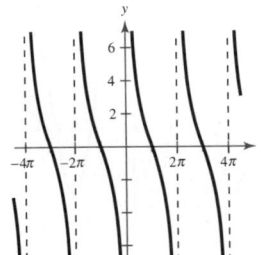

139.

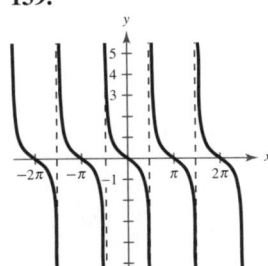

141.

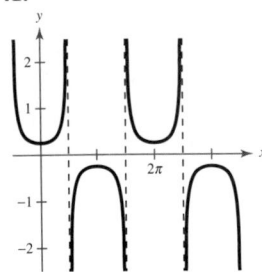

143.

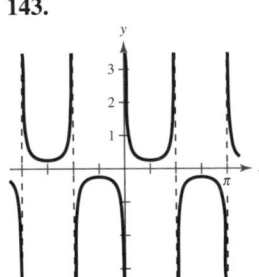

145.

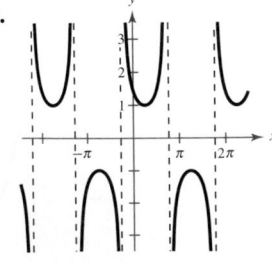

147.

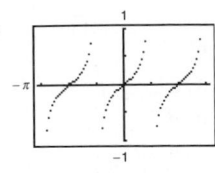

149.

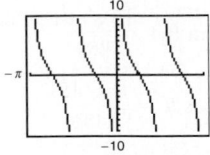

151.

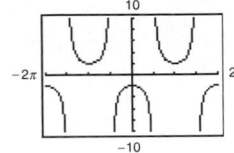

153.

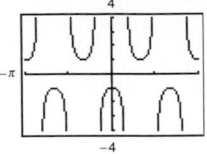

155.

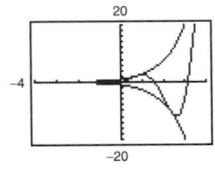

157.

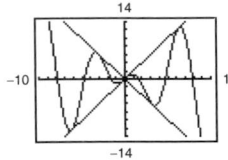

As $x \to \infty$, f oscillates
between $-\infty$ and ∞.

As $x \to \infty$, f oscillates
between $-2x$ and $2x$.

159. (a) $-\dfrac{\pi}{2}$ (b) Does not exist   **161.** (a) $\dfrac{\pi}{4}$ (b) $\dfrac{5\pi}{6}$

163. 1.14   **165.** -1.22 **167.** -1.49

169. 0.68   **171.** $\theta = \arcsin\left(\dfrac{x+3}{16}\right)$ **173.** $\dfrac{1}{\sqrt{2x - x^2}}$

175. $\dfrac{2\sqrt{4 - 2x^2}}{4 - x^2}$ **177.** 0.071 kilometer

179. 9.47 miles   **181.** $y = 3\cos\left(\dfrac{2\pi t}{15}\right)$

183. False. y is a function but is not one-to-one on
$30° \le \theta \le 150°$.

185. (a)

s	10	20	30
θ	0.0224	0.0894	0.1989

s	40	50	60
θ	0.3441	0.5105	0.6786

(b) θ is not a linear function of s.

Chapter Test   (page 349)

1. (a)

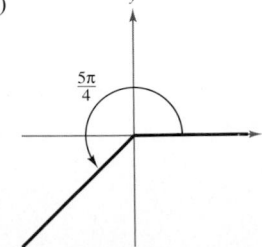

(b) Answers will vary.
    Sample answer:

$\dfrac{13\pi}{4}$, $-\dfrac{3\pi}{4}$

(c) $225°$

2. 2400 radians per minute

3. $\sin \theta = \dfrac{4\sqrt{17}}{17}$

$\cos \theta = -\dfrac{\sqrt{17}}{17}$

$\tan \theta = -4$

$\csc \theta = \dfrac{\sqrt{17}}{4}$

$\sec \theta = -\sqrt{17}$

$\cot \theta = -\dfrac{1}{4}$

4. $\sin \theta = \dfrac{7\sqrt{53}}{53}$

$\cos \theta = \dfrac{2\sqrt{53}}{53}$

$\csc \theta = \dfrac{\sqrt{53}}{7}$

$\sec \theta = \dfrac{\sqrt{53}}{2}$

$\cot \theta = \dfrac{2}{7}$

5. $\theta' = 75°$

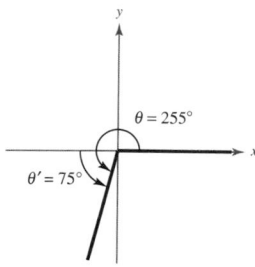

6. Quadrant III   **7.** $135°, 225°$ **8.** $1.33, 1.81$

9. $\sin \theta = \dfrac{4}{5}$

$\tan \theta = -\dfrac{4}{3}$

$\csc \theta = \dfrac{5}{4}$

$\sec \theta = -\dfrac{5}{3}$

$\cot \theta = -\dfrac{3}{4}$

10.

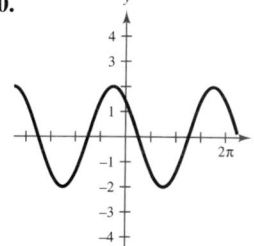

11.

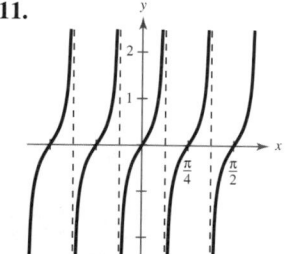

12.

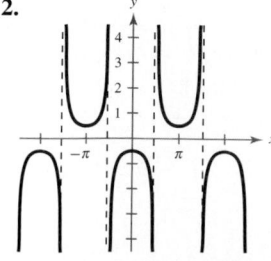

13.

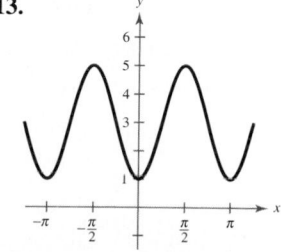

14.

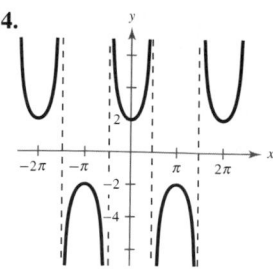

15.

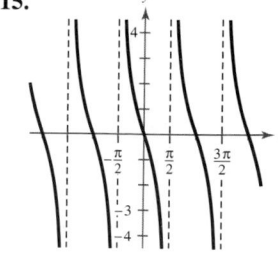

16.

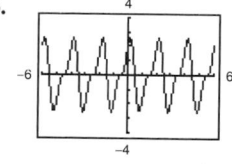

Period: 2

17.

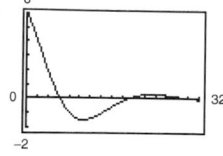

Not periodic

18. $a = -2, b = \dfrac{1}{2}, c = -\dfrac{\pi}{4}$ **19.** $\dfrac{\sqrt{5}}{2}$

20.

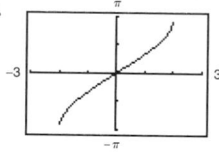

21.

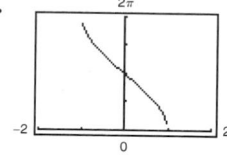

22.

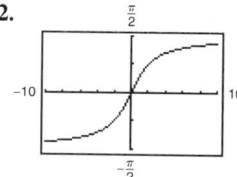

23. $214.51°$

Chapter 5

Section 5.1   (page 357)

1. $\tan x = \dfrac{\sqrt{3}}{3}$

$\csc x = 2$

$\sec x = \dfrac{2\sqrt{3}}{3}$

$\cot x = \sqrt{3}$

3. $\cos \theta = \dfrac{\sqrt{2}}{2}$

$\tan \theta = -1$

$\csc \theta = -\sqrt{2}$

$\cot \theta = -1$

CHAPTER 5

5. $\sin x = -\frac{7}{25}$

$\cos x = -\frac{24}{25}$

$\csc x = -\frac{25}{7}$

$\cot x = \frac{24}{7}$

7. $\cos \phi = -\frac{15}{17}$

$\tan \phi = -\frac{8}{15}$

$\csc \phi = \frac{17}{8}$

$\cot \phi = -\frac{15}{8}$

9. $\sin x = \frac{2}{3}$

$\cos x = -\frac{\sqrt{5}}{3}$

$\csc x = \frac{3}{2}$

$\sec x = -\frac{3\sqrt{5}}{5}$

$\cot x = -\frac{\sqrt{5}}{2}$

11. $\sin \theta = -\frac{2\sqrt{5}}{5}$

$\cos \theta = -\frac{\sqrt{5}}{5}$

$\csc \theta = -\frac{\sqrt{5}}{2}$

$\sec \theta = -\sqrt{5}$

$\cot \theta = \frac{1}{2}$

13. $\sin \theta = 0$

$\cos \theta = -1$

$\tan \theta = 0$

$\sec \theta = -1$

$\cot \theta$ is undefined.

15. d   **16.** a   **17.** b   **18.** f   **19.** e

20. c   **21.** b   **22.** c   **23.** f   **24.** a

25. e   **26.** d   **27.** $\cos x$ **29.** $\cos^2 \phi$

31. $\sec x$ **33.** 1   **35.** $\cot x$ **37.** $1 + \sin y$

39–49. Answers will vary.   **51.** $\cos^2 x$ **53.** $\cos x + 2$

55. $\sec^4 x$ **57.** $\sin^2 x - \cos^2 x$ **59.** $(\csc x - 1) \cot^2 x$

61. $1 + 2 \sin x \cos x$ **63.** $\cot^2 x$ **65.** $2 \csc^2 x$

67. $-\cot x$ **69.** $1 + \cos y$ **71.** $3(\sec x + \tan x)$

73.

x	0.2	0.4	0.6	0.8
y_1	0.1987	0.3894	0.5646	0.7174
y_2	0.1987	0.3894	0.5646	0.7174

x	1.0	1.2	1.4
y_1	0.8415	0.9320	0.9854
y_2	0.8415	0.9320	0.9854

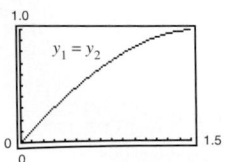

75.

x	0.2	0.4	0.6	0.8
y_1	1.2230	1.5085	1.8958	2.4650
y_2	1.2230	1.5085	1.8958	2.4650

x	1.0	1.2	1.4
y_1	3.4082	5.3319	11.6814
y_2	3.4082	5.3319	11.6814

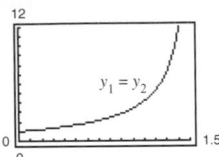

77. $\csc x$ **79.** $\tan x$ **81.** $5 \cos \theta$ **83.** $3 \tan \theta$

85. $3 \cos \theta$ **87.** $3 \sec \theta$ **89.** $3 \tan \theta$ **91.** $\sqrt{2} \cos \theta$

93. $0 \le \theta \le \pi$ **95.** $0 \le \theta < \frac{\pi}{2}, \frac{3\pi}{2} < \theta < 2\pi$

97. $\ln|\cot \theta|$ **99.** $\ln|(\cos x)(1 + \sin x)|$

101. The identity is not true when $\theta = \frac{7\pi}{6}$.

103. The identity is not true when $\theta = \frac{5\pi}{3}$.

105. The identity is not true when $\theta = \frac{7\pi}{4}$.

107. (a) and (b) Answers will vary.

109. (a) and (b) Answers will vary.

111. Answers will vary.   **113.** True, for all $\theta \ne n\pi$

115. 1, 1   **117.** ∞, 0

119. $\cos \theta = \pm \sqrt{1 - \sin^2 \theta}$

$\tan \theta = \pm \frac{\sin \theta}{\sqrt{1 - \sin^2 \theta}}$

$\csc \theta = \frac{1}{\sin \theta}$

$\sec \theta = \pm \frac{1}{\sqrt{1 - \sin^2 \theta}}$

$\cot \theta = \pm \frac{\sqrt{1 - \sin^2 \theta}}{\sin \theta}$

The sign depends on the choice of θ.

121. Answers will vary.

123. 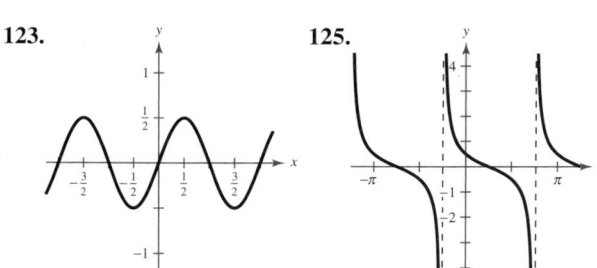 **125.**

Section 5.2 (page 365)

Vocabulary Check (page 365)

1. conditional **2.** identity **3.** $\cot u$ **4.** $\sin u$

5. $\tan u$ **6.** $\cos u$ **7.** $\cos^2 u$ **8.** $\cot u$

9. $-\sin u$ **10.** $\sec u$

1–9. Answers will vary.

11.

x	0.2	0.4	0.6	0.8
y_1	4.8348	2.1785	1.2064	0.6767
y_2	4.8348	2.1785	1.2064	0.6767

x	1.0	1.2	1.4
y_1	0.3469	0.1409	0.0293
y_2	0.3469	0.1409	0.0293

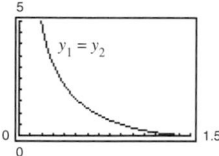

13.

x	0.2	0.4	0.6	0.8
y_1	4.8348	2.1785	1.2064	0.6767
y_2	4.8348	2.1785	1.2064	0.6767

x	1.0	1.2	1.4
y_1	0.3469	0.1409	0.0293
y_2	0.3469	0.1409	0.0293

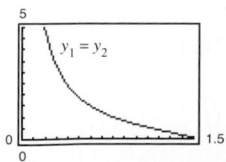

15.

x	0.2	0.4	0.6	0.8
y_1	5.0335	2.5679	1.7710	1.3940
y_2	5.0335	2.5679	1.7710	1.3940

x	1.0	1.2	1.4
y_1	1.1884	1.0729	1.0148
y_2	1.1884	1.0729	1.0148

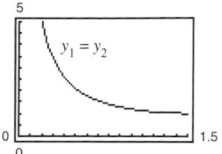

17.

x	0.2	0.4	0.6	0.8
y_1	5.1359	2.7880	2.1458	2.0009
y_2	5.1359	2.7880	2.1458	2.0009

x	1.0	1.2	1.4
y_1	2.1995	2.9609	5.9704
y_2	2.1995	2.9609	5.9704

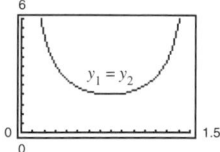

19. $\cot(-x) = -\cot(x)$

21–49. Answers will vary.

51. 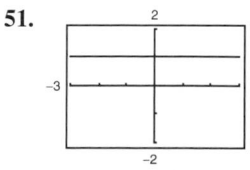 **53.**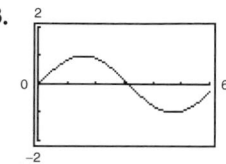

$y = 1$ $y = \sin x$

55. Answers will vary. **57.** Answers will vary.

59. 1 **61.** 2 **63–69.** Answers will vary.

71. $\mu = \tan \theta$, $W \neq 0$ **73.** True

75. False. $\sin^2\left(\dfrac{\pi}{4}\right) + \cos^2\left(\dfrac{\pi}{4}\right) \neq 1 + \tan^2\left(\dfrac{\pi}{4}\right)$

77. (a) Answers will vary. (b) No. Division by zero.

79. (a) Answers will vary. (b) Yes. Answers will vary.

81. $a \cos \theta$ **83.** $a \sec \theta$ **85.** $\sqrt{\tan^2 x} = |\tan x|$; $\dfrac{3\pi}{4}$

87. Answers will vary.   **89.** $f(x) = x^3 - x^2 + 64x - 64$

91. $f(x) = x^3 - 16x^2 + 85x - 148$

93. 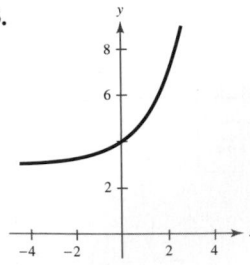   **95.**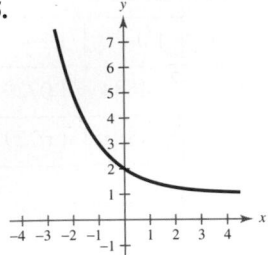

97. Quadrant II   **99.** Quadrant IV

Section 5.3   (page 376)

1–5. Answers will vary.   **7.** 30°, 150°   **9.** 120°, 240°

11. 45°, 225°   **13.** $\dfrac{5\pi}{6}, \dfrac{7\pi}{6}$ **15.** $\dfrac{3\pi}{4}, \dfrac{7\pi}{4}$

17. $\dfrac{5\pi}{6}, \dfrac{11\pi}{6}$ **19.** $\dfrac{7\pi}{6}, \dfrac{11\pi}{6}$ **21.** $\dfrac{\pi}{6}, \dfrac{7\pi}{6}$ **23.** $\dfrac{3\pi}{4}, \dfrac{7\pi}{4}$

25. $\dfrac{2\pi}{3} + 2n\pi, \dfrac{4\pi}{3} + 2n\pi$ **27.** $\dfrac{\pi}{6} + 2n\pi, \dfrac{11\pi}{6} + 2n\pi$

29. $\dfrac{\pi}{3} + n\pi, \dfrac{2\pi}{3} + n\pi$ **31.** $\dfrac{\pi}{3} + n\pi, \dfrac{2\pi}{3} + n\pi$

33. $\dfrac{\pi}{3} + n\pi, \dfrac{2\pi}{3} + n\pi$ **35.** $\dfrac{2\pi}{3}, \dfrac{5\pi}{3}$

37. $\dfrac{\pi}{4}, \dfrac{3\pi}{4}, \dfrac{5\pi}{4}, \dfrac{7\pi}{4}$ **39.** $0, \dfrac{\pi}{6}, \dfrac{5\pi}{6}, \pi, \dfrac{7\pi}{6}, \dfrac{11\pi}{6}$

41. $\dfrac{\pi}{3}, \pi, \dfrac{5\pi}{3}$ **43.** No solution   **45.** $\dfrac{\pi}{3}, \dfrac{5\pi}{3}$

47. $2.0344, 5.1760, \dfrac{\pi}{4}, \dfrac{5\pi}{4}$ **49.** 3.6652, 4.7124, 5.7596

51. 0.8614, 5.4218   **53.** 1.5708   **55.** 0.5236, 2.6180

57. (a)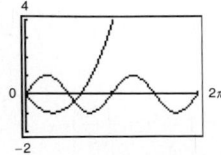

(b) $\sin 2x = x^2 - 2x$

(c) $(0, 0), (1.7757, -0.3984)$

59. (a)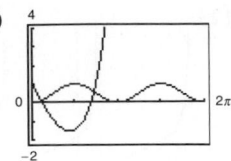

(b) $\sin^2 x = e^x - 4x$

(c) $(0.3194, 0.0986), (2.2680, 0.5878)$

61. $2\pi + 4n\pi$ **63.** $\dfrac{\pi}{8} + \dfrac{n\pi}{2}$ **65.** $\dfrac{2\pi}{3} + n\pi, \dfrac{5\pi}{6} + n\pi$

67. $\dfrac{\pi}{8} + \dfrac{n\pi}{4}$ **69.** $\dfrac{n\pi}{3}, \dfrac{\pi}{4} + n\pi$

71. $\dfrac{\pi}{2} + 4n\pi, \dfrac{7\pi}{2} + 4n\pi$ **73.** $x = -1, 3$

75. $x = \pm 2$ **77.** 1.1071, 4.2487   **79.** 0.8603, 3.4256

81. 0, 2.6779, 3.1416, 5.8195

83. 0.3398, 0.8481, 2.2935, 2.8018

85. $-1.154, 0.534$ **87.** 1.110

89. (a)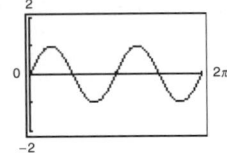

Maxima: $(0.7854, 1), (3.9270, 1)$

Minima: $(2.3562, -1), (5.4978, -1)$

(b) $\dfrac{\pi}{4}, \dfrac{3\pi}{4}, \dfrac{5\pi}{4}, \dfrac{7\pi}{4}$

91. (a)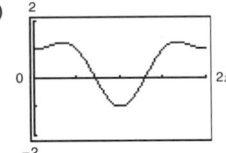

Maxima: $(1.0472, 1.25), (5.2360, 1.25)$

Minima: $(0, 1), (3.1416, -1), (6.2832, 1)$

(b) $0, \dfrac{\pi}{3}, \pi, \dfrac{5\pi}{3}, 2\pi$

93. (a)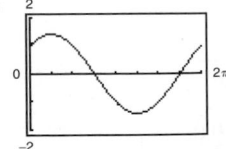

Maximum: $(0.7854, 1.4142)$

Minimum: $(3.9270, -1.4142)$

(b) $\dfrac{\pi}{4}, \dfrac{5\pi}{4}$

95. 1

97. (a) All real numbers x except $x = 0$

(b) y-axis symmetry; horizontal asymptote: $y = 1$

(c) Oscillates

(d) Infinite number of solutions

(e) Yes. 0.6366

99. April, May, June, July

101. 0.04 second, 0.43 second, 0.83 second

103. 36.87°, 53.13°

105. (a)

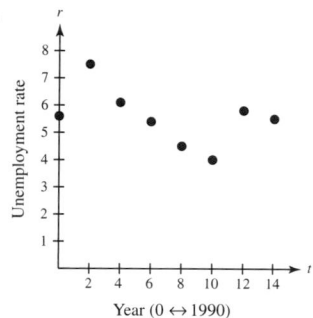

Year $(0 \leftrightarrow 1990)$

(b) By graphing the curves, it can be seen that model (1), $r = 124 \sin(0.47t + 0.40) + 5.45$, best fits the data.

(c) The constant terms gives the average unemployment rate of 5.45%.

(d) 13.37 years

(e) 2010

107. False

109. False. The range of the sine function does not include 3.4.

111. 2.164 rad   **113.** -0.007 rad

115. 24.25   **117.** 2290.38 feet

Section 5.4   (page 384)

Vocabulary Check   (page 384)

1. $\sin u \cos v - \cos u \sin v$

2. $\cos u \cos v - \sin u \sin v$ **3.** $\dfrac{\tan u + \tan v}{1 - \tan u \tan v}$

4. $\sin u \cos v + \cos u \sin v$

5. $\cos u \cos v + \sin u \sin v$ **6.** $\dfrac{\tan u - \tan v}{1 + \tan u \tan v}$

1. (a) $-\dfrac{1}{2}$ (b) $-\dfrac{3}{2}$ **3.** (a) $\dfrac{\sqrt{2} - \sqrt{6}}{4}$ (b) $\dfrac{1 + \sqrt{2}}{2}$

5. (a) $-\dfrac{\sqrt{2} + \sqrt{6}}{4}$ (b) $-\dfrac{\sqrt{2} + \sqrt{3}}{2}$

7. $\sin 105° = \dfrac{\sqrt{2} + \sqrt{6}}{4}$

$\cos 105° = \dfrac{\sqrt{2} - \sqrt{6}}{4}$

$\tan 105° = -2 - \sqrt{3}$

9. $\sin 195° = \dfrac{\sqrt{2} - \sqrt{6}}{4}$

$\cos 195° = -\dfrac{\sqrt{2} + \sqrt{6}}{4}$

$\tan 195° = 2 - \sqrt{3}$

11. $\sin \dfrac{11\pi}{12} = \dfrac{\sqrt{6} - \sqrt{2}}{4}$

$\cos \dfrac{11\pi}{12} = -\dfrac{\sqrt{2} + \sqrt{6}}{4}$

$\tan \dfrac{11\pi}{12} = -2 + \sqrt{3}$

13. $\sin\left(-\dfrac{\pi}{12}\right) = \dfrac{\sqrt{2} - \sqrt{6}}{4}$

$\cos\left(-\dfrac{\pi}{12}\right) = \dfrac{\sqrt{2} + \sqrt{6}}{4}$

$\tan\left(-\dfrac{\pi}{12}\right) = \sqrt{3} - 2$

15. $\sin 75° = \dfrac{\sqrt{2} + \sqrt{6}}{4}$

$\cos 75° = \dfrac{\sqrt{6} - \sqrt{2}}{4}$

$\tan 75° = 2 + \sqrt{3}$

17. $\sin(-225°) = \dfrac{\sqrt{2}}{2}$

$\cos(-225°) = -\dfrac{\sqrt{2}}{2}$

$\tan(-225°) = -1$

19. $\sin \dfrac{13\pi}{12} = \dfrac{\sqrt{2} - \sqrt{6}}{4}$

$\cos \dfrac{13\pi}{12} = -\dfrac{\sqrt{2} + \sqrt{6}}{4}$

$\tan \dfrac{13\pi}{12} = 2 - \sqrt{3}$

21. $\sin\left(-\dfrac{7\pi}{12}\right) = -\dfrac{\sqrt{2} + \sqrt{6}}{4}$

$\cos\left(-\dfrac{7\pi}{12}\right) = \dfrac{\sqrt{2} - \sqrt{6}}{4}$

$\tan\left(-\dfrac{7\pi}{12}\right) = 2 + \sqrt{3}$

23. $\cos 80°$ **25.** $\tan 239°$ **27.** $\sin 2.3$ **29.** $\cos \dfrac{16\pi}{63}$

31.

x	0.2	0.4	0.6	0.8
y_1	0.6621	0.7978	0.9017	0.9696
y_2	0.6621	0.7978	0.9017	0.9696

x	1.0	1.2	1.4
y_1	0.9989	0.9883	0.9384
y_2	0.9989	0.9883	0.9384

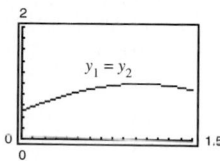

$y_1 = y_2$

33.

x	0.2	0.4	0.6	0.8
y_1	0.9605	0.8484	0.6812	0.4854
y_2	0.9605	0.8484	0.6812	0.4854

x	1.0	1.2	1.4
y_1	0.2919	0.1313	0.0289
y_2	0.2919	0.1313	0.0289

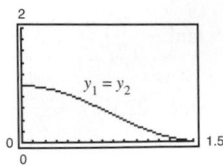

35. $-\frac{63}{65}$ **37.** $-\frac{63}{16}$ **39.** $\frac{36}{85}$ **41.** $\frac{13}{85}$ **43.** 1

45. $\dfrac{2x^2 - \sqrt{1 - x^2}}{\sqrt{4x^2 + 1}}$ **47.** 1   **49.** -1 **51.** $-\dfrac{1}{2}$

53. $\dfrac{\sqrt{3}}{3}$ **55.** 0   **57.** 0   **59.** $\dfrac{33}{65}$ **61.** $\dfrac{24}{25}$

63–69. Answers will vary.   **71.** $\dfrac{\pi}{2}$ **73.** $0, \dfrac{\pi}{3}, \pi, \dfrac{5\pi}{3}$

75. 0.7854, 5.4978   **77.** 0, 3.1416

79. Answers will vary.

81. False. $\cos(u \pm v) = \cos u \cos v \mp \sin u \sin v$

83. Answers will vary.   **85.** Answers will vary.

87. (a) $\sqrt{2}\sin\left(\theta + \dfrac{\pi}{4}\right)$ (b) $\sqrt{2}\cos\left(\theta - \dfrac{\pi}{4}\right)$

89. (a) $13\sin(3\theta + 0.3948)$ (b) $13\cos(3\theta - 1.1760)$

91. $2\cos\theta$ **93.** Answers will vary.

95. $u + v = w$. Answers will vary.   **97.** $(0, 19), (38, 0)$

99. $(0, 4), (2, 0), (7, 0)$ **101.** $\dfrac{\pi}{6}$ **103.** $\dfrac{\pi}{2}$

Section 5.5   (page 394)

Vocabulary Check   (page 394)

1. $2\sin u \cos u$ **2.** $\dfrac{1 + \cos 2u}{2}$ **3.** $\cos 2u$

4. $\tan\dfrac{u}{2}$ **5.** $\dfrac{2\tan u}{1 - \tan^2 u}$

6. $\frac{1}{2}[\cos(u - v) + \cos(u + v)]$ **7.** $\sin^2 u$

8. $\cos\dfrac{u}{2}$ **9.** $\frac{1}{2}[\sin(u + v) + \sin(u - v)]$

10. $2\sin\left(\dfrac{u + v}{2}\right)\cos\left(\dfrac{u - v}{2}\right)$

1. (a) $\frac{3}{5}$ (b) $\frac{4}{5}$ (c) $\frac{7}{25}$ (d) $\frac{24}{25}$
(e) $\frac{24}{7}$ (f) $\frac{25}{7}$ (g) $\frac{25}{24}$ (h) $\frac{7}{24}$

3. $0, 1.0472, 3.1416, 5.2360; 0, \dfrac{\pi}{3}, \pi, \dfrac{5\pi}{3}$

5. $0.2618, 1.3090, 3.4034, 4.4506; \dfrac{\pi}{12}, \dfrac{5\pi}{12}, \dfrac{13\pi}{12}, \dfrac{17\pi}{12}$

7. $0, 2.0944, 4.1888; 0, \dfrac{2\pi}{3}, \dfrac{4\pi}{3}$

9. $0, 1.5708, 3.1416, 4.7124; 0, \dfrac{\pi}{2}, \pi, \dfrac{3\pi}{2}$

11. $1.5708, 3.6652, 5.7596; \dfrac{\pi}{2}, \dfrac{7\pi}{6}, \dfrac{11\pi}{6}$

13. $\sin 2u = \frac{24}{25}$ **15.** $\sin 2u = \frac{4}{5}$
$\cos 2u = \frac{7}{25}$ $\cos 2u = \frac{3}{5}$
$\tan 2u = \frac{24}{7}$ $\tan 2u = \frac{4}{3}$

17. $\sin 2u = -\dfrac{4\sqrt{21}}{25}$

$\cos 2u = -\dfrac{17}{25}$

$\tan 2u = \dfrac{4\sqrt{21}}{17}$

19. $4\sin 2x$ **21.** $6\cos 2x$

23. $\frac{1}{8}(3 + 4\cos 2x + \cos 4x)$ **25.** $\frac{1}{8}(1 - \cos 4x)$

27. $\frac{1}{32}(2 + \cos 2x - 2\cos 4x - \cos 6x)$

29. $\frac{1}{2}(1 - \cos 4x)$ **31.** $\frac{1}{2}(1 + \cos x)$

33. $\frac{1}{8}(1 - \cos 8x)$ **35.** $\frac{1}{8}(3 - 4\cos x + \cos 2x)$

37. (a) $\dfrac{4\sqrt{17}}{17}$ (b) $\dfrac{\sqrt{17}}{17}$ (c) $\dfrac{1}{4}$ (d) $\dfrac{\sqrt{17}}{4}$

(e) $\sqrt{17}$ (f) 4   (g) $\dfrac{8}{17}$ (h) $\dfrac{2\sqrt{17}}{17}$

39. $\sin 15° = \dfrac{\sqrt{2 - \sqrt{3}}}{2}$ **41.** $\sin 112°\,30' = \dfrac{\sqrt{2 + \sqrt{2}}}{2}$

$\cos 15° = \dfrac{\sqrt{2 + \sqrt{3}}}{2}$ $\cos 112°\,30' = -\dfrac{\sqrt{2 - \sqrt{2}}}{2}$

$\tan 15° = 2 - \sqrt{3}$ $\tan 112°\,30' = -1 - \sqrt{2}$

43. $\sin\dfrac{\pi}{8} = \dfrac{\sqrt{2 - \sqrt{2}}}{2}$ **45.** $\sin\dfrac{3\pi}{8} = \dfrac{\sqrt{2 + \sqrt{2}}}{2}$

$\cos\dfrac{\pi}{8} = \dfrac{\sqrt{2 + \sqrt{2}}}{2}$ $\cos\dfrac{3\pi}{8} = \dfrac{\sqrt{2 - \sqrt{2}}}{2}$

$\tan\dfrac{\pi}{8} = \sqrt{2} - 1$ $\tan\dfrac{3\pi}{8} = \sqrt{2} + 1$

47. $\sin\dfrac{u}{2} = \dfrac{5\sqrt{26}}{26}$

$\cos\dfrac{u}{2} = \dfrac{\sqrt{26}}{26}$

$\tan\dfrac{u}{2} = 5$

49. $\sin\dfrac{u}{2} = \sqrt{\dfrac{89 - 5\sqrt{89}}{178}}$

$\cos\dfrac{u}{2} = -\sqrt{\dfrac{89 + 5\sqrt{89}}{178}}$

$\tan\dfrac{u}{2} = \dfrac{5 - \sqrt{89}}{8}$

51. $\sin\dfrac{u}{2} = \dfrac{3\sqrt{10}}{10}$

$\cos\dfrac{u}{2} = -\dfrac{\sqrt{10}}{10}$

$\tan\dfrac{u}{2} = -3$

53. $|\sin 3x|$ **55.** $-|\tan 4x|$ **57.** $\dfrac{\pi}{3}, \dfrac{5\pi}{3}$

59. $\dfrac{\pi}{3}, \pi, \dfrac{5\pi}{3}$ **61.** $3\left(\sin\dfrac{2\pi}{3} + \sin 0\right)$

63. $\dfrac{1}{2}(\sin 8\theta + \sin 2\theta)$ **65.** $5(\cos 60° + \cos 90°)$

67. $\dfrac{5}{2}(\cos 8\beta + \cos 2\beta)$ **69.** $\dfrac{1}{2}(\cos 2y - \cos 2x)$

71. $\dfrac{1}{2}(\sin 2\theta + \sin 2\pi)$ **73.** $2\cos 3\theta \sin 2\theta$

75. $2\cos 4x \cos 2x$ **77.** $2\cos\alpha \sin\beta$

79. $-2\sin\theta \sin\dfrac{\pi}{2}$ **81.** $\dfrac{\sqrt{2}}{2}$ **83.** $\dfrac{\sqrt{6}}{2}$

85. $0, \dfrac{\pi}{4}, \dfrac{\pi}{2}, \dfrac{3\pi}{4}, \pi, \dfrac{5\pi}{4}, \dfrac{3\pi}{2}, \dfrac{7\pi}{4}$ **87.** $\dfrac{\pi}{6}, \dfrac{5\pi}{6}$ **89.** $\dfrac{25}{169}$

91. $\dfrac{4}{13}$ **93–109.** Answers will vary.

111. $\dfrac{1 - \cos 2x}{2}$

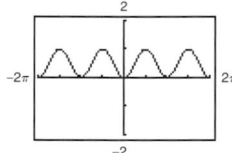

113. $\dfrac{3 + 4\cos 2x + \cos 4x}{8}$

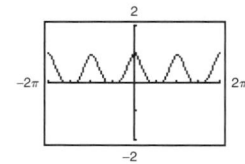

115. $2x\sqrt{1 - x^2}$ **117.** $1 - 2x^2$ **119.** $\dfrac{1 - x^2}{1 + x^2}$

121. (a)

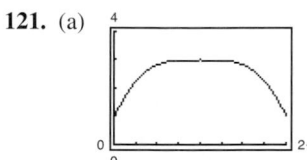

Maximum: $(3.1416, 3)$

(b) π

123. (a)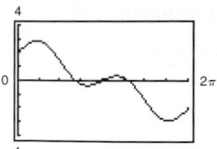

Minima: $(2.6078, -0.3486), (5.5839, -2.8642)$

Maxima: $(0.6993, 2.8642), (3.6754, 0.3486)$

(b) $0.6993, 2.6078, 3.6754, 5.5839$

125. (a) $x = 0, \dfrac{\pi}{3}, \pi, \dfrac{5\pi}{3}, 2\pi$

(b) $x = \arccos\dfrac{1 \pm \sqrt{33}}{8}, 2\pi - \arccos\dfrac{1 \pm \sqrt{33}}{8}$

127. (a) $r = \dfrac{1}{16}v_0{}^2 \sin\theta \cos\theta$

(b) 198.90 feet

(c) 98.73 feet per second

(d) $\theta = 45°$; Answers will vary.

129. $x = 2r(1 - \cos\theta)$

131. False. $\sin\dfrac{x}{2} = -\sqrt{\dfrac{1 - \cos x}{2}}$ for $\pi \le \dfrac{x}{2} \le 2\pi$.

133. (a) 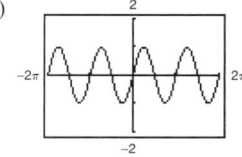   (b) $y = \sin 2x$

(c) Answers will vary.

135. Answers will vary.

137. (a) 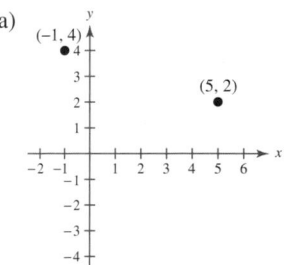   (b) $2\sqrt{10}$

(c) $(2, 3)$

139. (a) 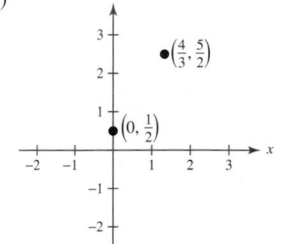   (b) $\dfrac{2\sqrt{13}}{3}$

(c) $\left(\dfrac{2}{3}, \dfrac{3}{2}\right)$

CHAPTER 5

141. (a) Complement: $35°$; supplement: $125°$

    (b) Complement: none; supplement: $18°$

143. (a) Complement: $\dfrac{4\pi}{9}$; supplement: $\dfrac{17\pi}{18}$

    (b) Complement: $\dfrac{\pi}{20}$; supplement: $\dfrac{11\pi}{20}$

145. 0.4667 rad

147.   **149.**

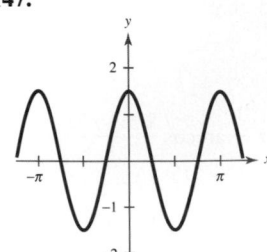 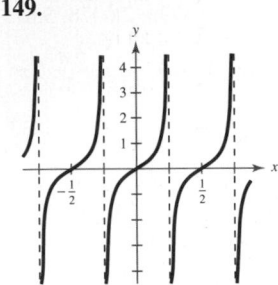

Review Exercises   (page 400)

1. $\sec x$ **3.** $\cos x$ **5.** $|\sin x|$ **7.** $\sec x$ **9.** $\sec x$

11. $\tan x = \dfrac{4}{3}$ **13.** $\cos x = \dfrac{\sqrt{2}}{2}$

    $\csc x = \dfrac{5}{4}$ $\tan x = -1$

                     $\csc x = -\sqrt{2}$

    $\sec x = \dfrac{5}{3}$ $\sec x = \sqrt{2}$

                     $\cot x = -1$

    $\cot x = \dfrac{3}{4}$

15. $\cos^2 x$ **17.** $1 + \cot\alpha$ **19.** 1 **21.** $\csc x$

23–35. Answers will vary.

37. $\dfrac{\pi}{6} + 2n\pi,\ \dfrac{5\pi}{6} + 2n\pi$ **39.** $\dfrac{\pi}{3} + 2n\pi,\ \dfrac{2\pi}{3} + 2n\pi$

41. $\dfrac{\pi}{6} + n\pi$ **43.** $\dfrac{\pi}{3} + n\pi,\ \dfrac{2\pi}{3} + n\pi$

45. $\dfrac{\pi}{6} + n\pi,\ \dfrac{5\pi}{6} + n\pi$ **47.** $n\pi$ **49.** $0,\ \dfrac{2\pi}{3},\ \dfrac{4\pi}{3}$

51. $0,\ \dfrac{\pi}{2},\ \pi$ **53.** $\dfrac{\pi}{8},\ \dfrac{3\pi}{8},\ \dfrac{9\pi}{8},\ \dfrac{11\pi}{8}$

55. $0,\ \dfrac{\pi}{8},\ \dfrac{3\pi}{8},\ \dfrac{5\pi}{8},\ \dfrac{7\pi}{8},\ \dfrac{9\pi}{8},\ \dfrac{11\pi}{8},\ \dfrac{13\pi}{8},\ \dfrac{15\pi}{8}$ **57.** $\dfrac{\pi}{2},\ \dfrac{3\pi}{2}$

59. $\dfrac{\pi}{12} + n\pi,\ \dfrac{5\pi}{12} + n\pi$ **61.** $\dfrac{\pi}{12} + \dfrac{n\pi}{6}$ **63.** $0,\ \pi$

65. $\arctan 2,\ \arctan(-5) + \pi,\ \arctan 2\ + \pi,$
    $\arctan(-5) + 2\pi$

67. $\sin 285° = -\dfrac{\sqrt{2}+\sqrt{6}}{4}$

    $\cos 285° = \dfrac{\sqrt{6}-\sqrt{2}}{4}$

    $\tan 285° = -2 - \sqrt{3}$

69. $\sin\dfrac{31\pi}{12} = \dfrac{\sqrt{2}+\sqrt{6}}{4}$

    $\cos\dfrac{31\pi}{12} = \dfrac{\sqrt{2}-\sqrt{6}}{4}$

    $\tan\dfrac{31\pi}{12} = -2 - \sqrt{3}$

71. $\sin 180°$ **73.** $\tan 35°$ **75.** $-\dfrac{117}{125}$

77. $\dfrac{3}{4}$ **79.** $-\dfrac{44}{125}$ **81.** 0 **83.** 0

85–89. Answers will vary.   **91.** $\dfrac{\pi}{4},\ \dfrac{7\pi}{4}$

93. $\sin 2u = \dfrac{20\sqrt{6}}{49}$ **95.** $\sin 2u = -\dfrac{36}{85}$

    $\cos 2u = -\dfrac{1}{49}$ $\cos 2u = \dfrac{77}{85}$

    $\tan 2u = -20\sqrt{6}$ $\tan 2u = -\dfrac{36}{77}$

97 and 99. Answers will vary.

101. $15°,\ 75°$ **103.** $\dfrac{1}{32}(10 - 15\cos 2x + 6\cos 4x - \cos 6x)$

105. $\dfrac{1}{8}(3 + 4\cos 4x + \cos 8x)$

107. $\sin 105° = \dfrac{\sqrt{2+\sqrt{3}}}{2}$

    $\cos 105° = -\dfrac{\sqrt{2-\sqrt{3}}}{2}$

    $\tan 105° = -2 - \sqrt{3}$

109. $\sin\dfrac{7\pi}{8} = \dfrac{\sqrt{2-\sqrt{2}}}{2}$ **111.** $\sin\dfrac{u}{2} = \dfrac{\sqrt{10}}{10}$

    $\cos\dfrac{7\pi}{8} = -\dfrac{\sqrt{2+\sqrt{2}}}{2}$ $\cos\dfrac{u}{2} = \dfrac{3\sqrt{10}}{10}$

    $\tan\dfrac{7\pi}{8} = 1 - \sqrt{2}$ $\tan\dfrac{u}{2} = \dfrac{1}{3}$

113. $\sin\dfrac{u}{2} = \dfrac{3\sqrt{14}}{14}$

    $\cos\dfrac{u}{2} = \dfrac{\sqrt{70}}{14}$

    $\tan\dfrac{u}{2} = \dfrac{3\sqrt{5}}{5}$

115. $-|\cos 4x|$ **117.** $V = \sin\dfrac{\theta}{2}\cos\dfrac{\theta}{2}$ cubic meters

119. $3\left(\sin \dfrac{\pi}{2} + \sin 0\right)$ **121.** $\dfrac{1}{2}(\cos \alpha - \cos 9\alpha)$

123. $2 \cos\left(\dfrac{9\theta}{2}\right) \cos\left(\dfrac{\theta}{2}\right)$ **125.** $2 \cos x \sin \dfrac{\pi}{4}$

127. $y = \dfrac{1}{2} \sqrt{10} \sin\left(8t - \arctan \dfrac{1}{3}\right)$ **129.** $\dfrac{\sqrt{10}}{2}$ feet

131. False. $\cos \dfrac{\theta}{2} > 0$ **133.** True

135. Answers will vary.   **137.** $y_3 = y_2 + 1$

Chapter Test   (page 403)

1. $\sin \theta = \dfrac{-3\sqrt{13}}{13}$ **2.** 1   **3.** 1

$\cos \theta = \dfrac{-2\sqrt{13}}{13}$

$\csc \theta = \dfrac{-\sqrt{13}}{3}$

$\sec \theta = \dfrac{-\sqrt{13}}{2}$

$\cot \theta = \dfrac{2}{3}$

4. $\csc \theta \sec \theta$ **5.** $0, \dfrac{\pi}{2} < \theta \le \pi, \dfrac{3\pi}{2} < \theta < 2\pi$

6.

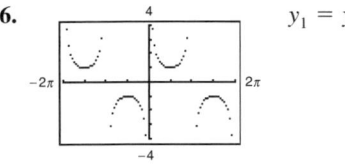

$y_1 = y_2$

7–12. Answers will vary.   **13.** $-2 - \sqrt{3}$

14. $\dfrac{1}{16}\left[\dfrac{10 - 15 \cos 2x + 6 \cos 4x - \cos 6x}{1 + \cos 2x}\right]$

15. $\tan 2\theta$ **16.** $2(\sin 6\theta + \sin 2\theta)$

17. $-2 \cos \dfrac{7\theta}{2} \sin \dfrac{\theta}{2}$ **18.** $0, \dfrac{3\pi}{4}, \pi, \dfrac{7\pi}{4}$

19. $\dfrac{\pi}{6}, \dfrac{\pi}{2}, \dfrac{5\pi}{6}, \dfrac{3\pi}{2}$ **20.** $\dfrac{\pi}{6}, \dfrac{5\pi}{6}, \dfrac{7\pi}{6}, \dfrac{11\pi}{6}$

21. $\dfrac{\pi}{6}, \dfrac{5\pi}{6}, \dfrac{3\pi}{2}$ **22.** $-2.938, -2.663, 1.170$

23. $\sin 2u = \dfrac{4}{5}$ **24.** $76.52°$

$\cos 2u = -\dfrac{3}{5}$

$\tan 2u = -\dfrac{4}{3}$

Chapter 6

Section 6.1   (page 414)

1. $C = 95°$, $b \approx 24.59$ inches, $c \approx 28.29$ inches

3. $A = 40°$, $a \approx 15.69$ centimeters, $b \approx 6.32$ centimeters

5. $C = 74°15'$, $a \approx 6.41$ kilometers, $c \approx 6.26$ kilometers

7. $B \approx 21.55°$, $C \approx 122.45°$, $c \approx 11.49$

9. $B = 60.9°$, $b \approx 19.32$, $c \approx 6.36$

11. $B \approx 18°13'$, $C \approx 51°32'$, $c \approx 40.05$

13. $B \approx 48.74°$, $C \approx 21.26°$, $c \approx 48.23$

15. No solution

17. Two solutions

   $B \approx 72.21°$, $C \approx 49.79°$, $c \approx 10.27$

   $B \approx 107.79°$, $C \approx 14.21°$, $c \approx 3.30$

19. 28.19 square units   **21.** 1782.32 square units

23. 2888.57 square units

25. (a)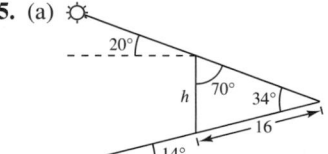

(b) $\dfrac{16}{\sin 70°} = \dfrac{h}{\sin 34°}$ (c) 9.52 meters

27. 240.03°

29. (a)

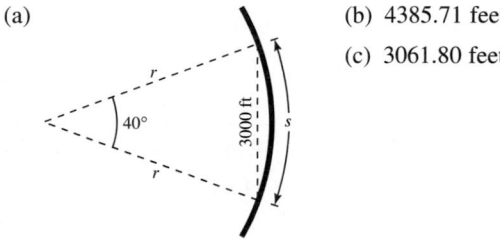

(b) 4385.71 feet

(c) 3061.80 feet

31. 15.53 kilometers from Colt Station; 42.43 kilometers from Pine Knob

CHAPTER 6

33. 16.08°

35. (a) $\alpha \approx 5.36°$

(b) $\beta = \arcsin\left(\dfrac{d \sin \theta}{58.36}\right)$

(c) $d = \sin(84.64 - \theta)\left[\dfrac{58.36}{\sin \theta}\right]$

(d)

θ	10°	20°	30°	40°	50°	60°
d	324.08	154.19	95.19	63.80	43.30	28.10

37. False. The triangle can't be solved if only three angles are known.

39. Yes, the Law of Sines can be used to solve a right triangle provided that at least one side and one angle are given or two sides are given. Answers will vary.

41. $C = 83°$, $b \approx 17.83$, $c \approx 22.46$

43. $\tan \theta = -\frac{12}{5}$; $\csc \theta = -\frac{13}{12}$; $\sec \theta = \frac{13}{5}$; $\cot \theta = -\frac{5}{12}$

45. $3(\sin 11\theta + \sin 5\theta)$

47. $\dfrac{3}{2}\left(\sin \dfrac{11\pi}{6} + \sin \dfrac{3\pi}{2}\right)$

Section 6.2 (page 421)

> ### Vocabulary Check (page 421)
>
> **1.** $c^2 = a^2 + b^2 - 2ab \cos C$ **2.** Heron's Area
>
> **3.** $\frac{1}{2}bh$, $\sqrt{s(s-a)(s-b)(s-c)}$

1. $A \approx 40.80°$, $B \approx 60.61°$, $C \approx 78.59°$

3. $A \approx 49.51°$, $B \approx 55.40°$, $C \approx 75.09°$

5. $A \approx 31.40°$, $C \approx 128.60°$, $b \approx 6.56$ millimeters

7. $A \approx 53°45'$, $C \approx 75°45'$, $b \approx 9.95$ feet

9. $A \approx 26.38°$, $B \approx 36.34°$, $C \approx 117.28°$

11. $B \approx 29.44°$, $C \approx 100.56°$, $a \approx 23.38$

13. $A \approx 36.87°$, $B \approx 53.13°$, $C = 90°$

15. $A \approx 103.52°$, $B \approx 38.24°$, $C \approx 38.24°$

17. $A \approx 154°14'$, $C \approx 17°31'$, $b \approx 8.58$

19. $A \approx 37°6'7''$, $C \approx 67°33'53''$, $b \approx 9.94$

	a	b	c	d	θ	ϕ
21.	4	8	11.64	4.96	30°	150°
23.	10	14	20	13.86	68.20°	111.80°
25.	15	16.96	25	20	77.22°	102.78°

27. 104.57 square inches **29.** 19.81

31. 0.27 square foot **33.** 15.52 **35.** 35.19

37.

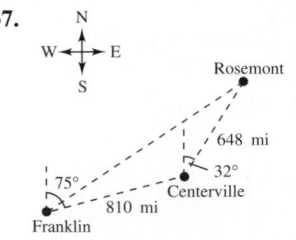

1357.85 miles, 236.01°

39.

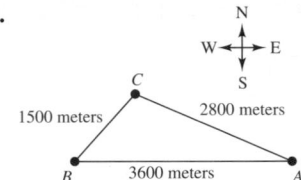

N 43.03° E, S 66.95° E

41. 43.27 miles **43.** $PQ \approx 9.43$, $QS \approx 5$, $RS \approx 12.81$

45. 18,617.66 square feet

47. (a) $49 = 2.25 + x^2 - 3x \cos \theta$

(b) $x = \frac{1}{2}\left(3 \cos \theta + \sqrt{9 \cos^2\theta + 187}\right)$

(c)

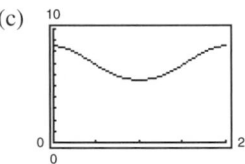

(d) 6 inches

49. False. A triangle cannot be formed with sides of lengths 10 feet, 16 feet, and 5 feet.

51. False. $s = \dfrac{a + b + c}{2}$ **53 and 55.** Proof.

57. Answers will vary. Yes; Answers will vary.

59. $-\dfrac{\pi}{2}$ **61.** $\dfrac{\pi}{3}$

Section 6.3 (page 433)

> ### Vocabulary Check (page 433)
>
> **1.** directed line segment **2.** initial, terminal
>
> **3.** magnitude **4.** vector
>
> **5.** standard position **6.** unit vector
>
> **7.** multiplication, addition **8.** resultant
>
> **9.** linear combination, horizontal, vertical

1. Answers will vary.

3. $\langle 4, 3 \rangle$, $\|\mathbf{v}\| = 5$ **5.** $\langle -3, 2 \rangle$, $\|\mathbf{v}\| = \sqrt{13}$

7. $\langle 0, 5 \rangle$, $\|\mathbf{v}\| = 5$ **9.** $\left\langle \frac{3}{5}, -\frac{3}{5} \right\rangle$, $\|\mathbf{v}\| = \frac{3\sqrt{2}}{5}$

11. $\left\langle \frac{7}{6}, \frac{9}{5} \right\rangle$, $\|\mathbf{v}\| = \frac{\sqrt{4141}}{30}$

13. 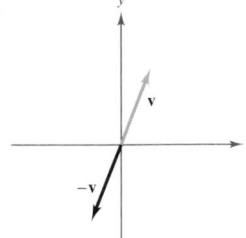   **15.**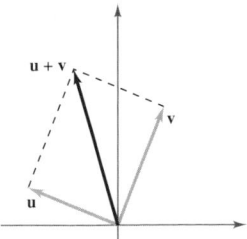

17. 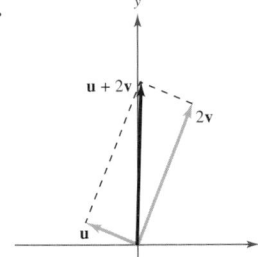   **19.**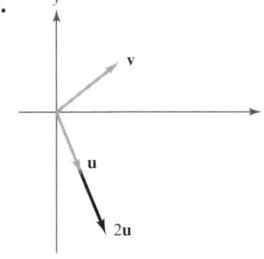

21. 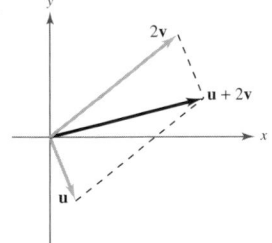   **23.**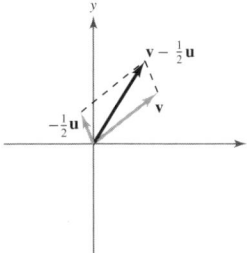

25. (a) $\langle 11, 3 \rangle$ (b) $\langle -3, 1 \rangle$ (c) $\langle -13, 1 \rangle$ (d) $\langle 23, 9 \rangle$

27. (a) $\langle -4, -4 \rangle$ (b) $\langle -8, -12 \rangle$ (c) $\langle -18, -28 \rangle$
   (d) $\langle -22, -28 \rangle$

29. (a) $3\mathbf{i} - 2\mathbf{j}$ (b) $-\mathbf{i} + 4\mathbf{j}$ (c) $-4\mathbf{i} + 11\mathbf{j}$
   (d) $6\mathbf{i} + \mathbf{j}$

31. $\mathbf{u} + \mathbf{v}$ **33.** $\mathbf{w} - \mathbf{v}$ **35.** $\langle 1, 0 \rangle$

37. $\left\langle -\frac{\sqrt{2}}{2}, \frac{\sqrt{2}}{2} \right\rangle$ **39.** $\left\langle -\frac{24}{25}, -\frac{7}{25} \right\rangle$ **41.** $\frac{4}{5}\mathbf{i} - \frac{3}{5}\mathbf{j}$

43. $\mathbf{j}$ **45.** $\frac{40\sqrt{61}}{61}\mathbf{i} + \frac{48\sqrt{61}}{61}\mathbf{j}$ **47.** $\frac{21}{5}\mathbf{i} + \frac{28}{5}\mathbf{j}$

49. $-8\mathbf{i}$ **51.** $7\mathbf{i} + 4\mathbf{j}$ **53.** $3\mathbf{i} + 8\mathbf{j}$

55. $\mathbf{v} = \left\langle 3, -\frac{3}{2} \right\rangle$

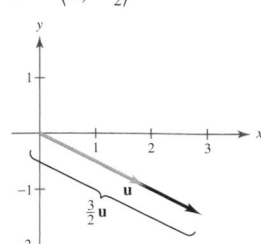

57. $\mathbf{v} = \langle 4, 3 \rangle$

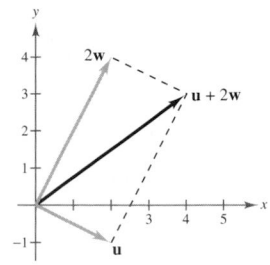

59. $\mathbf{v} = \left\langle \frac{7}{2}, -\frac{1}{2} \right\rangle$ **61.** $\|\mathbf{v}\| = 5$, $\theta = 30°$

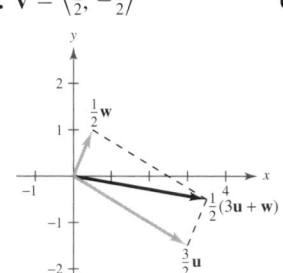

63. $\|\mathbf{v}\| = 6\sqrt{2}$, $\theta = 315°$ **65.** $\|\mathbf{v}\| = \sqrt{29}$, $\theta = 111.80°$

67. $\mathbf{v} = \langle 3, 0 \rangle$ **69.** $\mathbf{v} = \left\langle -\frac{3\sqrt{6}}{2}, \frac{3\sqrt{2}}{2} \right\rangle$

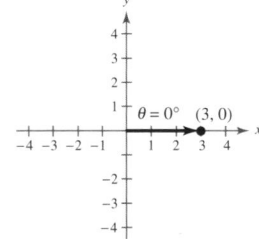   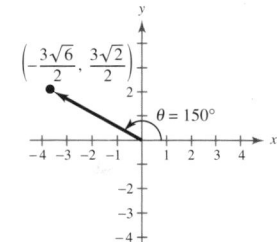

71. $\mathbf{v} = \left\langle \frac{\sqrt{10}}{5}, \frac{3\sqrt{10}}{5} \right\rangle$ **73.** $\left\langle \frac{5}{2}, \frac{10 + 5\sqrt{3}}{2} \right\rangle$

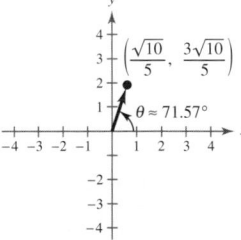

75. $\left\langle 10\sqrt{2} - 25\sqrt{3}, 25 + 10\sqrt{2} \right\rangle$ **77.** $90°$

79.

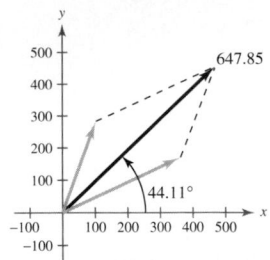

$\|\mathbf{v}\| \approx 647.85,\ \theta \approx 44.11°$

81. $62.72°$

83. Horizontal component: ≈ 53.62 feet per second

Vertical component: ≈ 45.00 feet per second

85. $T_{AC} \approx 3611.11$ pounds, $T_{BC} \approx 2169.51$ pounds

87. (a) $T = 3000 \sec \theta$; Domain: $0° \le \theta < 90°$

(b)

θ	$10°$	$20°$	$30°$
T	3046.28	3192.53	3464.10

θ	$40°$	$50°$	$60°$
T	3916.22	4667.17	6000

(c)

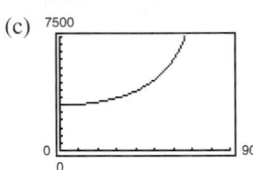

(d) The component in the direction of the motion of the barge decreases.

89. N $26.67°$ E, 130.35 kilometers per hour

91. (a) $12.10°$, 357.85 newtons

(b) $M = 10\sqrt{660\cos\theta + 709}$

$\alpha = \arctan\dfrac{15\sin\theta}{15\cos\theta + 22}$

(c)

θ	$0°$	$30°$	$60°$	$90°$
M	$370°$	$357.85°$	$322.34°$	$266.27°$
α	$0°$	$12.10°$	$23.77°$	$34.29°$

θ	$120°$	$150°$	$180°$
M	$194.68°$	$117.23°$	$70°$
α	$41.86°$	$39.78°$	$0°$

(d)

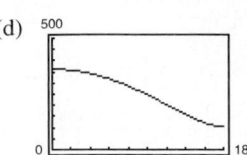 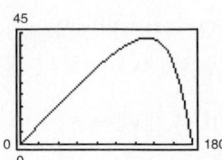

(e) For increasing θ, the two vectors tend to work against each other, resulting in a decrease in the magnitude of the resultant.

93. True   **95.** True. $a = b = 0$ **97.** True

99. True   **101.** True   **103.** False

105. (a) $0°$ (b) $180°$

(c) No. The magnitude is equal to the sum when the angle between the vectors is $0°$.

107. Proof.   **109.** $\langle 1, 3\rangle$ or $\langle -1, -3\rangle$

111. $12x^3y^7,\ x \neq 0,\ y \neq 0$ **113.** $48xy^2,\ x \neq 0$

115. 7.14×10^5 **117.** $7\cos\theta$ **119.** $10\csc\theta$

121. $\dfrac{\pi}{2} + n\pi,\ \pi + 2n\pi$ **123.** $\dfrac{\pi}{3} + 2n\pi,\ \dfrac{5\pi}{3} + 2n\pi$

Section 6.4   (page 445)

Vocabulary Check   (page 445)

1. dot product   **2.** $\dfrac{\mathbf{u}\cdot\mathbf{v}}{\|\mathbf{u}\|\,\|\mathbf{v}\|}$ **3.** orthogonal

4. $\left(\dfrac{\mathbf{u}\cdot\mathbf{v}}{\|\mathbf{v}\|^2}\right)\mathbf{v}$ **5.** $\left\|\mathrm{proj}_{\overrightarrow{PQ}}\,\mathbf{F}\right\|\,\|\overrightarrow{PQ}\|,\ \mathbf{F}\cdot\overrightarrow{PQ}$

1. 0   **3.** 14   **5.** 8, scalar   **7.** 4, scalar

9. $\langle -114, -114\rangle$, vector   **11.** 13   **13.** $5\sqrt{41}$

15. 4   **17.** $90°$ **19.** $70.56°$ **21.** $90°$ **23.** $\dfrac{5\pi}{12}$

25.   **27.**

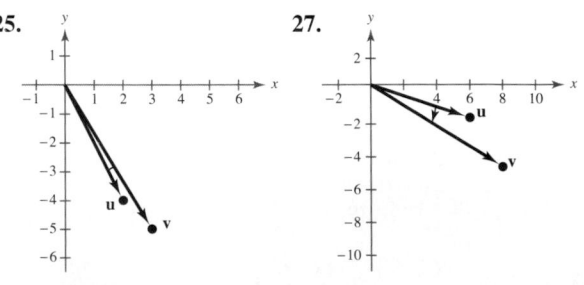

$\theta \approx 4.40°$ $\theta \approx 13.57°$

29. $26.57°, 63.43°, 90°$ **31.** $-162\sqrt{2}$ **33.** Parallel

35. Neither   **37.** Orthogonal   **39.** 3

41. 10   **43.** 0

45. $\frac{16}{17}\langle 4, 1\rangle$, $\mathbf{u} = \left\langle \frac{64}{17}, \frac{16}{17}\right\rangle + \left\langle -\frac{13}{17}, \frac{52}{17}\right\rangle$

47. $\frac{45}{229}\langle 2, 15\rangle$, $\mathbf{u} = \left\langle \frac{90}{229}, \frac{675}{229}\right\rangle + \left\langle -\frac{90}{229}, \frac{12}{229}\right\rangle$ **49. u**

51. 0   **53.** $\langle 3, -1\rangle, \langle -3, 1\rangle$ **55.** $-\frac{3}{4}\mathbf{i} - \frac{1}{2}\mathbf{j}, \frac{3}{4}\mathbf{i} + \frac{1}{2}\mathbf{j}$

57. 32

59. (a) 37,289; It is the total dollar value of the picture frames produced by the company.

(b) Multiply **v** by 1.02.

61. (a) Force $= 30,000 \sin d°$

(b)

d	0°	1°	2°	3°
Force	0	523.57	1046.98	1570.08

d	4°	5°	6°	7°
Force	2092.69	2614.67	3135.85	3656.08

d	8°	9°	10°
Force	4175.19	4693.03	5209.45

(c) 29,885.84 pounds

63. (a) $W = 15,691\dfrac{\sqrt{3}}{2}d$

(b)

d	0	200	400	800
Work	0	2,717,760.92	5,435,521.84	10,871,043.69

65. 21,650.64 foot-pounds   **67.** 725.05 foot-pounds

69. True. The zero vector is orthogonal to every vector.

71. Orthogonal. $\mathbf{u} \cdot \mathbf{v} = 0$

73. (a) **u** and **v** are parallel.   (b) **u** and **v** are orthogonal.

75 and 77. Proof

79. g is a horizontal shift of f four units to the right.

81. g is a vertical shift of f six units upward.

83. $-1 + 2i$ **85.** $15 + 12i$ **87.** 10   **89.** $\frac{47}{26} - \frac{27}{26}i$

91.
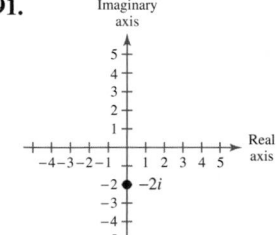

93.
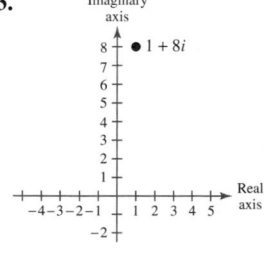

Section 6.5   (page 456)

Vocabulary Check   (page 456)

1. absolute value

2. trigonometric form, modulus, argument

3. DeMoivre's   4. nth root

1.
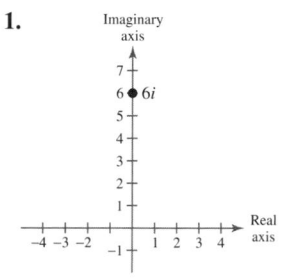

6

3.
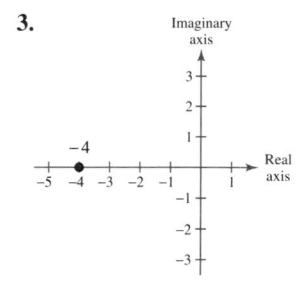

4

5.
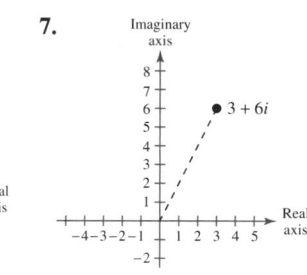

$4\sqrt{2}$

7.

$3\sqrt{5}$

9. $3\left(\cos\dfrac{\pi}{2} + i\sin\dfrac{\pi}{2}\right)$ **11.** $2(\cos \pi + i\sin \pi)$

13. $2\sqrt{2}\left(\cos\dfrac{5\pi}{4} + i\sin\dfrac{5\pi}{4}\right)$ **15.** $2\left(\cos\dfrac{11\pi}{6} + i\sin\dfrac{11\pi}{6}\right)$

17.
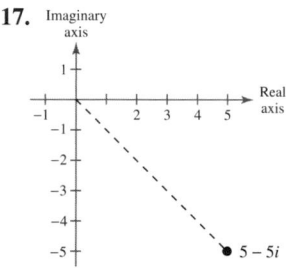

19.
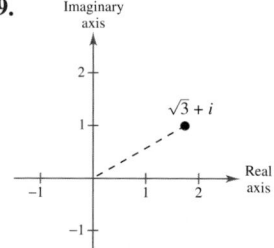

$5\sqrt{2}\left(\cos\dfrac{7\pi}{4} + i\sin\dfrac{7\pi}{4}\right)$ $2\left(\cos\dfrac{\pi}{6} + i\sin\dfrac{\pi}{6}\right)$

CHAPTER 6

21.

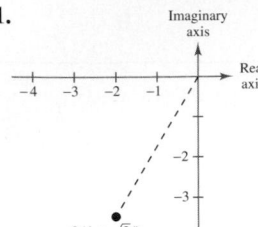

$$4\left(\cos \frac{4\pi}{3} + i \sin \frac{4\pi}{3}\right)$$

23.

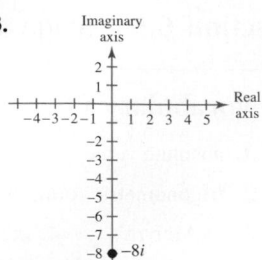

$$8\left(\cos \frac{3\pi}{2} + i \sin \frac{3\pi}{2}\right)$$

33.

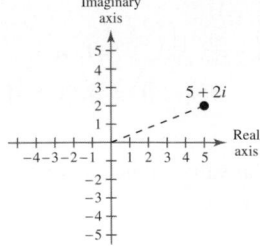

$$\sqrt{29}(\cos 0.381 + i \sin 0.381)$$

35.

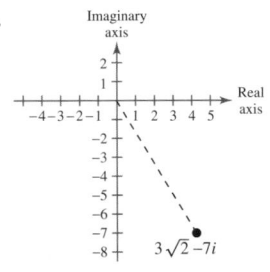

$$\sqrt{67}(\cos 5.257 + i \sin 5.257)$$

25.

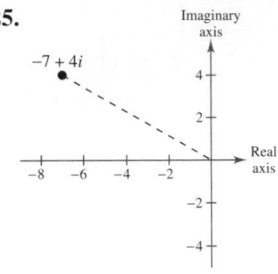

$$\sqrt{65}(\cos 2.622 + i \sin 2.622)$$

27.

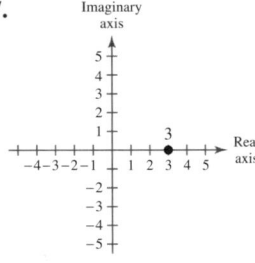

$$3(\cos 0 + i \sin 0)$$

29.

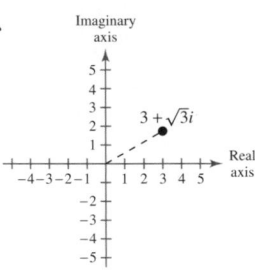

$$2\sqrt{3}\left(\cos \frac{\pi}{6} + i \sin \frac{\pi}{6}\right)$$

37.

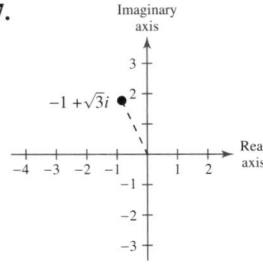

$$-1 + \sqrt{3}i$$

39.

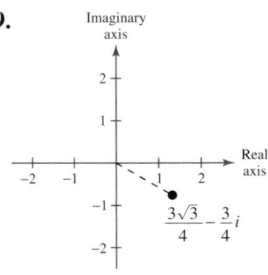

$$\frac{3\sqrt{3}}{4} - \frac{3}{4}i$$

31.

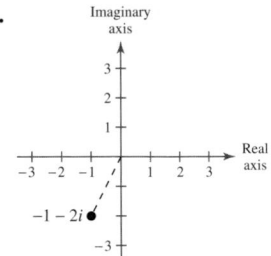

$$\sqrt{5}(\cos 4.249 + i \sin 4.249)$$

41.

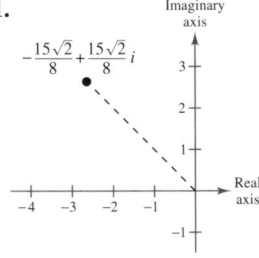

$$-\frac{15\sqrt{2}}{8} + \frac{15\sqrt{2}}{8}i$$

43.

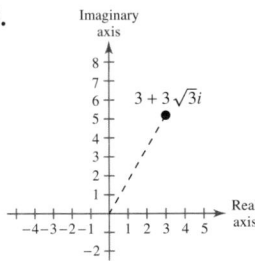

$$3 + 3\sqrt{3}i$$

45.

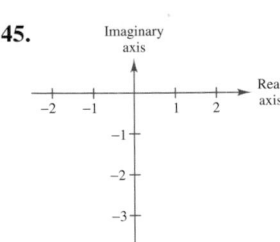

$-4i$

47.

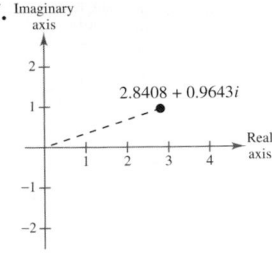

$2.8408 + 0.9643i$

49. $4.6985 + 1.7101i$ **51.** $4.7693 + 7.6324i$

53.

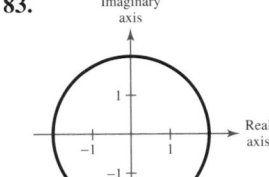

The absolute value of each is 1.

55. $12\left(\cos \dfrac{\pi}{2} + i \sin \dfrac{\pi}{2}\right)$ **57.** $\dfrac{10}{9}(\cos 200° + i \sin 200°)$

59. $\dfrac{11}{50}(\cos 130° + i \sin 130°)$ **61.** $\cos 30° + i \sin 30°$

63. $\dfrac{1}{2}(\cos 80° + i \sin 80°)$ **65.** $6(\cos 312° + i \sin 312°)$

67. (a) $2\sqrt{2}\left(\cos \dfrac{7\pi}{4} + i \sin \dfrac{7\pi}{4}\right)$

$\sqrt{2}\left(\cos \dfrac{\pi}{4} + i \sin \dfrac{\pi}{4}\right)$

(b) and (c) 4

69. (a) $2\sqrt{2}\left(\cos \dfrac{\pi}{4} + i \sin \dfrac{\pi}{4}\right)$

$\sqrt{2}\left(\cos \dfrac{7\pi}{4} + i \sin \dfrac{7\pi}{4}\right)$

(b) and (c) 4

71. (a) $2\left(\cos \dfrac{3\pi}{2} + i \sin \dfrac{3\pi}{2}\right)$

$\sqrt{2}\left(\cos \dfrac{\pi}{4} + i \sin \dfrac{\pi}{4}\right)$

(b) and (c) $2 - 2i$

73. (a) $2\left(\cos \dfrac{3\pi}{2} + i \sin \dfrac{3\pi}{2}\right)$

$2\left(\cos \dfrac{11\pi}{6} + i \sin \dfrac{11\pi}{6}\right)$

(b) and (c) $-2 - 2\sqrt{3}i$

75. (a) $2(\cos 0 + i \sin 0)$

$\sqrt{2}\left(\cos \dfrac{7\pi}{4} + i \sin \dfrac{7\pi}{4}\right)$

(b) and (c) $2 - 2i$

77. (a) $3\sqrt{2}\left(\cos \dfrac{\pi}{4} + i \sin \dfrac{\pi}{4}\right)$

$2\left(\cos \dfrac{5\pi}{3} + i \sin \dfrac{5\pi}{3}\right)$

(b) and (c) $\dfrac{3 - 3\sqrt{3}}{4} + \dfrac{3 + \sqrt{3}}{4}i$

79. (a) $5(\cos 0 + i \sin 0)$

$2\sqrt{2}\left(\cos \dfrac{\pi}{4} + i \sin \dfrac{\pi}{4}\right)$

(b) and (c) $\dfrac{5}{4} - \dfrac{5}{4}i$

81. (a) $4\left(\cos \dfrac{\pi}{2} + i \sin \dfrac{\pi}{2}\right)$

$\sqrt{2}\left(\cos \dfrac{3\pi}{4} + i \sin \dfrac{3\pi}{4}\right)$

(b) and (c) $2 - 2i$

83.

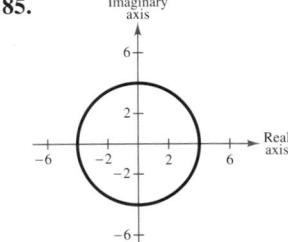

85.

87.

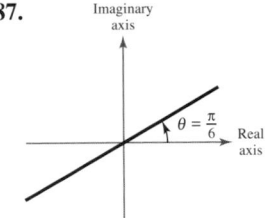

89.

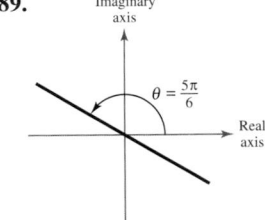

91. $-2 + 2i$ **93.** $-32i$ **95.** $-32\sqrt{3} + 32i$

97. $\dfrac{125}{2} + \dfrac{125\sqrt{3}}{2}i$ **99.** i **101.** $4.5386 - 15.3428i$

103. 256 **105.** $-597 - 122i$ **107.** $2048 + 2048\sqrt{3}i$

109. $\dfrac{9\sqrt{2}}{2} + \dfrac{9\sqrt{2}}{2}i$ **111.** Answers will vary.

113. (a) $2(\cos 30° + i \sin 30°)$; $2(\cos 150° + i \sin 150°)$;

$2(\cos 270° + i \sin 270°)$

(b) and (c) $8i$

115. (a) $\cos 0° + i \sin 0°$; $\cos 120° + i \sin 120°$;

$\cos 240° + i \sin 240°$

(b) and (c) 1

117. $1 + i, -1 - i$ **119.** $-\dfrac{\sqrt{6}}{2} + \dfrac{\sqrt{6}}{2}i, \dfrac{\sqrt{6}}{2} - \dfrac{\sqrt{6}}{2}i$

121. $-1.5538 + 0.6436i, 1.5538 - 0.6436i$

123. $\dfrac{\sqrt{6}}{2} + \dfrac{\sqrt{2}}{2}i, -\dfrac{\sqrt{6}}{2} - \dfrac{\sqrt{2}}{2}i$

125. (a) $\sqrt{5}(\cos 60° + i \sin 60°)$

$\sqrt{5}(\cos 240° + i \sin 240°)$

(b)

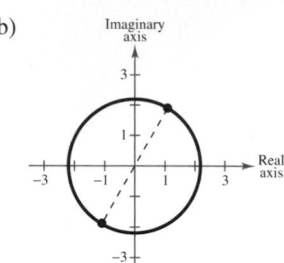

(c) $\dfrac{\sqrt{5}}{2} + \dfrac{\sqrt{15}}{2}i, -\dfrac{\sqrt{5}}{2} - \dfrac{\sqrt{15}}{2}i$

127. (a) $2\left(\cos \dfrac{\pi}{3} + i \sin \dfrac{\pi}{3}\right)$

$2\left(\cos \dfrac{5\pi}{6} + i \sin \dfrac{5\pi}{6}\right)$

$2\left(\cos \dfrac{4\pi}{3} + i \sin \dfrac{4\pi}{3}\right)$

$2\left(\cos \dfrac{11\pi}{6} + i \sin \dfrac{11\pi}{6}\right)$

(b)

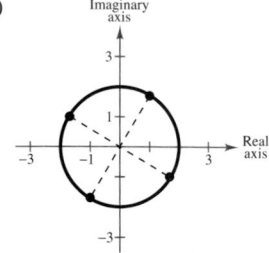

(c) $1 + \sqrt{3}i, -\sqrt{3} + i, -1 - \sqrt{3}i, \sqrt{3} - i$

129. (a) $3\left(\cos \dfrac{\pi}{2} + i \sin \dfrac{\pi}{2}\right)$

$3\left(\cos \dfrac{7\pi}{6} + i \sin \dfrac{7\pi}{6}\right)$

$3\left(\cos \dfrac{11\pi}{6} + i \sin \dfrac{11\pi}{6}\right)$

(b)

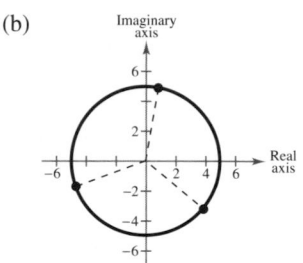

(c) $3i, -\dfrac{3\sqrt{3}}{2} - \dfrac{3}{2}i, \dfrac{3\sqrt{3}}{2} - \dfrac{3}{2}i$

131. (a) $5\left(\cos \dfrac{4\pi}{9} + i \sin \dfrac{4\pi}{9}\right)$

$5\left(\cos \dfrac{10\pi}{9} + i \sin \dfrac{10\pi}{9}\right)$

$5\left(\cos \dfrac{16\pi}{9} + i \sin \dfrac{16\pi}{9}\right)$

(b)

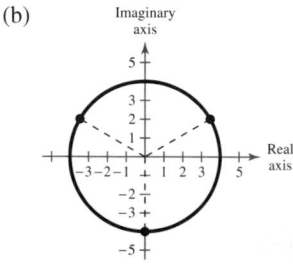

(c) $0.8682 + 4.9240i, -4.6985 - 1.7101i,$

$3.8302 - 3.2139i$

133. (a) $4\left(\cos \dfrac{\pi}{6} + i \sin \dfrac{\pi}{6}\right)$

$4\left(\cos \dfrac{5\pi}{6} + i \sin \dfrac{5\pi}{6}\right)$

$4\left(\cos \dfrac{3\pi}{2} + i \sin \dfrac{3\pi}{2}\right)$

(b)

(c) $2\sqrt{3} + 2i, -2\sqrt{3} + 2i, -4i$

135. (a) $\cos 0 + i \sin 0$

$$\cos \frac{2\pi}{5} + i \sin \frac{2\pi}{5}$$

$$\cos \frac{4\pi}{5} + i \sin \frac{4\pi}{5}$$

$$\cos \frac{6\pi}{5} + i \sin \frac{6\pi}{5}$$

$$\cos \frac{8\pi}{5} + i \sin \frac{8\pi}{5}$$

(b)

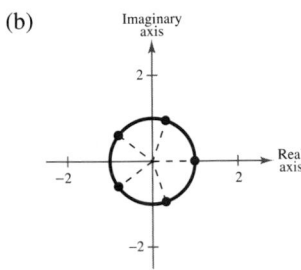

(c) $1, 0.3090 + 0.9511i, -0.8090 + 0.5878i,$
$-0.8090 - 0.5878i, 0.3090 - 0.9511i$

137. (a) $5\left(\cos \frac{\pi}{3} + i \sin \frac{\pi}{3}\right)$

$$5(\cos \pi + i \sin \pi)$$

$$5\left(\cos \frac{5\pi}{3} + i \sin \frac{5\pi}{3}\right)$$

(b)

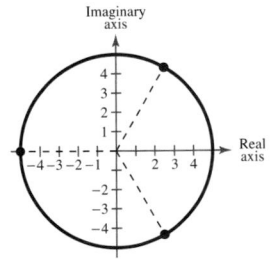

(c) $\dfrac{5}{2} + \dfrac{5\sqrt{3}}{2}i, -5, \dfrac{5}{2} - \dfrac{5\sqrt{3}}{2}i$

139. (a) $2\sqrt{2}\left(\cos \frac{3\pi}{20} + i \sin \frac{3\pi}{20}\right)$

$$2\sqrt{2}\left(\cos \frac{11\pi}{20} + i \sin \frac{11\pi}{20}\right)$$

$$2\sqrt{2}\left(\cos \frac{19\pi}{20} + i \sin \frac{19\pi}{20}\right)$$

$$2\sqrt{2}\left(\cos \frac{27\pi}{20} + i \sin \frac{27\pi}{20}\right)$$

$$2\sqrt{2}\left(\cos \frac{7\pi}{4} + i \sin \frac{7\pi}{4}\right)$$

(b)

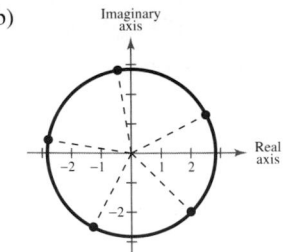

(c) $2.5201 + 1.2841i, -0.4425 + 2.7936i,$
$-2.7936 + 0.4425i, -1.2841 - 2.5201i,$
$2 - 2i$

141. $\cos \dfrac{\pi}{8} + i \sin \dfrac{\pi}{8}$

$$\cos \frac{5\pi}{8} + i \sin \frac{5\pi}{8}$$

$$\cos \frac{9\pi}{8} + i \sin \frac{9\pi}{8}$$

$$\cos \frac{13\pi}{8} + i \sin \frac{13\pi}{8}$$

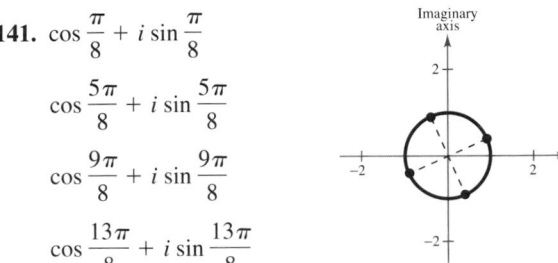

143. $3\left(\cos \dfrac{\pi}{5} + i \sin \dfrac{\pi}{5}\right)$

$$3\left(\cos \frac{3\pi}{5} + i \sin \frac{3\pi}{5}\right)$$

$$3(\cos \pi + i \sin \pi)$$

$$3\left(\cos \frac{7\pi}{5} + i \sin \frac{7\pi}{5}\right)$$

$$3\left(\cos \frac{9\pi}{5} + i \sin \frac{9\pi}{5}\right)$$

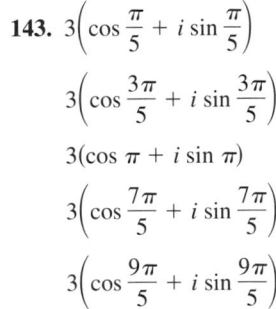

145. $2\left(\cos \dfrac{3\pi}{8} + i \sin \dfrac{3\pi}{8}\right)$

$$2\left(\cos \frac{7\pi}{8} + i \sin \frac{7\pi}{8}\right)$$

$$2\left(\cos \frac{11\pi}{8} + i \sin \frac{11\pi}{8}\right)$$

$$2\left(\cos \frac{15\pi}{8} + i \sin \frac{15\pi}{8}\right)$$

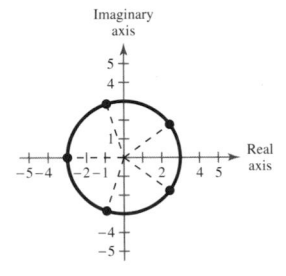

147. $\sqrt[6]{2}\left(\cos \dfrac{7\pi}{12} + i \sin \dfrac{7\pi}{12}\right)$

$$\sqrt[6]{2}\left(\cos \frac{5\pi}{4} + i \sin \frac{5\pi}{4}\right)$$

$$\sqrt[6]{2}\left(\cos \frac{23\pi}{12} + i \sin \frac{23\pi}{12}\right)$$

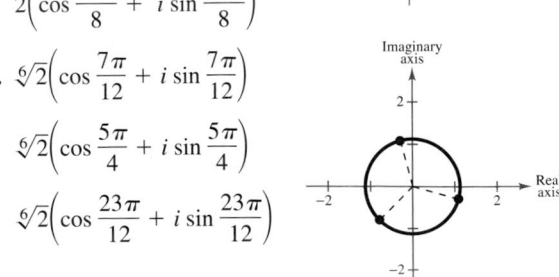

CHAPTER 6

149. $34 + 38i$ **151.** $\frac{3}{2} - \frac{1}{2}i$ **153.** $\frac{39}{34} + \frac{3}{34}i$

155. True   **157.** True   **159.** Answers will vary.

161. Answers will vary.

163. Maximum displacement: $16; t = 2$

165. Maximum displacement: $\frac{1}{8}; t = \frac{1}{24}$

167. $\dfrac{\pi}{2}, \dfrac{3\pi}{2}$ **169.** $\dfrac{5\pi}{6}, \dfrac{7\pi}{6}$

Review Exercises   (page 461)

1. $C = 98°, b \approx 23.13, c \approx 29.90$

3. $A = 50°, a \approx 19.83, b \approx 10.94$

5. $C = 74°15', a \approx 5.84, c \approx 6.48$

7. No solution

9. $A \approx 34.23°, C \approx 30.77°, c \approx 8.18$

11. Two solutions
$A \approx 60.52°, B \approx 69.48°, b \approx 26.90$
$A \approx 119.48°, B \approx 10.52°, b \approx 5.23$

13. 9.08   **15.** 221.34

17. 15.29 meters   **19.** 31.01 feet

21. $A \approx 82.82°, B \approx 41.41°, C \approx 55.77°$

23. $A \approx 15.29°, B \approx 20.59°, C \approx 144.11°$

25. $A \approx 13.19°, B \approx 20.98°, C \approx 145.83°$

27. $A \approx 86.38°, B \approx 28.62°, c \approx 22.70$

29. $A = 35°, C = 35°, b \approx 6.55$

31. $A \approx 41°41', C \approx 82°49', b \approx 15.37$

33. 4.29 feet, 12.63 feet   **35.** 1135.45 miles

37. 9.80 square units   **39.** 511.71 square units

41. $\langle 7, -5 \rangle$ **43.** $\langle 7, -7 \rangle$ **45.** $\langle -4, 4\sqrt{3} \rangle$

47.
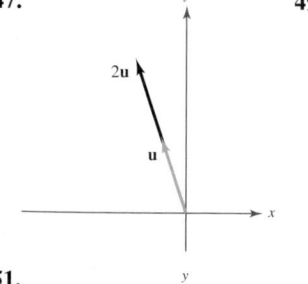

49.
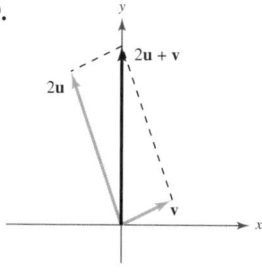

51.
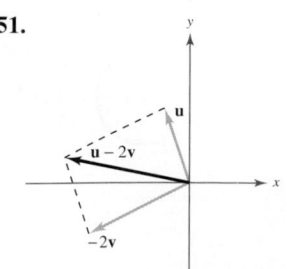

53. (a) $\langle -4, 3 \rangle$ (b) $\langle 2, -9 \rangle$ (c) $\langle -3, -9 \rangle$
(d) $\langle -11, -3 \rangle$

55. (a) $\langle -1, 6 \rangle$ (b) $\langle -9, -2 \rangle$ (c) $\langle -15, 6 \rangle$
(d) $\langle -17, 18 \rangle$

57. (a) $7\mathbf{i} + 2\mathbf{j}$ (b) $-3\mathbf{i} - 4\mathbf{j}$ (c) $6\mathbf{i} - 3\mathbf{j}$
(d) $20\mathbf{i} + \mathbf{j}$

59. (a) $3\mathbf{i} + 6\mathbf{j}$ (b) $5\mathbf{i} - 6\mathbf{j}$ (c) $12\mathbf{i}$ (d) $18\mathbf{i} + 12\mathbf{j}$

61. $\langle 30, 9 \rangle$ **63.** $\langle 74, -5 \rangle$

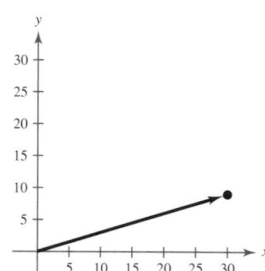
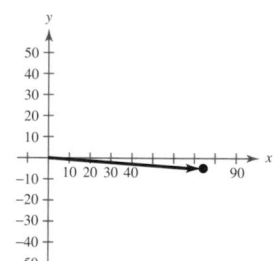

65. $\langle 0, -1 \rangle$ **67.** $\dfrac{\sqrt{29}}{29}\langle 5, -2 \rangle$ **69.** $9\mathbf{i} - 8\mathbf{j}$

71. $10\sqrt{2}\left[\left(\cos\dfrac{3\pi}{4}\right)\mathbf{i} + \left(\sin\dfrac{3\pi}{4}\right)\mathbf{j}\right]$

73.
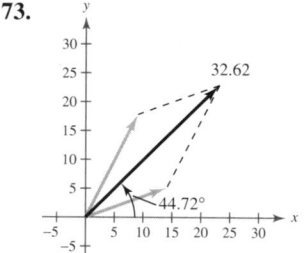

Magnitude: 32.62

Direction: 44.72°

75. 60°, 76.79 pounds   **77.** 180 pounds

79. 422.30 miles per hour, 130.41°   **81.** -20 **83.** 7

85. 25   **87.** -40 **89.** 2.802   **91.** $\dfrac{11\pi}{12}$

93.
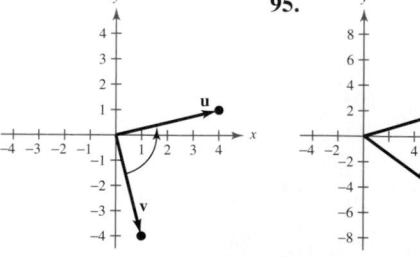

90°

95.

52.2°

97. Parallel   **99.** Neither   **101.** $\frac{1}{2}$ **103.** -1

105. $\frac{13}{17}\langle -4, -1 \rangle$, $\langle -\frac{52}{17}, -\frac{13}{17} \rangle + \langle -\frac{16}{17}, \frac{64}{17} \rangle$

107. $\frac{5}{2}\langle -1, 1 \rangle$, $\langle -\frac{5}{2}, \frac{5}{2} \rangle + \langle \frac{9}{2}, \frac{9}{2} \rangle$ **109.** 72,000 foot-pounds

111.

Imaginary axis

$-i$

1

113.

Imaginary axis

$7 - 5i$

$\sqrt{74}$

115. $2\sqrt{2}\left(\cos \dfrac{7\pi}{4} + i \sin \dfrac{7\pi}{4} \right)$

117. $2\left(\cos \dfrac{7\pi}{6} + i \sin \dfrac{7\pi}{6} \right)$ **119.** $2\left(\cos \dfrac{3\pi}{2} + i \sin \dfrac{3\pi}{2} \right)$

121. $10\left(\cos \dfrac{3\pi}{4} + i \sin \dfrac{3\pi}{4} \right)$ **123.** $4(\cos 240° + i \sin 240°)$

125. (a) $2\sqrt{2}\left(\cos \dfrac{7\pi}{4} + i \sin \dfrac{7\pi}{4} \right)$

$3\sqrt{2}\left(\cos \dfrac{\pi}{4} + i \sin \dfrac{\pi}{4} \right)$

(b) and (c) 12

127. (a) $\cos \dfrac{3\pi}{2} + i \sin \dfrac{3\pi}{2}$

$2\sqrt{2}\left(\cos \dfrac{\pi}{4} + i \sin \dfrac{\pi}{4} \right)$

(b) and (c) $2 - 2i$

129. (a) $3\sqrt{2}\left(\cos \dfrac{7\pi}{4} + i \sin \dfrac{7\pi}{4} \right)$

$2\sqrt{2}\left(\cos \dfrac{\pi}{4} + i \sin \dfrac{\pi}{4} \right)$

(b) and (c) $-\dfrac{3}{2}i$

131. $\dfrac{625}{2} + \dfrac{625\sqrt{3}}{2}i$ **133.** $2035 - 828i$

135. $\pm(0.3660 + 1.3660i)$ **137.** $-1 + i, 1 - i$

139. $\pm(0.6436 - 1.5538i)$

141. (a)

$3\left(\cos \dfrac{\pi}{4} + i \sin \dfrac{\pi}{4} \right)$

$3\left(\cos \dfrac{7\pi}{12} + i \sin \dfrac{7\pi}{12} \right)$

$3\left(\cos \dfrac{11\pi}{12} + i \sin \dfrac{11\pi}{12} \right)$

$3\left(\cos \dfrac{5\pi}{4} + i \sin \dfrac{5\pi}{4} \right)$

$3\left(\cos \dfrac{19\pi}{12} + i \sin \dfrac{19\pi}{12} \right)$

$3\left(\cos \dfrac{23\pi}{12} + i \sin \dfrac{23\pi}{12} \right)$

(b)

Imaginary axis

Real axis

(c) $2.1213 + 2.1213i$, $-0.7765 + 2.8978i$,

$-2.8978 + 0.7765i$, $-2.1213 - 2.1213i$,

$0.7765 - 2.8978i$, $2.8978 - 0.7765i$

143. (a) $2(\cos 0 + i \sin 0)$ (b)

$2\left(\cos \dfrac{2\pi}{3} + i \sin \dfrac{2\pi}{3} \right)$

$2\left(\cos \dfrac{4\pi}{3} + i \sin \dfrac{4\pi}{3} \right)$

Imaginary axis

Real axis

(c) $2, -1 + \sqrt{3}i, -1 - \sqrt{3}i$

145. $4\left(\cos \dfrac{\pi}{4} + i \sin \dfrac{\pi}{4} \right)$

$4\left(\cos \dfrac{3\pi}{4} + i \sin \dfrac{3\pi}{4} \right)$

$4\left(\cos \dfrac{5\pi}{4} + i \sin \dfrac{5\pi}{4} \right)$

$4\left(\cos \dfrac{7\pi}{4} + i \sin \dfrac{7\pi}{4} \right)$

Imaginary axis

Real axis

147. $2\left(\cos \dfrac{\pi}{2} + i \sin \dfrac{\pi}{2} \right)$

$2\left(\cos \dfrac{7\pi}{6} + i \sin \dfrac{7\pi}{6} \right)$

$2\left(\cos \dfrac{11\pi}{6} + i \sin \dfrac{11\pi}{6} \right)$

Imaginary axis

Real axis

149. True   **151.** Direction and magnitude

153. $z_1 z_2 = -4$, $\dfrac{z_1}{z_2} = -\dfrac{1}{4}z_1^{\,2}$

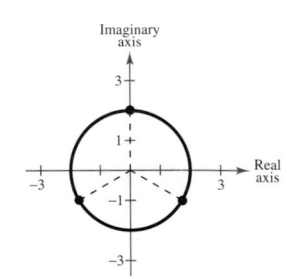

CHAPTER 6

Chapter Test (page 465)

1. $C = 46°$, $a \approx 13.07$, $b \approx 22.03$

2. $A \approx 22.33°$, $B \approx 49.46°$, $C \approx 108.21°$

3. $B \approx 40.11°$, $C = 104.89°$, $a \approx 7.12$

4. Two solutions

   $B \approx 41.10°$, $C \approx 113.90°$, $c \approx 38.94$

   $B \approx 138.90°$, $C \approx 16.10°$, $c \approx 11.81$

5. No solution   6. $B \approx 14.79°$, $C \approx 15.21°$, $c \approx 4.93$

7. 675 feet   8. 2337 square meters

9. $\mathbf{w} = \langle 12, 13 \rangle$, $\|\mathbf{w}\| \approx \sqrt{313}$

10. (a) $\langle -4, -20 \rangle$ (b) $\langle 6, 20 \rangle$ (c) $\langle 2, -12 \rangle$

11. (a) $\langle -4, -23 \rangle$ (b) $\langle 1, 27 \rangle$ (c) $\langle -9, -5 \rangle$

12. (a) $13\mathbf{i} + 17\mathbf{j}$ (b) $-17\mathbf{i} - 28\mathbf{j}$ (c) $-\mathbf{i} - 14\mathbf{j}$

13. (a) $-\mathbf{j}$ (b) $5\mathbf{i} + 9\mathbf{j}$ (c) $11\mathbf{i} + 17\mathbf{j}$

14. $\dfrac{\sqrt{65}}{65}\langle 7, 4 \rangle$ 15. $\left\langle \dfrac{18\sqrt{34}}{17}, -\dfrac{30\sqrt{34}}{17} \right\rangle$

16. $\theta \approx 14.87°$, 250.15 pounds   17. 3

18. $105.95°$ 19. No, because $\mathbf{u} \cdot \mathbf{v} = 24$, not 0.

20. $\left\langle \dfrac{185}{26}, \dfrac{37}{26} \right\rangle$, $\mathbf{u} = \left\langle \dfrac{185}{26}, \dfrac{37}{26} \right\rangle + \left\langle -\dfrac{29}{26}, \dfrac{145}{26} \right\rangle$

21. $z = 2\sqrt{2}\left(\cos\dfrac{3\pi}{4} + i \sin\dfrac{3\pi}{4} \right)$ 22. $-50 - 50\sqrt{3}\,i$

23. $-\dfrac{6561}{2} + \dfrac{6561\sqrt{3}}{2}\,i$ 24. $5832i$

25. $4\left(\cos\dfrac{\pi}{12} + i \sin\dfrac{\pi}{12} \right)$

   $4\left(\cos\dfrac{7\pi}{12} + i \sin\dfrac{7\pi}{12} \right)$

   $4\left(\cos\dfrac{13\pi}{12} + i \sin\dfrac{13\pi}{12} \right)$

   $4\left(\cos\dfrac{19\pi}{12} + i \sin\dfrac{19\pi}{12} \right)$

26. $5\left(\cos\dfrac{\pi}{8} + i \sin\dfrac{\pi}{8} \right)$

   $5\left(\cos\dfrac{5\pi}{8} + i \sin\dfrac{5\pi}{8} \right)$

   $5\left(\cos\dfrac{9\pi}{8} + i \sin\dfrac{9\pi}{8} \right)$

   $5\left(\cos\dfrac{13\pi}{8} + i \sin\dfrac{13\pi}{8} \right)$

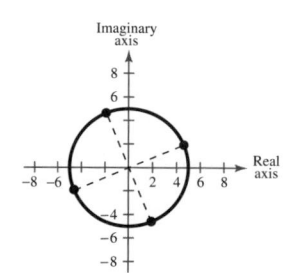

Cumulative Test for Chapters 4–6 (page 466)

1. (a)   (b) $210°$

   (c) $-\dfrac{5\pi}{6}$

   (d) $30°$

   (e) $\sin(-150°) = -\dfrac{1}{2}$

   $\cos(-150°) = -\dfrac{\sqrt{3}}{2}$

   $\tan(-150°) = \dfrac{\sqrt{3}}{3}$

   $\csc(-150°) = -2$

   $\sec(-150°) = -\dfrac{2\sqrt{3}}{3}$

   $\cot(-150°) = \sqrt{3}$

2. $146.1°$ 3. $\cos\theta = -\dfrac{5}{13}$

4. 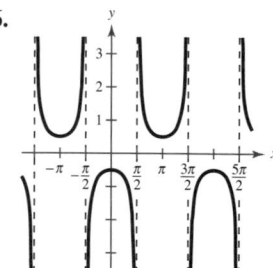   5.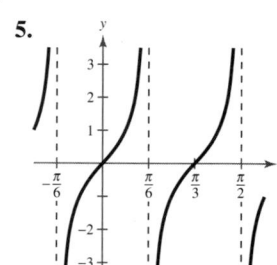

6.

7. $a = -3$, $b = \pi$, $c = 0$

8. $\dfrac{3}{5}$ 9. $-\dfrac{\sqrt{3}}{3}$ 10. $\dfrac{2x}{\sqrt{4x^2 + 1}}$ 11. $2\tan\theta$

12–14. Answers will vary.   15. $\dfrac{3\pi}{2} + 2n\pi$

16. $\dfrac{\pi}{6} + n\pi$, $\dfrac{5\pi}{6} + n\pi$ 17. $1.7646, 4.5186$

18.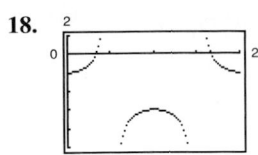

$$\frac{\pi}{3}, \frac{5\pi}{3}$$

19.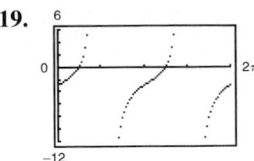

$$\frac{\pi}{4}, \frac{5\pi}{4}$$

20. $\dfrac{16}{63}$ **21.** $\dfrac{4}{3}$ **22.** $\dfrac{2\sqrt{5}}{5}$

23. $2\cos 6x \cos 2x$ **24–27.** Answers will vary.

28. $B \approx 14.89°$

$C \approx 119.11°$

$c \approx 17.00$

29. $B \approx 52.82°$

$C \approx 95.18°$

$a \approx 5.32$

30. $B = 55°$

$b \approx 20.14$

$c \approx 24.13$

31. $A \approx 26.07°$

$B \approx 33.33°$

$C \approx 120.60°$

32. 131.71 square inches   **33.** 94.10 square inches

34. $3\mathbf{i} + 5\mathbf{j}$ **35.** $\dfrac{\sqrt{5}}{5}\mathbf{i} - \dfrac{2\sqrt{5}}{5}\mathbf{j}$ **36.** -5 **37.** 1

38. $\left\langle -\frac{1}{13}, -\frac{5}{13} \right\rangle$; $\mathbf{u} = \left\langle \frac{105}{13}, -\frac{21}{13} \right\rangle + \left\langle -\frac{1}{13}, -\frac{5}{13} \right\rangle$

39. $3\sqrt{2}\left(\cos\dfrac{3\pi}{4} + i\sin\dfrac{3\pi}{4} \right)$ **40.** $-4\sqrt{3} + 4i$

41. $-12\sqrt{3} + 12i$

42. $1.4553 + 0.3436i, -1.4553 - 0.3436i$

43. $1, -\dfrac{1}{2} + \dfrac{\sqrt{3}}{2}i, -\dfrac{1}{2} - \dfrac{\sqrt{3}}{2}i$

44. $3\left(\cos\dfrac{\pi}{5} + i\sin\dfrac{\pi}{5} \right)$

$3\left(\cos\dfrac{3\pi}{5} + i\sin\dfrac{3\pi}{5} \right)$

$3(\cos\pi + i\sin\pi)$

$3\left(\cos\dfrac{7\pi}{5} + i\sin\dfrac{7\pi}{5} \right)$

$3\left(\cos\dfrac{9\pi}{5} + i\sin\dfrac{9\pi}{5} \right)$

45. 5 feet   **46.** $d = 4\sin\dfrac{\pi}{4}t$

47. $32.63°$, 543.88 kilometers per hour

48. $80.28°$

Chapter 7

Section 7.1   (page 481)

1. (a) No   (b) No   (c) No   (d) Yes

3. (a) No   (b) Yes   (c) No   (d) No

5. $(2, 2)$ **7.** $(2, 6), (-1, 3)$

9. $(0, 2), \left(\sqrt{3}, 2 - 3\sqrt{3} \right), \left(-\sqrt{3}, 2 + 3\sqrt{3} \right)$

11. $(4, 4)$ **13.** $(5, 5)$ **15.** $\left(\frac{1}{2}, 3 \right)$ **17.** $(1, 1)$

19. $\left(\frac{20}{3}, \frac{40}{3} \right)$ **21.** No solution   **23.** $(-2, 0), (3, 5)$

25. No real solution   **27.** $(0, 0), (1, 1), (-1, -1)$

29. $(4, 3)$ **31.** $\left(\frac{5}{2}, \frac{3}{2} \right)$ **33.** No real solution

35. $(3, 6), (-3, 0)$ **37.** $(4, -0.5)$ **39.** $(8, 3), (3, -2)$

41. $(\pm 1.540, 2.372)$ **43.** $(0, 1)$ **45.** $(2.318, 2.841)$

47. $(2.25, 5.5)$ **49.** $(0, -13), (\pm 12, 5)$ **51.** $(1, 2)$

53. $(-2, 0), \left(\frac{29}{10}, \frac{21}{10} \right)$ **55.** No real solution

57. $(0.25, 1.5)$ **59.** $(0.287, 1.751)$

61. $(0, 1), (1, 0)$ **63.** $\left(-4, -\frac{1}{4} \right), \left(\frac{1}{2}, 2 \right)$

65.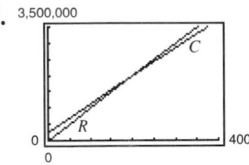

192 units; \$1,910,400

67.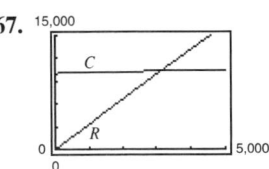

3133 units; \$10,308

69. (a)

Week	Animated	Horror
1	336	42
2	312	60
3	288	78
4	264	96
5	240	114
6	216	132
7	192	150
8	168	168
9	144	186
10	120	204
11	96	222
12	72	240

(b) and (c) $x = 8$

(d) The answers are the same.

(e) During week 8 the same number of animated and horror films were rented.

71. (a) $C = 35.45x + 16{,}000$

$R = 55.95x$

(b)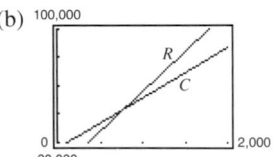

780 units

73. Sales greater than $11,667

75. (a) $\begin{cases} x + y = 20{,}000 \\ 0.065x + 0.085y = 1{,}600 \end{cases}$

(b)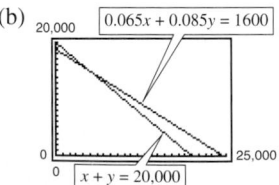

More invested at 6.5% means less invested at 8.5% and less interest.

(c) $5000

77. (a)

Year	Missouri	Tennessee
1990	5104	4875
1994	5294	5181
1998	5483	5487
2002	5673	5793
2006	5862	6099
2010	6052	6405

(b) From 1998 to 2010

(c) 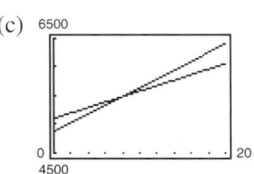   (d) $(7.87, 5477.01)$

$(7.87, 5477.01)$

(e) During 1997 the populations of Missouri and Tennessee were equal.

79. 6 meters $\times$ 9 meters   **81.** 8 miles $\times$ 12 miles

83. False. You can solve for either variable before back-substituting.

85. For a linear system, the result will be a contradictory equation such as $0 = N$, where N is a nonzero real number. For a nonlinear system, there may be an equation with imaginary roots.

87. (a) $y = x + 1$ (b) $y = 0$ (c) $y = -2$

89. $\begin{cases} y = x - 3 \\ 2y = x - 4 \end{cases}$ **91.** $2x + 7y - 45 = 0$

93. $y - 3 = 0$ **95.** $30x - 17y - 18 = 0$

97. Domain: All real numbers x except $x = 6$

Asymptotes: $y = 0, x = 6$

99. Domain: All real numbers x except $x = \pm 4$

Asymptotes: $y = 1, x = \pm 4$

101. Domain: All real numbers x

Asymptote: $y = 0$

Section 7.2   (page 491)

Vocabulary Check   (page 491)

1. method, elimination   **2.** equivalent

3. consistent, inconsistent

1. $(2, 1)$

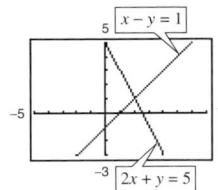

3. $(1, -1)$

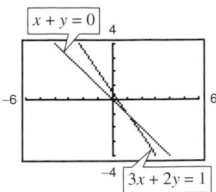

5. Inconsistent

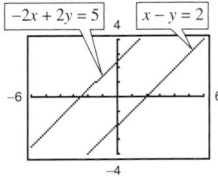

7. $\left(\frac{5}{2}, \frac{3}{4}\right)$ **9.** $(3, 4)$ **11.** $(4, -1)$ **13.** $\left(\frac{12}{7}, \frac{18}{7}\right)$

15. Inconsistent   **17.** b; One solution, consistent

19. c; One solution, consistent   **21.** $\left(-\frac{6}{35}, \frac{43}{35}\right)$

23. Inconsistent   **25.** All points on $6x + 8y - 1 = 0$

27. $(5, -2)$ **29.** $(7, 1)$

31. All points on $5x - 6y - 3 = 0$ **33.** $(101, 96)$

35. $\left(\frac{90}{31}, -\frac{67}{31}\right)$ **37.** $(-1, 1)$ **39.** $\left(1, \frac{1}{2}\right)$

41.

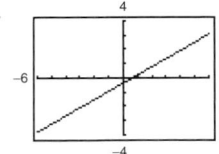

Consistent; $(5, 2)$

43.

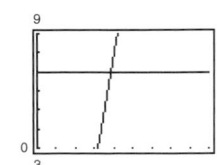

Inconsistent

45.

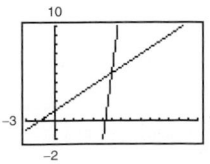

Consistent; all points
on $8x - 14y = 5$

47.

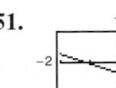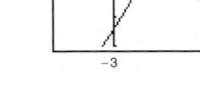

$(3.833, 7)$

49.

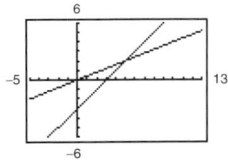

$(6, 5)$

51.

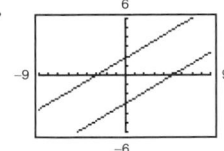

$(1, -0.667)$

53.

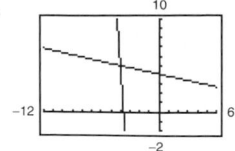

$(-4, 5)$

55. $(4, 1)$ **57.** $\left(-\frac{5}{3}, -\frac{11}{3}\right)$

59. $(6, -3)$ **61.** $\left(\frac{43}{6}, \frac{25}{6}\right)$

63. $\begin{cases} 3x + \frac{1}{2}y = 4 \\ x + 3y = 24 \end{cases}$ **65.** $\begin{cases} 2x + 2y = 11 \\ x - 4y = -7 \end{cases}$

67. $(80, 10)$ **69.** $(2{,}000{,}000, 100)$

71. Plane: 550 miles per hour; wind: 50 miles per hour

73. (a) $\begin{cases} 5.00A + 3.50C = 5087.50 \\ A + C = 1175 \end{cases}$

(b) $A = 650$, $C = 525$; Answers will vary.

(c) $A = 650$, $C = 525$

75. 9 oranges, 7 grapefruit

77. 185 movies, 125 video game cartridges

79. $y = 0.97x + 2.1$ **81.** $y = -2.5x + 5.54$

83. (a) and (b) $y = 14x + 19$

(c)

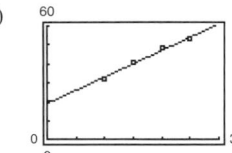

(d) 41.4 bushels per acre

85. True. A linear system can have only one solution, no solution, or infinitely many solutions.

87. $(39{,}600, 398)$. The solution is outside the viewing window.

89. No. Two lines can only intersect once, never, or infinitely many times.

91. $k = -4$ **93.** $u = -x \ln x$; $v = \ln x$

95. $x \le -\frac{22}{3}$ **97.** $-2 < x < 18$

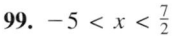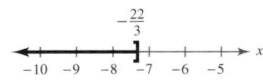   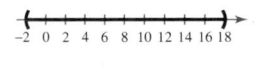

99. $-5 < x < \frac{7}{2}$ **101.** $\ln 6x$

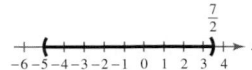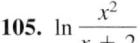

103. $\log_9 \dfrac{12}{x}$ **105.** $\ln \dfrac{x^2}{x + 2}$ **107.** Answers will vary.

CHAPTER 7

Section 7.3   (page 505)

1. (a) No   (b) Yes   (c) No   (d) No

3. (a) No   (b) No   (c) Yes   (d) No

5. $(2, -2, 2)$ **7.** $(3, 10, 2)$ **9.** $\left(\frac{1}{2}, -2, 2\right)$

11. $\begin{cases} x - 2y + 3z = 5 \\ \quad\quad y - 2z = 9 \\ 2x \quad\quad - 3z = 0 \end{cases}$

It removed the x-term from Equation 2.

13. $\begin{cases} x - 2y + z = 1 \\ \quad\quad 3y + z = -2 \\ 3x - y - 4z = 1 \end{cases}$

It removed the x-term from Equation 2.

15. $(1, 2, 3)$ **17.** $(-4, 8, 5)$ **19.** $(2, -3, -2)$

21. Inconsistent   **23.** $\left(1, -\frac{3}{2}, \frac{1}{2}\right)$

25. $(-3a + 10, 5a - 7, a)$ **27.** $(1, -1, 2)$

29. $(-a + 3, a + 1, a)$ **31.** Inconsistent

33. Inconsistent   **35.** $(-1, 1, 0)$

37. $(2a, 21a - 1, 8a)$ **39.** $\left(-\frac{3}{2}a + \frac{1}{2}, -\frac{2}{3}a + 1, a\right)$

41. $(-5a + 3, -a - 5, a)$ **43.** $\left(\frac{1}{2}, 0, \frac{1}{2}, \frac{3}{2}\right)$

45. $(1, -2, -1)$ **47.** $(1, -1, 2)$

49. $\begin{cases} 3x + y - z = 9 \\ x + 2y - z = 0 \\ -x + y + 3z = 1 \end{cases}$ **51.** $\begin{cases} x + 6y + 4z = 7 \\ 2x - 2y - 4z = 0 \\ -x + y + z = -\frac{7}{4} \end{cases}$

53.

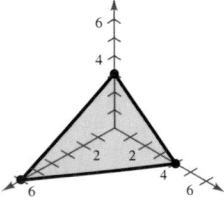

$(6, 0, 0), (0, 4, 0),$
$(0, 0, 3), (4, 0, 1)$

55.

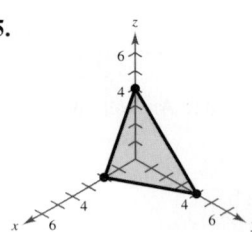

$(2, 0, 0), (0, 4, 0),$
$(0, 0, 4), (0, 2, 2)$

57. $\dfrac{A}{x} + \dfrac{B}{x - 14}$ **59.** $\dfrac{A}{x} + \dfrac{B}{x^2} + \dfrac{C}{x - 10}$

61. $\dfrac{A}{x - 5} + \dfrac{B}{(x - 5)^2} + \dfrac{C}{(x - 5)^3}$

63. $\dfrac{1}{2}\left(\dfrac{1}{x - 1} - \dfrac{1}{x + 1}\right)$ **65.** $\dfrac{1}{x} - \dfrac{1}{x + 1}$

67. $\dfrac{1}{x} - \dfrac{2}{2x + 1}$ **69.** $\dfrac{3}{2x - 1} - \dfrac{2}{x + 1}$

71. $-\dfrac{3}{x} - \dfrac{1}{x + 2} + \dfrac{5}{x - 2}$ **73.** $\dfrac{3}{x} - \dfrac{1}{x^2} + \dfrac{1}{x + 1}$

75. $\dfrac{3}{x} - \dfrac{3}{x - 3} + \dfrac{2}{(x - 3)^2}$

77. $2x - 7 + \dfrac{17}{x + 2} + \dfrac{1}{x + 1}$

79. $x + 3 + \dfrac{6}{x - 1} + \dfrac{4}{(x - 1)^2} + \dfrac{1}{(x - 1)^3}$

81. $\dfrac{3}{x} - \dfrac{2}{x - 4}$

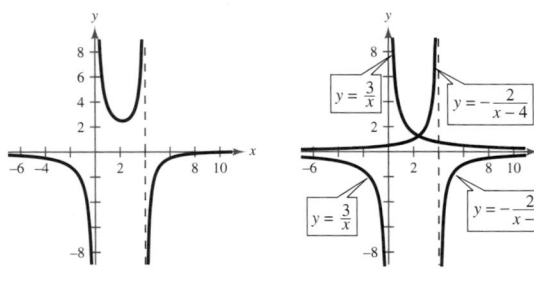

The vertical asymptotes are the same.

83. $s = -16t^2 + 144$ **85.** $s = -16t^2 - 32t + 500$

87. $y = \frac{1}{2}x^2 - 2x$ **89.** $y = x^2 - 6x + 8$

91. $x^2 + y^2 - 4x = 0$ **93.** $x^2 + y^2 + 6x - 8y = 0$

95. \$366,666.67 at 8%, \$316,666.67 at 9%, and \$91,666.67 at 10%

97. \$156,250 + 0.75g in certificates of deposit

\$125,000 in municipal bonds

\$218,750 - 1.75g in blue-chip stocks

g in growth stocks

99. 24 two-point field goals, 6 three-point field goals, 18 free throws

101. Four touchdowns, four extra-point kicks, one field goal

103. $I_1 = 1, I_2 = 2, I_3 = 1$

105. Four par-3 holes, 10 par-4 holes, four par-5 holes

107. $y = -\frac{5}{24}x^2 - \frac{3}{10}x + \frac{41}{6}$ **109.** $y = x^2 - x$

111. (a) $\begin{cases} 900a + 30b + c = 55 \\ 1600a + 40b + c = 105 \\ 2500a + 50b + c = 188 \end{cases}$

$y = 0.165x^2 - 6.55x + 103$

(b)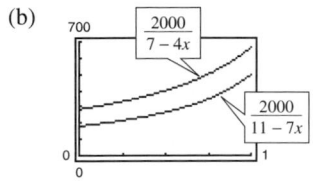

(c) 453 feet

113. (a) $\dfrac{2000}{7 - 4x} - \dfrac{2000}{11 - 7x}$, $0 \le x \le 1$

(b)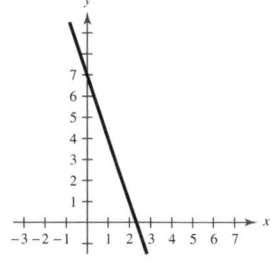

115. False. The leading coefficients are not all 1.

117. False. $\dfrac{A}{x + 10} + \dfrac{B}{x - 10} + \dfrac{C}{(x - 10)^2}$

119. $\dfrac{1}{2a}\left(\dfrac{1}{a + x} + \dfrac{1}{a - x}\right)$ **121.** $\dfrac{1}{a}\left(\dfrac{1}{y} + \dfrac{1}{a - y}\right)$

123. No. There are two arithmetic errors. The constant in the second equation should be -11 and the coefficient of z in the third equation should be 2.

125. $x = 5, y = 5, \lambda = -5$

127. $x = \dfrac{\sqrt{2}}{2}, y = \dfrac{1}{2}, \lambda = 1$ **129.**

$x = -\dfrac{\sqrt{2}}{2}, y = \dfrac{1}{2}, \lambda = 1$

$x = 0, y = 0, \lambda = 0$

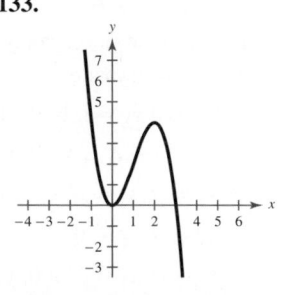

131.

133.

135. (a) $-4, 0, 3$

(b)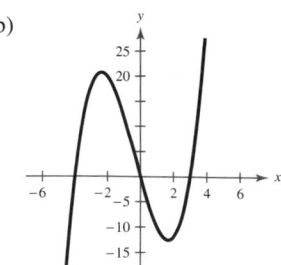

137. (a) $-4, -\dfrac{3}{2}, 3$

(b)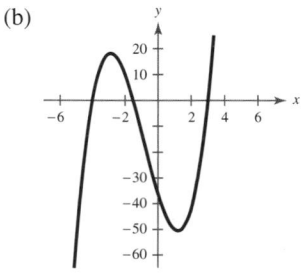

139.

x	-2	-1	0	1	2
y	16	8	4	2	1

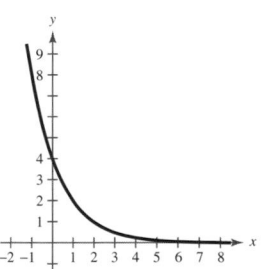

141.

x	-2	-1	0	1	2
y	-0.5	0	1	3	7

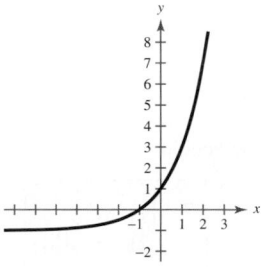

143. Answers will vary.

CHAPTER 7

Section 7.4 (page 521)

Vocabulary Check (page 521)

1. matrix   **2.** square

3. row matrix, column matrix   **4.** augmented matrix

5. coefficient matrix   **6.** row-equivalent

7. reduced row-echelon form

8. Gauss-Jordan elimination

1. 1×2 **3.** 3×1 **5.** 2×2

7. $\begin{bmatrix} 6 & -7 & \vdots & 11 \\ -2 & 5 & \vdots & -1 \end{bmatrix}$

9. $\begin{bmatrix} 1 & 10 & -2 & \vdots & 2 \\ 5 & -3 & 4 & \vdots & 0 \\ 2 & 1 & 0 & \vdots & 6 \end{bmatrix}$

11. $\begin{cases} 3x + 4y = 9 \\ x - y = -3 \end{cases}$ **13.** $\begin{cases} 9x + 12y + 3z = 0 \\ -2x + 18y + 5z = 10 \\ x + 7y - 8z = -4 \end{cases}$

15. $\begin{bmatrix} 1 & 4 & 3 \\ 0 & 2 & -1 \end{bmatrix}$

17. $\begin{bmatrix} 1 & 1 & 4 & -1 \\ 0 & 5 & -2 & 6 \\ 0 & 3 & 20 & 4 \end{bmatrix}, \begin{bmatrix} 1 & 1 & 4 & -1 \\ 0 & 1 & -\frac{2}{5} & \frac{6}{5} \\ 0 & 3 & 20 & 4 \end{bmatrix}$

19. Add -3 times R_2 to R_1.   **21.** Interchange R_1 and R_2.

23. Reduced row-echelon form

25. Not in row-echelon form   **27.** Not in row-echelon form

29. (a) $\begin{bmatrix} 1 & 2 & 3 \\ 0 & -5 & -10 \\ 3 & 1 & -1 \end{bmatrix}$ (b) $\begin{bmatrix} 1 & 2 & 3 \\ 0 & -5 & -10 \\ 0 & -5 & -10 \end{bmatrix}$

(c) $\begin{bmatrix} 1 & 2 & 3 \\ 0 & -5 & -10 \\ 0 & 0 & 0 \end{bmatrix}$ (d) $\begin{bmatrix} 1 & 2 & 3 \\ 0 & 1 & 2 \\ 0 & 0 & 0 \end{bmatrix}$

(e) $\begin{bmatrix} 1 & 0 & -1 \\ 0 & 1 & 2 \\ 0 & 0 & 0 \end{bmatrix}$

The matrix is in reduced row-echelon form.

31. (a) ```
*row+(-2,[A],1,2
)→[B]
   [[1 2 3]
    [0 -5 -10]
    [3 1 -1]]
```
(b) ```
*row+(-3,[B],1,3
)→[C]
   [[1 2 3 ]
    [0 -5 -10]
    [0 -5 -10]]
```

(c) ```
*row+(-1,[C],2,3
)→[D]
   [[1 2 3]
    [0 -5 -10]
    [0 0 0]]
```
(d) ```
*row(-1/5,[D],2)
→[E]
   [[1 2 3]
    [0 1 2]
    [0 0 0]]
```

(e) ```
*row+(-2,[E],2,1
)
   [[1 0 -1]
    [0 1 2]
    [0 0 0]]
```

The matrix is in reduced row-echelon form.

**33.** $\begin{bmatrix} 1 & 3 & \frac{3}{2} & 5 \\ 0 & 1 & \frac{3}{14} & 0 \\ 0 & 0 & 1 & -\frac{35}{12} \end{bmatrix}$    **35.** $\begin{bmatrix} 1 & -1 & -1 & 1 \\ 0 & 1 & 6 & 3 \\ 0 & 0 & 0 & 0 \end{bmatrix}$

**37.** $\begin{bmatrix} 1 & 0 & 0 \\ 0 & 1 & 0 \\ 0 & 0 & 1 \end{bmatrix}$    **39.** $\begin{bmatrix} 1 & 0 & -\frac{3}{7} & -\frac{8}{7} \\ 0 & 1 & -\frac{12}{7} & \frac{10}{7} \end{bmatrix}$

**41.** $\begin{cases} x - 2y = 4 \\ y = -3 \end{cases}$    **43.** $\begin{cases} x - y + 2z = 4 \\ y - z = 2 \\ z = -2 \end{cases}$
$(-2, -3)$
$(8, 0, -2)$

**45.** $(7, -5)$    **47.** $(-4, -8, 2)$    **49.** $(3, 2)$

**51.** Inconsistent    **53.** $(3, -2, 5, 0)$    **55.** $(4, -3, 2)$

**57.** $(2a + 1, 3a + 2, a)$    **59.** $(7, -3, 4)$

**61.** $(0, 2 - 4a, a)$    **63.** $(-5a, a, 3)$

**65.** Yes; $(-1, 1, -3)$    **67.** No    **69.** $y = x^2 + 2x + 5$

**71.** $y = 2x^2 - x + 1$    **73.** $f(x) = -9x^2 - 5x + 11$

**75.** $f(x) = x^3 - 2x^2 - 4x + 1$

**77.** \$150,000 at 7%, \$750,000 at 8%, \$600,000 at 10%

**79.** $I_1 = \frac{13}{10}, I_2 = \frac{11}{5}, I_3 = \frac{9}{10}$

**81.** (a) $y = -0.0400t^2 + 4.350t + 46.83$

(b)

(c) 2005: \$67.58; 2010: \$86.33; 2015: \$103.08

(d) Answers will vary.

**83.** $\begin{cases} 5.5g + 4.25s + 3.75m = 480 \\ -g + s + m = 0 \\ g + s + m = 100 \end{cases}$

50 pounds of glossy, 35 pounds of semi-gloss, 15 pounds of matte

**85.** (a) $x_1 = s, x_2 = t, x_3 = 600 - s,$
$x_4 = s - t, x_5 = 500 - t, x_6 = s, x_7 = t$

(b) $x_1 = 0, x_2 = 0, x_3 = 600, x_4 = 0, x_5 = 500,$
$x_6 = 0, x_7 = 0$

(c) $x_1 = 0, x_2 = -500, x_3 = 600, x_4 = 500,$
$x_5 = 1000, x_6 = 0, x_7 = -500$
```

87. False. It is a 2×4 matrix.

89. $\begin{cases} x + y + 7z = -1 \\ x + 2y + 11z = 0 \\ 2x + y + 10z = -3 \end{cases}$

(Answer is not unique.)

91. Gauss-Jordan elimination was not performed on the last column of the matrix.

93.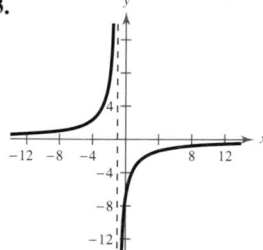

Asymptotes:
$x = -1, y = 0$

95.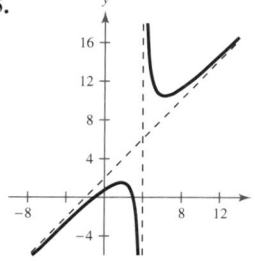

Asymptotes:
$x = 4, y = x + 2$

Section 7.5   (page 536)

Vocabulary Check   (page 536)

1. equal   **2.** scalars   **3.** zero, O **4.** identity
5. (a) iii   (b) i   (c) iv   (d) v   (e) ii
6. (a) ii   (b) iv   (c) i   (d) iii

1. $x = -4, y = 22$ **3.** $x = -1, y = 4, z = 6$

5. (a) $\begin{bmatrix} 3 & -2 \\ 1 & 7 \end{bmatrix}$ (b) $\begin{bmatrix} -1 & 0 \\ 3 & -9 \end{bmatrix}$
(c) $\begin{bmatrix} 3 & -3 \\ 6 & -3 \end{bmatrix}$ (d) $\begin{bmatrix} -1 & -1 \\ 8 & -19 \end{bmatrix}$

7. (a) $\begin{bmatrix} 9 & 5 \\ 1 & -2 \\ -3 & 15 \end{bmatrix}$ (b) $\begin{bmatrix} 7 & -7 \\ 3 & 8 \\ -5 & -5 \end{bmatrix}$
(c) $\begin{bmatrix} 24 & -3 \\ 6 & 9 \\ -12 & 15 \end{bmatrix}$ (d) $\begin{bmatrix} 22 & -15 \\ 8 & 19 \\ -14 & -5 \end{bmatrix}$

9. (a) $\begin{bmatrix} 5 & 5 & -2 & 4 & 4 \\ -5 & 10 & 0 & -4 & -7 \end{bmatrix}$
(b) $\begin{bmatrix} 3 & 5 & 0 & 2 & 4 \\ 7 & -6 & -4 & 2 & 7 \end{bmatrix}$
(c) $\begin{bmatrix} 12 & 15 & -3 & 9 & 12 \\ 3 & 6 & -6 & -3 & 0 \end{bmatrix}$
(d) $\begin{bmatrix} 10 & 15 & -1 & 7 & 12 \\ 15 & -10 & -10 & 3 & 14 \end{bmatrix}$

11. (a) Not possible   (b) Not possible
(c) $\begin{bmatrix} 18 & 0 & 9 \\ -3 & -12 & 0 \end{bmatrix}$ (d) Not possible

13. $\begin{bmatrix} -8 & -7 \\ 15 & -1 \end{bmatrix}$ **15.** $\begin{bmatrix} -24 & -4 & 12 \\ -12 & 32 & 12 \end{bmatrix}$

17. $\begin{bmatrix} -1 & 5 \\ 1 & -2 \end{bmatrix}$ **19.** $\begin{bmatrix} -14.646 & 21.306 \\ -41.546 & -69.137 \\ 78.117 & -32.064 \end{bmatrix}$

21. $\begin{bmatrix} -6 & -9 \\ -1 & 0 \\ 17 & -10 \end{bmatrix}$ **23.** $\begin{bmatrix} 3 & 3 \\ -\frac{1}{2} & 0 \\ -\frac{13}{2} & \frac{11}{2} \end{bmatrix}$

25. Not possible   **27.** $\begin{bmatrix} -2 & 51 \\ -8 & 33 \\ 0 & 27 \end{bmatrix}$

29. $\begin{bmatrix} 1 & 0 & 0 \\ 0 & 1 & 0 \\ 0 & 0 & \frac{7}{2} \end{bmatrix}$ **31.** $\begin{bmatrix} -15 & -5 & -25 & -45 \\ -18 & -6 & -30 & -54 \end{bmatrix}$

33. (a) $\begin{bmatrix} 0 & 15 \\ 8 & 11 \end{bmatrix}$ (b) $\begin{bmatrix} -3 & 2 \\ 39 & 14 \end{bmatrix}$ (c) $\begin{bmatrix} 11 & 6 \\ 15 & 14 \end{bmatrix}$

35. (a) $\begin{bmatrix} 0 & -10 \\ 10 & 0 \end{bmatrix}$ (b) $\begin{bmatrix} 0 & -10 \\ 10 & 0 \end{bmatrix}$ (c) $\begin{bmatrix} 8 & -6 \\ 6 & 8 \end{bmatrix}$

37. (a) $\begin{bmatrix} 7 & 7 & 14 \\ 8 & 8 & 16 \\ -1 & -1 & -2 \end{bmatrix}$ (b) $\begin{bmatrix} 13 \end{bmatrix}$ (c) Not possible

39. $\begin{bmatrix} 70 & -17 & 73 \\ 32 & 11 & 6 \\ 16 & -38 & 70 \end{bmatrix}$ **41.** $\begin{bmatrix} 151 & 25 & 48 \\ 516 & 279 & 387 \\ 47 & -20 & 87 \end{bmatrix}$

43. Not possible   **45.** $\begin{bmatrix} 5 & 8 \\ -4 & -16 \end{bmatrix}$ **47.** $\begin{bmatrix} -4 & 10 \\ 3 & 14 \end{bmatrix}$

49. (a) No   (b) Yes   (c) No   (d) No
51. (a) No   (b) Yes   (c) No   (d) No

53. (a) $\begin{bmatrix} -1 & 1 \\ -2 & 1 \end{bmatrix}\begin{bmatrix} x_1 \\ x_2 \end{bmatrix} = \begin{bmatrix} 4 \\ 0 \end{bmatrix}$ (b) $\begin{bmatrix} 4 \\ 8 \end{bmatrix}$

55. (a) $\begin{bmatrix} -2 & -3 \\ 6 & 1 \end{bmatrix}\begin{bmatrix} x_1 \\ x_2 \end{bmatrix} = \begin{bmatrix} -4 \\ -36 \end{bmatrix}$ (b) $\begin{bmatrix} -7 \\ 6 \end{bmatrix}$

57. (a) $\begin{bmatrix} 1 & -2 & 3 \\ -1 & 3 & -1 \\ 2 & -5 & 5 \end{bmatrix}\begin{bmatrix} x_1 \\ x_2 \\ x_3 \end{bmatrix} = \begin{bmatrix} 9 \\ -6 \\ 17 \end{bmatrix}$ (b) $\begin{bmatrix} 1 \\ -1 \\ 2 \end{bmatrix}$

59. (a) $\begin{bmatrix} 1 & -5 & 2 \\ -3 & 1 & -1 \\ 0 & -2 & 5 \end{bmatrix}\begin{bmatrix} x_1 \\ x_2 \\ x_3 \end{bmatrix} = \begin{bmatrix} -20 \\ 8 \\ -16 \end{bmatrix}$ (b) $\begin{bmatrix} -1 \\ 3 \\ -2 \end{bmatrix}$

61. (a) $\begin{bmatrix} 7 & -2 & 5 \\ -6 & 13 & -8 \\ 16 & 11 & -3 \end{bmatrix}$ (b) $\begin{bmatrix} 7 & -2 & 5 \\ -6 & 13 & -8 \\ 16 & 11 & -3 \end{bmatrix}$

The answers are the same.

CHAPTER 7

63. (a) $\begin{bmatrix} 26 & 11 & 0 \\ 11 & 20 & -3 \\ 11 & 14 & 0 \end{bmatrix}$ (b) $\begin{bmatrix} 26 & 11 & 0 \\ 11 & 20 & -3 \\ 11 & 14 & 0 \end{bmatrix}$

The answers are the same.

65. (a) $\begin{bmatrix} 25 & -34 & 28 \\ -53 & 34 & -7 \\ -76 & 30 & 21 \end{bmatrix}$ (b) $\begin{bmatrix} 25 & -34 & 28 \\ -53 & 34 & -7 \\ -76 & 30 & 21 \end{bmatrix}$

The answers are the same.

67. (a) and (b) $\begin{bmatrix} -1 & 10 & -4 \\ -5 & -1 & 0 \end{bmatrix}$

69. Not possible, undefined   **71.** Not possible, undefined

73. (a) and (b) $\begin{bmatrix} -6 & -12 & 12 \\ 6 & -6 & 0 \end{bmatrix}$

75. $\begin{bmatrix} -4 & 0 \\ 8 & 2 \end{bmatrix}$ **77.** $\begin{bmatrix} 84 & 60 & 30 \\ 42 & 120 & 84 \end{bmatrix}$

79. $[\$1037.50 \quad \$1400.00 \quad \$1012.50]$

The entries represent the total profit made at each outlet.

81. $\begin{bmatrix} \$15,770 & \$18,300 \\ \$26,500 & \$29,250 \\ \$21,260 & \$24,150 \end{bmatrix}$

The entries are the total wholesale and retail prices of the inventory at each outlet.

83. $\begin{bmatrix} 0.40 & 0.15 & 0.15 \\ 0.28 & 0.53 & 0.17 \\ 0.32 & 0.32 & 0.68 \end{bmatrix}$

P^2 represents the changes in party affiliations after two elections.

85. True. To add two matrices, you add corresponding entries.

87. Not possible   **89.** Not possible   **91.** 2×2

93. 2×3 **95.** Answers will vary.

97. Answers will vary.

99. $AC = BC = \begin{bmatrix} 2 & 3 \\ 2 & 3 \end{bmatrix}, A \neq B$

101. (a) $A^2 = \begin{bmatrix} -1 & 0 \\ 0 & -1 \end{bmatrix}, A^3 = \begin{bmatrix} -i & 0 \\ 0 & -i \end{bmatrix},$

$A^4 = \begin{bmatrix} 1 & 0 \\ 0 & 1 \end{bmatrix}$

The entries on the main diagonal are i^2 in A^2, i^3 in A^3, and i^4 in A^4.

(b) $B^2 = \begin{bmatrix} 1 & 0 \\ 0 & 1 \end{bmatrix}$

B^2 is the identity matrix.

103. (a) $A = \begin{bmatrix} 0 & 2 \\ 0 & 0 \end{bmatrix}, B = \begin{bmatrix} 0 & 2 & 3 \\ 0 & 0 & 4 \\ 0 & 0 & 0 \end{bmatrix}$

(Answers are not unique.)

(b) A^2 and B^3 are zero matrices.

(c) $A = \begin{bmatrix} 0 & 2 & 3 & 4 \\ 0 & 0 & 5 & 6 \\ 0 & 0 & 0 & 7 \\ 0 & 0 & 0 & 0 \end{bmatrix}$

A^4 is the zero matrix.

(d) A^n is the zero matrix.

105. $\ln\left(\dfrac{64}{\sqrt[3]{x^2+3}}\right)$ **107.** $\ln\left(\dfrac{\sqrt{x}(x+5)}{\sqrt{x-8}}\right)$

Section 7.6   (page 547)

Vocabulary Check   (page 547)

1. square   **2.** inverse   **3.** nonsingular, singular

1–9. Answers will vary.

11. $\begin{bmatrix} \frac{1}{2} & 0 \\ 0 & \frac{1}{3} \end{bmatrix}$ **13.** $\begin{bmatrix} -3 & 2 \\ -2 & 1 \end{bmatrix}$ **15.** Does not exist

17. $\begin{bmatrix} 1 & 1 & -1 \\ -3 & 2 & -1 \\ 3 & -3 & 2 \end{bmatrix}$ **19.** Does not exist

21. Does not exist   **23.** $\begin{bmatrix} -12 & -5 & -9 \\ -4 & -2 & -4 \\ -8 & -4 & -6 \end{bmatrix}$

25. $\frac{5}{11}\begin{bmatrix} 0 & -4 & 2 \\ -22 & 11 & 11 \\ 22 & -6 & -8 \end{bmatrix}$ **27.** $\begin{bmatrix} 1 & 0 & 1 & 0 \\ 0 & 1 & 0 & 1 \\ 2 & 0 & 1 & 0 \\ 0 & 1 & 0 & 2 \end{bmatrix}$

29. $\begin{bmatrix} \frac{1}{4} & \frac{1}{8} \\ -\frac{1}{4} & -\frac{5}{8} \end{bmatrix}$ **31.** $\frac{1}{59}\begin{bmatrix} 16 & 15 \\ -4 & 70 \end{bmatrix}$ **33.** $\begin{bmatrix} \frac{5}{13} & -\frac{3}{13} \\ \frac{1}{13} & \frac{2}{13} \end{bmatrix}$

35. $\begin{bmatrix} -1 & 0 \\ -\frac{3}{2} & -\frac{1}{2} \end{bmatrix}$ **37.** $k = 0$ **39.** $k = \frac{3}{5}$ **41.** $(5, 0)$

43. $(-8, -6)$ **45.** $(3, 8, -11)$ **47.** $(2, 1, 0, 0)$

49. $(2, -2)$

51. Not possible, because A is not invertible.

53. $(-4, -8)$ **55.** $(-1, 3, 2)$

57. $(0.3125t + 0.8125, 1.8175t + 0.6875, t)$

59. $(5, 0, -2, 3)$

61. (a) $(-3, 2)$ (b) $(-1, 1)$

(c) $B = \begin{bmatrix} -1 \\ 1 \\ 1 \end{bmatrix}$; they correspond to the same point.

(d) $A^{-1} = \begin{bmatrix} 1 & 0 & -2 \\ 0 & 1 & 1 \\ 0 & 0 & 1 \end{bmatrix}$

(e) $A^{-1}B = \begin{bmatrix} -3 \\ 2 \\ 1 \end{bmatrix}$ $A^{-1}B$ represents X.

63. (a) $(2, -4)$ (b) $(-1, 0)$

(c) $B = \begin{bmatrix} -1 \\ 0 \\ 1 \end{bmatrix}$; they correspond to the same point.

(d) $A^{-1} = \begin{bmatrix} 1 & 0 & 3 \\ 0 & 1 & -4 \\ 0 & 0 & 1 \end{bmatrix}$

(e) $A^{-1}B = \begin{bmatrix} 2 \\ -4 \\ 1 \end{bmatrix}$ $A^{-1}B$ represents X.

65. \$10,000 in AAA-rated bonds, \$5,000 in A-rated bonds, and \$10,000 in B-rated bonds

67. \$20,000 in AAA-rated bonds, \$15,000 in A-rated bonds, and \$30,000 in B-rated bonds

69. $I_1 = -3$ amperes, $I_2 = 8$ amperes, $I_3 = 5$ amperes

71. 0 muffins, 300 bones, 200 cookies

73. 100 muffins, 300 bones, 150 cookies

75. (a) $\begin{cases} 2f + 2.5h + 3s = 26 \\ f + h + s = 10 \\ h - s = 0 \end{cases}$

(b) $\begin{bmatrix} 2 & 2.5 & 3 \\ 1 & 1 & 1 \\ 0 & 1 & -1 \end{bmatrix} \begin{bmatrix} f \\ h \\ s \end{bmatrix} = \begin{bmatrix} 26 \\ 10 \\ 0 \end{bmatrix}$

(c) 2 pounds of French vanilla, 4 pounds of hazelnut, 4 pounds of Swiss chocolate

77. (a) $\begin{cases} 4a + 2b + c = 5343 \\ 9a + 3b + c = 5589 \\ 16a + 4b + c = 6309 \end{cases}$

(b) $A^{-1} = \begin{bmatrix} \frac{1}{2} & -1 & \frac{1}{2} \\ -\frac{7}{2} & 6 & -\frac{5}{2} \\ 6 & -8 & 3 \end{bmatrix}$

$y = 237t^2 - 939t + 6273$

(c)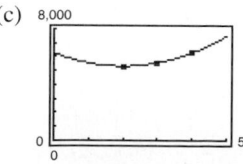

(d) 2005: 7,503,000 snowboarders

2010: 20,583,000 snowboarders

2015: 45,513,000 snowboarders

(e) Answers will vary.

79. True. $AA^{-1} = I = A^{-1}A$ **81.** Answers will vary.

83. $\dfrac{9}{2x + 6}$, $x \neq 0$ **85.** $\dfrac{x^2 + 2x - 13}{x(x - 2)}$, $x \neq \pm 3$

87. $\ln 3 \approx 1.099$ **89.** $\dfrac{e^{12/7}}{3} \approx 1.851$

91. Answers will vary.

Section 7.7   (page 556)

Vocabulary Check   (page 556)

1. determinant   **2.** minor   **3.** cofactor

4. expanding by cofactors   **5.** triangular

6. diagonal

1. 4   **3.** 16   **5.** 28   **7.** -24 **9.** 0

11. -9 **13.** -0.002

15. (a) $M_{11} = -5$, $M_{12} = 2$, $M_{21} = 4$, $M_{22} = 3$

(b) $C_{11} = -5$, $C_{12} = -2$, $C_{21} = -4$, $C_{22} = 3$

17. (a) $M_{11} = 10$, $M_{12} = -43$, $M_{13} = 2$, $M_{21} = -30$,

$M_{22} = 17$, $M_{23} = -6$, $M_{31} = 54$, $M_{32} = -53$,

$M_{33} = -34$

(b) $C_{11} = 10$, $C_{12} = 43$, $C_{13} = 2$, $C_{21} = 30$, $C_{22} = 17$,

$C_{23} = 6$, $C_{31} = 54$, $C_{32} = 53$, $C_{33} = -34$

19. (a) -75 (b) -75 **21.** (a) 170   (b) 170

23. -58 **25.** -168 **27.** 412   **29.** -60

31. -168 **33.** -336 **35.** 410

37. (a) -3 (b) -2 (c) $\begin{bmatrix} -2 & 0 \\ 0 & -3 \end{bmatrix}$ (d) 6

39. (a) 2   (b) -6 (c) $\begin{bmatrix} 1 & 4 & 3 \\ -1 & 0 & 3 \\ 0 & 2 & 0 \end{bmatrix}$ (d) -12

41. (a) -25 (b) -220

(c) $\begin{bmatrix} -7 & -16 & -1 & -28 \\ -4 & -14 & -11 & 8 \\ 13 & 4 & 4 & -4 \\ -2 & 3 & 2 & 2 \end{bmatrix}$ (d) 5500

43–47. Answers will vary.   **49.** $x = \pm 2$ **51.** $x = \pm\frac{3}{2}$

53. $x = 1 \pm \sqrt{2}$ **55.** $x = -4, -1$ **57.** $x = 1, \frac{1}{2}$

59. $x = 3$ **61.** $8uv - 1$ **63.** e^{5x} **65.** $1 - \ln x$

67. True

69. Answers will vary.

Sample answer: $A = \begin{bmatrix} 1 & 3 \\ -2 & 4 \end{bmatrix}$, $B = \begin{bmatrix} -4 & 0 \\ 3 & 5 \end{bmatrix}$

$|A + B| = -30$, $|A| + |B| = -10$

71. (a) 6   (b) $\begin{bmatrix} \frac{1}{3} & -\frac{1}{3} \\ \frac{1}{3} & \frac{1}{6} \end{bmatrix}$ (c) $\frac{1}{6}$ (d) They are reciprocals.

73. (a) 2   (b) $\begin{bmatrix} -4 & -5 & 1.5 \\ -1 & -1 & 0.5 \\ -1 & -1 & 0 \end{bmatrix}$

(c) $\frac{1}{2}$ (d) They are reciprocals.

75. (a) Columns 2 and 3 are interchanged.

(b) Rows 1 and 3 are interchanged.

77. (a) 5 is factored from the first row.

(b) 4 and 3 are factored from the second and third columns, respectively.

79. $(x - 2)(x - 1)$ **81.** $(2y - 3)^2$ **83.** $(2, -4)$

Section 7.8   (page 567)

Vocabulary Check   (page 567)

1. collinear   **2.** Cramer's Rule   **3.** cryptogram

4. uncoded, coded

1. $\frac{5}{2}$ **3.** $\frac{33}{8}$ **5.** 24   **7.** $x = 0, -\frac{16}{5}$ **9.** Collinear

11. Not collinear   **13.** $x = 3$ **15.** $(-3, -2)$

17. Not possible   **19.** $\left(\frac{32}{7}, \frac{30}{7}\right)$ **21.** $(-1, 3, 2)$

23. (a) and (b) $\left(0, -\frac{1}{2}, \frac{1}{2}\right)$ **25.** (a) and (b) $(1, -1, 2)$

27. (a) $y = -26.36t^2 + 241.5t + 784$

(b)

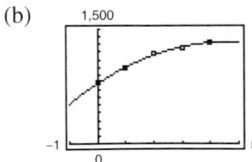

(c) Answers will vary.

29. Uncoded: $\begin{bmatrix} 3 & 1 & 12 \end{bmatrix}, \begin{bmatrix} 12 & 0 & 13 \end{bmatrix}, \begin{bmatrix} 5 & 0 & 20 \end{bmatrix},$
$\begin{bmatrix} 15 & 13 & 15 \end{bmatrix}, \begin{bmatrix} 18 & 18 & 15 \end{bmatrix}, \begin{bmatrix} 23 & 0 & 0 \end{bmatrix}$

Encoded: -68 21 35 -66 14 39 -115 35 60
-62 15 32 -54 12 27 23 -23 0

31. 38 63 51 -1 -14 -32 58 119 133 44 88 95

33. HAPPY NEW YEAR

35. YANKEES WIN WORLD SERIES

37. True. Cramer's Rule divides by the determinant.

39. Answers will vary.

41. $x + 4y - 19 = 0$ **43.** $2x - 7y - 27 = 0$

45.

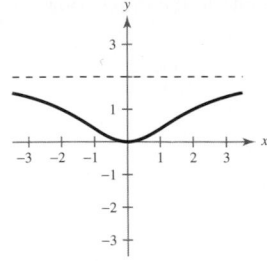

Asymptote: $y = 2$

Review Exercises   (page 570)

1. $(1, 1)$ **3.** $(5, 4)$ **5.** $(0, 0), (2, 8), (-2, 8)$

7. $(2, -0.5)$ **9.** $(0, 0), (4, -4)$

11. $(-1, 2), (0.67, 2.56)$ **13.** $(4, 4)$ **15.** $(-1, 1)$

17. 4762 units   **19.** 96 meters $\times$ 144 meters   **21.** $\left(\frac{5}{2}, 3\right)$

23. $\left(-\frac{1}{2}, \frac{4}{5}\right)$ **25.** $(0, 0)$ **27.** $\left(\frac{14}{5} + \frac{8}{5}a, a\right)$

29.

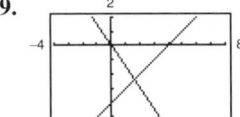

31.

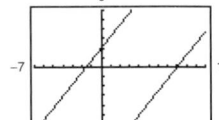

Consistent; $(1.6, -2.4)$ Inconsistent

33.

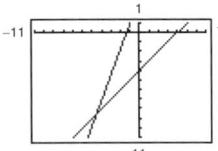

Consistent; $(-4.6, -8.6)$

35. $\left(\frac{500,000}{7}, \frac{159}{7}\right)$

37. 218.75 miles per hour; 193.75 miles per hour

39. $(4, 3, -2)$ **41.** $\left(\frac{38}{17}, \frac{40}{17}, -\frac{63}{17}\right)$

43. $(3a + 4, 2a + 5, a)$ **45.** $\left(-\frac{19}{6}, \frac{17}{12}, \frac{1}{3}\right)$

47. $(a - 4, a - 3, a)$

49.

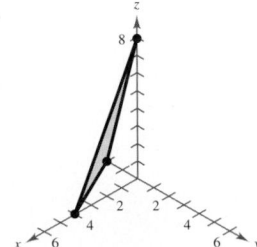

Sample answer: $(0, 0, 8), (0, -2, 0), (4, 0, 0), (1, -1, 2)$

51. (a) $\dfrac{3}{x+2} - \dfrac{4}{x+4}$ **53.** $\dfrac{1}{2}\left(\dfrac{3}{x-1} - \dfrac{x-3}{x^2+1}\right)$

55. $\dfrac{2x-1}{x^2+1} + \dfrac{-1}{x+2}$ **57.** $y = 2x^2 + x - 5$

59. Spray X: 10 gallons; spray Y: 5 gallons; spray Z: 12 gallons

61. 3×1 **63.** 1×1

65. $\begin{bmatrix} 3 & -10 & \vdots & 15 \\ 5 & 4 & \vdots & 22 \end{bmatrix}$

67. $\begin{bmatrix} 8 & -7 & 4 & \vdots & 12 \\ 3 & -5 & 2 & \vdots & 20 \\ 5 & 3 & -3 & \vdots & 26 \end{bmatrix}$

69. $\begin{cases} 5x + y + 7z = -9 \\ 4x + 2y \quad\;\; = 10 \\ 9x + 4y + 2z = 3 \end{cases}$

71. $\begin{bmatrix} 1 & 1 & 1 \\ 0 & 1 & 1 \\ 0 & 0 & 1 \end{bmatrix}$ **73.** $\begin{bmatrix} 1 & 0 & 3 & -2 \\ 0 & 1 & 4 & -3 \end{bmatrix}$

75. $\begin{bmatrix} 1 & 0 & 0 \\ 0 & 1 & 0 \\ 0 & 0 & 1 \end{bmatrix}$ **77.** $(10, -12)$ **79.** $(-0.2, 0.7)$

81. $\left(\frac{1}{2}, -\frac{1}{3}, 1\right)$ **83.** $(3a+1, -a, a)$ **85.** $(2, -3, 3)$

87. $\left(1, 2, \frac{1}{2}\right)$ **89.** $(3, 0, -4)$ **91.** $(2, 6, -10, -3)$

93. $x = 12, y = -7$ **95.** $x = 1, y = 11$

97. (a) $\begin{bmatrix} 17 & -17 \\ 13 & 2 \end{bmatrix}$ (b) $\begin{bmatrix} -3 & 23 \\ -15 & 8 \end{bmatrix}$

(c) $\begin{bmatrix} 28 & 12 \\ -4 & 20 \end{bmatrix}$ (d) $\begin{bmatrix} 37 & -57 \\ 41 & -4 \end{bmatrix}$

99. (a) $\begin{bmatrix} 6 & 5 & 8 \\ 1 & 7 & 8 \\ 5 & 1 & 4 \end{bmatrix}$ (b) $\begin{bmatrix} 6 & -5 & 6 \\ 9 & -9 & -4 \\ 1 & 3 & 2 \end{bmatrix}$

(c) $\begin{bmatrix} 24 & 0 & 28 \\ 20 & -4 & 8 \\ 12 & 8 & 12 \end{bmatrix}$ (d) $\begin{bmatrix} 6 & 15 & 10 \\ -7 & 23 & 20 \\ 9 & -1 & 6 \end{bmatrix}$

101. $\begin{bmatrix} -13 & -8 & 18 \\ 0 & 11 & -19 \end{bmatrix}$ **103.** $\begin{bmatrix} 9 & -7 \\ -9 & 4 \end{bmatrix}$

105. $\begin{bmatrix} 32 & -\frac{17}{2} & -\frac{3}{2} \\ 6 & 46 & 33 \end{bmatrix}$ **107.** $\begin{bmatrix} -14 & -4 \\ 7 & -17 \\ -17 & -2 \end{bmatrix}$

109. $\frac{1}{3}\begin{bmatrix} 9 & 2 \\ -4 & 11 \\ 10 & 0 \end{bmatrix}$ **111.** $\begin{bmatrix} 14 & -2 & 8 \\ 14 & -10 & 40 \\ 36 & -12 & 48 \end{bmatrix}$

113. $\begin{bmatrix} 11 & -6 \\ 18 & 39 \end{bmatrix}$ **115.** $\begin{bmatrix} 14 & -22 & 22 \\ 19 & -41 & 80 \\ 42 & -66 & 66 \end{bmatrix}$

117. $\begin{bmatrix} 1 & 17 \\ 12 & 36 \end{bmatrix}$

119. (a) $\begin{bmatrix} 438.20 & 47.40 \\ 612.10 & 66.20 \\ 717.28 & 77.20 \end{bmatrix}$

The dairy mart's sales and profits on milk for Friday, Saturday, and Sunday

(b) $\$190.80$

121. Answers will vary.

123. $\begin{bmatrix} 4 & -5 \\ 5 & -6 \end{bmatrix}$ **125.** $\begin{bmatrix} 13 & 6 & -4 \\ -12 & -5 & 3 \\ 5 & 2 & -1 \end{bmatrix}$

127. $\begin{bmatrix} \frac{1}{5} & \frac{1}{5} \\ \frac{1}{10} & -\frac{1}{15} \end{bmatrix}$ **129.** $\begin{bmatrix} 1 & 0 & 2 \\ 0 & 0 & -1 \\ 1 & 1 & 3 \end{bmatrix}$

131. $\begin{bmatrix} 1 & -1 \\ 4 & -\frac{7}{2} \end{bmatrix}$ **133.** $\begin{bmatrix} -\frac{1}{2} & \frac{1}{4} \\ \frac{1}{20} & \frac{1}{40} \end{bmatrix}$ **135.** $\left(1, -\frac{2}{5}\right)$

137. $(2, -1, -2)$ **139.** $(1, -2, 1, 0)$ **141.** $(-3, 1)$

143. $(1, 1, -2)$ **145.** -42 **147.** 550

149. (a) $M_{11} = 4, M_{12} = 7, M_{21} = -1, M_{22} = 2$

(b) $C_{11} = 4, C_{12} = -7, C_{21} = 1, C_{22} = 2$

151. (a) $M_{11} = 30, M_{12} = -12, M_{13} = -21, M_{21} = 20,$
$M_{22} = 19, M_{23} = 22, M_{31} = 5, M_{32} = -2,$
$M_{33} = 19$

(b) $C_{11} = 30, C_{12} = 12, C_{13} = -21, C_{21} = -20,$
$C_{22} = 19, C_{23} = -22, C_{31} = 5, C_{32} = 2, C_{33} = 19$

153. 130 **155.** -3 **157.** 279 **159.** -96

161. 16 **163.** 1.75 **165.** $\frac{13}{2}$ **167.** 48

169. Not collinear   **171.** $(1, 2)$ **173.** $(4, 7)$

175. $(-1, 4, 5)$ **177.** $(0, -2.4, -2.6)$

179. (a) and (b) $\left(\frac{53}{33}, -\frac{17}{33}, \frac{61}{66}\right)$

181. Uncoded:
$[9 \; 0 \; 8], [1 \; 22 \; 5], [0 \; 1 \; 0], [4 \; 18 \; 5], [1 \; 13 \; 0]$
Encoded: $-30, -2, 24, 38, 8, -51, 3, 0, -3, 32,$
$2, -39, 41, -2, -39$

183. I WILL BE BACK

185. THAT IS MY FINAL ANSWER

187. (a) $y = 0.41t + 11.74$

(b)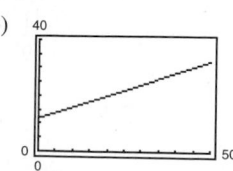

(c) and (d) $t \approx 20.15$ or 2010

189. True

191. An $n \times n$ matrix A has an inverse A^{-1} if $\det(A) \neq 0$.

Chapter Test   (page 576)

1. $(4, -2)$ **2.** $(0, -1), (1, 0), (2, 1)$

3. $(8, 5), (2, -1)$ **4.** $\left(\frac{28}{9}, -\frac{31}{9}\right)$ **5.** $\left(-\frac{2}{3}, -\frac{1}{2}, 1\right)$

6. $(1, 0, -2)$ **7.** $y = -\frac{1}{2}x^2 + x + 6$

8. $\dfrac{5}{x-1} + \dfrac{3}{(x-1)^2}$ **9.** $\dfrac{1}{x} + \dfrac{2}{x^2} - \dfrac{1}{x^2+1}$

10. $(-2a + 1.5, 2a + 1, a)$ **11.** $(5, 2, -6)$

12. (a) $\begin{bmatrix} 1 & 0 & 4 \\ -7 & -6 & -1 \\ 0 & 4 & 0 \end{bmatrix}$ (b) $\begin{bmatrix} 15 & 12 & 12 \\ -12 & -12 & 0 \\ 3 & 6 & 0 \end{bmatrix}$

    (c) $\begin{bmatrix} 7 & 4 & 12 \\ -18 & -16 & -2 \\ 1 & 10 & 0 \end{bmatrix}$ (d) $\begin{bmatrix} 36 & 20 & 4 \\ -28 & -24 & -4 \\ 10 & 8 & 2 \end{bmatrix}$

13. $\begin{bmatrix} -\frac{4}{3} & -\frac{5}{3} & 1 \\ -\frac{4}{3} & -\frac{8}{3} & 1 \\ \frac{1}{3} & \frac{2}{3} & 0 \end{bmatrix}, (-2, 3, -1)$ **14.** 67   **15.** -2

16. 30   **17.** $\left(1, -\frac{1}{2}\right)$

18. $x_1 = 700 - s - t, x_2 = 300 - s - t,$

    $x_3 = s, x_4 = 100 - t, x_5 = t$

Chapter 8

Section 8.1   (page 587)

Vocabulary Check   (page 587)

1. infinite sequence   **2.** terms   **3.** finite

4. recursively   **5.** factorial

6. summation notation

7. index, upper limit, lower limit   **8.** series

9. nth partial sum

1. 7, 9, 11, 13, 15   **3.** 2, 4, 8, 16, 32

5. $-\frac{1}{2}, \frac{1}{4}, -\frac{1}{8}, \frac{1}{16}, -\frac{1}{32}$ **7.** $2, \frac{3}{2}, \frac{4}{3}, \frac{5}{4}, \frac{6}{5}$

9. $\frac{1}{2}, \frac{2}{5}, \frac{3}{10}, \frac{4}{17}, \frac{5}{26}$ **11.** $0, 1, 0, \frac{1}{2}, 0$ **13.** $\frac{1}{2}, \frac{3}{4}, \frac{7}{8}, \frac{15}{16}, \frac{31}{32}$

15. $1, \dfrac{1}{2^{3/2}}, \dfrac{1}{3^{3/2}}, \dfrac{1}{4^{3/2}} = \dfrac{1}{8}, \dfrac{1}{5^{3/2}}$ **17.** $-1, \dfrac{1}{4}, -\dfrac{1}{9}, \dfrac{1}{16}, -\dfrac{1}{25}$

19. 3, 15, 35, 63, 99   **21.** -73 **23.** $\frac{100}{101}$ **25.** $\frac{64}{65}$

27.

29.

31.

33. 9, 15, 21, 27, 33, 39, 45, 51, 57, 63

35. $3, \frac{5}{2}, \frac{7}{3}, \frac{9}{4}, \frac{11}{5}, \frac{13}{6}, \frac{15}{7}, \frac{17}{8}, \frac{19}{9}, \frac{21}{10}$

37. 0, 2, 0, 2, 0, 2, 0, 2, 0, 2

39. c   **40.** b   **41.** d   **42.** a

43. $a_n = 3n - 2$ **45.** $a_n = n^2 - 1$ **47.** $a_n = \dfrac{n+1}{n+2}$

49. $a_n = \dfrac{(-1)^{n+1}}{2^n}$ **51.** $a_n = 1 + \dfrac{1}{n}$ **53.** $a_n = \dfrac{1}{n!}$

55. $a_n = (-1)^n + 2(1)^n = (-1)^n + 2$

57. 28, 24, 20, 16, 12   **59.** 3, 4, 6, 10, 18

61. 6, 8, 10, 12, 14; $a_n = 2n + 4$

63. 81, 27, 9, 3, 1; $a_n = \dfrac{243}{3^n}$ **65.** $1, 1, \dfrac{1}{2}, \dfrac{1}{6}, \dfrac{1}{24}$

67. $1, \dfrac{1}{3}, \dfrac{2}{5}, \dfrac{6}{7}, \dfrac{8}{3}$ **69.** $1, \dfrac{1}{2}, \dfrac{1}{24}, \dfrac{1}{720}, \dfrac{1}{40,320}$ **71.** $\dfrac{1}{12}$

73. 495   **75.** $n + 1$ **77.** $\dfrac{1}{2n(2n+1)}$ **79.** 35

81. 40   **83.** 30   **85.** $\frac{9}{5}$ **87.** 238   **89.** 30

91. 81   **93.** $\dfrac{47}{60}$ **95.** $\displaystyle\sum_{i=1}^{9} \dfrac{1}{3i} \approx 0.94299$

97. $\displaystyle\sum_{i=1}^{8} \left[2\left(\dfrac{i}{8}\right) + 3\right] = 33$ **99.** $\displaystyle\sum_{i=1}^{6} (-1)^{i+1} 3^i = -546$

101. $\displaystyle\sum_{i=1}^{20} \dfrac{(-1)^{i+1}}{i^2} \approx 0.821$ **103.** $\displaystyle\sum_{i=1}^{5} \dfrac{2^i - 1}{2^{i+1}} \approx 2.0156$

105. $\frac{75}{16}$ **107.** $-\frac{3}{2}$ **109.** (a) $\frac{3333}{5000}$ (b) $\frac{2}{3}$

111. (a) $\frac{1111}{10,000}$ (b) $\frac{1}{9}$

113. (a) $A_1 = \$5037.50, A_2 = \$5075.28,$

     $A_3 = \$5113.35, A_4 = \$5151.70,$

     $A_5 = \$5190.33, A_6 = \$5229.26,$

     $A_7 = \$5268.48, A_8 = \$5307.99$

    (b) $\$6741.74$

115. (a) $p_0 = 5500, p_n = (0.75)p_{n-1} + 500$

(b) 4625; 3969; 3477

(c) 2000 trout; Answers will vary. Sample answer: As time passes, the population of trout decreases at a decreasing rate. Because the population is growing smaller and still declines 25%, each time 25% is taken from a smaller number, there is a smaller decline in the number of trout.

117. (a) $a_0 = 50, a_n = a_{n-1}(1.005) + 50$

(b) \$616.78   (c) \$2832.26

119. (a)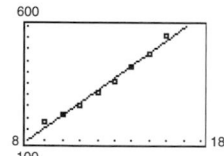

(b) 2010: \$12.25; 2015: \$12.64

(c) Answers will vary.   (d) 2008

121. (a)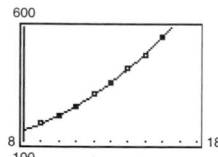

(b) Linear regression sequence:
$R_n = 54.58n - 336.3; 0.98656$

Quadratic regression sequence:
$R_n = 3.088n^2 - 22.62n + 130.0; 0.99919$

(c) Linear regression sequence:

Quadratic regression sequence:

(d) Quadratic regression sequence; the coefficient of determination for the quadratic regression sequence is closer to 1.

(e) 2010: \$912,850,000; 2015: \$1,494,560,000

(f) 2010

123. True   **125.** 1, 1, 2, 3, 5, 8, 13, 21, 34, 55, 89, 144;

$1, 2, \frac{3}{2}, \frac{5}{3}, \frac{8}{5}, \frac{13}{8}, \frac{21}{13}, \frac{34}{21}, \frac{55}{34}, \frac{89}{55}$

127. 1, 1, 2, 3, 5

129. $a_{n+1} = \frac{1}{2}a_n + \frac{\left(1 + \sqrt{5}\right)^n + \left(1 - \sqrt{5}\right)^n}{2^{n+1}}$

$a_{n+2} = \frac{3}{2}a_n + \frac{\left(1 + \sqrt{5}\right)^n + \left(1 - \sqrt{5}\right)^n}{2^{n+1}}$

131. $x, \frac{x^2}{2}, \frac{x^3}{6}, \frac{x^4}{24}, \frac{x^5}{120}$ **133.** $-\frac{x^3}{3}, \frac{x^5}{5}, -\frac{x^7}{7}, \frac{x^9}{9}, -\frac{x^{11}}{11}$

135. $-\frac{x^2}{2}, \frac{x^4}{24}, -\frac{x^6}{720}, \frac{x^8}{40,320}, -\frac{x^{10}}{3,628,800}$

137. $-x, \frac{x^2}{2}, -\frac{x^3}{6}, \frac{x^4}{24}, -\frac{x^5}{120}$

139. $x + 1, -\frac{(x+1)^2}{2}, \frac{(x+1)^3}{6}, -\frac{(x+1)^4}{24}, \frac{(x+1)^5}{120}$

141. $\frac{1}{4}, \frac{1}{12}, \frac{1}{24}, \frac{1}{40}, \frac{1}{60}; \frac{1}{2} - \frac{1}{2n+2}$

143. $\frac{1}{6}, \frac{1}{12}, \frac{1}{20}, \frac{1}{30}, \frac{1}{42}; \frac{1}{2} - \frac{1}{n+2}$

145. 0, ln 2, ln 3, ln 4, ln 5; ln($n!$)

147. (a) $\begin{bmatrix} 8 & 1 \\ -3 & 7 \end{bmatrix}$ (b) $\begin{bmatrix} -22 & -7 \\ 3 & -18 \end{bmatrix}$

(c) $\begin{bmatrix} 18 & 9 \\ 18 & 0 \end{bmatrix}$ (d) $\begin{bmatrix} 0 & 6 \\ 27 & 18 \end{bmatrix}$

149. (a) $\begin{bmatrix} -3 & -7 & 4 \\ 4 & 4 & 1 \\ 1 & 4 & 3 \end{bmatrix}$ (b) $\begin{bmatrix} 8 & 17 & -14 \\ -12 & -13 & -9 \\ -3 & -15 & -10 \end{bmatrix}$

(c) $\begin{bmatrix} -2 & 7 & -16 \\ 4 & 42 & 45 \\ 1 & 23 & 48 \end{bmatrix}$ (d) $\begin{bmatrix} 16 & 31 & 42 \\ 10 & 47 & 31 \\ 13 & 22 & 25 \end{bmatrix}$

Section 8.2   (page 598)

Vocabulary Check   (page 598)

1. arithmetic, common   **2.** $a_n = dn + c$

3. nth partial sum

1. Arithmetic sequence, $d = -2$

3. Arithmetic sequence, $d = -\frac{1}{2}$

5. Arithmetic sequence, $d = 8$

7. Arithmetic sequence, $d = 0.6$

9. 21, 34, 47, 60, 73

Arithmetic sequence, $d = 13$

11. $\frac{1}{2}, \frac{1}{3}, \frac{1}{4}, \frac{1}{5}, \frac{1}{6}$

Not an arithmetic sequence

13. 143, 136, 129, 122, 115

Arithmetic sequence, $d = -7$

15. 1, 5, 1, 5, 1

Not an arithmetic sequence

17. $a_n = -2 + 3n$ **19.** $a_n = 108 - 8n$

21. $a_n = \frac{13}{2} - \frac{5}{2}n$ **23.** $a_n = \frac{10}{3}n + \frac{5}{3}$

25. $a_n = 103 - 3n$ **27.** 5, 11, 17, 23, 29

29. $-10, -22, -34, -46, -58$ **31.** $-2, 2, 6, 10, 14$

33. 22.45, 20.725, 19, 17.275, 15.55

35. 15, 19, 23, 27, 31; $d = 4$; $a_n = 11 + 4n$

37. $\frac{3}{5}, \frac{1}{2}, \frac{2}{5}, \frac{3}{10}, \frac{1}{5}$; $d = -\frac{1}{10}$, $a_n = -\frac{1}{10}n + \frac{7}{10}$

39. 59   **41.** 18.6

43. 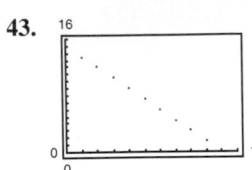   **45.**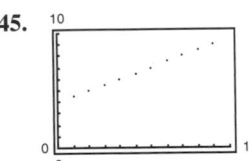

47. $-1, 3, 7, 11, 15, 19, 23, 27, 31, 35$

49. 19.25, 18.5, 17.75, 17, 16.25, 15.5, 14.75, 14, 13.25, 12.5

51. 1.55, 1.6, 1.65, 1.7, 1.75, 1.8, 1.85, 1.9, 1.95, 2

53. 110   **55.** -25 **57.** 2550   **59.** -4585

61. 620   **63.** 41   **65.** 4000   **67.** 1275

69. 25,250   **71.** 355   **73.** 129,250   **75.** 440

77. 2575   **79.** 14,268   **81.** 405 bricks

83. $150,000

85. (a) $a_n = 0.91n + 5.7$

(b)

Year	1997	1998	1999	2000
Sales (in billions of dollars)	12.1	13.0	13.9	14.8

Year	2001	2002	2003	2004
Sales (in billions of dollars)	15.7	16.6	17.5	18.4

The model fits the data.

(c) $122.0 billion

(d) $180.3 billion; Answers will vary.

87. True. Given a_1 and a_2, you know that $d = a_2 - a_1$. Hence, $a_n = a_1 + (n - 1)d$.

89. $x, 3x, 5x, 7x, 9x, 11x, 13x, 15x, 17x, 19x$ **91.** 4

93. (a) $-7, -4, -1, 2, 5, 8, 11$

$a_1 = -7, a_{n+1} = a_n + 3$

(b) 17, 23, 29, 35, 41, 47, 53, 59

$a_1 = 17, a_{n+1} = a_n + 6$

(c) Not possible

(d) 4, 7.5, 11, 14.5, 18, 21.5, 25, 28.5

$a_1 = 4, a_{n+1} = a_n + 3.5$

(e) Not possible

Answers will vary.

95. 20,100   **97.** 2601   **99.** $(1, 5, -1)$

101. 15 square units   **103.** Answers will vary.

Section 8.3   (page 607)

Vocabulary Check   (page 607)

1. geometric, common   **2.** $a_n = a_1 r^{n-1}$

3. $S_n = \sum_{i=1}^{n} a_1 r^{i-1} = a_1\left(\dfrac{1 - r^n}{1 - r}\right)$

4. geometric series   **5.** $S = \sum_{i=0}^{\infty} a_1 r^i = \dfrac{a_1}{1 - r}$

1. Geometric sequence, $r = 3$

3. Not a geometric sequence

5. Geometric sequence, $r = -\frac{1}{2}$

7. Geometric sequence, $r = 2$

9. Not a geometric sequence   **11.** 6, 18, 54, 162, 486

13. $1, \frac{1}{2}, \frac{1}{4}, \frac{1}{8}, \frac{1}{16}$ **15.** $5, -\frac{1}{2}, \frac{1}{20}, -\frac{1}{200}, \frac{1}{2000}$

17. $1, e, e^2, e^3, e^4$

19. 64, 32, 16, 8, 4; $r = \frac{1}{2}$;   **21.** 9, 18, 36, 72, 144; $r = 2$;

$a_n = 64\left(\frac{1}{2}\right)^{n-1}$ $a_n = 9(2)^{n-1}$

23. $6, -9, \frac{27}{2}, -\frac{81}{4}, \frac{243}{8}$; $r = -\frac{3}{2}$;

$a_n = 6\left(-\frac{3}{2}\right)^{n-1}$

25. $\dfrac{1}{128}$ **27.** $-\dfrac{2}{3^{10}}$ **29.** $500(1.02)^{13}$ **31.** $-\dfrac{2}{9}$

33. $a_n = 7(3)^{n-1}$; 45,927   **35.** $a_n = 5(6)^{n-1}$; 50,388,480

37. 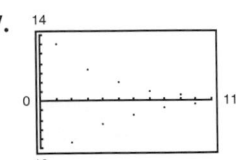   **39.**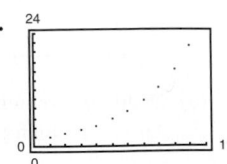

41. 8, 4, 6, 5

43.

n	S_n
1	16
2	24
3	28
4	30
5	31
6	31.5
7	31.75
8	31.875
9	31.9375
10	31.96875

45. 511   **47.** 43

49. 29,921.31   **51.** 6.4   **53.** 2092.60

55. $\sum_{n=1}^{7} 5(3)^{n-1}$ **57.** $\sum_{n=1}^{7} 2\left(-\frac{1}{4}\right)^{n-1}$ **59.** 50   **61.** $\frac{10}{3}$

63. Series does not have a finite sum because $\left|\frac{7}{3}\right| > 1$.

65. $\frac{1000}{89}$ **67.** $-\frac{30}{19}$ **69.** 32   **71.** $\frac{9}{4}$

73. $\frac{4}{11}$ **75.** $\frac{113}{90}$

77. (a) $1343.92   (b) $1346.86   (c) $1348.35

(d) $1349.35   (e) $1349.84

79. $6480.83   **81.** Answers will vary.

83. (a) $26,198.27   (b) $26,263.88

85. (a) $153,237.86   (b) $153,657.02

87. 126 square inches

89. (a) $T_n = 70(0.8)^n$ (b) 18.4°F; 4.8°F

(c)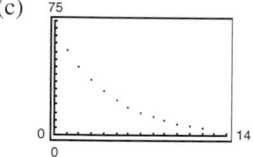

3.5 hours

91. $1600   **93.** $1250   **95.** $2181.82

97. (a) Option 1. Answers will vary. Sample answer: You make a cumulative amount of $157,689.86 from option 1 and $169,131.31 from option 2.

(b) Option 2. Answers will vary. Sample answer: You make about $33,114.39 from option 1 and $35,179.05 from option 2 the year prior to re-evaluation.

99. (a) 3208.53 feet; 2406.40 feet; 5614.93 feet total

(b) 5950 feet

101. False. A sequence is geometric if the ratios of consecutive terms are the same.

103. $3, \dfrac{3x}{2}, \dfrac{3x^2}{4}, \dfrac{3x^3}{8}, \dfrac{3x^4}{16}$ **105.** $100e^{8x}$

107. (a)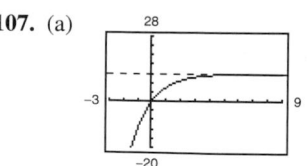

Horizontal asymptote: $y = 12$

Corresponds to the sum of the series

(b)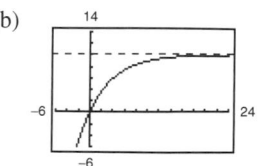

Horizontal asymptote: $y = 10$

Corresponds to the sum of the series

109. Divide the second term by the first to obtain the common ratio. The nth term is the first term times the common ratio raised to the $(n-1)$th power.

111. 45.65 miles per hour   **113.** -102

115. Answers will vary.

Section 8.4   (page 617)

1. $\dfrac{5}{(k+1)(k+2)}$ **3.** $\dfrac{2^{k+1}}{(k+2)!}$

5. $1 + 6 + 11 + \cdots + (5k - 4) + (5k + 1)$

7–19. Answers will vary.   **21.** 1,625,625   **23.** 572

25–41. Answers will vary.

43. 0, 3, 6, 9, 12

First differences: 3, 3, 3, 3

Second differences: 0, 0, 0

Linear

45. 3, 1, -2, -6, -11

First differences: $-2, -3, -4, -5$

Second differences: $-1, -1, -1$

Quadratic

47. 0, 1, 3, 6, 10

First differences: 1, 2, 3, 4

Second differences: 1, 1, 1

Quadratic

49. 2, 4, 6, 8, 10

First differences: 2, 2, 2, 2

Second differences: 0, 0, 0

Linear

51. $a_n = n^2 - 3n + 5, \ n \geq 1$ **53.** $a_n = \frac{1}{2}n^2 + n - 3$

55. (a) $a_n = 3(4)^{n-1}$

(b) $a_n = \frac{\sqrt{3}}{4}\left[1 + \frac{1}{3}\sum_{k=2}^{n}\left(\frac{4}{9}\right)^{k-2}\right]$ (c) $3\left(\frac{4}{3}\right)^{n-1}$

57. False. Not necessarily

59. False. It has $n - 2$ second differences.

61. $4x^4 - 4x^2 + 1$ **63.** $-64x^3 + 240x^2 - 300x + 125$

65. $7\sqrt{3}i$ **67.** $40\left(1 + \sqrt[3]{2}\right)$

Section 8.5   (page 624)

> **Vocabulary Check**   (page 624)
>
> **1.** binomial coefficients
>
> **2.** Binomial Theorem, Pascal's Triangle
>
> **3.** $_nC_r$ or $\binom{n}{r}$ **4.** expanding, binomial

1. 21   **3.** 1   **5.** 15,504   **7.** 14   **9.** 4950

11. 749,398   **13.** 4950   **15.** 31,125

17. $x^4 + 8x^3 + 24x^2 + 32x + 16$

19. $a^3 + 9a^2 + 27a + 27$

21. $y^4 - 8y^3 + 24y^2 - 32y + 16$

23. $x^5 + 5x^4y + 10x^3y^2 + 10x^2y^3 + 5xy^4 + y^5$

25. $729r^6 + 2916r^5s + 4860r^4s^2 + 4320r^3s^3 + 2160r^2s^4$
$\qquad + 576rs^5 + 64s^6$

27. $x^5 - 5x^4y + 10x^3y^2 - 10x^2y^3 + 5xy^4 - y^5$

29. $1 - 12x + 48x^2 - 64x^3$

31. $x^8 + 8x^6 + 24x^4 + 32x^2 + 16$

33. $x^{10} - 25x^8 + 250x^6 - 1250x^4 + 3125x^2 - 3125$

35. $x^8 + 4x^6y^2 + 6x^4y^4 + 4x^2y^6 + y^8$

37. $x^{18} - 6x^{15}y + 15x^{12}y^2 - 20x^9y^3 + 15x^6y^4 - 6x^3y^5 + y^6$

39. $\dfrac{1}{x^5} + \dfrac{5y}{x^4} + \dfrac{10y^2}{x^3} + \dfrac{10y^3}{x^2} + \dfrac{5y^4}{x} + y^5$

41. $\dfrac{16}{x^4} - \dfrac{32y}{x^3} + \dfrac{24y^2}{x^2} - \dfrac{8y^3}{x} + y^4$

43. $-512x^4 + 576x^3 - 240x^2 + 44x - 3$

45. $2x^4 - 24x^3 + 113x^2 - 246x + 207$

47. $-4x^6 - 24x^5 - 60x^4 - 83x^3 - 42x^2 - 60x + 20$

49. $61,440x^7$ **51.** $360x^3y^2$ **53.** $1,259,712x^2y^7$

55. $32,476,950,000x^4y^8$ **57.** 3,247,695   **59.** 180

61. $-489,888$ **63.** 210   **65.** 21   **67.** 6

69. $81t^4 - 216t^3v + 216t^2v^2 - 96tv^3 + 16v^4$

71. $32x^5 - 240x^4y + 720x^3y^2 - 1080x^2y^3$
$\qquad + 810xy^4 - 243y^5$

73. $x^2 + 20x^{3/2} + 150x + 500x^{1/2} + 625$

75. $x^2 - 3x^{4/3}y^{1/3} + 3x^{2/3}y^{2/3} - y$

77. $3x^2 + 3xh + h^2, \ h \neq 0$

79. $6x^5 + 15x^4h + 20x^3h^2 + 15x^2h^3 + 6xh^4 + h^5, \ h \neq 0$

81. $\dfrac{\sqrt{x+h} - \sqrt{x}}{h} = \dfrac{1}{\sqrt{x+h} + \sqrt{x}}, \ h \neq 0$

83. -4 **85.** $161 + 240i$ **87.** $2035 + 828i$

89. $-115 + 236i$ **91.** $-23 + 208\sqrt{3}i$

93. 1   **95.** $-\frac{1}{8}$ **97.** 1.172   **99.** 510,568.785

101.

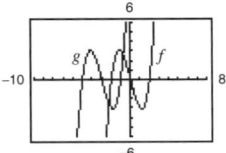

g is shifted three units to the left of f.

$g(x) = x^3 + 9x^2 + 23x + 15$

103.

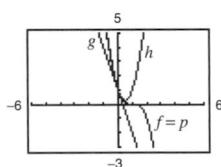

$p(x)$ is the expansion of $f(x)$.

105. 0.273   **107.** 0.171

109. (a) $g(t) = 0.064t^2 - 6.74t + 256.1, \ -15 \leq t \leq 3$

(b)

111. False. The correct term is $126,720x^4y^8$.

113. The first and last numbers in each row are 1. Every other number in each row is formed by adding the two numbers immediately above the number.

115. The terms of the expansion of $(x - y)^n$ alternate between positive and negative.

117 and 119. Answers will vary.

121. $g(x)$ is shifted eight units up from $f(x)$.

123. $g(x)$ is the reflection of $f(x)$ in the y-axis.

125. $\begin{bmatrix} 4 & -5 \\ 5 & -6 \end{bmatrix}$

Section 8.6   (page 634)

Vocabulary Check   (page 634)

1. Fundamental Counting Principle   **2.** permutation

3. $_nP_r = \dfrac{n!}{(n-r)!}$ **4.** distinguishable permutations

5. combinations

1. 6   **3.** 5   **5.** 3   **7.** 7   **9.** 120   **11.** 1024

13. (a) 900   (b) 648   **15.** 16,000,000

17. (a) 35,152   (b) 3902

19. (a) 100,000   (b) 20,000   **21.** (a) 720   (b) 48

23. 24   **25.** 336   **27.** 120   **29.** 27,907,200

31. 197,149,680   **33.** 120   **35.** 362,880

37. 11,880   **39.** 50,653

41. ABCD, ABDC, ACBD, ACDB, ADBC, ADCB,
BACD, BADC, CABD, CADB, DABC, DACB,
BCAD, BDAC, CBAD, CDAB, DBAC, DCAB,
BCDA, BDCA, CBDA, CDBA, DBCA, DCBA

43. 420   **45.** 2520   **47.** 10   **49.** 4   **51.** 1

53. 4845   **55.** 850,668

57. AB, AC, AD, AE, AF, BC, BD, BE, BF, CD, CE, CF, DE,
DF, EF

59. $4.418694268 \times 10^{16}$ **61.** 13,983,816   **63.** 36

65. 3744   **67.** (a) 495   (b) 210

69. 292,600   **71.** 5   **73.** 20

75. $n = 5$ or $n = 6$ **77.** $n = 10$

79. $n = 3$ **81.** $n = 2$ **83.** False.

85. For some calculators the answer is too large.

87. They are equal.   **89 and 91.** Answers will vary.

93. 8.303   **95.** 35   **97.** $(-2, -8)$ **99.** $(-1, 1)$

Section 8.7   (page 645)

Vocabulary Check   (page 645)

1. experiment, outcomes   **2.** sample space

3. probability   **4.** impossible, certain

5. mutually exclusive   **6.** independent

7. complement   **8.** (a) iii   (b) i   (c) iv   (d) ii

1. $\{(H, 1), (H, 2), (H, 3), (H, 4), (H, 5), (H, 6),$
$(T, 1), (T, 2), (T, 3), (T, 4), (T, 5), (T, 6)\}$

3. $\{ABC, ACB, BAC, BCA, CAB, CBA\}$

5. $\{(A, B), (A, C), (A, D), (A, E), (B, C), (B, D),$
$(B, E), (C, D), (C, E), (D, E)\}$

7. $\frac{3}{8}$ **9.** $\frac{7}{8}$ **11.** $\frac{3}{13}$ **13.** $\frac{4}{13}$ **15.** $\frac{5}{36}$ **17.** $\frac{11}{12}$

19. $\frac{1}{5}$ **21.** $\frac{2}{5}$ **23.** 0.25   **25.** $\frac{1}{3}$ **27.** 0.88   **29.** $\frac{7}{20}$

31. (a) 1.22 million   (b) 0.41   (c) 0.24   (d) 0.26

33. (a) 37.59 million   (b) 0.01   (c) 0.012

35. (a) 0.45   (b) 0.45   (c) 0.23

37. (a) $\frac{672}{1254}$ (b) $\frac{582}{1254}$ (c) $\frac{548}{1254}$

39. $P(\{\text{Taylor wins}\}) = \frac{1}{2}$
$P(\{\text{Moore wins}\}) = P(\{\text{Perez wins}\}) = \frac{1}{4}$

41. (a) $\frac{21}{1292} \approx 0.016$ (b) $\frac{225}{646} \approx 0.348$ (c) $\frac{49}{323} \approx 0.152$

43. (a) $\frac{1}{120}$ (b) $\frac{1}{24}$ **45.** (a) 0.008875   (b) $\frac{3}{8788}$

47. (a) 6.84×10^{-7} (b) 6.84×10^{-6}

49. (a) $\frac{5}{13}$ (b) $\frac{1}{2}$ (c) $\frac{4}{13}$ **51.** (a) $\frac{14}{55}$ (b) $\frac{12}{55}$ (c) $\frac{54}{55}$

53. 0.1024   **55.** (a) 0.9702   (b) 0.9998   (c) 0.0002

57. (a) $\dfrac{1}{15,625}$ (b) $\dfrac{4096}{15,625}$ (c) $\dfrac{11,529}{15,625}$

59. (a) $\dfrac{\pi}{4}$ (b) Answers will vary.   **61.** True

63. (a) As you consider successive people with distinct birthdays, the probabilities must decrease to take into account the birth dates already used. Because the birth dates of people are independent events, multiply the respective probabilities of distinct birthdays.

(b) $\frac{365}{365} \cdot \frac{364}{365} \cdot \frac{363}{365} \cdot \frac{362}{365}$

(c) Answers will vary.

(d) Q_n is the probability that the birthdays are *not* distinct, which is equivalent to at least two people having the same birthday.

(e)

n	10	15	20	23	30	40	50
P_n	0.88	0.75	0.59	0.49	0.29	0.11	0.03
Q_n	0.12	0.25	0.41	0.51	0.71	0.89	0.97

(f) 23

65. $x = \frac{11}{2}$ **67.** $x = -10$ **69.** $x = \ln 28 \approx 3.332$

71. $x = \frac{1}{6}e^4 \approx 9.100$ **73.** 60   **75.** 6,652,800

77. 15   **79.** 165

CHAPTER 8

Review Exercises   (page 651)

1. $\frac{2}{3}, \frac{4}{5}, \frac{8}{9}, \frac{16}{17}, \frac{32}{33}$ **3.** $-1, \frac{1}{2}, -\frac{1}{6}, \frac{1}{24}, -\frac{1}{120}$ **5.** $a_n = 5n$

7. $a_n = \dfrac{2}{2n-1}$ **9.** $9, 5, 1, -3, -7$ **11.** $\dfrac{1}{380}$

13. $(n+1)(n)$ **15.** 30   **17.** $\frac{205}{24}$ **19.** 51,005,000

21. 43,078   **23.** $\displaystyle\sum_{k=1}^{20} \frac{1}{2k} = 1.799$

25. $\displaystyle\sum_{k=1}^{9} \frac{k}{k+1} = 7.071$ **27.** (a) $\dfrac{1111}{2000}$ (b) $\dfrac{5}{9}$

29. (a) $\frac{15}{8}$ (b) 2

31. (a) 2512.50, 2525.06, 2537.69, 2550.38, 2563.13,
2575.94, 2588.82, 2601.77

(b) \$3051.99

33. Arithmetic sequence, $d = -2$

35. Arithmetic sequence, $d = \frac{1}{2}$

37. 3, 7, 11, 15, 19   **39.** 1, 4, 7, 10, 13

41. 35, 32, 29, 26, 23; $d = -3$;

$a_n = 38 - 3n$

43. 9, 16, 23, 30, 37; $d = 7$;

$a_n = 2 + 7n$

45. $a_n = 103 - 3n$; 1430   **47.** 80   **49.** 88

51. 25,250   **53.** (a) \$43,000   (b) \$192,500

55. Geometric sequence, $r = 2$

57. Geometric sequence, $r = -\frac{1}{3}$

59. $4, -1, \frac{1}{4}, -\frac{1}{16}, \frac{1}{64}$ **61.** $9, 6, 4, \frac{8}{3}, \frac{16}{9}$ or $9, -6, 4, -\frac{8}{3}, \frac{16}{9}$

63. $120, 40, \frac{40}{3}, \frac{40}{9}, \frac{40}{27}; r = \frac{1}{3}$;

$a_n = 120\left(\frac{1}{3}\right)^{n-1}$

65. $25, -15, 9, -\frac{27}{5}, \frac{81}{25}; r = -\frac{3}{5}$;

$a_n = 25\left(-\frac{3}{5}\right)^{n-1}$

67. $a_n = 16\left(-\frac{1}{2}\right)^{n-1}$; 10.67

69. $a_n = 100(1.05)^{n-1}$; 3306.60   **71.** 127   **73.** 3277

75. 1301.01   **77.** 24.85   **79.** 32   **81.** 12

83. (a) $a_t = 120,000(0.7)^t$ (b) \$20,168.40

85 and 87. Answers will vary.   **89.** 465   **91.** 4648

93. 5, 10, 15, 20, 25

First differences: 5, 5, 5, 5

Second differences: 0, 0, 0

Linear model

95. 16, 15, 14, 13, 12

First differences: $-1, -1, -1, -1$

Second differences: 0, 0, 0

Linear model

97. 45   **99.** 126   **101.** 20   **103.** 70

105. $x^4 + 20x^3 + 150x^2 + 500x + 625$

107. $a^5 - 20a^4b + 160a^3b^2 - 640a^2b^3 + 1280ab^4 - 1024b^5$

109. $1241 + 2520i$ **111.** 10

113. (a) 216   (b) 108   (c) 36   **115.** 45

117. 239,500,800   **119.** 4950   **121.** 999,000

123. 5040   **125.** 3,628,800   **127.** 15,504

129. $n = 3$ **131.** $\frac{1}{9}$

133. (a) 0.416   (b) 0.8   (c) 0.074   **135.** 0.0475

137. True. $\dfrac{(n+2)!}{n!} = \dfrac{(n+2)(n+1)n!}{n!} = (n+2)(n+1)$

139. (a) Each term is obtained by adding the same constant
(common difference) to the preceding term.

(b) Each term is obtained by multiplying the same
constant (common ratio) by the preceding term.

141. (a) Arithmetic. There is a constant difference between
consecutive terms.

(b) Geometric. Each term is a constant multiple of the
preceding term. In this case the common ratio is
greater than 1.

143. Each term of the sequence is defined using a previous
term or terms.

145. If n is even, the expressions are the same. If n is odd, the
expressions are negatives of each other.

Chapter Test   (page 655)

1. $1, -\frac{2}{3}, \frac{4}{9}, -\frac{8}{27}, \frac{16}{81}$ **2.** 12, 16, 20, 24, 28

3. $-x, \dfrac{x^2}{2}, -\dfrac{x^3}{3}, \dfrac{x^4}{4}, -\dfrac{x^5}{5}$

4. $-\dfrac{x^3}{6}, -\dfrac{x^5}{120}, -\dfrac{x^7}{5040}, -\dfrac{x^9}{362,880}, -\dfrac{x^{11}}{39,916,800}$

5. 7920   **6.** $\dfrac{1}{n+1}$ **7.** $2n$ **8.** $a_n = n^2 + 1$

9. $a_n = 5100 - 100n$ **10.** $a_n = 4\left(\frac{1}{2}\right)^{n-1}$

11. $\displaystyle\sum_{n=1}^{12} \frac{2}{3n+1}$ **12.** $\displaystyle\sum_{n=1}^{\infty} 2\left(\frac{1}{4}\right)^{n-1}$ **13.** 189

14. 28.80   **15.** $\frac{25}{7}$ **16.** Answers will vary.

17. $16a^4 - 160a^3b + 600a^2b^2 - 1000ab^3 + 625b^4$

18. 84   **19.** 1140   **20.** 72   **21.** 328,440

22. $n = 3$ **23.** 26,000   **24.** 12,650   **25.** $\frac{3}{26}$

26. $\frac{1}{462}$ **27.** (a) $\frac{1}{4}$ (b) $\frac{121}{3600}$ (c) $\frac{1}{60}$ **28.** 0.25

Chapter 9

Section 9.1   (page 667)

Vocabulary Check   (page 667)

1. conic section   **2.** locus   **3.** circle, center

4. parabola, directrix, focus   **5.** vertex

6. axis   **7.** tangent

1. $x^2 + y^2 = 18$ **3.** $(x - 3) + (y - 7)^2 = 53$

5. $(x + 3)^2 + (y + 1)^2 = 7$

7. Center: $(0, 0)$ **9.** Center: $(-2, 7)$
Radius: 7   Radius: 4

11. Center: $(1, 0)$ **13.** $x^2 + y^2 = 4$
Radius: $\sqrt{15}$ Center: $(0, 0)$
     Radius: 2

15. $x^2 + y^2 = \dfrac{3}{4}$
Center: $(0, 0)$
Radius: $\dfrac{\sqrt{3}}{2}$

17. $(x - 1)^2 + (y + 3)^2 = 1$ **19.** $\left(x + \dfrac{3}{2}\right)^2 + (y - 3)^2 = 1$
Center: $(1, -3)$ Center: $\left(-\dfrac{3}{2}, 3\right)$
Radius: 1   Radius: 1

21. Center: $(0, 0)$ **23.** Center: $(-2, -2)$
Radius: 4   Radius: 3

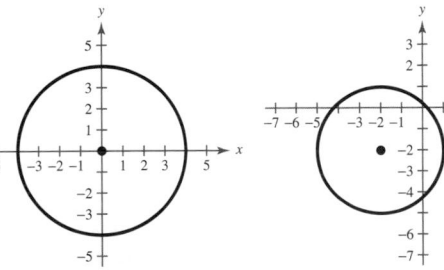

25. Center: $(7, -4)$ **27.** Center: $(-1, 0)$
Radius: 5   Radius: 6

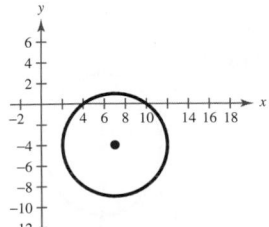   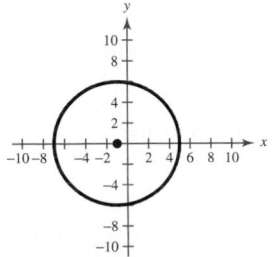

29. x-intercept: $(2, 0)$
y-intercepts: $\left(0, -3 \pm \sqrt{5}\right)$

31. x-intercepts: $\left(1 \pm 2\sqrt{2}, 0\right)$
y-intercepts: $(0, 9), (0, -3)$

33. x-intercept: $\left(6 \pm \sqrt{7}, 0\right)$
y-intercept: none

35. (a) $x^2 + y^2 = 6561$
(b) Yes
(c) 6 miles

37. e   **38.** b   **39.** d   **40.** f   **41.** a   **42.** c

43. $x^2 = \dfrac{3}{2}y$ **45.** $x^2 = -6y$ **47.** $y^2 = -8x$

49. $x^2 = 4y$ **51.** $y^2 = -8x$ **53.** $y^2 = 9x$

55. Vertex: $(0, 0)$ **57.** Vertex: $(0, 0)$
Focus: $\left(0, \dfrac{1}{2}\right)$ Focus: $\left(-\dfrac{3}{2}, 0\right)$
Directrix: $y = -\dfrac{1}{2}$ Directrix: $x = \dfrac{3}{2}$

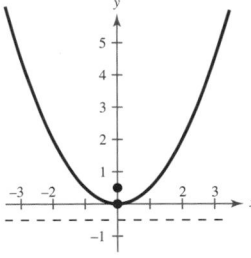   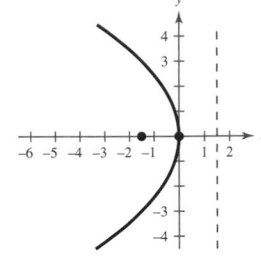

59. Vertex: $(0, 0)$ **61.** Vertex: $(-1, -3)$
Focus: $(0, -2)$ Focus: $(-1, -5)$
Directrix: $y = 2$ Directrix: $y = -1$

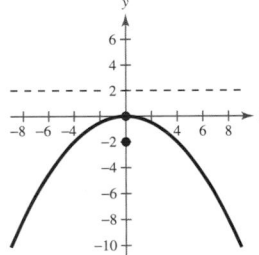   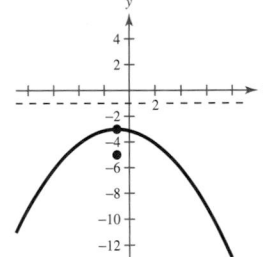

63. Vertex: $(-2, -3)$ **65.** Vertex: $\left(-\dfrac{3}{2}, 2\right)$
Focus: $(-4, -3)$ Focus: $\left(-\dfrac{3}{2}, 3\right)$
Directrix: $x = 0$ Directrix: $y = 1$

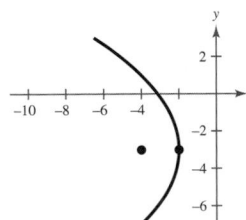   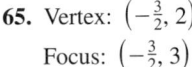 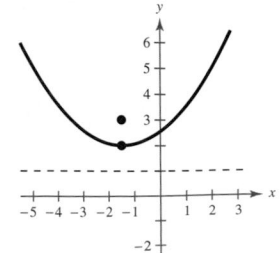

CHAPTER 9

67. Vertex: $(1, 1)$
Focus: $(1, 2)$
Directrix: $y = 0$

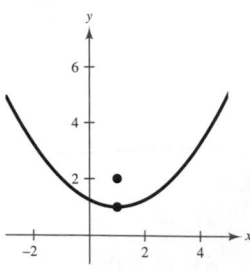

69. Vertex: $(-2, 1)$
Focus: $\left(-2, -\frac{1}{2}\right)$
Directrix: $y = \frac{5}{2}$

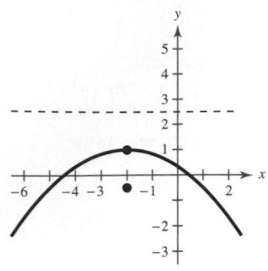

71. Vertex: $\left(\frac{1}{4}, -\frac{1}{2}\right)$
Focus: $\left(0, -\frac{1}{2}\right)$
Directrix: $x = \frac{1}{2}$

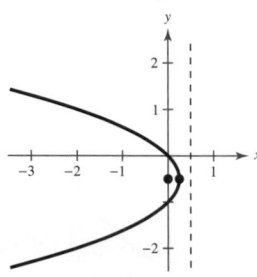

73. $(x - 3)^2 = -(y - 1)$ **75.** $y^2 = 2(x + 2)$
77. $(y - 2)^2 = -8(x - 5)$ **79.** $x^2 = 8(y - 4)$
81. $(y - 2)^2 = 8x$
83.

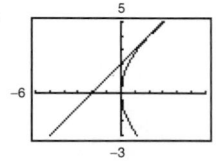

$(2, 4)$

85. $4x - y - 8 = 0$; $(2, 0)$ **87.** $4x - y + 2 = 0$; $\left(-\frac{1}{2}, 0\right)$
89.

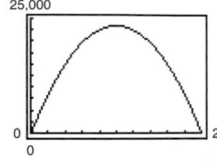

$x = 125$ televisions

91. (a) $y^2 = 6x$ (b) 2.67 inches

93. (a)

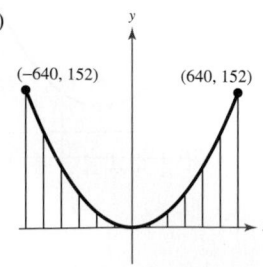

(b) $x^2 = \frac{51,200}{19}y$

(c)

x	0	200	400	500	600
y	0	14.844	59.375	92.773	133.59

95. $y^2 = 640x$

97. (a)

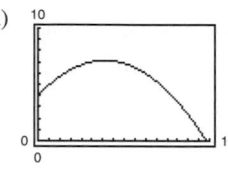

(b) Highest point: 7.125 feet
Distance: 15.687 feet

99. $y = \frac{3}{4}x - \frac{25}{4}$ **101.** $y = \frac{\sqrt{2}}{2}x - 3\sqrt{2}$

103. False. $x^2 + (y + 5)^2 = 25$ represents a circle with its center at $(0, -5)$ and a radius of 5.

105. False. A circle is a conic section.   **107.** True

109. The resulting surface has the property that all incoming rays parallel to the axis are reflected through the focus of the parabola. Graphical representations will vary.

111. $y = \sqrt{6(x + 1)} + 3$

113. Minimum: $(0.67, 0.22)$; maximum: $(-0.67, 3.78)$

115. Minimum: $(-0.79, 0.81)$

Section 9.2   (page 677)

Vocabulary Check   (page 677)

1. ellipse   **2.** major axis, center
3. minor axis   **4.** eccentricity

---

1. b   **2.** c   **3.** d   **4.** f   **5.** a   **6.** e
7. Center: $(0, 0)$
Vertices: $(\pm 8, 0)$
Foci: $\left(\pm \sqrt{55}, 0\right)$
Eccentricity: $\frac{\sqrt{55}}{8}$

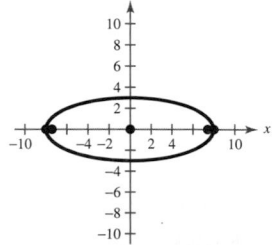

9. Center: $(4, -1)$
Vertices: $(4, 4), (4, -6)$
Foci: $(4, 2), (4, -4)$
Eccentricity: $\frac{3}{5}$

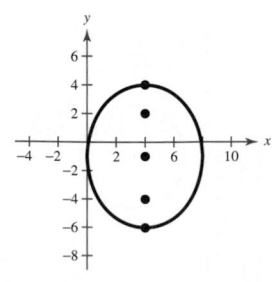

11. Center: $(-5, 1)$

Vertices:

$$\left(-\frac{7}{2}, 1\right), \left(-\frac{13}{2}, 1\right)$$

Foci: $\left(-5 \pm \frac{\sqrt{5}}{2}, 1\right)$

Eccentricity: $\frac{\sqrt{5}}{2}$

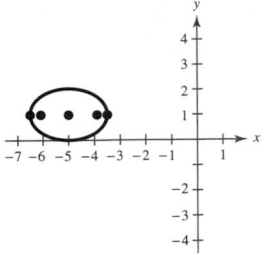

13. (a) $\dfrac{x^2}{36} + \dfrac{y^2}{4} = 1$ (c)

(b) Center: $(0, 0)$

Vertices: $(\pm 6, 0)$

Foci: $(\pm 4\sqrt{2}, 0)$

Eccentricity: $\dfrac{2\sqrt{2}}{3}$

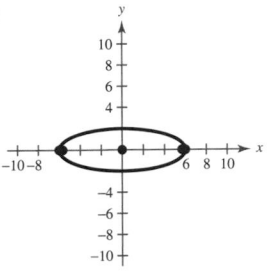

15. (a) $\dfrac{(x + 2)^2}{4} + \dfrac{(y - 3)^2}{9} = 1$

(b) Center: $(-2, 3)$

Vertices: $(-2, 6), (-2, 0)$

Foci: $(-2, 3 \pm \sqrt{5})$

Eccentricity: $\dfrac{\sqrt{5}}{3}$

(c)

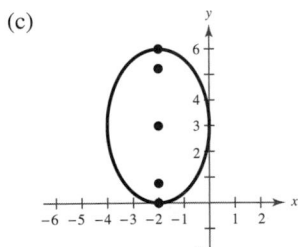

17. (a) $\dfrac{\left(x + \frac{3}{2}\right)^2}{4} + \dfrac{\left(y - \frac{5}{2}\right)^2}{12} = 1$

(b) Center: $\left(-\dfrac{3}{2}, \dfrac{5}{2}\right)$

Vertices: $\left(-\dfrac{3}{2}, \dfrac{5 \pm 4\sqrt{3}}{2}\right)$

Foci: $\left(-\dfrac{3}{2}, \dfrac{5}{2} \pm 2\sqrt{2}\right)$

Eccentricity: $\dfrac{\sqrt{6}}{3}$

(c)

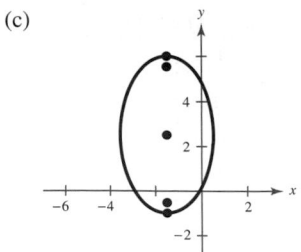

19. (a) $\dfrac{(x - 1)^2}{\frac{25}{16}} + (y + 1)^2 = 1$

(b) Center: $(1, -1)$

Vertices: $\left(\dfrac{9}{4}, -1\right), \left(-\dfrac{1}{4}, -1\right)$

Foci: $\left(\dfrac{7}{4}, -1\right), \left(\dfrac{1}{4}, -1\right)$

Eccentricity: $\dfrac{3}{5}$

(c)

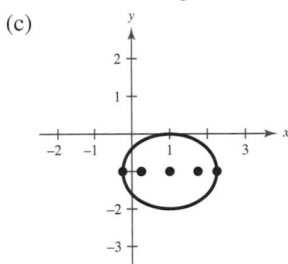

21. (a) $\dfrac{\left(x - \frac{1}{2}\right)^2}{5} + \dfrac{(y + 1)^2}{3} = 1$

(b) Center: $\left(\dfrac{1}{2}, -1\right)$

Vertices: $\left(\dfrac{1}{2} \pm \sqrt{5}, -1\right)$

Foci: $\left(\dfrac{1}{2} \pm \sqrt{2}, -1\right)$

Eccentricity: $\dfrac{\sqrt{10}}{5}$

(c)

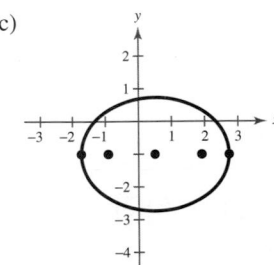

23. $\dfrac{x^2}{4} + \dfrac{y^2}{16} = 1$ **25.** $\dfrac{x^2}{9} + \dfrac{y^2}{5} = 1$ **27.** $\dfrac{x^2}{16} + \dfrac{y^2}{7} = 1$

CHAPTER 9

29. $\dfrac{x^2}{400/21} + \dfrac{y^2}{25} = 1$ **31.** $\dfrac{(x-2)^2}{1} + \dfrac{(y-3)^2}{9} = 1$

33. $\dfrac{(x-4)^2}{16} + \dfrac{(y-2)^2}{1} = 1$ **35.** $\dfrac{x^2}{308} + \dfrac{(y-4)^2}{324} = 1$

37. $\dfrac{(x-3)^2}{9} + \dfrac{(y-5)^2}{16} = 1$ **39.** $\dfrac{x^2}{16} + \dfrac{(y-4)^2}{12} = 1$

41. $\dfrac{\sqrt{5}}{3}$ **43.** $\dfrac{2\sqrt{2}}{3}$ **45.** $\dfrac{x^2}{2} + \dfrac{y^2}{9} = 1$

47. (a)

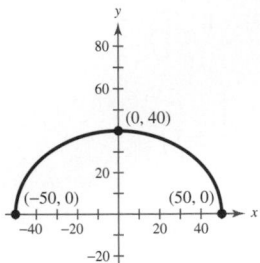

(b) $\dfrac{x^2}{2500} + \dfrac{y^2}{1600} = 1$ (c) 17.4 feet

49. $\left(\pm\sqrt{5}, 0\right)$; 6 feet   **51.** 40 units

53. $\dfrac{x^2}{4.88} + \dfrac{y^2}{1.39} = 1$ **55.** Answers will vary.

57.

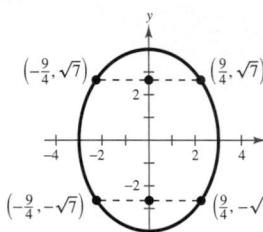

59.

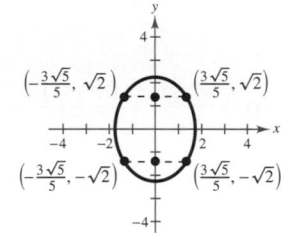

61. True

63. (a) $2a$

(b) The sum of the distances from the two fixed points is constant.

65. $\dfrac{(x-6)^2}{324} + \dfrac{(y-2)^2}{308} = 1$ **67.** Arithmetic

69. Geometric   **71.** 1093   **73.** 15.0990

Section 9.3   (page 687)

Vocabulary Check   (page 687)

1. hyperbola   **2.** branches

3. transverse axis, center   **4.** asymptotes

5. $Ax^2 + Cy^2 + Dx + Ey + F = 0$

1. b   **2.** c   **3.** a   **4.** d

5. Center: $(0, 0)$

Vertices: $(\pm 1, 0)$

Foci: $\left(\pm\sqrt{2}, 0\right)$

Asymptotes: $y = \pm x$

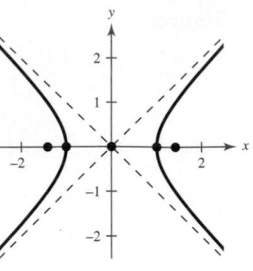

7. Center: $(0, 0)$

Vertices: $(0, \pm 1)$

Foci: $\left(0, \pm\sqrt{5}\right)$

Asymptotes: $y = \pm\tfrac{1}{2}x$

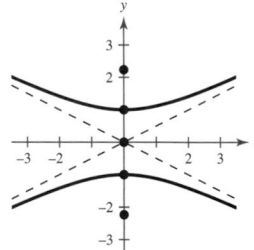

9. Center: $(0, 0)$

Vertices: $(0, \pm 5)$

Foci: $\left(0, \pm\sqrt{106}\right)$

Asymptotes: $y = \pm\tfrac{5}{9}x$

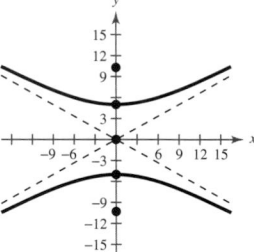

11. Center: $(1, -2)$

Vertices: $(3, -2), (-1, -2)$

Foci: $\left(1 \pm \sqrt{5}, -2\right)$

Asymptotes:

$y = -2 \pm \tfrac{1}{2}(x - 1)$

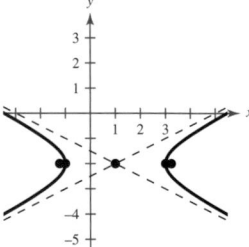

13. Center: $(1, -5)$

Vertices: $\left(1, -5 \pm \dfrac{1}{3}\right)$

Foci: $\left(1, -5 \pm \dfrac{\sqrt{13}}{6}\right)$

Asymptotes:

$y = -5 \pm \dfrac{2}{3}(x - 1)$

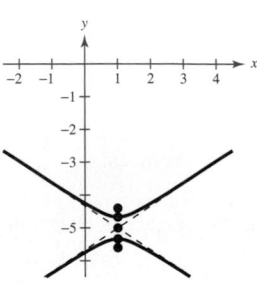

15. (a) $\dfrac{x^2}{9} - \dfrac{y^2}{4} = 1$ (c)

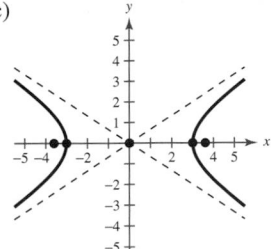

(b) Center: $(0, 0)$

Vertices: $(\pm 3, 0)$

Foci: $\left(\pm\sqrt{13}, 0\right)$

Asymptotes: $y = \pm\dfrac{2}{3}x$

17. (a) $\dfrac{x^2}{3} - \dfrac{y^2}{2} = 1$

(b) Center: $(0, 0)$

Vertices: $\left(\pm\sqrt{3}, 0\right)$

Foci: $\left(\pm\sqrt{5}, 0\right)$

Asymptotes: $y = \pm\dfrac{\sqrt{6}}{3}x$

(c)

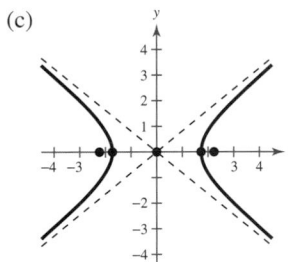

19. (a) $(x - 2)^2 - \dfrac{(y + 3)^2}{9} = 1$

(b) Center: $(2, -3)$

Vertices: $(3, -3), (1, -3)$

Foci: $\left(2 \pm \sqrt{10}, -3\right)$

Asymptotes: $y = -3 \pm 3(x - 2)$

(c)

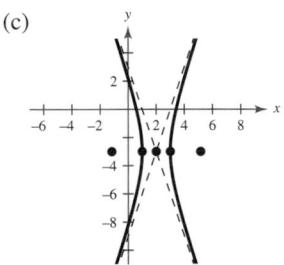

21. (a) $(x + 1)^2 - 9(y + 3)^2 = 0$

(b) It is a degenerate conic. The graph of this equation is two lines intersecting at $(-1, -3)$.

(c)

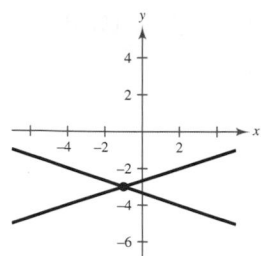

23. (a) $\dfrac{(y + 3)^2}{2} - \dfrac{(x - 1)^2}{18} = 1$

(b) Center: $(1, -3)$

Vertices: $\left(1, -3 \pm \sqrt{2}\right)$

Foci: $\left(1, -3 \pm 2\sqrt{5}\right)$

Asymptotes: $y = -3 \pm \dfrac{1}{3}(x - 1)$

(c)

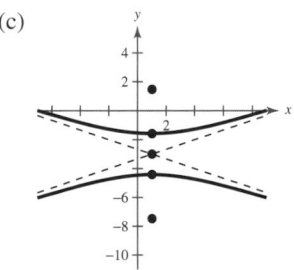

25. $\dfrac{y^2}{4} - \dfrac{x^2}{12} = 1$ **27.** $\dfrac{x^2}{1} - \dfrac{y^2}{25} = 1$

29. $\dfrac{17y^2}{1024} - \dfrac{17x^2}{64} = 1$ **31.** $\dfrac{(x - 4)^2}{4} - \dfrac{y^2}{12} = 1$

33. $\dfrac{(y - 5)^2}{16} - \dfrac{(x - 4)^2}{9} = 1$ **35.** $\dfrac{y^2}{9} - \dfrac{4(x - 2)^2}{9} = 1$

37. $\dfrac{(y - 2)^2}{4} - \dfrac{x^2}{4} = 1$ **39.** $\dfrac{(x - 2)^2}{1} - \dfrac{(y - 2)^2}{1} = 1$

41. $\dfrac{(x - 3)^2}{9} - \dfrac{(y - 2)^2}{4} = 1$

43. $\dfrac{x^2}{98,010,000} - \dfrac{y^2}{13,503,600} = 1$

45. (a) $x^2 - \dfrac{y^2}{27} = 1$ (b) 1.89 feet = 22.68 inches

47. $\left(12\sqrt{5} - 12, 0\right) \approx (14.83, 0)$ **49.** Ellipse

51. Hyperbola   **53.** Parabola   **55.** Circle

57. Parabola

59. True. For a hyperbola, $c^2 = a^2 + b^2$. The larger the ratio of b to a, the larger the eccentricity of the hyperbola, $e = c/a$.

61. False. If $D = E$ or $D = -E$, the graph is two intersecting lines. For example, the graph of $x^2 - y^2 - 2x + 2y = 0$ is two intersecting lines.

CHAPTER 9

63. Proof   **65.** Answers will vary.   **67.** Proof

69. $x^3 + x^2 + 2x - 6$ **71.** $x^2 - 2x + 1 + \dfrac{2}{x + 2}$

73. $x(x + 4)(x - 4)$ **75.** $2x(x - 6)^2$

77. $2(2x + 3)(4x^2 - 6x + 9)$

Section 9.4   (page 697)

Vocabulary Check   (page 697)

1. rotation, axes   **2.** invariant under rotation

3. discriminant

1. $(3, 0)$

3. $\dfrac{(y')^2}{2} - \dfrac{(x')^2}{2} = 1$ **5.** $(x')^2 - \dfrac{(y')^2}{1/3} = 1$

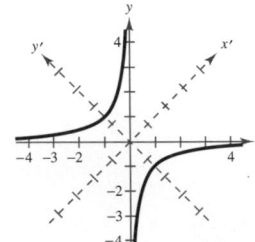
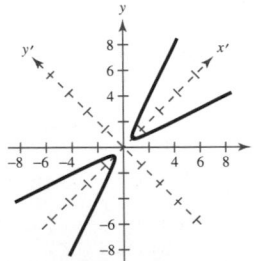

7. $\dfrac{\left(x' - 3\sqrt{2}\right)^2}{16} - \dfrac{\left(y' - \sqrt{2}\right)^2}{16} = 1$

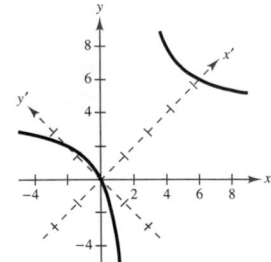

9. $\dfrac{(x')^2}{6} + \dfrac{(y')^2}{3/2} = 1$ **11.** $x' = -(y')^2$

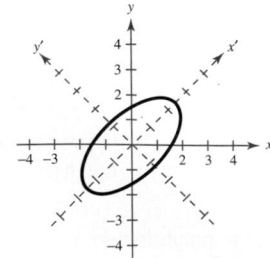
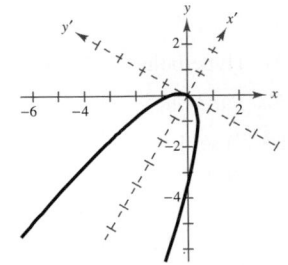

13. $y' = \frac{1}{6}(x')^2 - \frac{1}{3}x'$

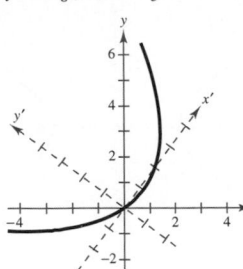

15.

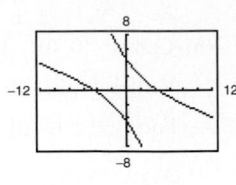

$\theta = 45°$

17.

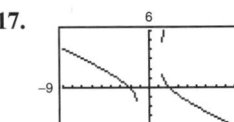

$\theta = 26.57°$

19.

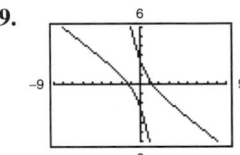

$\theta = 31.72°$

21. e   **22.** b   **23.** f   **24.** a   **25.** d   **26.** c

27. (a) Parabola

(b) $y = \dfrac{24x + 40 \pm \sqrt{3000x + 1600}}{18}$

(c)

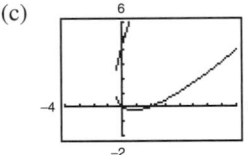

29. (a) Ellipse or circle

(b) $y = \dfrac{8x \pm \sqrt{-356x^2 + 1260}}{14}$

(c)

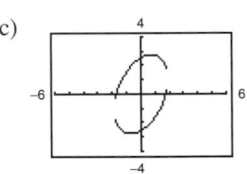

31. (a) Hyperbola

(b) $y = \dfrac{6x \pm \sqrt{56x^2 + 80x - 440}}{-10}$

(c)

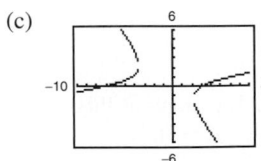

33. (a) Parabola

(b) $y = \dfrac{-4x + 1 \pm \sqrt{72x + 49}}{8}$

(c)
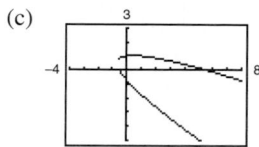

35.
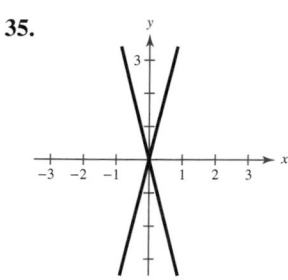

37.
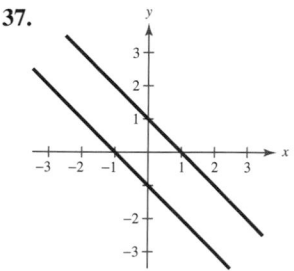

39. $\left(1, \sqrt{3}\right), \left(1, -\sqrt{3}\right)$ **41.** $(-8, 12)$

43. $(0, 8), (12, 8)$ **45.** $(0, 4)$

47. $\left(\sqrt{3}, -2\sqrt{3}\right), \left(-\sqrt{3}, 2\sqrt{3}\right)$ **49.** $(8, 0)$

51. $\left(-3, 0\right), \left(0, \tfrac{3}{2}\right)$

53. True. The discriminant will be greater than zero.

55.
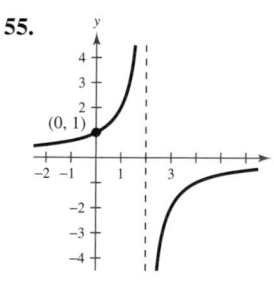

57.
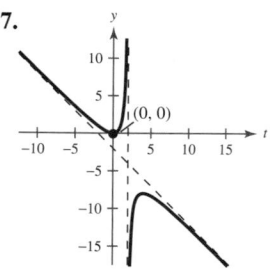

Intercept: $(0, 1)$ Intercept: $(0, 0)$

Asymptotes: Asymptotes:

$x = 2, y = 0$ $t = 2, y = -t - 2$

59. (a) $\begin{bmatrix} -15 & 9 \\ 25 & 7 \end{bmatrix}$ (b) $\begin{bmatrix} 12 & 30 \\ 3 & -20 \end{bmatrix}$ (c) $\begin{bmatrix} -5 & -18 \\ 12 & 19 \end{bmatrix}$

61. (a) $[45]$ (b) $\begin{bmatrix} 12 & -6 & 15 \\ -16 & 8 & -20 \\ 20 & -10 & 25 \end{bmatrix}$ (c) Not possible

63.
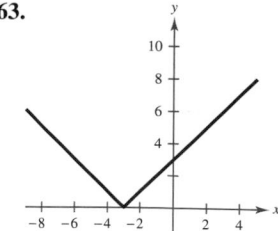

65.
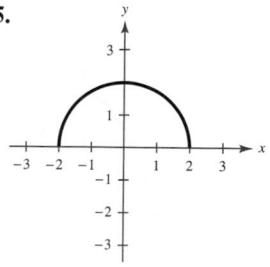

67.
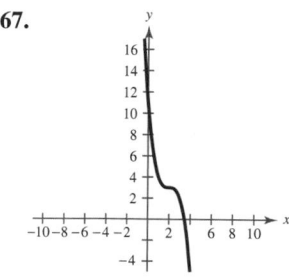

69.
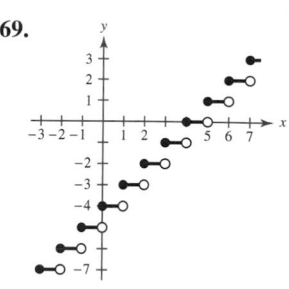

71. 45.11   **73.** 48.60

Section 9.5   (page 704)

Vocabulary Check   (page 704)

1. plane curve, parametric equations, parameter

2. orientation   **3.** eliminating, parameter

1. c   **2.** d   **3.** b   **4.** a   **5.** f   **6.** e

7. (a)

t	0	1	2	3	4
x	0	1	$\sqrt{2}$	$\sqrt{3}$	2
y	2	1	0	-1	-2

(b)
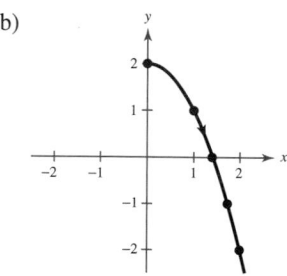

(c)
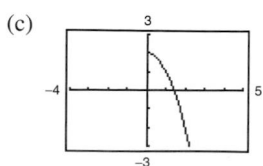

CHAPTER 9

(d) $y = 2 - x^2$

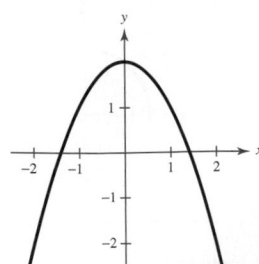

The graph is an entire parabola rather than just the right half.

9. b

11.

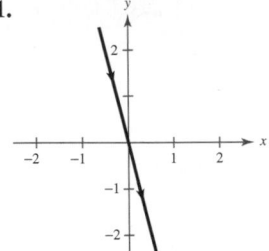

$y = -4x$

13.

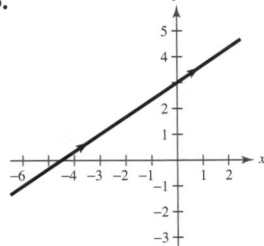

$y = \frac{2}{3}x + 3$

15.

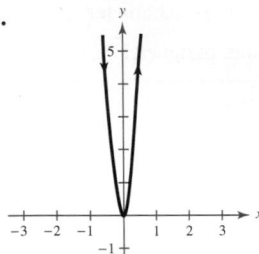

$y = 16x^2$

17.

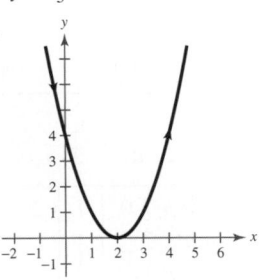

$y = (x - 2)^2$

19.

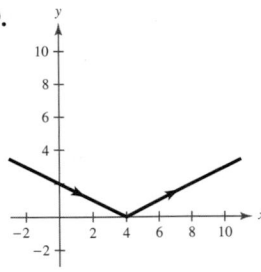

$y = \frac{1}{2}|x - 4|$

21.

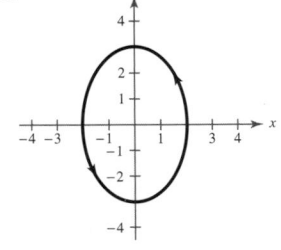

$\dfrac{x^2}{4} + \dfrac{y^2}{9} = 1$

23.

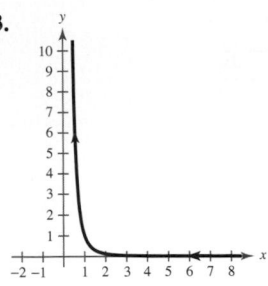

$y = x^{-3}, \; x > 0$

25.

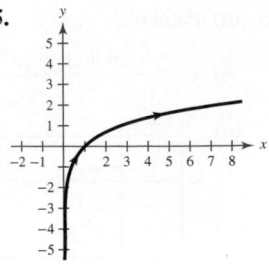

$y = \ln x$

27.

 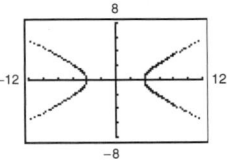

29.

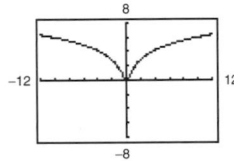

31.

33. Each curve represents a portion of the line $y = 2x + 1$.

Domain	Orientation
(a) $(-\infty, \infty)$	Left to right
(b) $[-1, 1]$	Depends on θ
(c) $(0, \infty)$	Right to left
(d) $(0, \infty)$	Left to right

35. $y - y_1 = \dfrac{y_2 - y_1}{x_2 - x_1}(x - x_1)$

37. $\dfrac{(x - h)^2}{a^2} + \dfrac{(y - k)^2}{b^2} = 1$

39. $x = 1 + 5t, \; y = 4 - 7t$

41. $x = 5 \cos \theta$
$\quad\; y = 3 \sin \theta$

43. $x = t, \; y = 5t - 3$
$\quad\; x = \frac{1}{5}t, \; y = t - 3$

45. $x = t, \; y = \dfrac{1}{t}$
$\quad\; x = t^3, \; y = \dfrac{1}{t^3}$

47. $x = t, \; y = 6t^2 - 5$
$\quad\; x = 2t, \; y = 24t^2 - 5$

49.

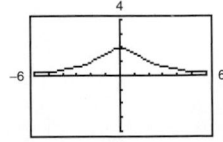

51. b   **52.** c   **53.** d   **54.** a

55. (a) $x = (146.67 \cos \theta)t$

$y = 3 + (146.67 \sin \theta)t - 16t^2$

(b) 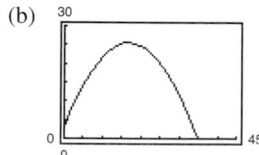   (c)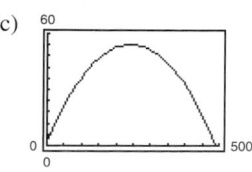

        No   Yes

(d) About $19.38°$

57. (a) $x = (v_0 \cos 35°)t$

$y = 7 + (v_0 \sin 35°)t - 16t^2$

(b) 54.09 feet per second

(c)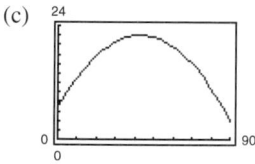

        22.04 feet

(d) 2.03 seconds

59. True. Both sets of parametric equations correspond to $y = x^2 + 1$.

61. False. $x = t^2$, $y = t$ does not correspond to y as a function of x.

63. Answers will vary. Sample answer:

$x = \cos \theta$

$y = -2 \sin \theta$

65. Even   **67.** Neither

Section 9.6   (page 711)

Vocabulary Check   (page 711)

1. pole   **2.** directed distance, directed angle

3. polar

1. $(0, 4)$ **3.** $\left(\dfrac{\sqrt{2}}{2}, \dfrac{\sqrt{2}}{2} \right)$

5.

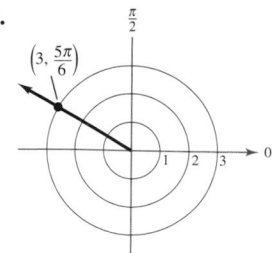

$\left(3, -\dfrac{7\pi}{6} \right), \left(-3, \dfrac{11\pi}{6} \right), \left(-3, -\dfrac{\pi}{6} \right)$

7.

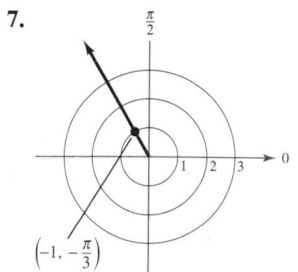

$\left(-1, \dfrac{5\pi}{3} \right), \left(1, \dfrac{2\pi}{3} \right), \left(1, -\dfrac{4\pi}{3} \right)$

9.

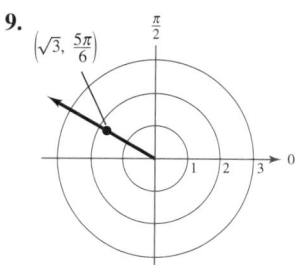

$\left(-\sqrt{3}, \dfrac{11\pi}{6} \right), \left(\sqrt{3}, -\dfrac{7\pi}{6} \right), \left(-\sqrt{3}, -\dfrac{\pi}{6} \right)$

11.

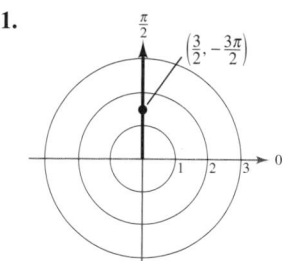

$\left(\dfrac{3}{2}, \dfrac{\pi}{2} \right), \left(-\dfrac{3}{2}, \dfrac{3\pi}{2} \right), \left(-\dfrac{3}{2}, -\dfrac{\pi}{2} \right)$

CHAPTER 9

13.

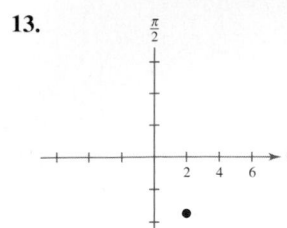

$\left(2, -2\sqrt{3}\right)$

15.

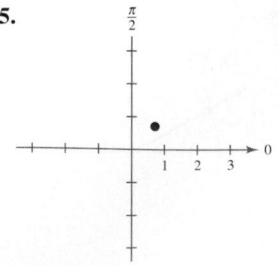

$\left(\dfrac{\sqrt{2}}{2}, \dfrac{\sqrt{2}}{2}\right)$

17.

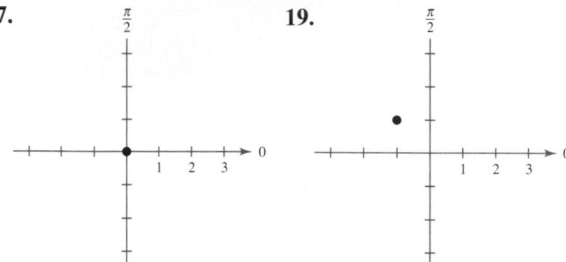

$(0, 0)$

19.

$(-1.004, 0.996)$

21. $(1.53, 1.29)$ **23.** $(-1.20, -4.34)$

25. $(-0.02, 2.50)$ **27.** $(-3.60, 1.97)$

29.

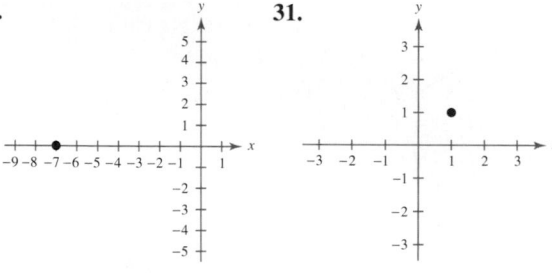

$(7, \pi), (-7, 0)$

31.

$\left(\sqrt{2}, \dfrac{\pi}{4}\right), \left(-\sqrt{2}, \dfrac{5\pi}{4}\right)$

33.

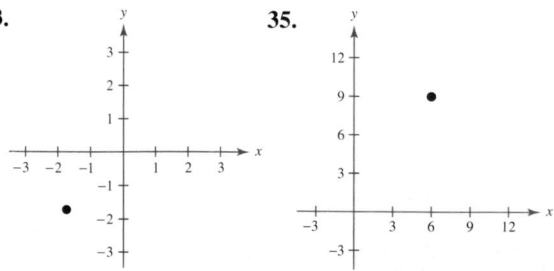

$\left(\sqrt{6}, \dfrac{5\pi}{4}\right), \left(-\sqrt{6}, \dfrac{\pi}{4}\right)$

35.

$(10.82, 0.98), (-10.82, 4.12)$

37. $(3.61, -0.59)$ **39.** $(2.65, 0.86)$ **41.** $(2.83, 0.49)$

43. $r = 3$ **45.** $r = 4 \csc \theta$ **47.** $r = 8 \sec \theta$

49. $r = -\dfrac{2}{3 \cos \theta - 6 \sin \theta}$ **51.** $r^2 = 8 \csc 2\theta$

53. $r^2 = 9 \cos 2\theta$ **55.** $r = 6 \cos \theta$ **57.** $r = 2a \cos \theta$

59. $r = \tan^2 \theta \sec \theta$ **61.** $x^2 + y^2 = 6y$

63. $y = \sqrt{3}x$ **65.** $y = -\dfrac{\sqrt{3}}{3}x$ **67.** $x = 0$

69. $x^2 + y^2 = 16$ **71.** $y = -3$

73. $(x^2 + y^2)^3 = x^2$ **75.** $(x^2 + y^2)^2 = 6x^2y - 2y^3$

77. $y^2 = 2x + 1$ **79.** $4x^2 - 5y^2 = 36y + 36$

81. The graph is a circle centered at the origin with a radius of 7; $x^2 + y^2 = 49$.

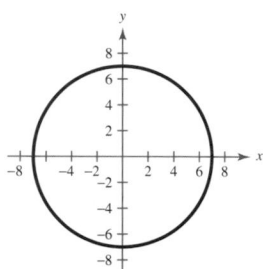

83. The graph consists of all points on the line that makes an angle of $\pi/4$ with the positive x-axis; $x - y = 0$.

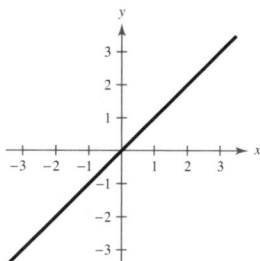

85. The graph is a vertical line through $(3, 0)$; $x - 3 = 0$.

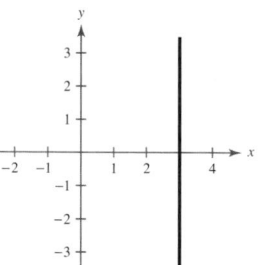

87. True. Because r is a directed distance, (r, θ) can be represented by $(-r, \theta \pm (2n + 1)\pi)$, so $|r| = |-r|$.

89. (a) Answers will vary.

(b) The points lie on a line passing through the pole.
$$d = \sqrt{r_1^2 + r_2^2 - 2r_1 r_2} = |r_1 - r_2|$$

(c) $d = \sqrt{r_1^2 + r_2^2}$ (Pythagorean Theorem)

Answers will vary.

(d) Answers will vary. The Distance Formula should give the same result in both cases.

91. $A \approx 30.68°$ **93.** $a \approx 16.16$ **95.** $A \approx 119.09°$
$B \approx 48.23°$ $b \approx 19.44$ $B \approx 25.91°$
$C \approx 101.09°$ $B \approx 86°$ $c \approx 5.25$

97. $(2, 3)$ **99.** $(0, 0, 0)$ **101.** $(2, -3, 3)$

103. Not collinear   **105.** Collinear

Section 9.7   (page 720)

Vocabulary Check   (page 720)

1. $\theta = \dfrac{\pi}{2}$ **2.** polar axis   **3.** convex limaçon

4. circle   **5.** lemniscate   **6.** cardioid

1. Rose curve   **3.** Lemniscate   **5.** Rose curve

7. a   **9.** c   **11.** Polar axis   **13.** $\theta = \dfrac{\pi}{2}$

15. $\theta = \dfrac{\pi}{2}$ **17.** Pole

19. Maximum: $|r| = 20$ **21.** Maximum: $|r| = 4$

Zero of r: $\theta = \dfrac{\pi}{2}$ Zeros of r: $\theta = \dfrac{\pi}{6}, \dfrac{\pi}{2}, \dfrac{5\pi}{6}$

23. 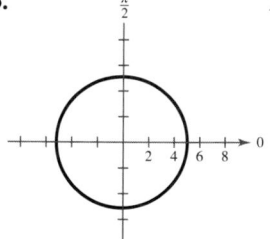   **25.**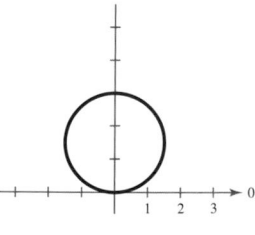

27. 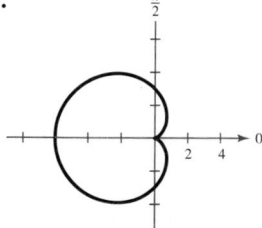   **29.**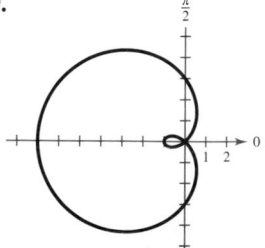

31. 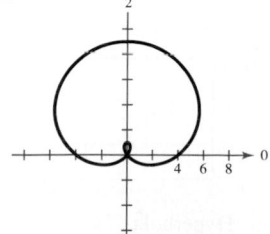   **33.**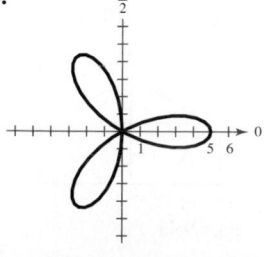

35. 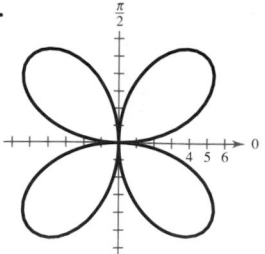   **37.**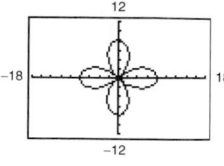

$0 \le \theta < 2\pi$
Answers will vary.

39. 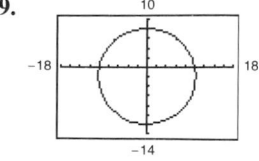   **41.**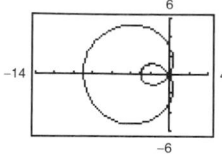

$0 \le \theta < 2\pi$
Answers will vary.

Answers will vary.

43. 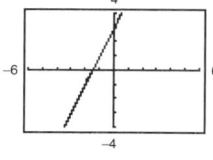   **45.**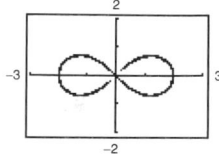

Answers will vary.   Answers will vary.

47. 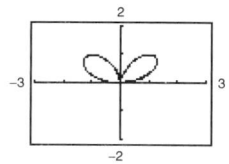   **49.**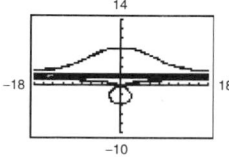

Answers will vary. $0 \le \theta < 2\pi$
Answers will vary.

51. 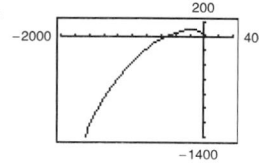   **53.**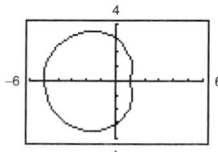

Answers will vary. $0 \le \theta < 2\pi$

55. 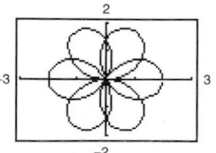   **57.**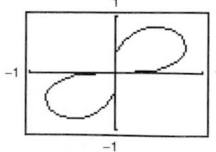

$0 \le \theta < 4\pi$ $0 \le \theta < \dfrac{\pi}{2}$

CHAPTER 9

59.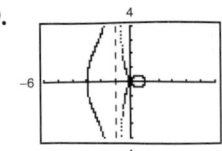

61.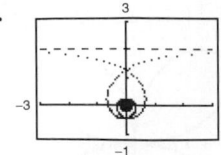

63. True

65. $n = -5$ $n = -4$

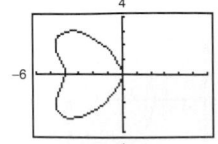   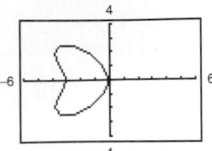

$n = -3$ $n = -2$

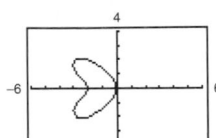   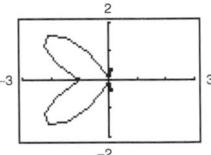

$n = -1$ $n = 0$

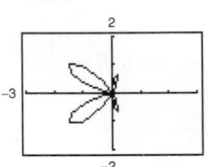   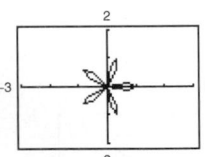

$n = 1$ $n = 2$

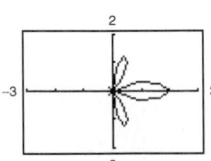   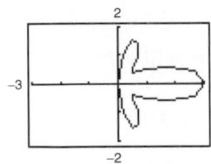

$n = 3$ $n = 4$

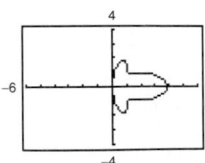   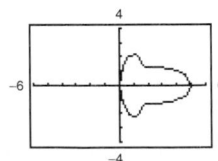

$n = 5$

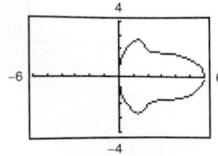

Negative values of n produce the heart-shaped curves; positive values of n produce the bell-shaped curves.

67. (a), (b), and (c) Answers will vary.

69. (a) $r = 4 \sin\left(\theta - \dfrac{\pi}{6}\right) \cos\left(\theta - \dfrac{\pi}{6}\right)$

(b) $r = -4 \sin\theta \cos\theta$

(c) $r = 4 \sin\left(\theta - \dfrac{2\pi}{3}\right) \cos\left(\theta - \dfrac{2\pi}{3}\right)$

(d) $r = 4 \sin\theta \cos\theta$

71.

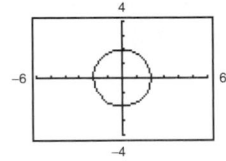   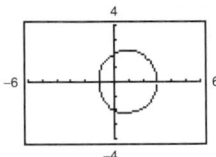

$k = 0$; circle $k = 1$; convex limaçon

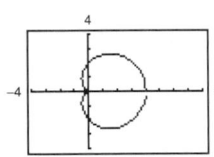   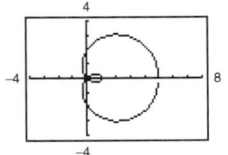

$k = 2$; cardioid $k = 3$; limaçon with inner loop

Section 9.8   (page 726)

Vocabulary Check   (page 726)

1. conic   **2.** eccentricity, e

3. (a) i   (b) iii   (c) ii

1. 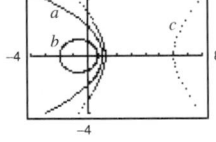   **3.**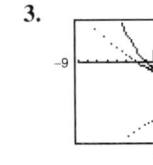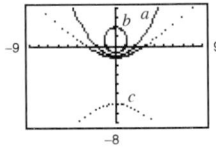

(a) Parabola   (a) Parabola

(b) Ellipse   (b) Ellipse

(c) Hyperbola   (c) Hyperbola

5. b   **6.** c   **7.** f   **8.** e   **9.** d   **10.** a

11. Parabola   **13.** Ellipse   **15.** Ellipse

17. Ellipse   **19.** Hyperbola

21. 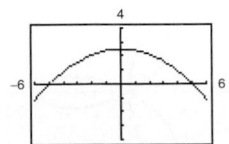   **23.**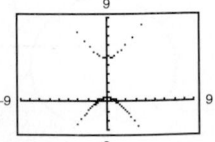

Parabola   Hyperbola

25.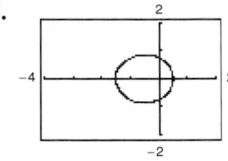

Ellipse

27.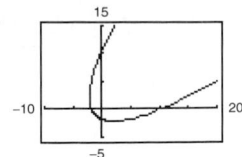

29.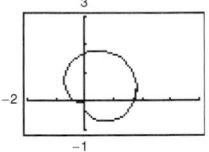

31.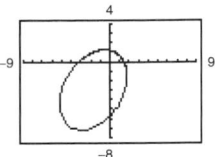

33. $r = \dfrac{1}{1 - \cos \theta}$ **35.** $r = \dfrac{1}{2 + \sin \theta}$

37. $r = \dfrac{2}{1 + 2 \cos \theta}$ **39.** $r = \dfrac{2}{1 - \sin \theta}$

41. $r = \dfrac{10}{1 - \cos \theta}$ **43.** $r = \dfrac{10}{3 + 2 \cos \theta}$

45. $r = \dfrac{20}{3 - 2 \cos \theta}$ **47.** $r = \dfrac{8}{3 + 5 \sin \theta}$

49. Answers will vary.

51. $r = \dfrac{9.2930 \times 10^7}{1 - 0.0167 \cos \theta}$

Perihelion: 9.1404×10^7 miles

Aphelion: 9.4508×10^7 miles

53. $r = \dfrac{7.7659 \times 10^8}{1 - 0.0484 \cos \theta}$

Perihelion: 7.4073×10^8 kilometers

Aphelion: 8.1609×10^8 kilometers

55. (a) $r_{\text{Neptune}} = \dfrac{4.4977 \times 10^9}{1 - 0.0086 \cos \theta}$

$r_{\text{Pluto}} = \dfrac{5.5404 \times 10^9}{1 - 0.2488 \cos \theta}$

(b) Neptune: Perihelion: 4.4593×10^9 kilometers

Aphelion: 4.5367×10^9 kilometers

Pluto: Perihelion: 4.4366×10^9 kilometers

Aphelion: 7.3754×10^9 kilometers

(c)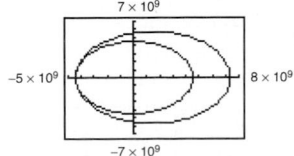

(d) Yes; because on average, Pluto is farther from the sun than Neptune.

(e) Using a graphing utility, it would appear that the orbits intersect. No, Pluto and Neptune will never collide because the orbits do not intersect in three-dimensional space.

57. False. The equation can be rewritten as

$$r = \dfrac{-4/3}{1 + \sin \theta}.$$

Because ep is negative, p must be negative and since p represents the distance between the pole and the directrix, the directrix has to be below the pole.

59. Answers will vary.   **61.** $r^2 = \dfrac{24{,}336}{169 - 25 \cos^2 \theta}$

63. $r^2 = \dfrac{400}{25 - 9 \cos^2 \theta}$ **65.** $r^2 = \dfrac{144}{25 \sin^2 \theta - 16}$

67. (a) Ellipse

(b) $r = \dfrac{4}{1 + 0.4 \cos \theta}$ is reflected about the line $\theta = \dfrac{\pi}{2}$.

$r = \dfrac{4}{1 - 0.4 \sin \theta}$ is rotated 90° counterclockwise.

69. Answers will vary.   **71.** $\dfrac{\pi}{6} + n\pi$

73. $\dfrac{\pi}{3} + n\pi, \dfrac{2\pi}{3} + n\pi$ **75.** $\dfrac{\pi}{2} + n\pi$

77. $\dfrac{\sqrt{2}}{10}$ **79.** $\dfrac{\sqrt{2}}{10}$ **81.** 220   **83.** 720

Review Exercises   (page 730)

1. $x^2 + y^2 = 25$ **3.** $(x - 2)^2 + (y - 4)^2 = 13$

5. $x^2 + y^2 = 36$ **7.** $\left(x - \tfrac{1}{2}\right)^2 + \left(y + \tfrac{3}{4}\right)^2 = 1$

Center: $(0, 0)$ Center: $\left(\tfrac{1}{2}, -\tfrac{3}{4}\right)$

Radius: 6   Radius: 1

9. 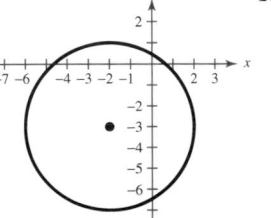   **11.** $\left(3 \pm \sqrt{6}, 0\right)$

Center: $(-2, -3)$

Radius: 4

13. Vertex: $(0, 0)$
   Focus: $(1, 0)$
   Directrix: $x = -1$

15. Vertex: $(0, 0)$
   Focus: $(-9, 0)$
   Directrix: $x = 9$

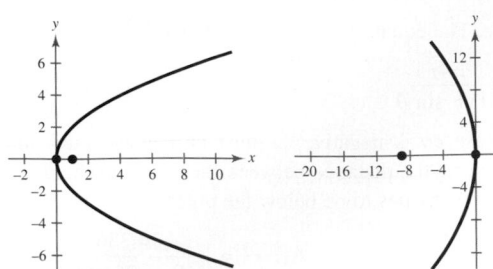

17. $y^2 = -24x$ **19.** $(y - 2)^2 = 12x$

21. $2x + y - 2 = 0; (1, 0)$ **23.** $8\sqrt{6}$ meters

25. Center: $(0, 0)$
   Vertices: $(0, \pm4)$
   Foci: $\left(0, \pm2\sqrt{3}\right)$
   Eccentricity: $\dfrac{\sqrt{3}}{2}$

27. Center: $(4, -4)$
   Vertices: $(4, -1), (4, -7)$
   Foci: $\left(4, -4 \pm \sqrt{3}\right)$
   Eccentricity: $\dfrac{\sqrt{3}}{3}$

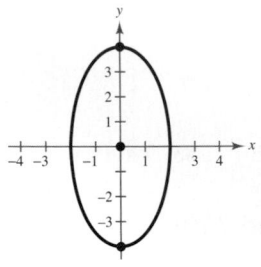

29. (a) $\dfrac{(x - 1)^2}{9} + \dfrac{(y + 4)^2}{16} = 1$

   (b) Center: $(1, -4)$
   Vertices: $(1, 0), (1, -8)$
   Foci: $\left(1, -4 \pm \sqrt{7}\right)$
   Eccentricity: $\dfrac{\sqrt{7}}{4}$

   (c)

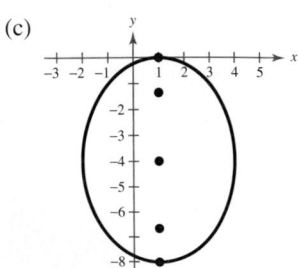

31. (a) $\dfrac{(x + 2)^2}{1/3} + \dfrac{(y - 7)^2}{1/8} = 1$

   (b) Center: $(-2, 7)$
   Vertices: $\left(-2 \pm \dfrac{\sqrt{3}}{3}, 7\right)$
   Foci: $\left(-2 \pm \dfrac{\sqrt{30}}{12}, 7\right)$
   Eccentricity: $\dfrac{\sqrt{30}}{4}$

   (c)

33. $\dfrac{x^2}{25} + \dfrac{y^2}{9} = 1$ **35.** $\dfrac{(x - 2)^2}{25} + \dfrac{y^2}{21} = 1$

37. The foci should be placed 3 feet on either side of the center at the same height as the pillars.

39. $e \approx 0.0543$

41. (a) $\dfrac{y^2}{4} - \dfrac{x^2}{5} = 1$

   (b) Center: $(0, 0)$
   Vertices: $(0, \pm2)$
   Foci: $(0, \pm3)$
   Eccentricity: $\dfrac{3}{2}$

   (c)

43. (a) $\dfrac{(x - 1)^2}{16} - \dfrac{(y + 1)^2}{9} = 1$

   (b) Center: $(1, -1)$
   Vertices: $(5, -1), (-3, -1)$
   Foci: $(6, -1), (-4, -1)$
   Eccentricity: $\dfrac{5}{4}$

(c)

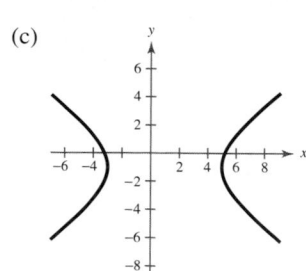

45. (a) $\dfrac{(x + 6)^2}{\frac{101}{2}} - \dfrac{(y - 1)^2}{202} = 1$

(b) Center: $(-6, 1)$

Vertices: $\left(-6 \pm \dfrac{\sqrt{202}}{2}, 1\right)$

Foci: $\left(-6 \pm \dfrac{\sqrt{1010}}{2}, 1\right)$

Eccentricity: $\sqrt{5}$

(c)

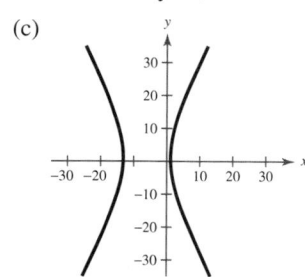

47. $\dfrac{x^2}{16} - \dfrac{y^2}{20} = 1$ **49.** $\dfrac{(x - 4)^2}{16/5} - \dfrac{y^2}{64/5} = 1$

51. ≈ 72 miles   **53.** Ellipse   **55.** Hyperbola

57. $\dfrac{(x')^2}{8} - \dfrac{(y')^2}{8} = 1$ **59.** $\dfrac{(x')^2}{3} + \dfrac{(y')^2}{2} = 1$

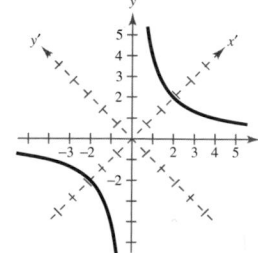   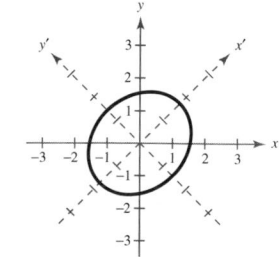

61. (a) Parabola

(b) $y = \dfrac{8x - 5 \pm \sqrt{(5 - 8x)^2 - 4(16x^2 - 10x)}}{2}$

(c)

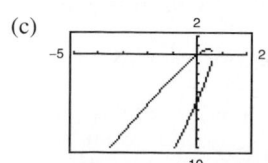

63. (a) Parabola

(b)
$$y = \dfrac{-(2x - 2\sqrt{2}) \pm \sqrt{(2x - 2\sqrt{2})^2 - 4(x^2 + 2\sqrt{2}x + 2)}}{2}$$

(c)

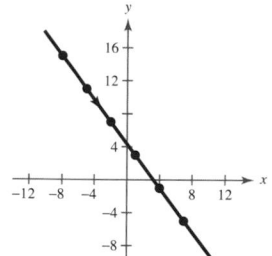

65. $(-10, 12)$

67.

t	-2	-1	0	1	2	3
x	-8	-5	-2	1	4	7
y	15	11	7	3	-1	-5

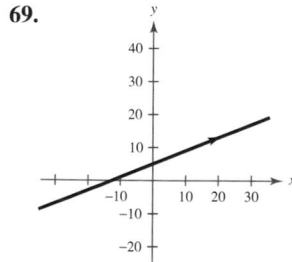

69.   **71.**

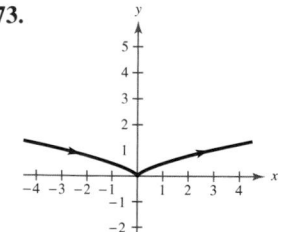   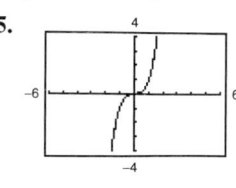

$y = \frac{2}{5}x + \frac{27}{5}$ $y = 4x - 11, \ x \geq 2$

73.   **75.**

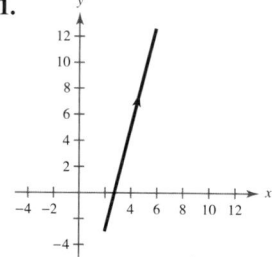

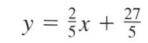

$y = \frac{1}{2}x^{2/3}$

77.   **79.**

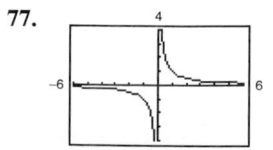   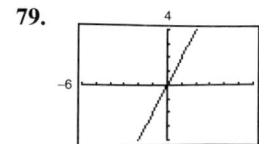

81. 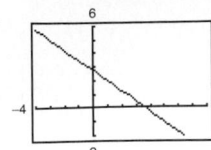   **83.**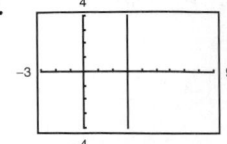

85.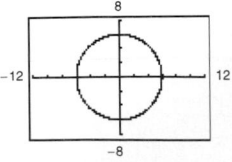

87. $x = t, y = 6t + 2$ **89.** $x = t, y = t^2 + 2$

   $x = 2t, y = 12t + 2$ $x = \frac{1}{2}t, y = \frac{1}{4}t^2 + 2$

91. $x = t, y = 5$ **93.** $x = -1 + 11t$

   $y = 6 - 6t$

95. 54.22 feet per second

97. 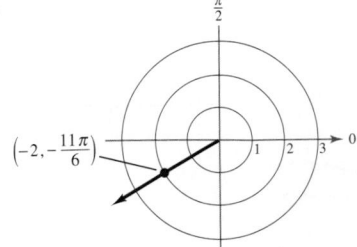   21.91 feet

99.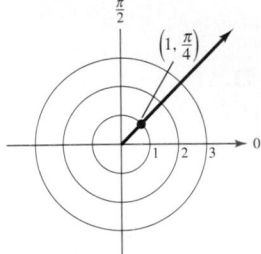

   $\left(1, -\frac{7\pi}{4}\right), \left(-1, \frac{5\pi}{4}\right), \left(-1, -\frac{3\pi}{4}\right)$

101.

   $\left(-2, \frac{\pi}{6}\right), \left(2, \frac{7\pi}{6}\right), \left(2, -\frac{5\pi}{6}\right)$

103.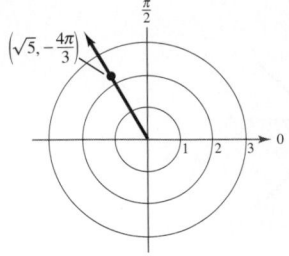

   $\left(\sqrt{5}, -\frac{2\pi}{3}\right), \left(-\sqrt{5}, \frac{\pi}{3}\right), \left(-\sqrt{5}, -\frac{5\pi}{3}\right)$

105. 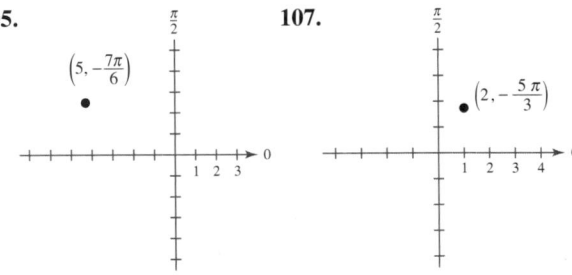   **107.**

   $\left(-\frac{5\sqrt{3}}{2}, \frac{5}{2}\right)$ $\left(1, \sqrt{3}\right)$

109. 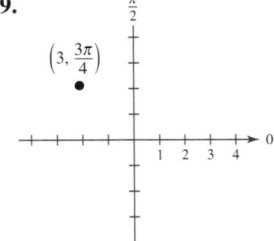   **111.**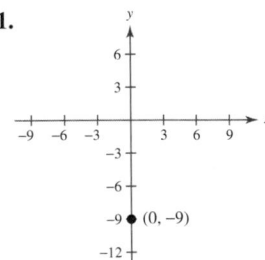

   $\left(-\frac{3\sqrt{2}}{2}, \frac{3\sqrt{2}}{2}\right)$ $\left(-9, \frac{\pi}{2}\right), \left(9, \frac{3\pi}{2}\right)$

113.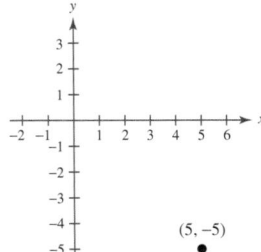

   $\left(-5\sqrt{2}, \frac{3\pi}{4}\right), \left(5\sqrt{2}, \frac{7\pi}{4}\right)$

115. $r = 3$ **117.** $r = 4\cos\theta$ **119.** $r^2 = 5\sec\theta\csc\theta$

121. $r^2 = \dfrac{1}{1 + 3\cos^2\theta}$ **123.** $x^2 + y^2 = 25$

125. $x^2 + y^2 = 3x$ **127.** $(x^2 + y^2)^2 - x^2 + y^2 = 0$

129. $y = -\dfrac{\sqrt{3}}{3}x$

131.

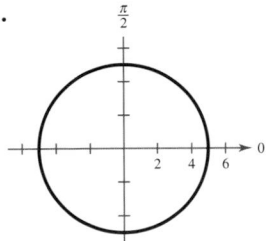

133.

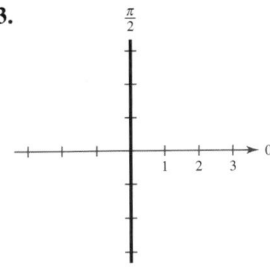

135.

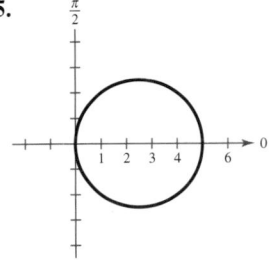

137. Dimpled limaçon

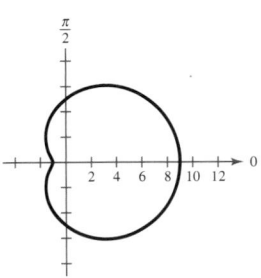

Symmetry: Polar axis

Maximum: $|r| = 9$

Zeros of r: None

139. Limaçon with inner loop

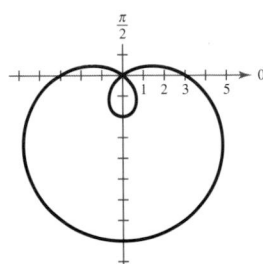

Symmetry: The line $\theta = \dfrac{\pi}{2}$

Maximum: $|r| = 8$

Zeros of r: $\theta \approx 0.64, 2.50$

141. Rose curve

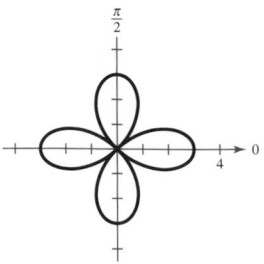

Symmetry: Pole, polar axis, and the line $\theta = \dfrac{\pi}{2}$

Maximum: $|r| = 3$

Zeros of r: $\theta = \dfrac{\pi}{4}, \dfrac{3\pi}{4}, \dfrac{5\pi}{4}, \dfrac{7\pi}{4}$

143. Lemniscate

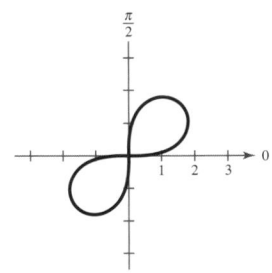

Symmetry: Pole

Maximum: $|r| = \sqrt{5}$

Zeros of r: $\theta = 0, \dfrac{\pi}{2}$

145. Parabola

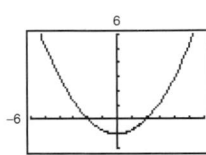

147. Ellipse

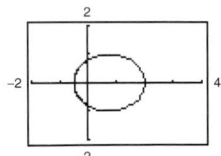

149. Ellipse

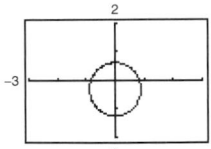

151. $r = \dfrac{4}{1 - \cos\theta}$

153. $r = \dfrac{5}{3 - 2\cos\theta}$

155. $r = \dfrac{1.512}{1 - 0.093\cos\theta}$

Perihelion: 1.383 astronomical units

Aphelion: 1.667 astronomical units

157. False. The equation of a hyperbola is a second-degree equation.

CHAPTER 9

159. (a) vertical translation   (b) horizontal translation

(c) reflection in the *y*-axis   (d) vertical shrink

161. 5; The ellipse gets closer and closer to circular and approaches a circle of radius 5.

163. (a) The time it takes to make one revolution is halved.

(b) The length of the major axis is increased by two units.

Chapter Test   (page 734)

1.
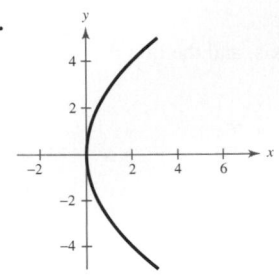

Vertex: $(0, 0)$

Focus: $(2, 0)$

2.
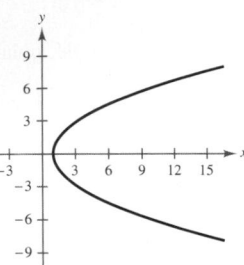

Vertex: $(1, 0)$

Focus: $(2, 0)$

3.
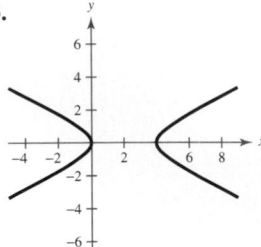

Vertices: $(0, 0), (4, 0)$

Foci: $\left(2 \pm \sqrt{5}, 0\right)$

4. $(y + 2)^2 = 8(x - 6)$
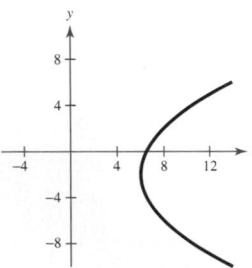

5. $\dfrac{(x + 6)^2}{16} + \dfrac{(y - 3)^2}{49} = 1$

6. $\dfrac{y^2}{9} - \dfrac{x^2}{4} = 1$

7.
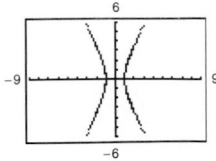

Answers will vary.

8. (a) $45°$

(b)
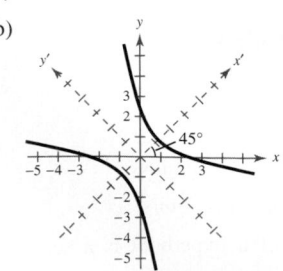

9. No real solution

10.
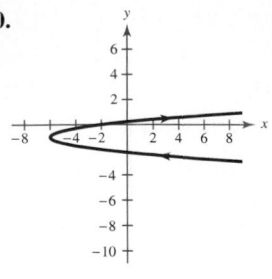

$(y + 1)^2 = \dfrac{1}{4}(x + 6)$

11.
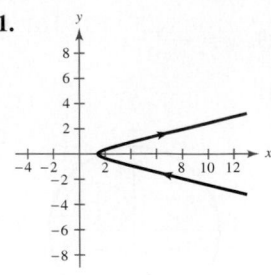

$\dfrac{x^2}{2} - \dfrac{y^2}{1/8} = 1,\ x \geq \sqrt{2}$

12.
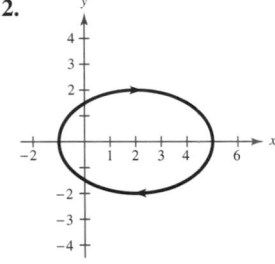

$\dfrac{(x - 2)^2}{9} + \dfrac{y^2}{4} = 1$

13. $x = t,\ y = t^2 + 10$

$x = 2t,\ y = 4t^2 + 10$

14. $x = 4 - t^2,\ y = t$

$x = 4 - 4t^2,\ y = 2t$

15. $x = \pm 2\sqrt{4 - t^2},\ y = t$

$x = \pm\sqrt{16 - t^2},\ y = \frac{1}{2}t$

16. $\left(-7, 7\sqrt{3}\right)$

17. $\left(2\sqrt{2}, \dfrac{7\pi}{4}\right); \left(2\sqrt{2}, -\dfrac{\pi}{4}\right), \left(-2\sqrt{2}, \dfrac{3\pi}{4}\right)$

18. $r = 12 \sin \theta$ **19.** $x^2 + (y - 1)^2 = 1$

20. Limaçon with inner loop   **21.** Parabola

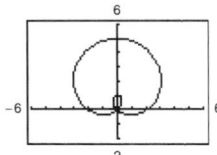 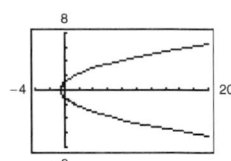

22. Hyperbola

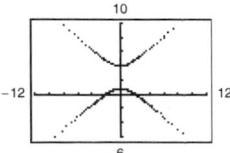

23. $r = \dfrac{4}{4 + \sin \theta}$ **24.** $r = \dfrac{10}{4 + 5 \sin \theta}$

25. Maximum: $|r| = 8$

Zeros of r: $\theta = \dfrac{\pi}{6}, \dfrac{\pi}{2}, \dfrac{5\pi}{6}$

Cumulative Test for Chapters 7–9 (page 735)

1. $(4, -3)$ **2.** $(8, 4), (2, -2)$

3. $\left(\frac{3}{5}, -4, -\frac{1}{5}\right)$ **4.** $(1, -4, -4)$

5. $\begin{bmatrix} -7 & -10 & -16 \\ -6 & 18 & 9 \\ -12 & 16 & 7 \end{bmatrix}$ **6.** $\begin{bmatrix} -18 & 15 & -14 \\ 28 & 11 & 34 \\ -20 & 52 & -1 \end{bmatrix}$

7. $\begin{bmatrix} 3 & -31 & 2 \\ 22 & 18 & 6 \\ 52 & -40 & 14 \end{bmatrix}$ **8.** $\begin{bmatrix} 5 & 36 & 31 \\ -36 & 12 & -36 \\ 16 & 0 & 18 \end{bmatrix}$

9. (a) $\begin{bmatrix} -175 & 37 & -13 \\ 95 & -20 & 7 \\ 14 & -3 & 1 \end{bmatrix}$ (b) 1 **10.** 22

11. (a) $\frac{1}{5}, -\frac{1}{7}, \frac{1}{9}, -\frac{1}{11}, \frac{1}{13}$ (b) 3, 6, 12, 24, 48 **12.** 135

13. $\frac{47}{52}$ **14.** 34.48 **15.** 66.67 **16.** $\frac{15}{8}$

17. $-\frac{5}{51}$ **18.** $\frac{8}{3}$ **19.** Answers will vary.

20. $x^4 + 12x^3 + 54x^2 + 108x + 81$

21. $32x^5 + 80x^4y^2 + 80x^3y^4 + 40x^2y^6 + 10xy^8 + y^{10}$

22. $x^6 - 12x^5y + 60x^4y^2 - 160x^3y^3 + 240x^2y^4$
$- 192xy^5 + 64y^6$

23. $6561a^8 - 69{,}984a^7b + 326{,}592a^6b^2 - 870{,}912a^5b^3$
$+ 1{,}451{,}520a^4b^4 - 1{,}548{,}288a^3b^5 + 1{,}032{,}192a^2b^6$
$- 393{,}216ab^7 + 65{,}536b^8$

24. 30 **25.** 120 **26.** 453,600 **27.** 151,200

28. Hyperbola **29.** Ellipse

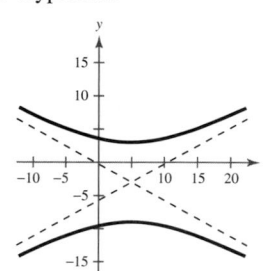 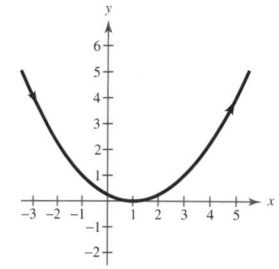...

30. Hyperbola **31.** Circle

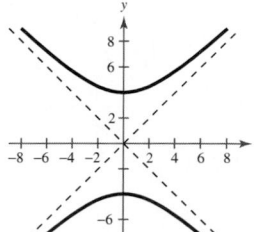 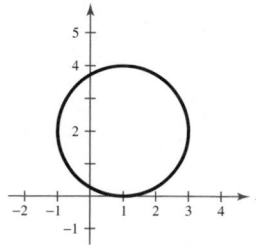

32. $(x - 2)^2 = -\frac{4}{3}(y - 3)$

33. $\dfrac{(x - 1)^2}{25} + \dfrac{(y - 4)^2}{4} = 1$ **34.** $\dfrac{(y + 4)^2}{4} - \dfrac{x^2}{16/3} = 1$

35.
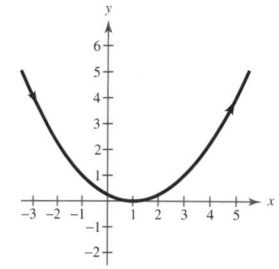

$\theta \approx 37.98°$

36. (a) and (b)
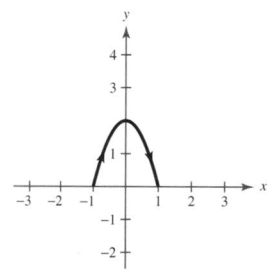

(c) $y = \dfrac{x^2 - 2x + 1}{4}$

37. (a) and (b)
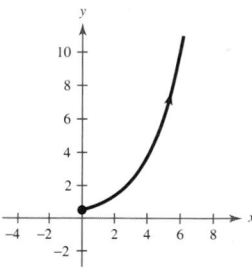

(c) $y = 2 - 2x^2, -1 \le x \le 1$

38. (a) and (b)

(c) $y = 0.5e^{0.5x}, \; x \ge 0$

39. $x = t, y = 3t - 2$ **40.** $x = \pm\sqrt{16 + t^2}, y = t$
$x = 2t, y = 6t - 2$ $x = \pm4\sqrt{1 + t^2}, y = 4t$

CHAPTER 9

41. $x = t, y = \dfrac{2}{t}$

$x = \dfrac{1}{t}, y = 2t$

42. $x = t, y = \dfrac{e^{2t}}{e^{2t} + 1}$

$x = 2t, y = \dfrac{e^{4t}}{e^{4t} + 1}$

47. $r = -\dfrac{1}{4 \sin \theta + 4 \cos \theta}$

48. $(x - 1)^2 + y^2 = 1$

49. $\dfrac{\left(x + \frac{10}{9}\right)^2}{\frac{64}{81}} - \dfrac{y^2}{\frac{4}{9}} = 1$

43.

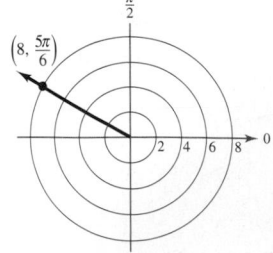

$\left(8, -\dfrac{7\pi}{6}\right), \left(-8, -\dfrac{\pi}{6}\right), \left(-8, \dfrac{11\pi}{6}\right)$

44.

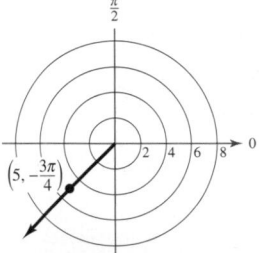

$\left(5, \dfrac{5\pi}{4}\right), \left(-5, -\dfrac{7\pi}{4}\right), \left(-5, \dfrac{\pi}{4}\right)$

45.

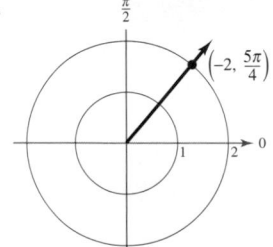

$\left(-2, -\dfrac{3\pi}{4}\right), \left(2, -\dfrac{7\pi}{4}\right), \left(2, \dfrac{\pi}{4}\right)$

46.

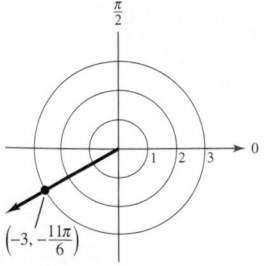

$\left(-3, \dfrac{\pi}{6}\right), \left(3, -\dfrac{5\pi}{6}\right), \left(3, \dfrac{7\pi}{6}\right)$

50. Circle

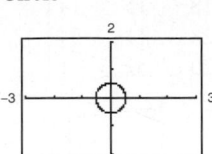

51. Dimpled limaçon

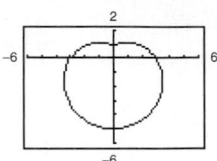

52. Limaçon with inner loop

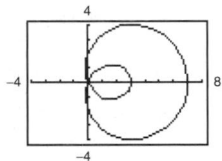

53. $701,303.32

54. $\frac{1}{4}$ **55.** $24\sqrt{2}$ meters

Chapter 10
Section 10.1   (page 747)

Vocabulary Check   (page 747)

1. three-dimensional

2. xy-plane, xz-plane, yz-plane   **3.** octants

4. Distance Formula   **5.** $\left(\dfrac{x_1 + x_2}{2}, \dfrac{y_1 + y_2}{2}, \dfrac{z_1 + z_2}{2}\right)$

6. sphere   **7.** surface, space   **8.** trace

1. A: $(-1, 4, 3)$ B: $(1, 3, -2)$, C: $(-3, 0, -2)$

3. A: $(-2, -1, 4)$ B: $(3, -2, 0)$, C: $(-2, 2, -3)$

5.

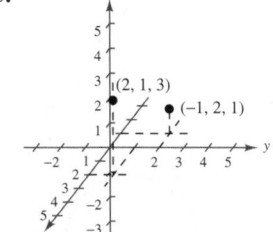

7.

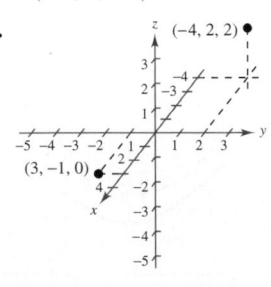

9.

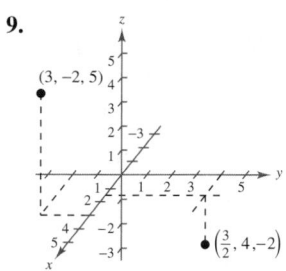

11. $(-3, 3, 4)$ **13.** $(10, 0, 0)$ **15.** Octant IV

17. Octants I, II, III, and IV   **19.** Octants II, IV, VI, and VIII

21. $\sqrt{189}$ units   **23.** $\sqrt{114}$ units   **25.** $\sqrt{110}$ units

27. $2\sqrt{5}, 3, \sqrt{29}$ **29.** $3, 6, 3\sqrt{5}$

31. $6, 6, 2\sqrt{10}$; isosceles triangle

33. $6, 6, 2\sqrt{10}$; isosceles triangle

35. $(0, -1, 7)$ **37.** $\left(1, 0, \frac{11}{2}\right)$ **39.** $\left(\frac{5}{2}, 2, 6\right)$

41. $(x - 3)^2 + (y - 2)^2 + (z - 4)^2 = 16$

43. $(x + 1)^2 + (y - 2)^2 + z^2 = 3$

45. $x^2 + (y - 4)^2 + (z - 3)^2 = 9$

47. $(x + 3)^2 + (y - 7)^2 + (z - 5)^2 = 25$

49. $\left(x - \frac{3}{2}\right)^2 + y^2 + (z - 3)^2 = \frac{45}{4}$

51. Center: $\left(\frac{5}{2}, 0, 0\right)$; radius: $\frac{5}{2}$

53. Center: $(2, -1, 0)$; radius: $\sqrt{5}$

55. Center: $(2, -1, 3)$; radius: 2

57. Center: $(-2, 0, 4)$; radius: 1

59. Center: $\left(1, \frac{1}{3}, 4\right)$; radius: 3

61. Center: $(1, -2, 0)$; radius: $\dfrac{\sqrt{21}}{2}$

63. Center: $\left(\frac{1}{3}, -1, 0\right)$; radius: 1

65.   **67.**

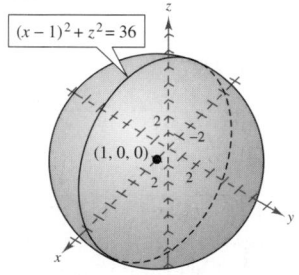   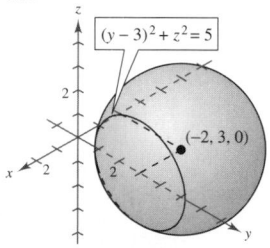

69.   **71.**

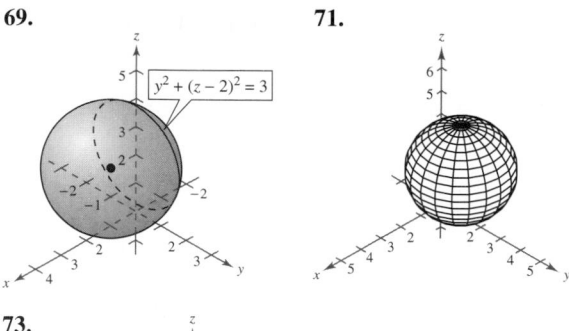

73.

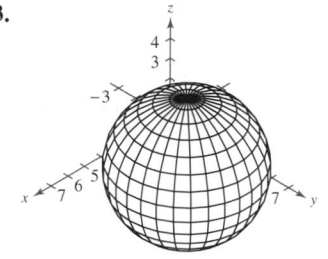

75. $(3, 3, 3)$ **77.** $x^2 + y^2 + z^2 = \dfrac{165^2}{4}$

79. False. z is the directed distance from the xy-plane to P.

81. $0; 0; 0$ **83.** A point or a circle

85. $(x_2, y_2, z_2) = (2x_m - x_1, 2y_m - y_1, 2z_m - z_1)$

87. $v = -\dfrac{3 \pm \sqrt{17}}{2}$ **89.** $x = \dfrac{5 \pm \sqrt{5}}{2}$

91. $y = -\dfrac{1 \pm \sqrt{10}}{2}$ **93.** $3\sqrt{2}, 315°$

95. $\sqrt{41}, 51.34°$ **97.** -7

99. $1, 2, 6, 15, 31$
First differences: $1, 4, 9, 16$
Second differences: $3, 5, 7$
Neither

101. $-1, 2, 5, 8, 11$
First differences: $3, 3, 3, 3$
Second differences: $0, 0, 0$
Linear

103. $(x + 5)^2 + (y - 1)^2 = 49$

105. $(y - 1)^2 = -12(x - 4)$

107. $\dfrac{(x - 3)^2}{9} + \dfrac{(y - 3)^2}{4} = 1$ **109.** $\dfrac{(x - 6)^2}{4} - \dfrac{y^2}{32} = 1$

Section 10.2   (page 755)

Vocabulary Check   (page 755)

1. zero   **2.** $\mathbf{v} = v_1\mathbf{i} + v_2\mathbf{j} + v_3\mathbf{k}$

3. component form   **4.** orthogonal   **5.** parallel

1. (a) $\langle -2, 3, 1 \rangle$

(b)

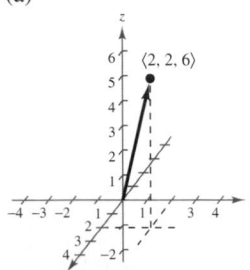

3. (a) $\langle 0, 0, -4 \rangle$

(b)

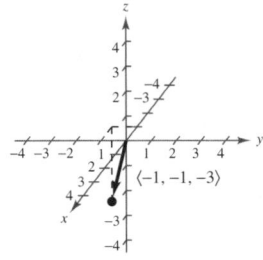

5. (a) $\langle 7, -5, 5 \rangle$ (b) $3\sqrt{11}$ (c) $\dfrac{\sqrt{11}}{33}\langle 7, -5, 5 \rangle$

7. (a) $\langle 2, 2, 0 \rangle$ (b) $2\sqrt{2}$ (c) $\dfrac{\sqrt{2}}{2}\langle 1, 1, 0 \rangle$

9. (a)

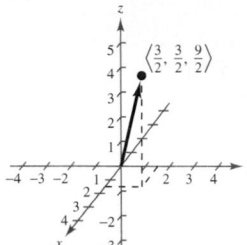

(b)

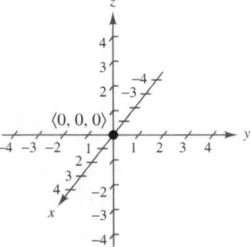

(c)

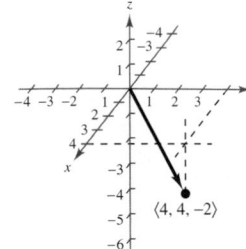

(d)

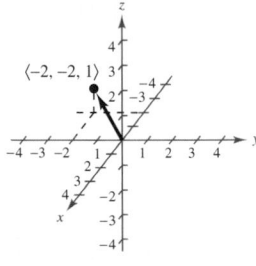

11. (a)

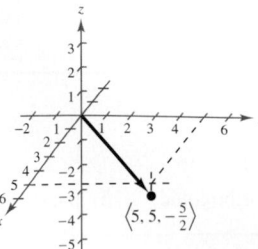

(b)

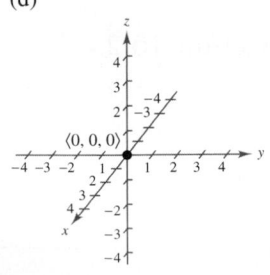

(c)

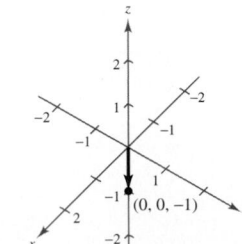

(d)

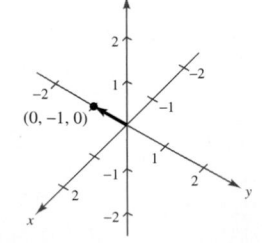

13. $\mathbf{z} = \langle -3, 7, 6 \rangle$ **15.** $\mathbf{z} = \left\langle \frac{1}{2}, 6, \frac{3}{2} \right\rangle$

17. $\mathbf{z} = \left\langle -\frac{5}{2}, 12, \frac{15}{2} \right\rangle$ **19.** $\mathbf{z} = \left\langle \frac{11}{2}, -\frac{5}{4}, -6 \right\rangle$

21. $9\sqrt{2}$ **23.** $\sqrt{21}$ **25.** $\sqrt{21}$ **27.** $\sqrt{74}$

29. $\sqrt{34}$ **31.** (a) $\frac{1}{13}(5\mathbf{i} - 12\mathbf{k})$ (b) $-\frac{1}{13}(5\mathbf{i} - 12\mathbf{k})$

33. (a) $\dfrac{\sqrt{74}}{74}(8\mathbf{i} + 3\mathbf{j} - \mathbf{k})$ (b) $-\dfrac{\sqrt{74}}{74}(8\mathbf{i} + 3\mathbf{j} - \mathbf{k})$

35. $\langle -26, 0, 48 \rangle$ **37.** ≈ 8.73 **39.** -4 **41.** 0

43. $\approx 124.45°$ **45.** $\approx 109.92°$ **47.** Parallel

49. Orthogonal   **51.** Neither   **53.** Orthogonal

55. Not collinear   **57.** Collinear

59. Right triangle. Answers will vary.

61. Acute triangle. Answers will vary.   **63.** $(3, 1, 7)$

65. $\left(6, \dfrac{5}{2}, -\dfrac{7}{4} \right)$ **67.** $\pm \dfrac{3\sqrt{14}}{14}$

69. $\langle 0, 2\sqrt{2}, 2\sqrt{2} \rangle$ or $\langle 0, 2\sqrt{2}, -2\sqrt{2} \rangle$

71. 10.91 pounds   **73.** True

75. (a)

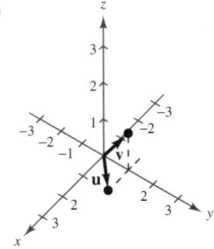

(b) Answers will vary.

(c) $a = b = 1$

(d) Answers will vary.

77. The angle between $\mathbf{u}$ and $\mathbf{v}$ is an obtuse angle.

Section 10.3   (page 762)

Vocabulary Check   (page 762)

1. cross product   **2.** $\mathbf{0}$ **3.** $\|\mathbf{u}\| \, \|\mathbf{v}\| \sin \theta$

4. triple scalar product

1. $-\mathbf{k}$

3. $-\mathbf{j}$

5. $\langle 1, 1, 1 \rangle$ **7.** $\langle 3, -3, -3 \rangle$ **9.** $\langle 0, 42, 0 \rangle$

11. $-7\mathbf{i} + 13\mathbf{j} + 16\mathbf{k}$ **13.** $-17\mathbf{i} + \mathbf{j} + 2\mathbf{k}$

15. $-\frac{7}{6}\mathbf{i} - \frac{7}{8}\mathbf{j}$ **17.** $-18\mathbf{i} - 6\mathbf{j}$ **19.** $-\mathbf{i} - 2\mathbf{j} - \mathbf{k}$

21. $\langle 10, -2, -4 \rangle$ **23.** $-6\mathbf{i} - 15\mathbf{j} - 6\mathbf{k}$

25. $-\frac{1}{4}\mathbf{i} - \frac{7}{10}\mathbf{j} - 2\mathbf{k}$ **27.** $\frac{\sqrt{166}}{166}\langle 9, 6, -7 \rangle$

29. $\frac{\sqrt{19}}{19}(\mathbf{i} - 3\mathbf{j} + 3\mathbf{k})$ **31.** $\frac{\sqrt{7602}}{7602}(-71\mathbf{i} - 44\mathbf{j} + 25\mathbf{k})$

33. $\frac{\sqrt{2}}{2}(\mathbf{i} - \mathbf{j})$ **35.** 1 **37.** $\sqrt{806}$ **39.** 14

41. (a) Answers will vary.

(b) $6\sqrt{10}$

(c) The parallelogram is not a rectangle.

43. $\frac{3\sqrt{13}}{2}$ **45.** $\frac{1}{2}\sqrt{4290}$ **47.** -16 **49.** 2

51. 2 **53.** 12 **55.** 84

57. (a) $T(p) = \frac{p}{2}\cos 40°$

(b)

p	15	20	25	30	35	40	45
T	5.75	7.66	9.58	11.49	13.41	15.32	17.24

59. True   **61 and 63.** Proofs

65. $-\frac{1}{2}$ **67.** $-\frac{1}{2}$

Section 10.4   (page 771)

Vocabulary Check   (page 771)

1. direction, $\dfrac{\overrightarrow{PQ}}{t}$ **2.** parametric equations

3. symmetric equations   **4.** normal

5. $a(x - x_1) + b(y - y_1) + c(z - z_1) = 0$

1. (a) $x = t, \ y = 2t, \ z = 3t$

(b) $x = \dfrac{y}{2} = \dfrac{z}{3}$

3. (a) $x = -4 + 3t, \ y = 1 + 8t, \ z = -6t$

(b) $\dfrac{x + 4}{3} = \dfrac{y - 1}{8} = \dfrac{z}{-6}$

5. (a) $x = 2 + 2t, \ y = -3 - 3t, \ z = 5 + t$

(b) $\dfrac{x - 2}{2} = \dfrac{y + 3}{-3} = z - 5$

7. (a) $x = 2 - t, \ y = 4t, \ z = 2 - 5t$

(b) $\dfrac{x - 2}{-1} = \dfrac{y}{4} = \dfrac{z - 2}{-5}$

9. (a) $x = -3 + 4t, \ y = 8 - 10t, \ z = 15 + t$

(b) $\dfrac{x + 3}{4} = \dfrac{y - 8}{-10} = z - 15$

11. (a) $x = 3 - 4t, \ y = 1, \ z = 2 + 3t$

(b) Not possible.

13. (a) $x = -\frac{1}{2} + 3t, \ y = 2 - 5t, \ z = \frac{1}{2} - t$

(b) $\dfrac{x + \frac{1}{2}}{3} = \dfrac{y - 2}{-5} = \dfrac{z - \frac{1}{2}}{-1}$

15.
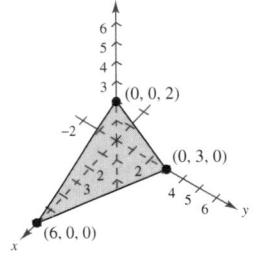
    17. $x - 2 = 0$

19. $-2x + y - 2z + 10 = 0$ **21.** $-x - 2y + z + 2 = 0$

23. $-3x - 9y + 7z = 0$ **25.** $6x - 2y - z - 8 = 0$

27. $y - 5 = 0$ **29.** $y - z + 2 = 0$

31. $7x + y - 11z - 5 = 0$ **33.** $x = 2, y = 3, z = 4 + t$

35. $x = 2 + 3t, \ y = 3 + 2t, \ z = 4 - t$

37. $x = 5 + 2t, \ y = -3 - t, \ z = -4 + 3t$

39. $x = 2 - t, \ y = 1 + t, \ z = 2 + t$

41. Orthogonal   **43.** Orthogonal

45. (a) $60.67°$ (b) $x = 2 - t, \ y = 8t, \ z = 7t$

47. (a) $77.83°$ (b) $x = 1 + 6t, \ y = t, \ z = 1 + 7t$

49.
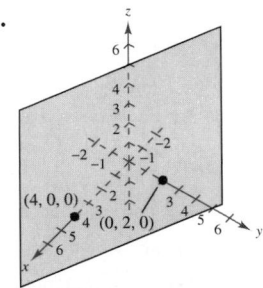
    51.

53.
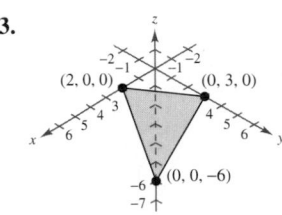

55. $\dfrac{8}{9}$ **57.** $\dfrac{2\sqrt{6}}{3}$ **59.** $88.45°$

61. False. Lines that do not intersect and are not in the same plane may not be parallel.

63. Parallel. $\langle 10, -18, 20 \rangle$ is a scalar multiple of $\langle -15, 27, -30 \rangle$.

65. $x^2 + y^2 = 100$ **67.** $x^2 + y^2 - 3x = 0$

69. $r = 7$ **71.** $r = 5 \csc \theta$

Review Exercises   (page 774)

1.

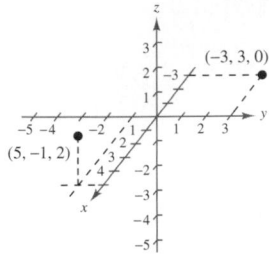

3. $(-5, 4, 0)$ **5.** $\sqrt{41}$

7. $\sqrt{29}, \sqrt{38}, \sqrt{67}$ **9.** $(0, -1, 0)$ **11.** $(1, 2, -9)$

13. $(x - 2)^2 + (y - 3)^2 + (z - 5)^2 = 1$

15. $(x - 1)^2 + (y - 5)^2 + (z - 2)^2 = 36$

17. Center: $(2, 3, 0)$; radius: 3

19. (a)   (b)

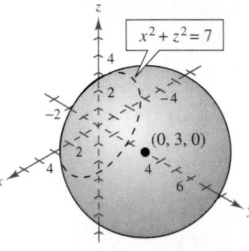

21. (a) $\langle 1, 4, -4 \rangle$ (b) $\sqrt{33}$ (c) $\dfrac{\sqrt{33}}{33}\langle 1, 4, -4 \rangle$

23. (a) $\langle -10, 6, 7 \rangle$ (b) $\sqrt{185}$ (c) $\dfrac{\sqrt{185}}{185}\langle -10, 6, 7 \rangle$

25. -9 **27.** 1 **29.** $90°$ **31.** $90°$

33. Orthogonal   **35.** Parallel   **37.** Not collinear

39. Collinear

41. A: 159.10 pounds of tension

B: 115.58 pounds of tension

C: 115.58 pounds of tension

43. $\langle -10, 0, -10 \rangle$

45. $\dfrac{\sqrt{7602}}{7602}(-71\mathbf{i} - 44\mathbf{j} + 25\mathbf{k})$

47. Answers will vary; $2\sqrt{43}$ **49.** 75

51. (a) $x = 3 + 6t, \ y = 11t, \ z = 2 + 4t$

(b) $\dfrac{x - 3}{6} = \dfrac{y}{11} = \dfrac{z - 2}{4}$

53. (a) $x = -1 + 4t, \ y = 3 + 3t, \ z = 5 - 6t$

(b) $\dfrac{x + 1}{4} = \dfrac{y - 3}{3} = \dfrac{z - 5}{-6}$

55. (a) $x = -2t, \ y = \dfrac{5}{2}t, \ z = t$ (b) $\dfrac{x}{-2} = \dfrac{y}{5/2} = z$

57. $-2x - 12y + 5z = 0$ **59.** $z - 2 = 0$

61.   **63.**

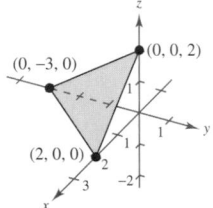
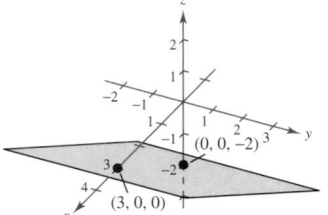

65. $\dfrac{\sqrt{110}}{110}$ **67.** $\dfrac{\sqrt{110}}{55}$ **69.** False. $\mathbf{u} \times \mathbf{v} = -(\mathbf{v} \times \mathbf{u})$

71 and 73. Answers will vary.

75. $\mathbf{u} \times \mathbf{v} = (u_2 v_3 - u_3 v_2)\mathbf{i} - (u_1 v_3 - u_3 v_1)\mathbf{j}$
$+ (u_1 v_2 - u_2 v_1)\mathbf{k}$

77. The magnitude of the cross product will increase by a factor of 4.

Chapter Test   (page 776)

1.

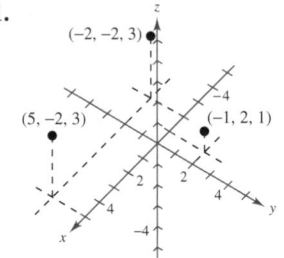

2. No. Answers will vary.   **3.** $(7, 1, 2)$

4. $(x - 7)^2 + (y - 1)^2 + (z - 2)^2 = 19$

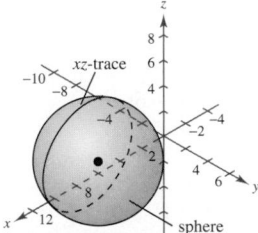

5. $\langle 2, 5, -10 \rangle$; $\sqrt{129}$ **6.** $\langle -3, -5, 8 \rangle$; $7\sqrt{2}$

7. $\mathbf{u} = \langle -2, 6, -6 \rangle$, $\mathbf{v} = \langle -12, 5, -5 \rangle$

8. (a) $\sqrt{194}$ (b) 84   (c) $\langle 0, 62, 62 \rangle$ **9.** $46.23°$

10. Answers will vary. Sample answer:

(a) $x = 8 - 2t$, $y = -2 + 6t$, $z = 5 - 6t$

(b) $\dfrac{x-8}{-2} = \dfrac{y+2}{6} = \dfrac{z-5}{-6}$

11. Neither   **12.** Orthogonal

13. Answers will vary; $2\sqrt{230}$

14. $27x + 4y + 32z + 33 = 0$ **15.** 200

16. 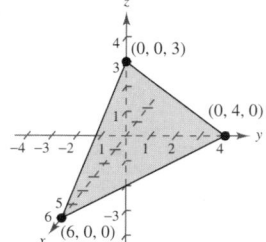   **17.**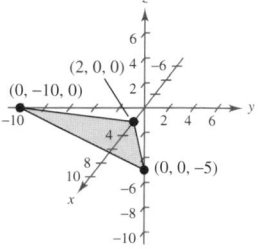

18. $\dfrac{4\sqrt{14}}{7}$

Chapter 11

Section 11.1   (page 788)

Vocabulary Check   (page 788)

1. limit   **2.** oscillates   **3.** direct substitution

1. (a) 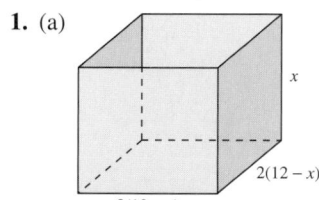   (b) Answers will vary.

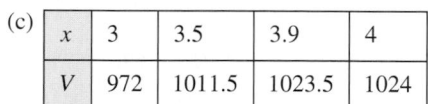

(c)

x	3	3.5	3.9	4
V	972	1011.5	1023.5	1024

x	4.1	4.5	5
V	1023.5	1012.5	980

$\displaystyle \lim_{x \to 4} V = 1024$

(d)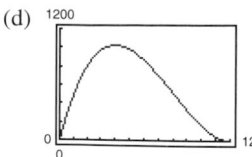

3.

x	1.9	1.99	1.999	2
$f(x)$	13.5	13.95	13.995	14

x	2.001	2.01	2.1
$f(x)$	14.005	14.05	14.5

14; Yes

5.

x	2.9	2.99	2.999	3
$f(x)$	0.1695	0.1669	0.1667	Error

x	3.001	3.01	3.1
$f(x)$	0.1666	0.1664	0.1639

0.1667; No

7.

x	-0.1	-0.01	-0.001	0
$f(x)$	1.9867	1.99987	1.9999987	Error

x	0.001	0.01	0.1
$f(x)$	1.9999987	1.99987	1.9867

2; No

9.

x	-0.1	-0.01	-0.001	0
$f(x)$	1.8127	1.9801	1.9980	Error

x	0.001	0.01	0.1
$f(x)$	2.0020	2.0201	2.2140

2; No

11.

x	0.9	0.99	0.999	1
$f(x)$	0.2564	0.2506	0.2501	Error

x	1.001	1.01	1.1
$f(x)$	0.2499	0.2494	0.2439

0.25

13.

x	-0.1	-0.01	-0.001	0
$f(x)$	0.2247	0.2237	0.2236	Error

x	0.001	0.01	0.1
$f(x)$	0.2236	0.2235	0.2225

0.2236

15.

x	-4.1	-4.01	-4.001	-4
$f(x)$	0.4762	0.4975	0.4998	Error

x	-3.999	-3.99	-3.9
$f(x)$	0.5003	0.5025	0.5263

0.5

17.

x	-0.1	-0.01	-0.001	0
$f(x)$	0.9983	0.99998	0.9999998	Error

x	0.001	0.01	0.1
$f(x)$	0.9999998	0.99998	0.9983

1

19.

x	-0.1	-0.01	-0.001	0
$f(x)$	-0.0997	-0.0100	-0.0010	Error

x	0.001	0.01	0.1
$f(x)$	0.0010	0.0100	0.0997

0

21.

x	-0.1	-0.01	-0.001	0
$f(x)$	0.9063	0.9901	0.9990	Error

x	0.001	0.01	0.1
$f(x)$	1.0010	1.0101	1.1070

1

23.

x	0.9	0.99	0.999	1
$f(x)$	2.2314	2.0203	2.0020	Error

x	1.001	1.01	1.1
$f(x)$	1.9980	1.9803	1.8232

2

25.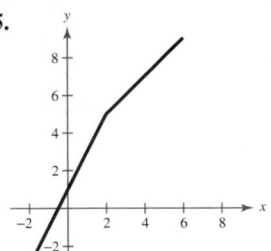

$\lim_{x \to 2} f(x) = 5$

27.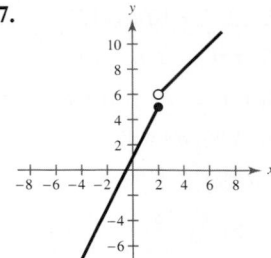

Limit does not exist.

29. 13

31. Limit does not exist; one-sided limits do not agree.

33. Limit does not exist; function oscillates between -2 and 2.

35. Limit does not exist; one-sided limits do not exist.

37.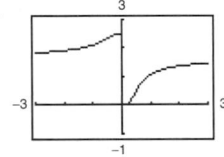

No

39.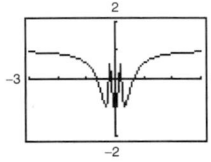

No

41.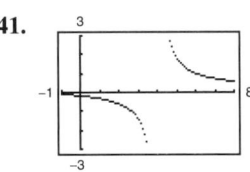

No

43.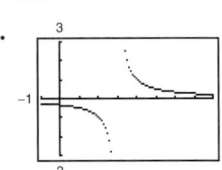

Yes

45.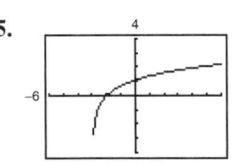

Yes

47. (a) -12 (b) 9   (c) $\frac{1}{2}$ (d) $\sqrt{3}$

49. (a) 8   (b) $\frac{3}{8}$ (c) 3   (d) $-\frac{61}{8}$

51. -15 **53.** 7   **55.** -3 **57.** $-\frac{9}{10}$ **59.** $\frac{7}{13}$

61. 1   **63.** $\frac{35}{3}$ **65.** e^3 **67.** 0   **69.** $\frac{\pi}{6}$

71. True   **73.** (a) and (b) Answers will vary.

75. (a) No. The function may approach different values from the right and left of 4.

(b) No. The function may approach 4 as x approaches 2, but the function could be undefined at $x = 2$.

77. $-\frac{1}{3}$, $x \neq 5$ **79.** $\frac{5x + 4}{5x + 2}$, $x \neq \frac{1}{3}$

81. $\dfrac{x^2 - 3x + 9}{x - 2}, \ x \neq -3$

Section 11.2   (page 798)

1. (a) 1   (b) 3   (c) 5

   $g_2(x) = -2x + 1$

3. (a) 2   (b) 0   (c) 0

   $g_2(x) = x(x + 1)$

5. $\dfrac{1}{12}$ **7.** 4   **9.** 12   **11.** 0   **13.** $\dfrac{1}{3}$ **15.** 3

17. $\dfrac{\sqrt{5}}{10}$ **19.** $\dfrac{1}{4}$ **21.** -1 **23.** Limit does not exist.

25. 0   **27.** 0

29.

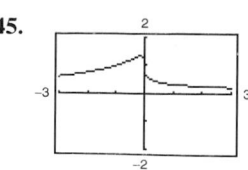

   0.29

31.

   1.00

33.

   80.00

35.

   -0.06

37.

   2.000

39.

   0.000

41.

   2.000

43.

   1.000

45.

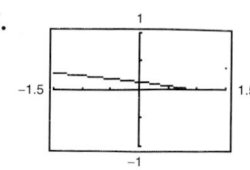

   0.333

47.

   0.135

49. (a) and (b) 0.50   (c) $\dfrac{1}{2}$

51. (a) and (b) -0.13 (c) $-\dfrac{1}{8}$

53.

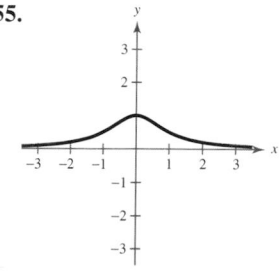

   Limit does not exist.

55.

   $\displaystyle\lim_{x \to 1} f(x) = \dfrac{1}{2}$

57.

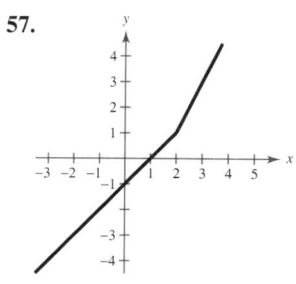

   $\displaystyle\lim_{x \to 2} f(x) = 1$

59.

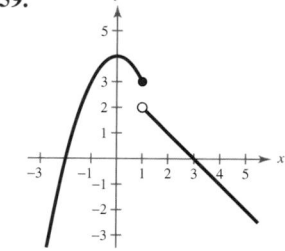

   Limit does not exist.

61.

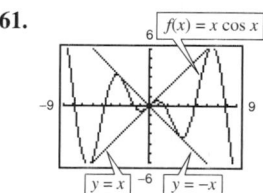

   $f(x) = x \cos x$

   $\displaystyle\lim_{x \to 0} f(x) = 0$

63.

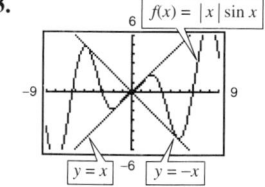

   $f(x) = |x| \sin x$

   $\displaystyle\lim_{x \to 0} f(x) = 0$

65.

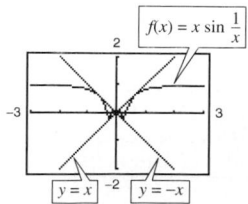

   $f(x) = x \sin \dfrac{1}{x}$

   $\displaystyle\lim_{x \to 0} f(x) = 0$

67. (a) Direct substitution; 0   (b) 1

CHAPTER 11

69. 3   **71.** $\dfrac{1}{2\sqrt{x}}$ **73.** $2x - 3$ **75.** $-\dfrac{1}{(x + 2)^2}$

77. -32 feet per second

79. (a)

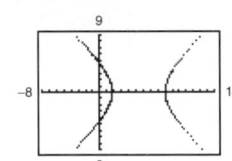

(b)

t	3	3.3	3.4	3.5	3.6	3.7	4
C	1.50	1.75	1.75	1.75	1.75	1.75	1.75

$\displaystyle \lim_{t \to 3.5} C(t) = 1.75$

(c)

t	2	2.5	2.9	3	3.1	3.5	4
C	1.25	1.50	1.50	1.50	1.75	1.75	1.75

Limit does not exist; the one-sided limits do not agree.

81. Answers will vary.   **83.** True.

85. (a) and (b) Answers will vary.

87. $x - 2y - 26 = 0$

89. Parabola   **91.** Hyperbola

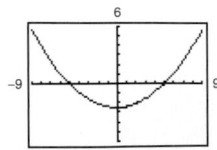

93. Parabola   **95.** Orthogonal

97. Parallel

Section 11.3   (page 808)

1. 0   **3.** $\frac{1}{2}$ **5.** 2   **7.** -2 **9.** -1 **11.** $\frac{1}{6}$

13. $m = -2x$; (a) 0   (b) 2

15. $m = -\dfrac{1}{(x + 4)^2}$; (a) $-\dfrac{1}{16}$ (b) $-\dfrac{1}{4}$

17. $m = \dfrac{1}{2\sqrt{x - 1}}$; (a) $\dfrac{1}{4}$ (b) $\dfrac{1}{6}$

19.   **21.**

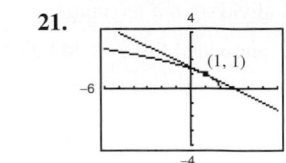

2   $-\dfrac{1}{2}$

23.   **25.** 0

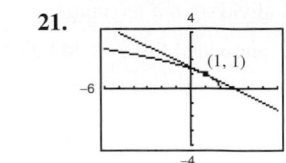

-1

27. $-\dfrac{1}{3}$ **29.** $-6x$ **31.** $-\dfrac{2}{x^3}$ **33.** $\dfrac{1}{2\sqrt{x - 4}}$

35. $-\dfrac{1}{(x + 2)^2}$ **37.** $-\dfrac{1}{2(x - 9)^{3/2}}$

39. (a) 4   **41.** (a) 1
(b) $y = 4x - 5$ (b) $y = x - 2$
(c)   (c)

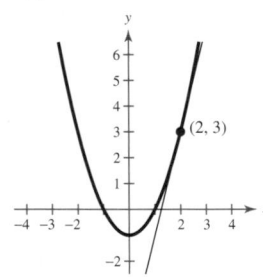
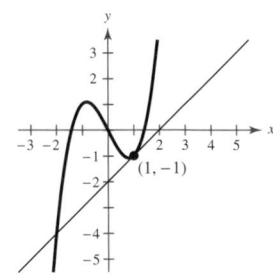

43. (a) $\frac{1}{4}$ **45.** (a) -1
(b) $y = \frac{1}{4}x + \frac{5}{4}$ (b) $y = -x - 3$
(c)   (c)

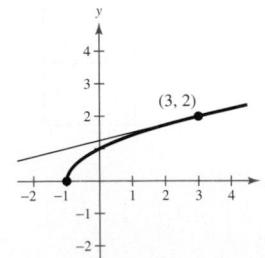
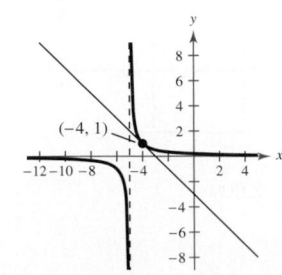

47.

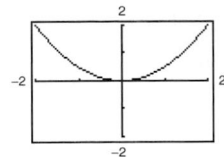

x	-2	-1.5	-1	-0.5	0
$f(x)$	2	1.125	0.5	0.125	0
$f'(x)$	-2	-1.5	-1	-0.5	0

x	0.5	1	1.5	2
$f(x)$	0.125	0.5	1.125	2
$f'(x)$	0.5	1	1.5	2

They appear to be the same.

49.

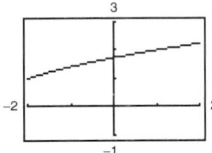

x	-2	-1.5	-1	-0.5	0
$f(x)$	1	1.225	1.414	1.581	1.732
$f'(x)$	0.5	0.408	0.354	0.316	0.289

x	0.5	1	1.5	2
$f(x)$	1.871	2	2.121	2.236
$f'(x)$	0.267	0.25	0.236	0.224

They appear to be the same.

51. $f'(x) = 2x - 4; (2, -1)$

53. $f'(x) = 9x^2 - 9; (-1, 6), (1, -6)$

55. $(-1, -1), (0, 0), (1, -1)$

57. $\left(\dfrac{\pi}{6}, \sqrt{3} + \dfrac{\pi}{6}\right), \left(\dfrac{5\pi}{6}, \dfrac{5\pi}{6} - \sqrt{3}\right)$

59. $(0, 0), (-2, 4e^{-2})$ **61.** $(e^{-1}, -e^{-1})$

63. (a) $P(t) = -0.63t^2 + 63.3t + 8448$

(b)

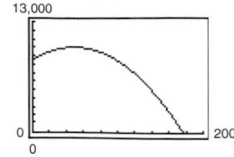

38; The population is increasing at approximately 38,000 people per year in 2020.

(c) $f'(x) = -1.26x + 63.3; f'(20) = 38.1$

(d) Answers will vary.

65. (a) $V'(r) = 4\pi r^2$ (b) ≈ 201.06

(c) Cubic inches per inch; The derivative is a formula for rate of change.

67. (a) $s'(t) = -32t + 64$ (b) 16 feet per second

(c) $t = 2$ seconds; Answers will vary.

(d) -96 feet per second

(e)

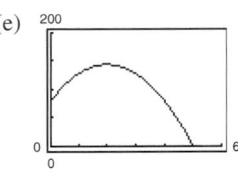

69. True. The graph of the derivative is a line, which is a one-to-one function.

71. b   **72.** a   **73.** d   **74.** c

75. Answers will vary. Sample answer: A sketch of any linear function with positive slope

77.

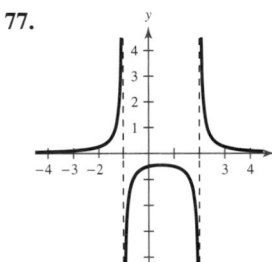

79.

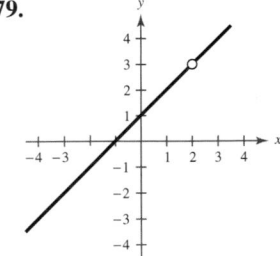

81. $\langle -2, 3, -1 \rangle$ **83.** $\langle 0, 0, -36 \rangle$

Section 11.4   (page 817)

Vocabulary Check   (page 817)

1. limit, infinity   **2.** converge   **3.** diverge

1. c   **2.** a   **3.** d   **4.** b   **5.** f   **6.** g   **7.** h

8. e   **9.** 0   **11.** -1 **13.** $\frac{5}{6}$ **15.** -4

17. Limit does not exist.   **19.** $\frac{4}{3}$ **21.** 2   **23.** -1

25. -4 **27.** -5

29. 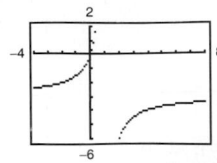   **31.**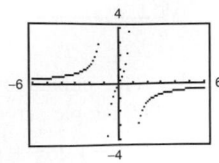

33.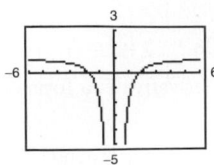

35. (a)

x	10^0	10^1	10^2	10^3
$f(x)$	-0.7321	-0.0995	-0.0100	-0.0010

x	10^4	10^5	10^6
$f(x)$	-1.0×10^{-4}	-1.0×10^{-5}	-1.0×10^{-6}

0

(b)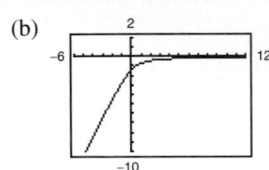

0

37. (a)

x	10^0	10^1	10^2	10^3
$f(x)$	-0.7082	-0.7454	-0.7495	-0.74995

x	10^4	10^5	10^6
$f(x)$	-0.749995	-0.7499995	-0.7500

(b)

-0.75

39. $1, \frac{3}{5}, \frac{2}{5}, \frac{5}{17}, \frac{3}{13}$ **41.** $\frac{1}{3}, \frac{2}{5}, \frac{3}{7}, \frac{4}{9}, \frac{5}{11}$

Limit: 0   Limit: $\frac{1}{2}$

43. $\frac{1}{5}, \frac{1}{2}, \frac{9}{11}, \frac{8}{7}, \frac{25}{17}$ **45.** 2, 3, 4, 5, 6

Limit does not exist.   Limit does not exist.

47. $-1, \frac{1}{2}, -\frac{1}{3}, \frac{1}{4}, -\frac{1}{5}$

Limit: 0

49.

n	10^0	10^1	10^2	10^3
a_n	2	1.55	1.505	1.5005

n	10^4	10^5	10^6
a_n	1.5001	1.5000	1.5000

1.5

$\lim_{n \to \infty} a_n = \frac{3}{2}$

51.

n	10^0	10^1	10^2	10^3
a_n	16	6.16	5.4136	5.3413

n	10^4	10^5	10^6
a_n	5.3341	5.3334	5.3333

5.33

$\lim_{n \to \infty} a_n = \frac{16}{3}$

53. (a) $\overline{C} = \dfrac{13.50x + 45,750}{x}$ (b) \$471; \$59.25

(c) \$13.50; As the number of PDAs produced gets very
large, the average cost approaches \$13.50.

55. (a)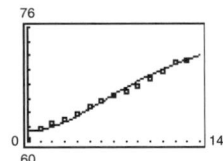

Answers will vary.

(b) 2004: 72,000,000; 2008: 73,800,000

(c) 78 million; As time passes, school enrollment in the
United States approaches 78 million.

(d) Answers will vary.

57. False. $y = \dfrac{x^2}{x + 1}$ does not have a horizontal asymptote.

59. True

61. Answers will vary. Sample answer: Let $f(x) = \dfrac{1}{x^2}$,

$g(x) = \dfrac{1}{x^2}$, and $c = 0$. Now $\lim\limits_{x \to 0} \dfrac{1}{x^2} = \infty$ and

$\lim\limits_{x \to 0} [f(x) - g(x)] = 0$.

63.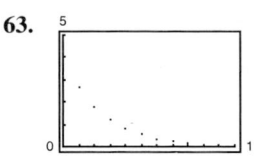

Converges to 0

65.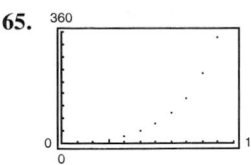

Diverges

67.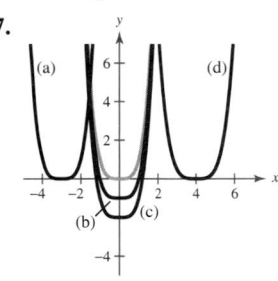

69. $x^2 + 2x + 1$ **71.** $x^3 + 5x^2 - 3 - \dfrac{2}{3x + 2}$

73. $-4, 5, 0, 0$ **75.** 3

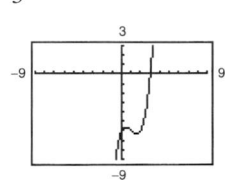

77. 60   **79.** 150

Section 11.5   (page 826)

Vocabulary Check   (page 826)

1. $\dfrac{n(n + 1)}{2}$ **2.** $\dfrac{n^2(n + 1)^2}{4}$ **3.** area

---

1. 420   **3.** 44,140   **5.** 44,140   **7.** 5850

9. (a) $S(n) = \dfrac{n^2 + 2n + 1}{4n^2}$

(b)

n	10^0	10^1	10^2	10^3	10^4
$S(n)$	1	0.3025	0.2550	0.2505	0.2501

(c) $\lim\limits_{n\to\infty} S(n) = \dfrac{1}{4}$

11. (a) $S(n) = \dfrac{2n^2 + 3n + 7}{2n^2}$

(b)

n	10^0	10^1	10^2	10^3	10^4
$S(n)$	6	1.185	1.0154	1.0015	1.0002

(c) $\lim\limits_{n\to\infty} S(n) = 1$

13. (a) $S(n) = \dfrac{14n^2 + 3n + 1}{6n^3}$

(b)

n	10^0	10^1	10^2	10^3	10^4
$S(n)$	3	0.2385	0.0234	0.0023	0.0002

(c) $\lim\limits_{n\to\infty} S(n) = 0$

15. (a) $S(n) = \dfrac{4n^2 - 3n - 1}{6n^2}$

(b)

n	10^0	10^1	10^2	10^3	10^4
$S(n)$	0	0.615	0.6617	0.6662	0.6666

(c) $\lim\limits_{n\to\infty} S(n) = \dfrac{2}{3}$

17. 14.25   **19.** 1.27

21.

n	4	8	20	50
Approximate area	18	21	22.8	23.52

23.

n	4	8	20	50
Approximate area	3.5156	2.8477	2.4806	2.3409

25.

n	4	8	20	50	100	∞
Area	40	38	36.8	36.32	36.16	36

27.

n	4	8	20	50	100	∞
Area	36	38	39.2	39.68	39.84	40

29.

n	4	8	20	50	100	∞
Area	14.25	14.81	15.13	15.25	15.29	$\frac{46}{3}$

31.

n	4	8	20	50	100	∞
Area	19	18.5	18.2	18.08	18.04	18

33. 3   **35.** 2   **37.** $\frac{10}{3}$ **39.** $\frac{17}{4}$ **41.** $\frac{3}{4}$ **43.** $\frac{51}{4}$

45.

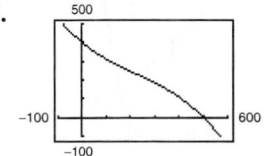

$\approx 105{,}208.33$ square feet

47. True   **49.** Answers will vary.   **51.** $n\pi$

53. $\langle 24, -30 \rangle$ **55.** $\sqrt{5} - 2$

Review Exercises   (page 829)

1.

x	2.9	2.99	2.999	3
$f(x)$	16.4	16.94	16.994	17

x	3.001	3.01	3.1
$f(x)$	17.006	17.06	17.6

17; Yes

3.

x	-0.1	-0.01	-0.001	0
$f(x)$	1.0517	1.0050	1.0005	Error

x	0.001	0.01	0.1
$f(x)$	0.9995	0.9950	0.9516

1; No

5. 2   **7.** 2   **9.** (a) 64   (b) 7   (c) 20   (d) $\frac{4}{5}$

11. 5   **13.** 77   **15.** $\frac{10}{3}$ **17.** -2 **19.** 0

21. $\frac{2}{e}$ **23.** $-\frac{\pi}{6}$ **25.** $-\frac{1}{4}$ **27.** $\frac{1}{15}$

29. $-\frac{1}{3}$ **31.** -1 **33.** $\frac{1}{4}$ **35.** $\frac{1}{4}$

37. (a) and (b) 0.17   **39.** Limit does not exist.

41. (a) and (b) 2   **43.** (a) and (b) 0.577

45.

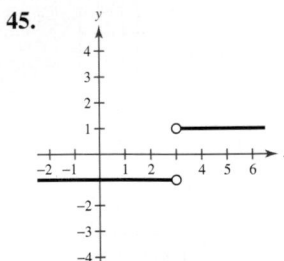

Limit does not exist.

47.

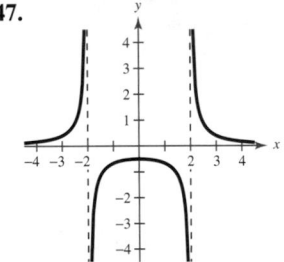

Limit does not exist.

49.

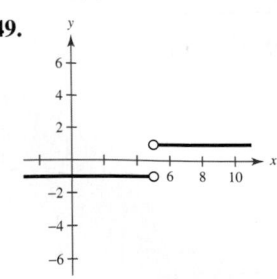

Limit does not exist.

51.

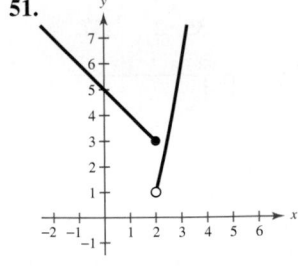

Limit does not exist.

53. $3 - 2x$ **55.** 2

57.

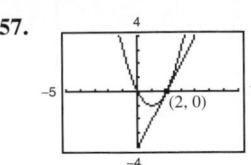

2

59.

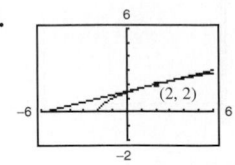

$\frac{1}{4}$

61.

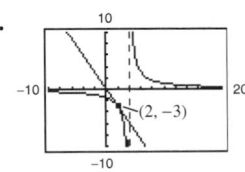

$-\frac{3}{2}$

63. $m = 2x - 4$; (a) -4 (b) 6

65. $m = -\dfrac{4}{(x - 6)^2}$; (a) -4 (b) -1

67. $f'(x) = 0$ **69.** $h'(x) = -\frac{1}{2}$ **71.** $g'(x) = 4x$

73. $f'(t) = \dfrac{1}{2\sqrt{t + 5}}$ **75.** $g'(s) = -\dfrac{4}{(s + 5)^2}$

77. $g'(x) = -\dfrac{1}{2(x + 4)^{3/2}}$ **79.** 2   **81.** 0

83. Limit does not exist.   **85.** 3

87. $-\frac{1}{9}, \frac{1}{14}, \frac{3}{19}, \frac{5}{24}, \frac{7}{29}$ **89.** $-1, \frac{1}{8}, -\frac{1}{27}, \frac{1}{64}, -\frac{1}{125}$

Limit: $\frac{2}{5}$ Limit: 0

91. $-\frac{1}{2}, -\frac{9}{8}, -\frac{7}{6}, -\frac{37}{32}, -\frac{57}{50}$

Limit: -1

93. (a) $S(n) = \dfrac{5n^2 + 9n + 4}{6n^2}$

(b)

n	10^0	10^1	10^2	10^3	10^4
$S(n)$	3	0.99	0.8484	0.8348	0.8335

(c) $\dfrac{5}{6}$

95. 675

97.

n	4	8	20	50
Approximate area	7.5	6.375	5.74	5.4944

99. 50   **101.** 15   **103.** 6   **105.** 34

107. (a) $y = (-3.376 \times 10^{-7})x^3 + (3.753 \times 10^{-4})x^2$
$- 0.168x + 132$

(b)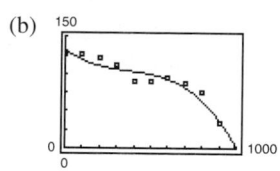

(c) 88,700 square feet

109. False. The limit of the rational function as x approaches ∞ does not exist.

Chapter Test   (page 832)

1.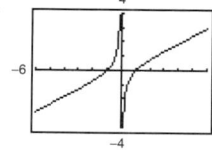

-0.75

$\lim_{x \to -2} f(x) = -\frac{3}{4}$

2.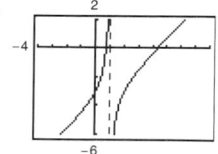

Limit does not exist.

3.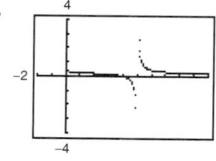

Limit does not exist.

4.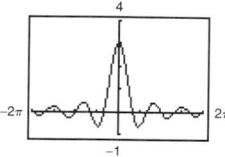

3.0000

5.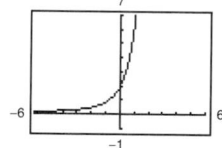

2.0000

6. (a) $m = 6x - 5; 7$ (b) $m = 6x^2 + 6; 12$

7. $f'(x) = -\frac{2}{5}$ **8.** $f'(x) = 4x + 4$

9. $f'(x) = -\dfrac{1}{(x + 3)^2}$ **10.** 0   **11.** -3

12. Limit does not exist.

13. $0, \frac{3}{4}, \frac{14}{19}, \frac{12}{17}, \frac{36}{53}$ **14.** $0, 1, 0, \frac{1}{2}, 0$

Limit: $\frac{1}{2}$ Limit: 0

15. 12.5   **16.** 8   **17.** $\frac{16}{3}$

18. (a) $y = 8.79x^2 - 6.2x - 0.4$ (b) 81.7 feet per second

Cumulative Test for Chapters 10 and 11 (page 833)

1. $(-6, 1, 3)$ **2.** $(0, -4, 0)$ **3.** $\sqrt{149}$

4. $3, 4, 5$ **5.** $\left(-1, 2, \frac{1}{2}\right)$

6. $(x - 2)^2 + (y - 2)^2 + (z - 4)^2 = 24$

7.

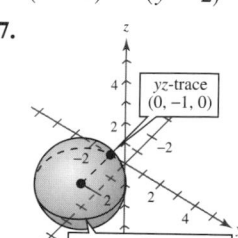

8. $\mathbf{u} \cdot \mathbf{v} = -38$

$\mathbf{u} \times \mathbf{v} = \langle -18, -6, -14 \rangle$

9. Neither   **10.** Orthogonal   **11.** Parallel   **12.** 12

13. (a) $x = -2 + 7t, y = 3 + 5t, z = 25t$

(b) $\dfrac{x + 2}{7} = \dfrac{y - 3}{5} = \dfrac{z}{25}$

14. $x = -1 + 2t, y = 2 - 4t, z = t$

15. $75x + 50y - 31z = 0$

16. 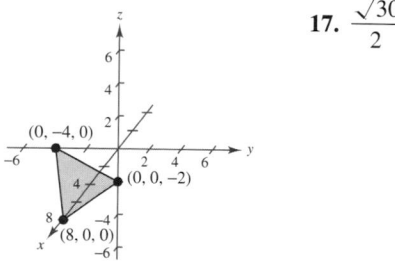   **17.** $\dfrac{\sqrt{30}}{2}$

18. 84.26°   **19.** 4   **20.** $-\frac{1}{3}$ **21.** $\frac{1}{14}$ **22.** $\frac{1}{4}$

23. -1 **24.** Limit does not exist.   **25.** $-\frac{1}{4}$ **26.** $\frac{1}{2}$

27. $\frac{1}{4}$ **28.** $m = -2x; 0$ **29.** $m = \frac{1}{2}(x + 3)^{-1/2}; \frac{1}{2}$

30. $m = -(x + 3)^{-2}; -\frac{1}{16}$ **31.** $m = 2x - 1; 1$

32. Limit does not exist.   **33.** -7 **34.** 3

35. 0   **36.** 0   **37.** Limit does not exist.

38. $-42,875$ **39.** 8190   **40.** 672,880

41. 10.5   **42.** 8.13   **43.** 2.69   **44.** 1.57   **45.** $\frac{5}{2}$

46. 64   **47.** 28   **48.** $\frac{76}{3}$ **49.** $\frac{16}{3}$ **50.** $\frac{3}{4}$

CHAPTER 11

Appendices

Appendix B.1 (page A32)

Vocabulary Check (page A32)

1. (a) iii (b) vi (c) i (d) iv (e) v (f) ii

2. Cartesian **3.** Distance Formula

4. Midpoint Formula

5. $(x - h)^2 + (y - k)^2 = r^2$, center, radius

1. A: $(2, 6)$; B: $(-6, -2)$; C: $(4, -4)$; D: $(-3, 2)$

3.

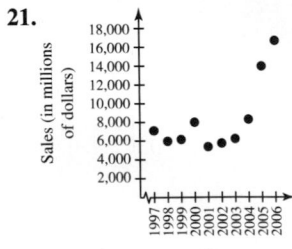

5.

7. $(-5, 4)$ **9.** $(-6, -6)$ **11.** Quadrant IV

13. Quadrant II **15.** Quadrant III or IV

17. Quadrant III **19.** Quadrant I or III

21.

23. 8 **25.** 5 **27.** 13 **29.** $\dfrac{\sqrt{277}}{6}$ **31.** $\sqrt{71.78}$

33. (a) 4, 3, 5 (b) $4^2 + 3^2 = 5^2$

35. (a) 10, 3, $\sqrt{109}$ (b) $10^2 + 3^2 = \left(\sqrt{109}\right)^2$

37. $\left(\sqrt{5}\right)^2 + \left(\sqrt{45}\right)^2 = \left(\sqrt{50}\right)^2$

39. Two equal sides of length $\sqrt{29}$

41. Opposite sides have equal lengths of $2\sqrt{5}$ and $\sqrt{85}$.

43. The diagonals are of equal length $\left(\sqrt{58}\right)$. The slope of the line between $(-5, 6)$ and $(0, 8)$ is $\frac{2}{5}$. The slope of the line between $(-5, 6)$ and $(-3, 1)$ is $-\frac{5}{2}$. The slopes are negative reciprocals, making them perpendicular lines, which form a right angle.

45. (a)

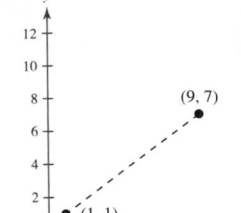

(b) 10
(c) $(5, 4)$

47. (a)

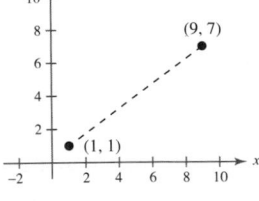

(b) 17
(c) $\left(0, \frac{5}{2}\right)$

49. (a)

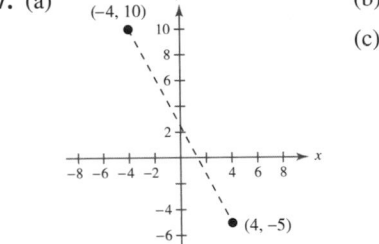

(b) $2\sqrt{10}$
(c) $(2, 3)$

51. (a)

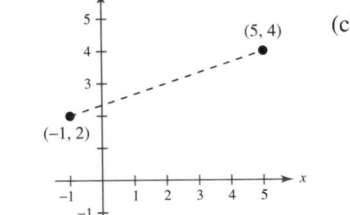

(b) $\dfrac{\sqrt{82}}{3}$
(c) $\left(-1, \dfrac{7}{6}\right)$

53. (a)

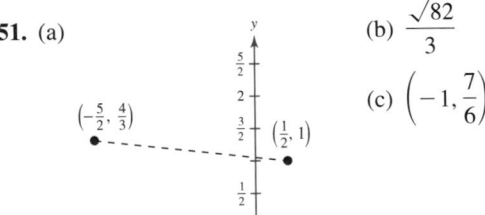

(b) $\sqrt{110.97}$
(c) $(1.25, 3.6)$

55. \$3,093.5 million

57. $(2x_m - x_1, 2y_m - y_1)$; (a) $(7, 0)$ (b) $(9, -3)$

59. $x^2 + y^2 = 9$ **61.** $(x - 2)^2 + (y + 1)^2 = 16$

63. $(x + 1)^2 + (y - 2)^2 = 5$

65. $(x - 3)^2 + (y - 4)^2 = 25$

67. $(x + 2)^2 + (y - 1)^2 = 1$

69. $(x - 3)^2 + (y + 6)^2 = 16$

71. $(x - 2)^2 + (y + 1)^2 = 16$

73. Center: $(0, 0)$ **75.** Center: $(1, -3)$

Radius $= 5$ Radius $= 2$

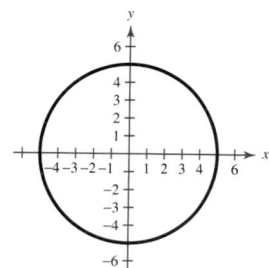 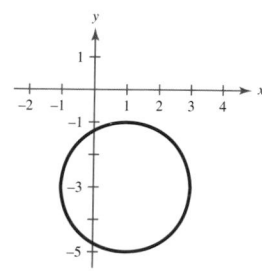

77. Center: $\left(\frac{1}{2}, \frac{1}{2}\right)$

Radius $= \frac{3}{2}$

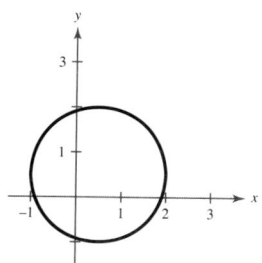

79. $(0, 1), (4, 2), (1, 4)$ **81.** $(-1, 5), (2, 8), (4, 5), (1, 2)$

83. 65

85. (a) Answers will vary. Sample answer: The number of artists elected each year seems to be nearly steady except for the first few years. Estimate: From 5 to 7 new members in 2007

(b) Answers will vary. Sample answer: The Rock and Roll Hall of Fame was opened in 1986.

87. $5\sqrt{74} \approx 43$ yards

89. (a)

N 1 unit : 8 mi (0, 64)
W—E ⊢—⊣ 4 P.M.
S   (0, 32)
         2 P.M.
(−24, 0)
2 P.M.
(−48, 0)
4 P.M.
Fisherman
Beach Lover

(b) 2 P.M.: 40 miles; 4 P.M.: 80 miles; The yachts are twice as far from each other at 4 P.M. as they were at 2 P.M.

91. The distance between $(2, 6)$ and $\left(2 + 2\sqrt{3}, 0\right)$ is $4\sqrt{3}$. The distance between $\left(2 + 2\sqrt{3}, 0\right)$ and $\left(2 - 2\sqrt{3}, 0\right)$ is $4\sqrt{3}$. The distance between $(2, 6)$ and $\left(2 - 2\sqrt{3}, 0\right)$ is $4\sqrt{3}$. Because the distance between each set of points is $4\sqrt{3}$, the sides connecting those points are all the same length, making the coordinates the vertices of an equilateral triangle.

93. False. You would have to use the Midpoint Formula 15 times.

95. False. It could be a rhombus.

97. No. The scales depend on the magnitudes of the quantities measured.

Appendix B.2   (page A43)

Vocabulary Check   (page A43)

1. solution point   **2.** graph   **3.** intercepts

1. (a) Yes   (b) Yes   **3.** (a) No   (b) Yes

5. (a) No   (b) Yes

7.

x	-2	0	$\frac{2}{3}$	1	2
y	-4	-1	0	$\frac{1}{2}$	2
Solution point	$(-2, -4)$	$(0, -1)$	$\left(\frac{2}{3}, 0\right)$	$\left(1, \frac{1}{2}\right)$	$(2, 2)$

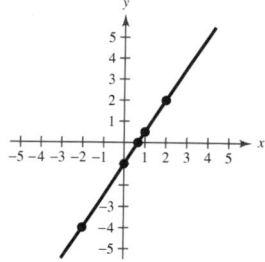

9. (a)

x	-2	-1	0	1	2
y	$-\frac{7}{2}$	$-\frac{13}{4}$	-3	$-\frac{11}{4}$	$-\frac{5}{2}$

APPENDICES

(b)

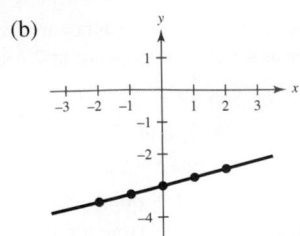

(c)

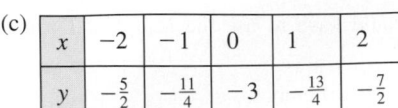

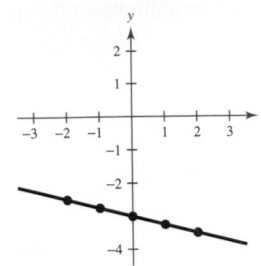

The lines have opposite slopes but the same y-intercept.

11. e **12.** f **13.** b **14.** d **15.** c **16.** a

17.

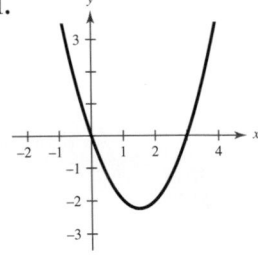

19.

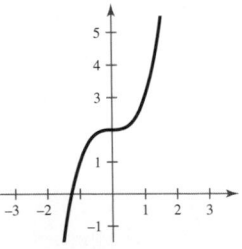

21.

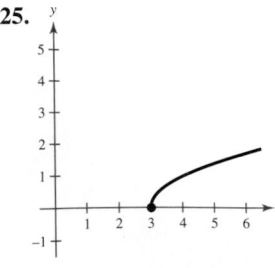

23.

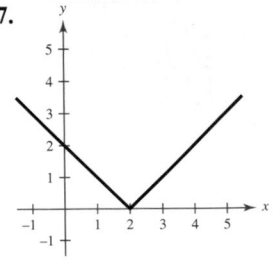

25.

27.

29.

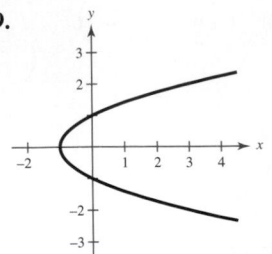

31.

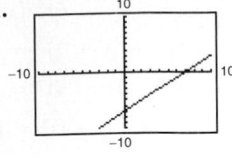

Intercepts: $(7, 0), (0, -7)$

33.

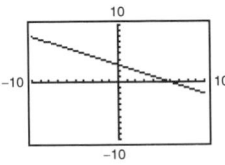

Intercepts: $(6, 0), (0, 3)$

35.

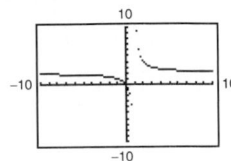

Intercept: $(0, 0)$

37.

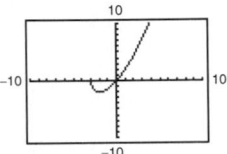

Intercepts: $(-3, 0), (0, 0)$

39.

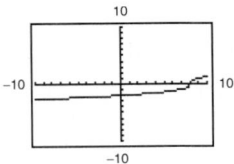

Intercepts: $(0, -2), (8, 0)$

41.

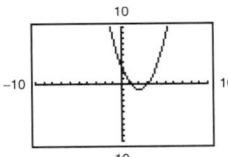

Intercepts: $(0, 3), (1, 0),$
$(3, 0)$

43.

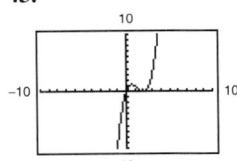

Intercepts: $(0, 0), (2, 0)$

45.

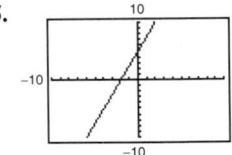

The standard setting gives a more complete graph.

47.

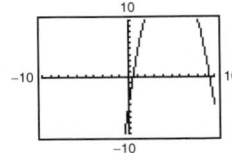

The specified setting gives a more complete graph.

49.
```
Xmin = -10
Xmax = 10
Xscl = 2
Ymin = -50
Ymax = 100
Yscl = 25
```

51.
```
Xmin = -5
Xmax = 1
Xscl = 1
Ymin = -3
Ymax = 1
Yscl = 1
```

53.
```
Xmin = -30
Xmax = 30
Xscl = 5
Ymin = -10
Ymax = 50
Yscl = 5
```

55. The graphs are identical. Distributive Property

57. The graphs are identical. Associative Property of Multiplication

59.

(a) $(2, 1.73)$

(b) $(-4, 3)$

61.

(a) $(-0.5, 2.47)$

(b) $(1, -4), (-1.65, -4)$

63. $y_1 = \sqrt{16 - x^2},\ y_2 = -\sqrt{16 - x^2}$

65. $y_1 = \sqrt{4 - (x - 1)^2} + 2,\ y_2 = -\sqrt{4 - (x - 1)^2} + 2$

67. a   **68.** c   **69.** b, c, d   **70.** c, d

71. (a)

(b) $109,000   (c) $178,000

73. (a)

Year	1995	1996	1997	1998	1999
Median sales price (in thousands of dollars)	123.4	127.1	131.5	136.5	142.3

Year	2000	2001	2002	2003	2004
Median sales price (in thousands of dollars)	148.6	155.5	163.0	171.1	179.6

The model fits the data fairly well.

(b)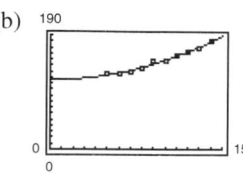

Answers will vary. Sample answer: Because the graphs overlap and have little divergence from one another, the model fits the data well.

(c) 2008: $218,155; 2010: $239,700; Answers will vary.

(d) 2000

75. (a)

(b) Answers will vary.

(c)

(d) $A \approx 5.4$ (e) $x = 3, w = 3$

77. False. $y = x^2 - 1$ has two x-intercepts.

79. Answers will vary. Sample answer: use the *ZoomFit* and *Zoom out* features as needed.

81. Answers will vary. Sample answer: $y = 250x + 1000$ could represent the amount of money in someone's checking account after time x if they deposited an initial $1000 and added $250 every x time increment (x could potentially stand for a month, for example).

APPENDICES

Appendix B.3   (page A59)

Vocabulary Check   (page A59)

1. equation   **2.** solve   **3.** extraneous
4. x-intercept, y-intercept   **5.** point of intersection

1. (a) Yes   (b) No   (c) No   (d) No
3. (a) Yes   (b) No   (c) No   (d) No
5. (a) No   (b) No   (c) No   (d) Yes
7. Identity   **9.** Identity   **11.** Conditional equation
13. $x = -\frac{96}{23}$; Answers will vary.
15. $x = \frac{20}{81}$; Answers will vary.   **17.** $x = 12$ **19.** $y = 1$
21. $y = -9$ **23.** $x = -10$ **25.** $z = -\frac{6}{5}$
27. $z = \frac{17}{48}$ **29.** $u = 10$ **31.** $x = 4$ **33.** $x = 5$
35. $x = \frac{11}{6}$ **37.** $x = \frac{5}{3}$ **39.** No solution
41. $(5, 0), (0, -5)$ **43.** $(-2, 0), (1, 0), (0, -2)$
45. $(-2, 0), (0, 0)$ **47.** No intercepts
49. $(-2, 0), (6, 0), (0, -2)$ **51.** $(1, 0), \left(0, \frac{1}{2}\right)$

53.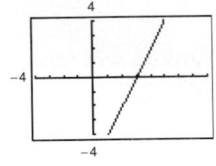
$(3, 0)$

55.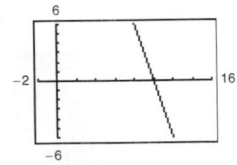
$(10, 0)$

57. $5(4 - 4) = 0$
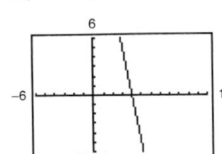

59. $(0)^3 - 6(0)^2 + 5(0) = 0$
$(5)^3 - 6(5)^2 + 5(5) = 0$
$(1)^3 - 6(1)^2 + 5(1) = 0$
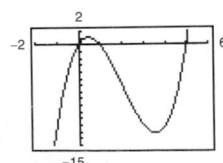

61. $\dfrac{1 + 2}{3} - \dfrac{1 - 1}{5} - 1 = 0$
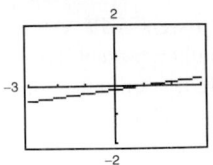

63. $\frac{12}{23}$ **65.** $\frac{89}{13}$ **67.** 6   **69.** -194 **71.** 3, 12
73. $1, -1.2$ **75.** $-\frac{3}{10}$ **77.** $0.5, -3, 3$ **79.** ± 1

81. -1.333 **83.** $-1, 7$ **85.** 11   **87.** $(1, 1)$
89. $(-1, 3), (2, 6)$ **91.** $(-1, 3)$ **93.** $(4, 1)$
95. $(1.449, 1.899), (-3.449, -7.899)$
97. $(0, 0), (-2, 8), (2, 8)$ **99.** $0, -\frac{1}{2}$ **101.** $4, -2$
103. $3, -\frac{1}{2}$ **105.** $2, -6$ **107.** $-a \pm b$ **109.** ± 7
111. $16, 8$ **113.** No solution
115. $\frac{1}{2} \pm \sqrt{3}$; $2.23, -1.23$ **117.** 2   **119.** $-8, 4$
121. $-3 \pm \sqrt{7}$ **123.** $1 \pm \dfrac{\sqrt{6}}{3}$ **125.** No solution
127. $\dfrac{-5 \pm \sqrt{89}}{4}$ **129.** $1 \pm \sqrt{3}$ **131.** $-4 \pm 2\sqrt{5}$
133. No solution   **135.** $\frac{2}{7}$ **137.** $-\dfrac{8}{3} \pm \dfrac{\sqrt{13}}{3}$
139. $1 \pm \sqrt{2}$ **141.** $6, -12$ **143.** 7   **145.** $\frac{1}{2} \pm \sqrt{3}$
147. $-\dfrac{1}{2}$ **149.** $\pm 2, 0$ **151.** $-3, 0$ **153.** $0, \pm\dfrac{3\sqrt{2}}{2}$
155. $\pm 1, \pm \sqrt{3}$ **157.** $3, 1, -1$ **159.** $\pm\frac{1}{2}, \pm 4$
161. $-\frac{1}{5}, -\frac{1}{3}$ **163.** $2, -\frac{3}{5}$ **165.** $\frac{1}{4}$ **167.** 26
169. 0   **171.** -256.5 **173.** 9   **175.** $-59, 69$
177. 1   **179.** $\dfrac{-3 \pm \sqrt{21}}{6}$ **181.** $2, -\dfrac{3}{2}$
183. $3, -2$ **185.** $\sqrt{3}, -3$

187. (a)   (b) and (c)
$x = 0, 3, -1$
(d) They are the same.

189. (a)   (b) and (c)
$x = 5, 6$
(d) They are the same.

191. (a)   (b) and (c)
$x = -1$
(d) They are the same.

193. (a)   (b) and (c)
$x = 1, -3$
(d) They are the same.

195. (a)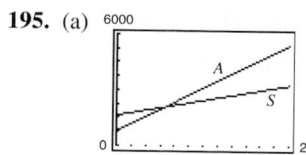

(6.6, 3386.5); The *x*-coordinate represents the year and the *y*-coordinate represents the population. In 1986, the populations were the same at about 3,386,500 people.

(b) (6.6, 3386.5); The point of intersection represents the point in time at which the populations of the two cities were the same.

(c) South Carolina increases by 45,200 people per year while Arizona increases by 128,200 people per year. Arizona has a greater population growth rate than South Carolina.

(d) South Carolina: 4,443,000 people

Arizona: 6,397,000 people

Answers will vary.

197. (a)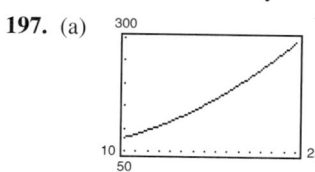

(b) 16.8°C   (c) 2.5

199. False. The lines could be identical.

201. $c = \frac{5}{8}$ **203.** (a) $0, -\frac{b}{a}$ (b) 0, 1

Appendix B.4   (page A72)

Vocabulary Check   (page A72)

1. negative   **2.** double   **3.** $-a \le x \le a$

4. $x \le -a, x \ge a$ **5.** zeros, undefined values

---

1. f   **2.** a   **3.** d   **4.** b   **5.** e   **6.** c

7. (a) Yes   (b) No   (c) Yes   (d) No

9. (a) No   (b) Yes   (c) Yes   (d) No

11. $x > -4$

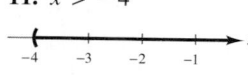

13. $x < -\frac{1}{2}$

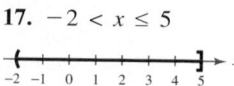

15. $x \ge 4$

17. $-2 < x \le 5$

19. $-\frac{9}{2} < x < \frac{15}{2}$

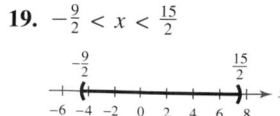

21.

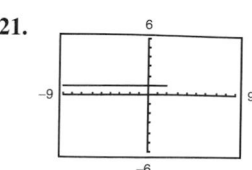

$x \le 2$

23.

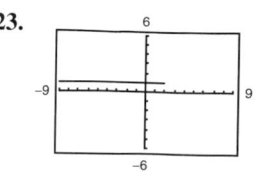

$x < 2$

25.

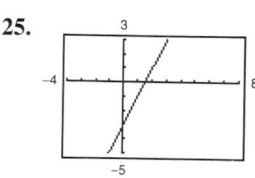

(a) $x \ge 2$

(b) $x \le \frac{3}{2}$

27.

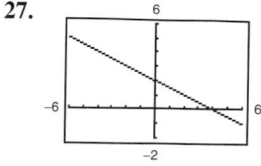

(a) $-2 \le x \le 4$

(b) $x \le 4$

29. $x < -2, x > 2$

31. $1 < x < 13$

33. $x < -28, x > 0$

35. $\frac{1}{4} < x < \frac{3}{4}$

37.

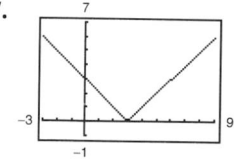

(a) $1 \le x \le 5$

(b) $x \le -1, x \ge 7$

39. $|x| \le 3$

41. $|x| > 3$ **43.** $|x - 7| \le 10$ **45.** $|x - 3| \ge 5$

47. Positive on: $(-\infty, -1) \cup (5, \infty)$

Negative on: $(-1, 5)$

49. Positive on: $\left(-\infty, \dfrac{2 - \sqrt{10}}{2}\right) \cup \left(\dfrac{2 + \sqrt{10}}{2}, \infty\right)$

Negative on: $\left(\dfrac{2 - \sqrt{10}}{2}, \dfrac{2 + \sqrt{10}}{2}\right)$

51. Positive on: $(-\infty, \infty)$

53. $(-7, 3)$

55. $(-\infty, -5], [1, \infty)$

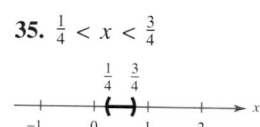

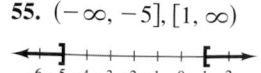

APPENDICES

57. $[-2, 0], [2, \infty)$

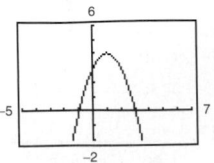

59. No solution

61. $(-3, -1), (\frac{3}{2}, \infty)$

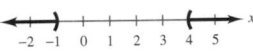

63. (a) $x = 1$
(b) $x \geq 1$
(c) $x > 1$

65.

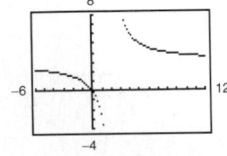

(a) $x \leq -1, \ x \geq 3$
(b) $0 \leq x \leq 2$

67. $(-\infty, -1), (0, 1)$

69. $(-\infty, -1), (4, \infty)$

71.

(a) $0 \leq x < 2$
(b) $2 < x \leq 4$

73. $[5, \infty)$ **75.** $(-\infty, \infty)$ **77.** $(-\infty, -2], [2, \infty)$

79. (a) 1994   (b) $(1990, 1994); (1994, 2004)$

81. (a) 10 seconds   (b) $(4, 6)$

83. (a)

(b) $(1991, 2000)$
(c) $1.28 < t < 10.09$
(d) No; Answers will vary.

85. $t \geq 2.11$; Answers will vary.

87. 20.70; Answers will vary.

89. $333\frac{1}{3}$ vibrations per second   **91.** $1.2 < t < 2.4$

93. (a) $A = 12 + 0.15x, B = 0.20x$

(b)

(c) Option B is the better option for monthly usage of up to 240 minutes. For more than 240 minutes, option A is the better option.

(d) Sample answer: I would choose option A because I normally use my cell phone more than 240 minutes per month.

95. False. $10 \geq -x$ **97.** a, b **99.** iv, ii, iii, i

Appendix B.5   (page A81)

Vocabulary Check   (page A81)

1. Statistics   **2.** Line plots   **3.** histogram
4. frequency distribution   **5.** bar graph
6. Line graphs

1. (a) \$2.569   (b) 0.19

3.

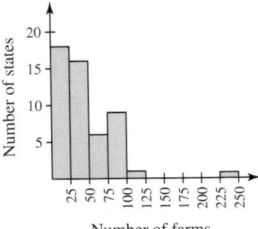

Quiz Scores

15

5. Sample answer:

Interval	Tally
$[0, 25)$	⫴⫴ ⫴⫴ ⫴⫴ ⫴⫴ ⫴⫴
$[25, 50)$	⫴⫴ ⫴⫴ ⫴⫴ │
$[50, 75)$	⫴⫴ │
$[75, 100)$	⫴⫴ ⫴⫴⫴
$[100, 125)$	│
$[125, 150)$	
$[150, 175)$	
$[175, 200)$	
$[200, 225)$	
$[225, 250)$	│

Number of farms
(in thousands)

7.

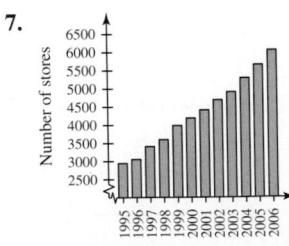

Year

Answers will vary. Sample answer: As time progresses from 1995 to 2006, the number of Wal-mart stores increases at a fairly constant rate.

9.

Year	1999	2000	2001
Difference in tuition charges (in dollars)	10,998	11,575	12,438

Year	2002	2003	2004
Difference in tuition charges (in dollars)	13,042	13,480	14,129

11.

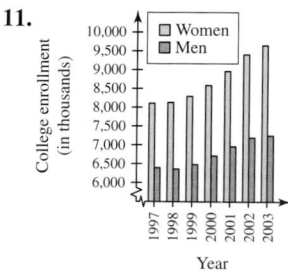

13. 118%   **15.** $2.59; January

17.

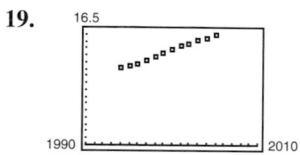

Answers will vary. Sample answer: As time progresses from 1995 to 2004, the number of women in the workforce increases at a fairly constant rate.

19.

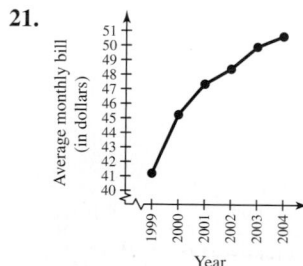

21.

Answers will vary. Sample answer: A line graph is best because the data are amounts that increased or decreased from year to year.

23.

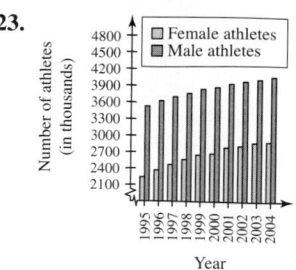

Answers will vary. Sample answer: A double bar graph is best because there are two different sets of data within the same time interval that do not deal primarily with increasing or decreasing behavior.

25. Line plots are useful for ordering small sets.

Histograms or bar graphs can be used to organize larger sets.

Line graphs are used to show trends over periods of time.

Appendix C.1   (page A91)

Vocabulary Check   (page A91)

1. measure, central tendency   **2.** modes, bimodal

3. variance, standard deviation   **4.** Quartiles

1. Mean: $\approx$ 8.86; median: 8; mode: 7

3. Mean: $\approx$ 10.29; median: 8; mode: 7

5. Mean: 9; median: 8; mode: 7

7. (a) The mean is sensitive to extreme values.

(b) Mean: $\approx$ 14.86; median: 14; mode: 13

Each is increased by 6.

(c) Each will increase by k.

9. Mean: 320; median: 320; mode: 320

11. (a) Jay: 199.67, Hank: 199.67, Buck: 229.33

(b) 209.56   (c) 202

13. Answers will vary. Sample answer: $\{4, 4, 10\}$

15. The median gives the most representative description.

17. (a) $\bar{x} = 12$; $\sigma \approx 2.83$ (b) $\bar{x} = 20$; $\sigma \approx 2.83$

(c) $\bar{x} = 12$; $\sigma \approx 1.41$ (d) $\bar{x} = 9$; $\sigma \approx 1.41$

19. $\bar{x} = 6$, $v = 10$, $\sigma \approx 3.16$

21. $\bar{x} = 2$, $v = \frac{4}{3}$, $\sigma \approx 1.15$ **23.** $\bar{x} = 4$, $v = 4$, $\sigma = 2$

25. $\bar{x} = 47$, $v = 226$, $\sigma \approx 15.03$ **27.** ≈ 3.42

29. ≈ 1.65 **31.** $\bar{x} = 12$ and $|x_i - 12| = 8$ for all x_i.

33. The mean will increase by 5, but the standard deviation will not change.

35. [179, 291]; [151, 319]
    [203, 267]; [187, 283]

37. (a) Upper quartile: 21.5   (b)
        Lower quartile: 13
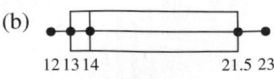

39. (a) Upper quartile: 51   (b)
        Lower quartile: 47
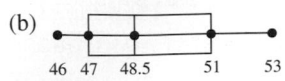

41.
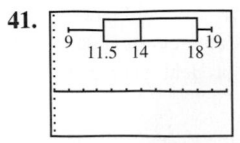
43.
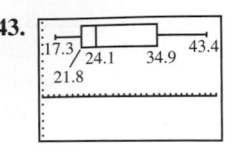

45.   Original design   New design
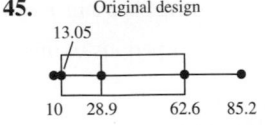
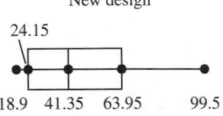

From the plots, you can see that the lifetimes of the sample units made by the new design are greater than the lifetimes of the sample units made by the original design. (The median lifetime increased by more than 12 months.)

Appendix C.2   (page A95)

1. $y = 1.6x + 7.5$ **3.** $y = 0.262x + 1.93$

Appendix D   (page A100)

Vocabulary Check   (page A100)

1. directly proportional   **2.** constant, variation

3. directly proportional   **4.** inverse   **5.** combined

6. jointly proportional

---

1. $y = \frac{12}{5}x$ **3.** $y = 205x$

5. Model: $y = \frac{33}{13}x$; 25.38 centimeters, 50.77 centimeters

7. $y = 0.0368x$; \$7360

9. (a) 0.05 meter   (b) $176\frac{2}{3}$ newtons

11.

x	2	4	6	8	10
$y = kx^2$	4	16	36	64	100

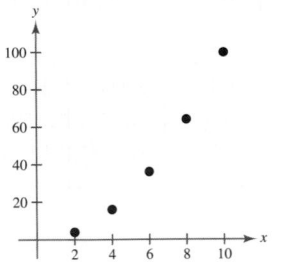

13.

x	2	4	6	8	10
$y = kx^2$	2	8	28	32	50

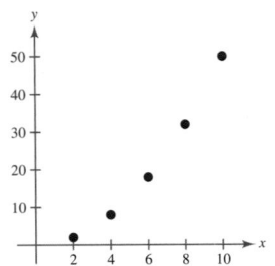

15. 0.61 mile per hour

17.

x	2	4	6	8	10
$y = k/x^2$	$\frac{1}{2}$	$\frac{1}{8}$	$\frac{1}{18}$	$\frac{1}{32}$	$\frac{1}{50}$

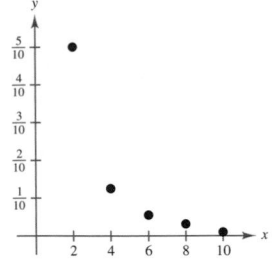

19.

x	2	4	6	8	10
$y = k/x^2$	$\frac{5}{2}$	$\frac{5}{8}$	$\frac{5}{18}$	$\frac{5}{32}$	$\frac{1}{10}$

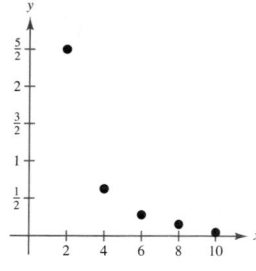

21. $y = \dfrac{5}{x}$ **23.** $y = -\dfrac{7}{10}x$ **25.** $A = kr^2$

27. $y = \dfrac{k}{x^2}$ **29.** $F = \dfrac{kg}{r^2}$ **31.** $P = \dfrac{k}{V}$

33. $R = k(T_e - T)$

35. The area of a triangle is jointly proportional to its base and height.

37. The volume of a sphere varies directly as the cube of its radius.

39. Average speed is directly proportional to the distance and inversely proportional to the time.

41. $A = \pi r^2$ **43.** $y = \dfrac{28}{x}$ **45.** $F = 14rs^3$

47. $z = \dfrac{2x^2}{3y}$ **49.** 506.27 feet   **51.** 1470 joules

53. The velocity is increased by 4.

55. (a)
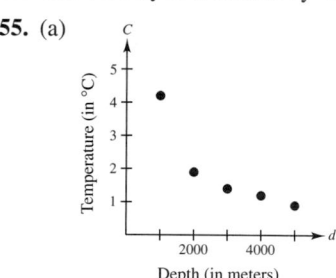

(b) Yes. $k_1 = 4200,\ k_2 = 3800,\ k_3 = 4200,$
$k_4 = 4800,\ k_5 = 4500$

(c) $C = \dfrac{4300}{d}$

(d)
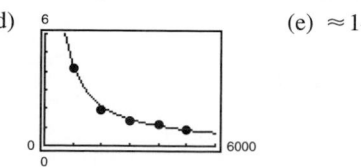

(e) ≈ 1433 meters

57. False. y will increase if k is positive and y will decrease if k is negative.

59. Inversely

Appendix E   (page A105)

1. 4   **3.** 7   **5.** 4   **7.** 20   **9.** 4   **11.** 3
13. 5   **15.** -10 **17.** No solution   **19.** $-\frac{6}{5}$
21. 9   **23.** $x < 2$ **25.** $x < 9$ **27.** $x \le -14$
29. $x > 10$ **31.** $x < 4$ **33.** $x < 3$ **35.** $x \ge 2$
37. $x \le -5$ **39.** $x < 6$ **41.** $x \ge 4$ **43.** $x \ge -4$

Appendix F.1   (page A113)

1. g   **2.** d   **3.** a   **4.** h   **5.** e   **6.** b
7. f   **8.** c

9. 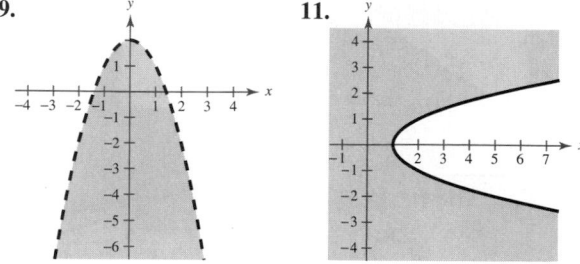   **11.**

13. 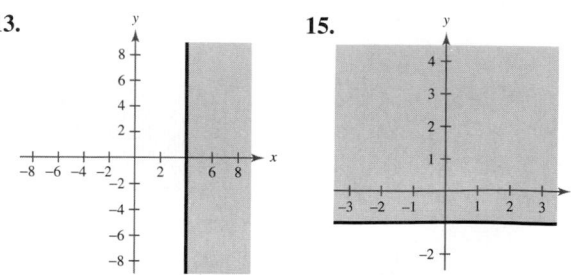   **15.**

17.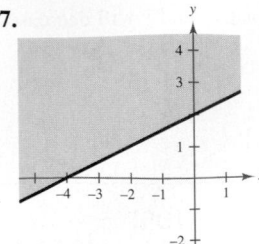

19.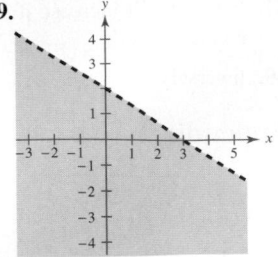

47.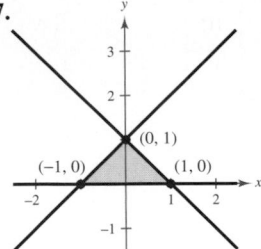

49.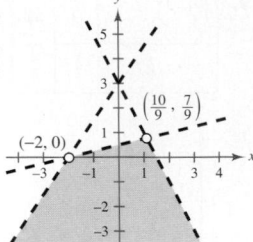

21.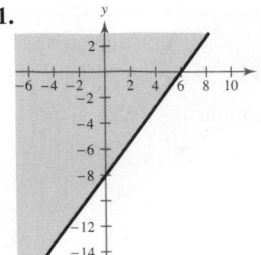

23.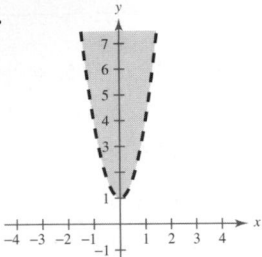

51.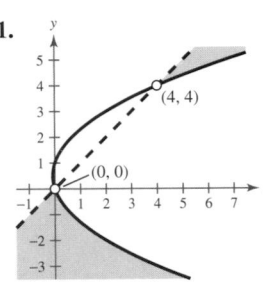

53.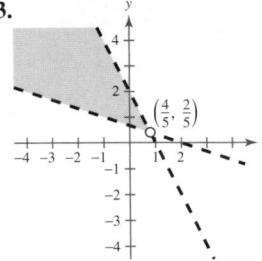

25.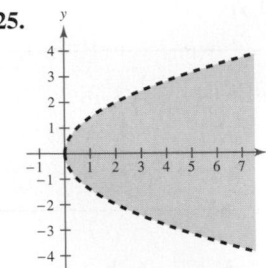

27.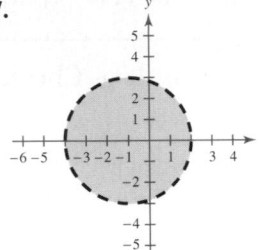

55.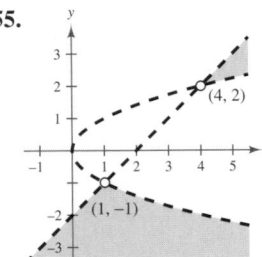

57.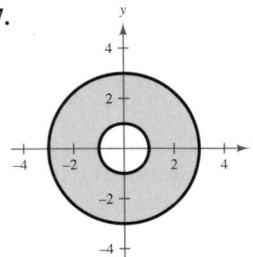

29.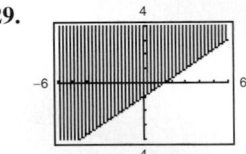

31.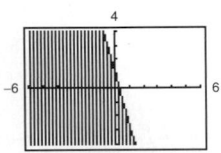

59.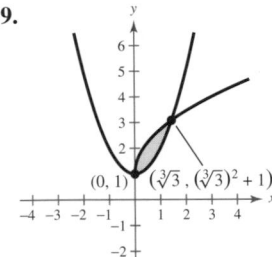

61.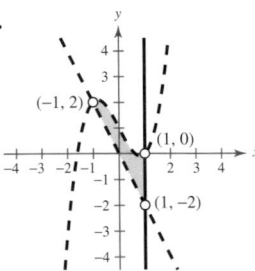

33.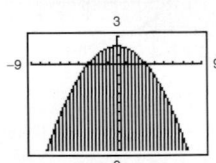

35.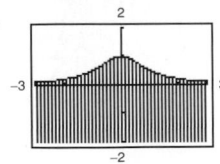

37.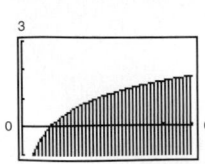

39.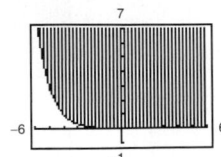

63.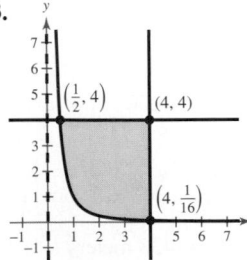

65. $\begin{cases} \frac{1}{4}x + \frac{1}{4}y < 1 \\ \quad x \geq 0 \\ \quad y \geq 0 \end{cases}$

41. $\dfrac{x}{3} + \dfrac{y}{2} > 1$ **43.** $x^2 + y^2 \leq 9$

45. (a) Yes   (b) No   (c) No   (d) No

67. $\begin{cases} y \leq 4 - x \\ y \leq 2 - \frac{1}{4}x \\ x \geq 0 \\ y \geq 0 \end{cases}$ **69.** $x^2 + (y - 2)^2 \leq 4$

71. $\begin{cases} 2 \le x \le 5 \\ 1 \le y \le 7 \end{cases}$ **73.** $\begin{cases} y \le \frac{3}{2}x \\ y \le -x + 5 \\ y \ge 0 \end{cases}$

75.

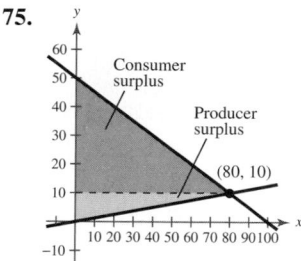

Consumer surplus: 1600

Producer surplus: 400

77.

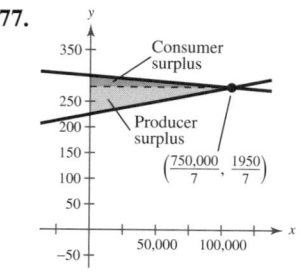

Consumer surplus: $\approx 1,147,959.18$

Producer surplus: $\approx 2,869,897.96$

79. (a) $\begin{cases} x + y \le 30,000 \\ x \ge 7500 \\ y \ge 7500 \\ x \ge 2y \end{cases}$ (b)

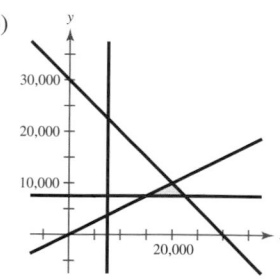

81. (a) $\begin{cases} 20x + 10y \ge 280 \\ 15x + 10y \ge 160 \\ 10x + 20y \ge 180 \\ x \ge 0 \\ y \ge 0 \end{cases}$ (b)

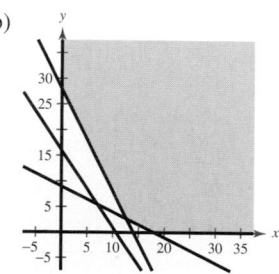

83. (a) $\begin{cases} xy \ge 500 \\ 2x + \pi y \ge 125 \\ x \ge 0 \\ y \ge 0 \end{cases}$ (b)

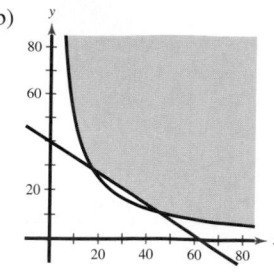

85. True   **87.** Test a point on either side.

Appendix F.2   (page A122)

Vocabulary Check   (page A122)

1. optimization   **2.** objective function

3. constraints, feasible solutions

1. Minimum at $(0, 0)$: 0   **3.** Minimum at $(0, 0)$: 0

Maximum at $(0, 6)$: 30   Maximum at $(6, 0)$: 60

5. Minimum at $(0, 0)$: 0   **7.** Minimum at $(0, 0)$: 0

Maximum at $(3, 4)$: 17   Maximum at $(4, 0)$: 20

9. Minimum at $(0, 0)$: 0

Maximum at $(60, 20)$: 740

11. Minimum at $(0, 0)$: 0

Maximum at any point on the line segment connecting $(60, 20)$ and $(30, 45)$: 2100

13.

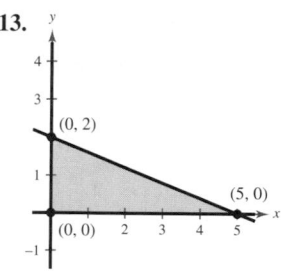

Minimum at $(0, 0)$: 0

Maximum at $(5, 0)$: 30

15.

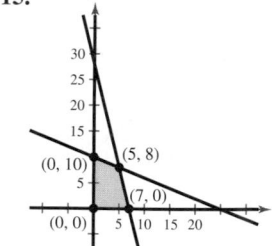

Minimum at $(0, 0)$: 0

Maximum at $(5, 8)$: 47

Minimum at $(0, 0)$: 0

Maximum at $(5, 8)$: 21

17.

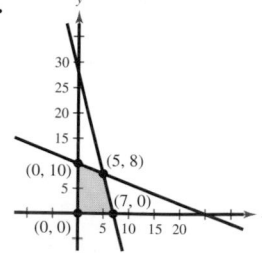

APPENDICES

19.

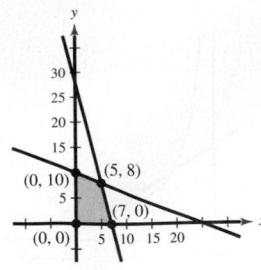

Minimum at any point on the line segment connecting $(0, 0)$ and $(0, 10)$: 0

Maximum at $(7, 0)$: 14

21.

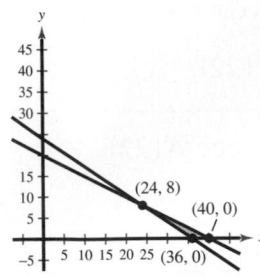

Minimum at $(24, 8)$: 104

Maximum at $(40, 0)$: 160

23.

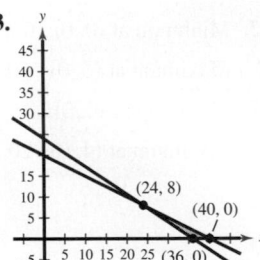

Minimum at $(36, 0)$: 36

Maximum at $(24, 8)$: 56

25.

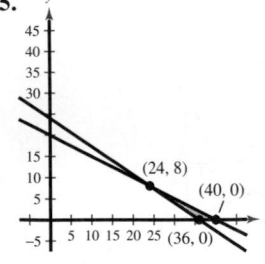

Minimum at any point on the line segment connecting $(24, 8)$ and $(36, 0)$: 72

Maximum at $(40, 0)$: 80

27. (a) and (b)   (c) $(3, 6)$

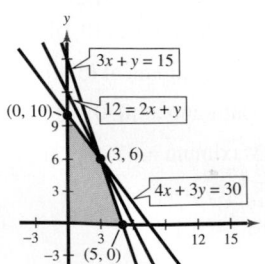

29. (a) and (b)   (c) $(0, 10)$

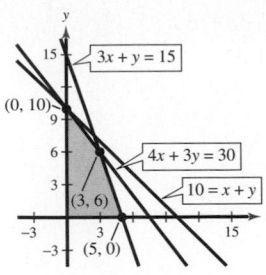

31.

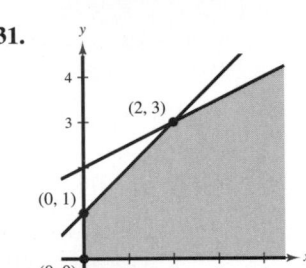

The constraints do not form a closed set of points. Therefore, $z = x + y$ is unbounded.

33.

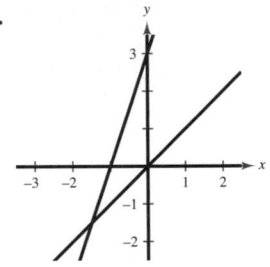

The feasible set is empty.

35. (a) Four audits, 32 tax returns

(b) Maximum revenue: $17,600

37. (a) Three bags of brand X, six bags of brand Y

(b) Minimum cost: $195

39. True   **41.** $z = x + 5y$ **43.** $z = 4x + y$

45. (a) $t > 9$ (b) $\frac{3}{4} < t < 9$

Index of Selected Applications

Go to this textbook's *Online Study Center* for a complete list of applications.

Biology and Life Sciences

Bacteria count, 57, 61
Carbon dioxide emissions, 245
Ecology, A100
Elk population, 164
Exercise, 573
Fish, 589
Forestry, 223
Health
   death rate due to heart disease, 626
Height of a person, 72
Heights and diameters of trees, A94
Logistic growth, A101
Pollution, 61
Predator-prey model, 317
Tree farm, 589
Wildlife, population of deer, 155
Yeast growth, 235

Business

Average cost, 163, 814, 818
Break-even analysis, 479, 482, 483,
   570
Earnings per share, Wal-Mart, 510
Gold prices, 177
Market research, 809
Multiplier effect, 609
Production, 549
Profit, 27, 102, 145, 173
Reimbursed expenses, 14
Rental demand, 14
Retail sales
   for lawn care, 568
   for lawn care products and services,
     84
   of prescription drugs, 29
Revenue, 28, 101, 114, 446, 539, 668
   California Pizza Kitchen, Inc., 590
   Costco Wholesale, A82
   Papa Johns Int'l., A34
   for a video rental store, 493
   Wendy's Int'l., Inc., A33
Sales, 60, 78, 83, 234, 296, 306, 342,
   347, 378, 493, 573, 599, A78
   Anheuser-Busch Companies, Inc.,
     590
   Apple Computer, Inc., A32
   AutoZone, Inc., 244
   Coca-Cola Enterprises, Inc., 599
   of college textbooks, 170
   of exercise equipment, 250

Family Dollar Stores, Inc. and
   Dollar General Corp., 493
   Goodyear Tire, 13
   of jogging and running shoes, 170
   Kohl's Corporation, 236
   Kraft Foods Inc., A29
Supply and demand, 492, 570, A111,
   A114

Chemistry and Physics

Astronomy
   eccentricities of planetary orbits,
     676
Automobile aerodynamics, 101
Beam load, A102
Boyle's Law, A101
Crystals, 748
Distinct vision, 154
Earthquake, 667
Effectiveness of a flu vaccine, 647
Electrical circuits, 275
Electrical engineering, 459
Energy, 130
Falling object, 599
Fluid flow, 40, A102
Free-falling object, 799
Free vibrations, A101
Friction, 367
Gas law, A98
Harmonic motion, 275, 306, 318, 334,
   335, 336, 341, 378, 386, 402,
   459, 467
Locating an explosion, 685
Mach number, 398
Machine design, 772
Mechanical design, 772
Meteorology, 14
   monthly normal temperatures for
     Santa Fe, New Mexico, 296
   normal daily high temperatures in
     Phoenix and Seattle, 379
   normal daily maximum temperatures
     and normal precipitation in
     Honolulu, Hawaii, 308
Newton's Law of Cooling, 236, A101
Newton's Law of Universal
   Gravitation, A101
Ocean temperatures, A102
Path of a baseball, 22
Path of a projectile, 670
Physics experiment, 154, A102

Planetary motion, 727, 731, 733
Projectile motion, 378, 397, 401, 670,
   699, 706
Radioactive decay, 192, 194, 229, 233,
   234, 247, 250, 252
Resistance, A101
Satellite orbit, 670, 679
Sound intensity, 205, 212, 235
Sound location, 688, 731
Standing waves, 385
Surface area of a tumor, 810
Television antenna dishes, 670
Temperature
   of a cup of water, 212, 244
   of water in an ice cube tray, 609
Tension, 435, 436, 463, 754, 756, 775
Torque, 763
Velocity, 431, 435
Vertical motion, 810
Work, 444, 446, 447, 463, A102

Construction

Architecture
   church window, 730
   fireplace arch, 678
   railroad underpass, 678
Bridge design, 415
Inclined plane, 287
Landau Building, 422
Machine shop calculations, 287
Railroad track, 397
Statuary Hall, 678
Surveying, 339, 422, 462
Suspension bridge, 669

Consumer

Communications, 40, 799, 800
Consumer awareness, 61, 634
DVD rentals, 482
Home mortgage, 205, 235, 236, 248
Income tax, A96
Mortgage payments, 589
Moving, A75
Pay rate, 14
Produce, 493
Repaying a loan, 14
Retail price of ground beef, A83
Salary options, 609
Selling prices of new homes, A92
Spending, A102
Taxes, A100

Geometry
Angle of elevation, 330, 338, 340, 415
Area of an inscribed rectangle, 173
Area of a parcel of land, 465, 827
Dimensions of a rectangular region, 162
Postal regulations, 27
Shadow length, 367, 416
Surface area of a sphere, A101
Volume of a balloon, 809
Volume of a right circular cylinder, A101
Volume of a sphere, 21, A101

Interest Rate
Finance, mortgage debt outstanding, 50
Investment portfolio, 508, 549, 570, 571

Miscellaneous
Ages of siblings, 61
Attaché case, 634
Batting order, 635
A boy or a girl?, 648
Coffee, 549
Course schedule, 634, 653
Cupcakes, 41
Cutting string, 602
Dominoes, 612
Elections, registered voters, 243
Ferris wheel, 306
Fibonacci sequence, 591
Flowers, 549
Gardening, 173
Gauss, 600
Geography, 749
IQ scores, 234
Koch snowflake, 618
Landscaper, 667
License plate numbers, 655
Maximum bench-press weight, A74
Moving, 340
Panoramic photo, 689
Paper, 525
Pendulum, 688
Photography, 329
PIN codes, 634, 648
Political party affiliation, 647
Radio stations, 634, 647
Riding in a car, 635
School locker, 635
Shoe sizes, 71, 77
Solar cooker, 669
Sphereflake, 609
Sports, 483, 508, 509, 653, 654, 655
    attendance at women's college basketball games, 164

baseball throw, A35
figure skating, 267
football pass, 732, A35
mean salary for professional football players, 79, 567
men's 400-meter free-style swimming event, 87
Telemarketing, 653
Tickets, 525
Tower of Hanoi, 618
Voting preference, 539
ZIP codes, 634, 635

Time and Distance
Airplane ascent, 339
Airplane speed, 490, 492, 570
Angular and linear speed of a car's wheels, 264
Angular speed
    of a DVD, 268
    spin balance machine, 268
Depth of the tide, 41, 72
Distance, 379, 415, 461
    a bungee jumper falls, 610
Flight path, 415
Height, 337
    of a football, 41
Location of a fire, 339
Navigation, 333, 339, 349, 419, 422, 432, 436, 461, 463, 467, 688, 731
Speed of fan blades, 268
Vertical motion, 504, 507

U.S. Demographics
Amount spent on movie theater admissions, 168
Annual salary for public school teachers, 524
Average salary for public school teachers, 220
Average wages, for instructional teacher aides, 590
Bachelor's degrees earned, 15
Broccoli consumption, 178
Cable television bills, 80
Cell phone calls, 79
Cellular communications employees, 22
Cellular phones, A84
College enrollment, A82
Defense budget, 179
Defense outlays, 155
Education
    associate's degrees, A79
    full-time faculty, 651
    master's degrees, 600

number of doctorate degrees, A74
preprimary enrollment, 590
Educational attainment, 646
Entertainment, 171
    amount spent on the Internet, 170
Grants, 98
High school athletics, A84
High school graduates, A81
Highway safety, 819
Hourly earnings for production workers, A83
Household credit market debt, 73, 75
International travelers to U.S., 550
Internet access, A84
Leisure time, A74
Manpower of the Department of Defense, 164
Medical costs, A62
Motor vehicles, 28
Number of FM radio stations, 170
Number of hospitals, 223
Population(s), 205
    65 and older, A77
    of California, 41
    of Colorado and Alabama, 480
    of Florida, 575
    of Italy, 251
    of Las Vegas, Nevada, 234, A74
    of Missouri and Tennessee, 483
    of New Jersey, 80, 809
    per square mile of the U.S., 195
    of Pittsburgh, Pennsylvania, 233, A74
    by race, 646
    of Reno, Nevada, 234
    of South Carolina, 72
    of South Carolina and Arizona, A62
    of the United States, 129, 243, 586
    of West Virginia, 40
    of Wyoming, 80
Population growth of California, 195
Price of gold, A93
Retail price of prescriptions, 524
SAT scores, A83
School enrollment, 818
Snowboarding participants, 550
Sports
    average salary of football players, 567
    female participants, 251
Teacher's salaries, 79
Transportation, 28, 72
Tuition, 484, A82
Tuition, room, and board charges, 494
U.S. households with color television, 550
U.S. Supreme Court, 635
Unemployment rate, 379

Index

A

Absolute value
  of a complex number, 448
  function, 19
  inequality, solution of, A66
Acute angle, 259
Addition
  of a complex number, 132
  matrix, 526
  vector, 426
    properties of, 428
    resultant of, 426
Additive identity
  for a complex number, 132
  for a matrix, 529
Additive inverse for a complex number, 132
Adjacent side of a right triangle, 277
Algebraic function, 2, 184, 293
Algebraic properties of the cross product, 758, 773, 777
Alternative definition of conic, 722
Alternative form of Law of Cosines, 417, 469
Alternative formula for standard deviation, A88
Amplitude of sine and cosine curves, 299
Analyzing graphs
  of functions, 82
  of polynomial functions, 172
  of quadratic functions, 172
Angle(s), 258
  acute, 259
  between two planes, 767
  between two vectors, 439, 471
    in space, 751, 773
  central, 259
  complementary, 260
  conversions between radians and degrees, 262, 343
  coterminal, 258
  degree, 261
  of depression, 282, 331
  directed, 707
  of elevation, 282, 331
  initial side, 258
  measure of, 259
  negative, 258
  obtuse, 259
  positive, 258
  radian, 259
  reference, 290
  of repose, 329

standard position, 258
supplementary, 260
terminal side, 258
vertex, 258
Angular speed, 263
Aphelion, 727
Apogee, 675
Arc length, 263
Arccosine function, 322
Arcsine function, 322
Arctangent function, 322
Area
  of an oblique triangle, 412, 460
  of a plane region, 824, 828
  of a region, 811
  of a triangle, 559
    formulas for, 420
Argument of a complex number, 449
Arithmetic combination, 51, 82
Arithmetic sequence, 592
  common difference of, 592
  nth term of, 593, 650
  recursion formula, 594
  sum of a finite, 595, 657
Associative Property of Addition
  for complex numbers, 133
  for matrices, 529
Associative Property of Matrix Multiplication, 533
Associative Property of Multiplication
  for complex numbers, 133
  for matrices, 529
Associative Property of Scalar Multiplication, for matrices, 529, 533
Astronomical unit, 725
Asymptote(s)
  horizontal, 147
  of a hyperbola, 682
  oblique, 159
  of a rational function, 148, 172
  slant, 159
  vertical, 147
Augmented matrix, 512
Average, A85
Axis
  imaginary, 135
  of a parabola, 92, 663
  polar, 707
  real, 135
  of symmetry, 92

B

Back-substitution, 475

Bar graph, A78
Base, natural, 188
Basic equation, 502
Basic limits, 785, 828
Bearings, 333
Bell-shaped curve, 229
Biconditional statement, 90
Bimodal, A85
Binomial, 619
  coefficient, 619
  expanding, 621
Binomial Theorem, 619, 650, 658
Bounded, sequence, 135
Box-and-whisker plot, A90
Branches of a hyperbola, 680
Break-even point, 479
Butterfly curve, 716

C

Calculus, 801
  area of a region, 811
  tangent line, 811
Cardioid, 718
Cartesian plane, A25
Center
  of a circle, 661, A30
  of an ellipse, 671
  of a hyperbola, 680
Central angle of a circle, 259
Certain event, 638
Change-of-base formula, 207, 246
Characteristics of a function from set A to set B, 16
Chebychev's Theorem, A89
Circle, 661, 718, A30
  center of, 661, A30
  central angle of, 259
  radius of, 661, A30
  of radius r, A29
  standard form of the equation of, 661, 729, A30
  unit, 269
Classification of conics by the discriminant, 694
Classifying a conic from its general equation, 686
Coded row matrices, 564
Coefficient
  binomial, 619
  correlation, 77
  matrix, 512, 534
Cofactor
  expanding by, 554
  of a matrix, 553

Cofunction identities, 352
Collinear points, 560, 753
    test for, 560
Column matrix, 511
Combination of n elements taken r at a
        time, 632
Combined variation, A98
Common difference, 592
Common logarithmic function, 197
Common ratio, 601
Commutative Property of Addition
    for complex numbers, 133
    for matrices, 529
Commutative Property of
        Multiplication, for
        complex numbers, 133
Complement
    of an event, 644
    probability of, 644
Complementary angles, 260
Completing the square, A54
Complex conjugates, 134
Complex number(s), 131
    absolute value of, 448
    addition of, 132
    additive identity, 132
    additive inverse, 132
    argument of, 449
    Associative Property of Addition,
        133
    Associative Property of
        Multiplication, 133
    Commutative Property of Addition,
        133
    Commutative Property of
        Multiplication, 133
    conjugate of, 141
    Distributive Property, 133
    equality of, 131
    imaginary part of, 131
    modulus of, 449
    nth root of, 453, 454, 460
    nth roots of unity, 455
    polar form, 449
    product of two, 450
    quotient of two, 450
    real part of, 131
    standard form of, 131
    subtraction of, 132
    trigonometric form of, 449
Complex plane, 135
Complex zeros
    occur in conjugate pairs, 141
    of polynomial functions, 172
Component form
    of a vector $\mathbf{v}$, 425
    of a vector in space, 750

Components, vector, 441, 442
Composition, 53, 82
Compound interest
    continuous compounding, 190
    formulas for, 190
Conclusion, 89
Conditional equation, 360, A47
Conditional statement, 89
Conditions under which limits do not
        exist, 784, 828
Conics (or conic sections), 660
    alternative definition, 722
    circle, 661
    classifying
        by the discriminant, 694
        from its general equation, 686
    degenerate, 660
    eccentricity of, 722
    ellipse, 671
    hyperbola, 680
    parabola, 663
    polar equations of, 722, 729, 739
Conjugate axis of a hyperbola, 682
Conjugate of a complex number, 134,
        141
Connected mode, A11
Consistent system of linear equations,
        487, 497
    dependent, 497
    independent, 497
Constant
    function, 32, 92
    matrix, 534
    of proportionality, A96
    spring, 79
    of variation, A96
Constraints, A116
Consumer surplus, A111
Continuous compounding, 190
Continuous function, 103, 699
Contradiction, proof by, 577
Contrapositive, 89
Converge, 815
Converse, 89
Conversions between degrees and
        radians, 262, 343
Convex limaçon, 718
Coordinate(s), A25
    polar, 707
    x-coordinate, A25
    y-coordinate, A25
Coordinate axes, reflection in, 45
Coordinate conversion, 708, 729
Coordinate planes, 742
    xy-plane, 742
    xz-plane, 742
    yz-plane, 742

Coordinate system
    polar, 707
    three-dimensional, 500, 742
Correlation
    coefficient, 77
    negative, 74
    positive, 74
Cosecant function, 270, 277
    of any angle, 288
    graph of, 312, 315, 343
Cosine curve, amplitude of, 299
Cosine function, 270, 277
    of any angle, 288
    common angles, 291
    domain of, 272
    graph of, 301, 315, 343
    inverse, 322
    period of, 300
    range of, 272
    special angles, 279
Cotangent function, 270, 277
    of any angle, 288
    graph of, 311, 315, 343
Coterminal angle, 258
Counterexample, 89
Counting problems, solving, 650
Cramer's Rule, 561, 562, 569
Critical numbers
    of a polynomial inequality, A67
    of a rational inequality, A70
Cross product of two vectors in space,
        757, 773
    algebraic properties of, 758, 773, 777
    determinant form, 757
    geometric properties of, 759, 778
Cryptogram, 564
Cubic function, 104
Cumulative sum feature, A2
Curve
    plane, 699
    rose, 717, 718
    sine, 297

D

Damping factor, 314
Data-defined function, 17
Decomposition of $N(x)/D(x)$ into
        partial fractions, 501
Decreasing function, 32
Definitions of trigonometric functions
        of any angle, 288
Degenerate conic, 660
Degree(s)
    of an angle, 261
    conversion to radians, 262, 343
Degree mode, A9
DeMoivre's Theorem, 452

Dependent system of linear equations, 497
Dependent variable, 18, 23
Derivative, 806, 828
Descartes's Rule of Signs, 124
Determinant
    of a matrix, 551, 554
    of a 2 × 2 matrix, 545, 551
Determinant feature, A2
Determinant form, 757
Diagonal
    matrix, 540, 555
    of a matrix, 555
    of a polygon, 636
Difference
    of functions, 51
    of limits, 785
    quotient, 23, 797, 803
    of vectors, 426
Differences
    first, 616
    second, 616
Differentiation, 326
Dimpled limaçon, 718
Direct substitution, 785
Direct variation, A96
    as an nth power, A97
Directed angle, 707
Directed distance, 707
Directed line segment, 424
    initial point, 424
    length of, 424
    magnitude, 424
    terminal point, 424
Direction angle of a vector, 430
Direction numbers, 764
Direction vector, 764
Directly proportional, A96
    to the nth power, A97
Directrix of a parabola, 663
Distance
    between a point and a plane, 770, 773
    directed, 707
Distance Formula, A27
    in space, 743, 773
Distinguishable permutations, 631
Distributive Property
    for complex numbers, 133
    for matrices, 529
Diverge, 815
Dividing out technique, 791
Division
    long, 116
    synthetic, 119
Division Algorithm, 117
Domain

of cosine function, 272
defined, 23
of a function, 16, 23, 82
implied, 20, A71
of a rational function, 146, 172
of sine function, 272
undefined, 23
Dot mode, A11
Dot product, 438
    properties of, 438, 471
    use, 460
    of vectors in space, 750, 773
Double inequality, A65
Double-angle formulas, 387, 399, 405
Doyle Log Rule, 415
Draw inverse feature, A2

E
Eccentricity
    of a conic, 722
    of an ellipse, 676, 722
    of a hyperbola, 684, 722
    of a parabola, 722
Elementary row operations
    for matrices, 513
    for systems of equations, 496
Elementary row operations features, A3
Eliminating the parameter, 702
Elimination
    Gaussian, 496, 569
        with back-substitution, 517, 569
    Gauss-Jordan, 518
    method of, 485, 486, 569
Ellipse, 671, 722
    center of, 671
    eccentricity of, 676, 722
    foci of, 671
    latus rectum of, 679
    major axis of, 671
    minor axis of, 671
    standard form of the equation of, 672, 729
    vertices of, 671
Entry of a matrix, 511
Equal matrices, 526
Equality
    of complex numbers, 131
    of vectors, 425
        in space, 750, 773
Equation(s), A36, A47
    basic, 502
    circle, standard form, 661, 729, A30
    conditional, 360, A47
    ellipse, standard form, 672, 729
    equivalent, A103
    exponential, 214
    graph of, A36

using a graphing utility, A38
hyperbola, standard form, 680, 729
identity, 360, A47
of a line
    general form, 8
    point-slope form, 5, 8, 82
    slope-intercept form, 7, 8
    summary of, 8
    symmetric, 764
    two-point form, 5
linear, A103
    in one variable, A47
logarithmic, 214
of a plane
    general form, 766
    standard form, 766, 773
parabola, standard form, 663, 729, 737
parametric, 699
position, 504, A71
solution of, A47
    graphical approximations, A50
solution point, A36
solving, A47
sphere, standard form, 744, 773
symmetric, 764
system of, 474
    in three variables, graph of, 500
Equivalent
    equations, A103
        generating, A103
    inequalities, A63, A104
        generating, A104
    systems, 486
Evaluating
    exponential functions, 246
    functions, 82
    logarithmic functions, 246
    trigonometric functions
        of any angle, 291, 343
Even function, 36
    test for, 36
    trigonometric, 273
Even and odd trigonometric functions, 273
Even/odd identities, 352
Event(s), 637
    certain, 638
    complement of, 644
        probability of, 644
    finding probabilities of, 650
    impossible, 638
    independent, 643
        probability of, 643
    mutually exclusive, 641
    probability of, 638
        the union of two, 641

Existence of an inverse function, 66
Existence of a limit, 795, 828
Existence theorems, 139
Expanding
    a binomial, 621
    by cofactors, 554
Experiment, 637
    outcome of, 637
    sample space of, 637
Exponential decay model, 225
Exponential equation, 214
    solving, 214, 246
Exponential function(s), 184, 186
    evaluating, 246
    f with base *a*, 184
    graph of, 246
    natural, 188
    one-to-one property, 214
Exponential growth model, 225
Exponentiating, 217
Extended principle of mathematical
        induction, 613
Extracting square roots, A54
Extraneous solution, A48
Extrema, 106
    maxima, 106
    minima, 106

F

Factor Theorem, 120, 180
Factorial, 582
Factoring, A54
Factors of a polynomial, 141, 181
Feasible solutions, A116
Fermat number, 611
Find linear models, 82
Find probabilities of events, 650
Find quadratic models for data, 172
Finding an inverse function, 67, 82
Finding an inverse matrix, 543, 569
Finding test intervals for a polynomial,
        A67
Finite
    sequence, 580
    series, 585
First differences, 616
Fitting a line to data, 74
Fitting nonlinear models to data, 246
Fixed point, 377
Focal chord of a parabola, 665
Focus (foci)
    of an ellipse, 671
    of a hyperbola, 680
    of a parabola, 663
Formula(s)
    for area of a triangle, 420
    change-of-base, 246

for compound interest, 190
double-angle, 387, 399, 405
half-angle, 390, 399
Heron's Area, 420, 460, 470
power-reducing, 389, 399, 405
product-to-sum, 391, 399
Quadratic, A54
reduction, 382
sum and difference, 380, 399, 404
summation, 820, 828
sum-to-product, 392, 399, 406
Fractal, 135
Fractal geometry, 135
Fraction
    partial, 501
        decomposition, 501
Frequency, 334, A76
    distribution, A77
Function(s), 16
    absolute value, 19
    algebraic, 184, 293
    arithmetic combination of, 51, 82
    characteristics of, 16
    common logarithmic, 197
    composition, 53, 82
    constant, 92
    continuous, 103, 699
    cosecant, 270, 277
    cosine, 270, 277
    cotangent, 270, 277
    cubic, 104
    damping factor, 314
    data-defined, 17
    decreasing, 32
    derivative of, 806, 828
    difference of, 51
    domain of, 16, 23, 82
    evaluating, 82
    even, 36
        test for, 36
    exponential, 184, 186
    extrema, 106
        maxima, 106
        minima, 106
    fixed point of, 377
    graph of, 30
        analyze, 82
    greatest integer, 35
    of half-angles, 387
    implied domain of, 20, A71
    increasing, 32
    input, 16
    inverse, 62, 63
        existence of, 66
    inverse cosine, 322
    inverse sine, 320, 322
    inverse tangent, 322

linear, 6, 92
logarithmic, 196, 199
of multiple angles, 387
name of, 18, 23
natural exponential, 188
natural logarithmic, 200
notation, 18, 23
objective, A116
odd, 36
    test for, 36
one-to-one, 66
output, 16
period of, 272
periodic, 272
piecewise-defined, 19
polynomial, 92, 104
product of, 51
quadratic, 92, 93
    graph of, 94
    maximum value of, 97
    minimum value of, 97
quotient of, 51
radical, 20
range of, 16, 23
rational, 146, 147
reciprocal, 147
relative maximum, 33
relative minimum, 33
secant, 270, 277
sine, 270, 277
square root, 20
squaring, 93
step, 35
sum of, 51
summary of terminology, 23
tangent, 270, 277
transcendental, 184, 293
trigonometric, 270, 277
value of, 18, 23
Vertical Line Test, 31
zero of, 106
Function mode, A9
Fundamental Counting Principle, 628
Fundamental Theorem of Algebra, 139
Fundamental trigonometric identities,
        280, 352

G

Gaussian elimination, 496, 569
    with back-substitution, 517, 569
Gaussian model, 225
Gauss-Jordan elimination, 518
General form
    of the equation of a line, 8
    of the equation of a plane, 766
Generating equivalent
    equations, A103

inequalities, A104
Geometric properties of the cross product, 759, 778
Geometric property of the triple scalar product, 761
Geometric sequence, 601
  common ratio of, 601
  nth term of, 602, 650
  sum of a finite, 604, 650, 657
Geometric series, 605
  sum of an infinite, 605, 650
Gnomon, 367
Graph(s), A36
  bar, A78
  of cosecant function, 312, 315, 343
  of cosine function, 301, 315, 343
  of cotangent function, 311, 315, 343
  of an equation, A36
    in three variables, 500
    using a graphing utility, A38
  of a function, 30
    analyzing, 82
  of an inequality, A63, A106
    in two variables, A106
  intercepts of, A37
  of inverse cosine function, 322
  of inverse sine function, 322
  of inverse tangent function, 322
  line, A79
  point of intersection, A52
  point-plotting method, A36
  of a rational function, 156
  of secant function, 312, 315, 343
  slope of, 803
  special polar, 718
  of sine function, 301, 315, 343
  of tangent function, 309, 315, 343
Graphical approximations of solutions of an equation, A50
Graphical interpretations of solutions, 487
Graphing
  exponential functions, 246
  logarithmic functions, 246
  method of, 474, 569
  parametric equations, 729
Graphing utility
  equation editor, A1
  features
    cumulative sum, A2
    determinant, A2
    draw inverse, A2
    elementary row operations, A3
    intersect, A5
    list, A2
    matrix, A6

maximum, A7
mean, A8
median, A8
minimum, A7
  $_nC_r$, A12
  $_nP_r$, A12
one-variable statistics, A12
reduced row-echelon, A15
regression, A13
row addition, A4
row multiplication and addition, A4
row swap, A3
row-echelon, A14
sequence, A15
shade, A15
statistical plotting, A16
store, A3
sum, A17
sum sequence, A17
table, A17
tangent, A18
trace, A19, A23
value, A19
zero or root, A22
zoom, A23
inverse matrix, A7
list editor, A5
matrix editor, A6
matrix operations, A6
mode settings, A9
  ask, A18
  connected, A11
  degree, A9
  dot, A11
  function, A9
  parametric, A9
  polar, A10
  radian, A9
  sequence, A10
  uses of, A1
  viewing window, A20
Greatest integer function, 35
Guidelines
  for graphing rational functions, 156
  for verifying trigonometric identities, 360

H

Half-angle formulas, 390, 399
Harmonic motion, simple, 334, 335, 343
Heron's Area Formula, 420, 460, 470
Histogram, A77
Horizontal asymptote, 147
Horizontal components of $\mathbf{v}$, 429
Horizontal line, 8

Horizontal Line Test, 66
Horizontal shift, 43, 82
Horizontal shrink, 47
  of a trigonometric function, 300
Horizontal stretch, 47
  of a trigonometric function, 300
Horizontal translation of a trigonometric function, 301
Human memory model, 202
Hyperbola, 680, 722
  asymptotes of, 682
  branches of, 680
  center of, 680
  conjugate axis of, 682
  eccentricity of, 684, 722
  foci of, 680
  standard form of the equation of, 680, 729
  transverse axis of, 680
  vertices of, 680
Hypotenuse of a right triangle, 277
Hypothesis, 89

I

Identity (identities), 360, A47
  guidelines for verifying trigonometric, 360
  matrix of order n, 533
If-then form, 89
Imaginary
  axis, 135
  number, pure, 131
  part of a complex number, 131
  unit i, 131
Implied domain, 20, A71
Impossible event, 638
Improper rational expression, 117
Inconsistent system of linear equations, 487, 497
Increasing annuity, 606
Increasing function, 32
Independent events, 643
  probability of, 643
Independent system of linear equations, 497
Independent variable, 18, 23
Indeterminate form, 792
Index of summation, 584
Indirect proof, 577
Inductive, 554
Inequality (inequalities)
  absolute value, solution of, A66
  double, A65
  equivalent, A63, A104
  graph of, A63, A106
  linear, 21, A64, A104
  properties of, A63, A64

satisfying, A63
solution of, A63, A106
solution set, A63
Infinite geometric series, 605
sum of, 605, 650
Infinite sequence, 580
Infinite series, 585, 650
Infinity, limit at, 811, 812, 828
rational functions, 814, 828
Initial point, 424
Initial side of an angle, 258
Input, 16
Integration, 326
Intercept(s), A37, A48
x-intercept, A48
y-intercept, A48
Intermediate Value Theorem, 111
Intersect feature, A5
Invariant under rotation, 694
Inverse, 89
Inverse function, 62, 63
cosine, 322
existence of, 66
finding, 67, 82
Horizontal Line Test, 66
sine, 320, 322
tangent, 322
Inverse of a matrix, 541
finding an, 543, 569
Inverse matrix with a graphing utility,
A7
Inverse properties
of logarithms, 197, 214
of natural logarithms, 200
of trigonometric functions, 324, 343
Inverse trigonometric functions, 322
Inverse variation, A98
Inversely proportional, A98
Invertible matrix, 542
Irreducible
over the rationals, 142
over the reals, 142

J
Joint variation, A99
Jointly proportional, A99

K
Kepler's Laws, 725
Key points
of the graph of a trigonometric
function, 298
intercepts, 298
maximum points, 298
minimum points, 298
Koch snowflake, 618

L
Lagrange multiplier, 510
Latus rectum
of an ellipse, 679
of a parabola, 665
Law of Cosines, 417, 460, 469
alternative form, 417, 469
standard form, 417, 469
Law of Sines, 408, 460, 468
Leading 1, 515
Leading Coefficient Test, 105
Least squares regression, A94
line, 75
Left-handed orientation, 742
Lemniscate, 718
Length of a directed line segment, 424
Limaçon, 715, 718
convex, 718
dimpled, 718
with inner loop, 718
Limit(s), 781
basic, 785, 828
converge, 815
difference of, 785
diverge, 815
evaluating
direct substitution, 785
dividing out technique, 791
rationalizing technique, 793
existence of, 795, 828
indeterminate form, 792
at infinity, 811, 812, 828
rational functions, 814, 828
nonexistence of, 784
one-sided, 795
of a polynomial function, 835
of polynomial and rational
functions, 787, 828
power of, 785
of a power function, 835
product of, 785
properties of, 785, 828
quotient of, 785
scalar multiple of, 785
of a sequence, 815, 828
sum of, 785
Line(s)
horizontal, 8
least squares regression, 75
parallel, 9
perpendicular, 9
slope of, 3, 82
secant, 803
in space, parametric equations of,
764, 773
tangent, 665, 801, 811
vertical, 8

Line graph, A79
Line plot, A76
Linear combination of vectors, 429
Linear equation, A103
general form, 8
in one variable, A47
point-slope form, 5, 8, 82
slope-intercept form, 7, 8
summary of, 8
two-point form, 5
Linear extrapolation, 6
Linear Factorization Theorem, 139,
181
Linear function, 6, 92
Linear inequality, 21, A64, A106
Linear interpolation, 6
Linear programming, A116
Linear speed, 263
List editor, A5
Locus, 660
Logarithm(s)
change-of-base formula, 207, 246
natural, properties of, 200, 208, 255
inverse, 200
one-to-one, 200
power, 208, 255
product, 208, 255
quotient, 208, 255
properties of, 197, 208, 246, 255
inverse, 197, 214
one-to-one, 197, 214
power, 208, 255
product, 208, 255
quotient, 208, 255
Logarithmic equation, 214
solving, 214, 246
exponentiating, 217
Logarithmic function, 196, 199
with base a, 196
common, 197
evaluating, 246
graph of, 246
natural, 200
Logarithmic model, 225
Logistic
curve, 230
growth model, 225
Long division of polynomials, 116
Lower bound, 125
Lower limit of summation, 584
Lower quartile, A90
Lower triangular matrix, 555

M
Magnitude
of a directed line segment, 424
of a vector, 425

in space, 750, 773
Main diagonal of a square matrix, 511
Major axis of an ellipse, 671
Mandelbrot Set, 135
Mathematical induction, 611
    extended principle of, 613
    Principle of, 612, 650
Mathematical model, 73
Matrix (matrices), 511
    addition, 526
        properties of, 529
    additive identity, 529
    area of a triangle, 559
    augmented, 512
    coded row, 564
    coefficient, 512, 534
    cofactor of, 553
    collinear points, 560
    column, 511
    constant, 534
    Cramer's Rule, 561, 562, 569
    cryptogram, 564
    determinant of, 545, 551, 554
    diagonal, 540, 555
    elementary row operations, 513
    entry of a, 511
    equal, 526
    identity, 533
    inverse of, 541
    invertible, 542
    leading 1, 515
    minor of, 553
    multiplication, 531
        properties of, 533
    nonsingular, 542
    operations, 569
    order of a, 511
    in reduced row-echelon form, 515
    representation of, 526
    row, 511
    in row-echelon form, 514, 515
    row-equivalent, 513
    scalar identity, 529
    scalar multiplication, 527
    singular, 542
    square, 511
    stochastic, 539
    triangular, 555
        lower, 555
        upper, 555
    uncoded row, 564
    zero, 529
Matrix editor, A6
Matrix feature, A6
Matrix operations with a graphing
    utility, A6
Maxima, 106

Maximum feature, A7
Maximum value of a quadratic
    function, 97
Mean, A85
Mean feature, A8
Measure of an angle, 259
Measure of central tendency, A85
    average, A85
    mean, A85
    median, A85
    mode, A85
Measure of dispersion, A86
    standard deviation, A87
    variance, A87
Median, A85
Median feature, A8
Method
    of elimination, 485, 486, 569
    of graphing, 474, 569
    of least squares, A94
    of substitution, 474, 569
Midpoint Formula, A28, A29
    in space, 743, 773
Midpoint of a line segment, A28
Minima, 106
Minimum feature, A7
Minimum value of a quadratic
    function, 97
Minor axis of an ellipse, 671
Minor of a matrix, 553
Minors and cofactors of a square
    matrix, 553
Mode, A85
Mode settings, A9
    connected, A11
    degree, A9
    dot, A11
    function, A9
    parametric, A9
    polar, A10
    radian, A9
    sequence, A10
Modulus of a complex number, 449
Multiplication
    matrix, 531
    scalar, 527
Multiplicative inverse for a matrix, 541
Multiplicity, 108
Multiplier effect, 609
Mutually exclusive events, 641

N
n factorial, 582
Name of a function, 18, 23
Natural base, 188
Natural exponential function, 188
Natural logarithm, properties of, 200,

208, 255
Natural logarithmic function, 200
${}_nC_r$ feature, A12
Negation, 89
Negative
    angle, 258
    correlation, 74
    of a vector, 426
Nonexistence of a limit, 784
Nonrigid transformation, 47, 82
Nonsingular matrix, 542
Nonsquare system of linear equations,
    499
Normal vector, 766
Normally distributed, 229
${}_nP_r$ feature, A12
nth partial sum of an arithmetic
    sequence, 596, 650
nth power, direct variation as, A97
nth root(s)
    of a complex number, 453, 454, 460
    of unity, 455
nth term
    of an arithmetic sequence, 593, 650
    of a geometric sequence, 602, 650
Number(s)
    complex, 131
    critical, A67, A70
    direction, 764
Number of combinations of n elements
    taken r at a time, 632
Number of permutations of n elements,
    629
    taken r at a time, 629, 630
Number of solutions of a linear system,
    497

O
Objective function, A116
Oblique asymptote, 159
Oblique triangle, 408
    area of, 412, 460
Obtuse angle, 259
Octant, 742
Odd function, 36
    test for, 36
    trigonometric, 273
One cycle of a sine curve, 297
One-sided limit, 795
One-to-one function, 66
One-to-one property
    of exponential functions, 214
    of logarithms, 197, 214
    of natural logarithms, 200
One-variable statistics feature, A12
Opposite side of a right triangle, 277
Optimal solution of a linear
    programming problem, A116
Optimization, A116

Order of a matrix, 511
Ordered pair, A25
Ordered triple, 495
Orientation of a curve, 700
Origin
    of a polar coordinate system, 707
    of a rectangular coordinate system, A25
    symmetry, 36
Orthogonal vectors, 440
    in space, 751
Outcome, 637
Output, 16

P

Parabola, 92, 663, 722
    axis of, 92, 663
    directrix of, 663
    eccentricity of, 722
    focal chord of, 665
    focus of, 663
    latus rectum of, 665
    reflective property of, 666
    standard form of the equation of, 663, 729, 737
    tangent line, 665
    vertex of, 92, 663
Parallel lines, 9
Parallel vectors in space, 752
Parallelogram law, 426
Parameter, 699
    eliminating the, 702
Parametric equation(s), 699
    graphing, 729
    of a line in space, 764, 773
Parametric mode, A9
Partial fraction, 501
    decomposition, 501
Partial sum of a series, 585, 650
Pascal's Triangle, 623
Performing matrix operations, 569
Perigee, 675
Perihelion, 727
Period
    of a function, 272
    of sine and cosine functions, 300
Periodic function, 272
Permutation, 629
    distinguishable, 631
    of n elements, 629
        taken r at a time, 629, 630
Perpendicular lines, 9, 10
Phase shift, 301
Piecewise-defined function, 19
Plane(s)
    angle between two, 767

general form of the equation of, 766
standard form of the equation of, 766, 773
trace of, 769
Plane curve, 699
    orientation of, 700
Point
    of diminishing returns, 114
    of equilibrium, 492, A111
    of intersection, 474, A52
Point-plotting method, A36
Point-slope form, 5, 8, 82
Polar axis, 707
Polar coordinate system, 707
Polar coordinates, 707
    conversion, 708, 729
    directed angle, 707
    directed distance, 707
    quick test for symmetry in, 715
    test for symmetry in, 714, 729
Polar equations of conics, 722, 729, 739
Polar form of a complex number, 449
Polar mode, A10
Pole, 707
Polynomial(s)
    factors of, 141, 181
    finding test intervals for, A67
    function, 92, 104
        analyze graphs of, 172
        complex zeros of, 172
        limit of, 787, 835
        rational zero of, 172
        real zeros of, 107, 172
        of x with degree n, 92
    inequality
        critical numbers, A67
        test intervals, A67
    long division of, 116
    prime factor, 142
    synthetic division, 119
Position equation, 504, A71
Positive
    angle, 258
    correlation, 74
Power function, limit of, 835
Power of a limit, 785
Power property
    of logarithms, 208, 255
    of natural logarithms, 208, 255
Power-reducing formulas, 389, 399, 405
Prime factor of a polynomial, 142
Principle of Mathematical Induction, 612, 650
Probability
    of a complement, 644

of an event, 638
of independent events, 643
of the union of two events, 641
Producer surplus, A111
Product
    of functions, 51
    of limits, 785
    of trigonometric functions, 387
    of two complex numbers, 450
Product property
    of logarithms, 208, 255
    of natural logarithms, 208, 255
Product-to-sum formulas, 391, 399
Projection of a vector, 442
Proof
    by contradiction, 577
    indirect, 577
    without words, 578
Proper rational expression, 117
Properties
    of the cross product
        algebraic, 758, 773, 777
        geometric, 759, 778
    of the dot product, 438, 471
    of inequalities, A63, A64
    of inverse trigonometric functions, 324, 343
    of limits, 785, 828
    of logarithms, 197, 208, 246, 255
        inverse, 197, 214
        one-to-one, 197, 214
        power, 208, 255
        product, 208, 255
        quotient, 208, 255
    of matrix addition and scalar multiplication, 529
    of matrix multiplication, 533
    of natural logarithms, 200, 208, 255
        inverse, 200
        one-to-one, 200
        power, 208, 255
        product, 208, 255
        quotient, 208, 255
    reflective property of a parabola, 666
    summation, 820, 828
    of sums, 585, 656
    of vector addition and scalar multiplication, 428
Pure imaginary number, 131
Pythagorean identities, 280, 352
Pythagorean Theorem, 350

Q

Quadrant, A25
Quadratic equation, solving
    by completing the square, A54

by extracting square roots, A54
by factoring, A54
using the Quadratic Formula, A54
Quadratic Formula, A54
Quadratic function, 92, 93
    analyzing graphs of, 172
    maximum value of, 97
    minimum value of, 97
    standard form of, 95
Quadratic type, A55
Quartile, A90
    lower, A90
    upper, A90
Quick test for symmetry in polar
        coordinates, 715
Quotient
    difference, 23, 797, 803
    of functions, 51
    of limits, 785
    of two complex numbers, 450
Quotient identities, 280, 352
Quotient property
    of logarithms, 208, 255
    of natural logarithms, 208, 255

R
Radian, 259
    conversion to degrees, 262, 343
Radian mode, A9
Radical function, 20
Radius of a circle, 661, A30
Random selection
    with replacement, 627
    without replacement, 627
Range, A76
    of a function, 16, 23
Rational expression(s)
    improper, 117
    proper, 117
    undefined values of, A70
    zero of, A70
Rational function, 146, 147
    asymptotes of, 148, 172
    domain of, 146, 172
    guidelines for graphing, 156
    limit of, 787
        at infinity, 814, 828
    sketching graphs of, 172
Rational Zero Test, 122
Rational zeros of polynomial functions,
        172
Rationalizing the denominator, 363
Rationalizing technique, 793
Real axis, 135
Real part of a complex number, 131
Real zeros of polynomial functions,
        107, 172

Reciprocal function, 147
Reciprocal identities, 280, 352
Rectangular coordinate system, A25
    origin of, A25
Recursion formula, 594
Recursively defined sequence, 582
Reduced row-echelon feature, A15
Reduced row-echelon form, 515
Reducible over the reals, 142
Reduction formulas, 382
Reference angle, 290
Reflection, 45, 82
        of a trigonometric function, 300
Reflective property of a parabola, 666
Regression feature, A13
Relation, 16
Relative maximum, 33
Relative minimum, 33
Remainder Theorem, 120, 180
Repeated zero, 108
Representation of matrices, 526
Resultant of vector addition, 426
Right triangle
    adjacent side, 277
    definitions of trigonometric
            functions, 277, 343
    hypotenuse of, 277
    opposite side of, 277
    right side of, 277
    solving, 281
Right-handed orientation, 742
Rigid transformation, 47
Root of a complex number, 453, 454,
        460,
Rose curve, 717, 718
Rotation of axes to eliminate an
        xy-term, 690, 738
Rotation invariants, 694
Row addition and row multiplication
        and addition features, A4
Row-echelon feature, A14
Row-echelon form, 495, 514, 515
    reduced, 515
Row-equivalent, 513
Row matrix, 511
Row multiplication feature, A4
Row swap feature, A3

S
Sample space, 637
Satisfying the inequality, A63
Scalar, 527
    identity, 529
Scalar multiple, 527
    of a limit, 785
    of a vector, 426
        in space, 750, 773

Scalar multiplication, 527
    properties of, 529
    of a vector, 426
        properties of, 428
Scatter plot, 73, A26
Scribner Log Rule, 415
Secant function, 270, 277
    of any angle, 288
    graph of, 312, 315, 343
Secant line, 803
Second differences, 616
Sequence, 580
    arithmetic, 592
    bounded, 135
    finite, 580
    first differences of, 616
    geometric, 601
    infinite, 580
    limit of, 815, 828
    recursively defined, 582
    second differences of, 616
    terms of, 580
    unbounded, 135
Sequence feature, A15
Sequence mode, A10
Series, 585
    finite, 585, 604
    geometric, 605
    infinite, 585, 650
        geometric, 605
Shade feature, A15
Sigma notation, 584
Sigmoidal curve, 230
Simple harmonic motion, 334, 335, 343
Sine curve, 297
    amplitude of, 299
    one cycle of, 297
Sine function, 270, 277
    of any angle, 288
    common angles, 291
    curve, 297
    domain of, 272
    graph of, 301, 315, 343
    inverse, 320, 322
    period of, 300
    range of, 272
    special angles, 279
Sines, cosines, and tangents of special
        angles, 279
Singular matrix, 542
Sketching the graph of an inequality in
        two variables, A106
Sketching graphs of rational functions,
        172
Slant asymptote, 159
Slope, 764
    of a graph, 803
    of a line, 3, 82

Slope-intercept form, 7, 8
Solid analytic geometry, 742
Solution
    of an absolute value inequality, A66
    of an equation, A47
        graphical approximations, A50
    extraneous, A48
    of an inequality, A63, A106
    set of an inequality, A63
    of a system of equations, 474
        graphical interpretations, 487
    of a system of inequalities, A108
Solution point, A36
Solving
    an absolute value inequality, A66
    counting problems, 650
    an equation, A47
    exponential and logarithmic
        equations, 214, 246
    an inequality, A63
    a linear programming problem,
        A116
    a quadratic equation, A54
        completing the square, A54
        extracting square roots, A54
        factoring, A54
        Quadratic Formula, A54
    right triangles, 281
    a system of equations, 474
        Gaussian elimination, 496, 569
            with back-substitution, 517,
            569
        Gauss-Jordan elimination, 518
        method of elimination, 485, 486,
            569
        method of graphing, 474, 569
        method of substitution, 474, 569
    trigonometric equations, 399
Space
    collinear points, 753
    Distance Formula, 743, 773
    Midpoint Formula, 743, 773
    surface in, 745
    vector in, 750
Special polar graphs, 718
Speed
    angular, 263
    linear, 263
Sphere, 744
    standard equation of, 744, 773
Spring constant, 79
Square matrix, 511
    determinant of, 554, 569
    main diagonal of, 511
    minors and cofactors of, 553
Square root(s)
    extracting, A54

function, 20
Square system of linear equations, 499
Square of trigonometric functions, 387
Squaring function, 93
Standard deviation, A87
    alternative formula for, A88
Standard form
    of a complex number, 131
    of the equation of a circle, 661,
        729, A30
    of the equation of an ellipse, 672,
        729
    of the equation of a hyperbola, 680,
        729
    of the equation of a parabola, 663,
        729, 737
    of the equation of a plane, 766, 773
    of the equation of a sphere, 744, 773
    of Law of Cosines, 417, 469
    of a quadratic function, 95
Standard position
    of an angle, 258
    of a vector, 425
Standard unit vector, 429
    notation in space, 750
Statistical plotting feature, A16
Step function, 35
Stochastic matrix, 539
Strategies for solving exponential and
        logarithmic equations, 214
Substitution, method of, 474, 569
Subtraction of a complex number, 132
Sum(s)
    of a finite arithmetic sequence, 595,
        657
    of a finite geometric sequence, 604,
        650, 657
    of functions, 51
    of an infinite geometric series, 605,
        650
    of limits, 785
    nth partial, of an arithmetic
            sequence, 596, 650
    partial, 585, 650
    of powers of integers, 615
    properties of, 585, 656
    of the squared differences, 75, 238,
        A94
    of vectors, 426
        in space, 750, 773
Sum and difference formulas, 380, 399,
        404
Sum feature, A17
Sum sequence feature, A17
Summary
    of equations of lines, 8
    of function terminology, 23

Summation
    formulas and properties, 820, 828
    index of, 584
    lower limit of, 584
    notation, 584
    upper limit of, 584
Sum-to-product formulas, 392, 399,
        406
Supplementary angles, 260
Surface in space, 745
    trace of, 746
Surplus
    consumer, A111
    producer, A111
Symmetric equations of a line, 764
Symmetry
    quick test for, in polar coordinates,
        715
    with respect to the origin, 36
    with respect to the x-axis, 36
    with respect to the y-axis, 36
    test for, in polar coordinates, 714,
        729
Synthetic division, 119
    using the remainder, 121
System of equations, 474
    equivalent, 486
    solution of, 474
    solving, 474
        Gaussian elimination, 496, 569
            with back-substitution, 517,
            569
        Gauss-Jordan elimination, 518
        method of elimination, 485, 486,
            569
        method of graphing, 474, 569
        method of substitution, 474, 569
    with a unique solution, 546
System of inequalities, solution of,
        A108
System of linear equations
    consistent, 487, 497
    dependent, 497
    elementary row operations, 496
    inconsistent, 487, 497
    independent, 497
    nonsquare, 499
    number of solutions, 497
    row-echelon form, 495
    square, 499

T

Table feature, A17
Tangent feature, A18
Tangent function, 270, 277
    of any angle, 288
    common angles, 291

graph of, 309, 315, 343
inverse, 322
special angles, 279
Tangent line, 801, 811
to a parabola, 665
Term of a sequence, 580
Terminal point, 424
Terminal side of an angle, 258
Test
for collinear points, 560
for even and odd functions, 36
for symmetry in polar coordinates, 714, 729
Test intervals, A67
Three-dimensional coordinate system, 500, 742
left-handed orientation, 742
right-handed orientation, 742
Tower of Hanoi, 618
Trace
of a plane, 769
of surface, 746
Trace feature, A19, A23
Transcendental function, 2, 184, 293
Transformation
nonrigid, 47, 82
rigid, 47
Transverse axis of a hyperbola, 680
Triangle
area of, 559
formulas for area of, 420
oblique, 408
area of, 412, 460
Triangular matrix, 555
lower, 555
upper, 555
Trigonometric equations, solving, 399
Trigonometric form of a complex number, 449
argument of, 449
modulus of, 449
Trigonometric function
of any angle, 288
evaluating, 291, 343
cosecant, 270, 277
cosine, 270, 277
cotangent, 270, 277
even and odd, 273
horizontal shrink of, 300
horizontal stretch of, 300
horizontal translation of, 301
inverse properties of, 324, 343
key points, 298
intercepts, 298
maximum points, 298
minimum points, 298
product of, 387

reflection of, 300
right triangle definitions of, 277, 343
secant, 270, 277
sine, 270, 277
square of, 387
tangent, 270, 277
unit circle definitions of, 270, 343
vertical translation of, 302
Trigonometric identities, 352
cofunction identities, 352
even/odd identities, 352
fundamental, 352
guidelines for verifying, 360
Pythagorean identities, 280, 352
quotient identities, 280, 352
reciprocal identities, 280, 352
verifying, 399
Trigonometric values of common angles, 291
Trigonometry, 258
Triple scalar product, 761
geometric property of, 761
Two-point form, 5

U
Unbounded sequence, 135
Uncoded row matrices, 564
Undefined values of a rational expression, A70
Unit circle, 269
definitions of trigonometric functions, 270, 343
Unit vector, 425, 428
in the direction of v, 428
in space
in the direction of v, 750, 773
standard, 429
Upper bound, 125
Upper limit of summation, 584
Upper and Lower Bound Rules, 125
Upper quartile, A90
Upper triangular matrix, 555
Uses of a graphing utility, A1
Using the dot product, 460
Using a graphing utility to graph an equation, A38
Using nonalgebraic models to solve real-life problems, 246
Using the remainder in synthetic division, 121
Using scatter plots, 82
Using vectors in the plane, 460

V
Value feature, A19
Value of a function, 18, 23
Variable

dependent, 18, 23
independent, 18, 23
Variance, A87
Variation
combined, A98
constant of, A96
direct, A96
inverse, A98
joint, A99
in sign, 125
Varies directly, A96
as the nth power, A97
Varies inversely, A98
Varies jointly, A99
Vector(s), 764
addition, 426
properties of, 428
resultant of, 426
angle between two, 439, 471
component form of, 425
components, 441, 442
difference of, 426
directed line segment of, 424
direction, 764
direction angle of, 430
dot product of, 438
properties of, 438, 471
use, 460
equality of, 425
horizontal component of, 429
linear combination of, 429
magnitude of, 425
negative of, 426
normal, 766
orthogonal, 440
parallelogram law, 426
in the plane, using, 460
projection, 442
resultant, 426
scalar multiple of, 426
scalar multiplication of, 426
properties of, 428
in space, 750
angle between two, 751, 773
component form, 750
cross product of, 757, 773
dot product of, 750, 773
equality of, 750, 773
magnitude of, 750, 773
orthogonal, 751
parallel, 752
scalar multiple of, 750, 773
standard unit, 750
sum of, 750, 773
unit, in the direction of v, 750, 773
zero, 750

standard position of, 425
sum of, 426
unit, 425, 428
    in the direction of **v**, 428
    standard, 429
v in the plane, 424
vertical component of, 429
zero, 425
Verifying trigonometric identities, 399
Vertex (vertices)
    of an angle, 258
    of an ellipse, 671
    of a hyperbola, 680
    of a parabola, 92, 663
Vertical asymptote, 147
Vertical components of **v**, 429
Vertical line, 8
Vertical Line Test, 31
Vertical shift, 43, 82

Vertical shrink, 47
Vertical stretch, 47
Vertical translation of a trigonometric
    function, 302
Viewing window, A20

W

With replacement, 627
Without replacement, 627
Work, 444

X

x-axis, A25
    symmetry, 36
x-coordinate, A25
x-intercept, A48
xy-plane, 742
xz-plane, 742

Y

y-axis, A25
    symmetry, 36
y-coordinate, A25
y-intercept, A48
yz-plane, 742

Z

Zero(s)
    of a function, 106
    matrix, 529
    of a polynomial function, 107
        real, 107
    of a rational expression, A70
    repeated, 108
    vector, 425
        in space, 750
Zero or root feature, A22
Zoom feature, A23

COMMON FORMULAS

Temperature

$$F = \frac{9}{5}C + 32$$

F = degrees Fahrenheit

C = degrees Celsius

Distance

$$d = rt$$

d = distance traveled

t = time

r = rate

Simple Interest

$$I = Prt$$

I = interest

P = principal

r = annual interest rate

t = time in years

Compound Interest

$$A = P\left(1 + \frac{r}{n}\right)^{nt}$$

A = balance

P = principal

r = annual interest rate

n = compoundings per year

t = time in years

Coordinate Plane: Midpoint Formula

$$\left(\frac{x_1 + x_2}{2}, \frac{y_1 + y_2}{2}\right)$$

midpoint of line segment joining (x_1, y_1) and (x_2, y_2)

Coordinate Plane: Distance Formula

$$d = \sqrt{(x_2 - x_1)^2 + (y_2 - y_1)^2}$$

d = distance between points (x_1, y_1) and (x_2, y_2)

Quadratic Formula

If $p(x) = ax^2 + bx + c$, $a \neq 0$ and $b^2 - 4ac \geq 0$, then the real zeros of p are

$$x = \frac{-b \pm \sqrt{b^2 - 4ac}}{2a}.$$

CONVERSIONS

Length and Area

1 foot = 12 inches	1 yard = 3 feet	1 meter = 1000 millimeters
1 mile = 5280 feet	1 mile = 1760 yards	1 centimeter ≈ 0.394 inch
1 kilometer = 1000 meters	1 meter = 100 centimeters	1 inch ≈ 2.540 centimeters
1 kilometer ≈ 0.621 mile	1 mile ≈ 1.609 kilometers	1 acre = 4840 square yards
1 meter ≈ 3.281 feet	1 meter ≈ 39.370 inches	1 square mile = 640 acres
1 foot ≈ 0.305 meter	1 foot ≈ 30.480 centimeters	

Volume

1 gallon = 4 quarts	1 quart = 2 pints	1 pint = 16 fluid ounces
1 gallon = 231 cubic inches	1 gallon ≈ 0.134 cubic foot	1 cubic foot ≈ 7.48 gallons
1 liter = 1000 milliliters	1 liter = 100 centiliters	
1 liter ≈ 1.057 quarts	1 liter ≈ 0.264 gallon	
1 gallon ≈ 3.785 liters	1 quart ≈ 0.946 liter	

Weight and Mass on Earth

1 ton = 2000 pounds	1 pound = 16 ounces	1 kilogram = 1000 grams
1 kilogram ≈ 2.205 pounds	1 pound ≈ 0.454 kilogram	1 gram ≈ 0.035 ounce

FORMULAS FROM GEOMETRY

Triangle

$h = a \sin \theta$

$\text{Area} = \dfrac{1}{2}bh$

Laws of Cosines:

$c^2 = a^2 + b^2 - 2ab \cos \theta$

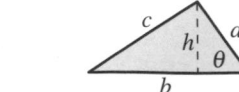

Right Triangle

Pythagorean Theorem:

$c^2 = a^2 + b^2$

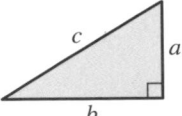

Equilateral Triangle

$h = \dfrac{\sqrt{3}s}{2}$

$\text{Area} = \dfrac{\sqrt{3}s^2}{4}$

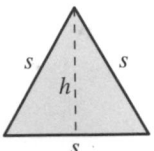

Parallelogram

$\text{Area} = bh$

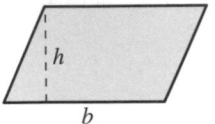

Trapezoid

$\text{Area} = \dfrac{h}{2}(a + b)$

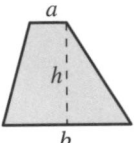 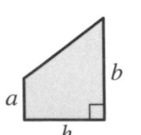

Circle

$\text{Area} = \pi r^2$

$\text{Circumference} = 2\pi r$

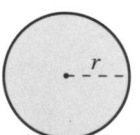

Sector of Circle

$\text{Area} = \dfrac{\theta r^2}{2}$

$s = r\theta$

(θ in radians)

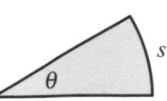

Circular Ring

$\text{Area} = \pi(R^2 - r^2)$

$\qquad = 2\pi pw$

(p = average radius,

w = width of ring)

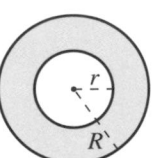 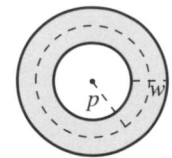

Ellipse

$\text{Area} = \pi ab$

$\text{Circumference} \approx 2\pi \sqrt{\dfrac{a^2 + b^2}{2}}$

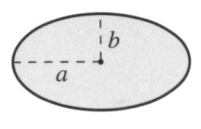

Cone

(A = area of base)

$\text{Volume} = \dfrac{Ah}{3}$

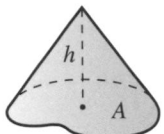

Right Circular Cone

$\text{Volume} = \dfrac{\pi r^2 h}{3}$

$\text{Lateral Surface Area} = \pi r \sqrt{r^2 + h^2}$

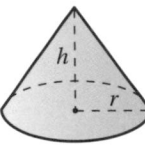

Frustum of Right Circular Cone

$\text{Volume} = \dfrac{\pi(r^2 + rR + R^2)h}{3}$

$\text{Lateral Surface Area} = \pi s(R + r)$

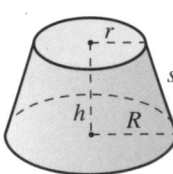

Right Circular Cylinder

$\text{Volume} = \pi r^2 h$

$\text{Lateral Surface Area} = 2\pi rh$

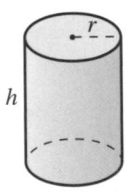

Sphere

$\text{Volume} = \dfrac{4}{3}\pi r^3$

$\text{Surface Area} = 4\pi r^2$

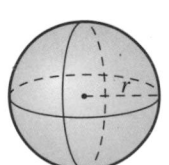